Applications

COLLEGE ALGEBRA and CALCULUS
An Applied Approach
Second Edition

Ron Larson

The Pennsylvania State University
The Behrend College

Anne V. Hodgkins

Phoenix College

With the assistance of David C. Falvo

The Pennsylvania State University
The Behrend College

BROOKS/COLE
CENGAGE Learning·

Australia • Brazil • Japan • Korea • Mexico • Singapore • Spain • United Kingdom • United States

BROOKS/COLE
CENGAGE Learning·

College Algebra and Calculus: An Applied Approach
Second Edition

Ron Larson
Anne V. Hodgkins

Vice President, Editorial Director: P.J. Boardman
Publisher: Richard Stratton
Senior Development Editor: Laura Wheel
Senior Editorial Assistant: Haeree Chang
Associate Media Editor: Andrew Coppola
Senior Marketing Manager: Barb Bartoszek
Marketing Coordinator: Michael Ledesma
Marketing Communications Manager: Mary Anne Payumo
Content Project Manager: Jill Quinn
Senior Art Director: Jill Ort Haskell
Manufacturing Planner: Doug Bertke
Rights Acquisition Specialist: Shalice Shah-Caldwell
Text Designer: Larson Texts, Inc.
Cover Designer: Larson Texts, Inc.
Compositor: Larson Texts, Inc.
Cover Image: Luis Stortini Sabor
          aka CVADRAT/Shutterstock.com

For product information and technology assistance, contact us at **Cengage Learning Customer & Sales Support, 1-800-354-9706.**
For permission to use material from this text or product, submit all requests online at **www.cengage.com/permissions.** Further permissions questions can be emailed to **permissionrequest@cengage.com.**

Library of Congress Control Number: 2011935271

Student Edition:
ISBN-13: 978-1-133-10506-0
ISBN-10: 1-133-10506-8

Brooks/Cole
20 Channel Center Street
Boston, MA 02210
USA

Cengage Learning is a leading provider of customized learning solutions with office locations around the globe, including Singapore, the United Kingdom, Australia, Mexico, Brazil, and Japan. Locate your local office at: **international.cengage.com/region**

Cengage Learning products are represented in Canada by Nelson Education, Ltd.

For your course and learning solutions, visit **www.cengage.com.** Purchase any of our products at your local college store or at our preferred online store *www.cengagebrain.com.*

Instructors: Please visit *login.cengage.com* and log in to access instructor-specific resources.

Printed in China
3 4 5 6 7 15 14 13 12

Contents

Appendices

*Available at the text-specific website *www.cengagebrain.com*

Preface

Welcome to the Second Edition of *College Algebra and Calculus: An Applied Approach*! We are always excited about a new edition, but with this edition, we are even more excited. We had a single goal in mind with this revision—to provide you with a book that is both real and relevant. This book has a bright business-oriented design that complements the multitude of business and life sciences applications found throughout.

The theme for the revision is **"IT'S ALL ABOUT YOU."** The pedagogy of the book is rock solid and is based on years of teaching, years of writing, and years of feedback from instructors and students. Please pay special attention to the study aids with a red **U**. These study aids will help you learn algebra and calculus, use technology, and prepare for tests. For an overview of these aids, check out ALGEBRA, CALCULUS & YOU on page 0.

In each exercise set, quiz, and test, be sure to notice the reference to **CalcChat.com**. At this free site, you can download a step-by-step solution to any odd-numbered exercise. Also, you can talk to a tutor, free, during the hours posted at the site (20 hours a week in the summer, 40 hours a week during the school year).

COLLEGE ALGEBRA and CALCULUS
An Applied Approach

IT'S ALL ABOUT
YOU

Second Edition

Ron Larson
Anne Hodgkins

2 Functions and Graphs

2.1 Graphs of Equations
2.2 Lines in the Plane
2.3 Linear Modeling and Direct Variation
2.4 Functions
2.5 Graphs of Functions
2.6 Transformations of Functions
2.7 The Algebra of Functions

Example 8 on page 189 shows how a scatter plot can be used to find a linear model that approximates the number of alternative-fueled vehicles in use in the United States for a given year.

157

New To This Edition

NEW Chapter Opener
Each *Chapter Opener* highlights a real-life problem from an example in the chapter, showing a graph related to the data and describing the math concept used to solve the problem.

NEW Section Opener
Each *Section Opener* highlights a real-life problem in the exercises, showing a graph for the situation with a description of how you will use the math of the section to solve the problem.

NEW SUMMARIZE
The *Summarize* feature at the end of each section helps you organize the lesson's key concepts into a concise summary, providing you with a valuable study tool.

NEW HOW DO YOU SEE IT? Exercise

The *How Do You See It?* exercise in each section presents a real-life problem that you will solve by visual inspection using the concepts learned in the lesson.

REVISED Exercise Sets

The exercise sets have been carefully and extensively examined to ensure they are rigorous, relevant, and cover all topics suggested by our users. The exercises have been reorganized and titled so you can better see the connections between examples and exercises. Multi-step, real-life exercises reinforce problem-solving skills and mastery of concepts by giving you the opportunity to apply the concepts in real-life situations.

Calc Chat

For the past several years, an independent website—**CalcChat.com**—has been maintained to provide free solutions to all odd-numbered problems in the text. Thousands of students have visited the site for practice and help with their homework. For this edition, information from **CalcChat.com**, including which solutions students accessed most often, was used to help guide the revision of the exercises.

Table of Contents Changes

We have moved Section 2.8 (Inverse Functions), to the beginning of Chapter 4 (Exponential and Logarithmic Functions) as the new Section 4.1. Based on feedback from users, old Section 12.2 (Partial Fractions and Logistic Growth) has been removed.

60. **HOW DO YOU SEE IT?** The graph shows the cost and revenue equations for a product.

$C = 0.5x + 4000$

$(10,000, 9000)$

$R = 0.9x$

(a) For what numbers of units sold is there a loss for the company?

(b) For what number of units sold does the company break even?

(c) For what numbers of units sold is there a profit for the company?

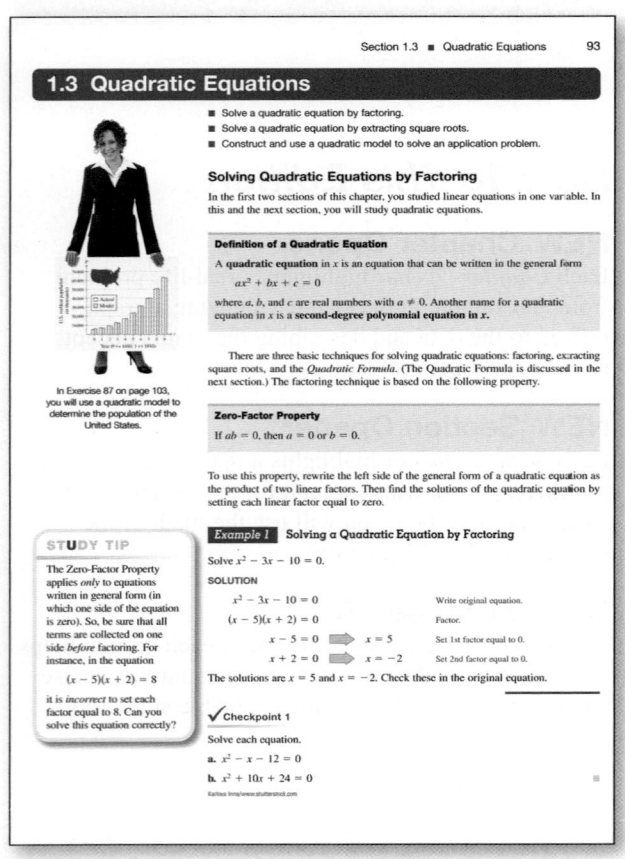

Trusted Features

Section Objectives

A bulleted list of learning objectives provides you the opportunity to preview what will be presented in the upcoming section.

Definitions and Theorems

All definitions and theorems are highlighted for emphasis and easy recognition.

Checkpoint

Paired with every example, the *Checkpoint* problems encourage immediate practice and check your understanding of the concepts presented in the example. Answers to all *Checkpoint* problems appear at the back of the text to reinforce understanding of the skill sets learned.

Business Capsule

Business Capsules appear at the end of selected sections. These capsules and their accompanying research project highlight business situations related to the mathematical concepts covered in the chapter.

STUDY TIP

These hints and tips can be used to reinforce or expand upon concepts, help you learn how to study mathematics, caution you about common errors, address special cases, or show alternative or additional steps to a solution of an example.

TECH TUTOR

The *Tech Tutor* gives suggestions for effectively using tools such as calculators, graphing calculators, and spreadsheet programs to help deepen your understanding of concepts, ease lengthy calculations, and provide alternate solution methods for verifying answers obtained by hand.

ALGEBRA TUTOR

The *Algebra Tutor* appears throughout each chapter and offers algebraic help at point of use. This support is revisited in a two-page algebra review at the end of the chapter, where additional details of example solutions with explanations are provided.

Business Capsule

CitiKitty, Inc. was founded in 2005 by 26-year-old Rebecca Rescate after she moved into a small apartment in New York City with no place to hide her cat's litter box. Finding no easy-to-use cat toilet training kit, she created one, and CitiKitty was born with an initial investment of $20,000. Today the company flourishes with an expanded product line. Revenues in 2010 reached $350,000.

83. Research Project Use your school's library, the Internet, or some other reference source to find information about the start-up costs of beginning a business, such as the example above. Write a short paper about the company.

71. Project: Number of Stores For a project analyzing the numbers of Tiffany & Co. stores from 2000 through 2009, visit this text's website at *www.cengagebrain.com.*
(Source: Tiffany & Co.)

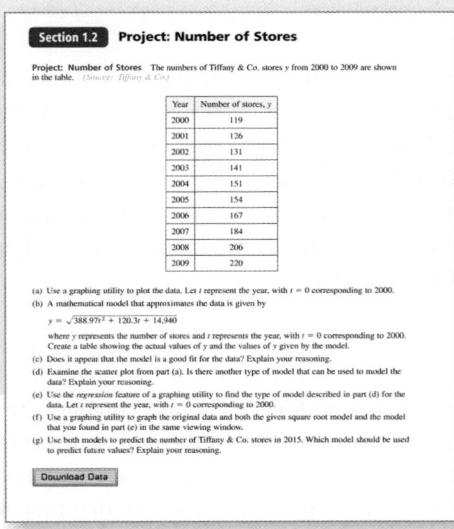

SKILLS WARM UP

The *Skills Warm Up* appears at the beginning of the exercise set for each section. These problems help you review previously learned skills that you will use in solving the section exercises.

Project

The projects at the end of selected sections involve in-depth applied exercises in which you will work with large, real-life data sets, often creating or analyzing models. These projects are offered online at *www.cengagebrain.com.*

Instructor Resources

Print

Complete Solutions Manual
ISBN-13: 978-1-133-62843-9
The *Complete Solutions Manual* provides worked-out solutions for all exercises in the text, including Checkpoints, Quiz Yourself, Test Yourself, and Tech Tutors.

Media

PowerLecture
ISBN-13: 978-1-133-10947-1
This comprehensive CD-ROM provides dynamic media tools designed to help you teach. PowerLecture includes Solution Builder, Diploma Computerized Testing, Microsoft® Powerpoint® lecture slides, and all art from the text.

Solution Builder
www.cengage.com/solutionbuilder
This online instructor database offers complete worked-out solutions of all exercises in the text. Solution Builder allows you to create customized, secure solutions printouts (in PDF format) matched exactly to the problems you assign in class.

Diploma Computerized Testing
Diploma is an easy-to-use assessment software containing hundreds of algorithmic questions derived from the text exercises. With Diploma, you can quickly create, customize, and deliver tests in both print and online formats. Diploma is available on the PowerLecture CD.

ENHANCED WebAssign
www.webassign.net
WebAssign's homework delivery system lets you deliver, collect, grade, and record assignments via the web. Enhanced WebAssign includes Cengage YouBook interactive eBook, Personal Study Plans, a Show My Work feature, Answer Evaluator, quizzes, videos, and more!

Cengage YouBook
YouBook is an interactive and customizable eBook! Containing all the content from the printed text, YouBook features a text edit tool that allows you to modify the textbook narrative as needed. With YouBook, you can quickly re-order entire sections and chapters or hide any content you don't teach to create an eBook that perfectly matches your syllabus. You can further customize the text by publishing web links. Additional media assets include: animated figures, video clips, highlighting, notes, and more! YouBook is available in Enhanced WebAssign.

CourseMate
www.cengagebrain.com
Interested in a simple way to complement your text and course content with study and practice materials? CourseMate brings course concepts to life with interactive learning, study, and exam preparation tools that support the printed textbook. CourseMate includes Engagement Tracker, a first-of-its-kind tool that monitors student engagement. Watch student comprehension soar as your class works with both the printed text and text-specific website.

Student Resources

Print

Student Solutions Manual
ISBN-13: 978-1-133-10860-3

The *Student Solutions Manual* provides complete worked-out solutions to all odd-numbered exercises in the text. In addition, the solutions of all Checkpoint, Quiz Yourself, Test Yourself, and Tech Tutor exercises are included.

Media

www.cengagebrain.com

Interested in a simple way to complement your text and course content with study and practice materials? CourseMate brings course concepts to life with interactive learning, study, and exam preparation tools that support the printed textbook. CourseMate includes: an interactive eBook, quizzes, flashcards, Excel guide, videos, and more!

www.webassign.net

Enhanced WebAssign is an online homework system that lets instructors deliver, collect, grade, and record assignments via the web. Enhanced WebAssign includes Cengage YouBook interactive eBook, Personal Study Plans, a Show My Work feature, Answer Evaluator, quizzes, videos, and more! Be sure to check with your instructor to find out if Enhanced WebAssign is required for your course.

CengageBrain.com

To access additional course materials and companion resources, please visit *www.cengagebrain.com*. At the CengageBrain.com home page, search for the ISBN of your title (from the back cover of your book) using the search box at the top of the page. This will take you to the product page where free companion resources can be found.

Acknowledgements

We would like to thank our colleagues who have helped us develop this program. Their encouragement, criticisms, and suggestions have been invaluable to us.

Reviewers of the Second Edition of College Algebra with Applications for Business and the Life Sciences

Victor Akatsa, *Chicago State University*

Victor Cifarelli, *UNC Charlotte*

Diane Fournier, *St. Cloud Technical and Community College*

Javier Garza, *Tarleton State University*

Jane Golden, *Hillsborough Community College*

Libin Mou, *Bradley University*

Dale Quinn, *Coastal Carolina University*

Kia Shafe, *Saddleback College*

Judith Wood, *College of Central Florida*

Reviewers of the Ninth Edition of Calculus: An Applied Approach

Nasri Abdel-Aziz, *State University of New York College of Environmental Sciences and Forestry*

Alejandro Acuna, *Central New Mexico Community College*

Dona Boccio, *Queensborough Community College*

George Bradley, *Duquesne University*

Andrea Marchese, *Pace University*

Benselamonyuy Ntatin, *Austin Peay State University*

Maijian Qian, *California State University, Fullerton*

Judy Smalling, *St. Petersburg College*

Eddy Stringer, *Tallahassee Community College*

Our thanks to Robert Hostetler, The Pennsylvania State University, The Behrend College, Bruce Edwards, University of Florida, and David Heyd, The Pennsylvania State University, The Behrend College, for their significant contributions to the previous edition of this text.

We would also like to thank the staff at Larson Texts, Inc. who assisted with proofreading the manuscript, preparing and proofreading the art package, and checking and typesetting the supplements.

On a personal level, we would like to thank our families, especially Deanna Gilbert Larson and Jay N. Torok, for their love, patience, and support. Also, a special thanks goes to R. Scott O'Neil.

If you have suggestions for improving this text, please feel free to write to us. Over the past two decades we have received many useful comments from both instructors and students, and we value these comments very highly.

Ron Larson, Ph.D.
Professor of Mathematics
Penn State University
www.RonLarson.com

Anne V. Hodgkins, Ed.D.
Mathematics Faculty
Phoenix College

ALGEBRA, CALCULUS & YOU

Every feature in this text is designed to help you learn calculus. Whenever you see a red **U**, pay special attention to the study aid. These study aids represent years of experience in teaching students *just like you*.   Ron Larson

STUDY TIP

The expressions for $f(g(x))$ and $g(f(x))$ are different in Example 5. In general, the composite of f with g is not the same as the composite of g with f.

The *Study Tips* occur at point of use throughout the text. They represent **common questions** that students ask me, **insights** into understanding concepts, and **alternative ways to look at concepts**. For instance, the *Study Tip* at the left provides insight into the importance of order when working with composite functions.

TECH TUTOR

If you have access to a symbolic differentiation utility, try using it to confirm the derivatives shown in this section.

The *Tech Tutors* give suggestions on how you can use various types of technology to help understand the material. This includes **graphing calculators**, **computer graphing programs**, and **spreadsheet programs** such as Excel. For instance, the *Tech Tutor* at the left points out that some calculators and some computer programs are capable of symbolic differentiation.

ALGEBRA TUTOR xy

For help in solving equations similar to the one in Example 2, see the review of solving equations on page 148.

Throughout years of teaching, I have found that the greatest stumbling block to success in calculus is a weakness in algebra. Each time you see an *Algebra Tutor*, please read it carefully. Then, flip ahead to the referenced page and give yourself a chance to enjoy a brief **algebra refresher**. It will be time well spent.

HOW DO YOU SEE IT?

The *How Do You See It?* question in each exercise set helps you **visually summarize concepts** without messy computations.

SUMMARIZE

The *Summarize* outline at the end of each section asks you to write each learning objective in **your own words**.

SKILLS WARM UP°

The *Skills Warm Up* exercises that precede each exercise set will help you **review previously learned skills**.

SUMMARY AND STUDY STRATEGIES

The *Summary and Study Strategies,* coupled with the Review Exercises are designed to help you organize your thoughts as you **prepare for a chapter test**.

QUIZ YOURSELF

The *Quiz Yourself* occurs midway in each chapter. Take each of these quizzes as you would **take a quiz in class**.

TEST YOURSELF

The *Test Yourself* occurs at the end of each chapter. All questions are answered so you can **check your progress**.

0

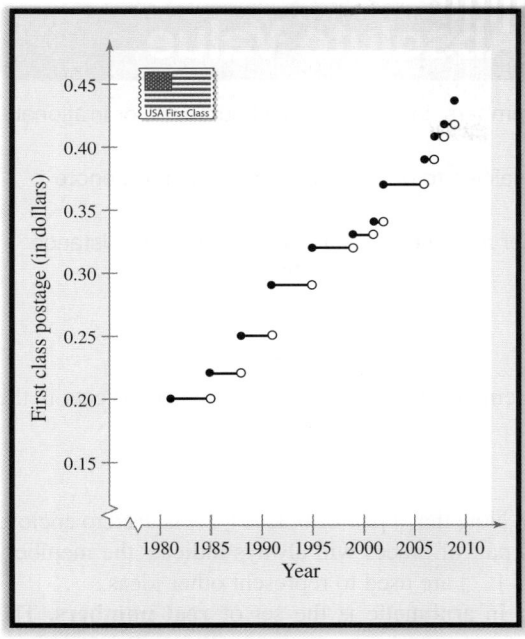

0 Fundamental Concepts of Algebra

Example 9 on page 26 shows how the compound interest formula can be used to determine the average annual rate of inflation for a first class postage stamp over a 26-year period.

R. Gino Santa Maria/Shutterstock.com
Kurhan/Shutterstock.com

1

0.1 Real Numbers: Order and Absolute Value

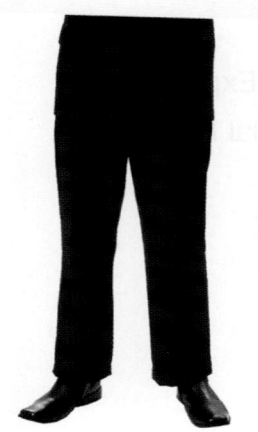

In Exercise 81 on page 9, you can use a number line to determine the range of temperatures for a city in a given month.

■ Classify real numbers as natural numbers, integers, rational numbers, or irrational numbers.

■ Use the real number line and inequalities to order real numbers and to denote subsets of real numbers.

■ Find the absolute value of a number and use absolute value to find the distance between two numbers.

Real Numbers

The formal term that is used in mathematics to refer to a collection of objects is the word **set.** For instance, the set

$$\{1, 2, 3\}$$

contains the three numbers 1, 2, and 3. Note that a pair of braces { } is used to enclose the members of the set. In this text, a pair of braces will always indicate the members of a set. Parentheses ( ) and brackets [ ] are used to represent other ideas.

The set of numbers that is used in arithmetic is the set of **real numbers.** The term *real* distinguishes real numbers from *imaginary* numbers. (You will learn about imaginary numbers in Chapter 3.)

A set A is called a **subset** of a set B if every member of A is also a member of B. Here are some examples.

- $\{1, 2, 3\}$ is a subset of $\{1, 2, 3, 4\}$.

- $\{0, 4\}$ is a subset of $\{0, 1, 2, 3, 4\}$.

- $\{0, 5\}$ is *not* a subset of $\{0, 1, 2, 3, 4\}$ because 5 is not a member of $\{0, 1, 2, 3, 4\}$.

One of the most commonly used subsets of real numbers is the set of **natural numbers** or **positive integers**

$$\{1, 2, 3, 4, . . .\}. \qquad \text{Set of positive integers}$$

Note that the three dots indicate that the pattern continues. For instance, the set also contains the numbers 5, 6, 7, and so on.

Positive integers can be used to describe many quantities that you encounter in everyday life. Here are some examples.

- You are taking four classes this term.

- You are paying $700 per month for rent.

- You are working 20 hours per week.

But even in everyday life, positive integers cannot describe some concepts accurately. For instance, you could have a zero balance in your checking account, or the temperature could be $-10°$ (10 degrees below zero). To describe such quantities, you need to expand the set of positive integers to include **zero** and the **negative integers.** The expanded set is called the set of **integers,** which can be written as shown.

$$\underbrace{\{. . . , -3, -2, -1,}_{\text{Negative integers}} \overset{\text{Zero}}{0,} \underbrace{1, 2, 3, . . .\}}_{\text{Positive integers}}$$

The set of integers is a subset of the set of real numbers. This means that every integer is a real number.

Even with the set of integers, there are still many quantities in everyday life that you cannot describe accurately. The costs of many items are not in whole dollar amounts, but in parts of dollars, such as $1.19 or $39.98. You might work $8\frac{1}{2}$ hours, or you might miss the first *half* of a movie. To describe such quantities, the set of integers is expanded to include **fractions**. The expanded set is called the set of **rational numbers.** Formally, a real number is called **rational** if it can be written as the ratio p/q of two integers, where $q \neq 0$. (The symbol $\neq$ means **not equal to.**) Here are some examples of rational numbers.

$$2 = \frac{2}{1} \qquad\qquad 0.333\ldots = \frac{1}{3}$$

$$0.125 = \frac{1}{8} \qquad\qquad 1.126126\ldots = \frac{125}{111}$$

Real numbers that cannot be written as the ratio of two integers are called **irrational.** For instance, the numbers

$$\sqrt{2} = 1.4142135\ldots$$

and

$$\pi = 3.1415926\ldots$$

are irrational numbers. The decimal representation of a rational number is either *terminating* or *repeating*. For instance, the decimal representation of

$$\frac{1}{4} = 0.25 \qquad\qquad \text{Terminating decimal}$$

is terminating, and the decimal representation of

$$\frac{4}{11} = 0.363636\ldots \qquad\qquad \text{Repeating decimal}$$
$$= 0.\overline{36}$$

is repeating. (The line over "36" indicates which digits repeat.)

The decimal representation of an irrational number neither terminates nor repeats. When you perform calculations using decimal representations of nonterminating decimals, you usually use a decimal approximation that has been **rounded** to a certain number of decimal places. For instance, rounded to four decimal places, the decimal approximations of $\frac{2}{3}$ and π are

$$\frac{2}{3} \approx 0.6667$$

and

$$\pi \approx 3.1416.$$

The symbol $\approx$ means **approximately equal to.**

The Venn diagram in Figure 0.1 shows the relationships between the real numbers and several commonly used subsets of the real numbers.

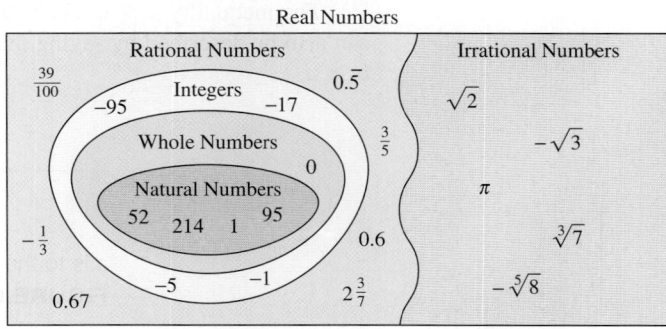

FIGURE 0.1

The Real Number Line and Ordering

The **real number line** can be used to represent the real numbers. It consists of a horizontal line with a point (the **origin**) labeled as 0 (zero). Points to the left of zero are associated with **negative numbers,** and points to the right of zero are associated with **positive numbers,** as shown in Figure 0.2. The real number zero is neither positive nor negative. So, when you want to talk about real numbers that might be positive *or* zero, you can use the term **nonnegative real numbers.**

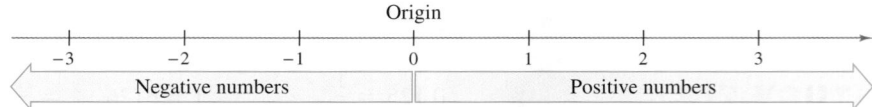

The Real Number Line

FIGURE 0.2

Each point on the real number line corresponds to exactly one real number, and each real number corresponds to exactly one point on the real number line, as shown in Figure 0.3. The number associated with a point on the real number line is the **coordinate** of the point.

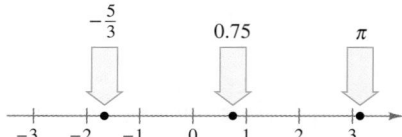

Every real number corresponds to a point on the real number line.

FIGURE 0.3

The real number line provides you with a way of comparing any two real numbers. For instance, if you choose any two (different) numbers on the real number line, one of the numbers must be to the left of the other number. The number to the left is **less than** the number to the right, and the number to the right is **greater than** the number to the left.

Definition of Order on the Real Number Line

If the real number a lies to the left of the real number b on the real number line, a is **less than** b, which is denoted by

$a < b$

as shown in Figure 0.4. This relationship can also be described by saying that b is **greater than** a and writing $b > a$.

The inequality $a \le b$ means that a is **less than or equal to** b. This relationship can also be described by saying that b is **greater than or equal to** a and writing $b \ge a$.

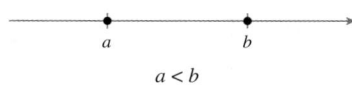

$a < b$

a is to the left of b.

FIGURE 0.4

The symbols <, >, ≤, and ≥ are called **inequality symbols.** Inequalities are useful in denoting subsets of real numbers, as shown in Examples 1 and 2.

Example 1   Interpreting Inequalities

a. The inequality $x \leq 2$ denotes all real numbers that are less than or equal to 2, as shown in Figure 0.5(a).

b. The inequality $-2 \leq x < 3$ means that $x \geq -2$ *and* $x < 3$. This **double inequality** denotes all real numbers between -2 and 3, including -2 but *not* including 3, as shown in Figure 0.5(b).

c. The inequality $x > -5$ denotes all real numbers that are greater than -5, as shown in Figure 0.5(c).

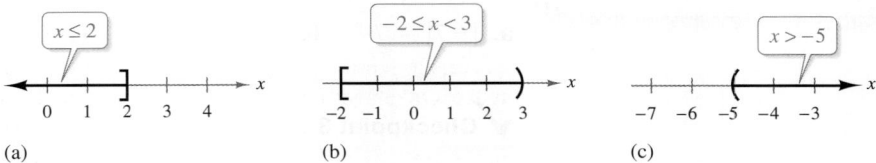

(a)   (b)   (c)

FIGURE 0.5

✓ Checkpoint 1

Give a verbal description of the subset of real numbers represented by $x \geq 7$. ▨

In Figure 0.5, notice that a bracket is used to *include* the endpoint of an interval and a parenthesis is used to *exclude* the endpoint.

Example 2   Inequalities and Subsets of Real Numbers

a. "c is nonnegative" means that c is greater than or equal to zero, which you can write as $c \geq 0$.

b. "b is at most 5" can be written as $b \leq 5$.

c. "d is negative" can be written as $d < 0$, and "d is greater than -3" can be written as $-3 < d$. Combining these two inequalities produces $-3 < d < 0$.

d. "x is positive" can be written as $0 < x$, and "x is not more than 6" can be written as $x \leq 6$. Combining these two inequalities produces $0 < x \leq 6$.

✓ Checkpoint 2

Use inequality notation to describe each subset of real numbers.

a. x is at least 5.

b. y is greater than 4, but no more than 11.

c. z is greater than or equal to -1 and less than or equal to 10. ▨

The following property of real numbers is called the **Law of Trichotomy.** As the "tri" in its name suggests, this law states that for any two real numbers a and b, precisely one of *three* relationships is possible.

$$a < b, \quad a = b, \quad \text{or} \quad a > b \qquad \text{Law of Trichotomy}$$

Absolute Value and Distance

The **absolute value** of a real number is its *magnitude*, or its value disregarding its sign. For instance, the absolute value of -3, written $|-3|$, has the value of 3.

Definition of Absolute Value

Let a be a real number. The **absolute value** of a, denoted by $|a|$, is

$$|a| = \begin{cases} a, & \text{if } a \geq 0 \\ -a, & \text{if } a < 0 \end{cases}.$$

Example 3   Finding Absolute Value

a. $|-7| = 7$ **b.** $|-4.8| = 4.8$ **c.** $-|-9| = -(9) = -9$

✓**Checkpoint 3**

Evaluate $|-12|$.

Example 4   Comparing Real Numbers

Place the correct symbol ($<$, $>$, or $=$) between the two real numbers.

a. $|-4|$ ⬚ $|4|$ **b.** $|-5|$ ⬚ 3 **c.** $-|-1|$ ⬚ $|-1|$

SOLUTION

a. $|-4| = |4|$, because both are equal to 4.

b. $|-5| > 3$, because $|-5| = 5$ and 5 is greater than 3.

c. $-|-1| < |-1|$, because $-|-1| = -1$ and $|-1| = 1$.

✓**Checkpoint 4**

Place the correct symbol ($<$, $>$, or $=$) between the two real numbers.

a. $-|-6|$ ⬚ $-|6|$ **b.** $-|5|$ ⬚ $|-5|$

Properties of Absolute Value

Let a and b be real numbers. Then the following properties are true.

1. $|a| \geq 0$ **2.** $|-a| = |a|$

3. $|ab| = |a| \, |b|$ **4.** $\left|\dfrac{a}{b}\right| = \dfrac{|a|}{|b|}, b \neq 0$

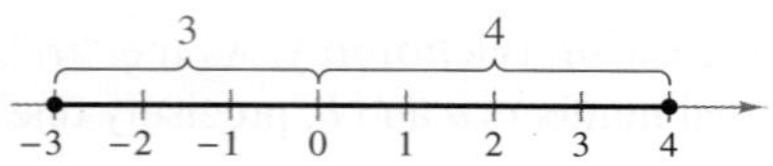

Absolute value can be used to define the distance between two numbers on the real number line. To see how this is done, consider the numbers -3 and 4, as shown in Figure 0.6. To find the distance between these two numbers, subtract either number from the other and then take the absolute value of the difference.

$$(\text{Distance between } -3 \text{ and } 4) = |-3 - 4| = |-7| = 7$$

The distance between -3 and 4 is 7.

FIGURE 0.6

Distance Between Two Numbers

Let a and b be real numbers. The **distance between a and b** is given by

$$\text{Distance} = |b - a| = |a - b|.$$

Example 5 **Distance Between Two Numbers**

a. The distance between 2 and -6 is $|2 - (-6)| = |8| = 8$.

b. "The distance between x and 2 is at least 3" can be written as $|x - 2| \geq 3$.

✓**Checkpoint 5**

Find the distance between -5 and 3.

 Example 6 **Budget Variance**

For each type of expense, a home health care company wants the absolute value of the difference between the actual and budgeted amounts to be less than or equal to $500 *and* less than or equal to 5% of the budgeted amount. By letting a represent the actual expenses and b the budgeted expenses, these restrictions can be written as

$$|a - b| \leq 500 \quad \text{and} \quad |a - b| \leq 0.05b.$$

For travel, office supplies, and wages, the company budgeted $12,500, $750, and $84,600. The actual amounts paid for these expenses were $12,872.56, $704.15, and $85,143.95. Are these amounts within budget restrictions?

SOLUTION Create a table.

	Budgeted Expense, b	Actual Expense, a	$|a - b|$	$0.05b$
Travel	$12,500	$12,872.56	$372.56	$625.00
Office supplies	$750	$704.15	$45.85	$37.50
Wages	$84,600	$85,143.95	$543.95	$4230.00

From this table, you can see that travel expenses pass both tests, so they are within budget restrictions. Office supply expenses pass the first test but fail the second test, so they are *not* within budget restrictions. Wage expenses fail the first test but pass the second test, so they are *not* within budget restrictions.

✓**Checkpoint 6**

In Example 6, the company budgeted $28,000 for medical supplies, but actually paid $30,100. Is this within budget restrictions?

SUMMARIZE   (Section 0.1)

1. Describe the commonly used subsets of the real numbers *(pages 2 and 3)*.

2. Describe how to use the real number line to denote subsets of real numbers *(pages 4 and 5)*. For examples of using the real number line to denote subsets of real numbers, see Examples 1 and 2.

3. State the absolute value of a real number *(page 6)*. For examples of using absolute value, see Examples 3, 4, 5, and 6.

Exercises 0.1

See www.CalcChat.com for worked-out solutions to odd-numbered exercises.

Classifying Real Numbers   In Exercises 1–6, determine which numbers in the set are (a) natural numbers, (b) integers, (c) rational numbers, and (d) irrational numbers.

1. $\left\{-9, -\frac{7}{2}, 5, \frac{2}{3}, \sqrt{2}, 0.1\right\}$
2. $\left\{\sqrt{5}, -7, -\frac{7}{3}, 0, 3.12, \frac{5}{4}\right\}$
3. $\left\{12, -13, 1, \sqrt{4}, \sqrt{6}, \frac{3}{2}\right\}$
4. $\left\{3, -1, \frac{1}{3}, \frac{6}{3}, -\frac{1}{2}\sqrt{2}, -7.5\right\}$
5. $\left\{\frac{8}{2}, -\frac{8}{3}, \sqrt{10}, -4, 9, 14.2\right\}$
6. $\left\{25, -17, \frac{12}{5}, \sqrt{9}, \sqrt{8}, -\sqrt{8}\right\}$

Writing Fractions as Decimals   In Exercises 7–10, use a calculator to find the decimal form of the rational number. If the number is a nonterminating decimal, then write the repeating pattern.

7. $\dfrac{2}{3}$
8. $\dfrac{9}{40}$
9. $\dfrac{14}{111}$
10. $\dfrac{49}{160}$

Identifying Points on the Real Number Line   In Exercises 11 and 12, approximate the two plotted numbers and place the correct symbol ($<$ or $>$) between them.

11.

12.

Comparing Numbers on the Real Number Line   In Exercises 13–18, plot the two real numbers on the real number line and place the appropriate inequality symbol ($<$ or $>$) between them.

13. $-6, 7$
14. $-4, -8$
15. $1, -3.5$
16. $4.3, 1$
17. $\frac{5}{6}, \frac{2}{3}$
18. $-\frac{8}{7}, -\frac{3}{7}$

Using a Calculator to Order Numbers   In Exercises 19–22, use a calculator to order the numbers from least to greatest.

19. $\dfrac{7}{2}, 2\sqrt{3}, 3.45, \dfrac{204}{60}, \dfrac{31}{9}$
20. $\dfrac{559}{500}, 1.12, \dfrac{\sqrt{5}}{2}, \dfrac{115}{99}, \dfrac{\pi}{3}$
21. $\dfrac{7071}{5000}, \dfrac{584}{413}, \sqrt{2}, \dfrac{47}{33}, \dfrac{127}{90}$
22. $\dfrac{26}{15}, \sqrt{3}, 1.73\overline{20}, \dfrac{381}{220}, \sqrt{10}-\sqrt{2}$

Writing an Inequality   In Exercises 23–26, write an inequality that describes the graph.

23.

24.

25.

26.

Interpreting Inequalities   In Exercises 27–36, give a verbal description of the subset of real numbers that is represented by the inequality, and sketch the subset on the real number line. *See Example 1.*

27. $x < 0$
28. $x < 2$
29. $x \le 5$
30. $x \ge -2$
31. $x > 3.5$
32. $x \ge 4.5$
33. $-2 < x < 2$
34. $0 \le x \le 5$
35. $-1 \le x < 0$
36. $0 < x \le 6$

Inequalities and Subsets of Real Numbers   In Exercises 37–44, use inequality notation to describe the subset of real numbers. *See Example 2.*

37. x is positive.
38. t is no more than 20.
39. y is greater than 5 and less than or equal to 12.
40. m is at least -3 and at most 8.
41. The graduate assistant's age A is at least 21.
42. The yield Y is no more than 42 bushels per acre.
43. In the last decade, the annual rate of inflation r was at least -0.5%, but no more than 3.8%.
44. A bank may charge up to $4.00 for an ATM transaction.

Finding Absolute Value   In Exercises 45–54, evaluate the expression. *See Example 3.*

45. $|-10|$
46. $|0|$
47. $-\dfrac{3}{2} - \left|-\dfrac{3}{2}\right|$
48. $-\left|\dfrac{1}{3}\right| - \left|-\dfrac{2}{3}\right|$
49. $-0.8|-2|$
50. $-6|-0.2|$
51. $\dfrac{-5}{|-5|}$
52. $\dfrac{|-4|}{-4}$
53. $|3 - \pi|$
54. $|4 - \pi|$

Comparing Real Numbers In Exercises 55–60, place the correct symbol (<, >, or =) between the two real numbers. *See Example 4.*

55. $|-7|$ $|7|$

56. -5 $-|5|$

57. $|-3|$ $-|-3|$

58. $-|-6|$ $|-6|$

59. $-|-2|$ $-|2|$

60. $-(-2)$ -2

The Distance Between Two Numbers In Exercises 61–70, find the distance between a and b. *See Example 5.*

61. $a = -1$ $b = 3$

$$\begin{array}{ccccc} & \bullet & & & \bullet \\ \hline -1 & 0 & 1 & 2 & 3 \end{array}$$

62. $a = -\frac{5}{2}$ $b = 0$

$$\begin{array}{cccc} & \bullet & & \bullet \\ \hline -3 & -2 & -1 & 0 \end{array}$$

63. $a = -4, b = -3$

64. $a = -1, b = 11$

65. $a = -7, b = 0$

66. $a = 3, b = 9$

67. $a = 126, b = 75$

68. $a = -126, b = -75$

69. $a = 2.22, b = -1.45$

70. $a = 9.34, b = -5.65$

Using Absolute Value Notation In Exercises 71–78, use absolute value notation to describe the sentence. *See Example 5.*

71. The distance between z and 6 is greater than 1.

72. The distance between x and $\frac{9}{2}$ is no more than 3.

73. The distance between x and -10 is at least 6.

74. The distance between z and 0 is less than 8.

75. y is at least six units from 0.

76. x is less than eight units from 0.

77. x is more than five units from m.

78. y is at most two units from a.

79. **Travel** While traveling on the Overseas Highway in the Florida Keys, you pass mile marker 108 at Jewfish Creek, and then mile marker 27 at Ramrod Key. How far do you travel between the mile markers?

80. **HOW DO YOU SEE IT?** Match each description with its graph. [The graphs are labeled (i) and (ii).] Which types of real numbers shown in the Venn diagram on page 3 may be included in a range of prices? a range of lengths? Explain.

(i)
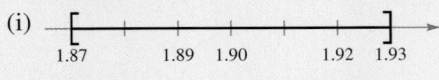

(ii)

    1.87 1.88 1.89 1.90 1.91 1.92 1.93

(a) An item is priced within $0.03 of $1.90.

(b) The distance between the prongs of an electric plug may not differ from 1.9 centimeters by more than 0.03 centimeter.

Temperature In Exercises 81 and 82, the record January temperatures (in degrees Fahrenheit) for a city are given. Find the distance between the numbers to determine the range of temperatures for January.

81. Milwaukee, Wisconsin: lowest: $-26°F$

                  highest: $63°F$

82. Fairbanks, Alaska: lowest: $-61°F$

                highest: $52°F$

Budget Variance In Exercises 83–86, the accounting department of an Internet start-up company checks whether various actual expenses differ from the budgeted expenses by more than $500 or by more than 5%. Complete the missing parts of the table. Then determine whether the actual expense passes the "budget variance test." *See Example 6.*

	Budgeted Expense, b	Actual Expense, a	$|a - b|$	0.05b
83.	$30,000	$29,123.45		
84.	$125,500	$126,347.85		
85.	$12,000	$11,735.68		
86.	$8300	$8632.59		

Quality Control In Exercises 87–92, the quality control inspector for a tire factory tests the rim diameters of various tires. The inspector rejects any tire that differs by more than 0.02 inch or by more than 0.12% of the expected diameter measure. Complete the missing parts of the table. Then determine whether the tire passes the test.

	Expected Diameter, b	Actual Diameter, a	$|a - b|$	0.0012b
87.	14 in.	13.998 in.		
88.	15 in.	15.012 in.		
89.	16 in.	15.973 in.		
90.	18 in.	18.027 in.		
91.	13 in.	13.022 in.		
92.	17 in.	16.992 in.		

93. **Think About It** Describe the real number values of u and v for which $|u + v|$ is *greater than*, *less than*, and *equal to* $|u| + |v|$.

94. **Think About It** Is the set of nonnegative real numbers the same as the set of positive real numbers? Explain.

95. **Writing** For what real numbers a is $|a| = -a$? Explain.

96. **Writing** Describe the differences among the sets of natural numbers, integers, rational numbers, and irrational numbers.

0.2 The Basic Rules of Algebra

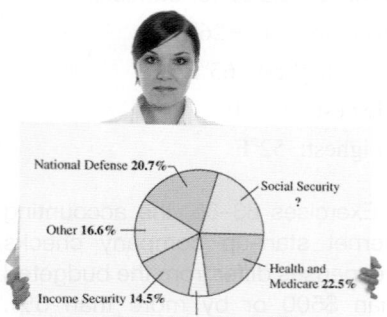

■ Identify the terms of, and evaluate, an algebraic expression.
■ Identify basic rules of algebra and perform operations on real numbers.
■ Develop an understanding of the properties of equality.
■ Use a calculator to evaluate an expression.

Algebraic Expressions

One of the basic characteristics of algebra is the use of letters (or combinations of letters) to represent numbers. The letters used to represent numbers are called **variables,** and combinations of letters and numbers are called **algebraic expressions.** Some examples of algebraic expressions are

$$5x, \quad 2x - 3, \quad \frac{4}{x^2 + 2}, \quad \text{and} \quad 7x + y.$$

Algebraic Expression

A collection of letters (called **variables**) and real numbers (called **constants**) that are combined using the operations of addition, subtraction, multiplication, and division is an **algebraic expression.** (Other operations can also be used to form an algebraic expression.)

In Exercise 69 on page 19, you will use a circle graph to analyze the expenses of the federal government.

The **terms** of an algebraic expression are those parts that are separated by addition. For example, the algebraic expression

$$x^2 - 5x + 8$$

has three terms: x^2, $-5x$, and 8. Note that $-5x$, rather than $5x$, is a term, because

$$x^2 - 5x + 8 = x^2 + (-5x) + 8.$$

The terms x^2 and $-5x$ are the **variable terms** of the expression, and 8 is the **constant term** of the expression. The numerical factor of a variable term is the **coefficient** of the variable term. For instance, the coefficient of the variable term $-5x$ is -5, and the coefficient of the variable term x^2 is 1.

Example 1   Identifying the Terms of an Algebraic Expression

Algebraic Expression	*Terms*
a. $4x - 3$	$4x, -3$
b. $2x + 4y - 5$	$2x, 4y, -5$
c. $-7 + 6p^3 - p^2$	$-7, 6p^3, -p^2$

✓ **Checkpoint 1**

Identify the terms of each algebraic expression.

a. $8 - 15x$

b. $4x^2 - 3y - 7$

c. $z^4 - 10z^3 + 1$

ALGEBRA TUTOR
xy

For help in evaluating the expressions in Example 2, see the *Chapter 0 Algebra Tutor* on page 62.

Example 2 Symbols of Grouping

a. $7 - 3(4 - 2) = 7 - 3(2)$
$$= 7 - 6$$
$$= 1$$

b. $(4 - 5) - (3 - 6) = (-1) - (-3)$
$$= -1 + 3$$
$$= 2$$

✓ Checkpoint 2

Simplify the expression $5(7 - 3) + 9$.

The **Substitution Principle** states, "If $a = b$, then a can be replaced by b in any expression involving a." You use this principle to **evaluate** an algebraic expression by substituting numerical values for each of the variables in the expression. In the first evaluation shown below, 3 is substituted for x in the expression $-3x + 5$.

TECH TUTOR

To evaluate the expression $3 + 4x$ for the x-values 2 and 5, use the *last entry* feature of a graphing utility.

1. Evaluate $3 + 4 \cdot 2$.

2. Press (2nd) [ENTRY] (recalls previous expression to the home screen).

3. Cursor to 2, replace 2 with 5, and press (ENTER).

Expression	Value of Variable	Substitution	Value of Expression
$-3x + 5$	$x = 3$	$-3(3) + 5$	$-9 + 5 = -4$
$3x^2 + 2x - 1$	$x = -1$	$3(-1)^2 + 2(-1) - 1$	$3 - 2 - 1 = 0$
$-2x(x + 4)$	$x = -2$	$-2(-2)(-2 + 4)$	$-2(-2)(2) = 8$
$\dfrac{1}{x - 2}$	$x = 2$	$\dfrac{1}{2 - 2}$	$\dfrac{1}{0}$ is undefined.

Example 3 Evaluating Algebraic Expressions

Evaluate each algebraic expression at $x = -2$ and $y = 3$.

a. $4y - 2x$ **b.** $5 + x^2$ **c.** $5 - y^2$

SOLUTION

a. When $x = -2$ and $y = 3$, the expression $4y - 2x$ has a value of
$$4(3) - 2(-2) = 12 + 4$$
$$= 16.$$

b. When $x = -2$, the expression $5 + x^2$ has a value of
$$5 + (-2)^2 = 5 + 4$$
$$= 9.$$

c. When $y = 3$, the expression $5 - y^2$ has a value of
$$5 - (3)^2 = 5 - 9$$
$$= -4.$$

✓ Checkpoint 3

Evaluate each algebraic expression at $x = 4$ and $y = -5$.

a. $7x - y$ **b.** $3y + x^2$ **c.** $x - y^2$

Basic Rules of Algebra

The four basic arithmetic operations are **addition, multiplication, subtraction,** and **division,** denoted by the symbols $+$, $\times$ or $\cdot$, $-$, and $\div$, respectively. Of these, addition and multiplication are considered to be the two primary arithmetic operations. Subtraction and division are defined as the inverse operations of addition and multiplication, as follows.

Subtraction: Add the opposite.

$$a - b = a + (-b)$$

Division: Multiply by the reciprocal.

If $b \neq 0$, then $a \div b = a\left(\dfrac{1}{b}\right) = \dfrac{a}{b}$.

In these definitions, $-b$ is called the **additive inverse** (or opposite) of b, and $1/b$ is called the **multiplicative inverse** (or reciprocal) of b. In place of $a \div b$, you can use the fractional form a/b. In this form, a is called the **numerator** of the fraction and b is called the **denominator.**

Basic Rules of Algebra

Let a, b, and c be real numbers, variables, or algebraic expressions.

Property	Example
Commutative Property of Addition	
$a + b = b + a$	$4x + x^2 = x^2 + 4x$
Commutative Property of Multiplication	
$ab = ba$	$(4 - x)x^2 = x^2(4 - x)$
Associative Property of Addition	
$(a + b) + c = a + (b + c)$	$(-x + 5) + x^2 = -x + (5 + x^2)$
Associative Property of Multiplication	
$(ab)c = a(bc)$	$(2x \cdot 3y)(8) = (2x)(3y \cdot 8)$
Distributive Property	
$a(b + c) = ab + ac$	$3x(5 + 2x) = 3x \cdot 5 + 3x \cdot 2x$
$(a + b)c = ac + bc$	$(y + 8)y = y \cdot y + 8 \cdot y$
Additive Identity Property	
$a + 0 = a$	$5y^2 + 0 = 5y^2$
Multiplicative Identity Property	
$a \cdot 1 = a$	$(4x^2)(1) = 4x^2$
Additive Inverse Property	
$a + (-a) = 0$	$5x^3 + (-5x^3) = 0$
Multiplicative Inverse Property	
$a \cdot \dfrac{1}{a} = 1, \quad a \neq 0$	$(x^2 + 4)\left(\dfrac{1}{x^2 + 4}\right) = 1$

Because subtraction is defined as "adding the opposite," the Distributive Property is also true for subtraction. For instance, the "subtraction form" of $a(b + c) = ab + ac$ is

$$a(b - c) = a[b + (-c)]$$
$$= ab + a(-c)$$
$$= ab - ac.$$

Example 4 Identifying the Basic Rules of Algebra

Identify the rule of algebra illustrated by each statement.

a. $(4x^2)5 = 5(4x^2)$

b. $(2y^3 + y) - (2y^3 + y) = 0$

c. $(4 + x^2) + 3x^2 = 4 + (x^2 + 3x^2)$

d. $(x - 5)7 + (x - 5)x = (x - 5)(7 + x)$

e. $2x \cdot \dfrac{1}{2x} = 1, \quad x \neq 0$

SOLUTION

a. This statement illustrates the Commutative Property of Multiplication.

b. This statement illustrates the Additive Inverse Property.

c. This statement illustrates the Associative Property of Addition.

d. This statement illustrates the Distributive Property in reverse order.

$$ab + ac = a(b + c) \qquad \text{Distributive Property}$$
$$(x - 5)7 + (x - 5)x = (x - 5)(7 + x)$$

e. This statement illustrates the Multiplicative Inverse Property. ——————

> **STUDY TIP**
>
> Note that in Example 4(e), it is important that x be a nonzero number. If x was allowed to be zero, then you would have 0 in the denominator, and the reciprocal of zero is undefined.

✓ Checkpoint 4

Identify the rule of algebra illustrated by each statement.

a. $3x^2 \cdot 1 = 3x^2$

b. $x^2 + 5 = 5 + x^2$

The following three lists summarize the basic properties of negation, zero, and fractions. When studying these lists, do not only *memorize* a verbal description of each property, but try to gain an understanding of why each property is valid.

Properties of Negation

Let a and b be real numbers, variables, or algebraic expressions.

Property	*Example*
1. $(-1)a = -a$	$(-1)7 = -7$
2. $-(-a) = a$	$-(-6) = 6$
3. $(-a)b = -(ab) = a(-b)$	$(-5)3 = -(5 \cdot 3) = 5(-3)$
4. $(-a)(-b) = ab$	$(-2)(-6) = 2 \cdot 6$
5. $-(a + b) = (-a) + (-b)$	$-(3 + 8) = (-3) + (-8)$

Be sure you see the difference between the opposite of a number and a negative number. If a is negative, then its opposite, $-a$, is positive. For instance, if $a = -5$, then $-a = -(-5) = 5$.

Properties of Zero

Let a and b be real numbers, variables, or algebraic expressions. Then the following properties are true.

1. $a + 0 = a$ and $a - 0 = a$

2. $a \cdot 0 = 0$

3. $\dfrac{0}{a} = 0, \quad a \neq 0$

4. $\dfrac{a}{0}$ is undefined.

5. Zero-Factor Property: If $ab = 0$, then $a = 0$ or $b = 0$.

The "or" in the Zero-Factor Property includes the possibility that both factors are zero. This is called an **inclusive or,** and it is the way the word "or" is always used in mathematics.

Properties of Fractions

Let a, b, c, and d be real numbers, variables, or algebraic expressions such that $b \neq 0$ and $d \neq 0$. Then the following properties are true.

1. *Equivalent fractions:* $\dfrac{a}{b} = \dfrac{c}{d}$ if and only if $ad = bc$.

2. *Rules of signs:* $-\dfrac{a}{b} = \dfrac{-a}{b} = \dfrac{a}{-b}$ and $\dfrac{-a}{-b} = \dfrac{a}{b}$

3. *Generate equivalent fractions:* $\dfrac{a}{b} = \dfrac{ac}{bc}, \quad c \neq 0$

4. *Add or subtract with like denominators:* $\dfrac{a}{b} \pm \dfrac{c}{b} = \dfrac{a \pm c}{b}$

5. *Add or subtract with unlike denominators:* $\dfrac{a}{b} \pm \dfrac{c}{d} = \dfrac{ad \pm bc}{bd}$

6. *Multiply fractions:* $\dfrac{a}{b} \cdot \dfrac{c}{d} = \dfrac{ac}{bd}$

7. *Divide fractions:* $\dfrac{a}{b} \div \dfrac{c}{d} = \dfrac{a}{b} \cdot \dfrac{d}{c} = \dfrac{ad}{bc}, \quad c \neq 0$

In Property 1 (equivalent fractions), the phrase "if and only if" implies the two statements below.

- If $\dfrac{a}{b} = \dfrac{c}{d}$, then $ad = bc$.

- If $ad = bc$, where $b \neq 0$ and $d \neq 0$, then $\dfrac{a}{b} = \dfrac{c}{d}$.

Example 5   **Properties of Zero and Properties of Fractions**

a. $x - \dfrac{0}{5} = x - 0 = x$ Properties 3 and 1 of zero

b. $\dfrac{x}{5} = \dfrac{3 \cdot x}{3 \cdot 5} = \dfrac{3x}{15}$ Generate equivalent fractions.

c. $\dfrac{x}{3} + \dfrac{2x}{5} = \dfrac{x \cdot 5 + 3 \cdot 2x}{15}$ Property 5 of fractions

$\qquad\qquad = \dfrac{11x}{15}$ Simplify.

d. $\dfrac{7}{x} \div \dfrac{3}{2} = \dfrac{7}{x} \cdot \dfrac{2}{3} = \dfrac{14}{3x}$ Divide fractions.

✓**Checkpoint 5**

Simplify the expression $\dfrac{x}{4} + \dfrac{2x}{3}$.

   If a, b, and c are integers such that $ab = c$, then a and b are **factors** or **divisors** of c. For example, 2 and 3 are factors of 6 because $2 \cdot 3 = 6$. A **prime number** is a positive integer that has exactly two factors: itself and 1. For example, 2, 3, 5, 7, and 11 are prime numbers, whereas 1, 4, 6, 8, 9, and 10 are not. The numbers 4, 6, 8, 9, and 10 are **composite** because they can be written as the products of two or more prime numbers. The number 1 is neither prime nor composite. The **Fundamental Theorem of Arithmetic** states that every positive integer greater than 1 is a prime number or can be written as the product of prime numbers in precisely one way (disregarding order). For instance, the *prime factorization* of 24 is

$24 = 2 \cdot 2 \cdot 2 \cdot 3.$

   When you are adding or subtracting fractions that have unlike denominators, you can use Property 4 of fractions by rewriting both of the fractions so that they have the same denominator. This is called the **least common denominator** method.

Example 6   **Adding and Subtracting Fractions**

Evaluate $\dfrac{2}{15} - \dfrac{5}{9} + \dfrac{4}{5}$.

SOLUTION   Begin by factoring the denominators to find the least common denominator (LCD). Use the LCD, 45, to rewrite the fractions and simplify.

$$\dfrac{2}{15} - \dfrac{5}{9} + \dfrac{4}{5} = \dfrac{2 \cdot 3}{15 \cdot 3} - \dfrac{5 \cdot 5}{9 \cdot 5} + \dfrac{4 \cdot 9}{5 \cdot 9}$$

$$= \dfrac{6 - 25 + 36}{45}$$

$$= \dfrac{17}{45}$$

✓**Checkpoint 6**

Evaluate $\dfrac{3}{4} + \dfrac{2}{3} - \dfrac{1}{2}$.

Equations and Calculators

An **equation** is a statement of equality between two expressions. So, the statement

$$a + b = c + d$$

means that the expressions

$$a + b \quad \text{and} \quad c + d$$

represent the same number. For instance, because $1 + 4$ and $3 + 2$ both represent the number 5, you can write

$$1 + 4 = 3 + 2.$$

Three important properties of equality follow.

Properties of Equality

Let a, b, and c be real numbers, variables, or algebraic expressions.

1. Reflexive: $a = a$

2. Symmetric: If $a = b$, then $b = a$.

3. Transitive: If $a = b$ and $b = c$, then $a = c$.

In algebra, you often rewrite expressions by making substitutions that are permitted under the Substitution Principle. Two important consequences of the Substitution Principle are the following rules.

1. If $a = b$, then $a + c = b + c$.

2. If $a = b$, then $ac = bc$.

The first rule allows you to add the same number to each side of an equation. The second allows you to multiply each side of an equation by the same number. The converses of these two rules are also true and are listed below.

1. If $a + c = b + c$, then $a = b$.

2. If $ac = bc$ and $c \neq 0$, then $a = b$.

So, you can also subtract the same number from each side of an equation as well as divide each side of an equation by the same nonzero number.

The table below shows keystrokes for several similar functions on a standard scientific calculator and a graphing calculator. These keystrokes may not be the same as those for your calculator. Consult your user's guide for specific keystrokes.

Graphing Calculator	Scientific Calculator
[ENTER]	[=]
[(−)]	[+/−]
[^]	[y^x]
[x^{-1}]	[1/x]

For example, you can evaluate the expressions 10^2 and 13^3 on a graphing calculator or a scientific calculator as shown.

Expression	Graphing Calculator	Scientific Calculator
10^2	10 [x^2] [ENTER]	10 [x^2] [=]
13^3	13 [^] 3 [ENTER]	13 [y^x] 3 [=]

Example 7   Using a Calculator

Scientific Calculator

Expression	Keystrokes	Display
a. $-12^2 - 100$	12 [x²] [+/−] [−] 100 [=]	−244
b. $24 \div 2^3$	24 [÷] 2 [yˣ] 3 [=]	3
c. $3(10 - 4^2) \div 2$	3 [×] [(] 10 [−] 4 [x²] [)] [÷] 2 [=]	−9
d. 37% of 40	.37 [×] 40 [=]	14.8

Graphing Calculator

Expression	Keystrokes	Display
a. $-12^2 - 100$	[(−)] 12 [x²] [−] 100 [ENTER]	−244
b. $24 \div 2^3$	24 [÷] 2 [∧] 3 [ENTER]	3
c. $3(10 - 4^2) \div 2$	3 [(] 10 [−] 4 [x²] [)] [÷] 2 [ENTER]	−9
d. 37% of 40	.37 [×] 40 [ENTER]	14.8

✓ Checkpoint 7

Write the keystrokes you can use to evaluate $6(8^3 - 481)$ on a graphing calculator or a scientific calculator. ■

When rounding decimals, look at the *decision digit* (the digit at the right of the last digit you want to keep). Round up when the decision digit is 5 or greater, and round down when it is 4 or less.

Example 8   Rounding Decimal Numbers

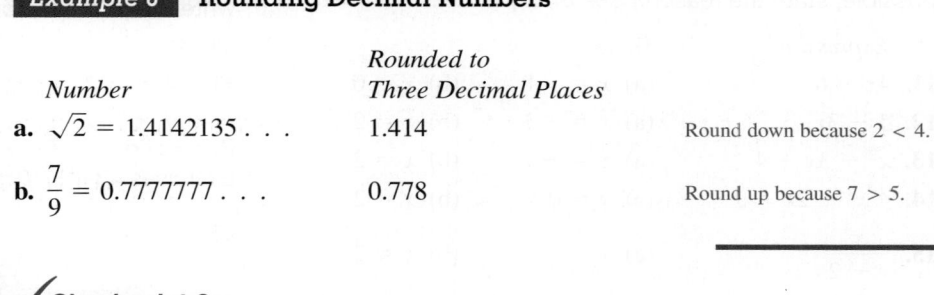

Number	Rounded to Three Decimal Places	
a. $\sqrt{2} = 1.4142135\ldots$	1.414	Round down because 2 < 4.
b. $\dfrac{7}{9} = 0.7777777\ldots$	0.778	Round up because 7 > 5.

✓ Checkpoint 8

Use a calculator to evaluate $4\left(\frac{2}{3} + \frac{4}{5}\right)$. Then round the result to two decimal places. ■

SUMMARIZE   (Section 0.2)

1. State the definition of an algebraic expression *(page 10)*. For an example of evaluating an algebraic expression, see Example 3.

2. State a basic rule of algebra, a property of negation, a property of zero, and a property of fractions *(pages 12–14)*. For examples that use these properties, see Examples 4, 5, and 6.

3. State the properties of equality *(page 16)*.

SKILLS WARM UP 0.2   The following warm-up exercises involve skills that were covered in earlier sections. You will use these skills in the exercise set for this section. For additional help, review Section 0.1.

In Exercises 1–4, place the correct inequality symbol ($<$ or $>$) between the two real numbers.

1. $-4 \quad \boxed{} \quad -2$ **2.** $0 \quad \boxed{} \quad -3$ **3.** $-\frac{11}{9} \quad \boxed{} \quad -\frac{13}{9}$ **4.** $-\frac{12}{33} \quad \boxed{} \quad \frac{11}{33}$

In Exercises 5–8, find the distance between the two real numbers.

5. $4, 6$ **6.** $-2, 2$ **7.** $0, -5$ **8.** $-1, 3$

Exercises 0.2

See www.CalcChat.com for worked-out solutions to odd-numbered exercises.

Identifying the Terms of an Algebraic Expression In Exercises 1–6, identify the terms of the algebraic expression. *See Example 1.*

1. $7x + 4$
2. $3x - 6$
3. $x^2 - 4x + 8$
4. $4x^3 + x - 5$
5. $2x^2 - 9x + 13$
6. $3x^4 + 2x^3 + x^2 - 1$

Symbols of Grouping In Exercises 7–10, simplify the expression. *See Example 2.*

7. $(2 - 7) + 3$
8. $-3 - (5 - 2)$
9. $(4 - 7)(-2)$
10. $-5(-2 - 6)$

Evaluating Algebraic Expressions In Exercises 11–16, evaluate the expression at each value of x. (If not possible, state the reason.) *See Example 3.*

Expression	Values	
11. $4x - 6$	(a) $x = -1$	(b) $x = 0$
12. $5 - 3x$	(a) $x = -3$	(b) $x = 2$
13. $x^2 - 3x + 4$	(a) $x = -2$	(b) $x = 2$
14. $-x^3 + 2x - 1$	(a) $x = 0$	(b) $x = 2$
15. $\dfrac{x}{x - 2}$	(a) $x = -2$	(b) $x = 2$
16. $\dfrac{x + 3}{x - 3}$	(a) $x = 3$	(b) $x = -3$

Evaluating Algebraic Expressions In Exercises 17–22, evaluate the expression at $x = 3$, $y = -2$, and $z = 4$. *See Example 3.*

17. $x + 3y + z$
18. $6z + 5x - 3y$
19. $x^2 - 5y + 4z$
20. $z^2 + 6y - x$
21. $\dfrac{x - y}{5z}$
22. $\dfrac{4z - 2y}{20x}$

Identifying the Basic Rules of Algebra In Exercises 23–38, identify the rule(s) of algebra illustrated by the statement. *See Example 4.*

23. $3 + 4 = 4 + 3$
24. $x + 9 = 9 + x$

25. $-15 + 15 = 0$
26. $(x + 2) - (x + 2) = 0$
27. $2(x + 3) = 2x + 6$
28. $(5 + 9)(6) = 5(6) + 9(6)$
29. $2\left(\frac{1}{2}\right) = 1$
30. $\dfrac{1}{h + 6}(h + 6) = 1, \quad h \neq -6$
31. $h + 0 = h$
32. $(z - 2) + 0 = z - 2$
33. $57 \cdot 1 = 57$
34. $1 \cdot (1 + x) = 1 + x$
35. $6 + (7 + 8) = (6 + 7) + 8$
36. $x + (y + 10) = (x + y) + 10$
37. $x(3y) = (x \cdot 3)y = (3x)y$
38. $\frac{1}{7}(7 \cdot 12) = \left(\frac{1}{7} \cdot 7\right)12 = 1 \cdot 12 = 12$

Writing the Prime Factorization In Exercises 39–42, write the prime factorization of the number.

39. 48
40. 24
41. 240
42. 150

Properties of Zero and Properties of Fractions In Exercises 43–48, simplify the expression. *See Example 5.*

43. $\dfrac{0}{7} + 3x$
44. $\dfrac{0}{12}(x + 1)$
45. $\dfrac{x}{4} + \dfrac{3x}{5}$
46. $\dfrac{x}{8} + \dfrac{5}{36}$
47. $\dfrac{8}{x} \div \dfrac{2}{3}$
48. $\dfrac{3x}{4} \div \dfrac{9}{2}$

Operations with Fractions In Exercises 49–56, perform the indicated operation(s). (Write fractional answers in simplest form.) *See Examples 5 and 6.*

49. $\dfrac{2 - 79}{-11}$
50. $\dfrac{27 - 35}{4}$
51. $\frac{5}{8} + \frac{1}{4} - \frac{5}{6}$
52. $\frac{10}{11} + \frac{6}{33} - \frac{13}{66}$
53. $\frac{2}{5} \cdot \frac{7}{8}$
54. $\left(-\frac{2}{3}\right) \cdot \frac{5}{8} \cdot \frac{3}{4}$
55. $\left(\frac{2}{3} \div 8\right) + \left(3 \cdot \frac{3}{12}\right)$
56. $\left(\frac{3}{5} \div 3\right) - \left(6 \cdot \frac{4}{8}\right)$

Using a Calculator In Exercises 57–64, use a calculator to evaluate the expression. (Round to two decimal places.) *See Examples 7 and 8.*

57. $3\left(-\frac{5}{12} + \frac{3}{8}\right)$

58. $2\left(-7 + \frac{1}{6}\right)$

59. $\dfrac{11.46 - 5.37}{3.91}$

60. $\dfrac{-8.31 + 4.83}{7.65}$

61. $-108 \div 3^3$

62. $-80 \div 2^5$

63. $5(18 - 2^3) \div 10$

64. $-6^2 - [7 + (-2)^3]$

Using a Calculator In Exercises 65–68, use a calculator to solve. *See Example 7.*

65. 35% of 68

66. 35% of 820

67. 125% of 37

68. 147% of 44

69. Federal Government Expenses The circle graph shows the types of expenses of the federal government in 2008. *(Source: U.S. Office of Management and Budget)*

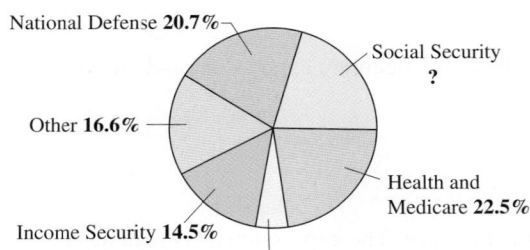

National Defense **20.7%**

Social Security **?**

Other **16.6%**

Health and Medicare **22.5%**

Income Security **14.5%**

Education and Veterans' Benefits **5.0%**

(a) What percent of the total expenses was spent on Social Security?

(b) The total 2008 expenses were $2,982,900,000,000. Find the amount spent for each category in the circle graph. (Round to the nearest billion dollars.)

70. Research Study The percent of people in a research study who have a particular health risk is 39.5%. The total number of people in the study is 12,857. How many people have the health risk?

71. Clinical Trial The percent of patients in a clinical trial of a cancer treatment showing a decrease in tumor size is 49.2%. There are 3445 patients in the trial. How many patients show a decrease in tumor size?

72. Calculator Keystrokes Write the expression that corresponds to each set of keystrokes.

(a) 5 ⊠ ⦇ 2.7 ⊟ 9.4 ⦈ ▣   Scientific

  5 ⦇ 2.7 ⊟ 9.4 ⦈ [ENTER]   Graphing

(b) 2 ⊠ ⦇ 4 [+/−] ⊞ 2 ⦈ ▣   Scientific

  2 ⦇ [(−)] 4 ⊞ 2 ⦈ [ENTER]   Graphing

The symbol 📟 indicates an exercise in which you are instructed to use graphing technology or a spreadsheet software program. The solutions of other exercises may also be facilitated by use of appropriate technology.

73. Pet Spending The data below show pet spending in the United States in 2009. Find the percent of total pet spending for each category. Then use a spreadsheet software program to make a circle graph for the percent data. *(Source: American Pet Products Association)*

Total pet spending (2009):	**$45.53 billion**
Food:	$17.56 billion
Vet care:	$12.04 billion
Supplies/OTC medicine:	$10.41 billion
Live animal purchases:	$2.16 billion
Grooming and boarding:	$3.36 billion

74. HOW DO YOU SEE IT? You earn $7.25 per hour for each of two work study jobs. You work a hours as an office assistant and b hours as an events coordinator. The area model represents the situation.

7.25 ⬚ + 7.25 ⬚ = 7.25 ⬚⬚

  a b a b

(a) What quantity does the area of the gold rectangle represent? the blue rectangle? the green rectangle?

(b) What rule of algebra does the model illustrate?

Business Capsule

PetSmart, the largest pet store chain in the U.S. and Canada, has grown by offering pet lodging services in 180 of their 1187 stores. PetsHotels provide supervised play areas with toys and slides, blankets, TV, and snacks. Special fee services (grooming, training, and phoning pet parents) are also provided. Sales for these services have increased an average of 19% each year since PetsHotels began in 2004.

75. Research Project Use your campus library, the Internet, or some other reference source to find information on a "special services" company experiencing strong growth, and write a brief report about the company.

0.3 Integer Exponents

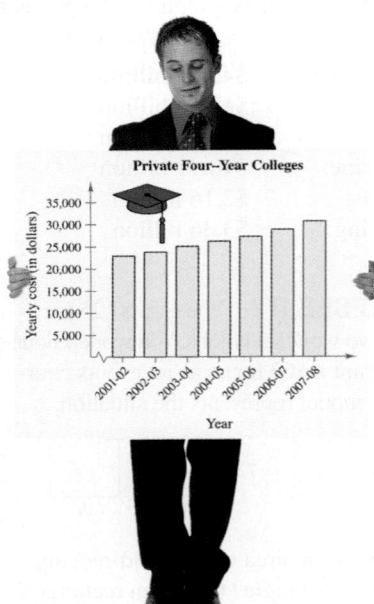

Private Four-Year Colleges

In Exercise 83 on page 28, you will use the compound interest formula to find the average rate of inflation for the yearly costs of attending a private four-year college in the United States.

■ Use properties of exponents.
■ Use scientific notation to represent real numbers and use a calculator to evaluate exponential expressions.
■ Use interest formulas to solve application problems.

Properties of Exponents

Repeated multiplication of a real number by itself can be written in **exponential form.** Here are some examples.

Repeated Multiplication	Exponential Form
$7 \cdot 7$	7^2
$a \cdot a \cdot a \cdot a \cdot a$	a^5
$(-4)(-4)(-4)$	$(-4)^3$
$(2x)(2x)(2x)(2x)$	$(2x)^4$

Exponential Notation

Let a be a real number, a variable, or an algebraic expression, and let n be a positive integer. Then

$$a^n = \underbrace{a \cdot a \cdot a \cdots a}_{n \text{ factors}}$$

where n is the **exponent** and a is the **base.** The expression a^n is read as "a to the nth **power**" or simply "a to the nth."

When multiplying exponential expressions with the same base, *add* exponents.

$$a^m \cdot a^n = a^{m+n} \qquad \text{Add exponents when multiplying.}$$

For instance, to multiply 2^2 and 2^3, you can write

$$2^2 \cdot 2^3 = \overbrace{(2 \cdot 2)}^{\substack{\text{Two} \\ \text{factors}}} \cdot \overbrace{(2 \cdot 2 \cdot 2)}^{\substack{\text{Three} \\ \text{factors}}} = \overbrace{2 \cdot 2 \cdot 2 \cdot 2 \cdot 2}^{\substack{\text{Five} \\ \text{factors}}}$$

$$= 2^{2+3} = 2^5.$$

On the other hand, when dividing exponential expressions with the same base, *subtract* exponents. That is,

$$\frac{a^m}{a^n} = a^{m-n}, \quad a \neq 0. \qquad \text{Subtract exponents when dividing.}$$

For instance, to divide 3^4 by 3^2, you can write

$$\frac{3^4}{3^2} = \frac{\overbrace{3 \cdot 3 \cdot 3 \cdot 3}^{\text{Four factors}}}{\underbrace{3 \cdot 3}_{\text{Two factors}}} = \frac{\overset{1}{\cancel{3}} \cdot \overset{1}{\cancel{3}} \cdot 3 \cdot 3}{\underset{1}{\cancel{3}} \cdot \underset{1}{\cancel{3}}} = \overbrace{3 \cdot 3}^{\text{Two factors}}$$

$$= 3^{4-2} = 3^2.$$

These and other properties of exponents are summarized on the following page.

Properties of Exponents

Let a and b be real numbers, variables, or algebraic expressions, and let m and n be integers. (Assume all denominators and bases are nonzero.)

Property	Example	
1. $a^m a^n = a^{m+n}$	$3^2 \cdot 3^4 = 3^{2+4} = 3^6$	Product of Powers
2. $\dfrac{a^m}{a^n} = a^{m-n}$	$\dfrac{x^7}{x^4} = x^{7-4} = x^3$	Quotient of Powers
3. $(ab)^m = a^m b^m$	$(5x)^3 = 5^3 x^3 = 125x^3$	Power of a Product
4. $\left(\dfrac{a}{b}\right)^m = \dfrac{a^m}{b^m}$	$\left(\dfrac{2}{x}\right)^3 = \dfrac{2^3}{x^3} = \dfrac{8}{x^3}$	Power of a Quotient
5. $(a^m)^n = a^{mn}$	$(y^3)^{-4} = y^{3(-4)} = y^{-12}$	Power of a Power
6. $a^{-n} = \dfrac{1}{a^n}$	$y^{-4} = \dfrac{1}{y^4}$	Definition of negative exponent
7. $a^0 = 1, \quad a \neq 0$	$(x^2 + 1)^0 = 1$	Definition of zero exponent
8. $\left(\dfrac{a}{b}\right)^{-n} = \left(\dfrac{b}{a}\right)^n, \quad a \neq 0, b \neq 0$	$\left(\dfrac{3}{2}\right)^{-3} = \left(\dfrac{2}{3}\right)^3$	
9. $|a^2| = |a|^2 = a^2$	$|2^2| = |2|^2 = 2^2$	

Notice that these properties of exponents apply for *all* integers m and n, not just positive integers. For instance, by the Quotient of Powers Property,

$$\frac{3^4}{3^{-5}} = 3^{4-(-5)} = 3^{4+5} = 3^9.$$

STUDY TIP

It is important to recognize the difference between exponential forms such as $(-2)^4$ and -2^4. In $(-2)^4$, the parentheses indicate that the exponent applies to the negative sign as well as to the 2, but in $-2^4 = -(2^4)$, the exponent applies only to the 2. Similarly, in $(5x)^3$, the parentheses indicate that the exponent applies to the 5 as well as to the x, whereas in $5x^3 = 5(x^3)$, the exponent applies only to the x.

Example 1   Using Properties of Exponents

a. $3^{-1} \cdot 3^4 = 3^{-1+4} = 3^3 = 27$

b. $\dfrac{5^6}{5^4} = 5^{6-4} = 5^2 = 25$

c. $5\left(\dfrac{2}{5}\right)^3 = 5 \cdot \dfrac{2^3}{5^3} = 5 \cdot 5^{-3} \cdot 2^3 = 5^{-2} \cdot 2^3 = \dfrac{2^3}{5^2} = \dfrac{8}{25}$

d. $(-5 \cdot 2^3)^2 = (-5)^2 \cdot (2^3)^2 = 25 \cdot 2^6 = 25 \cdot 64 = 1600$

e. $(-3ab^4)(4ab^{-3}) = -3(4)(a)(a)(b^4)(b^{-3}) = -12a^2 b$

f. $3a(-4a^2)^0 = 3a(1) = 3a, \quad a \neq 0$

g. $\left(\dfrac{5x^3}{y}\right)^2 = \dfrac{5^2(x^3)^2}{y^2} = \dfrac{25x^6}{y^2}$

✓ Checkpoint 1

Evaluate each expression.

a. $(2^4)^2$

b. $8^{-1} \cdot 4^3 \cdot 4^2$

c. $(-6)^0 + 3^{-1}$

Example 2   Rewriting with Positive Exponents

a. $x^{-1} = \dfrac{1}{x}$ Definition of negative exponent

b. $\dfrac{1}{3x^{-2}} = \dfrac{1(x^2)}{3}$ The exponent -2 applies only to x.

$= \dfrac{x^2}{3}$ Simplify.

c. $\dfrac{12a^3b^{-4}}{4a^{-2}b} = \dfrac{12a^3 \cdot a^2}{4b \cdot b^4}$ Definition of negative exponent

$= \dfrac{3a^5}{b^5}$ Product of Powers Property

d. $\left(\dfrac{3x^2}{y}\right)^{-2} = \dfrac{3^{-2}(x^2)^{-2}}{y^{-2}}$ Power of a Quotient and Power of a Product Properties

$= \dfrac{3^{-2}x^{-4}}{y^{-2}}$ Power of a Power Property

$= \dfrac{y^2}{3^2x^4}$ Definition of negative exponent

$= \dfrac{y^2}{9x^4}$ Simplify.

✓ **Checkpoint 2**

Rewrite $\left(\dfrac{3}{x^{-2}z^4}\right)^{-3}$ with positive exponents and simplify.

 Example 3   Ratio of Volume to Surface Area

The volume V and surface area S of a sphere are given by

$$V = \frac{4}{3}\pi r^3 \quad \text{and} \quad S = 4\pi r^2$$

where r is the radius of the sphere. A spherical weather balloon has a radius of 2 feet, as shown in Figure 0.7. Find the ratio of the volume to the surface area.

SOLUTION   To find the ratio, write the quotient V/S and simplify.

$$\frac{V}{S} = \frac{\frac{4}{3}\pi r^3}{4\pi r^2}$$

$$= \frac{\frac{4}{3}\pi 2^3}{4\pi 2^2}$$

$$= \frac{4\pi \cdot 2^3}{3 \cdot 4\pi \cdot 2^2}$$

$$= \frac{1}{3}(2) = \frac{2}{3}$$

✓ **Checkpoint 3**

Evaluate $\dfrac{\frac{5}{7}x^7}{25x^5}$ at $x = 7$.

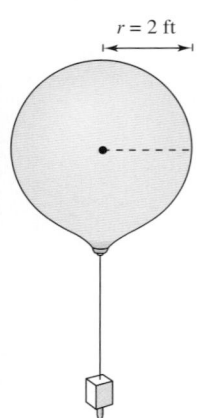

$r = 2$ ft

FIGURE 0.7

Scientific Notation and Calculators

Exponents provide an efficient way of writing and computing with very large (or very small) numbers. For instance, a drop of water contains more than 33 billion billion molecules—that is, 33 followed by 18 zeros.

$$33,000,000,000,000,000,000$$

It is convenient to write such numbers in **scientific notation.** This notation has the form

$$c \times 10^n$$

where $1 \le c < 10$ and n is an integer. So, the number of molecules in a drop of water can be written in scientific notation as

$$3.3 \times 10,000,000,000,000,000,000 = 3.3 \times 10^{19}.$$

The *positive* exponent 19 indicates that the number is *large* (10 or more) and that the decimal point has been moved 19 places. A *negative* exponent in scientific notation indicates that the number is *small* (less than 1). For instance, the mass (in grams) of one electron is approximately

$$9.0 \times 10^{-28} = 0.0000000000000000000000000009.$$

28 decimal places

Example 4 Converting to Scientific Notation

a. $0.0000572 = 5.72 \times 10^{-5}$ Number is less than 1.

b. $149,400,000 = 1.494 \times 10^8$ Number is greater than 10.

c. $32.675 = 3.2675 \times 10^1$ Number is greater than 10.

✓ **Checkpoint 4**

Write each number in scientific notation.

a. 6,318,000   **b.** 0.00345

Example 5 Converting to Decimal Notation

a. $3.125 \times 10^2 = 312.5$ Number is greater than 10.

b. $3.73 \times 10^{-6} = 0.00000373$ Number is less than 1.

c. $7.91 \times 10^5 = 791,000$ Number is greater than 10.

✓ **Checkpoint 5**

Write each number in decimal notation.

a. 4.28×10^5 **b.** 2.5×10^{-3}

Most calculators automatically use scientific notation when showing large (or small) numbers that exceed the display range. Try multiplying $86,500,000 \times 6000$. If your calculator follows standard conventions, then its display should be

5.19 11   or   5.19E11 .

This means that $c = 5.19$ and the exponent of 10 is $n = 11$, which implies that the number is 5.19×10^{11}. To *enter* numbers in scientific notation, your calculator should have an exponential entry key labeled EXP or EE.

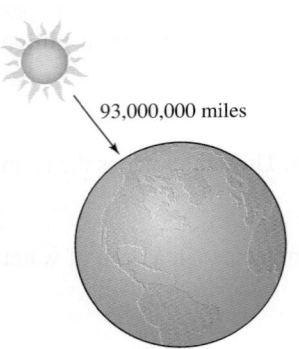

FIGURE 0.8

93,000,000 miles

Not drawn to scale

Example 6   The Speed of Light

The distance between Earth and the sun is approximately 93 million miles, as shown in Figure 0.8. How many minutes does it take for light to travel from the sun to Earth? Use the fact that light travels at a rate of approximately 186,000 miles per second.

SOLUTION   Use the formula *distance = (rate)(time)* to find the time.

$$\text{time} = \frac{\text{distance}}{\text{rate}} = \frac{93 \text{ million miles}}{186{,}000 \text{ miles per second}}$$

$$= \frac{9.3 \times 10^7 \text{ miles}}{1.86 \times 10^5 \text{ miles/second}}$$

$$= 5 \times 10^2 \text{ seconds}$$

$$= 500 \text{ seconds}$$

$$\approx 8.33 \text{ minutes}$$

Note that to convert 500 seconds to 8.33 minutes, you divide by 60 because there are 60 seconds in one minute.

✓ Checkpoint 6

The distance between Mars and the sun is approximately 142 million miles. How many minutes does it take for light to travel from the sun to Mars?

One of the most useful features of a calculator is its ability to evaluate exponential expressions. Consult your user's guide for specific keystrokes.

Example 7   Using a Calculator to Raise a Number to a Power

Scientific Calculator

Expression	Keystrokes	Display
a. $13^4 + 5$	13 y^x 4 $+$ 5 $=$	28566
b. $3^{-2} + 4^{-1}$	3 y^x 2 $+/-$ $+$ 4 y^x 1 $+/-$ $=$	0.361111111
c. $\dfrac{7^5 + 1}{7^5 - 1}$	(7 y^x 5 $+$ 1) $\div$ (7 y^x 5 $-$ 1) $=$	1.000119005

Graphing Calculator

Expression	Keystrokes	Display
a. $13^4 + 5$	13 $\wedge$ 4 $+$ 5 (ENTER)	28566
b. $3^{-2} + 4^{-1}$	3 $\wedge$ $(-)$ 2 $+$ 4 $\wedge$ $(-)$ 1 (ENTER)	.3611111111
c. $\dfrac{7^5 + 1}{7^5 - 1}$	(7 $\wedge$ 5 $+$ 1) $\div$ (7 $\wedge$ 5 $-$ 1) (ENTER)	1.000119005

TECH TUTOR

Make sure you include parentheses as needed when entering expressions in your calculator. Notice the use of parentheses in Example 7(c).

✓ Checkpoint 7

Use a calculator to evaluate each expression.

a. $5^{-3} - 2^{-4}$

b. $\dfrac{4^4 - 6}{2^5 + 18}$

Applications

One of the most familiar investment problems is finding the balance in a savings account. The formulas below can be used to find the balance in a savings account that earns simple interest or compound interest. You will learn more about interest in Chapter 4.

Balance in an Account

The balance A in an account that earns an annual interest rate r (in decimal form) for t years is given by one of the following.

$$A = P(1 + rt)$$ Simple interest

$$A = P\left(1 + \frac{r}{n}\right)^{nt}$$ Compound interest

In both formulas, P is the principal (or the initial deposit). In the formula for compound interest, n is the number of compoundings *per year*. Make sure you convert all units of time t to years. For instance, 6 months $= \frac{1}{2}$ year. So, $t = \frac{1}{2}$.

 Example 8   Finding the Balance in an Account

A family receives a tax refund of $6000 and wants to invest it for 5 years. Which savings plan will earn more money?

a. 2% simple annual interest

b. 1.5% annual interest compounded quarterly

SOLUTION

a. The balance after 5 years is

$$A = P(1 + rt)$$

$$= 6000[1 + 0.02(5)]$$

$$= \$6600.$$

b. The balance after 5 years is

$$A = P\left(1 + \frac{r}{n}\right)^{nt}$$

$$= 6000\left(1 + \frac{0.015}{4}\right)^{(4)(5)}$$

$$\approx \$6466.40.$$

Savings plan (a) will earn

$$6600 - 6000 = \$600$$

and savings plan (b) will earn

$$6466.40 - 6000 = \$466.40.$$

So, plan (a) will earn more money.

✓Checkpoint 8

In Example 8, how much money would the family earn in a savings plan with 3.4% annual interest compounded monthly?

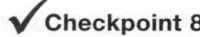

Yuri Arcurs/Shutterstock.com

The compound interest formula can also be used to determine the rate of inflation. To apply the formula, you must know the costs of an item in two different years, as demonstrated in Example 9.

 **Example 9**   **Finding the Rate of Inflation**

In 1984, the cost of a first class postage stamp was $0.20. By 2010, the cost had increased to $0.44, as shown in Figure 0.9. Find the average annual rate of inflation for a first class postage stamp over this 26-year period.   *(Source: U.S. Postal Service)*

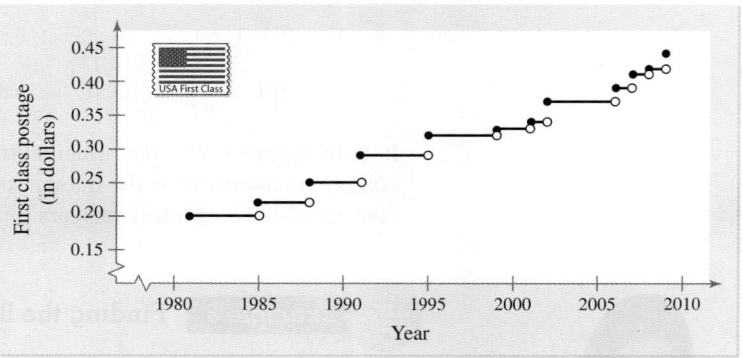

FIGURE 0.9

SOLUTION   To find the average annual rate of inflation, use the formula for compound interest with *annual* compounding. So, find the value of r that will make the following equation true.

$$A = P\left(1 + \frac{r}{n}\right)^{nt}$$

$$0.44 = 0.20(1 + r)^{26}$$

You can begin by guessing that the average annual rate of inflation was 4%. Entering $r = 0.04$ in the formula, you find that $0.20(1 + 0.04)^{26} \approx 0.5545$. Because this result is more than 0.44, try some smaller values of r. Finally, you can discover that

$$0.20(1 + 0.031)^{26} \approx 0.44.$$

So, the average annual rate of inflation for a first class postage stamp from 1984 to 2010 was about 3.1%.

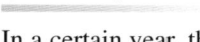**Checkpoint 9**

In a certain year, the registration fee for the Medical College Admission Test (MCAT) was $210. Four years later, the fee was $235. What was the average annual rate of inflation over this four-year period?

SUMMARIZE   (Section 0.3)

1. Make a list of the properties of exponents *(page 21)*. For examples that use these properties, see Examples 1, 2, and 3.

2. State the definition of scientific notation *(page 23)*. For examples of scientific notation, see Examples 4, 5, and 6.

3. State the formulas for simple and compound interest *(page 25)*. For applications of these formulas, see Examples 8 and 9.

In Exercises 1–10, perform the indicated operation(s) and simplify.

1. $\left(\frac{2}{3}\right)\left(\frac{3}{2}\right)$

2. $\left(\frac{1}{4}\right)(5)(4)$

3. $3\left(-\frac{2}{7}\right) + 11\left(-\frac{2}{7}\right)$

4. $11\left(-\frac{1}{4}\right) + \frac{5}{4}$

5. $\frac{1}{2} \div (-2)$

6. $-\frac{1}{3} \div \left(-\frac{1}{3}\right)$

7. $\frac{1}{7} + \frac{1}{3} - \frac{1}{21}$

8. $\frac{1}{3} + \frac{1}{2} - \frac{5}{6}$

9. $\frac{1}{12} - \left(\frac{1}{3} + \frac{1}{8}\right)$

10. $\left(\frac{1}{2} - \frac{1}{3}\right) \div \frac{1}{6}$

Exercises 0.3

See www.CalcChat.com for worked-out solutions to odd-numbered exercises.

Using Properties of Exponents In Exercises 1–14, evaluate the expression. Write fractional answers in simplest form. *See Example 1.*

1. $2^2 \cdot 2^4$

2. $3 \cdot 3^5$

3. $\dfrac{2^6}{2^3}$

4. $\dfrac{5^7}{5^5}$

5. $(3^3)^2$

6. $(2^5)^3$

7. -3^4

8. $(-3)^4$

9. $(2^3 \cdot 3^2)^2$

10. $(-3 \cdot 4^2)^3$

11. $\left(-\dfrac{3}{5}\right)^3\left(\dfrac{5}{3}\right)^2$

12. $\left(\dfrac{-5}{4}\right)^3\left(\dfrac{4}{5}\right)^2$

13. 3^0

14. $(-2)^0$

Using Properties of Exponents In Exercises 15–36, simplify the expression. *See Example 1.*

15. $(-5z)^3$

16. $(-2w)^5$

17. $(8x^4)(2x^3)$

18. $5x^4(x^2)$

19. $10(x^2)^2$

20. $2(4x^4)^3$

21. $(-z)^3(3z^4)$

22. $(6y^2)(2y^3)^3$

23. $\dfrac{25y^8}{15y^4}$

24. $\dfrac{10x^9}{4x^6}$

25. $\left(\dfrac{4}{y}\right)^3\left(\dfrac{3}{y}\right)^4$

26. $\left(\dfrac{5}{z}\right)^2\left(\dfrac{2}{z}\right)^3$

27. $\dfrac{7x^2}{x^3}$

28. $\dfrac{5z^5}{z^7}$

29. $\dfrac{x^2 \cdot x^n}{x^3 \cdot x^n}$

30. $\dfrac{x^n \cdot x^{2n}}{x^{3n}}$

31. $3^n \cdot 3^{2n}$

32. $2^m \cdot 2^{3m}$

33. $\dfrac{15(x+3)^3}{9(x+3)^2}$

34. $\dfrac{24(x-2)^5}{8(x-2)^4}$

35. $(2x^5)^0, \quad x \neq 0$

36. $(x+5)^0, \quad x \neq -5$

Rewriting with Positive Exponents In Exercises 37–50, rewrite the expression with positive exponents and simplify. *See Examples 1 and 2.*

37. $8 \cdot (-2)^{-2} \cdot 4^{-1}$

38. $6 \cdot 2^{-3} \cdot 3^{-1}$

39. $\left(\frac{1}{2}\right)^{-3}$

40. $\left(\frac{2}{3}\right)^{-3}$

41. $5^{-1} - 2^{-1}$

42. $4^{-1} - 2^{-2}$

43. $4x^{-2}$

44. $-7x^{-1}$

45. $\dfrac{9a^2b^{-2}}{3a^{-1}b}$

46. $\dfrac{10x^4y^{-4}}{5x^2y^{-2}}$

47. $\left(\dfrac{x}{10}\right)^{-1}$

48. $\left(\dfrac{y}{5}\right)^{-2}$

49. $\left(\dfrac{x^{-3}y^4}{5}\right)^{-3}$

50. $\left(\dfrac{2z^2}{y}\right)^{-2}$

Evaluating Expressions with Exponents In Exercises 51–54, evaluate the expression for the indicated value of *x*. *See Example 3.*

Expression	Value	Expression	Value
51. $\dfrac{x^4}{2x^2}$	$x = -6$	**52.** $4x^{-3}$	$x = 2$
53. $7x^{-2}$	$x = 4$	**54.** $8x^0 - (8x)^0$	$x = -7$

55. Discovery Using your calculator, find the values of $10^3, 10^2, 10^1, 10^0, 10^{-1}$, and 10^{-2}. What do you notice?

56. Error Analysis Describe the error.

$10^{-5} = 0.000001$

Converting to Scientific Notation In Exercises 57–62, write the number in scientific notation. *See Example 4.*

57. Land Area of Earth: 57,300,000 square miles

58. Water Area of Earth: 139,500,000 square miles

59. Light Year: 9,460,000,000,000 kilometers

60. Mass of a Bacterium: 0.0000000000000003 gram

61. Thickness of a Soap Bubble: 0.0000001 meter

62. One Micron (millionth of a meter): 0.00003937 inch

Converting to Decimal Notation   In Exercises 63–66, write the number in decimal notation. *See Example 5.*

63. Number of Air Sacs in the Lungs: 3.5×10^8

64. Temperature of the Core of the Sun:
1.5×10^7 degrees Celsius

65. Charge of an Electron: 1.602×10^{-19} coulomb

66. Width of a Human Hair: 9.0×10^{-5} meter

Operations with Scientific Notation   In Exercises 67 and 68, evaluate each expression without using a calculator. *See Example 6.*

67. (a) $(1.2 \times 10^7)(5 \times 10^{-3})$ (b) $\dfrac{6.0 \times 10^8}{3.0 \times 10^{-3}}$

68. (a) $(9.8 \times 10^{-2})(3 \times 10^7)$ (b) $\dfrac{9.0 \times 10^5}{4.5 \times 10^{-2}}$

Operations with Scientific Notation   In Exercises 69 and 70, write each number in scientific notation. Perform the operation and write your answer in scientific notation.

69. (a) $48,000,000,000(250,000,000)$

(b) $\dfrac{0.000000012}{0.0000064}$

70. (a) $0.00000034(0.00000006)$ (b) $\dfrac{18,000,000,000}{2,400,000}$

Using a Calculator to Raise a Number to a Power
In Exercises 71–76, use a calculator to evaluate the expression. *See Example 7.*

71. $11^5 + 49$

72. $17^3 + 5^3$

73. $2^{-3} + 5^{-1}$

74. $3^{-4} + 64^{-1}$

75. $\dfrac{4^4 + 1}{4^4 - 1}$

76. $\dfrac{5^4 - 1}{5^4 + 1}$

Using a Calculator   In Exercises 77–80, use a calculator to evaluate each expression. Write your answer in scientific notation. (Round to three decimal places.)

77. (a) $0.000345(8,900,000,000)$

(b) $\dfrac{67,000,000 + 93,000,000}{0.0052}$

78. (a) $0.000045(9,200,000)$

(b) $\dfrac{0.0000928 - 0.0000021}{0.0061}$

79. (a) $(9.3 \times 10^6)^3(6.1 \times 10^{-4})$ (b) $\dfrac{(2.414 \times 10^4)^6}{(1.68 \times 10^5)^5}$

80. (a) $(1.2 \times 10^2)^2(5.3 \times 10^{-5})$ (b) $\dfrac{(3.28 \times 10^{-6})^{10}}{(5.34 \times 10^{-3})^{22}}$

81. Balance in an Account   You deposit $10,000 in an account with an annual interest rate of 3.75%. Determine the balance in the account after 12 years when the interest is compounded (a) daily ($n = 365$), (b) weekly, (c) monthly, and (d) quarterly. How is the balance affected by the type of compounding?

82. **HOW DO YOU SEE IT?**   Each bar graph represents the value y of a $1000 investment after x years. One investment earns simple interest and the other earns compound interest. Which type of interest does each graph show? Explain.

(a)
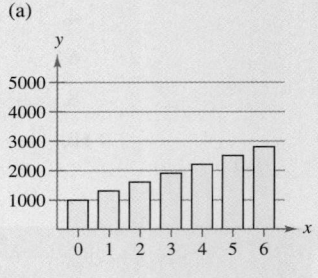

(b)
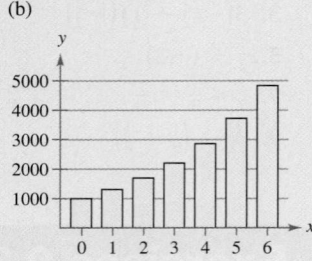

83. College Costs   The average yearly cost of tuition, fees, and room and board at private four-year colleges in the United States was $22,896 for the academic year 2001/2002 and $30,778 for the academic year 2007/2008. Find the average annual rate of inflation over this seven-year period. *(Source: U.S. National Center for Education Statistics)*

84. Becoming a Millionaire   The compound interest formula can be rewritten as

$$P = \frac{A}{(1 + r/n)^{nt}}.$$

Find the principal P that you need to invest on the day your child is born at 4.5% annual interest compounded quarterly to make your child a millionaire on his or her 21st birthday.

85. The Speed of Light   The closest distance from Earth to Neptune is 2.68 billion miles. At this distance, how long does it take for the image of Neptune to reach Earth, given that light travels about 186,000 miles per second?

86. Electron Microscopes   Electron microscopes provide greater magnification than traditional light microscopes by using focused beams of electrons instead of visible light. It is the extremely short wavelengths of the electron beams that make electron microscopes so powerful. The wavelength λ (in meters) of any object in motion is given by

$$\lambda = \frac{6.626 \times 10^{-34}}{mv}$$

where m is the mass (in kilograms) of the object and v is its velocity (in meters per second). Find the wavelength of an electron with a mass of 9.11×10^{-31} kilogram and a velocity of 5.9×10^6 meters per second.

(Submitted by Brian McIntyre, Senior Laboratory Engineer for the Optics Electron Microscopy Facility at the University of Rochester.)

0.4 Radicals and Rational Exponents

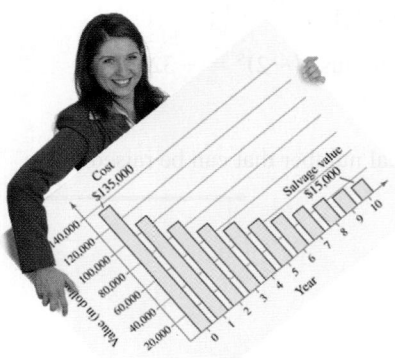

■ Evaluate expressions involving radicals and develop an understanding of the properties of radicals.

■ Simplify a radical and rationalize a denominator.

■ Use properties of rational exponents and use a calculator to evaluate a radical.

Radicals and Properties of Radicals

A **square root** of a number is defined as one of its two equal factors. For example, 5 is a square root of 25 because 5 is one of the two equal factors of 25. In a similar way, a **cube root** of a number is one of its three equal factors. Here are some examples.

Number	*Equal Factors*	*Root*	
$25 = (-5)^2$	$(-5)(-5)$	-5	(square root)
$-64 = (-4)^3$	$(-4)(-4)(-4)$	-4	(cube root)
$81 = 3^4$	$3 \cdot 3 \cdot 3 \cdot 3$	3	(fourth root)

In Exercise 100 on page 38, you will use a bar graph to find the annual depreciation rate of an ultrasound machine.

Definition of *n*th Root of a Number

Let a and b be real numbers and let n be a positive integer. If

$$a = b^n$$

then b is an ***n*th root of *a*.** If

$$n = 2$$

the root is a **square root,** and if

$$n = 3$$

the root is a **cube root.**

From this definition, you can see that some numbers have more than one *n*th root. For example, both 5 and -5 are square roots of 25. The following definition distinguishes between these two roots.

Principal *n*th Root of a Number

Let a be a real number that has at least one real *n*th root. **The principal *n*th root of *a*** is the *n*th root that has the same sign as a, and it is denoted by the **radical symbol**

$$\sqrt[n]{a}. \qquad \text{Principal } n\text{th root}$$

The positive integer n is the **index** (the plural of index is indexes or indices) of the radical, and the number a is the **radicand.**

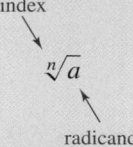

If $n = 2$, omit the index and write $\sqrt{a}$ rather than $\sqrt[2]{a}$.

| Example 1 | Evaluating Expressions Involving Radicals |

a. The principal square root of 121 is $\sqrt{121} = 11$ because $11^2 = 121$.

b. The principal cube root of $\frac{125}{64}$ is $\sqrt[3]{\frac{125}{64}} = \frac{5}{4}$ because $\left(\frac{5}{4}\right)^3 = \frac{5^3}{4^3} = \frac{125}{64}$.

c. The principal fifth root of -32 is $\sqrt[5]{-32} = -2$ because $(-2)^5 = -32$.

d. $-\sqrt{49} = -7$ because $7^2 = 49$.

e. $\sqrt[4]{-81}$ is not a real number because there is no real number that can be raised to the fourth power to produce -81.

✓**Checkpoint 1**

Evaluate each expression.

a. $\sqrt{\dfrac{1}{49}}$

b. $\sqrt[4]{625}$

c. $-\sqrt[3]{-8}$

From Example 1, you can make the following generalizations about nth roots of a real number.

1. If a is a positive real number and n is a positive *even* integer, then a has exactly two real nth roots, which are denoted by $\sqrt[n]{a}$ and $-\sqrt[n]{a}$.

2. If a is any real number and n is an *odd* integer, then a has only one (real) nth root. It is the principal nth root and is denoted by $\sqrt[n]{a}$.

3. If a is negative and n is an *even* integer, then a has no (real) nth root.

Integers such as 1, 4, 9, 16, 49, and 81 are called **perfect squares** because they have integer square roots. Similarly, integers such as 1, 8, 27, 64, and 125 are called **perfect cubes** because they have integer cube roots.

Properties of Radicals

Let a and b be real numbers, variables, or algebraic expressions such that the indicated roots are real numbers, and let m and n be positive integers. Then the following properties are true.

Property	*Example*
1. $\sqrt[n]{a^m} = \left(\sqrt[n]{a}\right)^m$	$\sqrt[3]{8^2} = \left(\sqrt[3]{8}\right)^2 = (2)^2 = 4$
2. $\sqrt[n]{a} \cdot \sqrt[n]{b} = \sqrt[n]{ab}$	$\sqrt{5} \cdot \sqrt{7} = \sqrt{5 \cdot 7} = \sqrt{35}$
3. $\dfrac{\sqrt[n]{a}}{\sqrt[n]{b}} = \sqrt[n]{\dfrac{a}{b}},\ b \neq 0$	$\dfrac{\sqrt[4]{27}}{\sqrt[4]{9}} = \sqrt[4]{\dfrac{27}{9}} = \sqrt[4]{3}$
4. $\sqrt[m]{\sqrt[n]{a}} = \sqrt[mn]{a}$	$\sqrt[3]{\sqrt{10}} = \sqrt[6]{10}$
5. $\left(\sqrt[n]{a}\right)^n = a$	$\left(\sqrt{3}\right)^2 = 3$
6. For n even, $\sqrt[n]{a^n} = |a|$.	$\sqrt{(-12)^2} = |-12| = 12$
For n odd, $\sqrt[n]{a^n} = a$.	$\sqrt[3]{(-12)^3} = -12$

A common special case of Property 6 is $\sqrt{a^2} = |a|$.

Simplifying Radicals

Radical Expressions in Simplest Form

An expression involving radicals is in **simplest form** when the following conditions are satisfied.

1. All possible factors have been removed from the radical.

2. All fractions have radical-free denominators (accomplished by a process called *rationalizing the denominator*).

3. The index of the radical has been reduced as far as possible.

To simplify a radical, factor the radicand into factors whose exponents are multiples of the index. The roots of these factors are written outside the radical, and the "leftover" factors make up the new radicand.

Example 2 Simplifying Even Roots

a. $\sqrt[4]{48} = \sqrt[4]{16 \cdot 3}$ Find largest fourth-power factor.

$\qquad\quad = \sqrt[4]{2^4 \cdot 3}$ Rewrite.

$\qquad\quad = 2\sqrt[4]{3}$ Find fourth root.

b. $\sqrt{75x^3} = \sqrt{25x^2 \cdot 3x}$ Find largest square factor.

$\qquad\quad = \sqrt{(5x)^2 \cdot 3x}$ Rewrite.

$\qquad\quad = 5x\sqrt{3x}, \quad x \geq 0$ Find root of perfect square.

c. $\sqrt[4]{(5x)^4} = |5x| = 5|x|$

d. $\sqrt{\dfrac{7x^2}{9}} = \dfrac{\sqrt{7x^2}}{\sqrt{9}} = \dfrac{\sqrt{7}|x|}{3}$

✓**Checkpoint 2**

Simplify $\sqrt{18x^5}$.

Example 3 Simplifying Odd Roots

a. $\sqrt[3]{24} = \sqrt[3]{8 \cdot 3}$ Find largest cube factor.

$\qquad\quad = \sqrt[3]{2^3 \cdot 3}$ Rewrite.

$\qquad\quad = 2\sqrt[3]{3}$ Find root of perfect cube.

b. $\sqrt[5]{32a^{11}} = \sqrt[5]{32a^{10} \cdot a}$ Find largest fifth-power factor.

$\qquad\quad = \sqrt[5]{(2a^2)^5 \cdot a}$ Rewrite.

$\qquad\quad = 2a^2\sqrt[5]{a}$ Find fifth root.

c. $\sqrt[3]{-40x^6} = \sqrt[3]{(-8x^6) \cdot 5}$ Find largest cube factor.

$\qquad\quad = \sqrt[3]{(-2x^2)^3 \cdot 5}$ Rewrite.

$\qquad\quad = -2x^2\sqrt[3]{5}$ Find root of perfect cube.

✓**Checkpoint 3**

Simplify $\sqrt[3]{54x^4}$.

Some fractions have radicals in the denominator. To **rationalize a denominator** of the form $a + b\sqrt{m}$, multiply the numerator and denominator by the **conjugate** $a - b\sqrt{m}$.

$$a + b\sqrt{m} \quad \text{and} \quad a - b\sqrt{m} \qquad \text{Conjugates}$$

When $a = 0$, the rationalizing factor of $\sqrt{m}$ is itself, $\sqrt{m}$.

Example 4   Rationalizing Single-Term Denominators

a. To rationalize the denominator of the following fraction, multiply *both* the numerator and the denominator by $\sqrt{3}$ to obtain

$$\frac{5}{2\sqrt{3}} = \frac{5}{2\sqrt{3}} \cdot \frac{\sqrt{3}}{\sqrt{3}}$$

$$= \frac{5\sqrt{3}}{2\sqrt{3^2}}$$

$$= \frac{5\sqrt{3}}{2(3)} = \frac{5\sqrt{3}}{6}.$$

b. To rationalize the denominator of the following fraction, multiply *both* the numerator and the denominator by $\sqrt[3]{5^2}$. Note how this eliminates the radical from the denominator by producing a perfect *cube* in the radicand.

$$\frac{2}{\sqrt[3]{5}} = \frac{2}{\sqrt[3]{5}} \cdot \frac{\sqrt[3]{5^2}}{\sqrt[3]{5^2}}$$

$$= \frac{2\sqrt[3]{5^2}}{\sqrt[3]{5^3}}$$

$$= \frac{2\sqrt[3]{25}}{5}$$

✓**Checkpoint 4**

Simplify $\dfrac{1}{\sqrt[3]{4}}$ by rationalizing the denominator. ■

Example 5   Rationalizing a Denominator with Two Terms

$$\frac{2}{3 + \sqrt{7}} = \frac{2}{3 + \sqrt{7}} \cdot \frac{3 - \sqrt{7}}{3 - \sqrt{7}} \qquad \text{Multiply numerator and denominator by conjugate.}$$

$$= \frac{2(3 - \sqrt{7})}{3^2 - (\sqrt{7})^2} \qquad \text{Multiply fractions.}$$

$$= \frac{2(3 - \sqrt{7})}{9 - 7} \qquad \text{Simplify.}$$

$$= \frac{\cancel{2}(3 - \sqrt{7})}{\cancel{2}} \qquad \text{Divide out common factors.}$$

$$= 3 - \sqrt{7} \qquad \text{Simplify.}$$

STUDY TIP

Don't confuse an expression such as $\sqrt{2} + \sqrt{7}$ with $\sqrt{2 + 7}$. In general, $\sqrt{x + y} \neq \sqrt{x} + \sqrt{y}$.

✓**Checkpoint 5**

Simplify $\dfrac{10}{4 - \sqrt{11}}$ by rationalizing the denominator. ■

Rational Exponents and Calculators

The definition below shows how radicals are used to define **rational exponents**. Until now, work with exponents has been restricted to integer exponents.

> **Definition of Rational Exponents**
>
> If a is a real number and n is a positive integer such that the principal nth root of a exists, then $a^{1/n}$ is defined to be $a^{1/n} = \sqrt[n]{a}$.
>
> If m is a positive integer that has no common factor with n, then $a^{m/n} = (a^{1/n})^m = \left(\sqrt[n]{a}\right)^m$ and $a^{m/n} = (a^m)^{1/n} = \sqrt[n]{a^m}$.

The properties of exponents mentioned in Section 0.3 also apply to rational exponents (provided the roots indicated by the denominators exist). Some of those properties are relisted here, with different examples.

> **Properties of Exponents**
>
> Let r and s be rational numbers, and let a and b be real numbers, variables, or algebraic expressions. If the roots indicated by the rational exponents exist, then the following properties are true.
>
> | *Property* | *Example* |
> |---|---|
> | **1.** $a^r a^s = a^{r+s}$ | $4^{1/2}(4^{1/3}) = 4^{5/6}$ |
> | **2.** $\dfrac{a^r}{a^s} = a^{r-s}, \quad a \neq 0$ | $\dfrac{x^2}{x^{1/2}} = x^{2-(1/2)} = x^{3/2}, x \neq 0$ |
> | **3.** $(ab)^r = a^r b^r$ | $(2x)^{1/2} = 2^{1/2}(x^{1/2})$ |
> | **4.** $\left(\dfrac{a}{b}\right)^r = \dfrac{a^r}{b^r}, \quad b \neq 0$ | $\left(\dfrac{x}{3}\right)^{1/3} = \dfrac{x^{1/3}}{3^{1/3}}$ |
> | **5.** $(a^r)^s = a^{rs}$ | $(x^3)^{1/3} = x$ |
> | **6.** $a^{-r} = \dfrac{1}{a^r}, \quad a \neq 0$ | $4^{-1/2} = \dfrac{1}{4^{1/2}} = \dfrac{1}{2}$ |
> | **7.** $\left(\dfrac{a}{b}\right)^{-r} = \left(\dfrac{b}{a}\right)^r, \quad a \neq 0, \quad b \neq 0$ | $\left(\dfrac{x}{4}\right)^{-1/2} = \left(\dfrac{4}{x}\right)^{1/2} = \dfrac{2}{x^{1/2}}$ |

Example 6   Simplifying with Rational Exponents

a. $(27)^{1/3} = \sqrt[3]{27}$

$\qquad\qquad\ = 3$

b. $(-32)^{-4/5} = \left(\sqrt[5]{-32}\right)^{-4} = (-2)^{-4} = \dfrac{1}{(-2)^4} = \dfrac{1}{16}$

c. $(-5x^{2/3})(3x^{-1/3}) = -15x^{(2/3)-(1/3)}$

$\qquad\qquad\qquad\qquad\ = -15x^{1/3}, \quad x \neq 0$

✓ **Checkpoint 6**

Simplify $(3^{1/2})(3^{3/2})$.

Example 7   Reducing the Index of a Radical

a. $\sqrt[6]{a^4} = a^{4/6} = a^{2/3} = \sqrt[3]{a^2}$

b. $\sqrt[3]{\sqrt{125}} = (125^{1/2})^{1/3}$ Rewrite with rational exponents.

$\qquad\qquad = (125)^{1/6}$ Multiply exponents.

$\qquad\qquad = (5^3)^{1/6}$ Rewrite base as perfect cube.

$\qquad\qquad = 5^{3/6}$ Multiply exponents.

$\qquad\qquad = 5^{1/2}$ Reduce exponent.

$\qquad\qquad = \sqrt{5}$ Rewrite as radical.

✓ Checkpoint 7

Use rational exponents to reduce the index of the radical $\sqrt[6]{2^3}$.   ■

Radical expressions can be combined (added or subtracted) if they are **like radicals**—that is, if they have the same index and radicand. For instance, $2\sqrt{3x}$ and $\frac{1}{2}\sqrt{3x}$ are like radicals, but $\sqrt[3]{3x}$ and $2\sqrt{3x}$ are not like radicals.

Example 8   Simplifying by Combining Like Radicals

a. $2\sqrt{48} + 3\sqrt{27} = 2\sqrt{16 \cdot 3} + 3\sqrt{9 \cdot 3}$ Find square factors.

$\qquad\qquad\quad = 2 \cdot 4\sqrt{3} + 3 \cdot 3\sqrt{3}$ Find square roots.

$\qquad\qquad\quad = 8\sqrt{3} + 9\sqrt{3}$ Simplify.

$\qquad\qquad\quad = 17\sqrt{3}$ Combine like terms.

b. $\sqrt[3]{16x} - \sqrt[3]{54x} = \sqrt[3]{8 \cdot 2x} - \sqrt[3]{27 \cdot 2x}$ Find cube factors.

$\qquad\qquad\quad = 2\sqrt[3]{2x} - 3\sqrt[3]{2x}$ Find cube roots.

$\qquad\qquad\quad = -\sqrt[3]{2x}$ Combine like terms.

✓ Checkpoint 8

Simplify the expression $\sqrt{25x} + \sqrt{x}$.   ■

You can use a calculator to evaluate radicals by using the square root key , the cube root key ⬚, or the xth root key ⬚. You can also use the exponential key ⬚ or ⬚. To use these keys, first convert the radical to exponential form.

Example 9   Evaluating a Cube Root with a Calculator

Two ways to evaluate $\sqrt[3]{25}$ using a calculator are shown below.

25 ⬚ ⬚ 1 ⬚ 3 ⬚ ⬚   Exponential key

⬚ 25 ⬚ ENTER   Cube root key

Most calculators will display 2.924017738. So, $\sqrt[3]{25} \approx 2.924$.

✓ Checkpoint 9

Use a calculator to evaluate $\sqrt[3]{18}$. Round to three decimal places.   ■

Example 10 **Evaluating Radicals with a Calculator**

a. Use the following keystroke sequence to evaluate $\sqrt[3]{-4}$.

4 [+/−] [y^x] [(] 1 [÷] 3 [)] [=]   Scientific

[(−)] 4 [∧] [(] 1 [÷] 3 [)] [ENTER]   Graphing

The display is -1.587401052. So, $\sqrt[3]{-4} \approx -1.587$.

b. Use the following keystroke sequence to evaluate $(1.4)^{-2/5}$.

1.4 [y^x] [(] 2 [÷] 5 [+/−] [)] [=]   Scientific

1.4 [∧] [(] [(−)] 2 [÷] 5 [)] [ENTER]   Graphing

The display is 0.874075175. So, $(1.4)^{-2/5} \approx 0.874$.

✓**Checkpoint 10**

Use a calculator to evaluate $6.7^{3/4}$. Round to three decimal places.

 Example 11 **Escape Velocity**

A rocket, launched vertically from Earth, has a velocity of 10,000 meters per second. The *escape velocity*, or the minimum velocity necessary for the rocket to escape the gravitational field of Earth, is

$$\sqrt{\frac{2(6.67 \times 10^{-11})(5.98 \times 10^{24})}{6.37 \times 10^{6}}}$$ meters per second.

Will the rocket escape Earth's gravitational field?

SOLUTION   The escape velocity is

$$\sqrt{\frac{2(6.67 \times 10^{-11})(5.98 \times 10^{24})}{6.37 \times 10^{6}}} \approx 11{,}190.7 \text{ meters per second.}$$

The velocity of 10,000 meters per second is less than the escape velocity of 11,190.7 meters per second. So, the rocket will not escape Earth's gravitational field.

✓**Checkpoint 11**

Will an object traveling at 10,000 meters per second exceed the escape velocity of Venus, which is

$$\sqrt{\frac{2(6.67 \times 10^{-11})(4.87 \times 10^{24})}{6.05 \times 10^{6}}}$$

meters per second?

SUMMARIZE   (Section 0.4)

1. State the definitions of the *n*th root and principal *n*th root of a number *(page 29)*. For an example of finding roots of numbers, see Example 1.

2. State what is meant by a radical expression in simplest form *(page 31)*. For examples of simplifying radical expressions, see Examples 2, 3, 4, and 5.

3. State the definition of rational exponents *(page 33)*. For examples of simplifying expressions with rational exponents, see Examples 6 and 7.

SKILLS WARM UP 0.4   The following warm-up exercises involve skills that were covered in earlier sections. You will use these skills in the exercise set for this section. For additional help, review Section 0.3.

In Exercises 1–10, simplify the expression.

1. $\left(\frac{1}{3}\right)\left(\frac{2}{3}\right)^2$

2. $3(-4)^2$

3. $(-2x)^3$

4. $(-2x^3)(-3x^4)$

5. $(7x^5)(4x)$

6. $(5x^4)(25x^2)^{-1}$

7. $\dfrac{12z^6}{4z^2}$

8. $\left(\dfrac{2x}{5}\right)^2\left(\dfrac{2x}{5}\right)^{-4}$

9. $\left(\dfrac{3y^2}{x}\right)^0, \quad x \neq 0,\ y \neq 0$

10. $[(x+2)^2(x+2)^3]^2$

Exercises 0.4

See www.CalcChat.com for worked-out solutions to odd-numbered exercises.

Finding the nth Roots of a Number   In Exercises 1–10, use a radical symbol to denote and evaluate the principal nth root (if it exists) of the number. Then find any other real nth roots of the number. *See Example 1.*

Number	n		Number	n
1. 4	2	**2.** 1		2
3. −8	3	**4.** 8		3
5. 27	3	**6.** −27		3
7. −1	4	**8.** −9		2
9. 64	6	**10.** 81		4

Evaluating an Expression Involving Radicals   In Exercises 11–20, evaluate the expression. *See Example 1.*

11. $\sqrt{9}$

12. $\sqrt[3]{64}$

13. $-\sqrt[3]{-27}$

14. $\sqrt[3]{0}$

15. $\dfrac{14}{\sqrt{49}}$

16. $\dfrac{\sqrt[4]{81}}{6}$

17. $\sqrt[4]{-16}$

18. $\sqrt[6]{-1}$

19. $\sqrt[3]{\dfrac{-27}{125}}$

20. $\sqrt[4]{\dfrac{16}{81}}$

Simplifying a Radical Expression   In Exercises 21–30, simplify the expression. *See Examples 2 and 3.*

21. $\sqrt[3]{-54}$

22. $\sqrt[4]{162}$

23. $\left(\sqrt[3]{-125}\right)^3$

24. $\sqrt[4]{562^4}$

25. $\sqrt[3]{16x^5}$

26. $\sqrt[5]{96x^5}$

27. $\sqrt{(4x^3)^2}$

28. $\sqrt[4]{(3x^2)^4}$

29. $\sqrt{\dfrac{z^2}{25}}$

30. $\sqrt[5]{\dfrac{64}{y^5}}$

Evaluating an Expression Involving Radicals   In Exercises 31–36, evaluate the expression at $x = 2$, $y = 3$, and $z = 5$.

31. $\sqrt{2xy^4z^2}$

32. $\sqrt{3x^2yz^6}$

33. $\sqrt[4]{16x^{-4}y^8z^4}$

34. $\sqrt[5]{243x^5y^{-5}z^{15}}$

35. $\sqrt[3]{z} \cdot \sqrt[3]{z^2}$

36. $\sqrt[5]{y^7} \cdot \sqrt[5]{y^3}$

Rationalizing a Denominator   In Exercises 37–44, rewrite the expression by rationalizing the denominator. Simplify your answer. *See Examples 4 and 5.*

37. $\dfrac{1}{\sqrt{5}}$

38. $\dfrac{5}{\sqrt{10}}$

39. $\dfrac{8}{\sqrt[3]{2}}$

40. $\dfrac{5}{\sqrt[3]{(5x)^2}}$

41. $\dfrac{2x}{5 - \sqrt{3}}$

42. $\dfrac{5x}{\sqrt{14} - 2}$

43. $\dfrac{3}{\sqrt{5} + \sqrt{6}}$

44. $\dfrac{5}{2\sqrt{10} - 5}$

Converting Between Forms   In Exercises 45–54, fill in the missing form.

Radical Form	Rational Exponent Form
45. $\sqrt{9} = 3$	
46. $\sqrt[3]{125} = 5$	
47.	$32^{1/5} = 2$
48.	$-(144^{1/2}) = -12$
49. $\sqrt[3]{-216} = -6$	
50. $\sqrt[5]{-243} = -3$	
51. $\left(\sqrt[4]{81}\right)^3 = 27$	
52. $\sqrt[4]{81^3} = 27$	
53.	$125^{2/3} = 25$
54.	$16^{5/4} = 32$

Simplifying with Rational Exponents In Exercises 55–70, simplify the expression. *See Example 6.*

55. $16^{1/2}$

56. $125^{1/3}$

57. $\left(\frac{16}{81}\right)^{-3/4}$

58. $\left(\frac{9}{4}\right)^{-1/2}$

59. $\left(-\frac{1}{64}\right)^{-1/3}$

60. $\left(-\frac{125}{27}\right)^{-1/3}$

61. $5^{1/2} \cdot 5^{3/2}$

62. $4^{1/3} \cdot 4^{5/3}$

63. $\dfrac{2^{3/2}}{2}$

64. $\dfrac{5^{1/2}}{5}$

65. $\dfrac{x^2}{x^{1/2} \cdot x^{3/2}}$

66. $\dfrac{x \cdot x^{1/2}}{x^{3/2}}$

67. $(x^6 x^3)^{1/3}$

68. $(x^3 x^{12})^{1/5}$

69. $(16x^8 y^4)^{3/4}$

70. $(27x^6 y^9)^{2/3}$

Reducing the Index of a Radical In Exercises 71–76, use rational exponents to reduce the index of the radical. *See Example 7.*

71. $\sqrt[4]{y^2}$

72. $\sqrt[9]{x^3}$

73. $\sqrt{\sqrt{32}}$

74. $\sqrt{\sqrt{8}}$

75. $\sqrt[8]{(x + 1)^6}$

76. $\sqrt[6]{(x + 2)^4}$

Simplifying by Combining Like Radicals In Exercises 77–82, simplify the expression. *See Example 8.*

77. $5\sqrt{x} - 3\sqrt{x}$

78. $3\sqrt{x + 1} + 10\sqrt{x + 1}$

79. $5\sqrt{50} + 3\sqrt{8}$

80. $2\sqrt{27} - \sqrt{75}$

81. $2\sqrt{4y} - 2\sqrt{9y}$

82. $2\sqrt{108x} + \sqrt{147x}$

Evaluating Radicals with a Calculator In Exercises 83–90, use a calculator to approximate the number. (Round to three decimal places.) *See Examples 9 and 10.*

83. $\sqrt[3]{45}$

84. $\sqrt{57}$

85. $6.7^{2/5}$

86. $18.4^{1.1}$

87. $(0.38)^{-0.6}$

88. $(5.25)^{-1/2}$

89. $\dfrac{3 - \sqrt{5}}{2}$

90. $\dfrac{-4 + \sqrt{12}}{4}$

91. Calculator Write the keystrokes used to evaluate

$$\frac{4 - \sqrt{7}}{3}$$

in one step on your calculator.

92. Calculator Write the keystrokes you can use to evaluate

$$\sqrt[3]{(-5)^5}$$

in one step on your calculator.

Comparing Radical Expressions In Exercises 93–96, complete the statement with <, =, or >.

93. $\sqrt{5} + \sqrt{3}$ ▨ $\sqrt{5 + 3}$

94. $\sqrt{3} - \sqrt{2}$ ▨ $\sqrt{3 - 2}$

95. $\sqrt{3} \cdot \sqrt[4]{3}$ ▨ $\sqrt[8]{3}$

96. $\sqrt{\dfrac{3}{11}}$ ▨ $\dfrac{\sqrt{3}}{\sqrt{11}}$

97. Geometry Find the dimensions of a square classroom that has 1100 square feet of floor space (see figure).

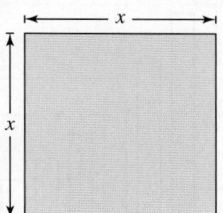

98. HOW DO YOU SEE IT? Package A is a cube with a volume of 500 cubic inches. Package B is a cube with a volume of 250 cubic inches. Is the length x of a side of package A greater than, less than, or equal to twice the length of a side of package B? Explain.

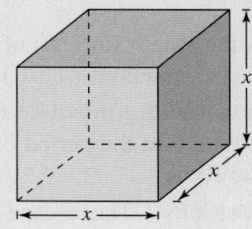

Declining Balances Depreciation In Exercises 99 and 100, find the annual depreciation rate r by using the declining balances formula

$$r = 1 - \left(\frac{S}{C}\right)^{1/n}$$

where n is the useful life of the item (in years), S is the salvage value (in dollars), and C is the original cost (in dollars).

99. A mammography machine with an original cost of $25,000 is depreciated over an eight-year period, as shown in the bar graph.

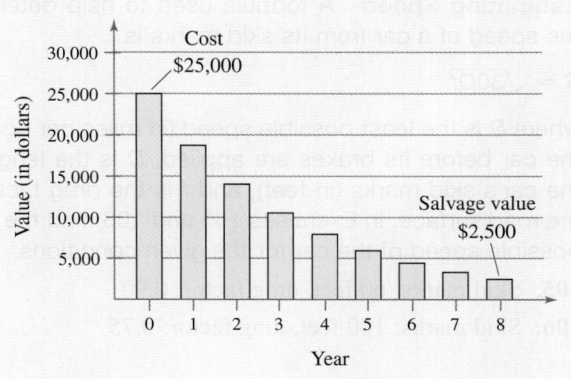

100. An ultrasound machine with an original cost of $135,000 is depreciated over a 10-year period, as shown in the bar graph.

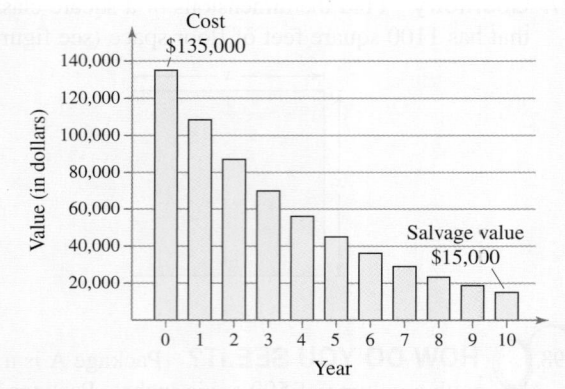

101. Period of a Pendulum   The period T (in seconds) of a pendulum is given by

$$T = 2\pi\sqrt{\frac{L}{32}}$$

where L is the length (in feet) of the pendulum. Find the period of a pendulum whose length is 4 feet.

102. Period of a Pendulum   Use the formula given in Exercise 101 to find the period of a pendulum whose length is 2.5 feet.

103. Escape Velocity   The escape velocity (in meters per second) on the moon is

$$\sqrt{\frac{2(6.67 \times 10^{-11})(7.36 \times 10^{22})}{1.74 \times 10^{6}}}.$$

A rocket, launched vertically from the moon, has a velocity of 2000 meters per second. Will the rocket escape the moon's gravitational field?

104. Erosion   A stream of water moving at the rate of v feet per second can carry particles of size

$0.03\sqrt{v}$ inches.

Find the size of the largest particle that can be carried by a stream flowing at the rate of $\frac{1}{2}$ foot per second.

Estimating Speed   A formula used to help determine the speed of a car from its skid marks is

$S = \sqrt{30Df}$

where S is the least possible speed (in miles per hour) of the car before its brakes are applied, D is the length of the car's skid marks (in feet), and f is the drag factor of the road surface. In Exercises 105 and 106, find the least possible speed of the car for the given conditions.

105. Skid marks: 60 feet, drag factor: 0.90

106. Skid marks: 100 feet, drag factor: 0.75

Wind Chill   A wind chill temperature is a measure of how cold it *feels* outside. The wind chill temperature W (in degrees Fahrenheit) is given by

$W = 35.75 + 0.6215T - 35.75v^{0.16} + 0.4275Tv^{0.16}$

where T is the actual temperature (in degrees Fahrenheit) and v is the wind speed (in miles per hour). In Exercises 107 and 108, find the wind chill temperature for the given conditions.   *(Source: NOAA's National Weather Service)*

107. Actual temperature: 30°F, wind speed: 20 mph

108. Actual temperature: 10°F, wind speed: 10 mph

Notes on a Musical Scale   In Exercises 109–111, find the frequency of the indicated note on a piano (see figure). The musical note A above middle C has a frequency of 440 vibrations per second. If we denote this frequency by F_1, then the frequency of the next higher note is given by $F_2 = F_1 \cdot 2^{1/12}$. Similarly, the frequency of the next note is given by $F_3 = F_2 \cdot 2^{1/12}$.

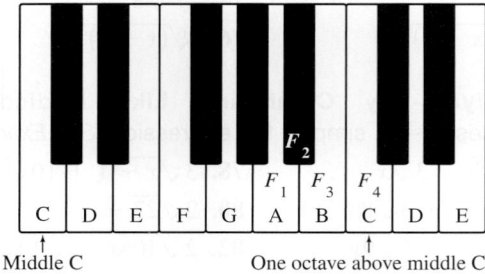

109. Find the frequency of the musical note B above middle C.

110. Find the frequency of the musical note C that is one octave above middle C.

111. Which note would you expect to have a higher frequency? Explain your reasoning.

(a) Musical note E one octave above the E above middle C

(b) Musical note D one octave above the D above middle C

112. Calculator Experiment   Enter any positive real number in your calculator and repeatedly take the square root. What real number does the display appear to be approaching?

113. Calculator Experiment   Square the real number $2/\sqrt{5}$ and note that the radical is eliminated from the denominator. Is this equivalent to rationalizing the denominator? Why or why not?

114. Think About It   How can you show that $a^0 = 1$, $a \neq 0$? (*Hint:* Use the property of exponents $a^m/a^n = a^{m-n}$.)

115. Think About It   Is it true that $\sqrt{4x^2} = 2x$ for every real number x? Explain.

116. Think About It   Explain why $\sqrt{2} + \sqrt{3} \neq \sqrt{5}$.

QUIZ YOURSELF

See www.CalcChat.com for worked-out solutions to odd-numbered exercises.

Take this quiz as you would take a quiz in class. When you are done, check your work against the answers given in the back of the book.

In Exercises 1 and 2, place the correct symbol ($<$, $>$, or $=$) between the two real numbers.

1. $-|-7|$ ▨ $|-7|$

2. $-(-3)$ ▨ $|-3|$

3. Determine which numbers in the set

$$\left\{-6, \sqrt{11}, \sqrt{1}, \tfrac{3}{4}, -\tfrac{1}{2}\right\}$$

are natural numbers, integers, rational numbers, and irrational numbers.

4. Use inequality notation to describe the subset of real numbers: x is positive or x is equal to zero.

5. Find the distance between the numbers -6 and 5.

6. Identify the terms of the algebraic expression $3x^2 - 7x + 2$.

In Exercises 7–10, perform the indicated operation(s). (Write fractional answers in simplest form.)

7. $-4 - 0 \cdot 7$

8. $\dfrac{31 - 5}{-2}$

9. $\dfrac{2}{3} \cdot \dfrac{5}{4} \cdot \dfrac{3}{7}$

10. $\dfrac{11}{15} \div \dfrac{3}{5}$

In Exercises 11 and 12, rewrite the expression with positive exponents and simplify.

11. $(-x)^3(2x^{-4})$

12. $\dfrac{5y^7}{15y^{-3}}$

13. A flea consumes 0.000014 liter of blood in one day. Write the number of liters in scientific notation.

14. You deposit $5000 in an account with an annual interest rate of 2.2%, compounded quarterly. Find the balance in the account after 5 years.

In Exercises 15 and 16, evaluate the expression.

15. $\dfrac{-\sqrt[4]{81}}{3}$

16. $\left(\sqrt[3]{-64}\right)^3$

In Exercises 17–19, simplify the expression.

17. $3^{1/2} \cdot 3^{3/2}$

18. $\sqrt[3]{81} - 4\sqrt[3]{3}$

19. $\sqrt[10]{12^5}$

20. Find the dimensions of a cube that has a volume of 10,648 cubic centimeters.

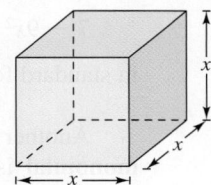

0.5 Polynomials and Special Products

■ Write a polynomial in standard form and perform operations with polynomials.
■ Use special products to multiply polynomials.
■ Use polynomials to solve application problems.

Polynomials and Operations with Polynomials

One of the simplest and most common types of algebraic expressions is a **polynomial.** Here are some examples.

$$2x + 5, \quad 3x^4 - 7x^2 + 2x + 4, \quad 5x^2y^2 - xy + 3$$

The first two are *polynomials in x* and the third is a *polynomial in x and y*. The terms of a polynomial in x have the form ax^k, where a is the **coefficient** and k is the **degree** of the term. Because a polynomial is defined as an algebraic sum, the coefficients take on the signs between the terms. For instance, the polynomial

$$2x^3 - 5x^2 + 1 = 2x^3 + (-5)x^2 + (0)x + 1$$

has coefficients 2, -5, 0, and 1.

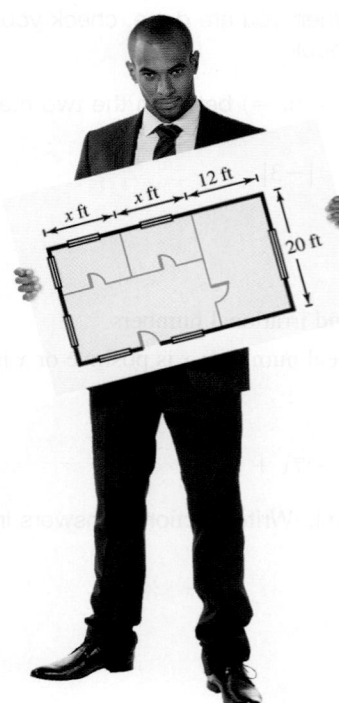

In Exercise 64 on page 47, you will find a polynomial that represents the total number of square feet in a floor plan.

Definition of a Polynomial in *x*

Let $a_n, \ldots, a_2, a_1, a_0$ be real numbers and let n be a *nonnegative integer.* A **polynomial in *x*** is an expression of the form

$$a_n x^n + \cdots + a_2 x^2 + a_1 x + a_0$$

where $a_n \neq 0$. The polynomial is of **degree** n, and the number a_n is the **leading coefficient.** The number a_0 is the **constant term.** The constant term is considered to have a degree of zero.

Note in the definition of a polynomial in x that the polynomial is written in descending powers of x. This is called the **standard form** of a polynomial.

Example 1   Rewriting a Polynomial in Standard Form

	Polynomial	*Standard Form*	*Degree*
a.	$4x^2 - 5x^3 - 2 + 3x$	$-5x^3 + 4x^2 + 3x - 2$	3
b.	$4 - 9x^2$	$-9x^2 + 4$	2
c.	8	$8 \;\; (8 = 8x^0)$	0

✓ **Checkpoint 1**

Rewrite the polynomial

$$7 - 9x^2 + 3x$$

in standard form and state its degree.

Another way to describe a polynomial is that it is the sum of *monomials*. A **monomial** is an expression of the form ax^k, where a is a real number and k is a nonnegative integer. The sum of two monomials is called a **binomial.** The sum of three monomials is called a **trinomial.**

A polynomial that has all zero coefficients is called the **zero polynomial,** denoted by 0. This particular polynomial is not considered to have a degree. For a polynomial in more than one variable, the *degree of a term* is the sum of the powers of the variables in the term. The *degree of the polynomial* is the highest degree of all its terms. For instance, the polynomial $5x^3y - x^2y^2 + 2xy - 5$ has two terms of degree 4, one term of degree 2, and one term of degree 0. The degree of the polynomial is 4.

Example 2 Identifying a Polynomial and Its Degree

a. $-2x^3 + x^2 + 3x - 2$ is a polynomial of degree 3.

b. $\sqrt{x^2 - 3x}$ is not a polynomial because the radical sign indicates a noninteger power of x.

c. $x^2 + 5x^{-1}$ is not a polynomial because of the negative exponent.

✓ Checkpoint 2

Determine whether the expression

$$\frac{2x + 5}{x}$$

is a polynomial. If it is, state its degree.

You can **add** and **subtract** polynomials in much the same way that you add and subtract real numbers—you simply add or subtract the *like terms* (terms having the same variables raised to the same powers) by adding or subtracting their coefficients. For instance, $-3x^2$ and $5x^2$ are like terms and their sum is given by

$$-3x^2 + 5x^2 = (-3 + 5)x^2 = 2x^2.$$

ALGEBRA TUTOR *xy*

For help in finding the sums and differences in Example 3, see the *Chapter 0 Algebra Tutor* on page 63.

Example 3 Sums and Differences of Polynomials

a. $(5x^3 - 7x^2 - 3) + (x^3 + 2x^2 - x + 8)$

$= (5x^3 + x^3) + (2x^2 - 7x^2) - x + (8 - 3)$ Group like terms.

$= 6x^3 - 5x^2 - x + 5$ Combine like terms.

b. $(7x^4 - x^2 - 4x + 2) - (3x^4 - 4x^2 + 3x)$

$= 7x^4 - x^2 - 4x + 2 - 3x^4 + 4x^2 - 3x$ Distributive Property

$= (7x^4 - 3x^4) + (4x^2 - x^2) + (-4x - 3x) + 2$ Group like terms.

$= 4x^4 + 3x^2 - 7x + 2$ Combine like terms.

✓ Checkpoint 3

Find the sum $(2x^2 + x + 3) + (4x + 1)$ and write the resulting polynomial in standard form.

A common mistake is to fail to change the sign of *each* term inside parentheses preceded by a minus sign. Here is an example.

$-(3x^4 - 4x^2 + 3x) = -3x^4 + 4x^2 - 3x$ Correct

$\overline{-(3x^4 - 4x^2 + 3x) = -3x^4 - 4x^2 + 3x}$ Common mistake

To find the **product** of two polynomials, you can use the left and right Distributive Properties. For example, treat $(5x + 7)$ as a single quantity and multiply $(3x - 2)$ by $(5x + 7)$, as shown.

$$(3x - 2)(5x + 7) = 3x(5x + 7) - 2(5x + 7)$$

$$= (3x)(5x) + (3x)(7) - (2)(5x) - (2)(7)$$

$$= 15x^2 + 21x - 10x - 14$$

Product of First terms	Product of Outer terms	Product of Inner terms	Product of Last terms

$$= 15x^2 + 11x - 14$$

You can use the four special products shown in the boxes above to write the product of two binomials in the FOIL form in just one step. This is called the **FOIL Method.**

Example 4   Using the FOIL Method

Use the FOIL Method to find the product of $(2x - 4)$ and $(x + 5)$.

SOLUTION

$$(2x - 4)(x + 5) = \overset{\text{F}}{2x^2} + \overset{\text{O}}{10x} - \overset{\text{I}}{4x} - \overset{\text{L}}{20}$$

$$= 2x^2 + 6x - 20$$

✓ Checkpoint 4

Find each product.

a. $(x + 5)(x + 9)$

b. $(3x + 1)(x - 1)$

When multiplying two polynomials, be sure to use the Distributive Property to multiply *each* term of one polynomial by *each* term of the other. The vertical format is a convenient way to multiply two polynomials.

Example 5   Using a Vertical Format to Multiply Polynomials

Multiply $(x^2 - 2x + 2)$ by $(x^2 + 2x + 2)$.

SOLUTION

$$\begin{array}{l} x^2 - 2x + 2 \\ \underline{x^2 + 2x + 2} \\ x^4 - 2x^3 + 2x^2 \\ 2x^3 - 4x^2 + 4x \\ 2x^2 - 4x + 4 \\ \hline x^4 + 0x^3 + 0x^2 + 0x + 4 = x^4 + 4 \end{array}$$

Standard form

Standard form

$x^2(x^2 - 2x + 2)$

$2x(x^2 - 2x + 2)$

$2(x^2 - 2x + 2)$

Combine like terms.

So, $(x^2 - 2x + 2)(x^2 + 2x + 2) = x^4 + 4$.

✓ Checkpoint 5

Multiply $(x^2 + x + 4)$ by $(x^2 - 3x + 1)$.

Special Products

Special Products

Let u and v be real numbers, variables, or algebraic expressions.

Special Product	*Example*

Sum and Difference of Two Terms

$(u + v)(u - v) = u^2 - v^2$ $(x + 4)(x - 4) = x^2 - 16$

Square of a Binomial

$(u + v)^2 = u^2 + 2uv + v^2$ $(x + 3)^2 = x^2 + 6x + 9$

$(u - v)^2 = u^2 - 2uv + v^2$ $(3x - 2)^2 = 9x^2 - 12x + 4$

Cube of a Binomial

$(u + v)^3 = u^3 + 3u^2v + 3uv^2 + v^3$ $(x + 2)^3 = x^3 + 6x^2 + 12x + 8$

$(u - v)^3 = u^3 - 3u^2v + 3uv^2 - v^3$ $(x - 1)^3 = x^3 - 3x^2 + 3x - 1$

Example 6 **Sum and Difference of Two Terms**

$$(5x + 9)(5x - 9) = (5x)^2 - 9^2$$
$$= 25x^2 - 81$$

✓**Checkpoint 6**

Find the product $(3 - x)(3 + x)$.

Example 7 **Square and Cube of a Binomial**

a. $(6x - 5)^2 = (6x)^2 - 2(6x)(5) + 5^2$
$$= 36x^2 - 60x + 25$$

b. $(3x + 2)^3 = (3x)^3 + 3(3x)^2(2) + 3(3x)(2)^2 + 2^3$
$$= 27x^3 + 54x^2 + 36x + 8$$

✓**Checkpoint 7**

Find each product.

a. $(x + 4)^2$ **b.** $(x - 3)^3$

Example 8 **The Product of Two Trinomials**

$$(x + y - 2)(x + y + 2) = [(x + y) - 2][(x + y) + 2]$$
$$= (x + y)^2 - 2^2$$
$$= x^2 + 2xy + y^2 - 4$$

✓**Checkpoint 8**

Find the product $(x + 5 - y)(x + 5 + y)$.

Applications

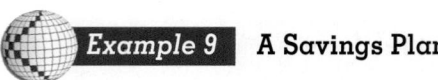 **Example 9**   A Savings Plan

At the same time each year for five consecutive years, you deposit money in an account that earns 7% interest, compounded annually. The deposit amounts are $1500, $1800, $2400, $2600, and $3000. After the last deposit, is there enough money to pay a $12,000 college tuition bill?

SOLUTION   Using the formula for compound interest, for *each* deposit you have

$$\text{Balance} = P\left(1 + \frac{r}{n}\right)^{nt}$$

$$= P(1 + 0.07)^{t}$$

$$= P(1.07)^{t}.$$

For the first deposit, $P = 1500$ and $t = 4$. For the second deposit, $P = 1800$ and $t = 3$, and so on. The balances for the five deposits are as follows.

Date	Deposit	Time in Account	Balance in Account
First Year	$1500	4 years	$1500(1.07)^4$
Second Year	$1800	3 years	$1800(1.07)^3$
Third Year	$2400	2 years	$2400(1.07)^2$
Fourth Year	$2600	1 year	$2600(1.07)$
Fifth Year	$3000	0 years	3000

By adding these five balances, you can find the total balance in the account to be

$$1500(1.07)^4 + 1800(1.07)^3 + 2400(1.07)^2 + 2600(1.07) + 3000.$$

Note that this expression is in polynomial form. By evaluating the expression, you can find the balance to be $12,701.03, as shown in Figure 0.10.

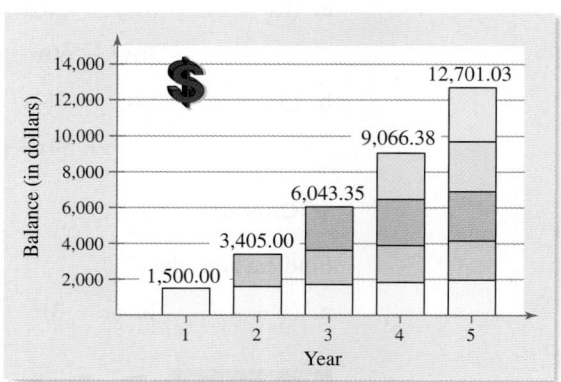

FIGURE 0.10

After the fifth deposit, there is enough money in the account to pay the college tuition bill.

**Checkpoint 9**

In Example 9, suppose the account earns 5% interest. What is the balance in the account after the last deposit?

Example 10 Geometry: Volume of a Box

An open box is made by cutting squares from the corners of a piece of metal that measures 16 inches by 20 inches and turning up the sides, as shown in Figure 0.11. The sides of the cut-out squares are all x inches long, so the box is x inches tall. Write an expression for the volume of the box. Then find the volume when $x = 1$, $x = 2$, and $x = 3$ inches.

SOLUTION

Verbal Model: Volume = Length · Width · Height

Labels: Height = x (inches)

Width = $16 - 2x$ (inches)

Length = $20 - 2x$ (inches)

Equation: Volume = $(20 - 2x)(16 - 2x)(x)$

$$= (320 - 72x + 4x^2)(x)$$

$$= 320x - 72x^2 + 4x^3$$

When $x = 1$ inch, the volume of the box is

Volume = $320(1) - 72(1)^2 + 4(1)^3$

$$= 252 \text{ cubic inches.}$$

When $x = 2$ inches, the volume of the box is

Volume = $320(2) - 72(2)^2 + 4(2)^3$

$$= 384 \text{ cubic inches.}$$

When $x = 3$ inches, the volume of the box is

Volume = $320(3) - 72(3)^2 + 4(3)^3$

$$= 420 \text{ cubic inches.}$$

FIGURE 0.11

✓ **Checkpoint 10**

In Example 10, suppose the original piece of metal is 10 inches by 12 inches. Write an expression for the volume of the box. Then find the volume when $x = 2$ and $x = 3$ inches.

SUMMARIZE   (Section 0.5)

1. State the definition of a polynomial in x and explain what is meant by the standard form of a polynomial *(page 40)*. For an example of rewriting polynomials in standard form, see Example 1.

2. Describe how to add and subtract polynomials *(page 41)*. For examples of adding and subtracting polynomials, see Example 3.

3. Describe the FOIL Method *(page 42)*. For an example that uses the FOIL Method, see Example 4.

4. Explain how to find the square and cube of a binomial *(page 43)*. For an example of squaring and cubing a binomial, see Example 7.

SKILLS WARM UP 0.5   The following warm-up exercises involve skills that were covered in earlier sections. You will use these skills in the exercise set for this section. For additional help, review Section 0.3.

In Exercises 1–10, perform the indicated operation(s).

1. $(7x^2)(6x)$

2. $(5x^3)(-9x^2)$

3. $(10z^3)(-2z^{-1})$

4. $(-3x^4)(5x^{-5})$

5. $(-3x^2)^3$

6. $-3(x^2)^3$

7. $\dfrac{27z^5}{12z^2}$

8. $\dfrac{32x^4}{20x}$

9. $\left(\dfrac{2x}{3}\right)^{-2}$

10. $\left(\dfrac{x^{-1}}{y^2}\right)^{-1}$

Exercises 0.5

See www.CalcChat.com for worked-out solutions to odd-numbered exercises.

Rewriting a Polynomial in Standard Form   In Exercises 1–6, rewrite the polynomial in standard form. Then identify the degree and leading coefficient of the polynomial. *See Example 1.*

1. $3x + 4x^2 + 2$

2. $x^2 - 4 - 3x^4$

3. $-8 + x^7$

4. $23 - x^3$

5. $1 - x + 6x^4 - 2x^5$

6. $7 + 8x$

Identifying a Polynomial and Its Degree   In Exercises 7–12, determine whether the algebraic expression is a polynomial. If it is, write the polynomial in standard form and state its degree. *See Example 2.*

7. $2x - 3x^3 + 8$

8. $2x^3 + x - 3x^{-1}$

9. $\dfrac{3x + 4}{x}$

10. $\dfrac{2x^2 + 5x - 3}{3}$

11. $w^2 - w^4 + 2w^3$

12. $\sqrt{y^2 - y^4}$

Evaluating a Polynomial   In Exercises 13–16, evaluate the polynomial for each value of x.

13. $4x + 5$
(a) $x = -2$ (b) $x = -1$
(c) $x = 0$ (d) $x = 3$

14. $-x^2 + 3$
(a) $x = -3$ (b) $x = -2$
(c) $x = 0$ (d) $x = 1$

15. $-2x^2 + 3x + 4$
(a) $x = -2$ (b) $x = -1$
(c) $x = 0$ (d) $x = 1$

16. $x^3 - 4x^2 + x$
(a) $x = -1$ (b) $x = 0$
(c) $x = 1$ (d) $x = 2$

A Sum or Difference of Polynomials   In Exercises 17–22, find the sum or difference and write the resulting polynomial in standard form. *See Example 3.*

17. $(6x + 5) - (8x + 15)$

18. $(3x^2 + 1) - (2x^2 - 2x + 3)$

19. $-(x^3 + 5) + (3x^3 - 4x)$

20. $-(5x^2 - 1) + (-3x^2 + 5)$

21. $(15x^2 - 6) - (-8x^3 - 14x^2 - 17)$

22. $(15x^4 - 18x - 19) - (13x^4 - 5x + 15)$

Multiplying Polynomials   In Exercises 23–52, find the product. *See Examples 4, 5, 6, 7, and 8.*

23. $3x(x^2 - 2x + 1)$

24. $z^2(2z^2 + 3z + 1)$

25. $-3x(-x)(3x - 7)$

26. $(2 - x^2)(-2x)(4x)$

27. $(x + 3)(x + 4)$

28. $(x - 5)(x + 10)$

29. $(3x - 5)(2x + 1)$

30. $(7x - 2)(4x - 3)$

31. $(x^2 + 2x)(x^2 + x - 3)$

32. $(3x - 4)(x^2 - 2x + 1)$

33. $(-x^2 + x - 5)(3x^2 + 4x + 1)$

34. $(x^2 + 3x + 2)(2x^2 - x + 4)$

35. $(x + 5)(x - 5)$

36. $(3x + 2)(3x - 2)$

37. $(x + 6)^2$

38. $(3x - 2)^2$

39. $(2x - 5y)^2$

40. $(5 - 8x)^2$

41. $[(x - 3) + y]^2$

42. $[(x + 1) - y]^2$

43. $(x + 1)^3$

44. $(x - 2)^3$

45. $(2x - y)^3$

46. $(3x + 2y)^3$

47. $(3y^2 - 1)(3y^2 + 1)$

48. $(3x^2 - 4y^2)(3x^2 + 4y^2)$

49. $(m - 3 + n)(m - 3 - n)$

50. $(x + y + 1)(x + y - 1)$

51. $(x^2 - x + 1)(x^2 + x + 1)$

52. $(x^2 + 3x - 2)(x^2 - 3x - 2)$

Operations with Polynomials   In Exercises 53–56, perform the indicated operation and write the result in standard form.

53. Find the product of $(3x - 7)$ and $(8x + 1)$.

54. Subtract $(3x^2 - 2)$ from $(5x^2 - 7x + 4)$.

55. Find the sum of $(4x^3 + 6)$ and $(6x^2 - 8x - 11)$.

56. Multiply $(x^2 - 2x - 4)$ by $(x + 3)$.

57. Error Analysis   Describe and correct the error.
$$\cancel{(x - 3)^2 = x^2 + 9}$$

58. Error Analysis Describe and correct the error.

$$\overline{x^2 - 6(x + 5) = x^2 - 6x + 30}$$

59. Compound Interest After 3 years, an investment of $1000 earning an interest rate r compounded annually will be worth $1000(1 + r)^3$ dollars. Write this expression as a polynomial in standard form.

60. Compound Interest After 2 years, an investment of $800 earning an interest rate r compounded annually will be worth $800(1 + r)^2$ dollars. Write this expression as a polynomial in standard form.

61. Savings Plan At the same time each year for five consecutive years, you deposit money in an account that earns annually compounded interest. The deposits are $1200, $1700, $1500, $2200, and $3000. Is there enough money in the account after the last deposit to pay a $10,000 college tuition bill when the rate of annually compounded interest is 4%? 3%? 2%?

62. Savings Plan You have an investment that pays an annual dividend. Each January for six consecutive years, you reinvest this dividend in an account that earns 4% interest, compounded annually. The dividends are shown in the table. Is there enough money in the account after the sixth deposit for a $7000 down payment on a car?

Year	Dividend
1	$920
2	$1000
3	$780
4	$1310
5	$1020
6	$1200

63. Federal Student Aid The total amount (in millions of dollars) of federal student aid disbursed each year from 2000 through 2007 can be approximated by

$$-164.14x^2 + 6777.8x + 43,161$$

where x represents the year, with $x = 0$ corresponding to 2000. Evaluate the polynomial at $x = 6$ and $x = 7$. Explain the meaning of your results in the context of the situation. *(Source: U.S. Department of Education)*

64. Geometry Find a polynomial that represents the total number of square feet in the floor plan.

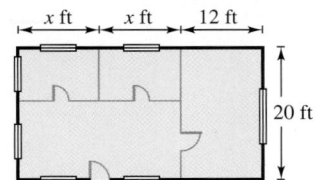

65. Geometry A box has a length of $(57 - 2x)$ inches, a width of $(39 - 2x)$ inches, and a height of x inches. Find the volume when $x = 4$, $x = 6$, and $x = 10$ inches. Which x-value gives the greatest volume?

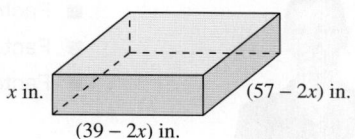

66. HOW DO YOU SEE IT? An open box has a length of $(52 - 2x)$ inches, a width of $(42 - 2x)$ inches, and a height of x inches, as shown.

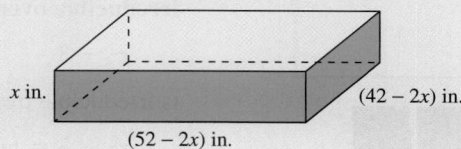

(a) Describe a way that the box could have been made from a rectangular piece of cardboard. Give the original dimensions of the cardboard.

(b) What degree is the polynomial that represents the volume of the box? Explain.

(c) Describe a procedure for finding the value of x (to the nearest tenth of an inch) that yields the maximum possible volume of the box.

67. Project: Population For a project involving the population of the United States from 1990 to 2009, visit this text's website at *www.cengagebrain.com.* *(Source: U.S. Census Bureau)*

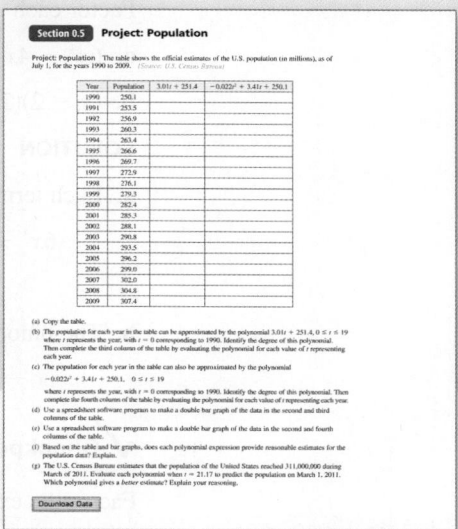

0.6 Factoring

- Factor a polynomial by factoring out common factors.
- Factor a polynomial in a special form.
- Factor a trinomial as the product of two binomials.
- Factor a polynomial by grouping.

Common Factors

The process of writing a polynomial as a product is called **factoring.** It is an important tool for solving equations and reducing fractional expressions.

A polynomial that cannot be factored using integer coefficients is called **prime** or **irreducible over the integers.** For instance, the polynomial

$$x^2 - 3$$

is irreducible over the integers. [Over the *real numbers*, it can be factored as

$$x^2 - 3 = \left(x + \sqrt{3}\right)\left(x - \sqrt{3}\right).]$$

A polynomial is **completely factored** when each of its factors is prime. For instance,

$$x^3 - x^2 + 4x - 4 = (x - 1)(x^2 + 4) \qquad \text{Completely factored}$$

is completely factored, but

$$x^3 - x^2 - 4x + 4 = (x - 1)(x^2 - 4) \qquad \text{Not completely factored}$$

is not completely factored. Its complete factorization is

$$x^3 - x^2 - 4x + 4 = (x - 1)(x + 2)(x - 2).$$

The simplest type of factoring involves a polynomial that can be written as the product of a monomial and another polynomial. To factor such a polynomial, you can use the Distributive Property in the *reverse* direction.

$$ab + ac = a(b + c) \qquad a \text{ is a common factor.}$$

Area $= (2x^2 - x - 3)$ ft^2 $(x + 1)$ ft

In Exercise 85 on page 54, you will factor a polynomial expression for the floor space of a room to find an expression for the length of the room.

Example 1   Factoring Out Common Factors

Factor each expression.

a. $6x^3 - 4x$

b. $(x - 2)(2x) + (x - 2)(3)$

SOLUTION

a. Each term of this polynomial has $2x$ as a common factor.

$$6x^3 - 4x = 2x(3x^2) - 2x(2)$$
$$= 2x(3x^2 - 2)$$

b. The binomial factor $(x - 2)$ is common to both terms.

$$(x - 2)(2x) + (x - 2)(3) = (x - 2)(2x + 3)$$

✓ Checkpoint 1

Factor the expression

$$(x + 1)^2 + 2x(x + 1).$$

Factoring Special Polynomial Forms

Factoring Special Polynomial Forms

Factored Form	*Example*
Difference of Two Squares	
$u^2 - v^2 = (u + v)(u - v)$	$9x^2 - 4 = (3x + 2)(3x - 2)$
Perfect Square Trinomial	
$u^2 + 2uv + v^2 = (u + v)^2$	$x^2 + 6x + 9 = (x + 3)^2$
$u^2 - 2uv + v^2 = (u - v)^2$	$x^2 - 6x + 9 = (x - 3)^2$
Sum or Difference of Two Cubes	
$u^3 + v^3 = (u + v)(u^2 - uv + v^2)$	$x^3 + 8 = (x + 2)(x^2 - 2x + 4)$
$u^3 - v^3 = (u - v)(u^2 + uv + v^2)$	$27x^3 - 1 = (3x - 1)(9x^2 + 3x + 1)$

STUDY TIP

In Example 2, note that the first step in factoring a polynomial is to check for common factors. Once the common factor is factored out, it is often possible to recognize patterns that were not obvious at first glance.

Example 2   Factoring Out a Common Factor First

Factor the expression $3 - 12x^2$.

SOLUTION

$$3 - 12x^2 = 3(1 - 4x^2) \qquad \text{3 is a common factor.}$$
$$= 3[1^2 - (2x)^2] \qquad \text{Difference of two squares}$$
$$= 3(1 + 2x)(1 - 2x) \qquad \text{Completely factored}$$

✓ **Checkpoint 2**

Factor the expression $x^3 - x$.

Example 3   Factoring the Difference of Two Squares

a. $(x + 2)^2 - y^2 = [(x + 2) + y][(x + 2) - y]$
$$= (x + 2 + y)(x + 2 - y)$$
$$= (x + y + 2)(x - y + 2)$$

b. The polynomial $16x^4 - 81$ can be factored by applying the difference of two squares formula twice.

$$16x^4 - 81 = (4x^2)^2 - 9^2$$
$$= (4x^2 + 9)(4x^2 - 9) \qquad \text{First application}$$
$$= (4x^2 + 9)[(2x)^2 - 3^2]$$
$$= (4x^2 + 9)(2x + 3)(2x - 3) \qquad \text{Second application}$$

✓ **Checkpoint 3**

Factor the expression $100 - 4y^2$.

A perfect square trinomial is the square of a binomial, and it has the following form. Note that the first and last terms of a perfect square trinomial are squares and the middle term is twice the product of u and v.

$$u^2 + 2uv + v^2 = (u + v)^2 \qquad \text{or} \qquad u^2 - 2uv + v^2 = (u - v)^2$$

Same sign   Same sign

Example 4   Factoring Perfect Square Trinomials

a. $16x^2 + 8x + 1 = (4x)^2 + 2(4x)(1) + 1^2$

$$= (4x + 1)^2$$

b. $x^2 - 10x + 25 = x^2 - 2(x)(5) + 5^2$

$$= (x - 5)^2$$

✔ **Checkpoint 4**

Factor each expression.

a. $4x^2 - 4x + 1$

b. $x^2 - 12x + 36$

The next two formulas show that sums and differences of cubes factor easily. Pay special attention to the signs of the terms.

Like signs   Like signs

$$u^3 + v^3 = (u + v)(u^2 - uv + v^2) \qquad u^3 - v^3 = (u - v)(u^2 + uv + v^2)$$

Unlike signs   Unlike signs

Example 5   Factoring the Sum and Difference of Cubes

Factor each expression.

a. $x^3 - 27$

b. $3x^3 + 192$

SOLUTION

a. $x^3 - 27 = x^3 - 3^3$ Rewrite 27 as 3^3.

$$= (x - 3)(x^2 + 3x + 9)$$ Factor.

b. $3x^3 + 192 = 3(x^3 + 64)$ 3 is a common factor.

$$= 3(x^3 + 4^3)$$ Rewrite 64 as 4^3.

$$= 3(x + 4)(x^2 - 4x + 16)$$ Factor.

✔ **Checkpoint 5**

Factor each expression.

a. $y^3 + 1$

b. $5z^3 - 40$

Trinomials with Binomial Factors

To factor a trinomial of the form $ax^2 + bx + c$, use the following pattern.

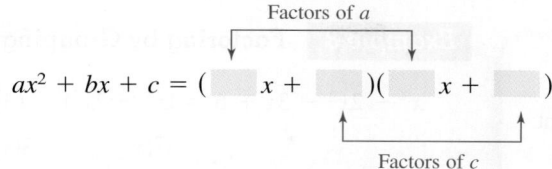

The goal is to find a combination of factors of a and c such that the outer and inner products add up to the middle term bx. For instance, for the trinomial $6x^2 + 17x + 5$ you can write

$$\overset{\text{F}\quad\text{O}\quad\text{I}\quad\text{L}}{(2x + 5)(3x + 1)} = 6x^2 + 2x + 15x + 5$$
$$\underset{\text{O + I}}{}$$
$$= 6x^2 + 17x + 5.$$

Note that the outer (O) and inner (I) products add up to $17x$.

Example 6   Factoring a Trinomial: Leading Coefficient Is 1

Factor the trinomial $x^2 - 7x + 12$.

SOLUTION   For this trinomial, you have $a = 1$, $b = -7$, and $c = 12$. Because b is negative and c is positive, both factors of 12 must be negative. That is, $12 = (-2)(-6)$, $12 = (-1)(-12)$, or $12 = (-3)(-4)$. So, the possible factorizations of $x^2 - 7x + 12$ are

$$(x - 2)(x - 6), \quad (x - 1)(x - 12), \quad \text{and} \quad (x - 3)(x - 4).$$

Testing the middle term, you find that $x^2 - 7x + 12 = (x - 3)(x - 4)$.

___

✓ Checkpoint 6

Factor the trinomial $x^2 + x - 6$.   ■

Example 7   Factoring a Trinomial: Leading Coefficient Is Not 1

Factor the trinomial $2x^2 + x - 15$.

SOLUTION   For this trinomial, you have $a = 2$ and $c = -15$, which means that the factors of -15 must have unlike signs. The eight possible factorizations are:

$$(2x - 1)(x + 15) \quad (2x + 1)(x - 15) \quad (2x - 3)(x + 5) \quad (2x + 3)(x - 5)$$

$$(2x - 5)(x + 3) \quad (2x + 5)(x - 3) \quad (2x - 15)(x + 1) \quad (2x + 15)(x - 1)$$

Testing the middle term, you find that $2x^2 + x - 15 = (2x - 5)(x + 3)$.

___

✓ Checkpoint 7

Factor the trinomial $2x^2 - 5x + 3$.   ■

Factoring by Grouping

Sometimes polynomials with more than three terms can be **factored by grouping.**

Example 8   Factoring by Grouping

$$x^3 - 2x^2 - 3x + 6 = (x^3 - 2x^2) - (3x - 6)$$ Group terms.
$$= x^2(x - 2) - 3(x - 2)$$ Factor groups.
$$= (x - 2)(x^2 - 3)$$ Distributive Property

✓ Checkpoint 8

Factor the polynomial $x^3 + x^2 + 5x + 5$.   ■

Factoring by grouping can save you some of the trial and error involved in factoring a trinomial. To factor a trinomial of the form $ax^2 + bx + c$ by grouping, rewrite the middle term using the sum of two factors of the product ac that add up to b. This technique is illustrated in Example 9.

Example 9   Factoring a Trinomial by Grouping

Use factoring by grouping to factor $2x^2 + 5x - 3$.

SOLUTION   In the trinomial $2x^2 + 5x - 3$, $a = 2$ and $c = -3$, so the product ac is -6. Notice that -6 factors as $(6)(-1)$, and $6 - 1 = 5 = b$. So, you can rewrite the middle term as $5x = 6x - x$.

$$2x^2 + 5x - 3 = 2x^2 + 6x - x - 3$$ Rewrite middle term.
$$= (2x^2 + 6x) - (x + 3)$$ Group terms.
$$= 2x(x + 3) - (x + 3)$$ Factor groups.
$$= (x + 3)(2x - 1)$$ Distributive Property

The trinomial factors as $2x^2 + 5x - 3 = (x + 3)(2x - 1)$.

✓ Checkpoint 9

Use factoring by grouping to factor $2x^2 + 5x - 12$.   ■

STUDY TIP

In some cases, several different groupings will work. In Example 8, a different grouping could have been used.

$$x^3 - 2x^2 - 3x + 6$$
$$= (x^3 - 3x) - (2x^2 - 6)$$
$$= x(x^2 - 3) - 2(x^2 - 3)$$
$$= (x^2 - 3)(x - 2)$$

SUMMARIZE   (Section 0.6)

1. Describe what it means to completely factor a polynomial *(page 48)*. For an example of factoring out common factors, see Example 1.

2. Make a list of the special polynomial forms of factoring *(page 49)*. For examples of factoring these special forms, see Examples 3, 4, and 5.

3. Describe how to factor a trinomial of the form $ax^2 + bx + c$ *(page 51)*. For examples of factoring trinomials of this form, see Examples 6 and 7.

4. Explain how to factor a polynomial by grouping *(page 52)*. For examples of factoring by grouping, see Examples 8 and 9.

In Exercises 1–10, find the product.

1. $3x(5x - 2)$

2. $-2y(y + 1)$

3. $(2x + 3)^2$

4. $(3x - 8)^2$

5. $(2x - 3)(x + 8)$

6. $(4 - 5z)(1 + z)$

7. $(2y + 1)(2y - 1)$

8. $(x + a)(x - a)$

9. $(x + 4)^3$

10. $(2x - 3)^3$

Exercises 0.6

Factoring Out Common Factors In Exercises 1–6, factor out the common factor. *See Example 1.*

1. $3x + 6$

2. $6y - 30$

3. $3x^3 - 6x$

4. $4x^3 - 6x^2 + 12x$

5. $3x(x - 5) + 8(x - 5)$

6. $4(5x - 4)^2 + x(5x - 4)$

Factoring the Difference of Two Squares In Exercises 7–14, factor the difference of two squares. *See Example 3.*

7. $x^2 - 36$

8. $x^2 - \frac{1}{9}$

9. $16x^2 - 9y^2$

10. $x^2 - 49y^2$

11. $(x - 1)^2 - 4$

12. $25 - (z + 5)^2$

13. $81x^4 - 1$

14. $x^4 - 256$

Factoring Out a Common Factor First In Exercises 15–20, factor the expression. *See Example 2.*

15. $3x^2 - 3$

16. $8x^2 - 18$

17. $8x^2 - 50y^2$

18. $48 - 27x^2$

19. $25(x - 1) - x^2(x - 1)$

20. $4x^2(x + 2) - 36(x + 2)$

Factoring a Perfect Square Trinomial In Exercises 21–26, factor the perfect square trinomial. *See Example 4.*

21. $x^2 - 4x + 4$

22. $x^2 + 10x + 25$

23. $4y^2 + 12y + 9$

24. $9x^2 - 12x + 4$

25. $y^2 - \frac{2}{3}y + \frac{1}{9}$

26. $z^2 + z + \frac{1}{4}$

Factoring the Sum or Difference of Cubes In Exercises 27–32, factor the sum or difference of cubes. *See Example 5.*

27. $x^3 - 8$

28. $x^3 - 27$

29. $y^3 + 125$

30. $y^3 + 1000$

31. $x^3 - \frac{8}{27}$

32. $x^3 + \frac{8}{125}$

Factoring a Trinomial In Exercises 33–46, factor the trinomial. *See Examples 6 and 7.*

33. $x^2 + x - 2$

34. $x^2 + 6x + 8$

35. $w^2 - 5w + 6$

36. $z^2 - z - 6$

37. $y^2 + y - 20$

38. $z^2 - 4z - 21$

39. $x^2 - 30x + 200$

40. $x^2 - 5x - 150$

41. $3x^2 - 5x + 2$

42. $2x^2 - x - 1$

43. $9x^2 - 3x - 2$

44. $12y^2 + 7y + 1$

45. $6x^2 + 37x + 6$

46. $5u^2 + 13u - 6$

Factoring by Grouping In Exercises 47–56, factor by grouping. *See Examples 8 and 9.*

47. $x^3 - x^2 + 2x - 2$

48. $x^3 + 5x^2 - 5x - 25$

49. $2x^3 - x^2 - 6x + 3$

50. $5x^3 - 10x^2 + 3x - 6$

51. $6 + 2y - 3y^3 - y^4$

52. $z^5 + 2z^3 + z^2 + 2$

53. $2x^2 + 3x - 2$

54. $3x^2 + 5x - 2$

55. $6x^2 - 7x - 10$

56. $8x^2 + 11x + 3$

Factoring Completely In Exercises 57–80, completely factor the expression. *See Examples 1–9.*

57. $4x^2 - 8x$

58. $12x^3 - 48x$

59. $y^3 - \frac{1}{9}y$

60. $x^3 - \frac{1}{4}x$

61. $3x^2 - 48$

62. $7y^2 - 63$

63. $x^2 - 2x + 1$

64. $9x^2 - 6x + 1$

65. $1 - 4x + 4x^2$

66. $16 + 6x - x^2$

67. $2y^3 - 7y^2 - 15y$

68. $3x^4 + x^3 - 10x^2$

69. $5x^2 - 4x + 3$

70. $7x^2 + 5x - 4$

71. $3x^3 + x^2 + 15x + 5$

72. $5 - x + 5x^2 - x^3$

73. $x^4 - 4x^3 + x^2 - 4x$

74. $3u - 2u^2 + 6 - u^3$

75. $25 - (x + 5)^2$

76. $(t - 1)^2 - 49$

77. $(x^2 + 1)^2 - 4x^2$

78. $(x^2 + 8)^2 - 36x^2$

79. $2t^3 - 16$

80. $3x^3 + 81$

Geometric Modeling   In Exercises 81–84, make a "geometric factoring model" to represent the given factorization. For instance, a factoring model for $2x^2 + 5x + 2 = (2x + 1)(x + 2)$ is shown below.

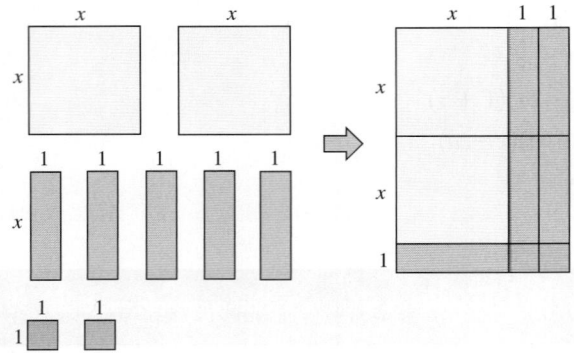 

81. $x^2 + 3x + 2 = (x + 2)(x + 1)$

82. $x^2 + 4x + 3 = (x + 3)(x + 1)$

83. $2x^2 + 7x + 3 = (2x + 1)(x + 3)$

84. $3x^2 + 7x + 2 = (3x + 1)(x + 2)$

85. **Geometry**   The room shown in the figure has a floor space of $(2x^2 - x - 3)$ square feet. The width of the room is $(x + 1)$ feet. What is the length?

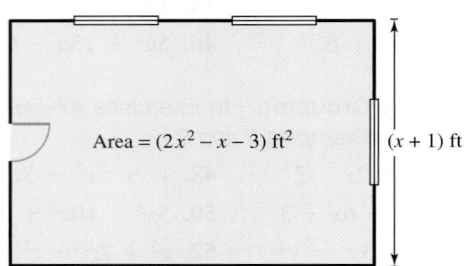

Area = $(2x^2 - x - 3)$ ft² $(x + 1)$ ft

86. **Geometry**   The room shown in the figure has a floor space of $(3x^2 + 8x + 4)$ square feet. The width of the room is $(x + 2)$ feet. What is the length?

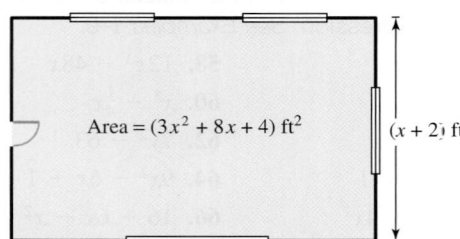

Area = $(3x^2 + 8x + 4)$ ft² $(x + 2)$ ft

87. **Choosing a Method**   Factor each trinomial. State whether you used factoring by grouping or factoring by trial and error.

   (a) $x^2 + 11x + 24$

   (b) $3x^2 + 7x - 20$

88. **Think About It**   Find all integers b for which $x^2 + bx + 24$ can be factored. Describe how you found these values of b.

89. **Think About It**   Find all integers $c > 0$ for which $x^2 + 8x + c$ can be factored. Describe how you found these values of c.

90. **Error Analysis**   Describe and correct the error.
   $$x^3 - 8 = (x - 2)(x^2 + 4).$$

91. **Think About It**   Describe two different ways to factor $2x^2 - 7x - 15$.

92. **HOW DO YOU SEE IT?**   The figure shows a large square with an area of a^2 that contains a smaller square with an area of b^2.

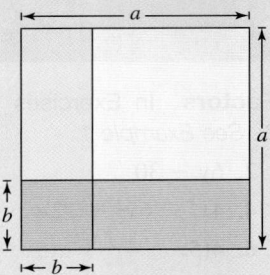

   (a) Describe the regions that represent $a^2 - b^2$. How can you rearrange these regions to show that
   $$a^2 - b^2 = (a - b)(a + b)?$$

   (b) How can you use the figure to show that
   $$(a - b)^2 = a^2 - 2ab + b^2?$$

   (c) Draw another figure to show that
   $$(a + b)^2 = a^2 + 2ab + b^2.$$
   Explain how the figure shows this.

93. **Geometric Modeling**   The figure shows a large cube with a volume of a^3 that contains a smaller cube with a volume of b^3. When the smaller cube is removed, the remaining solid has a volume of
   $$a^3 - b^3$$
   and consists of the three rectangular boxes labeled Box 1, Box 2, and Box 3. Explain how you can use the figure to obtain the factoring formula
   $$a^3 - b^3 = (a - b)(a^2 + ab + b^2).$$

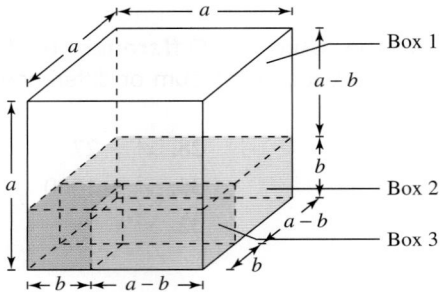

0.7 Fractional Expressions

In Exercise 72 on page 61, you will draw conclusions about the oxygen level of a pond based on a bar graph obtained using a rational model.

- Find the domain of an algebraic expression.
- Simplify a rational expression and perform operations with rational expressions.
- Simplify a complex fraction.

Domain of an Expression

The set of all real numbers for which an algebraic expression is defined is called the **domain** of the expression. For instance, the domain of

$$\frac{1}{x}$$

is all real numbers other than $x = 0$. Two algebraic expressions are **equivalent** if they have the same domain and yield the same values for all numbers in their domain. For instance, the expressions below are equivalent.

$$[(x + 1) + (x + 2)] \qquad 2x + 3$$

Example 1   Finding the Domain of an Algebraic Expression

a. The domain of the polynomial

$$2x^3 + 3x + 4$$

is the set of all real numbers. In fact, the domain of any polynomial is the set of all real numbers (unless the domain is specifically restricted).

b. The domain of the polynomial

$$x^2 + 5x + 2, \quad x > 0$$

is the set of positive real numbers, because the polynomial is specifically restricted to that set.

c. The domain of the radical expression

$$\sqrt{x}$$

is the set of nonnegative real numbers, because the square root of a negative number is not a real number.

d. The domain of the expression

$$\frac{x + 2}{x - 3}$$

is the set of all real numbers except $x = 3$, because the value $x = 3$ results in division by zero, which is undefined.

✓ Checkpoint 1

Find the domain of each expression.

a. $\dfrac{1}{x - 5}$

b. $1 - 2x - x^2$

c. $\sqrt{4x}$

Edyta Pawlowska /Shutterstock.com

Simplifying Rational Expressions

The quotient of two algebraic expressions is a **fractional expression.** Moreover, the quotient of two *polynomials* such as

$$\frac{1}{x}, \quad \frac{2x-1}{x+1}, \quad \text{or} \quad \frac{x^2-1}{x^2+1}$$

is a **rational expression.** Recall that a fraction is in simplest form if its numerator and denominator have no factors in common aside from ± 1. To write a fraction in simplest form, divide out common factors.

$$\frac{a \cdot \overset{1}{\cancel{c}}}{b \cdot \underset{1}{\cancel{c}}} = \frac{a}{b}, \quad b \neq 0, \quad c \neq 0$$

The key to success in simplifying rational expressions lies in your ability to *factor* polynomials.

Example 2   Simplifying a Rational Expression

Write the expression $\dfrac{x^2 + 4x - 12}{3x - 6}$ in simplest form.

SOLUTION

$$\frac{x^2 + 4x - 12}{3x - 6} = \frac{(x+6)(x-2)}{3(x-2)} \qquad \text{Factor completely.}$$

$$= \frac{(x+6)(x-\overset{1}{\cancel{2}})}{3(x-\underset{1}{\cancel{2}})} \qquad \text{Divide out common factors.}$$

$$= \frac{x+6}{3}, \quad x \neq 2 \qquad \text{Simplify.}$$

Note that the original expression is undefined when $x = 2$ (because division by zero is undefined). Because this is not obvious in the simplified expression, you must add the domain restriction $x \neq 2$ to the simplified expression to make it *equivalent* to the original expression.

✓Checkpoint 2

Write the expression $\dfrac{2x^2 - 2}{3x - 3}$ in simplest form.

In Example 2, do not make the mistake of trying to simplify further by dividing out *terms.*

$$\frac{x+6}{3} = \frac{x + \overset{2}{\cancel{6}}}{\underset{1}{\cancel{3}}} = x + 2$$

Remember that to simplify fractions, you divide out *factors*, not terms.

When simplifying rational expressions, be sure to factor each polynomial completely before concluding that the numerator and denominator have no factors in common. Moreover, changing the sign of a factor may allow further simplification, as demonstrated in part (b) of the next example.

Example 3 Simplifying Rational Expressions

STUDY TIP

In Example 3(b), note that when factoring completely, $4 - x = -(x - 4)$.

a. $\dfrac{x^3 - 4x}{x^2 + x - 2} = \dfrac{x(x + 2)(x - 2)}{(x + 2)(x - 1)} = \dfrac{x(x - 2)}{x - 1}, \quad x \neq -2$

Factor completely and divide out common factors.

b. $\dfrac{12 + x - x^2}{2x^2 - 9x + 4} = \dfrac{(4 - x)(3 + x)}{(2x - 1)(x - 4)} = \dfrac{-(x - 4)(3 + x)}{(2x - 1)(x - 4)}$

Factor completely and divide out common factors.

$= -\dfrac{3 + x}{2x - 1}, \quad x \neq 4$

Simplify.

✓ **Checkpoint 3**

Write the expression $\dfrac{3 - 2x - x^2}{2x^2 - 2}$ in simplest form.

To multiply or divide rational expressions, use the properties of fractions (see Section 0.2). Recall that to divide fractions you invert the divisor and multiply.

Example 4 Multiplying Rational Expressions

$\dfrac{6x^2 - 6x}{x^2 + 2x - 3} \cdot \dfrac{x^2 + x - 6}{2x}$

Original product

$= \dfrac{6x(x - 1)(x + 3)(x - 2)}{(x - 1)(x + 3)(2x)}$

Factor and multiply.

$= \dfrac{3(2x)(x - 1)(x + 3)(x - 2)}{(x - 1)(x + 3)(2x)}$

Divide out common factors.

$= 3(x - 2), \quad x \neq -3, x \neq 0, x \neq 1$

Simplify.

✓ **Checkpoint 4**

Multiply and simplify: $\dfrac{3}{x - 2} \cdot \dfrac{x - 2}{3x + 3}$.

Example 5 Dividing Rational Expressions

$\dfrac{2x}{3x - 12} \div \dfrac{x^2 - 2x}{x^2 - 6x + 8} = \dfrac{2x}{3x - 12} \cdot \dfrac{x^2 - 6x + 8}{x^2 - 2x}$

Invert and multiply.

$= \dfrac{(2x)(x - 2)(x - 4)}{(3)(x - 4)(x)(x - 2)}$

Factor and multiply.

$= \dfrac{(2x)(x - 2)(x - 4)}{(3)(x - 4)(x)(x - 2)}$

Divide out common factors.

$= \dfrac{2}{3}, \quad x \neq 0, x \neq 2, x \neq 4$

Simplify.

✓ **Checkpoint 5**

Divide and simplify: $\dfrac{4x + 4y}{5} \div \dfrac{x + y}{2}$.

To add or subtract rational expressions, use the least common denominator (LCD) method or the following basic property of fractions (see Section 0.2).

$$\frac{a}{b} \pm \frac{c}{d} = \frac{ad \pm bc}{bd}, \quad b \neq 0, d \neq 0$$

This property is efficient for adding or subtracting *two* fractions that have no common factors in their denominators.

Example 6   Adding Rational Expressions

$$\frac{x}{x-3} + \frac{2}{3x+4} = \frac{x(3x+4) + 2(x-3)}{(x-3)(3x+4)} \qquad \frac{a}{b} + \frac{c}{d} = \frac{ad+bc}{bd}$$

$$= \frac{3x^2 + 4x + 2x - 6}{(x-3)(3x+4)} \qquad \text{Distributive Property}$$

$$= \frac{3x^2 + 6x - 6}{(x-3)(3x+4)} \qquad \text{Combine like terms.}$$

$$= \frac{3(x^2 + 2x - 2)}{(x-3)(3x+4)} \qquad \text{Factor.}$$

✓**Checkpoint 6**

Subtract: $\dfrac{4}{x} - \dfrac{2x}{3}$. ■

For fractions with repeated factors in their denominators, the LCD method works well.

Example 7   Combining Rational Expressions: The LCD Method

Perform the indicated operations and simplify: $\dfrac{3}{x-1} - \dfrac{2}{x} + \dfrac{x+3}{x^2-1}$.

SOLUTION   Using the factored denominators $(x-1)$, x, and $(x+1)(x-1)$, you can see that the least common denominator is $x(x+1)(x-1)$.

$$\frac{3}{x-1} - \frac{2}{x} + \frac{x+3}{x^2-1}$$

$$= \frac{3(x)(x+1)}{x(x+1)(x-1)} - \frac{2(x+1)(x-1)}{x(x+1)(x-1)} + \frac{(x+3)(x)}{x(x+1)(x-1)}$$

$$= \frac{3(x)(x+1) - 2(x+1)(x-1) + (x+3)(x)}{x(x+1)(x-1)}$$

$$= \frac{3x^2 + 3x - 2x^2 + 2 + x^2 + 3x}{x(x+1)(x-1)}$$

$$= \frac{2x^2 + 6x + 2}{x(x+1)(x-1)} = \frac{2(x^2 + 3x + 1)}{x(x+1)(x-1)}$$

✓**Checkpoint 7**

Perform the indicated operations and simplify:

$$\frac{5}{x} + \frac{4}{x-1} - \frac{4}{x(x-1)}. \qquad ■$$

STUDY TIP

Recall that the least common denominator of two or more fractions consists of the product of all prime factors in the denominators, with each factor given the highest power of its occurrence in any denominator.

Complex Fractions

Fractional expressions with separate fractions in the numerator and/or the denominator are called **complex fractions.** Here are two examples.

$$\frac{\left(\dfrac{1}{x}\right)}{x^2 + 1} \quad \text{and} \quad \frac{\left(\dfrac{1}{x}\right)}{\left(\dfrac{1}{x^2 + 1}\right)}$$

A complex fraction can be simplified by combining the fractions in its numerator into a single fraction and then combining the fractions in its denominator into a single fraction. Then invert the denominator and multiply.

Example 8 **Simplifying a Complex Fraction**

$$\frac{\left(\dfrac{2}{x} - 3\right)}{\left(1 - \dfrac{1}{x-1}\right)} = \frac{\left[\dfrac{2 - 3(x)}{x}\right]}{\left[\dfrac{1(x-1) - 1}{x-1}\right]} \qquad \text{Combine fractions.}$$

$$= \frac{\left(\dfrac{2 - 3x}{x}\right)}{\left(\dfrac{x - 2}{x - 1}\right)} \qquad \text{Simplify.}$$

$$= \frac{2 - 3x}{x} \cdot \frac{x - 1}{x - 2} \qquad \text{Invert and multiply.}$$

$$= \frac{(2 - 3x)(x - 1)}{x(x - 2)}, \quad x \neq 1$$

✔ **Checkpoint 8**

Simplify the complex fraction $\dfrac{\left(\dfrac{x}{3} - 1\right)}{x - 3}$.

Another way to simplify the complex fraction in Example 8 is to multiply the numerator and denominator by the LCD of all fractions in its numerator and denominator, as shown below. After simplifying, the same result is obtained.

$$\frac{\left(\dfrac{2}{x} - 3\right)}{\left(1 - \dfrac{1}{x-1}\right)} = \frac{\left(\dfrac{2}{x} - 3\right)}{\left(1 - \dfrac{1}{x-1}\right)} \cdot \frac{x(x-1)}{x(x-1)} \qquad \text{LCD is } x(x - 1).$$

SUMMARIZE   (Section 0.7)

1. State the definition of the domain of an algebraic expression *(page 55)*. For an example of finding the domain of an algebraic expression, see Example 1.

2. State the definition of a rational expression and describe how to simplify a rational expression *(page 56)*. For examples of simplifying rational expressions, see Examples 2, 3, 4, 5, 6, and 7.

3. State the definition of a complex fraction *(page 59)*. For an example of simplifying a complex fraction, see Example 8.

In Exercises 1–10, completely factor the polynomial.

1. $5x^2 - 15x^3$

2. $16x^2 - 9$

3. $9x^2 - 6x + 1$

4. $9 + 12y + 4y^2$

5. $z^2 + 4z + 3$

6. $2x^2 - 30x + 100$

7. $3 + 8x - 3x^2$

8. $3x^2 - 46x + 15$

9. $s^3 + s^2 - 4s - 4$

10. $2x^3 - 128$

Exercises 0.7

See www.CalcChat.com for worked-out solutions to odd-numbered exercises.

Checking Domain Values   In Exercises 1–4, determine whether each value of x is in the domain of the expression.

1. $\dfrac{x + 2}{4x + 2}$ (a) $x = -\dfrac{1}{2}$ (b) $x = 2$

2. $\dfrac{2x + 3}{x - 4}$ (a) $x = -\dfrac{3}{2}$ (b) $x = 4$

3. $\sqrt{2x + 4}$ (a) $x = -2$ (b) $x = 2$

4. $\sqrt{3x - 9}$ (a) $x = -3$ (b) $x = 3$

Finding the Domain of an Algebraic Expression   In Exercises 5–12, find the domain of the expression. *See Example 1.*

5. $3x^2 - 4x + 7$

6. $6x^2 - 9, \ x > 0$

7. $\dfrac{1}{3 - x}$

8. $\dfrac{x + 1}{2x + 1}$

9. $\dfrac{x - 1}{x^2 - 4x}$

10. $\dfrac{4x + 3}{x^2 - 36}$

11. $\sqrt{x + 1}$

12. $\sqrt{x - 1}$

Equivalent Rational Expressions   In Exercises 13–18, find the missing factor and state any domain restrictions necessary to make the two fractions equivalent.

13. $\dfrac{5}{2x} = \dfrac{5()}{6x^2}$

14. $\dfrac{3}{4} = \dfrac{3()}{4(x + 1)}$

15. $\dfrac{x + 1}{x} = \dfrac{(x + 1)()}{x(x - 2)}$

16. $\dfrac{3y - 4}{y + 1} = \dfrac{(3y - 4)()}{y^2 - 1}$

17. $\dfrac{3x}{x - 3} = \dfrac{3x()}{x^2 - x - 6}$

18. $\dfrac{1 - z}{z^2} = \dfrac{(1 - z)()}{z^3 + z^2}$

Simplifying a Rational Expression   In Exercises 19–34, write the rational expression in simplest form. *See Examples 2 and 3.*

19. $\dfrac{15x^2}{10x}$

20. $\dfrac{24y^3}{56y^7}$

21. $\dfrac{2x}{4x + 4}$

22. $\dfrac{9x^2 + 9x}{2x + 2}$

23. $\dfrac{x - 5}{10 - 2x}$

24. $\dfrac{3 - x}{8x - 24}$

25. $\dfrac{x^2 - 25}{5 - x}$

26. $\dfrac{x^2 - 16}{4 - x}$

27. $\dfrac{x^3 + 5x^2 + 6x}{x^2 - 4}$

28. $\dfrac{x^3 - 2x^2 - 3x}{x^2 - 9}$

29. $\dfrac{y^2 - 7y + 12}{y^2 + 3y - 18}$

30. $\dfrac{x^2 + 8x - 20}{x^2 + 11x + 10}$

31. $\dfrac{2 - x + 2x^2 - x^3}{x - 2}$

32. $\dfrac{x^2 - 9}{x^3 + x^2 - 9x - 9}$

33. $\dfrac{z^3 - 27}{z^2 + 3z + 9}$

34. $\dfrac{y^3 - 2y^2 - 8y}{y^3 + 8}$

Multiplying or Dividing Rational Expressions   In Exercises 35–48, perform the indicated operations and simplify. *See Examples 4 and 5.*

35. $\dfrac{5}{x - 1} \cdot \dfrac{x - 1}{25(x - 2)}$

36. $\dfrac{x + 13}{x^3(3 - x)} \cdot \dfrac{x(x - 3)}{5}$

37. $\dfrac{(x - 9)(x + 7)}{x + 1} \cdot \dfrac{x}{9 - x}$

38. $\dfrac{(x + 5)(x - 3)}{x + 2} \cdot \dfrac{1}{(x + 5)(x + 2)}$

39. $\dfrac{r}{r - 1} \cdot \dfrac{r^2 - 1}{r^2}$

40. $\dfrac{4y - 16}{5y + 15} \cdot \dfrac{2y + 6}{4 - y}$

41. $\dfrac{t^2 - t - 6}{t^2 + 6t + 9} \cdot \dfrac{t + 3}{t^2 - 4}$

42. $\dfrac{y^3 - 8}{2y^3} \cdot \dfrac{4y}{y^2 - 5y + 6}$

43. $\dfrac{x^2 + x - 2}{x^3 + x^2} \cdot \dfrac{x}{x^2 + 3x + 2}$

44. $\dfrac{x^3 - 8}{x + 1} \cdot \dfrac{x^2 - 1}{x^3 - 3x^2 + 2x}$

45. $\dfrac{3(x + y)}{4} \div \dfrac{x + y}{2}$

46. $\dfrac{x + 2}{5(x - 3)} \div \dfrac{x - 2}{5(x - 3)}$

47. $\dfrac{x^2}{(x+1)^2} \div \dfrac{x}{(x+1)^3}$ **48.** $\dfrac{x^2-1}{x} \div \dfrac{(x-1)^2}{x}$

Finding the Least Common Denominator In Exercises 49–52, find the least common denominator of the expressions.

49. $\dfrac{10}{x+4}, \dfrac{x+1}{x^2-2x-24}$ **50.** $\dfrac{x}{x+2}, \dfrac{3}{x^2+5x+6}$

51. $\dfrac{1}{x^2}, \dfrac{1}{x-1}, \dfrac{1}{x^2-x}$ **52.** $\dfrac{1}{x}, \dfrac{1}{x^2+3x}, \dfrac{1}{x+3}$

Adding or Subtracting Rational Expressions In Exercises 53–62, perform the indicated operations and simplify. *See Examples 6 and 7.*

53. $\dfrac{4x}{x-2} + \dfrac{x}{x-2}$ **54.** $\dfrac{3x-2}{x+1} + \dfrac{2-x}{x+1}$

55. $\dfrac{3x}{x-4} - \dfrac{x}{4-x}$ **56.** $\dfrac{4x}{3-x} - \dfrac{5x}{x-3}$

57. $4 + \dfrac{3}{x-5}$ **58.** $\dfrac{4}{x+2} - 6$

59. $\dfrac{2}{x^2-4} - \dfrac{1}{x^2-3x+2}$ **60.** $\dfrac{x}{x^2+x-2} - \dfrac{1}{x+2}$

61. $-\dfrac{1}{x} + \dfrac{2}{x^2+1} + \dfrac{1}{x^3+x}$ **62.** $\dfrac{2}{x+1} + \dfrac{2}{x-1} + \dfrac{1}{x^2-1}$

Simplifying a Complex Fraction In Exercises 63–68, simplify the complex fraction. *See Example 8.*

63. $\dfrac{\left(\dfrac{x}{2}-1\right)}{(x-2)}$ **64.** $\dfrac{(x-3)}{\left(\dfrac{x}{4}-\dfrac{4}{x}\right)}$

65. $\dfrac{\left(\dfrac{1}{x}-\dfrac{1}{x+1}\right)}{\left(\dfrac{1}{x+1}\right)}$ **66.** $\dfrac{\left(\dfrac{5}{y}-\dfrac{6}{2y+1}\right)}{\left(\dfrac{5}{y}+4\right)}$

67. $\dfrac{\left(\sqrt{x}-\dfrac{1}{2\sqrt{x}}\right)}{\sqrt{x}}$ **68.** $\dfrac{\left(\dfrac{1}{\sqrt{2y}}+\sqrt{2y}\right)}{\sqrt{2y}}$

Monthly Payment In Exercises 69 and 70, use the formula for the approximate annual interest rate r of a monthly installment loan

$$r = \dfrac{\left[\dfrac{24(NM-P)}{N}\right]}{\left(P+\dfrac{NM}{12}\right)}$$

where N is the total number of payments, M is the monthly payment, and P is the amount financed.

69. (a) Approximate the annual interest rate r for a four-year car loan of $18,000 with monthly payments of $415.

(b) Simplify the expression for the annual interest rate r, and then rework part (a).

70. (a) Approximate the annual interest rate r for a five-year car loan of $20,000 with monthly payments of $415.

(b) Simplify the expression for the annual interest rate r, and then rework part (a).

71. Refrigeration Food at a temperature of 75°F is placed in a 40°F refrigerator. A model for the temperature T (in degrees Fahrenheit) of the food after t hours is

$$T = 10\left(\dfrac{4t^2+16t+75}{t^2+4t+10}\right), \quad t \geq 0.$$

Sketch a bar graph showing the temperature of the food at $t = 0, 1, 2, 3, 4,$ and 5 hours. According to the bar graph, does it seem likely that the temperature of the food will drop to 40°F within 7 hours of being placed in the refrigerator? Explain.

 **72.** **HOW DO YOU SEE IT?** The mathematical model

$$P = 100\left(\dfrac{t^2-t+1}{t^2+1}\right), \quad t \geq 0$$

gives the percent P of the normal level of oxygen in a pond, where t is the time (in weeks) after organic waste is dumped into the pond. The bar graph shows the situation. What conclusions can you draw from the bar graph?

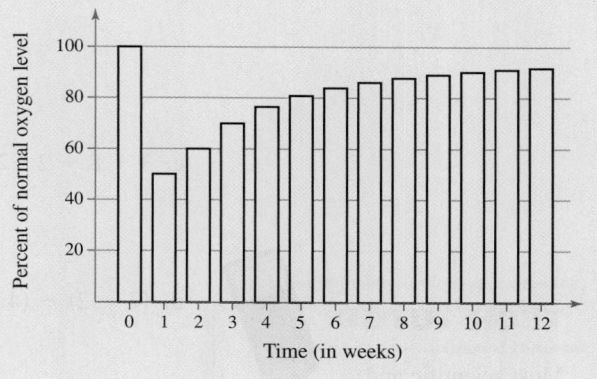

True or False? In Exercises 73 and 74, determine whether the statement is true or false. Justify your answer.

73. $\dfrac{x^{2n}-1^{2n}}{x^n-1^n} = x^n + 1^n$ **74.** $\dfrac{x^2-3x+2}{x-1} = x-2$

75. Writing In simplifying a rational expression, when is it necessary to indicate domain restrictions?

76. Think About It Is it true that

$$\dfrac{ax-b}{b-ax} = -1$$

for all nonzero numbers a and b? Explain.

ALGEBRA TUTOR

Order of Operations

Much of the algebra in this chapter involves evaluation of algebraic expressions. When you evaluate an algebraic expression, you need to know the priorities assigned to different operations. These priorities are called the *order of operations.*

1. Perform operations inside *symbols of grouping* or *absolute value symbols*, starting with the innermost symbol.

2. Evaluate all *exponential* expressions.

3. Perform all *multiplications* and *divisions* from left to right.

4. Perform all *additions* and *subtractions* from left to right.

Example 1   **Using Order of Operations**

Evaluate each expression.

a. $19 - 3 \cdot 6$ **b.** $20 - 2 \cdot 3^2 + 11$

c. $3 + 8 \div 2 \cdot 2$ **d.** $(7 - 2) - (4 - 5)$

e. $7 - [(5 \cdot 3) + 2^3]$ **f.** $16 + (12 - |4 - 8|)$

SOLUTION

a. $19 - 3 \cdot 6 = 19 - 18$ Multiply.

$= 1$ Subtract.

b. $20 - 2 \cdot 3^2 + 11 = 20 - 2 \cdot 9 + 11$ Evaluate exponential expression.

$= 20 - 18 + 11$ Multiply.

$= 2 + 11$ Subtract.

$= 13$ Add.

c. $3 + 8 \div 2 \cdot 2 = 3 + 4 \cdot 2$ Divide.

$= 3 + 8$ Multiply.

$= 11$ Add.

d. $(7 - 2) - (4 - 5) = 5 - (-1)$ Subtract inside parentheses.

$= 6$ Subtract.

e. $7 - [(5 \cdot 3) + 2^3] = 7 - [15 + 2^3]$ Multiply inside parentheses.

$= 7 - [15 + 8]$ Evaluate exponential expression.

$= 7 - 23$ Add inside brackets.

$= -16$ Subtract.

f. $16 + (12 - |4 - 8|) = 16 + (12 - |-4|)$ Subtract inside absolute value symbols.

$= 16 + (12 - 4)$ Evaluate absolute value.

$= 16 + 8$ Subtract inside parentheses.

$= 24$ Add.

TECH TUTOR

Most scientific and graphing calculators use the same order of operations listed above. Try entering the expressions in Example 1 into your calculator. Do you get the same results?

Simplifying Expressions

A second algebraic skill studied in this chapter is simplifying expressions by combining like terms.

Like terms have the same variables raised to the same powers. For instance, $5x^2$ and $7x^2$ are like terms. Constant terms, such as 8 and -3, are also like terms. An expression is said to be *simplified* when it contains no symbols of grouping and all like terms are combined. The Distributive Property allows you to combine like terms by adding coefficients. For instance,

$$5x^2 + 7x^2 = (5 + 7)x^2 = 12x^2.$$

Example 2   Simplifying Expressions by Combining Like Terms

Simplify each expression.

a. $-8x^2 + x^2 - 9x^2$

b. $7 - x + 20 + 4x - 1$

c. $4x^3 + 5x^2 - 9x^3 - 2x^2$

d. $x^2 - 8x + x + 6x^2 - 3x$

e. $12x - 10y - 5x + y$

f. $4(w + 9) - 5(w - 3)$

g. $3z + 1 - 6z^2 - z - 4 + 11z^2$

SOLUTION

a. $-8x^2 + x^2 - 9x^2 = (-8 + 1 - 9)x^2$ Distributive Property

$\qquad\qquad\qquad = -16x^2$ Add coefficients.

b. $7 - x + 20 + 4x - 1 = (-x + 4x) + (7 + 20 - 1)$ Group like terms.

$\qquad\qquad\qquad = 3x + 26$ Combine like terms.

c. $4x^3 + 5x^2 - 9x^3 - 2x^2 = (4x^3 - 9x^3) + (5x^2 - 2x^2)$ Group like terms.

$\qquad\qquad\qquad = -5x^3 + 3x^2$ Combine like terms.

d. $x^2 - 8x + x + 6x^2 - 3x = (x^2 + 6x^2) + (-8x + x - 3x)$ Group like terms.

$\qquad\qquad\qquad = 7x^2 - 10x$ Combine like terms.

e. $12x - 10y - 5x + y = (12x - 5x) + (-10y + y)$ Group like terms.

$\qquad\qquad\qquad = 7x - 9y$ Combine like terms.

f. $4(w + 9) - 5(w - 3) = 4w + 36 - 5w + 15$ Distributive Property

$\qquad\qquad\qquad = (4w - 5w) + (36 + 15)$ Group like terms.

$\qquad\qquad\qquad = -w + 51$ Combine like terms.

g. $3z + 1 - 6z^2 - z - 4 + 11z^2$

$\qquad = (-6z^2 + 11z^2) + (3z - z) + (1 - 4)$ Group like terms.

$\qquad = 5z^2 + 2z - 3$ Combine like terms.

SUMMARY AND STUDY STRATEGIES

After studying this chapter, you should have acquired the following skills. The exercise numbers are keyed to the Review Exercises that begin on page 66. Answers to odd-numbered Review Exercises are given in the back of the text.*

Section 0.1	Review Exercises
∎ Classify real numbers as natural numbers, integers, rational numbers, or irrational numbers.	*1, 2*
∎ Order real numbers.	*3, 4*
∎ Use and interpret inequality notation.	*5–10*
∎ Find the absolute value of a number and use absolute value to find the distance between two numbers.	*11–18*

Section 0.2

∎ Identify the terms of an algebraic expression.	*19, 20*
∎ Evaluate an algebraic expression.	*21, 22*
∎ Identify basic rules of algebra.	*23–26*
∎ Perform operations on real numbers.	*27–30*
∎ Use a calculator to evaluate an expression and round decimal numbers.	*31, 32*

Section 0.3

∎ Use properties of exponents to simplify expressions with exponents.

$$a^m a^n = a^{m+n} \qquad \frac{a^m}{a^n} = a^{m-n} \qquad (ab)^m = a^m b^m$$

$$\left(\frac{a}{b}\right)^m = \frac{a^m}{b^m} \qquad (a^m)^n = a^{mn} \qquad a^{-n} = \frac{1}{a^n}$$

$$a^0 = 1 \qquad \left(\frac{a}{b}\right)^{-n} = \left(\frac{b}{a}\right)^n \qquad |a^2| = |a|^2 = a^2 \qquad \textit{33–36}$$

∎ Use scientific notation.	*37–40*
∎ Use a calculator to evaluate exponential expressions.	*41, 42*
∎ Use interest formulas to solve application problems.	

Simple interest: $A = P(1 + rt)$

Compound interest: $A = P\left(1 + \dfrac{r}{n}\right)^{nt}$ *43, 44*

---

* A wide range of valuable study aids are available to help you master the material in this chapter. The *Student Solutions Manual* includes step-by-step solutions to all odd-numbered exercises to help you review and prepare. The student website at *www.cengagebrain.com* offers algebra help and a *Graphing Technology Guide*, which contains step-by-step commands and instructions for a wide variety of graphing calculators.

Section 0.4   Review Exercises

■ Simplify and evaluate expressions involving radicals.

$$\sqrt[n]{a^m} = \left(\sqrt[n]{a}\right)^m \qquad \sqrt[n]{a} \cdot \sqrt[n]{b} = \sqrt[n]{ab} \qquad \frac{\sqrt[n]{a}}{\sqrt[n]{b}} = \sqrt[n]{\frac{a}{b}}$$

$$\sqrt[m]{\sqrt[n]{a}} = \sqrt[mn]{a} \qquad \left(\sqrt[n]{a}\right)^n = a \quad \text{For } n \text{ even, } \sqrt[n]{a^n} = |a|. \text{ For } n \text{ odd, } \sqrt[n]{a^n} = a. \qquad \textit{45–52}$$

■ Rationalize a denominator by using its conjugate.   *53, 54*

■ Combine like radicals.   *55, 56*

■ Use properties of rational exponents.   *57–60*

■ Use a calculator to evaluate a radical.   *61, 62*

Section 0.5

■ Write a polynomial in standard form.   *63–72*

■ Add and subtract polynomials by combining like terms.   *63–66*

■ Multiply polynomials using the FOIL Method or a vertical format.   *67, 69, 70*

■ Use special products to multiply polynomials.

$$(u + v)(u - v) = u^2 - v^2 \qquad (u \pm v)^2 = u^2 \pm 2uv + v^2$$

$$(u \pm v)^3 = u^3 \pm 3u^2v + 3uv^2 \pm v^3 \qquad\qquad\qquad\qquad \textit{68, 71, 72}$$

■ Use polynomials to solve application problems.   *73, 74*

Section 0.6

■ Factor a polynomial by factoring out common factors.   *75, 76, 78, 80*

■ Factor a polynomial in a special form.

$$u^2 - v^2 = (u + v)(u - v) \qquad u^2 \pm 2uv + v^2 = (u \pm v)^2$$

$$u^3 \pm v^3 = (u \pm v)(u^2 \mp uv + v^2) \qquad\qquad\qquad\qquad \textit{75–78}$$

■ Factor a trinomial as the product of two binomials.   *79, 80*

■ Factor a polynomial by grouping.   *81, 82*

Section 0.7

■ Find the domain of an algebraic expression by finding values of the variable that
  make a denominator zero or a radicand negative.   *83–88*

■ Simplify a rational expression by dividing out common factors from the
  numerator and denominator.   *89–92*

■ Perform operations with rational expressions by using properties of fractions.   *93–98*

■ Simplify a complex fraction.   *99, 100*

Study Strategies

■ **Use the Skills Warm-Up Exercises**   Each exercise set in this text (except the set for Section 0.1) begins with a set
  of skills warm-up exercises. You should begin each homework session by quickly working through all of these exercises
  (all are answered in the back of the text). The "old" skills covered in these exercises are needed to master the "new"
  skills in the section exercise set. The skills warm-up exercises remind you that mathematics is cumulative—to be
  successful in this course, you must retain "old" skills.

■ **Use the Additional Study Aids**   The additional study aids were prepared specifically to help you master the concepts
  discussed in the text. They are the *Student Solutions Manual*, the *Graphing Technology Guide*, and the *Instructional DVD*.

Review Exercises

See www.CalcChat.com for worked-out solutions to odd-numbered exercises.

Classifying Real Numbers In Exercises 1 and 2, determine which numbers are (a) natural numbers, (b) integers, (c) rational numbers, and (d) irrational numbers.

1. $\left\{9, -7, -\frac{8}{9}, \frac{5}{2}, \sqrt{6}, 0.4\right\}$ **2.** $\left\{\sqrt{8}, -8, -\frac{10}{3}, 0, 5.2, \frac{3}{7}\right\}$

Comparing Numbers on the Real Number Line In Exercises 3 and 4, plot the two real numbers on the real number line and place the appropriate inequality sign ($<$ or $>$) between them.

3. $-4, -3$ **4.** $\frac{1}{5}, \frac{1}{6}$

Interpreting an Inequality In Exercises 5 and 6, give a verbal description of the subset of real numbers that is represented by the inequality, and sketch the subset on the real number line.

5. $x \le -6$ **6.** $x > 5$

Inequalities and Subsets of Real Numbers In Exercises 7–10, use inequality notation to describe the subset of real numbers.

7. x is nonnegative. **8.** x is at most 7.

9. x is greater than 2 and less than or equal to 5.

10. x is less than or equal to -2 or x is greater than 2.

Finding Absolute Value In Exercises 11 and 12, evaluate the expression.

11. $-|-14|$ **12.** $|-4 - 2|$

The Distance Between Two Numbers In Exercises 13–16, find the distance between a and b.

13. $a = -14, \quad b = -18$ **14.** $a = -1, \quad b = -5$

15. $a = 2, \quad b = -8$ **16.** $a = 10, \quad b = -7$

Using Absolute Value Notation In Exercises 17 and 18, use absolute value notation to describe the sentence.

17. The distance between x and 7 is at least 4.

18. The distance between x and -22 is no more than 10.

Identifying Terms In Exercises 19 and 20, identify the terms of the algebraic expression.

19. $5x^3 - 2x + 7$ **20.** $7x^5 + 4x^2 - 8$

Evaluating Algebraic Expressions In Exercises 21 and 22, evaluate the expression for each value of x.

21. $-4x^2 - 6x$ (a) $x = -1$ (b) $x = 0$

22. $12 - 5x^2$ (a) $x = -2$ (b) $x = 3$

Basic Rules of Algebra In Exercises 23–26, identify the rule of algebra illustrated by the statement.

23. $5(x^2 + x) = 5x^2 + 5x$

24. $(x + 6) + 3 = x + (6 + 3)$

25. $3x + 7 = 7 + 3x$ **26.** $(x^2 - 1)\left(\dfrac{1}{x^2 - 1}\right) = 1$

Properties of Zero and Properties of Fractions In Exercises 27–30, perform the indicated operation(s). (Write fractional answers in simplest form.)

27. $-3(0) - 2(4 - 5)$ **28.** $\dfrac{-3 + 3}{9} - 5(2 - 3)$

29. $\dfrac{1}{2} + \dfrac{1}{3} - \dfrac{1}{6}$ **30.** $\dfrac{5}{12} + \dfrac{3}{5}$

Using a Calculator In Exercises 31 and 32, use a calculator to evaluate the expression. (Round to two decimal places.)

31. $4\left(\dfrac{1}{6} - \dfrac{1}{7}\right)$ **32.** $-2 + 3\left(\dfrac{1}{2} - \dfrac{1}{3}\right)$

Using Properties of Exponents In Exercises 33–36, simplify the expression.

33. $\dfrac{(4x)^2}{2x}$ **34.** $(-x)^2(-3x)^3$

35. $\dfrac{10x^2}{2x^6}$ **36.** $2x(5x^2)^3$

Converting to Scientific Notation In Exercises 37 and 38, write the number in scientific notation.

37. $308,700,000$ **38.** 0.0002036

Converting to Decimal Notation In Exercises 39 and 40, write the number in decimal notation.

39. 7.28×10^6 **40.** 8.4×10^{-5}

Using a Calculator In Exercises 41 and 42, use a calculator to evaluate the expression. (Round to three decimal places.)

41. $1800(1 + 0.08)^{24}$ **42.** $50,000\left(1 + \dfrac{0.075}{12}\right)^{48}$

Finding the Balance in an Account In Exercises 43 and 44, you deposit $8000 in an account earning the interest described. Complete the table showing the balance over time.

Year	5	10	15	20	25
Balance					

43. Annual interest rate of 4.5% compounded monthly

44. Annual interest rate of 4% compounded quarterly

Converting Between Forms In Exercises 45 and 46, fill in the missing form.

Radical Form　　*Rational Exponent Form*

45. $\sqrt{16} = 4$

46. 　　　　　　　　　$16^{1/4} = 2$

Evaluating an Expression Involving Radicals In Exercises 47 and 48, evaluate the expression.

47. $\sqrt{169}$　　　　　　**48.** $\sqrt[3]{125}$

Simplifying a Radical Expression In Exercises 49–52, simplify the expression.

49. $\sqrt{4x^4}$　　　　　　**50.** $\sqrt[3]{54x^3}$

51. $\sqrt{6}\sqrt{2}$　　　　　**52.** $\sqrt{6}\sqrt{3}$

Rationalizing a Denominator In Exercises 53 and 54, rewrite the expression by rationalizing the denominator. Simplify your answer.

53. $\dfrac{1}{2 - \sqrt{3}}$　　　　　**54.** $\dfrac{2}{3 + \sqrt{5}}$

Simplifying Expressions In Exercises 55–58, simplify the expression.

55. $2\sqrt{x} - 5\sqrt{x}$　　　　**56.** $\sqrt{72} + \sqrt{128}$

57. $(64)^{-2/3}$　　　　　　**58.** $4^{1/3} \cdot 4^{5/3}$

Reducing the Index In Exercises 59 and 60, use rational exponents to reduce the index of the radical.

59. $\sqrt[4]{5^2}$　　　　　　**60.** $\sqrt[8]{x^4}$

Evaluating Radicals with a Calculator In Exercises 61 and 62, use a calculator to approximate the number. (Round your answer to three decimal places.)

61. $\sqrt{127}$　　　　　　**62.** $\sqrt[3]{52}$

Operations with Polynomials In Exercises 63–72, perform the indicated operation(s) and write the resulting polynomial in standard form.

63. $(x - 3) - (2x - 8)$　　**64.** $(x^2 + 2) + (2 - 4x)$

65. $x(x - 2) - 2(3x + 7)$　**66.** $2x(x + 1) + 3(x^2 - x)$

67. $(x + 1)(x - 2)$　　　　**68.** $(2x - 5)(2x + 5)$

69. $(x + 4)(x^2 - 4x + 16)$　**70.** $(x - 2)(x^2 + 6x + 9)$

71. $(x + 4)^2$　　　　　　**72.** $(2x + 1)^3$

73. Cell Sites The numbers of mobile telephone cell sites in the United States from 2005 through 2010 can be approximated by the polynomial $14,945.3x + 105,238$, where x represents the year, with $x = 5$ corresponding to 2005. Evaluate the polynomial at $x = 10$. Then interpret your result in the context of the situation. *(Source: CTIA-The Wireless Association)*

74. Mobile Home Prices The average sale price (in dollars) of a newly manufactured residential mobile home in the United States each year from 2004 through 2009 can be approximated by the polynomial

$$-707.14x^2 + 10,104.3x + 29,311$$

where x represents the year, with $x = 4$ corresponding to 2004. Evaluate the polynomial at $x = 4$, $x = 8$, and $x = 9$. Interpret your results in the context of the situation. *(Source: U.S. Census Bureau)*

Factoring Completely Exercises 75–82, completely factor the expression.

75. $4x^2 - 36$　　　　　　**76.** $x^3 - 16x$

77. $8x^3 - 125$　　　　　**78.** $2x^6 - 16x^3$

79. $x^2 - 4x - 5$　　　　　**80.** $-3x^2 - 6x + 3x^3$

81. $x^3 - 4x^2 - 2x + 8$　　**82.** $x^3 - 2x^2 - 9x + 18$

Finding the Domain of an Algebraic Expression In Exercises 83–86, find the domain of the expression.

83. $\dfrac{2x + 1}{x - 4}$　　　　　**84.** $\dfrac{x - 3}{7 - x}$

85. $2\sqrt{x + 1}$　　　　　　**86.** $\sqrt{x - 5}$

Equivalent Rational Expressions In Exercises 87 and 88, find the missing factor and state any domain restrictions necessary to make the two fractions equivalent.

87. $\dfrac{4}{3x} = \dfrac{4(\ \ \ \ )}{9x^2}$　　　　**88.** $\dfrac{5}{7} = \dfrac{5(\ \ \ \ )}{7(x + 2)}$

Simplifying a Rational Expression In Exercises 89–92, write the rational expression in simplest form.

89. $\dfrac{x^2 - 4}{2x + 4}$　　　　　**90.** $\dfrac{2x^2 + 4x}{2x}$

91. $\dfrac{x^3 - 9x}{x^3 - 4x^2 + 3x}$　　　**92.** $\dfrac{x^3 + 64}{x^2 - x - 20}$

Operations with Rational Expressions In Exercises 93–98, perform the indicated operation(s) and simplify.

93. $\dfrac{2x - 1}{x + 1} \cdot \dfrac{x^2 - 1}{2x^2 - 7x + 3}$　**94.** $\dfrac{x + 2}{x - 4} \div \dfrac{2x + 4}{8x}$

95. $\dfrac{x}{x - 1} + \dfrac{2x}{x - 2}$　　　**96.** $\dfrac{2}{x + 2} - \dfrac{3}{x - 2}$

97. $\dfrac{2}{x - 1} + \dfrac{4}{x + 1} + \dfrac{8}{x^2 - 1}$　**98.** $\dfrac{1}{x - 1} + \dfrac{2}{x} - \dfrac{1}{x^2 - x}$

Simplifying a Complex Fraction In Exercises 99 and 100, simplify the complex fraction.

99. $\dfrac{\left(\dfrac{x^2 - 1}{x}\right)}{\left[\dfrac{(x - 1)^2}{x}\right]}$　　　　**100.** $\dfrac{(x - 4)}{\left(\dfrac{x}{4} - \dfrac{4}{x}\right)}$

TEST YOURSELF

See www.CalcChat.com for worked-out solutions to odd-numbered exercises.

Take this test as you would take a test in class. When you are done, check your work against the answers given in the back of the book.

1. Evaluate the expression $-3x^2 - 5x$ at $x = -3$.

2. Complete the table at the left given that $4000 is deposited in an account with an annual interest rate of 3.5%, compounded monthly. What can you conclude from the table?

Year	Balance
5	
10	
15	
20	
25	

Table for 2

In Exercises 3–8, simplify the expression.

3. $8(-2x^2)^3$

4. $3\sqrt{x} - 7\sqrt{x}$

5. $5^{1/4} \cdot 5^{7/4}$

6. $\sqrt{48} - \sqrt{80}$

7. $\sqrt{12x^3}$

8. $\dfrac{2}{5 - \sqrt{7}}$

In Exercises 9 and 10, write the polynomial in standard form.

9. $(3x + 7)^2$

10. $3x(x + 5) - 2x(4x - 7)$

In Exercises 11–14, completely factor the expression.

11. $5x^2 - 80$

12. $4x^2 + 12x + 9$

13. $x^3 - 6x^2 - 3x + 18$

14. $x^3 + 2x^2 - 4x - 8$

15. Simplify: $\dfrac{x^2 - 16}{3x + 12}$.

16. Multiply and simplify: $\dfrac{3x - 5}{x + 3} \cdot \dfrac{x^2 + 7x + 12}{9x^2 - 25}$.

17. Add and simplify: $\dfrac{x}{x - 3} + \dfrac{3x}{x - 4}$.

18. Subtract and simplify: $\dfrac{3}{x + 5} - \dfrac{4}{x - 2}$.

In Exercises 19 and 20, find the domain of the expression.

19. $4\sqrt{x - 10}$

20. $\dfrac{3}{x + 9}$

21. Simplify the complex fraction $\dfrac{\left(\dfrac{2x - 9}{x - 1}\right)}{\left(\dfrac{3}{x - 1} + \dfrac{1 - x}{x + 2}\right)}$.

22. Movie Price The average price of a movie ticket in the United States each year from 2000 through 2009 can be approximated by the polynomial $0.223x + 5.35$, where x represents the year, with $x = 0$ corresponding to 2000. Evaluate the polynomial at $x = 0$ and $x = 9$. Then explain the meaning of your results in the context of the situation. *(Source: National Association of Theatre Owners)*

1 Equations and Inequalities

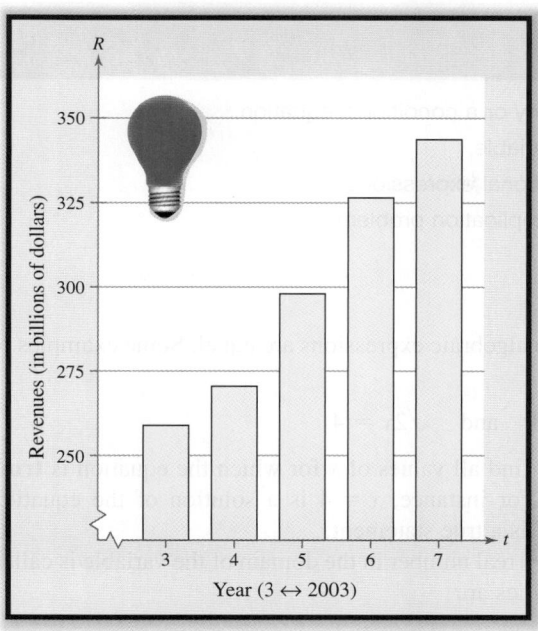

Example 9 on page 99 shows how a quadratic equation can be used to approximate revenues for the electric power industry in the United States for given years.

junjie/www.shutterstock.com
Kurhan/Shutterstock.com

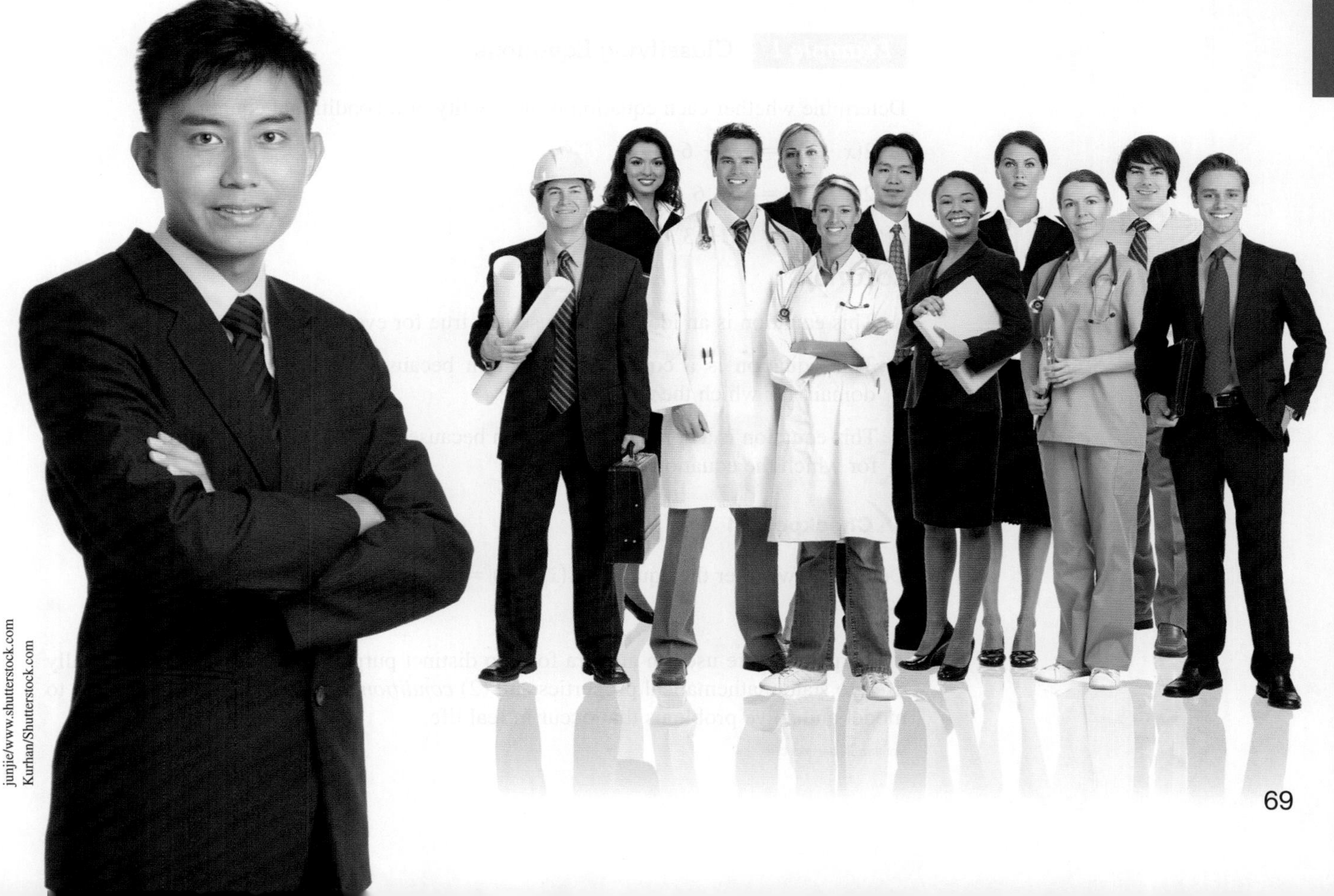

1.1 Linear Equations

■ Classify an equation as an identity or a conditional equation.

■ Solve a linear equation in one variable.

■ Solve an equation involving fractional expressions.

■ Use a linear model to solve an application problem.

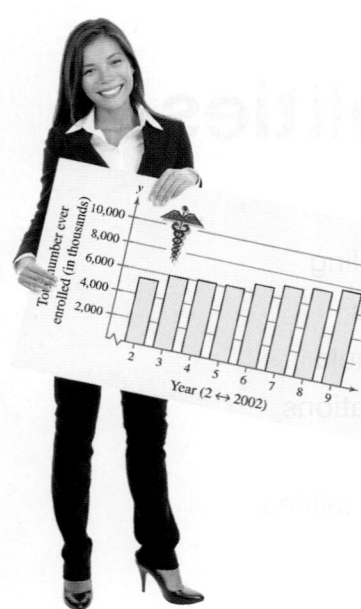

In Exercise 79 on page 79, you will use a linear model to determine the number of enrollees in a health insurance program.

Equations and Solutions

An **equation** is a statement that two algebraic expressions are equal. Some examples of equations in x are

$$3x - 5 = 7, \quad x^2 - x - 6 = 0, \quad \text{and} \quad \sqrt{2x} = 4.$$

To **solve** an equation in x means to find all values of x for which the equation is **true.** Such values are called **solutions.** For instance, $x = 4$ is a solution of the equation $3x - 5 = 7$, because $3(4) - 5 = 7$ is a true statement.

An equation that is true for *every* real number in the domain of the variable is called an **identity.** Two examples of identities are

$$x^2 - 9 = (x + 3)(x - 3) \quad \text{and} \quad \frac{x}{3x^2} = \frac{1}{3x}, \quad x \neq 0.$$

The first equation is an identity because it is a true statement for all real values of x. The second equation is an identity because it is true for all nonzero real values of x.

An equation that is true for just *some* (or even none) of the real numbers in the domain of the variable is called a **conditional equation.** For example, the equation $x^2 - 9 = 0$ is conditional because $x = 3$ and $x = -3$ are the only values in the domain that satisfy the equation.

Example 1   Classifying Equations

Determine whether each equation is an identity or a conditional equation.

a. $2(x + 3) = 2x + 6$

b. $2(x + 3) = x + 6$

c. $2(x + 3) = 2x + 3$

SOLUTION

a. This equation is an identity because it is true for every real value of x.

b. This equation is a conditional equation because $x = 0$ is the only value in the domain for which the equation is true.

c. This equation is a conditional equation because there are no real number values of x for which the equation is true.

✓ **Checkpoint 1**

Determine whether the equation $4(x + 1) = 4x + 4$ is an identity or a conditional equation.

    ■

Equations are used in algebra for two distinct purposes: (1) *identities* are usually used to state mathematical properties and (2) *conditional equations* are usually used to model and solve problems that occur in real life.

Linear Equations in One Variable

The most common type of conditional equation is a **linear equation.**

Definition of a Linear Equation

A **linear equation** in one variable x is an equation that can be written in the standard form

$$ax + b = 0$$

where a and b are real numbers with $a \neq 0$.

A linear equation in x has exactly one solution. To see this, consider the following steps. (Remember that $a \neq 0$.)

$$ax + b = 0 \qquad \text{Original equation}$$

$$ax = -b \qquad \text{Subtract } b \text{ from each side.}$$

$$x = -\frac{b}{a} \qquad \text{Divide each side by } a.$$

So, the equation $ax + b = 0$ has exactly one solution,

$$x = -\frac{b}{a}.$$

To solve a linear equation in x, isolate x by forming a sequence of **equivalent** (and usually simpler) equations, each having the same solution as the original equation. The operations that yield equivalent equations come from the basic rules of algebra reviewed in Section 0.2.

Forming Equivalent Equations

A given equation can be transformed into an equivalent equation by one or more of the following steps.

	Given Equation	Equivalent Equation
1. Remove symbols of grouping, combine like terms, or simplify one or both sides of the equation.	$2x - x = 4$ $3(x - 2) = 5$	$x = 4$ $3x - 6 = 5$
2. Add (or subtract) the same quantity to (from) *each* side of the equation.	$x + 1 = 6$ $x - 4 = 3$	$x = 5$ $x = 7$
3. Multiply (or divide) *each* side of the equation by the same *nonzero* quantity.	$2x = 6$ $\dfrac{x}{8} = 4$	$x = 3$ $x = 32$
4. Interchange the two sides of the equation.	$2 = x$	$x = 2$

The steps for solving a linear equation in x written in standard form are shown in Example 2.

ALGEBRA TUTOR xy

For help in solving equations similar to the one in Example 2, see the review of solving equations on page 148.

Example 2   Solving a Linear Equation

Solve $3x - 6 = 0$.

SOLUTION

$3x - 6 = 0$	Write original equation.
$3x = 6$	Add 6 to each side.
$x = 2$	Divide each side by 3.

✓**Checkpoint 2**

Solve each equation.

a. $5 + 5x = 15$ **b.** $10 - x = 11$ ■

After solving an equation, you should **check each solution** in the *original* equation. For instance, in Example 2, you can check that 2 is the solution by substituting 2 for x in the original equation

$$3x - 6 = 0.$$

CHECK

$3x - 6 = 0$	Write original equation.
$3(2) - 6 \overset{?}{=} 0$	Substitute 2 for x.
$6 - 6 = 0$	Solution checks. ✓

Example 3   Solving a Linear Equation

Solve $6(x - 1) + 4 = 3(7x + 1)$.

SOLUTION

$6(x - 1) + 4 = 3(7x + 1)$	Write original equation.
$6x - 6 + 4 = 21x + 3$	Distributive Property
$6x - 2 = 21x + 3$	Simplify.
$-15x = 5$	Add 2 to, and subtract $21x$ from, each side.
$x = -\frac{1}{3}$	Divide each side by -15.

The solution is $x = -\frac{1}{3}$.

STUDY TIP

You may think a solution to a problem looks easy when it is worked out in class, but you may not know where to begin when solving the problem on your own. Keep in mind that many problems involve some trial and error before a solution is found.

CHECK

$6(x - 1) + 4 = 3(7x + 1)$	Write original equation.
$6\left(-\frac{1}{3} - 1\right) + 4 \overset{?}{=} 3\left[7\left(-\frac{1}{3}\right) + 1\right]$	Substitute $-\frac{1}{3}$ for x.
$6\left(-\frac{4}{3}\right) + 4 \overset{?}{=} 3\left(-\frac{7}{3} + 1\right)$	Simplify.
$-8 + 4 \overset{?}{=} -7 + 3$	Simplify.
$-4 = -4$	Solution checks. ✓

✓**Checkpoint 3**

Solve each equation.

a. $3(x + 5) = x - 7$ **b.** $2(x + 2) + 6 = 4(2x - 3)$ ■

Some equations in one variable have *infinitely many solutions*. To recognize an equation of this type, perform the regular steps for solving the equation. If, when writing equivalent equations, you reach a statement that is true for all values in the domain of the variable, then the equation is an identity and has infinitely many solutions.

Example 4   An Equation with Infinitely Many Solutions

Solve $x + 4(x - 2) = 3x + 2(x - 4)$.

SOLUTION

$$x + 4(x - 2) = 3x + 2(x - 4) \qquad \text{Write original equation.}$$

$$x + 4x - 8 = 3x + 2x - 8 \qquad \text{Distributive Property}$$

$$5x - 8 = 5x - 8 \qquad \text{Simplify.}$$

$$-8 = -8 \qquad \text{Subtract } 5x \text{ from each side.}$$

Because the last equation is true for every real value of x, the original equation is an identity and you can conclude that it has infinitely many solutions.

✓ Checkpoint 4

Solve $x + 5 + 3(2x + 1) = 7x + 8$.

It is also possible for an equation in one variable to have *no solution*. When solving an equation of this type, you will reach a statement that is not true for any value of the variable.

Example 5   An Equation with No Solution

Solve $4x - 9 + 2(x + 8) = 1 + 6(x + 4)$.

SOLUTION

$$4x - 9 + 2(x + 8) = 1 + 6(x + 4) \qquad \text{Write original equation.}$$

$$4x - 9 + 2x + 16 = 1 + 6x + 24 \qquad \text{Distributive Property}$$

$$6x + 7 = 6x + 25 \qquad \text{Simplify.}$$

$$7 \neq 25 \qquad \text{Subtract } 6x \text{ from each side.}$$

Because the statement

$$7 = 25$$

is not true, you can conclude that the original equation has no solution.

✓ Checkpoint 5

Solve $1 + 4(x + 1) = 4(2 + x)$.

Equations in one variable with infinitely many solutions or no solution are not linear because they cannot be written in the standard form

$$ax + b = 0.$$

Note that a linear equation in x has exactly one solution.

Equations Involving Fractional Expressions

To solve an equation involving fractional expressions, you can multiply every term in the equation by the least common denominator (LCD) of the terms.

Use the *table* feature of your graphing utility to check the solution in Example 3. In the equation editor, enter the expression to the left of the equal sign in y_1 and enter the expression to the right of the equal sign in y_2, as follows.

$$y_1 = 6(x - 1) + 4$$

$$y_2 = 3(7x + 1)$$

Set the *table* feature to ASK mode. When you enter the solution $-\frac{1}{3}$ for x, both y_1 and y_2 are -4, as shown.

X	Y₁	Y₂
-.3333	-4	-4

X=-.333333333333

Similarly, a graphing utility can help you determine if a solution is extraneous. For instance, enter the equation from Example 7 into the graphing utility's equation editor. Then, use the *table* feature in ASK mode to enter -2 for x. You will see that the graphing utility displays ERROR in the y_2 column. So, the solution $x = -2$ is extraneous.

Example 6   An Equation Involving Fractional Expressions

$$\frac{x}{3} + \frac{3x}{4} = 2 \qquad \text{Original equation}$$

$$(12)\frac{x}{3} + (12)\frac{3x}{4} = (12)2 \qquad \text{Multiply each term by the least common denominator.}$$

$$4x + 9x = 24 \qquad \text{Simplify.}$$

$$13x = 24 \qquad \text{Combine like terms.}$$

$$x = \frac{24}{13} \qquad \text{Divide each side by 13.}$$

✓ **Checkpoint 6**

Solve $\dfrac{4x}{5} - \dfrac{x}{2} = 9$.   ■

When multiplying or dividing an equation by a *variable expression*, it is possible to introduce an **extraneous** solution—one that does not satisfy the original equation. In such cases, a check is especially important.

Example 7   An Equation with an Extraneous Solution

Solve $\dfrac{1}{x - 2} = \dfrac{3}{x + 2} - \dfrac{6x}{x^2 - 4}$.

SOLUTION   The least common denominator is $x^2 - 4 = (x + 2)(x - 2)$. Multiply each term by this LCD and simplify.

$$\frac{1}{x - 2} = \frac{3}{x + 2} - \frac{6x}{x^2 - 4} \qquad \text{Write original equation.}$$

$$\frac{1}{x - 2}(x + 2)(x - 2) = \frac{3}{x + 2}(x + 2)(x - 2) - \frac{6x}{x^2 - 4}(x + 2)(x - 2)$$

$$x + 2 = 3(x - 2) - 6x, \quad x \neq \pm 2 \qquad \text{Simplify.}$$

$$x + 2 = 3x - 6 - 6x \qquad \text{Distributive Property}$$

$$4x = -8 \qquad \text{Simplify.}$$

$$x = -2 \qquad \text{Extraneous solution}$$

By checking $x = -2$, you can see that it yields a denominator of zero for the fraction

$$\frac{3}{x + 2}.$$

So, $x = -2$ is extraneous, and the equation has *no solution*.   ───

✓ **Checkpoint 7**

Solve $\dfrac{1}{x - 4} + \dfrac{1}{x} = \dfrac{4}{x(x - 4)}$.   ■

An equation with a *single fraction* on each side can be cleared of denominators by **cross-multiplying,** which is equivalent to multiplying each side of the equation by the least common denominator and then simplifying.

Example 8   Cross-Multiplying to Solve an Equation

Solve $\dfrac{3y - 2}{2y + 1} = \dfrac{6y - 9}{4y + 3}$.

SOLUTION

$\dfrac{3y - 2}{2y + 1} = \dfrac{6y - 9}{4y + 3}$	Write original equation.
$(3y - 2)(4y + 3) = (6y - 9)(2y + 1)$	Cross-multiply.
$12y^2 + y - 6 = 12y^2 - 12y - 9$	Multiply.
$13y = -3$	Isolate y-term on left.
$y = -\dfrac{3}{13}$	Divide each side by 13.

The solution is $y = -\frac{3}{13}$. Check this in the original equation.

TECH TUTOR

When using the *equation editor* of a graphing utility, you must enter equations in terms of x. So, to enter an equation like the one shown in Example 8, replace y with x, as shown.

$y_1 = (3x - 2)/(2x + 1)$

$y_2 = (6x - 9)/(4x + 3)$

✓**Checkpoint 8**

Solve $\dfrac{3x - 6}{x + 10} = \dfrac{3}{4}$.

Example 9   Using a Calculator to Solve an Equation

Solve $\dfrac{1}{9.38} - \dfrac{3}{x} = \dfrac{5}{0.3714}$.

SOLUTION   To minimize roundoff error, solve for x in the equation before performing any calculations. The least common denominator is $(9.38)(0.3714)(x)$.

$$\frac{1}{9.38} - \frac{3}{x} = \frac{5}{0.3714}$$

$$(9.38)(0.3714)(x)\left(\frac{1}{9.38} - \frac{3}{x}\right) = (9.38)(0.3714)(x)\left(\frac{5}{0.3714}\right)$$

$$0.3714x - 3(9.38)(0.3714) = 5(9.38)(x), \quad x \neq 0$$

$$[0.3714 - 5(9.38)]x = 3(9.38)(0.3714)$$

$$x = \frac{3(9.38)(0.3714)}{0.3714 - 5(9.38)}$$

$$x \approx -0.225 \qquad \text{Round to three decimal places.}$$

The solution is $x \approx -0.225$. Check this in the original equation.

STUDY TIP

Because of roundoff error, a check of a decimal solution may not yield exactly the same values for each side of the original equation. The difference, however, should be quite small.

✓**Checkpoint 9**

Solve $\dfrac{5}{x} + \dfrac{1}{2.7} = \dfrac{4}{0.6}$.

Application

 Example 10   Hourly Earnings

The mean hourly earnings y (in dollars) of employees in the manufacturing industry in the United States from 2003 to 2008 can be modeled by the linear equation

$$y = 0.388t + 14.58, \quad 3 \le t \le 8$$

where t represents the year, with $t = 3$ corresponding to 2003. Use the model to estimate the year in which the mean hourly earnings were \$17.30.   *(Source: U.S. Bureau of Labor Statistics)*

SOLUTION To determine when the mean hourly earnings were \$17.30, solve the model for t when $y = 17.3$.

$y = 0.388t + 14.58$	Write original model.
$17.3 = 0.388t + 14.58$	Substitute 17.3 for y.
$2.72 = 0.388t$	Subtract 14.58 from each side.
$t = \dfrac{2.72}{0.388}$	Divide each side by 0.388.
$t \approx 7$	Use a calculator.

Because $t = 3$ corresponds to 2003, it follows that $t = 7$ corresponds to 2007. See Figure 1.1. So, the mean hourly earnings were \$17.30 in 2007. ────────

FIGURE 1.1

✓ **Checkpoint 10**

The mean hourly earnings y (in dollars) of employees in the trade, transportation, and utilities industry in the United States from 2000 to 2008 can be modeled by the linear equation

$$y = 0.350t + 13.29, \quad 0 \le t \le 8$$

where t represents the year, with $t = 0$ corresponding to 2000. Use the model to estimate the year in which the mean hourly earnings were \$14.35.   *(Source: U.S. Bureau of Labor Statistics)*

SUMMARIZE   (Section 1.1)

1. State the definitions of an identity and a conditional equation *(page 70)*. For an example of identifying identities and conditional equations, see Example 1.

2. State the definition of a linear equation and list the four steps that can be used to form an equivalent equation *(page 71)*. For examples of solving linear equations, see Examples 2 and 3.

3. Describe two methods for solving equations involving fractional expressions *(pages 74 and 75)*. For examples of solving equations involving fractional expressions, see Examples 6, 7, 8, and 9.

4. Describe a real-life example of how a linear equation can be used to analyze the mean hourly earnings of employees in a particular industry *(page 76, Example 10)*.

SKILLS WARM UP 1.1 The following warm-up exercises involve skills that were covered in earlier sections. You will use these skills in the exercise set for this section. For additional help, review Sections 0.5 and 0.7.

In Exercises 1–10, perform the indicated operations and simplify your answer.

1. $(2x - 4) - (5x + 6)$

2. $(3x - 5) + (2x - 7)$

3. $2(x + 1) - (x + 2)$

4. $-3(2x - 4) + 7(x + 2)$

5. $\dfrac{x}{2} + \dfrac{x}{7}$

6. $x - \dfrac{x}{4}$

7. $\dfrac{1}{x + 1} - \dfrac{1}{x}$

8. $\dfrac{5}{x} + \dfrac{3}{x}$

9. $\dfrac{4}{x} + \dfrac{3}{x - 2}$

10. $\dfrac{1}{x + 1} - \dfrac{1}{x - 1}$

Exercises 1.1

See www.CalcChat.com for worked-out solutions to odd-numbered exercises.

Classifying Equations In Exercises 1–6, determine whether the equation is an identity or a conditional equation. *See Example 1.*

1. $2(x - 1) = 2x - 2$

2. $3(x + 2) = 3x + 6$

3. $2(x - 1) = 3x + 4$

4. $3(x + 2) = 2x + 4$

5. $2(x + 1) = 2x + 1$

6. $3(x + 4) = 3x + 4$

Checking Solutions In Exercises 7–16, determine whether each value of *x* is a solution of the equation.

Equation *Values*

7. $5x - 3 = 3x + 5$ (a) $x = 0$ (b) $x = -5$
(c) $x = 4$ (d) $x = 10$

8. $7 - 3x = 5x - 17$ (a) $x = -3$ (b) $x = 0$
(c) $x = 8$ (d) $x = 3$

9. $3x^2 + 2x - 5 = 2x^2 - 2$ (a) $x = -3$ (b) $x = 1$
(c) $x = 4$ (d) $x = -5$

10. $5x^3 + 2x - 3 = 4x^3 + 2x - 11$
(a) $x = 2$ (b) $x = -2$
(c) $x = 0$ (d) $x = 10$

11. $\dfrac{5}{2x} - \dfrac{4}{x} = 3$ (a) $x = -\dfrac{1}{2}$ (b) $x = 4$

(c) $x = 0$ (d) $x = \dfrac{1}{4}$

12. $3 + \dfrac{1}{x + 2} = 4$ (a) $x = -1$ (b) $x = -2$
(c) $x = 0$ (d) $x = 5$

Equation *Values*

13. $(x + 5)(x - 3) = 20$ (a) $x = 3$ (b) $x = -2$
(c) $x = 0$ (d) $x = -7$

14. $(3x + 5)(2x - 7) = 0$ (a) $x = -\dfrac{5}{3}$ (b) $x = -\dfrac{2}{7}$
(c) $x = \dfrac{2}{3}$ (d) $x = \dfrac{3}{2}$

15. $\sqrt{2x - 3} = 3$ (a) $x = 6$ (b) $x = -3$
(c) $x = -\dfrac{1}{3}$ (d) $x = -2$

16. $\sqrt[3]{x - 8} = 3$ (a) $x = 2$ (b) $x = -5$
(c) $x = 35$ (d) $x = 8$

Solving an Equation In Exercises 17–32, solve the equation and check your solution. (Some equations have no solution.) *See Examples 2, 3, 4, and 5.*

17. $x + 10 = 15$

18. $9 - x = 13$

19. $7 - 2x = 15$

20. $7x + 2 = 16$

21. $8x - 5 = 3x + 10$

22. $7x + 3 = 3x - 13$

23. $5x - x = 4x - 4$

24. $3 - x = 4x - 5x$

25. $4x - 8 = -4(2 - x)$

26. $3(3 - x) = 9 - 3x$

27. $2(x + 5) - 7 = 3(x - 2)$

28. $2(13t - 15) + 3(t - 19) = 0$

29. $6[x - (2x + 3)] = 8 - 5x$

30. $3[2x - (x + 7)] = 5(x - 3)$

31. $x - 3(2x + 3) = 8 - 5x$

32. $9x - 10 = 5x + 2(2x - 5)$

An Equation Involving Fractional Expressions In Exercises 33–48, solve the equation and check your solution. (Some equations have no solution.) *See Examples 6, 7, and 8.*

33. $\dfrac{5x}{4} + \dfrac{1}{2} = x - \dfrac{1}{2}$

34. $\dfrac{x}{5} - \dfrac{x}{2} = 3$

35. $\dfrac{3}{2}(z + 5) = \dfrac{1}{4}(z + 24)$

36. $\dfrac{3x}{2} + \dfrac{1}{4}(x - 2) = 10$

37. $\dfrac{100 - 4u}{3} = \dfrac{5u + 6}{4} + 6$

38. $\dfrac{17 + y}{y} + \dfrac{32 + y}{y} = 100$

39. $\dfrac{5x - 4}{5x + 4} = \dfrac{2}{3}$

40. $\dfrac{10x + 3}{5x + 6} = \dfrac{1}{2}$

41. $10 - \dfrac{13}{x} = 4 + \dfrac{5}{x}$

42. $\dfrac{15}{x} - 4 = \dfrac{6}{x} + 3$

43. $\dfrac{1}{x - 3} + \dfrac{1}{x + 3} = \dfrac{10}{x^2 - 9}$

44. $\dfrac{1}{x - 2} + \dfrac{3}{x + 3} = \dfrac{4}{x^2 + x - 6}$

45. $\dfrac{6}{(x - 3)(x - 1)} = \dfrac{3}{x - 3} + \dfrac{4}{x - 1}$

46. $\dfrac{2}{(x - 4)(x - 2)} = \dfrac{1}{x - 4} + \dfrac{2}{x - 2}$

47. $\dfrac{x - 3}{2x + 1} = \dfrac{x - 5}{2x + 4}$

48. $\dfrac{2x - 3}{6x - 5} = \dfrac{x + 2}{3x + 4}$

Solving an Equation In Exercises 49–56, solve the equation and check your solution. (Some equations have no solution.)

49. $0.25x + 0.75(10 - x) = 3$

50. $0.60x + 0.40(100 - x) = 50$

51. $(x + 2)^2 + 5 = (x + 3)^2$

52. $(x + 1)^2 + 2(x - 2) = (x + 1)(x - 2)$

53. $(x + 2)^2 - x^2 = 4(x + 1)$

54. $4(x + 1) - 3x = x + 5$

55. $(2x + 1)^2 = 4(x^2 + x + 1)$

56. $(2x - 1)^2 = 4(x^2 - x + 6)$

The symbol ▨ indicates an exercise in which you are instructed to use graphing technology or a symbolic computer algebra system. The solutions of other exercises may also be facilitated by use of appropriate technology.

57. Error Analysis A student states that the solution of the equation

$$\frac{2}{x(x - 2)} + \frac{5}{x} = \frac{1}{x - 2}$$

is $x = 2$. Describe and correct the student's error.

58. Error Analysis A student states that the equation

$$-3(x + 2) = -3x + 6$$

is an identity. Describe and correct the student's error.

59. Writing Explain why a solution of an equation involving fractional expressions may be extraneous.

60. Writing Describe two methods you can use to check a solution of an equation involving fractional expressions.

61. Writing What is meant by "equivalent equations?" Give an example of two equivalent equations.

62. Think About It For what value(s) of b does the equation $7x + 3 = 7x + b$ have infinitely many solutions? no solution?

Using a Calculator to Solve an Equation In Exercises 63–68, use a calculator to solve the equation. (Round your solution to three decimal places.) *See Example 9.*

63. $0.275x + 0.725(500 - x) = 300$

64. $2.763 - 4.5(2.1x - 5.1432) = 6.32x + 5$

65. $\dfrac{x}{0.6321} + \dfrac{x}{0.0692} = 1000$

66. $(x + 5.62)^2 + 10.83 = (x + 7)^2$

67. $\dfrac{2}{7.398} - \dfrac{4.405}{x} = \dfrac{1}{x}$

68. $\dfrac{x}{2.625} + \dfrac{x}{4.875} = 1$

69. Think About It What method or methods would you recommend for checking the solutions to Exercises 63–68 using your graphing utility?

70. Think About It In Exercises 63–68, your answers are rounded to three decimal places. What effect does rounding have as you check a solution?

Roundoff Error In Exercises 71–74, evaluate the expression in two ways. (a) Calculate entirely on your calculator using appropriate parentheses, and then round the answer to two decimal places. (b) Round both the numerator and the denominator to two decimal places before dividing, and then round the final answer to two decimal places. (c) Then describe why roundoff error is introduced by the second method.

71. $\dfrac{1 + 0.73205}{1 - 0.73205}$

72. $\dfrac{1 + 0.86603}{1 - 0.86603}$

73. $\dfrac{333 + \dfrac{1.98}{0.74}}{4 + \dfrac{6.25}{3.15}}$

74. $\dfrac{1.73205 - 1.19195}{3 - (1.73205)(1.19195)}$

75. Personal Income The per capita personal income y (in dollars) in West Virginia from 2000 through 2009 can be approximated by the linear equation

$$y = 1114.4t + 21{,}952, \quad 0 \le t \le 9$$

where t represents the year, with $t = 0$ corresponding to 2000. Use the model to estimate the year in which the per capita personal income was about \$32,000. *(Source: Bureau of Economic Analysis)*

76. Annual Sales The annual sales S (in billions of dollars) of Oracle Corporation from 2004 through 2009 can be approximated by the linear equation

$$S = 2.903t - 1.98, \quad 4 \le t \le 9$$

where t represents the year, with $t = 4$ corresponding to 2004. Use the model to estimate the year in which Oracle's annual sales were about \$18 billion. *(Source: Oracle Corporation)*

Forensics In Exercises 77 and 78, use the following information. The relationship between the length of an adult's femur (thigh bone) and the height of the adult can be approximated by the linear equations

$$y = 0.432x - 10.44 \qquad \text{Female}$$
$$y = 0.449x - 12.15 \qquad \text{Male}$$

where y is the length of the femur in inches and x is the height of the adult in inches (see figure).

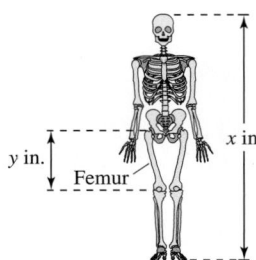

77. A crime scene investigator discovers a femur belonging to an adult human female. The bone is 18 inches long. Estimate the height of the female.

78. Officials search a forest for a missing man who is 6 feet 2 inches tall. They find an adult male femur that is 21 inches long. Is it possible that the femur belongs to the missing man?

79. Health Insurance The Children's Health Insurance Program (CHIP) tracks the total number of children ever enrolled in the program. For the years 2002 through 2009, the total numbers ever enrolled y (in thousands) are shown in the bar graph and can be approximated by

$$y = 322.5t + 4788, \quad 2 \le t \le 9$$

where t is the year, with $t = 2$ corresponding to 2002. *(Source: Centers for Medicare and Medicaid Services)*

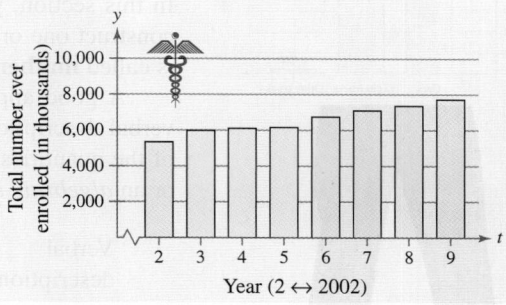

(a) In what year did the total number of CHIP enrollees exceed 7 million?

(b) Use the model to predict the year in which the total number of CHIP enrollees will reach 8.3 million.

80. HOW DO YOU SEE IT? From 1997 through 2006, the federal minimum wage was \$5.15 per hour. Adjusting for inflation, the federal minimum wage's value y (in 1996 dollars) during each of these years is shown in the bar graph. *(Source: U.S. Department of Labor)*

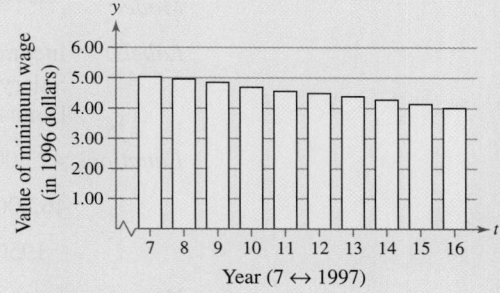

(a) In which year was the value of the minimum wage equal to \$4.85 in 1996 dollars?

(b) By 2009, the minimum wage increased to \$7.25, a value of \$5.30 in 1996 dollars. Suppose it had not increased. Use the linear relationship shown in the bar graph to predict the value (in 1996 dollars) of a 2009 minimum wage of \$5.15 per hour.

1.2 Mathematical Modeling

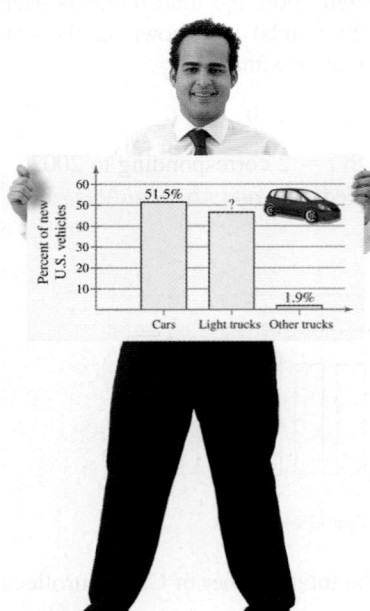

In Exercise 37 on page 89, you will use a mathematical model to determine the number of new cars that were sold or leased in 2009.

- Construct a mathematical model from a verbal model.
- Model and solve percent and distance problems.
- Model and solve mixture problems.
- Use common formulas to solve geometry problems.

Introduction to Problem Solving

In this section, you will use algebra to solve real-life problems. To do this, you will construct one or more equations that represent each real-life problem. This procedure is called **mathematical modeling.**

A good approach to mathematical modeling is to use two stages. First, use the verbal description of the problem to form a *verbal model.* Then, assign labels to each of the quantities in the verbal model and use the labels to form a *mathematical model* or an *algebraic equation.*

Verbal description	⟹	Verbal model	⟹	Algebraic equation

When you are trying to construct a verbal model, it is helpful to look for a *hidden equality*—a statement that two algebraic expressions are equal. These two expressions might be explicitly stated as being equal, or they might be known to be equal (based on prior knowledge or experience).

Example 1   Using a Verbal Model

You accept a job with an annual income of $47,800. This includes your salary and a $1000 year-end bonus. You are paid twice a month. What is your salary per pay period?

SOLUTION   Because there are 12 months in a year and you are paid twice a month, it follows that there are 24 pay periods during the year.

Verbal Model: Income for year = 24 pay periods · Salary per pay period + Bonus

Labels: Income for year = 47,800   (dollars)
         Salary per pay period = x (dollars)
         Bonus = 1000   (dollars)

Equation: $47,800 = 24x + 1000$

$$46,800 = 24x$$

$$1950 = x$$

Your salary per pay period is $1950.

CHECK

$47,800 = 24x + 1000$	Write original equation.
$47,800 \overset{?}{=} 24(1950) + 1000$	Substitute 1950 for x.
$47,800 = 47,800$	Solution checks. ✓

✓ Checkpoint 1

You accept a job with an annual income of $51,000. This includes your salary and a $1600 year-end bonus. You are paid weekly. What is your salary per pay period?   ■

Translating Key Words and Phrases

Key Words and Phrases	Verbal Description	Algebraic Statement
Consecutive		
Next, subsequent	Consecutive integers	$n, n + 1$
Addition		
Sum, plus, greater, increased by, more than, exceeds, total of	The sum of 5 and x Seven more than y	$5 + x$ $y + 7$
Subtraction		
Difference, minus, less than, decreased by, subtracted from, reduced by, the remainder	Four decreased by b Three less than z Five subtracted from w	$4 - b$ $z - 3$ $w - 5$
Multiplication		
Product, multiplied by, twice, times, percent of	Two times x	$2x$
Division		
Quotient, divided by, per	The quotient of x and 8	$\dfrac{x}{8}$

 Example 2 Constructing Mathematical Models

a. A salary of $50,700 is increased by 6%. Write an equation that represents the new salary.

Verbal Model: New salary $=$ 6%(Original salary) $+$ Original salary

Labels: Original salary $= 50{,}700$ (dollars)
       New salary $= S$ (dollars)
       Percent $= 0.06$ (percent in decimal form)

Equation: $S = 0.06(50{,}700) + 50{,}700$

b. An e-reader is marked down 10% to $225. Write an equation you can use to find the original price.

Verbal Model: Original price $-$ 10%(Original price) $=$ Sale price

Labels: Original price $= p$ (dollars)
       Sale price $= 225$ (dollars)
       Percent $= 0.1$ (percent in decimal form)

Equation: $p - 0.1p = 225$

✓**Checkpoint 2**

A bid price of $6800 for an auction item is increased by 5%. Write an equation that you can use to find the new bid price.

In Example 2, notice that part of the labeling process is to list the unit of measure for each labeled quantity. Developing this habit helps in checking the validity of a verbal model.

Using Mathematical Models

 **Example 3** **Finding the Percent of a Raise**

You accept a job that pays $12.50 an hour. You are told that after a two-month probationary period, your hourly wage will be increased to $13 an hour. What percent raise will you receive after the two-month period?

SOLUTION

Verbal Model: Raise = Percent · Old wage

Labels: Old wage = 12.5   (dollars)
           Raise = 13 − 12.5 = 0.5   (dollars)
           Percent = r (percent in decimal form)

Equation: $0.5 = r \cdot 12.5$

By solving this equation, you can find that you will receive a raise of

$$r = \frac{0.5}{12.5} = 0.04$$

or 4%.

✓ **Checkpoint 3**

You buy stock at $25 per share. You sell the stock at $30 per share. What is the percent increase of the stock's value?

 Example 4 **Finding the Percent of Employees**

The number of people employed by the U.S. federal government for a recent year is 2,730,040. The table shows the number of people employed in the executive branch of the federal government. What percent of employees in the U.S. federal government are in the executive branch? *(Source: U.S. Office of Personnel Management)*

	Number of Employees
Executive Office of the President	1717
Executive Departments	1,740,979
Independent Agencies	923,744

SOLUTION

Verbal Model: Executive branch employees = Percent · Federal government employees

Labels: Federal government employees = 2,730,040   (people)
           Executive branch employees = 2,666,440   (people)
           Percent = r (percent in decimal form)

Equation: $2,666,440 = r \cdot 2,730,040$

By solving this equation, you can find that the percent of federal government employees in the executive branch is $r = 2,666,440/2,730,040$, or about 97.7%.

✓ **Checkpoint 4**

Your income last year was $42,000. By the end of the year, you paid a total of $648 for parking fees. What percent of your income was paid for parking fees?

The President presides over the executive branch of the federal government, which employs more than 2.5 million people.

Example 5 **Finding the Dimensions of a Room**

A rectangular family room is twice as long as it is wide, and its perimeter is 84 feet. Find the dimensions of the family room.

SOLUTION For this problem, it helps to sketch a diagram, as shown in Figure 1.2.

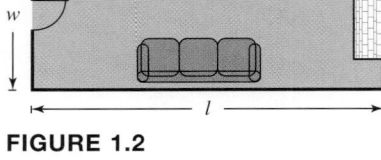

Verbal Model: $2 \cdot \text{Length} + 2 \cdot \text{Width} = \text{Perimeter}$

Labels: Perimeter $= 84$ (feet)
         Width $= w$ (feet)
         Length $= l = 2w$ (feet)

Equation: $2(2w) + 2w = 84$
          $4w + 2w = 84$
          $6w = 84$
          $w = 14$ feet
          $l = 2w = 2(14) = 28$ feet

FIGURE 1.2

The dimensions of the room are 14 feet by 28 feet.

✓**Checkpoint 5**

A rectangular driveway is three times as long as it is wide, and its perimeter is 120 feet. Find the dimensions of the driveway.

 **Example 6** **A Distance Problem**

A plane travels nonstop from New York to San Francisco, a distance of about 2600 miles. It takes 1.5 hours to fly from New York to Chicago, a distance of about 700 miles (see Figure 1.3). Assuming the plane flies at a constant speed, how long does the entire trip take? What time (EST) should the plane leave New York to arrive in San Francisco by 5 P.M. PST (8 P.M. EST)?

SOLUTION To solve this problem, use the formula that relates distance, rate, and time. That is, (distance) = (rate)(time). Because it took the plane 1.5 hours to travel a distance of about 700 miles, you can conclude that its rate (or speed) is

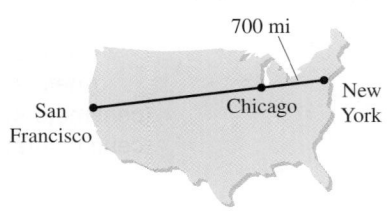

700 mi

San Francisco   Chicago   New York

FIGURE 1.3

$$\text{Rate} = \frac{\text{distance}}{\text{time}}$$

$$= \frac{700 \text{ miles}}{1.5 \text{ hours}} \approx 466.67 \text{ miles per hour.}$$

Because the entire trip is about 2600 miles, the time for the entire trip is

$$\text{Time} = \frac{\text{distance}}{\text{rate}}$$

$$= \frac{2600 \text{ miles}}{466.67 \text{ miles per hour}} \approx 5.57 \text{ hours.}$$

Because 0.57 hour represents about 34 minutes, you can conclude that the entire trip takes about 5 hours and 34 minutes. So, the plane must leave New York by 2:26 P.M. EST in order to arrive in San Francisco by 8 P.M. EST.

✓**Checkpoint 6**

A boat travels at full speed to an island 11 miles away. It takes 0.3 hour to travel the first 3 miles. How long does the entire trip take?

Another way to solve the distance problem in Example 6 is to use the concept of **ratio and proportion.** To do this, let x represent the time required to fly from New York to San Francisco, set up the following proportion, and solve for x.

$$\frac{\text{Time to San Francisco}}{\text{Time to Chicago}} = \frac{\text{Distance to San Francisco}}{\text{Distance to Chicago}}$$

$$\frac{x}{1.5} = \frac{2600}{700}$$

$$1.5 \cdot \frac{x}{1.5} = 1.5 \cdot \frac{2600}{700}$$

$$x \approx 5.57$$

Notice how ratio and proportion are used with a property from geometry to solve the problem in the following example.

Example 7   An Application Involving Similar Triangles

To determine the height of Petronas Tower 1 (in Kuala Lumpur, Malaysia), you measure the shadow cast by the building to be 113 meters long, as shown in Figure 1.4. Then you measure the shadow cast by a 100-centimeter post and find that its shadow is 25 centimeters long. Use this information to determine the height of Petronas Tower 1.

SOLUTION   To find the height of the tower, you can use a property from geometry that states that the ratios of corresponding sides of similar triangles are equal.

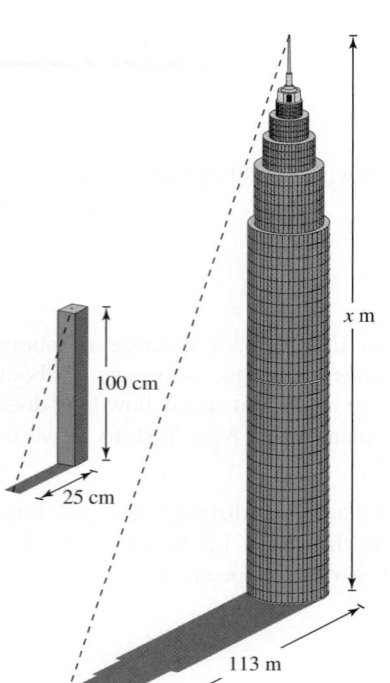

FIGURE 1.4

Verbal Model: $\dfrac{\text{Height of tower}}{\text{Length of tower's shadow}} = \dfrac{\text{Height of post}}{\text{Length of post's shadow}}$

Labels:
Height of tower $= x$ (meters)
Length of tower's shadow $= 113$ (meters)
Height of post $= 100$ (centimeters)
Length of post's shadow $= 25$ (centimeters)

Equation:

$$\frac{x}{113} = \frac{100}{25}$$

$$113 \cdot \frac{x}{113} = 113 \cdot \frac{100}{25}$$

$$x = 113 \cdot 4$$

$$x = 452 \text{ meters}$$

The Petronas Tower 1 is 452 meters high.

CHECK

$$\frac{x}{113} = \frac{100}{25} \qquad \text{Write original equation.}$$

$$\frac{452}{113} = \frac{100}{25} \qquad \text{Substitute 452 for } x.$$

$$4 = 4 \qquad \text{Solution checks. } ✓$$

✓ **Checkpoint 7**

A tree casts a shadow that is 24 feet long. At the same time, a four-foot tall mailbox casts a shadow that is 3 feet long. How tall is the tree?

Mixture Problems

The next example is called a **mixture problem** because it involves two different unknown quantities that are *mixed* in a specific way. Watch for a *hidden product* in the verbal model.

 Example 8 **A Simple Interest Problem**

You invested a total of $10,000 in accounts that earned $1\frac{1}{2}\%$ and $2\frac{1}{2}\%$ simple interest. In 1 year, the two accounts earned $207.50 in interest. How much did you invest in each account?

SOLUTION   The formula for simple interest is

$$I = Prt$$

where I is the interest, P is the principal, r is the annual interest rate (in decimal form), and t is the time in years.

Verbal Model:	Interest from $1\frac{1}{2}\%$	$+$	Interest from $2\frac{1}{2}\%$	$=$	Total interest

You can let x represent the amount invested at $1\frac{1}{2}\%$. Because the total amount invested at $1\frac{1}{2}\%$ and $2\frac{1}{2}\%$ is $10,000, you can let $10,000 - x$ represent the amount invested at $2\frac{1}{2}\%$.

Labels:
Amount invested at $1\frac{1}{2}\% = x$ (dollars)
Amount invested at $2\frac{1}{2}\% = 10,000 - x$ (dollars)
Interest from $1\frac{1}{2}\% = Prt = (x)(0.015)(1)$ (dollars)
Interest from $2\frac{1}{2}\% = Prt = (10,000 - x)(0.025)(1)$ (dollars)
Total interest $= 207.50$ (dollars)

Equation:
$$0.015x + 0.025(10,000 - x) = 207.50$$
$$0.015x + 250 - 0.025x = 207.50$$
$$-0.01x = -42.50$$
$$x = 4250$$

So, the amount invested at $1\frac{1}{2}\%$ is $4250 and the amount invested at $2\frac{1}{2}\%$ is

$$10,000 - x = 10,000 - 4250$$
$$= \$5750.$$

Check these results in the original statement of the problem.

CHECK

Interest from $1\frac{1}{2}\%$ Interest from $2\frac{1}{2}\%$ Total interest

$$0.015(4250) + 0.025(10,000 - 4250) \stackrel{?}{=} 207.50$$
$$63.75 + 143.75 \stackrel{?}{=} 207.50$$
$$207.50 = 207.50 \qquad \text{Solution checks.} ✓$$

✓ **Checkpoint 8**

You invested a total of $2500 in accounts that earned 1% and 2% simple interest. In 1 year, the two accounts earned a total of $38 in interest. How much did you invest in each account?

Common Formulas

Many common types of geometric, scientific, and investment problems use ready-made equations, called **formulas.** Knowing formulas such as those in the following lists will help you translate and solve a wide variety of real-life problems involving perimeter, area, volume, temperature, interest, and distance.

Common Formulas for Area, Perimeter, and Volume

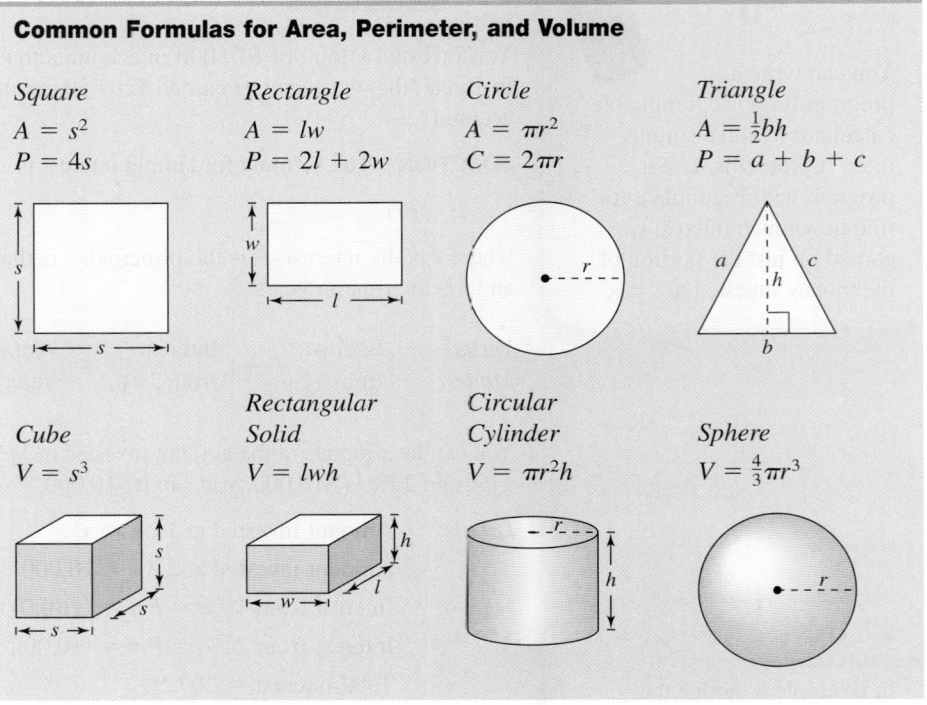

Square

$A = s^2$

$P = 4s$

Rectangle

$A = lw$

$P = 2l + 2w$

Circle

$A = \pi r^2$

$C = 2\pi r$

Triangle

$A = \frac{1}{2}bh$

$P = a + b + c$

Cube

$V = s^3$

Rectangular Solid

$V = lwh$

Circular Cylinder

$V = \pi r^2 h$

Sphere

$V = \frac{4}{3}\pi r^3$

Miscellaneous Common Formulas

Temperature: $F = \dfrac{9}{5}C + 32$

$C = \dfrac{5}{9}(F - 32)$

F = degrees Fahrenheit
C = degrees Celsius

Simple Interest: $I = Prt$ I = interest
P = principal
r = interest rate
t = time

Distance: $d = rt$ d = distance traveled
r = rate
t = time

When working with applied problems, you often need to rewrite common formulas. For instance, the formula

$$P = 2l + 2w$$

for the perimeter of a rectangle can be rewritten or solved for w to produce

$$w = \tfrac{1}{2}(P - 2l).$$

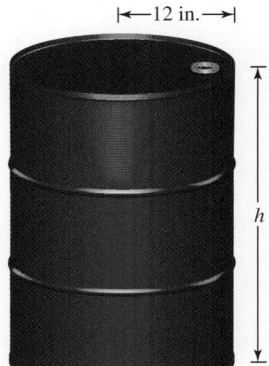

|←—12 in.—→|

h

FIGURE 1.5

Example 9 Using a Formula

A cylindrical shipping container has a volume of 15,607 cubic inches and a radius of 12 inches, as shown in Figure 1.5. Find the height of the container.

SOLUTION The formula for the *volume of a cylinder* is $V = \pi r^2 h$. To find the height of the container, solve for h.

$$h = \frac{V}{\pi r^2}$$

Then, using $V = 15{,}607$ and $r = 12$, find the height.

$$h = \frac{15{,}607}{\pi(12)^2} \qquad \text{Substitute 15,607 for } V \text{ and 12 for } r.$$

$$= \frac{15{,}607}{144\pi} \qquad \text{Simplify denominator.}$$

$$\approx 34.5 \qquad \text{Use a calculator.}$$

So, the height of the container is about 34.5 inches. You can use unit analysis to check that your answer is reasonable.

$$\frac{15{,}607 \text{ in.}^3}{144\pi \text{ in.}^2} \approx 34.5 \text{ in.}$$

✓ Checkpoint 9

One cubic foot of water fills a cylindrical pipe with a radius of 0.5 foot. What is the height of the pipe?

Strategy for Solving Word Problems

1. *Search* for the hidden equality—two expressions said to be equal or known to be equal. A sketch may be helpful.

2. *Write* a verbal model that equates these two expressions. Identify any *hidden* products.

3. *Assign* numbers to the known quantities and letters (or algebraic expressions) to the unknown quantities.

4. *Rewrite* the verbal model as an algebraic equation using the assigned labels.

5. *Solve* the resulting algebraic equation.

6. *Check* that the answer satisfies the word problem as stated. (Remember that "solving for x" or another variable may not completely answer the question.)

SUMMARIZE (Section 1.2)

1. Describe the procedure of mathematical modeling *(page 80)*. For examples of constructing and using mathematical models, see Examples 1, 2, 3, 4, 5, 6, and 7.

2. Describe a mixture problem *(page 85)*. For an example of a mixture problem, see Example 8.

3. State the definition of a formula *(page 86)*. For an example that uses a volume formula, see Example 9.

SKILLS WARM UP 1.2   The following warm-up exercises involve skills that were covered in earlier sections. You will use these skills in the exercise set for this section. For additional help, review Section 1.1.

In Exercises 1–10, solve the equation (if possible) and check your solution.

1. $3x - 42 = 0$

2. $64 - 16x = 0$

3. $2 - 3x = 14 + x$

4. $7 + 5x = 7x - 1$

5. $5[1 + 2(x + 3)] = 6 - 3(x - 1)$

6. $2 - 5(x - 1) = 2[x + 10(x - 1)]$

7. $\dfrac{x}{3} + \dfrac{x}{2} = \dfrac{1}{3}$

8. $\dfrac{2}{x} + \dfrac{2}{5} = 1$

9. $1 - \dfrac{2}{z} = \dfrac{z}{z + 3}$

10. $\dfrac{x}{x + 1} - \dfrac{1}{2} = \dfrac{4}{3}$

Exercises 1.2

See www.CalcChat.com for worked-out solutions to odd-numbered exercises.

Writing an Algebraic Expression In Exercises 1–6, (a) use the verbal description to write a verbal model, (b) assign labels to the quantities in the verbal model, and (c) use the labels to write an algebraic expression.

1. The sum of two consecutive natural numbers

2. The product of two natural numbers whose sum is 25

3. Distance Traveled The distance traveled in t hours by a car traveling at 50 miles per hour

4. Travel Time The travel time for a plane that is traveling at a rate of r miles per hour for 200 miles

5. Geometry The perimeter of a rectangle whose width is x and whose length is twice the width

6. Geometry The area of a triangle whose base is 20 inches and whose height is h inches

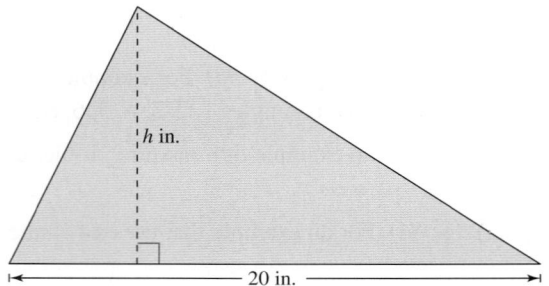

Writing an Algebraic Equation In Exercises 7–12, write an equation that represents the verbal description. *See Example 2.*

7. The sum of 5 and x equals 8.

8. The difference of n and 7 is 4.

9. The quotient of r and 2 is 9.

10. The product of x and 6 equals -9.

11. A salary of \$60,000 is increased by 8%.

12. A food processor is marked down 30% to \$105.

Constructing a Mathematical Model In Exercises 13–22, (a) use the verbal description to write a verbal model, (b) assign labels to the quantities in the verbal model, (c) use the labels to write a mathematical model, and (d) solve the problem. *See Examples 3 and 4.*

13. Find two consecutive numbers whose sum is 525.

14. Find three consecutive natural numbers whose sum is 804.

15. One positive number is five times another positive number. The difference of the two numbers is 148. Find the numbers.

16. One positive number is one-fifth of another number. The difference of the two numbers is 76. Find the numbers.

17. Salary Your weekly salary increases from \$440 to \$506. What percent is your raise?

18. Salary Your monthly salary increases from \$2200 to \$2376. What percent is your raise?

19. Discount Rate A satellite radio system for your car has been discounted by \$30. The sale price is \$119. What percent of the original list price is the discount?

20. Discount Rate The price of a shirt has been discounted by \$20. The sale price is \$29.95. What percent of the original list price is the discount?

21. Weekly Paycheck Your weekly paycheck is 12% *more* than your coworker's. Your two paychecks total $848. Find the amount of each paycheck.

22. Weekly Paycheck Your weekly paycheck is 12% *less* than your coworker's. Your two paychecks total $848. Find the amount of each paycheck.

Movie Series In Exercises 23–28, use the table, which shows Harry Potter movies in the order of their releases. Find the percent increase or decrease in the U.S. gross revenue of the two indicated movies. *(Source: Nash Information Services)*

Movie Title: Harry Potter and the . . .	U.S. gross revenue
Sorcerer's Stone (2001)	$317,557,891
Chamber of Secrets (2002)	$261,987,880
Prisoner of Azkaban (2004)	$249,538,952
Goblet of Fire (2005)	$290,013,036
Order of the Phoenix (2007)	$292,004,738
Half-Blood Prince (2009)	$301,959,197

23. *Sorcerer's Stone* (2001) to *Chamber of Secrets* (2002)

24. *Chamber of Secrets* (2002) to *Prisoner of Azkaban* (2004)

25. *Prisoner of Azkaban* (2004) to *Goblet of Fire* (2005)

26. *Goblet of Fire* (2005) to *Order of the Phoenix* (2007)

27. *Order of the Phoenix* (2007) to *Half-Blood Prince* (2009)

28. *Sorcerer's Stone* (2001) to *Half-Blood Prince* (2009)

Size Inflation In Exercises 29–32, use the following information. Restaurants tend to serve food in larger portions now than they have in the past. Several examples are shown in the table. Find the percent increase in size from the past to 2011 for the indicated food item.

Food or drink item	Past size	2011 size
Small soft drink	7 fl oz	16 fl oz
Small French fries	2.4 oz	2.5 oz
Large French fries	3.5 oz	5.4 oz
Pizza	10 in.	12 in.

29. Small soft drink

30. Small French fries

31. Large French fries

32. Pizza (Compare the areas.)

33. Comparing Calories A lunch consisting of a large deluxe cheeseburger, large fries, and a large soft drink contains 1560 calories. A lunch consisting of a small cheeseburger, small fries, and a small soft drink contains 680 calories. Find the percent change in calories from the larger to the smaller lunch.

34. Comparing Calories One slice (or one-eighth) of a 14-inch cheese pizza has 250 calories. One slice of a pepperoni pizza has 280 calories. Find the percent change in calories from a slice of the cheese pizza to a slice of the pepperoni pizza.

35. Salary You accept a new job with a starting salary of $35,000. You receive an 8% raise at the start of your second year, a 7.8% raise at the start of your third year, and a 9.4% raise at the start of your fourth year.

(a) Find your salary for the second year.

(b) Find your salary for the third year.

(c) Find your salary for the fourth year.

36. Salary You accept a new job with a starting salary of $48,000. You receive a 4% raise at the start of your second year, a 5.5% raise at the start of your third year, and an 11.4% raise at the start of your fourth year.

(a) Find your salary for the second year.

(b) Find your salary for the third year.

(c) Find your salary for the fourth year.

37. New Vehicle Sales In 2009, about 10,601,000 new motor vehicles (cars, light trucks, and other types of trucks) were sold or leased in the United States. The bar graph shows the percent of vehicles sold or leased by type. *(Source: U.S. Bureau of Economic Analysis)*

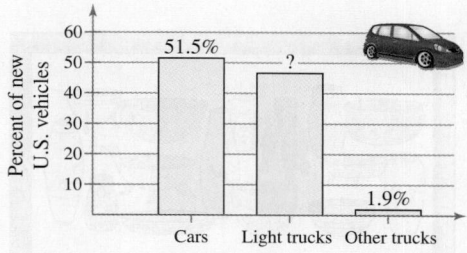

(a) Determine the percent of new motor vehicles that were light trucks.

(b) Determine the numbers of new cars, light trucks, and other types of trucks that were sold or leased in 2009.

(c) Determine the percent of new trucks that were light trucks.

38. Media Usage   It is projected that the average person will spend 3515 hours using various media in 2012. The bar graph shows these hours by media type (a person can use more than one type of media at a time). *(Source: Veronis Suhler Stevenson)*

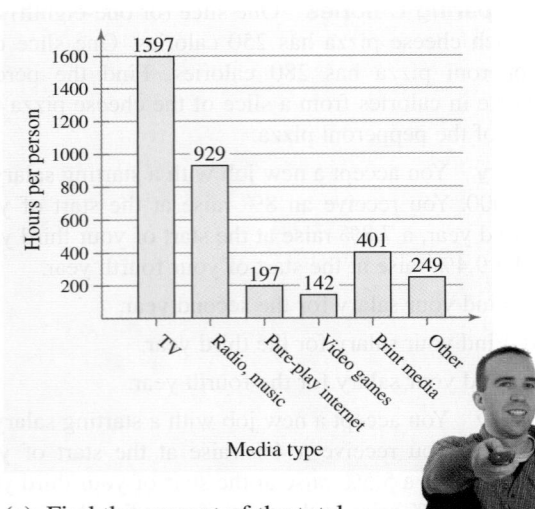

Media type

(a) Find the percent of the total media time spent reading print media.

(b) Find the percent of the total media time spent on something other than reading print media.

(c) Find the percent of the total media time spent playing video games and watching TV.

39. Geometry   A picture frame (see figure) has a total perimeter of 3 feet. The width of the frame is 0.62 times its length. Find the dimensions of the frame.

40. Geometry   A room is 1.5 times as long as it is wide, and its perimeter is 75 feet. Sketch a diagram of the room. Then find the dimensions of the room.

41. Simple Interest   You invest $2500 at 4.5% simple interest. How many years will it take for the investment to earn $1000 in interest?

42. Simple Interest   An investment earns $1075.20 interest over a seven-year period. What is the rate of simple interest on a $4800 principal investment?

43. Course Grade   To get an A in a course, you need an average of 90% or better on four tests that are worth 100 points each. Your scores on the first three tests were 87, 92, and 84. What must you score on the fourth test to get an A for the course?

44. Course Grade   To get an A in a course, you need an average of 90% or better on four tests. The first three tests are worth 100 points each and the fourth is worth 200 points. Your scores on the first three tests are 87, 92, and 84. What must you score on the fourth test to get an A for the course?

45. List Price   The price of a swimming pool has been discounted 15%. The sale price is $1200. Find the original list price of the swimming pool.

46. List Price   The price of a home theater system has been discounted 10%. The sale price is $499. Find the original list price of the system.

47. Wholesale Price   A store marks up an electric power drill 60% from its wholesale price. In a clearance sale, the price is discounted by 25%. The sale price is $21.60. What was the wholesale price of the power drill?

48. Wholesale Price   A store marks up a picture frame 80% from its wholesale price. In a clearance sale, the price is discounted by 40%. The sale price is $28.08. What was the wholesale price of the picture frame?

49. Weekly Salary   Due to economic factors, your employer has reduced your weekly wage by 15%. Before the reduction, your weekly salary was $425.

(a) What is your reduced salary?

(b) What percent raise must you receive to bring your weekly salary back up to $425? Explain why the percent raise is different from the percent reduction.

50. HOW DO YOU SEE IT?   The number of Internet users in the world increased by approximately the same amount from 2004 to 2007 as from 2007 to 2009, as shown in the bar graph. Was the *percent* increase from 2004 to 2007 approximately the same as from 2007 to 2009? Explain.   *(Source: ITU World Telecommunication)*

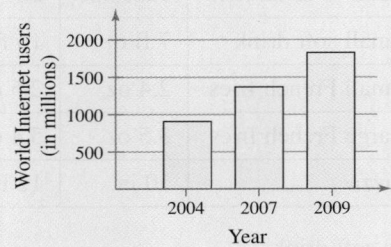

Year

51. Travel Time   You are driving to a college 150 miles from home. It takes 28 minutes to travel the first 30 miles. At this rate, how long is your entire trip?

52. Catch-Up Time   Students are traveling in two cars to a football game 135 miles away. One car travels at an average speed of 45 miles per hour. The second car starts $\frac{1}{2}$ hour later and travels at an average speed of 55 miles per hour. How long will it take the second car to catch up to the first car?

53. Height of a Tree   To determine the height of a tree, you measure its shadow and the shadow of a five-foot lamppost, as shown in the figure. How tall is the tree?

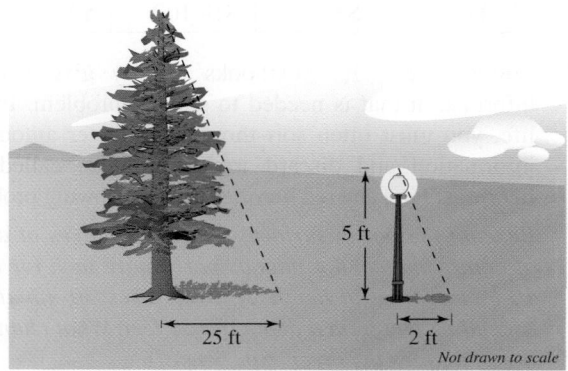

25 ft   2 ft

5 ft

Not drawn to scale

54. Height of a Building   To determine the height of a building, you measure the building's shadow and the shadow of a four-foot stake, as shown in the figure. How tall is the building?

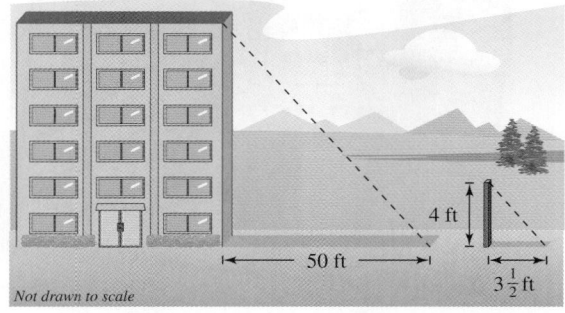

4 ft

50 ft $3\frac{1}{2}$ ft

Not drawn to scale

55. Radio Waves   Radio waves travel at the same speed as light, 3.0×10^8 meters per second. Find the time required for a radio wave to travel from mission control in Houston to NASA astronauts on the surface of the moon 3.84×10^8 meters away.

56. Distance to a Star   Find the distance (in miles) to a star that is 50 light years (distance traveled by light in 1 year) away. (Light travels at 186,000 miles per second.)

57. Investment Mix   You invest $15,000 in two funds paying 6.5% and 7.5% simple interest. The total annual interest is $1020. How much do you invest in each fund?

58. Investment Mix   You invest $30,000 in two funds paying 3% and $4\frac{1}{2}$% simple interest. The total annual interest is $1230. How much do you invest in each fund?

59. Stock Mix   You invest $5000 in two stocks. In one year, the value of stock A increases by 9.8% and the value of stock B increases by 6.2%. The total value of the stocks is now $5389.20. How much did you originally invest in each stock?

60. Stock Mix   You invest $4000 in two stocks. In one year, the value of stock A increases by 5.4% and the value of stock B increases by 12.8%. The total value of the stocks is now $4401. How much did you originally invest in each stock?

Production Limit   In Exercises 61 and 62, use the following information. *Variable costs* depend on the number of units produced. *Fixed costs* are the same regardless of how many units are produced. Find the greatest number of units the company can produce each month.

61. The company has fixed monthly costs of $15,000 and variable monthly costs of $8.75 per unit. The company has $90,000 available each month to cover costs.

62. The company has fixed monthly costs of $10,000 and variable monthly costs of $9.30 per unit. The company has $85,000 available each month to cover costs.

Rewriting Formulas   In Exercises 63–70, solve for the indicated variable.

63. Area of a Triangle
Solve for h in $A = \frac{1}{2}bh$.

64. Volume of a Rectangular Prism
Solve for l in $V = lwh$

65. Perimeter of a Rectangle
Solve for l in $P = 2l + 2w$.

66. Markup
Solve for C in $S = C + RC$.

67. Discount
Solve for L in $S = L - RL$.

68. Investment at Simple Interest
Solve for r in $A = P + Prt$.

69. Investment at Compound Interest
Solve for P in $A = P\left(1 + \dfrac{r}{n}\right)^{nt}$.

70. Area of a Trapezoid
Solve for b in $A = \frac{1}{2}(a + b)h$.

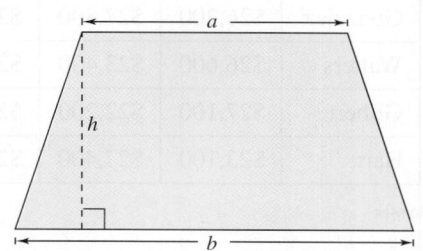

a

h

b

71. Length of a Tank The diameter of a cylindrical propane gas tank is 4 feet (see figure). The total volume of the tank is 603.2 cubic feet. Find the length of the tank.

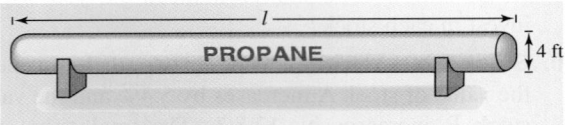

72. Water Depth A trough is 12 feet long, 3 feet deep, and 3 feet wide (see figure). Find the depth of the water when the trough contains 70 gallons of water. (1 gallon ≈ 0.13368 cubic foot.)

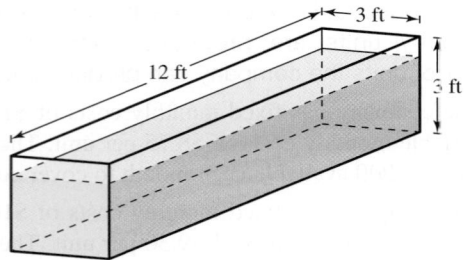

73. Mixture A 55-gallon barrel contains a mixture with a concentration of 40%. How much of this mixture must be withdrawn and replaced by 100% concentrate to bring the mixture up to 75% concentration? (See figure.)

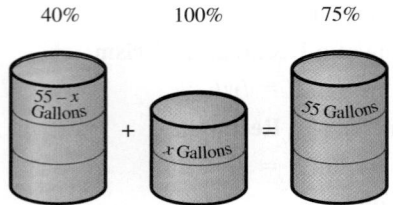

74. Fuel Mixture You have two gallons of a fuel mixture that is 32 parts gasoline and 1 part two-cycle engine oil. How much gasoline do you need to add to form a new mixture that is 40 parts gasoline and 1 part oil?

75. Monthly Sales The table shows the monthly sales of a sales team for the second quarter of the calendar year. Find the average monthly sales for each salesperson. Then find the team's average sales for each month.

Name	April	May	June
Williams	$25,000	$28,800	$21,000
Gonzalez	$26,200	$27,800	$29,500
Walters	$26,600	$23,400	$26,900
Gilbert	$27,100	$22,200	$29,000
Hart	$23,100	$27,400	$22,800

Courtesy of CitiKitty, Inc.

76. Monthly Sales The table shows the monthly sales of a sales team for the fourth quarter of the calendar year. Use a spreadsheet software program to find the average monthly sales for each salesperson. Then find the team's average sales for each month.

Name	October	November	December
Williams	$20,000	$25,100	$23,900
Gonzalez	$24,200	$23,600	$18,500
Walters	$31,900	$23,800	$18,400
Gilbert	$24,600	$23,100	$30,700
Hart	$32,400	$19,100	$28,600

77. Think About It Textbooks usually give just the information that is needed to solve a problem. In real life, you must often sort through a greater amount of information and discard irrelevant facts, called **red herrings.** Find any red herrings in the following problem.

Beneath the ocean surface, pressure changes at a rate of approximately 4.4 pounds per square inch for every 10-foot change in depth. A diver takes 30 minutes to ascend 25 feet from a depth of 150 feet. What change in pressure does the diver experience?

Business Capsule

CitiKitty, Inc. was founded in 2005 by 26-year-old Rebecca Rescate after she moved into a small apartment in New York City with no place to hide her cat's litter box. Finding no easy-to-use cat toilet training kit, she created one, and CitiKitty was born with an initial investment of $20,000. Today the company flourishes with an expanded product line. Revenues in 2010 reached $350,000.

78. Research Project Use your school's library, the Internet, or some other reference source to find information about the start-up costs of beginning a business, such as the example above. Write a short paper about the company.

1.3 Quadratic Equations

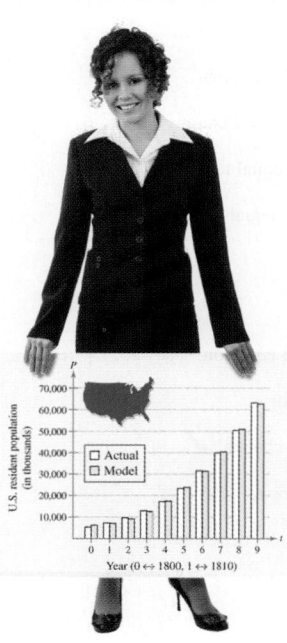

- Solve a quadratic equation by factoring.
- Solve a quadratic equation by extracting square roots.
- Construct and use a quadratic model to solve an application problem.

Solving Quadratic Equations by Factoring

In the first two sections of this chapter, you studied linear equations in one variable. In this and the next section, you will study quadratic equations.

> **Definition of a Quadratic Equation**
>
> A **quadratic equation** in x is an equation that can be written in the general form
>
> $$ax^2 + bx + c = 0$$
>
> where a, b, and c are real numbers with $a \neq 0$. Another name for a quadratic equation in x is a **second-degree polynomial equation in x.**

There are three basic techniques for solving quadratic equations: factoring, extracting square roots, and the *Quadratic Formula*. (The Quadratic Formula is discussed in the next section.) The factoring technique is based on the following property.

> **Zero-Factor Property**
>
> If $ab = 0$, then $a = 0$ or $b = 0$.

To use this property, rewrite the left side of the general form of a quadratic equation as the product of two linear factors. Then find the solutions of the quadratic equation by setting each linear factor equal to zero.

In Exercise 87 on page 103, you will use a quadratic model to determine the population of the United States.

In Exercise 87 on page 103,

STUDY TIP

The Zero-Factor Property applies *only* to equations written in general form (in which one side of the equation is zero). So, be sure that all terms are collected on one side *before* factoring. For instance, in the equation

$$(x - 5)(x + 2) = 8$$

it is *incorrect* to set each factor equal to 8. Can you solve this equation correctly?

Example 1 Solving a Quadratic Equation by Factoring

Solve $x^2 - 3x - 10 = 0$.

SOLUTION

$x^2 - 3x - 10 = 0$	Write original equation.
$(x - 5)(x + 2) = 0$	Factor.
$x - 5 = 0$ $\implies$ $x = 5$	Set 1st factor equal to 0.
$x + 2 = 0$ $\implies$ $x = -2$	Set 2nd factor equal to 0.

The solutions are $x = 5$ and $x = -2$. Check these in the original equation.

✓ **Checkpoint 1**

Solve each equation.

a. $x^2 - x - 12 = 0$

b. $x^2 + 10x + 24 = 0$

Karlova Irina/www.shutterstock.com

Example 2 **Solving a Quadratic Equation by Factoring**

Solve $6x^2 - 3x = 0$.

SOLUTION

$6x^2 - 3x = 0$		Write original equation.
$3x(2x - 1) = 0$		Factor out common factor.
$3x = 0$ ⟹ $x = 0$		Set 1st factor equal to 0.
$2x - 1 = 0$ ⟹ $x = \frac{1}{2}$		Set 2nd factor equal to 0.

The solutions are $x = 0$ and $x = \frac{1}{2}$.

CHECK

$6x^2 - 3x = 0$	Write original equation.
$6(0)^2 - 3(0) \overset{?}{=} 0$	Substitute 0 for x.
$0 - 0 = 0$	First solution checks. ✓
$6\left(\frac{1}{2}\right)^2 - 3\left(\frac{1}{2}\right) \overset{?}{=} 0$	Substitute $\frac{1}{2}$ for x.
$\frac{6}{4} - \frac{3}{2} = 0$	Second solution checks. ✓

✓ Checkpoint 2

Solve $4x^2 - 8x = 0$. ∎

If the two factors of a quadratic expression are the same, the corresponding solution of the quadratic equation is a **double** or **repeated** solution.

Example 3 **A Quadratic Equation with a Repeated Solution**

Solve $9x^2 - 6x = -1$.

SOLUTION

$9x^2 - 6x = -1$	Write original equation.
$9x^2 - 6x + 1 = 0$	Write in general form.
$(3x - 1)^2 = 0$	Factor.
$3x - 1 = 0$	Set repeated factor equal to 0.
$x = \frac{1}{3}$	Solution

The only solution is $x = \frac{1}{3}$.

CHECK

$9x^2 - 6x = -1$	Write original equation.
$9\left(\frac{1}{3}\right)^2 - 6\left(\frac{1}{3}\right) \overset{?}{=} -1$	Substitute $\frac{1}{3}$ for x.
$1 - 2 \overset{?}{=} -1$	Simplify.
$-1 = -1$	Solution checks. ✓

✓ Checkpoint 3

Solve $x^2 + 4x = -4$. ∎

TECH TUTOR

To check the solution in Example 3 with your graphing utility, you should first write the equation in general form.

$9x^2 - 6x + 1 = 0$

Then enter the expression $9x^2 - 6x + 1$ into y_1 of the equation editor. Now you can use the ASK mode of the *table* feature of your graphing utility to check the solution. For instructions on how to use the *table* feature, see Appendix A.

Extracting Square Roots

There is a shortcut for solving equations of the form $u^2 = d$, where $d > 0$. By factoring, you can see that this equation has two solutions.

$u^2 = d$	Write original equation.
$u^2 - d = 0$	Write in general form.
$\left(u + \sqrt{d}\right)\left(u - \sqrt{d}\right) = 0$	Factor.
$u + \sqrt{d} = 0 \implies u = -\sqrt{d}$	Set 1st factor equal to 0.
$u - \sqrt{d} = 0 \implies u = \sqrt{d}$	Set 2nd factor equal to 0.

Solving an equation of the form $u^2 = d$ without going through the steps of factoring is called **extracting square roots.**

Extracting Square Roots

The equation $u^2 = d$, where $d > 0$, has exactly two solutions:

$$u = \sqrt{d} \quad \text{and} \quad u = -\sqrt{d}.$$

These solutions can also be written as $u = \pm\sqrt{d}$.

Example 4   Extracting Square Roots

Solve $4x^2 = 12$.

SOLUTION

$4x^2 = 12$	Write original equation.
$x^2 = 3$	Divide each side by 4.
$x = \pm\sqrt{3}$	Extract square roots.

The solutions are $x = \sqrt{3}$ and $x = -\sqrt{3}$. Check these in the original equation.

✓ Checkpoint 4

Solve $2x^2 = 8$.

Example 5   Extracting Square Roots

Solve $(x - 3)^2 = 7$.

SOLUTION

$(x - 3)^2 = 7$	Write original equation.
$x - 3 = \pm\sqrt{7}$	Extract square roots.
$x = 3 \pm \sqrt{7}$	Add 3 to each side.

The solutions are $x = 3 \pm \sqrt{7}$. Check these in the original equation.

✓ Checkpoint 5

Solve $(x - 1)^2 = 16$.

Applications

Quadratic equations often occur in problems dealing with area. Here is a simple example.

A square room has an area of 144 square feet. Find the dimensions of the room.

To solve this problem, let x represent the length of each side of the room. Then use the formula for the area of a square to write and solve the equation

$$x^2 = 144$$

to conclude that each side of the room is 12 feet long. Note that although the equation $x^2 = 144$ has two solutions, $x = -12$ and $x = 12$, the negative solution makes no sense (for this problem), so you should choose the positive solution.

Example 6   Finding the Dimensions of a Room

A rectangular sunroom is 3 feet longer than it is wide (see Figure 1.6) and has an area of 154 square feet. Find the dimensions of the room.

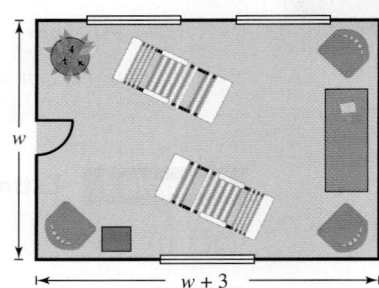

FIGURE 1.6

SOLUTION   Use the same type of problem-solving strategy that was presented in Section 1.2.

| *Verbal Model:* | $\dfrac{\text{Width}}{\text{of room}}$ $\cdot$ $\dfrac{\text{Length}}{\text{of room}}$ $=$ $\dfrac{\text{Area}}{\text{of room}}$ |

Labels:	Area of room $= 154$	(square feet)
	Width of room $= w$	(feet)
	Length of room $= w + 3$	(feet)

Equation:

$$w(w + 3) = 154$$
$$w^2 + 3w - 154 = 0$$
$$(w - 11)(w + 14) = 0$$
$$w - 11 = 0 \implies w = 11$$
$$w + 14 = 0 \implies w = -14$$

Choosing the positive value, you can conclude that the width is 11 feet and the length is $w + 3 = 14$ feet. Check this in the original statement of the problem, as shown.

CHECK

The length of 14 feet is 3 feet more than the width of 11 feet. ✓

The area of the sunroom is $(11)(14) = 154$ square feet. ✓

✓ Checkpoint 6

A rectangular kitchen is 8 feet longer than it is wide and has an area of 84 square feet. Find the dimensions of the kitchen.

Another application of quadratic equations involves an object that is falling (or is vertically projected into the air). The equation that gives the height of such an object is called a **position equation,** and on Earth's surface it has the form

$$s = -16t^2 + v_0t + s_0.$$

In this equation, s represents the height of the object (in feet), v_0 represents the initial velocity of the object (in feet per second), s_0 represents the initial height of the object (in feet), and t represents the time (in seconds).

 Example 7 **Falling Object**

A construction worker accidentally drops a wrench from a height of 235 feet and yells "Look out below!" (see Figure 1.7). Could a person at ground level hear this warning in time to get out of the way of the falling wrench?

SOLUTION Because sound travels at about 1100 feet per second, it follows that a person at ground level hears the warning within 1 second of the time the wrench is dropped. To set up a mathematical model for the height of the wrench, use the position equation

$$s = -16t^2 + v_0t + s_0.$$

Because the object is dropped rather than thrown, the initial velocity is $v_0 = 0$ feet per second. Moreover, because the initial height is $s_0 = 235$ feet, you have the following model.

$$s = -16t^2 + (0)t + 235$$
$$= -16t^2 + 235$$

After falling for 1 second, the height of the wrench is

$$-16(1)^2 + 235 = 219 \text{ feet.}$$

After falling for 2 seconds, the height of the wrench is

$$-16(2)^2 + 235 = 171 \text{ feet.}$$

To find the number of seconds it takes the wrench to hit the ground, let the height s be zero and solve the equation for t.

$s = -16t^2 + 235$	Write position equation.
$0 = -16t^2 + 235$	Substitute 0 for s.
$16t^2 = 235$	Add $16t^2$ to each side.
$t^2 = \dfrac{235}{16}$	Divide each side by 16.
$t = \dfrac{\sqrt{235}}{4}$	Extract positive square root.
$t \approx 3.83$	Use a calculator.

The wrench will take about 3.83 seconds to hit the ground. If the person hears the warning within 1 second after the wrench is dropped, the person still has almost 3 seconds to get out of the way.

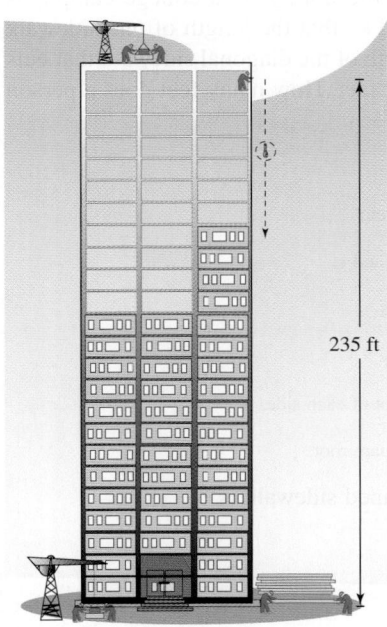

235 ft

FIGURE 1.7

✓**Checkpoint 7**

You drop a rock from a height of 144 feet. How long does it take the rock to hit the ground?

A third type of application using a quadratic equation involves the hypotenuse of a right triangle. Recall from geometry that the sides of a right triangle are related by a formula called the **Pythagorean Theorem.** This theorem states that if a and b are the lengths of the legs of the triangle and c is the length of the hypotenuse (see Figure 1.8),

$$a^2 + b^2 = c^2. \qquad \text{Pythagorean Theorem}$$

Notice how this formula is used in the next example.

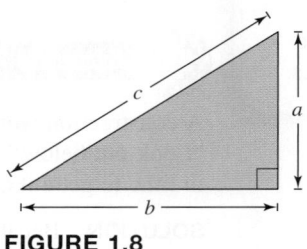

FIGURE 1.8

 Example 8 **Cutting Across the Lawn**

An L-shaped sidewalk from the athletic center to the library on a college campus is shown in Figure 1.9. The sidewalk was constructed so that the length of one sidewalk forming the L is twice as long as the other. The length of the diagonal sidewalk that cuts across the grounds between the two buildings is 32 feet. How many feet does a person save by walking on the diagonal sidewalk?

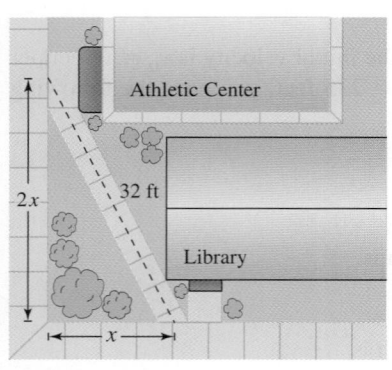

FIGURE 1.9

SOLUTION   Use the Pythagorean Theorem.

$$a^2 + b^2 = c^2 \qquad \text{Pythagorean Theorem}$$

$$x^2 + (2x)^2 = 32^2 \qquad \text{Substitute for } a, b, \text{ and } c.$$

$$5x^2 = 1024 \qquad \text{Combine like terms.}$$

$$x^2 = 204.8 \qquad \text{Divide each side by 5.}$$

$$x = \pm\sqrt{204.8} \qquad \text{Take the square root of each side.}$$

$$x = \sqrt{204.8} \qquad \text{Extract positive square root.}$$

The total distance covered by walking on the L-shaped sidewalk is

$$x + 2x = 3x$$

$$= 3\sqrt{204.8}$$

$$\approx 42.9 \text{ feet.}$$

Walking on the diagonal sidewalk saves a person about

$$42.9 - 32 = 10.9 \text{ feet.}$$

✓**Checkpoint 8**

In Example 8, suppose the length of one sidewalk forming the L is three times as long as the other. How many feet does a person save by walking on the 32-foot diagonal sidewalk?

A fourth type of application of a quadratic equation is one in which a quantity y is changing over time t according to a quadratic model. In the next example, y is exchanged with R because R is a better descriptor in the model.

Example 9 **Electric Power Revenues**

From 2003 to 2007, the yearly revenues R (in billions of dollars) for the electric power industry in the United States can be modeled by

$$R = 2.237t^2 + 239.12, \quad 3 \le t \le 7$$

where t represents the year, with $t = 3$ corresponding to 2003 (see Figure 1.10). Use the model to approximate the year that revenues were about \$325,000,000,000. *(Source: Edison Electric Institute)*

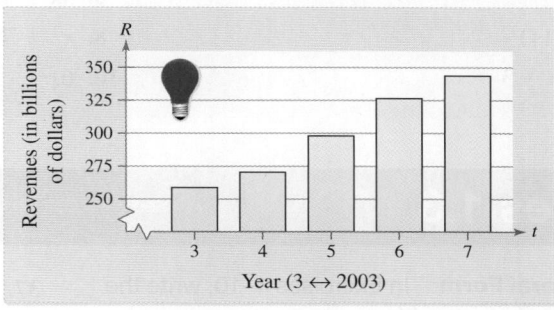

FIGURE 1.10

SOLUTION To solve this problem, let the yearly revenues R be 325 billion dollars and solve the equation for t.

$2.237t^2 + 239.12 = 325$	Substitute 325 for R.
$2.237t^2 = 85.88$	Subtract 239.12 from each side.
$t^2 \approx 38.391$	Divide each side by 2.237.
$t \approx \sqrt{38.391}$	Extract positive square root.
$t \approx 6.196$	Simplify.

The solution is $t \approx 6$. Because $t = 3$ represents 2003, you can conclude that, according to the model, revenues were about \$325 billion in the year 2006.

✓Checkpoint 9

Use the model in Example 9 to predict the year that revenues will be about \$500 billion.

SUMMARIZE (Section 1.3)

1. Describe how to solve a quadratic equation by factoring *(page 93)*. For examples of solving a quadratic equation by factoring, see Examples 1, 2, and 3.

2. Describe how to solve a quadratic equation by extracting square roots *(page 95)*. For examples of solving a quadratic equation by extracting square roots, see Examples 4 and 5.

3. Describe how a quadratic model can be used to represent data, such as the revenues of a particular industry *(page 99, Example 9)*.

SKILLS WARM UP 1.3   The following warm-up exercises involve skills that were covered in earlier sections. You will use these skills in the exercise set for this section. For additional help, review Sections 0.4 and 0.6.

In Exercises 1–4, simplify the expression.

1. $\sqrt{\frac{7}{50}}$

2. $\sqrt{32}$

3. $\sqrt{7^2 + 3 \cdot 7^2}$

4. $\sqrt{\frac{1}{4} + \frac{3}{8}}$

In Exercises 5–10, factor the expression.

5. $3x^2 + 5x$

6. $4x^2 - 25$

7. $16 - (x - 11)^2$

8. $x^2 + 7x - 18$

9. $10x^2 + 13x - 3$

10. $6x^2 - 73x + 12$

Exercises 1.3

See www.CalcChat.com for worked-out solutions to odd-numbered exercises.

Writing the General Form   In Exercises 1–10, write the quadratic equation in general form.

1. $2x^2 = 3 - 5x$

2. $4x^2 - 2x = 9$

3. $x^2 = 19x$

4. $10x^2 = 70$

5. $(x - 3)^2 = 5$

6. $12 - 3(x + 7)^2 = 0$

7. $x(x + 2) = 3x^2 + 1$

8. $x(x + 5) = 2(x + 5)$

9. $\dfrac{3x^2 - 10}{5} = 12x$

10. $\dfrac{x^2 - 7}{3} = 2x$

Solving a Quadratic Equation by Factoring   In Exercises 11–22, solve the quadratic equation by factoring. *See Examples 1, 2, and 3.*

11. $x^2 - 2x - 8 = 0$

12. $x^2 - 10x + 9 = 0$

13. $6x^2 + 3x = 0$

14. $9x^2 - 1 = 0$

15. $x^2 + 10x + 25 = 0$

16. $16x^2 + 56x + 49 = 0$

17. $3 + 5x - 2x^2 = 0$

18. $2x^2 = 19x + 33$

19. $x^2 + 4x = 12$

20. $x^2 + 4x = 21$

21. $-x^2 - 7x = 10$

22. $-x^2 + 8x = 12$

Extracting Square Roots   In Exercises 23–40, solve the quadratic equation by extracting square roots. When a solution is irrational, list both the exact solution *and* its approximation rounded to two decimal places. *See Examples 4 and 5.*

23. $x^2 = 16$

24. $x^2 = 144$

25. $x^2 = 7$

26. $x^2 = 27$

27. $3x^2 = 36$

28. $9x^2 = 25$

29. $(x - 12)^2 = 144$

30. $(x + 13)^2 = 225$

31. $(x + 2)^2 = 12$

32. $(x + 5)^2 = 20$

33. $12x^2 = 300$

34. $6x^2 = 250$

35. $5x^2 = 190$

36. $15x^2 = 620$

37. $3x^2 + 2(x^2 - 4) = 15$

38. $x^2 + 3(x^2 - 5) = 10$

39. $6x^2 - 3(x^2 + 1) = 23$

40. $2x^2 + 5(x^2 - 2) = 29$

Choosing a Method   In Exercises 41–62, solve the quadratic equation using any convenient method. *See Examples 1, 2, 3, 4, and 5.*

41. $x^2 = 64$

42. $7x^2 = 16$

43. $x^2 - 2x + 1 = 0$

44. $x^2 - 6x + 5 = 0$

45. $16x^2 - 9 = 0$

46. $11x^2 + 33x = 0$

47. $4x^2 - 12x + 9 = 0$

48. $x^2 - 14x + 49 = 0$

49. $(x + 4)^2 = 49$

50. $(x - 3)^2 = 36$

51. $4x = 4x^2 - 3$

52. $80 + 6x = 9x^2$

53. $50 + 5x = 3x^2$

54. $144 - 73x + 4x^2 = 0$

55. $12x = x^2 + 27$

56. $26x = 8x^2 + 15$

57. $50x^2 - 60x + 10 = 0$

58. $9x^2 + 12x + 3 = 0$

59. $(x + 3)^2 - 4 = 0$

60. $(x - 2)^2 - 9 = 0$

61. $(x + 1)^2 = x^2$

62. $(x + 1)^2 = 4x^2$

63. Writing   Consider the expression $(x + 2)^2$. How would you convince someone in your class that $(x + 2)^2 \neq x^2 + 4$? Give an argument based on the rules of algebra. Give an argument using your graphing utility.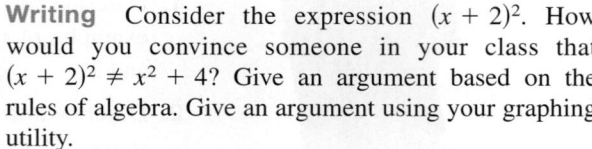

64. Writing   Consider the expression $\sqrt{a^2 + b^2}$. How would you convince someone in your class that $\sqrt{a^2 + b^2} \neq a + b$? Give an argument based on the rules of algebra or geometry. Give an argument using your graphing utility.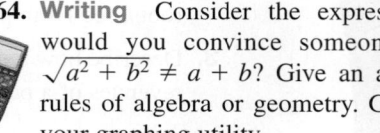

65. Geometry A one-story building is 14 feet longer than it is wide (see figure). The building has 1632 square feet of floor space. What are the dimensions of the building?

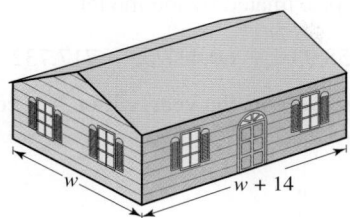

66. Geometry A billboard is 10 feet longer than it is high (see figure). The billboard has 336 square feet of advertising space. What are the dimensions of the billboard?

67. Geometry A triangular sign has a height that is equal to its base. The area of the sign is 4 square feet. Find the base and height of the sign.

68. Geometry The building lot shown in the figure has an area of 8000 square feet. What are the dimensions of the lot?

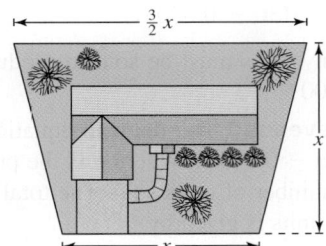

69. Geometry A rectangular garden that is 30 feet long and 20 feet wide is surrounded on all four sides by a rock path that is x feet wide. The total area of the garden and the rock path is 1200 square feet. What is the width of the path?

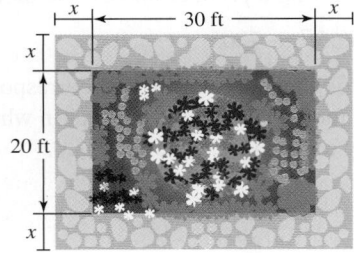

70. Geometry A rectangular pool is 30 feet wide and 40 feet long. It is surrounded on all four sides by a wooden deck that is x feet wide. The total area enclosed within the perimeter of the deck is 3000 square feet. What is the width of the deck?

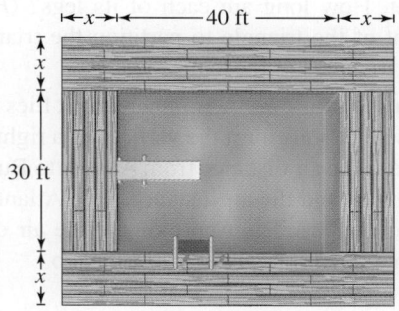

In Exercises 71–76, assume that air resistance is negligible, which implies that the position equation $s = -16t^2 + v_0 t + s_0$ is a reasonable model.

71. Falling Object A rock is dropped from the top of a 200-foot cliff that overlooks the ocean. How long will it take for the rock to hit the water?

72. Royal Gorge Bridge The Royal Gorge Bridge near Canon City, Colorado is one of the highest suspension bridges in the world. The bridge is 1053 feet above the Arkansas river. A rock is dropped from the bridge. How long does it take the rock to hit the water?

73. Olympic Diver The high-dive platform in the Olympics is 10 meters above the water. A diver wants to perform an armstand dive, which means she will drop to the water from a handstand position. How long will the diver be in the air? (*Hint:* 1 meter ≈ 3.2808 feet)

74. The Owl and the Mouse An owl is flying at a height of 100 feet and spots a mouse directly below. The owl folds up its wings and drops into a dive at the mouse. How long does the mouse have to escape?

75. Wind Resistance At the same time a skydiver jumps from an airplane 13,000 feet above the ground, a steel ball is dropped from the plane. Because of air resistance, it takes the skydiver 67 seconds to freefall to a height of 3000 feet, where the parachute opens. The steel ball has relatively no air resistance, so its height can be modeled by the position equation. How much faster does the ball reach a height of 3000 feet than the skydiver?

76. Wind Resistance At the same time a skydiver jumps from an airplane 8900 feet above the ground, a steel ball is dropped from the plane. Because of air resistance, it takes the skydiver 44 seconds to freefall to a height of 2500 feet, where the parachute opens. The steel ball has relatively no air resistance, so its height can be modeled by the position equation. How much faster does the ball reach a height of 2500 feet than the skydiver?

77. Geometry   The hypotenuse of an isosceles right triangle is 6 centimeters long. How long are the legs? (An isosceles right triangle is one whose two legs are of equal length.)

78. Geometry   An equilateral triangle has a height of 3 feet. How long are each of its legs? (*Hint:* Use the height of the triangle to partition the triangle into two right triangles of the same size.)

79. Flying Distance   A commercial jet flies to three cities whose locations form the vertices of a right triangle (see figure). The air distance from Atlanta to Buffalo is about 703 miles and the air distance from Atlanta to Chicago is about 583 miles. Approximate the air distance from Atlanta to Buffalo *by way of* Chicago.

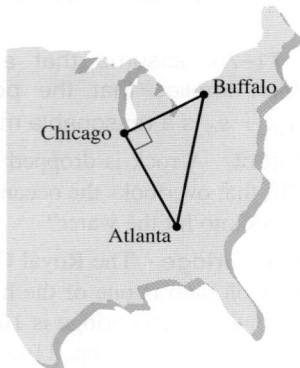

In Exercises 80 and 81, use the following information. The sum of the angles of a triangle is 180°. Also, if two angles of a triangle are equal, then the lengths of the sides opposite those angles are equal.

80. Depth of a Whale   The sonar on a research ship detects a whale that is 3000 feet from the ship. The angle formed by the ocean surface and a line from the ship to the whale is 45° (see figure). How deep is the whale?

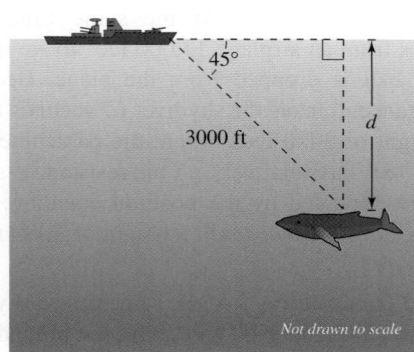

Not drawn to scale

81. Depth of a Whale Shark   A research ship is tracking the movements of a whale shark that is 700 meters from the ship. The angle formed by the ocean surface and a line from the ship to the whale shark is 45°. How deep is the whale shark?

82. Tennis   The winnings W (in dollars) of the men's singles champions of the Wimbledon Tennis Tournament from 2004 through 2009 can be approximated by the model

$$W = 8258.93t^2 - 60{,}437.5t + 717{,}732, \quad 4 \le t \le 9$$

where t represents the year, with $t = 4$ corresponding to 2004.   (*Source: Wimbledon Championships*)

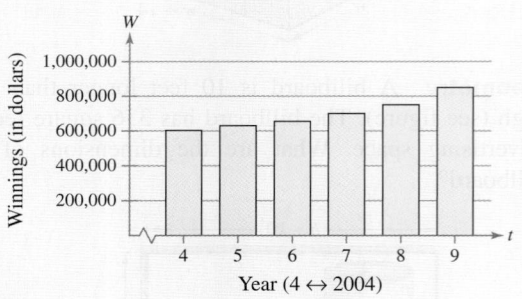

(a) According to the model, what is the first year the men's singles champion won $1 million or more?

(b) The Wimbledon men's singles champion won $1 million in 2010. Based on your answer in part (a), do you think this is a good model for making predictions after 2009? Explain.

83. Total Revenue   The demand equation for a product is $p = 36 - 0.0003x$, where p is the price per unit and x is the number of units sold. The total revenue R from selling x units is given by

$$R = xp = x(36 - 0.0003x).$$

How many units must be sold to produce a revenue of $1,080,000?

84. Total Revenue   The demand equation for a product is $p = 40 - 0.0005x$, where p is the price per unit and x is the number of units sold. The total revenue R from selling x units is given by

$$R = xp = x(40 - 0.0005x).$$

How many units must be sold to produce a revenue of $800,000?

85. Production Cost   A company determines that the average monthly cost C (in dollars) of raw materials for manufacturing a product line can be modeled by

$$C = 35.65t^2 + 7205, \quad t \ge 0$$

where t is the year, with $t = 0$ corresponding to 2010. Use the model to estimate the year in which the average monthly cost will be about $10,000.

86. Blue Oak   The blue oak tree, native to California, is known for its slow rate of growth. Fencing enclosures protect seedlings from herbivore damage and promote faster growth. The height H (in inches) of an enclosed blue oak tree is expected to change according to the model

$$H = 0.74t^2 + 25, \quad 0 \le t \le 5$$

where t represents the year, with $t = 0$ corresponding to 2010. Approximate the height of the tree in 2012. Then, use the model to approximate the year in which the height of the tree will be about 37 inches.

87. U.S. Population   The resident population P (in thousands) of the United States from 1800 to 1890 can be approximated by the model

$$P = 694.59t^2 + 6179, \quad 0 \le t \le 9$$

where t represents the year, with $t = 0$ corresponding to 1800, $t = 1$ corresponding to 1810, and so on (see figure). *(Source: U.S. Census Bureau)*

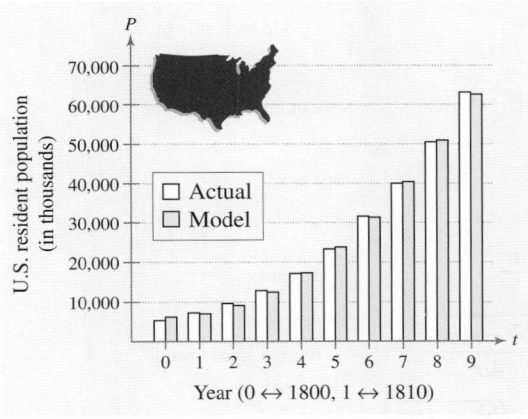

U.S. resident population (in thousands)

Year ($0 \leftrightarrow 1800$, $1 \leftrightarrow 1810$)

(a) Assume this model had continued to be valid up through the present time. In what year would the resident population of the United States have reached 250,000,000?

(b) Judging from the figure, would you say that this model is a good representation of the resident population through 1890?

(c) Use the model to predict the resident population in 2010. How does your answer compare with the actual 2010 resident population of approximately 310,000,000?

(d) The U.S. Census Bureau predicts that the resident population in 2050 will be 439,010,253. Would you say that the model supports this prediction? Explain.

88.   HOW DO YOU SEE IT?   Each figure shows a person or a ball about to go into motion through the air. Match each figure with the position equation that represents the situation. Without calculating, compare the terms of the equations to determine which situation involves the least amount of time in the air. Explain your reasoning.

(i)    (ii) 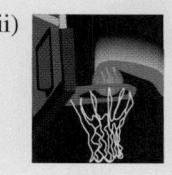   (iii)

(a) $s = -16t^2 + 33t$

(b) $s = -16t^2 + 10$

(c) $s = -16t^2 - 20t + 10$

89. Monthly Cost   A company determines that the average monthly cost C (in dollars) of staffing temporary positions can be modeled by

$$C = 135.47t^2 + 13{,}702, \quad t \ge 0$$

where t represents the year, with $t = 0$ corresponding to 2010. Use the model to predict the year in which the average monthly cost will be about $25,000.

90. Education   The enrollment E in an early childhood development program for a school district from 2000 to 2012 can be approximated by the model

$$E = 1.678t^2 + 1025, \quad 0 \le t \le 12$$

where t represents the year, with $t = 0$ corresponding to 2000. Use the model to approximate the year in which the early childhood enrollment was about 1160 children. Can you use the model to estimate early childhood enrollment for the year 1980? Explain.

91. Meteorology   The temperature T (in degrees Fahrenheit) during a certain day can be approximated by

$$T = 0.31t^2 + 32.9, \quad 7 \le t \le 15$$

where t represents the hour of the day, with $t = 7$ corresponding to 7 A.M. Use the model to approximate the time when the temperature was 85° F. Can you use this model to predict the temperature at 7 P.M.? Explain.

92. Electromedical Exports   From 2004 through 2009, the yearly exports E (in billions of dollars) of electromedical equipment from the United States can be approximated by

$$E = 0.119t^2 + 10.7, \quad 4 \le t \le 9$$

where t represents the year, with $t = 4$ corresponding to 2004. Use the model to estimate the year when the value of these exports was about $15 billion. *(Source: AeA)*

1.4 The Quadratic Formula

■ Develop the Quadratic Formula by completing the square.
■ Use the discriminant to determine the number of real solutions of a quadratic equation.
■ Solve a cuadratic equation using the Quadratic Formula.
■ Use the Quadratic Formula to solve an application problem.

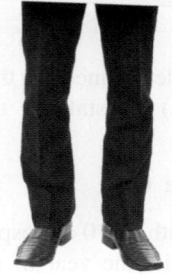

In Exercise 71 on page 112, you will use the Quadratic Formula to determine the year when only 60% of households still had VCRs.

Development of the Quadratic Formula

In Section 1.3, you studied two methods for solving quadratic equations. These two methods are efficient for special quadratic equations that are factorable or that can be solved by extracting square roots. There are, however, many quadratic equations that cannot be solved efficiently by either of these two techniques. Fortunately, there is a general formula that can be used to solve *any* quadratic equation. It is called the **Quadratic Formula.** This formula is derived using a process called **completing the square.**

$$ax^2 + bx + c = 0 \qquad \text{General form, } a \neq 0$$

$$ax^2 + bx = -c \qquad \text{Subtract } c \text{ from each side.}$$

$$x^2 + \frac{b}{a}x = -\frac{c}{a} \qquad \text{Divide each side by } a.$$

$$x^2 + \frac{b}{a}x + \left(\frac{b}{2a}\right)^2 = -\frac{c}{a} + \left(\frac{b}{2a}\right)^2 \qquad \text{Complete the square.}$$

$$\left(\text{half of } \frac{b}{a}\right)^2$$

$$\left(x + \frac{b}{2a}\right)^2 = \frac{b^2 - 4ac}{4a^2} \qquad \text{Simplify.}$$

$$x + \frac{b}{2a} = \pm\sqrt{\frac{b^2 - 4ac}{4a^2}} \qquad \text{Extract square roots.}$$

$$x = \frac{-b \pm \sqrt{b^2 - 4ac}}{2a} \qquad \text{Solutions}$$

The Quadratic Formula

The solutions of

$$ax^2 + bx + c = 0, \quad a \neq 0$$

are given by the **Quadratic Formula,**

$$x = \frac{-b \pm \sqrt{b^2 - 4ac}}{2a}.$$

To help you remember the Quadratic Formula, try to memorize a verbal statement of the rule, such as:

"The opposite of *b*, plus or minus the square root of *b* squared minus 4*ac*, all divided by 2*a*."

The Discriminant

In the Quadratic Formula, the quantity under the radical sign, $b^2 - 4ac$, is called the **discriminant** of the quadratic expression $ax^2 + bx + c$.

$$b^2 - 4ac \qquad \text{Discriminant}$$

It can be used to determine the number of real solutions of a quadratic equation.

Solutions of a Quadratic Equation

The solutions of a quadratic equation

$$ax^2 + bx + c = 0, \quad a \neq 0$$

can be classified by the discriminant, $b^2 - 4ac$, as follows.

1. If $b^2 - 4ac > 0$, then the equation has *two* distinct real solutions.

2. If $b^2 - 4ac = 0$, then the equation has *one* repeated real solution.

3. If $b^2 - 4ac < 0$, then the equation has *no* real solutions.

If the discriminant of a quadratic equation is negative, as in case 3 above, then its square root is imaginary (not a real number) and the Quadratic Formula yields two complex solutions. You will study complex solutions in Section 3.5.

Example 1   Using the Discriminant

Use the discriminant to determine the number of real solutions to each of the following quadratic equations.

a. $4x^2 - 20x + 25 = 0$ **b.** $13x^2 + 7x + 1 = 0$ **c.** $5x^2 = 8x$

SOLUTION

a. Using $a = 4$, $b = -20$, and $c = 25$, the discriminant is

$$b^2 - 4ac = (-20)^2 - 4(4)(25)$$

$$= 400 - 400 = 0.$$

Because $b^2 - 4ac = 0$, there is *one* repeated real solution.

b. Using $a = 13$, $b = 7$, and $c = 1$, the discriminant is

$$b^2 - 4ac = (7)^2 - 4(13)(1)$$

$$= 49 - 52 = -3.$$

Because $b^2 - 4ac < 0$, there are *no* real solutions.

c. In general form, this equation is $5x^2 - 8x = 0$, with $a = 5$, $b = -8$, and $c = 0$, which implies that the discriminant is

$$b^2 - 4ac = (-8)^2 - 4(5)(0)$$

$$= 64.$$

Because $b^2 - 4ac > 0$, there are *two* distinct real solutions.

✓ **Checkpoint 1**

Use the discriminant to determine the number of real solutions of $x^2 + 6x + 9 = 0$.

Using the Quadratic Formula

When using the Quadratic Formula, remember that *before* the formula can be applied, you must first write the quadratic equation in general form.

Example 2   Two Distinct Solutions

Solve $x^2 + 3x = 9$.

SOLUTION

$$x^2 + 3x = 9$$ Write original equation.

$$x^2 + 3x - 9 = 0$$ Write in general form.

$$x = \frac{-3 \pm \sqrt{(3)^2 - 4(1)(-9)}}{2(1)}$$ Quadratic Formula

$$x = \frac{-3 \pm \sqrt{45}}{2}$$ Simplify.

$$x = \frac{-3 \pm 3\sqrt{5}}{2}$$ Simplify.

The solutions are

$$x = \frac{-3 + 3\sqrt{5}}{2} \quad \text{and} \quad x = \frac{-3 - 3\sqrt{5}}{2}.$$

Check these in the original equation.

✓ Checkpoint 2

Solve $x^2 + 2x = 2$.

Example 3   One Repeated Solution

Solve $8x^2 - 24x + 18 = 0$.

SOLUTION   Begin by dividing each side by the common factor 2.

$$8x^2 - 24x + 18 = 0$$ Write original equation.

$$4x^2 - 12x + 9 = 0$$ Divide each side by 2.

$$x = \frac{-(-12) \pm \sqrt{(-12)^2 - 4(4)(9)}}{2(4)}$$ Quadratic Formula

$$x = \frac{12 \pm \sqrt{0}}{8}$$ Simplify.

$$x = \frac{3}{2}$$ Repeated solution

The only solution is

$$x = \tfrac{3}{2}.$$

Check this in the original equation.

✓ Checkpoint 3

Solve $27x^2 - 18x = -3$.

TECH TUTOR

You can write a program for a graphing utility to solve equations using the Quadratic Formula. Use a program to solve the equation in Example 2.

The discriminant in Example 3 is a perfect square (zero in this case), and you could have factored the quadratic as

$$4x^2 - 12x + 9 = 0$$

$$(2x - 3)^2 = 0$$

and concluded that the solution is $x = \frac{3}{2}$. Because factoring is easier than applying the Quadratic Formula, try factoring first when solving a quadratic equation. If factors cannot easily be found, then use the Quadratic Formula. For instance, try solving the quadratic equation

$$x^2 - x - 12 = 0$$

in two ways—by factoring and by the Quadratic Formula—to see that you get the same solutions either way.

When using a calculator with the Quadratic Formula, you should get in the habit of using the memory key to store intermediate steps. This will save steps and minimize roundoff error.

Example 4   Using a Calculator with the Quadratic Formula

Solve $16.3x^2 - 197.6x + 7.042 = 0$.

SOLUTION   In this case, $a = 16.3$, $b = -197.6$, $c = 7.042$, and you have

$$x = \frac{-(-197.6) \pm \sqrt{(-197.6)^2 - 4(16.3)(7.042)}}{2(16.3)}.$$

To evaluate these solutions, begin by calculating the square root of the discriminant, as follows.

Scientific Calculator Keystrokes

197.6 $\boxed{+/-}$ $\boxed{x^2}$ $\boxed{-}$ 4 $\boxed{\times}$ 16.3 $\boxed{\times}$ 7.042 $\boxed{=}$ $\boxed{\sqrt{}}$

Graphing Calculator Keystrokes

$\boxed{\sqrt{}}$ $\boxed{(}$ $\boxed{(}$ $\boxed{(-)}$ 197.6 $\boxed{)}$ $\boxed{x^2}$ $\boxed{-}$ 4 $\boxed{\times}$ 16.3 $\boxed{\times}$ 7.042 $\boxed{)}$ $\boxed{\text{ENTER}}$

In either case, the result is 196.434777. Storing this result and using the recall key, you can find the following two solutions.

$$x \approx \frac{197.6 + 196.434777}{2(16.3)} \approx 12.087 \qquad \text{Add stored value.}$$

$$x \approx \frac{197.6 - 196.434777}{2(16.3)} \approx 0.036 \qquad \text{Subtract stored value.}$$

✓ **Checkpoint 4**

Solve $4.7x^2 - 3.2x - 5.9 = 0$. ■

TECH TUTOR

Try to calculate the value of x in Example 4 by using additional parentheses instead of storing the intermediate result, 196.434777, in your calculator.

Applications

In Section 1.3, you studied four basic types of applications involving quadratic equations: area, falling bodies, the Pythagorean Theorem, and quadratic models. The solution to each of these types of problems can involve the Quadratic Formula. For instance, Example 5 shows how the Quadratic Formula can be used to analyze a quadratic model for a patient's blood oxygen level.

 Example 5   **Blood Oxygen Level**

Doctors treated a patient at an emergency room from 1:00 p.m. to 5:00 p.m. The patient's blood oxygen level L (in percent) during this time period can be modeled by

$$L = -0.25t^2 + 3.0t + 87, \quad 1 \le t \le 5$$

where t represents the time of day, with $t = 1$ corresponding to 1:00 p.m. Use the model to estimate the time when the patient's blood oxygen level was 95%.

SOLUTION   To find the hour when the patient's blood oxygen level was 95%, solve the equation

$$95 = -0.25t^2 + 3.0t + 87.$$

To begin, write the equation in general form.

$$-0.25t^2 + 3.0t - 8 = 0$$

Then apply the Quadratic Formula with $a = -0.25$, $b = 3.0$, and $c = -8$.

$$t = \frac{-3 \pm \sqrt{3^2 - 4(-0.25)(-8)}}{2(-0.25)}$$

$$= \frac{-3 \pm \sqrt{9 - 8}}{-0.5}$$

$$= \frac{-3 \pm \sqrt{1}}{-0.5}$$

$$= 4 \text{ or } 8$$

Of the two possible solutions, only $t = 4$ makes sense in the context of the problem because $t = 8$ is not in the domain of L. Because $t = 1$ corresponds to 1:00 P.M., it follows that $t = 4$ corresponds to 4:00 P.M. So, from the model you can conclude that the patient's blood oxygen level was 95% at about 4:00 P.M. Figure 1.11 shows the patient's oxygen level recorded every 30 minutes.

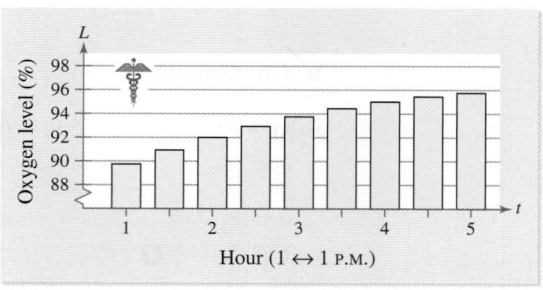

FIGURE 1.11

✓ Checkpoint 5

Use the model in Example 5 to estimate the time when the patient's blood oxygen level was 92%.

A pulse oximeter helps monitor the oxygen saturation of an individual's blood, as well as his or her heart rate.

In Section 1.3, you learned that the position equation for a falling object is of the form

$$s = -16t^2 + v_0t + s_0$$

where s is the height (in feet) of the object, v_0 is the initial velocity (in feet per second), t is the time (in seconds), and s_0 is the initial height (in feet). This equation is valid only for free-falling objects near Earth's surface. Because of differences in gravitational force, position equations are different on other planets or moons. The next example looks at a position equation for a falling object on our moon.

 **Example 6** **Throwing an Object on the Moon**

An astronaut standing on the surface of the moon throws a rock straight up at 27 feet per second from a height of 6 feet. The height s (in feet) of the rock is given by

$$s = -2.7t^2 + 27t + 6$$

where t is the time (in seconds). How much time elapses before the rock strikes the lunar surface?

SOLUTION Because s gives the height of the rock at any time t, you can find the time that the rock hits the surface of the moon by setting s equal to zero and solving for t.

$$-2.7t^2 + 27t + 6 = 0 \qquad \text{Substitute 0 for } s.$$

$$t = \frac{-27 \pm \sqrt{(27)^2 - 4(-2.7)(6)}}{2(-2.7)} \qquad \text{Quadratic Formula}$$

$$= \frac{-27 \pm \sqrt{793.8}}{-5.4} \qquad \text{Simplify.}$$

$$\approx 10.2 \text{ seconds} \qquad \text{Choose positive solution.}$$

So, about 10.2 seconds elapse before the rock hits the lunar surface.

✓ **Checkpoint 6**

In Example 6, suppose the rock is thrown straight up at 13 feet per second from a height of 4 feet. The height s (in feet) of the rock is given by

$$s = -2.7t^2 + 13t + 4.$$

How much time (in seconds) elapses before the rock strikes the lunar surface?

SUMMARIZE (Section 1.4)

1. State the definition of the discriminant of a quadratic expression and describe how it is used to determine the number of real solutions of a quadratic equation *(page 105)*. For an example of using the discriminant, see Example 1.

2. Describe how to solve a quadratic equation using the Quadratic Formula *(pages 106 and 107)*. For examples that use the Quadratic Formula, see Examples 2, 3, and 4.

3. Describe a real-life example that uses the Quadratic Formula *(pages 108 and 109, Examples 5 and 6)*.

SKILLS WARM UP 1.4 The following warm-up exercises involve skills that were covered in earlier sections. You will use these skills in the exercise set for this section. For additional help, review Sections 0.4 and 1.3.

In Exercises 1–4, simplify the expression.

1. $\sqrt{25 - 4(3)(-12)}$

2. $\sqrt{49 - 4(2)(3)}$

3. $\sqrt{12^2 - 4(3)(4)}$

4. $\sqrt{15^2 + 4(9)(12)}$

In Exercises 5–10, solve the quadratic equation by factoring.

5. $x^2 - x - 2 = 0$

6. $2x^2 + 3x - 9 = 0$

7. $x^2 - 4x = 5$

8. $2x^2 + 13x = 7$

9. $x(x - 7) = -10$

10. $x(x - 3) = 4$

Exercises 1.4

See www.CalcChat.com for worked-out solutions to odd-numbered exercises.

Using the Discriminant In Exercises 1–8, use the discriminant to determine the number of real solutions of the quadratic equation. *See Example 1.*

1. $x^2 - 4x + 4 = 0$

2. $2x^2 - x - 1 = 0$

3. $3x^2 + 4x + 1 = 0$

4. $x^2 + 2x + 4 = 0$

5. $2x^2 - 5x = -5$

6. $3 - 6x = -3x^2$

7. $\frac{1}{5}x^2 + \frac{6}{5}x - 8 = 0$

8. $\frac{1}{3}x^2 - 5x + 25 = 0$

Using the Quadratic Formula In Exercises 9–30, use the Quadratic Formula to solve the quadratic equation. *See Examples 2 and 3.*

9. $2x^2 + x - 1 = 0$

10. $2x^2 - x - 1 = 0$

11. $16x^2 + 8x - 3 = 0$

12. $25x^2 - 20x + 3 = 0$

13. $6x = 4 - x^2$

14. $2 + 2x = x^2$

15. $x^2 - 10x + 22 = 0$

16. $x^2 + 14x + 44 = 0$

17. $x^2 + 8x - 4 = 0$

18. $4x^2 - 4x - 4 = 0$

19. $12x - 9x^2 = -3$

20. $16x^2 + 22 = 40x$

21. $36x^2 + 24x = 7$

22. $3x + x^2 = 1$

23. $4x^2 + 4x = 7$

24. $16x^2 - 40x = -5$

25. $28x - 49x^2 = 4$

26. $9x^2 + 24x = -16$

27. $8t = 5 + 2t^2$

28. $25h^2 + 61 = -80h$

29. $(y - 5)^2 = 2y$

30. $(x + 6)^2 = -2x$

Using a Calculator In Exercises 31–36, use a calculator to solve the quadratic equation. (Round your answer to three decimal places.) *See Example 4.*

31. $5.1x^2 - 1.7x - 3.2 = 0$

32. $-0.067x^2 - 0.852x + 1.277 = 0$

33. $12.67x^2 + 31.55x + 8.09 = 0$

34. $422x^2 - 506x - 347 = 0$

35. $-0.003x^2 + 0.025x - 0.98 = 0$

36. $-0.005x^2 + 0.101x - 0.193 = 0$

Choosing a Method In Exercises 37–46, solve the quadratic equation using any convenient method. *See Examples 2, 3, and 4.*

37. $2x^2 + 7 = 2x^2 - x - 4$

38. $x^2 - 2x + 5 = x^2 - 5$

39. $4x^2 - 15 = 25$

40. $3x^2 - 16 = 38$

41. $x^2 - 2x - 1 = 0$

42. $x^2 + 3x - 4 = 0$

43. $(x - 1)^2 = 9$

44. $(x - 5)^2 = 16$

45. $3x^2 + 5x - 11 = 4(x - 2)$

46. $2x^2 + 4x - 9 = 2(x - 1)$

Writing Real-Life Problems In Exercises 47–50, solve the number problem *and* write a real-life problem that could be represented by this verbal model. For instance, an applied problem that could be represented by Exercise 47 is shown below.

The sum of the length and width of a one-story house is 100 feet. The house has 2500 square feet of floor space. What are the length and width of the house?

47. Find two numbers whose sum is 100 and whose product is 2500.

48. One number is 1 more than another number. The product of the two numbers is 72. Find the numbers.

49. One number is 1 more than another number. The sum of their squares is 113. Find the numbers.

50. One number is 2 more than another number. The product of the two numbers is 440. Find the numbers.

Cost Equation In Exercises 51–54, use the cost equation to find the number of units x that a manufacturer can produce for cost C. (Round your answer to the nearest positive integer.)

51. $C = 0.125x^2 + 20x + 5000, \quad C = \$14,000$

52. $C = 0.5x^2 + 15x + 5000, \quad C = \$11,500$

53. $C = 800 + 0.04x + 0.002x^2$, $C = \$1680$

54. $C = 312.5 - 10x + 0.4x^2$, $C = \$900$

55. Seating A lecture hall has 108 chairs arranged in rows with the same number of chairs in each row. You eliminate three rows by adding six chairs to each of the other rows. How many rows are there now? How many chairs are in each row?

56. Child Care A child care center has 200 feet of fencing to enclose two adjacent rectangular safe play areas (see figure).

(a) Write an expression for the total area A of the play areas in terms of x.

(b) Find the dimensions for which the enclosed area is 1600 square feet.

57. Geometry An open box is made from a square piece of material by cutting two-inch squares from the corners and turning up the sides (see figure). The volume of the finished box is 200 cubic inches. Find the size of the original piece of material.

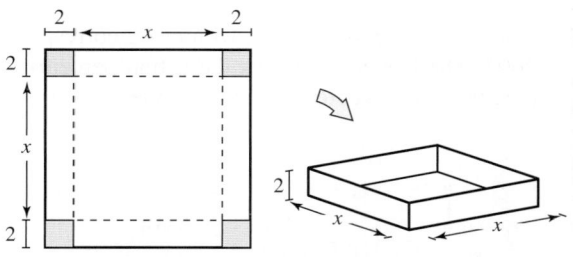

58. Geometry The swimming pool shown is 4 feet deep and has a volume of 1024 cubic feet. Find the dimensions of the base of the pool.

Not drawn to scale

59. Stratosphere You are at the Stratosphere Tower in Las Vegas. While riding the X-Scream tower ride, you accidently drop a coin from a height of 866 feet with an initial velocity of -5 feet per second. How long does it take the coin to hit the ground?

60. Sports You throw a baseball straight up into the air at a velocity of 80 feet per second. You release the baseball at a height of 5.5 feet and catch it when it falls back to a height of 5.5 feet.

(a) Use the position equation to write a mathematical model for the height of the baseball.

(b) Find the height of the baseball after 1 second, 2 seconds, and 3 seconds. What must have occurred sometime in the interval $2 \le t \le 3$? Explain.

(c) For how many seconds is the baseball in the air?

61. On the Moon An astronaut on the moon throws a rock straight upward into space. The height s (in feet) of the rock is given by

$$s = -2.7t^2 + 40t + 5$$

where the initial velocity is 40 feet per second, the initial height is 5 feet, and t is the time (in seconds). How long will it take the rock to hit the surface of the moon? Suppose a rock is thrown with the same initial velocity and height on Earth. How long will it take the rock to hit the ground?

62. Hot Air Balloon Two people are floating in a hot air balloon 200 feet above a lake. One person tosses out a coin with an initial velocity of 20 feet per second. One second later, the balloon is 20 feet higher and the other person drops another coin (see figure). The position equation for the first coin is $s = -16t^2 + 20t + 200$, and the position equation for the second coin is $s = -16t^2 + 220$. Which coin will hit the water first? (*Hint:* Remember that the first coin was tossed one second before the second coin was dropped.)

63. Comparing Gravity Does a rock thrown upward from an initial height of 6 feet with an initial velocity of 27 feet per second take longer to reach the ground on Earth or on the moon? (See Example 6.)

64. Comparing Gravity Does a rock thrown downward from an initial height of 6 feet with an initial velocity of −14 feet per second take longer to reach the ground on Earth or on the moon? (See Example 6.)

65. Aviation A commercial jet flies to three cities whose locations form the vertices of a right triangle (see figure). The total flight distance (from Oklahoma City to Austin to New Orleans and back to Oklahoma City) is approximately 1348 miles. It is 560 miles between Oklahoma City and New Orleans. Approximate the other two distances.

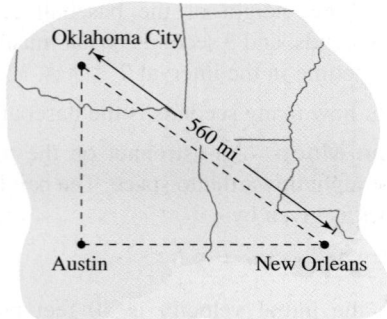

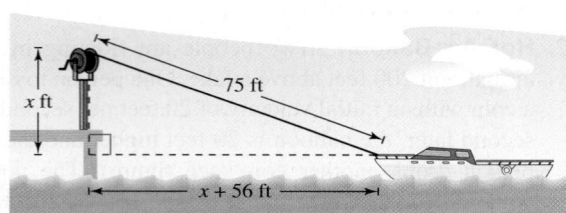

66. Distance from a Dock A windlass is used to tow a boat to a dock. The figure shows the boat when it is 75 feet from the windlass. How far is the boat from the dock?

67. Blood Oxygen Level Doctors treated a patient at an emergency room from 2:00 P.M. to 7:00 P.M. The patient's blood oxygen level L (in percent) during this time period can be modeled by

$$L = -0.270t^2 + 3.59t + 83.1, \quad 2 \le t \le 7$$

where t represents the time of day, with $t = 2$ corresponding to 2:00 P.M. Use the model to estimate the time (rounded to the nearest hour) when the patient's blood oxygen level was 93%.

68. Green Mountain Coffee Roasters The sales S (in millions of dollars) for Green Mountain Coffee Roasters from 2004 through 2009 can be approximated by

$$S = 31.61t^2 - 283.4t + 776, \quad 4 \le t \le 9$$

where t represents the year, with $t = 4$ corresponding to 2004. Use the model to estimate the year when sales were about $340 million. *(Source: Green Mountain Coffee Roasters, Inc.)*

69. Flying Speed Two planes leave simultaneously from the same airport, one flying due east and the other due south (see figure). The eastbound plane is flying 50 miles per hour faster than the southbound plane. After 3 hours the planes are 2440 miles apart. Find the speed of each plane.

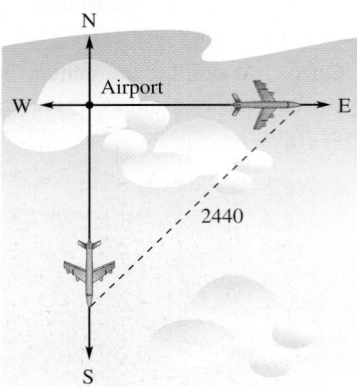

70. Flying Speed Two planes leave simultaneously from the same airport, one flying due east and the other due south. The eastbound plane is flying 100 miles per hour faster than the southbound plane. After 2 hours the planes are 1500 miles apart. Find the speed of each plane.

71. VCRs The percent P of homes in the United States with VCRs from 2002 through 2011 can be approximated by the model

$$P = -0.496t^2 + 3.01t + 87.2, \quad 2 \le t \le 11$$

where t represents the year, with $t = 2$ corresponding to 2002. The bar graph shows the actual percents and the percents represented by the model. *(Source: The Nielsen Company)*

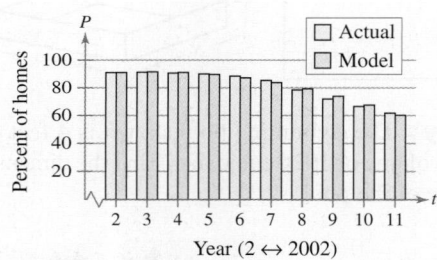

(a) Use the model to estimate the year when the percent of homes with VCRs was about 60%.

(b) Use the model to estimate the year when the percent of homes with VCRs was about 80%.

(c) Use the model to predict the last year that VCRs will exist in U.S. households. Would you say that the model is valid to represent data up to that year?

72. **HOW DO YOU SEE IT?** The bar graph shows the growth of a baby girl. A model that approximates her body length l (in centimeters) at m months of age is

$$l = -0.0813m^2 + 3.028m + 50.0, \quad 0 \le m \le 15.$$

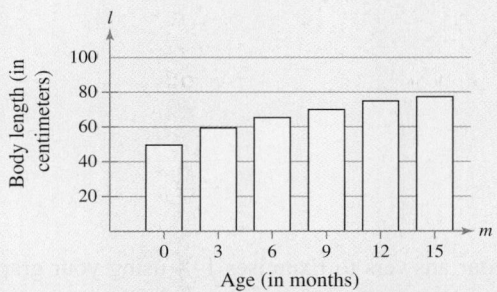

(a) During which 3-month period shown in the bar graph did the baby girl grow the most?

(b) For $l = 50$ centimeters, the model simplifies to

$$0 = -0.0813m^2 + 3.028m.$$

Is the discriminant positive, negative, or zero?

(c) Based on your answer in part (b) and the bar graph, will the original model work well for predicting the girl's height as she gets older? Explain.

73. **Biology** The metabolic rate of an ectothermic organism increases with increasing temperature within a certain range. The graph shows experimental data for the oxygen consumption C (in microliters per gram per hour) of a beetle at certain temperatures. The data can be approximated by the model

$$C = 0.45x^2 - 1.65x + 50.75, \quad 10 \le x \le 25$$

where x is the temperature in degrees Celsius.

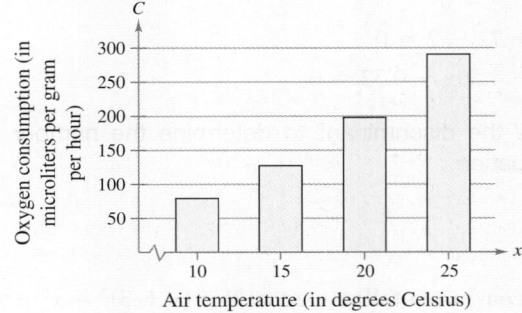

(a) The oxygen consumption is 150 microliters per gram per hour. What is the air temperature?

(b) The temperature is increased from 10°C to 20°C. The oxygen consumption is increased by approximately what factor?

74. **Total Revenue** The demand equation for a product is

$$p = 60 - 0.0004x$$

where p is the price per unit and x is the number of units sold. The total revenue R from selling x units is given by

$$R = xp.$$

How many units must be sold to produce a revenue of $220,000?

75. **Total Revenue** The demand equation for a product is

$$p = 50 - 0.0005x$$

where p is the price per unit and x is the number of units sold. The total revenue R from selling x units is given by

$$R = xp.$$

How many units must be sold to produce a revenue of $250,000?

76. **Think About It** When the Quadratic Formula is used to solve certain problems, such as the problem in Example 5 on page 108, why is only one solution used?

77. **Think About It** The discriminants of two quadratic equations are 5 and -10. Can the equations have the same solutions? Explain.

78. **Writing** List the methods you have learned for solving quadratic equations. Identify the type of quadratic equation that is most easily solved by each method. Give an example showing the solutions for each type.

79. **Project: Revenue** For a project involving the revenues for Buffalo Wild Wings from 2002 to 2010, visit this text's website at *www.cengagebrain.com*. (*Source: Buffalo Wild Wings, Inc.*)

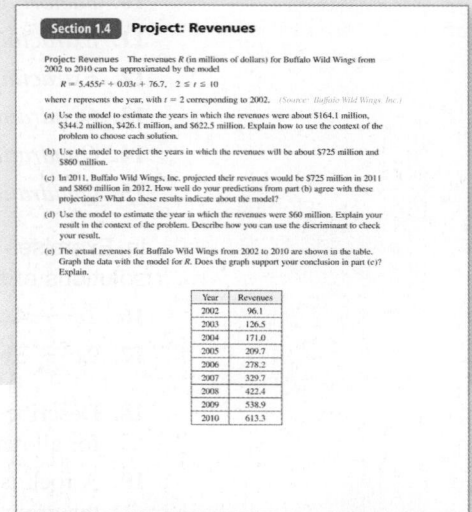

QUIZ YOURSELF

See www.CalcChat.com for worked-out solutions to odd-numbered exercises.

Take this quiz as you would take a quiz in class. When you are done, check your work against the answers given in the back of the book.

In Exercises 1–4, solve the equation and check your solution.

1. $3(x - 2) - 4(2x + 5) = 4$

2. $\dfrac{3x + 3}{5x - 2} = \dfrac{3}{4}$

3. $\dfrac{2}{x(x - 1)} + \dfrac{1}{x} = \dfrac{1}{x - 4}$

4. $(x + 3)^2 - x^2 = 6(x + 2)$

5. Describe how you can check your answers to Exercises 1–4 using your graphing utility.

In Exercises 6 and 7, use a calculator to solve the equation. (Round your solution to three decimal places.)

6. $\dfrac{x}{2.004} - \dfrac{x}{5.128} = 100$

7. $0.378x + 0.757(500 - x) = 215$

In Exercises 8 and 9, write an algebraic equation for the verbal description. Find the solution if possible and check.

8. A company has fixed costs of \$30,000 per month and variable costs of \$8.50 per unit manufactured. The company has \$200,000 available each month to cover monthly costs. How many units can the company manufacture?

9. The demand equation for a product is $p = 75 - 0.0002x$, where p is the price per unit and x is the number of units sold. The total revenue R from selling x units is given by $R = xp$. How many units must be sold to produce a revenue of \$300,000?

In Exercises 10–15, solve the quadratic equation by the indicated method.

10. *Factoring:* $3x^2 + 13x = 10$

11. *Extracting square roots:* $3x^2 = 15$

12. *Extracting square roots:* $(x + 3)^2 = 17$

13. *Quadratic Formula:* $2x + x^2 = 6$

14. *Quadratic Formula:* $3x^2 + 7x - 2 = 0$

15. *Quadratic Formula:* $3x^2 - 4.50x - 0.32 = 0$

In Exercises 16 and 17, use the discriminant to determine the number of real solutions of the quadratic equation.

16. $2x^2 - 4x + 9 = 0$

17. $9x^2 - 3x + \frac{1}{4} = 0$

18. Describe how you would convince a fellow student that $(x + 3)^2 = x^2 + 6x + 9$ for all real values of x.

19. A rock is dropped from a height of 324 feet. How long will it take the rock to hit the ground?

20. An open box has a square base and a height of 6 inches. The volume of the box is 486 cubic inches. Find the dimensions of the box.

1.5 Other Types of Equations

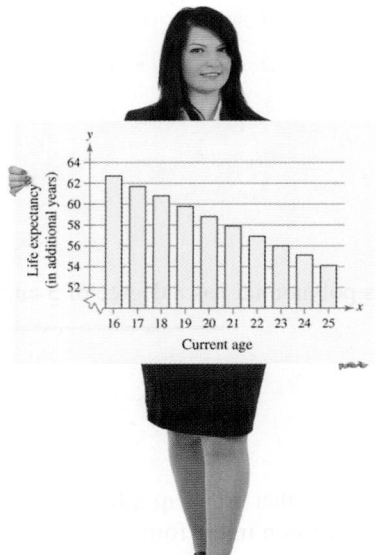

In Exercise 70 on page 124, you will solve an equation involving a radical to determine a person's age given his or her life expectancy.

■ Solve a polynomial equation by factoring.
■ Rewrite and solve an equation involving radicals or rational exponents.
■ Rewrite and solve an equation involving fractions or absolute value.
■ Construct and use a nonquadratic model to solve an application problem.

Polynomial Equations

In this section, you will extend the techniques for solving equations to nonlinear and nonquadratic equations. At this point in the text, you have three basic methods for solving nonlinear equations—*factoring, extracting square roots,* and the *Quadratic Formula.* So, the main goal of this section is to learn to rewrite nonlinear equations in a form to which you can apply one of these methods.

Example 1 Solving a Polynomial Equation by Factoring

Solve $3x^4 = 48x^2$.

SOLUTION The basic approach is first to write the polynomial equation in general form with zero on one side, then to factor the other side, and finally to set each factor equal to zero and solve.

$$3x^4 = 48x^2 \qquad \text{Write original equation.}$$
$$3x^4 - 48x^2 = 0 \qquad \text{Write in general form.}$$
$$3x^2(x^2 - 16) = 0 \qquad \text{Factor out common factor.}$$
$$3x^2(x + 4)(x - 4) = 0 \qquad \text{Difference of two squares}$$
$$3x^2 = 0 \implies x = 0 \qquad \text{Set 1st factor equal to 0.}$$
$$x + 4 = 0 \implies x = -4 \qquad \text{Set 2nd factor equal to 0.}$$
$$x - 4 = 0 \implies x = 4 \qquad \text{Set 3rd factor equal to 0.}$$

Check these solutions in the original equation.

CHECK

$$3x^4 = 48x^2 \qquad \text{Write original equation.}$$
$$3(0)^4 = 48(0)^2 \qquad \text{Substitute 0 for } x.$$
$$0 = 0 \qquad \text{0 checks. } \checkmark$$
$$3(-4)^4 = 48(-4)^2 \qquad \text{Substitute } -4 \text{ for } x.$$
$$768 = 768 \qquad -4 \text{ checks. } \checkmark$$
$$3(4)^4 = 48(4)^2 \qquad \text{Substitute 4 for } x.$$
$$768 = 768 \qquad 4 \text{ checks. } \checkmark$$

After checking, you can conclude that the solutions are

$$x = 0, x = -4, \text{ and } x = 4.$$

STUDY TIP

When solving an equation, avoid dividing each side by a common variable factor to simplify. You may lose solutions. For instance, if you divide each side by x^2 in Example 1, you lose the solution $x = 0$. Also, when solving an equation by factoring, be sure to set each variable factor equal to zero to find all of the possible solutions.

✓**Checkpoint 1**

Solve $3x^3 = 3x$.

Blaj Gabriel/www.shutterstock.com

Example 2   **Solving a Polynomial Equation by Factoring**

$x^3 - 3x^2 - 3x + 9 = 0$	Original equation
$x^2(x - 3) - 3(x - 3) = 0$	Group terms and factor.
$(x - 3)(x^2 - 3) = 0$	Distributive Property
$x - 3 = 0$ ⟹ $x = 3$	Set 1st factor equal to 0.
$x^2 - 3 = 0$ ⟹ $x = \pm\sqrt{3}$	Set 2nd factor equal to 0.

The solutions are

$$x = 3, \quad x = \sqrt{3}, \quad \text{and} \quad x = -\sqrt{3}.$$

Check these in the original equation. Notice that this polynomial has a degree of 3 and the equation has three solutions.

✓**Checkpoint 2**

Solve $x^3 - x^2 - 2x + 2 = 0$.

Occasionally, mathematical models involve equations that are of **quadratic type.** In general, an equation is of quadratic type if it can be written in the form

$$au^2 + bu + c = 0$$

where $a \neq 0$ and u is an algebraic expression.

Example 3   **Solving an Equation of Quadratic Type**

Solve $x^4 - 3x^2 + 2 = 0$.

SOLUTION   This equation is of quadratic type with $u = x^2$.

$$(x^2)^2 - 3(x^2) + 2 = 0$$
$$u^2 - 3u + 2 = 0$$

To solve, factor as the product of two second-degree polynomials.

$u^2 - 3u + 2 = 0$	Write equation.
$(u - 1)(u - 2) = 0$	Factor.
$(x^2 - 1)(x^2 - 2) = 0$	$u = x^2$
$(x + 1)(x - 1)(x^2 - 2) = 0$	Completely factor.
$x + 1 = 0$ ⟹ $x = -1$	Set 1st factor equal to 0.
$x - 1 = 0$ ⟹ $x = 1$	Set 2nd factor equal to 0.
$x^2 - 2 = 0$ ⟹ $x = \pm\sqrt{2}$	Set 3rd factor equal to 0.

The solutions are

$$x = -1, \quad x = 1, \quad x = \sqrt{2}, \quad \text{and} \quad x = -\sqrt{2}.$$

Check these in the original equation. Notice that this polynomial has a degree of 4 and the equation has four solutions.

✓**Checkpoint 3**

Solve $x^4 - 5x^2 + 4 = 0$.

STUDY TIP

When using the u-substitution, remember to substitute back to the original variable. If the original equation is in terms of x, the solutions should be in terms of x.

Solving Equations Involving Radicals

The steps involved in solving the remaining equations in this section will often introduce *extraneous solutions*, as discussed in Section 1.1. Operations such as squaring each side of an equation, raising each side of an equation to a rational power, or multiplying each side of an equation by a variable quantity all create this potential danger. So, when you use any of these operations, checking possible solutions is crucial.

Example 4 An Equation Involving a Radical

Solve $\sqrt{2x + 7} - x = 2$.

SOLUTION

$\sqrt{2x + 7} - x = 2$	Write original equation.
$\sqrt{2x + 7} = x + 2$	Isolate the square root.
$2x + 7 = x^2 + 4x + 4$	Square each side.
$0 = x^2 + 2x - 3$	Write in general form.
$0 = (x + 3)(x - 1)$	Factor.
$x + 3 = 0 \implies x = -3$	Set 1st factor equal to 0.
$x - 1 = 0 \implies x = 1$	Set 2nd factor equal to 0.

By checking these values, you can determine that the only solution is $x = 1$.

✓ **Checkpoint 4**

Solve $\sqrt{3x} - 6 = 0$.

Example 5 An Equation Involving a Rational Exponent

Solve $4x^{3/2} - 8 = 0$.

SOLUTION

$4x^{3/2} - 8 = 0$	Write original equation.
$4x^{3/2} = 8$	Add 8 to each side.
$x^{3/2} = 2$	Isolate $x^{3/2}$.
$x = 2^{2/3}$	Raise each side to the $\frac{2}{3}$ power.
$x \approx 1.587$	Round to three decimal places.

CHECK

$4x^{3/2} - 8 = 0$	Write original equation.
$4(2^{2/3})^{3/2} \overset{?}{=} 8$	Substitute $2^{2/3}$ for x.
$4(2) \overset{?}{=} 8$	Power of a Power Property
$8 = 8$	Solution checks. ✓

✓ **Checkpoint 5**

Solve $2x^{3/4} - 54 = 0$.

STUDY TIP

The basic technique used in Example 5 is to isolate the factor with the rational exponent and raise each side to the reciprocal power. In Example 4, this is equivalent to isolating the square root and squaring each side.

Equations Involving Fractions or Absolute Value

In Section 1.1, you learned how to solve equations involving fractions. Recall that the first step is to multiply each term of the equation by the least common denominator (LCD).

Example 6   An Equation Involving Fractions

Solve $\dfrac{2}{x} = \dfrac{3}{x-2} - 1$.

SOLUTION   For this equation, the LCD of the three terms is

$$x(x-2)$$

so begin by multiplying each term of the equation by this expression.

$$\dfrac{2}{x} = \dfrac{3}{x-2} - 1 \qquad\qquad \text{Write original equation.}$$

$$x(x-2)\dfrac{2}{x} = x(x-2)\dfrac{3}{x-2} - x(x-2)(1) \qquad \text{Multiply each term by LCD.}$$

$$2(x-2) = 3x - x(x-2), \quad x \neq 0, 2 \qquad \text{Simplify.}$$

$$2x - 4 = -x^2 + 5x \qquad\qquad \text{Simplify.}$$

$$x^2 - 3x - 4 = 0 \qquad\qquad \text{Write in general form.}$$

$$(x-4)(x+1) = 0 \qquad\qquad \text{Factor.}$$

$$x - 4 = 0 \quad\Longrightarrow\quad x = 4 \qquad \text{Set 1st factor equal to 0.}$$

$$x + 1 = 0 \quad\Longrightarrow\quad x = -1 \qquad \text{Set 2nd factor equal to 0.}$$

Notice that the values $x = 0$ and $x = 2$ are excluded from the domains of the fractions because they result in division by zero. So, both

$$x = 4 \quad \text{and} \quad x = -1$$

are possible solutions.

CHECK

$$\dfrac{2}{x} = \dfrac{3}{x-2} - 1 \qquad\qquad \text{Write original equation.}$$

$$\dfrac{2}{4} \stackrel{?}{=} \dfrac{3}{4-2} - 1 \qquad\qquad \text{Substitute 4 for } x.$$

$$\dfrac{1}{2} = \dfrac{3}{2} - 1 \qquad\qquad \text{4 checks. } \checkmark$$

$$\dfrac{2}{-1} \stackrel{?}{=} \dfrac{3}{-1-2} - 1 \qquad\qquad \text{Substitute } -1 \text{ for } x.$$

$$-2 = -1 - 1 \qquad\qquad -1 \text{ checks. } \checkmark$$

The solutions are

$$x = 4 \quad \text{and} \quad x = -1.$$

✓ Checkpoint 6

Solve $\dfrac{3}{x} + \dfrac{1}{x-2} = 2$.

To solve an equation involving absolute value, remember that the expression inside the absolute value signs can be positive or negative. This results in *two* separate equations, each of which must be solved. For instance, the equation

$$|x - 2| = 3$$

results in the two equations

$$x - 2 = 3 \quad \text{and} \quad -(x - 2) = 3$$

which implies that the original equation has two solutions:

$$x = 5 \quad \text{and} \quad x = -1.$$

When writing the negative expression, it is important to remember to place parentheses around the entire expression that is inside the absolute value signs. After you write the two equations, solve each one independently.

Example 7   An Equation Involving Absolute Value

Solve $|x^2 - 3x| = -4x + 6$.

SOLUTION   Because the variable expression inside the absolute value signs can be positive or negative, you must solve the following two equations.

First Equation

$x^2 - 3x = -4x + 6$	Use positive expression.
$x^2 + x - 6 = 0$	Write in general form.
$(x + 3)(x - 2) = 0$	Factor.
$x + 3 = 0 \implies x = -3$	Set 1st factor equal to 0.
$x - 2 = 0 \implies x = 2$	Set 2nd factor equal to 0.

Second Equation

$-(x^2 - 3x) = -4x + 6$	Use negative expression.
$x^2 - 7x + 6 = 0$	Write in general form.
$(x - 1)(x - 6) = 0$	Factor.
$x - 1 = 0 \implies x = 1$	Set 1st factor equal to 0.
$x - 6 = 0 \implies x = 6$	Set 2nd factor equal to 0.

CHECK

$|(-3)^2 - 3(-3)| = -4(-3) + 6$	-3 checks. ✓
$|(2)^2 - 3(2)| \neq -4(2) + 6$	2 does not check.
$|(1)^2 - 3(1)| = -4(1) + 6$	1 checks. ✓
$|(6)^2 - 3(6)| \neq -4(6) + 6$	6 does not check.

The only solutions are

$$x = -3 \quad \text{and} \quad x = 1.$$

✓ Checkpoint 7

Solve $|x^2 - 3| = 5x - 3$.

Applications

It would be virtually impossible to categorize all of the many different types of applications that involve nonlinear and nonquadratic models. However, from the few examples and exercises that follow, we hope you will gain some appreciation for the variety of applications that involve such models.

Example 8   Reduced Rates

A group of friends purchase a television sports package at a cost of $350. To lower the cost per person, the group invites three more people to contribute to the purchase. As a result, the cost per person decreases by $15. How many friends were in the original group that wanted to purchase the sports package?

SOLUTION   Begin the solution by creating a verbal model and assigning labels.

Verbal Model: Cost per person · Total number of people = Cost of package

Labels:
Cost of package $= 350$ (dollars)
Number of people in original group $= x$ (people)
Total number of people $= x + 3$ (people)
Original cost per person $= \dfrac{350}{x}$ (dollars per person)
Cost per person $= \dfrac{350}{x} - 15$ (dollars per person)

Equation:

$$\left(\frac{350}{x} - 15\right)(x + 3) = 350 \qquad \text{Original equation}$$

$$\left(\frac{350 - 15x}{x}\right)(x + 3) = 350 \qquad \text{Rewrite first factor.}$$

$$(350 - 15x)(x + 3) = 350x, \quad x \neq 0 \qquad \text{Multiply each side by } x.$$

$$350x + 1050 - 15x^2 - 45x = 350x \qquad \text{Multiply factors.}$$

$$-15x^2 - 45x + 1050 = 0 \qquad \text{Subtract } 350x \text{ from each side.}$$

$$x^2 + 3x - 70 = 0 \qquad \text{Divide each side by } -15.$$

$$(x + 10)(x - 7) = 0 \qquad \text{Factor.}$$

$$x + 10 = 0 \implies x = -10 \qquad \text{Set 1st factor equal to 0.}$$

$$x - 7 = 0 \implies x = 7 \qquad \text{Set 2nd factor equal to 0.}$$

Only the positive x-value makes sense in the context of the problem. So, there were 7 friends in the original group that wanted to purchase the sports package.

CHECK

$$\left(\frac{350}{7} - 15\right)(7 + 3) \stackrel{?}{=} 350 \qquad \text{Substitute 7 for } x.$$

$$(35)(10) \stackrel{?}{=} 350 \qquad \text{Simplify.}$$

$$350 = 350 \qquad \text{Solution checks. } \checkmark$$

✓ Checkpoint 8

In Example 8, suppose the group of friends invites two more people instead of three to contribute to the purchase. As a result, the cost per person decreases by $20. How many friends were in the original group that wanted to purchase the sports package? ■

Interest earned on a savings account is calculated by one of three basic methods: simple interest, interest compounded n times per year, and interest compounded continuously. The next example uses the formula for interest that is compounded n times per year,

$$A = P\left(1 + \frac{r}{n}\right)^{nt}.$$

In this formula, A is the balance in the account, P is the principal (or original deposit), r is the annual interest rate (in decimal form), n is the number of compoundings per year, and t is the time in years. In Chapter 4, you will study the derivation of this formula for compound interest.

 Example 9 **Compound Interest**

When you were born, your grandparents deposited $5000 in a savings account earning interest compounded quarterly. On your 25th birthday, the balance of the account is $25,062.59. What is the average annual interest rate of the account?

SOLUTION

Formula: $A = P\left(1 + \dfrac{r}{n}\right)^{nt}$

Labels:	Balance $= A = 25{,}062.59$	(dollars)
	Principal $= P = 5000$	(dollars)
	Time $= t = 25$	(years)
	Compoundings per year $= n = 4$	(compoundings)
	Annual interest rate $= r$	(percent in decimal form)

Equation:

$$25{,}062.59 = 5000\left(1 + \frac{r}{4}\right)^{4(25)}$$ Substitute.

$$\frac{25{,}062.59}{5000} = \left(1 + \frac{r}{4}\right)^{100}$$ Divide each side by 5000.

$$5.0125 \approx \left(1 + \frac{r}{4}\right)^{100}$$ Use a calculator.

$$(5.0125)^{1/100} \approx 1 + \frac{r}{4}$$ Raise each side to reciprocal power.

$$1.01625 \approx 1 + \frac{r}{4}$$ Use a calculator.

$$0.01625 \approx \frac{r}{4}$$ Subtract 1 from each side.

$$0.065 \approx r$$ Multiply each side by 4.

The average annual interest rate is about

$$0.065 = 6.5\%.$$

Check this in the original statement of the problem.

✓ **Checkpoint 9**

You placed $1000 in an account earning interest compounded monthly. After 3 years, the account balance is $1144.25. What is the annual interest rate of the account?

 **Example 10**   **Market Research**

The marketing department of a publishing company is asked to determine the price of a book. The department determines that the demand for the book depends on the price of the book according to the model

$$p = 40 - \sqrt{0.0001x + 1}, \quad 0 \le x \le 15{,}990{,}000$$

where p is the price per book (in dollars) and x is the number of books sold at the given price. For instance, in Figure 1.12, note that if the price was \$39, then (according to the model) no one would be willing to buy the book. On the other hand, if the price was \$17.60, 5 million copies could be sold. The marketing department set the price at \$12.95. How many copies can the publisher expect to sell?

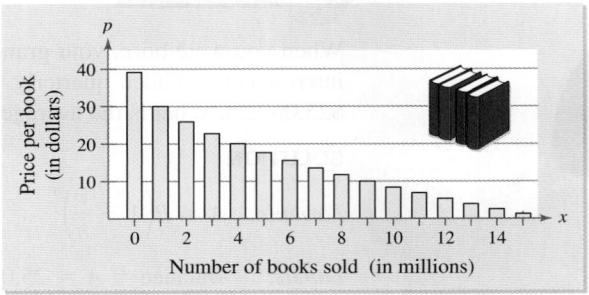

FIGURE 1.12

SOLUTION

$p = 40 - \sqrt{0.0001x + 1}$	Write given model.
$12.95 = 40 - \sqrt{0.0001x + 1}$	Set price at \$12.95.
$\sqrt{0.0001x + 1} = 27.05$	Isolate the radical.
$0.0001x + 1 = 731.7025$	Square each side.
$0.0001x = 730.7025$	Subtract 1 from each side.
$x = 7{,}307{,}025$	Divide each side by 0.0001.

So, by setting the book's price at \$12.95, the publisher can expect to sell about 7.3 million copies.

 **Checkpoint 10**

In Example 10, suppose the marketing department set the price at \$19.95. How many copies can the publisher expect to sell? ■

SUMMARIZE   (Section 1.5)

1. Describe how to solve a polynomial equation by factoring (*pages 115 and 116*). For examples of solving polynomial equations by factoring, see Examples 1, 2, and 3.

2. Describe how to solve an equation involving absolute value (*page 119*). For an example of solving an equation involving absolute value, see Example 7.

3. Describe a real-life example that uses a nonlinear or nonquadratic model (*pages 120–122, Examples 8, 9, and 10*).

In Exercises 1–10, find the real solution(s) of the equation.

1. $x^2 - 22x + 121 = 0$

2. $x(x - 20) + 3(x - 20) = 0$

3. $(x + 20)^2 = 625$

4. $5x^2 + x = 0$

5. $3x^2 + 4x - 4 = 0$

6. $12x^2 + 8x - 55 = 0$

7. $x^2 + 4x - 5 = 0$

8. $4x^2 + 4x - 15 = 0$

9. $x^2 - 3x + 1 = 0$

10. $x^2 - 4x + 2 = 0$

Exercises 1.5

Solving a Polynomial Equation In Exercises 1–20, find the real solution(s) of the polynomial equation. Check your solution(s). *See Examples 1, 2, and 3.*

1. $x^3 - 2x^2 - 3x = 0$

2. $20x^3 - 125x = 0$

3. $4x^4 - 36x^2 = 0$

4. $2x^4 - 15x^3 + 18x^2 = 0$

5. $x^4 - 81 = 0$

6. $x^6 - 64 = 0$

7. $5x^3 + 30x^2 + 45x = 0$

8. $9x^4 - 24x^3 + 16x^2 = 0$

9. $x^3 - 7x^2 - 4x + 28 = 0$

10. $x^3 + 2x^2 + 3x + 6 = 0$

11. $x^4 - x^3 + x - 1 = 0$

12. $x^4 + 2x^3 - 8x - 16 = 0$

13. $x^4 - 12x^2 + 11 = 0$

14. $x^4 - 29x^2 + 100 = 0$

15. $x^4 + 5x^2 - 36 = 0$

16. $x^4 - 4x^2 + 3 = 0$

17. $4x^4 - 65x^2 + 16 = 0$

18. $36t^4 + 29t^2 - 7 = 0$

19. $x^6 + 7x^3 - 8 = 0$

20. $x^6 + 3x^3 + 2 = 0$

An Equation Involving a Radical In Exercises 21–34, find the real solution(s) of the radical equation. Check your solution(s). *See Example 4.*

21. $\sqrt{2x} - 10 = 0$

22. $4\sqrt{x} - 3 = 0$

23. $\sqrt{x - 10} - 4 = 0$

24. $\sqrt{5 - x} - 3 = 0$

25. $\sqrt[3]{2x + 5} + 3 = 0$

26. $\sqrt[3]{3x + 1} - 5 = 0$

27. $2x + 9\sqrt{x} - 5 = 0$

28. $6x - 7\sqrt{x} - 3 = 0$

29. $x = \sqrt{11x - 30}$

30. $2x - \sqrt{15 - 4x} = 0$

31. $-\sqrt{26 - 11x} + 4 = x$

32. $x + \sqrt{31 - 9x} = 5$

33. $\sqrt{x + 1} - 3x = 1$

34. $\sqrt{2x + 1} + x = 7$

An Equation Involving a Rational Exponent In Exercises 35–40, find the real solution(s) of the equation involving rational exponents. Check your solution(s). *See Example 5.*

35. $(x - 5)^{2/3} - 16 = 0$

36. $(x + 3)^{4/3} - 16 = 0$

37. $(x + 3)^{3/2} - 8 = 0$

38. $(x^2 + 2)^{2/3} - 9 = 0$

39. $(x^2 - 5)^{2/3} = 16$

40. $(x^2 - x - 22)^{4/3} = 16$

An Equation Involving Fractions In Exercises 41–48, find the real solution(s) of the equation involving fractions. Check your solution(s). *See Example 6.*

41. $\dfrac{1}{x} = \dfrac{4}{x - 1} + 1$

42. $\dfrac{1}{x} - \dfrac{1}{x + 1} = 3$

43. $\dfrac{4}{x} - \dfrac{5}{3} = \dfrac{x}{6}$

44. $\dfrac{5}{x} - \dfrac{3}{2} = \dfrac{x}{5}$

45. $\dfrac{4}{x + 1} - \dfrac{3}{x + 2} = 1$

46. $\dfrac{x}{x^2 - 4} + \dfrac{1}{x + 2} = 3$

47. $\dfrac{x + 2}{x - 2} - \dfrac{x}{3} = 0$

48. $\dfrac{x + 1}{3} - \dfrac{x + 1}{x + 2} = 0$

An Equation Involving Absolute Value In Exercises 49–56, find the real solution(s) of the equation involving absolute value. Check your solution(s). *See Example 7.*

49. $|x + 1| = 2$

50. $|x - 2| = 3$

51. $|2x - 1| - 5 = 0$

52. $|3x + 2| - 7 = 0$

53. $|x| = x^2 + x - 3$

54. $|3x| = x^2 + 3x - 16$

55. $|x - 10| = x^2 - 10x$

56. $|x + 1| = x^2 - 5$

57. Error Analysis   Find the error(s) in the solution.

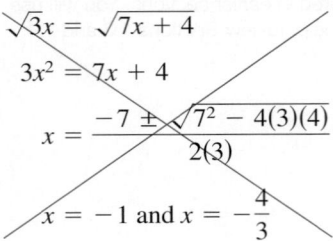

$$\sqrt{3}x = \sqrt{7x + 4}$$

$$3x^2 = 7x + 4$$

$$x = \frac{-7 \pm \sqrt{7^2 - 4(3)(4)}}{2(3)}$$

$$x = -1 \text{ and } x = -\frac{4}{3}$$

58. Error Analysis   Find the error(s) in the solution.

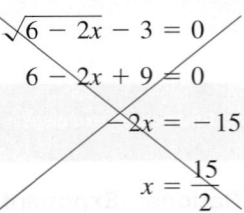

$$\sqrt{6 - 2x} - 3 = 0$$

$$6 - 2x + 9 = 0$$

$$-2x = -15$$

$$x = \frac{15}{2}$$

Using a Calculator   In Exercises 59–62, use a calculator to find the real solutions of the equation. (Round your answers to three decimal places.)

59. $3.2x^4 - 1.5x^2 - 2.1 = 0$

60. $7.08x^6 + 4.15x^3 - 9.6 = 0$

61. $1.8x - 6\sqrt{x} - 5.6 = 0$

62. $4x + 8\sqrt{x} + 3.6 = 0$

63. Sharing the Cost   A college charters a bus for $1700 to take a group of students to the Fiesta Bowl. When six more students join the trip, the cost per student decreases by $7.50. How many students were in the original group?

64. Sharing the Cost   Three students plan to share equally in the rent for an apartment. By adding a fourth person, the cost per person decreases by $125 per month. How much is the monthly rent for the apartment?

65. Compound Interest   A deposit of $3000 reaches a balance of $3422.91 after 6 years. The interest on the account is compounded monthly. What is the annual interest rate for this investment?

66. Compound Interest   A sales representative describes a "guaranteed investment fund" that is offered to new investors. By depositing $15,000 in the fund, you will be guaranteed to receive a total of at least $40,000 after 20 years. Assuming that interest is compounded quarterly, determine the fixed annual rates of interest that yield

(a) the minimum guarantee.

(b) a return of $48,000 after 20 years.

67. Borrowing Money   You borrow $300 from a friend and agree to pay the money back, plus $20 in interest, after 6 months. Assuming that the interest is compounded monthly, what annual interest rate are you paying?

68. Cash Advance   You take out a cash advance of $1000 on a credit card. After 2 months, you owe $1041.93. The interest is compounded monthly. What is the annual interest rate for this cash advance?

69. Airline Passengers   An airline offers daily flights between Chicago and Denver. The total monthly cost C (in millions of dollars) of these flights is modeled by

$$C = \sqrt{0.25x + 1}$$

where x is the number of passengers flying that month in thousands (see figure). The total cost of the flights for a month is 3.5 million dollars. Use the model to determine how many passengers flew that month.

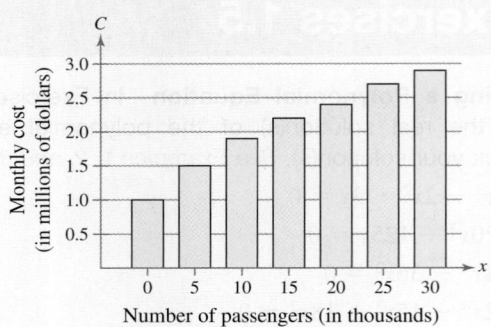

70. Life Expectancy   The life expectancy of a person who is 16 to 25 years old can be modeled by

$$y = \sqrt{1.216x^2 - 161.12x + 6197.8}, \quad 16 \le x \le 25$$

where y represents the number of additional years the person is expected to live and x represents the person's current age.   *(Source: U.S. National Center for Health Statistics)*

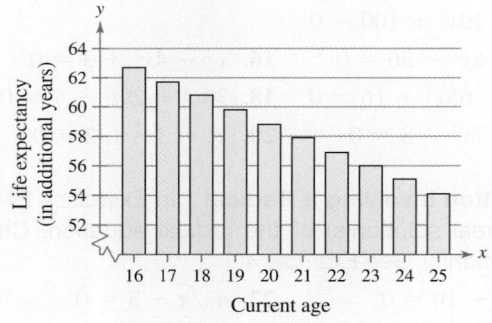

(a) Determine the life expectancies of persons who are 18, 20, and 22 years old.

(b) A person's life expectancy is 55 years. Use the model to determine the age of the person.

71. Life Expectancy The life expectancy of a person who is 50 to 65 years old can be modeled by

$$y = \sqrt{0.828x^2 - 135.71x + 5672.1}, \quad 50 \le x \le 65$$

where y represents the number of additional years the person is expected to live and x represents the person's current age. A person's life expectancy is 25 years. How old is the person? *(Source: U.S. National Center for Health Statistics)*

72. Fast-Food Industry The yearly revenues R (in billions of dollars) of fast-food restaurants in the United States from 2003 through 2009 can be approximated by

$$R = 125.7 + 4.69t - \frac{58.6}{t}, \quad 3 \le t \le 9$$

where t represents the year, with $t = 3$ corresponding to 2003 (see figure). Use the model to predict the year in which fast-food revenues will be about $172 billion. *(Source: National Restaurant Association)*

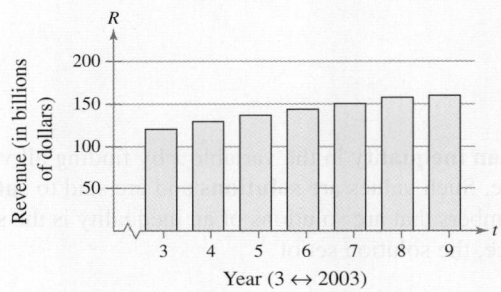

Year (3 ↔ 2003)

73. Market Research The demand equation for a product is modeled by

$$p = 40 - \sqrt{0.01x + 1}$$

where x is the number of units demanded per day and p is the price per unit. Find the demand when the price is set at $13.95. Explain why this model is valid only for $0 \le x \le 159,900$.

74. Power Line A power station is on one side of a river that is $\frac{1}{2}$ mile wide. A factory is 6 miles downstream on the other side of the river. It costs $18 per foot to run power lines over land and $24 per foot to run them under water. The project's cost is $616,877.27. Find the length x as labeled in the figure.

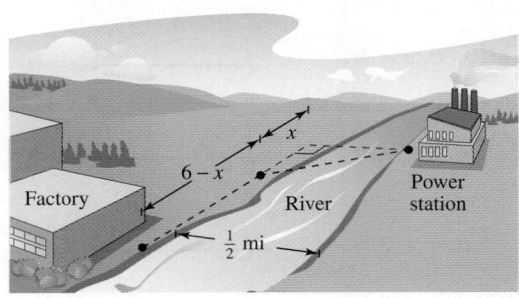

75. Flour Production A company weighs each 16-ounce bag of flour it produces. After production, any bag that does not weigh within 0.4 ounce of 16 ounces cannot be sold. Solve the equation $|x - 16| = 0.4$ to find the least and greatest acceptable weights of a 16-ounce bag of flour.

76. Sugar Production A company weighs each 80-ounce bag of sugar it produces. After production, any bag that does not weigh within 1.2 ounces of 80 ounces cannot be sold. Solve the equation $|x - 80| = 1.2$ to find the least and greatest acceptable weights of an 80-ounce bag of sugar.

77. Work Rate With only the cold water valve open, it takes 8 minutes to fill the tub of a washing machine. With both the hot and cold water valves open, it takes 5 minutes. The time it takes for the tub to fill with only the hot water valve open can be modeled by the equation

$$\frac{1}{8} + \frac{1}{t} = \frac{1}{5}$$

where t is the time (in minutes) for the tub to fill. How long does it take for the tub of the washing machine to fill with only the hot water valve open?

78. Community Service You and a friend volunteer to paint a house as a community service project. Working alone, you can paint the house in 28 hours. Your friend can paint the house in 25 hours working alone. How long will it take both of you, working together, to paint the house?

79. Discovery Look back at your solutions in Exercises 1–20 and recall what you learned in Sections 1.3 and 1.4 about the numbers of solutions of quadratic equations. Make a conjecture about how the number of solutions of a polynomial equation in general form is related to the degree of the polynomial.

 **80. HOW DO YOU SEE IT?** The figure shows a glass cube partially filled with water. The equation

$$x^2(x - 3) = 320$$

represents the situation.

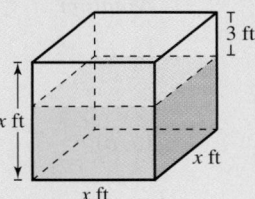

(a) What does the expression $x^2(x - 3)$ represent?

(b) Explain how you can use the given equation to find the capacity of the cube.

1.6 Linear Inequalities

- Write bounded and unbounded intervals using inequalities or interval notation.
- Develop an understanding of the properties of inequalities.
- Solve and graph a linear inequality.
- Solve an inequality involving absolute value.
- Construct and use a linear inequality to solve an application problem.

Introduction

Simple inequalities are used to *order* real numbers. The inequality symbols $<$, $\leq$, $>$, and $\geq$ are used to compare two numbers and to denote subsets of real numbers. For instance, the simple inequality

$$x \geq 3$$

denotes all real numbers x that are greater than or equal to 3.

In this section, you will expand your work with inequalities to include more involved statements such as

$$5x - 7 > 3x + 9$$

and

$$-3 \leq 6x - 1 < 3.$$

As with an equation, you **solve an inequality** in the variable x by finding all values of x for which the inequality is true. Such values are **solutions** and are said to **satisfy** the inequality. The set of all real numbers that are solutions of an inequality is the **solution set** of the inequality. For instance, the solution set of

$$x + 3 > 4$$

is all real numbers that are greater than 1.

The set of all points on the real number line that represent the solution set of an inequality is the **graph** of the inequality. Graphs of many types of inequalities consist of intervals on the real number line. The four different types of **bounded** intervals are summarized below.

In Exercise 85 on page 136, you will use an inequality to predict when the average National Football League player's salary will exceed $2.2 million.

Bounded Intervals on the Real Number Line

Let a and b be real numbers such that $a < b$. The following intervals on the real number line are **bounded.** The numbers a and b are the **endpoints** of each interval.

Notation	Interval Type	Inequality	Graph
$[a, b]$	Closed	$a \leq x \leq b$	
(a, b)	Open	$a < x < b$	
$[a, b)$		$a \leq x < b$	
$(a, b]$		$a < x \leq b$	

Note that a closed interval contains both of its endpoints and an open interval does not contain either of its endpoints. Often, the solution of an inequality is an interval on the real line that is **unbounded.** For instance, the interval consisting of all positive numbers is unbounded. The symbols ∞, **positive infinity,** and $-\infty$, **negative infinity,** do not represent real numbers. They are simply convenient symbols used to describe the unboundedness of an interval such as $(1, \infty)$.

Unbounded Intervals on the Real Number Line

Let a and b be real numbers. The following intervals on the real number line are **unbounded.**

Notation	Interval Type	Inequality	Graph
$[a, \infty)$		$x \geq a$	
(a, ∞)	Open	$x > a$	
$(-\infty, b]$		$x \leq b$	
$(-\infty, b)$	Open	$x < b$	
$(-\infty, \infty)$	Entire real line	$-\infty < x < \infty$	

Example 1   Intervals and Inequalities

Write an inequality to represent each interval. Then state whether the interval is bounded or unbounded.

a. $(-3, 5]$

b. $(-4, \infty)$

c. $[0, 2]$

d. $(-\infty, 0)$

SOLUTION

a. $(-3, 5]$ corresponds to $-3 < x \leq 5$.   Bounded

b. $(-4, \infty)$ corresponds to $x > -4$.   Unbounded

c. $[0, 2]$ corresponds to $0 \leq x \leq 2$.   Bounded

d. $(-\infty, 0)$ corresponds to $x < 0$.   Unbounded

✓ Checkpoint 1

Write an inequality to represent each interval. Then state whether the interval is bounded or unbounded.

a. $[2, 7)$

b. $(-\infty, 6)$

Properties of Inequalities

The procedures for solving linear inequalities in one variable are much like those for solving linear equations. To isolate the variable, you can make use of the **properties of inequalities.** These properties are similar to the properties of equality, but there are two important exceptions. When each side of an inequality is multiplied or divided by a negative number, the direction of the inequality symbol must be reversed. Here is an example.

$$-2 < 5$$ Original inequality

$$(-3)(-2) > (-3)(5)$$ Multiply each side by -3 and reverse the inequality symbol.

$$6 > -15$$ Simplify.

Notice in the example above that when the inequality symbol is not reversed, you obtain the false statement

$$6 < -15.$$ False statement

Two inequalities that have the same solution set are **equivalent.** For instance, the inequalities

$$x + 2 < 5 \quad \text{and} \quad x < 3$$

are equivalent. To obtain the second inequality from the first, you can subtract 2 from each side of the inequality. The following list describes operations that can be used to create equivalent inequalities.

Properties of Inequalities

Let a, b, c, and d be real numbers.

1. *Transitive Property*

$$a < b \text{ and } b < c \quad\Longrightarrow\quad a < c$$

2. *Addition of Inequalities*

$$a < b \text{ and } c < d \quad\Longrightarrow\quad a + c < b + d$$

3. *Addition of a Constant*

$$a < b \quad\Longrightarrow\quad a + c < b + c$$

4. *Multiplication by a Constant*

$$\text{For } c > 0, a < b \quad\Longrightarrow\quad ac < bc$$

$$\text{For } c < 0, a < b \quad\Longrightarrow\quad ac > bc$$ Reverse direction of inequality symbol.

Each of the properties above is true if the symbol $<$ is replaced by $\leq$ and the symbol $>$ is replaced by $\geq$. For instance, another form of the multiplication property would be as shown.

$$\text{For } c > 0, a \leq b \quad\Longrightarrow\quad ac \leq bc$$

$$\text{For } c < 0, a \leq b \quad\Longrightarrow\quad ac \geq bc$$

On your own, try to verify each of the properties of inequalities by using several examples with real numbers.

Solving a Linear Inequality

The simplest type of inequality to solve is a **linear inequality** in a single variable. For instance,

$$2x + 3 > 4$$

is a linear inequality in x.

As you read through the following examples, pay special attention to the steps in which the inequality symbol is reversed. Remember that when you multiply or divide by a negative number, you must reverse the inequality symbol.

ALGEBRA TUTOR xy

For help in solving inequalities similar to the one in Example 2, see the review of solving inequalities on page 149.

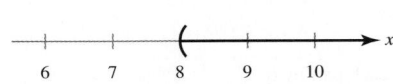

Solution Interval: $(8, \infty)$
FIGURE 1.13

Example 2   Solving a Linear Inequality

Solve $5x - 7 > 3x + 9$.

SOLUTION

$5x - 7 > 3x + 9$	Write original inequality.
$2x - 7 > 9$	Subtract $3x$ from each side.
$x > 8$	Add 7 to each side and then divide each side by 2.

The solution set is all real numbers that are greater than 8, which is denoted by $(8, \infty)$. The graph is shown in Figure 1.13.

✓**Checkpoint 2**

Solve $3x < 2x + 1$.

Checking the solution set of an inequality is not as simple as checking the solutions of an equation. You can, however, get an indication of the validity of a solution set by substituting a few convenient values of x to see whether the original inequality is satisfied. For instance, substitute several values of x that are greater than 8 into the original inequality in Example 2 and check to see if the inequality is satisfied.

Example 3   Solving a Linear Inequality

Solve $1 - \dfrac{3x}{2} \geq x - 4$.

SOLUTION

$1 - \dfrac{3x}{2} \geq x - 4$	Write original inequality.
$2 - 3x \geq 2x - 8$	Multiply each side by 2.
$2 - 5x \geq -8$	Subtract $2x$ from each side.
$-5x \geq -10$	Subtract 2 from each side.
$x \leq 2$	Divide each side by -5 and reverse inequality symbol.

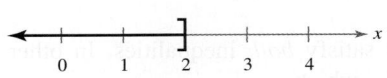

Solution Interval: $(-\infty, 2]$
FIGURE 1.14

The solution set is all real numbers that are less than or equal to 2, which is denoted by $(-\infty, 2]$. The graph is shown in Figure 1.14.

✓**Checkpoint 3**

Solve $-\dfrac{4x}{3} \leq 2 - x$. Then graph the solution set on the real number line.

Sometimes it is convenient to write two inequalities as a **double inequality.** For instance, you can write the two inequalities

$$-4 \le 5x - 2 \quad \text{and} \quad 5x - 2 < 7$$

more simply as

$$-4 \le 5x - 2 < 7.$$

This enables you to solve the two inequalities together, as demonstrated in Example 4.

Example 4   Solving a Double Inequality

Solve $-3 \le 6x - 1 < 3$.

SOLUTION   To solve a double inequality, you can isolate x as the middle term.

$-3 \le 6x - 1 < 3$	Write original inequality.
$-3 + 1 \le 6x - 1 + 1 < 3 + 1$	Add 1 to each part.
$-2 \le 6x < 4$	Simplify.
$\dfrac{-2}{6} \le \dfrac{6x}{6} < \dfrac{4}{6}$	Divide each part by 6.
$-\dfrac{1}{3} \le x < \dfrac{2}{3}$	Simplify.

The solution set is all real numbers that are greater than or equal to $-\frac{1}{3}$ and less than $\frac{2}{3}$. The interval notation for this solution set is $\left[-\frac{1}{3}, \frac{2}{3}\right)$. The graph of this solution set is shown in Figure 1.15.

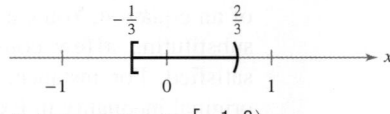

Solution Interval: $\left[-\frac{1}{3}, \frac{2}{3}\right)$

FIGURE 1.15

✓ Checkpoint 4

Solve $-1 < 3 - 2x \le 5$. Then graph the solution set on the real number line.   ■

The double inequality in Example 4 could have been solved in two parts as follows.

$-3 \le 6x - 1$	$\quad$ and $\quad$ $6x - 1 < 3$
$-2 \le 6x$	$6x < 4$
$-\dfrac{1}{3} \le x$	$x < \dfrac{2}{3}$

The solution set consists of all real numbers that satisfy *both* inequalities. In other words, the solution set is the set of all values of x for which

$$-\tfrac{1}{3} \le x < \tfrac{2}{3}.$$

When combining two inequalities to form a double inequality, be sure that the inequalities satisfy the Transitive Property. For instance, it is *incorrect* to combine the inequalities $3 < x$ and $x \le -1$ as $3 < x \le -1$. This is obviously wrong because 3 is not less than -1.

Inequalities Involving Absolute Value

Solving an Absolute Value Inequality

Let x be a variable or an algebraic expression and let a be a real number such that $a \geq 0$.

1. The solutions of $|x| < a$ are all values of x that lie between $-a$ and a.

   $$|x| < a \qquad \text{if and only if } -a < x < a.$$

2. The solutions of $|x| > a$ are all values of x that are less than $-a$ or greater than a.

   $$|x| > a \qquad \text{if and only if } x < -a \text{ or } x > a.$$

These rules are also valid if $<$ is replaced by $\leq$ and $>$ is replaced by $\geq$.

Example 5 Solving an Absolute Value Inequality

Solve $|x - 5| < 2$.

SOLUTION

$\lvert x - 5 \rvert < 2$	Write original inequality.
$-2 < x - 5 < 2$	Equivalent inequality
$-2 + 5 < x - 5 + 5 < 2 + 5$	Add 5 to each part.
$3 < x < 7$	Simplify.

The solution set consists of all real numbers that are greater than 3 and less than 7, which is denoted by $(3, 7)$. The graph is shown in Figure 1.16.

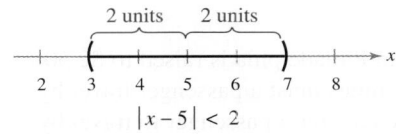

FIGURE 1.16

✓**Checkpoint 5**

Solve $|x + 2| \leq 7$. Then graph the solution set on the real number line. ■

Example 6 Solving an Absolute Value Inequality

Solve $|x + 3| \geq 7$.

SOLUTION

$\lvert x + 3 \rvert \geq 7$			Write original inequality.
$x + 3 \leq -7$	or	$x + 3 \geq 7$	Equivalent inequalities
$x + 3 - 3 \leq -7 - 3$		$x + 3 - 3 \geq 7 - 3$	Subtract 3 from each side.
$x \leq -10$		$x \geq 4$	Simplify.

The solution set is all real numbers that are less than or equal to -10 *or* greater than or equal to 4, which is denoted by

$$(-\infty, -10] \cup [4, \infty)$$

(see Figure 1.17). The symbol $\cup$ (union) means *or*.

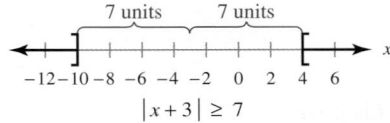

FIGURE 1.17

✓**Checkpoint 6**

Solve $|x + 1| > 3$. Then graph the solution set on the real number line. ■

Applications

 Example 7   **Comparing Taxi Fares**

The cost of traveling by taxi in Las Vegas is $3.20 for the initial charge and $2 for each mile driven. The cost in Philadelphia is $2.30 for the initial charge and $2.10 for each mile driven. How many miles must a passenger travel by taxi in Philadelphia before the cost is more than the cost for a passenger to travel by taxi in Las Vegas?

SOLUTION

Verbal Model:	Cost for taxi in Philadelphia	>	Cost for taxi in Las Vegas	

Labels: Miles driven $= m$ (miles)
Cost for taxi in Las Vegas $= 3.2 + 2m$ (dollars)
Cost for taxi in Philadelphia $= 2.3 + 2.1m$ (dollars)

Inequality: $2.3 + 2.1m > 3.2 + 2m$

$$0.1m > 0.9$$

$$m > 9$$

When a passenger travels more than 9 miles, the cost for a taxi in Philadelphia is more than the cost for a taxi in Las Vegas.

✓ Checkpoint 7

In Example 7, suppose the initial charge for a taxi in Philadelphia is raised to $2.50 and the cost per mile is raised to $2.20. How many miles must a passenger travel by taxi in Philadelphia before the cost is more than the cost for a passenger to travel by taxi in Las Vegas?

 Example 8   **Exercise Program**

A 225-pound man begins an exercise and diet program that is designed to reduce his weight by at least 2 pounds per week. Find the maximum number of weeks before the man's weight will reach his goal of 192 pounds.

SOLUTION

Verbal Model:	Desired weight	≤	Current weight	−	2 pounds per week	·	Number of weeks

Labels: Desired weight $= 192$ (pounds)
Current weight $= 225$ (pounds)
Number of weeks $= x$ (weeks)

Inequality: $192 \le 225 - 2x$

$$-33 \le -2x$$

$$16.5 \ge x$$

It will take at most $16\frac{1}{2}$ weeks for the man to reach his goal.

✓ Checkpoint 8

In Example 8, find the maximum number of weeks before the man's weight will reach 200 pounds.

 **Example 9**   **Accuracy of a Measurement**

You go to a candy store to buy chocolates that cost $9.89 per pound. The scale used in the store has a state seal of approval that indicates the scale is accurate to within one-half ounce. According to the scale, your purchase weighs one-half pound and costs $4.95. How much might you have been undercharged or overcharged due to an error in the scale?

SOLUTION   To solve this problem, let x represent the *true* weight of the candy. Because the scale is accurate to within one-half ounce (or $\frac{1}{32}$ pound), you can conclude that the absolute value of the difference of the exact weight (x) and the scale weight $\left(\frac{1}{2} \text{ pound}\right)$ is less than or equal to $\frac{1}{32}$ pound. That is,

$$\left| x - \frac{1}{2} \right| \leq \frac{1}{32}.$$

You can solve this inequality as shown.

$$-\frac{1}{32} \leq x - \frac{1}{2} \leq \frac{1}{32}$$

$$\frac{15}{32} \leq x \leq \frac{17}{32}$$

$$0.46875 \leq x \leq 0.53125$$

The graph of this solution set is shown in Figure 1.18. In other words, your "one-half" pound of candy could have weighed as little as 0.46875 pound (which would have cost $0.46875 \cdot \$9.89 \approx \$4.64$) or as much as 0.53125 pound (which would have cost $0.53125 \cdot \$9.89 \approx \$5.25$). So, you could have been undercharged by as much as $0.30 or overcharged by as much as $0.31.

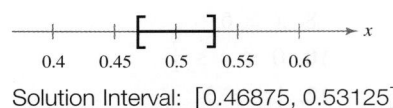

Solution Interval: [0.46875, 0.53125]
FIGURE 1.18

 **Checkpoint 9**

You go to a grocery store to buy ground beef that costs $3.96 per pound. The scale used in the store is accurate to within $\frac{1}{3}$ ounce $\left(\text{or } \frac{1}{48} \text{ pound}\right)$. According to the scale, your purchase weighs 7.5 pounds and costs $29.70. How much might you have been undercharged or overcharged due to an error in the scale?

SUMMARIZE   (Section 1.6)

1. Make a list of the different types of bounded and unbounded intervals on the real number line *(pages 126 and 127)*. For an example on bounded and unbounded intervals, see Example 1.

2. Make a list of the properties of inequalities *(page 128)*.

3. Describe how to solve a linear inequality *(pages 128 and 129)*. For examples of solving linear inequalities, see Examples 2 and 3.

4. State the definition of a double inequality *(page 130)*. For an example of solving a double inequality, see Example 4.

5. Describe how to solve an absolute value inequality *(page 131)*. For examples of solving absolute value inequalities, see Examples 5 and 6.

6. Describe a real-life example that uses an inequality *(pages 132 and 133, Examples 7, 8, and 9)*.

SKILLS WARM UP 1.6   The following warm-up exercises involve skills that were covered in earlier sections. You will use these skills in the exercise set for this section. For additional help, review Sections 0.1 and 0.2.

In Exercises 1–4, determine which of the two numbers is greater.

1. $-\frac{1}{2}, -7$

2. $-\frac{1}{3}, -\frac{1}{6}$

3. $-\pi, -3$

4. $-6, -\frac{13}{2}$

In Exercises 5–8, use inequality notation to describe the subset of real numbers.

5. x is nonnegative.

6. z is strictly between -3 and 10.

7. P is no more than 2.

8. W is at least 200.

In Exercises 9 and 10, evaluate the expression at each value of x.

9. $|x - 10|, x = 12, x = 3$

10. $|2x - 3|, x = \frac{3}{2}, x = 1$

Exercises 1.6

See www.CalcChat.com for worked-out solutions to odd-numbered exercises.

Intervals and Inequalities   In Exercises 1–6, write an inequality that represents the interval. Then state whether the interval is bounded or unbounded. *See Example 1.*

1. $[-1, 5]$

2. $(2, 10]$

3. $(11, \infty)$

4. $[-5, \infty)$

5. $(-\infty, -2)$

6. $(-\infty, 7]$

Matching   In Exercises 7–14, match the inequality with its graph. [The graphs are labeled (a), (b), (c), (d), (e), (f), (g), and (h).]

(a)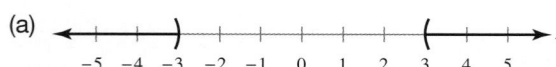

(b)

(c)

(d)

(e)

(f)

(g)

(h)

7. $x < 4$

8. $x \geq 6$

9. $-2 < x \leq 5$

10. $0 \leq x \leq \frac{7}{2}$

11. $|x| < 4$

12. $|x| > 3$

13. $|x - 5| > 2$

14. $|x + 6| < 3$

Checking Solutions   In Exercises 15–22, determine whether each value of x is a solution of the inequality.

15. $5x - 12 > 0$

    (a) $x = 3$ (b) $x = -3$ (c) $x = \frac{5}{2}$ (d) $x = \frac{3}{2}$

16. $4x - 5 < 3$

    (a) $x = 0$ (b) $x = 2$ (c) $x = -2$ (d) $x = 4$

17. $0 < \dfrac{x - 2}{4} < 2$

    (a) $x = 4$

    (b) $x = 10$

    (c) $x = 0$

    (d) $x = \frac{7}{2}$

18. $-1 < \dfrac{3 - x}{2} \leq 1$

    (a) $x = 0$

    (b) $x = -5$

    (c) $x = 1$

    (d) $x = 5$

19. $|x + 2| \leq 10$

    (a) $x = -15$

    (b) $x = -4$

    (c) $x = 1$

    (d) $x = 8$

20. $|x - 10| \geq 3$

    (a) $x = 13$

    (b) $x = -1$

    (c) $x = 14$

    (d) $x = 9$

21. $|2x - 3| < 15$

    (a) $x = -6$

    (b) $x = 0$

    (c) $x = 12$

    (d) $x = 7$

22. $|3x + 5| > 7$

    (a) $x = -5$

    (b) $x = -2$

    (c) $x = \frac{1}{3}$

    (d) $x = 10$

Solving an Inequality In Exercises 23–30, copy and complete the statement using the correct inequality symbol.

23. If $2x > 6$, then x ___?___ 3.

24. If $3x > 9$, then x ___?___ 3.

25. If $2x \leq -8$, then x ___?___ -4.

26. If $3x \leq -15$, then x ___?___ -5.

27. If $2 - 4x > -10$, then x ___?___ 3.

28. If $5 - 3x > -7$, then x ___?___ 4.

29. If $-\dfrac{2}{3}x \geq -6$, then x ___?___ 9.

30. If $-\dfrac{3}{4}x \geq -12$, then x ___?___ 16.

Solving a Linear Inequality In Exercises 31–44, solve the inequality. Then graph the solution set on the real number line. *See Examples 2 and 3.*

31. $\dfrac{3}{2}x \geq 9$

32. $\dfrac{2}{5}x > 7$

33. $-10x < 40$

34. $-6x > 15$

35. $\dfrac{3}{5}x - 7 < 8$

36. $\dfrac{5}{4}x + 1 \leq 11$

37. $2x + 7 < 3 + 4x$

38. $6x - 4 \leq 2 + 8x$

39. $2x - 1 \geq 5x + 1$

40. $3x + 1 \geq 2 + x$

41. $3(x + 2) + 7 < 2x - 5$

42. $2(x + 7) - 4 \geq 5(x - 3)$

43. $-3(x - 1) + 7 < 2x + 8$

44. $5 - 3x > -5(x + 4) + 6$

Solving a Double Inequality In Exercises 45–52, solve the inequality. Then graph the solution set on the real number line. *See Example 4.*

45. $3 \leq 2x - 1 < 7$

46. $3 > 1 - \dfrac{x}{2} > -3$

47. $-11 \leq 1 + 4(x + 1) < 13$

48. $-8 \leq 1 - 3(x - 2) < 13$

49. $-4 < \dfrac{x - 3}{3} < 4$

50. $0 \leq \dfrac{x + 3}{2} < 5$

51. $\dfrac{3}{4} > x + 1 > \dfrac{1}{4}$

52. $-\dfrac{1}{2} < x - 1 < \dfrac{3}{4}$

Solving an Absolute Value Inequality In Exercises 53–70, solve the inequality. Then graph the solution set on the real number line. *See Examples 5 and 6.*

53. $|x| < 6$

54. $|x| > 8$

55. $\left|\dfrac{x}{2}\right| > 3$

56. $|5x| > 10$

57. $|x + 3| < 5$

58. $|x - 7| < 6$

59. $|x - 5| < 0$

60. $|x - 5| \geq 0$

61. $|2x - 5| > 6$

62. $|2x - 20| \leq 4$

63. $2|x + 10| \geq 9$

64. $3|4 - 5x| \leq 9$

65. $\left|\dfrac{x - 3}{2}\right| \geq 5$

66. $\left|\dfrac{2x + 1}{2}\right| < 6$

67. $|9 - x| - 2 < -1$

68. $|x + 14| + 3 > 17$

69. $2|2x - 5| - 4 > 6$

70. $2|5 - 3x| + 7 < 21$

Using Absolute Value In Exercises 71–78, use absolute value notation to define the solution set.

71.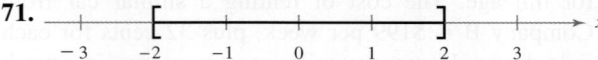

72. ![number line graph]

73. ![number line graph]

74. ![number line graph]

75. All real numbers at most 10 units from 12

76. All real numbers at least 5 units from 8

77. All real numbers whose distances from -3 are more than 5

78. All real numbers whose distances from -6 are no more than 7

79. Annual Operating Cost A utility company has a fleet of vans. The annual operating cost C per van is

$$C = 0.52m + 2500$$

where m is the number of miles traveled by a van in a year. What number of miles will yield an annual operating cost that is less than $12,000?

80. IQ Scores The admissions office of a college wants to determine whether there is a relationship between IQ scores x and grade-point averages y after the first year of school. An equation that models the data obtained by the admissions office is

$$y = 0.068x - 4.753.$$

Estimate the values of x that predict a grade-point average of at least 3.0.

81. Brand Name Drugs The average prices B (in dollars) of brand name prescription drugs from 2005 through 2009 can be approximated by

$$B = 14.516t + 24.65, \quad 5 \leq t \leq 9$$

where t represents the year, with $t = 5$ corresponding to 2005. Use the model to estimate the first year in which the average price exceeded $140. *(Source: National Association of Chain Drug Stores)*

82. Generic Drugs   The average prices G (in dollars) of generic prescription drugs from 2005 through 2009 can be approximated by

$$G = 2.649t + 15.00, \quad 5 \le t \le 9$$

where t represents the year, with $t = 5$ corresponding to 2005. Use the model to estimate the first year in which the average price exceeded \$30. *(Source: National Association of Chain Drug Stores)*

83. Comparative Shopping   The cost of renting a car from Company A is \$279 per week with no extra charge for mileage. The cost of renting a similar car from Company B is \$199 per week, plus 32 cents for each mile driven. How many miles must you drive in a week to make the rental fee for Company B greater than that for Company A?

84. Comparative Shopping   Your department sends its copying to a photocopy center, which bills your department \$0.08 per page. You are considering buying a departmental copier for \$2500. With your own copier, the cost per page would be \$0.025. The expected life of the copier is 4 years. How many copies must you make in the four-year period to justify purchasing the copier?

85. Football Salaries   The average National Football League player's salary S (in millions of dollars) each year from 1998 through 2009 can be approximated by

$$S = 0.1215t - 0.351, \quad 8 \le t \le 19$$

where t represents the year, with $t = 8$ corresponding to 1998 (see figure). *(Source: National Football League Players Association)*

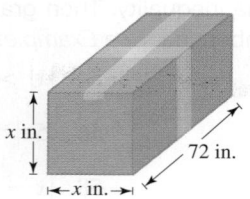

Year (8 ↔ 1998)

(a) Use the model to estimate the years when the average National Football League player's salary was less than \$1.5 million. Do your results agree with the bar graph? Explain.

(b) Use the model to predict the first year that the average National Football League player's salary will exceed \$2.2 million. Do your results seem reasonable? Explain.

86. Public College Enrollment   The projected public college enrollment E (in thousands) in the United States from 2012 to 2019 can be approximated by

$$E = 265.3t + 14{,}157, \quad 2 \le t \le 9$$

where t represents the year, with $t = 2$ corresponding to 2012 (see figure). *(Source: U.S. National Center for Education Statistics)*

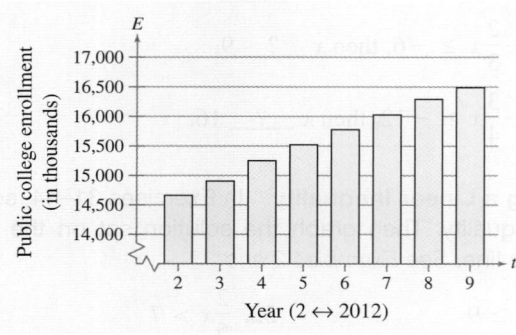

Year (2 ↔ 2012)

(a) Use the model to predict the first year that the public college enrollment will be greater than 16 million students. Do your results agree with the bar graph? Explain.

(b) Use the model to predict the first year that the public college enrollment will be greater than 17 million students. Do your results seem reasonable? Explain.

87. Weight Loss Program   A person enrolls in a weight loss program that guarantees a loss of at least $1\frac{1}{2}$ pounds per week. The person's weight at the beginning of the program is 180 pounds. Find the maximum number of weeks it will take for the person to attain a weight of 130 pounds.

88. Salary Increase   You accept a new job with a starting salary of \$34,800. You are told that you will receive an annual raise of at least \$1500. What is the maximum number of years you must work before your annual salary will be \$45,000?

89. Maximum Width   An overnight delivery company will not accept any package whose combined length and girth (perimeter of a cross section) exceeds 132 inches. The package shown is 72 inches long and has square cross sections. Describe the lengths of x for which the company will deliver the package.

90. Break-Even Analysis   The revenue R from selling x units of a product is

$$R = 139.95x.$$

The cost C of producing x units is

$$C = 97x + 850.$$

In order to obtain a profit, the revenue must be greater than the cost.

(a) Complete the table.

x	10	20	30	40	50	60
R						
C						

(b) For what values of x will this product return a profit?

91. Break-Even Analysis   The revenue R from selling x units of a product is $R = 25.95x$. The cost C of producing x units is

$$C = 13.95x + 125,000.$$

In order to obtain a profit, the revenue must be greater than the cost. For what values of x will the product return a profit?

92. HOW DO YOU SEE IT?   A company sells all the units of a product that it produces. The double bar graph shows the daily cost and revenue based on the numbers of units x produced.

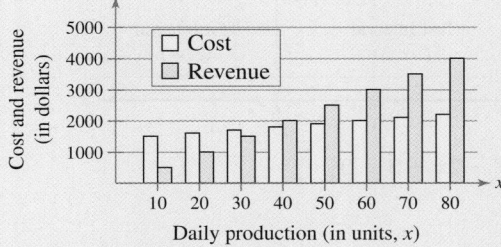

(a) Use the double bar graph to estimate the number of units the company must produce in a day to break even, with cost equal to revenue.

(b) Use your answer in part (a) to write an inequality that represents the daily numbers of units x for which the company receives a profit.

(c) Write an inequality that represents the daily numbers of units x for which the company receives a profit of at least $1000.

93. Simple Interest   For $1500 to grow to more than $1680 in 3 years, what must the simple interest rate be?

94. Simple Interest   For $2000 to grow to more than $2180 in 2 years, what must the simple interest rate be?

95. Geometry   You measure the side of a square as 10.4 inches with a possible error of $\frac{1}{16}$ inch. Using these measurements, determine the interval containing the possible areas of the square.

96. Geometry   You measure the side of a square as 24.2 centimeters with a possible error of 0.25 centimeter. Using these measurements, determine the interval containing the possible areas of the square.

97. Human Height   The heights h of two-thirds of a population satisfy the inequality

$$|h - 68.5| \le 2.7$$

where h is measured in inches. Determine the interval on the real number line in which these heights lie.

98. Time Study   A time study was conducted to determine the length of time required to perform a particular task in a manufacturing process. The times required by approximately two-thirds of the workers in the study satisfied the inequality

$$\left| \frac{t - 15.6}{1.9} \right| < 1$$

where t is time in minutes. Determine the interval on the real number line in which these times lie.

99. Humidity Control   The specifications for an electronic device state that it is to be operated in a room with relative humidity h defined by

$$|h - 50| \le 30.$$

What are the minimum and maximum relative humidities for the operation of this device?

100. Body Temperature   Physicians consider an adult's body temperature x (in degrees Fahrenheit) to be normal when it satisfies the inequality

$$|x - 98.6| \le 1.$$

Determine the range of temperatures that are considered to be normal.

101. Accuracy of Measurement   You buy six T-bone steaks that cost $8.99 per pound. The weight listed on the package is 5.72 pounds. The scale that weighed the package is accurate to within $\frac{1}{2}$ ounce. How much money might you have been undercharged or overcharged?

102. Daily Sales   A doughnut shop sells a dozen doughnuts for $7.95. Beyond the fixed costs (rent, utilities, and insurance) of $165 per day, it costs $1.45 for enough materials (flour, sugar, and so on) and labor to produce a dozen doughnuts. The daily profit from doughnut sales varies between $400 and $1200. Between what numbers of doughnuts (in dozens) do the daily sales vary?

1.7 Other Types of Inequalities

■ Use critical numbers to determine test intervals for a polynomial inequality.
■ Solve and graph a polynomial inequality.
■ Solve and graph a rational inequality.
■ Construct and use a polynomial inequality to solve an application problem.
■ Determine the domain of an expression involving a radical.

Finding Critical Numbers and Test Intervals

To solve a polynomial inequality such as

$$x^2 - 2x - 3 < 0$$

you can use the fact that a polynomial can change signs only at its **zeros.** The zeros are the x-values that make the polynomial equal to zero. Between two consecutive zeros, a polynomial must be entirely positive or entirely negative. This means that when the real zeros of a polynomial are put in order, they divide the real number line into intervals in which the polynomial has no sign changes. These zeros are the **critical numbers** of the inequality, and the resulting intervals are the **test intervals** for the inequality. For example, the polynomial above factors as

$$x^2 - 2x - 3 = (x + 1)(x - 3)$$

and has two zeros, $x = -1$ and $x = 3$. These zeros divide the real number line into three test intervals:

$$(-\infty, -1), \quad (-1, 3), \quad \text{and} \quad (3, \infty). \qquad \text{(See Figure 1.19.)}$$

So, to solve the inequality $x^2 - 2x - 3 < 0$, you need only test one value from each of these test intervals.

In Exercise 74 on page 147, you will use a quadratic inequality to make a prediction about corporate sales.

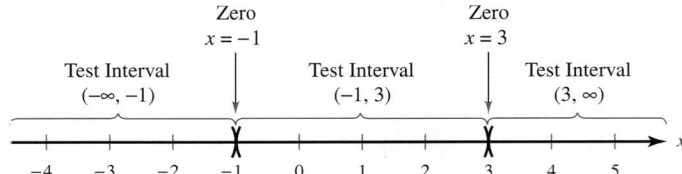

Three Test Intervals for $x^2 - 2x - 3 < 0$
FIGURE 1.19

Finding Test Intervals for a Polynomial

To determine the intervals on which the values of a polynomial are entirely negative or entirely positive, use the following steps.

1. Find all real zeros of the polynomial, and arrange the zeros in increasing order. These zeros are the critical numbers of the polynomial.

2. Use the critical numbers to determine the test intervals.

3. Choose one representative x-value in each test interval and evaluate the polynomial at that value. If the value of the polynomial is negative at the representative x-value, the polynomial will have negative values for *every* x-value in the interval. If the value of the polynomial is positive, the polynomial will have positive values for *every* x-value in the interval.

Polynomial Inequalities

Example 1 Solving a Polynomial Inequality

Solve $x^2 - x - 6 < 0$.

SOLUTION By factoring the quadratic as

$$x^2 - x - 6 = (x + 2)(x - 3)$$

you can see that the critical numbers are

$$x = -2 \quad \text{and} \quad x = 3. \qquad \text{Critical numbers}$$

The boundaries between the numbers that satisfy the inequality and the numbers that do not satisfy the inequality always occur at critical numbers. So, the polynomial's test intervals are

$$(-\infty, -2), \quad (-2, 3), \quad \text{and} \quad (3, \infty). \qquad \text{Test intervals}$$

In each test interval, choose a representative x-value and evaluate the polynomial.

Test Interval	x-Value	Polynomial Value	Conclusion
$(-\infty, -2)$	$x = -3$	$(-3)^2 - (-3) - 6 = 6$	Positive
$(-2, 3)$	$x = 0$	$(0)^2 - (0) - 6 = -6$	Negative
$(3, \infty)$	$x = 4$	$(4)^2 - (4) - 6 = 6$	Positive

From this, you can conclude that the polynomial is positive for all x-values in $(-\infty, -2)$ and $(3, \infty)$, and is negative for all x-values in $(-2, 3)$. This implies that the solution of the inequality

$$x^2 - x - 6 < 0$$

is the interval $(-2, 3)$, as shown in Figure 1.20.

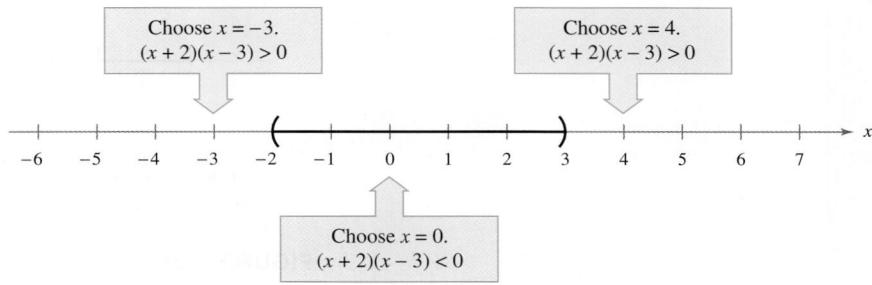

Choose $x = -3$.
$(x + 2)(x - 3) > 0$

Choose $x = 4$.
$(x + 2)(x - 3) > 0$

Choose $x = 0$.
$(x + 2)(x - 3) < 0$

FIGURE 1.20

✓ Checkpoint 1

Solve $x^2 + x - 2 < 0$.

As with linear inequalities, you can check a solution interval of a polynomial inequality by substituting x-values into the original inequality. For instance, to check the solution found in Example 1, try substituting several x-values from the interval $(-2, 3)$ into the inequality

$$x^2 - x - 6 < 0.$$

Regardless of which x-values you choose, the inequality will be satisfied.

In Example 1, the polynomial inequality was given in general form. Whenever this is not the case, begin the solution process by writing the inequality in general form—with the polynomial on one side and zero on the other.

Example 2   Solving a Polynomial Inequality

Solve $x^3 - 3x^2 > 10x$.

SOLUTION

$$x^3 - 3x^2 > 10x \qquad \text{Write original inequality.}$$

$$x^3 - 3x^2 - 10x > 0 \qquad \text{Write in general form.}$$

$$x(x - 5)(x + 2) > 0 \qquad \text{Factor.}$$

You can see that the critical numbers are $x = -2$, $x = 0$, and $x = 5$. So, the test intervals are $(-\infty, -2)$, $(-2, 0)$, $(0, 5)$, and $(5, \infty)$. In each test interval, choose a representative x-value and evaluate the polynomial.

Test Interval	x-Value	Polynomial Value	Conclusion
$(-\infty, -2)$	$x = -3$	$(-3)^3 - 3(-3)^2 - 10(-3) = -24$	Negative
$(-2, 0)$	$x = -1$	$(-1)^3 - 3(-1)^2 - 10(-1) = 6$	Positive
$(0, 5)$	$x = 2$	$(2)^3 - 3(2)^2 - 10(2) = -24$	Negative
$(5, \infty)$	$x = 6$	$(6)^3 - 3(6)^2 - 10(6) = 48$	Positive

From this, you can conclude that the inequality is satisfied on the open intervals $(-2, 0)$ and $(5, \infty)$. So, the solution set consists of all real numbers in the intervals $(-2, 0)$ and $(5, \infty)$, as shown in Figure 1.21.

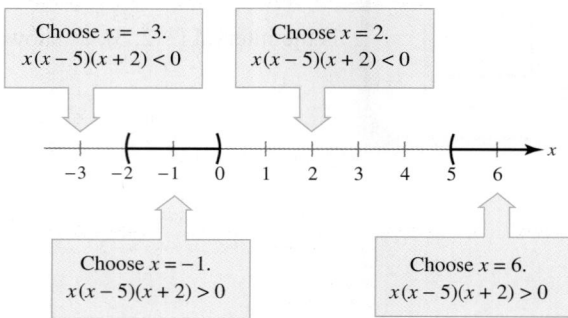

FIGURE 1.21

✓ **Checkpoint 2**

Solve $x^2 - 3x > -2$.

When solving a polynomial inequality, be sure to account for the type of inequality symbol it contains. For instance, in Example 2, note that the solution consisted of two *open* intervals because the original inequality contained a "greater than" symbol. If the original inequality had been

$$x^3 - 3x^2 \geq 10x$$

then the solution would have consisted of the *closed* interval $[-2, 0]$ and the interval $[5, \infty)$.

Each of the polynomial inequalities in Examples 1 and 2 has a solution set that consists of a single interval or the union of two intervals. When solving the exercises for this section, watch for some unusual solution sets, as illustrated in Example 3.

Example 3   Unusual Solution Sets

What is unusual about the solution set for each inequality?

a. $x^2 + 2x + 4 > 0$

The solution set for this inequality consists of the entire set of real numbers, $(-\infty, \infty)$. In other words, the value of the quadratic $x^2 + 2x + 4$ is positive for every real value of x.

b. $x^2 + 2x + 1 \leq 0$

The solution set for this inequality consists of the single real number $\{-1\}$, because the quadratic $x^2 + 2x + 1$ has one critical number, $x = -1$, and it is the only value that satisfies the inequality.

c. $x^2 + 3x + 5 < 0$

The solution set for this inequality is empty. In other words, the quadratic $x^2 + 3x + 5$ is *not* less than zero for any value of x.

d. $x^2 - 4x + 4 > 0$

The solution set for this inequality consists of all real numbers *except* the number 2. In interval notation, this solution can be written as $(-\infty, 2) \cup (2, \infty)$.

✓ **Checkpoint 3**

What is unusual about the solution set for each inequality?

a. $x^2 + x + 3 \leq 0$ **b.** $x^2 - 2x + 1 > 0$

TECH TUTOR

Most graphing utilities can graph inequalities. Consult your user's guide for specific instructions. Once you know how to graph an inequality, you can check solutions by graphing. (Make sure you use an appropriate viewing window that includes all of the critical numbers.) For example, the solution of

$$x^2 - 5x < 0$$

is the interval $(0, 5)$. When graphed, the solution occurs as an interval above the horizontal axis on the graphing utility, as shown in Figure 1.22. The graph does not indicate whether 0 and/or 5 are part of the solution. You must determine whether the endpoints are part of the solution based on the type of inequality.

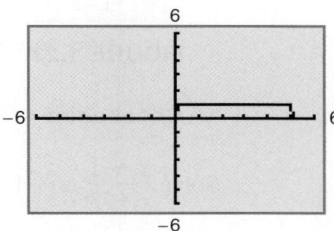

FIGURE 1.22

Rational Inequalities

The concepts of critical numbers and test intervals can be extended to inequalities involving rational expressions. Use the fact that the value of a rational expression can change sign only at its *zeros* (the x-values for which its numerator is zero) and its *undefined values* (the x-values for which its denominator is zero). These two types of numbers make up the **critical numbers** of a rational inequality.

Example 4   Solving a Rational Inequality

Solve $\dfrac{2x - 7}{x - 5} \le 3$.

SOLUTION

$$\frac{2x - 7}{x - 5} \le 3 \qquad \text{Write original inequality.}$$

$$\frac{2x - 7}{x - 5} - 3 \le 0 \qquad \text{Write in general form.}$$

$$\frac{2x - 7 - 3x + 15}{x - 5} \le 0 \qquad \text{Add fractions.}$$

$$\frac{-x + 8}{x - 5} \le 0 \qquad \text{Simplify.}$$

Critical numbers: $x = 5, x = 8$

Test intervals: $(-\infty, 5), (5, 8), (8, \infty)$

Test: Is $\dfrac{-x + 8}{x - 5} \le 0$?

After testing these intervals, as shown in Figure 1.23, you can see that the inequality is satisfied on the open intervals $(-\infty, 5)$ and $(8, \infty)$. Moreover, because

$$\frac{-x + 8}{x - 5} = 0$$

when $x = 8$, you can conclude that the solution set consists of all real numbers in the intervals $(-\infty, 5) \cup [8, \infty)$.

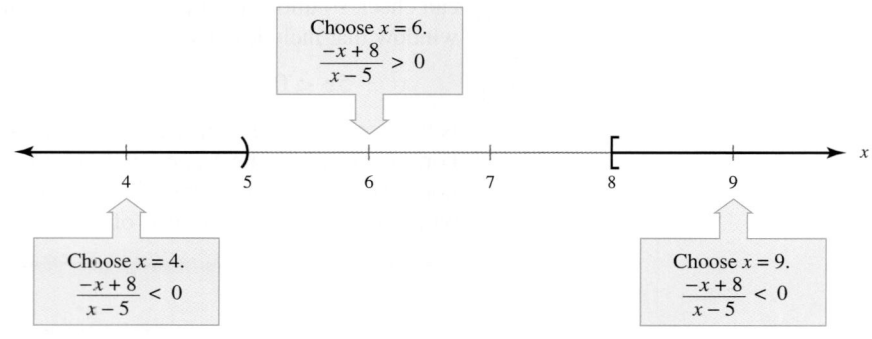

FIGURE 1.23

✓ Checkpoint 4

Solve $\dfrac{x - 1}{x - 3} \ge -1$.

Application

One common business application of inequalities involves profit, revenue, and cost. The formula that relates these three quantities is

Profit = Revenue − Cost

$$P = R - C.$$

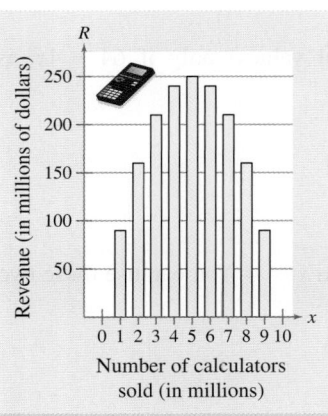

Example 5 Increasing the Profit for a Product

The marketing department of a calculator manufacturer has determined that the demand for a new model of calculator is given by

$$p = 100 - 10x, \quad 0 \le x \le 10 \qquad \text{Demand equation}$$

where p is the price per calculator (in dollars) and x is the number of calculators sold, in millions. (If this model is accurate, no one would be willing to pay $100 for the calculator. At the other extreme, the company couldn't *give* away more than 10 million calculators.) The revenue, in millions of dollars, for selling x million calculators is given by

$$R = xp = x(100 - 10x). \qquad \text{Revenue equation}$$

See Figure 1.24. The total cost of producing x million calculators is $10 per calculator plus a one-time development cost of $2,500,000. So, the total cost, in millions of dollars, is

$$C = 10x + 2.5. \qquad \text{Cost equation}$$

What prices can the company charge per calculator to obtain a profit of at least $190,000,000?

SOLUTION

Verbal Model: Profit = Revenue − Cost

Equation: $P = R - C$

$$P = 100x - 10x^2 - (10x + 2.5)$$

$$P = -10x^2 + 90x - 2.5$$

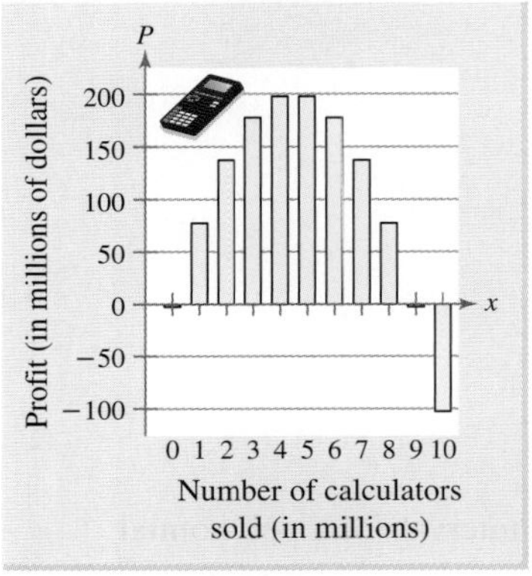

To answer the question, you must solve the inequality

$$-10x^2 + 90x - 2.5 \ge 190.$$

Using the techniques described in this section, you can find the solution set to be

$$3.5 \le x \le 5.5$$

as shown in Figure 1.25. The prices that correspond to these x-values are given by

$$100 - 10(3.5) \ge p \ge 100 - 10(5.5)$$

$$45 \le p \le 65.$$

The company can obtain a profit of $190,000,000 or better by charging at least $45 per calculator and at most $65 per calculator.

✓ Checkpoint 5

In Example 5, what prices can the company charge per calculator to obtain a profit of at least $160,000,000?

FIGURE 1.24

Revenue (in millions of dollars)

R

250
200
150
100
50

0 1 2 3 4 5 6 7 8 9 10 x

Number of calculators
sold (in millions)

FIGURE 1.25

Profit (in millions of dollars)

P

200
150
100
50
0
−50
−100

0 1 2 3 4 5 6 7 8 9 10

Number of calculators
sold (in millions)

Domain of a Radical Expression

Recall that the domain of an expression is the set of all x-values for which the expression is defined. You can use inequalities to find the domain of an expression that involves a radical, as shown in Example 6.

Example 6 | Finding the Domain of an Expression

Find the domain of the expression $\sqrt{64 - 4x^2}$.

SOLUTION   Because $\sqrt{64 - 4x^2}$ is defined (has real values) only if $64 - 4x^2$ is nonnegative, the domain is given by $64 - 4x^2 \geq 0$.

$$64 - 4x^2 \geq 0 \qquad \text{Write in general form.}$$

$$16 - x^2 \geq 0 \qquad \text{Divide each side by 4.}$$

$$(4 - x)(4 + x) \geq 0 \qquad \text{Factor.}$$

So, the inequality has two critical numbers: $x = -4$ and $x = 4$. You can use these two numbers to test the inequality, as follows.

Critical numbers: $x = -4, x = 4$

Test intervals: $(-\infty, -4), (-4, 4), (4, \infty)$

Test: Is $(4 - x)(4 + x) \geq 0$?

A test shows that $64 - 4x^2$ is greater than or equal to 0 in the *closed interval* $[-4, 4]$. So, the domain of the expression $\sqrt{64 - 4x^2}$ is the interval $[-4, 4]$, as shown in Figure 1.26.

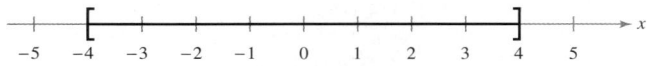

FIGURE 1.26

✓ Checkpoint 6

Find the domain of each expression.

a. $\sqrt{12 - 3x^2}$

b. $\sqrt[3]{x^2 - 2x - 8}$

SUMMARIZE   (Section 1.7)

1. Describe how to find the critical numbers and test intervals of a polynomial inequality *(page 138)*.

2. Describe how to solve a polynomial inequality *(pages 139–141)*. For examples of solving polynomial inequalities, see Examples 1, 2, and 3.

3. Describe how to solve a rational inequality *(page 142)*. For an example of solving a rational inequality, see Example 4.

4. Describe a real-life example of how a polynomial inequality can be used to analyze profit, revenue, and cost *(page 143, Example 5)*.

5. Describe how to find the domain of an expression involving a radical *(page 144)*. For an example of finding the domain of an expression involving a radical, see Example 6.

SKILLS WARM UP 1.7   The following warm-up exercises involve skills that were covered in earlier sections. You will use these skills in the exercise set for this section. For additional help, review Section 1.6.

In Exercises 1–10, solve the inequality.

1. $-\dfrac{y}{3} > 2$

2. $-6z < 27$

3. $-3 \le 2x + 3 < 5$

4. $-3x + 5 \ge 20$

5. $10 > 4 - 3(x + 1)$

6. $3 < 1 + 2(x - 4) < 7$

7. $2|x| \le 7$

8. $|x - 3| > 1$

9. $|x + 4| > 2$

10. $|2 - x| \le 4$

Exercises 1.7

See www.CalcChat.com for worked-out solutions to odd-numbered exercises.

Finding Test Intervals   In Exercises 1–6, find the test intervals of the inequality.

1. $x^2 - 25 < 0$

2. $x^2 - 6x + 8 > 0$

3. $2x^2 + 7x + 16 \ge 20$

4. $3x^2 - 26x + 25 \le 9$

5. $\dfrac{x - 3}{x - 1} < 2$

6. $\dfrac{x - 4}{2x + 3} \ge 1$

Solving a Polynomial Inequality   In Exercises 7–34, solve the inequality. Then graph the solution set on the real number line. *See Examples 1, 2, and 3.*

7. $x^2 \le 9$

8. $x^2 < 5$

9. $x^2 > 4$

10. $x^2 \ge 16$

11. $(x + 2)^2 < 25$

12. $(x + 6)^2 \le 8$

13. $x^2 + 4x + 4 \ge 9$

14. $x^2 - 6x + 9 < 16$

15. $x^2 + x < 6$

16. $x^2 + 2x > 3$

17. $3(x - 1)(x + 1) > 0$

18. $6(x + 2)(x - 1) < 0$

19. $x^2 + 2x - 3 < 0$

20. $x^2 - 4x - 1 > 0$

21. $x^2 - 8x + 16 > 0$

22. $x^2 + 6x + 9 > 0$

23. $4x^2 - 4x + 1 \le 0$

24. $4x^2 + 12x + 9 \le 0$

25. $x^2 + 4 > 0$

26. $x^2 + 9 \le 0$

27. $x^2 < -1$

28. $x^2 \ge -7$

29. $4x^3 - 6x^2 < 0$

30. $4x^3 - 12x^2 > 0$

31. $x^3 - 4x \ge 0$

32. $2x^3 - x^4 \le 0$

33. $x^3 - 2x^2 - x + 2 \ge 0$

34. $x^3 + 5x^2 - 4x - 20 \le 0$

Solving a Rational Inequality   In Exercises 35–44, solve the inequality. Then graph the solution set on the real number line. *See Example 4.*

35. $\dfrac{1}{x} > x$

36. $\dfrac{1}{x} < 4$

37. $\dfrac{x + 6}{x + 1} < 2$

38. $\dfrac{x + 12}{x + 2} \ge 3$

39. $\dfrac{3x - 5}{x - 5} > 4$

40. $\dfrac{5 + 7x}{1 + 2x} < 4$

41. $\dfrac{4}{x + 5} > \dfrac{1}{2x + 3}$

42. $\dfrac{5}{x - 6} > \dfrac{3}{x + 2}$

43. $\dfrac{1}{x - 3} \le \dfrac{9}{4x + 3}$

44. $\dfrac{1}{x} \ge \dfrac{1}{x + 3}$

Finding the Domain of an Expression   In Exercises 45–54, find the domain of the expression. *See Example 6.*

45. $\sqrt{x^2 - 9}$

46. $\sqrt{x^2 - 4}$

47. $\sqrt[4]{6 + x^2}$

48. $\sqrt{x^2 + 4}$

49. $\sqrt{81 - 4x^2}$

50. $\sqrt{147 - 3x^2}$

51. $\sqrt{x^2 - 7x + 10}$

52. $\sqrt{12 - x - x^2}$

53. $\sqrt{x^2 - 3x + 3}$

54. $\sqrt[4]{-x^2 + 2x - 2}$

Writing the Solution in Interval Notation   In Exercises 55–60, solve the inequality and write the solution set in interval notation.

55. $6x^3 - 10x^2 > 0$

56. $25x^3 - 10x^2 < 0$

57. $x^3 - 9x \le 0$

58. $4x^3 - x^4 \ge 0$

59. $(x - 1)^2(x + 2)^3 \ge 0$

60. $x^4(x - 3) \le 0$

Using a Calculator   In Exercises 61–66, use a calculator to solve the inequality. (Round each number in your answer to two decimal places.)

61. $0.4x^2 + 5.26 < 10.2$

62. $-1.3x^2 + 3.78 > 2.12$

63. $-0.5x^2 + 12.5x + 1.6 > 0$

64. $1.2x^2 + 4.8x + 3.1 < 5.3$

65. $\dfrac{1}{2.3x - 5.2} > 3.4$

66. $\dfrac{2}{3.1x - 3.7} > 5.8$

67. Geometry   A rectangular playing field with a perimeter of 100 meters is to have an area of at least 500 square meters (see figure). Describe the possible lengths l of the field.

68. Height of a Projectile   A projectile is fired straight upward from ground level with an initial velocity of 160 feet per second. During what time period will its height be less than 384 feet?

69. Company Profits   The revenue R and cost C for a product are given by

$$R = x(50 - 0.0002x)$$

and

$$C = 12x + 150{,}000$$

where R and C are measured in dollars and x represents the number of units produced and sold (see figure).

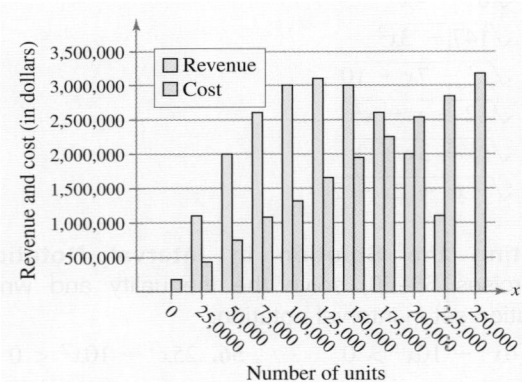

(a) How many units must be sold to obtain a profit of at least \$1,650,000?

(b) The demand equation for the product is

$$p = 50 - 0.0002x$$

where p is the price per unit. What prices will produce a profit of at least \$1,650,000?

(c) As the number of units increases, the revenue eventually decreases. After this point, at what number of units is the revenue approximately equal to the cost? How should this affect the company's decision about the level of production?

 **70.   HOW DO YOU SEE IT?**   The revenue R and cost C for a product are given by

$$R = x(75 - 0.0005x)$$

and

$$C = 30x + 250{,}000$$

where R and C are measured in dollars and x represents the number of units produced and sold. The figure shows the values of R and C at several production levels.

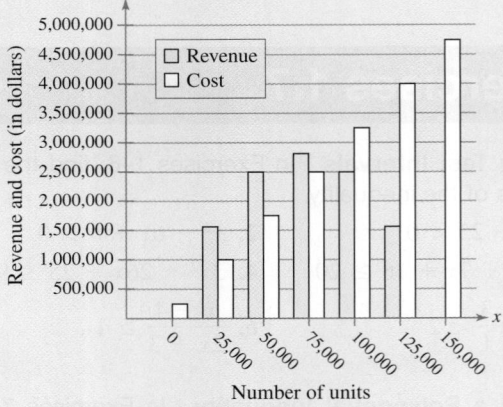

(a) What is the revenue at a production level of zero units?

(b) How many levels of production are there for which the company breaks even, with cost equal to revenue?

(c) Do the maximum revenue and the maximum profit occur at the same level of production? Explain.

(d) Describe the maximum profit shown by the graph and the corresponding level of production.

71. Compound Interest P dollars, invested at interest rate r compounded annually, increases to an amount A given by

$$A = P(1 + r)^3$$

in 3 years. For an investment of \$1000 to increase to an amount greater than \$1150 in 3 years, the interest rate must be greater than what percent?

72. Compound Interest P dollars, invested at interest rate r compounded annually, increases to an amount A given by

$$A = P(1 + r)^2$$

in 2 years. For an investment of \$2000 to increase to an amount greater than \$2200 in 2 years, the interest rate must be greater than what percent?

73. World Population The world population P (in millions) from 2000 to 2010 can be approximated by

$$P = 0.06t^2 + 75.9t + 6089, \quad 0 \le t \le 10$$

where t represents the year, with $t = 0$ corresponding to 2000 (see figure). *(Source: U.S. Census Bureau)*

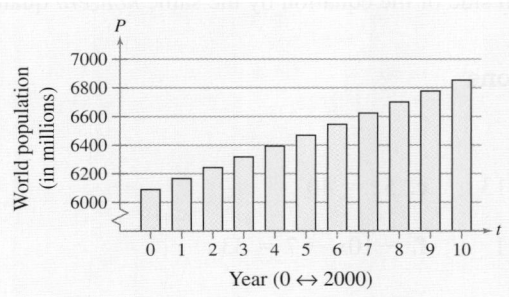

Year (0 ↔ 2000)

(a) Use the model to estimate the world population in 2004. Compare your answer with the value shown in the bar graph.

(b) Use the model to predict the first year in which the world population will exceed 7 billion. Does your answer seem reasonable?

74. Sales The total sales S (in billions of dollars) for Walgreen Company each year from 2000 through 2009 can be approximated by

$$S = 0.126t^2 + 3.68t + 20.9, \quad 0 \le t \le 9$$

where t represents the year, with $t = 0$ corresponding to 2000 (see figure). *(Source: Walgreen Company)*

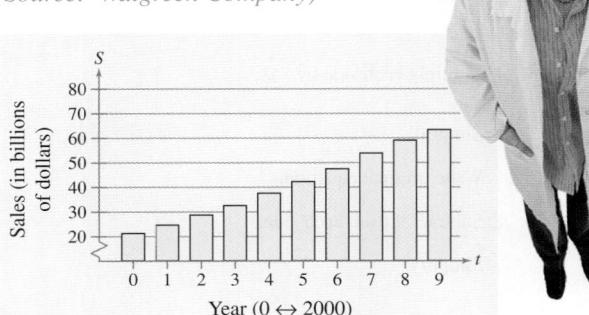

Year (0 ↔ 2000)

(a) Use the model to estimate Walgreen's sales in 2008. How well does your answer compare with the value shown in the bar graph?

(b) Use the model to predict the first year in which Walgreen's sales will be greater than $75 billion.

(c) Walgreen Company predicts yearly sales exceeding $85 billion between 2013 and 2015. Does the model support this prediction? Explain your reasoning.

75. Higher Education The average yearly dormitory charges C (in dollars) at public institutions of higher learning in the United States for the academic years 1997/1998 through 2007/2008 can be approximated by

$$C = 7.20t^2 - 9.7t + 1828, \quad 8 \le t \le 18$$

where t represents the year, with $t = 8$ corresponding to the academic year 1997/1998 (see figure). Use the model to predict the first academic year in which the average yearly dormitory charges will be greater than $6000. *(Source: U.S. National Center for Education Statistics)*

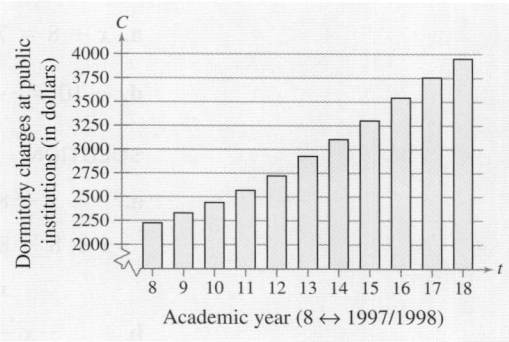

Academic year (8 ↔ 1997/1998)

76. Higher Education The average yearly dormitory charges C (in dollars) at private institutions of higher learning in the United States for the academic years 1996/1997 through 2007/2008 can be approximated by

$$C = 6.20t^2 + 26.3t + 2357, \quad 7 \le t \le 18$$

where t represents the year, with $t = 7$ corresponding to the academic year 1996/1997. Use the model to predict the first academic year in which the average yearly dormitory charges will be greater than $6000. *(Source: U.S. National Center for Education Statistics)*

77. Resistors When two resistors of resistances R_1 and R_2 are connected in parallel (see figure), the total resistance R satisfies the equation

$$\frac{1}{R} = \frac{1}{R_1} + \frac{1}{R_2}.$$

Find R_1 for a parallel circuit in which $R_2 = 2$ ohms and R must be at least 1 ohm.

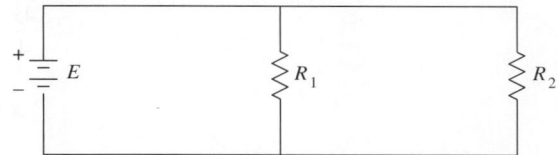

78. Think About It Explain why the domains of

$$\sqrt{x^2 - 7x + 12} \quad \text{and} \quad \sqrt[3]{x^2 - 7x + 12}$$

are different.

ALGEBRA TUTOR

Solving Equations

Much of the algebra in this chapter involves solving an equation in one variable. To solve a linear equation, you can add or subtract the same quantity from each side of the equation, or multiply or divide each side of the equation by the same *nonzero* quantity.

Example 1   Solving Equations

Solve each equation.

a. $x + 8 = 7$ **b.** $-x - 6 = 13$ **c.** $5x = 40$

d. $-10 = -\dfrac{x}{2}$ **e.** $9 + 4x = 11$ **f.** $-10x - 7 = 33$

SOLUTION

a. $x + 8 = 7$ — Write original equation.

$x + 8 - 8 = 7 - 8$ — Subtract 8 from each side.

$x = -1$ — Simplify.

b. $-x - 6 = 13$ — Write original equation.

$-x - 6 + 6 = 13 + 6$ — Add 6 to each side.

$-x = 19$ — Simplify.

$x = -19$ — Divide each side by -1.

c. $5x = 40$ — Write original equation.

$\dfrac{5x}{5} = \dfrac{40}{5}$ — Divide each side by 5.

$x = 8$ — Simplify.

d. $-10 = -\dfrac{x}{2}$ — Write original equation.

$-10 \cdot (-2) = -\dfrac{x}{2} \cdot (-2)$ — Multiply each side by -2.

$20 = x$ — Simplify.

e. $9 + 4x = 11$ — Write original equation.

$9 - 9 + 4x = 11 - 9$ — Subtract 9 from each side.

$4x = 2$ — Simplify.

$\dfrac{4x}{4} = \dfrac{2}{4}$ — Divide each side by 4.

$x = \dfrac{1}{2}$ — Simplify.

f. $-10x - 7 = 33$ — Write original equation.

$-10x - 7 + 7 = 33 + 7$ — Add 7 to each side.

$-10x = 40$ — Simplify.

$x = -4$ — Divide each side by -10.

> **STUDY TIP**
>
> Always remember to check your solution in the *original* equation.

Solving Inequalities

A second algebraic skill used extensively in this chapter is solving an inequality in one variable. This process is similar to solving a linear equation, except when you multiply or divide by a negative number. In this case, you reverse the inequality symbol.

Example 2 Solving Inequalities

Solve each inequality.

a. $x - 3 < 14$ **b.** $-x + 15 > 4$ **c.** $9x \geq 63$

d. $-20 < -\dfrac{x}{5}$ **e.** $6x - 1 \geq 9$ **f.** $-12x + 5 > 47$

SOLUTION

a.

$x - 3 < 14$	Write original inequality.
$x - 3 + 3 < 14 + 3$	Add 3 to each side.
$x < 17$	Simplify.

b.

$-x + 15 > 4$	Write original inequality.
$-x + 15 - 15 > 4 - 15$	Subtract 15 from each side.
$-x > -11$	Simplify.
$x < 11$	Divide each side by -1 and reverse inequality symbol.

c.

$9x \geq 63$	Write original inequality.
$\dfrac{9x}{9} \geq \dfrac{63}{9}$	Divide each side by 9.
$x \geq 7$	Simplify.

d.

$-20 < -\dfrac{x}{5}$	Write original inequality.
$-20 \cdot (-5) > -\dfrac{x}{5} \cdot (-5)$	Multiply each side by -5 and reverse inequality symbol.
$100 > x$	Simplify.

e.

$6x - 1 \geq 9$	Write original inequality.
$6x - 1 + 1 \geq 9 + 1$	Add 1 to each side.
$6x \geq 10$	Simplify.
$\dfrac{6x}{6} \geq \dfrac{10}{6}$	Divide each side by 6.
$x \geq \dfrac{5}{3}$	Simplify.

f.

$-12x + 5 > 47$	Write original inequality.
$-12x + 5 - 5 > 47 - 5$	Subtract 5 from each side.
$-12x > 42$	Simplify.
$x < -\dfrac{7}{2}$	Divide each side by -12 and reverse inequality symbol.

SUMMARY AND STUDY STRATEGIES

After studying this chapter, you should have acquired the following skills.
The exercise numbers are keyed to the Review Exercises that begin on page 152.
Answers to odd-numbered Review Exercises are given in the back of the text.*

Section 1.1	Review Exercises
■ Classify an equation as an identity or a conditional equation.	*1–4*
■ Determine whether a given value is a solution of an equation.	*5–8*
■ Solve a linear equation in one variable.	*9–12*
A linear equation can be written in the standard form: $ax + b = 0$.	
■ Recognize an equation with no solution or infinitely many solutions.	*13–16*
■ Solve an equation involving fractional expressions.	*17–20*
■ Use a calculator to solve an equation.	*21–24*
■ Use a linear model to solve an application problem.	*25, 26*

Section 1.2

	Review Exercises
■ Construct a mathematical model from a verbal model.	*27–30*
■ Model and solve percent and distance problems.	*31–36*
■ Model and solve mixture problems.	*37, 38*
■ Use common formulas to solve geometry and simple interest problems.	*39–42*

Square: $A = s^2$, $\quad$ Rectangle: $A = lw$,
$\qquad P = 4s$ $\qquad\qquad\qquad P = 2l + 2w$
Circle: $A = \pi r^2$, $\quad$ Triangle: $A = \frac{1}{2}bh$,
$\qquad C = 2\pi r$ $\qquad\qquad\qquad P = a + b + c$
Cube: $V = s^3$
Rectangular Solid: $V = lwh$
Circular Cylinder: $V = \pi r^2 h$
Sphere: $V = \frac{4}{3}\pi r^3$
Temperature: $F = \frac{9}{5}C + 32$
Simple Interest: $I = Prt$
Distance: $d = rt$

Section 1.3

	Review Exercises
■ Solve a quadratic equation by factoring.	*43–46*
A quadratic equation can be written in the general form: $ax^2 + bx + c = 0$.	
Zero-Factor Property: If $ab = 0$, then $a = 0$ or $b = 0$.	
■ Solve a quadratic equation by extracting square roots.	*47–50*
■ Construct and use a quadratic model to solve area problems, falling-object problems, right triangle problems, and other applications.	*51–54*

---

* A wide range of valuable study aids are available to help you master the material in this chapter.
The *Student Solutions Manual* includes step-by-step solutions to all odd-numbered exercises to
help you review and prepare. The student website at *www.cengagebrain.com* offers algebra help
and a *Graphing Technology Guide,* which contains step-by-step commands and instructions for a
wide variety of graphing calculators.

Section 1.4

■ Use the discriminant to determine the number of real solutions of a
quadratic equation.

If $b^2 - 4ac > 0$, then the equation has two distinct real solutions.

If $b^2 - 4ac = 0$, then the equation has one repeated real solution.

If $b^2 - 4ac < 0$, then the equation has no real solutions.

■ Solve a quadratic equation using the Quadratic Formula.

Quadratic Formula: $x = \dfrac{-b \pm \sqrt{b^2 - 4ac}}{2a}$

■ Use the Quadratic Formula to solve an application problem.

Section 1.5

■ Solve a polynomial equation by factoring.

■ Solve an equation of quadratic type.

■ Rewrite and solve an equation involving radicals or rational exponents.

■ Rewrite and solve an equation involving fractions or absolute value.

■ Construct and use a nonquadratic model to solve an application problem.

■ Solve a compound interest problem.

Section 1.6

■ Write bounded and unbounded intervals using inequalities or interval notation.

■ Solve and graph a linear inequality.

■ Solve and graph inequalities involving absolute value.

$|x| < a$ if and only if $-a < x < a$

$|x| > a$ if and only if $x < -a$ or $x > a$

■ Construct and use a linear inequality to solve an application problem.

Section 1.7

■ Use critical numbers to determine test intervals for a polynomial inequality.

■ Solve and graph a polynomial inequality.

■ Solve and graph a rational inequality.

■ Determine the domain of an expression involving a radical.

■ Construct and use a polynomial inequality to solve an application problem.

Study Strategies

■ **Check Your Answers**   Because of the number of steps involved in solving an
equation or inequality, there are many ways to make mistakes. So, always check
your answers. In some cases, you may even want to check your answers in more
than one way, just to be sure.

■ **Using Test Intervals**   Make sure that you understand how to use critical numbers
to determine test intervals for inequalities. The logic and mathematical reasoning
involved in this concept can be applied in many real-life situations.

Review Exercises

See www.CalcChat.com for worked-out solutions to odd-numbered exercises.

Classifying Equations In Exercises 1–4, determine whether the equation is an identity or a conditional equation.

1. $5(x - 3) = 2x + 9$ **2.** $3(x + 2) = 3x + 6$

3. $4(x + 2) = 4x + 8$ **4.** $2(x + 3) = 3x + 2$

Checking Solutions In Exercises 5–8, determine whether each value of x is a solution of the equation.

5. $5(x - 1) = 3x + 1$

  (a) $x = 0$ (b) $x = -1$ (c) $x = 3$ (d) $x = 2$

6. $3x^2 + 7x + 5 = x^2 + 9$

  (a) $x = 0$ (b) $x = \frac{1}{2}$ (c) $x = -4$ (d) $x = -1$

7. $6 + \dfrac{3}{x - 4} = 5$

  (a) $x = 5$ (b) $x = 0$ (c) $x = -2$ (d) $x = 7$

8. $\sqrt{x + 4} = 3$

  (a) $x = 0$ (b) $x = -1$ (c) $x = 5$ (d) $x = 12$

Solving an Equation In Exercises 9–16, solve the equation and check your solution. (Some equations have no solution.)

9. $x + 7 = 20$ **10.** $2x + 15 = 43$

11. $4(x + 3) - 3 = 2(4 - 3x) - 4$

12. $(x + 3) + 2(x - 4) = 5(x + 3)$

13. $5x + x = 6x + 3$ **14.** $4 - x = 2x - 3x + 1$

15. $8(x - 2) = 8x - 16$ **16.** $-(3 - 2x) = 2x - 3$

An Equation Involving Fractional Expressions In Exercises 17–20, solve the equation and check your solution.

17. $\dfrac{3x - 2}{5x - 1} = \dfrac{3}{4}$

18. $\dfrac{3}{x - 4} + \dfrac{8}{2x + 5} = \dfrac{11}{2x^2 - 3x - 20}$

19. $\dfrac{x}{x + 3} - \dfrac{4}{x + 3} + 2 = 0$

20. $7 - \dfrac{3}{x} = 8 + \dfrac{5}{x}$

Using a Calculator to Solve an Equation In Exercises 21–24, use a calculator to solve the equation. (Round your solution to three decimal places.)

21. $0.375x - 0.75(300 - x) = 200$

22. $0.235x + 2.6(-x - 4) = 30$

23. $\dfrac{x}{0.055} + \dfrac{x}{0.085} = 1$ **24.** $\dfrac{x}{0.0645} + \dfrac{x}{0.098} = 2$

25. **Stores** The numbers N of Safeway stores in operation from 2003 through 2009 can be approximated by the linear equation

$$N = -15.5t + 1859, \quad 3 \le t \le 9$$

where t represents the year, with $t = 3$ corresponding to 2003. Use the model to estimate the year in which the number of Safeway stores decreased to 1720. *(Source: Safeway, Inc.)*

26. **Federal Programs** The numbers of students S (in millions) participating in the school breakfast program from 2001 through 2009 can be approximated by the linear equation

$$S = 0.42t + 7.3, \quad 1 \le t \le 9$$

where t represents the year, with $t = 1$ corresponding to 2001. Use the model to estimate the year in which 9.4 million students participated in the school breakfast program. *(Source: U.S. Department of Agriculture, Food and Nutrition Service)*

Constructing a Mathematical Model In Exercises 27–30, use the verbal description to (a) write a verbal model, (b) assign labels to the quantities in the verbal model, (c) use the labels to write a mathematical model, and (d) solve the problem.

27. Three consecutive even integers have a sum of 42. Find the smallest of these integers.

28. One positive number is four times another number. The difference of the numbers is 51. Find the numbers.

29. **Annual Salary** Your annual salary is $28,900. You receive a 7% raise. What is your new annual salary?

30. **Salary** Your weekly salary increases from $550 to $594. What percent is your raise?

31. **List Price** The price of an outdoor barbecue grill is discounted 15%. The sale price is $139. Find the original list price of the grill.

32. **Discount Rate** The price of a home gym is discounted by $300. The sale price is $599.99. What percent of the original price is the discount?

33. **Travel Time** Two cars start at the same time at a given point and travel in the same direction at average speeds of 45 miles per hour and 50 miles per hour. After how long are the cars 10 miles apart?

34. **Exercise** Two bicyclists start at the same time at a given point and travel in the same direction at average speeds of 8 miles per hour and 10 miles per hour. After how long are the bicyclists 5 miles apart?

35. Box Office James Cameron's 1997 movie *Titanic* earned U.S. gross revenue of $600,788,188. Cameron's 2009 movie *Avatar* earned U.S. gross revenue of $760,507,625. Find the percent increase in the U.S. gross revenue from *Titanic* to *Avatar*. *(Source: Nash Information Services)*

36. Oil Imports The United States imported 1594 million barrels of crude oil from members of OPEC (Organization of Petroleum Exporting Countries) in 2009. Use the bar graph to determine the amount imported from each of the five top contributing countries. *(Source: U.S. Energy Information Administration)*

37. Mixture A car radiator contains 10 quarts of a 10% antifreeze solution. The car's owner wishes to create a 10-quart solution that is 30% antifreeze. How many quarts will have to be replaced with pure antifreeze?

38. Mixture A three-gallon acid solution contains 3% boric acid. How many gallons of 20% boric acid solution should be added to make a final solution that is 8% boric acid?

39. Geometry A volleyball court is twice as long as it is wide, and its perimeter is 177 feet. Find the dimensions of the volleyball court.

40. Geometry A room is 1.25 times as long as it is wide, and its perimeter is 90 feet. Find the dimensions of the room.

41. Simple Interest You deposit $500 in a savings account earning 4% simple interest. How much interest will you earn in 1 year?

42. Simple Interest You deposit $800 in a money market account. One year later the account balance is $814. What was the simple interest rate?

Solving a Quadratic Equation by Factoring In Exercises 43–46, solve the quadratic equation by factoring. Check your solutions.

43. $6x^2 = 5x + 4$

44. $-x^2 = 15x + 36$

45. $x^2 - 11x + 24 = 0$

46. $4 - 4x + x^2 = 0$

Extracting Square Roots In Exercises 47–50, solve the quadratic equation by extracting square roots. When a solution is irrational, list both the exact solution and its approximation rounded to two decimal places.

47. $x^2 = 11$

48. $16x^2 = 25$

49. $(x + 4)^2 = 18$

50. $(x - 1)^2 = 5$

51. Geometry A billboard is 12 feet longer than it is high. The billboard has 405 square feet of advertising space. What are the dimensions of the billboard?

52. Grand Canyon The Grand Canyon is 6000 feet deep at its deepest part. A rock is dropped over the deepest part of the canyon. How long does the rock take to hit the water in the Colorado River below?

53. Total Revenue The demand equation for a product is

$$p = 60 - 0.0001x$$

where p is the price per unit and x is the number of units sold. The total revenue R from selling x units is given by

$$R = xp = x(60 - 0.0001x).$$

How many units must be sold to produce a revenue of $8,000,000?

54. Depth of an Underwater Cable A ship's sonar locates a cable 2000 feet from the ship (see figure). The angle between the surface of the water and a line from the ship to the cable is 45°. How deep is the cable?

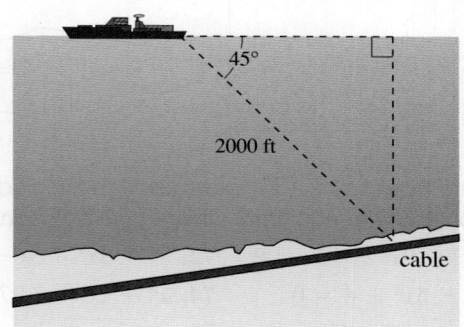

Using the Discriminant In Exercises 55–58, use the discriminant to determine the number of real solutions of the quadratic equation.

55. $x^2 + 11x + 24 = 0$

56. $x^2 + 5x + 12 = 0$

57. $3x^2 - 4x + 2 = 0$

58. $4x^2 + 12x + 9 = 0$

Using the Quadratic Formula In Exercises 59–64, use the Quadratic Formula to solve the quadratic equation. Check your solutions.

59. $x^2 - 12x + 30 = 0$

60. $5x^2 + 16x - 12 = 0$

61. $(y + 7)^2 = -5y$

62. $6x = 7 - 2x^2$

63. $x^2 + 6x - 3 = 0$

64. $10x^2 - 11x = 2$

Using a Calculator   In Exercises 65–68, use a calculator to solve the quadratic equation. (Round your answers to three decimal places.)

65. $3.6x^2 - 5.7x - 1.9 = 0$

66. $2.3x^2 + 6.6x - 3.9 = 0$

67. $34x^2 - 296x + 47 = 0$

68. $39x^2 + 75x - 21 = 0$

69. On the Moon   An astronaut standing on the edge of a cliff on the moon drops a rock over the cliff. The height s of the rock after t seconds is given by

$$s = -2.7t^2 + 200.$$

The rock's initial velocity is 0 feet per second and its initial height is 200 feet. Determine how long it will take the rock to hit the lunar surface. How long does it take a rock dropped from a height of 200 feet on Earth to hit the ground?

70. Geometry   An open box is made from a square piece of material by cutting three-inch squares from the corners and turning up the sides (see figure). The volume of the finished box is 363 cubic inches. Find the size of the original piece of material.

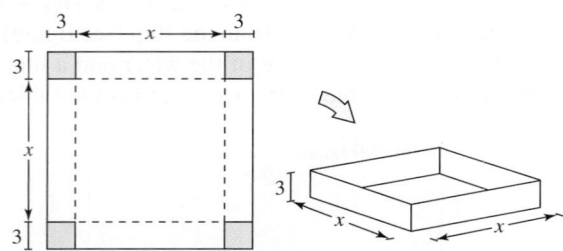

Solving Equations   In Exercises 71–84, find the real solution(s) of the equation. Check your solution(s).

71. $3x^3 - 9x^2 - 12x = 0$

72. $x^4 + 3x^3 - 5x - 15 = 0$

73. $x^4 - 5x^2 + 4 = 0$

74. $x^6 + 26x^3 - 27 = 0$

75. $2\sqrt{x} - 5 = 0$

76. $\sqrt{3x - 2} + x = 4$

77. $2\sqrt{x - 3} - 4 = 3x$

78. $\sqrt[3]{3x + 5} = 5$

79. $(x^2 - 5)^{2/3} = 9$

80. $(x^2 - 5x - 6)^{4/3} = 16$

81. $|5x + 4| = 11$

82. $|x^2 + 4x| - 2x = 8$

83. $\dfrac{5}{x + 1} + \dfrac{3}{x + 3} = 1$

84. $x + \dfrac{3}{x + 2} = 2$

85. Sharing the Cost   Three students are planning to share the expense of renting a condominium at a resort for 1 week. By adding a fourth person to the group, each person could save $75 in rental fees. How much is the rent for the week?

86. Sharing the Cost   A college charters a bus for $1800 to take a group to a museum. When four more students join the trip, the cost per student decreases by $5. How many students were in the original group?

87. Market Research   The demand equation for a product is given by

$$p = 45 - \sqrt{0.002x + 1}$$

where x is the number of units demanded per day and p is the price per unit. Find the demand when the price is set at $19.95.

88. Cash Advance   You take out a cash advance of $500 on a credit card. After 3 months, the amount you owe is $525.29. What is the annual percentage rate for this cash advance? (Assume that the interest is compounded monthly and that you have made no payments.)

Intervals and Inequalities   In Exercises 89–92, write an inequality that represents the interval. Then state whether the interval is bounded or unbounded.

89. $[-4, 7]$

90. $(6, 1000)$

91. $(-\infty, 1000]$

92. $(-10, \infty)$

Solving an Inequality   In Exercises 93–98, solve the inequality and graph the solution set on the real number line.

93. $3(x - 1) < 2x + 8$

94. $-5 \le 2 - 4(x + 2) \le 6$

95. $-3 < \dfrac{2x + 1}{4} < 3$

96. $-1 \le -5 - 3x < 4$

97. $|x + 10| + 3 < 5$

98. $|2x - 3| - 4 > 2$

99. Break-Even Analysis   The revenue R from selling x units of a product is

$$R = 89.95x.$$

The cost C of producing x units is

$$C = 35x + 2500.$$

For what values of x will this product return a profit?

100. Accuracy of Measurement   You buy a 16-inch gold chain that costs $9.95 per inch. The chain is measured accurately to within $\frac{1}{16}$ inch. How much money might you have been undercharged or overcharged?

Finding Test Intervals   In Exercises 101–104, find the test intervals of the inequality.

101. $x^2 - 1 < 0$

102. $x^2 - 3x - 4 \ge 0$

103. $\dfrac{x + 3}{x - 1} \le 2$

104. $\dfrac{x + 5}{x + 7} > 3$

Solving an Inequality   In Exercises 105–110, solve the inequality and graph the solution set on the real number line.

105. $5(x + 1)(x - 3) < 0$

106. $(x + 4)^2 \le 4$

107. $x^3 - 9x < 0$

108. $\dfrac{x + 5}{x + 8} \ge 2$

109. $\dfrac{2 + 3x}{4 - x} < 2$

110. $\dfrac{1}{x + 1} \geq \dfrac{1}{x + 5}$

Using a Calculator   In Exercises 111–114, use a calculator to solve the inequality. (Round each number in your answer to two decimal places.)

111. $-1.2x^2 + 4.76 > 1.32$

112. $3.5x^2 + 4.9x - 6.1 < 2.4$

113. $\dfrac{1}{3.7x - 6.1} > 2.9$

114. $\dfrac{3}{5.4x - 2.7} < 8.9$

Finding the Domain of an Expression   In Exercises 115–118, find the domain of the expression.

115. $\sqrt{x - 10}$

116. $\sqrt[4]{2x + 5}$

117. $\sqrt{x^2 - 15x + 54}$

118. $\sqrt{81 - 4x^2}$

119. Height of a Flare   A flare is fired straight upward from ground level with an initial velocity of 100 feet per second. During what time period will its height exceed 150 feet?

120. Path of a Soccer Ball   A soccer player kicks the ball toward the center of the goal. The path of the ball can be modeled by

$$y = -0.054x^2 + 1.43x$$

where x is the horizontal distance (in feet) from where the ball is kicked and y is the corresponding height (in feet).

(a) The soccer goal is 8 feet high. Write an inequality to determine the values of x for which the ball is low enough to go into the goal.

(b) Solve the inequality from part (a).

(c) The player kicks the ball 15 feet from the goal and the ball is not blocked. Does the player score a goal? Explain your reasoning.

121. Geometry   A rectangular field with a perimeter of 80 meters is to have an area of at least 380 square meters (see figure). Describe the possible lengths l of the field.

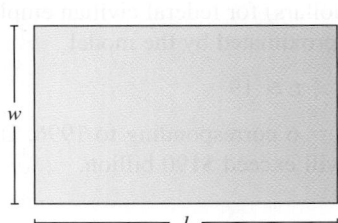

122. Compound Interest P dollars, invested at interest rate r compounded annually, increases to an amount A given by

$$A = P(1 + r)^5$$

in 5 years. An investment of $1000 increases to an amount greater than $1400 in 5 years. The interest rate must be greater than what percent?

123. Compound Interest P dollars, invested at an interest rate r compounded semiannually, increases to an amount

$$A = P(1 + r/2)^{2 \cdot 8}$$

in 8 years. An investment of $2000 increases to an amount greater than $2800 in 8 years. The interest rate must be greater than what percent?

124. Company Profits   The revenue R and cost C for a product are given by

$$R = x(75 - 0.0005x)$$

and

$$C = 25x + 100{,}000$$

where R and C are measured in dollars and x represents the number of units sold. How many units must be sold to obtain a profit of at least $500,000?

125. Price of a Product   In Exercise 124, the revenue equation

$$R = x(75 - 0.0005x)$$

implies that the demand equation is

$$p = 75 - 0.0005x$$

where p is the price per unit. What prices per unit can the company set to obtain a profit of at least $1,000,000?

126. Income   The total investment income I (in millions of dollars) of Aflac each year from 2001 through 2009 can be approximated by the model

$$I = 6.07t^2 + 90.3t + 1448, \quad 1 \leq t \leq 9$$

where t represents the year, with $t = 1$ corresponding to 2001.   *(Source: Aflac Incorporated)*

(a) Complete the table. Round each value of I to the nearest whole number.

t	1	3	5	9
I				

(b) Aflac expected to earn investment income of greater than $3 billion in 2010. Does the model support this? Explain.

(c) Aflac predicts that the yearly investment income will be at least $3.5 billion by 2015. Does the model support this prediction? Explain.

TEST YOURSELF   See www.CalcChat.com for worked-out solutions to odd-numbered exercises.

Take this test as you would take a test in class. When you are done, check your work against the answers given in the back of the book.

1. Solve the equation $3(x + 2) - 8 = 4(2 - 5x) + 7$.

2. Find the domain of (a) $\sqrt[3]{2x + 3}$ and (b) $\sqrt{9 - x^2}$.

3. In May, the total profit for a company was 8% less than it was in April. The total profit for the 2 months was $625,509.12. Find the profit for each month.

In Exercises 4–13, solve the equation. Check your solution(s).

4. *Factoring:* $6x^2 + 7x = 5$

5. *Factoring:* $12 + 5x - 2x^2 = 0$

6. *Extracting square roots:* $x^2 - 5 = 10$

7. *Quadratic Formula:* $(x + 5)^2 = -3x$

8. *Quadratic Formula:* $3x^2 - 11x = 2$

9. *Quadratic Formula:* $5.4x^2 - 3.2x - 2.5 = 0$

10. $|2x - 3| = 10$

11. $\sqrt{x - 3} + x = 5$

12. $x^4 - 10x^2 + 9 = 0$

13. $(x^2 - 9)^{2/3} = 9$

14. The demand equation for a product is $p = 40 - 0.0001x$, where p is the price per unit and x is the number of units sold. The total revenue R from selling x units is given by $R = xp$. How many units must be sold to produce a revenue of $2,000,000? Explain your reasoning.

In Exercises 15–18, solve the inequality and graph the solution set on the real number line.

15. $\dfrac{3x + 1}{5} < 2$

16. $|4 - 5x| \geq 24$

17. $\dfrac{x + 3}{x + 7} > 2$

18. $3x^3 - 12x \leq 0$

19. The revenue R and cost C for a product are given by

    $R = x(90 - 0.0004x)$

    and

    $C = 25x + 300,000$

    where R and C are measured in dollars and x represents the number of units sold. How many units must be sold to obtain a profit of at least $800,000?

20. The total payroll P (in billions of dollars) for federal civilian employees each year from 1996 through 2009 can be approximated by the model

    $P = 0.145t^2 + 0.79t + 107.4,\quad 6 \leq t \leq 19$

    where t represents the year, with $t = 6$ corresponding to 1996. Use the model to predict the year when the payroll will exceed $190 billion. *(Source: U.S. Office of Personnel Management)*

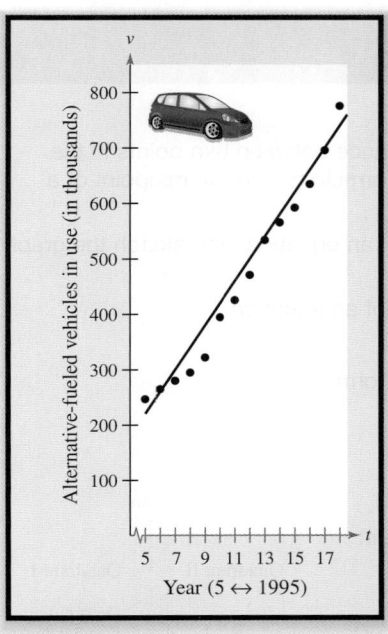

Year (5 ↔ 1995)

2 Functions and Graphs

Example 8 on page 189 shows how a scatter plot can be used to find a linear model that approximates the number of alternative-fueled vehicles in use in the United States for a given year.

GG Pro Photo/www.shutterstock.com
Kurhan/Shutterstock.com

157

2.1 Graphs of Equations

- Plot points in the Cartesian plane.
- Use the Distance Formula to find the distance between two points in the coordinate plane, and use the Midpoint Formula to find the midpoint of a line segment joining two points.
- Determine whether a point is a solution of an equation, and sketch the graph of an equation.
- Find the *x*- and *y*-intercepts of the graph of an equation.
- Determine the symmetry of a graph.
- Write the equation of a circle in standard form.

The Cartesian Plane

Just as you can represent real numbers by points on a real number line, you can represent ordered pairs of real numbers by points in a plane. This plane is called the **rectangular coordinate system,** or the **Cartesian plane,** named after the French mathematician René Descartes (1596–1650).

The Cartesian plane is formed by using two real number lines intersecting at right angles, as shown in Figure 2.1. The horizontal real number line is usually called the *x*-axis, and the vertical real number line is usually called the *y*-axis. The point of intersection of these two axes is the **origin,** and the two axes divide the plane into four parts called **quadrants.**

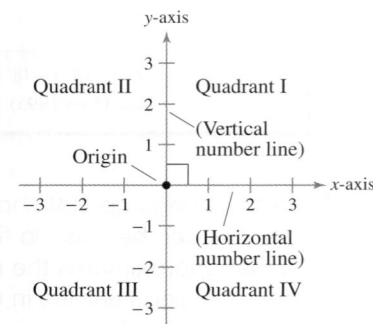

FIGURE 2.1

Each point in the plane corresponds to an **ordered pair** (x, y) of real numbers *x* and *y*, called the **coordinates** of the point. The *x*-coordinate represents the directed distance from the *y*-axis to the point, and the *y*-coordinate represents the directed distance from the *x*-axis to the point, as shown in Figure 2.2.

The notation (x, y) denotes both a point in the plane and an open interval on the real number line. The context will tell you which meaning is intended.

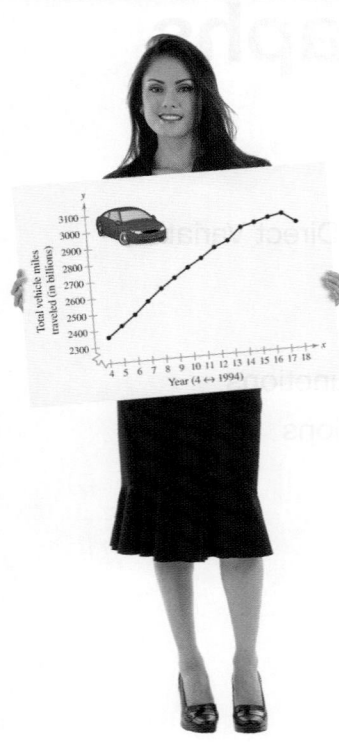

In Exercise 105 on page 170, you will use points plotted in a coordinate plane to determine the percent increase in miles traveled by vehicles over a specified period of time.

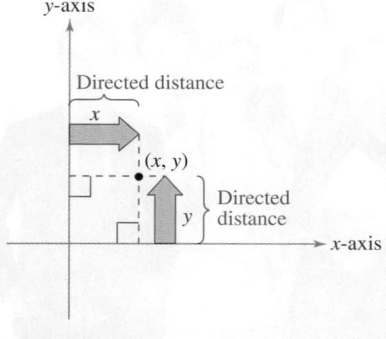

FIGURE 2.2

Example 1   **Plotting Points in the Cartesian Plane**

To plot the point $(-1, 2)$ in the Cartesian plane, imagine a vertical line through -1 on the *x*-axis and a horizontal line through 2 on the *y*-axis. The intersection of these two lines is the point $(-1, 2)$. Other points, such as $(3, 4)$, $(0, 0)$, $(3, 0)$, and $(-2, -3)$, can be plotted in a similar way, as shown in Figure 2.3.

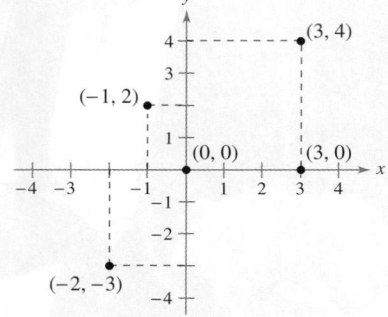

FIGURE 2.3

✓ **Checkpoint 1**

Plot the points $(0, -2)$ and $(-4, 1)$.

The Distance and Midpoint Formulas

ALGEBRA TUTOR

For help with identifying and plotting points in the Cartesian plane, see the *Chapter 2 Algebra Tutor* on pages 240 and 241.

Recall from the Pythagorean Theorem that, for a right triangle with hypotenuse of length c and legs of lengths a and b, you have

$$a^2 + b^2 = c^2$$

as shown in Figure 2.4. Note that the converse is also true. That is, if $a^2 + b^2 = c^2$, then the triangle is a right triangle.

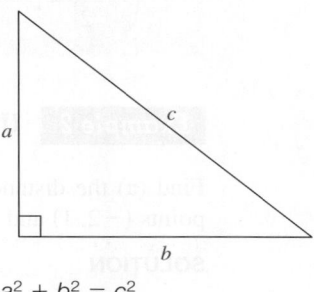

$$a^2 + b^2 = c^2$$

FIGURE 2.4

Suppose you want to determine the distance d between two points (x_1, y_1) and (x_2, y_2) that do not lie on the same horizontal or vertical line. With these two points, a right triangle can be formed, as shown in Figure 2.5.

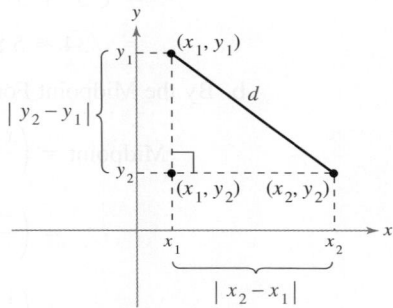

FIGURE 2.5

The length of the vertical side of the triangle is $|y_2 - y_1|$, and the length of the horizontal side is $|x_2 - x_1|$. By the Pythagorean Theorem, you can write

$$d^2 = |x_2 - x_1|^2 + |y_2 - y_1|^2$$
$$d = \sqrt{|x_2 - x_1|^2 + |y_2 - y_1|^2} \qquad \text{Choose positive square root.}$$
$$d = \sqrt{(x_2 - x_1)^2 + (y_2 - y_1)^2}.$$

The result is the **Distance Formula.**

The Distance Formula

The distance d between the points (x_1, y_1) and (x_2, y_2) in the coordinate plane is

$$d = \sqrt{(x_2 - x_1)^2 + (y_2 - y_1)^2}.$$

When using the Distance Formula, remember that assigning points to (x_1, y_1) and (x_2, y_2) is arbitrary.

The formula below shows how to find the *midpoint* of the line segment that joins two points.

The Midpoint Formula

The midpoint of the line segment joining the points (x_1, y_1) and (x_2, y_2) in the coordinate plane is

$$\left(\frac{x_1 + x_2}{2}, \frac{y_1 + y_2}{2} \right).$$

Example 2   Using the Distance and Midpoint Formulas

Find (a) the distance between, and (b) the midpoint of the line segment joining, the points $(-2, 1)$ and $(3, 4)$.

SOLUTION

Let $(x_1, y_1) = (-2, 1)$ and $(x_2, y_2) = (3, 4)$.

a. Apply the Distance Formula.

$$
\begin{aligned}
d &= \sqrt{(x_2 - x_1)^2 + (y_2 - y_1)^2} && \text{Distance Formula} \\
&= \sqrt{[3 - (-2)]^2 + (4 - 1)^2} && \text{Substitute for } x_1, x_2, y_1, \text{ and } y_2. \\
&= \sqrt{5^2 + 3^2} && \text{Simplify.} \\
&= \sqrt{34} \approx 5.83 && \text{Simplify. See Figure 2.6.}
\end{aligned}
$$

b. By the Midpoint Formula, you have

$$
\begin{aligned}
\text{Midpoint} &= \left(\frac{x_1 + x_2}{2}, \frac{y_1 + y_2}{2} \right) && \text{Midpoint Formula} \\
&= \left(\frac{-2 + 3}{2}, \frac{1 + 4}{2} \right) && \text{Substitute for } x_1, x_2, y_1, \text{ and } y_2. \\
&= \left(\frac{1}{2}, \frac{5}{2} \right). && \text{Simplify. See Figure 2.7.}
\end{aligned}
$$

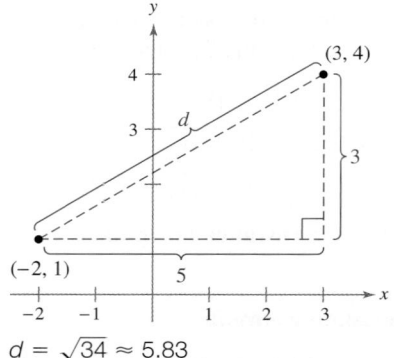

$d = \sqrt{34} \approx 5.83$

FIGURE 2.6

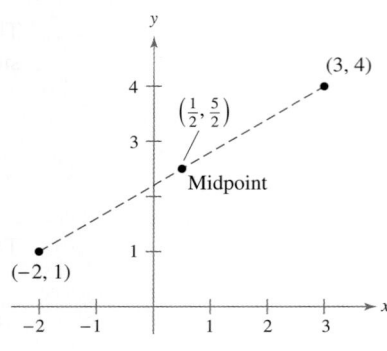

FIGURE 2.7

✓ Checkpoint 2

Find (a) the distance between, and (b) the midpoint of the line segment joining, the points $(-7, 5)$ and $(1, -1)$.

The Graph of an Equation

Frequently, a relationship between two quantities is written in the form of an equation. In the remainder of this section, you will study a procedure for sketching the graph of an equation. For an equation in the variables x and y, a point (a, b) is a **solution** if the substitutions $x = a$ and $y = b$ satisfy the equation.

Example 3   Solution of an Equation

Determine whether $(-1, 0)$ is a solution of the equation $y = 2x^2 - 4x - 6$.

SOLUTION

$$y = 2x^2 - 4x - 6 \qquad \text{Write original equation.}$$

$$0 \overset{?}{=} 2(-1)^2 - 4(-1) - 6 \qquad \text{Substitute } -1 \text{ for } x \text{ and } 0 \text{ for } y.$$

$$0 = 0 \qquad \text{Simplify.}$$

Both sides of the equation are equivalent. So, the point $(-1, 0)$ is a solution.

✓ Checkpoint 3

Determine whether $(-1, 3)$ is a solution of the equation $y = x + 4$. ■

Most equations have *infinitely* many solutions. The **graph of an equation** is the set of all points that are solutions of the equation. To sketch the graph of an equation, you can use the point-plotting method.

The Point-Plotting Method of Graphing

1. If possible, isolate one of the variables.

2. Construct a table of values showing several solution points.

3. Plot these points in a rectangular coordinate system.

4. Connect the points with a smooth curve or a line.

Step 4 of the point-plotting method can be difficult. For instance, how would you connect the four points in Figure 2.8? Without further information about the equation, any one of the three graphs in Figure 2.9 would be reasonable. These graphs show that with too few solution points, you can misrepresent the graph of an equation. Throughout this course, you will study many ways to improve your graphing techniques. For now, you should plot enough points to reveal the essential behavior of the graph. It is important to use negative values, zero, and positive values for x when constructing a table.

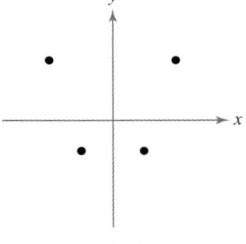

FIGURE 2.8

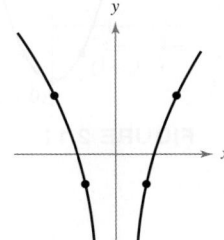 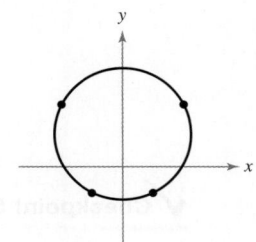 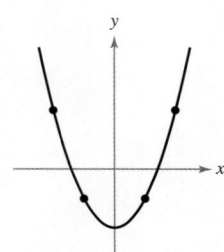

FIGURE 2.9

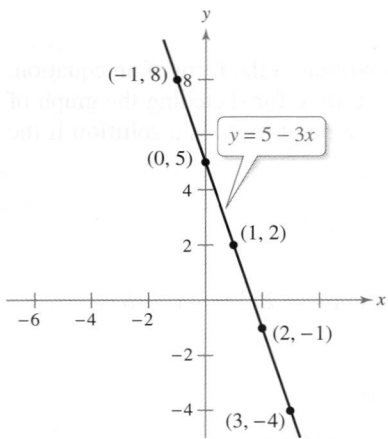

FIGURE 2.10

Example 4   Sketching the Graph of an Equation

Sketch the graph of $3x + y = 5$.

SOLUTION   First, rewrite the equation as $y = 5 - 3x$ with y isolated on the left. Next, construct a table of values by choosing several values of x and calculating the corresponding values of y.

x	-1	0	1	2	3
$y = 5 - 3x$	8	5	2	-1	-4

From the table, it follows that $(-1, 8)$, $(0, 5)$, $(1, 2)$, $(2, -1)$, and $(3, -4)$ are solution points of the equation. After plotting these points and connecting them, you can see that they appear to lie on a line, as shown in Figure 2.10.

✓ **Checkpoint 4**

Sketch the graph of each equation.

a. $y - x = 3$ **b.** $2x + y = -1$

Example 5   Sketching the Graph of an Equation

Sketch the graph of $y = x^2 - 2$.

SOLUTION   First, construct a table of values by choosing several convenient values of x and calculating the corresponding values of y.

x	-3	-2	-1	0	1	2	3
$y = x^2 - 2$	7	2	-1	-2	-1	2	7

Next, plot the corresponding solution points. Finally, connect the points with a smooth curve, as shown in Figure 2.11.

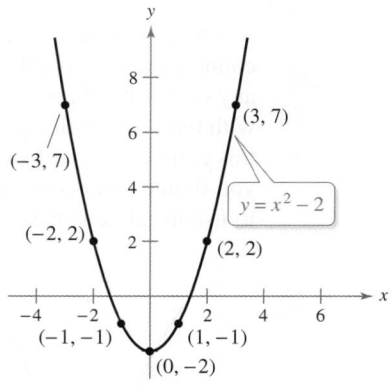

FIGURE 2.11

✓ **Checkpoint 5**

Sketch the graph of each equation.

a. $y = \frac{1}{4}x^2$ **b.** $y = 5 - x^2$

Intercepts of a Graph

When you are sketching a graph, points for which either the y-coordinate or the x-coordinate is zero are especially useful.

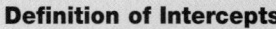

> **Definition of Intercepts**
>
> 1. The **x-intercepts** of a graph are the points at which the graph intersects the x-axis. To find the x-intercepts, let y equal zero and solve for x.
>
> 2. The **y-intercepts** of a graph are the points at which the graph intersects the y-axis. To find the y-intercepts, let x equal zero and solve for y.

Some texts denote the x-intercept as the x-coordinate of the point $(a, 0)$ rather than the point itself. Unless it is necessary to make a distinction, the term *intercept* will be used to mean either the point or the coordinate.

A graph may have no intercepts, one intercept, or several intercepts. For instance, consider the three graphs in Figure 2.12.

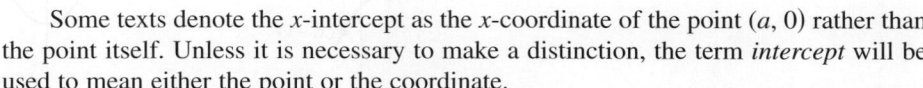

Three x-intercepts   No x-intercept   No intercepts
One y-intercept   One y-intercept

FIGURE 2.12

Example 6   **Finding x- and y-Intercepts**

Find the x- and y-intercepts of the graph of

$$y^2 - 3 = x.$$

SOLUTION   To find the x-intercept, let $y = 0$. This produces

$$-3 = x$$

which implies that the graph has one x-intercept, which occurs at

$$(-3, 0).   \text{x-intercept}$$

To find the y-intercept, let $x = 0$. This produces

$$y^2 - 3 = 0$$

which has two solutions: $y = \pm\sqrt{3}$. So, the graph has two y-intercepts, which occur at

$$\left(0, \sqrt{3}\right) \quad \text{and} \quad \left(0, -\sqrt{3}\right).   \text{y-intercepts}$$

See Figure 2.13.

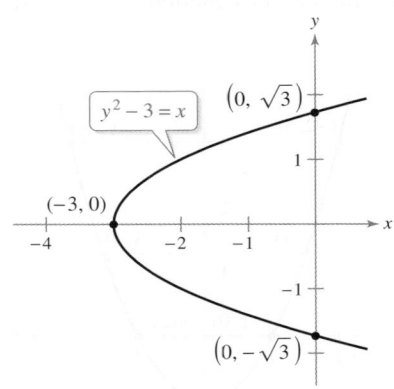
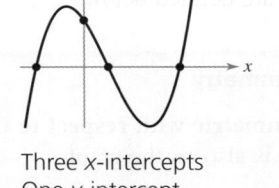

FIGURE 2.13

✓ **Checkpoint 6**

Find the x- and y-intercepts of the graph of each equation.

a. $y = 4x + 7$ **b.** $y = x^2 - 1$

Symmetry

Symmetry with respect to the x-axis means that if the Cartesian plane is folded along the x-axis, then the portion of the graph above the x-axis would coincide with the portion below the x-axis. Symmetry with respect to the y-axis can be described in a similar manner. Symmetry with respect to the origin means that the graph rotates $180°$ about the origin, it looks the same. (See Figure 2.14.)

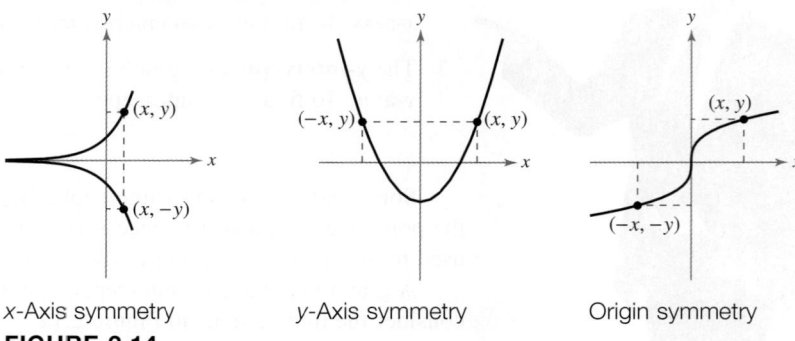

x-Axis symmetry   y-Axis symmetry   Origin symmetry

FIGURE 2.14

Knowing the symmetry of a graph *before* attempting to sketch it is helpful, because then you need only half as many solution points to sketch the graph. The three basic types of symmetry are defined below.

Definition of Symmetry

1. A graph is **symmetric with respect to the x-axis** if, whenever (x, y) is on the graph, $(x, -y)$ is also on the graph.

2. A graph is **symmetric with respect to the y-axis** if, whenever (x, y) is on the graph, $(-x, y)$ is also on the graph.

3. A graph is **symmetric with respect to the origin** if, whenever (x, y) is on the graph, $(-x, -y)$ is also on the graph.

You can apply this definition of symmetry to the graph of the equation $y = x^2 - 1$. Replacing x with $-x$ produces

$y = x^2 - 1$ Write original equation.

$y = (-x)^2 - 1$ Replace x with $-x$.

$y = x^2 - 1.$ Replacement yields equivalent equation.

Because the substitution did not change the equation, it follows that if (x, y) is a solution of the equation, then $(-x, y)$ must also be a solution. So, the graph of $y = x^2 - 1$ is symmetric with respect to the y-axis. By plotting the points in the table below, you can confirm that the graph is symmetric with respect to the y-axis, as shown in Figure 2.15.

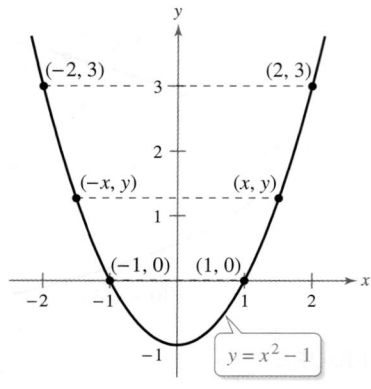

y-Axis Symmetry

FIGURE 2.15

x	-2	-1	1	2
y	3	0	0	3

Tests for Symmetry

1. The graph of an equation is symmetric with respect to the *x*-axis if replacing *y* with −*y* yields an equivalent equation.

2. The graph of an equation is symmetric with respect to the *y*-axis if replacing *x* with −*x* yields an equivalent equation.

3. The graph of an equation is symmetric with respect to the origin if replacing *x* with −*x* *and y* with −*y* yields an equivalent equation.

Example 7   Testing for Symmetry

Test $y = 2x^3$ for symmetry with respect to both axes and the origin.

SOLUTION

x-Axis:	$y = 2x^3$	Write original equation.
	$-y = 2x^3$	Replace *y* with −*y*. Result is *not* an equivalent equation.
y-Axis:	$y = 2x^3$	Write original equation.
	$y = 2(-x)^3$	Replace *x* with −*x*.
	$y = -2x^3$	Simplify. Result is *not* an equivalent equation.
Origin:	$y = 2x^3$	Write original equation.
	$-y = 2(-x)^3$	Replace *y* with −*y* and *x* with −*x*.
	$-y = -2x^3$	Simplify.
	$y = 2x^3$	Equivalent equation

Of the three tests for symmetry, the only one that is satisfied is the test for origin symmetry (see Figure 2.16).

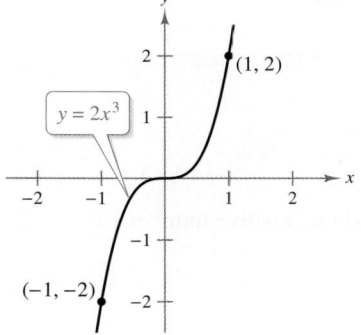

$y = 2x^3$

(1, 2)

(−1, −2)

FIGURE 2.16

✓ **Checkpoint 7**

Test $y = -4x^2$ for symmetry with respect to both axes and the origin.

Example 8   Using Symmetry as a Sketching Aid

Use symmetry to sketch the graph of $x - y^2 = 1$.

SOLUTION   Of the three tests for symmetry, the only one that is satisfied is the test for *x*-axis symmetry.

$x - y^2 = 1$	Write original equation.
$x - (-y)^2 = 1$	Replace *y* with −*y*.
$x - y^2 = 1$	Equivalent equation

So, the graph is symmetric with respect to the *x*-axis. To sketch the graph, plot the points above the *x*-axis and use symmetry to complete the graph, as shown in Figure 2.17.

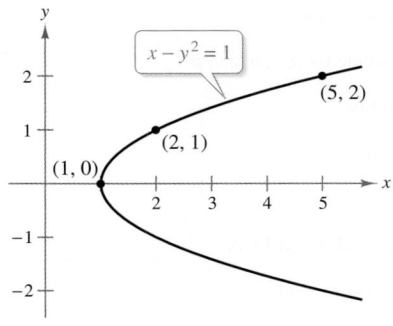

$x - y^2 = 1$

(5, 2)

(2, 1)

(1, 0)

FIGURE 2.17

✓ **Checkpoint 8**

Use symmetry to sketch the graph of each equation.

a. $y = |x|$ **b.** $y = x^3 + x$

The Equation of a Circle

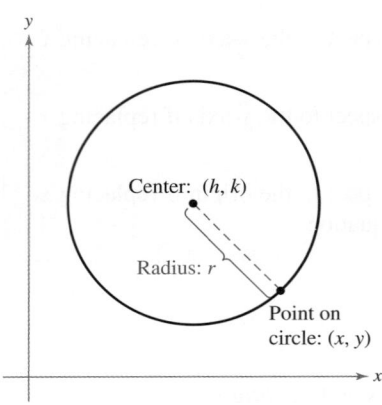

FIGURE 2.18

In this section, you have studied the point-plotting method and two additional concepts (intercepts and symmetry) that can be used to streamline the graphing procedure. Another graphing aid is *equation recognition*, which is the ability to recognize the general shape of a graph simply by looking at its equation.

Figure 2.18 shows a circle of *radius r* with *center* at the point (h, k). The point (x, y) is on this circle if and only if its distance from the center (h, k) is r. This means that a **circle** in the plane consists of all points (x, y) that are a given positive distance r from a fixed point (h, k). Using the Distance Formula, you can conclude that the point (x, y) lies on the circle if and only if

$$\sqrt{(x - h)^2 + (y - k)^2} = r.$$

By squaring each side of this equation, you obtain the **standard form of the equation of a circle.** For example, a circle with its center at the origin, $(h, k) = (0, 0)$, and radius $r = 4$ is given by

$$\sqrt{(x - 0)^2 + (y - 0)^2} = 4 \qquad \text{Substitute for } h, k, \text{ and } r.$$

$$\sqrt{x^2 + y^2} = 4 \qquad \text{Simplify.}$$

$$x^2 + y^2 = 16. \qquad \text{Square each side.}$$

STUDY TIP

The standard form of the equation of a circle with radius r whose center is at the *origin* is

$$x^2 + y^2 = r^2.$$

Standard Form of the Equation of a Circle

The **standard form of the equation of a circle** is

$$(x - h)^2 + (y - k)^2 = r^2.$$

The point (h, k) is called the **center** of the circle, and the positive number r is called the **radius** of the circle.

Example 9   Finding the Equation of a Circle

The point $(3, 4)$ lies on a circle whose center is at $(-1, 2)$, as shown in Figure 2.19. Write the standard form of the equation of this circle.

SOLUTION   The radius of the circle is the distance between the center $(-1, 2)$ and the point $(3, 4)$.

$$r = \sqrt{(x - h)^2 + (y - k)^2} \qquad \text{Distance Formula}$$

$$r = \sqrt{[3 - (-1)]^2 + (4 - 2)^2} \qquad \text{Substitute for } x, y, h, \text{ and } k.$$

$$= \sqrt{4^2 + 2^2} \qquad \text{Simplify.}$$

$$= \sqrt{16 + 4} \qquad \text{Simplify.}$$

$$= \sqrt{20} \qquad \text{Radius}$$

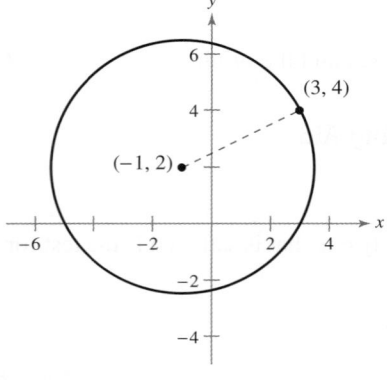

FIGURE 2.19

Using $(h, k) = (-1, 2)$ and $r = \sqrt{20}$, the equation of the circle is

$$(x - h)^2 + (y - k)^2 = r^2 \qquad \text{Equation of circle}$$

$$[x - (-1)]^2 + (y - 2)^2 = \left(\sqrt{20}\right)^2 \qquad \text{Substitute for } h, k, \text{ and } r.$$

$$(x - 1)^2 + (y - 2)^2 = 20. \qquad \text{Standard form}$$

✓ **Checkpoint 9**

The point $(4, 4)$ lies on a circle whose center is at $(0, 1)$. Write the standard form of the equation of this circle. ■

When you expand the standard equation in Example 9, you obtain the result shown.

$$(x + 1)^2 + (y - 2)^2 = 20 \qquad \text{Standard form}$$

$$x^2 + 2x + 1 + y^2 - 4y + 4 = 20 \qquad \text{Expand terms.}$$

$$x^2 + y^2 + 2x - 4y - 15 = 0 \qquad \text{General form}$$

The last equation is in the **general form of the equation of a circle,**

$$Ax^2 + Ay^2 + Dx + Ey + F = 0, \quad A \neq 0.$$

The general form is less useful than the standard form. For instance, it is not immediately apparent from the general form shown above that the center is $(-1, 2)$ and the radius is $\sqrt{20}$. To graph the equation of a circle, write the equation in standard form. You can do this by **completing the square,** as shown in Example 10.

STUDY TIP

Recall that to complete the square, you add the square of half the coefficient of the linear term to each side.

Example 10   **Completing the Square to Sketch a Circle**

Sketch the circle given by $4x^2 + 4y^2 + 20x - 16y + 37 = 0$.

SOLUTION   Begin by writing the original equation in standard form by completing the square for both the x-terms *and* the y-terms.

$$4x^2 + 4y^2 + 20x - 16y + 37 = 0 \qquad \text{Write original equation.}$$

$$x^2 + y^2 + 5x - 4y + \frac{37}{4} = 0 \qquad \text{Divide by 4.}$$

$$\left(x^2 + 5x + \boxed{}\right) + \left(y^2 - 4y + \boxed{}\right) = -\frac{37}{4} \qquad \text{Group terms.}$$

$$\left[x^2 + 5x + \left(\frac{5}{2}\right)^2\right] + (y^2 - 4y + 2^2) = -\frac{37}{4} + \frac{25}{4} + 4 \qquad \text{Complete the square.}$$

$$\left(x + \frac{5}{2}\right)^2 + (y - 2)^2 = 1 \qquad \text{Standard form}$$

So, the center of the circle is $\left(-\frac{5}{2}, 2\right)$ and the radius of the circle is 1. Using this information, you can sketch the circle, as shown in Figure 2.20. ━━━

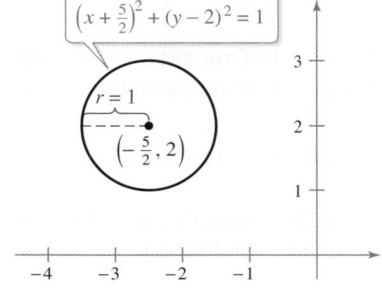

$\left(x + \frac{5}{2}\right)^2 + (y - 2)^2 = 1$

$r = 1$

$\left(-\frac{5}{2}, 2\right)$

FIGURE 2.20

✓**Checkpoint 10**

Sketch the circle given by $x^2 + y^2 - 2x - 4y + 1 = 0$.   ■

SUMMARIZE   (Section 2.1)

1. Describe how to plot a point in the Cartesian plane *(page 158)*. For an example of plotting points, see Example 1.

2. State the Distance Formula and the Midpoint Formula *(pages 159 and 160)*. For an example that uses these formulas, see Example 2.

3. Describe how to sketch the graph of an equation *(pages 161 and 162)*. For examples of graphing equations, see Examples 4 and 5.

4. State the definitions of x-intercept and y-intercept *(page 163)*. For an example of finding x- and y-intercepts, see Example 6.

5. State the standard form of the equation of a circle *(page 166)*. For an example of finding the standard form of the equation of a circle, see Example 9.

SKILLS WARM UP 2.1   The following warm-up exercises involve skills that were covered in earlier sections. You will use these skills in the exercise set for this section. For additional help, review Sections 0.2, 0.4, and 1.5.

In Exercises 1–6, simplify the expression.

1. $\sqrt{(2-6)^2 + [1-(-2)]^2}$

2. $\sqrt{(1-4)^2 + (-2-1)^2}$

3. $\dfrac{4 + (-2)}{2}$

4. $\dfrac{-1 + (-3)}{2}$

5. $\sqrt{18} + \sqrt{45}$

6. $\sqrt{12} + \sqrt{44}$

In Exercises 7–10, solve the equation.

7. $\sqrt{(4-x)^2 + (5-2)^2} = \sqrt{58}$

8. $\sqrt{(8-6)^2 + (y-5)^2} = 2\sqrt{5}$

9. $x^3 - 9x = 0$

10. $x^4 - 8x^2 + 16 = 0$

Exercises 2.1

See www.CalcChat.com for worked-out solutions to odd-numbered exercises.

Plotting Points in the Cartesian Plane   In Exercises 1 and 2, plot the points in the Cartesian plane. *See Example 1.*

1. $(-5, 3), (1, -1), (-2, -4), (2, 0), (1, -6)$

2. $(0, -4), (5, 1), (-3, 5), (2, -2), (-6, -1)$

Using the Distance and Midpoint Formulas   In Exercises 3–12, (a) plot the points, (b) find the distance between the points, and (c) find the midpoint of the line segment joining the points. *See Examples 1 and 2.*

3. $(2, -5), (-6, 1)$

4. $(1, 12), (6, 0)$

5. $(3, -11), (-12, -3)$

6. $(-7, 3), (2, -9)$

7. $(-1, 2), (5, 4)$

8. $(2, 10), (10, 2)$

9. $\left(\frac{1}{2}, 1\right), \left(-\frac{3}{2}, -5\right)$

10. $\left(-\frac{1}{3}, -\frac{1}{3}\right), \left(-\frac{1}{6}, -\frac{1}{2}\right)$

11. $(1.8, 7.5), (-2.5, 2.1)$

12. $(5.2, 6.4), (-2.7, 1.8)$

Finding the Hypotenuse   In Exercises 13–16, find the length of the hypotenuse in two ways: (a) using the Pythagorean Theorem and (b) using the Distance Formula.

13.
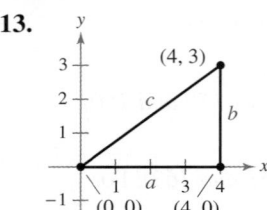

14.
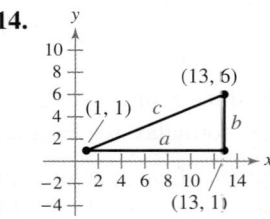

15.
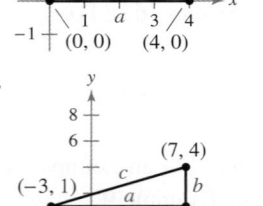

16.
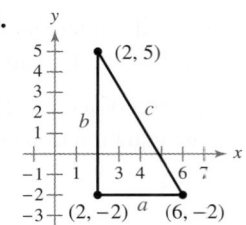

Using the Distance Formula   In Exercises 17 and 18, find the values of *x* for which the distance between the points is 15.

17. $(3, -4), (x, 5)$

18. $(x, 8), (-9, -4)$

Using the Distance Formula   In Exercises 19 and 20, find the values of *y* for which the distance between the points is 20.

19. $(-15, y), (-3, -7)$

20. $(6, -1), (-10, y)$

Solution of an Equation   In Exercises 21–24, determine whether each point is a solution of the equation. *See Example 3.*

Equation	Points	
21. $2x - 3y + 11 = 0$	(a) $(2, 5)$	(b) $(3, 2)$
22. $y = 2x^2 - 7x + 3$	(a) $(1, -1)$	(b) $(3, 0)$
23. $y = \sqrt{x - 5}$	(a) $(9, 2)$	(b) $(21, 4)$
24. $y = \dfrac{x + 1}{5 - x}$	(a) $\left(1, \frac{1}{2}\right)$	(b) $(0, 1)$

Sketching the Graph of an Equation   In Exercises 25–28, complete the table below for the given equation. Use the resulting solution points to sketch the graph of the equation. *See Examples 4 and 5.*

x	-3	-2	-1	0	1	2	3
y							

25. $y = \frac{3}{4}x - 1$

26. $\frac{3}{2}x + y = 2$

27. $y = x^2 - 4$

28. $y = \frac{1}{2}x^2$

Finding x- and y-Intercepts In Exercises 29–36, find the x- and y-intercepts of the graph of the equation. *See Example 6.*

29. $y = 2x - 1$

30. $2x = -y - 6$

31. $y = x^2 + x - 2$

32. $y = 4 - x^2$

33. $y = \sqrt{4 - x^2}$

34. $y = \sqrt{x^2 + 9}$

35. $2y - xy + 3x = 4$

36. $x^2y - x^2 + 4y = 0$

37. Think About It Use your knowledge of the Cartesian plane and intercepts to explain why you let y equal zero when you are finding the x-intercepts of the graph of an equation, and why you let x equal zero when you are finding the y-intercepts of the graph of an equation.

38. Think About It Is it possible for a graph to have no x-intercepts? no y-intercepts? no x-intercepts and no y-intercepts? Give examples to support your answers.

Testing for Symmetry In Exercises 39–50, test for symmetry with respect to both axes and the origin. *See Example 7.*

39. $x^4 - 2y = 0$

40. $y = x^4 - x^2 + 3$

41. $x - y^2 = 0$

42. $y^2 = x + 2$

43. $y = \sqrt{16 - x^2}$

44. $y = \sqrt{4 - x^2}$

45. $xy = 2$

46. $x^3y = 1$

47. $y = \dfrac{x}{x^2 - 4}$

48. $y = \dfrac{x}{x^2 + 1}$

49. $x^2 + y^2 = 25$

50. $x^2 + y^2 = 9$

Using Symmetry as a Sketching Aid In Exercises 51–54, use symmetry to complete the graph of the equation. *See Example 8.*

51. y-axis symmetry
$y = -x^2 + 4$

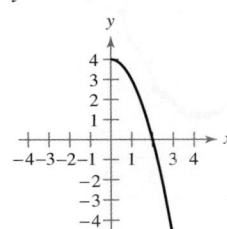

52. x-axis symmetry
$y^2 = -x + 4$

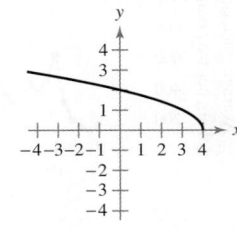

53. Origin symmetry
$y = -x^3 + x$

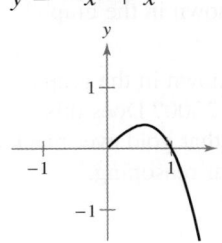

54. y-axis symmetry
$y = |x| - 2$

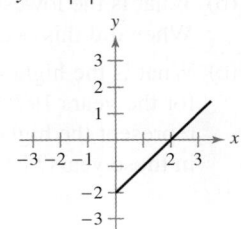

Matching In Exercises 55–60, match the equation with its graph. [The graphs are labeled (a), (b), (c), (d), (e), and (f).]

(a)
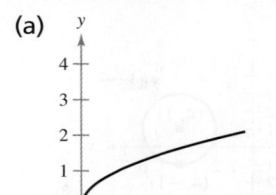

(b)
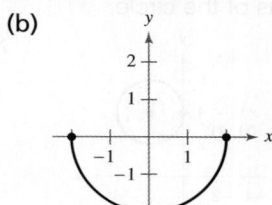

(c)
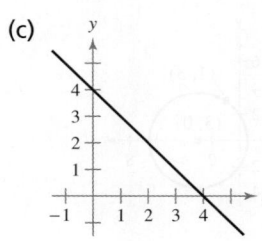

(d)
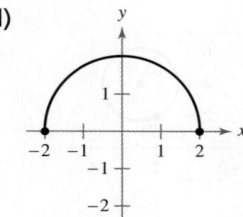

(e)
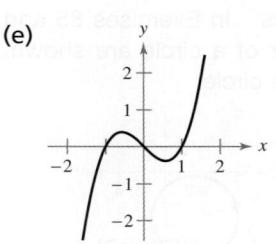

(f)
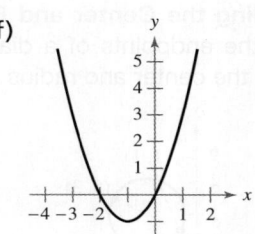

55. $y = 4 - x$

56. $y = \sqrt{4 - x^2}$

57. $y = x^2 + 2x$

58. $y = \sqrt{x}$

59. $y = x^3 - x$

60. $y = -\sqrt{4 - x^2}$

Sketching the Graph of an Equation In Exercises 61–80, sketch the graph of the equation. Identify any intercepts and test for symmetry.

61. $y = 5 - 3x$

62. $y = 2x - 3$

63. $y = 1 - x^2$

64. $y = x^2 - 1$

65. $y = x^2 - 4x + 3$

66. $y = -x^2 - 4x$

67. $y = x^3 + 2$

68. $y = x^3 - 1$

69. $y = \dfrac{8}{x^2 + 4}$

70. $y = \dfrac{4}{x^2 + 1}$

71. $y = \sqrt{x + 1}$

72. $y = \sqrt{1 - x}$

73. $y = \sqrt[3]{x}$

74. $y = \sqrt[3]{x + 1}$

75. $y = |x - 4|$

76. $y = |x| - 3$

77. $x = y^2 - 1$

78. $x = y^2 - 4$

79. $x^2 + y^2 = 4$

80. $x^2 + y^2 = 16$

Finding the Radius In Exercises 81–84, the center of a circle and a point on the circle are shown. Find the radius of the circle.

81.

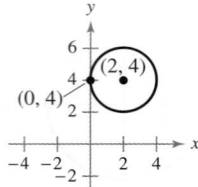

82.

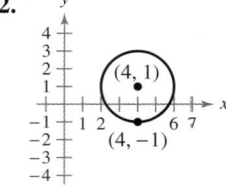

83.

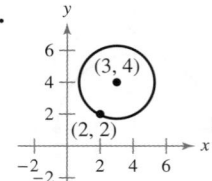

84.

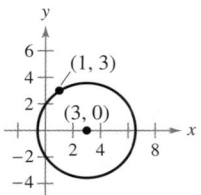

Finding the Center and Radius In Exercises 85 and 86, the endpoints of a diameter of a circle are shown. Find the center and radius of the circle.

85.

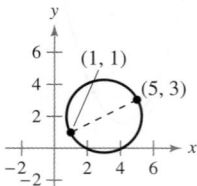

86.

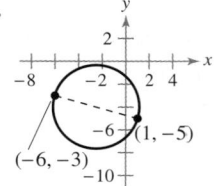

Finding the Equation of a Circle In Exercises 87–94, find the standard form of the equation of the circle. *See Example 9.*

87. Center: $(0, 0)$; radius: 3

88. Center: $(0, 0)$; radius: 5

89. Center: $(-4, 1)$; radius: $\sqrt{2}$

90. Center: $\left(0, \frac{1}{2}\right)$; radius: $\frac{2}{3}$

91. Center: $(-1, 2)$; point on circle: $(0, 0)$

92. Center: $(3, -2)$; point on circle: $(-1, 1)$

93. Endpoints of a diameter: $(-3, 4)$, $(5, -2)$

94. Endpoints of a diameter: $(-4, -1)$, $(4, 1)$

Completing the Square to Sketch a Circle In Exercises 95–102, write the equation of the circle in standard form. Then sketch the circle. *See Example 10.*

95. $x^2 + y^2 - 6x + 4y - 3 = 0$

96. $x^2 + y^2 - 2x + 6y - 15 = 0$

97. $x^2 + y^2 - 4x + 6y + 9 = 0$

98. $5x^2 + 5y^2 + 10x + 1 = 0$

99. $2x^2 + 2y^2 - 2x - 2y - 3 = 0$

100. $4x^2 + 4y^2 - 4x + 2y - 1 = 0$

101. $16x^2 + 16y^2 + 16x + 40y - 7 = 0$

102. $x^2 + y^2 - 4x + 2y + 3 = 0$

Using the Standard Form In Exercises 103 and 104, an equation of a circle is written in standard form. Indicate the coordinates of the center of the circle and determine the radius of the circle. Rewrite the equation of the circle in general form.

103. $(x - 3)^2 + (y + 1)^2 = 25$

104. $\left(x - \frac{1}{2}\right)^2 + (y - 2)^2 = 7$

105. Vehicle Distance The graph shows the total number of miles traveled by vehicles in the United States each year from 1994 through 2008. *(Source: U.S. Federal Highway Administration)*

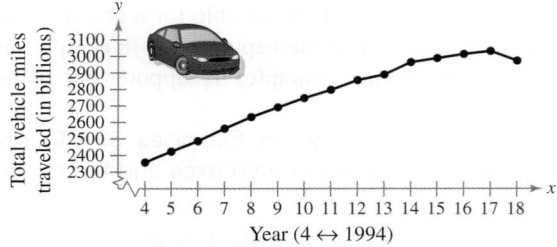

(a) Estimate the percent increase in miles traveled by vehicles from 1994 to 2000.

(b) Estimate the percent increase in miles traveled by vehicles from 2000 to 2008.

106. HOW DO YOU SEE IT? The graph shows the average prices of gold for the years 1975 through 2009. *(Sources: U.S. Bureau of Mines; U.S. Geological Survey)*

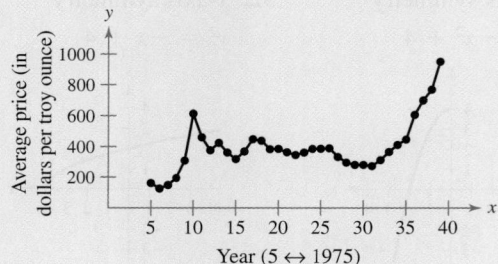

(a) What is the highest price shown in the graph? When did this occur?

(b) What is the lowest price shown in the graph? When did this occur?

(c) What is the highest price shown in the graph for the years 1975 through 2000? Does this represent the highest price that gold ever reached in those years? Explain your reasoning.

107. Population The population y (in millions of people) of North America from 1980 through 2050 can be modeled by

$$y = 5.3x + 482, \quad -20 \le x \le 50$$

where x represents the year, with $x = 50$ corresponding to 2050. *(Source: U.S. Census Bureau)*

(a) Find the y-intercept of the graph of the model. What does it represent in the given situation?

(b) Construct a table of values for $x = -20, -10, 0, 10, 20, 30, 40,$ and 50.

(c) Plot the solution points given by the table in part (b) and use the points to sketch the graph of the model.

108. Population The population y (in millions of people) of South America from 1980 through 2050 can be modeled by

$$y = -0.03x^2 + 5.0x + 351, \quad -20 \le x \le 50$$

where x represents the year, with $x = 50$ corresponding to 2050. *(Source: U.S. Census Bureau)*

(a) Find the y-intercept of the graph of the model. What does it represent in the given situation?

(b) Construct a table of values for $x = -20, -10, 0, 10, 20, 30, 40,$ and 50.

(c) Plot the solution points given by the table in part (b) and use the points to sketch the graph of the model.

109. Retail Stores The number y of stores operated by Family Dollar Stores from 2000 through 2010 can be approximated by the model

$$y = -24.91t^2 + 566.6t + 3624, \quad 0 \le t \le 10$$

where t represents the year, with $t = 0$ corresponding to 2000. *(Source: Family Dollar Stores, Inc.)*

(a) Sketch a graph of the model.

(b) Family Dollar Stores planned to have 7050 stores in operation by the end of 2011. Use the model to predict the number of stores in 2011. How well does the model support the company's plans?

(c) Complete the table by using the model to predict the number of stores in each year from 2012 through 2015.

Year, t	12	13	14	15
Number of stores				

(d) Extend the graph of the model in part (a) to include the years 2011 through 2015.

(e) Family Dollar Stores expects to have 7550 stores in operation sometime during the years 2013 through 2015. Based on your results in parts (c) and (d), what can you conclude about the model?

110. Observation Wheel The Singapore Flyer opened in Singapore in 2008 as the largest observation wheel in the world. Use the diagram to write an equation that models the circular shape of the wheel.

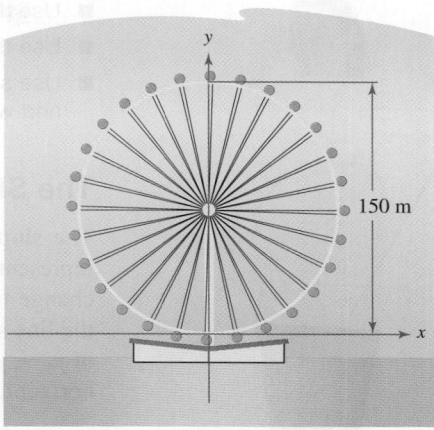

111. Spatial Learning The Morris water maze is an experiment for studying spatial learning in rats. A rat is repeatedly placed in a circular pool of water with a hidden platform that the rat uses to escape the water. The paths taken by the rat are studied as the rat uses visual cues to learn the location of the platform (see figure).

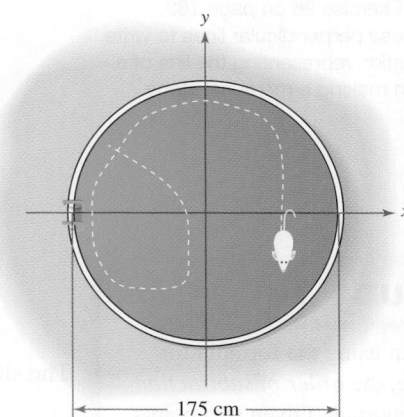

(a) Use the figure to write an equation that models the circular shape of the pool, using centimeters as units.

(b) Researchers track the time that a rat spends in each quadrant. The white dashed line shows the path taken by the rat. Determine the sequence of quadrants that the rat passed through.

(c) The platform is located at the point $(-40, 0)$ and the rat is at $(52, -40)$. How far is the rat from the platform? Round your answer to the nearest centimeter.

2.2 Lines in the Plane

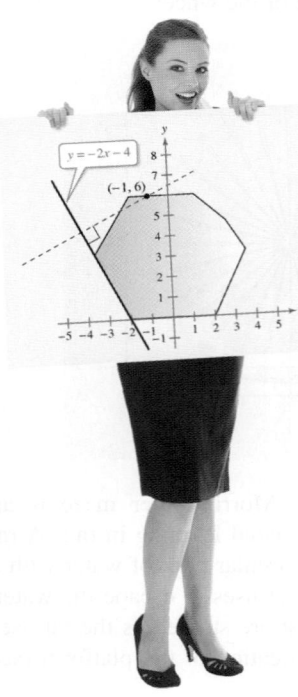

■ Find the slope of a line passing through two points.
■ Use the point-slope form to find the equation of a line.
■ Use the slope-intercept form to sketch the graph of a line.
■ Use slope to determine whether lines are parallel or perpendicular, and write the equation of a line parallel or perpendicular to a given line.

The Slope of a Line

The **slope** of a nonvertical line is a measure of the steepness of the line. The slope represents the number of units the line rises or falls vertically for each unit of horizontal change from left to right. For instance, consider the two points (x_1, y_1) and (x_2, y_2) on the line shown in Figure 2.21. As you move from left to right along this line, a change of $y_2 - y_1$ units in the vertical direction corresponds to a change of $x_2 - x_1$ units in the horizontal direction. That is,

$$y_2 - y_1 = \text{the change in } y$$

and

$$x_2 - x_1 = \text{the change in } x.$$

In Exercise 96 on page 182, you will use perpendicular lines to write an equation representing the line of a cut in making a memorial stone.

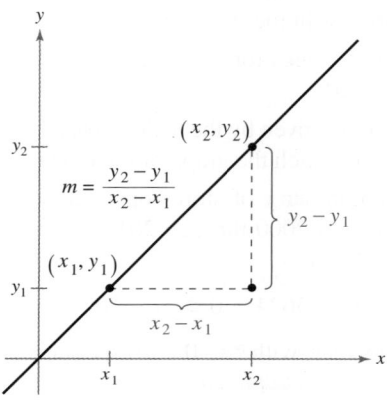

FIGURE 2.21

The slope of the line is defined as the quotient of these two changes.

Definition of the Slope of a Line

The **slope** m of the nonvertical line passing through the points (x_1, y_1) and (x_2, y_2) is

$$m = \frac{y_2 - y_1}{x_2 - x_1} = \frac{\text{change in } y}{\text{change in } x}$$

where $x_1 \neq x_2$.

The change in x is sometimes called the *run* and the change in y is sometimes called the *rise*.

In real-life problems, such as finding the steepness of a ramp or the increase in the value of a product, the slope of a line can be interpreted as either a *ratio* or a *rate*. If the x-axis and the y-axis have the same units of measure, then the slope has no units and is a *ratio*. If the x-axis and the y-axis have different units of measure, then the slope is a *rate* or *rate of change*. You will learn more about rates of change in Section 2.3.

STUDY TIP

When using the formula for slope, the *order of subtraction* is important. Given two points on a line, you can label either one of them as (x_1, y_1) and the other as (x_2, y_2). Once this is done, you must form the numerator and denominator using the same order of subtraction.

$$m = \frac{y_2 - y_1}{x_2 - x_1} \quad \text{Correct}$$

$$m = \frac{y_1 - y_2}{x_1 - x_2} \quad \text{Correct}$$

$$m = \frac{y_2 - y_1}{x_1 - x_2} \quad \text{Incorrect}$$

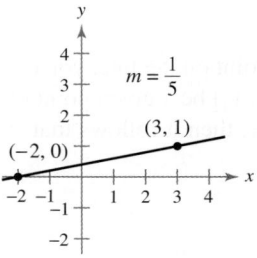

(a) If m is positive, the line rises from left to right.

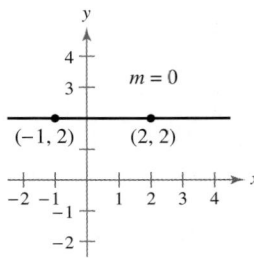

(b) If m is zero, the line is horizontal.

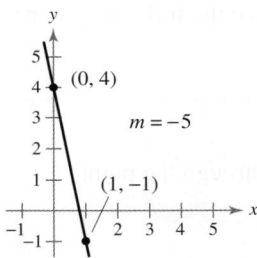

(c) If m is negative, the line falls from left to right.

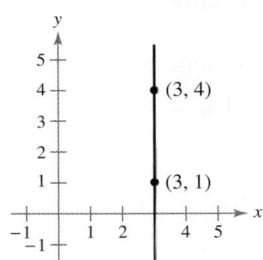

(d) If the line is vertical, the slope is undefined.

FIGURE 2.22

Example 1   **Finding the Slope of a Line Through Two Points**

Find the slope of the line passing through each pair of points.

a. $(-2, 0)$ and $(3, 1)$ **b.** $(-1, 2)$ and $(2, 2)$

c. $(0, 4)$ and $(1, -1)$ **d.** $(3, 4)$ and $(3, 1)$

SOLUTION

a. Letting $(x_1, y_1) = (-2, 0)$ and $(x_2, y_2) = (3, 1)$, you obtain a slope of

$$m = \frac{y_2 - y_1}{x_2 - x_1} \qquad \frac{\text{Difference in } y\text{-values}}{\text{Difference in } x\text{-values}}$$

$$= \frac{1 - 0}{3 - (-2)}$$

$$= \frac{1}{5}.$$

b. The slope of the line passing through $(-1, 2)$ and $(2, 2)$ is

$$m = \frac{2 - 2}{2 - (-1)}$$

$$= \frac{0}{3}$$

$$= 0.$$

c. The slope of the line passing through $(0, 4)$ and $(1, -1)$ is

$$m = \frac{-1 - 4}{1 - 0}$$

$$= \frac{-5}{1}$$

$$= -5.$$

d. The slope of the line passing through $(3, 4)$ and $(3, 1)$ is undefined. Applying the formula for slope, you have

$$m = \frac{1 - 4}{3 - 3} = \frac{-3}{0}. \qquad \text{Division by zero is undefined.}$$

Because division by zero is not defined, the slope of a vertical line is not defined.

The graphs of the four lines are shown in Figure 2.22.

✓ **Checkpoint 1**

Find the slope of the line passing through the points $(-1, 2)$ and $(3, 4)$.

From Example 1, you can make the following generalizations about the slope of a line.

Slope of a Line

1. A line with positive slope ($m > 0$) *rises* from left to right.

2. A line with negative slope ($m < 0$) *falls* from left to right.

3. A line with zero slope ($m = 0$) is *horizontal*.

4. A line with undefined slope is *vertical*.

The Point-Slope Form

When you know the slope of a line and the coordinates of one point on the line, you can find an equation of the line. For instance, in Figure 2.23, let (x_1, y_1) be a given point on the line whose slope is m. If (x, y) is *any other* point on the line, then it follows that

$$\frac{y - y_1}{x - x_1} = m.$$

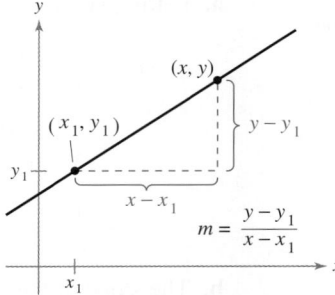

$$m = \frac{y - y_1}{x - x_1}$$

Any two points on a line can be used to determine the slope of the line.

FIGURE 2.23

ALGEBRA TUTOR
xy

For help with identifying and plotting points in the Cartesian plane, see the *Chapter 2 Algebra Tutor* on pages 240 and 241.

This equation in the variables x and y can be rewritten to produce the following **point-slope form** of the equation of a line.

Point-Slope Form of the Equation of a Line

The **point-slope form** of the equation of the line that passes through the point (x_1, y_1) and has a slope of m is

$$y - y_1 = m(x - x_1).$$

Example 2   The Point-Slope Form of the Equation of a Line

Find an equation of the line that passes through $(1, -2)$ and has a slope of 3.

SOLUTION   Use the point-slope form with $(x_1, y_1) = (1, -2)$ and $m = 3$.

$y - y_1 = m(x - x_1)$	Point-slope form
$y - (-2) = 3(x - 1)$	Substitute $y_1 = -2$, $x_1 = 1$, and $m = 3$.
$y + 2 = 3x - 3$	Simplify.
$y = 3x - 5$	Equation of line

The graph of this line is shown in Figure 2.24.

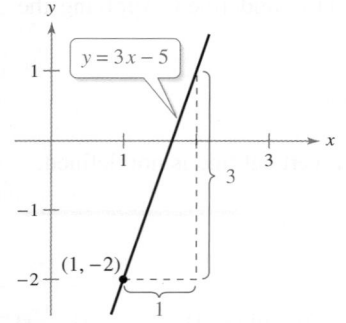

FIGURE 2.24

✓ Checkpoint 2

Find an equation of the line that passes through the given point and has the given slope.

a. $(2, 4)$, $m = -2$

b. $(-8, -3)$, $m = \frac{3}{2}$

c. $(5, 1)$, $m = 0$

The point-slope form can be used to find the equation of a line passing through two points (x_1, y_1) and (x_2, y_2). First, use the formula for the slope of a line passing through two points. Then, use the point-slope form to obtain

$$y - y_1 = \frac{y_2 - y_1}{x_2 - x_1}(x - x_1).$$

This is sometimes called the **two-point form** of the equation of a line.

Example 3 A Linear Model for Sales Prediction

During the first two quarters of the calendar year, Apple Inc. had sales of $13.5 billion and $15.7 billion, respectively.

a. Write a linear equation giving the sales y in terms of the quarter x.

b. Use the equation to predict the sales during the fourth quarter. Can you assume that sales will follow this linear pattern?   *(Source: Apple Inc.)*

SOLUTION

a. Let $(x_1, y_1) = (1, 13.5)$ and $(x_2, y_2) = (2, 15.7)$ be two points on the line. Use the two-point form to find an equation of the line.

$$y - 13.5 = \frac{15.7 - 13.5}{2 - 1}(x - 1) \qquad \text{Substitute for } x_1, y_1, x_2, \text{ and } y_2 \text{ in two-point form.}$$

$$y - 13.5 = 2.2(x - 1) \qquad \text{Simplify quotient.}$$

$$y = 2.2x + 11.3 \qquad \text{Equation of line}$$

b. Using the equation from part (a), the fourth-quarter sales ($x = 4$) should be

$$y = 2.2(4) + 11.3 \qquad \text{Substitute 4 for } x.$$

$$= \$20.1 \text{ billion.} \qquad \text{Simplify.}$$

See Figure 2.25. Without more data, you cannot assume that the sales pattern will be linear. Many factors, such as seasonal demand and past sales history, help to determine the sales pattern.

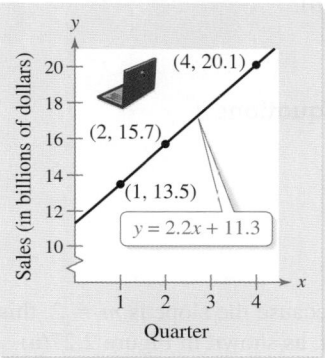

FIGURE 2.25

✓ Checkpoint 3

A company has sales of $1.2 million and $1.4 million in its first two years. Write a linear equation giving the sales y in terms of the year x. ■

The estimation method illustrated in Example 3 is called **linear extrapolation.** Note in Figure 2.26(a) that for linear extrapolation, the estimated point lies to the *right* of the given points. When the estimated point lies *between* two given points, the procedure is called **linear interpolation,** as shown in Figure 2.26(b).

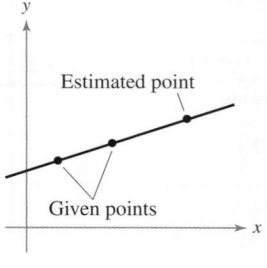

(a) Linear extrapolation

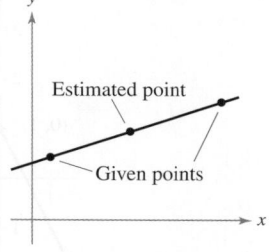

(b) Linear interpolation

FIGURE 2.26

Sketching Graphs of Lines

You have seen that for *finding the equation of a line* it is convenient to use the point-slope form. This formula, however, is not particularly useful for *sketching the graph of a line*. The form that is better suited to graphing linear equations is the **slope-intercept form** of the equation of a line. You can derive the slope-intercept form as shown.

$$y - y_1 = m(x - x_1) \qquad \text{Point-slope form}$$

$$y = mx - mx_1 + y_1 \qquad \text{Solve for } y.$$

$$y = mx + (y_1 - mx_1) \qquad \text{Commutative Property of Addition}$$

$$y = mx + b \qquad \text{Slope-intercept form } (b = y_1 - mx_1)$$

Slope-Intercept Form of the Equation of a Line

The graph of the equation

$$y = mx + b$$

is a line whose slope is m and whose y-intercept is $(0, b)$.

Example 4 Sketching the Graphs of Linear Equations

Sketch the graph of each linear equation.

a. $y = 2x + 1$ **b.** $y = 2$ **c.** $x + y = 2$

SOLUTION

a. Because $b = 1$, the y-intercept is $(0, 1)$. Moreover, because the slope is $m = 2$, this line *rises* two units for each unit it moves to the right, as shown in Figure 2.27(a).

b. By writing the equation $y = 2$ in the form $y = (0)x + 2$ you can see that the y-intercept is $(0, 2)$ and the slope is zero. A zero slope implies that the line is horizontal, as shown in Figure 2.27(b).

c. By writing the equation $x + y = 2$ in slope-intercept form $y = -x + 2$ you can see that the y-intercept is $(0, 2)$. Moreover, because the slope is $m = -1$, this line *falls* one unit for each unit it moves to the right, as shown in Figure 2.27(c).

✓ Checkpoint 4

Sketch the graph of each linear equation.

a. $y = 4x + 1$ **b.** $y - 2x = -3$

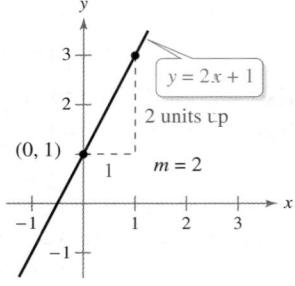

(a) When m is positive, the line rises from left to right.

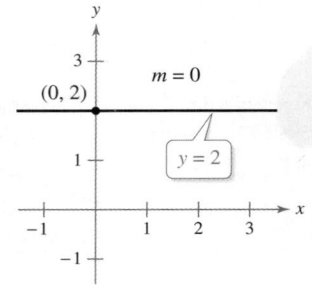

(b) When m is zero, the line is horizontal.

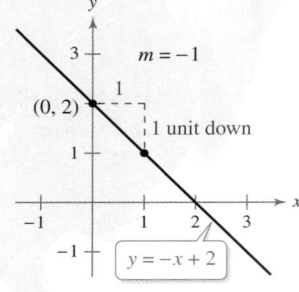

(c) When m is negative, the line falls from left to right.

FIGURE 2.27

From the slope-intercept form of the equation of a line, you can see that a horizontal line ($m = 0$) has an equation of the form

$$y = (0)x + b \quad \text{or} \quad y = b. \qquad \text{Horizontal line}$$

This is consistent with the fact that each point on a horizontal line through $(0, b)$ has a y-coordinate of b, as shown in Figure 2.28.

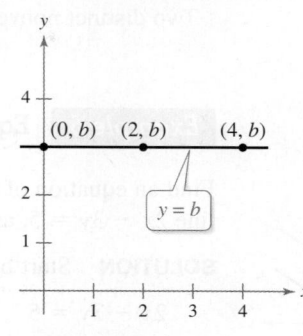

Horizontal Line
FIGURE 2.28

Similarly, each point on a vertical line through $(a, 0)$ has an x-coordinate of a, as shown in Figure 2.29. So, a vertical line has an equation of the form

$$x = a. \qquad \text{Vertical line}$$

This equation cannot be written in slope-intercept form because the slope of a vertical line is undefined. However, *every* line has an equation that can be written in the **general form**

$$Ax + By + C = 0 \qquad \text{General form}$$

where A and B are not *both* zero. If $A = 0$ (and $B \neq 0$), then the general equation can be reduced to the form $y = b$, which represents a horizontal line. If $B = 0$ (and $A \neq 0$), then the general equation can be reduced to the form $x = a$, which represents a vertical line.

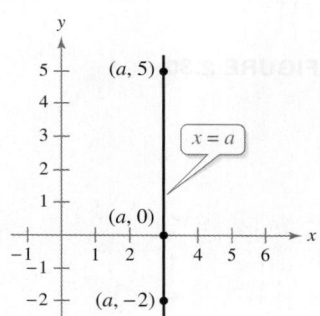

Vertical Line
FIGURE 2.29

Summary of Equations of Lines

1. General form: $Ax + By + C = 0$

2. Vertical line: $x = a$

3. Horizontal line: $y = b$

4. Slope-intercept form: $y = mx + b$

5. Point-slope form: $y - y_1 = m(x - x_1)$

Make sure you understand that one form of the equation of a line may be more efficient than another form for different problem situations. For instance, the point-slope form (or two-point form) is more efficient for problems in which two points are given. Note that another form of an equation of a line is the *two-intercept* or *intercept form*. This form is introduced in Exercises 59–64 in this section.

Parallel and Perpendicular Lines

The slope of a line is a convenient tool for determining whether two lines are parallel, perpendicular, or neither.

Parallel Lines

Two distinct nonvertical lines are **parallel** if and only if their slopes are equal.

Example 5   **Equations of Parallel Lines**

Find an equation of the line that passes through the point $(2, -1)$ and is parallel to the line $2x - 3y = 5$, as shown in Figure 2.30.

SOLUTION   Start by rewriting the given equation in slope-intercept form.

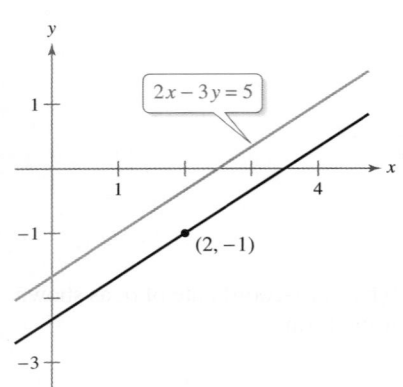

FIGURE 2.30

$$2x - 3y = 5 \qquad \text{Write original equation.}$$

$$-3y = -2x + 5 \qquad \text{Subtract } 2x \text{ from each side.}$$

$$y = \frac{2}{3}x - \frac{5}{3} \qquad \text{Write in slope-intercept form.}$$

So, the given line has a slope of

$$m = \frac{2}{3}.$$

Because any line parallel to the given line must also have a slope of $\frac{2}{3}$, the required line through $(2, -1)$ has the following equation.

$$y - y_1 = m(x - x_1) \qquad \text{Point-slope form}$$

$$y - (-1) = \frac{2}{3}(x - 2) \qquad \text{Substitute for } y_1, x_1, \text{ and } m.$$

$$y + 1 = \frac{2}{3}x - \frac{4}{3} \qquad \text{Simplify.}$$

$$y = \frac{2}{3}x - \frac{4}{3} - 1 \qquad \text{Solve for } y.$$

$$y = \frac{2}{3}x - \frac{7}{3} \qquad \text{Write in slope-intercept form.}$$

Notice the similarity between the slope-intercept form of the original equation and the slope-intercept form of the parallel equation.   ⎯⎯⎯⎯⎯

✓ Checkpoint 5

Find an equation of the line that passes through the given point and is parallel to the line $2y - 6x = 2$.

a. $(2, 4)$

b. $(-2, 0)$ ▪

You have seen that two nonvertical lines are parallel if and only if they have the same slope. Two nonvertical lines are *perpendicular* if and only if their slopes are negative reciprocals of each other. For instance, the lines

$$y = 2x \quad \text{and} \quad y = -\tfrac{1}{2}x$$

are perpendicular because one has a slope of $2 = \frac{2}{1}$ and the other has a slope of $-\frac{1}{2}$.

Perpendicular Lines

Two nonvertical lines are **perpendicular** if and only if their slopes are negative reciprocals of each other. That is,

$$m_1 = -\frac{1}{m_2}.$$

Example 6   Equations of Perpendicular Lines

Find an equation of the line that passes through the point $(2, -1)$ and is perpendicular to the line $2x - 3y = 5$, as shown in Figure 2.31.

SOLUTION   By rewriting the given equation in slope-intercept form

$$y = \frac{2}{3}x - \frac{5}{3}$$

you can see that the line has a slope of $\frac{2}{3}$. So, any line that is perpendicular to this line must have a slope of $-\frac{3}{2}$ because $-\frac{3}{2}$ is the negative reciprocal of $\frac{2}{3}$. The required line through the point $(2, -1)$ has the following equation.

$$y - y_1 = m(x - x_1) \qquad \text{Point-slope form}$$

$$y - (-1) = -\frac{3}{2}(x - 2) \qquad \text{Substitute for } y_1, x_1, \text{ and } m.$$

$$y + 1 = -\frac{3}{2}x + 3 \qquad \text{Simplify.}$$

$$y = -\frac{3}{2}x + 3 - 1 \qquad \text{Solve for } y.$$

$$y = -\frac{3}{2}x + 2 \qquad \text{Write in slope-intercept form.}$$

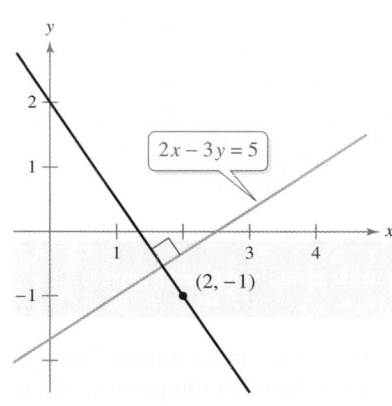

FIGURE 2.31

✓**Checkpoint 6**

Find an equation of the line that passes through the point $(-2, 12)$ and is perpendicular to the line $y = \frac{1}{4}x - 2$.

SUMMARIZE   (Section 2.2)

1. State the definition of the slope of a line and describe how to find the slope of a line passing through two points *(page 172)*. For an example of finding the slope of a line through two points, see Example 1.

2. State the point-slope form of the equation of a line *(page 174)*. For examples of using the point-slope form of the equation of a line, see Examples 2 and 3.

3. State the slope-intercept form of the equation of a line and describe how it is used to sketch the graph of a line *(page 176)*. For an example of using the slope-intercept form to sketch the graphs of lines, see Example 4.

4. Describe the slopes of two nonvertical parallel lines and the slopes of two nonvertical perpendicular lines *(pages 178 and 179)*. For examples of finding equations of parallel and perpendicular lines, see Examples 5 and 6.

SKILLS WARM UP 2.2   The following warm-up exercises involve skills that were covered in earlier sections. You will use these skills in the exercise set for this section. For additional help, review Sections 0.2 and 1.2.

In Exercises 1 and 2, perform the indicated operations.

1. $\dfrac{4 - (-4)}{-3 - (-1)}$

2. $\dfrac{-5 - 8}{0 - (-3)}$

3. Evaluate $-1/m$ when $m = 4/5$.

4. Evaluate $-1/m$ when $m = -3$.

In Exercises 5–10, solve for y in terms of x.

5. $2x - 3y = 6$

6. $4x + 2y = 0$

7. $y - (-4) = 3[x - (-1)]$

8. $y - 7 = \dfrac{2}{3}(x - 3)$

9. $y - (-1) = \dfrac{3 - (-1)}{2 - 4}(x - 4)$

10. $y - 5 = \dfrac{3 - 5}{0 - 2}(x - 2)$

Exercises 2.2

See www.CalcChat.com for worked-out solutions to odd-numbered exercises.

Finding the Slope of a Line   In Exercises 1–4, use the change in y and the change in x to find the slope of the line.

1.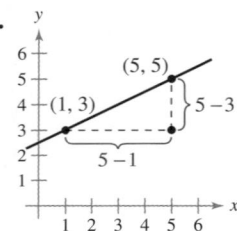

2.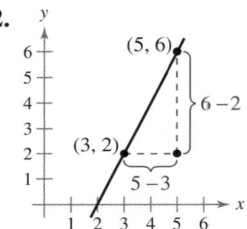

3.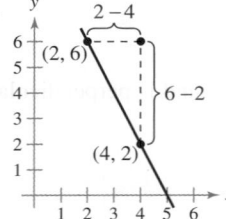

4.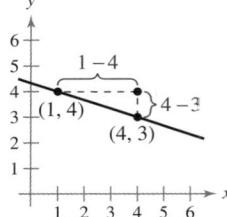

Using Slope   In Exercises 5 and 6, identify the line that has the indicated slope.

5. (a) $m = \dfrac{1}{3}$
   (b) m is undefined.
   (c) $m = -2$

6. (a) $m = 0$
   (b) $m = -\dfrac{3}{4}$
   (c) $m = 1$

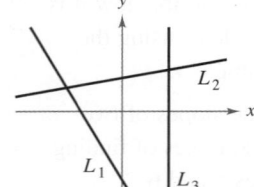

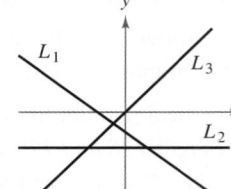

Sketching the Graphs of Lines   In Exercises 7 and 8, sketch the graphs of the lines through the point with the indicated slopes in the same coordinate plane.

Point	Slopes
7. $(-3, 4)$	(a) -2 (b) $\dfrac{2}{3}$
	(c) 0 (d) Undefined
8. $(-2, -5)$	(a) -1 (b) $\dfrac{3}{4}$
	(c) 0 (d) Undefined

Finding the Slope of a Line Through Two Points   In Exercises 9–14, plot the points and find the slope of the line passing through the points. *See Example 1.*

9. $(6, 9), (-4, -1)$

10. $(2, 4), (4, -4)$

11. $(-6, -1), (-6, 4)$

12. $(0, -10), (-4, 0)$

13. $\left(-\dfrac{1}{3}, 1\right), \left(-\dfrac{2}{3}, \dfrac{5}{6}\right)$

14. $\left(\dfrac{7}{8}, \dfrac{3}{4}\right), \left(\dfrac{5}{4}, -\dfrac{1}{4}\right)$

Using Slope   In Exercises 15–22, use the point on the line and the slope of the line to find three additional points through which the line passes. (There are many correct answers.)

Point	Slope
15. $(5, -2)$	$m = 0$
16. $(-3, 4)$	$m = 0$
17. $(2, -5)$	m is undefined.
18. $(-1, 3)$	m is undefined.
19. $(5, -6)$	$m = 1$
20. $(10, -6)$	$m = -1$
21. $(-6, -1)$	$m = \dfrac{1}{2}$
22. $(7, -5)$	$m = -\dfrac{2}{3}$

The Point-Slope Form of the Equation of a Line   In Exercises 23–32, find an equation of the line that passes through the point and has the indicated slope. Then sketch the graph of the line. *See Example 2.*

	Point	Slope
23.	$(7, 0)$	$m = 1$
24.	$(0, -4)$	$m = -1$
25.	$(-2, 0)$	$m = -4$
26.	$(1, 3)$	$m = 3$
27.	$(4, 0)$	$m = -\frac{1}{3}$
28.	$(-8, 3)$	$m = -\frac{1}{2}$
29.	$(-2, -7)$	$m = 0$
30.	$(-10, 4)$	$m = 0$
31.	$\left(4, \frac{5}{2}\right)$	$m = \frac{4}{3}$
32.	$\left(-\frac{1}{2}, \frac{3}{2}\right)$	$m = -3$

Writing an Equation of a Line   In Exercises 33–44, find an equation of the line passing through the points.

33. $(0, 0), (-1, -4)$ **34.** $(6, -1), (0, 0)$

35. $(7, -4), (-7, 3)$ **36.** $(4, 3), (-4, -4)$

37. $(-9, 11), (-9, 14)$ **38.** $(3, 5), (3, -2)$

39. $(-1, 7), (3, 7)$ **40.** $(3, -2), (-8, -2)$

41. $\left(2, \frac{1}{2}\right), \left(\frac{1}{2}, \frac{5}{4}\right)$ **42.** $(1, 1), \left(6, -\frac{2}{3}\right)$

43. $(1, 0.6), (-2, -0.6)$ **44.** $(-8, 0.6), (2, -2.4)$

45. Discovery   Use a graphing utility to graph each equation in the same viewing window. What is the y-intercept of each graph?

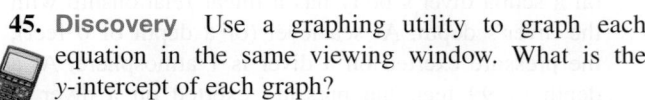

$$y_1 = x + 2 \qquad y_2 = \tfrac{1}{4}x + 2 \qquad y_3 = -x + 2$$

46. Discovery   In Exercise 45, what can you determine about each graph from the coefficient of x in its equation?

Sketching the Graph of a Linear Equation   In Exercises 47–56, find the slope and y-intercept (if possible) of the line specified by the equation. Then sketch the graph of the line. *See Example 4.*

47. $y = 2x - 1$ **48.** $y = 3 - x$

49. $4x - y - 6 = 0$ **50.** $2x + 3y - 9 = 0$

51. $8 - 3x = 0$ **52.** $2x + 5 = 0$

53. $7x + 6y - 30 = 0$ **54.** $x - y - 10 = 0$

55. $2y - 7 = 0$ **56.** $8 - 5y = 0$

57. Writing   A student is confused about why the slope of a vertical line is undefined. Describe how you would help this student understand the concept of undefined slope.

58. Writing   A student makes the comment, "I do not understand why a horizontal line has zero slope and how that is different from undefined or no slope." Describe how you would explain the concepts of zero slope and undefined slope and how they are different from each other.

The Intercept Form of the Equation of a Line   In Exercises 59–64, use the *intercept form* to find an equation of the line with the given intercepts. The intercept form of the equation of a line with intercepts $(a, 0)$ and $(0, b)$ is

$$\frac{x}{a} + \frac{y}{b} = 1, \quad a \neq 0, \quad b \neq 0.$$

59. x-intercept: $(1, 0)$ **60.** x-intercept: $(-3, 0)$
    y-intercept: $(0, -4)$ y-intercept: $(0, 4)$

61. x-intercept: $(-2, 0)$ **62.** x-intercept: $(5, 0)$
    y-intercept: $(0, -2)$ y-intercept: $(0, 1)$

63. x-intercept: $\left(-\frac{1}{6}, 0\right)$ **64.** x-intercept: $\left(-\frac{2}{3}, 0\right)$
    y-intercept: $\left(0, -\frac{2}{3}\right)$ y-intercept: $\left(0, \frac{1}{2}\right)$

65. Discovery   Use a graphing utility to graph the equation $y = x$ in a viewing window with a *square* setting, in which the horizontal and vertical tick marks have equal spacing. Then adjust the viewing window several times by increasing and decreasing the number of units on the y-axis. How does changing the viewing window affect the graph? Which setting best represents the slope of the line?

66. Discovery   Use a graphing utility to graph each equation in the same viewing window with a *square* setting (see Exercise 65).

$$y_1 = \tfrac{4}{7}x - 1 \qquad y_2 = \tfrac{4}{7}x + 2 \qquad y_3 = -\tfrac{7}{4}x + 8$$

(a) What is the geometric relationship between the two lines that have the same slope?

(b) In each pair of perpendicular lines, what is the numerical relationship between their slopes?

Parallel and Perpendicular Lines   In Exercises 67–74, the equations of two lines are given. Determine whether the lines L_1 and L_2 are parallel, perpendicular, or neither.

67. $L_1:\ y = 3x + 4;\ L_2:\ y = x - \frac{1}{4}$

68. $L_1:\ y = \frac{3}{4}x + 1;\ L_2:\ y = -\frac{4}{3}x + 3$

69. $L_1:\ 2x - y = 1;\ L_2:\ x + 2y = -1$

70. $L_1:\ x - 5y = -2;\ L_2:\ -3x + 15y = 6$

71. $L_1:\ x - 3y = -3;\ L_2:\ 2x - 6y = 6$

72. $L_1:\ 4x - y = -2;\ L_2:\ 8x - 2y = 6$

73. $L_1:\ 2x - 3y - 15 = 0;\ L_2:\ 3x + 2y + 8 = 0$

74. $L_1:\ x - 4y - 12 = 0;\ L_2:\ 3x - 4y - 8 = 0$

Parallel and Perpendicular Lines   In Exercises 75–82, determine whether the lines L_1 and L_2 passing through the indicated pairs of points are parallel, perpendicular, or neither.

75. $L_1:\ (-5, 0), (-2, 1);\ L_2:\ (0, 1), (3, 2)$

76. $L_1:\ (-1, 6), (1, 4);\ L_2:\ (3, -3), (6, -9)$

77. $L_1:\ (0, -1), (5, 9);\ L_2:\ (0, 3), (4, 1)$

78. $L_1:\ (3, 6), (-6, 0);\ L_2:\ (0, -1), \left(5, \frac{7}{3}\right)$

79. $L_1:\ (-2, -1), (1, 5);\ L_2:\ (1, 3), (5, -5)$

80. L_1: $(4, 8), (-4, 2)$; L_2: $(3, -5), \left(-1, \frac{1}{3}\right)$

81. L_1: $(-1, 7), (-6, 4)$; L_2: $(0, 1), (5, 4)$

82. L_1: $(-1, 3), (2, -5)$; L_2: $(3, 0), (2, -7)$

Equations of Parallel and Perpendicular Lines In Exercises 83–88, write equations of the lines passing through the given point that are (a) parallel to the given line and (b) perpendicular to the given line. *See Examples 5 and 6.*

	Point	*Line*
83.	$(6, 2)$	$y = 2x - 1$
84.	$(-5, 4)$	$y = -x + 8$
85.	$\left(\frac{1}{4}, -\frac{2}{3}\right)$	$2x - 3y = 5$
86.	$\left(\frac{7}{8}, \frac{3}{4}\right)$	$5x + 3y = 0$
87.	$(-1, 0)$	$y = -3$
88.	$(2, 5)$	$x = 4$

89. Wheelchair Ramp The maximum recommended slope of a wheelchair ramp is $\frac{1}{12}$. A business is installing a wheelchair ramp that rises 34 inches over a horizontal length of 30 feet. Is the ramp steeper than recommended? *(Source: Americans with Disabilities Act Handbook)*

90. HOW DO YOU SEE IT? Match the description of the situation with its graph. Then write the equation of the line.

(a) You are paying $10 per week to repay a $100 loan.

(b) An employee is paid $12.50 per hour plus $1.50 for each unit produced per hour.

(c) A sales representative receives $30 per day for food plus $0.51 for each mile traveled.

(d) A computer that was purchased for $800 depreciates $100 per year.

(i)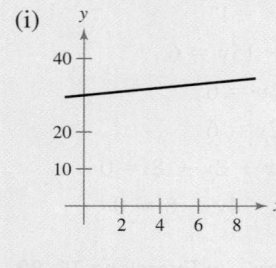

(ii)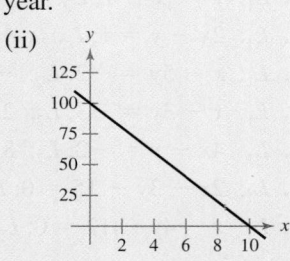

(iii)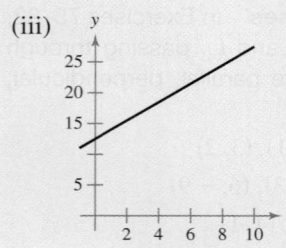

(iv)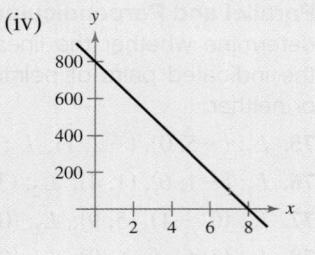

91. Temperature Find an equation of the line that gives the relationship between the temperature in degrees Celsius C and the temperature in degrees Fahrenheit F. Remember that water freezes at $0°$ Celsius ($32°$ Fahrenheit) and boils at $100°$ Celsius ($212°$ Fahrenheit).

92. Temperature Use the result of Exercise 91 to complete the table. Is there a temperature for which the Fahrenheit reading is the same as the Celsius reading? If so, what is it?

C		$-10°$	$10°$			$177°$
F	$0°$			$68°$	$90°$	

93. Sales During the first two quarters of the calendar year, a business had sales of $158,000 and $165,000, respectively. Write a linear equation giving the sales y in terms of the quarter x. Use the equation to predict the fourth quarter sales. Can you assume that sales will follow this linear pattern?

94. Retail Stores AutoZone operated 3483 stores in 2004. By 2007, AutoZone operated 4056 stores. Assuming the number of stores followed a linear pattern, how many stores would AutoZone have operated in 2010? The actual number of stores operated by AutoZone in 2010 was 4627. Was the actual increase in the number of stores approximately linear? *(Source: AutoZone)*

95. Scuba Diving The pressure (in atmospheres) exerted on a scuba diver's body has a linear relationship with the diver's depth. At sea level (or a depth of 0 feet), the pressure exerted on a diver is 1 atmosphere. At a depth of 99 feet, the pressure exerted on a diver is 4 atmospheres. Write a linear equation to describe the pressure p (in atmospheres) in terms of the depth d (in feet) below the surface of the sea. What is the rate of change in pressure with respect to depth? *(Source: PADI Open Water Diver Manual)*

96. Stone Cutting A stone cutter is making a 6-foot-tall memorial stone. The diagram shows coordinates labeled in feet. The stone cutter plans to cut through the point $(-1, 6)$ along a line that is perpendicular to one side of the stone, as indicated by the dashed line in the figure. Find an equation of the line of the cut.

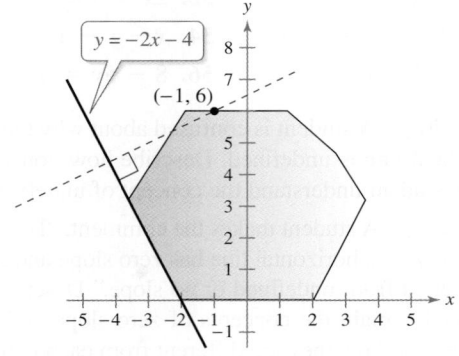

2.3 Linear Modeling and Direct Variation

- Use a mathematical model to approximate a set of data points.
- Construct a linear model to relate quantities that vary directly.
- Construct and use a linear model with slope as the rate of change.
- Use a scatter plot to find a linear model that fits a set of data.

Mathematical Models

The primary objective of applied mathematics is to find equations or **mathematical models** that describe real-world situations. In developing a mathematical model to represent actual data, you should strive for two (often conflicting) goals—accuracy and simplicity. That is, you want the model to be simple enough to be workable, yet accurate enough to produce meaningful results.

You have already studied some techniques for fitting models to data. For instance, in Section 2.2, you learned how to find the equation of a line that passes through two points. In this section, you will study other techniques for fitting models to data: *direct variation, rates of change,* and *linear regression.*

Example 1 A Mathematical Model

The weight of a puppy recorded every two months is shown in the table.

Age (in months)	2	4	6	8	10	12
Weight (in pounds)	24	45	67	93	117	130

A linear model that approximates the puppy's weight w (in pounds) in month t is

$$w = 11.03t + 2.1, \quad 2 \le t \le 12.$$

How closely does the model represent the data?

SOLUTION By graphing the data points with the linear model (see Figure 2.32), you can see that the model is a "good fit" for the actual data. The table shows how each actual weight w compares with the weight $w*$ given by the model.

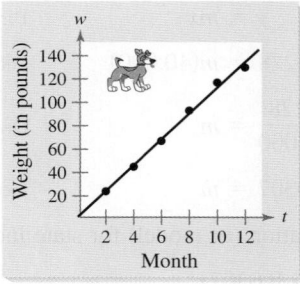

FIGURE 2.32

t	w	$w*$
2	24	24.16
4	45	46.22
6	67	68.28
8	93	90.34
10	117	112.4
12	130	134.46

In Exercise 41 on page 192, you will use data points from a scatter plot to create a linear model describing the number of autistic children receiving disability services.

✓ **Checkpoint 1**

In Example 1, what are the best and worst approximations given by the model?

Direct Variation

There are two basic types of linear models in x and y. The more general model has a y-intercept that is nonzero:

$$y = mx + b, \quad b \neq 0.$$

The simpler model,

$$y = mx$$

has a y-intercept that is zero. In the simpler model, y is said to **vary directly** as x, or to be **directly proportional** to x.

Direct Variation

The following statements are equivalent.

1. y **varies directly** as x.

2. y is **directly proportional** to x.

3. $y = mx$ for some nonzero constant m, where m is the **constant of variation** or the **constant of proportionality.**

 Example 2   State Income Tax

In Pennsylvania, state income tax is directly proportional to *taxable income*. For a taxable income of $40,000, the Pennsylvania state income tax is $1228. Find a mathematical model that gives the Pennsylvania state income tax in terms of taxable income.

SOLUTION

Verbal Model: $\dfrac{\text{State}}{\text{income tax}} = m \cdot \dfrac{\text{Taxable}}{\text{income}}$

Labels: State income tax $= y$ (dollars)
　　　　 Taxable income $= x$ (dollars)
　　　　 Income tax rate $= m$ (percent in decimal form)

Equation: $y = mx$

Find m by substituting the given information into the equation $y = mx$.

$$y = mx \qquad \text{Direct variation model}$$

$$1228 = m(40{,}000) \qquad \text{Substitute } y = 1228 \text{ and } x = 40{,}000.$$

$$\frac{1228}{40{,}000} = m \qquad \text{Divide each side by 40,000.}$$

$$0.0307 = m \qquad \text{Income tax rate}$$

An equation (or model) for state income tax in Pennsylvania is

$$y = 0.0307x.$$

So, Pennsylvania has a state income tax rate of 3.07% of taxable income. The graph of this equation is shown in Figure 2.33.

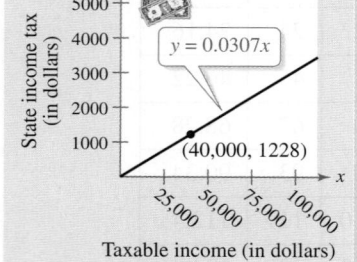

FIGURE 2.33

State income tax (in dollars) / Taxable income (in dollars)
$y = 0.0307x$
$(40{,}000, 1228)$

✓ Checkpoint 2

You buy a flash drive for $14.50 and pay sales tax of $0.87. The sales tax is directly proportional to the price. Find a mathematical model that gives the sales tax in terms of the price.

Most measurements in the English system and the metric system are directly proportional. The next example shows how to use a direct proportion to convert between centimeters per day and inches per day.

 Example 3 The English and Metric Systems

Bamboos are some of the fastest growing plants in the world. One type of bamboo has been measured to grow 47.6 inches, or 120.9 centimeters, per day. Use this information to find a mathematical model that relates inches per day to centimeters per day. *(Source: American Bamboo Society)*

SOLUTION Let y represent the growth rate in inches per day and let x represent the growth rate in centimeters per day. Then y and x are related by the equation

$$y = mx.$$

Use the fact that $y = 47.6$ when $x = 120.9$ to find the value of m.

$y = mx$	Direct variation model
$47.6 = m(120.9)$	Substitute $y = 47.6$ and $x = 120.9$.
$\dfrac{47.6}{120.9} = m$	Divide each side by 120.9.
$0.39371 \approx m$	Use a calculator.

So, the conversion factor from centimeters per day to inches per day is approximately 0.39371, and the model is

$$y = 0.39371x.$$

The graph of this equation is shown in Figure 2.34.

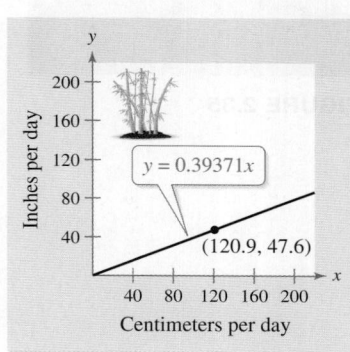

FIGURE 2.34

✓ **Checkpoint 3**

You buy an ice bucket with a capacity of 44 ounces, or 1.3 liters. Write a mathematical model that relates ounces to liters.

You can use the model from Example 3 to convert any growth rate in centimeters per day to inches per day, as shown in the table.

Centimeters per day	Inches per day
40	15.7
80	31.5
120	47.2
160	63.0
200	78.7

Rates of Change

A second common type of linear model is one that involves a known rate of change. In the linear equation

$$y = mx + b$$

you know that m represents the slope of the line. In real-life problems, the slope can often be interpreted as the **rate of change** of y with respect to x. Rates of change should always be listed in appropriate units of measure.

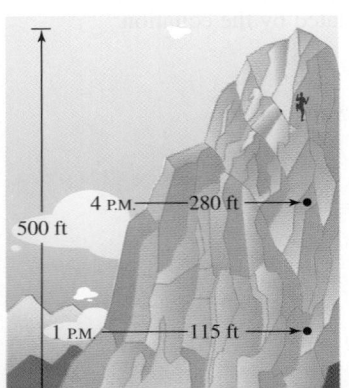

FIGURE 2.35

Example 4   Mountain Climbing

A mountain climber is climbing up a 500-foot cliff. At 1 P.M., the climber is 115 feet up the cliff. By 4 P.M., the climber has reached a height of 280 feet, as shown in Figure 2.35.

a. Find the average rate of change of the climber. Use this rate of change to find an equation that relates the height of the climber to the time.

b. Use the equation to estimate the time when the climber reaches the top of the cliff.

SOLUTION

a. Let y represent the climber's height on the cliff (in feet). Let $t = 1$ represent 1 P.M., and let t be measured in hours. Then the two points that represent the climber's two positions are

$$(t_1, y_1) = (1, 115) \quad \text{and} \quad (t_2, y_2) = (4, 280).$$

So, the average rate of change of the climber is

$$\text{Average rate of change} = \frac{y_2 - y_1}{t_2 - t_1}$$

$$= \frac{280 - 115}{4 - 1}$$

$$= 55 \text{ feet per hour.}$$

An equation that relates the height of the climber to the time is

$y - y_1 = m(t - t_1)$	Point-slope form
$y - 115 = 55(t - 1)$	Substitute $y_1 = 115$, $t_1 = 1$, and $m = 55$.
$y = 55t + 60.$	Linear model

If you had chosen to use the point (t_2, y_2) to determine the equation, then you would have obtained a different equation initially: $y - 280 = 55(t - 4)$. However, simplifying this equation yields the same linear model $y = 55t + 60$.

b. To estimate the time when the climber reaches the top of the cliff, let $y = 500$ and solve for t.

$500 = 55t + 60$	Substitute 500 for y.
$440 = 55t$	Subtract 60 from each side.
$8 = t$	Divide each side by 55.

Because $t = 8$ corresponds to 8 P.M., at the average rate of change, the climber will reach the top at 8 P.M.

✓Checkpoint 4

At what time does the climber in Example 4 reach a height of 390 feet?   ■

Example 5 **Population of San Francisco, California**

Between 2000 and 2009, the population of San Francisco, California increased at an average rate of approximately 4292 people per year. In 2000, the population was 776,733. Find a mathematical model that gives the population of San Francisco in terms of the year, and use the model to predict the population in 2012. *(Source: U.S. Census Bureau)*

SOLUTION Let y represent the population of San Francisco and let t represent the calendar year, with $t = 0$ corresponding to 2000. It is convenient to let $t = 0$ correspond to 2000 because you are given the population in 2000. Now, using the rate of change of 4292 people per year, you have

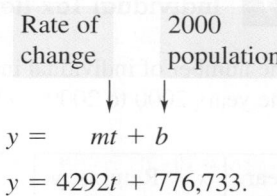

$$y = mt + b$$

$$y = 4292t + 776,733.$$

Using this model, you can predict the 2012 population to be

$$2012 \text{ population} = 4292(12) + 776,733$$

$$= 828,237.$$

The graph is shown in Figure 2.36.

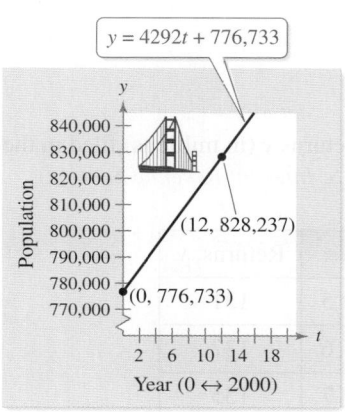

$$y = 4292t + 776,733$$

(12, 828,237)

(0, 776,733)

Year (0 ↔ 2000)

FIGURE 2.36

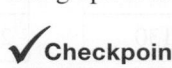 **Checkpoint 5**

Use the model in Example 5 to predict the population of San Francisco in 2014. ■

 Example 6 **Straight-Line Depreciation**

A racing team buys a $4750 welder that has a useful life of 10 years. The salvage value of the welder at the end of the 10 years is $400. Write a linear equation that describes the value of the welder throughout its useable life.

SOLUTION Let V represent the value of the welder (in dollars) at the end of year t. Because the value of the welder at the time of purchase is $4750, you can represent the initial value of the welder by the ordered pair $(t_1, V_1) = (0, 4750)$. You can represent the salvage value by the ordered pair $(t_2, V_2) = (10, 400)$. The slope of the line is

$$m = \frac{400 - 4750}{10 - 0}$$

$$= \frac{-4350}{10}$$

$$= -435$$

which represents the annual depreciation in *dollars per year*. Using the slope-intercept form, you can write the equation of the line.

$$V = -435t + 4750 \qquad \text{Slope-intercept form}$$

The graph of the equation is shown in Figure 2.37.

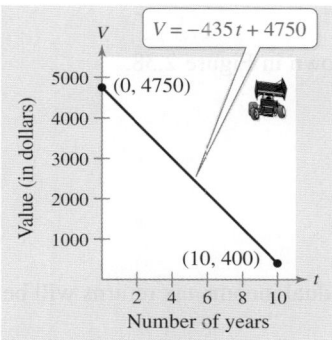

$$V = -435t + 4750$$

(0, 4750)

(10, 400)

Number of years

FIGURE 2.37

 **Checkpoint 6**

Write a linear equation to model the value of a new machine that costs $2300 and is worth $350 after 10 years. ■

Scatter Plots and Regression Analysis

Another type of linear modeling is a graphical approach that is commonly used in statistics. To find a mathematical model that approximates a set of actual data points, plot the points in a rectangular coordinate system. This collection of points is called a **scatter plot.** You can use the statistical features of a graphing utility to calculate the equation of the best-fitting line for the data in your scatter plot. The statistical method of fitting a line to a collection of points is called **linear regression.** A discussion of linear regression is beyond the scope of this text, but the program in most graphing utilities is easy to use and allows you to analyze linear data that may not be convenient to graph by hand.

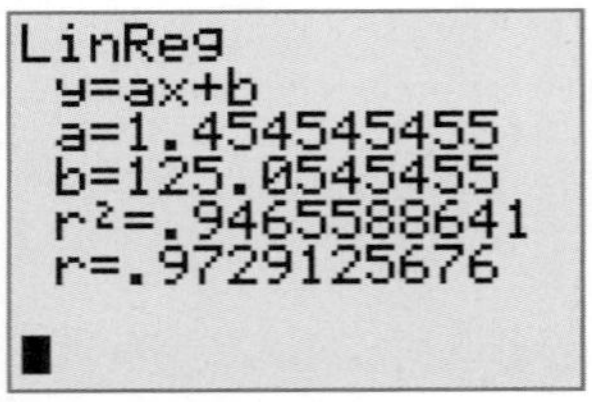

Example 7 Individual Tax Returns

The table shows the number of individual income tax returns y (in millions) filed in the United States in the years 2000 to 2009. *(Source: U.S. Internal Revenue Service)*

Year	x	Returns, y	Year	x	Returns, y
2000	0	125	2005	5	131
2001	1	127	2006	6	132
2002	2	129	2007	7	135
2003	3	130	2008	8	138
2004	4	130	2009	9	139

a. Use the *regression* feature of a graphing utility to find a linear model for the data. Let $x = 0$ represent 2000.

b. Use the graphing utility to graph the linear model along with a scatter plot of the data.

c. Use the linear model to predict the number of individual tax returns in 2011.

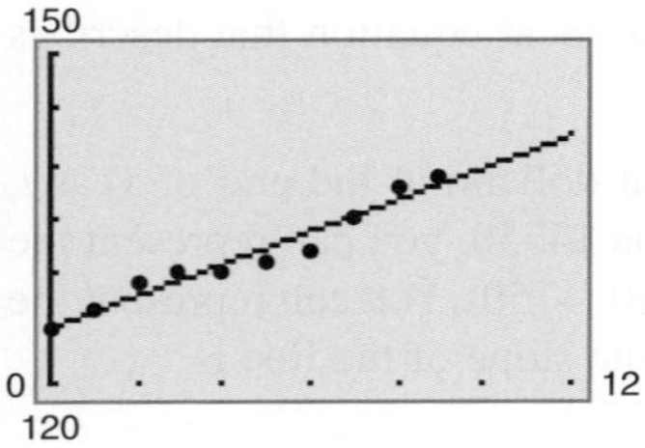

SOLUTION

a. Enter the data into the graphing utility. Then, using the *regression* feature of the graphing utility, you should obtain a linear model for the data that can be rounded to the following:

$$y = 1.45x + 125.1, \quad 0 \le x \le 9.$$

b. The graph of the equation and the scatter plot are shown in Figure 2.38.

c. Substitute $x = 11$ into the equation found in part (a).

$y = 1.45x + 125.1$ Write equation.

$y = 1.45(11) + 125.1$ Substitute 11 for x.

$y = 141.05$ Simplify.

So, according to the model, about 141,000,000 individual income tax returns will be filed in the United States in 2011.

FIGURE 2.38

✓ **Checkpoint 7**

Repeat Example 7 using only the data for the years 2004–2009.

Example 8 **Alternative-Fueled Vehicles**

The number of alternative-fueled vehicles v (in thousands) in use in the United States each year from 1995 through 2008 is shown in the table. Construct a scatter plot that represents the data and find a linear model that approximates the data. *(Source: U.S. Energy Information Administration)*

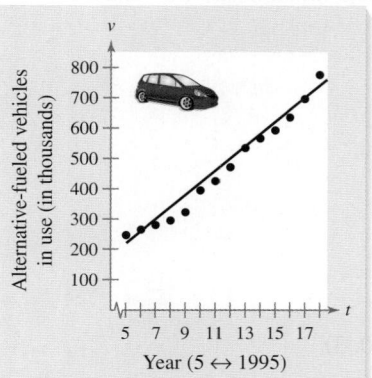

FIGURE 2.39

Year	1995	1996	1997	1998	1999	2000	2001
v	246.9	265.0	280.2	295.0	322.3	394.7	425.5

Year	2002	2003	2004	2005	2006	2007	2008
v	471.1	534.0	565.5	592.1	634.6	695.8	775.7

SOLUTION Let $t = 5$ represent 1995. The scatter plot of the data is shown in Figure 2.39. Draw a line on the scatter plot that approximates the data. To find an equation of the line, approximate two points on the line:

$$(t_1, v_1) = (7, 300)$$

and

$$(t_2, v_2) = (17, 700).$$

So, the slope of the line is

$$m \approx \frac{v_2 - v_1}{t_2 - t_1}$$

$$= \frac{700 - 300}{17 - 7}$$

$$= 40.$$

Using the point-slope form, you can determine that an equation of the line is

$$v - 300 = 40(t - 7) \qquad \text{Point-slope form}$$

$$v = 40t + 20. \qquad \text{Slope-intercept form}$$

To check this model, compare the actual v-values with the v-values given by the model.

✓ **Checkpoint 8**

Repeat Example 8 using only the data for 2003 through 2008. ■

SUMMARIZE (Section 2.3)

1. State the definition of a mathematical model *(page 183)*. For an example that uses a mathematical model, see Example 1.

2. Describe what it means when two quantities vary directly *(page 184)*. For examples of direct variation, see Examples 2 and 3.

3. Describe the relationship between slope and rate of change *(page 186)*. For examples that involve rates of change, see Examples 4, 5, and 6.

4. State the definition of linear regression *(page 188)*. For examples of linear regression, see Examples 7 and 8.

SKILLS WARM UP 2.3   The following warm-up exercises involve skills that were covered in earlier sections. You will use these skills in the exercise set for this section. For additional help, review Sections 2.1 and 2.2.

In Exercises 1–4, sketch the graph of the line.

1. $y = 2x$

2. $y = \dfrac{1}{2}x$

3. $y = 2x + 1$

4. $y = \dfrac{1}{2}x + 1$

In Exercises 5 and 6, find an equation of the line that has the given slope and y-intercept.

5. Slope: 1; y-intercept: $(0, 2)$

6. Slope: $\dfrac{3}{2}$; y-intercept: $(0, 3)$

In Exercises 7–10, find an equation of the line that passes through the two points.

7. $(1, 3)$ and $(6, 8)$

8. $(0, 4)$ and $(7, 10)$

9. $(1, 5.2)$ and $(5, 4.7)$

10. $(2, 6.5)$ and $(8, 3.6)$

Exercises 2.3   See www.CalcChat.com for worked-out solutions to odd-numbered exercises.

1. Dog Growth The weight of a puppy recorded every two months is shown in the table.

Age (in months)	2	4	6
Weight (in pounds)	21	44	63

Age (in months)	8	10	12
Weight (in pounds)	82	92	101

A linear model that approximates the puppy's weight w (in pounds) in month t is $w = 8.0t + 11$, $2 \le t \le 12$. Plot the actual data with the model. How closely does the model represent the data?

2. Non-Wage Earners The numbers of working-age civilians (in millions) in the United States who were not involved in the labor force from 2000 through 2009 are given by the ordered pairs.

(2000, 70.0)   (2001, 71.4)   (2002, 72.7)
(2003, 74.7)   (2004, 76.0)   (2005, 76.8)
(2006, 77.4)   (2007, 78.7)   (2008, 79.5)
(2009, 81.7)

A linear model that approximates the data is

$y = 1.22t + 70.4, \ 0 \le t \le 9$

where y is the number of civilians (in millions) and $t = 0$ represents 2000. Plot the actual data with the model. How closely does the model represent the data? *(Source: U.S. Bureau of Labor Statistics)*

3. Aeropostale Sales The yearly sales y (in millions of dollars) of Aeropostale from 2001 through 2009 are given by the ordered pairs.

(2001, 304.8)   (2002, 550.9)   (2003, 734.9)
(2004, 964.2)   (2005, 1204.3)   (2006, 1413.2)
(2007, 1590.9)   (2008, 1885.5)   (2009, 2230.1)

Use a graphing utility to create a scatter plot of the data. Let $x = 1$ represent 2001. Then use the *regression* feature of the graphing utility to find a best-fitting line for the data. Graph the model and the data together. How closely does the model represent the data? *(Source: Aeropostale, Inc.)*

4. Consumer Price Index For urban consumers of educational and communication materials, the Consumer Price Index y, giving the dollar amount equal to the buying power of $100 in December 1997, is shown for each year from 1998 through 2009 by the ordered pairs.

(1998, 100.3)   (1999, 101.2)   (2000, 102.5)
(2001, 105.2)   (2002, 107.9)   (2003, 109.8)
(2004, 111.6)   (2005, 113.7)   (2006, 116.8)
(2007, 119.6)   (2008, 123.6)   (2009, 127.4)

Use a graphing utility to create a scatter plot of the data. Let $x = 8$ represent 1998. Then use the *regression* feature of the graphing utility to find a best-fitting line for the data. Graph the model and the data together. How closely does the model represent the data? *(Source: U.S. Bureau of Labor Statistics)*

Direct Variation   In Exercises 5–10, y is proportional to x. Use the x- and y-values to find a linear model that relates y and x.

5. $x = 8$, $y = 3$

6. $x = 5$, $y = 9$

7. $x = 15$, $y = 300$

8. $x = 12$, $y = 204$

9. $x = 7$, $y = 3.2$

10. $x = 11$, $y = 1.5$

Direct Variation   In Exercises 11–14, write a linear model that relates the variables.

11. H varies directly as p; $H = 27$ when $p = 9$

12. s is proportional to t; $s = 32$ when $t = 4$

13. c is proportional to d; $c = 12$ when $d = 20$

14. r varies directly as s; $r = 25$ when $s = 40$

15. Simple Interest   The simple interest received from an investment is directly proportional to the amount of the investment. By investing \$2000 in a corporate bond, you obtain interest of \$120 at the end of 1 year. Find a mathematical model that gives the interest I at the end of 1 year in terms of the amount invested P.

16. Simple Interest   The simple interest received from an investment is directly proportional to the amount of the investment. By investing \$4000 in a municipal bond, you obtain interest of \$280 at the end of 1 year. Find a mathematical model that gives the interest I at the end of 1 year in terms of the amount invested P.

17. Property Tax   Your property tax is based on the assessed value of your property. (The assessed value is often lower than the actual value of the property.) A house that has an assessed value of \$165,000 has a property tax of \$2211.

(a) Find a mathematical model that gives the amount of property tax y in terms of the assessed value x of the property.

(b) Use the model to find the property tax on a house that has an assessed value of \$185,000.

18. State Sales Tax   An item that sells for \$145.99 has a sales tax of \$10.22.

(a) Find a mathematical model that gives the amount of sales tax y in terms of the retail price x.

(b) Use the model to find the sales tax on a purchase that has a retail price of \$540.50.

19. Miles and Kilometers   While driving, your speedometer indicates that your speed is 64 miles per hour, or 103 kilometers per hour.

(a) Use this information to find a mathematical model that relates miles per hour to kilometers per hour.

(b) Use the model to complete the table.

Kilometers per hour	40	60	80	100	120
Miles per hour					

20. Liters and Gallons   You are buying gasoline and notice that 14 gallons of gasoline is the same as 53 liters.

(a) Use this information to find a mathematical model that relates gallons to liters.

(b) Use the model to complete the table.

Gallons	5	10	20	25	30
Liters					

Writing a Linear Equation   In Exercises 21–26, you are given the 2012 value of a product *and* the rate at which the value is expected to change during the next 5 years. Use this information to write a linear equation that gives the dollar value V of the product in terms of the year. (Let $t = 12$ represent 2012.)

	2012 Value	*Rate*
21.	\$2540	\$140 increase per year
22.	\$156	\$4.50 increase per year
23.	\$20,400	\$2142 decrease per year
24.	\$45,000	\$2800 decrease per year
25.	\$154,000	\$10,780 increase per year
26.	\$245,000	\$5600 increase per year

27. Parachuting   After opening the parachute, the descent of a parachutist follows a linear model. At 2:08 P.M., the height of the parachutist is 7000 feet. At 2:10 P.M., the height is 4600 feet.

(a) Write a linear equation that gives the height h of the parachutist in terms of the time t. (Let $t = 0$ represent 2:08 P.M., and let t be measured in seconds.)

(b) Use the equation in part (a) to find the time when the parachutist will reach the ground.

28. Distance Traveled by a Car   You are driving at a constant speed. At 4:30 P.M., you drive by a sign that gives the distance to Montgomery, Alabama as 84 miles. At 4:59 P.M., you drive by another sign that gives the distance to Montgomery as 56 miles.

(a) Write a linear equation that gives your distance from Montgomery in terms of time t. (Let $t = 0$ represent 4:30 P.M. and let t be measured in minutes.)

(b) Use the equation in part (a) to find the time when you will reach Montgomery.

29. Straight-Line Depreciation   A business purchases a piece of equipment for \$875. After 5 years, the equipment will have no value. Write a linear equation giving the value V of the equipment during the 5 years.

30. Straight-Line Depreciation   A business purchases a piece of equipment for \$25,000. The equipment will be replaced in 10 years, at which time its salvage value is expected to be \$2000. Write a linear equation giving the value V of the equipment during the 10 years.

31. Hourly Wages A manufacturer pays its assembly line workers $11.50 per hour. In addition, workers receive a piecework rate of $0.75 per unit produced. Write a linear equation for the hourly wages W in terms of the number of units x produced per hour.

32. Sales Commission A salesperson receives a monthly salary of $2500 plus a commission of 7% of sales. Write a linear equation for the salesperson's monthly wage W in terms of the person's monthly sales S.

33. Hair Growth Rate A beautician cuts a customer's hair to a length of 4 inches. During the next year, the customer's hair grows at a rate of 0.5 inch per month.

(a) Write a linear equation giving the hair length H (in inches) in terms of the number of months t.

(b) The hair keeps growing at this constant rate. Predict the hair length in 1.5 years.

34. Pest Management The cost of implementing an invasive species management system in a forest is related to the area of the forest. It costs $630 to implement the system in a forest area of 10 acres. It costs $1070 in a forest area of 18 acres.

(a) Write a linear equation giving the cost of the invasive species management system in terms of the number of acres x of forest.

(b) Use the equation in part (a) to find the cost of implementing the system in a forest area of 30 acres.

Finding an Equation of a Line In Exercises 35–40, can the data be approximated by a linear model? If so, sketch the line that best approximates the data. Then find an equation of the line.

35.
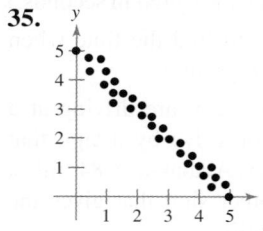

36.
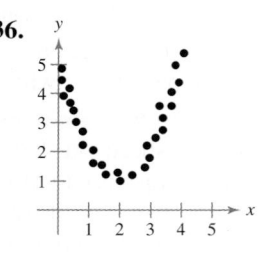

37.
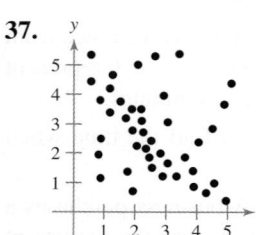

38.
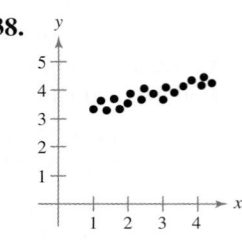

39.
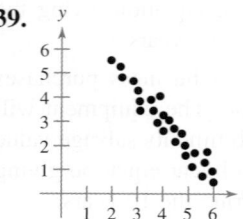

40.
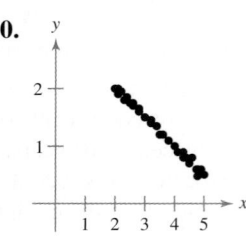

41. Autism The numbers of autistic children C (in thousands) receiving disability services for the years 2001 through 2009 are shown in the table. *(Source: U.S. Department of Education)*

Year	2001	2002	2003	2004	2005
Autistic children, C	98.6	118.8	141.1	166.5	193.8

Year	2006	2007	2008	2009
Autistic children, C	224.6	258.0	292.8	333.0

(a) Use a graphing utility to create a scatter plot of the data. Let $t = 1$ represent 2001. Do the data appear linear?

(b) Use the *regression* feature of the graphing utility to find a linear model for the data.

(c) State the slope of the graph of the linear model from part (b) and interpret its meaning in the context of the problem.

(d) Use the linear model to predict the numbers of autistic children receiving disability services in 2011 and 2012. Are your predictions reasonable?

42. Japan The population of Japan is expected to drop by 30% over the next 50 years as the percent of its citizens that are elderly increases. Projections for Japan's population through 2050 are shown in the table. *(Source: National Institute of Population and Social Security Research)*

Year, t	2005	2010	2020
Population, P (in millions)	127.7	127.5	124.1

Year, t	2030	2040	2050
Population, P (in millions)	117.6	109.3	100.6

(a) Use a graphing utility to create a scatter plot of the data. Let $t = 5$ represent 2005. Do the data appear linear?

(b) Use the *regression* feature of the graphing utility to find a linear model for the data.

(c) Identify the slope of the model from part (b) and interpret its meaning in the context of the problem.

(d) Use the linear model to predict the populations in 2025, 2035, and 2060. Are these predictions reasonable?

43. Yearly Revenue  The yearly revenues y (in millions of dollars) for California Pizza Kitchen for the years 2000 through 2009 are given by the ordered pairs. *(Source: California Pizza Kitchen, Inc.)*

(2000, 210.8)	(2001, 249.3)
(2002, 306.3)	(2003, 359.9)
(2004, 422.5)	(2005, 479.6)
(2006, 554.6)	(2007, 632.9)
(2008, 677.1)	(2009, 664.7)

(a) Use a graphing utility to create a scatter plot of the data. Let $t = 0$ represent 2000.

(b) Use two points on the scatter plot to find an equation of a line that approximates the data.

(c) Use the *regression* feature of the graphing utility to find a linear model for the data. Use this model and the model from part (b) to predict the revenues in 2010 and 2011.

(d) California Pizza Kitchen projected the revenues in 2010 and 2011 to be $645 million and $685 million, respectively. How close are these projections to the predictions from the models?

(e) California Pizza Kitchen also expects the yearly revenue to reach $850 million in 2013, 2014, or 2015. Do the models from parts (b) and (c) support this? Explain your reasoning.

44. Revenue per Share  The revenues per share of stock y (in dollars) for California Pizza Kitchen for the years 2000 through 2009 are given by the ordered pairs. *(Source: California Pizza Kitchen, Inc.)*

(2000, 7.86)	(2001, 9.02)
(2002, 10.90)	(2003, 12.66)
(2004, 14.64)	(2005, 16.26)
(2006, 19.16)	(2007, 22.32)
(2008, 28.37)	(2009, 27.47)

(a) Use a graphing utility to create a scatter plot of the data. Let $t = 0$ represent 2000.

(b) Use two points on the scatter plot to find an equation of a line that approximates the data.

(c) Use the *regression* feature of the graphing utility to find a linear model for the data. Use this model and the model from part (b) to predict the revenues per share in 2010 and 2011.

(d) California Pizza Kitchen projected the revenues per share in 2010 and 2011 to be $25.80 and $27.40, respectively. How close are these projections to the predictions from the models?

(e) California Pizza Kitchen also expects the revenue per share to reach $34.00 in 2013, 2014, or 2015. Do the models from parts (b) and (c) support this? Explain your reasoning.

45. Purchasing Power  The value y (in 1982 dollars) of each dollar received by producers in each of the years from 1994 through 2008 in the United States is represented by the ordered pairs. *(Source: U.S. Bureau of Labor Statistics)*

(1994, 0.797)	(1995, 0.782)
(1996, 0.762)	(1997, 0.759)
(1998, 0.765)	(1999, 0.752)
(2000, 0.725)	(2001, 0.711)
(2002, 0.720)	(2003, 0.698)
(2004, 0.673)	(2005, 0.642)
(2006, 0.623)	(2007, 0.600)
(2008, 0.565)	

(a) Use a spreadsheet software program to generate a scatter plot of the data. Let $t = 4$ represent 1994. Do the data appear linear?

(b) Use the *regression* feature of the spreadsheet software program to find a linear model for the data.

(c) Use the model to predict the value (in 1982 dollars) of 1 dollar received by producers in 2010 and in 2011. Discuss the reliability of your predictions based on your scatter plot and the graph of your linear model for the data.

46. Purchasing Power  The value y (in 1982–1984 dollars) of each dollar paid by consumers in each of the years from 1994 through 2008 in the United States is represented by the ordered pairs. *(Source: U.S. Bureau of Labor Statistics)*

(1994, 0.675)	(1995, 0.656)
(1996, 0.638)	(1997, 0.623)
(1998, 0.613)	(1999, 0.600)
(2000, 0.581)	(2001, 0.565)
(2002, 0.556)	(2003, 0.543)
(2004, 0.529)	(2005, 0.512)
(2006, 0.496)	(2007, 0.482)
(2008, 0.464)	

(a) Use a spreadsheet software program to generate a scatter plot of the data. Let $t = 4$ represent 1994. Do the data appear linear?

(b) Use the *regression* feature of the spreadsheet software program to find a linear model for the data.

(c) Use the model to predict the value (in 1982–1984 dollars) of 1 dollar paid by consumers in 2010 and in 2011. Discuss the reliability of your predictions based on your scatter plot and the graph of your linear model for the data.

47. Health Services   The numbers of employees E (in thousands) in the health services industry for the years 2004 through 2009 are shown in the table.   *(Source: U.S. Bureau of Labor Statistics)*

Year	2004	2005	2006
Employees, E	14,190	14,536	14,925

Year	2007	2008	2009
Employees, E	15,380	15,798	16,101

(a) Use a graphing utility to create a scatter plot of the data. Let $t = 4$ represent 2004. Do the data appear linear?

(b) Use the *regression* feature of the graphing utility to find a linear model for the data.

(c) Use the model to predict the numbers of employees in 2011 and 2013.

(d) Graph the linear model along with the scatter plot of the data. Comparing the data with the model, are the predictions in part (c) most likely high, low, or just about right? Explain your reasoning.

48. Health Care   The total yearly health care expenditures E (in billions of dollars) in the United States for the years 2000 through 2009 are shown in the table.   *(Source: U.S. Centers for Medicare and Medicaid Services)*

Year	2000	2001	2002	2003	2004
Expenditures, E	1353	1469	1602	1735	1855

Year	2005	2006	2007	2008	2009
Expenditures, E	1983	2113	2240	2339	2486

(a) Use a graphing utility to create a scatter plot of the data. Let $t = 0$ represent 2000. Do the data appear linear?

(b) Use the *regression* feature of the graphing utility to find a linear model for the data.

(c) Use the model to predict the health care expenditures in 2011 and 2012.

(d) Graph the linear model along with the scatter plot. The government's projections for 2011 and 2012 are $2703 billion and $2850 billion, respectively. Compare these projections with the predictions in part (c). Are the government's projections high, low, or just about right? Explain your reasoning.

49. Think About It   Annual data from three years are used to create linear models for the population and the yearly snowfall of Reno, Nevada. Which model is more likely to give better predictions for future years? Discuss the appropriateness of using only three data points in each situation.

50.  **HOW DO YOU SEE IT?**   Match the description of the situation with its graph. Find the slope of the line and describe how it is interpreted in the real-life situation. [The graphs are labeled (i), (ii), (iii), and (iv).]

(a) You pay $0.06 in sales tax for every dollar spent.

(b) A person pays $40 to reserve a banquet hall, plus $12 for each hour it is occupied.

(c) You pay $800 for a new refrigerator that depreciates $50 per year.

(d) You pay $34.99 per day to rent a car.

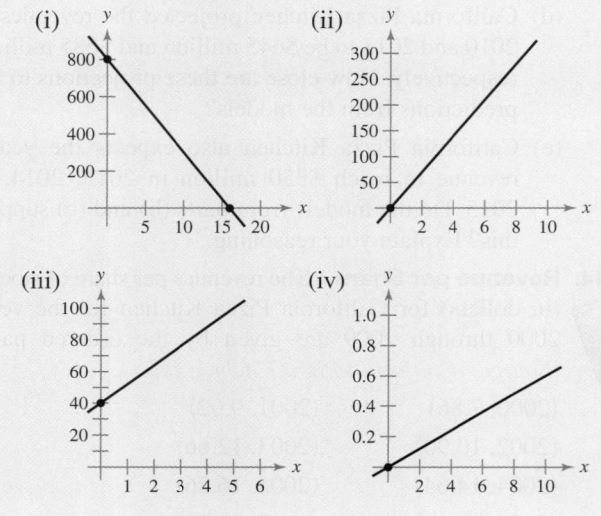

51. Think About It   You begin a dance video game with 800 points. You lose 50 points each time you incorrectly complete a dance move. Does this description match graph (i) in Exercise 50? Explain.

52. Think About It   You start with $12 and save $40 per week. Does this description match graph (iii) in Exercise 50? Explain.

53. Think About It   How can you use the rate of change of y with respect to x to determine whether the data points $(1, 10)$, $(2, 20)$, $(3, 30)$, $(4, 40)$, and $(5, 50)$ are linear?

54. Writing   Use your campus library, the Internet, or some other reference source to locate data that you think describes a linear relationship. Use a graphing utility to create a scatter plot of the data and find a linear model that represents the points. Interpret the slope and y-intercept in the context of the data. Write a summary of your findings.

2.4 Functions

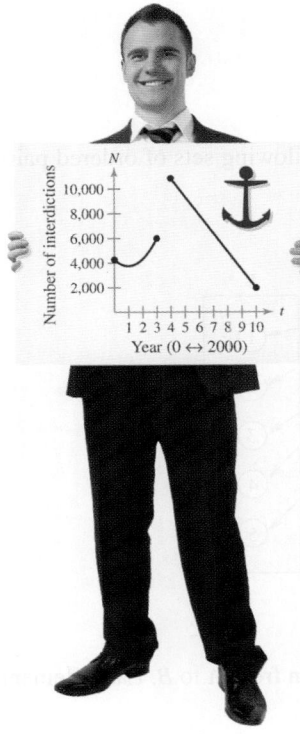

In Exercise 86 on page 206, you will use a piecewise-defined function to approximate the numbers of Coast Guard interdictions in various years.

originalpunkt/www.shutterstock.com

- ■ Determine whether a correspondence is a function, and use function notation.
- ■ Find the domain of a function.
- ■ Write a function that relates quantities in an application problem.

Introduction to Functions and Function Notation

Many everyday phenomena involve two quantities that are related to each other by some rule of correspondence. Here are some examples.

1. The simple interest I earned on $1000 for 1 year is related to the annual interest rate r by the formula $I = 1000r$.

2. The distance d traveled on a bicycle in 2 hours is related to the speed s of the bicycle by the formula $d = 2s$.

3. The area A of a circle is related to its radius r by the formula $A = \pi r^2$.

Not all correspondences between two quantities have simple mathematical formulas. For instance, people commonly match up athletes with jersey numbers and hours of the day with temperatures. In each of these cases, however, there is some rule of correspondence that matches each item from one set with exactly one item from a different set. Such a rule of correspondence is called a **function.**

> ### Definition of a Function
>
> A **function** f from a set A to a set B is a rule of correspondence that assigns to each element x in the set A exactly one element y in the set B. The set A is the **domain** (or set of inputs) of the function f, and the set B contains the **range** (or set of outputs).

To get a better idea of this definition, look at the function that relates the time of day to the temperature in Figure 2.40. This function can be represented by the following set of ordered pairs.

$$\{(1, 9°), (2, 13°), (3, 15°), (4, 15°), (5, 12°), (6, 4°)\}$$

In each ordered pair, the first coordinate (x-value) is the input and the second coordinate (y-value) is the output. In this example, note the following characteristics of a function.

1. Each element of A (the domain) must be matched with an element of B (the range).

2. Some elements of B may not be matched with any element of A.

3. Two or more elements of A may be matched with the same element of B.

4. An element of A cannot be matched with two different elements of B.

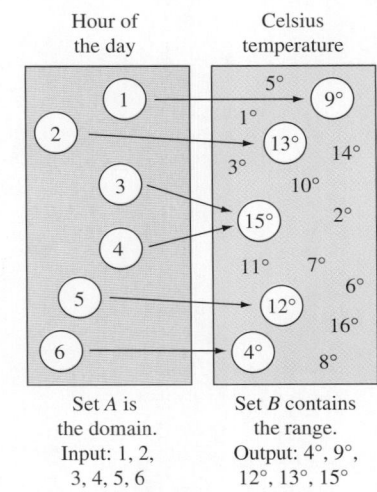

Hour of the day	Celsius temperature
Set A is the domain. Input: 1, 2, 3, 4, 5, 6	Set B contains the range. Output: 4°, 9°, 12°, 13°, 15°

Function from Set A to Set B

FIGURE 2.40

To determine whether a correspondence is a function, you must decide whether each element of the domain *A* is matched with exactly one element of the range *B*. When any element of *A* is matched with two or more elements of *B*, the correspondence is not a function. For example, people are not a function of their birthday month because many people are born in any given month.

Example 1   Testing for Functions

Let $A = \{a, b, c\}$ and $B = \{1, 2, 3, 4, 5\}$. Which of the following sets of ordered pairs or figures represent functions from set *A* to set *B*?

a. $\{(a, 2), (b, 3), (c, 4)\}$

b. $\{(a, 4), (b, 5)\}$

c.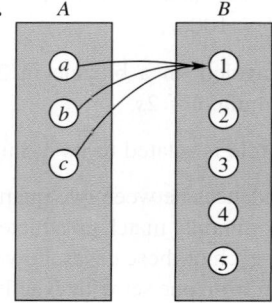

d.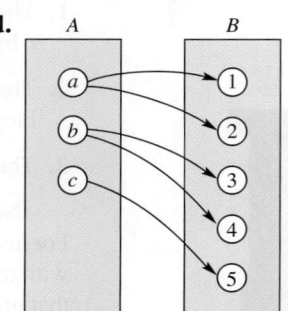

SOLUTION

a. This collection of ordered pairs *does* represent a function from *A* to *B*. Each element of *A* is matched with exactly one element of *B*.

b. This collection of ordered pairs *does not* represent a function from *A* to *B*. Not every element of *A* is matched with an element of *B*. The element *c* of *A* is not matched with an element of *B*.

c. This figure *does* represent a function from *A* to *B*. It does not matter that each element of *A* is matched with the same element of *B*.

d. This figure *does not* represent a function from *A* to *B*. The element *a* of *A* is matched with *two* elements of *B*. This is also true of the element *b*. _____

✓ Checkpoint 1

Let $A = \{a, b, c, d\}$ and $B = \{1, 3, 5, 7\}$. Decide whether each set of ordered pairs represents a function from set *A* to set *B*.

a. $\{(a, 1), (c, 5), (d, 7)\}$

b. $\{(a, 3), (b, 7), (c, 1), (d, 3)\}$

c. $\{(b, 3), (b, 5)\}$

Representing functions by sets of ordered pairs is a common practice in *discrete mathematics*. In algebra, however, it is more common to represent functions by equations or formulas involving two variables. For instance, the equation

$$y = x^2 \qquad \text{\textit{y} is a function of \textit{x}.}$$

represents the variable *y* as a function of the variable *x*. In this equation, *x* is the **independent variable** and *y* is the **dependent variable.** The domain of the function is the set of all values taken on by the independent variable *x*, and the range of the function is the set of all values taken on by the dependent variable *y*.

Example 2   **Testing for Functions Represented by Equations**

Which of the equations represent(s) y as a function of x?

a. $x^2 + y = 1$ **b.** $-x + y^2 = 1$

SOLUTION   To determine whether y is a function of x, try to solve for y in terms of x.

a. Solving for y yields

$$x^2 + y = 1 \qquad \text{Write original equation.}$$
$$y = 1 - x^2. \qquad \text{Solve for } y.$$

To each value of x there corresponds exactly one value of y. So, y *is* a function of x.

b. Solving for y yields

$$-x + y^2 = 1 \qquad \text{Write original equation.}$$
$$y^2 = 1 + x \qquad \text{Add } x \text{ to each side.}$$
$$y = \pm\sqrt{1 + x}. \qquad \text{Solve for } y.$$

The $\pm$ indicates that to a given value of x there correspond two values of y. So, y *is not* a function of x.

✓**Checkpoint 2**

Determine whether each equation represents y as a function of x.

a. $y - 2 = x^2$ **b.** $x^2 + y^2 = 25$ ■

When an equation is used to represent a function, it is convenient to name the function so that it can be referenced easily. For example, you know that the equation $y = 1 - x^2$ describes y as a function of x. Suppose you give this function the name "f." Then you can use the **function notation** shown below.

Input	Output	Equation
x	$f(x)$	$f(x) = 1 - x^2$

The symbol $f(x)$ is read as the **value of f at x** or simply f **of x.** The symbol $f(x)$ corresponds to the y-value for a given x. So, you can write $y = f(x)$. Keep in mind that f is the *name* of the function, whereas $f(x)$ is the *value* of the function at x. For instance, the function given by

$$f(x) = 3 - 2x$$

has *function values* denoted by $f(-1)$, $f(0)$, $f(2)$, and so on. To find these values, substitute the specified input values into the given equation.

For $x = -1$, $f(-1) = 3 - 2(-1)$

$$= 3 + 2$$
$$= 5.$$

For $x = 0$, $f(0) = 3 - 2(0)$

$$= 3 - 0$$
$$= 3.$$

For $x = 2$, $f(2) = 3 - 2(2)$

$$= 3 - 4$$
$$= -1.$$

STUDY TIP

Make sure you understand how function notation is written and how it is read. Frequently, $f(x)$ is misinterpreted as "f times x" rather than "f of x."

Although f is often used as a convenient function name and x is often used as the independent variable, you can use other letters. For instance,

$$f(x) = x^2 - 4x + 7, \quad f(t) = t^2 - 4t + 7, \quad \text{and} \quad g(s) = s^2 - 4s + 7$$

all define the same function. In fact, the role of the independent variable in a function is simply that of a "placeholder." Consequently, the function above could be described by the form

$$f() = ()^2 - 4() + 7.$$

Example 3   Evaluating a Function

Let $g(x) = -x^2 + 4x + 1$. Evaluate $g(x)$ at each specified value and simplify.

a. $g(2)$ **b.** $g(t)$ **c.** $g(x + 2)$

SOLUTION

a. Replace x with 2 in $g(x) = -x^2 + 4x + 1$.

$$g(2) = -(2)^2 + 4(2) + 1 = -4 + 8 + 1 = 5$$

b. Replace x with t.

$$g(t) = -(t)^2 + 4(t) + 1 = -t^2 + 4t + 1$$

c. Replace x with $x + 2$.

$$g(x + 2) = -(x + 2)^2 + 4(x + 2) + 1$$
$$= -(x^2 + 4x + 4) + 4x + 8 + 1$$
$$= -x^2 - 4x - 4 + 4x + 8 + 1$$
$$= -x^2 + 5$$

> **STUDY TIP**
>
> In Example 3(c), note that $g(x + 2)$ is not equal to $g(x) + g(2)$. In general, $g(u + v) \neq g(u) + g(v)$.

✓ **Checkpoint 3**

Let $h(x) = 2x^2 + x - 4$. Find $h(-1)$.   ■

A function defined by two or more equations over a specified domain is called a **piecewise-defined function.**

Example 4   A Piecewise-Defined Function

Evaluate the function when $x = -1, 0,$ and 1.

$$f(x) = \begin{cases} x^2 + 1, & x < 0 \\ x - 1, & x \geq 0 \end{cases}$$

SOLUTION For each x-value, only one part of the piecewise function is used. Because $x = -1$ is less than 0, use $f(x) = x^2 + 1$ to obtain

$$f(-1) = (-1)^2 + 1 = 2.$$

For $x = 0$, use $f(x) = x - 1$ to obtain $f(0) = (0) - 1 = -1$. For $x = 1$, use $f(x) = x - 1$ to obtain $f(1) = (1) - 1 = 0$. The graph of f is shown in Figure 2.41.

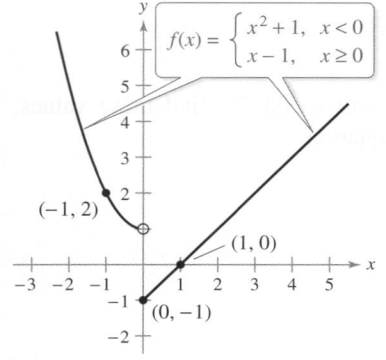

FIGURE 2.41

✓ **Checkpoint 4**

Evaluate the function in Example 4 when $x = -3$ and 3.   ■

Finding the Domain of a Function

The domain of a function can be described explicitly or it can be *implied* by the expression used to define the function. The **implied domain** is the set of all real numbers for which the expression is defined. For instance, the function given by

$$f(x) = \frac{1}{x^2 - 4}$$ Domain excludes x-values
   that result in division by zero.

has an implied domain that consists of all real x other than $x = \pm 2$. These two values are excluded from the domain because division by zero is undefined. Another common type of implied domain results from the restrictions needed to avoid even roots of negative numbers. For example, the function given by

$$f(x) = \sqrt{x}$$ Domain excludes x-values that result
   in even roots of negative numbers.

is defined only for $x \geq 0$. So, its implied domain is the interval $[0, \infty)$. In general, the domain of a function *excludes* values that would cause division by zero *or* result in the even root of a negative number.

Example 5   Finding the Domain of a Function

Find the domain of each function.

a. f: $\{(-3, 0), (-1, 4), (0, 2), (2, 2), (4, -1)\}$ **b.** $g(x) = \dfrac{1}{x + 5}$

c. Volume of a sphere: $V = \dfrac{4}{3}\pi r^3$ **d.** $h(x) = \sqrt{4 - x^2}$

e. $r(x) = \sqrt[3]{x + 3}$

SOLUTION

a. The domain of f consists of all first coordinates in the set of ordered pairs.

   Domain $= \{-3, -1, 0, 2, 4\}$

b. Excluding x-values that yield zero in the denominator, the domain of g is the set of all real numbers x such that $x \neq -5$.

c. Because this function represents the volume of a sphere, the values of the radius r must be positive. So, the domain is the set of all real numbers r such that $r > 0$.

d. This function is defined only for x-values for which

   $4 - x^2 \geq 0$.

   Using the methods described in Section 1.7, you can conclude that $-2 \leq x \leq 2$. So, the domain of h is the interval $[-2, 2]$.

e. Because the cube root of any real number is defined, the domain of r is the set of all real numbers, or $(-\infty, \infty)$.

✓ Checkpoint 5

Find the domain of the function $f(x) = 6 - x^3$.

In Example 5(c), note that the domain of a function may be implied by the physical context. For instance, from the equation $V = \frac{4}{3}\pi r^3$, you would have no reason to restrict r to positive values, but the physical context implies that a sphere cannot have a negative or zero radius.

Applications

 Example 6   **The Dimensions of a Container**

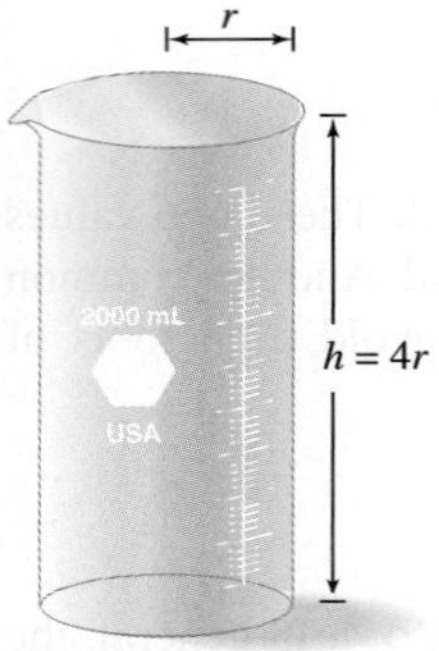

You are working with a cylindrical beaker in a chemistry lab experiment. The height of the beaker is 4 times the radius, as shown in Figure 2.42.

a. Write the volume of the beaker as a function of the radius r.

b. Write the volume of the beaker as a function of the height h.

SOLUTION

a. $V = \pi r^2 h$ Write volume formula.

  $= \pi r^2 (4r)$ Substitute $4r$ for h.

  $= 4\pi r^3$ V is a function of r.

b. $V = \pi \left(\dfrac{h}{4}\right)^2 h$ Substitute $\dfrac{h}{4}$ for r.

  $= \dfrac{\pi h^3}{16}$ V is a function of h.

FIGURE 2.42 (left margin, with beaker diagram showing r and $h = 4r$)

✓ **Checkpoint 6**

In Example 6, write the volume of the beaker as a function of the height h when the radius is twice the height.

 Example 7   **The Path of a Baseball**

A baseball is hit at a point 3 feet above home plate at a velocity of 100 feet per second and an angle of 45°. The path of the baseball is given by the function

$$y = -0.0032x^2 + x + 3$$

where y and x are measured in feet. Will the baseball clear a 10-foot fence located 300 feet from home plate?

SOLUTION   When $x = 300$, the height of the baseball is given by

$$y = -0.0032(300)^2 + 300 + 3 = 15 \text{ feet.}$$

Yes, the ball will clear the fence, as shown in Figure 2.43.

STUDY TIP

Notice that in Figure 2.43, the baseball is not at the point $(0, 0)$ when it is initially hit. This is because the baseball is hit at a point 3 feet above home plate. So, it is hit at the point $(0, 3)$.

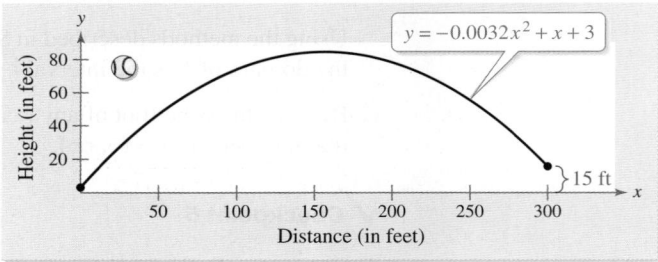

FIGURE 2.43

✓ **Checkpoint 7**

In Example 7, will the baseball clear a 35-foot fence located 280 feet from home plate?

 **Example 8** **Houses Sold**

The number H (in thousands) of new privately owned one-family houses sold in the United States increased in a quadratic pattern from 1998 through 2004. Then, in 2005, the pattern changed from a quadratic to a linear pattern (see Figure 2.44). These two patterns can be approximated by the function

$$H = \begin{cases} 15.036t^2 - 278.68t + 2159.3, & 8 \le t \le 14 \\ -238.20t + 4843.4, & 15 \le t \le 19 \end{cases}$$

with $t = 8$ corresponding to 1998. Use this function to approximate the total number of new privately owned one-family houses sold between 1998 and 2009. *(Source: U.S. Census Bureau and U.S. Department of Housing and Urban Development)*

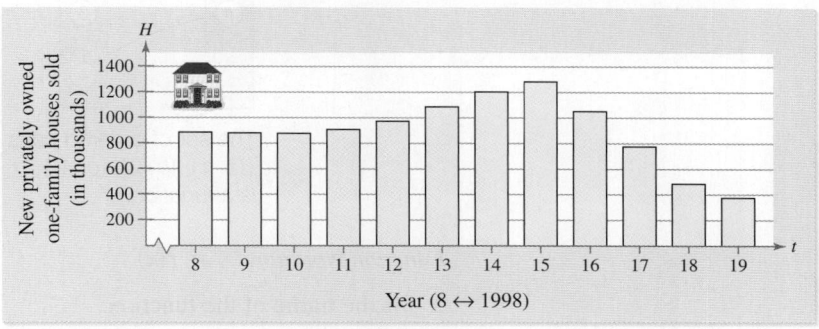

FIGURE 2.44

SOLUTION From 1998 through 2004, use the equation $H = 15.036t^2 - 278.68t + 2159.3$ to approximate the number of houses sold, as shown in the table.

t	8	9	10	11	12	13	14
H	892.2	869.1	876.1	913.2	980.3	1077.5	1204.8

From 2005 through 2009, use the equation $H = -238.20t + 4843.4$ to approximate the number of houses sold, as shown in the table.

t	15	16	17	18	19
H	1270.4	1032.2	794.0	555.8	317.6

To approximate the total number of houses sold from 1998 through 2009, you can add the amounts for the 12 years to obtain the sum 10,783.2. Because H is measured in thousands, you can conclude that the total number of new privately owned one-family houses sold between 1998 and 2009 was approximately 10,783,200.

✓ **Checkpoint 8**

The median sales price P (in thousands of dollars) of new privately owned one-family houses sold in the United States from 1999 through 2009 can be modeled by the function

$$P = \begin{cases} 8.66t + 82.3, & 9 \le t \le 13 \\ -4.680t^2 + 153.12t - 1004.5, & 14 \le t \le 19 \end{cases}$$

with $t = 9$ corresponding to 1999. Use the function to approximate the median sales price for each year from 1999 through 2009. *(Source: U.S. Census Bureau and U.S. Department of Housing and Urban Development)*

Summary of Function Terminology

Function: A **function** is a relationship between two variables such that to each value of the independent variable there corresponds exactly one value of the dependent variable.

For instance, let $A = \{a, b, c\}$ and $B = \{1, 2, 3, 4\}$.

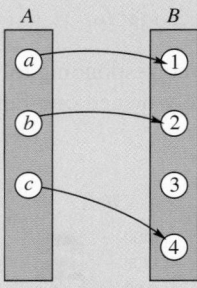

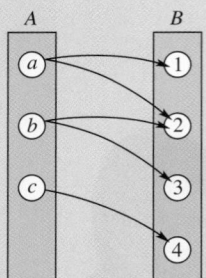

The set of ordered pairs $\{(a, 1), (b, 2), (c, 4)\}$ *is a function.*

The set of ordered pairs $\{(a, 1), (a, 2), (b, 2), (b, 3), (c, 4)\}$ *is not a function.*

Function Notation: $y = f(x)$

f is the **name** of the function.

y is the **dependent variable.**

x is the **independent variable.**

$f(x)$ is the **value of the function at x.**

Domain: The **domain** of a function is the set of all values (inputs) of the independent variable for which the function is defined. If x is in the domain of f, then f is said to be **defined** at x. If x is not in the domain of f, then f is said to be **undefined** at x.

Range: The **range** of a function is the set of all values (outputs) assumed by the dependent variable (that is, the set of all function values).

Implied Domain: If f is defined by an algebraic expression and the domain is not specified, then the **implied domain** consists of all real numbers for which the expression is defined.

SUMMARIZE   (Section 2.4)

1. State the definition of a function *(page 195)*. For examples of determining functions, see Examples 1 and 2.

2. Describe function notation *(pages 197 and 198)*. For examples of using function notation, see Examples 3 and 4.

3. State the definition of an implied domain of a function *(page 199)*. For an example of finding the domain of a function, see Example 5.

4. Describe a real-life example of how a function can be used to analyze the path of an object *(page 200, Example 7)*.

SKILLS WARM UP 2.4 The following warm-up exercises involve skills that were covered in earlier sections. You will use these skills in the exercise set for this section. For additional help, review Sections 0.2, 1.2, and 1.7.

In Exercises 1–4, simplify the expression.

1. $2(-3)^3 + 4(-3) - 7$

2. $4(-1)^2 - 5(-1) + 4$

3. $(x + 1)^2 + 3(x + 1) - 4 - (x^2 + 3x - 4)$

4. $(x - 2)^2 - 4(x - 2) - (x^2 - 4)$

In Exercises 5 and 6, solve for y in terms of x.

5. $2x + 5y - 7 = 0$

6. $y^2 = x^2$

In Exercises 7–10, solve the inequality.

7. $x^2 - 4 \geq 0$

8. $9 - x^2 \geq 0$

9. $x^2 + 2x + 1 \geq 0$

10. $x^2 - 3x + 2 \geq 0$

Exercises 2.4

See www.CalcChat.com for worked-out solutions to odd-numbered exercises.

Testing for Functions In Exercises 1–4, decide whether the set of figures represents a function from A to B.

$A = \{a, b, c\}$

$B = \{1, 2, 3, 4\}$

Give a reason for your answer. *See Example 1.*

1.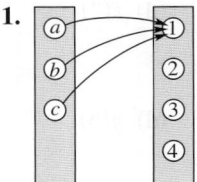

2.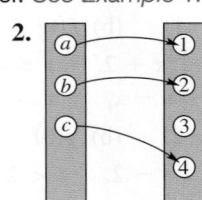

3.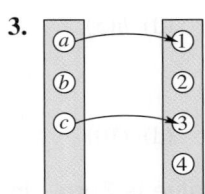

4.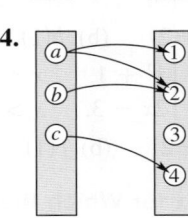

Testing for Functions In Exercises 5–8, decide whether the set of ordered pairs represents a function from A to B.

$A = \{0, 1, 2, 3\}$

$B = \{-2, -1, 0, 1, 2\}$

Give a reason for your answer. *See Example 1.*

5. $\{(0, 1), (1, -2), (2, 0), (3, 2)\}$

6. $\{(0, -1), (2, 2), (1, -2), (3, 0), (1, 1)\}$

7. $\{(0, 0), (1, 0), (2, 0), (3, 0)\}$

8. $\{(0, 2), (3, 0), (1, 1)\}$

Testing for Functions In Exercises 9–14, decide whether the set of ordered pairs represents a function from A to B.

$A = \{a, b, c\}$

$B = \{0, 1, 2, 3\}$

Give a reason for your answer. *See Example 1.*

9. $\{(a, 1), (c, 2), (c, 3), (b, 3)\}$

10. $\{(a, 3), (c, 1)\}$

11. $\{(0, a), (2, c), (3, b)\}$

12. $\{(a, 0), (c, 1), (b, 3)\}$

13. $\{(a, 1), (b, 2), (c, 3)\}$

14. $\{(c, 0), (b, 0), (a, 3)\}$

Testing for Functions Represented by Equations In Exercises 15–24, determine whether the equation represents y as a function of x. *See Example 2.*

15. $x^2 + y^2 = 4$

16. $x = y^2$

17. $x^2 + y = 4$

18. $y = \sqrt{x + 5}$

19. $2x + 3y = 4$

20. $5x - 7 = 3y$

21. $y^2 = x^2 - 1$

22. $x + y^2 = 4$

23. $x^2y - x^2 + 4y = 0$

24. $xy - y - x - 2 = 0$

Evaluating a Function In Exercises 25–28, fill in the blank and simplify. *See Example 3.*

25. $f(x) = 6 - 4x$

(a) $f(3) = 6 - 4(\boxed{})$

(b) $f(-7) = 6 - 4(\boxed{})$

(c) $f(t) = 6 - 4(\boxed{})$

(d) $f(c + 1) = 6 - 4(\boxed{})$

26. $f(s) = \dfrac{1}{s+1}$

(a) $f(4) = \dfrac{1}{(\quad) + 1}$ (b) $f(0) = \dfrac{1}{(\quad) + 1}$

(c) $f(4x) = \dfrac{1}{(\quad) + 1}$

(d) $f(x+1) = \dfrac{1}{(\quad) + 1}$

27. $g(x) = \dfrac{1}{x^2 - 2x}$

(a) $g(1) = \dfrac{1}{(\quad)^2 - 2(\quad)}$

(b) $g(-3) = \dfrac{1}{(\quad)^2 - 2(\quad)}$

(c) $g(t) = \dfrac{1}{(\quad)^2 - 2(\quad)}$

(d) $g(t+1) = \dfrac{1}{(\quad)^2 - 2(\quad)}$

28. $f(t) = \sqrt{25 - t^2}$

(a) $f(3) = \sqrt{25 - (\quad)^2}$

(b) $f(5) = \sqrt{25 - (\quad)^2}$

(c) $f(x+5) = \sqrt{25 - (\quad)^2}$

(d) $f(2x) = \sqrt{25 - (\quad)^2}$

Evaluating a Function   In Exercises 29–40, evaluate the function at each specified value of the independent variable and simplify. *See Example 3.*

29. $f(x) = 2x - 3$

(a) $f(1)$

(b) $f(-3)$

(c) $f(x-1)$

(d) $f\left(\tfrac{1}{4}\right)$

30. $g(y) = 7 - 3y$

(a) $g(0)$

(b) $g\left(\tfrac{7}{3}\right)$

(c) $g(s)$

(d) $g(s+2)$

31. $h(t) = t^2 - 2t$

(a) $h(2)$

(b) $h(-1)$

(c) $h(x+2)$

(d) $h(1.5)$

32. $k(b) = 2b^2 + 7b + 3$

(a) $k(0)$

(b) $k\left(-\tfrac{1}{2}\right)$

(c) $k(a)$

(d) $k(x+2)$

33. $V(r) = \dfrac{4}{3}\pi r^3$

(a) $V(3)$

(b) $V(0)$

(c) $V\left(\tfrac{3}{2}\right)$

(d) $V(2r)$

34. $A(s) = \dfrac{\sqrt{3}\,s^2}{4}$

(a) $A(1)$

(b) $A(0)$

(c) $A(2x)$

(d) $A(3)$

35. $f(y) = 3 - \sqrt{y}$

(a) $f(4)$

(b) $f(100)$

(c) $f(4x^2)$

(d) $f(0.25)$

36. $f(x) = \sqrt{x+3} - 2$

(a) $f(-3)$

(b) $f(1)$

(c) $f(x-3)$

(d) $f(x+4)$

37. $c(x) = \dfrac{1}{x^2 - 16}$

(a) $c(4)$

(b) $c(0)$

(c) $c(y+2)$

(d) $c(y-2)$

38. $q(t) = \dfrac{2t^2 + 3}{t^2}$

(a) $q(2)$

(b) $q(0)$

(c) $q(x)$

(d) $q(-x)$

39. $f(x) = \dfrac{|x|}{x}$

(a) $f(2)$

(b) $f(-2)$

(c) $f(x^2)$

(d) $f(x-1)$

40. $f(x) = |x| + 4$

(a) $f(2)$

(b) $f(-2)$

(c) $f(x^2)$

(d) $f(x+2)$

A Piecewise-Defined Function   In Exercises 41–44, evaluate the function at each specified value of the independent variable and simplify. *See Example 4.*

41. $f(x) = \begin{cases} 3x - 1, & x < 0 \\ 2x + 3, & x \geq 0 \end{cases}$

(a) $f(-1)$ (b) $f(0)$ (c) $f(-2)$ (d) $f(2)$

42. $g(x) = \begin{cases} \tfrac{1}{4}x + 2, & x \leq 4 \\ 8 - x, & x > 4 \end{cases}$

(a) $g(-2)$ (b) $g(0)$ (c) $g(4)$ (d) $g(6)$

43. $h(t) = \begin{cases} t^2 - 2, & t < 2 \\ 3, & t \geq 2 \end{cases}$

(a) $h(-2)$ (b) $h(1)$ (c) $h(2)$ (d) $h(5)$

44. $f(x) = \begin{cases} x^2 + 1, & x \leq 1 \\ 2x - 3, & x > 1 \end{cases}$

(a) $f(-2)$ (b) $f(1)$ (c) $f\left(\tfrac{3}{2}\right)$ (d) $f(0)$

Values of *x* for Which the Function Value is Zero   In Exercises 45–52, find all real values of *x* for which $f(x) = 0$.

45. $f(x) = 15 - 3x$

46. $f(x) = \dfrac{2x - 5}{3}$

47. $f(x) = x^2 - 9$

48. $f(x) = 2x^2 - 11x + 5$

49. $f(x) = x^3 - x$

50. $f(x) = x^3 - 3x^2 - 4x + 12$

51. $f(x) = \dfrac{3}{x-1} + \dfrac{4}{x-2}$

52. $f(x) = 3 + \dfrac{2}{x-1}$

Finding the Domain of a Function In Exercises 53–70, find the domain of the function. *See Example 5.*

53. $f: \{(0, 1), (3, 4), (5, 7), (8, 11), (9, 12)\}$

54. $g: \{(10, -4), (20, 1), (30, 6), (40, 9), (50, 13)\}$

55. $h: \{(-2, 7.4), (-1, 6.5), (0, 4.8), (1, 4.1), (2, 3.3)\}$

56. $s: \{(-5.7, 9.1), (-3.9, 5.1), (-2.9, 3.2), (-1.5, 1.1)\}$

57. $g(x) = 1 - 2x^2$

58. $f(x) = 5x^2 + 2x - 1$

59. $h(t) = \dfrac{4}{t}$

60. $s(y) = \dfrac{3y}{y + 5}$

61. $g(y) = \sqrt[3]{y - 10}$

62. $f(t) = \sqrt[3]{t + 4}$

63. $f(x) = \sqrt[4]{1 - x^2}$

64. $g(x) = \sqrt{x + 1}$

65. $g(x) = \dfrac{1}{x} - \dfrac{3}{x + 2}$

66. $h(x) = \dfrac{10}{x^2 - 2x}$

67. $f(x) = \dfrac{\sqrt{x + 1}}{x - 2}$

68. $f(s) = \dfrac{\sqrt{s - 1}}{s - 4}$

69. $f(x) = \dfrac{x - 4}{\sqrt{x}}$

70. $f(x) = \dfrac{x - 5}{\sqrt{x^2 - 9}}$

Writing a Function as a Set of Ordered Pairs In Exercises 71–74, the domain of *f* is the set

$A = \{-2, -1, 0, 1, 2\}.$

Write the function as a set of ordered pairs.

71. $f(x) = x^2$

72. $f(x) = \dfrac{2x}{x^2 + 1}$

73. $f(x) = \sqrt{x + 2}$

74. $f(x) = |x + 1|$

75. Think About It Consider

$f(x) = \sqrt{x - 2}$ and $g(x) = \sqrt[3]{x - 2}.$

Why are the domains of *f* and *g* different?

76. Think About It A student says that the domain of

$f(x) = \dfrac{\sqrt{x + 1}}{x - 3}$

is all real numbers except $x = 3$. Is the student correct? Explain.

77. Think About It Set *A* contains *n* elements and set *B* contains *m* elements. How many ordered pairs are there in a set that represents a function *f* from set *A* to set *B*?

78. Discovery Use a graphing utility to graph the functions

$f(x) = \sqrt{4 - x^2}$ and $g(x) = \sqrt{x^2 - 4}.$

What is the domain of each function? Do the domains of these two functions overlap? If so, for what values?

79. Volume of a Box An open box is to be made from a square piece of material 18 inches on a side by cutting equal squares from the corners and turning up the sides (see figure).

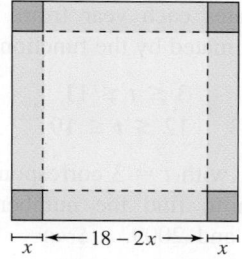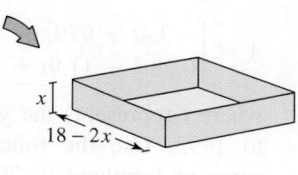

(a) Write the volume *V* of the box as a function of its height *x*.

(b) What is the domain of the function?

(c) Determine the volume of a box with a height of 4 inches.

80. Height of a Balloon A balloon carrying a transmitter ascends vertically from a point 2000 feet from the receiving station (see figure). Let *d* be the distance between the balloon and the receiving station. Write the height *h* of the balloon as a function of *d*. What is the domain of this function?

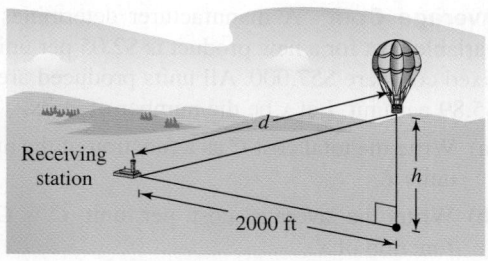

81. Path of a Ball A player throws a baseball toward a teammate. The height *y* (in feet) of the baseball is given by the function

$y = -\dfrac{1}{10}x^2 + 3x + 6$

where *x* is the horizontal distance (in feet) from where the ball is thrown. The teammate is 30 feet away and can reach to a maximum height of 7 feet. Can the teammate catch the ball without backing up? Explain.

82. Path of a Salmon   Part of the life cycle of a salmon is migration for reproduction. Salmon are anadromous fish. This means that they swim from the ocean to fresh water streams to lay their eggs. During migration, salmon must jump waterfalls to reach their destination. A migrating salmon initiates a jump 4 feet from a waterfall that is 3 feet high. Its path through the air is given by the function

$$h = -0.42x^2 + 2.52x$$

where h is the height (in feet) and x is the horizontal distance (in feet) from where the salmon leaves the water. Will the salmon clear the waterfall?

83. Farmland   The number of acres A (in millions) of farmland in the United States each year from 1993 through 2009 can be approximated by the function

$$A = \begin{cases} -3.4t + 979, & 3 \le t \le 11 \\ 0.28t^2 - 11.9t + 1043, & 12 \le t \le 19 \end{cases}$$

where t represents the year, with $t = 3$ corresponding to 1993. Use the function to find the numbers of acres of farmland in 2000 and 2009. *(Source: U.S. Department of Agriculture)*

84. Cost, Revenue, and Profit   A company produces a product for which the variable cost is $11.75 per unit and the fixed costs are $112,000. The product sells for $21.95 per unit. Let x be the number of units produced and sold.

(a) Add the variable cost and the fixed costs to write the total cost C as a function of the number of units x.

(b) Write the revenue R as a function of the number of units x.

(c) Use the formula $P = R - C$ to write the profit P as a function of the number of units x.

85. Average Cost   A manufacturer determines that the variable cost for a new product is $2.05 per unit and the fixed costs are $57,000. All units produced are sold for $5.89 per unit. Let x be the number of units.

(a) Write the total cost C as a function of the number of units x.

(b) Write the average cost per unit $\overline{C} = C/x$ as a function of x.

(c) Complete the table.

x	100	1000	10,000	100,000
$\overline{C}$				

(d) Write a paragraph analyzing the data in the table. What do you observe about the average cost per unit as x gets larger?

86. Migrant Interdictions   The U.S. Coast Guard interdicts undocumented migrants at sea to prevent them from entering the United States, its territories, and its possessions. The number N of migrants interdicted each year from 2000 through 2010 can be approximated by the function

$$N = \begin{cases} 556.50t^2 - 1096.5t + 4280, & 0 \le t \le 3 \\ -1481.9t + 16,793, & 4 \le t \le 10 \end{cases}$$

where t represents the year, with $t = 0$ corresponding to 2000 (see figure). *(Source: U.S. Coast Guard)*

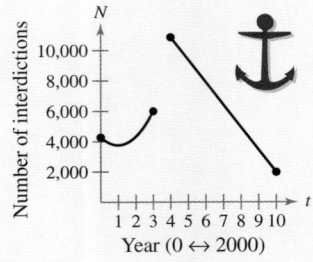

(a) Use the function to approximate the numbers of migrants interdicted in 2003, 2004, and 2010.

(b) Use the function to approximate the total number of migrants interdicted from 2000 through 2010.

87. Profits   Johnson & Johnson manufactures and sells health care products. The net profits P (in billions of dollars) of Johnson & Johnson for the years 2000 through 2009 are shown in the table. *(Source: Johnson & Johnson)*

Year	2000	2001	2002	2003	2004
Profit, P	4.80	5.89	6.81	8.10	9.30

Year	2005	2006	2007	2008	2009
Profit, P	10.55	11.13	12.09	12.95	12.91

(a) Use a graphing utility to create a scatter plot of the data. Let t represent the year, with $t = 0$ corresponding to 2000.

(b) Use the *regression* feature of the graphing utility to find a linear model and a quadratic model for the data.

(c) Use each model to approximate the net profits for the years 2000 through 2009. Compare the values generated by each model with the actual values shown in the table. Which model is a better fit? Explain.

88. Dividends   The dividends D (in dollars) per share declared by Coca-Cola for the years 1995 through 2010 are shown in the table.   *(Source: Coca-Cola Company)*

Year	Dividend, D	Year	Dividend, D
1995	0.44	2003	0.88
1996	0.50	2004	1.00
1997	0.56	2005	1.12
1998	0.60	2006	1.24
1999	0.64	2007	1.36
2000	0.68	2008	1.52
2001	0.72	2009	1.64
2002	0.80	2010	1.76

(a) Use a graphing utility to create a scatter plot of the data. Let t represent the year, with $t = 5$ corresponding to 1995.

(b) Use the *regression* feature of the graphing utility to find a linear model and a quadratic model for the data.

(c) Use the graphing utility to graph each model from part (b) with the data.

(d) Which model do you think better fits the data? Explain your reasoning.

(e) Use the model you selected in part (d) to predict the dividends per share in 2011 and 2015. Coca-Cola predicts the dividends per share to be about $1.88 in 2011 and to reach $2.48 by one of the years from 2013 to 2015. Do your predictions support those of Coca-Cola? Explain.

89. Ripples in a Pond   A stone is thrown into the middle of a calm pond, causing ripples to form in concentric circles. The radius r of the outermost ripple increases at the rate of 0.75 foot per second.

(a) Write a function for the radius r of the circle formed by the outermost ripple in terms of time t.

(b) Write a function for the area A enclosed by the outermost ripple. Complete the table.

Time, t	1	2	3	4	5
Radius, r (in feet)					
Area, A (in square feet)					

(c) Compare the ratios $A(2)/A(1)$ and $A(4)/A(2)$. What do you observe? Based on your observation, predict the area when $t = 8$. Verify by checking $t = 8$ in the area function.

90. HOW DO YOU SEE IT?   You pop the cork from a bottle at an outdoor event. The graph represents the height h of the cork after t seconds.

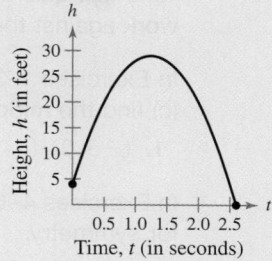

(a) Explain why h is a function of t.

(b) Approximate the height of the cork after 0.5 second and after 1.25 seconds.

(c) Approximate the domain of h.

(d) Is t a function of h? Explain.

Think About It   In Exercises 91 and 92, determine whether the statements use the word *function* in ways that are *mathematically* correct. Explain your reasoning.

91. (a) The sales tax on a purchased item is a function of the selling price.

(b) Intelligence is a function of education.

92. (a) The market value of a home is a function of its size.

(b) The speed at which a free-falling baseball strikes the ground is a function of its initial height.

93. Project: Expenditures   For a project involving the personal expenditures for Internet access in the United States from 2000 to 2009, visit this text's website at *www.cengagebrain.com*.   *(Source: Bureau of Economic Analysis)*

QUIZ YOURSELF

Take this quiz as you would take a quiz in class. When you are done, check your work against the answers given in the back of the book.

In Exercises 1–3, (a) plot the points, (b) find the distance between the points, and (c) find the midpoint of the line segment joining the points.

1. $(-3, 2), (4, -5)$ **2.** $(1.3, -4.5), (-3.7, 0.7)$ **3.** $(4, -2), \left(-1, -\frac{5}{2}\right)$

In Exercises 4–6, sketch the graph of the equation. Identify any intercepts and test for symmetry.

4. $y = 9 - x^2$ **5.** $y = x\sqrt{x + 4}$ **6.** $y = |x - 3|$

7. The point $(0, 3)$ lies on a circle whose center is at $(4, 5)$. Write the standard form of the equation of this circle.

8. Write the equation of the circle $x^2 + y^2 - 2x + 4y - 4 = 0$ in standard form. Then sketch the circle.

In Exercises 9–12, find an equation of the line that passes through the point and has the indicated slope. Then sketch the graph of the line.

	Point	*Slope*
9.	$(3, 5)$	$m = \frac{2}{3}$
10.	$(-2, 4)$	$m = 0$
11.	$(2, -3)$	m is undefined.
12.	$(-2, -5)$	$m = -2$

13. A city had a population of 233,134 in 2010 and 244,288 in 2012. Assuming the population followed a linear pattern, predict the population of the city in 2015.

14. Determine whether the equation $x^2 + y = 3$ represents y as a function of x.

In Exercises 15 and 16, let $A = \{6, 8, 10\}$ and $B = \{1, 2, 4, 16\}$. Decide whether the set of ordered pairs represents a function from A to B.

15. $\{(6, 1), (8, 1), (10, 2)\}$ **16.** $\{(6, 2), (10, 4)\}$

In Exercises 17 and 18, evaluate the function at each specified value of the independent variable and simplify.

17. $f(x) = 3(x + 2) - 4$
(a) $f(0)$ (b) $f(-3)$

18. $g(t) = 2t^3 - t^2$
(a) $g(1)$ (b) $g(-2)$

In Exercises 19 and 20, find the domain of the function.

19. $h(x) = \sqrt{x - 4}$ **20.** $f(x) = \dfrac{x}{x + 2}$

In Exercises 21 and 22, use the table, which shows the federal cost C (in millions of dollars) of the National School Lunch Program for each year from 2004 through 2009. *(Source: U.S. Department of Agriculture)*

21. Use a graphing utility to create a scatter plot of the data and use the regression feature to find a linear model and a quadratic model for the data. Let $t = 4$ represent 2004.

22. Use each model you found in Exercise 21 to predict the federal costs of the national school lunch program in 2010 and 2011.

23. Write the area A of a circle as a function of its circumference C.

Year	Cost, C (in millions of dollars)
2004	6663
2005	7055
2006	7389
2007	7707
2008	8265
2009	8873

Table for 21 and 22

2.5 Graphs of Functions

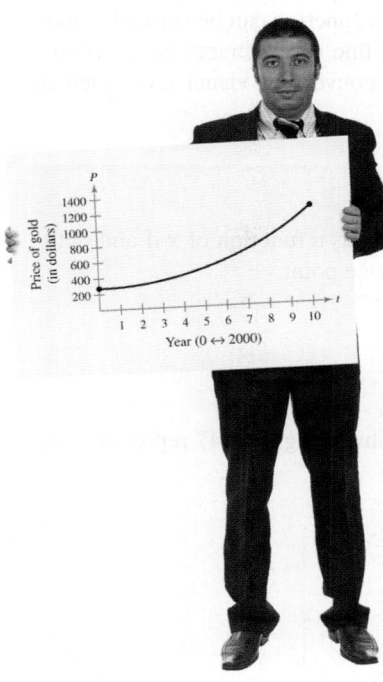

■ Find the domain and range of a function using its graph.
■ Identify the graph of a function using the Vertical Line Test.
■ Describe the increasing, decreasing, and constant behavior of a function.
■ Find the relative minima and relative maxima of the graph of a function.
■ Sketch the graph of a step function.
■ Classify a function as even or odd.
■ Identify six common graphs and use them to sketch the graphs of functions.

In Exercise 79 on page 218, you will use a graph to determine in which years the price of an ounce of gold was increasing or decreasing.

The Graph of a Function

In Section 2.4, you studied functions from an algebraic point of view. In this section, you will study functions from a graphical perspective.

The **graph of a function** f is the collection of ordered pairs $(x, f(x))$ such that x is in the domain of f. As you study this section, remember that x is the directed distance from the y-axis and $f(x)$ is the directed distance from the x-axis, as shown in Figure 2.45. If the graph of a function has an x-intercept at $(a, 0)$, then a is a **zero** of the function. In other words, the zeros of a function are the values of x for which $f(x) = 0$. For instance, the function given by $f(x) = x^2 - 4$ has two zeros: -2 and 2.

The **range** of a function (the set of values assumed by the dependent variable) is often easier to determine graphically than algebraically.

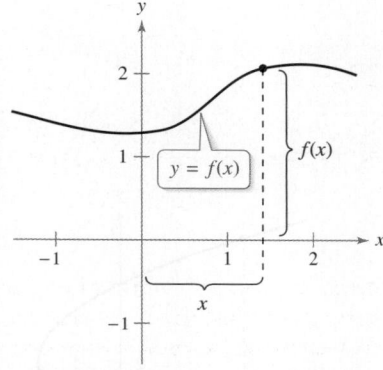

FIGURE 2.45

Example 1   Finding the Domain and Range of a Function

Use the graph of the function f, shown in Figure 2.46, to find (a) the domain of f, (b) the function values $f(-1)$ and $f(2)$, and (c) the range of f.

SOLUTION

a. Because the graph does not extend beyond $x = -1$ (on the left) and $x = 4$ (on the right), the domain of f is all x in the interval $[-1, 4]$.

b. Because $(-1, -5)$ is a point on the graph of f, it follows that

$$f(-1) = -5.$$

Similarly, because $(2, 4)$ is a point on the graph of f, it follows that

$$f(2) = 4.$$

c. Because the graph does not extend below $f(-1) = -5$ or above $f(2) = 4$, the range of f is the interval $[-5, 4]$.

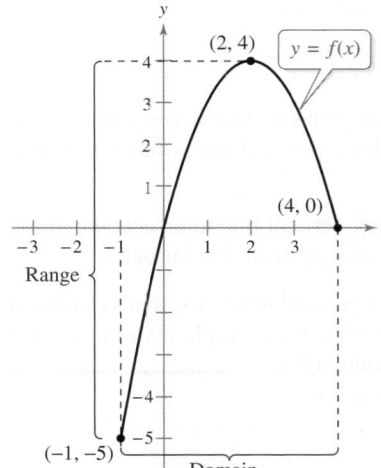

FIGURE 2.46

✓ Checkpoint 1

Use the graph of $f(x) = x^2 - 3$ to find the domain and range of f.

Vertical Line Test

By the definition of a function, at most one y-value corresponds to a given x-value. This means that the graph of a function cannot have two or more different points with the same x-coordinate, and no two points on the graph of a function can be vertically above or below each other. It follows, then, that a vertical line can intersect the graph of a function at most once. This observation provides a convenient visual test called the **Vertical Line Test** for functions.

Vertical Line Test for Functions

A set of points in a coordinate plane is the graph of y as a function of x if and only if no vertical line intersects the graph at more than one point.

Example 2   Using the Vertical Line Test

Use the Vertical Line Test to decide whether the graphs in Figure 2.47 represent y as a function of x.

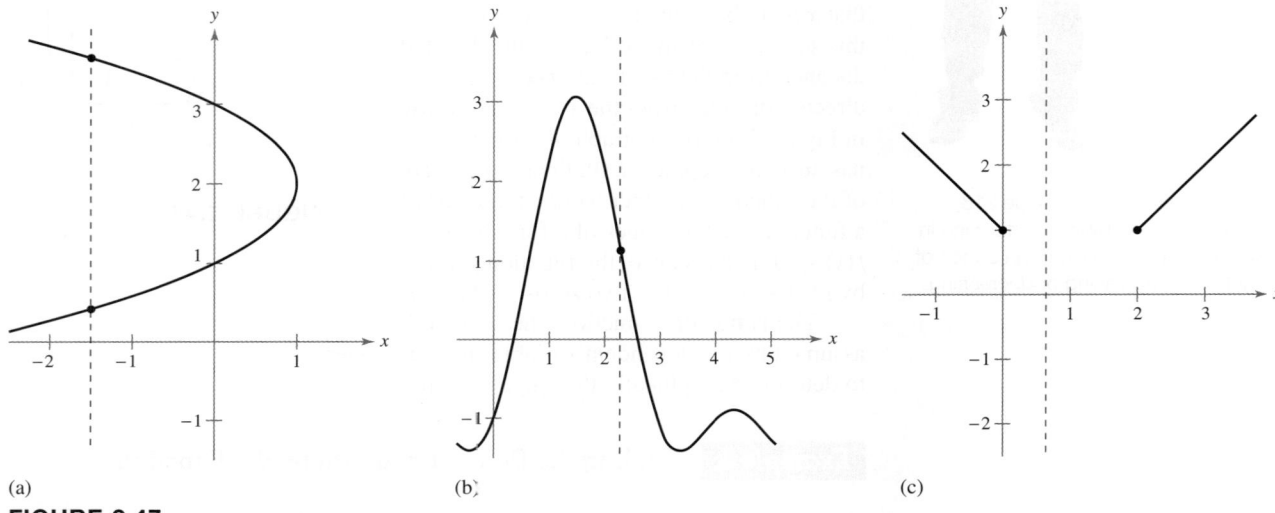

(a)

(b)

(c)

FIGURE 2.47

SOLUTION

a. This *is not* a graph of y as a function of x because you can find a vertical line that intersects the graph twice. That is, for a particular input x, there is more than one output y.

b. This *is* a graph of y as a function of x because every vertical line intersects the graph at most once. That is, for a particular input x, there is at most one output y.

c. This *is* a graph of y as a function of x. That is, for a particular input x, there is at most one output y. Note that if a vertical line does not intersect the graph, it simply means that the function is undefined for that particular value of x. ——————

✓ Checkpoint 2

Use the Vertical Line Test to decide whether the graph of each equation represents y as a function of x.

a. $x^2 + y = 2$ **b.** $y = 2x^3$

Increasing and Decreasing Functions

The more you know about the graph of a function, the more you know about the function itself. For example, consider the graph that is shown in Figure 2.48. As you move from *left to right*, this graph decreases, then is constant, and then increases.

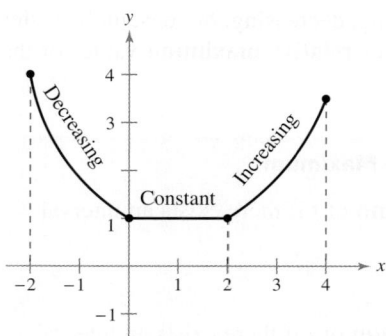

As *x* increases from left to right, notice how *y* decreases, then remains constant, and then increases.

FIGURE 2.48

Increasing, Decreasing, and Constant Functions

A function f is **increasing** on an interval if, for any x_1 and x_2 in the interval, $x_1 < x_2$ implies

$$f(x_1) < f(x_2).$$

A function f is **decreasing** on an interval if, for any x_1 and x_2 in the interval, $x_1 < x_2$ implies

$$f(x_1) > f(x_2).$$

A function f is **constant** on an interval if, for any x_1 and x_2 in the interval,

$$f(x_1) = f(x_2).$$

Example 3 **Describing Function Behavior**

Describe the increasing, decreasing, and constant behavior of each function shown in Figure 2.49.

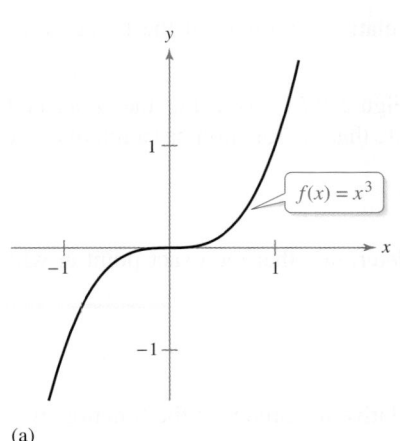

(a)

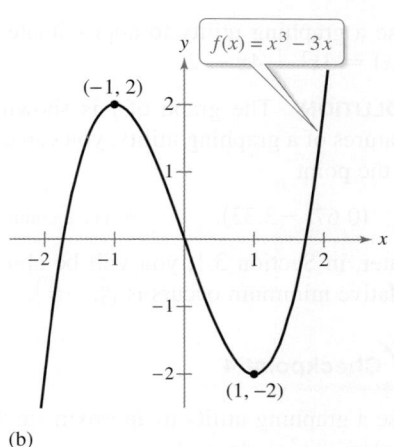

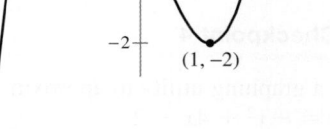

(b)

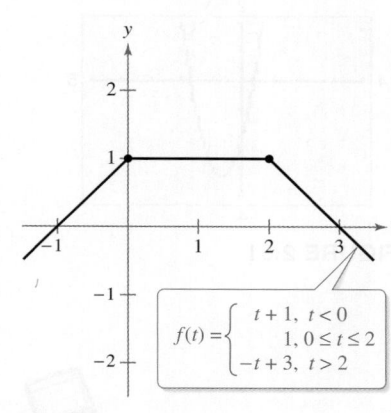

(c)

FIGURE 2.49

SOLUTION

a. This function is increasing over the entire real line.

b. This function is increasing on the interval $(-\infty, -1)$, decreasing on the interval $(-1, 1)$, and increasing on the interval $(1, \infty)$.

c. This function is increasing on the interval $(-\infty, 0)$, constant on the interval $(0, 2)$, and decreasing on the interval $(2, \infty)$.

✓ **Checkpoint 3**

Describe the increasing, decreasing, and constant behavior of the function $f(x) = x^2 + 3x$.

Make sure you understand that the behavior of functions (increasing, decreasing, or constant) is determined by the *y*-values, but that the intervals are written in terms of *x*.

Relative Minima and Relative Maxima

The points at which a function changes its increasing, decreasing, or constant behavior are helpful in determining the **relative minimum** or **relative maximum** values of the function.

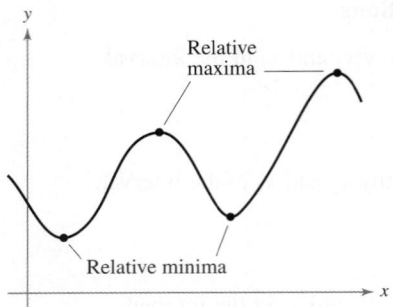

FIGURE 2.50

> ### Definition of Relative Minimum and Relative Maximum
>
> A function value $f(a)$ is called a **relative minimum** of f if there exists an interval (x_1, x_2) that contains a such that
>
> $$x_1 < x < x_2 \quad \text{implies} \quad f(a) \le f(x).$$
>
> A function value $f(a)$ is called a **relative maximum** of f if there exists an interval (x_1, x_2) that contains a such that
>
> $$x_1 < x < x_2 \quad \text{implies} \quad f(a) \ge f(x).$$

Figure 2.50 shows several examples of relative minima and relative maxima. In Section 3.1, you will study a technique for finding the *exact point* at which a second-degree polynomial function has a relative minimum or relative maximum. For the time being, however, you can use a graphing utility to find reasonable approximations of these points.

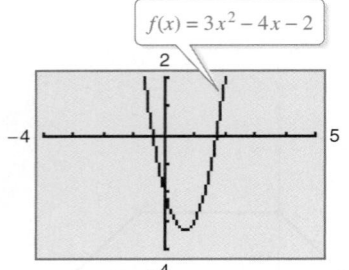

FIGURE 2.51

Example 4   **Approximating a Relative Minimum**

Use a graphing utility to approximate the relative minimum of the function given by $f(x) = 3x^2 - 4x - 2$.

SOLUTION   The graph of f is shown in Figure 2.51. By using the *zoom* and *trace* features of a graphing utility, you can estimate that the function has a relative minimum at the point

$(0.67, -3.33).$ Relative minimum

Later, in Section 3.1, you will be able to determine that the exact point at which the relative minimum occurs is $\left(\frac{2}{3}, -\frac{10}{3}\right)$.

 **Checkpoint 4**

Use a graphing utility to approximate the relative maximum of the function given by $f(x) = -x^2 + 4x - 2$.   ■

You can also use the *table* feature of a graphing utility to approximate numerically the relative minimum of the function in Example 4. Using a table that begins at 0.6 and increments the value of x by 0.01, you can approximate the minimum of $f(x) = 3x^2 - 4x - 2$ to be -3.33, which occurs at $(0.67, -3.33)$. A third way to find the relative minimum is to use the *minimum* feature of a graphing utility.

TECH TUTOR

For instructions on how to use the *table* feature and the *minimum* feature, see Appendix A.

TECH TUTOR

When using a graphing utility to estimate the *x*- and *y*-values of a relative minimum or relative maximum, the *zoom* feature will often produce graphs that are nearly flat. To overcome this problem, you can manually change the vertical setting of the viewing window. The graph will stretch vertically if the values of Ymin and Ymax are closer together.

Step Functions

The **greatest integer function** is denoted by $[\![x]\!]$ and is defined as

$$f(x) = [\![x]\!] = \text{the greatest integer less than or equal to } x.$$

The graph of this function is shown in Figure 2.52. Note that the graph of the greatest integer function jumps vertically one unit at each integer and is constant (a horizontal line segment) between each pair of consecutive integers. Because of the jumps in its graph, the greatest integer function is an example of a type of function called a **step function.** Some values of the greatest integer function are as follows.

$$[\![-1]\!] = -1 \qquad [\![-0.5]\!] = -1$$
$$[\![0]\!] = 0 \qquad [\![0.5]\!] = 0$$
$$[\![1]\!] = 1 \qquad [\![1.5]\!] = 1$$

The range of the greatest integer function is the set of all integers.

When using a graphing utility to graph a step function, set the utility to *dot* mode rather than *connected* mode.

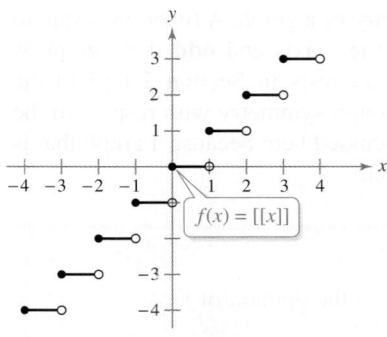

Greatest Integer Function
FIGURE 2.52

Example 5   **The Cost of a Machinery Rental**

The cost of renting a warehouse forklift is $200 for up to, but not including, the first day and $125 for each additional day (or portion of a day). The greatest integer function

$$C = 200 + 125[\![t]\!], \quad t > 0$$

can be used to model the cost of this rental, where C is the total cost of the rental (in dollars) and t is the length of the rental (in days).

a. Sketch the graph of this function.

b. How long can a company rent the forklift without spending more than $1000?

SOLUTION

a. For rentals up to, but not including, 1 day, the cost is $200. For rentals between 1 and 2 days, the cost is $325, for rentals between 2 and 3 days, the cost is $450, and so on, as shown in the table. Using these and other values, you can sketch the graph shown in Figure 2.53.

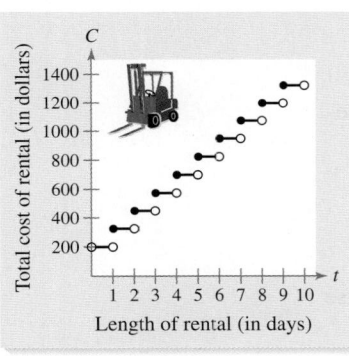

FIGURE 2.53

Length of rental, t	Cost of rental, C
$0 < t < 1$	$200
$1 \le t < 2$	$325
$2 \le t < 3$	$450
$\vdots$	$\vdots$
$9 \le t < 10$	$1325

b. From the graph, you can see that the company must limit the rental period to 7 days to avoid spending more than $1000.

✓**Checkpoint 5**

Repeat Example 5 when the cost of each additional day (or portion of a day) is $100.

Even and Odd Functions

In Section 2.1, you studied different types of symmetry of a graph. A function is said to be **even** if its graph is symmetric with respect to the y-axis and **odd** if its graph is symmetric with respect to the origin. The symmetry tests in Section 2.1 yield the following tests for even and odd functions. Even though symmetry with respect to the x-axis is introduced in Section 2.1, it will not be discussed here because a graph that is symmetric with respect to the x-axis is not a function.

Tests for Even and Odd Functions

A function given by $y = f(x)$ is even if, for each x in the domain of f,

$$f(-x) = f(x).$$

A function given by $y = f(x)$ is odd if, for each x in the domain of f,

$$f(-x) = -f(x).$$

Example 6   **Identifying Even and Odd Functions**

Decide whether each function is even, odd, or neither.

a. $g(x) = x^3 - x$ **b.** $h(x) = x^2 + 1$

SOLUTION

a. The function given by $g(x) = x^3 - x$ is odd because

$$g(-x) = (-x)^3 - (-x)$$
$$= -x^3 + x$$
$$= -(x^3 - x) = -g(x).$$

b. The function given by $h(x) = x^2 + 1$ is even because

$$h(-x) = (-x)^2 + 1$$
$$= x^2 + 1 = h(x).$$

The graphs of the two functions are shown in Figure 2.54.

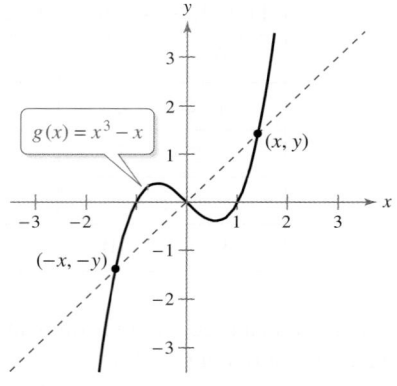

(a) Odd function (symmetric about origin)

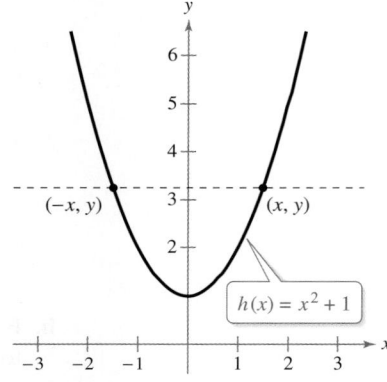

(b) Even function (symmetric about y-axis)

FIGURE 2.54

✓ **Checkpoint 6**

Decide whether the function $f(x) = -2x^2 + x - 1$ is even, odd, or neither.

Common Graphs

Figure 2.55 shows the graphs of six common functions. You need to be familiar with these graphs. They can be used as an aid when sketching other graphs. For instance, the graph of the absolute value function given by

$$f(x) = |x - 2|$$

is V-shaped.

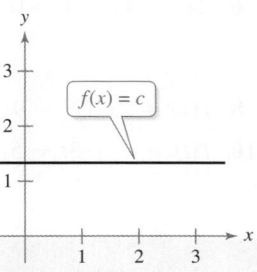

(a) Constant function

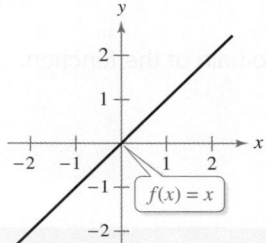

(b) Identity function

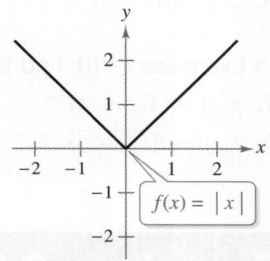

(c) Absolute value function

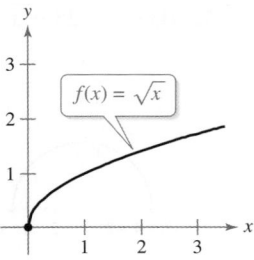

(d) Square root function

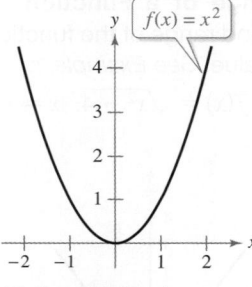

(e) Squaring function

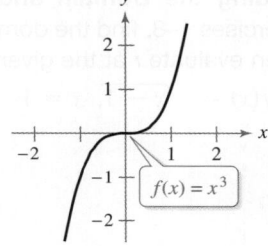

(f) Cubing function

FIGURE 2.55

SUMMARIZE   (Section 2.5)

1. State the definition of the graph of a function *(page 209)*. For an example of finding the domain and range of a function using its graph, see Example 1.

2. Describe the Vertical Line Test *(page 210)*. For an example that uses the Vertical Line Test, see Example 2.

3. Describe when a function is increasing, decreasing, or constant on an interval *(page 211)*. For an example of describing the increasing, decreasing, or constant behavior of a function, see Example 3.

4. State the definitions of relative minimum and relative maximum *(page 212)*. For an example of approximating a relative minimum using a graphing utility, see Example 4.

5. Describe the greatest integer function *(page 213)*. For an example that uses a greatest integer function, see Example 5.

6. State the definitions of an even function and an odd function *(page 214)*. For an example of identifying an even or odd function, see Example 6.

SKILLS WARM UP 2.5   The following warm-up exercises involve skills that were covered in earlier sections. You will use these skills in the exercise set for this section. For additional help, review Sections 1.4, 1.5, and 2.4.

1. Find $f(2)$ for $f(x) = -x^3 + 5x$.

2. Find $f(6)$ for $f(x) = x^2 - 6x$.

3. Find $f(-x)$ for $f(x) = 3/x$.

4. Find $f(-x)$ for $f(x) = x^2 + 3$.

In Exercises 5 and 6, solve the equation.

5. $x^3 - 16x = 0$

6. $2x^2 - 3x + 1 = 0$

In Exercises 7–10, find the domain of the function.

7. $g(x) = 4(x - 4)^{-1}$

8. $f(x) = 2x/(x^2 - 9x + 20)$

9. $h(t) = \sqrt[4]{5 - 3t}$

10. $f(t) = t^3 + 3t - 5$

Exercises 2.5

See www.CalcChat.com for worked-out solutions to odd-numbered exercises.

Finding the Domain and Range of a Function   In Exercises 1–8, find the domain and range of the function. Then evaluate f at the given x-value. *See Example 1.*

1. $f(x) = \sqrt{x - 1}$, $x = 1$

2. $f(x) = \sqrt{x^2 - 4}$, $x = -2$

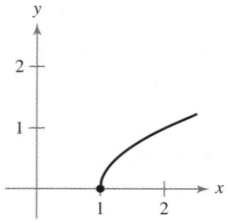

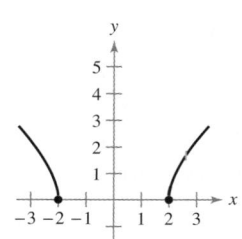

3. $f(x) = 4 - x^2$, $x = 0$

4. $f(x) = |x - 2|$, $x = 2$

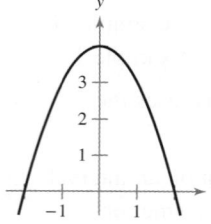

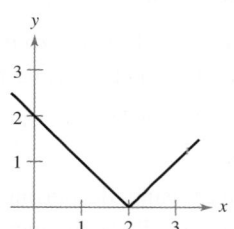

5. $f(x) = x^3 - 1$, $x = 0$

6. $f(x) = \dfrac{|x|}{x}$, $x = 5$

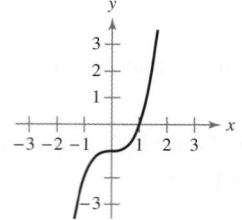

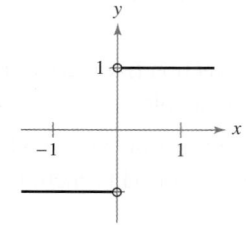

7. $f(x) = \sqrt{25 - x^2}$, $x = 0$ **8.** $f(x) = \sqrt{x^2 - 9}$, $x = 3$

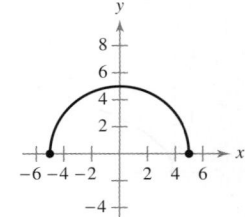

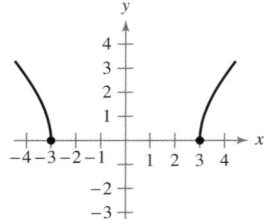

Using the Vertical Line Test   In Exercises 9–16, use the Vertical Line Test to decide whether y is a function of x. *See Example 2.*

9. $y = x^2$

10. $x^2 = xy - 1$

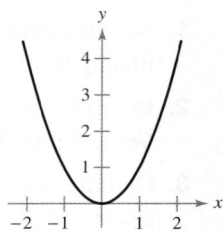

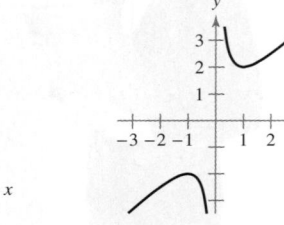

11. $x^2 + y^2 = 9$

12. $x - y^2 = 0$

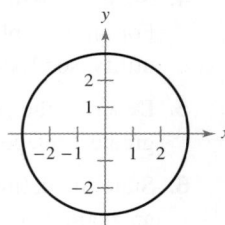

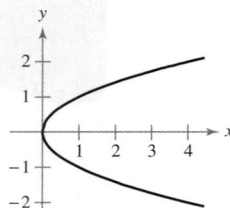

13. $y = \frac{1}{2}x^2$

14. $y = \frac{1}{4}x^3$

15. $x^2 + y^2 = 25$

16. $x = |y + 2|$

Describing Function Behavior In Exercises 17–28, describe the increasing, decreasing, and constant behavior of the function. Find the point or points where the behavior of the function changes. *See Example 3.*

17. $f(x) = 2x$

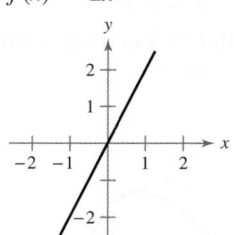

18. $f(x) = x^2 - 2x$

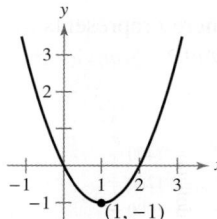

19. $f(x) = x^3 - 3x^2$

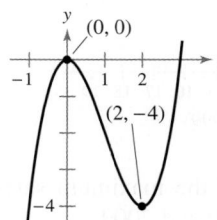

20. $f(x) = \sqrt{x^2 - 4}$

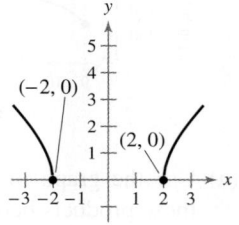

21. $f(x) = 3x^4 - 6x^2$

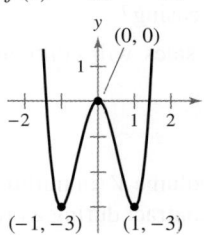

22. $f(x) = x^{2/3}$

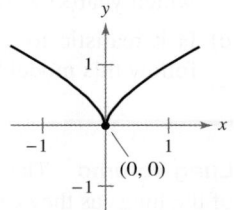

23. $y = x\sqrt{x + 3}$

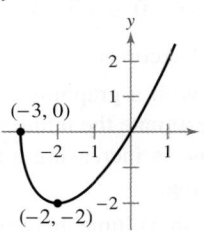

24. $y = |x + 1| + |x - 1|$

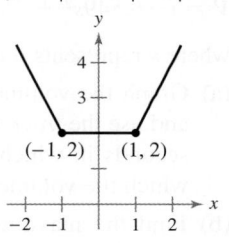

25. $f(x) = x^2 - 4x$

26. $f(x) = \frac{3}{2}x$

27. $f(x) = x^3 - 3x^2 + 2$

28. $f(x) = \sqrt{x^2 - 1}$

Approximating a Relative Minimum and Maximum In Exercises 29–34, use a graphing utility to graph the function, approximate the relative minimum or maximum of the function, and estimate the open intervals on which the function is increasing or decreasing. *See Example 4.*

29. $f(x) = x^2 - 4x + 1$

30. $f(x) = -x^2 + 6x + 3$

31. $f(x) = x^3 - 3x^2$

32. $f(x) = -x^3 + 3x + 1$

33. $f(x) = \frac{1}{4}(-4x^4 - 5x^3 + 10x^2 + 8x + 6)$

34. $f(x) = \frac{1}{4}(x^4 + x^3 - 10x^2 + 2x - 15)$

Using the Greatest Integer Function In Exercises 35–38, evaluate the function at each specified value of the independent variable.

35. $f(x) = [\![x]\!]$
    (a) $f(2)$ (b) $f(2.5)$
    (c) $f(-2.5)$ (d) $f(-4)$

36. $f(x) = [\![-x]\!]$
    (a) $f(3)$ (b) $f(6.1)$
    (c) $f(-5.9)$ (d) $f(-9)$

37. $f(x) = [\![x - 1.8]\!]$
    (a) $f(4)$ (b) $f(3.7)$
    (c) $f(-5.8)$ (d) $f(-6.3)$

38. $f(x) = [\![x + 0.3]\!]$
    (a) $f(2.9)$ (b) $f(4.6)$
    (c) $f(-2.3)$ (d) $f(-4.2)$

Identifying Even and Odd Functions In Exercises 39–44, decide whether the function is even, odd, or neither. *See Example 6.*

39. $f(x) = x^6 - 2x^2 + 3$

40. $f(t) = t^2 + 3t - 10$

41. $g(x) = x^3 - 5x$

42. $h(x) = x^3 + 3$

43. $f(x) = x\sqrt{4 - x^2}$

44. $g(s) = 4s^{2/3}$

Sketching Graphs of Functions In Exercises 45–58, sketch the graph of the function and determine whether the function is even, odd, or neither.

45. $f(x) = 3$ **46.** $g(x) = x$

47. $f(x) = 5 - 3x$

48. $h(x) = x^2 - 4$

49. $g(s) = \dfrac{s^3}{4}$ **50.** $f(t) = -t^4$

51. $f(x) = \sqrt{1 - x}$

52. $g(t) = \sqrt[3]{t - 1}$

53. $f(x) = x^{3/2}$

54. $f(x) = |x + 2|$

55. $f(x) = \begin{cases} x^2 + 1, & x \le 1 \\ 3x - 1, & x > 1 \end{cases}$ **56.** $f(x) = \begin{cases} 2x - 1, & x \le -1 \\ x^2 - 1, & x > -1 \end{cases}$

57. $f(x) = \begin{cases} x + 1, & x \le 0 \\ 4, & 0 < x \le 2 \\ 3x - 1, & x > 2 \end{cases}$

58. $f(x) = \begin{cases} 2x - 1, & x \le 1 \\ 3, & 1 < x \le 3 \\ 2x + 1, & x > 3 \end{cases}$

Graphing Functions In Exercises 59–64, use a graphing utility to graph the function and determine whether the function is even, odd, or neither.

59. $f(x) = x^2 - x^4$

60. $g(x) = 2x^3 + 1$

61. $h(x) = x^5 - 2x^3 + x$

62. $f(x) = x^5 - 2x^4 + x - 2$

63. $g(x) = 2 - x^6 - x^8$

64. $h(x) = x^9 + 3x^5 - x^3 + x$

Sketching Graphs of Functions In Exercises 65–78, sketch the graph of the function.

65. $f(x) = 4 - x$

66. $f(x) = 4x + 2$

67. $f(x) = x^2 - 9$

68. $f(x) = x^2 - 4x$

69. $f(x) = 1 - x^4$

70. $f(x) = x^4 - 4x^2$

71. $f(x) = \frac{1}{3}(3 + |x|)$

72. $f(x) = -1(1 + |x|)$

73. $f(x) = \sqrt{x + 3}$

74. $f(x) = \sqrt{x - 1}$

75. $f(x) = -[\![x]\!]$

76. $f(x) = 2[\![x]\!]$

77. $f(x) = [\![x - 1]\!]$

78. $f(x) = [\![x + 1]\!]$

79. Price of Gold The price P (in dollars) of an ounce of gold from 2000 through 2010 can be approximated by the model

$$P = 0.038214t^4 - 0.48703t^3 + 11.0120t^2 -$$
$$6.018t + 276.49, \quad 0 \le t \le 10$$

where t represents the year, with $t = 0$ corresponding to 2000. *(Source: World Gold Council)*

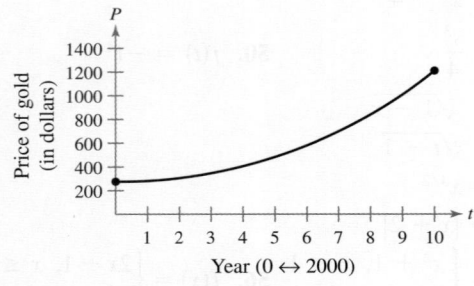

Year (0 ↔ 2000)

(a) Use the graph of P to find the maximum price of gold between 2000 and 2010.

(b) During which year(s) was the price decreasing? During which year(s) was the price increasing?

(c) Is it realistic to assume that the price of gold will continue to follow this model?

80. Sales The sales S (in billions of dollars) of petroleum and coal products from 1999 through 2009 can be approximated by the model

$$S = -1.28418t^4 + 67.9435t^3 - 1316.592t^2 +$$
$$11,169.74t - 34,707.3, \quad 9 \le t \le 19$$

where t represents the year, with $t = 9$ corresponding to 1999. *(Source: U.S. Census Bureau)*

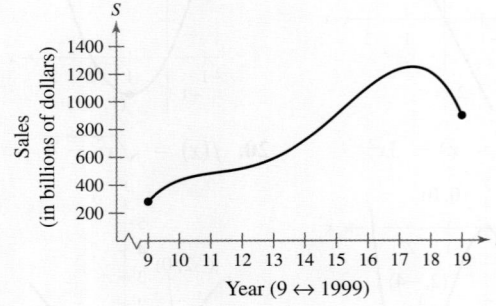

Year (9 ↔ 1999)

(a) Use the graph of S to find the maximum sales of these products between 1999 and 2009.

(b) During which year(s) were sales decreasing? During which year(s) were sales increasing?

(c) Is it realistic to assume that sales will continue to follow this model?

81. Lung Volume The change in volume V (in milliliters) of the lungs as they expand and contract during a breath can be approximated by the model

$$V = (-6.549s^2 + 26.20s - 3.8)^2, \quad 0 \le s \le 4$$

where s represents the number of seconds.

(a) Graph the volume function with a graphing utility and use the *trace* feature to estimate the numbers of seconds in which the volume is increasing and in which the volume is decreasing.

(b) Find the maximum change in volume between 0 and 4 seconds.

82. Maximum Profit The marketing department of a company estimates that the demand for a product is given by $p = 100 - 0.0001x$, where p is the price per unit and x is the number of units. The cost C of producing x units is given by

$$C = 350,000 + 30x$$

and the profit P for producing and selling x units is given by

$$P = R - C = xp - C.$$

Sketch the graph of the profit function and estimate the number of units that would produce a maximum profit. Verify your estimate using a graphing utility.

83. Maximum Profit The marketing department of a company estimates that the demand for a product is given by $p = 125 - 0.0002x$, where p is the price per unit and x is the number of units. The cost C of producing x units is given by

$$C = 225,000 + 80x$$

and the profit P for producing and selling x units is given by

$$P = R - C = xp - C.$$

Sketch the graph of the profit function and estimate the number of units that would produce a maximum profit. Verify your estimate using a graphing utility.

84. HOW DO YOU SEE IT? For the years 1995 through 2009, the book value B (in dollars) of a share of Wells Fargo stock can be approximated by the model

$$B = 0.0497t^2 - 0.163t + 3.53, \quad 5 \le t \le 19$$

where t represents the year, with $t = 5$ corresponding to 1995 (see figure). *(Source: Wells Fargo)*

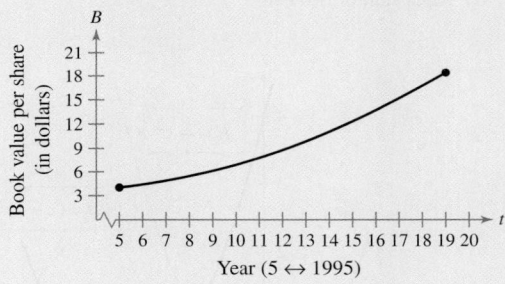

(a) Estimate the maximum book value per share from 1995 through 2009.

(b) Estimate the minimum book value per share from 1995 through 2009.

(c) Assume the book value B continues to follow the model through 2011. In which year is B at a maximum?

85. Cost of Overnight Delivery The cost of sending an overnight package from New York to Atlanta is $26.10 for up to, but not including, the first pound and $4.25 for each additional pound (or portion of a pound). A model for the total cost C of sending the package is $C = 26.1 + 4.25[\![x]\!]$, $x > 0$, where x is the weight of the package (in pounds). Sketch the graph of this function.

86. Cost of Overnight Delivery The cost of sending an overnight package from Los Angeles to Miami is $31.60 for up to, but not including, the first pound and $5.50 for each additional pound (or portion of a pound). A model for the total cost C of sending the package is $C = 31.6 + 5.5[\![x]\!]$, $x > 0$, where x is the weight of the package (in pounds). Sketch the graph of this function.

87. Strategic Reserve The total volume V (in millions of barrels) of the Strategic Oil Reserve in the United States from 1998 through 2009 can be approximated by

$$V = \begin{cases} -2.2292t^4 + 94.319t^3 - 1472.15t^2 + \\ \qquad 10,050.1t - 24,773, \quad 8 \le t \le 13 \\ 0.796t^3 - 38.13t^2 + 611.8t - 2601, \\ \qquad 14 \le t \le 19 \end{cases}$$

where t represents the year, with $t = 8$ corresponding to 1998. Sketch the graph of this function. *(Source: U.S. Energy Information Administration)*

88. Grade Level Salaries The 2010 salary S (in dollars) for federal employees at the Step 1 level can be approximated by the model

$$S = \begin{cases} 3109.2x + 13,012, & x = 1, 2, \ldots, 10 \\ 12,310.5x - 86,725, & x = 11, \ldots, 15 \end{cases}$$

where x represents the "GS" grade. Sketch a *bar graph* that represents this function. *(Source: U.S. Office of Personnel Management)*

89. Revenues The revenues of Symantec Corporation (in millions of dollars) from 2000 through 2009 are given by the ordered pairs. *(Source: Symantec Corporation)*

(2000, 853.6), (2001, 1071.4), (2002, 1406.9), (2003, 1870.1), (2004, 2582.8), (2005, 4143.4), (2006, 5199.4), (2007, 5874.4), (2008, 6149.9), (2009, 5985.0)

(a) Use the *regression* feature of a graphing utility to find a linear model for the data from 2000 through 2004. Let t represent the year, with $t = 0$ corresponding to 2000. Then determine the domain and range of the function.

(b) Use the *regression* feature of the graphing utility to find a quadratic model for the data from 2005 through 2009. Let t represent the year, with $t = 5$ corresponding to 2005. Then determine the domain and range of the function.

(c) Use your results from parts (a) and (b) to construct a piecewise model for all of the data. Use the graphing utility to graph the function.

(d) During which years did the revenues increase? During which year did the revenue decrease?

90. Think About It If f is an even function, determine whether g is even, odd, or neither. Explain.

(a) $g(x) = -f(x)$ (b) $g(x) = f(-x)$

(c) $g(x) = f(x) - 2$ (d) $g(x) = f(x - 2)$

Think About It In Exercises 91–94, find the coordinates of a second point on the graph of a function f when the given point is on the graph and the function is (a) even and (b) odd.

91. $\left(-\frac{3}{2}, 4\right)$ **92.** $\left(-\frac{5}{3}, -7\right)$

93. $(4, 9)$ **94.** $(5, -1)$

2.6 Transformations of Functions

- Use vertical and horizontal shifts to sketch graphs of functions.
- Use reflections to sketch graphs of functions.
- Use nonrigid transformations to sketch graphs of functions.

Vertical and Horizontal Shifts

Many functions have graphs that are simple transformations of the common graphs that are summarized on page 215. For example, you can obtain the graph of

$$h(x) = x^2 + 2$$

by shifting the graph of $f(x) = x^2$ *upward* two units, as shown in Figure 2.56. In function notation, h and f are related as shown below.

$$h(x) = x^2 + 2 = f(x) + 2 \qquad \text{Upward shift of two units}$$

Similarly, you can obtain the graph of

$$g(x) = (x - 2)^2$$

by shifting the graph of $f(x) = x^2$ to the *right* two units, as shown in Figure 2.57. In this case, the functions g and f have the following relationship.

$$g(x) = (x - 2)^2 = f(x - 2) \qquad \text{Right shift of two units}$$

In Exercise 65 on page 229, you will see how a transformation affects the graph of a company's profit function.

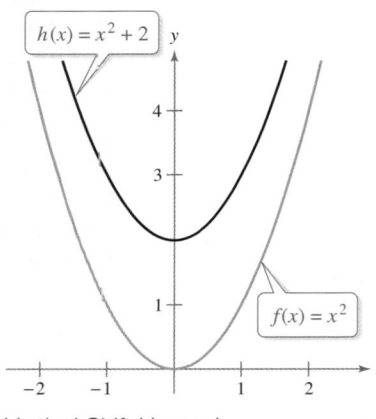

Vertical Shift Upward
FIGURE 2.56

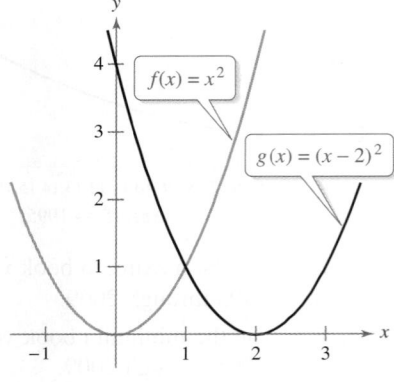

Horizontal Shift to the Right
FIGURE 2.57

The following list summarizes this discussion about vertical and horizontal shifts.

Vertical and Horizontal Shifts

Let c be a positive real number. **Vertical** and **horizontal shifts** of the graph of $y = f(x)$ are represented as follows.

1. Vertical shift c units **upward:** $h(x) = f(x) + c$
2. Vertical shift c units **downward:** $h(x) = f(x) - c$
3. Horizontal shift c units to the **right:** $h(x) = f(x - c)$
4. Horizontal shift c units to the **left:** $h(x) = f(x + c)$

Some graphs can be obtained from a combination of vertical and horizontal shifts, as demonstrated in Example 1(b). Vertical and horizontal shifts generate a *family of functions*, each with the same shape but at different locations in the plane.

Example 1   Shifts in the Graph of a Function

Use the graph of $f(x) = x^3$ to sketch the graph of each function.

a. $g(x) = x^3 + 1$

b. $h(x) = (x + 2)^3 + 1$

SOLUTION

a. Relative to the graph of $f(x) = x^3$, the graph of $g(x) = x^3 + 1$ is an upward shift of one unit, as shown in Figure 2.58(a).

b. Relative to the graph of $f(x) = x^3$, the graph of $h(x) = (x + 2)^3 + 1$ involves a left shift of two units *and* an upward shift of one unit, as shown in Figure 2.58(b).

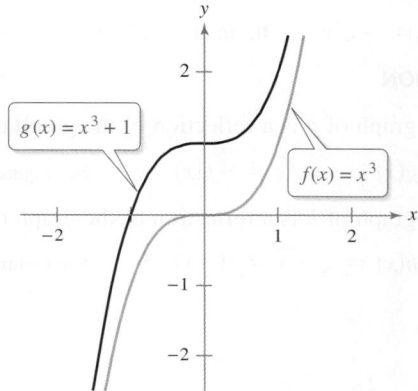

(a) Vertical shift: one unit upward

(b) Horizontal shift: two units left; Vertical shift: one unit upward

FIGURE 2.58

Note that the functions f, g, and h belong to the family of cubic functions.

✓ Checkpoint 1

Use the graph of $f(x) = \sqrt{x}$ to sketch the graph of

$$g(x) = \sqrt{x - 3} - 4.$$

TECH TUTOR

Graphing utilities are ideal tools for exploring transformations of functions. Try to predict how the graphs of g and h relate to the graph of f. Graph f, g, and h in the same viewing window to confirm your prediction.

a. $f(x) = x^2$, $g(x) = (x - 4)^2$, $h(x) = (x - 4)^2 + 3$

b. $f(x) = x^2$, $g(x) = (x + 1)^2$, $h(x) = (x + 1)^2 - 2$

c. $f(x) = x^2$, $g(x) = (x + 4)^2$, $h(x) = (x + 4)^2 + 2$

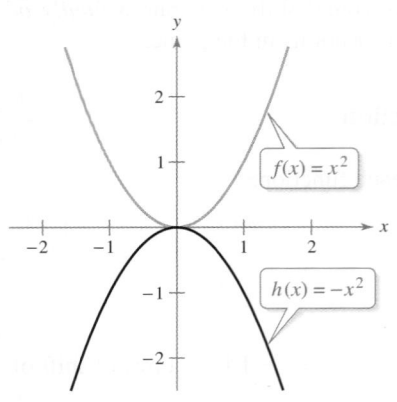

Reflection
FIGURE 2.59

Reflections

Another common type of transformation is a **reflection.** For instance, if you consider the *x*-axis to be a mirror, the graph of $h(x) = -x^2$ is the mirror image (or reflection) of the graph of $f(x) = x^2$, as shown in Figure 2.59.

Reflections in the Coordinate Axes

Reflections in the coordinate axes of the graph of $y = f(x)$ are represented as follows.

1. **Reflection in the *x*-axis:** $g(x) = -f(x)$

2. **Reflection in the *y*-axis:** $h(x) = f(-x)$

TECH TUTOR

Several models of graphing utilities have programs that will give you practice working with reflections, horizontal shifts, and vertical shifts. These programs will graph the function

$$y = R(x + H)^2 + V$$

where $R = \pm 1$, H is an integer between -6 and 6, and V is an integer between -3 and 3. Each time you run the program, different values of R, H, and V are possible. From the graph, you should be able to determine the values of R, H, and V.

Example 2   Reflections of the Graph of a Function

Compare the graph of each function with the graph of $f(x) = \sqrt{x}$.

a. $g(x) = -\sqrt{x}$ **b.** $h(x) = \sqrt{-x}$

SOLUTION

a. The graph of g is a reflection of the graph of f in the *x*-axis because

$$g(x) = -\sqrt{x} = -f(x). \qquad \text{See Figure 2.60(a).}$$

b. The graph of h is a reflection of the graph of f in the *y*-axis because

$$h(x) = \sqrt{-x} = f(-x). \qquad \text{See Figure 2.60(b).}$$

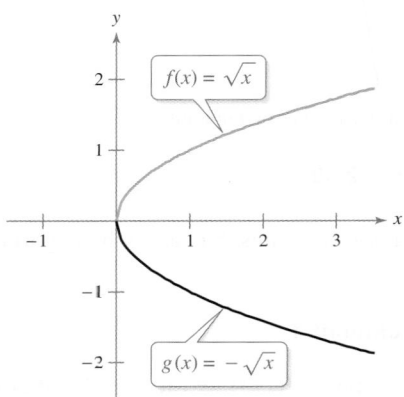

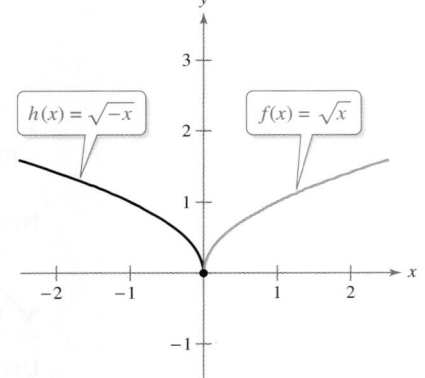

(a) Reflection in *x*-axis   (b) Reflection in *y*-axis
FIGURE 2.60

✓ **Checkpoint 2**

Compare the graph of each function with the graph of $f(x) = |x|$.

a. $g(x) = -|x|$ **b.** $h(x) = |-x|$

When sketching the graph of a function involving square roots, remember that the domain must be restricted to exclude numbers that make the radicand negative. For instance, here are the domains of the functions in Example 2.

Domain of $g(x) = -\sqrt{x}$: $x \geq 0$

Domain of $h(x) = \sqrt{-x}$: $x \leq 0$

> ### Example 3 Reflections and Shifts
>
> Use the graph of $f(x) = x^2$ to sketch the graph of each function.
>
> **a.** $g(x) = -(x - 3)^2$
>
> **b.** $h(x) = -x^2 + 2$
>
> **SOLUTION**
>
> **a.** To sketch the graph of $g(x) = -(x - 3)^2$, first shift the graph of $f(x) = x^2$ to the right three units. Then reflect the result in the x-axis. See Figure 2.61(a).
>
> **b.** To sketch the graph of $h(x) = -x^2 + 2$, first reflect the graph of $f(x) = x^2$ in the x-axis. Then shift the result upward two units. See Figure 2.61(b).

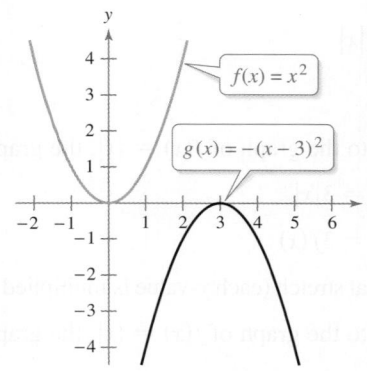

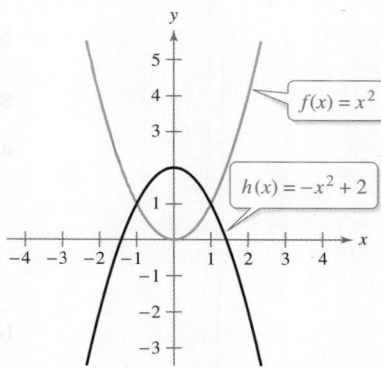

(a) Shift and then reflect in x-axis. (b) Reflect in x-axis and then shift.

FIGURE 2.61

✓ **Checkpoint 3**

Use the graph of $f(x) = x^3$ to sketch the graph of each function.

a. $g(x) = -(x + 2)^3$

b. $h(x) = -x^3 - 3$

> ### Example 4 Writing Equations from Graphs
>
> The graphs labeled g and h in Figure 2.62(a) are transformations of the graph of $f(x) = x^4$. Write an equation for each function.
>
> **SOLUTION** The graph of g is a reflection in the x-axis *followed* by a downward shift of two units of the graph of $f(x) = x^4$. So, the equation for g is $g(x) = -x^4 - 2$. The graph of h is a horizontal shift of one unit to the left *followed* by a reflection in the x-axis of the graph of $f(x) = x^4$. So, the equation for h is $h(x) = -(x + 1)^4$.

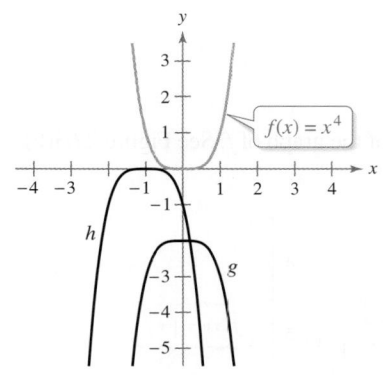

(a)

(b)

FIGURE 2.62

✓ **Checkpoint 4**

The graph labeled h in Figure 2.62(b) is a transformation of the graph of $f(x) = x^2$. Write an equation for the function h.

Can you think of another way to find an equation for g in Example 4? When you shift the graph of f upward two units and then reflect the graph in the x-axis, you get $g(x) = -(x^4 + 2)$. The Distributive Property yields $g(x) = -x^4 - 2$, which is the same equation obtained in Example 4.

Nonrigid Transformations

Horizontal shifts, vertical shifts, and reflections are **rigid** transformations because the basic shape of the graph is unchanged. These transformations change only the *position* of the graph in the *xy*-plane. A **nonrigid** transformation is one that causes a *distortion*— a change in the shape of the original graph. For instance, a nonrigid transformation of the graph of $y = f(x)$ is represented by $g(x) = cf(x)$, where the transformation is a **vertical stretch** if $c > 1$ and a **vertical shrink** if $0 < c < 1$.

Example 5   Nonrigid Transformations

Compare the graph of each function with the graph of $f(x) = |x|$.

a. $h(x) = 3|x|$

b. $g(x) = \dfrac{1}{3}|x|$

SOLUTION

a. Relative to the graph of $f(x) = |x|$, the graph of

$$h(x) = 3|x|$$
$$= 3f(x)$$

is a vertical stretch (each *y*-value is multiplied by 3) of the graph of *f*. See Figure 2.63(a).

b. Relative to the graph of $f(x) = |x|$, the graph of

$$g(x) = \frac{1}{3}|x|$$

$$= \frac{1}{3}f(x)$$

is a vertical shrink $\left(\text{each } y\text{-value is multiplied by } \tfrac{1}{3}\right)$ of the graph of *f*. See Figure 2.63(b).

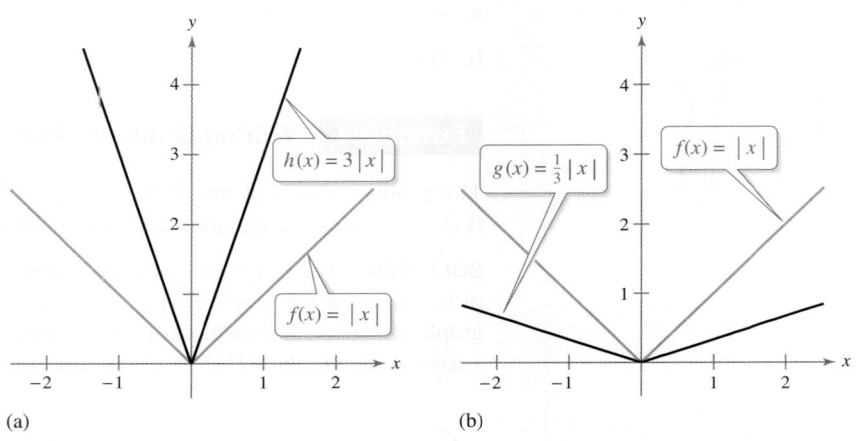

(a)   (b)

FIGURE 2.63

✓Checkpoint 5

Compare the graph of each function with the graph of $f(x) = \sqrt{x}$.

a. $g(x) = 4\sqrt{x}$

b. $h(x) = \dfrac{1}{4}\sqrt{x}$

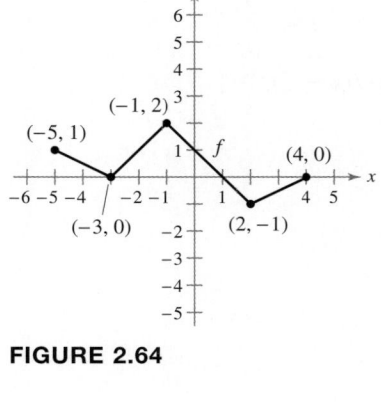

FIGURE 2.64

Example 6 Rigid and Nonrigid Transformations

Use the graph of f shown in Figure 2.64 to sketch each graph.

a. $g(x) = f(x - 2) + 1$

b. $h(x) = \frac{1}{2}f(x)$

SOLUTION

a. The graph of g is a horizontal shift to the right two units and a vertical shift upward one unit of the graph of f. The graph of g is shown in Figure 2.65(a).

b. The graph of h is a vertical shrink of the graph of f.

For $x = -5$, $h(-5) = \frac{1}{2}f(-5) = \frac{1}{2}(1) = \frac{1}{2}$.

For $x = -3$, $h(-3) = \frac{1}{2}f(-3) = \frac{1}{2}(0) = 0$.

For $x = -1$, $h(-1) = \frac{1}{2}f(-1) = \frac{1}{2}(2) = 1$.

For $x = 2$, $h(2) = \frac{1}{2}f(2) = \frac{1}{2}(-1) = -\frac{1}{2}$.

For $x = 4$, $h(4) = \frac{1}{2}f(4) = \frac{1}{2}(0) = 0$.

The graph of h is shown in Figure 2.65(b).

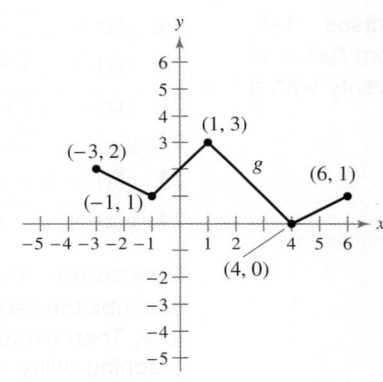

(a)

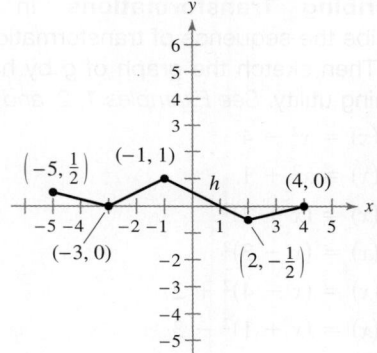

(b)

FIGURE 2.65

 **Checkpoint 6**

Use the graph of g shown in Figure 2.65(a) to sketch the graph of

$$p(x) = 2g(x) - 1.$$

SUMMARIZE (Section 2.6)

1. Describe how to shift the graph of $y = f(x)$ vertically and horizontally *(page 220)*. For an example of shifting the graph of a function, see Example 1.

2. Describe how to reflect the graph of $y = f(x)$ in the x-axis and in the y-axis *(page 222)*. For examples of reflecting and shifting the graph of a function, see Examples 2, 3, and 4.

3. State the definitions of a rigid transformation, a nonrigid transformation, a vertical stretch, and a vertical shrink *(page 224)*. For examples of rigid and nonrigid transformations, see Examples 5 and 6.

SKILLS WARM UP 2.6   The following warm-up exercises involve skills that were covered in earlier sections. You will use these skills in the exercise set for this section. For additional help, review Sections 1.4, 1.5, 2.4, and 2.5.

In Exercises 1 and 2, evaluate the function at the indicated value.

1. Find $f(3)$ for $f(x) = x^2 - 4x + 15$.

2. Find $f(-x)$ for $f(x) = 2x/(x - 3)$.

In Exercises 3 and 4, solve the equation.

3. $-x^3 + 10x = 0$

4. $3x^2 + 2x - 8 = 0$

In Exercises 5–10, sketch the graph of the function.

5. $f(x) = -2$

6. $f(x) = -x$

7. $f(x) = x + 5$

8. $f(x) = 2 - x$

9. $f(x) = 3x - 4$

10. $f(x) = 9x + 10$

Exercises 2.6   See www.CalcChat.com for worked-out solutions to odd-numbered exercises.

Describing Transformations In Exercises 1–8, describe the sequence of transformations from $f(x) = x^2$ to g. Then sketch the graph of g by hand. Verify with a graphing utility. *See Examples 1, 2, and 3.*

1. $g(x) = x^2 - 4$

2. $g(x) = x^2 + 1$

3. $g(x) = (x + 2)^2$

4. $g(x) = (x - 3)^2$

5. $g(x) = (x - 4)^2 + 2$

6. $g(x) = (x + 1)^2 - 3$

7. $g(x) = -x^2 + 1$

8. $g(x) = -(x - 2)^2$

Describing Transformations In Exercises 9–16, describe the sequence of transformations from $f(x) = |x|$ to g. Then sketch the graph of g by hand. Verify with a graphing utility. *See Examples 1, 2, and 3.*

9. $g(x) = |x| + 2$ **10.** $g(x) = |x| - 3$

11. $g(x) = |x - 1|$

12. $g(x) = |x + 4|$

13. $g(x) = -|x| + 3$

14. $g(x) = 5 - |x - 1|$

15. $g(x) = |x + 1| - 4$

16. $g(x) = |x - 2| - 1$

Describing Transformations In Exercises 17–24, describe the sequence of transformations from $f(x) = \sqrt{x}$ to g. Then sketch the graph of g by hand. Verify with a graphing utility. *See Examples 1, 2, 3, and 5.*

17. $g(x) = \sqrt{x - 3}$ **18.** $g(x) = \sqrt{x + 4}$

19. $g(x) = \sqrt{x - 3} + 1$

20. $g(x) = \sqrt{x + 5} - 2$

21. $g(x) = \sqrt{2x}$

22. $g(x) = \sqrt{2x} - 5$

23. $g(x) = 2 + \sqrt{-x - 4}$

24. $g(x) = \sqrt{-x} + 1$

Describing Transformations In Exercises 25–34, describe the sequence of transformations from $f(x) = \sqrt[3]{x}$ to y. Then sketch the graph of y by hand. Verify with a graphing utility. *See Examples 1, 2, 3, and 5.*

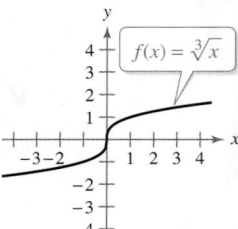

25. $y = \sqrt[3]{-x}$

26. $y = -\sqrt[3]{x}$

27. $y = \sqrt[3]{x} - 2$

28. $y = \sqrt[3]{x} + 3$

29. $y = 2 - \sqrt[3]{x + 1}$

30. $y = -\sqrt[3]{x - 1} - 4$

31. $y = 2\sqrt[3]{x + 1} - 1$

32. $y = 2\sqrt[3]{x - 2} + 1$

33. $y = \frac{1}{2}\sqrt[3]{x}$

34. $y = \frac{1}{2}\sqrt[3]{x} - 3$

Writing Equations from Graphs   In Exercises 35–42, identify the transformation shown in the graph and identify the associated common function. Write the equation of the graphed function. *See Example 4.*

35.

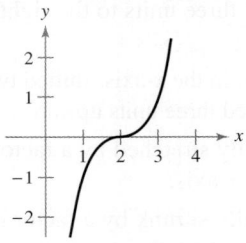

36.

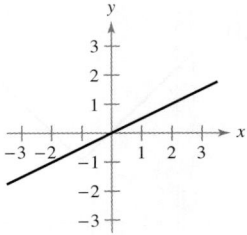

37.

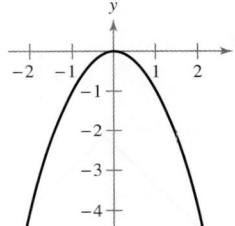

38.

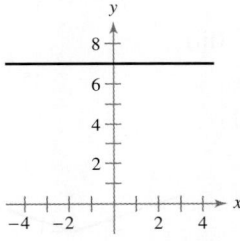

39.

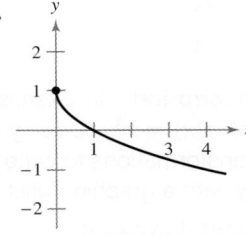

40.

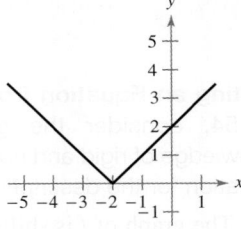

41.

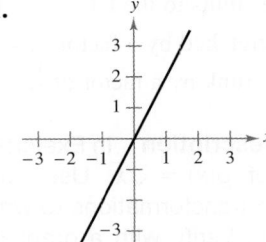

42.

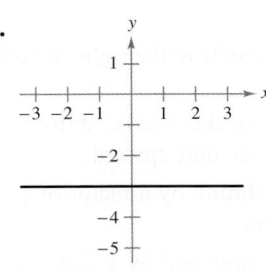

43. Comparing Graphs of Functions   Use a graphing utility to graph f for $c = -2$, 0, and 2 in the same viewing window.

(a) $f(x) = \frac{1}{2}x + c$ (b) $f(x) = \frac{1}{2}(x - c)$

(c) $f(x) = \frac{1}{2}(cx)$

In each case, compare the graph with the graph of $y = \frac{1}{2}x$.

44. Comparing Graphs of Functions   Use a graphing utility to graph f for $c = -2$, 0, and 2 in the same viewing window.

(a) $f(x) = x^3 + c$ (b) $f(x) = (x - c)^3$

(c) $f(x) = (x - 2)^3 + c$

In each case, compare the graph with the graph of $y = x^3$.

45. Writing Equations from Graphs   Use the graph of $f(x) = x^2$ to write equations for the functions whose graphs are shown. *See Example 4.*

(a)

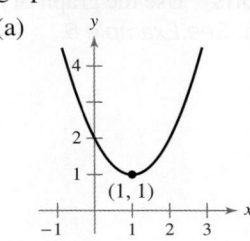

(b)

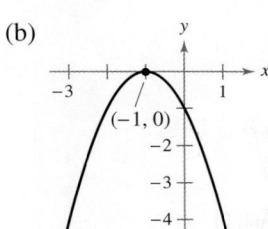

46. Writing Equations from Graphs   Use the graph of $f(x) = x^3$ to write equations for the functions whose graphs are shown. *See Example 4.*

(a)

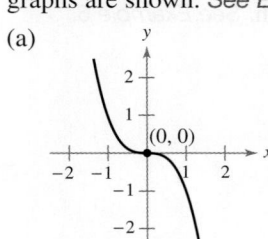

(b)

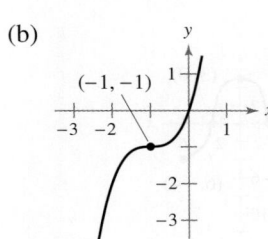

47. Sketching Transformations   Use the graph of f (see figure) to sketch each graph. *See Example 6.*

(a) $y = f(x) + 2$

(b) $y = -f(x)$

(c) $y = f(x - 2)$

(d) $y = f(x + 3)$

(e) $y = 2f(x)$

(f) $y = f(-x)$

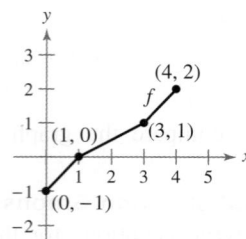

48. Sketching Transformations   Use the graph of f (see figure) to sketch each graph. *See Example 6.*

(a) $y = f(x) - 1$

(b) $y = f(x + 1)$

(c) $y = f(x - 1)$

(d) $y = -f(x - 2)$

(e) $y = f(-x)$

(f) $y = \frac{1}{2}f(x)$

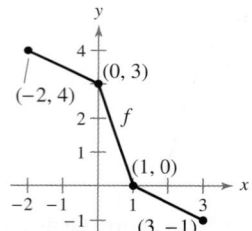

49. Sketching Transformations   Use the graph of f (see figure) to sketch each graph. *See Example 6.*

(a) $y = f(x - 5)$

(b) $y = -f(x) + 3$

(c) $y = \frac{1}{3}f(x)$

(d) $y = -f(x + 1)$

(e) $y = f(-x)$

(f) $y = f(x) - 5$

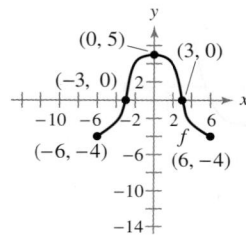

50. HOW DO YOU SEE IT?   Consider the graph of $f(x) = |x|$. Match the description of the transformation with its graph.

(a) The graph of f is shifted three units to the right and two units upward.

(b) The graph of f is reflected in the x-axis, shifted two units to the left, and shifted three units upward.

(c) The graph of f is vertically stretched by a factor of 4 and reflected in the x-axis.

(d) The graph of f is vertically shrunk by a factor of $\frac{1}{3}$ and shifted two units to the left.

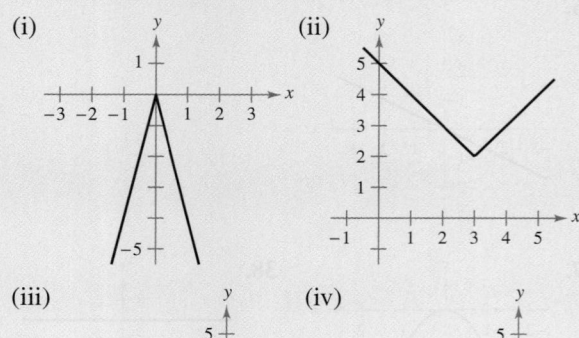

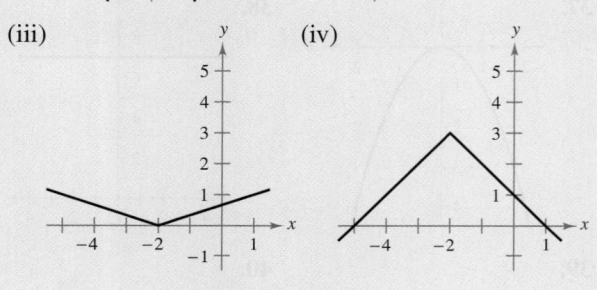

Writing an Equation from a Description   In Exercises 51–54, consider the graph of $f(x) = x^3$. Use your knowledge of rigid and nonrigid transformations to write an equation for the description. Verify with a graphing utility.

51. The graph of f is shifted two units downward.

52. The graph of f is shifted three units to the left.

53. The graph of f is vertically stretched by a factor of 4.

54. The graph of f is vertically shrunk by a factor of $\frac{1}{3}$.

Writing an Equation from a Description   In Exercises 55–58, consider the graph of $g(x) = \sqrt{x}$. Use your knowledge of rigid and nonrigid transformations to write an equation for the description. Verify with a graphing utility.

55. The graph of g is shifted four units to the right and three units downward.

56. The graph of g is reflected in the x-axis, shifted two units to the left, and shifted one unit upward.

57. The graph of g is vertically shrunk by a factor of $\frac{1}{2}$ and shifted three units to the right.

58. The graph of g is vertically stretched by a factor of 2, reflected in the x-axis, and shifted three units upward.

Writing an Equation from a Graph In Exercises 59–62, use the graph of $f(x) = x^3 - 3x^2$ to write an equation for the function g.

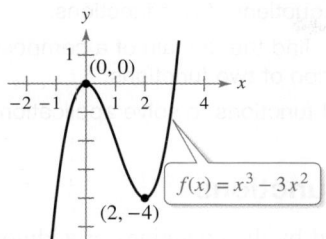

(0, 0)

$f(x) = x^3 - 3x^2$

(2, −4)

59.

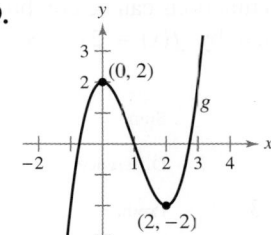

(0, 2)

g

(2, −2)

60.

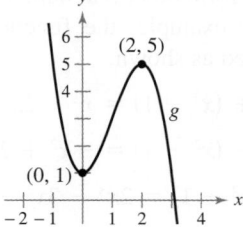

(2, 5)

g

(0, 1)

61.

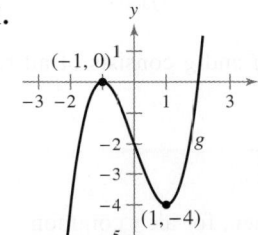

(−1, 0)

g

(1, −4)

62.

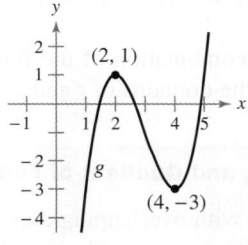

(2, 1)

g

(4, −3)

63. Exploration The point $(3, 9)$ on the graph of $f(x) = x^2$ has been shifted to the point $(4, 7)$ after a rigid transformation. Identify the shift and write the new function g in terms of f.

64. Exploration The point $(8, 2)$ on the graph of $f(x) = \sqrt[3]{x}$ has been shifted to the point $(5, 0)$ after a rigid transformation. Identify the shift and write the new function h in terms of f.

65. Profit A company's weekly profit P (in hundreds of dollars) from a product is given by the model

$$P(x) = 80 + 20x - 0.5x^2, \quad 0 \le x \le 20$$

where x is the amount (in hundreds of dollars) spent on advertising.

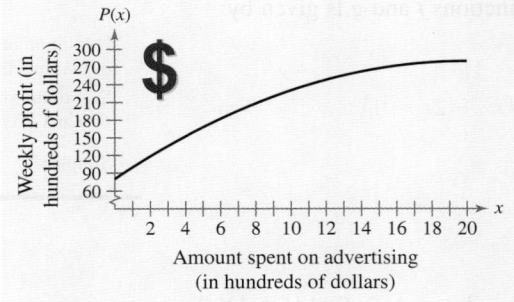

$P(x)$

Weekly profit (in hundreds of dollars)

300
270
240
210
180
150
120
90
60

2   4   6   8   10 12 14 16 18 20

Amount spent on advertising
(in hundreds of dollars)

(a) The company estimates that taxes and operating costs will increase by an average of $2500 per week during the next year. Rewrite the profit equation to reflect this expected decrease in profits. Identify the type of transformation applied to the graph of the equation.

(b) Use a graphing utility to graph the profit equation from part (a).

(c) Rewrite the original profit equation so that x measures advertising expenditures in dollars. [Find $P(x/100)$.] Identify the type of transformation applied to the graph of the equation.

66. Automobile Aerodynamics The number of horsepower H required to overcome wind drag on an automobile is approximated by

$$H(x) = 0.002x^2 + 0.005x - 0.029, \quad 10 \le x \le 100$$

where x is the speed of the car (in miles per hour).

(a) Use a graphing utility to graph the function.

(b) Rewrite the horsepower function so that x represents the speed in kilometers per hour. [Find $H(x/1.6)$.] Identify the type of transformation applied to the graph of the horsepower function.

67. Finance The amount M (in billions of dollars) of home mortgage debt outstanding in the United States from 1995 through 2009 can be approximated by the model

$$M(t) = 34.0t^2 + 3705, \quad 5 \le t \le 19$$

where t represents the year, with $t = 5$ corresponding to 1995. *(Source: Board of Governors of the Federal Reserve System)*

(a) Describe the sequence of transformations from $f(t) = t^2$ to M.

(b) Use a graphing utility to graph the function M.

(c) According to the model, when did the amount of debt exceed 13 trillion dollars?

(d) Rewrite the function so that $t = 5$ represents 2005. Identify the type of transformation applied to the graph of the home mortgage debt function M.

68. Exploration Use a graphing utility to graph the six functions below in the same viewing window. Describe any similarities and differences you observe among the graphs.

(a) $y = x$ (b) $y = x^2$

(c) $y = x^3$ (d) $y = x^4$

(e) $y = x^5$ (f) $y = x^6$

69. Reasoning Use the results of Exercise 68 to make a conjecture about the shapes of the graphs of $y = x^7$ and $y = x^8$. Use a graphing utility to verify your conjecture.

2.7 The Algebra of Functions

■ Find the sum, difference, product, and quotient of two functions.
■ Form the composition of two functions, find the domain of a composite function, and identify a function as the composition of two functions.
■ Use combinations and compositions of functions to solve application problems.

Arithmetic Combinations of Functions

Just as two real numbers can be combined by the operations of addition, subtraction, multiplication, and division to form other real numbers, two functions can be combined to create new functions. For example, the functions given by $f(x) = 2x - 3$ and $g(x) = x^2 - 1$ can be combined as shown.

$$f(x) + g(x) = (2x - 3) + (x^2 - 1) = x^2 + 2x - 4 \qquad \text{Sum}$$

$$f(x) - g(x) = (2x - 3) - (x^2 - 1) = -x^2 + 2x - 2 \qquad \text{Difference}$$

$$f(x)g(x) = (2x - 3)(x^2 - 1) = 2x^3 - 3x^2 - 2x + 3 \qquad \text{Product}$$

$$\frac{f(x)}{g(x)} = \frac{2x - 3}{x^2 - 1}, \quad x \neq \pm 1, \quad g(x) \neq 0 \qquad \text{Quotient}$$

The domain of an arithmetic combination of the functions f and g consists of all real numbers that are common to the domains of f and g.

In Exercise 56 on page 237, you will use a composite function to determine the trend of the profits for a company that owns two manufacturing plants.

Sum, Difference, Product, and Quotient of Functions

Let f and g be two functions with overlapping domains. Then, for all x common to both domains, the **sum, difference, product,** and **quotient** of f and g are defined as follows.

1. *Sum:* $(f + g)(x) = f(x) + g(x)$

2. *Difference:* $(f - g)(x) = f(x) - g(x)$

3. *Product:* $(fg)(x) = f(x) \cdot g(x)$

4. *Quotient:* $\left(\dfrac{f}{g}\right)(x) = \dfrac{f(x)}{g(x)}, \quad g(x) \neq 0$

Example 1   Finding the Sum of Two Functions

Given $f(x) = 2x + 1$ and $g(x) = x^2 + 2x - 1$, find

   $(f + g)(x)$.

SOLUTION   The sum of the functions f and g is given by

$$(f + g)(x) = f(x) + g(x) \qquad \text{Definition of sum of two functions}$$

$$= (2x + 1) + (x^2 + 2x - 1) \qquad \text{Substitute for } f(x) \text{ and } g(x).$$

$$= x^2 + 4x. \qquad \text{Simplify.}$$

✓ Checkpoint 1

Given $f(x) = x^2 - 4$ and $h(x) = x^2 + x + 3$, find $(f + h)(x)$.

Example 2   Finding the Difference of Two Functions

Given the functions

$$f(x) = 2x + 1 \quad \text{and} \quad g(x) = x^2 + 2x - 1$$

find $(f - g)(x)$. Then evaluate the difference when $x = 2$.

SOLUTION   The difference of the functions f and g is given by

$(f - g)(x) = f(x) - g(x)$	Definition of difference of two functions
$= (2x + 1) - (x^2 + 2x - 1)$	Substitute for $f(x)$ and $g(x)$.
$= -x^2 + 2.$	Simplify.

When $x = 2$, the value of this difference is

$$(f - g)(2) = -(2)^2 + 2 = -2.$$

✓ **Checkpoint 2**

Given $f(x) = x^2 - 4$ and $h(x) = x^2 + x + 3$, find $(f - h)(x)$. Then evaluate the difference when $x = 3$.   ■

In Examples 1 and 2, both f and g have domains that consist of all real numbers. So, the domains of $(f + g)$ and $(f - g)$ are also the set of all real numbers. Remember that any restrictions on the domains of f and g must be considered when forming the sum, difference, product, or quotient of f and g.

Example 3   The Quotient of Two Functions

Find the domains of $\left(\dfrac{f}{g}\right)(x)$ and $\left(\dfrac{g}{f}\right)(x)$ for the functions

$$f(x) = \sqrt{x} \quad \text{and} \quad g(x) = \sqrt{4 - x^2}.$$

SOLUTION   The quotient of f and g is given by

$\left(\dfrac{f}{g}\right)(x) = \dfrac{f(x)}{g(x)}$	Definition of quotient of two functions
$= \dfrac{\sqrt{x}}{\sqrt{4 - x^2}}$	Substitute for $f(x)$ and $g(x)$.

and the quotient of g and f is given by

$\left(\dfrac{g}{f}\right)(x) = \dfrac{g(x)}{f(x)}$	Definition of quotient of two functions
$= \dfrac{\sqrt{4 - x^2}}{\sqrt{x}}.$	Substitute for $g(x)$ and $f(x)$.

The domain of f is $[0, \infty)$ and the domain of g is $[-2, 2]$. The intersection of these two domains is $[0, 2]$, which implies that the domains of f/g and g/f are as follows. Notice that the domains differ slightly.

$$\text{Domain of } \frac{f}{g}: \ [0, 2) \qquad \text{Domain of } \frac{g}{f}: \ (0, 2]$$

✓ **Checkpoint 3**

Find the domains of $\left(\dfrac{f}{h}\right)(x)$ and $\left(\dfrac{h}{f}\right)(x)$ for $f(x) = x - 1$ and $h(x) = x - 3$.   ■

Composition of Functions

Another way to combine two functions is to form the **composition** of one with the other. To form a composition, the output of one function becomes the input for another function. For instance, when $f(x) = x^2$ and $g(x) = x + 1$, the composition of f with g is given by

$$f(g(x)) = f(x + 1)$$
$$= (x + 1)^2.$$

This composition is denoted as $f \circ g$ and is read as "f composed with g."

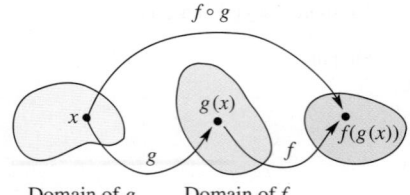

$f \circ g$

x $g(x)$ $f(g(x))$

g f

Domain of g Domain of f

FIGURE 2.66

Definition of the Composition of Two Functions

The **composition** of the functions f and g is given by

$$(f \circ g)(x) = f(g(x)).$$

The domain of $f \circ g$ is the set of all x in the domain of g such that $g(x)$ is in the domain of f. (See Figure 2.66.)

From the definition above, it follows that the domain of $f \circ g$ is always a subset of the domain of g, and the range of $f \circ g$ is always a subset of the range of f.

Example 4   Composition of Functions

Given $f(x) = x + 2$ and $g(x) = 4 - x^2$, find each composition.

a. $(f \circ g)(x)$

b. $(g \circ f)(x)$

SOLUTION

a. The composition of f with g is

$(f \circ g)(x) = f(g(x))$	Definition of $f \circ g$
$= f(4 - x^2)$	Definition of $g(x)$
$= (4 - x^2) + 2$	Definition of $f(x)$
$= -x^2 + 6.$	Simplify.

b. The composition of g with f is

$(g \circ f)(x) = g(f(x))$	Definition of $g \circ f$
$= g(x + 2)$	Definition of $f(x)$
$= 4 - (x + 2)^2$	Definition of $g(x)$
$= 4 - (x^2 + 4x + 4)$	Expand.
$= -x^2 - 4x.$	Simplify.

Note that, in this case, $(f \circ g)(x) \neq (g \circ f)(x)$.

✓ Checkpoint 4

Given $f(x) = x^2 - 2$ and $g(x) = x + 1$, find each composition.

a. $(f \circ g)(x)$

b. $(g \circ f)(x)$

Example 5   Finding the Domain of a Composite Function

Find the composition $(f \circ g)(x)$ for the functions given by

$$f(x) = x^2 - 9 \quad \text{and} \quad g(x) = \sqrt{9 - x^2}.$$

Then find the domain of $f \circ g$.

SOLUTION   The composition of the functions is

$$
\begin{aligned}
(f \circ g)(x) &= f(g(x)) \\
&= f\left(\sqrt{9 - x^2}\right) \\
&= \left(\sqrt{9 - x^2}\right)^2 - 9 \\
&= 9 - x^2 - 9 \\
&= -x^2.
\end{aligned}
$$

From this result, it might appear that the domain of the composition is the set of all real numbers. However, because the domain of f is the set of all real numbers and the domain of g is $[-3, 3]$, the domain of $f \circ g$ is $[-3, 3]$.

✓ Checkpoint 5

Find the composition $(f \circ g)(x)$ for the functions given by $f(x) = \sqrt{x}$ and $g(x) = 3 - x^2$. Then find the domain of $f \circ g$.   ■

In Examples 4 and 5, you formed the composition of two functions. To "decompose" a composite function, look for an "inner" function and an "outer" function. For instance, the function h given by

$$h(x) = (3x - 5)^3$$

is the composition of f with g, where $f(x) = x^3$ and $g(x) = 3x - 5$. That is,

$$h(x) = (3x - 5)^3 = [g(x)]^3 = f(g(x)).$$

In the function h, $g(x) = 3x - 5$ is the *inner* function and $f(x) = x^3$ is the *outer* function.

Example 6   Identifying a Composite Function

Write the function given by $h(x) = \dfrac{1}{(x - 2)^2}$ as a composition of two functions.

SOLUTION   One way to write h as a composition of two functions is to take the inner function to be $g(x) = x - 2$ and the outer function to be

$$f(x) = \frac{1}{x^2} = x^{-2}.$$

Then you can write

$$h(x) = \frac{1}{(x - 2)^2} = (x - 2)^{-2} = f(x - 2) = f(g(x)).$$

✓ Checkpoint 6

Write the function given by

$$h(x) = (x - 1)^2 + 2$$

as a composition of two functions.   ■

TECH TUTOR

In Example 5, the domain of the composite function is $[-3, 3]$. To convince yourself of this, use a graphing utility to graph

$$y = \left(\sqrt{9 - x^2}\right)^2 - 9$$

as shown in the figure below. Notice that the graphing utility does not extend the graph to the left of $x = -3$ or to the right of $x = 3$.

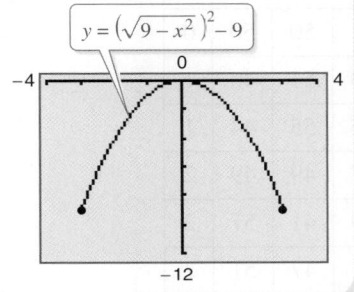

STUDY TIP

When writing a function as a composition of two functions, there can be many correct answers. For instance,

$$h(x) = (3x - 5)^3$$

can be written as the composition of f with g, where $f(x) = (x - 5)^3$ and $g(x) = 3x$.

Applications

Example 7 **Political Makeup of the U.S. Senate**

Consider three functions R, D, and I that represent the numbers of Republicans, Democrats, and Independents, respectively, in the U.S. Senate from 1977 to 2011. Sketch the graphs of R, D, and I and the sum of R, D, and I in the same coordinate plane. The numbers of senators from the three political parties immediately following the elections are shown below. *(Source: United States Senate)*

In recent years, no party has had a strong majority in the Senate, which can make it difficult to pass legislation.

Year	R	D	I
1977	38	61	1
1979	41	58	1
1981	53	46	1
1983	55	45	0
1985	53	47	0
1987	45	55	0
1989	45	55	0
1991	44	56	0
1993	43	57	0

Year	R	D	I
1995	52	48	0
1997	55	45	0
1999	55	45	0
2001	50	50	0
2003	51	48	1
2005	55	44	1
2007	49	49	2
2009	41	57	2
2011	47	51	2

SOLUTION The graphs of R, D, and I are shown in Figure 2.67. Note that the sum of R, D, and I is the constant function

$$R + D + I = 100.$$

This follows from the fact that the number of senators in the United States Senate is 100 (two from each state).

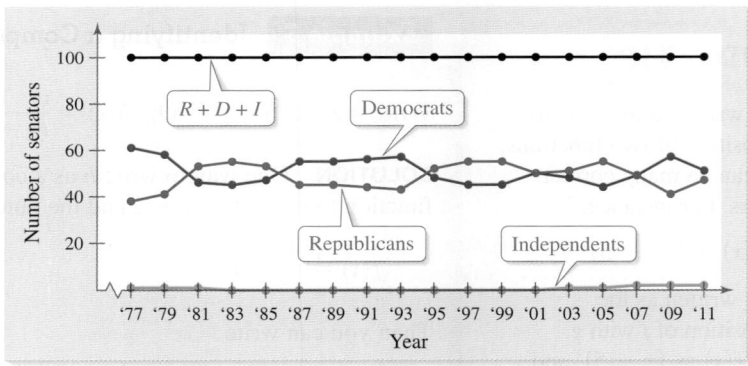

Numbers of U.S. Senators by Political Party
FIGURE 2.67

✓ Checkpoint 7

In Example 7, consider the function f given by

$$f = 100 - (R + D).$$

What does f represent in the context of the real-life situation?

Example 8   **Bacteria Count**

The number N of bacteria in a certain food is given by

$$N(T) = 20T^2 - 80T + 500, \quad 2 \le T \le 14$$

where T is the temperature of the food in degrees Celsius. When the food is removed from refrigeration, the temperature of the food is given by

$$T(t) = 4t + 2, \quad 0 \le t \le 3$$

where t is the time in hours. Find

a. the composition $N(T(t))$,

b. the number of bacteria in the food when $t = 2$ hours, and

c. how long the food can remain unrefrigerated before the bacteria count reaches 2000.

SOLUTION

a. $N(T(t)) = 20(4t + 2)^2 - 80(4t + 2) + 500$

$\qquad\qquad = 20(16t^2 + 16t + 4) - 320t - 160 + 500$

$\qquad\qquad = 320t^2 + 320t + 80 - 320t - 160 + 500$

$\qquad\qquad = 320t^2 + 420$

b. When $t = 2$, the number of bacteria is

$\qquad N(T(2)) = 320(2)^2 + 420$

$\qquad\qquad\qquad = 1280 + 420$

$\qquad\qquad\qquad = 1700.$

c. The bacteria count will reach $N = 2000$ when $320t^2 + 420 = 2000$. By solving this equation, you can determine that the bacteria count will reach 2000 when $t \approx 2.2$ hours. So, the food can remain unrefrigerated for about 2 hours and 12 minutes.

✓ **Checkpoint 8**

In Example 8, how long can the food remain unrefrigerated before the bacteria count reaches 1000?

SUMMARIZE   (Section 2.7)

1. Describe how to find the sum, difference, product, and quotient of two functions with overlapping domains *(page 230)*. For examples of these arithmetic combinations of two functions, see Examples 1, 2, and 3.

2. State the definition of the composition of two functions *(page 232)*. For examples that use the composition of two functions, see Examples 4, 5, and 6.

3. Describe real-life examples that use a combination of two functions *(pages 234 and 235, Examples 7 and 8)*.

SKILLS WARM UP 2.7   The following warm-up exercises involve skills that were covered in earlier sections. You will use these skills in the exercise set for this section. For additional help, review Section 0.7.

In Exercises 1–10, perform the indicated operations and simplify the result.

1. $\dfrac{1}{x} + \dfrac{1}{1-x}$

2. $\dfrac{2}{x+3} - \dfrac{2}{x-3}$

3. $\dfrac{3}{x-2} - \dfrac{2}{x(x-2)}$

4. $\dfrac{x}{x-5} + \dfrac{1}{3}$

5. $(x-1)\left(\dfrac{1}{\sqrt{x^2-1}}\right)$

6. $\left(\dfrac{x}{x^2-4}\right)\left(\dfrac{x^2-x-2}{x^2}\right)$

7. $(x^2-4) \div \left(\dfrac{x+2}{5}\right)$

8. $\left(\dfrac{x}{x^2+3x-10}\right) \div \left(\dfrac{x^2+3x}{x^2+6x+5}\right)$

9. $\dfrac{(1/x)+5}{3-(1/x)}$

10. $\dfrac{(x/4)-(4/x)}{x-4}$

Exercises 2.7

See www.CalcChat.com for worked-out solutions to odd-numbered exercises.

Graphing the Sum of Two Functions   In Exercises 1–4, use the graphs of f and g to graph $h(x) = (f+g)(x)$.

1.
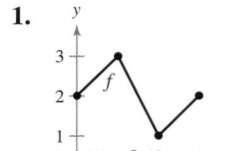

2.
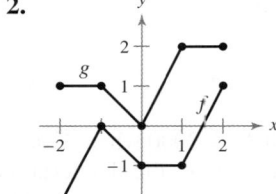

3.
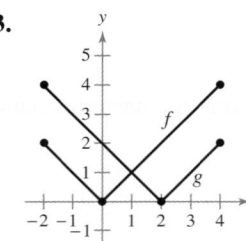

4.
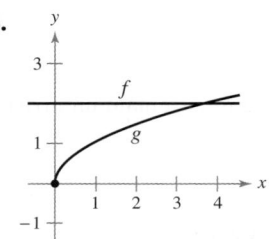

Finding Arithmetic Combinations of Functions   In Exercises 5–12, find (a) $(f+g)(x)$, (b) $(f-g)(x)$, (c) $(fg)(x)$, and (d) $(f/g)(x)$. What is the domain of f/g? *See Examples 1, 2, and 3.*

5. $f(x) = x+1$, $g(x) = x-1$

6. $f(x) = 2x-3$, $g(x) = 1-x$

7. $f(x) = 3x-1$, $g(x) = 1-x^2$

8. $f(x) = 2x+3$, $g(x) = x^2-1$

9. $f(x) = x^2+5$, $g(x) = \sqrt{1-x}$

10. $f(x) = x^2-4$, $g(x) = \sqrt{x-3}$

11. $f(x) = \dfrac{1}{x}$, $g(x) = \dfrac{1}{x^2}$

12. $f(x) = \dfrac{1}{x}$, $g(x) = \dfrac{2}{x^3}$

Evaluating Arithmetic Combinations of Functions   In Exercises 13–24, evaluate the function for

$$f(x) = 2x+1 \quad \text{and} \quad g(x) = x^2-2.$$

13. $(f+g)(3)$

14. $(f-g)(-2)$

15. $(f-g)(2t)$

16. $(f+g)(3t)$

17. $(fg)(-2)$

18. $(fg)(-6)$

19. $\left(\dfrac{f}{g}\right)(5)$

20. $\left(\dfrac{f}{g}\right)(0)$

21. $(f-g)(0)$

22. $(f+g)(1)$

23. $(4f)(2) - (2g)(4)$

24. $(2f)(5) + (3g)(-4)$

Composition of Functions   In Exercises 25–28, find (a) $f \circ g$, (b) $g \circ f$, and (c) $f \circ f$. *See Example 4.*

25. $f(x) = 3x$, $g(x) = 2x+5$

26. $f(x) = 2x-1$, $g(x) = 7-x$

27. $f(x) = x^2$, $g(x) = 3x+1$

28. $f(x) = x^3$, $g(x) = \dfrac{1}{x}$

Composition of Functions   In Exercises 29–36, find (a) $f \circ g$ and (b) $g \circ f$. *See Example 4.*

29. $f(x) = \frac{1}{3}x - 3, \quad g(x) = 3x + 1$

30. $f(x) = \frac{1}{2}x + 1, \quad g(x) = 2x + 3$

31. $f(x) = \sqrt{x + 4}, \quad g(x) = x^2 - 4$

32. $f(x) = \sqrt[3]{x - 1}, \quad g(x) = x^3 + 1$

33. $f(x) = \sqrt{x}, \quad g(x) = \sqrt{x}$

34. $f(x) = x^2, \quad g(x) = x^2$

35. $f(x) = |x|, \quad g(x) = x + 6$

36. $f(x) = x - 4, \quad g(x) = |x|$

Finding the Domain of a Composite Function   In Exercises 37–40, determine the domain of (a) f, (b) g, and (c) $f \circ g$. *See Example 5.*

37. $f(x) = x^2 + 3,$
$g(x) = \sqrt{x}$

38. $f(x) = \sqrt[3]{x + 1},$
$g(x) = x^3$

39. $f(x) = \dfrac{1}{x^2},$
$g(x) = x - 2$

40. $f(x) = \dfrac{5}{x^2 - 4},$
$g(x) = x + 3$

Evaluating Combinations of Functions   In Exercises 41–44, use the graphs of f and g to evaluate the functions.

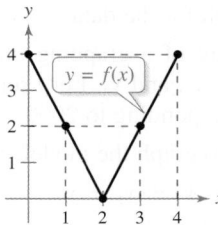
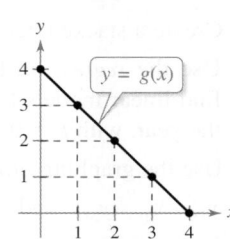

41. (a) $(f + g)(3)$ (b) $\left(\dfrac{f}{g}\right)(2)$

42. (a) $(f - g)(1)$ (b) $(fg)(4)$

43. (a) $(f \circ g)(2)$ (b) $(g \circ f)(2)$

44. (a) $(f \circ g)(0)$ (b) $(g \circ f)(3)$

Identifying a Composite Function   In Exercises 45–52, find two functions f and g such that $(f \circ g)(x) = h(x)$. (There are many correct answers.) *See Example 6.*

45. $h(x) = (2x + 1)^2$

46. $h(x) = (1 - x)^3$

47. $h(x) = \sqrt[3]{x^2 - 4}$

48. $h(x) = \sqrt{9 - x}$

49. $h(x) = \dfrac{1}{x + 2}$

50. $h(x) = \dfrac{4}{(5x + 2)^2}$

51. $h(x) = (x + 4)^2 + 2(x + 4)$

52. $h(x) = (x + 3)^{3/2}$

53. Stopping Distance   While driving at x miles per hour, you are required to stop quickly to avoid an accident. The distance the car travels (in feet) during your reaction time is given by $R(x) = \frac{3}{4}x$. The distance the car travels (in feet) while you are braking is given by $B(x) = \frac{1}{15}x^2$. Find the function that represents the total stopping distance T. (*Hint:* $T = R + B$.) Graph the functions $R, B,$ and T on the same set of coordinate axes for $0 \le x \le 60$.

54. Cost   The weekly cost C of producing x units in a manufacturing process is given by the function

$$C(x) = 70x + 800.$$

The number of units x produced in t hours is given by

$$x(t) = 40t.$$

Find and interpret $(C \circ x)(t)$.

55. Cost   The weekly cost C of producing x units in a manufacturing process is given by the function

$$C(x) = 50x + 495.$$

The number of units x produced in t hours is given by

$$x(t) = 30t.$$

Find and interpret $(C \circ x)(t)$.

56. Profits   A manufacturing company has two plants, one in New Jersey and another in California. From 2005 through 2012, the profits for the plant in New Jersey decreased according to the function

$$P_1 = 18.97 - 0.55t, \quad t = 5, 6, 7, 8, 9, 10, 11, 12$$

and the profits for the plant in California increased according to the function

$$P_2 = 15.85 + 0.67t, \quad t = 5, 6, 7, 8, 9, 10, 11, 12$$

where P_1 and P_2 represent the yearly profits (in millions of dollars) and t represents the year, with $t = 5$ corresponding to 2005 (see figure).

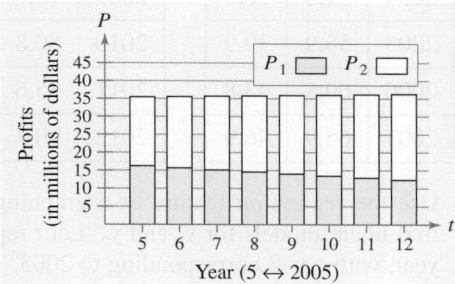

(a) Use the stacked bar graph to determine whether the company's total yearly profits for both plants decreased or increased from 2005 through 2012.

(b) Write a function that represents the total yearly profits for both plants from 2005 through 2012. Explain how the function supports your answer in part (a).

57. Sales You own two fast-food restaurants. From 2005 through 2012, the sales of one restaurant decreased according to the function

$$R_1 = 525 - 15.2t, \quad t = 5, 6, 7, 8, 9, 10, 11, 12$$

and the sales of the other restaurant increased according to the function

$$R_2 = 392 + 8.5t, \quad t = 5, 6, 7, 8, 9, 10, 11, 12$$

where R_1 and R_2 represent the yearly sales (in thousands of dollars) and t represents the year, with $t = 5$ corresponding to 2005 (see figure).

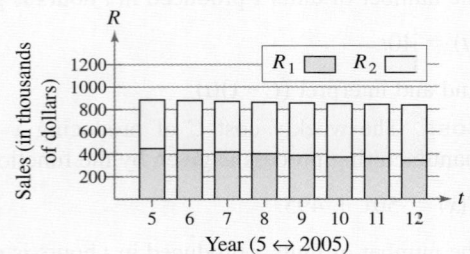

(a) Use the stacked bar graph to determine whether your total yearly sales of both restaurants decreased or increased from 2005 through 2012.

(b) Write a function that represents your total yearly sales of both restaurants from 2005 through 2012. Explain how the function supports your answer in part (a).

58. Cost, Revenue, and Profit The table shows the revenues y_1 (in thousands of dollars) and total costs y_2 (in thousands of dollars) for a sports memorabilia store for the years 2003 through 2012.

Year	y_1	y_2	Year	y_1	y_2
2003	46.3	32.9	2008	71.0	51.1
2004	51.3	36.5	2009	75.7	53.7
2005	55.9	39.9	2010	80.8	57.6
2006	60.8	43.8	2011	85.6	62.1
2007	65.9	46.9	2012	90.7	68.7

(a) Use the *regression* feature of a graphing utility to find linear models for y_1 and y_2. Let t represent the year, with $t = 3$ corresponding to 2003.

(b) Use the graphing utility to graph the models for

$$y_1, \quad y_2, \quad \text{and} \quad y_3 = y_1 - y_2$$

in the same viewing window. What does y_3 represent in the context of the problem? Predict the value of y_3 in 2014.

(c) Create a stacked bar graph for y_2 and y_3. What do the heights of the bars represent?

59. Employment The table shows the numbers of employees in the offices of health care practitioners from 2000 through 2009. The numbers of employees (in thousands) in the offices of physicians, dentists, and other health care practitioners are represented by y_1, y_2, and y_3, respectively.
(Source: U.S. Bureau of Labor Statistics)

Year	y_1	y_2	y_3
2000	1840	688	438
2001	1911	705	457
2002	1968	725	486
2003	2003	744	503
2004	2048	760	527
2005	2094	774	549
2006	2148	786	573
2007	2202	808	600
2008	2253	818	627
2009	2280	818	651

(a) Create a stacked bar graph for the data.

(b) Use the *regression* feature of a graphing utility to find linear models for y_1, y_2, and y_3. Let t represent the year, with $t = 0$ corresponding to 2000.

(c) Use the graphing utility to graph the models for

$$y_1, \quad y_2, \quad y_3, \quad \text{and} \quad y_4 = y_1 + y_2 + y_3$$

in the same viewing window. Use y_4 to predict the total number of employees in the offices of health care practitioners in 2011 and in 2013.

60. Bacteria Count A package of food is removed from a refrigerator. The number N of bacteria in the food is given by

$$N(T) = 10T^2 - 20T + 600, \quad 1 \le T \le 16$$

where T is the temperature (in degrees Celsius) of the food. The temperature of the food increases according to the function

$$T(t) = 3t + 1, \quad 0 \le t \le 5$$

where t is the time in hours.

(a) Find the composite function $N(T(t))$.

(b) Find the number of bacteria in the food when $t = 2$ hours.

(c) Find the time t when the bacteria count reaches 1500.

61. Ripples in Water A pebble is dropped into a calm pond, causing ripples in the form of concentric circles. The radius (in feet) of the outermost ripple is given by

$$r(t) = 0.6t$$

where t is time in seconds after the pebble strikes the water. The area of the outermost circle is given by the function

$$A(r) = \pi r^2.$$

Find and interpret

$$(A \circ r)(t).$$

62. HOW DO YOU SEE IT? The graphs labeled L_1, L_2, L_3, and L_4 represent four different pricing discounts, where p is the original price (in dollars) and S is the sale price (in dollars). Match each function with its graph. Describe the situations in parts (c) and (d).

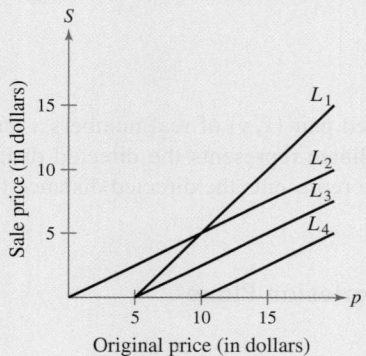

(a) $f(p)$: A 50% discount is applied.

(b) $g(p)$: A $5 discount is applied.

(c) $(g \circ f)(p)$

(d) $(f \circ g)(p)$

Price-Earnings Ratio The average annual price-earnings ratio for a corporation's stock is defined as the average price of the stock divided by the earnings per share. In Exercises 63 and 64, the average price of a corporation's stock is given as the function P, and the earnings per share is given as the function E. Find the price-earnings ratios, P/E, for the years 2005 through 2009.

63. DISH Network

Year	2005	2006	2007	2008	2009
P	$29.06	$31.92	$43.51	$25.15	$16.05
E	$1.74	$1.52	$1.68	$1.98	$1.42

(Source: DISH Network)

AP Photo/Jeff Chiu

64. Walt Disney

Year	2005	2006	2007	2008	2009
P	$26.72	$27.53	$34.18	$32.09	$22.75
E	$1.31	$1.61	$1.92	$2.26	$1.82

(Source: Walt Disney)

65. Think About It Find the domains of $(f/g)(x)$ and $(g/f)(x)$ for the functions $f(x) = \sqrt{x}$ and $g(x) = \sqrt{9 - x^2}$. Why do the two domains differ?

True or False? In Exercises 66 and 67, determine whether the statement is true or false. Justify your answer.

66. If $f(x) = x + 1$ and $g(x) = 6x$, then

$$(f \circ g)(x) = (g \circ f)(x).$$

67. Given two functions $f(x)$ and $g(x)$, you can calculate $(f \circ g)(x)$ if and only if the range of g is a subset of the domain of f.

Business Capsule

SunPower Corporation develops and manufactures solar-electric power products. SunPower's high-efficiency solar cells generate up to 50% more power than other solar technologies. SunPower's technology was developed by Dr. Richard Swanson and his students while he was Professor of Engineering at Stanford University. From 2005 to 2010, SunPower's revenues increased by almost 3000%.

68. Research Project Use your campus library, the Internet, or some other reference source to find information about an alternative-energy business experiencing strong growth similar to the example above. Write a brief report about the business.

ALGEBRA TUTOR

Identifying and Plotting Points in the Cartesian Plane

To be successful at graphing equations, you must be good at plotting points in the Cartesian plane. The Cartesian plane is formed by the intersection of a horizontal real number line called the *x*-axis and a vertical real number line called the *y*-axis. These axes intersect at a point called the origin and divide the plane into four quadrants, as shown below.

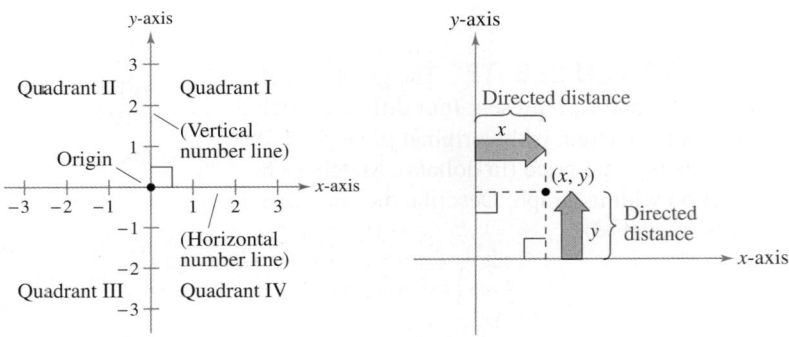

Each point in the plane corresponds to an ordered pair (x, y) of real numbers x and y, called the coordinates of the point. The *x*-coordinate represents the directed distance from the *y*-axis to the point, and the *y*-coordinate represents the directed distance from the *x*-axis to the point, as shown above.

Example 1   Identifying Points in the Cartesian Plane

Give the coordinates of each point.

a. *A*   **b.** *B*   **c.** *C*

d. *D*   **e.** *E*   **f.** *F*

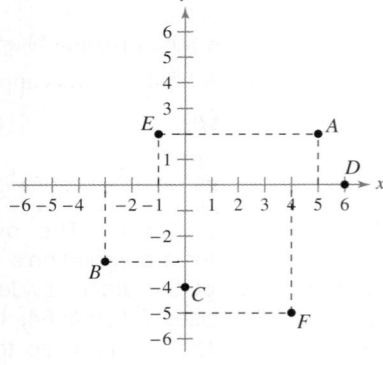

SOLUTION

a. Point *A* is 5 units to the right of the origin and 2 units up. So, the *x*-coordinate is 5 and the *y*-coordinate is 2. The coordinates are $(5, 2)$.

b. Point *B* is 3 units to the left of the origin and 3 units down. So, the *x*-coordinate is -3 and the *y*-coordinate is -3. The coordinates are $(-3, -3)$.

c. Point *C* is 4 units down from the origin. So, the *x*-coordinate is 0 and the *y*-coordinate is -4. The coordinates are $(0, -4)$.

d. Point *D* is 6 units to the right of the origin. So, the *x*-coordinate is 6 and the *y*-coordinate is 0. The coordinates are $(6, 0)$.

e. Point *E* is 1 unit to the left of the origin and 2 units up. So, the *x*-coordinate is -1 and the *y*-coordinate is 2. The coordinates are $(-1, 2)$.

f. Point *F* is 4 units to the right of the origin and 5 units down. So, the *x*-coordinate is 4 and the *y*-coordinate is -5. The coordinates are $(4, -5)$.

Example 2   **Plotting Points in the Cartesian Plane**

Plot each point in the Cartesian plane.

a. $P(7, -2)$ **b.** $Q(-1, -6)$

c. $R(0, 4)$ **d.** $S(-2, 0)$

e. $T(8, 3)$ **f.** $U(-3, 5)$

g. $V\left(-\frac{3}{2}, \frac{7}{2}\right)$ **h.** $W\left(4, \frac{1}{2}\right)$

i. $X(2.5, -6.5)$

SOLUTION

a. Start at the origin. The x-coordinate is 7, so move right 7 units. The y-coordinate is -2, so move down 2 units. Plot the point at $(7, -2)$ and label it P.

b. Start at the origin. The x-coordinate is -1, so move left 1 unit. The y-coordinate is -6, so move down 6 units. Plot the point at $(-1, -6)$ and label it Q.

c. Start at the origin. The y-coordinate is 4, so move up 4 units. Plot the point at $(0, 4)$ and label it R.

d. Start at the origin. The x-coordinate is -2, so move left 2 units. Plot the point at $(-2, 0)$ and label it S.

e. Start at the origin. The x-coordinate is 8, so move right 8 units. The y-coordinate is 3, so move up 3 units. Plot the point at $(8, 3)$ and label it T.

f. Start at the origin. The x-coordinate is -3, so move left 3 units. The y-coordinate is 5, so move up 5 units. Plot the point at $(-3, 5)$ and label it U.

g. Start at the origin. The x-coordinate is $-\frac{3}{2}$, so move left $\frac{3}{2}$ units. The y-coordinate is $\frac{7}{2}$, so move up $\frac{7}{2}$ units. Plot the point at $\left(-\frac{3}{2}, \frac{7}{2}\right)$ and label it V.

h. Start at the origin. The x-coordinate is 4, so move right 4 units. The y-coordinate is $\frac{1}{2}$, so move up $\frac{1}{2}$ unit. Plot the point at $\left(4, \frac{1}{2}\right)$ and label it W.

i. Start at the origin. The x-coordinate is 2.5, so move right 2.5 units. The y-coordinate is -6.5, so move down 6.5 units. Plot the point at $(2.5, -6.5)$ and label it X.

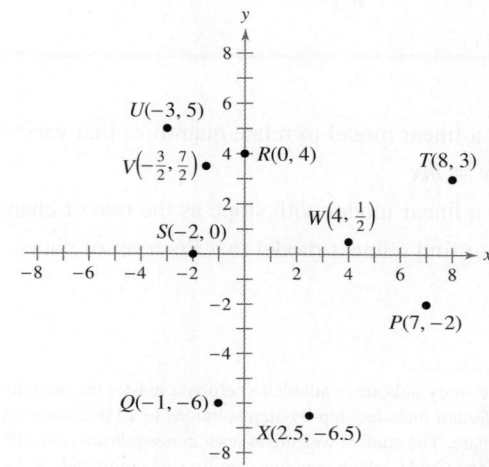

SUMMARY AND STUDY STRATEGIES

After studying this chapter, you should have acquired the following skills.
The exercise numbers are keyed to the Review Exercises that begin on page 244.
Answers to odd-numbered Review Exercises are given in the back of the text.*

Section 2.1

Review Exercises

- Plot points in the Cartesian plane, find the distance between two points, and find the midpoint of a line segment joining two points.

  $$d = \sqrt{(x_2 - x_1)^2 + (y_2 - y_1)^2}$$

  $$\text{Midpoint} = \left(\frac{x_1 + x_2}{2}, \frac{y_1 + y_2}{2}\right)$$

  1–6

- Determine whether a point is a solution of an equation.   *7, 8*
- Sketch the graph of an equation.   *9, 10*
- Find the x- and y-intercepts, and determine the symmetry, of the graph of an equation.   *11–16*
- Write the equation of a circle in standard form.   *17–20*

  $$(x - h)^2 + (y - k)^2 = r^2$$

Section 2.2

- Find the slope of a line passing through two points.   *21–24*

  $$m = \frac{y_2 - y_1}{x_2 - x_1}$$

- Use the point-slope form to find the equation of a line.   *25–28*

  $$y - y_1 = m(x - x_1)$$

- Use the slope-intercept form to sketch the graph of a line.   *29–32*

  $$y = mx + b$$

- Use slope to determine whether lines are parallel or perpendicular, and write the equation of a line parallel or perpendicular to a given line.   *33–42*

  Parallel lines: $m_1 = m_2$

  Perpendicular lines: $m_1 = -\dfrac{1}{m_2}$

Section 2.3

- Construct and use a linear model to relate quantities that vary directly.   *43–52*

  Direct variation: $y = mx$

- Construct and use a linear model with slope as the rate of change.   *53–55*
- Use a scatter plot to find a linear model that fits a set of data.   *56*

---

* A wide range of valuable study aids are available to help you master the material in this chapter.
  The *Student Solutions Manual* includes step-by-step solutions to all odd-numbered exercises to
  help you review and prepare. The student website at *www.cengagebrain.com* offers algebra help
  and a *Graphing Technology Guide*, which contains step-by-step commands and instructions for
  a wide variety of graphing calculators.

Section 2.4

	Review Exercises
■ Determine whether a correspondence is a function.	*57–62*
■ Use function notation, evaluate a function, and find the domain of a function.	*63–71*
■ Write a function that relates quantities in an application problem.	*72–74*

Section 2.5

■ Find the domain and range of a function using its graph.	*75–78*
■ Identify the graph of a function using the Vertical Line Test.	*79–82*
■ Describe the increasing and decreasing behavior of a function.	*75–78*
■ Find the relative minima and relative maxima of the graph of a function.	*75–78*
■ Sketch the graph of a step function.	*85, 86*
■ Classify a function as even or odd.	*75–78*

In an even function, $f(-x) = f(x)$.

In an odd function, $f(-x) = -f(x)$.

■ Identify six common graphs and use them to sketch the graphs of functions.	*83–90*

Section 2.6

■ Use vertical and horizontal shifts, reflections, and nonrigid transformations to sketch graphs of functions.	*91–98*

Vertical shifts: $h(x) = f(x) + c$ or $h(x) = f(x) - c$

Horizontal shifts: $h(x) = f(x - c)$ or $h(x) = f(x + c)$

Reflection in the *x*-axis: $h(x) = -f(x)$

Reflection in the *y*-axis: $h(x) = f(-x)$

Nonrigid transformation: $h(x) = cf(x)$

Section 2.7

■ Find the sum, difference, product, and quotient of two functions.	*99–104*
■ Form the composition of two functions and determine its domain.	*105–108*
■ Identify a function as the composition of two functions.	*109–112*
■ Use combinations and compositions of functions to solve application problems.	*113–115*

Study Strategies

- **To Memorize or Not To Memorize?**   When studying mathematics, you often need to memorize formulas, rules, and properties. The formulas that you use most often can become committed to memory through practice. Some formulas, however, are used infrequently or may be easily forgotten. When you are unsure of a formula, you may be able to *derive* it using other information that you know. For instance, if you forget the standard form of the equation of a circle, you can use the Distance Formula and properties of a circle to derive it, as shown on page 166. If you also forget the Distance Formula, you can depict the distance between two generic points graphically and use the Pythagorean Theorem to derive the formula, as shown on page 159.

- **Choose Convenient Values for Yearly Data**   When you work with data involving years, you may want to reassign simpler values to represent the years. For instance, you might represent the years 1993 through 2010 by the *x*-values 3 through 10. If you sketch a graph of these data, be sure to account for this in the *x*-axis title: Year (3 ↔ 1993).

Review Exercises

See www.CalcChat.com for worked-out solutions to odd-numbered exercises.

Using the Distance and Midpoint Formulas In Exercises 1–4, (a) plot the points, (b) find the distance between the points, and (c) find the midpoint of the line segment joining the points.

1. $(3, 2), (-3, -5)$

2. $(-9, 3), (5, 7)$

3. $(3.45, 6.55), (-1.06, -3.87)$

4. $(-6.7, -3.9), (5.1, 8.2)$

Using the Distance Formula In Exercises 5 and 6, find the values of x for which the distance between the points is 25.

5. $(10, 10), (x, -5)$

6. $(x, -5), (-15, 10)$

Solution of an Equation In Exercises 7 and 8, determine whether each point is a solution of the equation.

7. $y = 2x^2 - 7x - 15$

   (a) $(5, 0)$ (b) $(-2, 7)$

8. $y = \sqrt{16 - x^2}$

   (a) $(1, 5)$ (b) $(4, 0)$

Sketching the Graph of an Equation In Exercises 9 and 10, complete the table. Use the resulting solution points to sketch the graph of the equation.

9. $y = -\frac{1}{2}x + 2$

x	-2	0	2	3	4
y					

10. $y = x^2 - 3x$

x	-1	0	1	2	3
y					

Sketching the Graph of an Equation In Exercises 11–16, sketch the graph of the equation. Identify any intercepts and test for symmetry.

11. $y = x^2 + 3$

12. $y^2 = x$

13. $y = 3x - 4$

14. $y = \sqrt{9 - x}$

15. $y = x^3 + 1$

16. $y = |x - 3|$

Finding the Equation of a Circle In Exercises 17 and 18, find the standard form of the equation of the circle.

17. Center: $(-1, 2)$; radius: 6

18. Endpoints of a diameter: $(-2, -3), (4, 5)$

Completing the Square to Sketch a Circle In Exercises 19 and 20, write the equation of the circle in standard form. Then sketch the circle.

19. $x^2 + y^2 - 4x + 6y - 12 = 0$

20. $4x^2 + 4y^2 - 4x - 8y - 11 = 0$

Finding the Slope of a Line Through Two Points In Exercises 21–24, plot the points and find the slope of the line passing through the points.

21. $(3, 7), (2, -1)$

22. $(3, -2), (-1, -2)$

23. $(3, 4), (3, -2)$

24. $(-1, 5), (2, -3)$

The Point-Slope Form of the Equation of a Line In Exercises 25–28, find an equation of the line that passes through the point and has the indicated slope. Then sketch the graph of the line.

	Point	*Slope*
25.	$(0, -5)$	$m = \frac{3}{2}$
26.	$(3, 0)$	$m = -\frac{1}{3}$
27.	$(-2, 6)$	$m = 0$
28.	$(5, 4)$	m is undefined.

Sketching the Graph of a Linear Equation In Exercises 29–32, find the slope and y-intercept (if possible) of the line specified by the equation. Then sketch the graph of the line.

29. $8x - 4y + 16 = 0$

30. $3y - 2 = 0$

31. $17 - 5x = 10$

32. $16x + 12y - 24 = 0$

Parallel and Perpendicular Lines In Exercises 33–36, determine whether the lines L_1 and L_2 passing through the pairs of points are parallel, perpendicular, or neither.

33. L_1: $(0, 3), (-2, 1)$; L_2: $(-8, -3), (4, 9)$

34. L_1: $(-1, 0), (5, 5)$; L_2: $(2, 1), (8, 6)$

35. L_1: $(3, 6), (-1, -5)$; L_2: $(-2, 3), (4, 7)$

36. L_1: $(-1, 2), (-1, 4)$; L_2: $(7, 3), (4, 7)$

Equations of Parallel and Perpendicular Lines In Exercises 37–42, write equations of the lines passing through the given point that are (a) parallel to the given line and (b) perpendicular to the given line.

	Point	*Line*
37.	$(4, 3)$	$y = \frac{1}{2}x - 1$
38.	$(1, -4)$	$y = -2x + 3$
39.	$(3, -2)$	$5x - 4y = 8$

	Point	Line
40.	$(-8, 3)$	$2x + 3y = 5$
41.	$(-1, -2)$	$y = 2$
42.	$(0, 5)$	$x = -3$

Direct Variation In Exercises 43–46, y is proportional to x. Use the x- and y-values to find a linear model that relates y and x.

43. $x = 3, y = 7$ **44.** $x = 5, y = 7.5$

45. $x = 10, y = 3480$

46. $x = 14, y = 1.95$

Direct Variation In Exercises 47–50, write a linear model that relates the variables.

47. A varies directly as r; $A = 30$ when $r = 6$.

48. y varies directly as z; $y = 7$ when $z = 14$.

49. a is proportional to b; $a = 15$ when $b = 20$.

50. m varies directly as n; $m = 12$ when $n = 36$.

51. Property Tax The property tax in a city is based on the assessed value of the property. A house that has an assessed value of $150,000 has a property tax of $1260. Find a mathematical model that gives the amount of property tax y in terms of the assessed value of the property x. Use the model to find the property tax on a house that has an assessed value of $175,000.

52. Feet and Meters You are driving and you notice a billboard that indicates it is 1000 feet or 305 meters to the next restaurant of a national fast-food chain. Use this information to find a linear model that relates feet to meters. Use the model to complete the table.

Feet	20	50	100	120
Meters				

53. Fourth-Quarter Sales During the second and third quarters of the year, a business had sales of $2,950,000 and $3,150,000, respectively. Assume the growth of the sales follows a linear pattern. What will sales be during the fourth quarter?

54. Dollar Value The dollar value of a product in 2011 is $75 and the item is expected to increase in value at a rate of $4.75 per year. Write a linear equation that gives the dollar value of the product in terms of the year. Use this model to predict the dollar value of the product in 2013. (Let $t = 11$ represent 2011.)

55. Straight-Line Depreciation A small business purchases a piece of equipment for $135,000. After 10 years, the equipment will have to be replaced. Its salvage value at that time is expected to be $5500. Write a linear equation giving the value V of the equipment during the 10 years it will be used.

56. Sales The sales S (in millions of dollars) for Intuit Corporation for the years 2005 through 2010 are shown in the table. (*Source: Intuit Corporation*)

Year	Sales S (in millions of dollars)
2005	2079.9
2006	2342.3
2007	2672.9
2008	3071.0
2009	3182.5
2010	3455.0

(a) Use a graphing utility to create a scatter plot of the data. Let t represent the year, with $t = 5$ corresponding to 2005. Do the data appear linear?

(b) Use the *regression* feature of the graphing utility to find a linear model for the data.

(c) Use the linear model from part (b) to predict sales in 2011.

(d) Intuit Corporation predicts sales of $3820 million for 2011. Does your prediction from part (c) agree with that of Intuit Corporation? Which prediction do you think is more reasonable? Explain.

Testing for Functions Represented by Equations In Exercises 57–60, decide whether the equation represents y as a function of x.

57. $3x + y = 12$ **58.** $y^2 = x^2 - 9$

59. $y^2 = x + 3$ **60.** $x^2 + y^2 - 6x + 8y = 0$

Testing for Functions In Exercises 61 and 62, decide whether the set of ordered pairs represents a function from A to B.

$A = \{1, 2, 3\}$ $B = \{-3, -4, -7\}$

Give a reason for your answer.

61. $\{(1, -3), (2, -7), (3, -3)\}$

62. $\{(1, -4), (2, -3), (3, -9)\}$

Evaluating a Function In Exercises 63 and 64, evaluate the function at each specified value of the independent variable and simplify.

63. $f(x) = \sqrt{x + 4} - 5$

(a) $f(5)$ (b) $f(0)$ (c) $f(-4)$ (d) $f(x + 3)$

64. $f(x) = \begin{cases} 2x - 1, & x \le 1 \\ x^2 + 2, & x > 1 \end{cases}$

(a) $f(0)$ (b) $f(1)$ (c) $f(3)$ (d) $f(-4)$

Finding the Domain of a Function   In Exercises 65–70, find the domain of the function.

65. $f(x) = 2x^2 + 7x + 3$

66. $g(t) = 3t^2 + 1$

67. $h(x) = \sqrt{x + 5}$

68. $f(t) = \sqrt[3]{t - 3}$

69. $g(t) = \dfrac{\sqrt{t - 1}}{t - 4}$

70. $h(x) = \dfrac{2}{x^2 - 4}$

71. Reasoning   A student has difficulty understanding why the domains of

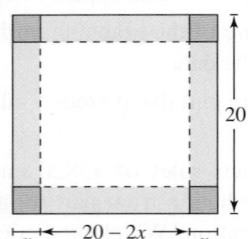

$$h(x) = \frac{x^2 - 4}{x} \quad \text{and} \quad k(x) = \frac{x}{x^2 - 4}$$

are different. How would you explain their respective domains algebraically? How could you use a graphing utility to explain their domains?

72. Volume of a Box   An open box is to be made from a square piece of material 20 inches on a side by cutting equal squares from the corners and turning up the sides (see figure).

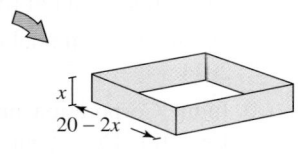

(a) Write the volume V of the box as a function of its height x.

(b) What is the domain of the function?

(c) Use a graphing utility to graph the function.

(d) Determine the volume of the box with a height of 3 inches.

73. Balance in an Account   A person deposits $6500 in an account that pays 5.75% interest compounded quarterly.

(a) Write the balance of the account in terms of the time t that the principal is left in the account.

(b) What is the domain of this function?

74. Vertical Motion   The velocity v (in feet per second) of a ball thrown vertically upward from ground level is given by

$$v(t) = -32t + 80$$

where t is the time (in seconds).

(a) Find the velocity when $t = 1$.

(b) Find the time when the ball reaches its maximum height. [*Hint:* Find the time when $v(t) = 0$.]

(c) Find the velocity when $t = 3$.

Describing Function Behavior   In Exercises 75–78, (a) determine the domain and range of the function, (b) determine the intervals over which the function is increasing, decreasing, or constant, (c) determine whether the function is even, odd, or neither, and (d) approximate any relative minimum or relative maximum values of the function.

75. $f(x) = x^2 + 1$

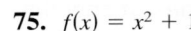
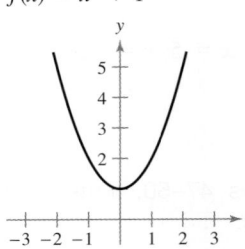

76. $f(x) = \sqrt{x^2 - 9}$

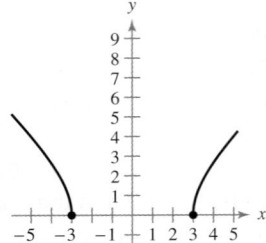

77. $f(x) = x^3 - 4x^2$

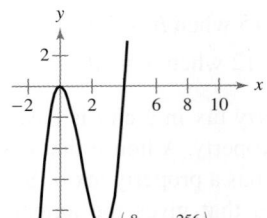

78. $f(x) = |x - 2|$

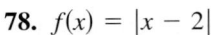
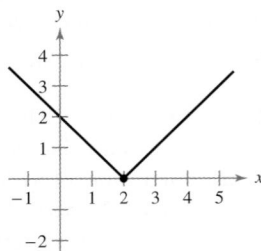

Using the Vertical Line Test   In Exercises 79–82, use the Vertical Line Test to decide whether y is a function of x.

79. $y = \frac{1}{2}x^2$

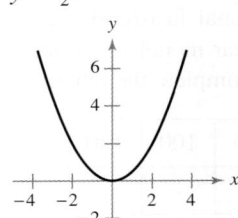

80. $y = \frac{1}{4}x^3$

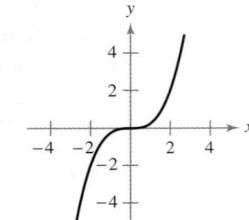

81. $x - y^2 = 1$

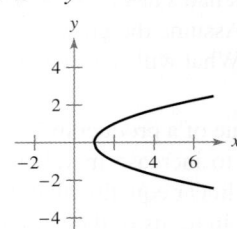

82. $x^2 + y^2 = 25$

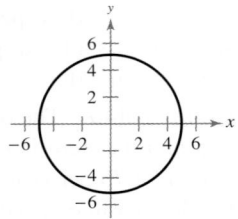

Sketching Graphs   In Exercises 83–90, sketch the graph of the function.

83. $f(x) = |x + 3|$

84. $g(x) = \sqrt{x^2 - 16}$

85. $h(x) = -[\![x]\!] + 1$

86. $f(x) = 2[\![x]\!] + 2$

87. $g(x) = \begin{cases} x + 2, & x < 0 \\ 2, & x = 0 \\ x^2 + 2, & x > 0 \end{cases}$

88. $g(x) = \begin{cases} 3x + 1, & x < -1 \\ x^2 - 3, & x \geq -1 \end{cases}$

89. $h(x) = x^2 - 3x$

90. $f(x) = \sqrt{9 - x^2}$

Describing Transformations In Exercises 91 and 92, describe the sequence of transformations from $f(x) = x^2$ to g. Then sketch the graph of g by hand. Verify with a graphing utility.

91. $g(x) = -(x - 1)^2 - 2$

92. $g(x) = -x^2 + 3$

Describing Transformations In Exercises 93 and 94, describe the sequence of transformations from $f(x) = \sqrt{x}$ to g. Then sketch the graph of g by hand. Verify with a graphing utility.

93. $g(x) = \sqrt{x - 2}$ **94.** $g(x) = \sqrt{x} + 2$

Describing Transformations In Exercises 95 and 96, describe the sequence of transformations from $f(x) = \sqrt[3]{x}$ to g. Then sketch the graph of g by hand. Verify with a graphing utility.

95. $g(x) = \sqrt[3]{x + 2}$ **96.** $g(x) = 2\sqrt[3]{x} + 1$

Identifying Transformations In Exercises 97 and 98, identify the transformation shown in the graph and identify the associated common function. Write the equation of the graphed function.

97.
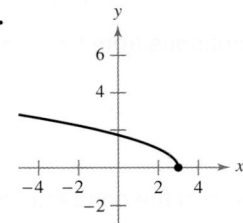

98.
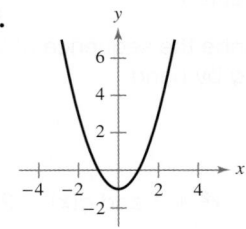

Finding Arithmetic Combinations of Functions In Exercises 99 and 100, find (a) $(f + g)(x)$, (b) $(f - g)(x)$, (c) $(fg)(x)$, and (d) $(f/g)(x)$. What is the domain of f/g?

99. $f(x) = 3x - 1$, $g(x) = x^2 + 2x$

100. $f(x) = 3x$, $g(x) = \sqrt{x^2 + 1}$

Evaluating Arithmetic Combinations of Functions In Exercises 101–104, evaluate the function for $f(x) = x^2 + 3x$ and $g(x) = 2x - 5$.

101. $(f + g)(2)$ **102.** $(f - g)(-1)$

103. $(fg)(3)$ **104.** $\left(\dfrac{f}{g}\right)(0)$

Composition of Functions In Exercises 105–108, find and determine the domains of (a) $f \circ g$ and (b) $g \circ f$.

105. $f(x) = x^2$, $g(x) = x + 3$

106. $f(x) = 2x - 5$, $g(x) = x^2 + 2$

107. $f(x) = \dfrac{1}{x}$, $g(x) = 3x + x^2$

108. $f(x) = \dfrac{1}{x^2}$, $g(x) = x^3$

Identifying a Composite Function In Exercises 109–112, find two functions f and g such that $(f \circ g)(x) = h(x)$. (There are many correct answers.)

109. $h(x) = (6x - 5)^2$

110. $h(x) = \sqrt[3]{x + 2}$

111. $h(x) = \dfrac{1}{(x - 1)^2}$

112. $h(x) = (x - 3)^3 + 2(x - 3)$

113. Area A square concrete foundation is prepared as a base for a large cylindrical aquatic tank that is to be used in ecology experiments (see figure).

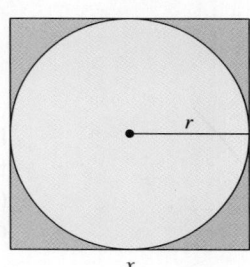

(a) Write the radius r of the tank as a function of the length x of the sides of the square.

(b) Write the area A of the circular base of the tank as a function of the radius r.

(c) Find and interpret $(A \circ r)(x)$.

114. Salary You are a sales representative for an automobile manufacturer. You are paid an annual salary plus a bonus of 4% of your sales over \$500,000. Consider the two functions given by

$$f(x) = x - 500,000$$

and

$$g(x) = 0.04x.$$

If x is greater than \$500,000, does $f(g(x))$ or $g(f(x))$ represent your bonus? Explain.

115. Bacteria The number N of bacteria in a petri dish is given by $N(T) = 8T^2 - 14T + 200$, $2 \leq T \leq 32$, where T is the temperature (in degrees Fahrenheit). The temperature is $T(t) = 2t + 2$, $0 \leq t \leq 15$, where t is the time in hours. Find and interpret $(N \circ T)(t)$.

TEST YOURSELF

See www.CalcChat.com for worked-out solutions to odd-numbered exercises.

Take this test as you would take a test in class. When you are done, check your work against the answers given in the back of the book.

In Exercises 1 and 2, find the distance between the points and the midpoint of the line segment joining the points.

1. $(-3, 2)$, $(5, -2)$

2. $(3.25, 7.05)$, $(-2.37, 1.62)$

3. Find the x- and y-intercepts of the graph of $y = (x + 5)(x - 3)$.

4. Describe the symmetry of the graph of

$$y = \frac{x}{x^2 - 4}.$$

5. Write the equation of the circle in standard form and sketch its graph.

$$x^2 + y^2 - 6x + 4y - 3 = 0$$

6. Find an equation of the line that passes through $(-3, 5)$ and has a slope of $\frac{2}{3}$.

In Exercises 7 and 8, decide whether the statement is true or false. Explain.

7. The equation $2x - 3y = 5$ represents y as a function of x.

8. If $A = \{3, 4, 5\}$ and $B = \{-1, -2, -3\}$, then the set $\{(3, -9), (4, -2), (5, -3)\}$ represents a function from A to B.

In Exercises 9 and 10, (a) find the domain and range of the function, (b) determine the intervals over which the function is increasing, decreasing, or constant, (c) determine whether the function is even or odd, and (d) approximate any relative minimum or relative maximum values of the function.

9. $f(x) = 2 - x^2$ (See figure.)

10. $g(x) = \sqrt{x^2 - 4}$ (See figure.)

In Exercises 11 and 12, describe the sequence of transformations from $f(x) = x^2$ to g. Then sketch the graph of g by hand.

11. $g(x) = -(x + 2)^2$

12. $g(x) = (x - 3)^2 + 4$

In Exercises 13–16, use $f(x) = x^2 + 1$ and $g(x) = 2x - 2$ to find the combination or composition.

13. $(f - g)(x)$

14. $(fg)(x)$

15. $(f \circ g)(x)$

16. $(g \circ f)(x)$

17. A business purchases a piece of equipment for $30,000. After 5 years, the equipment will be worth only $3500. Write a linear equation that gives the value V of the equipment during the 5 years.

18. The projected populations P (in millions) of children under the age of 5 in the United States for selected years from 2015 through 2050 are shown in the table. Use a graphing utility to create a scatter plot of the data and find a linear model for the data. Let t represent the year, with $t = 15$ corresponding to 2015. *(Source: U.S. Census Bureau)*

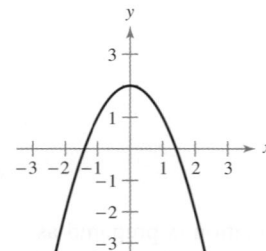

Figure for 9

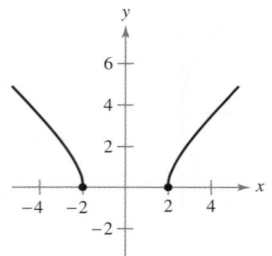

Figure for 10

Year	Population, P (in millions)
2015	22.1
2020	22.8
2025	23.5
2030	24.2
2035	25.1
2040	26.1
2045	27.2
2050	28.1

Table for 18

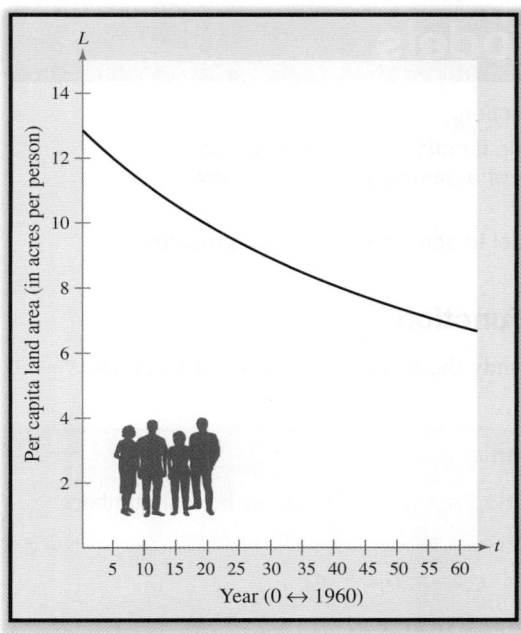

Example 8 on page 321 shows how a rational function can be used to model the per capita land area of the United States.

3 Polynomial and Rational Functions

249

3.1 Quadratic Functions and Models

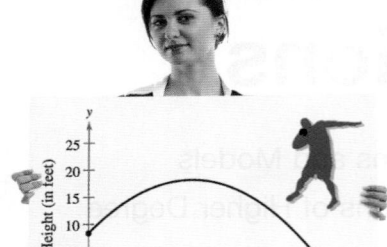

- Sketch the graph of a quadratic function.
- Use the standard form of a quadratic function to sketch the graph of a parabola and find the equation of a parabola given its vertex and a point on its graph.
- Construct and use a quadratic model to solve an application problem.

The Graph of a Quadratic Function

In this and the next section, you will study the graphs of polynomial functions.

In Exercise 56 on page 260, you will use a quadratic function to determine the distance traveled by a shot in a shot put event.

Definition of a Polynomial Function

Let n be a nonnegative integer and let $a_n, a_{n-1}, \ldots, a_2, a_1, a_0$ be real numbers with $a_n \neq 0$. The function given by

$$f(x) = a_n x^n + a_{n-1} x^{n-1} + \cdots + a_2 x^2 + a_1 x + a_0$$

is called a **polynomial function of x with degree n.**

For instance, each of the following functions is a polynomial function.

$$f(x) = 3x + 5 \qquad g(x) = \frac{1}{4}x^2 \qquad h(x) = 2x^3 + 7x^2 - 5x + 11$$

Polynomial functions are classified by degree. Recall that the degree of a polynomial is the highest degree of its terms. For instance, the polynomial function given by

$$f(x) = a, \quad a \neq 0 \qquad \text{Constant function}$$

has degree 0 and is called a **constant function.** In Chapter 2, you learned that the graph of this type of function is a horizontal line. The polynomial function given by

$$f(x) = ax + b, \quad a \neq 0 \qquad \text{Linear function}$$

has degree 1 and is called a **linear function.** In Chapter 2, you learned that the graph of the linear function given by $f(x) = ax + b$ is a line whose slope is a and whose y-intercept is $(0, b)$. In this section, you will study second-degree polynomial functions, which are called **quadratic functions.**

For instance, each of the following functions is a quadratic function.

$$f(x) = x^2 + 6x + 2 \qquad g(x) = 2(x + 1)^2 - 3 \qquad h(x) = (x - 2)(x + 1)$$

Definition of a Quadratic Function

Let a, b, and c be real numbers with $a \neq 0$. The function of x given by

$$f(x) = ax^2 + bx + c \qquad \text{Quadratic function}$$

is called a quadratic function.

The graph of a quadratic function is called a **parabola.** It is "∪"-shaped and can open upward or downward. Parabolas occur in many real-life applications, such as modeling the path of a baseball or the revenue and production costs of a company. You will learn about such applications later in this section.

All parabolas are symmetric with respect to a line called the **axis of symmetry,** or simply the **axis** of the parabola. The point at which the axis intersects the parabola is the **vertex** of the parabola, as shown in Figure 3.1. When the leading coefficient is positive, the graph of $f(x) = ax^2 + bx + c$ is a parabola that opens upward. When the leading coefficient is negative, the graph of $f(x) = ax^2 + bx + c$ is a parabola that opens downward. Later in this section, you will learn ways to find the coordinates of the vertex of a parabola.

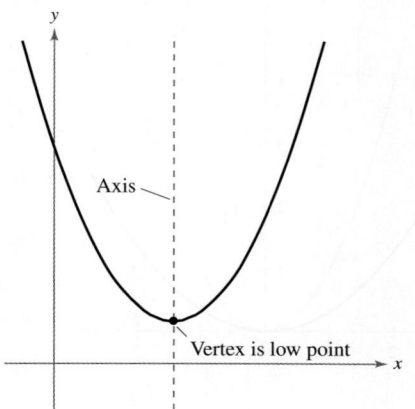

$a > 0$: Parabola opens upward. $a < 0$: Parabola opens downward.
FIGURE 3.1

The simplest type of quadratic function is

$$f(x) = ax^2.$$

Its graph is a parabola whose vertex is $(0, 0)$. When $a > 0$, the vertex is the point with the *minimum* y-value on the graph, and when $a < 0$, the vertex is the point with the *maximum* y-value on the graph, as shown in Figure 3.2.

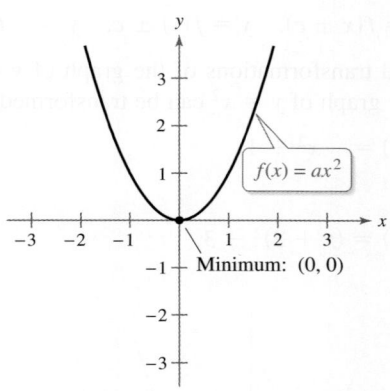
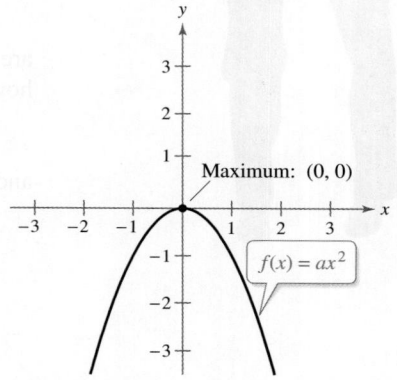

$a > 0$: Parabola opens upward. $a < 0$: Parabola opens downward.
FIGURE 3.2

When sketching the graph of

$$f(x) = ax^2$$

it is helpful to use the graph of $y = x^2$ as a reference, as discussed in Section 2.6. There you learned that when $a > 1$, the graph of $y = af(x)$ is a vertical stretch of the graph of $y = f(x)$. When $0 < a < 1$, the graph of $y = af(x)$ is a vertical shrink of the graph of $y = f(x)$. This is demonstrated again in Example 1.

| Example 1 | **Sketching the Graph of a Quadratic Function** |

a. Compared with the graph of $y = x^2$, each output of $f(x) = \frac{1}{3}x^2$ vertically "shrinks" the graph by a factor of $\frac{1}{3}$, creating the wider parabola shown in Figure 3.3(a).

b. Compared with the graph of $y = x^2$, each output of $g(x) = 2x^2$ vertically "stretches" the graph by a factor of 2, creating the narrower parabola shown in Figure 3.3(b).

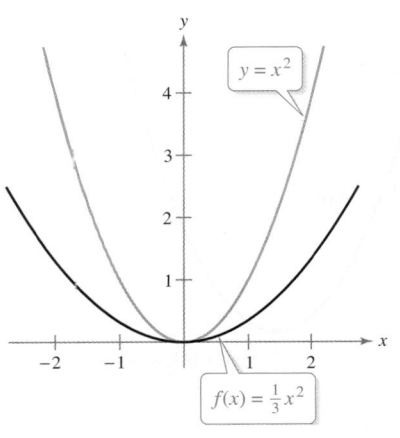

(a)

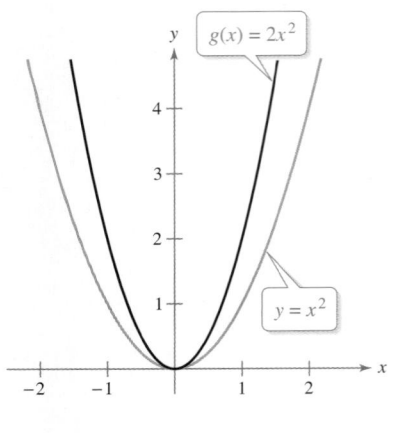

(b)

FIGURE 3.3

✓ **Checkpoint 1**

Sketch the graph of $f(x) = 4x^2$. Then compare this graph with the graph of $y = x^2$. ■

Note that the coefficient a determines how widely the parabola given by $f(x) = ax^2$ opens. When $|a|$ is small, the parabola opens more widely than when $|a|$ is large.

Recall from Section 2.6 that the graphs of

$$y = f(x \pm c), \quad y = f(x) \pm c, \quad y = -f(x), \quad \text{and} \quad y = f(-x)$$

are rigid transformations of the graph of $y = f(x)$. For instance, in Figure 3.4, notice how the graph of $y = x^2$ can be transformed to produce the graphs of

$$f(x) = -x^2 + 1$$

and

$$g(x) = (x + 2)^2 - 3.$$

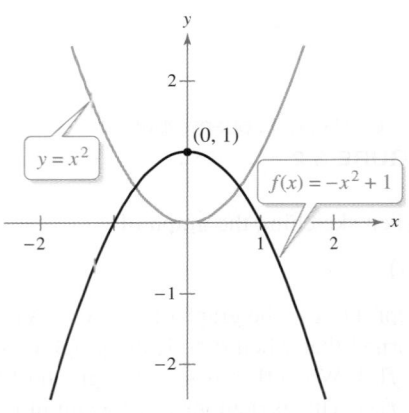

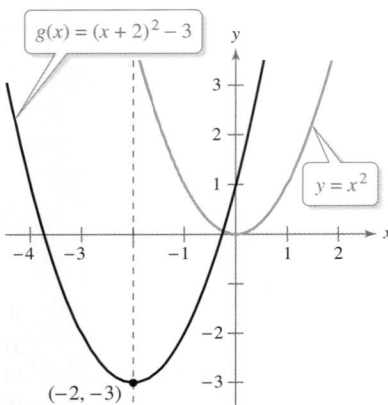

FIGURE 3.4

The Standard Form of a Quadratic Function

The **standard form** of a quadratic function is

$$f(x) = a(x - h)^2 + k.$$

This form is especially convenient for sketching a parabola because it identifies the vertex of the parabola.

ALGEBRA TUTOR xy

For help in factoring polynomial expressions, see the *Chapter 3 Algebra Tutor* on page 326.

> **Standard Form of a Quadratic Function**
>
> The quadratic function given by
>
> $$f(x) = a(x - h)^2 + k, \quad a \neq 0$$
>
> is said to be in **standard form.** The graph of f is a parabola whose axis is the vertical line $x = h$ and whose vertex is the point (h, k).
>
> When $a > 0$, the parabola opens upward.
>
> When $a < 0$, the parabola opens downward.

To write a quadratic function in standard form, you can use the process of *completing the square*, as illustrated in Example 2.

Example 2 Using Standard Form to Graph a Parabola

Sketch the graph of $f(x) = 2x^2 + 8x + 7$. Identify the vertex.

SOLUTION Begin by writing the quadratic function in standard form. The first step in completing the square is to factor out any coefficient of x^2 that is not 1.

$$\begin{aligned}
f(x) &= 2x^2 + 8x + 7 && \text{Write original function.} \\
&= (2x^2 + 8x) + 7 && \text{Group } x\text{-terms.} \\
&= 2(x^2 + 4x) + 7 && \text{Factor 2 out of } x\text{-terms.} \\
&= 2(x^2 + 4x + 4 - 4) + 7 && \text{Add and subtract 4 within parentheses to complete the square.}
\end{aligned}$$

$$\left(\tfrac{4}{2}\right)^2$$

After adding and subtracting 4 within the parentheses, regroup the terms to form a perfect square trinomial. The -4 can be removed from inside the parentheses. But, because of the 2 outside the parentheses, you must multiply -4 by 2 as shown below.

$$\begin{aligned}
f(x) &= 2(x^2 + 4x + 4) - 2(4) + 7 && \text{Regroup terms.} \\
&= 2(x^2 + 4x + 4) - 8 + 7 && \text{Simplify.} \\
&= 2(x + 2)^2 - 1 && \text{Standard form}
\end{aligned}$$

From this form, you can see that the graph of f is a parabola that opens upward with vertex $(-2, -1)$. This corresponds to a left shift of two units and a downward shift of one unit relative to the graph of $y = 2x^2$, as shown in Figure 3.5.

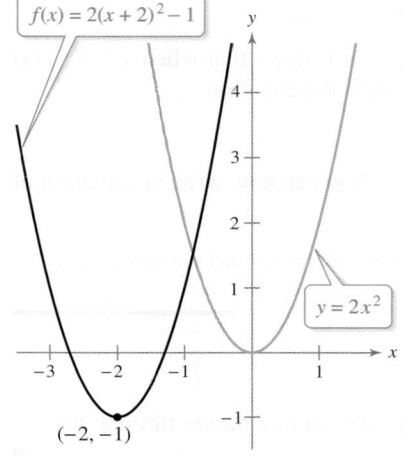

$f(x) = 2(x + 2)^2 - 1$

$y = 2x^2$

$(-2, -1)$

FIGURE 3.5

✓ Checkpoint 2

Sketch the graph of each quadratic function. Identify the vertex.

a. $f(x) = x^2 + 2x + 4$

b. $f(x) = 2x^2 - 12x + 20$

Example 3   Using Standard Form to Graph a Parabola

Sketch the graph of $f(x) = -x^2 + 6x - 8$. Identify the vertex.

SOLUTION   As in Example 2, begin by writing the quadratic function in standard form.

$$
\begin{aligned}
f(x) &= -x^2 + 6x - 8 && \text{Write original function.} \\
&= (-x^2 + 6x) - 8 && \text{Group } x\text{-terms.} \\
&= -(x^2 - 6x) - 8 && \text{Factor } -1 \text{ out of } x\text{-terms.} \\
&= -(x^2 - 6x + 9 - 9) - 8 && \text{Add and subtract 9 within parentheses to complete the square.} \\
& \quad\quad\quad\underset{(-6/2)^2}{\underline{}} \\
&= -(x^2 - 6x + 9) - (-9) - 8 && \text{Regroup terms.} \\
&= -(x - 3)^2 + 1 && \text{Standard form}
\end{aligned}
$$

So, the graph of f is a parabola that opens downward with vertex at $(3, 1)$, as shown in Figure 3.6.

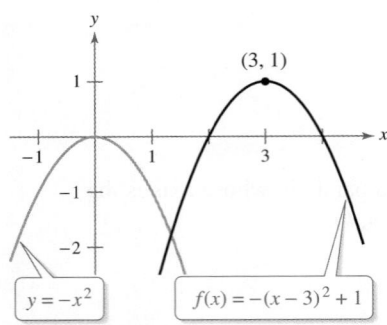

FIGURE 3.6

✓ **Checkpoint 3**

Sketch the graph of $f(x) = -3x^2 + 12x + 1$. Identify the vertex.   ■

Example 4   Finding an Equation of a Parabola

Find an equation of the parabola whose vertex is $(1, 2)$ and that passes through the point $(0, 0)$, as shown in Figure 3.7.

SOLUTION   Because the parabola has a vertex at $(h, k) = (1, 2)$, the equation must have the form

$$f(x) = a(x - 1)^2 + 2. \qquad \text{Standard form}$$

Because the parabola passes through the point $(0, 0)$, it follows that when $x = 0$, $f(x)$ must equal 0. Substitute 0 for x and 0 for $f(x)$ to obtain the equation

$$0 = a(0 - 1)^2 + 2.$$

By solving this equation for a, you find that $a = -2$. You can now write an equation of the parabola.

$$
\begin{aligned}
f(x) &= -2(x - 1)^2 + 2 && \text{Substitute for } a, h, \text{ and } k \text{ in standard form.} \\
&= -2x^2 + 4x && \text{Simplify.}
\end{aligned}
$$

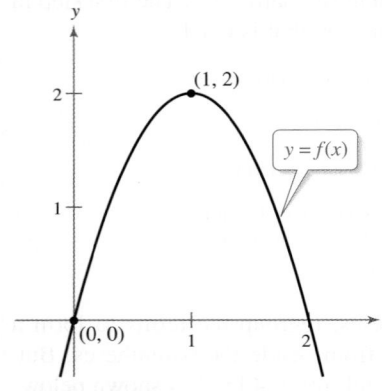

FIGURE 3.7

✓ **Checkpoint 4**

Find an equation of the parabola whose vertex is $(3, 4)$ and that passes through the point $(2, 5)$.   ■

To find the x-intercepts of the graph of $f(x) = ax^2 + bx + c$, you must solve the equation

$$ax^2 + bx + c = 0.$$

If the expression $ax^2 + bx + c$ does not factor, then use the Quadratic Formula to determine the x-intercepts. Remember, however, that a parabola may have no x-intercepts.

Applications

Many applications involve finding the maximum or minimum value of a quadratic function. By writing $f(x) = ax^2 + bx + c$ in standard form, you can determine that the x-coordinate of the vertex is $-b/2a$. The y-coordinate of the vertex can then be found by evaluating the function at $x = -b/2a$.

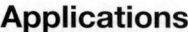 **Example 5** **The Maximum Height of a Baseball**

A baseball is hit 3 feet above the ground at a velocity of 100 feet per second and at an angle of 45° with respect to the ground. The path of the baseball is given by

$$f(x) = -0.0032x^2 + x + 3$$

where $f(x)$ is the height of the baseball (in feet) and x is the distance from home plate (in feet). What is the maximum height reached by the baseball?

SOLUTION For this quadratic function, you have

$$f(x) = ax^2 + bx + c$$

$$= -0.0032x^2 + x + 3.$$

So, $a = -0.0032$ and $b = 1$. Because the function has a maximum when $x = -b/2a$, the baseball reaches its maximum height when it is

$$x = -\frac{b}{2a} = -\frac{1}{2(-0.0032)} = 156.25 \text{ feet}$$

from home plate. At this distance, the maximum height is

$$f(156.25) = -0.0032(156.25)^2 + 156.25 + 3$$

$$= 81.125 \text{ feet.}$$

The path of the baseball is shown in Figure 3.8.

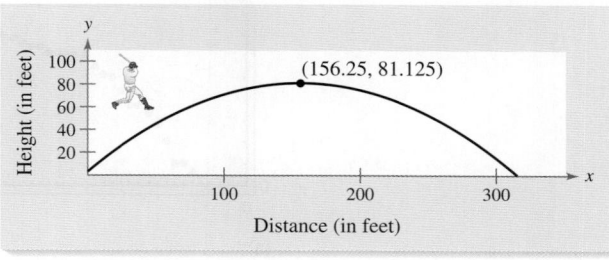

FIGURE 3.8

✓ **Checkpoint 5**

In Example 5, the baseball is hit at a velocity of 70 feet per second. The path of the baseball is given by

$$f(x) = -0.007x^2 + x + 4$$

where $f(x)$ is the height of the baseball (in feet) and x is the distance from home plate (in feet). What is the maximum height reached by the baseball? ■

    In Section 2.3, you plotted data points in the coordinate plane and estimated the best-fitting line. Fitting a quadratic model using this same process would be complicated. Most graphing utilities have a built-in statistical program that easily calculates the best-fitting quadratic model for a set of data points. Refer to the user's guide of your graphing utility for the required steps.

 TECH TUTOR

Your graphing utility may have *minimum* and *maximum* features that determine the minimum and maximum points of the graph of a function. You can use these features to find the vertex of a parabola. For instructions on how to use the *minimum* and *maximum* features, see Appendix A. Consult the user's guide for your graphing utility for more information.

 **Example 6**   **Fitting a Quadratic Function to Data**

Each year, grizzly bears in Yellowstone National Park feed on moths along the mountain slopes of the park. The bears dig into the mountains and eat the moths they uncover. The table shows the number N of confirmed moth sites in the Greater Yellowstone Ecosystem each year from 1986 to 2008. Use a graphing utility to plot the data and find the quadratic model that best fits the data. Find the vertex of the graph of the quadratic model and interpret its meaning in the context of the problem. Let $x = 6$ represent the year 1986.   *(Source: Yellowstone Grizzly Bear Investigations 2008)*

x	6	7	8	9	10	11	12	13	14	15	16	17
N	3	5	5	10	14	17	19	19	22	25	26	28

x	18	19	20	21	22	23	24	25	26	27	28
N	30	30	30	31	31	32	32	33	34	35	35

SOLUTION   Begin by entering the data into your graphing utility and displaying the scatter plot. From the scatter plot that is shown in Figure 3.9(a), you can see that the points have a parabolic trend. Use the *quadratic regression* feature to find the quadratic function that best fits the data. The quadratic equation that best fits the data is

$$N = -0.076x^2 + 4.04x - 19.5, \quad 6 \le x \le 28.$$

Graph the data and the equation in the same viewing window, as shown in Figure 3.9(b). By using the *maximum* feature of your graphing utility, you can see that the vertex of the graph is approximately (26.6, 34.2), as shown in Figure 3.9(c). The vertex corresponds to the year in which the number of confirmed moth sites was the greatest. So, in 2006, the number of confirmed moth sites reached a maximum.

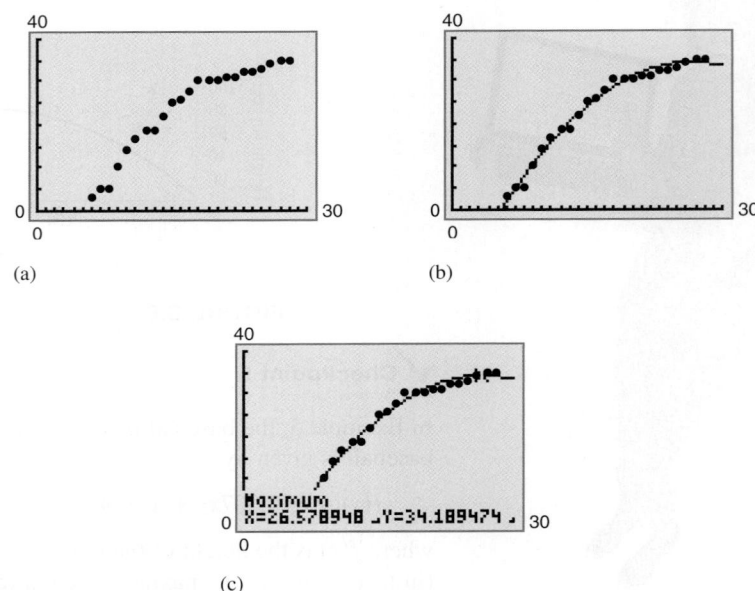

(a)

(b)

(c)

FIGURE 3.9

✓ **Checkpoint 6**

In Example 6, use the model to predict the number of confirmed moth sites in 2014. What do you notice?

TECH TUTOR

For instructions on how to use the *regression* feature, see Appendix A. Consult the user's guide for your graphing utility for more information.

Example 7   Charitable Contributions

The percent of a family's income that is given to charities is related to the income of the family. For families with annual incomes between $5000 and $100,000, the percent P can be modeled by

$$P(x) = 0.0014x^2 - 0.1529x + 5.855, \quad 5 \le x \le 100$$

where x is the annual income (in thousands of dollars). Use the model to estimate the income that corresponds to the minimum percent of income given to charities.

SOLUTION   One way to answer the question is to sketch the graph of the quadratic function, as shown in Figure 3.10. From this graph, it appears that the minimum percent corresponds to an income of about $55,000.

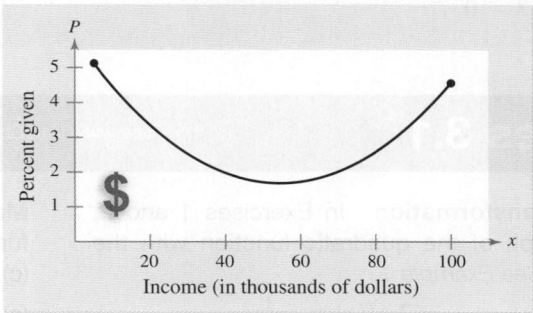

FIGURE 3.10

Another way to answer the question is to use the fact that the minimum point of the parabola occurs when $x = -b/2a$.

$$x = -\frac{b}{2a}$$

$$= -\frac{-0.1529}{2(0.0014)}$$

$$\approx 54.6$$

From this x-value, you can conclude that the minimum percent corresponds to an income of about $54,600.

 **Checkpoint 7**

A manufacturer has daily production costs C (in dollars per unit) of $C = 0.15x^2 - 9x + 700$, where x is the number of units produced. How many units should be produced each day to yield a minimum cost per unit?

SUMMARIZE   (Section 3.1)

1. State the definition of a quadratic function and describe its graph *(pages 250–252)*. For an example of sketching the graph of a quadratic function, see Example 1.

2. State the standard form of a quadratic function *(page 253)*. For examples that use the standard form of a quadratic function, see Examples 2, 3, and 4.

3. Describe a real-life example of how a quadratic function can be used to analyze the path of a moving object *(page 255, Example 5)*.

SKILLS WARM UP 3.1　The following warm-up exercises involve skills that were covered in earlier sections. You will use these skills in the exercise set for this section. For additional help, review Sections 1.3 and 1.4.

In Exercises 1–4, solve the quadratic equation by factoring.

1. $2x^2 + 11x - 6 = 0$　　　　　　　　　　**2.** $5x^2 - 12x - 9 = 0$

3. $3 + x - 2x^2 = 0$　　　　　　　　　　　**4.** $x^2 + 20x + 100 = 0$

In Exercises 5–10, use the Quadratic Formula to solve the quadratic equation.

5. $x^2 - 6x + 4 = 0$　　　　　　　　　　　**6.** $x^2 + 4x + 1 = 0$

7. $2x^2 - 16x + 25 = 0$　　　　　　　　　　**8.** $3x^2 + 30x + 74 = 0$

9. $x^2 + 3x + 1 = 0$

10. $x^2 + 3x - 3 = 0$

Exercises 3.1

See www.CalcChat.com for worked-out solutions to odd-numbered exercises.

Describing a Transformation　In Exercises 1 and 2, compare the graph of the quadratic function with the graph of $y = x^2$. *See Example 1.*

1. $f(x) = 5x^2$　　　　　　　**2.** $f(x) = -\frac{1}{4}x^2$

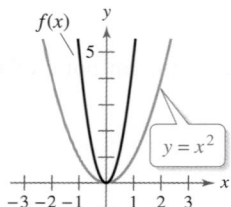　　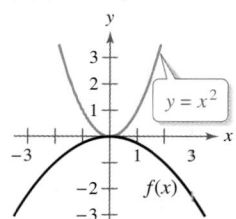

Sketching the Graph of a Quadratic Function　In Exercises 3–6, sketch the graph of the quadratic function and compare it with the graph of $y = x^2$. *See Example 1.*

3. $f(x) = 3x^2$

4. $f(x) = -2x^2$

5. $g(x) = -\frac{1}{2}x^2$

6. $g(x) = \frac{3}{4}x^2$

Describing a Transformation　In Exercises 7 and 8, compare the graph of the quadratic function with the graph of $y = x^2$.

7. $f(x) = -(x + 1)^2 + 1$　　**8.** $f(x) = 3(x - 2)^2 - 1$

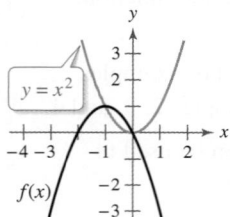　　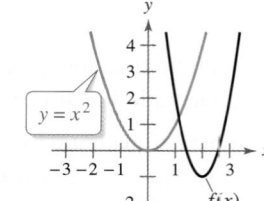

Matching　In Exercises 9–14, match the quadratic function with its graph. [The graphs are labeled (a), (b), (c), (d), (e), and (f).]

(a) 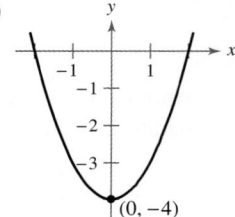　　(b)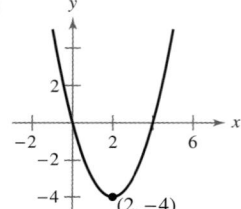

(c) 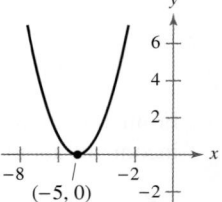　　(d)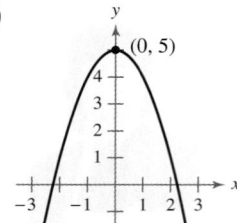

(e) 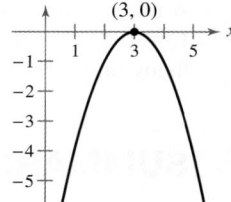　　(f)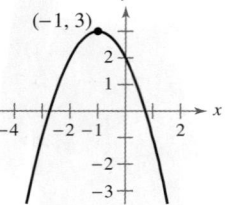

9. $f(x) = -(x - 3)^2$

10. $f(x) = (x + 5)^2$

11. $f(x) = x^2 - 4$

12. $f(x) = 5 - x^2$

13. $f(x) = -(x + 1)^2 + 3$

14. $f(x) = (x - 2)^2 - 4$

Using Standard Form to Graph a Parabola   In Exercises 15–30, write the quadratic function in standard form (if necessary) and sketch its graph. Identify the vertex. *See Examples 2 and 3.*

15. $f(x) = 16 - x^2$ **16.** $h(x) = x^2 - 9$

17. $f(x) = (x + 5)^2 - 6$ **18.** $f(x) = (x - 6)^2 + 3$

19. $g(x) = x^2 + 2x + 1$ **20.** $h(x) = x^2 - 4x + 2$

21. $f(x) = -(x^2 + 2x - 3)$

22. $f(x) = -(x^2 + 6x - 3)$

23. $f(x) = x^2 - x + \frac{5}{4}$ **24.** $f(x) = x^2 + 3x + \frac{1}{4}$

25. $f(x) = -x^2 + 2x + 5$

26. $f(x) = -x^2 - 4x + 1$

27. $h(x) = 4x^2 - 4x + 21$

28. $f(x) = 2x^2 - x + 1$

29. $f(x) = \frac{1}{4}(x^2 - 16x + 32)$

30. $g(x) = \frac{1}{2}(x^2 + 4x - 2)$

Finding an Equation of a Parabola   In Exercises 31–36, find an equation of the parabola. *See Example 4.*

31.

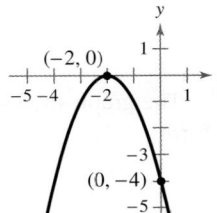

32.

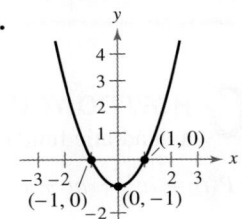

33.

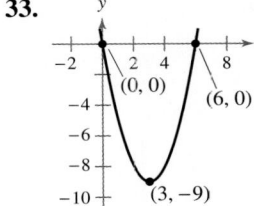

34.

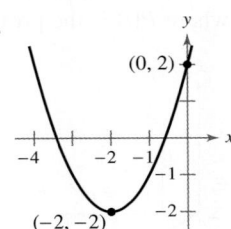

35.

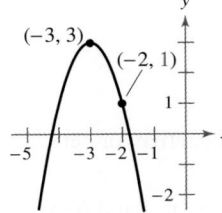

36.

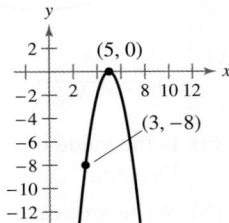

Finding an Equation of a Parabola   In Exercises 37–40, find an equation of the parabola that has the indicated vertex and whose graph passes through the given point. *See Example 4.*

37. Vertex: $(2, -1)$; point: $(4, -3)$

38. Vertex: $(-3, 5)$; point: $(-6, -1)$

39. Vertex: $(5, 12)$; point: $(7, 15)$

40. Vertex: $(-2, -2)$; point: $(-1, 0)$

Finding the Slope of a Line   In Exercises 41–46, find two quadratic functions whose graphs have the given x-intercepts. Find one function whose graph opens upward and another whose graph opens downward. (There are many correct answers.)

41. $(2, 0), (-1, 0)$

42. $(-4, 0), (0, 0)$

43. $(0, 0), (10, 0)$

44. $(4, 0), (8, 0)$

45. $(-3, 0), \left(-\frac{1}{2}, 0\right)$

46. $\left(-\frac{5}{2}, 0\right), (2, 0)$

47. Optimal Area   The perimeter of a rectangle is 200 feet. Let x represent the width of the rectangle. Write a quadratic function for the area of the rectangle in terms of its width. Find the vertex of the graph of the quadratic function and interpret its meaning in the context of the problem.

48. Optimal Area   The perimeter of a rectangle is 540 feet. Let x represent the width of the rectangle. Write a quadratic function for the area of the rectangle in terms of its width. Find the vertex of the graph of the quadratic function and interpret its meaning in the context of the problem.

49. Optimal Area   A rancher has 1200 feet of fencing with which to enclose two adjacent rectangular corrals (see figure). What measurements will produce a maximum enclosed area?

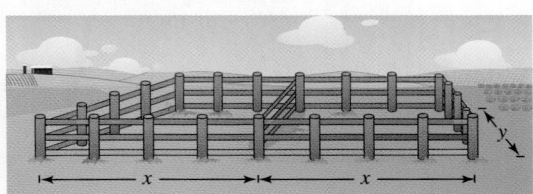

50. Optimal Area   An indoor physical-fitness room consists of a rectangular region with a semicircle on each end (see figure). The perimeter of the room is to be a 200-meter running track. What measurements will produce a maximum area of the rectangle?

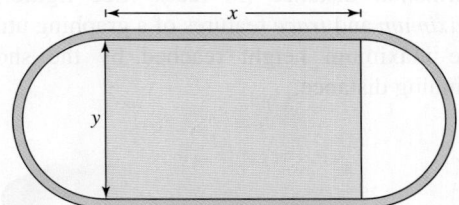

Optimal Revenue   In Exercises 51 and 52, find the number of units that produces a maximum revenue. The revenue R is measured in dollars and x is the number of units produced.

51. $R = 1000x - 0.02x^2$

52. $R = 80x - 0.0001x^2$

53. Optimal Cost   The daily production costs C (in dollars per unit) for a manufacturer of lighting fixtures are given by the quadratic function

$$C(x) = 800 - 10x + 0.25x^2$$

where x is the number of units produced. How many fixtures should be produced each day to yield a minimum cost per unit?

54. Optimal Profit   The profit P (in dollars) for a manufacturer of sound systems is given by the quadratic function

$$P(x) = -0.0003x^2 + 150x - 375,000$$

where x is the number of units produced. What production level will yield a maximum profit?

55. Maximum Height of a Diver   The path of a diver is given by

$$y = -\frac{4}{9}x^2 + \frac{24}{9}x + 10$$

where y is the height (in feet) and x is the horizontal distance from the end of the diving board (in feet). (See figure.) Use the *trace* or *maximum* feature of a graphing utility to find the maximum height of the diver.

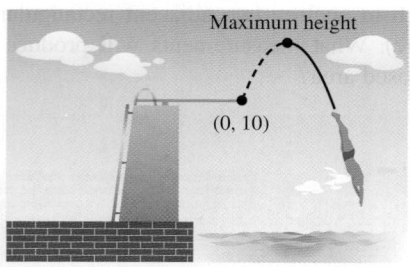

56. Maximum Height   The path of the winning shot at a shot put event can be approximated by the quadratic function

$$y = -0.011x^2 + 0.65x + 8.3$$

where y is the height of the shot (in feet) and x is the horizontal distance (in feet). (See figure.) Use the *maximum* and *trace* features of a graphing utility to find the maximum height reached by the shot and its winning distance.

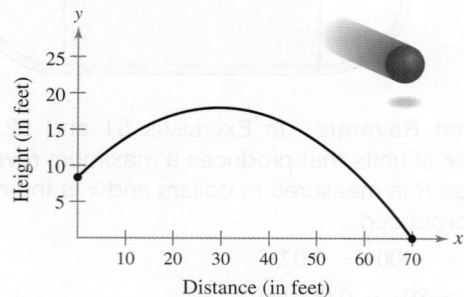

57. Revenue   The table shows the total yearly revenues R (in millions of dollars) for golf courses and country clubs in the United States from 2003 through 2009.   (*Source: U.S. Census Bureau*)

Year	2003	2004	2005	2006
Revenue, R	17,291	18,469	19,356	20,523

Year	2007	2008	2009
Revenue, R	21,195	21,044	20,326

(a) Use a graphing utility to create a scatter plot of the data. Let t represent the year, with $t = 3$ corresponding to 2003.

(b) Use the *regression* feature of the graphing utility to find a quadratic model for the data. Graph the model with the scatter plot from part (a).

(c) Use the graph of the model from part (b) to estimate when the yearly revenue was the greatest. Does this result agree with the actual data?

58. 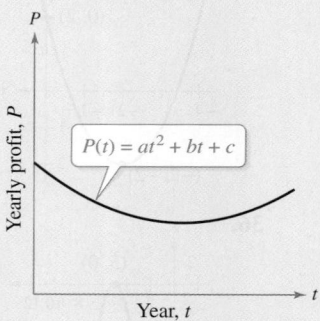   **HOW DO YOU SEE IT?**   The graph shows a quadratic function of the form

$$P(t) = at^2 + bt + c$$

which represents the yearly profits for a company, where $P(t)$ is the profit in year t.

(a) Is the value of a positive, negative, or zero? Explain.

(b) Write an expression in terms of a and b that represents the year t when the company made the least profit.

(c) The company made the same yearly profits in 2004 and 2012. Estimate the year in which the company made the least profit.

(d) Assume that the model is still valid today. Are the yearly profits currently increasing, decreasing, or constant? Explain.

59. Osteopathic Physicians The table shows the numbers N (in thousands) of female osteopathic physicians from 2001 through 2010. *(Source: American Osteopathic Association)*

Year	2001	2002	2003	2004	2005
Physicians, N	10.9	11.8	12.9	14.0	15.1

Year	2006	2007	2008	2009	2010
Physicians, N	16.4	17.7	19.1	20.8	22.5

(a) Use a graphing utility to create a scatter plot of the data. Let t represent the year, with $t = 1$ corresponding to 2001.

(b) Use the *regression* feature of the graphing utility to find a quadratic model for the data.

(c) Use the graphing utility to graph the model from part (b) in the same viewing window as the scatter plot of the data.

(d) Use the graph of the model from part (c) to predict the number of female osteopathic physicians in 2012.

60. Kidney Transplants The table shows the numbers K of kidney transplants performed in the United States in the years 1999 through 2009. *(Source: U.S. Department of Health and Human Services)*

Year	1999	2000	2001	2002
Transplants, K	12,455	13,258	14,152	14,741

Year	2003	2004	2005	2006
Transplants, K	15,129	16,000	16,481	17,094

Year	2007	2008	2009
Transplants, K	16,624	16,517	16,829

(a) Use a graphing utility to create a scatter plot of the data. Let t represent the year, with $t = 9$ corresponding to 1999.

(b) Use the *regression* feature of the graphing utility to find a quadratic model for the data.

(c) Use the graphing utility to graph the model from part (b) in the same viewing window as the scatter plot of the data.

(d) Use the graph of the model from part (c) to predict the number of kidney transplants performed in 2010. Does your answer seem reasonable? Explain.

61. Analyzing Profits Let x be the number of units (in tens of thousands) that a computer company produces and let $p(x)$ be the profit (in hundreds of thousands of dollars). The table shows the profits for different levels of production.

Units, x	2	4	6	8	10
Profit, $p(x)$	270.5	307.8	320.1	329.2	325.0

Units, x	12	14	16	18	20
Profit, $p(x)$	311.2	287.8	254.8	212.2	160.0

(a) Use a graphing utility to create a scatter plot of the data.

(b) Use the *regression* feature of the graphing utility to find a quadratic model for $p(x)$.

(c) Use the graphing utility to graph your model for $p(x)$ with the scatter plot of the data.

(d) Find the vertex of the graph of the model from part (c). Interpret its meaning in the context of the problem.

(e) With these data and this model, the profit begins to decrease. Discuss how it is possible for production to increase and profit to decrease.

62. Baseball Distance Let x be the angle (in degrees) at which a baseball is hit with no spin at an initial speed of 40 meters per second, and let $d(x)$ be the distance (in meters) the ball travels. The table shows the distances for the different angles at which the ball is hit. *(Source: The Physics of Sports)*

Angle, x	10	15	30	36	42
Distance, $d(x)$	58.3	79.7	126.9	136.6	140.6

Angle, x	44	45	48	54	60
Distance, $d(x)$	140.9	140.9	139.3	132.5	120.5

(a) Use a graphing utility to create a scatter plot of the data.

(b) Use the *regression* feature of the graphing utility to find a quadratic model for $d(x)$.

(c) Use the graphing utility to graph your model for $d(x)$ with the scatter plot of the data.

(d) Find the vertex of the graph of the model from part (c). Interpret its meaning in the context of the problem.

63. Verifying a Formula Write the quadratic function $f(x) = ax^2 + bx + c$ in standard form to verify that the vertex occurs at $(-b/2a, f(-b/2a))$.

3.2 Polynomial Functions of Higher Degree

■ Sketch a transformation of a monomial function.
■ Apply the Leading Coefficient Test to determine right-hand and left-hand behavior of graphs of polynomial functions.
■ Find the real zeros and sketch the graph of a polynomial function.
■ Use a polynomial model to solve an application problem.

Graphs of Polynomial Functions

The graphs of polynomial functions of degree greater than 2 are more complicated than those of degree 0, 1, or 2. However, using the characteristics presented in this section, together with point plotting, intercepts, and symmetry, you should be able to make reasonably accurate sketches *by hand*. Of course, a graphing utility makes the task easier.

In this section, you will study basic characteristics of the graphs of polynomial functions. The first characteristic is that the graph of a polynomial function is **continuous.** Essentially, this means that the graph of a polynomial function has no breaks, as shown in Figure 3.11(a). Functions with graphs that are not continuous are not polynomial functions, as shown in Figure 3.11(b).

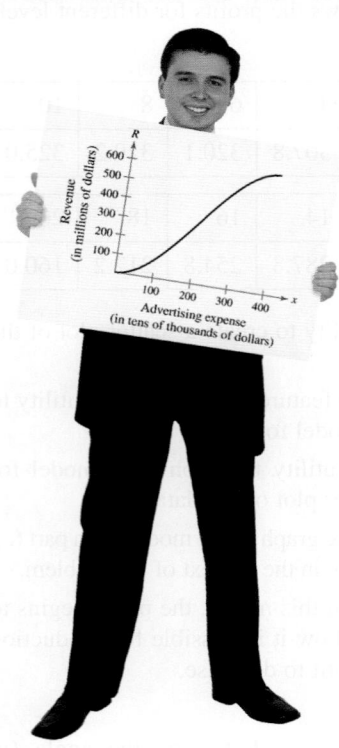

In Exercise 78 on page 271, you will use the graph of a cubic function to find the point of diminishing returns for the advertising expenses of a company.

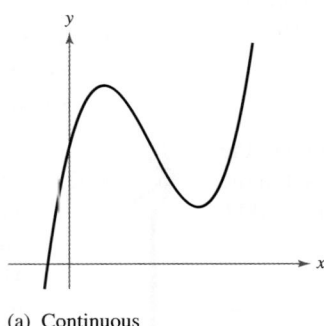

(a) Continuous

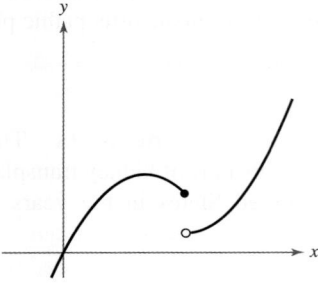

(b) Not continuous

FIGURE 3.11

The second characteristic is that the graph of a polynomial function has only smooth, rounded turns, as shown in Figure 3.12(a). A polynomial function cannot have a sharp turn. For instance, the function given by

$$f(x) = |x|$$

which has a sharp turn at the point $(0, 0)$, as shown in Figure 3.12(b), is not a polynomial function.

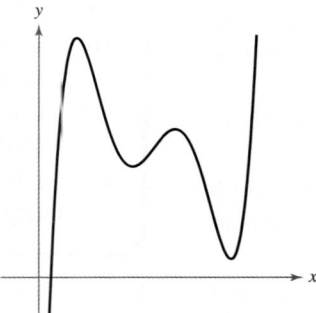

(a) Polynomial functions have smooth, rounded turns.

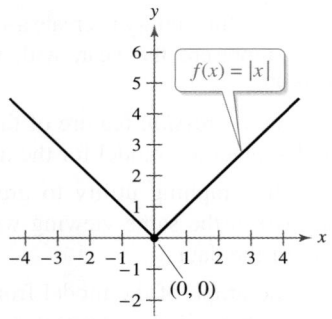

(b) Polynomial functions cannot have sharp turns.

FIGURE 3.12

The polynomial functions that have the simplest graphs are monomial functions of the form $f(x) = x^n$, where n is an integer greater than zero. From Figure 3.13, you can see that when n is *even*, the graph is similar to the graph of $f(x) = x^2$, and when n is *odd*, the graph is similar to the graph of $f(x) = x^3$. Moreover, the greater the value of n, the flatter the graph near the origin.

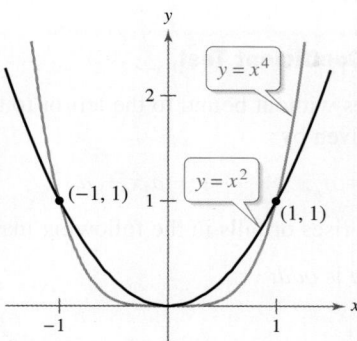
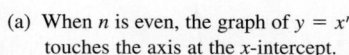
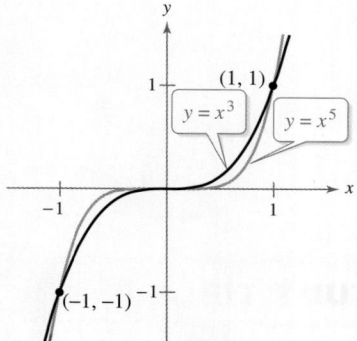

(a) When n is even, the graph of $y = x^n$ touches the axis at the x-intercept.

(b) When n is odd, the graph of $y = x^n$ crosses the axis at the x-intercept.

FIGURE 3.13

Example 1   Sketching Transformations of Monomial Functions

Sketch the graph of each function.

a. $f(x) = -x^5$

b. $h(x) = (x + 1)^4$

SOLUTION

a. The degree of $f(x) = -x^5$ is odd. So, its graph is similar to the graph of $y = x^3$. In Figure 3.14(a), note that the negative coefficient has the effect of reflecting the graph about the x-axis.

b. The graph of $h(x) = (x + 1)^4$ is a left shift, by one unit, of the graph of $y = x^4$, as shown in Figure 3.14(b).

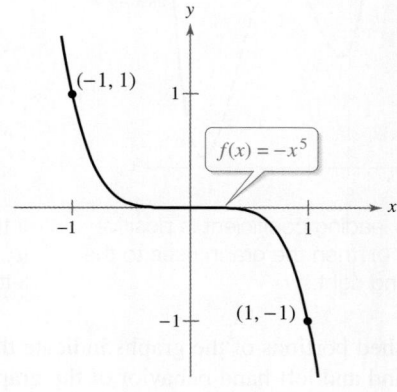

(a)

(b)

FIGURE 3.14

✓ **Checkpoint 1**

Sketch the graph of $f(x) = (x - 3)^3$.

The Leading Coefficient Test

In Example 1, note that both graphs eventually rise or fall without bound as x moves to the right. Whether the graph of a polynomial function eventually rises or falls can be determined by the function's degree (even or odd) and by its leading coefficient (positive or negative), as indicated in the **Leading Coefficient Test.**

Leading Coefficient Test

As x moves without bound to the left or to the right, the graph of the polynomial function given by

$$f(x) = a_n x^n + \cdots + a_1 x + a_0$$

eventually rises or falls in the following manner.

1. When n is *odd:*

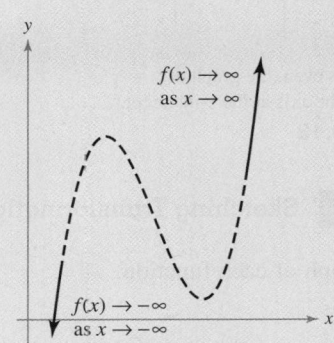 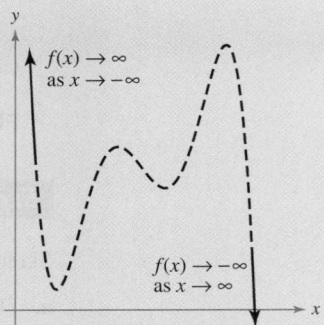

If the leading coefficient is positive ($a_n > 0$), then the graph falls to the left and rises to the right.

If the leading coefficient is negative ($a_n < 0$), then the graph rises to the left and falls to the right.

2. When n is *even:*

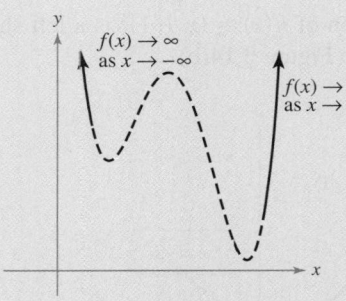 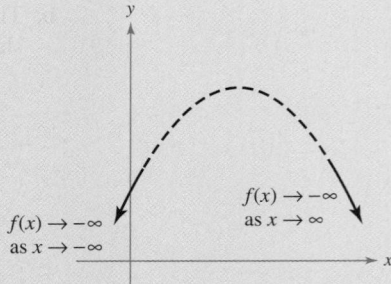

If the leading coefficient is positive ($a_n > 0$), then the graph rises to the left and right.

If the leading coefficient is negative ($a_n < 0$), then the graph falls to the left and right.

The dashed portions of the graphs indicate that the test determines *only* the right-hand and left-hand behavior of the graph.

As you continue to study polynomial functions and their graphs, you will notice that the degree of a polynomial plays an important role in determining other characteristics of the polynomial function and its graph.

Yuri Arcurs/www.shutterstock.com

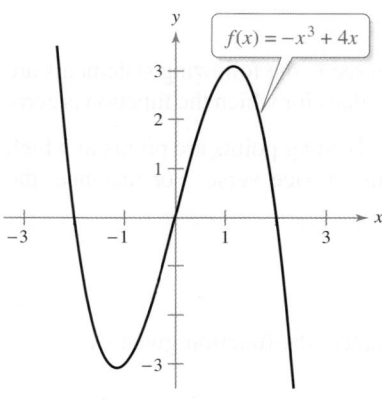

FIGURE 3.15

The functions in Example 3 are fourth- and fifth-degree polynomial functions, respectively. A fourth-degree polynomial function can also be referred to as a *quartic* function and a fifth-degree polynomial function as a *quintic* function.

Example 2   **Applying the Leading Coefficient Test**

Describe the right-hand and left-hand behavior of the graph of

$$f(x) = -x^3 + 4x.$$

SOLUTION   Because the degree is odd and the leading coefficient is negative, the graph rises to the left and falls to the right, as shown in Figure 3.15.   ———

✓ **Checkpoint 2**

Describe the right-hand and left-hand behavior of the graph of each function.

a. $f(x) = 3 + 2x + x^2$

b. $h(x) = -2x^4 + x$ ■

In Example 2, note that the Leading Coefficient Test tells you only whether the graph *eventually* rises or falls to the right or left. Other characteristics of the graph, such as intercepts, relative minima, and relative maxima, must be determined by other tests. For example, later you will use the number of real zeros of a polynomial function to determine how many times the graph of the function crosses the *x*-axis.

Example 3   **Applying the Leading Coefficient Test**

Describe the right-hand and left-hand behavior of the graph of each function.

a. $f(x) = x^4 - 5x^2 + 4$

b. $f(x) = x^5 - x$

SOLUTION

a. Because the degree is even and the leading coefficient is positive, the graph rises to the left and right, as shown in Figure 3.16(a).

b. Because the degree is odd and the leading coefficient is positive, the graph falls to the left and rises to the right, as shown in Figure 3.16(b).

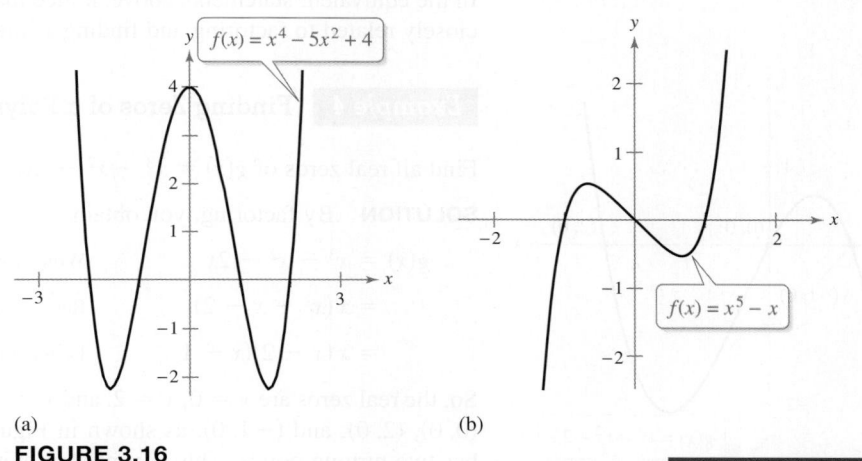

(a)   (b)

FIGURE 3.16

✓ **Checkpoint 3**

Describe the right-hand and left-hand behavior of the graph of

$$g(x) = -x^5 + 6x^2.$$ ■

Real Zeros of Polynomial Functions

It can be shown that for a polynomial function f of degree n, the following statements are true. Remember that the **zeros** of a function are the x-values for which the function is zero.

1. The graph of f has, at most, $n - 1$ *turning points*. Turning points are points at which the graph changes from increasing to decreasing, or vice versa. For instance, the graph of

$$f(x) = x^4 - 1$$

has at most $4 - 1 = 3$ turning points.

2. The function f has, at most, n real zeros. For instance, the function given by

$$f(x) = x^4 - 1$$

has at most $n = 4$ real zeros. (You will study this result in detail in Section 3.6 on the Fundamental Theorem of Algebra.)

Finding the zeros of polynomial functions is one of the most important problems in algebra. There is a strong interplay between graphical and algebraic approaches to this problem. Sometimes you can use information about the graph of a function to help find its zeros, and in other cases you can use information about the zeros of a function to help sketch its graph.

Real Zeros of Polynomial Functions

If f is a polynomial function and a is a real number, then the following statements are equivalent.

1. $x = a$ is a zero of the function f.

2. $x = a$ is a solution of the polynomial equation $f(x) = 0$.

3. $(x - a)$ is a factor of the polynomial $f(x)$.

4. $(a, 0)$ is an x-intercept of the graph of f.

In the equivalent statements above, notice that finding zeros of polynomial functions is closely related to factoring and finding x-intercepts.

Example 4   Finding Zeros of a Polynomial Function

Find all real zeros of $g(x) = x^3 - x^2 - 2x$.

SOLUTION   By factoring, you obtain

$$g(x) = x^3 - x^2 - 2x \qquad \text{Write original function.}$$
$$= x(x^2 - x - 2) \qquad \text{Remove common monomial factor.}$$
$$= x(x - 2)(x + 1). \qquad \text{Factor completely.}$$

So, the real zeros are $x = 0$, $x = 2$, and $x = -1$, and the corresponding x-intercepts are $(0, 0)$, $(2, 0)$, and $(-1, 0)$, as shown in Figure 3.17. Note that the graph in the figure has two turning points. This is consistent with the fact that the graph of a third-degree polynomial function can have *at most* $3 - 1 = 2$ turning points.

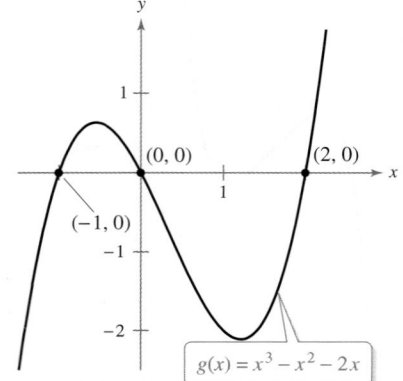

FIGURE 3.17

✓**Checkpoint 4**

Find all real zeros of $f(x) = x^2 - 4$.

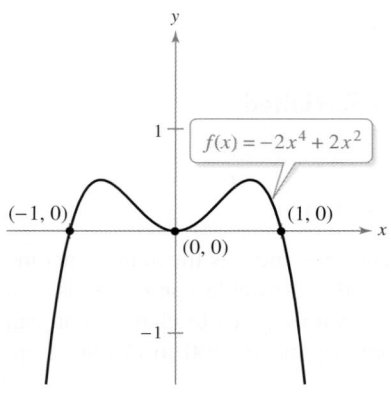

$f(x) = -2x^4 + 2x^2$

$(-1, 0)$ $(1, 0)$ $(0, 0)$

FIGURE 3.18

Example 5 **Finding Zeros of a Polynomial Function**

Find all real zeros of

$$f(x) = -2x^4 + 2x^2.$$

SOLUTION In this case, the polynomial factors as

$$f(x) = -2x^2(x^2 - 1) = -2x^2(x - 1)(x + 1).$$

So, the real zeros are $x = 0$, $x = 1$, and $x = -1$, and the corresponding x-intercepts are $(0, 0)$, $(1, 0)$, and $(-1, 0)$, as shown in Figure 3.18. Note that the graph in the figure has three turning points, which is consistent with the fact that the graph of a fourth-degree polynomial function can have *at most* $4 - 1 = 3$ turning points.

✓**Checkpoint 5**

Find all real zeros of $f(x) = x^3 - x$.

In Example 5, the real zero arising from $-2x^2 = 0$ is called a **repeated zero.** In general, a factor $(x - a)^k$ yields a repeated zero $x = a$ of **multiplicity** k. If k is odd, then the graph *crosses* the x-axis at $x = a$. If k is even, then the graph *touches* (but does not cross) the x-axis at $x = a$. This is illustrated in Figure 3.18.

Example 6 **Sketching the Graph of a Polynomial Function**

Sketch the graph of

$$f(x) = 3x^4 - 4x^3.$$

SOLUTION Because the leading coefficient is positive and the degree is even, you know that the graph eventually rises to the left and right, as shown in Figure 3.19(a). By factoring $f(x) = 3x^4 - 4x^3$ as $f(x) = x^3(3x - 4)$, you can see that the zeros of f are $x = 0$ and $x = \frac{4}{3}$ (both of odd multiplicity). So, the x-intercepts occur at $(0, 0)$ and $\left(\frac{4}{3}, 0\right)$. To sketch the graph by hand, find a few additional points, plot the points, and draw a continuous curve through the points to complete the graph, as shown in Figure 3.19(b). If you are unsure of the shape of a portion of a graph, then plot some additional points.

TECH TUTOR

Example 6 uses an algebraic approach to describe the graph of the function. A graphing utility is a valuable complement to this approach. Remember that when using a graphing utility, it is important that you find a viewing window that shows all important parts of the graph. For instance, the graph below shows the important parts of the graph of the function in Example 6.

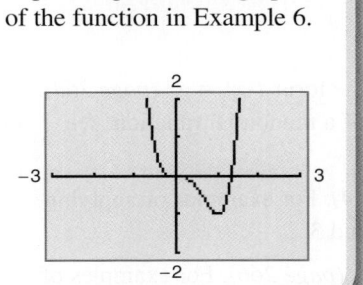

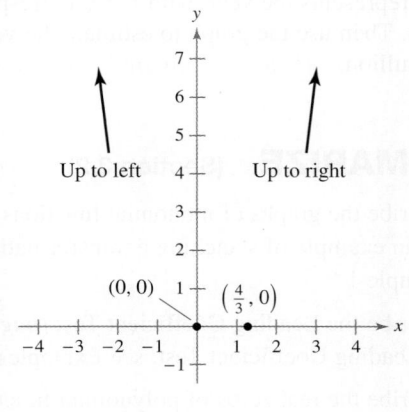

Up to left   Up to right

$(0, 0)$ $\left(\frac{4}{3}, 0\right)$

(a)

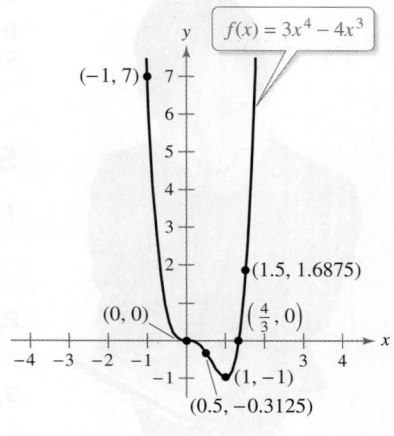

$f(x) = 3x^4 - 4x^3$

$(-1, 7)$

$(1.5, 1.6875)$

$(0, 0)$ $\left(\frac{4}{3}, 0\right)$

$(1, -1)$

$(0.5, -0.3125)$

(b)

FIGURE 3.19

✓**Checkpoint 6**

Sketch the graph of $h(x) = 2x^3 - 3x^2$.

Application

 Example 7   Charitable Contributions Revisited

Example 7 in Section 3.1 discussed the model

$$P(x) = 0.0014x^2 - 0.1529x + 5.855, \quad 5 \le x \le 100$$

where P is the percent of annual income given to charities and x is the annual income (in thousands of dollars). Note that this model gives the charitable contributions as a *percent* of annual income. To find the *amount* that a family gives to charity, you can multiply the given model by the income $1000x$ (and divide by 100 to change from percent to decimal form) to obtain

$$A(x) = 0.014x^3 - 1.529x^2 + 58.55x, \quad 5 \le x \le 100$$

where A represents the amount of charitable contributions (in dollars). Sketch the graph of the function A. Then use the graph to estimate the annual income of a family that gives $1000 per year to charities.

SOLUTION   Because the leading coefficient is positive and the degree is odd, you know that the graph eventually falls to the left and rises to the right. To sketch the graph by hand, find a few points, as shown in the table. Then plot the points and complete the graph, as shown in Figure 3.20.

x	5	25	45	65	86	100
$A(x)$	256.28	726.88	814.28	1190.48	2527.48	4565.00

FIGURE 3.20

From the graph, you can see that an annual contribution of $1000 corresponds to an annual income of about $59,000.

✓ Checkpoint 7

The net profits P (in millions of dollars) for the Colgate-Palmolive Company for the years 2000 to 2009 can be approximated by the function

$$P(t) = 2.864t^3 - 27.93t^2 + 156.0t + 1062$$

where t represents the year, with $t = 0$ corresponding to 2000. Sketch the graph of the function. Then use the graph to estimate the year in which the net profit was about $1600 million.   *(Source: Colgate-Palmolive Company)*   ■

SUMMARIZE   (Section 3.2)

1. Describe the graphs of monomial functions of the form $f(x) = x^n$ *(page 263)*. For an example of sketching a transformation of a monomial function, see Example 1.

2. Describe the Leading Coefficient Test *(page 264)*. For examples of applying the Leading Coefficient Test, see Examples 2 and 3.

3. Describe the real zeros of polynomial functions *(page 266)*. For examples of finding zeros of a polynomial function, see Examples 4 and 5.

SKILLS WARM UP 3.2

The following warm-up exercises involve skills that were covered in earlier sections. You will use these skills in the exercise set for this section. For additional help, review Sections 0.6, 1.3, 1.4, and 1.5.

In Exercises 1–6, factor the expression completely.

1. $12x^2 + 7x - 10$

2. $25x^3 - 60x^2 + 36x$

3. $12z^4 + 17z^3 + 5z^2$

4. $y^3 + 125$

5. $x^3 + 3x^2 - 4x - 12$

6. $x^3 + 2x^2 + 3x + 6$

In Exercises 7–10, find all real solutions of the equation.

7. $5x^2 + 8 = 0$

8. $x^2 - 6x + 4 = 0$

9. $4x^2 + 4x - 11 = 0$

10. $x^4 - 18x^2 + 81 = 0$

Exercises 3.2

See www.CalcChat.com for worked-out solutions to odd-numbered exercises.

Matching In Exercises 1–4, match the polynomial function with its graph. [The graphs are labeled (a), (b), (c), and (d).]

(a)
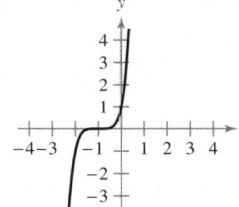

(b)
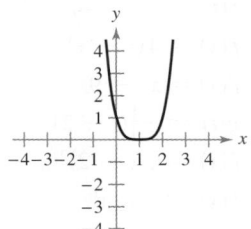

(c)
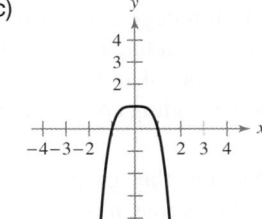

(d)
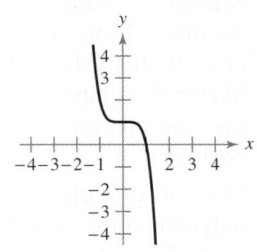

1. $f(x) = (x - 1)^4$

2. $f(x) = -x^5 + 1$

3. $f(x) = -x^4 + 1$

4. $f(x) = (x + 1)^5$

Sketching a Transformation of a Monomial Function In Exercises 5–8, use the graph of $y = x^3$ to sketch the graph of the function. *See Example 1.*

5. $f(x) = x^3 - 2$

6. $f(x) = (x + 3)^3$

7. $f(x) = (x + 1)^3 - 4$

8. $f(x) = -(x - 2)^3 + 2$

Sketching a Transformation of a Monomial Function In Exercises 9–12, use the graph of $y = x^4$ to sketch the graph of the function. *See Example 1.*

9. $f(x) = (x + 3)^4$

10. $f(x) = x^4 - 4$

11. $f(x) = 3 - x^4$

12. $f(x) = \frac{1}{2}(x - 1)^4$

Analyzing End Behavior In Exercises 13–20, (a) identify the degree of the function and state whether the degree is even or odd, (b) identify the leading coefficient and state whether it is positive or negative, (c) use a graphing utility to graph the function, and (d) describe the right-hand and left-hand behavior of the graph.

13. $y = 2x^5 + 2x^2 - 5x + 1$

14. $y = x^3 - 2x^2 - x + 1$

15. $y = -2x^5 - x^2 + 5x + 3$

16. $y = -x^3 + 5x - 2$

17. $y = 2x^2 + 3x - 4$

18. $y = x^4 - 3x^2 + 2x - 1$

19. $y = -x^6 - x^2 - 5x + 4$

20. $y = -x^2 + 3x + 4$

21. Discovery Using your results in Exercises 13–16, describe how the right-hand and left-hand behavior of the graph of a polynomial function of odd degree is related to the sign of the leading coefficient.

22. Discovery Using your results in Exercises 17–20, describe how the right-hand and left-hand behavior of the graph of a polynomial function of even degree is related to the sign of the leading coefficient.

Applying the Leading Coefficient Test In Exercises 23–32, describe the right-hand and left-hand behavior of the graph of the polynomial function. *See Examples 2 and 3.*

23. $f(x) = -x^3 + 1$

24. $f(x) = \frac{1}{3}x^3 + 5x$

25. $g(x) = 6 - 4x^2 + x - 3x^5$

26. $f(x) = 2x^5 - 5x + 7.5$

27. $f(x) = 4x^8 - 2$

28. $h(x) = 1 - x^6$

29. $f(x) = 2 + 5x - x^2 - x^3 + 2x^4$

30. $f(x) = \dfrac{3x^4 - 2x + 5}{4}$

31. $h(t) = -\frac{2}{3}(t^2 - 5t + 3)$

32. $f(s) = -\frac{7}{8}(s^3 + 5s^2 - 7s + 1)$

The Numbers of Turning Points and Real Zeros   In Exercises 33–36, determine (a) the maximum number of turning points of the graph of the function and (b) the maximum number of real zeros of the function.

33. $f(x) = x^2 - 4x + 1$ **34.** $f(x) = -3x^4 + 1$

35. $f(x) = -x^5 + x^4 - x$ **36.** $f(x) = 2x^3 + x^2 + 1$

Finding Zeros of a Polynomial Function   In Exercises 37–52, find all real zeros of the function algebraically. Then use a graphing utility to confirm your results. *See Examples 4 and 5.*

37. $f(x) = 9 - x^2$ **38.** $f(x) = x^2 - 25$

39. $h(t) = t^2 + 8t + 16$ **40.** $f(x) = x^2 - 12x + 36$

41. $f(x) = \frac{1}{3}x^2 + \frac{1}{3}x + \frac{2}{3}$

42. $f(x) = \frac{1}{2}x^2 + \frac{5}{2}x - \frac{3}{2}$

43. $g(x) = -5x^2 - 10x + 20$

44. $f(x) = 2x^2 + 4x + 6$

45. $f(t) = t^3 - 4t^2 + 4t$

46. $f(x) = x^4 - x^3 - 20x^2$

47. $g(t) = \frac{1}{2}t^4 - \frac{1}{2}$

48. $f(x) = \frac{1}{3} - \frac{1}{3}x^4$

49. $f(x) = 2x^4 - 2x^2 - 40$

50. $g(t) = t^5 - 6t^3 + 9t$

51. $f(x) = x^3 - 3x^2 + 2x - 6$

52. $f(x) = x^3 - 4x^2 - 25x + 100$

Matching   In Exercises 53–60, match the polynomial function with its graph. [The graphs are labeled (a), (b), (c), (d), (e), (f), (g), and (h).]

(a)
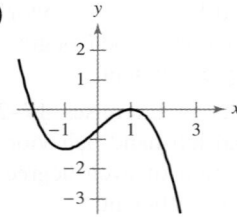

(b)
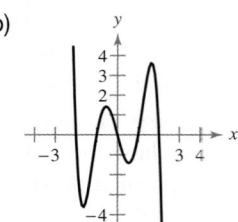

(c)
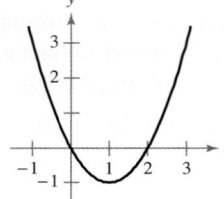

(d)
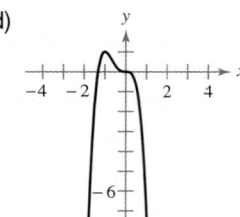

(e)
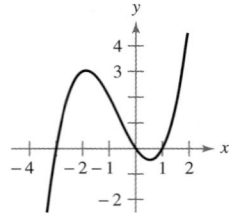

(f)
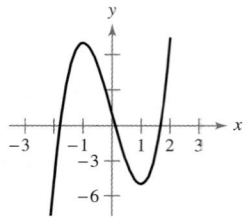

(g)
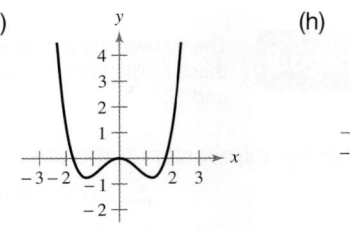

(h)
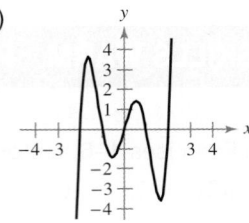

53. $f(x) = \frac{1}{2}(x^3 + 2x^2 - 3x)$

54. $f(x) = x^2 - 2x$ **55.** $f(x) = \frac{1}{3}x^4 - x^2$

56. $f(x) = -3x^4 - 4x^3$ **57.** $f(x) = 3x^3 - 9x + 1$

58. $f(x) = x^5 - 5x^3 + 4x$ **59.** $f(x) = -\frac{1}{3}x^3 + x - \frac{2}{3}$

60. $f(x) = -x^5 + 5x^3 - 4x$

Sketching the Graph of a Polynomial Function   In Exercises 61–72, analyze the graph of the function algebraically and use the results to sketch the graph *by hand*. Then use a graphing utility to confirm your sketch. *See Example 6.*

61. $f(x) = \frac{2}{3}x + 5$ **62.** $h(x) = -\frac{3}{4}x + 2$

63. $f(t) = \frac{1}{2}(t^2 - 4t - 1)$ **64.** $g(x) = -x^2 + 10x - 16$

65. $f(x) = 4x^2 - x^3$ **66.** $f(x) = 1 - x^3$

67. $f(x) = x^3 - 9x$ **68.** $f(x) = \frac{1}{4}x^3 - 2x^2$

69. $g(t) = -\frac{1}{4}(t - 2)^2(t + 2)^2$

70. $f(x) = x(x - 2)^2(x + 1)$

71. $f(x) = 1 - x^6$ **72.** $g(x) = 1 - (x + 1)^6$

73. Modeling Polynomials   Sketch the graph of a polynomial function that is of fourth degree, has a zero of multiplicity 2, and has a negative leading coefficient. Sketch the graph of another polynomial function with the same characteristics except the leading coefficient is positive.

74. Modeling Polynomials   Sketch the graph of a polynomial function that is of fifth degree, has a zero of multiplicity 2, and has a negative leading coefficient. Sketch the graph of another polynomial function with the same characteristics except the leading coefficient is positive.

75. Modeling Polynomials   Determine the equation of the fourth-degree polynomial function f whose graph is shown.

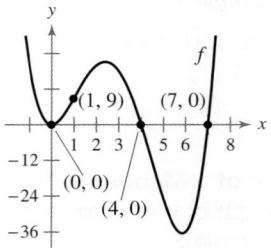

Figure for 75   Figure for 76

76. Modeling Polynomials   Determine the equation of the third-degree polynomial function g whose graph is shown.

77. Population The table shows the populations P (in millions) of foreign-born people living in the United States in 2009 and every 10 years from 1900 through 2000. *(Source: U.S. Census Bureau)*

Year	1900	1910	1920	1930
Population, P	10.3	13.5	13.9	14.2

Year	1940	1950	1960	1970
Population, P	11.6	10.3	9.7	9.7

Year	1980	1990	2000	2009
Population, P	14.1	19.8	31.1	38.5

(a) Use a graphing utility to create a scatter plot of the data. Let $t = 0$ and $t = 10$ correspond to 1900 and 1910, respectively.

(b) Use what you know about end behavior and the scatter plot from part (a) to predict the sign of the leading coefficient of a cubic model for P.

(c) Use the *regression* feature of the graphing utility to find a cubic model for P. Does your model agree with your answer from part (b)?

(d) Use the graphing utility to graph the model from part (c). Use the graph to predict the year in which the immigrant population will be about 45 million. Is your prediction reasonable?

78. Advertising Expenses The total revenue R (in millions of dollars) for a soft drink company is related to its advertising expenses by the function

$$R = \frac{1}{50,000}(-x^3 + 600x^2), \quad 0 \le x \le 400$$

where x is the amount spent on advertising (in tens of thousands of dollars). Use the graph of R to estimate the point on the graph at which the function is increasing most rapidly. This point is called the *point of diminishing returns* because any expenditure above this amount will yield less return per dollar invested in advertising.

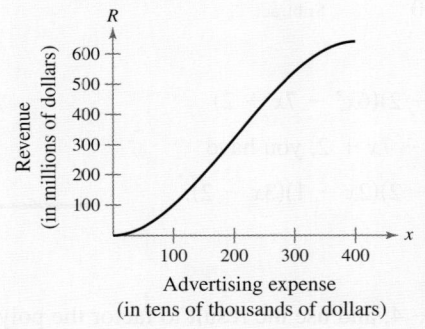

R

Revenue (in millions of dollars)

600
500
400
300
200
100

100 200 300 400 x

Advertising expense (in tens of thousands of dollars)

79. Advertising Expenses The total revenue R (in millions of dollars) for a hotel corporation is related to its advertising expenses by the function

$$R = -0.148x^3 + 4.89x^2 - 17.8x + 125, \quad 0 \le x \le 20$$

where x is the amount spent on advertising (in millions of dollars). Use a graphing utility to graph the function and estimate the point on the graph at which the function is increasing most rapidly. This point is called the *point of diminishing returns* because any expenditure above this amount will yield less return per dollar invested in advertising.

80. HOW DO YOU SEE IT? The scatter plot shows the number y of patients in a hospital at the start of each week x for five consecutive weeks. What is the least possible degree of a polynomial function whose graph passes through all five data points? Explain your reasoning.

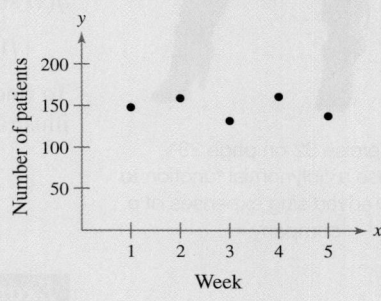

y

Number of patients

200
150
100
50

1 2 3 4 5 x

Week

81. Maximum Value You are making an open box with locking tabs from a square piece of material 24 inches on a side. You cut squares of the same size from the corners and fold along the dashed lines, as shown in the figure.

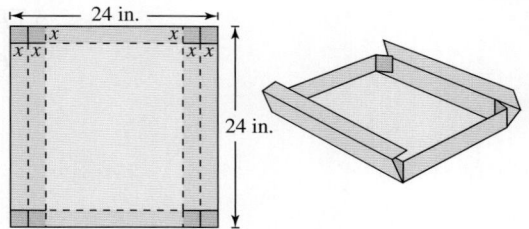

(a) Verify that the volume of the box is given by the function

$$V(x) = 8x(6 - x)(12 - x).$$

(b) Determine the domain of the function V.

(c) Sketch a graph of the function and estimate the value of x for which the volume $V(x)$ is maximum.

3.3 Polynomial Division

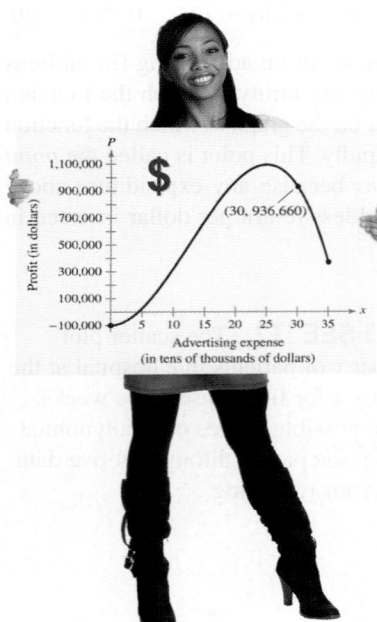

In Exercise 82 on page 281, you will use a polynomial function to estimate advertising expenses of a company.

■ Divide one polynomial by a second polynomial using long division.
■ Use synthetic division to divide two polynomials.
■ Use the Remainder Theorem and synthetic division to evaluate a polynomial.
■ Use the Factor Theorem to factor a polynomial.
■ Use polynomial division to solve an application problem.

Long Division of Polynomials

Suppose you are given the graph of

$$f(x) = 6x^3 - 19x^2 + 16x - 4.$$

Notice that a zero of f occurs at $x = 2$, as shown in Figure 3.21. Because $x = 2$ is a zero of the polynomial function f, you know that $(x - 2)$ is a factor of $f(x)$. This means that there exists a second-degree polynomial $q(x)$ such that

$$f(x) = (x - 2) \cdot q(x).$$

To find $q(x)$, you can use **long division,** as illustrated in Example 1.

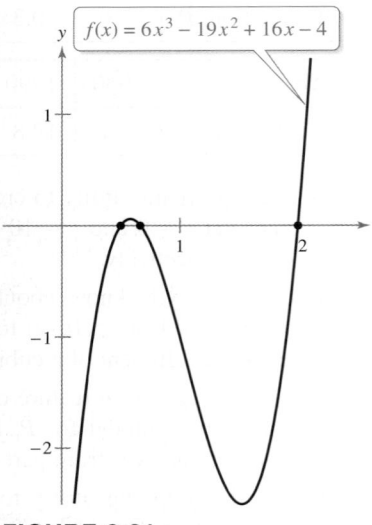

FIGURE 3.21

Example 1   **Long Division of Polynomials**

Divide the polynomial $6x^3 - 19x^2 + 16x - 4$ by $x - 2$, and use the result to factor the polynomial completely.

SOLUTION

$$
\begin{array}{r}
6x^2 - 7x + 2 \\
x - 2 \overline{)\, 6x^3 - 19x^2 + 16x - 4} \\
\underline{6x^3 - 12x^2} \\
-7x^2 + 16x \\
\underline{-7x^2 + 14x} \\
2x - 4 \\
\underline{2x - 4} \\
0
\end{array}
$$

Multiply $6x^2$ by $x - 2$.

Subtract and bring down $16x$.

Multiply $-7x$ by $x - 2$.

Subtract and bring down -4.

Multiply 2 by $x - 2$.

Subtract.

So, you can conclude that

$$6x^3 - 19x^2 + 16x - 4 = (x - 2)(6x^2 - 7x + 2)$$

and by factoring the quadratic $6x^2 - 7x + 2$, you have

$$6x^3 - 19x^2 + 16x - 4 = (x - 2)(2x - 1)(3x - 2).$$

✓ Checkpoint 1

Divide $x^3 - 6x^2 + 5x + 12$ by $x - 4$, and use the result to factor the polynomial completely.

Note that the factorization shown in Example 1 agrees with the graph in Figure 3.21 in that the three x-intercepts occur at

$$x = 2, \quad x = \tfrac{1}{2}, \quad \text{and} \quad x = \tfrac{2}{3}.$$

In Example 1, $x - 2$ is a factor of the polynomial

$$6x^3 - 19x^2 + 16x - 4$$

and the long division process produces a remainder of zero. Often, long division will produce a nonzero remainder. For instance, divide $x^2 + 3x + 5$ by $x + 1$, as shown.

$$
\begin{array}{r}
x + 2 \quad \longleftarrow \text{ Quotient}\\
x + 1 \,\overline{)\, x^2 + 3x + 5} \quad \longleftarrow \text{ Dividend}\\
\underline{x^2 + x}\\
2x + 5\\
\underline{2x + 2}\\
3 \quad \longleftarrow \text{ Remainder}
\end{array}
$$

Divisor $\Longrightarrow$

In fractional form, you can write this result as shown.

$$
\underbrace{\dfrac{\overbrace{x^2 + 3x + 5}^{\text{Dividend}}}{\underbrace{x + 1}_{\text{Divisor}}}}_{} = \overbrace{x + 2}^{\text{Quotient}} + \dfrac{\overset{\text{Remainder}}{3}}{\underbrace{x + 1}_{\text{Divisor}}}
$$

This implies that

$$x^2 + 3x + 5 = (x + 1)(x + 2) + 3 \qquad \text{Multiply each side by } (x + 1).$$

which illustrates the following well-known theorem called the **Division Algorithm.**

The Division Algorithm

If $f(x)$ and $d(x)$ are polynomials such that $d(x) \neq 0$, and the degree of $d(x)$ is less than or equal to the degree of $f(x)$, then there exist unique polynomials $q(x)$ and $r(x)$ such that

$$f(x) = d(x)q(x) + r(x)$$

Dividend | Quotient
Divisor | Remainder

where $r(x) = 0$ or the degree of $r(x)$ is less than the degree of $d(x)$. If the remainder $r(x)$ is zero, then $d(x)$ **divides evenly** into $f(x)$.

The Division Algorithm can also be written as

$$\dfrac{f(x)}{d(x)} = q(x) + \dfrac{r(x)}{d(x)}.$$

In the Division Algorithm, the rational expression

$$f(x)/d(x)$$

is **improper** because the degree of $f(x)$ is greater than or equal to the degree of $d(x)$. On the other hand, the rational expression

$$r(x)/d(x)$$

is **proper** because the degree of $r(x)$ is less than the degree of $d(x)$.

Before you apply the Division Algorithm, follow these steps.

1. Write the dividend and divisor in descending powers of the variable.

2. Insert placeholders with zero coefficients for missing powers of the variable.

Note how these steps are applied in the next example.

Example 2   Long Division of Polynomials

Divide $x^3 - 1$ by $x - 1$.

SOLUTION   Because there is no x^2-term or x-term in the dividend, you need to line up the subtraction by using zero coefficients (or leaving spaces) for the missing terms.

$$
\begin{array}{r}
x^2 + \ x + 1 \\
x - 1 \overline{)\ x^3 + 0x^2 + 0x - 1} \\
\end{array}
$$

$x^3 - \ x^2$	Insert $0x^2$ and $0x$.
$x^2 + 0x$	Multiply x^2 by $x - 1$.
$x^2 - \ x$	Subtract and bring down $0x$.
$x - 1$	Multiply x by $x - 1$.
$x - 1$	Subtract and bring down -1.
0	Multiply 1 by $x - 1$.
	Subtract.

> **STUDY TIP**
>
> You can check the result of a division problem by multiplying. For instance, in Example 2, try checking that
> $$(x - 1)(x^2 + x + 1)$$
> $$= x^3 - 1.$$

So, $x - 1$ divides evenly into $x^3 - 1$ and you can write

$$\frac{x^3 - 1}{x - 1} = x^2 + x + 1.$$

✓ Checkpoint 2

Divide $x^3 - 8$ by $x + 2$.

In each of the long division examples, the divisor has been a first-degree polynomial. The Division Algorithm works just as well with polynomial divisors of degree two or more, as shown in Example 3.

Example 3   Long Division of Polynomials

Divide $2x^4 + 4x^3 - 5x^2 + 3x - 2$ by $x^2 + 2x - 3$.

SOLUTION

$$
\begin{array}{r}
2x^2 \qquad\ + 1 \\
x^2 + 2x - 3 \overline{)\ 2x^4 + 4x^3 - 5x^2 + 3x - 2} \\
\end{array}
$$

$2x^4 + 4x^3 - 6x^2$	Multiply $2x^2$ by $x^2 + 2x - 3$.
$x^2 + 3x - 2$	Subtract and bring down $3x - 2$.
$x^2 + 2x - 3$	Multiply 1 by $x^2 + 2x - 3$.
$x + 1$	Subtract.

Note that the first subtraction eliminated two terms from the dividend. When this happens, the quotient skips a term. So, you can write

$$\frac{2x^4 + 4x^3 - 5x^2 + 3x - 2}{x^2 + 2x - 3} = 2x^2 + 1 + \frac{x + 1}{x^2 + 2x - 3}.$$

✓ Checkpoint 3

Divide $5x^4 + 10x^3 - 7x^2 + 28x - 39$ by $x^2 + 2x - 4$.

Synthetic Division

There is a nice shortcut for long division of polynomials when dividing by divisors of the form $x - k$. This shortcut is called **synthetic division.** The pattern for synthetic division of a cubic polynomial is shown below. (The pattern for higher-degree polynomials is similar.)

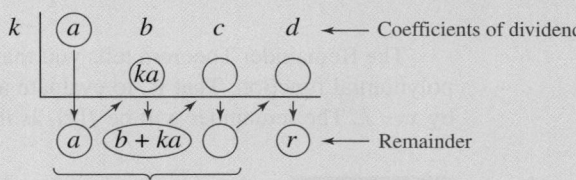

Synthetic Division (for a Cubic Polynomial)

To divide $ax^3 + bx^2 + cx + d$ by $x - k$, use the following pattern.

Vertical pattern: Add terms in columns.
Diagonal pattern: Multiply results by k.

This algorithm for synthetic division works only for divisors of the form $x - k$. Remember that

$$x + k = x - (-k).$$

Example 4 **Using Synthetic Division**

Use synthetic division to divide $x^4 - 10x^2 - 2x + 4$ by $x + 3$.

SOLUTION Set up the array as shown. Note that a zero is included for the missing x^3-term in the dividend.

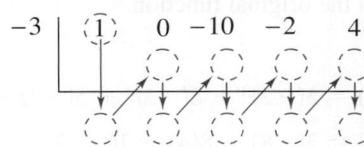

Then, use the synthetic division pattern by adding terms in columns and multiplying the results by -3.

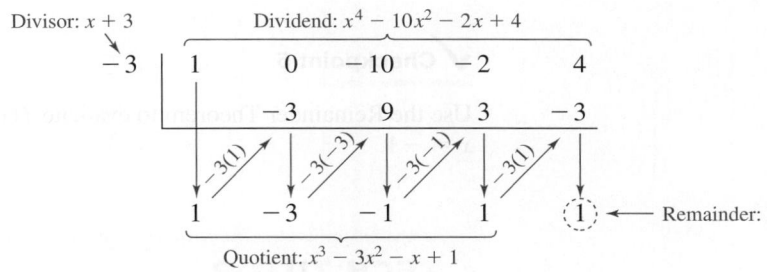

So, you have

$$\frac{x^4 - 10x^2 - 2x + 4}{x + 3} = x^3 - 3x^2 - x + 1 + \frac{1}{x + 3}.$$

✓ **Checkpoint 4**

Use synthetic division to divide $2x^3 - 7x^2 - 80$ by $x - 5$.

Andresr /Shutterstock.com

Remainder Theorem

The remainder obtained in the synthetic division process has an important interpretation, as described in the Remainder Theorem.

The Remainder Theorem

If a polynomial $f(x)$ is divided by $x - k$, then the remainder is

$$r = f(k).$$

The Remainder Theorem tells you that synthetic division can be used to evaluate a polynomial function. That is, to evaluate a polynomial function f at $x = k$, divide $f(x)$ by $x - k$. The remainder will be $f(k)$, as illustrated in Example 5.

Example 5   **Using the Remainder Theorem**

Use the Remainder Theorem to evaluate

$$f(x) = 3x^3 + 8x^2 + 5x - 7$$

when $x = -2$.

SOLUTION   Use synthetic division.

$$
\begin{array}{r|rrrr}
-2 & 3 & 8 & 5 & -7 \\
   &   & -6 & -4 & -2 \\
\hline
   & 3 & 2 & 1 & -9
\end{array}
$$

Because the remainder is $r = -9$, you can conclude that

$$f(-2) = -9.$$

This means that $(-2, -9)$ is a point on the graph of f. You can check this by substituting $x = -2$ in the original function.

CHECK

$$
\begin{aligned}
f(-2) &= 3(-2)^3 + 8(-2)^2 + 5(-2) - 7 \\
&= 3(-8) + 8(4) - 10 - 7 \\
&= -24 + 32 - 10 - 7 \\
&= -9
\end{aligned}
$$

✓ **Checkpoint 5**

Use the Remainder Theorem to evaluate $f(x) = 4x^3 + 6x^2 + 4x + 5$ when $x = -1$.

TECH TUTOR

Remember, you can also evaluate a function with your graphing utility by entering the function in the equation editor and using the *table* feature in ASK mode. For instructions on how to use the *table* feature, see Appendix A. Consult the user's guide for your graphing utility for more information.

Factor Theorem

Another important theorem is the Factor Theorem, which is stated below.

Factor Theorem

A polynomial $f(x)$ has a factor $(x - k)$ if and only if $f(k) = 0$.

You can think of the Factor Theorem as stating that if $(x - k)$ is a factor of $f(x)$, then $f(k) = 0$. Conversely, if $f(k) = 0$, then $(x - k)$ is a factor of $f(x)$.

Example 6   Factoring a Polynomial: Repeated Division

Show that $(x - 2)$ and $(x + 3)$ are factors of the polynomial

$$f(x) = 2x^4 + 7x^3 - 4x^2 - 27x - 18.$$

Then find the remaining factors of $f(x)$.

SOLUTION   Use synthetic division with the factor $(x - 2)$.

$$
\begin{array}{r|rrrrr}
2 & 2 & 7 & -4 & -27 & -18 \\
  &   & 4 & 22 & 36 & 18 \\
\hline
  & 2 & 11 & 18 & 9 & 0
\end{array}
$$

⟹ 0 remainder, so $f(2) = 0$ and $(x - 2)$ is a factor.

Take the result of this division and perform synthetic division again using the factor $(x + 3)$.

$$
\begin{array}{r|rrrr}
-3 & 2 & 11 & 18 & 9 \\
   &   & -6 & -15 & -9 \\
\hline
   & 2 & 5 & 3 & 0
\end{array}
$$

⟹ 0 remainder, so $f(-3) = 0$ and $(x + 3)$ is a factor.

Quadratic: $2x^2 + 5x + 3$

Because the resulting quadratic expression factors as

$$2x^2 + 5x + 3 = (2x + 3)(x + 1)$$

the complete factorization of $f(x)$ is

$$f(x) = (x - 2)(x + 3)(2x + 3)(x + 1).$$

Note that this factorization implies that f has four real zeros: 2, -3, $-\frac{3}{2}$, and -1. This is confirmed by the graph of f, which is shown in Figure 3.22.

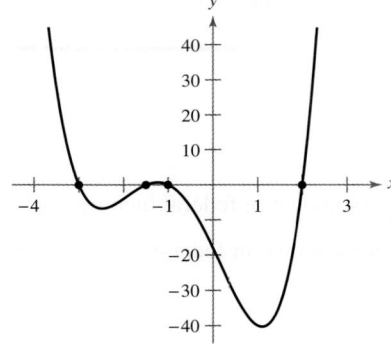

FIGURE 3.22

✓ Checkpoint 6

Show that $(x + 2)$ and $(x + 4)$ are factors of the polynomial $f(x) = x^4 + 6x^3 + 7x^2 - 6x - 8$. Then find the remaining factors of $f(x)$. ■

STUDY TIP

Try using more than one strategy to solve some of the exercises in this section. For instance, if you find that $x - k$ divides evenly into $f(x)$ (with no remainder), then try sketching the graph of f. You should see that $(k, 0)$ is an x-intercept.

Uses of the Remainder in Synthetic Division

The remainder r obtained in the synthetic division of $f(x)$ by $x - k$ provides the following information.

1. The remainder r gives the value of f at $x = k$. That is, $r = f(k)$.

2. If $r = 0$, then $(x - k)$ is a factor of $f(x)$.

3. If $r = 0$, then $(k, 0)$ is an x-intercept of the graph of f.

Application

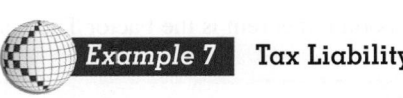 **Example 7**   **Tax Liability**

The 2009 federal income tax liability for an employee who was single and claimed no dependents is given by the function

$$y = 0.00000077x^2 + 0.096x - 1426.86, \quad 20{,}000 \le x \le 100{,}000$$

where y represents the tax liability (in dollars) and x represents the employee's yearly salary (in dollars) (see Figure 3.23).   *(Source: U.S. Department of the Treasury)*

a. Find a function that gives the tax liability as a *percent* of the yearly salary.

b. Graph the function in part (a). What conclusions can you make from the graph?

SOLUTION

a. Because the yearly salary is given by x and the tax liability is given by y, the percent (in decimal form) of yearly salary that a person owes in federal income tax is

$$
\begin{aligned}
P &= \frac{y}{x} \\
&= \frac{0.00000077x^2 + 0.096x - 1426.86}{x} \\
&= 0.00000077x + 0.096 - \frac{1426.86}{x}.
\end{aligned}
$$

b. The graph of the function P is shown in Figure 3.24. From the graph you can see that as a person's yearly salary increases, the percent that he or she must pay in federal income tax also increases.

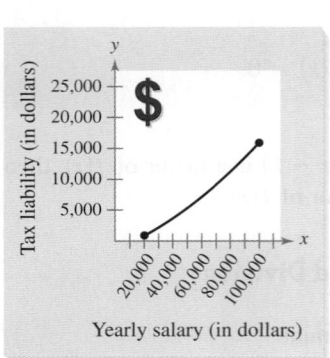

FIGURE 3.23

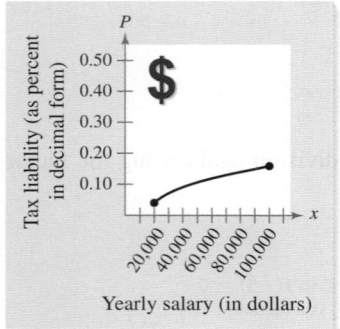

FIGURE 3.24

✓ **Checkpoint 7**

Use the function P from part (a) of Example 7.

a. What percent of a $50,000 yearly salary does a person owe in federal income tax?

b. What percent of a $100,000 yearly salary does a person owe in federal income tax?

SUMMARIZE   (Section 3.3)

1. Describe how to divide one polynomial by another polynomial using long division *(pages 272–274)*. For examples of long division of polynomials, see Examples 1, 2, and 3.

2. Describe the algorithm for synthetic division *(page 275)*. For an example of using synthetic division, see Example 4.

3. State the Remainder Theorem *(page 276)*. For an example of using the Remainder Theorem, see Example 5.

4. State the Factor Theorem *(page 277)*. For an example of using the Factor Theorem, see Example 6.

5. Describe a real-life example of how polynomial division is used to write a function that gives the tax liability as a percent of an employee's yearly salary *(page 278, Example 7)*.

SKILLS WARM UP 3.3   The following warm-up exercises involve skills that were covered in earlier sections. You will use these skills in the exercise set for this section. For additional help, review Sections 0.5 and 0.6.

In Exercises 1–4, write the expression in standard polynomial form.

1. $(x - 1)(x^2 + 2) + 5$

2. $(x^2 - 3)(2x + 4) + 8$

3. $(x^2 + 1)(x^2 - 2x + 3) - 10$

4. $(x + 6)(2x^3 - 3x) - 5$

In Exercises 5–10, factor the polynomial.

5. $x^2 - 4x + 3$

6. $8x^2 - 24x - 80$

7. $3x^2 + 2x - 5$

8. $9x^2 - 24x + 16$

9. $4x^3 - 10x^2 + 6x$

10. $6x^3 + 7x^2 + 2x$

Exercises 3.3

See www.CalcChat.com for worked-out solutions to odd-numbered exercises.

Long Division of Polynomials   In Exercises 1–22, use long division to divide. *See Examples 1, 2, and 3.*

	Dividend	Divisor
1.	$x^2 - x - 6$	$x - 3$
2.	$x^2 + 2x - 35$	$x + 7$
3.	$3x^2 - 7x + 4$	$x - 1$
4.	$5x^2 - 17x - 12$	$x - 4$
5.	$2x^2 + 10x + 12$	$x + 3$
6.	$2x^2 + x - 11$	$x + 5$
7.	$x^3 + 3x^2 + 5x + 3$	$x + 1$
8.	$x^3 - 6x^2 + 7x - 2$	$x - 1$
9.	$2x^3 + 6x^2 - x - 3$	$2x^2 - 1$
10.	$3x^3 - 12x^2 - 2x + 8$	$3x^2 - 2$
11.	$x^4 + 5x^3 + 6x^2 - x - 2$	$x + 2$
12.	$x^4 + 2x^3 - 3x^2 - 8x - 4$	$x^2 - 4$
13.	$7x + 3$	$x + 4$
14.	$8x - 5$	$2x + 3$
15.	$6x^3 + 10x^2 + x + 8$	$2x^2 + 1$
16.	$2x^3 - 8x^2 + 3x - 9$	$2x^2 + 3$
17.	$x^3 - 27$	$x^2 - 1$
18.	$x^3 - 9$	$x^2 + 1$
19.	$x^3 - 4x^2 + 5x - 2$	$x + 2$
20.	$x^3 - x^2 + 2x - 8$	$x - 2$
21.	$2x^5 - 8x^3 + 4x - 1$	$x^2 - 2x + 1$
22.	$x^4 + 3x^2 + 1$	$x^2 - 2x + 3$

Using Synthetic Division   In Exercises 23–42, use synthetic division to divide. *See Example 4.*

	Dividend	Divisor
23.	$x^2 + 6x - 7$	$x - 1$
24.	$x^2 - 4x - 21$	$x + 3$
25.	$2x^3 + 5x^2 - 7x + 20$	$x + 4$
26.	$3x^3 - 23x^2 - 12x + 32$	$x - 8$
27.	$4x^3 - 9x + 8x^2 - 18$	$x + 2$
28.	$9x^3 - 16x - 18x^2 + 32$	$x - 2$
29.	$-x^3 + 75x - 250$	$x + 10$
30.	$3x^3 - 16x^2 - 72$	$x - 6$
31.	$x^4 - 4x^3 - 7x^2 + 22x + 24$	$x + 3$
32.	$6x^4 - 15x^3 - 11x$	$x + 2$
33.	$10x^4 - 50x^3 - 800$	$x - 6$
34.	$x^5 - 13x^4 - 120x + 80$	$x + 3$
35.	x^5	$x + 1$
36.	$5x^3$	$x + 3$
37.	$-3x^4$	$x - 2$
38.	$2x^5$	$x + 3$
39.	$5 - 3x + 2x^2 - x^3$	$x + 1$
40.	$180x - x^4$	$x - 6$
41.	$4x^3 + 16x^2 - 23x - 15$	$x + \frac{1}{2}$
42.	$3x^3 - 4x^2 + 5$	$x - \frac{3}{2}$

Using the Remainder Theorem   In Exercises 43–50, write the function in the form

$$f(x) = (x - k)q(x) + r$$

for the given value of k, and demonstrate that $f(k) = r$.

43. $f(x) = x^3 + x^2 - 12x + 20, \quad k = 2$

44. $f(x) = x^3 - 2x^2 - 15x + 7, \quad k = -4$

45. $f(x) = 3x^3 + 2x^2 + 5x - 2, \quad k = \frac{1}{3}$

46. $f(x) = 4x^4 + 6x^3 + 4x^2 - 5x + 13, \quad k = -\frac{1}{2}$

47. $f(x) = x^3 + 2x^2 - 3x - 12, \quad k = \sqrt{3}$

48. $f(x) = x^3 + 3x^2 - 7x - 6, \quad k = -\sqrt{2}$

49. $f(x) = 2x^3 + x^2 - 14x - 10, \quad k = 1 + \sqrt{3}$

50. $f(x) = 3x^3 - 19x^2 + 27x - 7, \quad k = 3 - \sqrt{2}$

Using the Remainder Theorem   In Exercises 51–54, use the Remainder Theorem to find each function value. *See Example 5.*

51. $f(x) = 2x^5 - 3x^2 - 4x - 1$

   (a) $f(-2)$ (b) $f(-4)$

   (c) $f(1)$ (d) $f(3)$

52. $g(x) = x^6 - 4x^4 + 3x^2 + 2$

   (a) $g(2)$ (b) $g(-4)$

   (c) $g(7)$ (d) $g(-1)$

53. $f(x) = 2x^3 + 3x^2 + 8x - 14$

   (a) $f(2)$ (b) $f(-1)$

   (c) $f(1.1)$ (d) $f\left(\frac{1}{2}\right)$

54. $f(x) = 3x^4 - 7x^3 + 5x - 12$

   (a) $f(1)$ (b) $f(4)$

   (c) $f\left(\frac{1}{3}\right)$ (d) $f(-1.2)$

Factoring a Polynomial   In Exercises 55–60, (a) verify the given factors of $f(x)$, (b) find the remaining factor of $f(x)$, (c) use your results to write the complete factorization of $f(x)$, (d) list all real zeros of f, and (e) confirm your results by using a graphing utility to graph the function. *See Example 6.*

Function	Factors
55. $f(x) = x^3 - 12x - 16$	$(x + 2), (x - 4)$
56. $f(x) = x^3 - 28x - 48$	$(x + 4), (x - 6)$
57. $f(x) = 3x^3 + 10x^2 - 27x - 10$	$(3x + 1), (x - 2)$
58. $f(x) = 5x^3 - 11x^2 - 38x + 8$	$(5x - 1), (x - 4)$
59. $f(x) = x^3 + 2x^2 - 3x - 6$	$(x - \sqrt{3}), (x + 2)$
60. $f(x) = x^3 + 2x^2 - 2x - 4$	$(x - \sqrt{2}), (x + 2)$

61. Reasoning   You divide a polynomial by another polynomial. The remainder is zero. What conclusion(s) can you make?

62. Reasoning   A remainder obtained in a polynomial division by $x - k$ is zero. How is the divisor related to the graph of the dividend?

Finding Real Zeros   In Exercises 63–68, match the function with its graph and use the result to find all real solutions of $f(x) = 0$. [The graphs are labeled (a), (b), (c), (d), (e), and (f).]

(a) 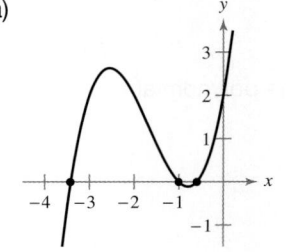   (b)

(c) 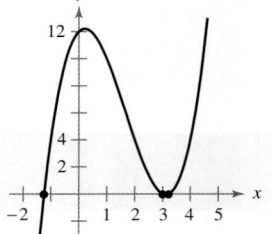   (d)

(e) 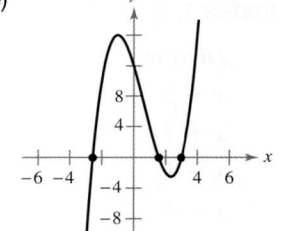   (f)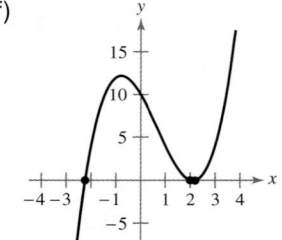

63. $f(x) = x^3 - 2x^2 - 7x + 12$

64. $f(x) = x^3 - x^2 - 5x + 2$

65. $f(x) = x^3 + 5x^2 + 6x + 2$

66. $f(x) = x^3 - 5x^2 + 2x + 12$

67. $f(x) = x^3 + 3x^2 - 5x - 15$

68. $f(x) = x^3 - 2x^2 - 5x + 10$

69. Modeling Polynomials   A third-degree polynomial function f has real zeros $-1, 2,$ and $\frac{10}{3}$. Find two different polynomial functions, one with a positive leading coefficient and one with a negative leading coefficient, that could be f. How many different polynomial functions are possible for f?

70. Modeling Polynomials   A fourth-degree polynomial function g has real zeros $-2, 0, 1,$ and 5. Find two different polynomial functions, one with a positive leading coefficient and one with a negative leading coefficient, that could be g. How many different polynomial functions are possible for g?

Simplifying a Rational Expression In Exercises 71–78, simplify the rational expression.

71. $\dfrac{x^3 - 10x^2 + 31x - 30}{x - 3}$

72. $\dfrac{x^3 + 15x^2 + 68x + 96}{x + 4}$

73. $\dfrac{6x^3 + x^2 - 21x - 10}{2x + 1}$

74. $\dfrac{3x^3 - 5x^2 - 34x + 24}{3x - 2}$

75. $\dfrac{x^4 - 5x^3 + 14x^2 - 120x}{x^2 + x + 20}$

76. $\dfrac{x^4 + x^3 + 3x^2 + 10x}{x^2 - x + 5}$

77. $\dfrac{x^4 + 4x^3 - 6x^2 - 36x - 27}{x^2 - 9}$

78. $\dfrac{x^4 + x^3 - 13x^2 - x + 12}{x^2 + x - 12}$

79. Examination Room A rectangular examination room in a veterinary clinic has a volume of

$$x^3 + 11x^2 + 34x + 32$$

cubic feet. The height of the room is $(x + 2)$ feet (see figure). Find the number of square feet of floor space in the examination room.

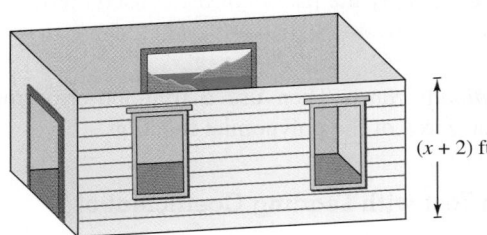

$(x + 2)$ ft

80. Veterinary Clinic A rectangular veterinary clinic has a volume of

$$x^3 + 55x^2 + 650x + 1784$$

cubic feet (the space in the attic is not included). The height of the clinic is $(x + 4)$ feet (see figure). Find the number of square feet of floor space *on the first floor* of the clinic.

$(x + 4)$ ft

VETERINARY CLINIC

81. Writing Briefly explain how to check polynomial division, and justify your answer. Give an example.

82. HOW DO YOU SEE IT? A company making fishing poles estimated that the profit P (in dollars) from selling a particular fishing pole was

$$P = -140.75x^3 + 5348.3x^2 - 76,560, \quad 0 \le x \le 35$$

where x was the advertising expense (in tens of thousands of dollars). For this fishing pole, the advertising expense was \$300,000 ($x = 30$) and the profit was \$936,660.

(a) From the graph shown in the figure, it appears that the company could have obtained the same profit by spending less on advertising. Use the graph to estimate another amount the company could have spent on advertising that would have produced the same profit.

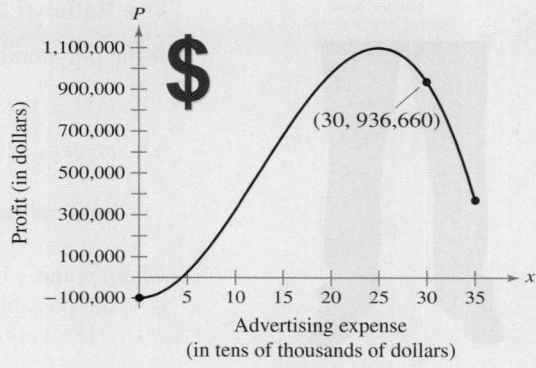

P

Profit (in dollars)

1,100,000
900,000
700,000
500,000
300,000
100,000
$-100,000$

(30, 936,660)

5   10   15   20   25   30   35 x

Advertising expense
(in tens of thousands of dollars)

(b) Explain how you could verify the amount from part (a) algebraically.

83. Profit A company that produces calculators estimated that the profit P (in dollars) from selling a particular model of calculator was

$$P = -152x^3 + 7545x^2 - 169,625, \quad 0 \le x \le 45$$

where x was the advertising expense (in tens of thousands of dollars). For this model of calculator, the advertising expense was \$400,000 ($x = 40$) and the profit was \$2,174,375.

(a) Use a graphing utility to graph the profit function.

(b) Use the graph from part (a) to estimate another amount the company could have spent on advertising that would have produced the same profit.

(c) Use synthetic division to confirm the result of part (b) algebraically.

Exploration In Exercises 84 and 85, find the constant c such that the denominator will divide evenly into the numerator.

84. $\dfrac{x^3 + 4x^2 - 3x + c}{x - 4}$

85. $\dfrac{x^5 - 2x^2 + x + c}{x + 2}$

3.4 Real Zeros of Polynomial Functions

- Find all possible rational zeros of a function using the Rational Zero Test.
- Approximate the real zeros of a polynomial function using the Intermediate Value Theorem.
- Approximate the real zeros of a polynomial function using a graphing utility.
- Apply techniques for approximating real zeros to solve an application problem.

The Rational Zero Test

The **Rational Zero Test** relates the possible rational zeros of a polynomial function (having integer coefficients) to the leading coefficient and to the constant term of the polynomial.

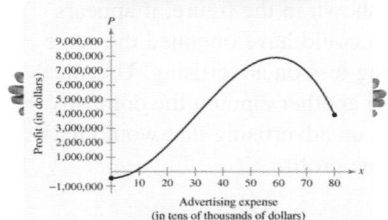

> ### The Rational Zero Test
>
> If the polynomial function given by
>
> $$f(x) = a_n x^n + a_{n-1} x^{n-1} + \cdots + a_2 x^2 + a_1 x + a_0$$
>
> has *integer* coefficients, then every rational zero of f has the form
>
> $$\text{Rational zero} = \frac{\text{a factor of the constant term } a_0}{\text{a factor of the leading coefficient } a_n} = \frac{p}{q}$$
>
> where p and q have no common factors other than 1. When the leading coefficient is 1, the possible rational zeros are simply the factors of the constant term.

In Exercise 68 on page 294, you will use a graph to determine how much money a company should spend on advertising to obtain a certain profit.

Make a list of *possible rational zeros*. Then use a trial-and-error method to determine which, if any, are actual zeros of the polynomial function.

Example 1   Rational Zero Test with Leading Coefficient of 1

Find the rational zeros of

$$f(x) = x^3 + x + 1.$$

SOLUTION   Because the leading coefficient is 1, the possible rational zeros are the factors of the constant term.

Possible rational zeros: 1 and -1

By testing these possible zeros, you can see that neither checks.

$$f(1) = (1)^3 + 1 + 1 = 3$$

$$f(-1) = (-1)^3 + (-1) + 1 = -1$$

So, you can conclude that the given function has *no* rational zeros. Note from the graph of f in Figure 3.25 that f does have one real zero (between -1 and 0). By the Rational Zero Test, you know that this real zero is *not* a rational number.

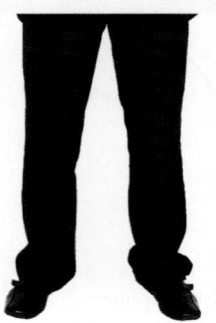

$$f(x) = x^3 + x + 1$$

FIGURE 3.25

✓ **Checkpoint 1**

Find the rational zeros of

$$f(x) = x^3 + 2x^2 + 2.$$

Example 2   Rational Zero Test with Leading Coefficient of 1

Find the rational zeros of

$$f(x) = x^4 - x^3 + x^2 - 3x - 6.$$

SOLUTION   Because the leading coefficient is 1, the possible rational zeros are the factors of the constant term.

Possible rational zeros: $\pm 1, \pm 2, \pm 3, \pm 6$

Test each possible rational zero. The test shows $x = -1$ and $x = 2$ are the only two rational zeros of the function.

✓**Checkpoint 2**

Find the rational zeros of

$$f(x) = x^4 + 2x^3 + x^2 - 4.$$ ∎

When the leading coefficient of a polynomial is not 1, the list of possible rational zeros can increase dramatically. In such cases, the search can be shortened in several ways.

- A programmable calculator can be used to speed up the calculations.

- A graph, created either by hand or with a graphing utility, can give a good estimate of the locations of the zeros.

- Synthetic division can be used to test the possible rational zeros.

TECH TUTOR

There are several ways to use your graphing utility to locate the zeros of a polynomial function after listing the possible rational zeros. You can use the *table* feature by setting the increments of x to the smallest difference between possible rational zeros, or use the *table* feature in ASK mode. In either case, the value in the function column will be 0 when x is a zero of the function. Another way to locate zeros is to graph the function. Be sure that your viewing window contains all the possible rational zeros.

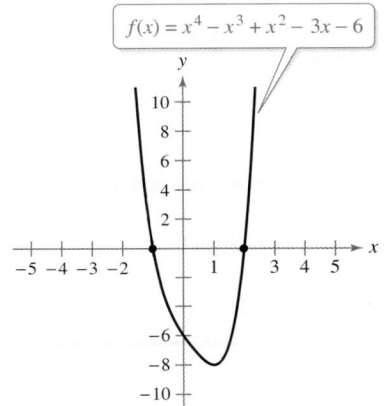

$$f(x) = x^4 - x^3 + x^2 - 3x - 6$$

FIGURE 3.26

To see how to use synthetic division to test the possible rational zeros, let's take another look at the function given by

$$f(x) = x^4 - x^3 + x^2 - 3x - 6$$

from Example 2. To test that $x = -1$ and $x = 2$ are zeros of f, you can apply synthetic division, as shown.

$$
\begin{array}{r|rrrrr}
-1 & 1 & -1 & 1 & -3 & -6 \\
   &   & -1 & 2 & -3 & 6 \\
\hline
   & 1 & -2 & 3 & -6 & 0
\end{array}
\qquad
\begin{array}{r|rrrrr}
2 & 1 & -2 & 3 & -6 \\
  &   & 2 & 0 & 6 \\
\hline
  & 1 & 0 & 3 & 0
\end{array}
$$

So, you have

$$f(x) = (x + 1)(x - 2)(x^2 + 3).$$

Because the factor $(x^2 + 3)$ produces no real zeros, you can conclude that $x = -1$ and $x = 2$ are the only *real* zeros of f. This is verified in the graph of f shown in Figure 3.26.

Finding the first zero is often the hardest part. After that, the search is simplified by using the lower-degree polynomial obtained in synthetic division. Once the lower-degree polynomial is quadratic, either factoring or the Quadratic Formula can be used to find the remaining zeros.

Example 3   Using the Rational Zero Test

Find the rational zeros of

$$f(x) = 2x^3 + 3x^2 - 8x + 3.$$

SOLUTION   The leading coefficient is 2 and the constant term is 3.

$$Possible\ rational\ zeros:\ \frac{\text{Factors of 3}}{\text{Factors of 2}} = \frac{\pm 1, \pm 3}{\pm 1, \pm 2} = \pm 1, \pm 3, \pm\frac{1}{2}, \pm\frac{3}{2}$$

By synthetic division, you can determine that $x = 1$ is a rational zero.

$$
\begin{array}{r|rrrr}
1 & 2 & 3 & -8 & 3 \\
  &   & 2 & 5 & -3 \\
\hline
  & 2 & 5 & -3 & 0
\end{array}
$$

So, $f(x)$ factors as

$$f(x) = (x - 1)(2x^2 + 5x - 3)$$
$$= (x - 1)(2x - 1)(x + 3)$$

and you can conclude that the rational zeros of f are $x = 1$, $x = \frac{1}{2}$, and $x = -3$.

✓**Checkpoint 3**

Find the rational zeros of $f(x) = 2x^3 + 5x^2 + x - 2$. ∎

Example 4   Using the Rational Zero Test

Find all the real zeros of

$$f(x) = 10x^3 - 15x^2 - 16x + 12.$$

SOLUTION   The leading coefficient is 10 and the constant term is 12.

$$Possible\ rational\ zeros:\ \frac{\text{Factors of 12}}{\text{Factors of 10}} = \frac{\pm 1, \pm 2, \pm 3, \pm 4, \pm 6, \pm 12}{\pm 1, \pm 2, \pm 5, \pm 10}$$

There are 32 possibilities. So, it is worth your time to stop and sketch a graph. From Figure 3.27, it looks like three reasonable choices would be $x = -\frac{6}{5}$, $x = \frac{1}{2}$, and $x = 2$. Testing these by synthetic division shows that only $x = 2$ checks. So, you have

$$f(x) = (x - 2)(10x^2 + 5x - 6).$$

Using the Quadratic Formula, you find that the two additional zeros are irrational numbers.

$$x = \frac{-5 + \sqrt{265}}{20} \approx 0.5639 \quad \text{and} \quad x = \frac{-5 - \sqrt{265}}{20} \approx -1.0639$$

You can conclude that the real zeros of f are $x = 2$, $x \approx 0.5639$, and $x \approx -1.0639$.

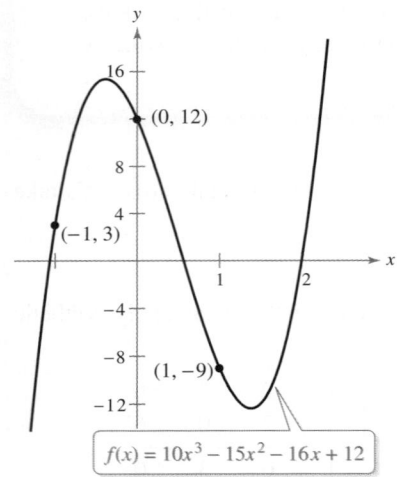

$f(x) = 10x^3 - 15x^2 - 16x + 12$

FIGURE 3.27

✓**Checkpoint 4**

Find the rational zero of $f(x) = 3x^3 + 2x^2 - 5x + 6$. ∎

The Intermediate Value Theorem

The next theorem, called the **Intermediate Value Theorem,** tells you of the existence of real zeros of polynomial functions. The theorem implies that if

$$(a, f(a)) \quad \text{and} \quad (b, f(b))$$

are two points on the graph of a polynomial function such that $f(a) \neq f(b)$, then for any number d between $f(a)$ and $f(b)$ there must be a number c between a and b such that $f(c) = d$. (See Figure 3.28.)

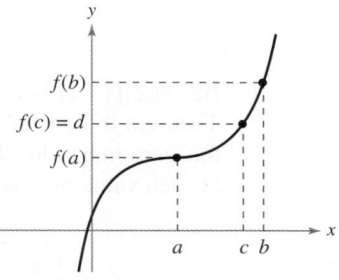

FIGURE 3.28

Intermediate Value Theorem

Let a and b be real numbers such that $a < b$. If f is a polynomial function such that $f(a) \neq f(b)$, then, in the interval $[a, b]$, f takes on every value between $f(a)$ and $f(b)$.

The Intermediate Value Theorem helps you locate the real zeros of a polynomial function in the following way. If you can find a value $x = a$ where a polynomial function is positive, and another value $x = b$ where it is negative, then you can conclude that the function has at least one real zero between these two values. For example, the function given by

$$f(x) = x^3 + x^2 + 1$$

is negative when $x = -2$ and positive when $x = -1$. So, it follows from the Intermediate Value Theorem that f must have a real zero somewhere between -2 and -1, as shown in Figure 3.29.

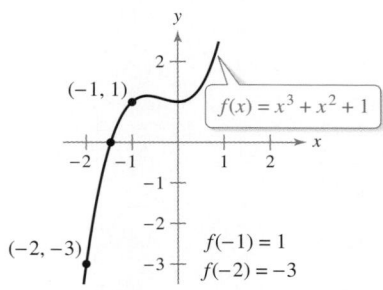

The function f must have a real zero somewhere between -2 and -1.

FIGURE 3.29

By continuing this line of reasoning, you can approximate any real zeros of a polynomial function to any desired level of accuracy. This concept is further demonstrated in Example 5.

Example 5   **Using the Intermediate Value Theorem**

Use the Intermediate Value Theorem to approximate a real zero of

$$f(x) = x^3 - x^2 + 1.$$

SOLUTION   Compute a few function values.

x	-2	-1	0	1
$f(x)$	-11	-1	1	1

Because $f(-1)$ is negative and $f(0)$ is positive, you can apply the Intermediate Value Theorem to conclude that the function has a zero between -1 and 0. To approximate this zero more closely, divide the interval $[-1, 0]$ into tenths and evaluate the function at each value of x. A few function values in the interval $[-1, 0]$ are shown below.

x	-0.9	-0.8	-0.7	-0.6	-0.5
$f(x)$	-0.539	-0.152	0.167	0.424	0.625

x	-0.4	-0.3	-0.2	-0.1
$f(x)$	0.776	0.883	0.952	0.989

You can see that

$$f(-0.8) = -0.152$$

and

$$f(-0.7) = 0.167.$$

So, f must have a zero between -0.8 and -0.7, as shown in Figure 3.30. By continuing this process, you can approximate this zero to any desired level of accuracy.

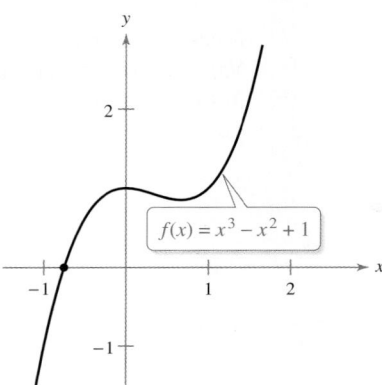

$f(x) = x^3 - x^2 + 1$

f has a zero between -0.8 and -0.7.
FIGURE 3.30

✓**Checkpoint 5**

Use the Intermediate Value Theorem to approximate a real zero of

$$f(x) = x^3 + x - 4.$$

Approximating Zeros Using a Graphing Utility

In this section, you will study two techniques for approximating the zeros of a polynomial function using a graphing utility. The first is called the **zoom-and-trace** technique.

STUDY TIP

To help you visually determine when you have zoomed in enough times to reach the desired level of accuracy, set the X-scale of the viewing window to the accuracy you need and zoom in repeatedly. For instance, to approximate the zero to the nearest hundredth, set the X-scale to 0.01.

Zoom-and-Trace Technique

To approximate a real zero of a function with a graphing utility, use the following steps.

1. Graph the function so that the real zero you want to approximate appears as an x-intercept on the screen.

2. Move the cursor near the x-intercept and use the *zoom* feature to zoom in to get a better look at the intercept.

3. Use the *trace* feature to find the x-values that occur just before and just after the x-intercept. If the difference between these values is sufficiently small, then use their average as the approximation. If not, continue zooming in until the approximation reaches the desired level of accuracy.

The amount that a graphing utility zooms in is determined by the *zoom factor*. The zoom factor is a positive number greater than or equal to 1 that gives the ratio of the larger screen to the smaller screen. For instance, if you zoom in with a zoom factor of 2, you will obtain a screen in which the x- and y-values are half their original values. This text uses a zoom factor of 4.

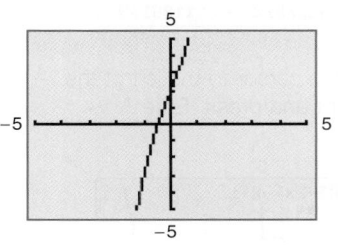

(a)

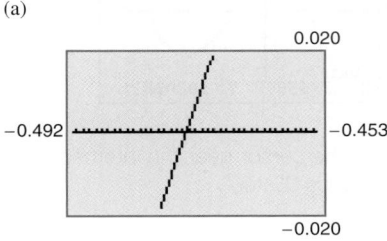

(b)
FIGURE 3.31

Example 6 Approximating a Zero of a Polynomial Function

Approximate a real zero of $f(x) = x^3 + 4x + 2$ to the nearest thousandth.

SOLUTION To begin, use a graphing utility to graph the function, as shown in Figure 3.31(a). Set the X-scale to 0.001 and zoom in several times until the tick marks on the x-axis become visible. The final screen should be similar to the one shown in Figure 3.31(b). At this point, you can use the *trace* feature to determine that the x-values just to the left and right of the x-intercept are

$$x \approx -0.4735 \quad \text{and} \quad x \approx -0.4733.$$

So, to the nearest thousandth, you can approximate the zero of the function to be

$$x \approx -0.473.$$

✓ Checkpoint 6

Approximate a real zero of $f(x) = 2x^3 - x + 3$ to the nearest thousandth.

STUDY TIP

Remember that when substituting a real zero that is an approximation into the original function, you usually obtain a result that is not exactly zero. For instance, in Example 6,

$$f(-0.473) \approx 0.002.$$

In Example 6, the cubic polynomial function has only one real zero. Remember that functions can also have two or more real zeros. In such cases, you can use the zoom-and-trace technique for each zero separately. For instance, the function given by $f(x) = x^3 - 4x^2 + x + 2$ has three real zeros, as shown in Figure 3.32. Using the zoom-and-trace technique for each real zero, you can approximate the real zeros to be

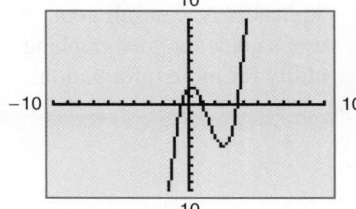

FIGURE 3.32

$$x \approx -0.562, \quad x = 1.000, \quad \text{and} \quad x \approx 3.562.$$

The second technique that can be used with some graphing utilities is to employ the graphing utility's *zero* or *root* feature. The name of this feature differs with different calculators. Consult your user's guide to determine if this feature is available.

Example 7   Approximating the Zeros of a Polynomial Function

Approximate the real zeros of

$$f(x) = x^3 - 2x^2 - x + 1.$$

SOLUTION   To begin, use a graphing utility to graph the function, as shown in the first screen in Figure 3.33. Notice that the graph has three *x*-intercepts. To approximate the leftmost intercept, find an appropriate viewing window and use the zero feature, as shown below. The calculator should display an approximation of $x \approx -0.8019377$, which is accurate to seven decimal places.

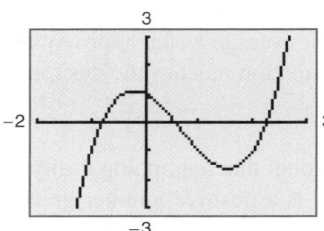

Find an appropriate viewing window, and then use the zero feature.

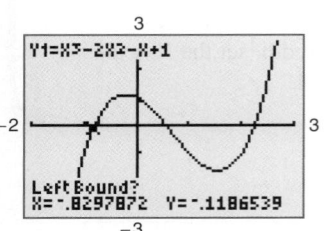

Move the cursor to the left of the intercept and press "Enter."

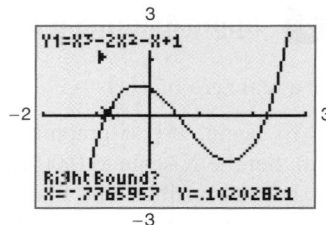

Move the cursor to the right of the intercept and press "Enter."

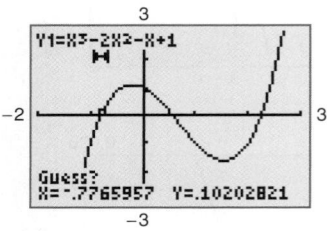

Move the cursor near the intercept and press "Enter."

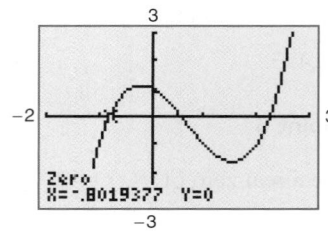

FIGURE 3.33

By repeating this process, you can determine that the other two zeros are $x \approx 0.555$ and $x \approx 2.247$.

✓ Checkpoint 7

Approximate the real zeros of

$$f(x) = x^3 - 4x^2 + 3x + 1.$$

TECH TUTOR

For instructions on how to use the *zoom*, *trace*, *zero*, and *root* features, see Appendix A. Consult your user's guide for your graphing utility for more information.

You may be wondering why so much time is spent in algebra trying to find the zeros of a function. The reason is that if you have a technique that will enable you to solve the equation $f(x) = 0$, you can use the same technique to solve the more general equation

$$f(x) = c$$

where c is any real number. This procedure is demonstrated in Example 8.

Example 8 ■ Solving the Equation $f(x) = c$

Find a value of x such that $f(x) = 30$ for the function given by

$$f(x) = x^3 - 4x + 4.$$

SOLUTION The graph of $f(x) = x^3 - 4x + 4$ is shown in Figure 3.34. Note from the graph that $f(x) = 30$ when x is about 3.5.

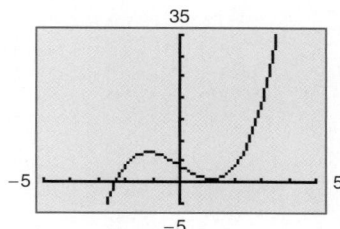

FIGURE 3.34

To use the zoom-and-trace technique to approximate this x-value more closely, consider the equation

$$x^3 - 4x + 4 = 30$$
$$x^3 - 4x - 26 = 0.$$

So, the *solutions* of the equation $f(x) = 30$ are precisely the same x-values as the *zeros* of

$$g(x) = x^3 - 4x - 26.$$

Using the graph of g, as shown in Figure 3.35, you can approximate the zero of g to be $x \approx 3.41$. You can check this value by substituting $x = 3.41$ into the original function.

$$f(3.41) = (3.41)^3 - 4(3.41) + 4$$
$$\approx 30.01 \ \checkmark$$

Remember that with decimal approximations, a check usually will not produce an exact value.

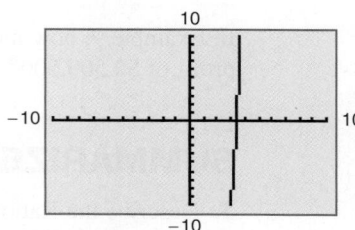

FIGURE 3.35

✓ Checkpoint 8

Find a value of x such that $f(x) = 20$ for the function given by

$$f(x) = x^3 + 4x^2 - 1.$$

Application

 Example 9 Profit and Advertising Expenses

A company that produces sports clothes estimates that the profit from selling a particular line of sportswear is given by

$$P = -0.014x^3 + 0.752x^2 - 40, \quad 0 \le x \le 50$$

where P is the profit (in tens of thousands of dollars) and x is the advertising expense (in tens of thousands of dollars). According to this model, how much money should the company spend on advertising to obtain a profit of $2,750,000?

SOLUTION From Figure 3.36, it appears that there are two different values of x between 0 and 50 that will produce a profit of $2,750,000.

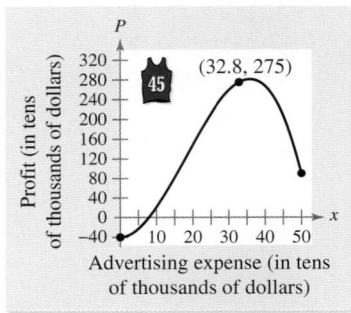

FIGURE 3.36

However, because of the context of the problem, it is clear that the better answer is the smaller of the two numbers. So, to solve the equation

$$-0.014x^3 + 0.752x^2 - 40 = 275$$

$$-0.014x^3 + 0.752x^2 - 315 = 0$$

find the zeros of the function $g(x) = -0.014x^3 + 0.752x^2 - 315$. Using the zoom-and-trace technique, you can find that the leftmost zero is $x \approx 32.8$. You can check this solution by substituting $x = 32.8$ into the original function.

$$P = -0.014(32.8)^3 + 0.752(32.8)^2 - 40 \approx 275$$

The company should spend about $328,000 on advertising for the line of sportswear.

✓**Checkpoint 9**

In Example 9, how much money should the company spend on advertising to obtain a profit of $2,500,000?

SUMMARIZE (Section 3.4)

1. Describe the Rational Zero Test *(page 282)*. For examples of using the Rational Zero Test, see Examples 1, 2, 3, and 4.

2. State the Intermediate Value Theorem *(page 285)*. For an example of using the Intermediate Value Theorem, see Example 5.

3. Describe the zoom-and-trace technique *(page 287)*. For an example of using the zoom-and-trace technique, see Example 6.

SKILLS WARM UP 3.4   The following warm-up exercises involve skills that were covered in earlier sections. You will use these skills in the exercise set for this section. For additional help, review Sections 1.5 and 3.3.

In Exercises 1 and 2, find a polynomial function with integer coefficients having the given zeros.

1. $-1, \frac{2}{3}, 3$

2. $-2, 0, \frac{3}{4}, 2$

In Exercises 3 and 4, use synthetic division to divide.

3. $\dfrac{x^5 - 9x^3 + 5x + 18}{x + 3}$

4. $\dfrac{3x^4 + 17x^3 + 10x^2 - 9x - 8}{x + \frac{2}{3}}$

In Exercises 5–8, use the given zero to find all the real zeros of f.

5. $f(x) = 2x^3 + 11x^2 + 2x - 4, \ x = \frac{1}{2}$

6. $f(x) = 6x^3 - 47x^2 - 124x - 60, \ x = 10$

7. $f(x) = 4x^3 - 13x^2 - 4x + 6, \ x = -\frac{3}{4}$

8. $f(x) = 10x^3 + 51x^2 + 48x - 28, \ x = \frac{2}{5}$

In Exercises 9 and 10, find all real solutions of the equation.

9. $x^4 - 3x^2 + 2 = 0$

10. $x^4 - 7x^2 + 12 = 0$

Exercises 3.4

See www.CalcChat.com for worked-out solutions to odd-numbered exercises.

Rational Zero Test with Leading Coefficient of 1   In Exercises 1–6, use the Rational Zero Test to list all possible rational zeros of f. Test each possible rational zero to determine whether it is an actual zero of f. *See Examples 1 and 2.*

1. $f(x) = x^3 + 2x^2 - 1$

2. $f(x) = x^3 + 3x^2 - x - 3$

3. $f(x) = x^3 - 4x^2 + 1$

4. $f(x) = x^4 + x^3 + 2x^2 - 1$

5. $f(x) = x^3 - 7x - 6$

6. $f(x) = x^4 + 2x^3 + 3x^2 + 2x - 8$

 Using the Rational Zero Test   In Exercises 7 and 8, use the Rational Zero Test to list all possible rational zeros of f. Then use a graphing utility to graph the function. Use the graph to help determine which of the possible rational zeros are actual zeros of the function.

7. $f(x) = x^3 + x^2 - 4x - 4$ **8.** $f(x) = 2x^4 - x^2 - 6$

Using the Rational Zero Test   In Exercises 9–12, use the Rational Zero Test to list all possible rational zeros of f. Use synthetic division to determine which of the possible rational zeros are actual zeros of the function. *See Example 3.*

9. $f(x) = x^3 - 9x^2 + 20x - 12$

10. $f(x) = x^3 - 9x^2 + 27x - 27$

11. $f(x) = 2x^3 - 3x^2 - 11x + 6$

12. $f(x) = 5x^3 - 7x^2 - 58x + 24$

Finding Rational Zeros   In Exercises 13–16, find the rational zeros of the polynomial function. *See Examples 1, 2, and 3.*

13. $f(x) = x^3 - \frac{3}{2}x^2 - \frac{23}{2}x + 6$

14. $f(x) = x^3 + 3x^2 - x - 3$

15. $f(x) = 4x^4 - 17x^2 + 4$

16. $f(x) = -2x^4 + 13x^3 - 21x^2 + 2x + 8$

Using the Rational Zero Test   In Exercises 17–24, find all the real zeros of the function. *See Example 4.*

17. $f(x) = x^3 - 6x^2 + 11x - 6$

18. $g(x) = x^3 - 4x^2 - x + 4$

19. $h(t) = t^3 + 12t^2 + 21t + 10$

20. $f(x) = x^3 - 4x^2 + 5x - 2$

21. $C(x) = 2x^3 + 3x^2 - 1$

22. $f(x) = 3x^3 - 19x^2 + 33x - 9$

23. $f(x) = x^4 - 11x^2 + 18$

24. $P(t) = t^4 - 19t^2 + 48$

Finding Real Solutions   In Exercises 25–30, find all real solutions of the polynomial equation.

25. $z^4 - z^3 - 2z - 4 = 0$

26. $x^4 - 13x^2 - 12x = 0$

27. $2y^4 + 7y^3 - 26y^2 + 23y - 6 = 0$

28. $2x^4 - 11x^3 - 6x^2 + 64x + 32 = 0$

29. $x^5 - x^4 - 3x^3 + 5x^2 - 2x = 0$

30. $x^5 - 7x^4 + 10x^3 + 14x^2 - 24x = 0$

Using the Rational Zero Test   In Exercises 31 and 32, (a) list all possible rational zeros of f, (b) sketch the graph of f so that some of the possible zeros in part (a) can be discarded, and (c) determine all the real zeros of f. *See Example 4.*

31. $f(x) = 32x^3 - 52x^2 + 17x + 3$

32. $f(x) = 4x^3 + 7x^2 - 11x - 18$

Using the Intermediate Value Theorem   In Exercises 33–36, use the Intermediate Value Theorem to show that the function has at least one zero in the interval $[a, b]$. (You do not have to approximate the zero.) *See Example 5.*

33. $f(x) = x^3 + 2x - 5$, $[1, 2]$

34. $f(x) = x^5 - 3x + 3$, $[-2, -1]$

35. $f(x) = x^4 - 3x^2 - 10$, $[2, 3]$

36. $f(x) = -x^3 + 2x^2 + 7x - 3$, $[3, 4]$

Using the Intermediate Value Theorem   In Exercises 37–40, use the Intermediate Value Theorem to approximate the zero of f in the interval $[a, b]$. Give your approximation to the nearest tenth. *See Example 5.*

37. $f(x) = x^3 + x - 1$, $[0, 1]$

38. $f(x) = x^5 + x + 1$, $[-1, 0]$

39. $f(x) = x^4 - 10x^2 - 11$, $[3, 4]$

40. $f(x) = -x^3 + 3x^2 + 9x - 2$, $[4, 5]$

Matching   In Exercises 41–46, match the function with its graph. Then approximate the real zeros of the function to three decimal places. [The graphs are labeled (a), (b), (c), (d), (e), and (f).]

(a)
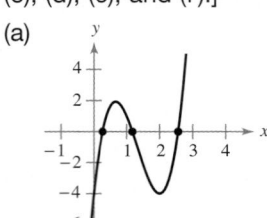

(b)
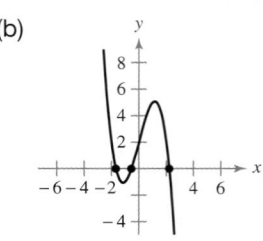

(c)
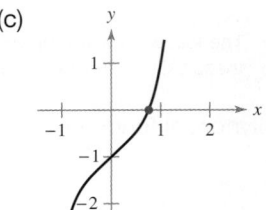

(d)
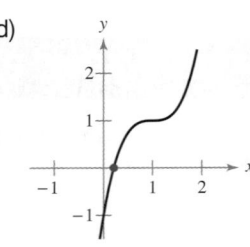

(e)
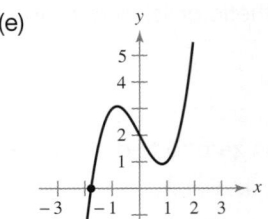

(f)
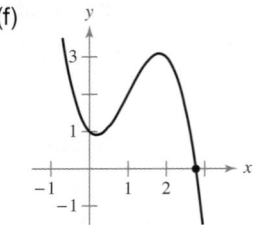

41. $f(x) = x^3 - 2x + 2$

42. $f(x) = x^5 + x - 1$

43. $f(x) = 2x^3 - 6x^2 + 6x - 1$

44. $f(x) = 5x^3 - 20x^2 + 20x - 4$

45. $f(x) = -x^3 + 3x^2 - x + 1$

46. $f(x) = -x^3 + 4x + 2$

 Approximating Zeros of a Polynomial Function In Exercises 47–50, use the *zoom* and *trace* features of a graphing utility to approximate the real zeros of f. Give your approximations to the nearest thousandth. *See Example 6.*

47. $f(x) = x^4 - x - 3$ **48.** $f(x) = 4x^3 + 14x - 8$

49. $f(x) = x^3 - 3.9x^2 + 4.79x - 1.881$

50. $f(x) = -x^3 + 2x^2 + 4x + 5$

 Approximating Zeros of a Polynomial Function In Exercises 51–54, use the *zero* or *root* feature of a graphing utility to approximate the real zeros of f. Give your approximations to the nearest thousandth. *See Example 7.*

51. $f(x) = x^4 + x - 3$

52. $f(x) = -x^4 + 2x^3 + 4$

53. $f(x) = 7x^4 - 42x^3 + 43x^2 + 216x - 324$

54. $f(x) = 3x^4 - 12x^3 + 27x^2 + 4x - 4$

Rational and Irrational Zeros   In Exercises 55–58, match the cubic function with the numbers of rational and irrational zeros.

(a) Rational zeros: 0; Irrational zeros: 1

(b) Rational zeros: 3; Irrational zeros: 0

(c) Rational zeros: 1; Irrational zeros: 2

(d) Rational zeros: 1; Irrational zeros: 0

55. $f(x) = x^3 - 1$ **56.** $f(x) = x^3 - 2$

57. $f(x) = x^3 - x$ **58.** $f(x) = x^3 - 2x$

59. Dimensions of a Box   An open box is to be made from a rectangular piece of material, 18 inches by 15 inches, by cutting equal squares from the corners and turning up the sides (see figure).

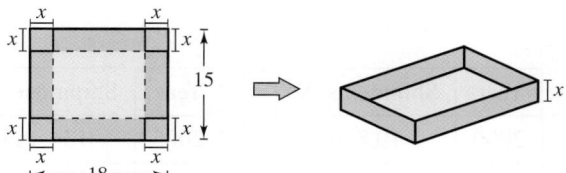

(a) Write the volume V of the box as a function of x. Determine the domain of the function.

(b) Sketch the graph of the function and approximate the dimensions of the box that yield a maximum volume.

(c) Find values of x such that $V = 108$. Which of these values is a physical impossibility in the construction of the box? Explain.

(d) What value of x should you use to make the tallest possible box with a volume of 108 cubic inches?

60. Dimensions of a Box   An open box is to be made from a rectangular piece of material, 16 inches by 12 inches, by cutting equal squares from the corners and turning up the sides (see figure).

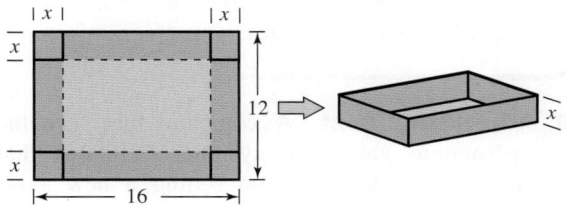

(a) Write the volume V of the box as a function of x. Determine the domain of the function.

(b) Sketch the graph of the function and approximate the dimensions of the box that yield a maximum volume.

(c) Find values of x such that $V = 120$. Which of these values is a physical impossibility in the construction of the box? Explain.

(d) What value of x should you use to make the tallest possible box with a volume of 120 cubic inches?

61. Dimensions of a Terrarium   A rectangular terrarium with a square cross section has a combined length and girth (perimeter of a cross section) of 108 inches (see figure). Find the dimensions of the terrarium, given that the volume is 11,664 cubic inches.

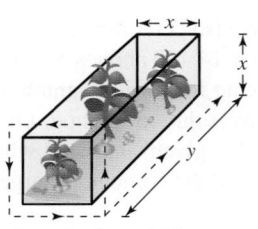

Figure for 61 and 62

62. Dimensions of a Terrarium   A rectangular terrarium has a combined length and girth (perimeter of a cross section) of 120 inches (see figure). Find the dimensions of the terrarium, given that the volume is 16,000 cubic inches.

63. Geometry   A bulk food storage bin with dimensions 2 feet by 3 feet by 4 feet needs to be increased in size to hold five times as much food as the current bin. (Assume each dimension is increased by the same amount.)

(a) Write a function that represents the volume V of the new bin.

(b) Find the dimensions of the new bin.

64. Medical Lab   A medical examiner wants to enlarge an existing medical lab such that the total area of the new lab is 1.5 times that of the original lab. The original lab's dimensions are 120 feet by 75 feet. The medical examiner wants to increase each dimension by the same amount.

(a) Write a function that represents the area A of the new lab.

(b) Find the dimensions of the new lab.

(c) The medical examiner wants to add a length to the sides of the lab that are 120 feet, and twice this length to the sides that are 75 feet, such that the total area of the new lab is 1.5 times that of the original lab. Repeat parts (a) and (b). Explain your results.

65. Medicine   The concentration C of a chemical in the bloodstream t hours after injection into muscle tissue is given by

$$C = \frac{3t^2 + t}{t^3 + 50}, \quad t \geq 0.$$

The concentration is greatest when

$$3t^4 + 2t^3 - 300t - 50 = 0.$$

Approximate this time to the nearest hundredth of an hour.

66. Transportation Cost   The transportation cost C (in thousands of dollars) of the components used in manufacturing prefabricated homes is given by

$$C = 100\left(\frac{200}{x^2} + \frac{x}{x + 30}\right), \quad x \geq 1$$

where x is the order size (in hundreds). The cost is a minimum when

$$3x^3 - 40x^2 - 2400x - 36,000 = 0.$$

Approximate the optimal order size to the nearest unit.

67. Online Sales The revenues per share R (in dollars) for Amazon.com for the years 2001 through 2010 are shown in the table. *(Source: Amazon.com)*

Year	Revenue per share, R	Year	Revenue per share, R
2001	8.37	2006	25.87
2002	10.14	2007	35.66
2003	13.05	2008	44.78
2004	17.16	2009	55.20
2005	20.41	2010	75.84

(a) Use a graphing utility to create a scatter plot of the data. Let t represent the year, with $t = 1$ corresponding to 2001.

(b) Use the *regression* feature of the graphing utility to find a linear model, a quadratic model, a cubic model, and a quartic model for the data.

(c) Use the graphing utility to graph each model separately with the data in the same viewing window. How well does each model fit the data?

(d) Use each model to predict the year in which the revenue per share will be about $100. Explain any differences in the predictions.

68. Advertising Cost A company that produces video games estimates that the profit P (in dollars) from selling a new game is given by

$$P = -82x^3 + 7250x^2 - 450{,}000, \quad 0 \le x \le 80$$

where x is the advertising expense (in tens of thousands of dollars). Using this model, how much money should the company spend on advertising to obtain a profit of $5,900,000?

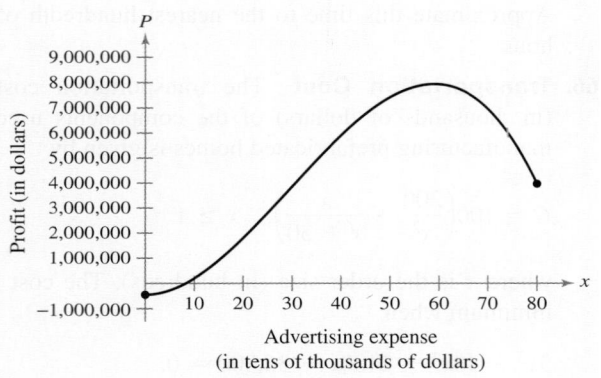

Advertising expense
(in tens of thousands of dollars)

69. Solar Energy Photovoltaic cells convert light energy into electricity. The domestic shipments S (in peak kilowatts) of photovoltaic cells and modules for the years 2000 through 2009 are shown in the table. *(Source: U.S. Energy Information Administration)*

Year	Shipments, S	Year	Shipments, S
2000	19,838	2005	134,465
2001	36,310	2006	206,511
2002	45,313	2007	280,475
2003	48,664	2008	524,252
2004	78,346	2009	601,133

(a) Use a spreadsheet software program to create a scatter plot of the data. Let t represent the year, with $t = 0$ corresponding to 2000.

(b) Use the *regression* feature of the spreadsheet software program to find a quadratic model and a cubic model for the data.

(c) Use each model to predict the year in which the shipments will be about 1,000,000 peak kilowatts. Then discuss the appropriateness of each model for predicting future values.

70. Advertising Cost A company that manufactures hydroponic gardening systems estimates that the profit P (in dollars) from selling a new system is given by

$$P = -35x^3 + 2700x^2 - 300{,}000, \quad 0 \le x \le 70$$

where x is the advertising expense (in tens of thousands of dollars). Using this model, how much money should the company spend on advertising to obtain a profit of $1,800,000?

71. Demand Function A company that produces cell phones estimates that the demand D for a new model of phone is given by

$$D = -x^3 + 54x^2 - 140x - 3000, \quad 10 \le x \le 50$$

where x is the price of the phone (in dollars).

(a) Use a graphing utility to graph D. Use the *trace* feature to determine the values of x for which the demand is 14,400 phones.

(b) You may also determine the values of x for which the demand is 14,400 phones by setting D equal to 14,400 and solving for x with a graphing utility. Discuss this alternative solution method. Of the solutions that lie within the given interval, what price would you recommend the company charge for the phones?

72. Demand Function A company that produces handheld organizers estimates that the demand D for a new model of organizer is given by

$$D = -0.005x^3 + 2.65x^2 - 70x - 2500,$$
$$50 \le x \le 500$$

where x is the price of the organizer (in dollars).

(a) Use a graphing utility to graph D. Use the *trace* feature to determine the values of x for which the demand will be 80,000 organizers.

(b) You may also determine the values of x for which the demand will be 80,000 organizers by setting D equal to 80,000 and solving for x with a graphing utility. Discuss this alternative solution method. Of the solutions that lie within the given interval, what price would you recommend the company charge for the new organizers?

73. Height of a Baseball A baseball is launched upward from ground level with an initial velocity of 48 feet per second, and its height h (in feet) is

$$h(t) = -16t^2 + 48t, \quad 0 \le t \le 3$$

where t is the time (in seconds). You are told the ball reaches a height of 64 feet. Is this possible?

74. Exploration Use a graphing utility to graph the function $f(x) = x^4 - 4x^2 + k$ for different values of k. Find the values of k such that the zeros of f satisfy the specified characteristics. (Some parts do not have unique answers.)

(a) Four real zeros

(b) Two real zeros and two complex roots

75. Reasoning Is it possible that a second-degree polynomial function with integer coefficients has one rational zero and one irrational zero? If so, give an example.

76. Graphical Reasoning The graph of one of the following functions is shown below. Identify the function shown in the graph. Explain why each of the others is not the correct function. Use a graphing utility to verify your result.

(a) $f(x) = x^2(x + 2)(x - 3.5)$

(b) $g(x) = (x + 2)(x - 3.5)$

(c) $h(x) = (x + 2)(x - 3.5)(x^2 + 1)$

(d) $k(x) = (x + 1)(x + 2)(x - 3.5)$

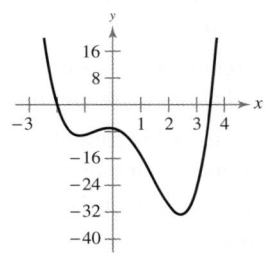

77. Reasoning Is it possible that a third-degree polynomial function with integer coefficients has one rational zero and two irrational zeros? If so, give an example.

78. HOW DO YOU SEE IT? Use the information in the table.

Interval	Value of $f(x)$
$(-\infty, -2)$	Positive
$(-2, 1)$	Negative
$(1, 4)$	Negative
$(4, \infty)$	Positive

(a) What are the three real zeros of the polynomial function f?

(b) What can be said about the behavior of the graph of f at $x = 1$?

(c) What is the least possible degree of f? Explain. Can the degree of f ever be odd? Explain.

(d) Is the leading coefficient of f positive or negative? Explain.

(e) Write an equation for f. (There are many correct answers.)

(f) Sketch a graph of the equation you wrote in part (e).

79. Project: Sales For a project involving the sales of the DVD *The Other Guys* for the first 8 weeks of its release, visit this text's website at *www.cengagebrain.com.* *(Source: Nash Information Services)*

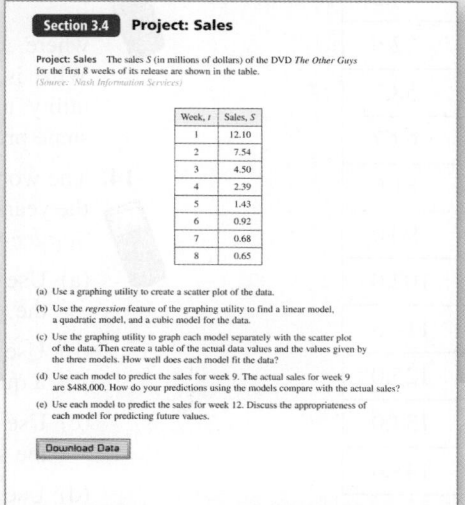

QUIZ YOURSELF

See www.CalcChat.com for worked-out solutions to odd-numbered exercises.

Take this quiz as you would take a quiz in class. When you are done, check your work against the answers given in the back of the book.

In Exercises 1 and 2, sketch the graph of the quadratic function. Identify the vertex and the intercepts.

1. $f(x) = (x + 1)^2 - 2$

2. $f(x) = 25 - x^2$

In Exercises 3 and 4, describe the right-hand and left-hand behavior of the graph of the polynomial function. Verify with a graphing utility.

3. $f(x) = -2x^3 + 7x^2 - 9$

4. $f(x) = x^4 + 7x^2 - 8$

5. Use synthetic division to evaluate $f(x) = 2x^4 + x^3 + 18x^2 - 4$ when $x = -3$.

In Exercises 6 and 7, write the function in the form $f(x) = (x - k)q(x) + r$ for the given value of k, and demonstrate that $f(k) = r$.

6. $f(x) = x^4 - 5x^2 + 4, \quad k = -1$

7. $f(x) = x^3 + 5x^2 - 2x - 24, \quad k = 2$

8. Simplify $\dfrac{2x^4 + 9x^3 - 32x^2 - 99x + 180}{x^2 + 2x - 15}$.

In Exercises 9–12, find all the real zeros of the function.

9. $f(x) = -2x^3 - 7x^2 + 10x + 35$

10. $f(x) = 4x^4 - 37x^2 + 9$

11. $f(x) = 3x^4 + 4x^3 - 3x - 4$

12. $f(x) = 2x^3 - 3x^2 + 2x - 3$

13. The profit P (in dollars) for a clothing company is

$$P = -95x^3 + 5650x^2 - 250{,}000, \quad 0 \le x \le 55$$

where x is the advertising expense (in tens of thousands of dollars). What is the profit for an advertising expense of \$450,000? Use a graphing utility to approximate another advertising expense that would yield the same profit.

14. The worldwide land areas A (in millions of hectares) of transgenic crops for the years 2000 through 2010 are shown in the table. *(Source: International Service for the Acquisition of Agri-Biotech Applications)*

(a) Use a graphing utility to create a scatter plot of the data. Let t represent the year, with $t = 0$ corresponding to 2000.

(b) Use the *regression* feature of the graphing utility to find a linear model, a quadratic model, a cubic model, and a quartic model for the data.

(c) Use the graphing utility to graph each model separately with the data in the same viewing window. How well does each model fit the data?

(d) Use each model to predict the year in which the land area will be about 170 million hectares. Explain any differences in the predictions.

Year	Area, A
2000	44.2
2001	52.6
2002	58.7
2003	67.7
2004	81.0
2005	90.0
2006	102.0
2007	114.3
2008	125.0
2009	134.0
2010	148.0

Table for 14

3.5 Complex Numbers

- Develop an understanding of the imaginary unit i and complex numbers.
- Perform operations with complex numbers and write the results in standard form.
- Find the complex conjugate of a complex number.
- Solve a polynomial equation that has complex solutions.
- Plot a complex number in the complex plane.

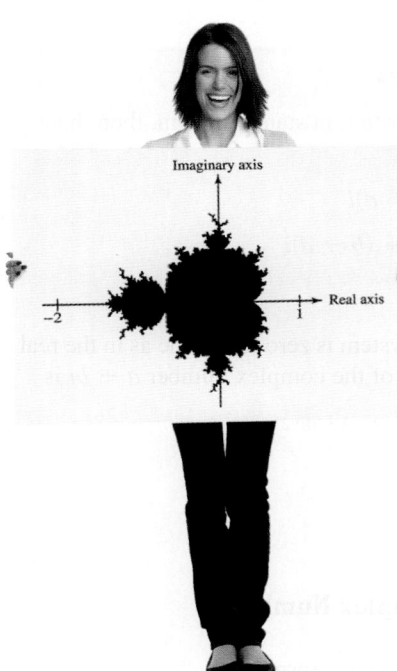

In Exercises 79–84 on page 305, you will determine whether various complex numbers are members of the Mandelbrot Set.

The Imaginary Unit i

Some quadratic equations have no real solutions. For instance, the quadratic equations

$$x^2 + 1 = 0 \quad \text{and} \quad x^2 = -5 \qquad \text{Equations with no real solutions}$$

have no real solutions because there is no real number x that can be squared to produce a negative number. To overcome this deficiency, mathematicians utilized an expanded system of numbers that used the **imaginary unit i,** which is defined as

$$i = \sqrt{-1} \qquad \text{Imaginary unit}$$

where $i^2 = -1$. By adding real numbers to real multiples of this imaginary unit, the set of **complex numbers** is obtained. Each complex number can be written in the **standard form $a + bi$.**

Definition of a Complex Number

Let a and b be real numbers. The number $a + bi$ is called a **complex number,** and it is said to be written in **standard form.** When $b = 0$, the number $a + bi = a$ is a real number. When $b \neq 0$, the number $a + bi$ is called an **imaginary number.** A number of the form bi, where $b \neq 0$, is called a **pure imaginary number.**

The set of real numbers is a subset of the set of complex numbers, as shown in Figure 3.37. This is true because every real number a can be written as a complex number using $b = 0$. That is, for every real number a, you can write $a = a + 0i$.

Complex numbers

Real numbers	Imaginary numbers
$3, -\frac{1}{2},$ $\sqrt{2}, \ 0$	$-2 + i$ Pure imaginary numbers $3i$

FIGURE 3.37

Equality of Complex Numbers

Two complex numbers $a + bi$ and $c + di$ written in standard form are **equal** to each other,

$$a + bi = c + di \qquad \text{Equality of two complex numbers}$$

if and only if $a = c$ and $b = d$.

Operations with Complex Numbers

To add (or subtract) two complex numbers, add (or subtract) the real and imaginary parts of the numbers separately.

Addition and Subtraction of Complex Numbers

If $a + bi$ and $c + di$ are two complex numbers written in standard form, then their sum and difference are defined as shown.

Sum: $(a + bi) + (c + di) = (a + c) + (b + d)i$

Difference: $(a + bi) - (c + di) = (a - c) + (b - d)i$

The **additive identity** in the complex number system is zero (the same as in the real number system). Furthermore, the **additive inverse** of the complex number $a + bi$ is

$$-(a + bi) = -a - bi. \qquad \text{Additive inverse}$$

So, you have

$$(a + bi) + (-a - bi) = 0 + 0i = 0.$$

Example 1   Adding and Subtracting Complex Numbers

Perform the operation(s) and write each result in standard form.

a. $(3 - i) + (2 + 3i)$

b. $2i + (-4 - 2i)$

c. $3 - (-2 + 3i) + (-5 + i)$

SOLUTION

a. $(3 - i) + (2 + 3i) = 3 - i + 2 + 3i$ Remove parentheses.

$= 3 + 2 - i + 3i$ Group like terms.

$= (3 + 2) + (-1 + 3)i$

$= 5 + 2i$ Write in standard form.

b. $2i + (-4 - 2i) = 2i - 4 - 2i$ Remove parentheses.

$= -4 + 2i - 2i$ Group like terms.

$= -4$ Write in standard form.

Note that the sum of two imaginary numbers can be a real number.

c. $3 - (-2 + 3i) + (-5 + i) = 3 + 2 - 3i - 5 + i$

$= 3 + 2 - 5 - 3i + i$

$= 0 - 2i$

$= -2i$

✓ Checkpoint 1

Perform the operation(s) and write each result in standard form.

a. $(4 + 7i) + (1 - 6i)$ **b.** $8i - (-10 + i) - (10 + 6i)$

Many of the properties of real numbers are valid for complex numbers as well. Here are some examples.

Associative Properties of Addition and Multiplication

Commutative Properties of Addition and Multiplication

Distributive Property of Multiplication Over Addition

Notice how these properties are used when two complex numbers are multiplied.

$$(a + bi)(c + di) = a(c + di) + bi(c + di) \qquad \text{Distributive Property}$$
$$= ac + (ad)i + (bc)i + (bd)i^2 \qquad \text{Distributive Property}$$
$$= ac + (ad)i + (bc)i + (bd)(-1) \qquad i^2 = -1$$
$$= ac - bd + (ad)i + (bc)i \qquad \text{Commutative Property}$$
$$= (ac - bd) + (ad + bc)i \qquad \text{Associative Property}$$

Rather than trying to memorize this multiplication rule, you should simply remember how the Distributive Property is used to multiply two complex numbers. The procedure is similar to multiplying two binomials and combining like terms (as in the FOIL Method).

Example 2   Multiplying Complex Numbers

Find each product.

a. $(i)(-6i)$ **b.** $(2 - i)(4 + 3i)$ **c.** $(3 + 2i)(3 - 2i)$ **d.** $(7 + 5i)^2$

SOLUTION

a. $(i)(-6i) = -6i^2$ Multiply.

$\qquad = -6(-1)$ $i^2 = -1$

$\qquad = 6$ Simplify.

b. $(2 - i)(4 + 3i) = 8 + 6i - 4i - 3i^2$ Distributive Property

$\qquad = 8 + 6i - 4i - 3(-1)$ $i^2 = -1$

$\qquad = 8 + 3 + 6i - 4i$ Group like terms.

$\qquad = 11 + 2i$ Write in standard form.

c. $(3 + 2i)(3 - 2i) = 9 - 6i + 6i - 4i^2$ Distributive Property

$\qquad = 9 - 4(-1)$ $i^2 = -1$

$\qquad = 9 + 4$ Simplify.

$\qquad = 13$ Write in standard form.

d. $(7 + 5i)^2 = 49 + 35i + 35i + 25i^2$ Distributive Property

$\qquad = 49 + 70i + 25i^2$ Simplify.

$\qquad = 49 + 70i + 25(-1)$ $i^2 = -1$

$\qquad = 49 - 25 + 70i$ Commutative Property

$\qquad = 24 + 70i$ Write in standard form.

 **Checkpoint 2**

Find each product.

a. $4(-2 + 3i)$ **b.** $(5 - 3i)^2$

Complex Conjugates

Notice in Example 2(c) that the product of two complex numbers can be a real number. This occurs with pairs of complex numbers of the form $a + bi$ and $a - bi$, called **complex conjugates.** In general, the product of two complex conjugates can be written as shown.

$$(a + bi)(a - bi) = a^2 - abi + abi - b^2i^2$$

$$= a^2 - b^2(-1) = a^2 + b^2$$

Complex conjugates can be used to write the quotient of $a + bi$ and $c + di$ in standard form, where c and d are not both zero. To do this, multiply the numerator and denominator by the complex conjugate of the denominator to obtain

$$\frac{a + bi}{c + di} = \frac{a + bi}{c + di}\left(\frac{c - di}{c - di}\right) = \frac{(ac + bd) + (bc - ad)i}{c^2 + d^2}.$$

TECH TUTOR

Some graphing utilities can perform operations with complex numbers. Consult the user's guide for your graphing utility for more information.

Example 3 **Writing Quotients of Complex Numbers in Standard Form**

Write each quotient in standard form.

a. $\dfrac{1}{1 + i}$ **b.** $\dfrac{2 + 3i}{4 - 2i}$

SOLUTION

a. $\dfrac{1}{1 + i} = \dfrac{1}{1 + i}\left(\dfrac{1 - i}{1 - i}\right)$ Multiply numerator and denominator by complex conjugate of denominator.

$= \dfrac{1 - i}{1^2 - i^2}$ Expand.

$= \dfrac{1 - i}{1 - (-1)}$ $i^2 = -1$

$= \dfrac{1 - i}{2}$ Simplify.

$= \dfrac{1}{2} - \dfrac{1}{2}i$ Write in standard form.

b. $\dfrac{2 + 3i}{4 - 2i} = \dfrac{2 + 3i}{4 - 2i}\left(\dfrac{4 + 2i}{4 + 2i}\right)$ Multiply numerator and denominator by complex conjugate of denominator.

$= \dfrac{8 + 4i + 12i + 6i^2}{16 - 4i^2}$ Expand.

$= \dfrac{8 - 6 + 16i}{16 + 4}$ $i^2 = -1$

$= \dfrac{2 + 16i}{20}$ Simplify.

$= \dfrac{1}{10} + \dfrac{4}{5}i$ Write in standard form.

✓ Checkpoint 3

Write $\dfrac{6 - 7i}{1 - 2i}$ in standard form.

Complex Solutions

When using the Quadratic Formula to solve a quadratic equation, you often obtain a result such as $\sqrt{-3}$, which is not a real number. By factoring out $i = \sqrt{-1}$, you can write this number in standard form.

$$\sqrt{-3} = \sqrt{3(-1)} = \sqrt{3}\,\sqrt{-1} = \sqrt{3}\,i$$

The number $\sqrt{3}\,i$ is called the *principal square root* of -3.

> **Principal Square Root of a Negative Number**
>
> If a is a positive number, the **principal square root** of the negative number $-a$ is defined as
>
> $$\sqrt{-a} = \sqrt{a}\,i.$$

Example 4 Writing Complex Numbers in Standard Form

a. $\sqrt{-3}\,\sqrt{-12} = \sqrt{3}\,i\sqrt{12}\,i = \sqrt{36}\,i^2 = 6(-1) = -6$

b. $\sqrt{-48} - \sqrt{-27} = \sqrt{48}\,i - \sqrt{27}\,i = 4\sqrt{3}\,i - 3\sqrt{3}\,i = \sqrt{3}\,i$

c. $\left(-1 + \sqrt{-3}\right)^2 = \left(-1 + \sqrt{3}\,i\right)^2$

$$= (-1)^2 - 2\sqrt{3}\,i + \left(\sqrt{3}\right)^2(i^2)$$

$$= 1 - 2\sqrt{3}\,i + 3(-1)$$

$$= -2 - 2\sqrt{3}\,i$$

✓**Checkpoint 4**

Write $4 + \sqrt{-9}$ in standard form.   ■

Example 5 Complex Solutions of a Quadratic Equation

Solve $3x^2 - 2x + 5 = 0$.

SOLUTION

$$x = \frac{-(-2) \pm \sqrt{(-2)^2 - 4(3)(5)}}{2(3)} \qquad \text{Quadratic Formula}$$

$$= \frac{2 \pm \sqrt{-56}}{6} \qquad \text{Simplify.}$$

$$= \frac{2 \pm 2\sqrt{14}\,i}{6} \qquad \text{Write in } i\text{-form.}$$

$$= \frac{1}{3} \pm \frac{\sqrt{14}}{3}\,i \qquad \text{Write in standard form.}$$

✓**Checkpoint 5**

Solve $x^2 + 3x + 4 = 0$.   ■

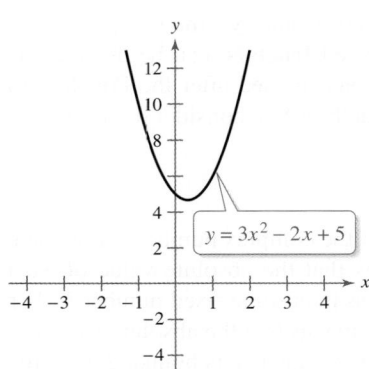

FIGURE 3.38

The graph of $f(x) = 3x^2 - 2x + 5$, shown in Figure 3.38, does not touch or cross the x-axis. This confirms that the equation in Example 5 has no real solution.

Applications

Most applications involving complex numbers are either theoretical (see the next section) or very technical, and so are not appropriate for inclusion in this text. However, to give you some idea of how complex numbers can be used in applications, a general description of their use in **fractal geometry** is presented.

To begin, consider a coordinate system called the **complex plane.** Just as every real number corresponds to a point on the real number line, every complex number corresponds to a point in the complex plane, as shown in Figure 3.39. In this figure, note that the vertical axis is the **imaginary axis** and the horizontal axis is the **real axis.** The point that corresponds to the complex number $a + bi$ is (a, b).

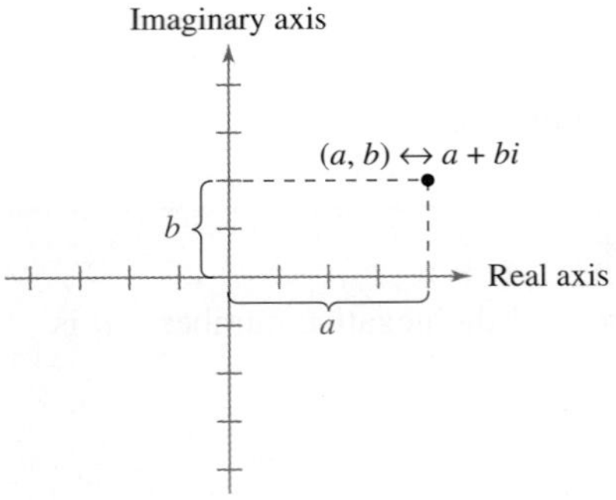

Imaginary axis

$(a, b) \leftrightarrow a + bi$

b

Real axis

a

FIGURE 3.39

Complex number		Ordered pair
$a + bi$	⇨	(a, b)

From Figure 3.39, you can see that i is called the imaginary unit because it is located one unit from the origin on the imaginary axis of the complex plane.

Example 6   Plotting Complex Numbers in the Complex Plane

Plot each complex number in the complex plane.

a. $2 + 3i$

b. $-1 + 2i$

c. 4

SOLUTION

a. To plot the complex number $2 + 3i$, move (from the origin) two units to the right on the real axis and then three units upward. See Figure 3.40. In other words, plotting the complex number $2 + 3i$ in the complex plane is comparable to plotting the point $(2, 3)$ in the Cartesian plane.

b. The complex number $-1 + 2i$ corresponds to the point $(-1, 2)$. See Figure 3.40.

c. The complex number 4 corresponds to the point $(4, 0)$. See Figure 3.40.

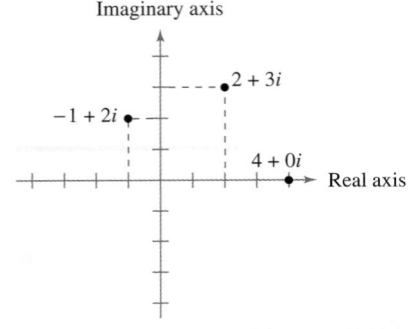

Imaginary axis

$2 + 3i$

$-1 + 2i$

$4 + 0i$

Real axis

FIGURE 3.40

✓Checkpoint 6

Plot each complex number in the complex plane.

a. $-3i$ **b.** $1 + 4i$ **c.** 1

In the hands of a person who understands "fractal geometry," the complex plane can become an easel on which stunning pictures, called **fractals,** can be drawn. The most famous such picture is called the **Mandelbrot Set,** named after the Polish-born mathematician Benoit Mandelbrot. To draw the Mandelbrot Set, consider the following sequence of numbers.

$$c, \quad c^2 + c, \quad (c^2 + c)^2 + c, \quad [(c^2 + c)^2 + c]^2 + c, \quad \ldots$$

The behavior of this sequence depends on the value of the complex number c. For some values of c, this sequence is **bounded,** which means that the absolute value of each number $\left(|a + bi| = \sqrt{a^2 + b^2}\right)$ in the sequence is less than some fixed number N. For other values of c, this sequence is **unbounded,** which means that the absolute values of the terms of the sequence become infinitely large. If the sequence is bounded, then the complex number c is in the Mandelbrot Set. If the sequence is unbounded, then the complex number c is not in the Mandelbrot Set.

| Example 7 | Members of the Mandelbrot Set |

Decide whether each complex number is in the Mandelbrot Set.

a. -2

b. i

c. $1 + i$

SOLUTION

a. For $c = -2$, the corresponding Mandelbrot sequence is

$$-2, \quad 2, \quad 2, \quad 2, \quad 2, \quad 2, \ldots$$

Because the sequence is bounded, the complex number -2 is in the Mandelbrot Set.

b. For $c = i$, the corresponding Mandelbrot sequence is

$$i, \quad -1 + i, \quad -i, \quad -1 + i, \quad -i, \quad -1 + i, \ldots$$

Because the sequence is bounded, the complex number i is in the Mandelbrot Set.

c. For $c = 1 + i$, the corresponding Mandelbrot sequence is

$$1 + i, \quad 1 + 3i, \quad -7 + 7i, \quad 1 - 97i, \quad -9407 - 193i,$$
$$88454401 + 3631103i, \ldots$$

Because the sequence is unbounded, the complex number $1 + i$ is *not* in the Mandelbrot Set.

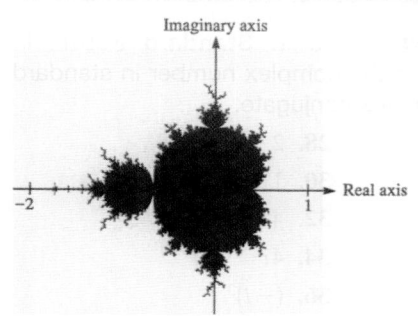

Imaginary axis

Real axis

−2   1

Mandelbrot Set
FIGURE 3.41

✓ **Checkpoint 7**

Decide whether -3 is in the Mandelbrot Set. Explain your reasoning. ■

With this definition, a picture of the Mandelbrot Set would have only two colors: one color for points that are in the set (the sequence is bounded) and one color for points that are outside the set (the sequence is unbounded). Figure 3.41 shows a black and yellow picture of the Mandelbrot Set. The points that are black are in the Mandelbrot Set and the points that are yellow are not.

SUMMARIZE   (Section 3.5)

1. State the definition of a complex number and describe how to add, subtract, and multiply complex numbers *(pages 297–299)*. For examples of performing operations with complex numbers, see Examples 1 and 2.

2. State the definition of a complex conjugate and describe how it is used to find the quotient of two complex numbers *(page 300)*. For an example of writing quotients of complex numbers in standard form, see Example 3.

3. State the definition of the principal square root of a negative number *(page 301)*. For examples that use the principal square root of a negative number, see Examples 4 and 5.

4. Describe how to plot a complex number in the complex plane *(page 302)*. For an example of plotting complex numbers in the complex plane, see Example 6.

SKILLS WARM UP 3.5   The following warm-up exercises involve skills that were covered in earlier sections. You will use these skills in the exercise set for this section. For additional help, review Sections 0.4 and 1.4.

In Exercises 1–8, simplify the expression.

1. $\sqrt{12}$

2. $\sqrt{500}$

3. $\sqrt{20} - \sqrt{5}$

4. $\sqrt{27} - \sqrt{243}$

5. $\sqrt{24}\sqrt{6}$

6. $2\sqrt{18}\sqrt{32}$

7. $\dfrac{1}{\sqrt{7}}$

8. $\dfrac{5}{\sqrt{5}}$

In Exercises 9 and 10, solve the quadratic equation.

9. $x^2 + x - 1 = 0$

10. $x^2 + 2x - 1 = 0$

Exercises 3.5   See www.CalcChat.com for worked-out solutions to odd-numbered exercises.

1. Discovery Write out the first 16 positive integer powers of i (i, i^2, i^3, . . . , i^{16}), and write each as i, $-i$, 1, or -1. What pattern do you observe?

2. Finding Powers of i Use the pattern you found in Exercise 1 to help you write each power of i as i, $-i$, 1, or -1.

   (a) i^{28} (b) i^{37} (c) i^{127} (d) i^{82}

Equality of Complex Numbers In Exercises 3–6, find the real numbers a and b such that the equation is true.

3. $a + bi = 7 + 12i$ **4.** $a + bi = -2 - 5i$

5. $(a + 3) + 2bi = 7 - 4i$

6. $(a + 6) + 2bi = 6 - 5i$

Adding, Subtracting, or Multiplying Complex Numbers In Exercises 7–26, perform the indicated operation(s) and write the result in standard form. *See Examples 1 and 2.*

7. $(-4 + 3i) + (6 - 2i)$

8. $(13 - 2i) + (-5 + 6i)$

9. $(12 + 5i) - (7 - i)$ **10.** $(3 + 2i) - (6 + 13i)$

11. $-\frac{5}{2}i + \left(\frac{5}{3} + \frac{11}{3}i\right)$ **12.** $-\frac{3}{4}i + \left(\frac{3}{5} + \frac{4}{5}i\right)$

13. $(2.7 + 4.3i) + 6.0 - (7.4 - 4.7i)$

14. $9.8 - (1.6 + 3.2i) + (-5.8 + 4.3i)$

15. $(3 + 4i)(3 - 4i)$ **16.** $(8 + 3i)(8 - 3i)$

17. $(2 + 3i)(1 - i)$ **18.** $(6 - 5i)(1 + i)$

19. $5i(4 - 6i)$ **20.** $-2i(7 + 9i)$

21. $(5 + 6i)^2$ **22.** $(3 - 7i)^2$

23. $(3 + 4i)^2 + (3 - 4i)^2$ **24.** $(2 - 5i)^2 - (2 + 5i)^2$

25. $\left(\sqrt{5} - \sqrt{3}\,i\right)\left(\sqrt{5} + \sqrt{3}\,i\right)$

26. $\left(\sqrt{14} + \sqrt{10}\,i\right)\left(\sqrt{14} - \sqrt{10}\,i\right)$

Writing Complex Numbers in Standard Form In Exercises 27–38, write the complex number in standard form and find its complex conjugate.

27. $9 + \sqrt{-16}$ **28.** $2 + \sqrt{-25}$

29. $-3 - \sqrt{-12}$ **30.** $1 - \sqrt{-8}$

31. -21 **32.** 45

33. $-6i + i^2$ **34.** $4i^2 - 2i^3$

35. $-5i^5$ **36.** $(-i)^3$

37. $\left(\sqrt{-6}\right)^2 + 3$ **38.** $\left(\sqrt{-4}\right)^2 - 5$

Writing Quotients of Complex Numbers in Standard Form In Exercises 39–50, write the quotient in standard form. *See Example 3.*

39. $\dfrac{3 - i}{3 + i}$ **40.** $\dfrac{8 - 5i}{1 - 3i}$

41. $\dfrac{5}{4 - 2i}$ **42.** $\dfrac{3}{1 + 2i}$

43. $\dfrac{7 + 10i}{2i}$ **44.** $\dfrac{8 + 15i}{3i}$

45. $\dfrac{1}{(2i)^3}$ **46.** $\dfrac{1}{(3i)^3}$

47. $\dfrac{4}{(1 - 2i)^3}$ **48.** $\dfrac{3}{(5 - 2i)^2}$

49. $\dfrac{(21 - 7i)(4 + 3i)}{2 - 5i}$ **50.** $\dfrac{(3 - i)(2 + 5i)}{4 + 3i}$

Writing Complex Numbers in Standard Form In Exercises 51–58, perform the indicated operation and write the result in standard form. *See Example 4.*

51. $\left(-2 + \sqrt{-8}\right) - \left(5 - \sqrt{-50}\right)$

52. $\left(5 + \sqrt{-18}\right) - \left(3 + \sqrt{-32}\right)$

53. $\sqrt{-3} \cdot \sqrt{-8}$ **54.** $\sqrt{-5} \cdot \sqrt{-10}$

55. $\left(\sqrt{-10}\right)^2$

56. $\left(\sqrt{-75}\right)^3$

57. $\left(2 - \sqrt{-8}\right)\left(8 + \sqrt{-6}\right)$

58. $\left(3 + \sqrt{-5}\right)\left(7 - \sqrt{-10}\right)$

Error Analysis In Exercises 59 and 60, a student has handed in the specified problem. Find the error(s) and discuss how to explain the error(s) to the student.

59. Write $\dfrac{5}{3 - 2i}$ in standard form.

$$\dfrac{5}{3 - 2i} \cdot \dfrac{3 + 2i}{3 + 2i} = \dfrac{15 + 10i}{9 - 4} = 3 + 2i$$

60. Multiply $\left(\sqrt{-4} + 3\right)\left(i - \sqrt{-3}\right)$.

$$\left(\sqrt{-4} + 3\right)\left(i - \sqrt{-3}\right)$$
$$= i\sqrt{-4} - \sqrt{-4}\sqrt{-3} + 3i - 3\sqrt{-3}$$
$$= -2i - \sqrt{12} + 3i - 3i\sqrt{3}$$
$$= \left(1 - 3\sqrt{3}\right)i - 2\sqrt{3}$$

Complex Solutions of a Quadratic Equation In Exercises 61–70, solve the quadratic equation. *See Example 5.*

61. $x^2 + 9 = 0$

62. $x^2 + 36 = 0$

63. $x^2 - 2x + 2 = 0$

64. $x^2 + 6x + 10 = 0$

65. $4x^2 + 16x + 17 = 0$

66. $9x^2 - 6x + 37 = 0$

67. $4x^2 + 16x + 20 = 0$

68. $9x^2 - 6x + 2 = 0$

69. $16t^2 - 4t + 3 = 0$

70. $5s^2 + 6s + 3 = 0$

71. Think About It Can a quadratic equation have one real solution and one complex imaginary solution? Explain. (*Hint:* Consider the Quadratic Formula.)

72. HOW DO YOU SEE IT? Which statement is true for the function shown in the graph? Explain. For each of the other statements, describe a translation of the graph that represents a function for which the statement is true.

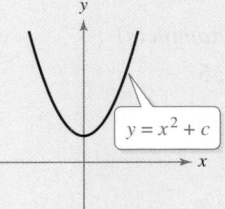

$y = x^2 + c$

(a) The function has two different real zeros.

(b) The function has one repeated real zero.

(c) The function has two complex imaginary zeros.

Plotting Complex Numbers in the Complex Plane In Exercises 73–78, plot the complex number in the complex plane. *See Example 6.*

73. 3

74. i

75. $-2 + i$

76. $-2 - 3i$

77. $1 - 2i$

78. $4 + 3i$

Members of the Mandelbrot Set In Exercises 79–84, decide whether the complex number is in the Mandelbrot Set. Explain your reasoning. *See Example 7.*

79. $c = 0$

80. $c = 2$

81. $c = 1$

82. $c = -1$

83. $c = \frac{1}{2}i$

84. $c = -i$

True or False? In Exercises 85 and 86, determine whether the statement is true or false. Justify your answer.

85. There is no complex number that is equal to its conjugate.

86. The conjugate of the sum of two complex numbers is equal to the sum of the conjugates of the two complex numbers.

Business Capsule

Fractal Graphics, established in Australia in 1992, built a world-class reputation as a leader in application of 3-D visualization technology to the interpretation of complex geoscientific models. In 2002, Fractal Graphics split to form the software development group Fractal Technologies Pty Ltd and the geological consulting group Fractal Geoscience. Five years later, Fractal Technologies and its software, FracSIS, were acquired by global mining technology leader Runge Limited. FracSIS combines geological, geochemical, and geophysical data with an interactive 3-D visualization environment.

87. Research Project Use your campus library, the Internet, or some other reference source to find information about a company that uses algorithms to generate 3-D images or gaming software. Write a report about the company.

3.6 The Fundamental Theorem of Algebra

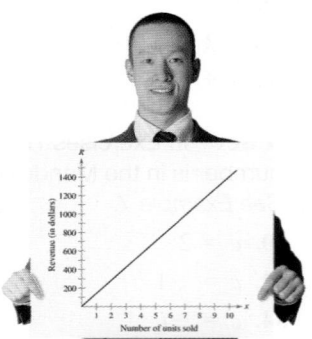

■ Use the Fundamental Theorem of Algebra and the Linear Factorization Theorem to write a polynomial as the product of linear factors.

■ Find a polynomial with real coefficients whose zeros are given.

■ Factor a polynomial over the rational, real, and complex numbers, and find all real and complex zeros of a polynomial function.

The Fundamental Theorem of Algebra

You have been using the fact that an nth-degree polynomial function can have at most n real zeros. In the complex number system, this statement can be improved. That is, in the complex number system, every nth-degree polynomial function has *precisely* n zeros. This important result is derived from the **Fundamental Theorem of Algebra,** first proved by the famous German mathematician Carl Friedrich Gauss (1777–1855).

> ### The Fundamental Theorem of Algebra
>
> If $f(x)$ is a polynomial of degree n, where $n > 0$, then f has at least one zero in the complex number system.

In Exercise 69 on page 313, you will use a graph to determine the price of a product that yields a specific revenue amount.

Using the Fundamental Theorem of Algebra and the equivalence of zeros and factors, you obtain the following theorem.

> ### Linear Factorization Theorem
>
> If $f(x)$ is a polynomial of degree n
>
> $$f(x) = a_n x^n + a_{n-1} x^{n-1} + \cdots + a_1 x + a_0$$
>
> where $n > 0$, then $f(x)$ has precisely n linear factors
>
> $$f(x) = a_n(x - c_1)(x - c_2) \cdots (x - c_n)$$
>
> where $c_1, c_2, \ldots, c_n$ are complex numbers and a_n is the leading coefficient of $f(x)$.

Note that neither the Fundamental Theorem of Algebra nor the Linear Factorization Theorem tells you *how* to find the zeros or factors of a polynomial. Such theorems are called **existence theorems.** To find the zeros of a polynomial function, you still must rely on the techniques developed in the earlier parts of the text.

Method	Section	Example(s)
Factoring	3.2	4, 5
Rational Zero Test	3.4	1–4
Intermediate Value Theorem	3.4	5
Graphing Utility	3.4	6, 7

Remember that the n zeros of a polynomial function can be real or complex, and they may be repeated. Example 1 illustrates several cases.

Example 1 Zeros of Polynomial Functions

Determine the number of zeros of each polynomial function. Then list the zeros.

a. $f(x) = x - 2$ **b.** $f(x) = x^2 - 6x + 9$

c. $f(x) = x^3 + 4x$ **d.** $f(x) = x^4 - 1$

SOLUTION

a. The first-degree polynomial function given by $f(x) = x - 2$ has exactly *one* zero: $x = 2$.

b. Counting multiplicity, the second-degree polynomial function given by

$$f(x) = x^2 - 6x + 9$$
$$= (x - 3)(x - 3)$$

has exactly *two* zeros: $x = 3$ and $x = 3$.

c. The third-degree polynomial function given by

$$f(x) = x^3 + 4x$$
$$= x(x - 2i)(x + 2i)$$

has exactly *three* zeros: $x = 0, x = 2i,$ and $x = -2i$.

d. The fourth-degree polynomial function given by

$$f(x) = x^4 - 1$$
$$= (x - 1)(x + 1)(x - i)(x + i)$$

has exactly *four* zeros: $x = 1, x = -1, x = i,$ and $x = -i$.

✓ **Checkpoint 1**

Determine the number of zeros of $f(x) = x^4 - 36$. Then list the zeros. ■

TECH TUTOR

Remember that when you use a graphing utility to locate the zeros of a function, the only zeros that appear as *x*-intercepts are the *real zeros*. Match the graphs below with the functions in Example 1. Which zeros appear on the graphs?

(a)
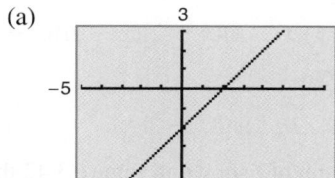

(b)
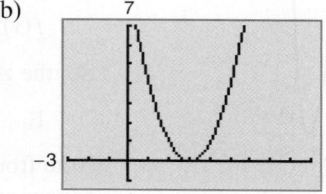

(c)
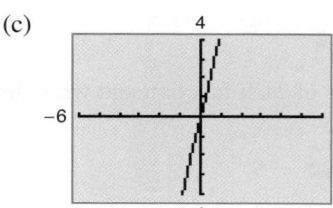

(d)
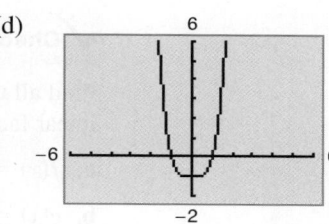

Example 2 shows how you can use the methods described in Sections 3.3 and 3.4 (the Rational Zero Test, synthetic division, and factoring) to find all the zeros of a polynomial function, including the complex zeros.

Example 2   Finding the Zeros of a Polynomial Function

Find all the zeros of

$$f(x) = x^5 + x^3 + 2x^2 - 12x + 8$$

and write the polynomial as the product of the linear factors.

SOLUTION   Using the Rational Zero Test, the possible rational zeros are

$$\pm 1, \quad \pm 2, \quad \pm 4, \quad \text{and} \quad \pm 8.$$

Synthetic division produces the results below.

$$
\begin{array}{r|rrrrrr}
1 & 1 & 0 & 1 & 2 & -12 & 8 \\
  &   & 1 & 1 & 2 & 4 & -8 \\
\hline
  & 1 & 1 & 2 & 4 & -8 & 0
\end{array}
$$
⇒ 1 is a zero.

$$
\begin{array}{r|rrrrr}
1 & 1 & 1 & 2 & 4 & -8 \\
  &   & 1 & 2 & 4 & 8 \\
\hline
  & 1 & 2 & 4 & 8 & 0
\end{array}
$$
⇒ 1 is a repeated zero.

$$
\begin{array}{r|rrrrr}
-2 & 1 &   & 2 & 4 & 8 \\
   &   &   & -2 & 0 & -8 \\
\hline
   & 1 & 0 & 4 & 0
\end{array}
$$
⇒ −2 is a zero.

So, you have

$$f(x) = x^5 + x^3 + 2x^2 - 12x + 8$$
$$= (x - 1)(x - 1)(x + 2)(x^2 + 4).$$

By factoring $x^2 + 4$ as the difference of two squares over the imaginary numbers

$$x^2 - (-4) = x^2 - \left(\sqrt{-4}\right)^2$$
$$= \left(x - \sqrt{-4}\right)\left(x + \sqrt{-4}\right)$$
$$= (x - 2i)(x + 2i)$$

you obtain

$$f(x) = (x - 1)(x - 1)(x + 2)(x - 2i)(x + 2i).$$

So, the zeros of f are

$$1, \quad 1, \quad -2, \quad 2i, \quad \text{and} \quad -2i.$$

Note from the graph of f shown in Figure 3.42 that the *real* zeros are the only ones that appear as x-intercepts.

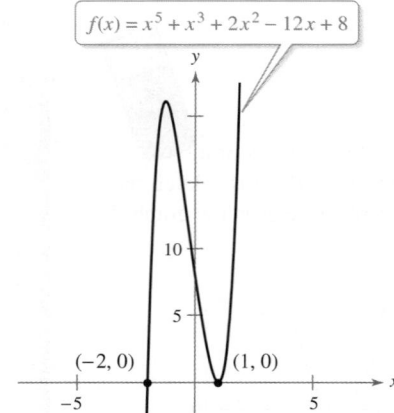

$f(x) = x^5 + x^3 + 2x^2 - 12x + 8$

(−2, 0)   (1, 0)

FIGURE 3.42

✓ Checkpoint 2

Find all the zeros of each function and write the polynomial as the product of linear factors.

a. $f(x) = x^4 + 8x^2 - 9$

b. $g(x) = x^5 - 5x^4 + 4x^3 + 4x^2 + 3x + 9$

Conjugate Pairs

In Example 2, note that the two imaginary zeros are **conjugates.** That is, they are of the form

$$a + bi \quad \text{and} \quad a - bi.$$

Complex Zeros Occur in Conjugate Pairs

Let f be a polynomial function that has *real coefficients*. If $a + bi$, where $b \neq 0$, is a zero of the function, then the conjugate $a - bi$ is also a zero of the function.

Be sure you see that this result is true only if the polynomial function has *real coefficients*. For instance, the result applies to the function given by

$$f(x) = x^2 + 1$$

but not to the function given by

$$g(x) = x - i.$$

You have been using the Rational Zero Test, synthetic division, and factoring to find the zeros of polynomial functions. The Linear Factorization Theorem enables you to reverse this process and find a polynomial function when its zeros are given.

Example 3   Finding a Polynomial Function with Given Zeros

Find a *fourth-degree* polynomial function with real coefficients that has

$$-1, \quad -1, \quad \text{and} \quad 3i$$

as zeros.

SOLUTION   Because $3i$ is a zero *and* the function is stated to have real coefficients, you know that the conjugate $-3i$ must also be a zero. So, the four zeros are

$$-1, \quad -1, \quad 3i, \quad \text{and} \quad -3i.$$

Then, using the Linear Factorization Theorem, $f(x)$ can be written as the product of linear factors, as shown.

$$f(x) = a(x + 1)(x + 1)(x - 3i)(x + 3i)$$

For simplicity, let $a = 1$. Then multiply the factors with real coefficients to get

$$(x + 1)(x + 1) = (x^2 + 2x + 1)$$

and multiply the complex conjugates to get

$$(x - 3i)(x + 3i) = (x^2 + 9).$$

So, you obtain the following fourth-degree polynomial function.

$$f(x) = (x^2 + 2x + 1)(x^2 + 9)$$
$$= x^4 + 2x^3 + 10x^2 + 18x + 9$$

✓ Checkpoint 3

Find a polynomial function with real coefficients that has the given zeros. (There are many correct answers.)

a. $-3, 3, 2i$

b. $1, 5 + i, 5 - i$

Factoring a Polynomial

The Linear Factorization Theorem shows that you can write any nth-degree polynomial as the product of n linear factors.

$$f(x) = a_n(x - c_1)(x - c_2)(x - c_3) \cdots (x - c_n)$$

However, this result includes the possibility that some of the values of c_i are complex. The following result implies that even if you do not want to get involved with "imaginary factors," you can still write $f(x)$ as the product of linear and/or quadratic factors.

Factors of a Polynomial

Every polynomial of degree $n > 0$ with real coefficients can be written as the product of linear and quadratic factors with real coefficients, where the quadratic factors have no real zeros.

A quadratic factor with no real zeros is said to be **irreducible over the reals.** Be sure you see that this is not the same as being *irreducible over the rationals*. For example, the quadratic $x^2 + 1 = (x - i)(x + i)$ is irreducible over the reals (and therefore over the rationals). On the other hand, the quadratic

$$x^2 - 2 = \left(x - \sqrt{2}\right)\left(x + \sqrt{2}\right)$$

is irreducible over the rationals, but it is *reducible* over the reals.

Example 4   Factoring a Polynomial

Complete parts (a)–(d) using the polynomial $x^4 - x^2 - 20$.

a. Write the polynomial as the product of factors that are irreducible over the *rationals*.

b. Write the polynomial as the product of linear factors and quadratic factors that are irreducible over the *reals*.

c. Write the polynomial in completely factored form.

d. How many of the zeros are rational, irrational, or imaginary?

SOLUTION

a. Factor the polynomial into the product of two quadratic polynomials.

$$x^4 - x^2 - 20 = (x^2 - 5)(x^2 + 4)$$

Both of these factors are irreducible over the rationals.

b. By factoring over the reals, you have

$$x^4 - x^2 - 20 = \left(x + \sqrt{5}\right)\left(x - \sqrt{5}\right)(x^2 + 4)$$

where the quadratic factor is irreducible over the reals.

c. In completely factored form, you have

$$x^4 - x^2 - 20 = \left(x + \sqrt{5}\right)\left(x - \sqrt{5}\right)(x - 2i)(x + 2i).$$

d. Using the completely factored form, you can conclude that there are no rational zeros, two irrational zeros $\left(\pm\sqrt{5}\right)$, and two imaginary zeros $(\pm 2i)$.

✓ **Checkpoint 4**

In Example 4, complete parts (a)–(d) using the polynomial $x^4 + x^2 - 12$.

> **Example 5** **Finding the Zeros of a Polynomial Function**

Find all the zeros of

$$f(x) = x^4 - 3x^3 + 6x^2 + 2x - 60$$

given that $1 + 3i$ is a zero of f.

SOLUTION   Because imaginary zeros occur in conjugate pairs, you know that $1 - 3i$ is also a zero of f. This means that both

$$[x - (1 + 3i)] \quad \text{and} \quad [x - (1 - 3i)]$$

are factors of $f(x)$. Multiplying these two factors produces

$$[x - (1 + 3i)][x - (1 - 3i)] = [(x - 1) - 3i][(x - 1) + 3i]$$
$$= (x - 1)^2 - 9i^2$$
$$= x^2 - 2x + 10.$$

Using long division, you can divide $x^2 - 2x + 10$ into $f(x)$ as shown.

$$
\begin{array}{r}
x^2 - x - 6 \\
x^2 - 2x + 10 \overline{\smash{)}\, x^4 - 3x^3 + 6x^2 + 2x - 60} \\
\underline{x^4 - 2x^3 + 10x^2} \\
-x^3 - 4x^2 + 2x \\
\underline{-x^3 + 2x^2 - 10x} \\
-6x^2 + 12x - 60 \\
\underline{-6x^2 + 12x - 60} \\
0
\end{array}
$$

So, you have

$$f(x) = (x^2 - 2x + 10)(x^2 - x - 6)$$
$$= (x^2 - 2x + 10)(x - 3)(x + 2)$$

and you can conclude that the zeros of f are $1 + 3i$, $1 - 3i$, 3, and -2.

✓ **Checkpoint 5**

Find all the zeros of $f(x) = 3x^3 - 5x^2 + 48x - 80$ given that $4i$ is a zero of f.

SUMMARIZE   *(Section 3.6)*

1. State the Fundamental Theorem of Algebra and the Linear Factorization Theorem *(page 306)*. For examples that apply these theorems, see Examples 1 and 2.

2. State the definition of conjugates *(page 309)*. For an example of using conjugates to find a polynomial function with given zeros, see Example 3.

3. Describe what it means for a quadratic factor to be irreducible over the reals and irreducible over the rationals *(page 310)*. For an example of writing a polynomial as the product of factors that are irreducible over the rationals, and as the product of linear factors and quadratic factors that are irreducible over the reals, see Example 4.

SKILLS WARM UP 3.6 The following warm-up exercises involve skills that were covered in earlier sections. You will use these skills in the exercise set for this section. For additional help, review Section 3.5.

In Exercises 1–4, write the complex number in standard form and find its complex conjugate.

1. $4 - \sqrt{-29}$ **2.** $-5 - \sqrt{-144}$ **3.** $-1 + \sqrt{-32}$ **4.** $6 + \sqrt{-1/4}$

In Exercises 5–10, perform the indicated operation(s) and write the result in standard form.

5. $(-3 + 6i) - (10 - 3i)$ **6.** $(12 - 4i) + 20i$

7. $(4 - 2i)(3 + 7i)$ **8.** $(2 - 5i)(2 + 5i)$

9. $\dfrac{1 + i}{1 - i}$ **10.** $(3 + 2i)^3$

Exercises 3.6

See www.CalcChat.com for worked-out solutions to odd-numbered exercises.

Zeros of Polynomial Functions In Exercises 1–6, determine the number of zeros of the polynomial function. *See Example 1.*

1. $f(x) = x - 7$ **2.** $g(x) = x^4 - 36$

3. $h(x) = -x^3 + 2x^2 - 5$ **4.** $f(t) = -2t^5 - 3t^3 + 1$

5. $f(x) = 6x - x^4$

6. $f(x) = 3 - 7x^2 - 5x^4 + 9x^6$

Finding the Zeros of a Polynomial Function In Exercises 7–36, find all the zeros of the function and write the polynomial as the product of linear factors. *See Example 2.*

7. $f(x) = x^2 + 25$ **8.** $f(x) = x^2 + 49$

9. $f(x) = x^2 - 12x + 26$ **10.** $f(x) = x^2 + 6x - 2$

11. $h(x) = x^2 - 4x + 1$ **12.** $g(x) = x^2 + 10x + 23$

13. $f(x) = x^4 - 81$ **14.** $f(t) = t^4 - 625$

15. $g(x) = x^3 + 5x$ **16.** $g(x) = x^3 + 7x$

17. $h(x) = x^3 - 11x^2 - 15x + 325$

18. $h(x) = x^3 - 3x^2 + 4x - 2$

19. $g(x) = x^3 - 6x^2 + 13x - 10$

20. $f(x) = x^3 - 2x^2 - 11x + 52$

21. $f(t) = t^3 - 3t^2 - 15t + 125$

22. $f(x) = x^3 + 8x^2 + 20x + 13$

23. $f(x) = x^3 + 24x^2 + 214x + 740$

24. $h(x) = x^3 - x + 6$

25. $h(x) = x^3 + 9x^2 + 27x + 35$

26. $f(s) = 2s^3 - 5s^2 + 12s - 5$

27. $f(x) = 16x^3 - 20x^2 - 4x + 15$

28. $f(x) = 9x^3 - 15x^2 + 11x - 5$

29. $f(x) = 5x^3 - 9x^2 + 28x + 6$

30. $g(x) = 3x^3 - 4x^2 + 8x + 8$

31. $g(x) = x^4 - 4x^3 + 8x^2 - 16x + 16$

32. $h(x) = x^4 + 6x^3 + 10x^2 + 6x + 9$

33. $f(x) = x^4 + 10x^2 + 9$

34. $f(x) = x^4 + 29x^2 + 100$

35. $f(t) = t^5 + 5t^4 - 7t^3 - 43t^2 - 8t - 48$

36. $g(x) = x^5 - 8x^4 + 28x^3 - 56x^2 + 64x - 32$

Finding a Polynomial Function with Given Zeros In Exercises 37–46, find a polynomial function with real coefficients that has the given zeros. (There are many correct answers.) *See Example 3.*

37. $2, i, -i$ **38.** $5, 2i, -2i$

39. $1, 2 + i, 2 - i$

40. $6, -5 + 2i, -5 - 2i$

41. $-4, 3i, -3i, 2i, -2i$

42. $2, 2, 2, 4i, -4i$

43. $-5, -5, 1 + \sqrt{3}i$ **44.** $0, 0, 4, 1 + i$

45. $\frac{2}{3}, -1, 3 + \sqrt{2}i$ **46.** $\frac{3}{4}, -2, -\frac{1}{2} + i$

Factoring a Polynomial In Exercises 47–50, write the polynomial (a) as the product of factors that are irreducible over the rationals, (b) as the product of linear and quadratic factors that are irreducible over the reals, and (c) in completely factored form. *See Example 4.*

47. $x^4 - 7x^2 - 8$

48. $x^4 - 6x^2 - 72$

49. $x^4 - 5x^3 + 4x^2 + x - 15$
   (*Hint:* One factor is $x^2 - 2x + 3$.)

50. $x^4 + x^3 + 8x^2 + 9x - 9$
   (*Hint:* One factor is $x^2 + 9$.)

Finding the Zeros of a Polynomial Function In Exercises 51–60, use the given zero of f to find all the zeros of f. See Example 5.

51. $f(x) = 2x^3 + 3x^2 + 50x + 75, \quad 5i$

52. $f(x) = 3x^3 - x^2 + 27x - 9, \quad 3i$

53. $f(x) = x^4 - 2x^3 + 37x^2 - 72x + 36, \quad 6i$

54. $f(x) = x^3 - 7x^2 - x + 87, \quad 5 + 2i$

55. $f(x) = 4x^3 + 23x^2 + 34x - 10, \quad -3 + i$

56. $f(x) = x^4 - 2x^3 + 22x^2 + 54x + 29, \quad 2 + 5i$

57. $f(x) = x^4 + 3x^3 - 5x^2 - 21x + 22, \quad -3 + \sqrt{2}i$

58. $f(x) = 2x^3 - 13x^2 + 34x - 35, \quad 2 - \sqrt{3}i$

59. $f(x) = 8x^3 - 14x^2 + 18x - 9, \quad \frac{1}{2}(1 - \sqrt{5}i)$

60. $f(x) = 25x^3 - 55x^2 - 54x - 18, \quad \frac{1}{5}(-2 + \sqrt{2}i)$

 **Graphical Analysis** In Exercises 61–64, (a) use the *zero* or *root* feature of a graphing utility to find the real zeros of the function and (b) find the exact values of the remaining zeros.

61. $f(x) = x^4 + 3x^3 - 5x^2 - 21x + 22$

62. $h(x) = x^3 + 4x^2 + 14x + 20$

63. $f(x) = 8x^3 - 14x^2 + 18x - 9$

64. $h(x) = 25x^3 - 55x^2 - 54x - 18$

65. **Graphical Reasoning** Solve $x^4 - 5x^2 + 4 = 0$.  Then use a graphing utility to graph

$y = x^4 - 5x^2 + 4$.

What is the connection between the solutions you found and the intercepts of the graph?

66. **Graphical Reasoning** Solve $x^4 + 5x^2 + 4 = 0$.  Then use a graphing utility to graph

$y = x^4 + 5x^2 + 4$.

What is the connection between the solutions you found and the intercepts of the graph?

67. **Graphical Analysis** Find a fourth-degree polynomial function that has (a) four real zeros, (b) two real zeros, and (c) no real zeros. Use a graphing utility to graph the functions and describe the similarities and differences among them.

68. **Graphical Analysis** Find a sixth-degree polynomial function that has (a) six real zeros, (b) four real zeros, (c) two real zeros, and (d) no real zeros. Use a graphing utility to graph the functions and describe the similarities and differences among them.

69. **Revenue** The demand equation for a stethoscope is given by $p = 140 - 0.0001x$, where p is the unit price (in dollars) and x is the number of units. The total revenue R (in dollars) obtained by producing and selling x units is given by the model $R = xp = -0.0001x^2 + 140x$ (see figure).

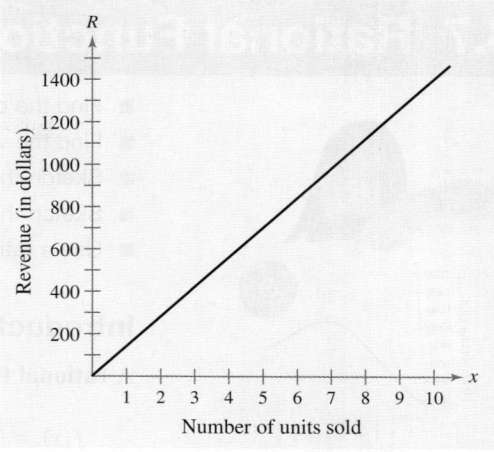

Write and solve an equation to determine a price that yields revenue of $50 million. Discuss the meaning of your solution in the context of the problem. Use a graphing utility to verify your results.

 70. **HOW DO YOU SEE IT?** Let f be a third-degree polynomial function with real coefficients. Two of the zeros of f are

$-\dfrac{2}{3}$ and $2 + 3i$.

The graph of f is shown below.

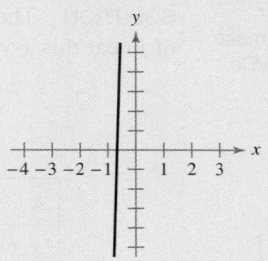

What is the third zero? Explain.

71. **Reasoning** Let f be a fourth-degree polynomial function with real coefficients. Three of the zeros of f are -1, $3 - 2i$, and $3 + 2i$. Explain why the fourth zero must be a real number.

72. **Reasoning** The imaginary number $2i$ is a zero of $f(x) = x^3 - 2ix^2 - 4x + 8i$, but the complex conjugate of $2i$ is not a zero of $f(x)$. Is this a contradiction of the conjugate pairs statement on page 309? Explain.

73. **Reasoning** Let f be a third-degree polynomial function with real coefficients. Explain how you know that f must have at least one zero that is a real number.

74. **Think About It** A student claims that the polynomial $x^4 - 7x^2 + 12$ may be factored over the rational numbers as $(x - \sqrt{3})(x + \sqrt{3})(x - 2)(x + 2)$. Do you agree with this claim? Explain your answer.

3.7 Rational Functions

■ Find the domain of a rational function.
■ Find the vertical and horizontal asymptotes of the graph of a rational function.
■ Sketch the graph of a rational function.
■ Sketch the graph of a rational function that has a slant asymptote.
■ Use a rational function model to solve an application problem.

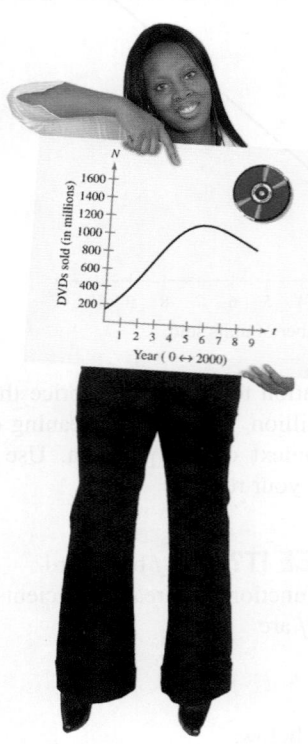

In Exercise 79 on page 324, you will use a rational function to make predictions about the sales of DVDs.

Introduction

A **rational function** is one that can be written in the form

$$f(x) = \frac{p(x)}{q(x)}$$

where $p(x)$ and $q(x)$ are polynomials, and $q(x)$ is not the zero polynomial. In this section, assume that $p(x)$ and $q(x)$ have no common factors. Unlike polynomial functions, whose domains consist of all real numbers, rational functions often have restricted domains. In general, the *domain* of a rational function of x includes all real numbers except x-values that make the denominator zero.

Example 1 Finding the Domain of a Rational Function

Find the domain of the rational function $f(x) = \dfrac{1}{x}$ and discuss the behavior of f near any excluded x-values.

SOLUTION The domain of f is all real numbers except $x = 0$. To determine the behavior of f near this x-value, evaluate $f(x)$ to the left and right of $x = 0$.

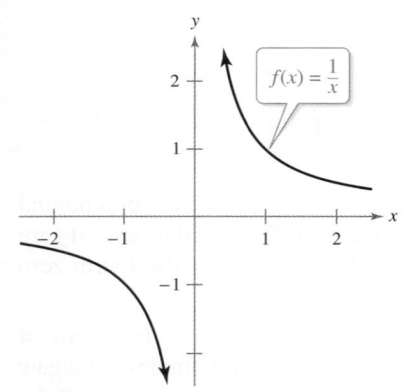

x approaches 0 from the left

x	-1	-0.5	-0.1	-0.01	-0.001	$\rightarrow 0$
$f(x)$	-1	-2	-10	-100	-1000	$\rightarrow -\infty$

x approaches 0 from the right

x	$0 \leftarrow$	0.001	0.01	0.1	0.5	1
$f(x)$	$\infty \leftarrow$	1000	100	10	2	1

Note that as x approaches 0 *from the left*, $f(x)$ decreases without bound. In contrast, as x approaches 0 *from the right*, $f(x)$ increases without bound. The graph of f is shown in Figure 3.43.

FIGURE 3.43

✓ Checkpoint 1

Find the domain of the rational function

$$f(x) = \frac{1}{x - 1}$$

and discuss the behavior of f near any excluded x-values.

Horizontal and Vertical Asymptotes

In Example 1, the behavior of f near $x = 0$ is denoted as shown.

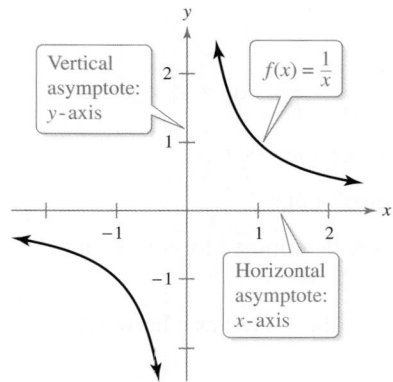

FIGURE 3.44

$$f(x) \to -\infty \text{ as } x \to 0^-$$

$f(x)$ decreases without bound as x approaches 0 from the left.

$$f(x) \to \infty \text{ as } x \to 0^+$$

$f(x)$ increases without bound as x approaches 0 from the right.

The line $x = 0$ is a **vertical asymptote** of the graph of f, as shown in Figure 3.44. In this figure, note that the graph of f also has a **horizontal asymptote**—the line $y = 0$. The behavior of f near $y = 0$ is denoted as shown.

$$f(x) \to 0 \text{ as } x \to -\infty$$

$f(x)$ approaches 0 as x decreases without bound.

$$f(x) \to 0 \text{ as } x \to \infty$$

$f(x)$ approaches 0 as x increases without bound.

Definitions of Vertical and Horizontal Asymptotes

1. The line $x = a$ is a **vertical asymptote** of the graph of f if

   $$f(x) \to \infty \quad \text{or} \quad f(x) \to -\infty$$

   as $x \to a$, either from the right or from the left.

2. The line $y = b$ is a **horizontal asymptote** of the graph of f if

   $$f(x) \to b$$

   as $x \to \infty$ or $x \to -\infty$.

The graph of a rational function can never intersect its vertical asymptote. It may or may not intersect its horizontal asymptote. In either case, the distance between the horizontal asymptote and the points on the graph must approach zero (as $x \to \infty$ or $x \to -\infty$). Figure 3.45 shows the horizontal and vertical asymptotes of the graphs of three rational functions.

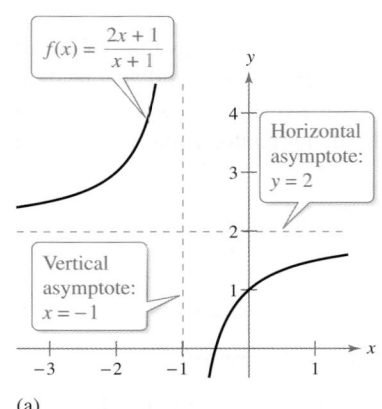

(a)

FIGURE 3.45

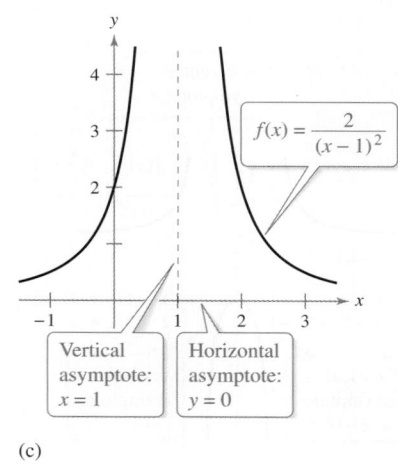

(b)   (c)

The graphs of $f(x) = 1/x$ in Figure 3.44 and

$$f(x) = \frac{2x + 1}{x + 1}$$

in Figure 3.45(a) are **hyperbolas.** You can learn more about hyperbolas in Appendix B.1.

Asymptotes of a Rational Function

Let f be the rational function given by

$$f(x) = \frac{p(x)}{q(x)} = \frac{a_n x^n + a_{n-1} x^{n-1} + \cdots + a_1 x + a_0}{b_m x^m + b_{m-1} x^{m-1} + \cdots + b_1 x + b_0}, \quad a_n \neq 0, b_m \neq 0.$$

Assume that $p(x)$ and $q(x)$ are polynomials with no common factors.

1. The graph of f has *vertical* asymptotes at the zeros of $q(x)$.

2. The graph of f has one or no *horizontal* asymptote determined by comparing the degrees of $p(x)$ and $q(x)$.

   a. When $n < m$, the graph of f has the line $y = 0$ (the x-axis) as a horizontal asymptote.

   b. When $n = m$, the graph of f has the line $y = a_n/b_m$ (ratio of the leading coefficients) as a horizontal asymptote.

   c. When $n > m$, the graph of f has no horizontal asymptote.

Example 2   Finding Horizontal and Vertical Asymptotes

Find all horizontal and vertical asymptotes of the graph of each rational function.

a. $f(x) = \dfrac{2x}{3x^2 + 1}$ **b.** $f(x) = \dfrac{2x^2}{x^2 - 1}$

SOLUTION

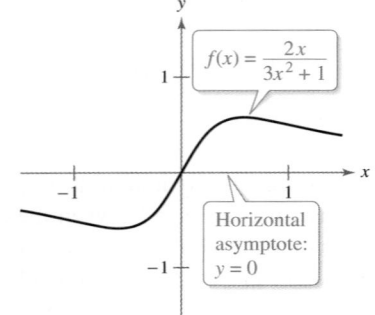

$f(x) = \dfrac{2x}{3x^2 + 1}$

Horizontal asymptote: $y = 0$

(a)

a. For this rational function, the degree of the numerator is *less than* the degree of the denominator, so the graph has the line $y = 0$ as a horizontal asymptote. To find any vertical asymptotes, set the denominator equal to zero and solve the resulting equation for x. Because the equation

$$3x^2 + 1 = 0$$

has no real solutions, you can conclude that the graph has no vertical asymptote. The graph of the function is shown in Figure 3.46(a).

b. For this rational function, the degree of the numerator is *equal* to the degree of the denominator. The leading coefficient of the numerator is 2 and the leading coefficient of the denominator is 1, so the graph has the line $y = 2$ as a horizontal asymptote. To find any vertical asymptotes, set the denominator equal to zero and solve the resulting equation for x.

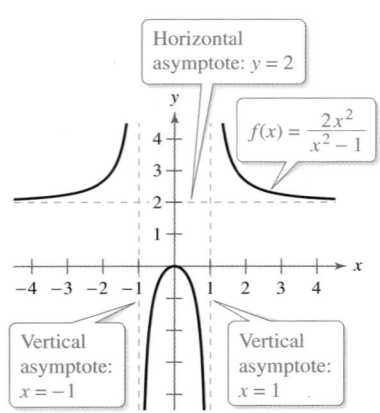

Horizontal asymptote: $y = 2$

$f(x) = \dfrac{2x^2}{x^2 - 1}$

Vertical asymptote: $x = -1$

Vertical asymptote: $x = 1$

(b)

FIGURE 3.46

$$x^2 - 1 = 0 \qquad \text{Set denominator equal to zero.}$$

$$(x + 1)(x - 1) = 0 \qquad \text{Factor.}$$

$$x + 1 = 0 \implies x = -1 \qquad \text{Set 1st factor equal to 0.}$$

$$x - 1 = 0 \implies x = 1 \qquad \text{Set 2nd factor equal to 0.}$$

This equation has two real solutions, $x = -1$ and $x = 1$, so the graph has the lines $x = -1$ and $x = 1$ as vertical asymptotes. The graph of the function is shown in Figure 3.46(b).

✓ Checkpoint 2

Find all horizontal and vertical asymptotes of the graph of $f(x) = \dfrac{x^2}{x^2 + 4}$.

Sketching the Graph of a Rational Function

Guidelines for Graphing Rational Functions

Let

$$f(x) = \frac{p(x)}{q(x)}$$

where $p(x)$ and $q(x)$ are polynomials with no common factors.

1. Find and plot the y-intercept (if any) by evaluating $f(0)$.

2. Find the zeros of the numerator (if any) by solving the equation $p(x) = 0$. Then plot the corresponding x-intercepts.

3. Find the zeros of the denominator (if any) by solving the equation $q(x) = 0$. Then sketch the corresponding vertical asymptotes.

4. Find and sketch the horizontal asymptote (if any) by using the rule for finding the horizontal asymptote of a rational function.

5. Test for symmetry.

6. Plot at least one point both *between and beyond* each x-intercept and vertical asymptote.

7. Use smooth curves to complete the graph between and beyond the vertical asymptotes.

Testing for symmetry can be useful, especially for simple rational functions. For example, the graph of $f(x) = 1/x$ is symmetric with respect to the origin, and the graph of $g(x) = 1/x^2$ is symmetric with respect to the y-axis.

STUDY TIP

Note that in the examples in this section, the vertical asymptotes are included in the table of additional points. This is done to emphasize numerically the behavior of the graph of the function.

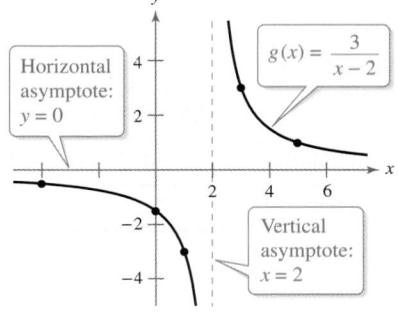

FIGURE 3.47

Example 3 Sketching the Graph of a Rational Function

Sketch the graph of $g(x) = \dfrac{3}{x - 2}$.

SOLUTION Begin by noting that the numerator and denominator have no common factors.

y-intercept:	$\left(0, -\frac{3}{2}\right)$, because $g(0) = -\frac{3}{2}$.
x-intercept:	None, numerator has no zeros.
Vertical asymptote:	$x = 2$, zero of denominator
Horizontal asymptote:	$y = 0$, degree of $p(x) <$ degree of $q(x)$

Additional points:

x	-4	1	2	3	5
$g(x)$	-0.5	-3	Undefined	3	1

By plotting the intercept, asymptotes, and a few additional points, you can obtain the graph shown in Figure 3.47. In the figure, note that the graph of g is a vertical stretch and a right shift of the graph of $y = 1/x$.

✓ Checkpoint 3

Sketch the graph of $f(x) = \dfrac{4}{x + 3}$.

Example 4   Sketching the Graph of a Rational Function

Sketch the graph of $f(x) = \dfrac{x}{x^2 - x - 2}$.

SOLUTION   Factor the denominator to determine more easily the zeros of the denominator.

$$f(x) = \frac{x}{x^2 - x - 2}$$

$$= \frac{x}{(x + 1)(x - 2)}$$

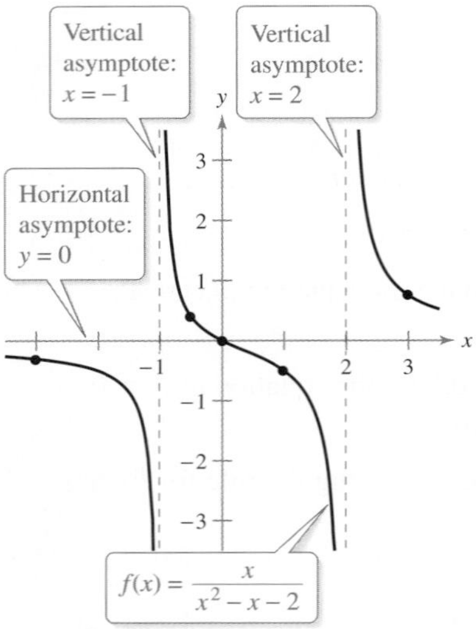

y-intercept: $(0, 0)$, because $f(0) = 0$.

x-intercept: $(0, 0)$

Vertical asymptotes: $x = -1$, $x = 2$, zeros of denominator

Horizontal asymptote: $y = 0$, degree of $p(x) <$ degree of $q(x)$

Additional points:

x	-3	-1	-0.5	1	2	3
$f(x)$	-0.3	Undefined	0.4	-0.5	Undefined	0.75

The graph is shown in Figure 3.48. Confirm the graph with your graphing utility.

Vertical asymptote: $x = -1$

Vertical asymptote: $x = 2$

Horizontal asymptote: $y = 0$

$f(x) = \dfrac{x}{x^2 - x - 2}$

FIGURE 3.48

✓ **Checkpoint 4**

Sketch the graph of $f(x) = \dfrac{3x}{x^2 + x - 6}$.

Example 5   Sketching the Graph of a Rational Function

Sketch the graph of $f(x) = \dfrac{2(x^2 - 9)}{x^2 - 4}$.

SOLUTION   By factoring the numerator and denominator, you have

$$f(x) = \frac{2(x^2 - 9)}{x^2 - 4}$$

$$= \frac{2(x - 3)(x + 3)}{(x - 2)(x + 2)}.$$

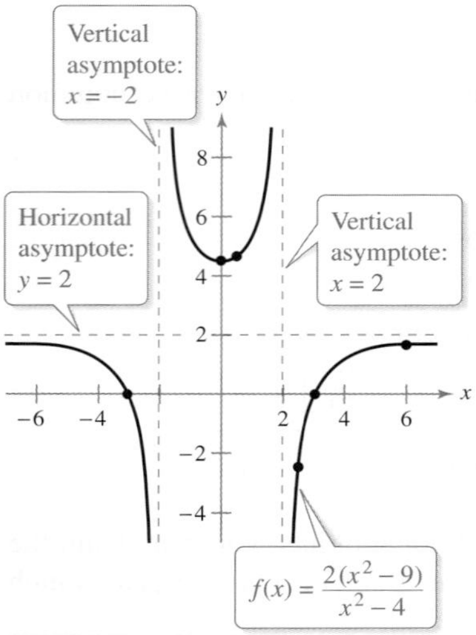

y-intercept: $\left(0, \frac{9}{2}\right)$, because $f(0) = \frac{9}{2}$.

x-intercepts: $(-3, 0)$ and $(3, 0)$

Vertical asymptotes: $x = -2$, $x = 2$, zeros of denominator

Horizontal asymptote: $y = 2$, degree of $p(x) =$ degree of $q(x)$

Symmetry: With respect to y-axis, because $f(-x) = f(x)$.

Additional points:

x	-2	0.5	2	2.5	6
$f(x)$	Undefined	4.67	Undefined	-2.44	1.6875

The graph is shown in Figure 3.49.

Vertical asymptote: $x = -2$

Horizontal asymptote: $y = 2$

Vertical asymptote: $x = 2$

$f(x) = \dfrac{2(x^2 - 9)}{x^2 - 4}$

FIGURE 3.49

✓ **Checkpoint 5**

Sketch the graph of $f(x) = \dfrac{5(x^2 - 1)}{x^2 - 9}$.

Slant Asymptotes

Consider a rational function whose denominator is of degree 1 or greater. If the degree of the numerator is exactly *one more* than the degree of the denominator, then the graph of the function has a **slant** (or **oblique**) **asymptote.** For example, the graph of

$$f(x) = \frac{x^2 - x}{x + 1}$$

has a slant asymptote, as shown in Figure 3.50. To find the equation of a slant asymptote, use long division. For instance, by dividing $x + 1$ into $x^2 - x$, you have

$$f(x) = \frac{x^2 - x}{x + 1} = \underbrace{x - 2}_{} + \frac{2}{x + 1}.$$

Slant asymptote
$(y = x - 2)$

As x increases or decreases without bound, the remainder term

$$\frac{2}{x + 1}$$

approaches 0, so the graph of f approaches the line $y = x - 2$, as shown in Figure 3.50.

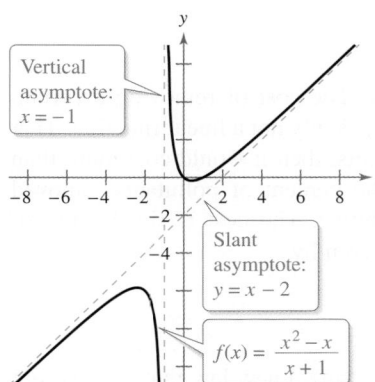

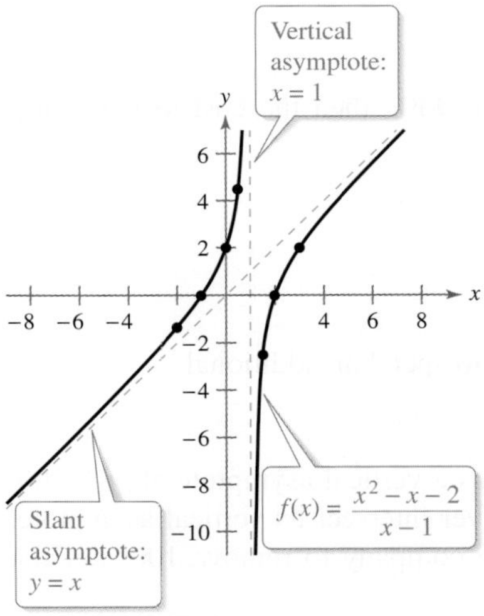

Vertical asymptote: $x = -1$

Slant asymptote: $y = x - 2$

$f(x) = \dfrac{x^2 - x}{x + 1}$

FIGURE 3.50

Example 6   A Rational Function with a Slant Asymptote

Sketch the graph of $f(x) = \dfrac{x^2 - x - 2}{x - 1}$.

SOLUTION   First write f in two different ways. Factoring the numerator enables you to recognize the x-intercepts.

$$f(x) = \frac{x^2 - x - 2}{x - 1}$$

$$= \frac{(x - 2)(x + 1)}{x - 1}$$

Then long division enables you to recognize that the line $y = x$ is a slant asymptote of the graph.

$$f(x) = \frac{x^2 - x - 2}{x - 1}$$

$$= x - \frac{2}{x - 1}$$

y-intercept: $(0, 2)$, because $f(0) = 2$.
x-intercepts: $(-1, 0)$ and $(2, 0)$
Vertical asymptote: $x = 1$, zero of denominator
Horizontal asymptote: None, degree of $p(x) >$ degree of $q(x)$
Slant asymptote: $y = x$
Additional points:

x	-2	0.5	1	1.5	3
$f(x)$	$-1.\overline{3}$	4.5	Undefined	-2.5	2

The graph is shown in Figure 3.51.

Vertical asymptote: $x = 1$

$f(x) = \dfrac{x^2 - x - 2}{x - 1}$

Slant asymptote: $y = x$

FIGURE 3.51

✓ Checkpoint 6

Sketch the graph of $f(x) = \dfrac{x^2 + 3x + 2}{x - 1}$.

Applications

There are many examples of asymptotic behavior in business and biology. For instance, the following example describes the asymptotic behavior related to the cost of removing smokestack emissions.

Example 7　Cost-Benefit Model

A utility company burns coal to generate electricity. The cost of removing a certain *percent* of the pollutants from the stack emissions is typically not a linear function. That is, if it costs C dollars to remove 25% of the pollutants, then it would cost more than $2C$ dollars to remove 50% of the pollutants. As the percent of pollutants removed approaches 100%, the cost tends to become prohibitive. The cost C (in dollars) of removing p percent of the smokestack pollutants is given by

$$C = \frac{80{,}000p}{100 - p}.$$

You are a member of a state legislature that is considering a new law that will require utility companies to remove 90% of the pollutants from their smokestack emissions. The current law requires 85% removal.

a. How much additional expense is the new law asking the utility company to incur?

b. According to the model, would it be possible to remove 100% of the smokestack pollutants?

SOLUTION

a. The graph of this function is shown in Figure 3.52. Note that the graph has a vertical asymptote at

$$p = 100. \qquad \text{Vertical asymptote}$$

Because the current law requires 85% removal, the current cost to the utility company is

$$C = \frac{80{,}000(85)}{100 - 85} \qquad \text{Substitute 85 for } p.$$

$$\approx \$453{,}333. \qquad \text{Use a calculator.}$$

If the new law increases the percent removal to 90%, then the cost to the utility company will be

$$C = \frac{80{,}000(90)}{100 - 90} \qquad \text{Substitute 90 for } p.$$

$$= \$720{,}000. \qquad \text{Use a calculator.}$$

The new law would require the utility company to spend an additional

$$\$720{,}000 - \$453{,}333 = \$266{,}667.$$

b. From Figure 3.52, you can see that the graph has a vertical asymptote at $p = 100$. Because the graph of a rational function can never intersect its vertical asymptote, you can conclude that it is not possible for the company to remove 100% of the pollutants from the stack emissions.

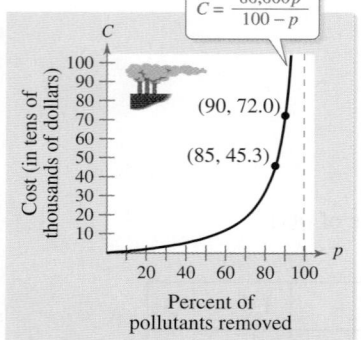

FIGURE 3.52

✓Checkpoint 7

In Example 7, a new law will require utility companies to remove 95% of the pollutants. Find the additional cost to the utility company.

 **Example 8** **Per Capita Land Area**

A model for the population P (in millions) of the United States from 1960 to 2009 is $P = 2.6174t + 176.112$, where t represents the year, with $t = 0$ corresponding to 1960. A model for the land area A (in millions of acres) of the United States from 1960 to 2009 is $A = 2263.960$. Construct a rational function for per capita land area L (in acres per person). Sketch a graph of the rational function. Use the model to predict the per capita land area in 2015. *(Source: U.S. Census Bureau)*

SOLUTION   The rational function for the per capita land area L is

$$L = \frac{A}{P} = \frac{2263.960}{2.6174t + 176.112}.$$

The graph of the function is shown in Figure 3.53. To find the per capita land area in 2015, substitute $t = 55$ into L.

$$L = \frac{2263.960}{2.6174t + 176.112} = \frac{2263.960}{2.6174(55) + 176.112}$$

$$= \frac{2263.960}{320.069}$$

$$\approx 7.07$$

The per capita land area will be approximately 7.1 acres per person in 2015.

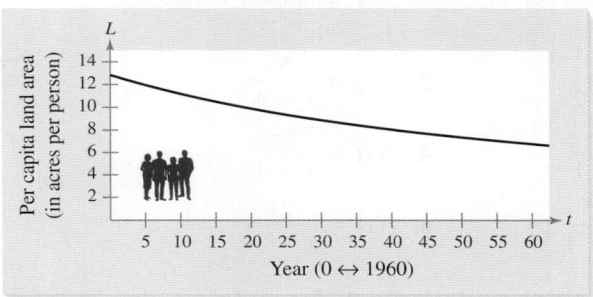

FIGURE 3.53

 **Checkpoint 8**

In Example 8, use the model to predict the per capita land area in 2020.

SUMMARIZE   (Section 3.7)

1. State the definition of a rational function and describe the domain of a rational function *(page 314)*. For an example of finding the domain of a rational function, see Example 1.

2. State the definitions of vertical and horizontal asymptotes *(page 315)*. For an example of finding vertical and horizontal asymptotes of rational functions, see Example 2.

3. State the guidelines for graphing a rational function *(page 317)*. For examples of sketching graphs of rational functions, see Examples 3, 4, and 5.

4. Describe when a rational function has a slant asymptote *(page 319)*. For an example of sketching the graph of a rational function with a slant asymptote, see Example 6.

SKILLS WARM UP 3.7

The following warm-up exercises involve skills that were covered in earlier sections. You will use these skills in the exercise set for this section. For additional help, review Sections 0.6 and 2.2.

In Exercises 1–6, factor the polynomial.

1. $x^2 - 4x$

2. $2x^3 - 6x$

3. $x^2 - 3x - 10$

4. $x^2 - 7x + 10$

5. $x^3 + 4x^2 + 3x$

6. $x^3 - 4x^2 - 2x + 8$

In Exercises 7–10, sketch the graph of the equation.

7. $y = 2$

8. $x = -1$

9. $y = x - 2$

10. $y = -x + 1$

Exercises 3.7

See www.CalcChat.com for worked-out solutions to odd-numbered exercises.

Finding the Domain of a Rational Function In Exercises 1–6, find the domain of the function f and discuss the behavior of f near any excluded x-values. *See Example 1.*

1. $f(x) = \dfrac{1}{x - 2}$

2. $f(x) = \dfrac{5}{x + 4}$

3. $f(x) = \dfrac{4x}{x + 1}$

4. $f(x) = \dfrac{5x}{x - 1}$

5. $f(x) = \dfrac{3x^2}{x^2 - 1}$

6. $f(x) = \dfrac{2x}{x^2 - 4}$

Finding Horizontal and Vertical Asymptotes In Exercises 7–14, find all horizontal and vertical asymptotes of the graph of the function. *See Example 2.*

7. $f(x) = \dfrac{3x}{x + 1}$

8. $f(x) = \dfrac{x}{x - 2}$

9. $f(x) = \dfrac{x - 7}{5 - x}$

10. $f(x) = \dfrac{1 - 5x}{1 + 2x}$

11. $f(x) = \dfrac{3x^2 + 1}{x^2 + 9}$

12. $f(x) = \dfrac{3x^2 + x - 5}{x^2 + 1}$

13. $f(x) = \dfrac{5}{(x + 4)^2}$

14. $f(x) = \dfrac{1}{(x - 1)^2}$

Matching In Exercises 15–20, match the function with its graph. [The graphs are labeled (a), (b), (c), (d), (e), and (f).]

(a)

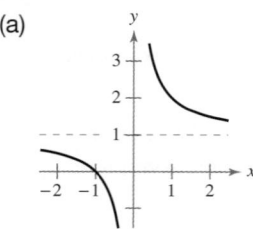

(b)

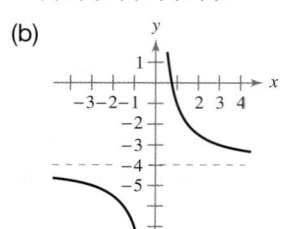

(c)

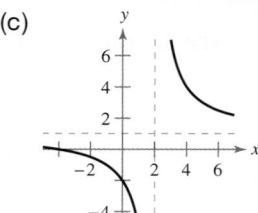

(d)

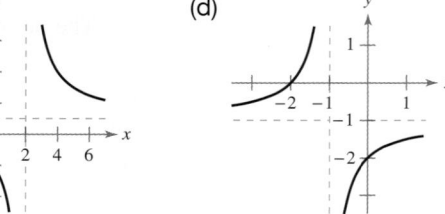

(e)

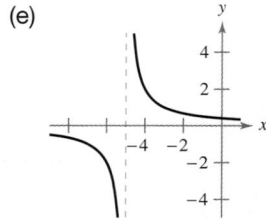

(f)

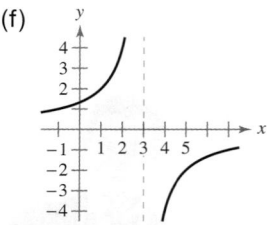

15. $f(x) = -\dfrac{4}{x - 3}$

16. $f(x) = \dfrac{2}{x + 5}$

17. $f(x) = \dfrac{x + 1}{x}$

18. $f(x) = \dfrac{3 - 4x}{x}$

19. $f(x) = \dfrac{x + 4}{x - 2}$

20. $f(x) = -\dfrac{x + 2}{x + 1}$

Describing a Transformation In Exercises 21–24, compare the graph of $f(x) = 1/x$ with the graph of g.

21. $g(x) = f(x) - 2 = \dfrac{1}{x} - 2$

22. $g(x) = f(x + 1) = \dfrac{1}{x + 1}$

23. $g(x) = -f(x) = -\dfrac{1}{x}$

24. $g(x) = -f(x + 1) = -\dfrac{1}{x + 1}$

Describing a Transformation In Exercises 25–28, compare the graph of $f(x) = 4/x^2$ with the graph of g.

25. $g(x) = f(x) + 3 = \dfrac{4}{x^2} + 3$

26. $g(x) = f(x - 1) = \dfrac{4}{(x - 1)^2}$

27. $g(x) = -f(x) = -\dfrac{4}{x^2}$

28. $g(x) = \dfrac{1}{8}f(x) = \dfrac{1}{2x^2}$

Describing a Transformation In Exercises 29–32, compare the graph of $f(x) = 8/x^3$ with the graph of g.

29. $g(x) = f(x) + 5 = \dfrac{8}{x^3} + 5$

30. $g(x) = f(x - 3) = \dfrac{8}{(x - 3)^3}$

31. $g(x) = -f(x) = -\dfrac{8}{x^3}$

32. $g(x) = \dfrac{1}{4}f(x) = \dfrac{2}{x^3}$

Sketching the Graph of a Rational Function In Exercises 33–58, sketch the graph of the rational function. To aid in sketching the graphs, check for intercepts, symmetry, vertical asymptotes, and horizontal asymptotes. *See Examples 3, 4, and 5.*

33. $f(x) = \dfrac{1}{x + 3}$

34. $f(x) = \dfrac{1}{x - 3}$

35. $f(x) = \dfrac{-1}{x + 1}$

36. $f(x) = \dfrac{-2}{x - 3}$

37. $f(x) = \dfrac{x + 4}{x - 5}$

38. $f(x) = \dfrac{x - 2}{x - 3}$

39. $f(x) = \dfrac{2 + x}{1 - x}$

40. $f(x) = \dfrac{3 - x}{2 - x}$

41. $f(t) = \dfrac{3t + 1}{t}$

42. $f(t) = \dfrac{1 - 2t}{t}$

43. $C(x) = \dfrac{5 + 2x}{1 + x}$

44. $P(x) = \dfrac{1 - 3x}{1 - x}$

45. $g(x) = \dfrac{1}{x + 2} + 2$

46. $h(x) = \dfrac{1}{x - 3} + 1$

47. $f(x) = \dfrac{1}{x^2} + 2$

48. $f(x) = 2 - \dfrac{3}{x^2}$

49. $h(x) = \dfrac{x^2}{x^2 - 9}$

50. $h(t) = \dfrac{3t^2}{t^2 - 4}$

51. $g(s) = \dfrac{s}{s^2 + 1}$

52. $g(x) = \dfrac{x}{x^2 + 3}$

53. $f(x) = \dfrac{x}{x^2 - 3x - 4}$

54. $f(x) = \dfrac{-x}{x^2 + x - 6}$

55. $f(x) = \dfrac{3x}{x^2 - x - 2}$

56. $f(x) = \dfrac{2x}{x^2 + x - 2}$

57. $g(x) = \dfrac{4(x^2 - 4)}{x^2 - 16}$

58. $g(x) = \dfrac{3(x^2 - 16)}{x^2 - 9}$

A Rational Function with a Slant Asymptote In Exercises 59–64, sketch the graph of the rational function. To aid in sketching the graphs, check for intercepts, vertical asymptotes, and slant asymptotes. *See Example 6.*

59. $f(x) = \dfrac{2x^2 + 1}{x}$

60. $g(x) = \dfrac{1 - x^2}{x}$

61. $h(x) = \dfrac{x^2}{x - 1}$

62. $f(x) = \dfrac{x^3}{x^2 - 1}$

63. $g(x) = \dfrac{x^3}{2x^2 - 8}$

64. $f(x) = \dfrac{x^3}{x^2 + 4}$

Writing a Rational Function In Exercises 65–68, write a rational function f that has the specified characteristics. (There are many correct answers.)

65. Vertical asymptote: None

Horizontal asymptote: $y = 2$

66. Vertical asymptotes: $x = 0, x = \dfrac{5}{2}$

Horizontal asymptote: $y = -3$

67. Vertical asymptotes: $x = -2, x = 1$

Horizontal asymptote: None

68. Vertical asymptote: $x = 3$

Horizontal asymptote: x-axis

 Graphical Analysis In Exercises 69–72, use a graphing utility to graph f and g in the same viewing window. (Notice that f has a common factor in the numerator and denominator.) Use the *trace* feature of the graphing utility to check the value of each function near any x-values excluded from its domain. Then, describe how the graphs of f and g are different.

69. $f(x) = \dfrac{(x - 3)(x + 3)}{x - 3}$, $g(x) = x + 3$

70. $f(x) = \dfrac{(x - 4)(x + 1)}{x - 4}$, $g(x) = x + 1$

71. $f(x) = \dfrac{x - 2}{x^2 - 4}$, $g(x) = \dfrac{1}{x + 2}$

72. $f(x) = \dfrac{x - 5}{x^2 - 4x - 5}$, $g(x) = \dfrac{1}{x + 1}$

73. Think About It You divide out the common factor $(x - 7)$ from the numerator and denominator of a rational function. How do you specify the domain of the new function so that both functions are the same?

74. HOW DO YOU SEE IT? For each statement below, determine which graph(s) show that the statement is incorrect.

(i)

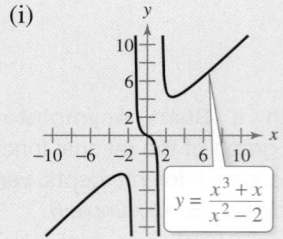

$y = \dfrac{x^3 + x}{x^2 - 2}$

(ii)

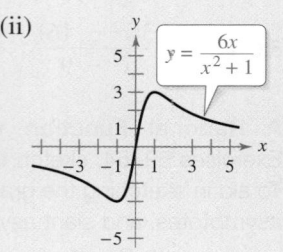

$y = \dfrac{6x}{x^2 + 1}$

(iii)

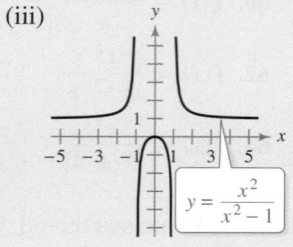

$y = \dfrac{x^2}{x^2 - 1}$

(iv)

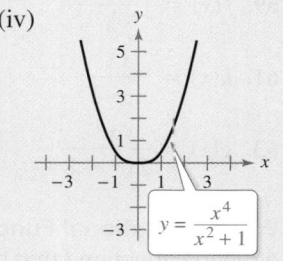

$y = \dfrac{x^4}{x^2 + 1}$

(a) Every rational function has a vertical asymptote.

(b) Every rational function has at least one asymptote.

(c) A rational function can have at most one vertical asymptote.

(d) The graph of a rational function with a slant asymptote cannot cross the slant asymptote.

75. Think About It Is it possible for a rational function to have all three types of asymptotes (vertical, horizontal, and slant)? Why or why not?

76. Think About It Is it possible for a rational function to have more than one horizontal asymptote? Why or why not?

77. Federal Drug Enforcement The cost C (in millions of dollars) for the federal government to seize p percent of an illegal drug as it enters the country is

$$C = \frac{528p}{100 - p}, \quad 0 \le p < 100.$$

(a) Find the costs of seizing 25% of the drug, 50% of the drug, and 75% of the drug.

(b) According to this model, is it possible to seize 100% of the drug? Explain.

78. Population of Deer The Game Commission introduces 100 deer into newly acquired state game lands. The population N of the herd is given by

$$N = \frac{25(4 + 2t)}{1 + 0.02t}, \quad t \ge 0, \text{ where } t \text{ is time (in years).}$$

(a) Find the populations when t is 5, 10, and 25.

(b) What is the limiting size of the herd as time progresses?

79. Retail Sales The table shows the numbers N (in millions) of DVDs sold in the years 2000 through 2009. The data can be modeled by

$$N = \frac{104.05t + 140.3}{0.026t^2 - 0.21t + 1}, \quad 0 \le t \le 9$$

where t is the year, with $t = 0$ corresponding to 2000 (see figure). *(Source: SNL Kagan)*

Year	DVDs, N	Year	DVDs, N
2000	146.6	2005	1072.4
2001	291.5	2006	1069.3
2002	507.5	2007	1063.4
2003	713.0	2008	1007.5
2004	976.6	2009	851.1

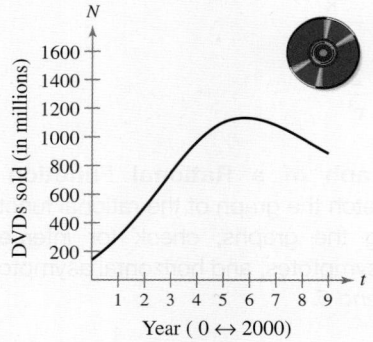

(a) Use a graphing utility to plot the data and graph the model in the same viewing window. How well does the model represent the data?

(b) Use the model to predict the numbers of DVDs sold in the years 2012, 2015, and 2020. Are the predictions reasonable?

(c) Determine the horizontal asymptote and explain its meaning in the context of the problem.

80. Average Cost The cost C (in dollars) of producing x basketballs is

$$C = 375,000 + 4x.$$

The average cost $\overline{C}$ per basketball is

$$\overline{C} = \frac{C}{x} = \frac{375,000 + 4x}{x}, \quad x > 0.$$

(a) Sketch the graph of the average cost function.

(b) Find the average costs of producing 1000, 10,000, and 100,000 basketballs.

(c) Find the horizontal asymptote and explain its meaning in the context of the problem.

81. Human Memory Model Psychologists have developed mathematical models to predict memory performance as a function of the number of trials n of a certain task. Consider the learning curve modeled by

$$P = \frac{0.6 + 0.95(n - 1)}{1 + 0.95(n - 1)}, \quad n > 0$$

where P is the percent of correct responses (in decimal form) after n trials.

(a) Complete the table.

n	1	2	3	4	5	6	7	8	9	10
P										

(b) According to this model, what is the limiting percent of correct responses as n increases?

82. Human Memory Model Consider the learning curve modeled by

$$P = \frac{0.55 + 0.87(n - 1)}{1 + 0.87(n - 1)}, \quad n > 0$$

where P is the percent of correct responses (in decimal form) after n trials.

(a) Complete the table.

n	1	2	3	4	5	6	7	8	9	10
P										

(b) According to this model, what is the limiting percent of correct responses as n increases?

83. Average Recycling Cost The cost C (in dollars) of recycling a waste product is

$$C = 450,000 + 6x, \quad x > 0$$

where x is the number of pounds of waste. The average recycling cost $\overline{C}$ per pound is

$$\overline{C} = \frac{C}{x} = \frac{450,000 + 6x}{x}, \quad x > 0.$$

(a) Use a graphing utility to graph $\overline{C}$.

(b) Find the average costs of recycling 10,000, 100,000, 1,000,000, and 10,000,000 pounds of waste. What can you conclude?

84. Drug Concentration The concentration C of a medication in the bloodstream t minutes after sublingual (under the tongue) application is given by

$$C(t) = \frac{3t - 1}{2t^2 + 5}, \quad t > 0.$$

(a) Use a graphing utility to graph the function. Estimate when the concentration is greatest.

(b) Does this function have a horizontal asymptote? If so, discuss the meaning of the asymptote in terms of the concentration of the medication.

85. Domestic Sales The U.S. domestic sales S of electricity (in millions of kilowatt-hours) from 1990 through 2008 can be modeled by

$$S = 27,904t + 906,950, \quad 0 \le t \le 18$$

where t represents the year, with $t = 0$ corresponding to 1990. The population P (in millions) of the United States from 1990 through 2008 can be modeled by

$$P = 3.023t + 250.90, \quad 0 \le t \le 18$$

where t represents the year, with $t = 0$ corresponding to 1990. *(Source: Edison Electric Institute and the U.S. Census Bureau)*

(a) Construct a rational function E to represent the per capita domestic sales of electricity.

(b) Use a graphing utility to graph the rational function E.

(c) Use the model to estimate the per capita domestic sales (in kilowatt-hours) of electricity in 2008.

86. Health Care Spending The total health care spending H (in millions of dollars) in the United States from 1990 through 2008 can be modeled by

$$H = 3258.97t^2 + 31,416.1t + 754,111, \quad 0 \le t \le 18$$

where t represents the year, with $t = 0$ corresponding to 1990. The population P (in millions) of the United States from 1990 through 2008 can be modeled by

$$P = 3.023t + 250.90, \quad 0 \le t \le 18$$

where t represents the year, with $t = 0$ corresponding to 1990. *(Sources: U.S. Centers for Medicare and Medicaid Services and the U.S. Census Bureau)*

(a) Construct a rational function S to represent the per capita health care spending.

(b) Use a graphing utility to graph the rational function S.

(c) Use the model to predict the per capita health care spending in 2012.

87. 3000-Meter Speed Skating The winning times for the women's 3000-meter speed skating race at the Olympics from 1960 through 2010 can be approximated by the quadratic model

$$y = 0.0237t^2 - 5.624t + 571.67, \quad 60 \le t \le 110$$

where y is the winning time (in seconds) and t represents the year, with $t = 60$ corresponding to 1960. *(Sources: DatabaseOlympics.com and Sports Reference LLC)*

(a) Use a graphing utility to graph the model.

(b) Use the model to predict the winning times in 2014 and 2018.

(c) Does this model have a horizontal asymptote? Do you think that a model for this type of data should have a horizontal asymptote?

ALGEBRA TUTOR

xy

Factoring Expressions

Much of the algebra in Chapter 3 involves factoring. On this page, you can review some of the basic techniques for factoring polynomial expressions. The simplest type of factoring involves recognizing common factors and factoring them out of the expression. Other types of factoring involve special polynomial forms. Several of these are listed below.

Difference of Two Squares: $u^2 - v^2 = (u + v)(u - v)$

Perfect Square Trinomial: $u^2 + 2uv + v^2 = (u + v)^2$
$$u^2 - 2uv + v^2 = (u - v)^2$$

Sum or Difference of Two Cubes: $u^3 + v^3 = (u + v)(u^2 - uv + v^2)$
$$u^3 - v^3 = (u - v)(u^2 + uv + v^2)$$

For trinomials of the form $ax^2 + bx + c$ that have binomial factors, you need to find a combination of factors of a and c such that the outer and inner products of the binomials add up to the middle term bx.

Example 1   Factoring Expressions

Completely factor the expression.

a. $6x^2 + 21x$ **b.** $36x^3 - 4x$

c. $4x^2 + 20x + 25$ **d.** $y^2 - 22y + 121$

e. $5z^3 + 40$ **f.** $x^2 - 12x + 27$

SOLUTION

STUDY TIP

Be sure to remember that a polynomial is completely factored when each of its factors is prime.

a. $6x^2 + 21x = 3x(2x) + 3x(7)$ $3x$ is a common factor.

$\qquad\qquad = 3x(2x + 7)$ Factor completely.

b. $36x^3 - 4x = 4x(9x^2 - 1)$ $4x$ is a common factor.

$\qquad\qquad = 4x[(3x)^2 - 1^2]$ $9x^2 - 1$ is a difference of two squares.

$\qquad\qquad = 4x(3x + 1)(3x - 1)$ Factor completely.

c. $4x^2 + 20x + 25 = (2x)^2 + 2(2x)(5) + 5^2$ Perfect square trinomial

$\qquad\qquad = (2x + 5)^2$ Factor completely.

d. $y^2 - 22y + 121 = y^2 - 2(y)(11) + 11^2$ Perfect square trinomial

$\qquad\qquad = (y - 11)^2$ Factor completely.

e. $5z^3 + 40 = 5(z^3 + 8)$ 5 is a common factor.

$\qquad\qquad = 5(z^3 + 2^3)$ $z^3 + 8$ is a sum of two cubes. Rewrite 8 as 2^3.

$\qquad\qquad = 5(z + 2)(z^2 - 2z + 4)$ Factor completely.

f. For $x^2 - 12x + 27$, you have $a = 1$, $b = -12$, and $c = 27$. Because b is negative and c is positive, both factors of 27 must be negative. That is, $27 = (-1)(-27)$, or $27 = (-3)(-9)$. So, the possible factorizations of $x^2 - 12x + 27$ are $(x - 1)(x - 27)$ and $(x - 3)(x - 9)$. Testing the middle term, you find that

$$x^2 - 12x + 27 = (x - 3)(x - 9).$$

Simplifying Radical Expressions

Performing operations with complex numbers involves several different algebraic skills. One such skill is simplifying expressions involving radicals. Here are some helpful simplification techniques.

1. To add or subtract radical expressions, the radicals must be like radicals (having the same index and the same radicand). The Distributive Property can then be used to combine like radicals.

2. To multiply or divide radical expressions, use the following properties.

   Let a and b be real numbers, variables, or algebraic expressions such that the indicated roots are real numbers, and let n be a positive integer.

$$\sqrt[n]{a} \cdot \sqrt[n]{b} = \sqrt[n]{ab} \qquad \frac{\sqrt[n]{a}}{\sqrt[n]{b}} = \sqrt[n]{\frac{a}{b}}, \quad b \neq 0$$

Example 2 Simplifying Expressions Involving Radicals

Simplify the expression.

a. $\sqrt{48}$ b. $\sqrt{7} + \sqrt{28}$ c. $\sqrt{125} - \sqrt{45}$

d. $\sqrt{5} \cdot \sqrt{20}$ e. $\sqrt{2}(8 - \sqrt{3})$ f. $(3 + \sqrt{5})^2$

SOLUTION

a. $\sqrt{48} = \sqrt{16 \cdot 3}$ Find largest square factor.

$\phantom{\sqrt{48}} = \sqrt{4^2 \cdot 3}$ Rewrite.

$\phantom{\sqrt{48}} = 4\sqrt{3}$ Find root of perfect square.

b. $\sqrt{7} + \sqrt{28} = \sqrt{7} + \sqrt{4 \cdot 7}$ Find largest square factor.

$\phantom{\sqrt{7} + \sqrt{28}} = \sqrt{7} + \sqrt{2^2 \cdot 7}$ Rewrite.

$\phantom{\sqrt{7} + \sqrt{28}} = \sqrt{7} + 2\sqrt{7}$ Find root of perfect square.

$\phantom{\sqrt{7} + \sqrt{28}} = (1 + 2)\sqrt{7}$ Distributive Property

$\phantom{\sqrt{7} + \sqrt{28}} = 3\sqrt{7}$ Add.

c. $\sqrt{125} - \sqrt{45} = \sqrt{25 \cdot 5} - \sqrt{9 \cdot 5}$ Find largest square factors.

$\phantom{\sqrt{125} - \sqrt{45}} = \sqrt{5^2 \cdot 5} - \sqrt{3^2 \cdot 5}$ Rewrite.

$\phantom{\sqrt{125} - \sqrt{45}} = 5\sqrt{5} - 3\sqrt{5}$ Find roots of perfect squares.

$\phantom{\sqrt{125} - \sqrt{45}} = (5 - 3)\sqrt{5}$ Distributive Property

$\phantom{\sqrt{125} - \sqrt{45}} = 2\sqrt{5}$ Subtract.

d. $\sqrt{5} \cdot \sqrt{20} = \sqrt{5 \cdot 20}$ Rewrite.

$\phantom{\sqrt{5} \cdot \sqrt{20}} = \sqrt{100}$ Multiply.

$\phantom{\sqrt{5} \cdot \sqrt{20}} = 10$ Simplify.

e. $\sqrt{2}(8 - \sqrt{3}) = \sqrt{2} \cdot 8 - \sqrt{2} \cdot \sqrt{3}$ Distributive Property

$\phantom{\sqrt{2}(8 - \sqrt{3})} = 8\sqrt{2} - \sqrt{2 \cdot 3}$ Rewrite.

$\phantom{\sqrt{2}(8 - \sqrt{3})} = 8\sqrt{2} - \sqrt{6}$ Multiply.

f. $(3 + \sqrt{5})^2 = 9 + 3\sqrt{5} + 3\sqrt{5} + 5$ Distributive Property

$\phantom{(3 + \sqrt{5})^2} = 14 + 6\sqrt{5}$ Combine like terms.

STUDY TIP

Be sure to remember that one of the conditions for a radical expression to be in simplest form is that all possible factors have been removed from the radical.

SUMMARY AND STUDY STRATEGIES

After studying this chapter, you should have acquired the following skills.
The exercise numbers are keyed to the Review Exercises that begin on page 330.
Answers to odd-numbered Review Exercises are given in the back of the text.*

Section 3.1

	Review Exercises
■ Sketch the graph of a quadratic function.	*1–4*
■ Find the equation of a parabola given its vertex and a point on its graph.	*5, 6*
■ Construct and use a quadratic model to solve an application problem.	*7–12*

Section 3.2

■ Sketch a transformation of a monomial function.	*13–20*
■ Apply the Leading Coefficient Test to determine right-hand and left-hand behavior of graphs of polynomial functions.	*21–24*

When n is odd and the leading coefficient is positive,

$$f(x) \to -\infty \text{ as } x \to -\infty \text{ and } f(x) \to \infty \text{ as } x \to \infty.$$

When n is odd and the leading coefficient is negative,

$$f(x) \to \infty \text{ as } x \to -\infty \text{ and } f(x) \to -\infty \text{ as } x \to \infty.$$

When n is even and the leading coefficient is positive,

$$f(x) \to \infty \text{ as } x \to -\infty \text{ and } f(x) \to \infty \text{ as } x \to \infty.$$

When n is even and the leading coefficient is negative,

$$f(x) \to -\infty \text{ as } x \to -\infty \text{ and } f(x) \to -\infty \text{ as } x \to \infty.$$

■ Find the real zeros and sketch the graph of a polynomial function.	*25–28*

Section 3.3

■ Divide one polynomial by a second polynomial using long division.	*29, 30*
■ Use synthetic division to divide two polynomials.	*31, 32, 39, 40*
■ Use the Remainder Theorem and synthetic division to evaluate a polynomial.	*33, 34*
■ Use the Factor Theorem to factor a polynomial.	*35, 36*
■ Simplify a rational expression using long division.	*37, 38*

Section 3.4

■ Find all possible rational zeros of a function using the Rational Zero Test.	*41, 42*
■ Find all the real zeros of a function.	*43–50*
■ Approximate the real zeros of a polynomial function using the Intermediate Value Theorem.	*51, 52*
■ Approximate the real zeros of a polynomial function using a graphing utility.	*53, 54*
■ Apply techniques for approximating real zeros to solve an application problem.	*55, 56*

---

* A wide range of valuable study aids are available to help you master the material in this chapter.
The *Student Solutions Manual* includes step-by-step solutions to all odd-numbered exercises to
help you review and prepare. The student website at *www.cengagebrain.com* offers algebra help
and a *Graphing Technology Guide*, which contains step-by-step commands and instructions for
a wide variety of graphing calculators.

Section 3.5   Review Exercises

■ Find the complex conjugate of a complex number.   *57–60*

■ Perform operations with complex numbers and write the results in standard form.   *61–74*

$$(a + bi) + (c + di) = (a + c) + (b + d)i$$
$$(a + bi) - (c + di) = (a - c) + (b - d)i$$
$$(a + bi)(c + di) = (ac - bd) + (ad + bc)i$$

■ Solve a polynomial equation that has complex solutions.   *75–78*

■ Plot a complex number in the complex plane.   *79, 80*

Section 3.6

■ Use the Fundamental Theorem of Algebra and the Linear Factorization   *81–86*
  Theorem to write a polynomial as the product of linear factors.

■ Find a polynomial with real coefficients whose zeros are given.   *87, 88*

■ Factor a polynomial over the rational, real, and complex numbers, and find   *89–94*
  all real and complex zeros of a polynomial function.

Section 3.7

■ Find the domain of a rational function.   *95–98*

■ Find the vertical and horizontal asymptotes of the graph of a rational function.   *95–98*

Let $f(x) = \dfrac{p(x)}{q(x)} = \dfrac{a_n x^n + a_{n-1}x^{n-1} + \cdots + a_1 x + a_0}{b_m x^m + b_{m-1}x^{m-1} + \cdots + b_1 x + b_0}$, $a_n \neq 0, b_m \neq 0$.

Assume that $p(x)$ and $q(x)$ are polynomials with no common factors.

1. The graph of f has *vertical* asymptotes at the zeros of $q(x)$.

2. The graph of f has one or no *horizontal* asymptote determined by comparing
   the degrees of $p(x)$ and $q(x)$.

   a. When $n < m$, the graph of f has the line $y = 0$ (the x-axis) as a horizontal asymptote.

   b. When $n = m$, the graph of f has the line $y = a_n/b_m$ (ratio of the leading coefficients)
      as a horizontal asymptote.

   c. When $n > m$, the graph of f has no horizontal asymptote.

■ Sketch the graph of a rational function, including graphs with slant asymptotes.   *99–104*

■ Use a rational function model to solve an application problem.   *105–108*

Study Strategies

■ **Use a Graphing Utility** A graphing calculator or graphing software for a computer can help you in this course in
  two important ways. As an *exploratory device*, a graphing utility helps you to learn concepts by comparing graphs of
  functions. For instance, sketching the graphs of $f(x) = x^3$ and $f(x) = -x^3$ shows that the negative coefficient has the
  effect of reflecting the graph in the x-axis. As a *problem-solving tool*, a graphing utility frees you from sketching
  complicated graphs by hand. The time you save can be spent using mathematics to solve real-life problems.

■ **Problem-Solving Strategies** When you get stuck trying to solve a real-life problem, consider the strategies below.

  1. *Draw a Diagram.* Draw a diagram representing the problem. Label all known values and unknown values on the diagram.

  2. *Solve a Simpler Problem.* Simplify the problem, or write several simple examples of the problem. For instance, if you
     are asked to find the dimensions that will produce a maximum area, try calculating the areas of several examples.

  3. *Rewrite the Problem in Your Own Words.* Rewriting a problem can help you understand it better.

  4. *Guess and Check.* Try guessing the answer, then check your guess in the statement of the original problem. By refining
     your guesses, you may be able to think of a general strategy for solving the problem.

Review Exercises

See www.CalcChat.com for worked-out solutions to odd-numbered exercises.

Using Standard Form to Graph a Parabola In Exercises 1–4, write the quadratic function in standard form (if necessary) and sketch its graph. Identify the vertex and intercepts.

1. $f(x) = (x + 3)^2 - 5$

2. $g(x) = -(x - 1)^2 + 3$

3. $h(x) = 3x^2 - 12x + 11$

4. $f(x) = \frac{1}{3}(x^2 + 5x - 4)$

Finding an Equation of a Parabola In Exercises 5 and 6, find an equation of the parabola that has the indicated vertex and whose graph passes through the given point.

5. Vertex: $(-5, -1)$; point: $(-2, 6)$

6. Vertex: $(2, 5)$; point: $(4, 7)$

7. **Optimal Area** The perimeter of a rectangular archaeological dig site is 500 feet. Let x represent the width of the dig site. Write a quadratic function for the area of the rectangle in terms of its width. Of all possible dig sites with perimeters of 500 feet, what are the measurements of the one with the greatest area?

8. **Optimal Revenue** Find the number of units that produces a maximum revenue R (in dollars) for

$$R = 900x - 0.015x^2$$

where x is the number of units produced.

9. **Optimal Cost** A manufacturer of retinal imaging systems has daily production costs C (in dollars per unit) of

$$C = 25{,}000 - 50x + 0.065x^2$$

where x is the number of units produced.

(a) Use a graphing utility to graph the cost function.

(b) Graphically estimate the number of units that should be produced to yield a minimum cost per unit.

(c) Explain how to confirm the result of part (b) algebraically.

10. **Optimal Profit** The profit P (in dollars) for an electronics company is given by

$$P = -0.00015x^2 + 155x - 450{,}000$$

where x is the number of units produced.

(a) Use a graphing utility to graph the profit function.

(b) Graphically estimate the number of units that should be produced to yield a maximum profit.

(c) Explain how to confirm the result of part (b) algebraically.

11. **Batting Distance** In a study, baseballs were batted at angles of x degrees above the horizontal, each with an initial speed of 40 meters per second and a backspin of 30 revolutions per second. The distances $d(x)$ (in meters) traveled by the balls are shown in the table. *(Source: The Physics of Sports)*

x	10	15	30	36	42	43
$d(x)$	61.2	83.0	130.4	139.4	143.2	143.3

x	44	45	48	54	60
$d(x)$	142.8	142.7	140.7	132.8	119.7

(a) Use a graphing utility to create a scatter plot of the data.

(b) Use the *regression* feature of the graphing utility to find a quadratic model for the data.

(c) Use the graphing utility to graph the model from part (b) in the same viewing window as the scatter plot of the data.

(d) Find the vertex of the graph of the model from part (c). Interpret its meaning in the context of the problem.

12. **Dormitory Room Costs** The average costs C (in dollars) of a college dormitory room for the years 2003 through 2010 are shown in the table. *(Source: Digest of Education Statistics)*

Year	2003	2004	2005	2006
Cost, C	3179	3359	3569	3804

Year	2007	2008	2009	2010
Cost, C	4019	4214	4446	4657

(a) Use a graphing utility to create a scatter plot of the data. Let t represent the year, with $t = 3$ corresponding to 2003.

(b) Use the *regression* feature of the graphing utility to find a quadratic model for the data.

(c) Use the graphing utility to graph the model from part (b) in the same viewing window as the scatter plot of the data.

(d) Use the graph of the model from part (c) to predict the average cost of a dormitory room in 2011.

Sketching a Transformation of a Monomial Function
In Exercises 13–16, use the graph of $y = x^5$ to sketch the graph of the function.

13. $f(x) = (x + 4)^5$ **14.** $f(x) = x^5 + 1$

15. $f(x) = 3 - \frac{1}{2}x^5$ **16.** $f(x) = 2(x + 3)^5$

Sketching a Transformation of a Monomial Function
In Exercises 17–20, use the graph of $y = x^6$ to sketch the graph of the function.

17. $f(x) = x^6 - 2$ **18.** $f(x) = -\frac{1}{4}x^6$

19. $f(x) = -\frac{1}{2}x^6 - 5$ **20.** $f(x) = -(x + 7)^6 + 2$

Applying the Leading Coefficient Test   In Exercises 21–24, describe the right-hand and left-hand behavior of the graph of the polynomial function.

21. $f(x) = \frac{1}{2}x^3 + 2x$ **22.** $f(x) = 5 + 4x^3 - x^5$

23. $f(x) = -x^6 + 3x^4 - x^2 + 6$

24. $f(x) = \frac{3}{4}(x^4 + 3x^2 + 2)$

Finding Zeros of a Polynomial Function   In Exercises 25–28, find all real zeros of the function algebraically and use the results to sketch the graph *by hand*. Then use a graphing utility to confirm your sketch.

25. $f(x) = 16 - x^2$ **26.** $f(x) = x^4 - 6x^2 + 8$

27. $f(x) = x^3 - 7x^2 + 10x$

28. $f(x) = x^3 - 6x^2 - 3x + 18$

Long Division of Polynomials   In Exercises 29 and 30, use long division to divide.

29. $(2x^3 - 5x^2 - x) \div (2x + 1)$

30. $(x^4 - 5x^3 + 10x^2 - 12) \div (x^2 - 2x + 4)$

Using Synthetic Division   In Exercises 31 and 32, use synthetic division to divide.

31. $(x^3 - 6x + 9) \div (x + 3)$

32. $(x^5 - x^4 + x^3 - 13x^2 + x + 6) \div (x - 2)$

Using Synthetic Division   In Exercises 33 and 34, use synthetic division to find each function value.

33. $f(x) = 6 + 2x^2 - 3x^3$ (a) $f(2)$ (b) $f(-1)$

34. $f(x) = 2x^4 + 3x^3 + 6$ (a) $f(\frac{1}{2})$ (b) $f(-1)$

Factoring a Polynomial   In Exercises 35 and 36, (a) verify the given factors of $f(x)$, (b) find the remaining factors of $f(x)$, (c) use your results to write the complete factorization of $f(x)$, (d) list all real zeros of f, and (e) confirm your results by using a graphing utility to graph the function.

Function	*Factors*
35. $f(x) = x^3 - 4x^2 - 11x + 30$	$(x - 5), (x + 3)$
36. $f(x) = 3x^3 + 23x^2 + 37x - 15$	$(3x - 1), (x + 5)$

Simplifying a Rational Expression   In Exercises 37 and 38, simplify the rational expression.

37. $\dfrac{x^3 + 9x^2 + 2x - 48}{x - 2}$

38. $\dfrac{x^4 + 5x^3 - 20x - 16}{x^2 - 4}$

39. Profit   The profit P (in dollars) from selling a motorcycle is given by

$$P = -42x^3 + 3000x^2 - 6000, \quad 0 \le x \le 65$$

where x is the advertising expense (in tens of thousands of dollars). For this motorcycle, the advertising expense was $600,000 ($x = 60$) and the profit was $1,722,000.

(a) Use a graphing utility to graph the function and use the result to find another advertising expense that would have produced the same profit.

(b) Use synthetic division to confirm the result of part (a) algebraically.

40. Profit   The profit P (in dollars) from selling a novel is given by

$$P = -150x^3 + 7500x^2 - 450,000, \quad 0 \le x \le 45$$

where x is the advertising expense (in tens of thousands of dollars). For this novel, the advertising expense was $400,000 ($x = 40$), and the profit was $1,950,000.

(a) Use a graphing utility to graph the function and use the result to find another advertising expense that would have produced the same profit.

(b) Use synthetic division to confirm the result of part (a) algebraically.

 Using the Rational Zero Test   In Exercises 41 and 42, use the Rational Zero Test to list all possible rational zeros of f. Then use a graphing utility to graph the function. Use the graph to help determine which of the possible rational zeros are actual zeros of the function.

41. $f(x) = -4x^3 + 8x^2 - 3x + 15$

42. $f(x) = 3x^4 + 4x^3 - 5x^2 + 10x - 8$

Using the Rational Zero Test   In Exercises 43–50, find all the real zeros of the function.

43. $f(x) = x^3 + 2x^2 - 5x - 6$

44. $g(x) = 2x^3 - 15x^2 + 24x + 16$

45. $h(x) = 3x^4 - 27x^2 + 60$

46. $f(x) = x^5 - 4x^3 + 3x$

47. $B(x) = 6x^3 - 19x^2 + 11x + 6$

48. $C(x) = 3x^4 + 3x^3 - 7x^2 - x + 2$

49. $p(x) = x^4 - x^3 - 2x - 4$

50. $q(x) = x^5 - 2x^4 + 2x^3 - 4x^2 - 3x + 6$

Using the Intermediate Value Theorem   In Exercises 51 and 52, use the Intermediate Value Theorem to approximate the zero of f in the interval $[a, b]$. Give your approximation to the nearest tenth.

51. $f(x) = x^3 - 4x + 3, \quad [-3, -2]$

52. $f(x) = x^5 + 5x^2 + x - 1, \quad [0, 1]$

 Approximating the Zeros of a Polynomial Function   In Exercises 53 and 54, use the *zero*, *root*, or *zoom* and *trace* features of a graphing utility to approximate the real zeros of f. Give your approximations to the nearest thousandth.

53. $f(x) = 5x^3 - 11x - 3$

54. $f(x) = 2x^4 - 9x^3 - 5x^2 + 10x + 12$

 55. Wholesale Revenue   The revenues R (in millions of dollars) for Costco Wholesale for the years 2001 through 2010 are shown in the table. *(Source: Costco Wholesale)*

Year	Revenue, R
2001	34,797
2002	38,762
2003	42,546
2004	48,107
2005	52,935
2006	60,151
2007	64,909
2008	72,483
2009	71,449
2010	77,946

(a) Use a graphing utility to create a scatter plot of the data. Let t represent the year, with $t = 1$ corresponding to 2001.

(b) Use the *regression* feature of the graphing utility to find a linear model, a quadratic model, and a quartic model for the data.

(c) Use the graphing utility to graph each model with the data. How well does each model fit the data?

(d) Use each model to predict the year in which the revenue will be about \$87 billion. Explain any differences in the predictions.

 56. Home Furnishing Sales   The sales S (in millions of dollars) for Bed Bath & Beyond for the years 2000 through 2009 are shown in the table. *(Source: Bed Bath & Beyond)*

Year	Sales, S	Year	Sales, S
2000	2396.7	2005	5809.6
2001	2928.0	2006	6617.4
2002	3665.2	2007	7048.9
2003	4478.0	2008	7208.3
2004	5147.7	2009	7828.8

(a) Use a graphing utility to create a scatter plot of the data. Let t represent the year, with $t = 0$ corresponding to 2000.

(b) Use the *regression* feature of the graphing utility to find a linear model, a quadratic model, and a quartic model for the data.

(c) Use the graphing utility to graph each model with the data. How well does each model fit the data?

(d) Use each model to predict the year in which the sales will be about \$8.8 billion. Explain any differences in the predictions.

Writing Complex Numbers in Standard Form   In Exercises 57–60, write the complex number in standard form and find its complex conjugate.

57. $\sqrt{-32}$ **58.** 12

59. $-3 + \sqrt{-16}$ **60.** $2 - \sqrt{-18}$

Adding, Subtracting, or Multiplying Complex Numbers   In Exercises 61–70, perform the indicated operation and write the result in standard form.

61. $(7 - 3i) + (-2 + 5i)$ **62.** $(14 + 8i) - (-1 - 2i)$

63. $(5 + 8i)(5 - 8i)$ **64.** $\left(\frac{1}{2} + \frac{3}{4}i\right)\left(\frac{1}{2} - \frac{3}{4}i\right)$

65. $-2i(4 - 5i)$ **66.** $-3(-2 + 4i)$

67. $(3 + 4i)^2$ **68.** $(2 - 5i)^2$

69. $(3 + 2i)^2 + (3 - 2i)^2$ **70.** $(1 + i)^2 - (1 - i)^2$

Writing Quotients in Standard Form   In Exercises 71–74, write the quotient in standard form.

71. $\dfrac{8 - i}{2 + i}$ **72.** $\dfrac{3 - 4i}{1 - 5i}$

73. $\dfrac{4 - 3i}{i}$ **74.** $\dfrac{2}{(1 + i)^2}$

Complex Solutions of a Quadratic Equation   In Exercises 75–78, solve the quadratic equation.

75. $2x^2 - x + 3 = 0$ **76.** $3x^2 + 6x + 11 = 0$

77. $4x^2 + 11x + 3 = 0$ **78.** $9x^2 - 2x + 5 = 0$

Plotting Complex Numbers in the Complex Plane
In Exercises 79 and 80, plot the complex number in the complex plane.

79. $-3 + 2i$ **80.** $-1 - 4i$

Finding the Zeros of a Polynomial Function   In Exercises 81–86, find all the zeros of the function and write the polynomial as the product of linear factors.

81. $f(x) = x^4 - 81$ **82.** $h(x) = x^4 + 17x^2 + 16$

83. $f(t) = t^3 + 5t^2 + 3t + 15$

84. $h(x) = 2x^3 - 5x^2 + 4x - 10$

85. $g(x) = 4x^3 - 8x^2 + 9x - 18$

86. $f(x) = x^5 - 2x^4 + x^3 - x^2 + 2x - 1$

Finding a Polynomial Function with Given Zeros   In Exercises 87 and 88, find a polynomial function with real coefficients that has the given zeros.

87. $1, 3i, -3i$ **88.** $1, -2, 1 - 3i, 1 + 3i$

Factoring a Polynomial   In Exercises 89 and 90, write the polynomial (a) as the product of factors that are irreducible over the rationals, (b) as the product of linear and quadratic factors that are irreducible over the *reals*, and (c) in completely factored form.

89. $x^4 + 5x^2 - 24$

90. $x^4 - 2x^3 - 2x^2 - 14x - 63$
    (*Hint:* One factor is $x^2 + 7$.)

Finding the Zeros of a Polynomial Function   In Exercises 91–94, use the given zero of f to find all the zeros of f.

91. $f(x) = 4x^3 - x^2 + 64x - 16,$ $-4i$

92. $f(x) = 50 - 75x + 2x^2 - 3x^3,$ $5i$

93. $f(x) = x^4 + 7x^3 + 24x^2 + 58x + 40,$ $-1 + 3i$

94. $f(x) = x^4 + 4x^3 + 8x^2 + 4x + 7,$ $-2 - \sqrt{3}\,i$

Finding the Domain and Asymptotes of a Rational Function   In Exercises 95–98, find the domain of the function and identify any horizontal or vertical asymptotes.

95. $f(x) = \dfrac{-3}{x + 2}$ **96.** $f(x) = \dfrac{3x^2 + 7x - 5}{x^2 + 1}$

97. $f(x) = \dfrac{2x^2}{x^2 - 9}$ **98.** $f(x) = \dfrac{3x}{x^2 + x - 6}$

Sketching the Graph of a Rational Function   In Exercises 99–102, sketch the graph of the rational function. To aid in sketching the graphs, check for intercepts, symmetry, vertical asymptotes, and horizontal asymptotes.

99. $P(x) = \dfrac{3 - x}{x + 2}$ **100.** $f(x) = \dfrac{4}{(x - 1)^2}$

101. $g(x) = \dfrac{1}{x^2 - 4} + 2$ **102.** $h(x) = \dfrac{-3x}{2x^2 + 3x - 5}$

Sketching the Graph of a Rational Function   In Exercises 103 and 104, find all asymptotes (vertical, horizontal, and/or slant) of the given function. Sketch the graph of f.

103. $f(x) = \dfrac{x^2 - 16}{x - 4}$ **104.** $f(x) = \dfrac{x^3}{x^2 - 5}$

105. Population of Fish   The Wildlife Commission introduces 60,000 game fish into a large lake. The population N (in thousands) of the fish is

$$N = \frac{20(3 + 5t)}{1 + 0.06t}, \quad t \geq 0$$

where t is time (in years).

(a) Find the populations when $t = 5$, 10, and 25.

(b) What is the limiting number of fish in the lake as time progresses?

106. Average Recycling Cost   The cost C (in dollars) of recycling a waste product is $C = 325{,}000 + 8.5x$, for $x > 0$, where x is the number of pounds of waste. The average recycling cost $\overline{C}$ per pound is

$$\overline{C} = \frac{C}{x} = \frac{325{,}000 + 8.5x}{x}, \quad x > 0.$$

(a) Sketch the graph of $\overline{C}$.

(b) Find the average costs of recycling 1000, 10,000, 100,000, and 1,000,000 pounds of waste. What can you conclude?

107. Smokestack Emissions   The cost C (in dollars) of removing p percent of the air pollutants in the stack emissions of a utility company that burns coal to generate electricity is

$$C = \frac{105{,}000p}{100 - p}, \quad 0 \leq p < 100.$$

(a) Find the cost of removing 25% of the pollutants.

(b) Find the cost of removing 60% of the pollutants.

108. Human Memory Model   Consider the learning curve modeled by

$$P = \frac{0.7 + 0.65(n - 1)}{1 + 0.65(n - 1)}, \quad n \geq 0$$

where P is the percent of correct responses (in decimal form) after n trials.

(a) Complete the table.

n	1	2	3	4	5	6	7	8	9	10
P										

(b) According to this model, what is the limiting percent of correct responses as n increases?

TEST YOURSELF

See www.CalcChat.com for worked-out solutions to odd-numbered exercises.

Take this test as you would take a test in class. When you are done, check your work against the answers given in the back of the book.

1. Sketch the graph of the quadratic function given by

   $$f(x) = -\tfrac{1}{2}(x - 1)^2 - 5.$$

   Identify the vertex and intercept(s).

2. Describe the right-hand and left-hand behavior of the graph of f.
   (a) $f(x) = 12x^3 - 5x^2 - 49x + 15$
   (b) $f(x) = 5x^4 - 3x^3 + 2x^2 + 11x + 12$

3. Simplify $\dfrac{x^4 + 4x^3 - 19x^2 - 106x - 120}{x^2 - 3x - 10}$.

4. List all possible rational zeros of

   $$f(x) = 4x^4 - 16x^3 + 3x^2 + 36x - 27.$$

   Use synthetic division to show that $x = -\tfrac{3}{2}$ and $x = \tfrac{3}{2}$ are zeros of f. Using these results, completely factor the polynomial.

5. The sales per share S (in dollars) for Hormel Foods for the years 2001 through 2010 are shown in the table at the left.   (Source: Hormel Foods Corporation)

   (a) Use a graphing utility to create a scatter plot of the data. Let t represent the year, with $t = 1$ corresponding to 2001.

   (b) Use the *regression* feature of the graphing utility to find a linear model, a quadratic model, and a quartic model for the data.

   (c) Use the graphing utility to graph each model separately with the data in the same viewing window. How well does each model fit the data?

   (d) Use each model to predict the year in which the sales per share will be about $35. Then discuss the appropriateness of each model for predicting future values.

Year	Sales per share, S
2001	14.87
2002	14.13
2003	15.15
2004	17.33
2005	19.64
2006	20.92
2007	22.82
2008	25.11
2009	24.45
2010	27.15

Table for 5

In Exercises 6–9, perform the indicated operation and write the result in standard form.

6. $(12 + 3i) + (4 - 5i)$

7. $(10 - 2i) - (3 + 8i)$

8. $\left(5 + \sqrt{-12}\right)\left(3 - \sqrt{-12}\right)$

9. $(4 + 3i)(2 - 5i)$

10. Write the quotient in standard form: $\dfrac{1 + i}{1 - i}$.

In Exercises 11 and 12, solve the quadratic equation.

11. $x^2 + 5x + 7 = 0$

12. $2x^2 - 5x + 11 = 0$

13. Find a polynomial function with real coefficients that has 2, 5, $3i$, and $-3i$ as zeros.

14. Find all the zeros of $f(x) = x^3 + 2x^2 + 5x + 10$, given that $\sqrt{5}i$ is a zero.

15. Sketch the graph of

    $$f(x) = \frac{3x}{x - 2}.$$

    Label any intercepts and asymptotes. What is the domain of f?

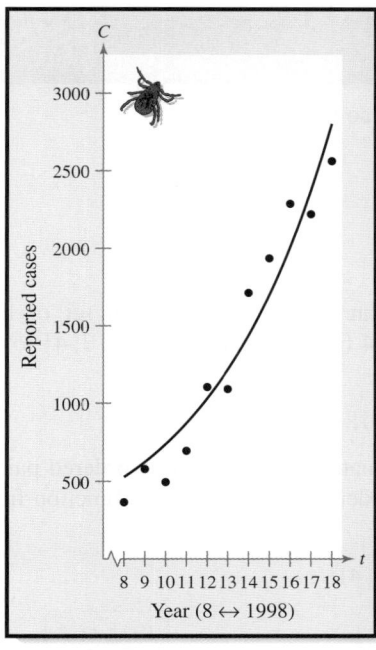

Example 1 on page 388 shows how an exponential growth function can be used to model the number of reported cases of Rocky Mountain spotted fever.

4 Exponential and Logarithmic Functions

Kurhan/Shutterstock.com
wavebreakmedia ltd/www.shutterstock.com

335

4.1 Inverse Functions

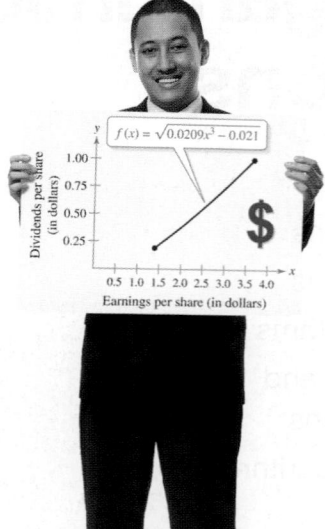

- Determine whether a function has an inverse function.
- Find the inverse function of a function.
- Graph a function and its inverse function.

Inverse Functions

Recall from Section 2.4 that a function can be represented by a set of ordered pairs. For instance, the function $f(x) = x + 4$ from the set $A = \{1, 2, 3, 4\}$ to the set $B = \{5, 6, 7, 8\}$ can be written as

$$f(x) = x + 4: \ \{(1, 5), (2, 6), (3, 7), (4, 8)\}.$$

By interchanging the first and second coordinates of each of these ordered pairs, you can form the **inverse function** of f, which is denoted by f^{-1}. It is a function from the set B to the set A and can be written as

$$f^{-1}(x) = x - 4: \ \{(5, 1), (6, 2), (7, 3), (8, 4)\}.$$

Note that the domain of f is equal to the range of f^{-1} and vice versa, as shown in Figure 4.1. Also note that the functions f and f^{-1} have the effect of "undoing" each other. In other words, when you form the composition of f with f^{-1} or the composition of f^{-1} with f, you obtain the identity function, as shown.

$$f(f^{-1}(x)) = f(x - 4)$$
$$= (x - 4) + 4 = x$$
$$f^{-1}(f(x)) = f^{-1}(x + 4)$$
$$= (x + 4) - 4 = x$$

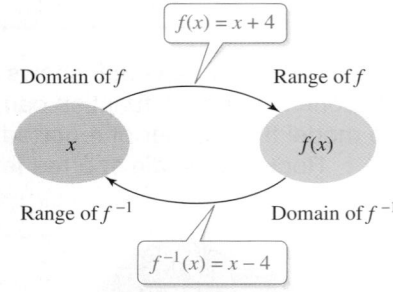

FIGURE 4.1

In Exercise 75 on page 345, you will use an inverse function to approximate the earnings per share in 2008 for Wal-Mart Stores.

Example 1   Finding Inverse Functions Informally

Find the inverse function of $f(x) = 4x$. Then verify that both $f(f^{-1}(x))$ and $f^{-1}(f(x))$ are equal to the identity function.

SOLUTION   The function $f(x) = 4x$ *multiplies* each input by 4. To "undo" this function, you need to *divide* each input by 4. So, the inverse function of $f(x) = 4x$ is

$$f^{-1}(x) = \frac{x}{4}.$$

You can verify that both $f(f^{-1}(x))$ and $f^{-1}(f(x))$ are equal to the identity function as shown.

$$f(f^{-1}(x)) = f\left(\frac{x}{4}\right) = 4\left(\frac{x}{4}\right) = x \qquad f^{-1}(f(x)) = f^{-1}(4x) = \frac{4x}{4} = x$$

✓ Checkpoint 1

Find the inverse function of $f(x) = \dfrac{x}{6}$. Then verify that both $f(f^{-1}(x))$ and $f^{-1}(f(x))$ are equal to the identity function.   ■

| Example 2 | Finding Inverse Functions Informally |

Find the inverse function of

$$f(x) = x - 6.$$

Then verify that both $f(f^{-1}(x))$ and $f^{-1}(f(x))$ are equal to the identity function.

SOLUTION   The function $f(x) = x - 6$ *subtracts* 6 from each input. To "undo" this function, you need to *add* 6 to each input. So, the inverse function of $f(x) = x - 6$ is

$$f^{-1}(x) = x + 6.$$

You can verify that both $f(f^{-1}(x))$ and $f^{-1}(f(x))$ are equal to the identity function as shown.

$$f(f^{-1}(x)) = f(x + 6) \qquad \text{Substitute } x + 6 \text{ for } f^{-1}(x).$$

$$= (x + 6) - 6 \qquad \text{Substitute } x + 6 \text{ into } f(x).$$

$$= x \qquad \text{Identity function}$$

$$f^{-1}(f(x)) = f^{-1}(x - 6) \qquad \text{Substitute } x - 6 \text{ for } f(x).$$

$$= (x - 6) + 6 \qquad \text{Substitute } x - 6 \text{ into } f^{-1}(x).$$

$$= x \qquad \text{Identity function}$$

✔ **Checkpoint 2**

Find the inverse function of $f(x) = x + 10$. Then verify that both $f(f^{-1}(x))$ and $f^{-1}(f(x))$ are equal to the identity function.   ■

The formal definition of inverse function is shown below.

Definition of Inverse Function

Let f and g be two functions such that

$$f(g(x)) = x \qquad \text{for every } x \text{ in the domain of } g$$

and

$$g(f(x)) = x \qquad \text{for every } x \text{ in the domain of } f.$$

Under these conditions, the function g is the **inverse function** of the function f. The function g is denoted by f^{-1} (read "f-inverse"). So,

$$f(f^{-1}(x)) = x \quad \text{and} \quad f^{-1}(f(x)) = x.$$

The domain of f must be equal to the range of f^{-1}, and the range of f must be equal to the domain of f^{-1}.

Do not be confused by the use of -1 to denote the inverse function f^{-1}. In this text, f^{-1} *always* refers to the inverse function of the function f and *not* to the reciprocal of $f(x)$. That is,

$$f^{-1}(x) \neq \frac{1}{f(x)}.$$

If the function g is the inverse function of the function f, then it must also be true that the function f is the inverse function of the function g. For this reason, you can say that the functions f and g are *inverse functions of each other.*

Example 3 **Verifying Inverse Functions**

Show that the following functions are inverse functions.

$$f(x) = 2x^3 - 1 \quad \text{and} \quad g(x) = \sqrt[3]{\frac{x + 1}{2}}$$

SOLUTION

$$f(g(x)) = f\left(\sqrt[3]{\frac{x + 1}{2}}\right) = 2\left(\sqrt[3]{\frac{x + 1}{2}}\right)^3 - 1$$

$$= 2\left(\frac{x + 1}{2}\right) - 1$$

$$= x + 1 - 1$$

$$= x$$

$$g(f(x)) = g(2x^3 - 1) = \sqrt[3]{\frac{(2x^3 - 1) + 1}{2}}$$

$$= \sqrt[3]{\frac{2x^3}{2}}$$

$$= \sqrt[3]{x^3}$$

$$= x$$

✓**Checkpoint 3**

Show that the following functions are inverse functions.

$$f(x) = x^3 + 6 \quad \text{and} \quad g(x) = \sqrt[3]{x - 6}$$

Example 4 **Verifying Inverse Functions**

Which of the functions given by

$$g(x) = \frac{x - 2}{5} \quad \text{and} \quad h(x) = \frac{5}{x} + 2$$

is the inverse function of $f(x) = \dfrac{5}{x - 2}$?

SOLUTION   By forming the composition of f with g, you can see that

$$f(g(x)) = f\left(\frac{x - 2}{5}\right) = \frac{5}{[(x - 2)/5] - 2} = \frac{25}{x - 12} \neq x.$$

Because this composition is not equal to the identity function x, it follows that g is *not* the inverse function of f. By forming the composition of f with h, you have

$$f(h(x)) = f\left(\frac{5}{x} + 2\right) = \frac{5}{[(5/x) + 2] - 2} = \frac{5}{5/x} = x.$$

So, it appears that h is the inverse function of f. You can confirm this result by showing that the composition of h with f is also equal to the identity function.

✓**Checkpoint 4**

Which of the functions given by $g(x) = \dfrac{x + 4}{3}$ and $h(x) = \dfrac{x}{3} + 4$ is the inverse function of $f(x) = 3x - 4$?

Finding Inverse Functions

For simple functions (such as the ones in Examples 1 and 2), you can find inverse functions by inspection. For more complicated functions, it is best to use the following guidelines. The key step in these guidelines is switching the roles of x and y. This step corresponds to the fact that inverse functions have ordered pairs with the coordinates reversed.

STUDY TIP

Note in Step 3 of the guidelines for finding inverse functions that it is possible for a function to have no inverse function. For instance, the function given by $f(x) = x^2$ has no inverse function.

Guidelines for Finding Inverse Functions

1. In the equation for $f(x)$, replace $f(x)$ with y.

2. Interchange the roles of x and y.

3. Solve the new equation for y. If the new equation does not represent y as a function of x, then the function f does not have an inverse function. If the new equation does represent y as a function of x, then continue to Step 4.

4. Replace y with $f^{-1}(x)$ in the new equation.

5. Verify that f and f^{-1} are inverse functions of each other by showing that the domain of f is equal to the range of f^{-1}, the range of f is equal to the domain of f^{-1}, and

$$f(f^{-1}(x)) = x = f^{-1}(f(x)).$$

Example 5 **Finding Inverse Functions**

Find the inverse function of

$$f(x) = \frac{5 - 3x}{2}.$$

SOLUTION

$f(x) = \dfrac{5 - 3x}{2}$	Write original function.
$y = \dfrac{5 - 3x}{2}$	Replace $f(x)$ with y.
$x = \dfrac{5 - 3y}{2}$	Interchange x and y.
$2x = 5 - 3y$	Multiply each side by 2.
$3y = 5 - 2x$	Isolate the y-term.
$y = \dfrac{5 - 2x}{3}$	Solve for y.
$f^{-1}(x) = \dfrac{5 - 2x}{3}$	Replace y with $f^{-1}(x)$.

Note that both f and f^{-1} have domains and ranges that consist of the entire set of real numbers. Check that

$$f(f^{-1}(x)) = x \quad \text{and} \quad f^{-1}(f(x)) = x.$$

✓ **Checkpoint 5**

Find the inverse function of $f(x) = 4x + 5$.

The Graph of an Inverse Function

The graphs of a function f and its inverse function f^{-1} are related to each other in the following way. If the point (a, b) lies on the graph of f, then the point (b, a) must lie on the graph of f^{-1}, and vice versa. This means that the graph of f^{-1} is a *reflection* of the graph of f in the line $y = x$, as shown in Figure 4.2.

TECH TUTOR

A graphing utility can help you check that the graphs of f and f^{-1} are reflections of each other in the line $y = x$. To do this, graph $y = f(x)$, $y = f^{-1}(x)$, and $y = x$ in the same viewing window, using a square setting.

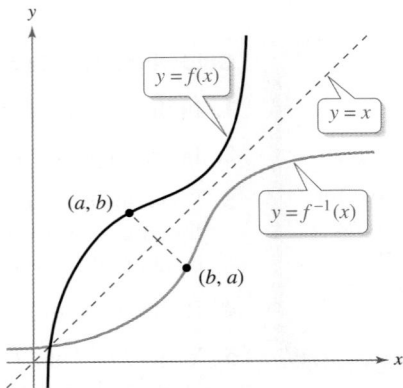

The graph of f^{-1} is a reflection of the graph of f in the line $y = x$.

FIGURE 4.2

Example 6   **The Graphs of f and f^{-1}**

Sketch the graphs of the inverse functions given by

$$f(x) = 2x - 3$$

and

$$f^{-1}(x) = \tfrac{1}{2}(x + 3)$$

in the same coordinate plane, and show that the graphs are reflections of each other in the line $y = x$.

SOLUTION   The graphs of f and f^{-1} are shown in Figure 4.3. Visually, it appears that the graphs are reflections of each other in the line $y = x$. You can further verify this reflective property by testing a few points on each graph. Note in the following list that if the point (a, b) is on the graph of f, then the point (b, a) is on the graph of f^{-1}.

Graph of $f(x) = 2x - 3$	Graph of $f^{-1}(x) = \tfrac{1}{2}(x + 3)$
$(0, -3)$	$(-3, 0)$
$(1, -1)$	$(-1, 1)$
$(2, 1)$	$(1, 2)$
$(3, 3)$	$(3, 3)$

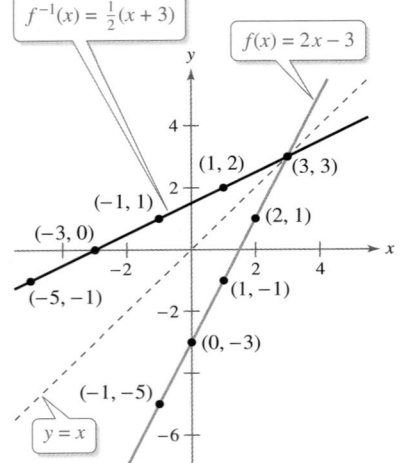

FIGURE 4.3

✓ **Checkpoint 6**

Sketch the graphs of the inverse functions given by

$$f(x) = \tfrac{2}{5}x + 2 \quad \text{and} \quad f^{-1}(x) = \tfrac{5}{2}x - 5$$

in the same coordinate plane, and show that the graphs are reflections of each other in the line $y = x$.

The Study Tip on page 339 mentioned that the function given by $f(x) = x^2$ has no inverse function. This means, *assuming the domain of f is the entire real line*, that the function given by $f(x) = x^2$ has no inverse function. If the domain of f is restricted to the nonnegative real numbers, however, then f does have an inverse function, as demonstrated in Example 7.

Example 7   The Graphs of f and f^{-1}

Sketch the graphs of the inverse functions given by

$$f(x) = x^2, \quad x \geq 0, \quad \text{and} \quad f^{-1}(x) = \sqrt{x}$$

in the same coordinate plane, and show that the graphs are reflections of each other in the line $y = x$.

SOLUTION   The graphs of f and f^{-1} are shown in Figure 4.4. Visually, it appears that the graphs are reflections of each other in the line $y = x$. You can further verify this reflective property by testing a few points on each graph. Note in the following list that if the point (a, b) is on the graph of f, then the point (b, a) is on the graph of f^{-1}.

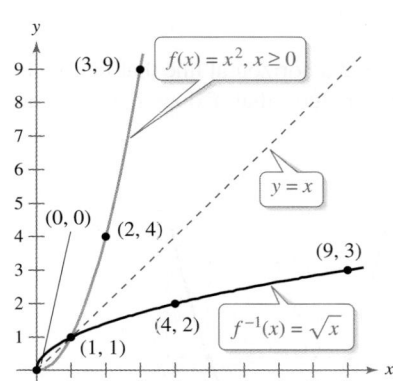

FIGURE 4.4

Graph of $f(x) = x^2, \quad x \geq 0$	*Graph of* $f^{-1}(x) = \sqrt{x}$
$(0, 0)$	$(0, 0)$
$(1, 1)$	$(1, 1)$
$(2, 4)$	$(4, 2)$
$(3, 9)$	$(9, 3)$

You can verify algebraically that the functions are inverse functions of each other by showing that $f(f^{-1}(x)) = x$ and $f^{-1}(f(x)) = x$ as shown.

$$f(f^{-1}(x)) = f\left(\sqrt{x}\right)$$
$$= \left(\sqrt{x}\right)^2$$
$$= x, \quad \text{if } x \geq 0$$
$$f^{-1}(f(x)) = f^{-1}(x^2)$$
$$= \sqrt{x^2}$$
$$= x, \quad \text{if } x \geq 0$$

✓**Checkpoint 7**

Sketch the graphs of the inverse functions given by

$$f(x) = x^2 + 3, \quad x \geq 0, \quad \text{and} \quad f^{-1}(x) = \sqrt{x - 3}$$

in the same coordinate plane, and show that the graphs are reflections of each other in the line $y = x$.   ■

The guidelines for finding the inverse function of a function include an *algebraic* test for determining whether a function has an inverse function. The reflective property of the graphs of inverse functions gives you a *geometric* test for determining whether a function has an inverse function. This test is called the **Horizontal Line Test** for inverse functions.

Horizontal Line Test for Inverse Functions

A function f has an inverse function if and only if no *horizontal* line intersects the graph of f at more than one point.

Example 8   **Applying the Horizontal Line Test**

Use the graph of f to determine whether the function has an inverse function.

a. $f(x) = x^3 - 1$ **b.** $f(x) = x^2 - 1$

SOLUTION

a. The graph of the function given by

$$f(x) = x^3 - 1$$

is shown in Figure 4.5(a). Because no horizontal line intersects the graph of f at more than one point, you can conclude that f does have an inverse function.

b. The graph of the function given by

$$f(x) = x^2 - 1$$

is shown in Figure 4.5(b). Because it is possible to find a horizontal line that intersects the graph of f at more than one point, you can conclude that f does not have an inverse function.

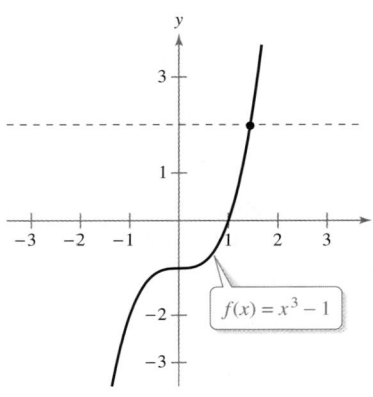
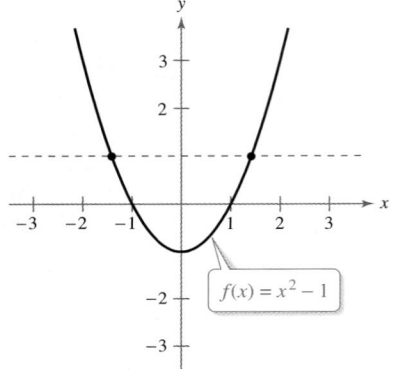

(a)   (b)

FIGURE 4.5

✓ **Checkpoint 8**

Use the graph of f to determine whether the function has an inverse function.

a. $f(x) = |x|$ **b.** $f(x) = \sqrt{x}$

SUMMARIZE   (Section 4.1)

1. State the definition of an inverse function *(page 337)*. For examples of finding inverse functions informally and verifying inverse functions, see Examples 1, 2, 3, and 4.

2. Describe how to find an inverse function *(page 339)*. For an example of finding an inverse function, see Example 5.

3. Describe the relationship between the graph of a function and the graph of its inverse function *(page 340)*. For examples of sketching the graphs of functions and their inverse functions, see Examples 6 and 7.

4. Describe the Horizontal Line Test for inverse functions *(page 341)*. For an example of applying the Horizontal Line Test, see Example 8.

The following warm-up exercises involve skills that were covered in earlier sections. You will use these skills in the exercise set for this section. For additional help, review Sections 0.2, 0.4, 1.2, and 2.4.

In Exercises 1–4, find the domain of the function.

1. $f(x) = \sqrt[3]{x + 1}$

2. $f(x) = \sqrt{x + 1}$

3. $g(x) = \dfrac{2}{x^2 - 2x}$

4. $h(x) = \dfrac{x}{3x + 5}$

In Exercises 5–8, simplify the expression.

5. $2\left(\dfrac{x + 5}{2}\right) - 5$

6. $7 - 10\left(\dfrac{7 - x}{10}\right)$

7. $\sqrt[3]{2\left(\dfrac{x^3}{2} - 2\right) + 4}$

8. $\sqrt[5]{(x + 2)^5} - 2$

In Exercises 9 and 10, solve for x in terms of y.

9. $y = \dfrac{2x - 6}{3}$

10. $y = \sqrt[3]{2x - 4}$

Exercises 4.1

Finding Inverse Functions   In Exercises 1–4, find the inverse function of the function f given by the set of ordered pairs.

1. $\{(1, 4), (2, 5), (3, 6), (4, 7)\}$

2. $\{(6, 2), (5, 3), (4, 4), (3, 5)\}$

3. $\{(-1, 1), (-2, 2), (-3, 3), (-4, 4)\}$

4. $\{(6, -2), (5, -3), (4, -4), (3, -5)\}$

Finding Inverse Functions Informally   In Exercises 5–8, find the inverse function informally. Verify that $f(f^{-1}(x)) = x$ and $f^{-1}(f(x)) = x$. *See Examples 1 and 2.*

5. $f(x) = 2x$

6. $f(x) = -\dfrac{x}{4}$

7. $f(x) = x - 5$

8. $f(x) = x + 7$

Verifying Inverse Functions   In Exercises 9–16, show that f and g are inverse functions by using the definition of inverse functions. *See Examples 3 and 4.*

9. $f(x) = 5x + 1, \quad g(x) = \dfrac{x - 1}{5}$

10. $f(x) = 3 - 4x, \quad g(x) = \dfrac{3 - x}{4}$

11. $f(x) = x^3, \quad g(x) = \sqrt[3]{x}$

12. $f(x) = \dfrac{1}{x}, \quad g(x) = \dfrac{1}{x}$

13. $f(x) = \sqrt{x - 4}, \quad g(x) = x^2 + 4, \quad x \geq 0$

14. $f(x) = 9 - x^2, \quad x \geq 0, \quad g(x) = \sqrt{9 - x}, \quad x \leq 9$

15. $f(x) = 1 - x^3, \quad g(x) = \sqrt[3]{1 - x}$

16. $f(x) = \dfrac{1}{1 + x}, \quad x \geq 0, \quad g(x) = \dfrac{1 - x}{x}, \quad 0 < x \leq 1$

Sketching Inverse Functions   In Exercises 17 and 18, use the graph of f to complete the table and to sketch the graph of f^{-1}.

17.

x	0	1	2	3	4
$f^{-1}(x)$					

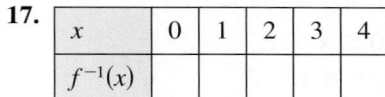
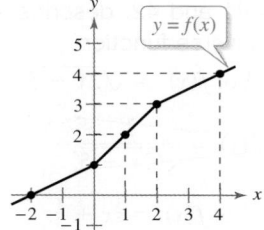

18.

x	0	2	4	6
$f^{-1}(x)$				

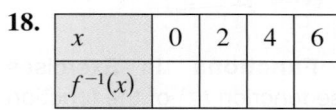
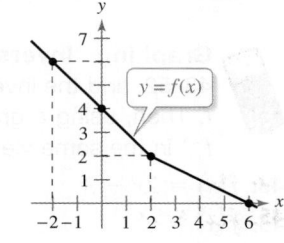

Finding Inverse Functions   In Exercises 19–34, determine whether the function has an inverse function. If it does, find its inverse function. *See Example 5.*

19. $f(x) = x^4$

20. $f(x) = \dfrac{1}{x^2}$

21. $g(x) = \dfrac{x}{8}$

22. $f(x) = 3x + 5$

23. $p(x) = -4$

24. $f(x) = \dfrac{3x + 4}{5}$

25. $f(x) = (x + 3)^2, \ x \geq -3$

26. $q(x) = (x - 5)^2$

27. $h(x) = \dfrac{1}{x}$

28. $f(x) = |x - 2|, \ x \leq 2$

29. $f(x) = \sqrt{2x + 3}$

30. $f(x) = \sqrt{x - 2}$

31. $g(x) = x^2 - x^4$

32. $f(x) = \dfrac{x^2}{x^2 + 1}$

33. $f(x) = 25 - x^2, \ x \leq 0$

34. $f(x) = 36 + x^2, \ x \leq 0$

The Graphs of f and f^{-1}   In Exercises 35–40, sketch the graphs of the inverse functions in the same coordinate plane and show that the graphs are reflections of each other in the line $y = x$. *See Examples 6 and 7.*

35. $f(x) = \dfrac{x}{3},$ $f^{-1}(x) = 3x$

36. $f(x) = x + 8,$ $f^{-1}(x) = x - 8$

37. $f(x) = 7x + 1,$ $f^{-1}(x) = \dfrac{x - 1}{7}$

38. $f(x) = \dfrac{4}{x},$ $f^{-1}(x) = \dfrac{4}{x}$

39. $f(x) = x^5 - 2,$ $f^{-1}(x) = \sqrt[5]{x + 2}$

40. $f(x) = \sqrt{4 - x^2}, \ 0 \leq x \leq 2,$
   $f^{-1}(x) = \sqrt{4 - x^2}, \ 0 \leq x \leq 2$

Error Analysis   In Exercises 41 and 42, describe and correct the error in finding the inverse function.

41. Find the inverse function f^{-1} of $f(x) = \sqrt{2x - 5}$.

$$f(x) = \sqrt{2x - 5}, \ \text{so} \ f^{-1}(x) = \dfrac{1}{\sqrt{2x - 5}}.$$

42. Find the inverse function f^{-1} of $f(x) = \frac{3}{5}x + \frac{1}{3}$.

$$f(x) = \tfrac{3}{5}x + \tfrac{1}{3}, \ \text{so} \ f^{-1}(x) = \tfrac{5}{3}x - 3.$$

 Graphing Inverse Functions   In Exercises 43–52, find the inverse function f^{-1} of the function f. Then, using a graphing utility, graph both f and f^{-1} in the same viewing window.

43. $f(x) = 2x - 3$

44. $f(x) = 5x + 2$

45. $f(x) = x^5$

46. $f(x) = x^3 + 1$

47. $f(x) = \sqrt{x}$

48. $f(x) = x^2, \ x \geq 0$

49. $f(x) = \sqrt{16 - x^2}, \ \ 0 \leq x \leq 4$

50. $f(x) = \dfrac{3}{x + 1}$

51. $f(x) = \sqrt[3]{x + 2}$

52. $f(x) = x^{3/5} - 2$

Applying the Horizontal Line Test   In Exercises 53–56, use the graph of f to determine whether the function has an inverse function. *See Example 8.*

53.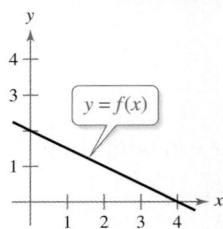

54.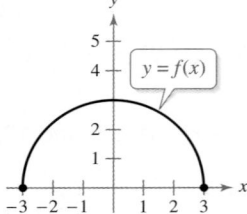

55.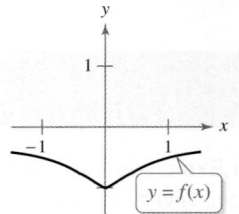

56.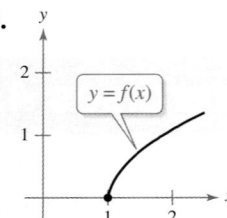

Applying the Horizontal Test   In Exercises 57–62, graph the function and use the Horizontal Line Test to determine whether the function has an inverse function.

57. $g(x) = \dfrac{5 - 2x}{3}$

58. $f(x) = 10$

59. $h(x) = |x - 5|$

60. $g(x) = (x - 3)^2$

61. $f(x) = -\sqrt{9 - x^2}$

62. $f(x) = (x - 1)^3$

Finding the Composition of Functions   In Exercises 63–66, use the functions given by

$$f(x) = x + 4 \quad \text{and} \quad g(x) = 2x - 6$$

to find the composition of functions.

63. $g^{-1} \circ f^{-1}$

64. $f^{-1} \circ g^{-1}$

65. $(f \circ g)^{-1}$

66. $(g \circ f)^{-1}$

Evaluating Functions   In Exercises 67–70, use the functions given by

$$f(x) = \dfrac{1}{8}x - 3 \quad \text{and} \quad g(x) = x^2$$

to find the value.

67. $(f^{-1} \circ g^{-1})(1)$

68. $(g^{-1} \circ f^{-1})(-3)$

69. $(f^{-1} \circ f^{-1})(6)$

70. $(g^{-1} \circ g^{-1})(-4)$

71. Profit A company's profit P for producing x units is given by $P(x) = 47x - 5736$. Find the inverse function $P^{-1}(x)$ and explain what it represents. Describe the domains of $P(x)$ and $P^{-1}(x)$.

72. HOW DO YOU SEE IT? The cost C for a business to make personalized T-shirts is given by

$$C(x) = 7.50x + 1500$$

where x represents the number of T-shirts.

(a) The graphs of $C(x)$ and $C^{-1}(x)$ are shown below. Match each function with its graph.

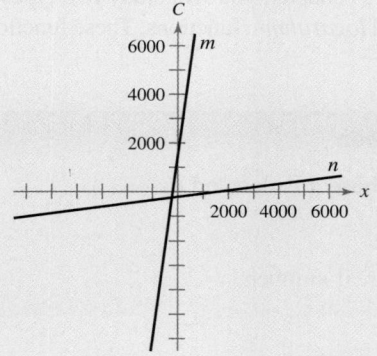

(b) Explain what $C(x)$ and $C^{-1}(x)$ represent in the context of the problem.

73. Movie Theaters The average prices of admission y (in dollars) to movie theaters for the years 2003 through 2010 are shown in the table. *(Source: National Association of Theater Owners)*

Year	2003	2004	2005	2006
Admission price, y	6.03	6.21	6.41	6.55

Year	2007	2008	2009	2010
Admission price, y	6.88	7.18	7.50	7.89

(a) Use a graphing utility to create a scatter plot of the data. Let t represent the year, with $t = 3$ corresponding to 2003.

(b) Use the *regression* feature of the graphing utility to find a linear model for the data.

(c) Algebraically find the inverse function of the model in part (b). Explain what this inverse function represents in a real-life context.

(d) Use the inverse function you found in part (c) to predict the year in which the average admission price to a movie theater will reach $9.00.

74. Lead Exposure A project is conducted to study the amount of lead accumulated in the bones of humans. The concentration L (in micrograms per gram of bone mineral) of lead found in the tibia of a man is measured every five years. The results are shown in the table.

Age	15	20	25	30	35	40
Lead, L	3.2	5.4	9.2	12.2	13.8	16.0

(a) Use a graphing utility to create a scatter plot of the data. Let x represent the age (in years) of the man.

(b) Use the *regression* feature of the graphing utility to find a linear model for the data.

(c) Algebraically find the inverse function of the model in part (b). Explain what this inverse function represents in a real-life context.

(d) Use the inverse function you found in part (c) to estimate the age of the man when the concentration of lead in his tibia reaches 25 micrograms per gram of bone mineral.

75. Earnings-Dividend Ratio From 2000 through 2009, the earnings per share for Wal-Mart Stores were approximately related to the dividends per share by the function

$$f(x) = \sqrt{0.0209x^3 - 0.021}, \quad 1.40 \le x \le 3.66$$

where f represents the dividends per share (in dollars) and x represents the earnings per share (in dollars). (See figure.)

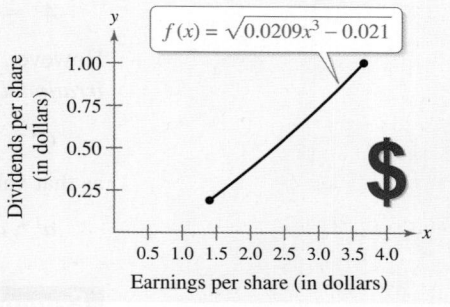

In 2008, Wal-Mart paid dividends of $0.93 per share. Find the inverse function of f and use the inverse function to approximate the earnings per share in 2008. *(Source: Wal-Mart Stores, Inc.)*

76. Diesel Mechanics The function given by

$$y = 0.03x^2 + 245.5, \quad 0 < x < 100$$

approximates the exhaust temperature y for a diesel engine in degrees Fahrenheit, where x is the percent load for the diesel engine. Solve the equation for x in terms of y and use the result to find the percent load for a diesel engine when the exhaust temperature is 410°F.

4.2 Exponential Functions

■ Evaluate an exponential expression.
■ Sketch the graph of an exponential function.
■ Evaluate and sketch the graph of the natural exponential function.
■ Use the compound interest formulas.
■ Use an exponential model to solve an application problem.

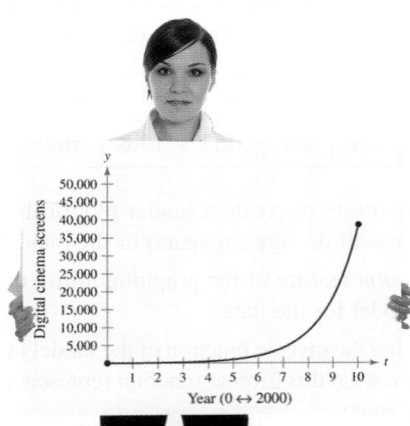

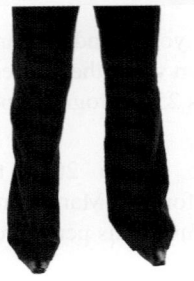

In Exercise 73 on page 356, you will use a natural exponential function to determine the number of digital cinema screens in the world.

Exponential Functions

So far, this text has dealt only with **algebraic functions,** which include polynomial functions and rational functions. In this chapter, you will study two types of nonalgebraic functions—*exponential* functions and *logarithmic* functions. These functions are examples of **transcendental functions.**

> **Definition of Exponential Function**
>
> The **exponential function** f **with base** a is denoted by
>
> $$f(x) = a^x$$
>
> where $a > 0$, $a \neq 1$, and x is any real number.

The base $a = 1$ is excluded because it yields

$$f(x) = 1^x = 1.$$

This is a constant function, not an exponential function.

You already know how to evaluate a^x for integer and rational values of x. For example, you know that

$$4^3 = 64 \quad \text{and} \quad 4^{1/2} = 2.$$

However, to evaluate 4^x for any real number x, you need to interpret forms with *irrational* exponents. For the purposes of this text, it is sufficient to think of

$$a^{\sqrt{2}} \quad \left(\text{where } \sqrt{2} \approx 1.414214\right)$$

as that value having the successively closer approximations

$$a^{1.4}, a^{1.41}, a^{1.414}, a^{1.4142}, a^{1.41421}, a^{1.414214}, \ldots.$$

| Example 1 | Evaluating an Exponential Expression |

Scientific Calculator

Number	*Keystrokes*	*Display*
$2^{-\pi}$	2 [y²] π [+/−] [=]	0.113314732

Graphing Calculator

Number	*Keystrokes*	*Display*
$2^{-\pi}$	2 [∧] [(−)] π [ENTER]	.1133147323

✓ **Checkpoint 1**

Use a calculator to evaluate $2.2^{\sqrt{3}}$. Round your result to three decimal places.

Graphs of Exponential Functions

The graphs of all exponential functions have similar characteristics, as shown in Examples 2, 3, and 4.

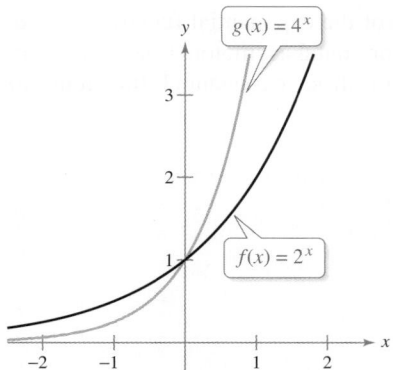

FIGURE 4.6

Example 2   Graphs of $y = a^x$

In the same coordinate plane, sketch the graph of each function.

a. $f(x) = 2^x$

b. $g(x) = 4^x$

SOLUTION   The table below lists some values for each function, and Figure 4.6 shows their graphs. Note that both graphs are increasing. Moreover, the graph of $g(x) = 4^x$ is increasing more rapidly than the graph of $f(x) = 2^x$.

x	-2	-1	0	1	2	3
$f(x) = 2^x$	$2^{-2} = \dfrac{1}{2^2} = \dfrac{1}{4}$	$2^{-1} = \dfrac{1}{2^1} = \dfrac{1}{2}$	$2^0 = 1$	$2^1 = 2$	$2^2 = 4$	$2^3 = 8$
$g(x) = 4^x$	$4^{-2} = \dfrac{1}{4^2} = \dfrac{1}{16}$	$4^{-1} = \dfrac{1}{4^1} = \dfrac{1}{4}$	$4^0 = 1$	$4^1 = 4$	$4^2 = 16$	$4^3 = 64$

✓ **Checkpoint 2**

Sketch the graph of $f(x) = 5^x$.   ■

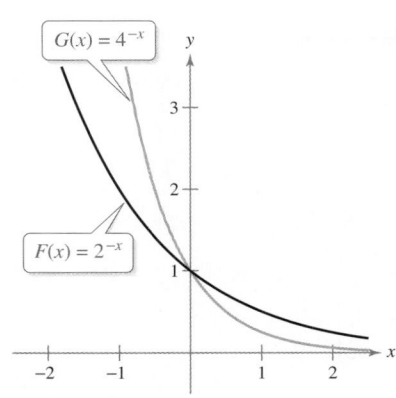

FIGURE 4.7

Example 3   Graphs of $y = a^{-x}$

In the same coordinate plane, sketch the graph of each function.

a. $F(x) = 2^{-x}$

b. $G(x) = 4^{-x}$

SOLUTION   The table below lists some values for each function, and Figure 4.7 shows their graphs. Note that both graphs are decreasing. Moreover, the graph of $G(x) = 4^{-x}$ is decreasing more rapidly than the graph of $F(x) = 2^{-x}$.

x	-3	-2	-1	0	1	2
$F(x) = 2^{-x}$	8	4	2	1	$\frac{1}{2}$	$\frac{1}{4}$
$G(x) = 4^{-x}$	64	16	4	1	$\frac{1}{4}$	$\frac{1}{16}$

✓ **Checkpoint 3**

Sketch the graph of $F(x) = 5^{-x}$.   ■

ALGEBRA TUTOR *xy*

For help with evaluating expressions involving exponents, see the *Chapter 4 Algebra Tutor* on page 400.

The tables in Examples 2 and 3 were evaluated by hand. You could use the *table* feature of a graphing utility to construct tables with more values.

In Example 3, note that the functions given by $F(x) = 2^{-x}$ and $G(x) = 4^{-x}$ can be rewritten with positive exponents.

$$F(x) = 2^{-x} = \left(\tfrac{2}{1}\right)^{-x} = \left(\tfrac{1}{2}\right)^x \quad \text{and} \quad G(x) = 4^{-x} = \left(\tfrac{4}{1}\right)^{-x} = \left(\tfrac{1}{4}\right)^x$$

Comparing the functions in Examples 2 and 3, observe that

$$F(x) = 2^{-x} = f(-x) \quad \text{and} \quad G(x) = 4^{-x} = g(-x).$$

Consequently, the graph of F is a reflection (in the y-axis) of the graph of f. The graphs of G and g have the same relationship.

The graphs in Figures 4.6 and 4.7 are typical of the exponential functions $y = a^x$ and $y = a^{-x}$. They have one y-intercept and one horizontal asymptote (the x-axis), and they are continuous. The basic characteristics of these exponential functions are summarized in Figures 4.8 and 4.9.

Characteristics of Exponential Functions

Graph of $y = a^x, a > 1$
- Domain: $(-\infty, \infty)$
- Range: $(0, \infty)$
- Intercept: $(0, 1)$
- Increasing
- x-axis is a horizontal asymptote $(a^x \to 0 \text{ as } x \to -\infty)$
- Continuous

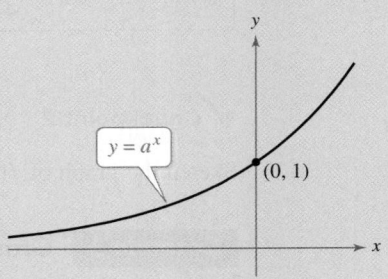

FIGURE 4.8

Graph of $y = a^{-x}, a > 1$
- Domain: $(-\infty, \infty)$
- Range: $(0, \infty)$
- Intercept: $(0, 1)$
- Decreasing
- x-axis is a horizontal asymptote $(a^{-x} \to 0 \text{ as } x \to \infty)$
- Continuous
- Reflection of graph of $y = a^x$ in y-axis

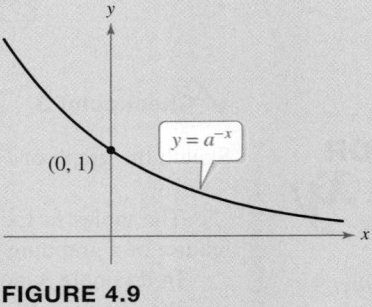

FIGURE 4.9

In the following example, notice how the graph of $y = a^x$ is used to sketch the graphs of functions of the form

$$f(x) = b \pm a^{x+c}.$$

Example 4 Transformations of Graphs of Exponential Functions

Each of the following graphs is a transformation of the graph of $f(x) = 3^x$, as shown in Figure 4.10.

a. Because $g(x) = 3^{x+1} = f(x + 1)$, the graph of g can be obtained by shifting the graph of f one unit to the *left*.

b. Because $h(x) = 3^x - 2 = f(x) - 2$, the graph of h can be obtained by shifting the graph of f *downward* two units.

c. Because $k(x) = -3^x = -f(x)$, the graph of k can be obtained by *reflecting* the graph of f in the x-axis.

d. Because $j(x) = 3^{-x} = f(-x)$, the graph of j can be obtained by *reflecting* the graph of f in the y-axis.

> **STUDY TIP**
>
> Notice in Example 4(b) that shifting the graph downward two units also shifts the horizontal asymptote of $f(x) = 3^x$ from the x-axis $(y = 0)$ to the line $y = -2$.

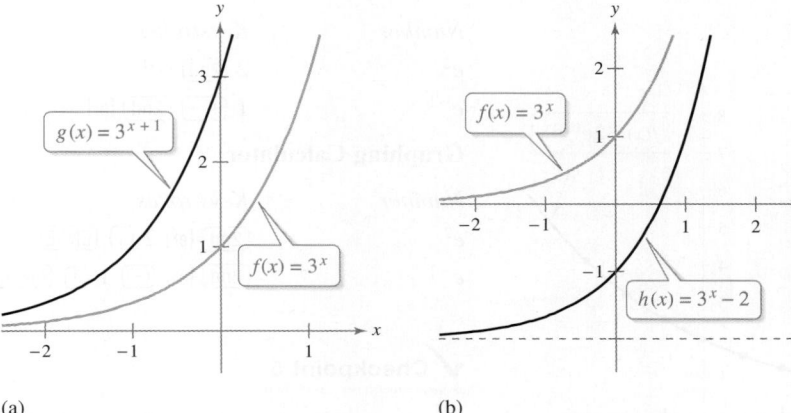

(a) (b)

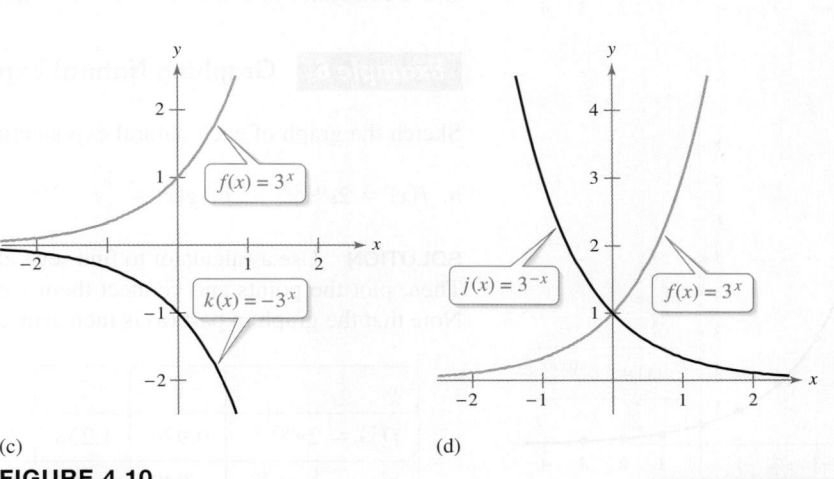

(c) (d)

FIGURE 4.10

✓ **Checkpoint 4**

Sketch the graph of

$$f(x) = 2^{x-1}.$$

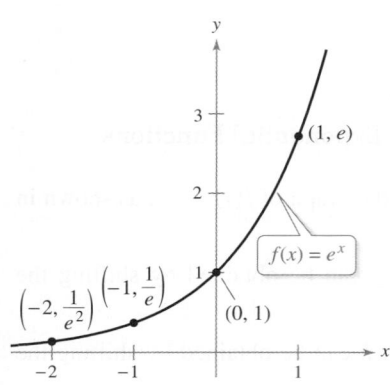

FIGURE 4.11

The Natural Base e

In many applications, the most convenient choice for a base is the irrational number

$$e = 2.718281828 . . .$$

called the **natural base.** The function given by

$$f(x) = e^x$$

is called the **natural exponential function.** Its graph is shown in Figure 4.11. The graph of the natural exponential function has the same basic characteristics as the graph of the exponential function given by $f(x) = a^x$ (see page 348). Be sure you see that for the exponential function given by $f(x) = e^x$, e is the constant $2.718281828 . . .$, whereas x is the variable.

Example 5   **Evaluating the Natural Exponential Function**

Use a calculator to evaluate the function given by $f(x) = e^x$ when $x = 2$ and $x = -1$.

SOLUTION

Scientific Calculator

Number	Keystrokes	Display
e^2	2 (2nd) [eˣ]	7.389056099
e^{-1}	1 (+/−) (2nd) [eˣ]	0.367879441

Graphing Calculator

Number	Keystrokes	Display
e^2	(2nd) [eˣ] 2 ( ) (ENTER)	7.389056099
e^{-1}	(2nd) [eˣ] (−) 1 ( ) (ENTER)	.3678794412

✓ **Checkpoint 5**

Use a calculator to evaluate $f(x) = e^x$ when $x = 6$ and $x = -3.5$.

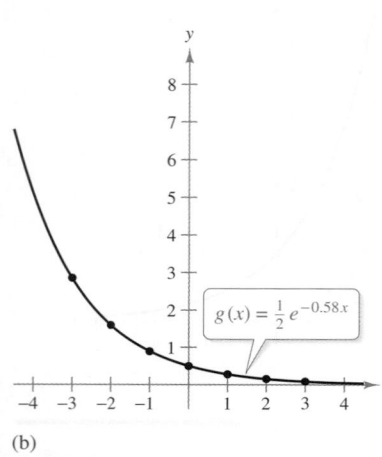

(a)

(b)

FIGURE 4.12

Example 6   **Graphing Natural Exponential Functions**

Sketch the graph of each natural exponential function.

a. $f(x) = 2e^{0.24x}$ **b.** $g(x) = \dfrac{1}{2}e^{-0.58x}$

SOLUTION   Use a calculator to find several points on each graph, as shown in the table. Then, plot the points and connect them with smooth curves, as shown in Figure 4.12. Note that the graph in part (a) is increasing, whereas the graph in part (b) is decreasing.

x	-3	-2	-1	0	1	2	3
$f(x) = 2e^{0.24x}$	0.974	1.238	1.573	2	2.542	3.232	4.109
$g(x) = \frac{1}{2}e^{-0.58x}$	2.849	1.595	0.893	0.5	0.280	0.157	0.088

✓ **Checkpoint 6**

Sketch the graph of $f(x) = e^{0.5x}$.

Compound Interest

One of the most familiar examples of exponential growth is that of an investment earning **continuously compounded interest.** The formula for the balance in an account that is compounded n times per year is

$$A = P\left(1 + \frac{r}{n}\right)^{nt}$$

where A is the balance in the account, P is the initial deposit, r is the annual interest rate (in decimal form), and t is the number of years. Using exponential functions, you will *develop* this formula and show how it leads to continuous compounding.

Suppose a principal P is invested at an annual interest rate r, compounded once a year. The principal at the end of the first year, P_1, is equal to the initial deposit P plus the interest earned, Pr. So,

$$P_1 = P + Pr.$$

This can be rewritten by factoring out P from each term as shown.

$$P_1 = P + Pr$$
$$= P(1 + r)$$

This pattern of multiplying the previous principal by $1 + r$ is then repeated each successive year, as shown below.

Year	Balance After Each Compounding
0	$P = P$
1	$P_1 = P(1 + r)$
2	$P_2 = P_1(1 + r) = P(1 + r)(1 + r) = P(1 + r)^2$
3	$P_3 = P_2(1 + r) = P(1 + r)^2(1 + r) = P(1 + r)^3$
$\vdots$	
t	$P_t = P(1 + r)^t$

To accommodate more frequent (quarterly, monthly, or daily) compounding of interest, let n be the number of compoundings per year and let t be the number of years. Then the rate per compounding is r/n and the account balance after t years is

$$A = P\left(1 + \frac{r}{n}\right)^{nt}. \qquad \text{Amount (balance) with } n \text{ compoundings per year}$$

When you let the number of compoundings n increase without bound, the process approaches what is called **continuous compounding.** In the formula for n compoundings per year, let $m = n/r$. This produces

$$A = P\left(1 + \frac{r}{n}\right)^{nt} \qquad \text{Amount with } n \text{ compoundings per year}$$

$$= P\left(1 + \frac{1}{m}\right)^{mrt} \qquad \text{Substitute } mr \text{ for } n \text{ and simplify.}$$

$$= P\left[\left(1 + \frac{1}{m}\right)^m\right]^{rt}. \qquad \text{Property of exponents}$$

As m increases without bound, it can be shown that

$$\left[1 + \left(\frac{1}{m}\right)\right]^m$$

approaches e. From this, you can conclude that the formula for continuous compounding is

$$A = Pe^{rt}. \qquad \text{Continuous compounding}$$

Formulas for Compound Interest

After t years, the balance A in an account with principal P and annual interest rate r (in decimal form) is given by the following formulas.

1. For n compoundings per year: $A = P\left(1 + \dfrac{r}{n}\right)^{nt}$

2. For continuous compounding: $A = Pe^{rt}$

Be sure that the annual interest rate is written in decimal form. For instance, 6% should be written as 0.06 when using compound interest formulas.

 **Example 7**   **Compound Interest**

You invest \$12,000 at an annual rate of 3%. Find the balance after 5 years when the interest is compounded (a) quarterly, (b) monthly, and (c) continuously. Which type of compounding earns the most money?

SOLUTION

a. For quarterly compounding, you have $n = 4$. So, in 5 years at 3%, the balance is

$$A = P\left(1 + \frac{r}{n}\right)^{nt} \qquad \text{Formula for compound interest}$$

$$= 12{,}000\left(1 + \frac{0.03}{4}\right)^{4(5)} \qquad \text{Substitute for } P, r, n, \text{ and } t.$$

$$\approx \$13{,}934.21. \qquad \text{Use a calculator.}$$

b. For monthly compounding, you have $n = 12$. So, in 5 years at 3%, the balance is

$$A = P\left(1 + \frac{r}{n}\right)^{nt} \qquad \text{Formula for compound interest}$$

$$= 12{,}000\left(1 + \frac{0.03}{12}\right)^{12(5)} \qquad \text{Substitute for } P, r, n, \text{ and } t.$$

$$\approx \$13{,}939.40. \qquad \text{Use a calculator.}$$

c. For continuous compounding, the balance is

$$A = Pe^{rt} \qquad \text{Formula for continuous compounding}$$

$$= 12{,}000e^{0.03(5)} \qquad \text{Substitute for } P, r, \text{ and } t.$$

$$\approx \$13{,}942.01. \qquad \text{Use a calculator.}$$

Continuous compounding yields more than quarterly and monthly compounding. This is typical of the two types of compounding. That is, for a given principal, interest rate, and time, continuous compounding will always yield a larger balance than compounding n times a year.

✓ **Checkpoint 7**

You invest \$6000 at an annual rate of 4%. Find the balance after 7 years when the interest is compounded

a. quarterly.

b. monthly.

c. continuously.

Another Application

 Example 8   **Radioactive Decay**

In 1986, a nuclear reactor accident occurred in Chernobyl in what was then the Soviet Union. The explosion spread highly toxic radioactive chemicals, such as plutonium, over hundreds of square miles, and the government evacuated the city and the surrounding area. Consider the model

$$P = 10\left(\frac{1}{2}\right)^{t/24,100}$$

which represents the amount of plutonium P that remains (from an initial amount of 10 pounds) after t years. Sketch the graph of this function over the interval from $t = 0$ to $t = 100,000$, where $t = 0$ represents 1986. How much of the 10 pounds of plutonium will remain in the year 2012? How much of the 10 pounds will remain after 100,000 years? Why is this city uninhabited?

SOLUTION   The graph of this function is shown in Figure 4.13. Plutonium has a *half-life* of about 24,100 years. That is, after 24,100 years, *half* of the original amount of plutonium will remain. After another 24,100 years, one-quarter of the original amount will remain, and so on. In the year 2012 ($t = 26$), there will still be

$$P = 10\left(\frac{1}{2}\right)^{26/24,100} \qquad \text{Substitute 26 for } t.$$

$$\approx 10\left(\frac{1}{2}\right)^{0.0010788} \qquad \text{Approximate exponent.}$$

$$\approx 9.993 \text{ pounds} \qquad \text{Use a calculator.}$$

of plutonium remaining. After 100,000 years, there will still be

$$P = 10\left(\frac{1}{2}\right)^{100,000/24,100} \qquad \text{Substitute 100,000 for } t.$$

$$\approx 10\left(\frac{1}{2}\right)^{4.149} \qquad \text{Approximate exponent.}$$

$$\approx 0.564 \text{ pound} \qquad \text{Use a calculator.}$$

of plutonium remaining. This city is uninhabited because much of the original amount of radioactive plutonium still remains.

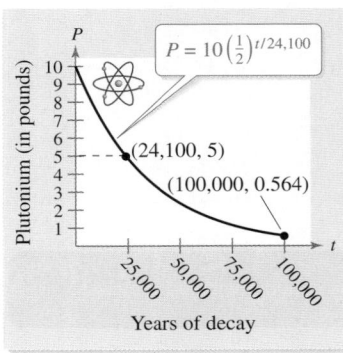

$$P = 10\left(\frac{1}{2}\right)^{t/24,100}$$

(24,100, 5)

(100,000, 0.564)

Plutonium (in pounds)

Years of decay

FIGURE 4.13

 **Checkpoint 8**

In Example 8, how much of the initial 10 pounds of plutonium will remain in the year 2086?

SUMMARIZE   (Section 4.2)

1. State the definition of an exponential function f with base a *(page 346)*. For an example of evaluating an exponential expression, see Example 1.

2. Describe the relationship between the graphs of the functions $f(x) = a^x$ and $g(x) = a^{-x}$, where $a > 1$ *(pages 347 and 348)*. For examples of graphing exponential functions, see Examples 2, 3, and 4.

3. State the definitions of the natural base and the natural exponential function *(page 350)*. For examples of evaluating and graphing natural exponential functions, see Examples 5 and 6.

SKILLS WARM UP 4.2   The following warm-up exercises involve skills that were covered in earlier sections. You will use these skills in the exercise set for this section. For additional help, review Sections 0.3 and 0.4.

In Exercises 1–12, use the properties of exponents to simplify the expression.

1. $5^{2x}(5^{-x})$

2. $3^{-x}(3^{3x})$

3. $\dfrac{4^{5x}}{4^{2x}}$

4. $\dfrac{10^{2x}}{10^x}$

5. $(4^x)^2$

6. $(4^{2x})^5$

7. $\left(\dfrac{2^x}{3^x}\right)^{-1}$

8. $(4^{6x})^{1/2}$

9. $(2^{3x})^{-1/3}$

10. $\left(\dfrac{3^{4x}}{5^{4x}}\right)^{1/4}$

11. $(16^x)^{1/4}$

12. $(27^x)^{1/3}$

Exercises 4.2

See www.CalcChat.com for worked-out solutions to odd-numbered exercises.

Evaluating an Exponential Expression   In Exercises 1–6, use a calculator to evaluate the expression. Round your result to three decimal places. *See Example 1.*

1. $(2.6)^{1.3}$

2. $(1.12)^{20.5}$

3. $100(1.03)^{-9/5}$

4. $1500(2^{-5/2})$

5. $6^{-\sqrt{2}}$

6. $1.3^{-\sqrt{5}}$

Matching   In Exercises 7–14, match the function with its graph. [The graphs are labeled (a), (b), (c), (d), (e), (f), (g), and (h).]

(a)
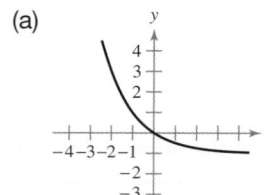

(b)
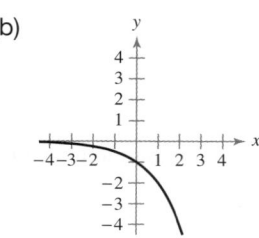

(c)
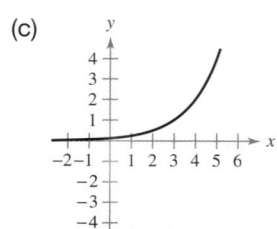

(d)
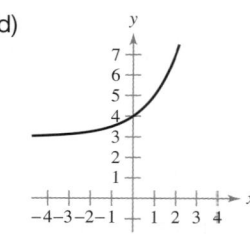

(e)
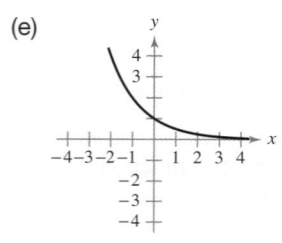

(f)
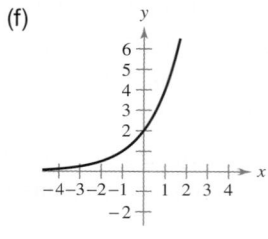

(g)
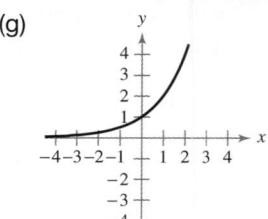

(h)
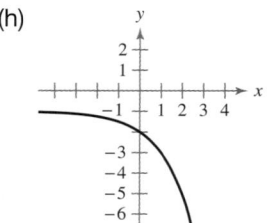

7. $f(x) = 2^x$

8. $f(x) = 2^{-x}$

9. $f(x) = -2^x$

10. $f(x) = -2^x - 1$

11. $f(x) = 2^x + 3$

12. $f(x) = 2^{-x} - 1$

13. $f(x) = 2^{x+1}$

14. $f(x) = 2^{x-3}$

Graphing an Exponential Function   In Exercises 15–26, sketch the graph of the function. *See Examples 2, 3, and 4.*

15. $g(x) = 4^x$

16. $f(x) = \left(\tfrac{3}{2}\right)^x$

17. $f(x) = 4^{-x}$

18. $h(x) = \left(\tfrac{3}{2}\right)^{-x}$

19. $h(x) = 4^{x-3}$

20. $g(x) = \left(\tfrac{3}{2}\right)^{x+2}$

21. $g(x) = 4^{-x} - 2$

22. $f(x) = \left(\tfrac{3}{2}\right)^{-x} + 2$

23. $h(x) = -4^x$

24. $g(x) = -\left(\tfrac{3}{2}\right)^x$

25. $y = 2^{-x^2}$

26. $y = 3^{-x^2}$

Evaluating the Natural Exponential Function   In Exercises 27–30, use a calculator to evaluate the function $f(x) = e^x$ for the given value of x. Round your result to three decimal places. *See Example 5.*

27. $x = 4$

28. $x = -5$

29. $x = -2.7$

30. $x = 0.3$

Graphing a Natural Exponential Function In Exercises 31–38, sketch the graph of the function. *See Example 6.*

31. $y = e^{-0.1x}$

32. $y = e^{0.2x}$

33. $f(x) = 2e^{0.12x}$

34. $f(x) = 3e^{-0.2x}$

35. $f(x) = e^{x+3}$

36. $h(x) = e^{x-2}$

37. $g(x) = 1 + e^{-x}$

38. $N(t) = 2 - e^t$

39. Think About It Is the natural exponential function also an exponential function with base a? Explain.

40. HOW DO YOU SEE IT?
The figure shows the graphs of $y = 2^x$, $y = e^x$, $y = 10^x$, $y = 2^{-x}$, $y = e^{-x}$, and $y = 10^{-x}$. Match each function with its graph. (The graphs are labeled A through F.) Explain your reasoning.

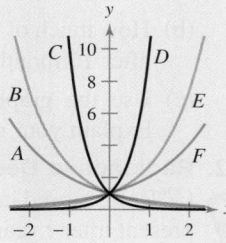

Compound Interest In Exercises 41–44, complete the table to find the balance A for P dollars invested at rate r for t years, compounded n times per year. *See Example 7.*

n	1	2	4	12	365	Continuous
A						

41. $P = \$5000$, $r = 8\%$, $t = 5$ years

42. $P = \$1000$, $r = 10\%$, $t = 10$ years

43. $P = \$2500$, $r = 5\%$, $t = 20$ years

44. $P = \$1000$, $r = 10\%$, $t = 40$ years

45. Compound Interest A bank offers two types of interest-bearing accounts. The first account pays 5% interest compounded quarterly. The second account pays 3% interest compounded continuously. Which account earns more money? Why?

46. Compound Interest A bank offers two types of interest-bearing accounts. The first account pays 6% interest compounded monthly. The second account pays 5% interest compounded continuously. Which account earns more money? Why?

Present Value The present value of money is the principal P you need to invest today so that it will grow to an amount A at the end of a specified time. The present value formula

$$P = A\left(1 + \frac{r}{n}\right)^{-nt}$$

is obtained by solving the compound interest formula

$$A = P\left(1 + \frac{r}{n}\right)^{nt}$$

for P. Recall that t is the number of years, r is the interest rate per year, and n is the number of compoundings per year. In Exercises 47–50, find the present value of amount A invested at rate r for t years, compounded n times per year.

47. $A = \$10,000$, $r = 6\%$, $t = 5$ years, $n = 4$

48. $A = \$50,000$, $r = 7\%$, $t = 10$ years, $n = 12$

49. $A = \$20,000$, $r = 8\%$, $t = 6$ years, $n = 4$

50. $A = \$1,000,000$, $r = 8\%$, $t = 20$ years, $n = 2$

Compound Interest In Exercises 51–54, complete the table to find the amount P that must be invested at rate r to obtain a balance of $A = \$100,000$ in t years.

t	1	10	20	30	40	50
P						

51. $r = 10\%$, compounded monthly

52. $r = 7\%$, compounded daily

53. $r = 9\%$, compounded continuously

54. $r = 5\%$, compounded continuously

Compound Interest In Exercises 55–58, you deposit a lump sum P in a trust fund on the day your child is born. The fund earns 6.5% interest compounded continuously. Find the amount P that will yield the given balance A on your child's 25th birthday.

55. $A = \$100,000$

56. $A = \$500,000$

57. $A = \$750,000$

58. $A = \$1,000,000$

Compound Interest In Exercises 59–62, you deposit a lump sum P in a trust fund on the day your grandchild is born. The fund earns 7.5% interest compounded continuously. Find the amount P that will yield the given balance A on your grandchild's 21st birthday.

59. $A = \$100,000$

60. $A = \$500,000$

61. $A = \$750,000$

62. $A = \$1,000,000$

63. Demand Function The demand function for a limited edition comic book is given by

$$p = 3000\left(1 - \frac{5}{5 + e^{-0.015x}}\right).$$

(a) Find the price p for a demand of $x = 75$ units.

(b) Find the price p for a demand of $x = 200$ units.

(c) Use a graphing utility to graph the demand function.

(d) Use the graph from part (c) to approximate the demand when the price is $100.

64. Demand Function The demand function for a home theater system is given by

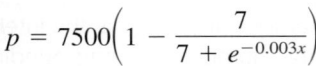

$$p = 7500\left(1 - \frac{7}{7 + e^{-0.003x}}\right).$$

(a) Find the price p for a demand of $x = 200$ units.

(b) Find the price p for a demand of $x = 900$ units.

(c) Use a graphing utility to graph the demand function.

(d) Use the graph from part (c) to approximate the demand when the price is $400.

65. Bacteria Growth The number of a certain type of bacteria increases according to the model

$$P(t) = 100e^{0.01896t}$$

where t is time (in hours).

(a) Find $P(0)$.   (b) Find $P(5)$.

(c) Find $P(10)$.   (d) Find $P(24)$.

66. Bacteria Growth As a result of a medical treatment, the number of a certain type of bacteria decreases according to the model

$$P(t) = 100e^{-0.685t}$$

where t is time (in hours).

(a) Find $P(0)$.   (b) Find $P(5)$.

(c) Find $P(10)$.   (d) Find $P(24)$.

67. Population Growth The population P of a town increases according to the model

$$P(t) = 4500e^{0.0272t}$$

where t represents the year, with $t = 0$ corresponding to 2000. Use the model to predict the population in each year.

(a) 2010   (b) 2012

(c) 2015   (d) 2020

68. Population Growth The population P of a small city increases according to the model

$$P(t) = 36{,}000e^{0.0156t}$$

where t represents the year, with $t = 0$ corresponding to 2000. Use the model to predict the population in each year.

(a) 2009   (b) 2011

(c) 2015   (d) 2018

69. Radioactive Decay Strontium-90 has a half-life of 29.1 years. The amount S of 100 kilograms of strontium-90 present after t years is given by

$$S = 100e^{-0.0238t}.$$

How much of the 100 kilograms will remain after 50 years?

70. Radioactive Decay Neptunium-237 has a half-life of 2.1 million years. The amount N of 200 kilograms of neptunium-237 present after t years is given by

$$N = 200e^{-0.00000033007t}.$$

How much of the 200 kilograms will remain after 20,000 years?

71. Radioactive Decay Five pounds of the element polonium (^{210}Po) is released in a nuclear accident. The amount of polonium P that is present after t months is given by $P = 5e^{-0.1507t}$.

(a) Use a graphing utility to graph this function over the interval from $t = 0$ to $t = 10$.

(b) How much of the 5 pounds of polonium will remain after 10 months?

(c) Use the graph to estimate the half-life of ^{210}Po. Explain your reasoning.

72. Radioactive Decay One hundred grams of radium (^{226}Ra) is stored in a container. The amount of radium R present after t years is given by $R = 100e^{-0.0004335t}$.

(a) Use a graphing utility to graph this function over the interval from $t = 0$ to $t = 10{,}000$.

(b) How much of the 100 grams of radium will remain after 10,000 years?

(c) Use the graph to estimate the half-life of ^{226}Ra. Explain your reasoning.

73. Digital Cinema Screens The numbers y of digital cinema screens in the world from 2000 through 2010 can be modeled by

$$y = 25.87e^{0.7323t}, \quad 0 \le t \le 10$$

where t represents the year, with $t = 0$ corresponding to 2000 (see figure).   *(Source: IHS Screen Digest)*

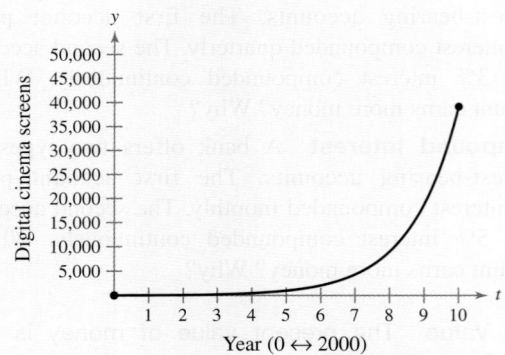

(a) Use the graph to estimate *graphically* the numbers of digital cinema screens in 2008 and 2010.

(b) Use the model to confirm *algebraically* the estimates obtained in part (a).

74. Revenue The revenue R (in millions of dollars) for Research in Motion each year from 2000 through 2009 can be modeled by

$$R = 162.53e^{0.5061t}, \quad 0 \le t \le 9$$

where t represents the year, with $t = 0$ corresponding to 2000 (see figure). *(Source: Research in Motion Limited)*

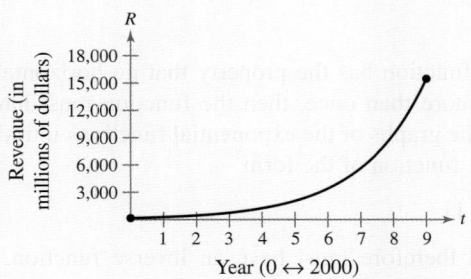

Year (0 ↔ 2000)

(a) Use the graph to estimate *graphically* the revenues for Research in Motion in 2003, 2007, and 2009.

(b) Use the model to confirm *algebraically* the estimates obtained in part (a).

75. Age at First Marriage For each year from 1980 through 2010, the median age A of an American woman at her first marriage can be modeled by

$$A = -45.10 + \frac{72.02}{1 + e^{-0.0523t - 2.6314}}, \quad 0 \le t \le 30$$

where t represents the year, with $t = 0$ corresponding to 1980 (see figure). *(Source: U.S. Census Bureau)*

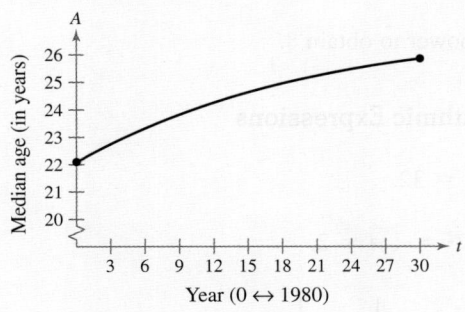

Year (0 ↔ 1980)

(a) Use the graph to estimate *graphically* the median age of an American woman at her first marriage in each of the years 1980, 1990, 2000, and 2010.

(b) Use the model to confirm *algebraically* the estimates obtained in part (a).

76. Prescriptions The numbers of mail-order prescriptions P (in millions) filled in the United States from 2002 through 2009 can be modeled by

$$P = 151 + \frac{89.24}{1 + e^{-0.0895t + 2.8739}}, \quad 2 \le t \le 9$$

where t represents the year, with $t = 2$ corresponding to 2002. *(Source: National Association of Chain Drug Stores)*

(a) Use a graphing utility to graph P for the years 2002 through 2009.

(b) Use the graph from part (a) to estimate the numbers of mail-order prescriptions filled in 2002, 2006, and 2009.

77. Gross Domestic Product The U.S. annual gross domestic product G (in billions of dollars) from 1985 through 2009 can be modeled by

$$G = 3355.4(1.0541)^t, \quad 5 \le t \le 29$$

where t represents the year, with $t = 5$ corresponding to 1985. *(Source: U.S. Bureau of Economic Analysis)*

(a) Use a graphing utility to graph G for the years 1985 through 2009.

(b) Use the graph from part (a) to estimate the gross domestic product in 1990, 2000, and 2009.

78. Writing Determine whether $e = \dfrac{271{,}801}{99{,}990}$. Justify your answer.

79. Project: Drug Concentration For a project involving the concentration of a drug in a patient's bloodstream, visit this text's website at *www.cengagebrain.com.*

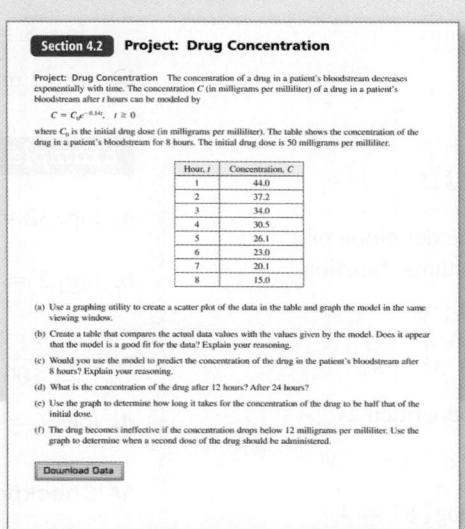

4.3 Logarithmic Functions

■ Recognize and evaluate a logarithmic function with base *a*.
■ Sketch the graph of a logarithmic function.
■ Recognize and evaluate the natural logarithmic function.
■ Use a logarithmic model to solve an application problem.

Logarithmic Functions

In Section 4.1, you learned that if a function has the property that no horizontal line intersects the graph of the function more than once, then the function must have an inverse function. By looking back at the graphs of the exponential functions introduced in Section 4.2, you will see that every function of the form

$$f(x) = a^x \text{ (where } a > 0 \text{ and } a \neq 1)$$

passes the Horizontal Line Test and therefore must have an inverse function. This inverse function is called the **logarithmic function with base *a*.**

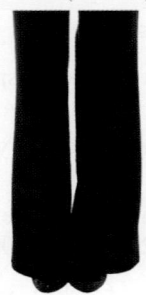

In Exercise 100 on page 367, you will use a natural logarithmic model to approximate the median age of the United States population.

Definition of Logarithmic Function

For $x > 0$, $a > 0$, and $a \neq 1$,

$$y = \log_a x \text{ if and only if } x = a^y.$$

The function given by

$$f(x) = \log_a x$$

is called the **logarithmic function with base *a*.**

The equations $y = \log_a x$ and $x = a^y$ are equivalent. The first equation is in logarithmic form and the second is in exponential form.

When evaluating logarithms, remember that *a logarithm is an exponent.* This means that $\log_a x$ is the exponent to which *a* must be raised to obtain *x*. For instance,

$$\log_2 8 = 3$$

because 2 must be raised to the third power to obtain 8.

STUDY TIP

By the definition of a logarithmic function,

$$\begin{array}{ccc} a & y & x \\ \searrow & \downarrow & \swarrow \\ & 3^4 = 81 & \end{array}$$

can be written as

$$\begin{array}{ccc} a & x & y \\ \searrow & \downarrow & \swarrow \\ \log_3 & 81 = 4. \end{array}$$

Example 1   Evaluating Logarithmic Expressions

a. $\log_2 32 = 5$ because $2^5 = 32.$

b. $\log_4 2 = \dfrac{1}{2}$ because $4^{1/2} = \sqrt{4} = 2.$

c. $\log_{10} \dfrac{1}{100} = -2$ because $10^{-2} = \dfrac{1}{10^2} = \dfrac{1}{100}.$

d. $\log_3 1 = 0$ because $3^0 = 1.$

✓**Checkpoint 1**

Evaluate each expression.

a. $\log_9 81$ **b.** $\log_7 \dfrac{1}{49}$

The logarithmic function with base 10 is called the **common logarithmic function.** On most calculators, this function is denoted by $\boxed{\text{LOG}}$.

Example 2   Evaluating Logarithmic Expressions on a Calculator

Scientific Calculator

Number	Keystrokes	Display
a. $\log_{10} 10$	10 $\boxed{\text{LOG}}$	1
b. $2 \log_{10} 2.5$	2.5 $\boxed{\text{LOG}}$ $\boxed{\times}$ 2 $\boxed{=}$	0.795880017
c. $\log_{10}(-2)$	2 $\boxed{+/-}$ $\boxed{\text{LOG}}$	ERROR

Graphing Calculator

Number	Keystrokes	Display
a. $\log_{10} 10$	$\boxed{\text{LOG}}$ 10 $\boxed{)}$ $\boxed{\text{ENTER}}$	1
b. $2 \log_{10} 2.5$	2 $\boxed{\text{LOG}}$ 2.5 $\boxed{)}$ $\boxed{\text{ENTER}}$	.7958800173
c. $\log_{10}(-2)$	$\boxed{\text{LOG}}$ $\boxed{(-)}$ 2 $\boxed{)}$ $\boxed{\text{ENTER}}$	ERROR

Many calculators display an error message (or a complex number) when you try to evaluate $\log_{10}(-2)$. This is because the domain of every logarithmic function is the set of *positive real numbers*. In other words, there is no real number power to which 10 can be raised to obtain -2.

✓**Checkpoint 2**

Use a calculator to evaluate the expression $\log_{10} 200$. Round your result to three decimal places.

The following properties follow directly from the definition of the logarithmic function with base a.

Properties of Logarithms

1. $\log_a 1 = 0$ because $a^0 = 1$.

2. $\log_a a = 1$ because $a^1 = a$.

3. $\log_a a^x = x$ and $a^{\log_a x} = x$ Inverse Properties

4. If $\log_a x = \log_a y$, then $x = y$.   One-to-One Property

---

Example 3   Using Properties of Logarithms

a. Solve the equation $\log_2 x = \log_2 3$ for x.

b. Solve the equation $\log_5 x = 1$ for x.

SOLUTION

a. Using the One-to-One Property (Property 4), you can conclude that $x = 3$.

b. Using Property 2, you can conclude that $x = 5$.

✓**Checkpoint 3**

Solve the equation $\log_4 1 = x$ for x.

Graphs of Logarithmic Functions

To sketch the graph of $y = \log_a x$, you can use the fact that the graphs of inverse functions are reflections of each other in the line $y = x$.

Example 4   Graphs of Exponential and Logarithmic Functions

In the same coordinate plane, sketch the graph of each function.

a. $f(x) = 2^x$

b. $g(x) = \log_2 x$

SOLUTION

a. For $f(x) = 2^x$, construct a table of values.

x	-2	-1	0	1	2	3
$f(x) = 2^x$	$\frac{1}{4}$	$\frac{1}{2}$	1	2	4	8

By plotting these points and connecting them with a smooth curve, you obtain the graph of f shown in Figure 4.14.

b. Because $g(x) = \log_2 x$ is the inverse function of $f(x) = 2^x$, the graph of g is obtained by plotting the points $(f(x), x)$ and connecting them with a smooth curve. The graph of g is a reflection of the graph of f in the line $y = x$, as shown in Figure 4.14.

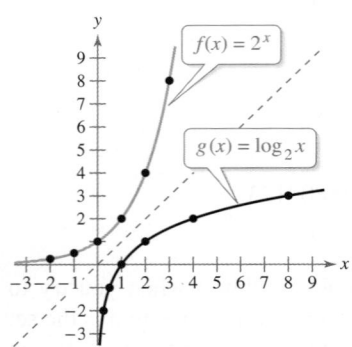

Inverse Functions

FIGURE 4.14

✓**Checkpoint 4**

In the same coordinate plane, sketch the graph of each function.

a. $f(x) = 4^x$

b. $g(x) = \log_4 x$

Before you can confirm the result of Example 4 with a graphing utility, you need to know how to enter $\log_2 x$. You will learn how to do this using the *change-of-base formula* discussed in Section 4.4.

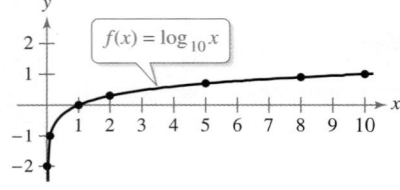

Common Logarithmic Function

FIGURE 4.15

Example 5   Sketching the Graph of a Logarithmic Function

Sketch the graph of the common logarithmic function given by $f(x) = \log_{10} x$.

SOLUTION   Begin by constructing a table of values. Note that some of the values can be obtained without a calculator by using the properties of logarithms and the definition of a logarithmic function. Others require a calculator. Next, plot the points and connect them with a smooth curve, as shown in Figure 4.15.

	Without Calculator				With Calculator		
x	$\frac{1}{100}$	$\frac{1}{10}$	1	10	2	5	8
$f(x) = \log_{10} x$	-2	-1	0	1	0.301	0.699	0.903

✓**Checkpoint 5**

Sketch the graph of the function given by $f(x) = 2 \log_{10} x$.

The nature of the graph in Figure 4.15 is typical of functions of the form $f(x) = \log_a x$, $a > 1$. They have one x-intercept and one vertical asymptote. Notice how slowly the graph rises for $x > 1$. The basic characteristics of logarithmic graphs are summarized below and shown in Figure 4.16. Note that the vertical asymptote occurs at $x = 0$, where $\log_a x$ is *undefined*.

Characteristics of Logarithmic Functions

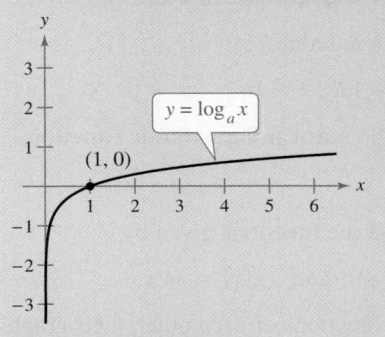

FIGURE 4.16

Graph of $y = \log_a x$, $a > 1$
- Domain: $(0, \infty)$
- Range: $(-\infty, \infty)$
- Intercept: $(1, 0)$
- Increasing
- Passes the Horizontal Line Test; therefore has an inverse function
- y-axis is a vertical asymptote ($\log_a x \to -\infty$ as $x \to 0^+$)
- Continuous
- Reflection of graph of $y = a^x$ about the line $y = x$

Remember that the domain of $y = \log_a x$ is the range of $y = a^x$ and the range of $y = \log_a x$ is the domain of $y = a^x$.

Example 6   Sketching the Graphs of Logarithmic Functions

The graph of each function below is similar to the graph of $f(x) = \log_{10} x$.

a. Because $g(x) = \log_{10}(x - 1) = f(x - 1)$, the graph of g can be obtained by shifting the graph of f one unit to the *right*. The domain of g is $(1, \infty)$ because $\log_{10}(x - 1)$ is defined only when $x - 1 > 0$. See Figure 4.17(a).

b. Because $h(x) = 2 + \log_{10} x = 2 + f(x)$, the graph of h can be obtained by shifting the graph of f two units *upward*. The domain of h is $(0, \infty)$ because $2 + \log_{10} x$ is defined only when $x > 0$. See Figure 4.17(b).

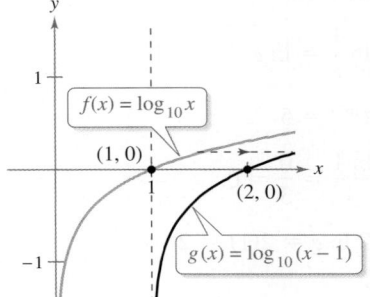

(a) Right shift of one unit
Domain of g: $(1, \infty)$

(b) Upward shift of two units
Domain of h: $(0, \infty)$

FIGURE 4.17

STUDY TIP

Notice in Example 6(a) that shifting the graph of $f(x)$ one unit to the right shifts the vertical asymptote from the y-axis ($x = 0$) to the line $x = 1$. It also shifts the x-intercept from $(1, 0)$ to $(2, 0)$.

✓ **Checkpoint 6**

Sketch the graph of $f(x) = \log_{10}(x + 3)$.

The Natural Logarithmic Function

By looking back at the graph of the natural exponential function introduced in Section 4.2, you will see that the graph of $f(x) = e^x$ passes the Horizontal Line Test. So, it has an inverse function. This inverse function is called the **natural logarithmic function** and is denoted by the special symbol ln x, read as "el en of x."

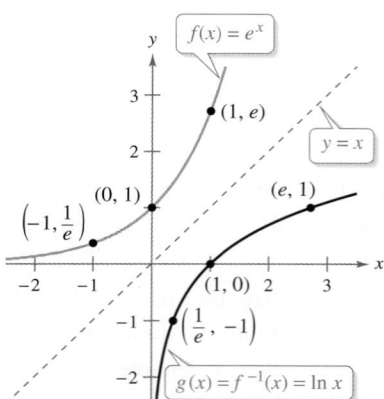

Inverse Functions

FIGURE 4.18

The Natural Logarithmic Function

The function defined by

$$f(x) = \log_e x = \ln x, \quad x > 0$$

is called the **natural logarithmic function.**

Because the functions given by

$$f(x) = e^x \quad \text{and} \quad g(x) = \ln x \qquad \text{Inverse functions}$$

are inverse functions of each other, their graphs are reflections of each other in the line $y = x$. This reflective property is illustrated in Figure 4.18. The four properties of logarithms listed on page 359 are also valid for natural logarithms.

Properties of Natural Logarithms

1. $\ln 1 = 0$ because $e^0 = 1$.

2. $\ln e = 1$ because $e^1 = e$.

3. $\ln e^x = x$ and $e^{\ln x} = x$ Inverse Properties

4. If $\ln x = \ln y$, then $x = y$.   One-to-One Property

 **Example 7**   **Using Properties of Natural Logarithms**

Evaluate each logarithmic expression.

a. $\ln \dfrac{1}{e}$ **b.** $e^{\ln 5}$ **c.** $\dfrac{\ln 1}{3}$ **d.** $2 \ln e$

SOLUTION

a. $\ln \dfrac{1}{e} = \ln e^{-1} = -1$ Inverse Property

b. $e^{\ln 5} = 5$ Inverse Property

c. $\dfrac{\ln 1}{3} = \dfrac{0}{3} = 0$ Property 1

d. $2 \ln e = 2(1) = 2$ Property 2

✓**Checkpoint 7**

Evaluate each logarithmic expression.

a. $\ln e^7$

b. $5 \ln 1$

c. $\frac{3}{4} \ln e$

On most calculators, the natural logarithm is denoted by $\boxed{\text{LN}}$, as illustrated in Example 8.

Example 8   Evaluating the Natural Logarithmic Function

Use a calculator to evaluate each expression.

a. ln 2   **b.** ln 0.3   **c.** ln e^2 **d.** ln(−1)

SOLUTION

Scientific Calculator

Number	Keystrokes	Display
a. ln 2	2 $\boxed{\text{LN}}$	0.69314718
b. ln 0.3	.3 $\boxed{\text{LN}}$	−1.203972804
c. ln e^2	2 $\boxed{\text{2nd}}$ [eˣ] $\boxed{\text{LN}}$	2
d. ln(−1)	1 $\boxed{+/-}$ $\boxed{\text{LN}}$	ERROR

Graphing Calculator

Number	Keystrokes	Display
a. ln 2	$\boxed{\text{LN}}$ 2 $\boxed{)}$ $\boxed{\text{ENTER}}$	.6931471806
b. ln 0.3	$\boxed{\text{LN}}$ .3 $\boxed{)}$ $\boxed{\text{ENTER}}$	−1.203972804
c. ln e^2	$\boxed{\text{LN}}$ $\boxed{\text{2nd}}$ [eˣ] 2 $\boxed{)}$ $\boxed{)}$ $\boxed{\text{ENTER}}$	2
d. ln(−1)	$\boxed{\text{LN}}$ $\boxed{(-)}$ 1 $\boxed{)}$ $\boxed{\text{ENTER}}$	ERROR

✓**Checkpoint 8**

Use a calculator to evaluate the expression ln 0.1. Round your result to three decimal places.

In Example 8, note that ln(−1) gives an error message on most calculators. This occurs because the domain of ln x is the set of *positive real numbers* (see Figure 4.18). So, ln(−1) is undefined.

Example 9   Finding the Domains of Natural Logarithmic Functions

Find the domain of each function.

a. $f(x) = \ln(x - 2)$ **b.** $g(x) = \ln(2 - x)$ **c.** $h(x) = \ln x^2$

SOLUTION

a. Because ln(x − 2) is defined only when $x - 2 > 0$, it follows that the domain of f is $(2, \infty)$.

b. Because ln(2 − x) is defined only when $2 - x > 0$, it follows that the domain of g is $(-\infty, 2)$. The graph of g is shown in Figure 4.19.

c. Because ln x^2 is defined only when $x^2 > 0$, it follows that the domain of h is all real numbers except $x = 0$.

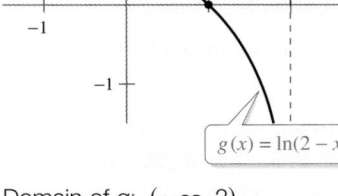

Domain of g: $(-\infty, 2)$

FIGURE 4.19

✓**Checkpoint 9**

Find the domain of the function given by $f(x) = \ln(x + 5)$.

Application

 Example 10   **Human Memory Model**

A group of students participating in a psychological experiment attended several lectures on a subject. Every month for a year after that, the students were tested to see how much of the material they remembered. The average scores $f(t)$ for the group are given by the *human memory model*

$$f(t) = 75 - 6 \ln(t + 1), \quad 0 \le t \le 12$$

where t is the time (in months). Based on the results of the experiment, how many months can a student wait before retaking the exam and still expect to score 60 or better? (Do not count portions of months.)

SOLUTION   To determine how many months a student can wait before retaking the exam and still expect to score 60 or better, use the model to create a table of values showing the scores for several months.

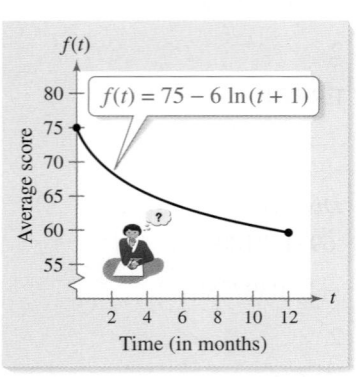

$f(t)$

$f(t) = 75 - 6 \ln(t + 1)$

Average score

Time (in months)

FIGURE 4.20

Month, t	0	1	2	3	4	5	6
Score, $f(t)$	75	70.84	68.41	66.68	65.34	64.25	63.32

Month, t	7	8	9	10	11	12
Score, $f(t)$	62.52	61.82	61.18	60.61	60.09	59.61

From the table, you can see that a student would need to retake the exam by the 11th month in order to score 60 or better. The graph of f is shown in Figure 4.20.

✓ **Checkpoint 10**

Biologists have found that an alligator's length l (in inches) can be approximated by the model

$$l = 27.1 \ln w - 32.8$$

where w is the weight (in pounds) of the alligator. Find the lengths of alligators for which $w = 150, 225, 380, 450,$ and 625 pounds. Round your results to the nearest tenth of an inch.

SUMMARIE   (Section 4.3)

1. State the definition of a logarithmic function with base a *(page 358)*. For examples of evaluating logarithmic expressions, see Examples 1 and 2.

2. Make a list of the properties of logarithms *(page 359)*. For an example that uses properties of logarithms, see Example 3.

3. Describe the relationship between the graphs of the functions $f(x) = a^x$ and $g(x) = \log_a x$, where $a > 1$ *(pages 360 and 361)*. For examples of graphing exponential and logarithmic functions, see Examples 4, 5, and 6.

4. State the definition of the natural logarithmic function and make a list of the properties of natural logarithms *(page 362)*. For examples of evaluating natural logarithms and using the properties of natural logarithms, see Examples 7 and 8.

SKILLS WARM UP 4.3 The following warm-up exercises involve skills that were covered in earlier sections. You will use these skills in the exercise set for this section. For additional help, review Sections 2.6 and 4.2.

In Exercises 1–4, determine the value of x.

1. $2^x = 8$ **2.** $4^x = 1$ **3.** $10^x = 0.1$ **4.** $e^x = e$

In Exercises 5 and 6, evaluate the expression. (Round the result to three decimal places.)

5. e^2 **6.** e^{-1}

In Exercises 7–10, describe how the graph of g is related to the graph of f.

7. $g(x) = f(x + 2)$ **8.** $g(x) = -f(x)$ **9.** $g(x) = -1 + f(x)$ **10.** $g(x) = f(-x)$

Exercises 4.3

See www.CalcChat.com for worked-out solutions to odd-numbered exercises.

Logarithmic Equations in Exponential Form In Exercises 1–6, match the logarithmic equation with its exponential form. [The exponential forms are labeled (a), (b), (c), (d), (e), and (f).]

1. $\log_4 16 = 2$ (a) $4^{1/2} = 2$

2. $\log_2 16 = 4$ (b) $2^{-4} = \frac{1}{16}$

3. $\log_2 \frac{1}{16} = -4$ (c) $4^2 = 16$

4. $\log_4 \frac{1}{16} = -2$ (d) $4^{-2} = \frac{1}{16}$

5. $\log_4 2 = \frac{1}{2}$ (e) $16^{1/2} = 4$

6. $\log_{16} 4 = \frac{1}{2}$ (f) $2^4 = 16$

Writing Logarithmic Equations In Exercises 7–12, write the exponential equation in logarithmic form. For example, the logarithmic form of $2^3 = 8$ is $\log_2 8 = 3$.

7. $4^4 = 256$ **8.** $7^3 = 343$

9. $81^{1/4} = 3$ **10.** $9^{3/2} = 27$

11. $6^{-2} = \frac{1}{36}$ **12.** $10^{-3} = 0.001$

Writing Exponential Equations In Exercises 13–20, write the logarithmic equation in exponential form. For example, the exponential form of $\log_5 125 = 3$ is $5^3 = 125$.

13. $\log_4 16 = 2$ **14.** $\log_{10} 1000 = 3$

15. $\log_2 \frac{1}{2} = -1$ **16.** $\log_3 \frac{1}{9} = -2$

17. $\log_5 0.2 = -1$ **18.** $\log_{10} 0.1 = -1$

19. $\log_{27} 3 = \frac{1}{3}$ **20.** $\log_8 2 = \frac{1}{3}$

Evaluating Logarithmic Expressions In Exercises 21–30, evaluate the expression without using a calculator. See Example 1.

21. $\log_3 9$ **22.** $\log_5 125$

23. $\log_2 \frac{1}{16}$ **24.** $\log_6 \frac{1}{36}$

25. $\log_8 2$ **26.** $\log_{64} 4$

27. $\log_7 7$ **28.** $\log_{12} 1$

29. $\log_{10} 0.0001$ **30.** $\log_{10} 100$

Evaluating Logarithmic Expressions on a Calculator In Exercises 31–36, use a calculator to evaluate the logarithm. Round your result to three decimal places. See Example 2.

31. $\log_{10} 345$ **32.** $\log_{10} 163$

33. $\log_{10} \frac{4}{5}$ **34.** $\log_{10} \frac{3}{4}$

35. $\log_{10} \sqrt{8}$ **36.** $\log_{10} \sqrt{3}$

Using Properties of Logarithms In Exercises 37–44, solve the equation for x. See Example 3.

37. $\log_5 5 = x$ **38.** $\log_6 6^2 = x$

39. $\log_2 2^{-1} = x$ **40.** $\log_4 4^3 = x$

41. $\log_2(x + 1) = \log_2 4$ **42.** $\log_2(x - 3) = \log_2 9$

43. $\log(2x + 1) = \log 15$ **44.** $\log(5x + 3) = \log 12$

Graphs of Exponential and Logarithmic Functions In Exercises 45–48, sketch the graphs of f and g in the same coordinate plane. See Example 4.

45. $f(x) = 7^x$, $g(x) = \log_7 x$

46. $f(x) = 5^x$, $g(x) = \log_5 x$

47. $f(x) = 6^x$, $g(x) = \log_6 x$

48. $f(x) = 10^x$, $g(x) = \log_{10} x$

Sketching the Graphs of Logarithmic Functions In Exercises 49–54, find the domain, vertical asymptote, and x-intercept of the logarithmic function. Then sketch its graph. See Examples 5 and 6.

49. $f(x) = \log_2 x$ **50.** $g(x) = \log_4 x$

51. $h(x) = \log_2(x + 4)$ **52.** $f(x) = \log_4(x - 3)$

53. $f(x) = -\log_2 x$ **54.** $h(x) = -\log_4(x - 1)$

Writing Logarithmic Equations In Exercises 55–58, write the exponential equation in logarithmic form.

55. $e^1 = e$

56. $e^4 = 54.5981...$

57. $e^x = 4$

58. $e^{-x} = 2$

Writing Exponential Equations In Exercises 59–62, write the logarithmic equation in exponential form.

59. $\ln e = 1$

60. $\ln \frac{1}{e} = -1$

61. $\ln 1 = 0$

62. $\ln \frac{1}{2} = -0.693...$

Using Properties of Natural Logarithms In Exercises 63–66, evaluate the expression without using a calculator. *See Example 7.*

63. $\ln e$

64. $\ln e^{10}$

65. $\ln e^{-4}$

66. $\ln \frac{1}{e^3}$

Evaluating the Natural Logarithmic Function In Exercises 67–72, use a calculator to evaluate the logarithm. Round your result to three decimal places. *See Example 8.*

67. $\ln 7$

68. $2 \ln 9$

69. $\ln 18.42$

70. $\ln 36.7$

71. $\ln \sqrt{6}$

72. $\ln \sqrt{10}$

Using Properties of Logarithms In Exercises 73–76, use the One-to-One Property to solve the equation for *x*.

73. $\ln(x + 2) = \ln 6$

74. $\ln(x - 4) = \ln 2$

75. $\ln(x^2 - 2) = \ln 23$

76. $\ln(x^2 - x) = \ln 6$

Graphs of Exponential and Logarithmic Functions In Exercises 77 and 78, sketch the graphs of *f* and *g* in the same coordinate plane.

77. $f(x) = e^x$, $g(x) = \ln x$ **78.** $f(x) = e^{2x}$, $g(x) = \frac{1}{2} \ln x$

Finding the Domains of Natural Logarithmic Functions In Exercises 79–82, find the domain, vertical asymptote, and *x*-intercept of the logarithmic function. Then sketch its graph. *See Example 9.*

79. $g(x) = \ln(-x)$

80. $f(x) = \ln(3 - x)$

81. $h(x) = \ln(x + 1)$

82. $f(x) = 3 + \ln x$

Matching In Exercises 83–86, match the function with its graph. [The graphs are labeled (a), (b), (c), and (d).]

(a)
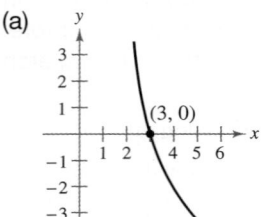

(b)
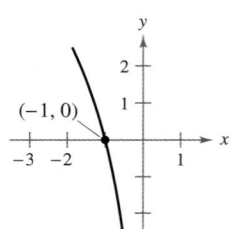

(c)
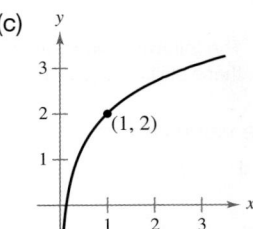

(d)
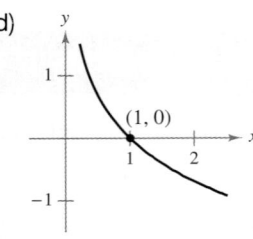

83. $f(x) = \ln x + 2$

84. $f(x) = -\ln x$

85. $f(x) = -3 \ln(x - 2)$

86. $f(x) = 4 \ln(-x)$

 Graphing Logarithmic Functions In Exercises 87–92, use a graphing utility to graph the function. Be sure to use an appropriate viewing window.

87. $f(x) = \log(x + 1)$

88. $f(x) = \log(x - 1)$

89. $f(x) = \ln(x - 1)$

90. $f(x) = \ln(x + 2)$

91. $f(x) = \ln x + 1$

92. $f(x) = 3 \ln x - 1$

93. Population Growth The population of a town will double in

$$t = \frac{8 \ln 3}{\ln 63 - \ln 45} \text{ years.}$$

Find *t*.

94. Work The work *W* (in foot-pounds) done in compressing a volume of 9 cubic feet at a pressure of 15 pounds per square inch to a volume of 3 cubic feet is $W = 19{,}440(\ln 9 - \ln 3)$. Find *W*.

 Skill Retention Model In Exercises 95 and 96, participants in an industrial psychology study were taught a simple mechanical task and tested monthly on this mechanical task for a period of 1 year. The average scores for the participants are given by the model $f(t) = 98 - 14 \log_{10}(t + 1)$, for $0 \le t \le 12$, where *t* is the time (in months).

95. Use a graphing utility to graph the function. Use the graph to discuss the domain and range of the function.

96. Think About It Based on the graph of *f*, do you think the study's participants practiced the simple mechanical task very often? Cite the behavior of the graph to justify your answer.

97. Human Memory Model Students in a seventh-grade class were given an exam. During the next 2 years, the same students were retested several times. The average score *g* can be approximated by the model

$$g(t) = 87 - 16 \log_{10}(t + 1), \quad 0 \le t \le 24$$

where *t* is the time (in months).

(a) What was the average score on the original exam?

(b) What was the average score after 6 months?

(c) When did the average score drop below 70?

98. **HOW DO YOU SEE IT?** The figure shows the graphs of $f(x) = 3^x$ and $g(x) = \log_3 x$. [The graphs are labeled m and n.]

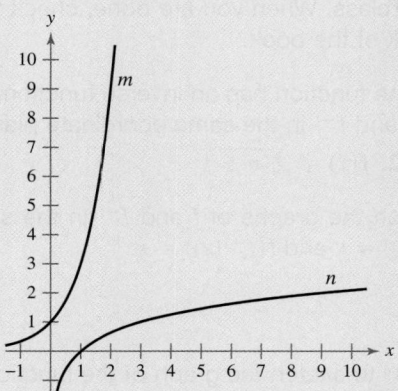

(a) Match each function with its graph.

(b) Given that $f(a) = b$, what is $g(b)$? Explain.

99. Monthly Payment The length t (in years) of a home mortgage of $150,000 at 8% interest can be approximated by the model

$$t = 12.542 \ln\left(\frac{x}{x - 1000}\right), \quad x > 1000$$

where x represents the monthly payment (in dollars) (see figure).

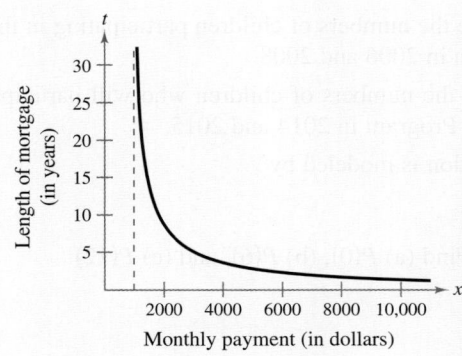

Monthly payment (in dollars)

(a) Use the model to approximate the length of a $150,000 mortgage at 8% interest when the monthly payment is $1100.65 and when the monthly payment is $1254.68.

(b) Approximate the total amount paid over the term of the mortgage with a monthly payment of $1100.65 and with a monthly payment of $1254.68.

(c) Approximate the total interest charge for a monthly payment of $1100.65 and for a monthly payment of $1254.68.

(d) What is the vertical asymptote of the model? Interpret its meaning in the context of the problem.

100. Median Age of U.S. Population The median age A of the U.S. population for each year from 1980 to 2050 can be approximated by the model

$$A = 16.15 - 0.036t + 5.956 \ln t, \quad 10 \le t \le 80$$

where t represents the year, with $t = 10$ corresponding to 1980 (see figure). *(Source: U.S. Census Bureau)*

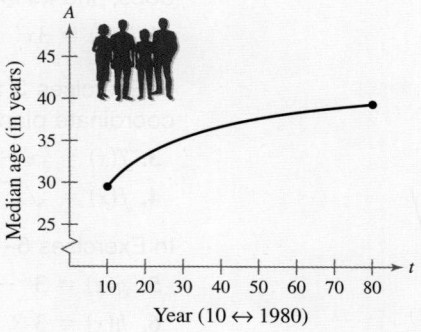

Year (10 ↔ 1980)

(a) Use the model to estimate the median age of the U.S. population in 1980.

(b) Use the model to estimate the median age of the U.S. population in 1990.

(c) Use the model to estimate the change in the median age of the U.S. population from 1980 to 2050.

(d) Use the model to project the change in the median age of the U.S. population from 1980 to 2050.

101. Investment Time A principal P, invested at 5.25% interest compounded continuously, increases to an amount that is K times the principal after t years, where t is given by

$$t = \frac{\ln K}{0.0525}.$$

(a) Complete the table.

K	1	2	4	6	8	10	12
t							

(b) Use the table in part (a) to graph the function.

102. Think About It The table of values was obtained by evaluating a function. Determine which of the statements may be true and which must be false.

x	1	2	8
y	0	1	3

(a) y is an exponential function of x.

(b) y is a logarithmic function of x.

(c) x is an exponential function of y.

(d) y is a linear function of x.

QUIZ YOURSELF

See www.CalcChat.com for worked-out solutions to odd-numbered exercises.

Take this quiz as you would take a quiz in class. When you are done, check your work against the answers given in the back of the book.

In Exercises 1 and 2, determine whether the function has an inverse function. If it does, find its inverse function and graph f and f^{-1} in the same coordinate plane.

1. $f(x) = 3x^2$

2. $f(x) = \sqrt[3]{x + 1}$

In Exercises 3 and 4, (a) find f^{-1}, (b) sketch the graphs of f and f^{-1} in the same coordinate plane, and (c) verify that $f^{-1}(f(x)) = x$ and $f(f^{-1}(x)) = x$.

3. $f(x) = \frac{1}{2}x - 3$

4. $f(x) = \sqrt{x + 1}$

In Exercises 5–8, use the graph of $f(x) = 3^x$ to sketch the graph of the function.

5. $g(x) = 3^x - 2$

6. $h(x) = 3^{-x}$

7. $k(x) = \log_3 x$

8. $j(x) = \log_3(x - 1)$

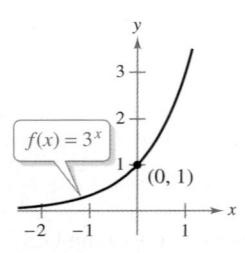

Figure for 5–8

9. For $P = \$10,000$, $r = 5.15\%$, and $t = 8$ years, find the balance in an account when interest is compounded (a) monthly and (b) continuously.

10. The numbers of children C (in millions) participating in the Federal School Breakfast Program from 2000 through 2009 can be approximated by the model

$$C = 7.50(1.04)^t, \quad 0 \le t \le 9$$

where t represents the year, with $t = 0$ corresponding to 2000. *(Source: U.S. Department of Agriculture)*

(a) Use the model to estimate the numbers of children participating in the Federal School Breakfast Program in 2006 and 2008.

(b) Use the model to predict the numbers of children who will participate in the Federal School Breakfast Program in 2014 and 2015.

11. The size of a bacteria population is modeled by

$$P(t) = 100e^{0.2154t}$$

where t is the time in hours. Find (a) $P(0)$, (b) $P(6)$, and (c) $P(12)$.

12. Use the demand function

$$p = 4000\left(1 - \frac{8}{8 + e^{-0.003x}}\right)$$

to find the price for a demand of $x = 500$ MP3 players.

In Exercises 13–16, evaluate the expression without using a calculator.

13. $\log_{10} 100$

14. $\ln e^4$

15. $\log_4 \frac{1}{16}$

16. $\ln 1$

17. Sketch the graphs of $f(x) = 3^x$ and $g(x) = \log_3 x$ in the same coordinate plane. Identify the domains of f and g. Discuss the special relationship between f and g that is shown by their graphs.

4.4 Properties of Logarithms

■ Evaluate a logarithm using the change-of-base formula.
■ Use properties of logarithms to evaluate or rewrite a logarithmic expression.
■ Use properties of logarithms to expand or condense a logarithmic expression.
■ Use logarithmic functions to model and solve real-life applications.

Change of Base

Most calculators have only two types of log keys, one for common logarithms (base 10) and one for natural logarithms (base e). In some situations, you must use logarithms with other bases. To evaluate logarithms with other bases, you can use the **change-of-base formula.**

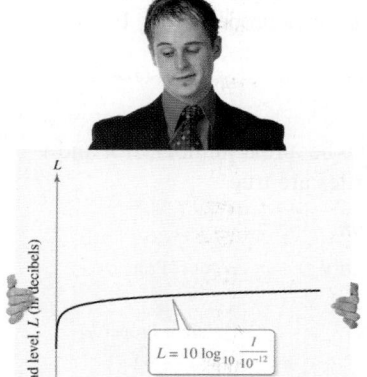

$$L = 10 \log_{10} \frac{I}{10^{-12}}$$

Sound level, L (in decibels)

Sound intensity, I
(in watts per square meter)

In Exercise 107 on page 376, you will use properties of logarithms to determine the difference in loudness between two sounds.

Change-of-Base Formula

Let a, b, and x be positive real numbers such that $a \neq 1$ and $b \neq 1$. Then $\log_a x$ can be converted to a different base as shown.

Base b	Base 10	Base e
$\log_a x = \dfrac{\log_b x}{\log_b a}$	$\log_a x = \dfrac{\log_{10} x}{\log_{10} a}$	$\log_a x = \dfrac{\ln x}{\ln a}$

One way to look at the change-of-base formula is that logarithms to base a are simply *constant multiples* of logarithms to base b. The constant multiplier is

$$\frac{1}{\log_b a}.$$

Example 1 Changing Bases Using Common Logarithms

a. $\log_4 30 = \dfrac{\log_{10} 30}{\log_{10} 4} \approx \dfrac{1.47712}{0.60206} \approx 2.4534$

b. $\log_2 14 = \dfrac{\log_{10} 14}{\log_{10} 2} \approx \dfrac{1.14613}{0.30103} \approx 3.8074$

Example 2 Changing Bases Using Natural Logarithms

a. $\log_4 30 = \dfrac{\ln 30}{\ln 4} \approx \dfrac{3.40120}{1.386294} \approx 2.4534$

b. $\log_2 14 = \dfrac{\ln 14}{\ln 2} \approx \dfrac{2.63906}{0.693147} \approx 3.8074$

✓ **Checkpoints 1 and 2**

Evaluate $\log_8 56$ using common logarithms and natural logarithms. Round your results to three decimal places.

Notice in Examples 1 and 2 that the result is the same whether common logarithms or natural logarithms are used in the change-of-base formula.

Properties of Logarithms

You know from the preceding section that the logarithmic function with base a is the *inverse function* of the exponential function with base a. So, it makes sense that the properties of exponents should have corresponding properties involving logarithms. For instance, the exponential property $a^0 = 1$ has the corresponding logarithmic property $\log_a 1 = 0$.

Properties of Logarithms

Let a be a positive number such that $a \neq 1$, and let n be a real number. If u and v are positive real numbers, then the following properties are true.

Logarithm with Base a	*Natural Logarithm*	
$\log_a(uv) = \log_a u + \log_a v$	$\ln(uv) = \ln u + \ln v$	Product Property
$\log_a \dfrac{u}{v} = \log_a u - \log_a v$	$\ln \dfrac{u}{v} = \ln u - \ln v$	Quotient Property
$\log_a u^n = n \log_a u$	$\ln u^n = n \ln u$	Power Property

STUDY TIP

There is no general property that can be used to rewrite $\log_a(u \pm v)$. Specifically, $\log_a(x + y)$ is *not* equal to $\log_a x + \log_a y$.

Example 3   Using Properties of Logarithms

Write each logarithm in terms of $\ln 2$ and $\ln 3$.

a. $\ln 6$ **b.** $\ln \dfrac{2}{27}$

SOLUTION

a. $\ln 6 = \ln(2 \cdot 3)$ Rewrite 6 as $2 \cdot 3$.

$\qquad = \ln 2 + \ln 3$ Product Property

b. $\ln \dfrac{2}{27} = \ln 2 - \ln 27$ Quotient Property

$\qquad\qquad = \ln 2 - \ln 3^3$ Rewrite 27 as 3^3.

$\qquad\qquad = \ln 2 - 3 \ln 3$ Power Property

✓**Checkpoint 3**

Write $\log_{10} \frac{25}{3}$ in terms of $\log_{10} 3$ and $\log_{10} 5$.

Example 4   Using Properties of Logarithms

Use the properties of logarithms to verify that $-\log_{10} \frac{1}{100} = \log_{10} 100$.

SOLUTION

$$-\log_{10} \frac{1}{100} = -\log_{10}(100^{-1}) \qquad \text{Definition of negative exponent}$$

$$= -(-1) \log_{10} 100 \qquad \text{Power Property}$$

$$= \log_{10} 100 \qquad \text{Simplify.}$$

✓**Checkpoint 4**

Use the properties of logarithms to verify that $-\ln \dfrac{2}{e} = 1 - \ln 2$.

Rewriting Logarithmic Expressions

The properties of logarithms are useful for rewriting logarithmic expressions in forms that simplify the operations of algebra. This is true because these properties convert complicated products, quotients, and exponential forms into simpler sums, differences, and products, respectively.

Example 5 — Expanding Logarithmic Expressions

Expand each logarithmic expression.

a. $\log_4 5x^3 y$ **b.** $\ln \dfrac{\sqrt{3x - 5}}{7}$

SOLUTION

a. $\log_4 5x^3 y = \log_4 5 + \log_4 x^3 + \log_4 y$ Product Property

$\qquad\qquad\quad = \log_4 5 + 3 \log_4 x + \log_4 y$ Power Property

b. $\ln \dfrac{\sqrt{3x - 5}}{7} = \ln \dfrac{(3x - 5)^{1/2}}{7}$ Rewrite using rational exponent.

$\qquad\qquad\quad = \ln(3x - 5)^{1/2} - \ln 7$ Quotient Property

$\qquad\qquad\quad = \dfrac{1}{2} \ln(3x - 5) - \ln 7$ Power Property

> **STUDY TIP**
>
> Be careful when rewriting the logarithmic expression $\log ax^n$. It should be rewritten as
>
> $$\log a + n \log x$$
>
> instead of $n \log ax$.

✓ **Checkpoint 5**

Expand the expression $\ln 2mn^2$. ■

Example 6 — Condensing Logarithmic Expressions

Condense each logarithmic expression.

a. $\frac{1}{2} \log_{10} x + 3 \log_{10}(x + 1)$ **b.** $2 \ln(x + 2) - \ln x$ **c.** $\frac{1}{3}[\log_2 x + \log_2(x + 1)]$

SOLUTION

a. $\frac{1}{2} \log_{10} x + 3 \log_{10}(x + 1) = \log_{10} x^{1/2} + \log_{10}(x + 1)^3$ Power Property

$\qquad\qquad\qquad\qquad\qquad = \log_{10}\left[\sqrt{x}(x + 1)^3\right]$ Product Property

b. $2 \ln(x + 2) - \ln x = \ln(x + 2)^2 - \ln x$ Power Property

$\qquad\qquad\qquad\qquad = \ln \dfrac{(x + 2)^2}{x}$ Quotient Property

c. $\frac{1}{3}[\log_2 x + \log_2(x + 1)] = \frac{1}{3}\{\log_2[x(x + 1)]\}$ Product Property

$\qquad\qquad\qquad\qquad\qquad = \log_2[x(x + 1)]^{1/3}$ Power Property

$\qquad\qquad\qquad\qquad\qquad = \log_2 \sqrt[3]{x(x + 1)}$ Rewrite with a radical.

> **STUDY TIP**
>
> Be careful when rewriting the logarithmic expression $\log x - \log y$. It should be rewritten as
>
> $$\log \dfrac{x}{y}$$
>
> instead of
>
> $$\dfrac{\log x}{\log y}.$$

✓ **Checkpoint 6**

Condense the expression $2 \log_{10}(x + 1) - 3 \log_{10}(x - 1)$. ■

When applying the properties of logarithms to a logarithmic function, be careful to check the domain of the function. For example, the domain of $f(x) = \ln x^2$ is all real $x \neq 0$, whereas the domain of $g(x) = 2 \ln x$ is all real $x > 0$.

Applications

One method of determining how the x- and y-values of a set of nonlinear data are related begins by taking the natural logarithm of each of the x- and y-values. If you graph the points $(\ln x, \ln y)$ and they fall in a straight line, then you can determine that the x- and y-values are related by the equation

$$\ln y = m \ln x$$

where m is the slope of the straight line.

 Example 7   Finding a Mathematical Model

The table shows the mean distance from the sun x and the period (the time it takes a planet to orbit the sun) y for each of the six planets that are closest to the sun. In the table, the mean distance is given in astronomical units (where Earth's mean distance is defined as 1.0), and the period is given in years. Find an equation that relates y and x.

Planet	Mean distance, x	Period, y
Mercury	0.387	0.241
Venus	0.723	0.615
Earth	1.000	1.000
Mars	1.524	1.881
Jupiter	5.203	11.863
Saturn	9.537	29.447

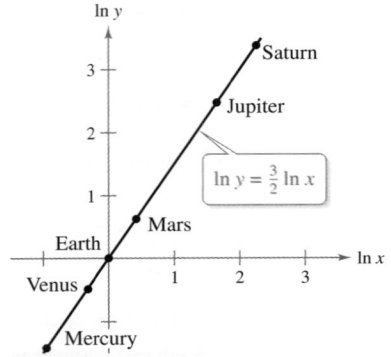

FIGURE 4.21

SOLUTION   The points in the table are plotted in Figure 4.21. From this figure, it is not clear how to find an equation that relates y and x. To solve this problem, take the natural logarithm of each of the x- and y-values in the table. This produces the following results.

Planet	Mercury	Venus	Earth	Mars	Jupiter	Saturn
$\ln x$	-0.949	-0.324	0.000	0.421	1.649	2.255
$\ln y$	-1.423	-0.486	0.000	0.632	2.473	3.383

Now, by plotting the points in the second table, you can see that all six of the points appear to lie in a line (see Figure 4.22). Choose any two points to determine the slope of the line. Using the two points $(0.421, 0.632)$ and $(0, 0)$, you can determine that the slope of the line is

$$m = \frac{0.632 - 0}{0.421 - 0} \approx 1.5 = \frac{3}{2}.$$

FIGURE 4.22

By the point-slope form, the equation of the line is $Y = \frac{3}{2}X$, where $Y = \ln y$ and $X = \ln x$. You can therefore conclude that $\ln y = \frac{3}{2} \ln x$.

✓ **Checkpoint 7**

Find a logarithmic equation that relates y and x.

x	1	2	3	4
y	1	1.414	1.732	2

 **Example 8** **Sound Intensity**

The level of sound L (in decibels) with an intensity of I (in watts per square meter) is given by

$$L = 10 \log_{10} \frac{I}{I_0}$$

where I_0 represents the faintest sound that can be heard by the human ear, and is approximately equal to 10^{-12} watt per square meter. You and your roommate are playing your stereos at the same time and at the same intensity. How much louder is the music when both stereos are playing than when just one stereo is playing?

SOLUTION Let L_1 represent the level of sound when one stereo is playing and let L_2 represent the level of sound when both stereos are playing. Using the formula for level of sound, you can express L_1 as

$$L_1 = 10 \log_{10} \frac{I}{10^{-12}}.$$

For L_2, multiply I by 2 as shown below

$$L_2 = 10 \log_{10} \frac{2I}{10^{-12}}$$

because L_2 represents the level of sound when *two* stereos are playing at the same intensity I. To determine the increase in loudness, subtract L_1 from L_2 as shown.

$$
\begin{aligned}
L_2 - L_1 &= 10 \log_{10} \frac{2I}{10^{-12}} - 10 \log_{10} \frac{I}{10^{-12}} \\
&= 10\left(\log_{10} \frac{2I}{10^{-12}} - \log_{10} \frac{I}{10^{-12}} \right) \\
&= 10\left(\log_{10} 2 + \log_{10} \frac{I}{10^{-12}} - \log_{10} \frac{I}{10^{-12}} \right) \\
&= 10 \log_{10} 2 \approx 3
\end{aligned}
$$

So, the music is about 3 decibels louder. Notice that the variable I drops out of the equation when it is simplified. This means that the loudness increases by 3 decibels when both stereos are playing at the same intensity, regardless of the individual intensities of the stereos.

✓ **Checkpoint 8**

Two sounds have intensities of $I_1 = 10^{-6}$ watt per square meter and $I_2 = 10^{-9}$ watt per square meter. Use the formula for the level of sound in Example 8 to find the difference in loudness between the two sounds.

SUMMARIZE (Section 4.4)

1. State the change-of-base formula *(page 369)*. For examples that use the change-of-base formula, see Examples 1 and 2.

2. Make a list of the properties of logarithms *(page 370)*. For examples that use properties of logarithms, see Examples 3 and 4.

3. Describe how the properties of logarithms are useful for rewriting logarithmic expressions *(page 371)*. For examples of expanding and condensing logarithmic expressions, see Examples 5 and 6.

SKILLS WARM UP 4.4 The following warm-up exercises involve skills that were covered in earlier sections. You will use these skills in the exercise set for this section. For additional help, review Sections 0.3, 0.4, and 4.3.

In Exercises 1–4, evaluate the expression without using a calculator.

1. $\log_7 49$

2. $\log_2 \dfrac{1}{32}$

3. $\ln \dfrac{1}{e^2}$

4. $\log_{10} 0.001$

In Exercises 5–8, simplify the expression.

5. $e^2 e^3$

6. $\dfrac{e^2}{e^3}$

7. $(e^2)^3$

8. $(e^2)^0$

In Exercises 9–12, rewrite the equation in exponential form.

9. $y = \dfrac{1}{x^2}$

10. $y = \sqrt{x}$

11. $\log_4 64 = 3$

12. $\log_{16} 4 = \dfrac{1}{2}$

Exercises 4.4

See www.CalcChat.com for worked-out solutions to odd-numbered exercises.

Changing the Base Using Common Logarithms In Exercises 1–4, evaluate the logarithm using common logarithms. Round your result to three decimal places. *See Example 1.*

1. $\log_5 8$

2. $\log_7 12$

3. $\log_4 30$

4. $\log_9 20$

Changing the Base Using Natural Logarithms In Exercises 5–8, evaluate the logarithm using natural logarithms. Round your result to three decimal places. *See Example 2.*

5. $\log_5 8$

6. $\log_7 12$

7. $\log_6 14$

8. $\log_8 18$

Changing the Base In Exercises 9–18, evaluate the logarithm. Round your result to three decimal places. *See Examples 1 and 2.*

9. $\log_2 6$

10. $\log_8 3$

11. $\log_{27} 35$

12. $\log_{19} 42$

13. $\log_{15} 1250$

14. $\log_{20} 1575$

15. $\log_5 \frac{1}{3}$

16. $\log_9 \frac{3}{5}$

17. $\log_{1/4} 10$

18. $\log_{1/3} 5$

Using Properties of Logarithms In Exercises 19–24, write the logarithm in terms of ln 2 and ln 5. *See Example 3.*

19. $\ln 10$

20. $\ln \frac{2}{5}$

21. $\ln \frac{5}{2}$

22. $\ln 20$

23. $\ln 100$

24. $\ln \frac{125}{16}$

Using Properties of Logarithms In Exercises 25–28, use the properties of logarithms to verify the statement. *See Example 4.*

25. $-\log_{10} \frac{11}{8} = \log_{10} \frac{8}{11}$

26. $-\ln \frac{3}{4} = \ln \frac{4}{3}$

27. $-2 \ln \dfrac{e}{7} = 2 \ln 7 - 2$

28. $-3 \log_{10} \frac{8}{100} = 6 - 3 \log_{10} 8$

Using Properties of Logarithms In Exercises 29–42, approximate the logarithm using the properties of logarithms, given $\log_b 2 \approx 0.3562$, $\log_b 3 \approx 0.5646$, and $\log_b 5 \approx 0.8271$.

29. $\log_b 10$

30. $\log_b 15$

31. $\log_b \frac{2}{3}$

32. $\log_b \frac{3}{5}$

33. $\log_b 8$

34. $\log_b 81$

35. $\log_b \sqrt{2}$

36. $\log_b \sqrt{5}$

37. $\log_b 40$

38. $\log_b 45$

39. $\log_b (2b)^{-2}$

40. $\log_b (3b^2)$

41. $\log_b \sqrt[3]{4b}$

42. $\log_b \sqrt[3]{3b}$

Using Properties of Logarithms In Exercises 43–50, use the properties of logarithms to simplify the logarithmic expression.

43. $\log_9 \frac{1}{18}$

44. $\log_5 \frac{1}{15}$

45. $\log_7 \sqrt{70}$

46. $\log_5 \sqrt{75}$

47. $\log_5 \frac{1}{250}$

48. $\log_{10} \frac{9}{300}$

49. $\ln(5e^6)$

50. $\ln \dfrac{6}{e^2}$

Error Analysis In Exercises 51–54, describe and correct the error.

51. $\log_2 5 = \dfrac{\ln 2}{\ln 5}$

52. $\ln 8x = \ln 8 \cdot \ln x$

53. $\ln 8 - \ln 2 = \dfrac{\ln 8}{\ln 2}$

54. $\log_{10} 5x^4 = 4 \log_{10} 5x$

Expanding a Logarithmic Expression In Exercises 55–76, use the properties of logarithms to expand the expression as a sum, difference, and/or multiple of logarithms. (Assume all variables are positive.) *See Example 5.*

55. $\log_2(4^3 \cdot 3^5)$

56. $\log_3(3^2 \cdot 4^2)$

57. $\log_3 4n$

58. $\log_6 6x$

59. $\log_5 \dfrac{x}{25}$

60. $\log_{10} \dfrac{y}{2}$

61. $\log_2 x^4$

62. $\log_2 z^{-3}$

63. $\ln \sqrt{z}$

64. $\ln \sqrt[3]{t}$

65. $\ln xyz$

66. $\ln \dfrac{xy}{z}$

67. $\ln \sqrt{a - 1}, \quad a > 1$

68. $\ln \sqrt[3]{y - 2}, \quad y > 2$

69. $\ln\left[\dfrac{(z - 1)^2}{z}\right]$

70. $\ln\left(\dfrac{x}{\sqrt{x^2 + 1}}\right)$

71. $\ln \dfrac{yz}{\sqrt[3]{z + 3}}$

72. $\log_9 \dfrac{\sqrt{y}}{7z^2}$

73. $\ln 5\sqrt[3]{\dfrac{x}{y}}$

74. $\ln 3\sqrt{\dfrac{x^2}{y^3}}$

75. $\ln \sqrt[4]{x^3(x^2 + 3)}$

76. $\ln \sqrt{x^2(x + 2)}$

Condensing a Logarithmic Expression In Exercises 77–92, condense the expression to the logarithm of a single quantity. *See Example 6.*

77. $\log_3 x + \log_3 5$

78. $\log_5 y + \log_5 x$

79. $\log_4 8 - \log_4 x$

80. $\log_{10} 4 - \log_{10} z$

81. $2 \log_{10}(x + 4)$

82. $-4 \log_{10} 2x$

83. $-\ln x - 3 \ln 6$

84. $2 \ln 8 + 5 \ln z$

85. $\frac{1}{3} \ln 5x - \ln(x + 1)$

86. $\frac{3}{2} \ln(z - 2) + \ln z$

87. $\log_8(x - 2) - \log_8(x + 2)$

88. $3 \log_7 x + 2 \log_7 y - 4 \log_7 z$

89. $2 \ln 4 + 2 \ln 3 - \frac{1}{2} \ln(x^2 + 1)$

90. $2 \ln(x^2 - 2) + \frac{3}{2} \ln t^6 - \frac{3}{4} \ln t^4$

91. $2 \ln x - [\ln(x + 2) + \ln(x - 2)]$

92. $4 \ln(x + 1) + 2 \ln(x - 1) - 3 \ln x$

93. Graphical Analysis Use a graphing utility to graph

$f(x) = \ln 5x \quad \text{and} \quad g(x) = \ln 5 + \ln x$

in the same viewing window. How do the two graphs compare? What property of logarithms do the graphs demonstrate?

94. Think About It Consider the functions below.

$$f(x) = \ln \dfrac{x}{2}, \quad g(x) = \dfrac{\ln x}{\ln 2}, \quad h(x) = \ln x - \ln 2$$

Which two functions have identical graphs? Verify your answer by using a graphing utility to graph all three functions in the same viewing window.

95. Graphical Analysis Use a graphing utility to graph

$f(x) = \log_{10} x^3 \quad \text{and} \quad g(x) = 3 \log_{10} x$

in the same viewing window. How do the two graphs compare? What property of logarithms is shown?

96. HOW DO YOU SEE IT?
The figure shows the graphs of $y = \ln x$, $y = \ln x^2$, $y = \ln 2x$, and $y = \ln 2$. Match each function with its graph. (The graphs are labeled A through D.) Explain your reasoning.

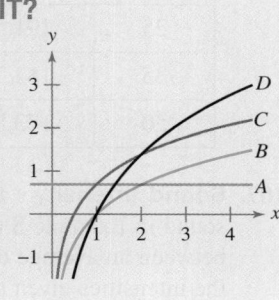

Using Inverse Properties In Exercises 97–102, find the exact value of the logarithmic expression without using a calculator.

97. $\log_4 \sqrt[3]{4}$

98. $\log_8 \sqrt[4]{8}$

99. $\ln \dfrac{1}{\sqrt{e}}$

100. $\ln \sqrt[4]{e^3}$

101. $\log_5 \frac{1}{125}$

102. $\log_7 \frac{49}{343}$

Finding a Mathematical Model In Exercises 103 and 104, find a logarithmic equation that relates y and x. Explain the steps used to find the equation. *See Example 7.*

103.

x	1	2	3	4	5	6
y	1	1.189	1.316	1.414	1.495	1.565

104.

x	1	2	3	4	5	6
y	1	1.587	2.080	2.520	2.924	3.302

105. Nail Length The approximate lengths and diameters (in inches) of common nails are shown in the table. Find a logarithmic equation that relates the diameter y of a common nail to its length x.

Length, x	Diameter, y	Length, x	Diameter, y
1	0.070	4	0.176
2	0.111	5	0.204
3	0.146	6	0.231

106. Galloping Speeds of Animals Four-legged animals run with two different types of motion: trotting and galloping. An animal that is trotting has at least one foot on the ground at all times, whereas an animal that is galloping has all four feet off the ground at some point in its stride. The number of strides per minute at which an animal breaks from a trot to a gallop depends on the weight of the animal. Use the table to find a logarithmic equation that relates an animal's lowest galloping speed y (in strides per minute) to its weight x (in pounds).

Weight, x	Galloping speed, y	Weight, x	Galloping speed, y
25	191.5	75	164.2
35	182.7	500	125.9
50	173.8	1000	114.2

107. Sound Intensity Use the equation for the level of sound in Example 8 to find the difference in loudness between an average office and a broadcast studio with the intensities given below.

Office: 1.26×10^{-7} watt per square meter

Broadcast studio: 3.16×10^{-10} watt per square meter

108. Sound Intensity Use the equation for the level of sound in Example 8 to find the difference in loudness between a bird singing and rustling leaves with the intensities given below.

Bird singing: 10^{-8} watt per square meter

Rustling leaves : 10^{-10} watt per square meter

109. Think About It For each property of logarithms given below, write a related property of exponents.

(a) $\log_a(uv) = \log_a u + \log_a v$

(b) $\log_a \dfrac{u}{v} = \log_a u - \log_a v$

(c) $\log_a u^n = n \log_a u$

110. Proof Complete the proof of the logarithmic property

$\log_a uv = \log_a u + \log_a v.$

Let $\log_a u = x$ and $\log_a v = y.$

$a^x = \boxed{}$ and $a^y = \boxed{}$ Rewrite in exponential form.

$u \cdot v = \boxed{} \cdot \boxed{} = a^{\boxed{}}$ Multiply and substitute for u and v.

$\boxed{} = x + y$ Rewrite in logarithmic form.

$\log_a uv = \boxed{} + \boxed{}$ Substitute for x and y.

111. Proof Complete the proof of the logarithmic property

$\log_a \dfrac{u}{v} = \log_a u - \log_a v.$

Let $\log_a u = x$ and $\log_a v = y.$

$a^x = \boxed{}$ and $a^y = \boxed{}$ Rewrite in exponential form.

$\dfrac{u}{v} = \dfrac{\boxed{}}{\boxed{}} = a^{\boxed{}}$ Divide and substitute for u and v.

$\boxed{} = x - y$ Rewrite in logarithmic form.

$\log_a \dfrac{u}{v} = \boxed{} - \boxed{}$ Substitute for x and y.

Homes of hospitality for expectant women.

www.maggiesplace.org

Business Capsule

Co-founded by five recent college graduates, Maggie's Place is a community of homes that provides hospitality for pregnant women who are alone or living on the streets. In homes shared with staff members, expectant mothers have their immediate needs met while being connected to community resources such as prenatal care, education programs, and low-cost housing. Maggie's Place opened its first home, the Magdalene House, on May 13, 2000 in Phoenix, Arizona. Today, there are five homes, and a resource and support center for women and children who have completed the Maggie's Place program.

112. Research Project Use your campus library, the Internet, or some other reference source to find information about a nonprofit group or company whose growth can be modeled by a logarithmic function. Write a brief report about the growth of the group or company.

4.5 Solving Exponential and Logarithmic Equations

■ Understand the strategies for solving exponential and logarithmic equations.
■ Solve an exponential equation.
■ Solve a logarithmic equation.
■ Use an exponential or a logarithmic model to solve an application problem.

Introduction

So far in this chapter, you have studied the definitions, graphs, and properties of exponential and logarithmic functions. In this section, you will study procedures for *solving equations* involving these exponential and logarithmic functions.

There are two basic strategies for solving exponential or logarithmic equations. The first is based on the One-to-One Properties and the second is based on the Inverse Properties. For $a > 0$ and $a \neq 1$, the following properties are true for all x and y for which $\log_a x$ and $\log_a y$ are defined.

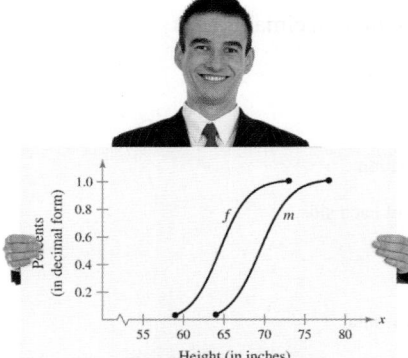

One-to-One Properties

$a^x = a^y$ if and only if $x = y$.

$\log_a x = \log_a y$ if and only if $x = y$.

Inverse Properties

$a^{\log_a x} = x$

$\log_a a^x = x$

In Exercise 110 on page 386, you will use exponential models to find the median heights of American males and females between the ages of 20 and 29 years old.

Example 1   Solving Simple Equations

Original Equation	*Rewritten Equation*	*Solution*	*Property*
a. $2^x = 32$	$2^x = 2^5$	$x = 5$	One-to-One
b. $\ln x - \ln 3 = 0$	$\ln x = \ln 3$	$x = 3$	One-to-One
c. $e^x = 7$	$\ln e^x = \ln 7$	$x = \ln 7$	Inverse
d. $\ln x = -3$	$e^{\ln x} = e^{-3}$	$x = e^{-3}$	Inverse
e. $\log_{10} x = -1$	$10^{\log_{10} x} = 10^{-1}$	$x = 10^{-1} = \frac{1}{10}$	Inverse

✓Checkpoint 1

Solve each equation for x.

a. $3^x = 81$

b. $\log_6 x = 3$

c. $\ln 5 - \ln x = 0$

Strategies for Solving Exponential and Logarithmic Equations

1. Rewrite the original equation in a form that allows the use of the One-to-One Property of exponential or logarithmic functions.

2. Rewrite an *exponential* equation in logarithmic form and apply the Inverse Property of logarithmic functions.

3. Rewrite a *logarithmic* equation in exponential form and apply the Inverse Property of exponential functions.

Solving Exponential Equations

TECH TUTOR

When solving an exponential or logarithmic equation, you can check your solution graphically by "graphing the left and right sides separately" and using the *intersect* feature of your graphing utility to determine the point of intersection. For instance, to check the solution of the equation in Example 2(a), graph

$$y_1 = 4^x \quad \text{and} \quad y_2 = 72$$

in the same viewing window, as shown below. Using the *intersect* feature, you can determine that the graphs intersect when $x \approx 3.085$, which confirms the solution found in Example 2(a).

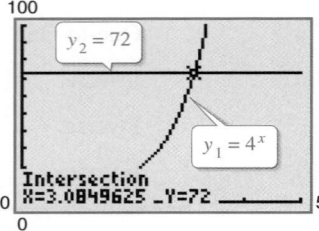

For instructions on how to use the *intersect* feature, see Appendix A. Consult the user's guide for your graphing utility for more information.

Example 2   Solving Exponential Equations

Solve each equation and approximate the result to three decimal places.

a. $4^x = 72$ **b.** $3(2^x) = 42$

SOLUTION

a.

$4^x = 72$	Write original equation.
$\log_4 4^x = \log_4 72$	Take log (base 4) of each side.
$x = \log_4 72$	Inverse Property
$x = \dfrac{\ln 72}{\ln 4} \approx 3.085$	Change-of-base formula

The solution is $x = \log_4 72 \approx 3.085$. Check this in the original equation.

b.

$3(2^x) = 42$	Write original equation.
$2^x = 14$	Divide each side by 3 to isolate the exponential expression.
$\log_2 2^x = \log_2 14$	Take log (base 2) of each side.
$x = \log_2 14$	Inverse Property
$x = \dfrac{\ln 14}{\ln 2} \approx 3.807$	Change-of-base formula

The solution is $x = \log_2 14 \approx 3.807$. Check this in the original equation.

✓ **Checkpoint 2**

Solve each equation and approximate the result to three decimal places.

a. $6^x = 84$ **b.** $62 - 10^x = 24$ ■

In Example 2(a), the exact solution is $x = \log_4 72$ and the approximate solution is $x \approx 3.085$. An exact answer is preferred when the solution is an intermediate step in a larger problem. For a final answer, an approximate solution in decimal form is easier to comprehend.

Example 3   Solving an Exponential Equation

$e^x + 5 = 60$	Original equation
$e^x = 55$	Subtract 5 from each side to isolate the exponential expression.
$\ln e^x = \ln 55$	Take natural log of each side.
$x = \ln 55$	Inverse Property
$x \approx 4.007$	Use a calculator.

The solution is $x = \ln 55 \approx 4.007$. Check this in the original equation.

✓ **Checkpoint 3**

Solve $2e^x - 7 = 23$ and approximate the result to three decimal places.   ■

Example 4 **Solving an Exponential Equation**

Solve $2(3^{2t-5}) - 4 = 11$ and approximate the result to three decimal places.

SOLUTION

$2(3^{2t-5}) - 4 = 11$	Write original equation.
$2(3^{2t-5}) = 15$	Add 4 to each side.
$3^{2t-5} = \dfrac{15}{2}$	Divide each side by 2.
$\log_3 3^{2t-5} = \log_3 \dfrac{15}{2}$	Take log (base 3) of each side.
$2t - 5 = \log_3 \dfrac{15}{2}$	Inverse Property
$2t = 5 + \log_3 7.5$	Add 5 to each side.
$t = \dfrac{5}{2} + \dfrac{1}{2}\log_3 7.5$	Divide each side by 2.
$t \approx 3.417$	Use a calculator.

The solution is $t = \frac{5}{2} + \frac{1}{2}\log_3 7.5 \approx 3.417$. Check this in the original equation.

STUDY TIP

Remember that to evaluate a logarithm such as $\log_3 7.5$, you need to use the change-of-base formula as shown.

$$\log_3 7.5 = \frac{\log_{10} 7.5}{\log_{10} 3} \approx 1.834$$

✓ **Checkpoint 4**

Solve $4(4^{2t-7}) + 14 = 110$ and approximate the result to three decimal places. ▪

When an equation involves two or more exponential expressions, you can still use a procedure similar to that demonstrated in Examples 2, 3, and 4. However, the algebra is a bit more complicated.

ALGEBRA TUTOR xy

For help with the algebra used to factor the left side of the equation in Example 5, see the *Chapter 4 Algebra Tutor* on page 401.

Example 5 **Solving an Exponential Equation of Quadratic Type**

$e^{2x} - 3e^x + 2 = 0$	Original equation
$(e^x)^2 - 3e^x + 2 = 0$	Write in quadratic form.
$(e^x - 2)(e^x - 1) = 0$	Factor.
$e^x - 2 = 0 \implies x = \ln 2$	Set 1st factor equal to 0.
$e^x - 1 = 0 \implies x = 0$	Set 2nd factor equal to 0.

The solutions are $x = \ln 2$ and $x = 0$. Check these in the original equation. Or, check by graphing $y = e^{2x} - 3e^x + 2$ using a graphing utility. The graph should have two x-intercepts: $x = \ln 2$ and $x = 0$, as shown in Figure 4.23.

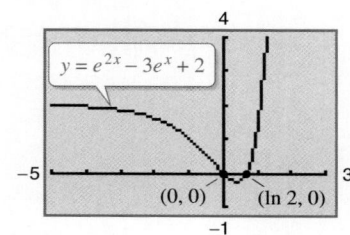

FIGURE 4.23

✓ **Checkpoint 5**

Solve $e^{2x} - 7e^x + 12 = 0$. ▪

Solving Logarithmic Equations

To solve a logarithmic equation such as

$$\ln x = 3 \qquad \text{Logarithmic form}$$

write the equation in exponential form as shown.

$$e^{\ln x} = e^3 \qquad \text{Exponentiate each side.}$$

$$x = e^3 \qquad \text{Exponential form}$$

This procedure is called **exponentiating** each side of an equation.

Example 6   **Solving Logarithmic Equations**

a. Solve $\ln(x + 5) = 1$.   **b.** Solve $2 \log_5 3x = 4$.

SOLUTION

a. $\ln(x + 5) = 1$ Write original equation.

$\qquad e^{\ln(x+5)} = e^1$ Exponentiate each side.

$\qquad\quad x + 5 = e$ Inverse Property

$\qquad\qquad x = e - 5$ Subtract 5 from each side.

The solution is $x = e - 5$. Check this in the original equation.

b. $2 \log_5 3x = 4$ Write original equation.

$\qquad \log_5 3x = 2$ Divide each side by 2.

$\qquad 5^{\log_5 3x} = 5^2$ Exponentiate each side (base 5).

$\qquad\quad 3x = 25$ Inverse Property

$\qquad\quad x = \dfrac{25}{3}$ Divide each side by 3.

The solution is $x = \frac{25}{3}$. Check this in the original equation. _____

 **Checkpoint 6**

Solve each logarithmic equation.

a. $\ln(x - 4) = 0$ **b.** $\log_3 2x = 4$ ▣

Example 7   **Solving a Logarithmic Equation**

Solve $\log_3(5x - 1) = \log_3(x + 7)$.

SOLUTION

$\qquad \log_3(5x - 1) = \log_3(x + 7)$ Write original equation.

$\qquad\qquad 5x - 1 = x + 7$ One-to-One Property

$\qquad\qquad\quad 4x = 8$ Add $-x$ and 1 to each side.

$\qquad\qquad\quad x = 2$ Divide each side by 4.

The solution is $x = 2$. Check this in the original equation. _____

 **Checkpoint 7**

Solve $\ln(3x + 2) = \ln(x + 8)$.   ▣

Ken Hurst/www.shutterstock.com

Example 8 Solving a Logarithmic Equation

Solve $5 + 2 \ln x = 4$ and approximate the result to three decimal places.

SOLUTION

$5 + 2 \ln x = 4$	Write original equation.
$2 \ln x = -1$	Subtract 5 from each side.
$\ln x = -\dfrac{1}{2}$	Divide each side by 2.
$e^{\ln x} = e^{-1/2}$	Exponentiate each side.
$x = e^{-1/2}$	Inverse Property
$x \approx 0.607$	Use a calculator.

The solution is $x = e^{-1/2} \approx 0.607$. Check this in the original equation.

✓ **Checkpoint 8**

Solve $4 + 3 \ln x = 16$ and approximate the result to three decimal places. ■

Because the domain of a logarithmic function generally does not include all real numbers, be sure to check for extraneous solutions of logarithmic equations.

Example 9 Checking for Extraneous Solutions

Solve $\log_{10} 5x + \log_{10}(x - 1) = 2$.

SOLUTION

$\log_{10} 5x + \log_{10}(x - 1) = 2$	Write original equation.
$\log_{10}[5x(x - 1)] = 2$	Product Property of logarithms
$10^{\log_{10}(5x^2 - 5x)} = 10^2$	Exponentiate each side (base 10).
$5x^2 - 5x = 100$	Inverse Property
$x^2 - x - 20 = 0$	Write in general form.
$(x - 5)(x + 4) = 0$	Factor.
$x - 5 = 0$	Set 1st factor equal to 0.
$x = 5$	Solution
$x + 4 = 0$	Set 2nd factor equal to 0.
$x = -4$	Solution

The solutions appear to be $x = 5$ and $x = -4$. However, when you check these in the original equation, you can see that $x = 5$ is the only solution.

✓ **Checkpoint 9**

Solve $\log_6 x + \log_6(x + 5) = 2$. ■

In Example 9, the domain of $\log_{10} 5x$ is $x > 0$ and the domain of $\log_{10}(x - 1)$ is $x > 1$, so the domain of the original equation is $x > 1$. Because the domain is all real numbers greater than 1, the solution $x = -4$ is extraneous.

TECH TUTOR

You can use a graphing utility to verify that the equation in Example 9 has $x = 5$ as its only solution. Graph

$$y_1 = \log_{10} 5x + \log_{10}(x - 1)$$

and

$$y_2 = 2$$

in the same viewing window, as shown below. It appears that the graphs intersect at one point. Use the *intersect* feature or the *zoom* and *trace* features to determine that $x = 5$ is the solution.

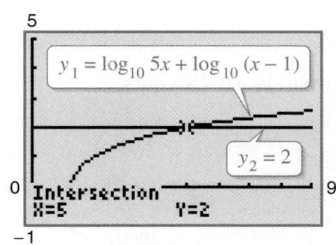

For instructions on how to use the *zoom* and *trace* features, see Appendix A. Consult the user's guide for your graphing utility for more information.

Applications

 Example 10   **Doubling and Tripling an Investment**

You deposit $500 in an account that pays 6.75% interest, compounded continuously.

a. How long will it take your money to double?

b. How long will it take your money to triple?

SOLUTION   Using the formula for compound interest with continuous compounding, you can find that the balance in the account is given by

$$A = Pe^{rt} = 500e^{0.0675t}.$$

a. To find the time required for the balance to double, let $A = 1000$ and solve the resulting equation for t.

$500e^{0.0675t} = 1000$	Substitute 1000 for A.
$e^{0.0675t} = 2$	Divide each side by 500.
$\ln e^{0.0675t} = \ln 2$	Take natural log of each side.
$0.0675t = \ln 2$	Inverse Property
$t = \dfrac{1}{0.0675} \ln 2$	Divide each side by 0.0675.
$t \approx 10.27$	Use a calculator.

The balance in the account will double after approximately 10.27 years.

b. To find the time required for the balance to triple, let $A = 1500$ and solve the resulting equation for t.

$500e^{0.0675t} = 1500$	Substitute 1500 for A.
$e^{0.0675t} = 3$	Divide each side by 500.
$\ln e^{0.0675t} = \ln 3$	Take natural log of each side.
$0.0675t = \ln 3$	Inverse Property
$t = \dfrac{1}{0.0675} \ln 3$	Divide each side by 0.0675.
$t \approx 16.28$	Use a calculator.

The balance in the account will triple after approximately 16.28 years.

Notice that it took 10.27 years to earn the first $500 and only 6.01 years to earn the second $500. This result is graphically demonstrated in Figure 4.24.

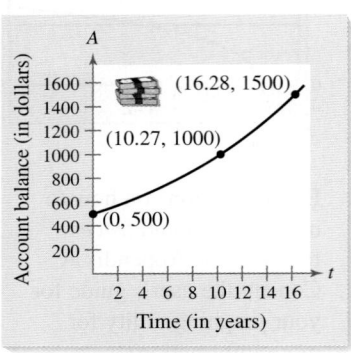

FIGURE 4.24

✓**Checkpoint 10**

In Example 10, how long will it take for the account balance to reach $600?

Example 11   Hospital Expenses

From 1997 through 2008, the yearly expenses y (in billions of dollars) of hospitals in the United States can be approximated by

$$y = 208.05\,e^{0.067t}$$

where t represents the year, with $t = 7$ corresponding to 1997 (see Figure 4.25). Use the model to estimate the year in which the expenses of hospitals reached about $610 billion. *(Source: Health Forum, An American Hospital Association Company)*

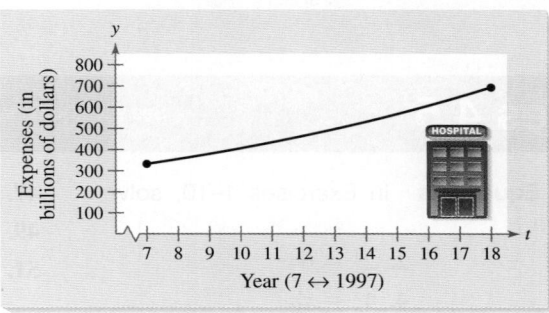

FIGURE 4.25

SOLUTION   To solve this problem, let $y = 610$ and solve the resulting equation for t.

$208.05e^{0.067t} = 610$	Substitute 610 for y.
$e^{0.067t} \approx 2.932$	Divide each side by 208.05.
$\ln e^{0.067t} \approx \ln 2.932$	Take natural log of each side.
$0.067t \approx 1.076$	Inverse Property
$t \approx 16$	Divide each side by 0.067.

The solution is $t \approx 16$. Because $t = 7$ represents 1997, you can conclude that, according to the model, expenses were about $610 billion in 2006.

**Checkpoint 11**

Use the model in Example 11 to estimate the year in which the expenses of hospitals reached about $435 billion.

SUMMARIZE   (Section 4.5)

1. State the One-to-One Properties and the Inverse Properties that can be used to solve simple exponential and logarithmic equations *(page 377)*. For an example of solving simple exponential and logarithmic equations, see Example 1.

2. Describe the strategies for solving exponential and logarithmic equations *(page 377)*. For examples of solving exponential equations, see Examples 2, 3, 4, and 5.

3. Describe what it means to exponentiate each side of an equation *(page 380)*. For examples of solving logarithmic equations, see Examples 6, 7, 8, and 9.

4. Describe a real-life example of how an exponential equation can be used to find the balance in an investment account *(page 382, Example 10)*.

SKILLS WARM UP 4.5

The following warm-up exercises involve skills that were covered in earlier sections. You will use these skills in the exercise set for this section. For additional help, review Sections 1.4 and 4.3.

In Exercises 1–6, solve for x.

1. $x \ln 2 = \ln 3$

2. $(x - 1) \ln 4 = 2$

3. $2xe^2 = e^3$

4. $4xe^{-1} = 8$

5. $x^2 - 4x + 5 = 0$

6. $2x^2 - 3x + 1 = 0$

In Exercises 7–10, simplify the expression.

7. $\log_{10} 10^x$

8. $\log_{10} 10^{2x}$

9. $\ln e^{2x}$

10. $\ln e^{-x^2}$

Exercises 4.5

See www.CalcChat.com for worked-out solutions to odd-numbered exercises.

Solving Simple Equations In Exercises 1–10, solve for x. See Example 1.

1. $5^x = 25$

2. $2^x = 64$

3. $7^x = \frac{1}{49}$

4. $4^x = \frac{1}{256}$

5. $4^{x-1} = 64$

6. $3^{x-1} = 27$

7. $\log_4 x = 3$

8. $\log_{10} x = -2$

9. $\log_5 5x = \log_5 2$

10. $\ln 2x = \ln 6$

Using Inverse Properties In Exercises 11–22, apply the Inverse Property of logarithmic or exponential functions to simplify the expression.

11. $\ln e^{x^2}$

12. $\ln e^{2x-1}$

13. $\log_{10} 10^x + 1$

14. $\log_{10} 10^{2x+3}$

15. $\log_5 5^{x^3} - 7$

16. $\log_8 8^{x^5} + 1$

17. $8 + e^{\ln x^3}$

18. $-1 + \ln e^{2x}$

19. $10^{\log_{10}(x+5)}$

20. $9^{\log_9(3x+7)}$

21. $2^{\log_2 x^2}$

22. $10^{\log_{10}(x^2+7x+10)}$

Solving Exponential Equations In Exercises 23–60, solve the exponential equation algebraically. Approximate the result to three decimal places. See Examples 2, 3, 4, and 5.

23. $3e^x = 9$

24. $5e^x = 20$

25. $2(3^x) = 16$

26. $3(4^x) = 81$

27. $e^x - 9 = 19$

28. $e^x + 4 = 18$

29. $3^{2x} = 80$

30. $6^{5x} = 3000$

31. $5^{-t/2} = 0.20$

32. $4^{-t/3} = 0.15$

33. $3^{x-1} = 28$

34. $2^{x-3} = 31$

35. $2^{3-x} = 565$

36. $8^{-2-x} = 431$

37. $8(10^{3x}) = 12$

38. $5(10^{2x}) = 7$

39. $3(5^{x-1}) = 21$

40. $8(3^{6-x}) = 40$

41. $e^{3x} = 12$

42. $e^{2x} = 50$

43. $500e^{-x} = 300$

44. $1000e^{-4x} = 75$

45. $7 - 2e^x = 6$

46. $-14 + 3e^x = 11$

47. $6(2^{3x-1}) - 7 = 9$

48. $8(4^{6-2x}) + 13 = 41$

49. $e^{2x} - 8e^x + 12 = 0$

50. $e^{2x} - 5e^x + 6 = 0$

51. $e^{2x} - 3e^x - 4 = 0$

52. $e^{2x} - 9e^x - 36 = 0$

53. $\dfrac{500}{100 - e^{x/2}} = 20$

54. $\dfrac{400}{1 + e^{-x}} = 350$

55. $\dfrac{3000}{2 + e^{2x}} = 2$

56. $\dfrac{119}{e^{6x} - 14} = 7$

57. $\left(1 + \dfrac{0.065}{365}\right)^{365t} = 4$

58. $\left(1 + \dfrac{0.075}{4}\right)^{4t} = 5$

59. $\left(1 + \dfrac{0.10}{12}\right)^{12t} = 2$

60. $\left(1 + \dfrac{0.0825}{26}\right)^{26t} = 9$

Solving Logarithmic Equations In Exercises 61–86, solve the logarithmic equation algebraically. Approximate the result to three decimal places. See Examples 6, 7, 8, and 9.

61. $\log_{10} x = 4$

62. $\log_{10} x = -5$

63. $\ln x = -3$

64. $\ln x = 6$

65. $\ln 2x = 2.4$

66. $\ln 4x = 1$

67. $\log_{10} 2x = 7$

68. $\log_{10} 3z = 2$

69. $4 \log_3(x + 1) = 12$

70. $5 \log_{10}(x - 2) = 10$

71. $3 \ln 5x = 10$

72. $2 \ln x = 7$

73. $\ln \sqrt{x + 2} = 1$

74. $\ln \sqrt{x - 8} = 5$

75. $7 + 3 \ln x = 5$

76. $2 - 6 \ln x = 10$

77. $\ln x - \ln(x + 1) = 2$

78. $\ln x - \ln(x + 2) = 3$

79. $\ln(x + 5) = \ln(x - 1) - \ln(x + 1)$

80. $\ln(x + 1) - \ln(x - 2) = \ln x$

81. $\log_2(2x - 3) = \log_2(x + 4)$

82. $\log_3(x + 8) = \log_3(3x + 2)$

83. $\log_{10}(x + 4) - \log_{10} x = \log_{10}(x + 2)$

84. $\log_{10} x + \log_{10}(x + 1) = \log_{10}(x + 3)$

85. $\log_4 x - \log_4(x - 1) = \frac{1}{2}$

86. $\log_3 x + \log_3(x - 8) = 2$

Rewriting Logarithmic Equations In Exercises 87–90, solve for y in terms of x.

87. $\ln y = \ln(2x + 1) + \ln 1$

88. $\ln y = 2 \ln x + \ln(x - 3)$

89. $\log_{10} y = 2 \log_{10}(x - 1) - \log_{10}(x + 2)$

90. $\log_{10}(y - 4) + \log_{10} x = 3 \log_{10} x$

 **Solving Equations Using a Graphing Utility** In Exercises 91–94, use a graphing utility to solve the equation. Approximate the result to three decimal places. Verify your result algebraically.

91. $2^x - 7 = 0$

92. $500 - 1500e^{-x/2} = 0$

93. $3 - \ln x = 0$

94. $10 - 4 \ln(x - 2) = 0$

Doubling an Investment In Exercises 95 and 96, find the time required for a $1000 investment to double at interest rate r, compounded continuously. *See Example 10.*

95. $r = 0.0725$

96. $r = 0.065$

Tripling an Investment In Exercises 97 and 98, find the time required for a $1000 investment to triple at interest rate r, compounded continuously. *See Example 10.*

97. $r = 0.0825$

98. $r = 0.0775$

99. Suburban Wildlife The number V of varieties of suburban nondomesticated wildlife in a community is approximated by the model

$$V = 15 \cdot 10^{0.02x}, \quad 0 \le x \le 36$$

where x is the number of months since the development of the community was completed. Use this model to approximate the number of months since the development was completed when $V = 60$.

100. Native Prairie Grasses The number A of varieties of native prairie grasses per acre within a farming region is approximated by the model

$$A = 10.5 \cdot 10^{0.04x}, \quad 0 \le x \le 24$$

where x is the number of months since the farming region was plowed. Use this model to approximate the number of months since the region was plowed using a test acre for which $A = 80$.

101. Demand Function The demand function for a special limited edition coin set is given by

$$p = 1000\left(1 - \frac{5}{5 + e^{-0.001x}}\right).$$

(a) Find the demand x for a price of $p = \$139.50$.

(b) Find the demand x for a price of $p = \$99.99$.

(c) Use a graphing utility to confirm graphically the results found in parts (a) and (b).

102. Demand Function The demand function for a hot tub spa is given by

$$p = 105,000\left(1 - \frac{3}{3 + e^{-0.002x}}\right).$$

(a) Find the demand x for a price of $p = \$25,000$.

(b) Find the demand x for a price of $p = \$21,000$.

(c) Use a graphing utility to confirm graphically the results found in parts (a) and (b).

103. Forest Yield The yield V (in millions of cubic feet per acre) for a forest at age t years is given by

$$V = 6.7e^{-48.1/t}, \quad t > 0.$$

(a) Use a graphing utility to find the time necessary to obtain a yield of 1.3 million cubic feet per acre.

(b) Use the graphing utility to find the time necessary to obtain a yield of 2 million cubic feet per acre.

104. Human Memory Model In a group project on learning theory, a mathematical model for the percent P (in decimal form) of correct responses after n trials was found to be

$$P = \frac{0.98}{1 + e^{-0.3n}}, \quad n \ge 0.$$

(a) After how many trials will 80% of the responses be correct? (That is, for what value of n will $P = 0.8$?)

(b) Use a graphing utility to graph the memory model and confirm the result found in part (a).

(c) Write a paragraph describing the memory model.

105. Drug Prescriptions 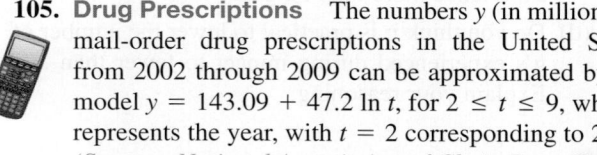 The numbers y (in millions) of mail-order drug prescriptions in the United States from 2002 through 2009 can be approximated by the model $y = 143.09 + 47.2 \ln t$, for $2 \le t \le 9$, where t represents the year, with $t = 2$ corresponding to 2002. *(Source: National Association of Chain Drug Stores)*

(a) Use a graphing utility to graph the model.

(b) Use the graphing utility to estimate the year when the number of mail-order drug prescriptions exceeded 200 million.

(c) Verify your answer to part (b) algebraically.

106. Air Transportation 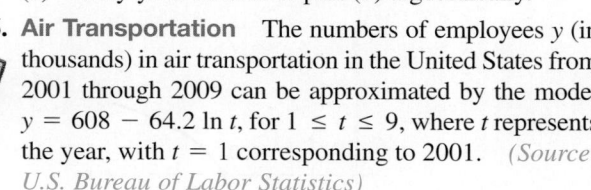 The numbers of employees y (in thousands) in air transportation in the United States from 2001 through 2009 can be approximated by the model $y = 608 - 64.2 \ln t$, for $1 \le t \le 9$, where t represents the year, with $t = 1$ corresponding to 2001. *(Source: U.S. Bureau of Labor Statistics)*

(a) Use a graphing utility to graph the model.

(b) Use the graphing utility to estimate the year in which the number of air transportation employees fell below 500,000.

(c) Verify your answer to part (b) algebraically.

107. Automobiles  Automobiles are designed with crumple zones that help protect their occupants. The crumple zones allow the occupants to move short distances when the automobiles come to abrupt stops. The greater the distance moved, the fewer g's the crash victims experience. (One g is equal to the acceleration due to gravity. For very short periods of time, humans have withstood as much as 40 g's.) In crash tests with vehicles moving at 90 kilometers per hour, analysts measured the numbers of g's experienced during deceleration by crash dummies that were permitted to move x meters during impact. The data are shown in the table.

x	0.2	0.4	0.6	0.8	1.0
g's	158	80	53	40	32

A model for these data is given by

$$y = -3.00 + 11.88 \ln x + \frac{36.94}{x}$$

where y is the number of g's.

(a) Use the model to expand the table by adding a row of y-values that correspond to the given x-values.

(b) Use a graphing utility to graph the data points and the model in the same viewing window. How do they compare?

(c) Use the model to estimate the least distance traveled during impact for which the passenger does not experience more than 30 g's.

(d) Do you think it is practical to lower the number of g's experienced during impact to fewer than 23? Explain your reasoning.

108.  **HOW DO YOU SEE IT?**   Solving $\log_3 x + \log_3(x - 8) = 2$ algebraically, the solutions appear to be $x = 9$ and $x = -1$. Use the graph of $y = \log_3 x + \log_3(x - 8) - 2$ to determine whether each value is an actual solution of the equation. Explain your reasoning.

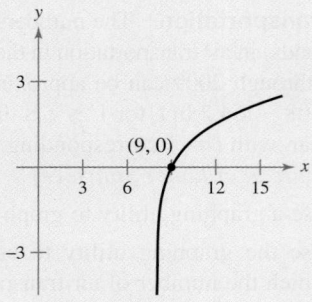

109. Think About It   Is it possible for a logarithmic equation to have more than one extraneous solution? Explain.

110. Average Heights   The percent m of American males (between 20 and 29 years old) who are less than x inches tall is approximated by

$$m = -0.018 + \frac{1.041}{1 + e^{-0.5325(x - 69.44)}}, \quad 64 \le x \le 78$$

and the percent f of American females (between 20 and 29 years old) who are less than x inches tall is approximated by

$$f = -0.013 + \frac{1.031}{1 + e^{-0.5604(x - 64.48)}}, \quad 59 \le x \le 73$$

where m and f are the percents (in decimal form) and x is the height (in inches) (see figure).   *(Source: U.S. National Center for Health Statistics)*

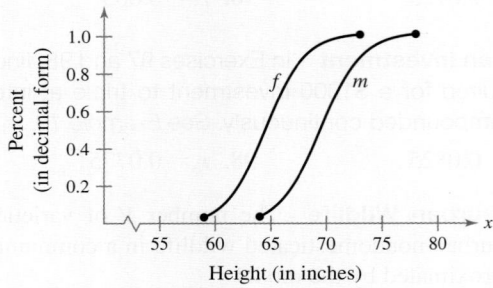

(a) What is the median height for each sex between 20 and 29 years old? (In other words, for what values of x are m and f equal to 0.5?)

(b) Write a paragraph describing each height model.

Writing Equations   In Exercises 111–114, rewrite each verbal statement as an equation. Then decide whether the statement is true or false. Justify your answer.

111. The logarithm of the product of two numbers is equal to the sum of the logarithms of the numbers.

112. The logarithm of the sum of two numbers is equal to the product of the logarithms of the numbers.

113. The logarithm of the difference of two numbers is equal to the difference of the logarithms of the numbers.

114. The logarithm of the quotient of two numbers is equal to the difference of the logarithms of the numbers.

115. Think About It   Are the times required for the investments in Exercises 95 and 96 to quadruple twice as long as the times for them to double? Give a reason for your answer and verify your answer algebraically.

4.6 Exponential and Logarithmic Models

- ■ Understand the shapes of the graphs of the most common types of mathematical models involving exponential and logarithmic functions.
- ■ Construct and use a model for exponential growth or exponential decay.
- ■ Use a Gaussian model to solve an application problem.
- ■ Use a logistic growth model to solve an application problem.
- ■ Use a logarithmic model to solve an application problem.
- ■ Choose an appropriate model involving exponential or logarithmic functions for a real-life situation.

In Exercise 43 on page 397, you will use the graph of a Gaussian model to determine the average IQ score for the students at a college.

Introduction

The five most common types of mathematical models involving exponential functions and logarithmic functions are shown below.

1. Exponential growth model: $y = ae^{bx}, \quad b > 0$

2. Exponential decay model: $y = ae^{-bx}, \quad b > 0$

3. Gaussian model: $y = ae^{-(x-b)^2/c}$

4. Logistic growth model: $y = \dfrac{a}{1 + be^{-rx}}$

5. Logarithmic models: $y = a + b \ln x$

$\qquad\qquad\qquad\qquad\quad y = a + b \log_{10} x$

The basic shape of the graph of each of these models is shown in Figure 4.26.

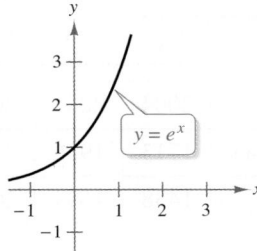
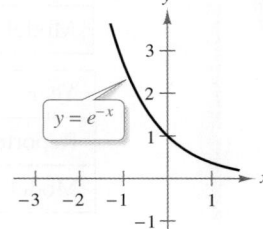
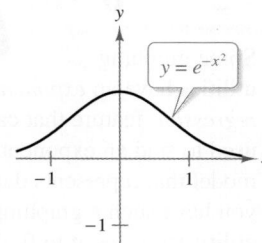
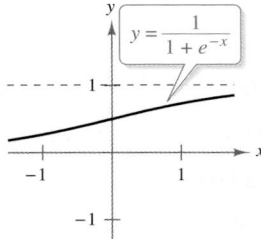
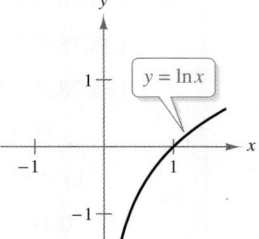
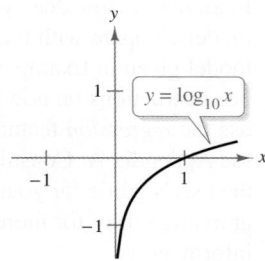

FIGURE 4.26

You can often gain quite a bit of insight into a situation modeled by an exponential or logarithmic function by identifying and interpreting the function's asymptotes. Use the graphs in Figure 4.26 to identify the asymptote(s) of the graph of each function.

Exponential Growth and Decay

 **Example 1**   Rocky Mountain Spotted Fever

The number of reported cases of Rocky Mountain spotted fever for each year from 1998 through 2008 are shown in the table. *(Source: Centers for Disease Control and Prevention)*

Year	1998	1999	2000	2001	2002	2003
Reported cases	365	579	495	695	1104	1091

Year	2004	2005	2006	2007	2008
Reported cases	1713	1936	2288	2221	2563

An exponential growth model that approximates these data is

$$C = 138.78e^{0.167t}, \quad 8 \le t \le 18$$

where C is the number of reported cases and t represents the year, with $t = 8$ corresponding to 1998. Compare the estimates given by the model with the values given by the Centers for Disease Control and Prevention. Use the model to predict the year in which the number of reported cases will reach 5500.

SOLUTION   The following table compares the two sets of data. The graph of the model and the original data values are shown in Figure 4.27.

Year	1998	1999	2000	2001	2002	2003
Reported cases	365	579	495	695	1104	1091
Model	528	624	737	871	1030	1217

Year	2004	2005	2006	2007	2008
Reported cases	1713	1936	2288	2221	2563
Model	1438	1699	2008	2373	2804

To find the year in which the number of reported cases will reach 5500, let $C = 5500$ in the model and solve for t.

$138.78e^{0.167t} = C$	Write original model.
$138.78e^{0.167t} = 5500$	Substitute 5500 for C.
$e^{0.167t} \approx 39.631$	Divide each side by 138.78.
$\ln e^{0.167t} \approx \ln 39.631$	Take natural log of each side.
$0.167t \approx 3.680$	Inverse Property
$t \approx 22$	Divide each side by 0.167.

The solution is $t \approx 22$. Because $t = 8$ represents 1998, you can conclude that, according to the model, the number of reported cases will reach 5500 in 2012.

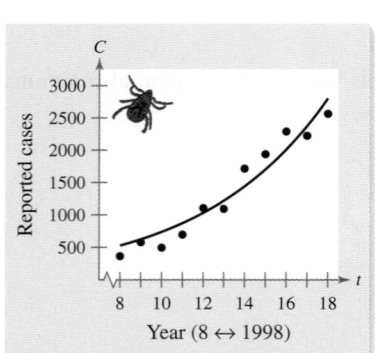

FIGURE 4.27

Reported cases
Year (8 ↔ 1998)

✓**Checkpoint 1**

Use the model in Example 1 to predict the year in which the number of reported cases will reach 7500.

The exponential model in Example 1 increases by the same percent each year. What is the annual percent increase for this exponential model?

In Example 1, you are given the exponential growth model. How could you find such a model if you are given only a set of data? To find an exponential growth model that fits a set of data, choose any two of the points and substitute them in the general exponential growth model $y = ae^{bx}$. This technique is demonstrated in Example 2.

Example 2 Finding an Exponential Growth Model

Find an exponential growth model whose graph passes through the points (0, 4453) and (7, 5024), as shown in Figure 4.28(a).

SOLUTION The general form of the model is

$$y = ae^{bx}.$$

From the fact that the graph passes through the point (0, 4453), you know that $y = 4453$ when $x = 0$. By substituting these values into the general model, you have

$$4453 = ae^0 \quad \Longrightarrow \quad a = 4453.$$

In a similar way, from the fact that the graph passes through the point (7, 5024), you know that $y = 5024$ when $x = 7$. Substitute these values into the model $y = ae^{bx}$ and solve for b.

$5024 = 4453e^{7b}$	Substitute for y, a, and x.
$1.12823 \approx e^{7b}$	Divide each side by 4453.
$\ln 1.12823 \approx \ln e^{7b}$	Take natural log of each side.
$0.12065 \approx 7b$	Inverse Property
$0.01724 \approx b$	Divide each side by 7.

So, the exponential growth model is

$$y = 4453e^{0.01724x}.$$

The graph of the model is shown in Figure 4.28(b).

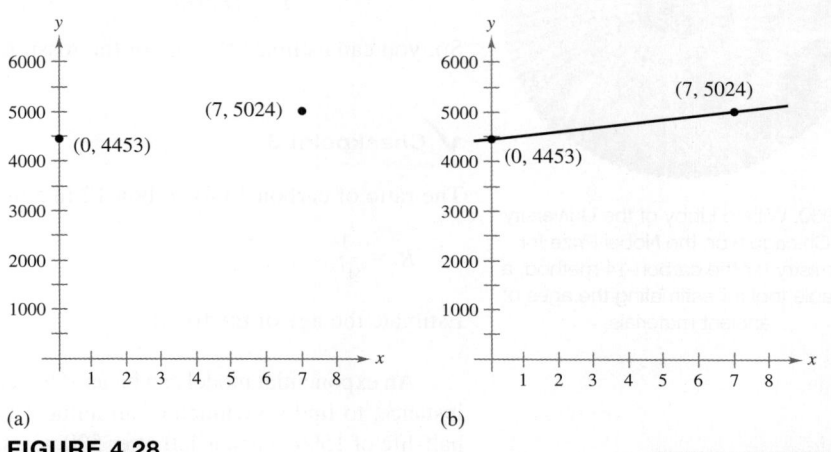

(a)   (b)

FIGURE 4.28

✓ Checkpoint 2

Find an exponential growth model whose graph passes through the points (0, 3) and (5, 8).

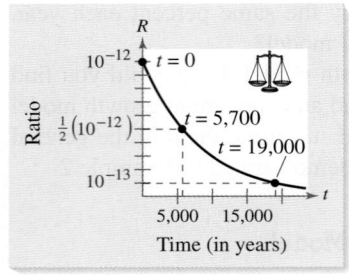

FIGURE 4.29

In living organic material, the ratio of the number of radioactive carbon isotopes (carbon 14) to the number of nonradioactive carbon isotopes (carbon 12) is about 1 to 10^{12}. When organic material dies, its carbon 12 content remains fixed, whereas its radioactive carbon 14 begins to decay with a half-life of about 5700 years. To estimate the age of dead organic material, scientists use the following formula, which denotes the ratio R of carbon 14 to carbon 12 present at any time t (in years).

$$R = \frac{1}{10^{12}} e^{-t/8223} \qquad \text{Carbon dating model}$$

In Figure 4.29, note that R decreases as the time t increases. Any material that is composed of carbon, such as wood, bone, hair, pottery, paper, and water, can be dated.

 **Example 3**   **Carbon Dating**

The ratio of carbon 14 to carbon 12 in a newly discovered fossil is

$$R = \frac{1}{10^{13}}.$$

Estimate the age of the fossil.

SOLUTION   In the carbon dating model, substitute for R and solve for t.

$\dfrac{1}{10^{12}} e^{-t/8223} = R$	Write original model.
$\dfrac{e^{-t/8223}}{10^{12}} = \dfrac{1}{10^{13}}$	Substitute $\dfrac{1}{10^{13}}$ for R.
$e^{-t/8223} = \dfrac{1}{10}$	Multiply each side by 10^{12}.
$\ln e^{-t/8223} = \ln \dfrac{1}{10}$	Take natural log of each side.
$-\dfrac{t}{8223} \approx -2.3026$	Inverse Property
$t \approx 18{,}934$	Multiply each side by -8223.

So, you can estimate the age of the fossil to be about 19,000 years.

✓**Checkpoint 3**

The ratio of carbon 14 to carbon 12 in a newly discovered fossil is

$$R = \frac{1}{9^{13}}.$$

Estimate the age of the fossil.

In 1960, Willard Libby of the University of Chicago won the Nobel Prize for Chemistry for the carbon 14 method, a valuable tool for estimating the ages of ancient materials.

An exponential model can be used to determine the *decay* of radioactive isotopes. For instance, to find how much of an initial 10 grams of radioactive radium (^{226}Ra), with a half-life of 1599 years, is left after 500 years, use the exponential decay model, as shown.

$$y = ae^{-bt} \implies \tfrac{1}{2}(10) = 10e^{-b(1599)} \implies \ln \tfrac{1}{2} = -1599b \implies -\frac{\ln \tfrac{1}{2}}{1599} = b$$

Using the value of b found above, $a = 10$, and $t = 500$, the amount left is

$$y = 10e^{-[-\ln(1/2)/1599](500)} \approx 8.05 \text{ grams.}$$

Gaussian Models

As mentioned at the beginning of this section, Gaussian models are of the form

$$y = ae^{-(x-b)^2/c}.$$

This type of model is commonly used in probability and statistics to represent populations that are **normally distributed.** For *standard* normal distributions, the model takes the form

$$y = \frac{1}{\sqrt{2\pi}} e^{-x^2/2}.$$

The graph of a Gaussian model is called a **bell-shaped curve.** Try to sketch the standard normal distribution curve with a graphing utility. Can you see why it is called a bell-shaped curve?

The **average value** of a population can be found from the bell-shaped curve by observing where the maximum *y*-value of the function occurs. The *x*-value corresponding to the maximum *y*-value of the function represents the average value of the independent variable, *x*.

 **Example 4**   **ACT Scores**

In 2010, the ACT science scores for high school graduates in the United States roughly followed a normal distribution given by

$$y = 0.0782e^{-(x-20.9)^2/52.02}, \quad 1 \le x \le 36$$

where *x* is the ACT science score. Sketch the graph of the function. From the graph, estimate the average ACT science score.   *(Source: ACT, Inc.)*

SOLUTION   The graph of the function is shown in Figure 4.30. On this bell-shaped curve, the *x*-value corresponding to the maximum value of the curve represents the average score. From the graph, you can estimate that the average ACT science score for high school graduates in 2010 was about 21.

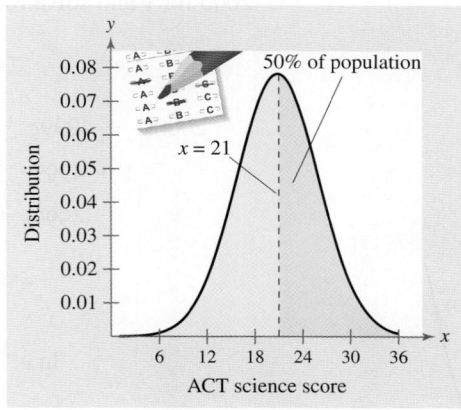

FIGURE 4.30

✓ Checkpoint 4

In 2010, the SAT critical reading scores for high school graduates in the United States roughly followed a normal distribution given by

$$y = 0.0036e^{-(x-501)^2/25,088}, \quad 200 \le x \le 800$$

where *x* is the SAT score for critical reading. Sketch the graph of the function. From the graph, estimate the average SAT critical reading score.   *(Source: The College Board)*

Logistic Growth Models

Some populations initially have rapid growth, followed by a declining rate of growth, as shown by the graph in Figure 4.31. One model for describing this type of growth pattern is the **logistic curve** given by the function

$$y = \frac{a}{1 + be^{-rx}}$$

where y is the population size and x is the time. An example is a bacteria culture that is initially allowed to grow under ideal conditions, followed by less favorable conditions that inhibit growth. A logistic growth curve is also called a **sigmoidal curve.**

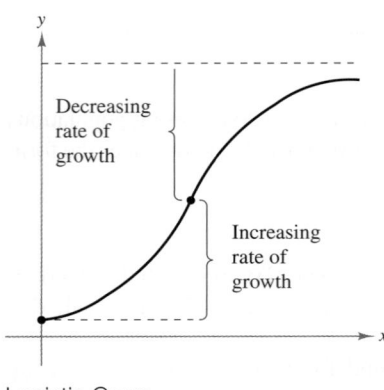

Logistic Curve
FIGURE 4.31

Example 5   Spread of a Virus

On a college campus of 5000 students, one student returned from vacation with a contagious flu virus. The spread of the virus through the student population is given by

$$y = \frac{5000}{1 + 4999e^{-0.8t}}, \quad t \geq 0$$

where y is the total number of students infected after t days.

a. How many students are infected after 5 days?

b. The college will cancel classes when 40% or more of the students become infected. After how many days will the college cancel classes?

SOLUTION

a. After 5 days, the number of students infected is

$$y = \frac{5000}{1 + 4999e^{-0.8(5)}} = \frac{5000}{1 + 4999e^{-4}} \approx 54.$$

b. Classes are cancelled when the number infected is $(0.40)(5000) = 2000$. So, substitute 2000 for y and solve for t as shown.

$$2000 = \frac{5000}{1 + 4999e^{-0.8t}}$$

$$2000(1 + 4999e^{-0.8t}) = 5000$$

$$1 + 4999e^{-0.8t} = 2.5$$

$$4999e^{-0.8t} = 1.5$$

$$e^{-0.8t} = \frac{1.5}{4999}$$

$$\ln e^{-0.8t} = \ln \frac{1.5}{4999}$$

$$-0.8t = \ln \frac{1.5}{4999}$$

$$t \approx 10.1$$

So, after about 10 days, 40% of the students will become infected, and the college will cancel classes. The graph of the function is shown in Figure 4.32.

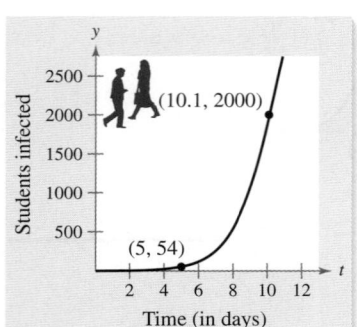

FIGURE 4.32

✓**Checkpoint 5**

In Example 5, how many days does it take for 25% of the students on campus to become infected?

Logarithmic Models

 Example 6   **Magnitudes of Earthquakes**

On the Richter scale, the magnitude R of an earthquake of intensity I per unit of area is given by $R = \log_{10}(I/I_0)$, where $I_0 = 1$ is the minimum intensity used for comparison. Find the intensity per unit of area for each earthquake. (Intensity is a measure of the wave energy of an earthquake.)

a. Chile, in 1960; $R = 9.5$ **b.** Southern Sumatra, Indonesia, in 2007; $R = 8.5$

SOLUTION

a. Because $I_0 = 1$ and $R = 9.5$,   **b.** For $R = 8.5$,

$$9.5 = \log_{10} I$$ $$8.5 = \log_{10} I$$

$$I = 10^{9.5} = 3,162,277,660.$$ $$I = 10^{8.5} = 316,227,766.$$

Note that an increase of 1 unit on the Richter scale (from 8.5 to 9.5) represents an intensity change by a factor of

$$\frac{3,162,277,660}{316,227,766} = 10.$$

In other words, the intensity of the earthquake in Chile in 1960 was 10 times greater than the intensity of the earthquake in Southern Sumatra, Indonesia in 2007.

✓ **Checkpoint 6**

In June 2010, an earthquake measuring 7.5 on the Richter scale occurred near the Nicobar Islands in the Indian Ocean. Find the intensity I per unit of area for this earthquake. (Let $I_0 = 1$.) How many times greater is the intensity of the 1960 Chile earthquake?

 Example 7   **pH Levels**

Acidity, or pH level, is a measure of the hydrogen ion concentration $[H^+]$ (measured in moles of hydrogen per liter) of a solution. Use the model given by

$$pH = -\log_{10}[H^+]$$

to determine the hydrogen ion concentration of milk of magnesia, which has a pH of 10.5.

SOLUTION

$pH = -\log_{10}[H^+]$	Write original model.
$10.5 = -\log_{10}[H^+]$	Substitute 10.5 for pH.
$-10.5 = \log_{10}[H^+]$	Multiply each side by -1.
$10^{-10.5} = 10^{\log_{10}[H^+]}$	Exponentiate each side (base 10).
$3.16 \times 10^{-11} = [H^+]$	Simplify.

So, the hydrogen ion concentration is 3.16×10^{-11} mole of hydrogen per liter.

✓ **Checkpoint 7**

Use the model in Example 7 to determine the hydrogen ion concentration of coffee, which has a pH of 5.0.

The severity of the destruction caused by an earthquake depends on its magnitude and duration. Earthquakes can destroy buildings, and can cause landslides and tsunamis.

Comparing Models

So far, you have been given the type of model to use for a data set. Now, you will use the general trends of the graphs of the five models presented in this section to choose appropriate models for real-life situations.

 **Example 8** Choosing an Appropriate Model

Decide whether to use an exponential growth model or a logistic growth model to represent each data set.

a.

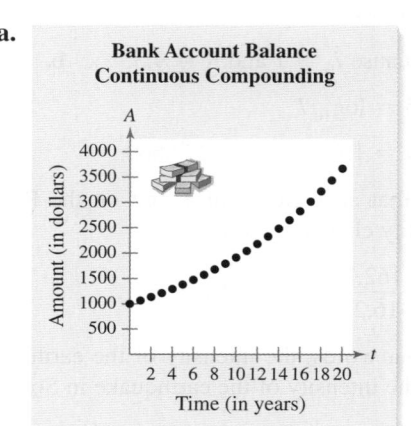

b.

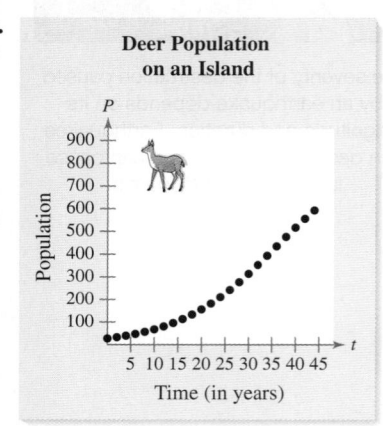

SOLUTION

a. As long as withdrawals and deposits are not made and the interest rate remains constant, the bank account balance will grow exponentially. So, an exponential growth model is an appropriate model.

b. The growth of the deer population will slow as the population approaches the carrying capacity of the island. So, a logistic growth model is an appropriate model.

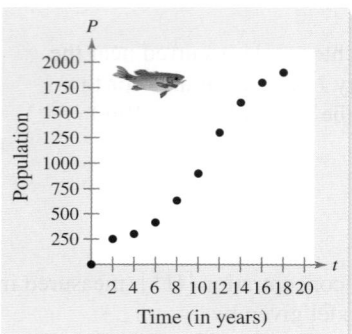

✓ **Checkpoint 8**

Decide whether to use an exponential growth model or a logistic growth model to represent the data for the fish population of a lake, shown in the figure at the left.

SUMMARIZE   (Section 4.6)

1. State the functions for an exponential growth model and an exponential decay model *(page 387)*. For examples of finding and using exponential growth and decay models, see Examples 1, 2, and 3.

2. State the function for a Gaussian model and describe its shape *(page 391)*. For an example of using a Gaussian model, see Example 4.

3. State the function for a logistic growth model *(page 392)*. For an example of using a logistic growth model, see Example 5.

4. State the functions for logarithmic models *(page 387)*. For examples of using logarithmic models, see Examples 6 and 7.

SKILLS WARM UP 4.6   The following warm-up exercises involve skills that were covered in earlier sections. You will use these skills in the exercise set for this section. For additional help, review Sections 4.2, 4.3, and 4.5.

In Exercises 1–6, sketch the graph of the equation.

1. $y = e^{0.1x}$

2. $y = e^{-0.25x}$

3. $y = e^{-x^2/5}$

4. $y = \dfrac{2}{1 + e^{-x}}$

5. $y = \log_{10} 2x$

6. $y = \ln 4x$

In Exercises 7 and 8, solve the equation algebraically.

7. $3e^{2x} = 7$

8. $4 \ln 5x = 14$

In Exercises 9 and 10, solve the equation graphically.

9. $2e^{-0.2x} = 0.002$

10. $6 \ln 2x = 12$

Exercises 4.6

See www.CalcChat.com for worked-out solutions to odd-numbered exercises.

Matching   In Exercises 1–6, match the function with its graph. [The graphs are labeled (a), (b), (c), (d), (e), and (f).]

(a)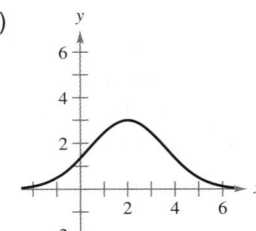

(b)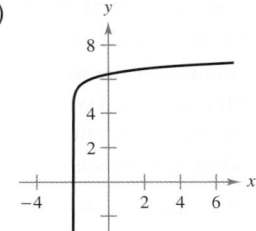

(c)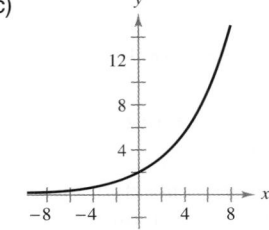

(d)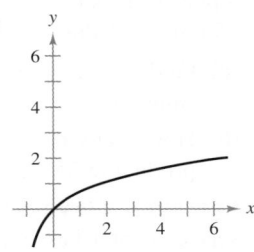

(e)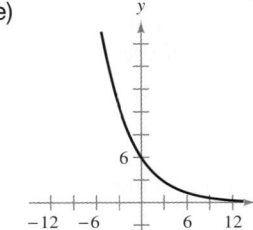

(f)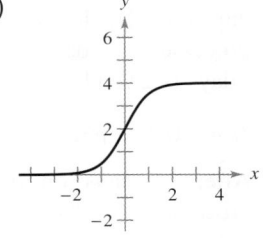

1. $y = 2e^{x/4}$

2. $y = 6e^{-x/4}$

3. $y = 6 + \log_{10}(x + 2)$

4. $y = 3e^{-(x-2)^2/5}$

5. $y = \ln(x + 1)$

6. $y = \dfrac{4}{1 + e^{-2x}}$

Classifying an Exponential Model   In Exercises 7–10, classify the model as an exponential growth model or an exponential decay model.

7. $y = 3e^{0.5t}$

8. $y = 2e^{-0.6t}$

9. $y = 20e^{-1.5t}$

10. $y = 4e^{0.07t}$

11. Population   The population P of a city is given by

$$P = 120{,}000e^{0.016t}$$

where t represents the year, with $t = 0$ corresponding to 2000. Sketch the graph of this equation. Use the model to predict the year in which the population of the city will reach 180,000.

12. Population   The population P of a city is given by

$$P = 240{,}360e^{0.012t}$$

where t represents the year, with $t = 0$ corresponding to 2000. Sketch the graph of this equation. Use the model to predict the year in which the population of the city will reach 300,000.

Finding an Exponential Growth or Decay Model   In Exercises 13–16, find the constants C and k such that the exponential function $y = Ce^{kt}$ passes through the points on the graph. See Example 2.

13.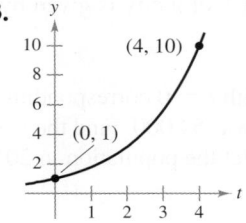

14.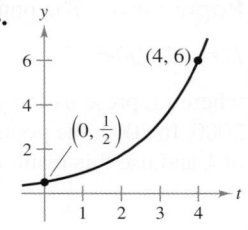

15.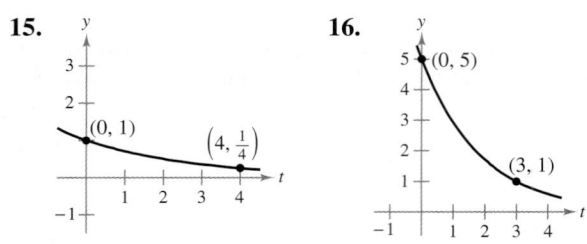

16.

Compound Interest   In Exercises 17–26, use the account balance formula $A = Pe^{rt}$ to complete the table for a savings account in which interest is compounded continuously.

Initial Investment	Annual % Rate	Time to Double	Amount After 10 Years
17. $5000	7%		
18. $1000	$9\frac{1}{4}\%$		
19. $500		10 yr	
20. $10,000		5 yr	
21. $1000			$2281.88
22. $2000			$3000
23.	11%		$19,205
24.	8%		$20,000
25. $5000			$11,127.70
26. $250			$600

Radioactive Decay   In Exercises 27–32, use the exponential decay model $y = ae^{-bt}$ to complete the table for the radioactive isotope.

Isotope	Half-Life (Years)	Initial Quantity	Amount After 1000 Years
27. ^{226}Ra	1599	4 g	
28. ^{226}Ra	1599		0.15 g
29. ^{14}C	5715		3.5 g
30. ^{14}C	5715	8 g	
31. ^{239}Pu	24,100		1.6 g
32. ^{239}Pu	24,100		0.38 g

33. Population   The population P of a city is given by

$$P = 25,000e^{kt}$$

where t represents the year, with $t = 0$ corresponding to 2000. In 1980, the population was 15,000. Find the value of k and use this result to predict the population in 2015.

34. Population   The population P of a city is given by

$$P = 52,500e^{kt}$$

where t represents the year, with $t = 0$ corresponding to 2000. In 2002, the population was 54,000. Find the value of k and use this result to predict the population in 2015.

35. Bacteria Growth   The number N of bacteria in a culture is given by the model $N = 100e^{kt}$, where t is the time (in hours), with $t = 0$ corresponding to the time when $N = 100$. When $t = 6$, there are 140 bacteria. How long does it take the bacteria population to double in size? To triple in size?

36. Bacteria Growth   The number N of bacteria in a culture is given by the model $N = 250e^{kt}$, where t is the time (in hours), with $t = 0$ corresponding to the time when $N = 250$. When $t = 10$, there are 320 bacteria. How long does it take the bacteria population to double in size? To triple in size?

37. Carbon Dating   The ratio of carbon 14 to carbon 12 in a piece of wood discovered in a cave is $R = 1/8^{14}$. Estimate the age of the piece of wood.

38. Carbon Dating   The ratio of carbon 14 to carbon 12 in a piece of paper buried in a tomb is $R = 1/13^{11}$. Estimate the age of the piece of paper.

39. Radioactive Decay   What percent of a present amount of radioactive cesium (^{137}Cs) will remain after 100 years? Use the fact that radioactive cesium has a half-life of 30 years.

40. Radioactive Decay   Find the half-life of radioactive iodine (^{131}I) if, after 20 days, 0.53 kilogram of an initial 3 kilograms remains.

41. Learning Curve   The management at a factory has found that the maximum number of units a worker can produce in a day is 40. The learning curve for the number of units N produced per day after a new employee has worked t days is given by

$$N = 40(1 - e^{kt}).$$

After 20 days on the job, a particular worker produced 25 units in 1 day.

(a) Find the learning curve for this worker (first find the value of k).

(b) How many days should pass before this worker is producing 35 units per day?

42. Learning Curve   The management at a customer service center has found that the maximum number of customer calls an employee can process effectively in a day is 90. The learning curve for the number N of calls processed per day after a new employee has worked t days is given by

$$N = 90(1 - e^{kt}).$$

After 15 days on the job, a particular employee processed 60 calls in 1 day.

(a) Find the learning curve for this worker (first find the value of k).

(b) How many days should pass before this employee will process 80 calls per day?

43. IQ Scores The IQ scores for the students at a college roughly follow the normal distribution

$$y = 0.0266e^{-(x-100)^2/450}, \quad 70 \le x \le 125$$

where x is the IQ score (see figure). Use the graph to estimate the average IQ score for the students.

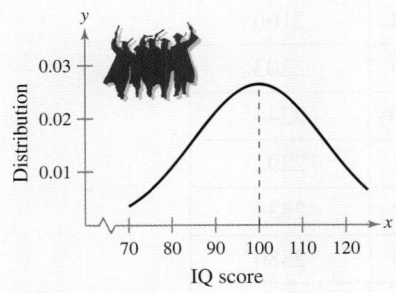

IQ score

44. Education The amount of time (in hours per week) a student spends at a math-tutoring center roughly follows the normal distribution

$$y = 0.7979e^{-(x-5.4)^2/0.5}, \quad 4 \le x \le 7$$

where x is the number of hours.

(a) Use a graphing utility to graph the function.

(b) From the graph in part (a), estimate the average number of hours per week a student spends at the tutoring center.

45. Stocking a Lake with Fish A lake is stocked with 500 fish, and the fish population P increases according to the logistic curve

$$P = \frac{10,000}{1 + 19e^{-t/5}}, \quad t \ge 0$$

where t is the time (in months).

(a) Use a graphing utility to graph the logistic curve.

(b) Find the fish population after 5 months.

(c) After how many months will the fish population reach 2000?

46. Endangered Species A conservation organization releases 100 animals of an endangered species into a game preserve. The organization believes that the preserve has a carrying capacity of 1000 animals and that the growth of the herd will be modeled by the logistic curve

$$p = \frac{1000}{1 + 9e^{-kt}}, \quad t \ge 0$$

where p is the number of animals and t is the time (in years). The herd size is 134 after 2 years. Find k. Then find the population after 5 years.

47. Think About It Does the graph of a logistic growth model have a relative maximum value?

48. HOW DO YOU SEE IT? Match the description of each situation with its graph. Explain your reasoning. [The graphs are labeled (i), (ii), (iii), and (iv).]

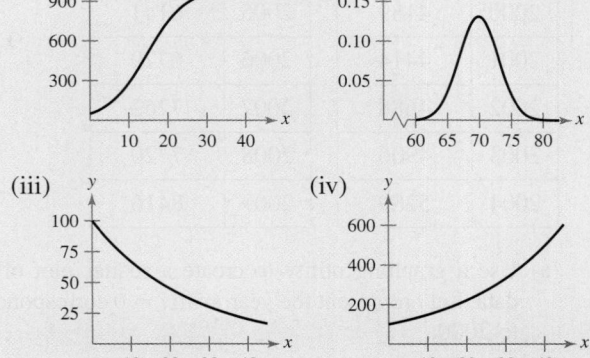

(a) The number of trout in a lake over time

(b) The amount of a radioactive substance over time

(c) The distribution of heights of men

(d) The balance of an interest-bearing account over time

49. Aged Population The table shows the projected U.S. populations P (in thousands) of people who are 85 years old or older for several years from 2020 through 2050.
(Source: U.S. Census Bureau)

Year	Population, P
2020	6597
2025	7239
2030	8745
2035	11,450
2040	14,198
2045	16,985
2050	19,041

(a) Use a graphing utility to create a scatter plot of the data. Let t represent the year, with $t = 20$ corresponding to 2020.

(b) Use the *regression* feature of the graphing utility to find an exponential model for the data. Use the Inverse Property $b = e^{\ln b}$ to rewrite the model as an exponential model in base e.

(c) Use the graphing utility to graph the exponential model in base e.

(d) Use the exponential model in base e to estimate the populations of people who are 85 years old or older in 2022 and 2042.

50. Number of Stores The table shows the number N of stores operated by Wal-Mart each year from 2000 through 2009. *(Source: Wal-Mart Stores, Inc.)*

Year	Stores, N	Year	Stores, N
2000	4189	2005	6141
2001	4414	2006	6779
2002	4688	2007	7262
2003	4906	2008	7720
2004	5289	2009	8416

(a) Use a graphing utility to create a scatter plot of the data. Let t represent the year, with $t = 0$ corresponding to 2000.

(b) Use the *regression* feature of the graphing utility to find an exponential model for the data. Use the Inverse Property $b = e^{\ln b}$ to rewrite the model as an exponential model in base e.

(c) Use the graphing utility to graph the exponential model in base e.

(d) Use the exponential model in base e to predict the numbers of stores Wal-Mart will operate in 2010 and 2012.

51. Island Population The table shows the population P of an island town every five years for forty years.

Year	Population, P	Year	Population, P
0	280	25	2340
5	390	30	2830
10	650	35	3100
15	1160	40	3240
20	1730		

(a) Use a graphing utility to create a scatter plot of the data. Let t represent the year.

(b) Based on the scatter plot, do you think the data can be better represented by an exponential growth model or a logistic growth model? Explain.

(c) Use the *regression* feature of the graphing utility to find a model of the type you chose in part (b) for the data.

(d) Use the graphing utility to graph the model with the scatter plot. Does the model fit the data well?

(e) Use the model to predict the population of the island town in year 50.

52. Population The populations P of Utah (in thousands) from 1997 through 2009 are shown in the table. *(Source: U.S. Census Bureau)*

Year	Population, P
1997	2120
1998	2166
1999	2203
2000	2244
2001	2291
2002	2334
2003	2380
2004	2439
2005	2500
2006	2584
2007	2664
2008	2727
2009	2785

(a) Use a graphing utility to create a scatter plot of the data. Let t represent the year, with $t = 7$ corresponding to 1997.

(b) Use the *regression* feature of the graphing utility to find an exponential model for the data. Use the Inverse Property

$$b = e^{\ln b}$$

to rewrite the model as an exponential model in base e.

(c) Use the *regression* feature of the graphing utility to find a linear model and a quadratic model for the data.

(d) Use the graphing utility to graph the exponential model in base e and the models in part (c) with the scatter plot.

(e) Use each model to predict the populations in 2010, 2012, and 2014. Do all of the models give reasonable predictions? Explain.

Earthquake Magnitudes In Exercises 53 and 54, use the Richter scale for measuring the magnitudes of earthquakes. *See Example 6.*

53. Find the magnitude R (on the Richter scale) of an earthquake of intensity I. (Let $I_0 = 1$.)

(a) $I = 80,500,000$

(b) $I = 48,275,000$

54. Find the intensity I of an earthquake measuring R on the Richter scale. (Let $I_0 = 1$.)

(a) In 2011, an earthquake in the Vanuatu Islands had a magnitude of $R = 6.3$.

(b) In 2011, an earthquake near the coast of Japan had a magnitude of $R = 9.0$.

Intensity of Sound   In Exercises 55 and 56, find the level of sound using the following information for determining sound intensity. The level of sound L (in decibels) of a sound with an intensity of I is given by

$$L = 10 \log_{10} \frac{I}{I_0}$$

where I_0 is an intensity of 10^{-12} watt per square meter, corresponding roughly to the faintest sound that can be heard by the human ear.

55. (a) $I = 10^{-10}$ watt per square meter (quiet room)

(b) $I = 10^{-5}$ watt per square meter (busy street corner)

56. (a) $I = 10^{-3}$ watt per square meter (loud car horn)

(b) $I \approx 10^0$ watt per square meter (threshold of pain)

pH Levels   In Exercises 57–60, use the model $\text{pH} = -\log_{10}[\text{H}^+]$. *See Example 7.*

57. Compute the hydrogen ion concentration $[\text{H}^+]$ of a solution for which $\text{pH} = 5.8$.

58. Compute the hydrogen ion concentration $[\text{H}^+]$ of a solution for which $\text{pH} = 7.3$.

59. A grape has a pH of 3.5, and baking soda has a pH of 8.0. The hydrogen ion concentration of the grape is how many times that of the baking soda?

60. The pH of a solution is decreased by one unit. The hydrogen ion concentration is increased by what factor?

61. Estimating the Time of Death   At 8:30 A.M., a coroner was called to the home of a person who had died during the night. The coroner assumed that the person had a normal body temperature of 98.6°F at death, and that the room temperature was a constant 70°F. From these two temperatures, the coroner was able to determine that the time elapsed since death and the body temperature are related by the formula

$$t = -10 \ln \frac{T - 70}{98.6 - 70}$$

where t is the time (in hours) elapsed since the person died, and T is the temperature (in degrees Fahrenheit) of the person's body. In order to estimate the time of death, the coroner took the person's temperature twice. At 9:00 A.M. the temperature was 85.7°F, and at 11:00 A.M. the temperature was 82.8°F. Use this formula to estimate the time of death of the person.

62. Thawing a Package of Steaks   You take a three-pound package of steaks out of the freezer at 11 A.M. and place it in the refrigerator. Will the steaks be thawed in time to be grilled at 6 P.M.? Assume that the refrigerator temperature is 40°F and that the freezer temperature is 0°F. Use the formula for Newton's Law of Cooling

$$t = -5.05 \ln \frac{T - 40}{0 - 40}$$

where t is the time in hours (with $t = 0$ corresponding to 11 A.M.) and T is the temperature of the package of steaks (in degrees Fahrenheit).

63. Worker's Productivity   The numbers n of units per day that a new worker can produce after t days on the job are listed in the table.

Days, t	5	10	15	20	25
Units, n	6	13	22	34	56

(a) Use a graphing utility to create a scatter plot of the data. Do the data fit an exponential model or a logarithmic model?

(b) Use the *regression* feature of the graphing utility to find the model of the type you chose in part (a).

(c) Use the graphing utility to graph the model with the data. Does the model fit the data well? Can you think of a better model to use for these data? Explain.

64. Chemical Reaction   The table shows the yield y (in milligrams) of a chemical reaction after x minutes.

Minutes, x	Yield, y
1	1.5
2	7.4
3	10.2
4	13.4
5	15.8
6	16.3
7	18.2
8	18.3

(a) Use a graphing utility to create a scatter plot of the data. Do the data fit an exponential model or a logarithmic model?

(b) Use the *regression* feature of the graphing utility to find the model of the type you chose in part (a).

(c) Use the graphing utility to graph the model with the data. Does the model fit the data well?

ALGEBRA TUTOR

Using Properties of Exponents

Much of the algebra in Chapter 4 involves evaluating expressions by using the properties of exponents. On this page, you can review some of these properties.

Let a and b be real numbers, variables, or algebraic expressions, and let m and n be integers. (Assume all denominators and bases are nonzero.)

1. $a^m a^n = a^{m+n}$ Product of Powers   **2.** $\dfrac{a^m}{a^n} = a^{m-n}$ Quotient of Powers

3. $(ab)^m = a^m b^m$ Power of a Product   **4.** $\left(\dfrac{a}{b}\right)^m = \dfrac{a^m}{b^m}$ Power of a Quotient

5. $(a^m)^n = a^{mn}$ Power of a Power   **6.** $a^{-n} = \dfrac{1}{a^n}$ Definition of negative exponent

7. $a^0 = 1, \quad a \neq 0$ Definition of zero exponent

8. $\left(\dfrac{a}{b}\right)^{-n} = \left(\dfrac{b}{a}\right)^n, \quad a \neq 0, b \neq 0$ **9.** $|a^2| = |a|^2 = a^2$

Example 1   Using Properties of Exponents

a. $(2^4)^3 = 2^{4 \cdot 3}$ Power of a Power

$\qquad = 2^{12}$ Multiply exponents.

$\qquad = 4096$ Simplify.

b. $10^{12} \cdot \dfrac{1}{10^{13}} = \dfrac{10^{12}}{10^{13}}$ Multiply.

$\qquad = 10^{12-13}$ Quotient of Powers

$\qquad = 10^{-1}$ Subtract exponents.

$\qquad = \dfrac{1}{10}$ Definition of negative exponent

c. $8^{-2} \cdot 4^0 = \dfrac{1}{8^2} \cdot 4^0$ Definition of negative exponent

$\qquad = \dfrac{1}{8^2} \cdot 1$ Definition of zero exponent

$\qquad = \dfrac{1}{64}$ Simplify.

d. $\left(\dfrac{2}{3}\right)^{4(2)} = \left(\dfrac{2}{3}\right)^8$ Multiply exponents.

$\qquad = \dfrac{2^8}{3^8}$ Power of a Quotient

$\qquad = \dfrac{256}{6561}$ Simplify.

e. $\left(\dfrac{3}{4}\right)^{-2} = \left(\dfrac{4}{3}\right)^2$ Property 8

$\qquad = \dfrac{16}{9}$ Power of a Quotient

STUDY TIP

Be sure to remember that these properties apply for all integers m and n, not just positive integers.

Factoring Trinomials

To solve equations that are of quadratic type, you must be good at factoring trinomials. Some examples are presented on this page. Remember that to factor a trinomial of the form $ax^2 + bx + c$, use the following pattern.

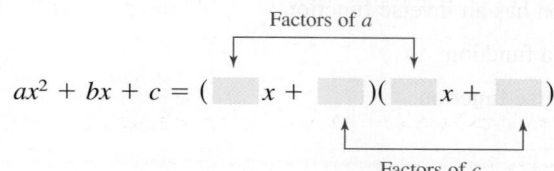

Factors of a

$$ax^2 + bx + c = (\quad x + \quad)(\quad x + \quad)$$

Factors of c

The goal is to find a combination of factors of a and c such that the outer and inner products add up to the middle term bx.

Example 2   Factoring Trinomials

Factor each trinomial.

a. $x^2 + 7x + 12$ **b.** $x^2 - 6x + 5$ **c.** $x^2 - 14x + 45$ **d.** $2x^2 + 11x - 21$

SOLUTION

a. For $x^2 + 7x + 12$, you have $a = 1$, $b = 7$, and $c = 12$. Because b and c are positive, both factors of 12 must be positive. That is, $12 = (1)(12)$, $12 = (2)(6)$, or $12 = (3)(4)$. So, the possible factorizations of $x^2 + 7x + 12$ are

$$(x + 1)(x + 12), \quad (x + 2)(x + 6), \quad \text{and} \quad (x + 3)(x + 4).$$

Testing the middle term, you can find the correct factorization to be

$$x^2 + 7x + 12 = (x + 3)(x + 4).$$

b. For $x^2 - 6x + 5$, you have $a = 1$, $b = -6$, and $c = 5$. Because b is negative and c is positive, both factors of 5 must be negative. That is, $5 = (-1)(-5)$. So, the factorization is

$$x^2 - 6x + 5 = (x - 1)(x - 5).$$

c. For $x^2 - 14x + 45$, you have $a = 1$, $b = -14$, and $c = 45$. Because b is negative and c is positive, both factors of 45 must be negative. That is, $45 = (-1)(-45)$, $45 = (-3)(-15)$, or $45 = (-5)(-9)$. So, the possible factorizations of $x^2 - 14x + 45$ are

$$(x - 1)(x - 45), \quad (x - 3)(x - 15), \quad \text{and} \quad (x - 5)(x - 9).$$

Testing the middle term, you can find the correct factorization to be

$$x^2 - 14x + 45 = (x - 5)(x - 9).$$

d. For $2x^2 + 11x - 21$, you have $a = 2$ and $c = -21$, which means that the factors of -21 must have unlike signs. The eight possible factorizations are shown below.

$$
\begin{array}{ll}
(2x - 1)(x + 21) & (2x + 1)(x - 21) \\
(2x - 3)(x + 7) & (2x + 3)(x - 7) \\
(2x - 7)(x + 3) & (2x + 7)(x - 3) \\
(2x - 21)(x + 1) & (2x + 21)(x - 1)
\end{array}
$$

Testing the middle term, you can find the correct factorization to be

$$2x^2 + 11x - 21 = (2x - 3)(x + 7).$$

SUMMARY AND STUDY STRATEGIES

After studying this chapter, you should have acquired the following skills.
The exercise numbers are keyed to the Review Exercises that begin on page 404.
Answers to odd-numbered Review Exercises are given in the back of the text.*

Section 4.1	Review Exercises
■ Determine whether a function has an inverse function.	*1–6*
■ Find the inverse function of a function.	*3–6*
■ Graph a function and its inverse function.	*3–6*

Section 4.2

■ Evaluate an exponential expression. *7–12*

■ Sketch the graph of an exponential function. *13–24*

Graph of $y = a^x$, $a > 1$

- Domain: $(-\infty, \infty)$, Range: $(0, \infty)$

- Intercept: $(0, 1)$

- Increasing

- x-axis is a horizontal asymptote: $(a^x \to 0$ as $x \to -\infty)$

- Continuous

Graph of $y = a^{-x}$, $a > 1$

- Domain: $(-\infty, \infty)$, Range: $(0, \infty)$

- Intercept: $(0, 1)$

- Decreasing

- x-axis is a horizontal asymptote: $(a^{-x} \to 0$ as $x \to \infty)$

- Continuous

- Reflection of graph of $y = a^x$ in y-axis

■ Use the compound interest formulas. *25–29*

For n compoundings per year: $A = P(1 + r/n)^{nt}$

For continuous compounding: $A = Pe^{rt}$

■ Use an exponential model to solve an application problem. *29, 30*

Section 4.3

■ Recognize and evaluate a logarithmic function. *31–44*

$y = \log_a x$ if and only if $x = a^y$ $\qquad y = \log_e x = \ln x$

■ Sketch the graph of a logarithmic function. *45–54*

Graph of $y = \log_a x$, $a > 1$

- Domain: $(0, \infty)$

- Range: $(-\infty, \infty)$

- Intercept: $(1, 0)$

- Increasing

- One-to-one; therefore has an inverse function

- y-axis is a vertical asymptote $(\log_a x \to -\infty$ as $x \to 0^+)$

- Continuous

- Reflection of graph of $y = a^x$ about the line $y = x$

■ Use a logarithmic model to solve an application problem. *55–58*

---

* A wide range of valuable study aids are available to help you master the material in this chapter.
  The *Student Solutions Manual* includes step-by-step solutions to all odd-numbered exercises to
  help you review and prepare. The student website at *www.cengagebrain.com* offers algebra help
  and a *Graphing Technology Guide*, which contains step-by-step commands and instructions for
  a wide variety of graphing calculators.

Section 4.4

- Evaluate a logarithm using the change-of-base formula.　*59–62*

$$\log_a x = \frac{\log_b x}{\log_b a}, \quad \log_a x = \frac{\log_{10} x}{\log_{10} a}, \quad \log_a x = \frac{\ln x}{\ln a}$$

- Use properties of logarithms to evaluate or rewrite a logarithmic expression.　*63–70*

$$\log_a(uv) = \log_a u + \log_a v \qquad\qquad \ln(uv) = \ln u + \ln v$$

$$\log_a \frac{u}{v} = \log_a u - \log_a v \qquad\qquad \ln \frac{u}{v} = \ln u - \ln v$$

$$\log_a u^n = n \log_a u \qquad\qquad\qquad \ln u^n = n \ln u$$

- Use properties of logarithms to expand or condense a logarithmic expression.　*71–82*
- Use logarithmic functions to model and solve real-life applications.　*83, 84*

Section 4.5

- Solve an exponential equation.　*85–90*
- Solve a logarithmic equation.　*91–98*
- Use an exponential or a logarithmic model to solve an application problem.　*99, 100*

Section 4.6

- Construct and use a model for exponential growth or exponential decay.　*101–107*

$$y = ae^{bx}, \quad b > 0 \qquad\qquad y = ae^{-bx}, \quad b > 0$$

- Use a Gaussian model to solve an application problem.　*108*

$$y = ae^{-(x-b)^2/c}$$

- Use a logistic growth model to solve an application problem.　*109*

$$y = \frac{a}{1 + be^{-rx}}$$

- Use a logarithmic model to solve an application problem.　*110, 111*

$$y = a + b \ln x, \quad y = a + b \log_{10} x$$

- Choose an appropriate model involving exponential or logarithmic functions for a real-life situation.　*112*

Study Strategies

- **Solve Problems Algebraically or Graphically**　When solving an exponential or logarithmic equation, you can use a variety of problem-solving strategies. For instance, you can solve the logarithmic equation $\ln(x + 4) - \ln x = 1$ *algebraically* or *graphically*.

　To solve the equation *algebraically*, you can use the properties of logarithms to rewrite the equation, exponentiate each side, use the Inverse Property, and solve the resulting equation to determine that $x \approx 2.328$.

　To solve the equation *graphically*, you can use a graphing utility to graph $y_1 = \ln(x + 4) - \ln x$ and $y_2 = 1$ in the same viewing window. Then use the *intersect* feature or the *zoom* and *trace* features to determine that the solution of the original equation is $x \approx 2.328$. (See figure.)

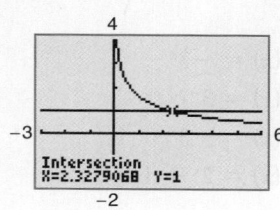

Review Exercises

See www.CalcChat.com for worked-out solutions to odd-numbered exercises.

Verifying Inverse Functions In Exercises 1 and 2, show that f and g are inverse functions of each other.

1. $f(x) = 3x + 5$, $g(x) = \dfrac{x - 5}{3}$

2. $f(x) = \sqrt[3]{x - 3}$, $g(x) = x^3 + 3$

Finding Inverse Functions In Exercises 3 and 4, determine whether the function has an inverse function. If it does, find its inverse function and graph f and f^{-1} in the same coordinate plane.

3. $f(x) = \dfrac{1}{x}$ **4.** $f(x) = \dfrac{x^2}{x^2 - 9}$

Finding and Sketching Inverse Functions In Exercises 5 and 6, (a) find f^{-1}, (b) sketch the graphs of f and f^{-1} in the same coordinate plane, and (c) verify that $f^{-1}(f(x)) = x$ and $f(f^{-1}(x)) = x$.

5. $f(x) = x^2$, $x \geq 0$ **6.** $f(x) = \sqrt[3]{x - 1}$

Evaluating an Exponential Expression In Exercises 7–12, use a calculator to evaluate the expression. Round your result to three decimal places.

7. 4^{12} **8.** $1.02^{-\pi}$

9. $2(100)^{-1/3}$ **10.** $7^{\sqrt{2}}$

11. $e^{3/4}$ **12.** $e^{-7.5}$

Matching In Exercises 13–16, match the function with its graph. [The graphs are labeled (a), (b), (c), and (d).]

(a)
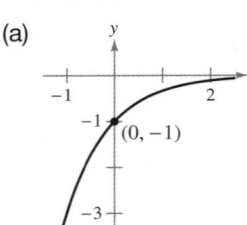

(b)
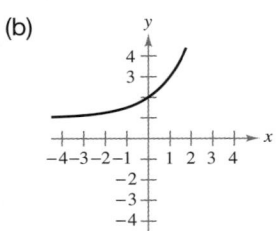

(c)
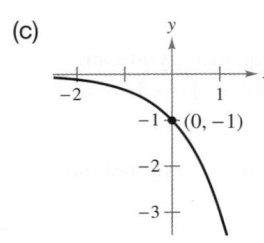

(d)
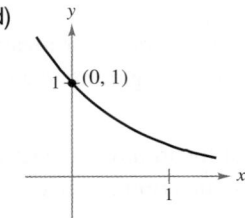

13. $f(x) = -3^x$

14. $f(x) = 3^{-x}$

15. $f(x) = -3^{-x}$

16. $f(x) = 2^x + 1$

Graphing Exponential Functions In Exercises 17–24, sketch the graph of the function.

17. $f(x) = 4^x$

18. $f(x) = 4^{x-1}$

19. $f(x) = \left(\tfrac{1}{2}\right)^x$

20. $f(x) = \left(\tfrac{1}{2}\right)^{x+1}$

21. $f(x) = 3e^{0.2x}$

22. $f(x) = 10e^{-0.1x}$

23. $f(x) = 3^{-x^2}$

24. $f(x) = 2^{1-x^2}$

Compound Interest In Exercises 25 and 26, complete the table to find the balance A for P dollars invested at rate r for t years, compounded n times per year.

n	1	2	4	12	365	Continuous
A						

25. $P = \$5000$, $r = 8.5\%$, $t = 12$ years

26. $P = \$8000$, $r = 6.75\%$, $t = 25$ years

Compound Interest In Exercises 27 and 28, complete the table to find the amount P that must be invested at rate r to obtain a balance of $A = \$200,000$ in t years.

t	1	10	20	30	40	50
P						

27. $r = 7.75\%$, compounded continuously

28. $r = 9.25\%$, compounded quarterly

29. Investment Plan You deposit $6000 in a fund that yields 5.75% interest, compounded continuously. How much money will be in the fund after 6 years?

30. Population The population P of a town increases according to the model

$$P(t) = 15,000e^{0.025t}$$

where t is the time in years, with $t = 8$ corresponding to 2008. Use the model to approximate the population in 2009 and 2011.

Writing Logarithmic Equations In Exercises 31–34, write the exponential equation in logarithmic form.

31. $4^3 = 64$ **32.** $25^{3/2} = 125$

33. $e^2 = 7.3890 \ldots$

34. $e^x = 8$

Writing Exponential Equations In Exercises 35–38, write the logarithmic equation in exponential form.

35. $\log_3 81 = 4$

36. $\log_5 0.2 = -1$

37. $\ln 1 = 0$

38. $\ln 4 = 1.3862\ldots$

Evaluating Logarithmic Expressions In Exercises 39–44, evaluate the expression without using a calculator.

39. $\log_2 32$

40. $\log_9 3$

41. $\ln e^7$

42. $\log_4 \frac{1}{4}$

43. $\ln e^{-1/2}$

44. $\ln 1$

Matching In Exercises 45–48, match the function with its graph. [The graphs are labeled (a), (b), (c), and (d).]

(a)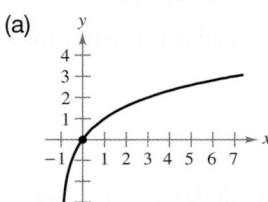

(b)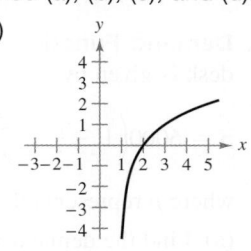

(c)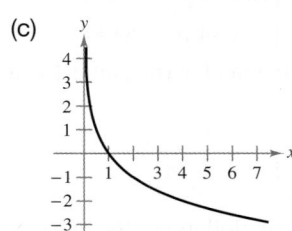

(d)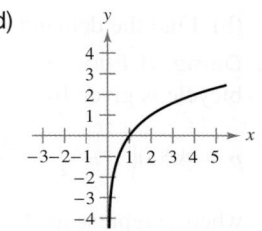

45. $f(x) = \log_2 x$ **46.** $f(x) = \log_2(x - 1)$

47. $f(x) = -\log_2 x$ **48.** $f(x) = \log_2(x + 1)$

Graphs of Exponential and Logarithmic Functions In Exercises 49 and 50, use the fact that f and g are inverse functions of each other to sketch their graphs in the same coordinate plane.

49. $f(x) = 10^x$, $g(x) = \log_{10} x$

50. $f(x) = e^x$, $g(x) = \ln x$

Sketching the Graphs of Logarithmic Functions In Exercises 51–54, find the domain, vertical asymptote, and x-intercept of the logarithmic function. Then sketch its graph.

51. $f(x) = \log_2(x - 3)$ **52.** $f(x) = 5 - 2\log_{10} x$

53. $g(x) = 2 \ln x$

54. $g(x) = \ln(4 - x)$

55. Human Memory Model Students in a sociology class were given an exam and then retested monthly for 6 months with an equivalent exam. The average score for the class is given by the human memory model

$$f(t) = 82 - 16 \log_{10}(t + 1), \quad 0 \le t \le 6$$

where t is the time (in months). How did the average score change over the six-month period?

56. Investment Time A principal P, invested at 5.85% interest compounded continuously, increases to an amount that is K times the principal after t years, where t is given by

$$t = \frac{\ln K}{0.0585}.$$

Complete the table and describe the result.

K	1	2	3	4	6	8	10
t							

57. Antler Spread The antler spread a (in inches) and shoulder height h (in inches) of an adult American elk are related by the model

$$h = 116 \log_{10}(a + 40) - 176.$$

(a) Approximate the shoulder height of an elk with an antler spread of 55 inches.

(b) Use a graphing utility to graph the model.

58. Snow Removal The number of miles s of roads cleared of snow is approximated by the model

$$s = 25 - \frac{13 \ln(h/12)}{\ln 3}, \quad 2 \le h \le 15$$

where h is the depth of the snow in inches.

(a) Use the model to find s when $h = 10$ inches.

(b) Use a graphing utility to graph the model.

Changing Bases In Exercises 59–62, evaluate the logarithm using the change-of-base formula. Evaluate each problem twice, once with common logarithms and once with natural logarithms. (Round your answer to three decimal places.)

59. $\log_3 10$ **60.** $\log_{1/4} 7$

61. $\log_{12} 200$

62. $\log_3 0.28$

Using Properties of Logarithms In Exercises 63–66, approximate the logarithm using the properties of logarithms, given $\log_b 2 \approx 0.3562$, $\log_b 3 \approx 0.5646$, and $\log_b 5 \approx 0.8271$.

63. $\log_b 6$ **64.** $\log_b \frac{4}{25}$

65. $\log_b \sqrt{3}$ **66.** $\log_b 30$

Using Inverse Properties In Exercises 67–70, find the *exact* value of the logarithm.

67. $\log_7 49$

68. $\log_6 \frac{1}{36}$

69. $\ln e^{3.2}$

70. $\ln \sqrt[5]{e^3}$

Expanding Logarithmic Expressions In Exercises 71–76, use the properties of logarithms to expand the expression as a sum, difference, and/or multiple of logarithms. (Assume that all variables are positive.)

71. $\log_{10} \dfrac{x}{y}$

72. $\log_{10} \dfrac{xy^3}{z^2}$

73. $\ln\left(x\sqrt{x-3}\right)$

74. $\ln \sqrt[3]{\dfrac{x^3}{y^2}}$

75. $\log_5(y-3)^4$

76. $\log_2 2xy^2 z$

Condensing Logarithmic Expressions In Exercises 77–82, condense the expression to the logarithm of a single quantity.

77. $\log_4 2 + \log_4 3$

78. $\ln y + 2 \ln z$

79. $\frac{1}{2} \ln x$

80. $4 \log_3 x + \log_3 y - 2 \log_3 z$

81. $\ln x - \ln(x-3) - \ln(x+1)$

82. $\log_{10}(x+2) + 2 \log_{10} x - 3 \log_{10}(x+4)$

83. Finding a Mathematical Model Find a logarithmic equation that relates y and x (see figure). Explain the steps used to find the equation.

x	1	2	3	4	5	6
y	1	2.520	4.327	6.350	8.550	10.903

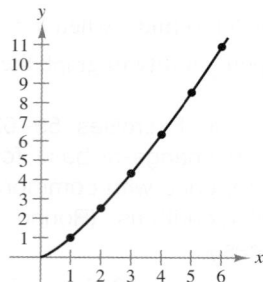

84. Human Memory Model Students in a learning theory study were given an exam and then retested monthly for 6 months with an equivalent exam. The average scores for the class are shown in the table, with $t = 1$ representing 1 month after the initial exam. Use the table to find a logarithmic equation that relates s and t.

Month, t	1	2	3	4	5	6
Score, s	87.9	79.7	74.8	71.3	68.6	66.5

Solving Exponential and Logarithmic Equations In Exercises 85–98, solve the equation. Approximate the result to three decimal places.

85. $e^x = 8$

86. $2e^{x+1} = 7$

87. $3^{4x-1} - 4 = 23$

88. $2^{3x+1} + 5 = 133$

89. $e^{2x} - 3e^x - 4 = 0$

90. $e^{2x} - 8e^x + 12 = 0$

91. $\ln 3x = 8.2$

92. $2 \log_3 4x = 15$

93. $-2 + \ln 5x = 0$

94. $\ln 4x^2 = 21$

95. $\ln x - \ln 3 = 2$

96. $\log_3 x - \log_3 4 = 5$

97. $\log_2 \sqrt[3]{x+1} = 1$

98. $\ln \sqrt{x+1} = 2$

99. Demand Function The demand x (in units) for a desk is given by

$$p = 6000\left(1 - \frac{5}{5 + e^{-0.004x}}\right)$$

where p represents the price (in dollars) of the desk.

(a) Find the demand x for a price of $p = \$500$.

(b) Find the demand x for a price of $p = \$400$.

100. Demand Function The demand x (in units) for a bicycle is given by

$$p = 4000\left(1 - \frac{3}{3 + e^{-0.004x}}\right)$$

where p represents the price (in dollars) of the bicycle.

(a) Find the demand x for a price of $p = \$700$.

(b) Find the demand x for a price of $p = \$400$.

Radioactive Decay In Exercises 101 and 102, complete the table for the radioactive isotope.

Isotope	Half-Life (Years)	Initial Quantity	Amount After 1000 Years
101. ^{14}C	5715	12 g	
102. ^{239}Pu	24,100		3.1 g

103. Population The population P of a city is given by

$$P = 185{,}000 e^{0.018t}$$

where t represents the year, with $t = 10$ corresponding to 2010.

(a) Use a graphing utility to graph this equation.

(b) Use the model to predict the year in which the population of the city will reach 270,000.

104. Population   The population P of a city is given by

$$P = 50,000e^{kt}$$

where t represents the year, with $t = 0$ corresponding to 2000. In 1990, the population was 34,500.

(a) Find the value of k and use this result to predict the population in the year 2025.

(b) Use a graphing utility to confirm the result of part (a).

105. Bacteria Growth   The number of bacteria N in a culture is given by the model $N = 250e^{kt}$, where t is the time (in hours), with $t = 0$ corresponding to the time when $N = 250$. When $t = 6$, there are 380 bacteria. How long does it take the bacteria population to double in size? To triple in size?

106. Bacteria Growth   The number of bacteria N in a culture is given by the model

$$N = 200e^{kt}$$

where t is the time (in hours), with $t = 0$ corresponding to the time when $N = 200$. When $t = 5$, there are 325 bacteria. How long does it take for the bacteria population to double in size? To triple in size?

107. Learning Curve   The management at a factory has found that the maximum number of units a worker can produce in a day is 50. The learning curve for the number of units N produced per day after a new employee has worked t days is given by

$$N = 50(1 - e^{kt}).$$

After 20 days on the job, a particular worker produced 31 units in 1 day.

(a) Find the learning curve for this worker.

(b) How many days should pass before this worker is producing 45 units per day?

108. Test Scores   The scores on a general aptitude test roughly follow a normal distribution given by

$$y = 0.0040e^{-[(x-300)^2]/20,000}, \quad 100 \le x \le 500.$$

Sketch the graph of this function. Estimate the average score on this test.

109. Wildlife Management   A state parks and wildlife department releases 100 deer into a wilderness area. The department believes that the carrying capacity of the area is 400 deer and that the growth of the herd will be modeled by the logistic curve

$$P = \frac{400}{1 + 3e^{-kt}}, \quad t \ge 0$$

where P is the number of deer and t is the time (in years).

(a) The herd size is 135 after 2 years. Find k.

(b) Find the populations after 5 years, after 10 years, and after 20 years.

110. Perrigo Sales   The annual sales S (in millions of dollars) for the Perrigo Company from 2004 through 2010 are shown in the table.   *(Source: Perrigo Company)*

Year	Sales, S
2004	898.2
2005	1024.1
2006	1366.8
2007	1447.4
2008	1822.1
2009	2006.9
2010	2268.9

(a) Use a graphing utility to create a scatter plot of the data. Let t represent the year, with $t = 4$ corresponding to 2004.

(b) Use the *regression* feature of the graphing utility to find an exponential model for the data. Use the Inverse Property $b = e^{\ln b}$ to rewrite the model as an exponential model in base e.

(c) Use the *regression* feature of the graphing utility to find a logarithmic model for the data.

(d) Use the exponential model in base e and the logarithmic model to predict sales in 2011. It is projected that sales in 2011 will be $2740 million. Do the predictions from the two models agree with this projection? Explain.

111. Thawing a Package of Steaks   You take a package of steaks out of a freezer at 10 A.M. and place it in the refrigerator. Will the steaks be thawed in time to be grilled at 6 P.M.? Assume that the refrigerator temperature is 40°F and the freezer temperature is 0°F. Use the formula

$$t = -3.95 \ln \frac{T - 40}{0 - 40}$$

where t is the time in hours (with $t = 0$ corresponding to 10 A.M.) and T is the temperature of the package of steaks (in degrees Fahrenheit).

112. Earthquake Magnitudes   On the Richter scale, the magnitude R of an earthquake of intensity I is given by

$$R = \log_{10} \frac{I}{I_0}$$

where $I_0 = 1$ is the minimum intensity used for comparison. Find the intensity per unit of area for each value of R.

(a) $R = 8.4$ (b) $R = 6.85$ (c) $R = 9.1$

TEST YOURSELF

See www.CalcChat.com for worked-out solutions to odd-numbered exercises.

Take this test as you would take a test in class. When you are done, check your work against the answers given in the back of the book.

In Exercises 1 and 2, determine whether the function has an inverse function. If it does, find the inverse function and graph f and f^{-1} in the same coordinate plane.

1. $f(x) = 5x - 7$ **2.** $f(x) = x^2 + 1$

In Exercises 3–6, sketch the graph of the function.

3. $y = 2^x$ **4.** $y = e^{-2x}$

5. $y = \ln x$ **6.** $y = \log_3(x - 1)$

In Exercises 7 and 8, students in a psychology class were given an exam and then retested monthly with an equivalent exam. The average score $f(t)$ for the class is given by the human memory model

$$f(t) = 87 - 15 \log_{10}(t + 1), \quad 0 \le t \le 18$$

where t is the time (in months).

7. What was the average score on the original exam? After 2 months? After 4 months?

8. Find the average scores after 6 months, 12 months, and 18 months. What might these scores indicate about human memory?

In Exercises 9–12, expand the logarithmic expression. (Assume all variables are positive).

9. $\ln \dfrac{x^2 y^3}{z}$ **10.** $\log_{10} 3xyz^2$

11. $\log_2\left(x \sqrt[3]{x - 2}\right)$ **12.** $\log_8 \sqrt[5]{x^2 + 1}$

In Exercises 13 and 14, condense the logarithmic expression.

13. $2 \ln x - 3 \ln y - \ln z$ **14.** $\frac{2}{3}(\log_{10} x + \log_{10} y)$

In Exercises 15–18, solve the equation. Approximate the result to three decimal places.

15. $2^{4x} = 21$ **16.** $e^{2x} - 8e^x + 12 = 0$

17. $\log_2(x + 1) - 7 = 0$ **18.** $\ln \sqrt{x + 2} = 3$

19. You deposit \$40,000 in a fund that pays 6.75% interest, compounded continuously. When will the balance be greater than \$120,000?

20. The population P of a city is given by

$$P = 85,000 e^{0.025t}$$

where t represents the year, with $t = 8$ corresponding to 2008. When will the city have a population of 125,000? Explain.

21. The number of bacteria N in a culture is given by

$$N = 100 e^{kt}$$

where t is the time (in hours), with $t = 0$ corresponding to the time when $N = 100$. When $t = 8$, $N = 175$. How long does it take the bacteria population to double?

22. Carbon 14 has a half-life of 5715 years. You have an initial quantity of 10 grams. How many grams will remain after 10,000 years? After 20,000 years?

23. Scientists release several bears on a small Alaskan island to start a bear population. Is the population likely to grow exponentially or logistically? Explain your reasoning.

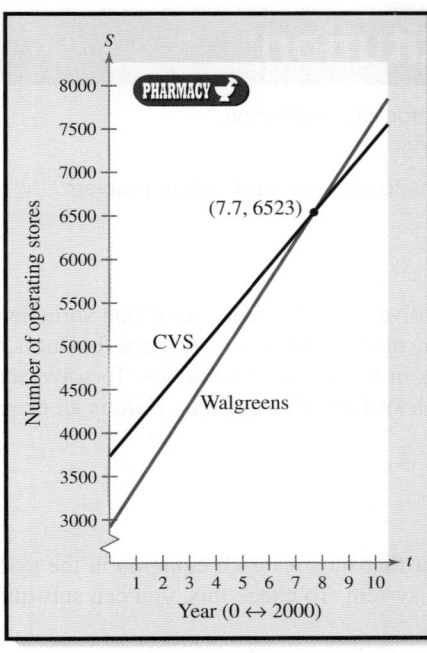

S

8000
7500
7000
6500 (7.7, 6523)
6000
5500
5000 CVS
4500
4000 Walgreens
3500
3000

Number of operating stores

1 2 3 4 5 6 7 8 9 10 t
Year (0 ↔ 2000)

Example 7 on page 416 shows how
a system of linear equations can be used
to estimate the year when the number
of operating stores of one pharmacy chain
exceeded the number of stores of
another chain.

5 Systems of Equations and Inequalities

5.1 Solving Systems Using Substitution

5.2 Solving Systems Using Elimination

5.3 Linear Systems in Three or More Variables

5.4 Systems of Inequalities

5.5 Linear Programming

5.1 Solving Systems Using Substitution

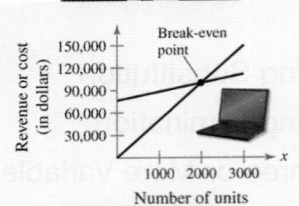

- Solve a system of equations by the method of substitution.
- Solve a system of equations graphically.
- Construct and use a system of equations to solve an application problem.

The Method of Substitution

So far, most problems in this text have involved either a function of one variable or a single equation in two variables. However, many problems in science, business, and engineering involve two or more equations in two or more variables. To solve such a problem, you need to find the solutions of a **system of equations.** Here is an example of a system of two equations in x and y.

$$\begin{cases} 2x + y = 5 & \text{Equation 1} \\ 3x - 2y = 4 & \text{Equation 2} \end{cases}$$

A **solution** of this system is an ordered pair that satisfies each equation in the system. The ordered pair $(2, 1)$ is a solution of this system. To check this, you can substitute 2 for x and 1 for y in *each* equation.

$2x + y = 5$	Write Equation 1.
$2(2) + 1 \overset{?}{=} 5$	Substitute 2 for x and 1 for y.
$4 + 1 \overset{?}{=} 5$	Multiply.
$5 = 5$	Solution checks in Equation 1. ✓

$3x - 2y = 4$	Write Equation 2.
$3(2) - 2(1) \overset{?}{=} 4$	Substitute 2 for x and 1 for y.
$6 - 2 \overset{?}{=} 4$	Multiply.
$4 = 4$	Solution checks in Equation 2. ✓

Finding the set of all solutions is called **solving the system of equations.** There are several different ways to solve systems of equations. In this chapter, you will study four of the most common techniques:

- *method of substitution,*
- *graphical approach,*
- *method of elimination,* and
- *Gaussian elimination.*

This section begins with the **method of substitution.**

Method of Substitution

1. *Solve* one of the equations for one variable in terms of the other.

2. *Substitute* the expression found in Step 1 into the other equation to obtain an equation in one variable.

3. *Solve* the equation obtained in Step 2.

4. *Back-substitute* the value found in Step 3 into the expression obtained in Step 1 to find the value of the other variable.

5. *Check* that the solution satisfies *each* of the original equations.

In Exercise 73 on page 419, you will use a system of equations to find the amount of sales a security software company needs to make in order to break even.

When using the method of substitution to solve a system of equations, it does not matter which variable you solve for first. You will obtain the same solution regardless. When making your choice, you should choose the variable that is easier to work with. For instance, solve for a variable that has a coefficient of 1 or −1 to avoid working with fractions.

Example 1 ■ Solving a System of Two Equations by Substitution

Solve the system of equations.

$$\begin{cases} x + y = 4 & \text{Equation 1} \\ x - y = 2 & \text{Equation 2} \end{cases}$$

SOLUTION   Begin by solving for y in Equation 1.

$$y = 4 - x \qquad \text{Revised Equation 1}$$

Next, substitute the expression $4 - x$ for y into Equation 2 and solve the resulting single-variable equation for x.

$x - y = 2$	Write Equation 2.
$x - (4 - x) = 2$	Substitute $4 - x$ for y.
$x - 4 + x = 2$	Distributive Property
$2x = 6$	Combine like terms.
$x = 3$	Divide each side by 2.

Finally, you can solve for y by *back-substituting* $x = 3$ into the equation $y = 4 - x$.

$y = 4 - x$	Write revised Equation 1.
$y = 4 - 3$	Substitute 3 for x.
$y = 1$	Solve for y.

The solution is the ordered pair $(3, 1)$.

CHECK

Substitute $(3, 1)$ into Equation 1:

$x + y = 4$	Write Equation 1.
$3 + 1 \overset{?}{=} 4$	Substitute for x and y.
$4 = 4$	Solution checks in Equation 1. ✓

Substitute $(3, 1)$ into Equation 2:

$x - y = 2$	Write Equation 2.
$3 - 1 \overset{?}{=} 2$	Substitute for x and y.
$2 = 2$	Solution checks in Equation 2. ✓

✓ Checkpoint 1

Solve the system of equations.

$$\begin{cases} x + y = 6 \\ x - y = 4 \end{cases}$$

The term *back-substitution* implies that you work *backwards*. First you solve for one of the variables, and then you substitute that value *back* into one of the equations in the system to find the value of the other variable.

TECH TUTOR

To check the solution in Example 1 with your graphing utility, graph $y_1 = -x + 4$ and $y_2 = x - 2$ in the same viewing window. Then use the *trace* feature to find the coordinates of the point of intersection. For instructions on how to use the *trace* feature, see Appendix A. Consult the user's guide for your graphing utility for more information.

STUDY TIP

Because many steps are required to solve a system of equations, it is easy to make errors in arithmetic. So, you should always check your solution by substituting it into *each* equation in the original system.

Example 2   Solving a System by Substitution

A total of $10,500 is invested in two funds paying 3% and 5% simple interest. (Recall that the formula for simple interest is $I = Prt$, where P is the principal, r is the annual interest rate, and t is the time.) The total annual interest is $465. How much is invested at each rate?

SOLUTION

Verbal Model:

$$\boxed{\text{Amount in 3\% fund}} + \boxed{\text{Amount in 5\% fund}} = \boxed{\text{Total investment}}$$

$$\boxed{\text{Interest for 3\% fund}} + \boxed{\text{Interest for 5\% fund}} = \boxed{\text{Total interest}}$$

Labels: Amount in 3% fund = x (dollars)
Interest for 3% fund = $0.03x$ (dollars)
Amount in 5% fund = y (dollars)
Interest for 5% fund = $0.05y$ (dollars)
Total investment = 10,500   (dollars)
Total interest = 465   (dollars)

System:
$$\begin{cases} x + y = 10{,}500 & \text{Equation 1} \\ 0.03x + 0.05y = 465 & \text{Equation 2} \end{cases}$$

To begin, it is convenient to multiply each side of Equation 2 by 100. This eliminates the need to work with decimals.

$$100(0.03x + 0.05y) = 100(465) \qquad \text{Multiply each side by 100.}$$

$$3x + 5y = 46{,}500 \qquad \text{Revised Equation 2}$$

To solve this system, you can begin by solving for x in Equation 1.

$$x = 10{,}500 - y \qquad \text{Revised Equation 1}$$

Then, substitute the expression $10{,}500 - y$ for x into revised Equation 2 and solve the resulting equation for y.

$$3x + 5y = 46{,}500 \qquad \text{Write revised Equation 2.}$$

$$3(10{,}500 - y) + 5y = 46{,}500 \qquad \text{Substitute } 10{,}500 - y \text{ for } x.$$

$$31{,}500 - 3y + 5y = 46{,}500 \qquad \text{Distributive Property}$$

$$2y = 15{,}000 \qquad \text{Combine like terms.}$$

$$y = 7500 \qquad \text{Divide each side by 2.}$$

Next, back-substitute $y = 7500$ into the equation $x = 10{,}500 - y$ to solve for x.

$$x = 10{,}500 - y \qquad \text{Write revised Equation 1.}$$

$$x = 10{,}500 - 7500 \qquad \text{Substitute 7500 for } y.$$

$$x = 3000 \qquad \text{Solve for } x.$$

The solution is (3000, 7500). So, $3000 is invested at 3% and $7500 is invested at 5%. Check this in the original system.

✓ Checkpoint 2

In Example 2, a total of $15,000 is invested in the same two funds. The total annual interest is $670. How much is invested at each rate? ▪

The equations in Examples 1 and 2 are linear. The method of substitution can also be used to solve systems in which one or both of the equations are nonlinear.

| Example 3 | Substitution: Two-Solution Case |

Solve the system of equations.

$$\begin{cases} x^2 - x - y = 1 & \text{Equation 1} \\ -x + y = -1 & \text{Equation 2} \end{cases}$$

SOLUTION   Begin by solving for y in Equation 2 to obtain $y = x - 1$. Next, substitute the expression $x - 1$ for y into Equation 1 and solve for x.

$x^2 - x - y = 1$	Write Equation 1.
$x^2 - x - (x - 1) = 1$	Substitute for y.
$x^2 - 2x + 1 = 1$	Simplify.
$x^2 - 2x = 0$	General form
$x(x - 2) = 0$	Factor.
$x = 0$	Set 1st factor equal to 0.
$x - 2 = 0 \implies x = 2$	Set 2nd factor equal to 0.

Back-substituting these values of x to solve for the corresponding values of y produces the two solutions $(0, -1)$ and $(2, 1)$. Check these solutions in the original system.

When using the method of substitution, you may encounter an equation that has no solution, as shown in Example 4.

| Example 4 | Substitution: No-Real-Solution Case |

Solve the system of equations.

$$\begin{cases} -x + y = 4 & \text{Equation 1} \\ x^2 + y = 3 & \text{Equation 2} \end{cases}$$

SOLUTION   Begin by solving for y in Equation 1 to obtain $y = x + 4$. Next, substitute the expression $x + 4$ for y into Equation 2 and solve for x.

$x^2 + y = 3$	Write Equation 2.
$x^2 + (x + 4) = 3$	Substitute $x + 4$ for y.
$x^2 + x + 1 = 0$	Simplify.
$x = \dfrac{-1 \pm \sqrt{1^2 - 4(1)(1)}}{2(1)}$	Use the Quadratic Formula.
$x = \dfrac{-1 \pm \sqrt{-3}}{2}$	Simplify.

Because the discriminant is negative, the equation $x^2 + x + 1 = 0$ has no (real) solution. So, this system has no (real) solution.

✓ **Checkpoints 3 and 4**

Solve each system of equations.

a. $\begin{cases} x^2 + 4x - y = 7 \\ 2x - y = -1 \end{cases}$

b. $\begin{cases} 2x^2 - y = 1 \\ x + y = -2 \end{cases}$

Graphical Approach to Finding Solutions

From Examples 1, 3, and 4, you can see that a system of two equations in two unknowns can have exactly one solution, more than one solution, or no solution. In practice, you can gain insight about the location and number of solutions of a system of equations by graphing each of the equations in the same coordinate plane. The solution(s) of the system correspond to the **point(s) of intersection** of the graphs. For instance, the graph of the system in Example 1 is two lines with a *single point* of intersection, as shown in Figure 5.1(a). The graph of the system in Example 3 is a parabola and a line with *two points* of intersection, as shown in Figure 5.1(b). The graph of the system in Example 4 is a line and a parabola that have *no points* of intersection, as shown in Figure 5.1(c).

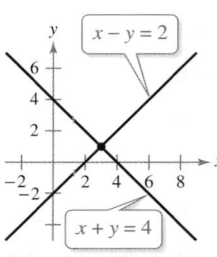
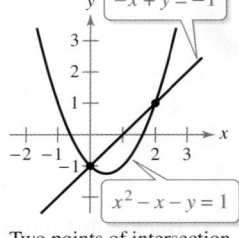
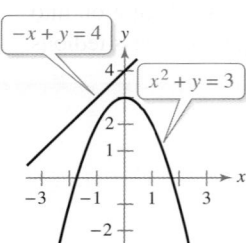

One point of intersection   Two points of intersection   No points of intersection

(a) One solution   (b) Two solutions   (c) No solution

FIGURE 5.1

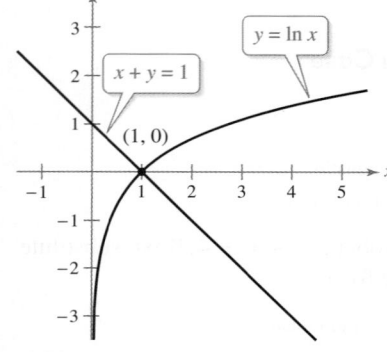

FIGURE 5.2

| Example 5 | Solving a System of Equations Graphically |

Solve the system of equations graphically.

$$\begin{cases} y = \ln x & \text{Equation 1} \\ x + y = 1 & \text{Equation 2} \end{cases}$$

SOLUTION   The graph of each equation is shown in Figure 5.2. From the graph, you can see that there is only one point of intersection. So, it appears that $(1, 0)$ is the solution point. You can confirm this by substituting 1 for x and 0 for y in *both* equations.

CHECK

Equation 1: $0 = \ln 1$ ✓

Equation 2: $1 + 0 = 1$ ✓

✓ **Checkpoint 5**

Solve the system of equations graphically.

$$\begin{cases} 2x + 2y = 8 \\ y = \ln \frac{1}{4}x \end{cases}$$

TECH TUTOR

Your graphing utility may have an *intersect* feature that approximates the point(s) of intersection of two graphs. Use the *intersect* feature to verify the solution of Example 5. For instructions on how to use the *intersect* feature, see Appendix A. Consult the user's guide for your graphing utility for more information.

Applications

The total cost C of producing x units of a product typically has two components—the initial cost and the cost per unit. When enough units have been sold so that the total revenue R equals the total cost C, the sales are said to have reached the **break-even point.** You will find that the break-even point corresponds to the point of intersection of the cost and revenue graphs.

Example 6   Break-Even Analysis

A shoe company invests $300,000 in equipment to produce cross-training shoes. Each pair of shoes costs $10 to produce and is sold for $75. How many pairs of shoes must be sold before the company breaks even?

SOLUTION   The total cost of producing x units is

$$\text{Total cost} = \text{Cost per unit} \cdot \text{Number of units} + \text{Initial cost}$$

$$C = 10x + 300,000.$$

The total revenue obtained by selling x units is

$$\text{Total revenue} = \text{Price per unit} \cdot \text{Number of units}$$

$$R = 75x.$$

Because the break-even point occurs when $R = C$, you have $C = 75x$, and the system of equations to solve is

$$\begin{cases} C = 10x + 300,000 & \text{Equation 1} \\ C = 75x & \text{Equation 2} \end{cases}.$$

Now, you can solve by substitution.

$$C = 10x + 300,000 \qquad \text{Write Equation 1.}$$
$$75x = 10x + 300,000 \qquad \text{Substitute } 75x \text{ for } C.$$
$$65x = 300,000 \qquad \text{Subtract } 10x \text{ from each side.}$$
$$x = \frac{300,000}{65} \qquad \text{Divide each side by 65.}$$
$$x \approx 4615 \qquad \text{Use a calculator.}$$

The company must sell about 4615 pairs of shoes to break even. Note in Figure 5.3 that sales less than the break-even point correspond to an overall loss, whereas sales greater than the break-even point correspond to a profit.

In 2009, the average price for cross-training shoes in the United States was $50.23.

(Source: National Sporting Goods Association)

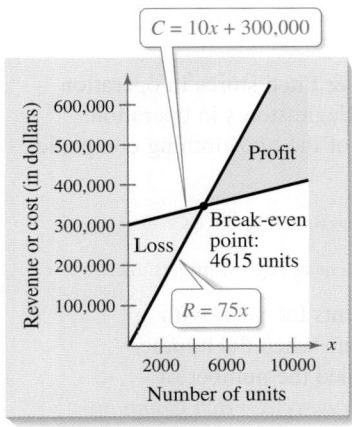

FIGURE 5.3

✓ Checkpoint 6

In Example 6, each pair of shoes costs $12 to produce. How many pairs of shoes must be sold before the business breaks even? ■

Another way to view the solution in Example 6 is to consider the profit function

$$P = R - C. \qquad \text{Profit function}$$

The break-even point occurs when the profit is 0, which is the same as saying that

$$R = C. \qquad \text{Profit is 0.}$$

 **Example 7**   **Operating Stores**

From 2000 through 2010, the number of Walgreens stores in operation increased more quickly than the number of CVS stores in operation. Models that represent the numbers of operating stores of the two pharmacy chains are given by

$$S = 362.24t + 3728.0 \qquad \text{CVS Caremark Corporation}$$

$$S = 467.94t + 2912.5 \qquad \text{Walgreen Company}$$

where S is the number of operating stores and t represents the year, with $t = 0$ corresponding to 2000. Use the models to estimate when the number of Walgreens stores in operation first exceeded the number of CVS stores in operation.   *(Source: CVS Caremark Corporation and Walgreen Company)*

SOLUTION   Both equations have already been solved for S in terms of t. So, substitute the value of S from the first equation into the second equation and solve for t.

$S = 467.94t + 2912.5$	Write second equation.
$362.24t + 3728.0 = 467.94t + 2912.5$	Substitute $362.24t + 3728.0$ for S.
$-105.7t + 3728.0 = 2912.5$	Subtract $467.94t$ from each side.
$-105.7t = -815.5$	Subtract 3728.0 from each side.
$t = \dfrac{-815.5}{-105.7}$	Divide each side by -105.7.
$t \approx 7.7$	Use a calculator.

Because the equations are equal when $t \approx 7.7$, you can conclude that the number of Walgreens stores first exceeded the number of CVS stores in 2007, as shown in Figure 5.4.

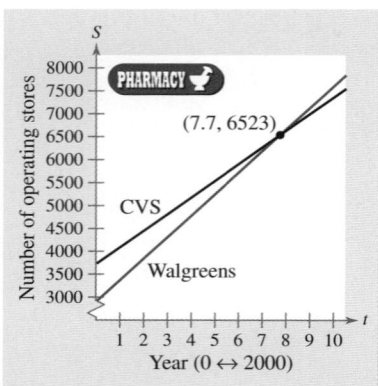

FIGURE 5.4

 **Checkpoint 7**

From 2002 through 2009, the number of Abercrombie & Fitch stores in operation increased more quickly than the number of American Eagle stores in operation. Models that represent the numbers of operating stores of the two clothing companies are given by

$$S = 76.81t + 469.5 \qquad \text{Abercrombie \& Fitch Company}$$

$$S = 52.14t + 634.7 \qquad \text{American Eagle Outfitters, Inc.}$$

where S is the number of operating stores and t represents the year, with $t = 2$ corresponding to 2002. Use the models to estimate when the number of Abercrombie & Fitch stores in operation first exceeded the number of American Eagle stores in operation.   *(Source: Abercrombie & Fitch Company and American Eagle Outfitters, Inc.)*

SUMMARIZE   (Section 5.1)

1. Describe how to solve a system of equations using the method of substitution *(page 410)*. For examples of solving systems of equations using the method of substitution, see Examples 1, 2, 3, and 4.

2. Describe how to solve a system of equations graphically *(page 414)*. For an example of solving a system of equations graphically, see Example 5.

3. State the definition of a break-even point *(page 415)*. For an example of using a system of equations to find a break-even point, see Example 6.

SKILLS WARM UP 5.1 The following warm-up exercises involve skills that were covered in earlier sections. You will use these skills in the exercise set for this section. For additional help, review Sections 0.5, 1.1, 1.3, and 2.1.

In Exercises 1–4, sketch the graph of the equation.

1. $y = -\frac{1}{3}x + 6$

2. $y = 2(x - 3)$

3. $x^2 + y^2 = 4$

4. $y = 5 - (x - 3)^2$

In Exercises 5–8, perform the indicated operations and simplify.

5. $(3x + 2y) - 2(x + y)$

6. $(-10u + 3v) + 5(2u - 8v)$

7. $x^2 + (x - 3)^2 + 6x$

8. $y^2 - (y + 1)^2 + 2y$

In Exercises 9 and 10, solve the equation.

9. $3x + (x - 5) = 15 + 4$

10. $y^2 + (y - 2)^2 = 2$

Exercises 5.1

See www.CalcChat.com for worked-out solutions to odd-numbered exercises.

Checking Solutions In Exercises 1–6, determine whether each ordered pair is a solution of the system of equations.

1. $\begin{cases} x + 4y = -3 \\ 5x - y = 6 \end{cases}$

(a) $(-1, -1)$

(b) $(1, -1)$

2. $\begin{cases} 2x - y = 2 \\ x + 3y = 8 \end{cases}$

(a) $(2, 1)$

(b) $(2, 2)$

3. $\begin{cases} 2x + 5y = -5 \\ 2x - y^2 = 1 \end{cases}$

(a) $(5, -3)$

(b) $(0, -1)$

4. $\begin{cases} 4x^2 + y = 3 \\ -x - y = 11 \end{cases}$

(a) $(-2, -9)$

(b) $(2, -13)$

5. $\begin{cases} y = -2e^x \\ 3x - y = 2 \end{cases}$

(a) $(-2, 0)$

(b) $(-1, 2)$

6. $\begin{cases} -\log_{10} x + 3 = y \\ \frac{1}{9}x + y = \frac{28}{9} \end{cases}$

(a) $(1, 3)$

(b) $\left(9, \frac{37}{9}\right)$

Solving a System by Substitution In Exercises 7–18, solve the system by the method of substitution. *See Example 1.*

7. $\begin{cases} 2x - y = -3 \\ -3x - 4y = -1 \end{cases}$

8. $\begin{cases} x + 2y = 1 \\ 5x - 4y = -23 \end{cases}$

9. $\begin{cases} 2x - y + 2 = 0 \\ 4x + y - 5 = 0 \end{cases}$

10. $\begin{cases} 6x - 3y - 4 = 0 \\ x + 2y - 4 = 0 \end{cases}$

11. $\begin{cases} x - y = 7 \\ 2x + y = 23 \end{cases}$

12. $\begin{cases} x - 2y = -2 \\ 3x - y = 6 \end{cases}$

13. $\begin{cases} 0.3x - 0.4y - 0.33 = 0 \\ 0.1x + 0.2y - 0.21 = 0 \end{cases}$

14. $\begin{cases} 1.5x + 0.8y = 2.3 \\ 0.3x - 0.2y = 0.1 \end{cases}$

15. $\begin{cases} \frac{1}{5}x + \frac{1}{2}y = 8 \\ x + y = 20 \end{cases}$

16. $\begin{cases} \frac{1}{2}x + \frac{3}{4}y = 10 \\ \frac{3}{2}x - y = 4 \end{cases}$

17. $\begin{cases} 6x + 5y = -3 \\ -x - \frac{5}{6}y = -7 \end{cases}$

18. $\begin{cases} -\frac{2}{3}x + y = 2 \\ 2x - 3y = 6 \end{cases}$

Solving a System by Substitution In Exercises 19–22, you are given the total annual interest earned from a total of $12,000 invested in two funds paying the given rates of simple interest. Write and solve a system of equations to find the amount invested at each rate. *See Example 2.*

	Annual Interest	Rate 1	Rate 2
19.	$500	3%	5%
20.	$630	4%	6%
21.	$396	2.8%	3.8%
22.	$254	1.75%	2.25%

Solving a System with a Nonlinear Equation In Exercises 23–32, solve the system by the method of substitution. *See Examples 3 and 4.*

23. $\begin{cases} y = 2x \\ y = x^2 - 1 \end{cases}$

24. $\begin{cases} x + y = 4 \\ x^2 - y = 2 \end{cases}$

25. $\begin{cases} 3x - 7y + 6 = 0 \\ x^2 - y^2 = 4 \end{cases}$

26. $\begin{cases} x^2 + y^2 = 25 \\ 2x + y = 10 \end{cases}$

27. $\begin{cases} x - 2y = 4 \\ x^2 - y = 0 \end{cases}$

28. $\begin{cases} x^2 + y^2 = 9 \\ x - y = -5 \end{cases}$

29. $\begin{cases} y = x^4 - 2x^2 + 1 \\ y = 1 - x^2 \end{cases}$

30. $\begin{cases} y = x^3 - 2x^2 + x - 1 \\ y = -x^2 + 3x - 1 \end{cases}$

31. $\begin{cases} xy - 2 = 0 \\ y = \sqrt{x - 1} \end{cases}$

32. $\begin{cases} xy = 3 \\ y = \sqrt{x - 2} \end{cases}$

Solving a System by Substitution In Exercises 33–38, solve the system by the method of substitution. Then use the graph to confirm your solution.

33. $\begin{cases} x + y = -1 \\ -2x + y = -7 \end{cases}$ 34. $\begin{cases} x - y = -5 \\ x + 2y = 4 \end{cases}$

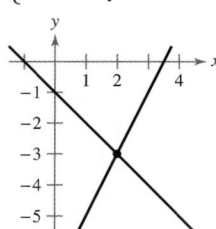

35. $\begin{cases} x - y = -3 \\ x^2 - y = -1 \end{cases}$ 36. $\begin{cases} y = x^3 - 3x^2 + 3 \\ 2x + y = 3 \end{cases}$

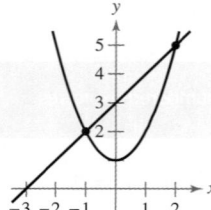   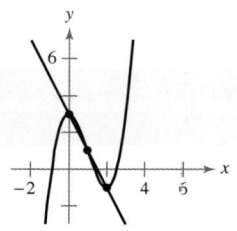

37. $\begin{cases} 3x + y = 4 \\ x^2 + y^2 = 16 \end{cases}$ 38. $\begin{cases} y = x^2 - 3x - 4 \\ y = -x^2 + 3x + 4 \end{cases}$

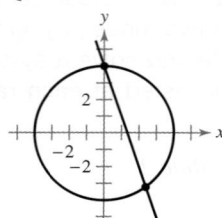   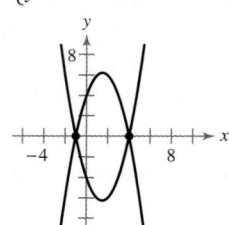

Solving a System of Equations Graphically In Exercises 39–50, solve the system graphically. See Example 5.

39. $\begin{cases} -x + 2y = 2 \\ 3x + y = 15 \end{cases}$ 40. $\begin{cases} x + y = 0 \\ 3x - 2y = 10 \end{cases}$

41. $\begin{cases} x - 3y = -2 \\ 5x + 3y = 17 \end{cases}$ 42. $\begin{cases} -x + 2y = 1 \\ x - y = 2 \end{cases}$

43. $\begin{cases} x + y = 4 \\ x^2 + y^2 - 4x = 0 \end{cases}$

44. $\begin{cases} -x + y = 3 \\ x^2 - 6x - 27 + y^2 = 0 \end{cases}$

45. $\begin{cases} x - y + 3 = 0 \\ x^2 - 4x + 7 = y \end{cases}$ 46. $\begin{cases} y^2 - 4x + 11 = 0 \\ -\frac{1}{2}x + y = -\frac{1}{2} \end{cases}$

47. $\begin{cases} 7x + 8y = 24 \\ x - 8y = 8 \end{cases}$ 48. $\begin{cases} x - y = 0 \\ 5x - 2y = 6 \end{cases}$

49. $\begin{cases} 3x - 2y = 0 \\ x^2 - y^2 = 4 \end{cases}$ 50. $\begin{cases} 2x - y + 3 = 0 \\ x^2 + y^2 - 4x = 0 \end{cases}$

Finding the Number of Solutions In Exercises 51–56, use a graphing utility to determine whether the system of equations has *one solution, two solutions,* or *no solution.*

51. $\begin{cases} y = -5x + 1 \\ y = x + 3 \end{cases}$ 52. $\begin{cases} -\frac{1}{2}x + y = -1 \\ 7x + y = 2 \end{cases}$

53. $\begin{cases} y = x^2 + 2x - 1 \\ y = 2x + 5 \end{cases}$ 54. $\begin{cases} x^2 + 3x + y = 4 \\ 3x + y = -5 \end{cases}$

55. $\begin{cases} y = x^2 + 3x + 7 \\ y = -x^2 - 3x + 1 \end{cases}$ 56. $\begin{cases} -10x + y = 2 \\ -10x + y = -3 \end{cases}$

Solving a System of Equations Graphically In Exercises 57–62, use a graphing utility to find the point(s) of intersection of the graphs. Then confirm your solution algebraically.

57. $\begin{cases} y = x^2 + 3x - 1 \\ y = -x^2 - 2x + 2 \end{cases}$ 58. $\begin{cases} y = -2x^2 + x - 1 \\ y = x^2 - 2x - 1 \end{cases}$

59. $\begin{cases} x - y + 3 = 0 \\ x^2 - 4x + 7 = y \end{cases}$ 60. $\begin{cases} x - y = 3 \\ x - y^2 = 1 \end{cases}$

61. $\begin{cases} y = e^x \\ x - y + 1 = 0 \end{cases}$ 62. $\begin{cases} x^2 + y^2 = 8 \\ y = x^2 + 4 \end{cases}$

Choosing a Solution Method In Exercises 63–66, solve the system graphically or algebraically. Explain your choice of method.

63. $\begin{cases} 4x - y = 0 \\ x^2 - y = -3 \end{cases}$ 64. $\begin{cases} x^2 + y^2 = 4 \\ x + y = 6 \end{cases}$

65. $\begin{cases} x^2 + y^2 = 25 \\ \frac{1}{2}x + y = \frac{5}{2} \end{cases}$ 66. $\begin{cases} y = \sqrt{x + 7} \\ y = x + 1 \end{cases}$

67. **Think About It** Create systems of two equations in two variables that have (a) one solution, (b) two solutions, and (c) no solution. (There are many correct answers.)

68. **HOW DO YOU SEE IT?** The cost C of producing x units and the revenue R obtained by selling x units are shown in the figure.

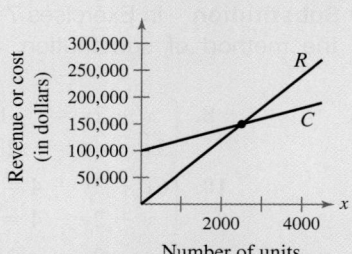

(a) Estimate the point of intersection. What does this point represent?

(b) Identify the area in the figure that corresponds to an overall loss. Identify the area that corresponds to a profit. Explain your reasoning.

Break-Even Analysis In Exercises 69–72, find the sales necessary to break even ($R = C$) for the cost C of producing x units and the revenue R obtained by selling x units. (Round your answer to the nearest whole unit.) *See Example 6.*

69. $C = 8650x + 250{,}000;\ R = 9950x$

70. $C = 5.5\sqrt{x} + 10{,}000;\ R = 3.29x$

71. $C = 2.65x + 350{,}000;\ R = 4.15x$

72. $C = 0.08x + 50{,}000;\ R = 0.25x$

73. Break-Even Analysis A security software company invests \$76,500 to produce software packages that will sell for \$49.95 each. Each software package can be produced for \$11.70. How many software packages must the company sell to break even?

74. Break-Even Analysis You invest \$3000 in a fishing lure business. A lure costs \$1.06 to produce and will be sold for \$5.86. How many lures must you sell to break even?

75. Comparing Populations From 2000 through 2009, the population of Alabama grew more slowly than that of Colorado. Models that represent the populations of the two states are given by

$$\begin{cases} P = 30.2t + 4420 & \text{Alabama} \\ P = 73.6t + 4332 & \text{Colorado} \end{cases}$$

where P is the population (in thousands) and t represents the year, with $t = 0$ corresponding to 2000. Use the models to estimate when the population of Colorado first exceeded the population of Alabama. *(Source: U.S. Census Bureau)*

76. Comparing Populations From 2000 through 2009, the population of Indiana grew more slowly than that of Arizona. Models that represent the populations of the two states are given by

$$\begin{cases} P = 37.6t + 6078 & \text{Indiana} \\ P = 168.5t + 5132 & \text{Arizona} \end{cases}$$

where P is the population (in thousands) and t represents the year, with $t = 0$ corresponding to 2000. Use the models to estimate when the population of Arizona first exceeded the population of Indiana. *(Source: U.S. Census Bureau)*

77. Body Mass Index Body mass index (BMI) is a measure of body fat based on height and weight. Models that represent the 85th percentile BMI for males and females, ages 9 to 20, are given by

$$\begin{cases} B = 0.77a + 11.7 & \text{Males} \\ B = 0.68a + 13.5 & \text{Females} \end{cases}$$

where B is the BMI (kg/m^2) and a represents the age, with $a = 9$ corresponding to 9 years old. Use a graphing utility to determine whether the BMI for males exceeds the BMI for females. *(Source: Centers for Disease Control and Prevention)*

78. Sales Per Share From 2000 through 2009, the sales per share of Ross Stores, Inc. grew faster than those of TJX Companies, Inc. Models that represent the sales per share of the two companies are given by

$$\begin{cases} S = 4.576t + 13.56 & \text{Ross Stores, Inc.} \\ S = 3.745t + 16.14 & \text{TJX Companies, Inc.} \end{cases}$$

where S is the sales per share (in dollars) and t represents the year, with $t = 0$ corresponding to 2000. Use a graphing utility to determine whether the sales per share of Ross Stores, Inc. will exceed the sales per share of TJX Companies, Inc. *(Source: Ross Stores, Inc. and TJX Companies, Inc.)*

79. Job Choices You are offered two jobs. Company A offers an annual salary of \$48,000 plus a year-end bonus of 2.5% of your total sales. Company B offers a salary of \$44,000 plus a year-end bonus of 6.5% of your total sales. What is the amount you must sell in one year to earn the same salary working for either company?

80. Camping You are choosing between camping outfitters. Outfitter A charges a reservation fee of \$150 plus a daily guide fee of \$70. Outfitter B charges a reservation fee of \$75 plus a daily guide fee of \$90. Estimate when the cost of Outfitter A equals the cost of Outfitter B.

81. Annual Sales The annual sales of food and beverage stores and general merchandise stores in the United States from 1992 through 2008 can be approximated by

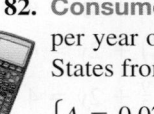

$$\begin{cases} S = 0.026t^3 - 0.36t^2 + 10.3t + 348 & \text{Food \& beverage} \\ S = -0.024t^3 + 1.12t^2 + 8.2t + 231 & \text{General merchandise} \end{cases}$$

where S is the annual sales (in billions of dollars) and t represents the year, with $t = 2$ corresponding to 1992. Use a graphing utility to determine when the sales of general merchandise stores first exceeded the sales of food and beverage stores. Do you think these models will continue to be accurate? Explain your reasoning. *(Source: U.S. Census Bureau)*

82. Consumer Spending The amount spent per person per year on video games and newspapers in the United States from 1996 through 2008 can be approximated by

$$\begin{cases} A = 0.077t^3 - 2.63t^2 + 31.1t - 99 & \text{Video games} \\ A = -0.10t^2 + 2.0t + 44 & \text{Newspapers} \end{cases}$$

where A is the amount spent (in dollars) and t represents the year, with $t = 6$ corresponding to 1996. Use a graphing utility to determine when the amount spent per person per year on video games first exceeded the amount spent per person per year on newspapers. Do you think these models will continue to be accurate? Explain your reasoning. *(Source: Veronis Suhler Stevenson)*

5.2 Solving Systems Using Elimination

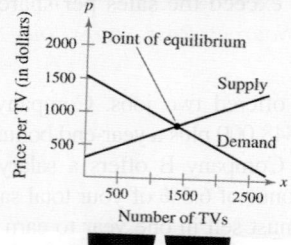

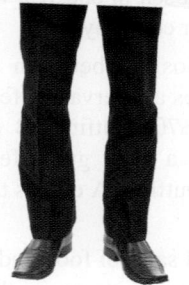

In Exercise 59 on page 430, you will use the method of elimination to solve a system of linear equations to find the equilibrium point that satisfies supply and demand equations.

■ Solve a linear system by the method of elimination.

■ Interpret the solution of a linear system graphically.

■ Construct and use a linear system to solve an application problem.

The Method of Elimination

In Section 5.1, you studied two methods for solving a system of equations: substitution and graphing. In this section, you will study a third method called the **method of elimination.** The key step in the method of elimination is to obtain, for one of the variables, coefficients that differ only in sign, so that *adding* the two equations eliminates this variable. Here is an example.

$$3x + 5y = 7 \qquad \text{Equation 1}$$
$$\underline{-3x - 2y = -1} \qquad \text{Equation 2}$$
$$3y = 6 \qquad \text{Add equations.}$$

Note that by adding the two equations, you eliminate the variable x and obtain a single equation in y. Solving this equation for y produces $y = 2$, which you can then back-substitute into one of the original equations to solve for x.

Example 1 | The Method of Elimination

Solve the system of linear equations.

$$\begin{cases} 3x + 2y = 4 & \text{Equation 1} \\ 5x - 2y = 8 & \text{Equation 2} \end{cases}$$

SOLUTION Because the coefficients of the y-terms differ only in sign, you can eliminate the y-terms by adding the two equations. This leaves you with a single equation in x.

$$3x + 2y = 4 \qquad \text{Write Equation 1.}$$
$$\underline{5x - 2y = 8} \qquad \text{Write Equation 2.}$$
$$8x = 12 \qquad \text{Add equations.}$$

Solving the equation $8x = 12$ produces $x = \frac{3}{2}$. By back-substituting $x = \frac{3}{2}$ into Equation 1, you can solve for y.

$$3x + 2y = 4 \qquad \text{Write Equation 1.}$$
$$3\left(\tfrac{3}{2}\right) + 2y = 4 \qquad \text{Substitute } \tfrac{3}{2} \text{ for } x.$$
$$\tfrac{9}{2} + 2y = 4 \qquad \text{Multiply.}$$
$$2y = -\tfrac{1}{2} \qquad \text{Subtract } \tfrac{9}{2} \text{ from each side.}$$
$$y = -\tfrac{1}{4} \qquad \text{Divide each side by 2.}$$

The solution is $\left(\tfrac{3}{2}, -\tfrac{1}{4}\right)$. Check this in the original system.

✓ Checkpoint 1

Solve the system of linear equations.

$$\begin{cases} 2x - 3y = 5 \\ 5x + 3y = 9 \end{cases}$$

STUDY TIP

It is important that you understand both the method of substitution and the method of elimination. In some cases, it is more efficient to use one method instead of the other. For instance, when two equations are written such that one variable is represented in terms of the other variable, substitution is a more efficient solution method.

To obtain coefficients (for one of the variables) that differ only in sign, you may need to multiply one or both of the equations by a suitable constant or constants, as demonstrated in Example 2.

Example 2   The Method of Elimination

Solve the system of linear equations.

$$\begin{cases} 2x - 3y = -7 \\ 3x + y = -5 \end{cases}$$ Equation 1

Equation 2

SOLUTION   For this system, you can obtain coefficients that differ only in sign by multiplying Equation 2 by 3. Then, by adding the two equations, you can eliminate the y-terms. This leaves you with a single equation in x.

$$2x - 3y = -7 \implies 2x - 3y = -7$$ Write Equation 1.

$$\underline{3x + y = -5} \implies \underline{9x + 3y = -15}$$ Multiply Equation 2 by 3.

$$11x = -22$$ Add equations.

Solving the equation $11x = -22$ produces $x = -2$. By back-substituting $x = -2$ into Equation 1, you can solve for y.

$$2x - 3y = -7$$ Write Equation 1.

$$2(-2) - 3y = -7$$ Substitute -2 for x.

$$-4 - 3y = -7$$ Multiply.

$$-3y = -3$$ Add 4 to each side.

$$y = 1$$ Divide each side by -3.

The solution is $(-2, 1)$.

CHECK

$$2(-2) - 3(1) \overset{?}{=} -7 \qquad\qquad 3(-2) + 1 \overset{?}{=} -5$$

$$-4 - 3 \overset{?}{=} -7 \qquad\qquad -6 + 1 \overset{?}{=} -5$$

$$-7 = -7 \checkmark \qquad\qquad\qquad -5 = -5 \checkmark$$

✓ Checkpoint 2

Solve the system of linear equations.

$$\begin{cases} 3x - 5y = -1 \\ x - 2y = -1 \end{cases}$$

In Example 2, the two systems of linear equations

$$\begin{cases} 2x - 3y = -7 \\ 3x + y = -5 \end{cases} \quad \text{and} \quad \begin{cases} 2x - 3y = -7 \\ 9x + 3y = -15 \end{cases}$$

are called **equivalent systems** because they have precisely the same solution set. The operations that can be performed on a system of linear equations to produce an equivalent system are:

(1) interchanging any two equations,

(2) multiplying an equation by a nonzero constant, and

(3) adding a multiple of one equation to any other equation in the system.

The Method of Elimination

To use the **method of elimination** to solve a system of two linear equations in x and y, use the following steps.

1. Examine the system to determine which variable is easiest to eliminate.

2. Obtain coefficients of x (or y) that differ only in sign by multiplying all terms of one or both equations by suitably chosen constants.

3. Add the equations to eliminate one variable and solve the resulting equation.

4. Back-substitute the value obtained in Step 3 into either of the original equations and solve for the other variable.

5. Check your solution in both of the original equations.

Example 3   The Method of Elimination

Solve the system of linear equations.

$$\begin{cases} 5x + 3y = 9 & \text{Equation 1} \\ 2x - 4y = 14 & \text{Equation 2} \end{cases}$$

SOLUTION   You can obtain coefficients of y that differ only in sign by multiplying Equation 1 by 4 and multiplying Equation 2 by 3.

$$5x + 3y = 9 \quad \Longrightarrow \quad 20x + 12y = 36 \qquad \text{Multiply Equation 1 by 4.}$$

$$\underline{2x - 4y = 14} \quad \Longrightarrow \quad \underline{6x - 12y = 42} \qquad \text{Multiply Equation 2 by 3.}$$

$$26x = 78 \qquad \text{Add equations.}$$

Solving the equation $26x = 78$ produces $x = 3$. By back-substituting $x = 3$ into Equation 2, you can solve for y.

$$2x - 4y = 14 \qquad \text{Write Equation 2.}$$

$$2(3) - 4y = 14 \qquad \text{Substitute 3 for } x.$$

$$y = -2 \qquad \text{Solve for } y.$$

The solution is $(3, -2)$.

CHECK

$$5(3) + 3(-2) \overset{?}{=} 9 \qquad\qquad 2(3) - 4(-2) \overset{?}{=} 14$$

$$15 + (-6) \overset{?}{=} 9 \qquad\qquad 6 - (-8) \overset{?}{=} 14$$

$$9 = 9 \checkmark \qquad\qquad 14 = 14 \checkmark$$

✓ **Checkpoint 3**

Solve the system of linear equations.

$$\begin{cases} -2x + 7y = -15 \\ 3x - 5y = 6 \end{cases}$$

Remember that you can check the solution of a system of equations graphically. For instance, to check the solution found in Example 3, graph both equations in the same viewing window, as shown in Figure 5.5. Notice that the two lines intersect at $(3, -2)$.

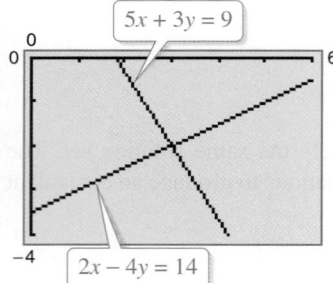

FIGURE 5.5

Example 4 illustrates a strategy for solving a system of linear equations that has decimal coefficients.

Example 4   A Linear System Having Decimal Coefficients

Solve the system of linear equations.

$$\begin{cases} 0.02x - 0.05y = -0.38 \\ 0.03x + 0.04y = 1.04 \end{cases}$$ Equation 1
Equation 2

SOLUTION   Because the coefficients in this system have two decimal places, you can begin by multiplying each equation by 100. This produces a system in which the coefficients are all integers.

$$\begin{cases} 2x - 5y = -38 \\ 3x + 4y = 104 \end{cases}$$ Revised Equation 1
Revised Equation 2

Now, to obtain coefficients of x that differ only in sign, multiply revised Equation 1 by 3 and revised Equation 2 by -2.

$$2x - 5y = -38 \implies 6x - 15y = -114$$ Multiply by 3.

$$\underline{3x + 4y = 104} \implies \underline{-6x - 8y = -208}$$ Multiply by -2.

$$-23y = -322$$ Add equations.

Solving the equation $-23y = -322$ produces $y = 14$. Now, back-substitute $y = 14$ into any of the original or revised equations of the system that contain the variable y. By back-substituting $y = 14$ into revised Equation 2, you can solve for x.

$$3x + 4y = 104$$ Write revised Equation 2.

$$3x + 4(14) = 104$$ Substitute 14 for y.

$$3x = 48$$ Subtract 56 from each side.

$$x = 16$$ Divide each side by 3.

The solution is $(16, 14)$. Check this in the original system.

✓ Checkpoint 4

Solve the system of linear equations.

$$\begin{cases} 0.03x + 0.04y = -0.13 \\ -0.04x + 0.05y = -0.24 \end{cases}$$

TECH TUTOR

The general solution of the linear system

$$\begin{cases} ax + by = c \\ dx + ey = f \end{cases}$$

is

$$x = (ce - bf)/(ae - bd) \quad \text{and} \quad y = (af - cd)/(ae - bd).$$

If $ae - bd = 0$, then the system does not have a unique solution. Several graphing utilities have programs for solving such a system. Try using one of these programs to check the solution of the system in Example 4.

Graphical Interpretation of Solutions

It is possible for a *general* system of equations to have

- exactly one solution,

- two or more solutions, or

- no solution.

If a system of *linear* equations has two different solutions, then it must have an *infinite* number of solutions. To see why this is true, consider the following graphical interpretations of systems of two linear equations in two variables. (Remember that the graph of a linear equation in two variables is a line.)

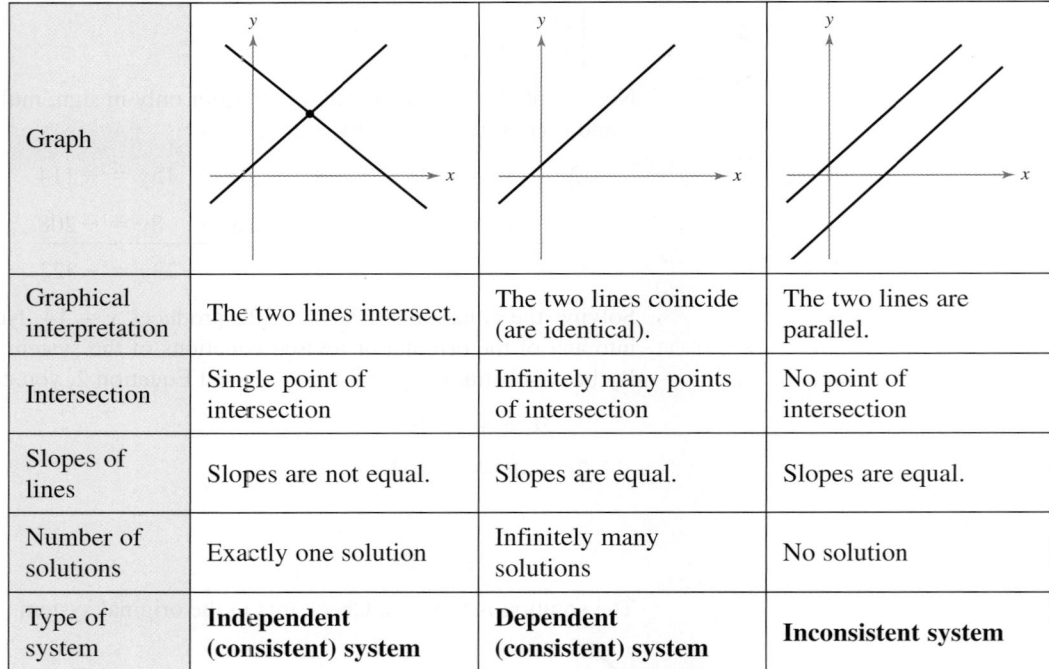

Graph			
Graphical interpretation	The two lines intersect.	The two lines coincide (are identical).	The two lines are parallel.
Intersection	Single point of intersection	Infinitely many points of intersection	No point of intersection
Slopes of lines	Slopes are not equal.	Slopes are equal.	Slopes are equal.
Number of solutions	Exactly one solution	Infinitely many solutions	No solution
Type of system	**Independent (consistent) system**	**Dependent (consistent) system**	**Inconsistent system**

A system of linear equations is *consistent* if it has at least one solution. A consistent system with exactly one solution is *independent*, whereas a consistent system with infinitely many solutions is *dependent*. A system is *inconsistent* if it has no solution.

From the graphs above, you can see that a comparison of the slopes and *y*-intercepts of two lines is helpful in determining the number of solutions of the corresponding system of equations. For instance:

- *Independent (consistent) systems* have lines with slopes that are not equal.

- *Dependent (consistent) systems* have lines with equal slopes and the same *y*-intercept.

- *Inconsistent systems* have lines with equal slopes, but different *y*-intercepts.

So, when solving a system of linear equations graphically, it is helpful to know the slope of each line. Writing each linear equation in the slope-intercept form

$$y = mx + b \qquad \text{Slope-intercept form}$$

enables you to identify the slopes quickly.

In Examples 5 and 6, note how you can use the method of elimination to determine when a linear system has no solution or infinitely many solutions.

$-2x + 4y = 1$

$x - 2y = 3$

No Solution

FIGURE 5.6

Example 5 The Method of Elimination: No-Solution Case

Solve the system of linear equations.

$$\begin{cases} x - 2y = 3 & \text{Equation 1} \\ -2x + 4y = 1 & \text{Equation 2} \end{cases}$$

SOLUTION Obtain coefficients that differ only in sign.

$$x - 2y = 3 \implies 2x - 4y = 6 \qquad \text{Multiply Equation 1 by 2.}$$

$$\underline{-2x + 4y = 1} \implies \underline{-2x + 4y = 1} \qquad \text{Write Equation 2.}$$

$$0 = 7 \qquad \text{False statement}$$

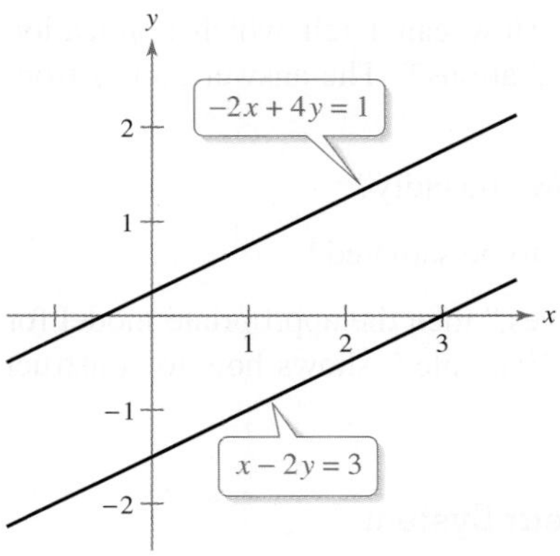

Because there are no values of x and y for which $0 = 7$, you can conclude that the system is inconsistent and has no solution. The graphs of the equations are shown in Figure 5.6. Note that the two lines have equal slopes, but different y-intercepts. So, the lines are parallel and have no point of intersection.

✓ **Checkpoint 5**

Solve the system of linear equations.

$$\begin{cases} -3x + 6y = 5 \\ x - 2y = 2 \end{cases}$$

In Example 5, note that the occurrence of a false statement, such as $0 = 7$, indicates that the system has no solution. In the next example, note that the occurrence of a statement that is true for all values of the variables, such as $0 = 0$, indicates that the system has infinitely many solutions.

Example 6 The Method of Elimination: Many-Solutions Case

Solve the system of linear equations.

$$\begin{cases} 2x - y = 1 & \text{Equation 1} \\ 4x - 2y = 2 & \text{Equation 2} \end{cases}$$

SOLUTION Obtain coefficients that differ only in sign.

$$2x - y = 1 \implies 2x - y = 1 \qquad \text{Write Equation 1.}$$

$$\underline{4x - 2y = 2} \implies \underline{-2x + y = -1} \qquad \text{Multiply Equation 2 by } -\frac{1}{2}.$$

$$0 = 0 \qquad \text{True statement}$$

Because the two equations are equivalent (have the same solution set), you can conclude that the system is consistent and has infinitely many solutions. The solution set consists of all points (x, y) lying on the line

$$2x - y = 1$$

as shown in Figure 5.7. To represent the solution set as an ordered pair, let $x = a$, where a is any real number. Then $y = 2a - 1$, and the solution set can be written as

$$(a, 2a - 1).$$

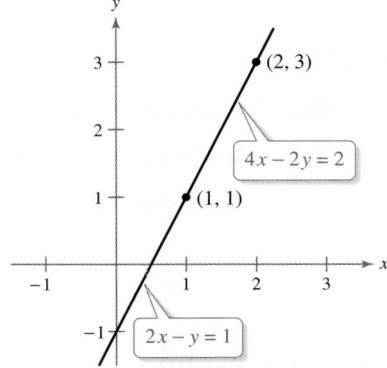

$(2, 3)$

$4x - 2y = 2$

$(1, 1)$

$2x - y = 1$

Infinitely Many Solutions

FIGURE 5.7

✓ **Checkpoint 6**

Solve the system of linear equations.

$$\begin{cases} -x - y = 5 \\ 4x + 4y = -20 \end{cases}$$

Applications

At this point, you may be asking the question, "How can I tell which application problems can be solved using a system of linear equations?" The answer comes from the following considerations.

1. Does the problem involve more than one unknown quantity?

2. Are there two (or more) equations or conditions to be satisfied?

If the answer to one or both of these questions is "yes," then the appropriate model for the problem may be a system of linear equations. Example 7 shows how to construct such a model.

Example 7   An Application of a Linear System

An airplane flying into a headwind travels the 2000-mile flying distance between Wilmington, Delaware and Tucson, Arizona in 4 hours and 24 minutes. On the return flight, the same distance is traveled in 4 hours. Find the airspeed of the plane and the speed of the wind, assuming that both remain constant.

SOLUTION   The two unknown quantities are the speeds of the wind and the plane. If r_1 is the airspeed of the plane and r_2 is the speed of the wind, then

$$r_1 - r_2 = \text{speed of the plane } against \text{ the wind}$$

$$r_1 + r_2 = \text{speed of the plane } with \text{ the wind}$$

as shown in Figure 5.8. Using the formula distance = (rate)(time) for these two speeds, you obtain the following equations.

$$2000 = (r_1 - r_2)\left(4 + \frac{24}{60}\right)$$

$$2000 = (r_1 + r_2)(4)$$

These two equations simplify as shown.

$$\begin{cases} 5000 = 11r_1 - 11r_2 & \text{Equation 1} \\ 500 = r_1 + r_2 & \text{Equation 2} \end{cases}$$

To solve this system by elimination, multiply Equation 2 by 11.

$$5000 = 11r_1 - 11r_2 \implies 5000 = 11r_1 - 11r_2 \qquad \text{Write Equation 1.}$$

$$\underline{500 = r_1 + r_2} \implies \underline{5500 = 11r_1 + 11r_2} \qquad \text{Multiply Equation 2 by 11.}$$

$$10{,}500 = 22r_1 \qquad \text{Add equations.}$$

Solving the equation $10{,}500 = 22r_1$ produces

$$r_1 = \frac{10{,}500}{22} = \frac{5250}{11} \approx 477.27.$$

By back-substituting $r_1 = 477.27$ into original Equation 2, you can solve for r_2.

$$500 = 477.27 + r_2 \qquad \text{Substitute 477.27 for } r_1.$$

$$22.73 = r_2 \qquad \text{Subtract 477.27 from each side.}$$

So, the airspeed of the plane is about 477.27 miles per hour and the speed of the wind is about 22.73 miles per hour. Check this solution in the original statement of the problem.

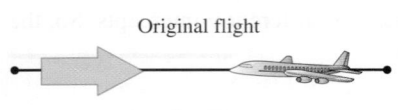

Original flight

$r_1 - r_2$

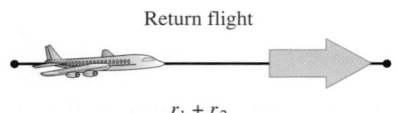

Return flight

$r_1 + r_2$

FIGURE 5.8

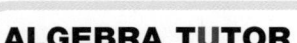

ALGEBRA TUTOR xy

For help with the algebra used to simplify the system of equations in Example 7, see the *Chapter 7 Algebra Tutor* on page 464.

 **Checkpoint 7**

In Example 7, the return flight takes 4 hours and 6 minutes. Find the airspeed of the plane and the speed of the wind, assuming that both remain constant.

In a free market, the demands for many products are related to the prices of the products. As the prices decrease, the demands by consumers increase and the amounts that producers are able or willing to supply decrease.

 Example 8 Finding the Point of Equilibrium

The demand and supply equations for a video game console are given by

$$p = 350 - 0.00003x \qquad \text{Demand equation}$$

$$p = 200 + 0.00003x \qquad \text{Supply equation}$$

where p is the price (in dollars) and x is the number of video game consoles. For how many units will the quantity demanded equal the quantity supplied? What price corresponds to this value?

SOLUTION   To obtain coefficients of p that differ only in sign, multiply the demand equation by -1.

$$p = 350 - 0.00003x \quad \Longrightarrow \quad -p = -350 + 0.00003x \qquad \text{Multiply demand equation by } -1.$$

$$p = 200 + 0.00003x \quad \Longrightarrow \quad \underline{p = 200 + 0.00003x} \qquad \text{Write supply equation.}$$

$$0 = -150 + 0.00006x \qquad \text{Add equations.}$$

By solving the equation $0 = -150 + 0.00006x$, you get $x = 2,500,000$. So, the quantity demanded equals the quantity supplied for 2,500,000 units (see Figure 5.9). The price that corresponds to this x-value is obtained by back-substituting $x = 2,500,000$ into either of the original equations. For instance, back-substituting into the demand equation produces

$$p = 350 - 0.00003x \qquad \text{Demand equation}$$

$$= 350 - 0.00003(2,500,000) \qquad \text{Substitute 2,500,000 for } x.$$

$$= \$275. \qquad \text{Simplify.}$$

Back-substitute $x = 2,500,000$ into the supply equation to see that you obtain the same price. The solution (2,500,000, 275) is called the *point of equilibrium*. The **point of equilibrium** is the price p and the number of units x that satisfy both the demand and supply equations.

✓**Checkpoint 8**

In Example 8, find the point of equilibrium when the supply equation is $p = 230 + 0.00003x$.

Figure 5.9 (margin):

$p = 350 - 0.00003x$

Demand

Supply
(2,500,000, 275)

$p = 200 + 0.00003x$

Price per console (in dollars)
Number of consoles

FIGURE 5.9

SUMMARIZE   (Section 5.2)

1. Describe how to solve a system of linear equations using the method of elimination *(pages 420–422)*. For examples of solving systems of linear equations using the method of elimination, see Examples 1, 2, 3, and 4.

2. Explain what it means for a system of linear equations to be an independent system, a dependent system, or an inconsistent system *(page 424)*. For examples of solving these types of linear systems using the method of elimination, see Examples 5 and 6.

3. State the definition of the point of equilibrium *(page 427)*. For an example of solving a system of supply and demand equations to find the point of equilibrium, see Example 8.

SKILLS WARM UP 5.2   The following warm-up exercises involve skills that were covered in earlier sections. You will use these skills in the exercise set for this section. For additional help, review Section 2.2.

In Exercises 1 and 2, sketch the graph of the equation.

1. $2x + y = 4$

2. $5x - 2y = 3$

In Exercises 3 and 4, find an equation of the line passing through the two points.

3. $(-1, 3), (4, 8)$

4. $(2, 6), (5, 1)$

In Exercises 5 and 6, determine the slope of the line.

5. $3x + 6y = 4$

6. $7x - 4y = 10$

In Exercises 7–10, determine whether the lines represented by the pair of equations are parallel, perpendicular, or neither.

7. $2x - 3y = -10$
    $3x + 2y = 11$

8. $4x - 12y = 5$
    $-2x + 6y = 3$

9. $5x + y = 2$
    $3x + 2y = 1$

10. $x - 3y = 2$
    $6x + 2y = 4$

Exercises 5.2

See www.CalcChat.com for worked-out solutions to odd-numbered exercises.

Solving a System by Elimination In Exercises 1–8, solve the system by the method of elimination. Then use the graph to confirm your solution. Copy the graph and label each line with the appropriate equation. *See Examples 1, 2, 3, 5, and 6.*

1. $\begin{cases} 3x - 2y = 2 \\ x + 2y = 6 \end{cases}$

2. $\begin{cases} -x + 3y = 2 \\ x - 4y = -4 \end{cases}$

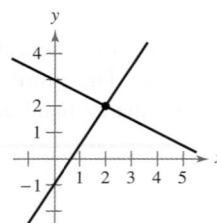

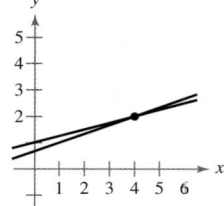

3. $\begin{cases} x - 4y = 2 \\ 2x + y = 4 \end{cases}$

4. $\begin{cases} 2x - y = 2 \\ 4x + 3y = 24 \end{cases}$

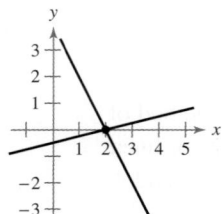

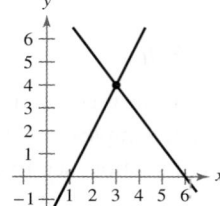

5. $\begin{cases} x - y = 1 \\ -2x + 2y = 5 \end{cases}$

6. $\begin{cases} 3x + 2y = 2 \\ 6x + 4y = 14 \end{cases}$

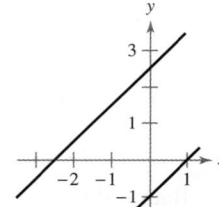

7. $\begin{cases} 3x - 2y = 6 \\ -6x + 4y = -12 \end{cases}$

8. $\begin{cases} 2x + 4y = 8 \\ 6x + 12y = 24 \end{cases}$

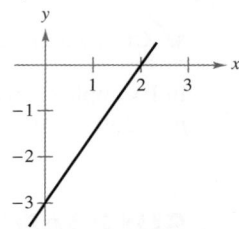

Solving a System by Elimination In Exercises 9–24, solve the system by the method of elimination. Then state whether the system is consistent or inconsistent. *See Examples 1, 2, 3, 5, and 6.*

9. $\begin{cases} x + 2y = 3 \\ x - 2y = 1 \end{cases}$

10. $\begin{cases} 2x - 3y = 4 \\ -2x - y = 4 \end{cases}$

11. $\begin{cases} 4x - 3y = 11 \\ -6x + 3y = 3 \end{cases}$

12. $\begin{cases} 3x - 5y = 2 \\ 2x + 5y = 13 \end{cases}$

13. $\begin{cases} 3x - y = 17 \\ 5x + 5y = -5 \end{cases}$

14. $\begin{cases} x + 7y = 12 \\ 3x - 5y = 10 \end{cases}$

15. $\begin{cases} 3x + 2y = 10 \\ 2x + 5y = 3 \end{cases}$ **16.** $\begin{cases} 8r + 16s = 20 \\ 16r + 50s = 55 \end{cases}$

17. $\begin{cases} 2u + v = 120 \\ u + 2v = 120 \end{cases}$ **18.** $\begin{cases} 5u + 6v = 24 \\ 3u + 5v = 18 \end{cases}$

19. $\begin{cases} 4b + 3m = 3 \\ 3b + 11m = 13 \end{cases}$ **20.** $\begin{cases} 3b + 3m = 7 \\ 3b + 5m = 3 \end{cases}$

21. $\begin{cases} \dfrac{x}{4} + \dfrac{y}{6} = 1 \\ x - y = 3 \end{cases}$ **22.** $\begin{cases} \dfrac{1}{6}x - \dfrac{2}{3}y = 3 \\ 3x + y = 15 \end{cases}$

23. $\begin{cases} \dfrac{x+3}{4} + \dfrac{y-1}{3} = 1 \\ x - y = 3 \end{cases}$ **24.** $\begin{cases} \dfrac{x-1}{2} + \dfrac{y+2}{3} = 4 \\ x - 2y = 5 \end{cases}$

Writing and Solving a System In Exercises 25–28, use the given statements to write a system of equations. Solve the system by the method of elimination.

25. The sum of a number x and a number y is 13. The difference of x and y is 3.

26. The sum of a number a and a number b is 43. The difference of a and b is -27.

27. The sum of twice a number r and a number s is 8. The difference of r and s is 7.

28. The difference of a number m and twice a number n is 1. The sum of two times m and n is 22.

Solving a System Having Decimal Coefficients In Exercises 29–34, solve the system by the method of elimination. Then state whether the system is consistent or inconsistent. *See Examples 4, 5, and 6.*

29. $\begin{cases} 6r - 5s = 3 \\ -1.2r + s = 0.5 \end{cases}$ **30.** $\begin{cases} 1.8x + 1.2y = 4 \\ 9x + 6y = 3 \end{cases}$

31. $\begin{cases} 2.5x - 3y = 1.5 \\ 10x - 12y = 6 \end{cases}$ **32.** $\begin{cases} 1.5x + 2y = 3.75 \\ 7.5x + 10y = 18.75 \end{cases}$

33. $\begin{cases} 0.05x - 0.03y = 0.21 \\ 0.07x + 0.02y = 0.16 \end{cases}$ **34.** $\begin{cases} 0.02x - 0.05y = -0.19 \\ 0.03x + 0.04y = 0.52 \end{cases}$

Choosing a Solution Method In Exercises 35–42, use any method to solve the system. Explain your choice of method.

35. $\begin{cases} 8x + 12y = 8 \\ 16x + 30y = 22 \end{cases}$ **36.** $\begin{cases} x - 9y = -7 \\ 5x + 9y = -35 \end{cases}$

37. $\begin{cases} -3x + 10y = -84 \\ y = x - 7 \end{cases}$ **38.** $\begin{cases} \frac{1}{8}x - \frac{3}{4}y = -3 \\ 4x + y = 29 \end{cases}$

39. $\begin{cases} y = e^x \\ 2x - 3y + 3 = 0 \end{cases}$ **40.** $\begin{cases} x + 5y = -11 \\ 8x - 6y = 50 \end{cases}$

41. $\begin{cases} 4x + 3y = 140 \\ 3x + 4y = 140 \end{cases}$ **42.** $\begin{cases} y = x^3 - 2x^2 \\ 10x - y = -7 \end{cases}$

43. Think About It Find a system of linear equations that has the given solution. (There are many correct answers.)

   (a) $(6, 0)$ (b) $(-3, 4)$ (c) $\left(-\frac{3}{4}, -7\right)$

44. HOW DO YOU SEE IT? Use the graphs of the two equations shown below.

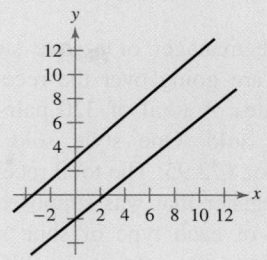

(a) Describe the graphs of the two equations.

(b) Can you conclude that the system of equations shown in the graph is inconsistent? Explain.

Recognizing a System of Equations In Exercises 45 and 46, the two lines appear to be parallel. Are they? Justify your answer by using the method of elimination to solve the system.

45. $\begin{cases} 200y - x = 200 \\ 199y - x = -198 \end{cases}$ **46.** $\begin{cases} 25x - 24y = 0 \\ 13x - 12y = 120 \end{cases}$

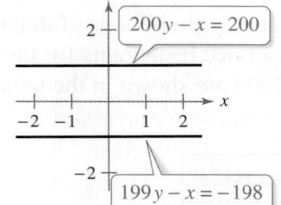

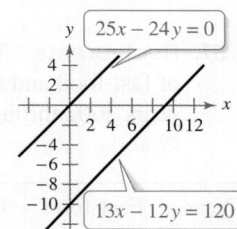

47. Airplane Speed An airplane flying into a headwind travels the 1800-mile flying distance between Los Angeles, California and South Bend, Indiana in 3 hours and 36 minutes. On the return flight, the distance is traveled in 3 hours. Find the airspeed of the plane and the speed of the wind, assuming that both remain constant.

48. Airplane Speed Two planes start from the same airport and fly in opposite directions. The second plane starts $\frac{1}{2}$ hour after the first plane, but its speed is 50 miles per hour faster. Find the airspeed of each plane if, 2 hours after the first plane departs, the planes are 2000 miles apart.

49. Acid Mixture Ten gallons of a 30% acid solution is obtained by mixing a 20% solution with a 50% solution. How much of each solution is required to obtain the specified concentration of the final mixture?

50. Fuel Mixture Five hundred gallons of 89-octane gasoline is obtained by mixing 87-octane gasoline with 92-octane gasoline. How much of each type of gasoline is required to obtain the specified mixture? (Octane ratings can be interpreted as percents.)

51. Investment Portfolio A total of $25,000 is invested in two corporate bonds that pay 4.75% and 5.5% simple interest. The total annual interest is $1315. How much is invested in each bond?

52. Shoe Sales You are the manager of a shoe store. On Sunday morning, you are going over the receipts for the previous week's sales. A total of 320 pairs of cross-training shoes were sold. One style sold for $56.95 and the other sold for $72.95. The total receipts were $21,024. The cash register that was supposed to keep track of the number of each type of shoe sold malfunctioned. Can you recover the information? If so, how many of each type were sold?

Finding the Point of Equilibrium In Exercises 53–56, find the point of equilibrium for the demand and supply equations. *See Example 8.*

Demand	*Supply*
53. $p = 56 - 0.0001x$	$p = 22 + 0.00001x$
54. $p = 60 - 0.00001x$	$p = 15 + 0.00004x$
55. $p = 140 - 0.00002x$	$p = 80 + 0.00001x$
56. $p = 400 - 0.0002x$	$p = 225 + 0.0005x$

57. Restaurants The sales y (in billions of dollars) of fast-food and full-service restaurants for the years 2002 through 2009 are shown in the table. *(Source: National Restaurant Association)*

Year	Fast-food	Full-service
2002	115.1	141.9
2003	120.5	148.3
2004	129.5	156.9
2005	136.9	165.2
2006	142.9	172.8
2007	150.8	179.1
2008	157.7	189.4
2009	160.0	182.0

(a) Use a spreadsheet software program to create a scatter plot of the data for fast-food sales and use the *regression* feature to find a linear model. Let x represent the year, with $x = 2$ corresponding to 2002. Repeat the procedure for the data for full-service sales.

(b) Assuming that the amounts for the given 8 years are representative of future years, will fast-food sales ever equal full-service sales? Discuss the appropriateness of using these models to determine whether fast-food sales will ever equal full-service sales.

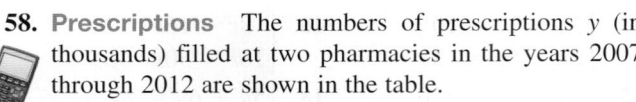

58. Prescriptions The numbers of prescriptions y (in thousands) filled at two pharmacies in the years 2007 through 2012 are shown in the table.

Year	Pharmacy A	Pharmacy B
2007	18.6	19.9
2008	19.2	20.4
2009	19.6	20.8
2010	20.0	21.1
2011	20.4	21.4
2012	21.3	22.0

(a) Use a spreadsheet software program to create a scatter plot of the data for pharmacy A and use the *regression* feature to find a linear model. Let x represent the year, with $x = 7$ corresponding to 2007. Repeat the procedure for the data for pharmacy B.

(b) Assuming that the amounts for the given 6 years are representative of future years, will the number of prescriptions filled at pharmacy A ever exceed the number of prescriptions filled at pharmacy B?

59. Supply and Demand The supply and demand equations for a small LCD television are given by

$$\begin{cases} p + 0.53x = 1542 & \text{Demand} \\ p - 0.37x = 300 & \text{Supply} \end{cases}$$

where p is the price (in dollars) and x represents the number of televisions. For how many units will the quantity demanded equal the quantity supplied? What price corresponds to this value?

60. Supply and Demand The supply and demand equations for a microscope are given by

$$\begin{cases} p + 0.85x = 650 & \text{Demand} \\ p - 0.4x = 75 & \text{Supply} \end{cases}$$

where p is the price (in dollars) and x represents the number of microscopes. For how many units will the quantity demanded equal the quantity supplied? What price corresponds to this value?

Fitting a Line to Data To find the least squares regression line $y = ax + b$ for a set of points (x_1, y_1), $(x_2, y_2), \ldots, (x_n, y_n)$, you can solve the following system for a and b. (If you are unfamiliar with summation notation, then look at the discussion in Section 7.1.)

$$\begin{cases} nb + \left(\displaystyle\sum_{i=1}^{n} x_i \right) a = \displaystyle\sum_{i=1}^{n} y_i \\ \left(\displaystyle\sum_{i=1}^{n} x_i \right) b + \left(\displaystyle\sum_{i=1}^{n} x_i^2 \right) a = \displaystyle\sum_{i=1}^{n} x_i y_i \end{cases}$$

In Exercises 61–64, the sums have been evaluated. Solve the given system of linear equations for a and b to find the least squares regression line for the points. Use a graphing utility to confirm the result.

61. $\begin{cases} 5b + 10a = 20.2 \\ 10b + 30a = 50.1 \end{cases}$

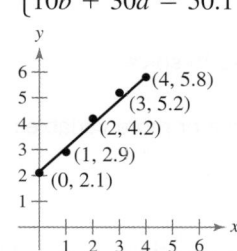

62. $\begin{cases} 5b + 10a = 11.7 \\ 10b + 30a = 25.6 \end{cases}$

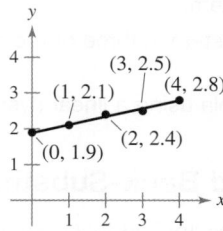

63. $\begin{cases} 7b + 21a = 35.1 \\ 21b + 91a = 114.2 \end{cases}$

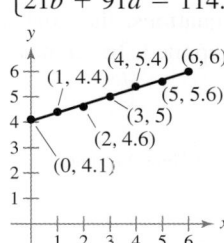

64. $\begin{cases} 6b + 15a = 23.6 \\ 15b + 55a = 48.8 \end{cases}$

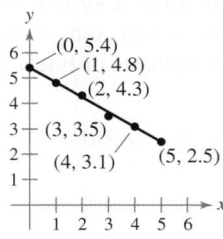

65. Atmosphere The concentration y (in parts per million) of carbon dioxide in the atmosphere is measured at the Mauna Loa Observatory in Hawaii. The greatest monthly carbon dioxide concentrations for the years 2006 through 2010 are shown in the table. *(Source: Scripps CO2 Program)*

Year	t	Concentration, y
2006	0	384.94
2007	1	386.43
2008	2	388.49
2009	3	390.18
2010	4	393.22

(a) Solve the following system for a and b to find the least squares regression line $y = at + b$ for the data. Let t represent the year, with $t = 0$ corresponding to 2006.

$\begin{cases} 5b + 10a = 1943.26 \\ 10b + 30a = 3906.83 \end{cases}$

(b) Use a graphing utility to graph the regression line and predict the greatest monthly carbon dioxide concentration in 2016.

(c) Use the *regression* feature of the graphing utility to find a linear model for the data. Compare this model with the one you found in part (a).

66. Stores in Operation The numbers of stores y operated by BJ's Wholesale Club, Inc. for the years 2005 through 2009 are shown in the table. *(Source: BJ's Wholesale Club, Inc.)*

Year	2005	2006	2007	2008	2009
t	0	1	2	3	4
Stores, y	165	172	177	180	187

(a) Solve the following system for a and b to find the least squares regression line $y = at + b$ for the data. Let t represent the year, with $t = 0$ corresponding to 2005.

$\begin{cases} 5b + 10a = 881 \\ 10b + 30a = 1814 \end{cases}$

(b) Use a graphing utility to graph the regression line and predict the number of stores in operation in 2011.

(c) Use the *regression* feature of the graphing utility to find a linear model for the data. Compare this model with the one you found in part (a).

Business Capsule

SAS is a leader in business software and services. Using SAS forecasting technologies, customers can accurately analyze and forecast processes that take place over time. SAS/ETS software contains popular forecasting methods such as regression analysis and trend extrapolation.

67. Research Project Use your campus library, the Internet, or some other reference source to find information about a company or small business that generates software which uses regression analysis to predict trends. Write a brief paper about the company or small business.

5.3 Linear Systems in Three or More Variables

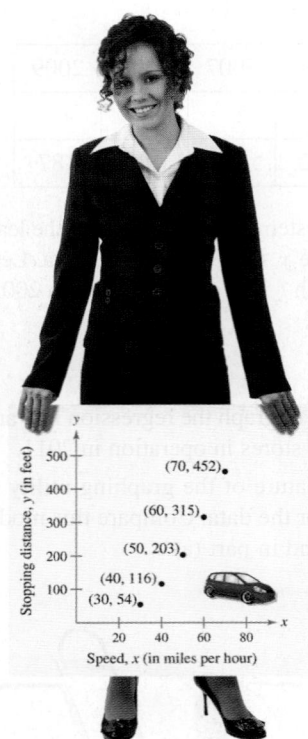

- Solve a linear system in row-echelon form using back-substitution.
- Use Gaussian elimination to solve a linear system.
- Solve a nonsquare linear system.
- Construct and use a linear system in three or more variables to solve an application problem.
- Find the equation of a parabola using a linear system in three or more variables.

Row-Echelon Form and Back-Substitution

The method of elimination can be applied to a system of linear equations in more than two variables. In fact, this method easily adapts to computer use for solving linear systems with dozens of variables.

When elimination is used to solve a system of linear equations, the goal is to rewrite the system in a form to which back-substitution can be applied. To see how this works, consider the following two systems of linear equations.

$$\begin{cases} x - 2y + 3z = 9 \\ -x + 3y = -4 \\ 2x - 5y + 5z = 17 \end{cases}$$ System of Three Linear Equations in Three Variables

$$\begin{cases} x - 2y + 3z = 9 \\ y + 3z = 5 \\ z = 2 \end{cases}$$ Equivalent System in Row-Echelon Form

The second system is said to be in **row-echelon form,** which means that it has a "stair-step" pattern with leading coefficients of 1. After comparing the two systems, it should be clear that it is easier to solve the second system.

In Exercise 69 on page 443, you will find the least squares regression parabola that models the stopping distance of an automobile based on its speed.

Example 1   Using Back-Substitution

Solve the system of linear equations.

$$\begin{cases} x - 2y + 3z = 9 & \text{Equation 1} \\ y + 3z = 5 & \text{Equation 2} \\ z = 2 & \text{Equation 3} \end{cases}$$

SOLUTION   From Equation 3, you know the value of z. To solve for y, substitute $z = 2$ into Equation 2 to obtain

$$y + 3(2) = 5 \implies y = -1.$$

Then substitute $y = -1$ and $z = 2$ into Equation 1 to obtain

$$x - 2(-1) + 3(2) = 9 \implies x = 1.$$

The solution is $x = 1$, $y = -1$, and $z = 2$, which can be written as the **ordered triple** $(1, -1, 2)$. Check this in the original system of equations.

✓ Checkpoint 1

Solve the system of linear equations.

$$\begin{cases} x - 3y - 5z = -12 \\ y + z = 1 \\ z = 4 \end{cases}$$

Gaussian Elimination

Two systems of equations are **equivalent** when they have the same solution set. To solve a system that is not in row-echelon form, first convert it to an *equivalent* system that is in row-echelon form by using one or more of the **elementary row operations** shown below. This process is called **Gaussian elimination,** after the German mathematician Carl Friedrich Gauss (1777–1855).

Elementary Row Operations for Systems of Equations

1. Interchange two equations.

2. Multiply one of the equations by a nonzero constant.

3. Add a multiple of one equation to another equation.

Example 2   **Using Elimination to Solve a System**

Solve the system of linear equations.

$$\begin{cases} x - 2y + 3z = 9 & \text{Equation 1} \\ -x + 3y \phantom{{}+ 3z} = -4 & \text{Equation 2} \\ 2x - 5y + 5z = 17 & \text{Equation 3} \end{cases}$$

SOLUTION   Because the leading coefficient of Equation 1 is 1, you can begin by keeping the x in the upper left position and eliminating the other x-terms from the first column.

$$\begin{cases} x - 2y + 3z = 9 \\ y + 3z = 5 \\ 2x - 5y + 5z = 17 \end{cases}$$

> Adding the first equation to the second equation produces a new second equation.

$$\begin{cases} x - 2y + 3z = 9 \\ y + 3z = 5 \\ {-y} - z = -1 \end{cases}$$

> Adding -2 times the first equation to the third equation produces a new third equation.

Now that all but the first x have been eliminated from the first column, work on the second column. (You need to eliminate y from the third equation.)

$$\begin{cases} x - 2y + 3z = 9 \\ y + 3z = 5 \\ 2z = 4 \end{cases}$$

> Adding the second equation to the third equation produces a new third equation.

Finally, you need a coefficient of 1 for z in the third equation.

$$\begin{cases} x - 2y + 3z = 9 \\ y + 3z = 5 \\ z = 2 \end{cases}$$

> Multiplying the third equation by $\frac{1}{2}$ produces a new third equation.

This is the same system that was solved in Example 1, and, as in that example, you can conclude that the solution is $x = 1$, $y = -1$, and $z = 2$. Check this in the original system of equations.

✓**Checkpoint 2**

Solve the system of linear equations.

$$\begin{cases} x + y + z = 6 \\ 2x - y + z = 3 \\ 3x \phantom{{}- y} - z = 0 \end{cases}$$

STUDY TIP

Elementary row operations involve a lot of arithmetic. So that you can go back and check your work, you should note the elementary row operations performed in each step next to the row you are changing.

The next example involves an inconsistent system—one that has no solution. The key to recognizing an inconsistent system is that at some stage in the elimination process, you obtain a false statement such as

$$0 = -2. \qquad \text{\small False statement}$$

Example 3   An Inconsistent System

Solve the system of linear equations.

$$\begin{cases} x - 3y + z = 1 & \text{\small Equation 1} \\ 2x - y - 2z = 2 & \text{\small Equation 2} \\ x + 2y - 3z = -1 & \text{\small Equation 3} \end{cases}$$

SOLUTION

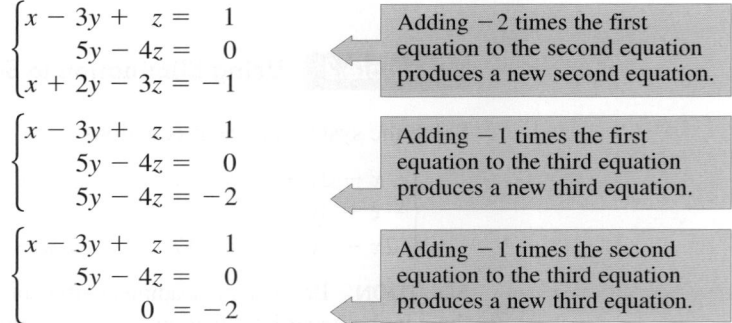

$$\begin{cases} x - 3y + z = 1 \\ 5y - 4z = 0 \\ x + 2y - 3z = -1 \end{cases}$$

Adding -2 times the first equation to the second equation produces a new second equation.

$$\begin{cases} x - 3y + z = 1 \\ 5y - 4z = 0 \\ 5y - 4z = -2 \end{cases}$$

Adding -1 times the first equation to the third equation produces a new third equation.

$$\begin{cases} x - 3y + z = 1 \\ 5y - 4z = 0 \\ 0 = -2 \end{cases}$$

Adding -1 times the second equation to the third equation produces a new third equation.

Because $0 = -2$ is a false statement, you can conclude that this system is inconsistent. So, it has no solution. Moreover, because this system is equivalent to the original system, you can conclude that the original system also has no solution. ————————

✓ Checkpoint 3

Solve the system of linear equations.

$$\begin{cases} 2x + y - z = 7 \\ x - 2y + 2z = -9 \\ 3x - y + z = 5 \end{cases}$$

As with a system of linear equations in two variables, the solution(s) of a system of linear equations in more than two variables must fall into one of three categories. Because an equation in three variables represents a plane in space, the possible solutions can be shown graphically. See Figure 5.10.

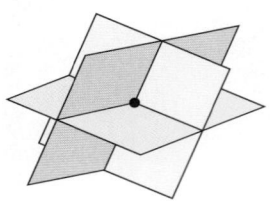

(a) Solution: one point

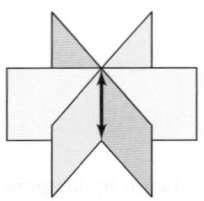

(b) Solution: one line

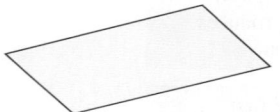

(c) Solution: one plane

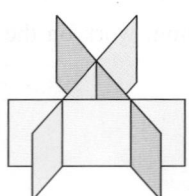

(d) Solution: none

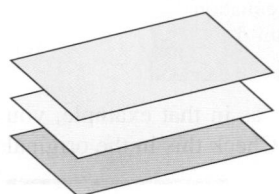

(e) Solution: none

FIGURE 5.10

The Number of Solutions of a Linear System

For a system of linear equations, exactly one of the following is true.

1. There is exactly one solution. [See Figure 5.10(a).]

2. There are infinitely many solutions. [See Figures 5.10(b) and (c).]

3. There is no solution. [See Figures 5.10(d) and (e).]

Recall from Section 5.2 that a system of linear equations is *consistent* if it has at least one solution. A consistent system with exactly one solution is *independent*, whereas a consistent system with infinitely many solutions is *dependent*. A system is *inconsistent* if it has no solution.

Example 4   A System with Infinitely Many Solutions

Solve the system of linear equations.

$$\begin{cases} x + y - 3z = -1 & \text{Equation 1} \\ y - z = 0 & \text{Equation 2} \\ -x + 2y = 1 & \text{Equation 3} \end{cases}$$

SOLUTION

$$\begin{cases} x + y - 3z = -1 \\ y - z = 0 \\ 3y - 3z = 0 \end{cases}$$

Adding the first equation to the third equation produces a new third equation.

$$\begin{cases} x + y - 3z = -1 \\ y - z = 0 \\ 0 = 0 \end{cases}$$

Adding -3 times the second equation to the third equation produces a new third equation.

This means that Equation 3 depends on Equations 1 and 2 in the sense that it gives us no additional information about the variables. Because $0 = 0$ is a true statement, you can conclude that this system has infinitely many solutions. So, the original system is equivalent to the system

$$\begin{cases} x + y - 3z = -1 \\ y - z = 0 \end{cases}.$$

In the equation $y - z = 0$, solve for y in terms of z to obtain $y = z$. Back-substituting for y into the equation $x + y - 3z = -1$ produces $x = 2z - 1$. Finally, letting $z = a$, where a is a real number, the solutions of the original system are all of the form

$$x = 2a - 1, \quad y = a, \quad \text{and} \quad z = a.$$

So, every ordered triple of the form

$$(2a - 1, a, a) \qquad a \text{ is a real number.}$$

is a solution of the system.

STUDY TIP

There are an infinite number of solutions to Example 4, but they are all of a specific form. By selecting, for instance, a-values of 0, 1, and 3, you can verify that $(-1, 0, 0)$, $(1, 1, 1)$, and $(5, 3, 3)$ are specific solutions. It is incorrect to say simply that the solution to Example 4 is "infinite." You must also specify the form of the solutions.

✓ **Checkpoint 4**

Solve the system of linear equations.

$$\begin{cases} 2x + y + 3z = 1 \\ 2x + 6y + 12z = 3 \\ 6x + 8y + 18z = 5 \end{cases}$$

In Example 4, there are other ways to write the same infinite set of solutions. For instance, the solutions could have been written as

$$\left(b, \tfrac{1}{2}(b + 1), \tfrac{1}{2}(b + 1)\right), \quad \text{where } b \text{ is a real number.}$$

This description produces the same set of solutions, as shown below.

Substitution	Solution
$a = 0$	$(2(0) - 1, 0, 0) = (-1, 0, 0)$
$b = -1$	$\left(-1, \tfrac{1}{2}(-1 + 1), \tfrac{1}{2}(-1 + 1)\right) = (-1, 0, 0)$
$a = 1$	$(2(1) - 1, 1, 1) = (1, 1, 1)$
$b = 1$	$\left(1, \tfrac{1}{2}(1 + 1), \tfrac{1}{2}(1 + 1)\right) = (1, 1, 1)$

In both cases, you obtain the same ordered triples. So, when comparing descriptions of an infinite solution set, keep in mind that there is more than one way to describe the set.

Nonsquare Systems

So far, each system of linear equations you have looked at has been **square,** which means that the number of equations is equal to the number of variables. In a **nonsquare** system, the number of equations differs from the number of variables. A system of linear equations cannot have a unique solution unless there are at least as many equations as there are variables in the system.

Example 5   A System with Fewer Equations than Variables

Solve the system of linear equations.

$$\begin{cases} x - 2y + z = 2 & \text{Equation 1} \\ 2x - y - z = 1 & \text{Equation 2} \end{cases}$$

SOLUTION   Begin by rewriting the system in row-echelon form.

$$\begin{cases} x - 2y + z = 2 \\ \quad\quad 3y - 3z = -3 \end{cases}$$

> Adding -2 times the first equation to the second equation produces a new second equation.

$$\begin{cases} x - 2y + z = 2 \\ \quad\quad y - z = -1 \end{cases}$$

> Multiplying the second equation by $\frac{1}{3}$ produces a new second equation.

Solving for y in terms of z, you obtain

$$y = z - 1.$$

Back-substitution into Equation 1 yields

$$x - 2(z - 1) + z = 2$$
$$x - 2z + 2 + z = 2$$
$$x - z = 0$$
$$x = z.$$

Finally, by letting $z = a$, where a is a real number, you have the solution

$$x = a, \quad y = a - 1, \quad \text{and} \quad z = a.$$

So, every ordered triple of the form

$$(a, a - 1, a) \qquad\qquad a \text{ is a real number.}$$

is a solution of the system. Because there were originally three variables and only two equations, the system cannot have a unique solution. ──────────

✓ Checkpoint 5

Solve the system of linear equations.

$$\begin{cases} 2x - 2y + 5z = 2 \\ 4x \quad\quad - z = 0 \end{cases}$$ ■

In Example 5, try choosing some values of a to obtain different solutions of the system, such as $(1, 0, 1)$, $(2, 1, 2)$, and $(3, 2, 3)$. Then check each of the solutions in the original system. For example, you can check the solution $(1, 0, 1)$ as shown.

Equation 1: $1 - 2(0) + 1 = 2$ ✓

Equation 2: $2(1) - 0 - 1 = 1$ ✓

Application

 Example 6   **Business Loans**

A company borrows $1,000,000. Some of the money is borrowed at 5%, some at 7%, some at 8%, and some at 10% simple annual interest. The amount borrowed at 8% is the same as the amount borrowed at 10%. How much is borrowed at each rate when the total annual interest is $79,000?

SOLUTION   Let x, y, z, and w represent the amounts borrowed at each rate. Because the total amount borrowed is $1,000,000, you can write the equation

$$x + y + z + w = 1,000,000.$$

A second equation can be derived from the fact that the total annual interest is $79,000.

$$0.05x + 0.07y + 0.08z + 0.10w = 79,000$$

Finally, because the amount borrowed at 8% is the same as the amount borrowed at 10%, you can write

$$z - w = 0.$$

These three equations make up the system shown.

$$\begin{cases} x + y + z + w = 1,000,000 & \text{Equation 1} \\ 0.05x + 0.07y + 0.08z + 0.10w = 79,000 & \text{Equation 2} \\ z - w = 0 & \text{Equation 3} \end{cases}$$

Using elimination, you find that the system has infinitely many solutions, which can be written as shown.

$$x = 2a - 450,000$$

$$y = -4a + 1,450,000$$

$$z = a$$

$$w = a$$

So, there are many different scenarios. One possible solution is to let $a = 250,000$. This yields the following scenario.

$$x = \$50,000 \text{ at } 5\%$$

$$y = \$450,000 \text{ at } 7\%$$

$$z = \$250,000 \text{ at } 8\%$$

$$w = \$250,000 \text{ at } 10\%$$

✓Checkpoint 6

In Example 6, the company borrows $1,200,000. How much is borrowed at each rate?

Remember that you can check solutions that are not unique by substituting several different values for a.

Finding the Equation of a Parabola

Example 7   Data Analysis: Curve Fitting

Find a quadratic equation $y = ax^2 + bx + c$ whose graph passes through the points $(-1, 3)$, $(1, 1)$, and $(2, 6)$.

SOLUTION   Because the graph of $y = ax^2 + bx + c$ passes through the points $(-1, 3)$, $(1, 1)$, and $(2, 6)$, you can substitute for x and y as shown below.

When $x = -1$ and $y = 3$: $a(-1)^2 + b(-1) + c = 3$

When $x = 1$ and $y = 1$: $a(1)^2 + b(1) + c = 1$

When $x = 2$ and $y = 6$: $a(2)^2 + b(2) + c = 6$

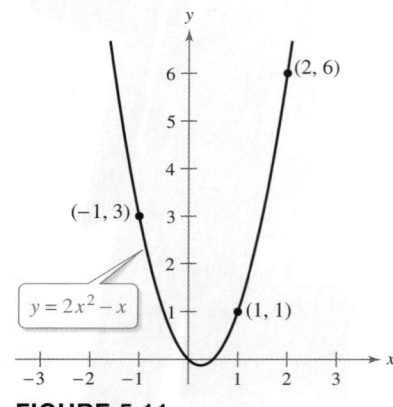

This produces the following system of linear equations.

$$\begin{cases} a - b + c = 3 & \text{Equation 1} \\ a + b + c = 1 & \text{Equation 2} \\ 4a + 2b + c = 6 & \text{Equation 3} \end{cases}$$

The solution of this system is

$$a = 2, \quad b = -1, \quad \text{and} \quad c = 0.$$

So, the equation of the parabola is

$$y = 2x^2 - x$$

as shown in Figure 5.11.

FIGURE 5.11

ALGEBRA TUTOR   *xy*

For help with the algebra used to solve the system of equations in Example 7, see the *Chapter 5 Algebra Tutor* on page 464.

✓ **Checkpoint 7**

Find a quadratic equation $y = ax^2 + bx + c$ whose graph passes through the points $(-1, 7)$, $(1, 3)$, and $(2, 7)$.

SUMMARIZE   (Section 5.3)

1. State the definition of row-echelon form *(page 432)*. For an example of solving a linear system in row-echelon form, see Example 1.

2. Describe the process of Gaussian elimination *(page 433)*. For examples that use Gaussian elimination to solve a linear system, see Examples 2, 3, and 4.

3. Describe the difference between square and nonsquare systems of linear equations *(page 436)*. For an example of solving a nonsquare linear system, see Example 5.

4. Describe a real-life example of how a linear system in three or more variables can be used to analyze the amount of money borrowed by a company *(page 437, Example 6)*.

5. Describe how a linear system in three or more variables can be used to find the equation of a parabola given three points on the parabola *(page 438)*. For an example of finding the equation of a parabola given three points on the parabola, see Example 7.

SKILLS WARM UP 5.3   The following warm-up exercises involve skills that were covered in earlier sections. You will use these skills in the exercise set for this section. For additional help, review Sections 1.2, 2.1, 5.1, and 5.2.

In Exercises 1–4, solve the system of linear equations.

1. $\begin{cases} x + y = 25 \\ \quad\ y = 15 \end{cases}$

2. $\begin{cases} 2x - 3y = \quad 4 \\ 6x \qquad\quad = -18 \end{cases}$

3. $\begin{cases} x + y = 32 \\ x - y = 24 \end{cases}$

4. $\begin{cases} 2r - \ s = \ 5 \\ \ r + 2s = 10 \end{cases}$

In Exercises 5–8, determine whether the ordered triple is a solution of the equation.

5. $5x - 3y + 4z = 2$

$(-1, -2, 1)$

6. $x - 2y + 12z = 9$

$(6, 3, 2)$

7. $2x - 5y + 3z = -9$

$(a - 2, a + 1, a)$

8. $-5x + y + z = 21$

$(a - 4, 4a + 1, a)$

In Exercises 9 and 10, solve for x in terms of a.

9. $x + 2y - 3z = 4$

$y = 1 - a, z = a$

10. $x - 3y + 5z = 4$

$y = 2a + 3, z = a$

Exercises 5.3

See www.CalcChat.com for worked-out solutions to odd-numbered exercises.

Matching In Exercises 1–4, match the system of equations with its solution. [The solutions are labeled (a), (b), (c), and (d).]

(a) $(-1, 0, 3)$ (b) $(6, 2, -2)$

(c) $(2, 1, -3)$ (d) $(4, -1, 5)$

1. $\begin{cases} -2x + 3y - 2z = \quad 5 \\ \ 3x - 4y + \ z = \ -1 \\ \quad x + 2y + 5z = -11 \end{cases}$

2. $\begin{cases} \ 5x + 2y - 4z = -17 \\ -8x + \ y - 5z = \ -7 \\ \ 4x + 3y - \ z = \ -7 \end{cases}$

3. $\begin{cases} 2x + 5y - \ 7z = \quad 36 \\ \ x + 6y - 10z = \quad 38 \\ \ x - 4y + \ 8z = -18 \end{cases}$

4. $\begin{cases} -x - 2y + 5z = \ 23 \\ -3x + \ y + 6z = \ 17 \\ \ 9x + 2y - 7z = \ -1 \end{cases}$

Identifying Row-Echelon Form In Exercises 5–8, determine whether the system of equations is in row-echelon form. Justify your answer.

5. $\begin{cases} x + 3y - 7z = -11 \\ \quad\ y - 2z = \ -3 \\ \qquad\quad z = \quad 2 \end{cases}$

6. $\begin{cases} x - y + 3z = -11 \\ \quad\ y + 8z = -12 \\ \qquad\quad z = \ -2 \end{cases}$

7. $\begin{cases} x - 9y + z = \ 22 \\ \quad\ 2y + z = -3 \\ \qquad\quad z = \ 1 \end{cases}$

8. $\begin{cases} x - \ y - 8z = \ 12 \\ \quad\ 2y - 2z = \quad 2 \\ \qquad\quad 7z = -7 \end{cases}$

Using Back-Substitution In Exercises 9 and 10, use back-substitution to solve the system of linear equations. *See Example 1.*

9. $\begin{cases} x - y + \ z = \quad 4 \\ \quad\ y + 2z = -6 \\ \qquad\quad z = -2 \end{cases}$

10. $\begin{cases} x - 2y + z = 8 \\ \quad\ y + z = 5 \\ \qquad\quad z = 2 \end{cases}$

Solving a System of Linear Equations In Exercises 11–36, solve the system of linear equations. *See Examples 2, 3, 4, and 5.*

11. $\begin{cases} 4x + \ y - 3z = \ 11 \\ 2x - 3y + 2z = \quad 9 \\ \ x + \ y + \ z = -3 \end{cases}$

12. $\begin{cases} \qquad 6y + 4z = -12 \\ 3x + 3y \qquad = \quad 9 \\ 2x \qquad - 3z = \quad 10 \end{cases}$

13. $\begin{cases} 3x \qquad + 2z = \ 13 \\ \ x + 2y + \ z = -5 \\ \qquad - 3y - \ z = \ 10 \end{cases}$

14. $\begin{cases} 2x + 3y + \ z = -4 \\ 2x - 4y + 3z = \ 18 \\ 3x - 2y + 2z = \quad 9 \end{cases}$

15. $\begin{cases} 3x - 2y + 4z = 1 \\ \ x + \ y - 2z = 3 \\ 2x - 3y + 6z = 8 \end{cases}$

16. $\begin{cases} 5x - \ 3y + 2z = 3 \\ 2x + \ 4y - \ z = 7 \\ \ x - 11y + 4z = 3 \end{cases}$

17. $\begin{cases} 3x + 3y + \ 5z = 1 \\ 3x + 5y + \ 9z = 0 \\ 5x + 9y + 17z = 0 \end{cases}$

18. $\begin{cases} 2x + \ y - \ z = \ 13 \\ \ x + 2y + \ z = \quad 2 \\ 8x - 3y + 4z = -2 \end{cases}$

19. $\begin{cases} x + 2y - 7z = -4 \\ 2x + y + z = 13 \\ 3x + 9y - 36z = -33 \end{cases}$

20. $\begin{cases} 2x + y - 3z = 4 \\ 4x + 2z = 10 \\ -2x + 3y - 13z = -8 \end{cases}$

21. $\begin{cases} x + 4z = 13 \\ 4x - 2y + z = 7 \\ 2x - 2y - 7z = -19 \end{cases}$ **22.** $\begin{cases} 4x - y + 5z = 11 \\ x + 2y - z = 5 \\ 5x - 8y + 13z = 7 \end{cases}$

23. $\begin{cases} x + 4z = 1 \\ x + y + 10z = 10 \\ 2x - y + 2z = -5 \end{cases}$ **24.** $\begin{cases} 3x - 2y - 6z = 4 \\ -3x + 2y + 6z = 1 \\ x - y - 5z = 3 \end{cases}$

25. $\begin{cases} 4x + 3y + 5z = 10 \\ 5x + 2y + 10z = 13 \\ 3x + y - 2z = -9 \end{cases}$ **26.** $\begin{cases} 2x + 5y = 25 \\ 3x - 2y + 4z = 1 \\ 4x - 3y + z = 9 \end{cases}$

27. $\begin{cases} 2x + 3y - z = 1 \\ x - 2y + z = 7 \\ 3x + y + 2z = 12 \end{cases}$ **28.** $\begin{cases} 2x + 3y = 0 \\ 4x + 3y - z = 0 \\ 8x + 3y + 3z = 0 \end{cases}$

29. $\begin{cases} 12x + 5y + z = 0 \\ 12x + 4y - z = 0 \end{cases}$ **30.** $\begin{cases} x - 2y + 5z = 2 \\ 3x + 2y - z = -2 \end{cases}$

31. $\begin{cases} x - 3y + 2z = 18 \\ 5x - 13y + 12z = 80 \end{cases}$

32. $\begin{cases} 2x - 3y + z = -2 \\ -4x + 9y = 7 \end{cases}$

33. $\begin{cases} 3x - 3y + 6z = 7 \\ -x + y - 2z = 3 \\ 2x + 3y - 4z = 8 \end{cases}$

34. $\begin{cases} 4x + 3y = 7 \\ x - 2y + z = 0 \\ -2x + 4y - 2z = 13 \end{cases}$

35. $\begin{cases} x + 3w = 4 \\ 2y - z - w = 0 \\ 3y - 2w = 1 \\ 2x - y + 4z = 5 \end{cases}$

36. $\begin{cases} x + y + z + w = 6 \\ 2x + 3y - w = 0 \\ -3x + 4y + z + 2w = 4 \\ x + 2y - z + w = 0 \end{cases}$

Think About It In Exercises 37–40, find two systems of linear equations that have the ordered triple as a solution. (There are many correct answers.)

37. $(3, -1, 2)$ **38.** $\left(-\frac{1}{2}, -2, 4\right)$

39. $(1, -5, -3)$ **40.** $\left(0, 2, \frac{1}{2}\right)$

Writing Ordered Triples In Exercises 41–44, write three ordered triples of the given form.

41. $\left(a, a + 2, \frac{1}{3}a\right)$ **42.** $(3a, 5 - a, a)$

43. $\left(\frac{1}{2}a, 4a, 7\right)$ **44.** $\left(-\frac{1}{4}a + 3, a, 8\right)$

Data Analysis: Curve Fitting In Exercises 45–48, find the quadratic equation

$$y = ax^2 + bx + c$$

whose graph passes through the points. *See Example 7.*

45. 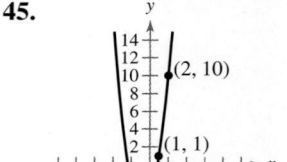   **46.**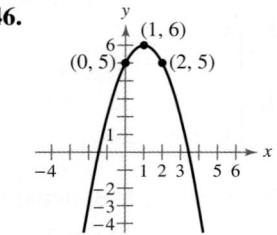

47. 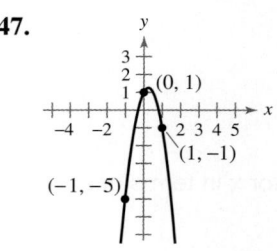   **48.**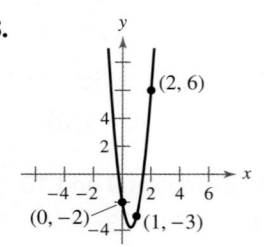

Data Analysis: Curve Fitting In Exercises 49–52, find the equation

$$x^2 + y^2 + Dx + Ey + F = 0$$

of the circle that passes through the points.

49. 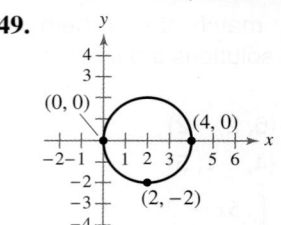   **50.**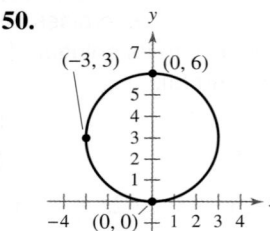

51. 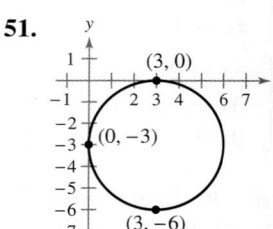   **52.**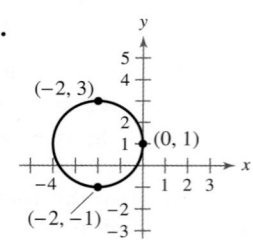

53. Investment A real estate company borrows $1,500,000. Some of the money is borrowed at 3%, some at 4%, and some at 6% simple annual interest. How much is borrowed at each rate when the total annual interest is $53,000 and the amount borrowed at 4% is the same as the amount borrowed at 6%?

54. Investment A clothing company borrows $700,000. Some of the money is borrowed at 8%, some at 9%, and some at 10% simple annual interest. How much is borrowed at each rate when the total annual interest is $60,500 and the amount borrowed at 8% is three times the amount borrowed at 10%?

55. Candles A candle company sells three types of candles for $15, $10, and $5 per unit. In one year, the total revenue for the three products was $550,000, which corresponded to the sale of 50,000 units. The company sold half as many units of the $15 candles as units of the $10 candles. How many units of each type of candle were sold?

56. Hair Products A hair product company sells three types of hair products for $30, $20, and $10 per unit. In one year, the total revenue for the three products was $800,000, which corresponded to the sale of 40,000 units. The company sold half as many units of the $30 product as units of the $20 product. How many units of each product were sold?

57. Crop Spraying A mixture of 5 gallons of chemical A, 8 gallons of chemical B, and 12 gallons of chemical C is required to kill a crop-destroying insect. Commercial spray X contains 1, 2, and 3 parts of these chemicals, respectively. Commercial spray Y contains only chemical C. Commercial spray Z contains chemicals A, B, and C in equal amounts. How much of each type of commercial spray is needed to obtain the desired mixture?

58. Acid Mixture A chemist needs 10 liters of a 25% acid solution. The solution is to be mixed from three solutions whose acid concentrations are

10%, 20%, and 50%.

How many liters of each solution should the chemist use to satisfy each of the following requirements?

(a) Use as little as possible of the 50% solution.

(b) Use as much as possible of the 50% solution.

(c) Use 2 liters of the 50% solution.

Investment Portfolio In Exercises 59 and 60, you have a total of $500,000 that is to be invested in certificates of deposit, municipal bonds, blue-chip stocks, and growth stocks. How much should be put in each type of investment? *See Example 6.*

59. The certificates of deposit pay 2.5% simple annual interest, and the municipal bonds pay 10% simple annual interest. Over a five-year period, you expect the blue-chip stocks to return 12% simple annual interest and the growth stocks to return 18% simple annual interest. You want a combined annual return of 10% and you also want to have only one-fourth of the portfolio invested in stocks.

60. The certificates of deposit pay 3% simple annual interest, and the municipal bonds pay 10% simple annual interest. Over a five-year period, you expect the blue-chip stocks to return 12% simple annual interest and the growth stocks to return 15% simple annual interest. You want a combined annual return of 10% and you also want to have only one-fourth of the portfolio invested in stocks.

Fitting a Parabola to Data To find the least squares regression parabola

$$y = ax^2 + bx + c$$

for a set of points

$$(x_1, y_1), (x_2, y_2), \ldots, (x_n, y_n)$$

you can solve the following system of linear equations for a, b, and c.

$$\begin{cases} nc + \left(\sum_{i=1}^{n} x_i\right)b + \left(\sum_{i=1}^{n} x_i^2\right)a = \sum_{i=1}^{n} y_i \\ \left(\sum_{i=1}^{n} x_i\right)c + \left(\sum_{i=1}^{n} x_i^2\right)b + \left(\sum_{i=1}^{n} x_i^3\right)a = \sum_{i=1}^{n} x_i y_i \\ \left(\sum_{i=1}^{n} x_i^2\right)c + \left(\sum_{i=1}^{n} x_i^3\right)b + \left(\sum_{i=1}^{n} x_i^4\right)a = \sum_{i=1}^{n} x_i^2 y_i \end{cases}$$

In Exercises 61–64, the sums have been evaluated. Solve the given system for a, b, and c to find the least squares regression parabola for the points. Use a graphing utility to confirm the result.

61. $\begin{cases} 5c + 10a = 15.5 \\ 10b = 6.3 \\ 10c + 34a = 32.1 \end{cases}$ **62.** $\begin{cases} 5c + 10a = 15.0 \\ 10b = 17.3 \\ 10c + 34a = 34.5 \end{cases}$

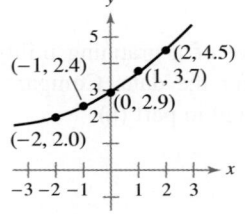

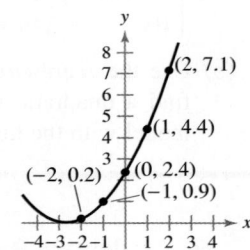

63. $\begin{cases} 6c + 3b + 19a = 23.9 \\ 3c + 19b + 27a = -7.2 \\ 19c + 27b + 115a = 48.8 \end{cases}$

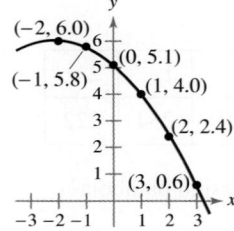

64. $\begin{cases} 6c + 3b + 19a = 13.1 \\ 3c + 19b + 27a = -2.6 \\ 19c + 27b + 115a = 29.0 \end{cases}$

65. Consumer Spending The annual amounts y (in billions of dollars) spent on hunting equipment and firearms in the United States from 2005 through 2009 are shown in the table. In the table, x represents the year, with $x = -2$ corresponding to 2005. *(Source: National Sporting Goods Association)*

Year, x	Amount, y
-2	3.6
-1	3.7
0	3.9
1	4.5
2	5.2

(a) Find the least squares regression parabola

$$y = ax^2 + bx + c$$

for the data by solving the system below.

$$\begin{cases} 5c \quad\;\; + 10a = 20.9 \\ \quad\;\; 10b \quad\quad = 4.0 \\ 10c \quad\;\; + 34a = 43.4 \end{cases}$$

(b) Use the *regression* feature of a graphing utility to find a quadratic model for the data. Compare the model with the model found in part (a).

66. Trademark Registration The annual numbers y (in thousands) of trademarks issued in the United States from 2005 through 2009 are shown in the table. In the table, x represents the year, with $x = -2$ corresponding to 2005. *(Source: U.S. Patent and Trademark Office)*

Year, x	-2	-1	0	1	2
Number, y	155	194	219	234	222

(a) Find the least squares regression parabola

$$y = ax^2 + bx + c$$

for the data by solving the system below.

$$\begin{cases} 5c \quad\;\; + 10a = 1024 \\ \quad\;\; 10b \quad\quad = 174 \\ 10c \quad\;\; + 34a = 1936 \end{cases}$$

(b) Use the *regression* feature of a graphing utility to find a quadratic model for the data. Compare the model with the model found in part (a).

67. Gardening The percents y of households in the United States involved in vegetable gardening from 2005 through 2009 are shown in the table. In the table, x represents the year, with $x = -1$ corresponding to 2005. *(Source: The National Gardening Association)*

Year, x	Percent, y
-1	25
0	22
1	22
2	23
3	27

(a) Find the least squares regression parabola $y = ax^2 + bx + c$ for the data by solving the system below.

$$\begin{cases} 5c + 5b + 15a = 119 \\ 5c + 15b + 35a = 124 \\ 15c + 35b + 99a = 382 \end{cases}$$

(b) Use the model found in part (a) to predict the percent of households involved in vegetable gardening in 2010.

68. Sales Per Share The sales per share y (in dollars) of Hershey's from 2005 through 2009 are shown in the table. In the table, x represents the year, with $x = -1$ corresponding to 2005. *(Source: The Hershey Company)*

Year, x	Sales per share, y
-1	20.11
0	21.47
1	21.79
2	22.61
3	23.24

(a) Find the least squares regression parabola $y = ax^2 + bx + c$ for the data by solving the system below.

$$\begin{cases} 5c + 5b + 15a = 109.22 \\ 5c + 15b + 35a = 116.62 \\ 15c + 35b + 99a = 341.50 \end{cases}$$

(b) Use the model found in part (a) to predict Hershey's sales per share in 2010 and 2011.

69. Automobile Stopping Distance Engineers tested the braking system of a new automobile. The scatter plot shows the stopping distances y (in feet) of the automobile for several speeds x (in miles per hour).

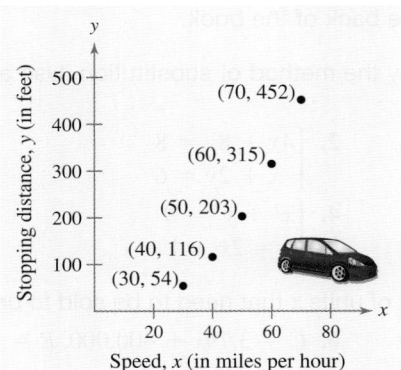

Speed, x (in miles per hour)

(a) Find the least squares regression parabola $y = ax^2 + bx + c$ for the data by solving the system below.

$$\begin{cases} 5c + 250b + 13{,}500a = 1140 \\ 250c + 13{,}500b + 775{,}000a = 66{,}950 \\ 13{,}500c + 775{,}000b + 46{,}590{,}000a = 4{,}090{,}500 \end{cases}$$

(b) Use the *regression* feature of a graphing utility to check your answer to part (a).

(c) Use the model found in part (a) to predict the stopping distance of the automobile when traveling at a speed of 75 miles per hour.

70. HOW DO YOU SEE IT? The number of sides x and the combined number of sides and diagonals y for each of three regular polygons are shown below. How can you create a system of linear equations to find an equation of the form $y = ax^2 + bx + c$ that represents the relationship between x and y for the three polygons?

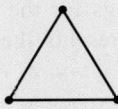
$x = 3, y = 3$

 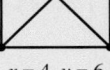
$x = 4, y = 6$

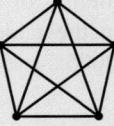
$x = 5, y = 10$

71. Geometry Find the equation of the form

$$y = ax^2 + bx + c$$

that represents the relationship between x and y for the three polygons shown in Exercise 70. Then sketch regular polygons with 6, 7, and 8 sides, and draw their diagonals. Does the equation represent the relationship between x and y for these polygons?

72. Think About It Is it possible for a square linear system to have no solution? Explain.

73. Think About It Is it possible for a square linear system to have infinitely many solutions? Explain.

74. Writing One solution for Exercise 30 is

$$(-a, 2a - 1, a).$$

A student gives

$$(b, -2b - 1, -b)$$

as a solution to the same exercise. Explain why both solutions are correct.

75. Think About It Find values of a, b, and c (if possible) such that the system of linear equations below has each number of solutions.

(a) A unique solution

(b) No solution

(c) Infinitely many solutions

$$\begin{cases} x + y = 2 \\ y + z = 2 \\ x + z = 2 \\ ax + by + cz = 0 \end{cases}$$

76. Writing When using Gaussian elimination to solve a system of linear equations, explain how you can recognize that the system has no solution. Give an example that illustrates your answer.

77. Project: Revenues per Share For a project involving the revenues per share of Amazon.com and Google Inc., visit this text's website at *www.cengagebrain.com.* (*Source: Amazon.com and Google Inc.*)

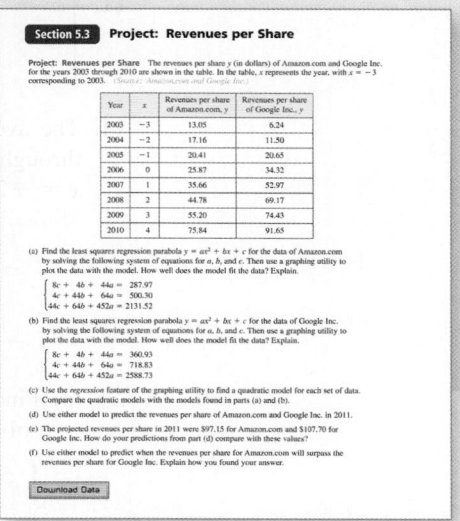

QUIZ YOURSELF

See www.CalcChat.com for worked-out solutions to odd-numbered exercises.

Take this quiz as you would take a quiz in class. When you are done, check your work against the answers given in the back of the book.

In Exercises 1–4, solve the system by the method of substitution. Use a graphing utility to verify your solution.

1. $\begin{cases} 3x + y = 11 \\ x - 2y = -8 \end{cases}$

2. $\begin{cases} 4x + 8y = 8 \\ x + 2y = 6 \end{cases}$

3. $\begin{cases} x + y = 4 \\ y = 2\sqrt{x} + 1 \end{cases}$

4. $\begin{cases} x^2 + y^2 = 9 \\ y = 2x + 1 \end{cases}$

In Exercises 5 and 6, find the number of units x that need to be sold to break even.

5. $C = 10.50x + 9000$, $R = 16.50x$

6. $C = 3.79x + 400,000$, $R = 4.59x$

In Exercises 7 and 8, solve the system by the method of elimination.

7. $\begin{cases} 2.5x - y = 6 \\ 3x + 4y = 2 \end{cases}$

8. $\begin{cases} \frac{1}{2}x + \frac{1}{3}y = 1 \\ x - 2y = -2 \end{cases}$

9. Find the point of equilibrium for the supply and demand equations. Verify your solution graphically.

Demand: $p = 50 - 0.002x$

Supply: $p = 20 + 0.004x$

10. The total numbers y (in millions) of Medicare enrollees in the years 2005 through 2009 are shown in the table at the left. Solve the following system for a and b to find the least squares regression line $y = at + b$ for the data. Let t represent the year, with $t = 0$ corresponding to 2005. *(Source: U.S. Centers for Medicare and Medicaid Services)*

$$\begin{cases} 5b + 10a = 221.4 \\ 10b + 30a = 451.2 \end{cases}$$

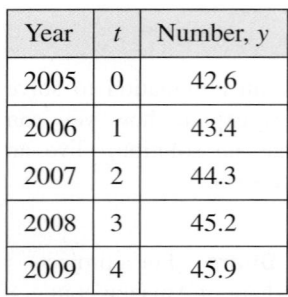

Year	t	Number, y
2005	0	42.6
2006	1	43.4
2007	2	44.3
2008	3	45.2
2009	4	45.9

Table for 10

In Exercises 11–13, solve the system of equations.

11. $\begin{cases} 2x + 3y - z = -7 \\ x \qquad + 3z = 10 \\ 2y + z = -1 \end{cases}$

12. $\begin{cases} x + y - 2z = 12 \\ 2x - y - z = 6 \\ y - z = 6 \end{cases}$

13. $\begin{cases} 3x + 2y + z = 17 \\ -x + y + z = 4 \\ x - y - z = 3 \end{cases}$

14. The average prices y (in dollars) of retail prescription drugs for the years 2005 through 2009 are shown in the table. In the table, x represents the year, with $x = -2$ corresponding to 2005. *(Source: National Association of Chain Drug Stores)*

Year, x	-2	-1	0	1	2
Average price, y	63.87	66.97	69.91	72.87	76.94

(a) Find the least squares regression parabola $y = ax^2 + bx + c$ for the data by solving the system below.

$$\begin{cases} 5c \qquad + 10a = 350.56 \\ 10b \qquad = 32.04 \\ 10c \qquad + 34a = 703.08 \end{cases}$$

(b) Use the model found in part (a) to predict the average prices of retail prescription drugs in 2010 and 2011.

5.4 Systems of Inequalities

- Sketch the graph of an inequality in two variables.
- Solve a system of inequalities.
- Construct and use a system of inequalities to solve an application problem.

The Graph of an Inequality

The following statements are inequalities in two variables:

$$3x - 2y < 6 \quad \text{and} \quad 2x^2 + 3y^2 \geq 6.$$

An ordered pair (a, b) is a **solution of an inequality** in x and y if the inequality is true when a and b are substituted for x and y, respectively. The **graph of an inequality** is the collection of all solutions of the inequality. To sketch the graph of an inequality, begin by sketching the graph of the *corresponding equation*. The graph of the equation will normally separate the plane into two or more regions. In each such region, one of the following must be true.

1. *All* points in the region are solutions of the inequality.

2. *No* point in the region is a solution of the inequality.

So, you can determine whether the points in an entire region satisfy the inequality simply by testing *one* point in the region. When possible, use test points that are convenient to substitute into the inequality, such as $(0, 0)$.

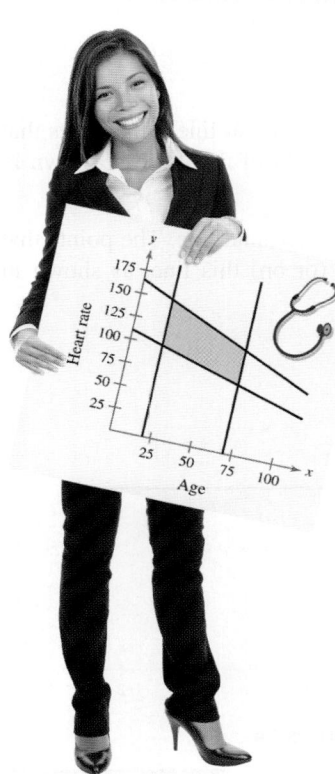

In Exercise 66 on page 453, you will use a graph to find the exercise target heart rate for individuals between the ages of 20 and 70 years old.

Sketching the Graph of an Inequality in Two Variables

1. Replace the inequality sign with an equal sign, and sketch the graph of the resulting equation. (Use a dashed line for $<$ or $>$ and a solid line for $\leq$ or $\geq$.)

2. Test one point in each of the regions formed by the graph in Step 1. If the point satisfies the inequality, then shade the entire region to denote that every point in the region satisfies the inequality.

Example 1   **Sketching the Graph of an Inequality**

Sketch the graph of the inequality $y \geq x^2 - 1$.

SOLUTION   The graph of the corresponding *equation* $y = x^2 - 1$ is a parabola, as shown in Figure 5.12. Test a point *above* the parabola $(0, 0)$ and a point *below* the parabola $(0, -2)$.

The points that satisfy the inequality are those lying above (or on) the parabola.

$$(0, 0): \ 0 \overset{?}{\geq} 0^2 - 1 \qquad\qquad (0, -2): \ -2 \overset{?}{\geq} 0^2 - 1$$

$$0 \geq -1 \quad \text{\small{(0, 0) is a solution.}} \qquad\qquad -2 \ngeq -1 \quad \text{\small{(0, -2) is not a solution.}}$$

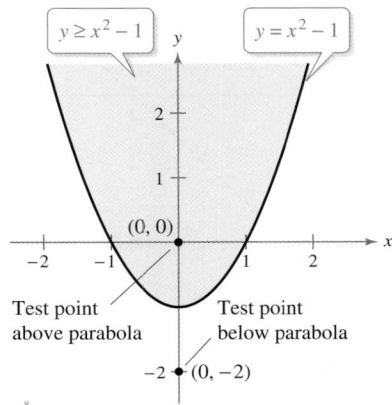

FIGURE 5.12

✓ **Checkpoint 1**

Sketch the graph of $y < x^2 + 2$.

The inequality in Example 1 is a nonlinear inequality in two variables. Most of the following examples involve **linear inequalities** such as $ax + by < c$ (a and b are not both zero). The graph of a linear inequality is a half-plane lying on one side of the line $ax + by = c$.

TECH TUTOR

A graphing utility can be used to graph an inequality or a system of inequalities. For instance, to graph $y \geq x^2 - 2$, enter $y = x^2 - 2$ and use the *shade* feature of the graphing utility to shade the part of the graph consisting of the points that satisfy the inequality. You should obtain the graph shown below. Consult the user's guide for your graphing utility for more information.

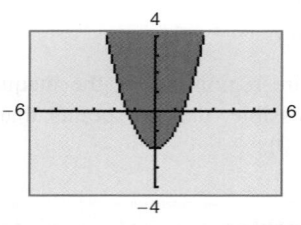

Example 2   Sketching the Graph of a Linear Inequality

Sketch the graph of each linear inequality.

a. $x > -2$ **b.** $y \leq 3$

SOLUTION

a. The graph of the corresponding equation $x = -2$ is a vertical line. The points that satisfy the inequality $x > -2$ are those lying to the right of this line, as shown in Figure 5.13.

b. The graph of the corresponding equation $y = 3$ is a horizontal line. The points that satisfy the inequality $y \leq 3$ are those lying below (or on) this line, as shown in Figure 5.14.

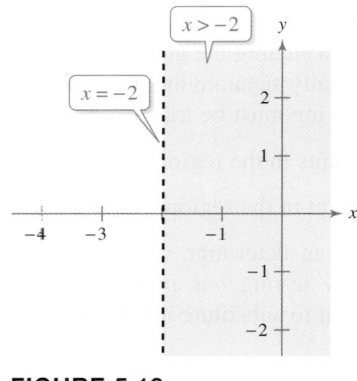

FIGURE 5.13   **FIGURE 5.14**

✓ Checkpoint 2

Sketch the graph of $x \geq -1$.

Example 3   Sketching the Graph of a Linear Inequality

Sketch the graph of $x - y < 2$.

SOLUTION   The graph of the corresponding equation $x - y = 2$ is a line, as shown in Figure 5.15. Because the origin $(0, 0)$ satisfies the inequality, the graph consists of the half-plane lying above the line. (Try checking a point below the line. Regardless of which point you choose, you will see that it does not satisfy the inequality.)

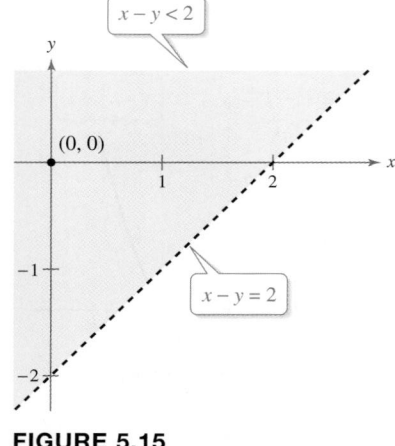

FIGURE 5.15

STUDY TIP

To graph a linear inequality, it can help to write the inequality in slope-intercept form. For instance, by writing $x - y < 2$ in the form

$$y > x - 2$$

you can see that the solution points lie above the line $x - y = 2$ (or $y = x - 2$), as shown in Figure 5.15.

✓ Checkpoint 3

Sketch the graph of $x + y \leq 1$.

Systems of Inequalities

A **solution** of a system of inequalities in x and y is a point (x, y) that satisfies each inequality in the system. To solve a system of inequalities in two variables, first sketch the graph of each individual inequality (on the same coordinate system) and then find the region that is *common* to every graph in the system. This region represents the **solution set** of the system. For systems of *linear* inequalities, it is helpful to find the vertices of the solution region.

Example 4 Solving a System of Inequalities

Solve the system of linear inequalities.

$$\begin{cases} x - y < & 2 & \text{Inequality 1} \\ x > & -2 & \text{Inequality 2} \\ y \leq & 3 & \text{Inequality 3} \end{cases}$$

SOLUTION To solve the system, sketch the graph of the solution set. The graphs of these inequalities are shown in Figures 5.15, 5.13, and 5.14 on page 446. The triangular region common to all three graphs can be found by superimposing the graphs on the same coordinate plane, as shown in Figure 5.16. To find the vertices of the region, solve the three systems of equations obtained by taking the *pairs* of equations representing the boundaries of the individual regions.

Vertex A: $(-2, -4)$
*Obtained by solving
the system*

$$\begin{cases} x - y = & 2 \\ x = & -2 \end{cases}$$

Vertex B: $(5, 3)$
*Obtained by solving
the system*

$$\begin{cases} x - y = 2 \\ y = 3 \end{cases}$$

Vertex C: $(-2, 3)$
*Obtained by solving
the system*

$$\begin{cases} x = -2 \\ y = & 3 \end{cases}$$

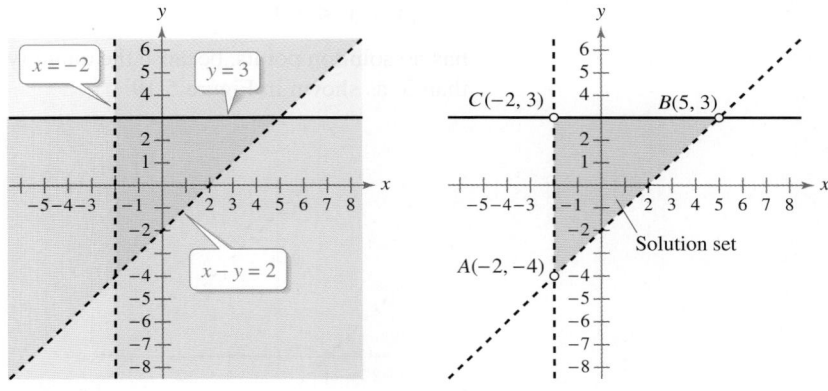

FIGURE 5.16

✓Checkpoint 4

Solve the system of linear inequalities.

$$\begin{cases} 2x - y > -3 \\ x \leq & 2 \\ y > & -2 \end{cases}$$

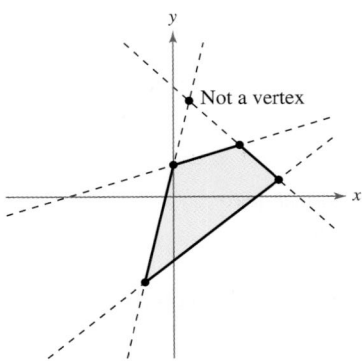

Boundary lines can intersect at a point that is not a vertex.
FIGURE 5.17

For the triangular region shown in Figure 5.16, each point of intersection of a pair of boundary lines corresponds to a vertex. With more complicated regions, two border lines can sometimes intersect at a point that is *not* a vertex of the region, as shown in Figure 5.17. To determine which points of intersection are actually vertices of the region, sketch the region and refer to your sketch as you find each point of intersection.

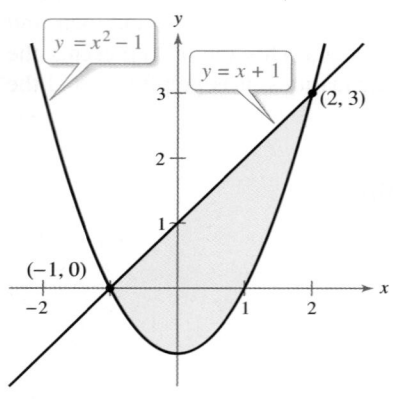

FIGURE 5.18

Example 5   Solving a System of Inequalities

Solve the system of inequalities.

$$\begin{cases} x^2 - y \le 1 & \text{Inequality 1} \\ -x + y \le 1 & \text{Inequality 2} \end{cases}$$

SOLUTION   To solve the system, sketch the graph of the solution set. As shown in Figure 5.18, the points that satisfy the inequality $x^2 - y \le 1$ are the points lying above (or on) the parabola given by $y = x^2 - 1$. The points that satisfy the inequality $-x + y \le 1$ are the points lying below (or on) the line given by $y = x + 1$. To find the points of intersection of the parabola and the line, solve the system of corresponding equations.

$$\begin{cases} x^2 - y = 1 \\ -x + y = 1 \end{cases}$$

Using the method of substitution, you can find the points of intersection to be $(-1, 0)$ and $(2, 3)$. The graph of the solution set of the system is shown in Figure 5.18.

✓ **Checkpoint 5**

Solve the system of inequalities.

$$\begin{cases} x^2 + y < 3 \\ x + y > -3 \end{cases}$$

When solving a system of inequalities, you should be aware that the system might have no solution. For instance, the system

$$\begin{cases} x + y > 3 \\ x + y < -1 \end{cases}$$

has no solution points, because the quantity $(x + y)$ cannot be less than -1 *and* greater than 3, as shown in Figure 5.19.

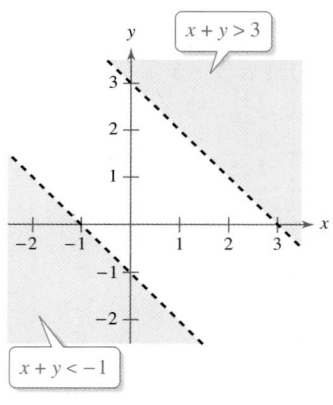

No Solution
FIGURE 5.19

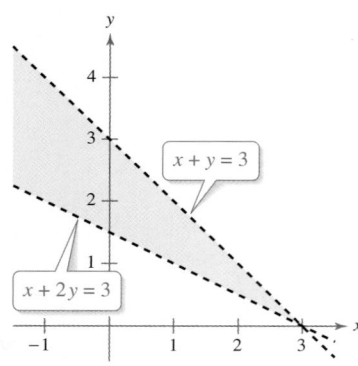

Unbounded Region
FIGURE 5.20

Another possibility is that the solution set of a system of inequalities can be unbounded. For instance, the solution set of

$$\begin{cases} x + y < 3 \\ x + 2y > 3 \end{cases}$$

forms an *infinite wedge*, as shown in Figure 5.20.

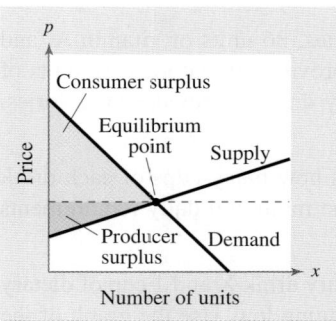

FIGURE 5.21

Applications

Example 8 in Section 5.2 discussed the *point of equilibrium* for a demand function and a supply function. The next example discusses two related concepts that economists call **consumer surplus** and **producer surplus.** As shown in Figure 5.21, the consumer surplus is defined as the area of the region that lies *below* the demand graph, *above* the horizontal line passing through the equilibrium point, and to the right of the *p*-axis. Similarly, the producer surplus is defined as the area of the region that lies *above* the supply graph, *below* the horizontal line passing through the equilibrium point, and to the right of the *p*-axis. The consumer surplus is a measure of the amount that consumers would have been willing to pay *above what they actually paid*, whereas the producer surplus is a measure of the amount that producers would have been willing to receive *below what they actually received.*

 Example 6 **Consumer and Producer Surpluses**

The demand and supply equations for a video game console are given by

$$\begin{cases} p = 350 - 0.00003x & \text{Demand equation} \\ p = 200 + 0.00003x & \text{Supply equation} \end{cases}$$

where *p* is the price (in dollars) and *x* is the number of video game consoles. Find the consumer surplus and producer surplus for these two equations.

SOLUTION In Example 8 in Section 5.2, you saw that the point of equilibrium for these equations is

$(2,500,000, 275)$.

So, the horizontal line passing through this point is $p = 275$. Now you can determine that the consumer surplus and producer surplus are the areas of the triangular regions given by the following systems of inequalities, respectively.

Consumer Surplus

$$\begin{cases} p \le 350 - 0.00003x \\ p \ge 275 \\ x \ge 0 \end{cases}$$

Producer Surplus

$$\begin{cases} p \ge 200 + 0.00003x \\ p \le 275 \\ x \ge 0 \end{cases}$$

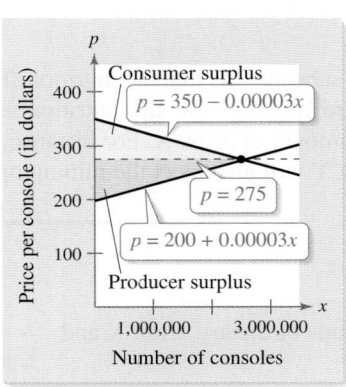

FIGURE 5.22

In Figure 5.22, you can see that the consumer and producer surpluses are defined as the areas of the shaded triangles. The base of the triangle representing the consumer surplus is 2,500,000 because the *x*-value of the point of equilibrium is 2,500,000. To find the height of this triangle, subtract the *p*-value of the point of equilibrium, 275, from the *p*-intercept of the demand equation, 350, to obtain 75. You can find the base and height of the triangle representing the producer surplus in a similar manner.

$$\text{Consumer surplus} = \tfrac{1}{2}(\text{base})(\text{height})$$

$$= \tfrac{1}{2}(2,500,000)(75)$$

$$= \$93,750,000$$

$$\text{Producer surplus} = \tfrac{1}{2}(\text{base})(\text{height})$$

$$= \tfrac{1}{2}(2,500,000)(75)$$

$$= \$93,750,000$$

✓ **Checkpoint 6**

In Example 6, the supply equation is given by $p = 230 + 0.00003x$ and the new point of equilibrium is $(2,000,000, 290)$. Find the consumer surplus and producer surplus.

 **Example 7** **Nutrition**

The liquid portion of a diet requires at least 300 calories, 36 units of vitamin A, and 90 units of vitamin C daily. A cup of dietary drink X provides 60 calories, 12 units of vitamin A, and 10 units of vitamin C. A cup of dietary drink Y provides 60 calories, 6 units of vitamin A, and 30 units of vitamin C.

a. Set up a system of linear inequalities that describes how many cups of each drink should be consumed each day to meet or exceed the minimum daily requirements for calories and vitamins.

b. A nutritionist normally gives a patient 6 cups of dietary drink X and 1 cup of dietary drink Y per day. Supplies of dietary drink X are running low. Use the graph of the system of linear inequalities to determine other combinations of drinks X and Y that can be given that will meet the minimum daily requirements.

SOLUTION

a. Begin by letting x represent the number of cups of drink X and letting y represent the number of cups of drink Y. To meet or exceed the minimum daily requirements, the following inequalities must be satisfied.

$$\begin{cases} 60x + 60y \geq 300 & \text{Calories} \\ 12x + 6y \geq 36 & \text{Vitamin A} \\ 10x + 30y \geq 90 & \text{Vitamin C} \\ x \geq 0 \\ y \geq 0 \end{cases}$$

The last two inequalities are included because x and y cannot be negative. The graph of this system of inequalities is shown in Figure 5.23. (This application is discussed again in Example 5 in Section 5.5.)

b. From Figure 5.23, there are many different possible substitutions that the nutritionist can make. Because supplies of dietary drink X are running low, the nutritionist should choose a combination that contains a small amount of drink X. For instance, 1 cup of dietary drink X and 4 cups of dietary drink Y will also meet the minimum daily requirements.

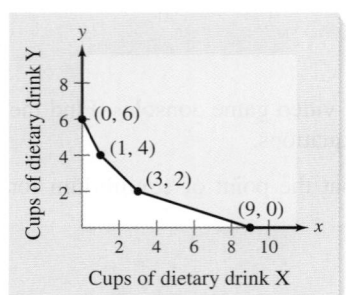

FIGURE 5.23

✓ **Checkpoint 7**

In Example 7, should the nutritionist give a patient 4 cups of dietary drink X and 1 cup of dietary drink Y? Explain.

SUMMARIZE (Section 5.4)

1. Describe how to sketch the graph of an inequality in two variables *(page 445)*. For examples of sketching the graph of an inequality, see Examples 1, 2, and 3.

2. Describe how to solve a system of inequalities *(page 447)*. For examples of solving a system of inequalities, see Examples 4 and 5.

3. State the definitions of consumer surplus and producer surplus *(page 449)*. For an example of finding consumer surplus and producer surplus, see Example 6.

4. Describe a real-life example of how a system of linear inequalities can be used to determine the amounts of dietary drinks that should be consumed each day to meet or exceed the minimum daily requirements *(page 450, Example 7)*.

SKILLS WARM UP 5.4

The following warm-up exercises involve skills that were covered in earlier sections. You will use these skills in the exercise set for this section. For additional help, review Sections 2.1, 2.2, 3.1, 5.1, and 5.2.

In Exercises 1–6, classify the graph of the equation as a line, a parabola, or a circle.

1. $x + y = 3$

2. $y = x^2 - 4$

3. $x^2 + y^2 = 9$

4. $y = -x^2 + 1$

5. $4x - y = 8$

6. $y^2 = 16 - x^2$

In Exercises 7–10, solve the system of equations.

7. $\begin{cases} x + 2y = 3 \\ 4x - 7y = -3 \end{cases}$

8. $\begin{cases} 2x - 3y = 4 \\ x + 5y = 2 \end{cases}$

9. $\begin{cases} x^2 + y = 5 \\ 2x - 4y = 0 \end{cases}$

10. $\begin{cases} x^2 + y^2 = 13 \\ x + y = 5 \end{cases}$

Exercises 5.4

See www.CalcChat.com for worked-out solutions to odd-numbered exercises.

Matching In Exercises 1–6, match the inequality with its graph. [The graphs are labeled (a), (b), (c), (d), (e), and (f).]

(a)
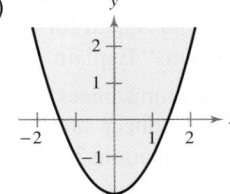

(b)
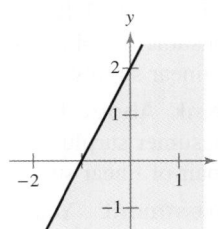

(c)
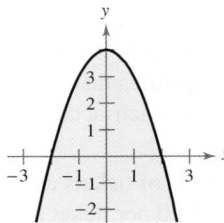

(d)
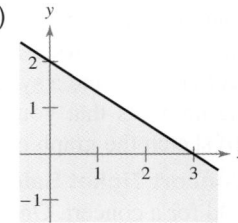

(e)
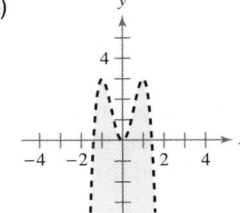

(f)
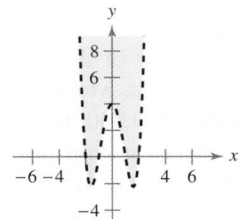

1. $2x + 3y \le 6$

2. $2x - y \ge -2$

3. $x^2 - y \le 2$

4. $y \le 4 - x^2$

5. $y > x^4 - 5x^2 + 4$

6. $3x^4 + y < 6x^2$

Sketching the Graph of an Inequality In Exercises 7–22, sketch the graph of the inequality. *See Examples 1, 2, and 3.*

7. $x \ge 2$

8. $x < 4$

9. $y + 2x^2 > 0$

10. $y^2 - x < 0$

11. $y > -1$

12. $y \le 3$

13. $y < 2 - x$

14. $y > 2x - 4$

15. $2y - x \ge 4$

16. $5x + 3y \ge -15$

17. $y < \ln x$

18. $y \ge -\ln x + 1$

19. $x^2 + y^2 \le 4$

20. $x^2 + y^2 > 4$

21. $y - (x + 2)^3 \le 0$

22. $y - (x - 3)^3 \ge 0$

Solving a System of Inequalities In Exercises 23–44, graph the solution set of the system of inequalities. *See Examples 4 and 5.*

23. $\begin{cases} x + y \le 2 \\ -x + y \le 2 \\ y \ge 0 \end{cases}$

24. $\begin{cases} 3x + 2y < 6 \\ x > 1 \\ y > 0 \end{cases}$

25. $\begin{cases} x + y \le 5 \\ x \ge 2 \\ y \ge 0 \end{cases}$

26. $\begin{cases} 2x + y \ge 2 \\ x \le 2 \\ y \le 1 \end{cases}$

27. $\begin{cases} -3x + 2y < 6 \\ x + 4y < -2 \\ 2x + y < 3 \end{cases}$

28. $\begin{cases} x - 7y > -36 \\ 5x + 2y > 5 \\ 6x - 5y > 6 \end{cases}$

29. $\begin{cases} x^2 + y \le 6 \\ x \ge -1 \\ y \ge 0 \end{cases}$

30. $\begin{cases} 2x^2 + y > 4 \\ x < -1 \\ y > -2 \end{cases}$

31. $\begin{cases} 2x + y > 2 \\ 6x + 3y < 2 \end{cases}$

32. $\begin{cases} 5x - 3y > -6 \\ 5x - 3y < -9 \end{cases}$

33. $\begin{cases} y \ge -3 \\ y \le 1 - x^2 \end{cases}$

34. $\begin{cases} y < -(x - 1)^2 \\ y > (x - 3)^2 - 4 \end{cases}$

35. $\begin{cases} x^2 + y^2 \le 16 \\ x^2 + y^2 < 1 \end{cases}$

36. $\begin{cases} x^2 + y^2 \le 25 \\ x^2 + y^2 \ge 9 \end{cases}$

37. $\begin{cases} x > y^2 \\ x < y + 2 \end{cases}$

38. $\begin{cases} x < 2y - y^2 \\ 0 < x + y \end{cases}$

39. $\begin{cases} y \le \sqrt{3x} + 1 \\ y \ge x + 1 \end{cases}$

40. $\begin{cases} y < \sqrt{2x} + 3 \\ y > x + 3 \end{cases}$

41. $\begin{cases} y < x^3 - 2x + 1 \\ y > -2x \\ x \le 1 \end{cases}$

42. $\begin{cases} x^2 + y \le 4 \\ y \ge 2x \\ x \ge -1 \end{cases}$

43. $\begin{cases} y \le e^x \\ y \ge \ln x \\ x \ge \frac{1}{2} \\ x \le 2 \end{cases}$

44. $\begin{cases} y \le e^{-x^2/2} \\ y \ge 0 \\ x \ge -1 \\ x \le 0 \end{cases}$

Geometry In Exercises 45–50, write a system of inequalities that corresponds to the solution set shown in the graph.

45. Parallelogram

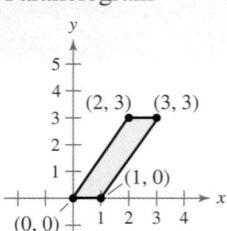

46. Rectangle

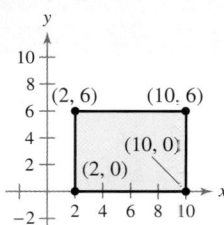

47. Triangle

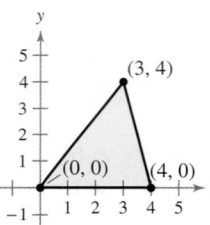

48. Triangle

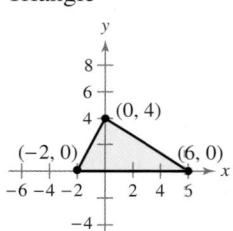

49. Sector of a circle

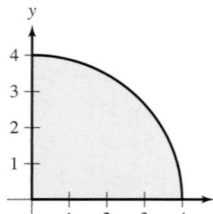

50. Sector of a circle

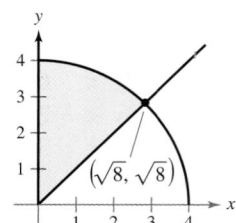

51. Furniture Production A furniture company produces tables and chairs. Each table requires 2 hours in the assembly center and $1\frac{1}{2}$ hours in the finishing center. Each chair requires $1\frac{1}{2}$ hours in the assembly center and $\frac{3}{4}$ hour in the finishing center. The company's assembly center is available 18 hours per day, and its finishing center is available 12 hours per day. Let x and y be the numbers of tables and chairs produced per day, respectively. (a) Find a system of inequalities describing all possible production levels, and (b) sketch the graph of the system.

52. Kayak Inventory A store sells two models of kayaks. Because of the demand, it is necessary to stock at least twice as many units of model A as units of model B. The costs to the store for the two models are $500 and $700, respectively. The management does not want more than $25,000 in kayak inventory at any one time, and it wants at least six model A kayaks and three model B kayaks in inventory at all times. (a) Find a system of inequalities describing all possible inventory levels, and (b) sketch the graph of the system.

Consumer and Producer Surpluses In Exercises 53–56, find the consumer surplus and producer surplus for the demand and supply equations. *See Example 6.*

Demand	Supply
53. $p = 56 - 0.0001x$	$p = 22 + 0.00001x$
54. $p = 60 - 0.00001x$	$p = 15 + 0.00004x$
55. $p = 140 - 0.00002x$	$p = 80 + 0.00001x$
56. $p = 600 - 0.0002x$	$p = 125 + 0.0006x$

57. Think About It Under what circumstances are the consumer surplus and producer surplus equal for a pair of linear supply and demand equations? Explain.

58. Think About It Under what circumstances is the consumer surplus greater than the producer surplus for a pair of linear supply and demand equations? Explain.

59. Investment You plan to invest up to $30,000 in two different interest-bearing accounts. Each account is to contain at least $7000. Moreover, one account should have at least twice the amount that is in the other account. (a) Find a system of inequalities that describes the amounts that you can invest in each account, and (b) sketch the graph of the system.

60. Concert Ticket Sales Two types of tickets are to be sold for a concert. One type costs $30 per ticket and the other type costs $40 per ticket. The promoter of the concert must sell at least 15,000 tickets, including at least 8000 of the $30 tickets and at least 4000 of the $40 tickets. Moreover, the gross receipts must total at least $500,000 in order for the concert to be held. (a) Find a system of inequalities describing the different numbers of tickets that must be sold, and (b) sketch the graph of the system.

61. Diet Supplement A dietitian designs a special diet supplement using two different foods. Each ounce of food X contains 20 units of calcium, 10 units of iron, and 15 units of vitamin B. Each ounce of food Y contains 15 units of calcium, 20 units of iron, and 20 units of vitamin B. The minimum daily requirements for the diet are 400 units of calcium, 250 units of iron, and 220 units of vitamin B. (a) Find a system of inequalities describing the different amounts of food X and food Y that the dietitian can use in the diet, and (b) sketch the graph of the system.

62. Diet Supplement A dietitian designs a special diet supplement using two different foods. Each ounce of food X contains 12 units of calcium, 10 units of iron, and 20 units of vitamin B. Each ounce of food Y contains 15 units of calcium, 20 units of iron, and 12 units of vitamin B. The minimum daily requirements for the diet are 300 units of calcium, 280 units of iron, and 300 units of vitamin B.

(a) Find a system of inequalities describing the different amounts of food X and food Y that the dietitian can use in the diet.

(b) Sketch the graph of the system.

(c) A nutritionist normally gives a patient 10 ounces of food X and 12 ounces of food Y per day. Supplies of food Y are running low. What other combinations of foods X and Y can be given to the patient to meet the minimum daily requirements?

63. Nesting Falcons  The numbers of nesting pairs y of falcons in a nature conservancy from 2006 through 2010 can be approximated by the linear model

$$y = 2.8t + 1.8, \quad 6 \le t \le 10$$

where t represents the year, with $t = 6$ corresponding to 2006.

(a) The *total* number of nesting pairs during this five-year period can be approximated by finding the area of the trapezoid represented by the system below.

$$\begin{cases} y \le 2.8t + 1.8 \\ y \ge 0 \\ t \ge 5.5 \\ t \le 10.5 \end{cases}$$

Graph this region using a graphing utility.

(b) Use the formula for the area of a trapezoid to approximate the total number of nesting pairs.

64. Sales The annual sales y (in millions of dollars) for WebMD Health from 2005 through 2010 can be approximated by the linear model

$$y = 69.50t - 169.5, \quad 5 \le t \le 10$$

where t represents the year, with $t = 5$ corresponding to 2005. *(Source: WebMD Health Corporation)*

(a) The *total* sales during this six-year period can be approximated by finding the area of the trapezoid represented by the system below.

$$\begin{cases} y \le 69.50t - 169.5 \\ y \ge 0 \\ t \ge 4.5 \\ t \le 10.5 \end{cases}$$

Graph this region using a graphing utility.

(b) Use the formula for the area of a trapezoid to approximate the total sales.

65. Writing Explain the difference between the graphs of the inequality $x \le 4$ on the real number line and on the rectangular coordinate system.

66.  **HOW DO YOU SEE IT?** A person's maximum heart rate is

$$220 - x$$

where x is the person's age in years for $20 \le x \le 70$. When a person exercises, it is recommended that the person strive for a heart rate that is at least 50% of the maximum and at most 75% of the maximum. The exercise target heart rate region can be approximated by the system of inequalities below, where y represents a person's heart rate. *(Source: American Heart Association)*

$$\begin{cases} y \ge 0.5(220 - x) \\ y \le 0.75(220 - x) \\ x \ge 20 \\ x \le 70 \end{cases}$$

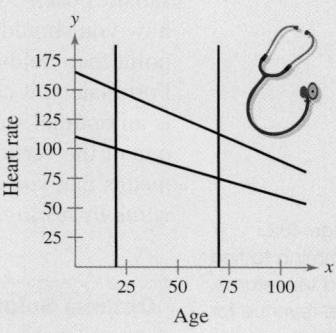

Using the graph of the system of inequalities above, find two solutions of the system and interpret the meanings of the solutions in the context of the problem.

67. Geometry Write a system of inequalities whose graphed solution set is a rectangle.

68. Geometry Write a system of inequalities whose graphed solution set is an isosceles triangle.

69. Graphical Reasoning Two concentric circles have radii x and y, where $y > x$. The area between the circles must be at least 10 square units.

(a) Find a system of inequalities describing the constraints on the circles.

(b) Use a graphing utility to graph the system of inequalities in part (a). Graph the line $y = x$ in the same viewing window.

(c) Identify the graph of the line in relation to the boundary of the inequality. Explain its meaning in the context of the problem.

5.5 Linear Programming

■ Use linear programming to minimize or maximize an objective function.

■ Use linear programming to optimize an application.

Linear Programming: A Graphical Approach

Many applications in business and economics involve a process called **optimization,** in which you are asked to find the minimum cost, the maximum profit, or the minimum use of resources. In this section, you will study an optimization strategy that applies systems of linear inequalities, called **linear programming.**

A two-dimensional linear programming problem consists of a linear **objective function** and a system of linear inequalities called **constraints.** The objective function gives the quantity that is to be maximized (or minimized), and the constraints determine the set of **feasible solutions.** For example, suppose you are asked to maximize the value of

$$z = ax + by \qquad \text{Objective function}$$

subject to a set of constraints that determines the region in Figure 5.24. Because every point in the region satisfies each constraint, it is not clear how you should go about finding the point that yields a maximum value of z. Fortunately, it can be shown that if there is an optimal solution, it must occur at one of the vertices of the region. This means that *you can find the maximum value by testing z at each of the vertices.*

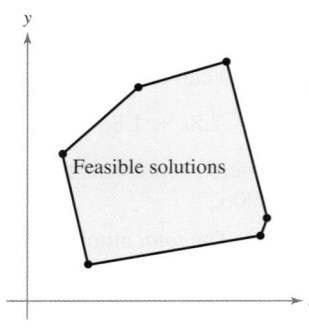

FIGURE 5.24

In Exercise 43 on page 462, you will use linear programming to find the numbers of audits and tax returns that will bring in an optimal revenue for an accounting firm.

Optimal Solution of a Linear Programming Problem

If a linear programming problem has a solution, then it must occur at a vertex of the set of feasible solutions. If the problem has more than one solution, then at least one solution must occur at a vertex of the set of feasible solutions. In either case, the value of the objective function is unique.

A linear programming problem can include hundreds, and sometimes even thousands, of variables. However, in this section, you will solve linear programming problems that involve only two variables. The guidelines for solving a linear programming problem in two variables are listed below.

Guidelines for Solving a Linear Programming Problem

1. Sketch the region corresponding to the system of constraints. (The points inside or on the boundary of the region are *feasible solutions.*)

2. Find the vertices of the region.

3. Test the objective function at each of the vertices and select the values of the variables that optimize the objective function. For a bounded region, both a minimum and a maximum value will exist. (For an unbounded region, *if* an optimal solution exists, then it will occur at a vertex.)

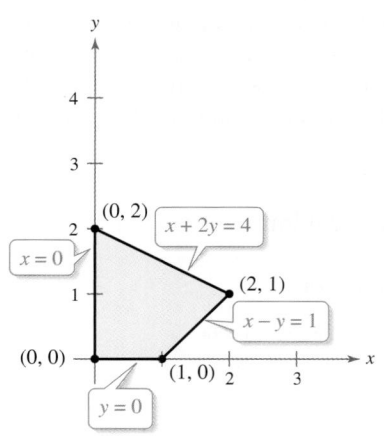

FIGURE 5.25

Example 1 Solving a Linear Programming Problem

Find the maximum value of

$$z = 3x + 2y \qquad \text{Objective function}$$

subject to the following constraints.

$$\left.\begin{array}{r} x \geq 0 \\ y \geq 0 \\ x + 2y \leq 4 \\ x - y \leq 1 \end{array}\right\} \qquad \text{Constraints}$$

SOLUTION The constraints form the region shown in Figure 5.25. At the four vertices of this region, the objective function has the values shown below.

At $(0, 0)$: $z = 3(0) + 2(0) = 0$

At $(1, 0)$: $z = 3(1) + 2(0) = 3$

At $(2, 1)$: $z = 3(2) + 2(1) = 8$ Maximum value of z

At $(0, 2)$: $z = 3(0) + 2(2) = 4$

So, the maximum value of z is 8, which occurs when $x = 2$ and $y = 1$.

✓ Checkpoint 1

Find the maximum value of

$$z = 2x + 3y$$

subject to the following constraints.

$$\left.\begin{array}{r} x \geq 0 \\ y \geq 0 \\ x + y \leq 3 \\ x - y \leq 2 \end{array}\right\}$$

STUDY TIP

Remember that a vertex of a region can be found using a system of linear equations. The system will consist of the equations of the lines passing through the vertex.

In Example 1, try testing some of the *interior* points of the region. You will see that the corresponding values of z are less than 8. Here are some examples.

At $(1, 1)$: $z = 3(1) + 2(1) = 5$

At $\left(1, \frac{1}{2}\right)$: $z = 3(1) + 2\left(\frac{1}{2}\right) = 4$

At $\left(\frac{1}{2}, \frac{3}{2}\right)$: $z = 3\left(\frac{1}{2}\right) + 2\left(\frac{3}{2}\right) = \frac{9}{2}$

At $\left(\frac{3}{2}, 1\right)$: $z = 3\left(\frac{3}{2}\right) + 2(1) = \frac{13}{2}$

To see why the maximum value of the objective function in Example 1 must occur at a vertex, consider writing the objective function in slope-intercept form

$$y = -\frac{3}{2}x + \frac{z}{2} \qquad \text{Family of lines}$$

where

$$\frac{z}{2}$$

is the y-intercept of the objective function. This equation represents a family of lines, each of slope $-\frac{3}{2}$. Of these infinitely many lines, you want the one that has the largest z-value while still intersecting the region determined by the constraints. In other words, of all the lines whose slope is $-\frac{3}{2}$, you want the one that has the largest y-intercept and intersects the given region, as shown in Figure 5.26. It should be clear that such a line will pass through one (or more) of the vertices of the region.

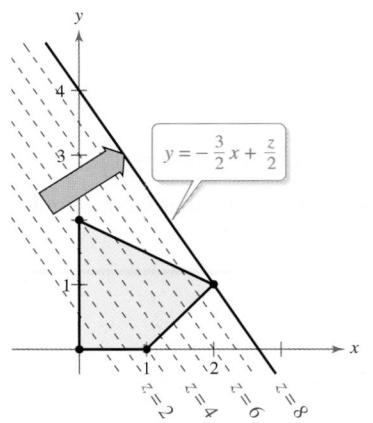

FIGURE 5.26

The guidelines for solving a linear programming problem will work whether the objective function is to be maximized or minimized. The steps used are precisely the same. In other words, once you have evaluated the objective function at the vertices of the set of feasible solutions, you simply choose the largest value as the maximum and the smallest value as the minimum. For instance, the same test used in Example 1 to find the maximum value of z can be used to conclude that the minimum value of z is 0, and that this value occurs at the vertex $(0, 0)$.

Example 2   Solving a Linear Programming Problem

Find (a) the maximum value and (b) the minimum value of

$$z = 4x + 6y \qquad \text{Objective function}$$

subject to the following constraints.

$$\left. \begin{aligned} x &\geq 0 \\ y &\geq 0 \\ -x + y &\leq 11 \\ x + y &\leq 27 \\ 2x + 5y &\leq 90 \end{aligned} \right\} \qquad \text{Constraints}$$

SOLUTION

a. The region bounded by the constraints is shown in Figure 5.27.

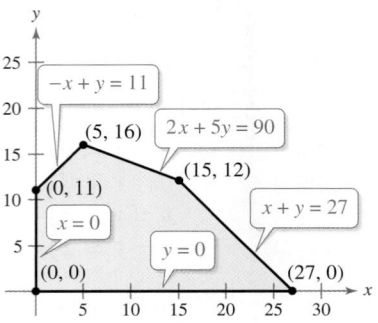

FIGURE 5.27

By testing the objective function at each vertex, you obtain the values shown below.

At $(0, 0)$: $z = 4(0) + 6(0) = 0$

At $(0, 11)$: $z = 4(0) + 6(11) = 66$

At $(5, 16)$: $z = 4(5) + 6(16) = 116$

At $(15, 12)$: $z = 4(15) + 6(12) = 132$ \qquad Maximum value of z

At $(27, 0)$: $z = 4(27) + 6(0) = 108$

So, the maximum value of z is 132, which occurs when $x = 15$ and $y = 12$.

b. Using the values of z at the vertices in part (a), you can conclude that the minimum value of z is 0, and that this value occurs when $x = 0$ and $y = 0$.

✓ **Checkpoint 2**

Find (a) the maximum value and (b) the minimum value of $z = 5x + 2y$ subject to the same constraints as in Example 2. ■

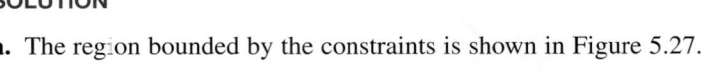

ALGEBRA TUTOR

For help with the algebra used to find the vertices of the region in Example 2, see the *Chapter 5 Algebra Tutor* on page 465.

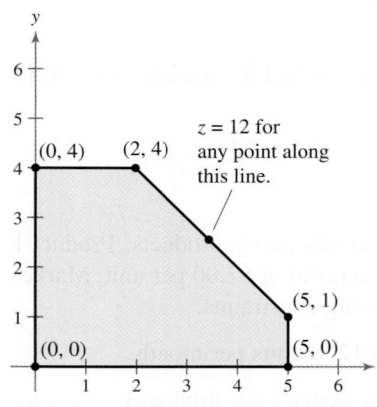

FIGURE 5.28

$z = 12$ for any point along this line.

It is possible for the maximum (or minimum) value in a linear programming problem to occur at *two* different vertices. For instance, at the vertices of the region shown in Figure 5.28, the objective function

$$z = 2x + 2y \qquad \text{Objective function}$$

has the values shown below.

At $(0, 0)$: $z = 2(0) + 2(0) = 0$

At $(0, 4)$: $z = 2(0) + 2(4) = 8$

At $(2, 4)$: $z = 2(2) + 2(4) = 12$ Maximum value of z

At $(5, 1)$: $z = 2(5) + 2(1) = 12$ Maximum value of z

At $(5, 0)$: $z = 2(5) + 2(0) = 10$

In this case, you can conclude that the objective function has a maximum value not only at the vertices $(2, 4)$ and $(5, 1)$; it also has a maximum value (of 12) at *any point on the line segment connecting these two vertices*. Note that the objective function

$$y = -x + \frac{1}{2}z$$

has the same slope as the line through the vertices $(2, 4)$ and $(5, 1)$.

Some linear programming problems have no optimal solution. This can occur when the region determined by the constraints is *unbounded*. This is shown in Example 3.

Example 3   An Unbounded Region

Find the maximum value of

$$z = 4x + 2y \qquad \text{Objective function}$$

where $x \geq 0$ and $y \geq 0$, subject to the following constraints.

$$\left. \begin{array}{r} x + 2y \geq 4 \\ 3x + y \geq 7 \\ -x + 2y \leq 7 \end{array} \right\} \qquad \text{Constraints}$$

SOLUTION The region determined by the constraints is shown in Figure 5.29. For this unbounded region, there is no maximum value of z. To see this, note that the point $(x, 0)$ lies in the region for all values of $x \geq 4$. By choosing x to be large, you can obtain values of

$$z = 4(x) + 2(0)$$

$$= 4x$$

that are as large as you want. So, there is no maximum value of z. For this problem, there *is* a minimum value of z, $z = 10$, which occurs at the vertex $(2, 1)$, as shown below.

At $(1, 4)$: $z = 4(1) + 2(4) = 12$

At $(2, 1)$: $z = 4(2) + 2(1) = 10$ Minimum value of z

At $(4, 0)$: $z = 4(4) + 2(0) = 16$

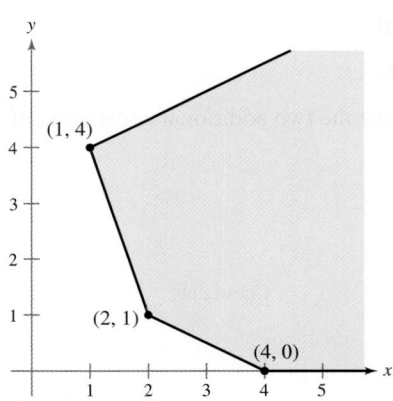

FIGURE 5.29

✓ **Checkpoint 3**

Find the maximum value of the objective function

$$z = x + 8y$$

where $x \geq 0$ and $y \geq 0$, subject to the same constraints as in Example 3.

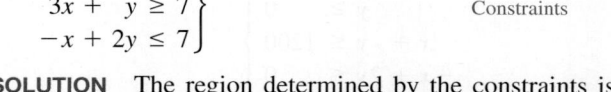

Applications

Example 4 shows how linear programming can be used to find the maximum profit in a business application.

Example 4   Optimal Profit

A manufacturer wants to maximize the profit for two laboratory products. Product I yields a profit of $1.50 per unit, and product II yields a profit of $2.00 per unit. Market tests and available resources have indicated the following constraints.

1. The combined production level should not exceed 1200 units per month.

2. The demand for product II is no more than half the demand for product I.

3. The production level of product I is less than or equal to 600 units plus three times the production level of product II.

What is the optimal production level for each product?

SOLUTION   Let x be the number of units of product I and let y be the number of units of product II. The objective function (for the combined profit) is given by

$$P = 1.5x + 2y. \qquad \text{Objective function}$$

The three constraints translate into the following linear inequalities.

1. $x + y \le 1200$ ⟹ $x + y \le 1200$

2. $y \le \frac{1}{2}x$ ⟹ $-x + 2y \le 0$

3. $x \le 3y + 600$ ⟹ $x - 3y \le 600$

Because neither x nor y can be negative, you also have the two additional constraints of $x \ge 0$ and $y \ge 0$.

So, the constraints are as shown.

$$\left.\begin{array}{rcl} x & \ge & 0 \\ y & \ge & 0 \\ x + y & \le & 1200 \\ -x + 2y & \le & 0 \\ x - 3y & \le & 600 \end{array}\right\} \qquad \text{Constraints}$$

Figure 5.30 shows the region determined by the constraints. To find the maximum profit, test the value of P at each vertex of the region.

At $(0, 0)$: $P = 1.5(0) + 2(0) = 0$

At $(800, 400)$: $P = 1.5(800) + 2(400) = 2000$ Maximum profit

At $(1050, 150)$: $P = 1.5(1050) + 2(150) = 1875$

At $(600, 0)$: $P = 1.5(600) + 2(0) = 900$

So, the maximum profit is $2000, and it occurs when the monthly production levels are 800 units of product I and 400 units of product II.

✓Checkpoint 4

In Example 4, the manufacturer improves the production of product I so that it yields a profit of $2.50 per unit. How would this improvement affect the optimal number of units the manufacturer should sell in order to obtain a maximum profit?   ▪

Example 5 shows how linear programming can be used to find the optimal cost in a real-life application.

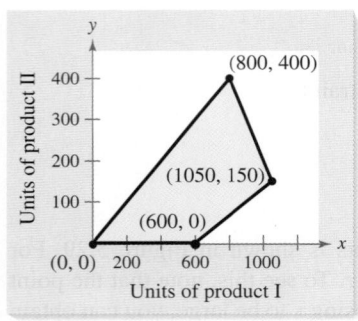

FIGURE 5.30

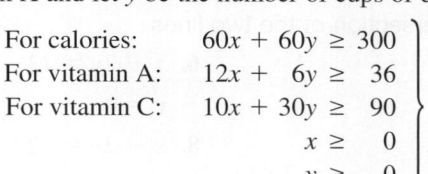

Example 5 **Optimal Cost**

The liquid portion of a diet requires at least 300 calories, 36 units of vitamin A, and 90 units of vitamin C daily. A cup of dietary drink X costs $0.12 and provides 60 calories, 12 units of vitamin A, and 10 units of vitamin C. A cup of dietary drink Y costs $0.15 and provides 60 calories, 6 units of vitamin A, and 30 units of vitamin C. How many cups of each drink should be consumed each day to obtain an optimal cost and still meet the daily requirements?

SOLUTION   As in Example 7 on page 450, let x be the number of cups of dietary drink X and let y be the number of cups of dietary drink Y.

$$\left.\begin{array}{lrl} \text{For calories:} & 60x + 60y & \geq 300 \\ \text{For vitamin A:} & 12x + 6y & \geq 36 \\ \text{For vitamin C:} & 10x + 30y & \geq 90 \\ & x & \geq 0 \\ & y & \geq 0 \end{array}\right\} \quad \text{Constraints}$$

The cost C is given by

$$C = 0.12x + 0.15y. \qquad \text{Objective function}$$

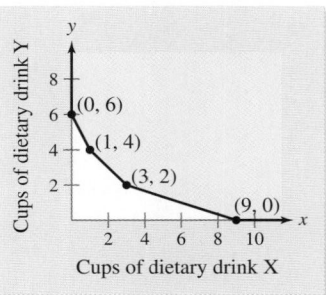

y

Cups of dietary drink Y

8
6 (0, 6)
4 (1, 4)
2 (3, 2)
(9, 0)
2 4 6 8 10 x

Cups of dietary drink X

FIGURE 5.31

The graph of the region corresponding to the constraints is shown in Figure 5.31. Because you want to incur as little cost as possible, you want to determine the *minimum* cost. To determine the minimum cost, test C at each vertex of the region, as shown below.

At $(0, 6)$: $C = 0.12(0) + 0.15(6) = 0.90$

At $(1, 4)$: $C = 0.12(1) + 0.15(4) = 0.72$

At $(3, 2)$: $C = 0.12(3) + 0.15(2) = 0.66$ Minimum value of C

At $(9, 0)$: $C = 0.12(9) + 0.15(0) = 1.08$

So, the minimum cost is $0.66 per day, and this cost occurs when 3 cups of dietary drink X and 2 cups of dietary drink Y are consumed each day. ————

 Checkpoint 5

In Example 5, a cup of dietary drink Y costs $0.11. How would this affect the number of cups of each drink that should be consumed each day to obtain an optimal cost and still meet the daily requirements?   ■

SUMMARIZE   (Section 5.5)

1. Describe the guidelines for solving a linear programming problem *(page 454)*. For an example of solving a linear programming problem, see Example 1.

2. Describe how to find the maximum and minimum values of an objective function of a linear programming problem *(page 456)*. For an example of maximizing and minimizing the objective function of a linear programming problem, see Example 2.

3. Describe a situation in which a linear programming problem has no optimal solution *(page 457)*. For an example of solving a linear programming problem that has no optimal solution, see Example 3.

4. Describe a real-life example of how linear programming can be used to find the maximum profit in a business application *(page 458, Example 4)*.

SKILLS WARM UP 5.5   The following warm-up exercises involve skills that were covered in earlier sections. You will use these skills in the exercise set for this section. For additional help, review Sections 2.2, 5.1, and 5.4.

In Exercises 1–4, sketch the graph of the linear equation.

1. $y + x = 3$

2. $y - x = 12$

3. $x = 0$

4. $y = 4$

In Exercises 5–8, find the point of intersection of the two lines.

5. $x + y = 4$
$\quad x = 0$

6. $x + 2y = 12$
$\quad y = 0$

7. $x + y = 4$
$\quad 2x + 3y = 9$

8. $x + 2y = 12$
$\quad 2x + y = 9$

In Exercises 9 and 10, sketch the graph of the inequality.

9. $2x + 3y \geq 18$

10. $4x + 3y \geq 12$

Exercises 5.5

See www.CalcChat.com for worked-out solutions to odd-numbered exercises.

Solving a Linear Programming Problem In Exercises 1–8, find the minimum and maximum values of the objective function and where they occur, subject to the indicated constraints. (For each exercise, the graph of the region determined by the constraints is provided.) *See Examples 1 and 2.*

1. *Objective function:*

$z = 3x + 4y$

Constraints:

$\quad x \geq 0$
$\quad y \geq 0$
$\quad x + y \leq 5$

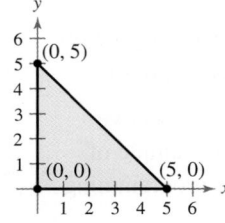

2. *Objective function:*

$z = 2x + 8y$

Constraints:

$\quad x \geq 0$
$\quad y \geq 0$
$\quad 2x + y \leq 4$

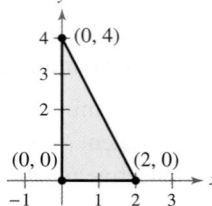

3. *Objective function:*

$z = 9x + 6y$

Constraints:

See Exercise 1.

4. *Objective function:*

$z = 7x + 3y$

Constraints:

See Exercise 2.

5. *Objective function:*

$z = 6x + 2y$

Constraints:

$\quad x \geq 0$
$\quad y \geq 0$
$\quad x + 3y \leq 15$
$\quad 4x + y \leq 16$

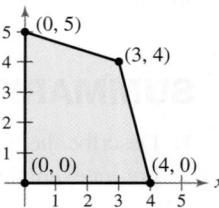

6. *Objective function:*

$z = 5x + 4y$

Constraints:

$\quad x \geq 0$
$\quad y \geq 0$
$\quad 2x + 3y \geq 6$
$\quad 3x - 2y \leq 9$
$\quad x + 5y \leq 20$

7. *Objective function:*

$z = 5x + 0.5y$

Constraints:

See Exercise 5.

8. *Objective function:*

$z = x + 6y$

Constraints:

See Exercise 6.

Solving a Linear Programming Problem In Exercises 9–20, sketch the region determined by the indicated constraints. Then find the minimum and maximum values of the objective function and where they occur, subject to the constraints. *See Examples 1, 2, and 3.*

9. *Objective function:*

$z = 6x + 10y$

Constraints:

$$x \geq 0$$
$$y \geq 0$$
$$3x + 5y \leq 15$$

10. *Objective function:*

$z = 7x + 8y$

Constraints:

$$x \geq 0$$
$$y \geq 0$$
$$x + 2y \leq 8$$

11. *Objective function:*

$z = 8x + 4y$

Constraints:

See Exercise 9.

12. *Objective function:*

$z = 10x + 3y$

Constraints:

See Exercise 10.

13. *Objective function:*

$z = 4x + 5y$

Constraints:

$$x \geq 0$$
$$y \geq 0$$
$$x + y \geq 8$$
$$3x + 5y \geq 30$$

14. *Objective function:*

$z = 4x + 5y$

Constraints:

$$x \geq 0$$
$$y \geq 0$$
$$x + y \leq 5$$
$$x + 2y \leq 6$$

15. *Objective function:*

$z = 2x + 7y$

Constraints:

See Exercise 13.

16. *Objective function:*

$z = 2x - y$

Constraints:

See Exercise 14.

17. *Objective function:*

$z = 3x + 6y$

Constraints:

$$x \geq 0$$
$$y \geq 0$$
$$x + 2y \leq 40$$
$$x + y \leq 30$$
$$2x + 3y \leq 65$$

18. *Objective function:*

$z = x$

Constraints:

$$x \geq 0$$
$$y \geq 0$$
$$2x + 3y \leq 60$$
$$2x + y \leq 28$$
$$4x + y \leq 48$$

19. *Objective function:*

$z = x + y$

Constraints:

See Exercise 17.

20. *Objective function:*

$z = y$

Constraints:

See Exercise 18.

Maximizing the Objective Function In Exercises 21–24, maximize the objective function subject to the constraints $3x + y \leq 15$, $4x + 3y \leq 30$, $x \geq 0$, and $y \geq 0$.

21. $z = 2x + y$

22. $z = 5x + y$

23. $z = 4x + 3y$

24. $z = 3x + y$

Maximizing the Objective Function In Exercises 25–28, maximize the objective function subject to the constraints $x + 5y \leq 25$, $x + y \leq 9$, $4x + 2y \leq 32$, $x \geq 0$, and $y \geq 0$.

25. $z = 6x + 7y$

26. $z = 10x + 8y$

27. $z = 15x + y$

28. $z = 5x + 11y$

Think About It In Exercises 29–36, find an objective function that has a maximum or minimum value at the indicated vertex of the constraint region shown. (There are many correct answers.)

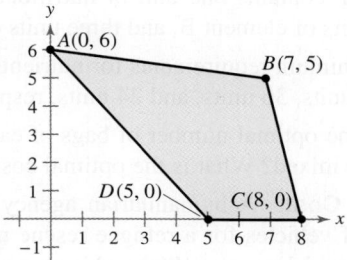

29. The maximum occurs at vertex A.

30. The maximum occurs at vertex B.

31. The minimum occurs at vertex C.

32. The minimum occurs at vertex D.

33. The maximum occurs at vertices A and B.

34. The maximum occurs at vertices B and C.

35. The minimum occurs at vertices A and D.

36. The minimum occurs at vertices C and D.

37. Optimal Profit A fruit grower raises crops A and B. The profit is \$185 per acre for crop A and \$245 per acre for crop B. Research and available resources indicate the following constraints.

- The fruit grower has 150 acres of land for raising the crops.

- It takes 1 day to trim an acre of crop A and 2 days to trim an acre of crop B, and there are 240 days per year available for trimming.

- It takes 0.3 day to pick an acre of crop A and 0.1 day to pick an acre of crop B, and there are 30 days per year available for picking.

What is the optimal acreage for each fruit? What is the optimal profit?

38. Optimal Profit The costs to a store for two models of Global Positioning System (GPS) receivers are \$80 and \$100. The \$80 model yields a profit of \$25 and the \$100 model yields a profit of \$30. Market tests and available resources indicate the following constraints.

- The merchant estimates that the total monthly demand will not exceed 200 units.

- The merchant does not want to invest more than \$18,000 in GPS receiver inventory.

What is the optimal inventory level for each model? What is the optimal profit?

39. Optimal Cost A farming cooperative mixes two brands of cattle feed. Brand X costs $30 per bag, and brand Y costs $25 per bag. Research and available resources have indicated the following constraints.

- Brand X contains two units of nutritional element A, two units of element B, and two units of element C.

- Brand Y contains one unit of nutritional element A, nine units of element B, and three units of element C.

- The minimum requirements for nutrients A, B, and C are 12 units, 36 units, and 24 units, respectively.

What is the optimal number of bags of each brand that should be mixed? What is the optimal cost?

40. Optimal Cost A humanitarian agency can use two models of vehicles for a refugee rescue mission. Each model A vehicle costs $1000 and each model B vehicle costs $1500. Mission strategies and objectives indicate the following constraints.

- A total of at least 20 vehicles must be used.

- A model A vehicle can hold 45 boxes of supplies. A model B vehicle can hold 30 boxes of supplies. The agency must deliver at least 690 boxes of supplies to the refugee camp.

- A model A vehicle can hold 20 refugees. A model B vehicle can hold 32 refugees. The agency must rescue at least 520 refugees.

What is the optimal number of vehicles of each model that should be used? What is the optimal cost?

41. Optimal Profit A manufacturer produces two models of bicycles. The times (in hours) required for assembling, painting, and packaging each model are shown in the table.

Process	Model A	Model B
Assembling	2	2.5
Painting	4	1
Packaging	1	0.75

The total times available for assembling, painting, and packaging are 4000 hours, 4800 hours, and 1500 hours, respectively. The profits per unit are $50 for model A and $75 for model B. What is the optimal production level for each model? What is the optimal profit?

42. Optimal Profit A company makes two models of doghouses. The times (in hours) required for assembling, painting, and packaging are shown in the table.

Process	Model A	Model B
Assembling	2.5	3
Painting	2	1
Packaging	0.75	1.25

The total times available for assembling, painting, and packaging are 4000 hours, 2500 hours, and 1500 hours, respectively. The profits per unit are $60 for model A and $75 for model B. What is the optimal production level for each model? What is the optimal profit?

43. Optimal Revenue An accounting firm charges $2500 for an audit and $350 for a tax return. Research and available resources have indicated the following constraints.

- The firm has 900 hours of staff time available each week.

- The firm has 155 hours of review time available each week.

- Each audit requires 75 hours of staff time and 10 hours of review time.

- Each tax return requires 12.5 hours of staff time and 2.5 hours of review time.

What numbers of audits and tax returns will bring in an optimal revenue?

44. Optimal Revenue The accounting firm in Exercise 43 lowers its charge for an audit to $2000. What numbers of audits and tax returns will bring in an optimal revenue?

45. Media Selection A company has budgeted a maximum of $4,800,000 for national advertising of an allergy medication. Each minute of television time costs $400,000 and each one-page newspaper ad costs $80,000. Each television ad is expected to be viewed by 20 million viewers, and each newspaper ad is expected to be seen by 5 million readers. The company's market research department recommends that at most 80% of the advertising budget be spent on television ads. What is the optimal amount that should be spent on each type of ad? What is the optimal total audience?

46. Optimal Profit A fruit juice company makes two drinks by blending apple and pineapple juices. The percents of apple juice and pineapple juice in each drink are shown in the table.

Mixture	Drink A	Drink B
Apple juice	30%	60%
Pineapple juice	70%	40%

There are 1000 liters of apple juice and 1500 liters of pineapple juice available. The profit for drink A is $0.70 per liter and the profit for drink B is $0.60 per liter. What is the optimal production level for each type of drink? What is the optimal profit?

47. Investments An investor has up to $250,000 to invest in two types of investments. Type A investments pay 7% annually and type B pay 12% annually. To have a well-balanced portfolio, the investor imposes the following conditions. At least one-fourth of the total portfolio is to be allocated to type A investments and at least one-fourth is to be allocated to type B investments. What is the optimal amount that should be invested in each type of investment? What is the optimal return?

48. Investments An investor has up to $450,000 to invest in two types of investments. Type A investments pay 8% annually and type B pay 14% annually. To have a well-balanced portfolio, the investor imposes the following conditions. At least one-half of the total portfolio is to be allocated to type A investments and at least one-fourth is to be allocated to type B investments. What is the optimal amount that should be invested in each type of investment? What is the optimal return?

49. Optimal Profit A company makes two models of a patio furniture set. The times for assembling, finishing, and packaging model A are 3 hours, 2.5 hours, and 0.6 hour, respectively. The times for model B are 2.75 hours, 1 hour, and 1.25 hours. The total times available for assembling, finishing, and packaging are 3000 hours, 2400 hours, and 1200 hours, respectively. The profit per unit for model A is $100 and the profit per unit for model B is $85. What is the optimal production level for each model? What is the optimal profit?

50. Optimal Profit A manufacturer produces two models of elliptical cross-training exercise machines. The times for assembling, finishing, and packaging model A are 3 hours, 3 hours, and 0.8 hour, respectively. The times for model B are 4 hours, 2.5 hours, and 0.4 hour. The total times available for assembling, finishing, and packaging are 6000 hours, 4200 hours, and 950 hours, respectively. The profits per unit are $300 for model A and $375 for model B. What is the optimal production level for each model? What is the optimal profit?

Recognizing Unusual Characteristics In Exercises 51–56, the given linear programming problem has an unusual characteristic. Sketch a graph of the solution region for the problem and describe the unusual characteristic. Find the maximum value of the objective function and where it occurs.

51. *Objective function:*

$z = 2.5x + y$

Constraints:

$x \geq 0$
$y \geq 0$
$3x + 5y \leq 15$
$5x + 2y \leq 10$

52. *Objective function:*

$z = x + y$

Constraints:

$x \geq 0$
$y \geq 0$
$-x + y \leq 1$
$-x + 2y \leq 4$

53. *Objective function:*

$z = -x + 2y$

Constraints:

$x \geq 0$
$y \geq 0$
$x \leq 10$
$x + y \leq 7$

54. *Objective function:*

$z = x + y$

Constraints:

$x \geq 0$
$y \geq 0$
$-x + y \leq 1$
$-3x + y \geq 3$

55. *Objective function:* $z = 3x + 4y$

Constraints:

$x \geq 0$
$y \geq 0$
$x + y \leq 1$
$2x + y \leq 4$

56. *Objective function:* $z = x + 2y$

Constraints:

$x \geq 0$
$y \geq 0$
$x + 2y \leq 4$
$2x + y \leq 4$

57. Reasoning An objective function has a minimum value at the vertex $(20, 0)$. Can you conclude that it also has a minimum value at the point $(0, 0)$?

58. 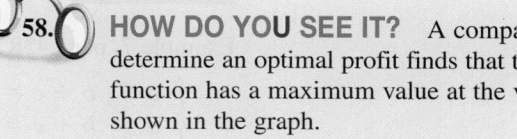 **HOW DO YOU SEE IT?** A company trying to determine an optimal profit finds that the objective function has a maximum value at the vertices shown in the graph.

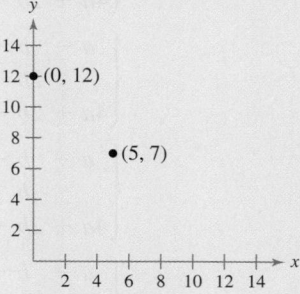

(a) Can you conclude that it also has a maximum value at the point $(3, 9)$? Explain.

(b) Can you conclude that it also has a maximum value at the point $(6, 6)$? Explain.

(c) Find two additional points that maximize the objective function.

59. Reasoning When solving a linear programming problem, you find that the objective function has a maximum value at more than one vertex. Can you assume that there are an infinite number of points that will produce the maximum value? Explain your reasoning.

ALGEBRA TUTOR

Simplifying and Solving Systems of Equations

On this page, you can review the algebra used in solving some of the examples in Chapter 5.

Example 1   Simplifying a System of Equations

Example 7, page 426

$$2000 = (r_1 - r_2)\left(4 + \frac{24}{60}\right) \quad \text{Equation 1} \qquad 2000 = (r_1 + r_2)(4) \quad \text{Equation 2}$$

$$2000 = (r_1 - r_2)\left(\frac{264}{60}\right) \quad \text{Add.} \qquad 500 = r_1 + r_2 \quad \begin{array}{l}\text{Divide each} \\ \text{side by 4.}\end{array}$$

$$\frac{5000}{11} = (r_1 - r_2) \qquad \begin{array}{l}\text{Multiply each} \\ \text{side by } \frac{60}{264}.\end{array}$$

$$5000 = 11(r_1 - r_2) \qquad \begin{array}{l}\text{Multiply each} \\ \text{side by 11.}\end{array}$$

$$5000 = 11r_1 - 11r_2 \qquad \text{Distributive Property}$$

Example 2   Solving a System of Equations

Example 7, page 438

$$\begin{cases} a - b + c = 3 \\ a + b + c = 1 \\ 4a + 2b + c = 6 \end{cases} \quad \begin{array}{l}\text{Equation 1} \\ \text{Equation 2} \\ \text{Equation 3}\end{array}$$

$$\begin{cases} a - b + c = 3 \\ 2b = -2 \\ 4a + 2b + c = 6 \end{cases}$$ Adding -1 times the first equation to the second equation produces a new second equation.

$$\begin{cases} a - b + c = 3 \\ b = -1 \\ 4a + 2b + c = 6 \end{cases}$$ Multiplying the second equation by $\frac{1}{2}$ produces a new second equation.

$$\begin{cases} a - b + c = 3 \\ b = -1 \\ 6b - 3c = -6 \end{cases}$$ Adding -4 times the first equation to the third equation produces a new third equation.

$$\begin{cases} a - b + c = 3 \\ b = -1 \\ -3c = 0 \end{cases}$$ Adding -6 times the second equation to the third equation produces a new third equation.

$$\begin{cases} a - b + c = 3 \\ b = -1 \\ c = 0 \end{cases}$$ Multiplying the third equation by $-\frac{1}{3}$ produces a new third equation.

To solve for a, substitute $b = -1$ and $c = 0$ into Equation 1 to obtain

$$a - (-1) + 0 = 3 \quad \Longrightarrow \quad a = 2.$$

So, the solution is $a = 2$, $b = -1$, and $c = 0$.

Finding Vertices of a Region

On this page, you can review the algebra used in finding the vertices of the region bounded by the constraints in Example 2 on page 456.

Example 3   Finding Vertices of a Region

Find the vertices of the region bounded by the constraints shown in the figure.

SOLUTION

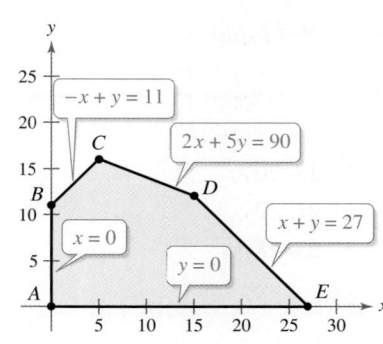

First, you know that A is at $(0, 0)$ because $x = 0$ and $y = 0$ intersect at the origin. To find vertex B, solve the system

$$\begin{cases} -x + y = 11 & \text{Equation 1} \\ x = 0 & \text{Equation 2} \end{cases}$$

by substituting $x = 0$ into Equation 1 and solving for y.

$$0 + y = 11 \quad \Longrightarrow \quad y = 11$$

So, B is at $(0, 11)$. To find vertex C, solve the system

$$\begin{cases} -x + y = 11 & \text{Equation 1} \\ 2x + 5y = 90 & \text{Equation 2} \end{cases}$$

$$\begin{cases} -x + y = 11 \\ 7y = 112 \end{cases}$$

Adding 2 times the first equation to the second equation produces a new second equation.

So, $y = 16$. By back-substituting $y = 16$ into Equation 1, you obtain

$$-x + 16 = 11 \quad \Longrightarrow \quad x = 5.$$

So, C is at $(5, 16)$. To find vertex D, solve the system

$$\begin{cases} 2x + 5y = 90 & \text{Equation 1} \\ x + y = 27 & \text{Equation 2} \end{cases}$$

First, solve for y in Equation 2 to get $y = 27 - x$. Then substitute the expression $27 - x$ for y into Equation 1 and solve for x.

$$2x + 5(27 - x) = 90 \qquad \text{Substitute } 27 - x \text{ for } y.$$

$$2x + 135 - 5x = 90 \qquad \text{Distributive Property}$$

$$-3x = -45 \qquad \text{Combine like terms.}$$

$$x = 15 \qquad \text{Divide each side by } -3.$$

By back-substituting $x = 15$ into Equation 2, you obtain

$$15 + y = 27 \quad \Longrightarrow \quad y = 12.$$

So, D is at $(15, 12)$. To find vertex E, solve the system

$$\begin{cases} x + y = 27 & \text{Equation 1} \\ y = 0 & \text{Equation 2} \end{cases}$$

by substituting $y = 0$ into Equation 1 and solving for x. You obtain

$$x + 0 = 27 \quad \Longrightarrow \quad x = 27.$$

So, E is at $(27, 0)$.

SUMMARY AND STUDY STRATEGIES

After studying this chapter, you should have acquired the following skills.
The exercise numbers are keyed to the Review Exercises that begin on page 467.
Answers to odd-numbered Review Exercises are given in the back of the text.*

Section 5.1

	Review Exercises
■ Solve a system of equations by the method of substitution.	*1–6*
■ Solve a system of equations graphically.	*7, 8, 12*
■ Construct and use a system of equations to solve an application problem.	*9–12*

Section 5.2

■ Solve a linear system by the method of elimination.	*13–20*
■ Interpret the solution of a linear system graphically.	*21, 22*
■ Construct and use a linear system to solve an application problem.	*23–28*

Section 5.3

■ Solve a linear system in row-echelon form using back-substitution.	*29, 30*
■ Use Gaussian elimination to solve a linear system.	*31–36*
■ Solve a nonsquare linear system.	*33, 34*
■ Find the equation of a parabola or a circle using a linear system in three or more variables.	*37–40*
■ Construct and use a linear system in three or more variables to solve an application problem.	*41–44*

Section 5.4

■ Sketch the graph of an inequality in two variables.	*45–50*
■ Solve a system of inequalities.	*51–58*
■ Construct and use a system of inequalities to solve an application problem.	*59–64*

Section 5.5

■ Use linear programming to minimize or maximize an objective function.	*65–72*
■ Use linear programming to optimize an application.	*73–77*

Study Strategies

■ **Units of Variables in Applied Problems**   When using systems of equations to solve real-life applications, be sure to keep track of the unit(s) assigned to each variable. This will allow you to write correctly each equation of the system based on the constraints given in the application.

---

* A wide range of valuable study aids are available to help you master the material in this chapter.
The *Student Solutions Manual* includes step-by-step solutions to all odd-numbered exercises to
help you review and prepare. The student website at *www.cengagebrain.com* offers algebra help
and a *Graphing Technology Guide*, which contains step-by-step commands and instructions for
a wide variety of graphing calculators.

Review Exercises

See www.CalcChat.com for worked-out solutions to odd-numbered exercises.

Solving a System by Substitution In Exercises 1–6, solve the system by the method of substitution.

1. $\begin{cases} x + 3y = 10 \\ 4x - 5y = -28 \end{cases}$

2. $\begin{cases} 3x - y - 13 = 0 \\ 4x + 3y - 26 = 0 \end{cases}$

3. $\begin{cases} \frac{1}{2}x + \frac{3}{5}y = -2 \\ 2x + y = 6 \end{cases}$

4. $\begin{cases} 1.3x + 0.9y = 7.5 \\ 0.4x - 0.5y = -0.8 \end{cases}$

5. $\begin{cases} x^2 + y^2 = 100 \\ x + 2y = 20 \end{cases}$

6. $\begin{cases} y = x^3 - 2x^2 - 2x - 3 \\ y = -x^2 + 4x - 3 \end{cases}$

 Solving a System of Equations Graphically In Exercises 7 and 8, use a graphing utility to find the point(s) of intersection of the graphs.

7. $\begin{cases} y = x^2 - 3x + 11 \\ y = -x^2 + 2x + 8 \end{cases}$

8. $\begin{cases} y = \sqrt{9 - x^2} \\ y = e^x + 1 \end{cases}$

9. **Break-Even Analysis** You invest $5000 in a greenhouse. The planter, potting soil, and seed for each plant costs $6.43, and the selling price is $12.68. How many of the potted plants do you need to sell to break even?

10. **Break-Even Analysis** You invest $50,000 to open a temporary employment agency. Each hour of labor that you contract costs you $13.72 and brings in revenue of $22.26. How many hours of labor must you contract to break even?

11. **Choice of Newscasts** Television stations A and B are competing for the 6 P.M. newscast audience. Station A is implementing a new newscast format for the 6 P.M. audience. Models that represent the numbers of 6 P.M. viewers each month for the two stations are given by

$\begin{cases} y = 950x + 10,000 & \text{Station A (new format)} \\ y = -875x + 18,000 & \text{Station B} \end{cases}$

where y represents the number of viewers and x represents the month, with $x = 1$ corresponding to the first month of station A's new format. Use the models to estimate when the number of viewers for station A's 6 P.M. newscast will exceed the number of viewers for station B's 6 P.M. newscast.

12. **Comparing Populations**  From 2000 through 2009, the population of Kentucky grew more slowly than that of South Carolina. Models that represent the populations of the two states are given by

$\begin{cases} P = 30.8t + 4035 & \text{Kentucky} \\ P = 61.6t + 3985 & \text{South Carolina} \end{cases}$

where P represents the population (in thousands) and t represents the year, with $t = 0$ corresponding to 2000. Use a graphing utility to estimate the year when the population of South Carolina first exceeded that of Kentucky. *(Source: U.S. Census Bureau)*

Solving a System by Elimination In Exercises 13–20, solve the system by the method of elimination.

13. $\begin{cases} 2x - 3y = 21 \\ 3x + y = 4 \end{cases}$

14. $\begin{cases} 3u + 5v = 9 \\ 12u + 10v = 22 \end{cases}$

15. $\begin{cases} 4x - 3y = 10 \\ 8x - 6y = 20 \end{cases}$

16. $\begin{cases} 3x + 4y = 18 \\ 6x + 8y = 18 \end{cases}$

17. $\begin{cases} 1.5x + 2.5y = 8.5 \\ 6x + 10y = 24 \end{cases}$

18. $\begin{cases} 1.25x - 2y = 3.5 \\ 5x - 8y = 14 \end{cases}$

19. $\begin{cases} \dfrac{x - 2}{3} + \dfrac{y + 3}{4} = 5 \\ 2x - y = 7 \end{cases}$

20. $\begin{cases} \dfrac{3}{5}x + \dfrac{2}{7}y = 10 \\ x + 2y = 38 \end{cases}$

Interpreting the Solution Graphically In Exercises 21 and 22, describe the graph of the solution of the linear system.

21. $\begin{cases} 2x + y = -1 \\ 3x - 2y = -5 \end{cases}$

22. $\begin{cases} x - 2y = -1 \\ -2x + 4y = 2 \end{cases}$

23. **Acid Mixture** Twelve gallons of a 25% acid solution is obtained by mixing a 10% solution with a 50% solution.

   (a) Write a system of equations that represents the problem and use a graphing utility to graph the equations in the same viewing window.

   (b) How much of each solution is required to obtain the specified concentration of the final mixture?

24. Acid Mixture Twenty gallons of a 30% acid solution is obtained by mixing a 12% solution with a 60% solution.

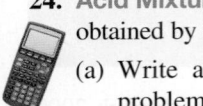

(a) Write a system of equations that represents the problem and use a graphing utility to graph the equations in the same viewing window.

(b) How much of each solution is required to obtain the specified concentration of the final mixture?

25. Fitting a Line to Data Find the least squares regression line $y = ax + b$ for the points shown in the graph below by solving the following system of linear equations for a and b.

$$\begin{cases} 5b + 10a = 17.8 \\ 10b + 30a = 45.7 \end{cases}$$

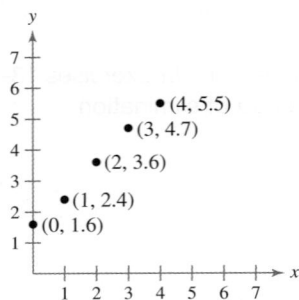

26. Ethanol Production The table shows the amount of corn y (in billions of bushels) used in the production of ethanol each year from 2005 through 2009. *(Source: U.S. Department of Agriculture)*

Year	2005	2006	2007	2008	2009
t	0	1	2	3	4
Amount, y	1.6	2.1	3.0	3.7	4.3

(a) Solve the system below for a and b to find the least squares regression line $y = at + b$ for the data. Let t represent the year, with $t = 0$ corresponding to 2005.

$$\begin{cases} 5b + 10a = 14.7 \\ 10b + 30a = 36.4 \end{cases}$$

(b) Use a graphing utility to graph the regression line and predict the amount of corn used to produce ethanol in 2010.

(c) Use the *regression* feature of the graphing utility to find a linear model for the data. Compare this model with the one you found in part (a).

Finding the Point of Equilibrium In Exercises 27 and 28, find the point of equilibrium for the demand and supply equations.

Demand   *Supply*

27. $p = 37 - 0.0002x$ $p = 22 + 0.00001x$

Demand   *Supply*

28. $p = 120 - 0.0001x$ $p = 45 + 0.0002x$

Using Back-Substitution In Exercises 29 and 30, use back-substitution to solve the system of linear equations.

29. $\begin{cases} x - 3y + 2z = 1 \\ y - 2z = 1 \\ z = 2 \end{cases}$

30. $\begin{cases} x + y - 4z = 6 \\ y + 2z = 2 \\ z = -4 \end{cases}$

Solving a System of Linear Equations In Exercises 31–36, solve the system of linear equations.

31. $\begin{cases} 2x + y + z = 6 \\ x - 4y - z = 3 \\ x + y + z = 4 \end{cases}$

32. $\begin{cases} x + 3y - z = 13 \\ 2x - 5z = 23 \\ 4x - y - 2z = 4 \end{cases}$

33. $\begin{cases} x + y + z = 10 \\ -2x + 3y + 4z = 22 \end{cases}$

34. $\begin{cases} 5x - 12y + 7z = 16 \\ 3x - 7y + 4z = 9 \end{cases}$

35. $\begin{cases} 2x + 6y - z = 1 \\ x - 3y + z = 2 \\ \frac{3}{2}x + \frac{3}{2}y = 6 \end{cases}$

36. $\begin{cases} x + y + z + w = 8 \\ 4y + 5z - 2w = 3 \\ 2x + 3y - z = -2 \\ 3x + 2y - 4w = -20 \end{cases}$

Data Analysis: Curve Fitting In Exercises 37 and 38, find the quadratic equation $y = ax^2 + bx + c$ whose graph passes through the points.

37.

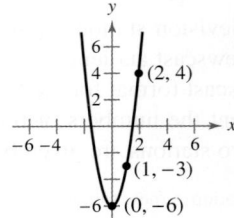

38.

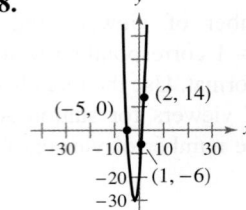

Data Analysis: Curve Fitting   In Exercises 39 and 40, find the equation $x^2 + y^2 + Dx + Ey + F = 0$ of the circle that passes through the points.

39.

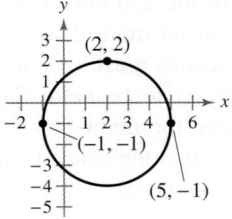

40.

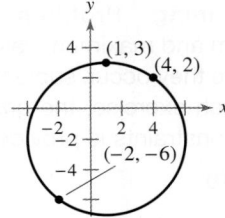

41. Investment Portfolio   You are creating an investment portfolio by investing a total of $500,000 in certificates of deposit, municipal bonds, blue-chip stocks, and growth stocks. The certificates of deposit pay 5% simple annual interest, and the municipal bonds pay 8% simple annual interest. Over a five-year period, you expect the blue-chip stocks to return 10% simple annual interest and the growth stocks to return 15% simple annual interest. You want to have two-fifths of your portfolio in stocks, and twice as much in certificates of deposit as in municipal bonds. How much should you allocate to each type of investment to have a combined return of 9.45%?

42. Investment   You receive $8580 a year in simple annual interest from three investments. The interest rates for the three investments are 6%, 8%, and 10%. The value of the 10% investment is two times that of the 6% investment, and the 8% investment is $1000 more than the 6% investment. What is the amount of each investment?

43. Fitting a Parabola to Data   Find the least squares regression parabola $y = ax^2 + bx + c$ for the points shown in the graph below by solving the following system of linear equations for a, b, and c.

$$\begin{cases} 5c \quad\;\; + 10a = \;\; 9.1 \\ \quad\;\; 10b \quad\quad\;\; = \;\; 8.0 \\ 10c \quad + 34a = 19.8 \end{cases}$$

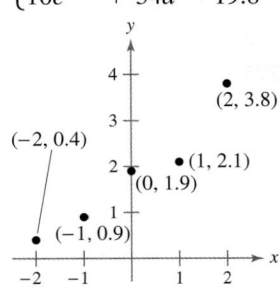

44. Sailboat Sales   The table shows the number y (in thousands) of new sailboats sold in the United States each year from 2005 through 2009. In the table, t represents the year, with $t = -1$ corresponding to 2005. *(Source: The Sailing Company's Annual Sailing Business Review)*

Year, t	Number, y
-1	14.4
0	12.9
1	11.8
2	9.3
3	5.4

(a) Use a graphing utility to create a scatter plot of the data.

(b) Solve the following system for a and b to find the least squares regression line $y = at + b$ for the data.

$$\begin{cases} 5b + \;\; 5a = 53.8 \\ 5b + 15a = 32.2 \end{cases}$$

(c) Solve the following system for a, b, and c to find the least squares regression parabola $y = at^2 + bt + c$ for the data.

$$\begin{cases} 5c + \;\; 5b + 15a = 53.8 \\ 5c + 15b + 35a = 32.2 \\ 15c + 35b + 99a = 112 \end{cases}$$

(d) Use the *regression* feature of the graphing utility to find linear and quadratic models for the data. Compare the models with the least squares regression models found in parts (b) and (c).

(e) Use the graphing utility to graph the linear and quadratic models with the scatter plot of the data. Use the models to predict the numbers of sailboats sold in 2010 and 2011. Discuss your results in the context of the situation.

Sketching the Graph of an Inequality   In Exercises 45–50, sketch the graph of the inequality.

45. $x \geq -5$ **46.** $y < -2$

47. $y \leq 5 + \frac{1}{2}x$ **48.** $3y - x \geq 7$

49. $y - 4x^2 > -1$ **50.** $y \leq \dfrac{3}{x^2 + 2}$

Solving a System of Inequalities   In Exercises 51–58, graph the solution set of the system of inequalities.

51. $\begin{cases} x + 4y < 8 \\ \quad\;\; x > 0 \\ \quad\;\; y > 0 \end{cases}$ **52.** $\begin{cases} 2x - y > \;\; 6 \\ \quad\;\; x < \;\; 6 \\ \quad\;\; y \geq -2 \end{cases}$

53. $\begin{cases} 3x - y > -4 \\ 2x + y > -1 \\ 7x + y < 4 \end{cases}$

54. $\begin{cases} x + y > 4 \\ 3x + y > 10 \\ x - y \geq 0 \end{cases}$

55. $\begin{cases} x^2 + y^2 \leq 9 \\ x^2 - x - 2 \leq y \end{cases}$

56. $\begin{cases} x^2 + y^2 \leq 4 \\ -2x^2 + 2 < y \end{cases}$

57. $\begin{cases} \ln x < y \\ y > -1 \\ x < 4 \end{cases}$

58. $\begin{cases} \ln x \geq y \\ -x + y < -2 \\ x > 2 \end{cases}$

Geometry In Exercises 59 and 60, write a system of inequalities that corresponds to the solution set shown in the graph.

59. Parallelogram

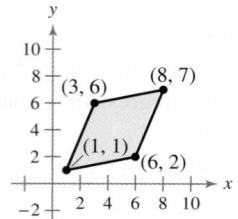

60. Triangle

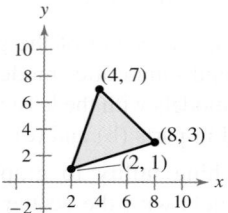

Consumer and Producer Surpluses In Exercises 61 and 62, find the consumer surplus and producer surplus for the demand and supply equations.

Demand	Supply
61. $p = 160 - 0.0001x$	$p = 70 + 0.0002x$
62. $p = 130 - 0.0002x$	$p = 30 + 0.0003x$

63. Movie Player Inventory A store sells two models of Blu-ray Disc™ players (BDPs). Because of the demand, it is necessary to stock at least twice as many units of model Y as units of model Z. The costs to the store for the two models are $100 and $150, respectively. The management does not want more than $4000 in BDP inventory at any one time, and it wants at least four model Y BDPs and two model Z BDPs in inventory at all times. Find a system of inequalities that describes all possible inventory levels. Sketch the graph of the system.

64. Concert Ticket Sales Two types of tickets are to be sold for a concert. One type costs $30 per ticket and the other type costs $50 per ticket. The promoter of the concert can sell up to 10,000 of the $50 tickets and up to 18,000 of the $30 tickets, but no more than 24,000 tickets in all. Also, the gross receipts must total at least $550,000 in order for the concert to be held. Find a system of inequalities describing the numbers of each type of ticket that can be sold so that the concert can be held. Sketch the graph of the system.

Solving a Linear Programming Problem In Exercises 65–68, find the minimum and maximum values of the objective function and where they occur, subject to the indicated constraints. (For each exercise, the graph of the region determined by the constraints is provided.)

65. Objective function: $z = 5x + 6y$

Constraints:
$$x \geq 0$$
$$y \geq 0$$
$$x + y \leq 8$$

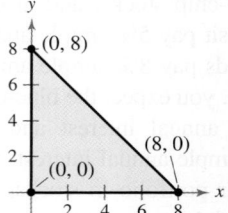

66. Objective function: $z = 15x + 12y$

Constraints:
$$x \geq 0$$
$$y \geq 0$$
$$x + 3y \leq 12$$
$$3x + 2y \leq 15$$

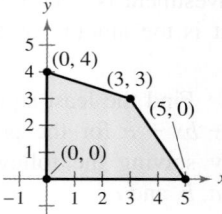

67. Objective function: $z = 8x + 10y$

Constraints:
$$0 \leq x \leq 50$$
$$0 \leq y \leq 35$$
$$4x + 5y \leq 275$$

68. *Objective function:* $z = 50x + 60y$

    Constraints:

$$x \geq 0$$
$$y \geq 0$$
$$3x + 4y \geq 1200$$
$$5x + 6y \leq 3000$$

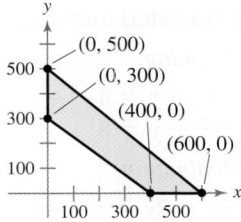

Solving a Linear Programming Problem In Exercises 69–72, sketch the region determined by the indicated constraints. Then find the minimum and maximum values of the objective function and where they occur, subject to the constraints.

69. *Objective function:* $z = 6x + 8y$

    Constraints:

$$x \geq 0$$
$$y \geq 0$$
$$x + 4y \leq 16$$
$$3x + 2y \leq 18$$

70. *Objective function:* $z = 5x + 8y$

    Constraints:

$$0 \leq x \leq 5$$
$$y \geq 0$$
$$x + 2y \leq 12$$
$$2x + 3y \leq 19$$

71. *Objective function:* $z = 8x + 3y$

    Constraints:

$$0 \leq x \leq 5$$
$$0 \leq y \leq 7$$
$$x + y \leq 9$$
$$3x + y \leq 17$$

72. *Objective function:* $z = 10x + 11y$

    Constraints:

$$x \geq 0$$
$$y \geq 0$$
$$2x + 5y \leq 30$$
$$x + y \geq 3$$
$$2x + y \leq 14$$

73. Optimal Profit A company makes two models of desks. The times (in hours) required for assembling, finishing, and packaging each model are shown in the table.

Process	Model A	Model B
Assembling	3.5	8
Finishing	2.5	2
Packaging	1.3	0.7

The total times available each month for assembling, finishing, and packaging are 5600 hours, 2000 hours, and 910 hours, respectively. The profits per unit are $100 for model A and $150 for model B. What is the optimal monthly production level for each model? What is the optimal monthly profit?

74. Optimal Profit A factory manufactures two television set models: an LCD model that yields $100 profit and a plasma model that yields a profit of $180. The times (in hours) required for assembling, testing, and packaging each model are shown in the table.

Process	LCD model	Plasma model
Assembling	2	5
Testing	1	2
Packaging	1	1

The total times available each month for assembling, testing, and packaging are 3000 hours, 1300 hours, and 1000 hours, respectively. What is the optimal monthly production level for each model? What is the optimal monthly profit?

75. Optimal Profit The costs to a merchant for two models of digital camcorders are $525 and $675. The $525 model yields a profit of $75 and the $675 model yields a profit of $125. The merchant estimates that the total monthly demand will not exceed 350 units. There should be no more than $206,250 in digital camcorder inventory. Find the number of units of each model that should be stocked in order to optimize profit. What is the optimal profit?

76. Optimal Profit The costs to a merchant for two models of home theater systems are $270 and $455. The $270 model yields a profit of $30 and the $455 model yields a profit of $45. The merchant estimates that the total monthly demand will not exceed 100 units. There should be no more than $36,250 in home theater system inventory. Find the number of units of each model that should be stocked in order to optimize profit. What is the optimal profit?

77. Optimal Revenue An accounting firm has 800 hours of staff time and 90 hours of review time available each week. The firm charges $2500 for an audit and $200 for a tax return. Each audit requires 100 hours of staff time and 10 hours of review time. Each tax return requires 10 hours of staff time and 2 hours of review time. What numbers of audits and tax returns will bring in an optimal revenue? What is the optimal revenue?

TEST YOURSELF   See www.CalcChat.com for worked-out solutions to odd-numbered exercises.

Take this test as you would take a test in class. When you are done, check your work against the answers given in the back of the book.

In Exercises 1–6, solve the system of equations using the indicated method.

1. *Substitution*

$$\begin{cases} 5x - 7y = -18 \\ 4x + 3y = 20 \end{cases}$$

2. *Substitution*

$$\begin{cases} x + y = 3 \\ x^2 + y = 9 \end{cases}$$

3. *Graphing*

$$\begin{cases} 2x + y = 4 \\ 2x^2 + y = 8 \end{cases}$$

4. *Graphing*

$$\begin{cases} 1.5x - 2.25y = 8 \\ 2.5x + 2y = 5.75 \end{cases}$$

5. *Elimination*

$$\begin{cases} 2x - 4y + z = 11 \\ x + 2y + 3z = 9 \\ 3y + 5z = 12 \end{cases}$$

6. *Elimination*

$$\begin{cases} 3x - 2y + z = 16 \\ 5x - z = 6 \\ 2x - y - z = 3 \end{cases}$$

7. A total of $80,000 is invested in two funds paying 2.3% and 3.1% simple interest. The total annual interest is $2000. How much is invested in each fund?

8. Find the point of equilibrium for a system that has a demand equation of $p = 49 - 0.0003x$ and a supply equation of $p = 33 + 0.00002x$.

9. The numbers y of adults (in millions) who participated in baking as a leisure activity in the years 2005 through 2009 are shown in the table at the left. In the table, x represents the year, with $x = 0$ corresponding to 2007. *(Source: Mediamark Research & Intelligence)*

(a) Find the least squares regression parabola $y = ax^2 + bx + c$ for the data by solving the system below.

$$\begin{cases} 5c + 10a = 239.6 \\ 10b = 49.5 \\ 10c + 34a = 487.5 \end{cases}$$

(b) Use the model found in part (a) to predict the number of adults who participated in baking as a leisure activity in 2010.

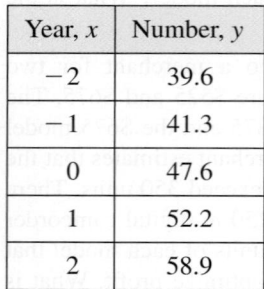

Year, x	Number, y
-2	39.6
-1	41.3
0	47.6
1	52.2
2	58.9

Table for 9

In Exercises 10–13, sketch the graph of the inequality.

10. $x \geq 0$

11. $y \geq 0$

12. $x + 3y \leq 12$

13. $3x + 2y \leq 15$

14. Sketch the solution set of the system of inequalities composed of the inequalities in Exercises 10–13.

15. Find the minimum and maximum values of the objective function $z = 6x + 7y$, subject to the constraints given in Exercises 10–13.

16. A manufacturer produces two models of exercise equipment. The times required for assembling, painting, and packaging each model are as follows.

• Assembling: 3.5 hours for model A; 8 hours for model B

• Painting: 2.5 hours for model A; 2 hours for model B

• Packaging: 1.3 hours for model A; 0.9 hour for model B

The total times available for assembling, painting, and packaging are 5600 hours, 2000 hours, and 900 hours, respectively. The profits per unit are $200 for model A and $275 for model B. What is the optimal production level for each model? What is the optimal profit? Explain your reasoning.

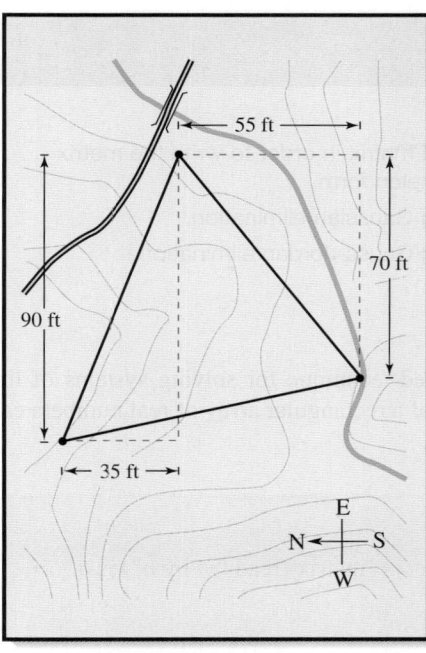

Exercise 16 on page 528 uses a determinant to find the number of square feet in a triangular tract of land.

6 Matrices and Determinants

Kurhan/Shutterstock.com
iofoto/www.shutterstock.com

473

6.1 Matrices and Linear Systems

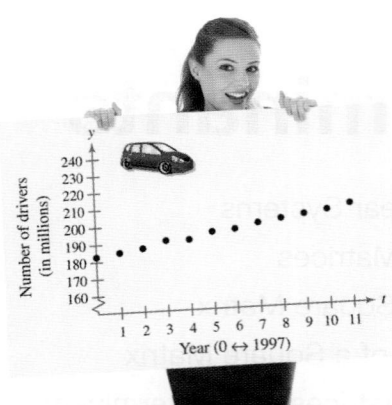

In Exercise 90 on page 486, you will find the least squares regression line for data shown in a graph by solving a system of equations using matrices.

■ Determine the dimension of a matrix.
■ Perform elementary row operations on a matrix in order to write the matrix in row-echelon form or reduced row-echelon form.
■ Solve a system of linear equations using Gaussian elimination.
■ Solve a system of linear equations using Gauss-Jordan elimination.

Matrices

In this section, you will study a streamlined technique for solving systems of linear equations. This technique involves the use of a rectangular array of real numbers called a **matrix.** The plural of matrix is **matrices.**

Definition of a Matrix

If m and n are positive integers, then an $m \times n$ matrix (read "m by n") is a rectangular array

$$
\left.
\begin{bmatrix}
a_{11} & a_{12} & a_{13} & \cdots & a_{1n} \\
a_{21} & a_{22} & a_{23} & \cdots & a_{2n} \\
a_{31} & a_{32} & a_{33} & \cdots & a_{3n} \\
\vdots & \vdots & \vdots & & \vdots \\
a_{m1} & a_{m2} & a_{m3} & \cdots & a_{mn}
\end{bmatrix}
\right\} m \text{ rows}
$$

$\underbrace{\phantom{a_{11} \quad a_{12} \quad a_{13} \quad \cdots \quad a_{1n}}}_{n \text{ columns}}$

in which each **entry**, a_{ij}, of the matrix is a number. An $m \times n$ matrix has m **rows** (horizontal lines) and n **columns** (vertical lines).

The entry in the ith row and jth column is denoted by the *double subscript* notation a_{ij}. That is, a_{21} refers to the entry in row 2, column 1. A matrix having m rows and n columns is said to be of **dimension** $m \times n$. If $m = n$, then the matrix is **square** of dimension $m \times m$ (or $n \times n$). For a square matrix, the entries $a_{11}, a_{22}, a_{33}, \ldots$ are the **main diagonal** entries.

Example 1   Dimensions of Matrices

a. The dimension of the matrix shown at the right is 1×4. $\begin{bmatrix} 1 & -3 & 0 & \frac{1}{2} \end{bmatrix}$

b. The dimension of the matrix shown at the right is 2×2. $\begin{bmatrix} 0 & 0 \\ 0 & 0 \end{bmatrix}$

c. The dimension of the matrix shown at the right is 3×2. $\begin{bmatrix} 5 & 0 \\ 2 & -2 \\ -7 & 4 \end{bmatrix}$

✓**Checkpoint 1**

Determine the dimension of the matrix.

$$\begin{bmatrix} 0 & 8 & 5 \\ 3 & 1 & -2 \end{bmatrix}$$

A matrix that has only one row [such as the matrix in Example 1(a)] is called a **row matrix,** and a matrix that has only one column is called a **column matrix.**

A matrix derived from a system of linear equations (each written in standard form with the constant term on the right) is the **augmented matrix** of the system. Moreover, the matrix derived from the coefficients of the system (but that does not include the constant terms) is the **coefficient matrix** of the system. Note in the matrices below the use of 0 for the coefficient of the missing y-variable in the third equation. Also note that the fourth column (the column of constant terms) in the augmented matrix is separated from the coefficients of the linear system by vertical dots.

$$\text{System:} \quad \begin{cases} x - 4y + 3z = 5 \\ -x + 3y - z = -3 \\ 2x - 4z = 6 \end{cases}$$

$$\text{Augmented Matrix:} \quad \begin{bmatrix} 1 & -4 & 3 & \vdots & 5 \\ -1 & 3 & -1 & \vdots & -3 \\ 2 & 0 & -4 & \vdots & 6 \end{bmatrix}$$

$$\text{Coefficient Matrix:} \quad \begin{bmatrix} 1 & -4 & 3 \\ -1 & 3 & -1 \\ 2 & 0 & -4 \end{bmatrix}$$

When forming either the coefficient matrix or the augmented matrix of a system, you should begin by vertically aligning the variables in the equations and using zeros for the coefficients of any missing variables.

Example 2 Writing an Augmented Matrix

Write the augmented matrix for the system of linear equations. What is the dimension of the augmented matrix?

$$\begin{cases} x + 3y = 9 \\ -y + 4z = -2 \\ x - 5z = 0 \end{cases}$$

SOLUTION Begin by writing the linear system and aligning the variables.

$$\begin{cases} x + 3y = 9 \\ -y + 4z = -2 \\ x - 5z = 0 \end{cases}$$

Next, use the coefficients and constant terms as the matrix entries. Include zeros for the coefficients of the missing variables.

$$\begin{matrix} R_1 \\ R_2 \\ R_3 \end{matrix} \begin{bmatrix} 1 & 3 & 0 & \vdots & 9 \\ 0 & -1 & 4 & \vdots & -2 \\ 1 & 0 & -5 & \vdots & 0 \end{bmatrix}$$

The augmented matrix has three rows and four columns, so it is a 3×4 matrix. The notation R_n is used to designate each row in the matrix. For instance, Row 1 is represented by R_1.

✓ Checkpoint 2

Write the augmented matrix for the system of linear equations. What is the dimension of the augmented matrix?

$$\begin{cases} 4x - 3y = -5 \\ -x + 3y = 12 \end{cases}$$

Elementary Row Operations

In Section 5.3, you studied three operations that can be used on a system of linear equations to produce an equivalent system.

1. Interchange two equations.

2. Multiply one of the equations by a nonzero constant.

3. Add a multiple of one equation to another equation.

In matrix terminology, these three operations correspond to **elementary row operations.** An elementary row operation on an augmented matrix of a given system of linear equations produces a new augmented matrix corresponding to a new (but equivalent) system of linear equations. Two matrices are **row-equivalent** when one can be obtained from the other by a sequence of elementary row operations.

Elementary Row Operations

1. Interchange two rows.

2. Multiply a row by a nonzero constant.

3. Add a multiple of a row to another row.

STUDY TIP

Elementary row operations involve a lot of arithmetic. So that you can go back and check your work, you should note the elementary row operation performed in each step next to the row you are changing, as shown in Example 3.

Example 3   **Elementary Row Operations**

a. Interchange the first and second rows.

Original Matrix

$$\begin{bmatrix} 0 & 1 & 3 & 4 \\ -1 & 2 & 0 & 3 \\ 2 & -3 & 4 & 1 \end{bmatrix}$$

New Row-Equivalent Matrix

$$\begin{matrix} R_2 \\ R_1 \end{matrix} \begin{bmatrix} -1 & 2 & 0 & 3 \\ 0 & 1 & 3 & 4 \\ 2 & -3 & 4 & 1 \end{bmatrix}$$

b. Multiply the first row by $\frac{1}{2}$.

Original Matrix

$$\begin{bmatrix} 2 & -4 & 6 & -2 \\ 1 & 3 & -3 & 0 \\ 5 & -2 & 1 & 2 \end{bmatrix}$$

New Row-Equivalent Matrix

$$\frac{1}{2}R_1 \rightarrow \begin{bmatrix} 1 & -2 & 3 & -1 \\ 1 & 3 & -3 & 0 \\ 5 & -2 & 1 & 2 \end{bmatrix}$$

c. Add -2 times the first row to the third row.

Original Matrix

$$\begin{bmatrix} 1 & 2 & -4 & 3 \\ 0 & 3 & -2 & -1 \\ 2 & 1 & 5 & -2 \end{bmatrix}$$

New Row-Equivalent Matrix

$$\begin{matrix} \\ \\ -2R_1 + R_3 \rightarrow \end{matrix} \begin{bmatrix} 1 & 2 & -4 & 3 \\ 0 & 3 & -2 & -1 \\ 0 & -3 & 13 & -8 \end{bmatrix}$$

✓**Checkpoint 3**

Identify the elementary row operation being performed to obtain the new row-equivalent matrix.

Original Matrix

$$\begin{bmatrix} 1 & 0 & 2 \\ 0 & 3 & 6 \end{bmatrix}$$

New Row-Equivalent Matrix

$$\begin{bmatrix} 1 & 0 & 2 \\ 0 & 1 & 2 \end{bmatrix}$$

Gaussian Elimination with Back-Substitution

In Example 2 in Section 5.3, you used Gaussian elimination with back-substitution to solve a system of linear equations. The next example demonstrates the matrix version of Gaussian elimination. The two methods are essentially the same. The basic difference is that with matrices you do not need to keep writing the variables.

TECH TUTOR

Most graphing utilities can perform elementary row operations on matrices. The screen below shows how one graphing utility displays the new row-equivalent matrix from Example 3(c).

```
*row+(-2,[A],1,3
)
  [[1  2  -4  3 ]
   [0  3  -2  -1]
   [0  -3  13  -8]]
```

Example 4   Comparing Linear Systems and Matrix Operations

Linear System   *Associated Augmented Matrix*

$$\begin{cases} x - 2y + 3z = 9 \\ -x + 3y = -4 \\ 2x - 5y + 5z = 17 \end{cases} \qquad \begin{bmatrix} 1 & -2 & 3 & \vdots & 9 \\ -1 & 3 & 0 & \vdots & -4 \\ 2 & -5 & 5 & \vdots & 17 \end{bmatrix}$$

Add the first equation to the second equation.

Add the first row to the second row $(R_1 + R_2)$.

$$\begin{cases} x - 2y + 3z = 9 \\ y + 3z = 5 \\ 2x - 5y + 5z = 17 \end{cases} \quad R_1 + R_2 \to \begin{bmatrix} 1 & -2 & 3 & \vdots & 9 \\ 0 & 1 & 3 & \vdots & 5 \\ 2 & -5 & 5 & \vdots & 17 \end{bmatrix}$$

Add -2 times the first equation to the third equation.

Add -2 times the first row to the third row $(-2R_1 + R_3)$.

$$\begin{cases} x - 2y + 3z = 9 \\ y + 3z = 5 \\ -y - z = -1 \end{cases} \quad -2R_1 + R_3 \to \begin{bmatrix} 1 & -2 & 3 & \vdots & 9 \\ 0 & 1 & 3 & \vdots & 5 \\ 0 & -1 & -1 & \vdots & -1 \end{bmatrix}$$

Add the second equation to the third equation.

Add the second row to the third row $(R_2 + R_3)$.

$$\begin{cases} x - 2y + 3z = 9 \\ y + 3z = 5 \\ 2z = 4 \end{cases} \quad R_2 + R_3 \to \begin{bmatrix} 1 & -2 & 3 & \vdots & 9 \\ 0 & 1 & 3 & \vdots & 5 \\ 0 & 0 & 2 & \vdots & 4 \end{bmatrix}$$

Multiply the third equation by $\frac{1}{2}$.

Multiply the third row by $\frac{1}{2}$.

$$\begin{cases} x - 2y + 3z = 9 \\ y + 3z = 5 \\ z = 2 \end{cases} \quad \frac{1}{2}R_3 \to \begin{bmatrix} 1 & -2 & 3 & \vdots & 9 \\ 0 & 1 & 3 & \vdots & 5 \\ 0 & 0 & 1 & \vdots & 2 \end{bmatrix}$$

At this point, you can use back-substitution to find that the solution is $x = 1$, $y = -1$, and $z = 2$, as shown in Example 1 in Section 5.3.

✓ Checkpoint 4

Write the system of equations represented by the augmented matrix. Use back-substitution to find the solution. (Use the variables x, y, and z.)

$$\begin{bmatrix} 1 & -2 & 5 & \vdots & 3 \\ 0 & 1 & 4 & \vdots & -3 \\ 0 & 0 & 1 & \vdots & 2 \end{bmatrix}$$

Remember that you can check a solution by substituting the values of x, y, and z into each equation in the original system. For instance, you can check the solution to Example 4 as shown.

Equation 1: $1 - 2(-1) + 3(2) = 9$ ✓

Equation 2: $-1 + 3(-1) = -4$ ✓

Equation 3: $2(1) - 5(-1) + 5(2) = 17$ ✓

The last matrix in Example 4 is said to be in *row-echelon form*. The term *echelon* refers to the stair-step pattern formed by the nonzero entries of the matrix. To be in this form, a matrix must have the following properties.

Row-Echelon Form and Reduced Row-Echelon Form

A matrix in **row-echelon form** has the following properties.

1. Any rows consisting entirely of zeros occur at the bottom of the matrix.

2. For each row that does not consist entirely of zeros, the first nonzero entry is 1 (called a **leading 1**).

3. For two successive (nonzero) rows, the leading 1 in the higher row is farther to the left than the leading 1 in the lower row.

A matrix in *row-echelon form* is in **reduced row-echelon form** when every column that has a leading 1 has zeros in every position above and below the leading 1.

Note that the row-echelon form of a matrix is not unique. That is, two different sequences of elementary row operations may yield different row-echelon forms. The *reduced* row-echelon form of a given matrix, however, is unique.

Example 5   Row-Echelon Form

Determine whether each matrix is in row-echelon form. If it is, determine whether the matrix is in reduced row-echelon form.

a. $\begin{bmatrix} 1 & 2 & -1 & 4 \\ 0 & 1 & 0 & 3 \\ 0 & 0 & 1 & -2 \end{bmatrix}$ **b.** $\begin{bmatrix} 1 & 2 & -1 & 2 \\ 0 & 0 & 0 & 0 \\ 0 & 1 & 2 & -4 \end{bmatrix}$

c. $\begin{bmatrix} 1 & -5 & 2 & -1 & 3 \\ 0 & 0 & 1 & 3 & -2 \\ 0 & 0 & 0 & 1 & 4 \\ 0 & 0 & 0 & 0 & 1 \end{bmatrix}$ **d.** $\begin{bmatrix} 1 & 0 & 0 & -1 \\ 0 & 1 & 0 & 2 \\ 0 & 0 & 1 & 3 \\ 0 & 0 & 0 & 0 \end{bmatrix}$

e. $\begin{bmatrix} 1 & 2 & -3 & 4 \\ 0 & 2 & 1 & -1 \\ 0 & 0 & 1 & -3 \end{bmatrix}$ **f.** $\begin{bmatrix} 0 & 1 & 0 & 5 \\ 0 & 0 & 1 & 3 \\ 0 & 0 & 0 & 0 \end{bmatrix}$

SOLUTION   The matrices in (a), (c), (d), and (f) are in row-echelon form. The matrices in (d) and (f) are in reduced row-echelon form because every column that has a leading 1 has zeros in every position above and below the leading 1. The matrix in (b) is not in row-echelon form because the row of all zeros does not occur at the bottom of the matrix. The matrix in (e) is not in row-echelon form because the first nonzero entry in row 2 is not 1.

✓ Checkpoint 5

Determine whether the matrix is in row-echelon form. If it is, determine whether the matrix is in reduced row-echelon form.

$$\begin{bmatrix} 1 & 0 & 3 & -2 \\ 0 & 1 & 0 & -3 \\ 0 & 0 & 1 & 4 \end{bmatrix}$$

Every matrix can be converted to a row-equivalent matrix that is in row-echelon form. For instance, in Example 5, you can change the matrix in part (e) to row-echelon form by multiplying its second row by $\frac{1}{2}$, as shown at the left.

Original Matrix

$$\begin{bmatrix} 1 & 2 & -3 & 4 \\ 0 & 2 & 1 & -1 \\ 0 & 0 & 1 & -3 \end{bmatrix}$$

Row-Echelon Form

$$\frac{1}{2}R_2 \rightarrow \begin{bmatrix} 1 & 2 & -3 & 4 \\ 0 & 1 & \frac{1}{2} & -\frac{1}{2} \\ 0 & 0 & 1 & -3 \end{bmatrix}$$

Gaussian elimination with back-substitution works well for solving systems of linear equations by hand or with a computer. For this algorithm, the order in which the elementary row operations are performed is important. You should operate from *left to right* by columns, using elementary row operations to obtain zeros in all entries directly below the leading 1's.

Example 6   Gaussian Elimination with Back-Substitution

Solve the system.

$$\begin{cases} y + z - 2w = -3 \\ x + 2y - z \quad\quad = 2 \\ 2x + 4y + z - 3w = -2 \\ x - 4y - 7z - w = -19 \end{cases}$$

SOLUTION

$$\begin{matrix} R_2 \\ R_1 \end{matrix} \begin{bmatrix} 1 & 2 & -1 & 0 & \vdots & 2 \\ 0 & 1 & 1 & -2 & \vdots & -3 \\ 2 & 4 & 1 & -3 & \vdots & -2 \\ 1 & -4 & -7 & -1 & \vdots & -19 \end{bmatrix}$$

Interchange R_1 and R_2 so first column has leading 1 in upper left corner.

$$\begin{matrix} \\ \\ -2R_1 + R_3 \rightarrow \\ -R_1 + R_4 \rightarrow \end{matrix} \begin{bmatrix} 1 & 2 & -1 & 0 & \vdots & 2 \\ 0 & 1 & 1 & -2 & \vdots & -3 \\ 0 & 0 & 3 & -3 & \vdots & -6 \\ 0 & -6 & -6 & -1 & \vdots & -21 \end{bmatrix}$$

Perform operations on R_3 and R_4 so first column has zeros below its leading 1.

$$\begin{matrix} \\ \\ \\ 6R_2 + R_4 \rightarrow \end{matrix} \begin{bmatrix} 1 & 2 & -1 & 0 & \vdots & 2 \\ 0 & 1 & 1 & -2 & \vdots & -3 \\ 0 & 0 & 3 & -3 & \vdots & -6 \\ 0 & 0 & 0 & -13 & \vdots & -39 \end{bmatrix}$$

Perform operations on R_4 so second column has zeros below its leading 1.

$$\begin{matrix} \\ \\ \tfrac{1}{3}R_3 \rightarrow \\ \\ \end{matrix} \begin{bmatrix} 1 & 2 & -1 & 0 & \vdots & 2 \\ 0 & 1 & 1 & -2 & \vdots & -3 \\ 0 & 0 & 1 & -1 & \vdots & -2 \\ 0 & 0 & 0 & -13 & \vdots & -39 \end{bmatrix}$$

Multiply R_3 by $\tfrac{1}{3}$ so third row has a leading 1.

$$\begin{matrix} \\ \\ \\ -\tfrac{1}{13}R_4 \rightarrow \end{matrix} \begin{bmatrix} 1 & 2 & -1 & 0 & \vdots & 2 \\ 0 & 1 & 1 & -2 & \vdots & -3 \\ 0 & 0 & 1 & -1 & \vdots & -2 \\ 0 & 0 & 0 & 1 & \vdots & 3 \end{bmatrix}$$

Multiply R_4 by $-\tfrac{1}{13}$ so fourth row has a leading 1.

The matrix is now in row-echelon form, and the corresponding system is

$$\begin{cases} x + 2y - z \quad\quad = 2 \\ y + z - 2w = -3 \\ z - w = -2 \\ w = 3 \end{cases}$$

Using back-substitution, you can determine that the solution is $x = -1$, $y = 2$, $z = 1$, and $w = 3$. Check this in the original system of equations. ▬▬▬

✓ Checkpoint 6

Solve the system.

$$\begin{cases} y + 2z - w = -5 \\ x - 3y - z \quad\quad = 0 \\ 2x - 6y + z + 3w = 6 \\ 4x + 4y - 2z + w = 1 \end{cases}$$

Gaussian Elimination with Back-Substitution

1. Write the augmented matrix of the system of linear equations.

2. Use elementary row operations to rewrite the augmented matrix in row-echelon form.

3. Write the system of linear equations corresponding to the matrix in row-echelon form, and use back-substitution to find the solution.

Remember that it is possible for a system to have no solution. In the elimination process, when you obtain a row with zeros except for the last entry, you can conclude that the system has no solution, or is inconsistent.

Example 7   A System with No Solution

Solve the system.

$$\begin{cases} x - y + 2z = 4 \\ x \quad\;\; + z = 6 \\ 2x - 3y + 5z = 4 \\ 3x + 2y - z = 1 \end{cases}$$

SOLUTION

$$\begin{bmatrix} 1 & -1 & 2 & \vdots & 4 \\ 1 & 0 & 1 & \vdots & 6 \\ 2 & -3 & 5 & \vdots & 4 \\ 3 & 2 & -1 & \vdots & 1 \end{bmatrix}$$ Write augmented matrix.

$$\begin{matrix} \\ -R_1 + R_2 \rightarrow \\ -2R_1 + R_3 \rightarrow \\ -3R_1 + R_4 \rightarrow \end{matrix} \begin{bmatrix} 1 & -1 & 2 & \vdots & 4 \\ 0 & 1 & -1 & \vdots & 2 \\ 0 & -1 & 1 & \vdots & -4 \\ 0 & 5 & -7 & \vdots & -11 \end{bmatrix}$$ Perform row operations.

$$\begin{matrix} \\ \\ R_2 + R_3 \rightarrow \\ \\ \end{matrix} \begin{bmatrix} 1 & -1 & 2 & \vdots & 4 \\ 0 & 1 & -1 & \vdots & 2 \\ 0 & 0 & 0 & \vdots & -2 \\ 0 & 5 & -7 & \vdots & -11 \end{bmatrix}$$ Perform row operations.

Note that the third row of this matrix consists of zeros except for the last entry. This means that the original system of linear equations is *inconsistent*. You can see why this is true by converting back to a system of linear equations.

$$\begin{cases} x - y + 2z = 4 \\ y - z = 2 \\ 0 = -2 \\ 5y - 7z = -11 \end{cases}$$

Because $0 = -2$ is a false statement, the system has no solution.

✓ **Checkpoint 7**

Solve the system.

$$\begin{cases} x + 2y - z = 3 \\ -x - y + 3z = 2 \\ 2x + 3y - 4z = 0 \\ 3x + 2y + z = -4 \end{cases}$$

Gauss-Jordan Elimination

With Gaussian elimination, elementary row operations are applied to a matrix to obtain a (row-equivalent) row-echelon form of the matrix. A second method of elimination, called **Gauss-Jordan elimination** after Carl Friedrich Gauss and Wilhelm Jordan (1842–1899), continues the reduction process until a *reduced* row-echelon form is obtained. This procedure is demonstrated in Example 8.

Example 8 Gauss-Jordan Elimination

Use Gauss-Jordan elimination to solve the system.

$$\begin{cases} x - 2y + 3z = 9 \\ -x + 3y = -4 \\ 2x - 5y + 5z = 17 \end{cases}$$

SOLUTION In Example 4, Gaussian elimination was used to obtain the row-echelon form

$$\begin{bmatrix} 1 & -2 & 3 & \vdots & 9 \\ 0 & 1 & 3 & \vdots & 5 \\ 0 & 0 & 1 & \vdots & 2 \end{bmatrix}.$$

Now, rather than using back-substitution, apply additional elementary row operations until you obtain a matrix in *reduced* row-echelon form. To do this, you must produce zeros above each of the leading 1's, as follows.

$$2R_2 + R_1 \rightarrow \begin{bmatrix} 1 & 0 & 9 & \vdots & 19 \\ 0 & 1 & 3 & \vdots & 5 \\ 0 & 0 & 1 & \vdots & 2 \end{bmatrix}$$

Perform operations on R_1 so second column has a zero above its leading 1.

$$\begin{matrix} -9R_3 + R_1 \rightarrow \\ -3R_3 + R_2 \rightarrow \end{matrix} \begin{bmatrix} 1 & 0 & 0 & \vdots & 1 \\ 0 & 1 & 0 & \vdots & -1 \\ 0 & 0 & 1 & \vdots & 2 \end{bmatrix}$$

Perform operations on R_1 and R_2 so third column has zeros above its leading 1.

The matrix is now in reduced row-echelon form. Converting back to a system of linear equations, you have

$$\begin{cases} x = 1 \\ y = -1. \\ z = 2 \end{cases}$$

An advantage of Gauss-Jordan elimination is that you can read the solution from the matrix in reduced row-echelon form.

✓ Checkpoint 8

Use Gauss-Jordan elimination to solve the system.

$$\begin{cases} x - 3y + 2z = 1 \\ -x - y + 3z = 4 \\ y - 2z = -5 \end{cases}$$

The elimination procedures described in this section sometimes result in fractional coefficients. For instance, in the elimination procedure for the system

$$\begin{cases} 2x - 5y + 5z = 17 \\ 3x - 2y + 3z = 11 \\ -3x + 3y = 6 \end{cases}$$

you may be inclined to multiply the first row by $\frac{1}{2}$ to produce a leading 1, which will result in working with fractional coefficients. You can sometimes avoid fractions by judiciously choosing the order in which you apply elementary row operations.

Example 9   **A System with an Infinite Number of Solutions**

Solve the system.

$$\begin{cases} 2x + 4y - 2z = 0 \\ 3x + 5y \quad\quad = 1 \end{cases}$$

SOLUTION

$$\begin{bmatrix} 2 & 4 & -2 & \vdots & 0 \\ 3 & 5 & 0 & \vdots & 1 \end{bmatrix}$$

$$\tfrac{1}{2}R_1 \rightarrow \begin{bmatrix} 1 & 2 & -1 & \vdots & 0 \\ 3 & 5 & 0 & \vdots & 1 \end{bmatrix}$$

$$-3R_1 + R_2 \rightarrow \begin{bmatrix} 1 & 2 & -1 & \vdots & 0 \\ 0 & -1 & 3 & \vdots & 1 \end{bmatrix}$$

$$-R_2 \rightarrow \begin{bmatrix} 1 & 2 & -1 & \vdots & 0 \\ 0 & 1 & -3 & \vdots & -1 \end{bmatrix}$$

$$-2R_2 + R_1 \rightarrow \begin{bmatrix} 1 & 0 & 5 & \vdots & 2 \\ 0 & 1 & -3 & \vdots & -1 \end{bmatrix}$$

The corresponding system of equations is

$$\begin{cases} x \quad\quad + 5z = \quad 2 \\ \quad y - 3z = -1 \end{cases}.$$

Solving for x and y in terms of z, you have $x = -5z + 2$ and $y = 3z - 1$. To write a solution of the system that does not use any of the three variables of the system, let a represent any real number and let $z = a$. Now, substitute a for z in the equations for x and y.

$$x = -5z + 2 = -5a + 2 \quad \text{and} \quad y = 3z - 1 = 3a - 1$$

So, the solution set has the form $(-5a + 2, 3a - 1, a)$, where a is a real number. Try substituting values for a to obtain a few solutions. Then check each solution in the original system of equations.

✓**Checkpoint 9**

Solve the system: $\begin{cases} \quad\quad y + 5z = -2 \\ -x + y - 4z = \quad 8 \end{cases}.$

SUMMARIZE   (Section 6.1)

1. State the definition of a matrix *(page 474)*. For examples of determining the dimensions of matrices, see Examples 1 and 2.

2. List the elementary row operations *(page 476)*. For examples of using elementary row operations, see Examples 3 and 4.

3. State the definitions of row-echelon form and reduced row-echelon form *(page 478)*. For an example of matrices in these forms, see Example 5.

4. Describe the procedure of Gaussian elimination with back-substitution *(page 479)*. For examples of using this procedure, see Examples 6 and 7.

5. Describe the procedure of Gauss-Jordan elimination *(page 481)*. For examples of using this procedure, see Examples 8 and 9.

SKILLS WARM UP 6.1

The following warm-up exercises involve skills that were covered in earlier sections. You will use these skills in the exercise set for this section. For additional help, review Sections 0.2. 5.1, and 5.3.

In Exercises 1–4, evaluate the expression.

1. $2(-1) - 3(5) + 7(2)$

2. $-4(-3) + 6(7) + 8(-3)$

3. $11\left(\frac{1}{2}\right) - 7\left(-\frac{3}{2}\right) - 5(2)$

4. $\frac{2}{3}\left(\frac{1}{2}\right) + \frac{4}{3}\left(-\frac{1}{3}\right)$

In Exercises 5 and 6, determine whether $x = 1$, $y = 3$, and $z = -1$ is a solution of the system.

5. $\begin{cases} 4x - 2y + 3z = -5 \\ x + 3y - z = 11 \\ -x + 2y = 5 \end{cases}$

6. $\begin{cases} -x + 2y + z = 4 \\ 2x - 3z = 5 \\ 3x + 5y - 2z = 21 \end{cases}$

In Exercises 7–10, use back-substitution to solve the system of linear equations.

7. $\begin{cases} 2x - 3y = 4 \\ y = 2 \end{cases}$

8. $\begin{cases} 5x + 4y = 0 \\ y = -3 \end{cases}$

9. $\begin{cases} x - 3y + z = 0 \\ y - 3z = 8 \\ z = 2 \end{cases}$

10. $\begin{cases} 2x - 5y + 3z = -2 \\ y - 4z = 0 \\ z = 1 \end{cases}$

Exercises 6.1

See www.CalcChat.com for worked-out solutions to odd-numbered exercises.

Dimensions of Matrices In Exercises 1–8, determine the dimension of the matrix. *See Example 1.*

1. $\begin{bmatrix} 0 & -3 & 0 \\ 9 & 2 & -7 \end{bmatrix}$

2. $\begin{bmatrix} -7 & 21 \end{bmatrix}$

3. $\begin{bmatrix} 6 & 4 & 1 \\ 8 & 3 & 0 \\ -1 & 2 & 1 \\ 1 & 5 & 4 \end{bmatrix}$

4. $\begin{bmatrix} 1 \\ 0 \\ 3 \\ 5 \\ 6 \end{bmatrix}$

5. $\begin{bmatrix} 33 & 45 \\ -9 & 20 \\ 12 & 15 \\ 16 & -2 \end{bmatrix}$

6. $\begin{bmatrix} 12 & -2 & 4 \\ -3 & 4 & 0 \\ -8 & 12 & 2 \end{bmatrix}$

7. $\begin{bmatrix} 2 & 7 & 11 & -3 \\ -1 & 10 & -5 & 0 \end{bmatrix}$

8. $\begin{bmatrix} -11 \end{bmatrix}$

Writing an Augmented Matrix In Exercises 9–14, write the augmented matrix for the system of linear equations. What is the dimension of the augmented matrix? *See Example 2.*

9. $\begin{cases} 4x - 3y = -5 \\ -x + 3y = 12 \end{cases}$

10. $\begin{cases} 7x + 4y = 22 \\ 5x - 9y = 15 \end{cases}$

11. $\begin{cases} x + 10y - 2z = 2 \\ 5x - 3y + 4z = 0 \\ 2x + y = 6 \end{cases}$

12. $\begin{cases} x - 3y + z = 1 \\ 4y = 0 \\ 7z = -5 \end{cases}$

13. $\begin{cases} 7x - 5y + z = 13 \\ 19x - 8z = 10 \end{cases}$

14. $\begin{cases} 9x + 2y - 3z = 20 \\ -25y + 11z = -5 \end{cases}$

Elementary Row Operations In Exercises 15–18, fill in the blank(s) to form the new row-equivalent matrix. *See Example 3.*

Original Matrix *New Row-Equivalent Matrix*

15. $\begin{bmatrix} 1 & 1 & 1 \\ 5 & -2 & 4 \end{bmatrix}$ $\begin{bmatrix} 1 & 1 & 1 \\ 0 & & -1 \end{bmatrix}$

16. $\begin{bmatrix} -3 & 3 & 12 \\ 18 & -8 & 4 \end{bmatrix}$ $\begin{bmatrix} 1 & -1 & \\ 18 & -8 & 4 \end{bmatrix}$

17. $\begin{bmatrix} 1 & 5 & 4 & -1 \\ 0 & 1 & -2 & 2 \\ 0 & 0 & 1 & -7 \end{bmatrix}$ $\begin{bmatrix} 1 & 0 & & \\ 0 & 1 & -2 & 2 \\ 0 & 0 & 1 & -7 \end{bmatrix}$

18. $\begin{bmatrix} 1 & 0 & 6 & 1 \\ 0 & -1 & 0 & 7 \\ 0 & 0 & -1 & 3 \end{bmatrix}$ $\begin{bmatrix} 1 & 0 & 6 & 1 \\ 0 & 1 & 0 & \\ 0 & 0 & 1 & \end{bmatrix}$

Elementary Row Operations In Exercises 19–22, identify the elementary row operation(s) being performed to obtain the new row-equivalent matrix. *See Example 3.*

Original Matrix *New Row-Equivalent Matrix*

19. $\begin{bmatrix} -2 & 5 & 1 \\ 3 & -1 & -8 \end{bmatrix}$ $\begin{bmatrix} 13 & 0 & -39 \\ 3 & -1 & -8 \end{bmatrix}$

20. $\begin{bmatrix} 3 & -1 & -4 \\ -4 & 3 & 7 \end{bmatrix}$ $\begin{bmatrix} 3 & -1 & -4 \\ 5 & 0 & -5 \end{bmatrix}$

21. $\begin{bmatrix} 0 & -1 & -5 & 5 \\ -1 & 3 & -7 & 6 \\ 4 & -5 & 1 & 3 \end{bmatrix}$ $\begin{bmatrix} -1 & 3 & -7 & 6 \\ 0 & -1 & -5 & 5 \\ 0 & 7 & -27 & 27 \end{bmatrix}$

Original Matrix   *New Row-Equivalent Matrix*

22. $\begin{bmatrix} -1 & -2 & 3 & -2 \\ 2 & -5 & 1 & -7 \\ 5 & 4 & -7 & 6 \end{bmatrix}$ $\begin{bmatrix} -1 & -2 & 3 & -2 \\ 0 & -9 & 7 & -11 \\ 0 & -6 & 8 & -4 \end{bmatrix}$

Writing a System of Linear Equations In Exercises 23–26, write the system of linear equations represented by the augmented matrix. (Use the variables x, y, z, and w.)

23. $\begin{bmatrix} 2 & 4 & \vdots & 6 \\ -1 & 3 & \vdots & -8 \end{bmatrix}$

24. $\begin{bmatrix} 7 & -2 & \vdots & 7 \\ -8 & 3 & \vdots & -3 \end{bmatrix}$

25. $\begin{bmatrix} 1 & 0 & 2 & \vdots & -10 \\ 0 & 3 & -1 & \vdots & 5 \\ 4 & 2 & 0 & \vdots & 3 \end{bmatrix}$

26. $\begin{bmatrix} 5 & 8 & 2 & 0 & \vdots & -1 \\ -2 & 15 & 5 & 1 & \vdots & 9 \\ 1 & 6 & -7 & 0 & \vdots & -3 \end{bmatrix}$

Using Back-Substitution In Exercises 27–30, write the system of linear equations represented by the augmented matrix. Use back-substitution to find the solution. (Use the variables x, y, z, and w.)

27. $\begin{bmatrix} 1 & -5 & \vdots & 6 \\ 0 & 1 & \vdots & -2 \end{bmatrix}$

28. $\begin{bmatrix} 1 & 2 & -1 & \vdots & 3 \\ 0 & 1 & -2 & \vdots & -3 \\ 0 & 0 & 1 & \vdots & 4 \end{bmatrix}$

29. $\begin{bmatrix} 1 & 3 & -1 & \vdots & 15 \\ 0 & 1 & 4 & \vdots & -12 \\ 0 & 0 & 1 & \vdots & -5 \end{bmatrix}$

30. $\begin{bmatrix} 1 & 2 & -2 & 0 & \vdots & -1 \\ 0 & 1 & 1 & 2 & \vdots & 9 \\ 0 & 0 & 1 & 0 & \vdots & 2 \\ 0 & 0 & 0 & 1 & \vdots & -3 \end{bmatrix}$

Comparing Linear Systems and Matrix Operations In Exercises 31 and 32, (a) perform the elementary row operations to solve the augmented matrix, (b) write and solve the system of linear equations represented by the augmented matrix, and (c) compare the two solution methods. Which do you prefer? *See Example 4.*

31. $\begin{bmatrix} -3 & 4 & \vdots & 22 \\ 6 & -4 & \vdots & -28 \end{bmatrix}$

(i) Add R_2 to R_1.

(ii) Add -2 times R_1 to R_2.

(iii) Multiply R_2 by $-\frac{1}{4}$.

(iv) Multiply R_1 by $\frac{1}{3}$.

32. $\begin{bmatrix} 7 & 13 & 1 & \vdots & -4 \\ -3 & -5 & -1 & \vdots & -4 \\ 3 & 6 & 1 & \vdots & -2 \end{bmatrix}$

(i) Add R_2 to R_1.

(ii) Multiply R_1 by $\frac{1}{4}$.

(iii) Add R_3 to R_2.

(iv) Add -3 times R_1 to R_3.

(v) Add -2 times R_2 to R_1.

33. Using a Graphing Utility Repeat steps (i) through (iv) for the matrix in Exercise 31 using a graphing utility.

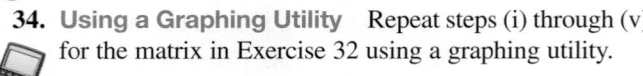

34. Using a Graphing Utility Repeat steps (i) through (v) for the matrix in Exercise 32 using a graphing utility.

Row-Echelon Form In Exercises 35–40, determine whether the matrix is in row-echelon form. If it is, determine if it is also in *reduced* row-echelon form. *See Example 5.*

35. $\begin{bmatrix} 1 & 0 & 0 & 0 \\ 0 & 1 & 1 & 5 \\ 0 & 0 & 0 & 0 \end{bmatrix}$ **36.** $\begin{bmatrix} 1 & 0 & 2 & 1 \\ 0 & 1 & -3 & 10 \\ 0 & 0 & 1 & 0 \end{bmatrix}$

37. $\begin{bmatrix} 2 & 0 & 4 & 0 \\ 0 & -1 & 3 & 6 \\ 0 & 0 & 1 & 5 \end{bmatrix}$ **38.** $\begin{bmatrix} 0 & 0 & 0 & 0 \\ 0 & 1 & 0 & 5 \\ 0 & 0 & 1 & 3 \end{bmatrix}$

39. $\begin{bmatrix} 1 & 3 & 0 & 0 & 0 & 0 \\ 0 & 0 & 1 & 8 & 1 & 0 \\ 0 & 0 & 0 & 0 & 1 & 1 \\ 0 & 0 & 0 & 0 & 1 & 1 \end{bmatrix}$

40. $\begin{bmatrix} 1 & 0 & 0 & 10 \\ 0 & 1 & 3 & 9 \\ 0 & 0 & 0 & 1 \\ 0 & 0 & 0 & 0 \end{bmatrix}$

Writing a Matrix in Row-Echelon Form In Exercises 41–44, write the matrix in row-echelon form. (*Note:* Row-echelon forms are not unique.)

41. $\begin{bmatrix} 1 & 2 & -1 & 5 \\ 3 & 2 & 1 & 11 \\ 4 & 8 & 1 & 10 \end{bmatrix}$ **42.** $\begin{bmatrix} 1 & 2 & -1 & 3 \\ 3 & 7 & -5 & 14 \\ -2 & -1 & -3 & 8 \end{bmatrix}$

43. $\begin{bmatrix} 1 & -1 & -1 & 1 \\ 5 & -4 & 1 & 8 \\ -6 & 8 & 18 & 0 \end{bmatrix}$ **44.** $\begin{bmatrix} 1 & -3 & 0 & -7 \\ -3 & 10 & 1 & 23 \\ 1 & 0 & 1 & 12 \\ 4 & -10 & 2 & -24 \end{bmatrix}$

Writing a Matrix in Reduced Row-Echelon Form In Exercises 45–50, write the matrix in reduced row-echelon form.

45. $\begin{bmatrix} 4 & 4 & 8 \\ 1 & 2 & 2 \\ -3 & 6 & -9 \end{bmatrix}$ **46.** $\begin{bmatrix} 1 & 3 & 2 \\ 5 & 15 & 9 \\ 2 & 6 & 10 \end{bmatrix}$

47. $\begin{bmatrix} 1 & 0 & 2 & 1 \\ 0 & 3 & -6 & 6 \\ 2 & 0 & 5 & -4 \\ 0 & 1 & 0 & 1 \end{bmatrix}$

48. $\begin{bmatrix} 1 & 2 & 3 & -5 \\ 1 & 2 & 4 & -9 \\ -2 & -4 & -4 & 3 \\ 4 & 8 & 11 & -14 \end{bmatrix}$

49. $\begin{bmatrix} 2 & 1 \\ 1 & 4 \\ -2 & -1 \end{bmatrix}$

50. $\begin{bmatrix} 1 & -3 \\ -1 & 8 \\ 0 & 4 \\ -2 & 10 \end{bmatrix}$

Interpreting Reduced Row-Echelon Form In Exercises 51–56, an augmented matrix that represents a system of linear equations (in variables x, y, and z) has been reduced using Gauss-Jordan elimination. Write the solution represented by the augmented matrix.

51. $\begin{bmatrix} 1 & 0 & \vdots & -4 \\ 0 & 1 & \vdots & 6 \end{bmatrix}$

52. $\begin{bmatrix} 1 & 0 & \vdots & 9 \\ 0 & 1 & \vdots & -3 \end{bmatrix}$

53. $\begin{bmatrix} 1 & 0 & 0 & \vdots & -4 \\ 0 & 1 & 0 & \vdots & -8 \\ 0 & 0 & 1 & \vdots & 2 \end{bmatrix}$

54. $\begin{bmatrix} 1 & 0 & 0 & \vdots & 3 \\ 0 & 1 & 0 & \vdots & -1 \\ 0 & 0 & 1 & \vdots & 0 \end{bmatrix}$

55. $\begin{bmatrix} 1 & 0 & 2 & \vdots & -4 \\ 0 & 1 & 1 & \vdots & 6 \\ 0 & 0 & 0 & \vdots & 0 \end{bmatrix}$

56. $\begin{bmatrix} 1 & 0 & 2 & \vdots & 9 \\ 0 & 1 & 5 & \vdots & -3 \\ 0 & 0 & 0 & \vdots & 0 \end{bmatrix}$

Gaussian Elimination with Back-Substitution In Exercises 57–74, use matrices to solve the system of equations (if possible). Use Gaussian elimination with back-substitution. *See Examples 6, 7, and 9.*

57. $\begin{cases} x + 2y = 7 \\ 2x + y = 8 \end{cases}$

58. $\begin{cases} 2x + 6y = 16 \\ 2x + 3y = 7 \end{cases}$

59. $\begin{cases} -3x + 5y = -22 \\ 3x + 4y = 4 \\ 4x - 8y = 32 \end{cases}$

60. $\begin{cases} x + 2y = 0 \\ x + y = 6 \\ 3x - 2y = 8 \end{cases}$

61. $\begin{cases} 8x - 4y = 7 \\ 5x + 2y = 1 \end{cases}$

62. $\begin{cases} x - 3y = 5 \\ -2x + 6y = -10 \end{cases}$

63. $\begin{cases} -x + 2y = 1.5 \\ 2x - 4y = 3 \end{cases}$

64. $\begin{cases} 2x - y = -0.1 \\ 3x + 2y = 1.6 \end{cases}$

65. $\begin{cases} x + 2y + z = 8 \\ 3x + 7y + 6z = 26 \end{cases}$

66. $\begin{cases} x + y + 4z = 5 \\ 2x + y - z = 9 \end{cases}$

67. $\begin{cases} 3x + 3y + 12z = 6 \\ x + y + 4z = 2 \\ 2x + 5y + 20z = 10 \\ -x + 2y + 8z = 4 \end{cases}$

68. $\begin{cases} 2x + 10y + 2z = 6 \\ x + 5y + 2z = 6 \\ x + 5y + z = 3 \\ -3x + 15y - 3z = -9 \end{cases}$

69. $\begin{cases} 4x + 12y - 7z - 20w = 22 \\ 3x + 9y - 5z - 28w = 30 \end{cases}$

70. $\begin{cases} x + 2y + 2z + 4w = 11 \\ 3x + 6y + 5z + 12w = 30 \end{cases}$

71. $\begin{cases} x + 2y = 0 \\ -x - y = 0 \end{cases}$

72. $\begin{cases} x + 2y = 0 \\ 2x + 4y = 0 \end{cases}$

73. $\begin{cases} x + y + z = 0 \\ 2x + 3y + z = 0 \\ 3x + 5y + z = 0 \end{cases}$

74. $\begin{cases} x - 2y + z + 3w = 0 \\ x - y + w = 0 \\ y - z + 2w = 0 \end{cases}$

Gauss-Jordan Elimination In Exercises 75–80, use matrices to solve the system of equations (if possible). Use Gauss-Jordan elimination. *See Examples 8 and 9.*

75. $\begin{cases} x - 3z = -2 \\ 3x + y - 2z = 5 \\ 2x + 2y + z = 4 \end{cases}$

76. $\begin{cases} 2x - y + 3z = 24 \\ 2y - z = 14 \\ 7x - 5y = 6 \end{cases}$

77. $\begin{cases} x + y - 5z = 3 \\ x - 2z = 1 \\ 2x - y - z = 0 \end{cases}$

78. $\begin{cases} 2x + 3z = 3 \\ 4x - 3y + 7z = 5 \\ 8x - 9y + 15z = 9 \end{cases}$

79. $\begin{cases} -x + y - z = -14 \\ 2x - y + z = 21 \\ 3x + 2y + z = 19 \end{cases}$

80. $\begin{cases} 2x + 2y - z = 2 \\ x - 3y + z = 28 \\ -x + y = 14 \end{cases}$

Comparing Solutions of Two Systems In Exercises 81–84, determine whether the two systems of linear equations yield the same solution. If so, find the solution using matrices.

81. (a) $\begin{cases} x - 2y + z = -6 \\ y - 5z = 16 \\ z = -3 \end{cases}$ (b) $\begin{cases} x + y - 2z = 6 \\ y + 3z = -8 \\ z = -3 \end{cases}$

82. (a) $\begin{cases} x - 3y + 4z = -11 \\ \quad\;\; y - z = -4 \\ \qquad\qquad z = 2 \end{cases}$ (b) $\begin{cases} x + 4y \quad\;\; = -11 \\ \quad\;\; y + 3z = 4 \\ \qquad\qquad z = 2 \end{cases}$

83. (a) $\begin{cases} x - 4y + 5z = 27 \\ \quad\;\; y - 7z = -54 \\ \qquad\qquad z = 8 \end{cases}$ (b) $\begin{cases} x - 6y + z = 15 \\ \quad\;\; y + 5z = 42 \\ \qquad\qquad z = 8 \end{cases}$

84. (a) $\begin{cases} x + 3y - z = 19 \\ \quad\;\; y + 6z = -18 \\ \qquad\qquad z = -4 \end{cases}$ (b) $\begin{cases} x - y + 3z = -15 \\ \quad\;\; y - 2z = 14 \\ \qquad\qquad z = -4 \end{cases}$

85. Breeding Facility A city zoo borrowed $2,000,000 at simple annual interest to construct a breeding facility. Some of the money was borrowed at 8%, some at 9%, and some at 12%. Use a system of linear equations to determine how much was borrowed at each rate if the total annual interest was $186,000 and the amount borrowed at 8% was twice the amount borrowed at 12%. Solve the system of linear equations using matrices.

86. Museum A natural history museum borrowed $2,000,000 at simple annual interest to purchase new exhibits. Some of the money was borrowed at 7%, some at 8.5%, and some at 9.5%. Use a system of linear equations to determine how much was borrowed at each rate if the total annual interest was $169,750 and the amount borrowed at 8.5% was four times the amount borrowed at 9.5%. Solve the system of linear equations using matrices.

87. Think About It You and a friend solve the following system of equations independently.

$$\begin{cases} 2x - 4y - 3z = 3 \\ \;\; x + 3y + z = -1 \\ 5x + \;\; y - 2z = 2 \end{cases}$$

You write your solution set as

$$(a, -a, 2a - 1)$$

where a is any real number. Your friend's solution set is

$$\left(\tfrac{1}{2}b + \tfrac{1}{2}, -\tfrac{1}{2}b - \tfrac{1}{2}, b\right)$$

where b is any real number. Are you both correct? Explain. If you let $a = 3$, what value of b must be selected so that you both have the same ordered triple?

88. HOW DO YOU SEE IT? Determine all values of a and b for which the augmented matrix has each given number of solutions.

$$\begin{bmatrix} 1 & 2 & \vdots & -4 \\ 0 & a & \vdots & b \end{bmatrix}$$

(a) Exactly one solution

(b) Infinitely many solutions

(c) No solution

89. Health and Wellness From 1998 through 2009, the numbers of new cases of a waterborne disease in a small city increased in a pattern that was approximately linear (see figure). Find the least squares regression line

$$y = at + b$$

for the data shown in the figure by solving the following system using matrices. Let t represent the year, with $t = 0$ corresponding to 1998.

$$\begin{cases} 12b + 66a = 831 \\ 66b + 506a = 5643 \end{cases}$$

Use the result to predict the number of new cases of the waterborne disease in 2012. Is the estimate reasonable? Explain.

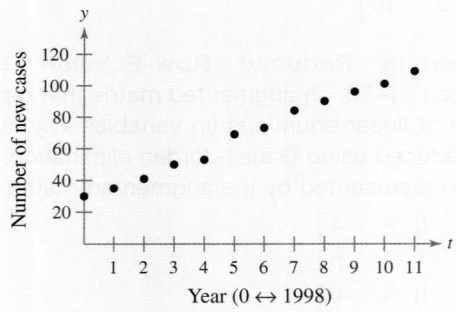

90. Drivers From 1997 through 2008, the numbers y (in millions) of drivers in the United States increased in a pattern that was approximately linear (see figure). Find the least squares regression line

$$y = at + b$$

for the data shown in the figure by solving the following system using matrices. Let t represent the year, with $t = 0$ corresponding to 1997.

$$\begin{cases} 12b + 66a = 2344 \\ 66b + 506a = 13{,}213 \end{cases}$$

Use the result to predict the number of drivers in 2013. Is the estimate reasonable? Explain. (*Source: Federal Highway Administration*)

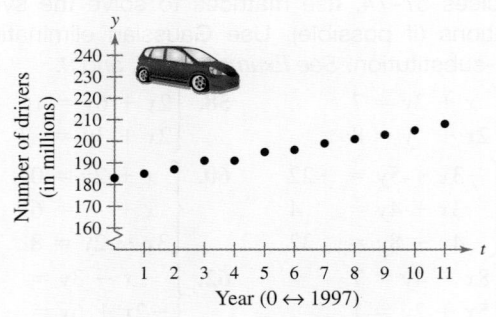

6.2 Operations with Matrices

- Determine whether two matrices are equal.
- Add or subtract two matrices and multiply a matrix by a scalar.
- Find the product of two matrices.
- Use matrix operations to solve an application problem.

Equality of Matrices

In Section 6.1, you used matrices to solve systems of linear equations. Matrices, however, can do much more than this. There is a rich mathematical theory of matrices, with numerous applications. It is standard mathematical convention to represent matrices in any of the following three ways.

1. A matrix can be denoted by an uppercase letter such as A, B, or C.

2. A matrix can be denoted by a representative element enclosed in brackets, such as $[a_{ij}]$, $[b_{ij}]$, or $[c_{ij}]$.

3. A matrix can be denoted by a rectangular array of numbers such as

$$A = [a_{ij}] = \begin{bmatrix} a_{11} & a_{12} & a_{13} & \cdots & a_{1n} \\ a_{21} & a_{22} & a_{23} & \cdots & a_{2n} \\ a_{31} & a_{32} & a_{33} & \cdots & a_{3n} \\ \vdots & \vdots & \vdots & & \vdots \\ a_{m1} & a_{m2} & a_{m3} & \cdots & a_{mn} \end{bmatrix}.$$

Two matrices

$$A = [a_{ij}] \quad \text{and} \quad B = [b_{ij}]$$

are **equal** when they have the same dimension $(m \times n)$ and all of their corresponding entries are equal.

In Exercise 89 on page 501, you will use matrix multiplication to find the numbers of subscribers that two competing satellite television providers have after a given period of time.

Example 1 Equality of Matrices

Solve for a_{11}, a_{12}, a_{21}, and a_{22} in the matrix equation.

$$\begin{bmatrix} a_{11} & a_{12} \\ a_{21} & a_{22} \end{bmatrix} = \begin{bmatrix} 2 & -1 \\ -3 & 0 \end{bmatrix}$$

SOLUTION Because two matrices are equal only when they have the same dimension and their corresponding entries are equal, you can conclude that

$$a_{11} = 2, \quad a_{12} = -1, \quad a_{21} = -3, \quad \text{and} \quad a_{22} = 0.$$

✓Checkpoint 1

Solve for a_{11}, a_{12}, a_{21}, and a_{22} in the matrix equation.

$$\begin{bmatrix} a_{11} & a_{12} \\ a_{21} & a_{22} \end{bmatrix} = \begin{bmatrix} 5 & 2 \\ -1 & 3 \end{bmatrix}$$

Be sure you see that for two matrices to be equal, they must have the same dimension and their corresponding entries must be equal. So,

$$\begin{bmatrix} 2 & -1 \\ \sqrt{4} & \frac{1}{2} \end{bmatrix} = \begin{bmatrix} 2 & -1 \\ 2 & 0.5 \end{bmatrix} \quad \text{but} \quad \begin{bmatrix} 2 & -1 \\ 3 & 4 \\ 0 & 0 \end{bmatrix} \neq \begin{bmatrix} 2 & -1 \\ 3 & 4 \end{bmatrix}.$$

Matrix Addition and Scalar Multiplication

You can **add** two matrices (of the same dimension) by adding their corresponding entries.

Definition of Matrix Addition

If $A = [a_{ij}]$ and $B = [b_{ij}]$ are matrices of dimension $m \times n$, then their **sum** is the $m \times n$ matrix given by

$$A + B = [a_{ij} + b_{ij}].$$

The sum of two matrices of different dimensions is undefined.

Example 2   Addition of Matrices

a. $\begin{bmatrix} -1 & 2 \\ 0 & 1 \end{bmatrix} + \begin{bmatrix} 1 & 3 \\ -1 & 2 \end{bmatrix} = \begin{bmatrix} -1+1 & 2+3 \\ 0+(-1) & 1+2 \end{bmatrix} = \begin{bmatrix} 0 & 5 \\ -1 & 3 \end{bmatrix}$

b. $\begin{bmatrix} 0 & 1 & -2 \\ 1 & 2 & 3 \end{bmatrix} + \begin{bmatrix} 0 & 0 & 0 \\ 0 & 0 & 0 \end{bmatrix} = \begin{bmatrix} 0 & 1 & -2 \\ 1 & 2 & 3 \end{bmatrix}$

c. $\begin{bmatrix} 1 \\ -3 \\ -2 \end{bmatrix} + \begin{bmatrix} -1 \\ 3 \\ 2 \end{bmatrix} = \begin{bmatrix} 0 \\ 0 \\ 0 \end{bmatrix}$

d. The sum of

$$A = \begin{bmatrix} 2 & 1 & 0 \\ 4 & 0 & -1 \\ 3 & -2 & 2 \end{bmatrix} \quad \text{and} \quad B = \begin{bmatrix} 0 & 1 \\ -1 & 3 \\ 2 & 4 \end{bmatrix}$$

is undefined because A is of dimension 3×3 and B is of dimension 3×2.

✓ Checkpoint 2

Evaluate the expression.

$$\begin{bmatrix} 2 & -7 \\ -1 & 3 \end{bmatrix} + \begin{bmatrix} 4 & 5 \\ -1 & -6 \end{bmatrix}$$

In operations with matrices, numbers are usually referred to as **scalars.** In this text, scalars will always be real numbers. You can multiply a matrix A by a scalar c by multiplying each entry in A by c, as shown below.

Scalar   Matrix

$$3 \begin{bmatrix} -1 & 2 \\ 6 & 5 \end{bmatrix} = \begin{bmatrix} 3(-1) & 3(2) \\ 3(6) & 3(5) \end{bmatrix} = \begin{bmatrix} -3 & 6 \\ 18 & 15 \end{bmatrix}$$

Definition of Scalar Multiplication

If $A = [a_{ij}]$ is an $m \times n$ matrix and c is a scalar, then the **scalar multiple** of A by c is the $m \times n$ matrix given by

$$cA = [ca_{ij}].$$

STUDY TIP

The order of operations for matrix expressions is similar to that for real numbers. In particular, you perform scalar multiplication before matrix addition and subtraction, as shown in Example 3(c).

The symbol $-A$ represents the negation of A, or the scalar product $(-1)A$. Moreover, if A and B are of the same dimension, then $A - B$ represents the sum of A and $(-1)B$. That is,

$$A - B = A + (-1)B.$$ Subtraction of matrices

Example 3   **Scalar Multiplication and Matrix Subtraction**

For the following matrices, find (a) $3A$, (b) $-A$, and (c) $3A - B$.

$$A = \begin{bmatrix} 2 & 2 & 4 \\ -3 & 0 & -1 \\ 2 & 1 & 2 \end{bmatrix} \quad \text{and} \quad B = \begin{bmatrix} 2 & 0 & 0 \\ 1 & -4 & 3 \\ -1 & 3 & 2 \end{bmatrix}$$

SOLUTION

a. $3A = 3\begin{bmatrix} 2 & 2 & 4 \\ -3 & 0 & -1 \\ 2 & 1 & 2 \end{bmatrix}$ Scalar multiplication

$$= \begin{bmatrix} 3(2) & 3(2) & 3(4) \\ 3(-3) & 3(0) & 3(-1) \\ 3(2) & 3(1) & 3(2) \end{bmatrix}$$ Multiply each entry by 3.

$$= \begin{bmatrix} 6 & 6 & 12 \\ -9 & 0 & -3 \\ 6 & 3 & 6 \end{bmatrix}$$ Simplify.

b. $-A = (-1)\begin{bmatrix} 2 & 2 & 4 \\ -3 & 0 & -1 \\ 2 & 1 & 2 \end{bmatrix}$ Definition of negation

$$= \begin{bmatrix} -2 & -2 & -4 \\ 3 & 0 & 1 \\ -2 & -1 & -2 \end{bmatrix}$$ Multiply each entry by -1.

c. $3A - B = \begin{bmatrix} 6 & 6 & 12 \\ -9 & 0 & -3 \\ 6 & 3 & 6 \end{bmatrix} - \begin{bmatrix} 2 & 0 & 0 \\ 1 & -4 & 3 \\ -1 & 3 & 2 \end{bmatrix}$ Perform scalar multiplication first.

$$= \begin{bmatrix} 4 & 6 & 12 \\ -10 & 4 & -6 \\ 7 & 0 & 4 \end{bmatrix}$$ Subtract corresponding entries.

TECH TUTOR

Most graphing utilities have the capability of performing matrix operations. Consult the user's guide for your graphing utility for more information. Try using a graphing utility to check the solutions to Example 3.

✓ **Checkpoint 3**

For the following matrices, find (a) $2A$ and (b) $2A - B$.

$$A = \begin{bmatrix} 2 & 4 & -1 \\ 0 & 1 & 3 \\ -3 & 2 & 5 \end{bmatrix} \quad \text{and} \quad B = \begin{bmatrix} 0 & 6 & 3 \\ 7 & -4 & 1 \\ 2 & 0 & -2 \end{bmatrix}$$

It is often convenient to rewrite the scalar multiple cA by factoring c out of every entry in the matrix. For instance, in the matrix below, the scalar $\frac{1}{2}$ has been factored out of the matrix.

$$\begin{bmatrix} \frac{1}{2} & -\frac{3}{2} \\ \frac{5}{2} & \frac{1}{2} \end{bmatrix} = \frac{1}{2}\begin{bmatrix} 1 & -3 \\ 5 & 1 \end{bmatrix}$$

The properties of matrix addition and scalar multiplication are similar to those of addition and multiplication of real numbers.

ALGEBRA TUTOR
xy

To review addition and multiplication properties of real numbers, see the *Chapter 6 Algebra Tutor* on page 530.

Properties of Matrix Addition and Scalar Multiplication

If A, B, and C are $m \times n$ matrices and c and d are scalars, then the following properties are true.

1. $A + B = B + A$ Commutative Property of Matrix Addition

2. $A + (B + C) = (A + B) + C$ Associative Property of Matrix Addition

3. $(cd)A = c(dA)$ Associative Property of Scalar Multiplication

4. $1A = A$ Scalar Identity Property

5. $c(A + B) = cA + cB$ Distributive Property

6. $(c + d)A = cA + dA$ Distributive Property

Note that the Associative Property of Matrix Addition allows you to write expressions such as $A + B + C$ without ambiguity, because you obtain the same sum regardless of how the matrices are grouped. In other words, you obtain the same sum whether you group $A + B + C$ as $(A + B) + C$ or as $A + (B + C)$. This same reasoning applies to sums of four or more matrices.

ALGEBRA TUTOR
xy

For more examples of the algebra involved in matrix addition and scalar multiplication, see the *Chapter 6 Algebra Tutor* on page 531.

Example 4 **Addition of More than Two Matrices**

By adding corresponding entries, you can obtain the sum of four matrices, as shown below.

$$\begin{bmatrix} 1 \\ 2 \\ -3 \end{bmatrix} + \begin{bmatrix} -1 \\ -1 \\ 2 \end{bmatrix} + \begin{bmatrix} 0 \\ 1 \\ 4 \end{bmatrix} + \begin{bmatrix} 2 \\ -3 \\ -2 \end{bmatrix} = \begin{bmatrix} 2 \\ -1 \\ 1 \end{bmatrix}$$

✓ **Checkpoint 4**

Add the matrices $\begin{bmatrix} 2 & -9 \\ 0 & 1 \end{bmatrix}$, $\begin{bmatrix} -1 & 4 \\ 7 & -3 \end{bmatrix}$, and $\begin{bmatrix} 0 & 6 \\ 5 & -2 \end{bmatrix}$.

STUDY TIP

In Example 5, you could add the two matrices first and then multiply the resulting matrix by 3. The result would be the same.

Example 5 **Using the Distributive Property**

$$3\left(\begin{bmatrix} -2 & 0 \\ 4 & 1 \end{bmatrix} + \begin{bmatrix} 4 & -2 \\ 3 & 7 \end{bmatrix} \right) = 3\begin{bmatrix} -2 & 0 \\ 4 & 1 \end{bmatrix} + 3\begin{bmatrix} 4 & -2 \\ 3 & 7 \end{bmatrix}$$

$$= \begin{bmatrix} -6 & 0 \\ 12 & 3 \end{bmatrix} + \begin{bmatrix} 12 & -6 \\ 9 & 21 \end{bmatrix}$$

$$= \begin{bmatrix} 6 & -6 \\ 21 & 24 \end{bmatrix}$$

✓ **Checkpoint 5**

Evaluate the expression using the Distributive Property.

$$4\left(\begin{bmatrix} -4 & 0 \\ 0 & 2 \end{bmatrix} - \begin{bmatrix} 2 & 1 \\ 3 & -6 \end{bmatrix} \right)$$

One important property of addition of real numbers is that the number 0 is the additive identity. That is, $c + 0 = c$ for any real number c. For matrices, a similar property holds. That is, if A is an $m \times n$ matrix and O is the $m \times n$ **zero matrix** consisting entirely of zeros, then $A + O = A$.

In other words, O is the **additive identity** for the set of all $m \times n$ matrices. For example, the following matrices are the additive identities for the sets of all 2×3 and 2×2 matrices, respectively.

$$O = \begin{bmatrix} 0 & 0 & 0 \\ 0 & 0 & 0 \end{bmatrix} \quad \text{and} \quad O = \begin{bmatrix} 0 & 0 \\ 0 & 0 \end{bmatrix}.$$

<u>2 × 3 zero matrix</u>   <u>2 × 2 zero matrix</u>

The algebra of real numbers and the algebra of matrices have many similarities. For example, compare the following solutions.

Real Numbers *(Solve for x.)*	*m × n Matrices* *(Solve for X.)*
$x + a = b$	$X + A = B$
$x + a + (-a) = b + (-a)$	$X + A + (-A) = B + (-A)$
$x + 0 = b - a$	$X + O = B - A$
$x = b - a$	$X = B - A$

This means that you can apply some of your knowledge of solving real number equations to solving matrix equations. It is often easier to complete the algebraic steps first, and then substitute the matrices in the equation, as shown in Example 6.

Example 6   Solving a Matrix Equation

Solve for X in the equation $3X + A = B$, where

$$A = \begin{bmatrix} 1 & -2 \\ 0 & 3 \end{bmatrix} \quad \text{and} \quad B = \begin{bmatrix} -3 & 4 \\ 2 & 1 \end{bmatrix}.$$

SOLUTION   Begin by solving the equation for X to obtain

$$3X + A = B$$
$$3X = B - A$$
$$X = \frac{1}{3}(B - A).$$

Now, using the matrices A and B, you have

$$X = \frac{1}{3}\left(\begin{bmatrix} -3 & 4 \\ 2 & 1 \end{bmatrix} - \begin{bmatrix} 1 & -2 \\ 0 & 3 \end{bmatrix}\right) \qquad \text{Substitute the matrices.}$$

$$= \frac{1}{3}\begin{bmatrix} -4 & 6 \\ 2 & -2 \end{bmatrix} \qquad \text{Subtract matrix } A \text{ from matrix } B.$$

$$= \begin{bmatrix} -\frac{4}{3} & 2 \\ \frac{2}{3} & -\frac{2}{3} \end{bmatrix}. \qquad \text{Multiply the resulting matrix by } \frac{1}{3}.$$

✓**Checkpoint 6**

Solve for X in the equation $2X - A = B$, where

$$A = \begin{bmatrix} 7 & 0 \\ -1 & 2 \end{bmatrix} \quad \text{and} \quad B = \begin{bmatrix} 3 & 1 \\ 2 & 4 \end{bmatrix}.$$

Matrix Multiplication

The third basic matrix operation is **matrix multiplication.** At first glance, the definition may seem unusual. You will see later, however, that this definition of the product of two matrices has many practical applications.

Definition of Matrix Multiplication

If $A = [a_{ij}]$ is an $m \times n$ matrix and $B = [b_{ij}]$ is an $n \times p$ matrix, then the **product** AB is an $m \times p$ matrix

$$AB = [c_{ij}]$$

where $c_{ij} = a_{i1}b_{1j} + a_{i2}b_{2j} + a_{i3}b_{3j} + \cdots + a_{in}b_{nj}.$

The definition of matrix multiplication indicates a *row-by-column* multiplication, where the entry in the ith row and jth column of the product AB is obtained by multiplying the entries in the ith row of A by the corresponding entries in the jth column of B and then adding the results. The general pattern for matrix multiplication is as follows.

$$
\begin{bmatrix}
a_{11} & a_{12} & a_{13} & \cdots & a_{1n} \\
a_{21} & a_{22} & a_{23} & \cdots & a_{2n} \\
a_{31} & a_{32} & a_{33} & \cdots & a_{3n} \\
\vdots & \vdots & \vdots & & \vdots \\
a_{i1} & a_{i2} & a_{i3} & \cdots & a_{in} \\
\vdots & \vdots & \vdots & & \vdots \\
a_{m1} & a_{m2} & a_{m3} & \cdots & a_{mn}
\end{bmatrix}
\begin{bmatrix}
b_{11} & b_{12} & \cdots & b_{1j} & \cdots & b_{1p} \\
b_{21} & b_{22} & \cdots & b_{2j} & \cdots & b_{2p} \\
b_{31} & b_{32} & \cdots & b_{3j} & \cdots & b_{3p} \\
\vdots & \vdots & & \vdots & & \vdots \\
b_{n1} & b_{n2} & \cdots & b_{nj} & \cdots & b_{np}
\end{bmatrix}
=
\begin{bmatrix}
c_{11} & c_{12} & \cdots & c_{1j} & \cdots & c_{1p} \\
c_{21} & c_{22} & \cdots & c_{2j} & \cdots & c_{2p} \\
\vdots & \vdots & & \vdots & & \vdots \\
c_{i1} & c_{i2} & \cdots & c_{ij} & \cdots & c_{ip} \\
\vdots & \vdots & & \vdots & & \vdots \\
c_{m1} & c_{m2} & \cdots & c_{mj} & \cdots & c_{mp}
\end{bmatrix}
$$

$$a_{i1}b_{1j} + a_{i2}b_{2j} + a_{i3}b_{3j} + \cdots + a_{in}b_{nj} = c_{ij}$$

Example 7 **Finding the Product of Two Matrices**

Find the product AB using $A = \begin{bmatrix} -1 & 3 \\ 4 & -2 \\ 5 & 0 \end{bmatrix}$ and $B = \begin{bmatrix} -3 & 2 \\ -4 & 1 \end{bmatrix}$.

SOLUTION First, note that the product AB is defined because the number of *columns* of A is equal to the number of *rows* of B. Moreover, the product AB has dimension 3×2. To find the entries of the product, multiply each row of A by each column of B.

$$AB = \begin{bmatrix} -1 & 3 \\ 4 & -2 \\ 5 & 0 \end{bmatrix} \begin{bmatrix} -3 & 2 \\ -4 & 1 \end{bmatrix}$$

$$= \begin{bmatrix} (-1)(-3) + (3)(-4) & (-1)(2) + (3)(1) \\ (4)(-3) + (-2)(-4) & (4)(2) + (-2)(1) \\ (5)(-3) + (0)(-4) & (5)(2) + (0)(1) \end{bmatrix}$$

$$= \begin{bmatrix} -9 & 1 \\ -4 & 6 \\ -15 & 10 \end{bmatrix}$$

✓ **Checkpoint 7**

Find the product AB using

$$A = \begin{bmatrix} -2 & 2 \\ 0 & 4 \\ 3 & -1 \end{bmatrix} \quad \text{and} \quad B = \begin{bmatrix} -3 & 0 \\ 1 & -4 \end{bmatrix}.$$

$$
\begin{array}{ccccc}
A & B & = & AB \\
m \times n & n \times p & & m \times p
\end{array}
$$

— Equal —
—Dimension of AB—

Be sure you understand that for the product of two matrices to be defined, the number of *columns* of the first matrix must equal the number of *rows* of the second matrix. That is, the middle two indices must be the same and the outside two indices give the dimension of the product, as shown at the left.

Example 8   Finding the Product of Two Matrices

Find the product AB using $A = \begin{bmatrix} 6 & 2 & 0 \\ 3 & -1 & 2 \\ 1 & 4 & 6 \end{bmatrix}$ and $B = \begin{bmatrix} 1 & 0 \\ 2 & 7 \\ -3 & 5 \end{bmatrix}$.

SOLUTION   Note that the dimension of A is 3×3 and the dimension of B is 3×2. So, the product AB is defined and is of dimension 3×2.

$$
AB = \begin{bmatrix} 6 & 2 & 0 \\ 3 & -1 & 2 \\ 1 & 4 & 6 \end{bmatrix} \begin{bmatrix} 1 & 0 \\ 2 & 7 \\ -3 & 5 \end{bmatrix}
$$

$$
= \begin{bmatrix} 6(1) + 2(2) + 0(-3) & 6(0) + 2(7) + 0(5) \\ 3(1) + (-1)(2) + 2(-3) & 3(0) + (-1)(7) + 2(5) \\ 1(1) + 4(2) + 6(-3) & 1(0) + 4(7) + 6(5) \end{bmatrix}
$$

$$
= \begin{bmatrix} 10 & 14 \\ -5 & 3 \\ -9 & 58 \end{bmatrix}
$$

✓ Checkpoint 8

Find the product AB using

$$
A = \begin{bmatrix} 0 & 4 & -3 \\ 2 & 1 & 7 \\ 3 & -2 & 1 \end{bmatrix} \quad \text{and} \quad B = \begin{bmatrix} -2 & 0 \\ 0 & -4 \\ 1 & 2 \end{bmatrix}.
$$

Example 9   Patterns in Matrix Multiplication

a. $\begin{bmatrix} 1 & 0 & 3 \\ 2 & -1 & -2 \end{bmatrix} \begin{bmatrix} -2 & 4 & 2 \\ 1 & 0 & 0 \\ -1 & 1 & -1 \end{bmatrix} = \begin{bmatrix} -5 & 7 & -1 \\ -3 & 6 & 6 \end{bmatrix}$

$\quad\quad\quad 2 \times 3 \quad\quad\quad\quad 3 \times 3 \quad\quad\quad\quad 2 \times 3$

b. $\begin{bmatrix} 3 & 4 \\ -2 & 5 \end{bmatrix} \begin{bmatrix} 1 & 0 \\ 0 & 1 \end{bmatrix} = \begin{bmatrix} 3 & 4 \\ -2 & 5 \end{bmatrix}$

$\quad\quad 2 \times 2 \quad\quad 2 \times 2 \quad\quad 2 \times 2$

c. The product AB for the following matrices is not defined.

$$
A = \begin{bmatrix} -2 & 1 \\ 1 & -3 \\ 1 & 4 \end{bmatrix} \quad \text{and} \quad B = \begin{bmatrix} -2 & 3 & 1 & 4 \\ 0 & 1 & -1 & 2 \\ 2 & -1 & 0 & 1 \end{bmatrix}
$$

$\quad\quad\quad\quad 3 \times 2 \quad\quad\quad\quad\quad\quad 3 \times 4$

✓ Checkpoint 9

Find AB, if possible, using $A = \begin{bmatrix} 2 & 0 & 1 \\ 6 & 1 & -3 \end{bmatrix}$ and $B = \begin{bmatrix} -5 & 3 \\ 3 & 0 \end{bmatrix}$.

Example 10   Patterns in Matrix Multiplication

a. $\begin{bmatrix} 1 & -2 \end{bmatrix} \begin{bmatrix} 2 \\ -1 \end{bmatrix} = \begin{bmatrix} 4 \end{bmatrix}$

$\quad\quad 1 \times 2 \quad\; 2 \times 1 \quad\; 1 \times 1$

b. $\begin{bmatrix} 2 \\ -1 \end{bmatrix} \begin{bmatrix} 1 & -2 \end{bmatrix} = \begin{bmatrix} 2 & -4 \\ -1 & 2 \end{bmatrix}$

$\quad\quad 2 \times 1 \quad\; 1 \times 2 \quad\quad 2 \times 2$

✓ **Checkpoint 10**

Find AB and BA using

$$A = \begin{bmatrix} 3 & -1 \end{bmatrix} \quad \text{and} \quad B = \begin{bmatrix} 1 \\ -3 \end{bmatrix}.$$

In Example 10, note that the two products are different. Even when AB and BA are defined, matrix multiplication is not, in general, commutative. That is, for most matrices, $AB \neq BA$.

Properties of Matrix Multiplication

Let A, B, and C be matrices and let c be a scalar.

1. $A(BC) = (AB)C$ Associative Property of Matrix Multiplication

2. $A(B + C) = AB + AC$ Left Distributive Property

3. $(A + B)C = AC + BC$ Right Distributive Property

4. $c(AB) = (cA)B = A(cB)$ Associative Property of Scalar Multiplication

---

Definition of the Identity Matrix

The $n \times n$ matrix that consists of 1's on its main diagonal and 0's elsewhere is called the **identity matrix of dimension $n \times n$** and is denoted by

$$I_n = \begin{bmatrix} 1 & 0 & 0 & \cdots & 0 \\ 0 & 1 & 0 & \cdots & 0 \\ 0 & 0 & 1 & \cdots & 0 \\ \vdots & \vdots & \vdots & & \vdots \\ 0 & 0 & 0 & \cdots & 1 \end{bmatrix}. \quad \text{Identity matrix}$$

Note that an identity matrix must be *square*. When the dimension is understood to be $n \times n$, you can denote I_n simply by I.

If A is an $n \times n$ matrix, then the identity matrix has the property that $AI_n = A$ and $I_nA = A$. For example,

$$\begin{bmatrix} 3 & -2 & 5 \\ 1 & 0 & 4 \\ -1 & 2 & -3 \end{bmatrix} \begin{bmatrix} 1 & 0 & 0 \\ 0 & 1 & 0 \\ 0 & 0 & 1 \end{bmatrix} = \begin{bmatrix} 3 & -2 & 5 \\ 1 & 0 & 4 \\ -1 & 2 & -3 \end{bmatrix} \quad AI = A$$

and

$$\begin{bmatrix} 1 & 0 & 0 \\ 0 & 1 & 0 \\ 0 & 0 & 1 \end{bmatrix} \begin{bmatrix} 3 & -2 & 5 \\ 1 & 0 & 4 \\ -1 & 2 & -3 \end{bmatrix} = \begin{bmatrix} 3 & -2 & 5 \\ 1 & 0 & 4 \\ -1 & 2 & -3 \end{bmatrix}. \quad IA = A$$

Applications

One application of matrix multiplication is the representation of a system of linear equations. Note how the system below can be written as the matrix equation $AX = B$ where A is the *coefficient matrix* of the system, B is the *constant matrix* of the system, and X is a column matrix.

System

$$\begin{cases} a_{11}x_1 + a_{12}x_2 + a_{13}x_3 = b_1 \\ a_{21}x_1 + a_{22}x_2 + a_{23}x_3 = b_2 \\ a_{31}x_1 + a_{32}x_2 + a_{33}x_3 = b_3 \end{cases}$$

Matrix Equation $AX = B$

$$\underbrace{\begin{bmatrix} a_{11} & a_{12} & a_{13} \\ a_{21} & a_{22} & a_{23} \\ a_{31} & a_{32} & a_{33} \end{bmatrix}}_{A} \times \underbrace{\begin{bmatrix} x_1 \\ x_2 \\ x_3 \end{bmatrix}}_{X} = \underbrace{\begin{bmatrix} b_1 \\ b_2 \\ b_3 \end{bmatrix}}_{B}$$

In Example 11, note that $[A \vdots B]$ represents the augmented matrix formed when matrix B is *adjoined* to matrix A. Also, $[I \vdots X]$ represents the reduced row-echelon form of the augmented matrix that yields the solution of the system.

Example 11   Solving a System of Linear Equations

Consider the system of linear equations.

$$\begin{cases} x_1 - 2x_2 + x_3 = -4 \\ x_2 + 2x_3 = 4 \\ 2x_1 + 3x_2 - 2x_3 = 2 \end{cases}$$

a. Write the system as a matrix equation $AX = B$.

b. Use Gauss-Jordan elimination on the augmented matrix $[A \vdots B]$ to solve for the matrix X.

SOLUTION

a. In matrix form $AX = B$, the system is written as

$$\underbrace{\begin{bmatrix} 1 & -2 & 1 \\ 0 & 1 & 2 \\ 2 & 3 & -2 \end{bmatrix}}_{\text{Coefficient matrix}} \begin{bmatrix} x_1 \\ x_2 \\ x_3 \end{bmatrix} = \underbrace{\begin{bmatrix} -4 \\ 4 \\ 2 \end{bmatrix}}_{\text{Constant matrix}}.$$

b. The augmented matrix is formed by adjoining matrix B to matrix A.

$$[A \vdots B] = \begin{bmatrix} 1 & -2 & 1 & \vdots & -4 \\ 0 & 1 & 2 & \vdots & 4 \\ 2 & 3 & -2 & \vdots & 2 \end{bmatrix}$$

Using Gauss-Jordan elimination, you can rewrite this matrix as

$$[I \vdots X] = \begin{bmatrix} 1 & 0 & 0 & \vdots & -1 \\ 0 & 1 & 0 & \vdots & 2 \\ 0 & 0 & 1 & \vdots & 1 \end{bmatrix}.$$

So, the solution of the matrix equation is

$$X = \begin{bmatrix} x_1 \\ x_2 \\ x_3 \end{bmatrix} = \begin{bmatrix} -1 \\ 2 \\ 1 \end{bmatrix}.$$

✓ **Checkpoint 11**

For the system of linear equations, (a) write the system as a matrix equation $AX = B$, and (b) use Gauss-Jordan elimination on $[A \vdots B]$ to solve for the matrix X.

$$\begin{cases} -2x_1 - 3x_2 = -4 \\ 6x_1 + x_2 = -36 \end{cases}$$

 **Example 12**   **Softball Team Expenses**

Two softball teams submit equipment lists to their sponsors, as shown in the table.

Equipment	Women's team	Men's team
Bats	12	15
Balls	45	38
Gloves	15	17

Each bat costs $90, each ball costs $6, and each glove costs $60. Use matrices to find the total cost of equipment for each team.

SOLUTION   The equipment lists E and the costs per item C can be written in matrix form as

$$E = \begin{bmatrix} 12 & 15 \\ 45 & 38 \\ 15 & 17 \end{bmatrix} \quad \text{and} \quad C = \begin{bmatrix} 90 & 6 & 60 \end{bmatrix}.$$

You can find the total cost of the equipment for each team using the product CE because the number of columns of C (3 columns) equals the number of rows of E (3 rows). The total cost of equipment for each team is given by

$$\begin{aligned} CE &= \begin{bmatrix} 90 & 6 & 60 \end{bmatrix} \begin{bmatrix} 12 & 15 \\ 45 & 38 \\ 15 & 17 \end{bmatrix} \\ &= \begin{bmatrix} 90(12) + 6(45) + 60(15) & 90(15) + 6(38) + 60(17) \end{bmatrix} \\ &= \begin{bmatrix} 2250 & 2598 \end{bmatrix}. \end{aligned}$$

So, the total cost of equipment for the women's team is $2250, and the total cost of equipment for the men's team is $2598. ────────────

✓**Checkpoint 12**

Repeat Example 12 when each bat costs $100, each ball costs $7, and each glove costs $65.

SUMMARIZE   (Section 6.2)

1. State the conditions under which two matrices are equal *(page 487)*. For an example of the equality of matrices, see Example 1.

2. State the definition of matrix addition *(page 488)*. For an example of matrix addition, see Example 2.

3. State the definition of scalar multiplication *(page 488)*. For an example of scalar multiplication, see Example 3.

4. List the properties of matrix addition and scalar multiplication *(page 490)*. For examples of using these properties, see Examples 4, 5, and 6.

5. State the definition of matrix multiplication *(page 492)*. For examples of matrix multiplication, see Examples 7, 8, 9, and 10.

SKILLS WARM UP 6.2   The following warm-up exercises involve skills that were covered in earlier sections. You will use these skills in the exercise set for this section. For additional help, review Sections 0.2 and 6.1.

In Exercises 1 and 2, evaluate the expression.

1. $-3\left(-\frac{5}{6}\right) + 10\left(-\frac{3}{4}\right)$

2. $-22\left(\frac{5}{2}\right) + 6(8)$

In Exercises 3 and 4, determine whether the matrix is in reduced row-echelon form.

3. $\begin{bmatrix} 0 & 1 & 0 & -5 \\ 1 & 0 & 3 & 2 \\ 0 & 0 & 1 & 0 \end{bmatrix}$

4. $\begin{bmatrix} 1 & 0 & 0 & 2 & 3 \\ 0 & 0 & 0 & 0 & 0 \\ 0 & 1 & 1 & 3 & 10 \end{bmatrix}$

In Exercises 5 and 6, write the augmented matrix for the system of linear equations.

5. $\begin{cases} -5x + 10y = 12 \\ 7x - 3y = 0 \end{cases}$

6. $\begin{cases} 10x + 15y - 9z = 42 \\ 6x - 5y = 0 \end{cases}$

In Exercises 7–10, solve the system of linear equations represented by the augmented matrix.

7. $\begin{bmatrix} 1 & 0 & \vdots & 0 \\ 0 & 1 & \vdots & 2 \end{bmatrix}$

8. $\begin{bmatrix} 1 & 0 & -1 & \vdots & 2 \\ 0 & 1 & 1 & \vdots & 3 \end{bmatrix}$

9. $\begin{bmatrix} 1 & 2 & 1 & \vdots & 0 \\ 0 & 0 & 1 & \vdots & -1 \\ 0 & 0 & 0 & \vdots & 0 \end{bmatrix}$

10. $\begin{bmatrix} 1 & -1 & 0 & \vdots & 3 \\ 0 & 1 & -2 & \vdots & 1 \\ 0 & 0 & 1 & \vdots & -1 \end{bmatrix}$

Exercises 6.2

See www.CalcChat.com for worked-out solutions to odd-numbered exercises.

Equality of Matrices   In Exercises 1–4, find x and y.
See Example 1.

1. $\begin{bmatrix} 4 & x \\ -1 & y \end{bmatrix} = \begin{bmatrix} 4 & -3 \\ -1 & 2 \end{bmatrix}$

2. $\begin{bmatrix} x & -7 \\ 9 & y \end{bmatrix} = \begin{bmatrix} 5 & -7 \\ 9 & -8 \end{bmatrix}$

3. $\begin{bmatrix} -4 & 3 \\ 6 & -1 \\ 8 & 2 \\ 5 & 9 \end{bmatrix} = \begin{bmatrix} x-2 & 3 \\ 6 & -1 \\ 8 & -x \\ 5 & 2y-1 \end{bmatrix}$

4. $\begin{bmatrix} x+2 & 8 & -3 \\ 1 & 2y & 2x \\ 7 & -2 & y+2 \end{bmatrix} = \begin{bmatrix} 2x+6 & 8 & -3 \\ 1 & 18 & -8 \\ 7 & -2 & 11 \end{bmatrix}$

Operations with Matrices   In Exercises 5–14, find (a) $A + B$, (b) $A - B$, (c) $6A$, and (d) $4A - 3B$. See *Examples 2 and 3.*

5. $A = \begin{bmatrix} 7 & -3 \end{bmatrix}$, $B = \begin{bmatrix} -4 & 2 \end{bmatrix}$

6. $A = \begin{bmatrix} -5 \\ -5 \end{bmatrix}$, $B = \begin{bmatrix} -1 \\ 10 \end{bmatrix}$

7. $A = \begin{bmatrix} 5 & -2 \\ 3 & 1 \end{bmatrix}$, $B = \begin{bmatrix} 3 & 1 \\ -2 & 6 \end{bmatrix}$

8. $A = \begin{bmatrix} 7 & 4 \\ -4 & 5 \end{bmatrix}$, $B = \begin{bmatrix} -3 & 1 \\ 8 & -4 \end{bmatrix}$

9. $A = \begin{bmatrix} 6 & -1 \\ 2 & 4 \\ -3 & 5 \end{bmatrix}$, $B = \begin{bmatrix} 1 & 4 \\ -1 & 5 \\ 1 & 10 \end{bmatrix}$

10. $A = \begin{bmatrix} 6 & 8 & -3 & 2 & 1 \\ -4 & 2 & 1 & 5 & -2 \end{bmatrix}$,
$B = \begin{bmatrix} 6 & 0 & 4 & -1 & 3 \\ 4 & 5 & -2 & 1 & 2 \end{bmatrix}$

11. $A = \begin{bmatrix} 2 & 2 & -1 \\ 1 & 1 & -2 \\ 1 & -1 & 3 \end{bmatrix}$, $B = \begin{bmatrix} 1 & 1 & -1 \\ -3 & 4 & 9 \\ 0 & -7 & 8 \end{bmatrix}$

12. $A = \begin{bmatrix} 3 \\ 2 \\ -1 \end{bmatrix}$, $B = \begin{bmatrix} -4 \\ 6 \\ 2 \end{bmatrix}$

13. $A = \begin{bmatrix} -13 & 24 & 20 & -11 \end{bmatrix}$,
$B = \begin{bmatrix} 15 & 16 & 10 & -12 \end{bmatrix}$

14. $A = \begin{bmatrix} 12 & 5 & -24 \\ -10 & 0 & 13 \end{bmatrix}$, $B = \begin{bmatrix} 17 & -6 & 0 \\ -1 & 15 & 18 \end{bmatrix}$

Evaluating an Expression In Exercises 15–24, evaluate the expression. *See Examples 4 and 5.*

15. $\begin{bmatrix} -5 & 0 \\ 3 & -6 \end{bmatrix} + \begin{bmatrix} 7 & 1 \\ -2 & -1 \end{bmatrix} + \begin{bmatrix} -10 & -8 \\ 14 & 6 \end{bmatrix}$

16. $\begin{bmatrix} 6 & 8 \\ -1 & 0 \end{bmatrix} + \begin{bmatrix} 0 & 5 \\ -3 & -1 \end{bmatrix} + \begin{bmatrix} -11 & -7 \\ 2 & -1 \end{bmatrix}$

17. $[4 \;\; -8] + [-3 \;\; 1] + [2 \;\; 0] + [9 \;\; 1]$

18. $\begin{bmatrix} 7 \\ -1 \\ 5 \\ -5 \end{bmatrix} + \begin{bmatrix} 0 \\ -4 \\ -9 \\ 3 \end{bmatrix} + \begin{bmatrix} 1 \\ 8 \\ 8 \\ 2 \end{bmatrix} + \begin{bmatrix} -3 \\ -5 \\ -7 \\ -1 \end{bmatrix}$

19. $5\left(\begin{bmatrix} 1 & -3 \\ -2 & -8 \end{bmatrix} + \begin{bmatrix} 10 & -7 \\ 5 & 4 \end{bmatrix} \right)$

20. $-7([-3 \;\; -1 \;\; 6] - [-5 \;\; 4 \;\; -4])$

21. $8\left(\begin{bmatrix} -4 & 0 & 1 \\ 0 & 2 & 3 \end{bmatrix} - \begin{bmatrix} 2 & 1 & -2 \\ 3 & -6 & 0 \end{bmatrix} \right)$

22. $\frac{1}{2}([5 \;\; -2 \;\; 4 \;\; 0] + [14 \;\; 6 \;\; -18 \;\; 9])$

23. $-3\left(\begin{bmatrix} 0 & -3 \\ 7 & 2 \end{bmatrix} + \begin{bmatrix} -6 & 3 \\ 8 & 1 \end{bmatrix} \right) - 2\begin{bmatrix} 4 & -4 \\ 7 & -9 \end{bmatrix}$

24. $-1\begin{bmatrix} 4 & 11 \\ -2 & -1 \\ 9 & 3 \end{bmatrix} + \frac{1}{6}\left(\begin{bmatrix} -5 & -1 \\ 3 & 4 \\ 0 & 13 \end{bmatrix} + \begin{bmatrix} 7 & 5 \\ -9 & -1 \\ 6 & -1 \end{bmatrix} \right)$

Operations with Matrices In Exercises 25–28, use the matrix capabilities of a graphing utility to evaluate the expression. Round your results to three decimal places, if necessary.

25. $\frac{3}{7}\begin{bmatrix} 2 & 5 \\ -1 & -4 \end{bmatrix} + 6\begin{bmatrix} -3 & 0 \\ 2 & 2 \end{bmatrix}$

26. $55\left(\begin{bmatrix} 14 & -11 \\ -22 & 19 \end{bmatrix} + \begin{bmatrix} -22 & 20 \\ 13 & 6 \end{bmatrix} \right)$

27. $-\begin{bmatrix} 3.211 & 6.829 \\ -1.004 & 4.914 \\ 0.055 & -3.889 \end{bmatrix} - \begin{bmatrix} -1.630 & -3.090 \\ 5.256 & 8.335 \\ -9.768 & 4.251 \end{bmatrix}$

28. $-12\left(\begin{bmatrix} 6 & 20 \\ 1 & -9 \\ -2 & 5 \end{bmatrix} + \begin{bmatrix} 14 & -15 \\ -8 & -6 \\ 7 & 0 \end{bmatrix} + \begin{bmatrix} -31 & -19 \\ 16 & 10 \\ 24 & -10 \end{bmatrix} \right)$

Solving a Matrix Equation In Exercises 29–32, solve for *X*, where

$A = \begin{bmatrix} -2 & -1 \\ 1 & 0 \\ 3 & -4 \end{bmatrix}$ and $B = \begin{bmatrix} 0 & 3 \\ 2 & 0 \\ -4 & -1 \end{bmatrix}$. *See Example 6.*

29. $X = 3A - 2B$

30. $2X = 2A - B$

31. $2X + 3A = B$

32. $2A + 4B = -2X$

Finding the Product of Two Matrices In Exercises 33–44, find *AB*, if possible. *See Examples 7, 8, 9, and 10.*

33. $A = [-3 \;\; 5], B = \begin{bmatrix} 4 \\ -4 \end{bmatrix}$

34. $A = \begin{bmatrix} 1 \\ 7 \end{bmatrix}, B = [8 \;\; -2]$

35. $A = \begin{bmatrix} -1 & 2 \\ 4 & -2 \end{bmatrix}, B = \begin{bmatrix} 3 \\ -1 \end{bmatrix}$

36. $A = [-5 \;\; 0], B = \begin{bmatrix} -3 & 1 \\ -1 & 4 \end{bmatrix}$

37. $A = \begin{bmatrix} 3 & -2 \\ 4 & 5 \\ 1 & -1 \end{bmatrix}, B = \begin{bmatrix} -1 & 4 & -2 & 5 \\ 2 & 1 & 3 & -1 \end{bmatrix}$

38. $A = \begin{bmatrix} 0 & -1 & 0 \\ 4 & 0 & 2 \\ 8 & -1 & 7 \end{bmatrix}, B = \begin{bmatrix} 2 & 1 \\ -3 & 4 \\ 1 & 6 \end{bmatrix}$

39. $A = \begin{bmatrix} -1 & 3 \\ 4 & -5 \\ 0 & 2 \end{bmatrix}, B = \begin{bmatrix} 1 & 2 \\ 0 & 7 \end{bmatrix}$

40. $A = \begin{bmatrix} 1 & 0 & 0 \\ 0 & 4 & 0 \\ 0 & 0 & -2 \end{bmatrix}, B = \begin{bmatrix} 3 & 0 & 0 \\ 0 & -1 & 0 \\ 0 & 0 & 5 \end{bmatrix}$

41. $A = \begin{bmatrix} 5 & 0 & 0 \\ 0 & -8 & 0 \\ 0 & 0 & 7 \end{bmatrix}, B = \begin{bmatrix} \frac{1}{5} & 0 & 0 \\ 0 & -\frac{1}{8} & 0 \\ 0 & 0 & \frac{1}{2} \end{bmatrix}$

42. $A = \begin{bmatrix} 6 \\ -2 \\ 1 \\ 6 \end{bmatrix}, B = [10 \;\; 12]$

43. $A = \begin{bmatrix} 0 & 1 & 0 \\ 3 & 0 & 2 \\ 5 & 0 & 0 \end{bmatrix}, B = \begin{bmatrix} 4 \\ -2 \\ 0 \\ 1 \end{bmatrix}$

44. $A = \begin{bmatrix} 1 & 0 & 3 & -2 & 4 \\ 6 & 13 & 8 & -17 & 10 \end{bmatrix}, B = \begin{bmatrix} 1 & 6 \\ 4 & 2 \end{bmatrix}$

Matrix Multiplication In Exercises 45–48, use the matrix capabilities of a graphing utility to find *AB*, if possible.

45. $A = \begin{bmatrix} 5 & 6 & -3 \\ -2 & 5 & 1 \\ 10 & -5 & 5 \end{bmatrix}, B = \begin{bmatrix} 1 & -1 & 2 \\ 8 & 1 & 4 \\ 4 & -2 & 9 \end{bmatrix}$

46. $A = \begin{bmatrix} 11 & -12 & 4 \\ 14 & 10 & 12 \\ 6 & -2 & 9 \end{bmatrix}, B = \begin{bmatrix} 12 & 10 \\ -5 & 12 \\ 15 & 16 \end{bmatrix}$

47. $A = \begin{bmatrix} -2 & 4 & 8 \\ 21 & 5 & 6 \\ 13 & 2 & 6 \end{bmatrix}, B = \begin{bmatrix} 2 & 0 \\ -7 & 15 \\ 32 & 14 \\ 0.5 & 1.6 \end{bmatrix}$

48. $A = \begin{bmatrix} 9 & 10 & -38 & 18 \\ 100 & -50 & 250 & 75 \end{bmatrix}$,

$B = \begin{bmatrix} 52 & -85 & 27 & 45 \\ 40 & -35 & 60 & 82 \end{bmatrix}$

Operations with Matrices   In Exercises 49–54, find (a) AB, (b) BA, and, if possible, (c) A^2. (*Note:* $A^2 = AA$)

49. $A = \begin{bmatrix} 1 & 2 \\ 4 & 2 \end{bmatrix}$, $B = \begin{bmatrix} 2 & -1 \\ -1 & 8 \end{bmatrix}$

50. $A = \begin{bmatrix} 2 & -1 \\ 1 & 4 \end{bmatrix}$, $B = \begin{bmatrix} 0 & 0 \\ 3 & -3 \end{bmatrix}$

51. $A = \begin{bmatrix} -1 & 2 & 3 \\ 4 & 1 & -1 \end{bmatrix}$, $B = \begin{bmatrix} 1 & 3 \\ -1 & -2 \\ 2 & 4 \end{bmatrix}$

52. $A = \begin{bmatrix} 1 & -1 & 7 \\ 2 & -1 & 8 \\ 3 & 1 & -1 \end{bmatrix}$, $B = \begin{bmatrix} 1 & 1 & 2 \\ 2 & 1 & 1 \\ 1 & -3 & 2 \end{bmatrix}$

53. $A = \begin{bmatrix} -4 & 2 & 3 \end{bmatrix}$, $B = \begin{bmatrix} 1 \\ 0 \\ 5 \end{bmatrix}$

54. $A = \begin{bmatrix} 3 & 2 & 1 & 0 \end{bmatrix}$, $B = \begin{bmatrix} 2 \\ 3 \\ 1 \\ 0 \end{bmatrix}$

Think About It   In Exercises 55–64, let matrices A, B, C, and D be dimensions 2×3, 2×3, 3×2, and 2×2, respectively. Determine whether it is possible to perform the operation(s). If so, give the dimension of the result.

55. $A + 2C$ **56.** $B - 3C$

57. AB **58.** BC

59. AD **60.** CD

61. $BC - D$ **62.** $CB - D$

63. $D(A - 3B)$ **64.** $(BC - D)A$

Solving a System of Linear Equations   In Exercises 65–72, (a) write the system of linear equations as a matrix equation $AX = B$, and (b) use Gauss-Jordan elimination on the augmented matrix $[A \,\vdots\, B]$ to solve for the matrix X. See Example 11.

65. $\begin{cases} -x + y = 4 \\ -2x + y = 0 \end{cases}$ **66.** $\begin{cases} 2x + 3y = 5 \\ x + 4y = 10 \end{cases}$

67. $\begin{cases} x + 2y = 3 \\ 3x - y = 2 \end{cases}$

68. $\begin{cases} -4x - 3y = -2 \\ x + y = -1 \end{cases}$

69. $\begin{cases} x - 4y + 5z = 3 \\ 2x + 5y + z = -24 \\ -x - y - z = 8 \end{cases}$

70. $\begin{cases} 2x - 4y + z = 0 \\ -x + 3y + z = 1 \\ x + y = 3 \end{cases}$

71. $\begin{cases} x - 2y + 3z = 9 \\ -x + 3y - z = -6 \\ 2x - 5y + 5z = 17 \end{cases}$ **72.** $\begin{cases} x + y - 3z = -1 \\ -x + 2y = 1 \\ x - y + z = 2 \end{cases}$

73. Manufacturing   A corporation that makes sunglasses has four factories, each of which manufactures two products. The number of units of product i produced at factory j in one day is represented by a_{ij} in the matrix

$$A = \begin{bmatrix} 100 & 120 & 60 & 40 \\ 140 & 160 & 200 & 80 \end{bmatrix}.$$

Find the production levels when production is increased by 10%. (*Hint:* Because an increase of 10% corresponds to 100% + 10%, multiply the matrix by 1.10.)

74. Manufacturing   A tire corporation has three factories, each of which manufactures two products. The number of units of product i produced at factory j in one day is represented by a_{ij} in the matrix

$$A = \begin{bmatrix} 80 & 120 & 140 \\ 40 & 100 & 80 \end{bmatrix}.$$

Find the production levels when production is decreased by 5%. (*Hint:* Because a decrease of 5% corresponds to 100% − 5%, multiply the matrix by 0.95.)

75. Vacation Packages   A vacation service has identified four resort hotels with a special all-inclusive package. The quoted room rates are for double and family (maximum of four people) occupancy for 5 days and 4 nights. The current rates for the two types of rooms at the four hotels are represented by the matrix A.

$$A = \begin{matrix} & \begin{matrix} \text{Hotel} \\ w \end{matrix} & \begin{matrix} \text{Hotel} \\ x \end{matrix} & \begin{matrix} \text{Hotel} \\ y \end{matrix} & \begin{matrix} \text{Hotel} \\ z \end{matrix} \\ \begin{bmatrix} 615 & 670 & 740 & 990 \\ 995 & 1030 & 1180 & 1105 \end{bmatrix} & & & \begin{matrix} \text{Double} \\ \text{Family} \end{matrix} \end{matrix} \Big\} \text{Occupancy}$$

Room rates are guaranteed not to increase by more than 12% by next season. What is the maximum rate per package per hotel?

76. HOW DO YOU SEE IT?   A fruit grower raises apples and peaches, which are shipped to three different outlets. The numbers of units of apples and peaches that are shipped to the three outlets are shown in the matrix A.

$$A = \begin{matrix} & \overbrace{\begin{matrix} \text{X} & \text{Y} & \text{Z} \end{matrix}}^{\text{Outlet}} \\ \begin{bmatrix} 125 & 100 & 75 \\ 100 & 175 & 125 \end{bmatrix} & \begin{matrix} \text{Apples} \\ \text{Peaches} \end{matrix} \end{matrix} \Big\} \text{Units shipped}$$

(a) The profit per unit of apples is $3.50 and the profit per unit of peaches is $6. Organize the profits per unit in a matrix B.

(b) The product BA is given by the matrix

$BA = \begin{bmatrix} \$1037.50 & \$1400.00 & \$1012.50 \end{bmatrix}$.

Explain what this matrix represents.

(c) Can you compute AB to find the profits? Explain.

77. Inventory Levels A company sells five different models of laptop computers through three retail outlets. The inventories of the five models at the three outlets are given by the matrix S.

Model

$$S = \begin{bmatrix} \overset{A}{3} & \overset{B}{2} & \overset{C}{2} & \overset{D}{3} & \overset{E}{0} \\ 0 & 2 & 3 & 4 & 3 \\ 4 & 2 & 1 & 3 & 2 \end{bmatrix} \begin{matrix} 1 \\ 2 \\ 3 \end{matrix} \text{Outlet}$$

The wholesale and retail prices for each model are given by the matrix T.

Price

$$T = \begin{bmatrix} \overset{\text{Wholesale}}{\$250} & \overset{\text{Retail}}{\$475} \\ \$400 & \$700 \\ \$575 & \$850 \\ \$650 & \$950 \\ \$760 & \$1200 \end{bmatrix} \begin{matrix} A \\ B \\ C \\ D \\ E \end{matrix} \text{Model}$$

(a) What is the total retail price of the inventory at Outlet 1?

(b) What is the total wholesale price of the inventory at Outlet 3?

(c) Compute the product ST and interpret the result in the context of the problem.

78. Labor/Wage Requirements A company that manufactures boats has the following labor-hour and wage requirements.

Labor-Hour Requirements (per boat)

Department

$$S = \begin{bmatrix} \overset{\text{Cutting}}{1.0 \text{ hour}} & \overset{\text{Assembly}}{0.5 \text{ hour}} & \overset{\text{Packaging}}{0.2 \text{ hour}} \\ 1.6 \text{ hours} & 1.0 \text{ hour} & 0.2 \text{ hour} \\ 2.5 \text{ hours} & 2.0 \text{ hours} & 0.4 \text{ hour} \end{bmatrix} \begin{matrix} \text{Small} \\ \text{Medium} \\ \text{Large} \end{matrix} \begin{matrix} \text{Boat} \\ \text{size} \end{matrix}$$

Wage Requirements (per hour)

Plant

$$T = \begin{bmatrix} \overset{A}{\$15} & \overset{B}{\$13} \\ \$12 & \$11 \\ \$11 & \$10 \end{bmatrix} \begin{matrix} \text{Cutting} \\ \text{Assembly} \\ \text{Packaging} \end{matrix} \text{Department}$$

(a) What is the labor cost for a medium boat at Plant B?

(b) What is the labor cost for a large boat at Plant A?

(c) Compute ST and interpret the result.

79. Exercise The numbers of calories burned by individuals of different body weights while performing different types of aerobic exercises for a one-hour time period are shown in the matrix A.

Calories burned

$$A = \begin{bmatrix} \overset{\substack{130\text{-lb} \\ \text{person}}}{472} & \overset{\substack{155\text{-lb} \\ \text{person}}}{563} \\ 590 & 704 \\ 177 & 211 \end{bmatrix} \begin{matrix} \text{Basketball} \\ \text{Jumping rope} \\ \text{Weight lifting} \end{matrix}$$

(a) A 130-pound person and a 155-pound person played basketball for 2 hours, jumped rope for 15 minutes, and lifted weights for 30 minutes. Organize the times spent exercising in a matrix B.

(b) Compute BA and interpret the result.

80. Exercise The numbers of calories burned by individuals of different body weights while performing different types of aerobic exercises for a 30-minute time period are shown in the matrix A.

Calories burned

$$A = \begin{bmatrix} \overset{\substack{180\text{-lb} \\ \text{person}}}{143} & \overset{\substack{205\text{-lb} \\ \text{person}}}{163} \\ 286 & 326 \\ 164 & 186 \end{bmatrix} \begin{matrix} \text{Calisthenics} \\ \text{Tennis} \\ \text{Water aerobics} \end{matrix}$$

(a) A 180-pound person and a 205-pound person performed calisthenics for 30 minutes, played tennis for 1 hour, and did water aerobics for 45 minutes. Organize the times spent exercising in a matrix B.

(b) Compute BA and interpret the result.

81. Contract Bonuses Professional athletes frequently have bonus or incentive clauses in their contracts. For example, a defensive football player might receive bonuses for defensive plays such as sacks, interceptions, and/or key tackles. In one contract, a sack is worth $2000, an interception is worth $1000, and a key tackle is worth $800. The table shows the numbers of sacks, interceptions, and key tackles for three players.

Player	Sacks	Interceptions	Key tackles
Player X	3	0	4
Player Y	1	2	5
Player Z	2	3	3

(a) Write a matrix D that represents the number of each type of defensive play i made by each player j using the data from the table. State what each entry d_{ij} of the matrix represents.

(b) Write a matrix B that represents the bonus amount received for each type of defensive play. State what each entry b_{ij} of the matrix represents.

(c) Find the product BD of the two matrices and state what each entry of matrix BD represents.

(d) Which player receives the largest bonus?

82. Health Care　The health care plans offered this year by a local manufacturing plant are as follows. For individuals, the comprehensive plan costs \$694.32, the HMO standard plan costs \$451.80, and the HMO Plus plan costs \$489.48. For families, the comprehensive plan costs \$1725.36, the HMO standard plan costs \$1187.76, and the HMO Plus plan costs \$1248.12. The plant expects the costs of the plans to change next year as follows. For individuals, the costs for the comprehensive, HMO standard, and HMO Plus plans will be \$683.91, \$463.10, and \$499.27, respectively. For families, the costs for the comprehensive, HMO standard, and HMO Plus plans will be \$1699.48, \$1217.45, and \$1273.08, respectively.

(a) Organize the information using two matrices A and B, where A represents the health care plan costs for this year and B represents the health care plan costs for next year. State what each entry of each matrix represents.

(b) Compute $A - B$ and interpret the result.

(c) The employees receive monthly paychecks from which the health care plan costs are deducted. Use the matrices from part (a) to write matrices that show how much will be deducted from each employee's paycheck this year and next year.

(d) Suppose instead that the costs of the health care plans increase by 4% next year. Write a matrix that shows the new monthly payments.

83. Voting Preference　The matrix

From

$$P = \begin{bmatrix} 0.6 & 0.1 & 0.1 \\ 0.2 & 0.7 & 0.1 \\ 0.2 & 0.2 & 0.8 \end{bmatrix} \begin{matrix} R \\ D \\ I \end{matrix} \Big\} \text{To}$$

with column labels R, D, I

is called a *stochastic matrix*. Each entry p_{ij} $(i \neq j)$ represents the proportion of the voting population that changes from Party i to Party j, and p_{ii} represents the proportion that remains loyal to the party from one election to the next. Use a graphing utility to find P^2. (This matrix gives the transition probabilities from the first election to the third.)

84. Voting Preference　Use a graphing utility to find P^3, P^4, P^5, P^6, P^7, and P^8 for the matrix given in Exercise 83. Can you detect a pattern?

Finding a Matrix　In Exercises 85 and 86, find a matrix B such that AB is the identity matrix. Is there more than one correct result?

85. $A = \begin{bmatrix} 1 & 3 \\ 1 & 2 \end{bmatrix}$　　**86.** $A = \begin{bmatrix} 2 & 1 \\ 5 & 2 \end{bmatrix}$

87. Think About It　If a, b, and c are real numbers such that $c \neq 0$ and $ac = bc$, then $a = b$. However, if A, B, and C are matrices such that $AC = BC$, then A is *not* necessarily equal to B. Illustrate this using the following matrices.

$$A = \begin{bmatrix} 1 & 2 & 3 \\ 0 & 5 & 4 \\ 3 & -2 & 1 \end{bmatrix}, B = \begin{bmatrix} 4 & -6 & 3 \\ 5 & 4 & 4 \\ -1 & 0 & 1 \end{bmatrix},$$

$$\text{and } C = \begin{bmatrix} 0 & 0 & 0 \\ 0 & 0 & 0 \\ 4 & -2 & 3 \end{bmatrix}$$

88. Think About It　If a and b are real numbers such that $ab = 0$, then $a = 0$ or $b = 0$. However, if A and B are matrices such that $AB = O$, then it is *not* necessarily true that $A = O$ or $B = O$. Illustrate this using the matrices $A = \begin{bmatrix} 3 & 3 \\ 4 & 4 \end{bmatrix}$ and $B = \begin{bmatrix} 1 & -1 \\ -1 & 1 \end{bmatrix}$. Find another example of two nonzero matrices whose product is the zero matrix.

89. Satellite Television　Two competing satellite television providers offer service to a city with 100,000 households. Gold Cable Company has 25,000 subscribers and Galaxy Cable Company has 30,000 subscribers. (The other 45,000 households do not subscribe.) The percent changes in subscriptions each year are shown in the matrix below.

Percent Changes

		From Gold	From Galaxy	From Non-subscriber
Percent Changes	To Gold	0.70	0.15	0.15
	To Galaxy	0.20	0.80	0.15
	To Nonsubscriber	0.10	0.05	0.70

Use matrix multiplication to find the number of subscribers each company will have in one year, in two years, and in three years. What is happening to the number of subscribers and nonsubscribers to each company? Explain.

90. Project: Sales of e-Readers　For a project involving the sales of e-readers, visit this textbook's website at *www.cengagebrain.com*.

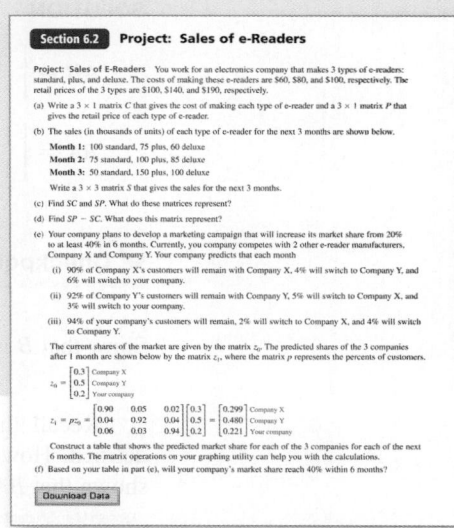

6.3 The Inverse of a Square Matrix

■ Verify that a matrix is the inverse of a given matrix.
■ Find the inverse of a matrix.
■ Find the inverse of a 2×2 matrix using a formula.
■ Use an inverse matrix to solve a system of linear equations.

The Inverse of a Matrix

This section further develops the algebra of matrices. To begin, consider the real number equation $ax = b$. To solve this equation for x, multiply each side of the equation by a^{-1} (provided $a \neq 0$).

$$ax = b$$
$$(a^{-1}a)x = a^{-1}b$$
$$(1)x = a^{-1}b$$
$$x = a^{-1}b$$

The number a^{-1} is called the *multiplicative inverse* of a because $a^{-1}a = 1$. The definition of the multiplicative inverse of a matrix is similar.

Definition of the Inverse of a Square Matrix

Let A be an $n \times n$ matrix and let I_n be the $n \times n$ identity matrix. If there exists a matrix A^{-1} such that

$$AA^{-1} = I_n = A^{-1}A$$

then A^{-1} is called the **inverse** of A. (The symbol A^{-1} is read "A inverse.")

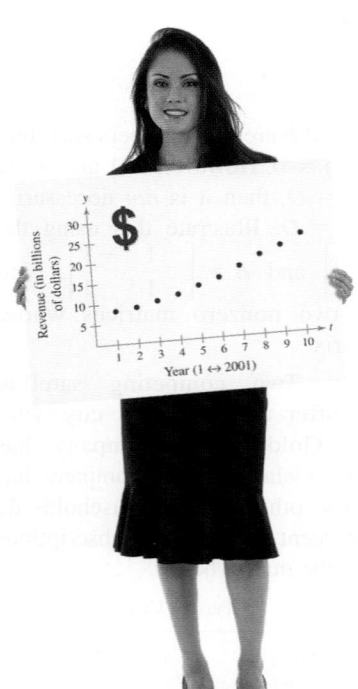

In Exercise 82 on page 511, you will use an inverse matrix to find an equation that models the annual revenues for a company.

Example 1 **The Inverse of a Matrix**

Show that B is the inverse of A, where

$$A = \begin{bmatrix} -1 & 2 \\ -1 & 1 \end{bmatrix} \quad \text{and} \quad B = \begin{bmatrix} 1 & -2 \\ 1 & -1 \end{bmatrix}.$$

SOLUTION To show that B is the inverse of A, show that $AB = I = BA$.

$$AB = \begin{bmatrix} -1 & 2 \\ -1 & 1 \end{bmatrix}\begin{bmatrix} 1 & -2 \\ 1 & -1 \end{bmatrix} = \begin{bmatrix} -1+2 & 2-2 \\ -1+1 & 2-1 \end{bmatrix} = \begin{bmatrix} 1 & 0 \\ 0 & 1 \end{bmatrix}$$

$$BA = \begin{bmatrix} 1 & -2 \\ 1 & -1 \end{bmatrix}\begin{bmatrix} -1 & 2 \\ -1 & 1 \end{bmatrix} = \begin{bmatrix} -1+2 & 2-2 \\ -1+1 & 2-1 \end{bmatrix} = \begin{bmatrix} 1 & 0 \\ 0 & 1 \end{bmatrix}$$

✓**Checkpoint 1**

Show that B is the inverse of A, where $A = \begin{bmatrix} 3 & 5 \\ -1 & -2 \end{bmatrix}$ and $B = \begin{bmatrix} 2 & 5 \\ -1 & -3 \end{bmatrix}$.   ■

Recall that it is not always true that $AB = BA$, even when both products are defined. However, if A and B are both square matrices and $AB = I_n$, then it can be shown that $BA = I_n$. So, in Example 1, you need only check that $AB = I_2$.

Finding Inverse Matrices

If a matrix A has an inverse, then A is called **invertible** (or **nonsingular**); otherwise, A is called **singular.** A nonsquare matrix cannot have an inverse. To see this, note that when A is of dimension $m \times n$ and B is of dimension $n \times m$ (where $m \neq n$), the products AB and BA are of different dimensions and therefore cannot be equal to each other. Not all square matrices have inverses (see the matrix at the bottom of page 505). When, however, a matrix does have an inverse, that inverse is unique. The next example shows how to use a system of equations to find an inverse.

Example 2   Finding the Inverse of a Matrix

Find the inverse of the matrix

$$A = \begin{bmatrix} 1 & 4 \\ -1 & -3 \end{bmatrix}.$$

SOLUTION   To find the inverse of A, try to solve the matrix equation $AX = I$ for X.

$$\overset{A}{\begin{bmatrix} 1 & 4 \\ -1 & -3 \end{bmatrix}} \overset{X}{\begin{bmatrix} x_{11} & x_{12} \\ x_{21} & x_{22} \end{bmatrix}} = \overset{I}{\begin{bmatrix} 1 & 0 \\ 0 & 1 \end{bmatrix}} \qquad \text{Write matrix equation.}$$

$$\begin{bmatrix} x_{11} + 4x_{21} & x_{12} + 4x_{22} \\ -x_{11} - 3x_{21} & -x_{12} - 3x_{22} \end{bmatrix} = \begin{bmatrix} 1 & 0 \\ 0 & 1 \end{bmatrix} \qquad \text{Multiply.}$$

Equating corresponding entries, you obtain the following two systems of linear equations.

$$\begin{cases} x_{11} + 4x_{21} = 1 \\ -x_{11} - 3x_{21} = 0 \end{cases} \qquad \begin{cases} x_{12} + 4x_{22} = 0 \\ -x_{12} - 3x_{22} = 1 \end{cases}$$

You can solve these systems using the methods learned in Chapter 5. From the first system you can determine that $x_{11} = -3$ and $x_{21} = 1$. From the second system you can determine that $x_{12} = -4$ and $x_{22} = 1$. So, the inverse of A is

$$X = A^{-1}$$

$$= \begin{bmatrix} -3 & -4 \\ 1 & 1 \end{bmatrix}.$$

You can use matrix multiplication to check this result.

CHECK

$$AA^{-1} = \begin{bmatrix} 1 & 4 \\ -1 & -3 \end{bmatrix} \begin{bmatrix} -3 & -4 \\ 1 & 1 \end{bmatrix}$$

$$= \begin{bmatrix} 1 & 0 \\ 0 & 1 \end{bmatrix} \checkmark$$

$$A^{-1}A = \begin{bmatrix} -3 & -4 \\ 1 & 1 \end{bmatrix} \begin{bmatrix} 1 & 4 \\ -1 & -3 \end{bmatrix}$$

$$= \begin{bmatrix} 1 & 0 \\ 0 & 1 \end{bmatrix} \checkmark$$

✓ Checkpoint 2

Find the inverse of the matrix

$$A = \begin{bmatrix} -4 & -1 \\ 5 & 1 \end{bmatrix}.$$

In Example 2, note that the two systems of linear equations have the *same coefficient matrix A*. Rather than solve the two systems represented by

$$\begin{bmatrix} 1 & 4 & \vdots & 1 \\ -1 & -3 & \vdots & 0 \end{bmatrix}$$

and

$$\begin{bmatrix} 1 & 4 & \vdots & 0 \\ -1 & -3 & \vdots & 1 \end{bmatrix}$$

separately, you can solve them simultaneously by adjoining the identity matrix to the coefficient matrix to obtain

$$\begin{array}{cc} A & \quad I \\ \begin{bmatrix} 1 & 4 & \vdots & 1 & 0 \\ -1 & -3 & \vdots & 0 & 1 \end{bmatrix}. \end{array}$$

This "doubly augmented" matrix can be represented as

$$[A \; \vdots \; I].$$

Then, by applying Gauss-Jordan elimination to this matrix, you can solve *both* systems with a single elimination process.

$$\begin{bmatrix} 1 & 4 & \vdots & 1 & 0 \\ -1 & -3 & \vdots & 0 & 1 \end{bmatrix}$$

$$R_1 + R_2 \rightarrow \begin{bmatrix} 1 & 4 & \vdots & 1 & 0 \\ 0 & 1 & \vdots & 1 & 1 \end{bmatrix}$$

$$-4R_2 + R_1 \rightarrow \begin{bmatrix} 1 & 0 & \vdots & -3 & -4 \\ 0 & 1 & \vdots & 1 & 1 \end{bmatrix}$$

So, from the "doubly augmented" matrix $[A \; \vdots \; I]$, you obtained the matrix $[I \; \vdots \; A^{-1}]$.

$$\begin{array}{cc} A & \quad I \\ \begin{bmatrix} 1 & 4 & \vdots & 1 & 0 \\ -1 & -3 & \vdots & 0 & 1 \end{bmatrix} \end{array} \implies \begin{array}{cc} I & \quad A^{-1} \\ \begin{bmatrix} 1 & 0 & \vdots & -3 & -4 \\ 0 & 1 & \vdots & 1 & 1 \end{bmatrix} \end{array}$$

This procedure (or algorithm) works for any square matrix that has an inverse.

Finding an Inverse Matrix

Let A be a square matrix of dimension $n \times n$.

1. Write the $n \times 2n$ matrix that consists of the given matrix A on the left and the $n \times n$ identity matrix I on the right to obtain

   $$[A \; \vdots \; I].$$

2. If possible, row reduce A to I using elementary row operations on the *entire* matrix

   $$[A \; \vdots \; I].$$

   The result will be the matrix

   $$[I \; \vdots \; A^{-1}].$$

   If this is not possible, then A is not invertible.

3. Check your work by multiplying to see that

   $$AA^{-1} = I = A^{-1}A.$$

Adrian Britton/www.shutterstock.com

TECH TUTOR

Most graphing utilities can find the inverse of a matrix by using the inverse key $\boxed{x^{-1}}$. The screen below shows how one graphing utility displays the inverse of a matrix. For specific instructions on how to use the inverse key to find the inverse of a matrix, consult the user's guide for your graphing utility.

```
[A]
        [[1   4 ]
         [-1  -3]]
[A]-1
        [[-3  -4]
         [1   1 ]]
```

Example 3 **Finding the Inverse of a Matrix**

Find the inverse of the matrix $A = \begin{bmatrix} 1 & -1 & 0 \\ 1 & 0 & -1 \\ 6 & -2 & -3 \end{bmatrix}$.

SOLUTION Begin by adjoining the identity matrix to A to form the matrix

$$[A \,\vdots\, I] = \begin{bmatrix} 1 & -1 & 0 & \vdots & 1 & 0 & 0 \\ 1 & 0 & -1 & \vdots & 0 & 1 & 0 \\ 6 & -2 & -3 & \vdots & 0 & 0 & 1 \end{bmatrix}.$$

Next, use elementary row operations to obtain the matrix $[I \,\vdots\, A^{-1}]$.

$$\begin{array}{l} -R_1 + R_2 \rightarrow \\ -6R_1 + R_3 \rightarrow \end{array} \begin{bmatrix} 1 & -1 & 0 & \vdots & 1 & 0 & 0 \\ 0 & 1 & -1 & \vdots & -1 & 1 & 0 \\ 0 & 4 & -3 & \vdots & -6 & 0 & 1 \end{bmatrix}$$

$$\begin{array}{l} R_2 + R_1 \rightarrow \\ \\ -4R_2 + R_3 \rightarrow \end{array} \begin{bmatrix} 1 & 0 & -1 & \vdots & 0 & 1 & 0 \\ 0 & 1 & -1 & \vdots & -1 & 1 & 0 \\ 0 & 0 & 1 & \vdots & -2 & -4 & 1 \end{bmatrix}$$

$$\begin{array}{l} R_3 + R_1 \rightarrow \\ R_3 + R_2 \rightarrow \end{array} \begin{bmatrix} 1 & 0 & 0 & \vdots & -2 & -3 & 1 \\ 0 & 1 & 0 & \vdots & -3 & -3 & 1 \\ 0 & 0 & 1 & \vdots & -2 & -4 & 1 \end{bmatrix}$$

So, the matrix A is invertible and its inverse is $A^{-1} = \begin{bmatrix} -2 & -3 & 1 \\ -3 & -3 & 1 \\ -2 & -4 & 1 \end{bmatrix}$.

CHECK Confirm this result by multiplying A and A^{-1} to obtain I.

$$AA^{-1} = \begin{bmatrix} 1 & -1 & 0 \\ 1 & 0 & -1 \\ 6 & -2 & -3 \end{bmatrix} \begin{bmatrix} -2 & -3 & 1 \\ -3 & -3 & 1 \\ -2 & -4 & 1 \end{bmatrix} = \begin{bmatrix} 1 & 0 & 0 \\ 0 & 1 & 0 \\ 0 & 0 & 1 \end{bmatrix} = I$$

✓**Checkpoint 3**

Find the inverse of the matrix $A = \begin{bmatrix} 1 & -2 & -1 \\ 0 & -1 & 2 \\ 1 & -2 & 0 \end{bmatrix}$. ▪

The process shown in Example 3 applies to any $n \times n$ matrix A. So, when A has an inverse, this process will find it. When using this process, if the matrix A does not reduce to the identity matrix, then A does not have an inverse.

For instance, consider the matrix A shown below. To confirm that A has no inverse, begin by adjoining the identity matrix to A.

$$A = \begin{bmatrix} 1 & 2 & 0 \\ 3 & -1 & 2 \\ -2 & 3 & -2 \end{bmatrix} \implies [A \,\vdots\, I] = \begin{bmatrix} 1 & 2 & 0 & \vdots & 1 & 0 & 0 \\ 3 & -1 & 2 & \vdots & 0 & 1 & 0 \\ -2 & 3 & -2 & \vdots & 0 & 0 & 1 \end{bmatrix}$$

Then use elementary row operations to obtain

$$\begin{bmatrix} 1 & 2 & 0 & \vdots & 1 & 0 & 0 \\ 0 & -7 & 2 & \vdots & -3 & 1 & 0 \\ 0 & 0 & 0 & \vdots & -2 & 1 & 1 \end{bmatrix}.$$

At this point in the elimination process, you can see that it is impossible to obtain the identity matrix I on the left. So, A is not invertible.

The Inverse of a 2 × 2 Matrix (Quick Method)

Using Gauss-Jordan elimination to find the inverse of a matrix works well (even as a computer technique) for matrices of dimensions 3×3 or greater. For 2×2 matrices, however, many people prefer to use a formula for the inverse rather than Gauss-Jordan elimination. This formula works *only* for 2×2 matrices. If A is a 2×2 matrix given by

$$A = \begin{bmatrix} a & b \\ c & d \end{bmatrix}$$

then A is invertible if and only if $ad - bc \neq 0$. Moreover, when $ad - bc \neq 0$, the inverse is given by

$$A^{-1} = \frac{1}{ad - bc} \begin{bmatrix} d & -b \\ -c & a \end{bmatrix}.$$ Formula for inverse of a 2 × 2 matrix

Try verifying this inverse by multiplication.

The denominator $ad - bc$ is called the **determinant** of the 2×2 matrix A. You will study determinants in the next section.

Example 4   Finding the Inverse of a 2 × 2 Matrix

If possible, find the inverse of each matrix.

a. $A = \begin{bmatrix} 3 & -1 \\ -2 & 2 \end{bmatrix}$ **b.** $B = \begin{bmatrix} 3 & -1 \\ -6 & 2 \end{bmatrix}$

SOLUTION

a. For matrix A, begin by applying the formula for the determinant of a 2×2 matrix to obtain

$$ad - bc = 3(2) - (-1)(-2) = 4.$$

Because this quantity is not zero, the matrix is invertible. The inverse is formed by interchanging the entries on the main diagonal, changing the signs of the other two entries, and multiplying by the scalar $\frac{1}{4}$.

$$A^{-1} = \frac{1}{ad - bc} \begin{bmatrix} d & -b \\ -c & a \end{bmatrix}$$ Formula for inverse of a 2 × 2 matrix

$$= \frac{1}{4} \begin{bmatrix} 2 & 1 \\ 2 & 3 \end{bmatrix}$$ Substitute for a, b, c, d, and the determinant.

$$= \begin{bmatrix} \frac{1}{4}(2) & \frac{1}{4}(1) \\ \frac{1}{4}(2) & \frac{1}{4}(3) \end{bmatrix}$$ Multiply by the scalar $\frac{1}{4}$.

$$= \begin{bmatrix} \frac{1}{2} & \frac{1}{4} \\ \frac{1}{2} & \frac{3}{4} \end{bmatrix}$$ Simplify.

b. For matrix B, you have

$$ad - bc = 3(2) - (-1)(-6)$$

$$= 0.$$

Because $ad - bc = 0$, B is not invertible.

✓Checkpoint 4

Find the inverse of the matrix

$$A = \begin{bmatrix} -2 & 3 \\ 4 & -1 \end{bmatrix}.$$

Systems of Linear Equations

You know that a system of linear equations can have exactly one solution, infinitely many solutions, or no solution. If the coefficient matrix A of a *square* system (a system that has the same number of equations as variables) is invertible, then the system has a unique solution, which is defined as follows.

A System of Equations with a Unique Solution

If A is an invertible matrix, then the system of linear equations represented by $AX = B$ has a unique solution given by $X = A^{-1}B$.

Example 5   **Solving a System of Equations Using an Inverse Matrix**

Use an inverse matrix to solve the system.

$$\begin{cases} 2x + 3y + z = -1 \\ 3x + 3y + z = 1 \\ 2x + 4y + z = -2 \end{cases}$$

SOLUTION   Begin by writing the system in the matrix form $AX = B$.

$$\begin{bmatrix} 2 & 3 & 1 \\ 3 & 3 & 1 \\ 2 & 4 & 1 \end{bmatrix} \begin{bmatrix} x \\ y \\ z \end{bmatrix} = \begin{bmatrix} -1 \\ 1 \\ -2 \end{bmatrix}$$

Next, use Gauss-Jordan elimination to find A^{-1}.

$$A^{-1} = \begin{bmatrix} -1 & 1 & 0 \\ -1 & 0 & 1 \\ 6 & -2 & -3 \end{bmatrix}$$

Finally, multiply B by A^{-1} on the left to obtain the solution.

$$X = A^{-1}B = \begin{bmatrix} -1 & 1 & 0 \\ -1 & 0 & 1 \\ 6 & -2 & -3 \end{bmatrix} \begin{bmatrix} -1 \\ 1 \\ -2 \end{bmatrix} = \begin{bmatrix} 2 \\ -1 \\ -2 \end{bmatrix}$$

So, the solution is $x = 2$, $y = -1$, and $z = -2$.

✓ Checkpoint 5

Use an inverse matrix to solve the system.

$$\begin{cases} -x + y + z = 4 \\ 2x - y - 3z = -7 \\ -2x + 3y + 2z = 10 \end{cases}$$

SUMMARIZE   (Section 6.3)

1. State the definition of the inverse of a square matrix *(page 502)*. For an example of showing that a matrix is the inverse of another matrix, see Example 1.

2. Explain how to find an inverse matrix *(page 504)*. For an example of finding an inverse matrix, see Example 3.

3. Give the formula for finding the inverse of a 2×2 matrix *(page 506)*. For an example of using this formula, see Example 4.

The following warm-up exercises involve skills that were covered in earlier sections. You will use these skills in the exercise set for this section. For additional help, review Sections 6.1 and 6.2.

In Exercises 1–8, perform the indicated matrix operation(s).

1. $4\begin{bmatrix} 1 & 6 \\ 0 & -4 \\ 12 & 2 \end{bmatrix}$

2. $\dfrac{1}{2}\begin{bmatrix} 11 & 10 & 48 \\ 1 & 0 & 16 \\ 0 & 2 & 8 \end{bmatrix}$

3. $\begin{bmatrix} 1 & -10 & 3 \\ 4 & 1 & 0 \end{bmatrix} - 2\begin{bmatrix} 3 & -4 & 8 \\ 0 & 7 & 1 \end{bmatrix}$

4. $\begin{bmatrix} 5 & 20 \\ -7 & 15 \end{bmatrix} - 3\begin{bmatrix} 6 & 3 \\ 4 & -2 \end{bmatrix}$

5. $\begin{bmatrix} 1 & -2 \\ -1 & 3 \end{bmatrix}\begin{bmatrix} 3 & 2 \\ 1 & 1 \end{bmatrix}$

6. $\begin{bmatrix} 1 & 0 \\ 0 & 1 \end{bmatrix}\begin{bmatrix} 6 & 5 \\ 3 & -2 \end{bmatrix}$

7. $\begin{bmatrix} 2 & 0 & 0 \\ 0 & -1 & 0 \\ 0 & 0 & 3 \end{bmatrix}\begin{bmatrix} \frac{1}{2} & 0 & 0 \\ 0 & -1 & 0 \\ 0 & 0 & \frac{1}{3} \end{bmatrix}$

8. $\begin{bmatrix} 1 & 2 & 3 \\ 3 & -1 & -2 \\ 3 & 1 & 1 \end{bmatrix}\begin{bmatrix} 1 & 1 & -1 \\ -9 & -8 & 11 \\ 6 & 5 & -7 \end{bmatrix}$

In Exercises 9 and 10, write the matrix in reduced row-echelon form.

9. $\begin{bmatrix} 3 & -2 & 1 & 0 \\ 4 & -3 & 0 & 1 \end{bmatrix}$

10. $\begin{bmatrix} 1 & 1 & 2 & 1 & 0 & 0 \\ -1 & 0 & 3 & 0 & 1 & 0 \\ 1 & 2 & 8 & 0 & 0 & 1 \end{bmatrix}$

Exercises 6.3

See www.CalcChat.com for worked-out solutions to odd-numbered exercises.

The Inverse of a Matrix   In Exercises 1–10, show that B is the inverse of A. *See Example 1.*

1. $A = \begin{bmatrix} 7 & 4 \\ 5 & 3 \end{bmatrix}$, $B = \begin{bmatrix} 3 & -4 \\ -5 & 7 \end{bmatrix}$

2. $A = \begin{bmatrix} -4 & 1 \\ -9 & 2 \end{bmatrix}$, $B = \begin{bmatrix} 2 & -1 \\ 9 & -4 \end{bmatrix}$

3. $A = \begin{bmatrix} 2 & -1 \\ 5 & -4 \end{bmatrix}$, $B = \begin{bmatrix} \frac{4}{3} & -\frac{1}{3} \\ \frac{5}{3} & -\frac{2}{3} \end{bmatrix}$

4. $A = \begin{bmatrix} 1 & -2 \\ 3 & -10 \end{bmatrix}$, $B = \begin{bmatrix} \frac{5}{2} & -\frac{1}{2} \\ \frac{3}{4} & -\frac{1}{4} \end{bmatrix}$

5. $A = \begin{bmatrix} -2 & 2 & 3 \\ 1 & -1 & 0 \\ 0 & 1 & 4 \end{bmatrix}$, $B = \dfrac{1}{3}\begin{bmatrix} -4 & -5 & 3 \\ -4 & -8 & 3 \\ 1 & 2 & 0 \end{bmatrix}$

6. $A = \begin{bmatrix} -1 & 0 & 2 \\ 1 & -2 & 0 \\ 1 & 0 & 3 \end{bmatrix}$, $B = \dfrac{1}{10}\begin{bmatrix} -6 & 0 & 4 \\ -3 & -5 & 2 \\ 2 & 0 & 2 \end{bmatrix}$

7. $A = \begin{bmatrix} 2 & -17 & 11 \\ -1 & 11 & -7 \\ 0 & 3 & -2 \end{bmatrix}$, $B = \begin{bmatrix} 1 & 1 & 2 \\ 2 & 4 & -3 \\ 3 & 6 & -5 \end{bmatrix}$

8. $A = \begin{bmatrix} -1 & 1 & -3 \\ 2 & -1 & 4 \\ -1 & 1 & -2 \end{bmatrix}$, $B = \begin{bmatrix} 2 & 1 & -1 \\ 0 & 1 & 2 \\ -1 & 0 & 1 \end{bmatrix}$

9. $A = \begin{bmatrix} 2 & 0 & 2 & 1 \\ 3 & 0 & 0 & 1 \\ -1 & 1 & -2 & 1 \\ 3 & -1 & 1 & 0 \end{bmatrix}$, $B = \dfrac{1}{3}\begin{bmatrix} -1 & 3 & -2 & -2 \\ -2 & 9 & -7 & -10 \\ 1 & 0 & -1 & -1 \\ 3 & -6 & 6 & 6 \end{bmatrix}$

10. $A = \begin{bmatrix} -1 & 1 & 0 & -1 \\ 1 & -1 & 2 & 0 \\ -1 & 1 & 2 & 0 \\ 0 & -1 & 1 & 1 \end{bmatrix}$, $B = \dfrac{1}{4}\begin{bmatrix} -4 & 1 & 1 & -4 \\ -4 & -1 & 3 & -4 \\ 0 & 1 & 1 & 0 \\ -4 & -2 & 2 & 0 \end{bmatrix}$

Finding the Inverse of a Matrix   In Exercises 11–14, find the inverse of matrix A by solving the matrix equation $AX = I$ for X. *See Example 2.*

11. $\begin{bmatrix} 1 & 2 \\ 3 & 7 \end{bmatrix}$

12. $\begin{bmatrix} -1 & 1 \\ -2 & 1 \end{bmatrix}$

13. $\begin{bmatrix} 8 & 4 \\ -2 & -2 \end{bmatrix}$

14. $\begin{bmatrix} 2 & 3 \\ 1 & 4 \end{bmatrix}$

Finding the Inverse of a Matrix In Exercises 15–32, find the inverse of the matrix (if it exists). *See Example 3.*

15. $\begin{bmatrix} 11 & 1 \\ -1 & 0 \end{bmatrix}$

16. $\begin{bmatrix} -7 & 33 \\ 4 & -19 \end{bmatrix}$

17. $\begin{bmatrix} 0 & 4 \\ -3 & 6 \end{bmatrix}$

18. $\begin{bmatrix} 2 & 3 \\ 6 & 9 \end{bmatrix}$

19. $\begin{bmatrix} 6 & 15 \\ 4 & 10 \end{bmatrix}$

20. $\begin{bmatrix} 2 & 4 \\ -3 & 0 \end{bmatrix}$

21. $\begin{bmatrix} 2 & 7 & 1 \\ -3 & -9 & 2 \end{bmatrix}$

22. $\begin{bmatrix} -2 & 5 \\ 6 & -15 \\ 0 & 1 \end{bmatrix}$

23. $\begin{bmatrix} 1 & 1 & 1 \\ 3 & 5 & 4 \\ 3 & 6 & 5 \end{bmatrix}$

24. $\begin{bmatrix} 1 & 2 & 2 \\ 3 & 7 & 9 \\ -1 & -4 & -7 \end{bmatrix}$

25. $\begin{bmatrix} 1 & 1 & 2 \\ 3 & 1 & 0 \\ -2 & 0 & 3 \end{bmatrix}$

26. $\begin{bmatrix} 3 & 2 & 2 \\ 2 & 2 & 2 \\ -4 & 4 & 3 \end{bmatrix}$

27. $\begin{bmatrix} 3 & 0 & 0 \\ 0 & -2 & 0 \\ 0 & 0 & 4 \end{bmatrix}$

28. $\begin{bmatrix} 2 & 0 & 0 \\ 0 & 3 & 0 \\ 0 & 0 & 5 \end{bmatrix}$

29. $\begin{bmatrix} 1 & 0 & 0 \\ 3 & 4 & 0 \\ 2 & 5 & 5 \end{bmatrix}$

30. $\begin{bmatrix} 1 & 0 & 0 \\ 3 & 0 & 0 \\ 2 & 5 & 5 \end{bmatrix}$

31. $\begin{bmatrix} 1 & 0 & 3 & 0 \\ 0 & 2 & 0 & 4 \\ 1 & 0 & 3 & 0 \\ 0 & 2 & 0 & 4 \end{bmatrix}$

32. $\begin{bmatrix} -1 & 0 & 1 & 0 \\ 0 & 2 & 0 & -1 \\ 2 & 0 & -1 & 0 \\ 0 & -1 & 0 & 1 \end{bmatrix}$

 Finding the Inverse of a Matrix In Exercises 33–40, use the matrix capabilities of a graphing utility to find the inverse of the matrix (if it exists).

33. $\begin{bmatrix} 1 & 2 & -1 \\ 3 & 7 & -10 \\ -5 & -7 & -15 \end{bmatrix}$

34. $\begin{bmatrix} 10 & 5 & -7 \\ -5 & 1 & 4 \\ 3 & 2 & -2 \end{bmatrix}$

35. $\begin{bmatrix} 0.1 & 0.2 & 0.3 \\ -0.3 & 0.2 & 0.2 \\ 0.5 & 0.4 & 0.4 \end{bmatrix}$

36. $\begin{bmatrix} 0.6 & 0 & -0.3 \\ 0.7 & -1 & 0.2 \\ 1 & 0 & -0.9 \end{bmatrix}$

37. $\begin{bmatrix} -\frac{1}{2} & \frac{3}{4} & \frac{1}{4} \\ 1 & 0 & -\frac{3}{2} \\ 0 & -1 & \frac{1}{2} \end{bmatrix}$

38. $\begin{bmatrix} -\frac{5}{6} & \frac{1}{3} & \frac{11}{6} \\ 0 & \frac{2}{3} & 2 \\ 1 & -\frac{1}{2} & -\frac{5}{2} \end{bmatrix}$

39. $\begin{bmatrix} 1 & -3 & 2 & -1 \\ 0 & 4 & -12 & 8 \\ 3 & 0 & 5 & -2 \\ 0 & -3 & 9 & -6 \end{bmatrix}$

40. $\begin{bmatrix} 1 & -2 & -1 & -2 \\ 3 & -5 & -2 & -3 \\ 2 & -5 & -2 & -5 \\ -1 & 4 & 4 & 11 \end{bmatrix}$

Finding the Inverse of a 2 × 2 Matrix In Exercises 41–46, use the formula on page 506 to find the inverse of the matrix (if it exists). *See Example 4.*

41. $\begin{bmatrix} 5 & -2 \\ 2 & 3 \end{bmatrix}$

42. $\begin{bmatrix} 7 & 12 \\ -8 & -5 \end{bmatrix}$

43. $\begin{bmatrix} -4 & -6 \\ 2 & 3 \end{bmatrix}$

44. $\begin{bmatrix} -12 & 3 \\ 5 & -2 \end{bmatrix}$

45. $\begin{bmatrix} \frac{7}{2} & -\frac{3}{4} \\ \frac{1}{5} & \frac{4}{5} \end{bmatrix}$

46. $\begin{bmatrix} -\frac{1}{4} & -\frac{2}{3} \\ \frac{1}{3} & \frac{8}{9} \end{bmatrix}$

Solving a System Using an Inverse Matrix In Exercises 47–50, use the inverse matrix found in Exercise 11 to solve the system of linear equations. *See Example 5.*

47. $\begin{cases} x + 2y = 0 \\ 3x + 7y = 1 \end{cases}$

48. $\begin{cases} x + 2y = -5 \\ 3x + 7y = -16 \end{cases}$

49. $\begin{cases} x + 2y = 8 \\ 3x + 7y = 26 \end{cases}$

50. $\begin{cases} x + 2y = -6 \\ 3x + 7y = -21 \end{cases}$

Solving a System Using an Inverse Matrix In Exercises 51–54, use the inverse matrix found in Exercise 14 to solve the system of linear equations. *See Example 5.*

51. $\begin{cases} 2x + 3y = 5 \\ x + 4y = 10 \end{cases}$

52. $\begin{cases} 2x + 3y = 0 \\ x + 4y = 3 \end{cases}$

53. $\begin{cases} 2x + 3y = 4 \\ x + 4y = 2 \end{cases}$

54. $\begin{cases} 2x + 3y = 1 \\ x + 4y = -2 \end{cases}$

Solving a System Using an Inverse Matrix In Exercises 55 and 56, use the inverse matrix found in Exercise 23 to solve the system of linear equations. *See Example 5.*

55. $\begin{cases} x + y + z = 0 \\ 3x + 5y + 4z = 5 \\ 3x + 6y + 5z = 2 \end{cases}$

56. $\begin{cases} x + y + z = -1 \\ 3x + 5y + 4z = 2 \\ 3x + 6y + 5z = 0 \end{cases}$

Solving a System Using an Inverse Matrix In Exercises 57 and 58, use the inverse matrix found in Exercise 40 to solve the system of linear equations. *See Example 5.*

57. $\begin{cases} x_1 - 2x_2 - x_3 - 2x_4 = 0 \\ 3x_1 - 5x_2 - 2x_3 - 3x_4 = 1 \\ 2x_1 - 5x_2 - 2x_3 - 5x_4 = -1 \\ -x_1 + 4x_2 + 4x_3 + 11x_4 = 2 \end{cases}$

58. $\begin{cases} x_1 - 2x_2 - x_3 - 2x_4 = 1 \\ 3x_1 - 5x_2 - 2x_3 - 3x_4 = -2 \\ 2x_1 - 5x_2 - 2x_3 - 5x_4 = 0 \\ -x_1 + 4x_2 + 4x_3 + 11x_4 = -3 \end{cases}$

Solving a System Using an Inverse Matrix In Exercises 59–66, use an inverse matrix to solve (if possible) the system of linear equations. *See Example 5.*

59. $\begin{cases} 3x + 4y = -2 \\ 5x + 3y = 4 \end{cases}$

60. $\begin{cases} 18x + 12y = 13 \\ 30x + 24y = 23 \end{cases}$

61. $\begin{cases} -0.4x + 0.8y = 1.6 \\ 2x - 4y = 5 \end{cases}$

62. $\begin{cases} 0.2x - 0.6y = 2.4 \\ -x + 1.4y = -8.8 \end{cases}$

63. $\begin{cases} -\frac{1}{4}x + \frac{3}{8}y = -2 \\ \frac{3}{2}x + \frac{3}{4}y = -12 \end{cases}$

64. $\begin{cases} \frac{5}{6}x - \phantom{\frac{8}{5}}y = -20 \\ \frac{4}{3}x - \frac{8}{5}y = -51 \end{cases}$

65. $\begin{cases} 4x - y + z = -5 \\ 2x + 2y + 3z = 10 \\ 5x - 2y + 6z = 1 \end{cases}$

66. $\begin{cases} 4x - 2y + 3z = -2 \\ 2x + 2y + 5z = 16 \\ 8x - 5y - 2z = 4 \end{cases}$

 Using a Graphing Utility   In Exercises 67 and 68, use the matrix capabilities of a graphing utility to solve (if possible) the system of linear equations.

67. $\begin{cases} 7x - 3y + 2w = 41 \\ -2x + y - w = -13 \\ 4x + z - 2w = 12 \\ -x + y - w = -8 \end{cases}$

68. $\begin{cases} 2x + 5y + w = 11 \\ x + 4y + 2z - 2w = -7 \\ 2x - 2y + 5z + w = 3 \\ x - 3w = -1 \end{cases}$

Writing a System of Equations   In Exercises 69 and 70, write a system of equations that has the given coefficient matrix and solution. Use an inverse matrix to verify that the system of equations has the given solution.

69. $\begin{bmatrix} 2 & 1 & 3 \\ 4 & 0 & -2 \\ 0 & 3 & 2 \end{bmatrix}$ $\begin{array}{l} x = 2 \\ y = -3 \\ z = 5 \end{array}$

70. $\begin{bmatrix} 1 & 0 & 2 \\ 1 & 1 & 1 \\ 2 & -1 & 0 \end{bmatrix}$ $\begin{array}{l} x = 5 \\ y = -2 \\ z = 1 \end{array}$

Bond Investment   In Exercises 71–74, you invest in AAA-rated bonds, A-rated bonds, and B-rated bonds. Your average yield is 9% on AAA bonds, 7% on A bonds, and 8% on B bonds. You invest twice as much in B bonds as in A bonds. The desired system of linear equations (where x, y, and z represent the amounts invested in AAA, A, and B bonds, respectively) is as follows.

$\begin{cases} x + y + z = \text{(total investment)} \\ 0.09x + 0.07y + 0.08z = \text{(annual return)} \\ 2y - z = 0 \end{cases}$

Use the inverse of the coefficient matrix of this system to find the amount invested in each type of bond for the given total investment and annual return.

71. Total investment = $35,000; annual return = $2950

72. Total investment = $50,000; annual return = $4180

73. Total investment = $36,000; annual return = $3040

74. Total investment = $45,000; annual return = $3770

Circuit Analysis   In Exercises 75 and 76, consider the circuit shown in the figure. The currents I_1, I_2, and I_3, in amperes, are the solution of the system of linear equations

$\begin{cases} 2I_1 + 4I_3 = E_1 \\ I_2 + 4I_3 = E_2 \\ I_1 + I_2 - I_3 = 0 \end{cases}$

where E_1 and E_2 are voltages. Use the inverse of the coefficient matrix of this system to find the unknown currents for the given voltages.

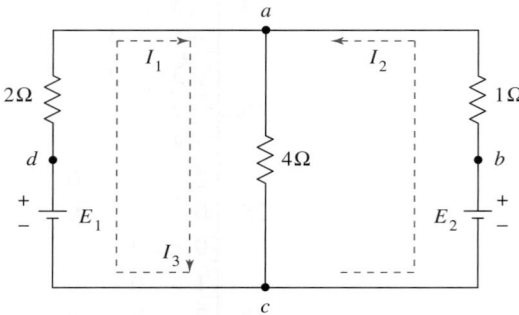

75. $E_1 = 28$ volts, $E_2 = 21$ volts

76. $E_1 = 24$ volts, $E_2 = 23$ volts

Raw Materials   In Exercises 77–80, find the numbers of bags of potting soil that a company can produce for seedlings, general potting, and hardwood plants with the given amounts of raw materials. The raw materials used in one bag of each type of potting soil are shown below.

	Sand	Loam	Peat moss
Seedlings	2 units	1 unit	1 unit
General	1 unit	2 units	1 unit
Hardwoods	2 units	2 units	2 units

77. 500 units of sand
   500 units of loam
   400 units of peat moss

78. 500 units of sand
   750 units of loam
   450 units of peat moss

79. 350 units of sand
   445 units of loam
   345 units of peat moss

80. 975 units of sand
   1050 units of loam
   725 units of peat moss

81. Floral Design A florist is creating 10 centerpieces for the tables at a wedding reception. The customer has a budget of $300 allocated for the centerpieces and wants each centerpiece to contain 12 flowers, with twice as many roses as irises and lilies combined. The florist uses the prices shown to determine the total cost to the customer.

$2.50 per rose

$4.00 per lily

$2.00 per iris

(a) Write a system of linear equations that represents the situation.

(b) Write a matrix equation that corresponds to your system.

(c) Solve your system of linear equations using an inverse matrix. Find the number of flowers of each type that the florist can use to create the 10 centerpieces.

82. DIRECTV® The figure shows the revenue y (in billions of dollars) for DIRECTV each year from 2001 through 2010. The least squares regression parabola $y = at^2 + bt + c$ for these data is found by solving the system

$$\begin{cases} 10c + 55b + 385a = 149.21 \\ 55c + 385b + 3025a = 969.73 \\ 385c + 3025b + 25{,}333a = 7440.01 \end{cases}$$

where t represents the year, with $t = 1$ corresponding to 2001. *(Source: DIRECTV)*

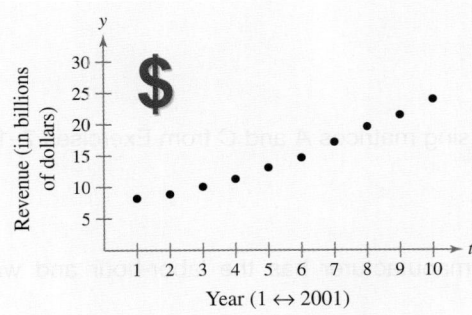

Year (1 ↔ 2001)

(a) Use a graphing utility to find an inverse matrix to solve this system, and find the equation of the least squares regression parabola.

(b) Use the result from part (a) to predict the revenue for DIRECTV in 2012.

(c) In 2011, DIRECTV projected a revenue of $28.9 billion for 2012. How does this compare with your prediction in part (b)? Do both amounts seem reasonable?

83. Discovery Let A, B, and C be the 2×2 matrices

$$A = \begin{bmatrix} 4 & 3 \\ -2 & 1 \end{bmatrix}, \; B = \begin{bmatrix} 1 & -2 \\ 3 & 4 \end{bmatrix}, \text{ and } C = \begin{bmatrix} 13 & 4 \\ 1 & 8 \end{bmatrix}.$$

(a) Find AB and BA. How is each product related to matrix C?

(b) Find C^{-1}, $A^{-1} \cdot B^{-1}$, and $B^{-1} \cdot A^{-1}$. Which of the resulting matrices are equal?

(c) Make a conjecture about the inverse of the product AB of two invertible matrices A and B.

 **84. HOW DO YOU SEE IT?** Let A be the 2×2 matrix

$$A = \begin{bmatrix} x & y \\ 0 & z \end{bmatrix}.$$

Use the determinant of A to state the conditions for which (a) A^{-1} exists and (b) $A^{-1} = A$.

Think About It In Exercises 85 and 86, find a value of k that makes the matrix invertible. (There are many correct answers.) Then find a value of k that makes the matrix singular.

85. $\begin{bmatrix} 4 & 3 \\ -2 & k \end{bmatrix}$ **86.** $\begin{bmatrix} 2k + 1 & 3 \\ -7 & 1 \end{bmatrix}$

True or False? In Exercises 87 and 88, determine whether the statement is true or false. Justify your answer.

87. There exists a matrix A such that $A = A^{-1}$.

88. Multiplication of a nonsingular matrix and its inverse is commutative.

89. Verifying a Formula Verify that the inverse of an invertible 2×2 matrix

$$A = \begin{bmatrix} a & b \\ c & d \end{bmatrix}$$

is given by $A^{-1} = \dfrac{1}{ad - bc} \begin{bmatrix} d & -b \\ -c & a \end{bmatrix}.$

90. Discovery Consider matrices of the form

$$A = \begin{bmatrix} a_{11} & 0 & 0 & 0 & \cdots & 0 \\ 0 & a_{22} & 0 & 0 & \cdots & 0 \\ 0 & 0 & a_{33} & 0 & \cdots & 0 \\ \vdots & \vdots & \vdots & \vdots & \cdots & \vdots \\ 0 & 0 & 0 & 0 & \cdots & a_{nn} \end{bmatrix}$$

(a) Write a 2×2 matrix and a 3×3 matrix of the form of A. Find the inverse of each.

(b) Use the result from part (a) to make a conjecture about the inverses of matrices of the form of A.

QUIZ YOURSELF

See www.CalcChat.com for worked-out solutions to odd-numbered exercises.

Take this quiz as you would take a quiz in class. When you are done, check your work against the answers given in the back of the book.

In Exercises 1 and 2, write a matrix of the given dimension.

1. 4×3

2. 3×1

In Exercises 3 and 4, write the augmented matrix for the system of linear equations.

3. $\begin{cases} 3x + 2y = -2 \\ 5x - y = 19 \end{cases}$

4. $\begin{cases} x \quad\;\; + 3z = -5 \\ x + 2y - z = 3 \\ 3x \quad\;\; + 4z = 0 \end{cases}$

5. Use Gaussian elimination with back-substitution to solve the augmented matrix found in Exercise 3.

6. Use Gauss-Jordan elimination to solve the augmented matrix found in Exercise 4.

In Exercises 7–12, use the following matrices to find the indicated matrix (if possible).

$$A = \begin{bmatrix} -1 & 4 \\ -2 & 6 \end{bmatrix}, \quad B = \begin{bmatrix} -1 & 2 & -3 \\ 2 & 0 & 5 \end{bmatrix}, \quad C = \begin{bmatrix} 0 & -1 \\ 3 & 1 \end{bmatrix}$$

7. $2A + 3C$

8. AB

9. $A - 3C$

10. C^2

11. A^{-1}

12. B^{-1}

In Exercises 13 and 14, solve for X using matrices A and C from Exercises 7–12.

13. $X = 3A - 2C$

14. $2X + 4A = 2C$

In Exercises 15–18, a hang glider manufacturer has the labor-hour and wage requirements indicated at the left.

15. What is the labor cost for model A at Plant 1?

16. What is the labor cost for model B at Plant 2?

17. What is the labor cost for model C at Plant 2?

18. Compute LW and interpret the result.

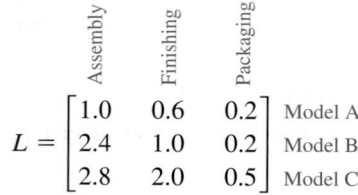

$$L = \begin{bmatrix} 1.0 & 0.6 & 0.2 \\ 2.4 & 1.0 & 0.2 \\ 2.8 & 2.0 & 0.5 \end{bmatrix} \begin{matrix} \text{Model A} \\ \text{Model B} \\ \text{Model C} \end{matrix}$$

Labor-Hour Requirements
(in hours per hang glider)

$$W = \begin{bmatrix} 15 & 12 \\ 10 & 11 \\ 9 & 8 \end{bmatrix} \begin{matrix} \text{Assembly} \\ \text{Finishing} \\ \text{Packaging} \end{matrix}$$

Wage Requirements
(in dollars per hour)

Matrices for 15–18

In Exercises 19 and 20, use an inverse matrix to solve the system of linear equations.

19. $\begin{cases} x - 3y = 10 \\ -2x + y = -10 \end{cases}$

20. $\begin{cases} 2x - y + z = 3 \\ 3x \quad\;\; - z = 15 \\ \quad\;\; 4y + 3z = -1 \end{cases}$

6.4 The Determinant of a Square Matrix

- Evaluate the determinant of a 2 × 2 matrix.
- Find the minors and cofactors of a matrix.
- Find the determinant of a square matrix.

The Determinant of a 2 × 2 Matrix

Every *square* matrix can be associated with a real number called its **determinant.** Determinants have many uses, and several will be discussed in this and the next section. The use of determinants is derived from special number patterns that occur when systems of linear equations are solved. For instance, the system

$$\begin{cases} a_1x + b_1y = c_1 \\ a_2x + b_2y = c_2 \end{cases}$$

has a solution given by

$$x = \frac{c_1b_2 - c_2b_1}{a_1b_2 - a_2b_1}$$

and

$$y = \frac{a_1c_2 - a_2c_1}{a_1b_2 - a_2b_1}$$

In Exercises 91–100 on page 520, you will use determinants to solve equations for *x*.

provided that $a_1b_2 - a_2b_1 \neq 0$. Note that the denominator of each fraction is the same. This denominator is called the *determinant* of the coefficient matrix of the system.

Coefficient Matrix	Determinant
$A = \begin{bmatrix} a_1 & b_1 \\ a_2 & b_2 \end{bmatrix}$	$\det(A) = a_1b_2 - a_2b_1$

The determinant of the matrix A can also be denoted by vertical bars on both sides of the matrix, as indicated in the following definition.

Definition of the Determinant of a 2 × 2 Matrix

The **determinant** of the matrix

$$A = \begin{bmatrix} a_1 & b_1 \\ a_2 & b_2 \end{bmatrix}$$

is given by

$$\det(A) = |A| = \begin{vmatrix} a_1 & b_1 \\ a_2 & b_2 \end{vmatrix} = a_1b_2 - a_2b_1.$$

In this text, $\det(A)$ and $|A|$ are used interchangeably to represent the determinant of A. Although vertical bars are also used to denote the absolute value of a real number, the context will show which use is intended.

A convenient method of remembering the formula for the determinant of a 2 × 2 matrix is shown in the following diagram.

$$\det(A) = \begin{vmatrix} a_1 & b_1 \\ a_2 & b_2 \end{vmatrix} = a_1b_2 - a_2b_1$$

Note that the determinant of a 2 × 2 matrix is the difference of the products of the two diagonals of the matrix.

In Example 1 you will see that the determinant of a matrix can be positive, zero, or negative.

TECH TUTOR

Try evaluating the determinant of A with your graphing utility.

$$A = \begin{bmatrix} 3 & -1 & 1 \\ 0 & 2 & 1 \end{bmatrix}$$

You should obtain an error message similar to the one shown below. The graphing utility displays this message because the matrix is not square.

```
ERR:INVALID DIM
1:Quit
2:Goto
```

Example 1   The Determinant of a 2 × 2 Matrix

Find the determinant of each matrix.

a. $A = \begin{bmatrix} 2 & -3 \\ 1 & 2 \end{bmatrix}$

b. $B = \begin{bmatrix} 2 & 1 \\ 4 & 2 \end{bmatrix}$

c. $C = \begin{bmatrix} 0 & 3 \\ 2 & 4 \end{bmatrix}$

SOLUTION   Use the formula $\det(A) = \begin{vmatrix} a_1 & b_1 \\ a_2 & b_2 \end{vmatrix} = a_1 b_2 - a_2 b_1$.

a. $\det(A) = \begin{vmatrix} 2 & -3 \\ 1 & 2 \end{vmatrix} = 2(2) - 1(-3) = 4 + 3 = 7$

b. $\det(B) = \begin{vmatrix} 2 & 1 \\ 4 & 2 \end{vmatrix} = 2(2) - 4(1) = 4 - 4 = 0$

c. $\det(C) = \begin{vmatrix} 0 & 3 \\ 2 & 4 \end{vmatrix} = 0(4) - 2(3) = 0 - 6 = -6$

✓ **Checkpoint 1**

Find the determinant of each matrix.

a. $A = \begin{bmatrix} 1 & 2 \\ 3 & -1 \end{bmatrix}$ **b.** $B = \begin{bmatrix} 5 & 0 \\ -4 & 2 \end{bmatrix}$ **c.** $C = \begin{bmatrix} 3 & 6 \\ 2 & 4 \end{bmatrix}$ ■

The determinant of a matrix of dimension 1×1 is defined simply as the entry of the matrix. For instance, if $A = [-2]$, then

$$\det(A) = -2.$$

TECH TUTOR

Most graphing utilities can evaluate the determinant of a matrix. Use a graphing utility to find the determinant of matrix A from Example 1. The result should be 7, as shown below. For specific keystrokes on how to use a graphing utility to evaluate the determinant of a matrix, consult the user's guide for your graphing utility.

```
det([A])
              7
```

Minors and Cofactors

To define the determinant of a square matrix of dimension 3×3 or higher, it is helpful to introduce the concepts of **minors** and **cofactors.**

Sign Pattern for Cofactors

$$\begin{bmatrix} + & - & + \\ - & + & - \\ + & - & + \end{bmatrix}$$

3×3 matrix

$$\begin{bmatrix} + & - & + & - \\ - & + & - & + \\ + & - & + & - \\ - & + & - & + \end{bmatrix}$$

4×4 matrix

$$\begin{bmatrix} + & - & + & - & + & \cdots \\ - & + & - & + & - & \cdots \\ + & - & + & - & + & \cdots \\ - & + & - & + & - & \cdots \\ + & - & + & - & + & \cdots \\ \vdots & \vdots & \vdots & \vdots & \vdots & \end{bmatrix}$$

$n \times n$ matrix

> ### Minors and Cofactors of a Square Matrix
>
> If A is a square matrix, then the **minor** M_{ij} of the entry a_{ij} is the determinant of the matrix obtained by deleting the ith row and jth column of A. The **cofactor** C_{ij} of the entry a_{ij} is given by
>
> $$C_{ij} = (-1)^{i+j} M_{ij}.$$

In the sign pattern for cofactors shown at the left, notice that *odd* positions (where $i + j$ is odd) have negative signs and *even* positions (where $i + j$ is even) have positive signs.

Example 2   Finding the Minors and Cofactors of a Matrix

Find all the minors and cofactors of $A = \begin{bmatrix} 0 & 2 & 1 \\ 3 & -1 & 2 \\ 4 & 0 & 1 \end{bmatrix}$.

SOLUTION   To find the minor M_{11}, delete the first row and first column of A and evaluate the determinant of the resulting matrix.

$$\begin{bmatrix} 0 & 2 & 1 \\ 3 & -1 & 2 \\ 4 & 0 & 1 \end{bmatrix}, \quad M_{11} = \begin{vmatrix} -1 & 2 \\ 0 & 1 \end{vmatrix} = -1(1) - 0(2) = -1$$

Similarly, to find M_{12}, delete the first row and second column.

$$\begin{bmatrix} 0 & 2 & 1 \\ 3 & -1 & 2 \\ 4 & 0 & 1 \end{bmatrix}, \quad M_{12} = \begin{vmatrix} 3 & 2 \\ 4 & 1 \end{vmatrix} = 3(1) - 4(2) = -5$$

Continuing this pattern, you obtain the following minors.

$$\begin{array}{lll} M_{11} = -1 & M_{12} = -5 & M_{13} = 4 \\ M_{21} = 2 & M_{22} = -4 & M_{23} = -8 \\ M_{31} = 5 & M_{32} = -3 & M_{33} = -6 \end{array}$$

Now, to find the cofactors, combine the minors above with the checkerboard pattern of signs for a 3×3 matrix shown at the upper left.

$$\begin{array}{lll} C_{11} = -1 & C_{12} = 5 & C_{13} = 4 \\ C_{21} = -2 & C_{22} = -4 & C_{23} = 8 \\ C_{31} = 5 & C_{32} = 3 & C_{33} = -6 \end{array}$$

✓ Checkpoint 2

Find all the minors and cofactors of

$$A = \begin{bmatrix} 1 & 2 & 3 \\ 0 & -1 & 5 \\ 2 & 1 & 4 \end{bmatrix}.$$

The Determinant of a Square Matrix

The definition below is called **inductive** because it uses determinants of matrices of dimension $(n - 1) \times (n - 1)$ to define determinants of matrices of dimension $n \times n$.

Determinant of a Square Matrix

If A is a square matrix (of dimension 2×2 or greater), then the determinant of A is the sum of the entries in any row (or column) of A multiplied by their respective cofactors. For instance, expanding along the first row yields

$$|A| = a_{11}C_{11} + a_{12}C_{12} + \cdots + a_{1n}C_{1n}.$$

Applying this definition to find a determinant is called **expanding by cofactors.**

Try checking that for a 2×2 matrix

$$A = \begin{bmatrix} a_1 & b_1 \\ a_2 & b_2 \end{bmatrix}$$

this definition of the determinant yields

$$|A| = a_1 b_2 - a_2 b_1$$

as previously defined.

Example 3 The Determinant of a Matrix of Dimension 3 × 3

Find the determinant of $A = \begin{bmatrix} 0 & 2 & 1 \\ 3 & -1 & 2 \\ 4 & 0 & 1 \end{bmatrix}$.

SOLUTION Note that this is the same matrix that was given in Example 2. There you found the cofactors of the entries in the first row to be

$$C_{11} = -1, \quad C_{12} = 5, \quad \text{and} \quad C_{13} = 4.$$

So, by the definition of a determinant, you have

$$\begin{aligned} |A| &= a_{11}C_{11} + a_{12}C_{12} + a_{13}C_{13} && \text{First-row expansion} \\ &= 0(-1) + 2(5) + 1(4) \\ &= 14. \end{aligned}$$

✓ Checkpoint 3

Find the determinant of

$$A = \begin{bmatrix} 3 & 0 & 2 \\ 1 & 2 & -1 \\ 2 & 0 & -4 \end{bmatrix}.$$

In Example 3, the determinant was found by expanding by the cofactors in the first row. You could have used any row or column. For instance, you could have expanded along the second row to obtain

$$\begin{aligned} |A| &= a_{21}C_{21} + a_{22}C_{22} + a_{23}C_{23} && \text{Second-row expansion} \\ &= 3(-2) + (-1)(-4) + 2(8) \\ &= 14. \end{aligned}$$

When expanding by cofactors, you do not need to find cofactors of zero entries, because zero times its cofactor is zero. So, the row (or column) containing the most zeros is usually the best choice for expansion by cofactors.

Example 4 — The Determinant of a Matrix of Dimension 4 × 4

Find the determinant of $A = \begin{bmatrix} 1 & -2 & 3 & 0 \\ -1 & 1 & 0 & 2 \\ 0 & 2 & 0 & 3 \\ 3 & 4 & 0 & 2 \end{bmatrix}$.

SOLUTION After inspecting this matrix, you can see that three of the entries in the third column are zeros. So, you can eliminate some of the work in the expansion by using the third column.

$$|A| = 3(C_{13}) + 0(C_{23}) + 0(C_{33}) + 0(C_{43})$$

Because C_{23}, C_{33}, and C_{43} have zero coefficients, you only need to find the cofactor C_{13}. To do this, delete the first row and third column of A and evaluate the determinant of the resulting matrix.

$$C_{13} = (-1)^{1+3} \begin{vmatrix} -1 & 1 & 2 \\ 0 & 2 & 3 \\ 3 & 4 & 2 \end{vmatrix} \qquad \text{Delete 1st row and 3rd column.}$$

$$= \begin{vmatrix} -1 & 1 & 2 \\ 0 & 2 & 3 \\ 3 & 4 & 2 \end{vmatrix} \qquad \text{Simplify.}$$

Expanding by minors in the second row yields

$$C_{13} = 0(-1)^3 \begin{vmatrix} 1 & 2 \\ 4 & 2 \end{vmatrix} + 2(-1)^4 \begin{vmatrix} -1 & 2 \\ 3 & 2 \end{vmatrix} + 3(-1)^5 \begin{vmatrix} -1 & 1 \\ 3 & 4 \end{vmatrix}$$

$$= 0 + 2(1)(-8) + 3(-1)(-7)$$

$$= 5.$$

So, you obtain $|A| = 3C_{13} = 3(5) = 15$.

✓ **Checkpoint 4**

Find the determinant of $A = \begin{bmatrix} 3 & 0 & 7 & 0 \\ 2 & 6 & 0 & 11 \\ 4 & 1 & 0 & 2 \\ 1 & 5 & 0 & 10 \end{bmatrix}$.

SUMMARIZE (Section 6.4)

1. State the definition of the determinant of a 2 × 2 matrix *(page 513)*. For an example of finding the determinant of a 2 × 2 matrix, see Example 1.

2. State the definition of minors and cofactors of a square matrix *(page 515)*. For an example of finding the minors and cofactors of a matrix, see Example 2.

3. State the definition of the determinant of a square matrix using expansion by cofactors *(page 516)*. For examples of finding determinants using this definition, see Examples 3 and 4.

SKILLS WARM UP 6.4   The following warm-up exercises involve skills that were covered in earlier sections. You will use these skills in the exercise set for this section. For additional help, review Sections 0.2, 0.3, and 6.2.

In Exercises 1–4, perform the indicated matrix operations.

1. $\begin{bmatrix} 1 & -2 \\ 0 & 3 \end{bmatrix} + \begin{bmatrix} 2 & 7 \\ 4 & -3 \end{bmatrix}$

2. $\begin{bmatrix} -2 & 5 \\ 3 & -2 \end{bmatrix} - \begin{bmatrix} 0 & -3 \\ 1 & 2 \end{bmatrix}$

3. $3\begin{bmatrix} 3 & -4 & 2 \\ 1 & 0 & -1 \\ 0 & 1 & -2 \end{bmatrix}$

4. $4\begin{bmatrix} 0 & 2 & 3 \\ -1 & 2 & 3 \\ -2 & 1 & -2 \end{bmatrix}$

In Exercises 5–10, perform the indicated arithmetic operations.

5. $[(1)(3) + (-3)(2)] - [(1)(4) + (3)(5)]$

6. $[(4)(4) + (-1)(-3)] - [(-1)(2) + (-2)(7)]$

7. $\dfrac{4(7) - 1(-2)}{(-5)(-2) - 3(4)}$

8. $\dfrac{3(6) - 2(7)}{6(-5) - 2(1)}$

9. $-5(-1)^2[6(-2) - 7(-3)]$

10. $4(-1)^3[3(6) - 2(7)]$

Exercises 6.4

See www.CalcChat.com for worked-out solutions to odd-numbered exercises.

Finding the Determinant of a Matrix   In Exercises 1–16, find the determinant of the matrix. *See Example 1.*

1. $[-5]$

2. $[6]$

3. $\begin{bmatrix} 1 & 3 \\ 2 & 7 \end{bmatrix}$

4. $\begin{bmatrix} -3 & 4 \\ -2 & 1 \end{bmatrix}$

5. $\begin{bmatrix} 5 & 6 \\ 2 & 3 \end{bmatrix}$

6. $\begin{bmatrix} -7 & -4 \\ 8 & 7 \end{bmatrix}$

7. $\begin{bmatrix} 9 & 3 \\ 12 & 4 \end{bmatrix}$

8. $\begin{bmatrix} -5 & -2 \\ 10 & 4 \end{bmatrix}$

9. $\begin{bmatrix} 2 & 1 \\ 3 & 4 \end{bmatrix}$

10. $\begin{bmatrix} -3 & 1 \\ 5 & 2 \end{bmatrix}$

11. $\begin{bmatrix} \frac{2}{3} & 0 \\ -1 & 6 \end{bmatrix}$

12. $\begin{bmatrix} 9 & -\frac{1}{4} \\ 8 & 0 \end{bmatrix}$

13. $\begin{bmatrix} -\frac{1}{2} & \frac{1}{3} \\ -6 & \frac{1}{3} \end{bmatrix}$

14. $\begin{bmatrix} \frac{2}{3} & \frac{4}{3} \\ -1 & -\frac{1}{3} \end{bmatrix}$

15. $\begin{bmatrix} 2.5 & 0.4 \\ -3.5 & -1.6 \end{bmatrix}$

16. $\begin{bmatrix} -0.1 & 7 \\ -0.8 & -5 \end{bmatrix}$

 Using a Graphing Utility   In Exercises 17–22, use the matrix capabilities of a graphing utility to find the determinant of the matrix.

17. $\begin{bmatrix} 5 & -3 & 2 \\ 7 & 5 & -7 \\ 0 & 6 & -1 \end{bmatrix}$

18. $\begin{bmatrix} 2 & 3 & 1 \\ 0 & 5 & -2 \\ 0 & 0 & -2 \end{bmatrix}$

19. $\begin{bmatrix} 0.1 & 0.3 & 0.2 \\ -0.3 & -0.2 & 0.1 \\ 1 & 2 & 3 \end{bmatrix}$

20. $\begin{bmatrix} 0.2 & -0.1 & -0.3 \\ 0.1 & -0.1 & 0.4 \\ -0.5 & -0.2 & -0.1 \end{bmatrix}$

21. $\begin{bmatrix} 0.9 & 0.7 & 0 \\ -0.1 & 0.3 & 1.3 \\ 2.2 & 4.2 & 6.1 \end{bmatrix}$

22. $\begin{bmatrix} 0.1 & 0.1 & -4.3 \\ 7.5 & 6.2 & 0.7 \\ 0.3 & 0.6 & -1.2 \end{bmatrix}$

Finding the Minors and Cofactors of a Matrix   In Exercises 23–30, find all the (a) minors and (b) cofactors of the matrix. *See Example 2.*

23. $\begin{bmatrix} 3 & 4 \\ 2 & -5 \end{bmatrix}$

24. $\begin{bmatrix} 11 & 0 \\ -3 & 2 \end{bmatrix}$

25. $\begin{bmatrix} 3 & 1 \\ -2 & -4 \end{bmatrix}$

26. $\begin{bmatrix} -6 & 5 \\ 7 & -2 \end{bmatrix}$

27. $\begin{bmatrix} 4 & 0 & 2 \\ -3 & 2 & 1 \\ 1 & -1 & 1 \end{bmatrix}$

28. $\begin{bmatrix} 1 & -1 & 0 \\ 3 & 2 & 5 \\ 4 & -6 & 4 \end{bmatrix}$

29. $\begin{bmatrix} 3 & -2 & 8 \\ 3 & 2 & -6 \\ -1 & 3 & 6 \end{bmatrix}$

30. $\begin{bmatrix} -2 & 9 & 4 \\ 7 & -6 & 0 \\ 6 & 7 & -6 \end{bmatrix}$

Finding the Determinant of a Matrix   In Exercises 31–36, find the determinant of the matrix by the method of expansion by cofactors. Expand along the indicated row or column. *See Examples 3 and 4.*

31. $\begin{bmatrix} 4 & 1 & -3 \\ 6 & 5 & -2 \\ -1 & 3 & -4 \end{bmatrix}$

   (a) Row 3

   (b) Column 2

32. $\begin{bmatrix} -3 & 4 & 2 \\ 6 & 3 & 1 \\ 4 & -7 & -8 \end{bmatrix}$

   (a) Row 2

   (b) Column 3

33. $\begin{bmatrix} 7 & 0 & -4 \\ 2 & -3 & 0 \\ 5 & 8 & 1 \end{bmatrix}$

(a) Row 1

(b) Column 3

34. $\begin{bmatrix} 10 & -5 & 5 \\ 30 & 0 & 10 \\ 0 & 10 & 1 \end{bmatrix}$

(a) Row 3

(b) Column 1

35. $\begin{bmatrix} 6 & 0 & -3 & 5 \\ 4 & 13 & 6 & -8 \\ -1 & 0 & 7 & 4 \\ 8 & 6 & 0 & 2 \end{bmatrix}$

(a) Row 2

(b) Column 2

36. $\begin{bmatrix} 10 & 8 & 3 & -7 \\ 4 & 0 & 5 & -6 \\ 0 & 3 & 2 & 7 \\ 1 & 0 & -3 & 2 \end{bmatrix}$

(a) Row 3

(b) Column 1

Finding the Determinant of a Matrix In Exercises 37–52, find the determinant of the matrix. Expand by cofactors along the row or column that appears to make the computations easiest. Use a graphing utility to confirm your result. *See Examples 3 and 4.*

37. $\begin{bmatrix} 1 & 4 & -2 \\ 3 & 2 & 0 \\ -1 & 4 & 3 \end{bmatrix}$

38. $\begin{bmatrix} 2 & -1 & 3 \\ 1 & 4 & 4 \\ 1 & 0 & 2 \end{bmatrix}$

39. $\begin{bmatrix} 2 & 4 & 6 \\ 0 & 3 & 1 \\ 0 & 0 & -5 \end{bmatrix}$

40. $\begin{bmatrix} -3 & 0 & 0 \\ 7 & 11 & 0 \\ 1 & 2 & 2 \end{bmatrix}$

41. $\begin{bmatrix} 2 & -1 & 0 \\ 4 & 2 & 1 \\ 4 & 2 & 1 \end{bmatrix}$

42. $\begin{bmatrix} -2 & 2 & 3 \\ 1 & -1 & 0 \\ 0 & 1 & 4 \end{bmatrix}$

43. $\begin{bmatrix} 1 & 4 & -2 \\ 3 & 6 & -6 \\ -2 & 1 & 4 \end{bmatrix}$

44. $\begin{bmatrix} -1 & 3 & 1 \\ 4 & 2 & 5 \\ -2 & 1 & 6 \end{bmatrix}$

45. $\begin{bmatrix} 0.3 & 0.2 & 0.2 \\ 0.2 & 0.2 & 0.2 \\ -0.4 & 0.4 & 0.3 \end{bmatrix}$

46. $\begin{bmatrix} 0.1 & 0.2 & 0.3 \\ -0.3 & 0.2 & 0.2 \\ 0.5 & 0.4 & 0.4 \end{bmatrix}$

47. $\begin{bmatrix} 6 & 3 & -7 \\ 0 & 0 & 0 \\ 4 & -6 & 3 \end{bmatrix}$

48. $\begin{bmatrix} 5 & 0 & 3 \\ -4 & 0 & 8 \\ 3 & 0 & -6 \end{bmatrix}$

49. $\begin{bmatrix} 3 & 6 & -5 & 4 \\ -2 & 0 & 6 & 0 \\ 1 & 1 & 2 & 2 \\ 0 & 3 & -1 & -1 \end{bmatrix}$

50. $\begin{bmatrix} 1 & 4 & 3 & 2 \\ -5 & 6 & 2 & 1 \\ 0 & 0 & 0 & 0 \\ 3 & -2 & 1 & 5 \end{bmatrix}$

51. $\begin{bmatrix} 3 & 2 & 4 & -1 & 5 \\ -2 & 0 & 1 & 3 & 2 \\ 1 & 0 & 0 & 4 & 0 \\ 6 & 0 & 2 & -1 & 0 \\ 3 & 0 & 5 & 1 & 0 \end{bmatrix}$

52. $\begin{bmatrix} 5 & 2 & 0 & 0 & -2 \\ 0 & 1 & 4 & 3 & 2 \\ 0 & 0 & 2 & 6 & 3 \\ 0 & 0 & 3 & 4 & 1 \\ 0 & 0 & 0 & 0 & 2 \end{bmatrix}$

Using a Graphing Utility In Exercises 53–56, use the matrix capabilities of a graphing utility to evaluate the determinant.

53. $\begin{vmatrix} 3 & 8 & -7 \\ 0 & -5 & 4 \\ 8 & 1 & 6 \end{vmatrix}$

54. $\begin{vmatrix} 5 & -8 & 0 \\ 9 & 7 & 4 \\ -8 & 7 & 1 \end{vmatrix}$

55. $\begin{vmatrix} 1 & -1 & 8 & 4 \\ 2 & 6 & 0 & -4 \\ 2 & 0 & 2 & 6 \\ 0 & 2 & 8 & 0 \end{vmatrix}$

56. $\begin{vmatrix} -2 & 0 & 0 & 0 & 0 \\ 0 & 3 & 0 & 0 & 0 \\ 0 & 0 & -1 & 0 & 0 \\ 0 & 0 & 0 & 2 & 0 \\ 0 & 0 & 0 & 0 & -4 \end{vmatrix}$

The Determinant of a Matrix Product In Exercises 57–64, find (a) $|A|$, (b) $|B|$, (c) AB, and (d) $|AB|$. What do you notice about $|AB|$?

57. $A = \begin{bmatrix} -1 & 0 \\ 0 & 3 \end{bmatrix}$, $B = \begin{bmatrix} 2 & 0 \\ 0 & -1 \end{bmatrix}$

58. $A = \begin{bmatrix} -2 & 1 \\ 4 & -2 \end{bmatrix}$, $B = \begin{bmatrix} 1 & 2 \\ 0 & -1 \end{bmatrix}$

59. $A = \begin{bmatrix} 4 & 0 \\ 3 & -2 \end{bmatrix}$, $B = \begin{bmatrix} -1 & 1 \\ -2 & 2 \end{bmatrix}$

60. $A = \begin{bmatrix} 5 & 4 \\ 3 & -1 \end{bmatrix}$, $B = \begin{bmatrix} 0 & 6 \\ 1 & -2 \end{bmatrix}$

61. $A = \begin{bmatrix} 0 & 1 & 2 \\ -3 & -2 & 1 \\ 0 & 4 & 1 \end{bmatrix}$, $B = \begin{bmatrix} 3 & -2 & 0 \\ 1 & -1 & 2 \\ 3 & 1 & 1 \end{bmatrix}$

62. $A = \begin{bmatrix} 3 & 2 & 0 \\ -1 & -3 & 4 \\ -2 & 0 & 1 \end{bmatrix}$, $B = \begin{bmatrix} -3 & 0 & 1 \\ 0 & 2 & -1 \\ -2 & -1 & 1 \end{bmatrix}$

63. $A = \begin{bmatrix} -1 & 2 & 1 \\ 1 & 0 & 1 \\ 0 & 1 & 0 \end{bmatrix}$, $B = \begin{bmatrix} -1 & 0 & 0 \\ 0 & 2 & 0 \\ 0 & 0 & 3 \end{bmatrix}$

64. $A = \begin{bmatrix} 2 & 0 & 1 \\ 1 & -1 & 2 \\ 3 & 1 & 0 \end{bmatrix}$, $B = \begin{bmatrix} 2 & -1 & 4 \\ 0 & 1 & 3 \\ 3 & -2 & 1 \end{bmatrix}$

Diagonal Matrices A **diagonal matrix** is a square matrix with all zero entries above and below its main diagonal. In Exercises 65–68, find the determinant of the diagonal matrix. What do you notice about your result?

65. $\begin{bmatrix} 9 & 0 \\ 0 & 5 \end{bmatrix}$

66. $\begin{bmatrix} -1 & 0 & 0 \\ 0 & 2 & 0 \\ 0 & 0 & 4 \end{bmatrix}$

67. $\begin{bmatrix} 5 & 0 & 0 \\ 0 & -5 & 0 \\ 0 & 0 & -3 \end{bmatrix}$

68. $\begin{bmatrix} 1 & 0 & 0 & 0 \\ 0 & 3 & 0 & 0 \\ 0 & 0 & -3 & 0 \\ 0 & 0 & 0 & 2 \end{bmatrix}$

Triangular Matrices   A **triangular matrix** is a square matrix with all zero entries either above or below its main diagonal. Such a matrix is **upper triangular** when it has all zeros below the main diagonal and **lower triangular** when it has all zeros above the main diagonal. A diagonal matrix is both upper and lower triangular. To find the determinant of a triangular matrix of any dimension, simply find the product of the entries on the main diagonal. In Exercises 69–74, state whether the matrix is *upper triangular*, *lower triangular*, or *diagonal*, and then find the determinant.

69. $\begin{bmatrix} 2 & 0 & 0 \\ 4 & -3 & 0 \\ 6 & 5 & 1 \end{bmatrix}$

70. $\begin{bmatrix} 1 & 1 & 5 \\ 0 & -1 & -4 \\ 0 & 0 & 5 \end{bmatrix}$

71. $\begin{bmatrix} 2 & 3 & -1 & -1 \\ 0 & -1 & -3 & 5 \\ 0 & 0 & -2 & 7 \\ 0 & 0 & 0 & -4 \end{bmatrix}$

72. $\begin{bmatrix} 4 & 0 & 0 & 0 \\ 6 & -5 & 0 & 0 \\ 1 & 3 & 2 & 0 \\ 1 & 2 & 7 & -1 \end{bmatrix}$

73. $\begin{bmatrix} -6 & 7 & 2 & 0 & 5 \\ 0 & -1 & 3 & 4 & -3 \\ 0 & 0 & -7 & 0 & 4 \\ 0 & 0 & 0 & -2 & 1 \\ 0 & 0 & 0 & 0 & -2 \end{bmatrix}$

74. $\begin{bmatrix} -2 & 0 & 0 & 0 & 0 \\ 0 & 3 & 0 & 0 & 0 \\ 0 & 0 & -1 & 0 & 0 \\ 0 & 0 & 0 & 2 & 0 \\ 0 & 0 & 0 & 0 & -4 \end{bmatrix}$

Think About It   In Exercises 75–80, find a 4×4 *upper triangular* matrix whose determinant is equal to the given value and a 4×4 *lower triangular* matrix whose determinant is equal to the given value. Use a graphing utility to confirm your results.

75. 6

76. -8

77. -18

78. -40

79. 28

80. 36

Choosing a Method   In Exercises 81–86, find the determinant of the matrix. Tell which method you used. (*Hint:* Use information from Exercises 65–74.)

81. $\begin{bmatrix} 2 & 1 & 3 \\ 7 & 3 & -2 \\ 4 & 1 & 1 \end{bmatrix}$

82. $\begin{bmatrix} 6 & -5 & 2 \\ 0 & 5 & -3 \\ 0 & 0 & 2 \end{bmatrix}$

83. $\begin{bmatrix} 3 & 0 & 0 \\ 4 & -2 & 0 \\ 5 & 4 & 3 \end{bmatrix}$

84. $\begin{bmatrix} 3 & 2 & -4 \\ -1 & 5 & -3 \\ 0 & 1 & 0 \end{bmatrix}$

85. $\begin{bmatrix} 2 & 0 & 0 & 0 \\ 0 & 5 & 0 & 0 \\ 0 & 0 & 8 & 0 \\ 0 & 0 & 0 & 1 \end{bmatrix}$

86. $\begin{bmatrix} -1 & 0 & 4 & -3 \\ 5 & 2 & 1 & 1 \\ 3 & -6 & 3 & 2 \\ -4 & -2 & 1 & 0 \end{bmatrix}$

Verifying an Equation   In Exercises 87–90, evaluate the determinant(s) to verify the equation.

87. $\begin{vmatrix} w & x \\ y & z \end{vmatrix} = -\begin{vmatrix} y & z \\ w & x \end{vmatrix}$

88. $\begin{vmatrix} w & cx \\ y & cz \end{vmatrix} = c\begin{vmatrix} w & x \\ y & z \end{vmatrix}$

89. $\begin{vmatrix} w & x \\ y & z \end{vmatrix} = \begin{vmatrix} w & x + cw \\ y & z + cy \end{vmatrix}$

90. $\begin{vmatrix} w & x \\ cw & cx \end{vmatrix} = 0$

Solving an Equation   In Exercises 91–100, solve for *x*.

91. $\begin{vmatrix} x & 2 \\ 1 & x \end{vmatrix} = 2$

92. $\begin{vmatrix} x & 4 \\ -1 & x \end{vmatrix} = 20$

93. $\begin{vmatrix} 2x & -3 \\ -2 & 2x \end{vmatrix} = 3$

94. $\begin{vmatrix} x & 2 \\ 4 & 9x \end{vmatrix} = 8$

95. $\begin{vmatrix} x & 1 \\ 2 & x - 2 \end{vmatrix} = -1$

96. $\begin{vmatrix} x - 2 & -1 \\ -3 & x \end{vmatrix} = 0$

97. $\begin{vmatrix} 2x & 1 \\ -1 & x - 1 \end{vmatrix} = x$

98. $\begin{vmatrix} x - 1 & x \\ x + 1 & 2 \end{vmatrix} = -8$

99. $\begin{vmatrix} 1 & 2 & x \\ -1 & 3 & 2 \\ 3 & -2 & 1 \end{vmatrix} = 0$

100. $\begin{vmatrix} 1 & x & -2 \\ 1 & 3 & 3 \\ 0 & 2 & -2 \end{vmatrix} = 0$

101. Discovery   Consider square matrices in which the entries are consecutive integers. An example of such a matrix is

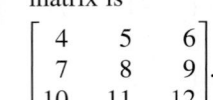

$$\begin{bmatrix} 4 & 5 & 6 \\ 7 & 8 & 9 \\ 10 & 11 & 12 \end{bmatrix}.$$

Use a graphing utility to find four determinants of matrices of this type. Make a conjecture based on your results. Then verify your conjecture.

102. HOW DO YOU SEE IT?   Explain why the determinant of the matrix is equal to zero.

(a) $\begin{bmatrix} 3 & 4 & -2 & 7 \\ 1 & 3 & -1 & 2 \\ 0 & 5 & 7 & 1 \\ 1 & 3 & -1 & 2 \end{bmatrix}$ (b) $\begin{bmatrix} 3 & 2 & -1 \\ -6 & -4 & 2 \\ 5 & -7 & 9 \end{bmatrix}$

(c) $\begin{bmatrix} 2 & -4 & 5 \\ 1 & -2 & 3 \\ 0 & 0 & 0 \end{bmatrix}$ (d) $\begin{bmatrix} 4 & -4 & 5 & 7 \\ 2 & -2 & 3 & 1 \\ 4 & -4 & 5 & 7 \\ 6 & 1 & -3 & -3 \end{bmatrix}$

True or False?   In Exercises 103 and 104, determine whether the statement is true or false. Justify your answer.

103. If a square matrix has an entire row of zeros, then the determinant will always be zero.

104. If two columns of a square matrix are the same, then the determinant of the matrix will be zero.

6.5 Applications of Matrices and Determinants

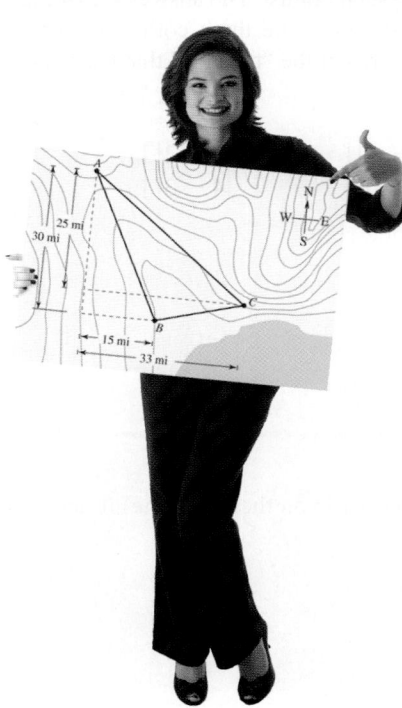

In Exercise 15 on page 527, you will use a determinant to find the area of a region of forest that has been infested with gypsy moths.

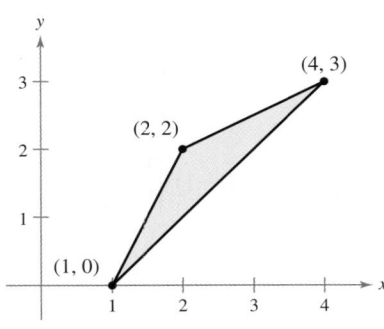

FIGURE 6.1

■ Find the area of a triangle using a determinant.
■ Use a determinant to test for collinear points and to find an equation of a line.
■ Encode and decode a cryptogram using a matrix.

Area of a Triangle

In this section, you will study some additional applications of matrices and determinants. The first application involves a formula for finding the area of a triangle whose vertices are given by three points on a rectangular coordinate system.

> ### Area of a Triangle
>
> The area of a triangle with vertices (x_1, y_1), (x_2, y_2), and (x_3, y_3) is given by
>
> $$\text{Area} = \pm\frac{1}{2}\begin{vmatrix} x_1 & y_1 & 1 \\ x_2 & y_2 & 1 \\ x_3 & y_3 & 1 \end{vmatrix}$$
>
> where the symbol ($\pm$) indicates that the appropriate sign should be chosen to yield a positive area.

Example 1　Finding the Area of a Triangle

Find the area of the triangle whose vertices are $(1, 0)$, $(2, 2)$, and $(4, 3)$, as shown in Figure 6.1.

SOLUTION　Let $(x_1, y_1) = (1, 0)$, $(x_2, y_2) = (2, 2)$, and $(x_3, y_3) = (4, 3)$. Then, to find the area of the triangle, evaluate the determinant

$$\begin{vmatrix} x_1 & y_1 & 1 \\ x_2 & y_2 & 1 \\ x_3 & y_3 & 1 \end{vmatrix} = \begin{vmatrix} 1 & 0 & 1 \\ 2 & 2 & 1 \\ 4 & 3 & 1 \end{vmatrix}$$

$$= 1(-1)^2\begin{vmatrix} 2 & 1 \\ 3 & 1 \end{vmatrix} + 0(-1)^3\begin{vmatrix} 2 & 1 \\ 4 & 1 \end{vmatrix} + 1(-1)^4\begin{vmatrix} 2 & 2 \\ 4 & 3 \end{vmatrix}$$

$$= 1(-1) + 0 + 1(-2)$$

$$= -3.$$

Using this value, you can conclude that the area of the triangle is

$$\text{Area} = -\frac{1}{2}\begin{vmatrix} 1 & 0 & 1 \\ 2 & 2 & 1 \\ 4 & 3 & 1 \end{vmatrix} \qquad \text{Choose } (-) \text{ so the area is positive.}$$

$$= -\frac{1}{2}(-3)$$

$$= \frac{3}{2} \text{ square units.}$$

✓ Checkpoint 1

Find the area of the triangle whose vertices are $(2, 1)$, $(3, 5)$, and $(10, 5)$.

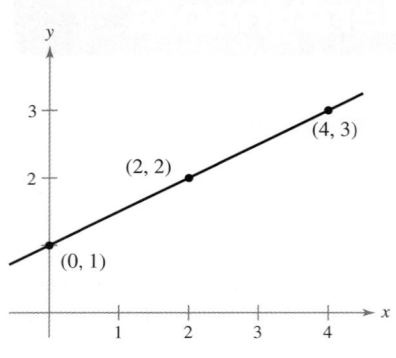

FIGURE 6.2

Lines in the Plane

Suppose the three points in Example 1 had been on the same line. What would have happened had the area formula been applied to three such points? The answer is that the determinant would have been zero. Consider, for instance, the three collinear points $(0, 1)$, $(2, 2)$, and $(4, 3)$, as shown in Figure 6.2. The area of the "triangle" that has these three points as vertices is

$$\frac{1}{2}\begin{vmatrix} 0 & 1 & 1 \\ 2 & 2 & 1 \\ 4 & 3 & 1 \end{vmatrix} = \frac{1}{2}\left[0(-1)^2\begin{vmatrix} 2 & 1 \\ 3 & 1 \end{vmatrix} + 1(-1)^3\begin{vmatrix} 2 & 1 \\ 4 & 1 \end{vmatrix} + 1(-1)^4\begin{vmatrix} 2 & 2 \\ 4 & 3 \end{vmatrix}\right]$$

$$= \frac{1}{2}[0(-1) - 1(-2) + 1(-2)]$$

$$= 0.$$

This result is generalized as follows.

Test for Collinear Points

Three points (x_1, y_1), (x_2, y_2), and (x_3, y_3) are collinear (lie on the same line) if and only if

$$\begin{vmatrix} x_1 & y_1 & 1 \\ x_2 & y_2 & 1 \\ x_3 & y_3 & 1 \end{vmatrix} = 0.$$

Example 2   Testing for Collinear Points

Determine whether the points $(-2, -2)$, $(1, 1)$, and $(7, 5)$ are collinear. (See Figure 6.3.)

SOLUTION   Letting $(x_1, y_1) = (-2, -2)$, $(x_2, y_2) = (1, 1)$, and $(x_3, y_3) = (7, 5)$, you have

$$\begin{vmatrix} x_1 & y_1 & 1 \\ x_2 & y_2 & 1 \\ x_3 & y_3 & 1 \end{vmatrix} = \begin{vmatrix} -2 & -2 & 1 \\ 1 & 1 & 1 \\ 7 & 5 & 1 \end{vmatrix}$$

$$= -2(-1)^2\begin{vmatrix} 1 & 1 \\ 5 & 1 \end{vmatrix} + (-2)(-1)^3\begin{vmatrix} 1 & 1 \\ 7 & 1 \end{vmatrix} + 1(-1)^4\begin{vmatrix} 1 & 1 \\ 7 & 5 \end{vmatrix}$$

$$= -2(-4) + 2(-6) + 1(-2)$$

$$= -6.$$

Because the value of this determinant is *not* zero, you can conclude that the three points *do not* lie on the same line. ──────────

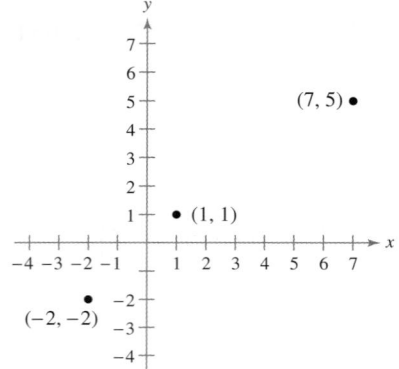

FIGURE 6.3

✓ Checkpoint 2

Determine whether the points

$$(-2, 4), \quad (3, 0), \quad \text{and} \quad (6, -4)$$

are collinear.

Another way to test for collinear points in Example 2 is to find the slope of the line between $(-2, -2)$ and $(1, 1)$ and the slope of the line between $(-2, -2)$ and $(7, 5)$. Try doing this. If the slopes are equal, then the points are collinear. If the slopes are not equal, then the points are not collinear.

The test for collinear points can be adapted for another use. Given two points on a rectangular coordinate system, you can find an equation of the line passing through the two points.

Two-Point Form of the Equation of a Line

An equation of the line passing through the distinct points (x_1, y_1) and (x_2, y_2) is given by

$$\begin{vmatrix} x & y & 1 \\ x_1 & y_1 & 1 \\ x_2 & y_2 & 1 \end{vmatrix} = 0.$$

Note that this method of finding the equation of a line works for all lines, including horizontal and vertical lines. (See Exercises 35–38.)

Example 3 **Finding an Equation of a Line**

Find an equation of the line passing through the two points $(2, 4)$ and $(-1, 3)$, as shown in Figure 6.4.

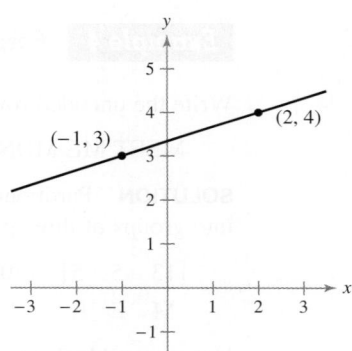

FIGURE 6.4

SOLUTION Let $(x_1, y_1) = (2, 4)$ and $(x_2, y_2) = (-1, 3)$. Applying the determinant formula for the equation of a line produces

$$\begin{vmatrix} x & y & 1 \\ 2 & 4 & 1 \\ -1 & 3 & 1 \end{vmatrix} = 0.$$

To evaluate this determinant, you can expand by cofactors along the first row to find an equation of the line.

$$x(-1)^2\begin{vmatrix} 4 & 1 \\ 3 & 1 \end{vmatrix} + y(-1)^3\begin{vmatrix} 2 & 1 \\ -1 & 1 \end{vmatrix} + 1(-1)^4\begin{vmatrix} 2 & 4 \\ -1 & 3 \end{vmatrix} = 0$$

$$x(1) - y(3) + (1)(10) = 0$$

$$x - 3y + 10 = 0$$

✓ **Checkpoint 3**

Find an equation of the line passing through the two points

$$(-3, -1) \quad \text{and} \quad (3, 5).$$

Cryptography

During World War II, Navajo soldiers created a code using their native language to send messages between batallions. Native words were assigned to represent characters in the English alphabet, and they created a number of expressions for important military terms, like iron-fish to mean submarine. Without the Navajo Code Talkers, the Second World War might have had a very different outcome.

A **cryptogram** is a message written according to a secret code. (The Greek word *kryptos* means "hidden.") Matrix multiplication can be used to **encode** and **decode** messages. To begin, you need to assign a number to each letter in the alphabet (with 0 assigned to a blank space), as follows.

0 = __	9 = I	18 = R
1 = A	10 = J	19 = S
2 = B	11 = K	20 = T
3 = C	12 = L	21 = U
4 = D	13 = M	22 = V
5 = E	14 = N	23 = W
6 = F	15 = O	24 = X
7 = G	16 = P	25 = Y
8 = H	17 = Q	26 = Z

Then the message is converted to numbers and partitioned into **uncoded row matrices,** each having n entries, as demonstrated in Example 4.

Example 4   Forming Uncoded Row Matrices

Write the uncoded row matrices of dimension 1×3 for the message

MEET ME MONDAY.

SOLUTION   Partitioning the message (including blank spaces, but ignoring punctuation) into groups of three produces the following uncoded row matrices.

$$\begin{matrix} [13 & 5 & 5] & [20 & 0 & 13] & [5 & 0 & 13] & [15 & 14 & 4] & [1 & 25 & 0] \\ M & E & E & T & & M & E & & M & O & N & D & A & Y \end{matrix}$$

Note that a blank space is used to fill out the last uncoded row matrix.

✓ Checkpoint 4

Write the uncoded row matrices of dimension 1×3 for the message

OWLS ARE NOCTURNAL.

To encode a message, choose an $n \times n$ invertible matrix A such as

$$A = \begin{bmatrix} 1 & -2 & 2 \\ -1 & 1 & 3 \\ 1 & -1 & -4 \end{bmatrix}$$

and multiply the uncoded row matrices by A to obtain **coded row matrices.** The uncoded matrix should be on the left, whereas the encoding matrix A should be on the right. Here is an example.

Uncoded Matrix	*Encoding Matrix A*	*Coded Matrix*
$[13 \quad 5 \quad 5]$	$\begin{bmatrix} 1 & -2 & 2 \\ -1 & 1 & 3 \\ 1 & -1 & -4 \end{bmatrix} =$	$[13 \quad -26 \quad 21]$

This technique is further illustrated in Example 5.

Example 5 Encoding a Message

Use the following matrix to encode the message MEET ME MONDAY.

$$A = \begin{bmatrix} 1 & -2 & 2 \\ -1 & 1 & 3 \\ 1 & -1 & -4 \end{bmatrix}$$

SOLUTION The coded row matrices are obtained by multiplying each of the uncoded row matrices found in Example 4 by the matrix A, as shown.

Uncoded Matrix	Encoding Matrix A	Coded Matrix

$$\begin{bmatrix} 13 & 5 & 5 \end{bmatrix} \begin{bmatrix} 1 & -2 & 2 \\ -1 & 1 & 3 \\ 1 & -1 & -4 \end{bmatrix} = \begin{bmatrix} 13 & -26 & 21 \end{bmatrix}$$

$$\begin{bmatrix} 20 & 0 & 13 \end{bmatrix} \begin{bmatrix} 1 & -2 & 2 \\ -1 & 1 & 3 \\ 1 & -1 & -4 \end{bmatrix} = \begin{bmatrix} 33 & -53 & -12 \end{bmatrix}$$

$$\begin{bmatrix} 5 & 0 & 13 \end{bmatrix} \begin{bmatrix} 1 & -2 & 2 \\ -1 & 1 & 3 \\ 1 & -1 & -4 \end{bmatrix} = \begin{bmatrix} 18 & -23 & -42 \end{bmatrix}$$

$$\begin{bmatrix} 15 & 14 & 4 \end{bmatrix} \begin{bmatrix} 1 & -2 & 2 \\ -1 & 1 & 3 \\ 1 & -1 & -4 \end{bmatrix} = \begin{bmatrix} 5 & -20 & 56 \end{bmatrix}$$

$$\begin{bmatrix} 1 & 25 & 0 \end{bmatrix} \begin{bmatrix} 1 & -2 & 2 \\ -1 & 1 & 3 \\ 1 & -1 & -4 \end{bmatrix} = \begin{bmatrix} -24 & 23 & 77 \end{bmatrix}$$

So, the sequence of coded row matrices is

$$\begin{bmatrix} 13 & -26 & 21 \end{bmatrix} \begin{bmatrix} 33 & -53 & -12 \end{bmatrix} \begin{bmatrix} 18 & -23 & -42 \end{bmatrix} \begin{bmatrix} 5 & -20 & 56 \end{bmatrix} \begin{bmatrix} -24 & 23 & 77 \end{bmatrix}.$$

Finally, removing the matrix notation produces the following cryptogram.

$$13 \;-26 \;21 \;33 \;-53 \;-12 \;18 \;-23 \;-42 \;5 \;-20 \;56 \;-24 \;23 \;77$$

✓ Checkpoint 5

Use the following matrix to encode the message OWLS ARE NOCTURNAL.

$$A = \begin{bmatrix} 1 & -1 & 0 \\ 1 & 0 & -1 \\ 6 & -2 & -3 \end{bmatrix}$$

For those who do not know the encoding matrix A, decoding the cryptogram found in Example 5 is difficult. But for an authorized receiver who knows the encoding matrix A, decoding is simple. The receiver only needs to multiply the coded row matrices by A^{-1} (on the right) to retrieve the uncoded row matrices. Here is an example.

$$\underbrace{\begin{bmatrix} 13 & -26 & 21 \end{bmatrix}}_{\text{Coded}} \underbrace{\begin{bmatrix} -1 & -10 & -8 \\ -1 & -6 & -5 \\ 0 & -1 & -1 \end{bmatrix}}_{A^{-1}} = \underbrace{\begin{bmatrix} 13 & 5 & 5 \end{bmatrix}}_{\text{Uncoded}}$$

The receiver could then easily refer to the number code chart on the previous page and translate $\begin{bmatrix} 13 & 5 & 5 \end{bmatrix}$ into the letters M E E.

Example 6   Decoding a Message

Use the inverse of the matrix $A = \begin{bmatrix} 1 & -2 & 2 \\ -1 & 1 & 3 \\ 1 & -1 & -4 \end{bmatrix}$ to decode the cryptogram.

$$13 \;\; -26 \;\; 21 \;\; 33 \;\; -53 \;\; -12 \;\; 18 \;\; -23 \;\; -42 \;\; 5 \;\; -20 \;\; 56 \;\; -24 \;\; 23 \;\; 77$$

SOLUTION   First, find the decoding matrix A^{-1} by using the techniques demonstrated in Section 6.3. Next, partition the message into groups of three to form the coded row matrices. Then multiply each coded row matrix by A^{-1} on the right to obtain the decoded row matrices.

Coded Matrix	Decoding Matrix A^{-1}	Decoded Matrix
$\begin{bmatrix} 13 & -26 & 21 \end{bmatrix}$	$\begin{bmatrix} -1 & -10 & -8 \\ -1 & -6 & -5 \\ 0 & -1 & -1 \end{bmatrix} =$	$\begin{bmatrix} 13 & 5 & 5 \end{bmatrix}$
$\begin{bmatrix} 33 & -53 & -12 \end{bmatrix}$	$\begin{bmatrix} -1 & -10 & -8 \\ -1 & -6 & -5 \\ 0 & -1 & -1 \end{bmatrix} =$	$\begin{bmatrix} 20 & 0 & 13 \end{bmatrix}$
$\begin{bmatrix} 18 & -23 & -42 \end{bmatrix}$	$\begin{bmatrix} -1 & -10 & -8 \\ -1 & -6 & -5 \\ 0 & -1 & -1 \end{bmatrix} =$	$\begin{bmatrix} 5 & 0 & 13 \end{bmatrix}$
$\begin{bmatrix} 5 & -20 & 56 \end{bmatrix}$	$\begin{bmatrix} -1 & -10 & -8 \\ -1 & -6 & -5 \\ 0 & -1 & -1 \end{bmatrix} =$	$\begin{bmatrix} 15 & 14 & 4 \end{bmatrix}$
$\begin{bmatrix} -24 & 23 & 77 \end{bmatrix}$	$\begin{bmatrix} -1 & -10 & -8 \\ -1 & -6 & -5 \\ 0 & -1 & -1 \end{bmatrix} =$	$\begin{bmatrix} 1 & 25 & 0 \end{bmatrix}$

So, the message is as follows.

$$\begin{bmatrix} 13 & 5 & 5 \end{bmatrix} \;\; \begin{bmatrix} 20 & 0 & 13 \end{bmatrix} \;\; \begin{bmatrix} 5 & 0 & 13 \end{bmatrix} \;\; \begin{bmatrix} 15 & 14 & 4 \end{bmatrix} \;\; \begin{bmatrix} 1 & 25 & 0 \end{bmatrix}$$
$$\text{M} \;\; \text{E} \;\; \text{E} \quad \text{T} \qquad \text{M} \;\; \text{E} \qquad \text{M} \;\; \text{O} \;\; \text{N} \quad \text{D} \;\; \text{A} \;\; \text{Y}$$

✓**Checkpoint 6**

Use the inverse of the matrix $A = \begin{bmatrix} 1 & -1 & 0 \\ 1 & 0 & -1 \\ 6 & -2 & -3 \end{bmatrix}$ to decode the cryptogram.

$$110, \; -39, \; -59, \; 25, \; -21, \; -3, \; 23, \; -18, \; -5,$$
$$47, \; -20, \; -24, \; 149, \; -56, \; -75, \; 87, \; -38, \; -37$$

SUMMARIZE   (Section 6.5)

1. State the formula for finding the area of a triangle using a determinant *(page 521)*. For an example of finding the area of a triangle, see Example 1.

2. State the test for collinear points *(page 522)*. For an example of testing for collinear points, see Example 2.

3. Explain how to use a determinant to find an equation of a line *(page 523)*. For an example of finding an equation of a line, see Example 3.

In Exercises 1–6, evaluate the determinant.

1. $\begin{vmatrix} 4 & 3 \\ -3 & -2 \end{vmatrix}$

2. $\begin{vmatrix} 10 & -20 \\ -1 & 2 \end{vmatrix}$

3. $\begin{vmatrix} 4 & 0 \\ -3 & -2 \end{vmatrix}$

4. $\begin{vmatrix} x & x^2 \\ 1 & 2x \end{vmatrix}$

5. $\begin{vmatrix} 4 & 0 & -2 \\ 3 & 1 & 2 \\ -8 & 0 & 6 \end{vmatrix}$

6. $\begin{vmatrix} 3 & 2 & 5 \\ 0 & 0 & -4 \\ -6 & 1 & 1 \end{vmatrix}$

In Exercises 7 and 8, find the inverse of the matrix.

7. $\begin{bmatrix} 1 & 3 \\ 2 & 7 \end{bmatrix}$

8. $\begin{bmatrix} 10 & 5 & -2 \\ -4 & -2 & 1 \\ 1 & 1 & 0 \end{bmatrix}$

In Exercises 9 and 10, perform the indicated matrix multiplication.

9. $\begin{bmatrix} 0.1 & 0.2 \\ 0.4 & 0.3 \end{bmatrix}\begin{bmatrix} 0.4 \\ 0.5 \end{bmatrix}$

10. $\begin{bmatrix} 2 & 5 \end{bmatrix}\begin{bmatrix} 1 & 2 \\ 1 & 2 \end{bmatrix}$

Exercises 6.5

Finding an Area   In Exercises 1–10, use a determinant to find the area of the triangle with the given vertices. *See Example 1.*

1.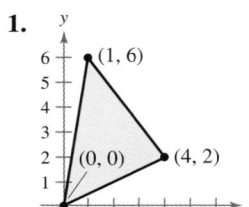

2.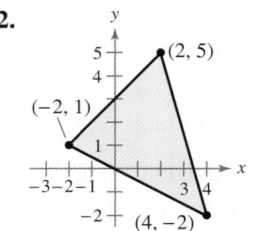

3.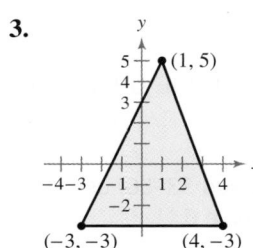

4.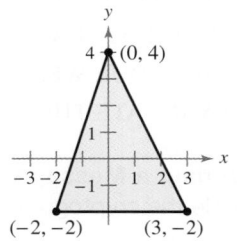

5. $\left(0, \frac{1}{2}\right), (4, 3), \left(\frac{5}{2}, 0\right)$

6. $(-4, -5), (6, 10), (6, -1)$

7. $(-2, 4), (2, 3), (-1, 5)$

8. $(0, -2), (-1, 4), (3, 5)$

9. $(-3, 5), (2, 6), (3, -5)$

10. $(-2, 4), (1, 5), (3, -2)$

Finding a Coordinate   In Exercises 11 and 12, find a value of y such that the triangle with the given vertices has an area of 4 square units.

11. $(-5, 1), (0, 2), (-2, y)$

12. $(-4, 2), (-3, 5), (-1, y)$

Finding a Coordinate   In Exercises 13 and 14, find a value of y such that the triangle with the given vertices has an area of 6 square units.

13. $(-2, -3), (1, -1), (-8, y)$

14. $(1, 0), (5, -3), (-3, y)$

15. Gypsy Moths   A large region of forest has been infested with gypsy moths. The region is roughly triangular, as shown in the figure. From the northernmost vertex A of the region, the distances to the other vertices are 30 miles south and 15 miles east (for vertex B), and 25 miles south and 33 miles east (for vertex C). Use a graphing utility to approximate the number of square miles in this region.

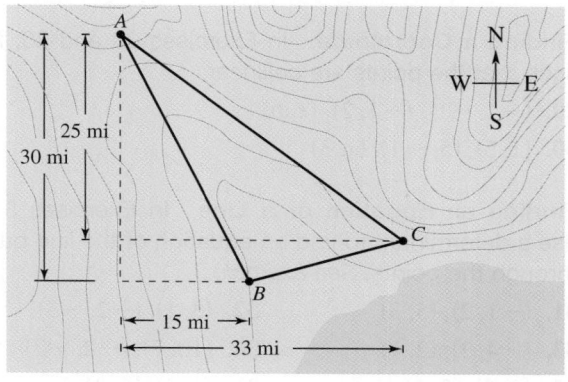

16. Botany A botanist is studying the plants growing in a triangular tract of land, as shown in the figure. To estimate the number of square feet in the tract, the botanist starts at one vertex, walks 70 feet east and 55 feet north to the second vertex, and then walks 90 feet west and 35 feet north to the third vertex. Use a graphing utility to determine how many square feet there are in the tract.

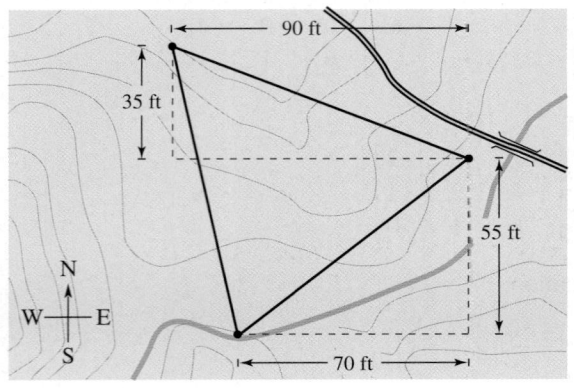

Testing for Collinear Points In Exercises 17–26, use a determinant to determine whether the points are collinear. *See Example 2.*

17. $(-4, -7), (0, -4), (4, -1)$

18. $(2, 4), (4, 5), (-2, 2)$

19. $(-1, -7), (0, -3), (1, 2)$

20. $(1, 7), (0, 4), (-1, 2)$

21. $(-2, -11), (4, 13), (2, 5)$

22. $(4, 3), (3, 1), (2, -1)$

23. $(-2, 3), (2, -1), (7, -4)$

24. $(-3, -4), (-1, -1), (5, 5)$

25. $\left(\frac{1}{2}, 3\right), \left(-\frac{3}{2}, -5\right), (0, 1)$ **26.** $\left(1, -\frac{5}{2}\right), (0, -3), (6, 0)$

Finding a Coordinate In Exercises 27 and 28, find y such that the points are collinear.

27. $(2, -5), (3, y), (5, -2)$ **28.** $(-6, 2), (-4, y), (-3, 5)$

Finding a Coordinate In Exercises 29 and 30, find x such that the points are collinear.

29. $(-4, -1), (-1, 2), (x, 6)$

30. $(1, 5), (5, -1), (x, 3)$

Finding an Equation of a Line In Exercises 31–40, use a determinant to find an equation of the line passing through the points. *See Example 3.*

31. $(-1, 2), (5, 3)$ **32.** $(3, 1), (-2, -5)$

33. $(-4, 3), (2, 1)$ **34.** $(10, 7), (-2, -7)$

35. $(0, 5), (2, 5)$ **36.** $(3, 3), (6, 3)$

37. $(-4, 5), (-4, -2)$ **38.** $(-1, 3), (-1, 0)$

39. $\left(-\frac{1}{2}, 3\right), \left(\frac{5}{2}, 1\right)$ **40.** $\left(\frac{2}{3}, 4\right), (6, 12)$

In the remaining exercises for this section, use the number code chart on page 524.

Encoding a Message In Exercises 41 and 42, find the uncoded 1×2 row matrices for the message. Then encode the message using the encoding matrix. *See Examples 4 and 5.*

Message	*Encoding Matrix*
41. COME HOME SOON	$\begin{bmatrix} 1 & 2 \\ 3 & 5 \end{bmatrix}$
42. HELP IS ON THE WAY	$\begin{bmatrix} -2 & 3 \\ -1 & 1 \end{bmatrix}$

Encoding a Message In Exercises 43 and 44, find the uncoded 1×3 row matrices for the message. Then encode the message using the encoding matrix. *See Examples 4 and 5.*

Message	*Encoding Matrix*
43. TEXT ME AT WORK	$\begin{bmatrix} 1 & -1 & 0 \\ 1 & 0 & -1 \\ -6 & 2 & 3 \end{bmatrix}$
44. MISSION ACCOMPLISHED	$\begin{bmatrix} 4 & 2 & 1 \\ -3 & -3 & -1 \\ 3 & 2 & 1 \end{bmatrix}$

Encoding a Message In Exercises 45–52, write a cryptogram for the message using the matrix

$$A = \begin{bmatrix} 1 & 2 & 2 \\ 3 & 7 & 9 \\ -1 & -4 & -7 \end{bmatrix}.$$

45. LANDING SUCCESSFUL

46. DOWNLOAD VIDEO   **47.** SEE YOU AT MOVIES

48. GOOD LUCK TOMORROW

49. CONGRATULATIONS

50. HEAD DUE WEST   **51.** PICK ME UP AT NOON

52. EVACUATE THE CITY

Decoding a Message In Exercises 53–56, use A^{-1} to decode the cryptogram. *See Example 6.*

53. $A = \begin{bmatrix} 1 & 2 \\ 3 & 5 \end{bmatrix}$

11, 21, 64, 112, 25, 50, 29, 53, 23, 46, 40, 75, 55, 92

54. $A = \begin{bmatrix} 2 & 3 \\ 3 & 4 \end{bmatrix}$

19, 26, 41, 57, 28, 42, 78, 109, 64, 87, 62, 83, 63, 87, 28, 42, 73, 102, 46, 69

55. $A = \begin{bmatrix} 4 & 2 & 1 \\ -3 & -3 & -1 \\ 3 & 2 & 1 \end{bmatrix}$

94, 35, 25, 44, 16, 10, 4, −10, 1, 27, 15, 9, 71, 43, 22

56. $A = \begin{bmatrix} 1 & -1 & 0 \\ 1 & 0 & -1 \\ -6 & 2 & 3 \end{bmatrix}$

9, −1, −9, 38, −19, −19, 28, −9, −19, −80, 25, 41, −64, 21, 31, −7, −4, 7

Decoding a Message In Exercises 57 and 58, use A^{-1} to decode the cryptogram. *See Example 6.*

$A = \begin{bmatrix} 1 & 2 & 2 \\ 3 & 7 & 9 \\ -1 & -4 & -7 \end{bmatrix}$

57. 20, 17, −15, −9, −44, −83, 64, 136, 157, 24, 31, 12, 4, −37, −102

58. −10, −57, −111, 74, 168, 209, 35, 75, 85, 16, 35, 42, 34, 55, 43

59. Decoding a Message The following cryptogram was encoded with a 2 × 2 matrix.

8, 21, −15, −10, −13, −13, 5, 10, 5, 25, 5, 19, −1, 6, 20, 40, −18, −18, 1, 16

The last word of the message is __RON. What is the message?

60. Decoding a Message The following cryptogram was encoded with a 2 × 2 matrix.

5, 2, 25, 11, −2, −7, −15, −15, 32, 14, −8, −13, 38, 19, −19, −19, 37, 16

The last word of the message is __SUE. What is the message?

61. Cryptography A code breaker intercepted the encoded message below.

45, −35, 38, −30, 18, −18, 35, −30, 81, −60, 42, −28, 75, −55, 2, −2, 22, −21, 15, −10

Let $A^{-1} = \begin{bmatrix} w & x \\ y & z \end{bmatrix}$. You know that $\begin{bmatrix} 45 & -35 \end{bmatrix}A^{-1} = \begin{bmatrix} 10 & 15 \end{bmatrix}$ and that $\begin{bmatrix} 38 & -30 \end{bmatrix}A^{-1} = \begin{bmatrix} 8 & 14 \end{bmatrix}$, where A^{-1} is the inverse of the encoding matrix A. Explain how you can find the values of $w, x, y,$ and z. Decode the message.

62. Cryptography Your biology professor gives you the encoded message below.

−204, 47, −231, 53, −265, 61, −223, 51, −9, 2, −117, 28, −117, 26, −166, 37, −265, 61, −145, 34, −112, 25, −76, 19

Let $A^{-1} = \begin{bmatrix} w & x \\ y & z \end{bmatrix}$. You know that $\begin{bmatrix} -204 & 47 \end{bmatrix}A^{-1} = \begin{bmatrix} 15 & 16 \end{bmatrix}$ and that $\begin{bmatrix} -231 & 53 \end{bmatrix}A^{-1} = \begin{bmatrix} 15 & 19 \end{bmatrix}$, where A^{-1} is the inverse of the encoding matrix A. Explain how you can find the values of $w, x, y,$ and z. Decode the message.

Kei Uesugi/Taxi/Getty Images

63. Reasoning Use a determinant to find the area of the triangle with vertices $(3, -1)$, $(7, -1)$, and $(7, 5)$. Confirm your answer by plotting the points in a coordinate plane and using the formula for the area of a triangle.

64. HOW DO YOU SEE IT? At this point in the text, you have learned several methods for finding an equation of a line that passes through two given points. Briefly describe the methods that can be used to find the equation of the line that passes through the two points shown. Discuss the advantages and disadvantages of each method.

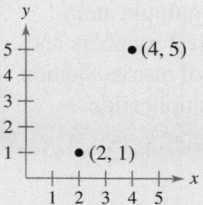

Business Capsule

Voltage Security, Inc. is a leader in secure business communications and data protection. The company provides the most scalable enterprise key management and encryption capabilities for securing data. Invented by Dr. Dan Boneh and Dr. Matt Franklin in 2001, Identity-Based Encryption or IBE is a breakthrough in cryptography. IBE enables users to simply use an identity, such as an e-mail address, to secure business communications.

65. Research Project Use your campus library, the Internet, or some other reference source to find information about a company that generates cryptographic software to secure data. Write a brief paper about such a company.

ALGEBRA TUTOR

xy

Basic Rules of Algebra

On this page, you can review some of the basic rules of algebra.

Let a, b, and c be real numbers, variables, or algebraic expressions.

Property

Commutative Property of Addition:	$a + b = b + a$
Commutative Property of Multiplication:	$ab = ba$
Associative Property of Addition:	$(a + b) + c = a + (b + c)$
Associative Property of Multiplication:	$(ab)c = a(bc)$
Distributive Property:	$a(b + c) = ab + ac$, $(a + b)c = ac + bc$
Additive Identity Property:	$a + 0 = a$
Multiplicative Identity Property:	$a \cdot 1 = a$
Additive Inverse Property:	$a + (-a) = 0$
Multiplicative Inverse Property:	$a \cdot \dfrac{1}{a} = 1, \quad a \neq 0$

STUDY TIP

Be sure you notice the similarities between the addition and multiplication properties of real numbers and the properties of matrix addition and scalar multiplication.

Example 1   **Identifying the Basic Rules of Algebra**

Identify the rule of algebra illustrated by each statement.

a. $(4 + x^3) + 5x^3 = 4 + (x^3 + 5x^3)$ **b.** $(6x^2)3 = 3(6x^2)$

c. $x^4 + 25 = 25 + x^4$ **d.** $7y \cdot 1 = 7y$

e. $(x^3 + 1)\left(\dfrac{1}{x^3 + 1}\right) = 1$ **f.** $8z^2 + 0 = 8z^2$

g. $(4w^2 + 7w) - (4w^2 + 7w) = 0$

SOLUTION

a. This statement illustrates the Associative Property of Addition. In other words, to form the sum $4 + x^3 + 5x^3$, it does not matter whether 4 and x^3, or x^3 and $5x^3$, are added first.

b. This statement illustrates the Commutative Property of Multiplication. In other words, you get the same result whether you multiply $6x^2$ by 3, or 3 by $6x^2$.

c. This statement illustrates the Commutative Property of Addition. In other words, you get the same result whether you add x^4 and 25, or 25 and x^4.

d. This statement illustrates the Multiplicative Identity Property. In other words, this property simply states that when any expression is multiplied by 1, the result is that expression.

e. This statement illustrates the Multiplicative Inverse Property. In other words, this property states that when any expression is multiplied by its reciprocal, the result is 1.

f. This statement illustrates the Additive Identity Property. In other words, this property simply states that when 0 is added to any expression, the result is that expression.

g. This statement illustrates the Additive Inverse Property. In terms of subtraction, this property simply states that when any expression is subtracted from itself, the result is 0.

Matrix Addition and Scalar Multiplication

Example 2 **Addition of Matrices**

$$\begin{bmatrix} -4 & 0 \\ 3 & 2 \end{bmatrix} + \begin{bmatrix} 5 & -1 \\ -2 & -3 \end{bmatrix} + \begin{bmatrix} 2 & 2 \\ 1 & -4 \end{bmatrix} = \begin{bmatrix} -4 + 5 + 2 & 0 + (-1) + 2 \\ 3 + (-2) + 1 & 2 + (-3) + (-4) \end{bmatrix}$$

$$= \begin{bmatrix} 3 & 1 \\ 2 & -5 \end{bmatrix}$$

Example 3 **Scalar Multiplication**

Evaluate each expression using two different methods.

a. $(2 + 5)\begin{bmatrix} 2 & -3 & -1 \\ 4 & 1 & -2 \\ -2 & 5 & 3 \end{bmatrix}$ **b.** $-2\left(4\begin{bmatrix} -3 & -1 \\ 5 & 4 \end{bmatrix}\right)$

SOLUTION

a. Method 1: $(2 + 5)\begin{bmatrix} 2 & -3 & -1 \\ 4 & 1 & -2 \\ -2 & 5 & 3 \end{bmatrix} = 2\begin{bmatrix} 2 & -3 & -1 \\ 4 & 1 & -2 \\ -2 & 5 & 3 \end{bmatrix} + 5\begin{bmatrix} 2 & -3 & -1 \\ 4 & 1 & -2 \\ -2 & 5 & 3 \end{bmatrix}$

$$= \begin{bmatrix} 4 & -6 & -2 \\ 8 & 2 & -4 \\ -4 & 10 & 6 \end{bmatrix} + \begin{bmatrix} 10 & -15 & -5 \\ 20 & 5 & -10 \\ -10 & 25 & 15 \end{bmatrix}$$

$$= \begin{bmatrix} 14 & -21 & -7 \\ 28 & 7 & -14 \\ -14 & 35 & 21 \end{bmatrix}$$

Method 2: $(2 + 5)\begin{bmatrix} 2 & -3 & -1 \\ 4 & 1 & -2 \\ -2 & 5 & 3 \end{bmatrix} = 7\begin{bmatrix} 2 & -3 & -1 \\ 4 & 1 & -2 \\ -2 & 5 & 3 \end{bmatrix}$

$$= \begin{bmatrix} 14 & -21 & -7 \\ 28 & 7 & -14 \\ -14 & 35 & 21 \end{bmatrix}$$

b. Method 1: $-2\left(4\begin{bmatrix} -3 & -1 \\ 5 & 4 \end{bmatrix}\right) = [-2(4)]\begin{bmatrix} -3 & -1 \\ 5 & 4 \end{bmatrix}$

$$= -8\begin{bmatrix} -3 & -1 \\ 5 & 4 \end{bmatrix}$$

$$= \begin{bmatrix} 24 & 8 \\ -40 & -32 \end{bmatrix}$$

Method 2: $-2\left(4\begin{bmatrix} -3 & -1 \\ 5 & 4 \end{bmatrix}\right) = -2\begin{bmatrix} -12 & -4 \\ 20 & 16 \end{bmatrix}$

$$= \begin{bmatrix} 24 & 8 \\ -40 & -32 \end{bmatrix}$$

Which method do you prefer for each expression in Example 3? Explain your reasoning.

SUMMARY AND STUDY STRATEGIES

After studying this chapter, you should have acquired the following skills.
The exercise numbers are keyed to the Review Exercises that begin on page 534.
Answers to odd-numbered Review Exercises are given in the back of the text.*

Section 6.1

Review Exercises

■ Determine the dimension of a matrix.   *1–4*

$$\begin{bmatrix} a_{11} & a_{12} & a_{13} & \cdots & a_{1n} \\ a_{21} & a_{22} & a_{23} & \cdots & a_{2n} \\ a_{31} & a_{32} & a_{33} & \cdots & a_{3n} \\ \vdots & \vdots & \vdots & & \vdots \\ a_{m1} & a_{m2} & a_{m3} & \cdots & a_{mn} \end{bmatrix} \Big\} \; m \text{ rows}$$

$$\underbrace{\qquad\qquad\qquad\qquad}_{n \text{ columns}}$$

A matrix having m rows and n columns is of dimension $m \times n$.

■ Perform elementary row operations on a matrix in order to write the matrix in   *5–8*
row-echelon form or reduced row-echelon form.

■ Solve a system of linear equations using Gaussian elimination or   *9–18*
Gauss-Jordan elimination.

Section 6.2

■ Determine whether two matrices are equal.   *19, 20*

■ Add or subtract two matrices and multiply a matrix by a scalar.   *21–34*

If $A = [a_{ij}]$ and $B = [b_{ij}]$ are $m \times n$ matrices and c is a scalar, then
$A + B = [a_{ij} + b_{ij}]$ and $cA = [ca_{ij}]$.

■ Find the product of two matrices.   *35–40*

If $A = [a_{ij}]$ is an $m \times n$ matrix and $B = [b_{ij}]$ is an $n \times p$ matrix, then AB is an
$m \times p$ matrix

$$AB = [c_{ij}]$$

where $c_{ij} = a_{i1}b_{1j} + a_{i2}b_{2j} + a_{i3}b_{3j} + \cdots + a_{in}b_{nj}$.

■ Use matrix operations to solve an application problem.   *41–44*

Section 6.3

■ Verify that a matrix is the inverse of a given matrix.   *45, 46*

■ Find the inverse of a matrix.   *47, 48*

■ Find the inverse of a 2×2 matrix using a formula.   *49, 50*

$$A^{-1} = \frac{1}{ad - bc}\begin{bmatrix} d & -b \\ -c & a \end{bmatrix}$$

■ Use an inverse matrix to solve a system of linear equations.   *51–60*

---

* A wide range of valuable study aids are available to help you master the material in this chapter.
  The *Student Solutions Manual* includes step-by-step solutions to all odd-numbered exercises to
  help you review and prepare. The student website at *www.cengagebrain.com* offers algebra help
  and a *Graphing Technology Guide*, which contains step-by-step commands and instructions for
  a wide variety of graphing calculators.

Section 6.4

■ Evaluate the determinant of a 2×2 matrix.

$$\det(A) = |A| = \begin{vmatrix} a_1 & b_1 \\ a_2 & b_2 \end{vmatrix} = a_1 b_2 - a_2 b_1$$

■ Find the minors and cofactors of a matrix.

■ Find the determinant of a square matrix.

$$|A| = a_{11}C_{11} + a_{12}C_{12} + \cdots + a_{1n}C_{1n}$$

Section 6.5

■ Find the area of a triangle using a determinant.

$$\text{Area} = \pm\frac{1}{2}\begin{vmatrix} x_1 & y_1 & 1 \\ x_2 & y_2 & 1 \\ x_3 & y_3 & 1 \end{vmatrix}$$

■ Use a determinant to test for collinear points.

$$\begin{vmatrix} x_1 & y_1 & 1 \\ x_2 & y_2 & 1 \\ x_3 & y_3 & 1 \end{vmatrix} = 0$$

■ Use a determinant to find an equation of a line.

$$\begin{vmatrix} x & y & 1 \\ x_1 & y_1 & 1 \\ x_2 & y_2 & 1 \end{vmatrix} = 0$$

■ Encode and decode a cryptogram using a matrix.

Study Strategies

■ **Variety of Approaches**   You can use a variety of approaches when finding the determinant of a square matrix.

1. For a 2×2 matrix, you can use the definition

$$|A| = \begin{vmatrix} a_1 & b_1 \\ a_2 & b_2 \end{vmatrix}$$

$$= a_1 b_2 - a_2 b_1.$$

2. For any square matrix (of dimension 2×2 or greater), you can use expansion by cofactors. Be sure you choose the row or column that makes the computations the easiest.

3. You can always use the matrix capabilities of a graphing utility.

■ **Using Technology**   Performing operations with matrices can be tedious. You can use a graphing utility to accomplish the following.

• Perform elementary row operations on matrices.

• Reduce matrices to row-echelon form and reduced row-echelon form.

• Add and subtract matrices.

• Multiply matrices.

• Multiply matrices by scalars.

• Find inverses of matrices.

• Solve systems of equations using matrices.

• Evaluate determinants of matrices.

Review Exercises

See www.CalcChat.com for worked-out solutions to odd-numbered exercises.

Dimensions of Matrices   In Exercises 1–4, determine the dimension of the matrix.

1. $[-4 \quad 0]$

2. $\begin{bmatrix} 9 & -1 & -2 \\ 0 & -5 & 3 \\ -4 & 10 & 6 \\ 5 & 5 & 1 \end{bmatrix}$

3. $\begin{bmatrix} 3 & 7 & 4 & -2 \\ 1 & 8 & 6 & 1 \end{bmatrix}$

4. $\begin{bmatrix} 5 \\ -1 \\ 2 \\ 4 \end{bmatrix}$

Writing a Matrix in Row-Echelon Form   In Exercises 5 and 6, write the matrix in row-echelon form.

5. $\begin{bmatrix} 1 & 3 & 0 & 2 \\ 3 & 10 & 1 & 8 \\ 2 & 3 & 3 & 10 \end{bmatrix}$

6. $\begin{bmatrix} 1 & 2 & -1 & 0 \\ -2 & -3 & 3 & 4 \\ 4 & 0 & 1 & 3 \end{bmatrix}$

Writing a Matrix in Reduced Row-Echelon Form   In Exercises 7 and 8, write the matrix in reduced row-echelon form.

7. $\begin{bmatrix} 1 & 2 & 3 \\ -2 & 0 & 2 \\ 2 & 1 & 2 \end{bmatrix}$

8. $\begin{bmatrix} 2 & 3 & 1 & -5 \\ 1 & 0 & 5 & 2 \\ -1 & 4 & 3 & 6 \\ 0 & -2 & 6 & -8 \end{bmatrix}$

Solving a System of Equations   In Exercises 9–16, use matrices to solve the system of equations (if possible). Use Gaussian elimination with back-substitution or Gauss-Jordan elimination.

9. $\begin{cases} 4x - 3y = 18 \\ x + y = 1 \end{cases}$

10. $\begin{cases} 2x + 4y = 16 \\ -x + 3y = 17 \end{cases}$

11. $\begin{cases} 2x + 3y - z = 13 \\ 3x + z = 8 \\ x - 2y + 3z = -4 \end{cases}$

12. $\begin{cases} 3x + 4y + 2z = 5 \\ 2x + 3y = 7 \\ 2y - 3z = 12 \end{cases}$

13. $\begin{cases} x + 2y + 2z = 10 \\ 2x + 3y + 5z = 20 \end{cases}$

14. $\begin{cases} 3x + 10y + 4z = 20 \\ x + 3y - 2z = 8 \end{cases}$

15. $\begin{cases} 2x + y - 3z = 4 \\ x + 2y + 2z = 10 \\ x - 2z = 12 \\ x + y + z = 6 \end{cases}$

16. $\begin{cases} 2x + 4y + 2z = 10 \\ x + 3z = 9 \\ 3x - 2y = 4 \\ x + y + z = 8 \end{cases}$

17. Biology A school district borrowed $200,000 at simple annual interest to upgrade microbiology equipment. Some of the money was borrowed at 8%, some at 10%, and some at 12%. Use a system of equations to determine how much was borrowed at each rate if the total annual interest was $20,000 and the amount borrowed at 10% was three times the amount borrowed at 8%. Solve the system using matrices.

18. Amusement Park An amusement park borrowed $650,000 at simple annual interest to renovate a roller coaster. Some of the money was borrowed at 8.5%, some at 9.5%, and some at 10%. Use a system of equations to determine how much was borrowed at each rate if the total annual interest was $58,250 and the amount borrowed at 8.5% was four times the amount borrowed at 10%. Solve the system using matrices.

Equality of Matrices   In Exercises 19 and 20, find x and y.

19. $\begin{bmatrix} -7 & -5 \\ 2 & -1 \end{bmatrix} = \begin{bmatrix} -7 & x + 1 \\ y & -1 \end{bmatrix}$

20. $\begin{bmatrix} 3 & x + 4 & -2 \\ 7y & 10 & y + 5 \\ -3 & 3x & 1 \end{bmatrix} = \begin{bmatrix} 3 & 2x - 1 & -2 \\ -21 & 10 & 2 \\ -3 & 15 & 1 \end{bmatrix}$

Operations with Matrices   In Exercises 21–26, find (a) $A + B$, (b) $A - B$, (c) $4A$, and (d) $5A - 2B$.

21. $A = \begin{bmatrix} -1 & 5 \\ 2 & 1 \end{bmatrix}$, $B = \begin{bmatrix} 4 & 2 \\ -6 & 3 \end{bmatrix}$

22. $A = [4 \quad 6 \quad -2], B = [3 \quad 7 \quad -9]$

23. $A = \begin{bmatrix} 1 & -4 \\ 0 & -1 \\ 4 & 2 \end{bmatrix}, B = \begin{bmatrix} 6 & 5 \\ 3 & 7 \\ -1 & 4 \end{bmatrix}$

24. $A = \begin{bmatrix} 1 & 0 & 2 \\ -1 & 3 & 5 \\ 2 & -2 & 3 \end{bmatrix}, B = \begin{bmatrix} 2 & 0 & 1 \\ 3 & -4 & 6 \\ 1 & 2 & -3 \end{bmatrix}$

25. $A = \begin{bmatrix} 1 & 3 & -2 & 6 \\ 0 & 1 & 3 & 2 \end{bmatrix}$

$B = \begin{bmatrix} 2 & 1 & 4 & -5 \\ 3 & -6 & 3 & -2 \end{bmatrix}$

26. $A = \begin{bmatrix} 3 \\ -2 \\ 3 \end{bmatrix}, B = \begin{bmatrix} -1 \\ 4 \\ 5 \end{bmatrix}$

Evaluating an Expression   In Exercises 27–30, evaluate the expression.

27. $\begin{bmatrix} 2 & 0 \\ -5 & -4 \end{bmatrix} + \begin{bmatrix} -10 & 3 \\ 3 & -5 \end{bmatrix} + \begin{bmatrix} -4 & -6 \\ 7 & 4 \end{bmatrix}$

28. $5\left(\begin{bmatrix} -6 & 13 \\ 4 & -8 \\ 7 & 10 \end{bmatrix} - \begin{bmatrix} -5 & 9 \\ -1 & -5 \\ 3 & 11 \end{bmatrix} \right)$

29. $\frac{1}{4}([3 \quad -1 \quad -2 \quad 18] + [5 \quad -3 \quad -10 \quad 2])$

30. $-2\begin{bmatrix} 1 & -3 \\ 2 & -1 \end{bmatrix} + 6\begin{bmatrix} -1 & 2 \\ -2 & -3 \end{bmatrix}$

Solving a Matrix Equation In Exercises 31–34, solve for X, where

$$A = \begin{bmatrix} 1 & -2 \\ 0 & 1 \\ 2 & 3 \end{bmatrix} \quad \text{and} \quad B = \begin{bmatrix} 0 & 1 \\ 1 & 1 \\ 3 & 5 \end{bmatrix}.$$

31. $X = 4A - 3B$ **32.** $X = 5B + 2A$

33. $2X - 3A = B$ **34.** $4X - 8B = 4A$

Find the Product of Two Matrices In Exercises 35–38, find AB, if possible.

35. $A = \begin{bmatrix} 1 & 4 \\ -2 & -1 \\ 3 & 2 \end{bmatrix}, B = \begin{bmatrix} -4 \\ 3 \end{bmatrix}$

36. $A = \begin{bmatrix} 3 \\ 2 \\ 4 \\ 6 \end{bmatrix}, B = \begin{bmatrix} 2 & 0 & -1 \end{bmatrix}$

37. $A = \begin{bmatrix} 4 & 0 & 0 \\ 0 & 3 & 0 \\ 0 & 0 & -2 \end{bmatrix}, B = \begin{bmatrix} \frac{1}{4} & 0 & 0 \\ 0 & \frac{1}{3} & 0 \\ 0 & 0 & -\frac{1}{2} \end{bmatrix}$

38. $A = \begin{bmatrix} 1 & 2 & 3 & 6 & -1 \\ 2 & 8 & 0 & 0 & 2 \end{bmatrix}, B = \begin{bmatrix} 3 & 2 \\ 4 & -1 \end{bmatrix}$

Operations with Matrices In Exercises 39 and 40, find (a) AB, (b) BA, and, if possible, (c) A^2. (*Note:* $A^2 = AA$.)

39. $A = \begin{bmatrix} 1 & -3 & 4 \end{bmatrix}, B = \begin{bmatrix} 2 \\ -2 \\ -1 \end{bmatrix}$

40. $A = \begin{bmatrix} 1 & 0 & 2 \\ 3 & 1 & -2 \\ 1 & 1 & 1 \end{bmatrix}, B = \begin{bmatrix} 2 & 0 & 0 \\ 1 & -2 & 1 \\ 5 & 4 & -2 \end{bmatrix}$

41. Factory Production A window corporation has four factories, each of which manufactures three products. The number of units of product i produced at factory j in one day is represented by a_{ij} in the matrix

$$A = \begin{bmatrix} 80 & 120 & 20 & 40 \\ 40 & 60 & 80 & 20 \\ 140 & 60 & 100 & 80 \end{bmatrix}.$$

Find the production levels when production is increased by 20%.

42. Factory Production An electronics manufacturer has three factories, each of which manufactures four products. The number of units of product i produced at factory j in one day is represented by a_{ij} in the matrix

$$A = \begin{bmatrix} 120 & 140 & 60 \\ 80 & 100 & 40 \\ 40 & 160 & 80 \\ 20 & 120 & 100 \end{bmatrix}.$$

Find the production levels when production is decreased by 10%.

43. Inventory Levels A company sells four different models of car sound systems through three retail outlets. The inventories of the four models at the three outlets are given by matrix S.

Model

$$S = \begin{bmatrix} 3 & 2 & 1 & 4 \\ 1 & 3 & 4 & 3 \\ 5 & 3 & 2 & 2 \end{bmatrix} \begin{matrix} 1 \\ 2 \\ 3 \end{matrix} \text{Outlet}$$

with columns labeled A, B, C, D.

The wholesale and retail prices of the four models are given by matrix T.

Price

	Wholesale	Retail	
	$350	$600	A
$T =$	$425	$705	B
	$300	$455	C
	$750	$1150	D

(Model)

(a) What is the total retail price of the inventory at Outlet 3?

(b) What is the total wholesale price of the inventory at Outlet 1?

(c) Compute ST and interpret the result.

44. Labor/Wage Requirements A company that manufactures racing bicycles has the following labor-hour and wage requirements.

Labor-Hour Requirements (per bicycle)

Department

	Cutting	Assembly	Packaging	
	0.9 hour	0.8 hour	0.2 hour	Basic
$S =$	1.5 hours	1.0 hour	0.4 hour	Light
	3.5 hours	3.0 hours	0.5 hour	Ultra-light

(Models)

Wage Requirements (per hour)

Plant

	A	B	
	$13.00	$14.00	Cutting
$T =$	$11.50	$9.50	Assembly
	$10.00	$9.00	Packaging

(Department)

(a) What is the labor cost for a light racing bicycle at Plant A?

(b) What is the labor cost for an ultra-light racing bicycle at Plant B?

(c) Compute ST and interpret the result.

The Inverse of a Matrix In Exercises 45 and 46, show that B is the inverse of A.

45. $A = \begin{bmatrix} 1 & 2 & 1 \\ 3 & 6 & 4 \\ 0 & 1 & 3 \end{bmatrix}, B = \begin{bmatrix} -14 & 5 & -2 \\ 9 & -3 & 1 \\ -3 & 1 & 0 \end{bmatrix}$

46. $A = \begin{bmatrix} 2 & 0 & 1 & 2 \\ 3 & 0 & 0 & 1 \\ -1 & 1 & 2 & 0 \\ 0 & -1 & 2 & 2 \end{bmatrix}$

$B = \dfrac{1}{9}\begin{bmatrix} -4 & 6 & 1 & 1 \\ 10 & -6 & 2 & -7 \\ -7 & 6 & 4 & 4 \\ 12 & -9 & -3 & -3 \end{bmatrix}$

Finding the Inverse of a Matrix In Exercises 47 and 48, find the inverse of the matrix.

47. $\begin{bmatrix} -1 & 0 & 0 \\ 0 & 2 & 0 \\ 0 & 0 & 4 \end{bmatrix}$ **48.** $\begin{bmatrix} 3 & 2 & 2 \\ 0 & 2 & 1 \\ 1 & 0 & 1 \end{bmatrix}$

Finding the Inverse of a 2 × 2 Matrix In Exercises 49 and 50, use the formula on page 506 to find the inverse of the matrix.

49. $\begin{bmatrix} 1 & 3 \\ 2 & 5 \end{bmatrix}$

50. $\begin{bmatrix} -2 & 1 \\ 4 & 3 \end{bmatrix}$

Solving a System Using an Inverse Matrix In Exercises 51 and 52, use the inverse matrix found in Exercise 49 to solve the system of linear equations.

51. $\begin{cases} x + 3y = 15 \\ 2x + 5y = 26 \end{cases}$ **52.** $\begin{cases} x + 3y = 7 \\ 2x + 5y = 11 \end{cases}$

Solving a System Using an Inverse Matrix In Exercises 53 and 54, use the inverse matrix found in Exercise 48 to solve the system of linear equations.

53. $\begin{cases} 3x + 2y + 2z = 13 \\ 2y + z = 4 \\ x + z = 5 \end{cases}$ **54.** $\begin{cases} 3x + 2y + 2z = 12 \\ 2y + z = 13 \\ x + z = 3 \end{cases}$

Solving a System Using an Inverse Matrix In Exercises 55 and 56, use an inverse matrix to solve the system of linear equations.

55. $\begin{cases} -3x + 10y = 8 \\ 5x - 17y = -13 \end{cases}$ **56.** $\begin{cases} -x + 4y - 2z = 12 \\ 2x - 9y + 5z = -25 \\ -x + 5y - 4z = 10 \end{cases}$

Raw Materials In Exercises 57 and 58, you are making three types of windshield washer fluid in chemistry class. Fluid X requires 9 cups of water, 1 cup of isopropyl alcohol, and 1 tablespoon of detergent. Fluid Y requires 10 cups of water, 3 cups of isopropyl alcohol, and 1 tablespoon of detergent. Fluid Z requires 14 cups of water, 2 cups of isopropyl alcohol, and 2 tablespoons of detergent. A system of linear equations (where x, y, and z represent fluids X, Y, and Z, respectively) is as follows.

$\begin{cases} 9x + 10y + 14z = \text{(cups of water)} \\ x + 3y + 2z = \text{(cups of isopropyl alcohol)} \\ x + y + 2z = \text{(tablespoons of detergent)} \end{cases}$

Use the inverse of the coefficient matrix of this system to find the numbers of units of fluids X, Y, and Z that you can produce with the given amounts of ingredients.

57. 240 cups of water
44 cups of isopropyl alcohol
28 tablespoons of detergent

58. 235 cups of water
41 cups of isopropyl alcohol
29 tablespoons of detergent

59. Operating Stores The numbers of PetSmart stores in operation from 2005 through 2009 increased in a pattern that was approximately parabolic. The least squares regression parabola $y = at^2 + bt + c$ for the data is found by solving the system

$\begin{cases} 5c + 10b + 30a = 5003 \\ 10c + 30b + 100a = 10{,}856 \\ 30c + 100b + 354a = 33{,}332 \end{cases}$

Let t represent the year, with $t = 0$ corresponding to 2005.   *(Source: PetSmart, Inc.)*

(a) Use a graphing utility to find an inverse matrix to solve the system, and find the equation of the least squares regression parabola.

(b) Use the result of part (a) to predict the number of stores in operation in 2010.

(c) The actual number of stores in operation in 2010 was about 1170. How does this compare with your prediction in part (b)?

60. Carnivorous Plants A Venus flytrap is grown in a greenhouse, and the size y (in millimeters) of its traps is measured at the end of each year for 5 years. The least squares regression parabola $y = at^2 + bt + c$ for the data is found by solving the system

$\begin{cases} 5c + 15b + 55a = 53.3 \\ 15c + 55b + 225a = 190.6 \\ 55c + 225b + 979a = 755.8 \end{cases}$

Let t represent the year, with $t = 1$ corresponding to the first year.

(a) Use a graphing utility to find an inverse matrix to solve the system, and find the equation of the least squares regression parabola.

(b) Use the result of part (a) to estimate the sizes of the traps after the first and third years.

(c) The actual sizes of the traps were 2.5 millimeters after the first year and 12.8 millimeters after the third year. How do these values compare with your estimates in part (b)?

Finding the Determinant of a Matrix In Exercises 61–64, find the determinant of the matrix.

61. $\begin{bmatrix} 8 & 4 \\ 3 & 2 \end{bmatrix}$

62. $\begin{bmatrix} 7 & 2 \\ 9 & -3 \end{bmatrix}$

63. $\begin{bmatrix} 5 & 2 \\ 0 & 0 \end{bmatrix}$

64. $\begin{bmatrix} 3 & 0 \\ 0 & -7 \end{bmatrix}$

Finding the Minors and Cofactors of a Matrix In Exercises 65–68, find all the (a) minors and (b) cofactors of the matrix.

65. $\begin{bmatrix} 2 & -1 \\ 7 & 4 \end{bmatrix}$

66. $\begin{bmatrix} 3 & 6 \\ 5 & -4 \end{bmatrix}$

67. $\begin{bmatrix} 3 & 2 & -1 \\ -2 & 5 & 0 \\ 1 & 8 & 6 \end{bmatrix}$

68. $\begin{bmatrix} 8 & 3 & 4 \\ 6 & 5 & -9 \\ -4 & 1 & 2 \end{bmatrix}$

Choosing a Method In Exercises 69–74, find the determinant of the matrix. Tell which method you used. (*Hint:* Use information from Exercises 69–74 on page 520.)

69. $\begin{bmatrix} 1 & 2 & 3 \\ 8 & 6 & 7 \\ 0 & 2 & -1 \end{bmatrix}$

70. $\begin{bmatrix} -2 & 3 & 3 \\ -1 & 0 & 5 \\ 1 & 2 & -1 \end{bmatrix}$

71. $\begin{bmatrix} 1 & 3 & 2 & 4 \\ 0 & -1 & 2 & 2 \\ 0 & 0 & 3 & 0 \\ 0 & 0 & 0 & 4 \end{bmatrix}$

72. $\begin{bmatrix} 2 & 0 & 0 & 0 \\ 3 & 4 & 0 & 0 \\ 5 & 1 & -2 & 0 \\ 6 & 3 & 1 & 1 \end{bmatrix}$

73. $\begin{bmatrix} -2 & 4 & 1 \\ 6 & 1 & 2 \\ 5 & 3 & 4 \end{bmatrix}$

74. $\begin{bmatrix} 4 & 7 & -1 \\ 2 & -3 & 4 \\ -5 & 1 & -1 \end{bmatrix}$

The Determinant of a Matrix Product In Exercises 75 and 76, find (a) $|A|$, (b) $|B|$, (c) AB, and (d) $|AB|$.

75. $A = \begin{bmatrix} -1 & 2 \\ -3 & -2 \end{bmatrix}$, $B = \begin{bmatrix} 4 & 0 \\ 1 & -3 \end{bmatrix}$

76. $A = \begin{bmatrix} 1 & 2 & -1 \\ 0 & 4 & 3 \\ -2 & 0 & 1 \end{bmatrix}$, $B = \begin{bmatrix} -3 & 4 & 0 \\ 1 & -1 & -2 \\ 0 & 1 & 1 \end{bmatrix}$

Finding an Area In Exercises 77–80, use a determinant to find the area of the triangle with the given vertices.

77.

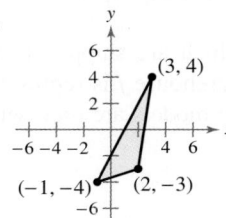

78.

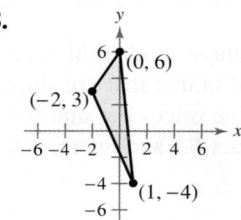

79. $(-1, 2), (3, 2), (-2, -3)$

80. $(-6, -1), (1, -3), (-4, -7)$

Testing for Collinear Points In Exercises 81–84, use a determinant to determine whether the points are collinear.

81. $(0, 3), (1, 5), (2, 8)$ **82.** $(2, 6), (-2, 3), (0, 5)$

83. $(-4, 1), (6, 6), (0, 3)$

84. $(-3, -1), (0, 5), (-4, -3)$

Finding an Equation of a Line In Exercises 85–88, use a determinant to find an equation of the line passing through the points.

85. $(-7, 3), (8, 2)$

86. $(5, -4), (-3, 2)$

87. $(2, 4), (2, -7)$

88. $(-5, -1), (1, -1)$

In the remaining exercises for this review, use the number code chart on page 524.

Encoding a Message In Exercises 89 and 90, find the uncoded row matrices for the message. Then encode the message using the encoding matrix.

Message	*Encoding Matrix*

89. TRANSMIT NOW $\begin{bmatrix} 2 & 3 \\ 3 & 4 \end{bmatrix}$

90. CALL AT MIDNIGHT $\begin{bmatrix} 1 & 2 & 2 \\ 3 & 7 & 9 \\ -1 & -4 & -7 \end{bmatrix}$

Decoding a Message In Exercises 91 and 92, use A^{-1} to decode the cryptogram.

91. $A = \begin{bmatrix} 1 & 2 \\ -1 & 3 \end{bmatrix}$

14, 53, −17, 96, 5, 10, 12, 64, 5, 10, 3, 11, 25, 50

92. $A = \begin{bmatrix} 1 & -1 & 0 \\ 1 & 0 & -1 \\ -6 & 2 & 3 \end{bmatrix}$

−14, −1, 10, −38, 2, 27, −94, 18, 57, 7, −11, −1, −96, 20, 57, −74, 23, 35, 17, −12, −5

93. Cryptography A family sends the encoded message below to a relative overseas.

−57, −13, 91, 26, 97, 29, −76, −19, 5, 5, −84, −21, 55, 16, −28, −7, 97, 28, −8, −2

Let $A^{-1} = \begin{bmatrix} w & x \\ y & z \end{bmatrix}$.

(a) You know that $\begin{bmatrix} -57 & -13 \end{bmatrix} A^{-1} = \begin{bmatrix} 23 & 5 \end{bmatrix}$ and that $\begin{bmatrix} 91 & 26 \end{bmatrix} A^{-1} = \begin{bmatrix} 0 & 13 \end{bmatrix}$, where A^{-1} is the inverse of the encoding matrix A. Explain how you can find the values of w, x, y, and z.

(b) Decode the message.

TEST YOURSELF

See www.CalcChat.com for worked-out solutions to odd-numbered exercises.

Take this test as you would take a test in class. When you are done, check your work against the answers given in the back of the book.

In Exercises 1 and 2, write the augmented matrix for the system of linear equations. What is the dimension of the augmented matrix?

1. $\begin{cases} 2x + y + 4z = 2 \\ x + 4y - z = 0 \\ -x + 3y + 3z = -1 \end{cases}$
2. $\begin{cases} 3x + 4y + 2z = 4 \\ 2x + 3y = -2 \\ 2y - 3z = -13 \end{cases}$

In Exercises 3–5, use matrices to solve the system of equations.

3. $\begin{cases} x + 2y + 3z = 16 \\ 5x + 4y - z = 22 \end{cases}$
4. $\begin{cases} x - 2y + z = 14 \\ y - 3z = 2 \\ z = -6 \end{cases}$
5. $\begin{cases} 2x - 3y + z = 14 \\ x + 2y = -4 \\ y - z = -4 \end{cases}$

In Exercises 6–13, use the matrices to find the indicated matrix.

$$A = \begin{bmatrix} 1 & 3 \\ 2 & 4 \end{bmatrix}, \quad B = \begin{bmatrix} 2 & -1 & 3 \\ 4 & 0 & 1 \end{bmatrix}, \quad C = \begin{bmatrix} 0 & -2 \\ 3 & 5 \end{bmatrix}, \quad D = \begin{bmatrix} 3 \\ 2 \\ -1 \end{bmatrix}, \quad E = \begin{bmatrix} 7 & -8 \\ 6 & 10 \end{bmatrix}$$

6. $2A + C$
7. CA
8. BD
9. C^2

10. $A + C + E$
11. $-4(A + E)$
12. $5(C - A)$
13. $3C + 2(A - E)$

In Exercises 14–16, find the inverse of the matrix.

14. $A = \begin{bmatrix} 2 & -1 \\ -3 & 4 \end{bmatrix}$
15. $A = \begin{bmatrix} 1 & 0 \\ 0 & 1 \end{bmatrix}$
16. $A = \begin{bmatrix} 3 & 4 & 2 \\ 2 & 3 & 0 \\ 0 & 2 & -3 \end{bmatrix}$

In Exercises 17–19, find the determinant of the matrix.

17. $\begin{bmatrix} -5 & 2 \\ 1 & 3 \end{bmatrix}$
18. $\begin{bmatrix} 3 & 2 & -1 \\ 1 & 0 & 2 \\ 4 & 5 & 2 \end{bmatrix}$
19. $\begin{bmatrix} 2 & 0 & 0 \\ 0 & 5 & 0 \\ 0 & 0 & -2 \end{bmatrix}$

20. Use the inverse matrix found in Exercise 16 to solve the system of linear equations in Exercise 2.

21. Find two nonzero matrices whose product is a zero matrix.

22. Use a determinant to find the area of the triangle whose vertices are $(-3, 1)$, $(0, 4)$, and $(5, 2)$.

23. Use a determinant to decide whether the points $(2, 1)$, $(-3, -14)$, and $(4, 7)$ are collinear.

24. Use a determinant to find an equation of the line passing through the points $(1, -2)$ and $(5, 2)$.

25. A manufacturer produces three models of a product, which are shipped to two warehouses. The number of units i that are shipped to warehouse j is represented by a_{ij} in matrix A below. The prices per unit for the three models are represented by matrix B. Find the product BA and interpret the result.

$$A = \begin{bmatrix} 1500 & 4000 \\ 3000 & 4500 \\ 5500 & 7000 \end{bmatrix}$$

$$B = \begin{bmatrix} \$55 & \$40 & \$33 \end{bmatrix}$$

7 Limits and Derivatives

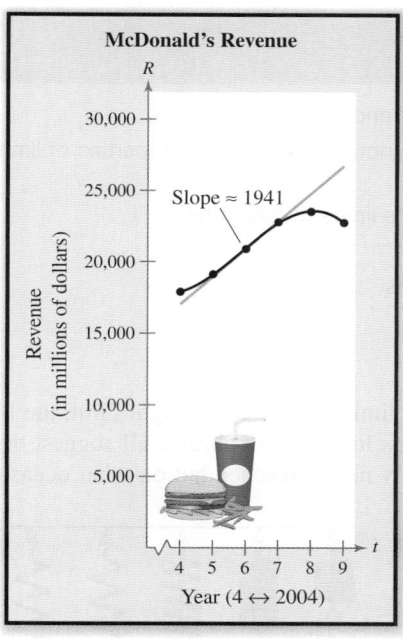

McDonald's Revenue
R
30,000
25,000
20,000
15,000
10,000
5,000
Revenue (in millions of dollars)
Slope ≈ 1941
t
4 5 6 7 8 9
Year (4 ↔ 2004)

Example 11 on page 582 shows how differentiation can be used to find the rate of change in a company's revenue.

David Gilder/www.shutterstock.com
Kurhan/Shutterstock.com

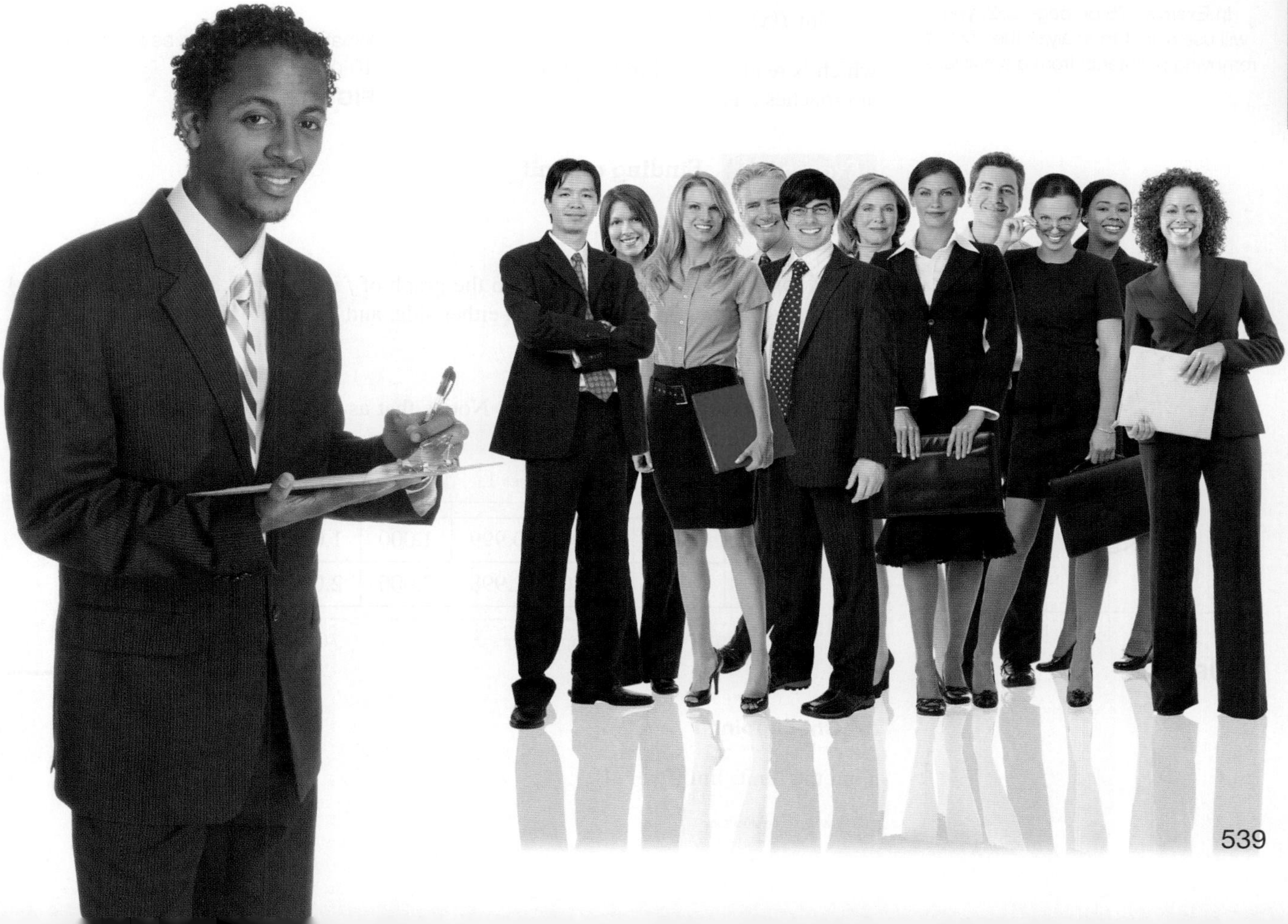

7.1 Limits

■ Find limits of functions graphically and numerically.

■ Understand the definition of the limit of a function and use the properties of limits to evaluate limits of functions.

■ Use different analytic techniques to evaluate limits of functions.

■ Evaluate one-sided limits.

■ Recognize unbounded behavior of functions.

The Limit of a Function

In everyday language, people refer to a speed limit, a wrestler's weight limit, the limit of one's endurance, or stretching a spring to its limit. These phrases all suggest that a limit is a bound, which on some occasions may not be reached but on other occasions may be reached or exceeded.

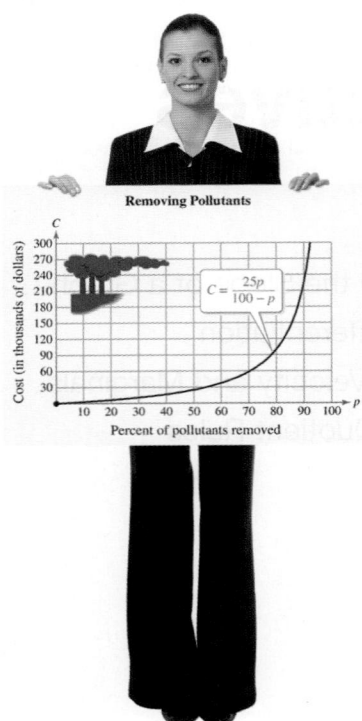

Removing Pollutants

$C = \dfrac{25p}{100 - p}$

In Exercise 75 on page 552, you will use a limit to analyze the cost of removing pollutants from a small lake.

Consider a spring that will break only when a weight of 10 pounds or more is attached. To determine how far the spring will stretch without breaking, you could attach increasingly heavier weights and measure the spring length s for each weight w, as shown in Figure 7.1. If the spring length approaches a value of L, then it is said that "the limit of s as w approaches 10 is L." A mathematical limit is much like the limit of a spring. The notation for a limit is

$$\lim_{x \to c} f(x) = L$$

which is read as "the limit of $f(x)$ as x approaches c is L."

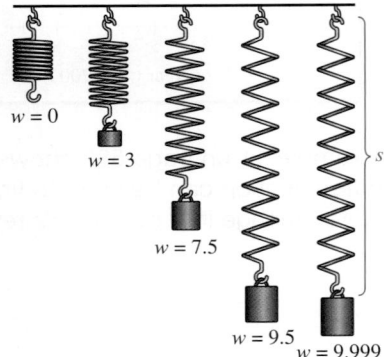

What is the limit of s as w approaches 10 pounds?

FIGURE 7.1

Example 1   Finding a Limit

Find the limit: $\lim_{x \to 1} (x^2 + 1)$.

SOLUTION   Let $f(x) = x^2 + 1$. From the graph of f in Figure 7.2, it appears that $f(x)$ approaches 2 as x approaches 1 from either side, and you can write

$$\lim_{x \to 1} (x^2 + 1) = 2.$$

The table yields the same conclusion. Notice that as x gets closer and closer to 1, $f(x)$ gets closer and closer to 2.

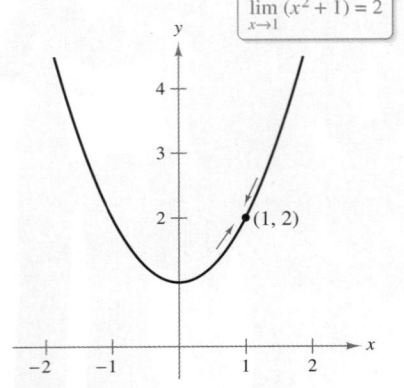

$\lim_{x \to 1} (x^2 + 1) = 2$

(1, 2)

FIGURE 7.2

	x approaches 1.				x approaches 1.		
x	0.900	0.990	0.999	1.000	1.001	1.010	1.100
$f(x)$	1.810	1.980	1.998	2.000	2.002	2.020	2.210
	$f(x)$ approaches 2.				$f(x)$ approaches 2.		

✓ **Checkpoint 1**

Find the limit: $\lim_{x \to 1} (2x + 4)$.

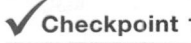

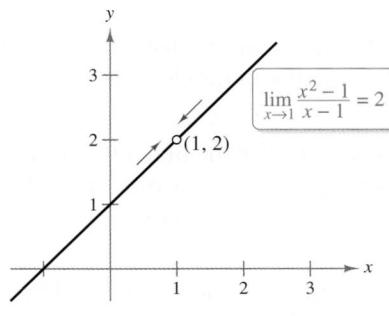

(a)

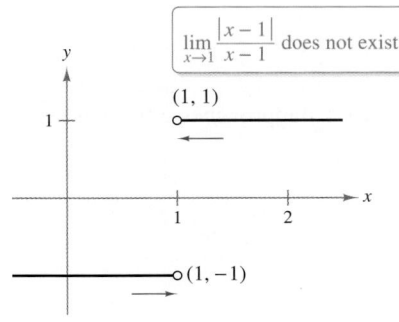

(b)

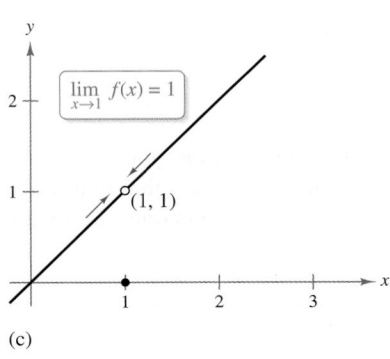

(c)

FIGURE 7.3

Example 2 Finding Limits Graphically and Numerically

Find the limit.

$$\lim_{x \to 1} f(x)$$

a. $f(x) = \dfrac{x^2 - 1}{x - 1}$ **b.** $f(x) = \dfrac{|x - 1|}{x - 1}$ **c.** $f(x) = \begin{cases} x, & x \neq 1 \\ 0, & x = 1 \end{cases}$

SOLUTION

a. From the graph of f in Figure 7.3(a), it appears that $f(x)$ approaches 2 as x approaches 1 from either side. A missing point is denoted by the open dot on the graph. This conclusion is reinforced by the table. Be sure you see that *it does not matter that $f(x)$ is undefined when $x = 1$*. The limit depends only on values of $f(x)$ near 1, not at 1.

	x approaches 1.				x approaches 1.		
x	0.900	0.990	0.999	1.000	1.001	1.010	1.100
f(x)	1.900	1.990	1.999	?	2.001	2.010	2.100

$f(x)$ approaches 2. $f(x)$ approaches 2.

b. From the graph of f in Figure 7.3(b), you can see that $f(x) = -1$ for all values to the left of $x = 1$ and $f(x) = 1$ for all values to the right of $x = 1$. So, $f(x)$ is approaching a different value from the left of $x = 1$ than it is from the right of $x = 1$. In such situations, you can conclude that *the limit does not exist*. This conclusion is reinforced by the table.

	x approaches 1.				x approaches 1.		
x	0.900	0.990	0.999	1.000	1.001	1.010	1.100
f(x)	-1.000	-1.000	-1.000	?	1.000	1.000	1.000

$f(x)$ approaches -1. $f(x)$ approaches 1.

c. From the graph of f in Figure 7.3(c), it appears that $f(x)$ approaches 1 as x approaches 1 from either side. This conclusion is reinforced by the table. It does not matter that $f(1) = 0$. The limit depends only on values of $f(x)$ near 1, not at 1.

	x approaches 1.				x approaches 1.		
x	0.900	0.990	0.999	1.000	1.001	1.010	1.100
f(x)	0.900	0.990	0.999	?	1.001	1.010	1.100

$f(x)$ approaches 1. $f(x)$ approaches 1.

✓ **Checkpoint 2**

Find the limit.

$$\lim_{x \to 2} f(x)$$

a. $f(x) = \dfrac{x^2 - 4}{x - 2}$ **b.** $f(x) = \dfrac{|x - 2|}{x - 2}$ **c.** $f(x) = \begin{cases} x^2, & x \neq 2 \\ 0, & x = 2 \end{cases}$

Definition of the Limit of a Function and Properties of Limits

There are three important ideas to learn from Examples 1 and 2.

1. Saying that the limit of $f(x)$ approaches L as x approaches c means that the value of $f(x)$ may be made *arbitrarily close* to the number L by choosing x closer and closer to c.

2. If $f(x)$ approaches the same number from *either side* of c, then the limit of $f(x)$ as x approaches c exists. However, if $f(x)$ approaches a different number from the right side of c than it does from the left side, then the limit of $f(x)$ as x approaches c *does not exist*. [See Example 2(b).]

3. The value of $f(x)$ when $x = c$ has no bearing on the existence or nonexistence of the limit of $f(x)$ as x approaches c. For instance, in Example 2(a), the limit of $f(x)$ exists as x approaches 1 even though the function f is not defined at $x = 1$.

Definition of the Limit of a Function

If $f(x)$ becomes arbitrarily close to a single number L as x approaches c from either side, then

$$\lim_{x \to c} f(x) = L$$

which is read as "the **limit** of $f(x)$ as x approaches c is L."

Many times the limit of $f(x)$ as x approaches c is simply $f(c)$, as shown in Example 1. Whenever the limit of $f(x)$ as x approaches c is

$$\lim_{x \to c} f(x) = f(c) \qquad \text{Substitute } c \text{ for } x.$$

the limit can be evaluated by **direct substitution.** (In the next section, you will learn that a function that has this property is *continuous at c.*) It is important that you learn to recognize the types of functions that have this property. Some basic ones are given in the following list.

Some Basic Limits

Let b and c be real numbers, and let n be a positive integer.

1. $\displaystyle\lim_{x \to c} b = b$　　2. $\displaystyle\lim_{x \to c} x = c$　　3. $\displaystyle\lim_{x \to c} x^n = c^n$　　4. $\displaystyle\lim_{x \to c} \sqrt[n]{x} = \sqrt[n]{c}$

In Property 4, if n is even, then c must be positive.

Example 3　Evaluating Basic Limits

a. $\displaystyle\lim_{x \to 2} 3 = 3$　　　　　**b.** $\displaystyle\lim_{x \to -4} x = -4$

c. $\displaystyle\lim_{x \to 2} x^3 = 2^3 = 8$　　**d.** $\displaystyle\lim_{x \to 9} \sqrt{x} = \sqrt{9} = 3$

✓ **Checkpoint 3**

Find the limit.

a. $\displaystyle\lim_{x \to 1} 5$　　**b.** $\displaystyle\lim_{x \to 6} x$　　**c.** $\displaystyle\lim_{x \to 5} x^2$　　**d.** $\displaystyle\lim_{x \to -8} \sqrt[3]{x}$

By combining the basic limits from the preceding page with the properties of limits shown below, you can find limits for a wide variety of algebraic functions.

Properties of Limits

Let b and c be real numbers, let n be a positive integer, and let f and g be functions with the following limits.

$$\lim_{x \to c} f(x) = L \quad \text{and} \quad \lim_{x \to c} g(x) = K$$

1. Scalar multiple: $\lim_{x \to c} [bf(x)] = bL$

2. Sum or difference: $\lim_{x \to c} [f(x) \pm g(x)] = L \pm K$

3. Product: $\lim_{x \to c} [f(x) \cdot g(x)] = LK$

4. Quotient: $\lim_{x \to c} \dfrac{f(x)}{g(x)} = \dfrac{L}{K}$, provided $K \neq 0$

5. Power: $\lim_{x \to c} [f(x)]^n = L^n$

6. Radical: $\lim_{x \to c} \sqrt[n]{f(x)} = \sqrt[n]{L}$

In Property 6, if n is even, then L must be positive.

TECH TUTOR

Symbolic computer algebra systems are capable of evaluating limits. Try using a computer algebra system to evaluate the limit given in Example 4.

Example 4   **Finding the Limit of a Polynomial Function**

Find the limit: $\lim_{x \to 2} (x^2 + 2x - 3)$.

SOLUTION

$$\begin{aligned}
\lim_{x \to 2} (x^2 + 2x - 3) &= \lim_{x \to 2} x^2 + \lim_{x \to 2} 2x - \lim_{x \to 2} 3 && \text{Apply Property 2.}\\
&= \lim_{x \to 2} x^2 + 2 \lim_{x \to 2} x - \lim_{x \to 2} 3 && \text{Apply Property 1.}\\
&= 2^2 + 2(2) - 3 && \text{Use direct substitution.}\\
&= 5
\end{aligned}$$

✓ **Checkpoint 4**

Find the limit: $\lim_{x \to 1} (2x^2 - x + 4)$.

In Example 4, note that the limit (as $x \to 2$) of the *polynomial function*

$$p(x) = x^2 + 2x - 3$$

is simply the value of p at $x = 2$.

$$\lim_{x \to 2} p(x) = p(2) = 2^2 + 2(2) - 3 = 4 + 4 - 3 = 5$$

This is an illustration of the following important result, which states that the limit of a polynomial function can be evaluated by direct substitution.

The Limit of a Polynomial Function

If p is a polynomial function and c is any real number, then

$$\lim_{x \to c} p(x) = p(c).$$

Techniques for Evaluating Limits

You have learned several techniques for evaluating limits. Another technique is shown in Example 5.

Example 5   **Finding the Limit of a Function**

Find the limit.

$$\lim_{x \to 1} \frac{x^3 - 1}{x - 1}$$

SOLUTION   Note that the numerator and denominator are zero when $x = 1$. This implies that $x - 1$ is a factor of both, and you can divide out this common factor.

$$\frac{x^3 - 1}{x - 1} = \frac{(x - 1)(x^2 + x + 1)}{x - 1} \qquad \text{Factor numerator.}$$

$$= \frac{(x - 1)(x^2 + x + 1)}{x - 1} \qquad \text{Divide out common factor.}$$

$$= x^2 + x + 1, \quad x \neq 1 \qquad \text{Simplify.}$$

So, the rational function $(x^3 - 1)/(x - 1)$ and the polynomial function $x^2 + x + 1$ agree for all values of x other than $x = 1$, and you can apply the Replacement Theorem.

$$\lim_{x \to 1} \frac{x^3 - 1}{x - 1} = \lim_{x \to 1} (x^2 + x + 1) = 1^2 + 1 + 1 = 3$$

Figure 7.4 illustrates this result graphically. Note that the two graphs are identical except that the graph of g contains the point $(1, 3)$, whereas this point is missing on the graph of f. (In the graph of f in Figure 7.4, the missing point is denoted by an open dot.)

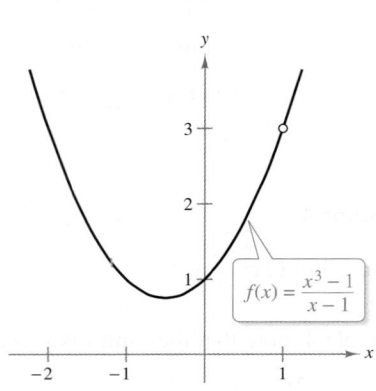
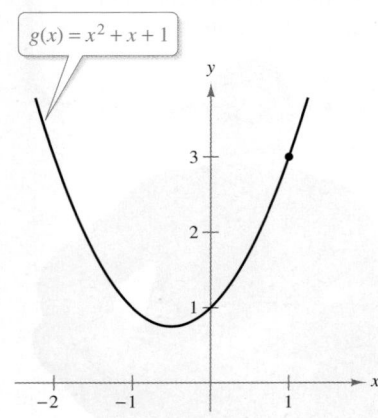

FIGURE 7.4

✓ **Checkpoint 5**

Find the limit.

$$\lim_{x \to 2} \frac{x^3 - 8}{x - 2}$$

The technique used to evaluate the limit in Example 5 is called the **dividing out** technique. The validity of this technique stems from the fact that when two functions agree at all but a single number c, the functions must have identical limit behavior at $x = c$. This technique is further demonstrated in the next example.

Example 6   Using the Dividing Out Technique

For $\lim\limits_{x \to -3} \dfrac{x^2 + x - 6}{x + 3}$, direct substitution fails because both the numerator and the denominator are zero when $x = -3$.

$$\lim_{x \to -3} \frac{x^2 + x - 6}{x + 3} \qquad \begin{array}{l} \longleftarrow \lim\limits_{x \to -3}(x^2 + x - 6) = 0 \\ \longleftarrow \lim\limits_{x \to -3}(x + 3) = 0 \end{array}$$

Because the limits of both the numerator and the denominator are zero when $x = -3$, they must have a *common factor* of $x + 3$. So, for all $x \neq -3$, you can divide out this factor and find the limit as shown.

$$\lim_{x \to -3} \frac{x^2 + x - 6}{x + 3} = \lim_{x \to -3} \frac{(x - 2)(x + 3)}{x + 3} \qquad \text{Factor numerator.}$$

$$= \lim_{x \to -3} \frac{(x - 2)\cancel{(x + 3)}}{\cancel{x + 3}} \qquad \text{Divide out common factor.}$$

$$= \lim_{x \to -3} (x - 2) \qquad \text{Simplify.}$$

$$= -3 - 2 \qquad \text{Direct substitution}$$

$$= -5 \qquad \text{Simplify.}$$

This result is shown graphically in Figure 7.5. Note that the graph of f coincides with the graph of $g(x) = x - 2$, except that the graph of f has a hole at $(-3, -5)$.

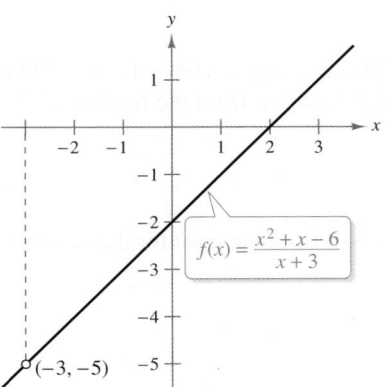

f is undefined when $x = -3$.
FIGURE 7.5

Graph label: $f(x) = \dfrac{x^2 + x - 6}{x + 3}$, point $(-3, -5)$.

✓**Checkpoint 6**

Find the limit: $\lim\limits_{x \to 3} \dfrac{x^2 + x - 12}{x - 3}$.

Example 7   Finding a Limit of a Function

Find the limit: $\lim\limits_{x \to 0} \dfrac{\sqrt{x + 1} - 1}{x}$.

SOLUTION   Direct substitution fails because both the numerator and the denominator are zero when $x = 0$. Rewrite the fraction by rationalizing the numerator.

$$\frac{\sqrt{x + 1} - 1}{x} = \left(\frac{\sqrt{x + 1} - 1}{x} \right)\left(\frac{\sqrt{x + 1} + 1}{\sqrt{x + 1} + 1} \right)$$

$$= \frac{(x + 1) - 1}{x(\sqrt{x + 1} + 1)}$$

$$= \frac{\cancel{x}}{\cancel{x}(\sqrt{x + 1} + 1)}$$

$$= \frac{1}{\sqrt{x + 1} + 1}, \quad x \neq 0$$

Now, using the Replacement Theorem, you can evaluate the limit as shown.

$$\lim_{x \to 0} \frac{\sqrt{x + 1} - 1}{x} = \lim_{x \to 0} \frac{1}{\sqrt{x + 1} + 1} = \frac{1}{\sqrt{0 + 1} + 1} = \frac{1}{1 + 1} = \frac{1}{2}$$

✓**Checkpoint 7**

Find the limit: $\lim\limits_{x \to 0} \dfrac{\sqrt{x + 4} - 2}{x}$.

STUDY TIP

When you try to evaluate a limit and both the numerator and denominator are zero, remember that you must rewrite the fraction so that the new denominator does not have 0 as its limit. One way to do this is to divide out common factors, as shown in Example 6. Another technique is to rationalize the numerator, as shown in Example 7.

One-Sided Limits

In Example 2(b), you saw that one way in which a limit can fail to exist is when a function approaches a different value from the left of c than it approaches from the right of c. This type of behavior can be described more concisely with the concept of a **one-sided limit.**

$$\lim_{x \to c^-} f(x) = L \qquad \text{Limit from the left}$$

$$\lim_{x \to c^+} f(x) = L \qquad \text{Limit from the right}$$

The first of these two limits is read as "the limit of $f(x)$ as x approaches c from the left is L." The second is read as "the limit of $f(x)$ as x approaches c from the right is L."

Example 8   Finding One-Sided Limits

Find the limit as $x \to 0$ from the left and the limit as $x \to 0$ from the right for the function

$$f(x) = \frac{|2x|}{x}.$$

SOLUTION   From the graph of f, shown in Figure 7.6, you can see that

$$f(x) = -2$$

for all $x < 0$. So, the limit from the left is

$$\lim_{x \to 0^-} \frac{|2x|}{x} = -2. \qquad \text{Limit from the left}$$

Because

$$f(x) = 2$$

for all $x > 0$, the limit from the right is

$$\lim_{x \to 0^+} \frac{|2x|}{x} = 2. \qquad \text{Limit from the right}$$

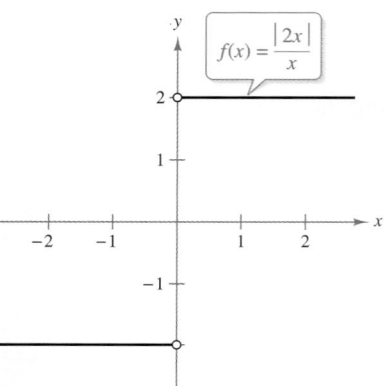

FIGURE 7.6

✓ Checkpoint 8

Find each limit.

a. $\displaystyle \lim_{x \to 2^-} \frac{|x - 2|}{x - 2}$

b. $\displaystyle \lim_{x \to 2^+} \frac{|x - 2|}{x - 2}$

In Example 8, note that the function approaches different limits from the left and from the right. In such cases, the limit of $f(x)$ as $x \to c$ does not exist. For the limit of a function to exist as $x \to c$, *both* one-sided limits must exist and must be equal.

Existence of a Limit

If f is a function and c and L are real numbers, then

$$\lim_{x \to c} f(x) = L$$

if and only if both the left and right limits are equal to L.

Example 9   Finding One-Sided Limits

Find the limit of $f(x)$ as x approaches 1.

$$f(x) = \begin{cases} 4 - x, & x < 1 \\ 4x - x^2, & x > 1 \end{cases}$$

SOLUTION   Remember that you are concerned about the value of f near $x = 1$ rather than at $x = 1$. So, for $x < 1$, $f(x)$ is given by

$$4 - x$$

and you can use direct substitution to obtain

$$\lim_{x \to 1^-} f(x) = \lim_{x \to 1^-} (4 - x) = 4 - 1 = 3.$$

For $x > 1$, $f(x)$ is given by

$$4x - x^2$$

and you can use direct substitution to obtain

$$\lim_{x \to 1^+} f(x) = \lim_{x \to 1^+} (4x - x^2) = 4(1) - 1^2 = 4 - 1 = 3.$$

Because both one-sided limits exist and are equal to 3, it follows that

$$\lim_{x \to 1} f(x) = 3.$$

The graph in Figure 7.7 confirms this conclusion.

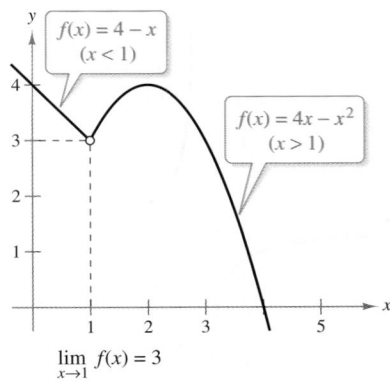

$f(x) = 4 - x$
$(x < 1)$

$f(x) = 4x - x^2$
$(x > 1)$

$$\lim_{x \to 1} f(x) = 3$$

FIGURE 7.7

✓ **Checkpoint 9**

Find the limit of $f(x)$ as x approaches 0.

$$f(x) = \begin{cases} x^2 + 1, & x < 0 \\ 2x + 1, & x > 0 \end{cases}$$

Example 10   Comparing One-Sided Limits

An overnight delivery service charges \$18 for the first pound and \$2 for each additional pound. Let x represent the weight of a parcel and let $f(x)$ represent the shipping cost.

$$f(x) = \begin{cases} 18, & 0 < x \leq 1 \\ 20, & 1 < x \leq 2 \\ 22, & 2 < x \leq 3 \end{cases}$$

Show that the limit of $f(x)$ as $x \to 2$ does not exist.

SOLUTION   The graph of f is shown in Figure 7.8. The limit of $f(x)$ as x approaches 2 from the left is

$$\lim_{x \to 2^-} f(x) = 20$$

whereas the limit of $f(x)$ as x approaches 2 from the right is

$$\lim_{x \to 2^+} f(x) = 22.$$

Because these one-sided limits are not equal, the limit of $f(x)$ as $x \to 2$ does not exist.

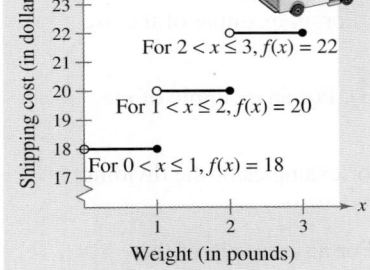

Overnight Delivery

For $2 < x \leq 3, f(x) = 22$

For $1 < x \leq 2, f(x) = 20$

For $0 < x \leq 1, f(x) = 18$

Shipping cost (in dollars)

Weight (in pounds)

FIGURE 7.8

✓ **Checkpoint 10**

Show that the limit of $f(x)$ as $x \to 1$ does not exist in Example 10.

Unbounded Behavior

Example 10 shows a limit that fails to exist because the limits from the left and right differ. Another important way in which a limit can fail to exist is when $f(x)$ increases or decreases without bound as x approaches c.

Example 11   An Unbounded Function

Find the limit (if possible): $\lim\limits_{x \to 2} \dfrac{3}{x - 2}$.

SOLUTION From Figure 7.9, you can see that $f(x)$ decreases without bound as x approaches 2 from the left and $f(x)$ increases without bound as x approaches 2 from the right. Symbolically, you can write this as

$$\lim_{x \to 2^-} \frac{3}{x - 2} = -\infty$$

and

$$\lim_{x \to 2^+} \frac{3}{x - 2} = \infty.$$

Because f is unbounded as x approaches 2, the limit does not exist.

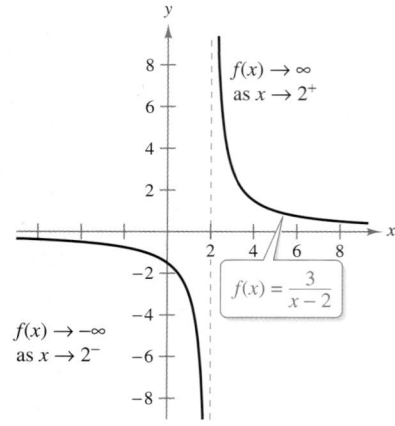

FIGURE 7.9

✓ Checkpoint 11

Find the limit (if possible): $\lim\limits_{x \to -2} \dfrac{5}{x + 2}$.

SUMMARIZE   (Section 7.1)

1. State the definition of the limit of a function *(page 542)*. For examples of limits, see Examples 1 and 2.

2. Make a list of the basic limits *(page 542)*. For examples of the basic limits, see Example 3.

3. Make a list of the properties of limits *(page 543)*. For an example of the use of these properties, see Example 4.

4. State the limit of a polynomial function *(page 543)*. For an example of the limit of a polynomial function, see Example 4.

5. Describe the dividing out technique *(page 544)*. For examples of the dividing out technique, see Examples 5 and 6.

6. Describe the rationalizing technique *(page 545)*. For an example of the rationalizing technique, see Example 7.

7. Describe a one-sided limit *(page 546)*. For examples of one-sided limits, see Examples 8, 9, and 10.

8. Describe the limit $\lim\limits_{x \to c} f(x)$ when $f(x)$ increases without bound as x approaches c *(page 548)*. For an example of an unbounded function, see Example 11.

SKILLS WARM UP 7.1   The following warm-up exercises involve skills that were covered in earlier sections. You will use these skills in the exercise set for this section. For additional help, review Sections 0.7, 2.4, and 2.5.

In Exercises 1–4, simplify the expression by factoring.

1. $\dfrac{2x^3 + x^2}{6x}$

2. $\dfrac{x^5 + 9x^4}{x^2}$

3. $\dfrac{x^2 - 3x - 28}{x - 7}$

4. $\dfrac{x^2 + 11x + 30}{x + 5}$

In Exercises 5–8, evaluate the expression and simplify.

5. $f(x) = x^2 - 3x + 3$

   (a) $f(-1)$ (b) $f(c)$ (c) $f(x + h)$

6. $f(x) = \begin{cases} 2x - 2, & x < 1 \\ 3x + 1, & x \geq 1 \end{cases}$

   (a) $f(-1)$ (b) $f(3)$ (c) $f(t^2 + 1)$

7. $f(x) = x^2 - 2x + 2$ $\dfrac{f(1 + h) - f(1)}{h}$

8. $f(x) = 4x$ $\dfrac{f(2 + h) - f(2)}{h}$

In Exercises 9–12, find the domain and range of the function and sketch its graph.

9. $h(x) = -\dfrac{5}{x}$

10. $g(x) = \sqrt{25 - x^2}$

11. $f(x) = |x - 3|$

12. $f(x) = \dfrac{|x|}{x}$

In Exercises 13 and 14, determine whether y is a function of x.

13. $9x^2 + 4y^2 = 49$

14. $2x^2y + 8x = 7y$

Exercises 7.1

See www.CalcChat.com for worked-out solutions to odd-numbered exercises.

Finding Limits Graphically   In Exercises 1–4, use the graph to find the limit. *See Examples 1 and 2.*

1.

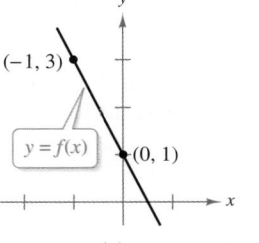

   (a) $\lim\limits_{x \to 0} f(x)$ (b) $\lim\limits_{x \to -1} f(x)$

2.

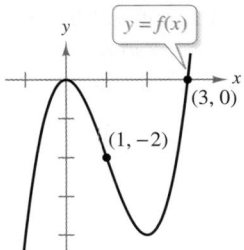

   (a) $\lim\limits_{x \to 1} f(x)$ (b) $\lim\limits_{x \to 3} f(x)$

3.

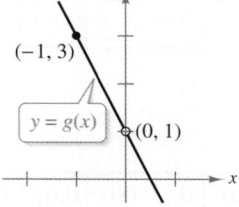

   (a) $\lim\limits_{x \to 0} g(x)$ (b) $\lim\limits_{x \to -1} g(x)$

4.

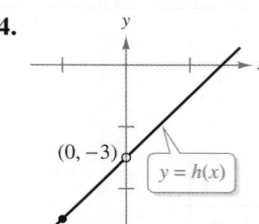

   (a) $\lim\limits_{x \to -2} h(x)$ (b) $\lim\limits_{x \to 0} h(x)$

Finding Limits Numerically In Exercises 5–12, complete the table and use the result to estimate the limit. Use a graphing utility to graph the function to confirm your result. *See Examples 1 and 2.*

5. $\lim\limits_{x \to 2} (2x + 5)$

x	1.9	1.99	1.999	2	2.001	2.01	2.1
$f(x)$				?			

6. $\lim\limits_{x \to 2} (x^2 - 3x + 1)$

x	1.9	1.99	1.999	2	2.001	2.01	2.1
$f(x)$				?			

7. $\lim\limits_{x \to 2} \dfrac{x - 2}{x^2 - 4}$

x	1.9	1.99	1.999	2	2.001	2.01	2.1
$f(x)$				?			

8. $\lim\limits_{x \to 2} \dfrac{x - 2}{x^2 - 3x + 2}$

x	1.9	1.99	1.999	2	2.001	2.01	2.1
$f(x)$				?			

9. $\lim\limits_{x \to 0} \dfrac{\sqrt{x + 1} - 1}{x}$

x	−0.1	−0.01	−0.001	0	0.001	0.01	0.1
$f(x)$				?			

10. $\lim\limits_{x \to 0} \dfrac{\sqrt{x + 2} - \sqrt{2}}{x}$

x	−0.1	−0.01	−0.001	0	0.001	0.01	0.1
$f(x)$				?			

11. $\lim\limits_{x \to -4} \dfrac{\dfrac{1}{x + 4} - \dfrac{1}{4}}{x}$

x	−4.1	−4.01	−4.001	−4
$f(x)$				?

x	−3.999	−3.99	−3.9
$f(x)$			

12. $\lim\limits_{x \to -2} \dfrac{\dfrac{1}{2} - \dfrac{1}{x + 2}}{2x}$

x	−2.1	−2.01	−2.001	−2
$f(x)$				?

x	−1.999	−1.99	−1.9
$f(x)$			

Evaluating Basic Limits In Exercises 13–20, find the limit. *See Example 3.*

13. $\lim\limits_{x \to 3} 6$

14. $\lim\limits_{x \to 5} 4$

15. $\lim\limits_{x \to -2} x$

16. $\lim\limits_{x \to 10} x$

17. $\lim\limits_{x \to 7} x^2$

18. $\lim\limits_{x \to 3} x^3$

19. $\lim\limits_{x \to 16} \sqrt{x}$

20. $\lim\limits_{x \to -1} \sqrt[3]{x}$

Operations with Limits In Exercises 21 and 22, find the limit of (a) $f(x) + g(x)$, (b) $f(x)g(x)$, and (c) $f(x)/g(x)$, as x approaches c.

21. $\lim\limits_{x \to c} f(x) = 3$

$\lim\limits_{x \to c} g(x) = 9$

22. $\lim\limits_{x \to c} f(x) = \frac{3}{2}$

$\lim\limits_{x \to c} g(x) = \frac{1}{2}$

Operations with Limits In Exercises 23 and 24, find the limit of (a) $\sqrt{f(x)}$, (b) $[3f(x)]$, and (c) $[f(x)]^2$, as x approaches c.

23. $\lim\limits_{x \to c} f(x) = 16$

24. $\lim\limits_{x \to c} f(x) = 9$

Using Properties of Limits In Exercises 25–36, find the limit using direct substitution. *See Examples 3 and 4.*

25. $\lim\limits_{x \to -3} (2x + 5)$

26. $\lim\limits_{x \to 0} (3x - 2)$

27. $\lim\limits_{x \to 1} (1 - x^2)$

28. $\lim\limits_{x \to 2} (-x^2 + x - 2)$

29. $\lim\limits_{x\to 3} \sqrt{x+6}$

30. $\lim\limits_{x\to 4} \sqrt[3]{x+4}$

31. $\lim\limits_{x\to -3} \dfrac{2}{x+2}$

32. $\lim\limits_{x\to -2} \dfrac{3x+1}{2-x}$

33. $\lim\limits_{x\to -2} \dfrac{x^2-1}{2x}$

34. $\lim\limits_{x\to 7} \dfrac{5x}{x+2}$

35. $\lim\limits_{x\to 5} \dfrac{\sqrt{x+11}+6}{x}$

36. $\lim\limits_{x\to 12} \dfrac{\sqrt{x-3}-2}{x}$

Finding Limits In Exercises 37–58, find the limit (if it exists). *See Examples 5, 6, 7, 9, and 11.*

37. $\lim\limits_{x\to -3} \dfrac{x^2-9}{x+3}$

38. $\lim\limits_{x\to -1} \dfrac{2x^2-x-3}{x+1}$

39. $\lim\limits_{x\to 2} \dfrac{2-x}{x^2-4}$

40. $\lim\limits_{t\to 1} \dfrac{t^2+t-2}{t^2-1}$

41. $\lim\limits_{x\to -2} \dfrac{x^3+8}{x+2}$

42. $\lim\limits_{x\to -1} \dfrac{x^3+1}{x+1}$

43. $\lim\limits_{\Delta x\to 0} \dfrac{2(x+\Delta x)-2x}{\Delta x}$

44. $\lim\limits_{\Delta x\to 0} \dfrac{4(x+\Delta x)-5-(4x-5)}{\Delta x}$

45. $\lim\limits_{\Delta t\to 0} \dfrac{(t+\Delta t)^2-5(t+\Delta t)-(t^2-5t)}{\Delta t}$

46. $\lim\limits_{\Delta t\to 0} \dfrac{(t+\Delta t)^2-4(t+\Delta t)+2-(t^2-4t+2)}{\Delta t}$

47. $\lim\limits_{x\to 4} \dfrac{\sqrt{x+5}-3}{x-4}$

48. $\lim\limits_{x\to 3} \dfrac{\sqrt{x+1}-2}{x-3}$

49. $\lim\limits_{x\to 0} \dfrac{\sqrt{x+5}-\sqrt{5}}{x}$

50. $\lim\limits_{x\to 0} \dfrac{\sqrt{x+2}-\sqrt{2}}{x}$

51. $\lim\limits_{x\to 2} f(x)$, where $f(x) = \begin{cases} 4-x, & x\neq 2 \\ 0, & x=2 \end{cases}$

52. $\lim\limits_{x\to 1} f(x)$, where $f(x) = \begin{cases} x^2+2, & x\neq 1 \\ 1, & x=1 \end{cases}$

53. $\lim\limits_{x\to 3} f(x)$, where $f(x) = \begin{cases} \frac{1}{3}x-2, & x\leq 3 \\ -2x+5, & x>3 \end{cases}$

54. $\lim\limits_{s\to 1} f(s)$, where $f(s) = \begin{cases} s, & s\leq 1 \\ 1-s, & s>1 \end{cases}$

55. $\lim\limits_{x\to -4} \dfrac{2}{x+4}$

56. $\lim\limits_{x\to 5} \dfrac{4}{x-5}$

57. $\lim\limits_{x\to 2} \dfrac{x-2}{x^2-4x+4}$

58. $\lim\limits_{t\to 4} \dfrac{t+4}{t^2-16}$

Finding One-Sided Limits In Exercises 59 and 60, use a graph to find the limit from the left and the limit from the right. *See Example 8.*

59. $\lim\limits_{x\to -3^-} \dfrac{|x+3|}{x+3}$

$\lim\limits_{x\to -3^+} \dfrac{|x+3|}{x+3}$

60. $\lim\limits_{x\to 6^-} \dfrac{|x-6|}{x-6}$

$\lim\limits_{x\to 6^+} \dfrac{|x-6|}{x-6}$

 Graphical, Numerical, and Analytic Analysis In Exercises 61–64, use a graphing utility to graph the function and estimate the limit. Use a table to reinforce your conclusion. Then find the limit by analytic methods.

61. $\lim\limits_{x\to 1^-} \dfrac{2}{x^2-1}$

62. $\lim\limits_{x\to 1^+} \dfrac{5}{1-x}$

63. $\lim\limits_{x\to -2^-} \dfrac{1}{x+2}$

64. $\lim\limits_{x\to 0^-} \dfrac{x+1}{x}$

Finding Limits Graphically In Exercises 65–70, use the graph to find the limit (if it exists).

(a) $\lim\limits_{x\to c^+} f(x)$ (b) $\lim\limits_{x\to c^-} f(x)$ (c) $\lim\limits_{x\to c} f(x)$

65.

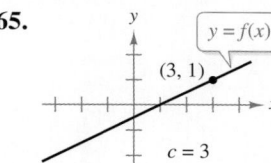

$y = f(x)$

$(3, 1)$

$c = 3$

66.

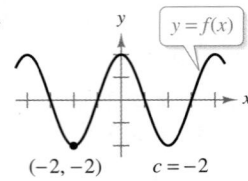

$(-2, -2)$ $c = -2$

67.

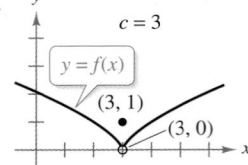

$c = 3$
$(3, 1)$
$(3, 0)$

68.

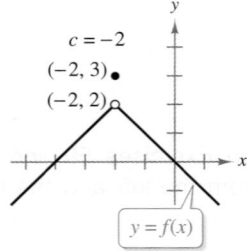

$c = -2$
$(-2, 3)$
$(-2, 2)$

69.

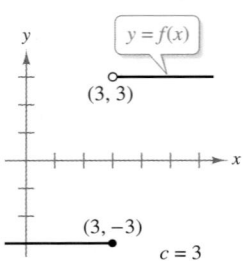

$(3, 3)$
$(3, -3)$
$c = 3$

70.

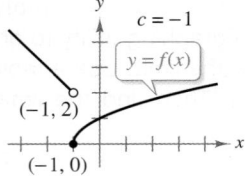

$c = -1$
$(-1, 2)$
$(-1, 0)$

 Estimating Limits In Exercises 71–74, use a graphing utility to estimate the limit (if it exists).

71. $\lim\limits_{x \to 2} \dfrac{x^2 - 5x + 6}{x^2 - 4x + 4}$

72. $\lim\limits_{x \to 1} \dfrac{x^2 + 6x - 7}{x^3 - x^2 + 2x - 2}$

73. $\lim\limits_{x \to -4} \dfrac{x^3 + 4x^2 + x + 4}{2x^2 + 7x - 4}$

74. $\lim\limits_{x \to -2} \dfrac{4x^3 + 7x^2 + x + 6}{3x^2 - x - 14}$

75. Environment The cost C (in thousands of dollars) of removing $p\%$ of the pollutants from the water in a small lake is given by

$$C = \frac{25p}{100 - p}, \quad 0 \le p < 100.$$

(a) Find the cost of removing 50% of the pollutants.

(b) What percent of the pollutants can be removed for $100 thousand?

(c) Evaluate $\lim\limits_{p \to 100^-} C$. Explain your results.

 **76.** **HOW DO YOU SEE IT?** The graph shows the cost C (in dollars) of making x photocopies at a copy shop.

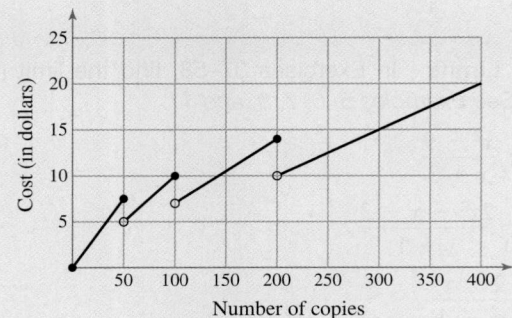

(a) Does $\lim\limits_{x \to 50} C$ exist? Explain your reasoning.

(b) Does $\lim\limits_{x \to 150} C$ exist? Explain your reasoning.

(c) You have to make 200 photocopies. Would it be better to make 200 or 201? Explain your reasoning.

 **77. Compound Interest** Consider a certificate of deposit that pays 10% (annual percentage rate) on an initial deposit of $1000. The balance A after 10 years is

$$A = 1000(1 + 0.1x)^{10/x}$$

where x is the length of the compounding period (in years).

(a) Use a graphing utility to graph A, where $0 \le x \le 1$.

(b) Use the *zoom* and *trace* features to estimate the balance for quarterly compounding and daily compounding.

(c) Use the *zoom* and *trace* features to estimate $\lim\limits_{x \to 0^+} A$. What do you think this limit represents? Explain your reasoning.

7.2 Continuity

■ Determine the continuity of functions.
■ Determine the continuity of functions on a closed interval.
■ Use the greatest integer function to model and solve real-life problems.
■ Use compound interest models to solve real-life problems.

Continuity

In mathematics, the term "continuous" has much the same meaning as it has in everyday use. To say that a function is continuous at

$$x = c$$

means that there is no interruption in the graph of f at c. That is, the graph of f

1. is unbroken at c.

2. has no holes, jumps, or gaps.

As simple as this concept may seem, its precise definition eluded mathematicians for many years. In fact, it was not until the early 1800's that a precise definition was finally developed.

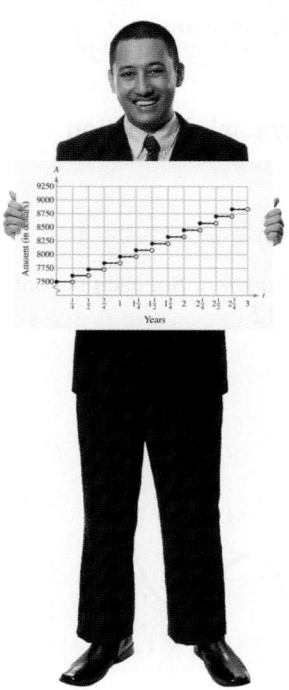

In Exercise 67 on page 562, you will examine the continuity of a function that represents an account balance.

Before looking at this definition, consider the function whose graph is shown in Figure 7.10. This figure identifies three values of x at which the function f is not continuous.

1. At $x = c_1$, $f(c_1)$ is not defined.

2. At $x = c_2$, $\lim_{x \to c_2} f(x)$ does not exist.

3. At $x = c_3$, $f(c_3) \neq \lim_{x \to c_3} f(x)$.

At all other points in the interval (a, b), the graph of f is uninterrupted, which implies that the function f is continuous at all other points in the interval (a, b).

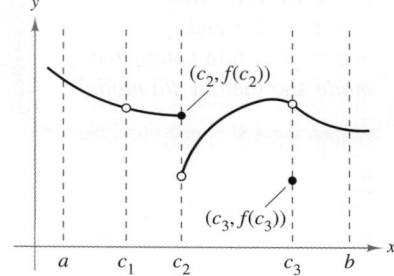

f is not continuous at $x = c_1, c_2, c_3$.

FIGURE 7.10

Definition of Continuity

Let c be a number in the interval (a, b), and let f be a function whose domain contains the interval (a, b). The function f is **continuous at the point c** when the following conditions are true.

1. $f(c)$ is defined.

2. $\lim_{x \to c} f(x)$ exists.

3. $\lim_{x \to c} f(x) = f(c)$.

If f is continuous at every point in the interval (a, b), then f is **continuous on the open interval (a, b).**

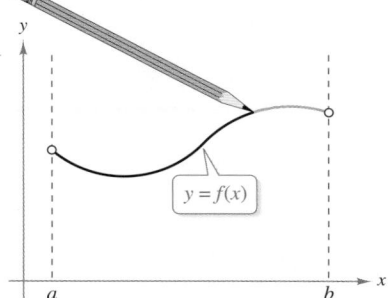

On the interval (a, b), the graph of f can be traced with a pencil.

FIGURE 7.11

Informally, you can say that a function is continuous on an interval when its graph on the interval can be traced using a pencil and paper without lifting the pencil from the paper, as shown in Figure 7.11.

TECH TUTOR

Most graphing utilities can draw graphs in two different modes: *connected mode* and *dot mode*. The *connected mode* works well as long as the function is continuous on the entire interval represented by the viewing window. If, however, the function is not continuous at one or more x-values in the viewing window, then the *connected mode* may try to "connect" parts of the graph that should not be connected. For instance, try graphing the function $y_1 = (x + 3)/(x - 2)$ in the viewing window $-8 \le x \le 8$ and $-6 \le y \le 6$ in *connected mode* and then in *dot mode*.

Continuity of Polynomial and Rational Functions

1. A polynomial function is continuous at every real number.

2. A rational function is continuous at every number in its domain.

Example 1   Determining Continuity of a Polynomial Function

Discuss the continuity of each function.

a. $f(x) = x^2 - 2x + 3$ **b.** $f(x) = x^3 - x$ **c.** $f(x) = x^4 - 2x^2 + 1$

SOLUTION   Each of these functions is a *polynomial function*. So, each is continuous on the entire real number line, as indicated in Figure 7.12.

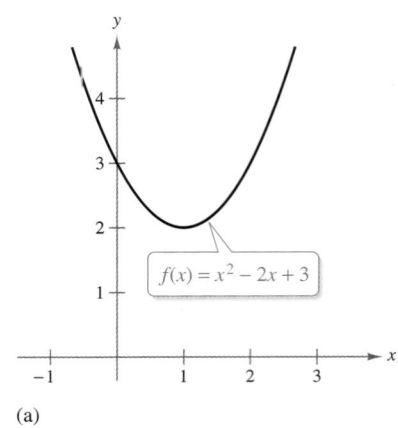

(a)

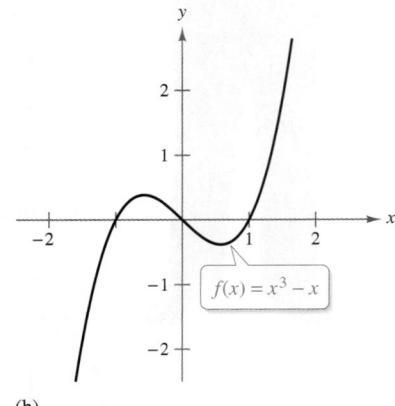

(b)

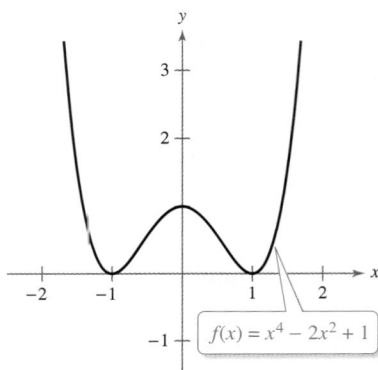

(c)

All three functions are continuous on $(-\infty, \infty)$.

FIGURE 7.12

✓**Checkpoint 1**

Discuss the continuity of each function.

a. $f(x) = x^2 + x + 1$ **b.** $f(x) = x^3 + x$ **c.** $f(x) = x^4$ ■

Polynomial functions are one of the most important types of functions used in calculus. Be sure you see from Example 1 that the graph of a polynomial function is continuous on the entire real number line and therefore has no holes, jumps, or gaps. Rational functions, on the other hand, need not be continuous on the entire real number line, as shown in Example 2.

Example 2   Determining Continuity of a Rational Function

Discuss the continuity of each function.

a. $f(x) = \dfrac{1}{x}$ **b.** $f(x) = \dfrac{x^2 - 1}{x - 1}$ **c.** $f(x) = \dfrac{1}{x^2 + 1}$

SOLUTION   Each of these functions is a rational function and is therefore continuous at every number in its domain.

a. The domain of $f(x) = 1/x$ consists of all real numbers except $x = 0$. So, this function is continuous on the intervals $(-\infty, 0)$ and $(0, \infty)$. [See Figure 7.13(a).]

b. The domain of $f(x) = (x^2 - 1)/(x - 1)$ consists of all real numbers except $x = 1$. So, this function is continuous on the intervals $(-\infty, 1)$ and $(1, \infty)$. [See Figure 7.13(b).]

c. The domain of $f(x) = 1/(x^2 + 1)$ consists of all real numbers. So, this function is continuous on the entire real number line. [See Figure 7.13(c).]

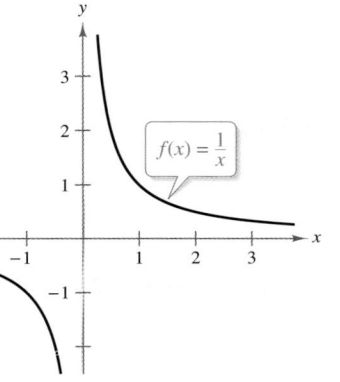

(a) Continuous on $(-\infty, 0)$ and $(0, \infty)$

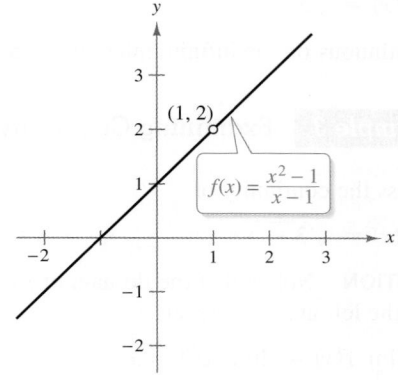
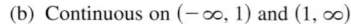

(b) Continuous on $(-\infty, 1)$ and $(1, \infty)$

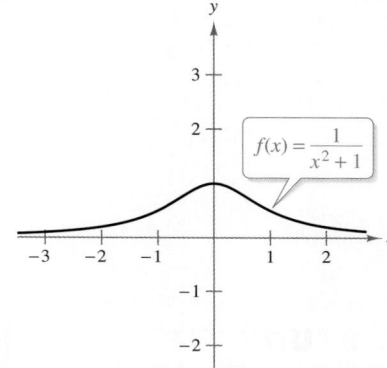

(c) Continuous on $(-\infty, \infty)$

FIGURE 7.13

✓**Checkpoint 2**

Discuss the continuity of each function.

a. $f(x) = \dfrac{1}{x - 1}$

b. $f(x) = \dfrac{x^2 - 4}{x - 2}$

c. $f(x) = \dfrac{1}{x^2 + 2}$

Consider an open interval I that contains a real number c. If a function f is defined on I (except possibly at c), and f is not continuous at c, then f is said to have a **discontinuity** at c. Discontinuities fall into two categories: **removable** and **nonremovable**. A discontinuity at c is called removable when f can be made continuous by appropriately defining (or redefining) $f(c)$. For instance, the function in Example 2(b) has a removable discontinuity at $(1, 2)$. To remove the discontinuity, all you need to do is redefine the function so that $f(1) = 2$.

A discontinuity at $x = c$ is nonremovable when the function cannot be made continuous at $x = c$ by defining or redefining the function at $x = c$. For instance, the function in Example 2(a) has a nonremovable discontinuity at $x = 0$.

Continuity on a Closed Interval

The intervals discussed in Examples 1 and 2 are open. To discuss continuity on a closed interval, you can use the concept of one-sided limits, as defined in Section 7.1.

Definition of Continuity on a Closed Interval

Let f be defined on a closed interval $[a, b]$. If f is continuous on the open interval (a, b) and

$$\lim_{x \to a^+} f(x) = f(a) \quad \text{and} \quad \lim_{x \to b^-} f(x) = f(b)$$

then f is **continuous on the closed interval** $[a, b]$. Moreover, f is **continuous from the right** at a and **continuous from the left** at b.

Similar definitions can be made to cover continuity on intervals of the form $(a, b]$ and $[a, b)$, or on infinite intervals. For instance, the function

$$f(x) = \sqrt{x}$$

is continuous on the infinite interval $[0, \infty)$.

Example 3 ■ Examining Continuity at an Endpoint

Discuss the continuity of

$$f(x) = \sqrt{3 - x}.$$

SOLUTION Notice that the domain of f is the set $(-\infty, 3]$. Moreover, f is continuous from the left at $x = 3$ because

$$\lim_{x \to 3^-} f(x) = \lim_{x \to 3^-} \sqrt{3 - x}$$
$$= \sqrt{3 - 3}$$
$$= 0$$
$$= f(3).$$

For all $x < 3$, the function f satisfies the three conditions for continuity. So, you can conclude that f is continuous on the interval $(-\infty, 3]$, as shown in Figure 7.14.

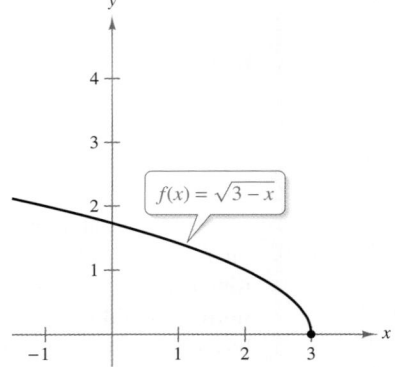

$f(x) = \sqrt{3 - x}$

FIGURE 7.14

STUDY TIP

When working with radical functions of the form

$$f(x) = \sqrt{g(x)}$$

remember that the domain of f coincides with the solution of $g(x) \geq 0$.

✓ **Checkpoint 3**

Discuss the continuity of

$$f(x) = \sqrt{x - 2}.$$

Example 4   **Examining Continuity on a Closed Interval**

Discuss the continuity of

$$g(x) = \begin{cases} 5 - x, & -1 \le x \le 2 \\ x^2 - 1, & 2 < x \le 3 \end{cases}.$$

SOLUTION   The polynomial functions

$$5 - x$$

and

$$x^2 - 1$$

are continuous on the intervals $[-1, 2]$ and $(2, 3]$, respectively. So, to conclude that g is continuous on the entire interval

$$[-1, 3]$$

you need only to check the behavior of g when $x = 2$. You can do this by taking the one-sided limits when $x = 2$.

$$\lim_{x \to 2^-} g(x) = \lim_{x \to 2^-} (5 - x) = 5 - 2 = 3 \qquad \text{Limit from the left}$$

and

$$\lim_{x \to 2^+} g(x) = \lim_{x \to 2^+} (x^2 - 1) = 2^2 - 1 = 3 \qquad \text{Limit from the right}$$

Because these two limits are equal,

$$\lim_{x \to 2} g(x) = g(2) = 3.$$

So, g is continuous at $x = 2$ and, consequently, it is continuous on the entire interval

$$[-1, 3].$$

The graph of g is shown in Figure 7.15.

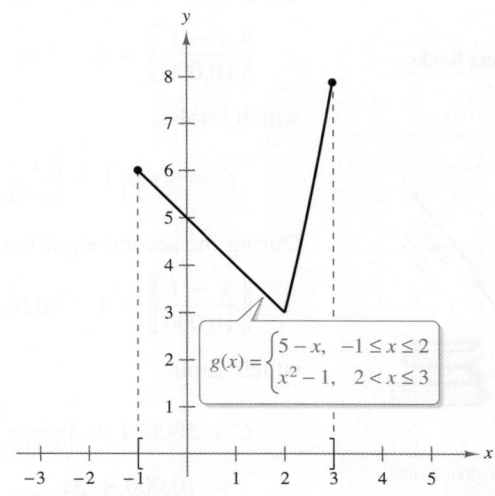

$$g(x) = \begin{cases} 5 - x, & -1 \le x \le 2 \\ x^2 - 1, & 2 < x \le 3 \end{cases}$$

FIGURE 7.15

✓**Checkpoint 4**

Discuss the continuity of

$$f(x) = \begin{cases} x + 2, & -1 \le x < 3 \\ 14 - x^2, & 3 \le x \le 5 \end{cases}.$$

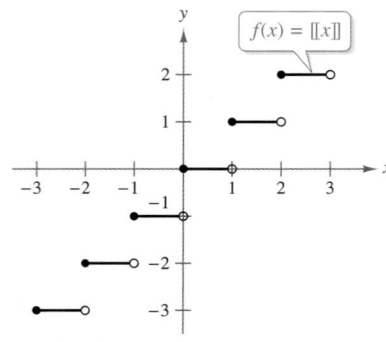

Greatest Integer Function
FIGURE 7.16

The Greatest Integer Function

Some functions that are used in business applications are **step functions.** For instance, the function in Example 10 in Section 7.1 is a step function. The **greatest integer function** is another example of a step function. This function is denoted by

$$[\![x]\!] = \text{greatest integer less than or equal to } x.$$

For example,

$$[\![-2.1]\!] = \text{greatest integer less than or equal to } -2.1 = -3$$
$$[\![-2]\!] = \text{greatest integer less than or equal to } -2 = -2$$
$$[\![1.5]\!] = \text{greatest integer less than or equal to } 1.5 = 1.$$

Note that the graph of the greatest integer function (Figure 7.16) jumps up one unit at each integer. This implies that the function is not continuous at each integer.

In real-life applications, the domain of the greatest integer function is often restricted to nonnegative values of x. In such cases, this function serves the purpose of **truncating** the decimal portion of x. For example, 1.345 is truncated to 1 and 3.57 is truncated to 3. That is,

$$[\![1.345]\!] = 1 \quad \text{and} \quad [\![3.57]\!] = 3.$$

 Example 5   Modeling a Cost Function

A bookbinding company produces 10,000 books in an eight-hour shift. The fixed cost *per shift* amounts to $5000, and the unit cost per book is $3. Using the greatest integer function, you can write the cost of producing x books as

$$C = 5000\left(1 + \left[\!\left[\frac{x-1}{10,000}\right]\!\right]\right) + 3x.$$

Sketch the graph of this cost function.

SOLUTION   Note that during the first eight-hour shift,

$$\left[\!\left[\frac{x-1}{10,000}\right]\!\right] = 0, \quad 1 \le x \le 10,000$$

which implies

$$C = 5000\left(1 + \left[\!\left[\frac{x-1}{10,000}\right]\!\right]\right) + 3x = 5000 + 3x.$$

During the second eight-hour shift,

$$\left[\!\left[\frac{x-1}{10,000}\right]\!\right] = 1, \quad 10,001 \le x \le 20,000$$

which implies

$$C = 5000\left(1 + \left[\!\left[\frac{x-1}{10,000}\right]\!\right]\right) + 3x$$
$$= 10,000 + 3x.$$

The graph of C is shown in Figure 7.17. Note the graph's discontinuities at $x = 10,000$, 20,000, and 30,000.

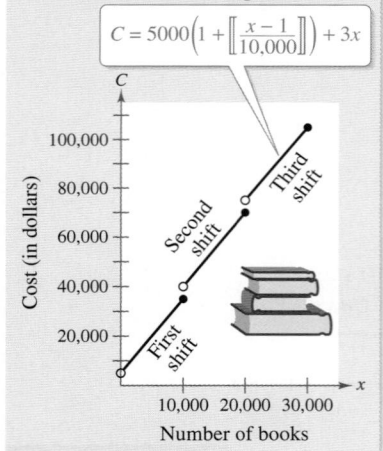

FIGURE 7.17

✓ Checkpoint 5

Use a graphing utility to graph the cost function in Example 5.

Extended Application: Compound Interest

Banks and other financial institutions differ on how interest is paid to an account. If the interest is added to the account so that future interest is paid on previously earned interest, then the interest is said to be **compounded.** For instance, a deposit of $10,000 is made in an account that pays 6% interest, compounded quarterly. Because the 6% is the annual interest rate, the quarterly rate is

$$\tfrac{1}{4}(0.06) = 0.015$$

or 1.5%. The balances during the first five quarters are shown below.

Quarter	Balance
1st	$10,000.00
2nd	$10,000.00 + (0.015)(10,000.00) = $10,150.00
3rd	$10,150.00 + (0.015)(10,150.00) = $10,302.25
4th	$10,302.25 + (0.015)(10,302.25) = $10,456.78
5th	$10,456.78 + (0.015)(10,456.78) = $10,613.63

 Example 6   **Graphing Compound Interest**

Sketch the graph of the balance in the account described above.

SOLUTION   Let A represent the balance in the account and let t represent the time, in years. You can use the greatest integer function to represent the balance, as shown.

$$A = 10,000(1 + 0.015)^{[\![4t]\!]}$$

From the graph shown in Figure 7.18, notice that the function has a discontinuity at each quarter. That is, A has discontinuities at

$$t = \frac{1}{4}, t = \frac{1}{2}, t = \frac{3}{4}, t = 1,$$

and

$$t = \frac{5}{4}.$$

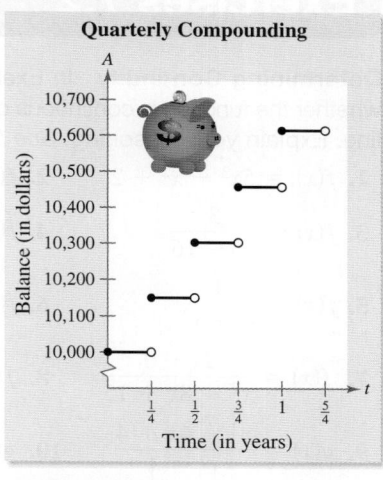

FIGURE 7.18

 **Checkpoint 6**

Write an equation that gives the balance of the account in Example 6 when the annual interest rate is 3%. Then sketch the graph of the equation.

SUMMARIZE   (Section 7.2)

1. State the definition of continuity *(page 553)*. For an example of a function that is continuous at every real number, see Example 1.

2. State the definition of continuity on a closed interval *(page 556)*. For an example of a function that is continuous on a closed interval, see Example 4.

3. State the definition of the greatest integer function *(page 558)*. For real-life examples of the greatest integer function, see Examples 5 and 6.

SKILLS WARM UP 7.2   The following warm-up exercises involve skills that were covered in a previous course or in earlier sections. You will use these skills in the exercise set for this section. For additional help, review Sections 0.7, 1.3, 1.5, and 7.1.

In Exercises 1–4, simplify the expression.

1. $\dfrac{x^2 + 6x + 8}{x^2 - 6x - 16}$

2. $\dfrac{x^2 - 5x - 6}{x^2 - 9x + 18}$

3. $\dfrac{2x^2 - 2x - 12}{4x^2 - 24x + 36}$

4. $\dfrac{x^3 - 16x}{x^3 + 2x^2 - 8x}$

In Exercises 5–8, solve for x.

5. $x^2 + 7x = 0$

6. $x^2 + 4x - 5 = 0$

7. $3x^2 + 8x + 4 = 0$

8. $x^3 + 5x^2 - 24x = 0$

In Exercises 9 and 10, find the limit.

9. $\lim\limits_{x \to 3} (2x^2 - 3x + 4)$

10. $\lim\limits_{x \to -2} (3x^3 - 8x + 7)$

Exercises 7.2

See www.CalcChat.com for worked-out solutions to odd-numbered exercises.

Determining Continuity In Exercises 1–10, determine whether the function is continuous on the entire real number line. Explain your reasoning. *See Examples 1 and 2.*

1. $f(x) = 5x^3 - x^2 + 2$

2. $f(x) = (x^2 - 1)^3$

3. $f(x) = \dfrac{3}{x^2 - 16}$

4. $f(x) = \dfrac{1}{9 - x^2}$

5. $f(x) = \dfrac{1}{4 + x^2}$

6. $f(x) = \dfrac{3x}{x^2 + 1}$

7. $f(x) = \dfrac{2x - 1}{x^2 - 8x + 15}$

8. $f(x) = \dfrac{x + 4}{x^2 - 6x + 5}$

9. $g(x) = \dfrac{x^2 - 4x + 4}{x^2 - 4}$

10. $g(x) = \dfrac{x^2 - 9x + 20}{x^2 - 16}$

Determining Continuity In Exercises 11–40, describe the interval(s) on which the function is continuous. Explain why the function is continuous on the interval(s). If the function has a discontinuity, identify the conditions of continuity that are not satisfied. *See Examples 1, 2, 3, 4, and 5.*

11. $f(x) = \dfrac{x^2 - 1}{x}$

12. $f(x) = \dfrac{1}{x^2 - 4}$

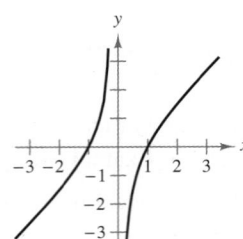

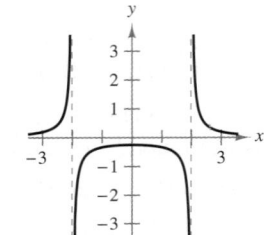

13. $f(x) = \dfrac{x^2 - 1}{x + 1}$

14. $f(x) = \dfrac{x^3 - 8}{x - 2}$

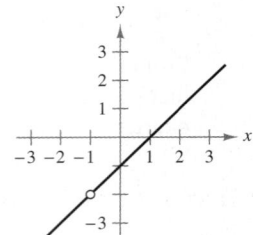

15. $f(x) = x^2 - 2x + 1$

16. $f(x) = 3 - 2x - x^2$

17. $f(x) = \dfrac{x}{x^2 - 1}$

18. $f(x) = \dfrac{x - 3}{x^2 - 9}$

19. $f(x) = \dfrac{x}{x^2 + 1}$

20. $f(x) = \dfrac{6}{x^2 + 3}$

21. $f(x) = \dfrac{x - 5}{x^2 - 9x + 20}$

22. $f(x) = \dfrac{x - 1}{x^2 + x - 2}$

23. $f(x) = \sqrt{4 - x}$

24. $f(x) = \sqrt{x - 1}$

25. $f(x) = \sqrt{x} + 2$

26. $f(x) = 3 - \sqrt{x}$

27. $f(x) = \begin{cases} -2x + 3, & -1 \le x \le 1 \\ x^2, & 1 < x \le 3 \end{cases}$

28. $f(x) = \begin{cases} \frac{1}{2}x + 1, & -3 \le x \le 2 \\ 3 - x, & 2 < x \le 4 \end{cases}$

29. $f(x) = \begin{cases} 3 + x, & x \le 2 \\ x^2 + 1, & x > 2 \end{cases}$

30. $f(x) = \begin{cases} x^2 - 4, & x \le 0 \\ 3x + 1, & x > 0 \end{cases}$

31. $f(x) = \dfrac{|x + 1|}{x + 1}$

32. $f(x) = \dfrac{|4 - x|}{4 - x}$

33. $f(x) = x\sqrt{x + 3}$

34. $f(x) = \dfrac{x + 1}{\sqrt{x}}$

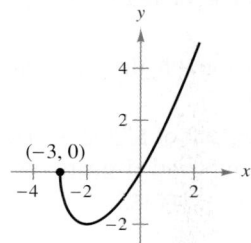

(−3, 0)

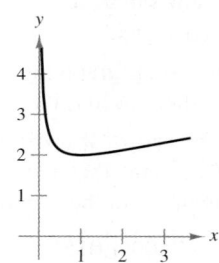

35. $f(x) = [\![2x]\!] + 1$

36. $f(x) = \dfrac{[\![x]\!]}{2} + x$

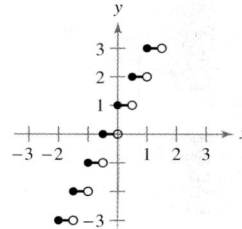

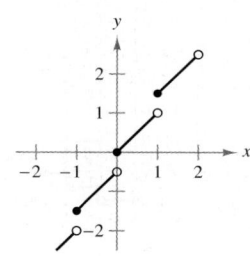

37. $f(x) = [\![x - 1]\!]$

38. $f(x) = x - [\![x]\!]$

39. $h(x) = f(g(x)),\quad f(x) = \dfrac{1}{\sqrt{x}},\quad g(x) = x - 1, x > 1$

40. $h(x) = f(g(x)),\quad f(x) = \dfrac{1}{x - 1},\quad g(x) = x^2 + 5$

Determining Continuity   In Exercises 41–46, sketch the graph of the function and describe the interval(s) on which the function is continuous.

41. $f(x) = \dfrac{x^2 - 16}{x - 4}$

42. $f(x) = \dfrac{2x^2 + x}{x}$

43. $f(x) = \dfrac{x + 4}{3x^2 - 12}$

44. $f(x) = \dfrac{x^3 + x}{x}$

45. $f(x) = \begin{cases} x^2 + 1, & x < 0 \\ x - 1, & x \ge 0 \end{cases}$

46. $f(x) = \begin{cases} x^2 - 4, & x \le 0 \\ 2x + 4, & x > 0 \end{cases}$

Determining Continuity on a Closed Interval   In Exercises 47–50, discuss the continuity of the function on the closed interval. If there are any discontinuities, determine whether they are removable.

Function	Interval
47. $f(x) = x^2 - 4x - 5$	$[-1, 5]$
48. $f(x) = \dfrac{5}{x^2 + 1}$	$[-2, 2]$
49. $f(x) = \dfrac{1}{x - 2}$	$[1, 4]$
50. $f(x) = \dfrac{x}{x^2 - 4x + 3}$	$[0, 4]$

Finding Discontinuities   In Exercises 51–56, use a graphing utility to graph the function. Use the graph to determine any x-value(s) at which the function is not continuous. Explain why the function is not continuous at the x-value(s).

51. $h(x) = \dfrac{1}{x^2 - x - 2}$

52. $k(x) = \dfrac{x - 4}{x^2 - 5x + 4}$

53. $f(x) = \begin{cases} 2x - 4, & x \le 3 \\ x^2 - 2x, & x > 3 \end{cases}$

54. $f(x) = \begin{cases} 3x - 1, & x \le 1 \\ x + 1, & x > 1 \end{cases}$

55. $f(x) = x - 2[\![x]\!]$

56. $f(x) = [\![2x - 1]\!]$

Making a Function Continuous   In Exercises 57 and 58, find the constant a (Exercise 57) and the constants a and b (Exercise 58) such that the function is continuous on the entire real number line.

57. $f(x) = \begin{cases} x^3, & x \le 2 \\ ax^2, & x > 2 \end{cases}$

58. $f(x) = \begin{cases} 2, & x \le -1 \\ ax + b, & -1 < x < 3 \\ -2, & x \ge 3 \end{cases}$

Writing   In Exercises 59 and 60, use a graphing utility to graph the function on the interval $[-4, 4]$. Does the graph of the function appear to be continuous on this interval? Is the function in fact continuous on $[-4, 4]$? Write a short paragraph about the importance of examining a function analytically as well as graphically.

59. $f(x) = \dfrac{x^2 + x}{x}$

60. $f(x) = \dfrac{x^3 - 8}{x - 2}$

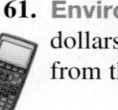

61. Environmental Cost   The cost C (in millions of dollars) of removing x percent of the pollutants emitted from the smokestack of a factory can be modeled by

$$C = \dfrac{2x}{100 - x}.$$

(a) What is the implied domain of C? Explain your reasoning.

(b) Use a graphing utility to graph the cost function. Is the function continuous on its domain? Explain your reasoning.

(c) Find the cost of removing 75% of the pollutants from the smokestack.

62. **HOW DO YOU SEE IT?** The graph shows the number of gallons G of gasoline in a person's car after t days.

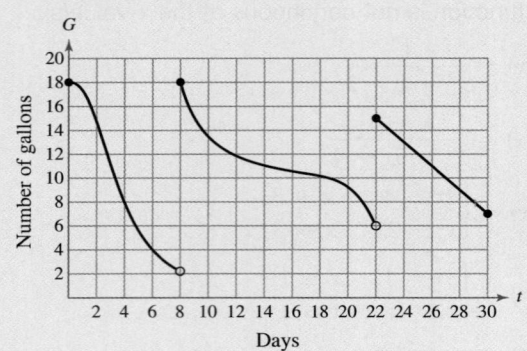

(a) On what days is the graph not continuous?

(b) What do you think happens on these days?

63. **Biology** The gestation period of rabbits is about 29 to 35 days. Therefore, the population of a form (rabbits' home) can increase dramatically in a short period of time. The table gives the population of a form, where t is the time in months and N is the rabbit population.

t	0	1	2	3	4	5	6
N	2	8	10	14	10	15	12

Graph the population as a function of time. Find any points of discontinuity in the function. Explain your reasoning.

64. **Owning a Franchise** You have purchased a franchise. You have determined a linear model for your revenue as a function of time. Is the model a continuous function? Would your actual revenue be a continuous function of time? Explain your reasoning.

65. **Consumer Awareness** The United States Postal Service first class mail rates for sending a letter are $0.44 for the first ounce and $0.20 for each additional ounce or fraction thereof up to 3.5 ounces. A model for the cost C (in dollars) of a first class mailing that weighs 3.5 ounces or less is given below. *(Source: United States Postal Service)*

$$C(x) = \begin{cases} 0.44, & 0 \le x \le 1 \\ 0.64, & 1 < x \le 2 \\ 0.84, & 2 < x \le 3 \\ 1.04, & 3 < x \le 3.5 \end{cases}$$

(a) Use a graphing utility to graph the function and then discuss its continuity. At what values is the function not continuous? Explain your reasoning.

(b) Find the cost of mailing a 2.5-ounce letter.

66. **Health Food** A co-op health food store charges $3.50 for the first pound of organically grown peanuts and $1.90 for each additional pound or fraction thereof.

(a) Use the greatest integer function to create a model for the cost C for x pounds of organically grown peanuts.

(b) Use a graphing utility to graph the function and then discuss its continuity.

67. **Compound Interest** A deposit of $7500 is made in an account that pays 6% compounded quarterly. The amount A in the account after t years is

$$A = 7500(1.015)^{[\![4t]\!]}, \quad t \ge 0.$$

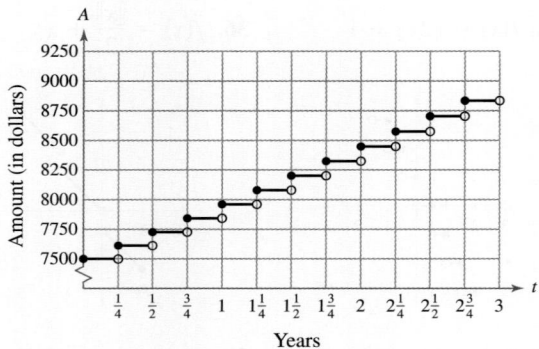

(a) Is the graph continuous? Explain your reasoning.

(b) What is the balance after 2 years?

(c) What is the balance after 7 years?

68. **Salary Contract** A union contract guarantees a 9% yearly increase for 5 years. For a current salary of $28,500, the salaries for the next 5 years are given by

$$S = 28,500(1.09)^{[\![t]\!]}$$

where $t = 0$ represents the present year.

(a) Use the greatest integer function of a graphing utility to graph the salary function and then discuss its continuity.

(b) Find the salary during the fifth year (when $t = 5$).

69. **Inventory Management** The number of units in inventory in a small company is

$$N = 25\left(2\left[\!\!\left[\frac{t+2}{2}\right]\!\!\right] - t\right), \quad 0 \le t \le 12$$

where the real number t is the time in months.

(a) Use the greatest integer function of a graphing utility to graph this function and then discuss its continuity.

(b) How often must the company replenish its inventory?

7.3 The Derivative and the Slope of a Graph

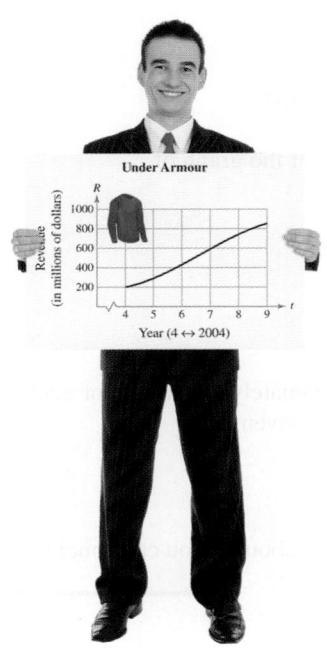

In Exercise 13 on page 572, you will estimate and interpret the slope of the graph of a revenue function.

- Identify tangent lines to a graph at a point.
- Approximate the slopes of tangent lines to graphs at points.
- Use the limit definition to find the slopes of graphs at points.
- Use the limit definition to find the derivatives of functions.
- Describe the relationship between differentiability and continuity.

Tangent Line to a Graph

Calculus is a branch of mathematics that studies rates of change of functions. In this course, you will learn that rates of change have many applications in real life. In Section 2.2, you learned how the slope of a line indicates the rate at which the line rises or falls. For a line, this rate (or slope) is the same at every point on the line. For graphs other than lines, the rate at which the graph rises or falls changes from point to point. For instance, in Figure 7.19, the parabola is rising more quickly at the point (x_1, y_1) than it is at the point (x_2, y_2). At the vertex (x_3, y_3), the graph levels off, and at the point (x_4, y_4), the graph is falling.

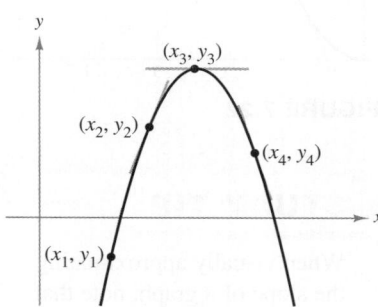

The slope of a nonlinear graph changes from one point to another.

FIGURE 7.19

To determine the rate at which a graph rises or falls at a *single point,* you can find the slope of the **tangent line** at the point. In simple terms, the tangent line to the graph of a function f at a point $P(x_1, y_1)$ is the line that best approximates the graph at that point, as shown in Figure 7.19. Figure 7.20 shows other examples of tangent lines.

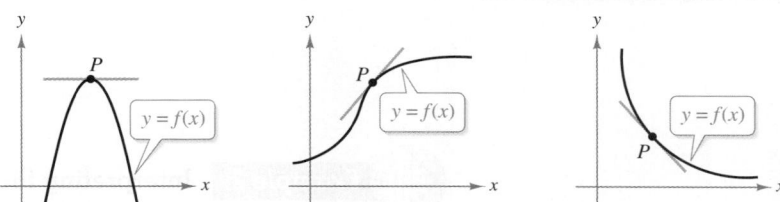

Tangent Line to a Graph at a Point

FIGURE 7.20

When Isaac Newton (1642–1727) was working on the "tangent line problem," he realized that it is difficult to define precisely what is meant by a tangent to a general curve. From geometry, you know that a line is tangent to a circle when the line intersects the circle at only one point, as shown in Figure 7.21. Tangent lines to a noncircular graph, however, can intersect the graph at more than one point. For instance, in the second graph in Figure 7.20, when the tangent line is extended, it intersects the graph at a point other than the point of tangency. In this section, you will see how the notion of a limit can be used to define a general tangent line.

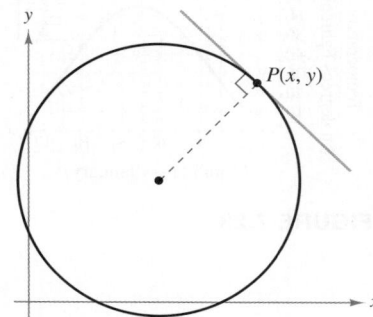

Tangent Line to a Circle

FIGURE 7.21

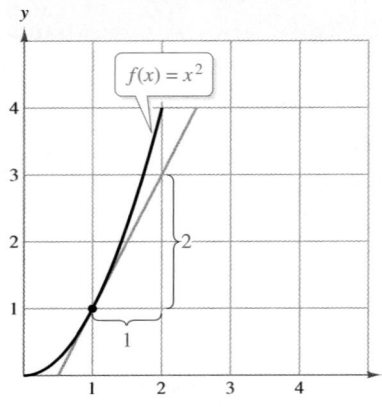

FIGURE 7.22

Slope of a Graph

Because a tangent line approximates the graph at a point, the problem of finding the slope of a graph at a point becomes one of finding the slope of the tangent line at the point.

Example 1 Approximating the Slope of a Graph

Use the graph in Figure 7.22 to approximate the slope of the graph of

$$f(x) = x^2$$

at the point $(1, 1)$.

SOLUTION From the graph of

$$f(x) = x^2$$

you can see that the tangent line at $(1, 1)$ rises approximately two units for each unit change in x. So, the slope of the tangent line at $(1, 1)$ is given by

$$\text{Slope} = \frac{\Delta y}{\Delta x} = \frac{\text{change in } y}{\text{change in } x} \approx \frac{2}{1} = 2.$$

Because the tangent line at the point $(1, 1)$ has a slope of about 2, you can conclude that the graph has a slope of about 2 at the point $(1, 1)$.

✓**Checkpoint 1**

Use the graph to approximate the slope of the graph of

$$f(x) = x^3$$

at the point $(1, 1)$.

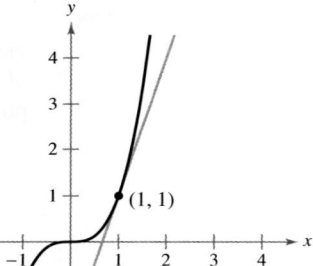

> **STUDY TIP**
>
> When visually approximating the slope of a graph, note that the scales on the horizontal and vertical axes may differ. When this happens (as it frequently does in applications), the slope of the tangent line is distorted, and you must be careful to account for the difference in scales.

Example 2 Interpreting Slope

Figure 7.23 graphically depicts the average monthly temperature (in degrees Fahrenheit) in Duluth, Minnesota. Estimate the slope of this graph at the indicated point and give a physical interpretation of the result. *(Source: National Oceanic and Atmospheric Administration)*

SOLUTION From the graph, you can see that the tangent line at the given point falls approximately 28 units for each two-unit change in x. So, you can estimate the slope at the given point to be

$$\text{Slope} = \frac{\Delta y}{\Delta x} = \frac{\text{change in } y}{\text{change in } x} \approx \frac{-28}{2} = -14 \text{ degrees per month.}$$

This means that you can expect the average daily temperatures in November to be about 14 degrees *lower* than the corresponding temperatures in October.

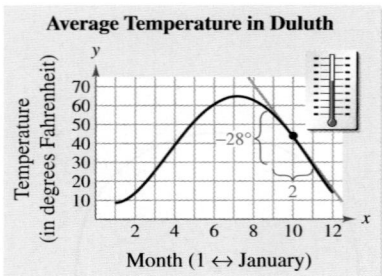

FIGURE 7.23

✓**Checkpoint 2**

In Figure 7.23, for which months do the slopes of the tangent lines appear to be positive? Negative? Interpret these slopes in the context of the problem.

Slope and the Limit Process

In Examples 1 and 2, you approximated the slope of a graph at a point by making a careful graph and then "eyeballing" the tangent line at the point of tangency. A more precise method of approximating the slope of a tangent line makes use of a **secant line** through the point of tangency and a second point on the graph, as shown in Figure 7.24. If $(x, f(x))$ is the point of tangency and

$$(x + \Delta x, f(x + \Delta x))$$

is a second point on the graph of f, then the slope of the secant line through the two points is

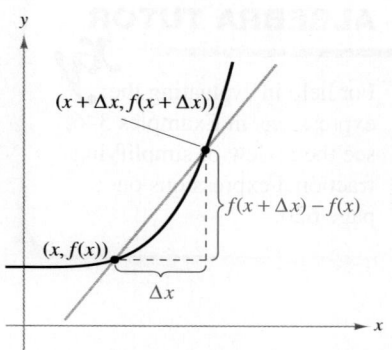

The Secant Line Through the Two Points $(x, f(x))$ and $(x + \Delta x, f(x + \Delta x))$

FIGURE 7.24

$$m = \frac{y_2 - y_1}{x_2 - x_1} \qquad \text{Formula for slope}$$

$$m_{\text{sec}} = \frac{f(x + \Delta x) - f(x)}{(x + \Delta x) - x} \qquad \frac{\text{Change in } y}{\text{Change in } x}$$

$$m_{\text{sec}} = \frac{f(x + \Delta x) - f(x)}{\Delta x}. \qquad \text{Slope of secant line}$$

The right side of this equation is called the **difference quotient.** The denominator Δx is the **change in x,** and the numerator is the **change in y.** The beauty of this procedure is that you obtain more and more accurate approximations of the slope of the tangent line by choosing points closer and closer to the point of tangency, as shown in Figure 7.25. Using the limit process, you can find the *exact* slope of the tangent line at $(x, f(x))$, which is also the slope of the graph of f at $(x, f(x))$.

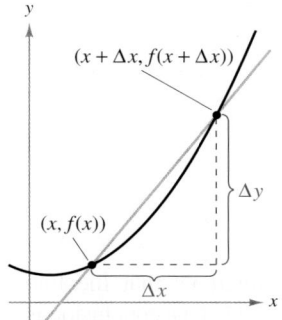
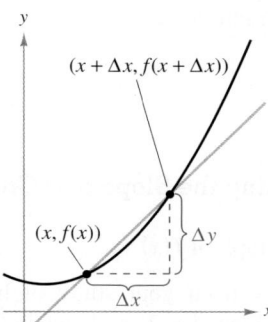
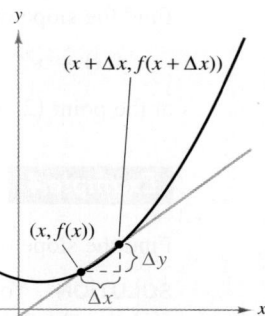
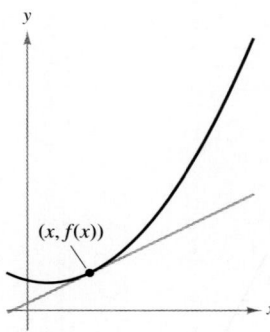

FIGURE 7.25   As Δx approaches 0, the secant lines approach the tangent line.

STUDY TIP

The variable Δx is used to represent the change in x in the definition of the slope of a graph. Other variables may also be used. Sometimes this definition is written as

$$m = \lim_{h \to 0} \frac{f(x + h) - f(x)}{h}.$$

Definition of the Slope of a Graph

The **slope** m of the graph of f at the point

$$(x, f(x))$$

is equal to the slope of the tangent line to the graph of f at $(x, f(x))$, and is given by

$$m = \lim_{\Delta x \to 0} m_{\text{sec}} = \lim_{\Delta x \to 0} \frac{f(x + \Delta x) - f(x)}{\Delta x}$$

provided this limit exists.

ALGEBRA TUTOR

For help in evaluating the expressions in Examples 3–6, see the review of simplifying fractional expressions on page 620.

Example 3 Finding Slope by the Limit Process

Find the slope of the graph of

$$f(x) = x^2$$

at the point $(-2, 4)$.

SOLUTION Begin by finding an expression that represents the slope of a secant line at the point $(-2, 4)$.

$$m_{\text{sec}} = \frac{f(-2 + \Delta x) - f(-2)}{\Delta x} \qquad \text{Set up difference quotient.}$$

$$= \frac{(-2 + \Delta x)^2 - (-2)^2}{\Delta x} \qquad \text{Use } f(x) = x^2.$$

$$= \frac{4 - 4\,\Delta x + (\Delta x)^2 - 4}{\Delta x} \qquad \text{Expand terms.}$$

$$= \frac{-4\,\Delta x + (\Delta x)^2}{\Delta x} \qquad \text{Simplify.}$$

$$= \frac{\Delta x(-4 + \Delta x)}{\Delta x} \qquad \text{Factor and divide out.}$$

$$= -4 + \Delta x, \quad \Delta x \neq 0 \qquad \text{Simplify.}$$

Next, take the limit of m_{sec} as $\Delta x \to 0$.

$$m = \lim_{\Delta x \to 0} m_{\text{sec}} = \lim_{\Delta x \to 0} (-4 + \Delta x) = -4 + 0 = -4$$

So, the graph of f has a slope of -4 at the point $(-2, 4)$, as shown in Figure 7.26.

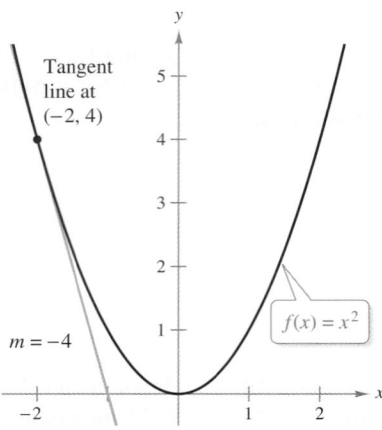

Tangent line at $(-2, 4)$

$m = -4$

$f(x) = x^2$

FIGURE 7.26

✓ Checkpoint 3

Find the slope of the graph of

$$f(x) = x^2$$

at the point $(2, 4)$.

Example 4 Finding the Slope of a Graph

Find the slope of the graph of $f(x) = -2x + 4$.

SOLUTION You know from your study of linear functions that the line given by $f(x) = -2x + 4$ has a slope of -2, as shown in Figure 7.27. This conclusion is consistent with the limit definition of slope.

$$m = \lim_{\Delta x \to 0} \frac{f(x + \Delta x) - f(x)}{\Delta x}$$

$$= \lim_{\Delta x \to 0} \frac{[-2(x + \Delta x) + 4] - (-2x + 4)}{\Delta x}$$

$$= \lim_{\Delta x \to 0} \frac{-2x - 2\,\Delta x + 4 + 2x - 4}{\Delta x}$$

$$= \lim_{\Delta x \to 0} \frac{-2\,\Delta x}{\Delta x}$$

$$= -2$$

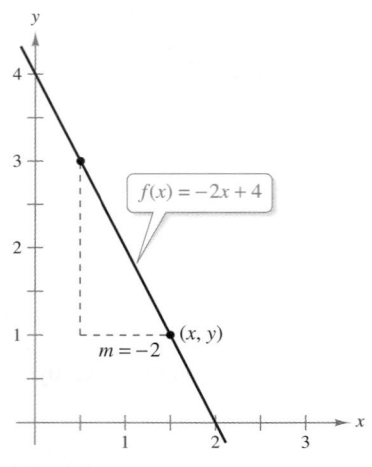

$f(x) = -2x + 4$

$m = -2$

(x, y)

FIGURE 7.27

✓ Checkpoint 4

Find the slope of the graph of $f(x) = 2x + 5$.

Example 5 Finding a Formula for the Slope of a Graph

Find a formula for the slope of the graph of $f(x) = x^2 + 1$. What are the slopes at the points $(-1, 2)$ and $(2, 5)$?

SOLUTION

$$
\begin{aligned}
m_{\text{sec}} &= \frac{f(x + \Delta x) - f(x)}{\Delta x} & \text{Set up difference quotient.} \\
&= \frac{[(x + \Delta x)^2 + 1] - (x^2 + 1)}{\Delta x} & \text{Use } f(x) = x^2 + 1. \\
&= \frac{x^2 + 2x \Delta x + (\Delta x)^2 + 1 - x^2 - 1}{\Delta x} & \text{Expand terms.} \\
&= \frac{2x \Delta x + (\Delta x)^2}{\Delta x} & \text{Simplify.} \\
&= \frac{\Delta x(2x + \Delta x)}{\Delta x} & \text{Factor and divide out.} \\
&= 2x + \Delta x, \quad \Delta x \neq 0 & \text{Simplify.}
\end{aligned}
$$

Next, take the limit of m_{sec} as $\Delta x \to 0$.

$$
\begin{aligned}
m &= \lim_{\Delta x \to 0} m_{\text{sec}} \\
&= \lim_{\Delta x \to 0} (2x + \Delta x) \\
&= 2x + 0 \\
&= 2x
\end{aligned}
$$

Using the formula $m = 2x$, you can find the slopes at the specified points. At $(-1, 2)$ the slope is $m = 2(-1) = -2$, and at $(2, 5)$ the slope is $m = 2(2) = 4$. The graph of f is shown in Figure 7.28.

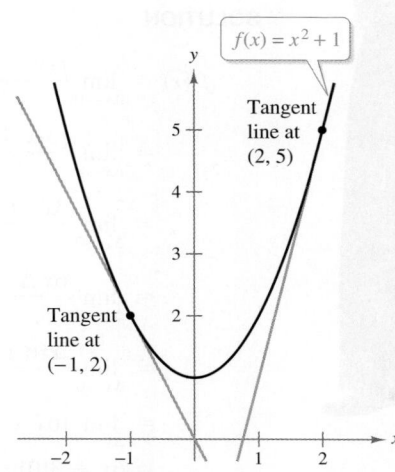

FIGURE 7.28

✓ Checkpoint 5

Find a formula for the slope of the graph of

$$f(x) = 4x^2 + 1.$$

What are the slopes at the points $(0, 1)$ and $(1, 5)$?

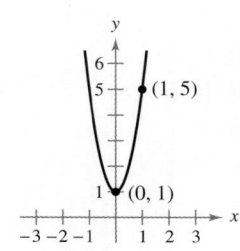

The Derivative of a Function

In Example 5, you started with the function

$$f(x) = x^2 + 1$$

and used the limit process to derive another function, $m = 2x$, that represents the slope of the graph of f at the point $(x, f(x))$. This derived function is called the **derivative** of f at x. It is denoted by $f'(x)$, which is read as "f prime of x."

STUDY TIP

The notation dy/dx is read as "the derivative of y with respect to x," and using limit notation, you can write

$$\frac{dy}{dx} = \lim_{\Delta x \to 0} \frac{\Delta y}{\Delta x}$$

$$= \lim_{\Delta x \to 0} \frac{f(x + \Delta x) - f(x)}{\Delta x}$$

$$= f'(x).$$

Definition of the Derivative

The **derivative of f at x** is given by

$$f'(x) = \lim_{\Delta x \to 0} \frac{f(x + \Delta x) - f(x)}{\Delta x}$$

provided this limit exists. A function is **differentiable** at x when its derivative exists at x. The process of finding derivatives is called **differentiation.**

In addition to $f'(x)$, other notations can be used to denote the derivative of $y = f(x)$. The most common are

$$\frac{dy}{dx}, \quad y', \quad \frac{d}{dx}[f(x)], \quad \text{and} \quad D_x[y].$$

Example 6   Finding a Derivative

Find the derivative of

$$f(x) = 3x^2 - 2x.$$

SOLUTION

$$f'(x) = \lim_{\Delta x \to 0} \frac{f(x + \Delta x) - f(x)}{\Delta x}$$

$$= \lim_{\Delta x \to 0} \frac{[3(x + \Delta x)^2 - 2(x + \Delta x)] - (3x^2 - 2x)}{\Delta x}$$

$$= \lim_{\Delta x \to 0} \frac{3x^2 + 6x\,\Delta x + 3(\Delta x)^2 - 2x - 2\,\Delta x - 3x^2 + 2x}{\Delta x}$$

$$= \lim_{\Delta x \to 0} \frac{6x\,\Delta x + 3(\Delta x)^2 - 2\,\Delta x}{\Delta x}$$

$$= \lim_{\Delta x \to 0} \frac{\Delta x(6x + 3\,\Delta x - 2)}{\Delta x}$$

$$= \lim_{\Delta x \to 0} (6x + 3\,\Delta x - 2)$$

$$= 6x + 3(0) - 2$$

$$= 6x - 2$$

So, the derivative of $f(x) = 3x^2 - 2x$ is

$$f'(x) = 6x - 2.$$

✓ Checkpoint 6

Find the derivative of

$$f(x) = x^2 - 5x.$$

In many applications, it is convenient to use a variable other than x as the independent variable. Example 7 shows a function that uses t as the independent variable.

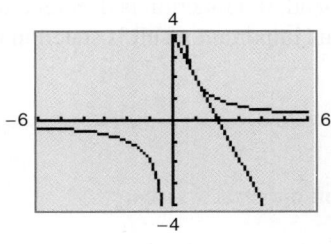
Example 7   Finding a Derivative

Find the derivative of y with respect to t for the function

$$y = \frac{2}{t}.$$

SOLUTION   Consider $y = f(t)$, and use the limit process as shown.

$$\frac{dy}{dt} = \lim_{\Delta t \to 0} \frac{f(t + \Delta t) - f(t)}{\Delta t} \qquad \text{Set up difference quotient.}$$

$$= \lim_{\Delta t \to 0} \frac{\dfrac{2}{t + \Delta t} - \dfrac{2}{t}}{\Delta t} \qquad \text{Use } f(t) = \frac{2}{t}.$$

$$= \lim_{\Delta t \to 0} \frac{\dfrac{2t - 2(t + \Delta t)}{t(t + \Delta t)}}{\Delta t} \qquad \text{Combine fractions in numerator.}$$

$$= \lim_{\Delta t \to 0} \frac{2t - 2t - 2\Delta t}{\Delta t(t)(t + \Delta t)} \qquad \text{Expand terms in numerator.}$$

$$= \lim_{\Delta t \to 0} \frac{-2\,\Delta t}{\Delta t(t)(t + \Delta t)} \qquad \text{Factor and divide out.}$$

$$= \lim_{\Delta t \to 0} \frac{-2}{t(t + \Delta t)} \qquad \text{Simplify.}$$

$$= \frac{-2}{t(t + 0)} \qquad \text{Direct substitution}$$

$$= -\frac{2}{t^2} \qquad \text{Simplify.}$$

So, the derivative of y with respect to t is

$$\frac{dy}{dt} = -\frac{2}{t^2}.$$

✓ **Checkpoint 7**

Find the derivative of y with respect to t for the function $y = 4/t$.   ■

Remember that the derivative of a function gives you a formula for finding the slope of the tangent line at any point on the graph of the function. For instance, in Example 7 the slope of the tangent line to the graph of f at the point $(1, 2)$ is given by

$$f'(1) = -\frac{2}{1^2} = -2.$$

To find the slopes of the graph at other points, substitute the t-coordinate of the point into the derivative, as shown below.

Point	t-Coordinate	Slope
$(2, 1)$	$t = 2$	$m = f'(2) = -\dfrac{2}{2^2} = -\dfrac{1}{2}$
$(-2, -1)$	$t = -2$	$m = f'(-2) = -\dfrac{2}{(-2)^2} = -\dfrac{1}{2}$

Differentiability and Continuity

Not every function is differentiable. Figure 7.29 shows some common situations in which a function will not be differentiable at a point—vertical tangent lines, discontinuities, and sharp turns in the graph. Each of the functions shown in Figure 7.29 is differentiable at every value of *x except x* = 0.

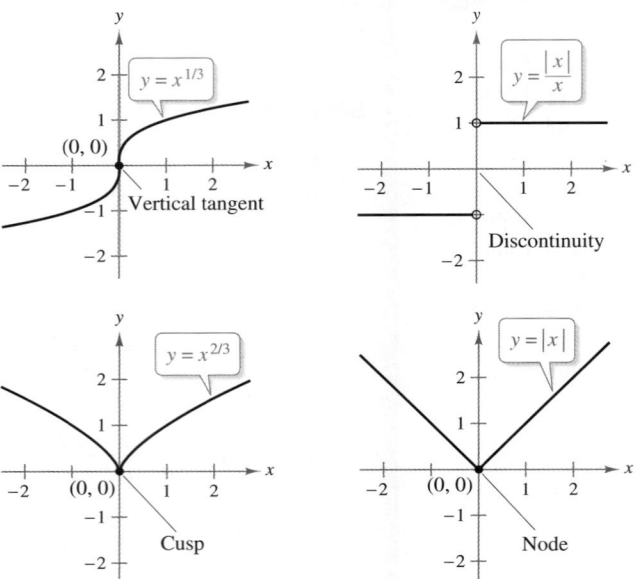

Functions That Are Not Differentiable at *x* = 0
FIGURE 7.29

In Figure 7.29, you can see that all but one of the functions are continuous at *x* = 0 but none are differentiable there. This shows that continuity is not a strong enough condition to guarantee differentiability. On the other hand, if a function is differentiable at a point, then it must be continuous at that point. This important result is stated in the following theorem.

Differentiability Implies Continuity

If a function *f* is differentiable at *x* = *c*, then *f* is continuous at *x* = *c*.

SUMMARIZE   (Section 7.3)

1. Describe a tangent line and how it can be used to approximate the slope of a graph at a point *(page 563)*. For an example of a tangent line, see Example 1.

2. State the definition of the slope of a graph using the limit process *(page 565)*. For examples of finding the slope of a graph using the limit process, see Examples 3, 4, and 5.

3. State the definition of the derivative of a function *(page 568)*. For examples of the derivative of a function, see Examples 6 and 7.

4. Describe the relationship between differentiability and continuity *(page 570)*. For an example showing that continuity does not guarantee differentiability, see Figure 7.29.

SKILLS WARM UP 7.3   The following warm-up exercises involve skills that were covered in earlier sections. You will use these skills in the exercise set for this section. For additional help, review Sections 2.2, 2.4, and 7.1.

In Exercises 1–4, find an equation of the line containing P and Q.

1. $P(2, 1)$, $Q(2, 4)$

2. $P(2, 2)$, $Q(-5, 2)$

3. $P(2, 0)$, $Q(3, -1)$

4. $P(3, 5)$, $Q(-1, -7)$

In Exercises 5–8, find the limit.

5. $\lim\limits_{\Delta x \to 0} \dfrac{2x\Delta x + (\Delta x)^2}{\Delta x}$

6. $\lim\limits_{\Delta x \to 0} \dfrac{3x^2\Delta x + 3x(\Delta x)^2 + (\Delta x)^3}{\Delta x}$

7. $\lim\limits_{\Delta x \to 0} \dfrac{1}{x(x + \Delta x)}$

8. $\lim\limits_{\Delta x \to 0} \dfrac{(x + \Delta x)^2 - x^2}{\Delta x}$

In Exercises 9–12, find the domain of the function.

9. $f(x) = 3x$

10. $f(x) = \dfrac{1}{x - 1}$

11. $f(x) = \dfrac{1}{5}x^3 - 2x^2 + \dfrac{1}{3}x - 1$

12. $f(x) = \dfrac{6x}{x^3 + x}$

Exercises 7.3

See www.CalcChat.com for worked-out solutions to odd-numbered exercises.

Sketching Tangent Lines In Exercises 1–6, trace the graph and sketch the tangent lines at (x_1, y_1) and (x_2, y_2).

Approximating the Slope of a Graph In Exercises 7–12, estimate the slope of the graph at the point (x, y). (Each square on the grid is 1 unit by 1 unit.) *See Example 1.*

1.
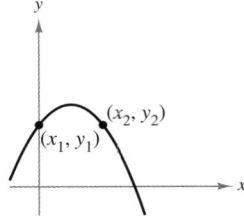

2.
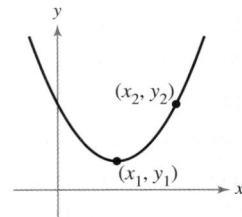

3.
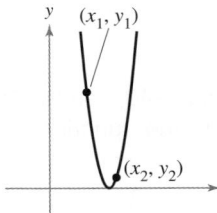

4.
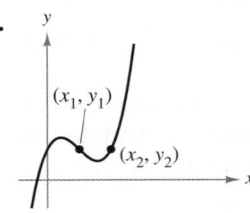

5.
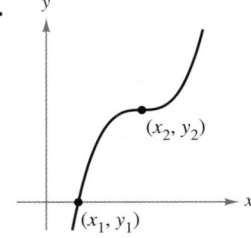

6.
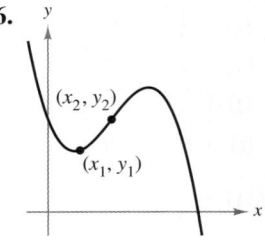

7.
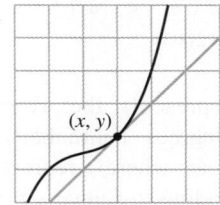

8.
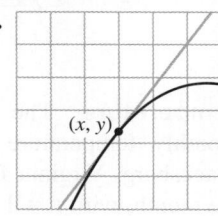

9.
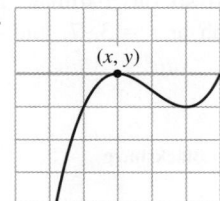

10.
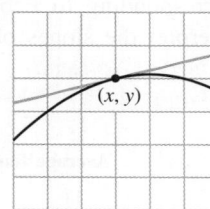

11.
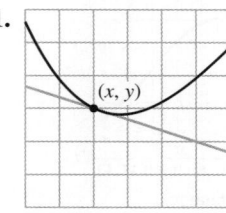

12.
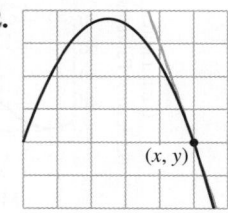

13. Revenue The graph represents the revenue R (in millions of dollars) for Under Armour from 2004 through 2009, where t represents the year, with $t = 4$ corresponding to 2004. Estimate and interpret the slopes of the graph for the years 2005 and 2007. *(Source: Under Armour, Inc.)*

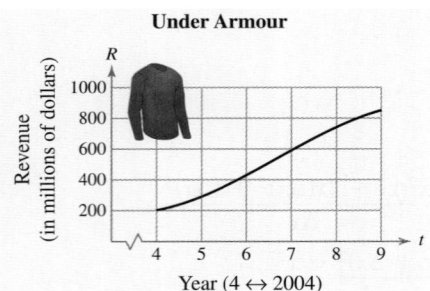

Under Armour

14. Sales The graph represents the sales S (in millions of dollars) for Scotts Miracle-Gro Company from 2003 through 2009, where t represents the year, with $t = 3$ corresponding to 2003. Estimate and interpret the slopes of the graph for the years 2006 and 2008. *(Source: Scotts Miracle-Gro Company)*

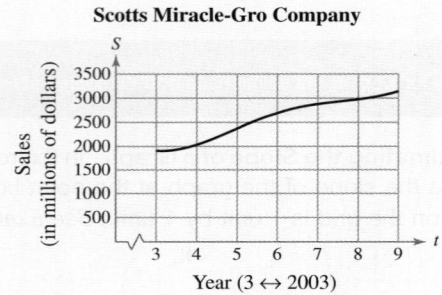

Scotts Miracle-Gro Company

15. Temperature The graph represents the average monthly temperature F (in degrees Fahrenheit) in Blacksburg, Virginia for one year, where t represents the month, with $t = 1$ corresponding to January, $t = 2$ corresponding to February, and so on. Estimate and interpret the slopes of the graph at $t = 3$, 7, and 10. *(Source: National Oceanic and Atmospheric Administration)*

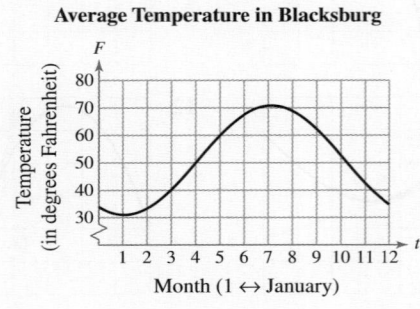

Average Temperature in Blacksburg

 **16.** **HOW DO YOU SEE IT?** Two long distance runners starting out side by side begin a 10,000-meter run. Their distances are given by $s = f(t)$ and $s = g(t)$, where s is measured in thousands of meters and t is measured in minutes.

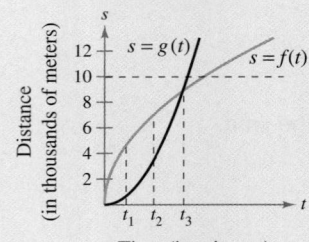

10,000-Meter Run

(a) Which runner is running faster at t_1?

(b) What conclusion can you make regarding their rates at t_2?

(c) What conclusion can you make regarding their rates at t_3?

(d) Which runner finishes the race first? Explain.

Finding the Slope of a Graph In Exercises 17–26, use the limit definition to find the slope of the tangent line to the graph of f at the given point. *See Examples 3, 4, and 5.*

17. $f(x) = -1$; $(0, -1)$

18. $f(x) = 6$; $(-2, 6)$

19. $f(x) = 8 - 3x$; $(2, 2)$

20. $f(x) = 6x + 3$; $(1, 9)$

21. $f(x) = 2x^2 - 3$; $(2, 5)$

22. $f(x) = 4 - x^2$; $(2, 0)$

23. $f(x) = x^3 - x$; $(2, 6)$

24. $f(x) = x^3 + 2x$; $(1, 3)$

25. $f(x) = 2\sqrt{x}$; $(4, 4)$

26. $f(x) = \sqrt{x + 1}$; $(8, 3)$

Finding a Derivative In Exercises 27–40, use the limit definition to find the derivative of the function. *See Examples 6 and 7.*

27. $f(x) = 3$

28. $f(x) = -2$

29. $f(x) = -5x$

30. $f(x) = 4x + 1$

31. $g(s) = \frac{1}{3}s + 2$

32. $h(t) = 6 - \frac{1}{2}t$

33. $f(x) = 4x^2 - 5x$

34. $f(x) = 2x^2 + 7x$

35. $h(t) = \sqrt{t - 1}$

36. $f(x) = \sqrt{x + 2}$

37. $f(t) = t^3 - 12t$

38. $f(t) = t^3 + t^2$

39. $f(x) = \dfrac{1}{x + 2}$

40. $g(s) = \dfrac{1}{s - 1}$

Finding an Equation of a Tangent Line In Exercises 41–48, use the limit definition to find an equation of the tangent line to the graph of f at the given point. Then verify your results by using a graphing utility to graph the function and its tangent line at the point.

41. $f(x) = \frac{1}{2}x^2;\ (2, 2)$ **42.** $f(x) = -x^2;\ (-1, -1)$

43. $f(x) = (x - 1)^2;\ (-2, 9)$

44. $f(x) = 2x^2 - 1;\ (0, -1)$

45. $f(x) = \sqrt{x} + 1;\ (4, 3)$ **46.** $f(x) = \sqrt{x + 2};\ (7, 3)$

47. $f(x) = \frac{1}{x};\ (1, 1)$ **48.** $f(x) = \frac{1}{x - 3};\ (2, -1)$

Finding an Equation of a Tangent Line In Exercises 49–52, find an equation of the line that is tangent to the graph of f and parallel to the given line.

Function	Line
49. $f(x) = -\frac{1}{4}x^2$	$x + y = 0$
50. $f(x) = x^2 - 7$	$2x + y = 0$
51. $f(x) = -\frac{1}{3}x^3$	$9x + y - 6 = 0$
52. $f(x) = x^2 - x$	$x + 2y - 6 = 0$

Determining Differentiability In Exercises 53–58, describe the x-values at which the function is differentiable. Explain your reasoning.

53. $y = |x + 3|$ **54.** $y = |x^2 - 9|$

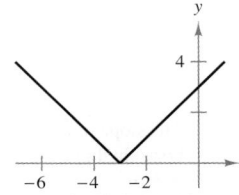   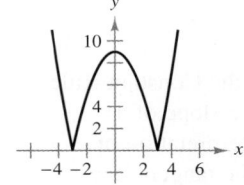

55. $y = (x - 3)^{2/3}$ **56.** $y = \sqrt{x - 1}$

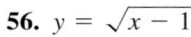

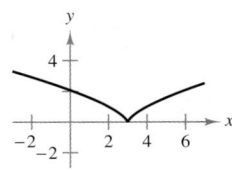   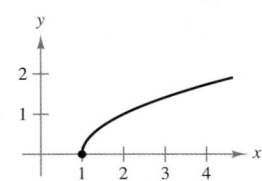

57. $y = \dfrac{x^2}{x^2 - 4}$ **58.** $y = \begin{cases} x^3 + 3, & x < 0 \\ x^3 - 3, & x \geq 0 \end{cases}$

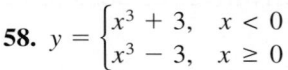

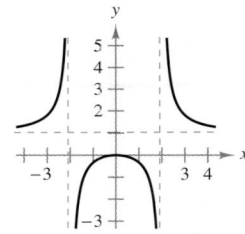   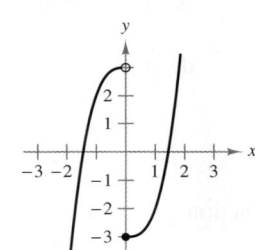

Writing a Function Using Derivatives In Exercises 59 and 60, identify a function f that has the given characteristics. Then sketch the function.

59. $f(0) = 2;\ f'(x) = -3$ for $-\infty < x < \infty$

60. $f(-2) = f(4) = 0;\ f'(1) = 0;$
$f'(x) < 0$ for $x < 1;\ f'(x) > 0$ for $x > 1$

 Graphical, Numerical, and Analytic Analysis In Exercises 61–64, use a graphing utility to graph f on the interval $[-2, 2]$. Complete the table by graphically estimating the slopes of the graph at the given points. Then evaluate the slopes analytically and compare your results with those obtained graphically.

x	-2	$-\frac{3}{2}$	-1	$-\frac{1}{2}$	0	$\frac{1}{2}$	1	$\frac{3}{2}$	2
$f(x)$									
$f'(x)$									

61. $f(x) = \frac{1}{4}x^3$ **62.** $f(x) = \frac{3}{4}x^2$

63. $f(x) = -\frac{1}{2}x^3$ **64.** $f(x) = -\frac{3}{2}x^2$

 Graphing a Function and Its Derivative In Exercises 65–68, find the derivative of the given function f. Then use a graphing utility to graph f and its derivative in the same viewing window. What does the x-intercept of the derivative indicate about the graph of f?

65. $f(x) = x^2 - 4x$ **66.** $f(x) = 2 + 6x - x^2$

67. $f(x) = x^3 - 3x$ **68.** $f(x) = x^3 - 6x^2$

True or False? In Exercises 69–72, determine whether the statement is true or false. If it is false, explain why or give an example that shows it is false.

69. The slope of the graph of $y = x^2$ is different at every point on the graph of f.

70. If a function is continuous at a point, then it is differentiable at that point.

71. If a function is differentiable at a point, then it is continuous at that point.

72. A tangent line to a graph can intersect the graph at more than one point.

73. Writing Use a graphing utility to graph the two functions

$$f(x) = x^2 + 1 \quad \text{and} \quad g(x) = |x| + 1$$

in the same viewing window. Use the *zoom* and *trace* features to analyze the graphs near the point $(0, 1)$. What do you observe? Which function is differentiable at this point? Write a short paragraph describing the geometric significance of differentiability at a point.

7.4 Some Rules for Differentiation

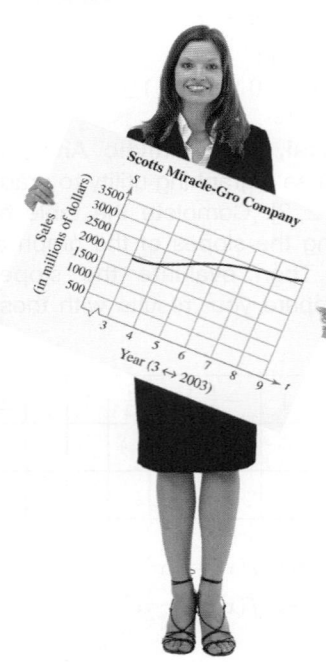

- Find the derivatives of functions using the Constant Rule.
- Find the derivatives of functions using the Power Rule.
- Find the derivatives of functions using the Constant Multiple Rule.
- Find the derivatives of functions using the Sum and Difference Rules.
- Use derivatives to answer questions about real-life situations.

The Constant Rule

In Section 7.3, you found derivatives by the limit process. This process is tedious, even for simple functions, but fortunately there are rules that greatly simplify differentiation. These rules allow you to calculate derivatives without the *direct* use of limits.

> **The Constant Rule**
>
> The derivative of a constant function is zero. That is,
>
> $$\frac{d}{dx}[c] = 0, \quad c \text{ is a constant.}$$

In Exercise 74 on page 585, you will use differentiation to find the rate of change in a company's sales.

PROOF Let $f(x) = c$. Then, by the limit definition of the derivative, you can write

$$f'(x) = \lim_{\Delta x \to 0} \frac{f(x + \Delta x) - f(x)}{\Delta x} = \lim_{\Delta x \to 0} \frac{c - c}{\Delta x} = \lim_{\Delta x \to 0} 0 = 0.$$

So, $\frac{d}{dx}[c] = 0$.

Note in Figure 7.30 that the Constant Rule is equivalent to saying that the slope of a horizontal line is zero. An interpretation of the Constant Rule says that the tangent line to a constant function is the function itself. For instance, the equation of the tangent line to $f(x) = 4$ at $x = -1$ is

$$y = 4.$$

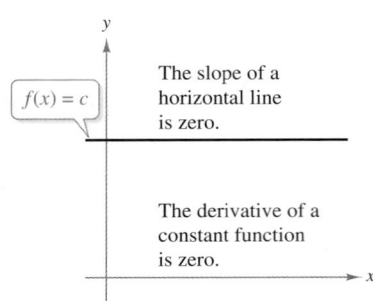

The slope of a horizontal line is zero.

The derivative of a constant function is zero.

FIGURE 7.30

Example 1   Finding Derivatives of Constant Functions

a. $\frac{d}{dx}[7] = 0$

b. If $f(x) = 0$, then $f'(x) = 0$.

c. If $y = 2$, then $\frac{dy}{dx} = 0$.

d. If $g(t) = -\frac{3}{2}$, then $g'(t) = 0$.

✓ **Checkpoint 1**

Find the derivative of each function.

a. $f(x) = -2$ **b.** $y = \pi$ **c.** $g(w) = \sqrt{5}$ **d.** $s(t) = 320.5$

The Power Rule

The binomial expansion process is used in proving a special case of the Power Rule.

$$(x + \Delta x)^2 = x^2 + 2x \, \Delta x + (\Delta x)^2$$

$$(x + \Delta x)^3 = x^3 + 3x^2 \, \Delta x + 3x(\Delta x)^2 + (\Delta x)^3$$

$$(x + \Delta x)^n = x^n + nx^{n-1} \, \Delta x + \underbrace{\frac{n(n-1)x^{n-2}}{2}(\Delta x)^2 + \cdots + (\Delta x)^n}_{(\Delta x)^2 \text{ is a factor of these terms.}}$$

The (Simple) Power Rule

$$\frac{d}{dx}[x^n] = nx^{n-1}, \quad n \text{ is any real number.}$$

PROOF This proof is limited to the case in which n is a positive integer. Let $f(x) = x^n$. Using the binomial expansion, you can write

$$f'(x) = \lim_{\Delta x \to 0} \frac{f(x + \Delta x) - f(x)}{\Delta x} \qquad \text{Definition of derivative}$$

$$= \lim_{\Delta x \to 0} \frac{(x + \Delta x)^n - x^n}{\Delta x}$$

$$= \lim_{\Delta x \to 0} \frac{x^n + nx^{n-1} \, \Delta x + \dfrac{n(n-1)x^{n-2}}{2}(\Delta x)^2 + \cdots + (\Delta x)^n - x^n}{\Delta x}$$

$$= \lim_{\Delta x \to 0} \left[nx^{n-1} + \frac{n(n-1)x^{n-2}}{2}(\Delta x) + \cdots + (\Delta x)^{n-1} \right]$$

$$= nx^{n-1} + 0 + \cdots + 0$$

$$= nx^{n-1}.$$

For the Power Rule, the case in which $n = 1$ is worth remembering as a separate differentiation rule. That is,

$$\frac{d}{dx}[x] = 1. \qquad \text{The derivative of } x \text{ is } 1.$$

This rule is consistent with the fact that the slope of the line given by $y = x$ is 1. (See Figure 7.31.)

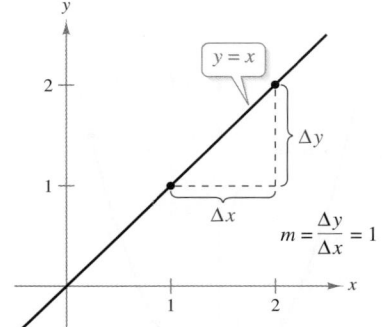

The slope of the line $y = x$ is 1.

FIGURE 7.31

Example 2   Applying the Power Rule

Original Function	Derivative
a. $f(x) = x^3$	$f'(x) = 3x^2$
b. $y = \dfrac{1}{x^2} = x^{-2}$	$\dfrac{dy}{dx} = (-2)x^{-3} = -\dfrac{2}{x^3}$
c. $g(t) = t$	$g'(t) = 1$

✓ Checkpoint 2

Find the derivative of each function.

a. $f(x) = x^4$ **b.** $y = \dfrac{1}{x^3}$ **c.** $g(w) = w^2$

In Example 2(b), note that *before* differentiating, you should rewrite $1/x^2$ as x^{-2}. Rewriting is the first step in *many* differentiation problems.

Original Function: $y = \dfrac{1}{x^2}$	$\Rightarrow$	Rewrite: $y = x^{-2}$	$\Rightarrow$	Differentiate: $\dfrac{dy}{dx} = (-2)x^{-3}$	$\Rightarrow$	Simplify: $\dfrac{dy}{dx} = -\dfrac{2}{x^3}$

Remember that the derivative of a function f is another function that gives the slope of the graph of f at any point at which f is differentiable. So, you can use the derivative to find slopes, as shown in Example 3.

Example 3   Finding the Slope of a Graph

Find the slopes of the graph of

$$f(x) = x^2$$

at $x = -2, -1, 0, 1,$ and 2.

SOLUTION   Begin by using the Power Rule to find the derivative of f.

$$f'(x) = 2x \qquad \text{Derivative}$$

You can use the derivative to find the slopes of the graph of f, as shown.

x-Value	Slope of Graph of f
$x = -2$	$m = f'(-2) = 2(-2) = -4$
$x = -1$	$m = f'(-1) = 2(-1) = -2$
$x = 0$	$m = f'(0) = 2(0) = 0$
$x = 1$	$m = f'(1) = 2(1) = 2$
$x = 2$	$m = f'(2) = 2(2) = 4$

The graph of f is shown in Figure 7.32.

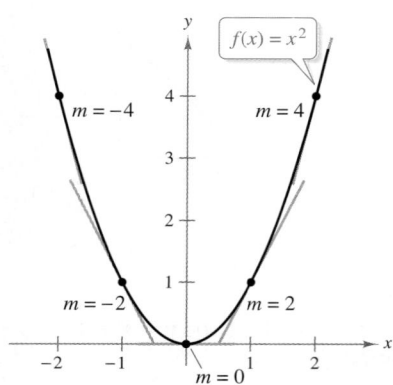

FIGURE 7.32

✓ Checkpoint 3

Find the slopes of the graph of

$$f(x) = x^3$$

at $x = -1, 0,$ and 1.

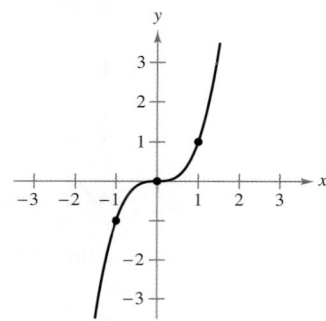

The Constant Multiple Rule

To prove the Constant Multiple Rule, the following property of limits is used.

$$\lim_{x \to a} cg(x) = c\left[\lim_{x \to a} g(x)\right]$$

The Constant Multiple Rule

If f is a differentiable function of x, and c is a real number, then

$$\frac{d}{dx}[cf(x)] = cf'(x), \quad c \text{ is a constant.}$$

PROOF Apply the definition of the derivative to produce

$$\frac{d}{dx}[cf(x)] = \lim_{\Delta x \to 0} \frac{cf(x + \Delta x) - cf(x)}{\Delta x} \qquad \text{Definition of derivative}$$

$$= \lim_{\Delta x \to 0} c\left[\frac{f(x + \Delta x) - f(x)}{\Delta x}\right]$$

$$= c\left[\lim_{\Delta x \to 0} \frac{f(x + \Delta x) - f(x)}{\Delta x}\right]$$

$$= cf'(x).$$

Informally, the Constant Multiple Rule states that constants can be factored out of the differentiation process.

$$\frac{d}{dx}[cf(x)] = c\frac{d}{dx}[f(x)] = cf'(x).$$

The usefulness of this rule is often overlooked, especially when the constant appears in the denominator, as shown below.

$$\frac{d}{dx}\left[\frac{f(x)}{c}\right] = \frac{d}{dx}\left[\frac{1}{c}f(x)\right]$$

$$= \frac{1}{c}\left(\frac{d}{dx}[f(x)]\right)$$

$$= \frac{1}{c}f'(x).$$

To use the Constant Multiple Rule efficiently, look for constants that can be factored out *before* differentiating. For example,

$$\frac{d}{dx}[5x^2] = 5\frac{d}{dx}[x^2] \qquad \text{Factor out 5.}$$

$$= 5(2x) \qquad \text{Differentiate.}$$

$$= 10x \qquad \text{Simplify.}$$

and

$$\frac{d}{dx}\left[\frac{x^2}{5}\right] = \frac{1}{5}\left(\frac{d}{dx}[x^2]\right) \qquad \text{Factor out } \tfrac{1}{5}.$$

$$= \frac{1}{5}(2x) \qquad \text{Differentiate.}$$

$$= \frac{2}{5}x. \qquad \text{Simplify.}$$

Example 4   **Using the Power and Constant Multiple Rules**

Find the derivative of (a) $y = 2x^{1/2}$ and (b) $f(t) = \dfrac{4t^2}{5}$.

SOLUTION

a. Using the Constant Multiple Rule and the Power Rule, you can write

$$\frac{dy}{dx} = \frac{d}{dx}[2x^{1/2}] = 2\underbrace{\frac{d}{dx}[x^{1/2}]}_{\text{Constant Multiple Rule}} = 2\underbrace{\left(\frac{1}{2}x^{-1/2}\right)}_{\text{Power Rule}} = x^{-1/2} = \frac{1}{\sqrt{x}}.$$

b. Begin by rewriting $f(t)$ as

$$f(t) = \frac{4t^2}{5} = \frac{4}{5}t^2.$$

Then, use the Constant Multiple Rule and the Power Rule to obtain

$$f'(t) = \frac{d}{dt}\left[\frac{4}{5}t^2\right] = \frac{4}{5}\left(\frac{d}{dt}[t^2]\right) = \frac{4}{5}(2t) = \frac{8}{5}t.$$

✓**Checkpoint 4**

Find the derivative of (a) $y = 4x^2$ and (b) $f(x) = 16x^{1/2}$.   ■

You may find it helpful to combine the Constant Multiple Rule and the Power Rule into one combined rule.

$$\frac{d}{dx}[cx^n] = cnx^{n-1}, \quad n \text{ is a real number, } c \text{ is a constant.}$$

For instance, in Example 4(b), you can apply this combined rule to obtain

$$\frac{d}{dt}\left[\frac{4}{5}t^2\right] = \left(\frac{4}{5}\right)(2)(t) = \frac{8}{5}t.$$

The three functions in the next example are simple, yet errors are frequently made in differentiating functions involving constant multiples of the first power of x. Keep in mind that

$$\frac{d}{dx}[cx] = c, \quad c \text{ is a constant.}$$

Example 5   **Applying the Constant Multiple Rule**

Original Function	*Derivative*
a. $y = -\dfrac{3x}{2}$	$y' = -\dfrac{3}{2}$
b. $y = 3\pi x$	$y' = 3\pi$
c. $y = -\dfrac{x}{2}$	$y' = -\dfrac{1}{2}$

✓**Checkpoint 5**

Find the derivative of (a) $y = \dfrac{t}{4}$ and (b) $y = -\dfrac{2x}{5}$.   ■

Parentheses can play an important role in the use of the Constant Multiple Rule and the Power Rule. In Example 6, be sure you understand the mathematical conventions involving the use of parentheses.

Example 6   Using Parentheses When Differentiating

Original Function	Rewrite	Differentiate	Simplify
a. $y = \dfrac{5}{2x^3}$	$y = \dfrac{5}{2}(x^{-3})$	$y' = \dfrac{5}{2}(-3x^{-4})$	$y' = -\dfrac{15}{2x^4}$
b. $y = \dfrac{5}{(2x)^3}$	$y = \dfrac{5}{8}(x^{-3})$	$y' = \dfrac{5}{8}(-3x^{-4})$	$y' = -\dfrac{15}{8x^4}$
c. $y = \dfrac{7}{3x^{-2}}$	$y = \dfrac{7}{3}(x^2)$	$y' = \dfrac{7}{3}(2x)$	$y' = \dfrac{14x}{3}$
d. $y = \dfrac{7}{(3x)^{-2}}$	$y = 63(x^2)$	$y' = 63(2x)$	$y' = 126x$

✓ Checkpoint 6

Find the derivative of each function.

a. $y = \dfrac{9}{4x^2}$

b. $y = \dfrac{9}{(4x)^2}$

When differentiating functions involving radicals, you should rewrite the function with rational exponents. For instance, you should rewrite

$$y = \sqrt[3]{x} \quad \text{as} \quad y = x^{1/3}$$

and you should rewrite

$$y = \dfrac{1}{\sqrt[3]{x^4}} \quad \text{as} \quad y = x^{-4/3}.$$

Example 7   Differentiating Radical Functions

Original Function	Rewrite	Differentiate	Simplify
a. $y = \sqrt{x}$	$y = x^{1/2}$	$y' = \left(\dfrac{1}{2}\right)x^{-1/2}$	$y' = \dfrac{1}{2\sqrt{x}}$
b. $y = \dfrac{1}{2\sqrt[3]{x^2}}$	$y = \dfrac{1}{2}x^{-2/3}$	$y' = \dfrac{1}{2}\left(-\dfrac{2}{3}\right)x^{-5/3}$	$y' = -\dfrac{1}{3x^{5/3}}$
c. $y = \sqrt{2x}$	$y = \sqrt{2}(x^{1/2})$	$y' = \sqrt{2}\left(\dfrac{1}{2}\right)x^{-1/2}$	$y' = \dfrac{1}{\sqrt{2x}}$

✓ Checkpoint 7

Find the derivative of each function.

a. $y = \sqrt{5x}$

b. $y = \sqrt[4]{x}$

The Sum and Difference Rules

To differentiate $y = 3x + 2x^3$, you would probably write

$$y' = 3 + 6x^2$$

without questioning your answer. The validity of differentiating a sum or difference of functions term by term is given by the Sum and Difference Rules.

The Sum and Difference Rules

The derivative of the sum or difference of two differentiable functions is the sum or difference of their derivatives.

$$\frac{d}{dx}[f(x) + g(x)] = f'(x) + g'(x) \qquad \text{Sum Rule}$$

$$\frac{d}{dx}[f(x) - g(x)] = f'(x) - g'(x) \qquad \text{Difference Rule}$$

PROOF   Let $h(x) = f(x) + g(x)$. Then, you can prove the Sum Rule as shown.

$$h'(x) = \lim_{\Delta x \to 0} \frac{h(x + \Delta x) - h(x)}{\Delta x} \qquad \text{Definition of derivative}$$

$$= \lim_{\Delta x \to 0} \frac{f(x + \Delta x) + g(x + \Delta x) - f(x) - g(x)}{\Delta x}$$

$$= \lim_{\Delta x \to 0} \frac{f(x + \Delta x) - f(x) + g(x + \Delta x) - g(x)}{\Delta x}$$

$$= \lim_{\Delta x \to 0} \left[\frac{f(x + \Delta x) - f(x)}{\Delta x} + \frac{g(x + \Delta x) - g(x)}{\Delta x} \right]$$

$$= \lim_{\Delta x \to 0} \frac{f(x + \Delta x) - f(x)}{\Delta x} + \lim_{\Delta x \to 0} \frac{g(x + \Delta x) - g(x)}{\Delta x}$$

$$= f'(x) + g'(x)$$

So,

$$\frac{d}{dx}[f(x) + g(x)] = f'(x) + g'(x).$$

The Difference Rule can be proved in a similar manner.

The Sum and Difference Rules can be extended to the sum or difference of any finite number of functions. For instance, if $y = f(x) + g(x) + h(x)$, then $y' = f'(x) + g'(x) + h'(x)$.

Example 8   **Using the Sum and Difference Rules**

Original Function	*Derivative*
a. $y = x^3 + 4x^2$	$y' = 3x^2 + 8x$
b. $f(x) = 3x^2 - 2x$	$f'(x) = 6x - 2$

✓ **Checkpoint 8**

Find the derivative of each function.

a. $f(x) = 2x^2 + 5x$ **b.** $y = x^4 - 2x$

STUDY TIP

Look back at Example 6 on page 568. Notice that the example asks for the derivative of the difference of two functions. Compare the result with the one obtained in Example 8(b) at the right.

With the differentiation rules listed in this section, you can differentiate any polynomial function.

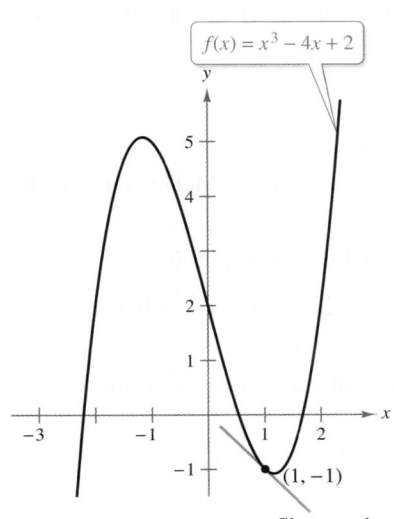

$f(x) = x^3 - 4x + 2$

Slope $= -1$

FIGURE 7.33

Example 9   Finding the Slope of a Graph

Find the slope of the graph of $f(x) = x^3 - 4x + 2$ at the point $(1, -1)$.

SOLUTION   The derivative of $f(x)$ is

$$f'(x) = 3x^2 - 4.$$

So, the slope of the graph of f at $(1, -1)$ is

$$\text{Slope} = f'(1) = 3(1)^2 - 4 = 3 - 4 = -1$$

as shown in Figure 7.33.

✓**Checkpoint 9**

Find the slope of the graph of $f(x) = x^2 - 5x + 1$ at the point $(2, -5)$.

Example 9 illustrates the use of the derivative for determining the shape of a graph. A rough sketch of the graph of $f(x) = x^3 - 4x + 2$ might lead you to think that the point $(1, -1)$ is a minimum point of the graph. After finding the slope at this point to be -1, however, you can conclude that the minimum point (where the slope is 0) is farther to the right. (You will study techniques for finding minimum and maximum points in Section 8.5.)

Example 10   Finding an Equation of a Tangent Line

Find an equation of the tangent line to the graph of

$$g(x) = -\frac{1}{2}x^4 + 3x^3 - 2x$$

at the point $\left(-1, -\frac{3}{2}\right)$.

SOLUTION   The derivative of $g(x)$ is $g'(x) = -2x^3 + 9x^2 - 2$, which implies that the slope of the graph at the point $\left(-1, -\frac{3}{2}\right)$ is

$$
\begin{aligned}
\text{Slope} &= g'(-1) \\
&= -2(-1)^3 + 9(-1)^2 - 2 \\
&= 2 + 9 - 2 \\
&= 9
\end{aligned}
$$

as shown in Figure 7.34. Using the point-slope form, you can write the equation of the tangent line at $\left(-1, -\frac{3}{2}\right)$ as shown.

$$y - \left(-\frac{3}{2}\right) = 9[x - (-1)] \qquad \text{Point-slope form}$$

$$y + \frac{3}{2} = 9x + 9 \qquad \text{Simplify.}$$

$$y = 9x + \frac{15}{2} \qquad \text{Equation of tangent line}$$

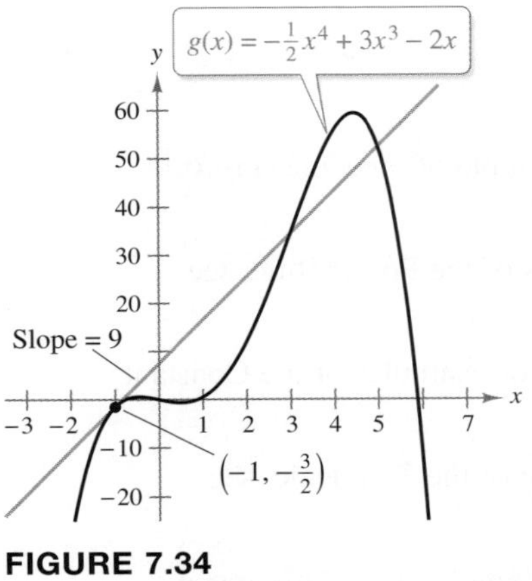

$g(x) = -\frac{1}{2}x^4 + 3x^3 - 2x$

Slope $= 9$

$\left(-1, -\frac{3}{2}\right)$

FIGURE 7.34

✓**Checkpoint 10**

Find an equation of the tangent line to the graph of $f(x) = -x^2 + 3x - 2$ at the point $(2, 0)$.

Application

There are many applications of the derivative that you will study in this textbook. In Example 11, you will use a derivative to find the rate of change of a company's revenue with respect to time.

 Example 11   **Modeling Revenue**

From 2004 through 2009, the revenue R (in millions of dollars) for McDonald's can be modeled by

$$R = -130.769t^3 + 2296.47t^2 - 11,493.5t + 35,493, \quad 4 \leq t \leq 9$$

where t represents the year, with $t = 4$ corresponding to 2004. At what rate was McDonald's revenue changing in 2006?   *(Source: McDonald's Corporation)*

SOLUTION   One way to answer this question is to find the derivative of the revenue model with respect to time.

$$\frac{dR}{dt} = -392.307t^2 + 4592.94t - 11,493.5, \quad 4 \leq t \leq 9$$

In 2006 (at $t = 6$), the rate of change of the revenue with respect to time is given by

$$\frac{dR}{dt} = -392.307(6)^2 + 4592.94(6) - 11,493.5 \approx 1941.$$

Because R is measured in millions of dollars and t is measured in years, it follows that the derivative dR/dt is measured in millions of dollars per year. So, at the end of 2006, McDonald's revenue was increasing at a rate of about $1941 million per year, as shown in Figure 7.35.

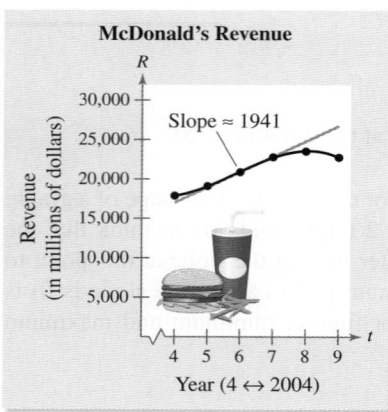

McDonald's Revenue

Slope ≈ 1941

Year (4 ↔ 2004)

FIGURE 7.35

✔ **Checkpoint 11**

From 2000 through 2010, the sales per share S (in dollars) for Microsoft Corporation can be modeled by

$$S = 0.0330t^2 + 0.208t + 2.13, \quad 0 \leq t \leq 10$$

where t represents the year, with $t = 0$ corresponding to 2000. At what rate was Microsoft's sales per share changing in 2002?   *(Source: Microsoft Corporation)*

SUMMARIZE   (Section 7.4)

1. State the Constant Rule *(page 574)*. For an example of the Constant Rule, see Example 1.

2. State the Power Rule *(page 575)*. For examples of the Power Rule, see Examples 2 and 3.

3. State the Constant Multiple Rule *(page 577)*. For examples of the Constant Multiple Rule, see Examples 4, 5, 6, and 7.

4. State the Sum Rule *(page 580)*. For an example of the Sum Rule, see Example 8.

5. State the Difference Rule *(page 580)*. For an example of the Difference Rule, see Example 8.

6. Describe a real-life example of how differentiation can be used to analyze the rate of change of a company's revenue *(page 582, Example 11)*.

SKILLS WARM UP 7.4   The following warm-up exercises involve skills that were covered in a previous course. You will use these skills in the exercise set for this section. For additional help, review Sections 0.3, 0.4, 1.3, and 1.5.

In Exercises 1 and 2, evaluate each expression when $x = 2$.

1. (a) $2x^2$ (b) $(2x)^2$ (c) $2x^{-2}$

2. (a) $\dfrac{1}{(3x)^2}$ (b) $\dfrac{1}{4x^3}$ (c) $\dfrac{(2x)^{-3}}{4x^{-2}}$

In Exercises 3–6, simplify the expression.

3. $4(3)x^3 + 2(2)x$

4. $\frac{1}{2}(3)x^2 - \frac{3}{2}x^{1/2}$

5. $\left(\frac{1}{4}\right)x^{-3/4}$

6. $\frac{1}{3}(3)x^2 - 2\left(\frac{1}{2}\right)x^{-1/2} + \frac{1}{3}x^{-2/3}$

In Exercises 7–10, solve the equation.

7. $3x^2 + 2x = 0$

8. $x^3 - x = 0$

9. $x^2 + 8x - 20 = 0$

10. $x^2 - 10x - 24 = 0$

Exercises 7.4

See www.CalcChat.com for worked-out solutions to odd-numbered exercises.

Finding Derivatives   In Exercises 1–24, find the derivative of the function. *See Examples 1, 2, 4, 5, and 8.*

1. $y = 3$

2. $f(x) = -8$

3. $y = x^5$

4. $f(x) = \dfrac{1}{x^5}$

5. $h(x) = 3x^3$

6. $h(x) = 2x^5$

7. $y = \dfrac{2x^3}{3}$

8. $g(t) = \dfrac{3t^2}{4}$

9. $f(x) = 4x$

10. $g(x) = 3x$

11. $y = 8 - x^3$

12. $y = t^2 - 6$

13. $f(x) = 4x^2 - 3x$

14. $g(x) = x^2 + 4x^3$

15. $f(t) = -3t^2 + 2t - 4$

16. $y = x^3 - 9x^2 + 2$

17. $s(t) = t^3 - 2t + 4$

18. $y = 2x^3 - x^2 + 3x - 1$

19. $g(x) = x^{2/3}$

20. $h(x) = x^{5/2}$

21. $y = 4t^{4/3}$

22. $f(x) = 10x^{1/2}$

23. $y = 4x^{-2} + 2x^2$

24. $s(t) = 4t^{-1} + 1$

Using Parentheses When Differentiating   In Exercises 25–30, find the derivative. *See Example 6.*

Function	Rewrite	Differentiate	Simplify
25. $y = \dfrac{2}{7x^4}$			
26. $y = \dfrac{2}{3x^2}$			
27. $y = \dfrac{1}{(4x)^3}$			
28. $y = \dfrac{\pi}{(3x)^2}$			
29. $y = \dfrac{4}{(2x)^{-5}}$			
30. $y = \dfrac{4x}{x^{-3}}$			

Differentiating Radical Functions   In Exercises 31–36, find the derivative. *See Example 7.*

Function	Rewrite	Differentiate	Simplify
31. $y = 6\sqrt{x}$			
32. $y = \dfrac{3\sqrt{x}}{4}$			
33. $y = \dfrac{1}{5\sqrt[5]{x}}$			
34. $y = \dfrac{3}{2\sqrt[4]{x^3}}$			
35. $y = \sqrt{3x}$			
36. $y = \sqrt[3]{6x^2}$			

Finding the Slope of a Graph   In Exercises 37–44, find the slope of the graph of the function at the given point. *See Examples 3 and 9.*

37. $y = x^{3/2}$

38. $y = x^{-1}$

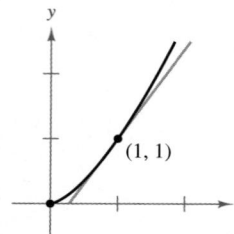

(1, 1)

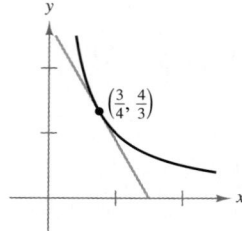

$\left(\frac{3}{4}, \frac{4}{3}\right)$

39. $f(t) = t^2;\ (4, 16)$

40. $f(x) = x^{-1/3};\ \left(8, \frac{1}{2}\right)$

41. $f(x) = 2x^3 + 8x^2 - x - 4;\ (-1, 3)$

42. $f(x) = 3x^4 - 5x^3 + 6x^2 - 10x;\ (1, -6)$

43. $f(x) = -\frac{1}{2}x(1 + x^2);\ (1, -1)$

44. $f(x) = 3(5 - x)^2;\ (5, 0)$

 Finding an Equation of a Tangent Line   In Exercises 45–50, (a) find an equation of the tangent line to the graph of the function at the given point, (b) use a graphing utility to graph the function and its tangent line at the point, and (c) use the *derivative* feature of a graphing utility to confirm your results. *See Example 10.*

45. $y = -2x^4 + 5x^2 - 3;\ (1, 0)$

46. $y = x^3 + x;\ (-1, -2)$

47. $f(x) = \sqrt[3]{x} + \sqrt[5]{x};\ (1, 2)$

48. $f(x) = \dfrac{1}{\sqrt[3]{x^2}} - x;\ (-1, 2)$

49. $y = 3x\left(x^2 - \dfrac{2}{x}\right);\ (2, 18)$

50. $y = (2x + 1)^2;\ (0, 1)$

Finding Derivatives   In Exercises 51–62, find $f'(x)$.

51. $f(x) = x^2 - \dfrac{4}{x} - 3x^{-2}$

52. $f(x) = x^2 - 3x - 3x^{-2} + 5x^{-3}$

53. $f(x) = x^2 - 2x - \dfrac{2}{x^4}$

54. $f(x) = x^2 + 4x + \dfrac{1}{x}$

55. $f(x) = x^{4/5} + x$

56. $f(x) = x^{1/3} - 1$

57. $f(x) = x(x^2 + 1)$

58. $f(x) = (x^2 + 2x)(x + 1)$

59. $f(x) = \dfrac{2x^3 - 4x^2 + 3}{x^2}$

60. $f(x) = \dfrac{2x^2 - 3x + 1}{x}$

61. $f(x) = \dfrac{4x^3 - 3x^2 + 2x + 5}{x^2}$

62. $f(x) = \dfrac{-6x^3 + 3x^2 - 2x + 1}{x}$

Finding Horizontal Tangent Lines   In Exercises 63–66, determine the point(s), if any, at which the graph of the function has a horizontal tangent line.

63. $y = -x^4 + 3x^2 - 1$

64. $y = x^3 + 3x^2$

65. $y = \frac{1}{2}x^2 + 5x$

66. $y = x^2 + 2x$

Exploring Relationships   In Exercises 67 and 68, (a) sketch the graphs of f and g, (b) find $f'(1)$ and $g'(1)$, (c) sketch the tangent line to each graph at $x = 1$, and (d) explain the relationship between f' and g'.

67. $f(x) = x^3$

$g(x) = x^3 + 3$

68. $f(x) = x^2$

$g(x) = 3x^2$

Exploring Relationships   In Exercises 69–72, the relationship between f and g is given. Explain the relationship between f' and g'.

69. $g(x) = f(x) + 6$

70. $g(x) = 2f(x)$

71. $g(x) = -5f(x)$

72. $g(x) = 3f(x) - 1$

73. Revenue   The revenue R (in millions of dollars) for Under Armour from 2004 through 2009 can be modeled by

$$R = -5.1509t^3 + 103.166t^2 - 526.15t + 985.4$$

where t is the year, with $t = 4$ corresponding to 2004. *(Source: Under Armour, Inc.)*

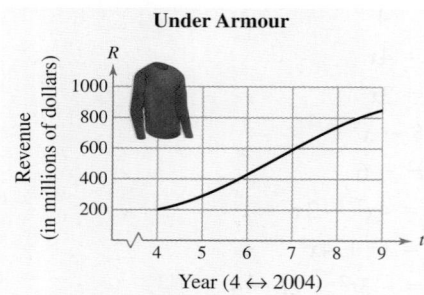

Under Armour

(a) Find the slopes of the graph for the years 2005 and 2007.

(b) Compare your results with those obtained in Exercise 13 in Section 7.3.

(c) Interpret the slope of the graph in the context of the problem.

74. Sales The sales S (in millions of dollars) for Scotts Miracle-Gro Company from 2003 through 2009 can be modeled by $S = 5.45682t^4 - 136.9359t^3 + 1219.018t^2 - 4294.73t + 7078.4$, where t is the year, with $t = 3$ corresponding to 2003. *(Source: Scotts Miracle-Gro Company)*

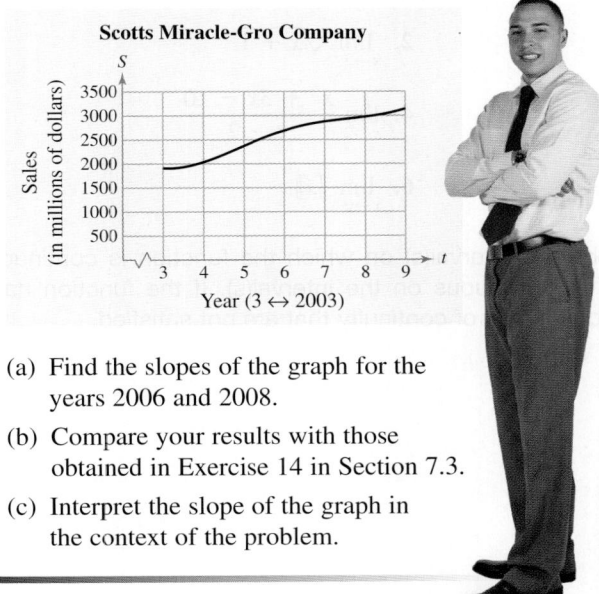

Scotts Miracle-Gro Company

Year (3 ↔ 2003)

(a) Find the slopes of the graph for the years 2006 and 2008.

(b) Compare your results with those obtained in Exercise 14 in Section 7.3.

(c) Interpret the slope of the graph in the context of the problem.

75. Psychology: Migraine Prevalence The graph illustrates the prevalence of migraine headaches in males and females in selected income groups. *(Source: Adapted from Sue/Sue/Sue, Understanding Abnormal Behavior, Seventh Edition)*

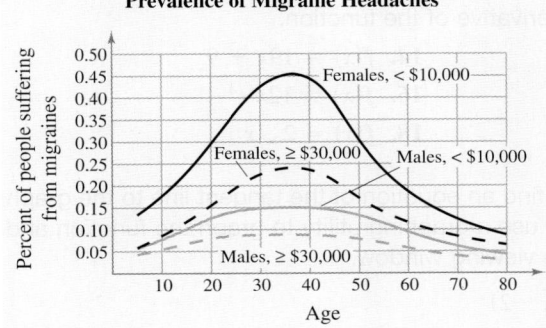

Prevalence of Migraine Headaches

(a) Write a short paragraph describing your general observations about the prevalence of migraines in females and males with respect to age group and income bracket.

(b) Describe the graph of the derivative of each curve, and explain the significance of each derivative. Include an explanation of the units of the derivatives, and indicate the time intervals in which the derivatives would be positive and negative.

76. HOW DO YOU SEE IT? The attendance for four high school basketball games is given by $s = f(t)$, and the attendance for four high school football games is given by $s = g(t)$, where $t = 1$ corresponds to the first game.

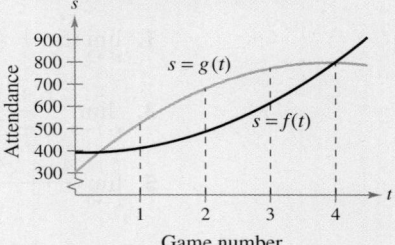

Attendance of High School Sports

Game number

(a) Which attendance rate, f' or g', is greater at game 1?

(b) What conclusion can you make regarding the attendance rates, f' and g', at game 3?

(c) What conclusion can you make regarding the attendance rates, f' and g', at game 4?

(d) Which sport do you think would have a greater attendance for game 5? Explain your reasoning.

77. Cost The marginal cost for manufacturing an electrical component is $7.75 per unit, and the fixed cost is $500. Write the cost C as a function of x, the number of units produced. Show that the derivative of this cost function is a constant and is equal to the marginal cost.

78. Political Fundraiser A politician raises funds by selling tickets to a dinner for $500. The politician pays $150 for each dinner and has fixed costs of $7000 to rent a dining hall and wait staff. Write the profit P as a function of x, the number of dinners sold. Show that the derivative of the profit function is a constant and is equal to the increase in profit from each dinner sold.

Finding Horizontal Tangent Lines In Exercises 79 and 80, use a graphing utility to graph f and f' over the given interval. Determine any points at which the graph of f has horizontal tangents.

Function	Interval
79. $f(x) = 4.1x^3 - 12x^2 + 2.5x$	$[0, 3]$
80. $f(x) = x^3 - 1.4x^2 - 0.96x + 1.44$	$[-2, 2]$

True or False? In Exercises 81 and 82, determine whether the statement is true or false. If it is false, explain why or give an example that shows it is false.

81. If $f'(x) = g'(x)$, then $f(x) = g(x)$.

82. If $f(x) = g(x) + c$, then $f'(x) = g'(x)$.

QUIZ YOURSELF

See www.CalcChat.com for worked-out solutions to odd-numbered exercises.

Take this quiz as you would take a quiz in class. When you are done, check your work against the answers given in the back of the book.

In Exercises 1–6, find the limit (if it exists).

1. $\lim\limits_{x \to 2} (5x + 4)$

2. $\lim\limits_{x \to 3^-} \sqrt{x + 1}$

3. $\lim\limits_{x \to -3} \dfrac{x + 1}{x + 3}$

4. $\lim\limits_{x \to 2} \dfrac{x^2 + 3x - 10}{x - 2}$

5. $\lim\limits_{x \to 0} \dfrac{4 - \sqrt{x + 16}}{x}$

6. $\lim\limits_{x \to 0^+} [\![x]\!]$

In Exercises 7–10, describe the interval(s) on which the function is continuous. Explain why the function is continuous on the interval(s). If the function has a discontinuity, identify the conditions of continuity that are not satisfied.

7. $f(x) = \dfrac{x}{x^2 + 2}$

8. $f(x) = \dfrac{x}{(x + 2)^2}$

9. $f(x) = \dfrac{x + 3}{x^2 + 2x - 3}$

10. $f(x) = \begin{cases} x^2, & x < 0 \\ x^3, & x \geq 0 \end{cases}$

In Exercises 11 and 12, use the limit definition to find the derivative of the function. Then find the slope of the tangent line to the graph of f at the given point.

11. $f(x) = -x + 2;\ (2, 0)$

12. $f(x) = \dfrac{4}{x};\ (1, 4)$

In Exercises 13–18, find the derivative of the function.

13. $f(x) = 12$

14. $f(x) = 19x + 9$

15. $f(x) = 5 - 3x^2$

16. $f(x) = 12x^{1/4}$

17. $f(x) = 4x^{-2}$

18. $f(x) = 2\sqrt{x}$

 In Exercises 19 and 20, find an equation of the tangent line to the graph of f at the given point. Then use a graphing utility to graph the function and the tangent line in the same viewing window.

19. $f(x) = 5x^2 + 6x - 1;\ (-1, -2)$

20. $f(x) = x^{4/3} + x;\ (0, 0)$

21. From 2003 through 2009, the sales per share S (in dollars) for Columbia Sportswear can be modeled by

$$S = -0.13556t^3 + 1.8682t^2 - 4.351t + 23.52, \quad 3 \leq t \leq 9$$

where t represents the year, with $t = 3$ corresponding to 2003. *(Source: Columbia Sportswear Company)*

(a) Find the rate of change of the sales per share with respect to the year.

(b) At what rate were the sales per share changing in 2004? in 2007? in 2008?

7.5 Rates of Change: Velocity and Marginals

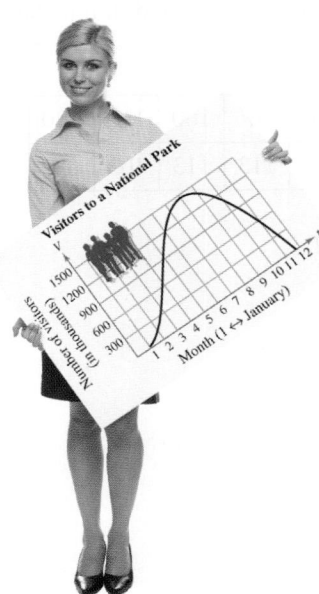

In Exercise 13 on page 598, you will use the graph of a function to estimate the rate of change of the number of visitors to a national park.

■ Find the average rates of change of functions over intervals.
■ Find the instantaneous rates of change of functions at points.
■ Find the marginal revenues, marginal costs, and marginal profits for products.

Average Rate of Change

In Sections 7.3 and 7.4, you studied the two primary applications of derivatives.

1. Slope The derivative of f is a function that gives the slope of the graph of f at a point $(x, f(x))$.

2. Rate of Change The derivative of f is a function that gives the rate of change of $f(x)$ with respect to x at the point $(x, f(x))$.

In this section, you will see that there are many real-life applications of rates of change. A few are velocity, acceleration, population growth rates, unemployment rates, production rates, and water flow rates. Although rates of change often involve change with respect to time, you can investigate the rate of change of one variable with respect to any other related variable.

When determining the rate of change of one variable with respect to another, you must be careful to distinguish between *average* and *instantaneous* rates of change. The distinction between these two rates of change is comparable to the distinction between the slope of the secant line through two points on a graph and the slope of the tangent line at one point on the graph.

STUDY TIP

In real-life problems, it is important to list the units of measure for a rate of change. The units for $\Delta y / \Delta x$ are "y-units" per "x-units." For example, if y is measured in miles and x is measured in hours, then $\Delta y / \Delta x$ is measured in *miles per hour*.

Definition of Average Rate of Change

If $y = f(x)$, then the **average rate of change** of y with respect to x on the interval $[a, b]$ is

$$\text{Average rate of change} = \frac{f(b) - f(a)}{b - a} = \frac{\Delta y}{\Delta x}.$$

Note that $f(a)$ is the value of the function at the *left* endpoint of the interval, $f(b)$ is the value of the function at the *right* endpoint of the interval, and $b - a$ is the width of the interval, as shown in Figure 7.36.

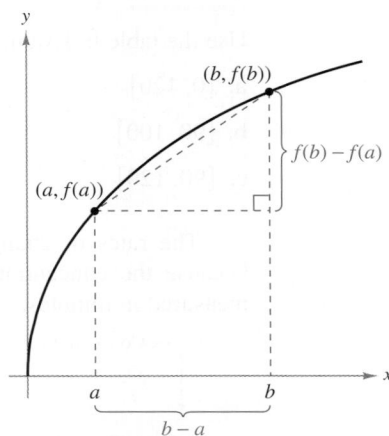

FIGURE 7.36

Example 1   Medicine

The concentration C (in milligrams per milliliter) of a drug in a patient's bloodstream is monitored over 10-minute intervals for 2 hours, where t is measured in minutes, as shown in the table.

t	0	10	20	30	40	50	60	70	80	90	100	110	120
C	0	2	17	37	55	73	89	103	111	113	113	103	68

Find the average rate of change of C over each interval.

a. $[0, 10]$

b. $[0, 20]$

c. $[100, 110]$

SOLUTION

a. For the interval $[0, 10]$, the average rate of change is

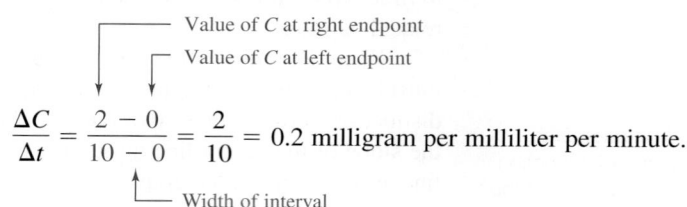

$$\frac{\Delta C}{\Delta t} = \frac{2 - 0}{10 - 0} = \frac{2}{10} = 0.2 \text{ milligram per milliliter per minute.}$$

b. For the interval $[0, 20]$, the average rate of change is

$$\frac{\Delta C}{\Delta t} = \frac{17 - 0}{20 - 0} = \frac{17}{20} = 0.85 \text{ milligram per milliliter per minute.}$$

c. For the interval $[100, 110]$, the average rate of change is

$$\frac{\Delta C}{\Delta t} = \frac{103 - 113}{110 - 100} = \frac{-10}{10} = -1 \text{ milligram per milliliter per minute.}$$

Notice in Figure 7.37 that the average rate of change is positive when the concentration increases and negative when the concentration decreases.

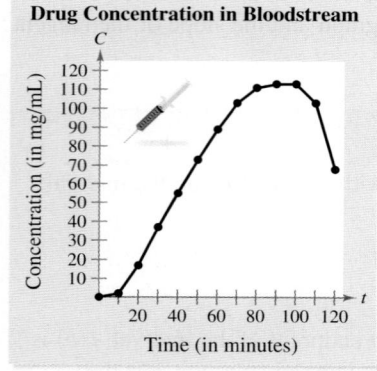

Drug Concentration in Bloodstream

Time (in minutes)

FIGURE 7.37

✓ **Checkpoint 1**

Use the table in Example 1 to find the average rate of change of C over each interval.

a. $[0, 120]$

b. $[90, 100]$

c. $[90, 120]$

The rates of change in Example 1 are in milligrams per milliliter per minute because the concentration is measured in milligrams per milliliter and the time is measured in minutes.

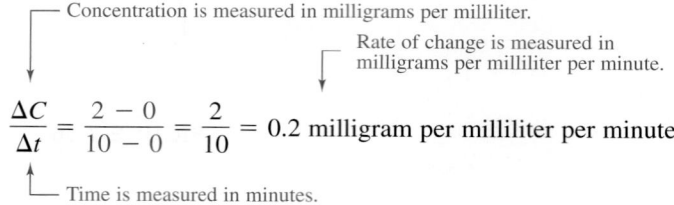

$$\frac{\Delta C}{\Delta t} = \frac{2 - 0}{10 - 0} = \frac{2}{10} = 0.2 \text{ milligram per milliliter per minute}$$

A common application of an average rate of change is to find the **average velocity** of an object that is moving in a straight line. That is,

$$\text{Average velocity} = \frac{\text{change in distance}}{\text{change in time}}.$$

This formula is demonstrated in Example 2.

Example 2 Finding an Average Velocity

A free-falling object is dropped from a height of 100 feet. *Neglecting air resistance,* the height h (in feet) of the object at time t (in seconds) is given by

$$h = -16t^2 + 100. \qquad \text{(See Figure 7.38.)}$$

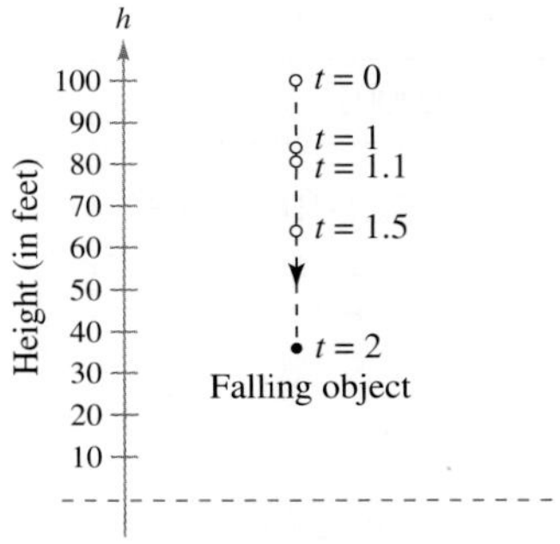

Find the average velocity of the object over each interval.

a. $[1, 2]$ **b.** $[1, 1.5]$ **c.** $[1, 1.1]$

SOLUTION You can use the position equation $h = -16t^2 + 100$ to determine the heights at

$$t = 1, 1.1, 1.5, \text{ and } 2$$

as shown in the table.

t (in seconds)	0	1	1.1	1.5	2
h (in feet)	100	84	80.64	64	36

a. For the interval $[1, 2]$, the object falls from a height of 84 feet to a height of 36 feet. So, the average velocity is

$$\frac{\Delta h}{\Delta t} = \frac{36 - 84}{2 - 1} = \frac{-48}{1} = -48 \text{ feet per second.}$$

b. For the interval $[1, 1.5]$, the average velocity is

$$\frac{\Delta h}{\Delta t} = \frac{64 - 84}{1.5 - 1} = \frac{-20}{0.5} = -40 \text{ feet per second.}$$

c. For the interval $[1, 1.1]$, the average velocity is

$$\frac{\Delta h}{\Delta t} = \frac{80.64 - 84}{1.1 - 1} = \frac{-3.36}{0.1} = -33.6 \text{ feet per second.}$$

✓ **Checkpoint 2**

The height h (in feet) of a free-falling object at time t (in seconds) is given by

$$h = -16t^2 + 180.$$

Find the average velocity of the object over each interval.

a. $[0, 1]$ **b.** $[1, 2]$ **c.** $[2, 3]$

Some falling objects have considerable air resistance. Other falling objects have negligible air resistance. When modeling a falling-body problem, you must decide whether to account for air resistance or neglect it.

FIGURE 7.38

STUDY TIP

In Example 2, the average velocities are negative because the object is moving downward.

Instantaneous Rate of Change and Velocity

Suppose in Example 2 you wanted to find the rate of change of h at the instant $t = 1$ second. Such a rate is called an **instantaneous rate of change.** You can approximate the instantaneous rate of change at $t = 1$ by calculating the average rate of change over smaller and smaller intervals of the form $[1, 1 + \Delta t]$, as shown in the table. From the table, it seems reasonable to conclude that the instantaneous rate of change of the height at $t = 1$ is -32 feet per second.

Δt approaches 0.

Δt	1	0.5	0.1	0.01	0.001	0.0001	0
$\dfrac{\Delta h}{\Delta t}$	-48	-40	-33.6	-32.16	-32.016	-32.0016	-32

$\dfrac{\Delta h}{\Delta t}$ approaches -32.

STUDY TIP

The limit in the definition of instantaneous rate of change is the same as the limit in the definition of the derivative of f at x. This is the second major interpretation of the derivative—as an *instantaneous rate of change in one variable with respect to another*. Recall that the first interpretation of the derivative is as the slope of the graph of f at x.

Definition of Instantaneous Rate of Change

The **instantaneous rate of change** (or simply **rate of change**) of $y = f(x)$ at x is the limit of the average rate of change on the interval

$$[x, x + \Delta x]$$

as Δx approaches 0.

$$\lim_{\Delta x \to 0} \frac{\Delta y}{\Delta x} = \lim_{\Delta x \to 0} \frac{f(x + \Delta x) - f(x)}{\Delta x}$$

If y is a distance and x is time, then the rate of change is a **velocity.**

Example 3　Finding an Instantaneous Rate of Change

Find the velocity of the object in Example 2 at $t = 1$.

SOLUTION　From Example 2, you know that the height of the falling object is given by

$$h = -16t^2 + 100. \qquad \text{Position function}$$

By taking the derivative of this position function, you obtain the velocity function.

$$h'(t) = -32t \qquad \text{Velocity function}$$

The velocity function gives the velocity at *any* time. So, at $t = 1$, the velocity is

$$h'(1) = -32(1)$$
$$= -32 \text{ feet per second.}$$

✓**Checkpoint 3**

The height of the object in Checkpoint 2 is given by

$$h = -16t^2 + 180.$$

Find the velocities of the object at

a. $t = 1.75$.

b. $t = 2$.

The general **position function** for a free-falling object, neglecting air resistance, is

$$h = -16t^2 + v_0 t + h_0 \qquad \text{Position function}$$

where h is the height (in feet), t is the time (in seconds), v_0 is the initial velocity (in feet per second), and h_0 is the initial height (in feet). Remember that the model assumes that positive velocities indicate upward motion and negative velocities indicate downward motion. The derivative

$$h' = -32t^2 + v_0 \qquad \text{Velocity function}$$

is the **velocity function.** The absolute value of the velocity is the **speed** of the object.

 Example 4   **Finding the Velocity of a Diver**

At time $t = 0$, a diver jumps from a diving board that is 32 feet high, as shown in Figure 7.39. Because the diver's initial velocity is 16 feet per second, the position of the diver is given by

$$h = -16t^2 + 16t + 32. \qquad \text{Position function}$$

a. When does the diver hit the water?

b. What is the diver's velocity at impact?

SOLUTION

a. To find the time at which the diver hits the water, let $h = 0$ and solve for t.

$$-16t^2 + 16t + 32 = 0 \qquad \text{Set } h \text{ equal to 0.}$$
$$-16(t^2 - t - 2) = 0 \qquad \text{Factor out common factor.}$$
$$-16(t + 1)(t - 2) = 0 \qquad \text{Factor.}$$
$$t = -1 \ \text{ or } \ t = 2 \qquad \text{Solve for } t.$$

The solution $t = -1$ does not make sense in the problem because it would mean that the diver hits the water 1 second before jumping. So, you can conclude that the diver hits the water at $t = 2$ seconds.

b. The velocity at time t is given by the derivative

$$h' = -32t + 16. \qquad \text{Velocity function}$$

The velocity at time $t = 2$ is

$$h' = -32(2) + 16 = -48 \text{ feet per second.}$$

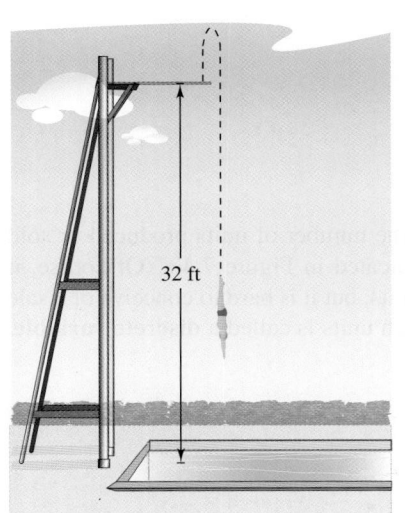

FIGURE 7.39

✓**Checkpoint 4**

At time $t = 0$, a diver jumps from a diving board that is 12 feet high with initial velocity 16 feet per second. The diver's position function is $h = -16t^2 + 16t + 12$.

a. When does the diver hit the water?

b. What is the diver's velocity at impact?

In Example 4, note that the diver's initial velocity is $v_0 = 16$ feet per second (upward) and the diver's initial height is $h_0 = 32$ feet.

Initial velocity is 16 feet per second.
Initial height is 32 feet.

$$h = -16t^2 + 16t + 32$$

Rates of Change in Economics: Marginals

Another important use of rates of change is in the field of economics. Economists refer to *marginal profit, marginal revenue,* and *marginal cost* as the rates of change of the profit, revenue, and cost with respect to x, the number of units produced or sold. An equation that relates these three quantities is

$$P = R - C$$

where P, R, and C represent the following quantities.

$$P = \text{total profit}, \quad R = \text{total revenue}, \quad \text{and} \quad C = \text{total cost}$$

The derivatives of these quantities are called the **marginal profit, marginal revenue,** and **marginal cost,** respectively.

$$\frac{dP}{dx} = \text{marginal profit}$$

$$\frac{dR}{dx} = \text{marginal revenue}$$

$$\frac{dC}{dx} = \text{marginal cost}$$

In many business and economics problems, the number of units produced or sold is restricted to nonnegative integer values, as indicated in Figure 7.40. (Of course, it could happen that a sale involves half or quarter units, but it is hard to conceive of a sale involving $\sqrt{2}$ units.) The variable that denotes such units is called a **discrete variable.**

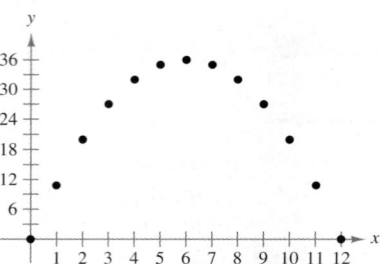

Function of a Discrete Variable

FIGURE 7.40

To analyze a function of a discrete variable x, you can temporarily assume that x is a **continuous variable** and is able to take on any real value in a given interval, as indicated in Figure 7.41. Then, you can use the methods of calculus to find the x-value that corresponds to the marginal revenue, maximum profit, minimum cost, or whatever is called for. Finally, you should round the solution to the nearest sensible x-value— cents, dollars, units, or days, depending on the context of the problem.

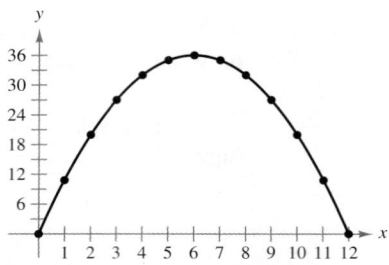

Function of a Continuous Variable

FIGURE 7.41

Example 5   Finding the Marginal Profit

The profit derived from selling x units of an alarm clock is given by

$$P = 0.0002x^3 + 10x.$$

a. Find the marginal profit for a production level of 50 units.

b. Compare the marginal profit with the actual gain in profit obtained by increasing the production level from 50 to 51 units.

SOLUTION

a. The profit is $P = 0.0002x^3 + 10x$. The marginal profit is given by the derivative

$$\frac{dP}{dx} = 0.0006x^2 + 10.$$

When $x = 50$, the marginal profit is

$$\frac{dP}{dx} = 0.0006(50)^2 + 10 \qquad \text{Substitute 50 for } x.$$
$$= 0.0006(2500) + 10$$
$$= 1.5 + 10$$
$$= \$11.50 \text{ per unit.} \qquad \text{Marginal profit for } x = 50$$

b. For $x = 50$, the actual profit is

$$P = 0.0002(50)^3 + 10(50) \qquad \text{Substitute 50 for } x.$$
$$= 0.0002(125,000) + 500$$
$$= 25 + 500$$
$$= \$525.00 \qquad \text{Actual profit for } x = 50$$

and for $x = 51$, the actual profit is

$$P = 0.0002(51)^3 + 10(51) \qquad \text{Substitute 51 for } x.$$
$$= 0.0002(132,651) + 510$$
$$\approx 26.53 + 510$$
$$= \$536.53. \qquad \text{Actual profit for } x = 51$$

So, the additional profit obtained by increasing production from 50 to 51 units is

$$536.53 - 525.00 = \$11.53. \qquad \text{Extra profit for one unit}$$

Note that the actual profit increase of \$11.53 (when x increases from 50 to 51 units) can be approximated by the marginal profit of \$11.50 per unit (when $x = 50$), as shown in Figure 7.42.

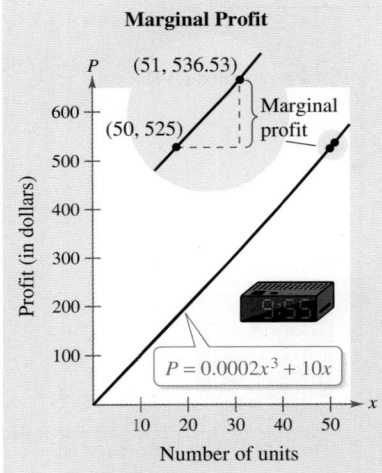

Marginal Profit

P (51, 536.53)

600 — Marginal
(50, 525) — profit

500

400

300

200

100

$P = 0.0002x^3 + 10x$

10   20   30   40   50

Number of units

Profit (in dollars)

FIGURE 7.42

✓ **Checkpoint 5**

Find the marginal profit in Example 5 for a production level of 100 units. Compare this with the actual gain in profit by increasing production from 100 to 101 units. ■

STUDY TIP

In Example 5, the marginal profit gives a good approximation of the actual change in profit because the graph of P is nearly straight over the interval $50 \le x \le 51$. You will study more about the use of marginals to approximate actual changes in Section 9.5.

The profit function in Example 5 is unusual in that the profit continues to increase as long as the number of units sold increases. In practice, it is more common to encounter situations in which sales can be increased only by lowering the price per item. Such reductions in price will ultimately cause the profit to decline.

The number of units x that consumers are willing to purchase at a given price per unit p is given by the **demand function**

$$p = f(x).$$ Demand function

The total revenue R is then related to the price per unit and the quantity demanded (or sold) by the equation

$$R = xp.$$ Revenue function

 Example 6   **Finding a Demand Function**

The table shows the numbers x (in millions) of prerecorded high-definition DVDs sold in the United States and the average unit prices p (in dollars) from 2006 through 2009. Use this information to find the demand function and the total revenue function. *(Source: SNL Kagan)*

Year	2006	2007	2008	2009
x	1.2	9.8	22.7	54.0
p	23.75	23.38	22.21	20.43

SOLUTION   Begin by making a scatter plot of the data using the ordered pairs (x, p), as shown in Figure 7.43. From the graph, it appears that a linear model would be a good fit for the data. To find a linear model for the demand function, use any two points, such as $(1.2, 23.75)$ and $(54.0, 20.43)$. The slope of the line through these points is

$$m = \frac{20.43 - 23.75}{54.0 - 1.2}$$

$$= \frac{-3.32}{52.8}$$

$$\approx -0.063.$$

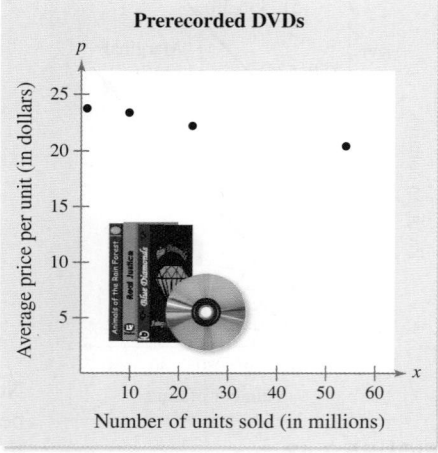

FIGURE 7.43

Using the point-slope form of a line, you can approximate the equation of the demand function to be $p = -0.063x + 23.83$. The total revenue function for prerecorded high-definition DVDs is

$$R = xp = x(-0.063x + 23.83) = -0.063x^2 + 23.83x.$$

 **Checkpoint 6**

Repeat Example 6 given that in 2010 an estimated 104.2 million prerecorded high-definition DVDs were sold at an average unit price of $17.88. To find a linear model for the demand function, use the points $(1.2, 23.75)$ and $(104.2, 17.88)$. *(Source: SNL Kagan)*

Example 7 **Finding the Marginal Revenue**

A fast-food restaurant has determined that the monthly demand for its hamburgers is given by

$$p = \frac{60,000 - x}{20,000}.$$

Figure 7.44 shows that as the price decreases, the quantity demanded increases. The table shows the demands for hamburgers at various prices.

x	60,000	50,000	40,000	30,000	20,000	10,000	0
p	$0.00	$0.50	$1.00	$1.50	$2.00	$2.50	$3.00

Find the increase in revenue per hamburger for monthly sales of 20,000 hamburgers. In other words, find the marginal revenue when $x = 20,000$.

SOLUTION Because the demand is given by

$$p = \frac{60,000 - x}{20,000}$$

and the revenue is given by $R = xp$, you have

$$R = xp \qquad \text{Formula for revenue}$$

$$= x\left(\frac{60,000 - x}{20,000}\right) \qquad \text{Substitute for } p.$$

$$= \frac{1}{20,000}(60,000x - x^2). \qquad \text{Revenue function}$$

By differentiating, you can find the marginal revenue to be

$$\frac{dR}{dx} = \frac{1}{20,000}(60,000 - 2x).$$

So, at $x = 20,000$, the marginal revenue is

$$\frac{dR}{dx} = \frac{1}{20,000}(60,000 - 2x) \qquad \text{Marginal revenue}$$

$$= \frac{1}{20,000}[60,000 - 2(20,000)] \qquad \text{Substitute 20,000 for } x.$$

$$= \frac{1}{20,000}(60,000 - 40,000) \qquad \text{Multiply.}$$

$$= \frac{1}{20,000}(20,000) \qquad \text{Subtract.}$$

$$= \$1 \text{ per unit.} \qquad \text{Marginal revenue when } x = 20,000$$

So, for monthly sales of 20,000 hamburgers, you can conclude that the increase in revenue per hamburger is $1.

✓**Checkpoint 7**

Find the revenue function and marginal revenue for a demand function of

$$p = 2000 - 4x.$$

Find the marginal revenue when $x = 250$.

Demand Function

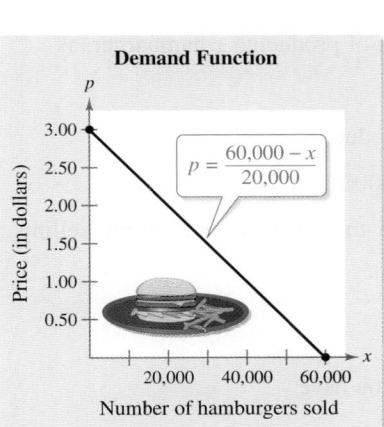

As the price decreases, more hamburgers are sold.

FIGURE 7.44

STUDY TIP

Writing a demand function in the form $p = f(x)$ is a convention used in economics. From a consumer's point of view, it might seem more reasonable to think that the quantity demanded is a function of the price. Mathematically, however, the two points of view are equivalent because a typical demand function is one-to-one and so has an inverse function. For instance, in Example 7, you could write the demand function as $x = 60,000 - 20,000p$.

Example 8 **Finding the Marginal Profit**

For the fast-food restaurant in Example 7, the cost of producing x hamburgers is

$$C = 5000 + 0.56x, \quad 0 \le x \le 50{,}000.$$

Find the profit and the marginal profit for each production level.

a. $x = 20{,}000$ **b.** $x = 24{,}400$ **c.** $x = 30{,}000$

SOLUTION From Example 7, you know that the total revenue from selling x hamburgers is

$$R = \frac{1}{20{,}000}(60{,}000x - x^2).$$

Because the total profit is given by $P = R - C$, you have

$$P = \frac{1}{20{,}000}(60{,}000x - x^2) - (5000 + 0.56x)$$

$$= 3x - \frac{x^2}{20{,}000} - 5000 - 0.56x$$

$$= 2.44x - \frac{x^2}{20{,}000} - 5000. \qquad \text{See Figure 7.45.}$$

So, the marginal profit is

$$\frac{dP}{dx} = 2.44 - \frac{x}{10{,}000}.$$

Using these formulas, you can compute the profit and marginal profit.

Production	Profit	Marginal Profit
a. $x = 20{,}000$	$P = \$23{,}800.00$	$2.44 - \dfrac{20{,}000}{10{,}000} = \$0.44$ per unit
b. $x = 24{,}400$	$P = \$24{,}768.00$	$2.44 - \dfrac{24{,}400}{10{,}000} = \$0.00$ per unit
c. $x = 30{,}000$	$P = \$23{,}200.00$	$2.44 - \dfrac{30{,}000}{10{,}000} = -\$0.56$ per unit

Profit Function

$P = 2.44x - \dfrac{x^2}{20{,}000} - 5000$

FIGURE 7.45

✓**Checkpoint 8**

From Example 8, compare the marginal profit when 10,000 units are produced with the actual increase in profit from 10,000 units to 10,001 units. ■

SUMMARIZE (Section 7.5)

1. State the definition of average rate of change *(page 587)*. For examples of average rate of change, see Examples 1 and 2.

2. State the definition of instantaneous rate of change *(page 590)*. For examples of an instantaneous rate of change, see Examples 3 and 4.

3. Describe a real-life example of how rates of change can be used in the field of economics *(pages 593–596, Examples 5, 6, 7, and 8)*.

SKILLS WARM UP 7.5

The following warm-up exercises involve skills that were covered in earlier sections. You will use these skills in the exercise set for this section. For additional help, review Sections 0.2, 7.3, and 7.4.

In Exercises 1–4, evaluate the expression.

1. $\dfrac{-63 - (-105)}{21 - 7}$

2. $\dfrac{-37 - 54}{16 - 3}$

3. $\dfrac{24 - 33}{9 - 6}$

4. $\dfrac{40 - 16}{18 - 8}$

In Exercises 5–12, find the derivative of the function.

5. $y = 4x^2 - 2x + 7$

6. $y = -3t^3 + 2t^2 - 8$

7. $s = -16t^2 + 24t + 30$

8. $y = -16x^2 + 54x + 70$

9. $A = \frac{1}{10}(-2r^3 + 3r^2 + 5r)$

10. $y = \frac{1}{9}(6x^3 - 18x^2 + 63x - 15)$

11. $y = 12x - \dfrac{x^2}{5000}$

12. $y = 138 + 74x - \dfrac{x^3}{10,000}$

Exercises 7.5

See www.CalcChat.com for worked-out solutions to odd-numbered exercises.

1. Research and Development The table shows the amounts A (in billions of dollars) spent on Research and Development in the United States from 1980 through 2008, where t is the year, with $t = 0$ corresponding to 1980. Approximate the average rate of change of A during each period. *(Source: U.S. National Science Foundation)*

(a) 1980–1985   (b) 1985–1990

(c) 1990–1995   (d) 1995–2000

(e) 2000–2005   (f) 1980–2008

(g) 1990–2008   (h) 2000–2008

t	0	1	2	3	4	5	6	7
A	63	72	81	90	102	115	120	126

t	8	9	10	11	12	13	14
A	134	142	152	161	165	166	169

t	15	16	17	18	19	20	21
A	184	197	212	226	245	267	277

t	22	23	24	25	26	27	28
A	276	288	299	322	347	373	398

2. Trade Deficit The graph shows the values I (in billions of dollars) of goods imported to the United States and the values E (in billions of dollars) of goods exported from the United States from 1980 through 2009. Approximate the average rates of change of I and E during each period. *(Source: U.S. International Trade Administration)*

(a) Imports: 1980–1990   (b) Exports: 1980–1990

(c) Imports: 1990–2000   (d) Exports: 1990–2000

(e) Imports: 2000–2009   (f) Exports: 2000–2009

(g) Imports: 1980–2009   (h) Exports: 1980–2009

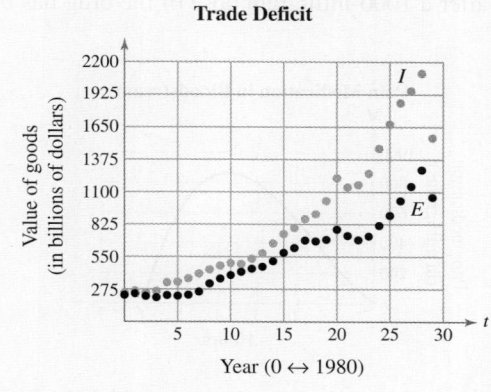

Finding Rates of Change   In Exercises 3–12, use a graphing utility to graph the function and find its average rate of change over the interval. Compare this rate with the instantaneous rates of change at the endpoints of the interval.

3. $f(t) = 3t + 5; [1, 2]$ **4.** $h(x) = 2 - x; [0, 2]$

5. $h(x) = x^2 - 4x + 2; [-2, 2]$

6. $f(x) = x^2 - 6x - 1; [-1, 3]$

7. $f(x) = 3x^{4/3}; [1, 8]$ **8.** $f(x) = x^{3/2}; [1, 4]$

9. $f(x) = \dfrac{1}{x}; [1, 4]$ **10.** $f(x) = \dfrac{1}{\sqrt{x}}; [1, 4]$

11. $g(x) = x^4 - x^2 + 2; [1, 3]$

12. $g(x) = x^3 - 1; [-1, 1]$

13. Consumer Trends   The graph shows the number of visitors V (in thousands) to a national park during a one-year period, where $t = 1$ represents January.

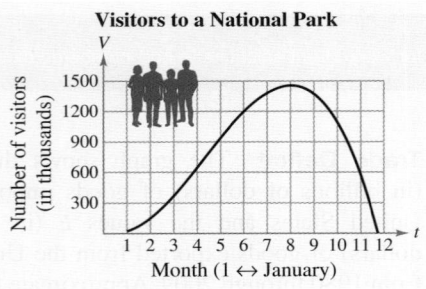

Visitors to a National Park

Month (1 ↔ January)

(a) Estimate the rate of change of V over the interval $[9, 12]$ and explain your results.

(b) Over what interval is the average rate of change approximately equal to the rate of change at $t = 8$? Explain your reasoning.

14. Medicine   The graph shows the estimated number of milligrams of a pain medication M in the bloodstream t hours after a 1000-milligram dose of the drug has been given.

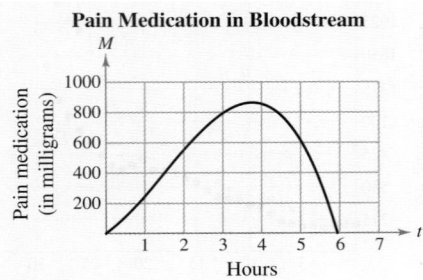

Pain Medication in Bloodstream

Hours

(a) Estimate the one-hour interval over which the average rate of change is the greatest.

(b) Over what interval is the average rate of change approximately equal to the rate of change at $t = 4$? Explain your reasoning.

15. Velocity   The height s (in feet) at time t (in seconds) of a ball thrown upward from the top of a building is given by

$$s = -16t^2 + 30t + 250.$$

Find the average velocity over each indicated interval and compare this velocity with the instantaneous velocity at the endpoints of the interval.

(a) $[0, 1]$ (b) $[1, 2]$ (c) $[2, 3]$ (d) $[3, 4]$

16. Chemistry: Wind Chill   At $0°$ Celsius, the heat loss H (in kilocalories per square meter per hour) from a person's body can be modeled by

$$H = 33\left(10\sqrt{v} - v + 10.45\right)$$

where v is the wind speed (in meters per second).

(a) Find $\dfrac{dH}{dv}$ and interpret its meaning in this situation.

(b) Find the rates of change of H when $v = 2$ and $v = 5$.

17. Velocity   The height s (in feet) at time t (in seconds) of a silver dollar dropped from the top of a building is given by $s = -16t^2 + 555$.

(a) Find the average velocity over the interval $[2, 3]$.

(b) Find the instantaneous velocities when $t = 2$ and $t = 3$.

(c) How long will it take the coin to hit the ground?

(d) Find the velocity of the coin when it hits the ground.

18. Velocity   A ball is thrown straight down from the top of a 210-foot building with an initial velocity of -18 feet per second.

(a) Find the position and velocity functions for the ball.

(b) Find the average velocity over the interval $[1, 2]$.

(c) Find the instantaneous velocities when $t = 1$ and $t = 2$.

(d) How long will it take the ball to hit the ground?

(e) Find the velocity of the ball when it hits the ground.

Marginal Cost   In Exercises 19–22, find the marginal cost for producing x units. (The cost is measured in dollars.)

19. $C = 205,000 + 9800x$

20. $C = 150,000 + 7x^3$

21. $C = 55,000 + 470x - 0.25x^2, \quad 0 \le x \le 940$

22. $C = 100\left(9 + 3\sqrt{x}\right)$

Marginal Revenue   In Exercises 23–26, find the marginal revenue for producing x units. (The revenue is measured in dollars.)

23. $R = 50x - 0.5x^2$ **24.** $R = 30x - x^2$

25. $R = -6x^3 + 8x^2 + 200x$

26. $R = 50(20x - x^{3/2})$

Marginal Profit In Exercises 27–30, find the marginal profit for producing x units. (The profit is measured in dollars.)

27. $P = -2x^2 + 72x - 145$

28. $P = -0.25x^2 + 2000x - 1,250,000$

29. $P = 0.0013x^3 + 12x$

30. $P = -0.5x^3 + 30x^2 - 164.25x - 1000$

31. **Marginal Cost** The cost C (in dollars) of producing x units of a product is given by

$$C = 3.6\sqrt{x} + 500.$$

(a) Find the additional cost when the production increases from 9 to 10 units.

(b) Find the marginal cost when $x = 9$.

(c) Compare the results of parts (a) and (b).

32. **Marginal Revenue** The revenue R (in dollars) from renting x apartments can be modeled by

$$R = 2x(900 + 32x - x^2).$$

(a) Find the additional revenue when the number of rentals is increased from 14 to 15.

(b) Find the marginal revenue when $x = 14$.

(c) Compare the results of parts (a) and (b).

33. **Marginal Profit** The profit P (in dollars) from selling x laptop computers is given by

$$P = -0.04x^2 + 25x - 1500.$$

(a) Find the additional profit when the sales increase from 150 to 151 units.

(b) Find the marginal profit when $x = 150$.

(c) Compare the results of parts (a) and (b).

34. **Marginal Profit** The profit P (in dollars) from selling x units of a product is given by

$$P = 36,000 + 2048\sqrt{x} - \frac{1}{8x^2}, \quad 150 \le x \le 275.$$

Find the marginal profit for each of the following sales.

(a) $x = 150$ (b) $x = 175$ (c) $x = 200$

(d) $x = 225$ (e) $x = 250$ (f) $x = 275$

35. **Population Growth** The population P (in thousands) of Japan from 1980 through 2010 can be modeled by

$$P = -15.56t^2 + 802.1t + 117,001$$

where t is the year, with $t = 0$ corresponding to 1980. *(Source: U.S. Census Bureau)*

(a) Evaluate P for $t = 0$, 5, 10, 15, 20, 25, and 30. Explain these values.

(b) Determine the population growth rate, dP/dt.

(c) Evaluate dP/dt for the same values as in part (a). Explain your results.

36. **Health** The temperature T (in degrees Fahrenheit) of a person during an illness can be modeled by the equation

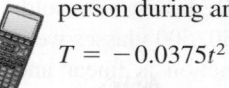

$$T = -0.0375t^2 + 0.3t + 100.4$$

where t is time in hours since the person started to show signs of a fever.

(a) Use a graphing utility to graph the function. Be sure to choose an appropriate window.

(b) Do the slopes of the tangent lines appear to be positive or negative? What does this tell you?

(c) Evaluate the function for $t = 0$, 4, 8, and 12.

(d) Find dT/dt and explain its meaning in this situation.

(e) Evaluate dT/dt for $t = 0$, 4, 8, and 12.

37. **Economics** Use the information in the table to find the models and answer the questions below.

Quantity produced and sold (Q)	Price (p)	Total revenue (TR)	Marginal revenue (MR)
0	160	0	—
2	140	280	130
4	120	480	90
6	100	600	50
8	80	640	10
10	60	600	−30

(a) Use the *regression* feature of a graphing utility to find a quadratic model that relates the total revenue (TR) to the quantity produced and sold (Q).

(b) Using derivatives, find a model for marginal revenue from the model you found in part (a).

(c) Calculate the marginal revenue for all values of Q using your model in part (b), and compare these values with the actual values given. How good is your model? *(Source: Adapted from Taylor, Economics, Fifth Edition)*

38. **Profit** The monthly demand function p and cost function C for x newspapers at a newsstand are given by

$$p = 5 - 0.001x \quad \text{and} \quad C = 35 + 1.5x.$$

(a) Find the monthly revenue R as a function of x.

(b) Find the monthly profit P as a function of x.

(c) Complete the table.

x	600	1200	1800	2400	3000
dR/dx					
dP/dx					
P					

39. Marginal Profit When the price of a glass of lemonade  at a lemonade stand was $1.75, 400 glasses were sold. When the price was lowered to $1.50, 500 glasses were sold. Assume that the demand function is linear and that the marginal and fixed costs are $0.10 and $25, respectively.

(a) Find the profit P as a function of x, the number of glasses of lemonade sold.

(b) Use a graphing utility to graph P, and comment about the slopes of P when $x = 300$ and $x = 700$.

(c) Find the marginal profits when 300 glasses of lemonade are sold and when 700 glasses of lemonade are sold.

40. Marginal Profit When the admission price for a  baseball game was $6 per ticket, 36,000 tickets were sold. When the price was raised to $7, only 33,000 tickets were sold. Assume that the demand function is linear and that the marginal and fixed costs for the ballpark owners are $0.20 and $85,000, respectively.

(a) Find the profit P as a function of x, the number of tickets sold.

(b) Use a graphing utility to graph P, and comment about the slopes of P when $x = 18,000$ and $x = 36,000$.

(c) Find the marginal profits when 18,000 tickets are sold and when 36,000 tickets are sold.

41. Fuel Cost A car is driven 15,000 miles a year and gets x miles per gallon. Assume that the average fuel cost is $2.95 per gallon.

(a) Find the annual cost of fuel C as a function of x.

(b) Find dC/dx and explain its meaning in this situation.

(c) Use the functions to complete the table.

x	10	15	20	25	30	35	40
C							
dC/dx							

(d) Who would benefit more from a 1 mile per gallon increase in fuel efficiency—the driver who gets 15 miles per gallon or the driver who gets 35 miles per gallon? Explain.

42. Gasoline Sales The number N of gallons of regular unleaded gasoline sold by a gasoline station at a price of p dollars per gallon is given by $N = f(p)$.

(a) Describe the meaning of $f'(2.959)$.

(b) Is $f'(2.959)$ usually positive or negative? Explain.

 43. Dow Jones Industrial Average The table shows the year-end closing prices p of the Dow Jones Industrial Average (DJIA) from 1995 through 2009, where t is the year, with $t = 5$ corresponding to 1995. *(Source: Dow Jones Industrial Average)*

t	5	6	7	8
p	5117.12	6448.26	7908.24	9181.43

t	9	10	11	12
p	11,497.12	10,786.85	10,021.50	8341.63

t	13	14	15	16
p	10,453.92	10,783.01	10,717.50	12,463.15

t	17	18	19
p	13,264.82	8776.39	10,428.05

(a) Determine the average rate of change in the value of the DJIA from 1995 through 2009.

(b) Estimate the instantaneous rate of change in 1998 by finding the average rate of change from 1996 to 2000.

(c) Estimate the instantaneous rate of change in 1998 by finding the average rate of change from 1997 to 1999.

(d) Compare your answers for parts (b) and (c). Which interval do you think produced the best estimate for the instantaneous rate of change in 1998?

 **44.** **HOW DO YOU SEE IT?** Many populations in nature exhibit logistic growth, which consists of four phases, as shown in the figure. Describe the rate of growth of the population in each phase, and give possible reasons as to why the rates might be changing from phase to phase. *(Source: Adapted from Levine/Miller, Biology: Discovering Life, Second Edition)*

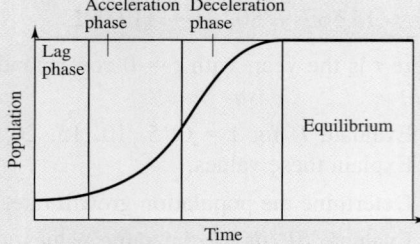

7.6 The Product and Quotient Rules

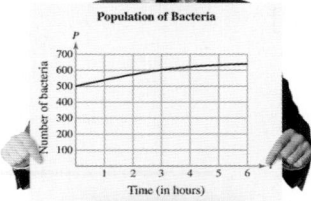

Population of Bacteria

■ Find the derivatives of functions using the Product Rule.
■ Find the derivatives of functions using the Quotient Rule.
■ Use derivatives to answer questions about real-life situations.

The Product Rule

In Section 7.4, you saw that the derivative of a sum or difference of two functions is simply the sum or difference of their derivatives. The rules for the derivative of a product or quotient of two functions are not as simple.

The Product Rule

The derivative of the product of two differentiable functions is equal to the first function times the derivative of the second plus the second function times the derivative of the first.

$$\frac{d}{dx}[f(x)g(x)] = f(x)g'(x) + g(x)f'(x)$$

In Exercise 63 on page 609, you will use the Quotient Rule to find the rate of change of a population of bacteria.

PROOF Some mathematical proofs, such as the proof of the Sum Rule, are straight-forward. Others involve clever steps that may not appear to follow clearly from a prior step. The proof below involves such a step—adding and subtracting the same quantity. (This step is shown in color.) Let $F(x) = f(x)g(x)$.

$$F'(x) = \lim_{\Delta x \to 0} \frac{F(x + \Delta x) - F(x)}{\Delta x}$$

$$= \lim_{\Delta x \to 0} \frac{f(x + \Delta x)g(x + \Delta x) - f(x)g(x)}{\Delta x}$$

$$= \lim_{\Delta x \to 0} \frac{f(x + \Delta x)g(x + \Delta x) - f(x + \Delta x)g(x) + f(x + \Delta x)g(x) - f(x)g(x)}{\Delta x}$$

$$= \lim_{\Delta x \to 0} \left[f(x + \Delta x) \frac{g(x + \Delta x) - g(x)}{\Delta x} + g(x) \frac{f(x + \Delta x) - f(x)}{\Delta x} \right]$$

$$= \lim_{\Delta x \to 0} f(x + \Delta x) \frac{g(x + \Delta x) - g(x)}{\Delta x} + \lim_{\Delta x \to 0} g(x) \frac{f(x + \Delta x) - f(x)}{\Delta x}$$

$$= \lim_{\Delta x \to 0} f(x + \Delta x) \cdot \lim_{\Delta x \to 0} \frac{g(x + \Delta x) - g(x)}{\Delta x} + \lim_{\Delta x \to 0} g(x) \cdot \lim_{\Delta x \to 0} \frac{f(x + \Delta x) - f(x)}{\Delta x}$$

$$= f(x)g'(x) + g(x)f'(x)$$

STUDY TIP

Rather than trying to remember the formula for the Product Rule, it can be more helpful to remember its verbal statement:

the first function times the derivative of the second plus the second function times the derivative of the first.

Example 1   Using the Product Rule

Find the derivative of $y = (3x - 2x^2)(5 + 4x)$.

SOLUTION   Using the Product Rule, you can write

$$\frac{dy}{dx} = \overbrace{(3x - 2x^2)}^{\text{First}} \overbrace{\frac{d}{dx}[5 + 4x]}^{\text{Derivative of second}} + \overbrace{(5 + 4x)}^{\text{Second}} \overbrace{\frac{d}{dx}[3x - 2x^2]}^{\text{Derivative of first}}$$

$$= (3x - 2x^2)(4) + (5 + 4x)(3 - 4x)$$

$$= (12x - 8x^2) + (15 - 8x - 16x^2)$$

$$= 15 + 4x - 24x^2.$$

✓ **Checkpoint 1**

Find the derivative of $y = (4x + 3x^2)(6 - 3x)$.

In general, the derivative of the product of two functions is not equal to the product of the derivatives of the two functions. To see this, compare the product of the derivatives of

$$f(x) = 3x - 2x^2 \quad \text{and} \quad g(x) = 5 + 4x$$

with the derivative found in Example 1.

In the next example, notice that the first step in differentiating is *rewriting the original function.*

TECH TUTOR

If you have access to a symbolic differentiation utility, try using it to confirm several of the derivatives in this section.

Example 2   Using the Product Rule

Find the derivative of $f(x) = \left(\dfrac{1}{x} + 1\right)(x - 1)$.

SOLUTION   Rewrite the function. Then use the Product Rule to find the derivative.

$$f(x) = \left(\frac{1}{x} + 1\right)(x - 1) \qquad\qquad \text{Write original function.}$$

$$= (x^{-1} + 1)(x - 1) \qquad\qquad \text{Rewrite function.}$$

$$f'(x) = (x^{-1} + 1)\frac{d}{dx}[x - 1] + (x - 1)\frac{d}{dx}[x^{-1} + 1] \qquad \text{Product Rule}$$

$$= (x^{-1} + 1)(1) + (x - 1)(-x^{-2})$$

$$= \frac{1}{x} + 1 - \frac{x - 1}{x^2}$$

$$= \frac{x + x^2 - x + 1}{x^2} \qquad\qquad \text{Write with common denominator.}$$

$$= \frac{x^2 + 1}{x^2} \qquad\qquad \text{Simplify.}$$

✓ **Checkpoint 2**

Find the derivative of

$$f(x) = \left(\frac{1}{x} + 1\right)(2x + 1).$$

You now have two differentiation rules that deal with products—the Constant Multiple Rule and the Product Rule. The difference between these two rules is that the Constant Multiple Rule is used when one of the factors is a constant

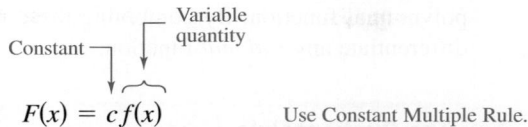

$$F(x) = c\,f(x) \qquad \text{Use Constant Multiple Rule.}$$

whereas the Product Rule is used when both of the factors are variable quantities

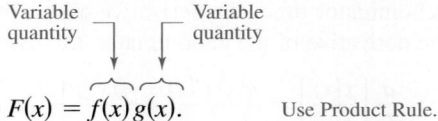

$$F(x) = f(x)\,g(x). \qquad \text{Use Product Rule.}$$

The next example compares these two rules.

Example 3   Comparing Differentiation Rules

Find the derivative of each function.

a. $y = 2x(x^2 + 3x)$

b. $y = 2(x^2 + 3x)$

SOLUTION

a. Because both factors are variable quantities, use the Product Rule.

$$y = 2x(x^2 + 3x)$$

$$\frac{dy}{dx} = (2x)\frac{d}{dx}[x^2 + 3x] + (x^2 + 3x)\frac{d}{dx}[2x] \qquad \text{Product Rule}$$

$$= (2x)(2x + 3) + (x^2 + 3x)(2)$$

$$= 4x^2 + 6x + 2x^2 + 6x$$

$$= 6x^2 + 12x$$

b. Because one of the factors is a constant, use the Constant Multiple Rule.

$$y = 2(x^2 + 3x)$$

$$\frac{dy}{dx} = 2\frac{d}{dx}[x^2 + 3x] \qquad \text{Constant Multiple Rule}$$

$$= 2(2x + 3)$$

$$= 4x + 6$$

> **STUDY TIP**
>
> You could calculate the derivative in Example 3(a) without the Product Rule. For instance,
>
> $$y = 2x(x^2 + 3x) = 2x^3 + 6x^2$$
>
> and
>
> $$\frac{dy}{dx} = 6x^2 + 12x.$$

✓ **Checkpoint 3**

Find the derivative of each function.

a. $y = 3x(2x^2 + 5x)$

b. $y = 3(2x^2 + 5x)$

The Product Rule can be extended to products that have more than two factors. For example, if f, g, and h are differentiable functions of x, then

$$\frac{d}{dx}[f(x)g(x)h(x)] = f'(x)g(x)h(x) + f(x)g'(x)h(x) + f(x)g(x)h'(x).$$

The Quotient Rule

In Section 7.4, you saw that by using the Constant Rule, the Power Rule, the Constant Multiple Rule, and the Sum and Difference Rules, you were able to differentiate any polynomial function. By combining these rules with the Quotient Rule, you can now differentiate any *rational* function.

The Quotient Rule

The derivative of the quotient of two differentiable functions is equal to the denominator times the derivative of the numerator minus the numerator times the derivative of the denominator, all divided by the square of the denominator.

$$\frac{d}{dx}\left[\frac{f(x)}{g(x)}\right] = \frac{g(x)f'(x) - f(x)g'(x)}{[g(x)]^2}, \quad g(x) \neq 0$$

PROOF   Begin by letting

$$F(x) = \frac{f(x)}{g(x)}.$$

As in the proof of the Product Rule, a key step in this proof is adding and subtracting the same quantity.

$$F'(x) = \lim_{\Delta x \to 0} \frac{F(x + \Delta x) - F(x)}{\Delta x}$$

$$= \lim_{\Delta x \to 0} \frac{\dfrac{f(x + \Delta x)}{g(x + \Delta x)} - \dfrac{f(x)}{g(x)}}{\Delta x}$$

$$= \lim_{\Delta x \to 0} \left[\left(\frac{g(x)f(x + \Delta x)}{g(x)g(x + \Delta x)} - \frac{f(x)g(x + \Delta x)}{g(x)g(x + \Delta x)}\right) \div \Delta x\right]$$

$$= \lim_{\Delta x \to 0} \left[\frac{g(x)f(x + \Delta x) - f(x)g(x + \Delta x)}{g(x)g(x + \Delta x)} \cdot \frac{1}{\Delta x}\right]$$

$$= \lim_{\Delta x \to 0} \frac{g(x)f(x + \Delta x) - f(x)g(x + \Delta x)}{\Delta x g(x)g(x + \Delta x)}$$

$$= \lim_{\Delta x \to 0} \frac{g(x)f(x + \Delta x) - f(x)g(x) + f(x)g(x) - f(x)g(x + \Delta x)}{\Delta x g(x)g(x + \Delta x)}$$

$$= \frac{\displaystyle\lim_{\Delta x \to 0} \frac{g(x)[f(x + \Delta x) - f(x)]}{\Delta x} - \lim_{\Delta x \to 0} \frac{f(x)[g(x + \Delta x) - g(x)]}{\Delta x}}{\displaystyle\lim_{\Delta x \to 0} [g(x)g(x + \Delta x)]}$$

$$= \frac{g(x)\left[\displaystyle\lim_{\Delta x \to 0} \frac{f(x + \Delta x) - f(x)}{\Delta x}\right] - f(x)\left[\displaystyle\lim_{\Delta x \to 0} \frac{g(x + \Delta x) - g(x)}{\Delta x}\right]}{\displaystyle\lim_{\Delta x \to 0} [g(x)g(x + \Delta x)]}$$

$$= \frac{g(x)f'(x) - f(x)g'(x)}{[g(x)]^2}$$

From the Quotient Rule, you can see that the derivative of a quotient is not, in general, the quotient of the derivatives. That is,

$$\frac{d}{dx}\left[\frac{f(x)}{g(x)}\right] \neq \frac{f'(x)}{g'(x)}.$$

ALGEBRA TUTOR *xy*

When applying the Quotient Rule, it is suggested that you enclose all factors and derivatives in symbols of grouping, such as parentheses. Also, pay special attention to the subtraction required in the numerator. For help in simplifying expressions like the one in Example 4, see the *Chapter 7 Algebra Tutor* on page 621, Example 2(d).

Example 4   Using the Quotient Rule

Find the derivative of $y = \dfrac{x - 1}{2x + 3}$.

SOLUTION   Apply the Quotient Rule, as shown.

$$\frac{dy}{dx} = \frac{(2x + 3)\dfrac{d}{dx}[x - 1] - (x - 1)\dfrac{d}{dx}[2x + 3]}{(2x + 3)^2}$$

$$= \frac{(2x + 3)(1) - (x - 1)(2)}{(2x + 3)^2}$$

$$= \frac{2x + 3 - 2x + 2}{(2x + 3)^2}$$

$$= \frac{5}{(2x + 3)^2}$$

✔ **Checkpoint 4**

Find the derivative of $y = \dfrac{x + 4}{5x - 2}$.

Example 5   Finding an Equation of a Tangent Line

Find an equation of the tangent line to the graph of

$$y = \frac{2x^2 - 4x + 3}{2 - 3x}$$

at $x = 1$.

SOLUTION   Apply the Quotient Rule, as shown.

$$\frac{dy}{dx} = \frac{(2 - 3x)\dfrac{d}{dx}[2x^2 - 4x + 3] - (2x^2 - 4x + 3)\dfrac{d}{dx}[2 - 3x]}{(2 - 3x)^2}$$

$$= \frac{(2 - 3x)(4x - 4) - (2x^2 - 4x + 3)(-3)}{(2 - 3x)^2}$$

$$= \frac{-12x^2 + 20x - 8 - (-6x^2 + 12x - 9)}{(2 - 3x)^2}$$

$$= \frac{-12x^2 + 20x - 8 + 6x^2 - 12x + 9}{(2 - 3x)^2}$$

$$= \frac{-6x^2 + 8x + 1}{(2 - 3x)^2}$$

When $x = 1$, the value of the function is $y = -1$ and the slope is $m = 3$. Using the point-slope form of a line, you can find the equation of the tangent line to be $y = 3x - 4$. The graph of the function and the tangent line is shown in Figure 7.46.

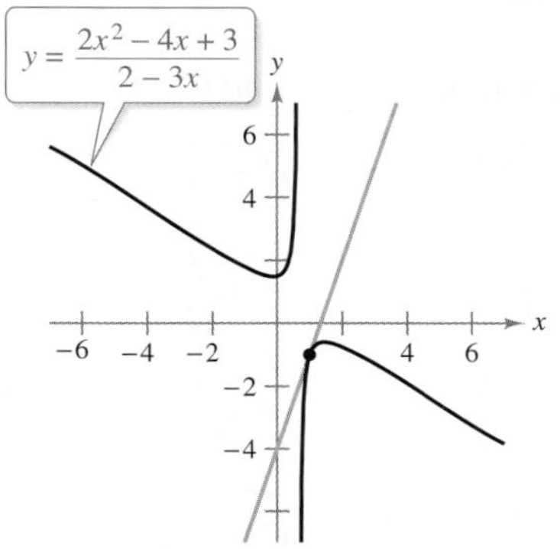

$y = \dfrac{2x^2 - 4x + 3}{2 - 3x}$

FIGURE 7.46

✔ **Checkpoint 5**

Find an equation of the tangent line to the graph of

$$y = \frac{x^2 - 4}{2x + 5}$$

at $x = 0$. Sketch the line tangent to the graph at $x = 0$.

STUDY TIP

Note in Example 6 that much of the work in obtaining the final form of the derivative occurs *after* applying the Quotient Rule. In general, direct application of differentiation rules often yields results that are not in simplified form. Note that two characteristics of simplified form are the absence of negative exponents and the combining of like terms.

Example 6   Rewriting Before Differentiating

Find the derivative of

$$y = \frac{3 - (1/x)}{x + 5}.$$

SOLUTION   Begin by rewriting the function. Then apply the Quotient Rule and simplify the result.

$$y = \frac{3 - (1/x)}{x + 5} \qquad \text{Write original function.}$$

$$= \frac{x[3 - (1/x)]}{x(x + 5)} \qquad \text{Multiply numerator and denominator by } x.$$

$$= \frac{3x - 1}{x^2 + 5x} \qquad \text{Rewrite.}$$

$$\frac{dy}{dx} = \frac{(x^2 + 5x)(3) - (3x - 1)(2x + 5)}{(x^2 + 5x)^2} \qquad \text{Apply Quotient Rule.}$$

$$= \frac{(3x^2 + 15x) - (6x^2 + 13x - 5)}{(x^2 + 5x)^2}$$

$$= \frac{-3x^2 + 2x + 5}{(x^2 + 5x)^2} \qquad \text{Simplify.}$$

✓**Checkpoint 6**

Find the derivative of $y = \dfrac{3 - (2/x)}{x + 4}$.

Not every quotient needs to be differentiated by the Quotient Rule. For instance, each quotient in the next example can be considered as the product of a constant and a function of x. In such cases, the Constant Multiple Rule is more efficient than the Quotient Rule.

STUDY TIP

To see the benefit of using the Constant Multiple Rule for some quotients, try using the Quotient Rule to differentiate the functions in Example 7. You should obtain the same results, but with more work.

Example 7   Using the Constant Multiple Rule

Original Function	Rewrite	Differentiate	Simplify
a. $y = \dfrac{x^2 + 3x}{6}$	$y = \dfrac{1}{6}(x^2 + 3x)$	$y' = \dfrac{1}{6}(2x + 3)$	$y' = \dfrac{1}{3}x + \dfrac{1}{2}$
b. $y = \dfrac{5x^4}{8}$	$y = \dfrac{5}{8}x^4$	$y' = \dfrac{5}{8}(4x^3)$	$y' = \dfrac{5}{2}x^3$
c. $y = \dfrac{-3(3x - 2x^2)}{7x}$	$y = -\dfrac{3}{7}(3 - 2x)$	$y' = -\dfrac{3}{7}(-2)$	$y' = \dfrac{6}{7}$
d. $y = \dfrac{9}{5x^2}$	$y = \dfrac{9}{5}(x^{-2})$	$y' = \dfrac{9}{5}(-2x^{-3})$	$y' = -\dfrac{18}{5x^3}$

✓**Checkpoint 7**

Find the derivative of each function.

a. $y = \dfrac{x^2 + 4x}{5}$ **b.** $y = \dfrac{3x^4}{4}$

Application

Example 8 Rate of Change of Systolic Blood Pressure

As blood moves from the heart through the major arteries out to the capillaries and back through the veins, the systolic blood pressure continuously drops. Consider a person whose systolic blood pressure P (in millimeters of mercury) is given by

$$P = \frac{25t^2 + 125}{t^2 + 1}, \quad 0 \le t \le 10$$

where t is measured in seconds. At what rate is the blood pressure changing 5 seconds after blood leaves the heart?

SOLUTION Begin by applying the Quotient Rule.

$$P = \frac{25t^2 + 125}{t^2 + 1} \qquad \text{Write original function.}$$

$$\frac{dP}{dt} = \frac{(t^2 + 1)(50t) - (25t^2 + 125)(2t)}{(t^2 + 1)^2} \qquad \text{Quotient Rule}$$

$$= \frac{50t^3 + 50t - 50t^3 - 250t}{(t^2 + 1)^2}$$

$$= -\frac{200t}{(t^2 + 1)^2} \qquad \text{Simplify.}$$

When $t = 5$, the rate of change is

$$\frac{dP}{dt} = -\frac{200(5)}{26^2} \approx -1.48 \text{ millimeters per second.}$$

So, the pressure is *dropping* at a rate of 1.48 millimeters per second at $t = 5$ seconds.

✓Checkpoint 8

In Example 8, find the rate at which systolic blood pressure is changing at each time shown in the table below. Describe the changes in blood pressure as the blood moves away from the heart.

t	0	1	2	3	4	5	6	7
$\dfrac{dP}{dt}$								

SUMMARIZE (Section 7.6)

1. State the Product Rule *(page 601)*. For examples of the Product Rule, see Examples 1, 2, and 3.

2. State the Quotient Rule *(page 604)*. For examples of the Quotient Rule, see Examples 4, 5, and 6.

3. Describe a real-life example of how the Quotient Rule can be used to analyze the rate of change of systolic blood pressure *(page 607, Example 8)*.

SKILLS WARM UP 7.6   The following warm-up exercises involve skills that were covered in a previous course or earlier sections. You will use these skills in the exercise set for this section. For additional help, review Sections 0.3, 0.6, 0.7, and 7.4.

In Exercises 1–10, simplify the expression.

1. $(x^2 + 1)(2) + (2x + 7)(2x)$

2. $(2x - x^3)(8x) + (4x^2)(2 - 3x^2)$

3. $x(4)(x^2 + 2)^3(2x) + (x^2 + 4)(1)$

4. $x^2(2)(2x + 1)(2) + (2x + 1)^4(2x)$

5. $\dfrac{(2x + 7)(5) - (5x + 6)(2)}{(2x + 7)^2}$

6. $\dfrac{(x^2 - 4)(2x + 1) - (x^2 + x)(2x)}{(x^2 - 4)^2}$

7. $\dfrac{(x^2 + 1)(2) - (2x + 1)(2x)}{(x^2 + 1)^2}$

8. $\dfrac{(1 - x^4)(4) - (4x - 1)(-4x^3)}{(1 - x^4)^2}$

9. $(x^{-1} + x)(2) + (2x - 3)(-x^{-2} + 1)$

10. $\dfrac{(1 - x^{-1})(1) - (x - 4)(x^{-2})}{(1 - x^{-1})^2}$

In Exercises 11–14, find $f'(2)$.

11. $f(x) = 3x^2 - x + 4$

12. $f(x) = -x^3 + x^2 + 8x$

13. $f(x) = \dfrac{1}{x}$

14. $f(x) = x^2 - \dfrac{1}{x^2}$

Exercises 7.6

See www.CalcChat.com for worked-out solutions to odd-numbered exercises.

Using the Product Rule In Exercises 1–10, use the Product Rule to find the derivative of the function. *See Examples 1, 2, and 3.*

1. $f(x) = (2x - 3)(1 - 5x)$

2. $g(x) = (x - 4)(x + 2)$

3. $f(x) = (6x - x^2)(4 + 3x)$

4. $f(x) = (x^2 + 1)(2x + 5)$

5. $f(x) = x(x^2 + 3)$

6. $f(x) = x^2(3x^3 - 1)$

7. $h(x) = \left(\dfrac{2}{x} - 3\right)(x^2 + 7)$

8. $f(x) = (3 - x)\left(\dfrac{4}{x^2} - 5\right)$

9. $g(x) = (x^2 - 4x + 3)(x - 2)$

10. $g(x) = (x^2 - 2x + 1)(x^3 - 1)$

Using the Quotient Rule In Exercises 11–20, use the Quotient Rule to find the derivative of the function. *See Examples 4 and 6.*

11. $h(x) = \dfrac{x}{x - 5}$

12. $h(x) = \dfrac{x^2}{x + 3}$

13. $f(t) = \dfrac{2t^2 - 3}{3t + 1}$

14. $f(x) = \dfrac{x + 1}{x - 1}$

15. $f(t) = \dfrac{t^2 - 1}{t + 4}$

16. $g(x) = \dfrac{4x - 5}{x^2 - 1}$

17. $f(x) = \dfrac{x^2 + 6x + 5}{2x - 1}$

18. $f(x) = \dfrac{4x^2 - x + 1}{x + 2}$

19. $f(x) = \dfrac{6 + (2/x)}{3x - 1}$

20. $f(x) = \dfrac{5 - (1/x^2)}{x + 2}$

Using the Constant Multiple Rule In Exercises 21–30, find the derivative of the function. *See Example 7.*

Original Function	Rewrite	Differentiate	Simplify
21. $f(x) = \dfrac{x^3 + 6x}{3}$			
22. $f(x) = \dfrac{3x^2}{7}$			
23. $y = \dfrac{x^2 + 2x}{3}$			
24. $y = \dfrac{4x^{3/2}}{x}$			
25. $y = \dfrac{7}{3x^3}$			
26. $y = \dfrac{4}{5x^2}$			
27. $y = \dfrac{4x^2 - 3x}{8\sqrt{x}}$			
28. $y = \dfrac{5(3x^2 + 5x)}{8x}$			
29. $y = \dfrac{x^2 - 4x + 3}{2(x - 1)}$			
30. $y = \dfrac{x^2 - 4}{4(x + 2)}$			

Finding Derivatives In Exercises 31–44, find the derivative of the function. State which differentiation rule(s) you used to find the derivative.

31. $f(x) = (x^3 - 3x)(2x^2 + 3x + 5)$

32. $h(t) = (t^5 - 1)(4t^2 - 7t - 3)$

33. $f(x) = \dfrac{x^3 + 3x + 2}{x^2 - 1}$

34. $f(x) = (x^5 - 3x)\left(\dfrac{1}{x^2}\right)$

35. $f(x) = \dfrac{x^2 - x - 20}{x + 4}$

36. $h(t) = \dfrac{t + 2}{t^2 + 5t + 6}$

37. $g(t) = (2t^3 - 1)^2$

38. $f(x) = \sqrt[3]{x}(x + 1)$

39. $g(s) = \dfrac{s^2 - 2s + 5}{\sqrt{s}}$

40. $f(x) = \dfrac{x + 1}{\sqrt{x}}$

41. $f(x) = \dfrac{(x - 2)(3x + 1)}{4x + 2}$

42. $f(x) = \dfrac{(x + 1)(2x - 7)}{2x + 1}$

43. $g(x) = \left(\dfrac{x - 3}{x + 4}\right)(x^2 + 2x + 1)$

44. $f(x) = (3x^3 + 4x)(x - 5)(x + 1)$

Finding an Equation of a Tangent Line In Exercises 45–52, find an equation of the tangent line to the graph of the function at the given point. Then use a graphing utility to graph the function and the tangent line in the same viewing window. *See Example 5.*

Function	*Point*
45. $f(x) = (5x + 2)(x^2 + x)$	$(-1, 0)$
46. $h(x) = (x^2 - 1)^2$	$(-2, 9)$
47. $f(x) = x^3(x^2 - 4)$	$(1, -3)$
48. $f(x) = \sqrt{x}(x - 3)$	$(9, 18)$
49. $f(x) = \dfrac{x - 2}{x + 1}$	$\left(1, -\dfrac{1}{2}\right)$
50. $f(x) = \dfrac{2x + 1}{x - 1}$	$(2, 5)$
51. $f(x) = \dfrac{(3x - 2)(6x + 5)}{2x - 3}$	$(-1, -1)$
52. $g(x) = \dfrac{(x + 2)(x^2 + x)}{x - 4}$	$(1, -2)$

Finding Horizontal Tangent Lines In Exercises 53–56, find the point(s), if any, at which the graph of f has a horizontal tangent line.

53. $f(x) = \dfrac{x^2}{x - 1}$

54. $f(x) = \dfrac{x^2}{x^2 + 1}$

55. $f(x) = \dfrac{x^4}{x^3 + 1}$

56. $f(x) = \dfrac{x^4 + 3}{x^2 + 1}$

 Graphing a Function and Its Derivative In Exercises 57–60, use a graphing utility to graph f and f' on the interval $[-2, 2]$.

57. $f(x) = x(x + 1)$

58. $f(x) = x^2(x + 1)$

59. $f(x) = x(x + 1)(x - 1)$

60. $f(x) = x^2(x + 1)(x - 1)$

Demand In Exercises 61 and 62, use the demand function to find the rate of change in the demand x for the given price p.

61. $x = 275\left(1 - \dfrac{3p}{5p + 1}\right)$, $p = \$4$

62. $x = 300 - p - \dfrac{2p}{p + 1}$, $p = \$3$

63. Population Growth A population of bacteria is introduced into a culture. The number of bacteria P can be modeled by

$$P = 500\left(1 + \dfrac{4t}{50 + t^2}\right)$$

where t is the time (in hours). Find the rate of change of the population at $t = 2$.

64. Quality Control The percent P of defective parts produced by a new employee t days after the employee starts work can be modeled by

$$P = \dfrac{t + 1750}{50(t + 2)}.$$

Find the rates of change of P at (a) $t = 1$ and (b) $t = 10$.

65. Environment The model

$$f(t) = \dfrac{t^2 - t + 1}{t^2 + 1}$$

measures the level of oxygen in a pond, where t is the time (in weeks) after organic waste is dumped into the pond. Find the rates of change of f with respect to t at (a) $t = 0.5$, (b) $t = 2$, and (c) $t = 8$. Interpret the meaning of these values.

66. Physical Science The temperature T (in degrees Fahrenheit) of food placed in a refrigerator is modeled by

$$T = 10\left(\frac{4t^2 + 16t + 75}{t^2 + 4t + 10}\right)$$

where t is the time (in hours). What is the initial temperature of the food? Find the rates of change of T with respect to t at (a) $t = 1$, (b) $t = 3$, (c) $t = 5$, and (d) $t = 10$. Interpret the meaning of these values.

67. Cost The cost C of producing x units of a product is given by $C = x^3 - 15x^2 + 87x - 73$ for $4 \le x \le 9$.

(a) Use a graphing utility to graph the marginal cost function and the average cost function, C/x, in the same viewing window.

(b) Find the point of intersection of the graphs of dC/dx and C/x. Does this point have any significance?

68.  **HOW DO YOU SEE IT?** The advertising manager for a new product determines that P percent of the potential market is aware of the product t weeks after the advertising campaign begins.

Market Awareness

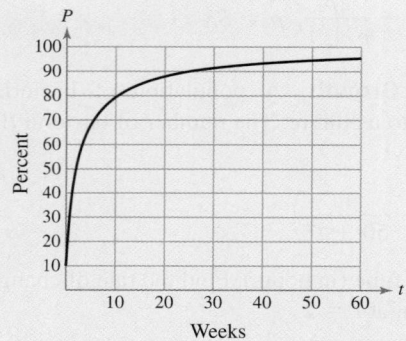

(a) What happens to the percent of people who are aware of the product in the long run?

(b) What happens to the rate of change of the percent of people who are aware of the product in the long run?

69. Inventory Replenishment The ordering and transportation cost C per unit (in thousands of dollars) of the components used in manufacturing a product is given by

$$C = 100\left(\frac{200}{x^2} + \frac{x}{x + 30}\right), \quad 1 \le x$$

where x is the order size (in hundreds). Find the rate of change of C with respect to x for each order size. What do these rates of change imply about increasing the size of an order? Of the given order sizes, which would you choose? Explain.

(a) $x = 10$ (b) $x = 15$ (c) $x = 20$

70. Managing a Store You are managing a store and have been adjusting the price of an item. You have found that you make a profit of $50 when 10 units are sold, $60 when 12 units are sold, and $65 when 14 units are sold.

(a) Use the *regression* feature of a graphing utility to find a quadratic model that relates the profit P to the number of units sold x.

(b) Use a graphing utility to graph P.

(c) Find the point on the graph at which the marginal profit is zero. Interpret this point in the context of the problem.

Using Relationships In Exercises 71–74, use the given information to find $f'(2)$.

$g(2) = 3$ and $g'(2) = -2$

$h(2) = -1$ and $h'(2) = 4$

71. $f(x) = 2g(x) + h(x)$ **72.** $f(x) = 3 - g(x)$

73. $f(x) = g(x)h(x)$ **74.** $f(x) = \dfrac{g(x)}{h(x)}$

Business Capsule

In 1978 Ben Cohen and Jerry Greenfield used their combined life savings of $8000 to convert an abandoned gas station in Burlington, Vermont into their first ice cream shop. Today, Ben & Jerry's Homemade Holdings, Inc. has almost 800 scoop shops in 25 countries. The company's three-part mission statement emphasizes product quality, economic reward, and a commitment to the community. Ben & Jerry's contributes a minimum of $1.8 million annually through corporate philanthropy that is primarily employee led.

75. Research Project Use your school's library, the Internet, or some other reference source to find information on a company that is noted for its philanthropy and community commitment. (One such business is described above.) Write a short paper about the company.

7.7 The Chain Rule

In Exercise 71 on page 619, you will use the General Power Rule to find the rate of change of the balance in an account.

■ Find derivatives using the Chain Rule.
■ Find derivatives using the General Power Rule.
■ Write derivatives in simplified form.
■ Use derivatives to answer questions about real-life situations.
■ Review the basic differentiation rules for algebraic functions.

The Chain Rule

In this section, you will study one of the most powerful rules of differential calculus—the **Chain Rule.** This differentiation rule deals with composite functions and adds versatility to the rules presented in Sections 7.4 and 7.6. For example, compare the functions below. Those on the left can be differentiated without the Chain Rule, whereas those on the right are best done with the Chain Rule.

Without the Chain Rule	*With the Chain Rule*
$y = x^2 + 1$	$y = \sqrt{x^2 + 1}$
$y = x + 1$	$y = (x + 1)^{-1/2}$
$y = 3x + 2$	$y = (3x + 2)^5$
$y = \dfrac{x + 5}{x^2 + 2}$	$y = \left(\dfrac{x + 5}{x^2 + 2}\right)^2$
$y = \dfrac{x + 1}{x}$	$y = \sqrt{\dfrac{x + 1}{x}}$

The Chain Rule

If $y = f(u)$ is a differentiable function of u, and $u = g(x)$ is a differentiable function of x, then $y = f(g(x))$ is a differentiable function of x, and

$$\frac{dy}{dx} = \frac{dy}{du} \cdot \frac{du}{dx}$$

or, equivalently,

$$\frac{d}{dx}[f(g(x))] = f'(g(x))g'(x).$$

Basically, the Chain Rule states that if y changes dy/du times as fast as u, and u changes du/dx times as fast as x, then y changes

$$\frac{dy}{du} \cdot \frac{du}{dx}$$

times as fast as x, as illustrated in Figure 7.47. One advantage of the

$$\frac{dy}{dx}$$

notation for derivatives is that it helps you remember differentiation rules, such as the Chain Rule. For instance, in the formula

$$\frac{dy}{dx} = \frac{dy}{du} \cdot \frac{du}{dx}$$

you can imagine that the du's divide out.

---

x

Input

Function g | Output

Rate of change of u with respect to x is $\dfrac{du}{dx}$.

$u = g(x)$

u

Input

Function f | Output

Rate of change of y with respect to u is $\dfrac{dy}{du}$.

$y = f(u) = f(g(x))$

Rate of change of y with respect to x is $\dfrac{dy}{dx} = \dfrac{dy}{du}\dfrac{du}{dx}$.

FIGURE 7.47

When applying the Chain Rule, it is helpful to think of the composite function $y = f(g(x))$ or $y = f(u)$ as having two parts—an *inside* and an *outside*—as illustrated below.

Inside

$$y = f(g(x)) = f(u)$$

Outside

The Chain Rule tells you that the derivative of $y = f(u)$ is the derivative of the outer function (at the inner function u) *times* the derivative of the inner function. That is,

$$y' = f'(u) \cdot u'.$$

Example 1 Decomposing Composite Functions

Write each function as the composition of two functions.

a. $y = \dfrac{1}{x + 1}$ **b.** $y = \sqrt{3x^2 - x + 1}$

SOLUTION   There is more than one correct way to decompose each function. One way for each is shown below.

$y = f(g(x))$	$u = g(x)$ *(inside)*	$y = f(u)$ *(outside)*
a. $y = \dfrac{1}{x + 1}$	$u = x + 1$	$y = \dfrac{1}{u}$
b. $y = \sqrt{3x^2 - x + 1}$	$u = 3x^2 - x + 1$	$y = \sqrt{u}$

✓ **Checkpoint 1**

Write each function as the composition of two functions, where $y = f(g(x))$.

a. $y = \dfrac{1}{\sqrt{x + 1}}$ **b.** $y = (x^2 + 2x + 5)^3$

Example 2 Using the Chain Rule

Find the derivative of $y = (x^2 + 1)^3$.

SOLUTION   To apply the Chain Rule, you need to identify the inside function u.

$$y = \overbrace{(x^2 + 1)}^{u}{}^3 = u^3$$

The inside function is $u = x^2 + 1$. By the Chain Rule, you can write the derivative as shown.

$$\frac{dy}{dx} = \overbrace{3(x^2 + 1)^2}^{\frac{dy}{du}}\overbrace{(2x)}^{\frac{du}{dx}} = 6x(x^2 + 1)^2$$

✓ **Checkpoint 2**

Find the derivative of $y = (x^3 + 1)^2$.

STUDY TIP

Try checking the result of Example 2 by expanding the function to obtain

$$y = x^6 + 3x^4 + 3x^2 + 1$$

and finding the derivative. Do you obtain the same answer?

The General Power Rule

The function in Example 2 illustrates one of the most common types of composite functions—a power function of the form

$$y = [u(x)]^n.$$

The rule for differentiating such functions is called the **General Power Rule,** and it is a special case of the Chain Rule.

The General Power Rule

If $y = [u(x)]^n$, where u is a differentiable function of x and n is a real number, then

$$\frac{dy}{dx} = n[u(x)]^{n-1}\frac{du}{dx}$$

or, equivalently,

$$\frac{d}{dx}[u^n] = nu^{n-1}u'.$$

PROOF   Apply the Chain Rule and the Simple Power Rule as shown.

$$\frac{dy}{dx} = \frac{dy}{du} \cdot \frac{du}{dx}$$

$$= \frac{d}{du}[u^n]\frac{du}{dx}$$

$$= nu^{n-1}\frac{du}{dx}$$

Example 3   Using the General Power Rule

Find the derivative of

$$y = (3x - 2x^2)^3.$$

SOLUTION   To apply the General Power Rule, you need to identify the inside function u.

$$y = \overbrace{(3x - 2x^2)}^{u}{}^3 = u^3$$

The inside function is

$$u = 3x - 2x^2.$$

So, by the General Power Rule,

$$\frac{dy}{dx} = \overset{n}{3}\,\overbrace{(3x - 2x^2)^2}^{u^{n-1}}\,\overbrace{\frac{d}{dx}[3x - 2x^2]}^{u'}$$

$$= 3(3x - 2x^2)^2(3 - 4x).$$

✓**Checkpoint 3**

Find the derivative of

$$y = (x^2 + 3x)^4.$$

Example 4   Finding an Equation of a Tangent Line

Find an equation of the tangent line to the graph of

$$y = \sqrt[3]{(x^2 + 4)^2}$$

at $x = 2$.

SOLUTION   Begin by rewriting the function in rational exponent form.

$$y = (x^2 + 4)^{2/3} \qquad \text{Rewrite original function.}$$

Then, using the inside function, $u = x^2 + 4$, apply the General Power Rule.

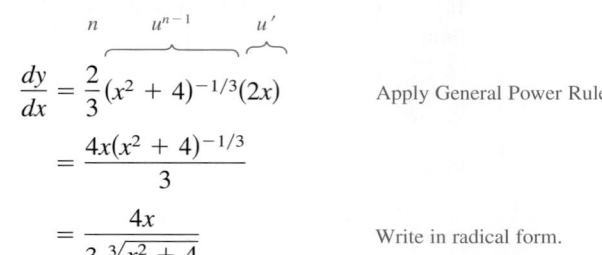

$$\frac{dy}{dx} = \frac{2}{3}(x^2 + 4)^{-1/3}(2x) \qquad \text{Apply General Power Rule.}$$

$$= \frac{4x(x^2 + 4)^{-1/3}}{3}$$

$$= \frac{4x}{3\sqrt[3]{x^2 + 4}} \qquad \text{Write in radical form.}$$

When $x = 2$, $y = 4$ and the slope of the line tangent to the graph at $(2, 4)$ is $\frac{4}{3}$. Using the point-slope form, you can find the equation of the tangent line to be $y = \frac{4}{3}x + \frac{4}{3}$. The graph of the function and the tangent line is shown in Figure 7.48.

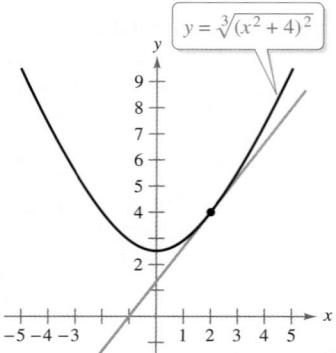

FIGURE 7.48

✓ **Checkpoint 4**

Find an equation of the tangent line to the graph of $y = \sqrt[3]{(x + 4)^2}$ at $x = 4$. ■

STUDY TIP

The derivative of a quotient can sometimes be found more easily with the General Power Rule than with the Quotient Rule. This is especially true when the numerator is a constant, as shown in Example 5.

Example 5   Differentiating a Quotient with a Constant Numerator

Find the derivative of

$$y = \frac{5}{(4x - 3)^2}.$$

SOLUTION   Begin by rewriting the function in rational exponent form.

$$y = 5(4x - 3)^{-2} \qquad \text{Rewrite original function.}$$

Then, using the inside function, $u = 4x - 3$, apply the General Power Rule.

$$\frac{dy}{dx} = 5(-2)(4x - 3)^{-3}(4) \qquad \text{Apply General Power Rule.}$$

Constant
Multiple Rule

$$= -40(4x - 3)^{-3} \qquad \text{Simplify.}$$

$$= \frac{-40}{(4x - 3)^3} \qquad \text{Write with positive exponent.}$$

✓ **Checkpoint 5**

Find the derivative of $y = \dfrac{4}{2x + 1}$. ■

Simplification Techniques

Throughout this chapter, writing derivatives in simplified form has been emphasized. The reason for this is that most applications of derivatives require a simplified form. The next two examples illustrate some useful simplification techniques.

ALGEBRA TUTOR xy

In Example 6, note that you subtract exponents when factoring. That is, when $(1 - x^2)^{-1/2}$ is factored out of $(1 - x^2)^{1/2}$, the *remaining* factor has an exponent of $\frac{1}{2} - \left(-\frac{1}{2}\right) = 1$. So, $(1 - x^2)^{1/2}$ is equal to the product of $(1 - x^2)^{-1/2}$ and $(1 - x^2)^{1}$. For help in simplifying expressions like the one in Example 6, see the *Chapter 7 Algebra Tutor* on pages 620 and 621.

Example 6 Simplifying by Factoring Out Least Powers

Find the derivative of $y = x^2\sqrt{1 - x^2}$.

SOLUTION

$$y = x^2\sqrt{1 - x^2} \qquad \text{Write original function.}$$
$$= x^2(1 - x^2)^{1/2} \qquad \text{Rewrite function.}$$
$$y' = x^2\frac{d}{dx}[(1 - x^2)^{1/2}] + (1 - x^2)^{1/2}\frac{d}{dx}[x^2] \qquad \text{Product Rule}$$
$$= x^2\left[\frac{1}{2}(1 - x^2)^{-1/2}(-2x)\right] + (1 - x^2)^{1/2}(2x) \qquad \text{General Power Rule}$$
$$= -x^3(1 - x^2)^{-1/2} + 2x(1 - x^2)^{1/2} \qquad \text{Simplify.}$$
$$= x(1 - x^2)^{-1/2}[-x^2(1) + 2(1 - x^2)] \qquad \text{Factor.}$$
$$= x(1 - x^2)^{-1/2}(2 - 3x^2) \qquad \text{Simplify.}$$
$$= \frac{x(2 - 3x^2)}{\sqrt{1 - x^2}} \qquad \text{Write in radical form.}$$

✓**Checkpoint 6**

Find and simplify the derivative of $y = x^2\sqrt{x^2 + 1}$.

STUDY TIP

In Example 7, try to find $f'(x)$ by applying the Quotient Rule to

$$f(x) = \frac{(3x - 1)^2}{(x^2 + 3)^2}.$$

Which method do you prefer?

Example 7 Differentiating a Quotient Raised to a Power

Find the derivative of

$$f(x) = \left(\frac{3x - 1}{x^2 + 3}\right)^2.$$

SOLUTION

$$f'(x) = \overset{n}{2}\overset{u^{n-1}}{\overbrace{\left(\frac{3x - 1}{x^2 + 3}\right)}}\overset{u'}{\overbrace{\frac{d}{dx}\left[\frac{3x - 1}{x^2 + 3}\right]}} \qquad \text{General Power Rule}$$
$$= \left[\frac{2(3x - 1)}{x^2 + 3}\right]\left[\frac{(x^2 + 3)(3) - (3x - 1)(2x)}{(x^2 + 3)^2}\right] \qquad \text{Quotient Rule}$$
$$= \frac{2(3x - 1)(3x^2 + 9 - 6x^2 + 2x)}{(x^2 + 3)^3} \qquad \text{Multiply.}$$
$$= \frac{2(3x - 1)(-3x^2 + 2x + 9)}{(x^2 + 3)^3} \qquad \text{Simplify.}$$

✓**Checkpoint 7**

Find the derivative of

$$f(x) = \left(\frac{x + 1}{x - 5}\right)^2.$$

Application

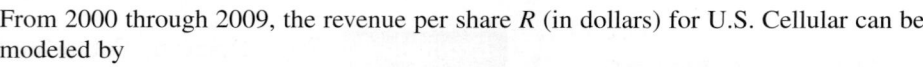 **Example 8**   **Finding Rates of Change**

From 2000 through 2009, the revenue per share R (in dollars) for U.S. Cellular can be modeled by

$$R = (-0.009t^2 + 0.39t + 4.3)^2, \quad 0 \le t \le 9$$

where t is the year, with $t = 0$ corresponding to 2000. Use the model to approximate the rates of change in the revenue per share in 2001, 2002, and 2005. Would U.S. Cellular stockholders have been satisfied with the performance of this stock from 2000 through 2009?   *(Source: U.S. Cellular)*

SOLUTION   The rate of change in R is given by the derivative dR/dt. You can use the General Power Rule to find the derivative.

$$\frac{dR}{dt} = 2(-0.009t^2 + 0.39t + 4.3)(-0.018t + 0.39)$$

$$= (-0.036t + 0.78)(-0.009t^2 + 0.39t + 4.3)$$

In 2001, the revenue per share was changing at a rate of

$$[-0.036(1) + 0.78][-0.009(1)^2 + 0.39(1) + 4.3] \approx \$3.48 \text{ per year.}$$

In 2002, the revenue per share was changing at a rate of

$$[-0.036(2) + 0.78][-0.009(2)^2 + 0.39(2) + 4.3] \approx \$3.57 \text{ per year.}$$

In 2005, the revenue per share was changing at a rate of

$$[-0.036(5) + 0.78][-0.009(5)^2 + 0.39(5) + 4.3] \approx \$3.62 \text{ per year.}$$

The graph of the revenue per share function R is shown in Figure 7.49. So, most stockholders would have been satisfied with the performance of this stock.

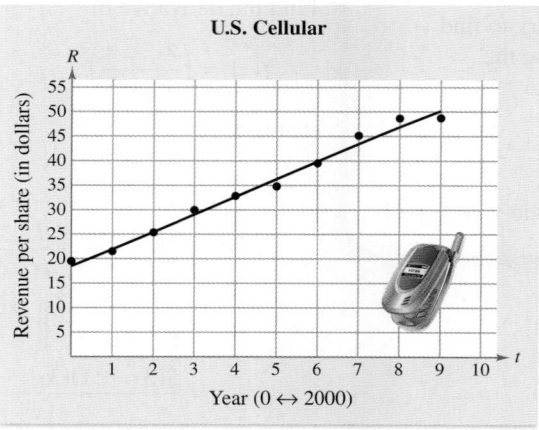

FIGURE 7.49

✓ Checkpoint 8

From 2000 through 2009, the sales per share S (in dollars) for Dollar Tree can be modeled by

$$S = (0.010t^2 + 0.27t + 3.1)^2, \quad 0 \le t \le 9$$

where t is the year, with $t = 0$ corresponding to 2000. Use the model to approximate the rate of change of sales per share in 2005.   *(Source: Dollar Tree, Inc.)*

Review of Basic Differentiation Rules

You now have all the rules you need to differentiate *any* algebraic function. For your convenience, they are summarized below.

Summary of Basic Differentiation Rules

Let u and v be differentiable functions of x.

1. Constant Rule $\dfrac{d}{dx}[c] = 0, \quad c$ is a constant.

2. Constant Multiple Rule $\dfrac{d}{dx}[cu] = c\,\dfrac{du}{dx}, \quad c$ is a constant.

3. Sum and Difference Rules $\dfrac{d}{dx}[u \pm v] = \dfrac{du}{dx} \pm \dfrac{dv}{dx}$

4. Product Rule $\dfrac{d}{dx}[uv] = u\,\dfrac{dv}{dx} + v\,\dfrac{du}{dx}$

5. Quotient Rule $\dfrac{d}{dx}\left[\dfrac{u}{v}\right] = \dfrac{v\,\dfrac{du}{dx} - u\,\dfrac{dv}{dx}}{v^2}$

6. Power Rules $\dfrac{d}{dx}[x^n] = nx^{n-1}$

$\dfrac{d}{dx}[u^n] = nu^{n-1}\,\dfrac{du}{dx}$

7. Chain Rule $\dfrac{dy}{dx} = \dfrac{dy}{du} \cdot \dfrac{du}{dx}$

SUMMARIZE   (Section 7.7)

1. State the Chain Rule *(page 611)*. For an example of the Chain Rule, see Example 2.

2. State the General Power Rule *(page 613)*. For examples of the General Power Rule, see Examples 3, 4, and 5.

3. Describe a real-life example of how the General Power Rule can be used to analyze the rate of change of a company's revenue per share *(page 616, Example 8)*.

4. Use the Summary of Basic Differentiation Rules to identify the differentiation rules illustrated by (a)–(f) below *(page 617)*.

(a) $\dfrac{d}{dx}[2x] = 2\,\dfrac{d}{dx}[x]$ (b) $\dfrac{d}{dx}[x^4] = 4x^3$

(c) $\dfrac{d}{dx}[8] = 0$ (d) $\dfrac{d}{dx}[x^2 + x] = \dfrac{d}{dx}[x^2] + \dfrac{d}{dx}[x]$

(e) $\dfrac{d}{dx}[x - x^3] = \dfrac{d}{dx}[x] - \dfrac{d}{dx}[x^3]$

(f) $\dfrac{d}{dx}[x(x + 1)] = (x)\,\dfrac{d}{dx}[x + 1] + (x + 1)\,\dfrac{d}{dx}[x]$

SKILLS WARM UP 7.7 The following warm-up exercises involve skills that were covered in a previous course. You will use these skills in the exercise set for this section. For additional help, review Sections 0.4 and 0.6.

In Exercises 1–6, rewrite the expression with rational exponents.

1. $\sqrt[5]{(1 - 5x)^2}$

2. $\sqrt[4]{(2x - 1)^3}$

3. $\dfrac{1}{\sqrt{4x^2 + 1}}$

4. $\dfrac{1}{\sqrt[3]{x - 6}}$

5. $\dfrac{\sqrt{x}}{\sqrt[3]{1 - 2x}}$

6. $\dfrac{\sqrt{(3 - 7x)^3}}{2x}$

In Exercises 7–10, factor the expression.

7. $3x^3 - 6x^2 + 5x - 10$

8. $5x\sqrt{x} - x - 5\sqrt{x} + 1$

9. $4(x^2 + 1)^2 - x(x^2 + 1)^3$

10. $-x^5 + 3x^3 + x^2 - 3$

Exercises 7.7

See www.CalcChat.com for worked-out solutions to odd-numbered exercises.

Decomposing Composite Functions In Exercises 1–6, identify the inside function, $u = g(x)$, and the outside function, $y = f(u)$. See Example 1.

$y = f(g(x))$	$u = g(x)$	$y = f(u)$
1. $y = (6x - 5)^4$		
2. $y = (x^2 - 2x + 3)^3$		
3. $y = \sqrt{5x - 2}$		
4. $y = \sqrt{1 - x^2}$		
5. $y = \dfrac{1}{3x + 1}$		
6. $y = \dfrac{1}{\sqrt{x^2 - 3}}$		

Using the Chain Rule In Exercises 7–12, find dy/du, du/dx, and dy/dx. See Example 2.

7. $y = u^2, u = 4x + 7$

8. $y = u^3, u = 3x^2 - 2$

9. $y = \sqrt{u}, u = 3 - x^2$

10. $y = 2\sqrt{u}, u = 5x + 9$

11. $y = u^{2/3}, u = 5x^4 - 2x$

12. $y = u^{-1}, u = x^3 + 2x^2$

Choosing a Differentiation Rule In Exercises 13–20, match the function with the rule that you would use to find the derivative *most efficiently*.

(a) Simple Power Rule

(b) Constant Rule

(c) General Power Rule

(d) Quotient Rule

13. $f(x) = \dfrac{2}{1 - x^3}$

14. $f(x) = \dfrac{2x}{1 - x^3}$

15. $f(x) = \sqrt[3]{8^2}$

16. $f(x) = \sqrt[3]{x^2}$

17. $f(x) = \dfrac{x^2 + 2}{x}$

18. $f(x) = \dfrac{\sqrt{x}}{x^3 + 2x - 5}$

19. $f(x) = \dfrac{2}{x - 2}$

20. $f(x) = \dfrac{5}{x^2 + 1}$

Using the General Power Rule In Exercises 21–36, use the General Power Rule to find the derivative of the function. See Examples 3 and 5.

21. $y = (2x - 7)^3$

22. $g(x) = (4 - 2x)^3$

23. $f(x) = (5x - x^2)^{3/2}$

24. $y = (2x^3 + 1)^2$

25. $h(x) = (6x - x^3)^2$

26. $f(x) = (4x - x^2)^3$

27. $f(t) = \sqrt{t + 1}$

28. $g(x) = \sqrt{5 - 3x}$

29. $s(t) = \sqrt{2t^2 + 5t + 2}$

30. $y = \sqrt[3]{3x^3 + 4x}$

31. $y = \sqrt[3]{9x^2 + 4}$

32. $y = 2\sqrt{4 - x^2}$

33. $f(x) = \dfrac{2}{(2 - 9x)^3}$

34. $g(x) = \dfrac{3}{\sqrt{(x^2 + 8x)^3}}$

35. $f(x) = \dfrac{1}{\sqrt{x^2 + 25}}$

36. $y = \dfrac{1}{\sqrt[3]{(4 - x^3)^4}}$

 Finding an Equation of a Tangent Line In Exercises 37–42, find an equation of the tangent line to the graph of *f* at the point $(2, f(2))$. Then use a graphing utility to graph the function and the tangent line in the same viewing window. See Example 4.

37. $f(x) = 2(x^2 - 1)^3$

38. $f(x) = 3(9x - 4)^4$

39. $f(x) = \sqrt{4x^2 - 7}$

40. $f(x) = x\sqrt{x^2 + 5}$

41. $f(x) = \sqrt{x^2 - 2x + 1}$

42. $f(x) = (4 - 3x^2)^{-2/3}$

 Using Technology In Exercises 43–46, use a symbolic differentiation utility to find the derivative of the function. Graph the function and its derivative in the same viewing window. Describe the behavior of the function when the derivative is zero.

43. $f(x) = \dfrac{\sqrt{x} + 1}{x^2 + 1}$

44. $f(x) = \sqrt{\dfrac{2x}{x + 1}}$

45. $f(x) = \sqrt{\dfrac{x + 1}{x}}$

46. $f(x) = \sqrt{x}(2 - x^2)$

Finding Derivatives In Exercises 47–62, find the derivative of the function. State which differentiation rule(s) you used to find the derivative.

47. $y = \dfrac{1}{4 - x^2}$

48. $s(t) = \dfrac{1}{t^2 + 3t - 1}$

49. $y = -\dfrac{4}{(t + 2)^2}$

50. $f(x) = \dfrac{3x}{(x^3 - 4)^2}$

51. $f(x) = (2x - 1)(9 - 3x^2)$

52. $y = x^3(7x + 2)$

53. $y = \dfrac{1}{\sqrt{x + 2}}$

54. $g(x) = \dfrac{3}{\sqrt[3]{x^3 - 1}}$

55. $f(x) = x(3x - 9)^3$

56. $f(x) = x^3(x - 4)^2$

57. $y = x\sqrt{2x + 3}$

58. $y = t\sqrt{t + 1}$

59. $y = t^2\sqrt{t - 2}$

60. $y = \sqrt{x}(x - 2)^2$

61. $y = \left(\dfrac{6 - 5x}{x^2 - 1}\right)^2$

62. $y = \left(\dfrac{4x^2}{3 - x}\right)^3$

Finding an Equation of a Tangent Line In Exercises 63–70, find an equation of the tangent line to the graph of the function at the given point. Then use a graphing utility to graph the function and the tangent line in the same viewing window.

63. $f(t) = \dfrac{36}{(3 - t)^2}$; $(0, 4)$

64. $y = (4x^3 + 3)^2$; $(-1, 1)$

65. $f(x) = \sqrt[5]{3x^3 + 4x}$; $(2, 2)$

66. $s(x) = \dfrac{1}{\sqrt{x^2 - 3x + 4}}$; $\left(3, \tfrac{1}{2}\right)$

67. $f(t) = (t^2 - 9)\sqrt{t + 2}$; $(-1, -8)$

68. $y = \dfrac{2x}{\sqrt{x + 1}}$; $(3, 3)$

69. $f(x) = \dfrac{x + 1}{\sqrt{2x - 3}}$; $(2, 3)$ **70.** $y = \dfrac{x}{\sqrt{25 + x^2}}$; $(0, 0)$

71. Compound Interest You deposit $1000 in an account with an annual interest rate of r (in decimal form) compounded monthly. At the end of 5 years, the balance A is

$$A = 1000\left(1 + \frac{r}{12}\right)^{60}.$$

Find the rates of change of A with respect to r when (a) $r = 0.08$, (b) $r = 0.10$, and (c) $r = 0.12$.

72. Biology The number N of bacteria in a culture after t days is modeled by

$$N = 400\left[1 - \frac{3}{(t^2 + 2)^2}\right].$$

Find the rate of change of N with respect to t when (a) $t = 0$, (b) $t = 1$, (c) $t = 2$, (d) $t = 3$, and (e) $t = 4$. (f) What can you conclude?

73. Depreciation The value V of a machine t years after it is purchased is inversely proportional to the square root of $t + 1$. The initial value of the machine is $10,000.

(a) Write V as a function of t.

(b) Find the rate of depreciation when $t = 1$.

(c) Find the rate of depreciation when $t = 3$.

74. HOW DO YOU SEE IT? The cost C (in dollars) of producing x units of a product is

$$C = 60x + 1350.$$

For one week, management determined that the number of units produced x at the end of t hours was

$$x = -1.6t^3 + 19t^2 - 0.5t - 1.$$

The graph shows the cost C in terms of the time t.

Cost of Producing a Product

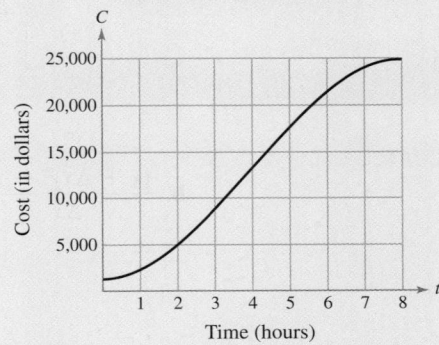

(a) Which is greater, the rate of change of the cost after 1 hour or the rate of change of the cost after 4 hours?

(b) Explain why the cost function is not increasing at a constant rate during the eight-hour shift.

75. Credit Card Rate The average annual rate r (in percent form) for commercial bank credit cards from 2003 through 2009 can be modeled by

$$r = \sqrt{2.8557t^4 - 72.792t^3 + 676.14t^2 - 2706t + 4096}$$

where t represents the year, with $t = 3$ corresponding to 2003. (*Source: Board of Governors of the Federal Reserve System*)

(a) Find the derivative of this model. Which differentiation rule(s) did you use?

(b) Use a graphing utility to graph the derivative on the interval $3 \le t \le 9$.

(c) Use the *trace* feature to find the year(s) during which the finance rate was changing the most.

(d) Use the *trace* feature to find the year(s) during which the finance rate was changing the least.

ALGEBRA TUTOR

Simplifying Algebraic Expressions

To be successful in using derivatives, you must be good at simplifying algebraic expressions. Here are some helpful simplification techniques.

1. Combine *like terms*. This may involve expanding an expression by multiplying factors.

2. Divide out *common factors* in the numerator and denominator of an expression.

3. Factor an expression.

4. Rationalize a denominator.

5. Add, subtract, multiply, or divide fractions.

TECH TUTOR

Symbolic algebra systems can simplify algebraic expressions. If you have access to such a system, try using it to simplify the expressions in this Algebra Tutor.

Example 1   Simplifying Fractional Expressions

a.
$$\frac{[3(x + \Delta x) + 5] - (3x + 5)}{\Delta x} = \frac{3x + 3\Delta x + 5 - 3x - 5}{\Delta x}$$ Multiply factors and remove parentheses.

$$= \frac{3\Delta x}{\Delta x}$$ Combine like terms.

$$= 3, \quad \Delta x \neq 0$$ Divide out common factors.

b.
$$\frac{(x + \Delta x)^2 - x^2}{\Delta x} = \frac{x^2 + 2x(\Delta x) + (\Delta x)^2 - x^2}{\Delta x}$$ Expand terms.

$$= \frac{2x(\Delta x) + (\Delta x)^2}{\Delta x}$$ Combine like terms.

$$= \frac{\Delta x(2x + \Delta x)}{\Delta x}$$ Factor.

$$= 2x + \Delta x, \quad \Delta x \neq 0$$ Divide out common factors.

c.
$$\frac{(x^2 - 1)(-2 - 2x) - (3 - 2x - x^2)(2)}{(x^2 - 1)^2}$$

$$= \frac{(-2x^2 - 2x^3 + 2 + 2x) - (6 - 4x - 2x^2)}{(x^2 - 1)^2}$$ Expand terms.

$$= \frac{-2x^2 - 2x^3 + 2 + 2x - 6 + 4x + 2x^2}{(x^2 - 1)^2}$$ Remove parentheses.

$$= \frac{-2x^3 + 6x - 4}{(x^2 - 1)^2}$$ Combine like terms.

d.
$$2\left(\frac{2x + 1}{3x}\right)\left[\frac{3x(2) - (2x + 1)(3)}{(3x)^2}\right]$$

$$= 2\left(\frac{2x + 1}{3x}\right)\left[\frac{6x - (6x + 3)}{(3x)^2}\right]$$ Multiply factors.

$$= \frac{2(2x + 1)(6x - 6x - 3)}{(3x)^3}$$ Multiply fractions and remove parentheses.

$$= \frac{2(2x + 1)(-3)}{3(9)x^3}$$ Combine like terms and factor.

$$= \frac{-2(2x + 1)}{9x^3}$$ Divide out common factors.

Example 2   Simplifying Expressions with Powers

Simplify each expression.

a. $(2x + 1)^2(6x + 1) + (3x^2 + x)(2)(2x + 1)(2)$

b. $(-1)(3x^2 - 2x)^{-2}(6x - 2)$

c. $(x)\left(\frac{1}{2}\right)(2x + 3)^{-1/2} + (2x + 3)^{1/2}(1)$

d. $\dfrac{x^2\left(\frac{1}{2}\right)(x^2 + 1)^{-1/2}(2x) - (x^2 + 1)^{1/2}(2x)}{x^4}$

SOLUTION

a. $(2x + 1)^2(6x + 1) + (3x^2 + x)(2)(2x + 1)(2)$

$= (2x + 1)[(2x + 1)(6x + 1) + (3x^2 + x)(2)(2)]$ Factor.

$= (2x + 1)[12x^2 + 8x + 1 + (12x^2 + 4x)]$ Multiply factors.

$= (2x + 1)(12x^2 + 8x + 1 + 12x^2 + 4x)$ Remove parentheses.

$= (2x + 1)(24x^2 + 12x + 1)$ Combine like terms.

b. $(-1)(3x^2 - 2x)^{-2}(6x - 2)$

$= \dfrac{(-1)(6x - 2)}{(3x^2 - 2x)^2}$ Rewrite as a fraction.

$= \dfrac{(-1)(2)(3x - 1)}{(3x^2 - 2x)^2}$ Factor.

$= \dfrac{-2(3x - 1)}{(3x^2 - 2x)^2}$ Multiply factors.

c. $(x)\left(\frac{1}{2}\right)(2x + 3)^{-1/2} + (2x + 3)^{1/2}(1)$

$= (2x + 3)^{-1/2}\left(\frac{1}{2}\right)[x + (2x + 3)(2)]$ Factor.

$= \dfrac{x + 4x + 6}{(2x + 3)^{1/2}(2)}$ Rewrite as a fraction.

$= \dfrac{5x + 6}{2(2x + 3)^{1/2}}$ Combine like terms.

d. $\dfrac{x^2\left(\frac{1}{2}\right)(x^2 + 1)^{-1/2}(2x) - (x^2 + 1)^{1/2}(2x)}{x^4}$

$= \dfrac{(x^3)(x^2 + 1)^{-1/2} - (x^2 + 1)^{1/2}(2x)}{x^4}$ Multiply factors.

$= \dfrac{(x^2 + 1)^{-1/2}(x)[x^2 - (x^2 + 1)(2)]}{x^4}$ Factor.

$= \dfrac{x[x^2 - (2x^2 + 2)]}{(x^2 + 1)^{1/2}x^4}$ Write with positive exponents.

$= \dfrac{x^2 - 2x^2 - 2}{(x^2 + 1)^{1/2}x^3}$ Divide out common factors and remove parentheses.

$= \dfrac{-x^2 - 2}{(x^2 + 1)^{1/2}x^3}$ Combine like terms.

STUDY TIP

All but one of the expressions in this Algebra Tutor are derivatives. Can you see what the original function is? Explain your reasoning.

SUMMARY AND STUDY STRATEGIES

After studying this chapter, you should have acquired the following skills.
The exercise numbers are keyed to the Review Exercises that begin on page 624.
Answers to odd-numbered Review Exercises are given in the back of the text.*

Section 7.1	Review Exercises
■ Use a table to estimate limits.	*1–4*
■ Determine whether limits exist. If they do, find the limits.	*5–22*

Section 7.2

■ Determine whether functions are continuous at a point, on an open interval, and on a closed interval.	*23–32*
■ Determine the constant such that *f* is continuous.	*33, 34*
■ Use analytic and graphical models of real-life data to solve real-life problems.	*35–38*

Section 7.3

■ Approximate the slope of the tangent line to a graph at a point.	*39–42*
■ Interpret the slope of a graph in a real-life setting.	*43, 44*
■ Use the limit definition to find the slope of a graph at a point and the derivative of a function.	*45–60*

$$f'(x) = \lim_{\Delta x \to 0} \frac{f(x + \Delta x) - f(x)}{\Delta x}$$

■ Use the graph of a function to recognize the *x*-values at which the function is differentiable.	*61–64*

Section 7.4

■ Use the Constant Rule for differentiation.	65, 66

$$\frac{d}{dx}[c] = 0$$

■ Use the Power Rule for differentiation.	67, 68

$$\frac{d}{dx}[x^n] = nx^{n-1}$$

■ Use the Constant Multiple Rule for differentiation.	69–72

$$\frac{d}{dx}[cf(x)] = cf'(x)$$

■ Use the Sum and Difference Rules for differentiation.	73–76

$$\frac{d}{dx}[f(x) \pm g(x)] = f'(x) \pm g'(x)$$

■ Use derivatives to find the slope of a graph.	77–80

---

* A wide range of valuable study aids are available to help you master the material in this chapter. The *Student Solutions Manual* includes step-by-step solutions to all odd-numbered exercises to help you review and prepare. The student website at *www.cengagebrain.com* offers algebra help and a *Graphing Technology Guide*, which contains step-by-step commands and instructions for a wide variety of graphing calculators.

Section 7.4 (continued)

Review Exercises

■ Use derivatives to write equations of tangent lines.

■ Use derivatives to answer questions about real-life situations.

Section 7.5

■ Find the average rate of change of a function over an interval and the instantaneous rate of change at a point.

Average rate of change: $\dfrac{f(b) - f(a)}{b - a}$;   Instantaneous rate of change: $\displaystyle\lim_{\Delta x \to 0} \dfrac{f(x + \Delta x) - f(x)}{\Delta x}$

■ Use derivatives to find the velocities of objects.

■ Find the marginal revenues, marginal costs, and marginal profits for products.

■ Use derivatives to answer questions about real-life situations.

Section 7.6

■ Use the Product Rule for differentiation.

$$\frac{d}{dx}[f(x)g(x)] = f(x)g'(x) + g(x)f'(x)$$

■ Use the Quotient Rule for differentiation.

$$\frac{d}{dx}\left[\frac{f(x)}{g(x)}\right] = \frac{g(x)f'(x) - f(x)g'(x)}{[g(x)]^2}$$

Section 7.7

■ Use the General Power Rule for differentiation.

$$\frac{d}{dx}[u^n] = nu^{n-1}u'$$

■ Use differentiation rules efficiently to find the derivative of any algebraic function, then simplify the result.

■ Use derivatives to answer questions about real-life situations.   (Sections 7.3–7.7)

Study Strategies

■ **Simplify Your Derivatives**   You may ask if you have to simplify your derivatives. The answer is "Yes, if you expect to use them." In the next chapter, you will see that almost all applications of derivatives require that the derivatives be written in simplified form. It is not difficult to see the advantage of a derivative in simplified form. Consider, for instance, the derivative of

$$f(x) = \frac{x}{\sqrt{x^2 + 1}}.$$

The "raw form" produced by the Quotient and Chain Rules

$$f'(x) = \frac{(x^2 + 1)^{1/2}(1) - (x)\left(\frac{1}{2}\right)(x^2 + 1)^{-1/2}(2x)}{\left(\sqrt{x^2 + 1}\right)^2}$$

is obviously much more difficult to use than the simplified form

$$f'(x) = \frac{1}{(x^2 + 1)^{3/2}}.$$

■ **List Units of Measure in Applied Problems**   When using derivatives in real-life applications, be sure to list the units of measure for each variable. For instance, if R is measured in dollars and t is measured in years, then the derivative dR/dt is measured in dollars per year.

Review Exercises

See www.CalcChat.com for worked-out solutions to odd-numbered exercises.

Finding Limits Numerically In Exercises 1–4, complete the table and use the result to estimate the limit. Use a graphing utility to graph the function to confirm your result.

1. $\lim\limits_{x \to 1} (4x - 3)$

x	0.9	0.99	0.999	1	1.001	1.01	1.1
$f(x)$				?			

2. $\lim\limits_{x \to 3} \dfrac{x - 3}{x^2 - 2x - 3}$

x	2.9	2.99	2.999	3	3.001	3.01	3.1
$f(x)$				?			

3. $\lim\limits_{x \to 0} \dfrac{\sqrt{x + 6} - 6}{x}$

x	−0.1	−0.01	−0.001	0	0.001	0.01	0.1
$f(x)$				?			

4. $\lim\limits_{x \to 7} \dfrac{\dfrac{1}{x - 7} - \dfrac{1}{7}}{x}$

x	6.9	6.99	6.999	7	7.001	7.01	7.1
$f(x)$				?			

Finding Limits In Exercises 5–22, find the limit (if it exists).

5. $\lim\limits_{x \to 3} 8$

6. $\lim\limits_{x \to 4} x^2$

7. $\lim\limits_{x \to 2} (5x - 3)$

8. $\lim\limits_{x \to 5} (2x + 4)$

9. $\lim\limits_{x \to -1} \dfrac{x + 3}{6x + 1}$

10. $\lim\limits_{t \to 3} \dfrac{t}{t + 5}$

11. $\lim\limits_{t \to 0} \dfrac{t^2 + 1}{t}$

12. $\lim\limits_{t \to 2} \dfrac{t + 1}{t - 2}$

13. $\lim\limits_{x \to -2} \dfrac{x + 2}{x^2 - 4}$

14. $\lim\limits_{x \to 3^-} \dfrac{x^2 - 9}{x - 3}$

15. $\lim\limits_{x \to 0^+} \left(x - \dfrac{1}{x} \right)$

16. $\lim\limits_{x \to 1/2} \dfrac{2x - 1}{6x - 3}$

17. $\lim\limits_{x \to 0} \dfrac{[1/(x - 2)] - 1}{x}$

18. $\lim\limits_{s \to 0} \dfrac{(1/\sqrt{1 + s}) - 1}{s}$

19. $\lim\limits_{x \to 0} f(x)$, where $f(x) = \begin{cases} x + 5, & x \ne 0 \\ 3, & x = 0 \end{cases}$

20. $\lim\limits_{x \to -2} f(x)$, where $f(x) = \begin{cases} \frac{1}{2}x + 5, & x < -2 \\ -x + 2, & x \ge -2 \end{cases}$

21. $\lim\limits_{\Delta x \to 0} \dfrac{(x + \Delta x)^3 - (x + \Delta x) - (x^3 - x)}{\Delta x}$

22. $\lim\limits_{\Delta x \to 0} \dfrac{1 - (x + \Delta x)^2 - (1 - x^2)}{\Delta x}$

Determining Continuity In Exercises 23–32, describe the interval(s) on which the function is continuous. Explain why the function is continuous on the interval(s). If the function has a discontinuity, identify the conditions of continuity that are not satisfied.

23. $f(x) = x + 6$

24. $f(x) = x^2 + 3x + 2$

25. $f(x) = \dfrac{1}{(x + 4)^2}$

26. $f(x) = \dfrac{x + 2}{x}$

27. $f(x) = \dfrac{3}{x + 1}$

28. $f(x) = \dfrac{x + 1}{2x + 2}$

29. $f(x) = [\![x + 3]\!]$

30. $f(x) = [\![x]\!] - 2$

31. $f(x) = \begin{cases} x, & x \le 0 \\ x + 1, & x > 0 \end{cases}$

32. $f(x) = \begin{cases} x, & x \le 0 \\ x^2, & x > 0 \end{cases}$

Making a Function Continuous In Exercises 33 and 34, find the constant a such that the function is continuous on the entire real number line.

33. $f(x) = \begin{cases} -x + 1, & x \le 3 \\ ax - 8, & x > 3 \end{cases}$

34. $f(x) = \begin{cases} x + 1, & x < 1 \\ 2x + a, & x \ge 1 \end{cases}$

35. Consumer Awareness The cost C (in dollars) of purchasing x bottles of vitamins at a whole foods store is shown below.

$$C(x) = \begin{cases} 5.99x, & 0 < x \le 5 \\ 4.99x, & 5 < x \le 10 \\ 3.99x, & 10 < x \le 15 \\ 2.99x, & x > 15 \end{cases}$$

(a) Use a graphing utility to graph the function and then discuss its continuity. At what values is the function not continuous? Explain your reasoning.

(b) Find the cost of purchasing 10 bottles.

36. Salary Contract A union contract guarantees a 10% salary increase yearly for 3 years. For a current salary of $28,000, the salaries S (in thousands of dollars) for the next 3 years are given by

$$S(t) = \begin{cases} 28.00, & 0 < t \le 1 \\ 30.80, & 1 < t \le 2 \\ 33.88, & 2 < t \le 3 \end{cases}$$

where $t = 0$ represents the present year. Does the limit of S exist as t approaches 2? Explain your reasoning.

37. Consumer Awareness A pay-as-you-go cellular phone charges $1 for the first minute and $0.10 for each additional minute or fraction thereof.

(a) Use the greatest integer function to create a model for the cost C of a phone call lasting t minutes.

(b) Use a graphing utility to graph the function and then discuss its continuity.

38. Recycling A recycling center pays $0.50 for each pound of aluminum cans. Twenty-four aluminum cans weigh one pound. A mathematical model for the amount A paid by the recycling center is

$$A = \frac{1}{2}\left[\!\left[\frac{x}{24}\right]\!\right]$$

where x is the number of cans.

(a) Use a graphing utility to graph the function and then discuss its continuity.

(b) How much does the recycling center pay out for 1500 cans?

Approximating the Slope of a Graph   In Exercises 39–42, estimate the slope of the graph at the point (x, y). (Each square on the grid is 1 unit by 1 unit.)

39.
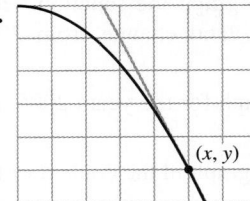

40.
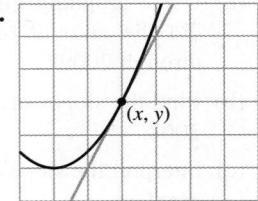

41.
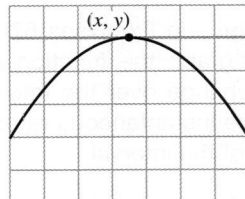

42.
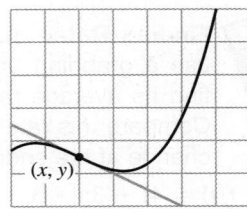

43. Consumer Trends   The graph shows the number of visitors V (in thousands) to a national park during a one-year period, where $t = 1$ corresponds to January. Estimate and interpret the slopes of the graph at $t = 1$, 8, and 12.

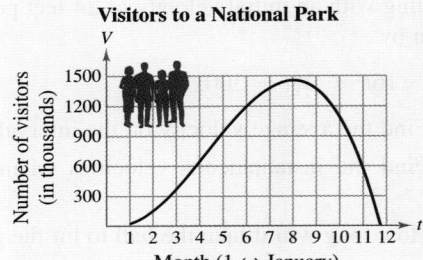
Visitors to a National Park

44. White-Water Rafting   Two white-water rafters leave a campsite simultaneously and start downstream on a 9-mile trip. Their distances from the campsite are given by $s = f(t)$ and $s = g(t)$, where s is measured in miles and t is measured in hours.

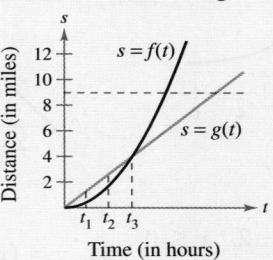
White-Water Rafting

(a) Which rafter is traveling at a greater rate at t_1?

(b) What can you conclude about their rates at t_2?

(c) What can you conclude about their rates at t_3?

(d) Which rafter finishes the trip first? Explain your reasoning.

Finding the Slope of a Graph   In Exercises 45–52, use the limit definition to find the slope of the tangent line to the graph of f at the given point.

45. $f(x) = -3x - 5;\ (-2, 1)$

46. $f(x) = 7x + 3;\ (-1, -4)$

47. $f(x) = x^2 - 4x;\ (1, -3)$

48. $f(x) = x^2 + 10;\ (2, 14)$

49. $f(x) = \sqrt{x + 9};\ (-5, 2)$

50. $f(x) = \sqrt{x - 1};\ (10, 3)$

51. $f(x) = \dfrac{1}{x - 5};\ (6, 1)$

52. $f(x) = \dfrac{1}{x + 4};\ (-3, 1)$

Finding a Derivative   In Exercises 53–60, use the limit definition to find the derivative of the function.

53. $f(x) = 9x + 1$

54. $f(x) = 1 - 4x$

55. $f(x) = -\frac{1}{2}x^2 + 2x$

56. $f(x) = 4 - x^2$

57. $f(x) = \sqrt{x - 5}$

58. $f(x) = \sqrt{x + 3}$

59. $f(x) = \dfrac{5}{x}$

60. $f(x) = \dfrac{1}{x + 4}$

Determining Differentiability In Exercises 61–64, determine the *x*-values at which the function is differentiable. Explain your reasoning.

61. $y = \dfrac{x + 1}{x - 1}$

62. $y = -|x| + 3$

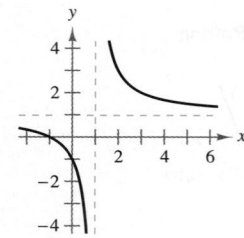

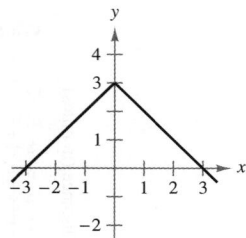

63. $y = \begin{cases} -x - 2, & x \le 0 \\ x^3 + 2, & x > 0 \end{cases}$

64. $y = (x + 1)^{2/3}$

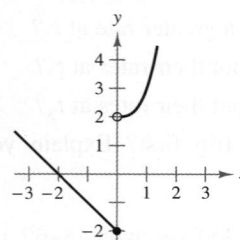

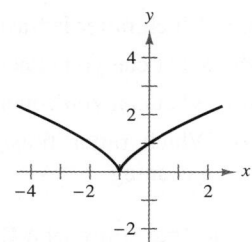

Finding Derivatives In Exercises 65–76, find the derivative of the function.

65. $y = -6$

66. $f(x) = 5$

67. $f(x) = x^3$

68. $h(x) = \dfrac{1}{x^4}$

69. $f(x) = 4x^2$

70. $g(t) = 6t^5$

71. $f(x) = \dfrac{2x^4}{5}$

72. $y = 3x^{2/3}$

73. $g(x) = 2x^4 + 3x^2$

74. $f(x) = 6x^2 - 4x$

75. $y = x^2 + 6x - 7$

76. $y = 2x^4 - 3x^3 + x$

Finding the Slope of a Graph In Exercises 77–80, find the slope of the graph of the function at the given point.

77. $f(x) = 2x^{-1/2};\ (4, 1)$

78. $y = \dfrac{3}{2x} + 3;\ \left(\dfrac{1}{2}, 6\right)$

79. $g(x) = x^3 - 4x^2 - 6x + 8;\ (-1, 9)$

80. $y = 2x^4 - 5x^3 + 6x^2 - x;\ (1, 2)$

 **Finding an Equation of a Tangent Line** In Exercises 81–84, (a) find an equation of the tangent line to the graph of the function at the given point, (b) use a graphing utility to graph the function and its tangent line at the point, and (c) use the *derivative* feature of a graphing utility to confirm your results.

81. $f(x) = 2x^2 - 3x + 1;\ (2, 3)$

82. $y = 11x^4 - 5x^2 + 1;\ (-1, 7)$

83. $f(x) = \sqrt{x} - \dfrac{1}{\sqrt{x}};\ (1, 0)$

84. $f(x) = -x^2 - 4x - 4;\ (-4, -4)$

85. Sales The annual sales S (in millions of dollars) for Tractor Supply Company for the years 2003 through 2009 can be modeled by

$$S = -0.7500t^4 + 13.278t^3 - 74.50t^2 + 440.2t + 523$$

where t is the year, with $t = 3$ corresponding to 2003. *(Source: Tractor Supply Company)*

(a) Use a graphing utility to graph the function.

(b) Find the slopes of the graph for the years 2004 and 2007.

(c) Interpret the slope of the graph in the context of the problem.

86. Farms The amount of farm land L (in millions of acres) in the United States for the years 2004 through 2009 can be modeled by

$$L = 0.0991t^3 + 1.512t^2 + 4.01t + 933.9$$

where t is the year, with $t = 4$ corresponding to 2004. *(Source: U.S. Department of Agriculture)*

(a) Use a graphing utility to graph the function. How did the amount of farm land change from 2004 through 2009?

(b) Find the slopes of the graph for the years 2005 and 2008.

(c) Interpret the slope of the graph in the context of the problem.

 Finding Rates of Change In Exercises 87–90, use a graphing utility to graph the function and find its average rate of change over the interval. Compare this rate with the instantaneous rates of change at the endpoints of the interval.

87. $f(t) = 4t + 3;\ [-3, 1]$

88. $f(x) = x^{2/3};\ [1, 8]$

89. $f(x) = x^2 + 3x - 4;\ [0, 1]$

90. $f(x) = x^3 + x;\ [-2, 2]$

91. Velocity The height s (in feet) at time t (in seconds) of a ball thrown upward from the top of a 300-foot building with an initial velocity of 24 feet per second is given by

$$s = -16t^2 + 24t + 300.$$

(a) Find the average velocity on the interval $[1, 2]$.

(b) Find the instantaneous velocities when $t = 1$ and $t = 3$.

(c) How long will it take the ball to hit the ground?

(d) Find the velocity of the ball when it hits the ground.

92. Velocity A rock is dropped from a tower on the Brooklyn Bridge, 276 feet above the East River. Let t represent the time in seconds.

(a) Find the position and velocity functions for the rock.

(b) Find the average velocity over the interval $[0, 2]$.

(c) Find the instantaneous velocities at $t = 2$ and $t = 3$.

(d) How long will it take the rock to hit the water?

(e) Find the velocity of the rock when it hits the water.

Marginal Cost In Exercises 93–96, find the marginal cost for producing x units. (The cost is measured in dollars.)

93. $C = 2500 + 320x$ **94.** $C = 3x^3 + 24,000$

95. $C = 370 + 2.55\sqrt{x}$ **96.** $C = 475 + 5.25x^{2/3}$

Marginal Revenue In Exercises 97–100, find the marginal revenue for producing x units. (The revenue is measured in dollars.)

97. $R = 150x - 0.6x^2$ **98.** $R = 150x - \frac{3}{4}x^2$

99. $R = -4x^3 + 2x^2 + 100x$ **100.** $R = 4x + 10\sqrt{x}$

Marginal Profit In Exercises 101 and 102, find the marginal profit for producing x units. (The profit is measured in dollars.)

101. $P = -0.0002x^3 + 6x^2 - x - 2000$

102. $P = -\frac{1}{15}x^3 + 4000x^2 - 120x - 144,000$

103. Marginal Profit The profit P (in dollars) from selling x units of a product is given by

$$P = -0.05x^2 + 20x - 1000.$$

(a) Find the additional profit when the sales increase from 100 to 101 units.

(b) Find the marginal profit when $x = 100$ units.

(c) Compare the results of parts (a) and (b).

104. Population Growth The population P (in millions) of Brazil from 1980 through 2010 can be modeled by

$$P = -0.007t^2 + 2.78t + 123.6$$

where t represents the year, with $t = 0$ corresponding to 1980. *(Source: U.S. Census Bureau)*

(a) Evaluate P for $t = 0, 5, 10, 15, 20, 25,$ and 30. Explain these values.

(b) Determine the population growth rate, dP/dt.

(c) Evaluate dP/dt for the same values as in part (a). Explain your results.

Finding Derivatives In Exercises 105–126, find the derivative of the function. State which differentiation rule(s) you used to find the derivative.

105. $f(x) = x^3(5 - 3x^2)$ **106.** $y = (3x^2 + 7)(x^2 - 2x)$

107. $y = (4x - 3)(x^3 - 2x^2)$

108. $s = \left(4 - \frac{1}{t^2}\right)(t^2 - 3t)$

109. $g(x) = \dfrac{x}{x + 3}$

110. $f(x) = \dfrac{2 - 5x}{3x + 1}$

111. $f(x) = \dfrac{6x - 5}{x^2 + 1}$

112. $f(x) = \dfrac{x^2 + x - 1}{x^2 - 1}$

113. $f(x) = (5x^2 + 2)^3$

114. $f(x) = \sqrt[3]{x^2 - 1}$

115. $h(x) = \dfrac{2}{\sqrt{x + 1}}$

116. $g(x) = \sqrt{x^6 - 12x^3 + 9}$

117. $g(x) = x\sqrt{x^2 + 1}$

118. $g(t) = \dfrac{t}{(1 - t)^3}$

119. $f(x) = x(1 - 4x^2)^2$

120. $f(x) = \left(x^2 + \dfrac{1}{x}\right)^5$

121. $h(x) = [x^2(2x + 3)]^3$

122. $f(x) = [(x - 2)(x + 4)]^2$

123. $f(x) = x^2(x - 1)^5$

124. $f(s) = s^3(s^2 - 1)^{5/2}$

125. $h(t) = \dfrac{\sqrt{3t + 1}}{(1 - 3t)^2}$

126. $g(x) = \dfrac{(3x + 1)^2}{(x^2 + 1)^2}$

127. Physical Science The temperature T (in degrees Fahrenheit) of food placed in a freezer can be modeled by

$$T = \dfrac{1300}{t^2 + 2t + 25}$$

where t is the time (in hours).

(a) Find the rates of change of T at $t = 1, 3, 5,$ and 10.

(b) Graph the model on a graphing utility and describe the rate at which the temperature is changing.

128. Forestry According to the *Doyle Log Rule*, the volume V (in board-feet) of a log of length L (in feet) and diameter D (in inches) at the small end is

$$V = \left(\dfrac{D - 4}{4}\right)^2 L.$$

Find the rates at which the volume is changing with respect to D for a 12-foot-long log whose smallest diameter is (a) 8 inches, (b) 16 inches, (c) 24 inches, and (d) 36 inches.

TEST YOURSELF

See www.CalcChat.com for worked-out solutions to odd-numbered exercises.

Take this test as you would take a test in class. When you are done, check your work against the answers given in the back of the book.

In Exercises 1–4, find the limit (if it exists).

1. $\displaystyle\lim_{x \to 0} \frac{x - 2}{x + 2}$

2. $\displaystyle\lim_{x \to 5} \frac{x + 5}{x - 5}$

3. $\displaystyle\lim_{x \to -3} \frac{x^2 + 2x - 3}{x^2 + 4x + 3}$

4. $\displaystyle\lim_{x \to 0} \frac{\sqrt{x + 9} - 3}{x}$

In Exercises 5–8, describe the interval(s) on which the function is continuous. Explain why the function is continuous on the interval(s). If the function has a discontinuity, identify the conditions of continuity that are not satisfied.

5. $f(x) = x^2 - 2x + 4$

6. $f(x) = \dfrac{x^2 - 16}{x - 4}$

7. $f(x) = \sqrt{5 - x}$

8. $f(x) = \begin{cases} 1 - x, & x < 1 \\ x - x^2, & x \geq 1 \end{cases}$

In Exercises 9 and 10, use the limit definition to find the derivative of the function. Then find the slope of the tangent line to the graph of f at the given point.

9. $f(x) = x^2 + 1; (2, 5)$

10. $f(x) = \sqrt{x} - 2; (4, 0)$

In Exercises 11–19, find the derivative of the function.

11. $f(t) = t^3 + 2t$

12. $f(x) = 4x^2 - 8x + 1$

13. $f(x) = x^{3/2}$

14. $f(x) = (x + 3)(x^2 + 2x)$

15. $f(x) = -3x^{-3}$

16. $f(x) = \sqrt{x}(5 + x)$

17. $f(x) = (3x^2 + 4)^2$

18. $f(x) = \sqrt{1 - 2x}$

19. $f(x) = \dfrac{(5x - 1)^3}{x}$

20. Find an equation of the tangent line to the graph of

$$f(x) = x - \frac{1}{x}$$

at the point $(1, 0)$. Then use a graphing utility to graph the function and the tangent line in the same viewing window.

21. The annual sales S (in billions of dollars) of CVS Caremark for the years 2004 through 2009 can be modeled by

$$S = -1.3241t^3 + 26.562t^2 - 155.81t + 314.3$$

where t represents the year, with $t = 4$ corresponding to 2004. *(Source: CVS Caremark Corporation)*

(a) Approximate the average rate of change for the years from 2005 through 2008.

(b) Find the instantaneous rates of change of the model for the years 2005 and 2008.

(c) Interpret the results of parts (a) and (b) in the context of the problem.

22. The monthly demand and cost functions for a product are given by

$$p = 1700 - 0.016x \quad \text{and} \quad C = 715{,}000 + 240x.$$

(a) Write the profit function for this product.

(b) Find the rate of change of the profit when the monthly sales are $x = 700$ units.

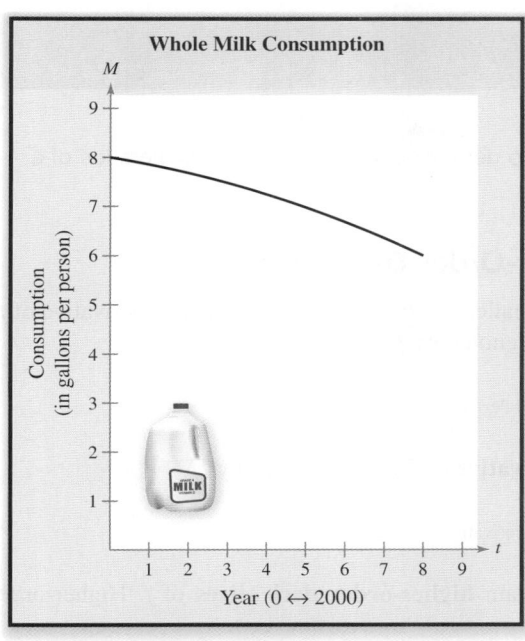

Whole Milk Consumption

M

Consumption (in gallons per person)

Year (0 ↔ 2000)

8 Applications of the Derivative

8.1 Higher-Order Derivatives

8.2 Implicit Differentiation

8.3 Related Rates

8.4 Increasing and Decreasing Functions

8.5 Extrema and the First-Derivative Test

8.6 Concavity and the Second-Derivative Test

Example 2 on page 653 shows how the derivative can be used to show that milk consumption decreased in the United States from 2000 through 2008.

8.1 Higher-Order Derivatives

■ Find higher-order derivatives.

■ Find and use a position function to determine the velocity and acceleration of a moving object.

Second, Third, and Higher-Order Derivatives

The "standard" derivative f' is often called the **first derivative** of f. The derivative of f' is the **second derivative** of f and is denoted by f''.

$$\frac{d}{dx}[f'(x)] = f''(x) \qquad \text{Second derivative}$$

The derivative of f'' is the **third derivative** of f and is denoted by f'''.

$$\frac{d}{dx}[f''(x)] = f'''(x) \qquad \text{Third derivative}$$

By continuing this process, you obtain **higher-order derivatives** of f. Higher-order derivatives are denoted as follows.

Position of a Ball

In Exercise 35 on page 635, you will use derivatives to find the velocity function and the acceleration function of a ball.

Notation for Higher-Order Derivatives				
1. 1st derivative:	y',	$f'(x)$,	$\dfrac{dy}{dx}$,	$\dfrac{d}{dx}[f(x)]$,	$D_x[y]$
2. 2nd derivative:	y'',	$f''(x)$,	$\dfrac{d^2y}{dx^2}$,	$\dfrac{d^2}{dx^2}[f(x)]$,	$D_x^2[y]$
3. 3rd derivative:	y''',	$f'''(x)$,	$\dfrac{d^3y}{dx^3}$,	$\dfrac{d^3}{dx^3}[f(x)]$,	$D_x^3[y]$
4. 4th derivative:	$y^{(4)}$,	$f^{(4)}(x)$,	$\dfrac{d^4y}{dx^4}$,	$\dfrac{d^4}{dx^4}[f(x)]$,	$D_x^4[y]$
5. nth derivative:	$y^{(n)}$,	$f^{(n)}(x)$,	$\dfrac{d^ny}{dx^n}$,	$\dfrac{d^n}{dx^n}[f(x)]$,	$D_x^n[y]$

Example 1   Finding Higher-Order Derivatives

Find the first five derivatives of

$$f(x) = 2x^4 - 3x^2.$$

SOLUTION

$f(x) = 2x^4 - 3x^2$	Write original function.
$f'(x) = 8x^3 - 6x$	First derivative
$f''(x) = 24x^2 - 6$	Second derivative
$f'''(x) = 48x$	Third derivative
$f^{(4)}(x) = 48$	Fourth derivative
$f^{(5)}(x) = 0$	Fifth derivative

✓ **Checkpoint 1**

Find the first four derivatives of $f(x) = 6x^3 - 2x^2 + 1$.

Example 2   Finding Higher-Order Derivatives

Find the value of $g'''(2)$ for the function

$$g(t) = -t^4 + 2t^3 + t + 4.$$

SOLUTION   Begin by differentiating three times.

$$g'(t) = -4t^3 + 6t^2 + 1 \qquad \text{First derivative}$$

$$g''(t) = -12t^2 + 12t \qquad \text{Second derivative}$$

$$g'''(t) = -24t + 12 \qquad \text{Third derivative}$$

Then, evaluate the third derivative of g at $t = 2$.

$$g'''(2) = -24(2) + 12$$

$$= -36 \qquad \text{Value of third derivative}$$

✓ **Checkpoint 2**

Find the value of $g'''(1)$ for $g(x) = x^4 - x^3 + 2x$.

Examples 1 and 2 show how to find higher-order derivatives of *polynomial* functions. Note that with each successive differentiation, the degree of the polynomial drops by one. Eventually, higher-order derivatives of polynomial functions degenerate to a constant function. Specifically, the nth-order derivative of an nth-degree polynomial function

$$f(x) = a_n x^n + a_{n-1} x^{n-1} + \cdots + a_1 x + a_0$$

is the constant function

$$f^{(n)}(x) = n! a_n$$

where $n! = 1 \cdot 2 \cdot 3 \cdots n$. Each derivative of order higher than n is the zero function.

TECH TUTOR

Higher-order derivatives of nonpolynomial functions can be difficult to find by hand. If you have access to a symbolic differentiation utility, try using it to find higher-order derivatives.

Example 3   Finding Higher-Order Derivatives

Find the first four derivatives of $y = x^{-1}$.

SOLUTION

$$y = x^{-1} = \frac{1}{x} \qquad \text{Write original function.}$$

$$y' = (-1)x^{-2} = -\frac{1}{x^2} \qquad \text{First derivative}$$

$$y'' = (-1)(-2)x^{-3} = \frac{2}{x^3} \qquad \text{Second derivative}$$

$$y''' = (-1)(-2)(-3)x^{-4} = -\frac{6}{x^4} \qquad \text{Third derivative}$$

$$y^{(4)} = (-1)(-2)(-3)(-4)x^{-5} = \frac{24}{x^5} \qquad \text{Fourth derivative}$$

✓ **Checkpoint 3**

Find the fourth derivative of

$$y = \frac{1}{x^2}.$$

Acceleration

In Section 7.5, you saw that the velocity of a free-falling object (neglecting air resistance) is given by the derivative of its position function. In other words, the rate of change of the position with respect to time is defined to be the velocity. In a similar way, the rate of change of the velocity with respect to time is defined to be the **acceleration** of the object.

$$s = f(t) \qquad \text{Position function}$$

$$\frac{ds}{dt} = f'(t) \qquad \text{Velocity function}$$

$$\frac{d^2s}{dt^2} = f''(t) \qquad \text{Acceleration function}$$

To find the position, velocity, or acceleration at a particular time t, substitute the given value of t into the appropriate function, as illustrated in Example 4.

 Example 4   Finding Acceleration

A ball is thrown upward from the top of a 160-foot cliff, as shown in Figure 8.1. The initial velocity of the ball is 48 feet per second, which implies that the position function is

$$s = -16t^2 + 48t + 160$$

where the time t is measured in seconds. Find the height, velocity, and acceleration of the ball at $t = 3$.

SOLUTION   Begin by differentiating to find the velocity and acceleration functions.

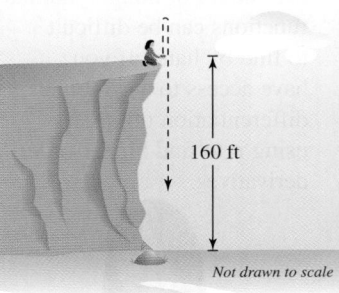

160 ft

Not drawn to scale

FIGURE 8.1

$$s = -16t^2 + 48t + 160 \qquad \text{Position function}$$

$$\frac{ds}{dt} = -32t + 48 \qquad \text{Velocity function}$$

$$\frac{d^2s}{dt^2} = -32 \qquad \text{Acceleration function}$$

To find the height, velocity, and acceleration at $t = 3$, substitute $t = 3$ into each of the functions above.

Height $= -16(3)^2 + 48(3) + 160 = 160$ feet

Velocity $= -32(3) + 48 = -48$ feet per second

Acceleration $= -32$ feet per second squared

✓ Checkpoint 4

A ball is thrown upward from the top of an 80-foot cliff with an initial velocity of 64 feet per second, which implies that the position function is

$$s = -16t^2 + 64t + 80$$

where the time t is measured in seconds. Find the height, velocity, and acceleration of the ball at $t = 2$.

The acceleration due to gravity on the surface of the moon is only about one-sixth that exerted on the surface of Earth.

In Example 4, notice that the acceleration of the ball is -32 feet per second squared at any time t. This constant acceleration is due to the gravitational force of Earth and is called the **acceleration due to gravity.** Note that the negative value indicates that the ball is being pulled *down*—toward Earth.

Although the acceleration exerted on a falling object is relatively constant near Earth's surface, it varies greatly throughout our solar system. Large planets exert a much greater gravitational pull than do small planets or moons. The next example describes the motion of a free-falling object on the moon.

Example 5   Finding Acceleration on the Moon

An astronaut standing on the surface of the moon throws a rock upward. The height s (in feet) of the rock is given by

$$s = -\frac{27}{10}t^2 + 27t + 6$$

where t is measured in seconds. How does the acceleration due to gravity on the moon compare with that on Earth?

SOLUTION

$$s = -\frac{27}{10}t^2 + 27t + 6 \qquad \text{Position function}$$

$$\frac{ds}{dt} = -\frac{27}{5}t + 27 \qquad \text{Velocity function}$$

$$\frac{d^2s}{dt^2} = -\frac{27}{5} \qquad \text{Acceleration function}$$

So, the acceleration at any time is

$$-\frac{27}{5} = -5.4 \text{ feet per second squared}$$

—about one-sixth of the acceleration due to gravity on Earth.   ▬▬▬▬▬▬

✓ Checkpoint 5

The position function on Earth, where s is measured in meters, t is measured in seconds, v_0 is the initial velocity in meters per second, and h_0 is the initial height in meters, is

$$s = -4.9t^2 + v_0 t + h_0.$$

An object is thrown upward with an initial velocity of 2.2 meters per second from an initial height of 3.6 meters. What is the acceleration due to gravity on Earth in meters per second squared?

The position function described in Example 5 neglects air resistance, which is appropriate because the moon has no atmosphere—and *no air resistance.* This means that the position function for any free-falling object on the moon is given by

$$s = -\frac{27}{10}t^2 + v_0 t + h_0$$

where s is the height (in feet), t is the time (in seconds), v_0 is the initial velocity (in feet per second), and h_0 is the initial height (in feet). For instance, the rock in Example 5 was thrown upward with an initial velocity of 27 feet per second and had an initial height of 6 feet. This position function is valid for all objects, whether heavy ones such as hammers or light ones such as feathers.

Example 6   **Finding Velocity and Acceleration**

The velocity v (in feet per second) of a certain automobile starting from rest is

$$v = \frac{80t}{t + 5}$$ Velocity function

where t is the time (in seconds). The positions of the automobile at 10-second intervals are shown in Figure 8.2. Find the velocity and acceleration of the automobile at 10-second intervals from $t = 0$ to $t = 60$.

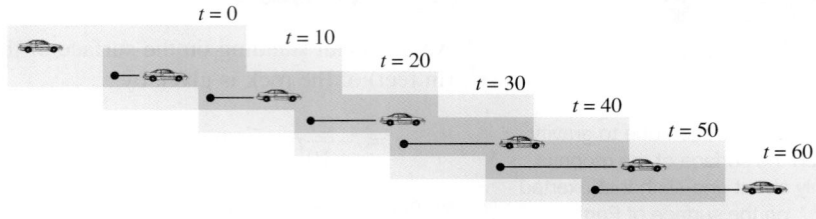

FIGURE 8.2

SOLUTION   To find the acceleration function, differentiate the velocity function.

$$\frac{dv}{dt} = \frac{(t + 5)(80) - (80t)(1)}{(t + 5)^2}$$ Apply Quotient Rule.

$$= \frac{400}{(t + 5)^2}$$ Acceleration function

t (seconds)	0	10	20	30	40	50	60
v (ft/sec)	0	53.3	64.0	68.6	71.1	72.7	73.8
$\dfrac{dv}{dt}$ (ft/sec²)	16	1.78	0.64	0.33	0.20	0.13	0.09

In the table, note that the acceleration approaches zero as the velocity levels off. This observation should agree with your experience—when riding in an accelerating automobile, you do not feel the velocity, but you do feel the acceleration. In other words, you feel changes in velocity.

 **Checkpoint 6**

Use a graphing utility to graph the velocity function and acceleration function in Example 6 in the same viewing window. Compare the graphs with the table in Example 6. As the velocity levels off, what does the acceleration approach?

SUMMARIZE   (Section 8.1)

1. State the meaning of each derivative listed below *(page 630)*. For examples of higher-order derivatives, see Examples 1, 2, and 3.

    (a) y'' (b) $f^{(4)}(x)$ (c) $\dfrac{d^3y}{dx^3}$

2. Describe a real-life example of how higher-order derivatives can be used to analyze the velocity and acceleration of an object *(page 632, Examples 4, 5, and 6)*.

SKILLS WARM UP 8.1   The following warm-up exercises involve skills that were covered in earlier sections. You will use these skills in the exercise set for this section. For additional help, review Sections 1.3, 1.4, 2.5, and 7.6.

In Exercises 1–4, solve the equation.

1. $-16t^2 + 24t = 0$

2. $-16t^2 + 80t + 224 = 0$

3. $-16t^2 + 128t + 320 = 0$

4. $-16t^2 + 9t + 1440 = 0$

In Exercises 5–8, find dy/dx.

5. $y = x^2(2x + 7)$ **6.** $y = (x^2 + 3x)(2x^2 - 5)$

7. $y = \dfrac{x^2}{2x + 7}$ **8.** $y = \dfrac{x^2 + 3x}{2x^2 - 5}$

In Exercises 9 and 10, find the domain and range of f.

9. $f(x) = x^2 - 4$

10. $f(x) = \sqrt{x - 7}$

Exercises 8.1

See www.CalcChat.com for worked-out solutions to odd-numbered exercises.

Finding Higher-Order Derivatives   In Exercises 1–12, find the second derivative of the function. *See Examples 1 and 3.*

1. $f(x) = 9 - 2x$

2. $f(x) = 4x + 15$

3. $f(x) = x^2 + 7x - 4$

4. $f(x) = 3x^2 + 4x$

5. $g(t) = \frac{1}{3}t^3 - 4t^2 + 2t$

6. $f(x) = -2x^5 + 3x^4 + 8x$

7. $f(t) = \dfrac{3}{4t^2}$

8. $g(t) = \dfrac{8}{t}$

9. $f(x) = 3(2 - x^2)^3$

10. $y = 4(x^2 + 5x)^3$

11. $f(x) = \dfrac{x + 1}{x - 1}$

12. $g(t) = -\dfrac{4}{(t + 2)^2}$

Finding Higher-Order Derivatives   In Exercises 13–18, find the third derivative of the function. *See Examples 1 and 3.*

13. $f(x) = x^5 - 3x^4$

14. $f(x) = x^4 - 2x^3$

15. $f(x) = 5x(x + 4)^3$

16. $f(x) = (x^3 - 6)^4$

17. $f(x) = \dfrac{3}{16x^2}$

18. $f(x) = -\dfrac{2}{x}$

Finding Higher-Order Derivatives   In Exercises 19–24, find the given value. *See Example 2.*

Function	Value
19. $g(t) = 5t^4 + 10t^2 + 3$	$g''(2)$
20. $f(x) = 9 - x^2$	$f''(-\sqrt{5})$
21. $f(x) = \sqrt{4 - x}$	$f'''(-5)$
22. $f(t) = \sqrt{2t + 3}$	$f'''(\frac{1}{2})$
23. $f(x) = (x^3 - 2x)^3$	$f''(1)$
24. $g(x) = (x^2 + 3x)^4$	$g''(-1)$

Finding Higher-Order Derivatives   In Exercises 25–30, find the higher-order derivative. *See Examples 1 and 3.*

Given	Derivative
25. $f'(x) = 2x^2$	$f''(x)$
26. $f''(x) = 20x^3 - 36x^2$	$f'''(x)$
27. $f'''(x) = 2\sqrt{x - 1}$	$f^{(4)}(x)$
28. $f'''(x) = 4x^{-4}$	$f^{(5)}(x)$
29. $f^{(4)}(x) = (x^2 + 1)^2$	$f^{(6)}(x)$
30. $f''(x) = 2x^2 + 7x - 12$	$f^{(5)}(x)$

Using Derivatives   In Exercises 31–34, find the second derivative and solve the equation $f''(x) = 0$.

31. $f(x) = x^3 - 9x^2 + 27x - 27$

32. $f(x) = (x + 2)(x - 2)(x + 3)(x - 3)$

33. $f(x) = x\sqrt{x^2 - 1}$

34. $f(x) = \dfrac{x}{x^2 + 3}$

35. Velocity and Acceleration   A ball is propelled straight upward from ground level with an initial velocity of 144 feet per second.

(a) Write the position, velocity, and acceleration functions of the ball.

(b) Find the height, velocity, and acceleration at $t = 3$.

(c) When is the ball at its highest point? How high is this point?

(d) How fast is the ball traveling when it hits the ground? How is this speed related to the initial velocity?

36. Velocity and Acceleration A brick becomes dislodged from the top of the Empire State Building (at a height of 1250 feet) and falls to the sidewalk below.

(a) Write the position, velocity, and acceleration functions of the brick.

(b) How long does it take the brick to hit the sidewalk?

(c) How fast is the brick traveling when it hits the sidewalk?

37. Velocity and Acceleration The velocity (in feet per second) of an automobile starting from rest is modeled by $ds/dt = 90t/(t + 10)$. Create a table showing the velocity and acceleration at 10-second intervals during the first minute of travel. What can you conclude?

38. Stopping Distance A car is traveling at a rate of 66 feet per second (45 miles per hour) when the brakes are applied. The position function for the car is given by $s = -8.25t^2 + 66t$, where s is measured in feet and t is measured in seconds. Use this function to complete the table showing the position, velocity, and acceleration for each given value of t. What can you conclude?

t	0	1	2	3	4
s					
$\dfrac{ds}{dt}$					
$\dfrac{d^2s}{dt^2}$					

39. Derivatives of Polynomial Functions Consider  the function $f(x) = x^2 - 6x + 6$.

(a) Use a graphing utility to graph f, f', and f'' in the same viewing window.

(b) What is the relationship among the degree of f and the degrees of its successive derivatives?

(c) Repeat parts (a) and (b) for $f(x) = 3x^3 - 9x$.

(d) In general, what is the relationship among the degree of a polynomial function and the degrees of its successive derivatives?

40. HOW DO YOU SEE IT? The graph shows the position, velocity, and acceleration functions of a particle. Identify each function. Explain your reasoning.

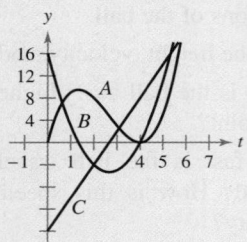

41. Modeling Data The table shows the revenues y (in millions of dollars) for eBay from 2004 to 2009, where t is the year, with $t = 4$ corresponding to 2004. *(Source: eBay Inc.)*

t	4	5	6	7	8	9
y	3271	4552	5970	7672	8541	8727

(a) Use a graphing utility to find a cubic model for the revenue $y(t)$ of eBay.

(b) Find the first and second derivatives of the function.

(c) Show that the revenue of eBay was increasing from 2005 to 2008.

(d) Find the year when the revenue was increasing at the greatest rate by solving $y''(t) = 0$.

42. Finding a Pattern Develop a general rule for

$$[x f(x)]^{(n)}$$

where f is a differentiable function of x.

True or False? In Exercises 43 and 44, determine whether the statement is true or false. If it is false, explain why or give an example that shows it is false.

43. If $y = f(x)g(x)$, then $y' = f'(x)g'(x)$.

44. If $f'(c)$ and $g'(c)$ are zero and $h(x) = f(x)g(x)$, then $h'(c) = 0$.

45. Project: Median Prices of U.S. Homes For a project analyzing the median prices of new privately owned homes in the United States from 2000 to 2009, visit this text's website at *www.cengagebrain.com.* *(Data Source: U.S. Census Bureau)*

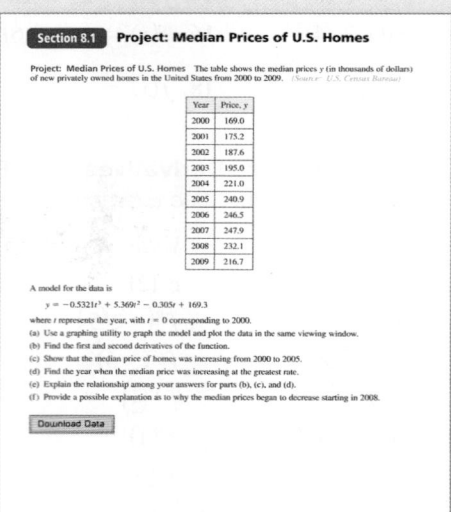

8.2 Implicit Differentiation

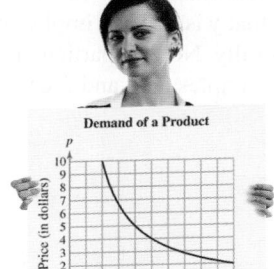

In Exercise 43 on page 643, you will use implicit differentiation to find the rate of change for a demand function.

- Find derivatives explicitly.
- Find derivatives implicitly.
- Use derivatives to answer questions about real-life situations.

Explicit and Implicit Functions

So far in this text, most functions involving two variables have been expressed in the **explicit form**

$$y = f(x). \qquad \text{Explicit form}$$

That is, one of the two variables has been explicitly given in terms of the other. For example, in the equation

$$y = 3x - 5$$

the variable y is explicitly written as a function of x. Some functions, however, are not given explicitly and are only implied by a given equation, as shown in Example 1.

Example 1   Finding a Derivative Explicitly

Find dy/dx for the equation

$$xy = 1.$$

SOLUTION   In this equation, y is **implicitly** defined as a function of x. One way to find dy/dx is first to solve the equation for y, then differentiate as usual.

$xy = 1$	Write original equation.
$y = \dfrac{1}{x}$	Solve for y.
$= x^{-1}$	Rewrite.
$\dfrac{dy}{dx} = -x^{-2}$	Differentiate with respect to x.
$= -\dfrac{1}{x^2}$	Simplify.

✓ Checkpoint 1

Find dy/dx for the equation

$$x^2 y = 1.$$

The procedure shown in Example 1 works well whenever you can easily write the given function explicitly. You cannot, however, use this procedure when you are unable to solve for y as a function of x. For instance, how would you find dy/dx for the equations

$$x^2 - 2y^3 + 4y = 2$$

and

$$x^2 + 2xy - y^3 = 5$$

where it is very difficult to express y as a function of x explicitly? To differentiate such equations, you can use a procedure called **implicit differentiation.**

Implicit Differentiation

To understand how to find dy/dx implicitly, you must realize that the differentiation is taking place *with respect to x*. This means that when you differentiate terms involving x alone, you can differentiate as usual. *But* when you differentiate terms involving y, you must apply the Chain Rule because you are assuming that y is defined implicitly as a differentiable function of x. Study the next example carefully. Note in particular how the Chain Rule is used to introduce the dy/dx factors in Examples 2(b) and 2(d).

Example 2   Applying the Chain Rule

Differentiate each expression with respect to x.

a. $3x^2$ **b.** $2y^3$

c. $x + 3y$ **d.** xy^2

SOLUTION

a. The only variable in this expression is x. So, to differentiate with respect to x, you can use the Simple Power Rule and the Constant Multiple Rule to obtain

$$\frac{d}{dx}[3x^2] = 6x.$$

b. This case is different. The variable in the expression is y, and yet you are asked to differentiate with respect to x. To do this, assume that y is a differentiable function of x and use the Chain Rule.

$$\frac{d}{dx}[2y^3] = \overset{c}{2} \ \overset{n}{(3)} \ \overset{u^{n-1}}{y^2} \ \overset{u'}{\frac{dy}{dx}} \qquad \text{Chain Rule}$$

$$= 6y^2 \frac{dy}{dx}$$

c. This expression involves both x and y. By the Sum Rule and the Constant Multiple Rule, you can write

$$\frac{d}{dx}[x + 3y] = 1 + 3\frac{dy}{dx}.$$

d. By the Product Rule and the Chain Rule, you can write

$$\frac{d}{dx}[xy^2] = x\frac{d}{dx}[y^2] + y^2\frac{d}{dx}[x] \qquad \text{Product Rule}$$

$$= x\left(2y\frac{dy}{dx}\right) + y^2(1) \qquad \text{Chain Rule}$$

$$= 2xy\frac{dy}{dx} + y^2.$$

✓ **Checkpoint 2**

Differentiate each expression with respect to x.

a. $4x^3$

b. $3y^2$

c. $x + 5y$

d. xy^3

Guidelines for Implicit Differentiation

Consider an equation involving x and y in which y is a differentiable function of x. You can use the steps below to find dy/dx.

1. Differentiate both sides of the equation *with respect to x*.

2. Write the result so that all terms involving dy/dx are on the left side of the equation and all other terms are on the right side of the equation.

3. Factor dy/dx out of the terms on the left side of the equation.

4. Solve for dy/dx by dividing both sides of the equation by the left-hand factor that does not contain dy/dx.

In Example 3, note that implicit differentiation can produce an expression for dy/dx that contains both x and y.

Example 3 **Using Implicit Differentiation**

Find dy/dx for the equation $y^3 + y^2 - 5y - x^2 = -4$.

SOLUTION

1. Differentiate both sides of the equation with respect to x.

$$\frac{d}{dx}[y^3 + y^2 - 5y - x^2] = \frac{d}{dx}[-4]$$

$$\frac{d}{dx}[y^3] + \frac{d}{dx}[y^2] - \frac{d}{dx}[5y] - \frac{d}{dx}[x^2] = \frac{d}{dx}[-4]$$

$$3y^2\frac{dy}{dx} + 2y\frac{dy}{dx} - 5\frac{dy}{dx} - 2x = 0$$

2. Collect the dy/dx terms on the left side of the equation and move all other terms to the right side of the equation.

$$3y^2\frac{dy}{dx} + 2y\frac{dy}{dx} - 5\frac{dy}{dx} = 2x$$

3. Factor dy/dx out of the left side of the equation.

$$\frac{dy}{dx}(3y^2 + 2y - 5) = 2x$$

4. Solve for dy/dx by dividing by $(3y^2 + 2y - 5)$.

$$\frac{dy}{dx} = \frac{2x}{3y^2 + 2y - 5}$$

✓**Checkpoint 3**

Find dy/dx for the equation $y^2 + x^2 - 2y - 4x = 4$. ■

To see how you can use an implicit derivative, consider the graph shown in Figure 8.3. The derivative found in Example 3 gives a formula for the slope of the tangent line at a point on this graph. For instance, the slope at the point $(1, -3)$ is

$$\frac{dy}{dx} = \frac{2(1)}{3(-3)^2 + 2(-3) - 5} = \frac{1}{8}.$$

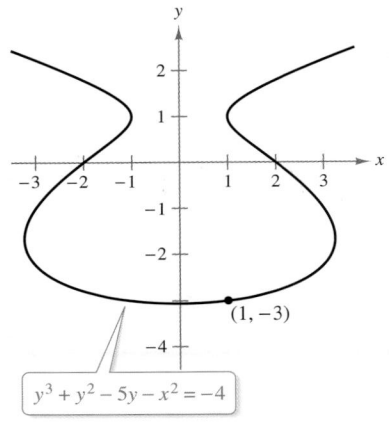

$y^3 + y^2 - 5y - x^2 = -4$

$(1, -3)$

FIGURE 8.3

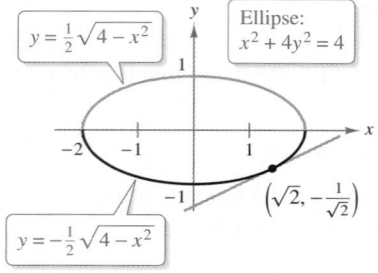

$y = \frac{1}{2}\sqrt{4 - x^2}$

Ellipse:
$x^2 + 4y^2 = 4$

$\left(\sqrt{2}, -\frac{1}{\sqrt{2}}\right)$

$y = -\frac{1}{2}\sqrt{4 - x^2}$

Slope of tangent line is $\frac{1}{2}$.

FIGURE 8.4

STUDY TIP

To see the benefit of implicit differentiation, try reworking Example 4 using the explicit function

$$y = -\frac{1}{2}\sqrt{4 - x^2}.$$

The graph of this function is the lower half of the ellipse.

Example 4   **Finding the Slope of a Graph Implicitly**

Find the slope of the tangent line to the ellipse given by $x^2 + 4y^2 = 4$ at the point $\left(\sqrt{2}, -1/\sqrt{2}\right)$, as shown in Figure 8.4.

SOLUTION

$$x^2 + 4y^2 = 4 \qquad \text{Write original equation.}$$

$$\frac{d}{dx}[x^2 + 4y^2] = \frac{d}{dx}[4] \qquad \text{Differentiate with respect to } x.$$

$$2x + 8y\frac{dy}{dx} = 0 \qquad \text{Implicit differentiation}$$

$$8y\frac{dy}{dx} = -2x \qquad \text{Subtract } 2x \text{ from each side.}$$

$$\frac{dy}{dx} = \frac{-2x}{8y} \qquad \text{Divide each side by } 8y.$$

$$\frac{dy}{dx} = -\frac{x}{4y} \qquad \text{Simplify.}$$

To find the slope at the given point, substitute $x = \sqrt{2}$ and $y = -1/\sqrt{2}$ into the derivative, as shown below.

$$\frac{dy}{dx} = -\frac{\sqrt{2}}{4\left(-1/\sqrt{2}\right)}$$

$$= \frac{1}{2}$$

✓ **Checkpoint 4**

Find the slope of the tangent line to the circle $x^2 + y^2 = 25$ at the point $(3, -4)$.   ▨

Example 5   **Finding the Slope of a Graph Implicitly**

Find the slope of the graph of $2x^2 - y^2 = 1$ at the point $(1, 1)$.

SOLUTION   Begin by finding dy/dx implicitly.

$$2x^2 - y^2 = 1 \qquad \text{Write original equation.}$$

$$4x - 2y\frac{dy}{dx} = 0 \qquad \text{Differentiate with respect to } x.$$

$$-2y\frac{dy}{dx} = -4x \qquad \text{Subtract } 4x \text{ from each side.}$$

$$\frac{dy}{dx} = \frac{2x}{y} \qquad \text{Divide each side by } -2y.$$

At the point $(1, 1)$, the slope of the graph is

$$\frac{dy}{dx} = \frac{2(1)}{1}$$

$$= 2$$

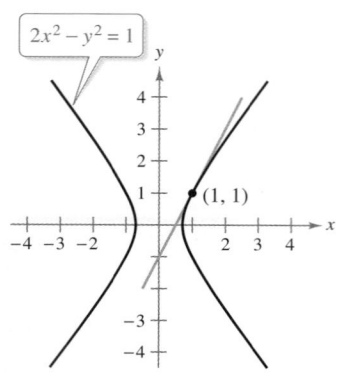

$2x^2 - y^2 = 1$

$(1, 1)$

Hyperbola

FIGURE 8.5

as shown in Figure 8.5. The graph is called a **hyperbola**.

✓ **Checkpoint 5**

Find the slope of the graph of $x^2 - 9y^2 = 16$ at the point $(5, 1)$.   ▨

Application

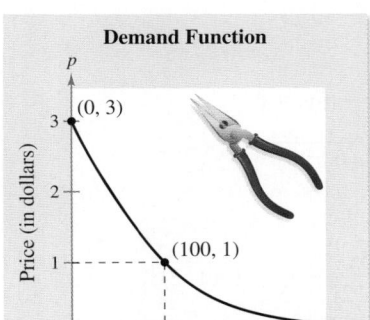

Demand Function

FIGURE 8.6

Example 6 **Using a Demand Function**

The demand function for a product is modeled by

$$p = \frac{3}{0.000001x^3 + 0.01x + 1}$$

where p is measured in dollars and x is measured in thousands of units, as shown in Figure 8.6. Find the rate of change of the demand x with respect to the price p when $x = 100$.

SOLUTION   To simplify the differentiation, begin by rewriting the function. Then, differentiate *with respect to p*.

$$p = \frac{3}{0.000001x^3 + 0.01x + 1}$$

$$0.000001x^3 + 0.01x + 1 = \frac{3}{p}$$

$$0.000003x^2 \frac{dx}{dp} + 0.01 \frac{dx}{dp} = -\frac{3}{p^2}$$

$$(0.000003x^2 + 0.01) \frac{dx}{dp} = -\frac{3}{p^2}$$

$$\frac{dx}{dp} = -\frac{3}{p^2(0.000003x^2 + 0.01)}$$

When $x = 100$, the price is

$$p = \frac{3}{0.000001(100)^3 + 0.01(100) + 1} = \$1.$$

So, when $x = 100$ and $p = 1$, the rate of change of the demand with respect to the price is

$$\frac{dx}{dp} = -\frac{3}{(1)^2[0.000003(100)^2 + 0.01]} = -75.$$

This means that when $x = 100$, the demand is dropping at the rate of 75 thousand units for each dollar increase in price.

✓ Checkpoint 6

The demand function for a product is given by

$$p = \frac{2}{0.001x^2 + x + 1}.$$

Find dx/dp implicitly.

SUMMARIZE   (Section 8.2)

1. State the guidelines for implicit differentiation *(page 639)*. For examples of implicit differentiation, see Examples 2, 3, 4, and 5.

2. Describe a real-life example of how implicit differentiation can be used to analyze the rate of change of a product's demand *(page 641, Example 6)*.

In Exercises 1–6, solve the equation for y.

1. $x - \dfrac{y}{x} = 2$

2. $\dfrac{4}{x - 3} = \dfrac{1}{y}$

3. $xy - x + 6y = 6$

4. $12 + 3y = 4x^2 + x^2y$

5. $x^2 + y^2 = 5$

6. $x = \pm\sqrt{6 - y^2}$

In Exercises 7–9, evaluate the expression at the given point.

7. $\dfrac{3x^2 - 4}{3y^2}$, $(2, 1)$

8. $\dfrac{x^2 - 2}{1 - y}$, $(0, -3)$

9. $\dfrac{5x}{3y^2 - 12y + 5}$, $(-1, 2)$

Exercises 8.2

See www.CalcChat.com for worked-out solutions to odd-numbered exercises.

Finding Derivatives   In Exercises 1–12, find dy/dx. See Examples 1 and 3.

1. $xy = 4$

2. $3x^2 - y = 8x$

3. $y^2 = 1 - x^2$, $0 \le x \le 1$

4. $y^3 = 4x^3 + 2x$

5. $x^2y^2 - 2x = 3$

6. $xy^2 + 4xy = 10$

7. $4y^2 - xy = 2$

8. $2xy^3 - x^2y = 2$

9. $\dfrac{xy - y^2}{y - x} = 1$

10. $\dfrac{2x + y}{x - 5y} = 1$

11. $\dfrac{2y - x}{y^2 - 3} = 5$

12. $\dfrac{4y^2}{y^2 - 9} = x^2$

Finding the Slope of a Graph Implicitly   In Exercises 13–26, find the slope of the graph of the function at the given point. See Examples 4 and 5.

Equation	Point
13. $x^2 + y^2 = 16$	$(0, 4)$
14. $x^2 - y^2 = 25$	$(5, 0)$
15. $y + xy = 4$	$(-5, -1)$
16. $16x^2 + 25y^2 = 400$	$(5, 4)$
17. $x^3 - xy + y^2 = 4$	$(0, -2)$
18. $x^2y + y^2x = -2$	$(2, -1)$
19. $x^3y^3 - y = x$	$(0, 0)$
20. $x^3 + y^3 = 6xy$	$\left(\dfrac{4}{3}, \dfrac{8}{3}\right)$
21. $x^{1/2} + y^{1/2} = 9$	$(16, 25)$
22. $x^{2/3} + y^{2/3} = 5$	$(8, 1)$
23. $\sqrt{xy} = x - 2y$	$(4, 1)$
24. $(x + y)^3 = x^3 + y^3$	$(-1, 1)$

Equation	Point
25. $y^2(x^2 + y^2) = 2x^2$	$(1, 1)$
26. $(x^2 + y^2)^2 = 8x^2y$	$(2, 2)$

Finding the Slope of a Graph Implicitly   In Exercises 27–32, find the slope of the graph at the given point. See Examples 4 and 5.

27. $3x^2 - 2y + 5 = 0$

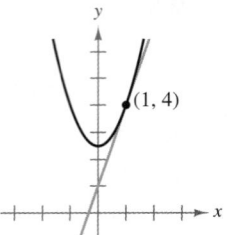

28. $4x^2 + 2y - 1 = 0$

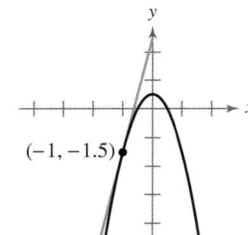

29. $x^2 + y^2 = 4$

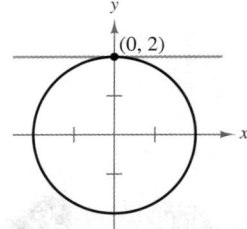

30. $4x^2 + 9y^2 = 36$

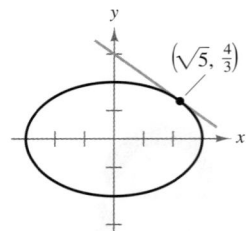

31. $(4 - x)y^2 = x^3$

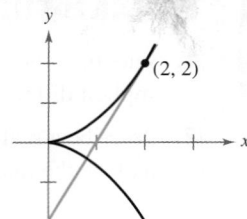

32. $x^2 - y^3 = 0$

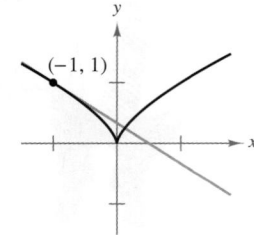

Finding Derivatives Implicitly and Explicitly In Exercises 33 and 34, find dy/dx implicitly and explicitly (the explicit functions are shown on the graph) and show that the results are equivalent. Use the graph to estimate the slope of the tangent line at the labeled point. Then verify your result analytically by evaluating dy/dx at the point.

33. $x - y^2 - 1 = 0$

34. $4y^2 - x^2 = 7$

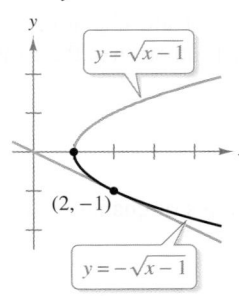

$y = \sqrt{x-1}$

$y = -\sqrt{x-1}$

$(2, -1)$

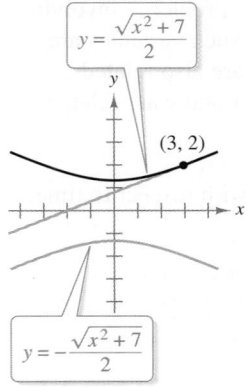

$y = \dfrac{\sqrt{x^2+7}}{2}$

$(3, 2)$

$y = -\dfrac{\sqrt{x^2+7}}{2}$

Finding an Equation of a Tangent Line In Exercises 35–42, find equations of the tangent lines to the graph at the given points. Use a graphing utility to graph the equation and the tangent lines in the same viewing window.

Equation	Points
35. $x^2 + y^2 = 100$	$(8, 6)$ and $(-6, 8)$
36. $x^2 + y^2 = 9$	$(0, 3)$ and $(2, \sqrt{5})$
37. $y^2 = 5x^3$	$(1, \sqrt{5})$ and $(1, -\sqrt{5})$
38. $4xy + x^2 = 5$	$(1, 1)$ and $(5, -1)$
39. $x^3 + y^3 = 8$	$(0, 2)$ and $(2, 0)$
40. $x + y^3 = 6xy^3 - 1$	$(-1, 0)$ and $(0, -1)$
41. $x^2 y - 8 = -4y$	$(-2, 1)$ and $(6, \tfrac{1}{5})$
42. $y^2 = \dfrac{x^3}{4 - x}$	$(2, 2)$ and $(2, -2)$

Demand In Exercises 43–46, find the rate of change of x with respect to p. *See Example 6.*

43. $p = \dfrac{2}{0.00001x^3 + 0.1x}, \quad x \geq 0$

44. $p = \dfrac{4}{0.000001x^2 + 0.05x + 1}, \quad x \geq 0$

45. $p = \sqrt{\dfrac{200 - x}{2x}}, \quad 0 < x \leq 200$

46. $p = \sqrt{\dfrac{500 - x}{2x}}, \quad 0 < x \leq 500$

47. Production Let x represent the units of labor and y the capital invested in a manufacturing process. When 135,540 units are produced, the relationship between labor and capital can be modeled by $100x^{0.75}y^{0.25} = 135,540$.

(a) Find the rate of change of y with respect to x when $x = 1500$ and $y = 1000$.

(b) The model used in this problem is called the *Cobb-Douglas production function*. Graph the model on a graphing utility and describe the relationship between labor and capital.

48.  **HOW DO YOU SEE IT?** The graph shows the demand function for a product.

Demand Function

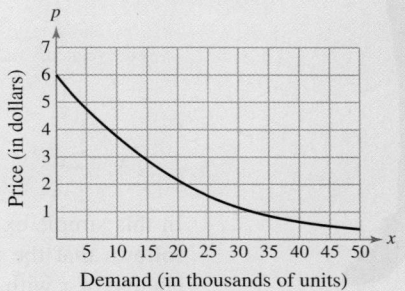

(a) What happens to the demand as the price increases?

(b) Over what interval is the rate of change of the demand with respect to the price decreasing?

49. Health: U.S. HIV/AIDS Epidemic The numbers (in thousands) of cases y of HIV/AIDS reported in the years 2004 through 2008 can be modeled by

$$y^2 - 1952.4 = 13.0345t^3 - 168.969t^2 + 465.66t$$

where t represents the year, with $t = 4$ corresponding to 2004. *(Source: U.S. Centers for Disease Control and Prevention)*

(a) Use a graphing utility to graph the model and describe the results.

(b) Use the graph to estimate the year during which the number of reported cases was decreasing at the greatest rate.

(c) Complete the table to estimate the year during which the number of reported cases was decreasing at the greatest rate. Compare this estimate with your answer in part (b).

t	4	5	6	7	8
y					
y'					

8.3 Related Rates

In Exercise 25 on page 650, you will use related rates to find the rate of change of the sales for a product.

■ Examine related variables.
■ Solve related-rate problems.

Related Variables

In this section, you will study problems involving variables that are changing with respect to time. If two or more such variables are related to each other, then their rates of change with respect to time are also related.

For instance, suppose that x and y are related by the equation

$$y = 2x.$$

If both variables are changing with respect to time, then their rates of change will also be related.

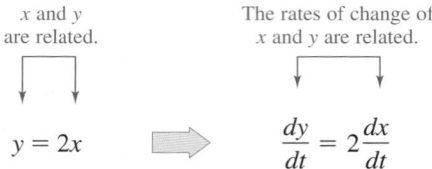

x and y are related.

The rates of change of x and y are related.

$$y = 2x \implies \frac{dy}{dt} = 2\frac{dx}{dt}$$

In this simple example, you can see that because y always has twice the value of x, it follows that the rate of change of y with respect to time is always twice the rate of change of x with respect to time.

Example 1   Examining Two Rates That Are Related

The variables x and y are differentiable functions of t and are related by the equation

$$y = x^2 + 3.$$

When $x = 1$, $dx/dt = 2$. Find dy/dt when $x = 1$.

SOLUTION   Use the Chain Rule to differentiate both sides of the equation with respect to t.

$$y = x^2 + 3 \qquad \text{Write original equation.}$$

$$\frac{d}{dt}[y] = \frac{d}{dt}[x^2 + 3] \qquad \text{Differentiate with respect to } t.$$

$$\frac{dy}{dt} = 2x\frac{dx}{dt} \qquad \text{Apply Chain Rule.}$$

When $x = 1$ and $dx/dt = 2$, you have

$$\frac{dy}{dt} = 2(1)(2)$$

$$= 4.$$

✓ **Checkpoint 1**

The variables x and y are differentiable functions of t and are related by the equation

$$y = x^3 + 2.$$

When $x = 1$, $dx/dt = 3$. Find dy/dt when $x = 1$.

Solving Related-Rate Problems

In Example 1, you were *given* the mathematical model.

Given equation: $y = x^2 + 3$

Given rate: $\dfrac{dx}{dt} = 2$ at $x = 1$

Find: $\dfrac{dy}{dt}$ at $x = 1$

In the next example, you must *create* a mathematical model from a verbal description.

 Example 2   Changing Area

A pebble is dropped into a calm pool of water, causing ripples in the form of concentric circles, as shown in the photo. The radius r of the outer ripple is increasing at a constant rate of 1 foot per second. When the radius is 4 feet, at what rate is the total area A of the disturbed water changing?

SOLUTION   The variables r and A are related by the equation for the area of a circle, $A = \pi r^2$. To solve this problem, use the fact that the rate of change of the radius is given by dr/dt.

Equation: $A = \pi r^2$

Given rate: $\dfrac{dr}{dt} = 1$ at $r = 4$

Find: $\dfrac{dA}{dt}$ at $r = 4$

Using this model, you can proceed as in Example 1.

$A = \pi r^2$	Write original equation.
$\dfrac{d}{dt}[A] = \dfrac{d}{dt}[\pi r^2]$	Differentiate with respect to t.
$\dfrac{dA}{dt} = 2\pi r \dfrac{dr}{dt}$	Apply Chain Rule.

When $r = 4$ and $dr/dt = 1$, you have

$\dfrac{dA}{dt} = 2\pi(4)(1) = 8\pi$ Substitute 4 for r and 1 for dr/dt.

When the radius is 4 feet, the area is changing at a rate of 8π square feet per second.

✓ Checkpoint 2

As in Example 2, a pebble is dropped into the pool, but this time the radius r of the outer ripple is increasing at a rate of 2 feet per second. At what rate is the total area changing when the radius is 3 feet?

In Example 2, note that the radius changes at a *constant* rate ($dr/dt = 1$ for all t), but the area changes at a *nonconstant* rate.

When $r = 1$ ft	When $r = 2$ ft	When $r = 3$ ft	When $r = 4$ ft
$\dfrac{dA}{dt} = 2\pi$ ft^2/sec	$\dfrac{dA}{dt} = 4\pi$ ft^2/sec	$\dfrac{dA}{dt} = 6\pi$ ft^2/sec	$\dfrac{dA}{dt} = 8\pi$ ft^2/sec

Total area increases as the outer radius increases.

The solution shown in Example 2 illustrates the steps for solving a related-rate problem.

Guidelines for Solving a Related-Rate Problem

1. Identify all *given* quantities and all quantities *to be determined*. If possible, make a sketch and label the quantities.

2. Write an equation that relates all variables whose rates of change are either given or to be determined.

3. Use the Chain Rule to differentiate both sides of the equation *with respect to time*.

4. *After* completing Step 3, substitute into the resulting equation all known values for the variables and their rates of change. Then solve for the required rate of change.

STUDY TIP

Be sure you notice the order of Steps 3 and 4 in the guidelines. Do not substitute the known values for the variables until after you have differentiated.

In Step 2 of the guidelines, note that you must write an equation that relates the given variables. To help you with this step, reference tables that summarize many common formulas are included in the appendices. For instance, the volume of a sphere of radius r is given by the formula

$$V = \frac{4}{3}\pi r^3$$

as listed in Appendix D.

The table below lists examples of the mathematical models for some common rates of change that can be used in the first step of the solution of a related-rate problem.

Verbal statement	Mathematical model
The velocity of a car after traveling for 1 hour is 50 miles per hour.	x = distance traveled $\dfrac{dx}{dt} = 50$ when $t = 1$
Water is being pumped into a swimming pool at a rate of 10 cubic feet per minute.	V = volume of water in pool $\dfrac{dV}{dt} = 10 \text{ ft}^3/\text{min}$
A population of bacteria is increasing at a rate of 2000 per hour.	x = number in population $\dfrac{dx}{dt} = 2000$ bacteria per hour
Revenue is increasing at a rate of $4000 per month.	R = revenue $\dfrac{dR}{dt} = 4000$ dollars per month
Profit is decreasing at a rate of $2500 per day.	P = profit $\dfrac{dP}{dt} = -2500$ dollars per day

Example 3   Analyzing a Profit Function

A company's profit P (in dollars) from selling x units of a product can be modeled by

$$P = 500x - \frac{1}{4}x^2. \qquad \text{Model for profit}$$

The sales are increasing at a rate of 10 units per day. Find the rate of change in the profit (in dollars per day) when 500 units have been sold.

SOLUTION   Because the sales are increasing at a rate of 10 units per day, you know that at time t the rate of change is $dx/dt = 10$. So, the problem can be stated as shown.

Given rate: $\dfrac{dx}{dt} = 10$

Find: $\dfrac{dP}{dt}$ when $x = 500$

To find the rate of change of the profit, use the model for profit that relates the profit P and the units of the product sold x.

Equation: $P = 500x - \dfrac{1}{4}x^2$

By differentiating both sides of the equation with respect to t, you obtain

$$\frac{d}{dt}[P] = \frac{d}{dt}\left[500x - \frac{1}{4}x^2\right] \qquad \text{Differentiate with respect to } t.$$

$$\frac{dP}{dt} = \left(500 - \frac{1}{2}x\right)\frac{dx}{dt}. \qquad \text{Apply Chain Rule.}$$

When $x = 500$ units and $dx/dt = 10$, the rate of change in the profit is

$$\frac{dP}{dt} = \left[500 - \frac{1}{2}(500)\right](10) = (500 - 250)(10) = 250(10) = \$2500 \text{ per day.}$$

The graph of the profit function (in terms of x) is shown in Figure 8.7.

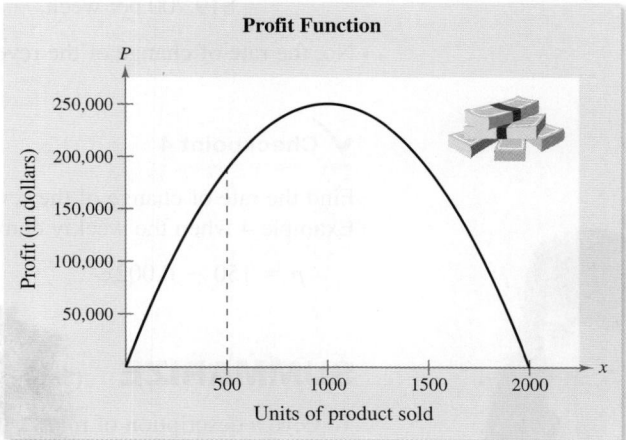

FIGURE 8.7

✓**Checkpoint 3**

Find the rate of change in profit (in dollars per day) when 50 units have been sold, sales have increased at a rate of 10 units per day, and $P = 200x - \frac{1}{2}x^2$.

Example 4   **Increasing Production**

A company is increasing the production of a product at the rate of 200 units per week. The weekly demand function is modeled by

$$p = 100 - 0.001x$$

where p is the price per unit and x is the number of units produced in a week. Find the rate of change of the revenue with respect to time when the weekly production is 2000 units. Will the rate of change of the revenue be greater than $20,000 per week?

SOLUTION   Because production is increasing at a rate of 200 units per week, you know that at time t the rate of change is $dx/dt = 200$. So, the problem can be stated as shown.

Given rate: $\dfrac{dx}{dt} = 200$

Find: $\dfrac{dR}{dt}$ when $x = 2000$

To find the rate of change of the revenue, you must find an equation that relates the revenue R and the number of units produced x.

Equation: $R = xp = x(100 - 0.001x) = 100x - 0.001x^2$

By differentiating both sides of the equation with respect to t, you obtain

$$R = 100x - 0.001x^2 \qquad \text{Write original equation.}$$

$$\frac{d}{dt}[R] = \frac{d}{dt}[100x - 0.001x^2] \qquad \text{Differentiate with respect to } t.$$

$$\frac{dR}{dt} = (100 - 0.002x)\frac{dx}{dt}. \qquad \text{Apply Chain Rule.}$$

Using $x = 2000$ and $dx/dt = 200$, you have

$$\frac{dR}{dt} = [100 - 0.002(2000)](200)$$

$$= \$19,200 \text{ per week.}$$

No, the rate of change of the revenue will not be greater than $20,000 per week.

✓ **Checkpoint 4**

Find the rate of change of the revenue with respect to time for the company in Example 4 when the weekly demand function is

$$p = 150 - 0.002x.$$

SUMMARIZE   (Section 8.3)

1. Give a description of related variables *(page 644)*. For an example of two related variables, see Example 1.

2. State the guidelines for solving a related-rate problem *(page 646)*. For examples of solving related-rate problems, see Examples 2, 3, and 4.

3. Describe a real-life example of how related rates can be used to analyze the rate of change of a company's revenue *(page 648, Example 4)*.

The following warm-up exercises involve skills that were covered in earlier sections. You will use these skills in the exercise set for this section. For additional help, review Sections 1.2 and 8.2.

In Exercises 1–6, write a formula for the given quantity.

1. Area of a circle

2. Volume of a sphere

3. Surface area of a cube

4. Volume of a cube

5. Volume of a cone

6. Area of a triangle

In Exercises 7–10, find dy/dx by implicit differentiation.

7. $x^2 + y^2 = 9$ **8.** $3xy - x^2 = 6$ **9.** $x^2 + 2y + xy = 12$ **10.** $x + xy^2 - y^2 = xy$

Exercises 8.3

Examining Two Rates That Are Related In Exercises 1–4, assume that x and y are both differentiable functions of t. Use the given values to find (a) dy/dt and (b) dx/dt. See Example 1.

Equation	Find	Given
1. $y = \sqrt{x}$	(a) $\dfrac{dy}{dt}$ when $x = 4$,	$\dfrac{dx}{dt} = 3$
	(b) $\dfrac{dx}{dt}$ when $x = 25$,	$\dfrac{dy}{dt} = 2$
2. $y = 2(x^2 - 3x)$	(a) $\dfrac{dy}{dt}$ when $x = 3$,	$\dfrac{dx}{dt} = 2$
	(b) $\dfrac{dx}{dt}$ when $x = 1$,	$\dfrac{dy}{dt} = 5$
3. $xy = 4$	(a) $\dfrac{dy}{dt}$ when $x = 8$,	$\dfrac{dx}{dt} = 10$
	(b) $\dfrac{dx}{dt}$ when $x = 1$,	$\dfrac{dy}{dt} = -6$
4. $x^2 + y^2 = 25$	(a) $\dfrac{dy}{dt}$ when $x = 3, y = 4$,	$\dfrac{dx}{dt} = 8$
	(b) $\dfrac{dx}{dt}$ when $x = 4, y = 3$,	$\dfrac{dy}{dt} = -2$

5. Area The radius r of a circle is increasing at a rate of 3 inches per minute. Find the rates of change of the area when (a) $r = 6$ inches and (b) $r = 24$ inches.

6. Volume The radius r of a sphere is increasing at a rate of 3 inches per minute. Find the rates of change of the volume when (a) $r = 6$ inches and (b) $r = 24$ inches.

7. Area Let A be the area of a circle of radius r that is changing with respect to time. If dr/dt is constant, is dA/dt constant? Explain your reasoning.

8. Volume Let V be the volume of a sphere of radius r that is changing with respect to time. If dr/dt is constant, is dV/dt constant? Explain your reasoning.

9. Volume A spherical balloon is inflated with gas at a rate of 10 cubic feet per minute. How fast is the radius of the balloon changing at the instant the radius is (a) 1 foot and (b) 2 feet?

10. Volume The radius r of a right circular cone is increasing at a rate of 2 inches per minute. The height h of the cone is related to the radius by

$$h = 3r.$$

Find the rates of change of the volume when (a) $r = 6$ inches and (b) $r = 24$ inches.

11. Cost, Revenue, and Profit A company that manufactures sport supplements calculates that its costs and revenue can be modeled by the equations

$$C = 125{,}000 + 0.75x \quad \text{and} \quad R = 250x - \frac{1}{10}x^2$$

where x is the number of units of sport supplements produced in 1 week. Production during one particular week is 1000 units and is increasing at a rate of 150 units per week. Find the rates at which the (a) cost, (b) revenue, and (c) profit are changing.

12. Cost, Revenue, and Profit A company that manufactures pet toys calculates that its costs and revenue can be modeled by the equations

$$C = 75{,}000 + 1.05x \quad \text{and} \quad R = 500x - \frac{x^2}{25}$$

where x is the number of toys produced in 1 week. Production during one particular week is 5000 toys and is increasing at a rate of 250 toys per week. Find the rates at which the (a) cost, (b) revenue, and (c) profit are changing.

13. Volume All edges of a cube are expanding at a rate of 3 centimeters per second. How fast is the volume changing when each edge is (a) 1 centimeter and (b) 10 centimeters?

14. Surface Area   All edges of a cube are expanding at a rate of 3 centimeters per second. How fast is the surface area changing when each edge is (a) 1 centimeter and (b) 10 centimeters?

15. Moving Point   A point is moving along the graph of $y = x^2$ such that dx/dt is 3 inches per second. Find dy/dt for (a) $x = -3$, (b) $x = 0$, (c) $x = 1$, and (d) $x = 3$.

16. Moving Point   A point is moving along the graph of $y = 1/(1 + x^2)$ such that dx/dt is 2 inches per second. Find dy/dt for (a) $x = -2$, (b) $x = 0$, (c) $x = 6$, and (d) $x = 10$.

17. Boating   A boat is pulled by a winch on a dock, and the winch is 12 feet above the deck of the boat (see figure). The winch pulls the rope at a rate of 4 feet per second. Find the speed of the boat when 13 feet of rope is out. What happens to the speed of the boat as it gets closer and closer to the dock?

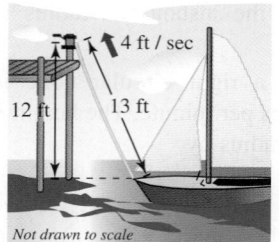

Figure for 17   Figure for 18

18. Shadow Length   A man 6 feet tall walks at a rate of 5 feet per second away from a light that is 15 feet above the ground (see figure).

  (a) When he is 10 feet from the base of the light, at what rate is the tip of his shadow moving?

  (b) When he is 10 feet from the base of the light, at what rate is the length of his shadow changing?

19. Air Traffic Control   An airplane flying at an altitude of 6 miles passes directly over a radar antenna (see figure). When the airplane is 10 miles away ($s = 10$), the radar detects that the distance s is changing at a rate of 240 miles per hour. What is the speed of the airplane?

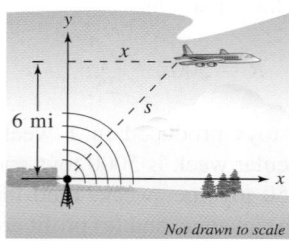   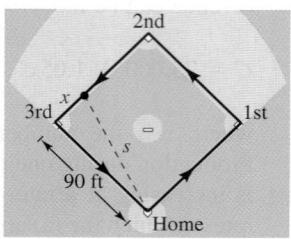

Figure for 19   Figure for 20

20. Baseball   A (square) baseball diamond has sides that are 90 feet long (see figure). A player 26 feet from third base is running at a speed of 30 feet per second. At what rate is the player's distance from home plate changing?

21. Air Traffic Control   An air traffic controller spots two airplanes at the same altitude converging to a point as they fly at right angles to each other. One airplane is 150 miles from the point and has a speed of 450 miles per hour. The other is 200 miles from the point and has a speed of 600 miles per hour.

  (a) At what rate is the distance between the planes changing?

  (b) How much time does the controller have to get one of the airplanes on a different flight path?

22. Advertising Costs   A retail sporting goods store estimates that weekly sales S and weekly advertising costs x are related by the equation $S = 2250 + 50x + 0.35x^2$. The current weekly advertising costs are $1500, and these costs are increasing at a rate of $125 per week. Find the current rate of change of weekly sales.

23. Environment   An accident at an oil drilling platform is causing a circular oil slick. The slick is 0.08 foot thick, and when the radius of the slick is 150 feet, the radius is increasing at the rate of 0.5 foot per minute. At what rate (in cubic feet per minute) is oil flowing from the site of the accident?

24. Profit   A company is increasing the production of a product at the rate of 25 units per week. The demand and cost functions for the product are given by $p = 50 - 0.01x$ and $C = 4000 + 40x - 0.02x^2$. Find the rate of change of the profit with respect to time when the weekly sales are $x = 800$ units. Use a graphing utility to graph the profit function, and use the *zoom* and *trace* features of the graphing utility to verify your result.

25. Sales   The profit for a product is increasing at a rate of $5600 per week. The demand and cost functions for the product are given by $p = 6000 - 25x$ and $C = 2400x + 5200$. Find the rate of change of sales with respect to time when the weekly sales are $x = 44$ units.

26. HOW DO YOU SEE IT?   The graph shows the demand and supply equations for a product, where x represents the number of units (in thousands) and p is the price (in dollars). Using the graph, (a) determine whether dp/dt is positive or negative given that dx/dt is negative, and (b) determine whether dx/dt is positive or negative given that dp/dt is positive.

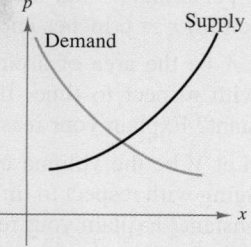

QUIZ YOURSELF

See www.CalcChat.com for worked-out solutions to odd-numbered exercises.

Take this quiz as you would take a quiz in class. When you are done, check your work against the answers given in the back of the book.

In Exercises 1–4, find the second derivative of the function. Simplify your result.

1. $f(x) = x^3 - x^2 + 2x - 1$

2. $h(x) = \dfrac{1}{\sqrt[3]{x - 2}}$

3. $g(x) = (x^2 + 1)^3$

4. $f(x) = \dfrac{x - 5}{2x + 5}$

In Exercises 5–7, find the given value.

Function	Value
5. $f(x) = \sqrt{x}$	$f''(4)$
6. $f(x) = x^5 - 4x^3 + \frac{3}{2}x^2 + 19$	$f^{(4)}(-1)$
7. $f(x) = \dfrac{1}{x}$	$f'''\left(\dfrac{1}{2}\right)$

8. An object is thrown upward from the top of an 800-foot building with an initial velocity of 80 feet per second. Find the height, the velocity, and the acceleration of the object when $t = 1$.

In Exercises 9–12, use implicit differentiation to find dy/dx.

9. $x^2 + 3y = x$

10. $\sqrt{y} = x^3$

11. $xy = x + y$

12. $y^3 + y - 2x^2y = 12$

13. Use implicit differentiation to find an equation of the tangent line to the graph of $-2xy + 3x^2 = 1$ at the point $(-1, -1)$. Use a graphing utility to graph the equation and the tangent line in the same viewing window.

In Exercises 14 and 15, use the given values to find dy/dt.

Equation	Given
14. $y = 2x^2 + 5$	$x = 1, \dfrac{dx}{dt} = \dfrac{1}{2}$
15. $x^2 - y^2 = \dfrac{16}{y}$	$x = \sqrt{12}, y = -4, \dfrac{dx}{dt} = 1$

16. A company that manufactures a type of automobile part calculates that its costs and revenue can be modeled by the equations

$$C = 200{,}000 + 0.95x$$

and

$$R = 300x - \tfrac{1}{75}x^2$$

where x is the number of parts produced in 1 week. If production in one particular week is 7500 parts and is increasing at a rate of 200 parts per week, find the rates of change of (a) the cost, (b) the revenue, and (c) the profit.

8.4 Increasing and Decreasing Functions

■ Test for increasing and decreasing functions.
■ Find the critical numbers of functions and find the open intervals on which functions are increasing or decreasing.
■ Use increasing and decreasing functions to model and solve real-life problems.

Increasing and Decreasing Functions

A function is **increasing** when its graph moves up as x moves to the right and **decreasing** when its graph moves down as x moves to the right. The following definition states this more formally.

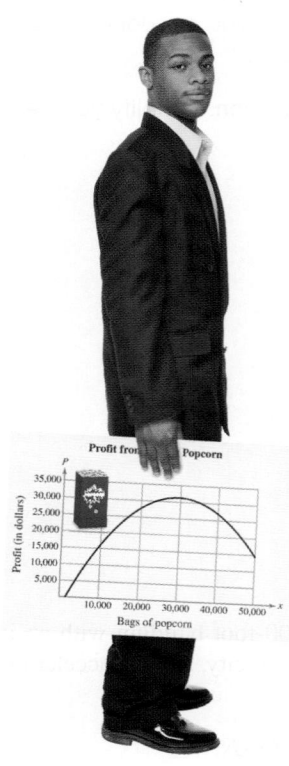

In Exercise 47 on page 660, you will use derivatives and critical numbers to find the intervals on which the profit from selling popcorn is increasing and decreasing.

Definition of Increasing and Decreasing Functions

A function f is **increasing** on an interval when, for any two numbers x_1 and x_2 in the interval,

$$x_2 > x_1 \quad \text{implies} \quad f(x_2) > f(x_1).$$

A function f is **decreasing** on an interval when, for any two numbers x_1 and x_2 in the interval,

$$x_2 > x_1 \quad \text{implies} \quad f(x_2) < f(x_1).$$

The function in Figure 8.8 is decreasing on the interval $(-\infty, a)$, constant on the interval (a, b), and increasing on the interval (b, ∞). Actually, from the definition of increasing and decreasing functions, the function shown in Figure 8.8 is decreasing on the interval $(-\infty, a]$ and increasing on the interval $[b, \infty)$. This text restricts the discussion to finding *open* intervals on which a function is increasing or decreasing.

The derivative of a function can be used to determine whether the function is increasing or decreasing on an interval.

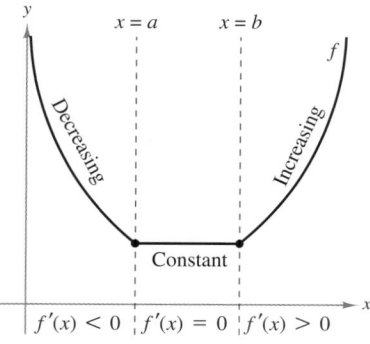

FIGURE 8.8

Test for Increasing and Decreasing Functions

Let f be differentiable on the interval (a, b).

1. If $f'(x) > 0$ for all x in (a, b), then f is increasing on (a, b).

2. If $f'(x) < 0$ for all x in (a, b), then f is decreasing on (a, b).

3. If $f'(x) = 0$ for all x in (a, b), then f is constant on (a, b).

STUDY TIP

The conclusions in the first two cases of testing for increasing and decreasing functions are valid even when $f'(x) = 0$ at a finite number of x-values in (a, b).

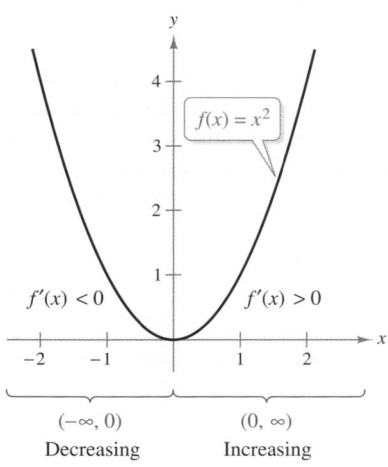

$f(x) = x^2$

$f'(x) < 0$ $f'(x) > 0$

$(-\infty, 0)$ $(0, \infty)$
Decreasing   Increasing

FIGURE 8.9

Example 1   Testing for Increasing and Decreasing Functions

Show that the function $f(x) = x^2$ is decreasing on the open interval $(-\infty, 0)$ and increasing on the open interval $(0, \infty)$.

SOLUTION   The derivative of f is

$$f'(x) = 2x.$$

On the open interval $(-\infty, 0)$, the fact that x is negative implies that $f'(x) = 2x$ is also negative. So, by the test for a decreasing function, you can conclude that f is *decreasing* on this interval. Similarly, on the open interval $(0, \infty)$, the fact that x is positive implies that $f'(x) = 2x$ is also positive. So, it follows that f is *increasing* on this interval, as shown in Figure 8.9.

✓**Checkpoint 1**

Show that the function $f(x) = x^4$ is decreasing on the open interval $(-\infty, 0)$ and increasing on the open interval $(0, \infty)$. ■

 Example 2   Modeling Consumption

From 2000 through 2008, the consumption M of whole milk in the United States (in gallons per person) can be modeled by

$$M = -0.015t^2 + 0.13t + 8.0, \quad 0 \le t \le 8$$

where $t = 0$ corresponds to 2000 (see Figure 8.10). Show that the consumption of whole milk was decreasing from 2000 to 2008. *(Source: U.S. Department of Agriculture)*

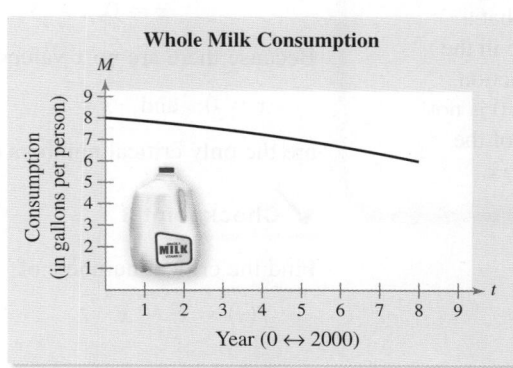

Whole Milk Consumption

Consumption (in gallons per person)

Year (0 ↔ 2000)

FIGURE 8.10

SOLUTION   The derivative of this model is $dM/dt = -0.030t + 0.13$. For the open interval $(0, 8)$, the derivative is negative. So, the function is decreasing, which implies that the consumption of whole milk was decreasing during the given time period.

✓**Checkpoint 2**

From 2003 through 2008, the consumption F of fresh fruit in the United States (in pounds per person) can be modeled by

$$F = -0.7674t^2 + 2.872t + 277.87, \quad 3 \le t \le 8$$

where $t = 3$ corresponds to 2003. Show that the consumption of fresh fruit was decreasing from 2003 to 2008. *(Source: U.S. Department of Agriculture)* ■

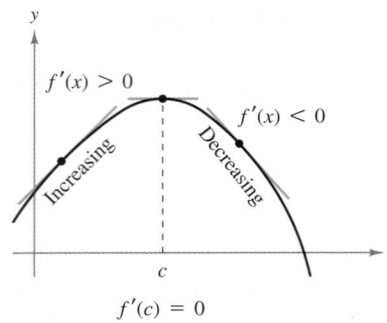

$f'(c) = 0$

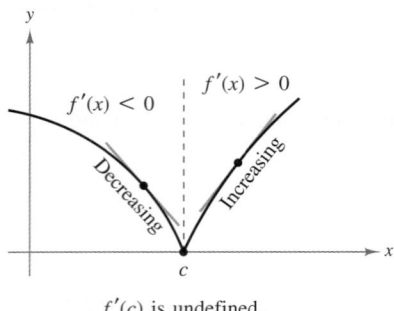

$f'(c)$ is undefined.

FIGURE 8.11

Critical Numbers and Their Use

In Example 1, you were given two intervals: one on which the function was decreasing and one on which it was increasing. Suppose you had been asked to determine these intervals. To do this, you could have used the fact that for a continuous function, $f'(x)$ can change signs only at x-values for which $f'(x) = 0$ or at x-values for which $f'(x)$ is undefined, as shown in Figure 8.11. These two types of numbers are called the **critical numbers** of f.

Definition of a Critical Number

If f is defined at c, then c is a critical number of f when $f'(c) = 0$ or when $f'(c)$ is undefined.

Example 3   **Finding Critical Numbers**

Find the critical numbers of

$$f(x) = 2x^3 - 9x^2.$$

SOLUTION   Begin by differentiating the function.

$$f(x) = 2x^3 - 9x^2 \qquad \text{Write original function.}$$
$$f'(x) = 6x^2 - 18x \qquad \text{Differentiate.}$$

To find the critical numbers of f, you must find all x-values for which $f'(x) = 0$ and all x-values for which $f'(x)$ is undefined.

$$6x^2 - 18x = 0 \qquad \text{Set } f'(x) \text{ equal to 0.}$$
$$6x(x - 3) = 0 \qquad \text{Factor.}$$
$$x = 0, x = 3 \qquad \text{Critical numbers}$$

Because there are no x-values for which f' is undefined, you can conclude that

$$x = 0 \quad \text{and} \quad x = 3$$

are the only critical numbers of f.

✓**Checkpoint 3**

Find the critical numbers of

$$f(x) = x^2 - x.$$

To determine the intervals on which a continuous function is increasing or decreasing, you can use the guidelines below.

Guidelines for Applying the Increasing/Decreasing Test

1. Find the derivative of f.

2. Locate the critical numbers of f and use these numbers to determine test intervals. That is, find all x for which $f'(x) = 0$ or $f'(x)$ is undefined.

3. Determine the sign of $f'(x)$ at one test value in each of the intervals.

4. Use the test for increasing and decreasing functions to decide whether f is increasing or decreasing on each interval.

Example 4   **Intervals on Which _f_ Is Increasing or Decreasing**

Find the open intervals on which the function is increasing or decreasing.

$$f(x) = x^3 - \frac{3}{2}x^2$$

SOLUTION   Begin by finding the derivative of f. Then set the derivative equal to zero and solve for the critical numbers.

$f'(x) = 3x^2 - 3x$	Differentiate original function.
$3x^2 - 3x = 0$	Set derivative equal to 0.
$3x(x - 1) = 0$	Factor.
$x = 0, x = 1$	Critical numbers

Because there are no x-values for which f' is undefined, it follows that $x = 0$ and $x = 1$ are the only critical numbers. So, the intervals that need to be tested are

$$(-\infty, 0), (0, 1), \quad \text{and} \quad (1, \infty). \qquad \text{Test intervals}$$

The table summarizes the testing of these three intervals.

Interval	$-\infty < x < 0$	$0 < x < 1$	$1 < x < \infty$
Test value	$x = -1$	$x = \frac{1}{2}$	$x = 2$
Sign of $f'(x)$	$f'(-1) = 6 > 0$	$f'\left(\frac{1}{2}\right) = -\frac{3}{4} < 0$	$f'(2) = 6 > 0$
Conclusion	Increasing	Decreasing	Increasing

The graph of f is shown in Figure 8.12. Note that the test values in the intervals were chosen for convenience—other x-values could have been used. ───────

The graph in the left margin:

$f(x) = x^3 - \frac{3}{2}x^2$

Increasing, Decreasing, Increasing $\left(1, -\frac{1}{2}\right)$, $(0, 0)$

FIGURE 8.12

✓**Checkpoint 4**

Find the open intervals on which the function $f(x) = x^3 - 12x$ is increasing or decreasing.

■

TECH TUTOR

You can use the *trace* feature of a graphing utility to confirm the result of Example 4. Begin by graphing the function, as shown below. Then use the *trace* feature and move the cursor from left to right. In intervals on which the function is increasing, note that the y-values increase as the x-values increase, whereas in intervals on which the function is decreasing, the y-values decrease as the x-values increase.

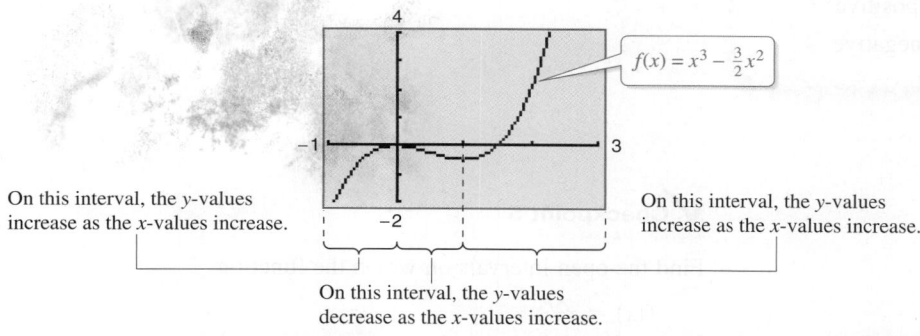

$f(x) = x^3 - \frac{3}{2}x^2$

On this interval, the y-values increase as the x-values increase.

On this interval, the y-values increase as the x-values increase.

On this interval, the y-values decrease as the x-values increase.

Not only is the function in Example 4 continuous on the entire real number line, it is also differentiable there. For such functions, the only critical numbers are those for which $f'(x) = 0$. The next example considers a continuous function that has *both* types of critical numbers—those for which $f'(x) = 0$ and those for which $f'(x)$ is undefined.

ALGEBRA TUTOR *xy*

For help on the algebra in Example 5, see Example 2(c) in the *Chapter 8 Algebra Tutor*, on page 681.

Example 5   **Intervals on Which *f* Is Increasing or Decreasing**

Find the open intervals on which the function

$$f(x) = (x^2 - 4)^{2/3}$$

is increasing or decreasing.

SOLUTION   Begin by finding the derivative of the function.

$$f'(x) = \frac{2}{3}(x^2 - 4)^{-1/3}(2x) \qquad \text{Differentiate.}$$

$$= \frac{4x}{3(x^2 - 4)^{1/3}} \qquad \text{Simplify.}$$

From this, you can see that the derivative is zero when $x = 0$ and the derivative is undefined when $x = \pm 2$. So, the critical numbers are

$$x = -2, \quad x = 0, \quad \text{and} \quad x = 2. \qquad \text{Critical numbers}$$

This implies that the test intervals are

$$(-\infty, -2), \quad (-2, 0), \quad (0, 2), \quad \text{and} \quad (2, \infty). \qquad \text{Test intervals}$$

The table summarizes the testing of these four intervals, and the graph of the function is shown in Figure 8.13.

STUDY TIP

To test the intervals in the table in Example 5, it is not necessary to *evaluate* $f'(x)$ at each test value—you only need to determine its sign. For instance, you can determine the sign of $f'(-3)$ as shown.

$$f'(-3) = \frac{4(-3)}{3(9 - 4)^{1/3}}$$

$$= \frac{\text{negative}}{\text{positive}}$$

$$= \text{negative}$$

Interval	$-\infty < x < -2$	$-2 < x < 0$	$0 < x < 2$	$2 < x < \infty$
Test value	$x = -3$	$x = -1$	$x = 1$	$x = 3$
Sign of $f'(x)$	$f'(-3) < 0$	$f'(-1) > 0$	$f'(1) < 0$	$f'(3) > 0$
Conclusion	Decreasing	Increasing	Decreasing	Increasing

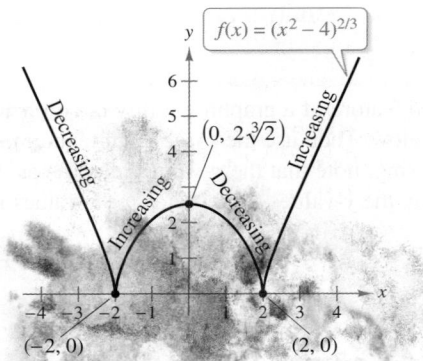

FIGURE 8.13

✓**Checkpoint 5**

Find the open intervals on which the function

$$f(x) = x^{2/3}$$

is increasing or decreasing.

The functions in Examples 1 through 5 are continuous on the entire real number line. If there are isolated x-values at which a function is not continuous, then these x-values should be used along with the critical numbers to determine the test intervals.

Example 6   Testing a Function That Is Not Continuous

The function

$$f(x) = \frac{x^4 + 1}{x^2}$$

is not continuous at $x = 0$. Because the derivative of f

$$f'(x) = \frac{2(x^4 - 1)}{x^3}$$

is zero at $x = \pm 1$, you should use the following numbers to determine the test intervals.

$x = -1, x = 1$ Critical numbers

$x = 0$ Discontinuity

After testing $f'(x)$, you can determine that f is decreasing on the intervals $(-\infty, -1)$ and $(0, 1)$, and increasing on the intervals $(-1, 0)$ and $(1, \infty)$, as shown in Figure 8.14.

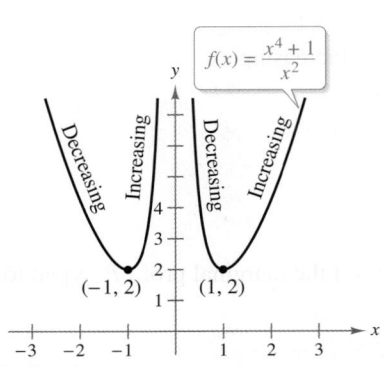

FIGURE 8.14

✓ Checkpoint 6

Find the open invervals on which the function $f(x) = \dfrac{x^2 + 1}{x}$ is increasing or decreasing.   ■

The converse of the test for increasing and decreasing functions is *not* true. For instance, it is possible for a function to be increasing on an interval even though its derivative is not positive at every point in the interval.

Example 7   Testing an Increasing Function

Show that $f(x) = x^3 - 3x^2 + 3x$ is increasing on the entire real number line.

SOLUTION   From the derivative of f

$$f'(x) = 3x^2 - 6x + 3 = 3(x - 1)^2$$

you can see that the only critical number is $x = 1$. So, the test intervals are $(-\infty, 1)$ and $(1, \infty)$. The table summarizes the testing of these two intervals. From Figure 8.15, you can see that f is increasing on the entire real number line, even though $f'(1) = 0$. To convince yourself of this, look back at the definition of an increasing function.

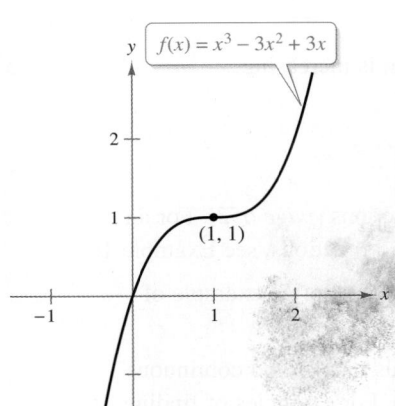

FIGURE 8.15

Interval	$-\infty < x < 1$	$1 < x < \infty$
Test value	$x = 0$	$x = 2$
Sign of $f'(x)$	$f'(0) = 3(-1)^2 > 0$	$f'(2) = 3(1)^2 > 0$
Conclusion	Increasing	Increasing

✓ Checkpoint 7

Show that $f(x) = -x^3 + 2$ is decreasing on the entire real number line.   ■

Application

Example 8   Profit Analysis

A national toy distributor determines the cost and revenue models for one of its games.

$$C = 2.4x - 0.0002x^2, \quad 0 \le x \le 6000$$
$$R = 7.2x - 0.001x^2, \quad 0 \le x \le 6000$$

Determine the interval on which the profit function is increasing.

SOLUTION   The profit for producing x games is

$$P = R - C$$
$$= (7.2x - 0.001x^2) - (2.4x - 0.0002x^2)$$
$$= 4.8x - 0.0008x^2.$$

To find the interval on which the profit is increasing, set the marginal profit P' equal to zero and solve for x.

$P' = 4.8 - 0.0016x$	Differentiate profit function.
$4.8 - 0.0016x = 0$	Set P' equal to 0.
$-0.0016x = -4.8$	Subtract 4.8 from each side.
$x = \dfrac{-4.8}{-0.0016}$	Divide each side by -0.0016.
$x = 3000$ games	Simplify.

On the interval $(0, 3000)$, P' is positive and the profit is *increasing*. On the interval $(3000, 6000)$, P' is negative and the profit is *decreasing*. The graphs of the cost, revenue, and profit functions are shown in Figure 8.16.

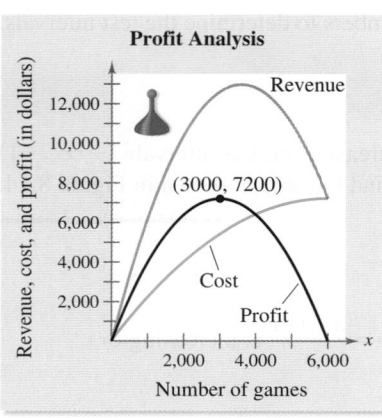

Profit Analysis

(3000, 7200)

Revenue

Cost

Profit

Number of games

Revenue, cost, and profit (in dollars)

FIGURE 8.16

✓ Checkpoint 8

A national distributor of pet toys determines the cost and revenue functions for one of its toys.

$$C = 1.2x - 0.0001x^2, \quad 0 \le x \le 6000$$
$$R = 3.6x - 0.0005x^2, \quad 0 \le x \le 6000$$

Determine the interval on which the profit function is increasing.

SUMMARIZE   (Section 8.4)

1. State the test for increasing and decreasing functions *(page 652)*. For an example of testing for increasing and decreasing functions, see Example 1.

2. State the definition of a critical number *(page 654)*. For an example of finding a critical number, see Example 3.

3. State the guidelines for determining the intervals on which a continuous function is increasing or decreasing *(page 654)*. For examples of finding the intervals on which a function is increasing or decreasing, see Examples 4, 5, and 7.

4. Describe a real-life example of how testing for increasing and decreasing functions can be used to analyze the profit of a company *(page 658, Example 8)*.

SKILLS WARM UP 8.4   The following warm-up exercises involve skills that were covered in a previous course or earlier sections. You will use these skills in the exercise set for this section. For additional help, review Sections 0.2, 1.3, 1.5, and 2.4.

In Exercises 1–4, solve the equation.

1. $x^2 = 8x$

2. $15x = \dfrac{5}{8}x^2$

3. $\dfrac{x^2 - 25}{x^3} = 0$

4. $\dfrac{2x}{\sqrt{1 - x^2}} = 0$

In Exercises 5–8, find the domain of the function.

5. $y = \dfrac{x + 3}{x - 3}$

6. $y = \dfrac{2}{\sqrt{1 - x}}$

7. $y = \dfrac{2x + 1}{x^2 - 3x - 10}$

8. $y = \dfrac{3x}{\sqrt{9 - 3x^2}}$

In Exercises 9–12, evaluate the expression when $x = -2, 0,$ and 2.

9. $-2(x + 1)(x - 1)$

10. $4(2x + 1)(2x - 1)$

11. $\dfrac{2x + 1}{(x - 1)^2}$

12. $\dfrac{-2(x + 1)}{(x - 4)^2}$

Exercises 8.4

See www.CalcChat.com for worked-out solutions to odd-numbered exercises.

Using Graphs   In Exercises 1–4, use the graph to estimate the open intervals on which the function is increasing or decreasing.

1. $f(x) = -(x + 1)^2$

2. $f(x) = \dfrac{x^3}{4} - 3x$

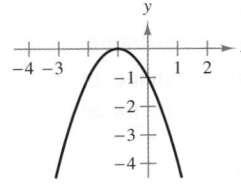
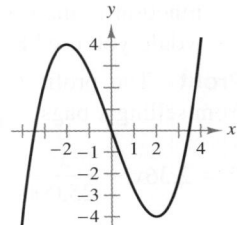

3. $f(x) = x^4 - 2x^2$

4. $f(x) = -(x^2 - 9)^{2/3}$

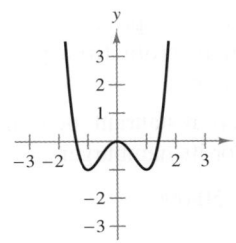
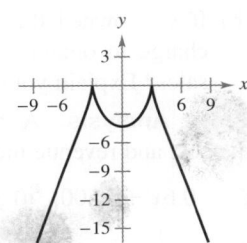

Finding Critical Numbers   In Exercises 5–10, find the critical numbers of the function. *See Example 3.*

5. $f(x) = 4x^2 - 6x$

6. $f(x) = 3x^2 + 10$

7. $y = x^4 + 4x^3 + 8$

8. $g(x) = 2x^2 - 54x$

9. $f(x) = \sqrt{x^2 - 4}$

10. $y = \dfrac{x}{x^2 + 16}$

Intervals on Which f Is Increasing or Decreasing   In Exercises 11–34, find the critical numbers and the open intervals on which the function is increasing or decreasing. Use a graphing utility to verify your results. *See Examples 4 and 5.*

11. $f(x) = 2x - 3$

12. $f(x) = 5 - 3x$

13. $y = x^2 - 6x$

14. $y = -x^2 + 2x$

15. $f(x) = -2x^2 + 4x + 3$

16. $f(x) = x^2 + 8x + 10$

17. $y = 3x^3 + 12x^2 + 15x$

18. $y = x^3 - 3x + 2$

19. $f(x) = x^4 - 2x^3$

20. $f(x) = \frac{1}{4}x^4 - 2x^2$

21. $g(x) = (x + 2)^2$

22. $y = (x - 2)^3$

23. $g(x) = -(x - 1)^2$

24. $y = x^3 - 6x^2$

25. $y = x^{1/3} + 1$

26. $y = x^{2/3} - 4$

27. $f(x) = \sqrt{x^2 - 1}$

28. $f(x) = \sqrt{9 - x^2}$

29. $g(x) = (x + 2)^{1/3}$

30. $g(x) = (x - 1)^{2/3}$

31. $f(x) = x\sqrt{x + 1}$

32. $h(x) = x\sqrt[3]{x - 1}$

33. $f(x) = \dfrac{x}{x^2 + 9}$

34. $f(x) = \dfrac{x^2}{x^2 + 4}$

Intervals on Which f Is Increasing or Decreasing In Exercises 35–42, find the critical numbers and the open intervals on which the function is increasing or decreasing. (*Hint:* Check for discontinuities.) Sketch the graph of the function. *See Example 6.*

35. $f(x) = \dfrac{x + 4}{x - 5}$

36. $f(x) = \dfrac{x}{x + 1}$

37. $f(x) = \dfrac{2x}{16 - x^2}$

38. $f(x) = \dfrac{x^2}{x^2 - 9}$

39. $y = \begin{cases} 4 - x^2, & x \le 0 \\ -2x, & x > 0 \end{cases}$

40. $y = \begin{cases} 3x + 1, & x \le 1 \\ 5 - x^2, & x > 1 \end{cases}$

41. $y = \begin{cases} 2x + 1, & x \le -1 \\ x^2 - 2, & x > -1 \end{cases}$

42. $y = \begin{cases} -x^3 + 1, & x \le 0 \\ -x^2 + 2x, & x > 0 \end{cases}$

43. Sales The sales S of Wal-Mart (in billions of dollars) from 2003 through 2009 can be modeled by

$$S = -1.598t^2 + 45.61t + 130.2, \quad 3 \le t \le 9$$

where t is the time in years, with $t = 3$ corresponding to 2003. Show that the sales were increasing from 2003 through 2009. *(Source: Wal-Mart Stores, Inc.)*

44. 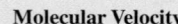 **HOW DO YOU SEE IT?** Plots of the relative numbers of N_2 (nitrogen) molecules that have a given velocity at each of three temperatures (in degrees Kelvin) are shown in the figure. Identify the differences in the average velocities (indicated by the peaks of the curves) for the three temperatures, and describe the intervals on which the velocity is increasing and decreasing for each of the three temperatures. *(Source: Adapted from Zumdahl, Chemistry, Seventh Edition)*

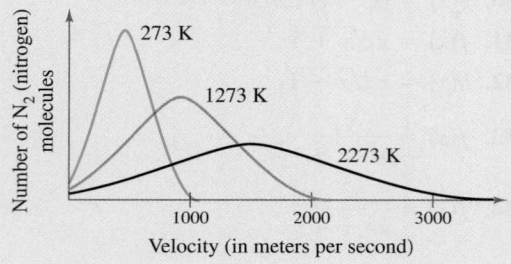

Molecular Velocity

45. Medical Degrees  The number y of medical degrees conferred in the United States from 1970 through 2008 can be modeled by

$$y = 0.692t^3 - 50.11t^2 + 1119.7t + 7894, \; 0 \le t \le 38$$

where t is the time in years, with $t = 0$ corresponding to 1970. *(Source: U.S. National Center for Education Statistics)*

(a) Use a graphing utility to graph the model. Then graphically estimate the years during which the model is increasing and the years during which it is decreasing.

(b) Use the test for increasing and decreasing functions to verify the result of part (a).

46. Cost  The ordering and transportation cost C (in hundreds of dollars) for an automobile dealership is modeled by

$$C = 10\left(\dfrac{1}{x} + \dfrac{x}{x + 3}\right), \quad x \ge 1$$

where x is the number of automobiles ordered.

(a) Find the intervals on which C is increasing or decreasing.

(b) Use a graphing utility to graph the cost function.

(c) Use the *trace* feature to determine the order sizes for which the cost is $900. Assuming that the revenue function is increasing for $x \ge 0$, which order size would you use? Explain your reasoning.

47. Profit The profit P (in dollars) made by a cinema from selling x bags of popcorn can be modeled by

$$P = 2.36x - \dfrac{x^2}{25{,}000} - 3500, \quad 0 \le x \le 50{,}000.$$

(a) Find the intervals on which P is increasing and decreasing.

(b) If you owned the cinema, what price would you charge to obtain a maximum profit from popcorn sales? Explain your reasoning.

48. Profit Analysis A fast-food restaurant determines the cost and revenue models for its hamburgers.

$$C = 0.6x + 7500, \quad 0 \le x \le 50{,}000$$

$$R = \dfrac{1}{20{,}000}(65{,}000x - x^2), \quad 0 \le x \le 50{,}000$$

(a) Write the profit function for this situation.

(b) Determine the intervals on which the profit function is increasing and decreasing.

(c) Determine how many hamburgers the restaurant needs to sell to obtain a maximum profit. Explain your reasoning.

8.5 Extrema and the First-Derivative Test

■ Recognize the occurrence of relative extrema of functions.
■ Use the First-Derivative Test to find the relative extrema of functions.
■ Find absolute extrema of continuous functions on a closed interval.
■ Find minimum and maximum values of real-life models and interpret the results in context.

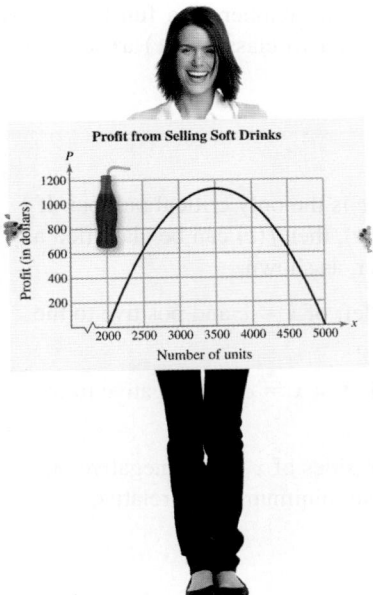

In Exercise 49 on page 669, you will use the First-Derivative Test to find the price of a soft drink that yields a maximum profit.

Relative Extrema

You have used the derivative to determine the intervals on which a function is increasing or decreasing. In this section, you will examine the points at which a function changes from increasing to decreasing, or vice versa. At such a point, the function has a **relative extremum.** (The plural of extremum is *extrema*.) The **relative extrema** of a function include the **relative minima** and **relative maxima** of the function. For instance, the function shown in Figure 8.17 has a relative maximum at the left point and a relative minimum at the right point.

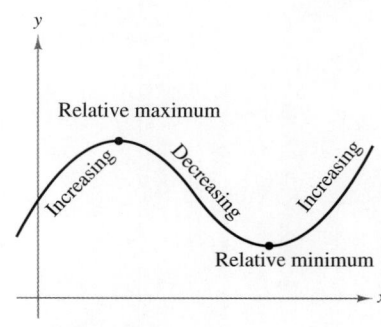

FIGURE 8.17

Definition of Relative Extrema

Let f be a function defined at c.

1. $f(c)$ is a **relative maximum** of f when there exists an interval (a, b) containing c such that $f(x) \leq f(c)$ for all x in (a, b).

2. $f(c)$ is a **relative minimum** of f when there exists an interval (a, b) containing c such that $f(x) \geq f(c)$ for all x in (a, b).

If $f(c)$ is a relative extremum of f, then the relative extremum is said to occur at $x = c$.

For a continuous function, the relative extrema must occur at critical numbers of the function, as shown in Figure 8.18.

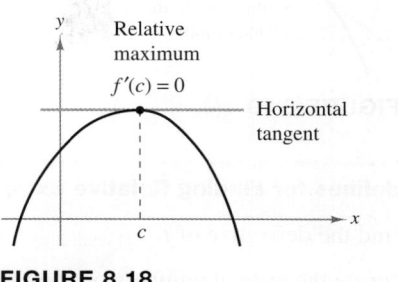

FIGURE 8.18

Occurrences of Relative Extrema

If f has a relative minimum or relative maximum at $x = c$, then c is a critical number of f. That is, either $f'(c) = 0$ or $f'(c)$ is undefined.

The First-Derivative Test

The discussion on the preceding page implies that in your search for relative extrema of a continuous function, you need to test only the critical numbers of the function. Once you have determined that c is a critical number of a function f, the **First-Derivative Test** for relative extrema enables you to classify $f(c)$ as a relative minimum, a relative maximum, or neither.

First-Derivative Test for Relative Extrema

Let f be continuous on the interval (a, b) in which c is the only critical number. If f is differentiable on the interval (except possibly at c), then $f(c)$ can be classified as a relative minimum, a relative maximum, or neither, as shown.

1. On the interval (a, b), if $f'(x)$ is negative to the left of $x = c$ and positive to the right of $x = c$, then $f(c)$ is a relative minimum.

2. On the interval (a, b), if $f'(x)$ is positive to the left of $x = c$ and negative to the right of $x = c$, then $f(c)$ is a relative maximum.

3. On the interval (a, b), if $f'(x)$ is positive on both sides of $x = c$ or negative on both sides of $x = c$, then $f(c)$ is neither a relative minimum nor a relative maximum.

A graphical interpretation of the First-Derivative Test is shown in Figure 8.19.

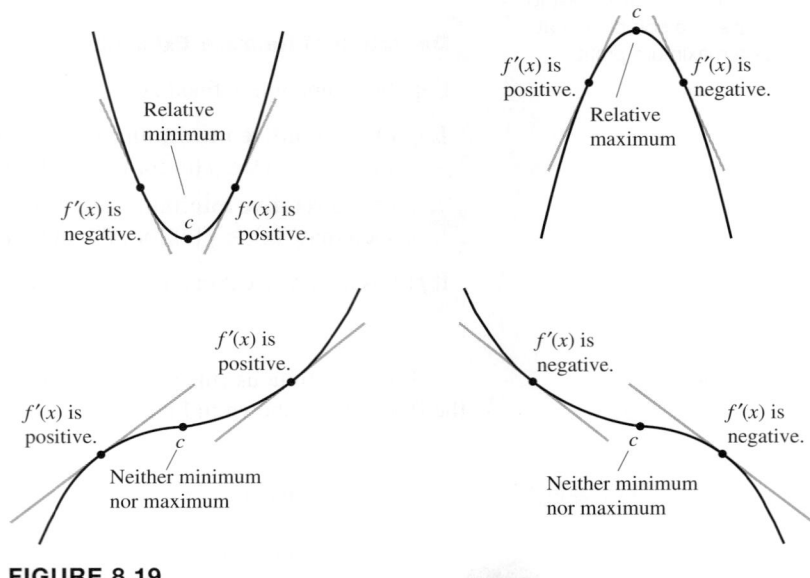

FIGURE 8.19

Guidelines for Finding Relative Extrema

1. Find the derivative of f.

2. Locate the critical numbers of f and use these numbers to determine the test intervals.

3. Test the sign of $f'(x)$ at an arbitrary number in each of the test intervals.

4. For each critical number c, use the First-Derivative Test to decide whether $f(c)$ is a relative minimum, a relative maximum, or neither.

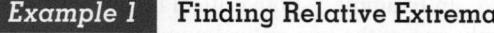

Example 1 **Finding Relative Extrema**

Find all relative extrema of the function

$$f(x) = 2x^3 - 3x^2 - 36x + 14.$$

SOLUTION Begin by finding the derivative of f.

$$f'(x) = 6x^2 - 6x - 36 \qquad \text{Differentiate.}$$

Next, find the critical numbers of f.

$$6x^2 - 6x - 36 = 0 \qquad \text{Set derivative equal to 0.}$$
$$6(x^2 - x - 6) = 0 \qquad \text{Factor out common factor.}$$
$$6(x - 3)(x + 2) = 0 \qquad \text{Factor.}$$
$$x = -2,\, x = 3 \qquad \text{Critical numbers}$$

Because $f'(x)$ is defined for all x, the only critical numbers of f are

$$x = -2 \quad \text{and} \quad x = 3. \qquad \text{Critical numbers}$$

Using these numbers, you can form the three test intervals

$$(-\infty, -2), (-2, 3), \quad \text{and} \quad (3, \infty). \qquad \text{Test intervals}$$

The testing of the three intervals is shown in the table.

Interval	$-\infty < x < -2$	$-2 < x < 3$	$3 < x < \infty$
Test value	$x = -3$	$x = 0$	$x = 4$
Sign of $f'(x)$	$f'(-3) = 36 > 0$	$f'(0) = -36 < 0$	$f'(4) = 36 > 0$
Conclusion	Increasing	Decreasing	Increasing

Using the First-Derivative Test, you can conclude that the critical number -2 yields a relative maximum [$f'(x)$ changes sign from positive to negative], and the critical number 3 yields a relative minimum [$f'(x)$ changes sign from negative to positive]. The graph of f is shown in Figure 8.20. The relative maximum is

$$f(-2) = 58$$

and the relative minimum is

$$f(3) = -67.$$

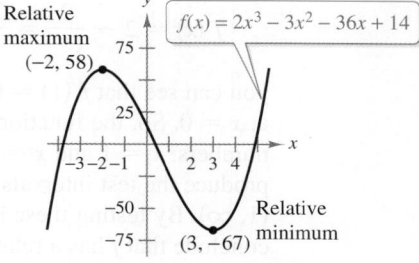

FIGURE 8.20

✓ Checkpoint 1

Find all relative extrema of

$$f(x) = 2x^3 - 6x + 1.$$

TECH TUTOR

Some graphing calculators have a special feature that allows you to find the minimum or maximum of a function on an interval. Consult the user's manual for information on the *minimum* and *maximum* features of your graphing utility.

In Example 1, both critical numbers yielded relative extrema. In the next example, only one of the two critical numbers yields a relative extremum.

ALGEBRA TUTOR *xy*

For help on the algebra in Example 2, see Example 2(b) in the *Chapter 8 Algebra Tutor*, on page 681.

Example 2   Finding Relative Extrema

Find all relative extrema of the function $f(x) = x^4 - x^3$.

SOLUTION   From the derivative of the function

$$f'(x) = 4x^3 - 3x^2 = x^2(4x - 3)$$

you can see that the function has only two critical numbers: $x = 0$ and $x = \frac{3}{4}$. These numbers produce the test intervals $(-\infty, 0)$, $\left(0, \frac{3}{4}\right)$, and $\left(\frac{3}{4}, \infty\right)$, which are tested in the table.

Interval	$-\infty < x < 0$	$0 < x < \frac{3}{4}$	$\frac{3}{4} < x < \infty$
Test value	$x = -1$	$x = \frac{1}{2}$	$x = 1$
Sign of $f'(x)$	$f'(-1) = -7 < 0$	$f'\left(\frac{1}{2}\right) = -\frac{1}{4} < 0$	$f'(1) = 1 > 0$
Conclusion	Decreasing	Decreasing	Increasing

By the First-Derivative Test, it follows that f has a relative minimum at $x = \frac{3}{4}$, as shown in Figure 8.21. The relative minimum is

$$f\left(\frac{3}{4}\right) = -\frac{27}{256}.$$

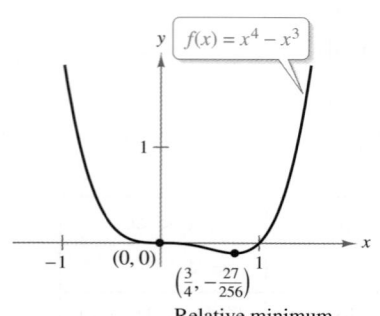

FIGURE 8.21

Note that the critical number $x = 0$ does not yield a relative extremum because $f'(x)$ is negative on both sides of $x = 0$.

✔**Checkpoint 2**

Find all relative extrema of $f(x) = x^4 - 4x^3$.

Example 3   Finding Relative Extrema

Find all relative extrema of the function

$$f(x) = 2x - 3x^{2/3}.$$

SOLUTION   From the derivative of the function

$$f'(x) = 2 - \frac{2}{x^{1/3}} = \frac{2(x^{1/3} - 1)}{x^{1/3}}$$

you can see that $f'(1) = 0$ and f' is undefined at $x = 0$. So, the function has two critical numbers: $x = 1$ and $x = 0$. These numbers produce the test intervals $(-\infty, 0)$, $(0, 1)$, and $(1, \infty)$. By testing these intervals, you can conclude that f has a relative maximum at $(0, 0)$ and a relative minimum at $(1, -1)$, as shown in Figure 8.22.

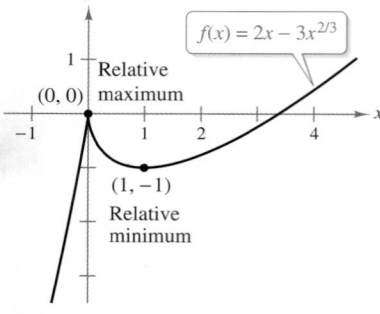

FIGURE 8.22

✔**Checkpoint 3**

Find all relative extrema of $f(x) = 3x^{2/3} - 2x$.

Absolute Extrema

The terms *relative minimum* and *relative maximum* describe the *local* behavior of a function. To describe the *global* behavior of the function on an entire interval, you can use the terms **absolute maximum** and **absolute minimum.**

Definition of Absolute Extrema

Let f be defined on an interval I containing c.

1. $f(c)$ is an **absolute minimum of** f on I when $f(c) \le f(x)$ for every x in I.

2. $f(c)$ is an **absolute maximum of** f on I when $f(c) \ge f(x)$ for every x in I.

The absolute minimum and absolute maximum values of a function on an interval are sometimes called simply the **minimum** and **maximum** of f on I.

Be sure that you understand the distinction between relative extrema and absolute extrema. For instance, in Figure 8.23, the function has a relative minimum that also happens to be an absolute minimum on the interval $[a, b]$. The relative maximum of f, however, is not the absolute maximum on the interval $[a, b]$. The next theorem points out that if a continuous function has a closed interval as its domain, then it *must* have both an absolute minimum and an absolute maximum on the interval. From Figure 8.23, note that these extrema can occur at the endpoints of the interval.

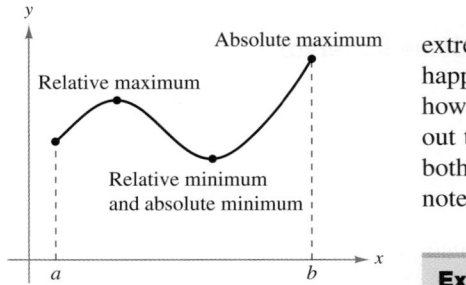

FIGURE 8.23

Extreme Value Theorem

If f is continuous on a closed interval $[a, b]$, then f has both a minimum value and a maximum value on $[a, b]$.

Although a continuous function has just one minimum and one maximum value on a closed interval, either of these values can occur for more than one x-value. For instance, on the interval $[-3, 3]$, the function

$$f(x) = 9 - x^2$$

has a minimum value of zero at $x = -3$ *and* at $x = 3$, as shown in Figure 8.24.

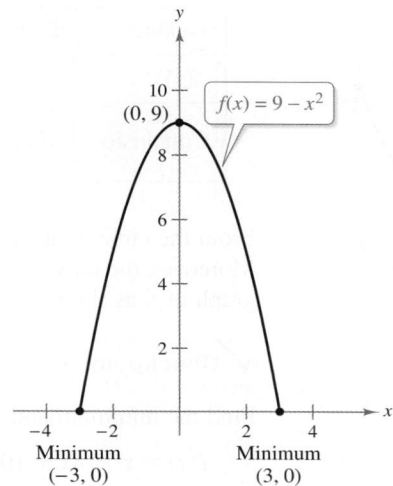

FIGURE 8.24

When looking for extrema of a function on a *closed* interval, remember that you must consider the values of the function at the endpoints as well as at the critical numbers of the function. You can use the guidelines below to find extrema on a closed interval.

Guidelines for Finding Extrema on a Closed Interval

To find the extrema of a continuous function f on a closed interval $[a, b]$, use the following steps.

1. Find the critical numbers of f in the open interval (a, b).

2. Evaluate f at each of its critical numbers in (a, b).

3. Evaluate f at each endpoint, a and b.

4. The least of these values is the minimum, and the greatest is the maximum.

Example 4   Finding Extrema on a Closed Interval

Find the minimum and maximum values of

$$f(x) = x^2 - 6x + 2$$

on the interval $[0, 5]$.

SOLUTION   Begin by differentiating the function.

$$f(x) = x^2 - 6x + 2 \qquad \text{Write original function.}$$
$$f'(x) = 2x - 6 \qquad \text{Differentiate.}$$

Next, find the critical numbers of f.

$$2x - 6 = 0 \qquad \text{Set derivative equal to 0.}$$
$$2x = 6 \qquad \text{Add 6 to each side.}$$
$$x = 3 \qquad \text{Solve for } x.$$

Because f' is defined for all x, you can conclude that the only critical number of f is $x = 3$. Because this number lies in the interval $[0, 5]$, you should evaluate f at this number *and* at the endpoints of the interval, as shown in the table.

x-value	Endpoint: $x = 0$	Critical number: $x = 3$	Endpoint: $x = 5$
$f(x)$	$f(0) = 2$	$f(3) = -7$	$f(5) = -3$
Conclusion	Maximum is 2.	Minimum is -7.	Neither maximum nor minimum

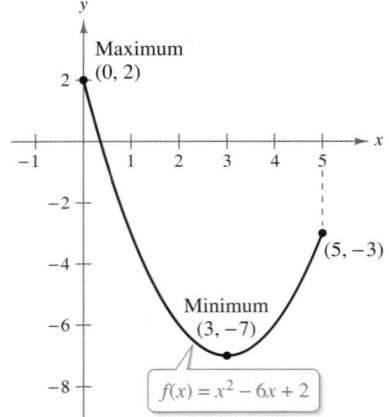

FIGURE 8.25

From the table, you can see that the minimum of f on the interval $[0, 5]$ is $f(3) = -7$. Moreover, the maximum of f on the interval $[0, 5]$ is $f(0) = 2$. This is confirmed by the graph of f, as shown in Figure 8.25.

✓ Checkpoint 4

Find the minimum and maximum values of

$$f(x) = x^2 - 8x + 10$$

on the interval $[0, 7]$. Sketch the graph of f and label the minimum and maximum values.

Application

Finding the minimum and maximum values of a function is one of the most common applications of calculus.

 Example 5   **Finding the Maximum Profit**

Recall the fast-food restaurant in Examples 7 and 8 in Section 7.5. The restaurant's profit function for hamburgers is given by

$$P = 2.44x - \frac{x^2}{20,000} - 5000, \quad 0 \le x \le 50,000.$$

Find the sales level that yields a maximum profit.

SOLUTION   To begin, find an equation for marginal profit.

$$\frac{dP}{dx} = 2.44 - \frac{x}{10,000} \qquad \text{Find marginal profit.}$$

Next, set the marginal profit equal to zero and solve for x.

$$2.44 - \frac{x}{10,000} = 0 \qquad \text{Set marginal profit equal to 0.}$$

$$-\frac{x}{10,000} = -2.44 \qquad \text{Subtract 2.44 from each side.}$$

$$x = 24,400 \text{ hamburgers} \qquad \text{Critical number}$$

In Figure 8.26, you can see that the critical number $x = 24,400$ corresponds to the sales level that yields a maximum profit. To find the maximum profit, substitute $x = 24,400$ into the profit function.

$$P = 2.44x - \frac{x^2}{20,000} - 5000$$

$$= 2.44(24,400) - \frac{(24,400)^2}{20,000} - 5000$$

$$= \$24,768$$

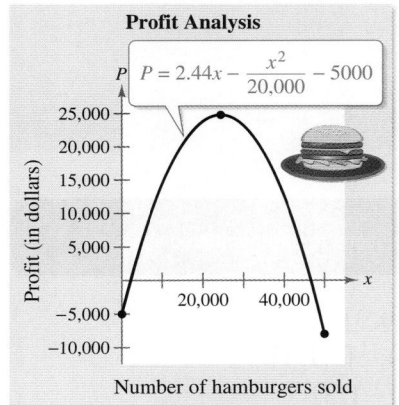

Profit Analysis

$P = 2.44x - \frac{x^2}{20,000} - 5000$

Number of hamburgers sold

FIGURE 8.26

✓**Checkpoint 5**

Verify the results of Example 5 by completing the table.

x (units)	24,000	24,200	24,300	24,400	24,500	24,600	24,800
P (profit)							

SUMMARIZE   (Section 8.5)

1. State the First-Derivative Test *(page 662)*. For examples in which the First-Derivative Test is used, see Examples 1, 2, and 3.

2. State the guidelines for finding extrema on a closed interval *(page 666)*. For an example of finding the extrema of a function on a closed interval, see Example 4.

3. Describe a real-life example of how the First-Derivative Test can be used to find the sales level that yields a maximum profit for a company *(page 667, Example 5)*.

SKILLS WARM UP 8.5   The following warm-up exercises involve skills that were covered in earlier sections. You will use these skills in the exercise set for this section. For additional help, review Sections 7.4, 7.6, and 8.4.

In Exercises 1–6, solve the equation $f'(x) = 0$.

1. $f(x) = 4x^4 - 2x^2 + 1$

2. $f(x) = \frac{1}{3}x^3 - \frac{3}{2}x^2 - 10x$

3. $f(x) = 5x^{4/5} - 4x$

4. $f(x) = \frac{1}{2}x^2 - 3x^{5/3}$

5. $f(x) = \dfrac{x + 4}{x^2 + 1}$

6. $f(x) = \dfrac{x - 1}{x^2 + 4}$

In Exercises 7–10, use $g(x) = -x^5 - 2x^4 + 4x^3 + 2x - 1$ to determine the sign of the derivative.

7. $g'(-4)$

8. $g'(0)$

9. $g'(1)$

10. $g'(3)$

In Exercises 11 and 12, decide whether the function is increasing or decreasing on the given interval.

11. $f(x) = 2x^2 - 11x - 6$, $(3, 6)$

12. $f(x) = x^3 + 2x^2 - 4x - 8$, $(-2, 0)$

Exercises 8.5

See www.CalcChat.com for worked-out solutions to odd-numbered exercises.

Finding Relative Extrema In Exercises 1–12, find all relative extrema of the function. *See Examples 1, 2, and 3.*

1. $f(x) = -2x^2 + 4x + 3$

2. $f(x) = x^2 + 8x + 10$

3. $f(x) = x^2 - 6x$

4. $f(x) = -4x^2 + 4x + 1$

5. $f(x) = x^4 - 12x^3$

6. $g(x) = \frac{1}{5}x^5 - x$

7. $h(x) = -(x + 4)^3$

8. $h(x) = 2(x - 3)^3$

9. $f(x) = x^3 - 6x^2 + 15$

10. $f(x) = x^4 - 32x + 4$

11. $f(x) = 6x^{2/3} + 4x$

12. $f(x) = 3x - 36x^{1/3}$

 Finding Relative Extrema In Exercises 13–18, use a graphing utility to find graphically all relative extrema of the function.

13. $f(x) = 2x - 6x^{2/3}$

14. $f(t) = (t - 1)^{1/3}$

15. $g(t) = t - \dfrac{1}{2t^2}$

16. $f(x) = x + \dfrac{1}{x}$

17. $f(x) = \dfrac{x}{x + 1}$

18. $h(x) = \dfrac{6}{x^2 + 2}$

Finding Extrema on a Closed Interval In Exercises 19–30, find the absolute extrema of the function on the closed interval. Use a graphing utility to verify your results. *See Example 4.*

19. $f(x) = 2(3 - x)$, $[-1, 2]$

20. $f(x) = \frac{1}{3}(2x + 5)$, $[0, 5]$

21. $f(x) = 5 - 2x^2$, $[0, 3]$

22. $f(x) = x^2 + 2x - 4$, $[-1, 1]$

23. $f(x) = x^3 - 3x^2$, $[-1, 3]$

24. $f(x) = x^3 - 12x$, $[0, 4]$

25. $h(s) = \dfrac{1}{3 - s}$, $[0, 2]$

26. $h(t) = \dfrac{t}{t - 2}$, $[3, 5]$

27. $g(t) = \dfrac{t^2}{t^2 + 3}$, $[-1, 1]$

28. $g(x) = 4\left(1 + \dfrac{1}{x} + \dfrac{1}{x^2}\right)$, $[-4, 5]$

29. $h(t) = (t - 1)^{2/3}$, $[-7, 2]$

30. $g(x) = (x^2 - 4)^{2/3}$, $[-6, 3]$

Determining Types of Extrema In Exercises 31–34, approximate the critical numbers of the function shown in the graph. Determine whether the function has a relative maximum, a relative minimum, an absolute maximum, an absolute minimum, or none of these at each critical number on the interval shown.

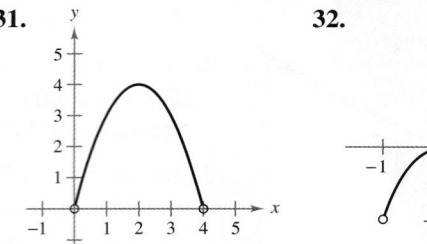

31.

32.

33.

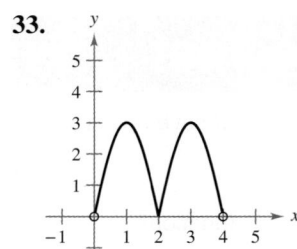

34.

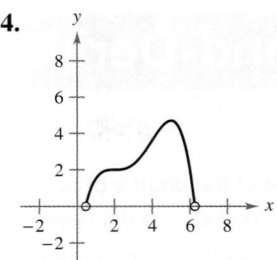

 Finding Extrema on a Closed Interval In Exercises 35–38, use a graphing utility to find graphically the absolute extrema of the function on the closed interval.

35. $f(x) = 0.4x^3 - 1.8x^2 + x - 3$, $[0, 5]$

36. $f(x) = 3.2x^5 + 5x^3 - 3.5x$, $[0, 1]$

37. $f(x) = \frac{4}{3}x\sqrt{3 - x}$, $[0, 3]$

38. $f(x) = 4\sqrt{x} - 2x + 1$, $[0, 6]$

Finding Absolute Extrema In Exercises 39–42, find the absolute extrema of the function on the interval $[0, \infty)$.

39. $f(x) = x^2 + \dfrac{16}{x}$

40. $f(x) = \dfrac{8}{x + 1}$

41. $f(x) = \dfrac{2x}{x^2 + 4}$

42. $f(x) = 8 - \dfrac{4x}{x^2 + 1}$

Creating the Graph of a Function In Exercises 43 and 44, graph a function on the closed interval $[-2, 5]$ having the given characteristics. (There are many correct answers.)

43. Absolute maximum at $x = -2$

Absolute minimum at $x = 1$

Relative maximum at $x = 3$

44. Relative minimum at $x = -1$

Critical number at $x = 0$, but no extrema

Absolute maximum at $x = 2$

Absolute minimum at $x = 5$

45. Population The resident populations P (in millions) of the United States from 1790 through 2010 can be modeled by $P = 0.000006t^3 + 0.005t^2 + 0.14t + 4.6$, $-10 \le t \le 210$, where $t = 0$ corresponds to 1800. *(Source: U.S. Census Bureau)*

(a) Make a conjecture about the maximum and minimum populations of the United States from 1790 to 2010.

(b) Analytically find the maximum and minimum populations over the interval.

(c) Write a brief paragraph comparing your conjecture with your results in part (b).

46.  **HOW DO YOU SEE IT?** The graph of the United States fertility rate shows the number of births per 1000 women in their lifetime according to the birth rate in that particular year. *(Source: U.S. National Center for Health Statistics)*

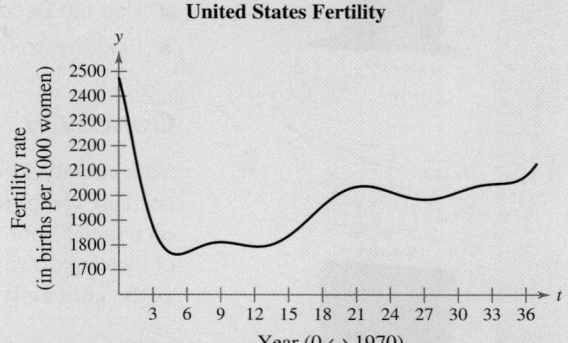

United States Fertility

(a) Around what year was the fertility rate the highest, and to how many births per 1000 women did this rate correspond?

(b) During which time periods was the fertility rate increasing most rapidly? most slowly?

(c) During which time periods was the fertility rate decreasing most rapidly? most slowly?

(d) Give some possible real-life reasons for fluctuations in the fertility rate.

47. Cost A retailer determines that the cost C of ordering and storing x units of a product can be modeled by

$$C = 3x + \frac{20{,}000}{x}, \quad 0 < x \le 200.$$

The delivery truck can carry at most 200 units per order. Find the order size that will minimize the cost. Use a graphing utility to verify your result.

48. Medical Science Coughing forces the trachea (windpipe) to contract, which in turn affects the velocity of the air passing through the trachea. The velocity of the air during coughing can be modeled by

$$v = k(R - r)r^2, \quad 0 \le r < R$$

where k is a constant, R is the normal radius of the trachea, and r is the radius during coughing. What radius r will produce the maximum air velocity?

49. Profit When soft drinks are sold for $1.00 per can at football games, approximately 6000 cans are sold. When the price is raised to $1.20 per can, the quantity demanded drops to 5600. The initial cost is $5000 and the cost per unit is $0.50. Assuming that the demand function is linear, what number of units and what price will yield a maximum profit?

8.6 Concavity and the Second-Derivative Test

Production Cost

In Exercise 69 on page 678, you will use the Second-Derivative Test to find the production level that will minimize the average cost per unit.

■ Determine the intervals on which the graphs of functions are concave upward or concave downward.

■ Find the points of inflection of the graphs of functions.

■ Use the Second-Derivative Test to find the relative extrema of functions.

■ Find the points of diminishing returns of input-output models.

Concavity

You already know that locating the intervals over which a function f increases or decreases helps to describe its graph. In this section, you will see that locating the intervals on which f' increases or decreases can determine where the graph of f is curving upward or curving downward. This property of curving upward or downward is defined formally as the **concavity** of the graph of the function.

Definition of Concavity

Let f be differentiable on an open interval I. The graph of f is

1. **concave upward** on I when f' is increasing on the interval.

2. **concave downward** on I when f' is decreasing on the interval.

In Figure 8.27, you can observe the following graphical interpretation of concavity.

1. A curve that is concave upward lies *above* its tangent line.

2. A curve that is concave downward lies *below* its tangent line.

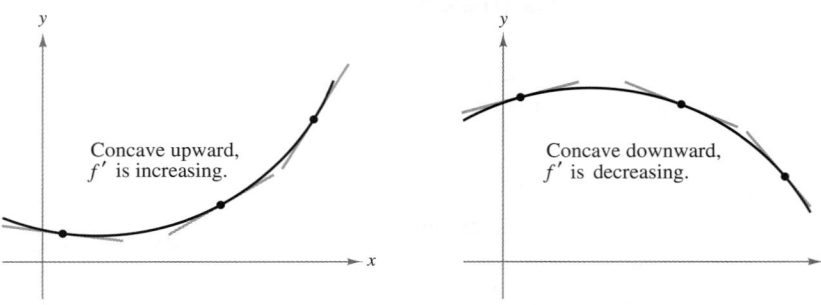

Concave upward, f' is increasing.

Concave downward, f' is decreasing.

FIGURE 8.27

To find the open intervals on which the graph of a function is concave upward or concave downward, you can use the second derivative of the function as follows.

Test for Concavity

Let f be a function whose second derivative exists on an open interval I.

1. If $f''(x) > 0$ for all x in I, then the graph of f is concave upward on I.

2. If $f''(x) < 0$ for all x in I, then the graph of f is concave downward on I.

Example 1   Determining Concavity

a. The graph of the function

$$f(x) = x^2 \qquad \text{Original function}$$

is concave upward on the entire real number line because its second derivative

$$f''(x) = 2 \qquad \text{Second derivative}$$

is positive for all x. (See Figure 8.28.)

b. The graph of the function

$$f(x) = \sqrt{x} \qquad \text{Original function}$$

is concave downward for $x > 0$ because its second derivative

$$f''(x) = -\frac{1}{4}x^{-3/2} \qquad \text{Second derivative}$$

is negative for all $x > 0$. (See Figure 8.29.)

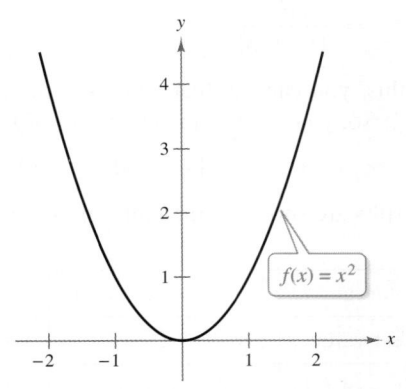

Concave Upward

FIGURE 8.28

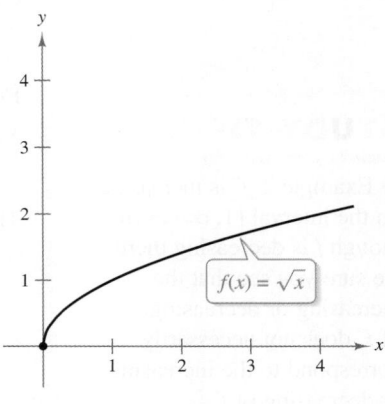

Concave Downward

FIGURE 8.29

 Checkpoint 1

Find the second derivative of f and discuss the concavity of its graph.

a. $f(x) = -2x^2$

b. $f(x) = -2\sqrt{x}$

For a *continuous* function f, you can find the open intervals on which the graph of f is concave upward and concave downward as follows. [When there are x-values at which the function is not continuous, these values should be used, along with the points at which $f''(x) = 0$ or $f''(x)$ is undefined, to form the test intervals.]

Guidelines for Applying the Concavity Test

1. Locate the x-values at which $f''(x) = 0$ or $f''(x)$ is undefined.

2. Use these x-values to determine the test intervals.

3. Test the sign of $f''(x)$ in each test interval.

Example 2   **Applying the Test for Concavity**

Determine the open intervals on which the graph of

$$f(x) = \frac{6}{x^2 + 3}$$

is concave upward or concave downward.

SOLUTION   Begin by finding the second derivative of f.

$$f(x) = 6(x^2 + 3)^{-1}$$ Rewrite original function.

$$f'(x) = 6(-1)(x^2 + 3)^{-2}(2x)$$ Chain Rule

$$= \frac{-12x}{(x^2 + 3)^2}$$ Simplify.

$$f''(x) = \frac{(x^2 + 3)^2(-12) - (-12x)(2)(x^2 + 3)(2x)}{(x^2 + 3)^4}$$ Quotient Rule

$$= \frac{-12(x^2 + 3) + 48x^2}{(x^2 + 3)^3}$$ Simplify.

$$= \frac{36(x^2 - 1)}{(x^2 + 3)^3}$$ Simplify.

From this, you can see that $f''(x)$ is defined for all real numbers and $f''(x) = 0$ when $x = \pm 1$. So, you can test the concavity of f by testing the intervals

$$(-\infty, -1), (-1, 1), \text{and} (1, \infty).$$ Test intervals

The results are shown in the table and in Figure 8.30.

Interval	$-\infty < x < -1$	$-1 < x < 1$	$1 < x < \infty$
Test value	$x = -2$	$x = 0$	$x = 2$
Sign of $f''(x)$	$f''(-2) > 0$	$f''(0) < 0$	$f''(2) > 0$
Conclusion	Concave upward	Concave downward	Concave upward

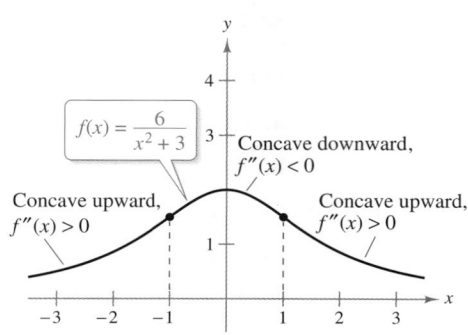

FIGURE 8.30

✓**Checkpoint 2**

Determine the intervals on which the graph of

$$f(x) = \frac{12}{x^2 + 4}$$

is concave upward or concave downward.

Points of Inflection

If the tangent line to a graph exists at a point at which the concavity changes, then the point is a **point of inflection.** Three examples of inflection points are shown in Figure 8.31. (Note that the third graph has a vertical tangent line at its point of inflection.)

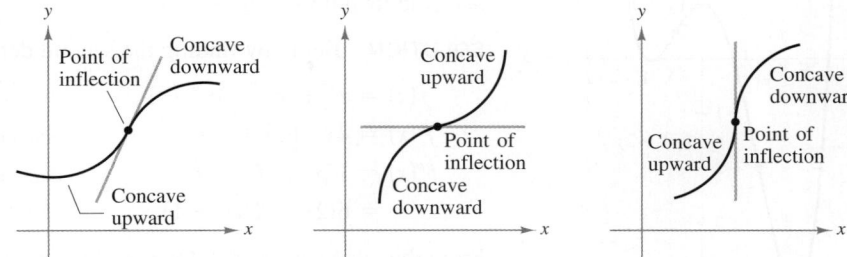

The graph *crosses* its tangent line at a point of inflection.
FIGURE 8.31

Definition of Point of Inflection

If the graph of a continuous function has a tangent line at a point where its concavity changes from upward to downward (or downward to upward), then the point is a **point of inflection.**

Because a point of inflection occurs where the concavity of a graph changes, it must be true that at such points the sign of f'' changes. So, to locate possible points of inflection, you need to determine only the values of x for which $f''(x) = 0$ or for which $f''(x)$ does not exist. This parallels the procedure for locating the relative extrema of f by determining the critical numbers of f.

Property of Points of Inflection

If $(c, f(c))$ is a point of inflection of the graph of f, then either $f''(c) = 0$ or $f''(c)$ is undefined.

Example 3   Finding a Point of Inflection

Discuss the concavity of the graph of $f(x) = 2x^3 + 1$ and find its point of inflection.

SOLUTION   Differentiating twice produces the following.

$$f(x) = 2x^3 + 1 \qquad \text{Write original function.}$$
$$f'(x) = 6x^2 \qquad \text{Find first derivative.}$$
$$f''(x) = 12x \qquad \text{Find second derivative.}$$

Setting $f''(x) = 0$, you can determine that the only possible point of inflection occurs at $x = 0$. After testing the intervals $(-\infty, 0)$ and $(0, \infty)$, you can determine that the graph is concave downward on $(-\infty, 0)$ and concave upward on $(0, \infty)$. Because the concavity changes at $x = 0$, you can conclude that the graph of f has a point of inflection at $(0, 1)$, as shown in Figure 8.32.

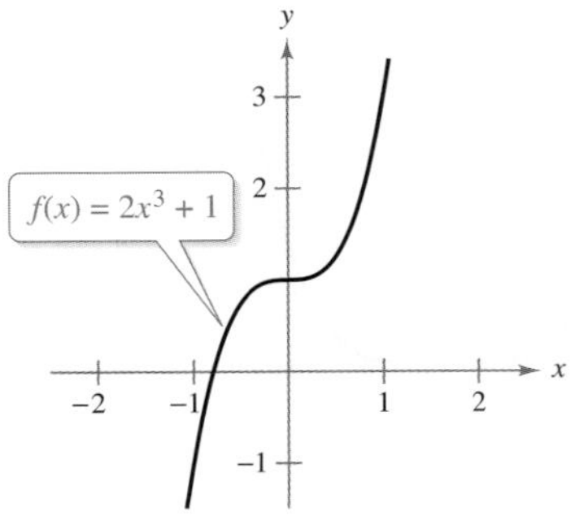

$f(x) = 2x^3 + 1$

FIGURE 8.32

 **Checkpoint 3**

Discuss the concavity of the graph of $f(x) = -x^3$ and find its point of inflection. ■

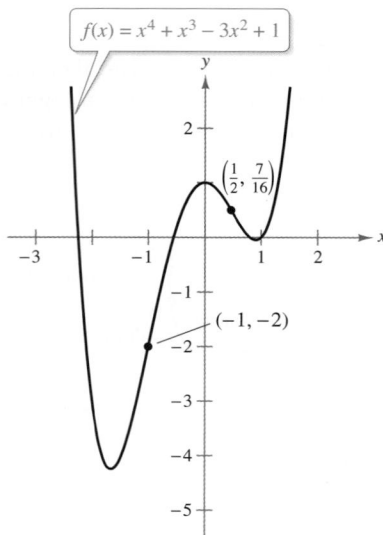

$f(x) = x^4 + x^3 - 3x^2 + 1$

Two Points of Inflection

FIGURE 8.33

Example 4   **Finding Points of Inflection**

Discuss the concavity of the graph of

$$f(x) = x^4 + x^3 - 3x^2 + 1$$

and find its points of inflection.

SOLUTION   Begin by finding the second derivative of f.

$f(x) = x^4 + x^3 - 3x^2 + 1$	Write original function.
$f'(x) = 4x^3 + 3x^2 - 6x$	Find first derivative.
$f''(x) = 12x^2 + 6x - 6$	Find second derivative.
$\quad\;\; = 6(2x - 1)(x + 1)$	Factor.

From this, you can see that the possible points of inflection occur at $x = \frac{1}{2}$ and $x = -1$. After testing the intervals $(-\infty, -1)$, $\left(-1, \frac{1}{2}\right)$, and $\left(\frac{1}{2}, \infty\right)$, you can determine that the graph is concave upward on $(-\infty, -1)$, concave downward on $\left(-1, \frac{1}{2}\right)$, and concave upward on $\left(\frac{1}{2}, \infty\right)$. Because the concavity changes at $x = -1$ and $x = \frac{1}{2}$, you can conclude that the graph of f has points of inflection at these x-values, as shown in Figure 8.33. The points of inflection are

$$(-1, -2) \quad \text{and} \quad \left(\frac{1}{2}, \frac{7}{16}\right).$$

✓ **Checkpoint 4**

Discuss the concavity of the graph of

$$f(x) = x^4 - 2x^3 + 1$$

and find its points of inflection.

It is possible for the second derivative to be zero at a point that is *not* a point of inflection. For example, compare the graphs of

$$f(x) = x^3 \quad \text{and} \quad g(x) = x^4$$

as shown in Figure 8.34. Both second derivatives are zero when $x = 0$, but only the graph of f has a point of inflection at $x = 0$. This shows that before concluding that a point of inflection exists at a value of x for which $f''(x) = 0$, you must test to be certain that the concavity actually changes at that point.

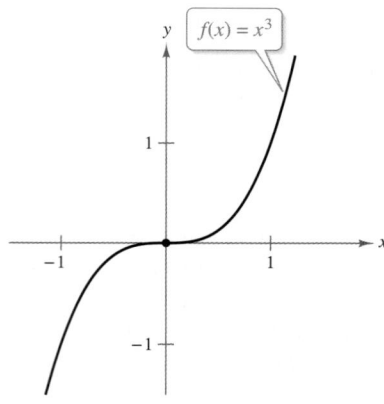

$f''(0) = 0$, and $(0, 0)$ is a point of inflection.

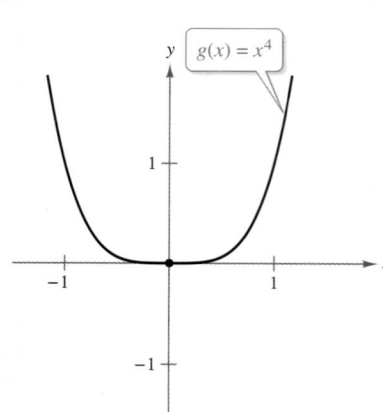

$g''(0) = 0$, but $(0, 0)$ is not a point of inflection.

FIGURE 8.34

The Second-Derivative Test

The second derivative can be used to perform a simple test for relative minima and relative maxima. If f is a function such that $f'(c) = 0$ and the graph of f is concave upward at $x = c$, then $f(c)$ is a relative minimum of f. Similarly, if f is a function such that $f'(c) = 0$ and the graph of f is concave downward at $x = c$, then $f(c)$ is a relative maximum of f, as shown in Figure 8.35.

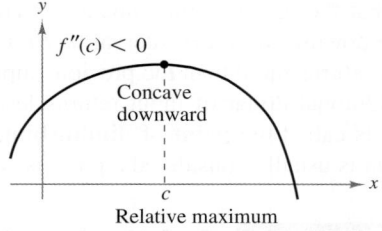
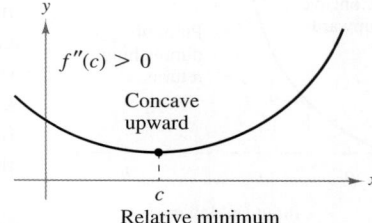

FIGURE 8.35

Second-Derivative Test

Let $f'(c) = 0$, and let f'' exist on an open interval containing c.

1. If $f''(c) > 0$, then $f(c)$ is a relative minimum.

2. If $f''(c) < 0$, then $f(c)$ is a relative maximum.

3. If $f''(c) = 0$, then the test fails. In such cases, you can use the First-Derivative Test to determine whether $f(c)$ is a relative minimum, a relative maximum, or neither.

Example 5 **Using the Second-Derivative Test**

Find the relative extrema of $f(x) = -3x^5 + 5x^3$.

SOLUTION Begin by finding the first derivative of f.

$$f'(x) = -15x^4 + 15x^2 = 15x^2(1 - x^2)$$

From this derivative, you can see that $x = 0$, $x = -1$, and $x = 1$ are the only critical numbers of f. Using the second derivative

$$f''(x) = -60x^3 + 30x = 30x(1 - 2x^2)$$

you can apply the Second-Derivative Test, as shown.

Point	$(-1, -2)$	$(0, 0)$	$(1, 2)$
Sign of $f''(x)$	$f''(-1) > 0$	$f''(0) = 0$	$f''(1) < 0$
Conclusion	Relative minimum	Test fails.	Relative maximum

Because the Second-Derivative Test fails at $(0, 0)$, you can use the First-Derivative Test and observe that f is positive on both sides of $x = 0$. So, $(0, 0)$ is neither a relative minimum nor a relative maximum. A test for concavity would show that $(0, 0)$ is a point of inflection. The graph of f is shown in Figure 8.36.

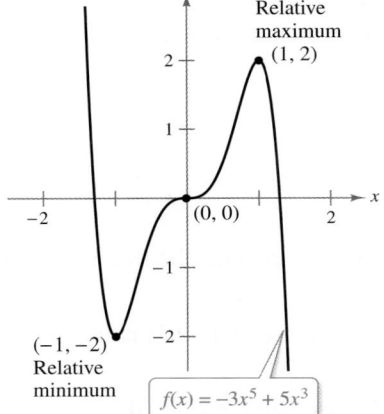

FIGURE 8.36

✓**Checkpoint 5**

Find all relative extrema of $f(x) = x^4 - 4x^3 + 1$.

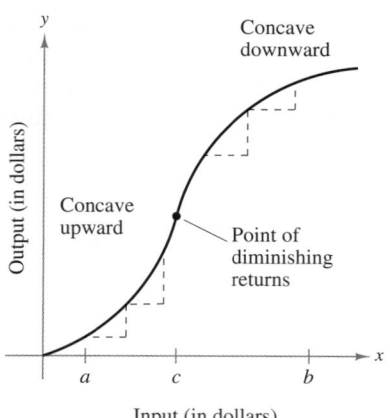

FIGURE 8.37

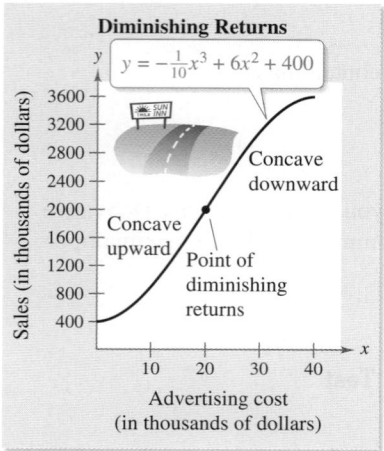

FIGURE 8.38

Extended Application: Diminishing Returns

In economics, the notion of concavity is related to the concept of **diminishing returns.** Consider a function

Output — ↓   ↓ — Input
$$y = f(x)$$

where x measures input (in dollars) and y measures output (in dollars). In Figure 8.37, notice that the graph of this function is concave upward on the interval (a, c) and is concave downward on the interval (c, b). On the interval (a, c), each additional dollar of input returns more than the previous input dollar. By contrast, on the interval (c, b), each additional dollar of input returns less than the previous input dollar. The point $(c, f(c))$ is called the **point of diminishing returns.** An increased investment beyond this point is usually considered a poor use of capital.

Example 6   **Exploring Diminishing Returns**

By increasing its advertising cost x (in thousands of dollars) for a product, a company discovers that it can increase the sales y (in thousands of dollars) according to the model

$$y = -\frac{1}{10}x^3 + 6x^2 + 400, \quad 0 \le x \le 40.$$

Find the point of diminishing returns for this product.

SOLUTION   Begin by finding the first and second derivatives.

$$y' = 12x - \frac{3x^2}{10} \qquad \text{First derivative}$$

$$y'' = 12 - \frac{3x}{5} \qquad \text{Second derivative}$$

The second derivative is zero only when $x = 20$. By testing for concavity on the intervals $(0, 20)$ and $(20, 40)$, you can conclude that the graph has a point of diminishing returns when $x = 20$, as shown in Figure 8.38. So, the point of diminishing returns for this product occurs when \$20,000 is spent on advertising. ────────

✓Checkpoint 6

Find the point of diminishing returns for the model below, where R is the revenue (in thousands of dollars) and x is the advertising cost (in thousands of dollars).

$$R = \frac{1}{20,000}(450x^2 - x^3), \quad 0 \le x \le 300$$

SUMMARIZE   (Section 8.6)

1. State the test for concavity *(page 670)*. For examples of applying the test for concavity, see Examples 1 and 2.

2. State the definition of point of inflection *(page 673)*. For examples of finding points of inflection, see Examples 3 and 4.

3. State the Second-Derivative Test *(page 675)*. For an example of using the Second-Derivative Test, see Example 5.

4. Describe a real-life example of how the second derivative can be used to find the point of diminishing returns for a product *(page 676, Example 6).*

SKILLS WARM UP 8.6 The following warm-up exercises involve skills that were covered in earlier sections. You will use these skills in the exercise set for this section. For additional help, review Sections 7.6, 7.7, 8.1, and 8.4.

In Exercises 1–6, find the second derivative of the function.

1. $f(x) = 4x^4 - 9x^3 + 5x - 1$

2. $g(s) = (s^2 - 1)(s^2 - 3s + 2)$

3. $g(x) = (x^2 + 1)^4$

4. $f(x) = (x - 3)^{4/3}$

5. $h(x) = \dfrac{4x + 3}{5x - 1}$

6. $f(x) = \dfrac{2x - 1}{3x + 2}$

In Exercises 7–10, find the critical numbers of the function.

7. $f(x) = 5x^3 - 5x + 11$

8. $f(x) = x^4 - 4x^3 - 10$

9. $g(t) = \dfrac{16 + t^2}{t}$

10. $h(x) = \dfrac{x^4 - 50x^2}{8}$

Exercises 8.6

See www.CalcChat.com for worked-out solutions to odd-numbered exercises.

Using Graphs In Exercises 1–4, state the signs of $f'(x)$ and $f''(x)$ on the interval $(0, 2)$.

1.
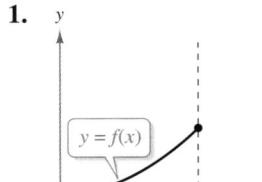

2.
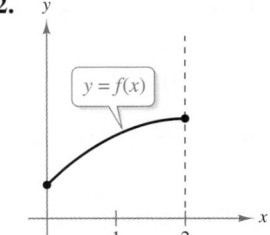

3.
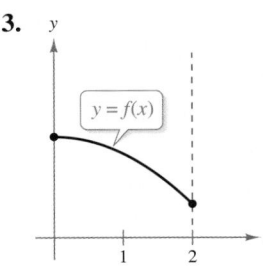

4.
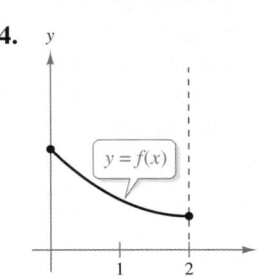

Applying the Test for Concavity In Exercises 5–12, determine the open intervals on which the graph of the function is concave upward or concave downward. *See Examples 1 and 2.*

5. $f(x) = -3x^2$

6. $f(x) = -5\sqrt{x}$

7. $y = -x^3 + 3x^2 - 2$

8. $y = -x^3 + 6x^2 - 9x - 1$

9. $f(x) = \dfrac{x^2 - 1}{2x + 1}$

10. $f(x) = \dfrac{x^2 + 4}{4 - x^2}$

11. $f(x) = \dfrac{24}{x^2 + 12}$

12. $f(x) = \dfrac{x^2}{x^2 + 1}$

Finding Points of Inflection In Exercises 13–20, discuss the concavity of the graph of the function and find the points of inflection. *See Examples 3 and 4.*

13. $f(x) = x^3 - 9x^2 + 24x - 18$

14. $f(x) = -4x^3 - 8x^2 + 32$

15. $f(x) = 2x^3 - 3x^2 - 12x + 5$

16. $f(x) = \frac{1}{2}x^4 + 2x^3$

17. $g(x) = 2x^4 - 8x^3 + 12x^2 + 12x$

18. $g(x) = x^5 + 5x^4 - 40x^2$

19. $f(x) = x(6 - x)^2$

20. $f(x) = (x - 1)^3(x - 5)$

Using the Second-Derivative Test In Exercises 21–34, find all relative extrema of the function. Use the Second-Derivative Test when applicable. *See Example 5.*

21. $f(x) = 6x - x^2$

22. $f(x) = 9x^2 - x^3$

23. $f(x) = x^3 - 5x^2 + 7x$

24. $f(x) = x^4 + 8x^3 - 6$

25. $f(x) = x^{2/3} - 3$

26. $f(x) = x + \dfrac{4}{x}$

27. $f(x) = \sqrt{x^2 + 1}$

28. $f(x) = \sqrt{2x^2 + 6}$

29. $f(x) = \sqrt{9 - x^2}$

30. $f(x) = \sqrt{4 - x^2}$

31. $f(x) = \dfrac{8}{x^2 + 2}$

32. $f(x) = \dfrac{x}{x^2 + 16}$

33. $f(x) = \dfrac{x}{x - 1}$

34. $f(x) = \dfrac{x}{x^2 - 1}$

Finding Relative Extrema   In Exercises 35–38, use a graphing utility to estimate graphically all relative extrema of the function.

35. $f(x) = 5 + 3x^2 - x^3$ **36.** $f(x) = x^3 - 6x^2 + 7$

37. $f(x) = \frac{1}{2}x^4 - \frac{1}{3}x^3 - \frac{1}{2}x^2$

38. $f(x) = -\frac{1}{3}x^5 - \frac{1}{2}x^4 + x$

Using the Second-Derivative Test   In Exercises 39–50, find all relative extrema and points of inflection. Then use a graphing utility to graph the function.

39. $f(x) = x^3 - 12x$ **40.** $f(x) = x^3 - 3x$

41. $g(x) = \sqrt{x} + \dfrac{4}{\sqrt{x}}$ **42.** $f(x) = x^3 - \frac{3}{2}x^2 - 6x$

43. $f(x) = \frac{1}{4}x^4 - 2x^2$ **44.** $f(x) = 2x^4 - 8x + 3$

45. $g(x) = (x - 2)(x + 1)^2$ **46.** $g(x) = (x - 6)(x + 2)^3$

47. $g(x) = x\sqrt{x + 3}$ **48.** $g(x) = x\sqrt{9 - x}$

49. $f(x) = \dfrac{4}{1 + x^2}$

50. $f(x) = \dfrac{2}{x^2 - 1}$

Creating a Function   In Exercises 51–54, sketch a graph of a function f having the given characteristics. (There are many correct answers.)

51. $f(2) = f(4) = 0$ **52.** $f(2) = f(4) = 0$
 $f'(x) < 0$ if $x < 3$ $f'(x) > 0$ if $x < 3$
 $f'(3) = 0$ $f'(3)$ is undefined.
 $f'(x) > 0$ if $x > 3$ $f'(x) < 0$ if $x > 3$
 $f''(x) > 0$ $f''(x) > 0,\ x \neq 3$

53. $f(0) = f(2) = 0$ **54.** $f(0) = f(2) = 0$
 $f'(x) > 0$ if $x < 1$ $f'(x) < 0$ if $x < 1$
 $f'(1) = 0$ $f'(1) = 0$
 $f'(x) < 0$ if $x > 1$ $f'(x) > 0$ if $x > 1$
 $f''(x) < 0$ $f''(x) > 0$

Using Graphs   In Exercises 55 and 56, use the graph to sketch the graph of f'. Find the intervals on which (a) $f'(x)$ is positive, (b) $f'(x)$ is negative, (c) f' is increasing, and (d) f' is decreasing. For each of these intervals, describe the corresponding behavior of f.

55.

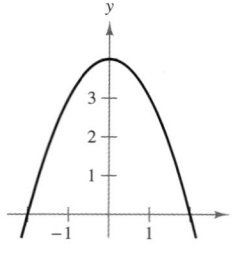

56.

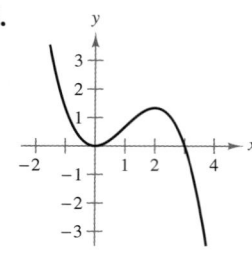

Using the First Derivative   In Exercises 57–60, you are given f'. Find the intervals on which (a) $f'(x)$ is increasing or decreasing and (b) the graph of f is concave upward or concave downward. (c) Find the x-values of the relative extrema and inflection points of f.

57. $f'(x) = 2x + 5$ **58.** $f'(x) = 3x^2 - 2$

59. $f'(x) = -x^2 + 2x - 1$ **60.** $f'(x) = x^2 + x - 6$

Point of Diminishing Returns   In Exercises 61 and 62, find the point of diminishing returns for the function. For each function, R is the revenue (in thousands of dollars) and x is the amount spent (in thousands of dollars) on advertising. Use a graphing utility to verify your results. See Example 6.

61. $R = \dfrac{1}{50,000}(600x^2 - x^3), \quad 0 \leq x \leq 400$

62. $R = -\dfrac{4}{9}x^3 + 4x^2 + 12, \quad 0 \leq x \leq 5$

Productivity   In Exercises 63 and 64, consider a college student who works from 7 P.M. to 11 P.M. assembling mechanical components. The number N of components assembled after t hours is given by the function. At what time is the student assembling components at the greatest rate?

63. $N = -0.12t^3 + 0.54t^2 + 8.22t, \quad 0 \leq t \leq 4$

64. $N = \dfrac{20t^2}{4 + t^2}, \quad 0 \leq t \leq 4$

Comparing a Function and Its Derivatives   In Exercises 65–68, use a graphing utility to graph f, f', and f'' in the same viewing window. Graphically locate the relative extrema and points of inflection of the graph of f. State the relationship between the behavior of f and the signs of f' and f''.

65. $f(x) = \frac{1}{2}x^3 - x^2 + 3x - 5, \quad [0, 3]$

66. $f(x) = -\frac{1}{20}x^5 - \frac{1}{12}x^2 - \frac{1}{3}x + 1, \quad [-2, 2]$

67. $f(x) = \dfrac{2}{x^2 + 1}, \quad [-3, 3]$

68. $f(x) = \dfrac{x^2}{x^2 + 1}, \quad [-3, 3]$

69. Average Cost   A manufacturer has determined that the total cost C (in dollars) of operating a factory is $C = 0.5x^2 + 10x + 7200$, where x is the number of units produced. At what level of production will the average cost per unit be minimized? (The average cost per unit is C/x.)

70. Inventory Cost   The cost C (in dollars) of ordering and storing x units is $C = 2x + 300,000/x$. What order size will produce a minimum cost?

71. Home Sales   The median sales price p (in thousands of dollars) of new single-family houses sold in the United States from 1995 through 2009 can be modeled by $p = -0.02812t^4 + 1.177t^3 - 17.02t^2 + 108.7t - 115$, for $5 \le t \le 19$, where t is the year, with $t = 5$ corresponding to 1995.   *(Source: U.S. Census Bureau)*

(a) Use a graphing utility to graph the model on the interval $[5, 19]$.

(b) Use the graph in part (a) to estimate the year corresponding to the absolute minimum sales price.

(c) Use the graph in part (a) to estimate the year corresponding to the absolute maximum sales price.

(d) During approximately which year was the rate of increase of the sales price the greatest? the least?

72. HOW DO YOU SEE IT?   The graph shows the Dow Jones Industrial Average y on Black Monday, October 19, 1987, where $t = 0$ corresponds to 9:30 A.M., when the market opens, and $t = 6.5$ corresponds to 4 P.M., the closing time.
(Source: Wall Street Journal)

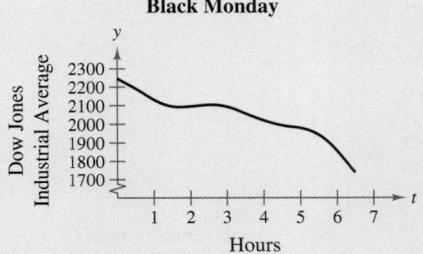

Black Monday

(a) Estimate the relative extrema and absolute extrema of the graph. Interpret your results in the context of the problem.

(b) Estimate the point of inflection of the graph on the interval $[1, 3]$. Interpret your result in the context of the problem.

73. Veteran Benefits   From 1995 through 2008, the number v of veterans (in thousands) receiving compensation and pension benefits for service in the armed forces can be modeled by

$v = -0.0687t^4 + 3.169t^3 - 45t^2 + 230.6t + 2950$

for $5 \le t \le 18$, where t is the year, with $t = 5$ corresponding to 1995.   *(Source: U.S. Department of Veterans Affairs)*

(a) Use a graphing utility to graph the model on the interval $[5, 18]$.

(b) Use the second derivative to determine the concavity of v.

(c) Find the point(s) of inflection of the graph of v.

(d) Interpret the meaning of the inflection point(s) of the graph of v.

74. Think About It   Let S represent monthly sales of a new digital audio player. Write a statement describing S' and S'' for each of the following.

(a) The rate of change of sales is increasing.

(b) Sales are increasing, but at a greater rate.

(c) The rate of change of sales is steady.

(d) Sales are steady.

(e) Sales are declining, but at a lower rate.

(f) Sales have bottomed out and have begun to rise.

Business Capsule

While working in New York City in 2004, Matthew Corrin noticed an abundance of fresh food bars and decided that if someone could successfully brand one, that person could create the "Starbucks of the fresh food business." With $275,000, he opened his first Freshii store in Toronto in 2005, and soon began developments for more. By the end of 2011, he will have 80–90 locations worldwide, with agreements signed for 400 additional stores in 25 cities and four countries. The mission of this eco-friendly chain is "to eliminate the excuse of people not eating fresh food because it isn't convenient," Corrin said.

75. Research Project   Use your school's library, the Internet, or some other reference source to research the financial history of a fast-growing company like the one discussed above. Gather data on the company's costs and revenues over a period of time, and use a graphing utility to graph a scatter plot of the data. Fit models to the data. Do the models appear to be concave upward or downward? Do they appear to be increasing or decreasing? Discuss the implications of your answers.

ALGEBRA TUTOR

Solving Equations

Much of the algebra in Chapter 8 involves simplifying algebraic expressions (see pages 620 and 621) and solving algebraic equations, as illustrated in the following examples. In Example 1, you can review some of the basic techniques for solving equations. In Example 2 on the next page, you can review some of the more complicated techniques for solving equations.

When solving an equation, remember that your basic goal is to isolate the variable on one side of the equation.

1. To solve a *linear equation*, you can add or subtract the same quantity from each side of the equation. You can also multiply or divide each side of the equation by the same *nonzero* quantity.

2. To solve a *quadratic equation*, you can take the square root of each side, use factoring, or use the Quadratic Formula.

3. To solve a *radical equation*, isolate the radical on one side of the equation and square each side of the equation.

Example 1   Solving Equations

Solve each equation.

a. $3x - 3 = 5x - 7$ **b.** $2x^2 = 10$

c. $2x^2 + 5x - 6 = 6$ **d.** $\sqrt{2x - 7} = 5$

SOLUTION

a. $3x - 3 = 5x - 7$ Write original (linear) equation.

$-3 = 2x - 7$ Subtract $3x$ from each side.

$4 = 2x$ Add 7 to each side.

$2 = x$ Divide each side by 2.

b. $2x^2 = 10$ Write original (quadratic) equation.

$x^2 = 5$ Divide each side by 2.

$x = \pm\sqrt{5}$ Take the square root of each side.

c. $2x^2 + 5x - 6 = 6$ Write original (quadratic) equation.

$2x^2 + 5x - 12 = 0$ Write in general form.

$(2x - 3)(x + 4) = 0$ Factor.

$2x - 3 = 0 \implies x = \tfrac{3}{2}$ Set first factor equal to zero.

$x + 4 = 0 \implies x = -4$ Set second factor equal to zero.

d. $\sqrt{2x - 7} = 5$ Write original (radical) equation.

$2x - 7 = 25$ Square each side.

$2x = 32$ Add 7 to each side.

$x = 16$ Divide each side by 2.

Example 2 Solving an Equation

Solve each equation.

a. $\dfrac{36(x^2 - 1)}{(x^2 + 3)^3} = 0$

b. $x^2(4x - 3) = 0$

c. $\dfrac{4x}{3(x^2 - 4)^{1/3}} = 0$

d. $g'(x) = 0$, where $g(x) = (x - 2)(x + 1)^2$

SOLUTION

a. $\dfrac{36(x^2 - 1)}{(x^2 + 3)^3} = 0$ Example 2, page 672

$\qquad 36(x^2 - 1) = 0$ A fraction is zero only if its numerator is zero.

$\qquad\qquad x^2 - 1 = 0$ Divide each side by 36.

$\qquad\qquad\quad\ x^2 = 1$ Add 1 to each side.

$\qquad\qquad\quad\ x = \pm 1$ Take the square root of each side.

b. $x^2(4x - 3) = 0$ Example 2, page 664

$\qquad\quad x^2 = 0 \implies x = 0$ Set first factor equal to zero.

$\qquad 4x - 3 = 0 \implies x = \frac{3}{4}$ Set second factor equal to zero.

c. $\dfrac{4x}{3(x^2 - 4)^{1/3}} = 0$ Example 5, page 656

$\qquad\qquad 4x = 0$ A fraction is zero only if its numerator is zero.

$\qquad\qquad\ x = 0$ Divide each side by 4.

d. $\qquad\qquad\qquad g(x) = (x - 2)(x + 1)^2$ Exercise 45, page 678

$(x - 2)(2)(x + 1) + (x + 1)^2(1) = 0$ Find derivative and set equal to zero.

$\qquad (x + 1)[2(x - 2) + (x + 1)] = 0$ Factor.

$\qquad\qquad (x + 1)(2x - 4 + x + 1) = 0$ Multiply factors.

$\qquad\qquad\qquad\ (x + 1)(3x - 3) = 0$ Combine like terms.

$\qquad\qquad\qquad\quad x + 1 = 0 \implies x = -1$ Set first factor equal to zero.

$\qquad\qquad\qquad\quad 3x - 3 = 0 \implies x = 1$ Set second factor equal to zero.

SUMMARY AND STUDY STRATEGIES

After studying this chapter, you should have acquired the following skills.
The exercise numbers are keyed to the Review Exercises that begin on page 684.
Answers to odd-numbered Review Exercises are given in the back of the text.*

Section 8.1   Review Exercises

■ Find higher-order derivatives.   *1–12, 16*

  1st derivative: y', $f'(x)$, $\dfrac{dy}{dx}$, $\dfrac{d}{dx}[f(x)]$, $D_x[y]$

  2nd derivative: y'', $f''(x)$, $\dfrac{d^2y}{dx^2}$, $\dfrac{d^2}{dx^2}[f(x)]$, $D_x^2[y]$

  nth derivative: $y^{(n)}$, $f^{(n)}(x)$, $\dfrac{d^n y}{dx^n}$, $\dfrac{d^n}{dx^n}[f(x)]$, $D_x^n[y]$

■ Find and use a position function to determine the velocity and acceleration   *13–15*
  of a moving object.

  Position function: $s = f(t)$

  Velocity function: $\dfrac{ds}{dt} = f'(t)$

  Acceleration function: $\dfrac{d^2s}{dt^2} = f''(t)$

Section 8.2

■ Find derivatives implicitly.   *17–20*
■ Use implicit differentiation to write equations of tangent lines.   *21–28*

Section 8.3

■ Solve related-rate problems.   *29–34*

Section 8.4

■ Find the critical numbers of a function.   *35–42*

  If f is defined at c, then c is a critical number of f when $f'(c) = 0$ or $f'(c)$ is undefined.

■ Find the open intervals on which a function is increasing or decreasing.   *43–46*

  f is increasing when $f'(x) > 0$.

  f is decreasing when $f'(x) < 0$.

■ Find intervals on which a real-life model is increasing or decreasing.   *47–50*

---

* A wide range of valuable study aids are available to help you master the material in this chapter.
  The *Student Solutions Manual* includes step-by-step solutions to all odd-numbered exercises to
  help you review and prepare. The student website at *www.cengagebrain.com* offers algebra help
  and a *Graphing Technology Guide*, which contains step-by-step commands and instructions for
  a wide variety of graphing calculators.

Section 8.5

- Use the First-Derivative Test to find the relative extrema of a function.   *51–60*

- Find the absolute extrema of a continuous function on a closed interval.   *61–68*

  If f is continuous on $[a, b]$, then f has both a minimum value and a maximum value on $[a, b]$. The x-value where each of these extrema occurs is a, b, or one of the critical numbers of f in (a, b).

- Find minimum and maximum values of a real-life model and interpret the   *69–72*
  results in context.

Section 8.6

- Find the open intervals on which the graph of a function is concave upward   *73–76*
  or concave downward.

  f is concave upward when $f''(x) > 0$.

  f is concave downward when $f''(x) < 0$.

- Find the points of inflection of the graph of a function.   *77–80, 87*

  If $(c, f(c))$ is a point of inflection of the graph of a function f, then either $f''(x) = 0$ or $f''(x)$ is undefined.

- Use the Second-Derivative Test to find the relative extrema of a function.   *81–86, 88*

- Find the point of diminishing returns of an input-output model.   *89, 90*

Study Strategies

- **Solve Problems Graphically, Analytically, and Numerically**   When analyzing the graph of a function, use a variety of problem-solving strategies. For instance, to analyze the graph of

  $$f(x) = x^3 - 4x^2 + 5x - 4$$

  you can begin *graphically*. That is, use a graphing utility to find a viewing window that shows the important characteristics of the graph. From the graph shown below, f appears to have one relative minimum, one relative maximum, and one point of inflection.

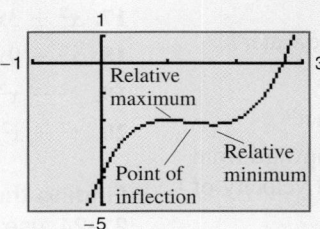

  Next, use calculus to *analyze* the graph. Because the derivative of f is

  $$f'(x) = 3x^2 - 8x + 5 = (3x - 5)(x - 1)$$

  the critical numbers of f are

  $$x = \tfrac{5}{3} \quad \text{and} \quad x = 1.$$

  By the First-Derivative Test, you can conclude that $x = \tfrac{5}{3}$ yields a relative minimum and $x = 1$ yields a relative maximum. Because

  $$f''(x) = 6x - 8$$

  you can conclude that $x = \tfrac{4}{3}$ yields a point of inflection. Finally, you can analyze the graph *numerically*. For instance, you can construct a table of values and observe that f is increasing on the interval $(-\infty, 1)$, decreasing on the interval $\left(1, \tfrac{5}{3}\right)$, and increasing on the interval $\left(\tfrac{5}{3}, \infty\right)$.

Review Exercises

See www.CalcChat.com for worked-out solutions to odd-numbered exercises.

Finding Higher-Order Derivatives   In Exercises 1–8, find the higher-order derivative.

Given	Derivative
1. $f(x) = 3x^2 + 7x + 1$	$f''(x)$
2. $f'(x) = 5x^4 - 6x^2 + 2x$	$f'''(x)$
3. $f'''(x) = -\dfrac{6}{x^4}$	$f^{(5)}(x)$
4. $f(x) = \sqrt{x}$	$f^{(4)}(x)$
5. $f'(x) = 7x^{5/2}$	$f''(x)$
6. $f(x) = x^2 + \dfrac{3}{x}$	$f''(x)$
7. $f''(x) = 6\sqrt[3]{x}$	$f'''(x)$
8. $f'''(x) = 20x^4 - \dfrac{2}{x^3}$	$f^{(5)}(x)$

Finding Higher-Order Derivatives   In Exercises 9–12, find the given value.

Function	Value
9. $f(x) = x^2 + 3x + 4$	$f''(1)$
10. $f(x) = \dfrac{1}{x}$	$f'''(3)$
11. $f(x) = \sqrt{16x + 9}$	$f'''(0)$
12. $f(x) = x^2(x - 2)^2$	$f''(-1)$

13. Athletics   A person dives from a 30-foot platform with an initial velocity of 5 feet per second (upward).

(a) Find the position function of the diver.

(b) How long will it take the diver to hit the water?

(c) What is the diver's velocity at impact?

(d) What is the diver's acceleration at impact?

14. Projectile Motion   An object is thrown upward from the top of a 96-foot building with an initial velocity of 80 feet per second.

(a) Write the position, velocity, and acceleration functions of the object.

(b) When will the object hit the ground?

(c) When is the velocity of the object zero?

(d) How high does the object go?

15. Velocity and Acceleration   The position function of a particle is given by

$$s = \frac{1}{t^2 + 2t + 1}$$

where s is the height (in feet) and t is the time (in seconds). Find the velocity and acceleration functions.

16. Modeling Data   The table shows the numbers y of new privately owned housing units (in thousands) started in the United States during the years 2004 through 2009, where t is the year, with $t = 4$ corresponding to 2004.   *(Source: U.S. Census Bureau)*

t	4	5	6	7	8	9
y	1956	2068	1801	1355	906	554

(a) Use a graphing utility to find a cubic model for the data.

(b) Use a graphing utility to graph the model and plot the data in the same viewing window. How well does the model fit the data?

(c) Find the first and second derivatives of the function.

(d) Show that the number of new privately owned housing units was decreasing from 2005 through 2009.

(e) Find the year in which the number was decreasing at the greatest rate by solving

$$y''(t) = 0.$$

(f) Explain the relationship among your answers for parts (c), (d), and (e).

Finding Derivatives   In Exercises 17–20, use implicit differentiation to find dy/dx.

17. $x^2 + 3xy + y^3 = 10$

18. $x^2 + 9xy + y^2 = 0$

19. $y^2 - x^2 + 8x - 9y - 1 = 0$

20. $y^2 + x^2 - 6y - 2x - 5 = 0$

Finding the Slope of a Graph Implicitly   In Exercises 21–24, use implicit differentiation to find the slope of the graph at the given point.

21. $5x^2 - 2y - 3 = 0$

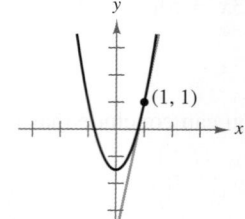

22. $3x^2 + 2y - 1 = 0$

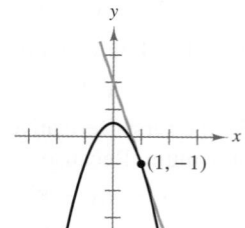

23. $x^2 + y^2 = 1$ **24.** $x^2 + 4y^2 = 16$

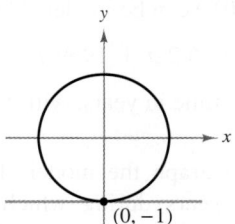

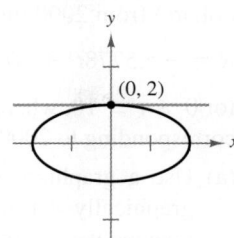

Finding an Equation of a Tangent Line   In Exercises 25–28, use implicit differentiation to find an equation of the tangent line to the graph of the function at the given point.

Equation	Point
25. $y^2 = x - y$	$(2, 1)$
26. $2\sqrt[3]{x} + 3\sqrt{y} = 10$	$(8, 4)$
27. $y^2 - 2x = xy$	$(1, 2)$
28. $y^3 - 2x^2y + 3xy^2 = -1$	$(0, -1)$

29. Area   The radius r of a circle is increasing at a rate of 2 inches per minute. Find the rate of change of the area at (a) $r = 3$ inches and (b) $r = 10$ inches.

30. Moving Point   A point is moving along the graph of $y = \sqrt{x}$ such that dx/dt is 3 centimeters per second. Find dy/dt for (a) $x = \frac{1}{4}$, (b) $x = 1$, and (c) $x = 4$.

31. Water Level   A swimming pool is 40 feet long, 20 feet wide, 4 feet deep at the shallow end, and 9 feet deep at the deep end (see figure). Water is being pumped into the pool at the rate of 10 cubic feet per minute. How fast is the water level rising when there is 4 feet of water in the deep end?

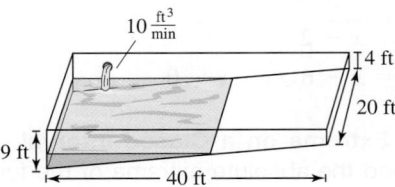

32. Electricity   The combined electrical resistance R of R_1 and R_2, connected in parallel, is given by

$$\frac{1}{R} = \frac{1}{R_1} + \frac{1}{R_2}$$

where R, R_1, and R_2 are measured in ohms. R_1 and R_2 are increasing at rates of 1 and 1.5 ohms per second, respectively. At what rate is R changing when $R_1 = 50$ ohms and $R_2 = 75$ ohms?

33. Water Level   A trough is 12 feet long and 3 feet across the top (see figure). Its ends are isosceles triangles with heights of 3 feet.

(a) If water is being pumped into the trough at 2 cubic feet per minute, how fast is the water level rising when h is 1 foot deep?

(b) If the water is rising at a rate of $\frac{3}{8}$ inch per minute when $h = 2$, determine the rate at which water is being pumped into the trough.

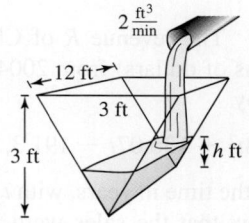

34. Profit   The demand and cost functions for a product can be modeled by

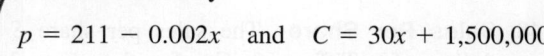

$$p = 211 - 0.002x \quad \text{and} \quad C = 30x + 1,500,000$$

where x is the number of units produced.

(a) Write the profit function for this product.

(b) Find the marginal profit when 80,000 units are produced.

(c) Graph the profit function on a graphing utility and use the graph to determine the price you would charge for the product. Explain your reasoning.

Finding Critical Numbers   In Exercises 35–42, find the critical numbers of the function.

35. $f(x) = -x^2 + 2x + 4$

36. $y = 3x^2 + 18x$ **37.** $y = 4x^3 - 108x$

38. $f(x) = x^4 - 8x^2 + 13$

39. $g(x) = (x - 1)^2(x - 3)$

40. $h(x) = \sqrt{x}(x - 3)$

41. $y = (x - 1)^{2/3}$ **42.** $y = 2x^{1/3} - 3$

Intervals on Which f Is Increasing or Decreasing   In Exercises 43–46, find the open intervals on which the function is increasing or decreasing. Verify your result with the graph of the function.

43. $f(x) = x^2 + x - 2$ **44.** $g(x) = (x + 2)^3$

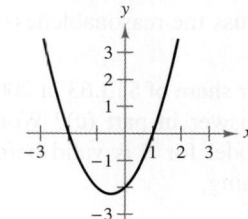

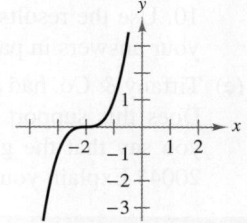

45. $h(x) = \dfrac{x^2 - 3x - 4}{x - 3}$ **46.** $f(x) = -x^3 + 3x^2 - 2$

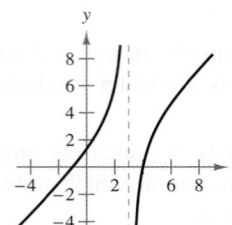

47. Revenue   The revenue R of Chipotle Mexican Grill (in millions of dollars) from 2004 through 2009 can be modeled by

$$R = 6.268t^2 + 136.07t - 191.3, \quad 4 \le t \le 9$$

where t is the time in years, with $t = 4$ corresponding to 2004. Show that the sales were increasing from 2004 through 2009.   *(Source: Chipotle Mexican Grill, Inc.)*

48. Sales Per Share   The sales per share S (in dollars) for Tiffany & Co. for the years 2004 through 2010 are shown in the table, where t is the time in years, with $t = 4$ corresponding to 2004. *(Source: Tiffany & Co.)*

Year, t	4	5	6	7
Sales per share, S	15.25	16.81	19.49	23.19

Year, t	8	9	10
Sales per share, S	23.09	21.45	24.30

(a) Use a graphing utility to create a scatter plot of the data.

(b) A model for the data is

$$S = 0.123295t^4 - 3.42977t^3 + 34.3254t^2$$
$$- 144.268t + 231.18, \quad 4 \le t \le 10.$$

Graph the model and the data in the same viewing window.

(c) Use the graph to estimate the periods of time when the sales per share were increasing and the periods of time when the sales per share were decreasing.

(d) Find $S'(t)$. Evaluate $S'(t)$ when $t = 4, 5, 6, \ldots,$ and 10. Use the results to discuss the reasonableness of your answers in part (c).

(e) Tiffany & Co. had sales per share of $13.63 in 2003. Does this support your answer in part (c)? Would you say that the given model for S is valid before 2004? Explain your reasoning.

49. Revenue   The revenue R of Cintas (in millions of dollars) from 2000 through 2010 can be modeled by

$$R = -5.5778t^3 + 67.524t^2 + 45.22t + 1969.2$$

for $0 \le t \le 10$, where t is the time in years, with $t = 0$ corresponding to 2000.   *(Source: Cintas Corporation)*

(a) Use a graphing utility to graph the model. Then graphically estimate the years during which the revenue was increasing and the years during which the revenue was decreasing.

(b) Use the test for increasing and decreasing functions to verify the result of part (a).

50. Meteorology   The monthly normal temperature T (in degrees Fahrenheit) for New York City can be modeled by

$$T = 0.0380t^4 - 1.092t^3 + 9.23t^2 - 19.6t + 44$$

where $1 \le t \le 12$ and $t = 1$ corresponds to January. *(Source: National Climatic Data Center)*

(a) Find the interval(s) on which the model is increasing.

(b) Find the interval(s) on which the model is decreasing.

(c) Interpret the results of parts (a) and (b).

(d) Use a graphing utility to graph the model.

Finding Relative Extrema   In Exercises 51–60, use the First-Derivative Test to find all relative extrema of the function. Use a graphing utility to verify your result.

51. $f(x) = 4x^3 - 6x^2 - 2$ **52.** $f(x) = \frac{1}{4}x^4 - 8x$

53. $g(x) = x^2 - 16x + 12$ **54.** $h(x) = 4 + 10x - x^2$

55. $h(x) = 2x^2 - x^4$ **56.** $s(x) = x^4 - 8x^2 + 3$

57. $f(x) = \dfrac{6}{x^2 + 1}$ **58.** $f(x) = \dfrac{2}{x^2 - 1}$

59. $h(x) = \dfrac{x^2}{x - 2}$

60. $g(x) = x - 6\sqrt{x}, \quad x > 0$

Finding Extrema on a Closed Interval   In Exercises 61–68, find the absolute extrema of the function on the closed interval. Use a graphing utility to verify your result.

61. $f(x) = x^2 + 5x + 6; \quad [-3, 0]$

62. $f(x) = x^4 - 2x^3; \quad [0, 2]$

63. $f(x) = x^3 - 12x + 1; \quad [-4, 4]$

64. $f(x) = x^3 + 2x^2 - 3x + 4; \quad [-3, 2]$

65. $f(x) = 2\sqrt{x} - x; \quad [0, 9]$

66. $f(x) = \dfrac{x}{\sqrt{x^2 + 1}}; \quad [0, 2]$

67. $f(x) = \dfrac{2x}{x^2 + 1}; \quad [-1, 2]$

68. $f(x) = \dfrac{8}{x} + x; \quad [1, 4]$

69. Surface Area A right circular cylinder of radius r and height h has a volume of 25 cubic inches (see figure). The total surface area of the cylinder in terms of r is given by

$$S = 2\pi r\left(r + \frac{25}{\pi r^2}\right).$$

Find the radius that will minimize the surface area. Use a graphing utility to verify your result.

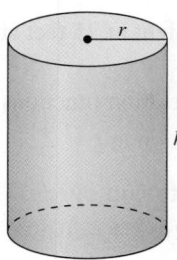

70. Profit The profit P (in dollars) made by a company from selling x tablet computers can be modeled by

$$P = 1.64x - \frac{x^2}{15,000} - 2500.$$

Find the number of units sold that will yield a maximum profit. What is the maximum profit?

71. Environment When organic waste is dumped into a pond, the decomposition of the waste consumes oxygen. A model for the oxygen level O (where 1 is the normal level) of a pond as waste material oxidizes is

$$O = \frac{t^2 - t + 1}{t^2 + 1}, \quad t \geq 0$$

where t is the time in weeks.

(a) When is the oxygen level lowest? What is this level?

(b) When is the oxygen level highest? What is this level?

(c) Describe the oxygen level as t increases.

72. Bloodstream The concentration C (in milligrams per milliliter) of a chemical in the bloodstream t hours after injection into muscle tissue can be modeled by

$$C = \frac{3t}{27 + t^3}, \quad t \geq 0.$$

(a) Complete the table and use it to approximate the time when the concentration reached a maximum.

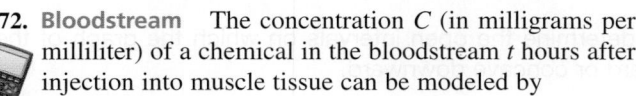

t	0	0.5	1	1.5	2	2.5	3
$C(t)$							

(b) Use a graphing utility to graph the concentration function. Use the *trace* feature to approximate the time when the concentration reached a maximum.

(c) Determine analytically the time when the concentration reached a maximum.

Applying the Test for Concavity In Exercises 73–76, determine the open intervals on which the graph of the function is concave upward or concave downward.

73. $f(x) = (x - 2)^3$ **74.** $h(x) = x^5 - 10x^2$

75. $g(x) = \frac{1}{4}(-x^4 + 8x^2 - 12)$

76. $h(x) = x^3 - 6x$

Finding Points of Inflection In Exercises 77–80, discuss the concavity of the graph of the function and find the points of inflection.

77. $f(x) = \frac{1}{2}x^4 - 4x^3$

78. $f(x) = \frac{1}{4}x^4 - 2x^2 - x$

79. $f(x) = x^3(x - 3)^2$

80. $f(x) = (x - 1)^2(x - 3)$

Using the Second-Derivative Test In Exercises 81–86, use the Second-Derivative Test to find all relative extrema of the function.

81. $f(x) = x^3 - 6x^2 + 12x$

82. $f(x) = x^4 - 32x^2 + 12$

83. $f(x) = x^5 - 5x^3$

84. $f(x) = x(x^2 - 3x - 9)$

85. $f(x) = 2x^2(1 - x^2)$

86. $f(x) = x - 4\sqrt{x + 1}$

87. Biology The growth of a red oak tree is approximated by the model

$$y = -0.003x^3 + 0.137x^2 + 0.458x - 0.839,$$

$$2 \leq x \leq 34$$

where y is the height of the tree in feet and x is its age in years. Find the age of the tree when it is growing most rapidly. Then use a graphing utility to graph the function and to verify your result. (*Hint:* Use the viewing window $2 \leq x \leq 34$ and $-10 \leq y \leq 60$.)

88. Medicine: Poiseuille's Law The speed of blood that is r centimeters from the center of an artery is modeled by

$$s(r) = c(R^2 - r^2), \quad c > 0$$

where c is a constant, R is the radius of the artery, and s is measured in centimeters per second. Show that the speed is a maximum at the center of an artery.

Point of Diminishing Returns In Exercises 89 and 90, find the point of diminishing returns for the function. For each function, R is the revenue (in thousands of dollars) and x is the amount spent (in thousands of dollars) on advertising. Use a graphing utility to verify your result.

89. $R = \frac{1}{1500}(150x^2 - x^3), \quad 0 \leq x \leq 100$

90. $R = -\frac{2}{3}(x^3 - 12x^2 - 6), \quad 0 \leq x \leq 8$

TEST YOURSELF

See www.CalcChat.com for worked-out solutions to odd-numbered exercises.

Take this test as you would take a test in class. When you are done, check your work against the answers given in the back of the book.

In Exercises 1–3, find the third derivative of the function.

1. $f(x) = 2x^2 + 3x + 1$ **2.** $f(x) = \sqrt{3-x}$ **3.** $f(x) = \dfrac{2x+1}{2x-1}$

4. A ball is thrown straight upward from a height of 75 feet above the ground with an initial velocity of 30 feet per second.

(a) Write the position, velocity, and acceleration functions of the ball.

(b) Find the height, velocity, and acceleration at $t = 2$.

In Exercises 5–7, use implicit differentiation to find dy/dx.

5. $x + xy = 6$ **6.** $y^2 + 2x - 2y + 1 = 0$ **7.** $x^2 - 2y^2 = 4$

8. The radius r of a right circular cylinder is increasing at a rate of 0.25 centimeter per minute. The height h of the cylinder is related to the radius by $h = 20r$. Find the rate of change of the volume when (a) $r = 0.5$ centimeter and (b) $r = 1$ centimeter.

In Exercises 9–11, find the critical numbers of the function and the open intervals on which the function is increasing or decreasing.

9. $f(x) = 3x^2 - 4$ **10.** $f(x) = x^3 - 12x$ **11.** $f(x) = (x-5)^4$

In Exercises 12–14, use the First-Derivative Test to find all relative extrema of the function.

12. $f(x) = \dfrac{1}{3}x^3 - 9x + 4$ **13.** $f(x) = 2x^4 - 4x^2 - 5$ **14.** $f(x) = \dfrac{5}{x^2+2}$

In Exercises 15–17, find the absolute extrema of the function on the closed interval. Use a graphing utility to verify your result.

15. $f(x) = x^2 + 6x + 8, \quad [-4, 0]$

16. $f(x) = 12\sqrt{x} - 4x, \quad [0, 5]$

17. $f(x) = \dfrac{6}{x} + \dfrac{x}{2}, \quad [1, 6]$

In Exercises 18 and 19, determine the open intervals on which the graph of the function is concave upward or concave downward.

18. $f(x) = x^5 - 80x^2$

19. $f(x) = \dfrac{20}{3x^2+8}$

In Exercises 20 and 21, discuss the concavity of the graph of the function and find the points of inflection.

20. $f(x) = x^4 + 6$

21. $f(x) = x^4 - 54x^2 + 230$

In Exercises 22 and 23, use the Second-Derivative Test to find all relative extrema of the function.

22. $f(x) = x^3 - 6x^2 - 36x + 50$

23. $f(x) = \frac{3}{5}x^5 - 9x^3$

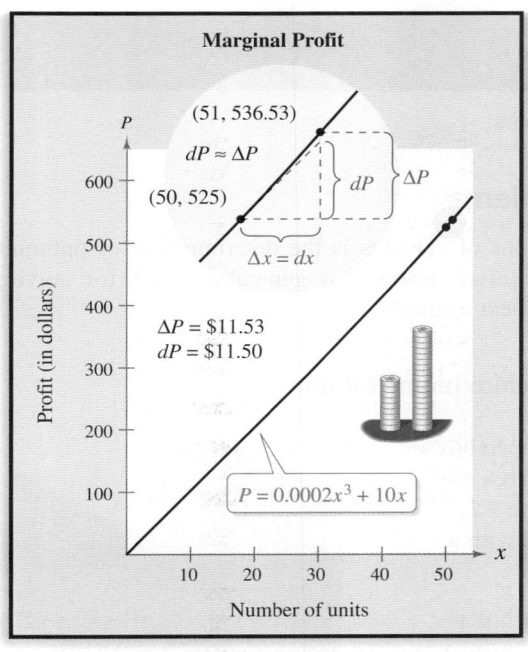

Marginal Profit

$dP \approx \Delta P$

$(51, 536.53)$

$(50, 525)$

$\Delta x = dx$

dP

ΔP

$\Delta P = \$11.53$
$dP = \$11.50$

$P = 0.0002x^3 + 10x$

Profit (in dollars)

Number of units

Example 3 on page 732 shows how to use
a differential to find the marginal profit for
a one-unit increase in production.

9 Further Applications of the Derivative

9.1 Optimization Problems

9.2 Business and Economics Applications

9.3 Asymptotes

9.4 Curve Sketching: A Summary

9.5 Differentials and Marginal Analysis

9.1 Optimization Problems

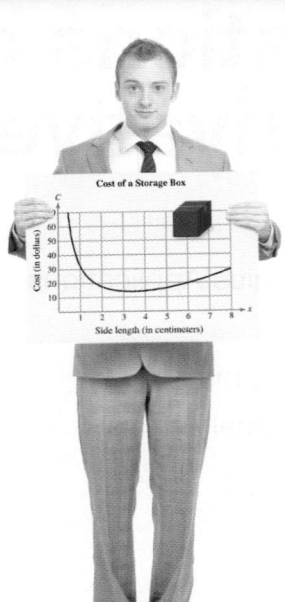

In Exercise 13 on page 696, you will use primary equations, secondary equations, and derivatives to find the dimensions of a box that will minimize the cost of making the box.

■ Solve real-life optimization problems.

Solving Optimization Problems

One of the most common applications of calculus is the determination of optimum (minimum or maximum) values. Before learning a general method for solving optimization problems, consider the next example.

 Example 1　**Finding the Maximum Volume**

A manufacturer wants to design an open box that has a square base and a surface area S of 108 square inches, as shown in Figure 9.1. What dimensions will produce a box with a maximum volume?

SOLUTION　Because the base of the box is square, the volume is

$$V = x^2h. \qquad \text{Primary equation}$$

This equation is called the **primary equation** because it gives a formula for the quantity to be optimized. The surface area of the box is

$$S = (\text{area of base}) + (\text{area of four sides})$$
$$108 = x^2 + 4xh. \qquad \text{Secondary equation}$$

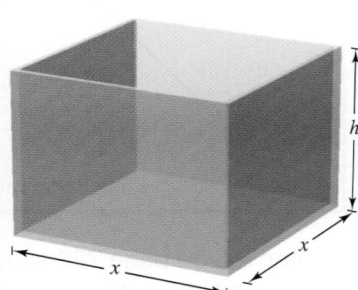

Open Box with Square Base:
$S = x^2 + 4xh = 108$
FIGURE 9.1

Because V is to be optimized, it helps to express V as a function of just one variable. To do this, solve the secondary equation for h in terms of x to obtain

$$h = \frac{108 - x^2}{4x}$$

and substitute for h in the primary equation.

$$V = x^2h = x^2\left(\frac{108 - x^2}{4x}\right) = 27x - \frac{1}{4}x^3 \qquad \text{Function of one variable}$$

Before finding which x-value yields a maximum value of V, you need to determine the *feasible domain* of the function. That is, what values of x make sense in this problem? Because x must be nonnegative and the area of the base ($A = x^2$) is at most 108, you can conclude that the feasible domain is

$$0 \le x \le \sqrt{108}. \qquad \text{Feasible domain}$$

Using the techniques described in the first three sections of this chapter, you can determine that $\left(\text{on the interval } 0 \le x \le \sqrt{108}\right)$ this function has an absolute maximum at $x = 6$ inches and $h = 3$ inches.

✓Checkpoint 1

Use a graphing utility to graph the volume function

$$V = 27x - \tfrac{1}{4}x^3$$

from Example 1 on $0 \le x \le \sqrt{108}$. Verify that the function has an absolute maximum at $x = 6$. What is the maximum volume?

ALGEBRA TUTOR xy

For help on the algebra in Example 1, see Example 1(c) in the *Chapter 9 Algebra Tutor*, on page 736.

In studying Example 1, be sure that you understand the basic question that it asks. Remember that you are not ready to begin solving an optimization problem until you have clearly identified the problem. Once you are sure you understand what is being asked, you are ready to begin considering a method for solving the problem.

For instance, in Example 1, you should realize that there are infinitely many open boxes having 108 square inches of surface area. To begin solving this problem, you might ask yourself which basic shape would seem to yield a maximum volume. Should the box be tall, squat, or nearly cubical? You might even try calculating a few volumes, as shown in Figure 9.2, to see if you can get a better feeling for what the optimum dimensions should be.

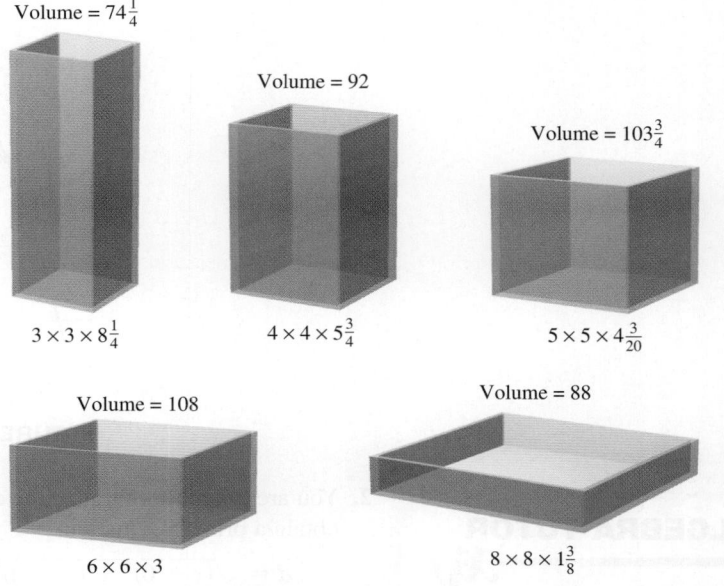

Volume = $74\frac{1}{4}$

Volume = 92

Volume = $103\frac{3}{4}$

$3 \times 3 \times 8\frac{1}{4}$ $4 \times 4 \times 5\frac{3}{4}$ $5 \times 5 \times 4\frac{3}{20}$

Volume = 108

Volume = 88

$6 \times 6 \times 3$ $8 \times 8 \times 1\frac{3}{8}$

Which box has the greatest volume?

FIGURE 9.2

There are several steps in the solution of Example 1. The first step is to sketch a diagram and identify all *known* quantities and all quantities *to be determined*. The second step is to write a primary equation for the quantity to be optimized. Then, a secondary equation is used to rewrite the primary equation as a function of one variable. Finally, calculus is used to determine the optimum value. These steps are summarized below.

STUDY TIP

When performing Step 5, remember that to determine the maximum or minimum value of a continuous function f on a closed interval, you need to compare the values of f at its critical numbers with the values of f at the endpoints of the interval. The greatest of these values is the desired maximum, and the least is the desired minimum.

Guidelines for Solving Optimization Problems

1. Identify all given quantities and all quantities to be determined. If possible, make a sketch.

2. Write a **primary equation** for the quantity that is to be maximized or minimized. (A summary of several common formulas is given in Appendix D.)

3. Reduce the primary equation to one having a single independent variable. This may involve the use of a **secondary equation** that relates the independent variables of the primary equation.

4. Determine the feasible domain of the primary equation. That is, determine the values for which the stated problem makes sense.

5. Determine the desired maximum or minimum value by the calculus techniques discussed in Sections 8.4 through 8.6.

Example 2   Finding a Minimum Distance

Find the points on the graph of

$$y = 4 - x^2$$

that are closest to $(0, 2)$.

SOLUTION

1. Figure 9.3 shows that there are two points at a minimum distance from $(0, 2)$.

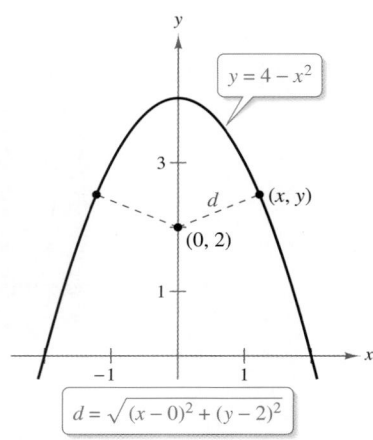

$y = 4 - x^2$

d ---- (x, y)

$(0, 2)$

$$d = \sqrt{(x - 0)^2 + (y - 2)^2}$$

FIGURE 9.3

ALGEBRA TUTOR *xy*

For help on the algebra in Example 2, see Example 1(b) in the *Chapter 9 Algebra Tutor*, on page 736.

2. You are asked to minimize the distance d. So, you can use the Distance Formula to obtain a primary equation.

$$d = \sqrt{(x - 0)^2 + (y - 2)^2} \qquad \text{Primary equation}$$

3. Using the secondary equation $y = 4 - x^2$, you can rewrite the primary equation as a function of a single variable.

$$d = \sqrt{x^2 + (4 - x^2 - 2)^2} \qquad \text{Substitute } 4 - x^2 \text{ for } y.$$
$$= \sqrt{x^2 + (2 - x^2)^2} \qquad \text{Simplify.}$$
$$= \sqrt{x^2 + 4 - 4x^2 + x^4} \qquad \text{Expand binomial.}$$
$$= \sqrt{x^4 - 3x^2 + 4} \qquad \text{Combine like terms.}$$

Because d is smallest when the expression under the radical is smallest, you simplify the problem by finding the minimum value of $f(x) = x^4 - 3x^2 + 4$.

4. The domain of f is the entire real number line.

5. To find the minimum value of $f(x)$, first find the critical numbers of f.

$$f'(x) = 4x^3 - 6x \qquad \text{Find derivative of } f.$$
$$0 = 4x^3 - 6x \qquad \text{Set derivative equal to 0.}$$
$$0 = 2x(2x^2 - 3) \qquad \text{Factor.}$$
$$x = 0, x = \sqrt{\tfrac{3}{2}}, x = -\sqrt{\tfrac{3}{2}} \qquad \text{Critical numbers}$$

The First-Derivative Test verifies that $x = 0$ yields a relative maximum, whereas both $\sqrt{3/2}$ and $-\sqrt{3/2}$ yield a minimum. So, the points closest to $(0, 2)$ are

$$\left(\sqrt{\tfrac{3}{2}}, \tfrac{5}{2}\right) \quad \text{and} \quad \left(-\sqrt{\tfrac{3}{2}}, \tfrac{5}{2}\right).$$

✓ **Checkpoint 2**

Find the points on the graph of $y = 4 - x^2$ that are closest to $(0, 3)$.

Example 3   Finding a Minimum Area

A rectangular page will contain 24 square inches of print. The margins at the top and bottom of the page are $1\frac{1}{2}$ inches wide. The margins on each side are 1 inch wide. What should the dimensions of the page be to minimize the amount of paper used?

SOLUTION

1. A diagram of the page is shown in Figure 9.4.

2. Letting A be the area to be minimized, the primary equation is

$$A = (x + 3)(y + 2). \qquad \text{Primary equation}$$

3. The printed area inside the margins is given by

$$24 = xy. \qquad \text{Secondary equation}$$

Solving this equation for y produces

$$y = \frac{24}{x}.$$

By substituting this result into the primary equation, you obtain

$$A = (x + 3)\left(\frac{24}{x} + 2\right) \qquad \text{Write as a function of one variable.}$$

$$= (x + 3)\left(\frac{24 + 2x}{x}\right) \qquad \text{Rewrite second factor as a single fraction.}$$

$$= \frac{2x^2}{x} + \frac{30x}{x} + \frac{72}{x} \qquad \text{Multiply and separate into terms.}$$

$$= 2x + 30 + \frac{72}{x}. \qquad \text{Simplify.}$$

4. Because x must be positive, the feasible domain is $x > 0$.

5. To find the minimum area, begin by finding the critical numbers of A.

$$\frac{dA}{dx} = 2 - \frac{72}{x^2} \qquad \text{Find derivative of } A.$$

$$0 = 2 - \frac{72}{x^2} \qquad \text{Set derivative equal to 0.}$$

$$-2 = -\frac{72}{x^2} \qquad \text{Subtract 2 from each side.}$$

$$x^2 = 36 \qquad \text{Simplify.}$$

$$x = \pm 6 \qquad \text{Critical numbers}$$

Because $x = -6$ is not in the feasible domain, you need to consider only the critical number $x = 6$. Using the First-Derivative Test, it follows that A is a minimum when $x = 6$. So, the dimensions of the page should be

$$x + 3 = 6 + 3 = 9 \text{ inches} \quad \text{by} \quad y + 2 = \frac{24}{6} + 2 = 6 \text{ inches.}$$

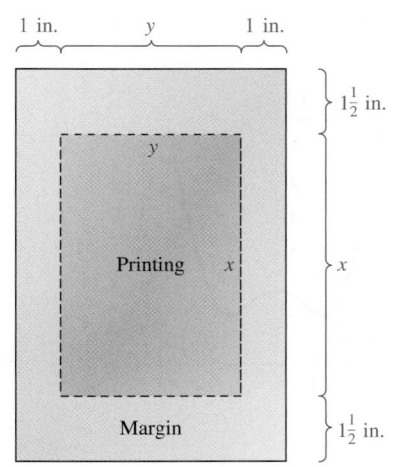

1 in.   y   1 in.

$1\frac{1}{2}$ in.

y

Printing   x

x

Margin $1\frac{1}{2}$ in.

$A = (x + 3)(y + 2)$

FIGURE 9.4

ALGEBRA TUTOR *xy*

For help on the algebra in Example 3, see Example 1(a) in the *Chapter 9 Algebra Tutor*, on page 736.

✓ **Checkpoint 3**

A rectangular page will contain 54 square inches of print. The margins at the top and bottom of the page are $1\frac{1}{2}$ inches wide. The margins on each side are 1 inch wide. What should the dimensions of the page be to minimize the amount of paper used? ■

As applications go, the examples described in this section are fairly simple, and yet the resulting primary equations are quite complicated. Real-life applications often involve equations that are at least as complex as the ones in the examples. Remember that one of the main goals of this course is to enable you to use the power of calculus to analyze equations that at first glance seem formidable.

Also remember that once you have found the primary equation, you can use the graph of the equation to help solve the problem. For instance, the graphs of the primary equations in Examples 1 through 3 are shown in Figure 9.5.

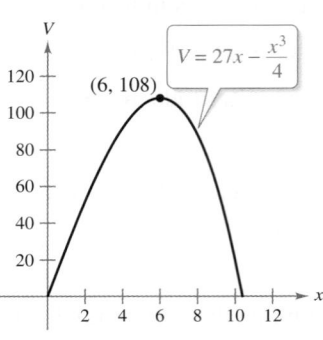

Example 1

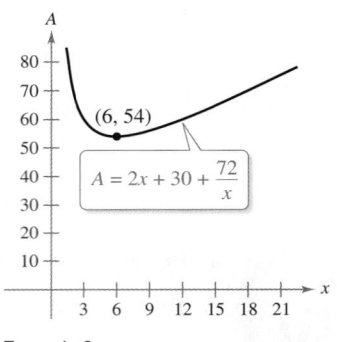

Example 2

Example 3

FIGURE 9.5

SUMMARIZE   (Section 9.1)

1. State what is meant by the primary equation of an optimization problem *(page 690)*. For examples of primary equations in optimization problems, see Examples 1, 2, and 3.

2. State what is meant by the feasible domain of a function *(page 690)*. For examples of feasible domains, see Examples 1, 2, and 3.

3. State what is meant by the secondary equation of an optimization problem *(page 691)*. For examples of secondary equations in optimization problems, see Examples 1, 2, and 3.

4. State the guidelines for solving optimization problems *(page 691)*. For examples of solving optimization problems, see Examples 2 and 3.

5. Describe a real-life example of how solving an optimization problem can be used to determine the dimensions of a page so that the amount of paper used is minimized *(page 693, Example 3)*.

SKILLS WARM UP 9.1 The following warm-up exercises involve skills that were covered in earlier sections. You will use these skills in the exercise set for this section. For additional help, review Sections 1.2 and 8.4.

In Exercises 1–4, write a formula for the written statement.

1. The sum of one number and half a second number is 12.

2. The product of one number and twice another is 24.

3. The area of a rectangle is 24 square units.

4. The distance between two points is 10 units.

In Exercises 5–10, find the critical numbers of the function.

5. $y = x^2 + 6x - 9$

6. $y = 2x^3 - x^2 - 4x$

7. $y = 5x + \dfrac{125}{x}$

8. $y = 3x + \dfrac{96}{x^2}$

9. $y = \dfrac{x^2 + 1}{x}$

10. $y = \dfrac{x}{x^2 + 9}$

Exercises 9.1

See www.CalcChat.com for worked-out solutions to odd-numbered exercises.

Maximum Area In Exercises 1 and 2, find the length and width of a rectangle that has the given perimeter and a maximum area.

1. Perimeter: 100 meters

2. Perimeter: P units

Minimum Perimeter In Exercises 3 and 4, find the length and width of a rectangle that has the given area and a minimum perimeter.

3. Area: 64 square feet

4. Area: A square centimeters

5. Maximum Area A rancher has 200 feet of fencing to enclose two adjacent rectangular corrals (see figure). What dimensions should be used so that the enclosed area will be a maximum?

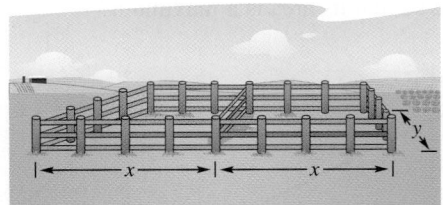

6. Minimum Dimensions A dairy farmer plans to enclose a rectangular pasture adjacent to a river. To provide enough grass for the herd, the pasture must contain 245,000 square meters. No fencing is required along the river. What dimensions will use the least amount of fencing?

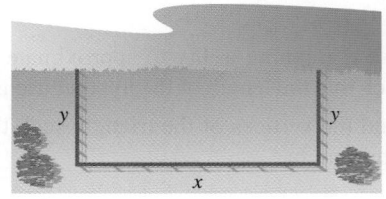

7. Maximum Volume

(a) Verify that each of the rectangular solids shown in the figure has a surface area of 150 square inches.

(b) Find the volume of each solid.

(c) Determine the dimensions of a rectangular solid (with a square base) of maximum volume if its surface area is 150 square inches.

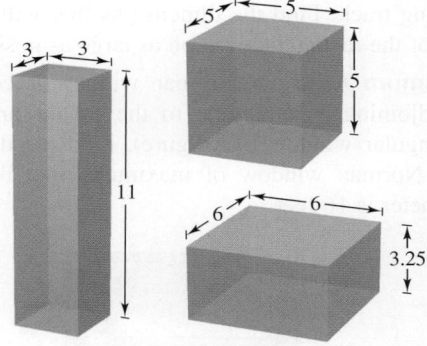

8. Maximum Volume A rectangular solid with a square base has a surface area of 337.5 square centimeters.

(a) Determine the dimensions that yield the maximum volume.

(b) Find the maximum volume.

9. Minimum Surface Area A rectangular solid with a square base has a volume of 8000 cubic inches.

(a) Determine the dimensions that yield the minimum surface area.

(b) Find the minimum surface area.

10. **HOW DO YOU SEE IT?**   The graph shows the profit P (in thousands of dollars) of a company in terms of its advertising cost x (in thousands of dollars).

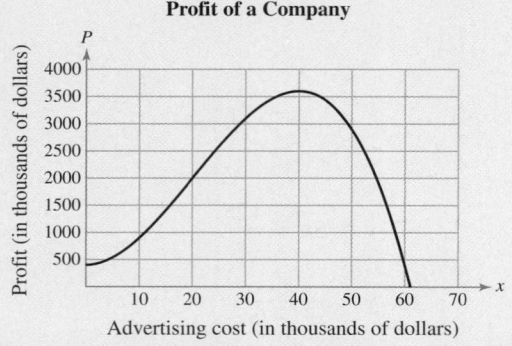

Profit of a Company

(a) Estimate the interval on which the profit is increasing.

(b) Estimate the interval on which the profit is decreasing.

(c) Estimate the amount of money the company should spend on advertising in order to yield a maximum profit.

(d) Estimate the point of diminishing returns.

11. Maximum Area   An indoor physical fitness room consists of a rectangular region with a semicircle on each end. The perimeter of the room is to be a 200-meter running track. Find the dimensions that will make the area of the rectangular region as large as possible.

12. Maximum Area   A Norman window is constructed by adjoining a semicircle to the top of an ordinary rectangular window (see figure). Find the dimensions of a Norman window of maximum area if the total perimeter is 16 feet.

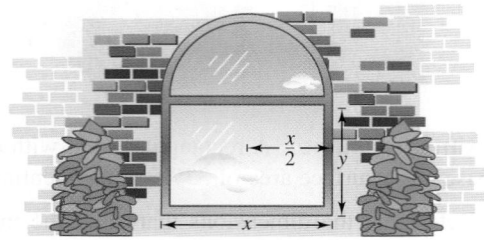

13. Minimum Cost   A storage box with a square base must have a volume of 80 cubic centimeters. The top and bottom cost $0.20 per square centimeter and the sides cost $0.10 per square centimeter. Find the dimensions that will minimize cost.

14. Minimum Surface Area   A net enclosure for golf practice is open at one end (see figure). The volume of the enclosure is $83\frac{1}{3}$ cubic meters. Find the dimensions that require the least amount of netting.

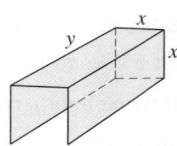

Figure for 14

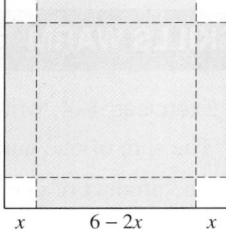

Figure for 15

15. Maximum Volume   An open box is to be made from a six-inch by six-inch square piece of material by cutting equal squares from the corners and turning up the sides (see figure). Find the volume of the largest box that can be made.

16. Maximum Volume   An open box is to be made from a three-foot by eight-foot rectangular piece of material by cutting equal squares from the corners and turning up the sides. Find the volume of the largest box that can be made in this manner.

17. Minimum Area   A rectangular page is to contain 36 square inches of print. The margins at the top and bottom and on each side are to be $1\frac{1}{2}$ inches. Find the dimensions of the page that will minimize the amount of paper used.

18. Minimum Area   A rectangular page is to contain 50 square inches of print. The margins at the top and bottom of the page are to be 2 inches wide. The margins on each side are to be 1 inch wide. Find the dimensions of the page that will minimize the amount of paper used.

19. Maximum Area   A rectangle is bounded by the x- and y-axes and the graph of

$$y = \tfrac{1}{2}(6 - x)$$

(see figure). What length and width should the rectangle have so that its area is a maximum?

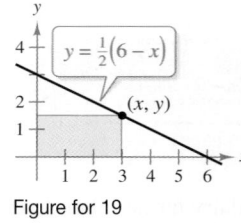

Figure for 19

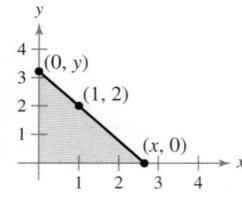

Figure for 20

20. Minimum Length and Minimum Area   A right triangle is formed in the first quadrant by the x- and y-axes and a line through the point $(1, 2)$ (see figure).

(a) Write the length L of the hypotenuse as a function of x.

(b) Use a graphing utility to approximate x graphically such that the length of the hypotenuse is a minimum.

(c) Find the vertices of the triangle such that its area is a minimum.

21. Maximum Area A rectangle is bounded by the x-axis and the semicircle $y = \sqrt{25 - x^2}$ (see figure). What length and width should the rectangle have so that its area is a maximum?

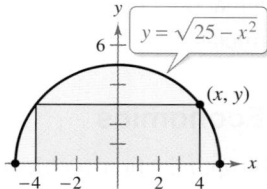

22. Maximum Area Find the dimensions of the largest rectangle that can be inscribed in a semicircle of radius r. (See Exercise 21.)

23. Minimum Surface Area You are designing a soft drink container that has the shape of a right circular cylinder. The container is supposed to hold 12 fluid ounces (1 fluid ounce is approximately 1.80469 cubic inches). Find the dimensions that will use a minimum amount of construction material.

24. Minimum Cost An energy drink container of the shape described in Exercise 23 must have a volume of 16 fluid ounces. The cost per square inch of constructing the top and bottom is twice the cost of constructing the lateral side. Find the dimensions that will minimize cost.

Finding a Minimum Distance In Exercises 25–28, find the points on the graph of the function that are closest to the given point. *See Example 2.*

25. $f(x) = x^2$, $\left(2, \frac{1}{2}\right)$ **26.** $f(x) = (x + 1)^2$, $(5, 3)$

27. $f(x) = \sqrt{x}$, $(4, 0)$ **28.** $f(x) = \sqrt{x - 8}$, $(12, 0)$

29. Maximum Volume A rectangular package to be sent by a postal service can have a maximum combined length and girth (perimeter of a cross section) of 108 inches. Find the dimensions of the package with maximum volume. Assume that the package's dimensions are x by x by y (see figure).

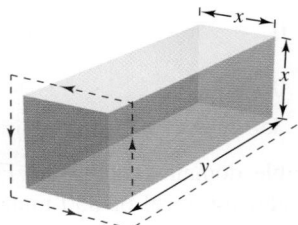

30. Minimum Surface Area A solid is formed by adjoining two hemispheres to the ends of a right circular cylinder. The total volume of the solid is 12 cubic inches. Find the radius of the cylinder that produces the minimum surface area.

31. Minimum Cost An industrial tank of the shape described in Exercise 30 must have a volume of 3000 cubic feet. The hemispherical ends cost twice as much per square foot of surface area as the sides. Find the dimensions that will minimize cost.

32. Minimum Time You are in a boat 2 miles from the nearest point on the coast. You are to go to point Q, located 3 miles down the coast and 1 mile inland (see figure). You can row at a rate of 2 miles per hour and you can walk at a rate of 4 miles per hour. Toward what point on the coast should you row in order to reach point Q in the least amount of time?

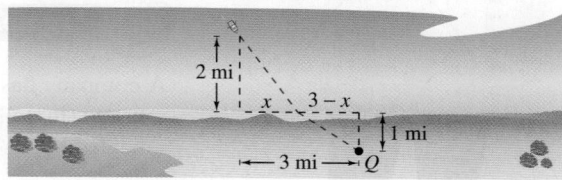

33. Minimum Area The sum of the circumference of a circle and the perimeter of a square is 16. Find the dimensions of the circle and square that produce a minimum total area.

34. Minimum Area The sum of the perimeters of an equilateral triangle and a square is 10. Find the dimensions of the triangle and square that produce a minimum total area.

35. Area Four feet of wire is to be used to form a square and a circle.

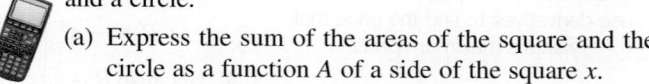

(a) Express the sum of the areas of the square and the circle as a function A of a side of the square x.

(b) What is the domain of A?

(c) Use a graphing utility to graph A on its domain.

(d) How much wire should be used for the square and how much for the circle in order to enclose the least total area? the greatest total area?

36. Maximum Yield A home gardener estimates that 16 apple trees will produce an average yield of 80 apples per tree. But because of the size of the garden, for each additional tree planted, the yield will decrease by four apples per tree. How many trees should be planted to maximize the total yield of apples? What is the maximum yield?

37. Farming A strawberry farmer will receive $30 per bushel of strawberries during the first week of harvesting. Each week after that, the value will drop $0.80 per bushel. The farmer estimates that there are approximately 120 bushels of strawberries in the fields, and that the crop is increasing at a rate of four bushels per week. When should the farmer harvest the strawberries to maximize their value? How many bushels of strawberries will yield the maximum value? What is the maximum value of the strawberries?

9.2 Business and Economics Applications

■ Solve business and economics optimization problems.

■ Find the price elasticity of demand for demand functions.

■ Recognize basic business terms and formulas.

Optimization in Business and Economics

The problems in this section are primarily optimization problems. So, the five-step procedure used in Section 9.1 is an appropriate strategy to follow.

 **Example 1**   **Finding the Maximum Revenue**

A company has determined that its total revenue (in dollars) for a product can be modeled by

$$R = -x^3 + 450x^2 + 52{,}500x$$

where x is the number of units produced (and sold). What production level will yield a maximum revenue?

SOLUTION

1. A sketch of the revenue function is shown in Figure 9.6.

2. The primary equation is the given revenue function.

$$R = -x^3 + 450x^2 + 52{,}500x$$

3. Because R is already given as a function of one variable, you do not need a secondary equation.

4. The feasible domain of the primary equation is

$$0 \le x \le 546. \qquad \text{Feasible domain}$$

This is determined by finding the x-intercepts of the revenue function, as shown in Figure 9.6.

5. To maximize the revenue, find the critical numbers.

$$\frac{dR}{dx} = -3x^2 + 900x + 52{,}500 = 0 \qquad \text{Set derivative equal to 0.}$$

$$-3(x - 350)(x + 50) = 0 \qquad \text{Factor.}$$

$$x = 350, \; x = -50 \qquad \text{Critical numbers}$$

The only critical number in the feasible domain is $x = 350$. From the graph of the function, you can see that the production level of 350 units corresponds to a maximum revenue.

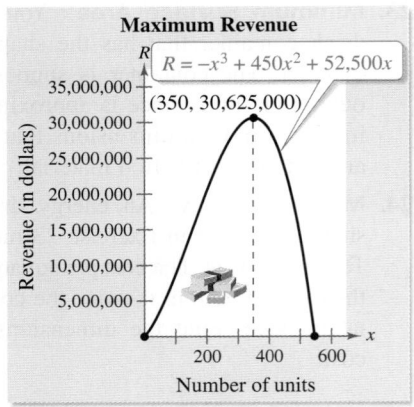

Maximum Revenue

$R = -x^3 + 450x^2 + 52{,}500x$

(350, 30,625,000)

Revenue (in dollars)

Number of units

Maximum revenue occurs when $dR/dx = 0$.

FIGURE 9.6

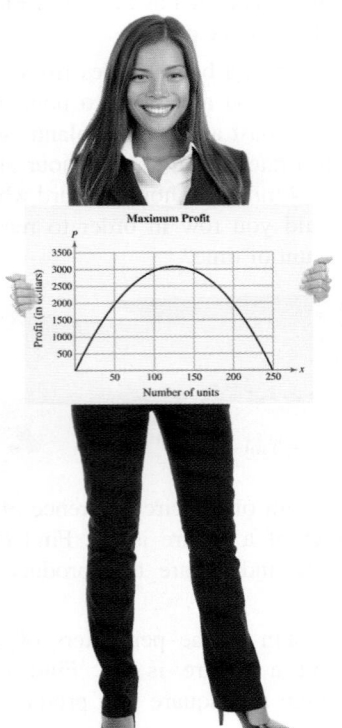

In Exercise 15 on page 705, you will use derivatives to find the price that yields a maximum profit.

✓ Checkpoint 1

Find the number of units that must be produced to maximize the revenue function $R = -x^3 + 150x^2 + 9375x$, where R is the total revenue (in dollars) and x is the number of units produced (and sold). What is the maximum revenue?

To study the effects of production levels on cost, one method economists use is the **average cost function** $\overline{C}$, which is defined as

$$\overline{C} = \frac{C}{x}$$ Average cost function

where $C = f(x)$ is the total cost function and x is the number of units produced.

 Example 2   **Finding the Minimum Average Cost**

A company estimates that the cost (in dollars) of producing x units of a product can be modeled by

$$C = 800 + 0.04x + 0.0002x^2.$$

Find the production level that minimizes the average cost per unit.

SOLUTION

1. C represents the total cost, x represents the number of units produced, and $\overline{C}$ represents the average cost per unit.

2. The primary equation is

$$\overline{C} = \frac{C}{x}.$$ Primary equation

3. Substituting the given equation for C produces

$$\overline{C} = \frac{800 + 0.04x + 0.0002x^2}{x}$$ Substitute for C.

$$= \frac{800}{x} + 0.04 + 0.0002x.$$ Function of one variable

4. The feasible domain of this function is

$$x > 0$$ Feasible domain

because the company cannot produce a negative number of units.

5. You can find the critical numbers as shown.

$$\frac{d\overline{C}}{dx} = -\frac{800}{x^2} + 0.0002 = 0$$ Set derivative equal to 0.

$$0.0002 = \frac{800}{x^2}$$

$$x^2 = \frac{800}{0.0002}$$ Multiply each side by x^2 and divide each side by 0.0002.

$$x^2 = 4,000,000$$

$$x = \pm 2000$$ Critical numbers

By choosing the positive value of x and sketching the graph of $\overline{C}$, as shown in Figure 9.7, you can see that a production level of $x = 2000$ minimizes the average cost per unit.

STUDY TIP

To see that $x = 2000$ corresponds to a minimum average cost in Example 2, try evaluating $\overline{C}$ for several values of x. For instance, when $x = 400$, the average cost per unit is $\overline{C} = \$2.12$, but when $x = 2000$, the average cost per unit is $\overline{C} = \$0.84$.

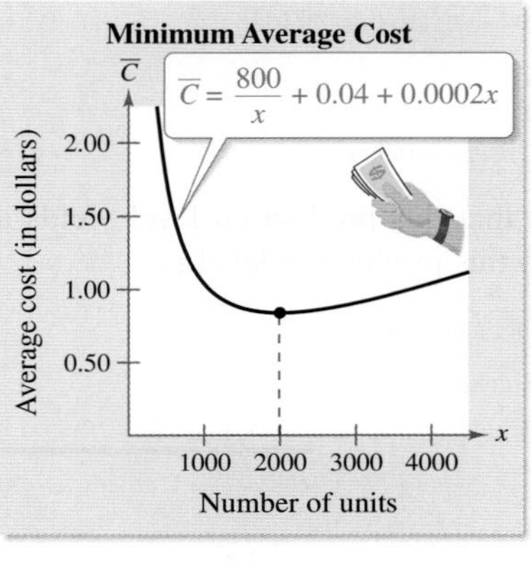

Minimum Average Cost

Minimum average cost occurs when $d\overline{C}/dx = 0$.

FIGURE 9.7

✓**Checkpoint 2**

Find the production level that minimizes the average cost per unit for the cost function

$$C = 400 + 0.05x + 0.0025x^2$$

where C is the cost (in dollars) of producing x units of a product.

Example 3   Finding the Maximum Revenue

A business sells 2000 units of a product per month at a price of $10 each. It can sell 250 more items per month for each $0.25 reduction in price. What price per unit will maximize the monthly revenue?

SOLUTION

1. Let x represent the number of units sold in a month, let p represent the price per unit, and let R represent the monthly revenue.

2. Because the revenue is to be maximized, the primary equation is

$$R = xp. \qquad \text{Primary equation}$$

3. A price of $p = \$10$ corresponds to $x = 2000$, and a price of $p = \$9.75$ corresponds to $x = 2250$. Using this information, you can use the two-point form to write the demand equation.

$$p - 10 = \frac{10 - 9.75}{2000 - 2250}(x - 2000) \qquad \text{Two-point form}$$

$$p - 10 = -0.001(x - 2000) \qquad \text{Simplify.}$$

$$p = -0.001x + 12 \qquad \text{Secondary equation}$$

Substituting this value into the revenue equation produces

$$R = x(-0.001x + 12) \qquad \text{Substitute for } p.$$

$$= -0.001x^2 + 12x. \qquad \text{Function of one variable}$$

4. The feasible domain of the revenue function is

$$0 \le x \le 12{,}000. \qquad \text{Feasible domain}$$

This is determined by finding the x-intercepts of the revenue function.

5. To maximize the revenue, find the critical numbers.

$$\frac{dR}{dx} = 12 - 0.002x = 0 \qquad \text{Set derivative equal to 0.}$$

$$-0.002x = -12$$

$$x = 6000 \qquad \text{Critical number}$$

From the graph of R in Figure 9.8, you can see that this production level yields a maximum revenue. The price that corresponds to this production level is

$$p = 12 - 0.001x \qquad \text{Demand function}$$

$$= 12 - 0.001(6000) \qquad \text{Substitute 6000 for } x.$$

$$= \$6. \qquad \text{Price per unit} \qquad \underline{\hspace{2cm}}$$

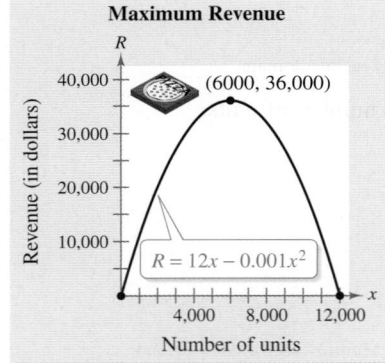

Maximum Revenue

(6000, 36,000)

$R = 12x - 0.001x^2$

Number of units

FIGURE 9.8

✓ **Checkpoint 3**

Find the price per unit that will maximize the monthly revenue for the business in Example 3 when it can sell only 200 more items per month for each $0.25 reduction in price.   ▨

In Example 3, the revenue function was written as a function of x. It could also have been written as a function of p. That is,

$$R = 1000(12p - p^2).$$

By finding the critical numbers of this function, you can determine that the maximum revenue occurs at $p = 6$.

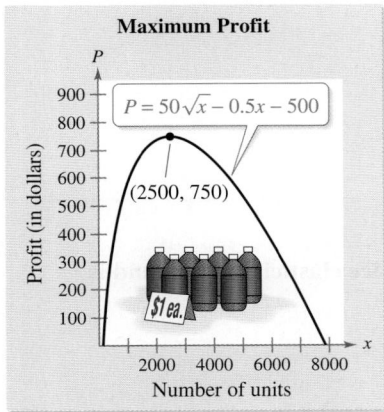

FIGURE 9.9

Example 4 Finding the Maximum Profit

The marketing department of a business has determined that the demand for a product can be modeled by $p = 50/\sqrt{x}$, where p is the price per unit (in dollars) and x is the number of units. The cost (in dollars) of producing x units is given by $C = 0.5x + 500$. What price will yield a maximum profit?

SOLUTION

1. Let R represent the revenue, P the profit, p the price per unit, x the number of units, and C the total cost of producing x units.

2. Because you are maximizing the profit, the primary equation is

$$P = R - C. \qquad \text{Primary equation}$$

3. Because the revenue is $R = xp$, you can write the profit function as

$$\begin{aligned}
P &= R - C \\
&= xp - (0.5x + 500) & \text{Substitute for } R \text{ and } C. \\
&= x\left(\frac{50}{\sqrt{x}}\right) - 0.5x - 500 & \text{Substitute for } p. \\
&= 50\sqrt{x} - 0.5x - 500. & \text{Function of one variable}
\end{aligned}$$

4. The feasible domain of the function is $127 < x \le 7872$. (When x is less than 127 or greater than or equal to 7872, the profit is negative.)

5. To maximize the profit, find the critical numbers.

$$\frac{dP}{dx} = \frac{25}{\sqrt{x}} - 0.5 = 0 \qquad \text{Set derivative equal to 0.}$$

$$\frac{25}{\sqrt{x}} = 0.5 \qquad \text{Add 0.5 to each side.}$$

$$50 = \sqrt{x} \qquad \text{Isolate } x\text{-term on one side.}$$

$$2500 = x \qquad \text{Critical number}$$

From the graph of the profit function shown in Figure 9.9, you can see that a maximum profit occurs at $x = 2500$. The price that corresponds to $x = 2500$ is

$$p = \frac{50}{\sqrt{x}} = \frac{50}{\sqrt{2500}} = \frac{50}{50} = \$1.00. \qquad \text{Price per unit}$$

✓ Checkpoint 4

Find the price that will maximize profit for the demand and cost functions

$$p = \frac{40}{\sqrt{x}} \quad \text{and} \quad C = 2x + 50$$

where p is the price per unit (in dollars), x is the number of units, and C is the cost (in dollars). ■

To find the maximum profit in Example 4, the equation $P = R - C$ was differentiated and set equal to zero. From the equation

$$\frac{dP}{dx} = \frac{dR}{dx} - \frac{dC}{dx} = 0$$

it follows that the maximum profit occurs when the marginal revenue is equal to the marginal cost, as shown in Figure 9.10.

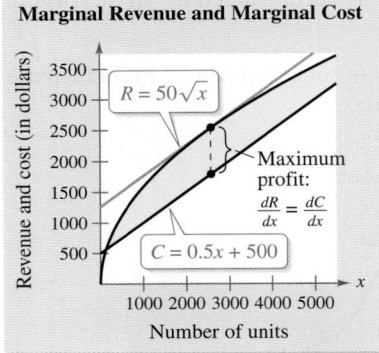

FIGURE 9.10

ALGEBRA TUTOR *xy*

For help on the algebra in Example 4, see Example 2(b) in the *Chapter 9 Algebra Tutor*, on page 737.

Price Elasticity of Demand

One way in which economists measure the responsiveness of consumers to a change in the price of a product is with **price elasticity of demand.** For example, a drop in the price of fresh tomatoes might result in a much greater demand for fresh tomatoes; such a demand is called **elastic.** On the other hand, the demand for items such as coffee and gasoline is relatively unresponsive to changes in price; the demand for such items is called **inelastic.**

More formally, the elasticity of demand is the percent change of a quantity demanded x, divided by the percent change in its price p. You can develop a formula for price elasticity of demand using the approximation

$$\frac{\Delta p}{\Delta x} \approx \frac{dp}{dx}$$

which is based on the definition of the derivative. Using this approximation, you can write

$$\text{Price elasticity of demand} = \frac{\text{rate of change in demand}}{\text{rate of change in price}}$$

$$= \frac{\Delta x / x}{\Delta p / p}$$

$$= \frac{p/x}{\Delta p / \Delta x}$$

$$\approx \frac{p/x}{dp/dx}.$$

Definition of Price Elasticity of Demand

If $p = f(x)$ is a differentiable function, then the **price elasticity of demand** is given by

$$\eta = \frac{p/x}{dp/dx}$$

where η is the lowercase Greek letter eta. For a given price, the demand is **elastic** when $|\eta| > 1$, the demand is **inelastic** when $|\eta| < 1$, and the demand has **unit elasticity** when $|\eta| = 1$.

Price elasticity of demand is related to the total revenue function, as indicated in Figure 9.11 and the list below.

1. If the demand is *elastic*, then a decrease in price is accompanied by an increase in unit sales sufficient to increase the total revenue.

2. If the demand is *inelastic*, then a decrease in price is not accompanied by an increase in unit sales sufficient to increase the total revenue.

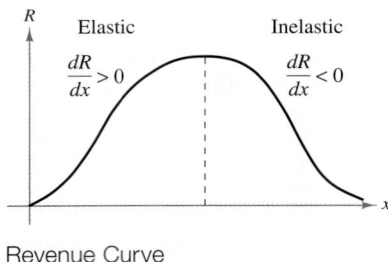

Revenue Curve
FIGURE 9.11

Demand Function of a Product

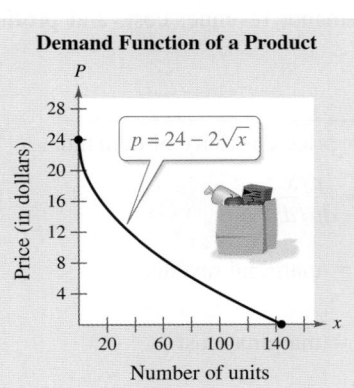

FIGURE 9.12

Example 5 Comparing Elasticity and Revenue

The demand function for a product is modeled by

$$p = 24 - 2\sqrt{x}, \quad 0 \le x \le 144$$

where p is the price per unit (in dollars) and x is the number of units. (See Figure 9.12.)

a. Determine when the demand is elastic, inelastic, and of unit elasticity.

b. Use the result of part (a) to describe the behavior of the revenue function.

SOLUTION

a. The price elasticity of demand is given by

$$\eta = \frac{p/x}{dp/dx} \qquad \text{Formula for price elasticity of demand}$$

$$= \frac{\dfrac{24 - 2\sqrt{x}}{x}}{\dfrac{-1}{\sqrt{x}}} \qquad \text{Substitute for } \frac{p}{x} \text{ and } \frac{dp}{dx}.$$

$$= \frac{\left(\dfrac{24 - 2\sqrt{x}}{x}\right)(-\sqrt{x})}{\left(\dfrac{-1}{\sqrt{x}}\right)(-\sqrt{x})} \qquad \text{Multiply numerator and denominator by } -\sqrt{x}.$$

$$= \frac{-24\sqrt{x} + 2x}{x} \qquad \text{Simplify.}$$

$$= -\frac{24\sqrt{x}}{x} + 2. \qquad \text{Rewrite as two fractions and simplify.}$$

The demand is of unit elasticity when $|\eta| = 1$. In the interval $[0, 144]$, the only solution of the equation

$$|\eta| = \left| -\frac{24\sqrt{x}}{x} + 2 \right| = 1 \qquad \text{Unit elasticity}$$

is $x = 64$. So, the demand is of unit elasticity when $x = 64$. For x-values in the interval $(0, 64)$,

$$|\eta| = \left| -\frac{24\sqrt{x}}{x} + 2 \right| > 1, \quad 0 < x < 64 \qquad \text{Elastic}$$

which implies that the demand is elastic when $0 < x < 64$. For x-values in the interval $(64, 144)$,

$$|\eta| = \left| -\frac{24\sqrt{x}}{x} + 2 \right| < 1, \quad 64 < x < 144 \qquad \text{Inelastic}$$

which implies that the demand is inelastic when $64 < x < 144$.

Revenue Function of a Product

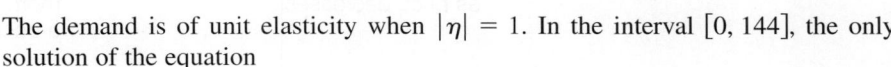

FIGURE 9.13

b. From part (a), you can conclude that the revenue function R is increasing on the open interval $(0, 64)$, is decreasing on the open interval $(64, 144)$, and is a maximum when $x = 64$, as indicated in Figure 9.13.

✓ Checkpoint 5

The demand function for a product is modeled by $p = 36 - 2\sqrt{x}, 0 \le x \le 324$, where p is the price per unit (in dollars) and x is the number of units. Determine when the demand is elastic, inelastic, and of unit elasticity.

Business Terms and Formulas

This section concludes with a summary of the basic business terms and formulas used in this section. A summary of the graphs of the demand, revenue, cost, and profit functions is shown in Figure 9.14.

Summary of Business Terms and Formulas

x = number of units produced (or sold)	η = price elasticity of demand
p = price per unit	$\quad = \dfrac{p/x}{dp/dx}$
R = total revenue from selling x units = xp	$\dfrac{dR}{dx}$ = marginal revenue
C = total cost of producing x units	$\dfrac{dC}{dx}$ = marginal cost
P = total profit from selling x units = $R - C$	$\dfrac{dP}{dx}$ = marginal profit
$\overline{C}$ = average cost per unit = $\dfrac{C}{x}$	

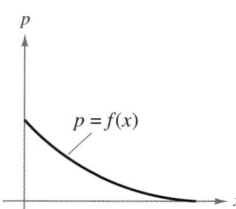

Demand Function
Quantity demanded increases as price decreases.

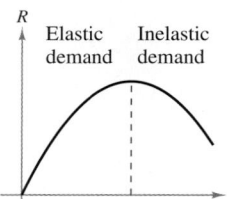

Revenue Function
The low prices required to sell more units eventually result in a decreasing revenue.

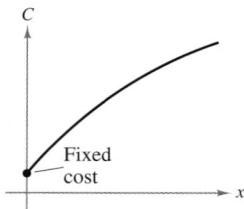

Cost Function
The total cost to produce x units includes the fixed cost.

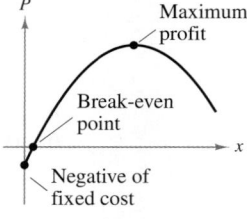

Profit Function
The break-even point occurs when $R = C$.

FIGURE 9.14

SUMMARIZE   (Section 9.2)

1. Describe a real-life example of how optimization can be used to find the maximum revenue for a product *(page 698, Example 1)*.

2. State the definition of the average cost function *(page 699)*. For an example of an average cost function, see Example 2.

3. State the definition of price elasticity of demand *(page 702)*. For an example of price elasticity of demand, see Example 5.

SKILLS WARM UP 9.2

The following warm-up exercises involve skills that were covered in a previous course or earlier sections. You will use these skills in the exercise set for this section. For additional help, review Sections 0.2, 0.3, 0.4, and 7.5.

In Exercises 1–4, evaluate the expression for $x = 150$.

1. $\left| -\dfrac{300}{x} + 3 \right|$

2. $\left| -\dfrac{600}{5x} + 2 \right|$

3. $\left| \dfrac{(20x^{-1/2})/x}{-10x^{-3/2}} \right|$

4. $\left| \dfrac{(4000/x^2)/x}{-8000x^{-3}} \right|$

In Exercises 5–10, find the marginal revenue, marginal cost, or marginal profit.

5. $C = 650 + 1.2x + 0.003x^2$

6. $P = 0.01x^2 + 11x$

7. $P = -0.7x^2 + 7x - 50$

8. $C = 1700 + 4.2x + 0.001x^3$

9. $R = 14x - \dfrac{x^2}{2000}$

10. $R = 3.4x - \dfrac{x^2}{1500}$

Exercises 9.2

See www.CalcChat.com for worked-out solutions to odd-numbered exercises.

Finding the Maximum Revenue In Exercises 1–4, find the number of units x that produces a maximum revenue R. *See Example 1.*

1. $R = 800x - 0.2x^2$
2. $R = 48x^2 - 0.02x^3$
3. $R = 400x - x^2$
4. $R = 30x^{2/3} - 2x$

Finding the Minimum Average Cost In Exercises 5–8, find the number of units x that produces the minimum average cost per unit $\overline{C}$. *See Example 2.*

5. $C = 0.125x^2 + 20x + 5000$
6. $C = 0.001x^3 + 5x + 250$
7. $C = 2x^2 + 255x + 5000$
8. $C = 0.02x^3 + 55x^2 + 1380$

Finding the Maximum Profit In Exercises 9–12, find the price that will maximize profit for the demand and cost functions, where p is the price, x is the number of units, and C is the cost. *See Example 4.*

Demand Function	Cost Function
9. $p = 90 - x$	$C = 100 + 30x$
10. $p = 70 - 0.01x$	$C = 8000 + 50x + 0.03x^2$
11. $p = 50 - 0.1\sqrt{x}$	$C = 35x + 500$
12. $p = \dfrac{24}{\sqrt{x}}$	$C = 0.4x + 600$

 Average Cost In Exercises 13 and 14, use the cost function to find the production level at which the average cost is a minimum. For this production level, show that the marginal cost and average cost are equal. Use a graphing utility to graph the average cost function and verify your results.

13. $C = 2x^2 + 5x + 18$ **14.** $C = x^3 - 6x^2 + 13x$

15. Maximum Profit A commodity has a demand function modeled by

$$p = 80 - 0.2x$$

and a total cost function modeled by

$$C = 30x + 40$$

where x is the number of units.

(a) What price yields a maximum profit?

(b) When the profit is maximized, what is the average cost per unit?

16. Maximum Profit A commodity has a demand function modeled by $p = 100 - 0.5x$, and a total cost function modeled by $C = 50x + 37.5$, where x is the number of units.

(a) What price yields a maximum profit?

(b) When the profit is maximized, what is the average cost per unit?

Maximum Profit In Exercises 17 and 18, find the amount s spent on advertising (in thousands of dollars) that maximizes the profit P (in thousands of dollars). Find the point of diminishing returns.

17. $P = -2s^3 + 35s^2 - 100s + 200$
18. $P = -0.1s^3 + 6s^2 + 400$

19. Maximum Profit The cost per unit of producing an MP3 player is $90. The manufacturer charges $150 per unit for orders of 100 or less. To encourage large orders, however, the manufacturer reduces the charge by $0.10 per player for each order in excess of 100 units. For instance, an order of 101 players would be $149.90 per player, an order of 102 players would be $149.80 per player, and so on. Find the largest order the manufacturer should allow to obtain a maximum profit.

20. Maximum Profit A real estate office handles a 50-unit apartment complex. When the rent is $580 per month, all units are occupied. For each $40 increase in rent, however, an average of one unit becomes vacant. Each occupied unit requires an average of $45 per month for service and repairs. What rent should be charged to obtain a maximum profit?

21. Maximum Revenue When a wholesaler sold a product at $40 per unit, sales were 300 units per week. After a price increase of $5, however, the average number of units sold dropped to 275 per week. Assuming that the demand function is linear, what price per unit will yield a maximum total revenue?

22. Maximum Profit Assume that the amount of money deposited in a bank is proportional to the square of the interest rate the bank pays on the money. Furthermore, the bank can reinvest the money at 12% simple interest. Find the interest rate the bank should pay to maximize its profit.

23. Minimum Cost A power station is on one side of a river that is 0.5 mile wide, and a factory is 6 miles downstream on the other side of the river (see figure). It costs $18 per foot to run overland power lines and $25 per foot to run underwater power lines. Write a cost function for running the power lines from the power station to the factory. Use a graphing utility to graph your function. Estimate the value of x that minimizes the cost. Explain your results.

24. Minimum Cost An offshore oil well is 1 mile off the coast. The oil refinery is 2 miles down the coast. Laying pipe in the ocean is twice as expensive as laying it on land. Find the most economical path for the pipe from the well to the oil refinery.

Minimum Cost In Exercises 25 and 26, find the speed v, in miles per hour, that will minimize costs on a 110-mile delivery trip. The cost per hour for fuel is C dollars, and the driver is paid W dollars per hour. (Assume there are no costs other than wages and fuel.)

25. Fuel cost: $C = \dfrac{v^2}{300}$

Driver: $W = \$12$

26. Fuel cost: $C = \dfrac{v^2}{500}$

Driver: $W = \$9.50$

Elasticity In Exercises 27–32, find the price elasticity of demand for the demand function at the indicated x-value. Is the demand elastic, inelastic, or of unit elasticity at the indicated x-value? Use a graphing utility to graph the revenue function, and identify the intervals of elasticity and inelasticity.

Demand Function	Quantity Demanded
27. $p = 600 - 5x$	$x = 60$
28. $p = 400 - 3x$	$x = 20$
29. $p = 5 - 0.03x$	$x = 100$
30. $p = 20 - 0.0002x$	$x = 30$
31. $p = \dfrac{500}{x + 2}$	$x = 23$
32. $p = \dfrac{500}{x^2} + 5$	$x = 5$

33. Elasticity The demand function for a product is modeled by

$$p = 20 - 0.02x, \quad 0 \le x \le 1000$$

where p is the price (in dollars) and x is the number of units.

(a) Determine when the demand is elastic, inelastic, and of unit elasticity.

(b) Use the result of part (a) to describe the behavior of the revenue function.

34. Elasticity The demand function for a product is given by $p = 800 - 4x$, $0 \le x \le 200$, where p is the price (in dollars) and x is the number of units.

(a) Determine when the demand is elastic, inelastic, and of unit elasticity.

(b) Use the result of part (a) to describe the behavior of the revenue function.

35. Minimum Cost The shipping and handling cost C of a manufactured product is modeled by

$$C = 4\left(\frac{25}{x^2} - \frac{x}{x - 10}\right), \quad 0 < x < 10$$

where C is measured in thousands of dollars and x is the number of units shipped (in hundreds). Use the *root* feature of a graphing utility to find the shipment size that minimizes the cost.

36. Minimum Cost The ordering and transportation cost C of the components used in manufacturing a product is modeled by

$$C = 8\left(\frac{2500}{x^2} - \frac{x}{x - 100}\right), \quad 0 < x < 100$$

where C is measured in thousands of dollars and x is the order size in hundreds. Use the *root* feature of a graphing utility to find the order size that minimizes the cost.

37. Revenue The demand for a car wash is $x = 900 - 45p$, where the current price is \$8. Can revenue be increased by lowering the price and thus attracting more customers? Use price elasticity of demand to determine your answer.

38. **HOW DO YOU SEE IT?** Match each graph with the function it best represents—a demand function, a revenue function, a cost function, or a profit function. Explain your reasoning. (The graphs are labeled *a–d*.)

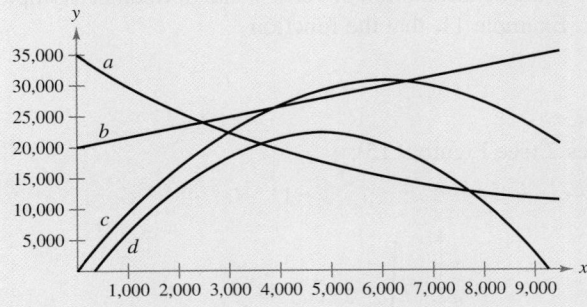

39. Sales The sales S (in millions of dollars per year) for The Clorox Company for the years 2001 through 2010 can be modeled by

$$S = -1.893t^3 + 41.03t^2 - 58.6t + 3972, \quad 1 \le t \le 10$$

where t represents the year, with $t = 1$ corresponding to 2001. *(Source: The Clorox Company)*

(a) During which year, from 2001 through 2010, were the company's sales increasing most rapidly?

(b) During which year were the sales increasing at the lowest rate?

(c) Find the rate of increase or decrease for each year in parts (a) and (b).

(d) Use a graphing utility to graph the sales function. Then use the *zoom* and *trace* features to confirm the results in parts (a), (b), and (c).

40. Sales The sales S (in billions of dollars) for Lockhead Martin Corporation from 2001 through 2010 can be modeled by

$$S = \frac{18.17 + 8.165t}{1 + 0.116t}, \quad 1 \le t \le 10$$

where t represents the year, with $t = 1$ corresponding to 2001. *(Source: Lockhead Martin Corporation)*

(a) During which year, from 2001 through 2010, were the company's sales the greatest? the least?

(b) During which year were the sales increasing at the greatest rate? decreasing at the greatest rate?

(c) Use a graphing utility to graph the revenue function. Then use the *zoom* and *trace* features to confirm the results in parts (a) and (b).

41. Demand A demand function is modeled by

$$x = \frac{a}{p^m}$$

where a is a constant and $m > 1$. Show that $\eta = -m$. In other words, show that a 1% increase in price results in an m% decrease in the quantity demanded.

42. Think About It Throughout this text, it is assumed that demand functions are decreasing. Can you think of a product that has an increasing demand function? That is, can you think of a product that becomes more in demand as its price increases? Explain your reasoning, and sketch a graph of the function.

Business Capsule

The website vWorker.com is a marketplace that links freelance workers with employers looking to outsource jobs. The site's founder, Ian Ippolito, noticed that more and more companies were outsourcing rather than hiring full-time employees. He saw the potential in this trend, borrowed \$5000 from his parents, and turned the capital into an online business in 2001. Today his company is known as vWorker.com—for virtual worker—and it connects more than 150,000 employers with over 300,000 workers worldwide. Posting and bidding are free, but vWorker takes a percentage of the workers' final earnings. Only eight years after the company's debut, annual revenues reached \$2.5 million in 2009.

43. Research Project Choose a company with an innovative product or service like the one described above. Use your school's library, the Internet, or some other reference source to research the history of the company. Collect data about the revenue that the product or service has generated, and find a mathematical model of the data. Summarize your findings.

9.3 Asymptotes

■ Find the vertical asymptotes of functions and find infinite limits.
■ Find the horizontal asymptotes of functions and find limits at infinity.
■ Use asymptotes to answer questions about real-life situations.

Vertical Asymptotes and Infinite Limits

In the first three sections of this chapter, you studied ways in which you can use calculus to help analyze the graph of a function. In this section, you will study another valuable aid to curve sketching: the determination of vertical and horizontal asymptotes.

Recall from Section 7.1, Example 11, that the function

$$f(x) = \frac{3}{x - 2}$$

is unbounded as x approaches 2 (see Figure 9.15).

In Exercise 61 on page 718, you will use limits at infinity to find the limit of an average cost function as the number of units produced increases.

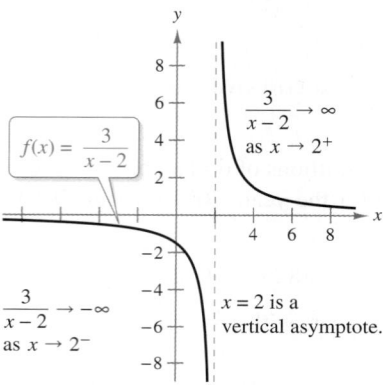

FIGURE 9.15

This type of behavior is described by saying that the line

$$x = 2 \qquad \text{Vertical asymptote}$$

is a **vertical asymptote** of the graph of f. The type of limit in which $f(x)$ approaches infinity (or negative infinity) as x approaches c from the left or from the right is an **infinite limit.** The infinite limits for the function $f(x) = 3/(x - 2)$ can be written as

$$\lim_{x \to 2^-} \frac{3}{x - 2} = -\infty$$

and

$$\lim_{x \to 2^+} \frac{3}{x - 2} = \infty.$$

Definition of Vertical Asymptote

If $f(x)$ approaches infinity (or negative infinity) as x approaches c from the right or from the left, then the line

$$x = c$$

is a **vertical asymptote** of the graph of f.

One of the most common instances of a vertical asymptote is the graph of a *rational function*—that is, a function of the form $f(x) = p(x)/q(x)$, where $p(x)$ and $q(x)$ are polynomials. If c is a real number such that $q(c) = 0$ and $p(c) \neq 0$, then the graph of f has a vertical asymptote at $x = c$. Example 1 shows four cases.

Example 1 Finding Infinite Limits

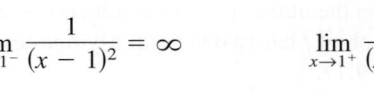 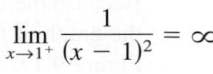

Limit from the left   *Limit from the right*

a. $\displaystyle\lim_{x \to 1^-} \frac{1}{x - 1} = -\infty$ $\displaystyle\lim_{x \to 1^+} \frac{1}{x - 1} = \infty$ See Figure 9.16(a).

b. $\displaystyle\lim_{x \to 1^-} \frac{-1}{x - 1} = \infty$ $\displaystyle\lim_{x \to 1^+} \frac{-1}{x - 1} = -\infty$ See Figure 9.16(b).

c. $\displaystyle\lim_{x \to 1^-} \frac{-1}{(x - 1)^2} = -\infty$ $\displaystyle\lim_{x \to 1^+} \frac{-1}{(x - 1)^2} = -\infty$ See Figure 9.16(c).

d. $\displaystyle\lim_{x \to 1^-} \frac{1}{(x - 1)^2} = \infty$ $\displaystyle\lim_{x \to 1^+} \frac{1}{(x - 1)^2} = \infty$ See Figure 9.16(d).

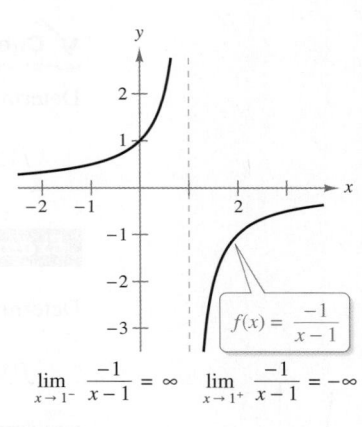

$\displaystyle\lim_{x \to 1^-} \frac{1}{x - 1} = -\infty$ $\displaystyle\lim_{x \to 1^+} \frac{1}{x - 1} = \infty$

(a)

$\displaystyle\lim_{x \to 1^-} \frac{-1}{x - 1} = \infty$ $\displaystyle\lim_{x \to 1^+} \frac{-1}{x - 1} = -\infty$

(b)

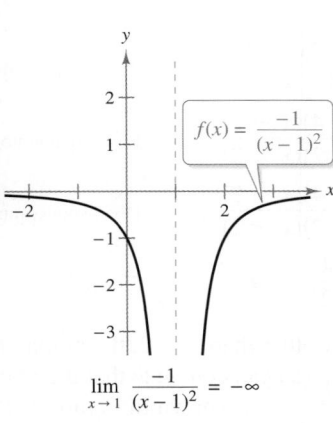 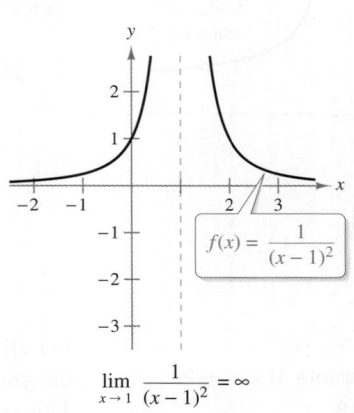

$\displaystyle\lim_{x \to 1} \frac{-1}{(x - 1)^2} = -\infty$

(c)

$\displaystyle\lim_{x \to 1} \frac{1}{(x - 1)^2} = \infty$

(d)

FIGURE 9.16

✓ Checkpoint 1

Find each limit.

a. $\displaystyle\lim_{x \to 2^-} \frac{1}{x - 2}$ **b.** $\displaystyle\lim_{x \to 2^+} \frac{1}{x - 2}$ **c.** $\displaystyle\lim_{x \to -3^-} \frac{1}{x + 3}$ **d.** $\displaystyle\lim_{x \to -3^+} \frac{1}{x + 3}$

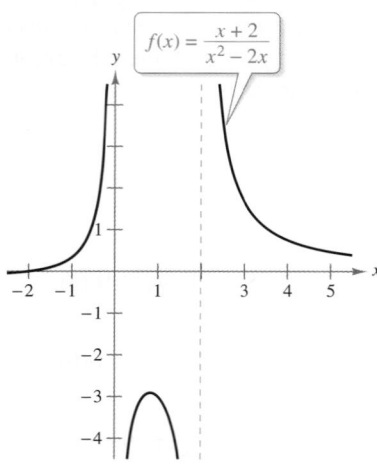

$f(x) = \dfrac{x+2}{x^2 - 2x}$

Vertical Asymptotes at $x = 0$ and $x = 2$

FIGURE 9.17

Each of the graphs in Example 1 has only one vertical asymptote. As shown in the next example, the graph of a rational function can have more than one vertical asymptote.

Example 2 **Finding Vertical Asymptotes**

Determine all vertical asymptotes of the graph of

$$f(x) = \frac{x+2}{x^2 - 2x}.$$

SOLUTION The possible vertical asymptotes correspond to the x-values for which the denominator is zero.

$x^2 - 2x = 0$	Set denominator equal to 0.
$x(x - 2) = 0$	Factor.
$x = 0, x = 2$	Zeros of denominator

Because the numerator of f is not zero at either of these x-values, you can conclude that the graph of f has two vertical asymptotes—one at $x = 0$ and one at $x = 2$, as shown in Figure 9.17.

✓**Checkpoint 2**

Determine all vertical asymptotes of the graph of

$$f(x) = \frac{x+4}{x^2 - 4x}.$$

Example 3 **Finding Vertical Asymptotes**

Determine all vertical asymptotes of the graph of

$$f(x) = \frac{x^2 + 2x - 8}{x^2 - 4}.$$

SOLUTION First factor the numerator and denominator. Then divide out common factors.

$f(x) = \dfrac{x^2 + 2x - 8}{x^2 - 4}$	Write original function.
$= \dfrac{(x + 4)(x - 2)}{(x + 2)(x - 2)}$	Factor numerator and denominator.
$= \dfrac{(x + 4)\cancel{(x - 2)}}{(x + 2)\cancel{(x - 2)}}$	Divide out common factors.
$= \dfrac{x + 4}{x + 2}, \quad x \neq 2$	Simplify.

For all values of x other than $x = 2$, the graph of this simplified function is the same as the graph of f. So, you can conclude that the graph of f has only one vertical asymptote. This occurs at $x = -2$, as shown in Figure 9.18.

✓**Checkpoint 3**

Determine all vertical asymptotes of the graph of

$$f(x) = \frac{x^2 + 4x + 3}{x^2 - 9}.$$

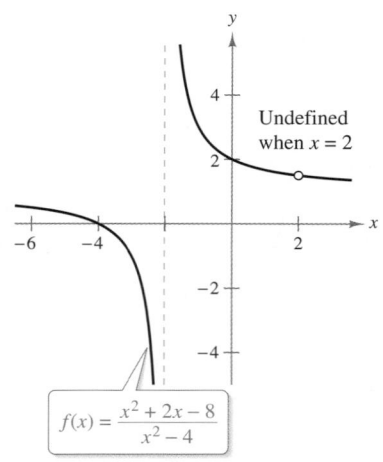

Undefined when $x = 2$

$f(x) = \dfrac{x^2 + 2x - 8}{x^2 - 4}$

Vertical Asymptote at $x = -2$

FIGURE 9.18

From Example 3, you know that the graph of

$$f(x) = \frac{x^2 + 2x - 8}{x^2 - 4}$$

has a vertical asymptote at $x = -2$. This implies that the limit of $f(x)$ as $x \to -2$ from the right (or from the left) is either ∞ or $-\infty$. But without looking at the graph, how can you determine that the limit from the left is *negative* infinity and the limit from the right is *positive* infinity? That is, why is the limit from the left

$$\lim_{x \to -2^-} \frac{x^2 + 2x - 8}{x^2 - 4} = -\infty \qquad \text{Limit from the left}$$

and why is the limit from the right

$$\lim_{x \to -2^+} \frac{x^2 + 2x - 8}{x^2 - 4} = \infty? \qquad \text{Limit from the right}$$

It is cumbersome to determine these limits analytically, and you may find the graphical method shown in Example 4 to be more efficient.

Example 4 Determining Infinite Limits

Find the limits.

$$\lim_{x \to 1^-} \frac{x^2 - 3x}{x - 1} \quad \text{and} \quad \lim_{x \to 1^+} \frac{x^2 - 3x}{x - 1}$$

SOLUTION Begin by considering the function

$$f(x) = \frac{x^2 - 3x}{x - 1}.$$

Because the denominator is zero when $x = 1$ and the numerator is not zero when $x = 1$, it follows that the graph of the function has a vertical asymptote at $x = 1$. This implies that each of the given limits is either ∞ or $-\infty$. To determine which, use a graphing utility to graph the function, as shown in Figure 9.19. From the graph, you can see that the limit from the left is positive infinity and the limit from the right is negative infinity. That is,

From the left, $f(x)$ approaches positive infinity.

From the right, $f(x)$ approaches negative infinity.

FIGURE 9.19

$$\lim_{x \to 1^-} \frac{x^2 - 3x}{x - 1} = \infty \qquad \text{Limit from the left}$$

and

$$\lim_{x \to 1^+} \frac{x^2 - 3x}{x - 1} = -\infty. \qquad \text{Limit from the right}$$

✓Checkpoint 4

Find the limits.

$$\lim_{x \to 2^-} \frac{x^2 - 4x}{x - 2} \quad \text{and} \quad \lim_{x \to 2^+} \frac{x^2 - 4x}{x - 2}$$

In Example 4, try evaluating $f(x)$ at x-values that are just barely to the left of 1. You will find that you can make the values of $f(x)$ arbitrarily large by choosing x sufficiently close to 1. For instance, $f(0.99999) \approx 199,999$.

TECH TUTOR

When you use a graphing utility to graph a function that has a vertical asymptote, the utility may try to connect separate branches of the graph. For instance, the figure below shows the graph of

$$f(x) = \frac{3}{x - 2}$$

on a graphing calculator.

This line is not part of the graph of the function.

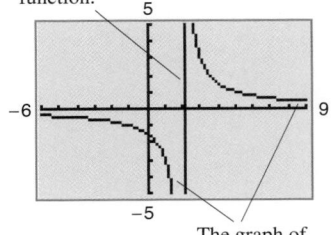

The graph of the function has two branches.

Horizontal Asymptotes and Limits at Infinity

Another type of limit, called a **limit at infinity,** specifies a finite value approached by a function as x increases (or decreases) without bound.

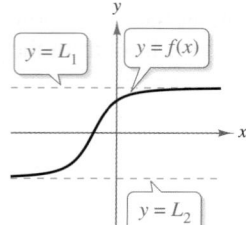

> **Definition of Horizontal Asymptote**
>
> If f is a function and L_1 and L_2 are real numbers, then the statements
>
> $$\lim_{x \to \infty} f(x) = L_1 \quad \text{and} \quad \lim_{x \to -\infty} f(x) = L_2$$
>
> denote **limits at infinity.** The lines $y = L_1$ and $y = L_2$ are **horizontal asymptotes** of the graph of f.

Figure 9.20 shows two ways in which the graph of a function can approach one or more horizontal asymptotes. Note that it is possible for the graph of a function to cross its horizontal asymptote.

Limits at infinity share many of the properties of limits discussed in Section 7.1. When finding horizontal asymptotes, you can use the property that

$$\lim_{x \to \infty} \frac{1}{x^r} = 0, \quad r > 0 \quad \text{and} \quad \lim_{x \to -\infty} \frac{1}{x^r} = 0, \quad r > 0.$$

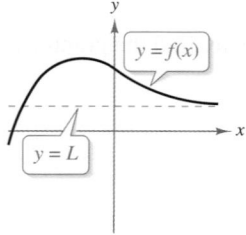

FIGURE 9.20

(The second limit assumes that x^r is defined when $x < 0$.)

Example 5   Finding Limits at Infinity

Find the limit: $\displaystyle \lim_{x \to \infty} \left(5 - \frac{2}{x^2} \right)$.

SOLUTION

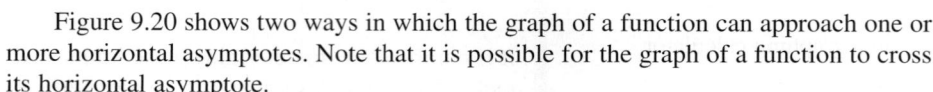

$$\lim_{x \to \infty} \left(5 - \frac{2}{x^2} \right) = \lim_{x \to \infty} 5 - \lim_{x \to \infty} \frac{2}{x^2} \qquad \lim_{x \to \infty} [f(x) - g(x)] = \lim_{x \to \infty} f(x) - \lim_{x \to \infty} g(x)$$

$$= \lim_{x \to \infty} 5 - 2 \left(\lim_{x \to \infty} \frac{1}{x^2} \right) \qquad \lim_{x \to \infty} cf(x) = c \lim_{x \to \infty} f(x)$$

$$= 5 - 2(0)$$

$$= 5$$

You can verify this limit by sketching the graph of

$$f(x) = 5 - \frac{2}{x^2}$$

as shown in Figure 9.21. Note that the graph has $y = 5$ as a horizontal asymptote to the right. By evaluating the limit

$$\lim_{x \to -\infty} \left(5 - \frac{2}{x^2} \right)$$

you can show that $y = 5$ is also a horizontal asymptote to the left.

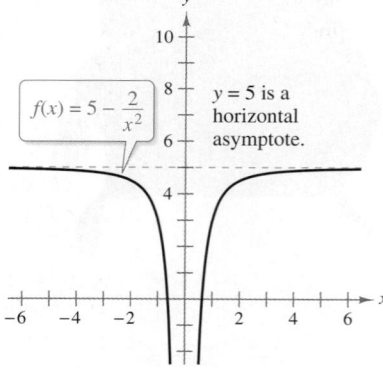

FIGURE 9.21

✓ Checkpoint 5

Find the limit: $\displaystyle \lim_{x \to \infty} \left(2 + \frac{5}{x^2} \right)$.

There is an easy way to determine whether the graph of a *rational* function has a horizontal asymptote. This shortcut is based on a comparison of the degrees of the numerator and denominator of the rational function.

Horizontal Asymptotes of Rational Functions

Let $f(x) = p(x)/q(x)$ be a rational function.

1. If the degree of the numerator is less than the degree of the denominator, then $y = 0$ is a horizontal asymptote of the graph of f (to the left and to the right).

2. If the degree of the numerator is equal to the degree of the denominator, then $y = a/b$ is a horizontal asymptote of the graph of f (to the left and to the right), where a and b are the leading coefficients of $p(x)$ and $q(x)$, respectively.

3. If the degree of the numerator is greater than the degree of the denominator, then the graph of f has no horizontal asymptote.

Example 6   Finding Horizontal Asymptotes

Find the horizontal asymptote of the graph of each function.

a. $y = \dfrac{-2x + 3}{3x^2 + 1}$ **b.** $y = \dfrac{-2x^2 + 3}{3x^2 + 1}$ **c.** $y = \dfrac{-2x^3 + 3}{3x^2 + 1}$

SOLUTION

a. Because the degree of the numerator is less than the degree of the denominator, $y = 0$ is a horizontal asymptote. [See Figure 9.22(a).]

b. Because the degree of the numerator is equal to the degree of the denominator, the line $y = -\frac{2}{3}$ is a horizontal asymptote. [See Figure 9.22(b).]

c. Because the degree of the numerator is greater than the degree of the denominator, the graph has no horizontal asymptote. [See Figure 9.22(c).]

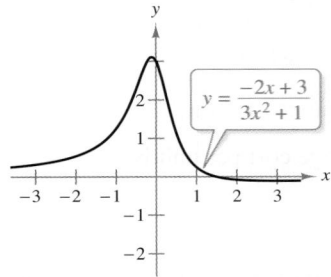

(a) $y = 0$ is a horizontal asymptote.

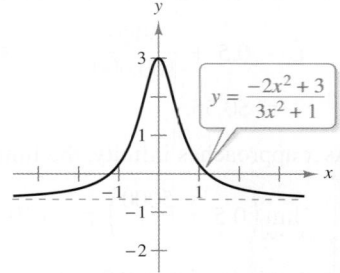

(b) $y = -\frac{2}{3}$ is a horizontal asymptote.

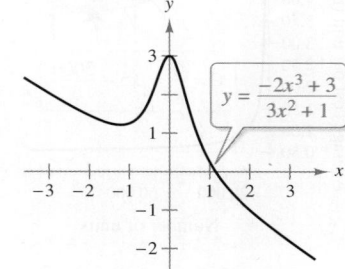

(c) No horizontal asymptote

FIGURE 9.22

✓ **Checkpoint 6**

Find the horizontal asymptote of the graph of each function.

a. $y = \dfrac{2x + 1}{4x^2 + 5}$ **b.** $y = \dfrac{2x^2 + 1}{4x^2 + 5}$ **c.** $y = \dfrac{2x^3 + 1}{4x^2 + 5}$

Some functions have two horizontal asymptotes: one to the right and one to the left (see Exercises 59 and 60).

Applications of Asymptotes

There are many examples of asymptotic behavior in real life. For instance, Example 7 describes the asymptotic behavior of an average cost function.

Example 7   Modeling Average Cost

A small business invests $5000 in a new product. In addition to this initial investment, the product will cost $0.50 per unit to produce.

a. Find the average cost per unit when 1000 units are produced.

b. Find the average cost per unit when 10,000 units are produced.

c. Find the average cost per unit when 100,000 units are produced.

d. What is the limit of the average cost as the number of units produced increases?

SOLUTION   From the given information, you can model the total cost C (in dollars) by

$$C = 0.5x + 5000 \qquad \text{Total cost function}$$

where x is the number of units produced. This implies that the average cost function is

$$\overline{C} = \frac{C}{x} = 0.5 + \frac{5000}{x}. \qquad \text{Average cost function}$$

a. When only 1000 units are produced, the average cost per unit is

$$\overline{C} = 0.5 + \frac{5000}{1000} \qquad \text{Substitute 1000 for } x.$$

$$= \$5.50. \qquad \text{Average cost for 1000 units}$$

b. When 10,000 units are produced, the average cost per unit is

$$\overline{C} = 0.5 + \frac{5000}{10,000} \qquad \text{Substitute 10,000 for } x.$$

$$= \$1.00. \qquad \text{Average cost for 10,000 units}$$

c. When 100,000 units are produced, the average cost per unit is

$$\overline{C} = 0.5 + \frac{5000}{100,000} \qquad \text{Substitute 100,000 for } x.$$

$$= \$0.55. \qquad \text{Average cost for 100,000 units}$$

d. As x approaches infinity, the limiting average cost per unit is

$$\lim_{x \to \infty} \left(0.5 + \frac{5000}{x} \right) = \$0.50.$$

As shown in Figure 9.23, this example points out one of the major problems faced by small businesses. That is, it is difficult to have competitively low prices when the production level is low.

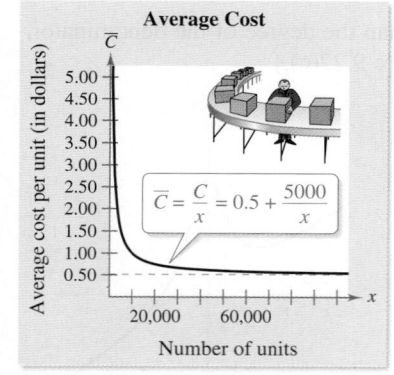

As $x \to \infty$, the average cost per unit approaches $0.50.

FIGURE 9.23

✓ Checkpoint 7

A small business invests $25,000 in a new product. In addition, the product will cost $0.75 per unit to produce. Find the cost function and the average cost function. What is the limit of the average cost function as the number of units produced increases? ■

In Example 7, suppose that the small business had made an initial investment of $50,000. How would this change the answers to the questions? Would it change the average cost of producing x units? Would it change the limiting average cost per unit?

Example 8 Modeling Smokestack Emission

A manufacturing plant has determined that the cost C (in dollars) of removing $p\%$ of the smokestack pollutants of its main smokestack is modeled by

$$C = \frac{80,000p}{100 - p}, \quad 0 \le p < 100.$$

What is the vertical asymptote of this function? What does the vertical asymptote mean to the plant owners?

SOLUTION   The graph of the cost function is shown in Figure 9.24. From the graph, you can see that $p = 100$ is the vertical asymptote. This means that as the plant attempts to remove higher and higher percents of the pollutants, the cost increases dramatically. For instance, the cost of removing 85% of the pollutants is

$$C = \frac{80,000(85)}{100 - 85} \approx \$453,333 \qquad \text{Cost for 85\% removal}$$

but the cost of removing 90% is

$$C = \frac{80,000(90)}{100 - 90} = \$720,000. \qquad \text{Cost for 90\% removal}$$

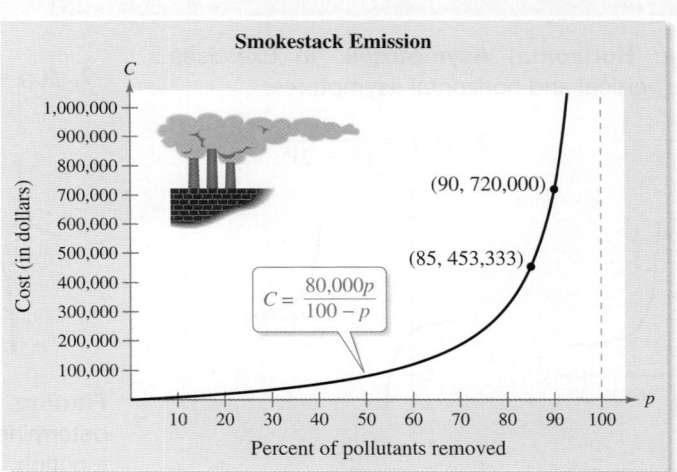

FIGURE 9.24

✓ **Checkpoint 8**

According to the cost function in Example 8, is it possible to remove 100% of the smokestack pollutants? Why or why not?

SUMMARIZE   *(Section 9.3)*

1. State the definition of vertical asymptote *(page 708)*. For examples of vertical asymptotes, see Examples 1, 2, and 3.

2. State the definition of horizontal asymptote *(page 712)*. For examples of horizontal asymptotes, see Example 6.

3. Describe a real-life example of how asymptotic behavior can be used to analyze the average cost for a new product *(page 714, Example 7)*.

SKILLS WARM UP 9.3   The following warm-up exercises involve skills that were covered in earlier sections. You will use these skills in the exercise set for this section. For additional help, review Sections 7.1, 7.5, and 9.2.

In Exercises 1–8, find the limit.

1. $\lim\limits_{x \to 2} (x + 1)$

2. $\lim\limits_{x \to -1} (3x + 4)$

3. $\lim\limits_{x \to -3} \dfrac{2x^2 + x - 15}{x + 3}$

4. $\lim\limits_{x \to 2} \dfrac{3x^2 - 8x + 4}{x - 2}$

5. $\lim\limits_{x \to 2^+} \dfrac{x^2 - 5x + 6}{x^2 - 4}$

6. $\lim\limits_{x \to 1^-} \dfrac{x^2 - 6x + 5}{x^2 - 1}$

7. $\lim\limits_{x \to 0^+} \sqrt{x}$

8. $\lim\limits_{x \to 1^+} \left(x + \sqrt{x - 1}\right)$

In Exercises 9–12, find the average cost and the marginal cost.

9. $C = 150 + 3x$

10. $C = 1900 + 1.7x + 0.002x^2$

11. $C = 0.005x^2 + 0.5x + 1375$

12. $C = 760 + 0.05x$

Exercises 9.3

See www.CalcChat.com for worked-out solutions to odd-numbered exercises.

Vertical and Horizontal Asymptotes In Exercises 1–8, find the vertical and horizontal asymptotes.

1. $f(x) = \dfrac{x^2 + 1}{x^2}$

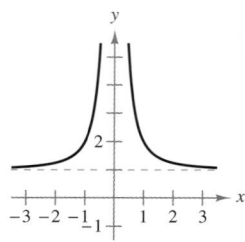

2. $f(x) = \dfrac{4}{(x - 2)^3}$

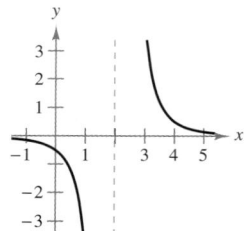

3. $f(x) = \dfrac{x^2 - 2}{x^2 - x - 2}$

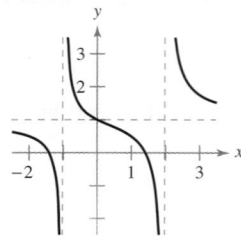

4. $y = \dfrac{x + 1}{x + 2}$

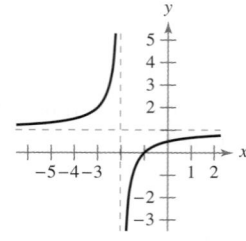

5. $f(x) = \dfrac{3x^2}{2(x^2 + 1)}$

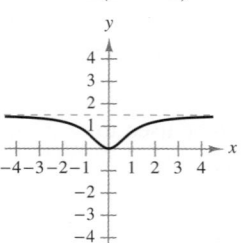

6. $f(x) = \dfrac{-4x}{x^2 + 4}$

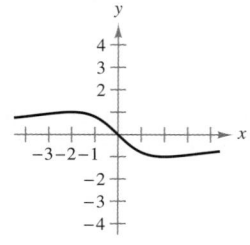

7. $f(x) = \dfrac{x^2 - 1}{2x^2 - 8}$

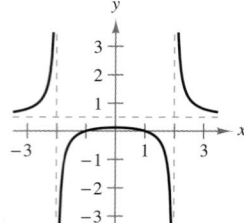

8. $f(x) = \dfrac{x^2 + 1}{x^3 - 8}$

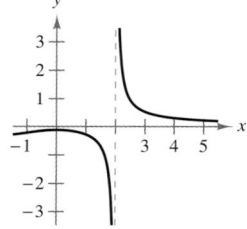

Finding Vertical Asymptotes In Exercises 9–14, determine all vertical asymptotes of the graph of the function. *See Examples 2 and 3.*

9. $f(x) = \dfrac{x - 3}{x^2 + 3x}$

10. $f(x) = \dfrac{x}{x^2 + 6x}$

11. $f(x) = \dfrac{x^2 - 8x + 15}{x^2 - 9}$

12. $f(x) = \dfrac{x^2 + 2x - 35}{x^2 - 25}$

13. $f(x) = \dfrac{2x^2 - x - 3}{2x^2 - 11x + 12}$

14. $f(x) = \dfrac{x^2 + x - 30}{4x^2 - 17x - 15}$

 Determining Infinite Limits In Exercises 15–20, use a graphing utility to find the limit. *See Example 4.*

15. $\lim\limits_{x \to 6^+} \dfrac{1}{(x - 6)^2}$

16. $\lim\limits_{x \to -2^-} \dfrac{1}{x + 2}$

17. $\lim\limits_{x \to 3^+} \dfrac{x - 4}{x - 3}$

18. $\lim\limits_{x \to 1^+} \dfrac{2 + x}{1 - x}$

19. $\lim\limits_{x \to -1^-} \dfrac{x^2 + 1}{x^2 - 1}$

20. $\lim\limits_{x \to 5^+} \dfrac{2x - 3}{x^2 - 25}$

Finding Limits at Infinity   In Exercises 21–24, find the limit. *See Example 5.*

21. $\lim\limits_{x\to\infty}\left(1+\dfrac{1}{x}\right)$

22. $\lim\limits_{x\to-\infty}\left(6-\dfrac{3}{x}\right)$

23. $\lim\limits_{x\to-\infty}\left(7+\dfrac{4}{x^2}\right)$

24. $\lim\limits_{x\to\infty}\left(10-\dfrac{8}{x^2}\right)$

Finding Horizontal Asymptotes   In Exercises 25–32, find the horizontal asymptote of the graph of the function. *See Example 6.*

25. $f(x)=\dfrac{4x-3}{2x+1}$

26. $f(x)=\dfrac{5x^2+1}{10x^3-3x^2+7}$

27. $f(x)=\dfrac{3x}{4x^2-1}$

28. $f(x)=\dfrac{2x^2-5x-12}{1-6x-8x^2}$

29. $f(x)=\dfrac{5x^2}{x+3}$

30. $f(x)=\dfrac{x^3-2x^2+3x+1}{x^2-3x+2}$

31. $f(x)=\dfrac{2x}{x-1}+\dfrac{3x}{x+1}$

32. $f(x)=\dfrac{2x^2}{x-1}+\dfrac{3x}{x+1}$

Using Horizontal Asymptotes   In Exercises 33–36, match the function with its graph. Use horizontal asymptotes as an aid. [The graphs are labeled (a)–(d).]

(a)
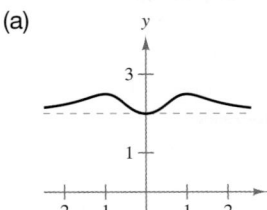

(b)
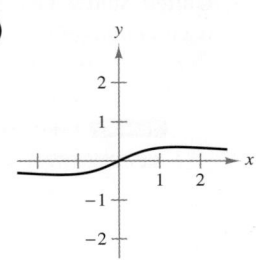

(c)
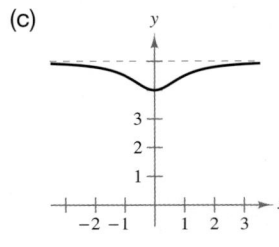

(d)
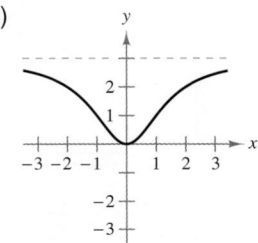

33. $f(x)=\dfrac{3x^2}{x^2+2}$

34. $f(x)=\dfrac{x}{x^2+2}$

35. $f(x)=2+\dfrac{x^2}{x^4+1}$

36. $f(x)=5-\dfrac{1}{x^2+1}$

Finding Limits at Infinity   In Exercises 37 and 38, find $\lim\limits_{x\to\infty} h(x)$, if possible.

37. $f(x)=5x^3-3$

  (a) $h(x)=\dfrac{f(x)}{x^2}$
  (b) $h(x)=\dfrac{f(x)}{x^3}$
  (c) $h(x)=\dfrac{f(x)}{x^4}$

38. $f(x)=3x^2+7$

  (a) $h(x)=\dfrac{f(x)}{x}$
  (b) $h(x)=\dfrac{f(x)}{x^2}$
  (c) $h(x)=\dfrac{f(x)}{x^3}$

Finding Limits at Infinity   In Exercises 39 and 40, find each limit, if possible.

39. (a) $\lim\limits_{x\to\infty}\dfrac{x^2+2}{x^3-1}$
 (b) $\lim\limits_{x\to\infty}\dfrac{x^2+2}{x^2-1}$
 (c) $\lim\limits_{x\to\infty}\dfrac{x^2+2}{x-1}$

40. (a) $\lim\limits_{x\to\infty}\dfrac{4-5x}{2x^3+6}$
 (b) $\lim\limits_{x\to\infty}\dfrac{4-5x}{2x+6}$
 (c) $\lim\limits_{x\to\infty}\dfrac{4-5x^2}{2x+6}$

 Estimating Limits at Infinity   In Exercises 41–44, use a graphing utility or spreadsheet software program to complete the table. Then use the result to estimate the limit of $f(x)$ as x approaches infinity.

x	10^0	10^1	10^2	10^3	10^4	10^5	10^6
$f(x)$							

41. $f(x)=\sqrt{x^3+6}-2x$

42. $f(x)=x-\sqrt{x(x-1)}$

43. $f(x)=\dfrac{x+1}{x\sqrt{x}}$

44. $f(x)=\dfrac{\sqrt{x}}{x^2+3}$

Sketching Graphs   In Exercises 45–60, sketch the graph of the equation. Use intercepts, extrema, and asymptotes as sketching aids.

45. $y=\dfrac{3x}{1-x}$

46. $y=\dfrac{x-3}{x-2}$

47. $f(x)=\dfrac{x^2}{x^2+9}$

48. $f(x)=\dfrac{x}{x^2+4}$

49. $g(x)=\dfrac{x^2}{x^2-16}$

50. $g(x)=\dfrac{x}{x^2-36}$

51. $y=1-\dfrac{3}{x^2}$

52. $y=1+\dfrac{1}{x}$

53. $f(x)=\dfrac{1}{x^2-x-2}$

54. $f(x)=\dfrac{x-2}{x^2-4x+3}$

55. $g(x)=\dfrac{x^2-x-2}{x-2}$

56. $g(x)=\dfrac{x^2-9}{x+3}$

57. $y=\dfrac{2x^2-6}{(x-1)^2}$

58. $y=\dfrac{x}{(x+1)^2}$

59. $y=\dfrac{x}{\sqrt{x^2+1}}$

60. $y=\dfrac{2x}{\sqrt{x^2+4}}$

61. Average Cost   The cost C (in dollars) of producing x units of a product is $C = 1.15x + 6000$.

(a) Find the average cost function $\overline{C}$.

(b) Find $\overline{C}$ when $x = 600$ and when $x = 6000$.

(c) Determine the limit of the average cost function as x approaches infinity. Interpret the limit in the context of the problem.

62. Average Cost   The cost C (in dollars) for a company to recycle x tons of material is $C = 1.25x + 10,500$.

(a) Find the average cost function $\overline{C}$.

(b) Find $\overline{C}$ when $x = 100$ and when $x = 1000$.

(c) Determine the limit of the average cost function as x approaches infinity. Interpret the limit in the context of the problem.

63. Average Profit   The cost C and revenue R functions (in dollars) for producing and selling x units of a product are $C = 34.5x + 15,000$ and $R = 69.9x$.

(a) Find the average profit function

$$\overline{P} = \frac{R - C}{x}.$$

(b) Find the average profits when x is 1000, 10,000, and 100,000.

(c) What is the limit of the average profit function as x approaches infinity? Explain your reasoning.

64. HOW DO YOU SEE IT?   The graph shows the temperature T (in degrees Fahrenheit) of an apple pie t seconds after it is removed from an oven.

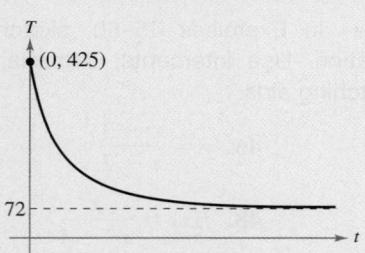

(a) Find $\lim\limits_{t \to 0^+} T$. What does this limit represent?

(b) Find $\lim\limits_{t \to \infty} T$. What does this limit represent?

65. Seizing Drugs   The cost C (in millions of dollars) for the federal government to seize $p\%$ of an illegal drug as it enters the country is modeled by

$$C = \frac{528p}{100 - p}, \quad 0 \le p < 100.$$

(a) Find the costs of seizing 25%, 50%, and 75%.

(b) Find the limit of C as $p \to 100^-$. Interpret the limit in the context of the problem. Use a graphing utility to verify your result.

66. Removing Pollutants   The cost C (in dollars) of removing $p\%$ of the air pollutants in the stack emission of a utility company that burns coal is modeled by

$$C = \frac{85,000p}{100 - p}, \quad 0 \le p < 100.$$

(a) Find the costs of removing 15%, 50%, and 95%.

(b) Find the limit of C as $p \to 100^-$. Interpret the limit in the context of the problem. Use a graphing utility to verify your result.

67. Learning Curve   Psychologists have developed mathematical models to predict performance P (the percent of correct responses in decimal form) as a function of n, the number of times a task is performed. One such model is

$$P = \frac{0.5 + 0.9(n - 1)}{1 + 0.9(n - 1)}, \quad n > 0.$$

(a) Use a spreadsheet software program to complete the table for the model.

n	1	2	3	4	5	6	7	8	9	10
P										

(b) Find the limit as n approaches infinity.

(c) Use a graphing utility to graph this learning curve, and interpret the graph in the context of the problem.

68. Project: Alternative-Fueled Vehicles
For a project analyzing the number of alternative-fueled vehicles in use in the United States, visit this text's website at *www.cengagebrain.com.* *(Source: U.S. Energy Information Administration)*

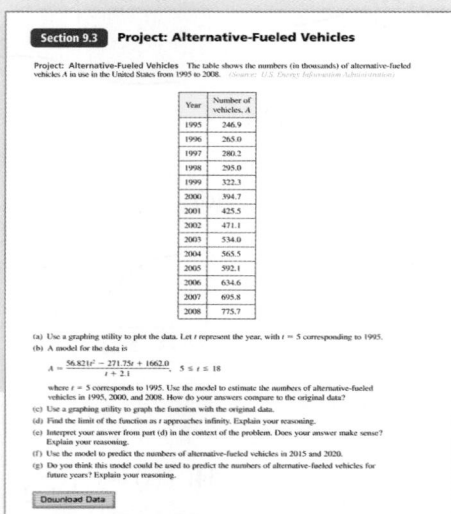

QUIZ YOURSELF

See www.CalcChat.com for worked-out solutions to odd-numbered exercises.

Take this quiz as you would take a quiz in class. When you are done, check your work against the answers given in the back of the book.

1. A gardener has 200 feet of fencing to enclose a rectangular garden adjacent to a river (see figure). No fencing is needed along the river.

   (a) What dimensions should be used so that the area of the garden will be a maximum?

   (b) Find the maximum area.

2. A rectangular page is to contain 48 square inches of print. The margins at the top and bottom of the page are to be 1 inch wide. The margins on each side are to be $\frac{3}{4}$ inch wide. Find the dimensions of the page that will minimize the amount of paper used.

Figure for 1

In Exercises 3 and 4, find the number of units x that produces the minimum average cost per unit $\overline{C}$.

3. $C = 0.06x^2 + 12x + 9600$

4. $C = 0.003x^3 + 8x + 2058$

In Exercises 5 and 6, find the price that will maximize profit for the cost and demand functions, where p is the price, x is the number of units, and C is the cost.

Cost Function	Demand Function
5. $C = 200 + 26x$	$p = 100 - x$
6. $C = 0.4x + 300$	$p = \dfrac{48}{\sqrt{x}}$

 In Exercises 7 and 8, (a) find the price elasticity of demand for the demand function at the indicated x-value, (b) determine whether the demand is elastic, inelastic, or of unit elasticity at the indicated x-value, (c) use a graphing utility to graph the revenue function, and (d) identify the intervals of elasticity and inelasticity.

Demand Function	Quantity Demanded
7. $p = 500 - 4x$	$x = 250$
8. $p = 15 - \sqrt{x}$	$x = 900$

In Exercises 9–14, find the limit.

9. $\displaystyle\lim_{x \to 5^+} \frac{5 - x}{x - 5}$

10. $\displaystyle\lim_{x \to 2^-} \frac{x^2}{x^2 + 2x - 8}$

11. $\displaystyle\lim_{x \to 0^+} \frac{x}{x^2 + 0.1x}$

12. $\displaystyle\lim_{x \to \infty} \frac{x}{3x + 2}$

13. $\displaystyle\lim_{x \to -\infty} \left(\frac{3}{x^2} - \frac{2}{x} - 1 \right)$

14. $\displaystyle\lim_{x \to \infty} \frac{x^2 - 9}{x + 3}$

In Exercises 15–17, find any vertical and horizontal asymptotes of the graph of the function.

15. $f(x) = \dfrac{2x + 1}{x - 1}$

16. $f(x) = \dfrac{3}{x^2 - 2x}$

17. $f(x) = \dfrac{x^2 - 4}{x - 3}$

9.4 Curve Sketching: A Summary

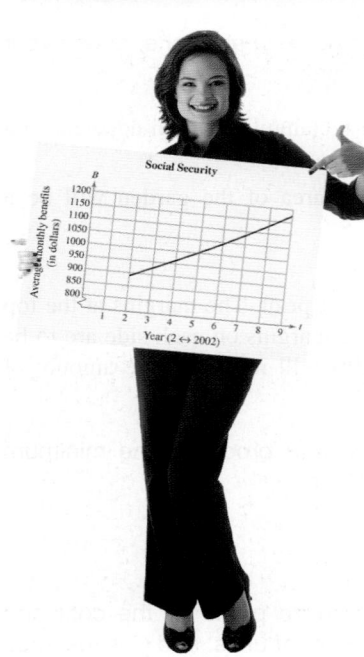

In Exercise 45 on page 728, you will analyze the graph of the Social Security average monthly benefits to determine whether the model is a good fit for the data.

■ Analyze the graphs of functions.

■ Recognize the graphs of simple polynomial functions.

Summary of Curve-Sketching Techniques

It would be difficult to overstate the importance of using graphs in mathematics. Descartes's introduction of analytic geometry contributed significantly to the rapid advances in calculus that began during the mid-seventeenth century.

So far, you have studied several concepts that are useful in analyzing the graph of a function.

• x-intercepts and y-intercepts	(Section 2.1)
• Domain and range	(Section 2.4)
• Continuity	(Section 7.2)
• Differentiability	(Section 7.3)
• Relative extrema	(Section 8.5)
• Concavity	(Section 8.6)
• Points of inflection	(Section 8.6)
• Vertical asymptotes	(Section 9.3)
• Horizontal asymptotes	(Section 9.3)

When you are sketching the graph of a function, either by hand or with a graphing utility, remember that you cannot normally show the *entire* graph. The decision as to which part of the graph to show is crucial. For instance, which of the viewing windows in Figure 9.25 better represents the graph of

$$f(x) = x^3 - 25x^2 + 74x - 20?$$

Figure 9.25(a) gives a more complete view of the graph, but the context of the problem might indicate that Figure 9.25(b) is better.

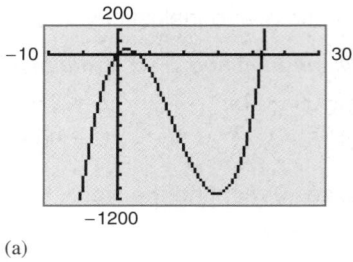

(a)

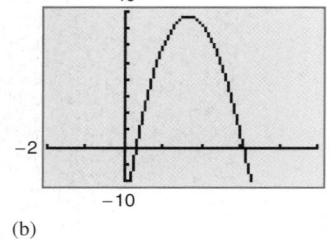

(b)

FIGURE 9.25

Guidelines for Analyzing the Graph of a Function

1. Determine the domain and range of the function. When the function models a real-life situation, consider the context.

2. Determine the intercepts and asymptotes of the graph.

3. Locate the x-values at which $f'(x)$ and $f''(x)$ are zero or undefined. Use the results to determine where the relative extrema and the points of inflection occur.

Example 1   Analyzing a Graph

Analyze the graph of

$$f(x) = x^3 + 3x^2 - 9x + 5.$$

SOLUTION   The y-intercept occurs at $(0, 5)$. Because this function factors as

$$f(x) = (x - 1)^2(x + 5) \qquad \text{Factored form}$$

the x-intercepts occur at $(-5, 0)$ and $(1, 0)$. The first derivative is

$$\begin{aligned} f'(x) &= 3x^2 + 6x - 9 \qquad \text{First derivative} \\ &= 3(x - 1)(x + 3). \qquad \text{Factored form} \end{aligned}$$

So, the critical numbers of f are $x = 1$ and $x = -3$. The second derivative of f is

$$\begin{aligned} f''(x) &= 6x + 6 \qquad \text{Second derivative} \\ &= 6(x + 1) \qquad \text{Factored form} \end{aligned}$$

which implies that the second derivative is zero when $x = -1$. By testing the values of $f'(x)$ and $f''(x)$, as shown in the table, you can see that f has one relative minimum, one relative maximum, and one point of inflection. The graph of f is shown in Figure 9.26.

	$f(x)$	$f'(x)$	$f''(x)$	Characteristics of graph
x in $(-\infty, -3)$		$+$	$-$	Increasing, concave downward
$x = -3$	32	0	$-$	Relative maximum
x in $(-3, -1)$		$-$	$-$	Decreasing, concave downward
$x = -1$	16	$-$	0	Point of inflection
x in $(-1, 1)$		$-$	$+$	Decreasing, concave upward
$x = 1$	0	0	$+$	Relative minimum
x in $(1, \infty)$		$+$	$+$	Increasing, concave upward

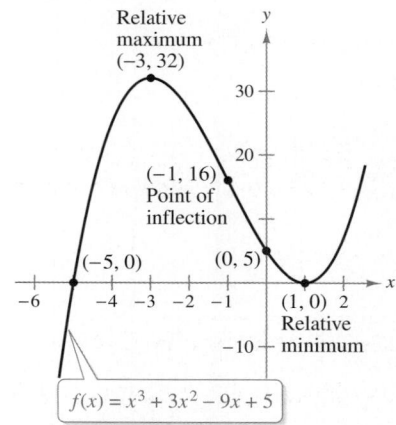

Relative maximum $(-3, 32)$

$(-1, 16)$ Point of inflection

$(-5, 0)$ $(0, 5)$

$(1, 0)$ Relative minimum

$$f(x) = x^3 + 3x^2 - 9x + 5$$

FIGURE 9.26

✓ Checkpoint 1

Analyze the graph of

$$f(x) = -x^3 + 3x^2 + 9x - 27.$$

Example 2   **Analyzing a Graph**

Analyze the graph of

$$f(x) = x^4 - 12x^3 + 48x^2 - 64x.$$

SOLUTION   One of the intercepts occurs at $(0, 0)$. Because this function factors as

$$f(x) = x(x^3 - 12x^2 + 48x - 64)$$
$$= x(x - 4)^3 \qquad \text{Factored form}$$

a second x-intercept occurs at $(4, 0)$. The first derivative is

$$f'(x) = 4x^3 - 36x^2 + 96x - 64 \qquad \text{First derivative}$$
$$= 4(x - 1)(x - 4)^2. \qquad \text{Factored form}$$

So, the critical numbers of f are $x = 1$ and $x = 4$. The second derivative of f is

$$f''(x) = 12x^2 - 72x + 96 \qquad \text{Second derivative}$$
$$= 12(x - 4)(x - 2) \qquad \text{Factored form}$$

which implies that the second derivative is zero when $x = 2$ and $x = 4$. By testing the values of $f'(x)$ and $f''(x)$, as shown in the table, you can see that f has one relative minimum and two points of inflection. The graph is shown in Figure 9.27.

	$f(x)$	$f'(x)$	$f''(x)$	Characteristics of graph
x in $(-\infty, 1)$		$-$	$+$	Decreasing, concave upward
$x = 1$	-27	0	$+$	Relative minimum
x in $(1, 2)$		$+$	$+$	Increasing, concave upward
$x = 2$	-16	$+$	0	Point of inflection
x in $(2, 4)$		$+$	$-$	Increasing, concave downward
$x = 4$	0	0	0	Point of inflection
x in $(4, \infty)$		$+$	$+$	Increasing, concave upward

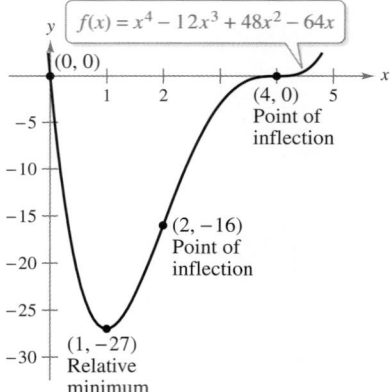

FIGURE 9.27

✓ **Checkpoint 2**

Analyze the graph of

$$f(x) = x^4 - 4x^3 + 5.$$

The fourth-degree polynomial function in Example 2 has one relative minimum and no relative maxima. In general, a polynomial function of degree n can have *at most* $n - 1$ relative extrema, and *at most* $n - 2$ points of inflection. Moreover, polynomial functions of even degree must have at least one relative extremum.

Example 3 Analyzing a Graph

Analyze the graph of

$$f(x) = \frac{x^2 - 2x + 4}{x - 2}.$$

SOLUTION The y-intercept occurs at $(0, -2)$. Using the Quadratic Formula on the numerator, you can see that there are no x-intercepts. Because the denominator is zero when $x = 2$ (and the numerator is not zero when $x = 2$), it follows that $x = 2$ is a vertical asymptote of the graph. There are no horizontal asymptotes because the degree of the numerator is greater than the degree of the denominator. The first derivative is

$$f'(x) = \frac{(x - 2)(2x - 2) - (x^2 - 2x + 4)}{(x - 2)^2}$$ First derivative

$$= \frac{x(x - 4)}{(x - 2)^2}.$$ Factored form

So, the critical numbers of f are $x = 0$ and $x = 4$. The second derivative is

$$f''(x) = \frac{(x - 2)^2(2x - 4) - (x^2 - 4x)(2)(x - 2)}{(x - 2)^4}$$ Second derivative

$$= \frac{(x - 2)(2x^2 - 8x + 8 - 2x^2 + 8x)}{(x - 2)^4}$$

$$= \frac{8}{(x - 2)^3}.$$ Factored form

Because the second derivative has no zeros and because $x = 2$ is not in the domain of the function, you can conclude that the graph has no points of inflection. By testing the values of $f'(x)$ and $f''(x)$, as shown in the table, you can see that f has one relative minimum and one relative maximum. The graph of f is shown in Figure 9.28.

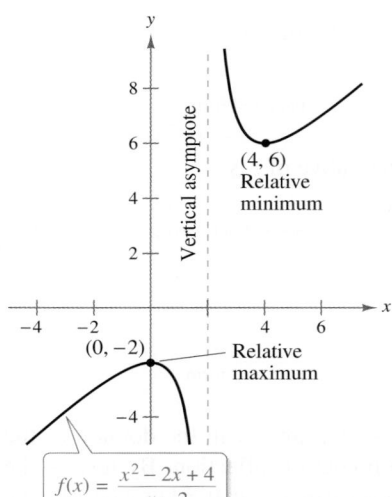

$f(x) = \dfrac{x^2 - 2x + 4}{x - 2}$

FIGURE 9.28

	$f(x)$	$f'(x)$	$f''(x)$	Characteristics of graph
x in $(-\infty, 0)$		$+$	$-$	Increasing, concave downward
$x = 0$	-2	0	$-$	Relative maximum
x in $(0, 2)$		$-$	$-$	Decreasing, concave downward
$x = 2$	Undef.	Undef.	Undef.	Vertical asymptote
x in $(2, 4)$		$-$	$+$	Decreasing, concave upward
$x = 4$	6	0	$+$	Relative minimum
x in $(4, \infty)$		$+$	$+$	Increasing, concave upward

✓ Checkpoint 3

Analyze the graph of

$$f(x) = \frac{x^2}{x - 1}.$$

Example 4   Analyzing a Graph

Analyze the graph of

$$f(x) = \frac{2(x^2 - 9)}{x^2 - 4}.$$

SOLUTION   Begin by writing the function in factored form.

$$f(x) = \frac{2(x - 3)(x + 3)}{(x - 2)(x + 2)} \qquad \text{Factored form}$$

The y-intercept is $\left(0, \frac{9}{2}\right)$, and the x-intercepts are $(-3, 0)$ and $(3, 0)$. The graph of f has vertical asymptotes at $x = \pm 2$ and a horizontal asymptote at $y = 2$. The first derivative is

$$f'(x) = \frac{2[(x^2 - 4)(2x) - (x^2 - 9)(2x)]}{(x^2 - 4)^2} \qquad \text{First derivative}$$

$$= \frac{2(2x^3 - 8x - 2x^3 + 18x)}{(x^2 - 4)^2} \qquad \text{Multiply.}$$

$$= \frac{20x}{(x^2 - 4)^2}. \qquad \text{Factored form}$$

So, the critical number of f is $x = 0$. The second derivative of f is

$$f''(x) = \frac{(x^2 - 4)^2(20) - (20x)(2)(x^2 - 4)(2x)}{(x^2 - 4)^4} \qquad \text{Second derivative}$$

$$= \frac{20(x^2 - 4)(x^2 - 4 - 4x^2)}{(x^2 - 4)^4}$$

$$= -\frac{20(3x^2 + 4)}{(x^2 - 4)^3}. \qquad \text{Factored form}$$

Because the second derivative has no zeros and $x = \pm 2$ are not in the domain of the function, you can conclude that the graph has no points of inflection. By testing the values of $f'(x)$ and $f''(x)$, as shown in the table, you can see that f has one relative minimum. The graph of f is shown in Figure 9.29.

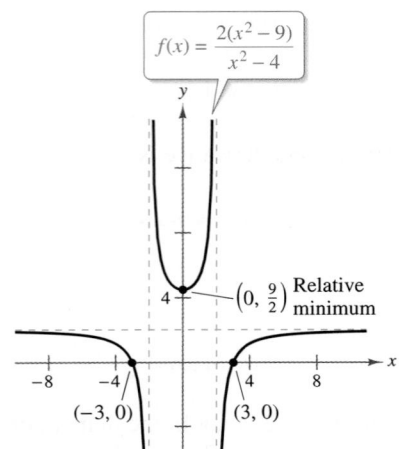

$f(x) = \dfrac{2(x^2 - 9)}{x^2 - 4}$

$\left(0, \frac{9}{2}\right)$ Relative minimum

$(-3, 0)$ $(3, 0)$

FIGURE 9.29

	$f(x)$	$f'(x)$	$f''(x)$	Characteristics of graph
x in $(-\infty, -2)$		$-$	$-$	Decreasing, concave downward
$x = -2$	Undef.	Undef.	Undef.	Vertical asymptote
x in $(-2, 0)$		$-$	$+$	Decreasing, concave upward
$x = 0$	$\frac{9}{2}$	0	$+$	Relative minimum
x in $(0, 2)$		$+$	$+$	Increasing, concave upward
$x = 2$	Undef.	Undef.	Undef.	Vertical asymptote
x in $(2, \infty)$		$+$	$-$	Increasing, concave downward

✓ Checkpoint 4

Analyze the graph of

$$f(x) = \frac{x^2 + 1}{x^2 - 1}.$$

TECH TUTOR

Some graphing utilities will not graph the function in Example 5 properly when the function is entered as

$$f(x) = 2x\wedge(5/3) - 5x\wedge(4/3).$$

To correct for this, you can enter the function as

$$f(x) = 2(\sqrt[3]{x})\wedge 5 - 5(\sqrt[3]{x})\wedge 4.$$

Try entering both functions into a graphing utility to see whether both functions produce correct graphs.

ALGEBRA TUTOR
xy

For help on the algebra in Example 5, see Example 2(a) in the *Chapter 9 Algebra Tutor*, on page 737.

| Example 5 | **Analyzing a Graph** |

Analyze the graph of

$$f(x) = 2x^{5/3} - 5x^{4/3}.$$

SOLUTION   Begin by writing the function in factored form.

$$f(x) = x^{4/3}(2x^{1/3} - 5) \qquad \text{Factored form}$$

One of the intercepts is $(0, 0)$. A second x-intercept occurs when $2x^{1/3} - 5 = 0$.

$$2x^{1/3} - 5 = 0$$
$$2x^{1/3} = 5$$
$$x^{1/3} = \frac{5}{2}$$
$$x = \left(\frac{5}{2}\right)^3$$
$$x = \frac{125}{8}$$

The first derivative is

$$f'(x) = \frac{10}{3}x^{2/3} - \frac{20}{3}x^{1/3} \qquad \text{First derivative}$$
$$= \frac{10}{3}x^{1/3}(x^{1/3} - 2). \qquad \text{Factored form}$$

So, the critical numbers of f are $x = 0$ and $x = 8$. The second derivative is

$$f''(x) = \frac{20}{9}x^{-1/3} - \frac{20}{9}x^{-2/3} \qquad \text{Second derivative}$$
$$= \frac{20}{9}x^{-2/3}(x^{1/3} - 1)$$
$$= \frac{20(x^{1/3} - 1)}{9x^{2/3}}. \qquad \text{Factored form}$$

So, possible points of inflection occur at $x = 1$ and when $x = 0$. By testing the values of $f'(x)$ and $f''(x)$, as shown in the table, you can see that f has one relative maximum, one relative minimum, and one point of inflection. The graph of f is shown in Figure 9.30.

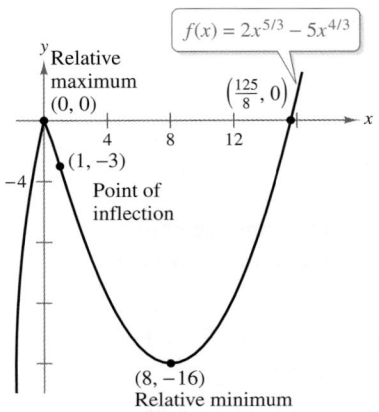

$$f(x) = 2x^{5/3} - 5x^{4/3}$$

FIGURE 9.30

	$f(x)$	$f'(x)$	$f''(x)$	Characteristics of graph
x in $(-\infty, 0)$		$+$	$-$	Increasing, concave downward
$x = 0$	0	0	Undef.	Relative maximum
x in $(0, 1)$		$-$	$-$	Decreasing, concave downward
$x = 1$	-3	$-$	0	Point of inflection
x in $(1, 8)$		$-$	$+$	Decreasing, concave upward
$x = 8$	-16	0	$+$	Relative minimum
x in $(8, \infty)$		$+$	$+$	Increasing, concave upward

✓**Checkpoint 5**

Analyze the graph of $f(x) = 2x^{3/2} - 6x^{1/2}$.

Summary of Simple Polynomial Graphs

A summary of the graphs of polynomial functions of degrees 0, 1, 2, and 3 is shown in Figure 9.31. Because of their simplicity, lower-degree polynomial functions are commonly used as mathematical models.

Constant function (degree 0): $y = a$

Horizontal line

Linear function (degree 1): $y = ax + b$

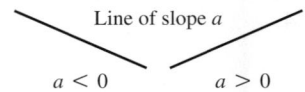

Line of slope a

$a < 0$ $a > 0$

Quadratic function (degree 2): $y = ax^2 + bx + c$

Parabola

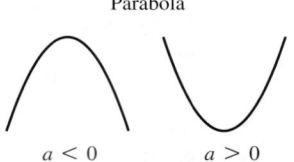

$a < 0$ $a > 0$

Cubic function (degree 3): $y = ax^3 + bx^2 + cx + d$

Cubic curve

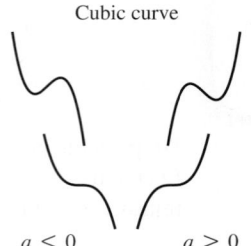

$a < 0$ $a > 0$

FIGURE 9.31

SUMMARIZE   (Section 9.4)

1. List the concepts you have learned that are useful in analyzing the graph of a function *(page 720)*. For an example that uses some of these concepts to analyze the graph of a function, see Example 1.

2. State the guidelines for analyzing the graph of a function *(page 720)*. For examples that use these guidelines, see Examples 3, 4, and 5.

3. State a general rule relating the degree *n* of a polynomial function with (a) the number of relative extrema and (b) the number of points of inflection *(page 723)*. For an example where this rule can be used to analyze the graph of a polynomial function, see Example 2.

SKILLS WARM UP 9.4   The following warm-up exercises involve skills that were covered in earlier sections. You will use these skills in the exercise set for this section. For additional help, review Sections 8.4 and 9.3.

In Exercises 1–4, find the vertical and horizontal asymptotes of the graph.

1. $f(x) = \dfrac{1}{x^2}$

2. $f(x) = \dfrac{8}{(x-2)^2}$

3. $f(x) = \dfrac{40x}{x+3}$

4. $f(x) = \dfrac{x^2 - 3}{x^2 - 4x + 3}$

In Exercises 5–10, determine the open intervals on which the function is increasing or decreasing.

5. $f(x) = x^2 + 4x + 2$

6. $f(x) = -x^2 - 8x + 1$

7. $f(x) = x^3 - 3x + 1$

8. $f(x) = \dfrac{-x^3 + x^2 - 1}{x^2}$

9. $f(x) = \dfrac{x-2}{x-1}$

10. $f(x) = -x^3 - 4x^2 + 3x + 2$

Exercises 9.4

See www.CalcChat.com for worked-out solutions to odd-numbered exercises.

Analyzing a Graph In Exercises 1–22, analyze and sketch the graph of the function. Label any intercepts, relative extrema, points of inflection, and asymptotes. *See Examples 1, 2, 3, 4, and 5.*

1. $y = -x^2 - 2x + 3$

2. $y = 2x^2 - 4x + 1$

3. $y = x^3 - 4x^2 + 6$

4. $y = -x^3 + x - 2$

5. $y = 2 - x - x^3$

6. $y = x^3 + 3x^2 + 3x + 2$

7. $y = 3x^4 + 4x^3$

8. $y = x^4 - 2x^2$

9. $y = x^4 - 8x^3 + 18x^2 - 16x + 5$

10. $y = x^4 - 4x^3 + 16x - 16$

11. $y = \dfrac{x^2 + 1}{x}$

12. $y = \dfrac{x+2}{x}$

13. $y = \dfrac{x^2 - 6x + 12}{x - 4}$

14. $y = \dfrac{x^2 + 4x + 7}{x + 3}$

15. $y = \dfrac{x^2 + 1}{x^2 - 9}$

16. $y = \dfrac{2x}{x^2 - 1}$

17. $y = 3x^{2/3} - x^2$

18. $y = x^{5/3} - 5x^{2/3}$

19. $y = x\sqrt{9 - x}$

20. $y = x\sqrt{4 - x^2}$

21. $y = \begin{cases} x^2 + 1, & x \le 0 \\ 1 - 2x, & x > 0 \end{cases}$

22. $y = \begin{cases} x^2 + 4, & x < 0 \\ 4 - x, & x \ge 0 \end{cases}$

 **Graphing a Function** In Exercises 23–36, use a graphing utility to graph the function. Choose a window that allows all relative extrema and points of inflection to be identified on the graph.

23. $y = 3x^3 - 9x + 1$

24. $y = -4x^3 + 6x^2$

25. $y = x^5 - 5x$

26. $y = (x - 1)^5$

27. $y = \dfrac{5 - 3x}{x - 2}$

28. $y = \dfrac{x}{x^2 + 1}$

29. $y = 1 - x^{2/3}$

30. $y = (1 - x)^{2/3}$

31. $y = x^{4/3}$

32. $y = x^{-1/3}$

33. $y = \dfrac{x}{\sqrt{x^2 - 4}}$

34. $y = \dfrac{x - 3}{x}$

35. $y = \dfrac{x^3}{x^3 - 1}$

36. $y = \dfrac{x^4}{x^4 - 1}$

Interpreting a Graph In Exercises 37–40, use the graph of f' or f'' to sketch the graph of f. (There are many correct answers.)

37.
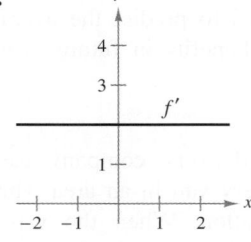

38.
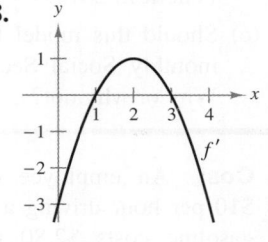

39.
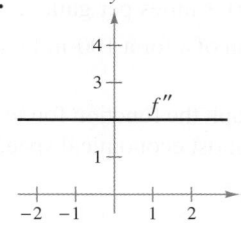

40.
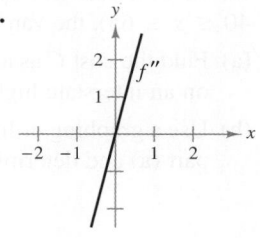

Sketching a Function In Exercises 41 and 42, sketch a graph of a function f having the given characteristics. (There are many correct answers.)

41. $f(-2) = f(0) = 0$

$f'(x) > 0$ if $x < -1$

$f'(x) < 0$ if $-1 < x < 0$

$f'(x) > 0$ if $x > 0$

$f'(-1) = f'(0) = 0$

42. $f(-1) = f(3) = 0$

$f'(1)$ is undefined.

$f'(x) < 0$ if $x < 1$

$f'(x) > 0$ if $x > 1$

$f''(x) < 0, x \neq 1$

$\lim\limits_{x \to \infty} f(x) = 4$

Creating a Function In Exercises 43 and 44, create a function whose graph has the given characteristics. (There are many correct answers.)

43. Vertical asymptote: $x = 5$

Horizontal asymptote: $y = 0$

44. Vertical asymptote: $x = -3$

Horizontal asymptote: None

 45. Social Security The table lists the average monthly Social Security benefits B (in dollars) for retired workers aged 62 and over from 2002 through 2009. A model for the data is

$$B = \frac{815.6 + 110.96t}{1 + 0.09t - 0.0033t^2}, \quad 2 \leq t \leq 9$$

where $t = 2$ corresponds to 2002. *(Source: U.S. Social Security Administration)*

t	2	3	4	5	6	7	8	9
B	895	922	955	1002	1044	1079	1153	1164

(a) Use a graphing utility to create a scatter plot of the data and graph the model in the same viewing window. How well does the model fit the data?

(b) Use the model to predict the average monthly benefit in 2014.

(c) Should this model be used to predict the average monthly Social Security benefits in future years? Why or why not?

46. Cost 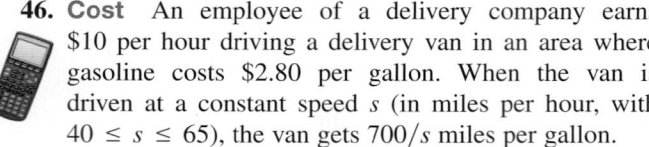 An employee of a delivery company earns $10 per hour driving a delivery van in an area where gasoline costs $2.80 per gallon. When the van is driven at a constant speed s (in miles per hour, with $40 \leq s \leq 65$), the van gets $700/s$ miles per gallon.

(a) Find the cost C as a function of s for a 100-mile trip on an interstate highway.

(b) Use a graphing utility to graph the function found in part (a) and determine the most economical speed.

47. Meteorology  The monthly average high temperature T (in degrees Fahrenheit) for Boston, Massachusetts can be modeled by

$$T = \frac{30.83 - 2.861t + 0.181t^2}{1 - 0.206t + 0.0139t^2}, \quad 1 \leq t \leq 12$$

where t is the month, with $t = 1$ corresponding to January. Use a graphing utility to graph the model and find all absolute extrema. Interpret the meaning of these values in the context of the problem. *(Source: National Climatic Data Center)*

 **48. HOW DO YOU SEE IT?** The graph shows a company's profits P for the years 1990 through 2010, where t is the year, with $t = 0$ corresponding to 2000.

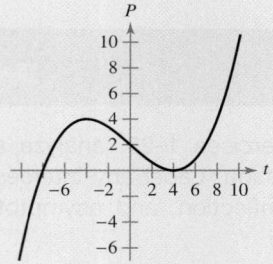

(a) For which values of t is P' zero? positive? negative? Interpret the meanings of these values in the context of the problem.

(b) For which values of t is P'' zero? positive? negative? Interpret the meanings of these values in the context of the problem.

Writing In Exercises 49 and 50, use a graphing utility to graph the function. Explain why there is no vertical asymptote when a superficial examination of the function may indicate that there should be one.

49. $h(x) = \dfrac{6 - 2x}{3 - x}$

50. $g(x) = \dfrac{x^2 + x - 2}{x - 1}$

51. Discovery Consider the function

$$f(x) = \frac{x^2 - 2x + 4}{x - 2}.$$

(a) Show that $f(x)$ can be rewritten as

$$f(x) = x + \frac{4}{x - 2}.$$

(b) Use a graphing utility to graph f and the line $y = x$. How do the two graphs compare as you zoom out?

(c) Use the results of part (b) to describe what is meant by a "slant asymptote."

9.5 Differentials and Marginal Analysis

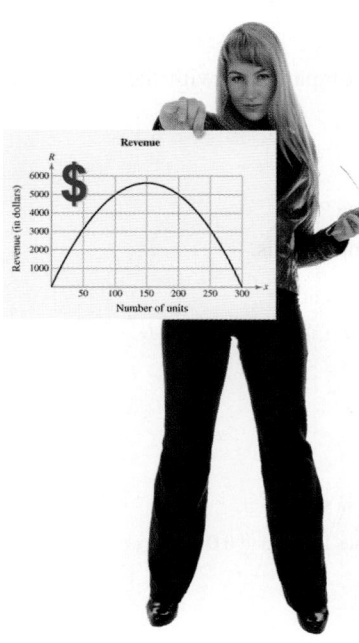

In Exercise 35 on page 735, you will use differentials to approximate the change in revenue for a one-unit increase in sales of a product.

■ Find the differentials of functions.
■ Use differentials in economics to approximate changes in revenue, cost, and profit.
■ Find the differential of a function using differentiation formulas.

Differentials

When the derivative was defined in Section 7.3 as the limit of the ratio $\Delta y/\Delta x$, it seemed natural to retain the quotient symbolism for the limit itself. So, the derivative of y with respect to x was denoted by

$$\frac{dy}{dx} = \lim_{\Delta x \to 0} \frac{\Delta y}{\Delta x}$$

even though dy/dx was not interpreted as the quotient of two separate quantities. In this section, you will see that the quantities dy and dx can be assigned meanings in such a way that their quotient, when $dx \neq 0$, is equal to the derivative of y with respect to x.

Definition of Differentials

Let $y = f(x)$ represent a differentiable function. The **differential of x** (denoted by dx) is any nonzero real number. The **differential of y** (denoted by dy) is $dy = f'(x)\,dx$.

In the definition of differentials, dx can have any nonzero value. In most applications, however, dx is chosen to be small, and this choice is denoted by $dx = \Delta x$.

One use of differentials is in approximating the change in $f(x)$ that corresponds to a change in x, as shown in Figure 9.32. This change is denoted by

$$\Delta y = f(x + \Delta x) - f(x). \qquad \text{Change in } y$$

In Figure 9.32, notice that as Δx gets smaller and smaller, the values of dy and Δy get closer and closer. That is, when Δx is small, $dy \approx \Delta y$. This **tangent line approximation** is the basis for most applications of differentials.

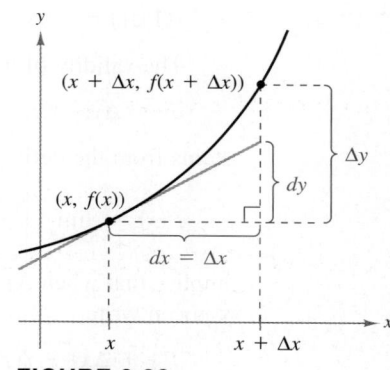

FIGURE 9.32

Note in Figure 9.32 that near the point of tangency, the graph of f is very close to the tangent line. This is the essence of the approximations used in this section. In other words, near the point of tangency, $dy \approx \Delta y$.

Example 1   Comparing Δy and dy

Consider the function given by

$$f(x) = x^2.$$

Find the value of dy when $x = 1$ and $dx = 0.01$. Compare this with the value of Δy when $x = 1$ and $\Delta x = 0.01$.

SOLUTION   Begin by finding the derivative of f.

$$f'(x) = 2x \qquad\qquad \text{Derivative of } f$$

When $x = 1$ and $dx = 0.01$, the value of the differential dy is

$dy = f'(x)\,dx$	Differential of y
$\quad = f'(1)(0.01)$	Substitute 1 for x and 0.01 for dx.
$\quad = 2(1)(0.01)$	Use $f'(x) = 2x$.
$\quad = 0.02.$	Simplify.

When $x = 1$ and $\Delta x = 0.01$, the value of Δy is

$\Delta y = f(x + \Delta x) - f(x)$	Change in y
$\quad = f(1.01) - f(1)$	Substitute 1 for x and 0.01 for Δx.
$\quad = (1.01)^2 - (1)^2$	
$\quad = 1.0201 - 1$	
$\quad = 0.0201.$	Simplify.

Note that $dy \approx \Delta y$, as shown in Figure 9.33.

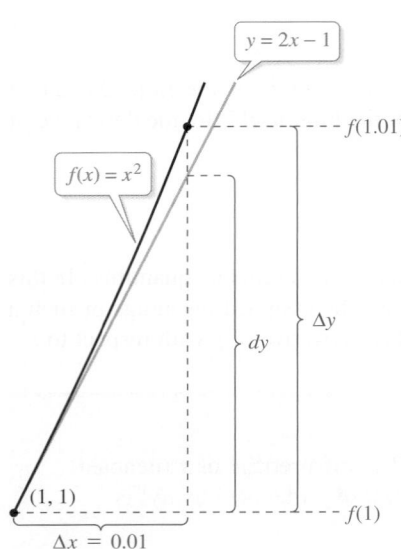

FIGURE 9.33

The labels on the figure: $y = 2x - 1$, $f(x) = x^2$, $f(1.01)$, Δy, dy, $(1, 1)$, $f(1)$, $\Delta x = 0.01$

✓**Checkpoint 1**

Find the value of dy when $x = 2$ and $dx = 0.01$ for $f(x) = x^4$. Compare this with the value of Δy when $x = 2$ and $\Delta x = 0.01$.

In Example 1, the tangent line to the graph of $f(x) = x^2$ at $x = 1$ is

$$y = 2x - 1 \quad \text{or} \quad g(x) = 2x - 1. \qquad \text{Tangent line to the graph of } f \text{ at } x = 1$$

For x-values near 1, this line is close to the graph of f, as shown in Figure 9.33. For instance,

$$f(1.01) = 1.01^2 = 1.0201 \quad \text{and} \quad g(1.01) = 2(1.01) - 1 = 1.02.$$

The validity of the approximation

$$dy \approx \Delta y, \quad dx \neq 0$$

stems from the definition of the derivative. That is, the existence of the limit

$$f'(x) = \lim_{\Delta x \to 0} \frac{f(x + \Delta x) - f(x)}{\Delta x}$$

implies that when Δx is close to zero, then $f'(x)$ is close to the difference quotient. So, you can write

$$\frac{f(x + \Delta x) - f(x)}{\Delta x} \approx f'(x)$$

$$f(x + \Delta x) - f(x) \approx f'(x)\,\Delta x$$

$$\Delta y \approx f'(x)\,\Delta x.$$

Substituting dx for Δx and dy for $f'(x)\,dx$ produces $\Delta y \approx dy$.

Marginal Analysis

Differentials are used in economics to approximate changes in revenue, cost, and profit. Let $R = f(x)$ be the total revenue for selling x units of a product. When the number of units increases by 1, the change in x is $\Delta x = 1$, and the change in R is

$$\Delta R = f(x + \Delta x) - f(x) \approx dR = \frac{dR}{dx}\, dx.$$

In other words, you can use the differential dR to approximate the change in the revenue that accompanies the sale of one additional unit. Similarly, the differentials dC and dP can be used to approximate the changes in cost and profit that accompany the sale (or production) of one additional unit.

 Example 2   Using Marginal Analysis

The demand function for a product is modeled by

$$p = 400 - x, \quad 0 \le x \le 400$$

where p is the price per unit (in dollars) and x is the number of units. Use differentials to approximate the change in revenue as sales increase from 149 units to 150 units. Compare this with the actual change in revenue.

SOLUTION   Begin by finding the revenue function. Because the demand is given by $p = 400 - x$, the revenue is

$$
\begin{aligned}
R &= xp & &\text{Formula for revenue} \\
&= x(400 - x) & &\text{Use } p = 400 - x. \\
&= 400x - x^2. & &\text{Multiply.}
\end{aligned}
$$

Next, find the marginal revenue, dR/dx.

$$\frac{dR}{dx} = 400 - 2x \qquad \text{Power Rule}$$

When $x = 149$ and $dx = \Delta x = 1$, the approximate change in the revenue is

$$
\begin{aligned}
\Delta R &\approx dR \\
&= \frac{dR}{dx}\, dx \\
&= (400 - 2x)\, dx \\
&= [400 - 2(149)](1) \\
&= \$102.
\end{aligned}
$$

When x increases from 149 to 150 and $R = f(x) = 400x - x^2$, the actual change in revenue is

$$
\begin{aligned}
\Delta R &= f(x + \Delta x) - f(x) \\
&= \left[400(150) - 150^2\right] - \left[400(149) - 149^2\right] \\
&= 37{,}500 - 37{,}399 \\
&= \$101.
\end{aligned}
$$

✓**Checkpoint 2**

The demand function for a product is modeled by $p = 200 - x$, $0 \le x \le 200$, where p is the price per unit (in dollars) and x is the number of units. Use differentials to approximate the change in revenue as sales increase from 89 to 90 units. Compare this with the actual change in revenue.

Example 3   **Using Marginal Analysis**

The profit (in dollars) derived from selling x units of an item is modeled by

$$P = 0.0002x^3 + 10x.$$

Use the differential dP to approximate the change in profit when the production level changes from 50 to 51 units. Compare this with the actual gain in profit obtained by increasing the production level from 50 to 51 units.

SOLUTION   The marginal profit is

$$\frac{dP}{dx} = 0.0006x^2 + 10.$$

When $x = 50$ and $dx = \Delta x = 1$, the approximate change in profit is

$$\Delta P \approx dP$$
$$= \frac{dP}{dx}\, dx$$
$$= (0.0006x^2 + 10)\, dx$$
$$= [0.0006(50)^2 + 10](1)$$
$$= \$11.50.$$

When x changes from 50 to 51 units and $P = f(x) = 0.0002x^3 + 10x$, the actual change in profit is

$$\Delta P = f(x + \Delta x) - f(x)$$
$$= [(0.0002)(51)^3 + 10(51)] - [(0.0002)(50)^3 + 10(50)]$$
$$\approx 536.53 - 525.00$$
$$= \$11.53.$$

These values are shown graphically in Figure 9.34.

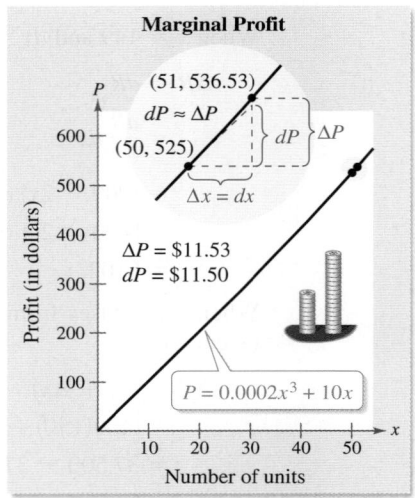

FIGURE 9.34

✓ **Checkpoint 3**

Use the differential dP to approximate the change in profit for the profit function in Example 3 when the production level changes from 40 to 41 units. Compare this with the actual gain in profit obtained by increasing the production level from 40 to 41 units.

STUDY TIP

Example 3 uses differentials to solve the same problem that was solved in Example 5 in Section 7.5. Look back at that solution. Which approach do you prefer?

Formulas for Differentials

You can use the definition of differentials to rewrite each differentiation rule in **differential form.**

Differential Forms of Differentiation Rules	
Constant Multiple Rule:	$d[cu] = c\,du$
Sum or Difference Rule:	$d[u \pm v] = du \pm dv$
Product Rule:	$d[uv] = u\,dv + v\,du$
Quotient Rule:	$d\left[\dfrac{u}{v}\right] = \dfrac{v\,du - u\,dv}{v^2}$
Constant Rule:	$d[c] = 0$
Power Rule:	$d[x^n] = nx^{n-1}\,dx$

The next example compares the derivatives and differentials of several simple functions.

Example 4   Finding Differentials

Function	Derivative	Differential
a. $y = x^2$	$\dfrac{dy}{dx} = 2x$	$dy = 2x\,dx$
b. $y = \dfrac{3x + 2}{5}$	$\dfrac{dy}{dx} = \dfrac{3}{5}$	$dy = \dfrac{3}{5}\,dx$
c. $y = 2x^2 - 3x$	$\dfrac{dy}{dx} = 4x - 3$	$dy = (4x - 3)\,dx$
d. $y = \dfrac{1}{x}$	$\dfrac{dy}{dx} = -\dfrac{1}{x^2}$	$dy = -\dfrac{1}{x^2}\,dx$

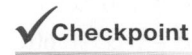 **Checkpoint 4**

Find the differential dy of each function.

a. $y = 4x^3$ **b.** $y = \dfrac{2x + 1}{3}$

c. $y = 3x^2 - 2x$ **d.** $y = \dfrac{1}{x^2}$

SUMMARIZE   (Section 9.5)

1. State the definition of differentials *(page 729)*. For an example of a differential, see Example 1.

2. Explain what is meant by marginal analysis *(page 731)*. For examples of marginal analysis, see Examples 2 and 3.

3. State the differential forms of the differentiation rules *(page 733)*. For an example that uses the differential forms of the differentiation rules, see Example 4.

SKILLS WARM UP 9.5   The following warm-up exercises involve skills that were covered in earlier sections. You will use these skills in the exercise set for this section. For additional help, review Sections 7.4 and 7.6.

In Exercises 1–12, find the derivative.

1. $C = 44 + 0.09x^2$

2. $C = 250 + 0.15x$

3. $R = x\left(1.25 + 0.02\sqrt{x}\right)$

4. $R = x(15.5 - 1.55x)$

5. $P = -0.03x^{1/3} + 1.4x - 2250$

6. $P = -0.02x^2 + 25x - 1000$

7. $A = \frac{1}{4}\sqrt{3}x^2$

8. $A = 6x^2$

9. $C = 2\pi r$

10. $P = 4w$

11. $S = 4\pi r^2$

12. $P = 2x + \sqrt{2}x$

In Exercises 13–16, write a formula for the quantity.

13. Area A of a circle of radius r

14. Area A of a square of side x

15. Volume V of a cube of edge x

16. Volume V of a sphere of radius r

Exercises 9.5

See www.CalcChat.com for worked-out solutions to odd-numbered exercises.

Comparing Δy and dy   In Exercises 1–6, compare the values of dy and Δy for the function. *See Example 1.*

Function	x-Value	Differential of x
1. $f(x) = 0.5x^3$	$x = 2$	$\Delta x = dx = 0.1$
2. $f(x) = 1 - 2x^2$	$x = 0$	$\Delta x = dx = -0.1$
3. $f(x) = x^4 + 1$	$x = -1$	$\Delta x = dx = 0.01$
4. $f(x) = 2x + 1$	$x = 1$	$\Delta x = dx = 0.01$
5. $f(x) = 3\sqrt{x}$	$x = 4$	$\Delta x = dx = 0.1$
6. $f(x) = 6x^{4/3}$	$x = -1$	$\Delta x = dx = 0.01$

Finding Differentials   In Exercises 7–12, let $x = 2$ and complete the table for the function.

$dx = \Delta x$	dy	Δy	$\Delta y - dy$	$\dfrac{dy}{\Delta y}$
1.000				
0.500				
0.100				
0.010				
0.001				

7. $y = x^2$

8. $y = x^5$

9. $y = \dfrac{1}{x^2}$

10. $y = \dfrac{1}{x}$

11. $y = \sqrt[4]{x}$

12. $y = \sqrt{x}$

Marginal Analysis   In Exercises 13–18, use differentials to approximate the change in cost, revenue, or profit corresponding to an increase in sales of one unit. For instance, in Exercise 13, approximate the change in cost as x increases from 12 units to 13 units. *See Examples 2 and 3.*

Function	x-Value
13. $C = 0.05x^2 + 4x + 10$	$x = 12$
14. $C = 0.025x^2 + 8x + 5$	$x = 10$
15. $R = 30x - 0.15x^2$	$x = 75$
16. $R = 50x - 1.5x^2$	$x = 15$
17. $P = -0.5x^3 + 2500x - 6000$	$x = 50$
18. $P = -x^2 + 60x - 100$	$x = 25$

Finding Differentials   In Exercises 19–28, find the differential dy. *See Example 4.*

19. $y = 6x^4$

20. $y = \dfrac{8 - 4x}{3}$

21. $y = 3x^2 - 4$

22. $y = 3x^{2/3}$

23. $y = (4x - 1)^3$

24. $y = (x^2 + 3)(2x + 4)^2$

25. $y = \dfrac{x + 1}{2x - 1}$

26. $y = \dfrac{x}{x^2 + 1}$

27. $y = \sqrt{9 - x^2}$

28. $y = \sqrt[3]{6x^2}$

Finding an Equation of a Tangent Line In Exercises 29–32, find an equation of the tangent line to the function at the given point. Then find the function values and the tangent line values at $f(x + \Delta x)$ and $y(x + \Delta x)$ for $\Delta x = -0.01$ and 0.01.

Function	Point
29. $f(x) = 2x^3 - x^2 + 1$	$(-2, -19)$
30. $f(x) = 3x^2 - 1$	$(2, 11)$
31. $f(x) = \dfrac{x}{x^2 + 1}$	$(0, 0)$
32. $f(x) = \sqrt{25 - x^2}$	$(3, 4)$

33. Profit The profit P for a company producing x units is

$$P = (500x - x^2) - \left(\frac{1}{2}x^2 - 77x + 3000\right).$$

(a) Use differentials to approximate the change in profit when the production level changes from 115 to 120 units.

(b) Compare this with the actual change in profit.

34. Revenue The revenue R for a company selling x units is $R = 900x - 0.1x^2$.

(a) Use differentials to approximate the change in revenue as the sales increase from 3000 units to 3100 units.

(b) Compare this with the actual change in revenue.

35. Demand The demand function for a product is modeled by $p = 75 - 0.25x$, where p is the price per unit (in dollars) and x is the number of units.

(a) Use differentials to approximate the change in revenue as sales increase from 7 units to 8 units.

(b) Repeat part (a) as sales increase from 70 units to 71 units.

36.  **HOW DO YOU SEE IT?** The graph shows the profit P (in dollars) from selling x units of an item. Use the graph to determine which is greater, the change in profit when the production level changes from 400 to 401 units or the change in profit when the production level changes from 900 to 901 units. Explain your reasoning.

Profit

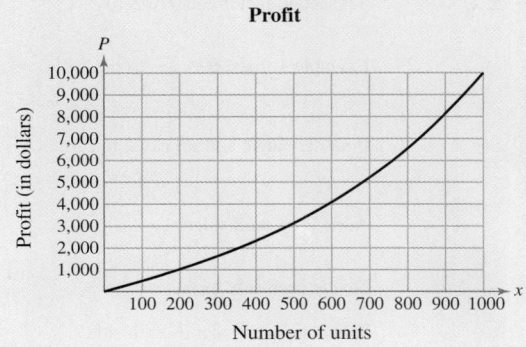

Number of units

37. Biology: Wildlife Management A state game commission introduces 50 deer into newly acquired state game lands. The population N of the herd can be modeled by

$$N = \frac{10(5 + 3t)}{1 + 0.04t}$$

where t is the time in years. Use differentials to approximate the change in the herd size from $t = 5$ to $t = 6$.

38. Medical Science The concentration C (in milligrams per milliliter) of a drug in a patient's bloodstream t hours after injection into muscle tissue is modeled by

$$C = \frac{3t}{27 + t^3}.$$

Use differentials to approximate the change in the concentration when t changes from $t = 1$ to $t = 1.5$.

39. Marginal Analysis A retailer has determined that the monthly sales x of a watch are 150 units when the price is \$50, but decrease to 120 units when the price is \$60. Assume that the demand is a linear function of the price. Find the revenue R as a function of x and approximate the change in revenue for a one-unit increase in sales when $x = 141$. Make a sketch showing dR and ΔR.

40. Marginal Analysis The demand x for a web camera is 30,000 units per month when the price is \$25 and 40,000 units when the price is \$20. The initial investment is \$275,000 and the cost per unit is \$17. Assume that the demand is a linear function of the price. Find the profit P as a function of x and approximate the change in profit for a one-unit increase in sales when $x = 28,000$. Make a sketch showing dP and ΔP.

Error Propagation In Exercises 41 and 42, use the following information. Given the error in a measurement (Δx), the *propagated error* (Δy) can be approximated by the differential dy. The ratio dy/y is the *relative error*, which corresponds to a *percentage error* of $dy/y \times 100\%$.

41. Area The side of a square measures 6 inches, with a possible error of $\pm\frac{1}{16}$ inch. Estimate the propagated error and the percentage error in computing the area of the square.

42. Volume The radius of a sphere measures 6 inches, with a possible error of ± 0.02 inch. Estimate the propagated error and the percentage error in computing the volume of the sphere.

True or False? In Exercises 43 and 44, determine whether the statement is true or false. If it is false, explain why or give an example that shows it is false.

43. If $y = x + c$, then $dy = dx$.

44. If $y = ax + b$, then $\Delta y/\Delta x = dy/dx$.

ALGEBRA TUTOR

Solving Equations

Example 1 on page 680 illustrates some of the basic techniques for solving equations. Example 2 on page 681 illustrates some of the more complicated techniques. In the examples that follow, you can further review some of the more complicated techniques for solving equations. Note in Example 2(c) that with an *absolute value* equation, the definition of absolute value is used to rewrite the equation as two equations.

When solving an equation, remember that your basic goal is to isolate the variable on one side of the equation. To do this, you use inverse operations. For instance, to isolate x in

$$x - 2 = 0$$

you add 2 to each side of the equation because *addition* is the inverse operation of *subtraction*. To isolate x in

$$\sqrt{x} = 2$$

you square each side of the equation because *squaring* is the inverse operation of *taking the square root*.

Example 1   Solving an Equation

Solve each equation.

a. $0 = 2 - \dfrac{72}{x^2}$ **b.** $0 = 2x(2x^2 - 3)$ **c.** $V' = 0$, where $V = 27x - \dfrac{1}{4}x^2$

SOLUTION

a. $0 = 2 - \dfrac{72}{x^2}$ Example 3, page 693

$-2 = -\dfrac{72}{x^2}$ Subtract 2 from each side.

$1 = \dfrac{36}{x^2}$ Divide each side by -2.

$x^2 = 36$ Multiply each side by x^2.

$x = \pm 6$ Take the square root of each side.

b. $0 = 2x(2x^2 - 3)$ Example 2, page 692

$2x = 0 \implies x = 0$ Set first factor equal to zero.

$2x^2 - 3 = 0 \implies x = \pm\sqrt{\dfrac{3}{2}}$ Set second factor equal to zero.

c. $V = 27x - \dfrac{1}{4}x^3$ Example 1, page 690

$27 - \dfrac{3}{4}x^2 = 0$ Find derivative and set equal to zero.

$27 = \dfrac{3}{4}x^2$ Add $\frac{3}{4}x^2$ to each side.

$36 = x^2$ Divide each side by $\frac{3}{4}$.

$\pm 6 = x$ Take the square root of each side.

Example 2   Solving an Equation

Solve each equation.

a. $\dfrac{20(x^{1/3} - 1)}{9x^{2/3}} = 0$

Example 5, page 725

$20(x^{1/3} - 1) = 0$

A fraction is zero only if its numerator is zero.

$x^{1/3} - 1 = 0$

Divide each side by 20.

$x^{1/3} = 1$

Add 1 to each side.

$x = 1$

Cube each side.

b. $\dfrac{25}{\sqrt{x}} - 0.5 = 0$

Example 4, page 701

$\dfrac{25}{\sqrt{x}} = 0.5$

Add 0.5 to each side.

$25 = 0.5\sqrt{x}$

Multiply each side by $\sqrt{x}$.

$50 = \sqrt{x}$

Divide each side by 0.5.

$2500 = x$

Square each side.

c. $\left| -\dfrac{24\sqrt{x}}{x} + 2 \right| = 1$

Example 5, page 703

First Equation

$-\dfrac{24\sqrt{x}}{x} + 2 = 1$

Use positive expression.

$-\dfrac{24\sqrt{x}}{x} = -1$

Subtract 2 from each side.

$24\sqrt{x} = x$

Multiply each side by $-x$.

$576x = x^2$

Square each side.

$0 = x(x - 576)$

Subtract $576x$ from each side and factor.

$x = 0$

Set first factor equal to zero (extraneous solution).

$x = 576$

Set second factor equal to zero.

Second Equation

$-\left(-\dfrac{24\sqrt{x}}{x} + 2 \right) = 1$

Use negative expression.

$\dfrac{24\sqrt{x}}{x} - 2 = 1$

Rewrite without parentheses.

$\dfrac{24\sqrt{x}}{x} = 3$

Add 2 to each side.

$24\sqrt{x} = 3x$

Multiply each side by x.

$576x = 9x^2$

Square each side.

$0 = 9x(x - 64)$

Subtract $576x$ from each side and factor.

$x = 0$

Set first factor equal to zero (extraneous solution).

$x = 64$

Set second factor equal to zero.

SUMMARY AND STUDY STRATEGIES

After studying this chapter, you should have acquired the following skills.
The exercise numbers are keyed to the Review Exercises that begin on page 739.
Answers to odd-numbered Review Exercises are given in the back of the text.*

Section 9.1

	Review Exercises
■ Solve real-life optimization problems.	*1–6*

Section 9.2

■ Solve business and economics optimization problems.	*7–12*
■ Find the price elasticity of demand for a demand function.	*13, 14*

Section 9.3

■ Find the vertical asymptotes of a function.	*15–18*
■ Find infinite limits.	*19–26*
■ Find the horizontal asymptotes of a function.	*27–32*
■ Use asymptotes to answer questions about real-life situations.	*33–36*

Section 9.4

■ Analyze the graph of a function.	*37–50*

Section 9.5

■ Use differentials to approximate changes in a function.	*51–54*
■ Use differentials in economics to approximate changes in cost, revenue, and profit.	*55–60*
■ Find the differential of a function using differentiation formulas.	*61–66*
■ Use differentials to approximate changes in real-life models.	*67–70*

Study Strategies

■ **Problem-Solving Strategies**   If you get stuck when trying to solve an optimization problem, consider the strategies below.

1. *Draw a Diagram.*   If feasible, draw a diagram that represents the problem. Label all known values and unknown values on the diagram.

2. *Solve a Simpler Problem.*   Simplify the problem, or write several simple examples of the problem. For instance, if you are asked to find the dimensions that will produce a maximum area, try calculating the areas of several examples.

3. *Rewrite the Problem in Your Own Words.*   Rewriting a problem can help you understand it better.

4. *Guess and Check.*   Try guessing the answer, then check your guess in the statement of the original problem. By refining your guesses, you may be able to think of a general strategy for solving the problem.

---

* A wide range of valuable study aids are available to help you master the material in this chapter. The *Student Solutions Manual* includes step-by-step solutions to all odd-numbered exercises to help you review and prepare. The student website at *www.cengagebrain.com* offers algebra help and a *Graphing Technology Guide*, which contains step-by-step commands and instructions for a wide variety of graphing calculators.

Review Exercises

See www.CalcChat.com for worked-out solutions to odd-numbered exercises.

1. **Minimum Perimeter** Find the length and width of a rectangle that has an area of 225 square meters and a minimum perimeter.

2. **Maximum Volume** A rectangular solid with a square base has a surface area of 432 square centimeters.

   (a) Determine the dimensions that yield the maximum volume.

   (b) Find the maximum volume.

3. **Maximum Volume** An open box is to be made from a 10-inch by 16-inch rectangular piece of material by cutting equal squares from the corners and turning up the sides (see figure). Find the volume of the largest box that can be made.

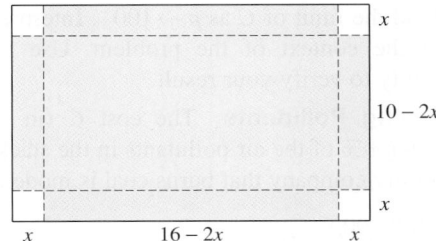

4. **Minimum Area** A rectangular page is to contain 108 square inches of print. The margins at the top and bottom of the page are to be $\frac{3}{4}$ inch wide. The margins on each side are to be 1 inch wide. Find the dimensions of the page that will minimize the amount of paper used.

5. **Minimum Length** Two posts, one 12 feet high and the other 28 feet high, stand 30 feet apart. They are to be secured by two wires, attached to a single stake, running from ground level to the top of each post (see figure). Where should the stake be placed to use the least amount of wire?

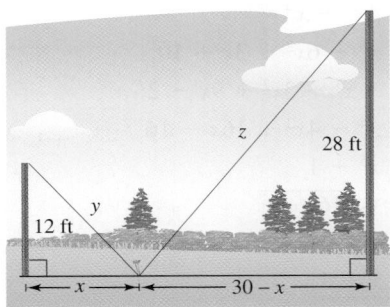

6. **Maximum Yield** A citrus grower estimates that 90 orange trees per acre will have an average yield of 700 oranges per tree. For each additional tree per acre, the yield will decrease by 25 oranges per tree.

   (a) How many trees should be planted per acre to maximize the yield of oranges?

   (b) What is the maximum yield per acre?

Finding the Maximum Revenue In Exercises 7 and 8, find the number of units x that produces a maximum revenue R.

7. $R = 450x - 0.25x^2$

8. $R = 36x^2 - 0.05x^3$

Finding the Minimum Average Cost In Exercises 9 and 10, find the number of units x that produces the minimum average cost per unit $\overline{C}$.

9. $C = 0.2x^2 + 10x + 4500$

10. $C = 0.03x^3 + 30x + 3840$

11. **Maximum Profit** A commodity has a demand function modeled by

    $$p = 36 - 4x$$

    and a total cost function modeled by

    $$C = 2x^2 + 6$$

    where x is the number of units.

    (a) What price yields a maximum profit?

    (b) When the profit is maximized, what is the average cost per unit?

12. **Maximum Profit** The profit P (in thousands of dollars) for a company in terms of the amount s spent on advertising (in thousands of dollars) can be modeled by

    $$P = -4s^3 + 72s^2 - 240s + 500.$$

    Find the amount of advertising that maximizes the profit. Find the point of diminishing returns.

13. **Elasticity** The demand function for a product is modeled by

    $$p = 60 - 0.04x, \quad 0 \le x \le 1500$$

    where p is the price (in dollars) and x is the number of units.

    (a) Determine when the demand is elastic, inelastic, and of unit elasticity.

    (b) Use the result of part (a) to describe the behavior of the revenue function.

14. **Elasticity** The demand function for a product is modeled by

    $$p = 960 - x, \quad 0 \le x \le 960$$

    where p is the price (in dollars) and x is the number of units.

    (a) Determine when the demand is elastic, inelastic, and of unit elasticity.

    (b) Use the result of part (a) to describe the behavior of the revenue function.

Finding Vertical Asymptotes In Exercises 15–18, determine all vertical asymptotes of the graph of the function.

15. $f(x) = \dfrac{x + 4}{x^2 + 7x}$

16. $f(x) = \dfrac{x - 1}{x^2 - 4}$

17. $f(x) = \dfrac{x^2 - 16}{2x^2 + 9x + 4}$

18. $f(x) = \dfrac{x^2 + 6x + 9}{x^2 - 5x - 24}$

 Determining Infinite Limits In Exercises 19–26, use a graphing utility to find the limit.

19. $\lim\limits_{x \to 0^+} \left(x - \dfrac{1}{x^3} \right)$

20. $\lim\limits_{x \to 0^-} \left(3 + \dfrac{1}{x} \right)$

21. $\lim\limits_{x \to -1^+} \dfrac{x^2 - 2x + 1}{x + 1}$

22. $\lim\limits_{x \to 3^-} \dfrac{3x^2 + 1}{x^2 - 9}$

23. $\lim\limits_{x \to \infty} \dfrac{2x^2}{3x^2 + 5}$

24. $\lim\limits_{x \to \infty} \dfrac{3x^2 - 2x + 3}{x + 1}$

25. $\lim\limits_{x \to -\infty} \dfrac{3x}{x^2 + 1}$

26. $\lim\limits_{x \to -\infty} \left(\dfrac{x}{x - 2} + \dfrac{2x}{x + 2} \right)$

Finding Horizontal Asymptotes In Exercises 27–32, find the horizontal asymptote of the graph of the function.

27. $f(x) = \dfrac{2x^2}{3x^2 + 5}$

28. $f(x) = \dfrac{3x^2 - 2x + 3}{x + 1}$

29. $f(x) = \dfrac{3x}{x^2 + 1}$

30. $f(x) = \dfrac{4}{x^2 + 1}$

31. $g(x) = \dfrac{3}{x} - 2$

32. $f(x) = \dfrac{x}{x - 2} + \dfrac{2x}{x + 2}$

33. Average Cost The cost C (in dollars) of producing x units of a product is

$$C = 0.75x + 4000.$$

(a) Find the average cost function $\overline{C}$.

(b) Find $\overline{C}$ when $x = 100$ and when $x = 1000$.

(c) Determine the limit of the average cost function as x approaches infinity. Interpret the limit in the context of the problem.

34. Average Cost The cost C (in dollars) of producing x units of a product is

$$C = 1.50x + 8000.$$

(a) Find the average cost function $\overline{C}$.

(b) Find $\overline{C}$ when $x = 1000$ and when $x = 10{,}000$.

(c) Determine the limit of the average cost function as x approaches infinity. Interpret the limit in the context of the problem.

35. Seizing Drugs The cost C (in millions of dollars) for the federal government to seize $p\%$ of an illegal drug as it enters the country is modeled by

$$C = \dfrac{250p}{100 - p}, \quad 0 \le p < 100.$$

(a) Find the costs of seizing 20%, 50%, and 90% of the drug.

(b) Find the limit of C as $p \to 100^-$. Interpret the limit in the context of the problem. Use a graphing utility to verify your result.

36. Removing Pollutants The cost C (in dollars) of removing $p\%$ of the air pollutants in the stack emission of a utility company that burns coal is modeled by

$$C = \dfrac{160{,}000p}{100 - p}, \quad 0 \le p < 100.$$

(a) Find the costs of seizing 25%, 50%, and 75% of the pollutants.

(b) Find the limit of C as $p \to 100^-$. Interpret the limit in the context of the problem. Use a graphing utility to verify your result.

Analyzing a Graph In Exercises 37–48, analyze and sketch the graph of the function. Label any intercepts, relative extrema, points of inflection, and asymptotes.

37. $f(x) = 4x - x^2$

38. $f(x) = 4x^3 - x^4$

39. $f(x) = x^3 - 6x^2 + 3x + 10$

40. $f(x) = -x^3 + 3x^2 + 9x - 2$

41. $f(x) = x^4 - 4x^3 + 16x - 16$

42. $f(x) = x^5 + 1$

43. $f(x) = x\sqrt{16 - x^2}$

44. $f(x) = x^2\sqrt{9 - x^2}$

45. $f(x) = \dfrac{x + 1}{x - 1}$

46. $f(x) = \dfrac{x - 1}{3x^2 + 1}$

47. $f(x) = 3x^{2/3} - 2x$

48. $f(x) = x^{4/5}$

49. Bacteria The data in the table show the number N of bacteria in a culture at time t, where t is measured in days.

t	1	2	3	4	5	6	7	8
N	25	200	804	1756	2296	2434	2467	2473

A model for these data is

$$N = \frac{24{,}670 - 35{,}153t + 13{,}250t^2}{100 - 39t + 7t^2}, \quad 1 \le t \le 8.$$

(a) Use a graphing utility to create a scatter plot of the data and graph the model in the same viewing window. How well does the model fit the data?

(b) Use the model to predict the number of bacteria in the culture after 10 days.

(c) Should this model be used to predict the number of bacteria in the culture after a few months? Why or why not?

50. Meteorology The monthly average high temperatures T (in degrees Fahrenheit) in New York City can be modeled by

$$T = \frac{31.6 - 1.822t + 0.0984t^2}{1 - 0.194t + 0.0131t^2}, \quad 1 \le t \le 12$$

where t is the month, with $t = 1$ corresponding to January. Use a graphing utility to graph the model and find all absolute extrema. Interpret the meaning of these values in the context of the problem. *(Source: National Climatic Data Center)*

Comparing Δy and dy In Exercises 51–54, compare the values of dy and Δy for the function.

Function	*x-Value*	*Differential of x*
51. $f(x) = 2x^2$	$x = 2$	$\Delta x = dx = 0.01$
52. $f(x) = x^4 + 3$	$x = 1$	$\Delta x = dx = 0.1$
53. $f(x) = 6x - x^3$	$x = 3$	$\Delta x = dx = 0.1$
54. $f(x) = 5x^{3/2}$	$x = 9$	$\Delta x = dx = 0.01$

Marginal Analysis In Exercises 55–60, use differentials to approximate the change in cost, revenue, or profit corresponding to an increase in sales of one unit. For instance, in Exercise 55, approximate the change in cost as x increases from 10 to 11.

Function	*x-Value*
55. $C = 40x^2 + 1225$	$x = 10$
56. $C = 1.5\sqrt[3]{x} + 500$	$x = 125$
57. $R = 6.25x + 0.4x^{3/2}$	$x = 225$
58. $R = 80x - 0.35x^2$	$x = 80$
59. $P = 0.003x^2 + 0.019x - 1200$	$x = 750$
60. $P = -0.2x^3 + 3000x - 7500$	$x = 50$

Finding Differentials In Exercises 61–66, find the differential dy.

61. $y = 0.5x^3$

62. $y = 7x^4 + 2x^2$

63. $y = (3x^2 - 2)^3$

64. $y = \sqrt{36 - x^2}$

65. $y = \dfrac{2 - x}{x + 5}$

66. $y = \dfrac{3x^2}{x - 4}$

67. Profit The profit P (in dollars) for a company producing x units is

$$P = -0.8x^2 + 324x - 2000.$$

(a) Use differentials to approximate the change in profit when the production level changes from 100 to 101 units.

(b) Compare this with the actual change in profit.

68. Demand The demand function for a product is modeled by

$$p = 108 - 0.2x$$

where p is the price per unit (in dollars) and x is the number of units.

(a) Use differentials to approximate the change in revenue as sales increase from 20 units to 21 units.

(b) Repeat part (a) when sales increase from 40 units to 41 units.

69. Physiology: Body Surface Area The body surface area (BSA) of a 180-centimeter-tall (about six-foot-tall) person is modeled by

$$B = 0.1\sqrt{5w}$$

where B is the BSA (in square meters) and w is the weight (in kilograms). Use differentials to approximate the change in the person's BSA when the person's weight changes from 90 kilograms to 95 kilograms.

70. Aquaculture The recommended daily percent p of biomass (plant matter) to be included in a fish's diet can be modeled by

$$p = 0.000235w^2 - 0.054w + 7.1$$

where w is the weight (in grams) of the fish. *(Source: Food and Agriculture Organization of the United Nations)*

(a) Use differentials to approximate the change in the recommended percent of biomass when the fish's weight changes from 10 grams to 20 grams.

(b) Use differentials to approximate the change in the recommended percent of biomass when the fish's weight changes from 40 grams to 60 grams.

TEST YOURSELF

See www.CalcChat.com for worked-out solutions to odd-numbered exercises.

Take this test as you would take a test in class. When you are done, check your work against the answers given in the back of the book.

In Exercises 1–4, find any vertical and horizontal asymptotes of the graph of the function.

1. $f(x) = \dfrac{2x - 5}{x - 1}$

2. $f(x) = \dfrac{2x}{x^2 + 3}$

3. $f(x) = \dfrac{3x + 2}{x - 5}$

4. $f(x) = \dfrac{x^2 + 2x + 3}{x^2 - 1}$

In Exercises 5–10, find the limit, if possible.

5. $\displaystyle\lim_{x \to 1^-} \dfrac{x + 1}{x - 1}$

6. $\displaystyle\lim_{x \to 2^+} \dfrac{x}{x^2 - 4}$

7. $\displaystyle\lim_{x \to -1^-} \dfrac{x^2 + 1}{x^2 - 1}$

8. $\displaystyle\lim_{x \to \infty} \left(\dfrac{3}{x} + 1\right)$

9. $\displaystyle\lim_{x \to \infty} \dfrac{3x^2 - 4x + 1}{x - 7}$

10. $\displaystyle\lim_{x \to -\infty} \dfrac{6x^2 + x - 5}{2x^2 - 5x}$

In Exercises 11–13, analyze and sketch the graph of the function. Label any intercepts, relative extrema, points of inflection, and asymptotes.

11. $y = -x^3 + 3x^2 + 9x - 2$

12. $y = x^5 - 5x$

13. $y = \dfrac{x}{x^2 - 4}$

In Exercises 14–16, find the differential dy.

14. $y = 5x^2 - 3$

15. $y = \dfrac{1 - x}{x + 3}$

16. $y = (x + 4)^3$

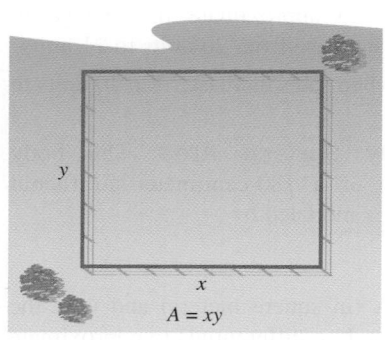

y

x

$A = xy$

Figure for 17

17. An ecologist has 500 meters of fencing to enclose a rectangular study plot (see figure). What dimensions will maximize the area of the enclosed study plot?

18. A rectangular solid with a square base has a volume of 8000 cubic inches.

(a) Determine the dimensions that yield the minimum surface area.

(b) Find the minimum surface area.

19. The demand function for a product is modeled by

$$p = 250 - 0.4x, \quad 0 \le x \le 625$$

where p is the price at which x units of the product are demanded by the market. Find the interval of inelasticity for the function.

20. Demand A store finds that the weekly demand for a product is modeled by

$$p = 20 - 0.05x$$

where p is the price per unit (in dollars) and x is the number of units. When x changes from 100 units to 101 units, what is the corresponding change in p? Compare the values of Δp and dp.

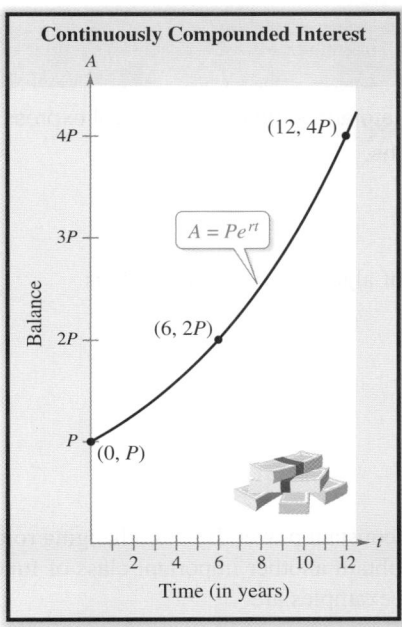

Continuously Compounded Interest
A
4P
3P
2P
P
Balance
(12, 4P)
A = Pe^rt
(6, 2P)
(0, P)
2
4
6
8
10
12
t
Time (in years)

Example 3 on page 789 shows how an exponential growth model can be used to find the annual interest rate of an account.

10 Derivatives of Exponential and Logarithmic Functions

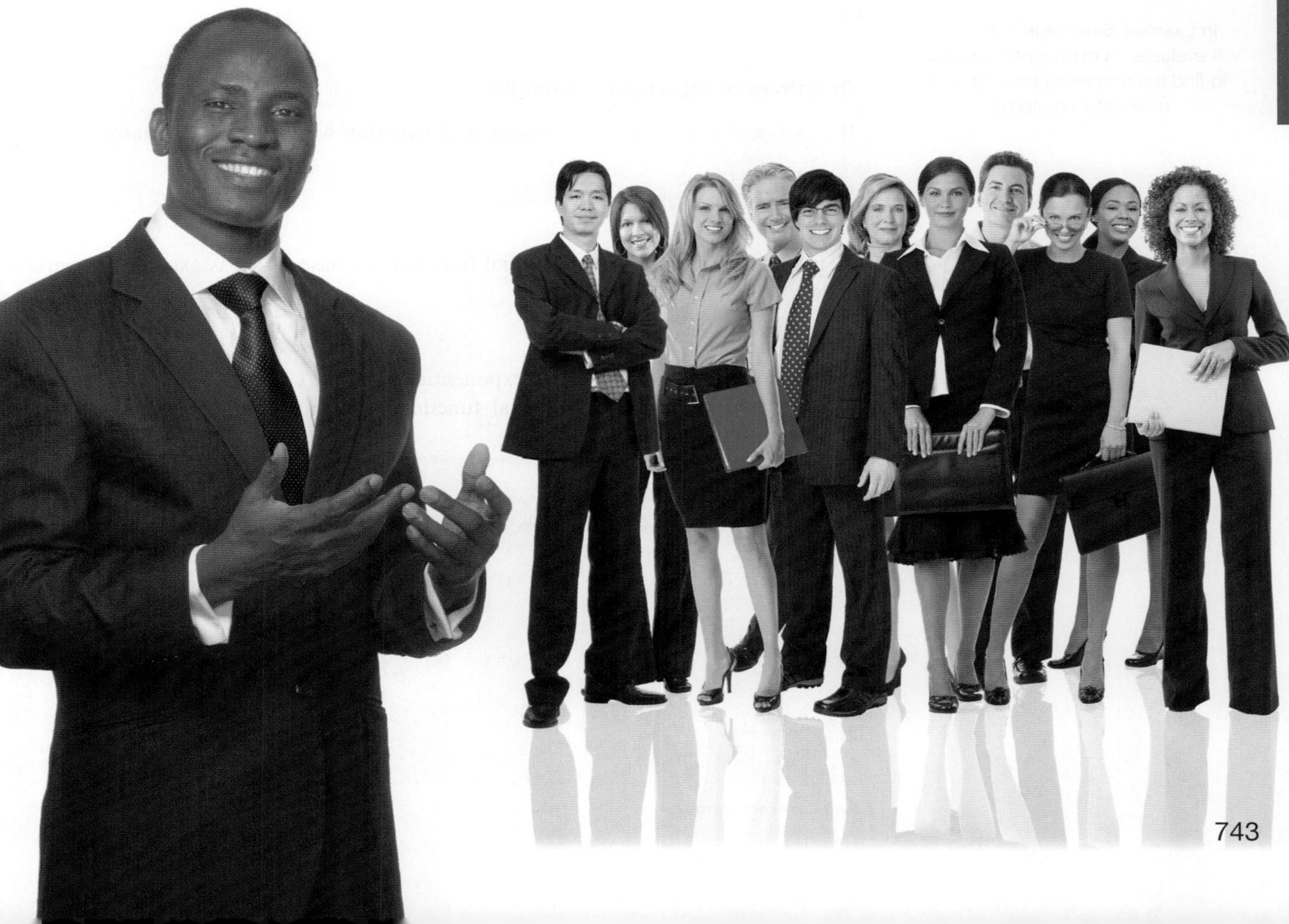

10.1 Exponential Functions

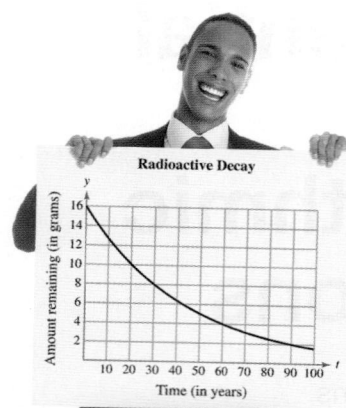

Radioactive Decay

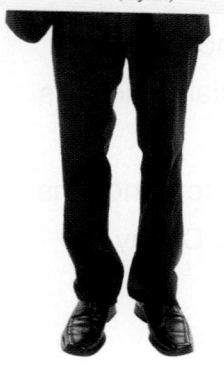

In Exercise 5 on page 748, you will evaluate an exponential function to find the remaining amount of a radioactive material.

■ Use the properties of exponents to evaluate and simplify exponential expressions.
■ Sketch the graphs of exponential functions.

Exponential Functions

You are already familiar with the behavior of algebraic functions such as

$$f(x) = x^2$$
$$g(x) = \sqrt{x} = x^{1/2}$$

and

$$h(x) = \frac{1}{x} = x^{-1}$$

each of which involves a variable raised to a constant power. By interchanging roles and raising a constant to a variable power, you obtain another important class of functions called **exponential functions.** Some simple examples are

$$f(x) = 2^x$$
$$g(x) = \left(\frac{1}{10}\right)^x = \frac{1}{10^x}$$

and

$$h(x) = 3^{2x} = 9^x.$$

In general, you can use any positive number $a \neq 1$ as the base of an exponential function.

Definition of Exponential Function

If $a > 0$ and $a \neq 1$, then the **exponential function** with base a is given by

$$f(x) = a^x.$$

In the definition of an exponential function, the base $a = 1$ is excluded because it yields

$$f(x) = 1^x = 1.$$

This is a constant function, not an exponential function.

When working with exponential functions, the properties of exponents, shown below, are useful.

Properties of Exponents

Let a and b be positive real numbers.

1. $a^0 = 1$ **2.** $a^x a^y = a^{x+y}$ **3.** $\dfrac{a^x}{a^y} = a^{x-y}$

4. $(a^x)^y = a^{xy}$ **5.** $(ab)^x = a^x b^x$ **6.** $\left(\dfrac{a}{b}\right)^x = \dfrac{a^x}{b^x}$

7. $a^{-x} = \dfrac{1}{a^x}$

Example 1 **Applying Properties of Exponents**

a. $(2^2)(2^3) = 2^{2+3} = 2^5 = 32$ Apply Property 2.

b. $(2^2)(2^{-3}) = 2^{2-3} = 2^{-1} = \dfrac{1}{2}$ Apply Properties 2 and 7.

c. $(3^2)^3 = 3^{2(3)} = 3^6 = 729$ Apply Property 4.

d. $\left(\dfrac{1}{3}\right)^{-2} = \left(\dfrac{3}{1}\right)^2 = \dfrac{3^2}{1^2} = 9$ Apply Properties 7 and 6.

e. $\dfrac{3^2}{3^3} = 3^{2-3} = 3^{-1} = \dfrac{1}{3}$ Apply Properties 3 and 7.

f. $(2^{1/2})(3^{1/2}) = [(2)(3)]^{1/2} = 6^{1/2} = \sqrt{6}$ Apply Property 5.

✓ **Checkpoint 1**

Simplify each expression using the properties of exponents.

a. $(3^2)(3^3)$ **b.** $(3^2)(3^{-1})$ **c.** $(2^3)^2$

d. $(1/2)^{-3}$ **e.** $2^2/2^3$ **f.** $(2^{1/2})(5^{1/2})$ ■

Although Example 1 demonstrates the properties of exponents with integer and rational exponents, it is important to realize that the properties hold for *all* real exponents. With a calculator, you can obtain approximations of a^x for any positive number a and any real number x. Here are some examples.

$$2^{-0.6} \approx 0.660, \qquad \pi^{0.75} \approx 2.360, \qquad (1.56)^{\sqrt{2}} \approx 1.876$$

 Example 2 **Dating Organic Material**

In living organic material, the ratio of radioactive carbon isotopes to the total number of carbon atoms is about 1 to 10^{12}. When organic material dies, its radioactive carbon isotopes begin to decay, with a half-life of about 5715 years. This means that after 5715 years, the ratio of isotopes to atoms will have decreased to one-half the original ratio; after a second 5715 years, the ratio will have decreased to one-fourth of the original; and so on. Figure 10.1 shows this decreasing ratio. The formula for the ratio R of carbon isotopes to carbon atoms is $R = (1/10^{12})(1/2)^{t/5715}$, where t is the time in years. Find the value of R for each period of time.

a. 10,000 years   **b.** 20,000 years   **c.** 25,000 years

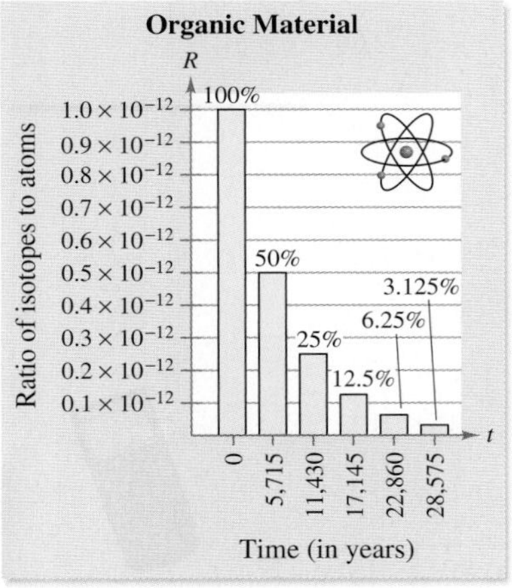

SOLUTION

a. $R = \left(\dfrac{1}{10^{12}}\right)\left(\dfrac{1}{2}\right)^{10,000/5715} \approx 2.973 \times 10^{-13}$ Ratio for 10,000 years

b. $R = \left(\dfrac{1}{10^{12}}\right)\left(\dfrac{1}{2}\right)^{20,000/5715} \approx 8.842 \times 10^{-14}$ Ratio for 20,000 years

c. $R = \left(\dfrac{1}{10^{12}}\right)\left(\dfrac{1}{2}\right)^{25,000/5715} \approx 4.821 \times 10^{-14}$ Ratio for 25,000 years

✓ **Checkpoint 2**

Use the formula for the ratio of carbon isotopes to carbon atoms in Example 2 to find the value of R for each period of time.

a. 5000 years   **b.** 15,000 years   **c.** 30,000 years   ■

Organic Material

Ratio of isotopes to atoms (vertical axis, R): 1.0×10^{-12}, 0.9×10^{-12}, 0.8×10^{-12}, 0.7×10^{-12}, 0.6×10^{-12}, 0.5×10^{-12}, 0.4×10^{-12}, 0.3×10^{-12}, 0.2×10^{-12}, 0.1×10^{-12}

Bar values: 100%, 50%, 25%, 12.5%, 6.25%, 3.125%

Time (in years) (horizontal axis, t): 0, 5,715, 11,430, 17,145, 22,860, 28,575

FIGURE 10.1

Graphs of Exponential Functions

The basic nature of the graph of an exponential function can be determined by the point-plotting method or by using a graphing utility.

Example 3   **Graphing Exponential Functions**

Sketch the graph of each exponential function.

a. $f(x) = 2^x$ **b.** $g(x) = \left(\frac{1}{2}\right)^x = 2^{-x}$ **c.** $h(x) = 3^x$

SOLUTION   To sketch these functions by hand, you can begin by constructing a table of values, as shown below.

x	-3	-2	-1	0	1	2	3	4
$f(x) = 2^x$	$\frac{1}{8}$	$\frac{1}{4}$	$\frac{1}{2}$	1	2	4	8	16
$g(x) = 2^{-x}$	8	4	2	1	$\frac{1}{2}$	$\frac{1}{4}$	$\frac{1}{8}$	$\frac{1}{16}$
$h(x) = 3^x$	$\frac{1}{27}$	$\frac{1}{9}$	$\frac{1}{3}$	1	3	9	27	81

The graphs of the three functions are shown in Figure 10.2. Note that the graphs of $f(x) = 2^x$ and $h(x) = 3^x$ are increasing, whereas the graph of $g(x) = 2^{-x}$ is decreasing.

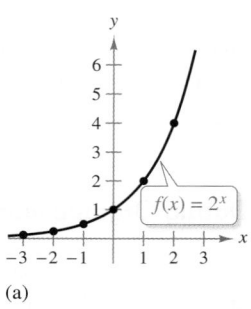

(a)

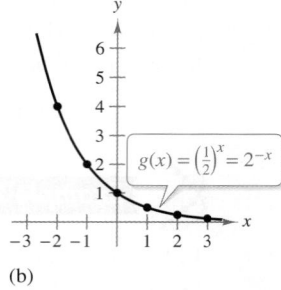

(b)

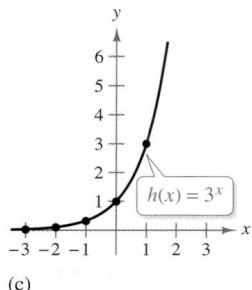

(c)

FIGURE 10.2

✓ **Checkpoint 3**

Sketch the graph of

$$f(x) = 5^x.$$

TECH TUTOR

Try graphing the functions

$$f(x) = 2^x \quad \text{and} \quad h(x) = 3^x$$

in the same viewing window, as shown at the right. From the display, you can see that the graph of h is increasing more rapidly than the graph of f.

The forms of the graphs in Figure 10.2 are typical of the graphs of the exponential functions $y = a^{-x}$ and $y = a^x$, where $a > 1$. The basic characteristics of such graphs are summarized in Figure 10.3.

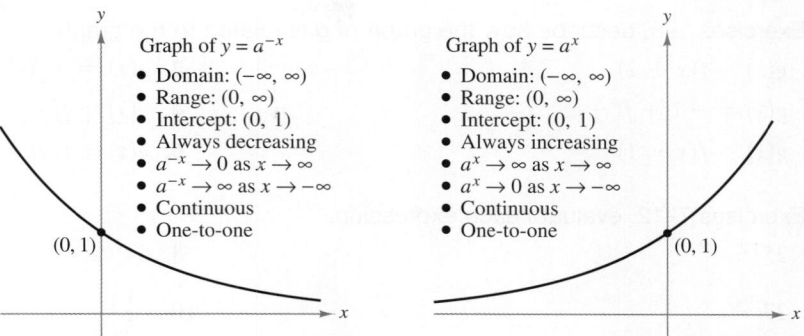

Graph of $y = a^{-x}$
- Domain: $(-\infty, \infty)$
- Range: $(0, \infty)$
- Intercept: $(0, 1)$
- Always decreasing
- $a^{-x} \to 0$ as $x \to \infty$
- $a^{-x} \to \infty$ as $x \to -\infty$
- Continuous
- One-to-one

Graph of $y = a^x$
- Domain: $(-\infty, \infty)$
- Range: $(0, \infty)$
- Intercept: $(0, 1)$
- Always increasing
- $a^x \to \infty$ as $x \to \infty$
- $a^x \to 0$ as $x \to -\infty$
- Continuous
- One-to-one

Characteristics of the Exponential Functions $y = a^{-x}$ and $y = a^x \ (a > 1)$

FIGURE 10.3

Example 4   Graphing an Exponential Function

Sketch the graph of $f(x) = 3^{-x} - 1$.

SOLUTION   Begin by creating a table of values, as shown below.

x	-2	-1	0	1	2
$f(x)$	$3^2 - 1 = 8$	$3^1 - 1 = 2$	$3^0 - 1 = 0$	$3^{-1} - 1 = -\frac{2}{3}$	$3^{-2} - 1 = -\frac{8}{9}$

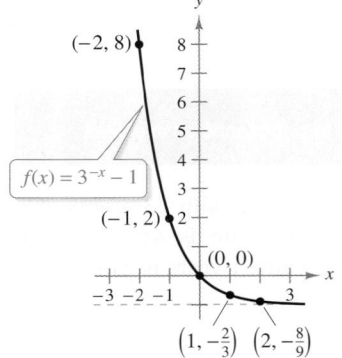

From the limit

$$\lim_{x \to \infty} (3^{-x} - 1) = \lim_{x \to \infty} 3^{-x} - \lim_{x \to \infty} 1$$

$$= \lim_{x \to \infty} \frac{1}{3^x} - \lim_{x \to \infty} 1$$

$$= 0 - 1$$

$$= -1$$

you can see that $y = -1$ is a horizontal asymptote of the graph. The graph is shown in Figure 10.4.

FIGURE 10.4

✓ **Checkpoint 4**

Sketch the graph of $f(x) = 2^{-x} + 1$. ■

SUMMARIZE   (Section 10.1)

1. State the definition of an exponential function *(page 744)*. For examples of exponential functions, see Example 3.

2. State the properties of exponents *(page 744)*. For examples of the properties of exponents, see Examples 1 and 2.

3. State the basic characteristics of the graphs of the exponential functions $y = a^{-x}$ and $y = a^x$ *(page 747)*. For an example of the graph of an exponential function, see Example 4.

SKILLS WARM UP 10.1   The following warm-up exercises involve skills that were covered in earlier sections. You will use these skills in the exercise set for this section. For additional help, review Sections 0.3, 0.4, 1.1, 1.3, and 2.6.

In Exercises 1–6, describe how the graph of g is related to the graph of f.

1. $g(x) = f(x + 2)$

2. $g(x) = -f(x)$

3. $g(x) = -1 + f(x)$

4. $g(x) = f(-x)$

5. $g(x) = f(x - 1)$

6. $g(x) = f(x) + 2$

In Exercises 7–12, evaluate each expression.

7. $25^{3/2}$

8. $64^{3/4}$

9. $27^{2/3}$

10. $\left(\dfrac{1}{5}\right)^3$

11. $\left(\dfrac{1}{8}\right)^{1/3}$

12. $\left(\dfrac{5}{8}\right)^2$

In Exercises 13–18, solve for x.

13. $2x - 6 = 4$

14. $3x + 1 = 5$

15. $(x + 4)^2 = 25$

16. $(x - 2)^2 = 8$

17. $x^2 + 4x - 5 = 0$

18. $2x^2 - 3x + 1 = 0$

Exercises 10.1

See www.CalcChat.com for worked-out solutions to odd-numbered exercises.

Applying Properties of Exponents   In Exercises 1–4, use the properties of exponents to simplify the expression. *See Example 1.*

1. (a) $(5^2)(5^3)$ (b) $(5^2)(5^{-3})$

(c) $(5^2)^2$ (d) 5^{-3}

2. (a) $\dfrac{5^3}{5^6}$ (b) $\left(\dfrac{1}{5}\right)^{-2}$

(c) $(8^{1/2})(2^{1/2})$ (d) $(32^{3/2})\left(\dfrac{1}{2}\right)^{3/2}$

3. (a) $\dfrac{5^3}{25^2}$ (b) $(9^{2/3})(3)(3^{2/3})$

(c) $[(25^{1/2})(5^2)]^{1/3}$ (d) $(8^2)(4^3)$

4. (a) $(4^3)(4^2)$ (b) $\left(\dfrac{1}{4}\right)^2(4^2)$

(c) $(4^6)^{1/2}$ (d) $[(8^{-1})(8^{2/3})]^3$

5. Radioactive Decay   Beginning with 16 grams of a radioactive element whose half-life is 30 years, the mass y (in grams) remaining after t years is given by

$$y = 16\left(\frac{1}{2}\right)^{t/30}, \quad t \geq 0.$$

How much of the initial mass remains after 90 years?

6. Radioactive Decay   Beginning with 23 grams of a radioactive element whose half-life is 45 years, the mass y (in grams) remaining after t years is given by

$$y = 23\left(\frac{1}{2}\right)^{t/45}, \quad t \geq 0.$$

How much of the initial mass remains after 150 years?

Graphing Exponential Functions   In Exercises 7–18, sketch the graph of the function. *See Examples 3 and 4.*

7. $f(x) = 6^x$

8. $f(x) = 4^x$

9. $f(x) = \left(\frac{1}{5}\right)^x = 5^{-x}$

10. $f(x) = \left(\frac{1}{4}\right)^x = 4^{-x}$

11. $y = 2^{x-1}$

12. $y = 4^x + 3$

13. $y = -2^x$

14. $y = -5^x$

15. $y = 3^{-x^2}$

16. $y = 2^{-x^2}$

17. $s(t) = \frac{1}{4}(3^{-t})$

18. $s(t) = 2^{-t} + 3$

19. Population Growth   The resident populations P (in millions) of the United States from 1995 through 2010 can be modeled by the exponential function

$$P(t) = 254.75(1.01)^t$$

where t is the time in years, with $t = 5$ corresponding to 1995. Use the model to estimate the populations in the years (a) 2013 and (b) 2020. *(Source: U.S. Census Bureau)*

20. Sales The sales S (in billions of dollars) for Walgreens from 2000 through 2010 can be modeled by the exponential function

$$S(t) = 22.52(1.125)^t$$

where t is the time in years, with $t = 0$ corresponding to 2000.

(a) Use the model to estimate the sales in 2014.

(b) Use the model to estimate the sales in 2018.

(Source: Walgreen Company)

21. Property Value A piece of property sells for $64,000. The value of the property doubles every 15 years. A model for the value V of the property t years after the date of purchase is

$$V(t) = 64,000(2)^{t/15}.$$

Use the model to approximate the value of the property (a) 5 years and (b) 20 years after it is purchased.

 **22.** **HOW DO YOU SEE IT?** Match the exponential function with its graph. Explain your reasoning. [The graphs are labeled (i), (ii), (iii), (iv), (v), and (vi).]

(i)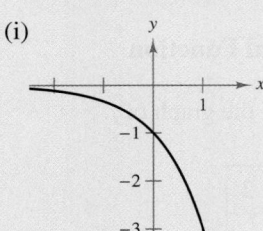

(ii)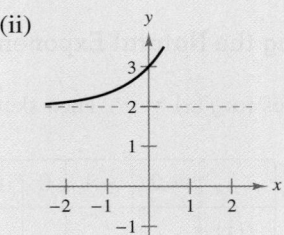

(iii)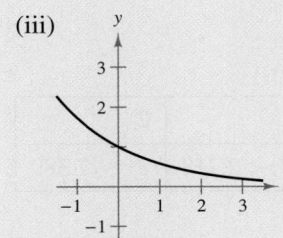

(iv)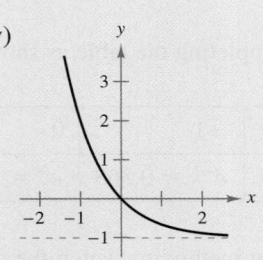

(v)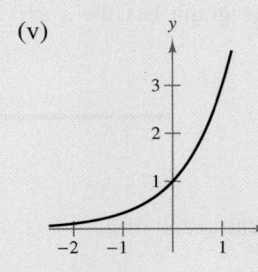

(vi)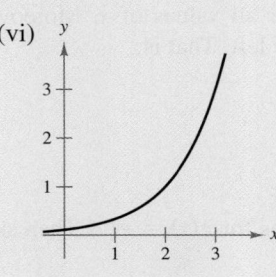

(a) $f(x) = 3^x$

(b) $f(x) = 3^{-x/2}$

(c) $f(x) = -3^x$

(d) $f(x) = 3^{x-2}$

(e) $f(x) = 3^{-x} - 1$

(f) $f(x) = 3^x + 2$

23. Inflation Rate With an annual rate of inflation of 4% over the next 10 years, the approximate cost C of goods or services during any year in the decade is given by

$$C(t) = P(1.04)^t, \ 0 \le t \le 10$$

where t is the time (in years) and P is the present cost. The price of an oil change for a car is presently $24.95. Estimate the price 10 years from now.

24. Inflation Rate Repeat Exercise 23 using an annual rate of inflation of 10% over the next 10 years. The approximate cost C of goods or services is given by

$$C(t) = P(1.10)^t, \ 0 \le t \le 10.$$

25. Depreciation A car sells for $28,000. The car depreciates such that each year it is worth $\frac{3}{4}$ of its value from the previous year. Find a model for the value V of the car after t years. Sketch a graph of the model and determine the value of the car 4 years after it is purchased.

26. Drug Concentration Immediately following an injection, the concentration of a drug in the bloodstream is 300 milligrams per milliliter. After t hours, the concentration is 75% of the level of the previous hour. Find a model for $C(t)$, the concentration of the drug after t hours. Sketch a graph of the model and determine the concentration of the drug after 8 hours.

 27. School Nurses For the years 2001 through 2008, the average salaries y (in dollars) of school nurses in the public school system in the United States are shown in the table. *(Source: Educational Research Service)*

Year	2001	2002	2003	2004
Salary	37,188	38,221	39,165	40,201

Year	2005	2006	2007	2008
Salary	40,520	41,746	43,277	46,025

A model for this data is given by

$$y = 35,963(1.0279)^t$$

where t represents the year, with $t = 1$ corresponding to 2001.

(a) Compare the actual salaries with those given by the model. How well does the model fit the data? Explain your reasoning.

(b) Use a graphing utility to graph the model.

(c) Use the *zoom* and *trace* features of a graphing utility to predict the year during which the average salary of school nurses will reach $54,000.

10.2 Natural Exponential Functions

- ■ Evaluate and graph functions involving the natural exponential function.
- ■ Solve compound interest problems.
- ■ Solve present value problems.

Natural Exponential Functions

In Section 10.1, exponential functions were introduced using an unspecified base a. In calculus, the most convenient (or natural) choice for a base is the irrational number e, whose decimal approximation is

$$e \approx 2.71828182846.$$

Although this choice of base may seem unusual, its convenience will become apparent as the rules for differentiating exponential functions are developed in Section 10.3. In that development, you will encounter the limit used in the definition of e.

Limit Definition of e

The irrational number e is defined to be the limit of $(1 + x)^{1/x}$ as $x \to 0$. That is,

$$\lim_{x \to 0} (1 + x)^{1/x} = e.$$

Example 1 Graphing the Natural Exponential Function

Complete the table of values for $f(x) = e^x$. Then sketch the graph of f.

x	-2	-1	0	1	2
$f(x)$					

SOLUTION Begin by completing the table as shown.

x	-2	-1	0	1	2
$f(x)$	$e^{-2} \approx 0.135$	$e^{-1} \approx 0.368$	$e^0 = 1$	$e^1 \approx 2.718$	$e^2 \approx 7.389$

Then use the point-plotting method to sketch the graph of f, as shown in Figure 10.5. Note that e^x is positive for all values of x. Moreover, the graph has the x-axis as a horizontal asymptote to the left. That is,

$$\lim_{x \to -\infty} e^x = 0.$$

✓**Checkpoint 1**

Complete the table of values for $g(x) = e^{-x}$. Then sketch the graph of g.

x	-2	-1	0	1	2
$g(x)$					

Las Vegas, Nevada

In Exercise 46 on page 758, you will evaluate a natural exponential function to find the population of Las Vegas, Nevada for several years.

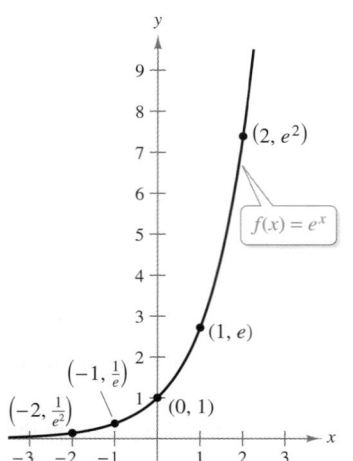

FIGURE 10.5

Exponential functions are often used to model the growth of a quantity or a population. When the quantity's growth *is not* restricted, an exponential model is often used. When the quantity's growth *is* restricted, the best model is often a **logistic growth model** of the form

$$f(t) = \frac{a}{1 + be^{-kt}}.$$

Graphs of both types of population growth models are shown in Figure 10.6. The graph of a logistic growth model is called a *logistic curve*.

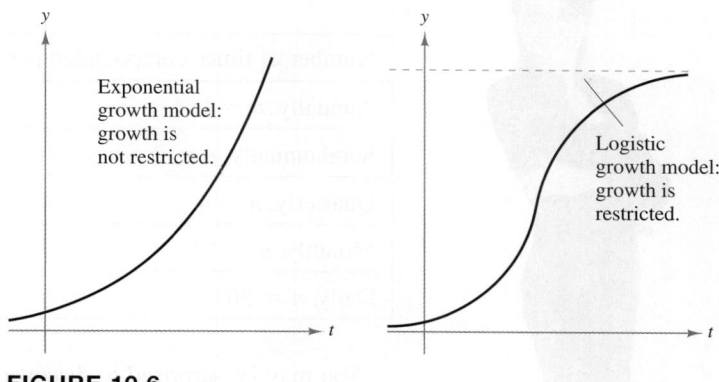

Exponential growth model: growth is not restricted.

Logistic growth model: growth is restricted.

FIGURE 10.6

 Example 2 **Modeling a Population**

A bacterial culture is growing according to the logistic growth model

$$y = \frac{1.25}{1 + 0.25e^{-0.4t}}, \quad t \geq 0$$

where y is the culture weight (in grams) and t is the time (in hours). Find the weight of the culture after 0 hours, 1 hour, and 10 hours. What is the limit of the model as t increases without bound?

SOLUTION The graph of the model is shown in Figure 10.7.

$$y = \frac{1.25}{1 + 0.25e^{-0.4(0)}} = 1 \text{ gram} \qquad \text{Weight when } t = 0$$

$$y = \frac{1.25}{1 + 0.25e^{-0.4(1)}} \approx 1.071 \text{ grams} \qquad \text{Weight when } t = 1$$

$$y = \frac{1.25}{1 + 0.25e^{-0.4(10)}} \approx 1.244 \text{ grams} \qquad \text{Weight when } t = 10$$

As t approaches infinity, the limit of y is

$$\lim_{t \to \infty} \frac{1.25}{1 + 0.25e^{-0.4t}} = \lim_{t \to \infty} \frac{1.25}{1 + (0.25/e^{0.4t})} = \frac{1.25}{1 + 0} = 1.25.$$

So, as t increases without bound, the weight of the culture approaches 1.25 grams.

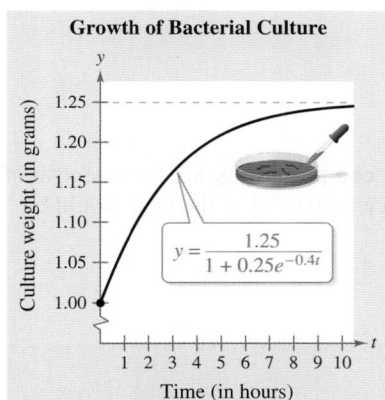

Growth of Bacterial Culture

Culture weight (in grams)

$$y = \frac{1.25}{1 + 0.25e^{-0.4t}}$$

Time (in hours)

When a culture is grown in a dish, the size of the dish and the available food limit the culture's growth.

FIGURE 10.7

✓Checkpoint 2

A bacterial culture is growing according to the model $y = 1.50/(1 + 0.2e^{-0.5t})$, $t \geq 0$, where y is the culture weight (in grams) and t is the time (in hours). Find the weight of the culture after 0 hours, 1 hour, and 10 hours. What is the limit of the model as t increases without bound?

Extended Application: Compound Interest

An amount of P dollars is deposited in an account at an annual interest rate of r (in decimal form). What is the balance after 1 year? The answer depends on the number of times the interest is compounded, according to the formula

$$A = P\left(1 + \frac{r}{n}\right)^n$$

where n is the number of compoundings per year. The balances for a deposit of $1000 at 8%, for various compounding periods, are shown in the table.

Number of times compounded per year, n	Balance (in dollars), A
Annually, $n = 1$	$A = 1000\left(1 + \frac{0.08}{1}\right)^1 = \$1080.00$
Semiannually, $n = 2$	$A = 1000\left(1 + \frac{0.08}{2}\right)^2 = \$1081.60$
Quarterly, $n = 4$	$A = 1000\left(1 + \frac{0.08}{4}\right)^4 \approx \$1082.43$
Monthly, $n = 12$	$A = 1000\left(1 + \frac{0.08}{12}\right)^{12} \approx \$1083.00$
Daily, $n = 365$	$A = 1000\left(1 + \frac{0.08}{365}\right)^{365} \approx \$1083.28$

You may be surprised to discover that as n increases, the balance A approaches a limit, as indicated in the following development. In this development, let

$$x = \frac{r}{n}.$$

Then $x \to 0$ as $n \to \infty$, and you have

$$A = \lim_{n \to \infty} P\left(1 + \frac{r}{n}\right)^n$$

$$= P \lim_{n \to \infty} \left(1 + \frac{r}{n}\right)^n$$

$$= P \lim_{n \to \infty} \left[\left(1 + \frac{r}{n}\right)^{n/r}\right]^r$$

$$= P\left[\lim_{x \to 0} (1 + x)^{1/x}\right]^r \qquad \text{Substitute } x \text{ for } r/n.$$

$$= Pe^r.$$

This limit is the balance after 1 year of **continuous compounding.** So, for a deposit of $1000 at 8%, compounded continuously, the balance at the end of the year would be

$$A = 1000e^{0.08}$$

$$\approx \$1083.29.$$

Summary of Compound Interest Formulas

Let P be the amount deposited, t the number of years, A the balance, and r the annual interest rate (in decimal form).

1. Compounded n times per year: $A = P\left(1 + \frac{r}{n}\right)^{nt}$

2. Compounded continuously: $A = Pe^{rt}$

The average interest rates paid by banks on savings accounts have varied greatly during the past 30 years. At times savings accounts have earned as much as 12% annual interest, and at times they have earned less than 1%. The next example shows how the annual interest rate can affect the balance of an account.

Example 3 Finding Account Balances

You are creating a trust fund for your newborn nephew. You plan to deposit $12,000 in an account, with instructions that the account be turned over to your nephew on his 25th birthday. Compare the balances in the account for each situation. Which account should you choose?

a. 7%, compounded continuously

b. 7%, compounded quarterly

c. 11%, compounded continuously

d. 11%, compounded quarterly

SOLUTION

a. $12{,}000e^{0.07(25)} \approx 69{,}055.23$ 7%, compounded continuously

b. $12{,}000\left(1 + \dfrac{0.07}{4}\right)^{4(25)} \approx 68{,}017.87$ 7%, compounded quarterly

c. $12{,}000e^{0.11(25)} \approx 187{,}711.58$ 11%, compounded continuously

d. $12{,}000\left(1 + \dfrac{0.11}{4}\right)^{4(25)} \approx 180{,}869.07$ 11%, compounded quarterly

The balances in the accounts for parts (a) and (c) are shown in Figure 10.8. Notice the dramatic difference between the balances at 7% and 11%. You should choose the account described in part (c) because it earns more money than the other accounts.

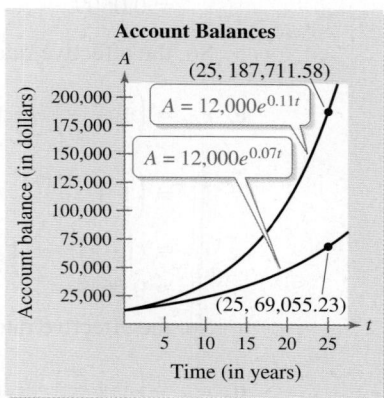

Account Balances

$A = 12{,}000e^{0.11t}$

$A = 12{,}000e^{0.07t}$

(25, 187,711.58)

(25, 69,055.23)

FIGURE 10.8

✓ Checkpoint 3

Find the balance in an account when $2000 is deposited for 10 years at an interest rate of 9%, compounded as follows. Compare the results and make a general statement about compounding.

a. quarterly   **b.** monthly

c. daily   **d.** continuously

In Example 3, note that the interest earned depends on the frequency with which the interest is compounded. The annual percentage rate is called the **stated rate** or **nominal rate.** However, the nominal rate does not reflect the actual rate at which interest is earned, which means that the compounding produced an **effective rate** that is larger than the nominal rate. In general, the effective rate corresponding to a nominal rate of r that is compounded n times per year is

$$\text{Effective rate} = r_{eff} = \left(1 + \frac{r}{n}\right)^n - 1.$$

 Example 4   **Finding the Effective Rate of Interest**

Find the effective rate of interest corresponding to a nominal rate of 6% per year compounded (a) annually, (b) semiannually, (c) quarterly, and (d) monthly.

SOLUTION

a. $r_{eff} = \left(1 + \dfrac{r}{n}\right)^n - 1$ Formula for effective rate of interest

$\quad = \left(1 + \dfrac{0.06}{1}\right)^1 - 1$ Substitute for r and n.

$\quad = 1.06 - 1$ Simplify.

$\quad = 0.06$

So, the effective rate is 6% per year.

b. $r_{eff} = \left(1 + \dfrac{r}{n}\right)^n - 1$ Formula for effective rate of interest

$\quad = \left(1 + \dfrac{0.06}{2}\right)^2 - 1$ Substitute for r and n.

$\quad = (1.03)^2 - 1$ Simplify.

$\quad = 0.0609$

So, the effective rate is 6.09% per year.

c. $r_{eff} = \left(1 + \dfrac{r}{n}\right)^n - 1$ Formula for effective rate of interest

$\quad = \left(1 + \dfrac{0.06}{4}\right)^4 - 1$ Substitute for r and n.

$\quad = (1.015)^4 - 1$ Simplify.

$\quad \approx 0.0614$

So, the effective rate is about 6.14% per year.

d. $r_{eff} = \left(1 + \dfrac{r}{n}\right)^n - 1$ Formula for effective rate of interest

$\quad = \left(1 + \dfrac{0.06}{12}\right)^{12} - 1$ Substitute for r and n.

$\quad = (1.005)^{12} - 1$ Simplify.

$\quad \approx 0.0617$

So, the effective rate is about 6.17% per year.

✓ **Checkpoint 4**

Repeat Example 4 using a nominal rate of 7%.

Present Value

In planning for the future, this problem often arises: "How much money P should be deposited now, at a fixed rate of interest r, in order to have a balance of A, t years from now?" The answer to this question is given by the **present value** of A.

To find the present value of a future investment, use the formula for compound interest as shown.

$$A = P\left(1 + \frac{r}{n}\right)^{nt} \qquad \text{Formula for compound interest}$$

Solving for P gives a present value of

$$P = \frac{A}{\left(1 + \dfrac{r}{n}\right)^{nt}} \quad \text{or} \quad P = \frac{A}{(1 + i)^N}$$

where $i = r/n$ is the interest rate per compounding period and $N = nt$ is the total number of compounding periods. You will learn another way to find the present value of a future investment in Section 12.1.

 ### Example 5 Finding Present Value

An investor is purchasing a 10-year certificate of deposit that pays an annual percentage rate of 8%, compounded monthly. How much should the person invest in order to obtain a balance of $15,000 at maturity?

SOLUTION Here, $A = 15,000$, $r = 0.08$, $n = 12$, and $t = 10$. Using the formula for present value, you obtain

$$P = \frac{15,000}{\left(1 + \dfrac{0.08}{12}\right)^{12(10)}} \qquad \text{Substitute for } A, r, n, \text{ and } t.$$

$$\approx 6757.85. \qquad \text{Simplify.}$$

So, the person should invest $6757.85 in the certificate of deposit.

✓ Checkpoint 5

How much money should be deposited in an account paying 6% interest compounded monthly in order to have a balance of $20,000 after 3 years?

SUMMARIZE   (Section 10.2)

1. State the limit definition of e *(page 750)*. For an example of a graph of a natural exponential function, see Example 1.

2. Describe a real-life example of how an exponential function can be used to model a population *(page 751, Example 2)*.

3. State the compound interest formulas for n compoundings per year and for continuous compounding *(page 752)*. For applications of these formulas, see Example 3.

4. State the formula for finding the effective rate of interest *(page 754)*. For an application of this formula, see Example 4.

5. State the formula for present value *(page 755)*. For an application of this formula, see Example 5.

SKILLS WARM UP 10.2   The following warm-up exercises involve skills that were covered in earlier sections. You will use these skills in the exercise set for this section. For additional help, review Sections 7.2 and 9.3.

In Exercises 1–4, discuss the continuity of the function.

1. $f(x) = \dfrac{3x^2 + 2x + 1}{x^2 + 1}$

2. $f(x) = \dfrac{x + 1}{x^2 - 4}$

3. $f(x) = \dfrac{x^2 - 6x + 5}{x^2 - 3}$

4. $g(x) = \dfrac{x^2 - 9x + 20}{x - 4}$

In Exercises 5–12, find the horizontal asymptote of the graph of the function.

5. $f(x) = \dfrac{25}{1 + 4x}$

6. $f(x) = \dfrac{16x}{3 + x^2}$

7. $f(x) = \dfrac{8x^3 + 2}{2x^3 + x}$

8. $f(x) = \dfrac{x}{2x}$

9. $f(x) = \dfrac{3}{2 + (1/x)}$

10. $f(x) = \dfrac{6}{1 + x^{-2}}$

11. $f(x) = 2^{-x}$

12. $f(x) = \dfrac{7}{1 + 5x}$

Exercises 10.2

See www.CalcChat.com for worked-out solutions to odd-numbered exercises.

Applying Properties of Exponents   In Exercises 1–4, use the properties of exponents to simplify the expression.

1. (a) $(e^3)(e^4)$ (b) $(e^3)^4$

   (c) $(e^3)^{-2}$ (d) e^0

2. (a) $\left(\dfrac{1}{e}\right)^{-2}$ (b) $\left(\dfrac{e^5}{e^2}\right)^{-1}$

   (c) $\dfrac{e^5}{e^3}$ (d) $\dfrac{1}{e^{-3}}$

3. (a) $(e^2)^{5/2}$ (b) $(e^2)(e^{1/2})$

   (c) $(e^{-2})^{-3}$ (d) $\dfrac{e^5}{e^{-2}}$

4. (a) $(e^{-3})^{2/3}$ (b) $\dfrac{e^4}{e^{-1/2}}$

   (c) $(e^{-2})^{-4}$ (d) $(e^{-4})(e^{-3/2})$

Matching   In Exercises 5–10, match the function with its graph. [The graphs are labeled (a)–(f).]

(a)
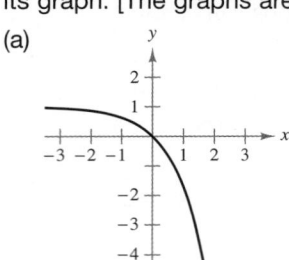

(b)
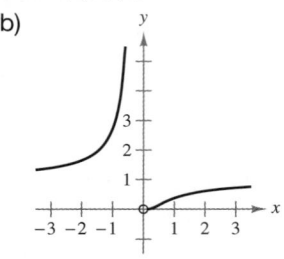

(c)
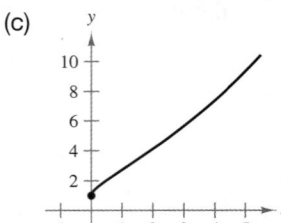

(d)
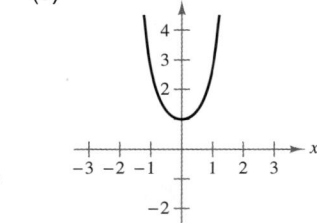

(e)
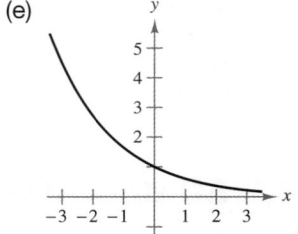

(f)
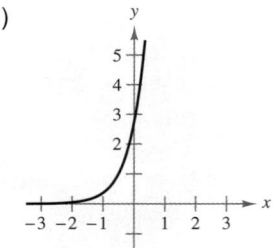

5. $f(x) = e^{2x+1}$ **6.** $f(x) = e^{-x/2}$

7. $f(x) = e^{x^2}$ **8.** $f(x) = e^{-1/x}$

9. $f(x) = e^{\sqrt{x}}$ **10.** $f(x) = -e^x + 1$

Graphing Natural Exponential Functions   In Exercises 11–16, sketch the graph of the function. *See Example 1.*

11. $f(x) = e^{-x/3}$ **12.** $f(x) = e^{2x}$

13. $g(x) = e^x - 2$ **14.** $h(x) = e^{-x} + 5$

15. $g(x) = e^{1-x}$ **16.** $j(x) = e^{-x+2}$

Graphing Functions   In Exercises 17–24, use a graphing utility to graph the function. Determine whether the function has any horizontal asymptotes and discuss the continuity of the function.

17. $N(t) = 500e^{-0.2t}$

18. $A(t) = 500e^{0.15t}$

19. $g(x) = \dfrac{2}{1 + e^{x^2}}$

20. $g(x) = \dfrac{10}{1 + e^{-x}}$

21. $f(x) = \dfrac{e^x + e^{-x}}{2}$

22. $f(x) = \dfrac{e^x - e^{-x}}{2}$

23. $f(x) = \dfrac{2}{1 + e^{1/x}}$

24. $f(x) = \dfrac{2}{1 + 2e^{-0.2x}}$

25. Graphing Exponential Functions   Use a graphing utility to graph $f(x) = e^x$ and the given function in the same viewing window. How are the two graphs related?

(a) $g(x) = e^{x-2}$ (b) $h(x) = -\frac{1}{2}e^x$

(c) $q(x) = e^x + 3$

26. Graphing Logistic Growth Functions   Use a graphing utility to graph the function. Describe the shape of the graph for very large and very small values of x.

(a) $f(x) = \dfrac{8}{1 + e^{-0.5x}}$ (b) $g(x) = \dfrac{8}{1 + e^{-0.5/x}}$

Finding Account Balances   In Exercises 27–30, complete the table to determine the balance A for P dollars invested at rate r for t years, compounded n times per year. *See Example 3.*

n	1	2	4	12	365	Continuous compounding
A						

27. $P = \$1000$, $r = 3\%$, $t = 10$ years

28. $P = \$2500$, $r = 2.5\%$, $t = 20$ years

29. $P = \$1000$, $r = 4\%$, $t = 20$ years

30. $P = \$2500$, $r = 5\%$, $t = 40$ years

Finding Present Value   In Exercises 31–34, complete the table to determine the amount of money P that should be invested at rate r to produce a final balance of $\$100,000$ in t years. *See Example 5.*

t	1	10	20	30	40	50
P						

31. $r = 4\%$, compounded continuously

32. $r = 3\%$, compounded continuously

33. $r = 5\%$, compounded monthly

34. $r = 6\%$, compounded daily

35. Trust Fund   On the day of a child's birth, a deposit of $\$20,000$ is made in a trust fund that pays 8% interest, compounded continuously. Determine the balance in this account on the child's 21st birthday.

36. Trust Fund   A deposit of $\$10,000$ is made in a trust fund that pays 7% interest, compounded continuously. It is specified that the balance will be given to the college from which the donor graduated after the money has earned interest for 50 years. How much will the college receive?

37. Effective Rate   Find the effective rate of interest corresponding to a nominal rate of 9% per year compounded (a) annually, (b) semiannually, (c) quarterly, and (d) monthly.

38. Effective Rate   Find the effective rate of interest corresponding to a nominal rate of 7.5% per year compounded (a) annually, (b) semiannually, (c) quarterly, and (d) monthly.

39. Present Value   How much should be deposited in an account paying 7.2% interest compounded monthly in order to have a balance of $\$8000$ after 3 years?

40. Present Value   How much should be deposited in an account paying 7.8% interest compounded monthly in order to have a balance of $\$21,000$ after 4 years?

41. Demand   The demand function for a product is modeled by

$$p = 5000\left(1 - \frac{4}{4 + e^{-0.002x}}\right).$$

Find the price p (in dollars) of the product when the quantity demanded is (a) $x = 100$ units and (b) $x = 500$ units. (c) What is the limit of the price as x increases without bound?

42. Demand   The demand function for a product is modeled by

$$p = 10,000\left(1 - \frac{3}{3 + e^{-0.001x}}\right).$$

Find the price p (in dollars) of the product when the quantity demanded is (a) $x = 1000$ units and (b) $x = 1500$ units. (c) What is the limit of the price as x increases without bound?

43. Probability   The average time between incoming calls at a switchboard is 3 minutes. If a call has just come in, the probability that the next call will come within the next t minutes is $P(t) = 1 - e^{-t/3}$. Find the probability of each situation.

(a) A call comes in within $\frac{1}{2}$ minute.

(b) A call comes in within 2 minutes.

(c) A call comes in within 5 minutes.

44. Consumer Awareness  An automobile gets 28 miles per gallon at speeds up to and including 50 miles per hour. At speeds greater than 50 miles per hour, the number of miles per gallon drops at the rate of 12% for each 10 miles per hour. If s is the speed (in miles per hour) and y is the number of miles per gallon, then $y = 28e^{0.6 - 0.012s}$, $s > 50$. Use this information and a spreadsheet to create a table showing the miles per gallon for $s = 50, 55, 60, 65,$ and 70. What can you conclude?

45. Federal Debt The federal debts D (in billions of dollars) of the United States at the end of each year from 2000 through 2009 are shown in the table. *(Source: U.S. Office of Management and Budget)*

Year	2000	2001	2002	2003	2004
Debt	5629	5770	6198	6760	7355

Year	2005	2006	2007	2008	2009
Debt	7905	8451	8951	9986	11,876

A model for these data is given by $y = 5364.1e^{0.0796t}$, where t represents the year, with $t = 0$ corresponding to 2000.

(a) How well does the model fit the data?

(b) Find a linear model for the data. How well does the linear model fit the data? Which model, exponential or linear, is a better fit?

(c) Use both models to predict the year in which the federal debt will exceed 18,000 billion dollars.

46. Population The populations P (in thousands) of Las Vegas, Nevada from 1960 through 2009 can be modeled by $P = 70.751e^{0.0451t}$, where t is the time in years, with $t = 0$ corresponding to 1960. *(Source: U.S. Census Bureau)*

(a) Find the populations in 1960, 1970, 1980, 1990, 2000, and 2009.

(b) Explain why the change in population from 1960 to 1970 is not the same as the change in population from 1980 to 1990.

(c) Use the model to estimate the population in 2020.

47. Biology The population y of a bacterial culture is modeled by the logistic growth function $y = 925/(1 + e^{-0.3t})$, where t is the time in days.

(a) Use a graphing utility to graph the model.

(b) Does the population have a limit as t increases without bound? Explain your answer.

(c) How would the limit change if the model were $y = 1000/(1 + e^{-0.3t})$? Explain your answer. Draw some conclusions about this type of model.

 **48. HOW DO YOU SEE IT?** The figure shows the graphs of $y = 2^x$, $y = e^x$, $y = 10^x$, $y = 2^{-x}$, $y = e^{-x}$, and $y = 10^{-x}$. Match each function with its graph. [The graphs are labeled (a)–(f).] Explain your reasoning.

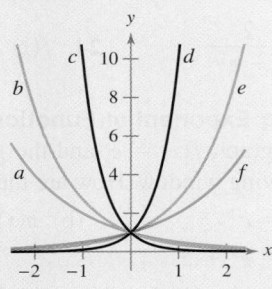

49. Learning Theory In a learning theory project, the proportion P of correct responses after n trials can be modeled by $P = 0.83/(1 + e^{-0.2n})$.

(a) Find the proportion of correct responses after 3 trials.

(b) Find the proportion of correct responses after 7 trials.

(c) Use a graphing utility to graph the model. Find the number of trials required for the proportion of correct responses to be 0.75.

(d) Does the proportion of correct responses have a limit as n increases without bound? Explain your reasoning.

50. Learning Theory In a typing class, the average number N of words per minute typed after t weeks of lessons can be modeled by $N = 95/(1 + 8.5e^{-0.12t})$.

(a) Find the average number of words per minute typed after 10 weeks.

(b) Find the average number of words per minute typed after 20 weeks.

(c) Use a graphing utility to graph the model. Find the number of weeks required to achieve an average of 70 words per minute.

(d) Does the number of words per minute have a limit as t increases without bound? Explain your reasoning.

51. Certificate of Deposit You want to invest $5000 in a certificate of deposit for 12 months. You are given the options below. Which would you choose? Explain.

(a) $r = 5.25\%$, quarterly compounding

(b) $r = 5\%$, monthly compounding

(c) $r = 4.75\%$, continuous compounding

10.3 Derivatives of Exponential Functions

■ Find the derivatives of natural exponential functions.
■ Use calculus to analyze the graphs of real-life functions that involve the natural exponential function.
■ Explore the normal probability density function.

Derivatives of Exponential Functions

In Section 10.2, it was stated that the most convenient base for exponential functions is the irrational number e. The convenience of this base stems primarily from the fact that the function

$$f(x) = e^x$$

is its own derivative. You will see that this is not true of other exponential functions of the form

$$y = a^x$$

where $a \neq e$. To verify that $f(x) = e^x$ is its own derivative, notice that the limit

$$\lim_{\Delta x \to 0} (1 + \Delta x)^{1/\Delta x} = e$$

implies that for small values of Δx,

$$e \approx (1 + \Delta x)^{1/\Delta x}$$

or

$$e^{\Delta x} \approx 1 + \Delta x.$$

This approximation is used in the following derivation.

$$
\begin{aligned}
f'(x) &= \lim_{\Delta x \to 0} \frac{f(x + \Delta x) - f(x)}{\Delta x} && \text{Definition of derivative} \\
&= \lim_{\Delta x \to 0} \frac{e^{x + \Delta x} - e^x}{\Delta x} && \text{Use } f(x) = e^x. \\
&= \lim_{\Delta x \to 0} \frac{e^x(e^{\Delta x} - 1)}{\Delta x} && \text{Factor numerator.} \\
&= \lim_{\Delta x \to 0} \frac{e^x[(1 + \Delta x) - 1]}{\Delta x} && \text{Substitute } 1 + \Delta x \text{ for } e^{\Delta x}. \\
&= \lim_{\Delta x \to 0} \frac{e^x(\Delta x)}{\Delta x} && \text{Divide out common factor.} \\
&= \lim_{\Delta x \to 0} e^x && \text{Simplify.} \\
&= e^x && \text{Evaluate limit.}
\end{aligned}
$$

When u is a function of x, you can apply the Chain Rule to obtain the derivative of e^u with respect to x. Both formulas are summarized below.

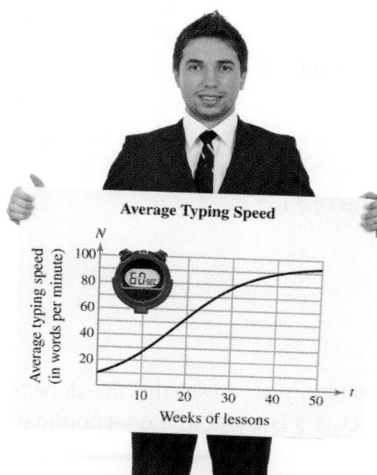

Average Typing Speed

In Exercise 48 on page 766, you will use the derivative of an exponential function to find the rate of change of the average typing speed after 5, 10, and 30 weeks of lessons.

Derivative of the Natural Exponential Function

Let u be a differentiable function of x.

1. $\dfrac{d}{dx}[e^x] = e^x$ 2. $\dfrac{d}{dx}[e^u] = e^u \dfrac{du}{dx}$

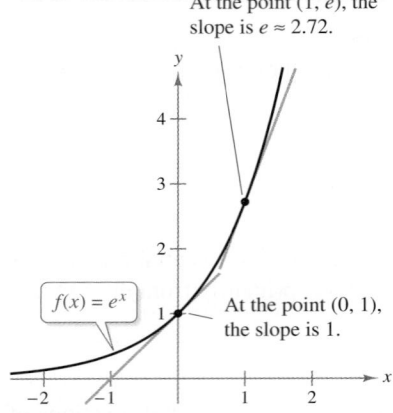

At the point $(1, e)$, the slope is $e \approx 2.72$.

$f(x) = e^x$

At the point $(0, 1)$, the slope is 1.

FIGURE 10.9

Example 1   Finding Slopes of Tangent Lines

Find the slopes of the tangent lines to

$$f(x) = e^x$$

at the points $(0, 1)$ and $(1, e)$. What conclusion can you make?

SOLUTION   Because the derivative of f is

$$f'(x) = e^x \qquad \text{Derivative}$$

it follows that the slope of the tangent line to the graph of f is

$$f'(0) = e^0 = 1 \qquad \text{Slope at point } (0, 1)$$

at the point $(0, 1)$ and

$$f'(1) = e^1 = e \qquad \text{Slope at point } (1, e)$$

at the point $(1, e)$, as shown in Figure 10.9. From this pattern, you can see that the slope of the tangent line to the graph of $f(x) = e^x$ at any point (x, e^x) is equal to the y-coordinate of the point.

✓ **Checkpoint 1**

Find the slopes of the tangent lines to $f(x) = 2e^x$ at the points $(0, 2)$ and $(1, 2e)$.   ■

Example 2   Differentiating Exponential Functions

Differentiate each function.

a. $f(x) = e^{2x}$ **b.** $f(x) = e^{-3x^2}$

c. $f(x) = 6e^{x^3}$ **d.** $f(x) = e^{-x}$

SOLUTION

a. Let $u = 2x$. Then $du/dx = 2$, and you can apply the Chain Rule.

$$f'(x) = e^u \frac{du}{dx} = e^{2x}(2) = 2e^{2x}$$

b. Let $u = -3x^2$. Then $du/dx = -6x$, and you can apply the Chain Rule.

$$f'(x) = e^u \frac{du}{dx} = e^{-3x^2}(-6x) = -6xe^{-3x^2}$$

c. Let $u = x^3$. Then $du/dx = 3x^2$, and you can apply the Chain Rule.

$$f'(x) = 6e^u \frac{du}{dx} = 6e^{x^3}(3x^2) = 18x^2e^{x^3}$$

d. Let $u = -x$. Then $du/dx = -1$, and you can apply the Chain Rule.

$$f'(x) = e^u \frac{du}{dx} = e^{-x}(-1) = -e^{-x}$$

✓ **Checkpoint 2**

Differentiate each function.

a. $f(x) = e^{3x}$ **b.** $f(x) = e^{-2x^3}$

c. $f(x) = 4e^{x^2}$ **d.** $f(x) = e^{-2x}$ ■

STUDY TIP

In Example 2, notice that when you differentiate an exponential function, the exponent does not change. For instance, the derivative of $f(x) = e^{3x}$ is $f'(x) = 3e^{3x}$. In both f and f', the exponent is $3x$.

The differentiation rules that you studied in Chapter 7 can be used with exponential functions, as shown in Example 3.

Example 3   Differentiating Exponential Functions

Differentiate each function.

a. $f(x) = 4e$ **b.** $f(x) = e^{2x-1}$

c. $f(x) = xe^x$ **d.** $f(x) = \dfrac{e^x - e^{-x}}{2}$

e. $f(x) = \dfrac{e^x}{x}$ **f.** $f(x) = xe^x - e^x$

SOLUTION

a. $f(x) = 4e$ Write original function.

   $f'(x) = 0$ Constant Rule

b. $f(x) = e^{2x-1}$ Write original function.

   $f'(x) = (e^{2x-1})(2)$ Chain Rule

   $= 2e^{2x-1}$ Simplify.

c. $f(x) = xe^x$ Write original function.

   $f'(x) = xe^x + e^x(1)$ Product Rule

   $= xe^x + e^x$ Simplify.

d. $f(x) = \dfrac{e^x - e^{-x}}{2}$ Write original function.

   $= \dfrac{1}{2}(e^x - e^{-x})$ Rewrite.

   $f'(x) = \dfrac{1}{2}[e^x - e^{-x}(-1)]$ Constant Multiple and Chain Rules

   $= \dfrac{1}{2}(e^x + e^{-x})$ Simplify.

e. $f(x) = \dfrac{e^x}{x}$ Write original function.

   $f'(x) = \dfrac{xe^x - e^x(1)}{x^2}$ Quotient Rule

   $= \dfrac{e^x(x-1)}{x^2}$ Simplify.

f. $f(x) = xe^x - e^x$ Write original function.

   $f'(x) = [xe^x + e^x(1)] - e^x$ Product and Difference Rules

   $= xe^x + e^x - e^x$

   $= xe^x$ Simplify.

✓ **Checkpoint 3**

Differentiate each function.

a. $f(x) = 9e$ **b.** $f(x) = e^{3x+1}$ **c.** $f(x) = x^2 e^x$

d. $f(x) = \dfrac{e^x + e^{-x}}{2}$ **e.** $f(x) = \dfrac{e^x}{x^2}$ **f.** $f(x) = x^2 e^x - e^x$

TECH TUTOR

If you have access to a symbolic differentiation utility, try using it to find the derivatives of the functions in Example 3.

@erics/www.shutterstock.com

Applications

In Chapter 8 and Chapter 9, you learned how to use derivatives to analyze the graphs of functions. The next example applies those techniques to a function composed of exponential functions. In the example, notice that

$$e^a = e^b$$

implies that $a = b$.

 Example 4   Analyzing a Catenary

When a telephone wire is hung between two poles, the wire forms a U-shaped curve called a **catenary.** For instance, the function

$$y = 30(e^{x/60} + e^{-x/60}), \quad -30 \le x \le 30$$

models the shape of a telephone wire strung between two poles that are 60 feet apart (x and y are measured in feet). Show that the lowest point on the wire is midway between the two poles. How much does the wire sag between the two poles?

SOLUTION   First, find the derivative of the function.

$y = 30(e^{x/60} + e^{-x/60})$	Write original function.
$y' = 30\left[e^{x/60}\left(\frac{1}{60}\right) + e^{-x/60}\left(-\frac{1}{60}\right)\right]$	Derivative
$= 30\left(\frac{1}{60}\right)(e^{x/60} - e^{-x/60})$	Factor out $\frac{1}{60}$.
$= \frac{1}{2}(e^{x/60} - e^{-x/60})$	Simplify.

To find the critical numbers, set the derivative equal to zero.

$\frac{1}{2}(e^{x/60} - e^{-x/60}) = 0$	Set derivative equal to 0.
$e^{x/60} - e^{-x/60} = 0$	Multiply each side by 2.
$e^{x/60} = e^{-x/60}$	Add $e^{-x/60}$ to each side.
$\dfrac{x}{60} = -\dfrac{x}{60}$	If $e^a = e^b$, then $a = b$.
$x = -x$	Multiply each side by 60.
$2x = 0$	Add x to each side.
$x = 0$	Divide each side by 2.

Using the First-Derivative Test, you can determine that the critical number $x = 0$ yields a relative minimum of the function. From the graph in Figure 10.10, you can see that this relative minimum is actually a minimum on the interval

$$[-30, 30].$$

So, you can conclude that the lowest point on the wire lies midway between the two poles. To find how much the wire sags between the two poles, you can compare its height at each pole with its height at the relative minimum.

$y = 30(e^{-30/60} + e^{-(-30)/60}) \approx 67.7$ feet	Height at left pole
$y = 30(e^{0/60} + e^{-(0)/60}) = 60$ feet	Height at relative minimum
$y = 30(e^{30/60} + e^{-(30)/60}) \approx 67.7$ feet	Height at right pole

From this, you can see that the wire sags about 7.7 feet. ▬▬▬▬▬▬

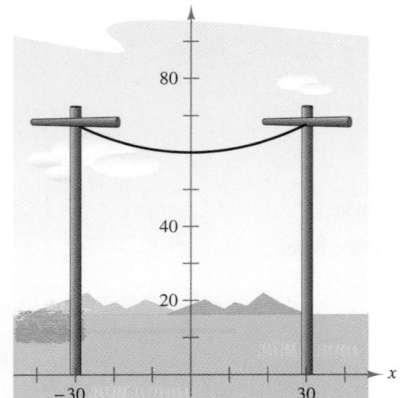

FIGURE 10.10

✓ Checkpoint 4

Use a graphing utility to graph the function in Example 4. Verify the minimum value. Use the information in the example to choose an appropriate viewing window. ▪

Example 5 **Finding a Maximum Revenue**

The demand function for a product is modeled by

$$p = 56e^{-0.000012x} \qquad \text{Demand function}$$

where p is the price per unit (in dollars) and x is the number of units. What price will yield a maximum revenue? What is the maximum revenue at this price?

SOLUTION The revenue function is

$$R = xp = 56xe^{-0.000012x}. \qquad \text{Revenue function}$$

To find the maximum revenue *analytically*, you would first find the marginal revenue

$$\frac{dR}{dx} = 56xe^{-0.000012x}(-0.000012) + e^{-0.000012x}(56).$$

You would then set dR/dx equal to zero

$$56xe^{-0.000012x}(-0.000012) + e^{-0.000012x}(56) = 0$$

and solve for x. At this point, you can see that the analytical approach is rather cumbersome. In this problem, it is easier to use a *graphical* approach. After experimenting to find a reasonable viewing window, you can obtain a graph of R that is similar to that shown in Figure 10.11. Using the *maximum* feature, you can conclude that the maximum revenue occurs when x is about 83,333 units. To find the price that corresponds to this production level, substitute $x \approx 83,333$ into the demand function.

$$p \approx 56e^{-0.000012(83,333)} \approx \$20.60.$$

So, a price of about $20.60 will yield a maximum revenue of

$$R \approx 56(83,333)e^{-0.000012(83,333)}$$
$$\approx \$1,716,771. \qquad \text{Maximum revenue}$$

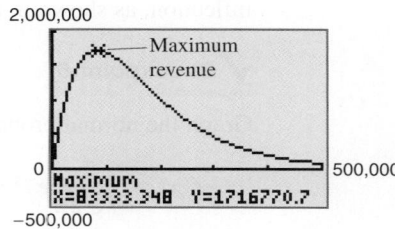

Use the *maximum* feature to approximate the x-value that corresponds to the maximum revenue.

FIGURE 10.11

 **Checkpoint 5**

The demand function for a product is modeled by

$$p = 50e^{-0.0000125x}$$

where p is the price per unit (in dollars) and x is the number of units. What price will yield a maximum revenue? What is the maximum revenue at this price?

Try solving Example 5 analytically. When you do this, you must solve the equation

$$56xe^{-0.000012x}(-0.000012) + e^{-0.000012x}(56) = 0.$$

Explain how you would solve this equation. What is the solution?

The Normal Probability Density Function

If you take a course in statistics or quantitative business analysis, you will spend quite a bit of time studying the characteristics and use of the **normal probability density function** given by

$$f(x) = \frac{1}{\sigma\sqrt{2\pi}}e^{-(x-\mu)^2/(2\sigma^2)}$$

where σ is the lowercase Greek letter sigma, and μ is the lowercase Greek letter mu. In this formula, σ represents the *standard deviation* of the probability distribution, and μ represents the *mean* of the probability distribution.

Example 6	**Exploring a Probability Density Function**

Show that the graph of the normal probability density function

$$f(x) = \frac{1}{\sqrt{2\pi}}e^{-x^2/2}$$

has points of inflection at $x = \pm 1$.

SOLUTION   Begin by finding the second derivative of the function.

$$f'(x) = \frac{1}{\sqrt{2\pi}}(-x)e^{-x^2/2} \qquad \text{First derivative}$$

$$f''(x) = \frac{1}{\sqrt{2\pi}}[(-x)(-x)e^{-x^2/2} + (-1)e^{-x^2/2}] \qquad \text{Second derivative}$$

$$= \frac{1}{\sqrt{2\pi}}(e^{-x^2/2})(x^2 - 1) \qquad \text{Simplify.}$$

By setting the second derivative equal to 0, you can determine that $x = \pm 1$. By testing the concavity of the graph, you can then conclude that these x-values yield points of inflection, as shown in Figure 10.12. _____

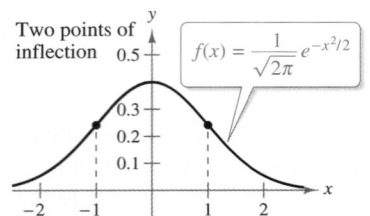

Two points of inflection

$$f(x) = \frac{1}{\sqrt{2\pi}}e^{-x^2/2}$$

The graph of the normal probability density function is bell-shaped.

FIGURE 10.12

✓ Checkpoint 6

Graph the normal probability density function

$$f(x) = \frac{1}{4\sqrt{2\pi}}e^{-x^2/32}$$

and approximate the points of inflection.   ■

SUMMARIZE   (Section 10.3)

1. State the derivative of the natural exponential function *(page 759)*. For examples of the derivative of the natural exponential function, see Examples 2 and 3.

2. Describe a real-life example of how a natural exponential function can be used to analyze the graph of a catenary *(page 762, Example 4)*.

3. Describe a real-life example of how a natural exponential function can be used to analyze a company's maximum revenue *(page 763, Example 5)*.

4. Describe a use of the natural exponential function in statistics *(page 764)*. For an example of the natural exponential function in statistics, see Example 6.

SKILLS WARM UP 10.3 The following warm-up exercises involve skills that were covered in a previous course or in earlier sections. You will use these skills in the exercise set for this section. For additional help, review Sections 0.6, 7.4, 7.6, and 8.5.

In Exercises 1–4, factor the expression.

1. $x^2e^x - \frac{1}{2}e^x$ **2.** $(xe^{-x})^{-1} + e^x$ **3.** $xe^x - e^{2x}$ **4.** $e^x - xe^{-x}$

In Exercises 5–8, find the derivative of the function.

5. $f(x) = \dfrac{3}{7x^2}$ **6.** $g(x) = 3x^2 - \dfrac{x}{6}$ **7.** $f(x) = (4x - 3)(x^2 + 9)$ **8.** $f(t) = \dfrac{t - 2}{\sqrt{t}}$

In Exercises 9 and 10, find the relative extrema of the function.

9. $f(x) = \frac{1}{8}x^3 - 2x$ **10.** $f(x) = x^4 - 2x^2 + 5$

Exercises 10.3

See www.CalcChat.com for worked-out solutions to odd-numbered exercises.

Differentiating Exponential Functions In Exercises 1–16, find the derivative of the function. *See Examples 2 and 3.*

1. $f(x) = 3e$

2. $f(x) = -5e$

3. $y = e^{5x}$

4. $y = e^{1-x}$

5. $y = e^{-x^2}$

6. $f(x) = e^{1/x}$

7. $f(x) = e^{-1/x^2}$

8. $g(x) = e^{\sqrt{x}}$

9. $f(x) = (x^2 + 1)e^{4x}$

10. $y = 4x^3e^{-x}$

11. $f(x) = \dfrac{2}{(e^x + e^{-x})^3}$

12. $f(x) = \dfrac{(e^x + e^{-x})^4}{2}$

13. $f(x) = \dfrac{e^x + 1}{e^x - 1}$

14. $f(x) = \dfrac{e^{2x}}{e^{2x} + 1}$

15. $y = xe^x - 4e^{-x}$

16. $y = x^2e^x - 2xe^x + 2e^x$

Finding the Slope of a Tangent Line In Exercises 17–20, find the slope of the tangent line to the exponential function at the point $(0, 1)$.

17. $y = e^{4x}$

18. $y = e^{x/2}$

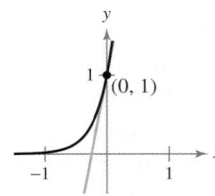   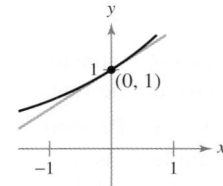

19. $y = e^{-3x}$

20. $y = e^{-x/2}$

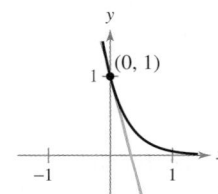   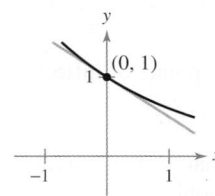

Finding an Equation of a Tangent Line In Exercises 21–26, find an equation of the tangent line to the graph of the function at the given point.

21. $y = e^{-2x+x^2}$, $(2, 1)$ **22.** $g(x) = e^{x^3}$, $\left(-1, \dfrac{1}{e}\right)$

23. $y = x^2e^{-x}$, $\left(2, \dfrac{4}{e^2}\right)$ **24.** $y = \dfrac{x}{e^{2x}}$, $\left(1, \dfrac{1}{e^2}\right)$

25. $y = (e^{2x} + 1)^3$, $(0, 8)$ **26.** $y = (e^{4x} - 2)^2$, $(0, 1)$

Finding Derivatives Implicitly In Exercises 27–30, find dy/dx implicitly.

27. $xe^y - 10x + 3y = 0$

28. $x^2y - e^y - 4 = 0$

29. $x^2e^{-x} + 2y^2 - xy = 0$

30. $e^{xy} + x^2 - y^2 = 10$

Finding Second Derivatives In Exercises 31–34, find the second derivative.

31. $f(x) = 2e^{3x} + 3e^{-2x}$ **32.** $f(x) = 5e^{-x} - 2e^{-5x}$

33. $f(x) = (1 + 2x)e^{4x}$ **34.** $f(x) = (3 + 2x)e^{-3x}$

Analyzing a Graph In Exercises 35–38, analyze and sketch the graph of the function. Label any relative extrema, points of inflection, and asymptotes.

35. $f(x) = \dfrac{1}{2 - e^{-x}}$ **36.** $f(x) = \dfrac{e^x - e^{-x}}{2}$

37. $f(x) = x^2e^{-x}$ **38.** $f(x) = xe^{-x}$

Solving Equations In Exercises 39–42, solve the equation for x.

39. $e^{-3x} = e$ **40.** $e^x = 1$

41. $e^{\sqrt{x}} = e^3$ **42.** $e^{-1/x} = e^{1/2}$

Depreciation In Exercises 43 and 44, the value V (in dollars) of an item is a function of the time t (in years).

(a) Sketch the function over the interval $[0, 10]$. Use a graphing utility to verify your graph.

(b) Find the rate of change of V when $t = 1$.

(c) Find the rate of change of V when $t = 5$.

(d) Use the values $(0, V(0))$ and $(10, V(10))$ to find the linear depreciation model for the item.

(e) Compare the exponential function and the model from part (d). What are the advantages of each?

43. $V = 15,000e^{-0.6286t}$ **44.** $V = 500,000e^{-0.2231t}$

45. Employment From 2000 through 2009, the numbers  y (in millions) of employed people in the United States can be modeled by

$$y = 136.855 - 0.5841t + 0.31664t^2 - 0.002166e^t$$

where t represents the year, with $t = 0$ corresponding to 2000. *(Source: U.S. Bureau of Labor Statistics)*

(a) Use a graphing utility to graph the model.

(b) Use the graph to estimate the rates of change in the number of employed people in 2000, 2004, and 2009.

(c) Confirm the results from part (b) analytically.

46. HOW DO YOU SEE IT? The yield y (in pounds per acre) of an orchard at age t (in years) is modeled by

$$y = 7955.6e^{-0.0458/t}.$$

The graph is shown below.

Orchard Yield

(a) What happens to the yield in the long run?

(b) What happens to the rate of change of the yield in the long run?

47. Compound Interest The balance A (in dollars) in a savings account is given by $A = 5000e^{0.08t}$, where t is measured in years. Find the rates at which the balance is changing when (a) $t = 1$ year, (b) $t = 10$ years, and (c) $t = 50$ years.

48. Learning Theory The average typing speed N (in words per minute) after t weeks of lessons is modeled by

$$N = \frac{95}{1 + 8.5e^{-0.12t}}.$$

Find the rates at which the typing speed is changing when (a) $t = 5$ weeks, (b) $t = 10$ weeks, and (c) $t = 30$ weeks.

49. Probability In a recent year, the mean SAT score for 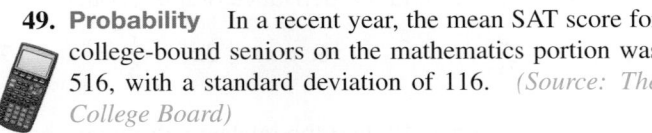 college-bound seniors on the mathematics portion was 516, with a standard deviation of 116. *(Source: The College Board)*

(a) Assuming the data can be modeled by a normal probability density function, find a model for these data.

(b) Use a graphing utility to graph the model. Be sure to choose an appropriate viewing window.

(c) Find the derivative of the model.

(d) Show that $f' > 0$ for $x < \mu$ and $f' < 0$ for $x > \mu$.

50. Probability A survey of a college freshman class has 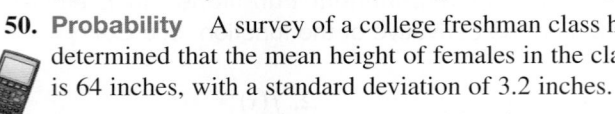 determined that the mean height of females in the class is 64 inches, with a standard deviation of 3.2 inches.

(a) Assuming the data can be modeled by a normal probability density function, find a model for these data.

(b) Use a graphing utility to graph the model. Be sure to choose an appropriate viewing window.

(c) Find the derivative of the model.

(d) Show that $f' > 0$ for $x < \mu$ and $f' < 0$ for $x > \mu$.

51. Normal Probability Density Function Use a 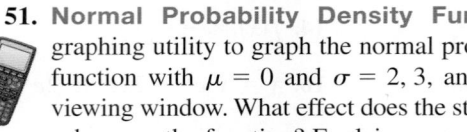 graphing utility to graph the normal probability density function with $\mu = 0$ and $\sigma = 2, 3$, and 4 in the same viewing window. What effect does the standard deviation σ have on the function? Explain your reasoning.

52. Normal Probability Density Function Use a graphing utility to graph the normal probability density function with

$$\sigma = 1 \quad \text{and} \quad \mu = -2, \quad 1, \quad \text{and} \quad 3$$

in the same viewing window. What effect does the mean μ have on the function? Explain your reasoning.

53. Normal Probability Density Function Use Example 6 as a model to show that the graph of the normal probability density function with $\mu = 0$

$$f(x) = \frac{1}{\sigma\sqrt{2\pi}}e^{-x^2/(2\sigma^2)}$$

has points of inflection at

$$x = \pm\sigma.$$

What is the maximum value of the function? Use a graphing utility to verify your answer by graphing the function for several values of σ.

QUIZ YOURSELF

See www.CalcChat.com for worked-out solutions to odd-numbered exercises.

Take this quiz as you would take a quiz in class. When you are done, check your work against the answers given in the back of the book.

In Exercises 1–8, use properties of exponents to simplify the expression.

1. $4^3(4^2)$

2. $\left(\dfrac{1}{6}\right)^{-3}$

3. $\dfrac{3^8}{3^5}$

4. $(5^{1/2})(3^{1/2})$

5. $(e^2)(e^5)$

6. $(e^{2/3})(e^3)$

7. $\dfrac{e^2}{e^{-4}}$

8. $(e^{-1})^{-3}$

In Exercises 9–14, sketch the graph of the function.

9. $f(x) = 3^x - 2$

10. $f(x) = 5^{-x} + 2$

11. $f(x) = 6^{x-3}$

12. $f(x) = e^{x+2}$

13. $f(x) = e^x + 3$

14. $f(x) = e^{-2x} + 1$

15. After t years, the remaining mass y (in grams) of an initial mass of 35 grams of a radioactive element whose half-life is 80 years is given by

$$y = 35\left(\frac{1}{2}\right)^{t/80}, \quad t \geq 0.$$

How much of the initial mass remains after 50 years?

16. With an annual rate of inflation of 4.5% over the next 10 years, the approximate cost C of goods or services during any year in the decade is given by

$$C(t) = P(1.045)^t, \quad 0 \leq t \leq 10$$

where t is the time (in years) and P is the present cost. The price of a baseball game ticket is presently $20. Estimate the price 10 years from now.

17. For $P = \$3000$, $r = 3.5\%$, and $t = 5$ years, find the balance in an account when interest is compounded (a) quarterly, (b) monthly, and (c) continuously.

18. How much should be deposited in an account paying 6% interest compounded monthly in order to have a balance of $14,000 after 5 years?

In Exercises 19–22, find the derivative of the function.

19. $y = e^{5x}$

20. $y = e^{x-4}$

21. $y = 5e^{x+2}$

22. $y = 3e^x - xe^x$

23. Determine an equation of the tangent line to

$$y = e^{-2x}$$

at the point $(0, 1)$.

24. Analyze and sketch the graph of

$$f(x) = 0.5x^2e^{-0.5x}.$$

Label any relative extrema, points of inflection, and asymptotes.

10.4 Logarithmic Functions

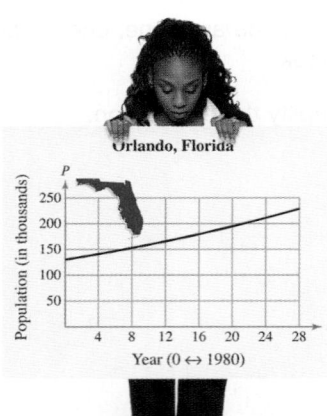

Orlando, Florida

In Exercise 77 on page 775, you will solve a natural exponential equation to predict when the population of Orlando, Florida will reach 300,000.

- Sketch the graphs of natural logarithmic functions.
- Use properties of logarithms to simplify, expand, and condense logarithmic expressions.
- Use inverse properties of exponential and logarithmic functions to solve exponential and logarithmic equations.
- Use properties of natural logarithms to answer questions about real-life situations.

The Natural Logarithmic Function

From your previous algebra courses, you should be somewhat familiar with logarithms. For instance, the **common logarithm** $\log_{10} x$ is defined as

$$\log_{10} x = b \quad \text{if and only if} \quad 10^b = x.$$

The base of common logarithms is 10. In calculus, the most useful base for logarithms is the number e.

Definition of the Natural Logarithmic Function

The **natural logarithmic function,** denoted by $\ln x$, is defined as

$$\ln x = b \quad \text{if and only if} \quad e^b = x.$$

$\ln x$ is read as "el en of x" or as "the natural log of x."

This definition implies that the natural logarithmic function and the natural exponential function are inverse functions. So, every logarithmic equation can be written in an equivalent exponential form, and every exponential equation can be written in logarithmic form. Here are some examples.

Logarithmic form:	*Exponential form:*
$\ln 1 = 0$	$e^0 = 1$
$\ln e = 1$	$e^1 = e$
$\ln \dfrac{1}{e} = -1$	$e^{-1} = \dfrac{1}{e}$
$\ln 2 \approx 0.693$	$e^{0.693} \approx 2$
$\ln 0.1 \approx -2.303$	$e^{-2.303} \approx 0.1$

Because the functions $f(x) = e^x$ and $g(x) = \ln x$ are inverse functions, their graphs are reflections of each other in the line

$$y = x.$$

This reflective property is illustrated in Figure 10.13. The figure also contains a summary of several properties of the graph of the natural logarithmic function.

   Notice that the domain of the natural logarithmic function is the set of *positive real numbers*—be sure you see that $\ln x$ is not defined for zero or for negative numbers. You can test this on your calculator. When you try evaluating

$$\ln(-1) \quad \text{or} \quad \ln 0$$

your calculator should indicate that the value is not a real number.

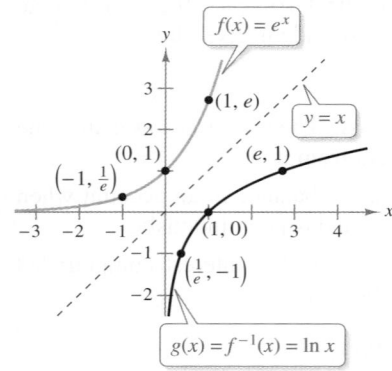

$g(x) = \ln x$

- Domain: $(0, \infty)$
- Range: $(-\infty, \infty)$
- Intercept: $(1, 0)$
- Always increasing
- $\ln x \to \infty$ as $x \to \infty$
- $\ln x \to -\infty$ as $x \to 0^+$
- Continuous
- One-to-one

FIGURE 10.13

Example 1   Graphing Logarithmic Functions

Sketch the graph of each function.

a. $f(x) = \ln(x + 1)$ **b.** $f(x) = 2 \ln(x - 2)$

SOLUTION

a. Because the natural logarithmic function is defined only for positive values, the domain of the function is $x + 1 > 0$, or

$$x > -1. \qquad \text{Domain}$$

To sketch the graph, begin by constructing a table of values, as shown below. Then plot the points in the table and connect them with a smooth curve, as shown in Figure 10.14(a).

x	-0.5	0	0.5	1	1.5	2
$\ln(x + 1)$	-0.693	0	0.405	0.693	0.916	1.099

b. The domain of this function is $x - 2 > 0$, or

$$x > 2. \qquad \text{Domain}$$

A table of values for the function is shown below, and its graph is shown in Figure 10.14(b).

x	2.5	3	3.5	4	4.5	5
$2 \ln(x - 2)$	-1.386	0	0.811	1.386	1.833	2.197

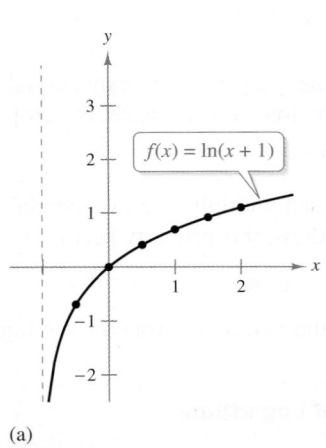

(a)

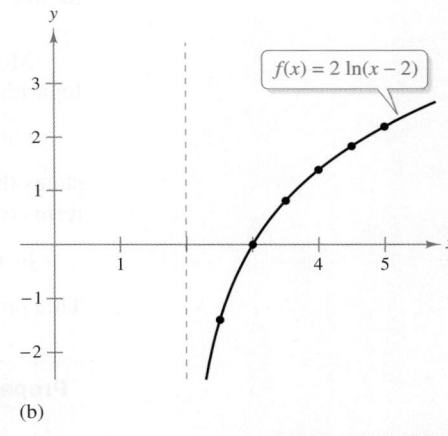

(b)

FIGURE 10.14

✓ Checkpoint 1

Complete the table and sketch the graph of

$$f(x) = \ln(x + 2).$$

x	-1.5	-1	-0.5	0	0.5	1
$f(x)$						

Properties of Logarithmic Functions

Recall from Section 4.1 that inverse functions have the property that

$$f(f^{-1}(x)) = x \quad \text{and} \quad f^{-1}(f(x)) = x.$$

The properties listed below follow from the fact that the natural logarithmic function and the natural exponential function are inverse functions.

Inverse Properties of Logarithms and Exponents

1. $\ln e^x = x$ **2.** $e^{\ln x} = x$

Example 2   **Applying Inverse Properties**

Simplify each expression.

a. $\ln e^{\sqrt{2}}$ **b.** $e^{\ln 3x}$

SOLUTION

a. Because $\ln e^x = x$, it follows that

$$\ln e^{\sqrt{2}} = \sqrt{2}.$$

b. Because $e^{\ln x} = x$, it follows that

$$e^{\ln 3x} = 3x.$$

✓ **Checkpoint 2**

Simplify each expression.

a. $\ln e^3$ **b.** $e^{\ln(x+1)}$

Most of the properties of exponential functions can be rewritten in terms of logarithmic functions. For instance, the property

$$e^x e^y = e^{x+y}$$

states that you can multiply two exponential expressions by adding their exponents. In terms of logarithms, this property becomes

$$\ln xy = \ln x + \ln y.$$

This property and two other properties of logarithms are summarized below.

Properties of Logarithms

1. $\ln xy = \ln x + \ln y$ **2.** $\ln \dfrac{x}{y} = \ln x - \ln y$ **3.** $\ln x^n = n \ln x$

STUDY TIP

There is no general property that can be used to rewrite $\ln(x + y)$. Specifically, $\ln(x + y)$ is not equal to $\ln x + \ln y$.

Rewriting a logarithm of a single quantity as the sum, difference, or multiple of logarithms is called *expanding* the logarithmic expression. The reverse procedure is called *condensing* a logarithmic expression.

Example 3   Expanding Logarithmic Expressions

Use the properties of logarithms to rewrite each expression as a sum, difference, or multiple of logarithms. (Assume $x > 0$ and $y > 0$.)

a. $\ln \dfrac{10}{9}$ **b.** $\ln \sqrt{x^2 + 1}$ **c.** $\ln \dfrac{xy}{5}$ **d.** $\ln[x^2(x + 1)]$

SOLUTION

a. $\ln \dfrac{10}{9} = \ln 10 - \ln 9$ Property 2

b. $\ln \sqrt{x^2 + 1} = \ln(x^2 + 1)^{1/2}$ Rewrite with rational exponent.

        $= \dfrac{1}{2}\ln(x^2 + 1)$ Property 3

c. $\ln \dfrac{xy}{5} = \ln(xy) - \ln 5$ Property 2

        $= \ln x + \ln y - \ln 5$ Property 1

d. $\ln[x^2(x + 1)] = \ln x^2 + \ln(x + 1)$ Property 1

        $= 2\ln x + \ln(x + 1)$ Property 3

✓ Checkpoint 3

Use the properties of logarithms to rewrite each expression as a sum, difference, or multiple of logarithms. (Assume $x > 0$ and $y > 0$.)

a. $\ln \dfrac{2}{5}$ **b.** $\ln \sqrt[3]{x + 2}$ **c.** $\ln \dfrac{x}{5y}$ **d.** $\ln x(x + 1)^2$

Example 4   Condensing Logarithmic Expressions

Use the properties of logarithms to rewrite each expression as the logarithm of a single quantity. (Assume $x > 0$ and $y > 0$.)

a. $\ln x + 2 \ln y$ **b.** $2 \ln(x + 2) - 3 \ln x$

SOLUTION

a. $\ln x + 2 \ln y = \ln x + \ln y^2$ Property 3

        $= \ln xy^2$ Property 1

b. $2 \ln(x + 2) - 3 \ln x = \ln(x + 2)^2 - \ln x^3$ Property 3

        $= \ln \dfrac{(x + 2)^2}{x^3}$ Property 2

✓ Checkpoint 4

Use the properties of logarithms to rewrite each expression as the logarithm of a single quantity. (Assume $x > 0$ and $y > 0$.)

a. $4 \ln x + 3 \ln y$ **b.** $\ln(x + 1) - 2 \ln(x + 3)$

TECH TUTOR

Try using a graphing utility to verify the results of Example 3(b). That is, try graphing the functions

$$y = \ln \sqrt{x^2 + 1}$$

and

$$y = \dfrac{1}{2} \ln(x^2 + 1).$$

Because these two functions are equivalent, their graphs should coincide.

Solving Exponential and Logarithmic Equations

To solve an exponential equation, first isolate the exponential expression. Then take the logarithm of each side of the equation and solve for the variable.

Example 5   Solving Exponential Equations

Solve each equation.

a. $e^x = 5$ **b.** $10 + e^{0.1t} = 14$

SOLUTION

a. $e^x = 5$ Write original equation.

$\ln e^x = \ln 5$ Take natural log of each side.

$x = \ln 5$ Inverse property: $\ln e^x = x$

b. $10 + e^{0.1t} = 14$ Write original equation.

$e^{0.1t} = 4$ Subtract 10 from each side.

$\ln e^{0.1t} = \ln 4$ Take natural log of each side.

$0.1t = \ln 4$ Inverse property: $\ln e^{0.1t} = 0.1t$

$t = 10 \ln 4$ Multiply each side by 10.

✔ **Checkpoint 5**

Solve each equation.

a. $e^x = 6$ **b.** $5 + e^{0.2t} = 10$

To solve a logarithmic equation, first isolate the logarithmic expression. Then exponentiate each side of the equation and solve for the variable.

Example 6   Solving Logarithmic Equations

Solve each equation.

a. $\ln x = 5$

b. $3 + 2 \ln x = 7$

SOLUTION

a. $\ln x = 5$ Write original equation.

$e^{\ln x} = e^5$ Exponentiate each side.

$x = e^5$ Inverse property: $e^{\ln x} = x$

b. $3 + 2 \ln x = 7$ Write original equation.

$2 \ln x = 4$ Subtract 3 from each side.

$\ln x = 2$ Divide each side by 2.

$e^{\ln x} = e^2$ Exponentiate each side.

$x = e^2$ Inverse property: $e^{\ln x} = x$

✔ **Checkpoint 6**

Solve each equation.

a. $\ln x = 4$

b. $4 + 5 \ln x = 19$

Application

 Example 7 **Finding Doubling Time**

You deposit P dollars in an account whose annual interest rate is r, compounded continuously. How long will it take for your balance to double?

SOLUTION The balance in the account after t years is $A = Pe^{rt}$. So, the balance will have doubled when $Pe^{rt} = 2P$. To find the "doubling time," solve this equation for t.

$Pe^{rt} = 2P$	Balance in account has doubled.
$e^{rt} = 2$	Divide each side by P.
$\ln e^{rt} = \ln 2$	Take natural log of each side.
$rt = \ln 2$	Inverse property: $\ln e^{rt} = rt$
$t = \dfrac{1}{r} \ln 2$	Divide each side by r.

From this result, you can see that the time it takes for the balance to double is inversely proportional to the interest rate r. The table shows the doubling times for several interest rates. Notice that the doubling time decreases as the rate increases. The relationship between doubling time and the interest rate is shown graphically in Figure 10.15.

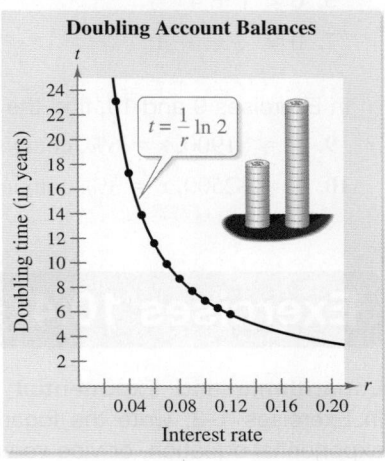

FIGURE 10.15

r	3%	4%	5%	6%	7%	8%	9%	10%	11%	12%
t	23.1	17.3	13.9	11.6	9.9	8.7	7.7	6.9	6.3	5.8

✓**Checkpoint 7**

Use the equation found in Example 7 to determine the amount of time it would take for your balance to double at an interest rate of 8.75%.

SUMMARIZE   (Section 10.4)

1. State the definition of the natural logarithmic function *(page 768)*. For an example of graphing logarithmic functions, see Example 1.

2. State the inverse properties of logarithms and exponents *(page 770)*. For an example of applying these properties, see Example 2.

3. State the properties of logarithms *(page 770)*. For examples of using these properties to expand and condense logarithmic expressions, see Examples 3 and 4.

4. Identify the properties of logarithms and exponents used to solve the exponential and logarithmic equations in Examples 5 and 6 *(page 772)*.

5. Describe a real-life example of how a logarithm is used to determine how long it will take for an account balance to double *(page 773, Example 7)*.

SKILLS WARM UP 10.4   The following warm-up exercises involve skills that were covered in a previous course or in earlier sections. You will use these skills in the exercise set for this section. For additional help, review Sections 1.6, 1.7, 4.1, and 10.2.

In Exercises 1–4, find the inverse function of f.

1. $f(x) = 5x$ **2.** $f(x) = x - 6$ **3.** $f(x) = 3x + 2$ **4.** $f(x) = \dfrac{3}{4}x - 9$

In Exercises 5–8, solve for x.

5. $0 < x + 4$ **6.** $0 < x^2 + 1$

7. $0 < \sqrt{x^2 - 1}$ **8.** $0 < x - 5$

In Exercises 9 and 10, find the balance in the account after 10 years.

9. $P = \$1900$, $r = 6\%$, compounded continuously

10. $P = \$2500$, $r = 3\%$, compounded continuously

Exercises 10.4

See www.CalcChat.com for worked-out solutions to odd-numbered exercises.

Logarithmic and Exponential Forms of Equations
In Exercises 1–8, write the logarithmic equation as an exponential equation, or vice versa.

1. $\ln 2 = 0.6931 \ldots$ **2.** $\ln 9 = 2.1972 \ldots$

3. $\ln 0.2 = -1.6094 \ldots$ **4.** $\ln 0.05 = -2.9957 \ldots$

5. $e^0 = 1$ **6.** $e^2 = 7.3891 \ldots$

7. $e^{-3} = 0.0498 \ldots$ **8.** $e^{0.25} = 1.2840 \ldots$

Matching   In Exercises 9–12, match the function with its graph. [The graphs are labeled (a)–(d).]

(a)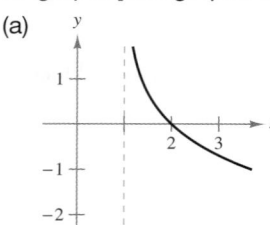

(b)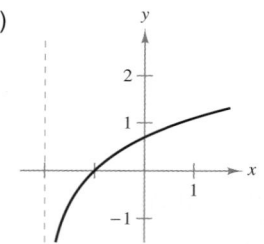

(c)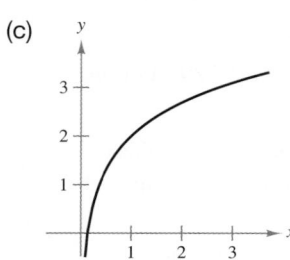

(d)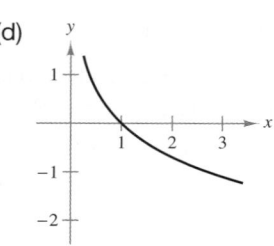

9. $f(x) = 2 + \ln x$

10. $f(x) = -\ln x$

11. $f(x) = \ln(x + 2)$

12. $f(x) = -\ln(x - 1)$

Graphing Logarithmic Functions   In Exercises 13–18, sketch the graph of the function. *See Example 1.*

13. $y = \ln(x - 1)$ **14.** $y = \ln|x|$

15. $y = \ln 2x$ **16.** $y = 5 + \ln x$

17. $y = 3 \ln x$ **18.** $y = \frac{1}{4} \ln x$

Applying Inverse Properties   In Exercises 19–24, apply the inverse properties of logarithmic and exponential functions to simplify the expression. *See Example 2.*

19. $\ln e^{x^2}$ **20.** $\ln e^{2x-1}$

21. $e^{\ln(5x+2)}$ **22.** $e^{\ln \sqrt{x}}$

23. $-1 + \ln e^{2x}$ **24.** $-8 + e^{\ln x^3}$

Expanding Logarithmic Expressions   In Exercises 25–34, use the properties of logarithms to rewrite the expression as a sum, difference, or multiple of logarithms. *See Example 3.*

25. $\ln \frac{2}{3}$ **26.** $\ln \frac{1}{5}$

27. $\ln xyz$ **28.** $\ln \dfrac{xy}{z}$

29. $\ln \sqrt[3]{2x + 7}$

30. $\ln \sqrt{\dfrac{x^3}{x + 1}}$

31. $\ln[z(z - 1)^2]$

32. $\ln\left(x \sqrt[3]{x^2 + 1}\right)$

33. $\ln \dfrac{3x(x + 1)}{(2x + 1)^2}$

34. $\ln \dfrac{2x}{\sqrt{x^2 - 1}}$

Inverse Functions   In Exercises 35–38, analytically show that the functions are inverse functions. Then use a graphing utility to show this graphically.

35. $f(x) = e^{2x}$

$g(x) = \ln \sqrt{x}$

36. $f(x) = e^x - 1$

$g(x) = \ln(x + 1)$

37. $f(x) = e^{2x-1}$

$g(x) = \frac{1}{2} + \ln \sqrt{x}$

38. $f(x) = e^{x/3}$

$g(x) = \ln x^3$

Using Properties of Logarithms   In Exercises 39 and 40, use the properties of logarithms and the fact that $\ln 2 \approx 0.6931$ and $\ln 3 \approx 1.0986$ to approximate the logarithm. Then use a calculator to confirm your approximation.

39. (a) $\ln 6$ (b) $\ln \frac{3}{2}$ (c) $\ln 81$ (d) $\ln \sqrt{3}$

40. (a) $\ln 0.25$ (b) $\ln 24$ (c) $\ln \sqrt[3]{12}$ (d) $\ln \frac{1}{72}$

Condensing Logarithmic Expressions   In Exercises 41–50, use the properties of logarithms to rewrite the expression as the logarithm of a single quantity. *See Example 4.*

41. $\ln(x - 2) - \ln(x + 2)$ **42.** $\ln(2x + 1) + \ln(2x - 1)$

43. $3 \ln x + 2 \ln y - 4 \ln z$ **44.** $4 \ln x + 6 \ln y - \ln z$

45. $4 \ln(x - 6) - \frac{1}{2} \ln(3x + 1)$

46. $2 \ln(5x + 3) + \frac{3}{2} \ln(x + 5)$

47. $3[\ln x + \ln(x + 3) - \ln(x + 4)]$

48. $\frac{1}{3}[2 \ln(x + 3) + \ln x - \ln(x^2 - 1)]$

49. $\frac{3}{2}[\ln x(x^2 + 1) - \ln(x + 1)]$

50. $2[\ln x + \frac{1}{4} \ln(x + 1)]$

Solving Exponential and Logarithmic Equations   In Exercises 51–72, solve for x or t. *See Examples 5 and 6.*

51. $e^{\ln x} = 4$ **52.** $e^{\ln x^2} - 9 = 0$

53. $e^{x+1} = 4$ **54.** $e^{-0.5x} = 0.075$

55. $300e^{-0.2t} = 700$ **56.** $400e^{-0.0174t} = 1000$

57. $4e^{2x-1} - 1 = 5$ **58.** $2e^{-x+1} - 5 = 9$

59. $\ln x = 0$ **60.** $2 \ln x = 4$

61. $\ln 2x = 2.4$ **62.** $\ln 4x = 1$

63. $3 + 4 \ln x = 15$ **64.** $6 + 3 \ln x = 8$

65. $\ln x - \ln(x - 6) = 3$ **66.** $\ln x + \ln(x + 2) = 0$

67. $5^{2x} = 15$ **68.** $2^{1-x} = 6$

69. $500(1.07)^t = 1000$ **70.** $400(1.06)^t = 1300$

71. $\left(1 + \dfrac{0.07}{12}\right)^{12t} = 3$ **72.** $\left(1 + \dfrac{0.06}{12}\right)^{12t} = 5$

Compound Interest   In Exercises 73 and 74, $3000 is invested in an account at interest rate r, compounded continuously. Find the time required for the amount to (a) double and (b) triple.

73. $r = 0.085$ **74.** $r = 0.045$

75. Compound Interest   A deposit of $1000 is made in an account that earns interest at an annual rate of 5%. How long will it take for the balance to double when the interest is compounded (a) annually, (b) monthly, (c) daily, and (d) continuously?

76. Compound Interest   Complete the table to determine the time t necessary for P dollars to triple when the interest is compounded continuously at rate r.

r	2%	4%	6%	8%	10%	12%	14%
t							

77. Population Growth   The population P (in thousands) of Orlando, Florida from 1980 through 2009 can be modeled by $P = 130e^{0.0205t}$, where $t = 0$ corresponds to 1980.   *(Source: U.S. Census Bureau)*

(a) What was the population of Orlando in 2009?

(b) In what year will Orlando have a population of 300,000?

78. Population Growth   The population P (in thousands) of Phoenix, Arizona from 1980 through 2009 can be modeled by

$$P = 788e^{0.0248t}$$

where $t = 0$ corresponds to 1980.   *(Source: U.S. Census Bureau)*

(a) What was the population of Phoenix in 2009?

(b) In what year will Phoenix have a population of 2,000,000?

Carbon Dating   In Exercises 79–82, you are given the ratio of carbon atoms in a fossil. Use the information to estimate the age of the fossil. In living organic material, the ratio of radioactive carbon isotopes to the total number of carbon atoms is about 1 to 10^{12}. (See Example 2 in Section 10.1.) When organic material dies, its radioactive carbon isotopes begin to decay, with a half-life of about 5715 years. So, the ratio R of carbon isotopes to carbon-14 atoms is modeled by $R = 10^{-12}\left(\frac{1}{2}\right)^{t/5715}$, where t is the time (in years) and $t = 0$ represents the time when the organic material died.

79. $R = 0.32 \times 10^{-12}$ **80.** $R = 0.27 \times 10^{-12}$

81. $R = 0.22 \times 10^{-12}$ **82.** $R = 0.13 \times 10^{-12}$

83. Learning Theory   Students in a mathematics class were given an exam and then retested monthly with equivalent exams. The average scores S (on a 100-point scale) for the class can be modeled by $S = 80 - 14 \ln(t + 1), 0 \le t \le 12$, where t is the time in months.

(a) What was the average score on the original exam?

(b) What was the average score after 4 months?

(c) After how many months was the average score 46?

84. **HOW DO YOU SEE IT?** The graph shows the percents of American males and females ages 20 and over who are no more than x inches tall. (*Source: National Center for Health Statistics*)

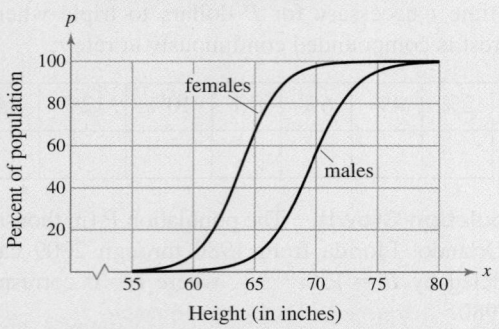

Heights of Americans Age 20 and Over

(a) Use the graph to determine the limit of each function as x approaches infinity. What do they mean?

(b) What is the median height of each sex?

85. Demand   The demand function for a product is given by

$$p = 5000\left(1 - \frac{4}{4 + e^{-0.002x}}\right)$$

where p is the price per unit (in dollars) and x is the number of units sold. Find the numbers of units sold for prices of (a) $p = \$200$ and (b) $p = \$800$.

86. Demand   The demand function for a product is given by

$$p = 10{,}000\left(1 - \frac{3}{3 + e^{-0.001x}}\right)$$

where p is the price per unit (in dollars) and x is the number of units sold. Find the numbers of units sold for prices of (a) $p = \$500$ and (b) $p = \$1500$.

87. Using a Property of Logarithms   Demonstrate that

$$\frac{\ln x}{\ln y} \neq \ln\frac{x}{y} = \ln x - \ln y$$

by using a spreadsheet to complete the table.

x	y	$\dfrac{\ln x}{\ln y}$	$\ln\dfrac{x}{y}$	$\ln x - \ln y$
1	2			
3	4			
10	5			
4	0.5			

88. Finding Limits and Relative Extrema   Use a spreadsheet to complete the table using

$$f(x) = \frac{\ln x}{x}.$$

x	1	5	10	10^2	10^4	10^6
$f(x)$						

(a) Use the table to estimate the limit: $\displaystyle\lim_{x\to\infty} f(x)$.

(b) Use a graphing utility to estimate the relative extrema of f.

Verifying Properties of Logarithms   In Exercises 89 and 90, use a graphing utility to verify that the functions are equivalent for $x > 0$.

89. $f(x) = \ln\dfrac{x^2}{4}$

$g(x) = 2\ln x - \ln 4$

90. $f(x) = \ln\sqrt{x(x^2 + 1)}$

$g(x) = \tfrac{1}{2}[\ln x + \ln(x^2 + 1)]$

True or False?   In Exercises 91–96, determine whether the statement is true or false given that $f(x) = \ln x$. If it is false, explain why or give an example that shows it is false.

91. $f(0) = 0$

92. $f(ax) = f(a) + f(x), \quad a > 0, x > 0$

93. $f(x - 2) = f(x) - f(2), \quad x > 2$

94. $\sqrt{f(x)} = \tfrac{1}{2}f(x)$

95. If $f(u) = 2f(v)$, then $v = u^2$.

96. If $f(x) < 0$, then $0 < x < 1$.

97. Finance   You are investing P dollars at an annual interest rate of r, compounded continuously, for t years. Which of the following options would you choose to get the highest value of the investment? Explain your reasoning.

(a) Double the amount you invest.

(b) Double your interest rate.

(c) Double the number of years.

98. Think About It   Are the times required for the investments in Exercises 75 and 76 to quadruple twice as long as the times for them to double? Give a reason for your answer and verify your answer algebraically.

99. Pursuit Curve   Use a graphing utility to graph

$$y = 10\ln\left(\frac{10 + \sqrt{100 - x^2}}{10}\right) - \sqrt{100 - x^2}$$

over the interval $(0, 10]$. This graph is called a *tractrix* or *pursuit curve*. Use your school's library, the Internet, or some other reference source to find information about a tractrix. Explain how such a curve can arise in a real-life setting.

10.5 Derivatives of Logarithmic Functions

In Exercise 73 on page 784, you will use the derivative of a logarithmic function to find the rate of change of a demand function.

■ Find the derivatives of natural logarithmic functions.
■ Find the derivatives of exponential and logarithmic functions involving other bases.

Derivatives of Logarithmic Functions

Implicit differentiation can be used to develop the derivative of the natural logarithmic function.

$$y = \ln x \qquad \text{Natural logarithmic function}$$

$$e^y = x \qquad \text{Write in exponential form.}$$

$$\frac{d}{dx}[e^y] = \frac{d}{dx}[x] \qquad \text{Differentiate with respect to } x.$$

$$e^y \frac{dy}{dx} = 1 \qquad \text{Chain Rule}$$

$$\frac{dy}{dx} = \frac{1}{e^y} \qquad \text{Divide each side by } e^y.$$

$$\frac{dy}{dx} = \frac{1}{x} \qquad \text{Substitute } x \text{ for } e^y.$$

This result and its Chain Rule version are summarized below.

Derivative of the Natural Logarithmic Function

Let u be a differentiable function of x.

1. $\dfrac{d}{dx}[\ln x] = \dfrac{1}{x}$ **2.** $\dfrac{d}{dx}[\ln u] = \dfrac{1}{u}\dfrac{du}{dx}$

Example 1   Differentiating a Logarithmic Function

Find the derivative of

$$f(x) = \ln 2x.$$

SOLUTION   Let $u = 2x$. Then $du/dx = 2$, and you can apply the Chain Rule as shown.

$$f'(x) = \frac{1}{u}\frac{du}{dx} \qquad \text{Chain Rule}$$

$$= \frac{1}{2x}(2)$$

$$= \frac{1}{x} \qquad \text{Simplify.}$$

✓ **Checkpoint 1**

Find the derivative of

$$f(x) = \ln 5x.$$

Example 2 Differentiating Logarithmic Functions

Find the derivative of each function.

a. $f(x) = \ln(2x^2 + 4)$ **b.** $f(x) = x \ln x$ **c.** $f(x) = \dfrac{\ln x}{x}$

SOLUTION

a. $f'(x) = \dfrac{1}{u}\dfrac{du}{dx}$ Chain Rule

$\qquad = \dfrac{1}{2x^2 + 4}(4x)$ $u = 2x^2 + 4, \, du/dx = 4x$

$\qquad = \dfrac{2x}{x^2 + 2}$ Simplify.

b. $f'(x) = x\dfrac{d}{dx}[\ln x] + (\ln x)\dfrac{d}{dx}[x]$ Product Rule

$\qquad = x\left(\dfrac{1}{x}\right) + (\ln x)(1)$

$\qquad = 1 + \ln x$ Simplify.

c. $f'(x) = \dfrac{x\dfrac{d}{dx}[\ln x] - (\ln x)\dfrac{d}{dx}[x]}{x^2}$ Quotient Rule

$\qquad = \dfrac{x\left(\dfrac{1}{x}\right) - \ln x}{x^2}$

$\qquad = \dfrac{1 - \ln x}{x^2}$ Simplify.

✓ **Checkpoint 2**

Find the derivative of each function.

a. $f(x) = \ln(x^2 - 4)$ **b.** $f(x) = x^2 \ln x$ **c.** $f(x) = -\dfrac{\ln x}{x^2}$ ■

Example 3 Rewriting Before Differentiating

STUDY TIP

When you are differentiating logarithmic functions, it is often helpful to use the properties of logarithms to rewrite the function *before* differentiating. To see the advantage of rewriting before differentiating, try using the Chain Rule to differentiate $f(x) = \ln\sqrt{x + 1}$ and compare your work with that shown in Example 3.

$f(x) = \ln\sqrt{x + 1}$ Original function

$\qquad = \ln(x + 1)^{1/2}$ Rewrite with rational exponent.

$\qquad = \dfrac{1}{2}\ln(x + 1)$ Property of logarithms

$f'(x) = \dfrac{1}{2}\left(\dfrac{1}{x + 1}\right)$ Differentiate.

$\qquad = \dfrac{1}{2(x + 1)}$ Simplify.

✓ **Checkpoint 3**

Find the derivative of

$f(x) = \ln\sqrt[3]{x + 1}.$ ■

Example 4 Rewriting Before Differentiating

Find the derivative of $f(x) = \ln[x(x^2 + 1)^2]$.

SOLUTION

$$
\begin{aligned}
f(x) &= \ln[x(x^2 + 1)^2] && \text{Write original function.}\\
&= \ln x + \ln(x^2 + 1)^2 && \text{Logarithmic properties}\\
&= \ln x + 2\ln(x^2 + 1) && \text{Logarithmic properties}\\
f'(x) &= \frac{1}{x} + 2\left(\frac{2x}{x^2 + 1}\right) && \text{Differentiate.}\\
&= \frac{1}{x} + \frac{4x}{x^2 + 1} && \text{Simplify.}
\end{aligned}
$$

✓ Checkpoint 4

Find the derivative of $f(x) = \ln[x^2\sqrt{x^2 + 1}]$.

Finding the derivative of the function in Example 4 without first rewriting would be a formidable task.

$$f'(x) = \frac{1}{x(x^2 + 1)^2} \frac{d}{dx}[x(x^2 + 1)^2]$$

You might try showing that this yields the same result obtained in Example 4, but be careful—the algebra is messy.

Example 5 Finding an Equation of a Tangent Line

Find an equation of the tangent line to the graph of $f(x) = 2 + 3x \ln x$ at the point $(1, 2)$.

SOLUTION Begin by finding the derivative of f.

$$
\begin{aligned}
f(x) &= 2 + 3x \ln x && \text{Write original function.}\\
f'(x) &= 3x\left(\frac{1}{x}\right) + (\ln x)(3) && \text{Differentiate.}\\
&= 3 + 3 \ln x && \text{Simplify.}
\end{aligned}
$$

The slope of the line tangent to the graph of f at $(1, 2)$ is

$$f'(1) = 3 + 3 \ln 1 = 3 + 3(0) = 3.$$

Using the point-slope form of a line, you can find the equation of the tangent line to be

$$y = 3x - 1.$$

The graph of the function and the tangent line are shown in Figure 10.16.

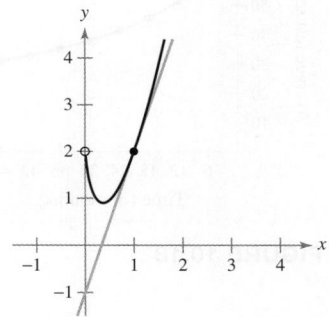

FIGURE 10.16

✓ Checkpoint 5

Find an equation of the tangent line to the graph of $f(x) = 4 \ln x$ at the point $(1, 0)$.

TECH TUTOR

A symbolic differentiation utility generally will not list the derivative of the logarithmic function in the form obtained in Example 4. Use a symbolic differentiation utility to find the derivative of the function in Example 4. Show that the two forms are equivalent by rewriting the answer obtained in Example 4.

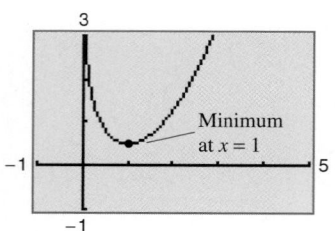

FIGURE 10.17

Example 6   Analyzing a Graph

Analyze the graph of the function $f(x) = \dfrac{x^2}{2} - \ln x$.

SOLUTION   From Figure 10.17, it appears that the function has a minimum at $x = 1$. To find the minimum analytically, find the critical numbers by setting the derivative of f equal to zero and solving for x.

$$f(x) = \frac{x^2}{2} - \ln x \qquad \text{Write original function.}$$

$$f'(x) = x - \frac{1}{x} \qquad \text{Differentiate.}$$

$$x - \frac{1}{x} = 0 \qquad \text{Set derivative equal to 0.}$$

$$x = \frac{1}{x} \qquad \text{Add } 1/x \text{ to each side.}$$

$$x^2 = 1 \qquad \text{Multiply each side by } x.$$

$$x = \pm 1 \qquad \text{Take square root of each side.}$$

Of these two possible critical numbers, only the positive one lies in the domain of f. By applying the First-Derivative Test, you can confirm that the function has a relative minimum at $x = 1$.

✓ **Checkpoint 6**

Determine the relative extrema of the function $f(x) = x - 2 \ln x$.   ■

 ### Example 7   Finding a Rate of Change

A group of 200 college students was tested every 6 months over a four-year period. The group was composed of students who took Spanish during the fall semester of their freshman year and did not take subsequent Spanish courses. The average test score p (in percent) is modeled by

$$p = 91.6 - 15.6 \ln(t + 1), \quad 0 \le t \le 48$$

where t is the time in months, as shown in Figure 10.18. At what rate was the average score changing after 1 year?

SOLUTION   The rate of change is

$$\frac{dp}{dt} = -\frac{15.6}{t + 1}.$$

The rate of change when $t = 12$ is

$$\frac{dp}{dt} = -\frac{15.6}{12 + 1} = -\frac{15.6}{13} = -1.2.$$

This means that the average score was decreasing at the rate of 1.2% per month.

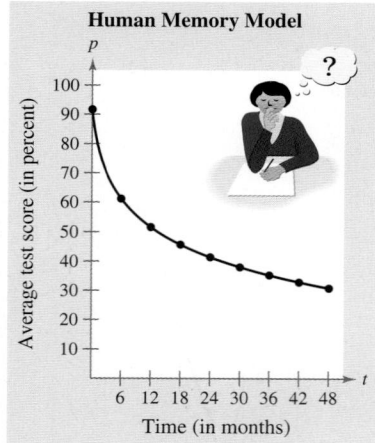

FIGURE 10.18

✓ **Checkpoint 7**

Suppose the average test score in Example 7 was modeled by

$$p = 92.3 - 16.9 \ln(t + 1), \quad 0 \le t \le 48$$

where t is the time in months. How would the rate at which the average test score was changing after 1 year compare with that of the model in Example 7?   ■

Other Bases

This chapter began with a definition of a general exponential function

$$f(x) = a^x$$

where a is a positive number such that $a \neq 1$. The corresponding **logarithm to the base a** is defined by

$$\log_a x = b \quad \text{if and only if} \quad a^b = x.$$

As with the natural logarithmic function, the domain of the logarithmic function to the base a is the set of positive numbers.

Example 8 Evaluating Logarithms

a. $\log_2 8 = 3$ $2^3 = 8$

b. $\log_{10} 100 = 2$ $10^2 = 100$

c. $\log_{10} \frac{1}{10} = -1$ $10^{-1} = \frac{1}{10}$

d. $\log_3 81 = 4$ $3^4 = 81$

✓ Checkpoint 8

Evaluate each logarithm without using a calculator.

a. $\log_2 16$ **b.** $\log_{10} \frac{1}{100}$ **c.** $\log_2 \frac{1}{32}$ **d.** $\log_5 125$ ■

Most calculators have only two logarithm keys—a natural logarithm key denoted by [LN] and a common logarithm key denoted by [LOG]. Logarithms to other bases can be evaluated with the following change-of-base formula.

$$\log_a x = \frac{\ln x}{\ln a} \qquad \text{Change-of-base formula}$$

Example 9 Changing Bases to Evaluate Logarithms

Use the change-of-base formula and a calculator to evaluate each logarithm.

a. $\log_2 3$

b. $\log_3 6$

c. $\log_2(-1)$

SOLUTION In each case, use the change-of-base formula and a calculator.

a. $\log_2 3 = \frac{\ln 3}{\ln 2} \approx 1.585$ $\log_a x = \frac{\ln x}{\ln a}$

b. $\log_3 6 = \frac{\ln 6}{\ln 3} \approx 1.631$ $\log_a x = \frac{\ln x}{\ln a}$

c. $\log_2(-1)$ is not defined.

✓ Checkpoint 9

Use the change-of-base formula and a calculator to evaluate each logarithm.

a. $\log_2 5$ **b.** $\log_3 18$ **c.** $\log_4 80$ **d.** $\log_{16} 0.25$ ■

To find derivatives of exponential or logarithmic functions to bases other than e, you can either convert to base e or use the differentiation rules shown below.

Other Bases and Differentiation

Let u be a differentiable function of x.

1. $\dfrac{d}{dx}[a^x] = (\ln a)a^x$ 2. $\dfrac{d}{dx}[a^u] = (\ln a)a^u \dfrac{du}{dx}$

3. $\dfrac{d}{dx}[\log_a x] = \left(\dfrac{1}{\ln a}\right)\dfrac{1}{x}$ 4. $\dfrac{d}{dx}[\log_a u] = \left(\dfrac{1}{\ln a}\right)\left(\dfrac{1}{u}\right)\dfrac{du}{dx}$

PROOF By definition, $a^x = e^{(\ln a)x}$. So, you can prove the first rule by letting $u = (\ln a)x$ and differentiating with base e to obtain

$$\frac{d}{dx}[a^x] = \frac{d}{dx}[e^{(\ln a)x}] = e^u \frac{du}{dx} = e^{(\ln a)x}(\ln a) = (\ln a)a^x.$$

 Example 10   Finding a Rate of Change

Radioactive carbon isotopes have a half-life of 5715 years. An object contains 1 gram of the isotopes. The amount A (in grams) that will be present after t years is

$$A = \left(\frac{1}{2}\right)^{t/5715}.$$

At what rate is the amount changing when $t = 10{,}000$ years?

SOLUTION   The derivative of A with respect to t is

$$\frac{dA}{dt} = \left(\ln \frac{1}{2}\right)\left(\frac{1}{2}\right)^{t/5715}\left(\frac{1}{5715}\right).$$

When $t = 10{,}000$, the rate at which the amount is changing is

$$\left(\ln \frac{1}{2}\right)\left(\frac{1}{2}\right)^{10,000/5715}\left(\frac{1}{5715}\right) \approx -0.000036$$

which implies that the amount of isotopes in the object is decreasing at the rate of 0.000036 gram per year.

✓**Checkpoint 10**

Use a graphing utility to graph the model in Example 10. Describe the rate at which the amount is changing as time t increases.   ◼

SUMMARIZE   (Section 10.5)

1. State the derivative of the natural logarithmic function *(page 777)*. For examples of the derivative of the natural logarithmic function, see Examples 1, 2, 3, and 4.

2. State the derivative of the logarithmic function to base a *(page 782)*. For an example of a derivative of a logarithmic function to base a, see Example 10.

SKILLS WARM UP 10.5   The following warm-up exercises involve skills that were covered in earlier sections. You will use these skills in the exercise set for this section. For additional help, review Sections 8.1, 8.2, and 10.4.

In Exercises 1–6, expand the logarithmic expression.

1. $\ln(x + 1)^2$

2. $\ln x(x + 1)$

3. $\ln \dfrac{x}{x + 1}$

4. $\ln\left(\dfrac{x}{x - 3}\right)^3$

5. $\ln \dfrac{4x(x - 7)}{x^2}$

6. $\ln x^3(x + 1)$

In Exercises 7 and 8, find dy/dx implicitly.

7. $y^2 + xy = 7$

8. $x^2y - xy^2 = 3x$

In Exercises 9 and 10, find the second derivative of f.

9. $f(x) = x^2(x + 1) - 3x^3$

10. $f(x) = -\dfrac{1}{x^2}$

Exercises 10.5

See www.CalcChat.com for worked-out solutions to odd-numbered exercises.

Differentiating a Logarithmic Function In Exercises 1–22, find the derivative of the function. *See Examples 1, 2, 3, and 4.*

1. $y = \ln x^2$

2. $f(x) = \ln 7x$

3. $y = \ln(x^2 + 3)$

4. $f(x) = \ln(1 - x^2)$

5. $y = \ln\sqrt{x - 4}$

6. $y = \ln(1 - x)^{3/2}$

7. $y = (\ln x)^4$

8. $y = (\ln x^2)^2$

9. $f(x) = 2x \ln x$

10. $y = \dfrac{\ln x}{x^2}$

11. $y = \ln\left(x\sqrt{x^2 - 1}\right)$

12. $y = \ln[x(2x + 3)^2]$

13. $y = \ln \dfrac{x}{x + 1}$

14. $y = \ln \dfrac{x}{x^2 + 1}$

15. $y = \ln \sqrt[3]{\dfrac{x - 1}{x + 1}}$

16. $y = \ln \sqrt{\dfrac{x + 1}{x - 1}}$

17. $y = \ln \dfrac{\sqrt{4 + x^2}}{x}$

18. $y = \ln \dfrac{(6 - x)^{3/2}}{x^{2/3}}$

19. $g(x) = e^{-x} \ln x$

20. $y = e^{x^2} \ln 4x^3$

21. $g(x) = \ln \dfrac{e^x + e^{-x}}{2}$

22. $f(x) = \ln \dfrac{1 + e^x}{1 - e^x}$

Evaluating Logarithms In Exercises 23–28, evaluate the logarithm without using a calculator. *See Example 8.*

23. $\log_5 25$

24. $\log_4 64$

25. $\log_3 \frac{1}{27}$

26. $\log_6 \frac{1}{36}$

27. $\log_7 49$

28. $\log_8 512$

Changing Bases to Evaluate Logarithms In Exercises 29–34, use the change-of-base formula and a calculator to evaluate the logarithm. *See Example 9.*

29. $\log_4 7$

30. $\log_6 10$

31. $\log_2 48$

32. $\log_5 12$

33. $\log_3 \frac{1}{2}$

34. $\log_7 \frac{2}{9}$

Differentiating Functions of Other Bases In Exercises 35–44, find the derivative of the function.

35. $y = 3^x$

36. $y = \left(\frac{1}{4}\right)^x$

37. $f(x) = \log_2 x$

38. $g(x) = \log_5 x$

39. $h(x) = 4^{2x-3}$

40. $f(x) = 10^{x^2}$

41. $y = \log_{10}(x^2 + 6x)$

42. $g(x) = \log_8(2x - 5)$

43. $y = x2^x$

44. $y = x3^{x+1}$

Finding an Equation of a Tangent Line In Exercises 45–52, find an equation of the tangent line to the graph of the function at the given point. *See Example 5.*

45. $y = \ln x^3$; $(1, 0)$

46. $y = \ln x^{5/2}$; $(1, 0)$

47. $y = x \ln x$; (e, e)

48. $y = \dfrac{\ln x}{x}$; $\left(e, \dfrac{1}{e}\right)$

49. $f(x) = \ln \dfrac{5(x + 2)}{x}$; $\left(-\dfrac{5}{2}, 0\right)$

50. $f(x) = \ln\left(x\sqrt{x + 3}\right)$; $\left(\frac{6}{5}, \frac{9}{10}\right)$

51. $y = \log_3 x$; $(27, 3)$

52. $g(x) = \log_{10} 2x$; $(5, 1)$

Finding Derivatives Implicitly   In Exercises 53–56, find dy/dx implicitly.

53. $x^2 - 3 \ln y + y^2 = 10$

54. $\ln xy + 5x = 30$

55. $4x^3 + \ln y^2 + 2y = 2x$

56. $4xy + \ln(x^2 y) = 7$

Finding an Equation of a Tangent Line   In Exercises 57 and 58, use implicit differentiation to find an equation of the tangent line to the graph of the function at the given point.

57. $x + y - 1 = \ln(x^2 + y^2);\ (1, 0)$

58. $y^2 + \ln(xy) = 2;\ (e, 1)$

Finding Higher-Order Derivatives   In Exercises 59–64, find the second derivative of the function.

59. $f(x) = x \ln \sqrt{x} + 2x$ **60.** $f(x) = 3 + 2 \ln x$

61. $f(x) = 2 + x^3 \ln x$ **62.** $f(x) = \dfrac{\ln x}{x^3} + x$

63. $f(x) = 5^x$ **64.** $f(x) = \log_{10} x$

65. Sound Intensity   The relationship between the number of decibels β and the intensity of a sound I (in watts per square centimeter) is given by

$$\beta = 10 \log_{10}\!\left(\dfrac{I}{10^{-16}}\right).$$

Find the rate of change in the number of decibels when the intensity is 10^{-4} watt per square centimeter.

66. Chemistry   The temperatures T (in °F) at which water boils at selected pressures p (in pounds per square inch) can be modeled by

$$T = 87.97 + 34.96 \ln p + 7.91 \sqrt{p}.$$

Find the rate of change of the temperature when the pressure is 60 pounds per square inch.

Analyzing a Graph   In Exercises 67–72, analyze and sketch the graph of the function. Label any relative extrema, points of inflection, and asymptotes. *See Example 6.*

67. $y = x - \ln x$

68. $y = \ln 2x - 2x^2$

69. $y = \dfrac{x}{\ln x}$

70. $y = \dfrac{\ln 5x}{x^2}$

71. $y = x^2 \ln \dfrac{x}{4}$

72. $y = (\ln x)^2$

Demand   In Exercises 73 and 74, find dx/dp for the demand function. Interpret this rate of change for a price of $10.

73. $x = \ln \dfrac{1000}{p}$

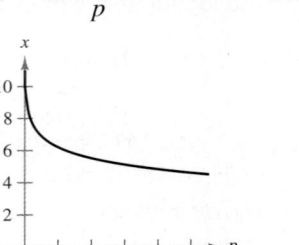

74. $x = \dfrac{500}{\ln(p^2 + 1)}$

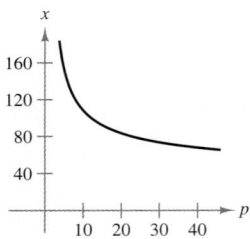

75. Demand   Solve the demand function in Exercise 73 for p. Use the result to find dp/dx. Then find the rate of change when $p = \$10$. What is the relationship between this derivative and dx/dp?

76. Demand   Solve the demand function in Exercise 74 for p. Use the result to find dp/dx. Then find the rate of change when $p = \$10$. What is the relationship between this derivative and dx/dp?

77. Minimum Average Cost   The cost of producing x units of a product is modeled by

$$C = 500 + 300x - 300 \ln x, \quad x \geq 1.$$

(a) Find the average cost function $\overline{C}$.

(b) Find the minimum average cost analytically. Use a graphing utility to confirm your result.

78. Minimum Average Cost   The cost of producing x units of a product is modeled by

$$C = 100 + 25x - 120 \ln x, \quad x \geq 1.$$

(a) Find the average cost function $\overline{C}$.

(b) Find the minimum average cost analytically. Use a graphing utility to confirm your result.

79. Consumer Trends   The numbers of employees E (in thousands) at outpatient care centers from 2004 through 2009 are shown in the table.

Year	2004	2005	2006	2007	2008	2009
Employees	451	473	493	512	533	543

The data can be modeled by $E = 287 + 116.7 \ln t$, where $t = 4$ corresponds to 2004. *(Source: U.S. Bureau of Labor Statistics)*

(a) Use a graphing utility to plot the data and graph E over the interval $[4, 9]$.

(b) At what rate were the numbers of employees changing in 2006?

80.  **HOW DO YOU SEE IT?** The graph shows the temperature T (in °C) of an object h hours after it is removed from a furnace.

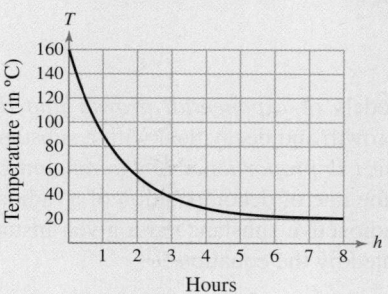

(a) Find $\displaystyle\lim_{h\to\infty} T$. What does this limit represent?

(b) When is the temperature changing most rapidly?

81. Home Mortgage The term t (in years) of a $200,000 home mortgage at 7.5% interest can be approximated by

$$t = -13.375 \ln \frac{x - 1250}{x}, \quad x > 1250$$

where x is the monthly payment in dollars.

(a) Use a graphing utility to graph the model.

(b) Use the model to approximate the term of a home mortgage for which the monthly payment is $1398.43. What is the total amount paid?

(c) Use the model to approximate the term of a home mortgage for which the monthly payment is $1611.19. What is the total amount paid?

(d) Find the instantaneous rate of change of t with respect to x when $x = 1398.43$ and $x = 1611.19$.

(e) Write a short paragraph describing the benefit of the higher monthly payment.

82. Earthquake Intensity On the Richter scale, the magnitude R of an earthquake of intensity I is given by

$$R = \frac{\ln I - \ln I_0}{\ln 10}$$

where I_0 is the minimum intensity used for comparison. Assume $I_0 = 1$.

(a) Find the intensity of the March 11, 2011 earthquake in Japan for which $R = 9.0$.

(b) Find the intensity of the January 12, 2010 earthquake in Haiti for which $R = 7.0$.

(c) Find the factor by which the intensity is increased when the value of R is doubled.

(d) Find dR/dI.

83. Learning Theory  Students in a learning theory study were given an exam and then retested monthly for 6 months with an equivalent exam. The data obtained in the study are shown in the table, where t is the time in months after the initial exam and s is the average score for the class.

t	1	2	3	4	5	6
s	84.2	78.4	72.1	68.5	67.1	65.3

(a) Use a graphing utility to find a logarithmic model for the average score s in terms of the time t.

(b) Use a graphing utility to plot the data and graph the model. How well does the model fit the data?

(c) Find the rate of change of s with respect to t when $t = 2$. Interpret the meaning of this rate of change in the context of the problem.

Business Capsule

While in college, Heikai Gani had a miserable experience trying to buy a new suit. With his friend Kyle Vucko, he created an online business model to deliver custom-tailored suits for men. With an initial investment of $800,000, Indochino.com was born in 2007. Today the company has seven-digit revenues with over 17,000 customers. Indochino offers a perfect-fit promise; they will pay for alterations at a local tailor or issue a full refund.

84. Research Project Use your school's library, the Internet, or some other reference source to research information about an e-commerce company, such as the one discussed above. Collect data about the company (sales over a 10-year period, for example) and find a mathematical model that represents the data.

10.6 Exponential Growth and Decay

■ Use exponential growth and decay to model real-life situations.

Exponential Growth and Decay

In this section, you will learn to create models of *exponential growth and decay*. Real-life situations that involve exponential growth and decay deal with a substance or population whose *rate of change at any time t is proportional to the amount of the substance present at that time*. For example, the rate of decomposition of a radioactive substance is proportional to the amount of radioactive substance at a given instant. In its simplest form, this relationship is represented by the equation

Rate of change of y ┐ is ┌ proportional to y.

$$\frac{dy}{dt} = ky.$$

In this equation, k is a constant and y is a function of t. The solution of this equation is shown below.

Population Growth of Bacteria

Time (in hours)

> **Exponential Growth and Decay**
>
> If y is a positive quantity whose rate of change with respect to time is proportional to the quantity present at any time t, then y is of the form
>
> $$y = Ce^{kt}$$
>
> where C is the **initial value** and k is the **constant of proportionality. Exponential growth** is indicated by $k > 0$ and **exponential decay** by $k < 0$.

In Exercise 23 on page 792, you will use exponential growth to find the time it takes a population of bacteria to double.

PROOF   Because the rate of change of y is proportional to y, you can write

$$\frac{dy}{dt} = ky.$$

You can see that $y = Ce^{kt}$ is a solution of this equation by differentiating to obtain $dy/dt = kCe^{kt}$ and substituting.

$y = Ce^{kt}$	Original equation
$\dfrac{dy}{dt} = kCe^{kt}$	Differentiate.
$= k(Ce^{kt})$	Rewrite.
$= ky$	Substitute y for Ce^{kt}.

> **STUDY TIP**
>
> In the model $y = Ce^{kt}$, C is called the "initial value" because, when $t = 0$,
>
> $$y = Ce^{k(0)}$$
> $$= C(1)$$
> $$= C.$$

Radioactive decay is measured in terms of **half-life,** the number of years required for half of the atoms in a sample of radioactive material to decay. The half-lives of some common radioactive isotopes are as shown.

Uranium (^{238}U)	4,470,000,000 years
Plutonium (^{239}Pu)	24,100 years
Carbon (^{14}C)	5,715 years
Radium (^{226}Ra)	1,599 years
Einsteinium (^{254}Es)	276 days
Nobelium (^{257}No)	25 seconds

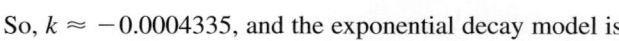 **Example 1**   **Modeling Radioactive Decay**

A sample contains 1 gram of radium. Will more than 0.5 gram of radium remain after 1000 years?

SOLUTION   Let y represent the mass (in grams) of the radium in the sample. Because the rate of decay is proportional to y, you can conclude that y is of the form $y = Ce^{kt}$, where t is the time in years. From the given information, you know that $y = 1$ when $t = 0$. Substituting these values into the model produces

$$1 = Ce^{k(0)} \qquad \text{Substitute 1 for } y \text{ and 0 for } t.$$

which implies that $C = 1$. Because radium has a half-life of 1599 years, you know that $y = \frac{1}{2}$ when $t = 1599$. Substituting these values into the model allows you to solve for k.

$y = e^{kt}$	Exponential decay model
$\frac{1}{2} = e^{k(1599)}$	Substitute $\frac{1}{2}$ for y and 1599 for t.
$\ln \frac{1}{2} = 1599k$	Take natural log of each side.
$\frac{1}{1599} \ln \frac{1}{2} = k$	Divide each side by 1599.

So, $k \approx -0.0004335$, and the exponential decay model is

$$y = e^{-0.0004335t}.$$

To find the amount of radium remaining in the sample after 1000 years, substitute $t = 1000$ into the model.

$$y = e^{-0.0004335(1000)} \approx 0.648 \text{ gram}$$

Yes, more than 0.5 gram of radium will remain after 1000 years. The graph of the model is shown in Figure 10.19.

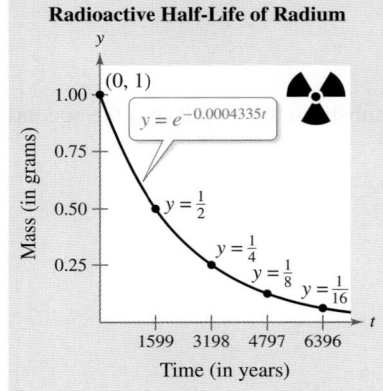

Radioactive Half-Life of Radium

$(0, 1)$

$y = e^{-0.0004335t}$

$y = \frac{1}{2}$

$y = \frac{1}{4}$

$y = \frac{1}{8}$

$y = \frac{1}{16}$

Mass (in grams)

Time (in years)

FIGURE 10.19

✓**Checkpoint 1**

Use the model in Example 1 to determine the number of years required for a one-gram sample of radium to decay to 0.4 gram.

Instead of approximating the value of k in Example 1, you could leave the value exact and obtain

$$y = e^{[(1/1599)\ln (1/2)]t}$$
$$= e^{\ln[(1/2)^{(t/1599)}]}$$
$$= \left(\frac{1}{2}\right)^{t/1599}.$$

This version of the model clearly shows the "half-life." When $t = 1599$, the value of y is $\frac{1}{2}$; when $t = 2(1599)$, the value of y is $\frac{1}{4}$; and so on.

Guidelines for Modeling Exponential Growth and Decay

1. Use the given information to write *two* sets of conditions involving y and t.

2. Substitute the given conditions into the model $y = Ce^{kt}$ and use the results to solve for the constants C and k. (When one of the conditions involves $t = 0$, substitute that value first to solve for C.)

3. Use the model $y = Ce^{kt}$ to answer the question.

Example 2 Modeling Population Growth

ALGEBRA TUTOR

xy

For help with the algebra in Example 2, see Example 1(c) in the *Chapter 10 Algebra Tutor* on page 794.

In a research experiment, a population of fruit flies is increasing in accordance with the exponential growth model. After 2 days, there are 100 flies, and after 4 days, there are 300 flies. How many flies will there be after 5 days?

SOLUTION   Let y be the number of flies at time t. From the given information, you know that $y = 100$ when $t = 2$ and $y = 300$ when $t = 4$. Substituting this information into the model $y = Ce^{kt}$ produces

$$100 = Ce^{2k} \quad \text{and} \quad 300 = Ce^{4k}.$$

To solve for k, solve for C in the first equation and substitute the result into the second equation.

$$300 = Ce^{4k} \qquad \text{Second equation}$$

$$300 = \left(\frac{100}{e^{2k}}\right)e^{4k} \qquad \text{Substitute } 100/e^{2k} \text{ for } C.$$

$$\frac{300}{100} = e^{2k} \qquad \text{Divide each side by 100.}$$

$$\ln 3 = 2k \qquad \text{Take natural log of each side.}$$

$$\frac{1}{2}\ln 3 = k \qquad \text{Solve for } k.$$

Using $k = \frac{1}{2}\ln 3 \approx 0.5493$, you can determine that

$$C \approx \frac{100}{e^{2(0.5493)}}$$

$$\approx 33.$$

So, the exponential growth model is

$$y = 33e^{0.5493t}$$

as shown in Figure 10.20. This implies that, after 5 days, the population is

$$y = 33e^{0.5493(5)}$$

$$\approx 514 \text{ flies.}$$

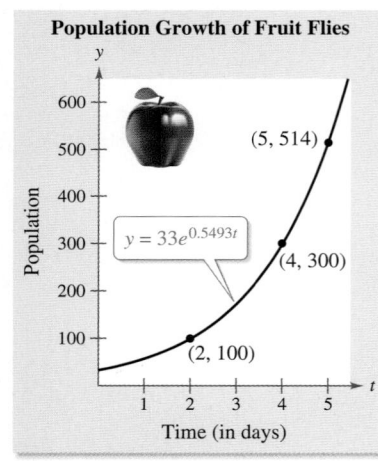

Population Growth of Fruit Flies

$y = 33e^{0.5493t}$

(5, 514)

(4, 300)

(2, 100)

Population

Time (in days)

FIGURE 10.20

✓ Checkpoint 2

Find the exponential growth model for a population of fruit flies for which there are 100 flies after 2 days and 400 flies after 4 days.

Example 3   Modeling Compound Interest

Money is deposited in an account for which the interest is compounded continuously. The balance in the account doubles in 6 years. What is the annual interest rate?

SOLUTION   The balance A in an account with continuously compounded interest is given by the exponential growth model

$$A = Pe^{rt} \qquad \text{Exponential growth model}$$

where P is the original deposit, r is the annual interest rate (in decimal form), and t is the time (in years). From the given information, you know that

$$A = 2P$$

when $t = 6$, as shown in Figure 10.21. Use this information to solve for r.

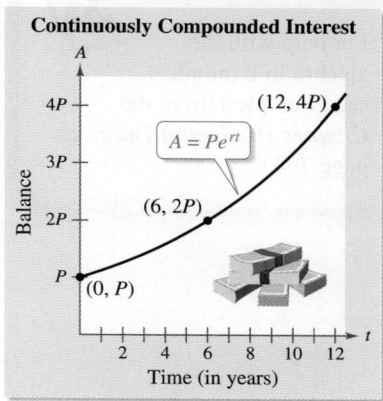

Continuously Compounded Interest

$A = Pe^{rt}$

FIGURE 10.21

$A = Pe^{rt}$	Exponential growth model
$2P = Pe^{r(6)}$	Substitute $2P$ for A and 6 for t.
$2 = e^{6r}$	Divide each side by P.
$\ln 2 = 6r$	Take natural log of each side.
$\frac{1}{6} \ln 2 = r$	Divide each side by 6.

So, the annual interest rate is

$$r = \frac{1}{6} \ln 2$$
$$\approx 0.1155$$

or about 11.55%.

✓ Checkpoint 3

Find the annual interest rate for an account whose balance doubles in 8 years and for which the interest is compounded continuously.

Each of the examples in this section uses the exponential growth model $y = Ce^{kt}$, in which the base is e. Exponential growth, however, can be modeled with any base. That is, the model

$$y = Ca^{bt}$$

also represents exponential growth. (To see this, note that the model can be written in the form $y = Ce^{(\ln a)bt}$.) In some real-life settings, bases other than e are more convenient. For instance, in Example 1, knowing that the half-life of radium is 1599 years, you can immediately write the exponential decay model as

$$y = \left(\frac{1}{2}\right)^{t/1599}.$$

Using this model, the amount of radium left in the sample after 1000 years is

$$y = \left(\frac{1}{2}\right)^{1000/1599}$$
$$\approx 0.648 \text{ gram}$$

which is the same answer obtained in Example 1.

Example 4 **Modeling Sales**

Four months after discontinuing advertising on national television, a manufacturer notices that sales have dropped from 100,000 MP3 players per month to 80,000. Using an exponential pattern of decline, what will the sales be after another 4 months?

SOLUTION   Let y represent the number of MP3 players, let t represent the time (in months), and consider the exponential decay model

$$y = Ce^{kt}.$$ Exponential decay model

From the given information, you know that $y = 100{,}000$ when $t = 0$. Using this information, you have

$$100{,}000 = Ce^0$$

which implies that $C = 100{,}000$. To solve for k, use the fact that $y = 80{,}000$ when $t = 4$.

$y = 100{,}000e^{kt}$	Exponential decay model
$80{,}000 = 100{,}000e^{k(4)}$	Substitute 80,000 for y and 4 for t.
$0.8 = e^{4k}$	Divide each side by 100,000.
$\ln 0.8 = 4k$	Take natural log of each side.
$\frac{1}{4}\ln 0.8 = k$	Divide each side by 4.

So, $k = \frac{1}{4}\ln 0.8 \approx -0.0558$, which means that the model is

$$y = 100{,}000e^{-0.0558t}.$$

After four more months ($t = 8$), you can expect sales to drop to

$$y = 100{,}000e^{-0.0558(8)}$$
$$\approx 64{,}000 \text{ MP3 players}$$

as shown in Figure 10.22.

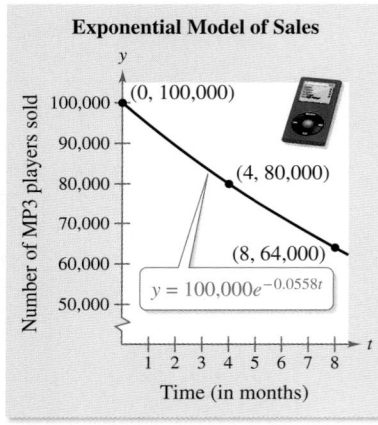

Exponential Model of Sales

(0, 100,000)
(4, 80,000)
(8, 64,000)
$y = 100{,}000e^{-0.0558t}$

Number of MP3 players sold

Time (in months)

FIGURE 10.22

✓ **Checkpoint 4**

Use the model in Example 4 to determine when sales will drop to 50,000 MP3 players. ■

SUMMARIZE   (Section 10.6)

1. State the model used for exponential growth and decay *(page 786)*. For examples of the use of this model, see Examples 1, 2, 3, and 4.

2. State the guidelines for modeling exponential growth and decay *(page 788)*. For examples of the use of these guidelines, see Examples 2, 3, and 4.

3. Describe a real-life example of an exponential decay model *(pages 787 and 790, Examples 1 and 4)*.

4. Describe a real-life example of an exponential growth model *(pages 788 and 789, Examples 2 and 3)*.

SKILLS WARM UP 10.6 The following warm-up exercises involve skills that were covered in earlier sections. You will use these skills in the exercise set for this section. For additional help, review Sections 10.3 and 10.4.

In Exercises 1–4, solve the equation for k.

1. $12 = 24e^{4k}$

2. $10 = 3e^{5k}$

3. $25 = 16e^{-0.01k}$

4. $22 = 32e^{-0.02k}$

In Exercises 5–8, find the derivative of the function.

5. $y = 32e^{0.23t}$

6. $y = 18e^{0.072t}$

7. $y = 24e^{-1.4t}$

8. $y = 25e^{-0.001t}$

In Exercises 9–12, simplify the expression.

9. $e^{\ln 4}$

10. $4e^{\ln 3}$

11. $e^{\ln(2x+1)}$

12. $e^{\ln(x^2+1)}$

Exercises 10.6

See www.CalcChat.com for worked-out solutions to odd-numbered exercises.

Modeling Exponential Growth and Decay In Exercises 1–6, find the exponential function $y = Ce^{kt}$ that passes through the two given points.

1. $y = Ce^{kt}$

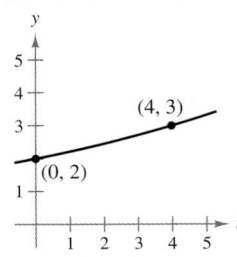

2. $y = Ce^{kt}$

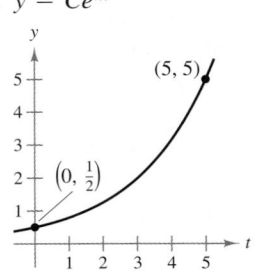

3. $y = Ce^{kt}$

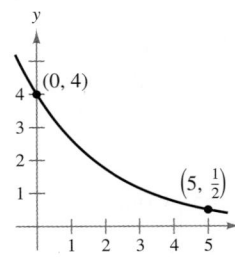

4. $y = Ce^{kt}$

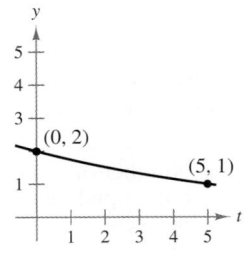

5. $y = Ce^{kt}$

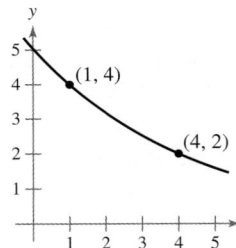

6. $y = Ce^{kt}$

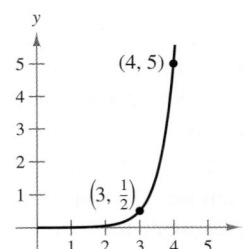

Determining Exponential Growth and Decay In Exercises 7–10, use the given information to write an exponential equation for y. Does the function represent exponential growth or exponential decay?

7. $\dfrac{dy}{dt} = 2y$, $y = 10$ when $t = 0$

8. $\dfrac{dy}{dt} = -\dfrac{2}{3}y$, $y = 20$ when $t = 0$

9. $\dfrac{dy}{dt} = -4y$, $y = 30$ when $t = 0$

10. $\dfrac{dy}{dt} = 5.2y$, $y = 18$ when $t = 0$

Modeling Radioactive Decay In Exercises 11–16, complete the table for each radioactive isotope. *See Example 1.*

Isotope	Half-life (in years)	Initial quantity	Amount after 1000 years	Amount after 10,000 years
11. ^{226}Ra	1599	10 grams		
12. ^{226}Ra	1599		1.5 grams	
13. ^{14}C	5715			2 grams
14. ^{14}C	5715	3 grams		
15. ^{239}Pu	24,100		2.1 grams	
16. ^{239}Pu	24,100			0.4 gram

17. Radioactive Decay What percent of a present amount of radioactive radium (^{226}Ra) will remain after 900 years?

18. Radioactive Decay   Find the half-life of a radioactive material for which 99.57% of the initial amount remains after 1 year.

19. Carbon Dating   Carbon-14 (^{14}C) dating assumes that the carbon dioxide on the Earth today has the same radioactive content as it did centuries ago. If this is true, then the amount of ^{14}C absorbed by a tree that grew several centuries ago should be the same as the amount of ^{14}C absorbed by a similar tree today. A piece of ancient charcoal contains only 15% as much of the radioactive carbon as a piece of modern charcoal. How long ago was the tree burned to make the ancient charcoal? (The half-life of ^{14}C is 5715 years.)

20. Carbon Dating   Repeat Exercise 19 for a piece of charcoal that contains 30% as much radioactive carbon as a modern piece.

Finding Exponential Models   In Exercises 21 and 22, find exponential models $y_1 = Ce^{k_1 t}$ and $y_2 = C(2)^{k_2 t}$ that pass through the two given points. Compare the values of k_1 and k_2. Briefly explain your results.

21. $(0, 5), (12, 20)$ **22.** $(0, 8), \left(20, \frac{1}{2}\right)$

23. Population Growth   The number of a certain type of bacteria increases continuously at a rate proportional to the number present. There are 150 bacteria at a given time and 450 bacteria 5 hours later.

(a) How many bacteria will there be 10 hours after the initial time?

(b) How long will it take for the population to double?

(c) Does the answer to part (b) depend on the starting time? Explain your reasoning.

24. HOW DO YOU SEE IT?   The graph shows the populations (in thousands) of Cleveland, Ohio and Atlanta, Georgia from 2000 through 2008 using exponential models. *(Source: U.S. Census Bureau)*

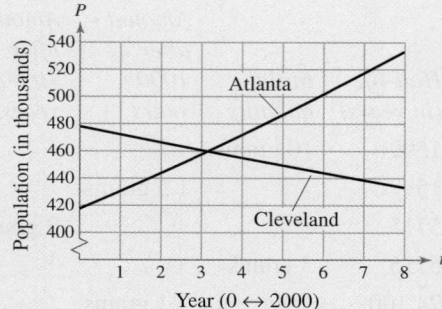

(a) Determine whether the population of each city is modeled by exponential growth or exponential decay. Explain your reasoning.

(b) Estimate the year when the two cities had the same population. What was this population?

Modeling Compound Interest   In Exercises 25–32, complete the table for an account in which interest is compounded continuously. *See Example 3.*

Initial investment	Annual rate	Time to double	Amount after 10 years	Amount after 25 years
25. $1000	12%			
26. $20,000	$10\frac{1}{2}$%			
27. $750		8 years		
28. $10,000		10 years		
29. $500			$1292.85	
30. $2000				$6008.33
31.	4.5%		$10,000.00	
32.	2%		$2000.00	

Finding Present Value   In Exercises 33 and 34, determine the principal P that must be invested at interest rate r, compounded continuously, so that $1,000,000 will be available for retirement in t years.

33. $r = 7.5\%, t = 40$

34. $r = 10\%, t = 25$

35. Effective Rate   The effective rate of interest r_{eff} is the annual rate that will produce the same interest per year as the nominal rate r.

(a) For a rate r (in decimal form) that is compounded n times per year, show that the effective rate r_{eff} (in decimal form) is

$$r_{eff} = \left(1 + \frac{r}{n}\right)^n - 1.$$

(b) For a rate r (in decimal form) that is compounded continuously, show that the effective rate r_{eff} (in decimal form) is $r_{eff} = e^r - 1$.

36. Effective Rate   Use the results of Exercise 35 to complete the table showing the effective rates for nominal rates of (a) $r = 5\%$, (b) $r = 6\%$, and (c) $r = 7\frac{1}{2}\%$.

Number of compoundings per year	4	12	365	Continuous
Effective yield				

37. Investment: Rule of 70   Verify that the time necessary for an investment to double in value is approximately $70/r$, where r is the annual interest rate entered as a percent.

38. Investment: Rule of 70   Use the Rule of 70 from Exercise 37 to approximate the times necessary for an investment to double in value when (a) $r = 10\%$ and (b) $r = 7\%$.

39. Depreciation A sports utility vehicle that costs $21,500 new has a book value of $13,600 after 2 years.

(a) Find a linear model for the value of the vehicle.

(b) Find an exponential model for the value of the vehicle.

(c) Find the book values of the vehicle after 1 year and after 4 years using each model.

(d) Use a graphing utility to graph the two models in the same viewing window. Which model depreciates faster in the first 2 years?

(e) Explain the advantages and disadvantages of using each model to a buyer and to a seller.

40. Population The table shows the populations P (in millions) of the United States from 1960 through 2010. *(Source: U.S. Census Bureau)*

Year	1960	1970	1980	1990	2000	2010
Population, P	181	205	228	250	282	309

(a) Use the 1960 and 1970 data to find an exponential model P_1. Let $t = 0$ represent 1960.

(b) Use a graphing utility to find an exponential model P_2 for all of the data. Let $t = 0$ represent 1960.

(c) Use a graphing utility to plot the data and graph both models in the same viewing window. Compare the actual data with the estimates from the models. Which model is more accurate?

41. Sales The cumulative sales S (in thousands of units) of a new product after it has been on the market for t years are modeled by $S = Ce^{k/t}$. During the first year, 5000 units were sold. The saturation point for the market is 30,000 units. That is, the limit of S as $t \to \infty$ is 30,000.

(a) Solve for C and k in the model.

(b) How many units will be sold after 5 years?

(c) Use a graphing utility to graph the sales function.

42. Sales The cumulative sales S (in thousands of units) of a new product after it has been on the market for t years are modeled by

$$S = 50(1 - e^{kt}).$$

During the first year, 8000 units were sold.

(a) Solve for k in the model.

(b) What is the saturation point for this product?

(c) How many units will be sold after 5 years?

(d) Use a graphing utility to graph the sales function.

43. Learning Curve The management of a factory finds that the maximum number of units a worker can produce in a day is 30. The learning curve for the number of units N produced per day after a new employee has worked for t days is modeled by

$$N = 30(1 - e^{kt}).$$

After 20 days on the job, a worker is producing 19 units in a day. How many days should pass before this worker is producing 25 units per day?

44. Learning Curve The management in Exercise 43 requires that a new employee be producing at least 20 units per day after 30 days on the job.

(a) Find a learning curve model that describes this minimum requirement.

(b) Find the number of days before a minimal achiever is producing 25 units per day.

45. Revenue A small business assumes that the demand function for one of its new products can be modeled by

$$p = Ce^{kx}.$$

When $p = \$45$, $x = 1000$ units, and when $p = \$40$, $x = 1200$ units.

(a) Solve for C and k in the model.

(b) Find the values of x and p that will maximize the revenue for this product.

46. Revenue Repeat Exercise 45 given that when $p = \$5$, $x = 300$ units, and when $p = \$4$, $x = 400$ units.

47. Project: Cell Phone Subscribers For a project analyzing the numbers of cell phone subscribers from 2000 to 2009, visit this text's website at *www.cengagebrain.com*. *(Source: CTIA-The Wireless Association)*

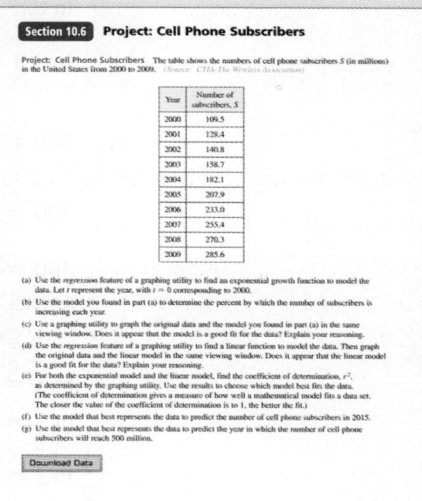

ALGEBRA TUTOR

Solving Exponential and Logarithmic Equations

To find the extrema or points of inflection of an exponential or logarithmic function, you must know how to solve exponential and logarithmic equations. A few examples are given on page 772. Some additional examples are presented in this Algebra Tutor.

As with all equations, remember that your basic goal is to isolate the variable on one side of the equation. To do this, you use inverse operations. For instance, to isolate x in

$$e^x = 7$$

take the natural log of each side of the equation and use the property $\ln e^x = x$. Similarly, to isolate x in

$$\ln x = 5$$

exponentiate each side of the equation and use the property $e^{\ln x} = x$.

Example 1   Solving Exponential Equations

Solve each exponential equation.

a. $25 = 5e^{7t}$ **b.** $80{,}000 = 100{,}000e^{k(4)}$ **c.** $300 = \left(\dfrac{100}{e^{2k}}\right)e^{4k}$

SOLUTION

a.

$25 = 5e^{7t}$	Write original equation.
$5 = e^{7t}$	Divide each side by 5.
$\ln 5 = \ln e^{7t}$	Take natural log of each side.
$\ln 5 = 7t$	Apply the property $\ln e^a = a$.
$\dfrac{1}{7}\ln 5 = t$	Divide each side by 7.

b.

$80{,}000 = 100{,}000e^{k(4)}$	Example 4, page 790
$0.8 = e^{4k}$	Divide each side by 100,000.
$\ln 0.8 = \ln e^{4k}$	Take natural log of each side.
$\ln 0.8 = 4k$	Apply the property $\ln e^a = a$.
$\dfrac{1}{4}\ln 0.8 = k$	Divide each side by 4.

c.

$300 = \left(\dfrac{100}{e^{2k}}\right)e^{4k}$	Example 2, page 788
$300 = (100)\dfrac{e^{4k}}{e^{2k}}$	Rewrite product.
$300 = 100e^{4k-2k}$	To divide powers, subtract exponents.
$300 = 100e^{2k}$	Simplify.
$3 = e^{2k}$	Divide each side by 100.
$\ln 3 = \ln e^{2k}$	Take natural log of each side.
$\ln 3 = 2k$	Apply the property $\ln e^a = a$.
$\dfrac{1}{2}\ln 3 = k$	Divide each side by 2.

Example 2   Solving Logarithmic Equations

Solve each logarithmic equation.

a. $\ln x = 2$ **b.** $5 + 2 \ln x = 4$

c. $2 \ln 3x = 4$ **d.** $\ln x - \ln(x - 1) = 1$

SOLUTION

a. $\ln x = 2$ Write original equation.

     $e^{\ln x} = e^2$ Exponentiate each side.

       $x = e^2$ Apply the property $e^{\ln a} = a$.

b. $5 + 2 \ln x = 4$ Write original equation.

       $2 \ln x = -1$ Subtract 5 from each side.

         $\ln x = -\dfrac{1}{2}$ Divide each side by 2.

       $e^{\ln x} = e^{-1/2}$ Exponentiate each side.

          $x = e^{-1/2}$ Apply the property $e^{\ln a} = a$.

c. $2 \ln 3x = 4$ Write original equation.

       $\ln 3x = 2$ Divide each side by 2.

     $e^{\ln 3x} = e^2$ Exponentiate each side.

        $3x = e^2$ Apply the property $e^{\ln a} = a$.

         $x = \dfrac{1}{3}e^2$ Divide each side by 3.

d. $\ln x - \ln(x - 1) = 1$ Write original equation.

       $\ln \dfrac{x}{x - 1} = 1$ $\ln m - \ln n = \ln(m/n)$

   $e^{\ln[x/(x-1)]} = e^1$ Exponentiate each side.

        $\dfrac{x}{x - 1} = e^1$ Apply the property $e^{\ln a} = a$.

           $x = ex - e$ Multiply each side by $x - 1$.

      $x - ex = -e$ Subtract ex from each side.

    $x(1 - e) = -e$ Factor.

           $x = \dfrac{-e}{1 - e}$ Divide each side by $1 - e$.

           $x = \dfrac{e}{e - 1}$ Simplify.

STUDY TIP

Because the domain of a logarithmic function generally does not include all real numbers, be sure to check for extraneous solutions.

SUMMARY AND STUDY STRATEGIES

After studying this chapter, you should have acquired the following skills.
The exercise numbers are keyed to the Review Exercises that begin on page 798.
Answers to odd-numbered Review Exercises are given in the back of the text.*

Section 10.1	Review Exercises

■ Use the properties of exponents to evaluate and simplify exponential   1, 2
expressions.

$$a^0 = 1, \quad a^x a^y = a^{x+y}, \quad \frac{a^x}{a^y} = a^{x-y}, \quad (a^x)^y = a^{xy}$$

$$(ab)^x = a^x b^x, \quad \left(\frac{a}{b}\right)^x = \frac{a^x}{b^x}, \quad a^{-x} = \frac{1}{a^x}$$

■ Sketch the graphs of exponential functions.   3–8
■ Use properties of exponents to answer questions about real-life situations.   9–12

Section 10.2

■ Use the properties of exponents to evaluate and simplify natural   13, 14
exponential expressions.
■ Sketch the graphs of natural exponential functions.   15–18
■ Solve compound interest problems.   19–24

$$A = P(1 + r/n)^{nt}, \quad A = Pe^{rt}$$

■ Solve effective rate of interest problems.   25, 26

$$r_{eff} = (1 + r/n)^n - 1$$

■ Solve present value problems.   27, 28

$$P = \frac{A}{(1 + r/n)^{nt}}$$

■ Answer questions involving the natural exponential function as a real-life model.   29–34

Section 10.3

■ Find the derivatives of natural exponential functions.   35–40

$$\frac{d}{dx}[e^x] = e^x, \quad \frac{d}{dx}[e^u] = e^u \frac{du}{dx}$$

■ Find equations of the tangent lines to the graphs of natural exponential functions.   41–44
■ Use calculus to analyze the graphs of functions that involve the natural   45–48
exponential function.

Section 10.4

■ Use the definition of the natural logarithmic function to write exponential   49–52
equations in logarithmic form, and vice versa.

$\ln x = b \quad$ if and only if $\quad e^b = x$.

---

* A wide range of valuable study aids are available to help you master the material in this chapter. The *Student Solutions Manual* includes step-by-step solutions to all odd-numbered exercises to help you review and prepare. The student website at *www.cengagebrain.com* offers algebra help and a *Graphing Technology Guide,* which contains step-by-step commands and instructions for a wide variety of graphing calculators.

Section 10.4 (continued)

Review Exercises

■ Sketch the graphs of natural logarithmic functions.   *53–56*

■ Use properties of logarithms to simplify, expand, and condense logarithmic   *57–66*
expressions.

$$\ln xy = \ln x + \ln y, \quad \ln \frac{x}{y} = \ln x - \ln y, \quad \ln x^n = n \ln x$$

■ Use inverse properties of exponential and logarithmic functions to solve   *67–80*
exponential and logarithmic equations.

$$\ln e^x = x, \quad e^{\ln x} = x$$

■ Use properties of natural logarithms to answer questions about real-life situations.   *81–84*

Section 10.5

■ Find the derivatives of natural logarithmic functions.   *85–98*

$$\frac{d}{dx}[\ln x] = \frac{1}{x}, \quad \frac{d}{dx}[\ln u] = \frac{1}{u}\frac{du}{dx}$$

■ Use the definition of logarithms to evaluate logarithmic expressions involving   *99–102*
other bases.

$$\log_a x = b \quad \text{if and only if} \quad a^b = x$$

■ Use the change-of-base formula to evaluate logarithmic expressions involving   *103–106*
other bases.

$$\log_a x = \frac{\ln x}{\ln a}$$

■ Find the derivatives of exponential and logarithmic functions involving other bases.   *107–112*

$$\frac{d}{dx}[a^x] = (\ln a)a^x, \quad \frac{d}{dx}[a^u] = (\ln a)a^u\frac{du}{dx}$$

$$\frac{d}{dx}[\log_a x] = \left(\frac{1}{\ln a}\right)\frac{1}{x}, \quad \frac{d}{dx}[\log_a u] = \left(\frac{1}{\ln a}\right)\left(\frac{1}{u}\right)\frac{du}{dx}$$

■ Use calculus to analyze the graphs of functions that involve the natural   *113–116*
logarithmic function.

■ Use calculus to answer questions about real-life situations.   *117, 118*

Section 10.6

■ Use exponential growth and decay to model real-life situations.   *119–132*

Study Strategies

■ **Classifying Differentiation Rules** Differentiation rules fall into two basic classes: (1) general rules that apply to all differentiable functions; and (2) specific rules that apply to special types of functions. At this point in the course, you have studied six general rules: the Constant Rule, the Constant Multiple Rule, the Sum Rule, the Difference Rule, the Product Rule, and the Quotient Rule. Although these rules were introduced in the context of algebraic functions, remember that they also can be used with exponential and logarithmic functions. You have also studied three specific rules: the Power Rule, the derivative of the natural exponential function, and the derivative of the natural logarithmic function. Each of these rules comes in two forms: the "simple" version, such as $D_x[e^x] = e^x$, and the Chain Rule version, such as $D_x[e^u] = e^u(du/dx)$.

■ **To Memorize or Not to Memorize?** When studying mathematics, you need to memorize some formulas and rules. Much of this will come from practice—the formulas that you use most often will be committed to memory. Some formulas, however, are used only infrequently. With these, it is helpful to be able to *derive* the formula from a *known* formula. For instance, knowing the Log Rule for differentiation and the change-of-base formula, $\log_a x = (\ln x)/(\ln a)$, allows you to derive the formula for the derivative of a logarithmic function to base a.

Review Exercises

See www.CalcChat.com for worked-out solutions to odd-numbered exercises.

Applying Properties of Exponents In Exercises 1 and 2, use the properties of exponents to simplify the expression.

1. (a) $(4^5)(4^2)$ (b) $(7^2)^3$

  (c) 2^{-4} (d) $\dfrac{3^8}{3^4}$

2. (a) $(5^4)(25^2)$ (b) $(9^{1/3})(3^{1/3})$

  (c) $\left(\dfrac{1}{3}\right)^{-3}$ (d) $(6^4)(6^{-5})$

Graphing Exponential Functions In Exercises 3–8, sketch the graph of the function.

3. $f(x) = 9^{x/2}$ **4.** $g(x) = 16^{3x/2}$

5. $f(t) = \left(\dfrac{1}{6}\right)^t$ **6.** $g(t) = \left(\dfrac{1}{3}\right)^{-t}$

7. $f(x) = \left(\dfrac{1}{2}\right)^{2x} + 4$ **8.** $g(x) = \left(\dfrac{2}{3}\right)^{2x} + 1$

9. Population Growth The resident populations P (in thousands) of Wisconsin from 2000 through 2009 can be modeled by the exponential function

$$P(t) = 5382(1.0057)^t$$

where t is the time in years, with $t = 0$ corresponding to 2000. Use the model to estimate the populations in the years (a) 2016 and (b) 2025. *(Source: U.S. Census Bureau)*

10. Revenue The revenues R (in millions of dollars) for Panera Bread Company from 2000 through 2009 can be modeled by the exponential function

$$R(t) = 163.82(1.2924)^t$$

where t is the time in years, with $t = 0$ corresponding to 2000. Use the model to estimate the sales in the years (a) 2014 and (b) 2017. *(Source: Panera Bread Company)*

11. Property Value Suppose that the value of a piece of property doubles every 12 years. If you buy the property for \$55,000, its value t years after the date of purchase should be

$$V(t) = 55{,}000(2)^{t/12}.$$

Use the model to approximate the value of the property (a) 4 years and (b) 25 years after it is purchased.

12. Inflation Rate Suppose the annual rate of inflation averages 2% over the next 10 years. With this rate of inflation, the approximate cost C of goods or services during any year in the decade will be given by

$$C(t) = P(1.02)^t, \quad 0 \le t \le 10$$

where t is time in years and P is the present cost. If the cost of a graphing calculator is presently \$80, estimate the cost 10 years from now.

Applying Properties of Exponents In Exercises 13 and 14, use the properties of exponents to simplify the expression.

13. (a) $(e^5)^2$ (b) $\dfrac{e^3}{e^5}$

  (c) $(e^4)(e^{3/2})$ (d) $(e^2)^{-4}$

14. (a) $(e^6)(e^{-3})$ (b) $(e^{-2})^{-5}$

  (c) $\left(\dfrac{e^6}{e^2}\right)^{-1}$ (d) $(e^3)^{4/3}$

Graphing Natural Exponential Functions In Exercises 15–18, sketch the graph of the function.

15. $f(x) = e^{-x} + 1$

16. $g(x) = e^{2x} - 1$

17. $f(x) = 1 - e^x$

18. $g(x) = 2 + e^{x-1}$

Finding Account Balances In Exercises 19–22, complete the table to determine the balance A for P dollars invested at rate r for t years, compounded n times per year.

n	1	2	4	12	365	Continuous compounding
A						

19. $P = \$1000$, $r = 4\%$, $t = 5$ years

20. $P = \$7000$, $r = 6\%$, $t = 20$ years

21. $P = \$3000$, $r = 3.5\%$, $t = 10$ years

22. $P = \$4500$, $r = 2\%$, $t = 25$ years

Comparing Account Balances In Exercises 23 and 24, \$2000 is deposited in an account. Decide which account, (a) or (b), will have the greater balance after 10 years.

23. (a) 5%, compounded continuously

    (b) 6%, compounded quarterly

24. (a) $6\frac{1}{2}\%$, compounded monthly

    (b) $6\frac{1}{4}\%$, compounded continuously

25. Effective Rate Find the effective rate of interest corresponding to a nominal rate of 6% per year compounded (a) annually, (b) semiannually, (c) quarterly, and (d) monthly.

26. Effective Rate Find the effective rate of interest corresponding to a nominal rate of 8.25% per year compounded (a) annually, (b) semiannually, (c) quarterly, and (d) monthly.

27. Present Value How much should be deposited in an account paying 5% interest compounded quarterly in order to have a balance of $12,000 three years from now?

28. Present Value How much should be deposited in an account paying 8% interest compounded monthly in order to have a balance of $20,000 five years from now?

29. Demand The demand function for a product is modeled by

$$p = 12,500 - \frac{10,000}{2 + e^{-0.001x}}.$$

Find the price p (in dollars) of the product when the quantity demanded is (a) $x = 1000$ units and (b) $x = 2500$ units. (c) What is the limit of the price as x increases without bound?

30. Demand The demand function for a product is modeled by

$$p = 8000\left(1 - \frac{5}{5 + e^{-0.002x}}\right).$$

Find the price p (in dollars) of the product when the quantity demanded is (a) $x = 1000$ units and (b) $x = 2500$ units. (c) What is the limit of the price as x increases without bound?

31. Profit The net profits P (in millions of dollars) of Medco Health Solutions from 2000 through 2009 are shown in the table.

Year	2000	2001	2002	2003	2004
Profit	216.8	256.6	361.6	425.8	481.6

Year	2005	2006	2007	2008	2009
Profit	602.0	729.8	912.0	1102.9	1280.3

A model for this data is given by $P = 223.89e^{0.1979t}$, where t represents the year, with $t = 0$ corresponding to 2000. *(Source: Medco Health Solutions, Inc.)*

(a) How well does the model fit the data?

(b) Find a linear model for the data. How well does the linear model fit the data? Which model, exponential or linear, is a better fit?

(c) Use both models to predict the net profit in 2015.

32. Population The populations P (in thousands) of Albuquerque, New Mexico from 2000 through 2009 can be modeled by $P = 450e^{0.019t}$, where t is the time in years, with $t = 0$ corresponding to 2000. *(Source: U.S. Census Bureau)*

(a) Find the populations in 2000, 2005, and 2009.

(b) Use the model to estimate the population in 2020.

33. Biology A lake is stocked with 500 fish, and the fish population P begins to increase according to the logistic growth model

$$P = \frac{10,000}{1 + 19e^{-t/5}}, \quad t \geq 0$$

where t is measured in months.

(a) Find the number of fish in the lake after 4 months.

(b) Use a graphing utility to graph the model. Find the number of months it takes for the population of fish to reach 4000.

(c) Does the population have a limit as t increases without bound? Explain your reasoning.

34. Medicine On a college campus of 5000 students, the spread of a flu virus through the student body is modeled by

$$P = \frac{5000}{1 + 4999e^{-0.8t}}, \quad t \geq 0$$

where P is the total number of infected people and t is the time, measured in days.

(a) Find the number of students infected after 5 days.

(b) Use a graphing utility to graph the model. Find the number of days it takes for 2000 students to become infected with the flu.

(c) According to this model, will all the students on campus become infected with the flu? Explain your reasoning.

Differentiating Exponential Functions In Exercises 35–40, find the derivative of the function.

35. $y = 4e^{x^2}$

36. $y = 4e^{\sqrt{x}}$

37. $y = \dfrac{x}{e^{2x}}$

38. $y = x^2 e^x$

39. $y = \dfrac{5}{1 + e^{2x}}$

40. $y = \dfrac{10}{1 - 2e^x}$

Finding an Equation of a Tangent Line In Exercises 41–44, find an equation of the tangent line to the graph of the function at the given point.

41. $y = e^{2-x}, \quad (2, 1)$

42. $y = e^{2x^2}, \quad (1, e^2)$

43. $y = x^2 e^{-x}, \quad \left(1, \dfrac{1}{e}\right)$

44. $y = xe^x - e^x, \quad (1, 0)$

Analyzing a Graph In Exercises 45–48, analyze and sketch the graph of the function. Label any relative extrema, points of inflection, and asymptotes.

45. $f(x) = x^3 e^x$

46. $f(x) = \dfrac{e^x}{x^2}$

47. $f(x) = \dfrac{1}{xe^x}$

48. $f(x) = \dfrac{x^2}{e^x}$

Logarithmic and Exponential Forms of Equations
In Exercises 49–52, write the logarithmic equation as an exponential equation, or vice versa.

49. $\ln 12 = 2.4849\ldots$

50. $\ln 0.6 = -0.5108\ldots$

51. $e^{1.5} = 4.4816\ldots$

52. $e^{-4} = 0.0183\ldots$

Graphing Logarithmic Expressions In Exercises 53–56, sketch the graph of the function.

53. $y = \ln(4 - x)$

54. $y = \ln x - 3$

55. $y = \ln \dfrac{x}{3}$

56. $y = -2 \ln x$

Expanding Logarithmic Expressions In Exercises 57–62, use the properties of logarithms to rewrite the expression as a sum, difference, or multiple of logarithms.

57. $\ln \sqrt{x^2(x - 1)}$

58. $\ln \sqrt[3]{x^2 - 1}$

59. $\ln \dfrac{x^2}{(x + 1)^3}$

60. $\ln \dfrac{x^2}{x^2 + 1}$

61. $\ln \left(\dfrac{1 - x}{3x} \right)^3$

62. $\ln \left(\dfrac{x - 1}{x + 1} \right)^2$

Condensing Logarithmic Expressions In Exercises 63–66, use the properties of logarithms to rewrite the expression as the logarithm of a single quantity.

63. $\ln(2x + 5) + \ln(x - 3)$

64. $\frac{1}{3} \ln(x^2 - 6) - 2 \ln(3x + 2)$

65. $4[\ln(x^3 - 1) + 2 \ln x - \ln(x - 5)]$

66. $\frac{1}{2}[\ln x + 3 \ln(x + 1) - \ln(x - 2)]$

Solving Exponential and Logarithmic Equations In Exercises 67–80, solve for x.

67. $e^{\ln x} = 3$

68. $e^{\ln(x + 2)} = 5$

69. $\ln x = 3$

70. $\ln 5x = 2$

71. $\ln 2x - \ln(3x - 1) = 0$

72. $\ln x - \ln(x + 1) = 2$

73. $\ln x + \ln(x - 3) = 0$

74. $2 \ln x + \ln(x - 2) = 0$

75. $e^{-1.386x} = 0.25$

76. $e^{-0.01x} - 5.25 = 0$

77. $e^{2x-1} - 6 = 0$

78. $4e^{2x-3} - 5 = 0$

79. $100(1.21)^x = 110$

80. $500(1.075)^{120x} = 100,000$

81. Compound Interest A deposit of $400 is made in an account that earns interest at an annual rate of 2.5%. How long will it take for the balance to double when the interest is compounded (a) annually, (b) monthly, (c) daily, and (d) continuously?

82. Hourly Earnings The average hourly wages w (in dollars) for private industry employees in the United States from 1990 through 2009 can be modeled by

$$w = 10.2e^{0.0315t}$$

where $t = 0$ corresponds to 1990. *(Source: U.S. Bureau of Labor Statistics)*

(a) What was the average hourly wage in 2000?

(b) In what year will the average hourly wage be $23?

83. Learning Theory Students in a psychology experiment were given an exam and then retested monthly with equivalent exams. The average scores S (on a 100-point scale) for the students can be modeled by

$$S = 75 - 6 \ln(t + 1), \quad 0 \le t \le 12$$

where t is the time in months.

(a) What was the average score on the original exam?

(b) What was the average score after 4 months?

(c) After how many months was the average score 60?

84. Demand The demand function for a product is given by

$$p = 8000 \left(1 - \frac{5}{5 + e^{-0.002x}} \right)$$

where p is the price per unit (in dollars) and x is the number of units sold. Find the numbers of units sold for prices of (a) $p = 200 and (b) $p = 800.

Differentiating a Logarithmic Function In Exercises 85–98, find the derivative of the function.

85. $f(x) = \ln 3x^2$

86. $y = \ln \sqrt{x}$

87. $y = \ln \dfrac{x(x - 1)}{x - 2}$

88. $y = \ln \dfrac{x^2}{x + 1}$

89. $f(x) = \ln e^{2x+1}$

90. $f(x) = \ln e^{x^2}$

91. $y = \dfrac{\ln x}{x^3}$

92. $y = \dfrac{x^2}{\ln x}$

93. $y = \ln(x^2 - 2)^{2/3}$

94. $y = \ln \sqrt[3]{x^3 + 1}$

95. $f(x) = \ln(x^2 \sqrt{x + 1})$

96. $f(x) = \ln \dfrac{x}{\sqrt{x + 1}}$

97. $y = \ln \dfrac{e^x}{1 + e^x}$

98. $y = \ln(e^{2x} \sqrt{e^{2x} - 1})$

Evaluating Logarithms   In Exercises 99–102, evaluate the logarithm without using a calculator.

99. $\log_6 36$

100. $\log_2 32$

101. $\log_{10} 1$

102. $\log_4 \frac{1}{64}$

Changing Bases to Evaluate Logarithms   In Exercises 103–106, use the change-of-base formula and a calculator to evaluate the logarithm.

103. $\log_5 13$

104. $\log_4 18$

105. $\log_{16} 64$

106. $\log_4 125$

Differentiating Functions of Other Bases   In Exercises 107–112, find the derivative of the function.

107. $y = 5^{2x+1}$

108. $y = 8^{x^3}$

109. $y = \log_3(2x - 1)$

110. $y = \log_{16}(x^2 - 3x)$

111. $y = \log_{10} \dfrac{3}{x}$

112. $y = \log_2 \dfrac{1}{x^2}$

Analyzing a Graph   In Exercises 113–116, analyze and sketch the graph of the function. Label any relative extrema, points of inflection, and asymptotes.

113. $y = \ln(x + 3)$

114. $y = \dfrac{8 \ln x}{x^2}$

115. $y = \ln \dfrac{10}{x + 2}$

116. $y = \ln \dfrac{x^2}{9 - x^2}$

117. Music   The numbers of download music singles D (in millions) from 2004 through 2009 can be modeled by

$$D = -1671.88 + 1282 \ln t$$

where $t = 4$ corresponds to 2004. Find the rates of change of the number of download music singles in 2005 and 2008. *(Source: Recording Industry Association of America)*

118. Minimum Average Cost   The cost of producing x units of a product is modeled by

$$C = 200 + 75x - 300 \ln x, \quad x \geq 1.$$

(a) Find the average cost function $\overline{C}$.

(b) Find the minimum average cost analytically. Use a graphing utility to confirm your result.

Modeling Exponential Growth and Decay   In Exercises 119 and 120, find the exponential function

$$y = Ce^{kt}$$

that passes through the two given points.

119. $(0, 3), (4, 1)$

120. $(1, 1), (5, 5)$

Modeling Radioactive Decay   In Exercises 121–126, complete the table for each radioactive isotope.

	Isotope	Half-life (in years)	Initial quantity	Amount after 1000 years	Amount after 10,000 years
121.	^{226}Ra	1599	8 grams		
122.	^{226}Ra	1599		0.7 gram	
123.	^{14}C	5715			6 grams
124.	^{14}C	5715	5 grams		
125.	^{239}Pu	24,100		2.4 grams	
126.	^{239}Pu	24,100			7.1 grams

Modeling Compound Interest   In Exercises 127–130, complete the table for an account in which interest is compounded continuously.

	Initial investment	Annual rate	Time to double	Amount after 10 years	Amount after 25 years
127.	$600	8%			
128.	$2000		7 years		
129.	$15,000			$18,321.04	
130.		4%		$11,934.60	

131. Medical Science   Soon after an injection, the concentration D (in milligrams per milliliter) of a drug in a patient's bloodstream is 500 milligrams per milliliter. After 6 hours, 50 milligrams per milliliter of the drug remains in the bloodstream.

(a) Find an exponential model for the concentration D after t hours.

(b) What is the concentration of the drug after 4 hours?

132. Population Growth   The number of a certain type of bacteria increases continuously at a rate proportional to the number present. After 2 hours, there are 200 bacteria, and after 4 hours, there are 300 bacteria.

(a) Find an exponential model given the population P after t hours.

(b) How many bacteria will there be after 7 hours?

(c) How long will it take for the population to double?

TEST YOURSELF

See www.CalcChat.com for worked-out solutions to odd-numbered exercises.

Take this test as you would take a test in class. When you are done, check your work against the answers given in the back of the book.

In Exercises 1–4, use the properties of exponents to simplify the expression.

1. $3^2(3^{-2})$

2. $\left(\dfrac{2^3}{2^{-5}}\right)^{-1}$

3. $(e^{1/2})(e^4)$

4. $(e^3)^4$

In Exercises 5–10, sketch the graph of the function.

5. $f(x) = 5^{x-2}$

6. $f(x) = 4^{-x}$

7. $f(x) = e^{x-3}$

8. $f(x) = 8 + \ln x^2$

9. $f(x) = \ln(x - 5)$

10. $f(x) = 0.5 \ln x$

In Exercises 11–13, use the properties of logarithms to rewrite the expression as a sum, difference, or multiple of logarithms.

11. $\ln \dfrac{3}{2}$

12. $\ln \sqrt{x + y}$

13. $\ln \dfrac{x + 1}{y}$

In Exercises 14–16, use the properties of logarithms to rewrite the expression as the logarithm of a single quantity.

14. $\ln y + \ln(x + 1)$

15. $3 \ln x - 2 \ln(x - 1)$

16. $\ln x + 4 \ln y - \frac{1}{2} \ln(z + 4)$

In Exercises 17–19, solve the equation.

17. $e^{x-1} = 9$

18. $10e^{2x+1} = 900$

19. $50(1.06)^x = 1500$

20. A deposit of \$500 is made in an account that earns interest at an annual rate of 4%. How long will it take for the balance to double when the interest is compounded (a) annually, (b) monthly, (c) daily, and (d) continuously?

In Exercises 21–24, find the derivative of the function.

21. $y = e^{-3x} + 5$

22. $y = 7e^{x+2} + 2x$

23. $y = \ln(3 + x^2)$

24. $y = \ln \dfrac{5x}{x + 2}$

25. The revenues R (in millions of dollars) of skiing facilities in the United States from 2000 through 2008 can be modeled by

$$R = 1548e^{0.0617t}$$

where $t = 0$ corresponds to 2000.   *(Source: U.S. Census Bureau)*

(a) Use this model to estimate the revenues in 2006.

(b) At what rate were the revenues changing in 2006?

26. What percent of a present amount of radioactive radium (^{226}Ra) will remain after 1200 years? (The half-life of ^{226}Ra is 1599 years.)

27. A population is growing continuously at the rate of 1.75% per year. Find the time necessary for the population to double in size.

11 Integration and Its Applications

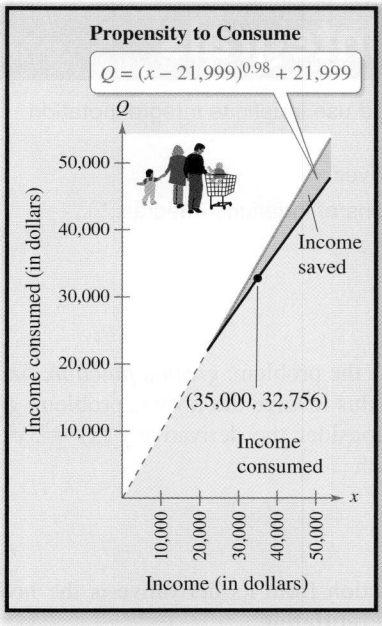

Propensity to Consume

$$Q = (x - 21{,}999)^{0.98} + 21{,}999$$

Income saved

(35,000, 32,756)

Income consumed

Example 8 on page 820 shows how integration can be used to analyze the marginal propensity to consume.

11.1 Antiderivatives and Indefinite Integrals

■ Understand the definition of antiderivative and use indefinite integral notation for antiderivatives.

■ Use basic integration rules to find antiderivatives.

■ Use initial conditions to find particular solutions of indefinite integrals.

■ Use antiderivatives to solve real-life problems.

Horry County, South Carolina

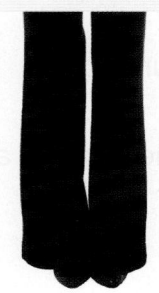

In Exercise 69 on page 813, you will use integration to find a model for the population of a county.

Antiderivatives

In Chapter 7, you were concerned primarily with the problem: *given a function, find its derivative.* Some important applications of calculus involve the inverse problem: *given a derivative, find the function.* For instance, consider the derivative $f'(x) = 3x^2$. To determine the function f, you might come up with

$$f(x) = x^3 \quad \text{because} \quad \frac{d}{dx}[x^3] = 3x^2.$$

This operation of determining the original function from its derivative is the inverse operation of differentiation. It is called **antidifferentiation.**

Definition of Antiderivative

A function F is an **antiderivative** of a function f when for every x in the domain of f, it follows that $F'(x) = f(x)$.

If $F(x)$ is an antiderivative of $f(x)$, then $F(x) + C$, where C is any constant, is also an antiderivative of $f(x)$. For example,

$$F(x) = x^3, \quad G(x) = x^3 - 5, \quad \text{and} \quad H(x) = x^3 + 0.3$$

are all antiderivatives of $3x^2$ because the derivative of each is $3x^2$. As it turns out, *all* antiderivatives of $3x^2$ are of the form $x^3 + C$. So, the process of antidifferentiation does not determine a single function, but rather a *family* of functions, each differing from the others by a constant.

The antidifferentiation process is also called **integration** and is denoted by

$$\int \qquad \text{Integral sign}$$

which is called an **integral sign.** The symbol

$$\int f(x)\, dx \qquad \text{Indefinite integral}$$

is the **indefinite integral** of $f(x)$, and it denotes the family of antiderivatives of $f(x)$. That is, if $F'(x) = f(x)$ for all x, then you can write

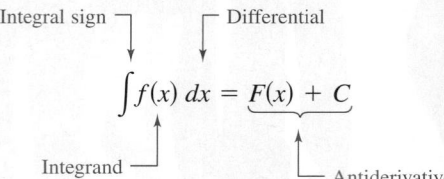

where $f(x)$ is the **integrand** and C is the **constant of integration.** The differential dx in the indefinite integral identifies the variable of integration. That is, the symbol $\int f(x)\, dx$ denotes the "antiderivative of f *with respect to x*" just as the symbol dy/dx denotes the "derivative of y *with respect to x.*"

STUDY TIP

In this text, the phrase "$F(x)$ is an antiderivative of $f(x)$" is used synonymously with "F is an antiderivative of f."

Finding Antiderivatives

The inverse relationship between the operations of integration and differentiation can be shown symbolically, as follows.

$$\frac{d}{dx}\left[\int f(x)\,dx\right] = f(x)$$
    Differentiation is the inverse of integration.

$$\int f'(x)\,dx = f(x) + C$$
    Integration is the inverse of differentiation.

This inverse relationship between integration and differentiation allows you to obtain integration formulas directly from differentiation formulas. The following summary lists the integration formulas that correspond to some of the differentiation formulas you have studied.

Basic Integration Rules

1. $\displaystyle\int k\,dx = kx + C, \quad k$ is a constant.   Constant Rule

2. $\displaystyle\int kf(x)\,dx = k\int f(x)\,dx$ Constant Multiple Rule

3. $\displaystyle\int [f(x) + g(x)]\,dx = \int f(x)\,dx + \int g(x)\,dx$ Sum Rule

4. $\displaystyle\int [f(x) - g(x)]\,dx = \int f(x)\,dx - \int g(x)\,dx$ Difference Rule

5. $\displaystyle\int x^n\,dx = \frac{x^{n+1}}{n+1} + C, \quad n \neq -1$ Simple Power Rule

Be sure you see that the Simple Power Rule has the restriction that n cannot be -1. So, you *cannot* use the Simple Power Rule to evaluate the integral

$$\int \frac{1}{x}\,dx.$$

To evaluate this integral, you need the Log Rule, which is described in Section 11.3.

Example 1   **Finding Indefinite Integrals**

Find each indefinite integral.

a. $\displaystyle\int \frac{1}{2}\,dx$ **b.** $\displaystyle\int 1\,dx$ **c.** $\displaystyle\int -5\,dt$

SOLUTION

a. $\displaystyle\int \frac{1}{2}\,dx = \frac{1}{2}x + C$ **b.** $\displaystyle\int 1\,dx = x + C$ **c.** $\displaystyle\int -5\,dt = -5t + C$

✓**Checkpoint 1**

Find each indefinite integral.

a. $\displaystyle\int 5\,dx$ **b.** $\displaystyle\int -1\,dr$ **c.** $\displaystyle\int 2\,dt$

Example 2   Finding an Indefinite Integral

$$\int 3x\, dx = 3 \int x\, dx \qquad \text{Constant Multiple Rule}$$

$$= 3 \int x^1\, dx \qquad \text{Rewrite } x \text{ as } x^1.$$

$$= 3\left(\frac{x^2}{2}\right) + C \qquad \text{Simple Power Rule with } n = 1$$

$$= \frac{3}{2}x^2 + C \qquad \text{Simplify.}$$

✓ **Checkpoint 2**

Find $\int 5x\, dx$.   ■

In finding indefinite integrals, a strict application of the basic integration rules tends to produce cumbersome constants of integration. For instance, in Example 2, you could have written

$$\int 3x\, dx = 3 \int x\, dx = 3\left(\frac{x^2}{2} + C\right) = \frac{3}{2}x^2 + 3C.$$

However, because C represents *any* constant, it is unnecessary to write $3C$ as the constant of integration. You can simply write $\frac{3}{2}x^2 + C$.

In Example 2, note that the general pattern of integration is similar to that of differentiation.

Original Integral: $\int 3x\, dx$	⇨	Rewrite: $3 \int x^1\, dx$	⇨	Integrate: $3\left(\frac{x^2}{2}\right) + C$	⇨	Simplify: $\frac{3}{2}x^2 + C$

Example 3   Rewriting Before Integrating

	Original Integral	Rewrite	Integrate	Simplify
a.	$\int \dfrac{1}{x^3}\, dx$	$\int x^{-3}\, dx$	$\dfrac{x^{-2}}{-2} + C$	$-\dfrac{1}{2x^2} + C$
b.	$\int \sqrt{x}\, dx$	$\int x^{1/2}\, dx$	$\dfrac{x^{3/2}}{3/2} + C$	$\dfrac{2}{3}x^{3/2} + C$

✓ **Checkpoint 3**

Find each indefinite integral.

a. $\int \dfrac{1}{x^2}\, dx$ **b.** $\int \sqrt[3]{x}\, dx$ ■

Remember that you can check your answer to an antidifferentiation problem by differentiating. For instance, in Example 3(b), you can confirm that $\frac{2}{3}x^{3/2} + C$ is the correct antiderivative by differentiating to obtain

$$\frac{d}{dx}\left[\frac{2}{3}x^{3/2} + C\right] = \left(\frac{2}{3}\right)\left(\frac{3}{2}\right)x^{1/2} = \sqrt{x}.$$

With the five basic integration rules, you can integrate *any* polynomial function, as demonstrated in the next example.

Example 4 **Integrating Polynomial Functions**

Find (a) $\int (x + 2)\, dx$ and (b) $\int (3x^4 - 5x^2 + x)\, dx$.

SOLUTION

a. $\int (x + 2)\, dx = \int x\, dx + \int 2\, dx$ Apply Sum Rule.

$$= \frac{x^2}{2} + C_1 + 2x + C_2 \qquad \text{Apply Simple Power and Constant Rules.}$$

$$= \frac{x^2}{2} + 2x + C \qquad\qquad C = C_1 + C_2$$

The second line in this solution is usually omitted.

b. $\int (3x^4 - 5x^2 + x)\, dx = 3\left(\frac{x^5}{5}\right) - 5\left(\frac{x^3}{3}\right) + \frac{x^2}{2} + C$

$$= \frac{3}{5}x^5 - \frac{5}{3}x^3 + \frac{1}{2}x^2 + C$$

✓**Checkpoint 4**

Find (a) $\int (x + 4)\, dx$ and (b) $\int (4x^3 - 5x + 2)\, dx$.

■

Example 5 **Rewriting Before Integrating**

Find $\int \dfrac{x - 1}{\sqrt{x}}\, dx$.

SOLUTION Begin by rewriting the quotient in the integrand as a difference. Then rewrite each term using rational exponents.

$\int \dfrac{x - 1}{\sqrt{x}}\, dx = \int \left(\dfrac{x}{\sqrt{x}} - \dfrac{1}{\sqrt{x}}\right) dx$ Rewrite as a difference.

$= \int (x^{1/2} - x^{-1/2})\, dx$ Rewrite using rational exponents.

$= \int x^{1/2}\, dx - \int x^{-1/2}\, dx$ Apply Difference Rule.

$= \dfrac{x^{3/2}}{3/2} - \dfrac{x^{1/2}}{1/2} + C$ Apply Simple Power Rule.

$= \dfrac{2}{3}x^{3/2} - 2x^{1/2} + C$ Simplify.

$= \dfrac{2}{3}\sqrt{x}(x - 3) + C$ Factor.

✓**Checkpoint 5**

Find $\int \dfrac{x + 2}{\sqrt{x}}\, dx$.

■

STUDY TIP

When integrating quotients, remember *not* to integrate the numerator and denominator separately. For instance, in Example 5, be sure you understand that

$$\int \frac{x - 1}{\sqrt{x}}\, dx = \frac{2}{3}\sqrt{x}(x - 3) + C$$

is not the same as

$$\frac{\int (x - 1)\, dx}{\int \sqrt{x}\, dx} = \frac{\frac{1}{2}x^2 - x + C_1}{\frac{2}{3}x\sqrt{x} + C_2}.$$

ALGEBRA TUTOR xy

For help on the algebra in Example 5, see Example 1(a) in the *Chapter 11 Algebra Tutor,* on page 858.

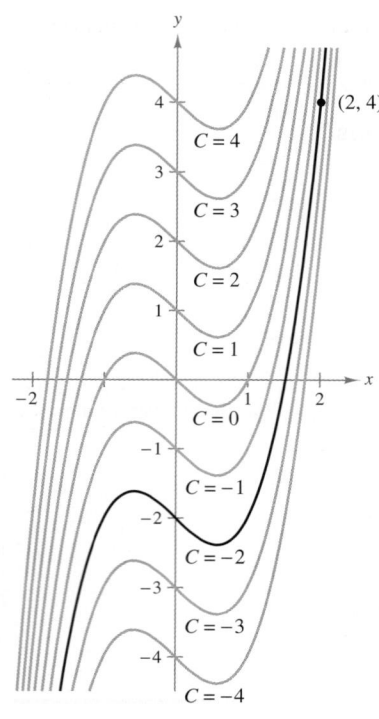

$F(x) = x^3 - x + C$

FIGURE 11.1

Particular Solutions

You have already seen that the equation $y = \int f(x)\, dx$ has many solutions, each differing from the others by a constant. This means that the graphs of any two antiderivatives of f are vertical translations of each other. For example, Figure 11.1 shows the graphs of several antiderivatives of the form

$$y = F(x) = \int (3x^2 - 1)\, dx = x^3 - x + C$$

for various integer values of C. Each of these antiderivatives is a solution of the *differential equation*

$$\frac{dy}{dx} = 3x^2 - 1.$$

A **differential equation** in x and y is an equation that involves x, y, and derivatives of y. The **general solution** of $dy/dx = 3x^2 - 1$ is $F(x) = x^3 - x + C$.

In many applications of integration, you are given enough information to determine a **particular solution.** To do this, you need to know the value of $F(x)$ for only one value of x. (This information is called an **initial condition.**) For example, in Figure 11.1, there is only one curve that passes through the point $(2, 4)$. To find this curve, use the information below.

$$F(x) = x^3 - x + C \qquad \text{General solution}$$
$$F(2) = 4 \qquad \text{Initial condition}$$

By using the initial condition in the general solution, you can determine that $F(2) = 2^3 - 2 + C = 4$, which implies that $C = -2$. So, the particular solution is

$$F(x) = x^3 - x - 2. \qquad \text{Particular solution}$$

Example 6   **Finding a Particular Solution**

Find the general solution of

$$F'(x) = 2x - 2$$

and find the particular solution that satisfies the initial condition $F(1) = 2$.

SOLUTION   Begin by integrating to find the general solution.

$$F(x) = \int (2x - 2)\, dx \qquad \text{Integrate } F'(x) \text{ to obtain } F(x).$$
$$= x^2 - 2x + C \qquad \text{General solution}$$

Using the initial condition $F(1) = 2$, you can write

$$F(1) = 1^2 - 2(1) + C = 2$$

which implies that $C = 3$. So, the particular solution is

$$F(x) = x^2 - 2x + 3. \qquad \text{Particular solution}$$

This solution is shown graphically in Figure 11.2. Note that each of the gray curves represents a solution of the equation $F'(x) = 2x - 2$. The black curve, however, is the only solution that passes through the point $(1, 2)$, which means that $F(x) = x^2 - 2x + 3$ is the only solution that satisfies the initial condition.

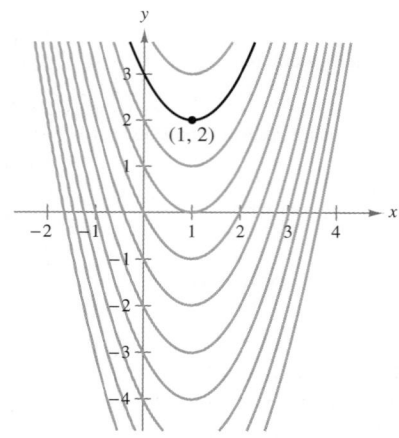

FIGURE 11.2

✓ Checkpoint 6

Find the general solution of $F'(x) = 4x + 2$, and find the particular solution that satisfies the initial condition $F(1) = 8$.

Applications

In Chapters 7 and 8, you used the general position function (neglecting air resistance) for a falling object

$$s(t) = -16t^2 + v_0 t + s_0$$

where $s(t)$ is the height (in feet) and t is the time (in seconds). In the next example, integration is used to *derive* this function.

 Example 7 **Deriving a Position Function**

A ball is thrown upward with an initial velocity of 64 feet per second from an initial height of 80 feet, as shown in Figure 11.3. Derive the position function giving the height s (in feet) as a function of the time t (in seconds). Will the ball be in the air for more than 5 seconds?

SOLUTION Let $t = 0$ represent the initial time. Then the two given conditions can be written as

$$s(0) = 80 \qquad \text{Initial height is 80 feet.}$$
$$s'(0) = 64. \qquad \text{Initial velocity is 64 feet per second.}$$

Because the acceleration due to gravity is -32 feet per second per second, you can integrate the acceleration function to find the velocity function, as shown.

$$s''(t) = -32 \qquad \text{Acceleration due to gravity}$$
$$s'(t) = \int -32 \, dt \qquad \text{Integrate } s''(t) \text{ to obtain } s'(t).$$
$$= -32t + C_1 \qquad \text{Velocity function}$$

Using the initial velocity, you can conclude that $C_1 = 64$. Next, integrate the velocity function to find the position function.

$$s'(t) = -32t + 64 \qquad \text{Velocity function}$$
$$s(t) = \int (-32t + 64) \, dt \qquad \text{Integrate } s'(t) \text{ to obtain } s(t).$$
$$= -16t^2 + 64t + C_2 \qquad \text{Position function}$$

Using the initial height, it follows that $C_2 = 80$. So, the position function is given by

$$s(t) = -16t^2 + 64t + 80. \qquad \text{Position function}$$

To find the time when the ball hits the ground, set the position function equal to 0 and solve for t.

$$-16t^2 + 64t + 80 = 0 \qquad \text{Set } s(t) \text{ equal to zero.}$$
$$-16(t + 1)(t - 5) = 0 \qquad \text{Factor.}$$
$$t = -1, \quad t = 5 \qquad \text{Solve for } t.$$

Because the time must be positive, you can conclude that the ball hits the ground 5 seconds after it is thrown. So, the ball is not in the air for more than 5 seconds.

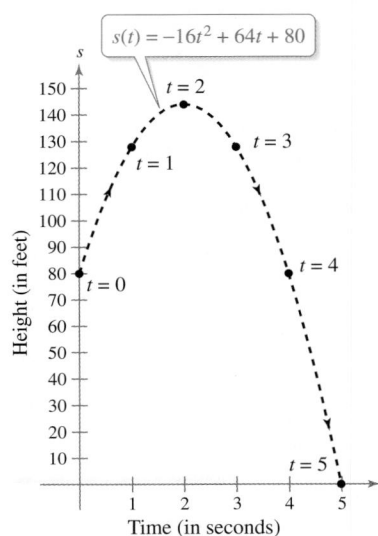

FIGURE 11.3

✓ **Checkpoint 7**

Derive the position function when a ball is thrown upward with an initial velocity of 32 feet per second from an initial height of 48 feet. When does the ball hit the ground? With what velocity does the ball hit the ground?

Example 8 **Finding a Cost Function**

The marginal cost of producing x units of a product is modeled by

$$\frac{dC}{dx} = 32 - 0.04x.$$ Marginal cost

It costs $50 to produce one unit. Find the total cost of producing 200 units.

SOLUTION   To find the cost function, integrate the marginal cost function.

$$C = \int (32 - 0.04x)\, dx$$ Integrate $\frac{dC}{dx}$ to obtain C.

$$= 32x - 0.04\left(\frac{x^2}{2}\right) + K$$

$$= 32x - 0.02x^2 + K$$ Cost function

To solve for K, use the initial condition $C = 50$ when $x = 1$.

$$50 = 32(1) - 0.02(1)^2 + K$$ Substitute 50 for C and 1 for x.

$$18.02 = K$$ Solve for K.

So, the total cost function is given by

$$C = 32x - 0.02x^2 + 18.02$$ Cost function

which implies that the cost of producing 200 units is

$$C = 32(200) - 0.02(200)^2 + 18.02$$

$$= \$5618.02.$$

 Checkpoint 8

The marginal cost function for producing x units of a product is modeled by

$$\frac{dC}{dx} = 28 - 0.02x.$$

It costs $40 to produce one unit. Find the total cost of producing 200 units.

STUDY TIP

In Example 8, note that K is used to represent the constant of integration rather than C. This is done to avoid confusion between the constant C and the cost function

$$C = 32x - 0.02x^2 + 18.02.$$

SUMMARIZE   (Section 11.1)

1. State the definition of antiderivative *(page 804)*. For examples of antiderivatives, see Examples 1, 2, 3, 4, and 5.

2. State the Constant Rule *(page 805)*. For an example of the Constant Rule, see Example 1.

3. State the Constant Multiple Rule *(page 805)*. For an example of the Constant Multiple Rule, see Example 2.

4. State the Sum Rule *(page 805)*. For an example of the Sum Rule, see Example 4.

5. State the Difference Rule *(page 805)*. For an example of the Difference Rule, see Example 5.

6. State the Simple Power Rule *(page 805)*. For examples of the Simple Power Rule, see Examples 2, 3, 4, and 5.

7. Describe a real-life example of how antidifferentiation can be used to find a cost function *(page 810, Example 8)*.

SKILLS WARM UP 11.1   The following warm-up exercises involve skills that were covered in a previous course or in earlier sections. You will use these skills in the exercise set for this section. For additional help, review Sections 0.2, 0.4, and 1.1.

In Exercises 1–6, rewrite the expression using rational exponents.

1. $\dfrac{\sqrt{x}}{x}$

2. $\sqrt[3]{2x}(2x)$

3. $\sqrt{5x^3} + \sqrt{x^5}$

4. $\dfrac{1}{\sqrt{x}} + \dfrac{1}{\sqrt[3]{x^2}}$

5. $\dfrac{(x+1)^3}{\sqrt{x+1}}$

6. $\dfrac{\sqrt{x}}{\sqrt[3]{x}}$

In Exercises 7–10, let $(x, y) = (2, 2)$, and solve the equation for C.

7. $y = x^2 + 5x + C$ **8.** $y = 3x^3 - 6x + C$ **9.** $y = -16x^2 + 26x + C$ **10.** $y = -\frac{1}{4}x^4 - 2x^2 + C$

Exercises 11.1

See www.CalcChat.com for worked-out solutions to odd-numbered exercises.

Integration and Differentiation   In Exercises 1–6, verify the statement by showing that the derivative of the right side is equal to the integrand on the left side.

1. $\displaystyle\int 4x\,dx = 2x^2 + C$

2. $\displaystyle\int 4x^3\,dx = x^4 + C$

3. $\displaystyle\int \left(-\dfrac{9}{x^4}\right)dx = \dfrac{3}{x^3} + C$

4. $\displaystyle\int \dfrac{4}{\sqrt{x}}\,dx = 8\sqrt{x} + C$

5. $\displaystyle\int \left(4x^3 - \dfrac{1}{x^2}\right)dx = x^4 + \dfrac{1}{x} + C$

6. $\displaystyle\int \left(1 - \dfrac{1}{\sqrt[3]{x^2}}\right)dx = x - 3\sqrt[3]{x} + C$

Finding Indefinite Integrals   In Exercises 7–18, find the indefinite integral. Check your result by differentiating. *See Examples 1 and 2.*

7. $\displaystyle\int du$

8. $\displaystyle\int dr$

9. $\displaystyle\int 6\,dx$

10. $\displaystyle\int -4\,dx$

11. $\displaystyle\int 7x\,dx$

12. $\displaystyle\int 2x\,dx$

13. $\displaystyle\int 5t^2\,dt$

14. $\displaystyle\int 3t^4\,dt$

15. $\displaystyle\int 5x^{-3}\,dx$

16. $\displaystyle\int 4y^{-2}\,dy$

17. $\displaystyle\int y^{3/2}\,dy$

18. $\displaystyle\int v^{-1/2}\,dv$

Rewriting Before Integrating   In Exercises 19–24, find the indefinite integral. *See Example 3.*

	Original Integral	Rewrite	Integrate	Simplify
19.	$\displaystyle\int \sqrt[3]{x^2}\,dx$			
20.	$\displaystyle\int \dfrac{1}{x^4}\,dx$			
21.	$\displaystyle\int \dfrac{1}{x\sqrt{x}}\,dx$			
22.	$\displaystyle\int x(x^2 + 3)\,dx$			
23.	$\displaystyle\int \dfrac{1}{2x^3}\,dx$			
24.	$\displaystyle\int \dfrac{1}{(3x)^2}\,dx$			

Finding Indefinite Integrals   In Exercises 25–36, find the indefinite integral. Check your result by differentiating. *See Examples 4 and 5.*

25. $\displaystyle\int (x + 3)\,dx$

26. $\displaystyle\int (5 - x)\,dx$

27. $\displaystyle\int (x^3 + 2)\,dx$

28. $\displaystyle\int (x^2 - 7)\,dx$

29. $\displaystyle\int (3x^3 - 6x^2 + 2)\,dx$

30. $\displaystyle\int (x^3 - 4x + 2)\,dx$

31. $\displaystyle\int (x^2 + 5x + 1)\,dx$

32. $\displaystyle\int (2x^4 - x^2 + 3)\,dx$

33. $\displaystyle\int \frac{2x^3 - 1}{x^3}\, dx$

34. $\displaystyle\int \frac{t^2 + 2}{t^2}\, dt$

35. $\displaystyle\int \frac{5x + 4}{\sqrt[3]{x}}\, dx$

36. $\displaystyle\int \frac{2x - 1}{\sqrt{x}}\, dx$

Interpreting a Graph   In Exercises 37–40, the graph of the derivative of a function is given. Sketch the graphs of *two* functions that have the given derivative. (There is more than one correct answer.)

37.

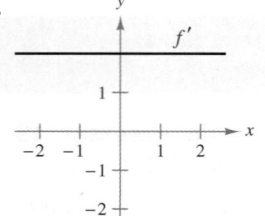

38.

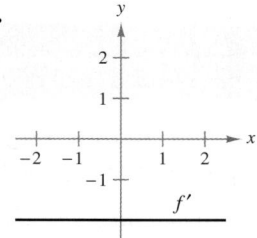

39.

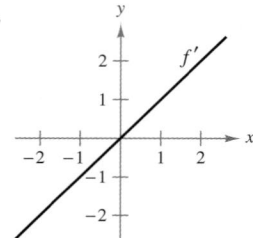

40.

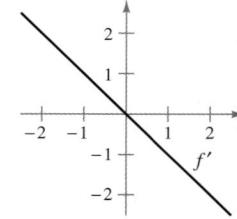

Finding Particular Solutions   In Exercises 41–48, find the particular solution that satisfies the differential equation and the initial condition. *See Example 6.*

41. $f'(x) = 4x;\ f(0) = 6$

42. $f'(x) = 9x^2;\ f(0) = -1$

43. $f'(x) = 2x + 4;\ f(-2) = 3$

44. $f'(x) = \frac{1}{5}x - 2;\ f(10) = -10$

45. $f'(x) = 10x - 12x^3;\ f(3) = 2$

46. $f'(x) = 2\sqrt{x};\ f(4) = 12$

47. $f'(x) = \dfrac{2 - x}{x^3},\ x > 0;\ f(2) = \dfrac{3}{4}$

48. $f'(x) = \dfrac{x^2 - 5}{x^2},\ x > 0;\ f(1) = 2$

Finding Particular Solutions   In Exercises 49–52, find a function *f* that satisfies the initial conditions.

49. $f''(x) = 2,\ f'(2) = 5,\ f(2) = 10$

50. $f''(x) = x^2,\ f'(0) = 6,\ f(0) = 3$

51. $f''(x) = x^{-2/3},\ f'(8) = 6,\ f(0) = 0$

52. $f''(x) = x^{-3/2},\ f'(1) = 2,\ f(9) = -4$

Finding a Cost Function   In Exercises 53–56, find the cost function for the given marginal cost and fixed cost. *See Example 8.*

Marginal Cost	Fixed Cost $(x = 0)$
53. $\dfrac{dC}{dx} = 85$	\$5500
54. $\dfrac{dC}{dx} = \dfrac{1}{50}x + 10$	\$1000
55. $\dfrac{dC}{dx} = \dfrac{1}{20\sqrt{x}} + 4$	\$750
56. $\dfrac{dC}{dx} = \dfrac{\sqrt[4]{x}}{10} + 10$	\$2300

Revenue and Demand   In Exercises 57 and 58, find the revenue and demand functions for the given marginal revenue. (Use the fact that $R = 0$ when $x = 0$.)

57. $\dfrac{dR}{dx} = 225 - 3x$

58. $\dfrac{dR}{dx} = 310 - 4x$

Profit   In Exercises 59–62, find the profit function for the given marginal profit and initial condition.

Marginal Profit	Initial Condition
59. $\dfrac{dP}{dx} = -18x + 1650$	$P(15) = \$22{,}725$
60. $\dfrac{dP}{dx} = -40x + 250$	$P(5) = \$650$
61. $\dfrac{dP}{dx} = -24x + 805$	$P(12) = \$8000$
62. $\dfrac{dP}{dx} = -30x + 920$	$P(8) = \$6500$

Vertical Motion   In Exercises 63–66, use $s''(t) = -32$ feet per second per second as the acceleration due to gravity. *See Example 7.*

63. The Grand Canyon is 6000 feet deep at the deepest part. A rock is dropped from this height. Express the height s (in feet) of the rock as a function of the time t (in seconds). How long will it take the rock to hit the canyon floor?

64. A ball is thrown upward with an initial velocity of 60 feet per second from an initial height of 16 feet. Express the height s (in feet) of the ball as a function of the time t (in seconds). How long will the ball be in the air?

65. With what initial velocity must an object be thrown upward from the ground to reach the height of the Washington Monument (550 feet)?

66. With what initial velocity must an object be thrown upward from a height of 5 feet to reach a maximum height of 230 feet?

67. Cost A company produces a product for which the marginal cost of producing x units is modeled by $dC/dx = 2x - 12$, and the fixed costs are $125.

(a) Find the total cost function and the average cost function.

(b) Find the total cost of producing 50 units.

(c) In part (b), how much of the total cost is fixed? How much is variable? Give examples of fixed costs associated with the manufacture of a product. Give examples of variable costs.

68. Tree Growth An evergreen nursery usually sells a certain shrub after 6 years of growth and shaping. The growth rate during those 6 years is approximated by $dh/dt = 1.5t + 5$, where t is the time (in years) and h is the height (in centimeters). The seedlings are 12 centimeters tall when planted ($t = 0$).

(a) Find the height function.

(b) How tall are the shrubs when they are sold?

69. Population Growth The growth rate of the population of Horry County in South Carolina from 1970 through 2009 can be modeled by

$$\frac{dP}{dt} = 158.80t + 1758.6$$

where t is the time in years, with $t = 0$ corresponding to 1970. The county's population was 263,868 in 2009. *(Source: U.S. Census Bureau)*

(a) Find the model for Horry County's population.

(b) Use the model to predict the population in 2015. Does your answer seem reasonable? Explain your reasoning.

70. HOW DO YOU SEE IT? The graph shows the rate of change of the revenue of a company from 1990 through 2010.

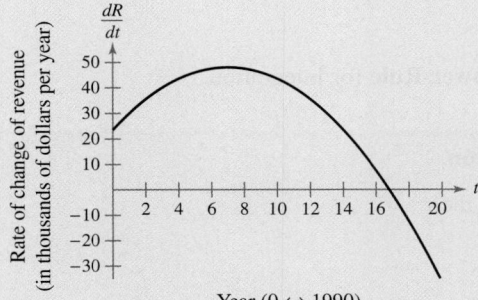

Year (0 ↔ 1990)

(a) Approximate the rate of change of the revenue in 1993. Explain your reasoning.

(b) Approximate the year when the revenue is maximum. Explain your reasoning.

71. Vital Statistics The rate of increase of the number of married couples M (in thousands) in the United States from 1980 through 2009 can be modeled by

$$\frac{dM}{dt} = -0.105t^2 + 14.02t + 217.8$$

where t is the time in years, with $t = 0$ corresponding to 1980. The number of married couples in 2009 was 60,844 thousand. *(Source: U.S. Census Bureau)*

(a) Find the model for the number of married couples in the United States.

(b) Use the model to predict the number of married couples in the United States in 2015. Does your answer seem reasonable? Explain your reasoning.

72. Internet Users The rate of growth of the number of Internet users I (in millions) in the world from 1991 through 2009 can be modeled by

$$\frac{dI}{dt} = 0.0556t^3 - 1.557t^2 + 25.70t - 59.2$$

where t is the time in years, with $t = 1$ corresponding to 1991. The number of Internet users in 2009 was 1833 million. *(Source: International Telecommunication Union)*

(a) Find the model for the number of Internet users in the world.

(b) Use the model to predict the number of Internet users in the world in 2015. Does your answer seem reasonable? Explain your reasoning.

73. Economics: Marginal Benefits and Costs The table gives the marginal benefit and marginal cost of producing x units of a product for a given company. (a) Plot the points in each column and use the *regression* feature of a graphing utility to find a linear model for marginal benefit and a quadratic model for marginal cost. (b) Use integration to find the benefit B and cost C equations. Assume $B(0) = 0$ and $C(0) = 425$. (c) Find the intervals in which the benefit exceeds the cost of producing x units. Make a recommendation for how many units the company should produce based on your findings. *(Source: Adapted from Taylor, Economics, Fifth Edition)*

Number of units	1	2	3	4	5
Marginal benefit	330	320	290	270	250
Marginal cost	150	120	100	110	120

Number of units	6	7	8	9	10
Marginal benefit	230	210	190	170	160
Marginal cost	140	160	190	250	320

11.2 Integration by Substitution and the General Power Rule

- Use the General Power Rule to find indefinite integrals.
- Use substitution to find indefinite integrals.
- Use the General Power Rule to solve real-life problems.

The General Power Rule

In Section 11.1, you used the Simple Power Rule

$$\int x^n \, dx = \frac{x^{n+1}}{n+1} + C, \quad n \neq -1$$

to find antiderivatives of functions expressed as powers of x alone. In this section, you will study a technique for finding antiderivatives of more complicated functions.

To begin, consider how you might find the antiderivative of

$$2x(x^2 + 1)^3.$$

Because you are hunting for a function whose derivative is $2x(x^2 + 1)^3$, you might discover the antiderivative as shown.

$$\frac{d}{dx}[(x^2 + 1)^4] = 4(x^2 + 1)^3(2x) \qquad \text{Use Chain Rule.}$$

$$\frac{d}{dx}\left[\frac{(x^2 + 1)^4}{4}\right] = (x^2 + 1)^3(2x) \qquad \text{Divide both sides by 4.}$$

$$\frac{(x^2 + 1)^4}{4} + C = \int 2x(x^2 + 1)^3 \, dx \qquad \text{Write in integral form.}$$

The key to this solution is the presence of the factor $2x$ in the integrand. In other words, this solution works because $2x$ is precisely the derivative of $(x^2 + 1)$. Letting $u = x^2 + 1$, you can write

$$\int \overbrace{(x^2 + 1)^3}^{u^3} \underbrace{2x \, dx}_{du} = \int u^3 \, du$$

$$= \frac{u^4}{4} + C.$$

This is an example of the **General Power Rule** for integration.

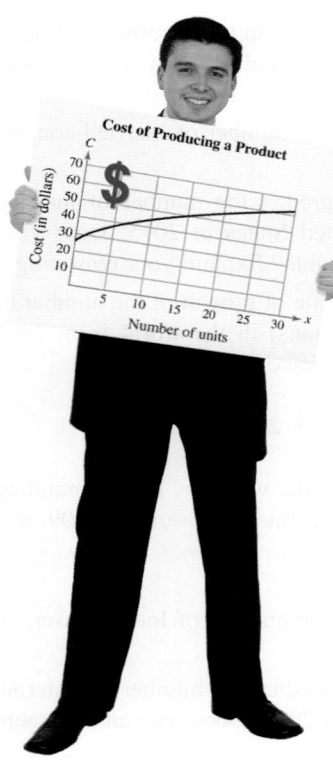

Cost of Producing a Product

In Exercise 49 on page 822, you will use integration to find a model for the cost of producing a product.

General Power Rule for Integration

If u is a differentiable function of x, then

$$\int u^n \frac{du}{dx} \, dx = \int u^n \, du$$

$$= \frac{u^{n+1}}{n+1} + C, \quad n \neq -1.$$

When using the General Power Rule, you must first identify a factor u of the integrand that is raised to a power. Then, you must show that its derivative du/dx is also a factor of the integrand. This is demonstrated in Example 1.

Example 1 **Applying the General Power Rule**

Find each indefinite integral.

a. $\int 3(3x - 1)^4 \, dx$ b. $\int (2x + 1)(x^2 + x) \, dx$

c. $\int 3x^2 \sqrt{x^3 - 2} \, dx$ d. $\int \frac{-4x}{(1 - 2x^2)^2} \, dx$

SOLUTION

a. $\int 3(3x - 1)^4 \, dx = \int \overbrace{(3x - 1)^4}^{u^n} \overbrace{(3)}^{\frac{du}{dx}} dx$ Let $u = 3x - 1$.

$= \frac{(3x - 1)^5}{5} + C$ General Power Rule

b. $\int (2x + 1)(x^2 + x) \, dx = \int \overbrace{(x^2 + x)}^{u^n} \overbrace{(2x + 1)}^{\frac{du}{dx}} dx$ Let $u = x^2 + x$.

$= \frac{(x^2 + x)^2}{2} + C$ General Power Rule

c. $\int 3x^2 \sqrt{x^3 - 2} \, dx = \int \overbrace{(x^3 - 2)^{1/2}}^{u^n} \overbrace{(3x^2)}^{\frac{du}{dx}} dx$ Let $u = x^3 - 2$.

$= \frac{(x^3 - 2)^{3/2}}{3/2} + C$ General Power Rule

$= \frac{2}{3}(x^3 - 2)^{3/2} + C$ Simplify.

d. $\int \frac{-4x}{(1 - 2x^2)^2} \, dx = \int \overbrace{(1 - 2x^2)^{-2}}^{u^n} \overbrace{(-4x)}^{\frac{du}{dx}} dx$ Let $u = 1 - 2x^2$.

$= \frac{(1 - 2x^2)^{-1}}{-1} + C$ General Power Rule

$= -\frac{1}{1 - 2x^2} + C$ Simplify.

STUDY TIP

Example 1(b) illustrates a case of the General Power Rule that is sometimes overlooked—when the power is $n = 1$. In this case, the rule takes the form

$\int u \frac{du}{dx} \, dx = \frac{u^2}{2} + C.$

✓ **Checkpoint 1**

Find each indefinite integral.

a. $\int (3x^2 + 6)(x^3 + 6x)^2 \, dx$ b. $\int 2x\sqrt{x^2 - 2} \, dx$

Remember that you can verify the result of an indefinite integral by differentiating the function. For instance, you can check the answer to Example 1(a) as follows.

$\frac{d}{dx}\left[\frac{(3x - 1)^5}{5} + C \right] = \left(\frac{1}{5} \right)(5)(3x - 1)^4(3)$ Apply Chain Rule.

$= 3(3x - 1)^4$ Simplify.

Many times, part of the derivative du/dx is missing from the integrand, and in *some* cases you can make the necessary adjustments to apply the General Power Rule.

Example 2   **Multiplying and Dividing by a Constant**

Find $\displaystyle\int x(3 - 4x^2)^2 \, dx.$

SOLUTION   Let $u = 3 - 4x^2$. To apply the General Power Rule, you need to create $du/dx = -8x$ as a factor of the integrand. You can accomplish this by multiplying and dividing by the constant -8.

$$\int x(3 - 4x^2)^2 \, dx = \int \left(-\frac{1}{8}\right)\overbrace{(3 - 4x^2)^2}^{u^n}\overbrace{(-8x)}^{\frac{du}{dx}} \, dx \qquad \text{Multiply and divide by } -8.$$

$$= -\frac{1}{8}\int (3 - 4x^2)^2(-8x) \, dx \qquad \text{Factor } -\tfrac{1}{8} \text{ out of integrand.}$$

$$= \left(-\frac{1}{8}\right)\left[\frac{(3 - 4x^2)^3}{3}\right] + C \qquad \text{General Power Rule}$$

$$= -\frac{(3 - 4x^2)^3}{24} + C \qquad \text{Simplify.}$$

✓**Checkpoint 2**

Find $\displaystyle\int x^3(3x^4 + 1)^2 \, dx.$

Example 3   **Multiplying and Dividing by a Constant**

Find $\displaystyle\int (x^2 + 2x)^3(x + 1) \, dx.$

SOLUTION   Let $u = x^2 + 2x$. To apply the General Power Rule, you need to create $du/dx = 2x + 2$ as a factor of the integrand. You can accomplish this by multiplying and dividing by the constant 2.

$$\int (x^2 + 2x)^3(x + 1) \, dx = \int \left(\frac{1}{2}\right)\overbrace{(x^2 + 2x)^3}^{u^n}\overbrace{(2)(x + 1)}^{\frac{du}{dx}} \, dx \qquad \text{Multiply and divide by 2.}$$

$$= \frac{1}{2}\int (x^2 + 2x)^3(2x + 2) \, dx \qquad \text{Rewrite integrand.}$$

$$= \frac{1}{2}\left[\frac{(x^2 + 2x)^4}{4}\right] + C \qquad \text{General Power Rule}$$

$$= \frac{1}{8}(x^2 + 2x)^4 + C \qquad \text{Simplify.}$$

✓**Checkpoint 3**

Find $\displaystyle\int (x^3 - 3x)^2(x^2 - 1) \, dx.$

Example 4   **A Failure of the General Power Rule**

Find $\int -8(3 - 4x^2)^2 \, dx$.

SOLUTION   Let $u = 3 - 4x^2$. To apply the General Power Rule, you must create $du/dx = -8x$ as a factor of the integrand. In Examples 2 and 3, this was done by multiplying and dividing by a constant, and then factoring that constant out of the integrand. This strategy doesn't work with variables. That is,

$$\int -8(3 - 4x^2)^2 \, dx \neq \frac{1}{x}\int (3 - 4x^2)^2(-8x) \, dx.$$

To find this indefinite integral, you can expand the integrand and use the Simple Power Rule.

$$\int -8(3 - 4x^2)^2 \, dx = \int (-72 + 192x^2 - 128x^4) \, dx$$

$$= -72x + 64x^3 - \frac{128}{5}x^5 + C$$

✓**Checkpoint 4**

Find $\int 2(3x^4 + 1)^2 \, dx$.

When an integrand contains an extra constant factor that is not needed as part of du/dx, you can simply move the factor outside the integral sign, as shown in the next example.

Example 5   **Applying the General Power Rule**

Find $\int 7x^2 \sqrt{x^3 + 1} \, dx$.

SOLUTION   Let $u = x^3 + 1$. Then you need to create $du/dx = 3x^2$ by multiplying and dividing by 3. The constant factor $\frac{7}{3}$ is not needed as part of du/dx, and can be moved outside the integral sign.

$$\int 7x^2 \sqrt{x^3 + 1} \, dx = \int 7x^2(x^3 + 1)^{1/2} \, dx \qquad \text{Rewrite with rational exponent.}$$

$$= \int \frac{7}{3}(x^3 + 1)^{1/2}(3x^2) \, dx \qquad \text{Multiply and divide by 3.}$$

$$= \frac{7}{3}\int (x^3 + 1)^{1/2}(3x^2) \, dx \qquad \text{Factor } \tfrac{7}{3} \text{ outside integral.}$$

$$= \frac{7}{3}\left[\frac{(x^3 + 1)^{3/2}}{3/2}\right] + C \qquad \text{General Power Rule}$$

$$= \frac{14}{9}(x^3 + 1)^{3/2} + C \qquad \text{Simplify.}$$

✓**Checkpoint 5**

Find $\int 5x \sqrt{x^2 - 1} \, dx$.

Substitution

The integration technique used in Examples 1, 2, 3, and 5 depends on your ability to recognize or create an integrand of the form

$$u^n \frac{du}{dx}.$$

With more complicated integrands, it is difficult to recognize the steps needed to fit the integrand to a basic integration formula. When this occurs, an alternative procedure called **substitution** or **change of variables** can be helpful. With this procedure, you completely rewrite the integral in terms of u and du. That is, if $u = f(x)$, then $du = f'(x)\,dx$, and the General Power Rule takes the form

$$\int u^n \frac{du}{dx}\,dx = \int u^n\,du. \qquad \text{General Power Rule}$$

Example 6   Integration by Substitution

Find $\displaystyle\int \sqrt{1 - 3x}\,dx$.

SOLUTION   Begin by letting $u = 1 - 3x$. Then, $du/dx = -3$ and $du = -3\,dx$. This implies that

$$dx = -\frac{1}{3}\,du$$

and you can find the indefinite integral as shown.

$$
\begin{aligned}
\int \sqrt{1 - 3x}\,dx &= \int (1 - 3x)^{1/2}\,dx && \text{Rewrite with rational exponent.}\\[2mm]
&= \int u^{1/2}\left(-\frac{1}{3}\,du\right) && \text{Substitute for } x \text{ and } dx.\\[2mm]
&= -\frac{1}{3}\int u^{1/2}\,du && \text{Factor } -\tfrac{1}{3} \text{ out of integrand.}\\[2mm]
&= \left(-\frac{1}{3}\right)\left(\frac{u^{3/2}}{3/2}\right) + C && \text{Apply Power Rule.}\\[2mm]
&= -\frac{2}{9}u^{3/2} + C && \text{Simplify.}\\[2mm]
&= -\frac{2}{9}(1 - 3x)^{3/2} + C && \text{Substitute } 1 - 3x \text{ for } u.
\end{aligned}
$$

You can check this result by differentiating.

$$
\begin{aligned}
\frac{d}{dx}\left[-\frac{2}{9}(1 - 3x)^{3/2} + C\right] &= \left(-\frac{2}{9}\right)\left(\frac{3}{2}\right)(1 - 3x)^{1/2}(-3)\\[2mm]
&= \left(-\frac{1}{3}\right)(-3)(1 - 3x)^{1/2}\\[2mm]
&= \sqrt{1 - 3x}
\end{aligned}
$$

✓ **Checkpoint 6**

Find $\displaystyle\int \sqrt{1 - 2x}\,dx$ by the method of substitution.

The basic steps for integration by substitution are outlined in the guidelines below.

Guidelines for Integration by Substitution

1. Let u be a function of x (usually part of the integrand).

2. Solve for x and dx in terms of u and du.

3. Convert the entire integral to u-variable form.

4. After integrating, rewrite the antiderivative as a function of x.

5. Check your answer by differentiating.

Example 7 Integration by Substitution

Find $\displaystyle\int x\sqrt{x^2 - 1}\, dx$.

SOLUTION Consider the substitution $u = x^2 - 1$, which produces

$$du = 2x\, dx.$$

To create $2x\, dx$ as part of the integral, multiply and divide by 2.

$$\int x\sqrt{x^2 - 1}\, dx = \frac{1}{2}\int \overbrace{(x^2 - 1)^{1/2}}^{u^{1/n}}\overbrace{2x\, dx}^{du} \qquad \text{Multiply and divide by 2.}$$

$$= \frac{1}{2}\int u^{1/2}\, du \qquad \text{Substitute for } x \text{ and } dx.$$

$$= \frac{1}{2}\left(\frac{u^{3/2}}{3/2}\right) + C \qquad \text{Apply Power Rule.}$$

$$= \frac{1}{3}u^{3/2} + C \qquad \text{Simplify.}$$

$$= \frac{1}{3}(x^2 - 1)^{3/2} + C \qquad \text{Substitute for } u.$$

You can check this result by differentiating.

$$\frac{d}{dx}\left[\frac{1}{3}(x^2 - 1)^{3/2} + C\right] = \frac{1}{3}\left(\frac{3}{2}\right)(x^2 - 1)^{1/2}(2x)$$

$$= \frac{1}{2}(2x)(x^2 - 1)^{1/2}$$

$$= x\sqrt{x^2 - 1}$$

✓ Checkpoint 7

Find $\displaystyle\int x\sqrt{x^2 + 4}\, dx$ by the method of substitution. ■

To become efficient at integration, you should learn to use *both* techniques discussed in this section. For simpler integrals, you should use pattern recognition and create du/dx by multiplying and dividing by an appropriate constant. For more complicated integrals, you should use a formal change of variables, as shown in Examples 6 and 7. For the integrals in this section's exercise set, try working several of the problems twice—once with pattern recognition and once using formal substitution.

Extended Application: Propensity to Consume

In 2009, the U.S. poverty level for a family of four was about $22,000. Families at or below the poverty level tend to consume 100% of their income—that is, they use all their income to purchase necessities such as food, clothing, and shelter. As income level increases, the average consumption tends to drop below 100%. For instance, a family earning $25,000 may be able to save $500 and so consume only $24,500 (98%) of their income. As the income increases, the ratio of consumption to savings tends to decrease. The rate of change of consumption with respect to income is called the **marginal propensity to consume.** *(Source: U.S. Census Bureau)*

 **Example 8** **Analyzing Consumption**

For a family of four in 2009, the marginal propensity to consume income x (in dollars) can be modeled by

$$\frac{dQ}{dx} = \frac{0.98}{(x - 21,999)^{0.02}}, \quad x \geq 22,000$$

where Q represents the income consumed (in dollars). Use the model to estimate the amount consumed by a family of four whose 2009 income was $35,000.

SOLUTION Begin by integrating dQ/dx to find a model for the consumption Q.

$$\begin{aligned} Q &= \int \frac{0.98}{(x - 21,999)^{0.02}} \, dx & &\text{Integrate } \frac{dQ}{dx} \text{ to obtain } Q. \\ &= \int 0.98(x - 21,999)^{-0.02} \, dx & &\text{Rewrite.} \\ &= (x - 21,999)^{0.98} + C & &\text{General Power Rule} \end{aligned}$$

To solve for C, use the initial condition that $Q = 22,000$ when $x = 22,000$.

$$22,000 = (22,000 - 21,999)^{0.98} + C$$
$$22,000 = 1 + C$$
$$21,999 = C$$

So, you can use the model $Q = (x - 21,999)^{0.98} + 21,999$ to estimate that a family of four with an income of $x = 35,000$ consumed about

$$Q = (35,000 - 21,999)^{0.98} + 21,999 \approx \$32,756.$$

The graph of Q is shown in Figure 11.4.

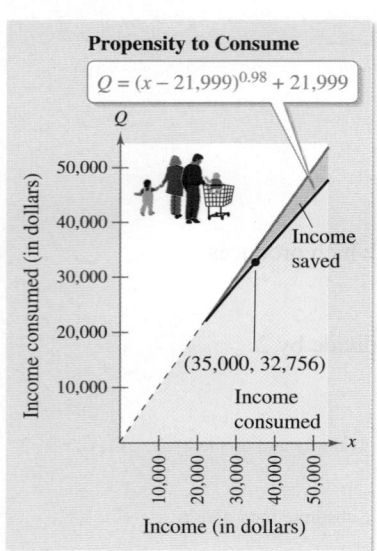

Propensity to Consume

$Q = (x - 21,999)^{0.98} + 21,999$

Income consumed (in dollars)

Income saved

(35,000, 32,756)

Income consumed

Income (in dollars)

FIGURE 11.4

✓ Checkpoint 8

According to the model in Example 8, at what income level would a family of four consume $32,000?

SUMMARIZE (Section 11.2)

1. State the General Power Rule for integration *(page 814)*. For examples of the General Power Rule, see Examples 1, 2, 3, and 5.

2. List the guidelines for integration by substitution *(page 819)*. For examples of integration by substitution, see Examples 6 and 7.

3. Describe a real-life example of how the General Power Rule can be used to analyze the marginal propensity to consume *(page 820, Example 8)*.

SKILLS WARM UP 11.2

The following warm-up exercises involve skills that were covered in earlier sections. You will use these skills in the exercise set for this section. For additional help, review Section 11.1.

In Exercises 1–9, find the indefinite integral.

1. $\int (2x^3 + 1)\, dx$

2. $\int (x^{1/2} + 3x - 4)\, dx$

3. $\int \frac{1}{x^2}\, dx$

4. $\int \frac{1}{3t^3}\, dt$

5. $\int (1 + 2t)t^{3/2}\, dt$

6. $\int \sqrt{x}(2x - 1)\, dx$

7. $\int \frac{5x^3 + 2}{x^2}\, dx$

8. $\int \frac{2x^2 - 5}{x^4}\, dx$

9. $\int \frac{8x^2 + 3}{\sqrt{x}}\, dx$

Exercises 11.2

See www.CalcChat.com for worked-out solutions to odd-numbered exercises.

Finding u and du/dx In Exercises 1–8, identify u and du/dx for the integral $\int u^n (du/dx)\, dx$.

1. $\int (5x^2 + 1)^2(10x)\, dx$

2. $\int (3 - 4x^2)^3(-8x)\, dx$

3. $\int \sqrt{1 - x^2}(-2x)\, dx$

4. $\int 3x^2\sqrt{x^3 + 1}\, dx$

5. $\int \left(4 + \frac{1}{x^2}\right)^5\left(\frac{-2}{x^3}\right) dx$

6. $\int \frac{1}{(1 + 2x)^2}(2)\, dx$

7. $\int (1 + \sqrt{x})^3\left(\frac{1}{2\sqrt{x}}\right) dx$

8. $\int (4 - \sqrt{x})^2\left(\frac{-1}{2\sqrt{x}}\right) dx$

Applying the General Power Rule In Exercises 9–34, find the indefinite integral. Check your result by differentiating. *See Examples 1, 2, 3, and 5.*

9. $\int (x - 1)^4\, dx$

10. $\int (x - 3)^{5/2}\, dx$

11. $\int (1 + 2x)^4(2)\, dx$

12. $\int (x^2 - 1)^3(2x)\, dx$

13. $\int (x^2 + 3x)(2x + 3)\, dx$

14. $\int (x^3 + 6x)^2(3x^2 + 6)\, dx$

15. $\int \sqrt{4x^2 - 5}(8x)\, dx$

16. $\int \sqrt[3]{1 - 2x^2}(-4x)\, dx$

17. $\int \frac{6x}{(3x^2 - 5)^4}\, dx$

18. $\int \frac{-12x^2}{(1 - 4x^3)^2}\, dx$

19. $\int x^2(2x^3 - 1)^4\, dx$

20. $\int x(1 - 2x^2)^3\, dx$

21. $\int t\sqrt{t^2 + 6}\, dt$

22. $\int t^4\sqrt[3]{t^5 - 9}\, dt$

23. $\int \frac{x^5}{(4 - x^6)^3}\, dx$

24. $\int \frac{x^2}{(x^3 - 1)^2}\, dx$

25. $\int (x^2 - 6x)^4(x - 3)\, dx$

26. $\int (4x^3 + 8x)^3(3x^2 + 2)\, dx$

27. $\int \frac{x + 1}{(x^2 + 2x - 3)^2}\, dx$

28. $\int \frac{x - 2}{\sqrt{x^2 - 4x + 3}}\, dx$

29. $\int 5x\sqrt[3]{1 - x^2}\, dx$

30. $\int 9x^3\sqrt{x^4 + 2}\, dx$

31. $\int \frac{6x}{(1 + x^2)^3}\, dx$

32. $\int \frac{4x + 6}{(x^2 + 3x + 7)^3}\, dx$

33. $\int \frac{-3}{\sqrt{2t + 3}}\, dt$

34. $\int \frac{3x^2}{\sqrt{1 - x^3}}\, dx$

Integration by Substitution In Exercises 35–42, use formal substitution to find the indefinite integral. Check your result by differentiating. *See Examples 6 and 7.*

35. $\int 12x(6x^2 - 1)^3\, dx$

36. $\int 3x^2(1 - x^3)^2\, dx$

37. $\int \sqrt[3]{4x + 3}\, dx$

38. $\int t\sqrt{t^2 + 1}\, dt$

39. $\int \frac{x}{\sqrt{x^2 + 25}}\, dx$

40. $\int \frac{3}{\sqrt{2x + 1}}\, dx$

41. $\int \frac{x^2 + 1}{\sqrt{x^3 + 3x + 4}}\, dx$

42. $\int \frac{x^2 + 3}{\sqrt[3]{x^3 + 9x}}\, dx$

Comparing Methods In Exercises 43–46, (a) perform the integration in two ways: once using the Simple Power Rule and once using the General Power Rule. (b) Explain the difference in the results. (c) Which method do you prefer? Explain your reasoning.

43. $\int (x - 1)^2\, dx$

44. $\int (3 - x)^2\, dx$

45. $\int x(x^2 - 1)^2\, dx$

46. $\int x(2x^2 + 1)^2\, dx$

47. Finding an Equation of a Function Find the equation of the function f whose graph passes through the point $(2, 10)$ and whose derivative is

$$f'(x) = 2x(4x^2 - 10)^2.$$

48. Finding an Equation of a Function Find the equation of the function f whose graph passes through the point $\left(0, \frac{7}{3}\right)$ and whose derivative is

$$f'(x) = x\sqrt{1 - x^2}.$$

49. Cost The marginal cost of a product is modeled by

$$\frac{dC}{dx} = \frac{4}{\sqrt{x + 1}}$$

where x is the number of units. When $x = 15$, $C = 50$.

(a) Find the cost function.

(b) Find the cost of producing 50 units.

50. Cost The marginal cost of a product is modeled by

$$\frac{dC}{dx} = \frac{12}{\sqrt[3]{12x + 1}}$$

where x is the number of units. When $x = 13$, $C = 100$.

(a) Find the cost function.

(b) Find the cost of producing 30 units.

Supply In Exercises 51 and 52, find the supply function $x = f(p)$ that satisfies the initial conditions.

51. $\dfrac{dx}{dp} = p\sqrt{p^2 - 25}$

$x = 600$ when $p = \$13$

52. $\dfrac{dx}{dp} = \dfrac{10}{\sqrt{p - 3}}$

$x = 100$ when $p = \$3$

Demand In Exercises 53 and 54, find the demand function $x = f(p)$ that satisfies the initial conditions.

53. $\dfrac{dx}{dp} = -\dfrac{6000p}{(p^2 - 16)^{3/2}}$

$x = 5000$ when $p = \$5$

54. $\dfrac{dx}{dp} = -\dfrac{400}{(0.02p - 1)^3}$

$x = 10{,}000$ when $p = \$100$

55. Gardening An evergreen nursery usually sells a type of shrub after 5 years of growth and shaping. The growth rate during those 5 years is approximated by

$$\frac{dh}{dt} = \frac{17.6t}{\sqrt{17.6t^2 + 1}}$$

where t is the time (in years) and h is the height (in inches). The seedlings are 6 inches tall when planted $(t = 0)$.

(a) Find the height function.

(b) How tall are the shrubs when they are sold?

56. **HOW DO YOU SEE IT?** The graph shows the rate of change of the revenue of a company from 1990 through 2010.

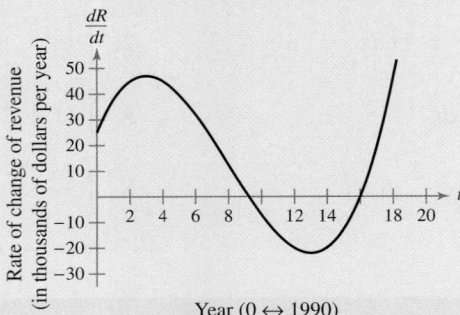

(a) Approximate the rate of change of the revenue in 2007. Explain your reasoning.

(b) Is $R(7) - R(6) > 0$? Explain your reasoning.

(c) Approximate the years in which the graph of the revenue is concave upward and the years in which it is concave downward. Approximate the years of any points of inflection.

Marginal Propensity to Consume In Exercises 57 and 58, (a) use the marginal propensity to consume, dQ/dx, to write Q as a function of x, where x is the income (in dollars) and Q is the income consumed (in dollars). Assume that families who have annual incomes of $\$25{,}000$ or less consume 100% of their income. (b) Use the result of part (a) and a spreadsheet to complete the table showing the income consumed and the income saved, $x - Q$, for various incomes. (c) Use a graphing utility to represent graphically the income consumed and saved. *See Example 8.*

x	25,000	50,000	100,000	150,000
Q				
$x - Q$				

57. $\dfrac{dQ}{dx} = \dfrac{0.95}{(x - 24{,}999)^{0.05}}$, $x \geq 25{,}000$

58. $\dfrac{dQ}{dx} = \dfrac{0.93}{(x - 24{,}999)^{0.07}}$, $x \geq 25{,}000$

Integration Using Technology In Exercises 59 and 60, use a symbolic integration utility to find the indefinite integral. Verify the result by differentiating.

59. $\displaystyle\int \frac{1}{\sqrt{x} + \sqrt{x + 1}}\, dx$

60. $\displaystyle\int \frac{x}{\sqrt{3x + 2}}\, dx$

11.3 Exponential and Logarithmic Integrals

■ Use the Exponential Rule to find indefinite integrals.
■ Use the Log Rule to find indefinite integrals.

Using the Exponential Rule

Each of the differentiation rules for exponential functions has a corresponding integration rule.

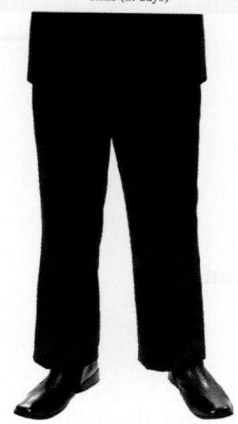

Population of Bacteria

In Exercise 51 on page 829, you will use integration to find a model for a population of bacteria.

Integrals of Exponential Functions

Let u be a differentiable function of x.

$$\int e^x \, dx = e^x + C \qquad \text{Simple Exponential Rule}$$

$$\int e^u \frac{du}{dx} \, dx = \int e^u \, du = e^u + C \qquad \text{General Exponential Rule}$$

Example 1 Integrating Exponential Functions

Find each indefinite integral.

a. $\displaystyle \int 2e^x \, dx$ **b.** $\displaystyle \int 2e^{2x} \, dx$ **c.** $\displaystyle \int (e^x + x) \, dx$

SOLUTION

a. $\displaystyle \int 2e^x \, dx = 2 \int e^x \, dx$ Constant Multiple Rule

$\qquad\qquad = 2e^x + C$ Simple Exponential Rule

b. $\displaystyle \int 2e^{2x} \, dx = \int e^{2x}(2) \, dx$ Let $u = 2x$, then $\dfrac{du}{dx} = 2$.

$\qquad\qquad = \int e^u \frac{du}{dx} \, dx$ Substitute u and $\dfrac{du}{dx}$.

$\qquad\qquad = e^u + C$ General Exponential Rule

$\qquad\qquad = e^{2x} + C$ Substitute for u.

c. $\displaystyle \int (e^x + x) \, dx = \int e^x \, dx + \int x \, dx$ Sum Rule

$\qquad\qquad = e^x + \dfrac{x^2}{2} + C$ Simple Exponential and Power Rules

You can check each of these results by differentiating. For instance, in part (a),

$$\frac{d}{dx}[2e^x + C] = 2e^x.$$

✓ Checkpoint 1

Find each indefinite integral.

a. $\displaystyle \int 3e^x \, dx$ **b.** $\displaystyle \int 5e^{5x} \, dx$ **c.** $\displaystyle \int (e^x - x) \, dx$

Example 2   **Integrating an Exponential Function**

Find $\int e^{3x+1}\, dx$.

SOLUTION   Let $u = 3x + 1$; then $du/dx = 3$. You can introduce the missing factor of 3 in the integrand by multiplying and dividing by 3.

$$\int e^{3x+1}\, dx = \frac{1}{3}\int e^{3x+1}(3)\, dx \qquad \text{Multiply and divide by 3.}$$

$$= \frac{1}{3}\int e^{u}\frac{du}{dx}\, dx \qquad \text{Substitute } u \text{ and } \frac{du}{dx}.$$

$$= \frac{1}{3}e^{u} + C \qquad \text{General Exponential Rule}$$

$$= \frac{1}{3}e^{3x+1} + C \qquad \text{Substitute for } u.$$

✓**Checkpoint 2**

Find $\int e^{2x+3}\, dx$.

Example 3   **Integrating an Exponential Function**

Find $\int 5xe^{-x^2}\, dx$.

SOLUTION   Let $u = -x^2$; then $du/dx = -2x$. You can create the factor $-2x$ in the integrand by multiplying and dividing by -2.

$$\int 5xe^{-x^2}\, dx = \int \left(-\frac{5}{2}\right)e^{-x^2}(-2x)\, dx \qquad \text{Multiply and divide by } -2.$$

$$= -\frac{5}{2}\int e^{-x^2}(-2x)\, dx \qquad \text{Factor } -\tfrac{5}{2} \text{ out of the integrand.}$$

$$= -\frac{5}{2}\int e^{u}\frac{du}{dx}\, dx \qquad \text{Substitute } u \text{ and } \frac{du}{dx}.$$

$$= -\frac{5}{2}e^{u} + C \qquad \text{General Exponential Rule}$$

$$= -\frac{5}{2}e^{-x^2} + C \qquad \text{Substitute for } u.$$

✓**Checkpoint 3**

Find $\int 4xe^{x^2}\, dx$.

Remember that you cannot introduce a missing *variable* in the integrand. For instance, you cannot find $\int e^{x^2}\, dx$ by multiplying and dividing by $2x$ and then factoring $1/(2x)$ out of the integrand. That is,

$$\int e^{x^2}\, dx \neq \frac{1}{2x}\int e^{x^2}(2x)\, dx.$$

Using the Log Rule

When the Power Rules for integration were introduced in Sections 11.1 and 11.2, you saw that they work for powers other than $n = -1$.

$$\int x^n \, dx = \frac{x^{n+1}}{n+1} + C, \quad n \neq -1 \qquad \text{Simple Power Rule}$$

$$\int u^n \frac{du}{dx} \, dx = \int u^n \, du = \frac{u^{n+1}}{n+1} + C, \quad n \neq -1 \qquad \text{General Power Rule}$$

The Log Rule for integration allows you to integrate functions of the form $\int x^{-1} \, dx$ and $\int u^{-1} \, du$.

STUDY TIP

Notice the absolute values in the Log Rule. For those special cases in which u or x cannot be negative, you can omit the absolute value. For instance, in Example 4(b), it is not necessary to write the antiderivative as $\ln|x^2| + C$ because x^2 cannot be negative.

Log Rule for Integration

Let u be a differentiable function of x.

$$\int \frac{1}{x} \, dx = \ln|x| + C \qquad \text{Simple Log Rule}$$

$$\int \frac{du/dx}{u} \, dx = \int \frac{1}{u} \, du = \ln|u| + C \qquad \text{General Log Rule}$$

You can verify each of these rules by differentiating. For instance, to verify that $d/dx[\ln|x|] = 1/x$, notice that

$$\frac{d}{dx}[\ln x] = \frac{1}{x} \quad \text{and} \quad \frac{d}{dx}[\ln(-x)] = \frac{-1}{-x} = \frac{1}{x}.$$

Example 4 Using the Log Rule for Integration

Find each indefinite integral.

a. $\displaystyle\int \frac{4}{x} \, dx$ **b.** $\displaystyle\int \frac{2x}{x^2} \, dx$ **c.** $\displaystyle\int \frac{3}{3x+1} \, dx$

SOLUTION

a. $\displaystyle\int \frac{4}{x} \, dx = 4\int \frac{1}{x} \, dx$ \qquad Constant Multiple Rule

$\qquad\qquad = 4\ln|x| + C$ \qquad Simple Log Rule

b. $\displaystyle\int \frac{2x}{x^2} \, dx = \int \frac{du/dx}{u} \, dx$ \qquad Let $u = x^2$; then $\dfrac{du}{dx} = 2x$.

$\qquad\qquad = \ln|u| + C$ \qquad General Log Rule

$\qquad\qquad = \ln x^2 + C$ \qquad Substitute for u.

c. $\displaystyle\int \frac{3}{3x+1} \, dx = \int \frac{du/dx}{u} \, dx$ \qquad Let $u = 3x + 1$; then $\dfrac{du}{dx} = 3$.

$\qquad\qquad = \ln|u| + C$ \qquad General Log Rule

$\qquad\qquad = \ln|3x + 1| + C$ \qquad Substitute for u.

✓ Checkpoint 4

Find each indefinite integral.

a. $\displaystyle\int \frac{2}{x} \, dx$ **b.** $\displaystyle\int \frac{3x^2}{x^3} \, dx$ **c.** $\displaystyle\int \frac{2}{2x+1} \, dx$

| Example 5 | Using the Log Rule for Integration |

Find $\int \dfrac{1}{2x - 1}\, dx$.

SOLUTION   Let $u = 2x - 1$; then $du/dx = 2$. You can create the necessary factor of 2 in the integrand by multiplying and dividing by 2.

$$\int \frac{1}{2x - 1}\, dx = \frac{1}{2} \int \frac{2}{2x - 1}\, dx \qquad \text{Multiply and divide by 2.}$$

$$= \frac{1}{2} \int \frac{du/dx}{u}\, dx \qquad \text{Substitute } u \text{ and } \frac{du}{dx}.$$

$$= \frac{1}{2} \ln|u| + C \qquad \text{General Log Rule}$$

$$= \frac{1}{2} \ln|2x - 1| + C \qquad \text{Substitute for } u.$$

✓ **Checkpoint 5**

Find $\int \dfrac{1}{4x + 1}\, dx$.   ▪

| Example 6 | Using the Log Rule for Integration |

Find $\int \dfrac{6x}{x^2 + 1}\, dx$.

SOLUTION   Let $u = x^2 + 1$; then

$$\frac{du}{dx} = 2x.$$

You can create the necessary factor of $2x$ in the integrand by factoring a 3 out of the integrand.

$$\int \frac{6x}{x^2 + 1}\, dx = 3 \int \frac{2x}{x^2 + 1}\, dx \qquad \text{Factor 3 out of integrand.}$$

$$= 3 \int \frac{du/dx}{u}\, dx \qquad \text{Substitute } u \text{ and } \frac{du}{dx}.$$

$$= 3 \ln|u| + C \qquad \text{General Log Rule}$$

$$= 3 \ln(x^2 + 1) + C \qquad \text{Substitute for } u.$$

✓ **Checkpoint 6**

Find $\int \dfrac{3x}{x^2 + 4}\, dx$.   ▪

ALGEBRA TUTOR xy

For help on the algebra at the right, see Example 2(d) in the *Chapter 11 Algebra Tutor*, on page 859.

Integrals to which the Log Rule can be applied are often given in disguised form. For instance, when a rational function has a numerator of degree greater than or equal to that of the denominator, you should use long division to rewrite the integrand. Here is an example.

$$\int \frac{x^2 + 6x + 1}{x^2 + 1}\, dx = \int \left(1 + \frac{6x}{x^2 + 1}\right) dx$$

$$= x + 3 \ln(x^2 + 1) + C$$

The next example summarizes some additional situations in which it is helpful to rewrite the integrand in order to recognize the antiderivative.

ALGEBRA TUTOR

For help on the algebra in Example 7, see Example 2(a)–(c) in the *Chapter 11 Algebra Tutor*, on page 859.

Example 7   Rewriting Before Integrating

Find each indefinite integral.

a. $\int \dfrac{3x^2 + 2x - 1}{x^2}\, dx$ **b.** $\int \dfrac{1}{1 + e^{-x}}\, dx$ **c.** $\int \dfrac{x^2 + x + 1}{x - 1}\, dx$

SOLUTION

a. Begin by rewriting the integrand as the sum of three fractions.

$$\int \frac{3x^2 + 2x - 1}{x^2}\, dx = \int \left(\frac{3x^2}{x^2} + \frac{2x}{x^2} - \frac{1}{x^2} \right) dx$$

$$= \int \left(3 + \frac{2}{x} - \frac{1}{x^2} \right) dx$$

$$= 3x + 2 \ln|x| + \frac{1}{x} + C$$

b. Begin by rewriting the integrand by multiplying and dividing by e^x.

$$\int \frac{1}{1 + e^{-x}}\, dx = \int \left(\frac{e^x}{e^x} \right) \frac{1}{1 + e^{-x}}\, dx$$

$$= \int \frac{e^x}{e^x + 1}\, dx$$

$$= \ln(e^x + 1) + C$$

c. Begin by dividing the numerator by the denominator.

$$\int \frac{x^2 + x + 1}{x - 1}\, dx = \int \left(x + 2 + \frac{3}{x - 1} \right) dx$$

$$= \frac{x^2}{2} + 2x + 3 \ln|x - 1| + C$$

✓**Checkpoint 7**

Find each indefinite integral.

a. $\int \dfrac{4x^2 - 3x + 2}{x^2}\, dx$ **b.** $\int \dfrac{2}{e^{-x} + 1}\, dx$ **c.** $\int \dfrac{x^2 + 2x + 4}{x + 1}\, dx$

SUMMARIZE   (Section 11.3)

1. State the Simple Exponential Rule *(page 823)*. For an example of the Simple Exponential Rule, see Example 1.

2. State the General Exponential Rule *(page 823)*. For examples of the General Exponential Rule, see Examples 2 and 3.

3. State the Simple Log Rule *(page 825)*. For an example of the Simple Log Rule, see Example 4.

4. State the General Log Rule *(page 825)*. For examples of the General Log Rule, see Examples 5 and 6.

In Exercises 1–4, use long division to rewrite the quotient.

1. $\dfrac{x^2 + 4x + 2}{x + 2}$

2. $\dfrac{x^2 - 6x + 9}{x - 4}$

3. $\dfrac{x^3 + 4x^2 - 30x - 4}{x^2 - 4x}$

4. $\dfrac{x^4 - x^3 + x^2 + 15x + 2}{x^2 + 5}$

In Exercises 5–8, find the indefinite integral.

5. $\displaystyle\int \left(x^3 + \dfrac{1}{x^2}\right) dx$

6. $\displaystyle\int \dfrac{x^2 + 2x}{x} \, dx$

7. $\displaystyle\int \dfrac{x^3 + 4}{x^2} \, dx$

8. $\displaystyle\int \dfrac{x + 3}{x^3} \, dx$

Exercises 11.3

Integrating Exponential Functions In Exercises 1–12, use the Exponential Rule to find the indefinite integral. *See Examples 1, 2, and 3.*

1. $\displaystyle\int 2e^{2x} \, dx$

2. $\displaystyle\int -3e^{-3x} \, dx$

3. $\displaystyle\int e^{4x} \, dx$

4. $\displaystyle\int e^{-0.25x} \, dx$

5. $\displaystyle\int e^{5x-3} \, dx$

6. $\displaystyle\int e^{-x-1} \, dx$

7. $\displaystyle\int 9xe^{-x^2} \, dx$

8. $\displaystyle\int 3xe^{0.5x^2} \, dx$

9. $\displaystyle\int 5x^2 e^{x^3} \, dx$

10. $\displaystyle\int -3x^3 e^{-2x^4} \, dx$

11. $\displaystyle\int (2x + 1)e^{x^2+x} \, dx$

12. $\displaystyle\int (x - 4)e^{x^2-8x} \, dx$

Using the Log Rule for Integration In Exercises 13–30, use the Log Rule to find the indefinite integral. *See Examples 4, 5, and 6.*

13. $\displaystyle\int \dfrac{1}{x + 1} \, dx$

14. $\displaystyle\int \dfrac{1}{x - 5} \, dx$

15. $\displaystyle\int \dfrac{5}{5x + 2} \, dx$

16. $\displaystyle\int \dfrac{4}{4x - 7} \, dx$

17. $\displaystyle\int \dfrac{1}{3 - 2x} \, dx$

18. $\displaystyle\int \dfrac{1}{6x - 5} \, dx$

19. $\displaystyle\int \dfrac{2}{3x + 5} \, dx$

20. $\displaystyle\int \dfrac{5}{2x - 1} \, dx$

21. $\displaystyle\int \dfrac{x}{x^2 + 1} \, dx$

22. $\displaystyle\int \dfrac{x^2}{3 - x^3} \, dx$

23. $\displaystyle\int \dfrac{x^2}{x^3 + 1} \, dx$

24. $\displaystyle\int \dfrac{x}{x^2 + 4} \, dx$

25. $\displaystyle\int \dfrac{x + 3}{x^2 + 6x + 7} \, dx$

26. $\displaystyle\int \dfrac{x^2 + 2x + 3}{x^3 + 3x^2 + 9x + 1} \, dx$

27. $\displaystyle\int \dfrac{1}{x \ln x} \, dx$

28. $\displaystyle\int \dfrac{1}{x(\ln x)^2} \, dx$

29. $\displaystyle\int \dfrac{e^{-x}}{1 - e^{-x}} \, dx$

30. $\displaystyle\int \dfrac{e^x}{1 + e^x} \, dx$

Finding Indefinite Integrals In Exercises 31–46, use any basic integration formula or formulas to find the indefinite integral. State which integration formula(s) you used to find the integral.

31. $\displaystyle\int \dfrac{x^3 - 8x}{2x^2} \, dx$

32. $\displaystyle\int \dfrac{x - 1}{4x} \, dx$

33. $\displaystyle\int \dfrac{8x^3 + 3x^2 + 6}{x^3} \, dx$

34. $\displaystyle\int \dfrac{2x^3 - 6x^2 - 5x}{x^2} \, dx$

35. $\displaystyle\int \dfrac{e^{2x} + 2e^x + 1}{e^x} \, dx$

36. $\displaystyle\int \dfrac{e^{5x} - 3e^{3x} + e^x}{e^{3x}} \, dx$

37. $\displaystyle\int e^x \sqrt{1 - e^x} \, dx$

38. $\displaystyle\int (6x + e^x)\sqrt{3x^2 + e^x} \, dx$

39. $\displaystyle\int \dfrac{1 + e^{-x}}{1 + xe^{-x}} \, dx$

40. $\displaystyle\int \dfrac{2(e^x - e^{-x})}{(e^x + e^{-x})^2} \, dx$

41. $\displaystyle\int \dfrac{5}{e^{-5x} + 7} \, dx$

42. $\displaystyle\int \dfrac{3}{1 + e^{-3x}} \, dx$

43. $\displaystyle\int \frac{x^2 + 2x + 5}{x - 1}\, dx$

44. $\displaystyle\int \frac{x^3 - 36x + 3}{x + 6}\, dx$

45. $\displaystyle\int \frac{x - 3}{x + 3}\, dx$

46. $\displaystyle\int \frac{x^2 + x + 1}{x^2 + 1}\, dx$

Finding an Equation of a Function In Exercises 47–50, find the equation of the function f whose graph passes through the given point.

47. $f'(x) = \dfrac{1}{x^2}e^{2/x};\ (4, 6)$

48. $f'(x) = \dfrac{2}{1 + e^{-x}};\ (0, 3)$

49. $f'(x) = \dfrac{x^2 + 4x + 3}{x - 1};\ (2, 4)$

50. $f'(x) = \dfrac{x^3 - 4x^2 + 3}{x - 3};\ (4, -1)$

51. Biology A population P of bacteria is growing at the rate of

$$\frac{dP}{dt} = \frac{3000}{1 + 0.25t}$$

where t is the time (in days). When $t = 0$, the population is 1000.

(a) Find a model for the population.

(b) What is the population after 3 days?

(c) After how many days will the population be 12,000?

52. Biology Because of an insufficient oxygen supply, the trout population P in a lake is dying. The population's rate of change can be modeled by

$$\frac{dP}{dt} = -125e^{-t/20}$$

where t is the time (in days). When $t = 0$, the population is 2500.

(a) Find a model for the population.

(b) What is the population after 15 days?

(c) How long will it take for the entire trout population to die?

53. Demand The marginal price for the demand of a product can be modeled by

$$\frac{dp}{dx} = 0.1e^{-x/500}$$

where x is the quantity demanded. When the demand is 600 units, the price p is \$30.

(a) Find the demand function.

(b) Use a graphing utility to graph the demand function. Does price increase or decrease as demand increases?

(c) Use the *zoom* and *trace* features of the graphing utility to find the quantity demanded when the price is \$22.

54. **HOW DO YOU SEE IT?** The graph shows the rate of change of the revenue of a company from 1990 through 2010.

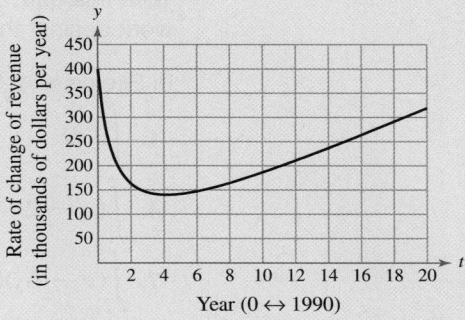

Year (0 ↔ 1990)

(a) Approximate the rate of change of the revenue in 2009. Explain your reasoning.

(b) Approximate the year when the rate of change of the revenue is the greatest. Explain your reasoning.

(c) Approximate the year when the revenue is maximum. Explain your reasoning.

55. Revenue The rate of change in revenue for Cablevision from 2002 through 2009 can be modeled by

$$\frac{dR}{dt} = 320.1e^{0.0993t}$$

where R is the revenue (in millions of dollars) and t is the time (in years), with $t = 2$ corresponding to 2002. In 2007, the revenue for Cablevision was \$6484.5 million. *(Source: Cablevision Systems Corporation)*

(a) Find a model for the revenue of Cablevision.

(b) Find Cablevision's revenue in 2009.

56. Revenue The rate of change in revenue for Under Armour from 2004 through 2009 can be modeled by

$$\frac{dR}{dt} = 13.897t + \frac{284.653}{t}$$

where R is the revenue (in millions of dollars) and t is the time (in years), with $t = 4$ corresponding to 2004. In 2008, the revenue for Under Armour was \$725.2 million. *(Source: Under Armour, Inc.)*

(a) Find a model for the revenue of Under Armour.

(b) Find Under Armour's revenue in 2006.

True or False In Exercises 57 and 58, determine whether the statement is true or false. If it is false, explain why or give an example that shows it is false.

57. $(\ln x)^{1/2} = \frac{1}{2}(\ln x)$

58. $\displaystyle\int \frac{1}{x}\, dx = \ln|ax| + C,\ a \neq 0$

Take this quiz as you would take a quiz in class. When you are done, check your work against the answers given in the back of the book.

In Exercises 1–9, find the indefinite integral. Check your result by differentiation.

1. $\displaystyle\int 3\,dx$ **2.** $\displaystyle\int 10x\,dx$ **3.** $\displaystyle\int \frac{1}{x^5}\,dx$

4. $\displaystyle\int (x^2 - 2x + 15)\,dx$ **5.** $\displaystyle\int (6x + 1)^3(6)\,dx$ **6.** $\displaystyle\int x(5x^2 - 2)^4\,dx$

7. $\displaystyle\int (x^2 - 5x)(2x - 5)\,dx$ **8.** $\displaystyle\int \frac{3x^2}{(x^3 + 3)^3}\,dx$ **9.** $\displaystyle\int \sqrt{5x + 2}\,dx$

In Exercises 10 and 11, find the particular solution that satisfies the differential equation and initial condition.

10. $f'(x) = 16x; f(0) = 1$ **11.** $f'(x) = 9x^2 + 4; f(1) = 5$

12. The marginal cost function for producing x units of a product is modeled by

$$\frac{dC}{dx} = 16 - 0.06x.$$

It costs \$25 to produce one unit. Find (a) the cost function C (in dollars), (b) the fixed cost (when $x = 0$), and (c) the total cost of producing 500 units.

13. Find the equation of the function f whose graph passes through the point $(0, 1)$ and whose derivative is

$$f'(x) = 2x^2 + 1.$$

14. The number of bolts B produced by a foundry changes according to the model

$$\frac{dB}{dt} = \frac{250t}{\sqrt{t^2 + 36}}, \quad 0 \le t \le 40$$

where t is the time (in hours). Find the number of bolts produced in (a) 8 hours and (b) 40 hours.

In Exercises 15–17, use the Exponential Rule to find the indefinite integral.

15. $\displaystyle\int 5e^{5x+4}\,dx$ **16.** $\displaystyle\int 3x^2 e^{x^3}\,dx$ **17.** $\displaystyle\int (x - 3)e^{x^2 - 6x}\,dx$

In Exercises 18–20, use the Log Rule to find the indefinite integral.

18. $\displaystyle\int \frac{2}{2x - 1}\,dx$ **19.** $\displaystyle\int \frac{1}{3 - 8x}\,dx$ **20.** $\displaystyle\int \frac{x}{3x^2 + 4}\,dx$

21. The rate of change in sales for Advance Auto Parts from 2001 through 2009 can be modeled by

$$\frac{dS}{dt} = 26.32t + \frac{848.99}{t}$$

where S is the sales (in millions) and t is the time (in years), with $t = 1$ corresponding to 2001. In 2001, the sales for Advance Auto Parts were \$2517.6 million. *(Source: Advance Auto Parts, Inc.)*

(a) Find a model for the sales of Advance Auto Parts.

(b) Find the sales for Advance Auto Parts in 2008.

11.4 Area and the Fundamental Theorem of Calculus

■ Understand the relationship between area and definite integrals.
■ Evaluate definite integrals using the Fundamental Theorem of Calculus.
■ Use definite integrals to solve marginal analysis problems.
■ Find the average values of functions over closed intervals.
■ Use properties of even and odd functions to help evaluate definite integrals.
■ Find the amounts of annuities.

One- to Four-Family Homes

In Exercise 79 on page 842, you will use integration to find a model for the mortgage debt outstanding for one- to four-family homes.

Area and Definite Integrals

From your study of geometry, you know that area is a number that defines the size of a bounded region. For simple regions, such as rectangles, triangles, and circles, area can be found using geometric formulas.

In this section, you will learn how to use calculus to find the areas of nonstandard regions, such as the region R shown in Figure 11.5.

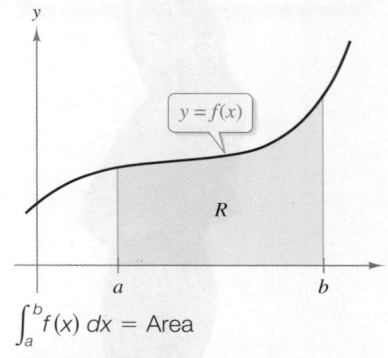

$$\int_a^b f(x)\, dx = \text{Area}$$

FIGURE 11.5

Definition of a Definite Integral

Let f be nonnegative and continuous on the closed interval $[a, b]$. The area of the region bounded by the graph of f, the x-axis, and the lines $x = a$ and $x = b$ is denoted by

$$\text{Area} = \int_a^b f(x)\, dx.$$

The expression $\int_a^b f(x)\, dx$ is called the **definite integral** from a to b, where a is the **lower limit of integration** and b is the **upper limit of integration.**

FIGURE 11.6

Example 1 Evaluating a Definite Integral Using a Geometric Formula

The definite integral

$$\int_0^2 2x\, dx$$

represents the area of the region bounded by the graph of $f(x) = 2x$, the x-axis, and the line $x = 2$, as shown in Figure 11.6. The region is triangular, with a height of 4 units and a base of 2 units. Using the formula for the area of a triangle, you have

$$\int_0^2 2x\, dx = \frac{1}{2}(\text{base})(\text{height}) = \frac{1}{2}(2)(4) = 4.$$

✓**Checkpoint 1**

Evaluate the definite integral using a geometric formula. Illustrate your answer with an appropriate sketch.

$$\int_0^3 4x\, dx$$

The Fundamental Theorem of Calculus

Consider the function A, which denotes the area of the region shown in Figure 11.7.

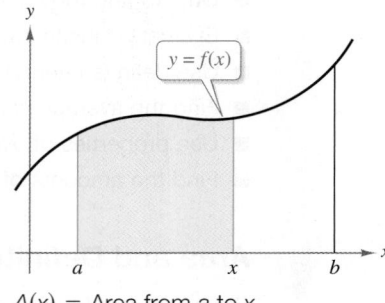

$A(x) =$ Area from a to x
FIGURE 11.7

To discover the relationship between A and f, let x increase by an amount Δx. This increases the area by ΔA. Let $f(m)$ and $f(M)$ denote the minimum and maximum values of f on the interval $[x, x + \Delta x]$.

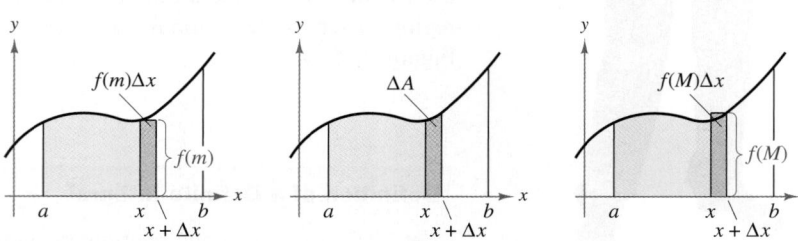

FIGURE 11.8

As indicated in Figure 11.8, you can write the inequality below.

$$f(m)\,\Delta x \leq \quad \Delta A \quad \leq f(M)\,\Delta x \qquad \text{See Figure 11.8.}$$

$$f(m) \leq \quad \frac{\Delta A}{\Delta x} \quad \leq f(M) \qquad \text{Divide each term by } \Delta x.$$

$$\lim_{\Delta x \to 0} f(m) \leq \lim_{\Delta x \to 0} \frac{\Delta A}{\Delta x} \leq \lim_{\Delta x \to 0} f(M) \qquad \text{Take limit of each term.}$$

$$f(x) \leq \quad A'(x) \quad \leq f(x) \qquad \text{Definition of derivative of } A(x)$$

So, $f(x) = A'(x)$, and $A(x) = F(x) + C$, where $F'(x) = f(x)$. Because $A(a) = 0$, it follows that $C = -F(a)$. So, $A(x) = F(x) - F(a)$, which implies that

$$A(b) = \int_a^b f(x)\,dx = F(b) - F(a).$$

This equation tells you that *if you can find an antiderivative for f,* then you can use the antiderivative to evaluate the definite integral $\int_a^b f(x)\,dx$. This result is called the **Fundamental Theorem of Calculus.**

The Fundamental Theorem of Calculus

If f is nonnegative and continuous on the closed interval $[a, b]$, then

$$\int_a^b f(x)\,dx = F(b) - F(a)$$

where F is any function such that $F'(x) = f(x)$ for all x in $[a, b]$.

Guidelines for Using the Fundamental Theorem of Calculus

1. The Fundamental Theorem of Calculus describes a way of *evaluating* a definite integral, not a procedure for finding antiderivatives.

2. In applying the Fundamental Theorem, it is helpful to use the notation

$$\int_a^b f(x)\, dx = F(x)\Big]_a^b = F(b) - F(a).$$

For instance, to evaluate $\int_1^3 x^3\, dx$, you can write

$$\int_1^3 x^3\, dx = \frac{x^4}{4}\Big]_1^3$$
$$= \frac{3^4}{4} - \frac{1^4}{4}$$
$$= 20.$$

3. The constant of integration C can be dropped because

$$\int_a^b f(x)\, dx = \left[F(x) + C \right]_a^b$$
$$= [F(b) + C] - [F(a) + C]$$
$$= F(b) - F(a) + C - C$$
$$= F(b) - F(a).$$

In the development of the Fundamental Theorem of Calculus, f was assumed to be nonnegative on the closed interval $[a, b]$. As such, the definite integral was defined as an area. Now, with the Fundamental Theorem, the definition can be extended to include functions that are negative on all or part of the closed interval $[a, b]$. Specifically, if f is *any* function that is continuous on a closed interval $[a, b]$, then the **definite integral** of $f(x)$ from a to b is defined to be

$$\int_a^b f(x)\, dx = F(b) - F(a)$$

where F is an antiderivative of f. Remember that definite integrals do not necessarily represent areas and can be negative, zero, or positive.

STUDY TIP

Be sure you see the distinction between indefinite and definite integrals. The *indefinite integral*

$$\int f(x)\, dx$$

denotes a *family of functions*, each of which is an antiderivative of f, whereas the *definite integral*

$$\int_a^b f(x)\, dx$$

is a *number*.

Properties of Definite Integrals

Let f and g be continuous on the closed interval $[a, b]$.

1. $\displaystyle \int_a^b k f(x)\, dx = k \int_a^b f(x)\, dx, \quad k$ is a constant.

2. $\displaystyle \int_a^b [f(x) \pm g(x)]\, dx = \int_a^b f(x)\, dx \pm \int_a^b g(x)\, dx$

3. $\displaystyle \int_a^b f(x)\, dx = \int_a^c f(x)\, dx + \int_c^b f(x)\, dx, \quad a < c < b$

4. $\displaystyle \int_a^a f(x)\, dx = 0$

5. $\displaystyle \int_a^b f(x)\, dx = -\int_b^a f(x)\, dx$

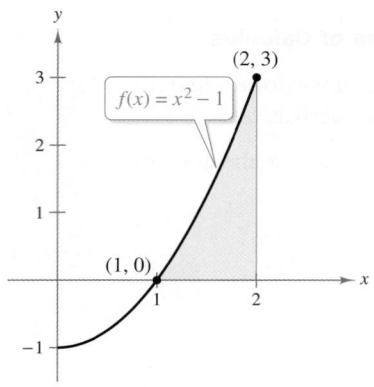

Area $= \displaystyle\int_1^2 (x^2 - 1)\, dx$

FIGURE 11.9

STUDY TIP

It is easy to make errors in signs when evaluating definite integrals. To avoid such errors, enclose the values of the antiderivative at the upper and lower limits of integration in separate sets of parentheses, as shown in Example 2.

Example 2   Finding Area by the Fundamental Theorem

Find the area of the region bounded by the x-axis and the graph of

$$f(x) = x^2 - 1, \quad 1 \le x \le 2.$$

SOLUTION   Note that $f(x) \ge 0$ on the interval $1 \le x \le 2$, as shown in Figure 11.9. So, you can represent the area of the region by a definite integral. To find the area, use the Fundamental Theorem of Calculus.

$$\begin{aligned}
\text{Area} &= \int_1^2 (x^2 - 1)\, dx && \text{Definition of definite integral} \\
&= \left[\frac{x^3}{3} - x\right]_1^2 && \text{Find antiderivative.} \\
&= \left(\frac{2^3}{3} - 2\right) - \left(\frac{1^3}{3} - 1\right) && \text{Apply Fundamental Theorem.} \\
&= \frac{2}{3} - \left(-\frac{2}{3}\right) \\
&= \frac{4}{3} && \text{Simplify.}
\end{aligned}$$

So, the area of the region is $\frac{4}{3}$ square units.

✓ **Checkpoint 2**

Find the area of the region bounded by the x-axis and the graph of

$$f(x) = x^2 + 1, \quad 2 \le x \le 3.$$

Example 3   Evaluating a Definite Integral

Evaluate the definite integral

$$\int_0^1 (4t + 1)^2\, dt$$

and sketch the region whose area is represented by the integral.

SOLUTION

$$\begin{aligned}
\int_0^1 (4t + 1)^2\, dt &= \frac{1}{4}\int_0^1 (4t + 1)^2(4)\, dt && \text{Multiply and divide by 4.} \\
&= \frac{1}{4}\left[\frac{(4t + 1)^3}{3}\right]_0^1 && \text{Find antiderivative.} \\
&= \frac{1}{4}\left[\left(\frac{5^3}{3}\right) - \left(\frac{1}{3}\right)\right] && \text{Apply Fundamental Theorem.} \\
&= \frac{1}{4}\left(\frac{124}{3}\right) \\
&= \frac{31}{3} && \text{Simplify.}
\end{aligned}$$

The region is shown in Figure 11.10.

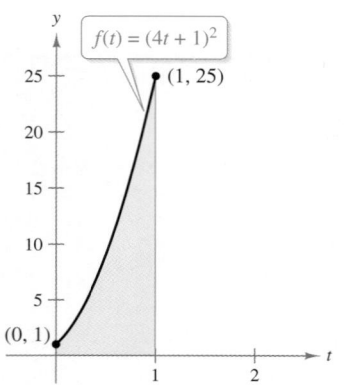

FIGURE 11.10

✓ **Checkpoint 3**

Evaluate $\displaystyle\int_0^1 (2t + 3)^3\, dt$.

Example 4 Evaluating Definite Integrals

Evaluate each definite integral.

a. $\displaystyle\int_0^3 e^{2x}\,dx$ **b.** $\displaystyle\int_1^2 \frac{1}{x}\,dx$ **c.** $\displaystyle\int_1^4 -3\sqrt{x}\,dx$

SOLUTION

a. $\displaystyle\int_0^3 e^{2x}\,dx = \frac{1}{2}e^{2x}\Big]_0^3 = \frac{1}{2}(e^6 - e^0) \approx 201.21$

b. $\displaystyle\int_1^2 \frac{1}{x}\,dx = \ln x\Big]_1^2 = \ln 2 - \ln 1 = \ln 2 \approx 0.69$

c. $\displaystyle\int_1^4 -3\sqrt{x}\,dx = -3\int_1^4 x^{1/2}\,dx$ Rewrite with rational exponent.

$\displaystyle\qquad = -3\left[\frac{x^{3/2}}{3/2}\right]_1^4$ Find antiderivative.

$\displaystyle\qquad = -2x^{3/2}\Big]_1^4$

$\displaystyle\qquad = -2(4^{3/2} - 1^{3/2})$ Apply Fundamental Theorem.

$\displaystyle\qquad = -2(8 - 1)$

$\displaystyle\qquad = -14$ Simplify.

✓ **Checkpoint 4**

Evaluate each definite integral.

a. $\displaystyle\int_0^1 e^{4x}\,dx$ **b.** $\displaystyle\int_2^5 -\frac{1}{x}\,dx$

Example 5 Interpreting Absolute Value

Evaluate $\displaystyle\int_0^2 |2x - 1|\,dx$.

SOLUTION   The region represented by the definite integral is shown in Figure 11.11. From the definition of absolute value, you can write

$$|2x - 1| = \begin{cases} -(2x - 1), & x < \frac{1}{2} \\ 2x - 1, & x \ge \frac{1}{2} \end{cases}.$$

Using Property 3 of definite integrals, rewrite the integral as two definite integrals.

$$\int_0^2 |2x - 1|\,dx = \int_0^{1/2} -(2x - 1)\,dx + \int_{1/2}^2 (2x - 1)\,dx$$

$$= \left[-x^2 + x\right]_0^{1/2} + \left[x^2 - x\right]_{1/2}^2$$

$$= \left(-\frac{1}{4} + \frac{1}{2}\right) - (0 + 0) + (4 - 2) - \left(\frac{1}{4} - \frac{1}{2}\right)$$

$$= \frac{5}{2}$$

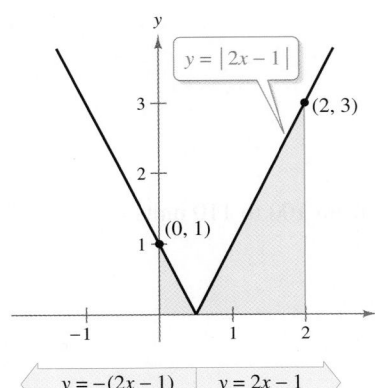

$y = -(2x - 1)$ $y = 2x - 1$

FIGURE 11.11

✓ **Checkpoint 5**

Evaluate $\displaystyle\int_0^5 |x - 2|\,dx$.

Marginal Analysis

You have already studied *marginal analysis* in the context of derivatives and differentials (Sections 7.5 and 9.5). There, you were given a cost, revenue, or profit function, and you used the derivative to approximate the additional cost, revenue, or profit obtained by selling one additional unit. In this section, you will examine the reverse process. That is, you will be given the marginal cost, marginal revenue, or marginal profit and you will use a definite integral to find the exact increase or decrease in cost, revenue, or profit obtained by selling one or several additional units.

For instance, you are asked to find the additional revenue obtained by increasing sales from x_1 to x_2 units. When you know the revenue function R, you can find the additional revenue by subtracting $R(x_1)$ from $R(x_2)$. When you don't know R, you can use the marginal revenue function dR/dx to find the additional revenue by using a definite integral.

$$\int_{x_1}^{x_2} \frac{dR}{dx}\, dx = R(x_2) - R(x_1)$$

 Example 6   **Analyzing a Profit Function**

The marginal profit for a product is modeled by

$$\frac{dP}{dx} = -0.0005x + 12.2.$$

a. Find the change in profit when sales increase from 100 to 101 units.

b. Find the change in profit when sales increase from 100 to 110 units.

SOLUTION

a. The change in profit obtained by increasing sales from 100 to 101 units is

$$\int_{100}^{101} \frac{dP}{dx}\, dx = \int_{100}^{101} (-0.0005x + 12.2)\, dx$$
$$= \left[-0.00025x^2 + 12.2x \right]_{100}^{101}$$
$$\approx \$12.15.$$

b. The change in profit obtained by increasing sales from 100 to 110 units is

$$\int_{100}^{110} \frac{dP}{dx}\, dx = \int_{100}^{110} (-0.0005x + 12.2)\, dx$$
$$= \left[-0.00025x^2 + 12.2x \right]_{100}^{110}$$
$$\approx \$121.48.$$

✓**Checkpoint 6**

The marginal profit for a product is modeled by

$$\frac{dP}{dx} = -0.0002x + 14.2.$$

a. Find the change in profit when sales increase from 100 to 101 units.

b. Find the change in profit when sales increase from 100 to 110 units.

Average Value

The *average value* of a function on a closed interval is defined below.

> **Definition of the Average Value of a Function**
>
> If f is continuous on $[a, b]$, then the **average value** of f on $[a, b]$ is
>
> $$\text{Average value of } f \text{ on } [a, b] = \frac{1}{b - a}\int_a^b f(x)\, dx.$$

In Section 9.2, you studied the effects of production levels on cost using an average cost function. In the next example, you will study the effects of time on cost by using integration to find the average cost.

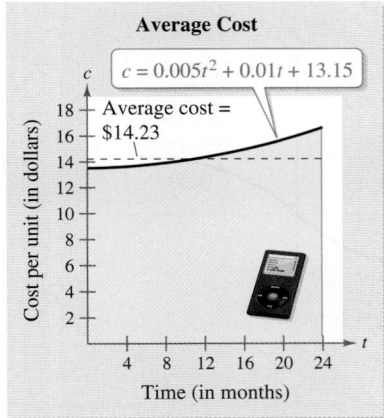

Average Cost

$c = 0.005t^2 + 0.01t + 13.15$

Average cost = $14.23

FIGURE 11.12

Example 7 Finding the Average Cost

The cost per unit c of producing MP3 players over a two-year period is modeled by

$$c = 0.005t^2 + 0.01t + 13.15, \quad 0 \le t \le 24$$

where t is the time (in months). Approximate the average cost per unit over the two-year period.

SOLUTION   The average cost can be found by integrating c over the interval $[0, 24]$.

$$\text{Average cost per unit} = \frac{1}{24}\int_0^{24}(0.005t^2 + 0.01t + 13.15)\, dt$$

$$= \frac{1}{24}\left[\frac{0.005t^3}{3} + \frac{0.01t^2}{2} + 13.15t\right]_0^{24}$$

$$= \frac{1}{24}(341.52)$$

$$= \$14.23 \qquad \text{(See Figure 11.12.)}$$

✓ Checkpoint 7

Find the average cost per unit over a two-year period when the cost per unit c of inline skates is given by

$$c = 0.005t^2 + 0.02t + 12.5, \quad 0 \le t \le 24$$

where t is the time (in months).

You can use a spreadsheet, as shown at the left, to check the reasonableness of the average value found in Example 7. The spreadsheet assumes that one unit is produced each month, beginning with $t = 0$ and ending with $t = 24$. So, when $t = 0$, the cost is

$$c = 0.005(0)^2 + 0.01(0) + 13.15$$

$$= \$13.15$$

and when $t = 1$ the cost is

$$c = 0.005(1)^2 + 0.01(1) + 13.15$$

$$= \$13.165$$

and so on. Note in the spreadsheet that the cost increases each month, and the average of the 25 costs is $14.25. So, you can conclude that the result of Example 7 is reasonable.

	A	B
1	t	$c = 0.005t^2 + 0.01t + 13.15$
2	0	13.15
3	1	13.165
4	2	13.19
5	3	13.225
6	4	13.27
7	5	13.325
8	6	13.39
9	7	13.465
10	8	13.55
11	9	13.645
12	10	13.75
13	11	13.865
14	12	13.99
15	13	14.125
16	14	14.27
17	15	14.425
18	16	14.59
19	17	14.765
20	18	14.95
21	19	15.145
22	20	15.35
23	21	15.565
24	22	15.79
25	23	16.025
26	24	16.27
27		
28	**Sum**	356.25
29	**Average**	14.25

Even and Odd Functions

Several common functions have graphs that are symmetric with respect to the y-axis or the origin, as shown in Figure 11.13. If the graph of f is symmetric with respect to the y-axis, as in Figure 11.13(a), then

$$f(-x) = f(x) \qquad \text{Even function}$$

and f is called an **even** function. If the graph of f is symmetric with respect to the origin, as in Figure 11.13(b), then

$$f(-x) = -f(x) \qquad \text{Odd function}$$

and f is called an **odd** function.

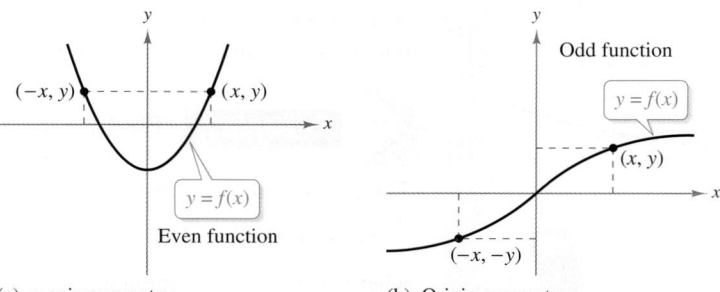

(a) y-axis symmetry   (b) Origin symmetry

FIGURE 11.13

Integration of Even and Odd Functions

1. If f is an *even* function, then $\displaystyle \int_{-a}^{a} f(x)\, dx = 2 \int_{0}^{a} f(x)\, dx.$

2. If f is an *odd* function, then $\displaystyle \int_{-a}^{a} f(x)\, dx = 0.$

Example 8   **Integrating Even and Odd Functions**

Evaluate each definite integral.

a. $\displaystyle \int_{-2}^{2} x^2\, dx$ **b.** $\displaystyle \int_{-2}^{2} x^3\, dx$

SOLUTION

a. Because $f(x) = x^2$ is an even function,

$$\int_{-2}^{2} x^2\, dx = 2 \int_{0}^{2} x^2\, dx = 2 \left[\frac{x^3}{3} \right]_{0}^{2} = 2 \left(\frac{8}{3} - 0 \right) = \frac{16}{3}.$$

b. Because $f(x) = x^3$ is an odd function,

$$\int_{-2}^{2} x^3\, dx = 0.$$

✓ **Checkpoint 8**

Evaluate each definite integral.

a. $\displaystyle \int_{-1}^{1} x^4\, dx$ **b.** $\displaystyle \int_{-1}^{1} x^5\, dx$

Annuity

A sequence of equal payments made at regular time intervals over a period of time is called an **annuity.** Some examples of annuities are payroll savings plans, monthly home mortgage payments, and individual retirement accounts. The **amount of an annuity** is the sum of the payments plus the interest earned.

Amount of an Annuity

If c represents a continuous income function in dollars per year (where t is the time in years), r represents the interest rate compounded continuously, and T represents the term of the annuity in years, then the **amount of an annuity** is

$$\text{Amount of an annuity} = e^{rT} \int_0^T c(t)e^{-rt}\, dt.$$

 Example 9 **Finding the Amount of an Annuity**

You deposit \$2000 each year for 15 years in an individual retirement account (IRA) paying 5% interest. How much will you have in your IRA after 15 years?

SOLUTION   The income function for your deposit is

$$c(t) = 2000.$$

So, the amount of the annuity after 15 years will be

$$\begin{aligned}
\text{Amount of an annuity} &= e^{rT} \int_0^T c(t)e^{-rt}\, dt \\
&= e^{(0.05)(15)} \int_0^{15} 2000e^{-0.05t}\, dt \\
&= 2000e^{0.75} \left[-\frac{e^{-0.05t}}{0.05} \right]_0^{15} \\
&\approx \$44{,}680.00.
\end{aligned}$$

✓**Checkpoint 9**

You deposit \$1000 each year in a savings account paying 4% interest. How much will be in the account after 10 years?

SUMMARIZE   (Section 11.4)

1. State the definition of a definite integral *(page 831)*. For an example of a definite integral, see Example 1.

2. State the Fundamental Theorem of Calculus *(page 832)*. For examples of the Fundamental Theorem of Calculus, see Examples 2 and 3.

3. State the properties of definite integrals *(page 833)*. For examples of these properties, see Examples 4 and 5.

4. State the definition of the average value of a function *(page 837)*. For an example of finding the average value of a function, see Example 7.

5. State the rules for integrating even and odd functions *(page 838)*. For an example of integrating even and odd functions, see Example 8.

In Exercises 1–4, find the indefinite integral.

1. $\int (3x + 7)\, dx$ **2.** $\int \left(x^{3/2} + 2\sqrt{x}\right) dx$ **3.** $\int \frac{1}{5x}\, dx$ **4.** $\int e^{-6x}\, dx$

In Exercises 5–8, integrate the marginal function.

5. $\dfrac{dC}{dx} = 0.02x^{3/2} + 29{,}500$

6. $\dfrac{dR}{dx} = 9000 + 2x$

7. $\dfrac{dP}{dx} = 25{,}000 - 0.01x$

8. $\dfrac{dC}{dx} = 0.03x^2 + 4600$

Exercises 11.4

Evaluating a Definite Integral Using a Geometric Formula In Exercises 1–6, sketch the region whose area is represented by the definite integral. Then use a geometric formula to evaluate the integral. *See Example 1.*

1. $\displaystyle\int_0^2 3\, dx$ **2.** $\displaystyle\int_0^3 4\, dx$

3. $\displaystyle\int_0^4 x\, dx$ **4.** $\displaystyle\int_0^3 \frac{x}{3}\, dx$

5. $\displaystyle\int_{-3}^3 \sqrt{9 - x^2}\, dx$ **6.** $\displaystyle\int_0^2 \sqrt{4 - x^2}\, dx$

Using Properties of Definite Integrals In Exercises 7 and 8, use the values $\int_0^5 f(x)\, dx = 6$ and $\int_0^5 g(x)\, dx = 2$ to evaluate the definite integral.

7. (a) $\displaystyle\int_0^5 [f(x) + g(x)]\, dx$ (b) $\displaystyle\int_0^5 [f(x) - g(x)]\, dx$

(c) $\displaystyle\int_0^5 -4f(x)\, dx$ (d) $\displaystyle\int_0^5 [f(x) - 3g(x)]\, dx$

8. (a) $\displaystyle\int_0^5 2g(x)\, dx$ (b) $\displaystyle\int_5^0 f(x)\, dx$

(c) $\displaystyle\int_5^5 f(x)\, dx$ (d) $\displaystyle\int_0^5 [f(x) - f(x)]\, dx$

Finding Area by the Fundamental Theorem In Exercises 9–16, find the area of the region. *See Example 2.*

9. $y = x - x^2$

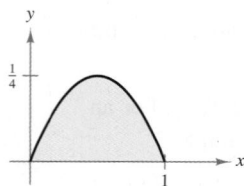

10. $y = 1 - x^4$

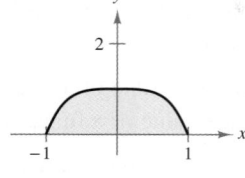

11. $y = \dfrac{1}{x^2}$

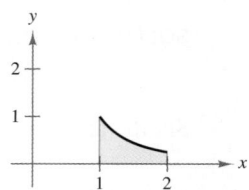

12. $y = \dfrac{2}{\sqrt{x}}$

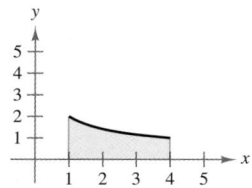

13. $y = 3e^{-x/2}$

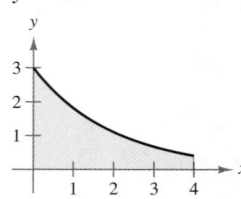

14. $y = 2e^{x/4}$

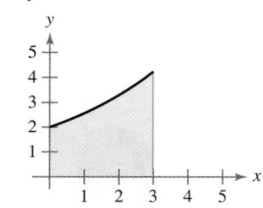

15. $y = \dfrac{x^2 + 4}{x}$

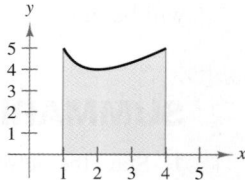

16. $y = \dfrac{x - 2}{x}$

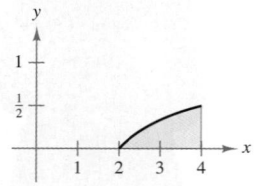

Evaluating a Definite Integral In Exercises 17–38, evaluate the definite integral. *See Examples 3 and 4.*

17. $\displaystyle\int_0^1 2x\, dx$ **18.** $\displaystyle\int_2^7 3v\, dv$

19. $\displaystyle\int_{-1}^0 (x - 2)\, dx$ **20.** $\displaystyle\int_2^5 (-3x + 4)\, dx$

21. $\displaystyle\int_{-1}^1 (3t + 4)^2\, dt$ **22.** $\displaystyle\int_0^1 (1 - 2x)^2\, dx$

23. $\int_0^3 (x-2)^3 \, dx$

24. $\int_1^3 (x-3)^4 \, dx$

25. $\int_{-1}^1 \left(\sqrt[3]{t} - 2 \right) dt$

26. $\int_1^4 \left(\sqrt{x} + x \right) dx$

27. $\int_{-1}^0 (t^{1/3} - t^{2/3}) \, dt$

28. $\int_0^4 (x^{1/2} + x^{1/4}) \, dx$

29. $\int_2^8 \frac{3}{x} \, dx$

30. $\int_1^4 -\frac{4}{x} \, dx$

31. $\int_0^4 \frac{1}{\sqrt{2x+1}} \, dx$

32. $\int_0^2 \frac{x}{\sqrt{1+2x^2}} \, dx$

33. $\int_1^2 e^{1-x} \, dx$

34. $\int_{-1}^1 (e^x - e^{-x}) \, dx$

35. $\int_0^1 e^{2x} \sqrt{e^{2x}+1} \, dx$

36. $\int_0^1 \frac{e^{-x}}{\sqrt{e^{-x}+1}} \, dx$

37. $\int_0^2 \frac{x}{1+4x^2} \, dx$

38. $\int_0^1 \frac{e^{2x}}{e^{2x}+1} \, dx$

Interpreting Absolute Value In Exercises 39–42, evaluate the definite integral. *See Example 5.*

39. $\int_{-2}^1 |4x| \, dx$

40. $\int_{-1}^3 \left| \frac{x}{3} \right| dx$

41. $\int_2^8 |3x-9| \, dx$

42. $\int_0^3 |2x-3| \, dx$

Area of a Region In Exercises 43–46, find the area of the region bounded by the graphs of the equations. Use a graphing utility to verify your results.

43. $y = 3x^2 + 1$, $y = 0$, $x = 0$, and $x = 2$

44. $y = 1 + \sqrt{x}$, $y = 0$, $x = 0$, and $x = 4$

45. $y = 4/x$, $y = 0$, $x = 1$, and $x = 3$

46. $y = e^x$, $y = 0$, $x = 0$, and $x = 2$

Marginal Analysis In Exercises 47–52, find the change in cost C, revenue R, or profit P, for the given marginal. In each case, assume that the number of units x increases by 3 from the specified value of x. *See Example 6.*

Marginal	Number of Units, x
47. $\dfrac{dC}{dx} = 2.25$	$x = 100$
48. $\dfrac{dC}{dx} = \dfrac{20{,}000}{x^2}$	$x = 10$
49. $\dfrac{dR}{dx} = 48 - 3x$	$x = 12$
50. $\dfrac{dR}{dx} = 75\left(20 + \dfrac{900}{x}\right)$	$x = 500$
51. $\dfrac{dP}{dx} = \dfrac{400 - x}{150}$	$x = 200$
52. $\dfrac{dP}{dx} = 12.5\left(40 - 3\sqrt{x}\right)$	$x = 125$

Average Value of a Function In Exercises 53–60, find the average value of the function on the interval. Then find all x-values in the interval for which the function is equal to its average value.

53. $f(x) = 6x$; $[1, 3]$

54. $f(x) = x^3$; $[0, 2]$

55. $f(x) = 4 - x^2$; $[-2, 2]$

56. $f(x) = x - 2\sqrt{x}$; $[0, 4]$

57. $f(x) = 2e^x$; $[-1, 1]$

58. $f(x) = e^{x/4}$; $[0, 4]$

59. $f(x) = \dfrac{3}{x+2}$; $[1, 5]$

60. $f(x) = \dfrac{1}{(x-3)^2}$; $[0, 2]$

Integrating Even and Odd Functions In Exercises 61–64, evaluate the definite integral using the properties of even and odd functions. *See Example 8.*

61. $\int_{-1}^1 3x^4 \, dx$

62. $\int_{-2}^2 (x^3 - 4x) \, dx$

63. $\int_{-1}^1 (2t^5 - 2t) \, dt$

64. $\int_{-2}^2 \left(\frac{1}{2}t^4 + 1 \right) dt$

65. Using Properties of Definite Integrals Use the value $\int_0^1 x^2 \, dx = \frac{1}{3}$ to evaluate each definite integral. Explain your reasoning.

(a) $\int_{-1}^0 x^2 \, dx$

(b) $\int_{-1}^1 x^2 \, dx$

(c) $\int_0^1 -x^2 \, dx$

66. Using Properties of Definite Integrals Use the value $\int_0^4 x^3 \, dx = 64$ to evaluate each definite integral. Explain your reasoning.

(a) $\int_{-4}^0 x^3 \, dx$

(b) $\int_{-4}^4 x^3 \, dx$

(c) $\int_0^4 2x^3 \, dx$

Finding the Amount of an Annuity In Exercises 67–70, find the amount of an annuity with income function $c(t)$, interest rate r, and term T. *See Example 9.*

67. $c(t) = \$250$, $r = 8\%$, $T = 6$ years

68. $c(t) = \$500$, $r = 7\%$, $T = 4$ years

69. $c(t) = \$1500$, $r = 2\%$, $T = 10$ years

70. $c(t) = \$2000$, $r = 3\%$, $T = 15$ years

Capital Accumulation In Exercises 71–74, you are given the rate of investment dI/dt. Find the capital accumulation over a five-year period by evaluating the definite integral

$$\text{Capital accumulation} = \int_0^5 \frac{dI}{dt}\, dt$$

where t is the time (in years).

71. $\dfrac{dI}{dt} = 500$

72. $\dfrac{dI}{dt} = 100t$

73. $\dfrac{dI}{dt} = 500\sqrt{t+1}$

74. $\dfrac{dI}{dt} = \dfrac{12{,}000t}{(t^2+2)^2}$

75. Cost The total cost of purchasing a piece of equipment and maintaining it for x years can be modeled by

$$C = 5000\left(25 + 3\int_0^x t^{1/4}\, dt\right).$$

Find the total cost after (a) 1 year, (b) 5 years, and (c) 10 years.

76. Depreciation A company purchases a new machine for which the rate of depreciation can be modeled by

$$\frac{dV}{dt} = 10{,}000(t-6), \quad 0 \le t \le 5$$

where V is the value of the machine after t years. Set up and evaluate the definite integral that yields the total loss of value of the machine over the first 3 years.

77. Compound Interest A deposit of $2250 is made in a savings account at an annual interest rate of 6%, compounded continuously. Find the average balance in the account during the first 5 years.

78. **HOW DO YOU SEE IT?** A college graduate has two job offers. The starting salary for each is $32,000, and after 8 years of service each will pay $54,000. The salary increase for each offer is shown in the figure. From a strictly monetary viewpoint, which is the better offer? Explain your reasoning.

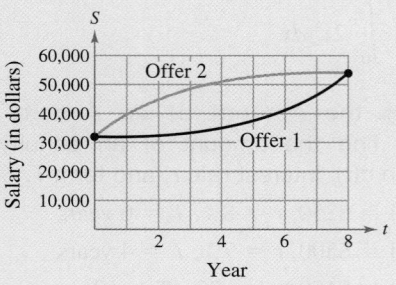

79. Mortgage Debt The rate of change of mortgage debt outstanding for one- to four-family homes in the United States from 2000 through 2009 can be modeled by

$$\frac{dM}{dt} = 547.56t - 69.459t^2 + 331.258e^{-t}$$

where M is the mortgage debt outstanding (in billions of dollars) and t is the year, with $t = 0$ corresponding to 2000. In 2000, the mortgage debt outstanding in the United States was $5107 billion. *(Source: Board of Governors of the Federal Reserve System)*

(a) Write a model for the debt as a function of t.

(b) What was the average mortgage debt outstanding for 2000 through 2009?

Business Capsule

After losing her job as an account executive in 1985, Avis Yates Rivers used $2500 to start a word processing business from the basement of her home. In 1996, as a spin-off, Ms. Yates Rivers established Technology Concepts Group. Today, this Somerset, New Jersey-based firm provides a wide range of information technology solutions. Revenues in 2010 were about $3 million, and the company projects revenues of $30 million in 2011 with the acquisition of equipment leasing expertise. Ms. Yates Rivers has become a nationally recognized leader, speaker, and advocate for minority- and women-owned small businesses.

80. Research Project Use your school's library, the Internet, or some other reference source to research a small company similar to the one described above. Describe the impact of different factors, such as start-up capital and market conditions, on a company's revenue.

11.5 The Area of a Region Bounded by Two Graphs

■ Find the areas of regions bounded by two graphs.
■ Find consumer and producer surpluses.
■ Use the areas of regions bounded by two graphs to solve real-life problems.

Area of a Region Bounded by Two Graphs

With a few modifications, you can extend the use of definite integrals from finding the area of a region *under a graph* to finding the area of a region *bounded by two graphs*. To see how this is done, consider the region bounded by the graphs of

$$f, \quad g, \quad x = a, \quad \text{and} \quad x = b$$

as shown in Figure 11.14. If the graphs of both f and g lie above the x-axis, then you can interpret the area of the region between the graphs as the area of the region under the graph of g subtracted from the area of the region under the graph of f, as shown in Figure 11.14.

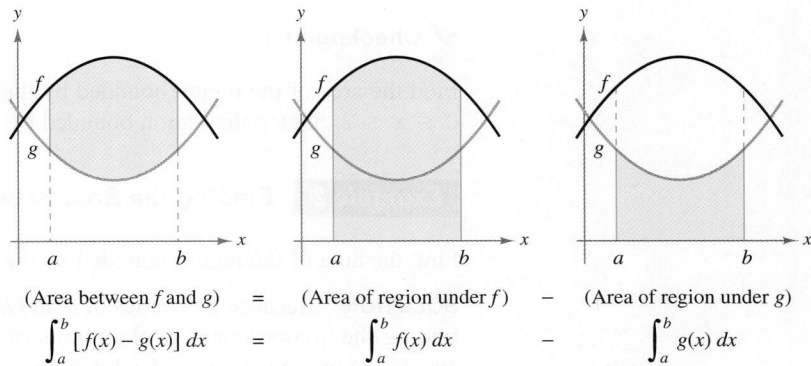

(Area between f and g) = (Area of region under f) − (Area of region under g)

$$\int_a^b [f(x) - g(x)]\, dx \quad = \quad \int_a^b f(x)\, dx \quad - \quad \int_a^b g(x)\, dx$$

FIGURE 11.14

Although Figure 11.14 depicts the graphs of f and g lying above the x-axis, this is not necessary, and the same integrand

$$[f(x) - g(x)]$$

can be used as long as both functions are continuous and $g(x) \le f(x)$ on the interval $[a, b]$.

Area of a Region Bounded by Two Graphs

If f and g are continuous on $[a, b]$ and $g(x) \le f(x)$ for all x in $[a, b]$, then the area of the region bounded by the graphs of

$$f, \quad g, \quad x = a, \quad \text{and} \quad x = b$$

(see Figure 11.15) is given by

$$A = \int_a^b [f(x) - g(x)]\, dx.$$

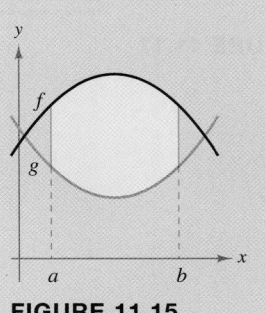

FIGURE 11.15

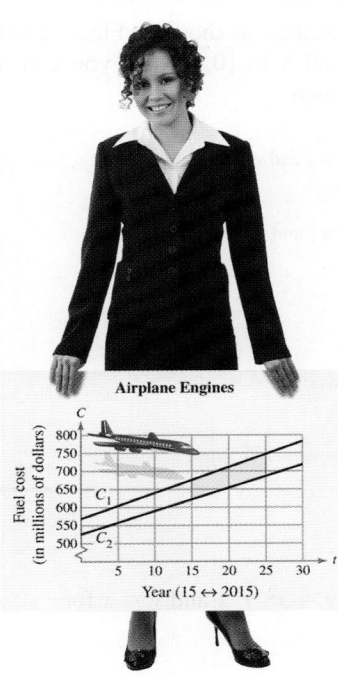

Airplane Engines

In Exercise 49 on page 850, you will use integration to find the amount saved on fuel costs by switching to more efficient airplane engines.

Example 1   **Finding the Area Bounded by Two Graphs**

Find the area of the region bounded by the graphs of $y = x^2 + 2$ and $y = x$ for $0 \le x \le 1$.

SOLUTION   Begin by sketching the graphs of both functions, as shown in Figure 11.16. From the figure, you can see that $x \le x^2 + 2$ for all x in $[0, 1]$. So, you can let $f(x) = x^2 + 2$ and $g(x) = x$. Then find the area as shown.

$$\text{Area} = \int_a^b [f(x) - g(x)]\, dx \qquad \text{Area between } f \text{ and } g$$

$$= \int_0^1 [(x^2 + 2) - (x)]\, dx \qquad \text{Substitute for } f \text{ and } g.$$

$$= \int_0^1 (x^2 - x + 2)\, dx$$

$$= \left[\frac{x^3}{3} - \frac{x^2}{2} + 2x \right]_0^1 \qquad \text{Find antiderivative.}$$

$$= \frac{11}{6} \text{ square units} \qquad \text{Apply Fundamental Theorem.}$$

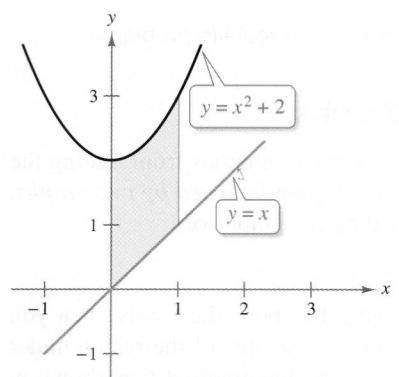

FIGURE 11.16

✔ **Checkpoint 1**

Find the area of the region bounded by the graphs of $y = x^2 + 1$ and $y = x$ for $0 \le x \le 2$. Sketch the region bounded by the graphs.   ■

Example 2   **Finding the Area Between Intersecting Graphs**

Find the area of the region bounded by the graphs of $y = 2 - x^2$ and $y = x$.

SOLUTION   Because the values of a and b are not given, you must determine them by finding the x-coordinates of the points of intersection of the two graphs. To do this, equate the two functions and solve for x.

$$2 - x^2 = x \qquad \text{Equate functions.}$$

$$-x^2 - x + 2 = 0 \qquad \text{Write in general form.}$$

$$-(x + 2)(x - 1) = 0 \qquad \text{Factor.}$$

$$x = -2,\; x = 1 \qquad \text{Solve for } x.$$

So, $a = -2$ and $b = 1$. In Figure 11.17, you can see that the graph of $f(x) = 2 - x^2$ lies above the graph of $g(x) = x$ for all x in the interval $[-2, 1]$.

$$\text{Area} = \int_a^b [f(x) - g(x)]\, dx \qquad \text{Area between } f \text{ and } g$$

$$= \int_{-2}^1 [(2 - x^2) - (x)]\, dx \qquad \text{Substitute for } f \text{ and } g.$$

$$= \int_{-2}^1 (-x^2 - x + 2)\, dx$$

$$= \left[-\frac{x^3}{3} - \frac{x^2}{2} + 2x \right]_{-2}^1 \qquad \text{Find antiderivative.}$$

$$= \frac{9}{2} \text{ square units} \qquad \text{Apply Fundamental Theorem.}$$

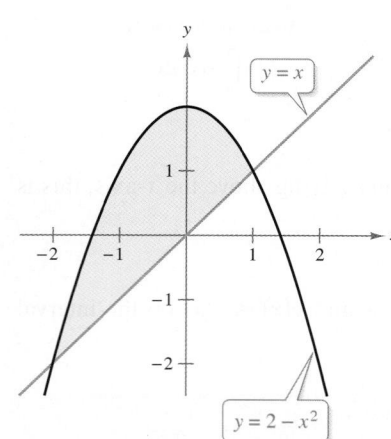

FIGURE 11.17

✔ **Checkpoint 2**

Find the area of the region bounded by the graphs of $y = 3 - x^2$ and $y = 2x$.   ■

Example 3 Finding an Area Below the *x*-Axis

Find the area of the region bounded by the graph of

$$y = x^2 - 3x - 4$$

and the *x*-axis.

SOLUTION Begin by finding the *x*-intercepts of the graph. To do this, set the function equal to zero and solve for *x*.

$x^2 - 3x - 4 = 0$	Set function equal to 0.
$(x - 4)(x + 1) = 0$	Factor.
$x = 4, x = -1$	Solve for *x*.

From Figure 11.18, you can see that $x^2 - 3x - 4 \le 0$ for all *x* in the interval $[-1, 4]$.

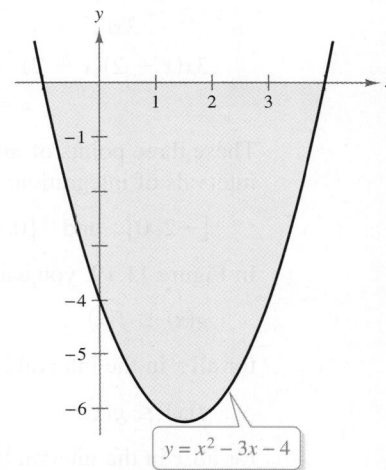

FIGURE 11.18

So, you can let

$$f(x) = 0 \quad \text{and} \quad g(x) = x^2 - 3x - 4$$

and find the area as shown.

$$\text{Area} = \int_a^b [f(x) - g(x)]\, dx \qquad \text{Area between } f \text{ and } g$$

$$= \int_{-1}^4 [(0) - (x^2 - 3x - 4)]\, dx \qquad \text{Substitute for } f \text{ and } g.$$

$$= \int_{-1}^4 (-x^2 + 3x + 4)\, dx$$

$$= \left[-\frac{x^3}{3} + \frac{3x^2}{2} + 4x \right]_{-1}^4 \qquad \text{Find antiderivative.}$$

$$= \frac{125}{6} \text{ square units} \qquad \text{Apply Fundamental Theorem.}$$

TECH TUTOR

Most graphing utilities can display regions that are bounded by two graphs. For instance, to graph the region in Example 3, set the viewing window to $-1 \le x \le 4$ and $-7 \le y \le 1$. Consult your user's manual for specific keystrokes on how to shade the graph. You should obtain the graph below.

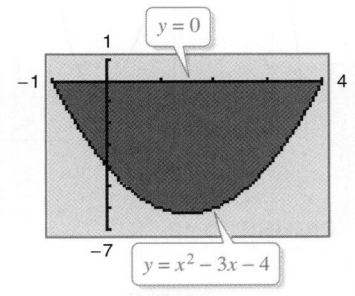

✓ Checkpoint 3

Find the area of the region bounded by the graph of

$$y = x^2 - x - 2$$

and the *x*-axis.

Sometimes two graphs intersect at more than two points. To determine the area of the region bounded by two such graphs, you must find *all* points of intersection and check to see which graph is above the other in each interval determined by the points.

Example 4   Using Multiple Points of Intersection

Find the area of the region bounded by the graphs of

$$f(x) = 3x^3 - x^2 - 10x \quad \text{and} \quad g(x) = -x^2 + 2x.$$

SOLUTION   To find the points of intersection of the two graphs, set the functions equal to each other and solve for x.

$f(x) = g(x)$	Set $f(x)$ equal to $g(x)$.
$3x^3 - x^2 - 10x = -x^2 + 2x$	Substitute for $f(x)$ and $g(x)$.
$3x^3 - 12x = 0$	Write in general form.
$3x(x^2 - 4) = 0$	
$3x(x - 2)(x + 2) = 0$	Factor.
$x = 0, x = 2, x = -2$	Solve for x.

These three points of intersection determine two intervals of integration:

$$[-2, 0] \quad \text{and} \quad [0, 2].$$

In Figure 11.19, you can see that

$$g(x) \le f(x)$$

for all x in the interval $[-2, 0]$, and that

$$f(x) \le g(x)$$

for all x in the interval $[0, 2]$. So, you must use two integrals to determine the area of the region bounded by the graphs of f and g: one for the interval $[-2, 0]$ and one for the interval $[0, 2]$.

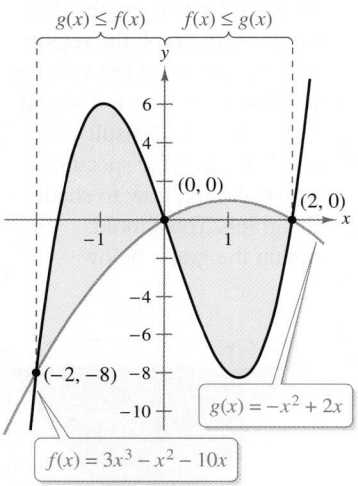

FIGURE 11.19

STUDY TIP

It is easy to make an error when calculating areas such as the one in Example 4. To check your solution, make a sketch of the region on graph paper and then use the grid on the graph paper to approximate the area. Try doing this with the graph shown in Figure 11.19. Is your approximation close to 24 square units?

$$\text{Area} = \int_{-2}^{0} [f(x) - g(x)]\, dx + \int_{0}^{2} [g(x) - f(x)]\, dx$$

$$= \int_{-2}^{0} (3x^3 - 12x)\, dx + \int_{0}^{2} (-3x^3 + 12x)\, dx$$

$$= \left[\frac{3x^4}{4} - 6x^2 \right]_{-2}^{0} + \left[-\frac{3x^4}{4} + 6x^2 \right]_{0}^{2}$$

$$= (0 - 0) - (12 - 24) + (-12 + 24) - (0 + 0)$$

$$= 24$$

So, the region has an area of 24 square units.

✓**Checkpoint 4**

Find the area of the region bounded by the graphs of

$$f(x) = x^3 + 2x^2 - 3x \quad \text{and} \quad g(x) = x^2 + 3x.$$

Sketch a graph of the region.

Consumer Surplus and Producer Surplus

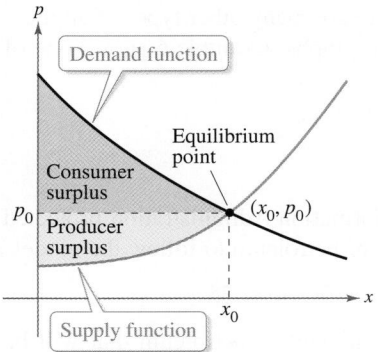

FIGURE 11.20

In Section 5.2, you learned that a demand function relates the price of a product to the consumer demand. You also learned that a supply function relates the price of a product to producers' willingness to supply the product. The point (x_0, p_0) at which a demand function $p = D(x)$ and a supply function $p = S(x)$ intersect is the equilibrium point.

Economists call the area of the region bounded by the graph of the demand function, the horizontal line $p = p_0$, and the vertical line $x = 0$ the **consumer surplus,** as shown in Figure 11.20. Consumer surplus is the difference between the amount consumers would be willing to pay and the actual amount paid for a product. The area of the region bounded by the graph of the supply function, the horizontal line $p = p_0$, and the vertical line $x = 0$ is called the **producer surplus,** as shown in Figure 11.20. Producer surplus is the difference between the amount a producer receives for selling a product and the minimum price needed to get the producer to supply the product.

Example 5   Finding Surpluses

The demand and supply functions for a product are modeled by

Demand: $p = -0.36x + 9$ and   *Supply:* $p = 0.14x + 2$

where p is the price (in dollars) and x is the number of units (in millions). Find the consumer and producer surpluses for this product.

SOLUTION   By equating the demand and supply functions, you can determine that the point of equilibrium occurs when $x = 14$ (million) and the price is $3.96 per unit.

$$\text{Consumer surplus} = \int_0^{14} (\text{demand function} - \text{price}) \, dx$$

$$= \int_0^{14} [(-0.36x + 9) - 3.96] \, dx$$

$$= \left[-0.18x^2 + 5.04x \right]_0^{14}$$

$$= 35.28$$

The consumer surplus is $35.28.

$$\text{Producer surplus} = \int_0^{14} (\text{price} - \text{supply function}) \, dx$$

$$= \int_0^{14} [3.96 - (0.14x + 2)] \, dx$$

$$= \left[-0.07x^2 + 1.96x \right]_0^{14}$$

$$= 13.72$$

The producer surplus is $13.72. The consumer surplus and producer surplus are shown in Figure 11.21.

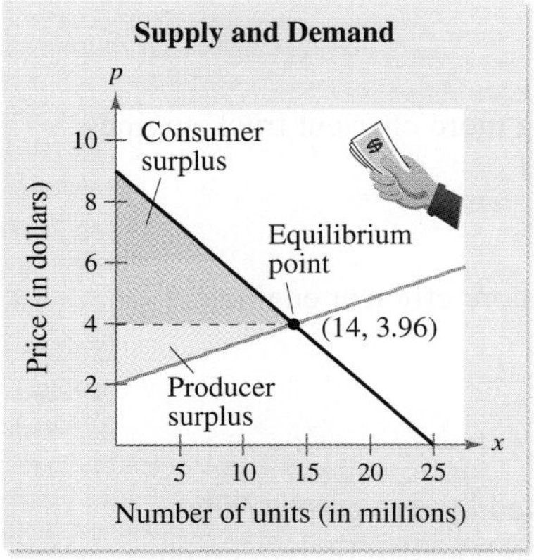

Supply and Demand

Consumer surplus

Equilibrium point

(14, 3.96)

Producer surplus

Price (in dollars)

Number of units (in millions)

FIGURE 11.21

✓Checkpoint 5

The demand and supply functions for a product are modeled by

Demand: $p = -0.2x + 8$ and   *Supply:* $p = 0.1x + 2$

where p is the price (in dollars) and x is the number of units (in millions). Find the consumer and producer surpluses for this product.

Application

In addition to consumer and producer surpluses, there are many other types of applications involving the area of a region bounded by two graphs. Example 6 shows one of these applications.

 Example 6   **Petroleum Consumption**

In the *Annual Energy Outlook*, the U.S. Energy Information Administration projected the consumption C (in quadrillions of Btu per year) of petroleum to follow the model

$$C_1 = 0.00078t^3 - 0.0445t^2 + 0.917t + 35.49, \quad 15 \le t \le 35$$

where $t = 15$ corresponds to 2015. Determine the amount of petroleum that will be saved when the actual consumption follows the model

$$C_2 = 0.0067t^2 - 0.211t + 40.95, \quad 15 \le t \le 35.$$

SOLUTION   The petroleum saved can be represented as the area of the region between the graphs of C_1 and C_2, as shown in Figure 11.22.

$$\text{Petroleum saved} = \int_{15}^{35} (C_1 - C_2)\, dt$$

$$= \int_{15}^{35} (0.00078t^3 - 0.0512t^2 + 1.128t - 5.46)\, dt$$

$$= \left[\frac{0.00078}{4}t^4 - \frac{0.0512}{3}t^3 + \frac{1.128}{2}t^2 - 5.46t \right]_{15}^{35}$$

$$\approx 63.42$$

So, about 63.42 quadrillion Btu of petroleum would be saved. ▬▬▬▬▬▬

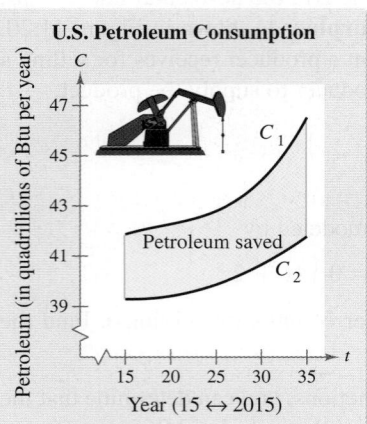

FIGURE 11.22

U.S. Petroleum Consumption

Petroleum (in quadrillions of Btu per year)

C

47

45

43

41

39

Petroleum saved

C_1

C_2

15 20 25 30 35 t

Year (15 ↔ 2015)

✓ **Checkpoint 6**

The projected fuel cost C (in millions of dollars per year) for a trucking company from 2012 through 2024 is

$$C_1 = 2.21t + 5.6, \quad 12 \le t \le 24$$

where $t = 12$ corresponds to 2012. After purchasing more efficient truck engines, the company expects fuels costs to follow the model

$$C_2 = 2.04t + 4.7, \quad 12 \le t \le 24.$$

How much money will the company save with the more efficient engines?

SUMMARIZE   (Section 11.5)

1. State the definition of the area of a region bounded by two graphs (*page 843*). For examples of finding the area of a region bounded by two graphs, see Examples 1, 2, 3, and 4.

2. Describe a real-life example of how finding the area of a region bounded by two graphs can be used to find the consumer and producer surpluses for a product (*page 847, Example 5*).

3. Describe a real-life example of how finding the area of a region bounded by two graphs can be used to analyze petroleum consumption (*page 848, Example 6*).

In Exercises 1–4, simplify the expression.

1. $(-x^2 + 4x + 3) - (x + 1)$

2. $(-2x^2 + 3x + 9) - (-x + 5)$

3. $(-x^3 + 3x^2 - 1) - (x^2 - 4x + 4)$

4. $(3x + 1) - (-x^3 + 9x + 2)$

In Exercises 5–8, find the points of intersection of the graphs.

5. $f(x) = x^2 - 4x + 4$, $g(x) = 4$

6. $f(x) = -3x^2$, $g(x) = 6 - 9x$

7. $f(x) = x^2$, $g(x) = -x + 6$

8. $f(x) = \frac{1}{2}x^3$, $g(x) = 2x$

Exercises 11.5

Finding the Area Bounded by Two Graphs In Exercises 1–8, find the area of the region. *See Examples 1, 2, 3, and 4.*

1. $f(x) = x^2 - 6x$
$g(x) = 0$

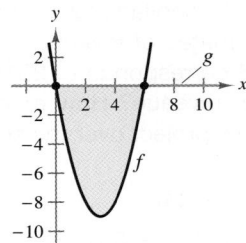

2. $f(x) = x^2 + 2x + 1$
$g(x) = 2x + 5$

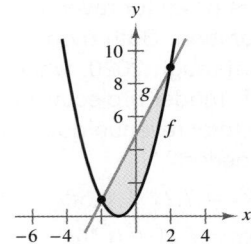

3. $f(x) = x^2 - 4x + 3$
$g(x) = -x^2 + 2x + 3$

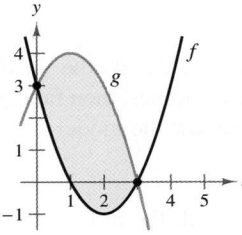

4. $f(x) = x^2$
$g(x) = x^3$

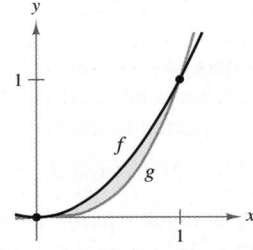

5. $f(x) = e^x - 1$
$g(x) = 0$

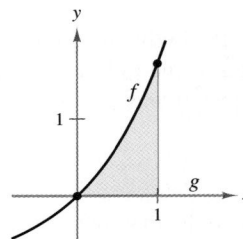

6. $f(x) = -x + 3$
$g(x) = \frac{2}{x}$

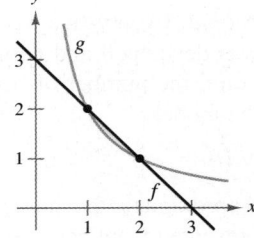

7. $f(x) = 3(x^3 - x)$
$g(x) = 0$

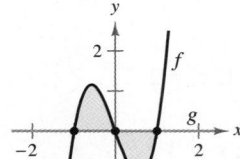

8. $f(x) = (x - 1)^3$
$g(x) = x - 1$

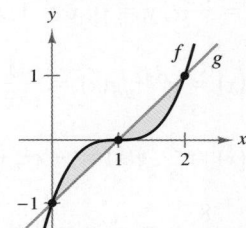

Finding the Region In Exercises 9–12, the integrand of the definite integral is a difference of two functions. Sketch the graph of each function and shade the region whose area is represented by the integral.

9. $\displaystyle\int_0^4 \left[(x + 1) - \tfrac{1}{2}x\right] dx$

10. $\displaystyle\int_{-1}^1 \left[(1 - x^2) - (x^2 - 1)\right] dx$

11. $\displaystyle\int_{-2}^2 \left[2x^2 - (x^4 - 2x^2)\right] dx$

12. $\displaystyle\int_{-4}^0 \left[(x - 6) - (x^2 + 5x - 6)\right] dx$

Think About It In Exercises 13 and 14, determine which value best approximates the area of the region bounded by the graphs of f and g. Make your selection based on a sketch of the region and not by performing any calculations.

13. $f(x) = x + 1$, $g(x) = (x - 1)^2$

(a) -2 (b) 2 (c) 10 (d) 4 (e) 8

14. $f(x) = 2 - \dfrac{1}{2}x$, $g(x) = 2 - \sqrt{x}$

(a) 1 (b) 6 (c) -3 (d) 3 (e) 4

Finding the Area Bounded by Two Graphs In Exercises 15–30, sketch the region bounded by the graphs of the functions and find the area of the region. *See Examples 1, 2, 3, and 4.*

15. $y = \dfrac{1}{x^2}, y = 0, x = 1, x = 5$

16. $y = 2x - 3, y = x^2 + 2x + 1, x = -2, x = 1$

17. $y = x^2 - 4x + 3, y = 3 + 4x - x^2$

18. $y = 4 - x^2, y = x^2$

19. $y = x^2 - 1, y = -x + 2, x = 0, x = 1$

20. $y = -x^3 + 3, y = x, x = -1, x = 1$

21. $f(x) = \sqrt[3]{x}, g(x) = x$

22. $f(x) = \sqrt{3x} + 1, g(x) = x + 1$

23. $f(x) = x^3 + 4x^2, g(x) = x + 4$

24. $f(x) = 1 - x, g(x) = x^4 - x$

25. $y = xe^{-x^2}, y = 0, x = 0, x = 1$

26. $y = \dfrac{e^{1/x}}{x^2}, y = 0, x = 1, x = 3$

27. $f(x) = e^{0.5x}, g(x) = -\dfrac{1}{x}, x = 1, x = 2$

28. $f(x) = \dfrac{1}{x}, g(x) = -e^x, x = \dfrac{1}{2}, x = 1$

29. $y = \dfrac{8}{x}, y = x^2, y = 0, x = 1, x = 4$

30. $y = x^2 - 2x + 1, y = x^2 - 10x + 25, y = 0$

Writing Integrals In Exercises 31–34, use a graphing utility to graph the region bounded by the graphs of the functions. Write the definite integral that represents the area of the region. (*Hint:* Multiple integrals may be necessary.)

31. $f(x) = 2x, g(x) = 4 - 2x, h(x) = 0$

32. $f(x) = x(x^2 - 3x + 3), g(x) = x^2$

33. $y = \dfrac{4}{x}, y = x, x = 1, x = 4$

34. $y = x^3 - 4x^2 + 1, y = x - 3$

Finding Area In Exercises 35–38, use a graphing utility to graph the region bounded by the graphs of the functions. Find the area of the region by hand.

35. $f(x) = x^2 - 4x, g(x) = 0$

36. $f(x) = 3 - 2x - x^2, g(x) = 0$

37. $f(x) = x^2 + 2x + 1, g(x) = x + 1$

38. $f(x) = -x^2 + 4x + 2, g(x) = x + 2$

Area of a Region In Exercises 39 and 40, use integration to find the area of the triangular region having the given vertices.

39. $(0, 0), (4, 0), (4, 4)$

40. $(0, 0), (4, 0), (6, 4)$

Consumer and Producer Surpluses In Exercises 41–46, find the consumer and producer surpluses by using the demand and supply functions, where p is the price (in dollars) and x is the number of units (in millions). *See Example 5.*

	Demand Function	Supply Function
41.	$p = 50 - 0.5x$	$p = 0.125x$
42.	$p = 300 - x$	$p = 100 + x$
43.	$p = 200 - 0.4x$	$p = 100 + 1.6x$
44.	$p = 975 - 23x$	$p = 42x$
45.	$p = 42 - 0.015x^2$	$p = 0.01x^2 + 2$
46.	$p = 62 - 0.3x$	$p = 0.002x^2 + 12$

Revenue In Exercises 47 and 48, two models, R_1 and R_2, are given for revenue (in billions of dollars) for a large corporation. Both models are estimates of revenues for 2015 through 2020, where $t = 15$ corresponds to 2015. Which model projects the greater revenue? How much more total revenue does that model project over the six-year period?

47. $R_1 = 7.21 + 0.58t, R_2 = 7.21 + 0.45t$

48. $R_1 = 7.21 + 0.26t + 0.02t^2, R_2 = 7.21 + 0.1t + 0.01t^2$

49. Fuel Cost The projected fuel cost C (in millions of dollars) for an airline from 2015 through 2025 is

$$C_1 = 568.5 + 7.15t$$

where $t = 15$ corresponds to 2015. If the airline purchases more efficient airplane engines, then fuel cost is expected to decrease and to follow the model

$$C_2 = 525.6 + 6.43t.$$

How much can the airline save with the more efficient engines? Explain your reasoning.

50. Health An epidemic was spreading such that t weeks after its outbreak it had infected

$$N_1(t) = 0.1t^2 + 0.5t + 150, \quad 0 \le t \le 50$$

people. Twenty-five weeks after the outbreak, a vaccine was developed and administered to the public. At that point, the number of people infected was governed by the model

$$N_2(t) = -0.2t^2 + 6t + 200.$$

Approximate the number of people that the vaccine prevented from becoming ill during the epidemic.

51. Consumer Trends For the years 1998 through 2008, the per capita consumption C of all fruit (in pounds) in the United States can be modeled by

$$C(t) = \begin{cases} -0.443t^2 + 5.02t + 277.7, & 8 \le t \le 12 \\ -0.775t^2 + 18.73t + 170.5, & 12 < t \le 18 \end{cases}$$

where t is the year, with $t = 8$ corresponding to 1998. *(Source: U.S. Department of Agriculture)*

(a) Use a graphing utility to graph this model.

(b) Suppose the fruit consumption from 2003 through 2008 had continued to follow the model for 1998 through 2002. How many more or fewer pounds of fruit would have been consumed from 2003 through 2008?

52. HOW DO YOU SEE IT? A state legislature is debating two proposals for eliminating the annual budget deficits by the year 2020. The rate of decrease of the deficits for each proposal is shown in the figure.

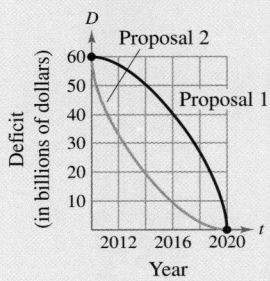

(a) What does the area between the two graphs represent?

(b) From the viewpoint of minimizing the cumulative state deficit, which is the better proposal? Explain your reasoning.

53. Cost, Revenue, and Profit The revenue from a manufacturing process (in millions of dollars) is projected to follow the model $R = 100$ for 10 years. Over the same period of time, the cost (in millions of dollars) is projected to follow the model $C = 60 + 0.2t^2$, where t is the time (in years). Approximate the profit over the 10-year period.

54. Cost, Revenue, and Profit Repeat Exercise 53 for revenue and cost models given by $R = 100 + 0.08t$ and $C = 60 + 0.2t^2$. Did the profit increase or decrease? Explain why.

55. Consumer and Producer Surpluses Factory orders for an air conditioner are about 6000 units per week when the price is $331 and about 8000 units per week when the price is $303. The supply function is given by $p = 0.0275x$. Find the consumer and producer surpluses. (Assume the demand function is linear.)

56. Consumer and Producer Surpluses Repeat Exercise 55 with a demand of about 6000 units per week when the price is $325 and about 8000 units per week when the price is $300. Find the consumer and producer surpluses. (Assume the demand function is linear.)

57. Lorenz Curve Economists use *Lorenz curves* to illustrate the distribution of income in a country. Letting x represent the percent of families in a country and y the percent of total income, the model $y = x$ would represent a country in which each family had the same income. The Lorenz curve, $y = f(x)$, represents the actual income distribution. The area between these two models, for $0 \le x \le 100$, indicates the "income inequality" of a country. In 2009, the Lorenz curve for the United States could be modeled by

$$y = (0.00061x^2 + 0.0224x + 1.666)^2, \quad 0 \le x \le 100$$

where x is measured from the poorest to the wealthiest families. Find the income inequality for the United States in 2009.   *(Source: U.S. Census Bureau)*

58. Income Distribution Using the Lorenz curve in Exercise 57 and a spreadsheet, complete the table, which lists the percent of total income earned by each quintile in the United States in 2009.

Quintile	Lowest	2nd	3rd	4th	Highest
Percent					

59. Project: Social Security For a project analyzing the receipts and expenditures for the Old-Age and Survivors Insurance Trust Fund (Social Security Trust Fund) from 1990 through 2009, visit this text's website at *www.cengagebrain.com.* *(Source: Social Security Administration)*

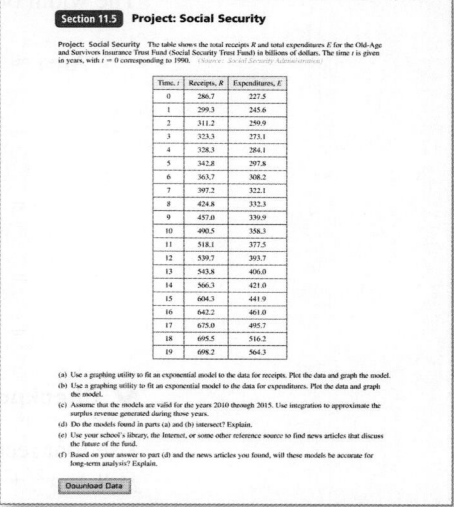

11.6 The Definite Integral as the Limit of a Sum

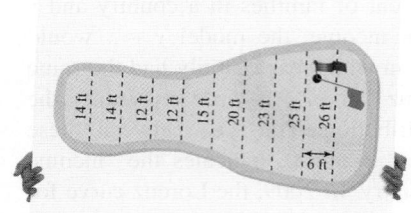

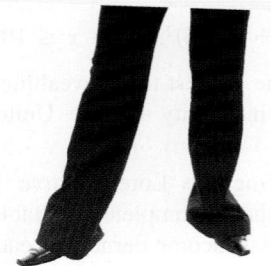

In Exercise 28 on page 857, you will use the Midpoint Rule to estimate the surface area of a golf green.

■ Use the Midpoint Rule to approximate definite integrals.
■ Understand the definite integral as the limit of a sum.

The Midpoint Rule

In Section 11.4, you learned that you cannot use the Fundamental Theorem of Calculus to evaluate a definite integral unless you can find an antiderivative of the integrand. When you cannot find an antiderivative of an integrand, you can use an approximation technique. One such technique, the **Midpoint Rule,** is demonstrated in Example 1.

Example 1 Approximating the Area of a Plane Region

Use the five rectangles in Figure 11.23 to approximate the area of the region bounded by the graph of $f(x) = -x^2 + 5$, the x-axis, and the lines $x = 0$ and $x = 2$.

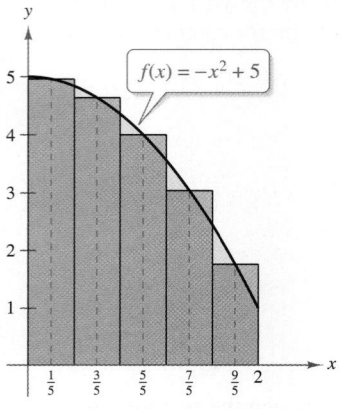

FIGURE 11.23

SOLUTION   You can find the heights of the five rectangles by evaluating f at the midpoint of each of the following intervals.

$$\left[0, \frac{2}{5}\right], \quad \left[\frac{2}{5}, \frac{4}{5}\right], \quad \left[\frac{4}{5}, \frac{6}{5}\right], \quad \left[\frac{6}{5}, \frac{8}{5}\right], \quad \left[\frac{8}{5}, \frac{10}{5}\right]$$

The width of each rectangle is $\frac{2}{5}$. So, the sum of the five areas is

$$\text{Area} \approx \frac{2}{5}f\left(\frac{1}{5}\right) + \frac{2}{5}f\left(\frac{3}{5}\right) + \frac{2}{5}f\left(\frac{5}{5}\right) + \frac{2}{5}f\left(\frac{7}{5}\right) + \frac{2}{5}f\left(\frac{9}{5}\right)$$

$$= \frac{2}{5}\left[f\left(\frac{1}{5}\right) + f\left(\frac{3}{5}\right) + f\left(\frac{5}{5}\right) + f\left(\frac{7}{5}\right) + f\left(\frac{9}{5}\right)\right]$$

$$= \frac{2}{5}\left(\frac{124}{25} + \frac{116}{25} + \frac{100}{25} + \frac{76}{25} + \frac{44}{25}\right)$$

$$= \frac{920}{125}$$

$$= 7.36.$$

✓ Checkpoint 1

Use four rectangles to approximate the area of the region bounded by the graph of $f(x) = x^2 + 1$, the x-axis, $x = 0$, and $x = 2$.

For the region in Example 1, you can find the exact area with a definite integral. That is,

$$\text{Area} = \int_0^2 (-x^2 + 5)\,dx = \frac{22}{3} \approx 7.33.$$

The approximation procedure used in Example 1 is the **Midpoint Rule.** You can use the Midpoint Rule to approximate *any* definite integral—not just those representing areas. The basic steps are summarized below.

Guidelines for Using the Midpoint Rule

To approximate the definite integral $\int_a^b f(x)\,dx$ with the Midpoint Rule, use the steps below.

1. Divide the interval $[a, b]$ into n subintervals, each of width

$$\Delta x = \frac{b - a}{n}.$$

2. Find the midpoint of each subinterval.

$$\text{Midpoints} = \{x_1, x_2, x_3, \ldots, x_n\}$$

3. Evaluate f at each midpoint and form the sum as shown.

$$\int_a^b f(x)\,dx \approx \frac{b - a}{n}[f(x_1) + f(x_2) + f(x_3) + \cdots + f(x_n)]$$

An important characteristic of the Midpoint Rule is that the approximation tends to improve as n increases. The table below shows the approximations for the area of the region described in Example 1 for various values of n. For example, when $n = 10$, the Midpoint Rule yields

$$\int_0^2 (-x^2 + 5)\,dx \approx \frac{2}{10}\left[f\left(\frac{1}{10}\right) + f\left(\frac{3}{10}\right) + \cdots + f\left(\frac{19}{10}\right)\right]$$
$$= 7.34.$$

n	5	10	15	20	25	30
Approximation	7.3600	7.3400	7.3363	7.3350	7.3344	7.3341

Note that as n increases, the approximation gets closer and closer to the exact value of the integral, which was found to be

$$\frac{22}{3} \approx 7.3333.$$

STUDY TIP

In Example 1, the Midpoint Rule is used to approximate an integral whose exact value can be found with the Fundamental Theorem of Calculus. This was done to illustrate the accuracy of the rule. In practice, of course, you would use the Midpoint Rule to approximate the values of definite integrals for which you cannot find an antiderivative. Examples 2 and 3 illustrate such integrals.

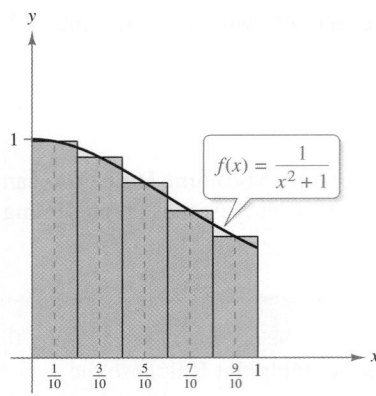

FIGURE 11.24

Example 2   Using the Midpoint Rule

Use the Midpoint Rule with $n = 5$ to approximate the area of the region bounded by the graph of

$$f(x) = \frac{1}{x^2 + 1}$$

the x-axis, and the lines $x = 0$ and $x = 1$.

SOLUTION   The region is shown in Figure 11.24. With $n = 5$, the interval $[0, 1]$ is divided into five subintervals.

$$\left[0, \frac{1}{5}\right], \quad \left[\frac{1}{5}, \frac{2}{5}\right], \quad \left[\frac{2}{5}, \frac{3}{5}\right], \quad \left[\frac{3}{5}, \frac{4}{5}\right], \quad \left[\frac{4}{5}, 1\right]$$

The midpoints of these intervals are $\frac{1}{10}, \frac{3}{10}, \frac{5}{10}, \frac{7}{10}$, and $\frac{9}{10}$. Because each subinterval has a width of $\Delta x = (1 - 0)/5 = \frac{1}{5}$, you can approximate the value of the definite integral as shown.

$$\int_0^1 \frac{1}{x^2 + 1} \, dx \approx \frac{1}{5}\left(\frac{1}{1.01} + \frac{1}{1.09} + \frac{1}{1.25} + \frac{1}{1.49} + \frac{1}{1.81}\right)$$

$$\approx 0.786$$

The actual area of this region is $\pi/4 \approx 0.785$. So, the approximation is off by about 0.001.

✔**Checkpoint 2**

Use the Midpoint Rule with $n = 4$ to approximate the area of the region bounded by the graph of $f(x) = 1/(x^2 + 2)$, the x-axis, and the lines $x = 0$ and $x = 1$.   ■

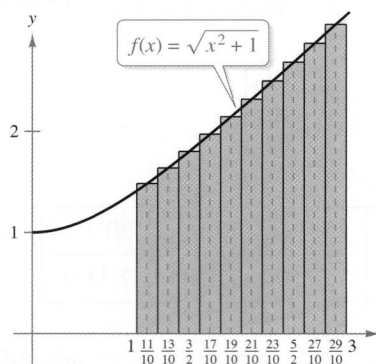

FIGURE 11.25

Example 3   Using the Midpoint Rule

Use the Midpoint Rule with $n = 10$ to approximate the area of the region bounded by the graph of $f(x) = \sqrt{x^2 + 1}$, the x-axis, and the lines $x = 1$ and $x = 3$.

SOLUTION   The region is shown in Figure 11.25. After dividing the interval $[1, 3]$ into 10 subintervals, you can determine that the midpoints of these intervals are

$$\frac{11}{10}, \quad \frac{13}{10}, \quad \frac{3}{2}, \quad \frac{17}{10}, \quad \frac{19}{10}, \quad \frac{21}{10}, \quad \frac{23}{10}, \quad \frac{5}{2}, \quad \frac{27}{10}, \quad \text{and} \quad \frac{29}{10}.$$

Because each subinterval has a width of $\Delta x = (3 - 1)/10 = \frac{1}{5}$, you can approximate the value of the definite integral as shown.

$$\int_1^3 \sqrt{x^2 + 1} \, dx \approx \frac{1}{5}\left[\sqrt{(1.1)^2 + 1} + \sqrt{(1.3)^2 + 1} + \cdots + \sqrt{(2.9)^2 + 1}\right]$$

$$\approx 4.504$$

It can be shown that the actual area is

$$\frac{1}{2}\left[3\sqrt{10} + \ln\left(3 + \sqrt{10}\right) - \sqrt{2} - \ln\left(1 + \sqrt{2}\right)\right] \approx 4.505.$$

So, the approximation is off by about 0.001.

✔**Checkpoint 3**

Use the Midpoint Rule with $n = 4$ to approximate the area of the region bounded by the graph of $f(x) = \sqrt{x^2 - 1}$, the x-axis, and the lines $x = 2$ and $x = 4$.   ■

STUDY TIP

The Midpoint Rule is necessary for solving certain real-life problems, such as measuring irregular areas like bodies of water (see Exercise 27).

The Definite Integral as the Limit of a Sum

Consider the closed interval $[a, b]$, divided into n subintervals whose midpoints are x_i and whose widths are $\Delta x = (b - a)/n$. In this section, you have seen that the midpoint approximation

$$\int_a^b f(x)\, dx \approx f(x_1)\, \Delta x + f(x_2)\, \Delta x + f(x_3)\, \Delta x + \cdots + f(x_n)\, \Delta x$$

$$= [f(x_1) + f(x_2) + f(x_3) + \cdots + f(x_n)]\Delta x$$

becomes better and better as n increases. In fact, the limit of this sum as n approaches infinity is exactly equal to the definite integral. That is,

$$\int_a^b f(x)\, dx = \lim_{n\to\infty} [f(x_1) + f(x_2) + f(x_3) + \cdots + f(x_n)]\Delta x.$$

It can be shown that this limit is valid as long as x_i is *any* point in the *i*th interval.

Example 4   **Approximating a Definite Integral**

Use the Midpoint Rule program in Appendix H or a symbolic integration utility to approximate the definite integral

$$\int_0^1 e^{-x^2}\, dx.$$

SOLUTION   Using the Midpoint Rule program (see Figure 11.26), you can complete the following table.

n	10	20	30	40	50
Approximation	0.7471	0.7469	0.7469	0.7468	0.7468

From the table, it appears that

$$\int_0^1 e^{-x^2}\, dx \approx 0.7468.$$

Using a symbolic integration utility, the value of the integral is approximately 0.7468241328.

```
UPPER LIMIT
?1
N DIVISIONS
?50
APPROXIMATION
           .7468363957
                   Done
```

FIGURE 11.26

✓ **Checkpoint 4**

Use the Midpoint Rule program in Appendix H or a symbolic integration utility to approximate the definite integral

$$\int_0^1 e^{x^2}\, dx.$$

SUMMARIZE   (Section 11.6)

1. Describe how to approximate the area of a region using rectangles (*page 852, Example 1*).

2. State the guidelines for using the Midpoint Rule (*page 853*). For examples of using these guidelines, see Examples 2 and 3.

3. State the definite integral as the limit of a sum (*page 855*).

SKILLS WARM UP 11.6

The following warm-up exercises involve skills that were covered in a previous course or in earlier sections. You will use these skills in the exercise set for this section. For additional help, review Sections 2.1 and 9.3.

In Exercises 1–6, find the midpoint of the interval.

1. $\left[0, \frac{1}{3}\right]$

2. $\left[\frac{1}{10}, \frac{2}{10}\right]$

3. $\left[\frac{3}{20}, \frac{4}{20}\right]$

4. $\left[1, \frac{7}{6}\right]$

5. $\left[2, \frac{31}{15}\right]$

6. $\left[\frac{26}{9}, 3\right]$

In Exercises 7–10, find the limit.

7. $\lim\limits_{x \to \infty} \dfrac{2x^2 + 4x - 1}{3x^2 - 2x}$

8. $\lim\limits_{x \to \infty} \dfrac{4x + 5}{7x - 5}$

9. $\lim\limits_{x \to \infty} \dfrac{x - 7}{x^2 + 1}$

10. $\lim\limits_{x \to \infty} \dfrac{5x^3 + 1}{x^3 + x^2 + 4}$

Exercises 11.6

See www.CalcChat.com for worked-out solutions to odd-numbered exercises.

Approximating the Area of a Plane Region In Exercises 1–6, use the rectangles to approximate the area of the region. Compare your result with the exact area obtained using a definite integral. *See Example 1.*

1. $f(x) = -2x + 3, [0, 1]$ **2.** $f(x) = \dfrac{1}{x}, [1, 5]$

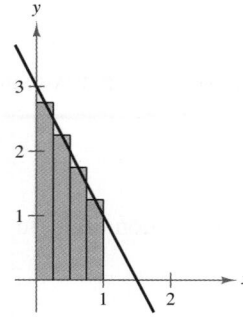

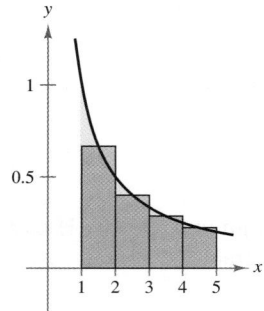

3. $f(x) = \sqrt{x}, [0, 1]$ **4.** $f(x) = 1 - x^2, [-1, 1]$

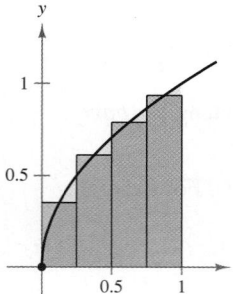

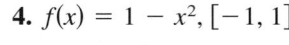

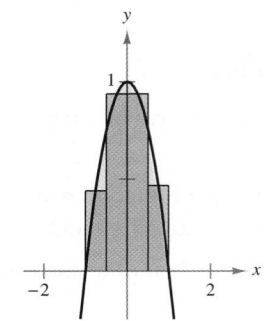

5. $f(x) = x^3 + 1, [0, 1]$ **6.** $f(x) = e^{-x/2}, [0, 3]$

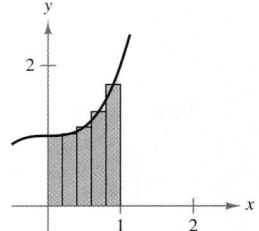

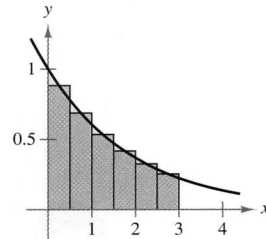

Using the Midpoint Rule In Exercises 7–12, use the Midpoint Rule with $n = 5$ to approximate the area of the region bounded by the graph of f and the x-axis over the interval. Sketch the region. *See Examples 2 and 3.*

Function	Interval
7. $f(x) = x^2$	$[1, 6]$
8. $f(x) = 4 - x^2$	$[-1, 0]$
9. $f(x) = x\sqrt{x + 4}$	$[0, 1]$
10. $f(x) = (x^2 + 1)^{2/3}$	$[0, 5]$
11. $f(x) = \dfrac{8}{x^2 + 1}$	$[-5, 5]$
12. $f(x) = \dfrac{5x}{x + 1}$	$[0, 2]$

Using the Midpoint Rule   In Exercises 13–18, use the Midpoint Rule with $n = 4$ to approximate the area of the region bounded by the graph of f and the x-axis over the interval. Sketch the region. *See Examples 2 and 3.*

Function	Interval
13. $f(x) = 2x^2$	$[1, 3]$
14. $f(x) = x^2 - x^3$	$[0, 1]$
15. $f(x) = (x^2 - 4)^2$	$[-2, 2]$
16. $f(x) = \sqrt{x^2 + 3}$	$[-1, 1]$
17. $f(x) = \dfrac{1}{(x^2 + 1)^2}$	$[-1, 3]$
18. $f(x) = \dfrac{2}{x^2 + 1}$	$[0, 2]$

 Approximating a Definite Integral   In Exercises 19–22, use the Midpoint Rule program in Appendix H or a symbolic integration utility to approximate the definite integral. If you use a Midpoint Rule program, complete the table. *See Example 4.*

n	10	20	30	40	50
Approximation					

19. $\displaystyle\int_0^4 \sqrt{2 + 3x^2}\, dx$

20. $\displaystyle\int_0^2 \frac{5}{x^3 + 1}\, dx$

21. $\displaystyle\int_1^3 x\sqrt[3]{x + 1}\, dx$

22. $\displaystyle\int_1^6 \frac{4}{\sqrt{1 + x^2}}\, dx$

 Making a Closer Approximation   In Exercises 23–26, use the Midpoint Rule program in Appendix H to approximate the definite integral. How large must n be to obtain an approximation that is correct to within 0.01?

23. $\displaystyle\int_0^4 (2x^2 + 3)\, dx$

24. $\displaystyle\int_1^2 (x^3 - 1)\, dx$

25. $\displaystyle\int_1^4 \frac{1}{x + 1}\, dx$

26. $\displaystyle\int_1^2 \sqrt{x + 2}\, dx$

27. Surface Area   Use the Midpoint Rule to estimate the surface area of the pond shown in the figure.

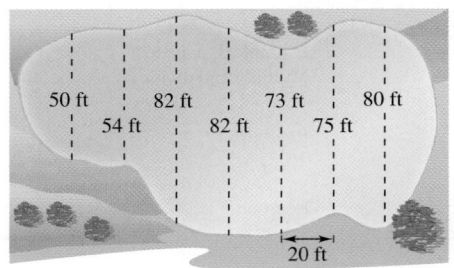

28. Surface Area   Use the Midpoint Rule to estimate the surface area of the golf green shown in the figure.

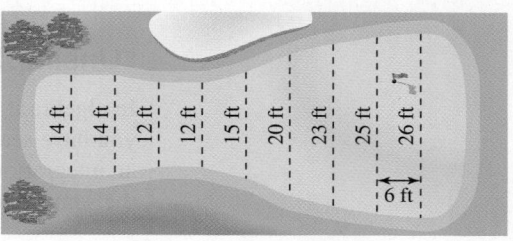

29. Surface Area   Use the Midpoint Rule to estimate the surface area of the oil spill shown in the figure.

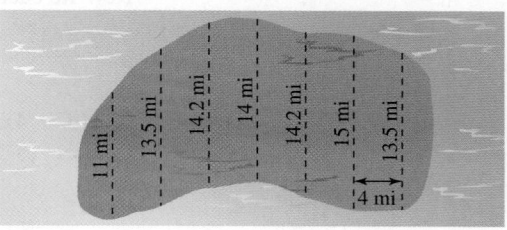

30. HOW DO YOU SEE IT?   The graph shows three areas representing awake time, REM (rapid eye movement) sleep time, and non-REM sleep time, over a typical individual's lifetime. *(Source: Adapted from Bernstein/Clarke-Stewart/Roy/Wickens, Psychology, Seventh Edition)*

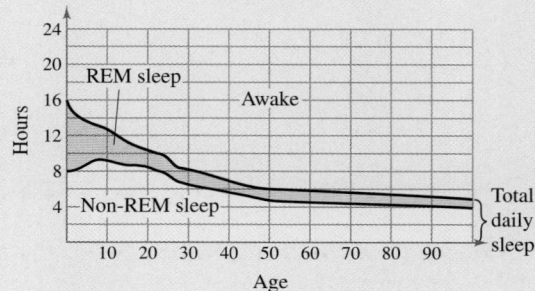

(a) Make generalizations about the amount of total sleep time (REM and non-REM) an individual gets as he or she gets older.

(b) How would you use the Midpoint Rule to estimate the amount of REM sleep time an individual gets between birth and age 10?

31. Numerical Approximation   Use the Midpoint Rule with $n = 4$ to approximate π, where

$$\pi = \int_0^1 \frac{4}{1 + x^2}\, dx.$$

Then use a graphing utility to evaluate the definite integral. Compare your results.

ALGEBRA TUTOR

"Unsimplifying" an Algebraic Expression

In algebra it is often helpful to write an expression in its simplest form. In this chapter, you have seen that the reverse is often true in integration. That is, to fit an integrand to an integration formula, it often helps to "unsimplify" the expression. To do this, you use the same algebraic rules, but your goal is different. Here are some examples.

Example 1   Rewriting Algebraic Expressions

Rewrite each algebraic expression as indicated in the example.

a. $\dfrac{x-1}{\sqrt{x}}$

b. $x(3 - 4x^2)^2$

c. $7x^2\sqrt{x^3 + 1}$

d. $5xe^{-x^2}$

SOLUTION

a. $\dfrac{x-1}{\sqrt{x}} = \dfrac{x}{\sqrt{x}} - \dfrac{1}{\sqrt{x}}$ Example 5, page 807
Rewrite as two fractions.

$\qquad = \dfrac{x^1}{x^{1/2}} - \dfrac{1}{x^{1/2}}$ Rewrite with rational exponents.

$\qquad = x^{1 - 1/2} - x^{-1/2}$ Properties of exponents

$\qquad = x^{1/2} - x^{-1/2}$ Simplify exponent.

b. $x(3 - 4x^2)^2 = \dfrac{-8}{-8}x(3 - 4x^2)^2$ Example 2, page 816
Multiply and divide by -8.

$\qquad = \left(-\dfrac{1}{8}\right)(-8)x(3 - 4x^2)^2$ Regroup.

$\qquad = \left(-\dfrac{1}{8}\right)(3 - 4x^2)^2(-8x)$ Regroup.

c. $7x^2\sqrt{x^3 + 1} = 7x^2(x^3 + 1)^{1/2}$ Example 5, page 817
Rewrite with rational exponent.

$\qquad = \dfrac{3}{3}(7x^2)(x^3 + 1)^{1/2}$ Multiply and divide by 3.

$\qquad = \dfrac{7}{3}(3x^2)(x^3 + 1)^{1/2}$ Regroup.

$\qquad = \dfrac{7}{3}(x^3 + 1)^{1/2}(3x^2)$ Regroup.

d. $5xe^{-x^2} = \dfrac{-2}{-2}(5x)e^{-x^2}$ Example 3, page 824
Multiply and divide by -2.

$\qquad = \left(-\dfrac{5}{2}\right)(-2x)e^{-x^2}$ Regroup.

$\qquad = \left(-\dfrac{5}{2}\right)e^{-x^2}(-2x)$ Regroup.

Example 2   Rewriting Algebraic Expressions

Rewrite each algebraic expression.

a. $\dfrac{3x^2 + 2x - 1}{x^2}$

b. $\dfrac{1}{1 + e^{-x}}$

c. $\dfrac{x^2 + x + 1}{x - 1}$

d. $\dfrac{x^2 + 6x + 1}{x^2 + 1}$

SOLUTION

a. $\dfrac{3x^2 + 2x - 1}{x^2} = \dfrac{3x^2}{x^2} + \dfrac{2x}{x^2} - \dfrac{1}{x^2}$ Example 7(a), page 827
Rewrite as separate fractions.

$$= 3 + \dfrac{2}{x} - x^{-2}$$ Properties of exponents

$$= 3 + 2\left(\dfrac{1}{x}\right) - x^{-2}$$ Regroup.

b. $\dfrac{1}{1 + e^{-x}} = \left(\dfrac{e^x}{e^x}\right)\dfrac{1}{1 + e^{-x}}$ Example 7(b), page 827
Multiply and divide by e^x.

$$= \dfrac{e^x}{e^x + e^x(e^{-x})}$$ Multiply.

$$= \dfrac{e^x}{e^x + e^{x-x}}$$ Property of exponents

$$= \dfrac{e^x}{e^x + e^0}$$ Simplify exponent.

$$= \dfrac{e^x}{e^x + 1}$$ $e^0 = 1$

c. $\dfrac{x^2 + x + 1}{x - 1} = x + 2 + \dfrac{3}{x - 1}$ Example 7(c), page 827
Use long division as shown below.

$$
\begin{array}{r}
x + 2 \\
x - 1 \overline{)\, x^2 + x + 1} \\
\underline{x^2 - x } \\
2x + 1 \\
\underline{2x - 2} \\
3
\end{array}
$$

d. $\dfrac{x^2 + 6x + 1}{x^2 + 1} = 1 + \dfrac{6x}{x^2 + 1}$ Bottom of page 826
Use long division as shown below.

$$
\begin{array}{r}
1 \\
x^2 + 1 \overline{)\, x^2 + 6x + 1} \\
\underline{x^2 + 1} \\
6x
\end{array}
$$

SUMMARY AND STUDY STRATEGIES

After studying this chapter, you should have acquired the following skills.
The exercise numbers are keyed to the Review Exercises that begin on page 862.
Answers to odd-numbered Review Exercises are given in the back of the text.*

Section 11.1

Review Exercises

■ Use basic integration rules to find indefinite integrals.   *1–14*

$$\int k \, dx = kx + C$$

$$\int kf(x) \, dx = k \int f(x) \, dx$$

$$\int [f(x) + g(x)] \, dx = \int f(x) \, dx + \int g(x) \, dx$$

$$\int [f(x) - g(x)] \, dx = \int f(x) \, dx - \int g(x) \, dx$$

$$\int x^n \, dx = \frac{x^{n+1}}{n+1} + C, \quad n \neq -1$$

■ Use initial conditions to find particular solutions of indefinite integrals.   *15–18*

■ Use antiderivatives to solve real-life problems.   *19, 20*

Section 11.2

■ Use the General Power Rule or integration by substitution to find indefinite integrals.   *21–32*

$$\int u^n \frac{du}{dx} \, dx = \int u^n \, du = \frac{u^{n+1}}{n+1} + C, \quad n \neq -1$$

■ Use the General Power Rule or integration by substitution to solve real-life problems.   *33, 34*

Section 11.3

■ Use the Exponential and Log Rules to find indefinite integrals.   *35–46*

$$\int e^x \, dx = e^x + C \qquad\qquad \int \frac{1}{x} \, dx = \ln|x| + C$$

$$\int e^u \frac{du}{dx} \, dx = \int e^u \, du = e^u + C \qquad \int \frac{du/dx}{u} \, dx = \int \frac{1}{u} \, du = \ln|u| + C$$

Section 11.4

■ Find the areas of regions using a geometric formula.   *47–50*

■ Use properties of definite integrals.   *51, 52*

■ Find the areas of regions bounded by the graph of a function and the x-axis.   *53–58*

---

* A wide range of valuable study aids are available to help you master the material in this chapter.
The *Student Solutions Manual* includes step-by-step solutions to all odd-numbered exercises to
help you review and prepare. The student website at *www.cengagebrain.com* offers algebra help
and a *Graphing Technology Guide,* which contains step-by-step commands and instructions for a
wide variety of graphing calculators.

Section 11.4 (continued)   Review Exercises

- Use the Fundamental Theorem of Calculus to evaluate definite integrals.   *59–70*

$$\int_a^b f(x)\, dx = F(x)\Big]_a^b = F(b) - F(a), \quad \text{where} \quad F'(x) = f(x)$$

- Find average values of functions over closed intervals.   *71–76*

$$\text{Average value} = \frac{1}{b-a}\int_a^b f(x)\, dx$$

- Use properties of even and odd functions to help evaluate definite integrals.   *77–80*

  Even function: $f(-x) = f(x)$

  If f is an *even* function, then $\displaystyle\int_{-a}^a f(x)\, dx = 2\int_0^a f(x)\, dx.$

  Odd function: $f(-x) = -f(x)$

  If f is an *odd* function, then $\displaystyle\int_{-a}^a f(x)\, dx = 0.$

- Find amounts of annuities.   *81, 82*
- Use definite integrals to solve marginal analysis problems.   *83, 84*
- Use average values to solve real-life problems.   *85, 86*

Section 11.5

- Find areas of regions bounded by two graphs.   *87–94*

$$A = \int_a^b [f(x) - g(x)]\, dx$$

- Find consumer and producer surpluses.   *95–98*
- Use the areas of regions bounded by two graphs to solve real-life problems.   *99–102*

Section 11.6

- Use the Midpoint Rule to approximate values of definite integrals.   *103–112*

$$\int_a^b f(x)\, dx \approx \frac{b-a}{n}[f(x_1) + f(x_2) + f(x_3) + \cdots + f(x_n)]$$

- Use the Midpoint Rule to solve real-life problems.   *113*

Study Strategies

- **Indefinite and Definite Integrals**   When evaluating integrals, remember that an indefinite integral is a *family of antiderivatives*, each differing by a constant C, whereas a definite integral is a *number*.

- **Checking Antiderivatives by Differentiating**   When finding an antiderivative, remember that you can check your result by differentiating. For example, you can confirm that the antiderivative

$$\int (3x^3 - 4x)\, dx = \frac{3}{4}x^4 - 2x^2 + C \quad \text{is correct by differentiating to obtain} \quad \frac{d}{dx}\left[\frac{3}{4}x^4 - 2x^2 + C\right] = 3x^3 - 4x.$$

  Because the derivative is equal to the original integrand, you know that the antiderivative is correct.

- **Grouping Symbols and the Fundamental Theorem**   When using the Fundamental Theorem of Calculus to evaluate a definite integral, you can avoid sign errors by using grouping symbols. Here is an example.

$$\int_1^3 (x^3 - 9x)\, dx = \left[\frac{x^4}{4} - \frac{9x^2}{2}\right]_1^3 = \left[\frac{3^4}{4} - \frac{9(3^2)}{2}\right] - \left[\frac{1^4}{4} - \frac{9(1^2)}{2}\right] = \frac{81}{4} - \frac{81}{2} - \frac{1}{4} + \frac{9}{2} = -16$$

Review Exercises

See www.CalcChat.com for worked-out solutions to odd-numbered exercises.

Finding Indefinite Integrals In Exercises 1–14, find the indefinite integral. Check your result by differentiation.

1. $\displaystyle\int 16\,dx$

2. $\displaystyle\int -9\,dx$

3. $\displaystyle\int \frac{3}{5}x\,dx$

4. $\displaystyle\int 6x\,dx$

5. $\displaystyle\int 3x^2\,dx$

6. $\displaystyle\int 8x^3\,dx$

7. $\displaystyle\int (2x^2 + 5x)\,dx$

8. $\displaystyle\int (5 - 6x^2)\,dx$

9. $\displaystyle\int \frac{2}{3\sqrt[3]{x}}\,dx$

10. $\displaystyle\int 6x^{5/2}\,dx$

11. $\displaystyle\int \left(\sqrt[3]{x^4} + 3x\right)dx$

12. $\displaystyle\int \left(\frac{4}{\sqrt{x}} + \sqrt{x}\right)dx$

13. $\displaystyle\int \frac{2x^4 - 1}{\sqrt{x}}\,dx$

14. $\displaystyle\int \frac{1 - 3x}{x^2}\,dx$

Finding Particular Solutions In Exercises 15–18, find the particular solution that satisfies the differential equation and the initial condition.

15. $f'(x) = 12x;\ f(0) = -3$

16. $f'(x) = 3x + 1;\ f(2) = 6$

17. $f'(x) = 3x^2 - 8x;\ f(1) = 12$

18. $f'(x) = \sqrt{x};\ f(9) = 4$

19. Vertical Motion An object is projected upward from the ground with an initial velocity of 80 feet per second. Express the height s (in feet) of the object as a function of the time t (in seconds). How long will the object be in the air? (Use $s''(t) = -32$ feet per second per second as the acceleration due to gravity.)

20. Revenue A company produces a new product for which the rate of change of the revenue can be modeled by

$$\frac{dR}{dt} = 0.675t^{3/2}, \quad 0 \le t \le 225$$

where t is the time (in weeks). When $t = 0$, $R = 0$.

(a) Find a model for the revenue function.

(b) What is the revenue after 20 weeks?

(c) When will the weekly revenue be $27,000?

Applying the General Power Rule In Exercises 21–32, find the indefinite integral. Check your result by differentiation.

21. $\displaystyle\int (x + 4)^3\,dx$

22. $\displaystyle\int (x - 6)^{4/3}\,dx$

23. $\displaystyle\int (5x + 1)^4(5)\,dx$

24. $\displaystyle\int (x^3 + 1)^2(3x^2)\,dx$

25. $\displaystyle\int (1 + 5x)^2\,dx$

26. $\displaystyle\int (6x - 2)^4\,dx$

27. $\displaystyle\int x^2(3x^3 + 1)^2\,dx$

28. $\displaystyle\int x(1 - 4x^2)^3\,dx$

29. $\displaystyle\int \frac{x^2}{(2x^3 - 5)^3}\,dx$

30. $\displaystyle\int \frac{x^2}{(x^3 - 4)^2}\,dx$

31. $\displaystyle\int \frac{1}{\sqrt{5x - 1}}\,dx$

32. $\displaystyle\int \frac{4x}{\sqrt{1 - 3x^2}}\,dx$

33. Production The rate of change of the output of a small sawmill is modeled by

$$\frac{dP}{dt} = 2t(0.001t^2 + 0.5)^{1/4}, \quad 0 \le t \le 40$$

where t is the time (in hours) and P is the output (in board-feet). Find the numbers of board-feet produced in (a) 6 hours and (b) 12 hours.

34. Cost The marginal cost for a catering service to cater to x people can be modeled by

$$\frac{dC}{dx} = \frac{5x}{\sqrt{x^2 + 1000}}.$$

When $x = 225$, the cost C (in dollars) is $1136.06. Find the costs of catering to (a) 500 people and (b) 1000 people.

Using the Exponential and Log Rules In Exercises 35–46, use the Exponential Rule or the Log Rule to find the indefinite integral.

35. $\displaystyle\int 4e^{4x}\, dx$

36. $\displaystyle\int 3e^{-3x}\, dx$

37. $\displaystyle\int e^{-5x}\, dx$

38. $\displaystyle\int e^{6x}\, dx$

39. $\displaystyle\int 7xe^{3x^2}\, dx$

40. $\displaystyle\int (2t - 1)e^{t^2 - t}\, dt$

41. $\displaystyle\int \frac{1}{x - 6}\, dx$

42. $\displaystyle\int \frac{1}{1 - 4x}\, dx$

43. $\displaystyle\int \frac{4}{6x - 1}\, dx$

44. $\displaystyle\int \frac{5}{2x + 3}\, dx$

45. $\displaystyle\int \frac{x^2}{1 - x^3}\, dx$

46. $\displaystyle\int \frac{x - 4}{x^2 - 8x}\, dx$

Evaluating a Definite Integral Using a Geometric Formula In Exercises 47–50, sketch the region whose area is given by the definite integral. Then use a geometric formula to evaluate the integral.

47. $\displaystyle\int_0^3 2\, dx$

48. $\displaystyle\int_0^6 \frac{x}{2}\, dx$

49. $\displaystyle\int_0^4 (4 - x)\, dx$

50. $\displaystyle\int_{-4}^4 \sqrt{16 - x^2}\, dx$

51. Using Properties of Definite Integrals Given

$$\int_2^6 f(x)\, dx = 10 \quad \text{and} \quad \int_2^6 g(x)\, dx = 3$$

evaluate the definite integral.

(a) $\displaystyle\int_2^6 [f(x) + g(x)]\, dx$

(b) $\displaystyle\int_2^6 [f(x) - g(x)]\, dx$

(c) $\displaystyle\int_2^6 [2f(x) - 3g(x)]\, dx$

(d) $\displaystyle\int_2^6 5f(x)\, dx$

52. Using Properties of Definite Integrals Given

$$\int_0^3 f(x)\, dx = 4 \quad \text{and} \quad \int_3^6 f(x)\, dx = -1$$

evaluate the definite integral.

(a) $\displaystyle\int_0^6 f(x)\, dx$

(b) $\displaystyle\int_6^3 f(x)\, dx$

(c) $\displaystyle\int_4^4 f(x)\, dx$

(d) $\displaystyle\int_3^6 -10f(x)\, dx$

Finding Area by the Fundamental Theorem In Exercises 53–58, find the area of the region.

53. $f(x) = 4 - x^2$

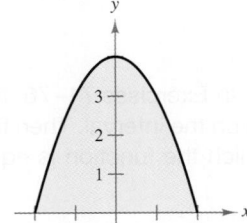

54. $f(x) = 9 - x^2$

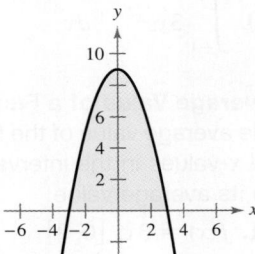

55. $f(x) = \dfrac{2}{x + 1}$

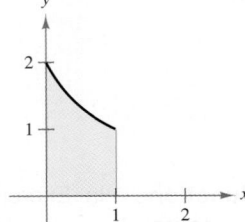

56. $f(x) = \dfrac{4}{\sqrt{x}}$

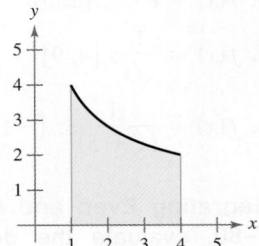

57. $f(x) = 2e^{x/2}$

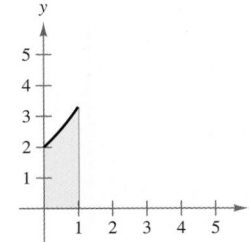

58. $f(x) = \dfrac{x - 1}{x}$

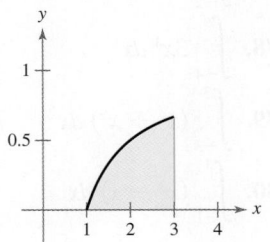

Evaluating a Definite Integral   In Exercises 59–70, use the Fundamental Theorem of Calculus to evaluate the definite integral.

59. $\displaystyle\int_0^4 (2 + x)\, dx$

60. $\displaystyle\int_{-1}^1 (t^2 + 2)\, dt$

61. $\displaystyle\int_{-1}^1 (4t^3 - 2t)\, dt$

62. $\displaystyle\int_{-2}^2 (x^4 + 2x^2 - 5)\, dx$

63. $\displaystyle\int_{-2}^0 (x + 2)^3\, dx$

64. $\displaystyle\int_2^4 (2x - 3)^2\, dx$

65. $\displaystyle\int_0^3 \frac{1}{\sqrt{1 + x}}\, dx$

66. $\displaystyle\int_3^6 \frac{x}{3\sqrt{x^2 - 8}}\, dx$

67. $\displaystyle\int_3^9 \frac{5}{x}\, dx$

68. $\displaystyle\int_1^2 \left(\frac{1}{x^2} - \frac{1}{x^3}\right)\, dx$

69. $\displaystyle\int_0^{\ln 5} e^{x/5}\, dx$

70. $\displaystyle\int_{-1}^1 3xe^{x^2 - 1}\, dx$

Average Value of a Function   In Exercises 71–76, find the average value of the function on the interval. Then find all x-values in the interval for which the function is equal to its average value.

71. $f(x) = 3x;\ [0, 2]$

72. $f(x) = x^2 + 2;\ [-3, 3]$

73. $f(x) = -2e^x;\ [0, 3]$

74. $f(x) = e^{5-x};\ [2, 5]$

75. $f(x) = \dfrac{1}{\sqrt{x}};\ [4, 9]$

76. $f(x) = \dfrac{1}{(x + 5)^2};\ [-1, 6]$

Integrating Even and Odd Functions   In Exercises 77–80, evaluate the definite integral by using the properties of even and odd functions.

77. $\displaystyle\int_{-2}^2 6x^5\, dx$

78. $\displaystyle\int_{-4}^4 3x^4\, dx$

79. $\displaystyle\int_{-3}^3 (x^4 + x^2)\, dx$

80. $\displaystyle\int_{-1}^1 (x^3 - x)\, dx$

Finding the Amount of an Annuity   In Exercises 81 and 82, find the amount of an annuity with income function $c(t)$, interest rate r, and term T.

81. $c(t) = \$3000,\ r = 6\%,\ T = 5$ years

82. $c(t) = \$1200,\ r = 7\%,\ T = 8$ years

83. Cost   The marginal cost of serving an additional typical client at a law firm can be modeled by

$$\frac{dC}{dx} = 675 + 0.5x$$

where x is the number of clients. Find the change in cost C (in dollars) when x increases from 50 to 51 clients.

84. Profit   The marginal profit obtained by selling x dollars of automobile insurance can be modeled by

$$\frac{dP}{dx} = 0.4\left(1 - \frac{5000}{x}\right), \quad x \geq 5000.$$

Find the change in the profit P (in dollars) when x increases from \$75,000 to \$100,000.

85. Compound Interest   A deposit of \$500 is made in a savings account at an annual interest rate of 4%, compounded continuously. Find the average balance in the account during the first 2 years.

86. Revenue   The rate of change in revenue for Texas Roadhouse from 2003 through 2009 can be modeled by

$$\frac{dR}{dt} = -11.5000t^2 + 142.140t - 294.91$$

where R is the revenue (in millions of dollars) and t is the time in years, with $t = 3$ corresponding to 2003. In 2006, the revenue for Texas Roadhouse was \$597.1 million. *(Source: Texas Roadhouse, Inc.)*

(a) Find the model for the revenue of Texas Roadhouse.

(b) What was the average revenue of Texas Roadhouse for 2003 through 2009?

Finding the Area Bounded by Two Graphs   In Exercises 87–94, sketch the region bounded by the graphs of the functions and find the area of the region.

87. $y = \dfrac{1}{x^3},\ y = 0,\ x = 1,\ x = 3$

88. $y = x^2 + 4x - 5,\ y = 4x - 1$

89. $y = (x - 3)^2,\ y = 8 - (x - 3)^2$

90. $y = 4 - x,\ y = x^2 - 5x + 8,\ x = 0$

91. $y = \dfrac{4}{\sqrt{x + 1}},\ y = 0,\ x = 0,\ x = 8$

92. $y = \sqrt{x}(1 - x),\ y = 0$

93. $y = x,\ y = x^3$

94. $y = x^3 - 4x,\ y = -x^2 - 2x$

Consumer and Producer Surpluses In Exercises 95–98, find the consumer and producer surpluses by using the demand and supply functions, where p is the price (in dollars) and x is the number of units (in millions).

Demand Function	Supply Function
95. $p = 36 - 0.35x$	$p = 0.05x$
96. $p = 200 - 0.2x$	$p = 50 + 1.3x$
97. $p = 250 - x$	$p = 150 + x$
98. $p = 500 - x$	$p = 1.25x + 162.5$

99. Revenue For the years 2015 through 2020, two models, R_1 and R_2, used to project the revenue (in millions of dollars) for a company are

$$R_1 = 24.3 + 8.24t$$

and

$$R_2 = 21.6 + 9.36t$$

where $t = 15$ corresponds to 2015. Which model projects the greater revenue? How much more total revenue does that model project over the six-year period?

100. Sales For the years 2000 through 2009, the sales (in millions of dollars) for Men's Wearhouse can be modeled by

$$R = \begin{cases} 23.596t^2 - 41.55t + 1310.5, & 0 \le t \le 6 \\ 38.7t^2 - 720.7t + 5261.2, & 6 < t \le 9 \end{cases}$$

where t is the year, with $t = 0$ corresponding to 2000. *(Source: Men's Wearhouse, Inc.)*

(a) Use a graphing utility to graph this model.

(b) Suppose the sales from 2007 through 2009 had continued to follow the model for 2000 through 2006. How much more or less would the sales have been for Men's Wearhouse?

101. Cost, Revenue, and Profit The revenue from a manufacturing process (in millions of dollars) is projected to follow the model

$$R = 70$$

for 10 years. Over the same period of time, the cost (in millions of dollars) is projected to follow the model

$$C = 30 + 0.3t^2$$

where t is the time (in years). Approximate the profit over the 10-year period.

102. Cost, Revenue, and Profit Repeat Exercise 101 for revenue and cost models given by

$$R = 70 + 0.1t$$

and

$$C = 30 + 0.3t^2.$$

Did the profit increase or decrease? Explain why.

Approximating the Area of a Plane Region In Exercises 103 and 104, use the rectangles to approximate the area of the region. Compare your result with the exact area obtained using a definite integral.

103. $f(x) = \dfrac{x}{3}$, $[0, 3]$ **104.** $f(x) = x^2 + 1$, $[0, 1]$

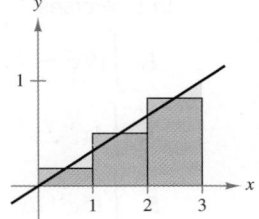   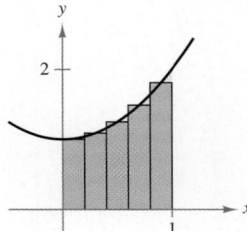

Using the Midpoint Rule In Exercises 105–108, use the Midpoint Rule with $n = 4$ to approximate the area of the region bounded by the graph of f and the x-axis over the interval. Sketch the region.

Function	Interval
105. $f(x) = x^2$	$[0, 2]$
106. $f(x) = 2x - x^3$	$[0, 1]$
107. $f(x) = (x^2 - 1)^2$	$[-1, 1]$
108. $f(x) = \dfrac{3x}{x + 2}$	$[0, 4]$

Using the Midpoint Rule In Exercises 109–112, use the Midpoint Rule with $n = 6$ to approximate the area of the region bounded by the graph of f and the x-axis over the interval. Sketch the region.

Function	Interval
109. $f(x) = x + 3$	$[0, 3]$
110. $f(x) = 9 - x^2$	$[-3, 3]$
111. $f(x) = x\sqrt{x + 1}$	$[0, 2]$
112. $f(x) = \dfrac{3}{x^2 + 1}$	$[-6, 6]$

113. Surface Area Use the Midpoint Rule to estimate the surface area of the swamp shown in the figure.

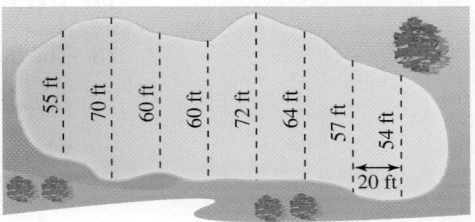

TEST YOURSELF

See www.CalcChat.com for worked-out solutions to odd-numbered exercises.

Take this test as you would take a test in class. When you are done, check your work against the answers given in the back of the book.

In Exercises 1–6, find the indefinite integral.

1. $\int (9x^2 - 4x + 13)\, dx$

2. $\int (x + 1)^2\, dx$

3. $\int 4x^3 \sqrt{x^4 - 7}\, dx$

4. $\int \dfrac{5x - 6}{\sqrt{x}}\, dx$

5. $\int 15e^{3x}\, dx$

6. $\int \dfrac{3}{4x - 1}\, dx$

In Exercises 7 and 8, find the particular solution that satisfies the differential equation and initial condition.

7. $f'(x) = 6x - 5; f(-1) = 6$

8. $f'(x) = e^x + 1; f(0) = 1$

In Exercises 9–14, evaluate the definite integral.

9. $\int_0^1 16x\, dx$

10. $\int_{-3}^3 (3 - 2x)\, dx$

11. $\int_{-1}^1 (x^3 + x^2)\, dx$

12. $\int_{-1}^2 \dfrac{2x}{\sqrt{x^2 + 1}}\, dx$

13. $\int_0^3 e^{4x}\, dx$

14. $\int_{-2}^3 \dfrac{1}{x + 3}\, dx$

15. The rate of change in sales of PetSmart from 2000 through 2009 can be modeled by

$$\frac{dS}{dt} = 226.912e^{0.1013t}$$

where S is the sales (in millions of dollars) and t is the time (in years), with $t = 0$ corresponding to 2000. In 2004, the sales of PetSmart were \$3363.5 million. *(Source: PetSmart, Inc.)*

(a) Find the model for the sales of PetSmart.

(b) What were the average sales for 2000 through 2009?

In Exercises 16 and 17, sketch the region bounded by the graphs of the functions and find the area of the region.

16. $f(x) = 6,\ g(x) = x^2 - x - 6$

17. $f(x) = \sqrt[3]{x},\ g(x) = x^2$

18. The demand and supply functions for a product are modeled by

Demand: $p = -0.625x + 10$ and   *Supply:* $p = 0.25x + 3$

where p is the price (in dollars) and x is the number of units (in millions). Find the consumer and producer surpluses for this product.

In Exercises 19 and 20, use the Midpoint Rule with $n = 4$ to approximate the area of the region bounded by the graph of f and the x-axis over the interval. Compare your result with the exact area. Sketch the region.

19. $f(x) = 3x^2,\ \ [0, 1]$

20. $f(x) = x^2 + 1,\ \ [-1, 1]$

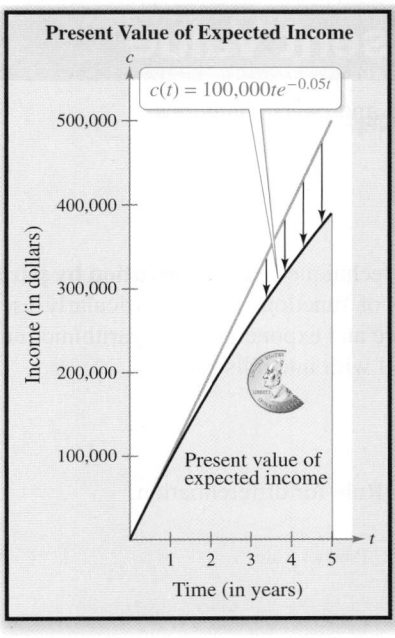

Present Value of Expected Income
$c(t) = 100{,}000te^{-0.05t}$
c
500,000
400,000
300,000
200,000
100,000
Income (in dollars)
Present value of
expected income
t
1 2 3 4 5
Time (in years)

Example 7 on page 874 shows how integration by parts can be used to find the present value of a company's future income.

12 Techniques of Integration

Dean Mitchell/www.shutterstock.com
Kurhan/www.shutterstock.com

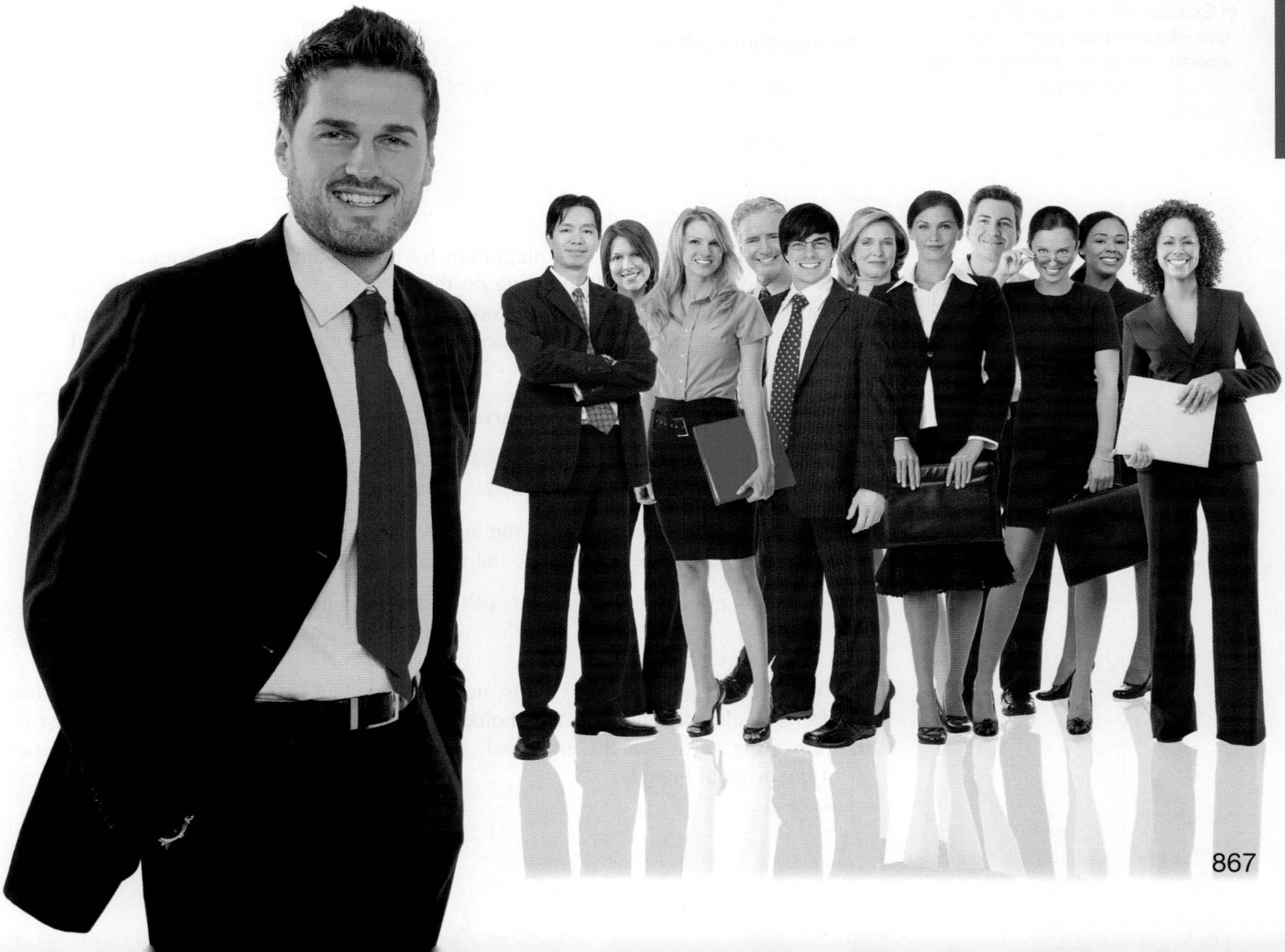

12.1 Integration by Parts and Present Value

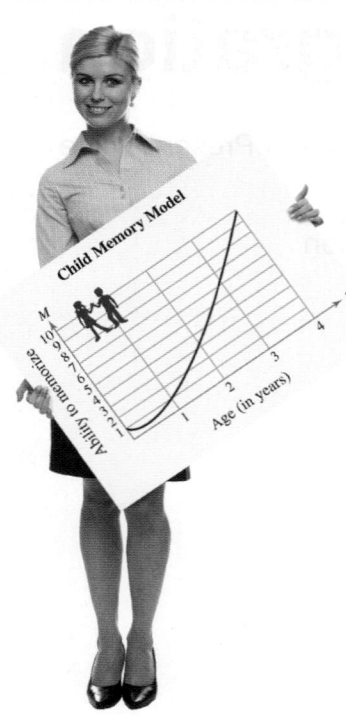

In Exercise 65 on page 876, you will use integration by parts to find the average value of a memory model for children.

■ Use integration by parts to find indefinite and definite integrals.
■ Find the present value of future income.

Integration by Parts

In this section, you will study an integration technique called **integration by parts.** This technique can be applied to a wide variety of functions and is particularly useful for integrands involving the products of algebraic and exponential or logarithmic functions. For instance, integration by parts works well with integrals such as

$$\int x^2 e^x \, dx \quad \text{and} \quad \int x \ln x \, dx.$$

Integration by parts is based on the Product Rule for differentiation.

$$\frac{d}{dx}[uv] = u\frac{dv}{dx} + v\frac{du}{dx} \qquad \text{Product Rule}$$

$$uv = \int u\frac{dv}{dx}\,dx + \int v\frac{du}{dx}\,dx \qquad \text{Integrate each side.}$$

$$uv = \int u\,dv + \int v\,du \qquad \text{Write in differential form.}$$

$$\int u\,dv = uv - \int v\,du \qquad \text{Rewrite.}$$

> **Integration by Parts**
>
> Let u and v be differentiable functions of x.
> $$\int u\,dv = uv - \int v\,du$$

Note that the formula for integration by parts expresses the original integral in terms of another integral. Depending on the choices of u and dv, it may be easier to evaluate the second integral than the original one. Because the choices of u and dv are critical in the integration by parts process, the following guidelines are provided.

> **Guidelines for Integration by Parts**
>
> 1. Try letting dv be the most complicated portion of the integrand that fits a basic integration rule. Then u will be the remaining factor(s) of the integrand.
>
> 2. Try letting u be the portion of the integrand whose derivative is a function simpler than u. Then dv will be the remaining factor(s) of the integrand.
>
> Note that dv always includes the dx of the original integrand.

When using integration by parts, note that you can first choose dv or first choose u. After you choose, however, the choice of the other factor is determined—it must be the remaining portion of the integrand. Also note that dv must contain the differential dx of the original integral.

Example 1   Integration by Parts

Find $\int xe^x \, dx$.

SOLUTION   To apply integration by parts, you must rewrite the original integral in the form $\int u \, dv$. That is, you must break $xe^x \, dx$ into two factors—one "part" representing u and the other "part" representing dv. There are several ways to do this.

$$\int \underbrace{(x)}_{u}\underbrace{(e^x \, dx)}_{dv} \qquad \int \underbrace{(e^x)}_{u}\underbrace{(x \, dx)}_{dv} \qquad \int \underbrace{(1)}_{u}\underbrace{(xe^x \, dx)}_{dv} \qquad \int \underbrace{(xe^x)}_{u}\underbrace{(dx)}_{dv}$$

The guidelines on the preceding page suggest the first option because $dv = e^x \, dx$ is the most complicated portion of the integrand that fits a basic integration formula *and* because the derivative of $u = x$ is simpler than x.

$$dv = e^x \, dx \qquad \Longrightarrow \qquad v = \int dv = \int e^x \, dx = e^x$$

$$u = x \qquad \Longrightarrow \qquad du = dx$$

Next, you can apply the integration by parts formula as shown.

$$\int u \, dv = uv - \int v \, du \qquad \text{Integration by parts formula}$$

$$\int xe^x \, dx = xe^x - \int e^x \, dx \qquad \text{Substitute.}$$

$$= xe^x - e^x + C \qquad \text{Integrate } \int e^x \, dx.$$

You can check this result by differentiating.

$$\frac{d}{dx}[xe^x - e^x + C] = xe^x + e^x(1) - e^x = xe^x$$

✓ Checkpoint 1

Find $\int xe^{2x} \, dx$.

In Example 1, notice that you do not need to include a constant of integration when solving $v = \int e^x \, dx = e^x$. To see why this is true, try replacing e^x by $e^x + C_1$ in the solution.

$$\int xe^x \, dx = x(e^x + C_1) - \int (e^x + C_1) \, dx$$

$$= xe^x + C_1 x - e^x - C_1 x + C$$

$$= xe^x - e^x + C$$

After integrating, you can see that the terms involving C_1 subtract out.

TECH TUTOR

If you have access to a symbolic integration utility, try using it to solve several of the exercises in this section. Note that the form of the integral may be slightly different from what you obtain when solving the exercise by hand.

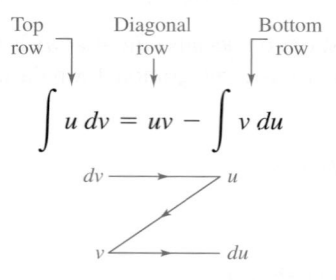

Example 2   Integration by Parts

Find $\displaystyle\int x^2 \ln x\, dx$.

SOLUTION   In this case, x^2 is more easily integrated than $\ln x$. Furthermore, the derivative of $\ln x$ is simpler than $\ln x$. So, you should choose $dv = x^2\, dx$.

$$dv = x^2\, dx \qquad\Longrightarrow\qquad v = \int dv = \int x^2\, dx = \frac{x^3}{3}$$

$$u = \ln x \qquad\Longrightarrow\qquad du = \frac{1}{x}\, dx$$

Next, apply the integration by parts formula.

$$\int u\,dv = uv - \int v\,du \qquad\qquad \text{Integration by parts formula}$$

$$\int x^2 \ln x\, dx = \frac{x^3}{3}\ln x - \int \left(\frac{x^3}{3}\right)\!\left(\frac{1}{x}\right)dx \qquad\qquad \text{Substitute.}$$

$$= \frac{x^3}{3}\ln x - \frac{1}{3}\int x^2\, dx \qquad\qquad \text{Simplify.}$$

$$= \frac{x^3}{3}\ln x - \frac{x^3}{9} + C \qquad\qquad \text{Integrate.}$$

✓ **Checkpoint 2**

Find $\displaystyle\int x \ln x\, dx$. ■

Example 3   Integrating by Parts with a Single Factor

Find $\displaystyle\int \ln x\, dx$.

SOLUTION   This integrand is unusual because it has only one factor. In such cases, you should choose $dv = dx$ and choose u to be the single factor.

$$dv = dx \qquad\Longrightarrow\qquad v = \int dv = \int dx = x$$

$$u = \ln x \qquad\Longrightarrow\qquad du = \frac{1}{x}\, dx$$

Next, apply the integration by parts formula.

$$\int u\,dv = uv - \int v\,du \qquad\qquad \text{Integration by parts formula}$$

$$\int \ln x\, dx = x \ln x - \int (x)\!\left(\frac{1}{x}\right)dx \qquad\qquad \text{Substitute.}$$

$$= x \ln x - \int dx \qquad\qquad \text{Simplify.}$$

$$= x \ln x - x + C \qquad\qquad \text{Integrate.}$$

✓ **Checkpoint 3**

Find $\displaystyle\int \ln 2x\, dx$. ■

Example 4   **Using Integration by Parts Repeatedly**

Find $\int x^2 e^x \, dx$.

SOLUTION   The factors x^2 and e^x are both easy to integrate. Notice, however, that the derivative of x^2 becomes simpler, whereas the derivative of e^x does not. So, you should let $u = x^2$ and let $dv = e^x \, dx$.

$$dv = e^x \, dx \qquad \Longrightarrow \qquad v = \int dv = \int e^x \, dx = e^x$$

$$u = x^2 \qquad \Longrightarrow \qquad du = 2x \, dx$$

Next, apply the integration by parts formula.

$$\int x^2 e^x \, dx = x^2 e^x - \int 2xe^x \, dx \qquad \text{First application of integration by parts}$$

This first use of integration by parts has succeeded in simplifying the original integral, but the integral on the right still doesn't fit a basic integration rule. To evaluate that integral, you can apply integration by parts again. This time, let $u = 2x$ and $dv = e^x \, dx$.

$$dv = e^x \, dx \qquad \Longrightarrow \qquad v = \int dv = \int e^x \, dx = e^x$$

$$u = 2x \qquad \Longrightarrow \qquad du = 2 \, dx$$

Next, apply the integration by parts formula.

$$\int x^2 e^x \, dx = x^2 e^x - \int 2xe^x \, dx \qquad \text{First application of integration by parts}$$

$$= x^2 e^x - \left(2xe^x - \int 2e^x \, dx \right) \qquad \text{Second application of integration by parts}$$

$$= x^2 e^x - 2xe^x + 2e^x + C \qquad \text{Integrate.}$$

$$= e^x(x^2 - 2x + 2) + C \qquad \text{Simplify.}$$

You can confirm this result by differentiating.

$$\frac{d}{dx}[e^x(x^2 - 2x + 2) + C] = e^x(2x - 2) + (x^2 - 2x + 2)(e^x)$$

$$= 2xe^x - 2e^x + x^2 e^x - 2xe^x + 2e^x$$

$$= x^2 e^x$$

✓ **Checkpoint 4**

Find $\int x^3 e^x \, dx$.   ■

When making repeated applications of integration by parts, you need to be careful not to interchange the substitutions in successive applications. For instance, in Example 4, the first substitution was $dv = e^x \, dx$ and $u = x^2$. If, in the second application, you had switched the substitution to $dv = 2x \, dx$ and $u = e^x$, you would have obtained

$$\int x^2 e^x \, dx = x^2 e^x - \int 2xe^x \, dx$$

$$= x^2 e^x - \left(x^2 e^x - \int x^2 e^x \, dx \right)$$

$$= \int x^2 e^x \, dx$$

thereby undoing the previous integration and returning to the *original* integral.

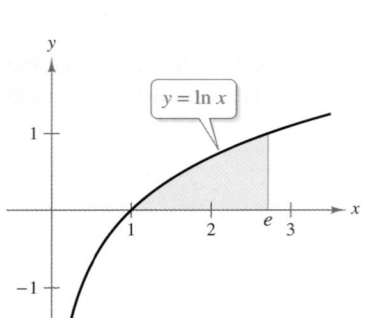

FIGURE 12.1

Example 5   **Evaluating a Definite Integral**

Evaluate $\displaystyle\int_1^e \ln x \, dx$.

SOLUTION   Integration by parts was used to find the antiderivative of $\ln x$ in Example 3. Using this result, you can evaluate the definite integral as shown.

$$\int_1^e \ln x \, dx = \Big[x \ln x - x \Big]_1^e \qquad \text{Use result of Example 3.}$$

$$= (e \ln e - e) - (1 \ln 1 - 1) \qquad \text{Apply Fundamental Theorem.}$$

$$= (e - e) - (0 - 1)$$

$$= 1 \qquad \text{Simplify.}$$

The area represented by this definite integral is shown in Figure 12.1.

ALGEBRA TUTOR xy

For help on the algebra in Example 5, see Example 1 in the *Chapter 12 Algebra Tutor*, on page 904.

✓ **Checkpoint 5**

Evaluate $\displaystyle\int_0^1 x^2 e^x \, dx$.   ▪

Before starting the exercises in this section, remember that it is not enough to know *how* to use the various integration techniques. You also must know *when* to use them. Integration is first and foremost a problem of recognition—recognizing which formula or technique to apply to obtain an antiderivative. Often, a slight alteration of an integrand will necessitate the use of a different integration technique. Here are some examples.

Integral	*Technique*	*Antiderivative*
$\displaystyle\int x \ln x \, dx$	Integration by parts	$\dfrac{x^2}{2} \ln x - \dfrac{x^2}{4} + C$
$\displaystyle\int \dfrac{\ln x}{x} \, dx$	Power Rule: $\displaystyle\int u^n \dfrac{du}{dx} \, dx$	$\dfrac{(\ln x)^2}{2} + C$
$\displaystyle\int \dfrac{1}{x \ln x} \, dx$	Log Rule: $\displaystyle\int \dfrac{1}{u} \dfrac{du}{dx} \, dx$	$\ln|\ln x| + C$

As you gain experience in using integration by parts, your skill in determining u and dv will improve. The following summary lists several common integrals with suggestions for the choices of u and dv.

Summary of Common Integrals Using Integration by Parts

1. For integrals of the form

$$\int x^n e^{ax} \, dx$$

   let $u = x^n$ and $dv = e^{ax} \, dx$. (See Examples 1 and 4.)

2. For integrals of the form

$$\int x^n \ln x \, dx$$

   let $u = \ln x$ and $dv = x^n \, dx$. (See Examples 2 and 3.)

Present Value

Recall from Section 10.2 that the present value of a future payment is the amount that would have to be deposited today to produce the future payment. What is the present value of a future payment of $1000 one year from now? Because of inflation, $1000 today buys more than $1000 will buy a year from now. The definition below considers only the effect of inflation.

Present Value

If c represents a continuous income function in dollars per year and the annual rate of inflation is r, then the actual total income over t_1 years is

$$\text{Actual income over } t_1 \text{ years} = \int_0^{t_1} c(t)\, dt$$

and its **present value** is

$$\text{Present value} = \int_0^{t_1} c(t)e^{-rt}\, dt.$$

Ignoring inflation, the equation for present value also applies to an interest-bearing account, where the annual interest rate r is compounded continuously and c is an income function in dollars per year.

 Example 6   Finding Present Value

You have just won $1,000,000 in a state lottery. You will be paid an annuity of $50,000 a year for 20 years. When the annual rate of inflation is 6%, what is the present value of this income?

SOLUTION   The income function for your winnings is given by $c(t) = 50,000$. So,

$$\text{Actual income} = \int_0^{20} 50,000\, dt$$

$$= \left[50,000t \right]_0^{20}$$

$$= \$1,000,000.$$

Because you do not receive this entire amount now, its present value is

$$\text{Present value} = \int_0^{20} 50,000e^{-0.06t}\, dt$$

$$= \left[\frac{50,000}{-0.06}e^{-0.06t} \right]_0^{20}$$

$$\approx \$582,338.$$

This present value represents the amount that the state must deposit now to cover your payments over the next 20 years. This shows why state lotteries are so profitable—for the states!

✓ **Checkpoint 6**

Find the present value of the income from the lottery ticket in Example 6 when the annual rate of inflation is 7%.

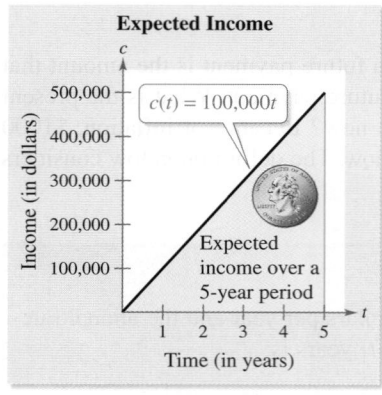

Expected Income

Income (in dollars)

$c(t) = 100,000t$

Expected income over a 5-year period

Time (in years)

(a)

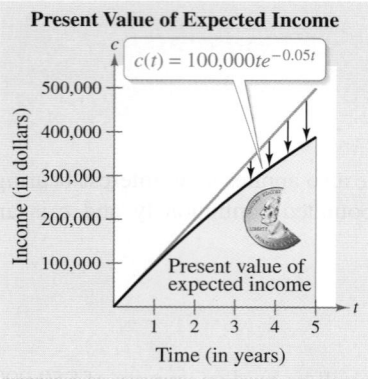

Present Value of Expected Income

Income (in dollars)

$c(t) = 100,000te^{-0.05t}$

Present value of expected income

Time (in years)

(b)

FIGURE 12.2

Example 7   **Finding Present Value**

A company expects its income during the next 5 years to be given by

$$c(t) = 100,000t, \quad 0 \le t \le 5. \qquad \text{See Figure 12.2(a).}$$

Assuming an annual inflation rate of 5%, can the company claim that the present value of this income is at least $1 million?

SOLUTION   The present value is

$$\text{Present value} = \int_0^5 100,000te^{-0.05t}\, dt = 100,000\int_0^5 te^{-0.05t}\, dt.$$

Using integration by parts, let $dv = e^{-0.05t}\, dt$.

$$dv = e^{-0.05t}\, dt \quad \Longrightarrow \quad v = \int dv = \int e^{-0.05t}\, dt = -20e^{-0.05t}$$

$$u = t \quad \Longrightarrow \quad du = dt$$

This implies that

$$\int te^{-0.05t}\, dt = -20te^{-0.05t} + 20\int e^{-0.05t}\, dt$$

$$= -20te^{-0.05t} - 400e^{-0.05t}$$

$$= -20e^{-0.05t}(t + 20).$$

So, the present value is

$$\text{Present value} = 100,000\int_0^5 te^{-0.05t}\, dt \qquad \text{See Figure 12.2(b).}$$

$$= 100,000\left[-20e^{-0.05t}(t + 20) \right]_0^5$$

$$\approx \$1,059,961.$$

Yes, the company can claim that the present value of its expected income during the next 5 years is at least $1 million.

✓**Checkpoint 7**

A company expects its income during the next 10 years to be given by $c(t) = 20,000t$, for $0 \le t \le 10$. Assuming an annual inflation rate of 5%, what is the present value of this income?

SUMMARIZE   (Section 12.1)

1. State the integration by parts formula *(page 868)*. For examples of using this formula, see Examples 1, 2, 3, 4, and 7.

2. State the guidelines for integration by parts *(page 868)*. For an example of using these guidelines, see Example 1.

3. Give a summary of the common integrals using integration by parts *(page 872)*. For examples of these common integrals, see Examples 1, 2, 3, and 4.

4. Describe a real-life example of how integration by parts can be used to find the present value of an annuity *(page 873, Example 6)*.

SKILLS WARM UP 12.1

The following warm-up exercises involve skills that were covered in earlier sections. You will use these skills in the exercise set for this section. For additional help, review Sections 10.3, 10.5, and 11.5.

In Exercises 1–6, find $f'(x)$.

1. $f(x) = \ln(x + 1)$

2. $f(x) = \ln(x^2 - 1)$

3. $f(x) = e^{x^3}$

4. $f(x) = e^{-x^2}$

5. $f(x) = x^2 e^x$

6. $f(x) = xe^{-2x}$

In Exercises 7–10, find the area between the graphs of f and g.

7. $f(x) = -x^2 + 4, \ g(x) = x^2 - 4$

8. $f(x) = -x^2 + 2, \ g(x) = 1$

9. $f(x) = 4x, \ g(x) = x^2 - 5$

10. $f(x) = x^3 - 3x^2 + 2, \ g(x) = x - 1$

Exercises 12.1

See www.CalcChat.com for worked-out solutions to odd-numbered exercises.

Setting Up Integration by Parts In Exercises 1–4, identify u and dv for finding the integral using integration by parts. (Do not evaluate the integral.)

1. $\displaystyle\int xe^{3x} \, dx$

2. $\displaystyle\int x^2 e^{3x} dx$

3. $\displaystyle\int x \ln 2x \, dx$

4. $\displaystyle\int \ln 4x \, dx$

Integration by Parts In Exercises 5–16, use integration by parts to find the indefinite integral. *See Examples 1, 2, 3, and 4.*

5. $\displaystyle\int xe^{3x} \, dx$

6. $\displaystyle\int xe^{-x} \, dx$

7. $\displaystyle\int x^3 \ln x \, dx$

8. $\displaystyle\int x^4 \ln x \, dx$

9. $\displaystyle\int \ln 2x \, dx$

10. $\displaystyle\int \ln x^2 \, dx$

11. $\displaystyle\int x^2 e^{-x} \, dx$

12. $\displaystyle\int x^2 e^{2x} \, dx$

13. $\displaystyle\int \sqrt{x} \ln x \, dx$

14. $\displaystyle\int x^2 \sqrt{x - 3} \, dx$

15. $\displaystyle\int 2x^2 e^x \, dx$

16. $\displaystyle\int \frac{2x}{e^x} \, dx$

Finding Indefinite Integrals In Exercises 17–38, find the indefinite integral. (*Hint:* Integration by parts is not required for all the integrals.)

17. $\displaystyle\int e^{4x} \, dx$

18. $\displaystyle\int e^{-2x} \, dx$

19. $\displaystyle\int xe^{4x} \, dx$

20. $\displaystyle\int xe^{-2x} \, dx$

21. $\displaystyle\int \frac{x}{e^{x/4}} \, dx$

22. $\displaystyle\int \frac{1}{2} x^3 e^x \, dx$

23. $\displaystyle\int t \ln(t + 1) \, dt$

24. $\displaystyle\int (x - 1)e^x \, dx$

25. $\displaystyle\int \frac{e^{1/t}}{t^2} \, dt$

26. $\displaystyle\int \frac{1}{x(\ln x)^3} \, dx$

27. $\displaystyle\int x(\ln x)^2 \, dx$

28. $\displaystyle\int \ln 3x \, dx$

29. $\displaystyle\int \frac{(\ln x)^2}{x} \, dx$

30. $\displaystyle\int \frac{1}{x \ln 3x} \, dx$

31. $\displaystyle\int \frac{\ln x}{x^2} \, dx$

32. $\displaystyle\int \frac{\ln 2x}{x^2} \, dx$

33. $\displaystyle\int x\sqrt{x - 1} \, dx$

34. $\displaystyle\int \frac{x}{\sqrt{x - 1}} \, dx$

35. $\displaystyle\int x(x + 1)^2 \, dx$

36. $\displaystyle\int \frac{x}{\sqrt{2 + 3x}} \, dx$

37. $\displaystyle\int \frac{xe^{2x}}{(2x + 1)^2} \, dx$

38. $\displaystyle\int \frac{x^3 e^{x^2}}{(x^2 + 1)^2} \, dx$

Evaluating Definite Integrals In Exercises 39–46, use integration by parts to evaluate the definite integral. *See Example 5.*

39. $\displaystyle\int_1^e x^5 \ln x\, dx$

40. $\displaystyle\int_1^e 2x \ln x\, dx$

41. $\displaystyle\int_0^1 \ln(1 + 2x)\, dx$

42. $\displaystyle\int_0^4 \frac{x}{e^{x/2}}\, dx$

43. $\displaystyle\int_0^8 x\sqrt{x+1}\, dx$

44. $\displaystyle\int_0^{12} \frac{x}{\sqrt{x+4}}\, dx$

45. $\displaystyle\int_1^2 x^2 e^x\, dx$

46. $\displaystyle\int_0^2 \frac{x^2}{e^{3x}}\, dx$

Area of a Region In Exercises 47–52, find the area of the region bounded by the graphs of the equations. Use a graphing utility to verify your results.

47. $y = x^3 e^x$, $y = 0$, $x = 0$, $x = 2$

48. $y = (x^2 - 1)e^x$, $y = 0$, $x = -1$, $x = 1$

49. $y = \frac{1}{9}xe^{-x/3}$, $y = 0$, $x = 0$, $x = 3$

50. $y = x^{-3}\ln x$, $y = 0$, $x = e$

51. $y = x^2 \ln x$, $y = 0$, $x = 1$, $x = e$

52. $y = \dfrac{\ln x}{x^2}$, $y = 0$, $x = 1$, $x = e$

Verifying Formulas In Exercises 53 and 54, use integration by parts to verify the formula.

53. $\displaystyle\int x^n \ln x\, dx = \frac{x^{n+1}}{(n+1)^2}[-1 + (n+1)\ln x] + C$,

$n \neq -1$

54. $\displaystyle\int x^n e^{ax}\, dx = \frac{x^n e^{ax}}{a} - \frac{n}{a}\int x^{n-1}e^{ax}\, dx$, $n > 0$

Using Formulas In Exercises 55–58, use the results of Exercises 53 and 54 to find the indefinite integral.

55. $\displaystyle\int x^2 e^{5x}\, dx$

56. $\displaystyle\int xe^{-3x}\, dx$

57. $\displaystyle\int x^{-4}\ln x\, dx$

58. $\displaystyle\int x^{3/2}\ln x\, dx$

 Integration Using Technology In Exercises 59–62, use a symbolic integration utility to evaluate the integral.

59. $\displaystyle\int_0^2 t^3 e^{-4t}\, dt$

60. $\displaystyle\int_1^4 (x^2 + 4)\ln x\, dx$

61. $\displaystyle\int_0^5 x^4(25 - x^2)^{3/2}\, dx$

62. $\displaystyle\int_1^e x^9 \ln x\, dx$

63. Demand A manufacturing company forecasts that the demand x (in units) for its product over the next 10 years can be modeled by

$$x = 500(20 + te^{-0.1t}), \quad 0 \le t \le 10$$

where t is the time in years.

(a) Use a graphing utility to decide whether the company is forecasting an increase or a decrease in demand over the decade.

(b) Find the total demand over the next 10 years.

(c) Find the average annual demand during the 10-year period.

64. Capital Campaign The board of trustees of a college is planning a five-year capital gifts campaign to raise money for the college. The goal is to have an annual gift income I that is modeled by

$$I = 2000(375 + 68te^{-0.2t}), \quad 0 \le t \le 5$$

where t is the time in years.

(a) Use a graphing utility to decide whether the board of trustees expects the gift income to increase or decrease over the five-year period.

(b) Find the expected total gift income over the five-year period.

(c) Determine the average annual gift income over the five-year period.

65. Memory Model A model for the ability M of a child to memorize, measured on a scale from 0 to 10, is

$$M = 1 + 1.6t \ln t, \quad 0 < t \le 4$$

where t is the child's age in years.

(a) Find the average value of this model between the child's first and second birthdays.

(b) Find the average value of this model between the child's third and fourth birthdays.

66. Revenue A company sells a seasonal product. The revenue R (in dollars) generated by sales of the product can be modeled by

$$R = 410.5t^2 e^{-t/30} + 25{,}000, \quad 0 \le t \le 365$$

where t is the time in days.

(a) Find the average daily revenue during the first quarter, which is given by $0 \le t \le 90$.

(b) Find the average daily revenue during the fourth quarter, which is given by $274 \le t \le 365$.

(c) Find the total daily revenue during the year.

Finding Present Value In Exercises 67–72, find the present value of the income c (in dollars) over t_1 years at the given annual inflation rate r. *See Examples 6 and 7.*

67. $c = 5000$, $r = 4\%$, $t_1 = 4$ years

68. $c = 450$, $r = 4\%$, $t_1 = 10$ years

69. $c = 100{,}000 + 4000t$, $r = 5\%$, $t_1 = 10$ years

70. $c = 30{,}000 + 500t$, $r = 7\%$, $t_1 = 6$ years

71. $c = 1000 + 50e^{t/2}$, $r = 6\%$, $t_1 = 4$ years

72. $c = 5000 + 25te^{t/10}$, $r = 6\%$, $t_1 = 10$ years

73. Present Value You have just won $1,500,000 in a state lottery. You will be paid an annuity of $100,000 a year for 15 years. When the annual rate of inflation is 5%, what is the present value of this income?

74. Present Value You have just won $65,000,000 in a lottery. You will be paid an annuity of $2,500,000 a year for 26 years. When the annual rate of inflation is 3%, what is the present value of this income?

75. Present Value A company expects its income c during the next 4 years to be modeled by

$$c = 150{,}000 + 75{,}000t, \quad 0 \le t \le 4.$$

(a) Find the actual income for the business over the 4 years.

(b) Assuming an annual inflation rate of 4%, what is the present value of this income?

76. Present Value A professional athlete signs a three-year contract in which the earnings c can be modeled by $c = 500{,}000 + 125{,}000t$, where t represents the year.

(a) Find the actual value of the athlete's contract.

(b) Assuming an annual inflation rate of 3%, what is the present value of the contract?

77. Present Value A professional athlete signs a four-year contract in which the earnings c can be modeled by $c = 3{,}000{,}000 + 750{,}000t$, where t represents the year.

(a) Find the actual value of the athlete's contract.

(b) Assuming an annual inflation rate of 5%, what is the present value of the contract?

78.  **HOW DO YOU SEE IT?** The graphs of two equations show the expected income and the present value of the expected income for a company. Which graph represents the expected income and which graph represents the present value of the expected income? Explain your reasoning.

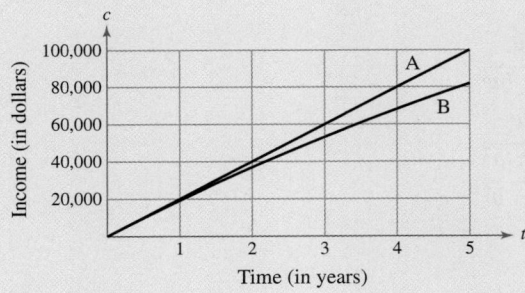

Time (in years)

Future Value In Exercises 79 and 80, find the future value of the income (in dollars) given by $f(t)$ over t_1 years at annual interest rate r. If the function f represents a continuous investment over a period of t_1 years at an annual interest rate r (compounded continuously), then the future value of the investment is given by

$$\text{Future value} = e^{rt_1} \int_0^{t_1} f(t)e^{-rt}\, dt.$$

79. $f(t) = 3000$, $r = 8\%$, $t_1 = 10$ years

80. $f(t) = 3000e^{0.05t}$, $r = 10\%$, $t_1 = 5$ years

81. Finance: Future Value Use the equation from Exercises 79 and 80 to calculate the following. *(Source: Adapted from Garman/Forgue, Personal Finance, Eighth Edition)*

(a) The future value of $1200 saved each year for 10 years earning 7% interest

(b) A person who wishes to invest $1200 each year finds one investment choice that is expected to pay 9% interest per year and another, riskier choice that may pay 10% interest per year. What is the difference in return (future value) if the investment is made for 15 years?

82. College Tuition Fund Assume your grandparents had continuously invested in a college fund according to the model

$$f(t) = 400t$$

for 18 years, at an annual interest rate of 7%.

(a) In 2010, the total cost of attending The Pennsylvania State University for 1 year was estimated to be $26,276. Will the fund have grown enough to allow you to cover 4 years of expenses at The Pennsylvania State University? *(Source: The Pennsylvania State University)*

(b) In 2010, the total cost of attending The Ohio State University for 1 year was estimated to be $23,604. Will the fund have grown enough to allow you to cover 4 years of expenses at The Ohio State University? *(Source: The Ohio State University)*

83. Midpoint Rule Use a program similar to the Midpoint Rule program in Appendix H with $n = 10$ to approximate

$$\int_1^4 \frac{4}{\sqrt{x} + \sqrt[3]{x}}\, dx.$$

84. Midpoint Rule Use a program similar to the Midpoint Rule program in Appendix H with $n = 12$ to approximate the area of the region bounded by the graphs of

$$y = \frac{10}{\sqrt{x}e^x}, \quad y = 0, \quad x = 1, \quad \text{and} \quad x = 4.$$

12.2 Integration Tables

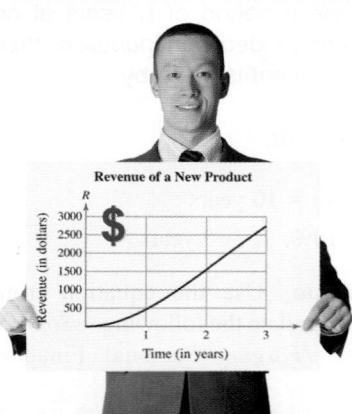

Revenue of a New Product

In Exercise 59 on page 884, you will use a formula from the integration table in Appendix G to find the total revenue of a new product during its first 2 years.

■ Use integration tables to find indefinite and definite integrals.
■ Use reduction formulas to find indefinite integrals.
■ Use integration tables to solve real-life problems.

Integration Tables

You have studied several integration techniques that can be used with the basic integration formulas. Certainly these techniques and formulas do not cover every possible method for finding an antiderivative, but they do cover most of the important ones.

In this section, you will expand the list of integration formulas to form a table of integrals. As you add new integration formulas to the basic list, two effects occur. On one hand, it becomes increasingly difficult to memorize, or even become familiar with, the entire list of formulas. On the other hand, with a longer list you need fewer techniques for fitting an integral to one of the formulas on the list. The procedure of integrating by means of a long list of formulas is called **integration by tables.** (The table in Appendix G constitutes only a partial listing of integration formulas. Much longer lists exist, some of which contain several hundred formulas.)

Integration by tables should not be considered a trivial task. It requires considerable thought and insight, and it often requires substitution. Many people find a table of integrals to be a valuable supplement to the integration techniques discussed in this text. As you gain competence in the use of integration tables, you will improve in the use of the various integration techniques. In doing so, you should find that a combination of techniques and tables is the most versatile approach to integration.

Each integration formula in Appendix G can be developed using one or more of the techniques you have studied. You should try to verify several of the formulas. For instance, Formula 17

$$\int \frac{\sqrt{a + bu}}{u}\, du = 2\sqrt{a + bu} + a \int \frac{1}{u\sqrt{a + bu}}\, du \qquad \text{Formula 17}$$

can be verified using integration by parts, Formula 39

$$\int \frac{1}{1 + e^u}\, du = u - \ln(1 + e^u) + C \qquad \text{Formula 39}$$

can be verified using substitution, and Formula 44

$$\int (\ln u)^2\, du = u[2 - 2\ln u - (\ln u)^2] + C \qquad \text{Formula 44}$$

can be verified using integration by parts twice.

In the table of integrals in Appendix G, the formulas have been classified according to the form of the integrand. Several of the forms are listed below.

- Forms involving u^n
- Forms involving $a + bu$
- Forms involving $\sqrt{a + bu}$
- Forms involving $u^2 - a^2$
- Forms involving $\sqrt{u^2 \pm a^2}$
- Forms involving $\sqrt{a^2 - u^2}$
- Forms involving e^u
- Forms involving $\ln u$

TECH TUTOR

Throughout this section, remember that a symbolic integration utility can be used instead of integration tables. If you have access to such a utility, try using it to find the indefinite integrals in Examples 1 and 2.

Example 1 **Using Integration Tables**

Find $\int \dfrac{x}{\sqrt{x-1}}\, dx$.

SOLUTION   Because the expression inside the radical is linear, you should consider forms involving $\sqrt{a + bu}$, as in Formula 19.

$$\int \frac{u}{\sqrt{a+bu}}\, du = -\frac{2(2a - bu)}{3b^2}\sqrt{a + bu} + C \qquad \text{Formula 19}$$

Using this formula, let $a = -1$, $b = 1$, and $u = x$. Then $du = dx$, and you obtain

$$\int \frac{x}{\sqrt{x-1}}\, dx = -\frac{2(-2 - x)}{3}\sqrt{x - 1} + C \qquad \begin{array}{l}\text{Substitute values}\\ \text{of } a, b, \text{ and } u.\end{array}$$

$$= \frac{2}{3}(2 + x)\sqrt{x - 1} + C. \qquad \text{Simplify.}$$

✔ **Checkpoint 1**

Use the integration table in Appendix G to find

$$\int \frac{x}{\sqrt{2 + x}}\, dx.$$

Example 2 **Using Integration Tables**

Find $\int x\sqrt{x^4 - 9}\, dx$.

SOLUTION   Because it is not clear which formula to use, you can begin by letting $u = x^2$ and $du = 2x\, dx$. With these substitutions, you can write the integral as shown.

$$\int x\sqrt{x^4 - 9}\, dx = \frac{1}{2}\int \sqrt{(x^2)^2 - 9}\,(2x)\, dx \qquad \text{Multiply and divide by 2.}$$

$$= \frac{1}{2}\int \sqrt{u^2 - 9}\, du \qquad \text{Substitute } u \text{ and } du.$$

Now, it appears that you can use Formula 23.

$$\int \sqrt{u^2 - a^2}\, du = \frac{1}{2}\left(u\sqrt{u^2 - a^2} - a^2 \ln\left|u + \sqrt{u^2 - a^2}\right|\right) + C$$

Letting $a = 3$, you obtain

$$\int x\sqrt{x^4 - 9}\, dx = \frac{1}{2}\int \sqrt{u^2 - a^2}\, du$$

$$= \frac{1}{2}\left[\frac{1}{2}\left(u\sqrt{u^2 - a^2} - a^2 \ln\left|u + \sqrt{u^2 - a^2}\right|\right)\right] + C$$

$$= \frac{1}{4}\left(x^2\sqrt{x^4 - 9} - 9 \ln\left|x^2 + \sqrt{x^4 - 9}\right|\right) + C.$$

✔ **Checkpoint 2**

Use the integration table in Appendix G to find

$$\int \frac{\sqrt{x^2 + 16}}{x}\, dx.$$

Example 3 **Using Integration Tables**

Find $\displaystyle\int \frac{1}{x\sqrt{x+1}}\,dx$.

SOLUTION Considering forms involving $\sqrt{a+bu}$, where $a=1$, $b=1$, and $u=x$, you can use Formula 15.

$$\int \frac{1}{u\sqrt{a+bu}}\,du = \frac{1}{\sqrt{a}}\ln\left|\frac{\sqrt{a+bu}-\sqrt{a}}{\sqrt{a+bu}+\sqrt{a}}\right| + C, \quad a>0$$

So,

$$\int \frac{1}{x\sqrt{x+1}}\,dx = \int \frac{1}{u\sqrt{a+bu}}\,du$$

$$= \frac{1}{\sqrt{a}}\ln\left|\frac{\sqrt{a+bu}-\sqrt{a}}{\sqrt{a+bu}+\sqrt{a}}\right| + C$$

$$= \ln\left|\frac{\sqrt{x+1}-1}{\sqrt{x+1}+1}\right| + C.$$

✓ **Checkpoint 3**

Use the integration table in Appendix G to find $\displaystyle\int \frac{1}{x^2-4}\,dx$. ■

Example 4 **Using Integration Tables**

Evaluate $\displaystyle\int_0^2 \frac{x}{1+e^{-x^2}}\,dx$.

SOLUTION Of the forms involving e^u, Formula 39

$$\int \frac{1}{1+e^u}\,du = u - \ln(1+e^u) + C$$

seems most appropriate. To use this formula, let $u = -x^2$ and $du = -2x\,dx$.

$$\int \frac{x}{1+e^{-x^2}}\,dx = -\frac{1}{2}\int \frac{1}{1+e^{-x^2}}(-2x)\,dx$$

$$= -\frac{1}{2}\int \frac{1}{1+e^u}\,du$$

$$= -\frac{1}{2}[u - \ln(1+e^u)] + C$$

$$= -\frac{1}{2}[-x^2 - \ln(1+e^{-x^2})] + C$$

$$= \frac{1}{2}[x^2 + \ln(1+e^{-x^2})] + C$$

So, the value of the definite integral is

$$\int_0^2 \frac{x}{1+e^{-x^2}}\,dx = \frac{1}{2}\Big[x^2 + \ln(1+e^{-x^2})\Big]_0^2 \approx 1.66. \qquad \text{See Figure 12.3.}$$

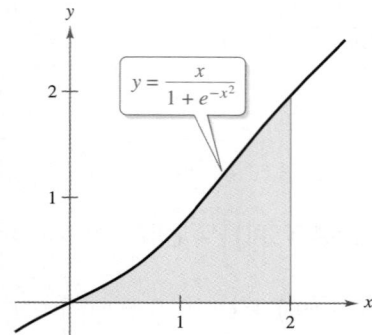

FIGURE 12.3

✓ **Checkpoint 4**

Use the integration table in Appendix G to evaluate $\displaystyle\int_0^1 \frac{x^2}{1+e^{x^3}}\,dx$. ■

Reduction Formulas

Several of the formulas in the integration table have the form

$$\int f(x)\,dx = g(x) + \int h(x)\,dx$$

where the right side contains an integral. Such integration formulas are called **reduction formulas** because they reduce the original integral to the sum of a function and a simpler integral.

ALGEBRA TUTOR
xy

For help on the algebra in Example 5, see Example 3 in the *Chapter 12 Algebra Tutor*, on page 905.

Example 5   Using a Reduction Formula

Find $\int x^2 e^x\,dx$.

SOLUTION   Using Formula 38

$$\int u^n e^u\,du = u^n e^u - n\int u^{n-1} e^u\,du$$

you can let $u = x$ and $n = 2$. Then $du = dx$, and you can write

$$\int x^2 e^x\,dx = x^2 e^x - 2\int x e^x\,dx.$$

Then, using Formula 37

$$\int u e^u\,du = (u - 1)e^u + C$$

you can write

$$\int x^2 e^x\,dx = x^2 e^x - 2\int x e^x\,dx$$
$$= x^2 e^x - 2(x - 1)e^x + C$$
$$= x^2 e^x - 2x e^x + 2e^x + C$$
$$= e^x(x^2 - 2x + 2) + C.$$

You can check this result by differentiating.

$$\frac{d}{dx}[e^x(x^2 - 2x + 2) + C] = e^x(2x - 2) + (x^2 - 2x + 2)(e^x)$$
$$= 2x e^x - 2e^x + x^2 e^x - 2x e^x + 2e^x$$
$$= x^2 e^x$$

✓**Checkpoint 5**

Use the integration table in Appendix G to find the indefinite integral $\int (\ln x)^2\,dx$.   ■

TECH TUTOR

You have now studied two ways to find the indefinite integral in Example 5. Example 5 uses an integration table, and Example 4 in Section 12.1 uses integration by parts. A third way would be to use a symbolic integration utility.

Application

Integration can be used to find the probability that an event will occur. In such an application, the real-life situation is modeled by a *probability density function f*, and the probability that x will lie between a and b is represented by

$$P(a \le x \le b) = \int_a^b f(x)\,dx.$$

The probability $P(a \le x \le b)$ must be a number between 0 and 1.

 Example 6 **Finding a Probability**

A psychologist finds that the probability that a participant in a memory experiment will recall between a and b percent (in decimal form) of the material is

$$P(a \le x \le b) = \int_a^b \frac{1}{e-2} x^2 e^x\,dx, \quad 0 \le a \le b \le 1.$$

Find the probability that a randomly chosen participant will recall between 0% and 87.5% of the material.

SOLUTION You can use the Constant Multiple Rule to rewrite the integral as

$$\frac{1}{e-2} \int_a^b x^2 e^x\,dx.$$

Note that the integrand is the same as the one in Example 5. Use the result of Example 5 to find the probability with $a = 0$ and $b = 0.875$.

$$\frac{1}{e-2} \int_0^{0.875} x^2 e^x\,dx = \frac{1}{e-2} \left[e^x(x^2 - 2x + 2) \right]_0^{0.875}$$
$$\approx 0.608$$

So, the probability is about 60.8%, as indicated in Figure 12.4.

y-axis graph:
$y = \dfrac{1}{e-2} x^2 e^x$

Area ≈ 0.608

FIGURE 12.4

✓ Checkpoint 6

Use Example 6 to find the probability that a participant will recall between 0% and 62.5% of the material.

SUMMARIZE   (Section 12.2)

1. Describe what is meant by integration by tables *(page 878)*. For examples of integration by tables, see Examples 1, 2, 3, and 4.

2. Describe what is meant by a reduction formula *(page 881)*. For an example of a reduction formula, see Example 5.

3. Describe a real-life example of how integration by tables can be used to analyze the results of a memory experiment *(page 882, Example 6)*.

SKILLS WARM UP 12.2   The following warm-up exercises involve skills that were covered in a previous course or in earlier sections. You will use these .skills in the exercise set for this section. For additional help, review Sections 0.5 and 12.1.

In Exercises 1–4, expand the expression.

1. $(x + 4)^2$

2. $(x - 1)^2$

3. $\left(x + \frac{1}{2}\right)^2$

4. $\left(x - \frac{1}{3}\right)^2$

In Exercises 5 and 6, use integration by parts to find the indefinite integral.

5. $\displaystyle\int 2xe^x\, dx$

6. $\displaystyle\int 3x^2 \ln x\, dx$

Exercises 12.2

See www.CalcChat.com for worked-out solutions to odd-numbered exercises.

Using Integration Tables In Exercises 1–8, use the indicated formula from the integration table in Appendix G to find the indefinite integral. *See Examples 1, 2, and 3.*

1. $\displaystyle\int \frac{x}{(2 + 3x)^2}\, dx$, Formula 4

2. $\displaystyle\int \frac{1}{x(2 + 3x)^2}\, dx$, Formula 11

3. $\displaystyle\int \frac{x}{\sqrt{2 + 3x}}\, dx$, Formula 19

4. $\displaystyle\int \frac{4}{x^2 - 9}\, dx$, Formula 21

5. $\displaystyle\int \frac{2x}{\sqrt{x^4 - 9}}\, dx$, Formula 27

6. $\displaystyle\int x^2\sqrt{x^2 + 9}\, dx$, Formula 24

7. $\displaystyle\int x^3 e^{x^2}\, dx$, Formula 37

8. $\displaystyle\int \frac{x}{1 + e^{x^2}}\, dx$, Formula 39

Using Integration Tables In Exercises 9–36, use the integration table in Appendix G to find the indefinite integral. *See Examples 1, 2, 3, and 5.*

9. $\displaystyle\int \frac{1}{x(1 + x)}\, dx$

10. $\displaystyle\int \frac{1}{x(1 + x)^2}\, dx$

11. $\displaystyle\int \frac{1}{x\sqrt{x^2 + 9}}\, dx$

12. $\displaystyle\int \frac{1}{\sqrt{x^2 - 1}}\, dx$

13. $\displaystyle\int \frac{1}{x\sqrt{4 - x^2}}\, dx$

14. $\displaystyle\int \frac{\sqrt{x^2 - 9}}{x^2}\, dx$

15. $\displaystyle\int 3x \ln 3x\, dx$

16. $\displaystyle\int (\ln 5x)^2\, dx$

17. $\displaystyle\int \frac{6x}{1 + e^{3x^2}}\, dx$

18. $\displaystyle\int \frac{1}{1 + e^x}\, dx$

19. $\displaystyle\int x^2\sqrt{3 + x}\, dx$

20. $\displaystyle\int \frac{x}{x^4 - 9}\, dx$

21. $\displaystyle\int \frac{t^2}{(2 + 3t)^3}\, dt$

22. $\displaystyle\int \frac{\sqrt{3 + 4t}}{t}\, dt$

23. $\displaystyle\int \frac{1}{x\sqrt{3 + 4x}}\, dx$

24. $\displaystyle\int \sqrt{3 + x^2}\, dx$

25. $\displaystyle\int \frac{x^2}{1 + x}\, dx$

26. $\displaystyle\int \frac{1}{1 + e^{2x}}\, dx$

27. $\displaystyle\int \frac{x^2}{(3 + 2x)^5}\, dx$

28. $\displaystyle\int \frac{1}{x^2\sqrt{x^2 - 4}}\, dx$

29. $\displaystyle\int \frac{1}{x^2\sqrt{1 - x^2}}\, dx$

30. $\displaystyle\int \frac{2x}{(1 - 3x)^2}\, dx$

31. $\displaystyle\int 4x^2 \ln 2x\, dx$

32. $\displaystyle\int xe^{x^2}\, dx$

33. $\displaystyle\int \frac{x^2}{(3x - 5)^2}\, dx$

34. $\displaystyle\int \frac{1}{2x^2(2x - 1)^2}\, dx$

35. $\displaystyle\int \frac{\ln x}{x(4 + 3\ln x)}\, dx$

36. $\displaystyle\int (\ln x)^3\, dx$

Using Integration Tables In Exercises 37–44, use the integration table in Appendix G to evaluate the definite integral. *See Example 4.*

37. $\displaystyle\int_0^1 \frac{x}{\sqrt{1 + x}}\, dx$

38. $\displaystyle\int_0^5 \frac{x}{\sqrt{5 + 2x}}\, dx$

39. $\displaystyle\int_0^5 \frac{x}{(4 + x)^2}\, dx$

40. $\displaystyle\int_2^4 \frac{x^2}{(3x - 5)}\, dx$

41. $\displaystyle\int_0^4 \frac{6}{1 + e^{0.5x}}\, dx$

42. $\displaystyle\int_2^4 \sqrt{3 + x^2}\, dx$

43. $\displaystyle\int_1^2 x^3 \ln x^2\, dx$

44. $\displaystyle\int_0^3 \frac{x}{(1 + 3x)^4}\, dx$

Area of a Region In Exercises 45–50, use the integration table in Appendix G to find the exact area of the region bounded by the graphs of the equations. Use a graphing utility to verify your results.

45. $y = \dfrac{1}{(16 - x^2)^{3/2}}$, $y = 0$, $x = -2$, $x = 2$

46. $y = \dfrac{2}{1 + e^{4x}}$, $y = 0$, $x = 0$, $x = 1$

47. $y = \dfrac{1}{9x^2(2 + 3x)}$, $y = 0$, $x = 1$, $x = 2$

48. $y = \dfrac{-e^x}{1 - e^{2x}}$, $y = 0$, $x = 1$, $x = 2$

49. $y = x^2\sqrt{x^2 + 4}$, $y = 0$, $x = \sqrt{5}$

50. $y = x \ln x^2$, $y = 0$, $x = 4$

Finding Indefinite Integrals Using Two Methods In Exercises 51–54, find the indefinite integral (a) using the integration table in Appendix G and (b) using integration by parts.

51. $\displaystyle \int \ln \frac{x}{3}\, dx$

52. $\displaystyle \int 4xe^{4x}\, dx$

53. $\displaystyle \int \frac{x}{\sqrt{7x - 3}}\, dx$

54. $\displaystyle \int 7x \ln 7x\, dx$

55. Probability The probability of recalling between a and b percent (in decimal form) of the material learned in a memory experiment is modeled by

$$P(a \le x \le b) = \int_a^b \frac{75}{14}\left(\frac{x}{\sqrt{4 + 5x}}\right) dx,$$

$$0 \le a \le b \le 1.$$

What are the probabilities of recalling (a) between 40% and 80% and (b) between 0% and 50% of the material?

56. Probability The probability of finding between a and b percent iron (in decimal form) in ore samples is modeled by

$$P(a \le x \le b) = \int_a^b 2x^3 e^{x^2}\, dx, \quad 0 \le a \le b \le 1.$$

What are the probabilities of finding (a) between 0% and 25% and (b) between 50% and 100% iron in a sample?

 Population Growth In Exercises 57 and 58, use a graphing utility to graph the growth function. Use the integration table in Appendix G to find the average value of the growth function over the interval, where N is the size of a population and t is the time in days.

57. $N = \dfrac{5000}{1 + e^{4.8 - 1.9t}}$, $[0, 2]$

58. $N = \dfrac{375}{1 + e^{4.20 - 0.25t}}$, $[21, 28]$

59. Revenue The revenue (in dollars) for a new product is modeled by $R = 10{,}000[1 - 1/(1 + 0.1t^2)^{1/2}]$, where t is the time in years. Estimate the total revenue of the product over its first 2 years on the market.

60. HOW DO YOU SEE IT? The graph shows the rate of change of the sales of a new product.

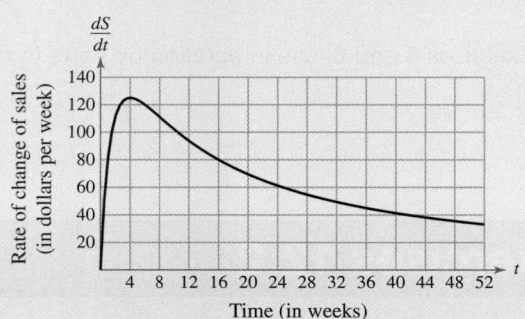

(a) Approximate the rate of change of the sales after 16 weeks. Explain your reasoning.

(b) Approximate the weeks for which the sales are increasing. Explain your reasoning.

61. Consumer and Producer Surpluses The demand and supply functions for a product are modeled by

Demand: $p = 60/\sqrt{x^2 + 81}$, *Supply:* $p = x/3$

where p is the price (in dollars) and x is the number of units (in millions). Find the consumer and producer surpluses for this product.

62. Project: Purchasing Power of the Dollar For a project analyzing the purchasing power of the dollar from 1983 through 2009, visit this text's website at *www.cengagebrain.com*. *(Source: U.S. Bureau of Labor Statistics)*

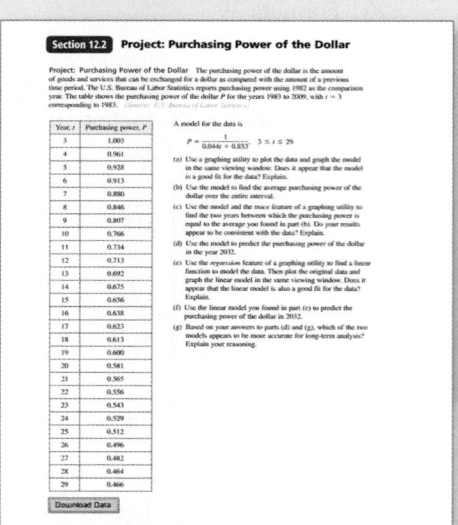

QUIZ YOURSELF

Take this quiz as you would take a quiz in class. When you are done, check your work against the answers given in the back of the book.

In Exercises 1–6, use integration by parts to find the indefinite integral.

1. $\displaystyle\int xe^{5x}\,dx$

2. $\displaystyle\int \ln x^3\,dx$

3. $\displaystyle\int (x+1)\ln x\,dx$

4. $\displaystyle\int x\sqrt{x+3}\,dx$

5. $\displaystyle\int x\ln\sqrt{x}\,dx$

6. $\displaystyle\int x^2 e^{-2x}\,dx$

7. A manufacturing company forecasts that the demand x (in units) for its product over the next 5 years can be modeled by

$$x = 1000(45 + 20te^{-0.5t})$$

where t is the time in years.

  (a) Find the total demand over the next 5 years.

  (b) Find the average annual demand during the 5-year period.

8. A small business expects its income c during the next 7 years to be given by

$$c(t) = 32{,}000t, \quad 0 \le t \le 7.$$

  (a) Find the actual income for the business over the 7 years.

  (b) Assuming an annual inflation rate of 3.3%, what is the present value of this income?

In Exercises 9–14, use the integration table in Appendix G to find the indefinite integral.

9. $\displaystyle\int \frac{x}{1+2x}\,dx$

10. $\displaystyle\int \frac{1}{x(0.1+0.2x)}\,dx$

11. $\displaystyle\int \frac{\sqrt{x^2-16}}{x^2}\,dx$

12. $\displaystyle\int \frac{1}{x\sqrt{4+9x}}\,dx$

13. $\displaystyle\int \frac{2x}{1+e^{4x^2}}\,dx$

14. $\displaystyle\int 2x(x^2+1)e^{x^2+1}\,dx$

15. The revenue (in millions of dollars) for a new product is modeled by

$$R = \sqrt{144t^2 + 400}$$

where t is the time in years.

  (a) Estimate the total revenue of the product over its first 3 years on the market.

  (b) Estimate the total revenue of the product over its first 6 years on the market.

In Exercises 16–21, evaluate the definite integral.

16. $\displaystyle\int_{-2}^{0} xe^{x/2}\,dx$

17. $\displaystyle\int_{1}^{2} 5x\ln x\,dx$

18. $\displaystyle\int_{0}^{8} \frac{x}{\sqrt{x+8}}\,dx$

19. $\displaystyle\int_{1}^{e} (\ln x)^2\,dx$

20. $\displaystyle\int_{2}^{3} \frac{1}{x^2\sqrt{9-x^2}}\,dx$

21. $\displaystyle\int_{4}^{6} \frac{2x}{x^4-4}\,dx$

12.3 Numerical Integration

■ Use the Trapezoidal Rule to approximate definite integrals.
■ Use Simpson's Rule to approximate definite integrals.
■ Analyze the sizes of the errors when approximating definite integrals with the Trapezoidal Rule and Simpson's Rule.

The Trapezoidal Rule

In Section 11.6, you studied one technique for approximating the value of a *definite* integral—the Midpoint Rule. In this section, you will study two other approximation techniques: the **Trapezoidal Rule** and **Simpson's Rule.**

To develop the Trapezoidal Rule, consider a function f that is nonnegative and continuous on the closed interval $[a, b]$. To approximate the area represented by

$$\int_a^b f(x)\,dx$$

partition the interval into n subintervals, each of width

$$\Delta x = \frac{b - a}{n}. \qquad \text{Width of each subinterval}$$

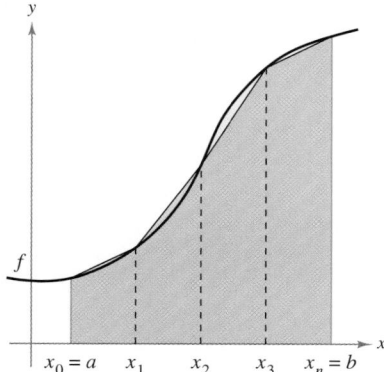

The area of the region can be approximated using four trapezoids.

FIGURE 12.5

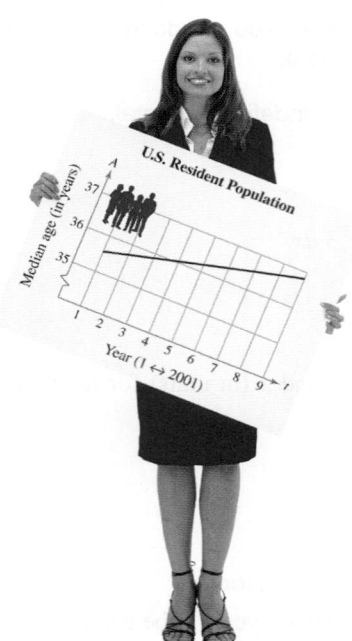

In Exercise 43 on page 893, you will use Simpson's Rule to find the average median age of the U.S. resident population from 2001 through 2009.

Next, form n trapezoids, as shown in Figure 12.5. As you can see in Figure 12.6, the area of the first trapezoid is

$$\text{Area of first trapezoid} = \left(\frac{b - a}{n}\right)\left[\frac{f(x_0) + f(x_1)}{2}\right].$$

The areas of the other trapezoids follow a similar pattern, and the sum of the n areas is

$$\begin{aligned}
\text{Area} &= \left(\frac{b - a}{n}\right)\left[\frac{f(x_0) + f(x_1)}{2} + \frac{f(x_1) + f(x_2)}{2} + \cdots + \frac{f(x_{n-1}) + f(x_n)}{2}\right] \\
&= \left(\frac{b - a}{2n}\right)[f(x_0) + f(x_1) + f(x_1) + f(x_2) + \cdots + f(x_{n-1}) + f(x_n)] \\
&= \left(\frac{b - a}{2n}\right)[f(x_0) + 2f(x_1) + 2f(x_2) + \cdots + 2f(x_{n-1}) + f(x_n)].
\end{aligned}$$

Although this development assumes f to be continuous *and* nonnegative on $[a, b]$, the resulting formula is valid as long as f is continuous on $[a, b]$.

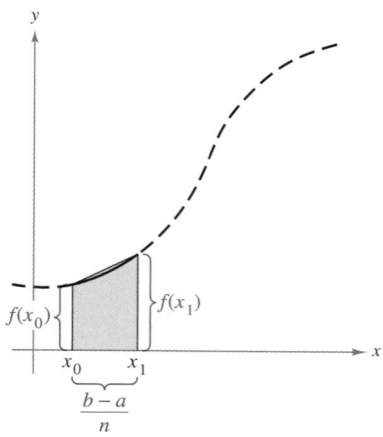

FIGURE 12.6

The Trapezoidal Rule

If f is continuous on $[a, b]$, then

$$\int_a^b f(x)\,dx \approx \left(\frac{b - a}{2n}\right)[f(x_0) + 2f(x_1) + \cdots + 2f(x_{n-1}) + f(x_n)].$$

Note that the coefficients in the Trapezoidal Rule have the following pattern.

$$1 \quad 2 \quad 2 \quad 2 \ldots 2 \quad 2 \quad 1$$

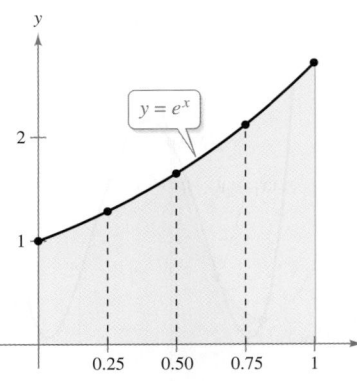

Four Subintervals
FIGURE 12.7

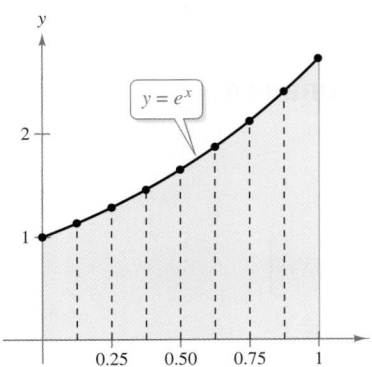

Eight Subintervals
FIGURE 12.8

Example 1   Using the Trapezoidal Rule

Use the Trapezoidal Rule to approximate $\int_0^1 e^x \, dx$. Compare the results for $n = 4$ and $n = 8$.

SOLUTION   When $n = 4$, the width of each subinterval is

$$\frac{1 - 0}{4} = \frac{1}{4}$$

and the endpoints of the subintervals are

$$x_0 = 0, \quad x_1 = \frac{1}{4}, \quad x_2 = \frac{1}{2}, \quad x_3 = \frac{3}{4}, \quad \text{and} \quad x_4 = 1$$

as indicated in Figure 12.7. So, by the Trapezoidal Rule,

$$\int_0^1 e^x \, dx = \frac{1}{8}(e^0 + 2e^{0.25} + 2e^{0.5} + 2e^{0.75} + e^1)$$

$$\approx 1.7272. \qquad \text{Approximation using } n = 4$$

When $n = 8$, the width of each subinterval is

$$\frac{1 - 0}{8} = \frac{1}{8}$$

and the endpoints of the subintervals are

$$x_0 = 0, \quad x_1 = \frac{1}{8}, \quad x_2 = \frac{1}{4}, \quad x_3 = \frac{3}{8}, \quad x_4 = \frac{1}{2},$$

$$x_5 = \frac{5}{8}, \quad x_6 = \frac{3}{4}, \quad x_7 = \frac{7}{8}, \quad \text{and} \quad x_8 = 1$$

as indicated in Figure 12.8. So, by the Trapezoidal Rule,

$$\int_0^1 e^x \, dx = \frac{1}{16}(e^0 + 2e^{0.125} + 2e^{0.25} + \cdots + 2e^{0.875} + e^1)$$

$$\approx 1.7205. \qquad \text{Approximation using } n = 8$$

Of course, for *this particular* integral, you could have found an antiderivative and used the Fundamental Theorem of Calculus to find the exact value of the definite integral. The exact value is

$$\int_0^1 e^x \, dx = e - 1 \qquad \text{Exact value}$$

which is approximately 1.718282.

✓ Checkpoint 1

Use the Trapezoidal Rule with $n = 4$ to approximate

$$\int_0^1 e^{2x} \, dx.$$

■

TECH TUTOR

A symbolic integration utility can be used to evaluate a definite integral. Use a symbolic integration utility to approximate the integral $\int_0^1 e^{x^2} \, dx$.

There are two important points that should be made concerning the Trapezoidal Rule. First, the approximation tends to become more accurate as n increases. For instance, in Example 1, when $n = 16$, the Trapezoidal Rule yields an approximation of 1.7188. Second, although you could have used the Fundamental Theorem of Calculus to evaluate the integral in Example 1, this theorem cannot be used to evaluate an integral as simple as $\int_0^1 e^{x^2} \, dx$. Yet the Trapezoidal Rule can be easily applied to estimate this integral.

Simpson's Rule

One way to view the Trapezoidal Rule is to say that f is approximated by a first-degree polynomial on each subinterval. In Simpson's Rule, f is approximated by a second-degree polynomial on each subinterval.

To develop Simpson's Rule, partition the interval $[a, b]$ into an *even number n* of subintervals, each of width

$$\Delta x = \frac{b - a}{n}.$$

On the subinterval $[x_0, x_2]$, approximate the function f by the second-degree polynomial $p(x)$ that passes through the points

$$(x_0, f(x_0)), \quad (x_1, f(x_1)), \quad \text{and} \quad (x_2, f(x_2))$$

as shown in Figure 12.9. The Fundamental Theorem of Calculus can be used to show that

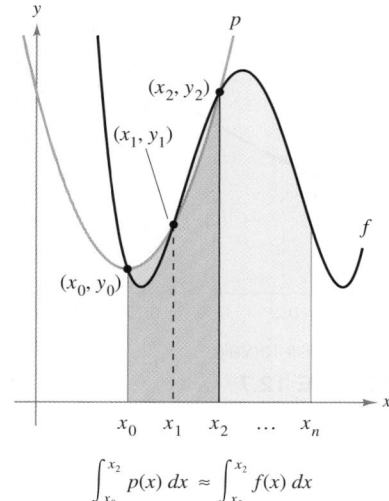

$$\int_{x_0}^{x_2} p(x)\, dx \approx \int_{x_0}^{x_2} f(x)\, dx$$

FIGURE 12.9

$$\int_{x_0}^{x_2} f(x)\, dx \approx \int_{x_0}^{x_2} p(x)\, dx$$

$$= \left(\frac{x_2 - x_0}{6}\right)\left[p(x_0) + 4p\left(\frac{x_0 + x_2}{2}\right) + p(x_2)\right]$$

$$= \frac{2[(b - a)/n]}{6}[p(x_0) + 4p(x_1) + p(x_2)]$$

$$= \left(\frac{b - a}{3n}\right)[f(x_0) + 4f(x_1) + f(x_2)].$$

Repeating this process on the subintervals $[x_{i-2}, x_i]$ produces

$$\int_a^b f(x)\, dx \approx \left(\frac{b - a}{3n}\right)[f(x_0) + 4f(x_1) + f(x_2) + f(x_2) + 4f(x_3) +$$

$$f(x_4) + \cdots + f(x_{n-2}) + 4f(x_{n-1}) + f(x_n)].$$

By grouping like terms, you can obtain the approximation shown below, which is known as Simpson's Rule. This rule is named after the English mathematician Thomas Simpson (1710–1761).

Simpson's Rule (n Is Even)

If f is continuous on $[a, b]$, then

$$\int_a^b f(x)\, dx \approx \left(\frac{b - a}{3n}\right)[f(x_0) + 4f(x_1) + 2f(x_2) + 4f(x_3) +$$

$$\cdots + 4f(x_{n-1}) + f(x_n)].$$

Note that the coefficients in Simpson's Rule have the following pattern.

$$1 \quad 4 \quad 2 \quad 4 \quad 2 \quad 4 \ldots 4 \quad 2 \quad 4 \quad 1$$

The Trapezoidal Rule and Simpson's Rule are necessary for solving certain real-life problems, such as approximating the present value of an income. You will see such problems in the exercise set for this section.

In Example 1, the Trapezoidal Rule was used to estimate the value of

$$\int_0^1 e^x \, dx.$$

The next example uses Simpson's Rule to approximate the same integral.

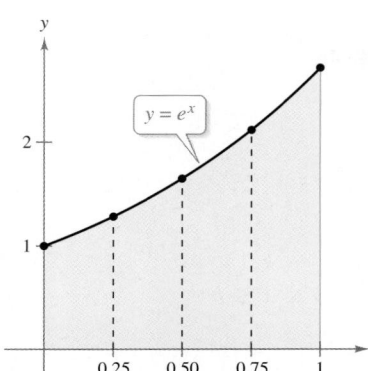

y = e^x

Four Subintervals

FIGURE 12.10

Example 2   Using Simpson's Rule

Use Simpson's Rule to approximate

$$\int_0^1 e^x \, dx.$$

Compare the results for $n = 4$ and $n = 8$.

SOLUTION   When $n = 4$, the width of each subinterval is

$$\frac{1 - 0}{4} = \frac{1}{4}$$

and the endpoints of the subintervals are

$$x_0 = 0, \quad x_1 = \frac{1}{4}, \quad x_2 = \frac{1}{2}, \quad x_3 = \frac{3}{4}, \quad \text{and} \quad x_4 = 1$$

as indicated in Figure 12.10. So, by Simpson's Rule

$$\int_0^1 e^x \, dx = \frac{1}{12}(e^0 + 4e^{0.25} + 2e^{0.5} + 4e^{0.75} + e^1)$$

$$\approx 1.718319. \qquad \text{Approximation using } n = 4$$

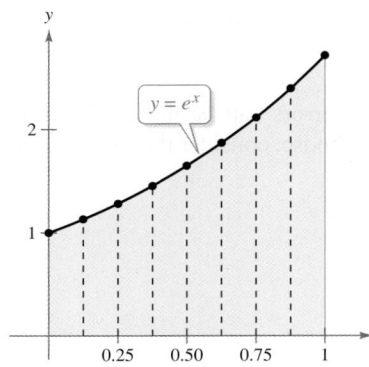

y = e^x

Eight Subintervals

FIGURE 12.11

When $n = 8$, the width of each subinterval is $(1 - 0)/8 = \frac{1}{8}$ and the endpoints of the subintervals are

$$x_0 = 0, \quad x_1 = \frac{1}{8}, \quad x_2 = \frac{1}{4}, \quad x_3 = \frac{3}{8}, \quad x_4 = \frac{1}{2},$$

$$x_5 = \frac{5}{8}, \quad x_6 = \frac{3}{4}, \quad x_7 = \frac{7}{8}, \quad \text{and} \quad x_8 = 1$$

as indicated in Figure 12.11. So, by Simpson's Rule

$$\int_0^1 e^x \, dx = \frac{1}{24}(e^0 + 4e^{0.125} + 2e^{0.25} + \cdots + 4e^{0.875} + e^1)$$

$$\approx 1.718284. \qquad \text{Approximation using } n = 8$$

Recall that the exact value of this integral is

$$\int_0^1 e^x \, dx = e - 1 \qquad \text{Exact value}$$

which is approximately

$$1.718282. \qquad \text{Approximate value}$$

So, with only eight subintervals, you obtained an approximation that is correct to the nearest 0.000002—an impressive result.

✓ Checkpoint 2

Use Simpson's Rule with $n = 4$ to approximate

$$\int_0^1 e^{2x} \, dx.$$

Error Analysis

A program for several models of graphing utilities that uses Simpson's Rule to approximate the definite integral $\int_a^b f(x)\,dx$ can be found in Appendix H.

TECH TUTOR

In Examples 1 and 2, you were able to calculate the exact value of the integral and compare that value with the approximations to see how good they were. In practice, you need to have a different way of telling how good an approximation is: such a way is provided in the next result.

Errors in the Trapezoidal Rule and Simpson's Rule

The errors E in approximating

$$\int_a^b f(x)\,dx$$

are as shown.

Trapezoidal Rule: $\quad |E| \le \dfrac{(b-a)^3}{12n^2}\big[\max|f''(x)|\big], \quad a \le x \le b$

Simpson's Rule: $\quad |E| \le \dfrac{(b-a)^5}{180n^4}\big[\max|f^{(4)}(x)|\big], \quad a \le x \le b$

This result indicates that the errors generated by the Trapezoidal Rule and Simpson's Rule have upper bounds dependent on the extreme values of

$$f''(x) \quad \text{and} \quad f^{(4)}(x)$$

in the interval $[a, b]$. Furthermore, the bounds for the errors can be made arbitrarily small by *increasing n*. To determine what value of n to choose, consider the steps below.

Trapezoidal Rule

1. Find $f''(x)$.

2. Find the maximum of $|f''(x)|$ on the interval $[a, b]$.

3. Set up the inequality

$$|E| \le \frac{(b-a)^3}{12n^2}\big[\max|f''(x)|\big].$$

4. For an error less than ϵ, solve for n in the inequality

$$\frac{(b-a)^3}{12n^2}\big[\max|f''(x)|\big] < \epsilon.$$

5. Partition $[a, b]$ into n subintervals and apply the Trapezoidal Rule.

Simpson's Rule

1. Find $f^{(4)}(x)$.

2. Find the maximum of $|f^{(4)}(x)|$ on the interval $[a, b]$.

3. Set up the inequality

$$|E| \le \frac{(b-a)^5}{180n^4}\big[\max|f^{(4)}(x)|\big].$$

4. For an error less than ϵ, solve for n in the inequality

$$\frac{(b-a)^5}{180n^4}\big[\max|f^{(4)}(x)|\big] < \epsilon.$$

5. Partition $[a, b]$ into n subintervals and apply Simpson's Rule.

ALGEBRA TUTOR xy

For help on the algebra in Example 3, see Example 4 in the Chapter 12 *Algebra Tutor*, on page 905.

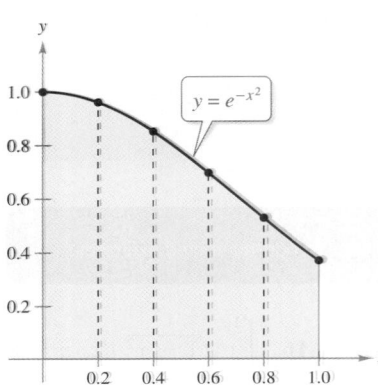

FIGURE 12.12

Example 3   **The Approximate Error in the Trapezoidal Rule**

Use the Trapezoidal Rule to estimate the value of $\int_0^1 e^{-x^2}\,dx$ such that the error in the approximation of the integral is less than 0.01.

SOLUTION

1. Begin by finding the second derivative of $f(x) = e^{-x^2}$.

$$f(x) = e^{-x^2}$$
$$f'(x) = -2xe^{-x^2}$$
$$f''(x) = 4x^2e^{-x^2} - 2e^{-x^2}$$
$$= 2e^{-x^2}(2x^2 - 1)$$

2. f'' has only one critical number in the interval $[0, 1]$, and the maximum value of $|f''(x)|$ on this interval is $|f''(0)| = 2$.

3. The error E using the Trapezoidal Rule is bounded by

$$|E| \le \frac{(b-a)^3}{12n^2}(2) = \frac{1}{12n^2}(2) = \frac{1}{6n^2}.$$

4. To ensure that the approximation has an error of less than 0.01, you should choose n such that

$$\frac{1}{6n^2} < 0.01.$$

Solving for n, you can determine that n must be 5 or more.

5. Partition $[0, 1]$ into five subintervals, as shown in Figure 12.12. Then apply the Trapezoidal Rule to obtain

$$\int_0^1 e^{-x^2}\,dx = \frac{1}{10}\left(\frac{1}{e^0} + \frac{2}{e^{0.04}} + \frac{2}{e^{0.16}} + \frac{2}{e^{0.36}} + \frac{2}{e^{0.64}} + \frac{1}{e^1}\right) \approx 0.744.$$

So, with an error less than 0.01, you know that

$$0.734 \le \int_0^1 e^{-x^2}\,dx \le 0.754.$$

✓ Checkpoint 3

Use the Trapezoidal Rule to estimate the value of

$$\int_0^1 \sqrt{1 + x^2}\,dx$$

such that the error in the approximation of the integral is less than 0.01.

SUMMARIZE (Section 12.3)

1. State the Trapezoidal Rule *(page 886)*. For an example of the Trapezoidal Rule, see Example 1.

2. State Simpson's Rule *(page 888)*. For an example of Simpson's Rule, see Example 2.

3. State the approximate errors in the Trapezoidal Rule and Simpson's Rule *(page 890)*. For an example of using the approximate error in the Trapezoidal Rule, see Example 3.

SKILLS WARM UP 12.3   The following warm-up exercises involve skills that were covered in a previous course or in earlier sections. You will use these skills in the exercise set for this section. For additional help, review Sections 1.7, 7.4, 8.1, 8.5, 10.3, and 10.5.

In Exercises 1–6, find the indicated derivative.

1. $f(x) = \dfrac{1}{x},\ f''(x)$

2. $f(x) = \ln(2x + 1),\ f^{(4)}(x)$

3. $f(x) = 2 \ln x,\ f^{(4)}(x)$

4. $f(x) = x^3 - 2x^2 + 7x - 12,\ f''(x)$

5. $f(x) = e^{2x},\ f^{(4)}(x)$

6. $f(x) = e^{x^2},\ f''(x)$

In Exercises 7 and 8, find the absolute maximum of f on the interval.

7. $f(x) = -x^2 + 6x + 9,\ [0, 4]$

8. $f(x) = \dfrac{8}{x^3},\ [1, 2]$

In Exercises 9 and 10, solve for n.

9. $\dfrac{1}{4n^2} < 0.001$

10. $\dfrac{1}{16n^4} < 0.0001$

Exercises 12.3

See www.CalcChat.com for worked-out solutions to odd-numbered exercises.

Using the Trapezoidal Rule and Simpson's Rule   In Exercises 1–10, use the Trapezoidal Rule and Simpson's Rule to approximate the value of the definite integral for the indicated value of n. Compare these results with the exact value of the definite integral. Round your answers to four decimal places. *See Examples 1 and 2.*

1. $\displaystyle\int_0^2 x^2\, dx,\ n = 4$

2. $\displaystyle\int_0^1 \left(\dfrac{x^2}{2} + 1\right) dx,\ n = 4$

3. $\displaystyle\int_0^2 e^{-4x}\, dx,\ n = 8$

4. $\displaystyle\int_1^3 (4 - x^2)\, dx,\ n = 4$

5. $\displaystyle\int_1^2 \dfrac{1}{x}\, dx,\ n = 8$

6. $\displaystyle\int_1^2 \dfrac{1}{x^2}\, dx,\ n = 4$

7. $\displaystyle\int_0^4 \sqrt{x}\, dx,\ n = 8$

8. $\displaystyle\int_0^8 \sqrt[3]{x}\, dx,\ n = 8$

9. $\displaystyle\int_0^1 xe^{3x^2}\, dx,\ n = 4$

10. $\displaystyle\int_0^2 x\sqrt{x^2 + 1}\, dx,\ n = 4$

Using the Trapezoidal Rule and Simpson's Rule   In Exercises 11–20, approximate the value of the definite integral using (a) the Trapezoidal Rule and (b) Simpson's Rule for the indicated value of n. Round your answers to three decimal places. *See Examples 1 and 2.*

11. $\displaystyle\int_0^1 \dfrac{1}{1 + x^2}\, dx,\ n = 4$

12. $\displaystyle\int_0^4 \dfrac{8}{x^2 + 3}\, dx,\ n = 4$

13. $\displaystyle\int_0^4 \dfrac{1}{\sqrt[3]{x^2 + 1}}\, dx,\ n = 8$

14. $\displaystyle\int_0^2 \dfrac{1}{\sqrt{1 + x^3}}\, dx,\ n = 4$

15. $\displaystyle\int_0^2 \sqrt{1 + x^3}\, dx,\ n = 4$

16. $\displaystyle\int_0^1 \sqrt{1 - x^2}\, dx,\ n = 8$

17. $\displaystyle\int_0^1 e^{x^2}\, dx,\ n = 8$

18. $\displaystyle\int_0^2 e^{-x^2}\, dx,\ n = 4$

19. $\displaystyle\int_0^3 \dfrac{1}{2 - 2x + x^2}\, dx,\ n = 6$

20. $\displaystyle\int_0^3 \dfrac{x}{2 + x + x^2}\, dx,\ n = 6$

 Present Value   In Exercises 21 and 22, use the Simpson's Rule program in Appendix H with $n = 8$ to approximate the present value of the income $c(t)$ over t_1 years at the given annual interest rate r. Then use the integration capabilities of a graphing utility to approximate the present value. Compare the results. (Present value is defined in Section 12.1.)

21. $c(t) = 6000 + 200\sqrt{t},\ r = 7\%,\ t_1 = 4$

22. $c(t) = 200{,}000 + 15{,}000\sqrt[3]{t},\ r = 10\%,\ t_1 = 8$

 Marginal Analysis   In Exercises 23 and 24, use the Simpson's Rule program in Appendix H with $n = 4$ to approximate the change in revenue from the marginal revenue function dR/dx. In each case, assume that the number of units sold x increases from 14 to 16.

23. $\dfrac{dR}{dx} = 5\sqrt{8000 - x^3}$

24. $\dfrac{dR}{dx} = 50\sqrt{x}\sqrt{20 - x}$

Probability In Exercises 25–28, use the Simpson's Rule program in Appendix H with $n = 6$ to approximate the indicated normal probability. The standard normal probability density function is

$$f(x) = \frac{1}{\sqrt{2\pi}} e^{-x^2/2}.$$

If x is chosen at random from a population with this density, then the probability that x lies in the interval $[a, b]$ is

$$P(a \le x \le b) = \int_a^b f(x)\, dx.$$

25. $P(0 \le x \le 1)$ **26.** $P(0 \le x \le 2)$

27. $P(0 \le x \le 4)$ **28.** $P(0 \le x \le 1.5)$

Surveying In Exercises 29 and 30, use the Simpson's Rule program in Appendix H to estimate the number of square feet of land in the lot, where x and y are measured in feet, as shown in the figures. In each case, the land is bounded by a stream and two straight roads.

29.

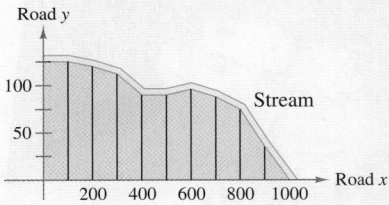

x	0	100	200	300	400	500
y	125	125	120	112	90	90

x	600	700	800	900	1000
y	95	88	75	35	0

30.

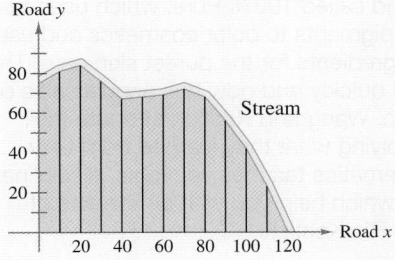

x	0	10	20	30	40	50	60
y	75	81	84	76	67	68	69

x	70	80	90	100	110	120
y	72	68	56	42	23	0

Error Analysis In Exercises 31–34, use the error formulas to find bounds for the error in approximating the definite integral using (a) the Trapezoidal Rule and (b) Simpson's Rule. Let $n = 4$.

31. $\displaystyle\int_0^2 (x^2 + 2x)\, dx$ **32.** $\displaystyle\int_0^1 \frac{1}{x + 1}\, dx$

33. $\displaystyle\int_0^1 e^{x^3}\, dx$ **34.** $\displaystyle\int_0^1 e^{2x^2}\, dx$

Error Analysis In Exercises 35–38, use the error formulas to find n such that the error in the approximation of the definite integral is less than 0.0001 using (a) the Trapezoidal Rule and (b) Simpson's Rule. *See Example 3.*

35. $\displaystyle\int_0^2 x^4\, dx$ **36.** $\displaystyle\int_1^3 \frac{1}{x}\, dx$

37. $\displaystyle\int_1^3 e^{2x}\, dx$ **38.** $\displaystyle\int_3^5 \ln x\, dx$

Using Simpson's Rule In Exercises 39–42, use the Simpson's Rule program in Appendix H with $n = 100$ to approximate the definite integral.

39. $\displaystyle\int_1^4 x\sqrt{x + 4}\, dx$

40. $\displaystyle\int_1^4 x^2\sqrt{x + 4}\, dx$

41. $\displaystyle\int_2^5 10xe^{-x}\, dx$

42. $\displaystyle\int_2^5 10x^2e^{-x}\, dx$

43. Median Age The table shows the median ages of the U.S. resident population for the years 2001 through 2009. *(Source: U.S. Census Bureau)*

Year	2001	2002	2003	2004	2005
Median age	35.5	35.7	35.9	36.0	36.2

Year	2006	2007	2008	2009
Median age	36.3	36.5	36.7	36.8

(a) Use Simpson's Rule to estimate the average median age over the time period.

(b) A model for the data is

$$A = 35.4 + 0.16t - 0.000004e^t, \quad 1 \le t \le 9$$

where A is the median age and t is the year, with $t = 1$ corresponding to 2001. Use integration to find the average median age over the time period.

(c) Compare the results of parts (a) and (b).

44. Electricity The table shows the residential prices of electricity (in cents per kilowatt-hour) for the years 2001 through 2009. *(Source: U.S. Energy Information Administration)*

Year	2001	2002	2003	2004	2005
Price	8.58	8.44	8.72	8.95	9.45

Year	2006	2007	2008	2009
Price	10.40	10.65	11.26	11.55

(a) Use Simpson's Rule to estimate the average residential price of electricity over the time period.

(b) A model for the data is

$$E = 8.4 - 1.39t + 0.291t^2 - 0.0160t^3 + 1.27097\sqrt{t}$$

for $1 \leq t \leq 9$, where E is the residential price of electricity (in cents per kilowatt-hour) and t is the year, with $t = 1$ corresponding to 2001. Use integration to find the average residential price of electricity over the time period.

(c) Compare the results of parts (a) and (b).

45. Medicine A body assimilates a 12-hour cold tablet at a rate modeled by $dC/dt = 8 - \ln(t^2 - 2t + 4)$, $0 \leq t \leq 12$, where dC/dt is measured in milligrams per hour and t is the time in hours. Use Simpson's Rule with $n = 8$ to estimate the total amount of the drug absorbed into the body during the 12 hours.

46. HOW DO YOU SEE IT? The graph shows the weekly revenue (in thousands of dollars) for a company.

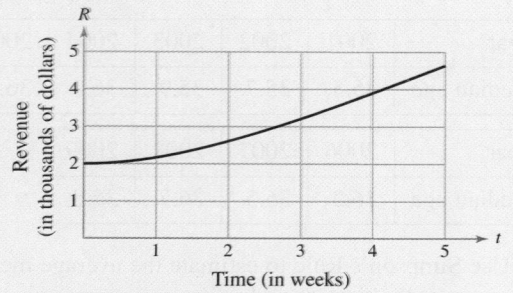

Time (in weeks)

(a) Which gives a more accurate approximation of the total weekly revenue for the first 4 weeks using the Trapezoidal Rule, $n = 8$ or $n = 16$?

(b) Which gives a more accurate approximation of the total weekly revenue for the first 4 weeks, the Trapezoidal Rule with $n = 8$ or Simpson's Rule with $n = 8$?

47. Consumer Trends The rate of change in the number of subscribers S to a newly introduced magazine is modeled by

$$\frac{dS}{dt} = 1000t^2 e^{-t}, \quad 0 \leq t \leq 6$$

where t is the time in years. Use Simpson's Rule with $n = 12$ to estimate the total increase in the number of subscribers during the first 6 years.

48. Using Simpson's Rule Prove that Simpson's Rule is exact when used to approximate the integral of a cubic polynomial function, and demonstrate the result for

$$\int_0^1 x^3 \, dx$$

with $n = 2$.

Business Capsule

Susie Wang and Ric Kostick graduated in 2002 from the University of California at Berkeley with degrees in mathematics. Together they launched a cosmetics brand called 100% Pure, which uses fruit and vegetable pigments to color cosmetics and uses only organic ingredients for the purest skin care. The company grew quickly and now has annual sales of over $15 million. Wang and Kostick attribute their success to applying what they learned from their studies. "Mathematics teaches you logic, discipline, and accuracy, which help you with all aspects of daily life," says Ric Kostick.

49. Research Project Use your school's library, the Internet, or some other reference source to research the opportunity cost of attending graduate school for 2 years to receive a Masters of Business Administration (MBA) degree rather than working for 2 years with a bachelor's degree. Write a short paper describing these costs.

12.4 Improper Integrals

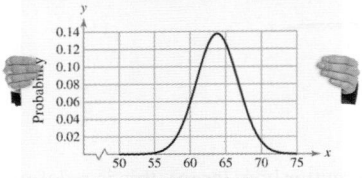

Height of 30- to 39-year-old American Women

In Exercise 27 on page 903, you will evaluate an improper integral to determine the probability that a 30- to 39-year-old woman is 6 feet or taller.

■ Recognize improper integrals.
■ Evaluate improper integrals with infinite limits of integration.
■ Use improper integrals to solve real-life problems.
■ Find the present value of a perpetuity.

Improper Integrals

The definition of a definite integral

$$\int_a^b f(x)\, dx$$

requires that the interval $[a, b]$ be finite. Furthermore, the Fundamental Theorem of Calculus, by which you have been evaluating definite integrals, requires that f be continuous on $[a, b]$. Some integrals do not satisfy these requirements because of one of the conditions below.

1. One or both of the limits of integration are infinite.

2. The function f has an infinite discontinuity in the interval $[a, b]$.

Integrals having either of these characteristics are called **improper integrals.** In this section, you will study integrals where one or both limits of integration are infinite. For instance, the integral

$$\int_0^\infty e^{-x}\, dx$$

is improper because one limit of integration is infinite, as indicated in Figure 12.13.

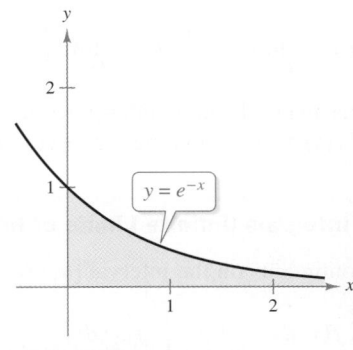

FIGURE 12.13

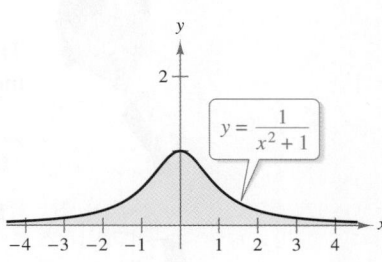

FIGURE 12.14

Similarly, the integral

$$\int_{-\infty}^\infty \frac{1}{x^2 + 1}\, dx$$

is improper because both limits of integration are infinite, as indicated in Figure 12.14.
The integrals

$$\int_1^5 \frac{1}{\sqrt{x - 1}}\, dx \quad \text{and} \quad \int_{-2}^2 \frac{1}{(x + 1)^2}\, dx$$

are improper because their integrands have an **infinite discontinuity**—that is, they approach infinity somewhere in the interval of integration. Evaluating an integral whose integrand has an infinite discontinuity is beyond the scope of this text.

Integrals with Infinite Limits of Integration

To see how to evaluate an improper integral, consider the integral shown in Figure 12.15.

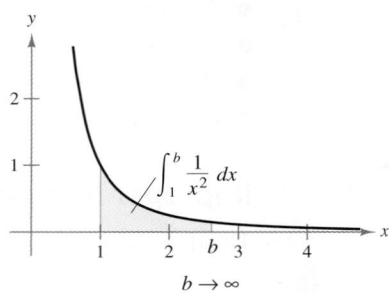

FIGURE 12.15

As long as b is a real number that is greater than 1 (no matter how large), this is a definite integral whose value is

$$\int_1^b \frac{1}{x^2} \, dx = \left[-\frac{1}{x} \right]_1^b = -\frac{1}{b} + 1 = 1 - \frac{1}{b}.$$

The table shows the values of this integral for several values of b.

b	2	5	10	100	1000	10,000
$\int_1^b \frac{1}{x^2} \, dx = 1 - \frac{1}{b}$	0.5000	0.8000	0.9000	0.9900	0.9990	0.9999

From this table, it appears that the value of the integral is approaching a limit as b increases without bound. This limit is denoted by the *improper integral* shown below.

$$\int_1^\infty \frac{1}{x^2} \, dx = \lim_{b \to \infty} \int_1^b \frac{1}{x^2} \, dx = \lim_{b \to \infty} \left(1 - \frac{1}{b} \right) = 1$$

This improper integral can be interpreted as the area of the *unbounded* region between the graph of $f(x) = 1/x^2$ and the x-axis (to the right of $x = 1$).

Improper Integrals (Infinite Limits of Integration)

1. If f is continuous on the interval $[a, \infty)$, then

$$\int_a^\infty f(x) \, dx = \lim_{b \to \infty} \int_a^b f(x) \, dx.$$

2. If f is continuous on the interval $(-\infty, b]$, then

$$\int_{-\infty}^b f(x) \, dx = \lim_{a \to -\infty} \int_a^b f(x) \, dx.$$

3. If f is continuous on the interval $(-\infty, \infty)$, then

$$\int_{-\infty}^\infty f(x) \, dx = \int_{-\infty}^c f(x) \, dx + \int_c^\infty f(x) \, dx$$

where c is any real number.

In the first two cases, if the limit exists, then the improper integral **converges;** otherwise, the improper integral **diverges.** In the third case, the integral on the left diverges when either one of the integrals on the right diverges.

Example 1   Evaluating an Improper Integral

Determine the convergence or divergence of $\int_1^\infty \frac{1}{x}\,dx$.

SOLUTION   Begin by applying the definition of an improper integral.

$$\int_1^\infty \frac{1}{x}\,dx = \lim_{b\to\infty}\int_1^b \frac{1}{x}\,dx \qquad \text{Definition of improper integral}$$

$$= \lim_{b\to\infty}\left[\ln x\right]_1^b \qquad \text{Find antiderivative.}$$

$$= \lim_{b\to\infty}(\ln b - 0) \qquad \text{Apply Fundamental Theorem.}$$

$$= \infty \qquad \text{Evaluate limit.}$$

Because the limit is infinite, the improper integral diverges.

✓ Checkpoint 1

Determine the convergence or divergence of each improper integral.

a. $\int_1^\infty \frac{1}{x^3}\,dx$ **b.** $\int_1^\infty \frac{1}{\sqrt{x}}\,dx$

As you begin to work with improper integrals, you will find that integrals that appear to be similar can have very different values. For instance, consider the two improper integrals

$$\int_1^\infty \frac{1}{x}\,dx = \infty \qquad \text{Divergent integral}$$

and

$$\int_1^\infty \frac{1}{x^2}\,dx = 1. \qquad \text{Convergent integral}$$

The first integral diverges and the second converges to 1. Graphically, this means that the areas shown in Figure 12.16 are very different. The region lying between the graph of

$$y = \frac{1}{x}$$

and the *x*-axis (for $x \ge 1$) has an *infinite* area, and the region lying between the graph of

$$y = \frac{1}{x^2}$$

and the *x*-axis (for $x \ge 1$) has a *finite* area.

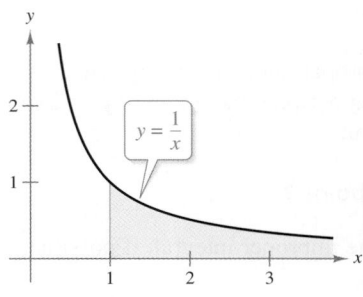

Diverges (infinite area)   Converges (finite area)

FIGURE 12.16

ALGEBRA TUTOR *xy*

For help on the algebra
in Example 2, see Example
2(a) in the *Chapter 12 Algebra
Tutor*, on page 904.

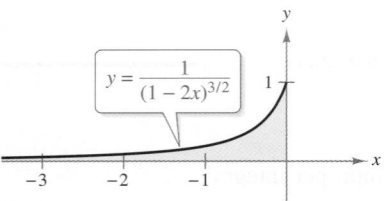

$$y = \frac{1}{(1 - 2x)^{3/2}}$$

FIGURE 12.17

| *Example 2* | **Evaluating an Improper Integral** |

Evaluate the improper integral.

$$\int_{-\infty}^{0} \frac{1}{(1 - 2x)^{3/2}} \, dx$$

SOLUTION   Begin by applying the definition of an improper integral.

$$\int_{-\infty}^{0} \frac{1}{(1 - 2x)^{3/2}} \, dx = \lim_{a \to -\infty} \int_{a}^{0} \frac{1}{(1 - 2x)^{3/2}} \, dx \qquad \text{Definition of improper integral}$$

$$= \lim_{a \to -\infty} \left[\frac{1}{\sqrt{1 - 2x}} \right]_{a}^{0} \qquad \text{Find antiderivative.}$$

$$= \lim_{a \to -\infty} \left(1 - \frac{1}{\sqrt{1 - 2a}} \right) \qquad \text{Apply Fundamental Theorem.}$$

$$= 1 - 0 \qquad \text{Evaluate limit.}$$

$$= 1 \qquad \text{Simplify.}$$

So, the improper integral converges to 1. As shown in Figure 12.17, this implies that the region lying between the graph of $y = 1/(1 - 2x)^{3/2}$ and the x-axis (for $x \leq 0$) has an area of 1 square unit.

✓**Checkpoint 2**

Evaluate the improper integral, if possible.

$$\int_{-\infty}^{0} \frac{1}{(x - 1)^2} \, dx$$

ALGEBRA TUTOR *xy*

For help on the algebra
in Example 3, see Example
2(b) in the *Chapter 12 Algebra
Tutor*, on page 904.

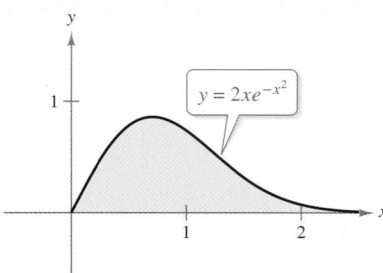

$$y = 2xe^{-x^2}$$

FIGURE 12.18

| *Example 3* | **Evaluating an Improper Integral** |

Evaluate the improper integral.

$$\int_{0}^{\infty} 2xe^{-x^2} \, dx$$

SOLUTION   Begin by applying the definition of an improper integral.

$$\int_{0}^{\infty} 2xe^{-x^2} \, dx = \lim_{b \to \infty} \int_{0}^{b} 2xe^{-x^2} \, dx \qquad \text{Definition of improper integral}$$

$$= \lim_{b \to \infty} \left[-e^{-x^2} \right]_{0}^{b} \qquad \text{Find antiderivative.}$$

$$= \lim_{b \to \infty} (-e^{-b^2} + 1) \qquad \text{Apply Fundamental Theorem.}$$

$$= 0 + 1 \qquad \text{Evaluate limit.}$$

$$= 1 \qquad \text{Simplify.}$$

So, the improper integral converges to 1. As shown in Figure 12.18, this implies that the region lying between the graph of $y = 2xe^{-x^2}$ and the x-axis (for $x \geq 0$) has an area of 1 square unit.

✓**Checkpoint 3**

Evaluate the improper integral, if possible.

$$\int_{-\infty}^{0} e^{2x} \, dx$$

Application

In Section 10.3, you studied the graph of the *normal probability density function*

$$f(x) = \frac{1}{\sigma\sqrt{2\pi}}e^{-(x-\mu)^2/(2\sigma^2)}.$$

This function is used in statistics to represent a population that is normally distributed with a mean of μ and a standard deviation of σ. Specifically, when an outcome x is chosen at random from the population, the probability that x will have a value between a and b is

$$P(a \le x \le b) = \int_a^b \frac{1}{\sigma\sqrt{2\pi}}e^{-(x-\mu)^2/(2\sigma^2)}dx.$$

As shown in Figure 12.19, the probability $P(-\infty < x < \infty)$ is

$$P(-\infty < x < \infty) = \int_{-\infty}^{\infty} \frac{1}{\sigma\sqrt{2\pi}}e^{-(x-\mu)^2/(2\sigma^2)}dx = 1.$$

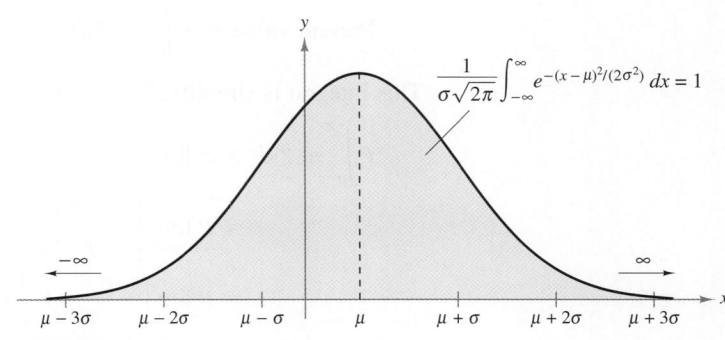

FIGURE 12.19

Example 4 Finding a Probability

The mean height of American men (from 20 to 29 years old) is 69 inches, and the standard deviation is 3 inches. A 20- to 29-year-old man is chosen at random from the population. What is the probability that he is 6 feet tall or taller? *(Source: U.S. National Center for Health Statistics)*

SOLUTION Note that the mean and standard deviation are given in inches and the height of the man chosen at random is given in feet. To calculate the probability, you need to use the same units for these quantities. Because it is easier to convert feet to inches, use 72 inches (1 foot = 12 inches) for the man's height. So, the probability can be written as $P(72 \le x < \infty)$. Using a mean of $\mu = 69$ and a standard deviation of $\sigma = 3$, the probability $P(72 \le x < \infty)$ is given by the improper integral

$$P(72 \le x < \infty) = \int_{72}^{\infty} \frac{1}{3\sqrt{2\pi}}e^{-(x-69)^2/18}dx.$$

Using a symbolic integration utility, you can approximate the value of this integral to be 0.158. So, the probability that the man is 6 feet tall or taller is about 15.8%.

Many professional basketball players are over $6\frac{1}{2}$ feet tall. When a man is chosen at random from the population, the probability that he is $6\frac{1}{2}$ feet tall or taller is less than half of one percent.

✓ Checkpoint 4

Use Example 4 to find the probability that a 20- to 29-year-old man chosen at random from the population is 6 feet 6 inches tall or taller.

Present Value of a Perpetuity

Recall from Section 12.1 that for an interest-bearing account, the present value over t_1 years is

$$\text{Present value} = \int_0^{t_1} c(t)e^{-rt}\, dt$$

where c represents a continuous income function (in dollars per year) and the annual interest rate r is compounded continuously. If the size of an annuity's payment is a constant number of dollars P, then $c(t)$ is equal to P and the present value is

$$\text{Present value} = \int_0^{t_1} Pe^{-rt}\, dt = P\int_0^{t_1} e^{-rt}\, dt. \qquad \begin{array}{l}\text{Present value of an annuity}\\ \text{with payment } P\end{array}$$

Consider an annuity, such as a scholarship fund, that pays the same amount each year *forever*. Because the annuity continues indefinitely, the number of years t_1 approaches infinity. Such an annuity is called a **perpetual annuity** or a **perpetuity.** This situation can be represented by the following improper integral.

$$\text{Present value} = P\int_0^{\infty} e^{-rt}\, dt \qquad \begin{array}{l}\text{Present value of a perpetuity}\\ \text{with payment } P\end{array}$$

This integral is simplified as follows.

$$P\int_0^{\infty} e^{-rt}\, dt = P\lim_{b\to\infty} \int_0^{b} e^{-rt}\, dt \qquad \text{Definition of improper integral}$$

$$= P\lim_{b\to\infty} \left[-\frac{e^{-rt}}{r}\right]_0^{b} \qquad \text{Find antiderivative.}$$

$$= P\lim_{b\to\infty} \left(-\frac{e^{-rb}}{r} + \frac{1}{r}\right) \qquad \text{Apply Fundamental Theorem.}$$

$$= P\left(0 + \frac{1}{r}\right) \qquad \text{Evaluate limit.}$$

$$= \frac{P}{r} \qquad \text{Simplify.}$$

So, the improper integral converges to P/r. As shown in Figure 12.20, this implies that the region lying between the graph of

$$y = Pe^{-rt}$$

and the t-axis for $t \geq 0$ has an area equal to the annual payment P divided by the annual interest rate r.

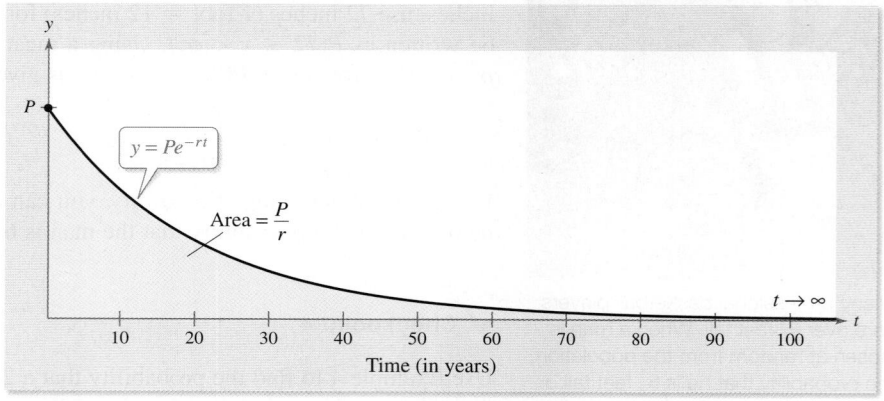

FIGURE 12.20

The present value of a perpetuity is defined as follows.

Present Value of a Perpetuity

If P represents the size of each annual payment in dollars and the annual interest rate is r (compounded continuously), then the present value of a perpetuity is

$$\text{Present value} = P \int_0^\infty e^{-rt}\, dt = \frac{P}{r}.$$

This definition is useful in determining the amount of money needed to start an endowment, such as a scholarship fund, as shown in Example 5.

 Example 5   **Finding Present Value**

You want to start a scholarship fund at your alma mater. You plan to give one $9000 scholarship annually beginning one year from now, and you have at most $120,000 to start the fund. You also want the scholarship to be given out indefinitely. Assuming an annual interest rate of 8% (compounded continuously), do you have enough money for the scholarship fund?

SOLUTION   To answer this question, you must find the present value of the scholarship fund. Because the scholarship is to be given out each year indefinitely, the time period is infinite. The fund is a perpetuity with $P = 9000$ and $r = 0.08$. The present value is

$$\begin{aligned}
\text{Present value} &= \frac{P}{r} \\
&= \frac{9000}{0.08} \\
&= 112{,}500.
\end{aligned}$$

The amount you need to start the scholarship fund is $112,500. Yes, you have enough money to start the scholarship fund.

✓ Checkpoint 5

In Example 5, do you have enough money to start a scholarship fund that pays $10,000 annually? Explain why or why not.

SUMMARIZE   (Section 12.4)

1. Describe the different types of improper integrals *(page 895)*. For examples of evaluating improper integrals, see Examples 1, 2, and 3.

2. Define the term *converges* as it applies to improper integrals *(page 896)*. For examples of improper integrals that converge, see Examples 2 and 3.

3. Define the term *diverges* as it applies to improper integrals *(page 896)*. For an example of an improper integral that diverges, see Example 1.

4. Describe a real-life example of how an improper integral can be used to find a probability *(page 899, Example 4)*.

5. Describe a real-life example of how an improper integral can be used to find the present value of a perpetuity *(page 901, Example 5)*.

SKILLS WARM UP 12.4   The following warm-up exercises involve skills that were covered in earlier sections. You will use these skills in the exercise set for this section. For additional help, review Sections 7.1, 10.1, and 10.4.

In Exercises 1–6, find the limit.

1. $\lim\limits_{x \to 2}(2x + 5)$

2. $\lim\limits_{x \to 1}\left(\dfrac{1}{x} + 2x^2\right)$

3. $\lim\limits_{x \to -4}\dfrac{x + 4}{x^2 - 16}$

4. $\lim\limits_{x \to 0}\dfrac{x^2 - 2x}{x^3 + 3x^2}$

5. $\lim\limits_{x \to 1}\dfrac{1}{\sqrt{x - 1}}$

6. $\lim\limits_{x \to -3}\dfrac{x^2 + 2x - 3}{x + 3}$

In Exercises 7–10, evaluate the expression (a) when $x = b$ and (b) when $x = 0$.

7. $\dfrac{4}{3}(2x - 1)^3$

8. $\dfrac{1}{x - 5} + \dfrac{3}{(x - 2)^2}$

9. $\ln(5 - 3x^2) - \ln(x + 1)$

10. $e^{3x^2} + e^{-3x^2}$

Exercises 12.4

See www.CalcChat.com for worked-out solutions to odd-numbered exercises.

Determining Whether an Integral Is Improper   In Exercises 1–6, decide whether the integral is improper. Explain your reasoning.

1. $\displaystyle\int_0^1 \dfrac{dx}{3x - 2}$

2. $\displaystyle\int_1^3 \dfrac{dx}{x^2}$

3. $\displaystyle\int_0^1 \dfrac{2x - 5}{x^2 - 5x + 6}\,dx$

4. $\displaystyle\int_1^\infty x^2\,dx$

5. $\displaystyle\int_0^5 e^{-x}\,dx$

6. $\displaystyle\int_{-\infty}^\infty \dfrac{1}{x^2 + 3}\,dx$

Evaluating an Improper Integral   In Exercises 7–20, determine whether the improper integral diverges or converges. Evaluate the integral if it converges. *See Examples 1, 2, and 3.*

7. $\displaystyle\int_1^\infty \dfrac{1}{x^2}\,dx$

8. $\displaystyle\int_1^\infty \dfrac{1}{\sqrt[3]{x}}\,dx$

9. $\displaystyle\int_0^\infty e^{x/3}\,dx$

10. $\displaystyle\int_0^\infty \dfrac{5}{e^{2x}}\,dx$

11. $\displaystyle\int_5^\infty \dfrac{x}{\sqrt{x^2 - 16}}\,dx$

12. $\displaystyle\int_5^\infty \dfrac{1}{\sqrt{2x - 1}}\,dx$

13. $\displaystyle\int_{-\infty}^0 e^{-x}\,dx$

14. $\displaystyle\int_{-\infty}^{-1} \dfrac{1}{x^2}\,dx$

15. $\displaystyle\int_1^\infty \dfrac{e^{\sqrt{x}}}{\sqrt{x}}\,dx$

16. $\displaystyle\int_{-\infty}^0 \dfrac{x}{x^2 + 1}\,dx$

17. $\displaystyle\int_{-\infty}^\infty 2xe^{-3x^2}\,dx$

18. $\displaystyle\int_{-\infty}^\infty x^2 e^{-x^3}\,dx$

19. $\displaystyle\int_4^\infty \dfrac{1}{x(\ln x)^3}\,dx$

20. $\displaystyle\int_{-\infty}^0 \dfrac{1}{\sqrt[3]{1 - x}}\,dx$

Area of a Region   In Exercises 21–26, find the area of the unbounded shaded region.

21. $y = e^{-x}$

22. $y = e^{x/4}$

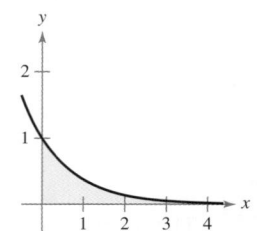

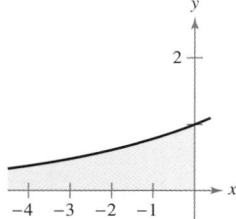

23. $y = -\dfrac{1}{x^3}$

24. $y = \dfrac{5}{\sqrt{4 - x}}$

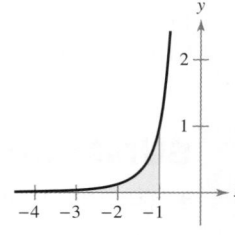

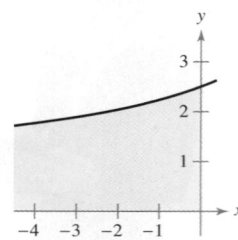

25. $y = \dfrac{6x}{x^2 + 1}$

26. $y = \dfrac{16x}{x^2 + 4}$

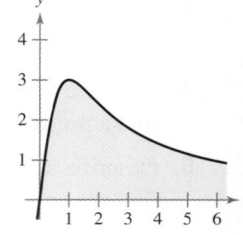

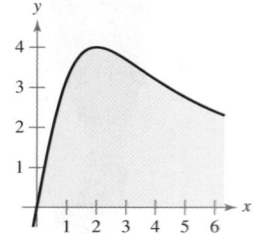

27. Women's Height  The mean height of American women (from 30 to 39 years old) is 63.8 inches, and the standard deviation is 2.9 inches. Use a symbolic integration utility or a graphing utility to find the probability that a 30- to 39-year-old woman chosen at random is

(a) between 5 and 6 feet tall.

(b) 5 feet 8 inches or taller.

(c) 6 feet or taller.

(Source: U.S. National Center for Health Statistics)

28. HOW DO YOU SEE IT? The graph shows the probability density function for a car brand that has a mean fuel efficiency of 26 miles per gallon and a standard deviation of 2.4 miles per gallon.

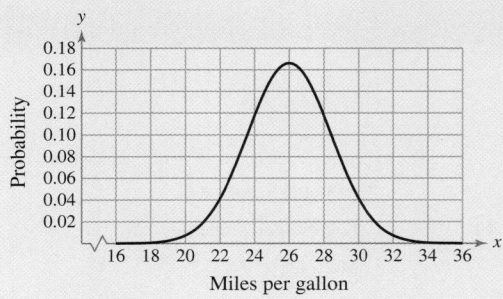

Miles per gallon

(a) Which is greater, the probability of choosing a car at random that gets between 26 and 28 miles per gallon or the probability of choosing a car at random that gets between 22 and 24 miles per gallon?

(b) Which is greater, the probability of choosing a car at random that gets between 20 and 22 miles per gallon or the probability of choosing a car at random that gets at least 30 miles per gallon?

29. Quality Control  A company manufactures wooden yardsticks. The lengths of the yardsticks are normally distributed with a mean of 36 inches and a standard deviation of 0.2 inch. Use a symbolic integration utility or a graphing utility to find the probability that a yardstick chosen at random is

(a) longer than 35.5 inches.

(b) longer than 35.9 inches.

30. Quality Control  A company manufactures compact fluorescent light bulbs. The life spans of the light bulbs are normally distributed with a mean of 9000 hours and a standard deviation of 500 hours. Use a symbolic integration utility or a graphing utility to find the probability that a light bulb chosen at random has a life span that is

(a) between 8000 and 10,000 hours.

(b) 11,000 hours or longer.

Endowment In Exercises 31 and 32, determine the amount of money required to set up a charitable endowment that pays the amount P each year indefinitely for the annual interest rate r compounded continuously. *See Example 5.*

31. $P = \$5000, r = 7.5\%$

32. $P = \$12,000, r = 6\%$

33. Scholarship Fund You want to start a scholarship fund at your alma mater. You plan to give one $18,000 scholarship annually beginning one year from now, and you have at most $400,000 to start the fund. You also want the scholarship to be given out indefinitely. Assuming an annual interest rate of 5% compounded continuously, do you have enough money for the scholarship fund?

34. Charitable Foundation A charitable foundation wants to help schools buy computers. The foundation plans to donate $35,000 each year to one school beginning one year from now, and the foundation has at most $500,000 to start the fund. The foundation wants the donation to be given out indefinitely. Assuming an annual interest rate of 8% compounded continuously, does the foundation have enough money to fund the donation?

35. Present Value A business is expected to yield a continuous flow of profit at the rate of $500,000 per year. Assuming an annual interest rate of 9% compounded continuously, what is the present value of the business

(a) for 20 years?

(b) forever?

36. Present Value A farm is expected to yield a continuous flow of profit at the rate of $75,000 per year. Assuming an annual interest rate of 8% compounded continuously, what is the present value of the farm

(a) for 20 years?

(b) forever?

Capitalized Cost In Exercises 37–40, the capitalized cost C of an asset is given by

$$C = C_0 + \int_0^n c(t)e^{-rt}\,dt$$

where C_0 is the original investment, t is the time in years, r is the annual interest rate compounded continuously, and $c(t)$ is the annual cost of maintenance (in dollars). Find the capitalized cost of an asset (a) for 5 years, (b) for 10 years, and (c) forever.

37. $C_0 = \$650,000, c(t) = 25,000, r = 10\%$

38. $C_0 = \$800,000, c(t) = 30,000, r = 4\%$

39. $C_0 = \$300,000, c(t) = 15,000t, r = 6\%$

40. $C_0 = \$650,000, c(t) = 25,000(1 + 0.08t), r = 12\%$

ALGEBRA TUTOR

Algebra and Integration Techniques

Integration techniques involve many different algebraic skills. For a definite integral, you need a variety of algebraic skills to apply the Fundamental Theorem of Calculus and evaluate the resulting expression. Study the examples in this Algebra Tutor. Be sure that you understand the algebra used in each step.

Example 1   Evaluating an Expression

Evaluate the expression

$$(e \ln e - e) - (1 \ln 1 - 1).$$

SOLUTION   Recall that

$$\ln e = 1 \quad \text{because} \quad e^1 = e$$

and

$$\ln 1 = 0 \quad \text{because} \quad e^0 = 1.$$

$$
\begin{aligned}
(e \ln e - e) &- (1 \ln 1 - 1) && \text{Example 5, page 872} \\
&= [e(1) - e] - [1(0) - 1] && \text{Logarithmic properties} \\
&= (e - e) - (0 - 1) && \text{Multiply.} \\
&= 0 - (-1) && \text{Simplify.} \\
&= 1 && \text{Simplify.}
\end{aligned}
$$

Example 2   Evaluating Expressions

Find the limit.

a. $\displaystyle\lim_{a \to -\infty} \left(1 - \frac{1}{\sqrt{1 - 2a}} \right)$

b. $\displaystyle\lim_{b \to \infty} \left(-e^{-b^2} + 1 \right)$

SOLUTION

a. $\displaystyle\lim_{a \to -\infty} \left(1 - \frac{1}{\sqrt{1 - 2a}} \right)$ Example 2, page 898

$$
\begin{aligned}
&= \lim_{a \to -\infty} 1 - \lim_{a \to -\infty} \left(\frac{1}{\sqrt{1 - 2a}} \right) && \lim_{x \to -\infty} [f(x) - g(x)] = \lim_{x \to -\infty} f(x) - \lim_{x \to -\infty} g(x) \\
&= 1 - 0 && \text{Evaluate limits.} \\
&= 1 && \text{Simplify.}
\end{aligned}
$$

b. $\displaystyle\lim_{b \to \infty} \left(-e^{-b^2} + 1 \right)$ Example 3, page 898

$$
\begin{aligned}
&= \lim_{b \to \infty} \left(-e^{-b^2} \right) + \lim_{b \to \infty} 1 && \lim_{x \to \infty} [f(x) + g(x)] = \lim_{x \to \infty} f(x) + \lim_{x \to \infty} g(x) \\
&= \lim_{b \to \infty} \left(\frac{1}{-e^{b^2}} \right) + \lim_{b \to \infty} 1 && \text{Rewrite with positive exponent.} \\
&= 0 + 1 && \text{Evaluate limits.} \\
&= 1 && \text{Simplify.}
\end{aligned}
$$

Example 3   Algebra and Integration Techniques

Simplify the expression

$$x^2 e^x - 2(x - 1)e^x.$$

SOLUTION

$$
\begin{aligned}
& x^2 e^x - 2(x - 1)e^x && \text{Example 5, page 881} \\
&= x^2 e^x - 2(xe^x - e^x) && \text{Multiply factors.} \\
&= x^2 e^x - 2xe^x + 2e^x && \text{Multiply factors.} \\
&= e^x(x^2 - 2x + 2) && \text{Factor.}
\end{aligned}
$$

Example 4   Solving a Rational Inequality

Solve the rational inequality

$$\frac{1}{6n^2} < 0.01$$

for n, where n is a positive integer.

SOLUTION

$$
\begin{aligned}
& \frac{1}{6n^2} < 0.01 && \text{Example 3, page 891} \\[2mm]
& \frac{1}{6n^2} < \frac{1}{100} && \text{Rewrite decimal as a fraction.} \\[2mm]
& \frac{100}{6n^2} < 1 && \text{Multiply each side by 100.} \\[2mm]
& 100 < 6n^2 && \text{Multiply each side by } 6n^2 \ (n > 0). \\[2mm]
& \frac{100}{6} < n^2 && \text{Divide each side by 6.} \\[2mm]
& \frac{50}{3} < n^2 && \text{Simplify.} \\[2mm]
& \sqrt{\frac{50}{3}} < n && \text{Take positive square root of each side } (n > 0).
\end{aligned}
$$

Because n is a positive integer and

$$\sqrt{\frac{50}{3}} \approx 4.08$$

n must be 5 or more. You can check this result using a graphing utility. Let $y_1(x) = 1/6x^2$ and $y_2(x) = 0.01$. Then use the *intersect* feature (see figure) to determine that $x \approx 4.08$. So, the solution found algebraically is correct.

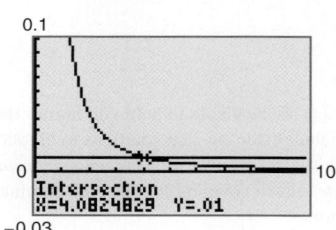

SUMMARY AND STUDY STRATEGIES

After studying this chapter, you should have acquired the following skills.
The exercise numbers are keyed to the Review Exercises that begin on page 908.
Answers to odd-numbered Review Exercises are given in the back of the text.*

Section 12.1

	Review Exercises
■ Use integration by parts to find indefinite and definite integrals.	*1–12*

$$\int u\, dv = uv - \int v\, du$$

For integrals of the form

$$\int x^n e^{ax}\, dx$$

let $u = x^n$ and $dv = e^{ax}\, dx$.

For integrals of the form

$$\int x^n \ln x\, dx$$

let $u = \ln x$ and $dv = x^n\, dx$.

■ Find the present value of future income.   *13–18*

$$\text{Actual income over } t_1 \text{ years} = \int_0^{t_1} c(t)\, dt$$

$$\text{Present value} = \int_0^{t_1} c(t)e^{-rt}\, dt$$

Section 12.2

■ Use integration tables to find indefinite integrals.	*19–32*
■ Use integration tables to solve real-life problems.	*33, 34*

Section 12.3

■ Use the Trapezoidal Rule and Simpson's Rule to approximate definite integrals.   *35–46*

Trapezoidal Rule:

$$\int_a^b f(x)\, dx \approx \left(\frac{b-a}{2n}\right)[f(x_0) + 2f(x_1) + \cdots + 2f(x_{n-1}) + f(x_n)]$$

Simpson's Rule:

$$\int_a^b f(x)\, dx \approx \left(\frac{b-a}{3n}\right)[f(x_0) + 4f(x_1) + 2f(x_2) + 4f(x_3) + \cdots + 4f(x_{n-1}) + f(x_n)]$$

---

* A wide range of valuable study aids are available to help you master the material in this chapter.
The *Student Solutions Manual* includes step-by-step solutions to all odd-numbered exercises to
help you review and prepare. The student website at *www.cengagebrain.com* offers algebra help
and a *Graphing Technology Guide*, which contains step-by-step commands and instructions for a
wide variety of graphing calculators.

Section 12.3   (continued)   Review Exercises

■ Analyze the sizes of the errors when approximating definite integrals with the   *47–50*
Trapezoidal Rule and Simpson's Rule.

Errors in the Trapezoidal Rule:

$$|E| \le \frac{(b-a)^3}{12n^2}[\max|f''(x)|], \quad a \le x \le b$$

Errors in Simpson's Rule:

$$|E| \le \frac{(b-a)^5}{180n^4}[\max|f^{(4)}(x)|], \quad a \le x \le b$$

Section 12.4

■ Evaluate improper integrals with infinite limits of integration.   *51–56*

$$\int_a^\infty f(x)\, dx = \lim_{b \to \infty} \int_a^b f(x)\, dx$$

$$\int_{-\infty}^b f(x)\, dx = \lim_{a \to -\infty} \int_a^b f(x)\, dx$$

$$\int_{-\infty}^\infty f(x)\, dx = \int_{-\infty}^c f(x)\, dx + \int_c^\infty f(x)\, dx$$

■ Find the area of an unbounded region.   *57–60*

■ Find the present value of a perpetuity.   *61–64*

$$\text{Present value} = P\int_0^\infty e^{-rt}\, dt = \frac{P}{r}$$

Study Strategies

■ **Use a Variety of Approaches**   To be efficient at finding antiderivatives, you need to use a variety of approaches.

1. Check to see whether the integral fits one of the basic integration formulas—you should have these formulas memorized.

2. Try an integration technique such as substitution or integration by parts to rewrite the integral in a form that fits one of the basic integration formulas.

3. Use a table of integrals.

4. Use a symbolic integration utility.

■ **Use Numerical Integration**   When solving a definite integral, remember that you cannot apply the Fundamental Theorem of Calculus unless you can find an antiderivative of the integrand. This is not always possible—even with a symbolic integration utility. In such cases, you can use a numerical technique such as the Midpoint Rule, the Trapezoidal Rule, or Simpson's Rule to approximate the value of the integral.

Review Exercises

See www.CalcChat.com for worked-out solutions to odd-numbered exercises.

Integration by Parts   In Exercises 1–8, use integration by parts to find the indefinite integral.

1. $\displaystyle\int \frac{\ln x}{\sqrt{x}}\,dx$

2. $\displaystyle\int x \ln 4x\,dx$

3. $\displaystyle\int (x + 1)e^x\,dx$

4. $\displaystyle\int xe^{-3x}\,dx$

5. $\displaystyle\int x\sqrt{x - 5}\,dx$

6. $\displaystyle\int \frac{x}{\sqrt{x + 8}}\,dx$

7. $\displaystyle\int 2x^2 e^{2x}\,dx$

8. $\displaystyle\int (\ln x)^3\,dx$

Evaluating Definite Integrals   In Exercises 9–12, use integration by parts to evaluate the definite integral.

9. $\displaystyle\int_1^e 6x \ln x\,dx$

10. $\displaystyle\int_0^4 \ln(1 + 3x)\,dx$

11. $\displaystyle\int_0^1 \frac{x}{e^{x/4}}\,dx$

12. $\displaystyle\int_0^2 x^2 e^{3x}\,dx$

Finding Present Value   In Exercises 13–16, find the present value of the income given by c (in dollars) over t_1 years at the given annual inflation rate r.

13. $c = 20{,}000$, $r = 4\%$, $t_1 = 5$ years

14. $c = 10{,}000 + 1500t$, $r = 6\%$, $t_1 = 10$ years

15. $c = 24{,}000t$, $r = 5\%$, $t_1 = 10$ years

16. $c = 20{,}000 + 100e^{t/2}$, $r = 5\%$, $t_1 = 5$ years

17. Present Value   A company expects its income c during the next 4 years to be modeled by

$$c = 200{,}000 + 50{,}000t, \quad 0 \le t \le 4.$$

(a) Find the actual income for the business over the 4 years.

(b) Assuming an annual inflation rate of 6%, what is the present value of this income?

18. Present Value   A company expects its income c during the next 7 years to be modeled by

$$c = 400{,}000 + 175{,}000t, \quad 0 \le t \le 7.$$

(a) Find the actual income for the business over the 7 years.

(b) Assuming an annual inflation rate of 4%, what is the present value of this income?

Using Integration Tables   In Exercises 19–22, use the indicated formula from the integration table in Appendix G to find the indefinite integral.

19. $\displaystyle\int \frac{x^2}{2 + 3x}\,dx$, Formula 6

20. $\displaystyle\int \frac{1}{1 + e^{6x}}\,dx$, Formula 40

21. $\displaystyle\int \sqrt{x^2 - 16}\,dx$, Formula 23

22. $\displaystyle\int x^5 \ln x\,dx$, Formula 43

Using Integration Tables   In Exercises 23–32, use the integration table in Appendix G to find the indefinite integral.

23. $\displaystyle\int \frac{x}{(2 + 3x)^2}\,dx$

24. $\displaystyle\int \frac{x}{\sqrt{2 + 3x}}\,dx$

25. $\displaystyle\int \frac{\sqrt{x^2 + 25}}{x}\,dx$

26. $\displaystyle\int \frac{1}{x(4 + 3x)}\,dx$

27. $\displaystyle\int \frac{1}{x^2 - 4}\,dx$

28. $\displaystyle\int (\ln 3x)^2\,dx$

29. $\displaystyle\int \frac{x}{\sqrt{1 + x}}\,dx$

30. $\displaystyle\int \frac{1}{x^2\sqrt{16 - x^2}}\,dx$

31. $\displaystyle\int \frac{\sqrt{1 + x}}{x}\,dx$

32. $\displaystyle\int \frac{1}{(x^2 - 9)^2}\,dx$

33. Probability   The probability of recalling between a and b percent (in decimal form) of the material learned in a memory experiment is modeled by

$$P(a \le x \le b) = \int_a^b \frac{96}{11}\left(\frac{x}{\sqrt{9 + 16x}}\right)dx,$$

$$0 \le a \le b \le 1.$$

What are the probabilities of recalling (a) between 0% and 80% and (b) between 0% and 50% of the material?

34. Probability   The probability of locating between a and b percent of the oil and gas deposits (in decimal form) in a region is modeled by

$$P(a \le x \le b) = \int_a^b 1.5x^2 e^{x^{1.5}}\,dx, \quad 0 \le a \le b \le 1.$$

What are the probabilities of locating (a) between 40% and 60% and (b) between 0% and 50% of the deposits?

Using the Trapezoidal Rule and Simpson's Rule   In Exercises 35–40, use the Trapezoidal Rule and Simpson's Rule to approximate the value of the definite integral for the indicated value of *n*. Compare these results with the exact value of the definite integral. Round your answers to four decimal places.

35. $\int_{1}^{3} \frac{1}{x^2}\, dx,\ n = 4$

36. $\int_{0}^{2} (x^2 + 1)\, dx,\ n = 8$

37. $\int_{1}^{2} \frac{1}{x^3}\, dx,\ n = 8$

38. $\int_{1}^{2} x^3\, dx,\ n = 4$

39. $\int_{0}^{4} e^{-x/2}\, dx,\ n = 4$

40. $\int_{0}^{8} \sqrt{x + 3}\, dx,\ n = 8$

Using the Trapezoidal Rule and Simpson's Rule   In Exercises 41–46, approximate the value of the definite integral using (a) the Trapezoidal Rule and (b) Simpson's Rule for the indicated value of *n*. Round your answers to three decimal places.

41. $\int_{1}^{2} \frac{1}{1 + \ln x}\, dx,\ n = 4$

42. $\int_{0}^{2} \frac{1}{\sqrt{1 + x^3}}\, dx,\ n = 8$

43. $\int_{0}^{1} \frac{x^{3/2}}{2 - x^2}\, dx,\ n = 4$

44. $\int_{0}^{1} e^{x^2}\, dx,\ n = 6$

45. $\int_{0}^{8} \frac{3}{x^2 + 2}\, dx,\ n = 8$

46. $\int_{0}^{1} \sqrt{1 - x}\, dx,\ n = 4$

Error Analysis   In Exercises 47 and 48, use the error formulas to find bounds for the error in approximating the definite integral using (a) the Trapezoidal Rule and (b) Simpson's Rule.

47. $\int_{0}^{2} e^{2x}\, dx,\ n = 4$ **48.** $\int_{2}^{4} \frac{1}{x - 1}\, dx,\ n = 8$

Error Analysis   In Exercises 49 and 50, use the error formulas to find *n* such that the error in the approximation of the definite integral is less than 0.0001 using (a) the Trapezoidal Rule and (b) Simpson's Rule.

49. $\int_{0}^{3} x^2\, dx$ **50.** $\int_{0}^{5} e^{x/5}\, dx$

Evaluating an Improper Integral   In Exercises 51–56, determine whether the improper integral diverges or converges. Evaluate the integral if it converges.

51. $\int_{-\infty}^{-1} \frac{1}{x^5}\, dx$ **52.** $\int_{1}^{\infty} \frac{1}{\sqrt{x}}\, dx$

53. $\int_{-\infty}^{0} \frac{1}{\sqrt[3]{8 - x}}\, dx$ **54.** $\int_{0}^{\infty} e^{-2x}\, dx$

55. $\int_{1}^{\infty} \frac{\ln x}{x}\, dx$ **56.** $\int_{0}^{\infty} \frac{e^x}{1 + e^x}\, dx$

Area of a Region   In Exercises 57–60, find the area of the unbounded shaded region.

57. $y = e^{-x/4}$ **58.** $y = \dfrac{2x}{x^2 + 2}$

59. $y = 4xe^{-2x^2}$ **60.** $y = \dfrac{3}{(1 - 3x)^{2/3}}$

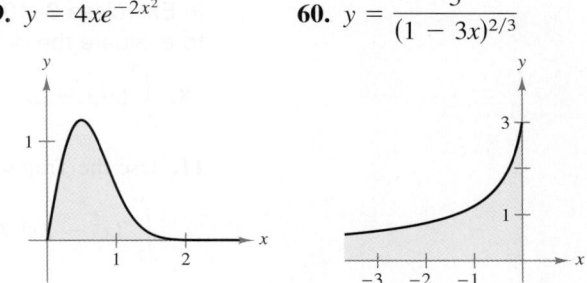

Endowment   In Exercises 61 and 62, determine the amount of money required to set up a charitable endowment that pays the amount *P* each year indefinitely for the annual interest rate *r* compounded continuously.

61. $P = \$8000,\ r = 3\%$

62. $P = \$15{,}000,\ r = 5\%$

63. Scholarship Fund   You want to start a scholarship fund at your alma mater. You plan to give one $21,000 scholarship annually beginning one year from now and you have at most $325,000 to start the fund. You also want the scholarship to be given out indefinitely. Assuming an annual interest rate of 7% compounded continuously, do you have enough money for the scholarship fund?

64. Present Value   You are considering buying a franchise that yields a continuous income stream of $100,000 per year. Assuming an annual interest rate of 6% compounded continuously, what is the present value of the franchise (a) for 15 years and (b) forever?

TEST YOURSELF

See www.CalcChat.com for worked-out solutions to odd-numbered exercises.

Take this test as you would take a test in class. When you are done, check your work against the answers given in the back of the book.

In Exercises 1–3, use integration by parts to find the indefinite integral.

1. $\displaystyle\int xe^{x+1}\,dx$ **2.** $\displaystyle\int 9x^2 \ln x\,dx$ **3.** $\displaystyle\int x^2\,e^{-x/3}\,dx$

4. The revenue R (in millions of dollars) for P. F. Chang's China Bistro from 2001 through 2009 can be modeled by

$$R = 295.1 + 147.66\sqrt{t}\ln t, \quad 1 \le t \le 9$$

where t is the year, with $t = 1$ corresponding to 2001.   *(Source: P. F. Chang's China Bistro)*

(a) Find the total revenue for the years 2001 through 2009.

(b) Find the average revenue for the years 2001 through 2009.

In Exercises 5–7, use the integration table in Appendix G to find the indefinite integral.

5. $\displaystyle\int \frac{x}{(7 + 2x)^2}\,dx$ **6.** $\displaystyle\int \frac{3x^2}{1 + e^{x^3}}\,dx$ **7.** $\displaystyle\int \frac{2x^3}{\sqrt{1 + 5x^2}}\,dx$

In Exercises 8–10, use integration by parts or the integration table in Appendix G to evaluate the definite integral.

8. $\displaystyle\int_0^1 \ln(3 - 2x)\,dx$ **9.** $\displaystyle\int_3^6 \frac{x}{\sqrt{x - 2}}\,dx$ **10.** $\displaystyle\int_{-3}^{-1} \frac{\sqrt{x^2 + 16}}{x}\,dx$

11. Use the Trapezoidal Rule with $n = 4$ to approximate

$$\int_2^5 (x^2 - 2x)\,dx.$$

Compare your result with the exact value of the definite integral.

12. Use Simpson's Rule with $n = 4$ to approximate

$$\int_0^1 9xe^{3x}\,dx.$$

Compare your result with the exact value of the definite integral.

In Exercises 13–15, determine whether the improper integral diverges or converges. Evaluate the integral if it converges.

13. $\displaystyle\int_0^{\infty} e^{-3x}\,dx$ **14.** $\displaystyle\int_0^9 \frac{2}{\sqrt{x}}\,dx$ **15.** $\displaystyle\int_{-\infty}^0 \frac{1}{(4x - 1)^{2/3}}\,dx$

16. A magazine publisher offers two subscription plans. Plan A is a one-year subscription for $19.95. Plan B is a lifetime subscription (lasting indefinitely) for $149.

(a) A subscriber considers using plan A indefinitely. Assuming an annual inflation rate of 4%, find the present value of the money the subscriber will spend using plan A.

(b) Based on your answer to part (a), which plan should the subscriber use? Explain.

13 Functions of Several Variables

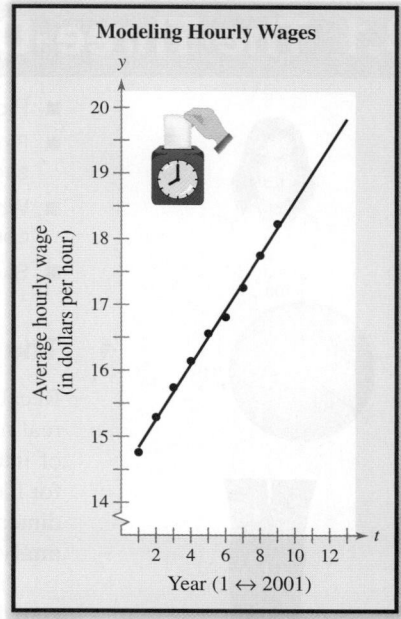

Modeling Hourly Wages
y
20
19
18
17
16
15
14
Average hourly wage (in dollars per hour)
t
2 4 6 8 10 12
Year (1 ↔ 2001)

Example 3 on page 969 shows how least squares regression analysis can be used to find the best-fitting line that models hourly wages for production workers in manufacturing industries.

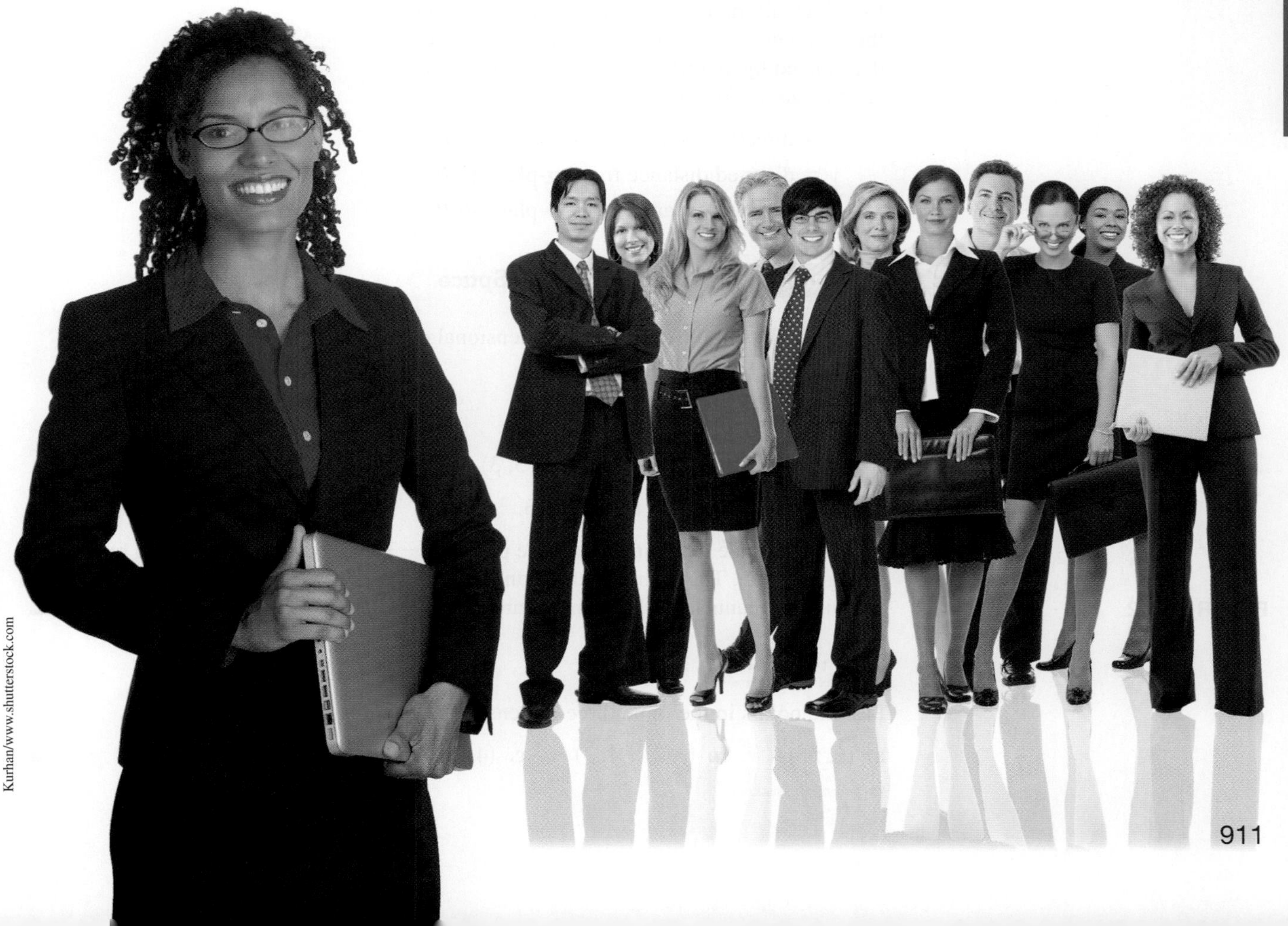

13.1 The Three-Dimensional Coordinate System

■ Plot points in space.
■ Find distances between points in space and find midpoints of line segments in space.
■ Write the standard forms of the equations of spheres and find the centers and radii of spheres.
■ Sketch the coordinate plane traces of surfaces.

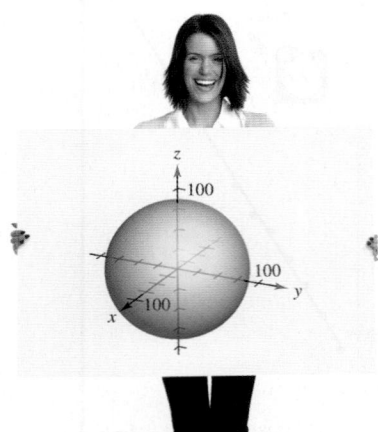

In Exercise 57 on page 918, you will model the shape of a spherical building using the standard equation of a sphere.

The Three-Dimensional Coordinate System

Recall from Section 2.1 that the Cartesian plane is determined by two perpendicular real number lines called the x-axis and the y-axis. These axes together with their point of intersection (the origin) allow you to develop a two-dimensional coordinate system for identifying points in a plane. To identify a point in space, you must introduce a third dimension to the model. The geometry of this three-dimensional model is called **solid analytic geometry.**

You can construct a **three-dimensional coordinate system** by passing a z-axis perpendicular to both the x- and y-axes at the origin. Figure 13.1 shows the positive portion of each coordinate axis. Taken as pairs, the axes determine three **coordinate planes:** the **xy-plane,** the **xz-plane,** and the **yz-plane.** These three coordinate planes separate the three-dimensional coordinate system into eight **octants.** The first octant is the one for which all three coordinates are positive. In this three-dimensional system, a point P in space is determined by an ordered triple (x, y, z), where x, y, and z are as follows.

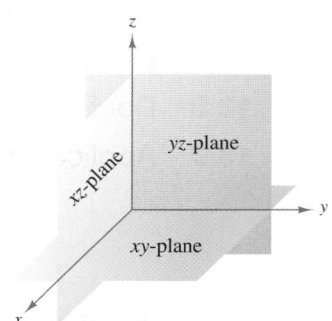

FIGURE 13.1

$$x = \text{directed distance from } yz\text{-plane to } P$$
$$y = \text{directed distance from } xz\text{-plane to } P$$
$$z = \text{directed distance from } xy\text{-plane to } P$$

Example 1   Plotting Points in Space

Plot the points in the same three-dimensional coordinate system.

a. $(2, -3, 3)$ **b.** $(-2, 6, 2)$

c. $(1, 4, 0)$ **d.** $(2, 2, -3)$

SOLUTION   To plot the point $(2, -3, 3)$, notice that

$$x = 2, \quad y = -3, \quad \text{and} \quad z = 3.$$

To help visualize the point, locate the point $(2, -3)$ in the xy-plane (denoted by a cross in Figure 13.2). The point lies three units above the cross. You can plot the other points in a similar manner, as shown in Figure 13.2.

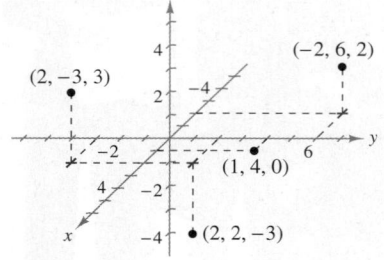

FIGURE 13.2

✓ Checkpoint 1

Plot the points in the same three-dimensional coordinate system.

a. $(2, 5, 1)$ **b.** $(-2, -4, 3)$ **c.** $(4, 0, -5)$

The Distance and Midpoint Formulas

Many of the formulas established for the two-dimensional coordinate system can be extended to three dimensions. For example, to find the distance between two points in space, you can use the Pythagorean Theorem twice, as shown in Figure 13.3. By doing this, you will obtain the formula for the distance between two points in space.

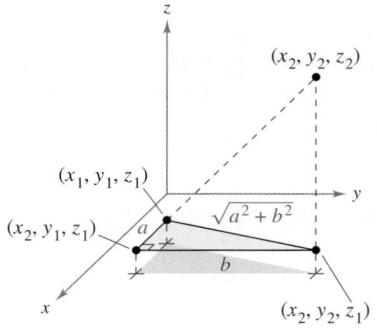

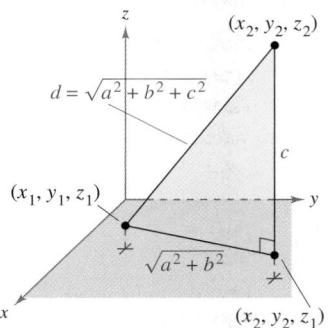

FIGURE 13.3

Distance Formula in Space

The distance between the points (x_1, y_1, z_1) and (x_2, y_2, z_2) is

$$d = \sqrt{(x_2 - x_1)^2 + (y_2 - y_1)^2 + (z_2 - z_1)^2}.$$

Example 2 Finding the Distance Between Two Points

Find the distance between $(1, 0, 2)$ and $(2, 4, -3)$.

SOLUTION

$$
\begin{aligned}
d &= \sqrt{(x_2 - x_1)^2 + (y_2 - y_1)^2 + (z_2 - z_1)^2} && \text{Write Distance Formula.}\\
&= \sqrt{(2 - 1)^2 + (4 - 0)^2 + (-3 - 2)^2} && \text{Substitute.}\\
&= \sqrt{1 + 16 + 25} && \text{Simplify.}\\
&= \sqrt{42} && \text{Simplify.}
\end{aligned}
$$

✓**Checkpoint 2**

Find the distance between $(2, 3, -1)$ and $(0, 5, 3)$. ■

Notice the similarity between the Distance Formula in the plane and the Distance Formula in space. The Midpoint Formulas in the plane and in space are also similar.

Midpoint Formula in Space

The midpoint of the line segment joining the points (x_1, y_1, z_1) and (x_2, y_2, z_2) is

$$\text{Midpoint} = \left(\frac{x_1 + x_2}{2}, \frac{y_1 + y_2}{2}, \frac{z_1 + z_2}{2} \right).$$

Example 3 Using the Midpoint Formula

Find the midpoint of the line segment joining

$$(5, -2, 3) \quad \text{and} \quad (0, 4, 4).$$

SOLUTION Using the Midpoint Formula, the midpoint is

$$\left(\frac{5 + 0}{2}, \frac{-2 + 4}{2}, \frac{3 + 4}{2} \right) = \left(\frac{5}{2}, 1, \frac{7}{2} \right)$$

as shown in Figure 13.4.

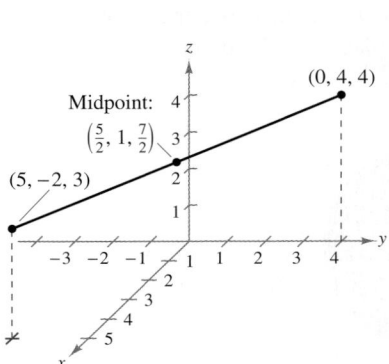

FIGURE 13.4

✓**Checkpoint 3**

Find the midpoint of the line segment joining

$$(3, -2, 0) \quad \text{and} \quad (-8, 6, -4).$$ ■

The Equation of a Sphere

A **sphere** with center at (h, k, l) and radius r is defined to be the set of all points (x, y, z) such that the distance between (x, y, z) and (h, k, l) is r, as shown in Figure 13.5. Using the Distance Formula, this condition can be written as

$$\sqrt{(x - h)^2 + (y - k)^2 + (z - l)^2} = r.$$

By squaring both sides of this equation, you obtain the standard equation of a sphere.

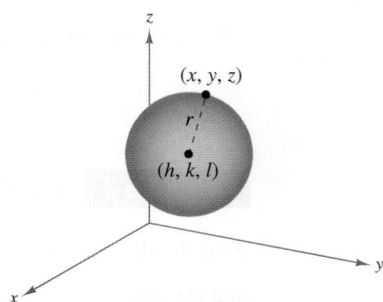

Sphere: Radius r, Center (h, k, l)

FIGURE 13.5

Standard Equation of a Sphere

The **standard equation of a sphere** with center at (h, k, l) and radius r is

$$(x - h)^2 + (y - k)^2 + (z - l)^2 = r^2.$$

Example 4   **Finding the Equation of a Sphere**

Find the standard equation of the sphere with center at $(2, 4, 3)$ and radius 3. Does this sphere intersect the xy-plane?

SOLUTION

$$(x - h)^2 + (y - k)^2 + (z - l)^2 = r^2 \qquad \text{Write standard equation.}$$
$$(x - 2)^2 + (y - 4)^2 + (z - 3)^2 = 3^2 \qquad \text{Substitute.}$$
$$(x - 2)^2 + (y - 4)^2 + (z - 3)^2 = 9 \qquad \text{Simplify.}$$

In Figure 13.6, note that the center of the sphere lies three units above the xy-plane. The sphere has a radius of 3, so it must intersect the xy-plane—at the point $(2, 4, 0)$.

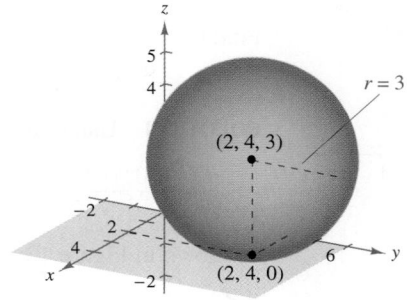

FIGURE 13.6

✓ **Checkpoint 4**

Find the standard equation of the sphere with center at $(4, 3, 2)$ and radius 5.

Example 5   **Finding the Equation of a Sphere**

Find the equation of the sphere that has the points $(3, -2, 6)$ and $(-1, 4, 2)$ as endpoints of a diameter.

SOLUTION   By the Midpoint Formula, the center of the sphere is

$$(h, k, l) = \left(\frac{3 + (-1)}{2}, \frac{-2 + 4}{2}, \frac{6 + 2}{2} \right)$$ Apply Midpoint Formula.

$$= (1, 1, 4).$$ Simplify.

By the Distance Formula, the radius is

$$r = \sqrt{(3 - 1)^2 + (-2 - 1)^2 + (6 - 4)^2}$$ Apply Distance Formula.

$$= \sqrt{4 + 9 + 4}$$ Simplify.

$$= \sqrt{17}.$$ Simplify.

So, the standard equation of the sphere is

$$(x - h)^2 + (y - k)^2 + (z - l)^2 = r^2$$ Write formula for a sphere.

$$(x - 1)^2 + (y - 1)^2 + (z - 4)^2 = 17.$$ Substitute.

✓**Checkpoint 5**

Find the equation of the sphere that has the points $(-2, 5, 7)$ and $(4, 1, -3)$ as endpoints of a diameter.

Example 6   **Finding the Center and Radius of a Sphere**

Find the center and radius of the sphere whose equation is

$$x^2 + y^2 + z^2 - 2x + 4y - 6z + 8 = 0.$$

SOLUTION   You can obtain the standard equation of the sphere by completing the square. To do this, begin by grouping terms with the same variable. Then add "the square of half the coefficient of each linear term" to each side of the equation. So, to complete the square of $(x^2 - 2x)$, add $\left[\frac{1}{2}(-2) \right]^2 = 1$ to each side. To complete the square of $(y^2 + 4y)$, add $\left[\frac{1}{2}(4) \right]^2 = 4$ to each side. To complete the square of $(z^2 - 6z)$, add $\left[\frac{1}{2}(-6) \right]^2 = 9$ to each side.

$$x^2 + y^2 + z^2 - 2x + 4y - 6z + 8 = 0$$

$$(x^2 - 2x + \boxed{}) + (y^2 + 4y + \boxed{}) + (z^2 - 6z + \boxed{}) = -8$$

$$(x^2 - 2x + 1) + (y^2 + 4y + 4) + (z^2 - 6z + 9) = -8 + 1 + 4 + 9$$

$$(x - 1)^2 + (y + 2)^2 + (z - 3)^2 = 6$$

So, the center of the sphere is $(1, -2, 3)$, and its radius is $\sqrt{6}$, as shown in Figure 13.7.

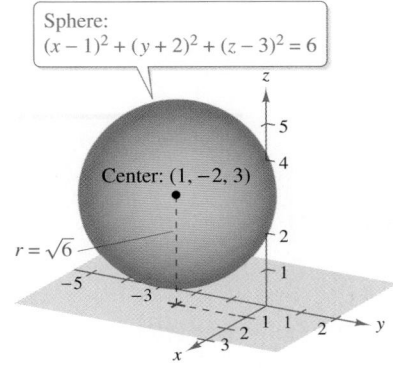

Sphere:
$(x - 1)^2 + (y + 2)^2 + (z - 3)^2 = 6$

Center: $(1, -2, 3)$

$r = \sqrt{6}$

FIGURE 13.7

✓**Checkpoint 6**

Find the center and radius of the sphere whose equation is

$$x^2 + y^2 + z^2 + 6x - 8y + 2z - 10 = 0.$$

Note in Example 6 that the points satisfying the equation of the sphere are "surface points," not "interior points." In general, the collection of points satisfying an equation involving x, y, and z is called a **surface in space.**

Traces of Surfaces

Finding the intersection of a surface with one of the three coordinate planes (or with a plane parallel to one of the three coordinate planes) helps visualize the surface. Such an intersection is called a **trace** of the surface. For example, the *xy*-trace of a surface consists of all points that are common to both the surface *and* the *xy*-plane. Similarly, the *xz*-trace of a surface consists of all points that are common to both the surface and the *xz*-plane.

Example 7   Finding a Trace of a Surface

Sketch the *xy*-trace of the sphere given by $(x - 3)^2 + (y - 2)^2 + (z + 4)^2 = 5^2$.

SOLUTION   To find the *xy*-trace of this surface, use the fact that every point in the *xy*-plane has a *z*-coordinate of zero. By substituting $z = 0$ into the original equation, the resulting equation will represent the intersection of the surface with the *xy*-plane.

$(x - 3)^2 + (y - 2)^2 + (z + 4)^2 = 5^2$	Write original equation.
$(x - 3)^2 + (y - 2)^2 + (0 + 4)^2 = 25$	Let $z = 0$ to find *xy*-trace.
$(x - 3)^2 + (y - 2)^2 + 16 = 25$	Simplify.
$(x - 3)^2 + (y - 2)^2 = 9$	Subtract 16 from each side.
$(x - 3)^2 + (y - 2)^2 = 3^2$	Equation of circle

From this equation, you can see that the *xy*-trace is a circle of radius 3, as shown in Figure 13.8.

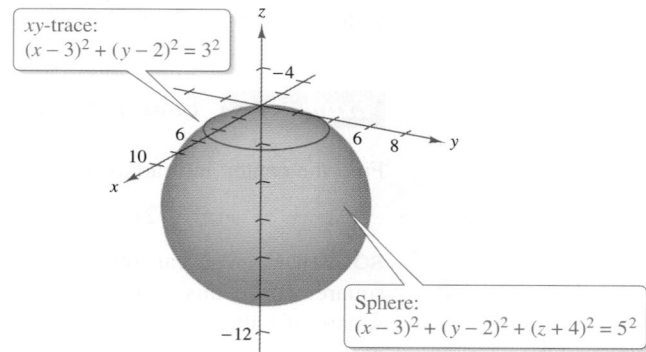

FIGURE 13.8

✓ **Checkpoint 7**

Find the equation of the *xy*-trace of the sphere given by

$$(x + 1)^2 + (y - 2)^2 + (z + 3)^2 = 5^2.$$

SUMMARIZE   (Section 13.1)

1. State the Distance Formula in space *(page 913)*. For an example of the Distance Formula in space, see Example 2.

2. State the Midpoint Formula in space *(page 913)*. For an example of the Midpoint Formula in space, see Example 3.

3. State the standard equation of a sphere *(page 914)*. For examples of finding equations of spheres, see Examples 4 and 5.

4. Explain what is meant by the trace of a surface *(page 916)*. For an example of finding the trace of a surface, see Example 7.

SKILLS WARM UP 13.1   The following warm-up exercises involve skills that were covered in earlier sections. You will use these skills in the exercise set for this section. For additional help, review Section 2.1.

In Exercises 1–4, find the distance between the points.

1. $(5, 1), (3, 5)$ **2.** $(2, 3), (-1, -1)$ **3.** $(-5, 4), (-5, -4)$ **4.** $(-3, 6), (-3, -2)$

In Exercises 5–8, find the midpoint of the line segment connecting the points.

5. $(2, 5), (6, 9)$ **6.** $(-1, -2), (3, 2)$ **7.** $(-6, 0), (6, 6)$ **8.** $(-4, 3), (2, -1)$

In Exercises 9 and 10, write the standard form of the equation of the circle.

9. Center: $(2, 3)$; radius: 2 **10.** Endpoints of a diameter: $(4, 0), (-2, 8)$

Exercises 13.1

See www.CalcChat.com for worked-out solutions to odd-numbered exercises.

Plotting Points in Space   In Exercises 1–4, plot the points in the same three-dimensional coordinate system. *See Example 1.*

1. $(2, 1, 3), (-1, 2, 1), (3, -2, 5), \left(\frac{3}{2}, 4, -2\right)$

2. $(-5, -2, 2), (5, -2, -2), (1, 3, 1), (-2, 4, -3)$

3. $(0, 4, -5), (4, 0, 5), \left(-2, \frac{1}{2}, 0\right), \left(-\frac{1}{2}, 3, 1\right)$

4. $(-3, 0, -1), (2, -1, 1), (-1, -3, -2), (1, 3, 4)$

Finding Points in Space   In Exercises 5–8, find the coordinates of the point.

5. The point is located three units behind the yz-plane, four units to the right of the xz-plane, and five units above the xy-plane.

6. The point is located seven units in front of the yz-plane, two units to the left of the xz-plane, and one unit below the xy-plane.

7. The point is located on the x-axis, 10 units in front of the yz-plane.

8. The point is located in the yz-plane, three units to the right of the xz-plane, and two units above the xy-plane.

9. Think About It   What is the z-coordinate of any point in the xy-plane?

10. Think About It   What is the y-coordinate of any point in the xz-plane?

Finding the Distance Between Two Points   In Exercises 11–14, find the distance between the two points. *See Example 2.*

11. $(4, 1, 5), (8, 2, 6)$

12. $(-4, -1, 1), (2, -1, 5)$

13. $(-1, -5, 7), (-3, 4, -4)$

14. $(8, -2, 2), (8, -2, 4)$

Using the Midpoint Formula   In Exercises 15–18, find the midpoint of the line segment joining the two points. *See Example 3.*

15. $(6, -4, 2), (-2, 1, 3)$ **16.** $(4, 0, -6), (8, 8, 20)$

17. $(-5, -2, 5), (6, 3, -7)$ **18.** $(0, -2, 5), (4, 2, 7)$

Using the Midpoint Formula   In Exercises 19–22, find (x, y, z).

19.

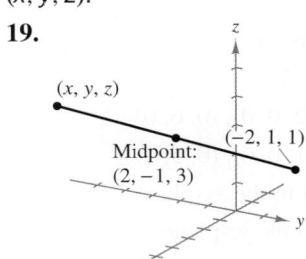

20.

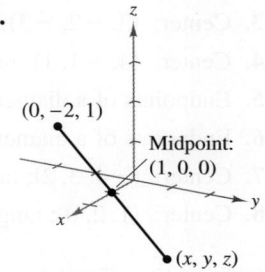

21.

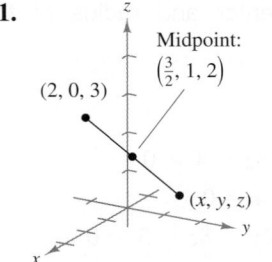

22.

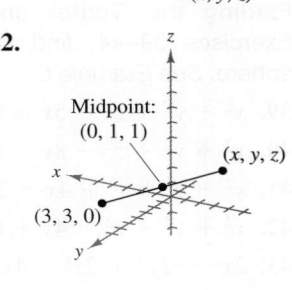

Identifying Triangles   In Exercises 23–26, find the lengths of the sides of the triangle with the given vertices, and determine whether the triangle is a right triangle, an isosceles triangle, or neither.

23. $(0, 0, 0), (2, 2, 1), (2, -4, 4)$

24. $(5, 3, 4), (7, 1, 3), (3, 5, 3)$

25. $(-1, 0, -2), (-1, 5, 2), (-3, -1, 1)$

26. $(5, 0, 0), (0, 2, 0), (0, 0, -3)$

27. Think About It The triangle in Exercise 23 is translated five units downward along the z-axis. Determine the coordinates of the translated triangle.

28. Think About It The triangle in Exercise 24 is translated three units to the right along the y-axis. Determine the coordinates of the translated triangle.

Finding the Equation of a Sphere In Exercises 29–38, find the standard equation of the sphere. *See Examples 4 and 5.*

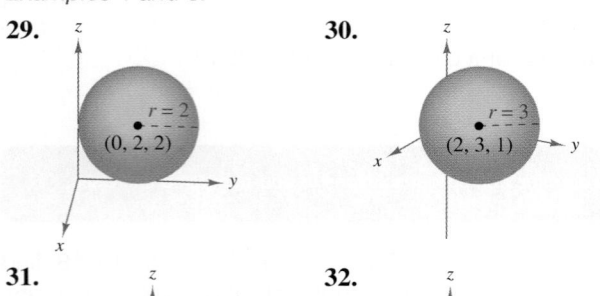

29.   **30.**

31.   **32.**

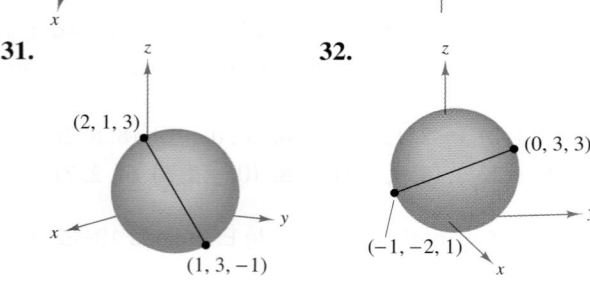

33. Center: $(3, -2, -3)$; radius: 4

34. Center: $(4, -1, 1)$; radius: 5

35. Endpoints of a diameter: $(2, 0, 0)$, $(0, 6, 0)$

36. Endpoints of a diameter: $(1, 0, 0)$, $(0, 5, 0)$

37. Center: $(-4, 3, 2)$; tangent to the xy-plane

38. Center: $(1, 2, 0)$; tangent to the yz-plane

Finding the Center and Radius of a Sphere In Exercises 39–44, find the center and radius of the sphere. *See Example 6.*

39. $x^2 + y^2 + z^2 - 5x = 0$

40. $x^2 + y^2 + z^2 - 8y = 0$

41. $x^2 + y^2 + z^2 + 4x - 2y + 8z - 4 = 0$

42. $x^2 + y^2 + z^2 - 4y + 6z + 4 = 0$

43. $2x^2 + 2y^2 + 2z^2 - 4x - 12y - 8z + 3 = 0$

44. $4x^2 + 4y^2 + 4z^2 - 8x + 16y + 11 = 0$

Finding the Trace of a Surface In Exercises 45–48, sketch the xy-trace of the sphere. *See Example 7.*

45. $(x - 1)^2 + (y - 3)^2 + (z - 2)^2 = 25$

46. $(x + 1)^2 + (y + 2)^2 + (z - 2)^2 = 16$

47. $x^2 + y^2 + z^2 - 6x - 10y + 6z + 30 = 0$

48. $x^2 + y^2 + z^2 - 4y + 2z - 60 = 0$

Finding the Trace of a Surface In Exercises 49–52, sketch the yz-trace of the sphere. *See Example 7.*

49. $x^2 + (y + 3)^2 + z^2 = 25$

50. $(x + 2)^2 + (y - 3)^2 + z^2 = 9$

51. $x^2 + y^2 + z^2 - 4x - 4y - 6z - 12 = 0$

52. $x^2 + y^2 + z^2 - 6x - 10y + 6z + 30 = 0$

Finding the Trace of a Surface In Exercises 53–56, sketch the trace of the intersection of each plane with the given sphere.

53. $x^2 + y^2 + z^2 = 25$

    (a) $z = 3$ (b) $x = 4$

54. $x^2 + y^2 + z^2 = 169$

    (a) $x = 5$ (b) $y = 12$

55. $x^2 + y^2 + z^2 - 4x - 6y + 9 = 0$

    (a) $x = 2$ (b) $y = 3$

56. $x^2 + y^2 + z^2 - 8x - 6z + 16 = 0$

    (a) $x = 4$ (b) $z = 3$

57. Architecture A spherical building has a diameter of 165 feet. The center of the building is placed at the origin of a three-dimensional coordinate system. What is the equation of the sphere that models the shape of the building?

58. HOW DO YOU SEE IT? Crystals are classified according to their symmetry.

(a) Crystals shaped like cubes are classified as isometric. The vertices of an isometric crystal mapped onto a three-dimensional coordinate system are shown in the figure. Determine (x, y, z).

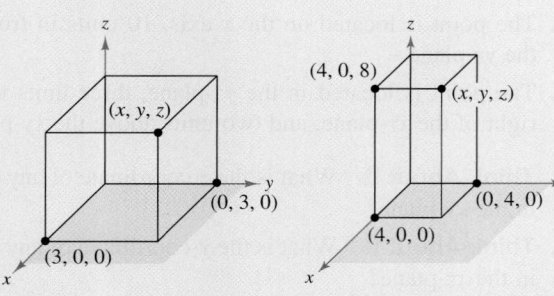

Figure for (a)   Figure for (b)

(b) Crystals shaped like rectangular prisms are classified as tetragonal. The vertices of a tetragonal crystal mapped onto a three-dimensional coordinate system are shown in the figure. Determine (x, y, z).

13.2 Surfaces in Space

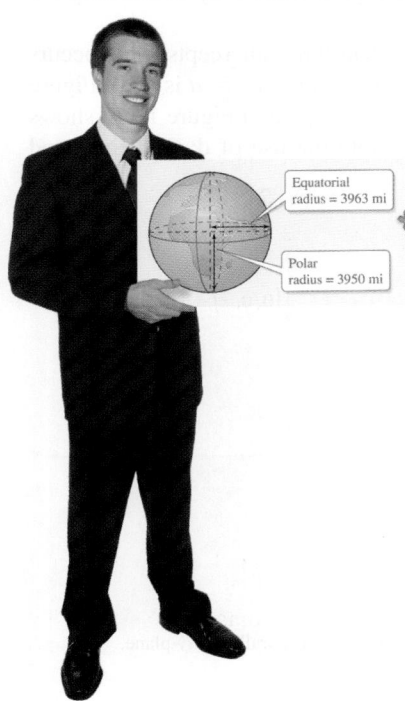

- Sketch planes in space.
- Draw planes in space with different numbers of intercepts.
- Classify quadric surfaces in space.

Equations of Planes in Space

In Section 13.1, you studied one type of surface in space—a sphere. In this section, you will study a second type—a plane in space. The **general equation of a plane** in space is

$$ax + by + cz = d. \qquad \text{General equation of a plane}$$

Note the similarity of this equation to the general equation of a line in the plane. In fact, when you intersect the plane represented by this equation with each of the three coordinate planes, you will obtain traces that are lines, as shown in Figure 13.9.

In Figure 13.9, the points where the plane intersects the three coordinate axes are the x-, y-, and z-intercepts of the plane. By connecting these three points, you can form a triangular region, which helps you visualize the plane in space.

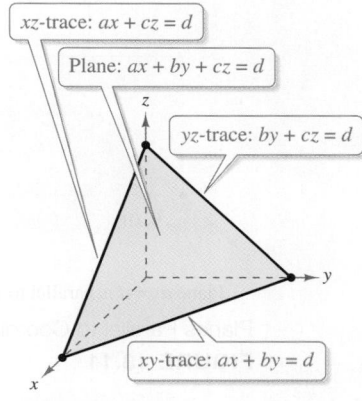

FIGURE 13.9

In Exercise 49 on page 927, you will use the dimensions of Earth to write an equation of an ellipsoid that models its shape.

Example 1 Sketching a Plane in Space

Find the x-, y-, and z-intercepts of the plane given by

$$3x + 2y + 4z = 12.$$

Then sketch the plane.

SOLUTION To find the x-intercept, let both y and z be zero.

$$3x + 2(0) + 4(0) = 12 \qquad \text{Substitute 0 for } y \text{ and } z.$$
$$3x = 12 \qquad \text{Simplify.}$$
$$x = 4 \qquad \text{Solve for } x.$$

So, the x-intercept is $(4, 0, 0)$. To find the y-intercept, let x and z be zero and conclude that $y = 6$. So, the y-intercept is $(0, 6, 0)$. Similarly, by letting x and y be zero, you can determine that $z = 3$ and that the z-intercept is $(0, 0, 3)$. Figure 13.10 shows the triangular portion of the plane formed by connecting the three intercepts

$$(4, 0, 0), \quad (0, 6, 0), \quad \text{and} \quad (0, 0, 3).$$

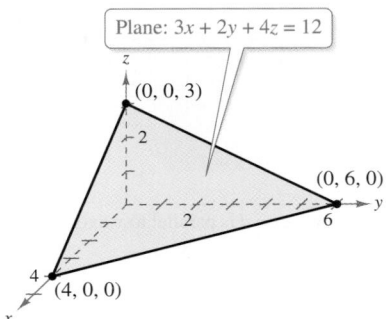

Sketch Made by Connecting Intercepts: $(4, 0, 0)$, $(0, 6, 0)$, $(0, 0, 3)$
FIGURE 13.10

✓Checkpoint 1

Find the x-, y-, and z-intercepts of the plane given by

$$2x + 4y + z = 8.$$

Then sketch the plane.

Drawing Planes in Space

The planes shown in Figures 13.9 and 13.10 have three intercepts. When this occurs, you can draw the plane by sketching the triangular region formed by connecting the three intercepts.

It is possible for a plane in space to have fewer than three intercepts. This occurs when one or more of the coefficients in the equation $ax + by + cz = d$ is zero. Figure 13.11 shows some planes in space that have only one intercept, and Figure 13.12 shows some that have only two intercepts. In each figure, note the use of dashed lines and shading to give the illusion of three dimensions.

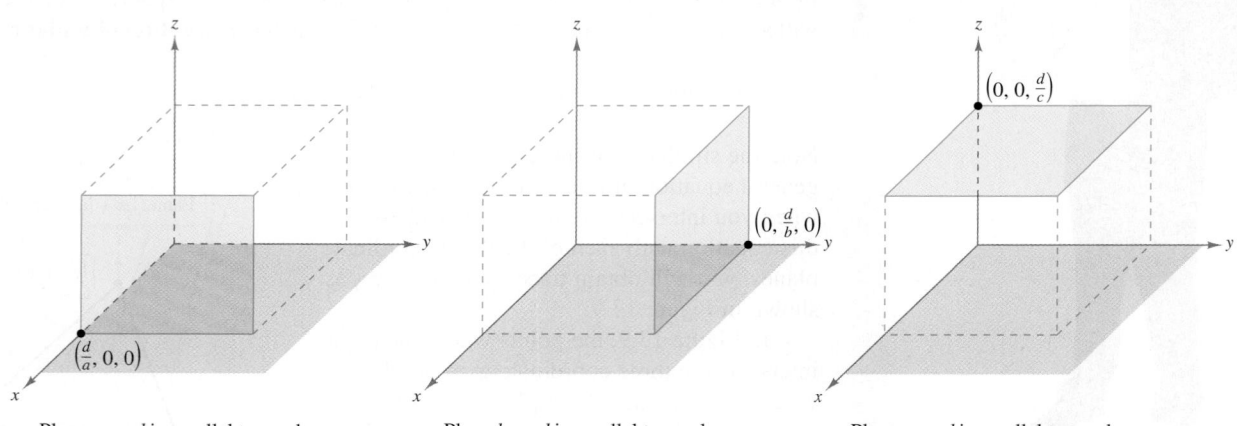

Plane $ax = d$ is parallel to yz-plane.　Plane $by = d$ is parallel to xz-plane.　Plane $cz = d$ is parallel to xy-plane.

Planes Parallel to Coordinate Planes

FIGURE 13.11

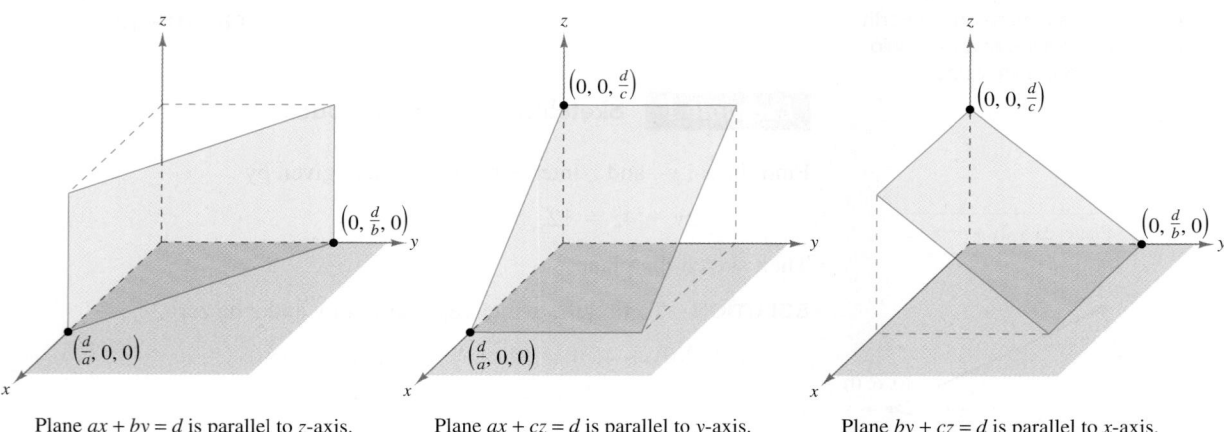

Plane $ax + by = d$ is parallel to z-axis.　Plane $ax + cz = d$ is parallel to y-axis.　Plane $by + cz = d$ is parallel to x-axis.

Planes Parallel to Coordinate Axes

FIGURE 13.12

When an equation of a plane has a missing variable, such as

$$2x + z = 1 \qquad \text{See Figure 13.13.}$$

the plane must be *parallel to the axis* represented by the missing variable, as shown in Figure 13.12. When two variables are missing from an equation of a plane, the plane is *parallel to the coordinate plane* represented by the missing variables, as shown in Figure 13.11.

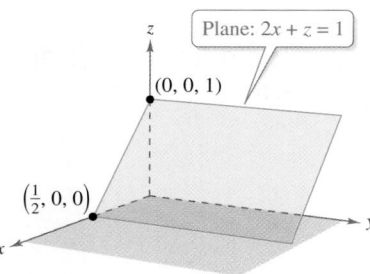

Plane $2x + z = 1$ is parallel to the y-axis.

FIGURE 13.13

Quadric Surfaces

A third common type of surface in space is a **quadric surface.** Quadric surfaces are the three-dimensional analogs of conic sections. The equation of a quadric surface in space is a second-degree equation in three variables, such as

$$Ax^2 + By^2 + Cz^2 + Dx + Ey + Fz + G = 0.$$ Second-degree equation

There are six basic types of quadric surfaces.

1. Elliptic cone   2. Elliptic paraboloid

3. Hyperbolic paraboloid   4. Ellipsoid

5. Hyperboloid of one sheet   6. Hyperboloid of two sheets

The six types are summarized on the next two pages. Notice that each surface is pictured with two types of three-dimensional sketches. The computer-generated sketches use traces with hidden lines to give the illusion of three dimensions. The artist-rendered sketches use shading to create the same illusion.

All of the quadric surfaces on the next two pages are centered at the origin and have axes along the coordinate axes. Moreover, only one of several possible orientations of each surface is shown. When the surface has a different center or is oriented along a different axis, its standard equation will change accordingly. For instance, the ellipsoid

$$\frac{x^2}{1^2} + \frac{y^2}{3^2} + \frac{z^2}{2^2} = 1$$

has $(0, 0, 0)$ as its center, but the ellipsoid

$$\frac{(x-2)^2}{1^2} + \frac{(y+1)^2}{3^2} + \frac{(z-4)^2}{2^2} = 1$$

has $(2, -1, 4)$ as its center. A computer-generated graph of the first ellipsoid is shown in Figure 13.14.

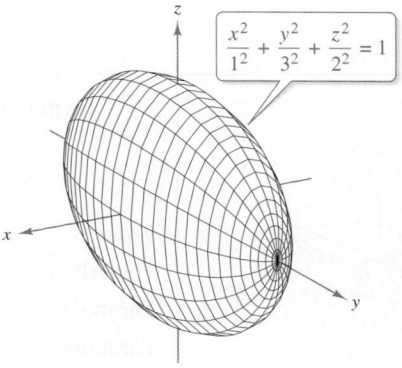

$$\frac{x^2}{1^2} + \frac{y^2}{3^2} + \frac{z^2}{2^2} = 1$$

FIGURE 13.14

TECH TUTOR

If you have access to a three-dimensional graphing utility, try using it to graph the surface in Figure 13.14. When you do this, you will discover that sketching surfaces in space is not a simple task—even with a graphing utility.

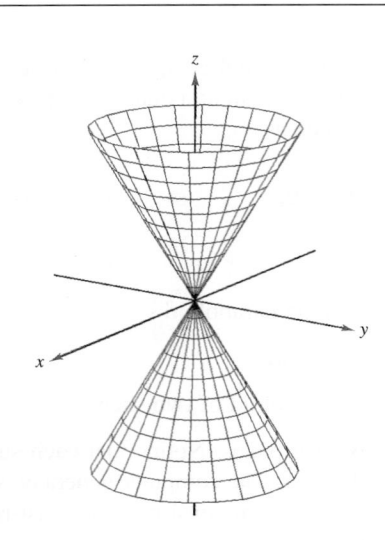

Elliptic Cone

$$\frac{x^2}{a^2} + \frac{y^2}{b^2} - \frac{z^2}{c^2} = 0$$

Trace	*Plane*
Ellipse	Parallel to xy-plane
Hyperbola	Parallel to xz-plane
Hyperbola	Parallel to yz-plane

The axis of the cone corresponds to the variable whose coefficient is negative. The traces in the coordinate planes parallel to this axis are intersecting lines.

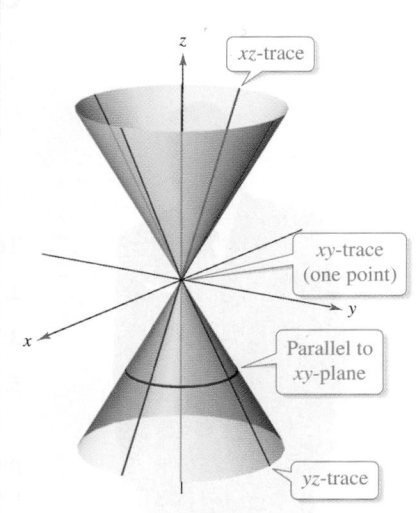

Elliptic Paraboloid

$$z = \frac{x^2}{a^2} + \frac{y^2}{b^2}$$

Trace	*Plane*
Ellipse	Parallel to xy-plane
Parabola	Parallel to xz-plane
Parabola	Parallel to yz-plane

The axis of the paraboloid corresponds to the variable raised to the first power.

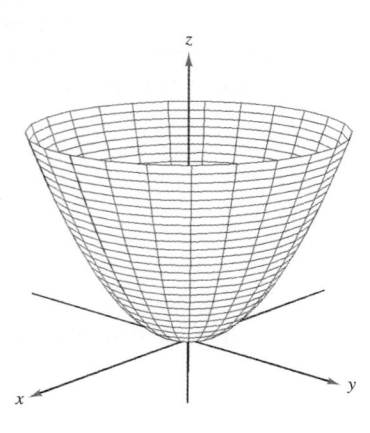

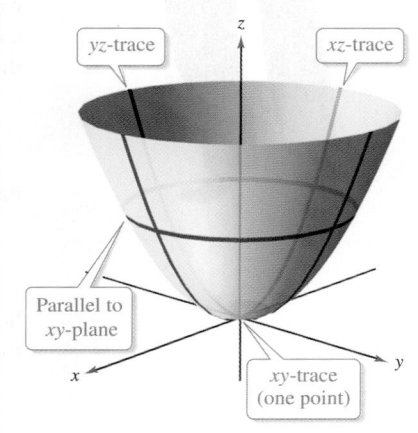

Hyperbolic Paraboloid

$$z = \frac{y^2}{b^2} - \frac{x^2}{a^2}$$

Trace	*Plane*
Hyperbola	Parallel to xy-plane
Parabola	Parallel to xz-plane
Parabola	Parallel to yz-plane

The axis of the paraboloid corresponds to the variable raised to the first power.

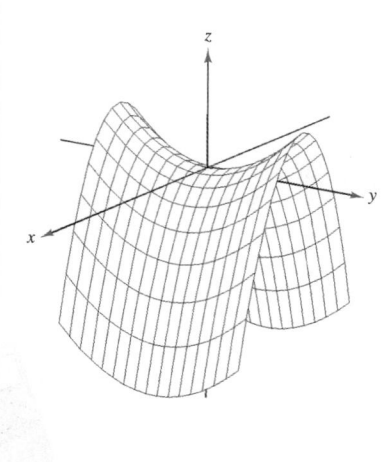

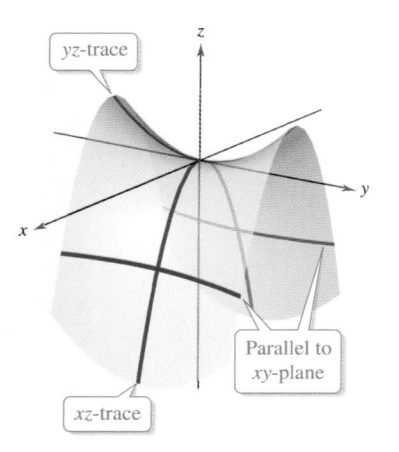

Ellipsoid

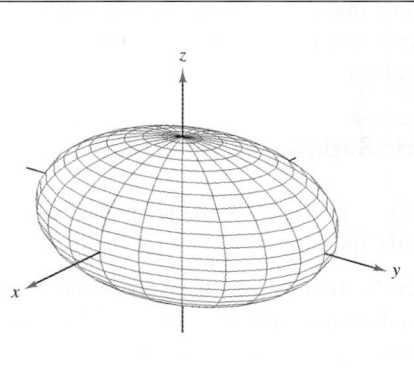

$$\frac{x^2}{a^2} + \frac{y^2}{b^2} + \frac{z^2}{c^2} = 1$$

Trace	Plane
Ellipse	Parallel to xy-plane
Ellipse	Parallel to xz-plane
Ellipse	Parallel to yz-plane

The surface is a sphere when the coefficients a, b, and c are equal and nonzero.

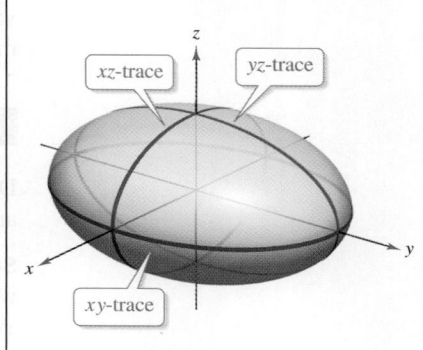

Hyperboloid of One Sheet

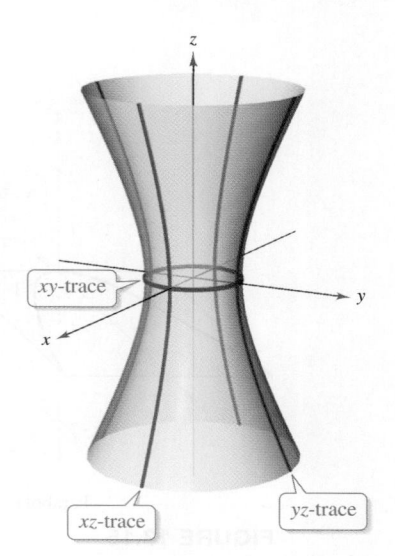

$$\frac{x^2}{a^2} + \frac{y^2}{b^2} - \frac{z^2}{c^2} = 1$$

Trace	Plane
Ellipse	Parallel to xy-plane
Hyperbola	Parallel to xz-plane
Hyperbola	Parallel to yz-plane

The axis of the hyperboloid corresponds to the variable whose coefficient is negative.

Hyperboloid of Two Sheets

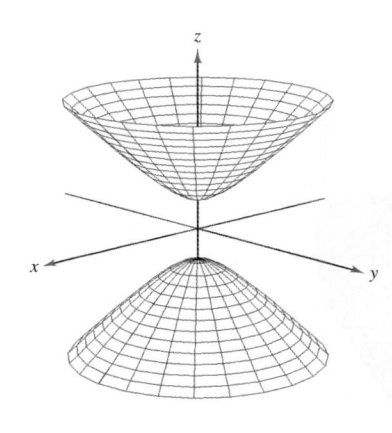

$$\frac{z^2}{c^2} - \frac{x^2}{a^2} - \frac{y^2}{b^2} = 1$$

Trace	Plane
Ellipse	Parallel to xy-plane
Hyperbola	Parallel to xz-plane
Hyperbola	Parallel to yz-plane

The axis of the hyperboloid corresponds to the variable whose coefficient is positive. There is no trace in the coordinate plane perpendicular to this axis.

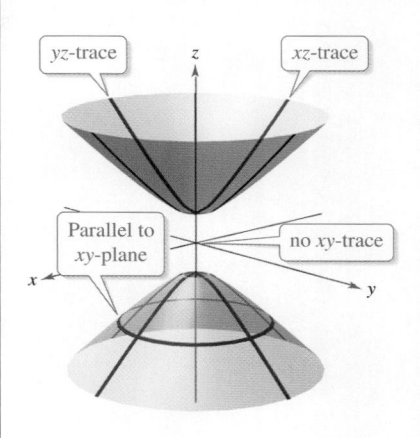

When classifying quadric surfaces, note that the two types of paraboloids have one variable raised to the first power. The other four types of quadric surfaces have equations that are of second degree in *all* three variables.

Example 2   Classifying a Quadric Surface

Describe the traces of the surface given by $x - y^2 - z^2 = 0$ in the xy-plane, the xz-plane, and the plane given by $x = 1$. Then classify the surface.

SOLUTION   Because x is raised only to the first power, the surface is a paraboloid whose axis is the x-axis. In standard form, the equation is $x = y^2 + z^2$. The traces in the xy-plane, the xz-plane, and the plane given by $x = 1$ are as shown.

Trace in xy-plane $(z = 0)$:	$x = y^2$	Parabola
Trace in xz-plane $(y = 0)$:	$x = z^2$	Parabola
Trace in plane $x = 1$:	$y^2 + z^2 = 1$	Circle

These three traces are shown in Figure 13.15.

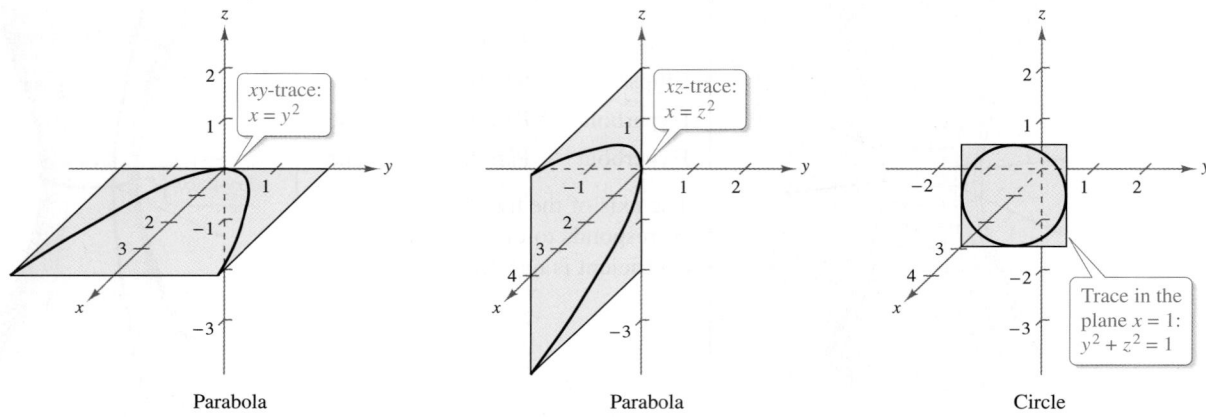

Parabola   Parabola   Circle

FIGURE 13.15

From the traces, you can see that the surface is an elliptic (or circular) paraboloid, as shown in Figure 13.16.

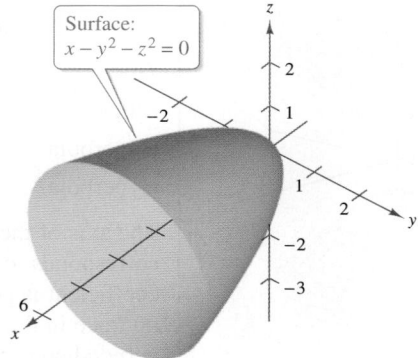

Elliptic Paraboloid
FIGURE 13.16

✓ Checkpoint 2

Describe the traces of the surface given by $x^2 + y^2 - z^2 = 1$ in the xy-plane, the yz-plane, the xz-plane, and the plane given by $z = 3$. Then classify the surface. ▪

Example 3   **Classifying Quadric Surfaces**

Classify the surface given by each equation.

a. $x^2 - 4y^2 - 4z^2 - 4 = 0$

b. $x^2 + 4y^2 + z^2 - 4 = 0$

SOLUTION

a. The equation $x^2 - 4y^2 - 4z^2 - 4 = 0$ can be written in standard form as

$$\frac{x^2}{4} - y^2 - z^2 = 1. \qquad \text{Standard form}$$

From the standard form, you can see that the graph is a **hyperboloid of two sheets**, with the x-axis as its axis, as shown in Figure 13.17(a).

b. The equation $x^2 + 4y^2 + z^2 - 4 = 0$ can be written in standard form as

$$\frac{x^2}{4} + y^2 + \frac{z^2}{4} = 1. \qquad \text{Standard form}$$

From the standard form, you can see that the graph is an **ellipsoid**, as shown in Figure 13.17(b).

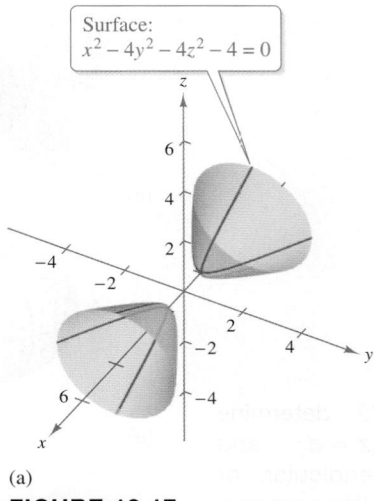

Surface:
$x^2 - 4y^2 - 4z^2 - 4 = 0$

Surface:
$x^2 + 4y^2 + z^2 - 4 = 0$

(a)   (b)

FIGURE 13.17

✓ **Checkpoint 3**

Classify the surface given by each equation.

a. $4x^2 + 9y^2 - 36z = 0$

b. $36x^2 + 16y^2 - 144z^2 = 0$

SUMMARIZE   (Section 13.2)

1. State the general equation of a plane in space *(page 919)*. For an example of sketching a plane in space, see Example 1.

2. List the six basic types of quadric surfaces *(page 921)*. For examples of classifying quadric surfaces, see Examples 2 and 3.

SKILLS WARM UP 13.2   The following warm-up exercises involve skills that were covered in earlier sections. You will use these skills in the exercise set for this section. For additional help, review Sections 2.1 and 13.1.

In Exercises 1–4, find the x- and y-intercepts of the function.

1. $3x + 4y = 12$

2. $6x + y = -8$

3. $-2x + y = -2$

4. $-x - y = 5$

In Exercises 5 and 6, write the equation of the sphere in standard form.

5. $16x^2 + 16y^2 + 16z^2 = 4$

6. $9x^2 + 9y^2 + 9z^2 = 36$

Exercises 13.2

See www.CalcChat.com for worked-out solutions to odd-numbered exercises.

Sketching a Plane in Space In Exercises 1–12, find the intercepts and sketch the graph of the plane. *See Example 1.*

1. $4x + 2y + 6z = 12$

2. $3x + 6y + 2z = 6$

3. $3x + 3y + 5z = 15$

4. $x + y + z = 3$

5. $2x - y + 3z = 4$

6. $2x - y + z = 4$

7. $z = 8$

8. $x = 5$

9. $y + z = 5$

10. $x + 2y = 4$

11. $x + z = 6$

12. $x - 3z = 3$

Comparing Planes In Exercises 13–22, determine whether the planes $a_1x + b_1y + c_1z = d_1$ and $a_2x + b_2y + c_2z = d_2$ are parallel, perpendicular, or neither. The planes are parallel when there exists a nonzero constant k such that $a_1 = ka_2$, $b_1 = kb_2$, and $c_1 = kc_2$, and are perpendicular when $a_1a_2 + b_1b_2 + c_1c_2 = 0$.

13. $5x - 3y + z = 4,\ x + 4y + 7z = 1$

14. $3x + y - 4z = 3,\ -9x - 3y + 12z = 4$

15. $x - 5y - z = 1,\ 5x - 25y - 5z = -3$

16. $x + 3y + 2z = 6,\ 4x - 12y + 8z = 24$

17. $x + 2y = 3,\ 4x + 8y = 5$

18. $x + 3y + z = 7,\ x - 5z = 0$

19. $2x + y = 3,\ 3x - 5z = 0$

20. $2x - z = 1,\ 4x + y + 8z = 10$

21. $x = 3,\ z = -1$

22. $x = -2,\ y = 4$

Matching In Exercises 23–28, match the equation with its graph. Then classify the quadric surface. [The graphs are labeled (a)–(f).]

(a)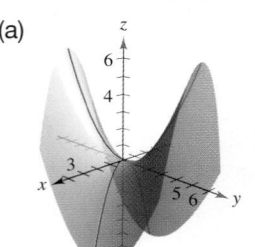

(b)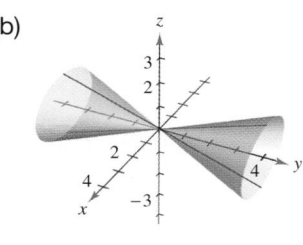

(c)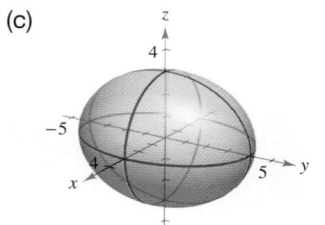

(d)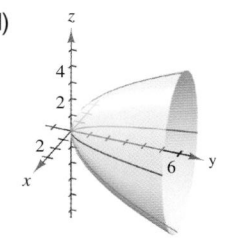

(e)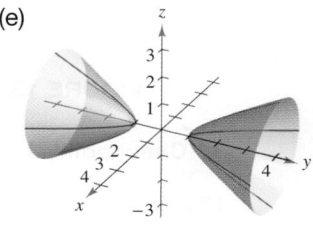

(f)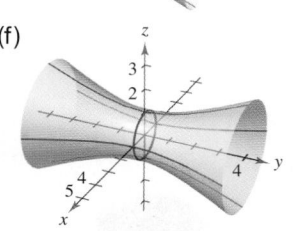

23. $\dfrac{x^2}{9} + \dfrac{y^2}{16} + \dfrac{z^2}{9} = 1$

24. $15x^2 - 4y^2 + 15z^2 = -4$

25. $4x^2 - y^2 + 4z^2 = 4$

26. $y^2 = 4x^2 + 9z^2$

27. $4x^2 - 4y + z^2 = 0$

28. $4x^2 - y^2 + 4z = 0$

Classifying a Quadric Surface In Exercises 29–34, describe the traces of the surface in the given planes. Then classify the surface. *See Example 2.*

29. $z = x^2 - y^2$

    (a) *xy*-plane   (b) $x = 3$ (c) *xz*-plane

30. $y = x^2 + z^2$

    (a) *xy*-plane   (b) $y = 1$ (c) *yz*-plane

31. $\dfrac{x^2}{4} + y^2 + z^2 = 1$

    (a) *xy*-plane   (b) *xz*-plane   (c) *yz*-plane

32. $y^2 + z^2 - x^2 = 1$

    (a) *xy*-plane   (b) *xz*-plane   (c) *yz*-plane

33. $z^2 - \dfrac{x^2}{9} - \dfrac{y^2}{16} = 1$

    (a) *xz*-plane   (b) $x = 2$ (c) $z = 4$

34. $y^2 + \dfrac{z^2}{4} - x^2 = 0$

    (a) $y = -1$ (b) $z = 4$ (c) *yz*-plane

Classifying a Quadric Surface In Exercises 35–48, classify the quadric surface. *See Example 3.*

35. $x^2 + \dfrac{y^2}{4} + z^2 = 1$ **36.** $z^2 = x^2 + \dfrac{y^2}{4}$

37. $25x^2 + 25y^2 - z^2 = 5$ **38.** $z = 4x^2 + y^2$

39. $x^2 - y^2 + z = 0$ **40.** $z^2 - x^2 - \dfrac{y^2}{4} = 1$

41. $x^2 - y + z^2 = 0$ **42.** $9x^2 + 4y^2 - 8z^2 = 72$

43. $z^2 = 9x^2 + y^2$ **44.** $\dfrac{x^2}{9} + \dfrac{y^2}{16} + \dfrac{z^2}{16} = 1$

45. $2x^2 - y^2 + 2z^2 = -4$

46. $4y = x^2 + z^2$

47. $3z = -y^2 + x^2$

48. $z^2 = 2x^2 + 2y^2$

49. Physical Science Because of the forces caused by its rotation, Earth is actually an oblate ellipsoid rather than a sphere. The equatorial radius is 3963 miles and the polar radius is 3950 miles. Find an equation of the ellipsoid. Assume that the center of Earth is at the origin and the *xy*-trace ($z = 0$) corresponds to the equator.

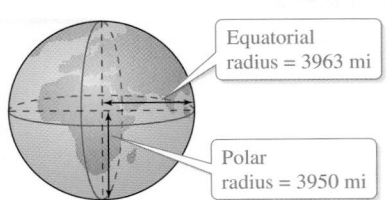

Equatorial radius = 3963 mi

Polar radius = 3950 mi

50. HOW DO YOU SEE IT? Each elliptic paraboloid below represents a statue that is modeled by the quadric surface $z = x^2 + y^2$. Match each of the four graphs with the point in space from which the statue is viewed. The four points are $(0, 0, 20)$, $(0, 20, 0)$, $(20, 0, 0)$, and $(10, 10, 20)$.

(a) 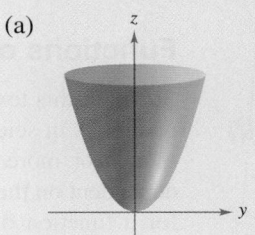   (b)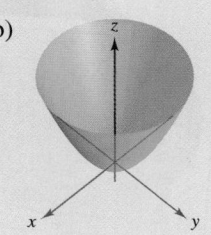

(c) 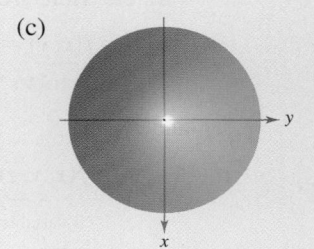   (d)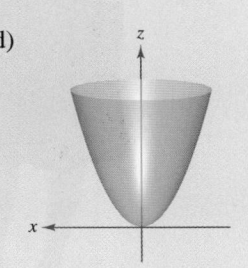

51. Modeling Data Personal consumption expenditures (in billions of dollars) for several types of recreation from 2004 through 2009 are shown in the table, where *x* is the expenditures on amusement parks and campgrounds, *y* is the expenditures on live entertainment (excluding sports), and *z* is the expenditures on spectator sports.

Year	2004	2005	2006	2007	2008	2009
x	33.1	34.9	37.4	40.6	43.0	41.8
y	13.2	13.8	14.9	15.0	15.4	14.5
z	15.5	16.3	17.8	19.5	20.5	20.7

A model for the data in the table is given by $-0.62x + 0.41y + z = 0.38$. *(Source: U.S. Bureau of Economic Analysis)*

(a) Complete a fourth row of the table using the model to approximate *z* for the given values of *x* and *y*. Compare the approximations with the actual values of *z*.

(b) According to this model, increases in expenditures of recreation types *y* and *z* would correspond to what kind of change in expenditures of recreation type *x*?

13.3 Functions of Several Variables

- Evaluate functions of several variables.
- Find the domains and ranges of functions of two variables.
- Read contour maps and sketch level curves of functions of two variables.
- Use functions of several variables to answer questions about real-life situations.

Functions of Several Variables

So far in this text, you have studied functions of a single independent variable. Many quantities in science, business, and technology, however, are functions not of one, but of two or more variables. For instance, the demand function for a product is often dependent on the price *and* the advertising, rather than on the price alone. The notation for a function of two or more variables is similar to that for a function of a single variable. Here are two examples.

$$z = f(\underbrace{x, y}_{\text{2 variables}}) = x^2 + xy \qquad \text{Function of two variables}$$

and

$$w = f(\underbrace{x, y, z}_{\text{3 variables}}) = x + 2y - 3z \qquad \text{Function of three variables}$$

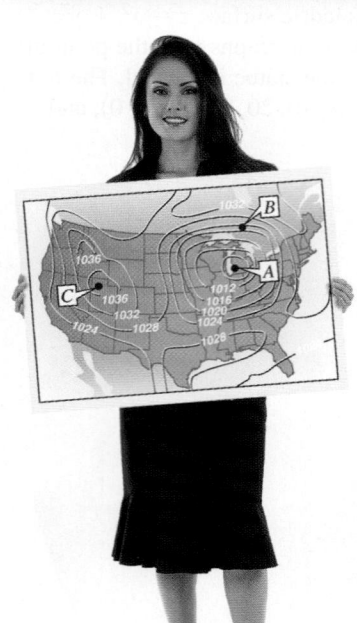

In Exercise 49 on page 935, you will read a weather map and identify areas of high and low pressure.

Definition of a Function of Two Variables

Let D be a set of ordered pairs of real numbers. If to each ordered pair (x, y) in D there corresponds a unique real number $f(x, y)$, then f is called a **function of x and y.** The set D is the **domain** of f, and the corresponding set of values for $f(x, y)$ is the **range** of f. Functions of three, four, or more variables are defined similarly.

For the function given by

$$z = f(x, y)$$

x and y are called the **independent variables** and z is called the **dependent variable.**

Example 1 Evaluating Functions of Several Variables

a. For $f(x, y) = 2x^2 - y^2$, you can evaluate $f(2, 3)$ as shown.

$$f(2, 3) = 2(2)^2 - (3)^2 = 8 - 9 = -1$$

b. For $f(x, y, z) = e^x(y + z)$, you can evaluate $f(0, -1, 4)$ as shown.

$$f(0, -1, 4) = e^0(-1 + 4) = (1)(3) = 3$$

✓ Checkpoint 1

Find the indicated function values.

a. For $f(x, y) = x^2 + 2xy$, find $f(2, -1)$.

b. For $f(x, y, z) = \dfrac{2x^2z}{y^3}$, find $f(-3, 2, 1)$.

The Domain and Range of a Function of Two Variables

A function of two variables can be represented graphically as a surface in space by letting

$$z = f(x, y). \qquad \text{Function of two variables}$$

When sketching the graph of a function of x and y, remember that even though the graph is three-dimensional, the domain of the function is two-dimensional—it consists of the points in the xy-plane for which the function is defined. As with functions of a single variable, unless specifically restricted, the domain of a function of two variables is assumed to be the set of all points (x, y) for which the defining equation has meaning. In other words, to each point (x, y) in the domain of f there corresponds a point (x, y, z) on the surface, and conversely, to each point (x, y, z) on the surface there corresponds a point (x, y) in the domain of f.

Example 2   Finding the Domain and Range of a Function

Find the domain and range of the function

$$f(x, y) = \sqrt{64 - x^2 - y^2}.$$

SOLUTION   Because no restrictions are given, the domain is assumed to be the set of all points for which the defining equation makes sense.

$$64 - x^2 - y^2 \geq 0 \qquad \text{Quantity inside radical must be nonnegative.}$$
$$-x^2 - y^2 \geq -64 \qquad \text{Subtract 64 from each side.}$$
$$x^2 + y^2 \leq 64 \qquad \text{Multiply each side by } -1 \text{ and reverse the inequality symbol.}$$

So, the domain of f is the set of all points that lie on or inside the circle given by

$$x^2 + y^2 \leq 8^2 \qquad \text{Domain of the function}$$

as shown in Figure 13.18. The range of f is the set

$$0 \leq z \leq 8. \qquad \text{Range of the function}$$

As shown in Figure 13.19, the graph of the function is a hemisphere.

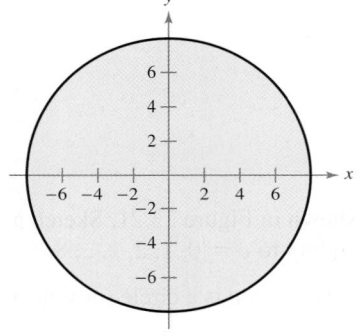

Domain of $f(x, y) = \sqrt{64 - x^2 - y^2}$

FIGURE 13.18

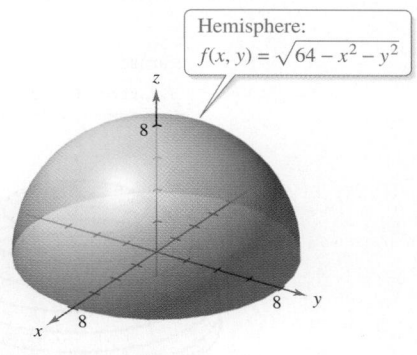

Hemisphere:
$f(x, y) = \sqrt{64 - x^2 - y^2}$

Domain: $x^2 + y^2 \leq 64$
Range: $0 \leq z \leq 8$

FIGURE 13.19

✓ Checkpoint 2

Consider the function

$$f(x, y) = \sqrt{9 - x^2 - y^2}.$$

a. Find the domain of f.

b. Find the range of f.

Contour Maps and Level Curves

A **contour map** of a surface is created by *projecting* traces, taken in evenly spaced planes that are parallel to the xy-plane, onto the xy-plane. Each projection is a **level curve** of the surface.

Contour maps are used to create weather, topographical, and population density maps. For instance, Figure 13.20(a) shows a graph of a "mountain and valley" surface given by $z = f(x, y)$. Each of the level curves in Figure 13.20(b) represents the intersection of the surface $z = f(x, y)$ with a plane $z = c$, where $c = 828, 830, \ldots, 854$.

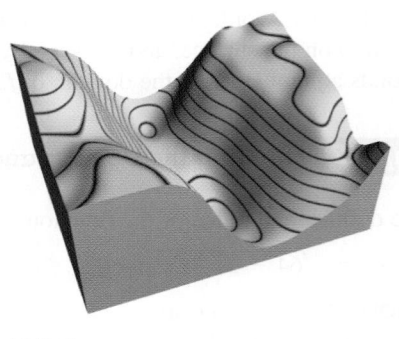

(a) Surface

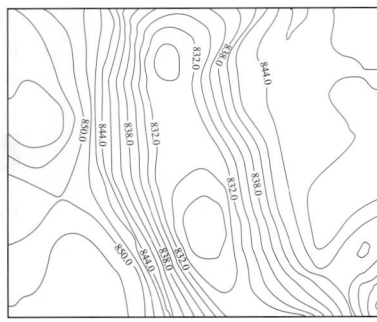

(b) Contour map

FIGURE 13.20

Example 3 Sketching a Contour Map

The hemisphere given by $f(x, y) = \sqrt{64 - x^2 - y^2}$ is shown in Figure 13.21. Sketch a contour map of this surface using level curves corresponding to $c = 0, 1, 2, \ldots, 8$.

SOLUTION For each value of c, the equation given by $f(x, y) = c$ is a circle (or point) in the xy-plane. For instance, when $c_1 = 0$, the level curve is

$$x^2 + y^2 = 8^2 \qquad \text{Circle of radius 8}$$

which is a circle of radius 8. Figure 13.22 shows the nine level curves for the hemisphere.

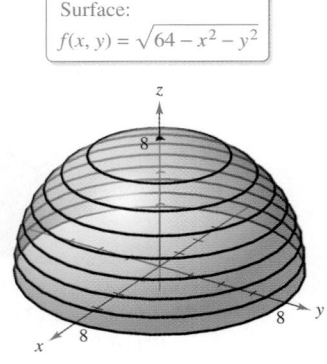

Surface:
$f(x, y) = \sqrt{64 - x^2 - y^2}$

Hemisphere

FIGURE 13.21

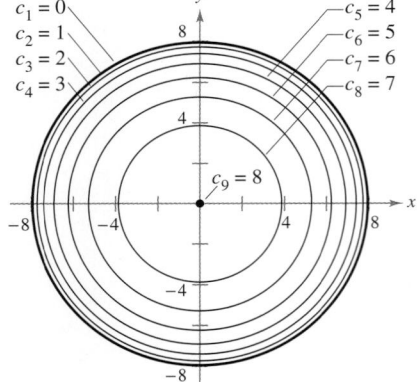

Contour map

FIGURE 13.22

✓ Checkpoint 3

Describe the level curves of $f(x, y) = \sqrt{9 - x^2 - y^2}$. Sketch the level curves for $c = 0, 1, 2,$ and 3.

Applications

The **Cobb-Douglas production function** is used in economics to represent the numbers of units produced by varying amounts of labor and capital. Let x represent the number of units of labor and let y represent the number of units of capital. Then, the number of units produced is modeled by

$$f(x, y) = Cx^a y^{1-a}$$

where C and a are constants, with $0 < a < 1$.

 Example 4 **Using a Production Function**

A manufacturer estimates that its production (measured in units of a product) can be modeled by $f(x, y) = 100x^{0.6}y^{0.4}$, where the labor x is measured in person-hours and the capital y is measured in thousands of dollars.

a. What is the production level when $x = 1000$ and $y = 500$?

b. What is the production level when $x = 2000$ and $y = 1000$?

c. How does doubling the amounts of labor and capital from part (a) to part (b) affect the production?

SOLUTION

a. When $x = 1000$ and $y = 500$, the production level is

$$f(1000, 500) = 100(1000)^{0.6}(500)^{0.4} \approx 75{,}786 \text{ units.}$$

b. When $x = 2000$ and $y = 1000$, the production level is

$$f(2000, 1000) = 100(2000)^{0.6}(1000)^{0.4} \approx 151{,}572 \text{ units.}$$

c. When the amounts of labor and capital are doubled, the production level also doubles. In Exercise 44, you are asked to show that this is characteristic of the Cobb-Douglas production function.

A contour map of this function is shown in Figure 13.23. Note that the level curves occur at increments of 10,000.

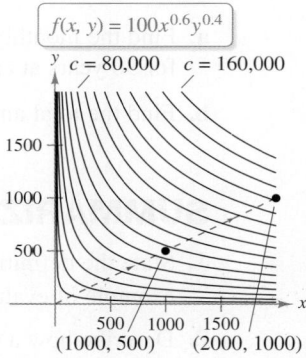

Level Curves (at Increments of 10,000)
FIGURE 13.23

✓**Checkpoint 4**

Use the Cobb-Douglas production function in Example 4 to find the production levels when $x = 1500$ and $y = 1000$ and when $x = 1000$ and $y = 1500$. Use your results to determine which variable has a greater influence on production.

Example 5   **Finding Monthly Payments**

The monthly payment M for an installment loan of P dollars taken out over t years at an annual interest rate of r is given by

$$M = f(P, r, t) = \frac{\dfrac{Pr}{12}}{1 - \left[\dfrac{1}{1 + (r/12)}\right]^{12t}}.$$

a. Find the monthly payment for a home mortgage of $100,000 taken out for 30 years at an annual interest rate of 7%.

b. Find the monthly payment for a car loan of $22,000 taken out for 5 years at an annual interest rate of 8%.

SOLUTION

a. When $P = \$100{,}000$, $r = 0.07$, and $t = 30$, the monthly payment is

$$M = f(100{,}000, 0.07, 30)$$

$$= \frac{\dfrac{(100{,}000)(0.07)}{12}}{1 - \left[\dfrac{1}{1 + (0.07/12)}\right]^{12(30)}}$$

$$\approx \$665.30.$$

b. When $P = \$22{,}000$, $r = 0.08$, and $t = 5$, the monthly payment is

$$M = f(22{,}000, 0.08, 5)$$

$$= \frac{\dfrac{(22{,}000)(0.08)}{12}}{1 - \left[\dfrac{1}{1 + (0.08/12)}\right]^{12(5)}}$$

$$\approx \$446.08.$$

✓**Checkpoint 5**

a. Find the monthly payment M for a home mortgage of $100,000 taken out for 30 years at an annual interest rate of 3%.

b. Find the total amount of money you will pay for the mortgage.

SUMMARIZE   (Section 13.3)

1. State the definition of a function of two variables *(page 928)*. For an example of evaluating a function of two variables, see Example 1.

2. Describe how a contour map of a surface is created *(page 930)*. For an example of sketching a contour map, see Example 3.

3. State the Cobb-Douglas production function *(page 931)*. For an example of using the Cobb-Douglas production function, see Example 4.

4. Describe a real-life example of how a function of several variables can be used to find the monthly payment for a loan *(page 932, Example 5)*.

In Exercises 1–4, evaluate the function when $x = -3$.

1. $f(x) = 5 - 2x$ **2.** $f(x) = -x^2 + 4x + 5$ **3.** $y = \sqrt{4x^2 - 3x + 4}$ **4.** $y = \sqrt[3]{34 - 4x + 2x^2}$

In Exercises 5–8, find the domain of the function.

5. $f(x) = 5x^2 + 3x - 2$ **6.** $g(x) = \dfrac{1}{2x} - \dfrac{2}{x + 3}$ **7.** $h(y) = \sqrt{y - 5}$ **8.** $f(y) = \sqrt{y^2 - 5}$

In Exercises 9 and 10, evaluate the expression.

9. $(476)^{0.65}$ **10.** $(251)^{0.35}$

Exercises 13.3

Evaluating Functions of Several Variables   In Exercises 1–14, find the function values. *See Example 1.*

1. $f(x, y) = \dfrac{x}{y}$

  (a) $f(3, 2)$ (b) $f(-1, 4)$ (c) $f(30, 5)$

  (d) $f(5, y)$ (e) $f(x, 2)$ (f) $f(5, t)$

2. $f(x, y) = 4 - x^2 - 4y^2$

  (a) $f(0, 0)$ (b) $f(0, 1)$ (c) $f(2, 3)$

  (d) $f(1, y)$ (e) $f(x, 0)$ (f) $f(t, 1)$

3. $f(x, y) = xe^y$

  (a) $f(5, 0)$ (b) $f(3, 2)$ (c) $f(2, -1)$

  (d) $f(5, y)$ (e) $f(x, 2)$ (f) $f(t, t)$

4. $g(x, y) = \ln|x + y|$

  (a) $g(2, 3)$ (b) $g(5, 6)$ (c) $g(e, 0)$

  (d) $g(0, 1)$ (e) $g(2, -3)$ (f) $g(e, e)$

5. $h(x, y, z) = \dfrac{xy}{z}$

  (a) $h(2, 3, 9)$ (b) $h(1, 0, 1)$

6. $f(x, y, z) = \sqrt{x + y + z}$

  (a) $f(0, 5, 4)$ (b) $f(6, 8, -3)$

7. $V(r, h) = \pi r^2 h$

  (a) $V(3, 10)$ (b) $V(5, 2)$

8. $F(r, N) = 500\left(1 + \dfrac{r}{12}\right)^N$

  (a) $F(0.09, 60)$ (b) $F(0.14, 240)$

9. $A(P, r, t) = P\left[\left(1 + \dfrac{r}{12}\right)^{12t} - 1\right]\left(1 + \dfrac{12}{r}\right)$

  (a) $A(100, 0.10, 10)$ (b) $A(275, 0.0925, 40)$

10. $A(P, r, t) = Pe^{rt}$

  (a) $A(500, 0.10, 5)$ (b) $A(1500, 0.12, 20)$

11. $f(x, y) = \displaystyle\int_x^y (2t - 3)\, dt$

  (a) $f(1, 2)$ (b) $f(1, 4)$

12. $g(x, y) = \displaystyle\int_x^y \dfrac{1}{t}\, dt$

  (a) $g(4, 1)$ (b) $g(6, 3)$

13. $f(x, y) = x^2 - 2y$

  (a) $f(x + \Delta x, y)$ (b) $\dfrac{f(x, y + \Delta y) - f(x, y)}{\Delta y}$

14. $f(x, y) = 3xy + y^2$

  (a) $f(x + \Delta x, y)$ (b) $\dfrac{f(x, y + \Delta y) - f(x, y)}{\Delta y}$

Finding the Domain and Range of a Function   In Exercises 15–30, find the domain and range of the function. *See Example 2.*

15. $f(x, y) = \sqrt{16 - x^2 - y^2}$

16. $z = \sqrt{4 - x^2 - y^2}$

17. $f(x, y) = x^2 + y^2$

18. $f(x, y) = x^2 + y^2 - 1$

19. $f(x, y) = e^{x/y}$

20. $f(x, y) = ye^{1/x}$

21. $g(x, y) = \ln(4 - x - y)$

22. $f(x, y) = \ln(x + y)$

23. $z = \sqrt{9 - 3x^2 - y^2}$

24. $z = \sqrt{4 - x^2 - 4y^2}$

25. $z = \dfrac{y}{x}$

26. $f(x, y) = \dfrac{x}{y}$

27. $f(x, y) = \dfrac{1}{xy}$

28. $g(x, y) = \dfrac{1}{x - y}$

29. $h(x, y) = x\sqrt{y}$

30. $f(x, y) = \sqrt{xy}$

Matching   In Exercises 31–34, match the graph of the surface with one of the contour maps. [The contour maps are labeled (a)–(d).]

(a)

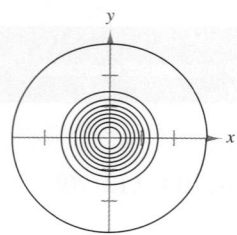

(b)

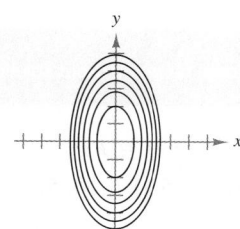

(c)

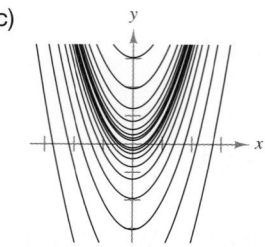

(d)

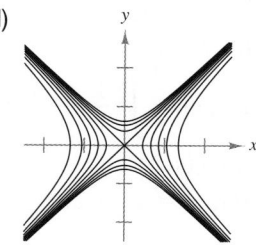

31. $f(x, y) = x^2 + \dfrac{y^2}{4}$ **32.** $f(x, y) = e^{1 - x^2 + y^2}$

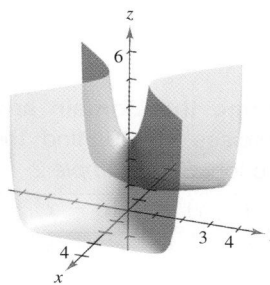

33. $f(x, y) = e^{1 - x^2 - y^2}$ **34.** $f(x, y) = \ln|y - x^2|$

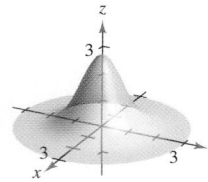

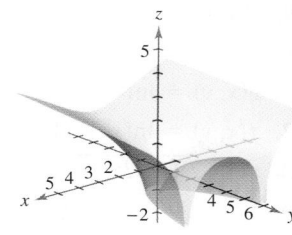

Sketching a Contour Map   In Exercises 35–42, describe the level curves of the function. Sketch a contour map of the surface using level curves for the given c-values. *See Example 3.*

Function	c-Values
35. $z = x + y$	$c = -1, 0, 2, 4$
36. $z = 6 - 2x - 3y$	$c = 0, 2, 4, 6, 8, 10$
37. $z = \sqrt{25 - x^2 - y^2}$	$c = 0, 1, 2, 3, 4, 5$
38. $f(x, y) = x^2 + y^2$	$c = 0, 2, 4, 6, 8$
39. $f(x, y) = xy$	$c = \pm 1, \pm 2, \ldots, \pm 6$
40. $z = e^{xy}$	$c = 1, 2, 3, 4, \tfrac{1}{2}, \tfrac{1}{3}, \tfrac{1}{4}$
41. $f(x, y) = \dfrac{x}{x^2 + y^2}$	$c = \pm\tfrac{1}{2}, \pm 1, \pm\tfrac{3}{2}, \pm 2$
42. $f(x, y) = \ln(x - y)$	$c = 0, \pm\tfrac{1}{2}, \pm 1, \pm\tfrac{3}{2}, \pm 2$

43. Cobb-Douglas Production Function   A manufacturer estimates the Cobb-Douglas production function to be given by

$$f(x, y) = 100x^{0.75}y^{0.25}.$$

Estimate the production level when $x = 1500$ and $y = 1000$.

44. Cobb-Douglas Production Function   Use the Cobb-Douglas production function (Example 4) to show that when both the number of units of labor and the number of units of capital are doubled, the production level is also doubled.

45. Profit   A sporting goods manufacturer produces regulation soccer balls at two plants. The costs of producing x_1 units at location 1 and x_2 units at location 2 are given by

$$C_1(x_1) = 0.02x_1^2 + 4x_1 + 500$$

and

$$C_2(x_2) = 0.05x_2^2 + 4x_2 + 275$$

respectively. If the product sells for \$50 per unit, then the profit function for the product is given by

$$P(x_1, x_2) = 50(x_1 + x_2) - C_1(x_1) - C_2(x_2).$$

Find (a) $P(250, 150)$, (b) $P(300, 200)$, and (c) $P(600, 400)$.

46. Queuing Model   The average amount of time that a customer waits in line for service is given by

$$W(x, y) = \dfrac{1}{x - y}, \quad y < x$$

where y is the average arrival rate and x is the average service rate (x and y are measured in the number of customers per hour). Evaluate W at each point.

(a) $(15, 10)$ (b) $(12, 9)$ (c) $(12, 6)$ (d) $(4, 2)$

47. Investment  In 2011, an investment of $2000 was made in a bond earning 10% compounded annually. The investor pays tax at rate R, and the annual rate of inflation is I. In the year 2021, the value V of the bond in constant 2011 dollars is given by

$$V(I, R) = 2000\left[\frac{1 + 0.10(1 - R)}{1 + I}\right]^{10}.$$

Use this function of two variables and a spreadsheet to complete the table.

Tax Rate	Inflation Rate		
	0	0.03	0.05
0			
0.28			
0.35			

48. Investment  A principal of $5000 is deposited in a savings account that earns an interest rate of r (in decimal form), compounded continuously. The amount $A(r, t)$ after t years is

$$A(r, t) = 5000e^{rt}.$$

Use this function of two variables and a spreadsheet to complete the table.

Rate	Number of Years			
	5	10	15	20
0.02				
0.04				
0.06				
0.08				

49. Meteorology Meteorologists measure the atmospheric pressure in millibars. From these observations they create weather maps on which the curves of equal atmospheric pressure (isobars) are drawn (see figure). On the map, the closer the isobars, the higher the wind speed. Match points A, B, and C with (a) highest pressure, (b) lowest pressure, and (c) highest wind velocity.

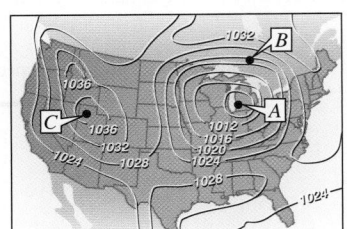

50.  **HOW DO YOU SEE IT?** The contour map of the Southern Hemisphere shown in the figure was computer generated using data collected by satellite instrumentation. Color is used to show the "ozone hole" in Earth's atmosphere. The purple and blue areas represent the lowest levels of ozone, and the green areas represent the highest levels. *(Source: National Aeronautics and Space Administration)*

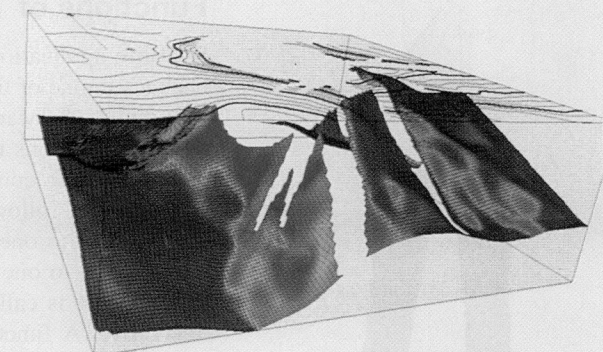

(a) Describe the areas that have the lowest levels of ozone.

(b) Do the level curves correspond to equally spaced ozone levels? Explain.

51. Earnings per Share The earnings per share z (in dollars) for Apple from 2005 through 2010 can be modeled by $z = 0.379x - 0.135y - 3.45$, where x is the sales (in billions of dollars) and y is the shareholder's equity (in billions of dollars). *(Source: Apple Inc.)*

(a) Find the earnings per share when $x = 20$ and $y = 10$.

(b) Which of the two variables in this model has the greater influence on the earnings per share? Explain.

52. Shareholder's Equity The shareholder's equity z (in millions of dollars) for Sketchers from 2001 through 2009 can be modeled by $z = 0.175x - 0.772y - 275$, where x is the sales (in millions of dollars) and y is the total assets (in millions of dollars). *(Source: Sketchers U.S.A. Inc.)*

(a) Find the shareholder's equity when $x = 1000$ and $y = 500$.

(b) Which of the two variables in this model has the greater influence on shareholder's equity? Explain.

53. Monthly Payments You are taking out a home mortgage for $120,000, and you are given the options below. Find the monthly payment and the total amount of money you will pay for each mortgage. Which option would you choose? Explain your reasoning.

(a) A fixed annual rate of 8%, over a term of 20 years.

(b) A fixed annual rate of 7%, over a term of 30 years.

(c) A fixed annual rate of 7%, over a term of 15 years.

13.4 Partial Derivatives

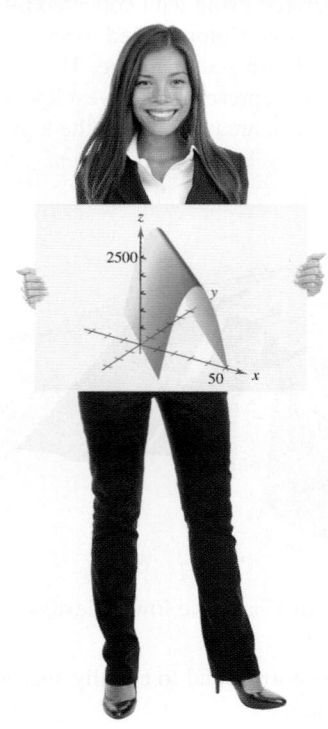

■ Find the first partial derivatives of functions of two variables.

■ Find the slopes of surfaces in the *x*- and *y*-directions and use partial derivatives to answer questions about real-life situations.

■ Find the partial derivatives of functions of several variables.

■ Find higher-order partial derivatives.

Functions of Two Variables

Real-life applications of functions of several variables are often concerned with how changes in one of the variables will affect the values of the functions. For instance, an economist who wants to determine the effect of a tax increase on the economy might make calculations using different tax rates while holding all other variables, such as unemployment, constant.

You can follow a similar procedure to find the rate of change of a function f with respect to one of its independent variables. That is, you find the derivative of f with respect to one independent variable while holding the other variable(s) constant. This process is called **partial differentiation,** and each derivative is called a **partial derivative.** A function of several variables has as many partial derivatives as it has independent variables.

In Exercise 60 on page 945, you will use partial derivatives to find the marginal revenues of a pharmaceutical corporation at two locations that produce the same medicine.

Partial Derivatives of a Function of Two Variables

If $z = f(x, y)$, then the **first partial derivatives of f with respect to x and y** are the functions $\partial z/\partial x$ and $\partial z/\partial y$, defined as shown.

$$\frac{\partial z}{\partial x} = \lim_{\Delta x \to 0} \frac{f(x + \Delta x, y) - f(x, y)}{\Delta x} \qquad y \text{ is held constant.}$$

$$\frac{\partial z}{\partial y} = \lim_{\Delta y \to 0} \frac{f(x, y + \Delta y) - f(x, y)}{\Delta y} \qquad x \text{ is held constant.}$$

This definition indicates that if $z = f(x, y)$, then to find $\partial z/\partial x$, you *consider y to be constant* and differentiate with respect to x. Similarly, to find $\partial z/\partial y$, you *consider x to be constant* and differentiate with respect to y.

STUDY TIP

The notation $\partial z/\partial x$ is read as "the partial derivative of z with respect to x," and $\partial z/\partial y$ is read as "the partial derivative of z with respect to y."

Example 1   Finding Partial Derivatives

Find $\partial z/\partial x$ and $\partial z/\partial y$ for the function $z = 3x - x^2y^2 + 2x^3y$.

SOLUTION

$$\frac{\partial z}{\partial x} = 3 - 2xy^2 + 6x^2y \qquad \text{Hold } y \text{ constant and differentiate with respect to } x.$$

$$\frac{\partial z}{\partial y} = -2x^2y + 2x^3 \qquad \text{Hold } x \text{ constant and differentiate with respect to } y.$$

✓ **Checkpoint 1**

Find $\dfrac{\partial z}{\partial x}$ and $\dfrac{\partial z}{\partial y}$ for $z = 2x^2 - 4x^2y^3 + y^4$.

Notation for First Partial Derivatives

The first partial derivatives of $z = f(x, y)$ are denoted by

$$\frac{\partial z}{\partial x} = f_x(x, y) = z_x = \frac{\partial}{\partial x}[f(x, y)]$$

and

$$\frac{\partial z}{\partial y} = f_y(x, y) = z_y = \frac{\partial}{\partial y}[f(x, y)].$$

The values of the first partial derivatives at the point (a, b) are denoted by

$$\frac{\partial z}{\partial x}\bigg|_{(a, b)} = f_x(a, b)$$

and

$$\frac{\partial z}{\partial y}\bigg|_{(a, b)} = f_y(a, b).$$

TECH TUTOR

Symbolic differentiation utilities can be used to find partial derivatives of a function of two variables. Try using a symbolic differentiation utility to find the first partial derivatives of the function in Example 2.

Example 2   Finding and Evaluating Partial Derivatives

Find the first partial derivatives of

$$f(x, y) = xe^{x^2y}$$

and evaluate each at the point $(1, \ln 2)$.

SOLUTION   To find the first partial derivative with respect to x, hold y constant and differentiate using the Product Rule.

$$f_x(x, y) = x\frac{\partial}{\partial x}[e^{x^2y}] + e^{x^2y}\frac{\partial}{\partial x}[x] \qquad \text{Apply Product Rule.}$$

$$= xe^{x^2y}(2xy) + e^{x^2y} \qquad \text{y is held constant.}$$

$$= e^{x^2y}(2x^2y + 1) \qquad \text{Simplify.}$$

At the point $(1, \ln 2)$, the value of this derivative is

$$f_x(1, \ln 2) = e^{(1)^2(\ln 2)}[2(1)^2(\ln 2) + 1] \qquad \text{Substitute for x and y.}$$

$$= 2(2 \ln 2 + 1) \qquad \text{Simplify.}$$

$$\approx 4.773. \qquad \text{Use a calculator.}$$

To find the first partial derivative with respect to y, hold x constant and differentiate to obtain

$$f_y(x, y) = xe^{x^2y}(x^2) \qquad \text{Apply Constant Multiple Rule.}$$

$$= x^3e^{x^2y}. \qquad \text{Simplify.}$$

At the point $(1, \ln 2)$, the value of this derivative is

$$f_y(1, \ln 2) = (1)^3e^{(1)^2(\ln 2)} \qquad \text{Substitute for x and y.}$$

$$= 2. \qquad \text{Simplify.}$$

✓**Checkpoint 2**

Find the first partial derivatives of

$$f(x, y) = x^2y^3$$

and evaluate each at the point $(1, 2)$.

Graphical Interpretation of Partial Derivatives

Earlier in the text, you studied graphical interpretations of the derivative of a function of a single variable. There, you found that $f'(x_0)$ represents the slope of the tangent line to the graph of $y = f(x)$ at the point (x_0, y_0). The partial derivatives of a function of two variables also have useful graphical interpretations. Consider the function

$$z = f(x, y). \qquad \text{Function of two variables}$$

As shown in Figure 13.24(a), the graph of this function is a surface in space. If $y = y_0$, then

$$z = f(x, y_0) \qquad \text{Function of one variable}$$

is a function of one variable. The graph of this function is the curve that is the intersection of the plane $y = y_0$ and the surface $z = f(x, y)$. On this curve, the partial derivative

$$f_x(x, y_0) \qquad \text{Slope in } x\text{-direction}$$

represents the slope in the plane $y = y_0$, as shown in Figure 13.24(a). Similarly, if $x = x_0$, then

$$z = f(x_0, y) \qquad \text{Function of one variable}$$

is a function of one variable. Its graph is the intersection of the plane $x = x_0$ and the surface $z = f(x, y)$. On this curve, the partial derivative

$$f_y(x_0, y) \qquad \text{Slope in } y\text{-direction}$$

represents the slope in the plane $x = x_0$, as shown in Figure 13.24(b).

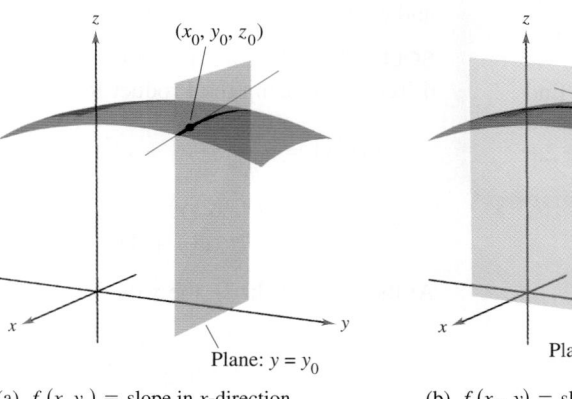

(a) $f_x(x, y_0)$ = slope in x-direction

(b) $f_y(x_0, y)$ = slope in y-direction

FIGURE 13.24

Informally, $f_x(x_0, y_0)$ and $f_y(x_0, y_0)$ at the point (x_0, y_0, z_0) denote the **slopes of the surface in the x- and y-directions,** respectively.

Guidelines for Finding the Slopes of a Surface at a Point

Let (x_0, y_0, z_0) be a point on the surface of

$$z = f(x, y).$$

1. Find the partial derivatives of f with respect to x and y.

2. The slope of the x-direction at (x_0, y_0, z_0) is $f_x(x_0, y_0)$.

3. The slope of the y-direction at (x_0, y_0, z_0) is $f_y(x_0, y_0)$.

Example 3   **Finding Slopes in the x- and y-Directions**

Find the slopes of the surface given by

$$f(x, y) = -\frac{x^2}{2} - y^2 + \frac{25}{8}$$

at the point $\left(\frac{1}{2}, 1, 2\right)$ in

a. the x-direction.

b. the y-direction.

SOLUTION

a. To find the slope in the x-direction, hold y constant and differentiate with respect to x to obtain

$$f_x(x, y) = -x. \qquad \text{Partial derivative with respect to } x$$

At the point $\left(\frac{1}{2}, 1, 2\right)$, the slope in the x-direction is

$$f_x\left(\frac{1}{2}, 1\right) = -\frac{1}{2} \qquad \text{Slope in } x\text{-direction}$$

as shown in Figure 13.25(a).

b. To find the slope in the y-direction, hold x constant and differentiate with respect to y to obtain

$$f_y(x, y) = -2y. \qquad \text{Partial derivative with respect to } y$$

At the point $\left(\frac{1}{2}, 1, 2\right)$, the slope in the y-direction is

$$f_y\left(\frac{1}{2}, 1\right) = -2 \qquad \text{Slope in } y\text{-direction}$$

as shown in Figure 13.25(b).

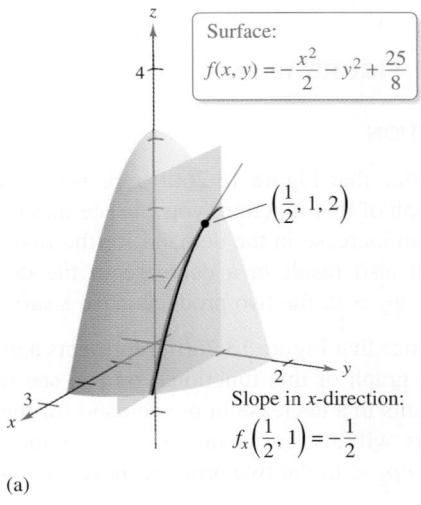

Slope in x-direction:
$$f_x\left(\frac{1}{2}, 1\right) = -\frac{1}{2}$$

(a)

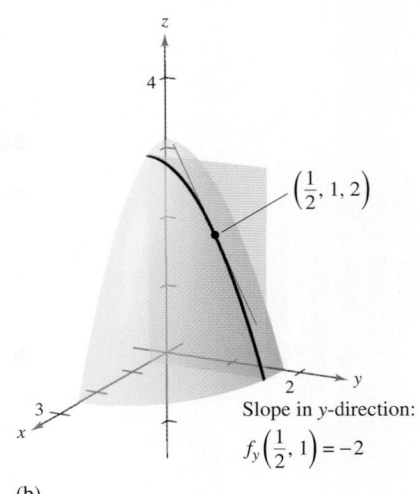

Slope in y-direction:
$$f_y\left(\frac{1}{2}, 1\right) = -2$$

(b)

FIGURE 13.25

✓**Checkpoint 3**

Find the slopes of the surface given by

$$f(x, y) = 4x^2 + 9y^2 + 36$$

at the point $(1, -1, 49)$ in

a. the x-direction.

b. the y-direction.

In 2010, Subway was chosen as the number one franchise by *Entrepreneur* magazine. In early 2011, Subway had more than 34,000 franchises worldwide. What type of product would be complementary to a Subway sandwich? What type of product would be a substitute?

Consumer products in the same market or in related markets can be classified as **complementary** or **substitute products.** When two products have a complementary relationship, an increase in the sale of one product will be accompanied by an increase in the sale of the other product. For instance, Blu-ray™ players and Blu-ray™ discs have a complementary relationship.

When two products have a substitute relationship, an increase in the sale of one product will be accompanied by a decrease in the sale of the other product. For instance, Blu-ray Disc™ players and DVD players both compete in the same home entertainment market, and you would expect a drop in the price of one to be a deterrent to the sale of the other.

Example 4   **Examining Demand Functions**

The demand functions for two products are represented by

$$x_1 = f(p_1, p_2) \quad \text{and} \quad x_2 = g(p_1, p_2)$$

where p_1 and p_2 are the prices per unit for the two products, and x_1 and x_2 are the numbers of units sold. The graphs of two different demand functions for x_1 are shown in Figure 13.26. Use them to classify the products as complementary or substitute products.

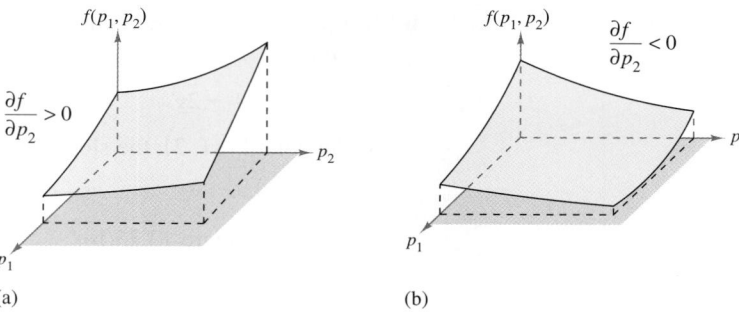

(a)   (b)

FIGURE 13.26

SOLUTION

a. Notice that Figure 13.26(a) represents the demand for the *first product*. From the graph of this function, you can see that for a fixed price p_1, an increase in p_2 results in an increase in the demand for the first product. Remember that an increase in p_2 will also result in a decrease in the demand for the second product. So, when $\partial f / \partial p_2 > 0$, the two products have a *substitute* relationship.

b. Notice that Figure 13.26(b) represents a different demand for the *first product*. From the graph of this function, you can see that for a fixed price p_1, an increase in p_2 results in a decrease in the demand for the first product. Remember that an increase in p_2 will also result in a decrease in the demand for the second product. So, when $\partial f / \partial p_2 < 0$, the two products have a *complementary* relationship.

✓Checkpoint 4

Determine if the demand functions below describe a complementary or a substitute product relationship.

$$x_1 = 100 - 2p_1 + 1.5p_2$$
$$x_2 = 145 + \tfrac{1}{2}p_1 - \tfrac{3}{4}p_2$$

Functions of Three Variables

The concept of a partial derivative can be extended naturally to functions of three or more variables. For instance, the function

$$w = f(x, y, z)$$ Function of three variables

has three partial derivatives, each of which is formed by considering two of the variables to be constant. That is, to define the partial derivative of w with respect to x, consider y *and* z to be constant and write

$$\frac{\partial w}{\partial x} = f_x(x, y, z) = \lim_{\Delta x \to 0} \frac{f(x + \Delta x, y, z) - f(x, y, z)}{\Delta x}.$$

To define the partial derivative of w with respect to y, consider x *and* z to be constant and write

$$\frac{\partial w}{\partial y} = f_y(x, y, z) = \lim_{\Delta y \to 0} \frac{f(x, y + \Delta y, z) - f(x, y, z)}{\Delta y}.$$

To define the partial derivative of w with respect to z, consider x *and* y to be constant and write

$$\frac{\partial w}{\partial z} = f_z(x, y, z) = \lim_{\Delta z \to 0} \frac{f(x, y, z + \Delta z) - f(x, y, z)}{\Delta z}.$$

Example 5 Finding Partial Derivatives of a Function

Find the three partial derivatives of the function

$$w = xe^{xy + 2z}.$$

SOLUTION   Holding y and z constant, you obtain

$$\frac{\partial w}{\partial x} = x\frac{\partial}{\partial x}[e^{xy + 2z}] + e^{xy + 2z}\frac{\partial}{\partial x}[x]$$ Apply Product Rule.

$$= x(e^{xy + 2z})(y) + e^{xy + 2z}(1)$$ Hold y and z constant.

$$= (xy + 1)e^{xy + 2z}.$$ Simplify.

Holding x and z constant, you obtain

$$\frac{\partial w}{\partial y} = x\frac{\partial}{\partial y}[e^{xy + 2z}]$$ Apply Constant Multiple Rule.

$$= x(e^{xy + 2z})(x)$$ Hold x and z constant.

$$= x^2 e^{xy + 2z}.$$ Simplify.

Holding x and y constant, you obtain

$$\frac{\partial w}{\partial z} = x\frac{\partial}{\partial z}[e^{xy + 2z}]$$ Apply Constant Multiple Rule.

$$= x(e^{xy + 2z})(2)$$ Hold x and y constant.

$$= 2xe^{xy + 2z}.$$ Simplify.

✓Checkpoint 5

Find the three partial derivatives of the function

$$w = x^2 y \ln(xz).$$ ■

In Example 5, the Product Rule is used only when finding the partial derivative with respect to x. For $\partial w/\partial y$ and $\partial w/\partial z$, x is considered to be constant, so the Constant Multiple Rule is used.

TECH TUTOR

A symbolic differentiation utility can be used to find the partial derivatives of a function of three or more variables. Try using a symbolic differentiation utility to find the partial derivative $f_y(x, y, z)$ for the function in Example 5.

Higher-Order Partial Derivatives

As with ordinary derivatives, it is possible to take second-, third-, and higher-order partial derivatives of a function of several variables, provided such derivatives exist. Higher-order derivatives are denoted by the order in which the differentiation occurs. For instance, there are four different ways to find a second partial derivative of $z = f(x, y)$.

1. $\dfrac{\partial}{\partial x}\left(\dfrac{\partial f}{\partial x}\right) = \dfrac{\partial^2 f}{\partial x^2} = f_{xx}$ Differentiate twice with respect to x.

2. $\dfrac{\partial}{\partial y}\left(\dfrac{\partial f}{\partial y}\right) = \dfrac{\partial^2 f}{\partial y^2} = f_{yy}$ Differentiate twice with respect to y.

3. $\dfrac{\partial}{\partial y}\left(\dfrac{\partial f}{\partial x}\right) = \dfrac{\partial^2 f}{\partial y \partial x} = f_{xy}$ Differentiate first with respect to x and then with respect to y.

4. $\dfrac{\partial}{\partial x}\left(\dfrac{\partial f}{\partial y}\right) = \dfrac{\partial^2 f}{\partial x \partial y} = f_{yx}$ Differentiate first with respect to y and then with respect to x.

The third and fourth cases are **mixed partial derivatives.** Notice that with the two types of notation for mixed partials, different conventions are used for indicating the order of differentiation. For instance, the partial derivative

$$\frac{\partial}{\partial y}\left(\frac{\partial f}{\partial x}\right) = \frac{\partial^2 f}{\partial y \partial x} \qquad \text{Right-to-left order}$$

indicates differentiation with respect to x first, but the partial derivative

$$(f_y)_x = f_{yx} \qquad \text{Left-to-right order}$$

indicates differentiation with respect to y first. To remember this, note that in each case you differentiate first with respect to the variable "nearest" f.

Example 6   Finding Second Partial Derivatives

Find the second partial derivatives of

$$f(x, y) = 3xy^2 - 2y + 5x^2y^2$$

and determine the value of $f_{xy}(-1, 2)$.

SOLUTION   Begin by finding the first partial derivatives.

$$f_x(x, y) = 3y^2 + 10xy^2 \qquad\qquad f_y(x, y) = 6xy - 2 + 10x^2y$$

Then, differentiating with respect to x and y produces

$$f_{xx}(x, y) = 10y^2, \qquad\qquad f_{yy}(x, y) = 6x + 10x^2,$$

$$f_{xy}(x, y) = 6y + 20xy, \qquad\qquad f_{yx}(x, y) = 6y + 20xy.$$

Finally, the value of $f_{xy}(x, y)$ at the point $(-1, 2)$ is

$$f_{xy}(-1, 2) = 6(2) + 20(-1)(2) = 12 - 40 = -28.$$

✓ Checkpoint 6

Find the second partial derivatives of

$$f(x, y) = 4x^2y^2 + 2x + 4y^2.$$

Notice in Example 6 that the two mixed partials are equal. It can be shown that when a function has continuous second partial derivatives, then the order in which the partial derivatives are taken is irrelevant.

A function of two variables has two first partial derivatives and four second partial derivatives. For a function of three variables, there are three first partials

$$f_x, \quad f_y, \quad \text{and} \quad f_z$$

and nine second partials

$$f_{xx}, \quad f_{xy}, \quad f_{xz}, \quad f_{yx}, \quad f_{yy}, \quad f_{yz}, \quad f_{zx}, \quad f_{zy}, \quad \text{and} \quad f_{zz}$$

of which six are mixed partials. To find partial derivatives of order three and higher, follow the same pattern used to find second partial derivatives. For instance, if $z = f(x, y)$, then

$$z_{xxx} = \frac{\partial}{\partial x}\left(\frac{\partial^2 f}{\partial x^2}\right) = \frac{\partial^3 f}{\partial x^3} \quad \text{and} \quad z_{xxy} = \frac{\partial}{\partial y}\left(\frac{\partial^2 f}{\partial x^2}\right) = \frac{\partial^3 f}{\partial y \partial x^2}.$$

Example 7   Finding Second Partial Derivatives

Find the second partial derivatives of

$$f(x, y, z) = ye^x + x \ln z.$$

SOLUTION   Begin by finding the first partial derivatives.

$$f_x(x, y, z) = ye^x + \ln z, \quad f_y(x, y, z) = e^x, \quad f_z(x, y, z) = \frac{x}{z}$$

Then, differentiate with respect to x, y, and z to find the nine second partial derivatives.

$$f_{xx}(x, y, z) = ye^x, \quad f_{xy}(x, y, z) = e^x, \quad f_{xz}(x, y, z) = \frac{1}{z}$$

$$f_{yx}(x, y, z) = e^x, \quad f_{yy}(x, y, z) = 0, \quad f_{yz}(x, y, z) = 0$$

$$f_{zx}(x, y, z) = \frac{1}{z}, \quad f_{zy}(x, y, z) = 0, \quad f_{zz}(x, y, z) = -\frac{x}{z^2}$$

✓ **Checkpoint 7**

Find the second partial derivatives of $f(x, y, z) = xe^y + 2xz + y^2$.

SUMMARIZE   (Section 13.4)

1. State the definition of partial derivatives of a function of two variables *(page 936)*. For an example of finding the partial derivatives of a function of two variables, see Example 1.

2. Describe the notation used for first partial derivatives *(page 937)*. For examples of this notation, see Examples 1 and 2.

3. State the guidelines for finding the slopes of a surface at a point *(page 938)*. For an example of finding slopes, see Example 3.

4. Describe a real-life example of how partial derivatives can be used to examine the demand functions of two products *(page 940, Example 4)*.

5. Explain how to find the partial derivatives of a function of three variables *(page 941)*. For an example of finding the partial derivatives of a function of three variables, see Example 5.

6. List the different ways to find the second partial derivatives of a function of two variables *(page 942)*. For an example of finding the second partial derivatives of a function of two variables, see Example 6.

In Exercises 1–8, find the derivative of the function.

1. $f(x) = \sqrt{x^2 + 3}$

2. $g(x) = (3 - x^2)^3$

3. $g(t) = te^{2t+1}$

4. $f(x) = e^{2x}\sqrt{1 - e^{2x}}$

5. $f(x) = \ln(3 - 2x)$

6. $u(t) = \ln\sqrt{t^3 - 6t}$

7. $g(x) = \dfrac{5x^2}{(4x - 1)^2}$

8. $f(x) = \dfrac{(x + 2)^3}{(x^2 - 9)^2}$

In Exercises 9 and 10, evaluate the derivative at the point (2, 4).

9. $f(x) = x^2 e^{x-2}$

10. $g(x) = x\sqrt{x^2 - x + 2}$

Exercises 13.4

Finding Partial Derivatives   In Exercises 1–14, find the first partial derivatives. *See Example 1.*

1. $z = 3x + 5y - 1$

2. $z = x^2 - 2y$

3. $f(x, y) = 3x - 6y^2$

4. $f(x, y) = x + 4y^{3/2}$

5. $f(x, y) = \dfrac{x}{y}$

6. $f(x, y) = \dfrac{xy}{x^2 + y^2}$

7. $f(x, y) = \sqrt{x^2 + y^2}$

8. $z = x\sqrt{y}$

9. $z = x^2 e^{2y}$

10. $z = xe^{x+y}$

11. $h(x, y) = e^{-(x^2 + y^2)}$

12. $g(x, y) = e^{x/y}$

13. $z = \ln\dfrac{x + y}{x - y}$

14. $g(x, y) = \ln(x^2 + y^2)$

Finding and Evaluating Partial Derivatives   In Exercises 15–22, find the first partial derivatives and evaluate each at the given point. *See Example 2.*

Function	Point
15. $f(x, y) = 3x^2 + xy - y^2$	(2, 1)
16. $f(x, y) = x^2 - 3xy + y^2$	(1, −1)
17. $f(x, y) = e^{3xy}$	(0, 4)
18. $f(x, y) = e^x y^2$	(0, 2)
19. $f(x, y) = \dfrac{xy}{x - y}$	(2, −2)
20. $f(x, y) = \dfrac{4xy}{\sqrt{x^2 + y^2}}$	(1, 0)
21. $f(x, y) = \ln(3x + 5y)$	(1, 0)
22. $f(x, y) = \ln\sqrt{xy}$	(−1, −1)

Finding Slopes in the x- and y-Directions   In Exercises 23–26, find the slopes of the surface at the given point in (a) the x-direction and (b) the y-direction. *See Example 3.*

23. $z = xy$

(1, 2, 2)

24. $z = \sqrt{25 - x^2 - y^2}$

(3, 0, 4)

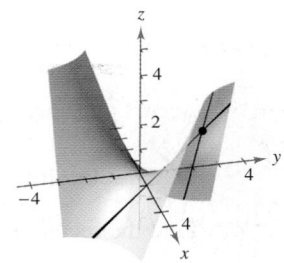

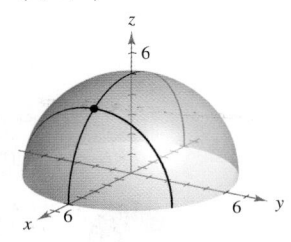

25. $z = 4 - x^2 - y^2$

(1, 1, 2)

26. $z = x^2 - y^2$

(−2, 1, 3)

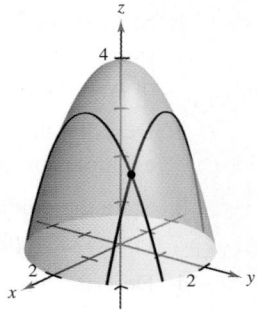

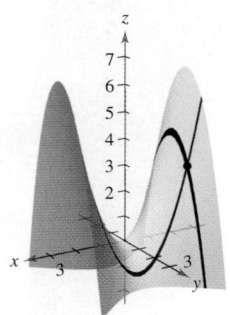

Finding Partial Derivatives   In Exercises 27–30, find the first partial derivatives. *See Example 5.*

27. $w = xy^2 z^4$

28. $w = x^3 yz^2$

29. $w = \dfrac{2z}{x + y}$

30. $w = \dfrac{xy}{x + y + z}$

Finding and Evaluating Partial Derivatives In Exercises 31–38, find the first partial derivatives and evaluate each at the given point.

Function	Point
31. $w = 2xz^2 + 3xyz - 6y^2z$	$(1, -1, 2)$
32. $w = 3x^2y - 5xyz + 10yz^2$	$(3, 4, -2)$
33. $w = \sqrt{x^2 + y^2 + z^2}$	$(2, -1, 2)$
34. $w = \sqrt{3x^2 + y^2 - 2z^2}$	$(1, -2, 1)$
35. $w = y^3z^2e^{2x^2}$	$\left(\frac{1}{2}, -1, 2\right)$
36. $w = xye^{z^2}$	$(2, 1, 0)$
37. $w = \ln(5x + 2y^3 - 3z)$	$(4, 1, -1)$
38. $w = \ln\sqrt{x^2 + y^2 + z^2}$	$(3, 0, 4)$

Using First Partial Derivatives In Exercises 39–42, find values of x and y such that $f_x(x, y) = 0$ and $f_y(x, y) = 0$ simultaneously.

39. $f(x, y) = x^2 + 4xy + y^2 - 4x + 16y + 3$

40. $f(x, y) = 3x^3 - 12xy + y^3$

41. $f(x, y) = \dfrac{1}{x} + \dfrac{1}{y} + xy$

42. $f(x, y) = \ln(x^2 + y^2 + 1)$

Finding Second Partial Derivatives In Exercises 43–50, find the four second partial derivatives. *See Example 6.*

43. $z = x^3 - 4y^2$ **44.** $z = 2x^2 + y^5$

45. $z = x^2 - 2xy + 3y^2$ **46.** $z = y^3 - 4xy^2 - 1$

47. $z = (3x^4 - 2y^3)^3$ **48.** $z = \sqrt{9 - x^2 - y^2}$

49. $z = \dfrac{x^2 - y^2}{2xy}$ **50.** $z = \dfrac{x}{x + y}$

Finding and Evaluating Second Partial Derivatives In Exercises 51–54, find the four second partial derivatives and evaluate each at the given point.

Function	Point
51. $f(x, y) = x^4 - 3x^2y^2 + y^2$	$(1, 0)$
52. $f(x, y) = x^3 + 2xy^3 - 3y$	$(3, 2)$
53. $f(x, y) = y^3e^{x^2}$	$(1, -1)$
54. $f(x, y) = x^2e^y$	$(-1, 0)$

Finding Second Partial Derivatives In Exercises 55–58, find the nine second partial derivatives. *See Example 7.*

55. $w = x^2 - 3xy + 4yz + z^3$

56. $w = x^2y^3 + 2xyz - 3yz$

57. $w = \dfrac{4xz}{x + y}$ **58.** $w = \dfrac{xy}{x + y + z}$

59. Marginal Cost A company manufactures two models of bicycles: a mountain bike and a racing bike. The cost function for producing x mountain bikes and y racing bikes is given by

$$C = 10\sqrt{xy} + 149x + 189y + 675.$$

(a) Find the marginal costs ($\partial C/\partial x$ and $\partial C/\partial y$) when $x = 120$ and $y = 160$.

(b) When additional production is required, which model of bicycle results in the cost increasing at a higher rate? How can this be determined from the cost model?

60. Marginal Revenue A pharmaceutical corporation has two locations that produce the same over-the-counter medicine. If x_1 and x_2 are the numbers of units produced at location 1 and location 2, respectively, then the total revenue for the product is given by

$$R = 200x_1 + 200x_2 - 4x_1^2 - 8x_1x_2 - 4x_2^2.$$

When $x_1 = 4$ and $x_2 = 12$, find

(a) the marginal revenue for location 1, $\partial R/\partial x_1$.

(b) the marginal revenue for location 2, $\partial R/\partial x_2$.

61. Marginal Productivity Consider the Cobb-Douglas production function

$$f(x, y) = 200x^{0.7}y^{0.3}.$$

When $x = 1000$ and $y = 500$, find

(a) the marginal productivity of labor, $\partial f/\partial x$.

(b) the marginal productivity of capital, $\partial f/\partial y$.

62. Marginal Productivity Repeat Exercise 61 for the production function given by $f(x, y) = 100x^{0.75}y^{0.25}$.

Complementary and Substitute Products In Exercises 63 and 64, determine whether the demand functions describe complementary or substitute product relationships. Using the notation of Example 4, let x_1 and x_2 be the demands for two products whose prices are p_1 and p_2, respectively. *See Example 4.*

63. $x_1 = 150 - 2p_1 - \frac{5}{2}p_2, \quad x_2 = 350 - \frac{3}{2}p_1 - 3p_2$

64. $x_1 = 150 - 2p_1 + 1.8p_2, \quad x_2 = 350 + \frac{3}{4}p_1 - 1.9p_2$

65. Expenditures The expenditures z (in billions of dollars) for spectator sports from 2004 through 2009 can be modeled by

$$z = 0.62x - 0.41y + 0.38$$

where x is the expenditures on amusement parks and campgrounds, and y is the expenditures on live entertainment (excluding sports), both in billions of dollars. *(Source: U.S. Bureau of Economic Analysis)*

(a) Find $\partial z/\partial x$ and $\partial z/\partial y$.

(b) Interpret the partial derivatives in the context of the problem.

66. Shareholder's Equity The shareholder's equity z (in millions of dollars) for Skechers from 2001 through 2009 can be modeled by

$$z = 0.175x - 0.772y - 275$$

where x is the sales (in millions of dollars) and y is the total assets (in millions of dollars). *(Source: Skechers U.S.A. Inc.)*

(a) Find $\partial z/\partial x$ and $\partial z/\partial y$.

(b) Interpret the partial derivatives in the context of the problem.

67. Psychology Early in the twentieth century, an intelligence test called the *Stanford-Binet Test* (more commonly known as the *IQ test*) was developed. In this test, an individual's mental age M is divided by the individual's chronological age C and the quotient is multiplied by 100. The result is the individual's IQ.

$$IQ(M, C) = \frac{M}{C} \times 100$$

Find the partial derivatives of IQ with respect to M and with respect to C. Evaluate the partial derivatives at the point $(12, 10)$ and interpret the result. *(Source: Adapted from Bernstein/Clark-Stewart/Roy/Wickens, Psychology, Fourth Edition)*

68. HOW DO YOU SEE IT? Use the graph of the surface to determine the sign of each partial derivative. Explain your reasoning.

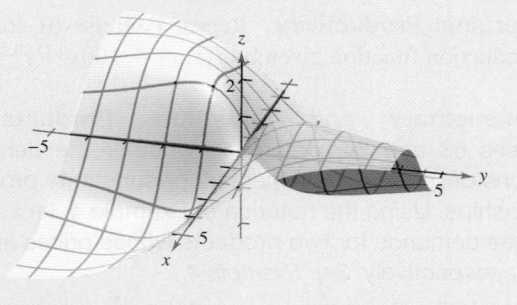

(a) $f_x(4, 1)$ (b) $f_y(4, 1)$

(c) $f_x(-1, -2)$ (d) $f_y(-1, -2)$

69. Investment The value of an investment of $1000 earning 10% interest compounded annually is

$$V(I, R) = 1000\left[\frac{1 + 0.10(1 - R)}{1 + I}\right]^{10}$$

where I is the annual rate of inflation and R is the tax rate for the person making the investment. Calculate $V_I(0.03, 0.28)$ and $V_R(0.03, 0.28)$. Determine whether the tax rate or the rate of inflation is the greater "negative" influence on the growth of the investment.

70. Think About It Let N be the number of applicants to a university, p the charge for food and housing at the university, and t the tuition. Suppose that N is a function of p and t such that $\partial N/\partial p < 0$ and $\partial N/\partial t < 0$. How would you interpret the fact that both partials are negative?

71. Marginal Utility The utility function $U = f(x, y)$ is a measure of the utility (or satisfaction) derived by a person from the consumption of two products x and y. Suppose the utility function is given by $U = -5x^2 + xy - 3y^2$.

(a) Determine the marginal utility of product x.

(b) Determine the marginal utility of product y.

(c) When $x = 2$ and $y = 3$, should a person consume one more unit of product x or one more unit of product y? Explain your reasoning.

(d) Use a three-dimensional graphing utility to graph the function. Interpret the marginal utilities of products x and y graphically.

Business Capsule

In 1996, twin sisters Izzy and Coco Tihanyi started Surf Diva, a surf school and apparel company for women and girls, in La Jolla, California. To advertise their business, they would donate surf lessons and give the surf report on local radio stations in exchange for air time. Today, they have schools and surf camps in Los Angeles, San Diego, and Costa Rica. Their clothing line can be found in the Surf Diva Boutique as well as other surf and specialty shops, sporting goods stores, and airport gift shops.

72. Research Project Use your school's library, the Internet, or some other reference source to research a company that increased the demand for its product by creative advertising. Write a paper about the company. Use graphs to show how a change in demand is related to a change in the marginal utility of a product or service.

13.5 Extrema of Functions of Two Variables

- Understand the relative extrema of functions of two variables.
- Use the First-Partials Test to find the relative extrema of functions of two variables.
- Use the Second-Partials Test to find the relative extrema of functions of two variables.
- Use relative extrema to answer questions about real-life situations.

Relative Extrema

Earlier in the text, you learned how to use derivatives to find the relative minimum and relative maximum values of a function of a single variable. In this section, you will learn how to use partial derivatives to find the relative minimum and relative maximum values of a function of two variables.

In Exercise 43 on page 954, you will find the dimensions of a rectangular package of maximum volume that can be sent by a shipping company.

Relative Extrema of a Function of Two Variables

Let f be a function defined on a region containing (x_0, y_0). The function f has a **relative maximum** at (x_0, y_0) when there is a circular region R centered at (x_0, y_0) such that

$$f(x, y) \le f(x_0, y_0) \qquad \text{f has a relative maximum at (x_0, y_0).}$$

for all (x, y) in R. The function f has a **relative minimum** at (x_0, y_0) when there is a circular region R centered at (x_0, y_0) such that

$$f(x, y) \ge f(x_0, y_0) \qquad \text{f has a relative minimum at (x_0, y_0).}$$

for all (x, y) in R.

To say that f has a relative maximum at (x_0, y_0) means that the point (x_0, y_0, z_0) is at least as high as all nearby points on the graph of $z = f(x, y)$. Similarly, f has a relative minimum at (x_0, y_0) when (x_0, y_0, z_0) is at least as low as all nearby points on the graph. (See Figure 13.27.)

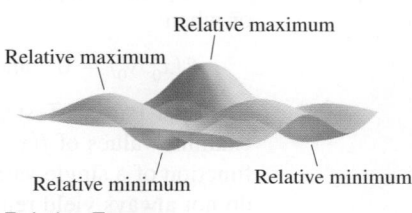

Relative Extrema

FIGURE 13.27

As in single-variable calculus, you need to distinguish between relative extrema and absolute extrema of a function of two variables. The number $f(x_0, y_0)$ is an absolute maximum of f in the region R when it is greater than or equal to all other function values in the region. (An absolute minimum of f in a region is defined similarly.) For instance, the function

$$f(x, y) = -(x^2 + y^2)$$

is a paraboloid, opening downward, with vertex at $(0, 0, 0)$. (See Figure 13.28.) The number

$$f(0, 0) = 0$$

is an absolute maximum of the function over the entire xy-plane.

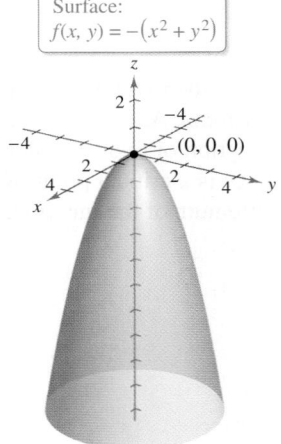

Surface:
$f(x, y) = -(x^2 + y^2)$

f has an absolute maximum at $(0, 0, 0)$.

FIGURE 13.28

The First-Partials Test for Relative Extrema

To locate the relative extrema of a function of two variables, you can use a procedure that is similar to the First-Derivative Test used for functions of a single variable.

First-Partials Test for Relative Extrema

If f has a relative extremum at (x_0, y_0) on an open region R in the xy-plane, and the first partial derivatives of f exist in R, then

$$f_x(x_0, y_0) = 0$$

and

$$f_y(x_0, y_0) = 0$$

as shown in Figure 13.29.

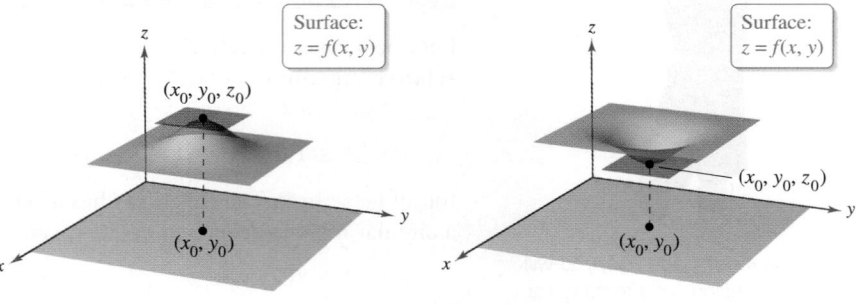

Relative maximum   Relative minimum

FIGURE 13.29

An *open* region in the xy-plane is similar to an open interval on the real number line. For instance, the region R consisting of the interior of the circle $x^2 + y^2 = 1$ is an open region. If the region R consists of the interior of the circle *and* the points on the circle, then it is a *closed* region.

A point (x_0, y_0) is a **critical point** of f when $f_x(x_0, y_0)$ or $f_y(x_0, y_0)$ is undefined or when

$$f_x(x_0, y_0) = 0 \quad \text{and} \quad f_y(x_0, y_0) = 0. \qquad \text{Critical point}$$

The First-Partials Test states that if the first partial derivatives exist, then you need only examine values of $f(x, y)$ at critical points to find the relative extrema. As is true for a function of a single variable, however, the critical points of a function of two variables do not always yield relative extrema. For instance, the point $(0, 0)$ is a critical point of the surface shown in Figure 13.30, but $f(0, 0)$ is not a relative extremum of the function. Such points are called **saddle points** of the function.

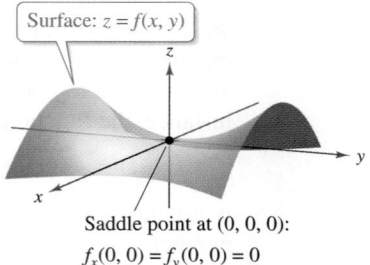

Saddle point at $(0, 0, 0)$:
$$f_x(0, 0) = f_y(0, 0) = 0$$

FIGURE 13.30

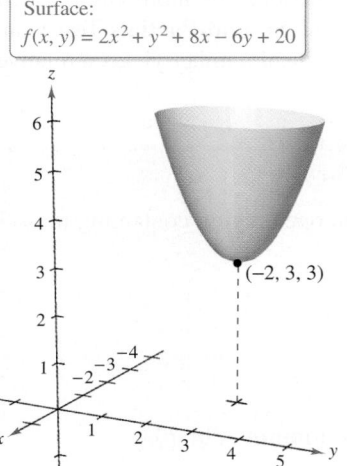

Surface:
$f(x, y) = 2x^2 + y^2 + 8x - 6y + 20$

$(-2, 3, 3)$

FIGURE 13.31

Example 1   **Finding Relative Extrema**

Find the relative extrema of

$$f(x, y) = 2x^2 + y^2 + 8x - 6y + 20.$$

SOLUTION   Begin by finding the first partial derivatives of f.

$$f_x(x, y) = 4x + 8 \quad \text{and} \quad f_y(x, y) = 2y - 6$$

Because these partial derivatives are defined for all points in the xy-plane, the only critical points are those for which both first partial derivatives are zero. To locate these points, set $f_x(x, y)$ and $f_y(x, y)$ equal to 0, and solve the resulting system of equations.

$$4x + 8 = 0 \qquad \text{Set } f_x(x, y) \text{ equal to 0.}$$
$$2y - 6 = 0 \qquad \text{Set } f_y(x, y) \text{ equal to 0.}$$

The solution of this system is $x = -2$ and $y = 3$. So, the point $(-2, 3)$ is the only critical number of f. From the graph of the function, shown in Figure 13.31, you can see that this critical point yields a relative minimum of the function. So, the function has only one relative extremum, which is

$$f(-2, 3) = 3. \qquad \text{Relative minimum}$$

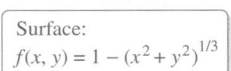 **Checkpoint 1**

Find the relative extrema of

$$f(x, y) = x^2 + 2y^2 + 16x - 8y + 8.$$

Example 1 shows a relative minimum occurring at one type of critical point—the type for which both $f_x(x, y)$ and $f_y(x, y)$ are zero. The next example shows a relative maximum that occurs at the other type of critical point—the type for which either $f_x(x, y)$ or $f_y(x, y)$ is undefined.

Example 2   **Finding Relative Extrema**

Find the relative extrema of

$$f(x, y) = 1 - (x^2 + y^2)^{1/3}.$$

SOLUTION   Begin by finding the first partial derivatives of f.

$$f_x(x, y) = -\frac{2x}{3(x^2 + y^2)^{2/3}} \quad \text{and} \quad f_y(x, y) = -\frac{2y}{3(x^2 + y^2)^{2/3}}$$

These partial derivatives are defined for all points in the xy-plane *except* the point $(0, 0)$. So, $(0, 0)$ is a critical point of f. Moreover, this is the only critical point, because there are no other values of x and y for which either partial derivative is undefined or for which both partial derivatives are zero. From the graph of the function, shown in Figure 13.32, you can see that this critical point yields a relative maximum of the function. So, the function has only one relative extremum, which is

$$f(0, 0) = 1. \qquad \text{Relative maximum}$$

Surface:
$f(x, y) = 1 - (x^2 + y^2)^{1/3}$

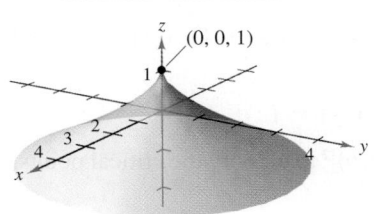

$(0, 0, 1)$

$f_x(x, y)$ and $f_y(x, y)$ are undefined at $(0, 0)$.

FIGURE 13.32

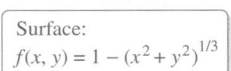 **Checkpoint 2**

Find the relative extrema of

$$f(x, y) = \sqrt{1 - \frac{x^2}{16} - \frac{y^2}{4}}.$$

The Second-Partials Test for Relative Extrema

For functions such as those in Examples 1 and 2, you can determine the *types* of extrema at the critical points by sketching the graph of the function. For more complicated functions, a graphical approach is not so easy to use. The **Second-Partials Test** is an analytical test that can be used to determine whether a critical number yields a relative minimum, a relative maximum, or neither.

Second-Partials Test for Relative Extrema

Let f have continuous second partial derivatives on an open region containing (a, b) for which

$$f_x(a, b) = 0 \quad \text{and} \quad f_y(a, b) = 0.$$

To test for relative extrema of f, consider the quantity

$$d = f_{xx}(a, b) f_{yy}(a, b) - [f_{xy}(a, b)]^2.$$

1. If $d > 0$ and $f_{xx}(a, b) > 0$, then f has a **relative minimum** at (a, b).

2. If $d > 0$ and $f_{xx}(a, b) < 0$, then f has a **relative maximum** at (a, b).

3. If $d < 0$, then $(a, b, f(a, b))$ is a **saddle point**.

4. The test gives no information when $d = 0$.

Note in the Second-Partials Test that if $d > 0$, then $f_{xx}(a, b)$ and $f_{yy}(a, b)$ must have the same sign. So, you can replace $f_{xx}(a, b)$ with $f_{yy}(a, b)$ in the first two parts of the test.

ALGEBRA TUTOR xy

For help in solving the system of equations

$$y - x^3 = 0$$
$$x - y^3 = 0$$

in Example 3, see Example 1(a) in the *Chapter 13 Algebra Tutor*, on page 988.

Example 3   **Applying the Second-Partials Test**

Find the relative extrema and saddle points of $f(x, y) = xy - \frac{1}{4}x^4 - \frac{1}{4}y^4$.

SOLUTION   Begin by finding the critical points of f. Because

$$f_x(x, y) = y - x^3 \quad \text{and} \quad f_y(x, y) = x - y^3$$

are defined for all points in the xy-plane, the only critical points are those for which both first partial derivatives are zero. By solving the equations

$$y - x^3 = 0 \quad \text{and} \quad x - y^3 = 0$$

simultaneously, you can determine that the critical points are $(1, 1)$, $(-1, -1)$, and $(0, 0)$. Furthermore, because

$$f_{xx}(x, y) = -3x^2, \quad f_{yy}(x, y) = -3y^2, \quad \text{and} \quad f_{xy}(x, y) = 1$$

you can use the quantity $d = f_{xx}(a, b) f_{yy}(a, b) - [f_{xy}(a, b)]^2$ to classify the critical points, as shown.

Critical Point	d	$f_{xx}(x, y)$	Conclusion
$(1, 1)$	$(-3)(-3) - 1 = 8$	-3	Relative maximum
$(-1, -1)$	$(-3)(-3) - 1 = 8$	-3	Relative maximum
$(0, 0)$	$(0)(0) - 1 = -1$	0	Saddle point

The graph of f is shown in Figure 13.33.

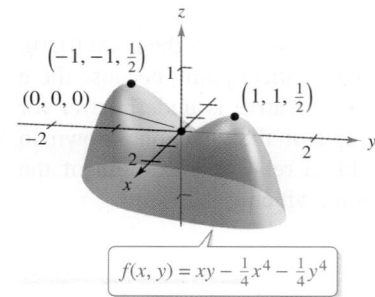

$(-1, -1, \frac{1}{2})$

$(0, 0, 0)$

$(1, 1, \frac{1}{2})$

$f(x, y) = xy - \frac{1}{4}x^4 - \frac{1}{4}y^4$

FIGURE 13.33

✓ **Checkpoint 3**

Find the relative extrema and saddle points of $f(x, y) = \dfrac{y^2}{16} - \dfrac{x^2}{4}$.

Applications

 Example 4 **Finding a Maximum Profit**

A company makes two substitute products whose demand functions are given by

$$x_1 = 200(p_2 - p_1)$$ Demand for product 1
$$x_2 = 500 + 100p_1 - 180p_2$$ Demand for product 2

where p_1 and p_2 are the prices per unit (in dollars) and x_1 and x_2 are the numbers of units sold. The costs of producing the two products are \$0.50 and \$0.75 per unit, respectively. Find the prices that will yield a maximum profit.

SOLUTION The cost function is

$$C = 0.5x_1 + 0.75x_2$$ Write cost function.
$$= 0.5(200)(p_2 - p_1) + 0.75(500 + 100p_1 - 180p_2)$$ Substitute.
$$= 375 - 25p_1 - 35p_2.$$ Simplify.

The revenue function is

$$R = p_1 x_1 + p_2 x_2$$ Write revenue function.
$$= p_1(200)(p_2 - p_1) + p_2(500 + 100p_1 - 180p_2)$$ Substitute.
$$= -200p_1{}^2 - 180p_2{}^2 + 300p_1 p_2 + 500p_2.$$ Simplify.

This implies that the profit function is

$$P = R - C$$ Write profit function.
$$= -200p_1{}^2 - 180p_2{}^2 + 300p_1 p_2 + 500p_2 - (375 - 25p_1 - 35p_2)$$
$$= -200p_1{}^2 - 180p_2{}^2 + 300p_1 p_2 + 25p_1 + 535p_2 - 375.$$

Next, find the first partial derivatives of P.

$$\frac{\partial P}{\partial p_1} = -400p_1 + 300p_2 + 25 \qquad \frac{\partial P}{\partial p_2} = 300p_1 - 360p_2 + 535$$

By setting the first partial derivatives equal to zero and solving the equations

$$-400p_1 + 300p_2 + 25 = 0$$
$$300p_1 - 360p_2 + 535 = 0$$

simultaneously, you can conclude that the solution is $p_1 \approx \$3.14$ and $p_2 \approx \$4.10$. From the graph of P shown in Figure 13.34, you can see that this critical number yields a maximum. So, the maximum profit is

$$P(p_1, p_2) \approx P(3.14, 4.10) = \$761.48.$$

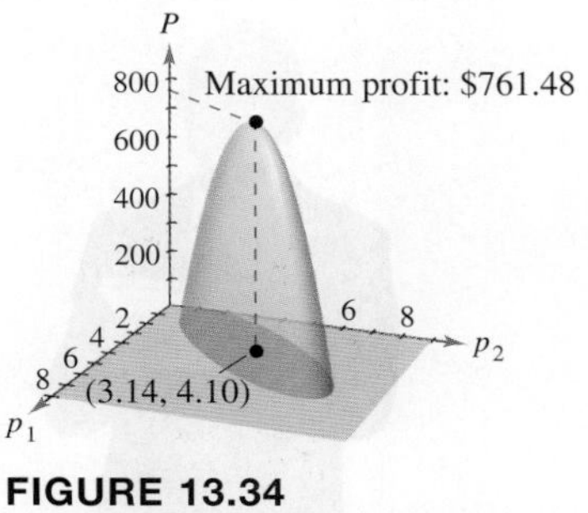

FIGURE 13.34

✓ **Checkpoint 4**

Find the prices that will yield a maximum profit for the products in Example 4 when the costs of producing the two products are \$0.75 and \$0.50 per unit, respectively. ■

In Example 4, to convince yourself that the maximum profit is \$761.48, try substituting other prices, such as $p_1 = \$2$ and $p_2 = \$3$, into the profit function. For each pair of prices, you will obtain a profit that is less than \$761.48.

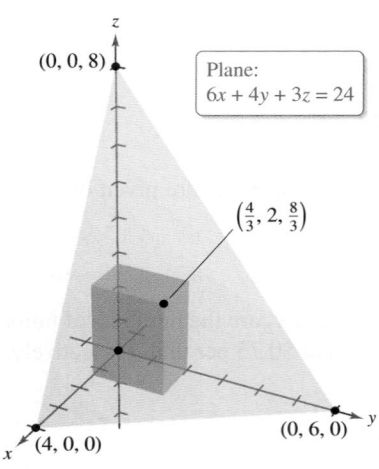

(0, 0, 8)

Plane:
$6x + 4y + 3z = 24$

$\left(\frac{4}{3}, 2, \frac{8}{3}\right)$

(4, 0, 0)

(0, 6, 0)

FIGURE 13.35

ALGEBRA TUTOR xy

For help in solving the system of equations

$$y(24 - 12x - 4y) = 0$$
$$x(24 - 6x - 8y) = 0$$

in Example 5, see Example 2(a) in the *Chapter 13 Algebra Tutor*, on page 989.

| Example 5 | **Finding a Maximum Volume** |

A rectangular box is resting on the xy-plane with one vertex at the origin. The opposite vertex lies in the plane

$$6x + 4y + 3z = 24$$

as shown in Figure 13.35. Find the maximum volume of such a box.

SOLUTION   Let x, y, and z represent the length, width, and height of the box. Because one vertex of the box lies in the plane given by $6x + 4y + 3z = 24$ or

$$z = \tfrac{1}{3}(24 - 6x - 4y) \qquad \text{Solve for } z.$$

you can write the volume of the box as a function of two variables.

$$
\begin{aligned}
V &= xyz & \text{Volume} = (\text{width})(\text{length})(\text{height})\\
&= xy\left(\tfrac{1}{3}\right)(24 - 6x - 4y) & \text{Substitute for } z.\\
&= \tfrac{1}{3}(24xy - 6x^2 y - 4xy^2) & \text{Simplify.}
\end{aligned}
$$

Next, find the first partial derivatives of V.

$$
\begin{aligned}
V_x &= \tfrac{1}{3}(24y - 12xy - 4y^2) & \text{Partial with respect to } x\\
&= \tfrac{1}{3}y(24 - 12x - 4y) & \text{Factor.}\\
V_y &= \tfrac{1}{3}(24x - 6x^2 - 8xy) & \text{Partial with respect to } y\\
&= \tfrac{1}{3}x(24 - 6x - 8y) & \text{Factor.}
\end{aligned}
$$

By solving the equations

$$
\begin{aligned}
\tfrac{1}{3}y(24 - 12x - 4y) &= 0 & \text{Set } V_x \text{ equal to 0.}\\
\tfrac{1}{3}x(24 - 6x - 8y) &= 0 & \text{Set } V_y \text{ equal to 0.}
\end{aligned}
$$

simultaneously, you can conclude that the solutions are $(0, 0)$, $(0, 6)$, $(4, 0)$, and $\left(\frac{4}{3}, 2\right)$. Using the Second-Partials Test, you can determine that the maximum volume occurs when the width is $x = \frac{4}{3}$ and the length is $y = 2$. For these values, the height of the box is

$$z = \tfrac{1}{3}\left[24 - 6\left(\tfrac{4}{3}\right) - 4(2)\right] = \tfrac{8}{3}.$$

So, the maximum volume is

$$V = xyz = \left(\tfrac{4}{3}\right)(2)\left(\tfrac{8}{3}\right) = \tfrac{64}{9} \text{ cubic units.}$$

✓**Checkpoint 5**

Find the maximum volume of a box that is resting on the xy-plane with one vertex at the origin and the opposite vertex in the plane $2x + 4y + z = 8$.

SUMMARIZE   (Section 13.5)

1. State the definition of relative extrema of a function of two variables (*page 947*). For examples of relative extrema, see Examples 1 and 2.

2. State the First-Partials Test for relative extrema (*page 948*). For examples of using the First-Partials Test, see Examples 1 and 2.

3. State the Second-Partials Test for relative extrema (*page 950*). For examples of using the Second-Partials Test, see Examples 3 and 5.

4. Describe a real-life example of how relative extrema can be used to find a company's maximum profit (*page 951, Example 4*).

SKILLS WARM UP 13.5 The following warm-up exercises involve skills that were covered in earlier sections. You will use these skills in the exercise set for this section. For additional help, review Sections 5.1, 5.2, and 13.4.

In Exercises 1–8, solve the system of equations.

1. $\begin{cases} 5x = 15 \\ 3x - 2y = 5 \end{cases}$

2. $\begin{cases} \frac{1}{2}y = 3 \\ -x + 5y = 19 \end{cases}$

3. $\begin{cases} x + y = 5 \\ x - y = -3 \end{cases}$

4. $\begin{cases} x + y = 8 \\ 2x - y = 4 \end{cases}$

5. $\begin{cases} 2x - y = 8 \\ 3x - 4y = 7 \end{cases}$

6. $\begin{cases} 2x - 4y = 14 \\ 3x + y = 7 \end{cases}$

7. $\begin{cases} x^2 + x = 0 \\ 2yx + y = 0 \end{cases}$

8. $\begin{cases} 3y^2 + 6y = 0 \\ xy + x + 2 = 0 \end{cases}$

In Exercises 9–14, find all first and second partial derivatives of the function.

9. $z = 4x^3 - 3y^2$

10. $z = 2x^5 - y^3$

11. $z = x^4 - \sqrt{xy} + 2y$

12. $z = 2x^2 - 3xy + y^2$

13. $z = ye^{xy^2}$

14. $z = xe^{xy}$

Exercises 13.5

See www.CalcChat.com for worked-out solutions to odd-numbered exercises.

Applying the Second-Partials Test In Exercises 1–18, find the critical points, relative extrema, and saddle points of the function. *See Examples 1, 2, and 3.*

1. $f(x, y) = x^2 - y^2 + 4x - 8y - 11$

2. $f(x, y) = x^2 + y^2 + 2x - 6y + 6$

3. $f(x, y) = \sqrt{x^2 + y^2 + 1}$

4. $f(x, y) = \sqrt{25 - (x - 2)^2 - y^2}$

5. $f(x, y) = (x - 1)^2 + (y - 3)^2$

6. $f(x, y) = 9 - (x - 3)^2 - (y + 2)^2$

7. $f(x, y) = 2x^2 + 2xy + y^2 + 2x - 3$

8. $f(x, y) = -x^2 - 5y^2 + 8x - 10y - 13$

9. $f(x, y) = -5x^2 + 4xy - y^2 + 16x + 10$

10. $f(x, y) = x^2 + 6xy + 10y^2 - 4y + 4$

11. $f(x, y) = 3x^2 + 2y^2 - 6x - 4y + 16$

12. $f(x, y) = -3x^2 - 2y^2 + 3x - 4y + 5$

13. $f(x, y) = -x^3 + 4xy - 2y^2 + 1$

14. $f(x, y) = x^2 - 3xy - y^2$

15. $f(x, y) = \frac{1}{2}xy$

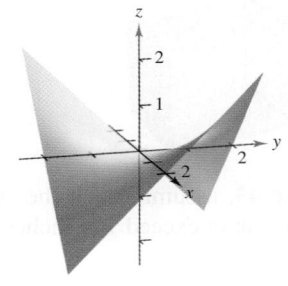

16. $f(x, y) = x + y + 2xy - x^2 - y^2$

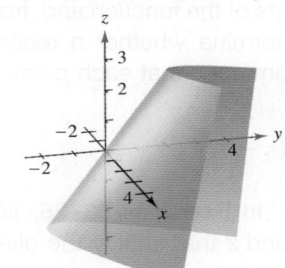

17. $f(x, y) = (x + y)e^{1 - x^2 - y^2}$

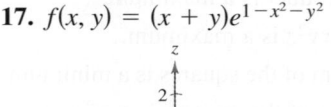
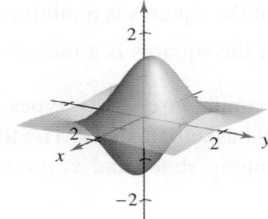

18. $f(x, y) = 3e^{-(x^2 + y^2)}$

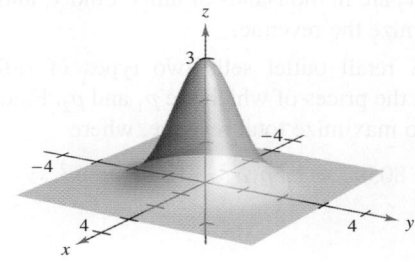

Think About It   In Exercises 19–24, determine whether there is a relative maximum, a relative minimum, a saddle point, or insufficient information to determine the nature of the function $f(x, y)$ at the critical point (x_0, y_0).

19. $f_{xx}(x_0, y_0) = 9$, $f_{yy}(x_0, y_0) = 4$, $f_{xy}(x_0, y_0) = 6$

20. $f_{xx}(x_0, y_0) = -3$, $f_{yy}(x_0, y_0) = -8$, $f_{xy}(x_0, y_0) = 2$

21. $f_{xx}(x_0, y_0) = -9$, $f_{yy}(x_0, y_0) = 6$, $f_{xy}(x_0, y_0) = 10$

22. $f_{xx}(x_0, y_0) = 25$, $f_{yy}(x_0, y_0) = 8$, $f_{xy}(x_0, y_0) = 10$

23. $f_{xx}(x_0, y_0) = 5$, $f_{yy}(x_0, y_0) = 5$, $f_{xy}(x_0, y_0) = 3$

24. $f_{xx}(x_0, y_0) = 8$, $f_{yy}(x_0, y_0) = 7$, $f_{xy}(x_0, y_0) = 9$

Analyzing a Function   In Exercises 25–30, find the critical points, relative extrema, and saddle points of the function. List the critical points for which the Second-Partials Test fails.

25. $f(x, y) = (xy)^2$

26. $f(x, y) = \sqrt{x^2 + y^2}$

27. $f(x, y) = x^3 + y^3$

28. $f(x, y) = x^3 + y^3 - 3x^2 + 6y^2 + 3x + 12y + 7$

29. $f(x, y) = x^{2/3} + y^{2/3}$

30. $f(x, y) = (x^2 + y^2)^{2/3}$

Analyzing a Function of Three Variables   In Exercises 31 and 32, find the critical points of the function and, from the form of the function, determine whether a relative maximum or a relative minimum occurs at each point.

31. $f(x, y, z) = (x - 1)^2 + (y + 3)^2 + z^2$

32. $f(x, y, z) = 6 - [x(y + 2)(z - 1)]^2$

Finding Positive Numbers   In Exercises 33–36, find three positive numbers x, y, and z that satisfy the given conditions.

33. The sum is 45 and the product is a maximum.

34. The sum is 32 and $P = xy^2z$ is a maximum.

35. The sum is 60 and the sum of the squares is a minimum.

36. The sum is 2 and the sum of the squares is a minimum.

37. Revenue   A company manufactures two types of sneakers: running shoes and basketball shoes. The total revenue from x_1 units of running shoes and x_2 units of basketball shoes is

$$R = -5x_1^2 - 8x_2^2 - 2x_1x_2 + 42x_1 + 102x_2$$

where x_1 and x_2 are in thousands of units. Find x_1 and x_2 so as to maximize the revenue.

38. Revenue   A retail outlet sells two types of riding lawn mowers, the prices of which are p_1 and p_2. Find p_1 and p_2 so as to maximize total revenue, where

$$R = 515p_1 + 805p_2 + 1.5p_1p_2 - 1.5p_1^2 - p_2^2.$$

Revenue   In Exercises 39 and 40, find p_1 and p_2, the prices per unit (in dollars), so as to maximize the total revenue

$$R = x_1p_1 + x_2p_2$$

where x_1 and x_2 are the numbers of units sold, for a retail outlet that sells two competitive products with the given demand functions. *See Example 4.*

39. $x_1 = 1000 - 2p_1 + p_2$, $x_2 = 1500 + 2p_1 - 1.5p_2$

40. $x_1 = 1000 - 4p_1 + 2p_2$, $x_2 = 900 + 4p_1 - 3p_2$

41. Profit   A corporation manufactures a product for a high-performance automobile engine at two locations. The cost of producing x_1 units at location 1 is

$$C_1 = 0.05x_1^2 + 15x_1 + 5400$$

and the cost of producing x_2 units at location 2 is

$$C_2 = 0.03x_2^2 + 15x_2 + 6100.$$

The demand function for the product is

$$p = 225 - 0.4(x_1 + x_2)$$

and the total revenue function is

$$R = [225 - 0.4(x_1 + x_2)](x_1 + x_2).$$

Find the production levels at the two locations that will maximize the profit

$$P = R - C_1 - C_2.$$

42. Profit   A corporation manufactures candles at two locations. The cost of producing x_1 units at location 1 is

$$C_1 = 0.02x_1^2 + 4x_1 + 500$$

and the cost of producing x_2 units at location 2 is

$$C_2 = 0.05x_2^2 + 4x_2 + 275.$$

The candles sell for $15 per unit. Find the quantity that should be produced at each location to maximize the profit

$$P = 15(x_1 + x_2) - C_1 - C_2.$$

43. Volume   Find the dimensions of a rectangular package of maximum volume that may be sent by a shipping company, assuming that the sum of the length and the girth (perimeter of a cross section) cannot exceed 96 inches.

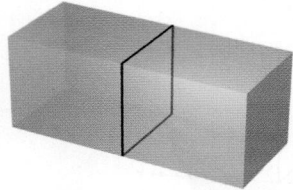

44. Volume   Repeat Exercise 43, assuming that the sum of the length and the girth cannot exceed 144 inches.

45. Cost A manufacturer makes an open-top wooden crate having a volume of 18 cubic feet. Material costs are $0.20 per square foot for the base and $0.15 per square foot for the sides. Find the dimensions that minimize the cost of each crate. What is the minimum cost?

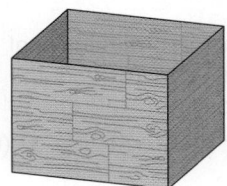

46. Cost A home improvement contractor is painting the walls and ceiling of a rectangular room. The volume of the room is 1584 cubic feet. The cost of wall paint is $0.06 per square foot and the cost of ceiling paint is $0.11 per square foot. Find the room dimensions that will minimize the cost of the paint. What is the minimum cost of the paint?

47. Cost An automobile manufacturer has determined that its annual labor and equipment cost (in millions of dollars) can be modeled by

$$C(x, y) = 2x^2 + 3y^2 - 15x - 20y + 4xy + 39$$

where x is the amount spent per year on labor and y is the amount spent per year on equipment (both in millions of dollars). Find the values of x and y that minimize the annual labor and equipment cost. What is this cost?

48. Medicine In order to treat a certain bacterial infection, a combination of two drugs is being tested. Studies have shown that the duration of the infection in laboratory tests can be modeled by

$$D(x, y) = x^2 + 2y^2 - 18x - 24y + 2xy + 120$$

where x is the dosage of the first drug and y is the dosage of the second drug (both in hundreds of milligrams). Find the amount of each drug necessary to minimize the duration of the infection.

49. Biology A lake is to be stocked with smallmouth and largemouth bass. Let x represent the number of smallmouth bass and let y represent the number of largemouth bass. The weight of each fish is dependent on the population densities. After a six-month period, the weight of a single smallmouth bass is given by

$$W_1 = 3 - 0.002x - 0.001y$$

and the weight of a single largemouth bass is given by

$$W_2 = 4.5 - 0.004x - 0.005y.$$

Assuming that no fish die during the six-month period, how many smallmouth and largemouth bass should be stocked in the lake so that the *total* weight T of bass in the lake is a maximum?

50.  **HOW DO YOU SEE IT?** The figure shows the level curves for a function $f(x, y)$. What, if anything, can be said about f at the points $A, B, C,$ and D? Explain your reasoning.

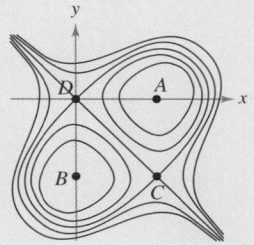

51. Hardy-Weinberg Law Common blood types are determined genetically by the three alleles A, B, and O. (An allele is any of a group of possible mutational forms of a gene.) A person whose blood type is AA, BB, or OO is homozygous. A person whose blood type is AB, AO, or BO is heterozygous. The Hardy-Weinberg Law states that the proportion P of heterozygous individuals in any given population is modeled by

$$P(p, q, r) = 2pq + 2pr + 2qr$$

where p represents the percent of allele A in the population, q represents the percent of allele B in the population, and r represents the percent of allele O in the population. Use the fact that $p + q + r = 1$ (the sum of the three must equal 100%) to show that the maximum proportion of heterozygous individuals in any population is $\frac{2}{3}$.

52. Shannon Diversity Index One way to measure species diversity is to use the Shannon diversity index H. A habitat consists of three species A, B, and C, and its Shannon diversity index is

$$H = -x \ln x - y \ln y - z \ln z$$

where x is the percent of species A in the habitat, y is the percent of species B in the habitat, and z is the percent of species C in the habitat. Use the fact that $x + y + z = 1$ (the sum of the three must equal 100%) to show that the maximum value of H occurs when

$$x = y = z = \frac{1}{3}.$$

What is the maximum value of H?

True or False? In Exercises 53 and 54, determine whether the statement is true or false. If it is false, explain why or give an example that shows it is false.

53. A saddle point always occurs at a critical point.

54. If $f(x, y)$ has a relative maximum at (x_0, y_0, z_0), then $f_x(x_0, y_0) = f_y(x_0, y_0) = 0$.

QUIZ YOURSELF

See www.CalcChat.com for worked-out solutions to odd-numbered exercises.

Take this quiz as you would take a quiz in class. When you are done, check your work against the answers given in the back of the book.

In Exercises 1–3, (a) plot the points in a three-dimensional coordinate system, (b) find the distance between the points, and (c) find the coordinates of the midpoint of the line segment joining the points.

1. $(1, 3, 2), (-1, 2, 0)$ **2.** $(-1, 3, 4), (5, 1, -6)$ **3.** $(0, -3, 3), (3, 0, -3)$

In Exercises 4 and 5, find the standard equation of the sphere.

4. Center: $(2, -1, 3)$; radius: 4

5. Endpoints of a diameter: $(0, 3, 1), (2, 5, -5)$

6. Find the center and radius of the sphere whose equation is

$$x^2 + y^2 + z^2 - 8x - 2y - 6z - 23 = 0.$$

In Exercises 7–9, find the intercepts and sketch the graph of the plane.

7. $2x + 3y + z = 6$ **8.** $x - 2z = 4$ **9.** $y = 3$

In Exercises 10–12, classify the quadric surface.

10. $\dfrac{x^2}{4} + \dfrac{y^2}{9} + \dfrac{z^2}{16} = 1$ **11.** $z^2 - x^2 - y^2 = 25$ **12.** $81z - 9x^2 - y^2 = 0$

In Exercises 13–15, find $f(1, 0)$ and $f(4, -1)$.

13. $f(x, y) = x - 9y^2$ **14.** $f(x, y) = \sqrt{4x^2 + y}$ **15.** $f(x, y) = \ln(x - 2y)$

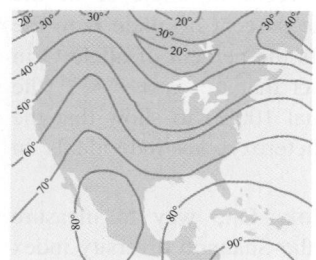

Figure for 16

16. The contour map shows level curves of equal temperature (isotherms), measured in degrees Fahrenheit, across North America on a spring day. Use the map to find the approximate range of temperatures in

(a) the Great Lakes region.

(b) the United States.

(c) Mexico.

In Exercises 17–20, find the first partial derivatives and evaluate each at the point $(-2, 3)$.

17. $f(x, y) = x^2 + 2y^2 - 3x - y + 1$ **18.** $f(x, y) = \dfrac{3x - y^2}{x + y}$

19. $f(x, y) = x^3 e^{2y}$ **20.** $f(x, y) = \ln(2x + 7y)$

In Exercises 21 and 22, find the critical points, relative extrema, and saddle points of the function.

21. $f(x, y) = 3x^2 + y^2 - 2xy - 6x + 2y$

22. $f(x, y) = -x^3 + 4xy - 2y^2 + 1$

23. A company manufactures two types of wood burning stoves: a freestanding model and a fireplace-insert model. The total cost (in thousands of dollars) for producing x freestanding stoves and y fireplace-insert stoves can be modeled by

$$C(x, y) = \tfrac{1}{16}x^2 + y^2 - 10x - 40y + 820.$$

Find the values of x and y that minimize the total cost. What is this cost?

13.6 Lagrange Multipliers

- ■ Understand the Method of Lagrange Multipliers.
- ■ Use Lagrange multipliers to solve constrained optimization problems.

In Exercise 37 on page 464, you will use Lagrange multipliers to find the dimensions that will minimize the cost of fencing in two corrals.

Lagrange Multipliers with One Constraint

In Example 5 in Section 13.5, you were asked to find the dimensions of the rectangular box of maximum volume that would fit in the first octant beneath the plane

$$6x + 4y + 3z = 24$$

as shown again in Figure 13.36. Another way of stating this problem is to say that you are asked to find the maximum of

$$V = xyz \qquad \text{Objective function}$$

subject to the constraint

$$6x + 4y + 3z - 24 = 0. \qquad \text{Constraint}$$

This type of problem is called a **constrained optimization** problem. In Section 13.5, you answered this question by solving for z in the constraint equation and then rewriting V as a function of two variables.

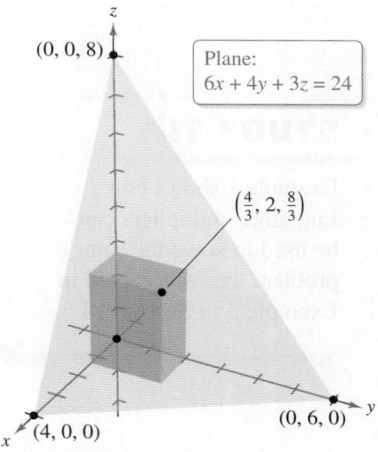

FIGURE 13.36

In this section, you will study a different (and often better) way to solve constrained optimization problems. This method involves the use of variables called **Lagrange multipliers,** named after the French mathematician Joseph Louis Lagrange (1736–1813).

Method of Lagrange Multipliers

If $f(x, y)$ has a maximum or minimum subject to the constraint $g(x, y) = 0$, then it will occur at one of the critical numbers of the function F defined by

$$F(x, y, \lambda) = f(x, y) - \lambda g(x, y).$$

The variable λ (the lowercase Greek letter lambda) is called a **Lagrange multiplier.** To find the minimum or maximum of f, use the following steps.

1. Solve the following system of equations.

$$F_x(x, y, \lambda) = 0 \qquad F_y(x, y, \lambda) = 0 \qquad F_\lambda(x, y, \lambda) = 0$$

2. Evaluate f at each solution point obtained in the first step. The greatest value yields the maximum of f subject to the constraint $g(x, y) = 0$, and the least value yields the minimum of f subject to the constraint $g(x, y) = 0$.

When using the Method of Lagrange Multipliers for functions of three variables, F has the form

$$F(x, y, z, \lambda) = f(x, y, z) - \lambda g(x, y, z).$$

The system of equations used in Step 1 is

$$F_x(x, y, z, \lambda) = 0 \quad F_y(x, y, z, \lambda) = 0 \quad F_z(x, y, z, \lambda) = 0 \quad F_\lambda(x, y, z, \lambda) = 0.$$

The Method of Lagrange Multipliers gives you a way of finding critical points but does not tell you whether these points yield minima, maxima, or neither. To make this distinction, you must rely on the context of the problem.

Constrained Optimization Problems

Example 1   Using Lagrange Multipliers

Find the maximum of

$$V = xyz \qquad \text{Objective function}$$

subject to the constraint

$$6x + 4y + 3z - 24 = 0. \qquad \text{Constraint}$$

SOLUTION   First, let $f(x, y, z) = xyz$ and $g(x, y, z) = 6x + 4y + 3z - 24$. Then, define a new function F as

$$F(x, y, z, \lambda) = f(x, y, z) - \lambda g(x, y, z)$$
$$= xyz - \lambda(6x + 4y + 3z - 24).$$

To find the critical numbers of F, begin by finding the partial derivatives of F with respect to x, y, z, and λ. Then, set the partial derivatives equal to zero.

$$F_x(x, y, z, \lambda) = yz - 6\lambda \qquad\Longrightarrow\qquad yz - 6\lambda = 0$$
$$F_y(x, y, z, \lambda) = xz - 4\lambda \qquad\Longrightarrow\qquad xz - 4\lambda = 0$$
$$F_z(x, y, z, \lambda) = xy - 3\lambda \qquad\Longrightarrow\qquad xy - 3\lambda = 0$$
$$F_\lambda(x, y, z, \lambda) = -6x - 4y - 3z + 24 \qquad\Longrightarrow\qquad -6x - 4y - 3z + 24 = 0$$

Solving for λ in the first equation produces

$$yz - 6\lambda = 0 \qquad\Longrightarrow\qquad \lambda = \frac{yz}{6}.$$

Substituting for λ in the second and third equations produces the following.

$$xz - 4\left(\frac{yz}{6}\right) = 0 \qquad\Longrightarrow\qquad y = \frac{3}{2}x$$

$$xy - 3\left(\frac{yz}{6}\right) = 0 \qquad\Longrightarrow\qquad z = 2x$$

Next, substitute for y and z in the equation $F_\lambda(x, y, z, \lambda) = 0$ and solve for x.

$$F_\lambda(x, y, z, \lambda) = 0$$
$$-6x - 4y - 3z + 24 = 0$$
$$-6x - 4\left(\tfrac{3}{2}x\right) - 3(2x) + 24 = 0$$
$$-18x = -24$$
$$x = \tfrac{4}{3}$$

Using this x-value, you can conclude that the critical values are $x = \frac{4}{3}$, $y = 2$, and $z = \frac{8}{3}$, which implies that the maximum is

$$V = xyz \qquad\qquad \text{Write objective function.}$$

$$= \left(\frac{4}{3}\right)(2)\left(\frac{8}{3}\right) \qquad\qquad \text{Substitute values of } x, y, \text{ and } z.$$

$$= \frac{64}{9} \text{ cubic units.} \qquad\qquad \text{Maximum volume}$$

✓ **Checkpoint 1**

Find the maximum volume of $V = xyz$ subject to the constraint

$$2x + 4y + z - 8 = 0.$$

STUDY TIP

Example 1 shows how Lagrange multipliers can be used to solve the same problem that was solved in Example 5 in Section 13.5.

ALGEBRA TUTOR *xy*

The most difficult aspect of many Lagrange multiplier problems is the complicated algebra needed to solve the system of equations arising from

$$F(x, y, \lambda) = f(x, y) - \lambda g(x, y).$$

There is no general way to proceed in every case, so you should study the examples carefully, and refer to the *Chapter 13 Algebra Tutor* on pages 988 and 989.

Example 2 Finding a Maximum Production Level

A manufacturer's production is modeled by the Cobb-Douglas function

$$f(x, y) = 100x^{3/4}y^{1/4} \qquad \text{Objective function}$$

where x represents the units of labor (at \$150 per unit) and y represents the units of capital (at \$250 per unit). The total costs for labor and capital cannot exceed \$50,000. Will the manufacturer's maximum production level exceed 16,000 units?

SOLUTION Because total labor and capital expenses cannot exceed \$50,000, the constraint is

$$150x + 250y = 50,000 \qquad \text{Constraint}$$
$$150x + 250y - 50,000 = 0. \qquad \text{Write in general form.}$$

To find the maximum production level, begin by writing the function

$$F(x, y, \lambda) = 100x^{3/4}y^{1/4} - \lambda(150x + 250y - 50,000).$$

Then find the partial derivatives of F with respect to x, y, and λ.

$$F_x(x, y, \lambda) = 75x^{-1/4}y^{1/4} - 150\lambda$$
$$F_y(x, y, \lambda) = 25x^{3/4}y^{-3/4} - 250\lambda$$
$$F_\lambda(x, y, \lambda) = -150x - 250y + 50,000$$

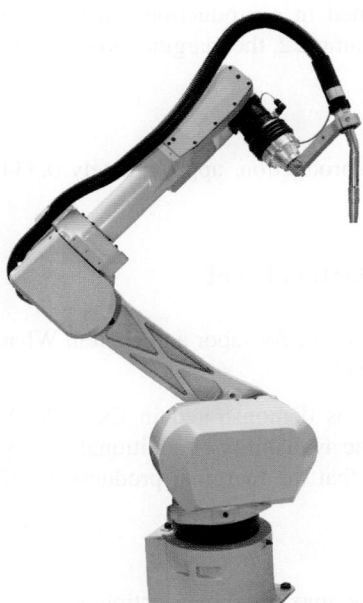

For some industrial applications, a simple robot can cost more than a year's wages and benefits for one employee. So, manufacturers must carefully balance the amount of money spent on labor and capital.

Next, set the partial derivatives equal to zero to obtain the following system of equations.

$$75x^{-1/4}y^{1/4} - 150\lambda = 0 \qquad \text{Equation 1}$$
$$25x^{3/4}y^{-3/4} - 250\lambda = 0 \qquad \text{Equation 2}$$
$$-150x - 250y + 50,000 = 0 \qquad \text{Equation 3}$$

By solving for λ in the first equation

$$75x^{-1/4}y^{1/4} - 150\lambda = 0 \qquad \text{Equation 1}$$
$$\lambda = \tfrac{1}{2}x^{-1/4}y^{1/4} \qquad \text{Solve for } \lambda.$$

and substituting for λ in Equation 2, you obtain

$$25x^{3/4}y^{-3/4} - 250\left(\tfrac{1}{2}\right)x^{-1/4}y^{1/4} = 0 \qquad \text{Substitute in Equation 2.}$$
$$25x - 125y = 0 \qquad \text{Multiply by } x^{1/4}y^{3/4}.$$
$$x = 5y. \qquad \text{Solve for } x.$$

So, $x = 5y$. By substituting for x in Equation 3, you obtain

$$-150(5y) - 250y + 50,000 = 0 \qquad \text{Substitute in Equation 3.}$$
$$-1000y = -50,000 \qquad \text{Simplify.}$$
$$y = 50 \qquad \text{Solve for } y.$$

When $y = 50$ units of capital, it follows that $x = 5(50) = 250$ units of labor. So, the maximum production level is

$$f(250, 50) = 100(250)^{3/4}(50)^{1/4} \qquad \text{Substitute for } x \text{ and } y.$$
$$\approx 16,719 \text{ units.} \qquad \text{Maximum production level}$$

You can conclude that the maximum production level will exceed 16,000 units.

TECH TUTOR

You can use a spreadsheet to solve constrained optimization problems. Try using a spreadsheet to solve the problem in Example 2. (Consult the user's manual of a spreadsheet software program for specific instructions on how to solve a constrained optimization problem.)

✓ **Checkpoint 2**

In Example 2, suppose that each labor unit costs \$200 and each capital unit costs \$250. Find the maximum production level when labor and capital cannot exceed \$50,000.

Economists call the Lagrange multiplier obtained in a production function the **marginal productivity of money.** For instance, in Example 2, the marginal productivity of money when $x = 250$ and $y = 50$ is

$$\lambda = \tfrac{1}{2}x^{-1/4}y^{1/4} = \tfrac{1}{2}(250)^{-1/4}(50)^{1/4} \approx 0.334.$$

This means that for each additional dollar spent on production, approximately 0.334 additional unit of the product can be produced.

 Example 3   **Finding a Maximum Production Level**

The manufacturer in Example 2 now has $70,000 available for labor and capital. What is the maximum number of units that can be produced?

SOLUTION   You could rework the entire problem, as demonstrated in Example 2. However, because the only change in the problem is the availability of additional money to spend on labor and capital, you can use the fact that the marginal productivity of money is

$$\lambda \approx 0.334.$$

Because an additional $20,000 is available and the maximum production level in Example 2 was 16,719 units, you can conclude that the maximum production level is now

$$16{,}719 + (0.334)(20{,}000) \approx 23{,}400 \text{ units.}$$

Try using the procedure demonstrated in Example 2 to confirm this result.

✓**Checkpoint 3**

The manufacturer in Example 2 now has $80,000 available for labor and capital. What is the maximum number of units that can be produced?

TECH TUTOR

You can use a three-dimensional graphing utility to confirm graphically the results of Examples 2 and 3. Begin by graphing the surface $f(x, y) = 100x^{3/4}y^{1/4}$. Then graph the vertical plane given by $150x + 250y = 50{,}000$. As shown below, the maximum production level corresponds to the highest point on the intersection of the surface and the plane.

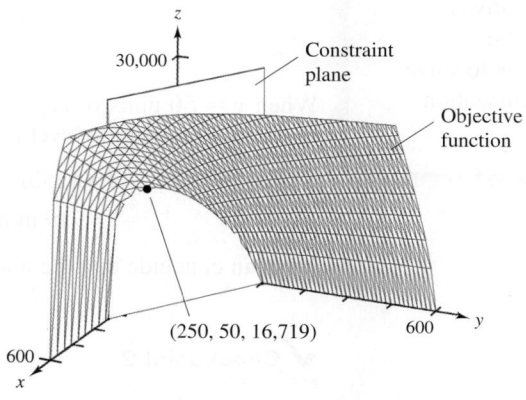

STUDY TIP

The constrained optimization problem in Example 4 on this page is represented graphically in Figure 13.37. The graph of the objective function is a paraboloid and the graph of the constraint is a vertical plane. In the "unconstrained" optimization problem on page 951, the maximum profit occurred at the vertex of the paraboloid. In this "constrained" problem, however, the maximum profit corresponds to the highest point on the curve that is the intersection of the paraboloid and the vertical "constraint" plane.

In Example 4 in Section 13.5, you found the maximum profit for two substitute products whose demand functions are given by

$$x_1 = 200(p_2 - p_1) \qquad \text{Demand for product 1}$$
$$x_2 = 500 + 100p_1 - 180p_2. \qquad \text{Demand for product 2}$$

With this model, the total demand, $x_1 + x_2$, is completely determined by the prices p_1 and p_2. In many real-life situations, this assumption is too simplistic; regardless of the prices of the substitute brands, the annual total demands for some products, such as toothpaste, are relatively constant. In such situations, the total demand is **limited,** and variations in price do not affect the total demand as much as they affect the market share of the substitute brands.

Example 4   Finding a Maximum Profit

A company makes two substitute products whose demand functions are given by

$$x_1 = 200(p_2 - p_1) \qquad \text{Demand for product 1}$$
$$x_2 = 500 + 100p_1 - 180p_2 \qquad \text{Demand for product 2}$$

where p_1 and p_2 are the prices per unit (in dollars) and x_1 and x_2 are the numbers of units sold. The costs of producing the two products are \$0.50 and \$0.75 per unit, respectively. The total demand is limited to 200 units per year. Find the prices that will yield a maximum profit.

SOLUTION   From Example 4 in Section 13.5, the profit function is modeled by

$$P = -200p_1^2 - 180p_2^2 + 300p_1p_2 + 25p_1 + 535p_2 - 375.$$

The total demand for the two products is

$$x_1 + x_2 = 200(p_2 - p_1) + 500 + 100p_1 - 180p_2$$
$$= 200p_2 - 200p_1 + 500 + 100p_1 - 180p_2$$
$$= -100p_1 + 20p_2 + 500.$$

Because the total demand is limited to 200 units, the constraint is

$$-100p_1 + 20p_2 + 500 = 200. \qquad \text{Constraint}$$

Using Lagrange multipliers, you can determine that the maximum profit occurs when $p_1 \approx \$3.94$ and $p_2 \approx \$4.69$. This corresponds to an annual profit of about \$712.21.

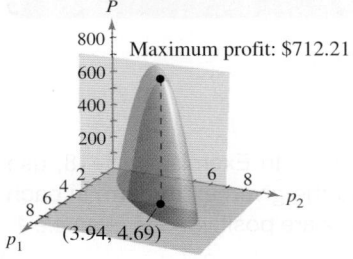

Maximum profit: \$712.21

(3.94, 4.69)

FIGURE 13.37

✓ Checkpoint 4

In Example 4, find the prices that will yield a maximum profit when the total demand is limited to 250 units per year.

SUMMARIZE   (Section 13.6)

1. Explain the Method of Lagrange Multipliers *(page 957)*. For an example of how to use Lagrange multipliers, see Example 1.

2. Describe a real-life example of using Lagrange multipliers to find a manufacturer's maximum production level *(page 959, Example 2)*.

3. Describe a real-life example of using Lagrange multipliers to find a company's maximum profit *(page 961, Example 4)*.

SKILLS WARM UP 13.6   The following warm-up exercises involve skills that were covered in earlier sections. You will use these skills in the exercise set for this section. For additional help, review Sections 5.1, 5.2, 5.3, and 13.4.

In Exercises 1–6, solve the system of linear equations.

1. $\begin{cases} 4x - 6y = 3 \\ 2x + 3y = 2 \end{cases}$

2. $\begin{cases} 6x - 6y = 5 \\ -3x - y = 1 \end{cases}$

3. $\begin{cases} 5x - y = 25 \\ x - 5y = 15 \end{cases}$

4. $\begin{cases} 4x - 9y = 5 \\ -x + 8y = -2 \end{cases}$

5. $\begin{cases} 2x - y + z = 3 \\ 2x + 2y + z = 4 \\ -x + 2y + 3z = -1 \end{cases}$

6. $\begin{cases} -x - 4y + 6z = -2 \\ x - 3y - 3z = 4 \\ 3x + y + 3z = 0 \end{cases}$

In Exercises 7–10, find all first partial derivatives.

7. $f(x, y) = x^2y + xy^2$

8. $f(x, y) = 25(xy + y^2)^2$

9. $f(x, y, z) = x(x^2 - 2xy + yz)$

10. $f(x, y, z) = z(xy + xz + yz)$

Exercises 13.6

See www.CalcChat.com for worked-out solutions to odd-numbered exercises.

Using Lagrange Multipliers   In Exercises 1–12, use Lagrange multipliers to find the given extremum. In each case, assume that x and y are positive. *See Example 1.*

1. Maximize $f(x, y) = xy$

   Constraint: $x + y = 10$

2. Maximize $f(x, y) = xy$

   Constraint: $x + 3y = 6$

3. Minimize $f(x, y) = x^2 + y^2$

   Constraint: $x + y - 8 = 0$

4. Minimize $f(x, y) = x^2 + y^2$

   Constraint: $-2x - 4y + 5 = 0$

5. Maximize $f(x, y) = x^2 - y^2$

   Constraint: $2y - x^2 = 0$

6. Minimize $f(x, y) = x^2 - y^2$

   Constraint: $x - 2y + 6 = 0$

7. Maximize $f(x, y) = 2x + 2xy + y$

   Constraint: $2x + y = 100$

8. Minimize $f(x, y) = 3x + y + 10$

   Constraint: $x^2y = 6$

9. Maximize $f(x, y) = \sqrt{6 - x^2 - y^2}$

   Constraint: $x + y - 2 = 0$

10. Minimize $f(x, y) = \sqrt{x^2 + y^2}$

    Constraint: $2x + 4y - 15 = 0$

11. Maximize $f(x, y) = e^{xy}$

    Constraint: $x^2 + y^2 - 8 = 0$

12. Minimize $f(x, y) = 2x + y$

    Constraint: $xy = 32$

Using Lagrange Multipliers   In Exercises 13–18, use Lagrange multipliers to find the given extremum. In each case, assume that x, y, and z are positive. *See Example 1.*

13. Minimize $f(x, y, z) = 2x^2 + 3y^2 + 2z^2$

    Constraint: $x + y + z - 24 = 0$

14. Maximize $f(x, y, z) = xyz$

    Constraint: $x + y + z - 6 = 0$

15. Minimize $f(x, y, z) = x^2 + y^2 + z^2$

    Constraint: $x + y + z = 1$

16. Minimize $f(x, y) = x^2 - 8x + y^2 - 12y + 48$

    Constraint: $x + y = 8$

17. Maximize $f(x, y, z) = x + y + z$

    Constraint: $x^2 + y^2 + z^2 = 1$

18. Maximize $f(x, y, z) = x^2y^2z^2$

    Constraint: $x^2 + y^2 + z^2 = 1$

Finding Positive Numbers   In Exercises 19–22, find three positive numbers x, y, and z that satisfy the given conditions.

19. The sum is 60 and the product is a maximum.

20. The sum is 80 and $P = x^2yz$ is a maximum.

21. The sum is 120 and the sum of the squares is a minimum.

22. The sum is 36 and the sum of the cubes is a minimum.

Finding Distance In Exercises 23–26, find the minimum distance from the curve or surface to the given point. (*Hint:* Start by minimizing the square of the distance.)

23. Line: $x + y = 6$, $(0, 0)$

    Minimize $d^2 = x^2 + y^2$

24. Circle: $(x - 4)^2 + y^2 = 4$, $(0, 10)$

    Minimize $d^2 = x^2 + (y - 10)^2$

25. Plane: $x + y + z = 1$, $(2, 1, 1)$

    Minimize $d^2 = (x - 2)^2 + (y - 1)^2 + (z - 1)^2$

26. Cone: $z = \sqrt{x^2 + y^2}$, $(4, 0, 0)$

    Minimize $d^2 = (x - 4)^2 + y^2 + z^2$

27. **Volume** A rectangular box is resting on the xy-plane with one vertex at the origin. The opposite vertex lies in the plane $2x + 3y + 5z = 90$. Find the dimensions that maximize the volume. (*Hint:* Maximize $V = xyz$ subject to the constraint $2x + 3y + 5z - 90 = 0$.)

28. **Volume** Find the dimensions of the rectangular package of largest volume subject to the constraint that the sum of the length and the girth cannot exceed 108 inches (see figure). (*Hint:* Maximize $V = xyz$ subject to the constraint $x + 2y + 2z = 108$.)

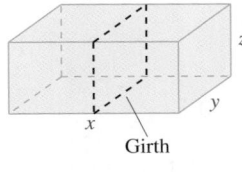

Girth

Figure for 28

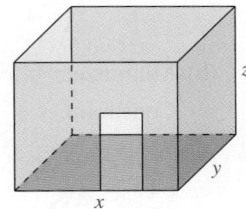

Figure for 29

29. **Cost** In redecorating an office, the cost for new carpeting is $3 per square foot and the cost of wallpapering a wall is $1 per square foot. Find the dimensions of the largest office that can be redecorated for $1296 (see figure). (*Hint:* Maximize $V = xyz$ subject to $3xy + 2xz + 2yz = 1296$.)

30. **Cost** A cargo container (in the shape of a rectangular solid) must have a volume of 480 cubic feet. Find the dimensions of the container that has a minimum cost, if the bottom will cost $5 per square foot to construct and the sides and top will cost $3 per square foot to construct.

31. **Cost** A manufacturer has an order for 1000 units of fine paper that can be produced at two locations. Let x_1 and x_2 be the numbers of units produced at the two locations. The cost function is modeled by

    $$C = 0.25x_1^2 + 25x_1 + 0.05x_2^2 + 12x_2.$$

    Find the number of units that should be produced at each location to minimize the cost.

32. **Cost** A manufacturer has an order for 2000 units of all-terrain vehicle tires that can be produced at two locations. Let x_1 and x_2 be the numbers of units produced at the two plants. The cost function is modeled by

    $$C = 0.25x_1^2 + 10x_1 + 0.15x_2^2 + 12x_2.$$

    Find the number of units that should be produced at each location to minimize the cost.

33. **Production** The production function for a company is given by

    $$f(x, y) = 100x^{0.25}y^{0.75}$$

    where x is the number of units of labor (at $48 per unit) and y is the number of units of capital (at $36 per unit). The total cost for labor and capital cannot exceed $100,000.

    (a) Find the maximum production level for this manufacturer.

    (b) Find the marginal productivity of money.

    (c) Use the marginal productivity of money to find the maximum number of units that can be produced when $125,000 is available for labor and capital.

    (d) Use the marginal productivity of money to find the maximum number of units that can be produced when $350,000 is available for labor and capital.

34. **Production** Repeat Exercise 33 for the production function given by

    $$f(x, y) = 100x^{0.6}y^{0.4}.$$

35. **Least-Cost Rule** The production function for a company is given by

    $$f(x, y) = 100x^{0.7}y^{0.3}$$

    where x is the number of units of labor (at $50 per unit) and y is the number of units of capital (at $100 per unit). Management sets a production goal of 20,000 units.

    (a) Find the numbers of units of labor and capital needed to meet the production goal while minimizing the cost.

    (b) Show that the conditions of part (a) are met when

    $$\frac{\text{Marginal productivity of labor}}{\text{Marginal productivity of capital}} = \frac{\text{Unit price of labor}}{\text{Unit price of capital}}.$$

    This proportion is called the *Least-Cost Rule* (or *Equimarginal Rule*).

36. **Least-Cost Rule** Repeat Exercise 35 for the production function given by

    $$f(x, y) = 100x^{0.4}y^{0.6}.$$

37. Construction A rancher plans to use an existing stone wall and the side of a barn as a boundary for two adjacent rectangular corrals (see figure). Fencing for the perimeter costs $10 per foot. To separate the corrals, a fence that costs $4 per foot will divide the region. The total area of the two corrals is to be 6000 square feet.

(a) Find the dimensions that will minimize the cost of the fencing.

(b) What is the minimum cost?

38. Office Space Partitions will be used in an office to form four equal work areas with a total area of 360 square feet (see figure). The partitions that are x feet long cost $100 per foot and the partitions that are y feet long cost $120 per foot.

(a) Find the dimensions x and y that will minimize the cost of the partitions.

(b) What is the minimum cost?

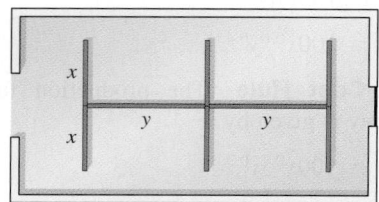

39. Biology A microbiologist must prepare a culture medium in which to grow a certain type of bacteria. The proportion of salt in this medium is given by

$$S = 12xyz$$

where x, y, and z are the amounts (in liters) of the three nutrient solutions to be mixed in the medium. For the bacteria to grow, the medium must be 13% salt. Nutrient solutions x, y, and z cost $1, $2, and $3 per liter, respectively. How much of each nutrient solution should be used to minimize the cost of the culture medium?

40. Biology Repeat Exercise 39 for a salt-content model given by

$$S = 0.01x^2y^2z^2.$$

41. Nutrition The number of grams of your favorite ice cream can be modeled by

$$G(x, y, z) = 0.05x^2 + 0.16xy + 0.25z^2$$

where x is the number of fat grams, y is the number of carbohydrate grams, and z is the number of protein grams. Find the maximum number of grams of ice cream you can eat without consuming more than 400 calories. Assume that there are 9 calories per fat gram, 4 calories per carbohydrate gram, and 4 calories per protein gram.

42. HOW DO YOU SEE IT? The graphs show the constraint and several level curves of the objective function. Use the graph to approximate the indicated extrema.

(a) Maximize $z = xy$

Constraint: $2x + y = 4$

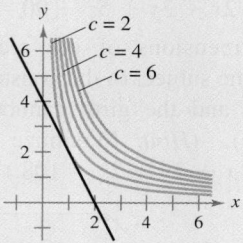

(b) Minimize $z = x^2 + y^2$

Constraint: $x + y - 4 = 0$

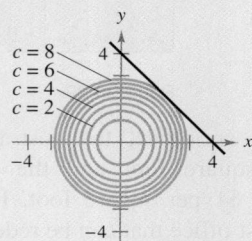

43. Advertising A private golf club is determining how to spend its $2700 advertising budget. The club knows from prior experience that the number of responses A is given by

$$A = 0.0001t^2pr^{1.5}$$

where t is the number of cable television ads, p is the number of newspaper ads, and r is the number of radio ads. A cable television ad costs $30, a newspaper ad costs $12, and a radio ad costs $15.

(a) How much should be spent on each type of advertising to obtain the maximum number of responses? (Assume the golf club uses each type of advertising.)

(b) What is the maximum number of responses expected?

13.7 Least Squares Regression Analysis

In Exercise 14 on page 970, you will find the least squares regression line that models the demand of a tool at a hardware store in terms of the price.

■ Find the sum of the squared errors for mathematical models.
■ Find the least squares regression lines for data.

Measuring the Accuracy of a Mathematical Model

When seeking a mathematical model to fit data, the goals are simplicity and accuracy. For instance, a simple linear model for the points shown in Figure 13.38(a) is

$$f(x) = 1.9x - 5. \qquad \text{Linear model}$$

Figure 13.38(b), however, shows that by choosing the slightly more complicated quadratic model

$$g(x) = 0.20x^2 - 0.7x + 1 \qquad \text{Quadratic model}$$

you can obtain significantly greater accuracy.*

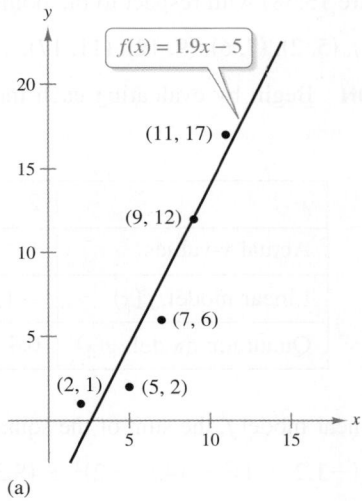

(a)

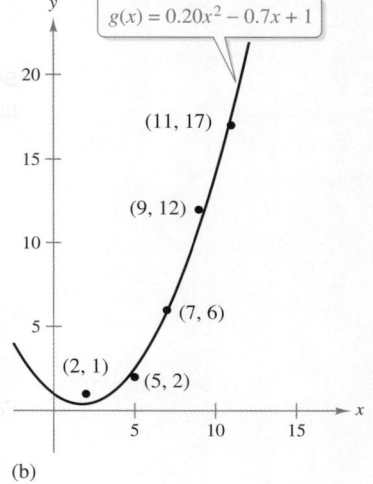

(b)

FIGURE 13.38

To measure how well the model

$$y = f(x)$$

fits a collection of points, sum the squares of the differences between the actual y-values and the model's y-values. This sum is called the **sum of the squared errors** and is denoted by S. Graphically, S can be interpreted as the sum of the squares of the vertical distances between the graph of f and the given points in the plane, as shown in Figure 13.39. If the model is a perfect fit, then

$$S = 0.$$

However, when a perfect fit is not feasible, you should use a model that minimizes S.

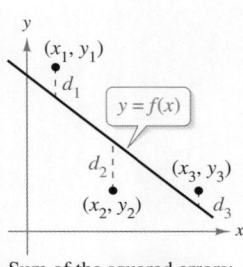

Sum of the squared errors:
$S = d_1^2 + d_2^2 + d_3^2$

FIGURE 13.39

---

* An analytic method for finding a quadratic model for a collection of data is not given in this text. You can perform this task using a graphing utility or a spreadsheet software program that has a built-in program for finding the least squares regression quadratic.

Definition of the Sum of the Squared Errors

The **sum of the squared errors** for the model $y = f(x)$ with respect to the points

$$(x_1, y_1), (x_2, y_2), \ldots, (x_n, y_n)$$

is given by

$$S = [f(x_1) - y_1]^2 + [f(x_2) - y_2]^2 + \cdots + [f(x_n) - y_n]^2.$$

Example 1   Finding the Sum of the Squared Errors

Find the sum of the squared errors for the linear model

$$f(x) = 1.9x - 5 \qquad \text{Linear model}$$

and the quadratic model

$$g(x) = 0.20x^2 - 0.7x + 1 \qquad \text{Quadratic model}$$

(see Figure 13.38) with respect to the points

$$(2, 1), (5, 2), (7, 6), (9, 12), (11, 17).$$

SOLUTION   Begin by evaluating each model at the given x-values, as shown in the table.

x	2	5	7	9	11
Actual y-values	1	2	6	12	17
Linear model, $f(x)$	-1.2	4.5	8.3	12.1	15.9
Quadratic model, $g(x)$	0.4	2.5	5.9	10.9	17.5

For the linear model f, the sum of the squared errors is

$$S = (-1.2 - 1)^2 + (4.5 - 2)^2 + (8.3 - 6)^2 + (12.1 - 12)^2 + (15.9 - 17)^2$$
$$= 17.6.$$

Similarly, the sum of the squared errors for the quadratic model g is

$$S = (0.4 - 1)^2 + (2.5 - 2)^2 + (5.9 - 6)^2 + (10.9 - 12)^2 + (17.5 - 17)^2$$
$$= 2.08.$$

✓ Checkpoint 1

Find the sum of the squared errors for the linear model

$$f(x) = 2.9x - 6 \qquad \text{Linear model}$$

and the quadratic model

$$g(x) = 0.20x^2 + 0.5x - 1 \qquad \text{Quadratic model}$$

with respect to the points

$$(2, 1), (4, 5), (6, 9), (8, 16), (10, 24).$$

Then decide which model is a better fit.   ■

In Example 1, note that the sum of the squared errors for the quadratic model is less than the sum of the squared errors for the linear model, which confirms that the quadratic model is a better fit.

Least Squares Regression Line

The sum of the squared errors can be used to determine which of several models is the best fit for a collection of data. In general, if the sum of the squared errors of f is less than the sum of the squared errors of g, then f is said to be a better fit for the data than g. In regression analysis, you consider all possible models of a certain type. The one that is defined to be the best-fitting model is the one with the least sum of the squared errors. Example 2 shows how to use the optimization techniques described in Section 13.5 to find the best-fitting linear model for a collection of data.

Example 2   Finding the Best Linear Model

Find the values of a and b such that the linear model

$$f(x) = ax + b$$

has a minimum sum of the squared errors for the points

$$(-3, 0), (-1, 1), (0, 2), (2, 3).$$

SOLUTION   The sum of the squared errors is

$$S = [f(x_1) - y_1]^2 + [f(x_2) - y_2]^2 + [f(x_3) - y_3]^2 + [f(x_4) - y_4]^2$$
$$= (-3a + b - 0)^2 + (-a + b - 1)^2 + (b - 2)^2 + (2a + b - 3)^2$$
$$= 14a^2 - 4ab + 4b^2 - 10a - 12b + 14.$$

To find the values of a and b for which S is a minimum, you can use the techniques described in Section 13.5. That is, find the partial derivatives of S.

$$\frac{\partial S}{\partial a} = 28a - 4b - 10 \qquad \text{Differentiate with respect to } a.$$

$$\frac{\partial S}{\partial b} = -4a + 8b - 12 \qquad \text{Differentiate with respect to } b.$$

Next, set each partial derivative equal to zero.

$$28a - 4b - 10 = 0 \qquad \text{Set } \partial S/\partial a \text{ equal to 0.}$$
$$-4a + 8b - 12 = 0 \qquad \text{Set } \partial S/\partial b \text{ equal to 0.}$$

The solution of this system of linear equations is

$$a = \frac{8}{13} \quad \text{and} \quad b = \frac{47}{26}.$$

So, the best-fitting linear model for the given points is

$$f(x) = \frac{8}{13}x + \frac{47}{26}.$$

The graph of this model is shown in Figure 13.40.

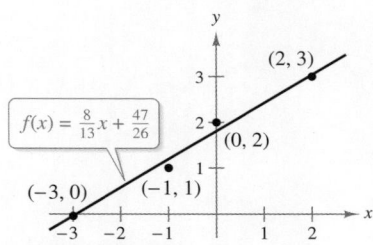

$$f(x) = \frac{8}{13}x + \frac{47}{26}$$

FIGURE 13.40

ALGEBRA TUTOR

xy

For help in solving the system of equations in Example 2, see Example 2(b) in the *Chapter 13 Algebra Tutor*, on page 989.

✓ Checkpoint 2

Find the values of a and b such that the linear model

$$f(x) = ax + b$$

has a minimum sum of the squared errors for the points

$$(-2, 0), (0, 2), (2, 5), (4, 7).$$

The line in Example 2 is called the **least squares regression line** for the given data. The solution shown in Example 2 can be generalized to find a formula for the least squares regression line. Consider the linear model

$$f(x) = ax + b$$

and the points

$$(x_1, y_1), (x_2, y_2), \ldots, (x_n, y_n).$$

The sum of the squared errors is

$$S = [f(x_1) - y_1]^2 + [f(x_2) - y_2]^2 + \cdots + [f(x_n) - y_n]^2$$
$$= (ax_1 + b - y_1)^2 + (ax_2 + b - y_2)^2 + \cdots + (ax_n + b - y_n)^2.$$

To minimize S, set the partial derivatives $\partial S/\partial a$ and $\partial S/\partial b$ equal to zero and solve for a and b. The results are summarized below.

The Least Squares Regression Line

The **least squares regression line** for the points

$$(x_1, y_1), (x_2, y_2), \ldots, (x_n, y_n)$$

is $f(x) = ax + b$, where

$$a = \frac{n \sum_{i=1}^{n} x_i y_i - \sum_{i=1}^{n} x_i \sum_{i=1}^{n} y_i}{n \sum_{i=1}^{n} x_i^2 - \left(\sum_{i=1}^{n} x_i\right)^2} \quad \text{and} \quad b = \frac{1}{n}\left(\sum_{i=1}^{n} y_i - a \sum_{i=1}^{n} x_i\right).$$

The summation notation

$$\sum_{i=1}^{n} x_i$$

where Σ is the Greek letter sigma, is used to indicate the sum of the numbers

$$x_1 + x_2 + \cdots + x_n.$$

Similarly,

$$\sum_{i=1}^{n} x_i y_i = x_1 y_1 + x_2 y_2 + \ldots + x_n y_n, \quad \sum_{i=1}^{n} x_i^2 = x_1^2 + x_2^2 + \ldots + x_n^2,$$

and so on.

In the formula for the least squares regression line, note that if the x-values are symmetrically spaced about zero, then

$$\sum_{i=1}^{n} x_i = 0$$

and the formulas for a and b simplify to

$$a = \frac{n \sum_{i=1}^{n} x_i y_i}{n \sum_{i=1}^{n} x_i^2} \quad \text{and} \quad b = \frac{1}{n} \sum_{i=1}^{n} y_i.$$

Note also that only the *development* of the least squares regression line involves partial derivatives. The *application* of this formula is a matter of computing the values of a and b. This task is performed much more simply on a calculator or a computer than by hand.

 **Example 3** **Modeling Hourly Wages**

The average hourly wages y (in dollars per hour) for production workers in manufacturing industries from 2001 through 2009 are shown in the table. Find the least squares regression line for the data and use the result to estimate the average hourly wage in 2013. *(Source: U.S. Bureau of Labor Statistics)*

Year	2001	2002	2003	2004	2005	2006	2007	2008	2009
y	14.76	15.29	15.74	16.14	16.56	16.81	17.26	17.75	18.23

SOLUTION   Let t represent the year, with $t = 1$ corresponding to 2001. Then, you need to find the linear model that best fits the points

$(1, 14.76)$, $(2, 15.29)$, $(3, 15.74)$, $(4, 16.14)$, $(5, 16.56)$, $(6, 16.81)$, $(7, 17.26)$, $(8, 17.75)$, $(9, 18.23)$.

Using a calculator with a built-in least squares regression program, you can determine that the best-fitting line is

$y = 0.416t + 14.42.$ Best-fitting line

With this model, you can estimate the 2013 average hourly wage, using $t = 13$, to be

$y = 0.416(13) + 14.42$

$= 19.828$

$\approx \$19.83$ per hour.

This result is shown graphically in Figure 13.41.

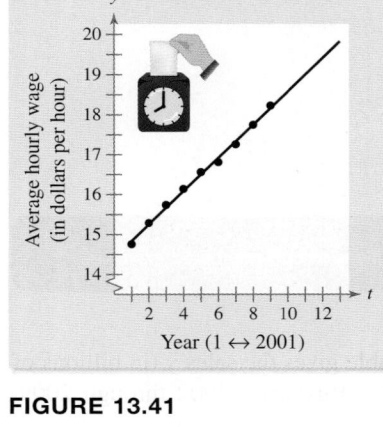

Modeling Hourly Wages

FIGURE 13.41

✓ **Checkpoint 3**

The numbers of cellular phone subscribers y (in thousands) for the years 2000 through 2009 are shown in the table. Find the least squares regression line for the data and use the result to estimate the number of subscribers in 2013. Let t represent the year, with $t = 0$ corresponding to 2000. *(Source: CTIA-The Wireless Association)*

Year	2000	2001	2002	2003	2004
y	190,478	128,375	140,767	158,722	182,140

Year	2005	2006	2007	2008	2009
y	207,896	233,041	255,396	270,334	285,646

SUMMARIZE   (Section 13.7)

1. State the definition of the sum of the squared errors *(page 966)*. For an example of finding the sum of the squared errors, See Example 1.

2. State the definition of the least squares regression line *(page 968)*. For an example of finding the least squares regression line, see Example 2.

3. Describe a real-life example of modeling hourly wages using the least squares regression line *(page 969, Example 3)*.

The following warm-up exercises involve skills that were covered in a previous course or in earlier sections. You will use these skills in the exercise set for this section. For additional help, review Sections 0.2 and 13.4

In Exercises 1 and 2, evaluate the expression.

1. $(2.5 - 1)^2 + (3.25 - 2)^2 + (4.1 - 3)^2$

2. $(1.1 - 1)^2 + (2.08 - 2)^2 + (2.95 - 3)^2$

In Exercises 3 and 4, find the partial derivatives of S.

3. $S = a^2 + 6b^2 - 4a - 8b - 4ab + 6$

4. $S = 4a^2 + 9b^2 - 6a - 4b - 2ab + 8$

In Exercises 5–10, evaluate the sum.

5. $\displaystyle\sum_{i=1}^{5} i$

6. $\displaystyle\sum_{i=1}^{6} 2i$

7. $\displaystyle\sum_{i=1}^{4} \frac{1}{i}$

8. $\displaystyle\sum_{i=1}^{3} i^2$

9. $\displaystyle\sum_{i=1}^{6} (2 - i)^2$

10. $\displaystyle\sum_{i=1}^{5} (30 - i^2)$

Exercises 13.7

See www.CalcChat.com for worked-out solutions to odd-numbered exercises.

Finding the Sum of the Squared Errors In Exercises 1–4, find the sum of the squared errors for the linear model $f(x)$ and the quadratic model $g(x)$ using the given points. *See Example 1.*

1. $f(x) = 1.6x + 6$, $g(x) = 0.29x^2 + 2.2x + 6$
   $(-3, 2), (-2, 2), (-1, 4), (0, 6), (1, 8)$

2. $f(x) = -0.7x + 2$, $g(x) = 0.06x^2 - 0.7x + 1$
   $(-3, 4), (-1, 2), (1, 1), (3, 0)$

3. $f(x) = -3.3x + 11$, $g(x) = -1.25x^2 + 0.5x + 10$
   $(0, 10), (1, 9), (2, 6), (3, 0)$

4. $f(x) = 2.0x - 3$, $g(x) = 0.14x^2 + 1.3x - 3$
   $(-1, -4), (1, -3), (2, 0), (4, 5), (6, 9)$

Finding the Least Squares Regression Line In Exercises 5–8, find the least squares regression line for the given points. Then plot the points and sketch the regression line. *See Example 2.*

5. $(-2, -1), (0, 0), (2, 3)$

6. $(-3, 0), (-1, 1), (1, 1), (3, 2)$

7. $(-2, 4), (-1, 1), (0, -1), (1, -3)$

8. $(-5, -3), (-4, -2), (-2, -1), (-1, 1)$

 Finding the Least Squares Regression Line In Exercises 9–12, use the regression capabilities of a graphing utility or a spreadsheet to find the least squares regression line for the given points.

9. $(-4, -1), (-2, 0), (2, 4), (4, 5)$

10. $(-5, 1), (1, 3), (2, 3), (2, 5)$

11. $(0, 6), (4, 3), (5, 0), (8, -4), (10, -5)$

12. $(-10, 10), (-5, 8), (3, 6), (7, 4), (5, 0)$

 13. Sales The table gives the sales y (in billions of dollars) for Best Buy from 2002 through 2009. *(Source: Best Buy Company, Inc.)*

Year	2002	2003	2004	2005
Sales, y	20.9	24.5	27.4	30.8

Year	2006	2007	2008	2009
Sales, y	35.9	40.0	45.0	49.7

(a) Use the regression capabilities of a graphing utility or a spreadsheet to find the least squares regression line for the data. Let $t = 2$ represent 2002.

(b) Estimate the sales in 2014.

(c) In what year will the sales be $85 billion?

14. Demand A hardware retailer wants to know the demand y for a tool as a function of price x. The monthly sales for four different prices of the tool are listed in the table.

Price, x	$25	$30	$35	$40
Demand, y	82	75	67	55

(a) Use the regression capabilities of a graphing utility or a spreadsheet to find the least squares regression line for the data.

(b) Estimate the demand when the price is $32.95.

(c) What price will create a demand of 83 tools?

15. Agriculture  An agronomist used four test plots to determine the relationship between the wheat yield y (in bushels per acre) and the amount of fertilizer x (in pounds per acre). The results are shown in the table.

Fertilizer, x	100	150	200	250
Yield, y	35	44	50	56

(a) Use the regression capabilities of a graphing utility or a spreadsheet to find the least squares regression line for the data.

(b) Estimate the yield for a fertilizer application of 160 pounds per acre.

16. **HOW DO YOU SEE IT?** Match the regression equation with the appropriate graph. Explain your reasoning. (Note that the x- and y-axes are broken.)

(a) $y = 0.22x - 7.5$ (b) $y = -0.35x + 11.5$

(c) $y = 0.09x + 19.8$ (d) $y = -1.29x + 89.8$

(i) 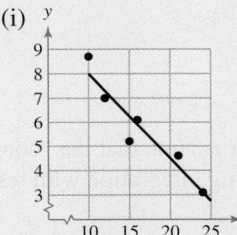   (ii)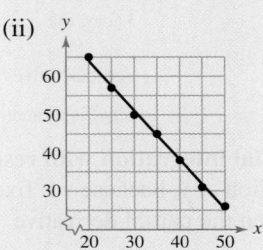

(iii) 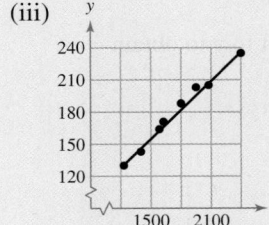   (iv)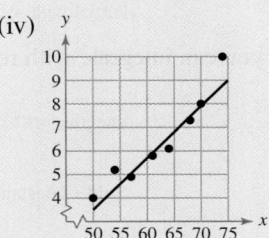

Determining Correlation In Exercises 17–22, plot the points and determine whether the data have positive, negative, or no linear correlation (see figures below). Then use a graphing utility to find the value of r and confirm your result. The number r is called the *correlation coefficient*. It is a measure of how well the model fits the data. Correlation coefficients vary between -1 and 1, and the closer $|r|$ is to 1, the better the model.

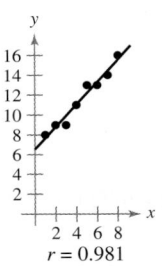
$r = 0.981$
Positive correlation

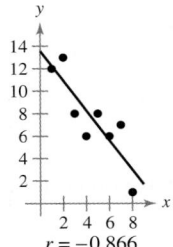
$r = -0.866$
Negative correlation

$r = 0.190$
No correlation

17. $(1, 4), (2, 6), (3, 8), (4, 11), (5, 13), (6, 15)$

18. $(1, 7.5), (2, 7), (3, 7), (4, 6), (5, 5), (6, 4.9)$

19. $(1, 3), (2, 6), (3, 2), (4, 3), (5, 9), (6, 1)$

20. $(0.5, 2), (0.75, 1.75), (1, 3), (1.5, 3.2), (2, 3.7), (2.6, 4)$

21. $(1, 36), (2, 10), (3, 0), (4, 4), (5, 16), (6, 36)$

22. $(0.5, 9), (1, 8.5), (1.5, 7), (2, 5.5), (2.5, 5), (3, 3.5)$

True or False? In Exercises 23–28, determine whether the statement is true or false. If it is false, explain why or give an example that shows it is false.

23. Data that are modeled by

$$y = 3.29x - 4.17$$

have a negative correlation.

24. Data that are modeled by

$$y = -0.238x + 25$$

have a negative correlation.

25. When the correlation coefficient is $r \approx -0.98781$, the model is a good fit.

26. A correlation coefficient of $r \approx 0.201$ implies that the data have no correlation.

27. A linear regression model with a positive correlation coefficient will have a slope that is greater than 0.

28. When the correlation coefficient for a linear regression model is close to -1, the regression line cannot be used to describe the data.

29. Project: Financial Data For a project analyzing the revenues per share, revenues, and shareholder's equity of McDonald's from 2000 through 2009, visit the text's website at *www.cengagebrain.com.* *(Source: McDonald's Corporation)*

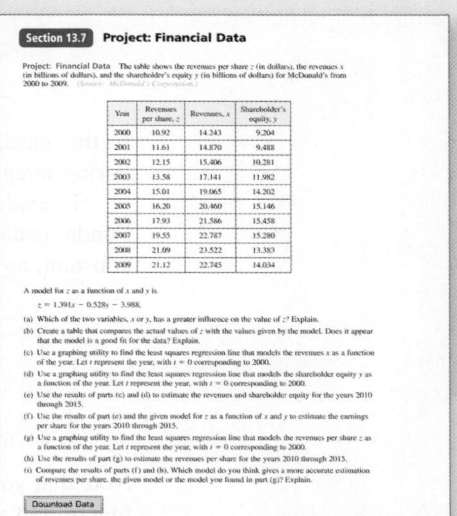

13.8 Double Integrals and Area in the Plane

■ Evaluate double integrals.
■ Use double integrals to find the areas of regions.

Double Integrals

In Section 13.4, you learned that it is meaningful to differentiate functions of several variables with respect to one variable while holding the other variable(s) constant. You can *integrate* functions of several variables by a similar procedure. For instance, if you are given the partial derivative

$$f_x(x, y) = 2xy \qquad \text{Partial with respect to } x$$

then, by holding y constant, you can integrate with respect to x to obtain

$$
\begin{aligned}
f(x, y) &= \int f_x(x, y)\, dx &&\text{Integrate with respect to } x. \\
&= \int 2xy\, dx &&\text{Hold } y \text{ constant.} \\
&= y \int 2x\, dx &&\text{Factor out constant } y. \\
&= y(x^2) + C(y) &&\text{Antiderivative of } 2x \text{ is } x^2. \\
&= x^2 y + C(y). &&C(y) \text{ is a function of } y.
\end{aligned}
$$

This procedure is called **partial integration with respect to x.** Note that the "constant of integration" $C(y)$ is a function of y, because y is fixed during integration with respect to x. Similarly, if you are given the partial derivative

$$f_y(x, y) = x^2 + 2 \qquad \text{Partial with respect to } y$$

then, by holding x constant, you can integrate with respect to y to obtain

$$
\begin{aligned}
f(x, y) &= \int f_y(x, y)\, dy &&\text{Integrate with respect to } y. \\
&= \int (x^2 + 2)\, dy &&\text{Hold } x \text{ constant.} \\
&= (x^2 + 2)\int dy &&\text{Factor out constant } x^2 + 2. \\
&= (x^2 + 2)(y) + C(x) &&\text{Antiderivative of 1 is } y. \\
&= x^2 y + 2y + C(x). &&C(x) \text{ is a function of } x.
\end{aligned}
$$

In this case, the "constant of integration" $C(x)$ is a function of x, because x is fixed during integration with respect to y.

To evaluate a definite integral of a function of several variables, you can apply the Fundamental Theorem of Calculus to one variable while holding the other variable(s) constant, as shown.

$$\int_1^{2y} 2xy\, dx = x^2 y \Big]_1^{2y} = (2y)^2 y - (1)^2 y = 4y^3 - y$$

x is the variable of integration and y is fixed. | Replace x by the limits of integration. | The result is a function of y.

Note that you omit the constant of integration, just as you do for a definite integral of a function of one variable.

In Exercise 29 on page 979, you will use a double integral to find the area of a region.

Example 1 Finding Partial Integrals

a. $\int_1^x (2x^2y^{-2} + 2y)\, dy = \left[\dfrac{-2x^2}{y} + y^2\right]_1^x$ Hold x constant.

$$= \left(\dfrac{-2x^2}{x} + x^2\right) - \left(\dfrac{-2x^2}{1} + 1\right)$$

$$= 3x^2 - 2x - 1$$

b. $\int_y^{5y} \sqrt{x - y}\, dx = \left[\dfrac{2}{3}(x - y)^{3/2}\right]_y^{5y}$ Hold y constant.

$$= \dfrac{2}{3}[(5y - y)^{3/2} - (y - y)^{3/2}]$$

$$= \dfrac{16}{3}y^{3/2}$$

✓ **Checkpoint 1**

Find each partial integral.

a. $\int_1^x (4xy + y^3)\, dy$ **b.** $\int_y^{y^2} \dfrac{1}{x + y}\, dx$ ■

In Example 1(a), note that the definite integral defines a function of x and can *itself* be integrated. An "integral of an integral" is called a **double integral.** With a function of two variables, there are two types of double integrals.

$$\int_a^b \int_{g_1(x)}^{g_2(x)} f(x, y)\, dy\, dx = \int_a^b \left[\int_{g_1(x)}^{g_2(x)} f(x, y)\, dy\right] dx$$

$$\int_a^b \int_{g_1(y)}^{g_2(y)} f(x, y)\, dx\, dy = \int_a^b \left[\int_{g_1(y)}^{g_2(y)} f(x, y)\, dx\right] dy$$

STUDY TIP

Notice that the difference between the two types of double integrals is the order in which the integration is performed, $dy\, dx$ or $dx\, dy$.

TECH TUTOR

A symbolic integration utility can be used to evaluate double integrals. To do this, you need to enter the integrand, then integrate twice—once with respect to one of the variables and then with respect to the other variable. Use a symbolic integration utility to evaluate the double integral in Example 2.

Example 2 Evaluating a Double Integral

$$\int_1^2 \int_0^x (2xy + 3)\, dy\, dx = \int_1^2 \left[\int_0^x (2xy + 3)\, dy\right] dx$$

$$= \int_1^2 \left[xy^2 + 3y\right]_0^x dx$$

$$= \int_1^2 (x^3 + 3x)\, dx$$

$$= \left[\dfrac{x^4}{4} + \dfrac{3x^2}{2}\right]_1^2$$

$$= \left(\dfrac{2^4}{4} + \dfrac{3(2^2)}{2}\right) - \left(\dfrac{1^4}{4} + \dfrac{3(1^2)}{2}\right)$$

$$= \dfrac{33}{4}$$

✓ **Checkpoint 2**

Evaluate $\int_1^2 \int_0^x (5x^2y - 2)\, dy\, dx.$ ■

Finding Area with a Double Integral

One of the simplest applications of a double integral is finding the area of a plane region. For instance, consider the region R that is bounded by

$$a \leq x \leq b \quad \text{and} \quad g_1(x) \leq y \leq g_2(x)$$

as shown in Figure 13.42. Using the techniques described in Section 11.5, you know that the area of R is

$$\int_a^b \left[g_2(x) - g_1(x)\right] dx. \qquad \text{Area of } R$$

This same area is also given by the double integral

$$\int_a^b \int_{g_1(x)}^{g_2(x)} dy \, dx \qquad \text{Area of } R$$

because

$$\int_a^b \int_{g_1(x)}^{g_2(x)} dy \, dx = \int_a^b \left[y \right]_{g_1(x)}^{g_2(x)} dx = \int_a^b \left[g_2(x) - g_1(x)\right] dx.$$

Figure 13.43 shows the two basic types of plane regions whose areas can be determined by a double integral.

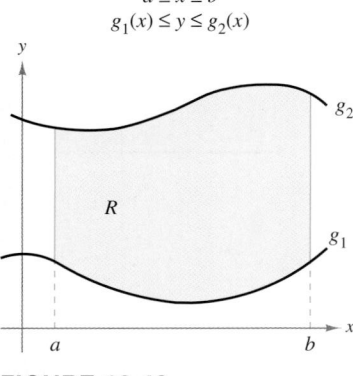

Region is bounded by
$a \leq x \leq b$
$g_1(x) \leq y \leq g_2(x)$

FIGURE 13.42

Determining Area in the Plane by Double Integrals

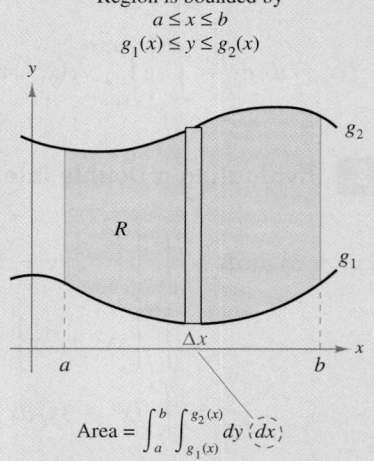

Region is bounded by
$a \leq x \leq b$
$g_1(x) \leq y \leq g_2(x)$

$$\text{Area} = \int_a^b \int_{g_1(x)}^{g_2(x)} dy \, dx$$

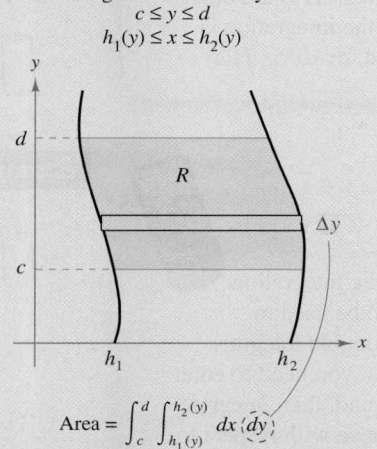

Region is bounded by
$c \leq y \leq d$
$h_1(y) \leq x \leq h_2(y)$

$$\text{Area} = \int_c^d \int_{h_1(y)}^{h_2(y)} dx \, dy$$

FIGURE 13.43

Example 3   **Finding Area with a Double Integral**

Use a double integral to find the area of the rectangular region shown in Figure 13.44.

SOLUTION   The bounds for x are $1 \le x \le 5$ and the bounds for y are $2 \le y \le 4$. So, the area of the region is

$$\int_1^5 \int_2^4 dy\, dx = \int_1^5 \left[y \right]_2^4 dx \qquad \text{Integrate with respect to } y.$$

$$= \int_1^5 (4 - 2)\, dx \qquad \text{Apply Fundamental Theorem of Calculus.}$$

$$= \int_1^5 2\, dx \qquad \text{Simplify.}$$

$$= \left[2x \right]_1^5 \qquad \text{Integrate with respect to } x.$$

$$= 10 - 2 \qquad \text{Apply Fundamental Theorem of Calculus.}$$

$$= 8 \text{ square units.} \qquad \text{Simplify.}$$

You can confirm this by noting that the rectangle measures two units by four units.

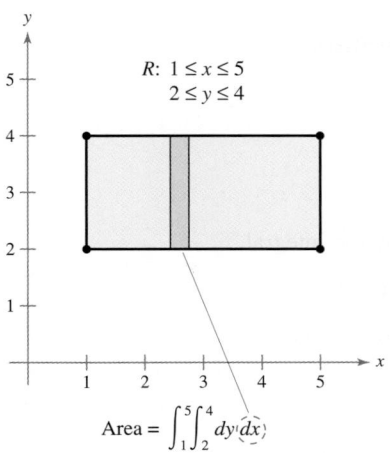

$R: 1 \le x \le 5$
$2 \le y \le 4$

$\text{Area} = \int_1^5 \int_2^4 dy\, dx$

FIGURE 13.44

✓**Checkpoint 3**

Use a double integral to find the area of the rectangular region shown in Example 3 by integrating with respect to x and then with respect to y.

Example 4   **Finding Area with a Double Integral**

Use a double integral to find the area of the region bounded by the graphs of

$$y = x^2 \quad \text{and} \quad y = x^3.$$

SOLUTION   As shown in Figure 13.45, the two graphs intersect when $x = 0$ and $x = 1$. Choosing x to be the outer variable, the bounds for x are $0 \le x \le 1$. On the interval $0 \le x \le 1$, the region is bounded above by $y = x^2$ and below by $y = x^3$. So, the bounds for y are

$$x^3 \le y \le x^2.$$

This implies that the area of the region is

$$\int_0^1 \int_{x^3}^{x^2} dy\, dx = \int_0^1 \left[y \right]_{x^3}^{x^2} dx \qquad \text{Integrate with respect to } y.$$

$$= \int_0^1 (x^2 - x^3)\, dx \qquad \text{Apply Fundamental Theorem of Calculus.}$$

$$= \left[\frac{x^3}{3} - \frac{x^4}{4} \right]_0^1 \qquad \text{Integrate with respect to } x.$$

$$= \frac{1}{3} - \frac{1}{4} \qquad \text{Apply Fundamental Theorem of Calculus.}$$

$$= \frac{1}{12} \text{ square unit.} \qquad \text{Simplify.}$$

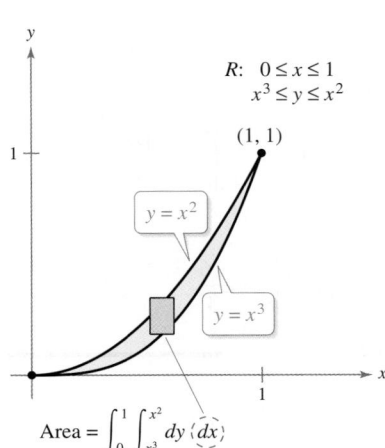

$R: 0 \le x \le 1$
$x^3 \le y \le x^2$

$(1, 1)$

$y = x^2$

$y = x^3$

$\text{Area} = \int_0^1 \int_{x^3}^{x^2} dy\, dx$

FIGURE 13.45

✓**Checkpoint 4**

Use a double integral to find the area of the region bounded by the graphs of

$$y = 2x \quad \text{and} \quad y = x^2.$$

In setting up double integrals, the most difficult task is likely to be determining the correct limits of integration. This can be simplified by making a sketch of the region R and identifying the appropriate bounds for x and y.

Example 5   Changing the Order of Integration

For the double integral

$$\int_0^2 \int_{y^2}^4 dx\, dy$$

a. sketch the region R whose area is represented by the integral,

b. rewrite the integral so that x is the outer variable, and

c. show that both orders of integration yield the same value.

SOLUTION

a. From the limits of integration, you know that

$$y^2 \le x \le 4 \qquad \text{Inner limits of integration}$$

which means that the region R is bounded on the left by the parabola $x = y^2$ and on the right by the line $x = 4$. Furthermore, because

$$0 \le y \le 2 \qquad \text{Outer limits of integration}$$

you know that the region lies above the x-axis, as shown in Figure 13.46.

b. If you interchange the order of integration so that x is the outer variable, then x will have constant bounds of integration given by

$$0 \le x \le 4. \qquad \text{Outer limits of integration}$$

By solving for y in the equation $x = y^2$, you can conclude that the bounds for y are

$$0 \le y \le \sqrt{x} \qquad \text{Inner limits of integration}$$

as shown in Figure 13.47. So, with x as the outer variable, the integral can be written as

$$\int_0^4 \int_0^{\sqrt{x}} dy\, dx.$$

c. Integrating with respect to x, you have

$$\int_0^2 \int_{y^2}^4 dx\, dy = \int_0^2 \left[x \right]_{y^2}^4 dy = \int_0^2 (4 - y^2)\, dy = \left[4y - \frac{y^3}{3} \right]_0^2 = \frac{16}{3}.$$

Integrating with respect to y, you have

$$\int_0^4 \int_0^{\sqrt{x}} dy\, dx = \int_0^4 \left[y \right]_0^{\sqrt{x}} dx = \int_0^4 \sqrt{x}\, dx = \left[\frac{2}{3} x^{3/2} \right]_0^4 = \frac{16}{3}.$$

So, both orders of integration yield the same value.

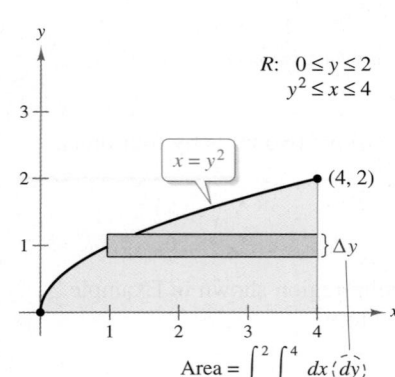

R: $0 \le y \le 2$
       $y^2 \le x \le 4$

$x = y^2$

$(4, 2)$

$\} \Delta y$

$\text{Area} = \int_0^2 \int_{y^2}^4 dx\, (dy)$

FIGURE 13.46

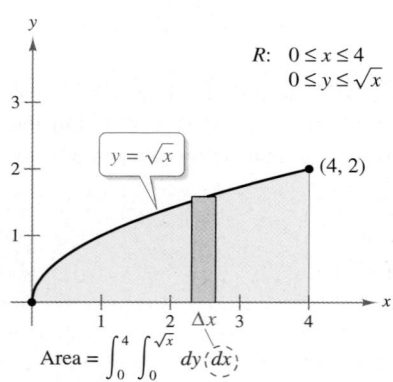

R: $0 \le x \le 4$
       $0 \le y \le \sqrt{x}$

$y = \sqrt{x}$

$(4, 2)$

Δx

$\text{Area} = \int_0^4 \int_0^{\sqrt{x}} dy\, (dx)$

FIGURE 13.47

✓ Checkpoint 5

For the double integral $\displaystyle\int_0^2 \int_{2y}^4 dx\, dy$,

a. sketch the region R whose area is represented by the integral,

b. rewrite the integral so that x is the outer variable, and

c. show that both orders of integration yield the same result.

Example 6 **Finding Area with a Double Integral**

Use a double integral to calculate the area denoted by

$$\int_R\int dA$$

where R is the region bounded by $y = x$ and $y = x^2 - x$.

SOLUTION Begin by sketching the region R, as shown in Figure 13.48. From the sketch, you can see that vertical rectangles of width dx are more convenient than horizontal ones. So, x is the outer variable of integration and its constant bounds are $0 \le x \le 2$. This implies that the bounds for y are $x^2 - x \le y \le x$, and the area is given by

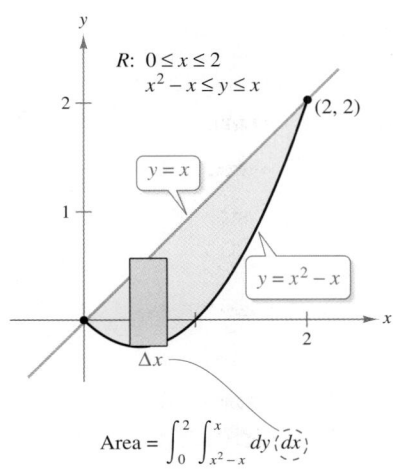

R: $0 \le x \le 2$
$x^2 - x \le y \le x$

(2, 2)

$y = x$

$y = x^2 - x$

Δx

Area $= \int_0^2 \int_{x^2-x}^x dy\,(dx)$

FIGURE 13.48

$$\int_R\int dA = \int_0^2 \int_{x^2-x}^x dy\,dx \qquad \text{Substitute bounds for region.}$$

$$= \int_0^2 \Big[\,y\,\Big]_{x^2-x}^x dx \qquad \text{Integrate with respect to } y.$$

$$= \int_0^2 \big[x - (x^2 - x)\big]\,dx \qquad \text{Apply Fundamental Theorem of Calculus.}$$

$$= \int_0^2 (2x - x^2)\,dx \qquad \text{Simplify.}$$

$$= \Big[x^2 - \frac{x^3}{3}\Big]_0^2 \qquad \text{Integrate with respect to } x.$$

$$= 4 - \frac{8}{3} \qquad \text{Apply Fundamental Theorem of Calculus.}$$

$$= \frac{4}{3} \text{ square units.} \qquad \text{Simplify.}$$

✓**Checkpoint 6**

Use a double integral to calculate the area denoted by

$$\int_R\int dA$$

where R is the region bounded by $y = 2x + 3$ and $y = x^2$.

As you are working the exercises for this section, you should be aware that the primary uses of double integrals will be discussed in Section 13.9. Double integrals by way of areas in the plane have been introduced so that you can gain practice in finding the limits of integration. When setting up a double integral, remember that your first step should be to sketch the region R. After doing this, you have two choices of integration orders: $dx\,dy$ or $dy\,dx$.

SUMMARIZE   (Section 13.8)

1. Describe a procedure for finding a partial integral with respect to one variable *(page 972)*. For an example of finding a partial integral with respect to x or to y, see Example 1.

2. Explain how to determine the area of a region in the plane using a double integral *(page 974)*. For examples of finding area using a double integral, see Examples 3, 4, and 5.

SKILLS WARM UP 13.8

The following warm-up exercises involve skills that were covered in earlier sections. You will use these skills in the exercise set for this section. For additional help, review Sections 11.2–11.5.

In Exercises 1–12, evaluate the definite integral.

1. $\int_0^1 dx$

2. $\int_0^2 3\, dy$

3. $\int_1^4 2x^2\, dx$

4. $\int_0^1 2x^3\, dx$

5. $\int_1^2 (x^3 - 2x + 4)\, dx$

6. $\int_0^2 (4 - y^2)\, dy$

7. $\int_1^2 \frac{2}{7x^2}\, dx$

8. $\int_1^4 \frac{2}{\sqrt{x}}\, dx$

9. $\int_0^2 \frac{2x}{x^2 + 1}\, dx$

10. $\int_2^e \frac{1}{y - 1}\, dy$

11. $\int_0^2 xe^{x^2+1}\, dx$

12. $\int_0^1 e^{-2y}\, dy$

In Exercises 13–16, sketch the region bounded by the graphs of the equations.

13. $y = x$, $y = 0$, $x = 3$

14. $y = x$, $y = 3$, $x = 0$

15. $y = 4 - x^2$, $y = 0$, $x = 0$

16. $y = x^2$, $y = 4x$

Exercises 13.8

See www.CalcChat.com for worked-out solutions to odd-numbered exercises.

Finding Partial Integrals In Exercises 1–10, find the partial integral. *See Example 1.*

1. $\int_0^x (2x - y)\, dy$

2. $\int_0^y (5x + 8y)\, dx$

3. $\int_x^{x^2} \frac{y}{x}\, dy$

4. $\int_1^{2y} \frac{y}{x}\, dx$

5. $\int_2^y (6x^2y + y^2)\, dx$

6. $\int_4^x (xy^3 + 4y)\, dy$

7. $\int_{x^3}^{\sqrt{x}} (x^2 + 3y^2)\, dy$

8. $\int_{-\sqrt{1-y^2}}^{\sqrt{1-y^2}} (x^2 + y^2)\, dx$

9. $\int_1^{e^y} \frac{y \ln x}{x}\, dx$

10. $\int_y^3 \frac{xy}{\sqrt{x^2 + 1}}\, dx$

19. $\int_0^1 \int_0^{3x} (3x^2 + 3y^2 + 1)\, dy\, dx$

20. $\int_0^1 \int_y^{2y} (1 + 2x^2 + 2y^2)\, dx\, dy$

21. $\int_0^1 \int_0^x \sqrt{1 - x^2}\, dy\, dx$

22. $\int_0^4 \int_0^x \frac{2}{x^2 + 1}\, dy\, dx$

23. $\int_0^\infty \int_0^\infty e^{-(x+y)/2}\, dy\, dx$

24. $\int_0^\infty \int_0^\infty xye^{-(x^2+y^2)}\, dx\, dy$

Evaluating a Double Integral In Exercises 11–24, evaluate the double integral. *See Example 2.*

11. $\int_0^1 \int_0^2 (x + y)\, dy\, dx$

12. $\int_0^2 \int_0^2 (6 - x^2)\, dy\, dx$

13. $\int_0^3 \int_0^4 xy\, dx\, dy$

14. $\int_{-1}^1 \int_{-2}^2 (x^2 - y^2)\, dy\, dx$

15. $\int_0^2 \int_0^{6x^2} x^3\, dy\, dx$

16. $\int_0^2 \int_{3y^2-6y}^{2y-y^2} 3y\, dx\, dy$

17. $\int_0^1 \int_0^y (x + y)\, dx\, dy$

18. $\int_0^2 \int_0^{\sqrt{1-y^2}} -5xy\, dx\, dy$

Finding Area with a Double Integral In Exercises 25–30, use a double integral to find the area of the specified region. *See Example 3.*

25.

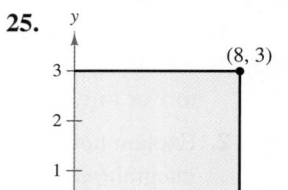

26.

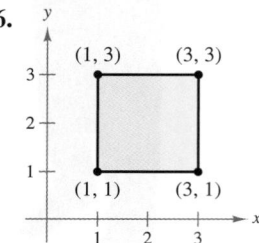

27.

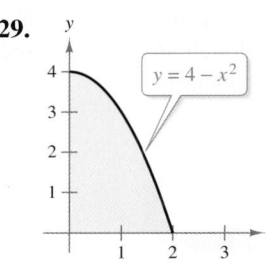

28.

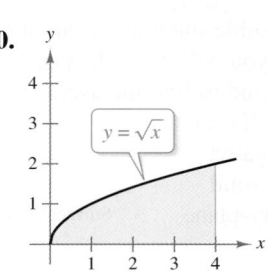

29.

30.

Finding Area with a Double Integral In Exercises 31–36, use a double integral to find the area of the region bounded by the graphs of the equations. *See Example 4.*

31. $y = 9 - x^2$, $y = 0$

32. $y = x^{3/2}$, $y = x$

33. $2x - 3y = 0$, $x + y = 5$, $y = 0$

34. $xy = 9$, $y = x$, $y = 0$, $x = 9$

35. $y = x$, $y = 2x$, $x = 2$

36. $y = 4 - x^2$, $y = x + 2$

Changing the Order of Integration In Exercises 37–44, sketch the region R whose area is given by the double integral. Then change the order of integration and show that both orders yield the same value. *See Example 5.*

37. $\int_0^1 \int_0^2 dy\,dx$

38. $\int_1^2 \int_2^4 dx\,dy$

39. $\int_0^1 \int_{2y}^2 dx\,dy$

40. $\int_0^4 \int_0^{\sqrt{x}} dy\,dx$

41. $\int_0^2 \int_{x/2}^1 dy\,dx$

42. $\int_0^4 \int_{\sqrt{x}}^2 dy\,dx$

43. $\int_0^1 \int_{y^2}^{\sqrt[3]{y}} dx\,dy$

44. $\int_{-2}^2 \int_0^{4-y^2} dx\,dy$

45. Think About It Explain why you need to change the order of integration to evaluate the double integral. Then evaluate the double integral.

(a) $\int_0^3 \int_y^3 e^{x^2} dx\,dy$ \qquad (b) $\int_0^2 \int_x^2 e^{-y^2} dy\,dx$

46. HOW DO YOU SEE IT? Complete the double integrals so that each one represents the area of the region R (see figure).

(a) Area $= \iint dx\,dy$ \qquad (b) Area $= \iint dy\,dx$

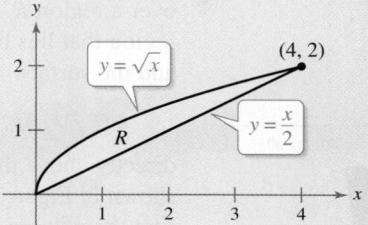

Evaluating a Double Integral In Exercises 47–54, use a symbolic integration utility to evaluate the double integral.

47. $\int_0^1 \int_0^2 e^{-x^2 - y^2} dx\,dy$

48. $\int_0^2 \int_{x^2}^{2x} (x^3 + 3y^2) dy\,dx$

49. $\int_1^2 \int_0^x e^{xy} dy\,dx$

50. $\int_1^2 \int_y^{2y} \ln(x + y) dx\,dy$

51. $\int_0^1 \int_x^1 \sqrt{1 - x^2} dy\,dx$

52. $\int_0^3 \int_0^{x^2} \sqrt{x}\sqrt{1 + x} dy\,dx$

53. $\int_0^2 \int_{\sqrt{4-x^2}}^{4-x^2/4} \frac{xy}{x^2 + y^2 + 1} dy\,dx$

54. $\int_0^4 \int_0^y \frac{2}{(x + 1)(y + 1)} dx\,dy$

True or False? In Exercises 55 and 56, determine whether the statement is true or false. If it is false, explain why or give an example that shows it is false.

55. $\int_{-1}^1 \int_{-2}^2 y\,dy\,dx = \int_{-1}^1 \int_{-2}^2 y\,dx\,dy$

56. $\int_2^5 \int_1^6 x\,dy\,dx = \int_1^6 \int_2^5 x\,dx\,dy$

13.9 Applications of Double Integrals

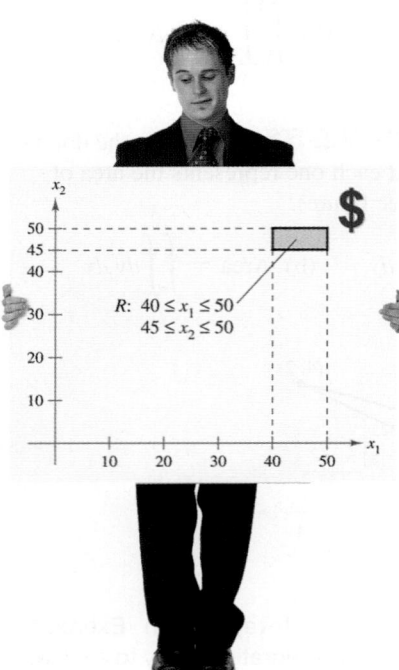

■ Use double integrals to find the volumes of solids.
■ Use double integrals to find the average values of real-life models.

Volume of a Solid Region

In Section 13.8, you used double integrals as an alternative way to find the area of a plane region. In this section, you will study the primary uses of double integrals: to find the volume of a solid region and to find the average value of a function.

Consider a function $z = f(x, y)$ that is continuous and nonnegative over a region R. Let S be the solid region that lies between the xy-plane and the surface

$$z = f(x, y)$$

directly above the region R, as shown in Figure 13.49. You can find the volume of S by integrating

$$f(x, y)$$

over the region R.

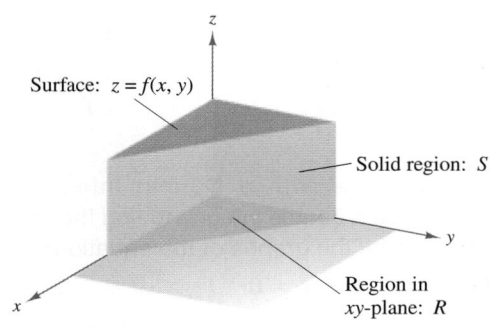

FIGURE 13.49

In Exercise 31 on page 987, you will use a double integral to find the average weekly profit of a company.

Determining Volume with Double Integrals

If R is a bounded region in the xy-plane and f is continuous and nonnegative over R, then the **volume of the solid** region between the surface

$$z = f(x, y)$$

and R is given by the double integral

$$\int_R \int f(x, y) \, dA$$

where $dA = dx \, dy$ or $dA = dy \, dx$.

You can use the following guidelines when finding the volume of a solid.

Guidelines for Finding the Volume of a Solid

1. Write the equation of the surface in the form

$$z = f(x, y)$$

and sketch the solid region.

2. Sketch the region R in the xy-plane and determine the order and limits of integration.

3. Evaluate the double integral

$$\int_R \int f(x, y) \, dA$$

using the order and limits determined in the second step.

| *Example 1* | **Finding the Volume of a Solid** |

Find the volume of the solid region bounded in the first octant by the plane

$$z = 2 - x - 2y.$$

SOLUTION

1. The equation of the surface is already in the form $z = f(x, y)$. A graph of the solid region is shown in Figure 13.50.

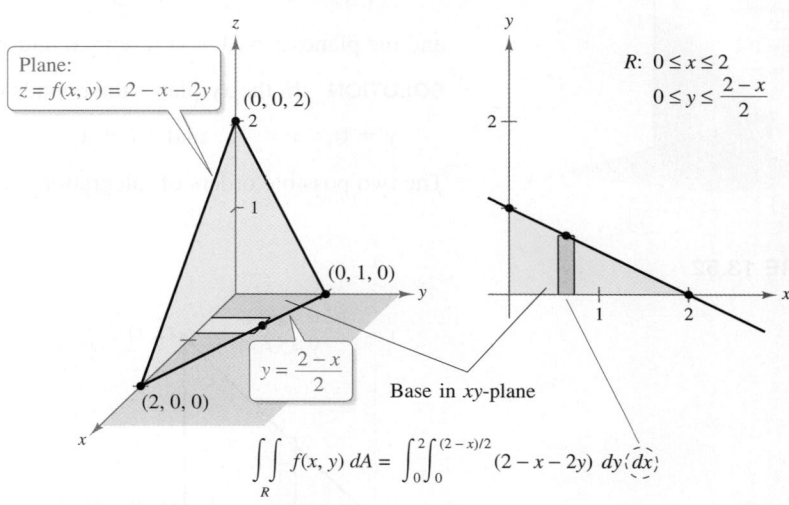

Plane:
$z = f(x, y) = 2 - x - 2y$ $(0, 0, 2)$

$(0, 1, 0)$

$y = \dfrac{2 - x}{2}$

$(2, 0, 0)$ Base in xy-plane

$R: 0 \le x \le 2$
$0 \le y \le \dfrac{2 - x}{2}$

$$\iint_R f(x, y)\, dA = \int_0^2 \int_0^{(2-x)/2} (2 - x - 2y)\ dy\ dx$$

FIGURE 13.50

2. Sketch the region R in the xy-plane. In Figure 13.50, you can see that the region R is bounded by the lines $x = 0$, $y = 0$, and $y = \frac{1}{2}(2 - x)$. One way to set up the double integral is to choose x as the outer variable. With that choice, the bounds for x are $0 \le x \le 2$ and the bounds for y are $0 \le y \le \frac{1}{2}(2 - x)$.

3. The volume of the solid region is

$$
\begin{aligned}
V &= \int_0^2 \int_0^{(2-x)/2} (2 - x - 2y)\ dy\ dx \\
&= \int_0^2 \left[(2 - x)y - y^2 \right]_0^{(2-x)/2} dx \\
&= \int_0^2 \left\{ (2 - x)\left(\frac{1}{2}\right)(2 - x) - \left[\frac{1}{2}(2 - x) \right]^2 \right\} dx \\
&= \frac{1}{4} \int_0^2 (2 - x)^2\ dx \\
&= \frac{1}{4} \left[-\frac{1}{3}(2 - x)^3 \right]_0^2 \\
&= \frac{2}{3}\ \text{cubic unit.}
\end{aligned}
$$

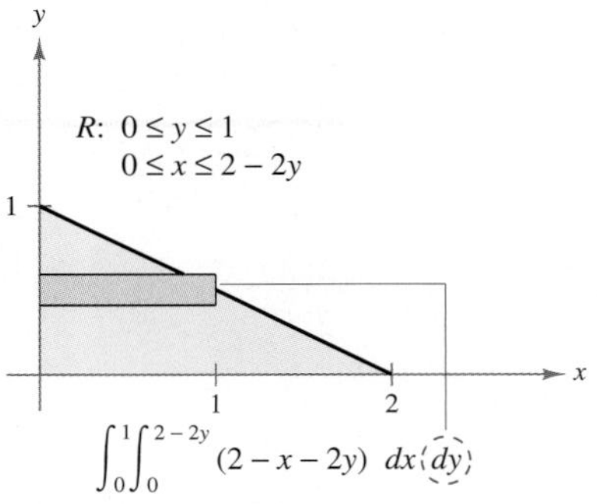

✓ **Checkpoint 1**

Find the volume of the solid region bounded in the first octant by the plane
$z = 4 - 2x - y$.

Example 1 uses $dy\ dx$ as the order of integration. The other order, $dx\ dy$, as indicated in Figure 13.51, produces the same result. Try verifying this.

$R: 0 \le y \le 1$
$0 \le x \le 2 - 2y$

$$\int_0^1 \int_0^{2-2y} (2 - x - 2y)\ dx\ dy$$

FIGURE 13.51

In Example 1, the problem could be solved with either order of integration. Moreover, had you used the order $dx\,dy$, you would have obtained a double integral of comparable difficulty. There are, however, some occasions in which one order of integration is much more convenient than the other. Example 2 shows such a case.

Example 2   Comparing Different Orders of Integration

Find the volume of the solid region bounded by the surface

$$f(x, y) = e^{-x^2} \qquad \text{Surface}$$

and the planes $z = 0$, $y = 0$, $y = x$, and $x = 1$, as shown in Figure 13.52.

SOLUTION   In the xy-plane, the bounds of region R are the lines

$$y = 0, \quad x = 1, \quad \text{and} \quad y = x.$$

The two possible orders of integration are indicated in Figure 13.53.

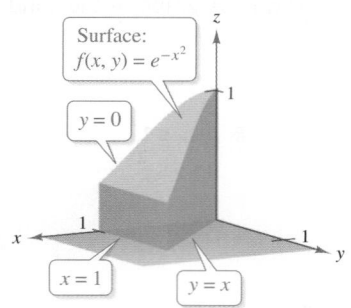

Surface:
$f(x, y) = e^{-x^2}$
$y = 0$
$x = 1$
$y = x$

FIGURE 13.52

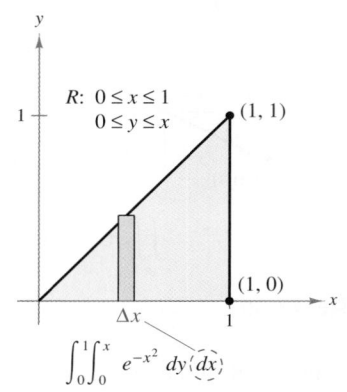

$R:\ 0 \le x \le 1$
$0 \le y \le x$
(1, 1)
(1, 0)
Δx

$$\int_0^1 \int_0^x e^{-x^2}\,dy\,dx$$

$R:\ 0 \le y \le 1$
$y \le x \le 1$
(1, 1)
Δy
(1, 0)

$$\int_0^1 \int_y^1 e^{-x^2}\,dx\,dy$$

FIGURE 13.53

By setting up the corresponding integrals, you can see that the order $dy\,dx$ produces an integral that is easier to evaluate than the order $dx\,dy$.

$$V = \int_0^1 \int_0^x e^{-x^2}\,dy\,dx$$

$$= \int_0^1 \left[e^{-x^2} y \right]_0^x dx$$

$$= \int_0^1 x e^{-x^2}\,dx$$

$$= \left[-\frac{1}{2} e^{-x^2} \right]_0^1$$

$$= -\frac{1}{2}\left(\frac{1}{e} - 1 \right)$$

$$\approx 0.316 \text{ cubic unit}$$

TECH TUTOR

Use a symbolic integration utility to evaluate the double integral in Example 2.

✓ Checkpoint 2

Find the volume under the surface

$$f(x, y) = e^{x^2}$$

bounded by the xz-plane and the planes $y = 2x$ and $x = 1$.

In the guidelines for finding the volume of a solid given at the beginning of this section, the first step suggests that you sketch the three-dimensional solid region. This is a good suggestion, but it is not always feasible and is not as important as making a sketch of the two-dimensional region R.

Example 3   Finding the Volume of a Solid

Find the volume of the solid bounded above by the surface

$$f(x, y) = 6x^2 - 2xy$$

and below by the plane region R shown in Figure 13.54.

SOLUTION   Because the region R is bounded by the parabola

$$y = 3x - x^2$$

and the line

$$y = x$$

the limits for y are $x \leq y \leq 3x - x^2$. The limits for x are $0 \leq x \leq 2$, and the volume of the solid is

$$
\begin{aligned}
V &= \int_0^2 \int_x^{3x-x^2} (6x^2 - 2xy)\, dy\, dx \\
&= \int_0^2 \left[6x^2 y - xy^2 \right]_x^{3x-x^2} dx \\
&= \int_0^2 \left[(18x^3 - 6x^4 - 9x^3 + 6x^4 - x^5) - (6x^3 - x^3) \right] dx \\
&= \int_0^2 (4x^3 - x^5)\, dx \\
&= \left[x^4 - \frac{x^6}{6} \right]_0^2 \\
&= \frac{16}{3} \text{ cubic units.}
\end{aligned}
$$

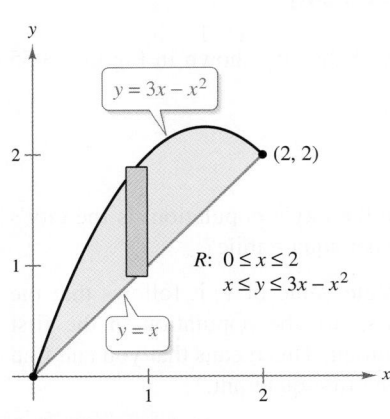

y = 3x − x²

(2, 2)

R: 0 ≤ x ≤ 2
    x ≤ y ≤ 3x − x²

y = x

FIGURE 13.54

✔ Checkpoint 3

Find the volume of the solid bounded above by the surface

$$f(x, y) = 4x^2 + 2xy$$

and below by the plane region bounded by $y = x^2$ and $y = 2x$.

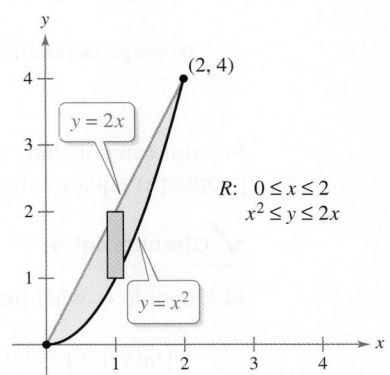

(2, 4)

y = 2x

R: 0 ≤ x ≤ 2
    x² ≤ y ≤ 2x

y = x²

A *population density function*

$$p = f(x, y)$$

is a model that describes the density (in people per square unit) of a region. To find the population of a region R, evaluate the double integral

$$\int_R \int f(x, y) \, dA.$$

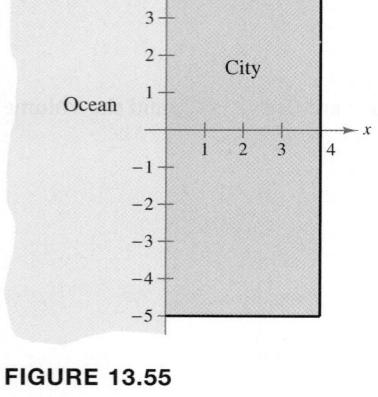

Example 4　Finding the Population of a City

The population density (in people per square mile) of the city shown in Figure 13.55 can be modeled by

$$f(x, y) = \frac{50,000}{x + |y| + 1}$$

where x and y are measured in miles. Approximate the city's population. Is the city's average population density less than 10,000 people per square mile?

SOLUTION　Because the model involves the absolute value of y, it follows that the population density is symmetrical about the x-axis. So, the population in the first quadrant is equal to the population in the fourth quadrant. This means that you can find the total population by doubling the population in the first quadrant.

$$\begin{aligned}
\text{Population} &= 2 \int_0^4 \int_0^5 \frac{50,000}{x + y + 1} \, dy \, dx \\
&= 100,000 \int_0^4 \int_0^5 \frac{1}{x + y + 1} \, dy \, dx \\
&= 100,000 \int_0^4 \left[\ln(x + y + 1) \right]_0^5 dx \\
&= 100,000 \int_0^4 \left[\ln(x + 6) - \ln(x + 1) \right] dx \\
&= 100,000 \left[(x + 6) \ln(x + 6) - (x + 6) - (x + 1) \ln(x + 1) + (x + 1) \right]_0^4 \\
&= 100,000 \left[(x + 6) \ln(x + 6) - (x + 1) \ln(x + 1) - 5 \right]_0^4 \\
&= 100,000 \left[10 \ln(10) - 5 \ln(5) - 5 - 6 \ln(6) + 5 \right] \\
&\approx 422,810 \text{ people}
\end{aligned}$$

So, the city's population is about 422,810. Because the city covers a region 4 miles wide and 10 miles long, its area is 40 square miles. So, the average population density is

$$\text{Average population density} = \frac{422,810}{40}$$

$$\approx 10,570 \text{ people per square mile.}$$

So, you can conclude that the city's average population density is not less than 10,000 people per square mile.

✓ Checkpoint 4

In Example 4, what integration technique was used to integrate

$$\int \left[\ln(x + 6) - \ln(x + 1) \right] dx?$$

FIGURE 13.55

Average Value of a Function over a Region

> ### Average Value of a Function over a Region
>
> If f is integrable over the plane region R with area A, then its **average value** over R is
>
> $$\text{Average value} = \frac{1}{A} \int_R \int f(x, y) \, dA.$$

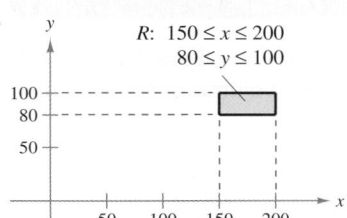 **Example 5** **Finding Average Profit**

A manufacturer determines that the profit for selling x units of one product and y units of a second product is modeled by

$$P = -(x - 200)^2 - (y - 100)^2 + 5000.$$

The weekly sales for product 1 vary between 150 and 200 units, and the weekly sales for product 2 vary between 80 and 100 units. Estimate the average weekly profit for the two products.

SOLUTION Because $150 \le x \le 200$ and $80 \le y \le 100$, you can estimate the weekly profit to be the average of the profit function over the rectangular region shown in Figure 13.56. Because the area of this rectangular region is $(50)(20) = 1000$, it follows that the average profit V is

$$V = \frac{1}{1000} \int_{150}^{200} \int_{80}^{100} \left[-(x - 200)^2 - (y - 100)^2 + 5000 \right] dy \, dx$$

$$= \frac{1}{1000} \int_{150}^{200} \left[-(x - 200)^2 y - \frac{(y - 100)^3}{3} + 5000y \right]_{80}^{100} dx$$

$$= \frac{1}{1000} \int_{150}^{200} \left[-20(x - 200)^2 - \frac{292{,}000}{3} \right] dx$$

$$= \frac{1}{3000} \left[-20(x - 200)^3 + 292{,}000x \right]_{150}^{200}$$

$$\approx \$4033.$$

✓ Checkpoint 5

Find the average value of $f(x, y) = 4 - \frac{1}{2}x - \frac{1}{2}y$ over the region $0 \le x \le 2$ and $0 \le y \le 2$.

R: $150 \le x \le 200$
       $80 \le y \le 100$

FIGURE 13.56

SUMMARIZE   (Section 13.9)

1. State the volume of a solid region using double integrals *(page 980)*. For examples of finding the volume of a solid, see Examples 1, 2, and 3.

2. Give the guidelines for finding the volume of a solid *(page 980)*. For examples of using these guidelines, see Examples 1, 2, and 3.

3. Describe a real-life example of how a double integral can be used to find a city's population *(page 984, Example 4)*.

4. State the average value of a function over a region *(page 985)*. For an example of finding the average value of a function, see Example 5.

In Exercises 1–4, sketch the region that is described.

1. $0 \le x \le 2, \ 0 \le y \le 1$

2. $1 \le x \le 3, \ 2 \le y \le 3$

3. $0 \le x \le 4, \ 0 \le y \le 2x - 1$

4. $0 \le x \le 2, \ 0 \le y \le x^2$

In Exercises 5–10, evaluate the double integral.

5. $\displaystyle\int_0^1 \int_1^2 dy \, dx$

6. $\displaystyle\int_0^3 \int_1^3 dx \, dy$

7. $\displaystyle\int_0^1 \int_0^x x \, dy \, dx$

8. $\displaystyle\int_0^4 \int_1^y y \, dx \, dy$

9. $\displaystyle\int_1^3 \int_x^{x^2} 2 \, dy \, dx$

10. $\displaystyle\int_0^1 \int_x^{-x^2+2} dy \, dx$

Exercises 13.9

See www.CalcChat.com for worked-out solutions to odd-numbered exercises.

Finding the Volume of a Solid In Exercises 1–8, sketch the region of integration in the xy-plane and evaluate the double integral. *See Example 1.*

1. $\displaystyle\int_0^2 \int_0^1 (3x + 4y) \, dy \, dx$ **2.** $\displaystyle\int_0^3 \int_0^1 (2x + 6y) \, dy \, dx$

3. $\displaystyle\int_{-1}^1 \int_0^{\sqrt{1-x^2}} x^2 y \, dy \, dx$ **4.** $\displaystyle\int_0^2 \int_0^{4-x^2} xy^2 \, dy \, dx$

5. $\displaystyle\int_0^1 \int_{y^2}^y (x^2 + y^2) \, dx \, dy$ **6.** $\displaystyle\int_0^6 \int_{y/2}^3 (x + y) \, dx \, dy$

7. $\displaystyle\int_{-a}^a \int_{-\sqrt{a^2-x^2}}^{\sqrt{a^2-x^2}} dy \, dx$ **8.** $\displaystyle\int_0^a \int_0^{\sqrt{a^2-x^2}} dy \, dx$

Comparing Different Orders of Integration In Exercises 9–12, set up the integral for both orders of integration and use the more convenient order to evaluate the integral over the region R. *See Example 2.*

9. $\displaystyle\int_R \int xy \, dA$

R: rectangle with vertices at $(0, 0), (0, 5), (3, 5), (3, 0)$

10. $\displaystyle\int_R \int x \, dA$

R: semicircle bounded by $y = \sqrt{25 - x^2}$ and $y = 0$

11. $\displaystyle\int_R \int \frac{y}{x^2 + y^2} \, dA$

R: triangle bounded by $y = x, y = 2x, x = 2$

12. $\displaystyle\int_R \int \frac{y}{1 + x^2} \, dA$

R: region bounded by $y = 0, y = \sqrt{x}, x = 4$

Finding the Volume of a Solid In Exercises 13–20, use a double integral to find the volume of the specified solid. *See Example 3.*

13.

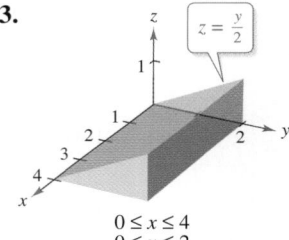

$0 \le x \le 4$
$0 \le y \le 2$

14.

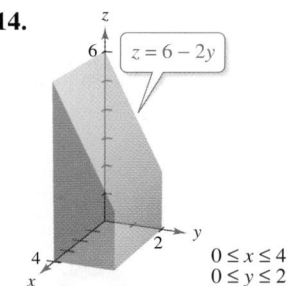

$0 \le x \le 4$
$0 \le y \le 2$

15.

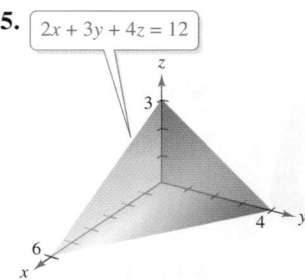

16.

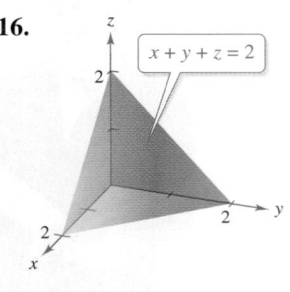

17.

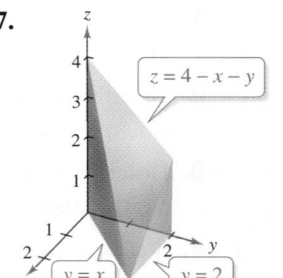

18.

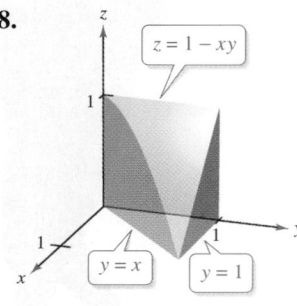

19.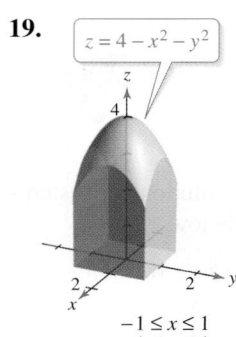
$z = 4 - x^2 - y^2$
$-1 \le x \le 1$
$-1 \le y \le 1$

20.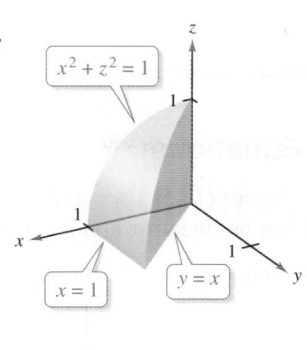
$x^2 + z^2 = 1$
$x = 1$
$y = x$

Finding the Volume of a Solid   In Exercises 21–24, use a double integral to find the volume of the solid bounded by the graphs of the equations.

21. $z = xy$, $z = 0$, $y = 2x$, $y = 0$, $x = 0$, $x = 3$

22. $z = x$, $z = 0$, $y = x$, $y = 0$, $x = 0$, $x = 4$

23. $z = 9 - x^2$, $z = 0$, $y = x + 2$, $y = 0$, $x = 0$, $x = 2$

24. $z = x + y$, $x^2 + y^2 = 4$ (first octant)

25. Population Density   The population density (in people per square mile) for a coastal town can be modeled by

$$f(x, y) = \frac{120,000}{(2 + x + y)^3}$$

where x and y are measured in miles. What is the population inside the rectangular area defined by the vertices

$(0, 0)$, $(2, 0)$, $(0, 2)$,   and $(2, 2)$?

26. Population Density   The population density (in people per square mile) for a coastal town on an island can be modeled by

$$f(x, y) = \frac{5000xe^y}{1 + 2x^2}$$

where x and y are measured in miles. What is the population inside the rectangular area defined by the vertices

$(0, 0)$, $(4, 0)$, $(0, -2)$,   and $(4, -2)$?

Average Value of a Function over a Region   In Exercises 27–30, find the average value of $f(x, y)$ over the region R. *See Example 5.*

27. $f(x, y) = y$

R: rectangle with vertices $(0, 0)$, $(5, 0)$, $(5, 3)$, $(0, 3)$

28. $f(x, y) = xy$

R: rectangle with vertices $(0, 0)$, $(4, 0)$, $(4, 2)$, $(0, 2)$

29. $f(x, y) = x^2 + y^2$

R: square with vertices $(0, 0)$, $(2, 0)$, $(2, 2)$, $(0, 2)$

30. $f(x, y) = e^{x+y}$

R: triangle with vertices $(0, 0)$, $(0, 1)$, $(1, 1)$

31. Average Weekly Profit   A firm's weekly profit (in dollars) in marketing two products is given by

$$P = 192x_1 + 576x_2 - x_1^2 - 5x_2^2 - 2x_1x_2 - 5000$$

where x_1 and x_2 represent the numbers of units of each product sold weekly. Estimate the average weekly profit when x_1 varies between 40 and 50 units and x_2 varies between 45 and 50 units.

32. Average Weekly Profit   After a change in marketing, the weekly profit of the firm in Exercise 31 is given by

$$P = 200x_1 + 580x_2 - x_1^2 - 5x_2^2 - 2x_1x_2 - 7500.$$

Estimate the average weekly profit when x_1 varies between 55 and 65 units and x_2 varies between 50 and 60 units.

33. Average Revenue   A company sells two products whose demand functions are given by

$$x_1 = 500 - 3p_1 \quad \text{and} \quad x_2 = 750 - 2.4p_2.$$

So, the total revenue is given by

$$R = x_1p_1 + x_2p_2.$$

Estimate the average revenue when price p_1 varies between \$50 and \$75 and price p_2 varies between \$100 and \$150.

34.  **HOW DO YOU SEE IT?**   The figure below shows Erie County, New York. Let $f(x, y)$ represent the total annual snowfall at the point (x, y) in the county, where R is the county. Interpret each of the following.

(a) $\displaystyle \int_R \int f(x, y) \, dA$

(b) $\displaystyle \frac{\int_R \int f(x, y) \, dA}{\int_R \int dA}$

35. Average Production   The Cobb-Douglas production function for an automobile manufacturer is

$$f(x, y) = 100x^{0.6}y^{0.4}$$

where x is the number of units of labor and y is the number of units of capital. Estimate the average production level when the number of units of labor x varies between 200 and 250 and the number of units of capital y varies between 300 and 325.

36. Average Production   Repeat Exercise 35 for the production function given by $f(x, y) = x^{0.25}y^{0.75}$.

ALGEBRA TUTOR

Solving Systems of Equations

Three of the sections in this chapter (13.5, 13.6, and 13.7) involve solutions of systems of equations. These systems can be linear or nonlinear, as shown below.

Nonlinear System in Two Variables

$$\begin{cases} 4x + 3y = 6 \\ x^2 - y = 4 \end{cases}$$

Linear System in Three Variables

$$\begin{cases} -x + 2y + 4z = 2 \\ 2x - y + z = 0 \\ 6x + 2z = 3 \end{cases}$$

There are many techniques for solving a system of linear equations. Two of the more common ones are listed here.

1. *Substitution:* Solve for one of the variables in one of the equations and substitute the value into another equation.

2. *Elimination:* Add multiples of one equation to a second equation to eliminate a variable in the second equation.

Example 1   Solving Systems of Equations

Solve each system of equations.

a. $\begin{cases} y - x^3 = 0 \\ x - y^3 = 0 \end{cases}$ b. $\begin{cases} -400p_1 + 300p_2 = -25 \\ 300p_1 - 360p_2 = -535 \end{cases}$

SOLUTION

a. Example 3, page 950

$$\begin{cases} y - x^3 = 0 \\ x - y^3 = 0 \end{cases}$$ Equation 1 / Equation 2

$y = x^3$	Solve for y in Equation 1.
$x - (x^3)^3 = 0$	Substitute x^3 for y in Equation 2.
$x - x^9 = 0$	$(x^m)^n = x^{mn}$
$x(x - 1)(x + 1)(x^2 + 1)(x^4 + 1) = 0$	Factor.
$x = 0$	Set factors equal to zero.
$x = 1$	Set factors equal to zero.
$x = -1$	Set factors equal to zero.

b. Example 4, page 951

$$\begin{cases} -400p_1 + 300p_2 = -25 \\ 300p_1 - 360p_2 = -535 \end{cases}$$ Equation 1 / Equation 2

$p_2 = \frac{1}{12}(16p_1 - 1)$	Solve for p_2 in Equation 1.
$300p_1 - 360\left(\frac{1}{12}\right)(16p_1 - 1) = -535$	Substitute for p_2 in Equation 2.
$300p_1 - 30(16p_1 - 1) = -535$	Multiply factors.
$-180p_1 = -565$	Combine like terms.
$p_1 = \frac{113}{36} \approx 3.14$	Divide each side by -180.
$p_2 = \frac{1}{12}\left[16\left(\frac{113}{36}\right) - 1\right]$	Find p_2 by substituting p_1.
$p_2 \approx 4.10$	Solve for p_2.

Example 2   Solving Systems of Equations

Solve each system of equations.

a. $\begin{cases} y(24 - 12x - 4y) = 0 \\ x(24 - 6x - 8y) = 0 \end{cases}$

b. $\begin{cases} 28a - 4b = 10 \\ -4a + 8b = 12 \end{cases}$

SOLUTION

a. Example 5, page 952

Before solving this system of equations, factor 4 out of the first equation and factor 2 out of the second equation.

$\begin{cases} y(24 - 12x - 4y) = 0 & \quad \text{Original Equation 1} \\ x(24 - 6x - 8y) = 0 & \quad \text{Original Equation 2} \end{cases}$

$\begin{cases} y(4)(6 - 3x - y) = 0 & \quad \text{Factor 4 out of Equation 1.} \\ x(2)(12 - 3x - 4y) = 0 & \quad \text{Factor 2 out of Equation 2.} \end{cases}$

$\begin{cases} y(6 - 3x - y) = 0 & \quad \text{Equation 1} \\ x(12 - 3x - 4y) = 0 & \quad \text{Equation 2} \end{cases}$

In each equation, either factor can be 0, so you obtain four different linear systems. For the first system, substitute $y = 0$ into the second equation to obtain $x = 4$.

$\begin{cases} y = 0 \\ 12 - 3x - 4y = 0 \end{cases}$ $(4, 0)$ is a solution.

You can solve the second system by the method of elimination.

$\begin{cases} 6 - 3x - y = 0 \\ 12 - 3x - 4y = 0 \end{cases}$ $\left(\frac{4}{3}, 2\right)$ is a solution.

The third system is already solved.

$\begin{cases} y = 0 \\ x = 0 \end{cases}$ $(0, 0)$ is a solution.

You can solve the last system by substituting $x = 0$ into the first equation to obtain $y = 6$.

$\begin{cases} 6 - 3x - y = 0 \\ \quad\quad x = 0 \end{cases}$ $(0, 6)$ is a solution.

b. Example 2, page 967

$\begin{cases} 28a - 4b = 10 & \quad \text{Equation 1} \\ -4a + 8b = 12 & \quad \text{Equation 2} \end{cases}$

$\begin{aligned} -2a + 4b &= 6 & &\text{Divide Equation 2 by 2.} \\ 26a \quad\quad &= 16 & &\text{Add new equation to Equation 1.} \\ a &= \tfrac{8}{13} & &\text{Divide each side by 26.} \\ 28\left(\tfrac{8}{13}\right) - 4b &= 10 & &\text{Substitute for } a \text{ in Equation 1.} \\ b &= \tfrac{47}{26} & &\text{Solve for } b. \end{aligned}$

SUMMARY AND STUDY STRATEGIES

After studying this chapter, you should have acquired the following skills.
The exercise numbers are keyed to the Review Exercises that begin on page 992.
Answers to odd-numbered Review Exercises are given in the back of the text.*

Section 13.1   Review Exercises

- Plot points in space.   *1, 2*
- Find the distance between two points in space.   *3, 4*

$$d = \sqrt{(x_2 - x_1)^2 + (y_2 - y_1)^2 + (z_2 - z_1)^2}$$

- Find the midpoint of a line segment in space.   *5, 6*

$$\text{Midpoint} = \left(\frac{x_1 + x_2}{2}, \frac{y_1 + y_2}{2}, \frac{z_1 + z_2}{2} \right)$$

- Write the standard forms of the equations of spheres.   *7–10*

$$(x - h)^2 + (y - k)^2 + (z - l)^2 = r^2$$

- Find the centers and radii of spheres.   *11, 12*
- Sketch the coordinate plane traces of spheres.   *13, 14*

Section 13.2

- Sketch planes in space.   *15–18*
- Classify quadric surfaces in space.   *19–26*

Section 13.3

- Evaluate functions of several variables.   *27, 28*
- Find the domains and ranges of functions of two variables.   *29–32*
- Sketch level curves of functions of two variables.   *33–36*
- Use functions of several variables to answer questions about real-life situations.   *37–40*

Section 13.4

- Find the first partial derivatives of functions of several variables. *41–50*
- Find the slopes of surfaces in the *x*- and *y*-directions.   *51–54*
- Find the second partial derivatives of functions of several variables. *55–60*
- Use partial derivatives to answer questions about real-life situations. *61, 62*

Section 13.5

- Find the relative extrema of functions of two variables.   *63–70*
- Use relative extrema to answer questions about real-life situations.   *71, 72*

---

* A wide range of valuable study aids are available to help you master the material in this chapter.
The *Student Solutions Manual* includes step-by-step solutions to all odd-numbered exercises to
help you review and prepare. The student website at *www.cengagebrain.com* offers algebra help
and a *Graphing Technology Guide*, which contains step-by-step commands and instructions for
a wide variety of graphing calculators.

Section 13.6

Section 13.7

$$a = \left[n \sum_{i=1}^{n} x_i y_i - \sum_{i=1}^{n} x_i \sum_{i=1}^{n} y_i \right] \bigg/ \left[n \sum_{i=1}^{n} x_i^2 - \left(\sum_{i=1}^{n} x_i \right)^2 \right], \quad b = \frac{1}{n} \left(\sum_{i=1}^{n} y_i - a \sum_{i=1}^{n} x_i \right)$$

Section 13.8

Section 13.9

$$\text{Volume} = \int_R \int f(x, y) \, dA$$

$$\text{Average value} = \frac{1}{A} \int_R \int f(x, y) \, dA$$

Study Strategies

■ **Comparing Two Dimensions with Three Dimensions**   Many of the formulas and techniques in this chapter are generalizations of formulas and techniques used in earlier chapters of the text. Here are several examples.

Two-Dimensional Coordinate System	Three-Dimensional Coordinate System
Distance Formula	*Distance Formula*
$d = \sqrt{(x_2 - x_1)^2 + (y_2 - y_1)^2}$	$d = \sqrt{(x_2 - x_1)^2 + (y_2 - y_1)^2 + (z_2 - z_1)^2}$
Midpoint Formula	*Midpoint Formula*
$\text{Midpoint} = \left(\dfrac{x_1 + x_2}{2}, \dfrac{y_1 + y_2}{2} \right)$	$\text{Midpoint} = \left(\dfrac{x_1 + x_2}{2}, \dfrac{y_1 + y_2}{2}, \dfrac{z_1 + z_2}{2} \right)$
Equation of Circle	*Equation of Sphere*
$(x - h)^2 + (y - k)^2 = r^2$	$(x - h)^2 + (y - k)^2 + (z - l)^2 = r^2$
Equation of Line	*Equation of Plane*
$ax + by = c$	$ax + by + cz = d$
Derivative of $y = f(x)$	*Partial Derivative of $z = f(x, y)$*
$\dfrac{dy}{dx} = \lim_{\Delta x \to 0} \dfrac{f(x + \Delta x) - f(x)}{\Delta x}$	$\dfrac{\partial z}{\partial x} = \lim_{\Delta x \to 0} \dfrac{f(x + \Delta x, y) - f(x, y)}{\Delta x}$
Area of Region	*Volume of Region*
$A = \displaystyle\int_a^b f(x) \, dx$	$V = \displaystyle\int_R \int f(x, y) \, dA$

Review Exercises

See www.CalcChat.com for worked-out solutions to odd-numbered exercises.

Plotting Points in Space   In Exercises 1 and 2, plot the points in the same three-dimensional coordinate system.

1. $(2, -1, 4), (-1, 3, -3), (-2, -2, 1), (3, 1, 2)$

2. $(1, -2, -3), (-4, -3, 5), \left(4, \frac{5}{2}, 1\right), (-2, 2, 2)$

Finding the Distance Between Two Points   In Exercises 3 and 4, find the distance between the two points.

3. $(1, 0, 2), (3, 5, 8)$

4. $(-4, 1, 5), (1, 3, 7)$

Using the Midpoint Formula   In Exercises 5 and 6, find the midpoint of the line segment joining the two points.

5. $(2, 6, 4), (-4, 2, 8)$

6. $(5, 0, 7), (-1, -2, 9)$

Finding the Equation of a Sphere   In Exercises 7–10, find the standard equation of the sphere.

7. Center: $(0, 1, 0)$; radius: 5

8. Center: $(4, -5, 3)$; radius: 10

9. Diameter endpoints: $(3, -4, -1), (1, 0, -5)$

10. Diameter endpoints: $(3, 4, 0), (5, 8, 2)$

Finding the Center and Radius of a Sphere   In Exercises 11 and 12, find the center and radius of the sphere.

11. $x^2 + y^2 + z^2 - 8x + 4y - 6z - 20 = 0$

12. $x^2 + y^2 + z^2 + 4y - 10z - 7 = 0$

Finding the Trace of a Surface   In Exercises 13 and 14, sketch the *xy*-trace of the sphere.

13. $(x + 2)^2 + (y - 1)^2 + (z - 3)^2 = 25$

14. $(x - 1)^2 + (y + 3)^2 + (z - 6)^2 = 72$

Sketching a Plane in Space   In Exercises 15–18, find the intercepts and sketch the graph of the plane.

15. $x + 2y + 3z = 6$ **16.** $2y + z = 4$

17. $3x - 6z = 12$ **18.** $4x - y + 2z = 8$

Classifying a Quadric Surface   In Exercises 19–26, classify the quadric surface.

19. $x^2 + y^2 + z^2 - 2x + 4y - 6z + 5 = 0$

20. $16x^2 + 16y^2 - 9z^2 = 0$

21. $x^2 + \frac{y^2}{16} + \frac{z^2}{9} = 1$

22. $x^2 - \frac{y^2}{16} - \frac{z^2}{9} = 1$

23. $z = \frac{x^2}{9} + y^2$

24. $-4x^2 + y^2 + z^2 = 4$

25. $z = \sqrt{x^2 + y^2}$

26. $z = x^2 - \frac{y^2}{4}$

Evaluating Functions of Several Variables   In Exercises 27 and 28, find the function values.

27. $f(x, y) = xy^2$

    (a) $f(2, 3)$ (b) $f(0, 1)$

    (c) $f(-5, 7)$ (d) $f(-2, -4)$

28. $f(x, y) = \frac{x^2}{y}$

    (a) $f(6, 9)$ (b) $f(8, 4)$ (c) $f(t, 2)$ (d) $f(r, r)$

Finding the Domain and Range of a Function   In Exercises 29–32, find the domain and range of the function.

29. $f(x, y) = \sqrt{1 - x^2 - y^2}$

30. $f(x, y) = x^2 + y^2 - 3$

31. $f(x, y) = e^{xy}$

32. $f(x, y) = \frac{1}{x + y}$

Sketching a Contour Map   In Exercises 33–36, describe the level curves of the function. Sketch a contour map of the surface using level curves for the given *c*-values.

Function	c-Values
33. $z = 10 - 2x - 5y$	$c = 0, 2, 4, 5, 10$
34. $z = \sqrt{9 - x^2 - y^2}$	$c = 0, 1, 2, 3$
35. $z = (xy)^2$	$c = 1, 4, 9, 12, 16$
36. $z = y - x^2$	$c = 0, \pm 1, \pm 2$

37. Meteorology   The contour map shown below represents the average yearly precipitation for Oklahoma.   *(Source: National Climatic Data Center)*

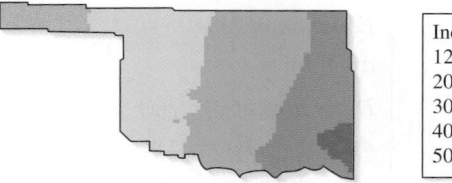 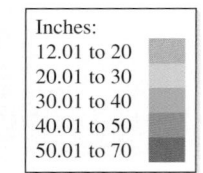

Inches:
12.01 to 20
20.01 to 30
30.01 to 40
40.01 to 50
50.01 to 70

(a) Do the level curves correspond to equally spaced levels of precipitation? Explain.

(b) Describe how to obtain a more detailed contour map.

38. Chemistry The acidity of rainwater is measured in units called pH, and smaller pH values are increasingly acidic. The map shows the curves of equal pH and gives evidence that downwind of heavily industrialized areas, the acidity has been increasing. Using the level curves on the map, determine the direction of the prevailing winds in the northeastern United States.

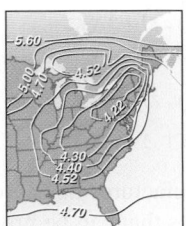

39. Earnings per Share The earnings per share z (in dollars) for Hewlett-Packard from 2003 through 2010 can be modeled by

$$z = -4.51 + 0.046x + 0.060y$$

where x is the sales (in billions of dollars) and y is the shareholder's equity (in billions of dollars). *(Source: Hewlett-Packard Company)*

(a) Find the earnings per share when $x = 100$ and $y = 40$.

(b) Which of the two variables in this model has the greater influence on the earnings per share? Explain.

40. Shareholder's Equity The shareholder's equity z (in billions of dollars) for Wal-Mart from 2000 through 2010 can be modeled by

$$z = 1.54 + 0.116x + 0.122y$$

where x is the net sales (in billions of dollars) and y is the total assets (in billions of dollars). *(Source: Wal-Mart Stores, Inc.)*

(a) Find the shareholder's equity when $x = 300$ and $y = 130$.

(b) Which of the two variables in this model has the greater influence on shareholder's equity? Explain.

Finding Partial Derivatives In Exercises 41–50, find the first partial derivatives.

41. $f(x, y) = x^2y + 3xy + 2x - 5y$

42. $f(x, y) = 4xy + xy^2 - 3x^2y$

43. $z = \dfrac{x^2}{y^2}$

44. $z = (xy + 2x + 4y)^2$

45. $f(x, y) = \ln(5x + 4y)$

46. $f(x, y) = \ln\sqrt{2x + 3y}$

47. $f(x, y) = xe^y + ye^x$

48. $f(x, y) = x^2e^{-2y}$

49. $w = xyz^2$

50. $w = 3xy - 5xz + 2yz$

Finding Slopes in the x- and y-Directions In Exercises 51–54, find the slopes of the surface at the given point in (a) the x-direction and (b) the y-direction.

51. $z = 3xy$
$(-2, -3, 18)$

52. $z = y^2 - x^2$
$(1, 2, 3)$

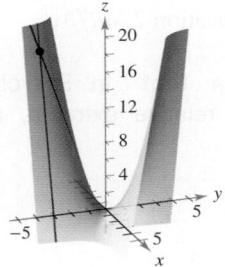
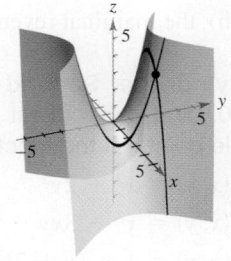

53. $z = 8 - x^2 - y^2$
$(1, 1, 6)$

54. $z = \sqrt{100 - x^2 - y^2}$
$(0, 6, 8)$

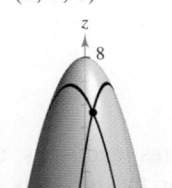
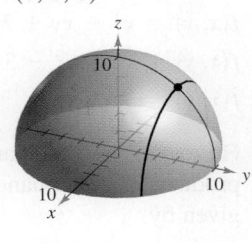

Finding Second Partial Derivatives In Exercises 55–60, find all second partial derivatives.

55. $f(x, y) = 3x^2 - xy + 2y^3$

56. $f(x, y) = \dfrac{y}{x + y}$

57. $f(x, y) = \sqrt{1 + x + y}$

58. $f(x, y) = x^2e^{-y^2}$

59. $f(x, y, z) = xy + 5x^2yz^3 - 3y^3z$

60. $f(x, y, z) = \dfrac{3yz}{x + z}$

61. Marginal Cost A company manufactures two models of skis: cross-country skis and downhill skis. The cost function for producing x pairs of cross-country skis and y pairs of downhill skis is given by

$$C = 15(xy)^{1/3} + 99x + 139y + 2293.$$

(a) Find the marginal costs ($\partial C/\partial x$ and $\partial C/\partial y$) when $x = 500$ and $y = 250$.

(b) When additional production is required, which model of skis results in the cost increasing at a higher rate? How can this be determined from the cost model?

62. Marginal Revenue   At a baseball stadium, souvenir hats are sold at two locations. If x_1 and x_2 are the numbers of baseball hats sold at location 1 and location 2, respectively, then the total revenue for the hats is modeled by

$$R = 15x_1 + 16x_2 - \frac{1}{10}x_1^2 - \frac{1}{10}x_2^2 - \frac{1}{100}x_1x_2.$$

When $x_1 = 50$ and $x_2 = 40$, find

(a) the marginal revenue for location 1, $\partial R/\partial x_1$.

(b) the marginal revenue for location 2, $\partial R/\partial x_2$.

Applying the Second-Partials Test   In Exercises 63–70, find the critical points, relative extrema, and saddle points of the function.

63. $f(x, y) = x^2 + 2y^2$

64. $f(x, y) = x^3 - 3xy + y^2$

65. $f(x, y) = 1 - (x + 2)^2 + (y - 3)^2$

66. $f(x, y) = e^x - x + y^2$

67. $f(x, y) = x^3 + y^2 - xy$

68. $f(x, y) = y^2 + xy + 3y - 2x + 5$

69. $f(x, y) = x^3 + y^3 - 3x - 3y + 2$

70. $f(x, y) = -x^2 - y^2$

71. Revenue   A company manufactures and sells two products. The demand functions for the products are given by

$$p_1 = 100 - x_1 \quad \text{and} \quad p_2 = 200 - 0.5x_2$$

where p_1 and p_2 are the prices per unit (in dollars) and x_1 and x_2 are the numbers of units sold. The total revenue function is given by

$$R = x_1p_1 + x_2p_2.$$

Find x_1 and x_2 so as to maximize revenue.

72. Profit   A company manufactures a product at two locations. The cost of producing x_1 units at location 1 is

$$C_1 = 0.03x_1^2 + 4x_1 + 300$$

and the cost of producing x_2 units at location 2 is

$$C_2 = 0.05x_2^2 + 7x_2 + 175.$$

The product sells for $10 per unit. Find the quantity that should be produced at each location to maximize the profit

$$P = 10(x_1 + x_2) - C_1 - C_2.$$

Using Lagrange Multipliers   In Exercises 73–78, use Lagrange multipliers to find the given extremum. In each case, assume that the variables are positive.

73. Maximize $f(x, y) = 2xy.$

Constraint: $2x + y = 12$

74. Maximize $f(x, y) = 2x + 3xy + y.$

Constraint: $x + 2y = 29$

75. Minimize $f(x, y) = x^2 + y^2.$

Constraint: $x + y = 4$

76. Minimize $f(x, y) = 3x^2 - y^2.$

Constraint: $2x - 2y + 5 = 0$

77. Maximize $f(x, y, z) = xyz.$

Constraint: $x + 2y + z - 4 = 0$

78. Maximize $f(x, y, z) = x^2z + yz.$

Constraint: $2x + y + z = 5$

79. Cost   A manufacturer has an order for 1000 units of wooden benches that can be produced at two locations. Let x_1 and x_2 be the numbers of units produced at the two locations. The cost function is modeled by

$$C = 0.25x_1^2 + 10x_1 + 0.15x_2^2 + 12x_2.$$

Use Lagrange multipliers to find the number of units that should be produced at each location to minimize the cost.

80. Production   The production function for a manufacturer is given by

$$f(x, y) = 4x + xy + 2y$$

where x is the number of units of labor (at $20 per unit) and y is the number of units of capital (at $4 per unit). The total cost for labor and capital cannot exceed $2000. Use Lagrange multipliers to find the maximum production level for this manufacturer.

Finding the Least Squares Regression Line   In Exercises 81 and 82, find the least squares regression line for the given points. Then plot the points and sketch the regression line.

81. $(-2, -3), (-1, -1), (1, 2), (3, 2)$

82. $(-3, -1), (-2, -1), (0, 0), (1, 1), (2, 1)$

83. Demand   A store manager wants to know the demand y for a digital camera as a function of price x. The monthly sales for four different prices of the digital camera are listed in the table.

Price, x	$80	$90	$100	$110
Demand, y	140	117	91	63

(a) Use the regression capabilities of a graphing utility or a spreadsheet to find the least squares regression line for the data.

(b) Estimate the demand when the price is $85.

(c) What price will create a demand of 200 cameras?

84. Work Force  The number of men x (in millions) and the number of women y (in millions) in the labor force from 2001 through 2010 are shown in the table. *(Source: U.S. Bureau of Labor Statistics)*

Year	2001	2002	2003	2004	2005
Men, x	76.9	77.5	78.2	79.0	80.0
Women, y	66.8	67.4	68.3	68.4	69.3

Year	2006	2007	2008	2009	2010
Men, x	81.3	82.1	82.5	82.1	82.0
Women, y	70.2	71.0	71.8	72.0	71.9

(a) Use the regression capabilities of a graphing utility or a spreadsheet to find the least squares regression line for the data.

(b) Estimate the number of women in the labor force when there are 80 million men in the labor force.

Evaluating a Double Integral In Exercises 85–88, evaluate the double integral.

85. $\int_0^1 \int_0^{1+x} (4x - 2y)\, dy\, dx$

86. $\int_{-3}^3 \int_0^4 (x - y^2)\, dx\, dy$

87. $\int_1^2 \int_1^{2y} \frac{x}{y^2}\, dx\, dy$

88. $\int_0^4 \int_0^{\sqrt{16-x^2}} 2x\, dy\, dx$

Finding Area with a Double Integral In Exercises 89–92, use a double integral to find the area of the region bounded by the graphs of the equations.

89. $y = 9 - x^2,\ y = 5$

90. $y = \frac{4}{x},\ y = 0,\ x = 1,\ x = 4$

91. $y = \sqrt{x + 3},\ y = \frac{1}{3}x + 1$

92. $y = x^2 - 2x - 2,\ y = -x$

Finding the Volume of a Solid In Exercises 93–96, use a double integral to find the volume of the specified solid.

93. **94.**

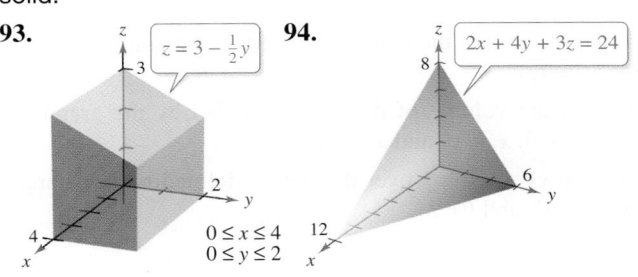

95. **96.**

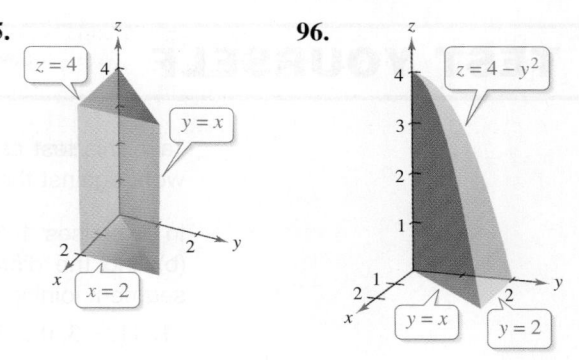

Finding the Volume of a Solid In Exercises 97 and 98, use a double integral to find the volume of the solid bounded by the graphs of the equations.

97. $z = (xy)^2,\ z = 0,\ y = 0,\ y = 4,\ x = 0,\ x = 4$

98. $z = x + y,\ z = 0,\ x = 0,\ x = 3,\ y = x,\ y = 0$

Average Value of a Function over a Region In Exercises 99 and 100, find the average value of $f(x, y)$ over the region R.

99. $f(x, y) = xy$

R: rectangle with vertices $(0, 0)$, $(4, 0)$, $(4, 3)$, $(0, 3)$

100. $f(x, y) = x^2 + 2xy + y^2$

R: rectangle with vertices $(0, 0)$, $(2, 0)$, $(2, 5)$, $(0, 5)$

101. Average Weekly Profit A firm's weekly profit (in dollars) in marketing two products is given by

$P = 150x_1 + 400x_2 - x_1^2 - 5x_2^2 - 2x_1x_2 - 3000$

where x_1 and x_2 represent the numbers of units of each product sold weekly. Estimate the average weekly profit when x_1 varies between 30 and 40 units and x_2 varies between 40 and 50 units.

102. Average Revenue A company sells two products whose demand functions are given by

$x_1 = 500 - 2.5p_1$ and $x_2 = 750 - 3p_2$.

So, the total revenue is given by

$R = x_1p_1 + x_2p_2$.

Estimate the average revenue when price p_1 varies between \$25 and \$50 and price p_2 varies between \$75 and \$125.

103. Real Estate The value of real estate (in dollars per square foot) for a city is given by

$f(x, y) = 0.003x^{2/3}y^{3/4}$

where x and y are measured in feet. What is the average value of real estate inside the rectangular area defined by the vertices $(0, 0)$, $(5280, 0)$, $(5280, 3960)$, and $(0, 3960)$?

TEST YOURSELF

Take this test as you would take a test in class. When you are done, check your work against the answers given in the back of the book.

In Exercises 1–3, (a) plot the points in a three-dimensional coordinate system, (b) find the distance between the points, and (c) find the midpoint of the line segment joining the points.

1. $(1, -3, 0), (3, -1, 0)$ **2.** $(-2, 2, 3), (-4, 0, 2)$ **3.** $(3, -7, 2), (5, 11, -6)$

4. Find the center and radius of the sphere whose equation is

$$x^2 + y^2 + z^2 - 20x + 10y - 10z + 125 = 0.$$

In Exercise 5–7, classify the quadric surface.

5. $4x^2 + 2y^2 - z^2 = 16$ **6.** $36x^2 + 9y^2 - 4z^2 = 0$

7. $4x^2 - y^2 - 16z = 0$

In Exercises 8–10, find $f(3, 3)$ and $f(1, 4)$.

8. $f(x, y) = x^2 + xy + 1$ **9.** $f(x, y) = \dfrac{x + 2y}{3x - y}$

10. $f(x, y) = xy \ln \dfrac{x}{y}$

In Exercises 11 and 12, find the first partial derivatives and evaluate each at the point $(10, -1)$.

11. $f(x, y) = 3x^2 + 9xy^2 - 2$ **12.** $f(x, y) = x\sqrt{x + y}$

In Exercises 13 and 14, find the critical points, relative extrema, and saddle points of the function.

13. $f(x, y) = 3x^2 + 4y^2 - 6x + 16y - 4$

14. $f(x, y) = 4xy - x^4 - y^4$

15. The production function for a company is given by

$$f(x, y) = 60x^{0.7}y^{0.3}$$

where x is the number of units of labor (at \$42 per unit) and y is the number of units of capital (at \$144 per unit). The total cost for labor and capital cannot exceed \$240,000. Use Lagrange multipliers to find the maximum production level for this manufacturer.

16. Find the least squares regression line for the points $(1, 2), (3, 3), (6, 4), (8, 6)$, and $(11, 7)$.

In Exercises 17 and 18, evaluate the double integral.

17. $\displaystyle\int_0^1 \int_x^1 (30x^2y - 1)\, dy\, dx$ **18.** $\displaystyle\int_0^{\sqrt{e-1}} \int_0^{2y} \frac{1}{y^2 + 1}\, dx\, dy$

19. Use a double integral to find the area of the region bounded by the graphs of $y = 3$ and $y = x^2 - 2x + 3$ (see figure).

20. Use a double integral to find the volume of the solid bounded by the graphs of $z = 8 - 2x, z = 0, y = 0, y = 3, x = 0$, and $x = 4$.

21. Find the average value of $f(x, y) = x^2 + y$ over the region defined by a rectangle with vertices $(0, 0), (1, 0), (1, 3)$, and $(0, 3)$.

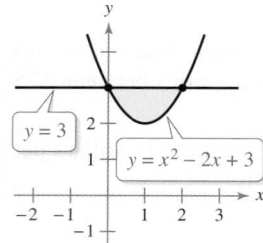

Figure for 19

Answers to Selected Exercises

Chapter 0

Section 0.1   *(page 8)*

1. (a) Natural: $\{5\}$
(b) Integer: $\{-9, 5\}$
(c) Rational: $\left\{-9, -\frac{7}{2}, 5, \frac{2}{3}, 0.1\right\}$
(d) Irrational: $\left\{\sqrt{2}\right\}$

3. (a) Natural: $\left\{12, 1, \sqrt{4}\right\}$ *(Note:* $\sqrt{4} = 2$)
(b) Integer: $\left\{12, -13, 1, \sqrt{4}\right\}$
(c) Rational: $\left\{12, -13, 1, \sqrt{4}, \frac{3}{2}\right\}$
(d) Irrational: $\left\{\sqrt{6}\right\}$

5. (a) Natural: $\left\{\frac{8}{2}, 9\right\}$ *(Note:* $\frac{8}{2} = 4$)
(b) Integer: $\left\{\frac{8}{2}, -4, 9\right\}$
(c) Rational: $\left\{\frac{8}{2}, -\frac{8}{3}, -4, 9, 14.2\right\}$
(d) Irrational: $\left\{\sqrt{10}\right\}$

7. $0.\overline{6}$ **9.** $0.\overline{126}$ **11.** $-1 < 2.5$
13. $-6 < 7$ **15.** $1 > -3.5$

17. $\frac{5}{6} > \frac{2}{3}$ **19.** $\frac{204}{60}, \frac{31}{9}, 3.45, 2\sqrt{3}, \frac{7}{2}$

21. $\frac{127}{90}, \frac{584}{413}, \frac{7071}{5000}, \sqrt{2}, \frac{47}{33}$ **23.** $x \le 4$ **25.** $0 < x \le 3$
27. $x < 0$ denotes all negative real numbers.   **29.** $x \le 5$ denotes all real numbers less than or equal to 5.

31. $x > 3.5$ denotes all real numbers greater than 3.5.

33. $-2 < x < 2$ denotes all real numbers greater than -2 and less than 2.

35. $-1 \le x < 0$ denotes all real numbers greater than or equal to -1 and less than 0.

37. $x > 0$ **39.** $5 < y \le 12$ **41.** $A \ge 21$
43. $-0.5\% \le r \le 3.8\%$ **45.** 10   **47.** -3 **49.** -1.6
51. -1 **53.** $\pi - 3$ **55.** $|-7| = |7|$
57. $|-3| > -|-3|$ **59.** $-|-2| = -|2|$ **61.** 4
63. 1   **65.** 7   **67.** 51   **69.** 3.67   **71.** $|z - 6| > 1$
73. $|x + 10| \ge 6$ **75.** $|y - 0| \ge 6 \Rightarrow |y| \ge 6$
77. $|x - m| > 5$ **79.** 81 miles   **81.** 89°F

$|a - b|$	$0.05b$	*Passes Budget Variance Test*
83. $876.55	$1500	No
85. $264.32	$600	Yes

$|a - b|$	$0.0012b$	*Passes Quality Control Test*
87. 0.002	0.0168	Yes
89. 0.027	0.0192	No
91. 0.022	0.0156	No

93. When u and v have the same sign, $|u + v| = |u| + |v|$. When u and v have different signs, $|u + v| < |u| + |v|$.

95. $a \le 0$; If the original value of a is negative, then $|a|$ results in a positive number. Because a is negative, the expression $|a| = a$ states that $|a|$ is equal to a negative number, which can never happen. So, if a is originally negative, $|a|$ must equal $-a$, which is a positive value.

Section 0.2   *(page 18)*

> **Skills Warm Up**   *(page 18)*
>
> **1.** $-4 < -2$ **2.** $0 > -3$ **3.** $-\frac{11}{9} > -\frac{13}{9}$
> **4.** $-\frac{12}{33} < \frac{11}{33}$ **5.** $|6 - 4| = 2$ **6.** $|2 - (-2)| = 4$
> **7.** $|0 - (-5)| = 5$ **8.** $|3 - (-1)| = 4$

1. $7x, 4$ **3.** $x^2, -4x, 8$ **5.** $2x^2, -9x, 13$ **7.** -2
9. 6   **11.** (a) -10 (b) -6 **13.** (a) 14   (b) 2
15. (a) $\frac{1}{2}$ (b) Undefined. You cannot divide by zero.
17. 1   **19.** 35   **21.** $\frac{1}{4}$ **23.** Commutative (addition)
25. Inverse (addition)   **27.** Distributive Property
29. Inverse (multiplication)   **31.** Identity (addition)
33. Identity (multiplication)   **35.** Associative (addition)
37. $x(3y) = (x \cdot 3)y$ Associative (multiplication)
$= (3x)y$ Commutative (multiplication)

39. $2^4 \cdot 3$ **41.** $2^4 \cdot 3 \cdot 5$ **43.** $3x$ **45.** $\dfrac{17x}{20}$ **47.** $\dfrac{12}{x}$
49. 7   **51.** $\frac{1}{24}$ **53.** $\frac{7}{20}$ **55.** $\frac{5}{6}$ **57.** -0.13
59. 1.56   **61.** -4 **63.** 5   **65.** 23.8   **67.** 46.25
69. (a) 20.7%
(b) National Defense: about $617 billion
Other: about $495 billion
Income Security: about $433 billion
Education and Veterans' Benefits: about $149 billion
Health and Medicare: about $671 billion
Social Security: about $617 billion
71. ≈ 1695 patients
73. Food: 38.6%   **75.** Answers will vary.
Vet care: 26.4%
Supplies/OTC medicine: 22.9%
Live animal purchases: 4.7%
Grooming and boarding: 7.4%

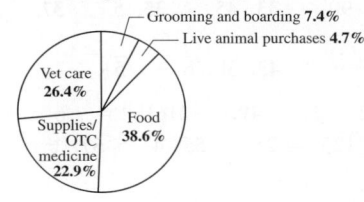

Section 0.3   *(page 27)*

Skills Warm Up   *(page 27)*

1. 1   **2.** 5   **3.** -4 **4.** $-\frac{3}{2}$ **5.** $-\frac{1}{4}$ **6.** 1

7. $\frac{3}{7}$ **8.** 0   **9.** $-\frac{3}{8}$ **10.** 1

1. 64   **3.** 8   **5.** 729   **7.** -81 **9.** 5184   **11.** $-\frac{3}{5}$

13. 1   **15.** $-125z^3$ **17.** $16x^7$ **19.** $10x^4$ **21.** $-3z^7$

23. $\dfrac{5y^4}{3}$ **25.** $\dfrac{5184}{y^7}$ **27.** $\dfrac{7}{x}$ **29.** $\dfrac{1}{x}$ **31.** 27^n

33. $\dfrac{5x}{3} + 5$ **35.** $1, x \neq 0$ **37.** $\dfrac{1}{2}$ **39.** 8   **41.** $-\dfrac{3}{10}$

43. $\dfrac{4}{x^2}$ **45.** $\dfrac{3a^3}{b^3}$ **47.** $\dfrac{10}{x}$ **49.** $\dfrac{125x^9}{y^{12}}$ **51.** 18

53. $\frac{7}{16}$ **55.** $10^3 = 1000$

$\qquad\qquad 10^2 = 100$ As the exponents decrease

$\qquad\qquad 10^1 = 10$ by 1, the answers decrease

$\qquad\qquad 10^0 = 1$ by a factor of 10.

$\qquad\qquad 10^{-1} = 0.1$

$\qquad\qquad 10^{-2} = 0.01$

57. 5.73×10^7 square miles   **59.** 9.46×10^{12} kilometers

61. 1×10^{-7} meter   **63.** 350,000,000 air sacs

65. 0.0000000000000000001602 coulomb

67. (a) 6.0×10^4 (b) 2.0×10^{11}

69. (a) $(4.8 \times 10^{10})(2.5 \times 10^8) = 1.2 \times 10^{19}$

$\qquad$ (b) $\dfrac{1.2 \times 10^{-8}}{6.4 \times 10^{-6}} = 1.875 \times 10^{-3}$

71. 161,100   **73.** 0.325   **75.** 1.007843137

77. (a) 3.071×10^6 (b) 3.077×10^{10}

79. (a) 4.907×10^{17} (b) 1.479

81. (a) \$15,682.76   (b) \$15,680.58

$\qquad$ (c) \$15,672.12   (d) \$15,650.28

As the number of compoundings per year decreases, the balance in the account also decreases.

83. About 4.32%   **85.** About 4 hours

Section 0.4   *(page 36)*

Skills Warm Up   *(page 36)*

1. $\frac{4}{27}$ **2.** 48   **3.** $-8x^3$ **4.** $6x^7$ **5.** $28x^6$

6. $\frac{1}{5}x^2$ **7.** $3z^4$ **8.** $\dfrac{25}{4x^2}$ **9.** 1   **10.** $(x+2)^{10}$

1. $\sqrt{4} = \pm 2$ **3.** $\sqrt[3]{-8} = -2$ **5.** $\sqrt[3]{27} = 3$

7. $\sqrt[4]{-1}$ is not a real number.   **9.** $\sqrt[6]{64} = \pm 2$ **11.** 3

13. 3   **15.** 2   **17.** $\sqrt[4]{-16}$ is not a real number.

19. $-\frac{3}{5}$ **21.** $-3\sqrt[3]{2}$ **23.** -125 **25.** $2x\sqrt[3]{2x^2}$

27. $4x^3$ **29.** $\dfrac{z}{5}$ **31.** 90   **33.** 45   **35.** 5   **37.** $\dfrac{\sqrt{5}}{5}$

39. $4\sqrt[3]{4}$ **41.** $\dfrac{x(5+\sqrt{3})}{11}$ **43.** $3(\sqrt{6} - \sqrt{5})$

45. $9^{1/2} = 3$ **47.** $\sqrt[5]{32} = 2$ **49.** $(-216)^{1/3} = -6$

51. $81^{3/4} = 27$ **53.** $\sqrt[3]{125^2} = 25$ **55.** 4   **57.** $\frac{27}{8}$

59. -4 **61.** 25   **63.** $2^{1/2}$ **65.** 1   **67.** x^3

69. $8x^6y^3$ **71.** $\sqrt{y}$ **73.** $2\sqrt[4]{2}$ **75.** $\sqrt[4]{(x+1)^3}$

77. $2\sqrt{x}$ **79.** $31\sqrt{2}$ **81.** $-2\sqrt{y}$ **83.** 3.557

85. 2.140   **87.** 1.787   **89.** 0.382

91. ⬛ 4 ⊟ ⬛ 7 ⬛ ⬛ ⬛ ÷ 3   **93.** $\sqrt{5} + \sqrt{3} > \sqrt{5+3}$

95. $\sqrt{3} \cdot \sqrt[4]{3} > \sqrt[8]{3}$ **97.** $10\sqrt{11}$ feet $\times\ 10\sqrt{11}$ feet

99. 25.01%   **101.** ≈ 2.221 seconds   **103.** No

105. ≈ 40.2 miles per hour   **107.** $\approx 17.4°$F

109. ≈ 494 vibrations per second

111. a; Higher notes have higher frequencies.

113. $\left(2/\sqrt{5}\right)^2 = \frac{4}{5}$. This is not the same as rationalizing the denominator, which gives the result

$\qquad 2/\sqrt{5} \cdot \sqrt{5}/\sqrt{5} = \left(2\sqrt{5}\right)/5.$

115. No; $\sqrt{4x^2} \geq 0$, whereas $2x$ can be less than zero.

Quiz Yourself   *(page 39)*

1. $-|-7| < |-7|$ **2.** $-(-3) = |-3|$

3. Natural: $\left\{\sqrt{1}\right\}$

$\qquad$ Integer: $\left\{-6, \sqrt{1}\right\}$

$\qquad$ Rational: $\left\{-6, \sqrt{1}, \frac{3}{4}, -\frac{1}{2}\right\}$

$\qquad$ Irrational: $\left\{\sqrt{11}\right\}$

4. $x \geq 0$ **5.** $|5 - (-6)| = 11$ **6.** $3x^2, -7x, 2$

7. -4 **8.** -13 **9.** $\dfrac{5}{14}$ **10.** $\dfrac{11}{9}$ **11.** $-\dfrac{2}{x}$

12. $\dfrac{y^{10}}{3}$ **13.** 1.4×10^{-5} **14.** \$5579.71

15. -1 **16.** -64 **17.** 9   **18.** $-\sqrt[3]{3}$ **19.** $2\sqrt{3}$

20. 22 cm $\times$ 22 cm $\times$ 22 cm

Section 0.5   *(page 46)*

Skills Warm Up   *(page 46)*

1. $42x^3$ **2.** $-45x^5$ **3.** $-20z^2$ **4.** $-15x^{-1}$

5. $-27x^6$ **6.** $-3x^6$ **7.** $\frac{9}{4}z^3, z \neq 0$

8. $\frac{8}{5}x^3, x \neq 0$ **9.** $\dfrac{9}{4x^2}$ **10.** xy^2

1. $4x^2 + 3x + 2$

$\qquad$ Degree: 2; leading coefficient: 4

3. $x^7 - 8$

$\qquad$ Degree: 7; leading coefficient: 1

5. $-2x^5 + 6x^4 - x + 1$

$\qquad$ Degree: 5; leading coeficient: -2

7. Polynomial, $-3x^3 + 2x + 8$, degree 3

9. Not a polynomial

11. Polynomial, $-w^4 + 2w^3 + w^2$, degree 4

13. (a) -3 (b) 1   (c) 5   (d) 17

15. (a) -10 (b) -1 (c) 4   (d) 5

17. $-2x - 10$ **19.** $2x^3 - 4x - 5$ **21.** $8x^3 + 29x^2 + 11$

23. $3x^3 - 6x^2 + 3x$ **25.** $9x^3 - 21x^2$ **27.** $x^2 + 7x + 12$

29. $6x^2 - 7x - 5$ **31.** $x^4 + 3x^3 - x^2 - 6x$

33. $-3x^4 - x^3 - 12x^2 - 19x - 5$ **35.** $x^2 - 25$

37. $x^2 + 12x + 36$ **39.** $4x^2 - 20xy + 25y^2$

41. $x^2 + 2xy + y^2 - 6x - 6y + 9$

43. $x^3 + 3x^2 + 3x + 1$ **45.** $8x^3 - 12x^2y + 6xy^2 - y^3$

47. $9y^4 - 1$ **49.** $m^2 - n^2 - 6m + 9$ **51.** $x^4 + x^2 + 1$

53. $24x^2 - 53x - 7$ **55.** $4x^3 + 6x^2 - 8x - 5$

57. The middle term was omitted when squaring the binomial.
$(x - 3)^2 = x^2 - 2(x)(3) + 3^2 = x^2 - 6x + 9$

59. $1000r^3 + 3000r^2 + 3000r + 1000$ **61.** Yes; yes; no

63. 77,918.76; 82,562.74; In the years 2006 and 2007, the total amounts of federal student aid disbursed were approximately $77,918,760,000 and $82,562,740,000.

65. $x = 4$ inches: $V = 6076$ cubic inches
$x = 6$ inches: $V = 7290$ cubic inches
$x = 10$ inches: $V = 7030$ cubic inches
$x = 6$ inches produces the greatest volume.

67. Answers will vary.

Section 0.6   *(page 53)*

Skills Warm Up   *(page 53)*

1. $15x^2 - 6x$ **2.** $-2y^2 - 2y$ **3.** $4x^2 + 12x + 9$

4. $9x^2 - 48x + 64$ **5.** $2x^2 + 13x - 24$

6. $-5z^2 - z + 4$ **7.** $4y^2 - 1$ **8.** $x^2 - a^2$

9. $x^3 + 12x^2 + 48x + 64$

10. $8x^3 - 36x^2 + 54x - 27$

1. $3(x + 2)$ **3.** $3x(x^2 - 2)$ **5.** $(x - 5)(3x + 8)$

7. $(x + 6)(x - 6)$ **9.** $(4x + 3y)(4x - 3y)$

11. $(x + 1)(x - 3)$ **13.** $(9x^2 + 1)(3x + 1)(3x - 1)$

15. $3(x + 1)(x - 1)$ **17.** $2(2x + 5y)(2x - 5y)$

19. $(5 - x)(5 + x)(x - 1)$ **21.** $(x - 2)^2$ **23.** $(2y + 3)^2$

25. $\left(y - \frac{1}{3}\right)^2$ **27.** $(x - 2)(x^2 + 2x + 4)$

29. $(y + 5)(y^2 - 5y + 25)$ **31.** $\left(x - \frac{2}{3}\right)\left(x^2 + \frac{2}{3}x + \frac{4}{9}\right)$

33. $(x + 2)(x - 1)$ **35.** $(w - 2)(w - 3)$

37. $(y + 5)(y - 4)$ **39.** $(x - 20)(x - 10)$

41. $(3x - 2)(x - 1)$ **43.** $(3x + 1)(3x - 2)$

45. $(6x + 1)(x + 6)$ **47.** $(x - 1)(x^2 + 2)$

49. $(2x - 1)(x^2 - 3)$ **51.** $(2 - y^3)(3 + y)$

53. $(2x - 1)(x + 2)$ **55.** $(x - 2)(6x + 5)$ **57.** $4x(x - 2)$

59. $y\left(y - \frac{1}{3}\right)\left(y + \frac{1}{3}\right)$ **61.** $3(x + 4)(x - 4)$ **63.** $(x - 1)^2$

65. $(1 - 2x)^2$ **67.** $y(2y + 3)(y - 5)$ **69.** Prime

71. $(3x + 1)(x^2 + 5)$ **73.** $x(x - 4)(x^2 + 1)$

75. $-x(x + 10)$ **77.** $(x + 1)^2(x - 1)^2$

79. $2(t - 2)(t^2 + 2t + 4)$

81.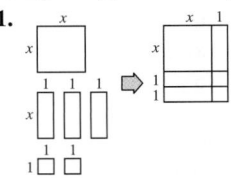

83. 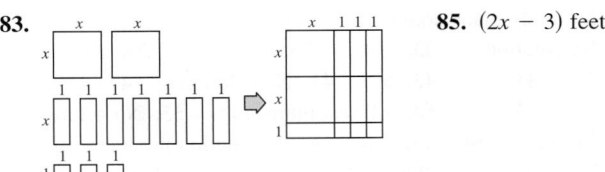   **85.** $(2x - 3)$ feet

87. (a) $(x + 8)(x + 3)$; factoring by grouping
(b) $(3x - 5)(x + 4)$; factoring by trial and error

89. $c = \{7, 12, 15, 16\}$; Answers will vary.

91. Answers will vary. Sample answer:
(1) Find a combination of factors of 2 and -15 such that the outer and inner products add up to the middle term $-7x$.
$2x^2 - 7x - 15 = (2x + 3)(x - 5)$
(2) Rewrite -7 as the sum of two factors of the product $2(-15)$. Then factor by grouping.
$$2x^2 - 7x - 15 = 2x^2 - 10x + 3x - 15$$
$$= 2x(x - 5) + 3(x - 5)$$
$$= (x - 5)(2x + 3)$$

93. Box 1: $V = (a - b)a^2$
Box 2: $V = (a - b)ab$
Box 3: $V = (a - b)b^2$
Multiplying $(a - b)$ by each term of $(a^2 + ab + b^2)$ produces the volumes of the three boxes.

Section 0.7   *(page 60)*

Skills Warm Up   *(page 60)*

1. $5x^2(1 - 3x)$ **2.** $(4x + 3)(4x - 3)$

3. $(3x - 1)^2$ **4.** $(2y + 3)^2$ **5.** $(z + 3)(z + 1)$

6. $2(x - 5)(x - 10)$ **7.** $(3 - x)(1 + 3x)$

8. $(3x - 1)(x - 15)$ **9.** $(s + 1)(s + 2)(s - 2)$

10. $2(x - 4)(x^2 + 4x + 16)$

1. (a) No   (b) Yes   **3.** (a) Yes   (b) Yes

5. All real numbers   **7.** All real numbers x except $x = 3$

9. All real numbers except $x = 0$ and $x = 4$

11. All real numbers greater than or equal to -1

13. $3x$, $x \neq 0$ **15.** $x - 2$, $x \neq 2$, $x \neq 0$

17. $x + 2$, $x \neq -2$, $x \neq 3$ **19.** $\dfrac{3x}{2}$, $x \neq 0$

21. $\dfrac{x}{2(x + 1)}$ **23.** $-\dfrac{1}{2}$, $x \neq 5$

25. $-(x + 5)$, $x \neq 5$ **27.** $\dfrac{x(x + 3)}{x - 2}$, $x \neq -2$

29. $\dfrac{y - 4}{y + 6}$, $y \neq 3$ **31.** $-1 - x^2$, $x \neq 2$ **33.** $z - 3$

35. $\dfrac{1}{5(x - 2)}$, $x \neq 1$ **37.** $-\dfrac{x(x + 7)}{x + 1}$, $x \neq 9$

39. $\dfrac{r + 1}{r}$, $r \neq 1$ **41.** $\dfrac{t - 3}{(t + 3)(t - 2)}$, $t \neq -2$

43. $\dfrac{x - 1}{x(x + 1)^2}$, $x \neq -2$ **45.** $\dfrac{3}{2}$, $x \neq -y$

47. $x(x + 1)$, $x \neq -1, 0$ **49.** $(x - 6)(x + 4)$

51. $x^2(x - 1)$ **53.** $\dfrac{5x}{x - 2}$ **55.** $\dfrac{4x}{x - 4}$ **57.** $\dfrac{4x - 17}{x - 5}$

59. $\dfrac{x - 4}{(x + 2)(x - 2)(x - 1)}$ **61.** $\dfrac{2 - x}{x^2 + 1}$, $x \neq 0$

63. $\dfrac{1}{2}$, $x \neq 2$ **65.** $\dfrac{1}{x}$, $x \neq -1$ **67.** $\dfrac{2x - 1}{2x}$, $x > 0$

69. (a) 4.88%   (b) $\dfrac{288(NM - P)}{N(12P + NM)}$; 4.88%

71.

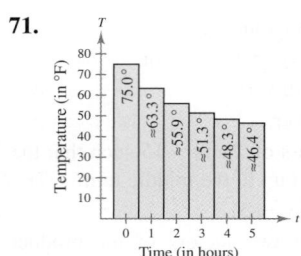

No

73. False. The domain of the left-hand side is $x^n \neq 1$.

75. Answers will vary.

Review Exercises   *(page 66)*

1. (a) Natural: $\{9\}$
   (b) Integer: $\{9, -7\}$
   (c) Rational: $\left\{9, -7, -\frac{8}{9}, \frac{5}{2}, 0.4\right\}$
   (d) Irrational: $\left\{\sqrt{6}\right\}$

3. $-4 < -3$

5. $x \leq -6$ denotes all real numbers less than or equal to -6.

7. $x \geq 0$ **9.** $2 < x \leq 5$ **11.** -14

13. 4   **15.** 10   **17.** $|x - 7| \geq 4$

19. $5x^3, -2x, 7$ **21.** (a) 2   (b) 0

23. Distributive Property   **25.** Commutative (addition)

27. 2   **29.** $\frac{2}{3}$ **31.** 0.10   **33.** $8x, x \neq 0$ **35.** $\frac{5}{x^4}$

37. 3.087×10^8 **39.** 7,280,000   **41.** 11,414.125

43.

Year	5	10	15
Balance	$10,014.37	$12,535.94	$15,692.44

Year	20	25
Balance	$19,643.73	$24,589.94

45. $16^{1/2} = 4$ **47.** 13   **49.** $2x^2$ **51.** $2\sqrt{3}$

53. $2 + \sqrt{3}$ **55.** $-3\sqrt{x}$ **57.** $\frac{1}{16}$ **59.** $\sqrt{5}$

61. 11.269   **63.** $-x + 5$ **65.** $x^2 - 8x - 14$

67. $x^2 - x - 2$ **69.** $x^3 + 64$ **71.** $x^2 + 8x + 16$

73. 254,691; In 2010, the number of mobile telephone sites in the United States was 254,691.

75. $4(x + 3)(x - 3)$ **77.** $(2x - 5)(4x^2 + 10x + 25)$

79. $(x + 1)(x - 5)$ **81.** $(x^2 - 2)(x - 4)$

83. All real numbers except $x = 4$

85. All real numbers where $x \geq -1$ **87.** $3x, x \neq 0$

89. $\frac{x - 2}{2}, x \neq -2$ **91.** $\frac{x + 3}{x - 1}, x \neq 0, 3$

93. $\frac{x - 1}{x - 3}, x \neq -1, \frac{1}{2}$ **95.** $\frac{3x^2 - 4x}{(x - 1)(x - 2)}$

97. $\frac{6}{x - 1}, x \neq -1$ **99.** $\frac{x + 1}{x - 1}, x \neq 0$

Test Yourself   *(page 68)*

1. -12

2.

Year	5	10	15
Balance	$4763.77	$5673.38	$6756.67

Year	20	25
Balance	$8046.81	$9583.29

The longer you leave the $4000 in the account, the more money you earn.

3. $-64x^6$ **4.** $-4\sqrt{x}$ **5.** 25   **6.** $4\left(\sqrt{3} - \sqrt{5}\right)$

7. $2x\sqrt{3x}$ **8.** $\frac{5 + \sqrt{7}}{9}$ **9.** $9x^2 + 42x + 49$

10. $-5x^2 + 29x$ **11.** $5(x + 4)(x - 4)$ **12.** $(2x + 3)^2$

13. $(x^2 - 3)(x - 6)$ **14.** $(x + 2)^2(x - 2)$

15. $\frac{1}{3}(x - 4), x \neq -4$ **16.** $\frac{x + 4}{3x + 5}, x \neq -3, \frac{5}{3}$

17. $\frac{4x^2 - 13x}{(x - 3)(x - 4)}$ **18.** $-\frac{x + 26}{(x + 5)(x - 2)}$

19. All real numbers greater than or equal to 10

20. All real numbers except $x = -9$

21. $\frac{2x^2 - 5x - 18}{5 + 5x - x^2}, x \neq 1, -2$

22. $x = 0$: 5.35, $x = 9$: 7.36
In 2000 and 2009, the average prices of a movie ticket in the United States were $5.35 and $7.36, respectively.

Chapter 1

Section 1.1   *(page 77)*

> **Skills Warm Up**   *(page 77)*
>
> **1.** $-3x - 10$ **2.** $5x - 12$ **3.** x **4.** $x + 26$
>
> **5.** $\frac{9x}{14}$ **6.** $\frac{3x}{4}$ **7.** $-\frac{1}{x(x + 1)}$ **8.** $\frac{8}{x}$
>
> **9.** $\frac{7x - 8}{x(x - 2)}$ **10.** $-\frac{2}{x^2 - 1}$

1. Identity   **3.** Conditional equation

5. Conditional equation

7. (a) No   (b) No   (c) Yes   (d) No

9. (a) Yes   (b) Yes   (c) No   (d) No

11. (a) Yes   (b) No   (c) No   (d) No

13. (a) No   (b) No   (c) No   (d) Yes

15. (a) Yes   (b) No   (c) No   (d) No

17. 5   **19.** -4 **21.** 3   **23.** No solution

25. Infinitely many solutions   **27.** 9   **29.** -26

31. No solution   **33.** -4 **35.** $z = -\frac{6}{5}$ **37.** 10

39. 4   **41.** 3   **43.** 5   **45.** No solution   **47.** $x = 1$

49. 9   **51.** 0   **53.** All real numbers   **55.** No solution

57. Because substituting 2 for x in the equation produces division by zero, $x = 2$ cannot be a solution to the equation.

59. Extraneous solutions may arise when a fractional expression is multiplied by factors involving the variable.

61. Equivalent equations have the same solutions.

Example: $2x - 6 = 0$ and $x - 3 = 0$ both have the solution $x = 3$.

63. $x \approx 138.889$ **65.** $x \approx 62.372$ **67.** $x \approx 19.993$

69. Use the *table* feature in ASK mode, or use the scientific calculator part of the graphing utility.

71. (a) 6.46   (b) 6.41

(c) There is a roundoff error because decimals in the original problem have 5 decimal places.

73. (a) 56.09   (b) 56.13

(c) There is a roundoff error because the fractions in the numerator and denominator have infinitely many decimal places.

75. 2009   **77.** 65.8 in.   **79.** (a) 2007   (b) 2010

Section 1.2   *(page 88)*

Skills Warm Up   *(page 88)*

1. 14   **2.** 4   **3.** -3 **4.** 4   **5.** -2

6. 1   **7.** $\frac{2}{5}$ **8.** $\frac{10}{3}$ **9.** 6   **10.** $-\frac{11}{5}$

---

1. (a) One natural number + Next consecutive natural number

(b) One natural number $= x$

Next consecutive natural number $= x + 1$

(c) $x + (x + 1) = 2x + 1$

3. (a) Distance $= t$ hours $\times$ Car speed

(b) Hours $= t$

Car speed $= 50$ mi/h

(c) $50t$

5. (a) Perimeter $= 2(\text{Width}) + 2(\text{Length})$

(b) Width $= x$

Length $= 2x$

(c) $2x + 2(2x) = 6x$

7. $5 + x = 8$ **9.** $\frac{r}{2} = 9$ **11.** $60{,}000 + 0.08(60{,}000)$

13. (a) Sum $=$ One number + Next consecutive number

(b) One number $= x$

Next consecutive number $= x + 1$

Sum $= 525$

(c) $525 = x + (x + 1)$ (d) 262, 263

15. (a) $148 =$ One positive number $-$ Another positive number

(b) One positive number $= x$

Another positive number $= 5x$

(c) $148 = 5x - x$ (d) 37, 185

17. (a) Raise $=$ Percent $\times$ Old wage

(b) Old wage $= \$440$

Raise $= \$66$

Percent $= r$

(c) $66 = 440r$ (d) 15%

19. (a) Change in price $=$ Percent $\times$ Original price

(b) Change in price $= \$30$

Percent $= r$

Original price $= \$149$

(c) $30 = 149r$ (d) $\approx 20.13\%$

21. (a) Total paycheck $=$ Your weekly paycheck + Your coworker's paycheck

(b) Total paycheck $= \$848$

Your weekly paycheck $= 1.12x$

Your coworker's paycheck $= x$

(c) $848 = 2.12x$

(d) \$400, \$448

23. $\approx 17.5\%$ decrease   **25.** $\approx 16.2\%$ increase

27. $\approx 3.4\%$ increase   **29.** $\approx 128.57\%$

31. $\approx 54.29\%$ **33.** $\approx 56.41\%$ decrease

35. (a) \$37,800   (b) \$40,748.40   (c) \$44,578.75

37. (a) 46.6%

(b) Cars: 5,459,515

Light trucks: 4,940,066

Other trucks: 201,419

(c) $\approx 96.1\%$

39. ≈ 0.57 ft $\times$ 0.93 ft   **41.** ≈ 8.9 years

43. 97 or greater   **45.** \$1411.76   **47.** \$18

49. (a) \$361.25

(b) Percent increase needed: $\approx 17.65\%$

A higher percent increase is needed because you are taking a percentage of a smaller number.

51. $2\frac{1}{3}$ hours   **53.** 62.5 feet   **55.** 1.28 seconds

57. \$10,500 at 6.5% and \$4500 at 7.5%

59. Stock A: \$2200, Stock B: \$2800

61. 8571 units per month   **63.** $h = \dfrac{2A}{b}$ **65.** $l = \dfrac{P - 2w}{2}$

67. $L = \dfrac{S}{1 - R}$ **69.** $P = \dfrac{A}{\left(1 + \dfrac{r}{n}\right)^{nt}}$

71. ≈ 48 ft   **73.** ≈ 32.1 gallons

75. Williams: $\approx \$24{,}933$

Gonzalez: $\approx \$27{,}833$

Walters: $\approx \$25{,}633$

Gilbert: $\approx \$26{,}100$

Hart: $\approx \$24{,}433$

Team average: April: \$25,600, May: \$25,920, June: \$25,840

77. "takes 30 minutes"; "from a depth of 150 feet"

Section 1.3   *(page 100)*

Skills Warm Up   *(page 100)*

1. $\dfrac{\sqrt{14}}{10}$ **2.** $4\sqrt{2}$ **3.** 14   **4.** $\dfrac{\sqrt{10}}{4}$

5. $x(3x + 5)$ **6.** $(2x - 5)(2x + 5)$

7. $-(x - 7)(x - 15)$ **8.** $(x - 2)(x + 9)$

9. $(5x - 1)(2x + 3)$ **10.** $(6x - 1)(x - 12)$

---

1. $2x^2 + 5x - 3 = 0$ **3.** $x^2 - 19x = 0$

5. $x^2 - 6x + 4 = 0$ **7.** $2x^2 - 2x + 1 = 0$

9. $3x^2 - 60x - 10 = 0$ **11.** $4, -2$ **13.** $0, -\frac{1}{2}$

15. -5 **17.** $3, -\frac{1}{2}$ **19.** $2, -6$ **21.** $-2, -5$

23. ± 4 **25.** $\pm\sqrt{7} \approx \pm 2.65$ **27.** $\pm 2\sqrt{3} \approx \pm 3.46$

29. 0, 24

31. $-2 + 2\sqrt{3} \approx 1.46$
$-2 - 2\sqrt{3} \approx -5.46$

33. ±5 **35.** $\pm\sqrt{38} \approx \pm6.16$ **37.** $\pm\dfrac{\sqrt{115}}{5} \approx \pm2.14$

39. $\pm\dfrac{\sqrt{78}}{3} \approx \pm2.94$ **41.** ±8 **43.** 1 **45.** $\pm\dfrac{3}{4}$

47. $\dfrac{3}{2}$ **49.** $3, -11$ **51.** $\dfrac{3}{2}, -\dfrac{1}{2}$ **53.** $5, -\dfrac{10}{3}$

55. $9, 3$ **57.** $\dfrac{1}{5}, 1$ **59.** $-1, -5$ **61.** $-\dfrac{1}{2}$

63. Algebra argument:

$(x + 2)^2 = (x + 2)(x + 2)$ Definition of exponent
$\qquad\quad = x^2 + 2x + 2x + 4$ FOIL
$\qquad\quad = x^2 + 4x + 4$ Combine like terms.

So, $(x + 2)^2 \neq x^2 + 4$.

Graphing utility argument:

(1) Let $y_1 = (x + 2)^2$ and $y_2 = x^2 + 4$. Use the table feature with an arbitrary value of x (but not $x = 0$). The table will show that the values of y_1 are not the same as the values of y_2.

(2) Use the scientific calculator portion of the graphing utility to show that if $x = 5$, $(5 + 2)^2 = 49$ and $5^2 + 4 = 29$. So, $(x + 2)^2$ is not equal to $x^2 + 4$.

65. 34 feet × 48 feet

67. Base: $2\sqrt{2}$ feet
Height: $2\sqrt{2}$ feet

69. 5 feet   **71.** ≈ 3.54 seconds   **73.** ≈ 1.43 seconds

75. 42 seconds faster   **77.** ≈ 4.24 centimeters

79. 976 miles   **81.** ≈ 494.97 meters   **83.** 60,000 units

85. 2019 $(t \approx 8.85)$

87. (a) 1987 $(t \approx 18.74)$

(b) Yes; the model is a good representation through 1890.

(c) 2010: 312,493.19 thousand
They are close.

(d) Yes

89. 2019 $(t \approx 9.13)$

91. 1 P.M. $(t \approx 12.96)$
No; For 7 P.M. $(t = 19)$, the model yields a value of $T = 144.81°F$, which is extremely high and unreasonable.

Section 1.4   *(page 110)*

Skills Warm Up   *(page 110)*

1. 13   **2.** 5   **3.** $4\sqrt{6}$ **4.** $3\sqrt{73}$
5. $2, -1$ **6.** $\dfrac{3}{2}, -3$ **7.** $5, -1$ **8.** $\dfrac{1}{2}, -7$
9. $5, 2$ **10.** $4, -1$

1. One real solution   **3.** Two real solutions
5. No real solutions   **7.** Two real solutions
9. $\dfrac{1}{2}, -1$ **11.** $\dfrac{1}{4}, -\dfrac{3}{4}$ **13.** $-3 \pm \sqrt{13}$ **15.** $5 \pm \sqrt{3}$
17. $-4 \pm 2\sqrt{5}$ **19.** $\dfrac{2}{3} \pm \dfrac{\sqrt{7}}{3}$ **21.** $-\dfrac{1}{3} \pm \dfrac{\sqrt{11}}{6}$
23. $-\dfrac{1}{2} \pm \sqrt{2}$ **25.** $\dfrac{2}{7}$ **27.** $2 \pm \dfrac{\sqrt{6}}{2}$ **29.** $6 \pm \sqrt{11}$
31. $x \approx 0.976, -0.643$ **33.** $-0.290, -2.200$
35. No real solution   **37.** -11 **39.** $\pm\sqrt{10}$
41. $1 \pm \sqrt{2}$ **43.** $-2, 4$

45. $\dfrac{-1 \pm \sqrt{37}}{6}$ **47.** 50, 50; Real-life problems will vary.

49. 7, 8, or $-8, -7$; Real-life problems will vary.

51. 200 units   **53.** 653 units

55. 6 rows, 18 chairs per row

57. 14 in. × 14 in.   **59.** ≈ 7.2 seconds

61. Moon: ≈ 14.9 seconds   **63.** Moon
Earth: ≈ 2.6 seconds

65. Oklahoma City to Austin: 354.45 mi
Austin to New Orleans: 433.55 mi

67. 4:00 P.M. $(t \approx 3.9)$

69. Southbound: ≈ 550 miles per hour
Eastbound: ≈ 600 miles per hour

71. (a) 2011 $(t \approx 11)$ (b) 2008 $(t \approx 8)$
(c) 2016 $(t \approx 16.6)$; yes

73. (a) $\approx 16.8°C$ (b) ≈ 2.5

75. 5279 units or 94,721 units

77. No; the equation that has a discriminant of 5 has two distinct real solutions, while the equation that has a discriminant of -10 has no real solutions.

79. Answers will vary.

Quiz Yourself   *(page 114)*

1. $x = -6$ **2.** $x = 6$ **3.** $x = -2$ **4.** No solution

5. Use the *table* feature in ASK mode or the scientific calculator portion of the graphing utility.

6. 328.954   **7.** 431.398

8. $8.50x + 30,000 = 200,000$; 20,000 units

9. $300,000 = x(75 - 0.0002x)$; 4044 units or 370,956 units

10. $x = \dfrac{2}{3}, -5$ **11.** $x = \pm\sqrt{5}; x \approx \pm2.24$

12. $x = -3 \pm \sqrt{17}; x \approx -7.12, 1.12$

13. $x = -1 \pm \sqrt{7}; x \approx -3.65, 1.65$

14. $x = \dfrac{-7 \pm \sqrt{73}}{6}; x \approx -2.59, 0.26$

15. $x \approx 1.568, -0.068$ **16.** No real solutions

17. One real solution

18. Answers will vary. Sample answer: Use the FOIL method $[(x + 3)^2 = (x + 3)(x + 3) = x^2 + 6x + 9]$, use the *table* feature of your graphing utility, or use the scientific calculator portion of your graphing utility to evaluate the solution.

19. 4.5 seconds   **20.** 9 in. × 9 in. × 6 in.

Section 1.5   *(page 123)*

Skills Warm Up   *(page 123)*

1. 11   **2.** $20, -3$ **3.** $5, -45$ **4.** $0, -\dfrac{1}{5}$
5. $\dfrac{2}{3}, -2$ **6.** $\dfrac{11}{6}, -\dfrac{5}{2}$ **7.** $1, -5$ **8.** $\dfrac{3}{2}, -\dfrac{5}{2}$
9. $\dfrac{3 \pm \sqrt{5}}{2}$ **10.** $2 \pm \sqrt{2}$

1. $3, -1, 0$ **3.** $0, \pm3$ **5.** ±3 **7.** $-3, 0$ **9.** $\pm2, 7$
11. ±1 **13.** $\pm\sqrt{11}, \pm1$ **15.** ±2 **17.** $\pm\dfrac{1}{2}, \pm4$
19. $1, -2$ **21.** 50   **23.** 26   **25.** -16 **27.** $\dfrac{1}{4}$
29. $6, 5$ **31.** $2, -5$ **33.** 0 **35.** $-59, 69$ **37.** 1
39. $\pm\sqrt{69}$ **41.** -1 **43.** $-12, 2$ **45.** $1, -3$

47. $6, -1$ **49.** $1, -3$ **51.** $3, -2$

53. $\sqrt{3}, -3$ **55.** $10, -1$

57. The quadratic equation was not written in general form before the values of a, b, and c were substituted in the Quadratic Formula. The general form for this equation is $3x^2 - 7x - 4 = 0$ ($a = 3$, $b = -7$, and $c = -4$), and the only solution is

$$x = \frac{-(-7) \pm \sqrt{(-7)^2 - 4(3)(-4)}}{2(3)}.$$

59. $x \approx \pm 1.038$ **61.** $x \approx 16.756$ **63.** 34 **65.** 2.2%

67. $\approx 12.98\%$ **69.** $45{,}000$ passengers **71.** 57 years old

73. $67{,}760$ units; It does not make sense for demand x or price p to be less than zero.

75. Least acceptable weight: 15.6 ounces
Greatest acceptable weight: 16.4 ounces

77. 13 minutes, 20 seconds

79. A polynomial of degree n has at most n real solutions.

Section 1.6 *(page 134)*

Skills Warm Up *(page 134)*

1. $-\frac{1}{2}$ **2.** $-\frac{1}{6}$ **3.** -3 **4.** -6 **5.** $x \geq 0$

6. $-3 < z < 10$ **7.** $P \leq 2$ **8.** $W \geq 200$

9. $2, 7$ **10.** $0, 1$

1. $-1 \leq x \leq 5$; Bounded **3.** $x > 11$; Unbounded

5. $x < -2$; Unbounded

7. c **8.** h **9.** f **10.** e

11. g **12.** a **13.** b **14.** d

15. (a) Yes (b) No (c) Yes (d) No

17. (a) Yes (b) No (c) No (d) Yes

19. (a) No (b) Yes (c) Yes (d) Yes

21. (a) No (b) Yes (c) No (d) Yes

23. If $2x > 6$, then $x > 3$. **25.** If $2x \leq -8$, then $x \leq -4$.

27. If $2 - 4x > -10$, then $x < 3$.

29. If $-\frac{2}{3}x \geq -6$, then $x \leq 9$.

31. $x \geq 6$

33. $x > -4$

35. $x < 25$

37. $x > 2$

39. $x \leq -\frac{2}{3}$

41. $x < -18$

43. $x > \frac{2}{5}$

45. $2 \leq x < 4$

47. $-4 \leq x < 2$

49. $-9 < x < 15$

51. $-\frac{3}{4} < x < -\frac{1}{4}$

53. $-6 < x < 6$

55. $x < -6$ or $x > 6$

57. $-8 < x < 2$

59. No solution

61. $x < -\frac{1}{2}$ or $x > \frac{11}{2}$

63. $x \leq -\frac{29}{2}$ or $x \geq -\frac{11}{2}$

65. $x \leq -7$ or $x \geq 13$

67. $8 < x < 10$

69. $x < 0$ or $x > 5$

71. $|x| \leq 2$ **73.** $|x - 9| \geq 3$ **75.** $|x - 12| \leq 10$

77. $|x + 3| > 5$ **79.** Less than $18{,}269.23$ miles

81. 2008 ($t > 7.9$) **83.** More than 250 miles

85. (a) 1998 to 2005 ($t < 15.23$); yes; According to the bar graph, the average salary was less than $\$1.5$ million from 1998 to 2005.

(b) 2011; Yes; The average salary has been increasing.

87. $33\frac{1}{3}$ weeks **89.** ≤ 15 inches

91. $\geq 10{,}417$ units **93.** Greater than 4%

95. $[\approx 106.864, \approx 109.464]$

97. $[65.8, 71.2]$

99. Minimum $= 20\%$; Maximum $= 80\%$

101. $\approx \$0.28$

Section 1.7 *(page 145)*

Skills Warm Up *(page 145)*

1. $y < -6$ **2.** $z > -\frac{9}{2}$ **3.** $-3 \leq x < 1$

4. $x \leq -5$ **5.** $-3 < x$ **6.** $5 < x < 7$

7. $-\frac{7}{2} \leq x \leq \frac{7}{2}$ **8.** $x < 2, x > 4$

9. $x < -6, x > -2$ **10.** $-2 \leq x \leq 6$

1. $(-\infty, -5), (-5, 5), (5, \infty)$

3. $(-\infty, -4), \left(-4, \frac{1}{2}\right), \left(\frac{1}{2}, \infty\right)$

5. $(-\infty, -1), (-1, 1), (1, \infty)$

7. $[-3, 3]$

9. $(-\infty, -2) \cup (2, \infty)$

11. $(-7, 3)$

13. $(-\infty, -5] \cup [1, \infty)$

15. $(-3, 2)$

17. $(-\infty, -1) \cup (1, \infty)$

19. $(-3, 1)$

21. $(-\infty, 4) \cup (4, \infty)$

23. $\frac{1}{2}$

25. All real numbers **27.** No solution

29. $(-\infty, 0) \cup (0, \frac{3}{2})$

31. $[-2, 0] \cup [2, \infty)$

33. $[-1, 1] \cup [2, \infty)$

35. $(-\infty, -1) \cup (0, 1)$

37. $(-\infty, -1) \cup (4, \infty)$

39. $(5, 15)$

41. $\left(-5, -\frac{3}{2}\right) \cup (-1, \infty)$

43. $\left(-\frac{3}{4}, 3\right) \cup [6, \infty)$

45. $(-\infty, -3] \cup [3, \infty)$ **47.** All real numbers
49. $\left[-\frac{9}{2}, \frac{9}{2}\right]$ **51.** $(-\infty, 2) \cup (5, \infty)$ **53.** All real numbers
55. $\left(\frac{5}{3}, \infty\right)$ **57.** $(-\infty, -3] \cup [0, 3]$ **59.** $[-2, \infty)$
61. $(-3.51, 3.51)$ **63.** $(-0.13, 25.13)$ **65.** $(2.26, 2.39)$
67. Between about 13.8 meters and about 36.2 meters
69. (a) $90{,}000 \le x \le 100{,}000$ (b) $\$30 \le p \le \$32$
  (c) About 185,967 units; The company should produce a
  maximum of about 185,967 units.
71. 4.8%
73. (a) 6393.56 million; It matches the value shown in the bar
  graph.
  (b) 2012; yes
75. 2014/2015   **77.** $R_1 \ge 2$ ohms

Review Exercises   *(page 152)*

1. Conditional equation   **3.** Identity
5. (a) No   (b) No   (c) Yes   (d) No
7. (a) No   (b) No   (c) No   (d) No
9. 13   **11.** $-\frac{1}{2}$ **13.** No solution
15. All real numbers   **17.** $-\frac{5}{3}$ **19.** $-\frac{2}{3}$
21. 377.778   **23.** 0.033   **25.** 2009
27. (a) Sum = First even integer + Second even integer
            + Third even integer
  (b) Sum = 42
    First even integer = x
    Second even integer = $x + 2$
    Third even integer = $x + 4$
  (c) $x + (x + 2) + (x + 4) = 42$
  (d) 12
29. (a) New salary = 7%(Annual salary) + Annual salary
  (b) New salary = S (dollars)
    Annual salary = 28,900 (dollars)
    Percent = 0.07 (percent in decimal form)
  (c) $S = 0.07(28{,}900) + 28{,}900$
  (d) \$30,923
31. \$163.53   **33.** 2 hours   **35.** $\approx 26.58\%$
37. $2\frac{2}{9}$ quarts   **39.** 29.5 feet × 59 feet   **41.** \$20
43. $-\frac{1}{2}, \frac{4}{3}$ **45.** 3, 8   **47.** $\pm\sqrt{11}, \approx \pm 3.32$

49. $-4 + 3\sqrt{2} \approx 0.24$
  $-4 - 3\sqrt{2} \approx -8.24$
51. 15 feet × 27 feet
53. 200,000 units or 400,000 units
55. Two real solutions   **57.** No real solutions
59. $6 \pm \sqrt{6}$ **61.** $\dfrac{-19 \pm \sqrt{165}}{2}$ **63.** $-3 \pm 2\sqrt{3}$
65. $1.866, -0.283$ **67.** $8.544, 0.162$
69. Moon: ≈ 8.61 seconds
  Earth: ≈ 3.54 seconds
71. $0, -1, 4$ **73.** $\pm 2, \pm 1$ **75.** $\frac{25}{4}$ **77.** No solution
79. $\pm 4\sqrt{2}$ **81.** $-3, \frac{7}{5}$ **83.** $2 \pm \sqrt{19}$ **85.** \$900
87. $\approx 313{,}251$ units   **89.** $-4 \le x \le 7$; bounded
91. $x \le 1000$; unbounded
93. $x < 11$ **95.** $-\frac{13}{2} < x < \frac{11}{2}$

97. $-12 < x < -8$ **99.** $x > 45$ units

101. $(-\infty, -1), (-1, 1), (1, \infty)$
103. $(-\infty, 1), (1, 5), (5, \infty)$
105. $(-1, 3)$ **107.** $(-\infty, -3) \cup (0, 3)$

109. $\left(-\infty, \frac{6}{5}\right) \cup (4, \infty)$

111. $(-1.69, 1.69)$ **113.** $(1.65, 1.74)$ **115.** $[10, \infty)$
117. $(-\infty, 6] \cup [9, \infty)$
119. Between 2.5 and 3.75 seconds
121. Between about 15.5 meters and about 24.5 meters
123. Greater than 4.25%
125. Between about \$41.34 and about \$58.66

Test Yourself   *(page 156)*

1. $\frac{17}{23}$ **2.** (a) All real numbers   (b) $-3 \le x \le 3$
3. April: \$325,786.00   **4.** $-\frac{5}{3}, \frac{1}{2}$ **5.** $4, -\frac{3}{2}$
  May: \$299,723.12
6. $\pm\sqrt{15}$ **7.** $\dfrac{-13 \pm \sqrt{69}}{2}$ **8.** $\dfrac{11 \pm \sqrt{145}}{6}$
9. $1.038, -0.446$ **10.** $-\frac{7}{2}, \frac{13}{2}$
11. 4 (7 is extraneous.)
12. $-1, 1, -3, 3$ **13.** $-6, 6$
14. Selling either 341,421 units or 58,579 units will produce a
  revenue of \$2,000,000.
15. $x < 3$ **16.** $x \le -4$ or $x \ge \frac{28}{5}$

17. $(-11, -7)$ **18.** $(-\infty, -2] \cup [0, 2]$

19. Between 19,189 units and 143,311 units   **20.** 2012

Chapter 2

Section 2.1   *(page 168)*

Skills Warm Up   *(page 168)*

1. 5   **2.** $3\sqrt{2}$ **3.** 1   **4.** -2
5. $3(\sqrt{2} + \sqrt{5})$ **6.** $2(\sqrt{3} + \sqrt{11})$ **7.** $-3, 11$
8. 9, 1   **9.** 0, ± 3 **10.** ± 2

1.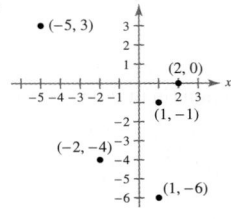

3. (a)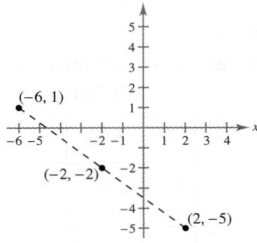

(b) 10   (c) $(-2, -2)$

5. (a)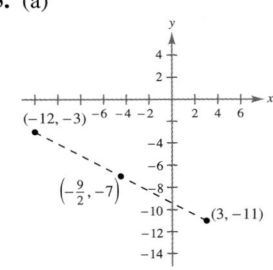

(b) 17   (c) $\left(-\frac{9}{2}, -7\right)$

7. (a)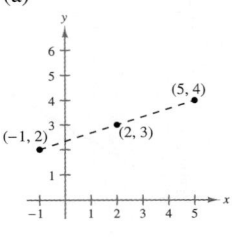

(b) $2\sqrt{10}$ (c) $(2, 3)$

9. (a)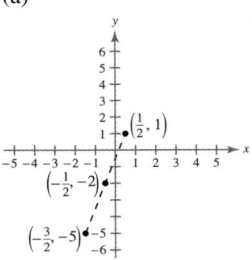

(b) $2\sqrt{10}$
(c) $\left(-\frac{1}{2}, -2\right)$

11. (a)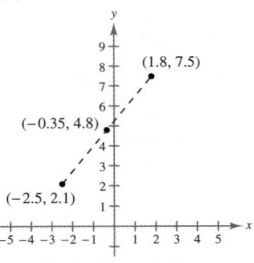

(b) $\sqrt{47.65}$
(c) $(-0.35, 4.8)$

13. 5   **15.** $\sqrt{109}$ **17.** $x = 15, -9$ **19.** $y = 9, -23$
21. (a) Yes (b) No   **23.** (a) Yes   (b) Yes
25.

x	-3	-2	-1
y	-3.25	-2.5	-1.75

x	0	1	2	3
y	-1	-0.25	0.5	1.25

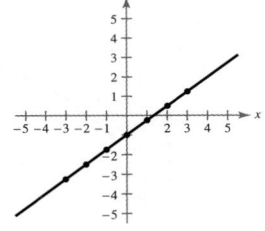

27.

x	-3	-2	-1	0	1	2	3
y	5	0	-3	-4	-3	0	5

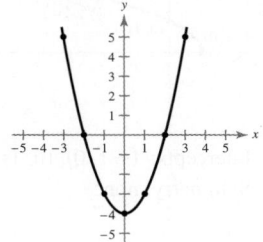

29. $\left(\frac{1}{2}, 0\right), (0, -1)$ **31.** $(-2, 0), (1, 0), (0, -2)$
33. $(-2, 0), (2, 0), (0, 2)$ **35.** $\left(\frac{4}{3}, 0\right), (0, 2)$
37. Every ordered pair on the x-axis has a y-coordinate of zero $[(x, 0)]$, so to find an x-intercept we let $y = 0$. Similarly, every ordered pair on the y-axis has an x-coordinate of zero $[(0, y)]$, so to find a y-intercept we let $x = 0$.
39. y-axis symmetry   **41.** x-axis symmetry
43. y-axis symmetry   **45.** Origin symmetry
47. Origin symmetry   **49.** x-axis, y-axis, and origin symmetry

51. 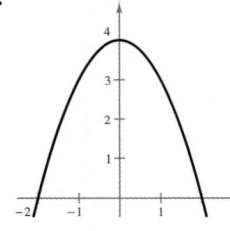   **53.**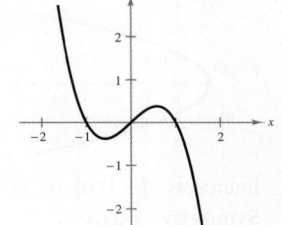

55. c   **56.** d   **57.** f   **58.** a   **59.** e   **60.** b
61. 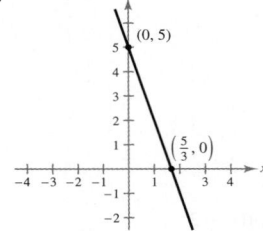   **63.**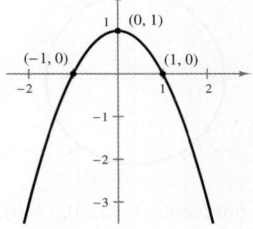

Intercepts:
$\left(\frac{5}{3}, 0\right), (0, 5)$
Symmetry: none

Intercepts:
$(-1, 0), (1, 0), (0, 1)$
Symmetry: y-axis

65. 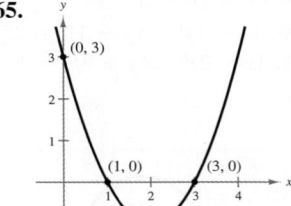   **67.**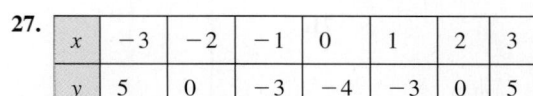

Intercepts:
$(3, 0), (1, 0), (0, 3)$
Symmetry: none

Intercepts:
$\left(-\sqrt[3]{2}, 0\right), (0, 2)$
Symmetry: none

CHAPTER 2

69.

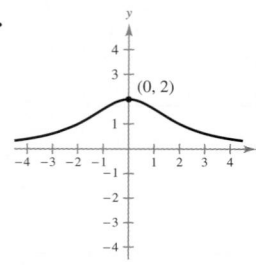

Intercept: $(0, 2)$
Symmetry: y-axis

71.

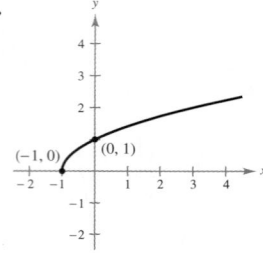

Intercepts: $(-1, 0), (0, 1)$
Symmetry: none

73.

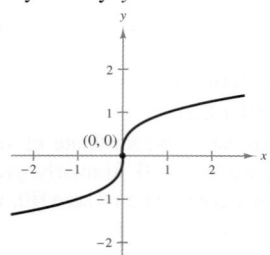

Intercept: $(0, 0)$
Symmetry: origin

75.

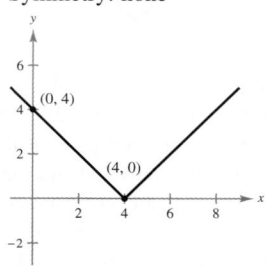

Intercepts: $(4, 0), (0, 4)$
Symmetry: none

77.

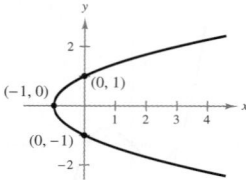

Intercepts: $(-1, 0), (0, 1), (0, -1)$
Symmetry: x-axis

79.

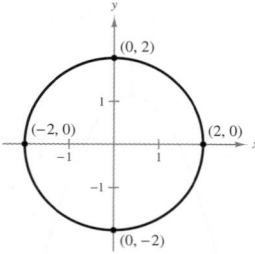

Intercepts: $(-2, 0), (2, 0), (0, 2), (0, -2)$
Symmetry: x-axis, y-axis, origin

81. Radius: 2   **83.** Radius: $\sqrt{5}$
85. Center: $(3, 2)$; Radius: $\sqrt{5}$
87. $x^2 + y^2 = 9$ **89.** $(x + 4)^2 + (y - 1)^2 = 2$
91. $(x + 1)^2 + (y - 2)^2 = 5$ **93.** $(x - 1)^2 + (y - 1)^2 = 25$
95. $(x - 3)^2 + (y + 2)^2 = 16$ **97.** $(x - 2)^2 + (y + 3)^2 = 4$

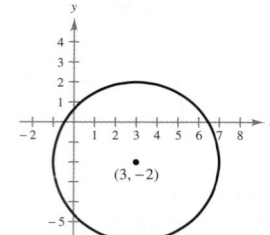

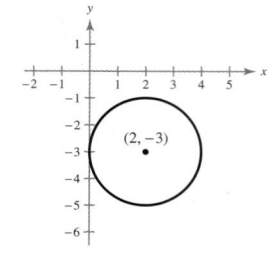

99. $\left(x - \frac{1}{2}\right)^2 + \left(y - \frac{1}{2}\right)^2 = 2$ **101.** $\left(x + \frac{1}{2}\right)^2 + \left(y + \frac{5}{4}\right)^2 = \frac{9}{4}$

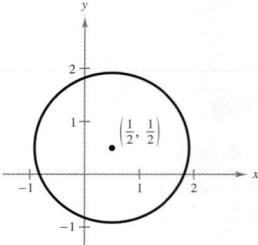

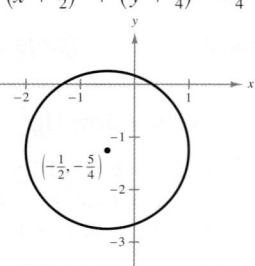

103. Center: $(3, -1)$; Radius: 5
$x^2 + y^2 - 6x + 2y - 15 = 0$
105. (a) $\approx 17\%$ (b) $\approx 8\%$
107. (a) $(0, 482)$; It represents the population (in millions of people) of North America in 2000.

(b)

x	-20	-10	0	10
y	376	429	482	535

x	20	30	40	50
y	588	641	694	747

(c)

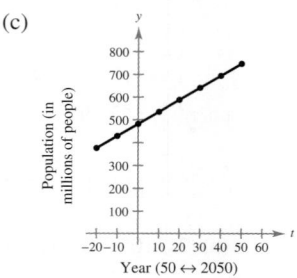

109. (a)

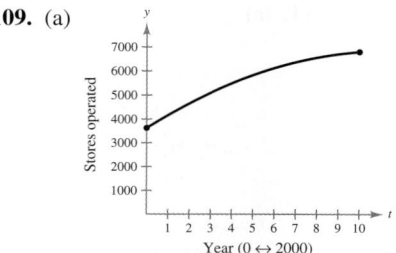

(b) 2011: about 6842; The model's prediction for 2011 is not close to the plan of Family Dollar Stores.

(c)

Year, t	12	13	14	15
Number of stores	6836	6780	6674	6518

(d)

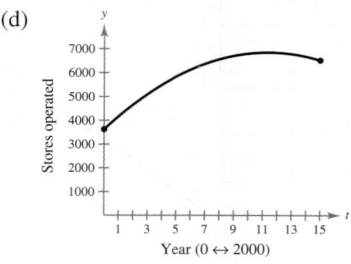

(e) The model does not support the company's expectations.

2013: 6780

2014: 6674

2015: 6518

The graph of the model begins to decrease in 2011.

111. (a) $x^2 + y^2 = 7656.25$

(b) Quadrant II, Quadrant III, Quadrant II, Quadrant I, Quadrant IV

(c) 100 cm

Section 2.2   *(page 180)*

Skills Warm Up   *(page 180)*

1. -4 **2.** $-\frac{13}{3}$ **3.** $-\frac{5}{4}$ **4.** $\frac{1}{3}$ **5.** $y = \frac{2}{3}x - 2$

6. $y = -2x$ **7.** $y = 3x - 1$ **8.** $y = \frac{2}{3}x + 5$

9. $y = -2x + 7$ **10.** $y = x + 3$

1. $\frac{1}{2}$ **3.** -2 **5.** (a) L_2 (b) L_3 (c) L_1

7.

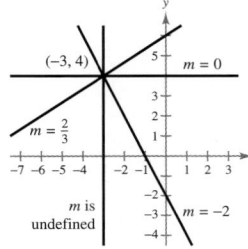

9.

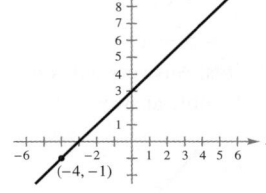

$m = 1$

11.

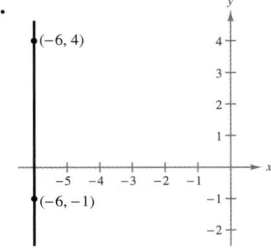

m is undefined.

13.

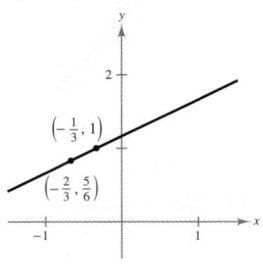

$m = \frac{1}{2}$

15. Answers will vary.

Sample answer: $(3, -2), (-1, -2), (0, -2)$

17. Answers will vary.

Sample answer: $(2, -3), (2, -7), (2, 9)$

19. Answers will vary.

Sample answer: $(6, -5), (7, -4), (8, -3)$

21. Answers will vary.

Sample answer: $(2, 3), (-4, 0), (4, 4)$

23. $x - y - 7 = 0$ **25.** $4x + y + 8 = 0$

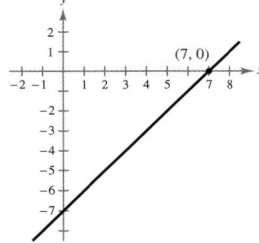

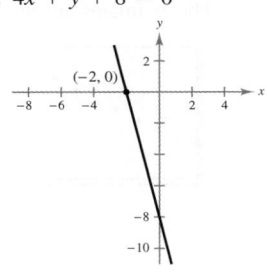

27. $x + 3y - 4 = 0$ **29.** $y + 7 = 0$

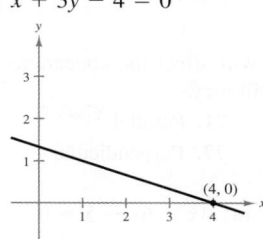

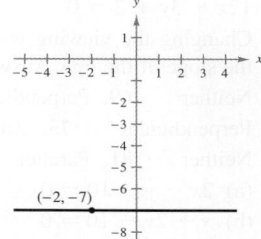

31. $8x - 6y - 17 = 0$ **33.** $y = 4x$

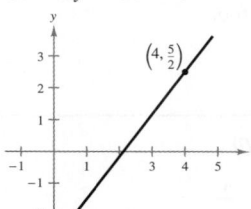

35. $x + 2y + 1 = 0$ **37.** $x + 9 = 0$ **39.** $y - 7 = 0$

41. $x + 2y - 3 = 0$ **43.** $2x - 5y + 1 = 0$

45. $(0, 2)$ **47.** $m = 2, (0, -1)$

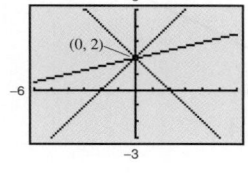

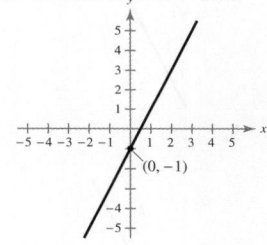

49. $m = 4, (0, -6)$ **51.** m is undefined; no y-intercept

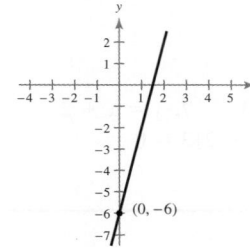

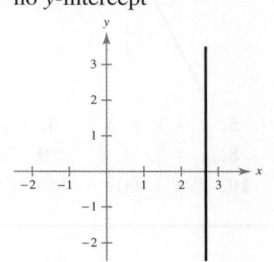

53. $m = -\frac{7}{6}, (0, 5)$ **55.** $m = 0, \left(0, \frac{7}{2}\right)$

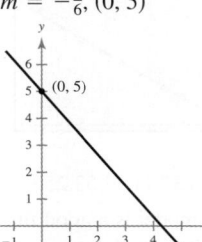

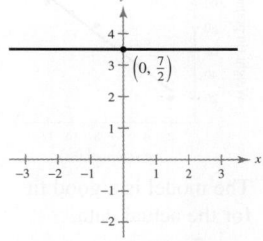

57. Answers will vary. Sample answer: You could graph a vertical line and pick two convenient points on the line to find the slope. Regardless of the points selected, the slope will have zero in the denominator. Division by zero is not possible, so the slope does not exist.

CHAPTER 2

59. $4x - y - 4 = 0$ **61.** $x + y + 2 = 0$

63. $12x + 3y + 2 = 0$

65. Changing the viewing window will affect the appearance of the slope of the line. Answers will vary.

67. Neither   **69.** Perpendicular   **71.** Parallel

73. Perpendicular   **75.** Parallel   **77.** Perpendicular

79. Neither   **81.** Parallel

83. (a) $2x - y - 10 = 0$ **85.** (a) $4x - 6y - 5 = 0$

    (b) $x + 2y - 10 = 0$ (b) $36x + 24y + 7 = 0$

87. (a) $y = 0$ (b) $x + 1 = 0$ **89.** Yes $\left(m = \frac{17}{180}\right)$

91. $F = \frac{9}{5}C + 32$ **93.** $y = 7000x + 151{,}000$; $179{,}000$; No

95. $p = \frac{1}{33}d + 1$; $\frac{1}{33}$ atmosphere per foot

Section 2.3   *(page 190)*

Skills Warm Up   *(page 190)*

1.
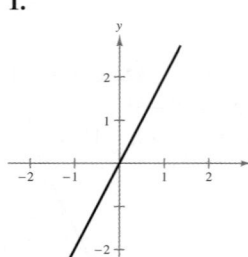

2.
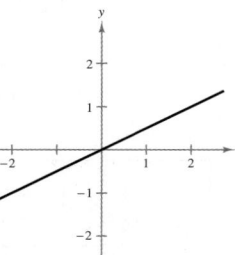

3.
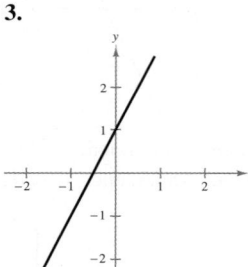

4.
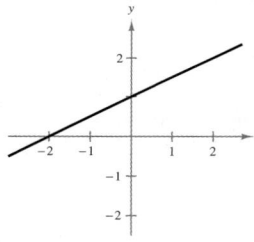

5. $y = x + 2$ **6.** $y = \frac{3}{2}x + 3$ **7.** $y = x + 2$

8. $y = \frac{6}{7}x + 4$ **9.** $5x + 40y - 213 = 0$

10. $29x + 60y - 448 = 0$

1.
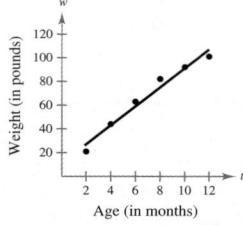

The model is a good fit for the actual data.

3.
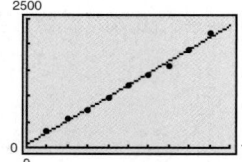

$y = 231.1x + 53.256$

The model is a good fit for the actual data.

5. $y = \frac{3}{8}x$ **7.** $y = 20x$ **9.** $y = \frac{3.2}{7}x$ or $y = \frac{16}{35}x$

11. $H = 3p$ **13.** $c = \frac{3}{5}d$ **15.** $I = 0.06P$

17. (a) $y = 0.0134x$ (b) \$2479

19. (a) $M = \frac{64}{103}K$

(b)

km/h	40	60	80	100	120
mi/h	24.85	37.28	49.71	62.14	74.56

21. $V = 140t + 860$, $12 \le t \le 17$

23. $V = -2142t + 46{,}104$, $12 \le t \le 17$

25. $V = 10{,}780t + 24{,}640$, $12 \le t \le 17$

27. (a) $h = 7000 - 20t$ (b) 2:13:50 P.M.

29. $V = 875 - 175t$, $0 \le t \le 5$

31. $W = 0.75x + 11.50$ **33.** (a) $H = 0.5t + 4$ (b) 13 in.

35. Yes; Answers will vary.   **37.** No

Sample answer:

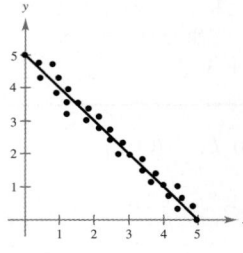

$y = 5 - x$

39. Yes; Answers will vary.

Sample answer:

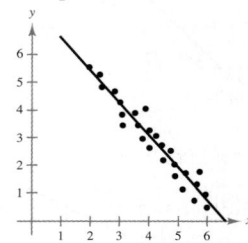

$y = -1.18x + 7.76$

41. (a)
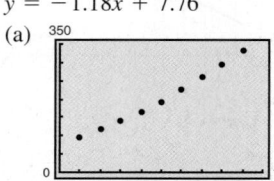

(b) $C = 29.19t + 57.1$

Yes

(c) Slope: 29.19

The number of autistic children receiving disability services is increasing by about 29 thousand children per year.

(d) 2011: 378 thousand children

2012: 407 thousand children

The estimates are reasonable.

43. (a)
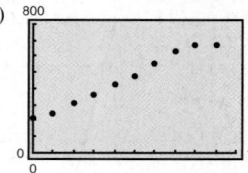

(b) Answers will vary. Sample answer:

$y = 58.1t + 210.8$

(c) $y = 56.69t + 200.7$

Using linear regression equation:

2010: $767.6 million

2011: $824.29 million

Using equation from part (b):

2010: $791.8 million

2011: $849.9 million

(d) The projections made by California Pizza Kitchen are lower than the predictions given by the models.

(e) Yes; Using the linear regression equation, the yearly revenue is expected to reach $937.67 million by 2013. Using the equation from part (b), the yearly revenue is expected to reach $966.1 million by 2013.

45. (a)

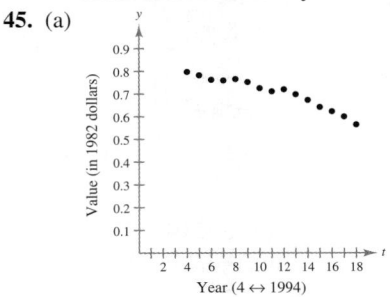

Year (4 ↔ 1994)

The data appear to be approximately linear.

(b) $y = -0.0152t + 0.873$

(c) 2010: 0.569; 2011: 0.5538

Because the data follow a linear pattern from 1994 to 2008, you can assume that the estimates for 2010 and 2011 are reliable.

47. (a)

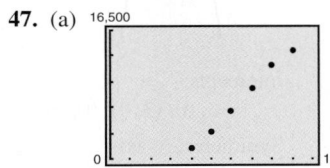

The data appear to be approximately linear.

(b) $E = 394.17t + 12,592.9$

(c) 2011: 16,928,770 employees

2013: 17,717,110 employees

(d)

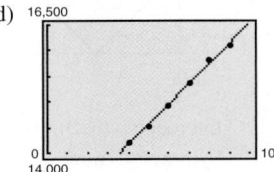

The predictions are most likely going to be just about right because the model is a good fit for the actual data.

49. The model for population, because population tends to change at a consistent rate, whereas snowfall can be quite different from year to year. You should use more than three data points to ensure that the data can be represented accurately by a linear model.

51. Yes; When you start, you have zero wrong moves ($x = 0$) and 800 points ($y = 800$). So, 800 is the y-intercept. Then you lose 50 points for each wrong move. This is a slope of -50.

53. A known rate of change corresponds to the slope of a linear model, so determine whether the rate of change of y with respect to x is constant. The data points show a 10-unit increase in y for each 1-unit increase in x, so the rate of change is constant and the data points are linear.

Section 2.4   *(page 203)*

Skills Warm Up   *(page 203)*

1. -73 **2.** 13   **3.** $2(x + 2)$ **4.** $-8(x - 2)$

5. $y = \frac{7}{5} - \frac{2}{5}x$ **6.** $y = \pm x$ **7.** $x \le -2, x \ge 2$

8. $-3 \le x \le 3$ **9.** All real numbers

10. $x \le 1, x \ge 2$

1. This is a function from A to B because each element of A is matched with an element of B.

3. Not a function; The relationship does not match the element b of A with an element of B.

5. This is a function from A to B because each element of A is matched with an element of B.

7. This is a function from A to B because each element of A is matched with an element of B.

9. Not a function; The relationship assigns two elements of B to the element c of A.

11. Not a function from A to B; The relationship defines a function from B to A.

13. This is a function from A to B because each element of A is matched with an element of B.

15. Not a function   **17.** Function   **19.** Function

21. Not a function   **23.** Function

25. (a) -6 (b) 34   (c) $6 - 4t$ (d) $2 - 4c$

27. (a) -1 (b) $\dfrac{1}{15}$ (c) $\dfrac{1}{t^2 - 2t}$ (d) $\dfrac{1}{t^2 - 1}$

29. (a) -1 (b) -9 (c) $2x - 5$ (d) $-\frac{5}{2}$

31. (a) 0   (b) 3   (c) $x^2 + 2x$ (d) -0.75

33. (a) 36π (b) 0   (c) $\dfrac{9\pi}{2}$ (d) $\dfrac{32\pi r^3}{3}$

35. (a) 1   (b) -7 (c) $3 - 2|x|$ (d) 2.5

37. (a) Undefined   (b) $-\dfrac{1}{16}$

(c) $\dfrac{1}{y^2 + 4y - 12}$, $y \ne -6, 2$

(d) $\dfrac{1}{y^2 - 4y - 12}$, $y \ne -2, 6$

39. (a) 1   (b) -1 (c) 1   (d) $\dfrac{|x - 1|}{x - 1}$

41. (a) -4 (b) 3   (c) -7 (d) 7

43. (a) 2   (b) -1 (c) 3   (d) 3   **45.** 5   **47.** ±3

49. $0, 1, -1$ **51.** $\frac{10}{7}$ **53.** $\{0, 3, 5, 8, 9\}$

55. $\{-2, -1, 0, 1, 2\}$ **57.** All real numbers x

59. All real numbers except $t = 0$ **61.** All real numbers y

63. $-1 \le x \le 1$ **65.** All real numbers except $x = 0, -2$

67. All real numbers $x \ge -1$ except $x = 2$ **69.** $x > 0$

71. $\{(-2, 4), (-1, 1), (0, 0), (1, 1), (2, 4)\}$

73. $\{(-2, 0), (-1, 1), (0, \sqrt{2}), (1, \sqrt{3}), (2, 2)\}$

75. The domain of $f(x) = \sqrt{x - 2}$ is all real numbers $x \geq 2$ because an even root of a negative number is not a real number. The domain of $g(x) = \sqrt[3]{x - 2}$ is all real numbers. f and g have different domains because an odd root of a negative number is a real number, but an even root of a negative number is not a real number.

77. n ordered pairs

79. (a) $V = x(18 - 2x)^2$ (b) Domain: $0 < x < 9$
(c) 400 cubic inches

81. Yes $[y(30) = 6]$

83. 2000: 945 million acres; 2009: 917.98 million acres

85. (a) $C = 57,000 + 2.05x$ (b) $\overline{C} = \dfrac{57,000}{x} + 2.05$

(c)

x	100	1000	10,000	100,000
$\overline{C}$	572.05	59.05	7.75	2.62

(d) Answers will vary. Sample answer: The average cost per unit decreases as x gets larger.

87. (a)

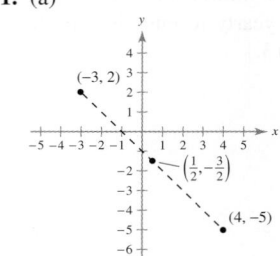

(b) Linear model: $P = 0.965t + 5.11$
Quadratic model: $P = -0.0456t^2 + 1.375t + 4.56$

(c)

Year	P (Actual)	P (Linear)	P (Quadratic)
2000	4.80	5.11	4.56
2001	5.89	6.08	5.89
2002	6.81	7.04	7.13
2003	8.10	8.01	8.27
2004	9.30	8.97	9.33
2005	10.55	9.94	10.30
2006	11.13	10.90	11.17
2007	12.09	11.87	11.95
2008	12.95	12.83	12.64
2009	12.91	13.80	13.24

The quadratic model is a better fit because its values of P tend to be closer to the actual values than those of the linear model.

89. (a) $r(t) = 0.75t$ (b) $A = 0.5625\pi t^2$

Time, t	1	2	3	4	5
Radius, r (in feet)	0.75	1.5	2.25	3	3.75
Area, A (in square feet)	1.767	7.069	15.904	28.274	44.179

(c) $\dfrac{A(2)}{A(1)} = \dfrac{7.069}{1.767} = 4.000$, $\dfrac{A(4)}{A(2)} = \dfrac{28.274}{7.069} = 4.000$
Predicted area when $t = 8$: 113.096 square feet
Calculated area when $t = 8$: 113.097 square feet

91. (a) Correct; Each price has exactly one tax amount.
(b) Incorrect; For any given value of education, there is not exactly one specific value of intelligence.

93. Answers will vary.

Quiz Yourself   *(page 208)*

1. (a)

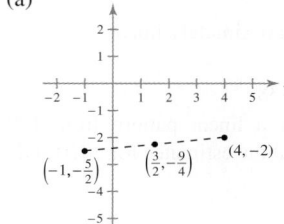

(b) $7\sqrt{2}$
(c) $\left(\frac{1}{2}, -\frac{3}{2}\right)$

2. (a)

(b) $\sqrt{52.04}$
(c) $(-1.2, -1.9)$

3. (a)

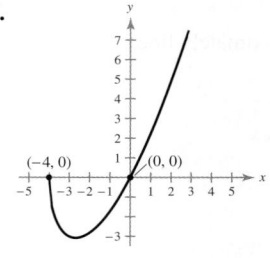

(b) $\dfrac{\sqrt{101}}{2}$ (c) $\left(\dfrac{3}{2}, -\dfrac{9}{4}\right)$

4.

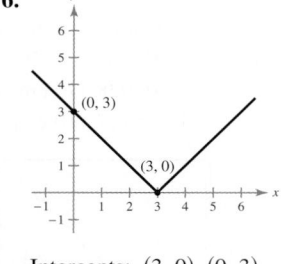

Intercepts:
$(-3, 0), (3, 0), (0, 9)$
Symmetry: y-axis

5.

Intercepts: $(-4, 0), (0, 0)$
Symmetry: none

6.

Intercepts: $(3, 0), (0, 3)$
Symmetry: none

7. $(x - 4)^2 + (y - 5)^2 = 20$
8. $(x - 1)^2 + (y + 2)^2 = 9$

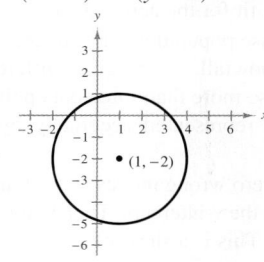

9. $2x - 3y + 9 = 0$

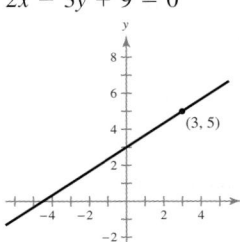

10. $y - 4 = 0$

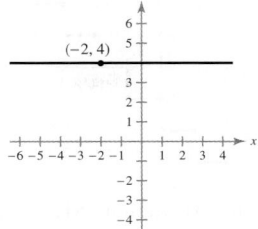

11. $x - 2 = 0$

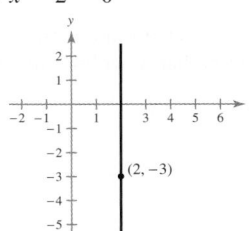

12. $2x + y + 9 = 0$

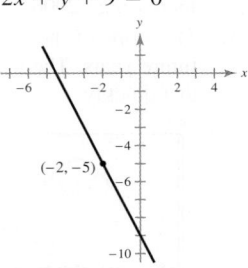

13. 261,019   **14.** y is a function of x.

15. The collection of ordered pairs does represent a function from A to B.

16. The collection of ordered pairs does not represent a function from A to B.

17. (a) 2   (b) -7 **18.** (a) 1   (b) -20

19. $x \geq 4$ **20.** All real numbers except $x = -2$

21.

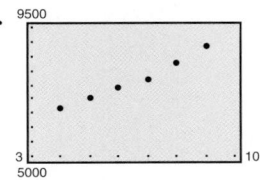

Linear: $y = 428.5t + 4873$

Quadratic: $y = 35.29t^2 - 30.2t + 6261$

22. Linear: Quadratic:

    2010: \$9158 million   2010: \$9488 million

    2011: \$9586.5 million   2011: \$10,198.89 million

23. $A = \dfrac{C^2}{4\pi}$

Section 2.5   *(page 216)*

1. Domain: $[1, \infty)$; Range: $[0, \infty)$; 0

3. Domain: $(-\infty, \infty)$; Range: $(-\infty, 4]$; 4

5. Domain: $(-\infty, \infty)$; Range: $(-\infty, \infty)$; -1

7. Domain: $[-5, 5]$; Range: $[0, 5]$; 5

9. Function   **11.** Not a function   **13.** Function

15. Not a function   **17.** Increasing on $(-\infty, \infty)$; No change

19. Increasing on $(-\infty, 0)$ and $(2, \infty)$, decreasing on $(0, 2)$; behavior changes at $(0, 0)$ and $(2, -4)$.

21. Increasing on $(-1, 0)$ and $(1, \infty)$, decreasing on $(-\infty, -1)$ and $(0, 1)$; behavior changes at $(-1, -3)$, $(0, 0)$, and $(1, -3)$.

23. Increasing on $(-2, \infty)$, decreasing on $(-3, -2)$; behavior changes at $(-2, -2)$.

25. Increasing on $(2, \infty)$, decreasing on $(-\infty, 2)$; behavior changes at $(2, -4)$.

27. Increasing on $(-\infty, 0)$ and $(2, \infty)$, decreasing on $(0, 2)$; behavior changes at $(0, 2)$ and $(2, -2)$.

29.

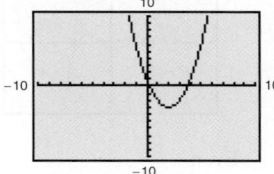

Minimum: $(2, -3)$

Increasing: $(2, \infty)$

Decreasing: $(-\infty, 2)$

31.

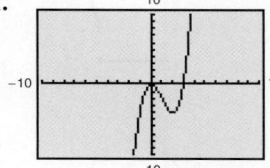

Relative maximum: $(0, 0)$

Relative minimum: $(2, -4)$

Increasing: $(-\infty, 0)$, $(2, \infty)$

Decreasing: $(0, 2)$

33.

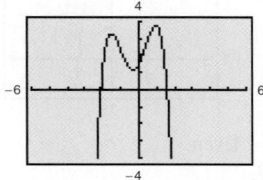

Relative maxima: $(-1.54, 3.29)$, $(0.95, 3.77)$

Relative minimum: $(-0.34, 1.14)$

Increasing: $(-\infty, -1.54)$, $(-0.34, 0.95)$

Decreasing: $(-1.54, -0.34)$, $(0.95, \infty)$

35. (a) 2   (b) 2   (c) -3 (d) -4

37. (a) 2   (b) 1   (c) -8 (d) -9

39. Even   **41.** Odd   **43.** Odd

45.

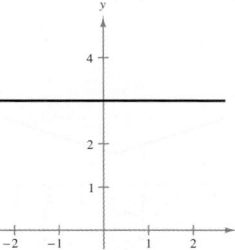

Even

47.

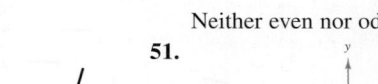

Neither even nor odd

49.

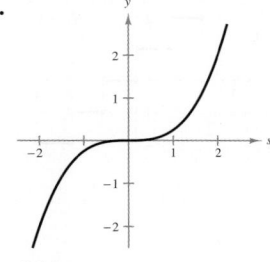

Odd

51.

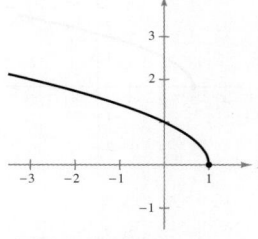

Neither even nor odd

53.

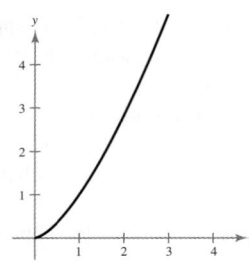

Neither even nor odd

55.

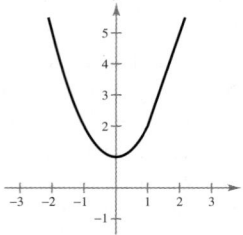

Neither even nor odd

57.

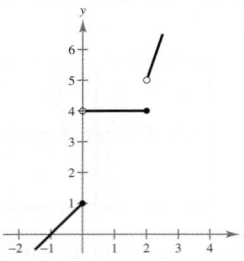

Neither even nor odd

59.

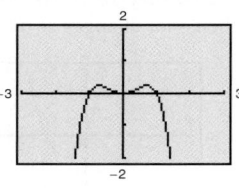

Even

61.

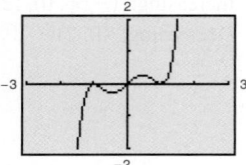

Odd

63.

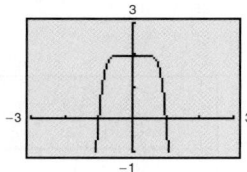

Even

65.

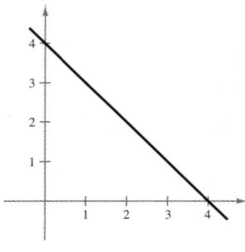

67.

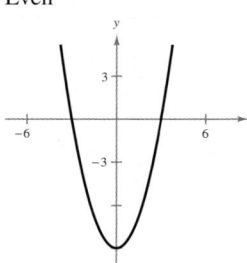

69.

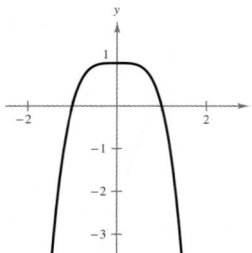

71.

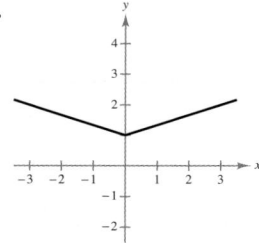

73.

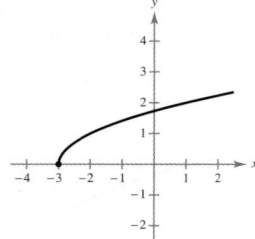

75.

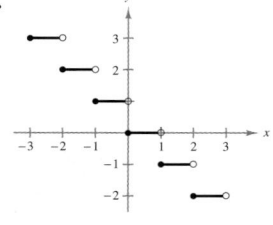

77.

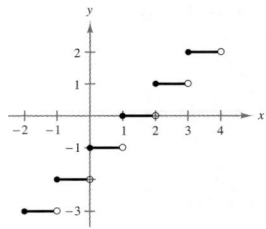

79. (a) Maximum: $1212.62
   (b) Decreasing: Price did not decrease during 2000–2010.
       Increasing: 2000–2010
   (c) The slope of the graph seems to increase quite rapidly in future years. The price of gold probably will not increase that quickly.

81. (a)

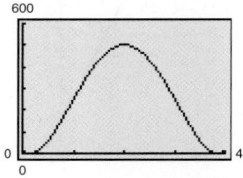

   Increasing: 0 seconds to 2 seconds
   Decreasing: 2 seconds to 4 seconds
   (b) Maximum change in volume: ≈ 501.9 milliliters

83.

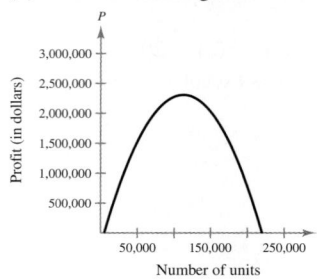

Approximately 112,500 units

85.

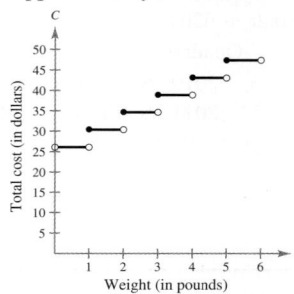

87.

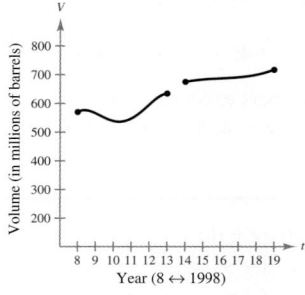

89. (a) $y = 425.71t + 705.54$
Domain: $(0, 4)$
Range: $(705.54, 2408.38)$

(b) $y = -202.95t^2 + 3304.67t - 7311.8$
Domain: $(5, 9)$
Range: $(4137.8, 6140.3)$

(c) $y = \begin{cases} 425.71t + 705.54, & 0 \le t \le 4 \\ -202.95t^2 + 3304.67t - 7311.8, & 5 \le t \le 9 \end{cases}$

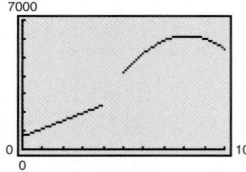

(d) The revenues increased from 2000 to 2008 and decreased from 2008 to 2009.

91. (a) $\left(\frac{3}{2}, 4\right)$ (b) $\left(\frac{3}{2}, -4\right)$

93. (a) $(-4, 9)$ (b) $(-4, -9)$

Section 2.6   *(page 226)*

Skills Warm Up   *(page 226)*

1. 12   **2.** $\dfrac{-2x}{-x-3}$ **3.** $0, \pm\sqrt{10}$ **4.** $\dfrac{4}{3}, -2$

5. $f(x) = -2$ **6.** $f(x) = -x$

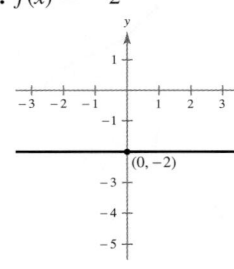

7. $f(x) = x + 5$ **8.** $f(x) = 2 - x$

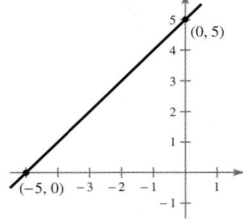
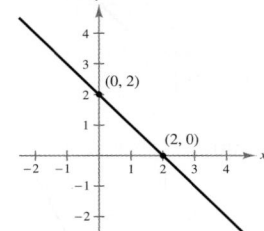

9. $f(x) = 3x - 4$ **10.** $f(x) = 9x + 10$

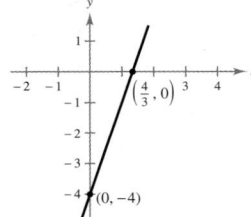
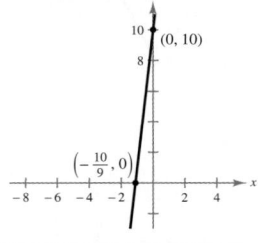

1. Shifted four units downward   **3.** Shifted two units to the left

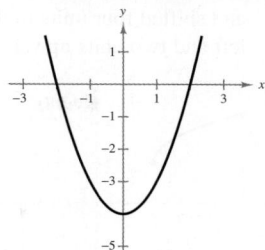

5. Shifted two units upward and four units to the right   **7.** Reflected about the *x*-axis and shifted one unit upward

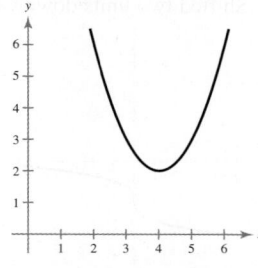

9. Shifted two units upward   **11.** Shifted one unit to the right

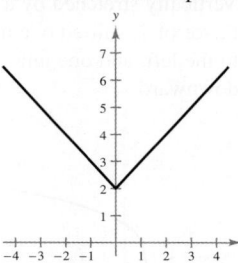
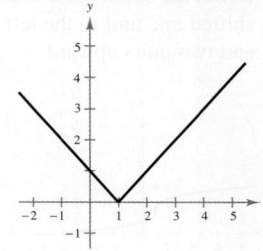

13. Reflected about the *x*-axis and shifted three units upward   **15.** Shifted one unit to the left and four units downward

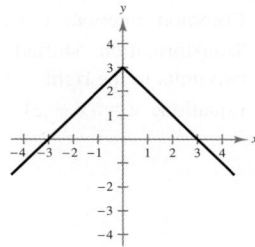
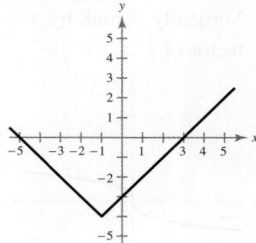

17. Shifted three units to the right   **19.** Shifted three units to the right and one unit upward

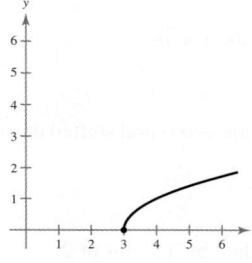
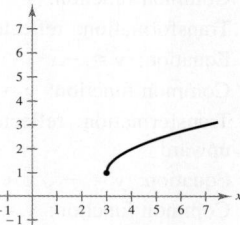

21. Vertically stretched by a factor of $\sqrt{2}$

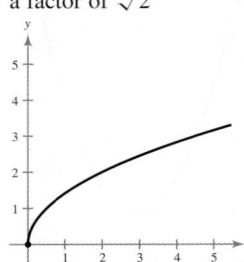

23. Reflected about the y-axis and shifted four units to the left and two units upward

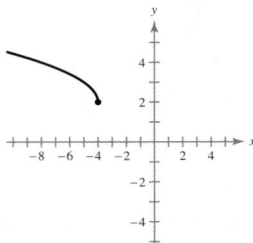

25. Reflected about the y-axis   **27.** Shifted two units downward

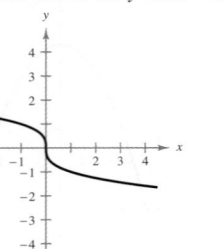

29. Reflected about the x-axis, shifted one unit to the left, and two units upward

31. Vertically stretched by a factor of 2, shifted one unit to the left, and one unit downward

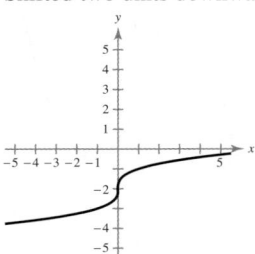

33. Vertically shrunk by a factor of $\frac{1}{2}$

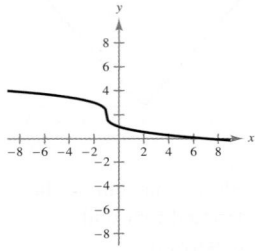

35. Common function: $y = x^3$
Transformation: shifted two units to the right
Equation: $y = (x - 2)^3$

37. Common function: $y = x^2$
Transformation: reflected about the x-axis
Equation: $y = -x^2$

39. Common function: $y = \sqrt{x}$
Transformation: reflected about the x-axis and shifted one unit upward
Equation: $y = -\sqrt{x} + 1$

41. Common function: $y = x$
Transformation: vertically stretched by a factor of 2
Equation: $y = 2x$

43. (a) Vertical shift of two units

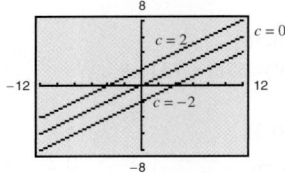

(b) Horizontal shift of two units

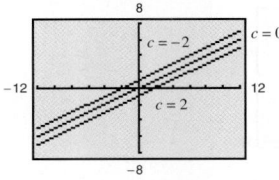

(c) Slope of the function changes

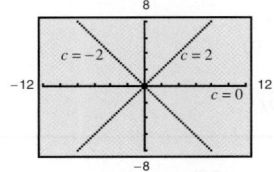

45. (a) $g(x) = (x - 1)^2 + 1$ (b) $g(x) = -(x + 1)^2$

47. (a)

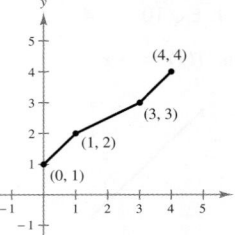

(b)

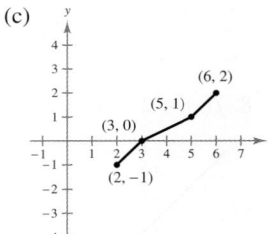

(c)

(d)

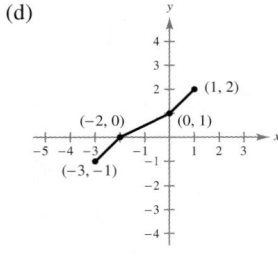

(e)

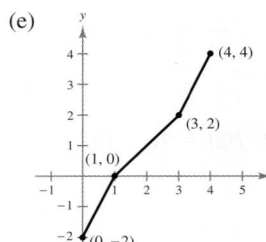

(f)

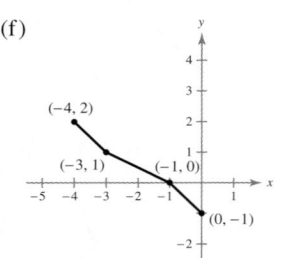

49. (a)

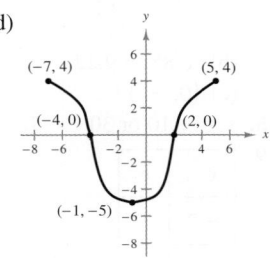

(b)

69. $y = x^7$ will be an odd function and its graph will be similar in shape to the graphs of $y = x^3$ and $y = x^5$. $y = x^8$ will be an even function and its graph will be similar in shape to the graphs of $y = x^2$, $y = x^4$, and $y = x^6$.

Section 2.7   *(page 236)*

Skills Warm Up   *(page 236)*

1. $\dfrac{1}{x(1-x)}$ **2.** $-\dfrac{12}{(x+3)(x-3)}$ **3.** $\dfrac{3x-2}{x(x-2)}$

4. $\dfrac{4x-5}{3(x-5)}$ **5.** $\dfrac{\sqrt{x^2-1}}{x+1}$, $x \neq 1$

6. $\dfrac{x+1}{x(x+2)}$, $x \neq 2$ **7.** $5(x-2)$, $x \neq -2$

8. $\dfrac{x+1}{(x-2)(x+3)}$, $x \neq -5, -1, 0$

9. $\dfrac{1+5x}{3x-1}$, $x \neq 0$ **10.** $\dfrac{x+4}{4x}$, $x \neq 4$

(c)

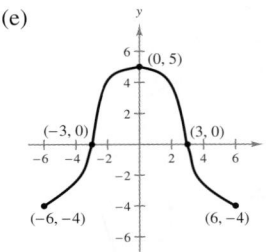

(d)

(e)

(f)

51. $y = x^3 - 2$ **53.** $y = 4x^3$
55. $h(x) = \sqrt{x-4} - 3$ **57.** $h(x) = \frac{1}{2}\sqrt{x-3}$
59. $g(x) = x^3 - 3x^2 + 2$ **61.** $g(x) = (x+1)^3 - 3(x+1)^2$
63. Shifted one unit to the right and two units downward
$g(x) = (x-1)^2 - 2$
65. (a) $P(x) = 80 + 20x - 0.5x^2 - 25$
$P(x) = 55 + 20x - 0.5x^2$, $0 \leq x \leq 20$
Shifted 25 units down

(b)

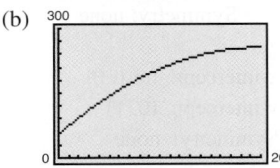

(c) $P\left(\dfrac{x}{100}\right) = 80 + \dfrac{x}{5} - 0.00005x^2$,
$0 \leq x \leq 2000$ (x in dollars)
Horizontal stretch
67. (a) Vertically stretched by a factor of 34, shifted 3705 units upward

(b)

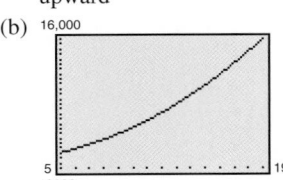

(c) 2007
(d) $M(t) = 34.0(t+10)^2 + 3705$;
To make a horizontal shift 10 years backward (10 units to the left), add 10 to t.

1.

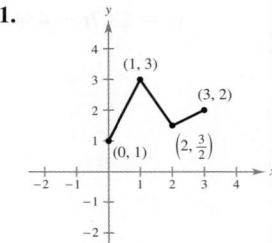

3.

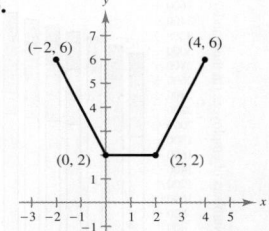

5. (a) $2x$ (b) 2 (c) $x^2 - 1$
(d) $\dfrac{x+1}{x-1}$; Domain: $(-\infty, 1) \cup (1, \infty)$
7. (a) $3x - x^2$ (b) $x^2 + 3x - 2$ (c) $-3x^3 + x^2 + 3x - 1$
(d) $\dfrac{3x-1}{1-x^2}$; Domain: $(-\infty, -1) \cup (-1, 1) \cup (1, \infty)$
9. (a) $x^2 + \sqrt{1-x} + 5$ (b) $x^2 - \sqrt{1-x} + 5$
(c) $x^2\sqrt{1-x} + 5\sqrt{1-x}$
(d) $\dfrac{x^2+5}{\sqrt{1-x}}$; Domain: $(-\infty, 1)$
11. (a) $\dfrac{x+1}{x^2}$ (b) $\dfrac{x-1}{x^2}$ (c) $\dfrac{1}{x^3}$
(d) $x, x \neq 0$; Domain: $(-\infty, 0) \cup (0, \infty)$
13. 14 **15.** $-4t^2 + 4t + 3$ **17.** -6 **19.** $\frac{11}{23}$ **21.** 3
23. -8 **25.** (a) $6x + 15$ (b) $6x + 5$ (c) $9x$
27. (a) $9x^2 + 6x + 1$ (b) $3x^2 + 1$ (c) x^4
29. (a) $x - \frac{8}{3}$ (b) $x - 8$
31. (a) $x, x \geq 0$ (b) $x, x \geq -4$ **33.** (a) $\sqrt[4]{x}$ (b) $\sqrt[4]{x}$
35. (a) $|x+6|$ (b) $|x| + 6$
37. (a) All real numbers, or $(-\infty, \infty)$
(b) $x \geq 0$, or $[0, \infty)$ (c) $x \geq 0$, or $[0, \infty)$
39. (a) All real numbers except $x = 0$, or $(-\infty, 0) \cup (0, \infty)$
(b) All real numbers, or $(-\infty, \infty)$
(c) All real numbers except $x = 2$, or $(-\infty, 2) \cup (2, \infty)$
41. (a) 3 (b) 0 **43.** (a) 0 (b) 4
45. Answers will vary.   **47.** Answers will vary.
Sample answer: Sample answer:
$f(x) = x^2, g(x) = 2x + 1$ $f(x) = \sqrt[3]{x}, g(x) = x^2 - 4$

CHAPTER 2

49. Answers will vary.
Sample answer:

$f(x) = \dfrac{1}{x}, g(x) = x + 2$

51. Answers will vary.
Sample answer:

$f(x) = x^2 + 2x,$
$g(x) = x + 4$

53. $T = \dfrac{3}{4}x + \dfrac{1}{15}x^2$

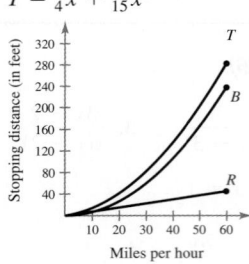

55. $(C \circ x)(t) = 1500t + 495$
$C \circ x$ represents the cost of producing x units in t hours.

57. (a) Total sales decreased.
(b) $R_1 + R_2 = 917 - 6.7t$
The function is linear with a negative slope, so its value decreases for increasing values of t.

59. (a)

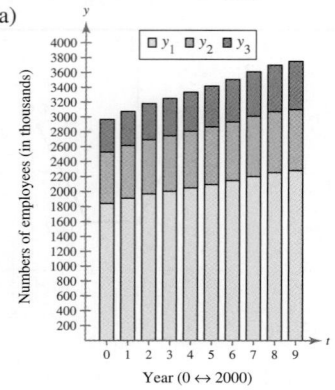

(b) $y_1 = 48.5t + 1856$
$y_2 = 15.2t + 694$
$y_3 = 23.7t + 434$

(c)

2011: 3,945,400 employees
2013: 4,120,200 employees

61. $(A \circ r)(t) = \pi(0.6t)^2 = 0.36\pi t^2$
$A \circ r$ represents the area of the circle at time t.

63.

Year	2005	2006	2007	2008	2009
P/E	16.7	21.0	25.9	12.7	11.3

65. Domain of $(f/g)(x)$: all real numbers $0 \le x < 3$, or $[0, 3)$
Domain of $(g/f)(x)$: all real numbers $0 < x \le 3$, or $(0, 3]$
The two domains differ because if $x = 3$, $(f/g)(x)$ is undefined (division by zero), and if $x = 0$, $(g/f)(x)$ is undefined (division by zero).

67. True. The range of g must be a subset of the domain of f for $(f \circ g)(x)$ to be defined.

Review Exercises   *(page 244)*

1. (a)

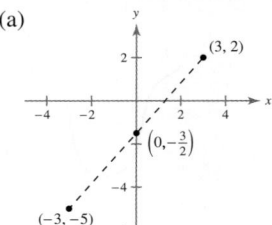

(b) $\sqrt{85} \approx 9.22$
(c) $\left(0, -\dfrac{3}{2}\right)$

3. (a)

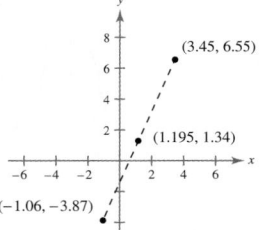

(b) $\sqrt{128.9165} \approx 11.35$
(c) $(1.195, 1.34)$

5. $x = -10$ or 30

7. (a) Yes   (b) Yes

9.

x	y
-2	3
0	2
2	1
3	$\dfrac{1}{2}$
4	0

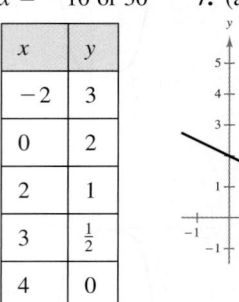

$y = -\dfrac{1}{2}x + 2$

11.

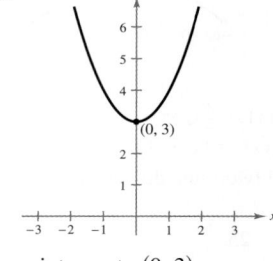

y-intercept: $(0, 3)$
Symmetry: y-axis

13.

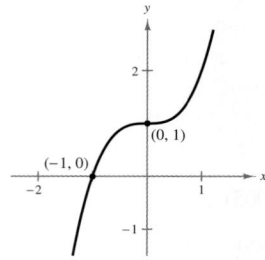

x-intercept: $\left(\dfrac{4}{3}, 0\right)$
y-intercept: $(0, -4)$
Symmetry: none

15.

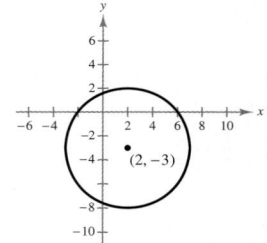

x-intercept: $(-1, 0)$
y-intercept: $(0, 1)$
Symmetry: none

17. $(x + 1)^2 + (y - 2)^2 = 36$

19. $(x - 2)^2 + (y + 3)^2 = 25$

21.

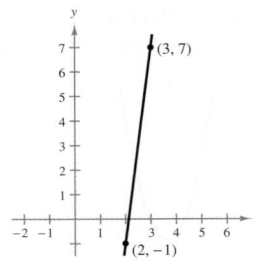

$m = 8$

23.

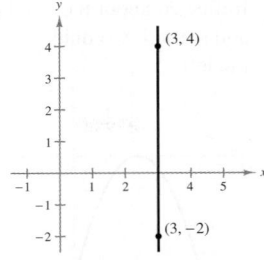

m is undefined.

25. $3x - 2y - 10 = 0$

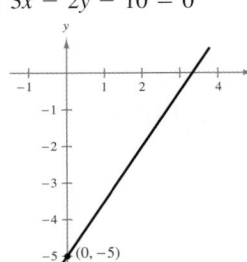

27. $y = 6$

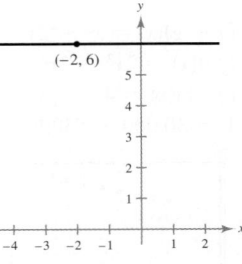

29. Slope: 2
y-intercept: $(0, 4)$

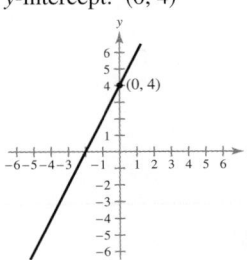

31. Slope: undefined
y-intercept: none

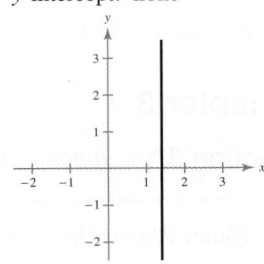

33. Parallel   **35.** Neither
37. (a) $y = \frac{1}{2}x + 1$ (b) $y = -2x + 11$
39. (a) $5x - 4y - 23 = 0$ (b) $4x + 5y - 2 = 0$
41. (a) $y = -2$ (b) $x = -1$
43. $y = \frac{7}{3}x$ **45.** $y = 348x$ **47.** $A = 5r$ **49.** $a = \frac{3}{4}b$
51. $y = 0.0084x$; \$1470   **53.** \$3,350,000
55. $V = -12,950t + 135,000,\ 0 \le t \le 10$
57. y is a function of x.   **59.** y is not a function of x.
61. Function; Every element of A is assigned to an element of B.
63. (a) -2 (b) -3 (c) -5 (d) $\sqrt{x + 7} - 5$
65. All real numbers x **67.** $x \ge -5$
69. $1 \le t < 4, t > 4$
71. The domain of $h(x)$ is all real numbers except $x = 0$, because division by zero is undefined. The domain of $k(x)$ is all real numbers except $x = -2$ and $x = 2$, because if $x = 2$ or $x = -2$, then $x^2 - 4$ equals zero, and division by zero is undefined. When a graphing utility and the *table* feature are used, $h(0)$ results in an error, and $k(-2)$ and $k(2)$ also result in errors.
73. (a) $B = 6500\left(1 + \dfrac{0.0575}{4}\right)^{4t}$ (b) $t \ge 0$
75. (a) Domain: all real numbers
Range: $[1, \infty)$
(b) Decreasing: $(-\infty, 0)$
Increasing: $(0, \infty)$
(c) Even   (d) Minimum: $(0, 1)$

77. (a) Domain: all real numbers
Range: all real numbers
(b) Decreasing: $\left(0, \frac{8}{3}\right)$
Increasing: $(-\infty, 0) \cup \left(\frac{8}{3}, \infty\right)$
(c) Neither
(d) Relative minimum: $\left(\frac{8}{3}, -\frac{256}{27}\right)$
Relative maximum: $(0, 0)$
79. y is a function of x.   **81.** y is not a function of x.

83.

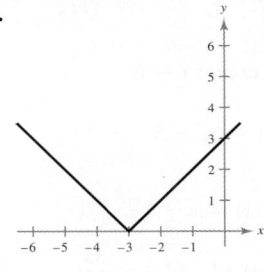

85.

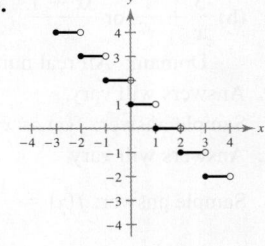

87.

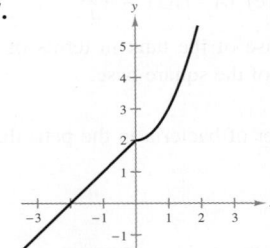

89.

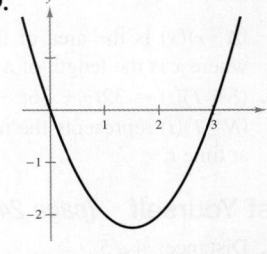

91. Reflected about the x-axis and shifted two units downward and one unit to the right

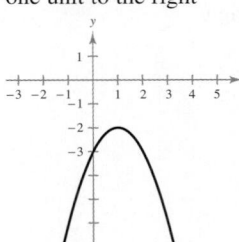

93. Shifted two units to the right

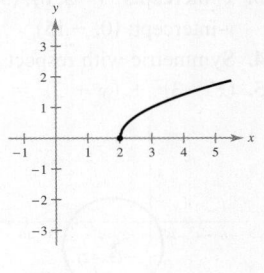

95. Shifted two units to the left

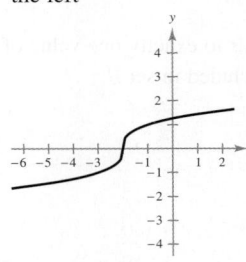

97. Common function: $y = \sqrt{x}$
Transformation: reflected about the y-axis and shifted three units to the right
Equation: $y = \sqrt{3 - x}$ or
$y = \sqrt{-(x - 3)}$

99. (a) $(f + g)(x) = x^2 + 5x - 1$
(b) $(f - g)(x) = -x^2 + x - 1$
(c) $(fg)(x) = 3x^3 + 5x^2 - 2x$
(d) $(f/g)(x) = \dfrac{3x - 1}{x^2 + 2x}$
Domain of f/g: $x < -2,\ -2 < x < 0, x > 0$, or
$(-\infty, -2) \cup (-2, 0) \cup (0, \infty)$

CHAPTER 2

101. 9   **103.** 18

105. (a) $x^2 + 6x + 9$

Domain: All real numbers

(b) $x^2 + 3$

Domain: All real numbers

107. (a) $\dfrac{1}{3x + x^2}$

Domain: All real numbers except $x = 0$ and $x = -3$

(b) $\dfrac{3}{x} + \dfrac{1}{x^2}$, or $\dfrac{3x + 1}{x^2}$

Domain: All real numbers except $x = 0$

109. Answers will vary.

Sample answer: $f(x) = x^2$, $g(x) = 6x - 5$

111. Answers will vary.

Sample answer: $f(x) = \dfrac{1}{x^2}$, $g(x) = x - 1$

113. (a) $r = \dfrac{x}{2}$ (b) $A = \pi r^2$ (c) $(A \circ r)(x) = \dfrac{\pi x^2}{4}$

$(A \circ r)(x)$ is the area of the base of the tank in terms of x, where x is the length of a side of the square base.

115. $(N \circ T)(t) = 32t^2 + 36t + 204$

$(N \circ T)(t)$ represents the number of bacteria in the petri dish at time t.

Test Yourself   *(page 248)*

1. Distance: $4\sqrt{5}$

Midpoint: $(1, 0)$

2. Distance: ≈ 7.81

Midpoint: $(0.44, 4.335)$

3. x-intercepts: $(-5, 0)$, $(3, 0)$

y-intercept: $(0, -15)$

4. Symmetric with respect to the origin

5. $(x - 3)^2 + (y + 2)^2 = 16$

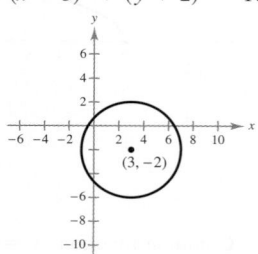

6. $2x - 3y + 21 = 0$

7. True. Each value of x corresponds to exactly one value of y.

8. False. The element -9 is not included in set B.

9. (a) Domain: All real numbers

Range: $(-\infty, 2]$

(b) Decreasing: $(0, \infty)$

Increasing: $(-\infty, 0)$

(c) Even

(d) Maximum: $(0, 2)$

10. (a) Domain: $(-\infty, -2] \cup [2, \infty)$

Range: $[0, \infty)$

(b) Decreasing: $(-\infty, -2)$

Increasing: $(2, \infty)$

(c) Even

(d) Minima: $(-2, 0)$, $(2, 0)$

11. Reflected about the x-axis and shifted two units to the left

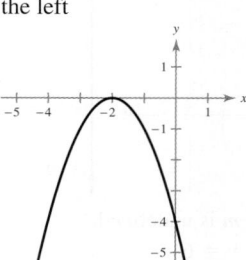

12. Shifted three units to the right and four units upward

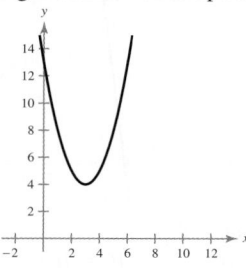

13. $(f - g)(x) = x^2 - 2x + 3$

14. $(fg)(x) = 2x^3 - 2x^2 + 2x - 2$

15. $(f \circ g)(x) = 4x^2 - 8x + 5$ **16.** $(g \circ f)(x) = 2x^2$

17. $V = 30,000 - 5300t$

18.

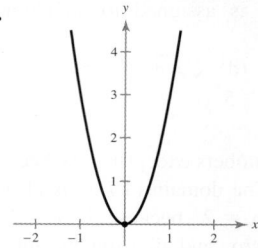

$P = 0.17t + 19.3$

Chapter 3

Section 3.1   *(page 258)*

Skills Warm Up   *(page 258)*

1. $\frac{1}{2}, -6$ **2.** $-\frac{3}{5}, 3$ **3.** $\frac{3}{2}, -1$ **4.** -10

5. $3 \pm \sqrt{5}$ **6.** $-2 \pm \sqrt{3}$ **7.** $4 \pm \dfrac{\sqrt{14}}{2}$

8. $-5 \pm \dfrac{\sqrt{3}}{3}$ **9.** $-\dfrac{3}{2} \pm \dfrac{\sqrt{5}}{2}$ **10.** $-\dfrac{3}{2} \pm \dfrac{\sqrt{21}}{2}$

1. Compared with the graph of $y = x^2$, each output of $f(x) = 5x^2$ vertically stretches the graph by a factor of 5.

3.

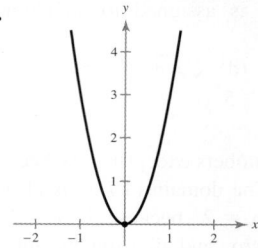

The graph of $f(x) = 3x^2$ is a vertical stretch of the graph of $y = x^2$ by a factor of 3.

5.

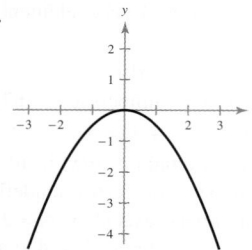

The graph of $g(x) = -\frac{1}{2}x^2$ is a reflection in the x-axis of the graph of $y = x^2$ and a vertical shrink by a factor of $\frac{1}{2}$.

7. The graph of f is the graph of $y = x^2$ reflected in the x-axis, shifted to the left 1 unit and shifted upward 1 unit.

9. e   **10.** c   **11.** a   **12.** d   **13.** f   **14.** b

15. $f(x) = -x^2 + 16$
Vertex: $(0, 16)$

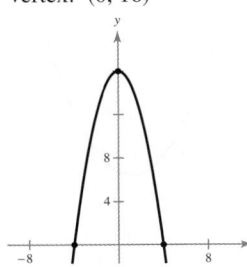

17. Vertex: $(-5, -6)$

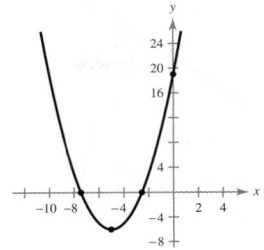

19. $g(x) = (x + 1)^2$
Vertex: $(-1, 0)$

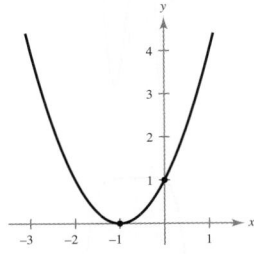

21. $f(x) = -(x + 1)^2 + 4$
Vertex: $(-1, 4)$

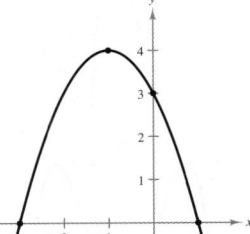

23. $f(x) = \left(x - \frac{1}{2}\right)^2 + 1$
Vertex: $\left(\frac{1}{2}, 1\right)$

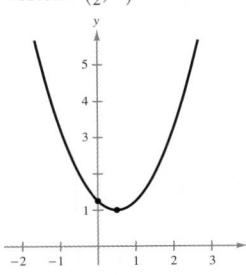

25. $f(x) = -(x - 1)^2 + 6$
Vertex: $(1, 6)$

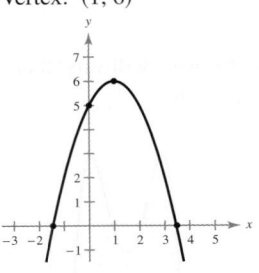

27. $h(x) = 4\left(x - \frac{1}{2}\right)^2 + 20$
Vertex: $\left(\frac{1}{2}, 20\right)$

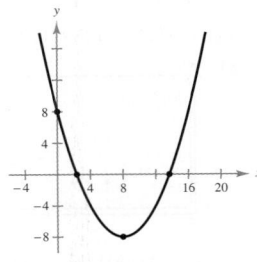

29. $f(x) = \frac{1}{4}(x - 8)^2 - 8$
Vertex: $(8, -8)$

31. $y = -(x + 2)^2$　　**33.** $y = (x - 3)^2 - 9$
35. $y = -2(x + 3)^2 + 3$
37. $y = -\frac{1}{2}(x - 2)^2 - 1$　　**39.** $y = \frac{3}{4}(x - 5)^2 + 12$
41. Answers will vary.　　**43.** Answers will vary.
Sample answer:　　　　　Sample answer:
$f(x) = x^2 - x - 2$　　　　$f(x) = x^2 - 10x$
$g(x) = -x^2 + x + 2$　　　$g(x) = -x^2 + 10x$
45. Answers will vary.　　**47.** $A = 100x - x^2$; $(50, 2500)$;
Sample answer:　　　　　The rectangle has the greatest
$f(x) = 2x^2 + 7x + 3$　　　area ($A = 2500$ square feet)
$g(x) = -2x^2 - 7x - 3$　　when its width is 50 feet.

49. $x = 150$ feet, $y = 200$ feet; 300 feet $\times$ 200 feet
51. 25,000 units　　**53.** 20 fixtures　　**55.** 14 feet
57. (a) and (b)

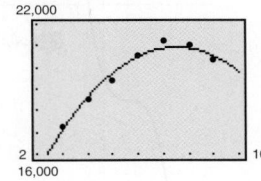

$$R = -186.43t^2 + 2811.9t + 10,329$$
(c) 2007 ($t \approx 7.5$); Yes, this agrees with the actual data.

59. (a)

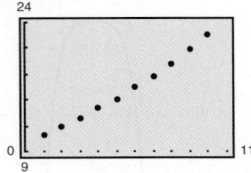

(b) $N = 0.047t^2 + 0.76t + 10.1$
(c)　　　　　　　　　　　　　　　　(d) 25,988 physicians

61. (a)

(b) $p(x) = -1.276x^2 + 21.63x + 236.5$
(c)

(d) $\approx (8.5, 328.2)$; Producing about 85,000 units yields the maximum profit, about \$32,820,000.
(e) Sample answer: Production costs may be growing faster than revenue, so profit decreases.

63. $f(x) = a\left(x + \dfrac{b}{2a}\right)^2 + \dfrac{4ac - b^2}{4a}$

Section 3.2　*(page 269)*

Skills Warm Up　*(page 269)*

1. $(3x - 2)(4x + 5)$　　**2.** $x(5x - 6)^2$
3. $z^2(12z + 5)(z + 1)$　　**4.** $(y + 5)(y^2 - 5y + 25)$
5. $(x + 3)(x + 2)(x - 2)$　　**6.** $(x + 2)(x^2 + 3)$
7. No real solution　　**8.** $3 \pm \sqrt{5}$　　**9.** $-\frac{1}{2} \pm \sqrt{3}$
10. ± 3

1. b   **2.** d   **3.** c   **4.** a

5. 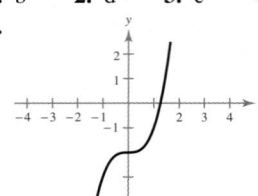   **7.**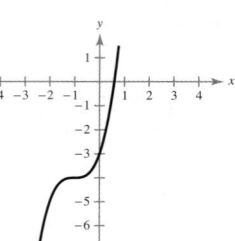

9. 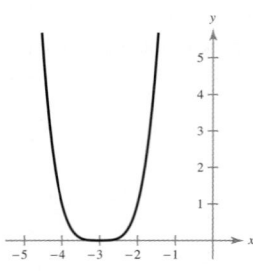   **11.**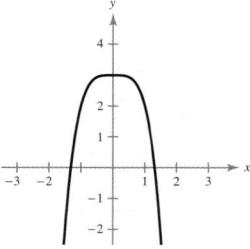

13. (a) 5; odd
(b) 2; positive
(c)
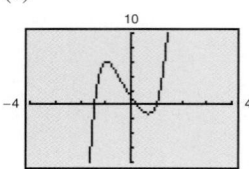
(d) Falls to the left, rises to the right

15. (a) 5; odd
(b) −2; negative
(c)
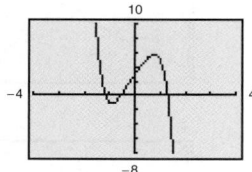
(d) Rises to the left, falls to the right

17. (a) 2; even
(b) 2; positive
(c)
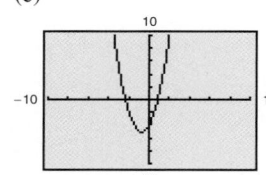
(d) Rises to the left and to the right

19. (a) 6; even
(b) −1; negative
(c)
(d) Falls to the left and to the right

21. When the leading coefficient is positive, the graph falls to the left and rises to the right. When the leading coefficient is negative, the graph rises to the left and falls to the right.

23. Rises to the left
Falls to the right

25. Rises to the left
Falls to the right

27. Rises to the left
Rises to the right

29. Rises to the left
Rises to the right

31. Falls to the left; Falls to the right

33. (a) 1   (b) 2   **35.** (a) 4   (b) 5   **37.** ±3

39. −4   **41.** No real zeros   **43.** −1 ± √5   **45.** 2, 0

47. ±1   **49.** ±√5   **51.** 3   **53.** e   **54.** c

55. g   **56.** d   **57.** f   **58.** h   **59.** a   **60.** b

61. 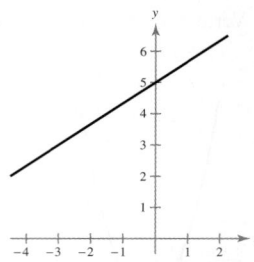   **63.**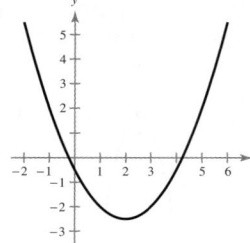

65. 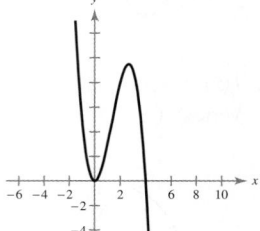   **67.**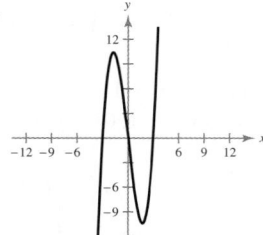

69. 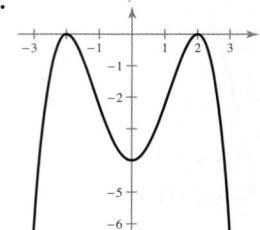   **71.**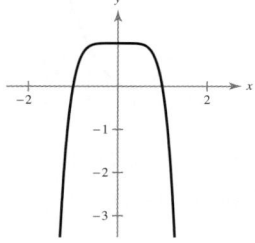

73. Answers will vary. Sample answers:
$a_n < 0$ $a_n > 0$
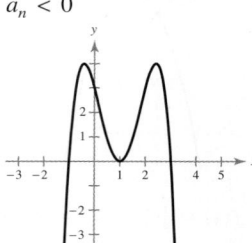   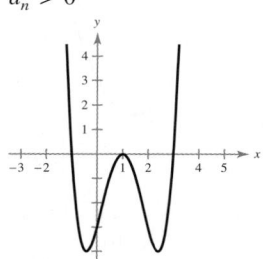

75. $f(x) = \frac{1}{2}(x^4 - 11x^3 + 28x^2)$

77. (a) 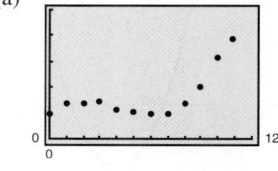   (b) Positive

(c) $P = 0.00011t^3 - 0.013t^2 + 0.37t + 10.8$
The model agrees with the prediction from part (b).

(d)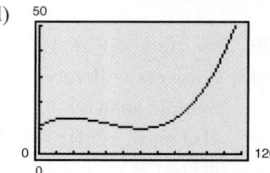

2012 ($t \approx 112.8$); The prediction seems reasonable.

79.

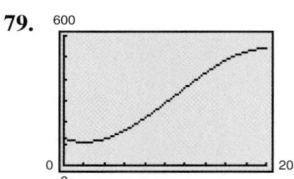

(11, 323.9)

81. (a) Answers will vary.   (b) Domain: $0 < x < 6$

(c)

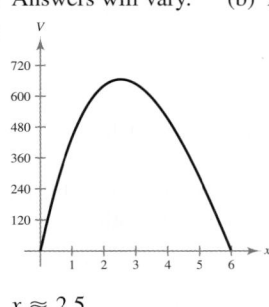

$x \approx 2.5$

Section 3.3   *(page 279)*

Skills Warm Up   *(page 279)*

1. $x^3 - x^2 + 2x + 3$ **2.** $2x^3 + 4x^2 - 6x - 4$
3. $x^4 - 2x^3 + 4x^2 - 2x - 7$
4. $2x^4 + 12x^3 - 3x^2 - 18x - 5$
5. $(x - 3)(x - 1)$ **6.** $8(x + 2)(x - 5)$
7. $(3x + 5)(x - 1)$ **8.** $(3x - 4)^2$
9. $2x(x - 1)(2x - 3)$ **10.** $x(3x + 2)(2x + 1)$

1. $x + 2$ **3.** $3x - 4$ **5.** $2x + 4$ **7.** $x^2 + 2x + 3$

9. $x + 3$ **11.** $x^3 + 3x^2 - 1$ **13.** $7 - \dfrac{25}{x + 4}$

15. $3x + 5 - \dfrac{2x - 3}{2x^2 + 1}$ **17.** $x + \dfrac{x - 27}{x^2 - 1}$

19. $x^2 - 6x + 17 - \dfrac{36}{x + 2}$

21. $2x^3 + 4x^2 - 2x - 8 - \dfrac{10x - 7}{x^2 - 2x + 1}$ **23.** $x + 7$

25. $2x^2 - 3x + 5$ **27.** $4x^2 - 9$ **29.** $-x^2 + 10x - 25$

31. $x^3 - 7x^2 + 14x - 20 + \dfrac{84}{x + 3}$

33. $10x^3 + 10x^2 + 60x + 360 + \dfrac{1360}{x - 6}$

35. $x^4 - x^3 + x^2 - x + 1 - \dfrac{1}{x + 1}$

37. $-3x^3 - 6x^2 - 12x - 24 - \dfrac{48}{x - 2}$

39. $-x^2 + 3x - 6 + \dfrac{11}{x + 1}$ **41.** $4x^2 + 14x - 30$

43. $f(x) = (x - 2)(x^2 + 3x - 6) + 8; f(2) = 8$
45. $f(x) = \left(x - \tfrac{1}{3}\right)(3x^2 + 3x + 6); f\left(\tfrac{1}{3}\right) = 0$
47. $f(x) = \left(x - \sqrt{3}\right)\left[x^2 + \left(2 + \sqrt{3}\right)x + 2\sqrt{3}\right] - 6;$
    $f\left(\sqrt{3}\right) = -6$

49. $f(x) = \left(x - 1 - \sqrt{3}\right)\left[2x^2 + \left(3 + 2\sqrt{3}\right)x + \left(-5 + 5\sqrt{3}\right)\right]; f\left(1 + \sqrt{3}\right) = 0$
51. (a) -69 (b) -2081 (c) -6 (d) 446
53. (a) 30   (b) -21 (c) 1.092   (d) -9
55. (a) Proof   (b) $x + 2$
    (c) $f(x) = (x - 4)(x + 2)^2$ (d) $4, -2$
    (e)

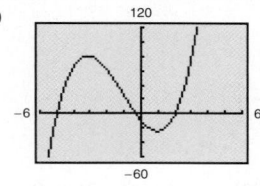

57. (a) Proof   (b) $x + 5$
    (c) $f(x) = (3x + 1)(x - 2)(x + 5)$ (d) $-\tfrac{1}{3}, 2, -5$
    (e)

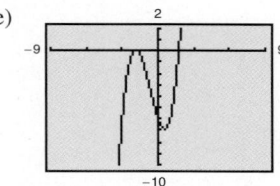

59. (a) Proof   (b) $x + \sqrt{3}$
    (c) $f(x) = \left(x - \sqrt{3}\right)(x + 2)\left(x + \sqrt{3}\right)$
    (d) $\sqrt{3}, -2, -\sqrt{3}$
    (e)

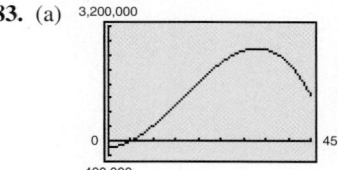

61. The second polynomial is a factor of the first polynomial.

63. e; 3, $\dfrac{-1 \pm \sqrt{17}}{2}$ **64.** d; -2, $\dfrac{3 \pm \sqrt{5}}{2}$

65. a; -1, $-2 \pm \sqrt{2}$ **66.** c; 3, $1 \pm \sqrt{5}$
67. b; -3, $\pm\sqrt{5}$ **68.** f; 2, $\pm\sqrt{5}$
69. Answers will vary. Sample answer:
    $f(x) = 3x^3 - 13x^2 + 4x + 20$
    $f(x) = -3x^3 + 13x^2 - 4x - 20$
    Infinitely many polynomial functions
71. $x^2 - 7x + 10$ **73.** $3x^2 - x - 10$ **75.** $x^2 - 6x$
77. $x^2 + 4x + 3$ **79.** $x^2 + 9x + 16$ ft^2
81. Multiply the divisor and the quotient, then add the remainder to obtain the dividend.
83. (a)

(b) Another nonnegative real solution is $x \approx 25.0366$, representing an advertising expense of $250,366.
(c) The answer is the same as part (b).
85. $c = 42$

Section 3.4   *(page 291)*

Skills Warm Up   *(page 291)*

1. Answers will vary. Sample answer:
$f(x) = 3x^3 - 8x^2 - 5x + 6$

2. Answers will vary. Sample answer:
$f(x) = 4x^4 - 3x^3 - 16x^2 + 12x$

3. $x^4 - 3x^3 + 5 + \dfrac{3}{x + 3}$

4. $3x^3 + 15x^2 - 9 - \dfrac{2}{x + (2/3)}$

5. $\frac{1}{2}, -3 \pm \sqrt{5}$ **6.** $10, -\frac{2}{3}, -\frac{3}{2}$ **7.** $-\frac{3}{4}, 2 \pm \sqrt{2}$

8. $\frac{2}{5}, -\frac{7}{2}, -2$ **9.** $\pm\sqrt{2}, \pm1$ **10.** $\pm2, \pm\sqrt{3}$

---

1. $\pm1; f(1) = 2, f(-1) = 0$
Actual zero of f: -1

3. $\pm1; f(1) = -2, f(-1) = -4$
Actual zero of f: none

5. $\pm1, \pm2, \pm3, \pm6;$
$f(1) = -12, f(-1) = 0, f(2) = -12, f(-2) = 0,$
$f(3) = 0, f(-3) = -12, f(6) = 168, f(-6) = -180$
Actual zeros of f: $-2, -1, 3$

7. Possible: $\pm1, \pm2, \pm4$
Actual: $-1, \pm2$

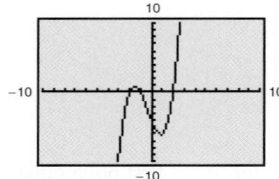

9. Possible: $\pm1, \pm2, \pm3, \pm4, \pm6, \pm12$
Actual: $1, 2, 6$

11. Possible: $\pm1, \pm2, \pm3, \pm6, \pm\frac{1}{2}, \pm\frac{3}{2}$
Actual: $-2, 3, \frac{1}{2}$

13. $-3, \frac{1}{2}, 4$ **15.** $\pm\frac{1}{2}, \pm2$ **17.** $1, 2, 3$ **19.** $-1, -10$

21. $\frac{1}{2}, -1$ **23.** $\pm3, \pm\sqrt{2}$ **25.** $-1, 2$ **27.** $-6, \frac{1}{2}, 1$

29. $-2, 0, 1$

31. (a) $\pm1, \pm3, \pm\frac{1}{2}, \pm\frac{3}{2}, \pm\frac{1}{4}, \pm\frac{3}{4}, \pm\frac{1}{8}, \pm\frac{3}{8}, \pm\frac{1}{16}, \pm\frac{3}{16}, \pm\frac{1}{32}, \pm\frac{3}{32}$
(b)   (c) $1, \frac{3}{4}, -\frac{1}{8}$

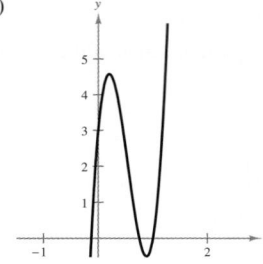

33. $f(1) = -2, f(2) = 7$ **35.** $f(2) = -6, f(3) = 44$

37. Real zero ≈ 0.7 **39.** Real zero ≈ 3.3

41. e; -1.769 **42.** c; 0.755 **43.** d; 0.206

44. a; $0.266, 1.175, 2.559$ **45.** f; 2.769

46. b; $-1.675, -0.539, 2.214$ **47.** $-1.164, 1.453$

49. $0.900, 1.100, 1.900$ **51.** $-1.453, 1.164$

53. $-2.177, 1.563$ **55.** d   **56.** a   **57.** b   **58.** c

59. (a) $V = x(18 - 2x)(15 - 2x)$
Domain: $0 < x < 7.5$
(b)

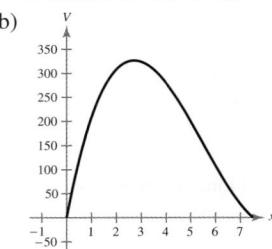

Approximate dimensions:
2.72 inches × 12.56 inches × 9.56 inches
(c) $x \approx 0.448, 6, \approx 10.052$
A value of $x \approx 10.052$ inches is impossible because it would yield a negative length and width.
(d) $x = 6$

61. 18 inches × 18 inches × 36 inches

63. (a) $V = x^3 + 9x^2 + 26x + 24 = 120$
(b) 4 feet by 5 feet by 6 feet

65. 4.49 hours

67. (a)

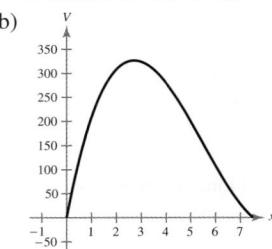

(b) Linear: $R = 6.923t - 7.43$
Quadratic: $R = 0.8884t^2 - 2.849t + 12.12$
Cubic: $R = 0.08806t^3 - 0.5646t^2 + 3.852t + 4.56$
Quartic:
$R = 0.012630t^4 - 0.18979t^3 + 1.4688t^2 - 1.705t + 8.90$
(c) Linear   Quadratic

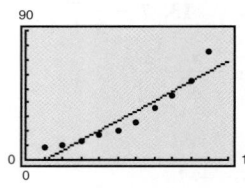   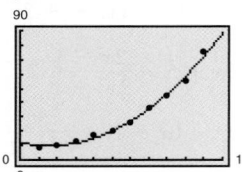

Cubic   Quartic

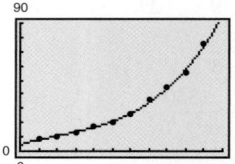   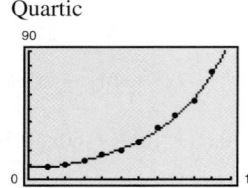

The nonlinear models fit the data well.
(d) Linear: 2015 ($t \approx 15.52$)
Quadratic: 2011 ($t \approx 11.68$)
Cubic: 2011 ($t \approx 11.18$)
Quartic: 2010 ($t \approx 10.99$)
Answers will vary. Sample answer: The higher the degree of the model, the faster the graph climbs after 2010, so the quartic model makes the earliest prediction of when the revenue per share will reach $100. Based on the increasing trend of the last few data points, the year (2015) predicted by the linear model seems too late.

69. (a)

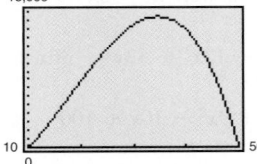

(b) Quadratic: $S = 11,001.49t^2 - 36,269.6t + 47,202$

Cubic: $S = 735.402t^3 + 1073.57t^2$
$$- 2367.6t + 28,670$$

(c) Quadratic: 2011 ($t \approx 11.1$)

Cubic: 2010 ($t \approx 10.6$)

Answers will vary. Sample answer: Based on the trend of the data, both models seem appropriate for predicting future values.

71. (a)

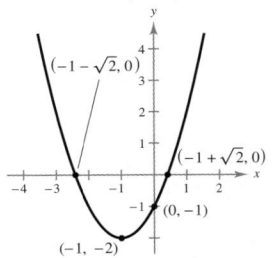

30 and ≈ 38.91

(b) You can solve

$$-x^3 + 54x^2 - 140x - 3000 = 14,400$$

by rewriting the equation as

$$y = -x^3 + 54x^2 - 140x - 17,400.$$

Using the *table* feature of a graphing utility, you can approximate the solutions to be $x \approx -14.91$, $x = 30$, and $x \approx 38.91$. The company should charge \$38.91 to generate greater revenue.

73. No; setting $h = 64$ and solving the resulting equation yields imaginary roots.

75. No   **77.** Yes. Answers will vary.

Sample answer: $f(x) = (x + 1)(x^2 - 5)$

79. Answers will vary.

Quiz Yourself   *(page 296)*

1.

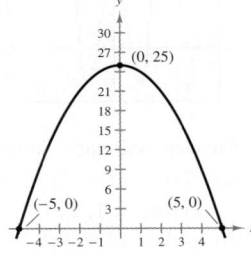

Vertex: $(-1, -2)$

Intercepts:

$\left(-1 \pm \sqrt{2}, 0\right), (0, -1)$

2.

Vertex: $(0, 25)$

Intercepts:

$(\pm 5, 0), (0, 25)$

3. Rises to the left   **4.** Rises to the left

Falls to the right   Rises to the right

5. 293   **6.** $f(x) = (x + 1)(x^3 - x^2 - 4x + 4) + 0$; $f(-1) = 0$

7. $f(x) = (x - 2)(x^2 + 7x + 12) + 0$; $f(2) = 0$

8. $2x^2 + 5x - 12$ **9.** $\pm\sqrt{5}, -\frac{7}{2}$ **10.** $\pm 3, \pm\frac{1}{2}$

11. $1, -\frac{4}{3}$ **12.** $\frac{3}{2}$ **13.** $P = \$2,534,375; \$337,600$

14. (a)

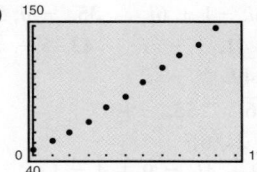

(b) Linear: $A = 10.53t + 39.9$

Quadratic: $A = 0.210t^2 + 8.43t + 43.0$

Cubic: $A = -0.0386t^3 + 0.789t^2 + 6.22t + 44.4$

Quartic:
$$A = 0.0062t^4 - 0.1633t^3 + 1.568t^2 + 4.66t + 44.9$$

(c) Linear   Quadratic

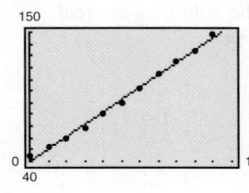

Cubic   Quartic

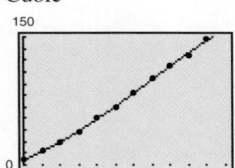

Each model could be considered a good fit, but the cubic and quartic models appear to fit the data a little better than the linear and quadratic models.

(d) Linear: 2012 ($t \approx 12.36$)

Quadratic: 2011 ($t \approx 11.67$)

Cubic: 2012 ($t \approx 12.49$)

Quartic: 2011 ($t \approx 11.79$)

Answers will vary. Sample answer: The linear and cubic predictions are close to each other and the quadratic and quartic predictions are close to each other. The quadratic and quartic models give earlier predictions because their graphs rise faster in the years after 2010 than the graphs of the linear and cubic models.

Section 3.5   *(page 304)*

Skills Warm Up   *(page 304)*

1. $2\sqrt{3}$ **2.** $10\sqrt{5}$ **3.** $\sqrt{5}$ **4.** $-6\sqrt{3}$

5. 12   **6.** 48   **7.** $\dfrac{\sqrt{7}}{7}$ **8.** $\sqrt{5}$

9. $-\dfrac{1}{2} \pm \dfrac{\sqrt{5}}{2}$ **10.** $-1 \pm \sqrt{2}$

CHAPTER 3

1. $i, -1, -i, 1, i, -1, -i, 1, i, -1, -i, 1, i, -1, -i, 1$
$i^{4n} = 1, i^{4n+1} = i, i^{4n+2} = -1, i^{4n+3} = -i, n$ is an integer.
3. $a = 7, b = 12$ **5.** $a = 4, b = -2$ **7.** $2 + i$
9. $5 + 6i$ **11.** $\frac{5}{3} + \frac{7}{6}i$ **13.** $1.3 + 9i$ **15.** 25
17. $5 + i$ **19.** $30 + 20i$ **21.** $-11 + 60i$ **23.** -14
25. 8 **27.** $9 + 4i; 9 - 4i$ **29.** $-3 - 2\sqrt{3}i; -3 + 2\sqrt{3}i$
31. $-21; -21$ **33.** $-1 - 6i; -1 + 6i$ **35.** $-5i; 5i$
37. $-3; -3$ **39.** $\frac{4}{5} - \frac{3}{5}i$ **41.** $1 + \frac{1}{2}i$ **43.** $5 - \frac{7}{2}i$
45. $\frac{1}{8}i$ **47.** $-\frac{44}{125} - \frac{8}{125}i$ **49.** $\frac{35}{29} + \frac{595}{29}i$
51. $-7 + 7\sqrt{2}i$ **53.** $-2\sqrt{6}$ **55.** -10
57. $\left(16 + 4\sqrt{3}\right) + \left(-16\sqrt{2} + 2\sqrt{6}\right)i$
59. Error: $(3 - 2i)(3 + 2i) = 9 - 4i^2 = 9 + 4 = 13$
(not $9 - 4 = 5$)
61. $\pm 3i$ **63.** $1 \pm i$ **65.** $-2 \pm \frac{1}{2}i$ **67.** $-2 \pm i$
69. $\dfrac{1}{8} \pm \dfrac{\sqrt{11}}{8}i$
71. No. Sample answer: When the discriminant of the Quadratic Formula is negative, there are two complex imaginary solutions to the equation. Otherwise, the solutions are real.
73. 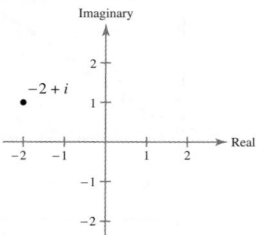   **75.**
77.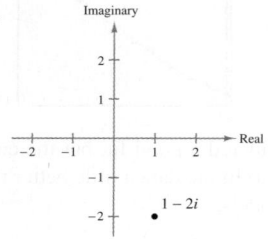
79. The complex number 0 is in the Mandelbrot Set because, for $c = 0$, the corresponding Mandelbrot sequence is $0, 0, 0, 0, 0, 0,$ which is bounded.
81. The complex number 1 is not in the Mandelbrot Set because, for $c = 1$, the corresponding Mandelbrot sequence is $1, 2, 5, 26, 677, 458,330, \ldots,$ which is unbounded.
83. The complex number $\frac{1}{2}i$ is in the Mandelbrot Set because, for $c = \frac{1}{2}i$, the corresponding Mandelbrot sequence is $\frac{1}{2}i,$
$-\frac{1}{4} + \frac{1}{2}i, -\frac{3}{16} + \frac{1}{4}i, -\frac{7}{256} + \frac{13}{32}i, -\frac{10,767}{65,536} + \frac{1957}{4096}i,$
$-\frac{864,513,055}{4,294,967,296} + \frac{46,037,845}{134,217,728}i,$ which is bounded.
85. False. If the complex number is real, it equals its conjugate.
87. Answers will vary.

Section 3.6   *(page 312)*

Skills Warm Up   *(page 312)*

1. $4 - \sqrt{29}i, 4 + \sqrt{29}i$ **2.** $-5 - 12i, -5 + 12i$
3. $-1 + 4\sqrt{2}i, -1 - 4\sqrt{2}i$ **4.** $6 + \frac{1}{2}i, 6 - \frac{1}{2}i$
5. $-13 + 9i$ **6.** $12 + 16i$ **7.** $26 + 22i$
8. 29 **9.** i **10.** $-9 + 46i$

1. 1 **3.** 3 **5.** 4 **7.** $\pm 5i; (x + 5i)(x - 5i)$
9. $6 \pm \sqrt{10}; \left(x - 6 - \sqrt{10}\right)\left(x - 6 + \sqrt{10}\right)$
11. $2 \pm \sqrt{3}; \left(x - 2 - \sqrt{3}\right)\left(x - 2 + \sqrt{3}\right)$
13. $\pm 3, \pm 3i; (x - 3)(x + 3)(x - 3i)(x + 3i)$
15. $0, \pm \sqrt{5}i; x\left(x - \sqrt{5}i\right)\left(x + \sqrt{5}i\right)$
17. $-5, 8 \pm i; (x + 5)(x - 8 + i)(x - 8 - i)$
19. $2, 2 \pm i; (x - 2)(x - 2 + i)(x - 2 - i)$
21. $-5, 4 \pm 3i; (t + 5)(t - 4 + 3i)(t - 4 - 3i)$
23. $-10, -7 \pm 5i; (x + 10)(x + 7 - 5i)(x + 7 + 5i)$
25. $-5, -2 \pm \sqrt{3}i; (x + 5)\left(x + 2 - \sqrt{3}i\right)\left(x + 2 + \sqrt{3}i\right)$
27. $-\frac{3}{4}, 1 \pm \frac{1}{2}i; (4x + 3)(2x - 2 + i)(2x - 2 - i)$
29. $-\frac{1}{5}, 1 \pm \sqrt{5}i; (5x + 1)\left(x - 1 + \sqrt{5}i\right)\left(x - 1 - \sqrt{5}i\right)$
31. $2, \pm 2i; (x - 2)^2(x + 2i)(x - 2i)$
33. $\pm i, \pm 3i; (x + i)(x - i)(x + 3i)(x - 3i)$
35. $-4, 3, \pm i; (t + 4)^2(t - 3)(t + i)(t - i)$
37. Answers will vary. Sample answer: $x^3 - 2x^2 + x - 2$
39. Answers will vary. Sample answer: $x^3 - 5x^2 + 9x - 5$
41. Answers will vary.
Sample answer: $x^5 + 4x^4 + 13x^3 + 52x^2 + 36x + 144$
43. Answers will vary.
Sample answer: $x^4 + 8x^3 + 9x^2 - 10x + 100$
45. Answers will vary.
Sample answer: $3x^4 - 17x^3 + 25x^2 + 23x - 22$
47. (a) $(x^2 - 8)(x^2 + 1)$ (b) $\left(x - 2\sqrt{2}\right)\left(x + 2\sqrt{2}\right)(x^2 + 1)$
(c) $\left(x - 2\sqrt{2}\right)\left(x + 2\sqrt{2}\right)(x - i)(x + i)$
49. (a) $(x^2 - 2x + 3)(x^2 - 3x - 5)$
(b) $(x^2 - 2x + 3)\left(x - \dfrac{3 + \sqrt{29}}{2}\right)\left(x - \dfrac{3 - \sqrt{29}}{2}\right)$
(c) $\left(x - 1 + \sqrt{2}i\right)\left(x - 1 - \sqrt{2}i\right)$
$\left(x - \dfrac{3 + \sqrt{29}}{2}\right)\left(x - \dfrac{3 - \sqrt{29}}{2}\right)$
51. $\pm 5i, -\frac{3}{2}$ **53.** $\pm 6i, 1$ **55.** $-3 \pm i, \frac{1}{4}$
57. $1, 2, -3 \pm \sqrt{2}i$ **59.** $\dfrac{3}{4}, \dfrac{1}{2} \pm \dfrac{\sqrt{5}}{2}i$
61. (a) $1, 2$ (b) $-3 \pm \sqrt{2}i$
63. (a) 0.75 (b) $\dfrac{1 \pm \sqrt{5}i}{2}$
65. $\pm 1, \pm 2$. The x-intercepts occur at the solutions of the equation.
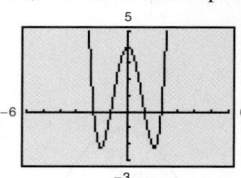
67. Answers will vary. Sample answer:
(a) $f(x) = x^4 - 7x^3 + 17x^2 - 17x + 6$
Zeros: $1, 1, 2, 3$
(b) $g(x) = x^4 - 3x^3 + 3x^2 - 3x + 2$
Zeros: $1, 2, \pm i$

(c) $h(x) = x^4 + 5x^2 + 4$
    Zeros: $\pm i, \pm 2i$

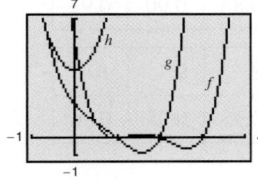

Similarities: f, g, and h all rise to the left and rise to the right.

Differences: f, g, and h have different numbers of x-intercepts.

69. $-0.0001x^2 + 140x = 50,000,000$
$$x = 700,000 \pm 100,000i$$

The solutions are imaginary, so there is no quantity of units x or price p that will yield a revenue of \$50 million.

71. Imaginary zeros of polynomials with real coefficients occur in conjugate pairs. So, a fourth-degree polynomial with real coefficients cannot have exactly three imaginary zeros.

73. Imaginary zeros occur only in conjugate pairs. Because f has three zeros, one or all of them must be real numbers.

Section 3.7   (page 322)

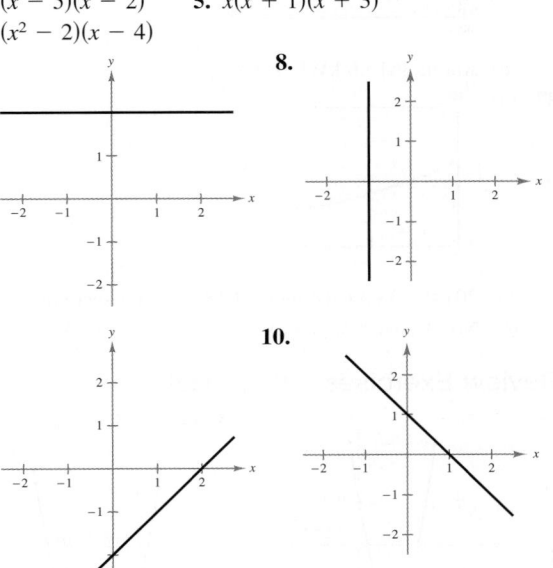

Skills Warm Up   (page 322)

1. $x(x - 4)$ **2.** $2x(x^2 - 3)$ **3.** $(x - 5)(x + 2)$
4. $(x - 5)(x - 2)$ **5.** $x(x + 1)(x + 3)$
6. $(x^2 - 2)(x - 4)$
7.
8.
9.
10.

1. Domain: All $x \neq 2$
$f(x) \to -\infty$ as $x \to 2^-$
$f(x) \to \infty$ as $x \to 2^+$

3. Domain: All $x \neq -1$
$f(x) \to \infty$ as $x \to -1^-$
$f(x) \to -\infty$ as $x \to -1^+$

5. Domain: All $x \neq \pm 1$
$f(x) \to -\infty$ as $x \to -1^-$ or $x \to 1^-$
$f(x) \to \infty$ as $x \to -1^+$ or $x \to 1^+$

7. Horizontal asymptote: $y = 3$; Vertical asymptote: $x = -1$

9. Horizontal asymptote: $y = -1$; Vertical asymptote: $x = 5$
11. Horizontal asymptote: $y = 3$; Vertical asymptote: None
13. Horizontal asymptote: $y = 0$; Vertical asymptote: $x = -4$
15. f   **16.** e   **17.** a   **18.** b   **19.** c   **20.** d
21. $g(x)$ shifts downward two units.
23. $g(x)$ is a reflection about the x-axis.
25. $g(x)$ shifts upward three units.
27. $g(x)$ is a reflection about the x-axis.
29. $g(x)$ shifts upward five units.
31. $g(x)$ is a reflection about the x-axis.

33.
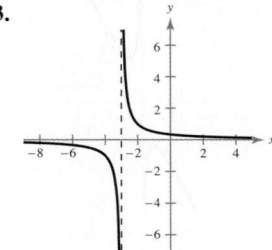
35.
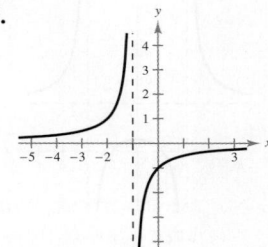

37.
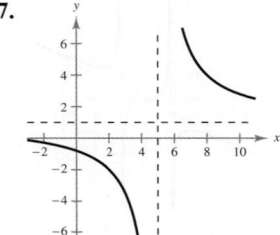
39.
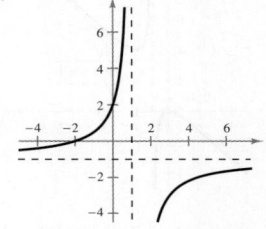

41.
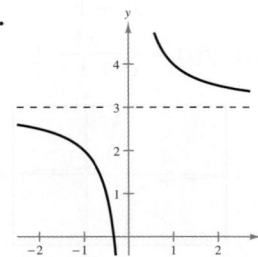
43.
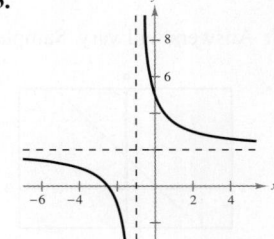

45.
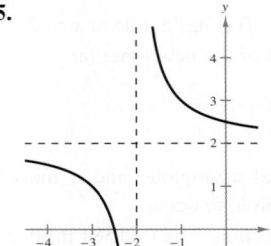
47.
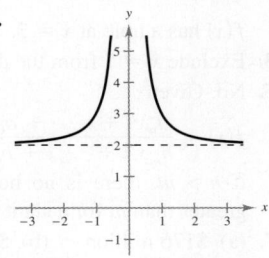

49.
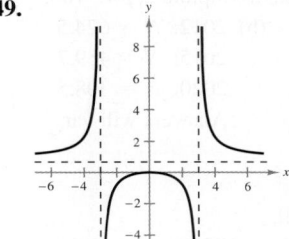
51.
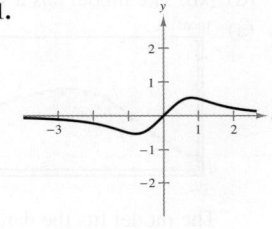

53.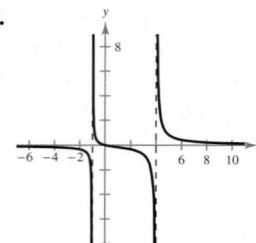

55.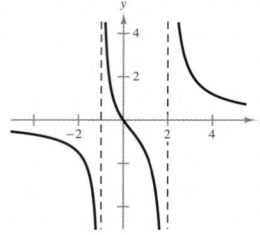

57.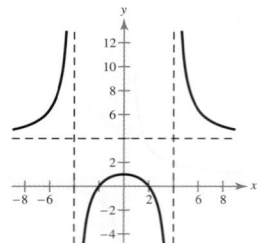

59.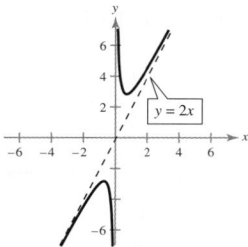

61.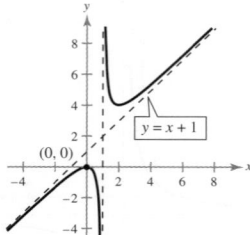

63.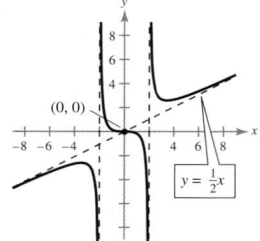

65. Answers will vary. Sample answer: $f(x) = \dfrac{2x^2}{x^2 + 1}$

67. Answers will vary. Sample answer: $f(x) = \dfrac{x^3}{x^2 + x - 2}$

69.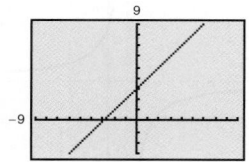

71.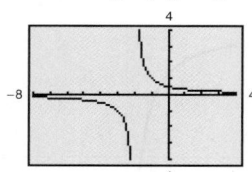

$f(x)$ has a hole at $x = 3$. $f(x)$ has a hole at $x = 2$.

73. Exclude $x = 7$ from the domain of the new function.

75. No. Given

$$f(x) = \frac{a_n x^n + \cdots + a_0}{b_m x^m + \cdots + b_0}$$

if $n > m$, there is no horizontal asymptote, and n must be greater than m for a slant asymptote to occur.

77. (a) \$176 million   (b) \$528 million   (c) \$1584 million

(d) No. The model has a vertical asymptote at $p = 100$.

79. (a)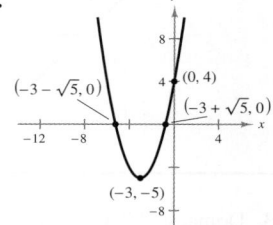

(b) 2012: $N \approx 624.5$
    2015: $N \approx 459.7$
    2020: $N \approx 308.5$
    Answers will vary.

The model fits the data well.

(c) $N = 0$; Over time, the number of DVDs sold will decrease and approach 0.

81. (a)

n	1	2	3	4	5
P	0.60	0.79	0.86	0.90	0.92

n	6	7	8	9	10
P	0.93	0.94	0.95	0.95	0.96

(b) 100%

83. (a)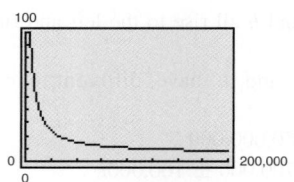

(b)

x	10,000	100,000	1,000,000	10,000,000
$\overline{C}$	\$51	\$10.50	\$6.45	\$6.05

Eventually, the average recycling cost per pound will approach the horizontal asymptote of \$6.

85. (a) $E = \dfrac{27,904t + 906,950}{3.023t + 250.90}$

(b)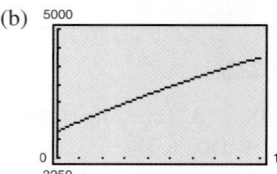

(c) About 4615.6 kWh/person

87. (a)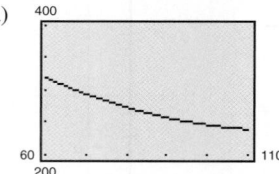

(b) 2014: 238.54 seconds; 2018: 238.04 seconds

(c) No; Answers will vary.

Review Exercises   *(page 330)*

1.

Vertex: $(-3, -5)$
Intercepts:
$(0, 4), \left(-3 \pm \sqrt{5}, 0\right)$

3. $h(x) = 3(x - 2)^2 - 1$

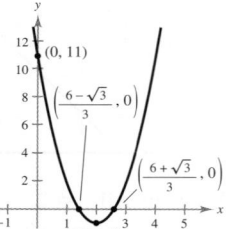

Vertex: $(2, -1)$
Intercepts:
$(0, 11), \left(\dfrac{6 \pm \sqrt{3}}{3}, 0\right)$

5. $f(x) = \frac{7}{9}(x + 5)^2 - 1$

7. $A(x) = 250x - x^2$; 125 feet $\times$ 125 feet

9. (a) 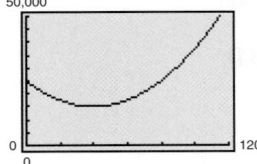 (b) About 385 units

(c) Write the equation of the quadratic function in standard form. The vertex is the minimum cost.

11. (a)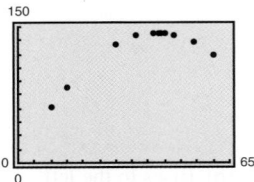

(b) $d(x) = -0.077x^2 + 6.59x + 2.4$

(c)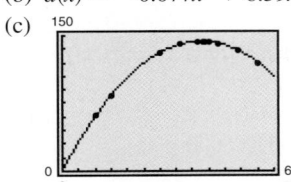

(d) $(42.8, 143.4)$; The vertex gives the angle $42.8°$, which results in the greatest distance of 143.4 meters.

13. 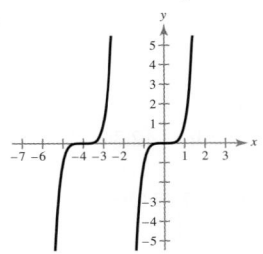   **15.**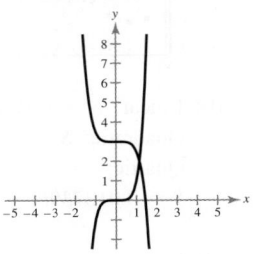

17. 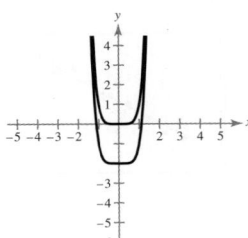   **19.**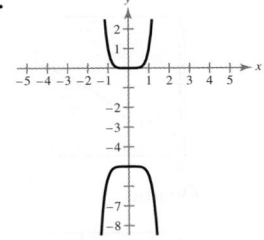

21. Falls to the left   **23.** Falls to the left
Rises to the right   Falls to the right

25. ± 4 **27.** $0, 2, 5$

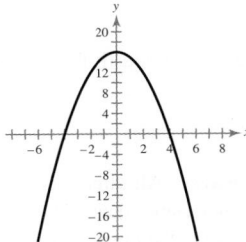

29. $x^2 - 3x + 1 - \dfrac{1}{2x + 1}$ **31.** $x^2 - 3x + 3$

33. (a) -10 (b) 11

35. (a) Proof   (b) $x - 2$
(c) $(x - 5)(x + 3)(x - 2)$ (d) $-3, 2, 5$
(e)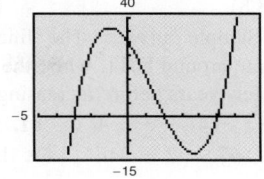

37. $x^2 + 11x + 24$

39. (a) 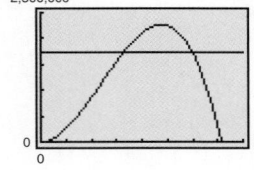   (b) Answers will vary.

$\approx \$325,167$

41. $\pm 1, \pm 3, \pm 5, \pm 15, \pm\frac{1}{2}, \pm\frac{3}{2}, \pm\frac{5}{2}, \pm\frac{15}{2}, \pm\frac{1}{4}, \pm\frac{3}{4}, \pm\frac{5}{4}, \pm\frac{15}{4}$

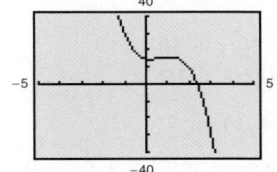

From the graph: $x \approx 2.357$, so the zero is not rational.
43. $-3, -1, 2$ **45.** $\pm 2, \pm\sqrt{5}$ **47.** $-\frac{1}{3}, \frac{3}{2}, 2$
49. $-1, 2$ **51.** $x \approx -2.3$ **53.** $-1.321, -0.283, 1.604$
55. (a)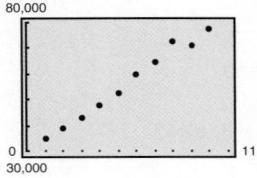

(b) Linear: $R = 4996.7t + 28,927$
Quadratic: $R = -36.14t^2 + 5394.3t + 28,132$
Quartic: $R = 3.4725t^4 - 134.788t^3 + 1486.42t^2$
$\qquad\qquad - 577.4t + 34,333$

(c) Linear   Quadratic
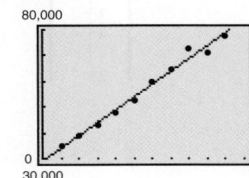   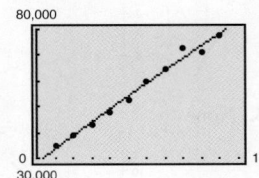

Quartic
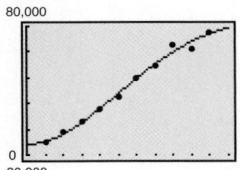

Each model could be considered a good fit for the data.

(d) Linear: 2011 ($t \approx 11.6$)

Quadratic: 2011 ($t \approx 11.9$)

Quartic: 2018 ($t \approx 18.8$)

Answers will vary. Sample answer: The linear and quadratic predictions are around 2011, while the quartic model levels off for a few years before increasing.

57. $4\sqrt{2}i; -4\sqrt{2}i$ **59.** $-3 + 4i; -3 - 4i$ **61.** $5 + 2i$

63. 89   **65.** $-10 - 8i$ **67.** $-7 + 24i$ **69.** 10

71. $3 - 2i$ **73.** $-3 - 4i$

75. $\dfrac{1 \pm \sqrt{23}i}{4}$ **77.** $\dfrac{-11 \pm \sqrt{73}}{8}$

79.

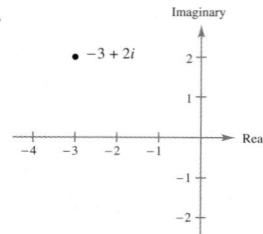

81. $\pm 3, \pm 3i; (x - 3)(x + 3)(x - 3i)(x + 3i)$

83. $-5, \pm\sqrt{3}i; (t + 5)(t - \sqrt{3}i)(t + \sqrt{3}i)$

85. $2, \pm\dfrac{3i}{2}; (x - 2)(2x - 3i)(2x + 3i)$

87. Answers will vary. Sample answer: $x^3 - x^2 + 9x - 9$

89. (a) $(x^2 + 8)(x^2 - 3)$ (b) $(x^2 + 8)(x - \sqrt{3})(x + \sqrt{3})$

(c) $(x + 2\sqrt{2}i)(x - 2\sqrt{2}i)(x - \sqrt{3})(x + \sqrt{3})$

91. $\pm 4i, \dfrac{1}{4}$ **93.** $-1 \pm 3i, -1, -4$

95. Domain: All $x \neq -2$

Vertical asymptote: $x = -2$

Horizontal asymptote: $y = 0$

97. Domain: All $x \neq \pm 3$

Vertical asymptotes: $x = -3, x = 3$

Horizontal asymptote: $y = 2$

99. 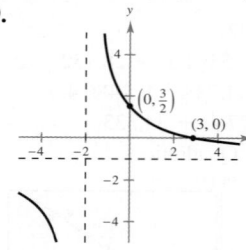   **101.**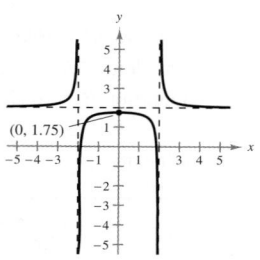

103. None

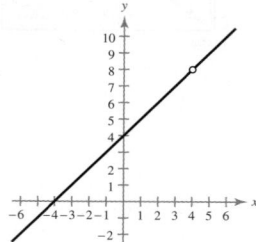

105. (a) 430,769 fish; 662,500 fish; 1,024,000 fish

(b) 1,666,667 fish

107. (a) \$35,000   (b) \$157,500

Test Yourself   *(page 334)*

1.

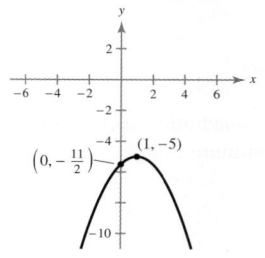

Vertex: $(1, -5)$

Intercept: $\left(0, -\dfrac{11}{2}\right)$

2. (a) Falls to the left   (b) Rises to the left

     Rises to the right   Rises to the right

3. $x^2 + 7x + 12$

4. $\pm 1, \pm 3, \pm 9, \pm 27, \pm\frac{1}{4}, \pm\frac{1}{2}, \pm\frac{3}{4}, \pm\frac{3}{2}, \pm\frac{9}{4}, \pm\frac{9}{2}, \pm\frac{27}{4}, \pm\frac{27}{2}$

$f\left(\frac{3}{2}\right) = 0$

$f\left(-\frac{3}{2}\right) = 0$

$f(x) = \left(x - \frac{3}{2}\right)\left(x + \frac{3}{2}\right)(x - 1)(x - 3)$

5. (a)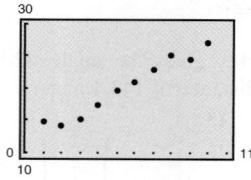

(b) Linear: $S = 1.517t + 11.81$

Quadratic: $S = 0.024t^2 + 1.253t + 12.34$

Quartic:

$S = 0.010739t^4 - 0.26789t^3 + 2.2749t^2$

$- 5.879t + 18.74$

(c) Linear   Quadratic

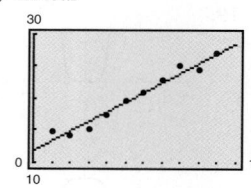 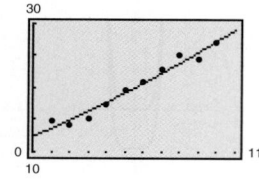

Quartic

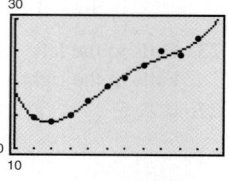

Each model could be considered a good fit for the data.

(d) Linear: 2015 ($t \approx 15.29$)

Quadratic: 2014 ($t \approx 14.21$)

Quartic: 2011 ($t \approx 11.9$)

Answers will vary. Sample answer: All three models give predictions that reflect the increasing data. The quartic model increases most rapidly, while the linear and quadratic models increase at slower rates.

6. $16 - 2i$ **7.** $7 - 10i$ **8.** $27 - 4\sqrt{3}i$ **9.** $23 - 14i$

10. i **11.** $x = \dfrac{-5 \pm \sqrt{3}i}{2}$ **12.** $x = \dfrac{5 \pm 3\sqrt{7}i}{4}$

13. Answers will vary. Sample answer:
$x^4 - 7x^3 + 19x^2 - 63x + 90$

14. $\pm\sqrt{5}i, -2$

15.

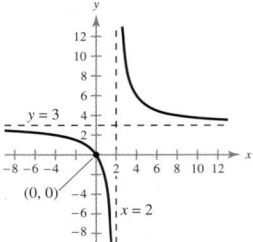

Domain: All $x \neq 2$

Chapter 4

Section 4.1   *(page 343)*

Skills Warm Up   *(page 343)*

1. All real numbers   **2.** $[-1, \infty)$

3. All real numbers except $x = 0, 2$

4. All real numbers except $x = -\frac{5}{3}$ **5.** x **6.** x

7. x **8.** x **9.** $x = \dfrac{3}{2}y + 3$ **10.** $x = \dfrac{y^3}{2} + 2$

1. $f^{-1} = \{(4, 1), (5, 2), (6, 3), (7, 4)\}$

3. $f^{-1} = \{(1, -1), (2, -2), (3, -3), (4, -4)\}$

5. $f^{-1}(x) = \frac{1}{2}x$ **7.** $f^{-1}(x) = x + 5$

9. $f(g(x)) = 5\left(\dfrac{x - 1}{5}\right) + 1 = x;\ g(f(x)) = \dfrac{(5x + 1) - 1}{5} = x$

11. $f(g(x)) = \left(\sqrt[3]{x}\right)^3 = x;\ g(f(x)) = \sqrt[3]{x^3} = x$

13. $f(g(x)) = \sqrt{(x^2 + 4) - 4} = x;\ g(f(x)) = \left(\sqrt{x - 4}\right)^2 + 4 = x$

15. $f(g(x)) = 1 - \left(\sqrt[3]{1 - x}\right)^3 = x;\ g(f(x)) = \sqrt[3]{1 - (1 - x^3)} = x$

17.

x	0	1	2	3	4
$f^{-1}(x)$	-2	0	1	2	4

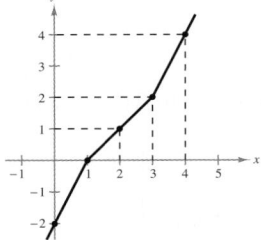

19. f doesn't have an inverse function.   **21.** $g^{-1}(x) = 8x$

23. p doesn't have an inverse function.

25. $f^{-1}(x) = \sqrt{x} - 3,\ x \geq 0$

27. $h^{-1}(x) = \dfrac{1}{x}$ **29.** $f^{-1}(x) = \dfrac{x^2 - 3}{2},\ x \geq 0$

31. g doesn't have an inverse function.

33. $f^{-1}(x) = -\sqrt{25 - x},\ x \leq 25$

35.

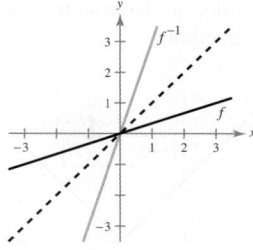
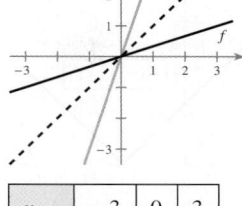
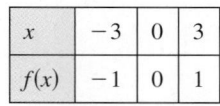

x	-3	0	3
$f(x)$	-1	0	1

x	-1	0	1
$f^{-1}(x)$	-3	0	3

37.

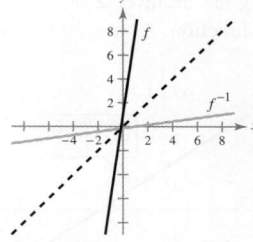

x	-1	0	1
$f(x)$	-6	1	8

x	-6	1	8
$f^{-1}(x)$	-1	0	1

39.

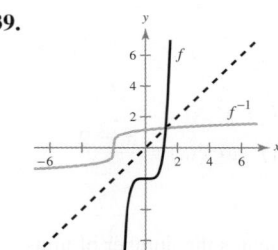

x	-1	0	1
$f(x)$	-3	-2	-1

x	-3	-2	-1
$f^{-1}(x)$	-1	0	1

41. Error: f^{-1} does not mean to take the reciprocal of $f(x)$.
$$f^{-1}(x) = \dfrac{x^2 + 5}{2}$$

43. $f^{-1}(x) = \dfrac{x + 3}{2}$ **45.** $f^{-1}(x) = \sqrt[5]{x}$

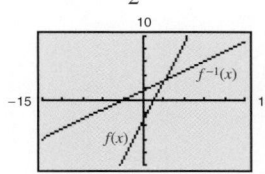
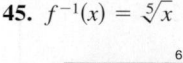
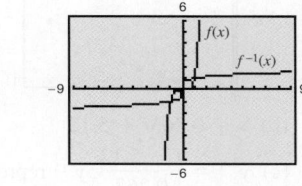

47. $f^{-1}(x) = x^2$,   **49.** $f^{-1}(x) = \sqrt{16 - x^2}$,
$\quad x \geq 0$ $0 \leq x \leq 4$

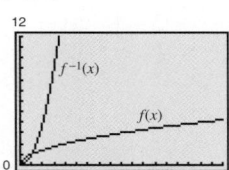
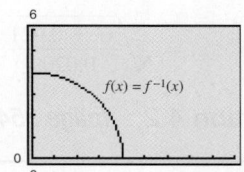

51. $f^{-1}(x) = x^3 - 2$

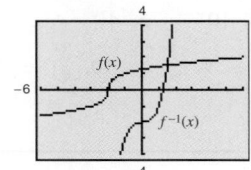

53. Because f is one-to-one, f has an inverse function.

55. f doesn't have an inverse function because two x-values share the same y-value.

57. *g* has an inverse function.

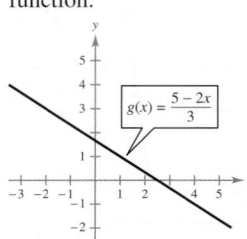

59. *h* doesn't have an inverse function.

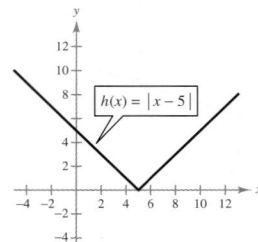

61. *f* doesn't have an inverse function.

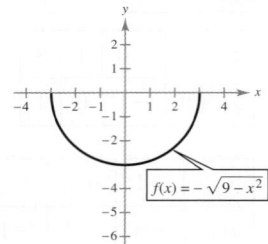

63. $(g^{-1} \circ f^{-1})(x) = \dfrac{x + 2}{2}$ **65.** $(f \circ g)^{-1}(x) = \dfrac{x + 2}{2}$

67. 32   **69.** 600

71. $P^{-1}(x) = \dfrac{x + 5736}{47}$; $P^{-1}(x)$ computes the number of units that must be sold to obtain a profit of x.
Domain of P: $[0, \infty)$
Domain of P^{-1}: $[-5736, \infty)$

73. (a)

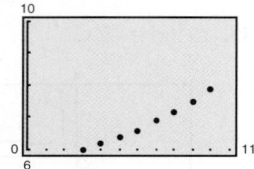

(b) $y = 0.263t + 5.12$

(c) $y^{-1} = \dfrac{t - 5.12}{0.263}$; y^{-1} represents the year in which the average price is t dollars.

(d) 2014 ($y^{-1} \approx 14.75$)

75. $f^{-1}(x) = \sqrt[3]{\dfrac{x^2 + 0.021}{0.0209}}$; \$3.49

Section 4.2   *(page 354)*

1. 3.463   **3.** 94.818   **5.** 0.079   **7.** g   **8.** e
9. b   **10.** h   **11.** d   **12.** a   **13.** f   **14.** c

15.

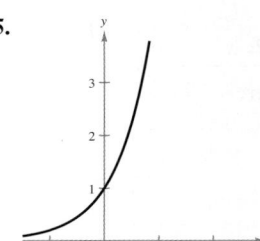

17.

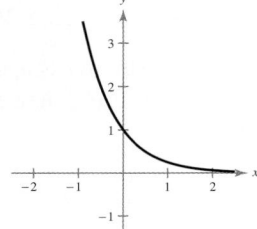

19.

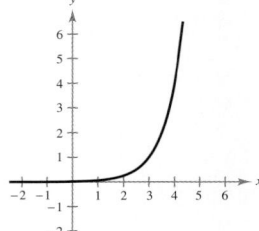

21.

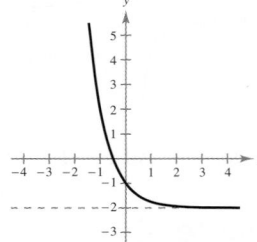

23.

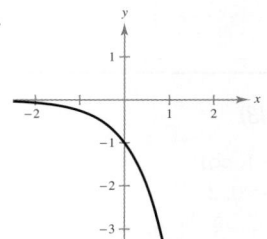

25.

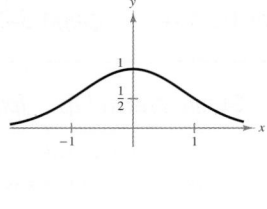

27. 54.598   **29.** 0.067

31.

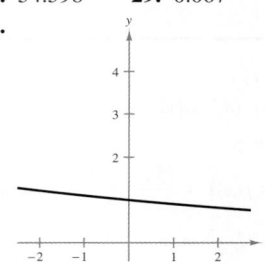

33.

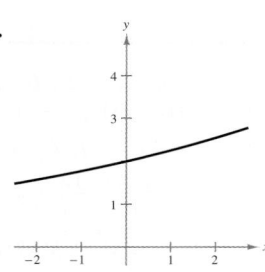

35.

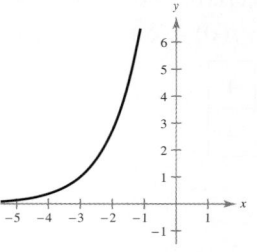

37.

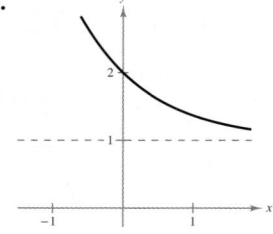

39. Yes. For the natural exponential function, $f(x) = e^x$, the base is $a = e$.

41.

n	1	2	4
A	\$7346.64	\$7401.22	\$7429.74

n	12	365	Continuous
A	\$7449.23	\$7458.80	\$7459.12

43.

n	1	2	4
A	\$6633.24	\$6712.66	\$6753.71

n	12	365	Continuous
A	\$6781.60	\$6795.24	\$6795.70

45. The account paying 5% interest compounded quarterly earns more money. Even though the interest is compounded less frequently, the higher interest rate yields a higher return.

47. \$7424.70   **49.** \$12,434.43

51.

t	1	10	20
P	\$90,521.24	\$36,940.70	\$13,646.15

t	30	40	50
P	\$5040.98	\$1862.17	\$687.90

53.

t	1	10	20
P	\$91,393.12	\$40,656.97	\$16,529.89

t	30	40	50
P	\$6720.55	\$2732.37	\$1110.90

55. \$19,691.17   **57.** \$147,683.76
59. \$20,700.76   **61.** \$155,255.66
63. (a) \$182.91   (b) \$29.58
  (c)   (d) ≈ 117 books

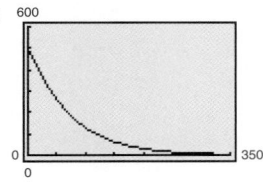

65. (a) 100   (b) ≈ 110   (c) ≈ 121   (d) ≈ 158
67. (a) 5907   (b) 6237   (c) 6767   (d) 7753
69. ≈ 30.42 kilograms
71. (a)   (b) ≈ 1.11 pounds

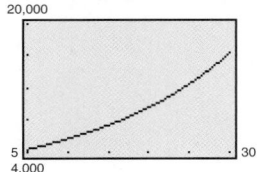

  (c) On the graph, when $P = 2.5$, $t \approx 4.6$ months.
73. (a) 9000 screens, 39,000 screens
  (b) 9059 screens, 39,186 screens
75. (a) 22, 24, 25, 26   (b) 22.1, 24.0, 25.1, 25.9
77. (a)

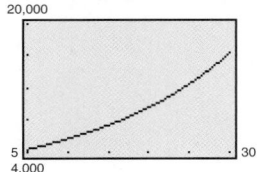

  (b) \$5682.8 billion, \$9624.6 billion, \$15,463.8 billion

79. Answers will vary.

Section 4.3   *(page 365)*

1. c   **2.** f   **3.** b   **4.** d   **5.** a   **6.** e
7. $\log_4 256 = 4$ **9.** $\log_{81} 3 = \frac{1}{4}$ **11.** $\log_6 \frac{1}{36} = -2$
13. $4^2 = 16$ **15.** $2^{-1} = \frac{1}{2}$ **17.** $5^{-1} = 0.2$
19. $27^{1/3} = 3$ **21.** 2   **23.** −4   **25.** $\frac{1}{3}$ **27.** 1
29. −4   **31.** 2.538   **33.** −0.097   **35.** 0.452   **37.** 1
39. −1   **41.** 3   **43.** 7
45.   **47.**

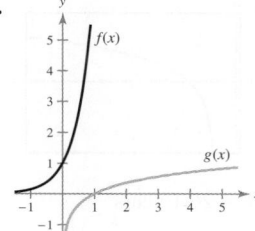
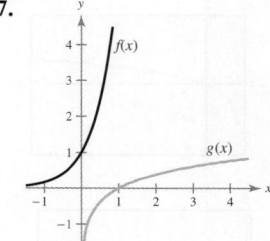

49. Domain: $(0, \infty)$ **51.** Domain: $(-4, \infty)$
  Asymptote: $x = 0$ Asymptote: $x = -4$
  x-intercept: $(1, 0)$ x-intercept: $(-3, 0)$

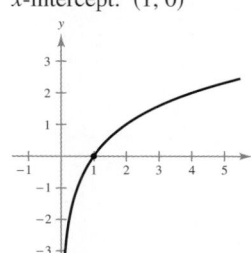
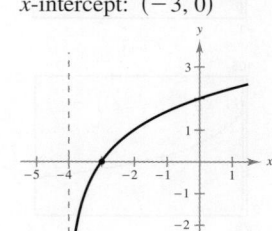

53. Domain: $(0, \infty)$ **55.** $\ln e = 1$
  Asymptote: $x = 0$
  x-intercept: $(1, 0)$

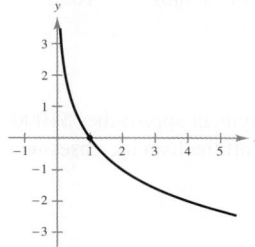

57. $\ln 4 = x$ **59.** $e^1 = e$ **61.** $e^0 = 1$ **63.** 1   **65.** -4
67. 1.946   **69.** 2.913   **71.** 0.896   **73.** 4   **75.** ± 5
77.

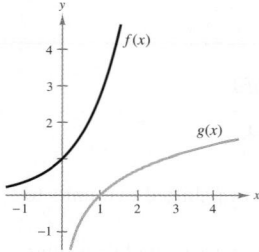

79. Domain: $(-\infty, 0)$
Asymptote: $x = 0$
x-intercept: $(-1, 0)$

81. Domain: $(-1, \infty)$
Asymptote: $x = -1$
x-intercept: $(0, 0)$

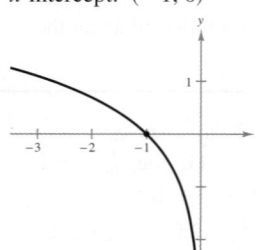   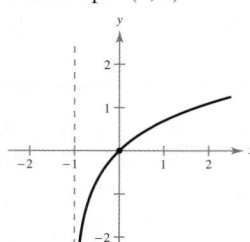

83. c   **84.** d   **85.** a   **86.** b
87.   **89.**

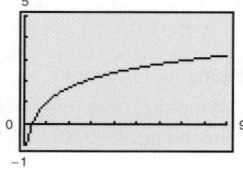

91.   **93.** $t \approx 26.1$ years

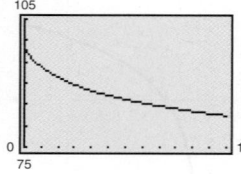

95.

The domain $0 \le t \le 12$ covers the period of 12 months or 1 year. The scores range from the average score on the original exam to the average score after 12 months, so the range is $82.40 \le f(t) \le 98$

97. (a) 87   (b) ≈ 73.5 (c) In 11 months ($t \approx 10.6$)
99. (a) 30 years; 20 years
(b) \$396,234; \$301,123
(c) \$246,234; \$151,123
(d) $x = 1000$; As the monthly payment approaches \$1000, the length of time it takes to pay off the loan increases without bound.

101. (a)

K	1	2	4	6	8	10	12
t	0	13.2	26.4	34.1	39.6	43.9	47.3

(b)

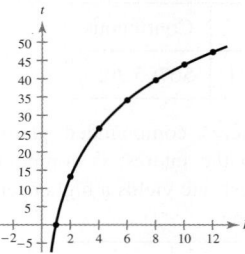

Quiz Yourself   *(page 368)*

1. f does not have an inverse function.
2. Yes; $f^{-1}(x) = x^3 - 1$

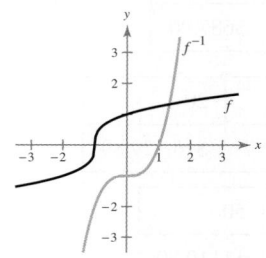

3. (a) $f^{-1}(x) = 2(x + 3)$
(b)

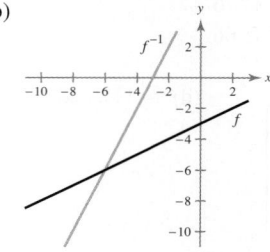

(c) $f^{-1}(f(x)) = 2\left(\frac{1}{2}x - 3 + 3\right) = x$
$f(f^{-1}(x)) = \frac{1}{2}[2(x + 3)] - 3 = x$

4. (a) $f^{-1}(x) = x^2 - 1, \ x \ge 0$
(b)

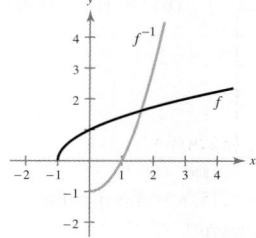

(c) $f^{-1}(f(x)) = \left(\sqrt{x + 1}\right)^2 - 1 = x$
$f(f^{-1}(x)) = \sqrt{x^2 - 1 + 1} = x, \ x \ge 0$

5.

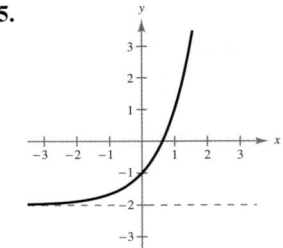

6.

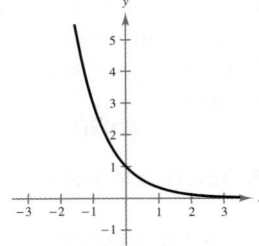

7.

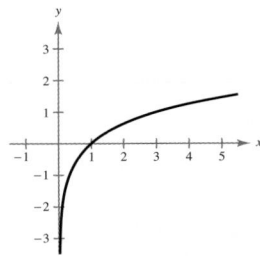

8.

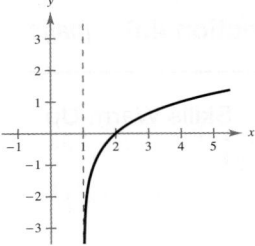

9. (a) Monthly: $15,085.04   (b) Continuously: $15,098.34

10. (a) 2006: 9.49 million; 2008: 10.26 million

(b) 2014: 12.99 million; 2015: 13.51 million

11. (a) 100   (b) ≈ 364 (c) ≈ 1326

12. $108.54   **13.** 2   **14.** 4   **15.** -2 **16.** 0

17.

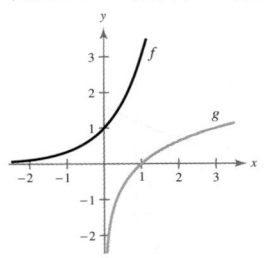

Domain of $f(x)$: $(-\infty, \infty)$
Domain of $g(x)$: $(0, \infty)$
The graphs are reflections of each other about the line $y = x$.

Section 4.4   *(page 374)*

Skills Warm Up   *(page 374)*

1. 2   **2.** -5 **3.** -2 **4.** -3 **5.** e^5

6. $\dfrac{1}{e}$ **7.** e^6 **8.** 1   **9.** $y = x^{-2}$

10. $y = x^{1/2}$ **11.** $4^3 = 64$ **12.** $16^{1/2} = 4$

1. $\dfrac{\log_{10} 8}{\log_{10} 5} \approx 1.292$ **3.** $\dfrac{\log_{10} 30}{\log_{10} 4} \approx 2.453$

5. $\dfrac{\ln 8}{\ln 5} \approx 1.292$ **7.** $\dfrac{\ln 14}{\ln 6} \approx 1.473$ **9.** 2.585

11. 1.079   **13.** 2.633   **15.** -0.683 **17.** -1.661

19. $\ln 5 + \ln 2$ **21.** $\ln 5 - \ln 2$ **23.** $2(\ln 5 + \ln 2)$

25. $-\log_{10} \frac{11}{8} = \log_{10}\left(\frac{11}{8}\right)^{-1} = \log_{10}\frac{8}{11}$

27. $-2\ln\frac{e}{7} = \ln\left(\frac{e}{7}\right)^{-2} = \ln\left(\frac{7}{e}\right)^2 = 2(\ln 7 - \ln e) = 2\ln 7 - 2$

29. 1.1833   **31.** -0.2084 **33.** 1.0686   **35.** 0.1781

37. 1.8957   **39.** -2.7124 **41.** 0.5708

43. $-1 - \log_9 2$ **45.** $\frac{1}{2} + \frac{1}{2}\log_7 10$

47. $-3 - \log_5 2$ **49.** $6 + \ln 5$

51. The change-of-base formula was used incorrectly. The formula is $\log_a x = \dfrac{\ln x}{\ln a}$. So, $\log_2 5 = \dfrac{\ln 5}{\ln 2}$.

53. The Quotient Property was used incorrectly. The Quotient Property is $\ln \dfrac{u}{v} = \ln u - \ln v$. So, $\ln 8 - \ln 2 = \ln \dfrac{8}{2} = \ln 4$.

55. $6 + 5\log_2 3$ **57.** $\log_3 4 + \log_3 n$ **59.** $\log_5 x - 2$

61. $4\log_2 x$ **63.** $\frac{1}{2}\ln z$ **65.** $\ln x + \ln y + \ln z$

67. $\frac{1}{2}\ln(a - 1)$ **69.** $2\ln(z - 1) - \ln z$

71. $\ln y + \ln z - \frac{1}{3}\ln(z + 3)$ **73.** $\ln 5 + \frac{1}{3}(\ln x - \ln y)$

75. $\frac{3}{4}\ln x + \frac{1}{4}\ln(x^2 + 3)$ **77.** $\log_3 5x$ **79.** $\log_4 \dfrac{8}{x}$

81. $\log_{10}(x + 4)^2$ **83.** $\ln \dfrac{1}{216x}$ **85.** $\ln \dfrac{\sqrt[3]{5x}}{x + 1}$

87. $\log_8 \dfrac{x - 2}{x + 2}$ **89.** $\ln \dfrac{144}{\sqrt{x^2 + 1}}$ **91.** $\ln \dfrac{x^2}{(x + 2)(x - 2)}$

93.

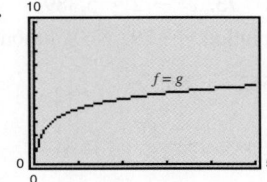

The two graphs are the same. The property is $\log_a(uv) = \log_a u + \log_a v$.

95.

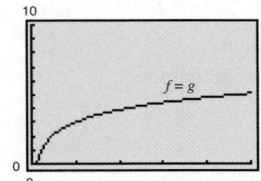

The two graphs are the same. The graphs demonstrate the Power Property for logarithms.

97. $\frac{1}{3}$ **99.** $-\frac{1}{2}$ **101.** -3 **103.** $\ln y = \frac{1}{4}\ln x$

105. $\ln y = \frac{2}{3}\ln x + \ln 0.070$ **107.** ≈ 26 decibels

109. (a) $a^m a^n = a^{m+n}$ (b) $\dfrac{a^m}{a^n} = a^{m-n}$ (c) $(a^m)^n = a^{mn}$

111. Let $\log_a u = x$ and $\log_a v = y$.

$a^x = u$ and $a^y = v$

$\dfrac{u}{v} = \dfrac{a^x}{a^y} = a^{x-y}$

$\log_a a^{x-y} = x - y$

$\log_a \dfrac{u}{v} = \log_a u - \log_a v$

Section 4.5   *(page 384)*

Skills Warm Up   *(page 384)*

1. $\dfrac{\ln 3}{\ln 2}$ **2.** $1 + \dfrac{2}{\ln 4}$ **3.** $\dfrac{e}{2}$ **4.** $2e$

5. $2 \pm i$ **6.** $\frac{1}{2}, 1$ **7.** x **8.** $2x$

9. $2x$ **10.** $-x^2$

1. 2   **3.** -2 **5.** 4   **7.** 64   **9.** $\frac{2}{5}$ **11.** x^2

13. $x + 1$ **15.** $x^3 - 7$ **17.** $8 + x^3$ **19.** $x + 5$

21. x^2 **23.** $\ln 3 \approx 1.099$ **25.** $\log_3 8 \approx 1.893$

27. $\ln 28 \approx 3.332$ **29.** $\frac{1}{2}\log_3 80 \approx 1.994$ **31.** 2

CHAPTER 4

33. $\log_3 28 + 1 \approx 4.033$ **35.** $3 - \log_2 565 \approx -6.142$

37. $\frac{1}{3}\log_{10}\frac{3}{2} \approx 0.059$ **39.** $\log_5 7 + 1 \approx 2.209$

41. $\frac{1}{3}\ln 12 \approx 0.828$ **43.** $\ln\frac{5}{3} \approx 0.511$

45. $\ln\frac{1}{2} \approx -0.693$ **47.** $\frac{1}{3} + \frac{1}{3}\log_2\frac{8}{3} \approx 0.805$

49. $\ln 6 \approx 1.792, \ln 2 \approx 0.693$ **51.** $\ln 4 \approx 1.386$

53. $2\ln 75 \approx 8.635$ **55.** $\frac{1}{2}\ln 1498 \approx 3.656$

57. $\dfrac{\ln 4}{365\ln\left(1 + \dfrac{0.065}{365}\right)} \approx 21.330$

59. $\dfrac{\ln 2}{12\ln\left(1 + \dfrac{0.10}{12}\right)} \approx 6.960$ **61.** 10,000

63. $e^{-3} \approx 0.050$ **65.** $\dfrac{e^{2.4}}{2} \approx 5.512$ **67.** 5,000,000

69. 26   **71.** $\frac{1}{5}e^{10/3} \approx 5.606$ **73.** $e^2 - 2 \approx 5.389$

75. $e^{-2/3} \approx 0.513$ **77.** No solution   **79.** No solution

81. 7   **83.** $\dfrac{-1 + \sqrt{17}}{2} \approx 1.562$ **85.** 2

87. $y = 2x + 1, x > -\dfrac{1}{2}$ **89.** $y = \dfrac{(x-1)^2}{x+2}, x \geq 1$

91. 2.807   **93.** 20.086   **95.** ≈ 9.56 years

97. ≈ 13.32 years   **99.** 30 months

101. (a) ≈ 210 coin sets   (b) ≈ 588 coin sets

(c)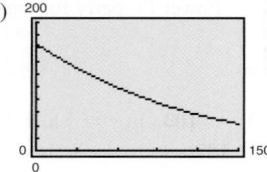

103. (a) ≈ 29.3 years   (b) ≈ 39.8 years

105. (a) 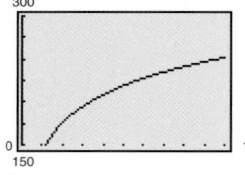   (b) and (c) 2003

107. (a)

x	y
0.2	162.6
0.4	78.5
0.6	52.5
0.8	40.5
1.0	33.9

(b)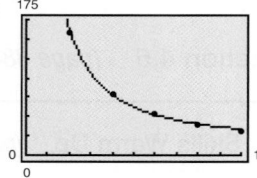

The model is a good fit for the data.

(c) 1.197 meters

(d) No. To reduce the g's to fewer than 23 requires a crumple zone of more than 2.27 meters, a length that exceeds the front width of most cars.

109. Yes. See Exercise 79.

111. $\log_b uv = \log_b u + \log_b v$
True by the Product Property in Section 4.4.

113. $\log_b(u - v) = \log_b u - \log_b v$
False.
$1.95 \approx \log(100 - 10) \neq \log 100 - \log 10 = 1$

115. Yes. Time to double: $t = \dfrac{\ln 2}{r}$.

Time to quadruple: $t = \dfrac{\ln 4}{r} = 2\left(\dfrac{\ln 2}{r}\right)$

Section 4.6   *(page 395)*

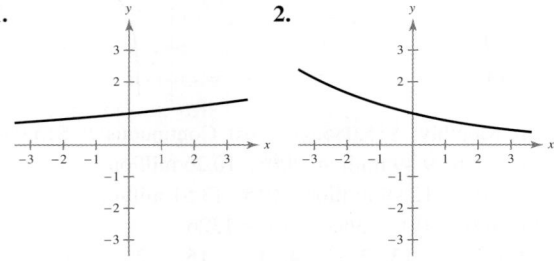

Skills Warm Up   *(page 395)*

1. **2.** **3.** **4.** **5.** **6.**

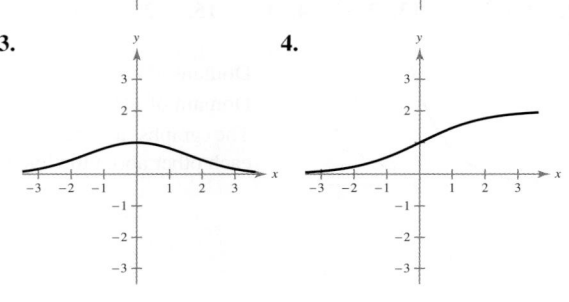
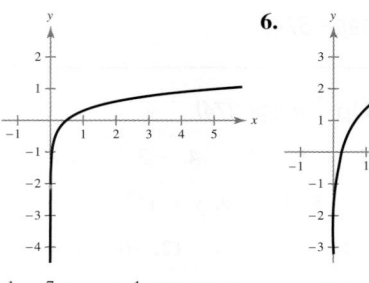

7. $\frac{1}{2}\ln\frac{7}{3}$ **8.** $\frac{1}{5}e^{7/2}$ **9.** ≈ 34.539 **10.** ≈ 3.695

1. c   **2.** e   **3.** b   **4.** a   **5.** d   **6.** f

7. Exponential growth   **9.** Exponential decay

11. 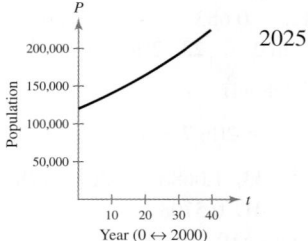   2025

13. $C = 1, k = \frac{1}{4}\ln 10$ **15.** $C = 1, k = \frac{1}{4}\ln\frac{1}{4}$

	Initial Investment	Annual % Rate	Time to Double	Amount After 10 Years
17.	$5000	7%	9.90 years	$10,068.76
19.	$500	6.93%	10 years	$1000.00
21.	$1000	8.25%	8.40 years	$2281.88
23.	$6392.79	11%	6.30 years	$19,205.00
25.	$5000	8%	8.66 years	$11,127.70

	Isotope	Half-Life (Years)	Initial Quantity	Amount After 1000 Years
27.	^{226}Ra	1599	4 g	2.59 g
29.	^{14}C	5715	3.95 g	3.5 g
31.	^{239}Pu	24,100	1.65 g	1.6 g

33. $k = -\frac{1}{20} \ln \frac{3}{5};\ \approx 36{,}671$ people

35. ≈ 12.36 hours to double

≈ 19.59 hours to triple

37. 12,180 years   **39.** $\approx 9.92\%$

41. (a) $N = 40(1 - e^{-0.049t})$ (b) ≈ 42 days   **43.** 100

45. (a) 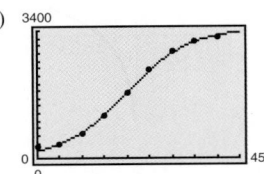   (b) ≈ 1252 fish

(c) ≈ 7.8 months

47. No, because the function is always increasing.

49. (a)

(b) $P = 2922.32(1.0391)^t;\ P = 2922.32e^{0.0384t}$

(c)   (d) 2022: $\approx 6{,}801{,}736$ people

2042: $\approx 14{,}660{,}808$ people

51. (a)

(b) Logistic growth model; The data points seem to be leveling off in the later years as though approaching an asymptote.

(c) $P = \dfrac{3455}{1 + 16.51e^{-0.1415t}}$

(d)   (e) 3407

The model is a good fit.

53. (a) ≈ 7.906 (b) ≈ 7.684

55. (a) 20 decibels   (b) 70 decibels

57. $\approx 1.585 \times 10^{-6}$ **59.** $\approx 31{,}623$ **61.** 3:00 A.M.

63.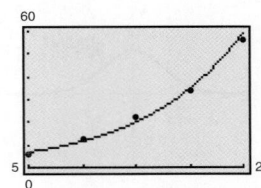

The data fit an exponential model.

Exponential model: $n = 3.9405e^{0.1086t}$

A logistic growth model would be more appropriate for this data because after the initial rapid growth in productivity, the worker's production rate will eventually level off.

Review Exercises   *(page 404)*

1. $f(g(x)) = 3\left(\dfrac{x - 5}{3}\right) + 5 = x$

$g(f(x)) = \dfrac{3x + 5 - 5}{3} = x$

$f(g(x)) = x = g(f(x))$, so f and g are inverse functions of each other.

3. $f^{-1}(x) = \dfrac{1}{x}$ (f is its own inverse function.)

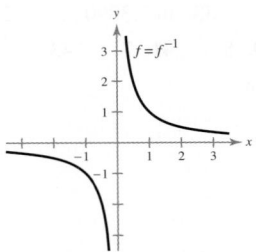

5. (a) $f^{-1}(x) = \sqrt{x},\ x \geq 0$

(b)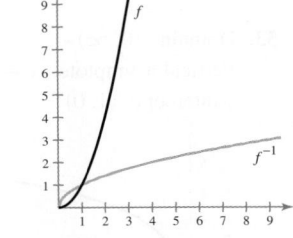

(c) $f^{-1}(f(x)) = \sqrt{x^2} = x$

$f(f^{-1}(x)) = (\sqrt{x})^2 = x$

7. 16,777,216   **9.** 0.431   **11.** 2.117

13. c   **14.** d   **15.** a   **16.** b

17.   **19.**

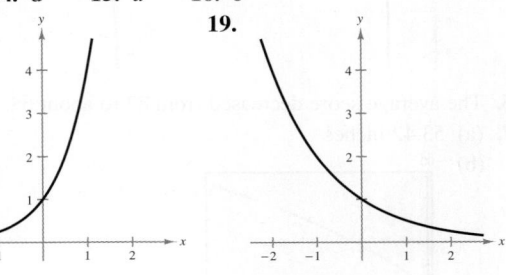

21.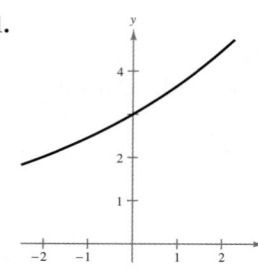

23.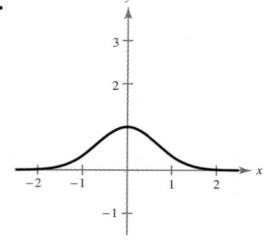

25.

n	1	2	4
A	\$13,308.43	\$13,576.74	\$13,718.59

n	12	365	Continuous
A	\$13,816.21	\$13,864.33	\$13,865.97

27.

t	1	10	20
P	\$185,085.40	\$92,140.76	\$42,449.59

t	30	40	50
P	\$19,556.69	\$9009.84	\$4150.87

29. \$8471.94   **31.** $\log_4 64 = 3$ **33.** $\ln 7.3890 \ldots = 2$
35. $3^4 = 81$ **37.** $e^0 = 1$ **39.** 5   **41.** 7   **43.** $-\frac{1}{2}$
45. d   **46.** b   **47.** c   **48.** a
49.

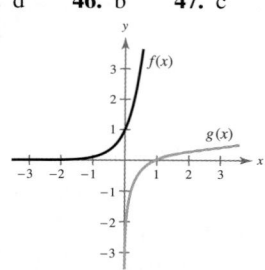

51. Domain: $(3, \infty)$
Vertical asymptote: $x = 3$
x-intercept: $(4, 0)$

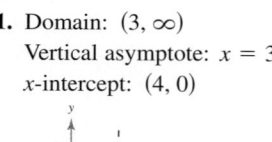

53. Domain: $(0, \infty)$
Vertical asymptote: $x = 0$
x-intercept: $(1, 0)$

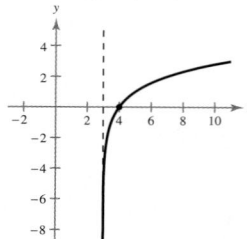 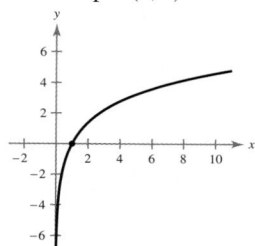

55. The average score decreased from 82 to about 68.
57. (a) 53.42 inches
(b)

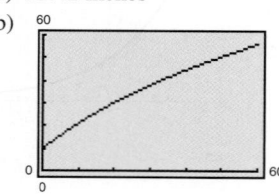

59. 2.096   **61.** 2.132   **63.** 0.9208   **65.** 0.2823
67. 2   **69.** 3.2   **71.** $\log_{10} x - \log_{10} y$
73. $\ln x + \frac{1}{2}\ln(x - 3)$ **75.** $4 \log_5(y - 3)$ **77.** $\log_4 6$
79. $\ln \sqrt{x}$ **81.** $\ln \dfrac{x}{(x - 3)(x + 1)}$ **83.** $\ln y = \dfrac{4}{3} \ln x$
85. $\ln 8 \approx 2.079$ **87.** 1   **89.** $\ln 4 \approx 1.386$
91. $\frac{1}{3}e^{8.2} \approx 1213.650$ **93.** $\frac{1}{5}e^2 \approx 1.478$
95. $3e^2 \approx 22.167$ **97.** 7
99. (a) ≈ 197 desks   (b) ≈ 257 desks   **101.** 10.63 g
103. (a)

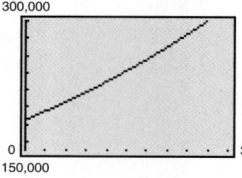

(b) 2021 ($t \approx 21.00$)
105. ≈ 9.93 hours to double
≈ 15.74 hours to triple
107. (a) $N = 50(1 - e^{-0.04838t})$ (b) 48 days
109. (a) $k \approx 0.2121$
(b) ≈ 196 deer; ≈ 294 deer; ≈ 383 deer
111. Yes

Test Yourself   *(page 408)*

1. Yes; $f^{-1}(x) = \dfrac{x + 7}{5}$

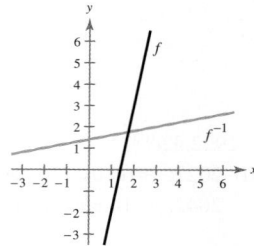

2. The function does not have an inverse function.

3. 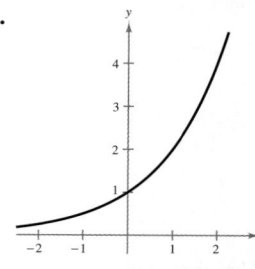   **4.**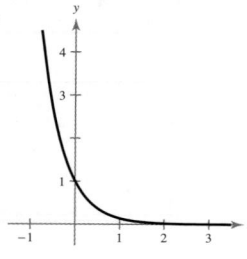

5. 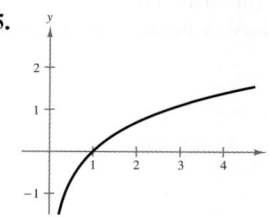   **6.**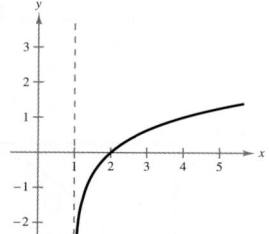

7. 87; ≈ 79.8; ≈ 76.5

8. After 6 months: ≈ 74.3

After 12 months: ≈ 70.3

After 18 months: ≈ 67.8

Human memory diminishes slowly over time.

9. $2 \ln x + 3 \ln y - \ln z$

10. $\log_{10} 3 + \log_{10} x + \log_{10} y + 2 \log_{10} z$

11. $\log_2 x + \frac{1}{3} \log_2(x - 2)$ **12.** $\frac{1}{5} \log_8(x^2 + 1)$

13. $\ln \dfrac{x^2 y^3}{z}$ **14.** $\log_{10} \sqrt[3]{x^2 y^2}$ **15.** $\dfrac{\log_2 21}{4} \approx 1.098$

16. $\ln 6 \approx 1.792$, $\ln 2 \approx 0.693$

17. 127   **18.** $e^6 - 2 \approx 401.429$ **19.** ≈ 16.3 years

20. About 2015 ($t \approx 15.4$). Because the exponent on e is positive, the population of the city is growing.

21. ≈ 9.9 hours   **22.** ≈ 2.97 grams; ≈ 0.88 gram

23. Answers will vary. Sample answer:

The growth of the bear population will slow down as the population approaches the carrying capacity of the island. So, the population will grow logistically.

Chapter 5

Section 5.1   *(page 417)*

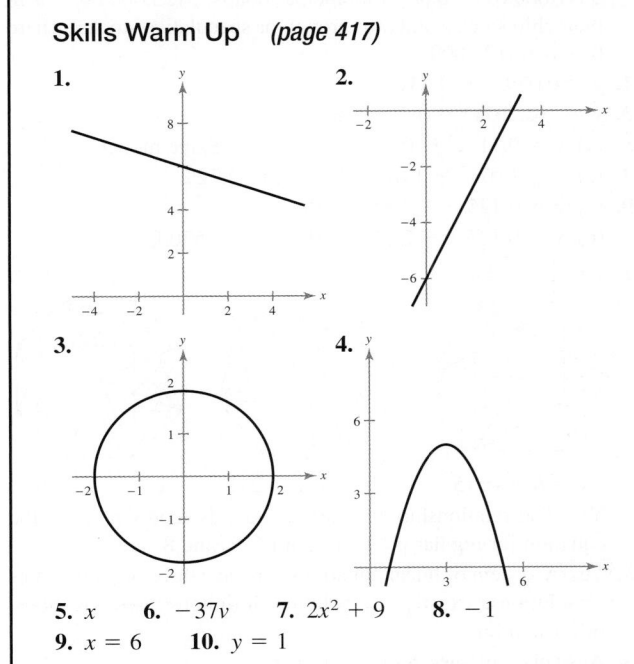

Skills Warm Up   *(page 417)*

1.

2.

3.

4.

5. x **6.** $-37v$ **7.** $2x^2 + 9$ **8.** -1

9. $x = 6$ **10.** $y = 1$

1. (a) No   (b) Yes   **3.** (a) Yes   (b) No

5. (a) No   (b) No   **7.** $(-1, 1)$ **9.** $\left(\frac{1}{2}, 3\right)$ **11.** $(10, 3)$

13. $(1.5, 0.3)$ **15.** $\left(\frac{20}{3}, \frac{40}{3}\right)$ **17.** No solution

19. \$5000 at 3%, \$7000 at 5%

21. \$6000 at 2.8%, \$6000 at 3.8%

23. $\left(1 + \sqrt{2}, 2 + 2\sqrt{2}\right)$, $\left(1 - \sqrt{2}, 2 - 2\sqrt{2}\right)$

25. $\left(\frac{29}{10}, \frac{21}{10}\right)$, $(-2, 0)$ **27.** No solution   **29.** $(0, 1)$, $(\pm 1, 0)$

31. $(2, 1)$ **33.** $(2, -3)$ **35.** $(-1, 2)$, $(2, 5)$

37. $(0, 4)$, $\left(\frac{12}{5}, -\frac{16}{5}\right)$ **39.** $(4, 3)$ **41.** $\left(\frac{5}{2}, \frac{3}{2}\right)$

43. $(2, 2)$, $(4, 0)$ **45.** $(1, 4)$, $(4, 7)$ **47.** $\left(4, -\frac{1}{2}\right)$

49. No solution   **51.** One solution   **53.** Two solutions

55. No solution   **57.** $\left(\frac{1}{2}, \frac{3}{4}\right)$, $(-3, -1)$ **59.** $(1, 4)$, $(4, 7)$

61. $(0, 1)$ **63.** $(1, 4)$, $(3, 12)$ **65.** $(-3, 4)$, $(5, 0)$

67. Answers will vary. Sample answers:

(a) $\begin{cases} x + 2y = 0 \\ 2x - 4y = 3 \end{cases}$ (b) $\begin{cases} 2x^2 - y = -1 \\ x + y = 4 \end{cases}$

(c) $\begin{cases} 3x + y = 8 \\ 3x + y = -2 \end{cases}$

69. 192 units   **71.** 233,333 units

73. 2000 software packages   **75.** 2002 ($t \approx 2.03$)

77. Yes, at age 20.   **79.** \$100,000

81. According to the graphs, sales of general merchandise stores first exceeded sales of food and beverage stores in 2003 ($t \approx 13.37$). Both models eventually decrease and become negative. So, it is unlikely that these models will continue to be accurate.

Section 5.2   *(page 428)*

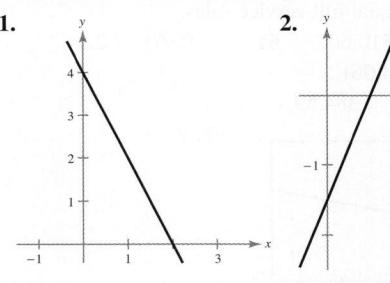

Skills Warm Up   *(page 428)*

1.

2.

3. $x - y + 4 = 0$ **4.** $5x + 3y - 28 = 0$

5. $-\frac{1}{2}$ **6.** $\frac{7}{4}$ **7.** Perpendicular   **8.** Parallel

9. Neither parallel nor perpendicular

10. Perpendicular

1. $(2, 2)$ **3.** $(2, 0)$ **5.** Inconsistent

7. $(2a, 3a - 3)$, where a is any real number

9. $\left(2, \frac{1}{2}\right)$; consistent   **11.** $(-7, -13)$; consistent

13. $(4, -5)$; consistent   **15.** $(4, -1)$; consistent

17. $(40, 40)$; consistent   **19.** $\left(-\frac{6}{35}, \frac{43}{35}\right)$; consistent

21. $\left(\frac{18}{5}, \frac{3}{5}\right)$; consistent   **23.** $\left(\frac{19}{7}, -\frac{2}{7}\right)$; consistent

25. $\begin{cases} x + y = 13 \\ x - y = 3 \end{cases}$; $(8, 5)$ **27.** $\begin{cases} 2r + s = 8 \\ r - s = 7 \end{cases}$; $(5, -2)$

29. No solution; inconsistent

31. $\left(a, \frac{5}{6}a - \frac{1}{2}\right)$, where a is any real number; consistent

33. $\left(\frac{90}{31}, -\frac{67}{31}\right)$; consistent   **35.** $\left(-\frac{1}{2}, 1\right)$ **37.** $(-2, -9)$

39. $(-0.874, 0.417)$, $(0, 1)$ **41.** $(20, 20)$

43. Answers will vary. Sample answers:

(a) $\begin{cases} 3x + y = 18 \\ x + y = 6 \end{cases}$ (b) $\begin{cases} -2x + 8y = 38 \\ 4x + y = -8 \end{cases}$

(c) $\begin{cases} 8x - y = 1 \\ -4x + 4y = -25 \end{cases}$

45. No. The solution of the system is (79,400, 398).

47. 550 miles per hour; 50 miles per hour

49. $6\frac{2}{3}$ gallons of 20% solution

$3\frac{1}{3}$ gallons of 50% solution

51. $8000 at 4.75%, $17,000 at 5.5%

53. $x \approx 309,091$ units; $p \approx \$25.09$

55. $x = 2,000,000$ units; $p = \$100.00$

57. (a) Fast-food   Full-service

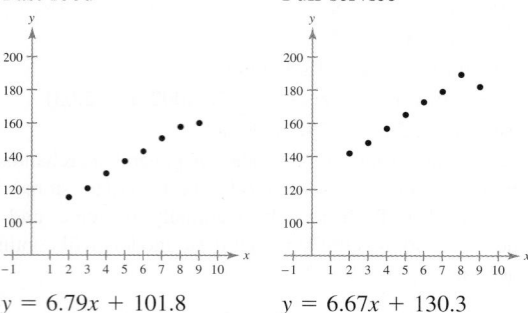

$y = 6.79x + 101.8$ $y = 6.67x + 130.3$

(b) In the near future, fast-food sales will not equal full-service sales. According to the models, fast-food sales will equal full-service sales in the year 2237. These models, therefore, are not appropriate to use for predicting when fast-food sales will equal full-service sales.

59. 1380 units at $810.60   **61.** $y = 0.97x + 2.1$

63. $y = 0.318x + 4.061$

65. (a) $y = 2.031t + 384.59$

(b)

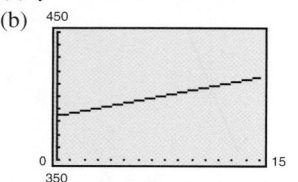

404.9

(c) You obtain the same model: $y = 2.031t + 384.59$.

67. Answers will vary.

Section 5.3   *(page 439)*

Skills Warm Up   *(page 439)*

1. (10, 15)   **2.** $\left(-3, -\frac{10}{3}\right)$ **3.** (28, 4)   **4.** (4, 3)

5. Not a solution   **6.** Not a solution   **7.** Solution

8. Solution   **9.** $5a + 2$ **10.** $a + 13$

---

1. c   **2.** a   **3.** b   **4.** d

5. Yes. The system has a "stair-step" pattern with leading coefficients of 1.

7. No. The system has a "stair-step" pattern, but not all of its leading coefficients are 1.

9. $(4, -2, -2)$ **11.** $(2, -3, -2)$ **13.** $(-1, -6, 8)$

15. Inconsistent   **17.** $\left(1, -\frac{3}{2}, \frac{1}{2}\right)$

19. $(-3a + 10, 5a - 7, a)$ **21.** $\left(-4a + 13, -\frac{15}{2}a + \frac{45}{2}, a\right)$

23. Inconsistent   **25.** $(-3, 4, 2)$ **27.** $(3, -1, 2)$

29. $\left(\frac{3}{4}a, -2a, a\right)$ **31.** $(-5a + 3, -a - 5, a)$

33. Inconsistent   **35.** (1, 1, 1, 1)

37. Answers will vary. Sample answer:

$$\begin{cases} 2x - y + z = 9 \\ y + z = 1, \\ z = 2 \end{cases} \begin{cases} -x + 2y - 4z = -13 \\ x + y + z = 4 \\ x + z = 5 \end{cases}$$

39. Answers will vary. Sample answer:

$$\begin{cases} x - y + z = 3 \\ y - z = -2, \\ z = -3 \end{cases} \begin{cases} x + 3y + 4z = -26 \\ 4x - y - 5z = 24 \\ x + 2y = -9 \end{cases}$$

41. Answers will vary.   **43.** Answers will vary.

Sample answer: Sample answer:

$a = 1: \left(1, 3, \frac{1}{3}\right)$ $a = -4: (-2, -16, 7)$

$a = 2: \left(2, 4, \frac{2}{3}\right)$ $a = 2: (1, 8, 7)$

$a = 0: (0, 2, 0)$ $a = 6: (3, 24, 7)$

45. $y = 2x^2 + 3x - 4$ **47.** $y = -4x^2 + 2x + 1$

49. $x^2 + y^2 - 4x = 0$ **51.** $x^2 + y^2 - 6x + 6y + 9 = 0$

53. $1,100,000 at 3%   **55.** 15,000 units of $15 candles

$200,000 at 4%   30,000 units of $10 candles

$200,000 at 6%   5000 units of $5 candles

57. 18 gallons of spray X

1 gallon of spray Y

6 gallons of spray Z

59. Invest $33,333.33 + 0.8a$ in certificates of deposit, $341,666.67 - 0.8a$ in municipal bonds, $125,000.00 - a$ in blue-chip stocks, and a in growth or speculative stocks, where $0 \le a \le 125,000$.

61. $y = 0.079x^2 + 0.63x + 2.9$

63. $y = -0.207x^2 - 0.89x + 5.1$

65. (a) $y = 0.114x^2 - 0.40x + 4.0$ (b) Same model

67. (a) $y = 1.07x^2 - 1.6x + 22$ (b) 32.72%

69. (a) $y = 0.125x^2 - 2.55x + 18$

(b) $y = 0.125x^2 - 2.55x + 18$ (c) ≈ 530 ft

71. $y = \frac{1}{2}x^2 - \frac{1}{2}x$

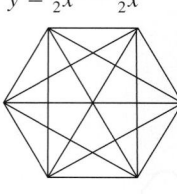 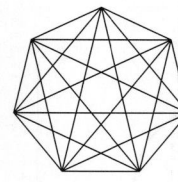 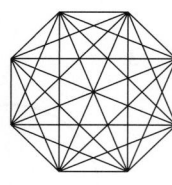

$x = 6, y = 15$ $x = 7, y = 21$ $x = 8, y = 28$

Yes. The relationship between x and y is represented by the equation for regular polygons with 6, 7, and 8 sides.

73. Yes. A system of linear equations can have three possible types of solutions: exactly one solution, infinitely many solutions, or no solution.

75. Answers will vary. Sample answer:

(a) $a = 4, b = -2, c = -2$

(b) $a = 3, b = 2, c = 1$

(c) Not possible

77. Answers will vary.

Quiz Yourself   *(page 444)*

1. (2, 5)   **2.** No solution   **3.** (1, 3)

4. $\left(-\frac{2}{5} \pm \frac{2\sqrt{11}}{5}, \frac{1}{5} \pm \frac{4\sqrt{11}}{5}\right)$ **5.** 1500 units

6. 500,000 units   **7.** $(2, -1)$ **8.** $\left(1, \frac{3}{2}\right)$

9. $x = 5000$ units
   $p = \$40$
10. $y = 0.84t + 42.6$ **11.** $(1, -2, 3)$
12. Answers will vary.
   Sample answer: $(a + 6, a + 6, a)$, a is any real number.
13. Inconsistent
14. (a) $y = 0.1400x^2 + 3.204x + 69.83$
   (b) 2010: \$80.70
      2011: \$84.89

Section 5.4   *(page 451)*

Skills Warm Up   *(page 451)*

1. Line   **2.** Parabola   **3.** Circle   **4.** Parabola
5. Line   **6.** Circle   **7.** $(1, 1)$ **8.** $(2, 0)$
9. $(2, 1), \left(-\frac{5}{2}, -\frac{5}{4}\right)$ **10.** $(2, 3), (3, 2)$

1. d   **2.** b   **3.** a   **4.** c   **5.** f   **6.** e

7. 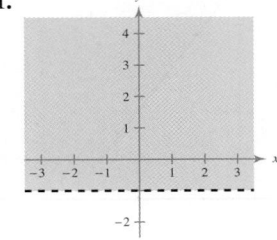   **9.**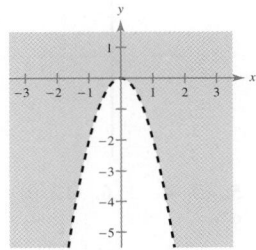

11. 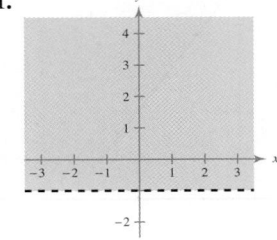   **13.**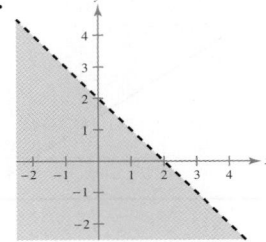

15. 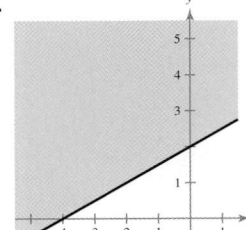   **17.**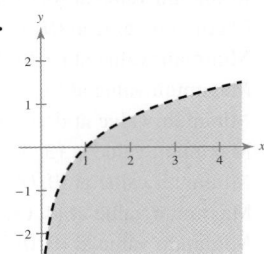

19. 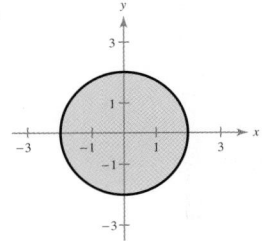   **21.**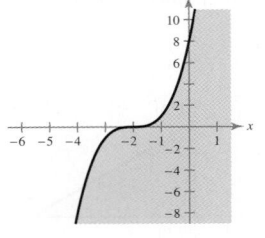

23. 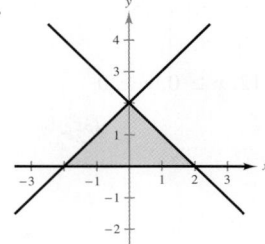   **25.**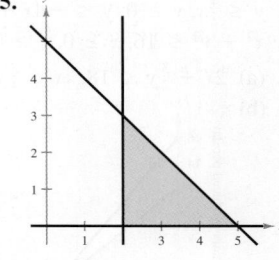

27. 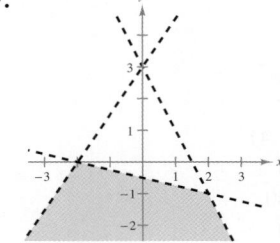   **29.**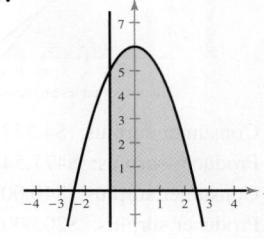

31. 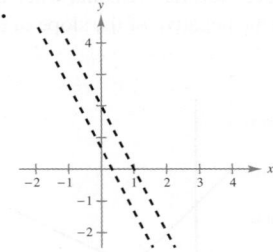   **33.**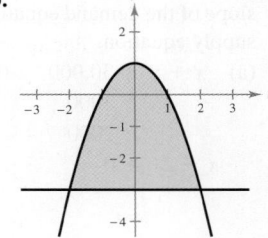

35. 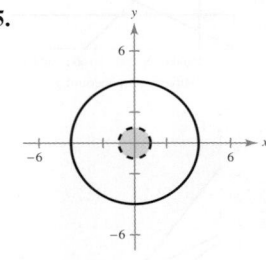   **37.**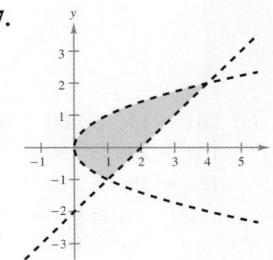

39. 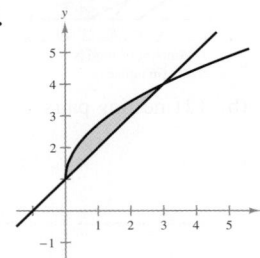   **41.**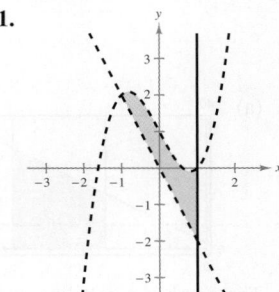

43.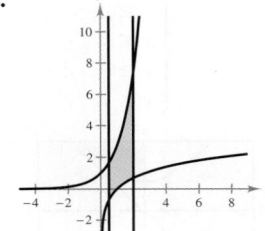

45. $y \le \frac{3}{2}x, y \ge \frac{3}{2}x - \frac{3}{2}, 0 \le y \le 3$

CHAPTER 5

47. $y \leq \frac{4}{3}x, y \geq 0, y \leq -4x + 16$

49. $x^2 + y^2 \leq 16, x \geq 0, y \geq 0$

51. (a) $2x + \frac{3}{2}y \leq 18, \frac{3}{2}x + \frac{3}{4}y \leq 12, x \geq 0, y \geq 0$

(b)

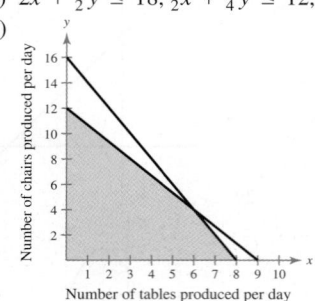

Number of tables produced per day

53. Consumer surplus: \$4,777,001.41
Producer surplus: \$477,545.60

55. Consumer surplus: \$40,000,000
Producer surplus: \$20,000,000

57. The consumer surplus and producer surplus are equal when the slope of the demand equation is the negative of the slope of the supply equation.

59. (a) $x + y \leq 30,000$ (b)
$x \geq 7000$
$y \geq 7000$
$x - 2y \geq 0$

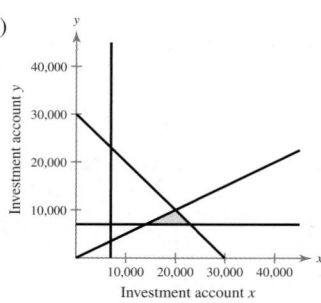

61. (a) $20x + 15y \geq 400$ (b)
$10x + 20y \geq 250$
$15x + 20y \geq 220$
$x \geq 0$
$y \geq 0$

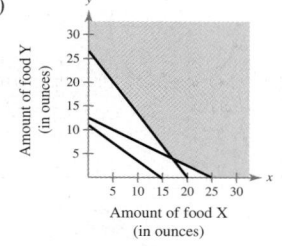

63. (a)   (b) 121 nesting pairs

65. The graph is a half-line on the real number line; on the rectangular coordinate system, the graph is a half-plane.

67. Answers will vary. Sample answer:
$$\begin{cases} x \geq 1 \\ x \leq 4 \\ y \geq 1 \\ y \leq 4 \end{cases}$$

69. (a) $\begin{cases} \pi y^2 - \pi x^2 \geq 10 \\ y > x \\ x > 0 \end{cases}$ (b)

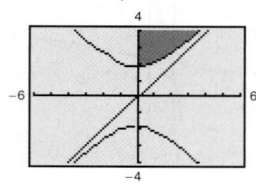

(c) The line is an asymptote to the boundary. The larger the circles, the closer the radii can be with the constraint still satisfied.

Section 5.5   (page 460)

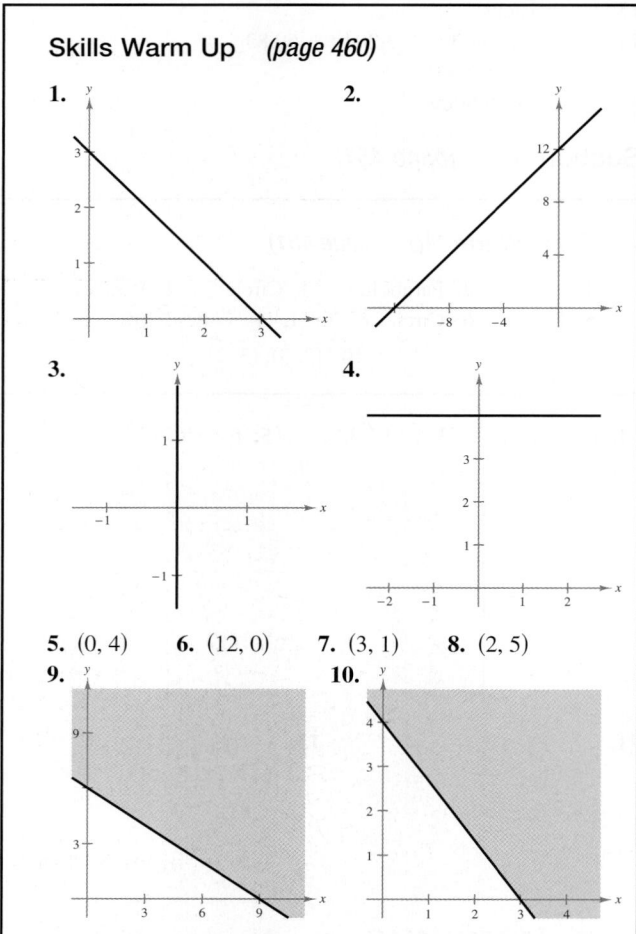

Skills Warm Up   (page 460)

1. 2. 3. 4.

5. $(0, 4)$ **6.** $(12, 0)$ **7.** $(3, 1)$ **8.** $(2, 5)$

9. 10.

1. Minimum value at $(0, 0)$: 0
Maximum value at $(0, 5)$: 20

3. Minimum value at $(0, 0)$: 0
Maximum value at $(5, 0)$: 45

5. Minimum value at $(0, 0)$: 0
Maximum value at $(3, 4)$: 26

7. Minimum value at $(0, 0)$: 0
Maximum value at $(4, 0)$: 20

9. Minimum value at $(0, 0)$: 0
Maximum value at any point on the line segment connecting the points $(5, 0)$ and $(0, 3)$: 30

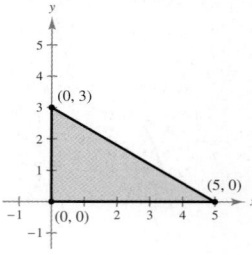

11. Minimum value at $(0, 0)$: 0
Maximum value at $(5, 0)$: 40
Same graph as in Answer 9

13. Minimum value at $(5, 3)$: 35
No maximum value

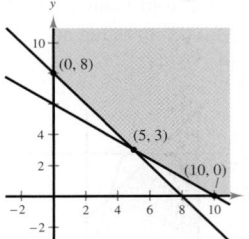

15. Minimum value at $(10, 0)$: 20
No maximum value
Same graph as in Answer 13

17. Minimum value at $(0, 0)$: 0
Maximum value at any point on the line segment connecting
the points $(0, 20)$ and $(10, 15)$: 120

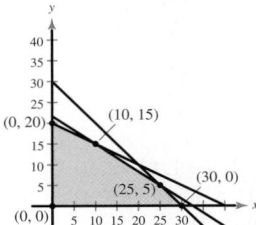

19. Minimum value at $(0, 0)$: 0
Maximum value at any point on the line segment connecting
the points $(25, 5)$ and $(30, 0)$: 30
Same graph as in Answer 17

21. Maximum value at $(3, 6)$: 12

23. Maximum value at any point on the line segment connecting
the points $(0, 10)$ and $(3, 6)$: 30

25. Maximum value at $(5, 4)$: 58

27. Maximum value at $(8, 0)$: 120

29. Answers will vary.
Sample answer:
$z = 2x + 15y$

31. Answers will vary.
Sample answer:
$z = -x$

33. Answers will vary.
Sample answer:
$z = x + 7y$

35. Answers will vary.
Sample answer:
$z = 6x + 5y$

37. Crop A: 60 acres
Crop B: 90 acres
$33,150

39. Brand X: 3 bags
Brand Y: 6 bags
$240

41. Model A: 0 bicycles
Model B: 1600 bicycles
$120,000

43. 12 audits and 0 tax returns

45. Television: None
Newspaper: $4,800,000
300 million people

47. Type A: $62,500
Type B: $187,500
$26,875

49. Model A: 929 units
Model B: 77 units
$99,445

51. z has a maximum value
(of 5) at any point on the
line segment connecting
the vertices $(2, 0)$ and $\left(\frac{20}{19}, \frac{45}{19}\right)$.

53. The constraint $x \le 10$ is
extraneous. The maximum
value of z, $z = 14$, occurs
at $(0, 7)$.

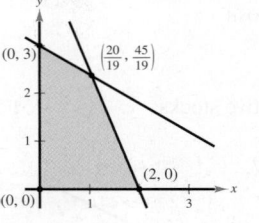
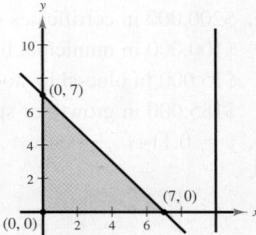

55. The constraint $2x + y \le 4$ is extraneous. The maximum value
of z, $z = 4$, occurs at $(0, 1)$.

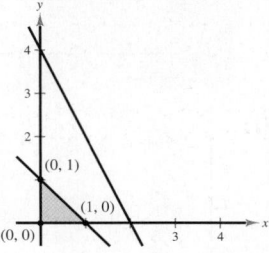

57. No. You need more information about the feasible region to
determine whether the minimum also occurs at $(0, 0)$.

59. Yes; The objective function also has the maximum value at
any point on the line segment connecting the two vertices,
so there are an infinite number of points that produce the
maximum value.

Review Exercises   *(page 467)*

1. $(-2, 4)$ **3.** $(8, -10)$ **5.** $(8, 6), (0, 10)$

7. $(1, 9), (1.5, 8.75)$

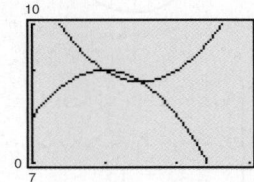

9. 800 plants   **11.** During the fourth month of the new format

13. $(3, -5)$ **15.** $\left(a, \frac{4}{3}a - \frac{10}{3}\right)$, where a is any real number

17. Inconsistent   **19.** $(8, 9)$

21. The graph is a point. Solution: $(-1, 1)$

23. (a) $\begin{cases} 0.1x + 0.5y = 0.25(12) \\ x + y = 12 \end{cases}$

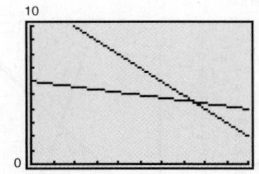

(b) 7.5 gallons of 10% solution
4.5 gallons of 50% solution

25. $y = 1.01x + 1.54$ **27.** $x = 71{,}429$ units, $p = \$22.71$

29. $(12, 5, 2)$ **31.** $(2, -1, 3)$

33. $\left(\frac{1}{5}a + \frac{8}{5}, -\frac{6}{5}a + \frac{42}{5}, a\right)$, a is any real number

35. Inconsistent   **37.** $y = 2x^2 + x - 6$

39. $x^2 + y^2 - 4x + 2y - 4 = 0$

41. $200,000 in certificates of deposit
$100,000 in municipal bonds
$15,000 in blue-chip stocks
$185,000 in growth or speculative stocks

43. $y = 0.114x^2 + 0.80x + 1.6$

45. 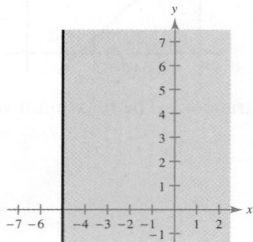   **47.**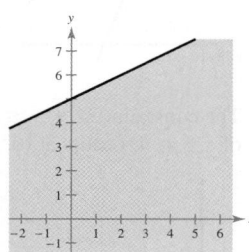

49. 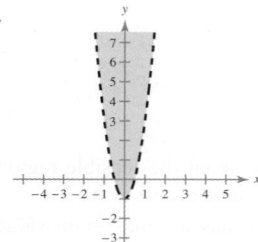   **51.**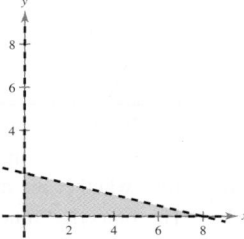

53. 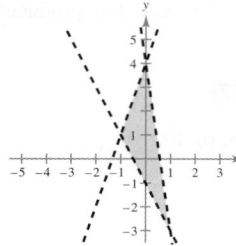   **55.**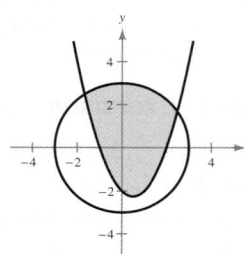

57.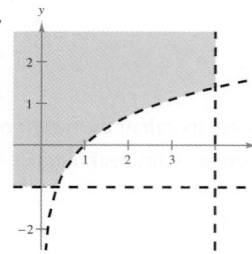

59. $\begin{cases} y \le \frac{5}{2}x - \frac{3}{2} \\ y \le \frac{1}{5}x + \frac{27}{5} \\ y \ge \frac{5}{2}x - 13 \\ y \ge \frac{1}{5}x + \frac{4}{5} \end{cases}$

61. Consumer surplus: $4,500,000
Producer surplus: $9,000,000

63. $\begin{cases} x \ge 2y \\ 100x + 150y \le 4000 \\ x \ge 4 \\ y \ge 2 \end{cases}$

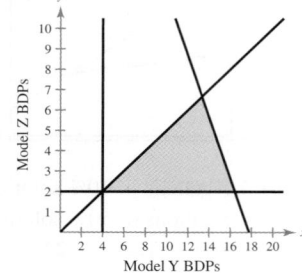

65. Minimum value at $(0, 0)$: 0
Maximum value at $(0, 8)$: 48

67. Minimum value at $(0, 0)$: 0
Maximum value at any point on the line segment connecting $(25, 35)$ and $(50, 15)$: 550

69. Minimum value at $(0, 0)$: 0   **71.** Minimum value at $(0, 0)$: 0
Maximum value at   Maximum value at
$(4, 3)$: 48 $(4, 5)$: 47

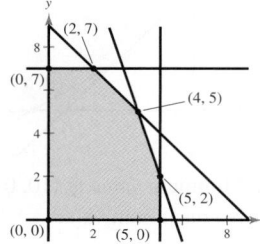

73. 369 units of model A
538 units of model B
$117,600

75. $525 model: 0 units
$675 model: 305 units
$38,125

77. 8 audits
0 tax returns
$20,000

Test Yourself   *(page 472)*

1. $(2, 4)$ **2.** $(-2, 5), (3, 0)$ **3.** $(-1, 6), (2, 0)$

4. $(3.36, -1.32)$ **5.** $(2, -1, 3)$ **6.** $(2, -3, 4)$

7. $60,000 at 2.3%   **8.** $(50,000, 34)$
$20,000 at 3.1%

9. (a) $y = 0.593x^2 + 4.95x + 46.7$ (b) ≈ 66.9 million adults

10. 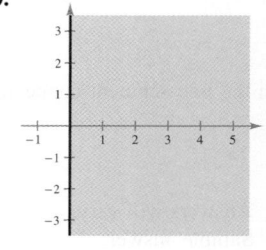   **11.**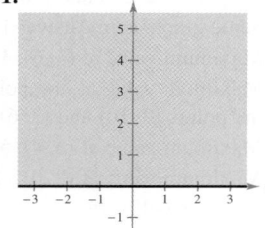

12. 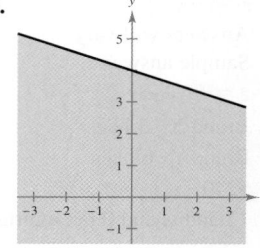   **13.**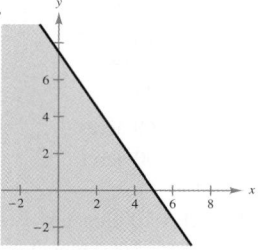

14.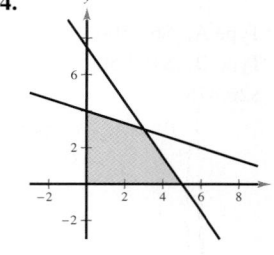

15. Minimum value at $(0, 0)$: 0
Maximum value at $(3, 3)$: 39

16. Model A: 297 units; model B: 570 units

$216,150; The profit model is $P = 200x + 275y$ with the constraints

$$\begin{cases} 3.5x + 8y \le 5600 \\ 2.5x + 2y \le 2000 \\ 1.3x + 0.9y \le 900 \\ \quad\quad x \ge 0 \\ \quad\quad y \ge 0 \end{cases}$$

and the maximum profit occurs at (297, 570).

Chapter 6

Section 6.1   *(page 483)*

> ### Skills Warm Up   *(page 483)*
>
> **1.** -3 **2.** 30   **3.** 6   **4.** $-\frac{1}{9}$ **5.** Solution
>
> **6.** Not a solution   **7.** (5, 2)   **8.** $\left(\frac{12}{5}, -3\right)$
>
> **9.** (40, 14, 2)   **10.** $\left(\frac{15}{2}, 4, 1\right)$

1. 2×3 **3.** 4×3 **5.** 4×2 **7.** 2×4

9. $\begin{bmatrix} 4 & -3 & \vdots & -5 \\ -1 & 3 & \vdots & 12 \end{bmatrix}$ **11.** $\begin{bmatrix} 1 & 10 & -2 & \vdots & 2 \\ 5 & -3 & 4 & \vdots & 0 \\ 2 & 1 & 0 & \vdots & 6 \end{bmatrix}$

2×3 3×4

13. $\begin{bmatrix} 7 & -5 & 1 & \vdots & 13 \\ 19 & 0 & -8 & \vdots & 10 \end{bmatrix}$ **15.** $\begin{bmatrix} 1 & 1 & 1 \\ 0 & -7 & -1 \end{bmatrix}$

2×4

17. $\begin{bmatrix} 1 & 0 & 14 & -11 \\ 0 & 1 & -2 & 2 \\ 0 & 0 & 1 & -7 \end{bmatrix}$ **19.** Add 5 times R_2 to R_1.

21. Interchange R_1 and R_2. Add 4 times R_1 to R_3.

23. $\begin{cases} 2x + 4y = 6 \\ -x + 3y = -8 \end{cases}$ **25.** $\begin{cases} x \quad + 2z = -10 \\ \quad 3y - z = 5 \\ 4x + 2y \quad = 3 \end{cases}$

27. $\begin{cases} x - 5y = 6 \\ \quad y = -2 \end{cases}$ **29.** $\begin{cases} x + 3y - z = 15 \\ \quad y + 4z = -12 \\ \quad z = -5 \end{cases}$

$(-4, -2)$ $(-14, 8, -5)$

31. (a) (i) $\begin{bmatrix} 3 & 0 & \vdots & -6 \\ 6 & -4 & \vdots & -28 \end{bmatrix}$ (ii) $\begin{bmatrix} 3 & 0 & \vdots & -6 \\ 0 & -4 & \vdots & -16 \end{bmatrix}$

(iii) $\begin{bmatrix} 3 & 0 & \vdots & -6 \\ 0 & 1 & \vdots & 4 \end{bmatrix}$ (iv) $\begin{bmatrix} 1 & 0 & \vdots & -2 \\ 0 & 1 & \vdots & 4 \end{bmatrix}$

$(-2, 4)$

(b) $\begin{cases} -3x + 4y = 22 \\ 6x - 4y = -28 \end{cases}$

$(-2, 4)$

(c) Answers will vary.

33. (i)
```
row+([A],2,1)→[B]
]
     [[3 0 -6 ]
      [6 -4 -28]]
```
(ii)
```
*row+(-2,[B],1,2
)→[C]
     [[3 0  -6 ]
      [0 -4 -16]]
```

(iii)
```
*row(-1/4,[C],2)
→[D]
     [[3 0 -6]
      [0 1 4 ]]
```
(iv)
```
*row(1/3,[D],1)→
[E]
     [[1 0 -2]
      [0 1 4 ]]
```

35. Reduced row-echelon form   **37.** Not in row-echelon form

39. Not in row-echelon form

41. $\begin{bmatrix} 1 & 2 & -1 & 5 \\ 0 & 1 & -1 & 1 \\ 0 & 0 & 1 & -2 \end{bmatrix}$ **43.** $\begin{bmatrix} 1 & -1 & -1 & 1 \\ 0 & 1 & 6 & 3 \\ 0 & 0 & 0 & 0 \end{bmatrix}$

45. $\begin{bmatrix} 1 & 0 & 0 \\ 0 & 1 & 0 \\ 0 & 0 & 1 \end{bmatrix}$ **47.** $\begin{bmatrix} 1 & 0 & 0 & 0 \\ 0 & 1 & 0 & 0 \\ 0 & 0 & 1 & 0 \\ 0 & 0 & 0 & 1 \end{bmatrix}$

49. $\begin{bmatrix} 1 & 0 \\ 0 & 1 \\ 0 & 0 \end{bmatrix}$ **51.** $(-4, 6)$ **53.** $(-4, -8, 2)$

55. $(-2a - 4, -a + 6, a)$ **57.** (3, 2)   **59.** (4, -2)

61. $\left(\frac{1}{2}, -\frac{3}{4}\right)$ **63.** Inconsistent   **65.** $(5a + 4, -3a + 2, a)$

67. $(0, 2 - 4a, a)$ **69.** $(-3b + 96a + 100, b, 52a + 54, a)$

71. (0, 0)   **73.** $(-2a, a, a)$ **75.** (4, -3, 2)

77. $(2a + 1, 3a + 2, a)$ **79.** (7, -3, 4)

81. Yes; $(-1, 1, -3)$ **83.** No

85. $1,200,000 was borrowed at 8%, $200,000 was borrowed at 9%, and $600,000 was borrowed at 12%.

87. Both are correct. Because there are infinitely many ordered triples that are solutions to this system, a solution can be written in many different ways. If $a = 3$, the ordered triple is $(3, -3, 5)$. You obtain the same triple when $b = 5$.

89. $y = 7.5t + 28$

133 new cases; Because the data values increased in a linear pattern, this estimate seems reasonable.

Section 6.2   *(page 497)*

> ### Skills Warm Up   *(page 497)*
>
> **1.** -5 **2.** -7
>
> **3.** Not in reduced row-echelon form
>
> **4.** Not in reduced row-echelon form
>
> **5.** $\begin{bmatrix} -5 & 10 & \vdots & 12 \\ 7 & -3 & \vdots & 0 \end{bmatrix}$
>
> **6.** $\begin{bmatrix} 10 & 15 & -9 & \vdots & 42 \\ 6 & -5 & 0 & \vdots & 0 \end{bmatrix}$
>
> **7.** (0, 2)   **8.** $(2 + a, 3 - a, a)$
>
> **9.** $(1 - 2a, a, -1)$ **10.** $(2, -1, -1)$

1. $x = -3, y = 2$ **3.** $x = -2, y = 5$

5. (a) $\begin{bmatrix} 3 & -1 \end{bmatrix}$ (b) $\begin{bmatrix} 11 & -5 \end{bmatrix}$

(c) $\begin{bmatrix} 42 & -18 \end{bmatrix}$ (d) $\begin{bmatrix} 40 & -18 \end{bmatrix}$

7. (a) $\begin{bmatrix} 8 & -1 \\ 1 & 7 \end{bmatrix}$ (b) $\begin{bmatrix} 2 & -3 \\ 5 & -5 \end{bmatrix}$

(c) $\begin{bmatrix} 30 & -12 \\ 18 & 6 \end{bmatrix}$ (d) $\begin{bmatrix} 11 & -11 \\ 18 & -14 \end{bmatrix}$

9. (a) $\begin{bmatrix} 7 & 3 \\ 1 & 9 \\ -2 & 15 \end{bmatrix}$ (b) $\begin{bmatrix} 5 & -5 \\ 3 & -1 \\ -4 & -5 \end{bmatrix}$

(c) $\begin{bmatrix} 36 & -6 \\ 12 & 24 \\ -18 & 30 \end{bmatrix}$ (d) $\begin{bmatrix} 21 & -16 \\ 11 & 1 \\ -15 & -10 \end{bmatrix}$

11. (a) $\begin{bmatrix} 3 & 3 & -2 \\ -2 & 5 & 7 \\ 1 & -8 & 11 \end{bmatrix}$ (b) $\begin{bmatrix} 1 & 1 & 0 \\ 4 & -3 & -11 \\ 1 & 6 & -5 \end{bmatrix}$

(c) $\begin{bmatrix} 12 & 12 & -6 \\ 6 & 6 & -12 \\ 6 & -6 & 18 \end{bmatrix}$ (d) $\begin{bmatrix} 5 & 5 & -1 \\ 13 & -8 & -35 \\ 4 & 17 & -12 \end{bmatrix}$

13. (a) $[2 \quad 40 \quad 30 \quad -23]$ (b) $[-28 \quad 8 \quad 10 \quad 1]$
(c) $[-78 \quad 144 \quad 120 \quad -66]$ (d) $[-97 \quad 48 \quad 50 \quad -8]$

15. $\begin{bmatrix} -8 & -7 \\ 15 & -1 \end{bmatrix}$ **17.** $[12 \quad -6]$ **19.** $\begin{bmatrix} 55 & -50 \\ 15 & -20 \end{bmatrix}$

21. $\begin{bmatrix} -48 & -8 & 24 \\ -24 & 64 & 24 \end{bmatrix}$ **23.** $\begin{bmatrix} 10 & 8 \\ -59 & 9 \end{bmatrix}$

25. $\begin{bmatrix} -17.143 & 2.143 \\ 11.571 & 10.286 \end{bmatrix}$ **27.** $\begin{bmatrix} -1.581 & -3.739 \\ -4.252 & -13.249 \\ 9.713 & -0.362 \end{bmatrix}$

29. $\begin{bmatrix} -6 & -9 \\ -1 & 0 \\ 17 & -10 \end{bmatrix}$ **31.** $\begin{bmatrix} 3 & 3 \\ -\frac{1}{2} & 0 \\ -\frac{13}{2} & \frac{11}{2} \end{bmatrix}$ **33.** $[-32]$

35. $\begin{bmatrix} -5 \\ 14 \end{bmatrix}$ **37.** $\begin{bmatrix} -7 & 10 & -12 & 17 \\ 6 & 21 & 7 & 15 \\ -3 & 3 & -5 & 6 \end{bmatrix}$ **39.** $\begin{bmatrix} -1 & 19 \\ 4 & -27 \\ 0 & 14 \end{bmatrix}$

41. $\begin{bmatrix} 1 & 0 & 0 \\ 0 & 1 & 0 \\ 0 & 0 & \frac{7}{2} \end{bmatrix}$ **43.** Not possible

45. $\begin{bmatrix} 41 & 7 & 7 \\ 42 & 5 & 25 \\ -10 & -25 & 45 \end{bmatrix}$ **47.** Not possible

49. (a) $\begin{bmatrix} 0 & 15 \\ 6 & 12 \end{bmatrix}$ (b) $\begin{bmatrix} -2 & 2 \\ 31 & 14 \end{bmatrix}$ (c) $\begin{bmatrix} 9 & 6 \\ 12 & 12 \end{bmatrix}$

51. (a) $\begin{bmatrix} 3 & 5 \\ 1 & 6 \end{bmatrix}$ (b) $\begin{bmatrix} 11 & 5 & 0 \\ -7 & -4 & -1 \\ 14 & 8 & 2 \end{bmatrix}$ (c) Not possible

53. (a) $[11]$ (b) $\begin{bmatrix} -4 & 2 & 3 \\ 0 & 0 & 0 \\ -20 & 10 & 15 \end{bmatrix}$ (c) Not possible

55. Cannot perform operation.   **57.** Cannot perform operation.
59. Cannot perform operation.   **61.** 2×2 **63.** 2×3

65. (a) $\begin{bmatrix} -1 & 1 \\ -2 & 1 \end{bmatrix}\begin{bmatrix} x \\ y \end{bmatrix} = \begin{bmatrix} 4 \\ 0 \end{bmatrix}$ (b) $\begin{bmatrix} 4 \\ 8 \end{bmatrix}$

67. (a) $\begin{bmatrix} 1 & 2 \\ 3 & -1 \end{bmatrix}\begin{bmatrix} x \\ y \end{bmatrix} = \begin{bmatrix} 3 \\ 2 \end{bmatrix}$ (b) $\begin{bmatrix} 1 \\ 1 \end{bmatrix}$

69. (a) $\begin{bmatrix} 1 & -4 & 5 \\ 2 & 5 & 1 \\ -1 & -1 & -1 \end{bmatrix}\begin{bmatrix} x \\ y \\ z \end{bmatrix} = \begin{bmatrix} 3 \\ -24 \\ 8 \end{bmatrix}$ (b) $\begin{bmatrix} -4 \\ -3 \\ -1 \end{bmatrix}$

71. (a) $\begin{bmatrix} 1 & -2 & 3 \\ -1 & 3 & -1 \\ 2 & -5 & 5 \end{bmatrix}\begin{bmatrix} x \\ y \\ z \end{bmatrix} = \begin{bmatrix} 9 \\ -6 \\ 17 \end{bmatrix}$ (b) $\begin{bmatrix} 1 \\ -1 \\ 2 \end{bmatrix}$

73. $\begin{bmatrix} 110 & 132 & 66 & 44 \\ 154 & 176 & 220 & 88 \end{bmatrix}$

75.   Hotel w Hotel x Hotel y Hotel z
$\begin{bmatrix} 688.80 & 750.40 & 828.80 & 1108.80 \\ 1114.40 & 1153.60 & 1321.60 & 1237.60 \end{bmatrix}\begin{matrix} \text{Double} \\ \text{Family} \end{matrix}\Big\}$ Occupancy

77. (a) \$7375   (b) \$5845
(c)   Wholesale   Retail
$ST = \begin{bmatrix} \$4650 & \$7375 \\ \$7405 & \$11,350 \\ \$5845 & \$9400 \end{bmatrix}\begin{matrix} 1 \\ 2 \\ 3 \end{matrix}\Big\}$ Outlet

ST represents the wholesale and retail prices of the laptop computer inventories at the three outlets.

79. (a) $B = [2 \quad 0.25 \quad 0.5]$
(b) $\overbrace{\text{Calories burned}}$

130-lb person	155-lb person

$BA = [1180 \qquad 1407.5]$
BA represents the total calories burned by each person.

81. (a)   X   Y   Z
$D = \begin{bmatrix} 3 & 1 & 2 \\ 0 & 2 & 3 \\ 4 & 5 & 3 \end{bmatrix}\begin{matrix} \text{Sacks} \\ \text{Interceptions} \\ \text{Key tackles} \end{matrix}$
Each entry d_{ij} represents the number of each type of defensive play made by each player.
(b)   Sacks   Interceptions   Tackles
$B = [\$2000 \qquad \$1000 \qquad \$800]$
Each entry b_{ij} represents the bonus received for each type of play.
(c)   X   Y   Z
$BD = [\$9200 \quad \$8000 \quad \$9400]$
Each entry represents the bonus each player will receive.
(d) Player Z

83. $\begin{bmatrix} 0.40 & 0.15 & 0.15 \\ 0.28 & 0.53 & 0.17 \\ 0.32 & 0.32 & 0.68 \end{bmatrix}$ **85.** $\begin{bmatrix} -2 & 3 \\ 1 & -1 \end{bmatrix}$; only one correct matrix

87. $AC = \begin{bmatrix} 12 & -6 & 9 \\ 16 & -8 & 12 \\ 4 & -2 & 3 \end{bmatrix} = BC$, but $A \neq B$.

89. One year:
Gold subscribers: 28,750
Galaxy subscribers: 35,750
Nonsubscribers: 35,500
Multiply the original matrix by the 3×1 matrix
$\begin{bmatrix} 25,000 \\ 30,000 \\ 45,000 \end{bmatrix}$
which represents the current numbers of subscribers for each company and the number of nonsubscribers.

Two years:

Gold subscribers: 30,813

Galaxy subscribers: 39,675

Nonsubscribers: 29,513

Multiply the original matrix by the 3×1 matrix

$$\begin{bmatrix} 28{,}750 \\ 35{,}750 \\ 35{,}500 \end{bmatrix}$$

which represents the numbers of subscribers for each company and the number of nonsubscribers after 1 year.

Three years:

Gold subscribers: 31,947

Galaxy subscribers: 42,330

Nonsubscribers: 25,724

Multiply the original matrix by the 3×1 matrix

$$\begin{bmatrix} 30{,}813 \\ 39{,}675 \\ 29{,}513 \end{bmatrix}$$

which represents the numbers of subscribers for each company and the number of nonsubscribers after 2 years.

The number of subscribers to each company is increasing each year. The number of nonsubscribers is decreasing each year.

Section 6.3 *(page 508)*

Skills Warm Up *(page 508)*

1. $\begin{bmatrix} 4 & 24 \\ 0 & -16 \\ 48 & 8 \end{bmatrix}$ **2.** $\begin{bmatrix} \frac{11}{2} & 5 & 24 \\ \frac{1}{2} & 0 & 8 \\ 0 & 1 & 4 \end{bmatrix}$

3. $\begin{bmatrix} -5 & -2 & -13 \\ 4 & -13 & -2 \end{bmatrix}$ **4.** $\begin{bmatrix} -13 & 11 \\ -19 & 21 \end{bmatrix}$

5. $\begin{bmatrix} 1 & 0 \\ 0 & 1 \end{bmatrix}$ **6.** $\begin{bmatrix} 6 & 5 \\ 3 & -2 \end{bmatrix}$ **7.** $\begin{bmatrix} 1 & 0 & 0 \\ 0 & 1 & 0 \\ 0 & 0 & 1 \end{bmatrix}$

8. $\begin{bmatrix} 1 & 0 & 0 \\ 0 & 1 & 0 \\ 0 & 0 & 1 \end{bmatrix}$ **9.** $\begin{bmatrix} 1 & 0 & \vdots & 3 & -2 \\ 0 & 1 & \vdots & 4 & -3 \end{bmatrix}$

10. $\begin{bmatrix} 1 & 0 & 0 & \vdots & -6 & -4 & 3 \\ 0 & 1 & 0 & \vdots & 11 & 6 & -5 \\ 0 & 0 & 1 & \vdots & -2 & -1 & 1 \end{bmatrix}$

1–9. $AB = I$ and $BA = I$ **11.** $\begin{bmatrix} 7 & -2 \\ -3 & 1 \end{bmatrix}$

13. $\begin{bmatrix} \frac{1}{4} & \frac{1}{2} \\ -\frac{1}{4} & -1 \end{bmatrix}$ **15.** $\begin{bmatrix} 0 & -1 \\ 1 & 11 \end{bmatrix}$ **17.** $\begin{bmatrix} \frac{1}{2} & -\frac{1}{3} \\ \frac{1}{4} & 0 \end{bmatrix}$

19. Does not exist **21.** Does not exist

23. $\begin{bmatrix} 1 & 1 & -1 \\ -3 & 2 & -1 \\ 3 & -3 & 2 \end{bmatrix}$ **25.** $\frac{1}{2}\begin{bmatrix} -3 & 3 & 2 \\ 9 & -7 & -6 \\ -2 & 2 & 2 \end{bmatrix}$

27. $\begin{bmatrix} \frac{1}{3} & 0 & 0 \\ 0 & -\frac{1}{2} & 0 \\ 0 & 0 & \frac{1}{4} \end{bmatrix}$ **29.** $\begin{bmatrix} 1 & 0 & 0 \\ -\frac{3}{4} & \frac{1}{4} & 0 \\ \frac{7}{20} & -\frac{1}{4} & \frac{1}{5} \end{bmatrix}$

31. Does not exist

33. $\begin{bmatrix} -175 & 37 & -13 \\ 95 & -20 & 7 \\ 14 & -3 & 1 \end{bmatrix}$ **35.** $\frac{1}{11}\begin{bmatrix} 0 & -20 & 10 \\ -110 & 55 & 55 \\ 110 & -30 & -40 \end{bmatrix}$

37. $\begin{bmatrix} -12 & -5 & -9 \\ -4 & -2 & -4 \\ -8 & -4 & -6 \end{bmatrix}$ **39.** Does not exist

41. $\frac{1}{19}\begin{bmatrix} 3 & 2 \\ -2 & 5 \end{bmatrix}$ **43.** Does not exist

45. $\frac{1}{59}\begin{bmatrix} 16 & 15 \\ -4 & 70 \end{bmatrix}$ **47.** $(-2, 1)$ **49.** $(4, 2)$

51. $(-2, 3)$ **53.** $(2, 0)$ **55.** $(3, 8, -11)$

57. $(2, 1, 0, 0)$ **59.** $(2, -2)$ **61.** Inconsistent

63. $(-4, -8)$ **65.** $(-1, 3, 2)$ **67.** $(5, 0, -2, 3)$

69. $\begin{cases} 2x + y + 3z = 16 \\ 4x - 2z = -2 \\ 3y + 2z = 1 \end{cases}$

71. AAA bonds: $20,000 **73.** AAA bonds: $21,000

A bonds: $5000   A bonds: $5000

B bonds: $10,000   B bonds: $10,000

75. $I_1 = 4$ amperes, $I_2 = 1$ ampere, $I_3 = 5$ amperes

77. 100 bags of potting soil for seedlings

100 bags of potting soil for general potting

100 bags of potting soil for hardwood plants

79. 5 bags of potting soil for seedlings

100 bags of potting soil for general potting

120 bags of potting soil for hardwood plants

81. (a) $\begin{cases} 2.5x + 4y + 2z = 300 \\ x + y + z = 120 \\ -x + 2y + 2z = 0 \end{cases}$

(b) $\begin{bmatrix} 2.50 & 4 & 2 \\ 1 & 1 & 1 \\ -1 & 2 & 2 \end{bmatrix}\begin{bmatrix} x \\ y \\ z \end{bmatrix} = \begin{bmatrix} 300 \\ 120 \\ 0 \end{bmatrix}$

(c) 80 roses, 10 lilies, 30 irises

83. (a) $AB = \begin{bmatrix} 13 & 4 \\ 1 & 8 \end{bmatrix}$, $BA = \begin{bmatrix} 8 & 1 \\ 4 & 13 \end{bmatrix}$

$AB = C$ and in the product BA, the diagonal entries are the reverse of those in C.

(b) $C^{-1} = \begin{bmatrix} \frac{2}{25} & -\frac{1}{25} \\ -\frac{1}{100} & \frac{13}{100} \end{bmatrix}$

$A^{-1} \cdot B^{-1} = \begin{bmatrix} \frac{13}{100} & -\frac{1}{100} \\ -\frac{2}{25} & \frac{2}{25} \end{bmatrix}$

$B^{-1} \cdot A^{-1} = \begin{bmatrix} \frac{2}{25} & -\frac{1}{25} \\ -\frac{1}{100} & \frac{13}{100} \end{bmatrix}$

$C^{-1} = B^{-1} \cdot A^{-1}$

(c) The inverse of a product AB is equal to the inverse of B times the inverse of A.

85. Answers will vary. Sample answer:
If $k = 3$, then

$$\begin{bmatrix} 4 & 3 \\ -2 & 3 \end{bmatrix}^{-1} = \begin{bmatrix} \frac{1}{6} & -\frac{1}{6} \\ \frac{1}{9} & \frac{2}{9} \end{bmatrix}.$$

If $k = -\frac{3}{2}$, the matrix is singular.
87. True. The inverse of I_n is I_n.   **89.** Answers will vary.

Quiz Yourself   *(page 512)*

1. Any matrix with four rows and three columns
2. Any matrix with three rows and one column

3. $\begin{bmatrix} 3 & 2 & \vdots & -2 \\ 5 & -1 & \vdots & 19 \end{bmatrix}$ **4.** $\begin{bmatrix} 1 & 0 & 3 & \vdots & -5 \\ 1 & 2 & -1 & \vdots & 3 \\ 3 & 0 & 4 & \vdots & 0 \end{bmatrix}$

5. $(2.769, -5.154)$ **6.** $(4, -2, -3)$ **7.** $\begin{bmatrix} -2 & 5 \\ 5 & 15 \end{bmatrix}$

8. $\begin{bmatrix} 9 & -2 & 23 \\ 14 & -4 & 36 \end{bmatrix}$ **9.** $\begin{bmatrix} -1 & 7 \\ -11 & 3 \end{bmatrix}$ **10.** $\begin{bmatrix} -3 & -1 \\ 3 & -2 \end{bmatrix}$

11. $\begin{bmatrix} 3 & -2 \\ 1 & -\frac{1}{2} \end{bmatrix}$ **12.** Not possible   **13.** $\begin{bmatrix} -3 & 14 \\ -12 & 16 \end{bmatrix}$

14. $\begin{bmatrix} 2 & -9 \\ 7 & -11 \end{bmatrix}$ **15.** \$22.80   **16.** \$41.40   **17.** \$59.60

18.

$$LW = \begin{bmatrix} \$22.80 & \$20.20 \\ \$47.80 & \$41.40 \\ \$66.50 & \$59.60 \end{bmatrix} \begin{matrix} A \\ B \\ C \end{matrix} \Big\} \text{Model}$$

Plant 1   Plant 2

LW represents the total labor costs for each model at each plant.
19. $(4, -2)$ **20.** $(4, 2, -3)$

Section 6.4   *(page 518)*

> ### Skills Warm Up   *(page 518)*
>
> **1.** $\begin{bmatrix} 3 & 5 \\ 4 & 0 \end{bmatrix}$ **2.** $\begin{bmatrix} -2 & 8 \\ 2 & -4 \end{bmatrix}$
>
> **3.** $\begin{bmatrix} 9 & -12 & 6 \\ 3 & 0 & -3 \\ 0 & 3 & -6 \end{bmatrix}$ **4.** $\begin{bmatrix} 0 & 8 & 12 \\ -4 & 8 & 12 \\ -8 & 4 & -8 \end{bmatrix}$
>
> **5.** -22 **6.** 35 **7.** -15 **8.** $-\frac{1}{8}$
> **9.** -45 **10.** -16

1. -5 **3.** 1 **5.** 3 **7.** 0 **9.** 5 **11.** 4
13. $\frac{11}{6}$ **15.** -2.6 **17.** 248 **19.** 0.14 **21.** -0.838
23. (a) $M_{11} = -5, M_{12} = 2, M_{21} = 4, M_{22} = 3$
    (b) $C_{11} = -5, C_{12} = -2, C_{21} = -4, C_{22} = 3$
25. (a) $M_{11} = -4, M_{12} = -2, M_{21} = 1, M_{22} = 3$
    (b) $C_{11} = -4, C_{12} = 2, C_{21} = -1, C_{22} = 3$
27. (a) $M_{11} = 3, M_{12} = -4, M_{13} = 1, M_{21} = 2, M_{22} = 2,$
    $M_{23} = -4, M_{31} = -4, M_{32} = 10, M_{33} = 8$
    (b) $C_{11} = 3, C_{12} = 4, C_{13} = 1, C_{21} = -2, C_{22} = 2,$
    $C_{23} = 4, C_{31} = -4, C_{32} = -10, C_{33} = 8$

29. (a) $M_{11} = 30, M_{12} = 12, M_{13} = 11, M_{21} = -36,$
    $M_{22} = 26, M_{23} = 7, M_{31} = -4,$
    $M_{32} = -42, M_{33} = 12$
    (b) $C_{11} = 30, C_{12} = -12, C_{13} = 11, C_{21} = 36,$
    $C_{22} = 26, C_{23} = -7, C_{31} = -4,$
    $C_{32} = 42, C_{33} = 12$
31. (a) -99 (b) -99 **33.** (a) -145 (b) -145
35. (a) 170 (b) 170 **37.** -58 **39.** -30 **41.** 0
43. 0 **45.** -0.002 **47.** 0 **49.** -108 **51.** 412
53. -126 **55.** -336

57. (a) -3 (b) -2 (c) $\begin{bmatrix} -2 & 0 \\ 0 & -3 \end{bmatrix}$ (d) 6

59. (a) -8 (b) 0 (c) $\begin{bmatrix} -4 & 4 \\ 1 & -1 \end{bmatrix}$ (d) 0

61. (a) -21 (b) -19 (c) $\begin{bmatrix} 7 & 1 & 4 \\ -8 & 9 & -3 \\ 7 & -3 & 9 \end{bmatrix}$ (d) 399

63. (a) 2 (b) -6 (c) $\begin{bmatrix} 1 & 4 & 3 \\ -1 & 0 & 3 \\ 0 & 2 & 0 \end{bmatrix}$ (d) -12

65. 45; The determinant is the product of the matrix's diagonal entries.
67. 75; The determinant is the product of the matrix's diagonal entries.
69. Lower triangular; -6 **71.** Upper triangular; -16
73. Upper triangular; -168
75. Matrices will vary. The determinant of each matrix is the product of the entries on the main diagonal, which in this case equals 6.
77. Matrices will vary. The determinant of each matrix is the product of the entries on the main diagonal, which in this case equals -18.
79. Matrices will vary. The determinant of each matrix is the product of the entries on the main diagonal, which in this case equals 28.
81. -20 **83.** -18 **85.** 80
87–89. Answers will vary.   **91.** ± 2 **93.** $\pm\frac{3}{2}$
95. $1 \pm \sqrt{2}$ **97.** $\frac{1}{2}, 1$ **99.** 3
101. Answers will vary. For an $n \times n$ matrix ($n > 2$) with consecutive integer entries, the determinant appears to be 0.
103. True. If an entire row is zeros, then each cofactor in the expansion is multiplied by zero.

Section 6.5   *(page 527)*

> ### Skills Warm Up   *(page 527)*
>
> **1.** 1 **2.** 0 **3.** -8 **4.** x^2 **5.** 8 **6.** 60
>
> **7.** $\begin{bmatrix} 7 & -3 \\ -2 & 1 \end{bmatrix}$ **8.** $\begin{bmatrix} 1 & 2 & -1 \\ -1 & -2 & 2 \\ 2 & 5 & 0 \end{bmatrix}$
>
> **9.** $\begin{bmatrix} 0.14 \\ 0.31 \end{bmatrix}$ **10.** $[7 \quad 14]$

1. 11 **3.** 28 **5.** $\frac{33}{8}$ **7.** $\frac{5}{2}$ **9.** 28

11. $y = \frac{16}{5}$ or $y = 0$ **13.** $y = -3$ or $y = -11$
15. 307.5 square miles   **17.** Collinear   **19.** Not collinear
21. Collinear   **23.** Not collinear   **25.** Collinear
27. $y = -4$ **29.** $x = 3$ **31.** $x - 6y + 13 = 0$
33. $x + 3y - 5 = 0$ **35.** $y = 5$ **37.** $x = -4$
39. $2x + 3y - 8 = 0$
41. $[3\ 15][13\ 5][0\ 8][15\ 13][5\ 0][19\ 15][15\ 14]$
   48, 81, 28, 51, 24, 40, 54, 95, 5, 10, 64, 113, 57, 100
43. $[20\ 5\ 24][20\ 0\ 13][5\ 0\ 1][20\ 0\ 23]\ [15\ 18\ 11]$
   -119, 28, 67, -58, 6, 39, -1, -3, 3, -118, 26, 69, -33,
   7, 15
45. 1, -25, -65, 17, 15, -9, -12, -62, -119,
   27, 51, 48, 43, 67, 48, 57, 111, 117
47. 29, 53, 48, 60, 115, 120, 20, 38, 35, 7, -12, -51, 72, 148,
   165, 62, 143, 181
49. 34, 55, 43, 60, 136, 169, 71, 139, 145, 52, 106, 119, 38, 52, 23
51. 40, 83, 92, -2, -30, -69, -16, -74, -137, 15, 28, 25, 6,
   -16, -58, 46, 79, 67
53. HAPPY NEW YEAR   **55.** SOUND ALL CLEAR
57. SEND MORE MONEY   **59.** MEET ME TONIGHT RON
61. Because $[45\ -35]\begin{bmatrix} w & x \\ y & z \end{bmatrix} = [10\ 15]$

and $[38\ -30]\begin{bmatrix} w & x \\ y & z \end{bmatrix} = [8\ 14]$,

you can solve $\begin{bmatrix} 45 & -35 \\ 38 & -30 \end{bmatrix}\begin{bmatrix} w & x \\ y & z \end{bmatrix} = \begin{bmatrix} 10 & 15 \\ 8 & 14 \end{bmatrix}$;

JOHN RETURN TO BASE
63. 12 square units   **65.** Answers will vary.

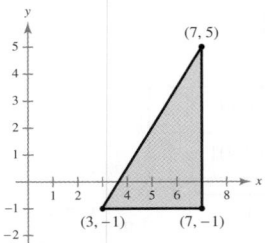

$A = \frac{1}{2}(4)(6) = 12$ square units

Review Exercises *(page 534)*

1. 1×2 **3.** 2×4
5. $\begin{bmatrix} 1 & 3 & 0 & 2 \\ 0 & 1 & 1 & 2 \\ 0 & 0 & 1 & 2 \end{bmatrix}$ **7.** $\begin{bmatrix} 1 & 0 & 0 \\ 0 & 1 & 0 \\ 0 & 0 & 1 \end{bmatrix}$ **9.** $(3, -2)$
11. $(3, 2, -1)$ **13.** $(10 - 4a, a, a)$ **15.** Inconsistent
17. \$40,000 was borrowed at 8%, \$120,000 was borrowed at 10%, and \$40,000 was borrowed at 12%.
19. $x = -6, y = 2$
21. (a) $\begin{bmatrix} 3 & 7 \\ -4 & 4 \end{bmatrix}$ (b) $\begin{bmatrix} -5 & 3 \\ 8 & -2 \end{bmatrix}$
  (c) $\begin{bmatrix} -4 & 20 \\ 8 & 4 \end{bmatrix}$ (d) $\begin{bmatrix} -13 & 21 \\ 22 & -1 \end{bmatrix}$
23. (a) $\begin{bmatrix} 7 & 1 \\ 3 & 6 \\ 3 & 6 \end{bmatrix}$ (b) $\begin{bmatrix} -5 & -9 \\ -3 & -8 \\ 5 & -2 \end{bmatrix}$

  (c) $\begin{bmatrix} 4 & -16 \\ 0 & -4 \\ 16 & 8 \end{bmatrix}$ (d) $\begin{bmatrix} -7 & -30 \\ -6 & -19 \\ 22 & 2 \end{bmatrix}$
25. (a) $\begin{bmatrix} 3 & 4 & 2 & 1 \\ 3 & -5 & 6 & 0 \end{bmatrix}$ (b) $\begin{bmatrix} -1 & 2 & -6 & 11 \\ -3 & 7 & 0 & 4 \end{bmatrix}$
  (c) $\begin{bmatrix} 4 & 12 & -8 & 24 \\ 0 & 4 & 12 & 8 \end{bmatrix}$ (d) $\begin{bmatrix} 1 & 13 & -18 & 40 \\ -6 & 17 & 9 & 14 \end{bmatrix}$
27. $\begin{bmatrix} -12 & -3 \\ 5 & -5 \end{bmatrix}$ **29.** $[2\ -1\ -3\ 5]$ **31.** $\begin{bmatrix} 4 & -11 \\ -3 & 1 \\ -1 & -3 \end{bmatrix}$
33. $\frac{1}{2}\begin{bmatrix} 3 & -5 \\ 1 & 4 \\ 9 & 14 \end{bmatrix}$ **35.** $\begin{bmatrix} 8 \\ 5 \\ -6 \end{bmatrix}$ **37.** $\begin{bmatrix} 1 & 0 & 0 \\ 0 & 1 & 0 \\ 0 & 0 & 1 \end{bmatrix}$
39. (a) $[4]$ (b) $\begin{bmatrix} 2 & -6 & 8 \\ -2 & 6 & -8 \\ -1 & 3 & -4 \end{bmatrix}$ (c) Not possible
41. $\begin{bmatrix} 96 & 144 & 24 & 48 \\ 48 & 72 & 96 & 24 \\ 168 & 72 & 120 & 96 \end{bmatrix}$
43. (a) \$8325 (b) \$5200
  (c)   Wholesale   Retail
    $ST = \begin{bmatrix} \$5200 & \$8265 \\ \$5075 & \$7985 \\ \$5125 & \$8325 \end{bmatrix}\begin{matrix} 1 \\ 2 \\ 3 \end{matrix}$ } Outlet

   ST represents the wholesale and retail values of the car sound system inventory at each outlet.
45. $AB = I$ and $BA = I$
47. $\begin{bmatrix} -1 & 0 & 0 \\ 0 & \frac{1}{2} & 0 \\ 0 & 0 & \frac{1}{4} \end{bmatrix}$ **49.** $\begin{bmatrix} -5 & 3 \\ 2 & -1 \end{bmatrix}$ **51.** $(3, 4)$
53. $\left(2, \frac{1}{2}, 3\right)$ **55.** $(-6, -1)$
57. 10 units of fluid X, 8 units of fluid Y, 5 units of fluid Z
59. (a) $\begin{bmatrix} \frac{31}{35} & -\frac{27}{35} & \frac{1}{7} \\ -\frac{27}{35} & \frac{87}{70} & -\frac{2}{7} \\ \frac{1}{7} & -\frac{2}{7} & \frac{1}{14} \end{bmatrix}$
   $y = -6.14t^2 + 109.6t + 818$
  (b) 1213 stores   (c) The estimate is too high.
61. 4   **63.** 0
65. (a) $M_{11} = 4, M_{12} = 7, M_{21} = -1, M_{22} = 2$
  (b) $C_{11} = 4, C_{12} = -7, C_{21} = 1, C_{22} = 2$
67. (a) $M_{11} = 30, M_{12} = -12, M_{13} = -21,$
    $M_{21} = 20, M_{22} = 19, M_{23} = 22,$
    $M_{31} = 5, M_{32} = -2, M_{33} = 19$
  (b) $C_{11} = 30, C_{12} = 12, C_{13} = -21,$
    $C_{21} = -20, C_{22} = 19, C_{23} = -22,$
    $C_{31} = 5, C_{32} = 2, C_{33} = 19$
69. 44; Answers will vary.   **71.** -12; Answers will vary.
73. -39; Answers will vary.
75. (a) 8 (b) -12 (c) $\begin{bmatrix} -2 & -6 \\ -14 & 6 \end{bmatrix}$ (d) -96
77. 10   **79.** 10   **81.** Not collinear   **83.** Collinear
85. $x + 15y - 38 = 0$ **87.** $x = 2$
89. $[20\ 18][1\ 14][19\ 13][9\ 20][0\ 14][15\ 23]$
   94, 132, 44, 59, 77, 109, 78, 107, 42, 56, 99, 137

CHAPTER 6

91. SEIZE THE DAY

93. (a) Because $\begin{bmatrix} -57 & -13 \end{bmatrix} \begin{bmatrix} w & x \\ y & z \end{bmatrix} = \begin{bmatrix} 23 & 5 \end{bmatrix}$ and

$\begin{bmatrix} 91 & 26 \end{bmatrix} \begin{bmatrix} w & x \\ y & z \end{bmatrix} = \begin{bmatrix} 0 & 13 \end{bmatrix}$, you can solve

$\begin{bmatrix} -57 & -13 \\ 91 & 26 \end{bmatrix} \begin{bmatrix} w & x \\ y & z \end{bmatrix} = \begin{bmatrix} 23 & 5 \\ 0 & 13 \end{bmatrix}$.

(b) WE MISS YOU BIG BOB

Test Yourself   *(page 538)*

1. $\begin{bmatrix} 2 & 1 & 4 & \vdots & 2 \\ 1 & 4 & -1 & \vdots & 0 \\ -1 & 3 & 3 & \vdots & -1 \end{bmatrix}$; 3×4

2. $\begin{bmatrix} 3 & 4 & 2 & \vdots & 4 \\ 2 & 3 & 0 & \vdots & -2 \\ 0 & 2 & -3 & \vdots & -13 \end{bmatrix}$; 3×4 **3.** $\left(\frac{7}{3}a - \frac{10}{3}, -\frac{8}{3}a + \frac{29}{3}, a \right)$

4. $(-12, -16, -6)$ **5.** $(2, -3, 1)$ **6.** $\begin{bmatrix} 2 & 4 \\ 7 & 13 \end{bmatrix}$

7. $\begin{bmatrix} -4 & -8 \\ 13 & 29 \end{bmatrix}$ **8.** $\begin{bmatrix} 1 \\ 11 \end{bmatrix}$ **9.** $\begin{bmatrix} -6 & -10 \\ 15 & 19 \end{bmatrix}$

10. $\begin{bmatrix} 8 & -7 \\ 11 & 19 \end{bmatrix}$ **11.** $\begin{bmatrix} -32 & 20 \\ -32 & -56 \end{bmatrix}$ **12.** $\begin{bmatrix} -5 & -25 \\ 5 & 5 \end{bmatrix}$

13. $\begin{bmatrix} -12 & 16 \\ 1 & 3 \end{bmatrix}$ **14.** $\frac{1}{5} \begin{bmatrix} 4 & 1 \\ 3 & 2 \end{bmatrix}$ **15.** $\begin{bmatrix} 1 & 0 \\ 0 & 1 \end{bmatrix}$

16. $\frac{1}{5} \begin{bmatrix} -9 & 16 & -6 \\ 6 & -9 & 4 \\ 4 & -6 & 1 \end{bmatrix}$ **17.** -17 **18.** -23 **19.** -20

20. $(2, -2, 3)$

21. Matrices will vary. Sample answer:

$\begin{bmatrix} 2 & -2 \\ -2 & 2 \end{bmatrix} \times \begin{bmatrix} 3 & 3 \\ 3 & 3 \end{bmatrix} = \begin{bmatrix} 0 & 0 \\ 0 & 0 \end{bmatrix}$

22. $\frac{21}{2}$ square units   **23.** Collinear   **24.** $-x + y + 3 = 0$
25. $BA = \begin{bmatrix} \$384{,}000 & \$631{,}000 \end{bmatrix}$

BA represents the total value of each product at each warehouse.

Chapter 7

Section 7.1   *(page 549)*

Skills Warm Up   *(page 549)*

1. $\frac{1}{3}x^2 + \frac{1}{6}x$ **2.** $x^2(x + 9)$ **3.** $x + 4$ **4.** $x + 6$
5. (a) 7   (b) $c^2 - 3c + 3$
(c) $x^2 + 2xh + h^2 - 3x - 3h + 3$
6. (a) -4 (b) 10   (c) $3t^2 + 4$ **7.** h **8.** 4
9. Domain: $(-\infty, 0) \cup (0, \infty)$
Range: $(-\infty, 0) \cup (0, \infty)$

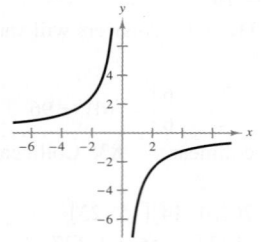

10. Domain: $[-5, 5]$ **11.** Domain: $(-\infty, \infty)$
Range: $[0, 5]$ Range: $[0, \infty)$

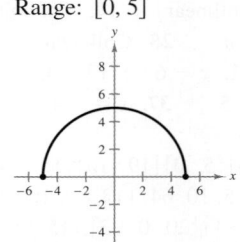 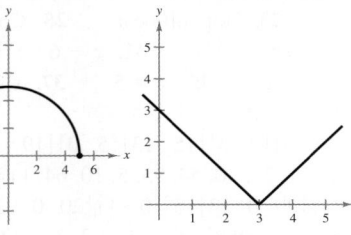

12. Domain: **13.** y is not a function of x.
$(-\infty, 0) \cup (0, \infty)$ **14.** y is a function of x.
Range: $-1, 1$

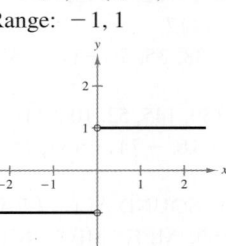

1. (a) 1   (b) 3   **3.** (a) 1   (b) 3

5.

x	1.9	1.99	1.999	2
$f(x)$	8.8	8.98	8.998	?

x	2.001	2.01	2.1
$f(x)$	9.002	9.02	9.2

$\lim\limits_{x \to 2}(2x + 5) = 9$

7.

x	1.9	1.99	1.999	2
$f(x)$	0.2564	0.2506	0.2501	?

x	2.001	2.01	2.1
$f(x)$	0.2499	0.2494	0.2439

$\lim\limits_{x \to 2} \dfrac{x - 2}{x^2 - 4} = \dfrac{1}{4}$

9.

x	-0.1	-0.01	-0.001	0
$f(x)$	0.5132	0.5013	0.5001	?

x	0.001	0.01	0.1
$f(x)$	0.4999	0.4988	0.4881

$\lim\limits_{x \to 0} \dfrac{\sqrt{x + 1} - 1}{x} = 0.5$

11.

x	-4.1	-4.01	-4.001	-4
$f(x)$	2.5	25	250	?

x	-3.999	-3.99	-3.9
$f(x)$	-250	-25	-2.5

The limit does not exist.

13. 6   **15.** −2   **17.** 49   **19.** 4

21. (a) 12   (b) 27   (c) $\frac{1}{3}$

23. (a) 4   (b) 48   (c) 256

25. −1   **27.** 0   **29.** 3   **31.** −2   **33.** $-\frac{3}{4}$ **35.** 2

37. −6   **39.** $-\frac{1}{4}$ **41.** 12   **43.** 2   **45.** $2t − 5$

47. $\frac{1}{6}$ **49.** $\dfrac{1}{2\sqrt{5}}$ **51.** 2   **53.** −1

55. Limit does not exist.   **57.** Limit does not exist.

59. $\displaystyle\lim_{x\to-3^-}\frac{|x+3|}{x+3}=-1,\ \lim_{x\to-3^+}\frac{|x+3|}{x+3}=1$

61.

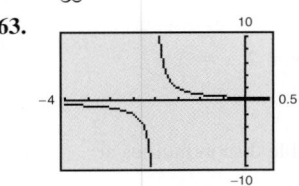

x	0	0.5	0.9	0.99
$f(x)$	−2	−2.67	−10.53	−100.5

x	0.999	0.9999	1
$f(x)$	−1000.5	−10,000.5	Undefined

$-\infty$

63.

x	−3	−2.5	−2.1	−2.01
$f(x)$	−1	−2	−10	−100

x	−2.001	−2.0001	−2
$f(x)$	−1000	−10,000	Undefined

$-\infty$

65. (a) 1   (b) 1   (c) 1

67. (a) 0   (b) 0   (c) 0

69. (a) 3   (b) −3   (c) Limit does not exist.

71.

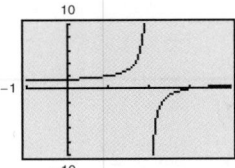

73.

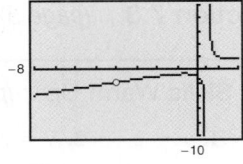

Limit does not exist. $-\frac{17}{9}\approx-1.8889$

75. (a) $25 thousand   (b) 80%

(c) ∞; The cost function increases without bound as x approaches 100 from the left. Therefore, according to the model, it is not possible to remove 100% of the pollutants.

77. (a)

(b) For $x = 0.25$, $A \approx \$2685.06$.

For $x = \frac{1}{365}$, $A \approx \$2717.91$.

(c) $\displaystyle\lim_{x\to0^+}1000(1+0.1x)^{10/x}=1000e\approx\$2718.28$; continuous compounding

Section 7.2   *(page 560)*

Skills Warm Up   *(page 560)*

1. $\dfrac{x+4}{x-8}$ **2.** $\dfrac{x+1}{x-3}$ **3.** $\dfrac{x+2}{2(x-3)}$ **4.** $\dfrac{x-4}{x-2}$

5. $x = 0, -7$ **6.** $x = -5, 1$ **7.** $x = -\frac{2}{3}, -2$

8. $x = 0, 3, -8$ **9.** 13   **10.** −1

1. Continuous; The function is a polynomial.

3. Not continuous $(x \ne \pm 4)$

5. Continuous; The rational function's domain is the set of real numbers.

7. Not continuous $(x \ne 3 \text{ and } x \ne 5)$

9. Not continuous $(x \ne \pm 2)$

11. $(-\infty, 0)$ and $(0, \infty)$; Explanations will vary. There is a discontinuity at $x = 0$, because $f(0)$ is not defined.

13. $(-\infty, -1)$ and $(-1, \infty)$; Explanations will vary. There is a discontinuity at $x = -1$, because $f(-1)$ is not defined.

15. $(-\infty, \infty)$; Explanations will vary.

17. $(-\infty, -1), (-1, 1)$, and $(1, \infty)$; Explanations will vary. There are discontinuities at $x = \pm 1$, because $f(\pm 1)$ is not defined.

19. $(-\infty, \infty)$; Explanations will vary.

21. $(-\infty, 4), (4, 5)$, and $(5, \infty)$; Explanations will vary. There are discontinuities at $x = 4$ and $x = 5$, because $f(4)$ and $f(5)$ are not defined.

23. $(-\infty, 4]$; Explanations will vary.

25. $[0, \infty)$; Explanations will vary.

27. $[-1, 3]$; Explanations will vary.

29. $(-\infty, \infty)$; Explanations will vary.

31. $(-\infty, -1)$ and $(-1, \infty)$; Explanations will vary. There is a discontinuity at $x = -1$, because $f(-1)$ is not defined.

33. $[-3, \infty)$; Explanations will vary.

35. Continuous on all intervals $\left(\dfrac{c}{2}, \dfrac{c}{2}+\dfrac{1}{2}\right)$, where c is an integer. Explanations will vary. There are discontinuities at $x = \dfrac{c}{2}$, where c is an integer, because $\displaystyle\lim_{x\to c}f\left(\dfrac{c}{2}\right)$ does not exist.

37. Continuous on all intervals $(c, c + 1)$, where c is an integer. Explanations will vary. There are discontinuities at $x = c$, where c is an integer, because $\displaystyle\lim_{x\to c^-}f(c)$ does not exist.

39. $(1, \infty)$; Explanations will vary.

CHAPTER 7

41.

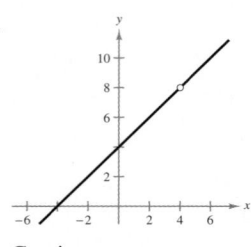

Continuous on
$(-\infty, 4)$ and $(4, \infty)$

43.

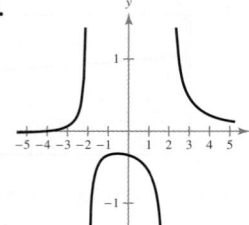

Continuous on $(-\infty, -2)$,
$(-2, 2)$, and $(2, \infty)$

45.

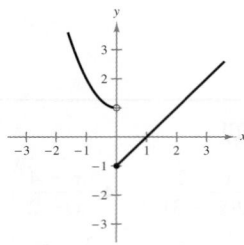

Continuous on $(-\infty, 0)$ and $(0, \infty)$

47. Continuous　　**49.** Nonremovable discontinuity at $x = 2$

51.

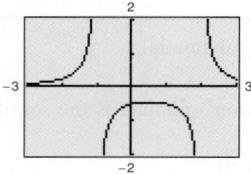

Not continuous at $x = 2$ and $x = -1$ because $f(-1)$ and $f(2)$
are not defined.

53.

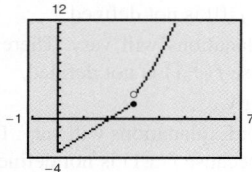

Not continuous at $x = 3$ because $\lim\limits_{x \to 3} f(3)$ does not exist.

55.

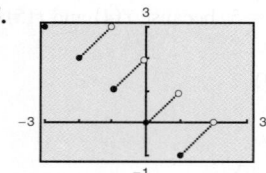

Not continuous at all integers c because $\lim\limits_{x \to c} f(c)$ does not exist.

57. $a = 2$

59.

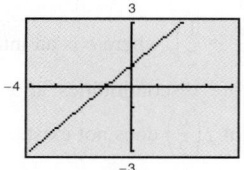

The graph of $f(x) = \dfrac{x^2 + x}{x}$ appears to be continuous on $[-4, 4]$,
but f is not continuous at $x = 0$. Explanations will vary.

61. (a) $[0, 100)$; Explanations will vary.

(b)

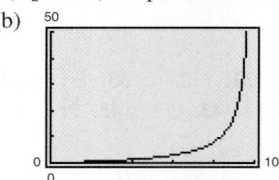

Continuous; Explanations will vary.

(c) $6 million

63.

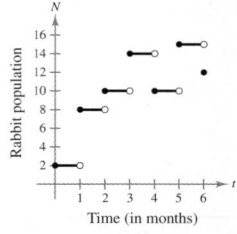

There are nonremovable discontinuities at $t = 1, 2, 3, 4, 5$,
and 6.

65. (a)

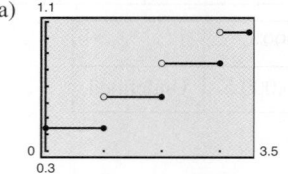

Discontinuities at $x = 1$, $x = 2$, $x = 3$
Explanations will vary.

(b) $0.84

67. (a) The graph has nonremovable discontinuities at
$t = \frac{1}{4}, \frac{1}{2}, \frac{3}{4}, 1, \frac{5}{4}, \dots$

(b) $8448.69

(c) $11,379.17

69. (a)

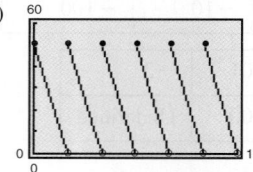

Nonremovable discontinuities at $t = 2, 4, 6, 8, \dots$;
N is not continuous at $t = 2, 4, 6, 8, \dots$.

(b) The company must replenish its inventory every two months.

Section 7.3　*(page 571)*

Skills Warm Up　*(page 571)*

1. $x = 2$　　**2.** $y = 2$　　**3.** $y = -x + 2$

4. $y = 3x - 4$　　**5.** $2x$　　**6.** $3x^2$　　**7.** $\dfrac{1}{x^2}$

8. $2x$　　**9.** $(-\infty, \infty)$　　**10.** $(-\infty, 1) \cup (1, \infty)$

11. $(-\infty, \infty)$　　**12.** $(-\infty, 0) \cup (0, \infty)$

1.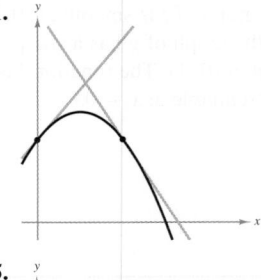

3.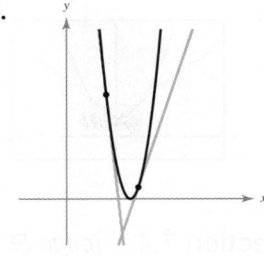

5.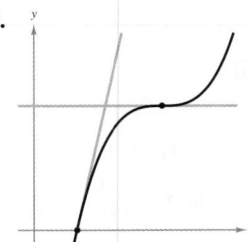

7. $m = 1$ **9.** $m = 0$ **11.** $m = -\frac{1}{3}$

13. 2005: $m \approx 119$

2007: $m \approx 161$

The slope is the rate of change of revenue at the given point in time.

15. $t = 3$: $m \approx 9$

$t = 7$: $m \approx 0$

$t = 10$: $m \approx -10$

The slope is the rate of change of the average temperature at the given point in time.

17. $f'(x) = 0$ **19.** $f'(x) = -3$ **21.** $f'(x) = 4x$

$f'(0) = 0$ $f'(2) = -3$ $f'(2) = 8$

23. $f'(x) = 3x^2 - 1$ **25.** $f'(x) = \dfrac{1}{\sqrt{x}}$

$f'(2) = 11$

$f'(4) = \dfrac{1}{2}$

27. $f(x) = 3$

$f(x + \Delta x) = 3$

$f(x + \Delta x) - f(x) = 0$

$\dfrac{f(x + \Delta x) - f(x)}{\Delta x} = 0$

$\lim\limits_{\Delta x \to 0} \dfrac{f(x + \Delta x) - f(x)}{\Delta x} = 0$

29. $f(x) = -5x$

$f(x + \Delta x) = -5x - 5\Delta x$

$f(x + \Delta x) - f(x) = -5\Delta x$

$\dfrac{f(x + \Delta x) - f(x)}{\Delta x} = -5$

$\lim\limits_{\Delta x \to 0} \dfrac{f(x + \Delta x) - f(x)}{\Delta x} = -5$

31. $g(s) = \dfrac{1}{3}s + 2$

$g(s + \Delta s) = \dfrac{1}{3}s + \dfrac{1}{3}\Delta s + 2$

$g(s + \Delta s) - g(s) = \dfrac{1}{3}\Delta s$

$\dfrac{g(s + \Delta s) - g(s)}{\Delta s} = \dfrac{1}{3}$

$\lim\limits_{\Delta s \to 0} \dfrac{g(s + \Delta s) - g(s)}{\Delta s} = \dfrac{1}{3}$

33. $f(x) = 4x^2 - 5x$

$f(x + \Delta x) = 4x^2 + 8x\Delta x + 4(\Delta x)^2 - 5x - 5\Delta x$

$f(x + \Delta x) - f(x) = 8x\Delta x + 4(\Delta x)^2 - 5\Delta x$

$\dfrac{f(x + \Delta x) - f(x)}{\Delta x} = 8x + 4\Delta x - 5$

$\lim\limits_{\Delta x \to 0} \dfrac{f(x + \Delta x) - f(x)}{\Delta x} = 8x - 5$

35. $h(t) = \sqrt{t - 1}$

$h(t + \Delta t) = \sqrt{t + \Delta t - 1}$

$h(t + \Delta t) - h(t) = \sqrt{t + \Delta t - 1} - \sqrt{t - 1}$

$\dfrac{h(t + \Delta t) - h(t)}{\Delta t} = \dfrac{1}{\sqrt{t + \Delta t - 1} + \sqrt{t - 1}}$

$\lim\limits_{\Delta t \to 0} \dfrac{h(t + \Delta t) - h(t)}{\Delta t} = \dfrac{1}{2\sqrt{t - 1}}$

37. $f(t) = t^3 - 12t$

$f(t + \Delta t) = t^3 + 3t^2\Delta t + 3t(\Delta t)^2 + (\Delta t)^3 - 12t - 12\Delta t$

$f(t + \Delta t) - f(t) = 3t^2\Delta t + 3t(\Delta t)^2 + (\Delta t)^3 - 12\Delta t$

$\dfrac{f(t + \Delta t) - f(t)}{\Delta t} = 3t^2 + 3t\Delta t + (\Delta t)^2 - 12$

$\lim\limits_{\Delta t \to 0} \dfrac{f(t + \Delta t) - f(t)}{\Delta t} = 3t^2 - 12$

39. $f(x) = \dfrac{1}{x + 2}$

$f(x + \Delta x) = \dfrac{1}{x + \Delta x + 2}$

$f(x + \Delta x) - f(x) = \dfrac{-\Delta x}{(x + \Delta x + 2)(x + 2)}$

$\dfrac{f(x + \Delta x) - f(x)}{\Delta x} = \dfrac{-1}{(x + \Delta x + 2)(x + 2)}$

$\lim\limits_{\Delta x \to 0} \dfrac{f(x + \Delta x) - f(x)}{\Delta x} = -\dfrac{1}{(x + 2)^2}$

41. $y = 2x - 2$ **43.** $y = -6x - 3$

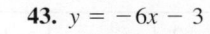
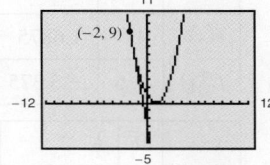

45. $y = \dfrac{x}{4} + 2$ **47.** $y = -x + 2$

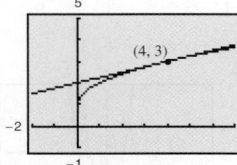
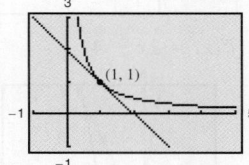

49. $y = -x + 1$ **51.** $y = -9x + 18,\ y = -9x - 18$

53. y is differentiable for all $x \neq -3$.

At $(-3, 0)$ the graph has a node.

55. y is differentiable for all $x \neq 3$.

At $(3, 0)$ the graph has a cusp.

57. y is differentiable for all $x = \pm 2$.

The function is not defined at $x = \pm 2$.

CHAPTER 7

59. $f(x) = -3x + 2$

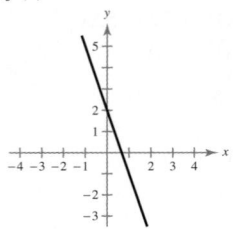

61.

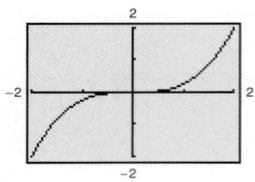

$f'(x) = \frac{3}{4}x^2$

x	-2	$-\frac{3}{2}$	-1	$-\frac{1}{2}$
$f(x)$	-2	-0.8438	-0.25	-0.0313
$f'(x)$	3	1.6875	0.75	0.1875

x	0	$\frac{1}{2}$	1	$\frac{3}{2}$	2
$f(x)$	0	0.0313	0.25	0.8438	2
$f'(x)$	0	0.1875	0.75	1.6875	3

63.

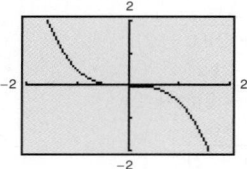

$f'(x) = -\frac{3}{2}x^2$

x	-2	$-\frac{3}{2}$	-1	$-\frac{1}{2}$
$f(x)$	4	1.6875	0.5	0.0625
$f'(x)$	-6	-3.375	-1.5	-0.375

x	0	$\frac{1}{2}$	1	$\frac{3}{2}$	2
$f(x)$	0	-0.0625	-0.5	-1.6875	-4
$f'(x)$	0	-0.375	-1.5	-3.375	-6

65. $f'(x) = 2x - 4$ **67.** $f'(x) = 3x^2 - 3$

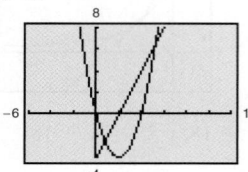 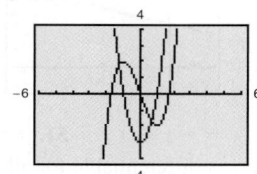

The x-intercept of the derivative indicates a point of horizontal tangency for f.

The x-intercepts of the derivative indicate points of horizontal tangency for f.

69. True   **71.** True

73.

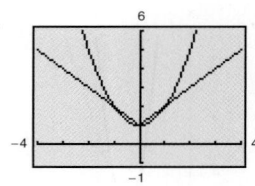

The graph of f is smooth at $(0, 1)$, but the graph of g has a sharp point at $(0, 1)$. The function f is differentiable at $x = 0$.

Section 7.4   *(page 583)*

Skills Warm Up   *(page 583)*

1. (a) 8 (b) 16 (c) $\frac{1}{2}$

2. (a) $\frac{1}{36}$ (b) $\frac{1}{32}$ (c) $\frac{1}{64}$ **3.** $4x(3x^2 + 1)$

4. $\frac{3}{2}x^{1/2}(x^{3/2} - 1)$ **5.** $\frac{1}{4x^{3/4}}$ **6.** $x^2 - \frac{1}{x^{1/2}} + \frac{1}{3x^{2/3}}$

7. $0, -\frac{2}{3}$ **8.** $0, \pm1$ **9.** $-10, 2$ **10.** $-2, 12$

---

1. 0 **3.** $5x^4$ **5.** $9x^2$ **7.** $2x^2$ **9.** 4 **11.** $-3x^2$

13. $8x - 3$ **15.** $-6t + 2$ **17.** $3t^2 - 2$

19. $\frac{2}{3\sqrt[3]{x}}$ **21.** $\frac{16}{3}t^{1/3}$ **23.** $-\frac{8}{x^3} + 4x$

25. Function: $y = \dfrac{2}{7x^4}$

Rewrite: $y = \dfrac{2}{7}x^{-4}$

Differentiate: $y' = -\dfrac{8}{7}x^{-5}$

Simplify: $y' = -\dfrac{8}{7x^5}$

27. Function: $y = \dfrac{1}{(4x)^3}$

Rewrite: $y = \dfrac{1}{64}x^{-3}$

Differentiate: $y' = -\dfrac{3}{64}x^{-4}$

Simplify: $y' = -\dfrac{3}{64x^4}$

29. Function: $y = \dfrac{4}{(2x)^{-5}}$

Rewrite: $y = 128x^5$

Differentiate: $y' = 128(5)x^4$

Simplify: $y' = 640x^4$

31. Function: $y = 6\sqrt{x}$

Rewrite: $y = 6x^{1/2}$

Differentiate: $y' = 6\left(\dfrac{1}{2}\right)x^{-1/2}$

Simplify: $y' = \dfrac{3}{\sqrt{x}}$

33. Function: $y = \dfrac{1}{5\sqrt[5]{x}}$

Rewrite: $y = \dfrac{1}{5}x^{-1/5}$

Differentiate: $y' = \dfrac{1}{5}\left(-\dfrac{1}{5}\right)x^{-6/5}$

Simplify: $y' = -\dfrac{1}{25\sqrt[5]{x^6}}$

35. Function: $y = \sqrt{3x}$

Rewrite: $y = (3x)^{1/2}$

Differentiate: $y' = \left(\dfrac{1}{2}\right)(3x)^{-1/2}(3)$

Simplify: $y' = \dfrac{3}{2\sqrt{3x}}$

37. $\dfrac{3}{2}$ **39.** 8   **41.** -11 **43.** -2

45. (a) $y = 2x - 2$

(b) and (c)

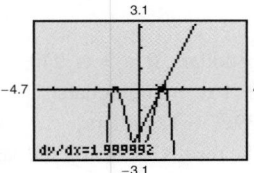

47. (a) $y = \dfrac{8}{15}x + \dfrac{22}{15}$

(b) and (c)

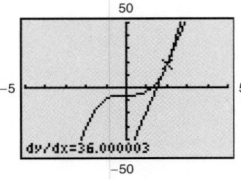

49. (a) $y = 36x - 54$

(b) and (c)

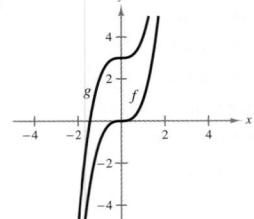

51. $2x + \dfrac{4}{x^2} + \dfrac{6}{x^3}$ **53.** $2x - 2 + \dfrac{8}{x^5}$ **55.** $\dfrac{4}{5x^{1/5}} + 1$

57. $3x^2 + 1$ **59.** $\dfrac{2x^3 - 6}{x^3}$ **61.** $\dfrac{4x^3 - 2x - 10}{x^3}$

63. $(0, -1), \left(-\dfrac{\sqrt{6}}{2}, \dfrac{5}{4}\right), \left(\dfrac{\sqrt{6}}{2}, \dfrac{5}{4}\right)$ **65.** $(-5, -12.5)$

67. (a)

(b) $f'(1) = g'(1) = 3$

(c)

(d) $f' = g' = 3x^2$ for every value of x.

69. $f'(x) = g'(x)$ **71.** $-5f'(x) = g'(x)$

73. (a) 2005: $m \approx 119.2$; 2007: $m \approx 161$

(b) These results are close to the estimates in Exercise 13 in Section 7.3.

(c) The slope of the graph at time t is the rate at which sales are increasing in millions of dollars per year.

75. (a) The men and women who seem to suffer most from migraines are those between 30 and 40 years old. More females than males suffer from migraines. Fewer people whose incomes are greater than or equal to $30,000 suffer from migraines than people whose incomes are less than $10,000.

(b) The derivatives are positive up to approximately 37 years old and negative after about 37 years of age. The percent of adults suffering from migraines increases up to about 37 years old, then decreases. The units of the derivative are percents of adults suffering from migraines per year.

77. $C = 7.75x + 500$

$C' = 7.75$, which equals the marginal cost.

79.

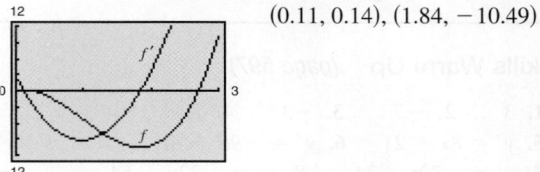

$(0.11, 0.14), (1.84, -10.49)$

81. False. Let $f(x) = x$ and $g(x) = x + 1$.

Quiz Yourself   *(page 586)*

1. 14   **2.** 2   **3.** Limit does not exist.   **4.** 7   **5.** $-\dfrac{1}{8}$

6. 0   **7.** $(-\infty, \infty)$; Explanations will vary.

8. $(-\infty, -2), (-2, \infty)$; There is a discontinuity at $x = -2$ because $f(-2)$ is not defined.

9. $(-\infty, 1), (1, \infty)$; There is a discontinuity at $x = 1$ because $f(1)$ is not defined.

10. $(-\infty, 0), (0, \infty)$; There is a discontinuity at $x = 0$ because $f(0)$ is not defined.

11. $f(x) = -x + 2$

$f(x + \Delta x) = -x - \Delta x + 2$

$f(x + \Delta x) - f(x) = -\Delta x$

$\dfrac{f(x + \Delta x) - f(x)}{\Delta x} = -1$

$\displaystyle\lim_{\Delta x \to 0} \dfrac{f(x + \Delta x) - f(x)}{\Delta x} = -1$

$f'(x) = -1$

$f'(2) = -1$

12. $f(x) = \dfrac{4}{x}$

$f(x + \Delta x) = \dfrac{4}{x + \Delta x}$

$f(x + \Delta x) - f(x) = -\dfrac{4\Delta x}{x(x + \Delta x)}$

$\dfrac{f(x + \Delta x) - f(x)}{\Delta x} = -\dfrac{4}{x(x + \Delta x)}$

$\displaystyle\lim_{\Delta x \to 0} \dfrac{f(x + \Delta x) - f(x)}{\Delta x} = -\dfrac{4}{x^2}$

$f'(x) = -\dfrac{4}{x^2}$

$f'(1) = -4$

13. $f'(x) = 0$ **14.** $f'(x) = 19$ **15.** $f'(x) = -6x$

16. $f'(x) = \dfrac{3}{x^{3/4}}$ **17.** $f'(x) = -\dfrac{8}{x^3}$ **18.** $f'(x) = \dfrac{1}{\sqrt{x}}$

19. $y = -4x - 6$ **20.** $y = x$

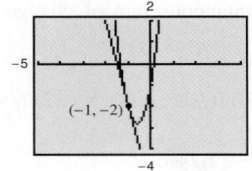

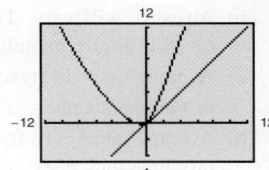

CHAPTER 7

21. (a) $\dfrac{dS}{dt} = -0.40668t^2 + 3.7364t - 4.351$

(b) 2004: \$4.08772/yr
2007: \$1.87648/yr
2008: $-\$0.48732$/yr

Section 7.5   *(page 597)*

Skills Warm Up   *(page 597)*

1. 3   **2.** -7 **3.** -3 **4.** 2.4
5. $y' = 8x - 2$ **6.** $y' = -9t^2 + 4t$
7. $s' = -32t + 24$ **8.** $y' = -32x + 54$
9. $A' = -\frac{3}{5}r^2 + \frac{3}{5}r + \frac{1}{2}$ **10.** $y' = 2x^2 - 4x + 7$
11. $y' = 12 - \dfrac{x}{2500}$ **12.** $y' = 74 - \dfrac{3x^2}{10{,}000}$

---

1. (a) \$10.4 billion/yr   (b) \$7.4 billion/yr
(c) \$6.4 billion/yr   (d) \$16.6 billion/yr
(e) \$11 billion/yr   (f) \$11.96 billion/yr
(g) \$13.67 billion/yr   (h) \$16.38 billion/yr

3.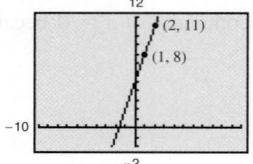
Average rate: 3
Instantaneous rates:
$f'(1) = f'(2) = 3$

5.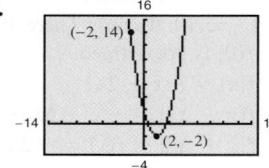
Average rate: -4
Instantaneous rates:
$h'(-2) = -8, h'(2) = 0$

7.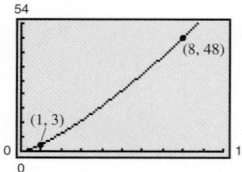
Average rate: $\frac{45}{7}$
Instantaneous rates:
$f'(1) = 4, f'(8) = 8$

9.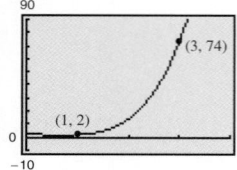
Average rate: $-\frac{1}{4}$
Instantaneous rates:
$f'(1) = -1, f'(4) = -\frac{1}{16}$

11.
Average rate: 36
Instantaneous rates:
$g'(1) = 2, g'(3) = 102$

13. (a) -450
The number of visitors to the park is decreasing at an average rate of 450 thousand people per month from September to December.
(b) Answers will vary. The instantaneous rate of change at $t = 8$ is approximately 0.

15. (a) Average rate: 14 ft/sec
Instantaneous rates: $s'(0) = 30$ ft/sec; $s'(1) = -2$ ft/sec
(b) Average rate: -18 ft/sec
Instantaneous rates: $s'(1) = -2$ ft/sec
$s'(2) = -34$ ft/sec

(c) Average rate: -50 ft/sec
Instantaneous rates: $s'(2) = -34$ ft/sec
$s'(3) = -66$ ft/sec
(d) Average rate: -82 ft/sec
Instantaneous rates: $s'(3) = -66$ ft/sec
$s'(4) = -98$ ft/sec

17. (a) -80 ft/sec
(b) $s'(2) = -64$ ft/sec; $s'(3) = -96$ ft/sec
(c) $\dfrac{\sqrt{555}}{4} \approx 5.89$ sec
(d) $-8\sqrt{555} \approx -188.5$ ft/sec

19. 9800 dollars   **21.** $470 - 0.5x$ dollars, $0 \le x \le 940$
23. $50 - x$ dollars   **25.** $-18x^2 + 16x + 200$ dollars
27. $-4x + 72$ dollars   **29.** $0.0039x^2 + 12$ dollars
31. (a) \$0.58   (b) \$0.60   (c) The results are nearly the same.
33. (a) \$12.96   (b) \$13.00
(c) The results are nearly the same.

35. (a) $P(0) = 117{,}001{,}000$ people
$P(5) = 120{,}622{,}500$ people
$P(10) = 123{,}466{,}000$ people
$P(15) = 125{,}531{,}500$ people
$P(20) = 126{,}819{,}000$ people
$P(25) = 127{,}328{,}500$ people
$P(30) = 127{,}060{,}000$ people
The population is growing from 1980 to 2005. It then begins to decline.

(b) $\dfrac{dP}{dt} = -31.12t + 802.1$

(c) $P'(0) = 802{,}100$ people per year
$P'(5) = 646{,}500$ people per year
$P'(10) = 490{,}900$ people per year
$P'(15) = 335{,}300$ people per year
$P'(20) = 179{,}700$ people per year
$P'(25) = 24{,}100$ people per year
$P'(30) = -131{,}500$ people per year
The rate of growth is decreasing.

37. (a) $TR = -10Q^2 + 160Q$
(b) $(TR)' = MR = -20Q + 160$
(c)

Q	0	2	4	6	8	10
Model	160	120	80	40	0	-40
Table	–	130	90	50	10	-30

Answers will vary.

39. (a) $P = -0.0025x^2 + 2.65x - 25$
(b)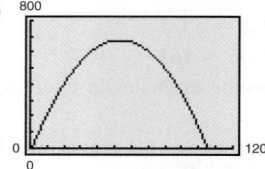
When $x = 300$, slope is positive.
When $x = 700$, slope is negative.
(c) $P'(300) = 1.15$
$P'(700) = -0.85$

41. (a) $C = \dfrac{44{,}250}{x}$

(b) $\dfrac{dC}{dx} = \dfrac{-44{,}250}{x^2}$

This is the rate of change of fuel cost.

(c)

x	10	15	20	25
C	4425.00	2950.00	2212.50	1770.00
dC/dx	-442.5	-196.67	-110.63	-70.80

x	30	35	40
C	1475.00	1264.29	1106.25
dC/dx	-49.17	-36.12	-27.66

(d) The driver who gets 15 mi/gal; Explanations will vary.

43. (a) Average rate of change from 1995 to 2009:

$\dfrac{\Delta p}{\Delta t} = \dfrac{10{,}428.05 - 5117.12}{19 - 5} \approx 379.35$ dollars per year

(b) Average rate of change from 1996 to 2000:

$\dfrac{\Delta p}{\Delta t} = \dfrac{10{,}786.85 - 6448.26}{10 - 6} \approx \$1084.65$

(c) Average rate of change from 1997 to 1999:

$\dfrac{\Delta p}{\Delta t} = \dfrac{11{,}497.12 - 7908.24}{9 - 7} \approx \$1794.44$

(d) The average rate of change from 1997 to 1999 is a better estimate because the data are closer to the year in question.

Section 7.6   *(page 608)*

Skills Warm Up   *(page 608)*

1. $2(3x^2 + 7x + 1)$ **2.** $4x^2(6 - 5x^2)$

3. $8x^2(x^2 + 2)^3 + (x^2 + 4)$

4. $(2x)(2x + 1)[2x + (2x + 1)^3]$

5. $\dfrac{23}{(2x + 7)^2}$ **6.** $-\dfrac{x^2 + 8x + 4}{(x^2 - 4)^2}$

7. $-\dfrac{2(x^2 + x - 1)}{(x^2 + 1)^2}$ **8.** $\dfrac{4(3x^4 - x^3 + 1)}{(1 - x^4)^2}$

9. $\dfrac{4x^3 - 3x^2 + 3}{x^2}$ **10.** $\dfrac{x^2 - 2x + 4}{(x - 1)^2}$

11. 11 **12.** 0 **13.** $-\frac{1}{4}$ **14.** $\frac{17}{4}$

1. $f'(x) = (2x - 3)(-5) + (2)(1 - 5x) = -20x + 17$

3. $f'(x) = (6x - x^2)(3) + (6 - 2x)(4 + 3x)$
$= -9x^2 + 28x + 24$

5. $f'(x) = x(2x) + 1(x^2 + 3) = 3x^2 + 3$

7. $h'(x) = \left(\dfrac{2}{x} - 3\right)(2x) + \left(-\dfrac{2}{x^2}\right)(x^2 + 7) = -\dfrac{14}{x^2} - 6x + 2$

9. $g'(x) = (x^2 - 4x + 3)(1) + (2x - 4)(x - 2)$
$= 3x^2 - 12x + 11$

11. $h'(x) = \dfrac{(x - 5)(1) - (x)(1)}{(x - 5)^2} = -\dfrac{5}{(x - 5)^2}$

13. $f'(t) = \dfrac{(3t + 1)(4t) - (2t^2 - 3)3}{(3t + 1)^2} = \dfrac{6t^2 + 4t + 9}{(3t + 1)^2}$

15. $f'(t) = \dfrac{(t + 4)(2t) - (t^2 - 1)(1)}{(t + 4)^2} = \dfrac{t^2 + 8t + 1}{(t + 4)^2}$

17. $f'(x) = \dfrac{(2x - 1)(2x + 6) - (x^2 + 6x + 5)(2)}{(2x - 1)^2}$
$= \dfrac{2x^2 - 2x - 16}{(2x - 1)^2}$

19. $f'(x) = \dfrac{(3x - 1)\left(-\dfrac{2}{x^2}\right) - \left(6 + \dfrac{2}{x}\right)(3)}{(3x - 1)^2}$
$= \dfrac{-18x^2 - 12x + 2}{x^2(3x - 1)^2}$

21. Function: $f(x) = \dfrac{x^3 + 6x}{3}$

Rewrite: $f(x) = \dfrac{x^3}{3} + 2x$

Differentiate: $f'(x) = x^2 + 2$

Simplify: $f'(x) = x^2 + 2$

23. Function: $y = \dfrac{x^2 + 2x}{3}$

Rewrite: $y = \dfrac{1}{3}(x^2 + 2x)$

Differentiate: $y' = \dfrac{1}{3}(2x + 2)$

Simplify: $y' = \dfrac{2}{3}(x + 1)$

25. Function: $y = \dfrac{7}{3x^3}$

Rewrite: $y = \dfrac{7}{3}x^{-3}$

Differentiate: $y' = -7x^{-4}$

Simplify: $y' = -\dfrac{7}{x^4}$

27. Function: $y = \dfrac{4x^2 - 3x}{8\sqrt{x}}$

Rewrite: $y = \dfrac{1}{2}x^{3/2} - \dfrac{3}{8}x^{1/2},\ x \neq 0$

Differentiate: $y' = \dfrac{3}{4}x^{1/2} - \dfrac{3}{16}x^{-1/2}$

Simplify: $y' = \dfrac{3}{4}\sqrt{x} - \dfrac{3}{16\sqrt{x}}$

29. Function: $y = \dfrac{x^2 - 4x + 3}{2(x - 1)}$

Rewrite: $y = \dfrac{1}{2}(x - 3),\ x \neq 1$

Differentiate: $y' = \dfrac{1}{2}(1),\ x \neq 1$

Simplify: $y' = \dfrac{1}{2},\ x \neq 1$

31. $10x^4 + 12x^3 - 3x^2 - 18x - 15$; Product Rule

33. $\dfrac{x^4 - 6x^2 - 4x - 3}{(x^2 - 1)^2}$; Quotient Rule   **35.** 1; Power Rule

37. $12t^2(2t^3 - 1)$; Product Rule

39. $\dfrac{3s^2 - 2s - 5}{2s^{3/2}}$; Quotient Rule

41. $\dfrac{12x^2 + 12x - 2}{(4x + 2)^2}$; Quotient Rule

43. $\dfrac{2x^3 + 11x^2 - 8x - 17}{(x + 4)^2}$; Quotient Rule

45. $y = 3x + 3$ **47.** $y = -7x + 4$

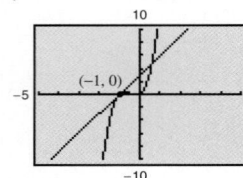   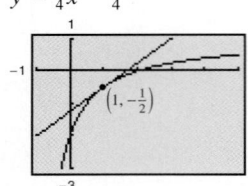

49. $y = \frac{3}{4}x - \frac{5}{4}$ **51.** $y = \frac{31}{5}x + \frac{26}{5}$

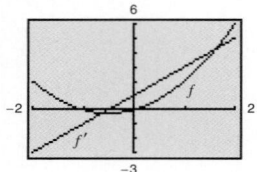   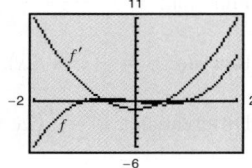

53. $(0, 0)$, $(2, 4)$ **55.** $(0, 0)$, $\left(\sqrt[3]{-4}, -2.117\right)$

57.   **59.**

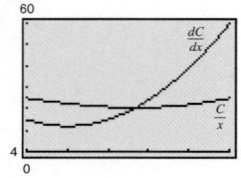

61. $-\$1.87$/unit   **63.** 31.55 bacteria/hr

65. (a) -0.480/wk   (b) 0.120/wk   (c) 0.015/wk

Each rate in parts (a), (b), and (c) is the rate at which the level of oxygen in the pond is changing at that particular time.

67. (a)

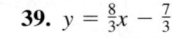

(b) At $x = 6.683$, $\dfrac{C}{x} = \dfrac{dC}{dx} \approx 20.50$.

So, the point of intersection is $(6.683, 20.50)$. At this point the average cost is at a minimum.

69. (a) -38.125 (b) -10.37 (c) -3.80

Increasing the order size reduces the cost per item; Choices and explanations will vary.

71. $f'(2) = 0$ **73.** $f'(2) = 14$ **75.** Answers will vary.

Section 7.7   *(page 618)*

Skills Warm Up   *(page 618)*

1. $(1 - 5x)^{2/5}$ **2.** $(2x - 1)^{3/4}$ **3.** $(4x^2 + 1)^{-1/2}$

4. $(x - 6)^{-1/3}$ **5.** $x^{1/2}(1 - 2x)^{-1/3}$

6. $(2x)^{-1}(3 - 7x)^{3/2}$ **7.** $(x - 2)(3x^2 + 5)$

8. $(x - 1)(5\sqrt{x} - 1)$ **9.** $(x^2 + 1)^2(4 - x - x^3)$

10. $(3 - x^2)(x - 1)(x^2 + x + 1)$

$y = f(g(x))$	$u = g(x)$	$y = f(u)$
1. $y = (6x - 5)^4$	$u = 6x - 5$	$y = u^4$
3. $y = \sqrt{5x - 2}$	$u = 5x - 2$	$y = \sqrt{u}$
5. $y = (3x + 1)^{-1}$	$u = 3x + 1$	$y = u^{-1}$

7. $\dfrac{dy}{du} = 2u$ **9.** $\dfrac{dy}{du} = \dfrac{1}{2\sqrt{u}}$

$\dfrac{du}{dx} = 4$ $\dfrac{du}{dx} = -2x$

$\dfrac{dy}{dx} = 32x + 56$ $\dfrac{dy}{dx} = -\dfrac{x}{\sqrt{3 - x^2}}$

11. $\dfrac{dy}{du} = \dfrac{2}{3u^{1/3}}$

$\dfrac{du}{dx} = 20x^3 - 2$

$\dfrac{dy}{dx} = \dfrac{40x^3 - 4}{3\sqrt[3]{5x^4 - 2x}}$

13. c   **15.** b   **17.** a   **19.** c   **21.** $6(2x - 7)^2$

23. $\dfrac{3\sqrt{5x - x^2}\,(5 - 2x)}{2}$ **25.** $6x(6 - x^2)(2 - x^2)$

27. $\dfrac{1}{2\sqrt{t + 1}}$ **29.** $\dfrac{4t + 5}{2\sqrt{2t^2 + 5t + 2}}$ **31.** $\dfrac{6x}{(9x^2 + 4)^{2/3}}$

33. $\dfrac{54}{(2 - 9x)^4}$ **35.** $\dfrac{-x}{(25 + x^2)^{3/2}}$

37. $y = 216x - 378$ **39.** $y = \frac{8}{3}x - \frac{7}{3}$

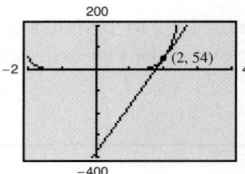   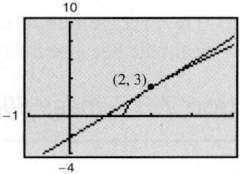

41. $y = x - 1$

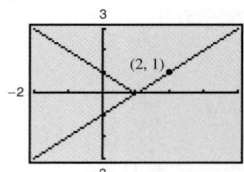

43. $f'(x) = \dfrac{1 - 3x^2 - 4x^{3/2}}{2\sqrt{x}(x^2 + 1)^2}$ **45.** $f'(x) = -\dfrac{\sqrt{(x + 1)/x}}{2x(x + 1)}$

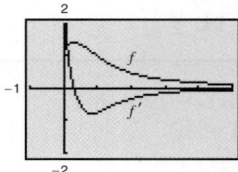   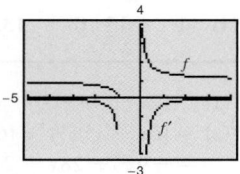

The zero of $f'(x)$ corresponds to the point on the graph of $f(x)$ where the tangent line is horizontal. $f'(x)$ has no zeros.

In Exercises 47–61, the differentiation rule(s) used may vary. A sample answer is provided.

47. $\dfrac{2x}{(4 - x^2)^2}$; Chain Rule   **49.** $\dfrac{8}{(t + 2)^3}$; Chain Rule

51. $-6(3x^2 - x - 3)$; Product Rule

53. $-\dfrac{1}{2(x+2)^{3/2}}$; Power Rule

55. $27(x-3)^2(4x-3)$; Product Rule and Chain Rule

57. $\dfrac{3(x+1)}{\sqrt{2x+3}}$; Product Rule and Chain Rule

59. $\dfrac{t(5t-8)}{2\sqrt{t-2}}$; Product Rule and Chain Rule

61. $\dfrac{2(6-5x)(5x^2-12x+5)}{(x^2-1)^3}$; Chain Rule and Quotient Rule

63. $y=\frac{8}{3}t+4$

65. $y=\frac{1}{2}x+1$

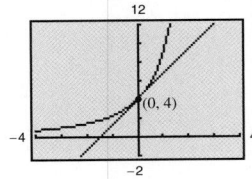

67. $y=-6t-14$

69. $y=-2x+7$

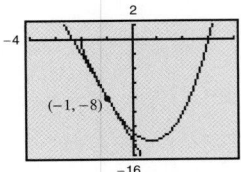
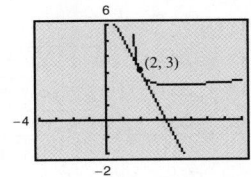

71. (a) \$74.00 per 1%   (b) \$81.59 per 1%   (c) \$89.94 per 1%

73. (a) $V=\dfrac{10,000}{\sqrt{t+1}}$

(b) About $-\$1767.77$ per year   (c) $-\$625.00$ per year

75. (a) $r'(t)=\dfrac{11.4228t^3-218.376t^2+1352.28t-2706}{2\sqrt{2.8557t^4-72.792t^3+676.14t^2-2706t+4096}}$;

Chain Rule

(b)

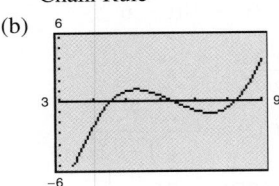

(c) $t=3$ (d) $t\approx4.52,\ t\approx6.36,\ t\approx8.24$

Review Exercises   *(page 624)*

1.

x	0.9	0.99	0.999	1	1.001	1.01	1.1
$f(x)$	0.6	0.96	0.996	?	1.004	1.04	1.4

$\lim\limits_{x\to1}(4x-3)=1$

3.

x	-0.1	-0.01	-0.001	0
$f(x)$	35.71	355.26	3550.71	?

x	0.001	0.01	0.1
$f(x)$	-3550.31	-354.85	-35.30

$\lim\limits_{x\to0}\dfrac{\sqrt{x+6}-6}{x}$ does not exist.

5. 8   **7.** 7   **9.** $-\frac{2}{5}$ **11.** Limit does not exist.

13. $-\frac{1}{4}$ **15.** $-\infty$ **17.** Limit does not exist.

19. 5   **21.** $3x^2-1$

23. $(-\infty,\infty)$; For any c on the real number line, $F(c)$ is defined, $\lim\limits_{x\to c}f(x)$ exists, and $\lim\limits_{x\to c}f(x)=f(c)$.

25. $(-\infty,-4)$ and $(-4,\infty)$; $f(-4)$ is undefined.

27. $(-\infty,-1)$ and $(-1,\infty)$; $f(-1)$ is undefined.

29. Continuous on all intervals $(c,c+1)$, where c is an integer; $\lim\limits_{x\to c}f(c)$ does not exist.

31. $(-\infty,0)$ and $(0,\infty)$; $\lim\limits_{x\to0}f(x)$ does not exist.   **33.** $a=2$

35. (a)

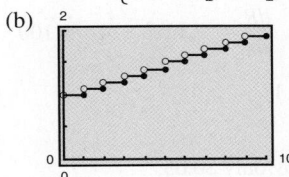

Explanations will vary. The function is defined for all values of x greater than zero. The function is discontinuous when $x=5$, $x=10$, and $x=15$.

(b) \$49.90

37. (a) $C(t)=\begin{cases}1+0.1[\![t]\!], & t>0,\ t\text{ not an integer}\\1+0.1[\![t-1]\!], & t>0,\ t\text{ an integer}\end{cases}$

(b)

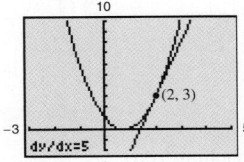

C is not continuous at $t=1,2,3,\ldots$.

39. -2 **41.** 0

43. Answers will vary. Sample answer:

$t=1$: slope ≈65 thousand visitors/month; The number of visitors to the national park is increasing at about 65,000 visitors per month in January.

$t=8$: slope ≈0 visitors/month; The number of visitors to the national park is neither increasing nor decreasing in August.

$t=12$: slope ≈-1000 thousand visitors/month; The number of visitors to the national park is decreasing at about 1,000,000 visitors per month in December.

45. -3 **47.** -2 **49.** $\frac{1}{4}$ **51.** -1 **53.** 9

55. $-x+2$ **57.** $\dfrac{1}{2\sqrt{x-5}}$ **59.** $-\dfrac{5}{x^2}$

61. All values except $x=1$; The function is not defined at $x=1$.

63. All values except $x=0$; A function is not differentiable at a discontinuity.

65. 0   **67.** $3x^2$ **69.** $8x$ **71.** $\dfrac{8x^3}{5}$ **73.** $8x^3+6x$

75. $2x+6$ **77.** -0.125 **79.** 5

81. (a) $y=5x-7$ **83.** (a) $y=x-1$

(b) and (c)   (b) and (c)

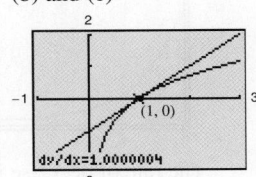

85. (a)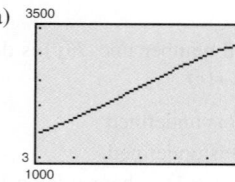

(b) 2004: $m \approx 290$
2007: $m \approx 320$

(c) The slope shows the rate at which sales were increasing or decreasing in a particular year.

87.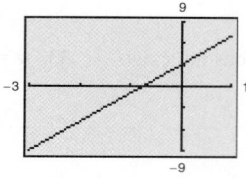

Average rate of change: 4
Instantaneous rate of change at $t = -3$: 4
Instantaneous rate of change at $t = 1$: 4

89.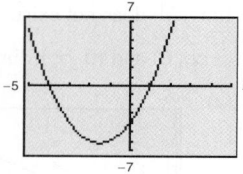

Average rate of change: 4
Instantaneous rate of change at $x = 0$: 3
Instantaneous rate of change at $x = 1$: 5

91. (a) -24 ft/sec   (b) $t = 1$: -8 ft/sec; $t = 3$: -72 ft/sec
(c) 5.14 sec   (d) -140.5 ft/sec

93. $\dfrac{dC}{dx} = 320$ **95.** $\dfrac{dC}{dx} = \dfrac{1.275}{\sqrt{x}}$

97. $\dfrac{dR}{dx} = -1.2x + 150$ **99.** $\dfrac{dR}{dx} = -12x^2 + 4x + 100$

101. $\dfrac{dP}{dx} = -0.0006x^2 + 12x - 1$

103. (a) \$9.95   (b) \$10
(c) Parts (a) and (b) differ by only \$0.05.

In Exercises 105–125, the differentiation rule(s) used may vary. A sample answer is provided.

105. $15x^2(1 - x^2)$; Power Rule

107. $16x^3 - 33x^2 + 12x$; Product Rule

109. $\dfrac{3}{(x + 3)^2}$
Quotient Rule

111. $\dfrac{2(3 + 5x - 3x^2)}{(x^2 + 1)^2}$
Quotient Rule

113. $30x(5x^2 + 2)^2$
Chain Rule

115. $-\dfrac{1}{(x + 1)^{3/2}}$
Quotient Rule

117. $\dfrac{2x^2 + 1}{\sqrt{x^2 + 1}}$
Product Rule

119. $80x^4 - 24x^2 + 1$
Product Rule

121. $18x^5(x + 1)(2x + 3)^2$
Chain Rule

123. $x(x - 1)^4(7x - 2)$
Product Rule

125. $\dfrac{3(9t + 5)}{2\sqrt{3t + 1}(1 - 3t)^3}$
Quotient Rule

127. (a) $t = 1$: -6.63 $t = 3$: -6.5
       $t = 5$: -4.33 $t = 10$: -1.36
(b)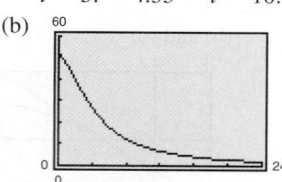

The rate of decrease is approaching zero.

Test Yourself   *(page 628)*

1. -1 **2.** Limit does not exist.   **3.** 2   **4.** $\frac{1}{6}$

5. $(-\infty, \infty)$; A polynomial function is continuous at every real number.

6. $(-\infty, 4)$ and $(4, \infty)$; Explanations will vary. There is a discontinuity at $x = 4$ because $f(4)$ is not defined.

7. $(-\infty, 5]$; Explanations will vary.

8. $(-\infty, \infty)$; Explanations will vary.

9. $f(x) = x^2 + 1$
$f(x + \Delta x) = x^2 + 2x\Delta x + \Delta x^2 + 1$
$f(x + \Delta x) - f(x) = 2x\Delta x + \Delta x^2$
$\dfrac{f(x + \Delta x) - f(x)}{\Delta x} = 2x + \Delta x$
$\lim\limits_{\Delta x \to 0} \dfrac{f(x + \Delta x) - f(x)}{\Delta x} = 2x$
$f'(x) = 2x$
$f'(2) = 4$

10. $f(x) = \sqrt{x} - 2$
$f(x + \Delta x) = \sqrt{x + \Delta x} - 2$
$f(x + \Delta x) - f(x) = \sqrt{x + \Delta x} - \sqrt{x}$
$\dfrac{f(x + \Delta x) - f(x)}{\Delta x} = \dfrac{1}{\sqrt{x + \Delta x} + \sqrt{x}}$
$\lim\limits_{\Delta x \to 0} \dfrac{f(x + \Delta x) - f(x)}{\Delta x} = \dfrac{1}{2\sqrt{x}}$
$f'(x) = \dfrac{1}{2\sqrt{x}}$
$f'(4) = \dfrac{1}{4}$

11. $f'(t) = 3t^2 + 2$ **12.** $f'(x) = 8x - 8$

13. $f'(x) = \dfrac{3\sqrt{x}}{2}$ **14.** $f'(x) = 3x^2 + 10x + 6$

15. $f'(x) = \dfrac{9}{x^4}$ **16.** $f'(x) = \dfrac{5 + x}{2\sqrt{x}} + \sqrt{x}$

17. $f'(x) = 36x^3 + 48x$ **18.** $f'(x) = -\dfrac{1}{\sqrt{1 - 2x}}$

19. $f'(x) = \dfrac{(10x + 1)(5x - 1)^2}{x^2} = 250x - 75 + \dfrac{1}{x^2}$

20. $y = 2x - 2$

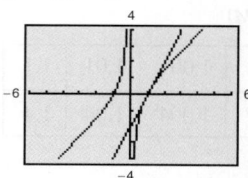

21. (a) \$18.69 billion/yr
(b) 2005: \$10.50 billion/yr; 2008: \$14.95 billion/yr
(c) The annual sales of CVS Caremark from 2005 to 2008 increased on average by about \$18.69 billion/yr, and the instantaneous rates of change for 2005 and 2008 are \$10.50 billion/yr and \$14.95 billion/yr, respectively.

22. (a) $P = -0.016x^2 + 1460x - 715,000$ (b) \$1437.60

Chapter 8

Section 8.1   *(page 635)*

Skills Warm Up   *(page 635)*

1. $t = 0, \frac{3}{2}$ **2.** $t = -2, 7$ **3.** $t = -2, 10$

4. $t = \dfrac{9 \pm 3\sqrt{10{,}249}}{32}$ **5.** $\dfrac{dy}{dx} = 6x^2 + 14x$

6. $\dfrac{dy}{dx} = 8x^3 + 18x^2 - 10x - 15$

7. $\dfrac{dy}{dx} = \dfrac{2x(x+7)}{(2x+7)^2}$ **8.** $\dfrac{dy}{dx} = -\dfrac{6x^2 + 10x + 15}{(2x^2-5)^2}$

9. Domain: $(-\infty, \infty)$ **10.** Domain: $[7, \infty)$
   Range: $[-4, \infty)$ Range: $[0, \infty)$

1. 0   **3.** 2   **5.** $2t - 8$ **7.** $\dfrac{9}{2t^4}$

9. $18(2 - x^2)(5x^2 - 2)$ **11.** $\dfrac{4}{(x-1)^3}$ **13.** $60x^2 - 72x$

15. $120x + 360$ **17.** $-\dfrac{9}{2x^5}$ **19.** 260   **21.** $-\dfrac{1}{648}$

23. 12   **25.** $4x$ **27.** $\dfrac{1}{\sqrt{x-1}}$ **29.** $12x^2 + 4$

31. $f''(x) = 6(x - 3) = 0$ at $x = 3$.

33. $f''(x) = \dfrac{x(2x^2 - 3)}{(x^2-1)^{3/2}} = 0$ at $x = \pm\dfrac{\sqrt{6}}{2}$.

35. (a) $s(t) = -16t^2 + 144t$ (b) $s(3) = 288$ ft
   $v(t) = -32t + 144$ $v(3) = 48$ ft/sec
   $a(t) = -32$ $a(3) = -32$ ft/sec²

(c) 4.5 sec; 324 ft

(d) $v(9) = -144$ ft/sec, which is the same speed as the initial velocity.

37.

t	0	10	20	30	40	50	60
$\dfrac{ds}{dt}$	0	45	60	67.5	72	75	77.1
$\dfrac{d^2s}{dt^2}$	9	2.25	1	0.56	0.36	0.25	0.18

As time increases, velocity increases and acceleration decreases.

39. (a)

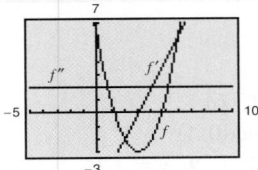

(b) The degree decreases by 1 for each successive derivative.

(c)

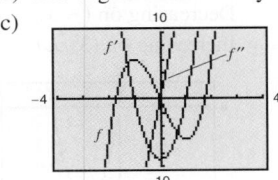

(d) The degree decreases by 1 for each successive derivative.

41. (a) $y = -68.991t^3 + 1208.34t^2 - 5445.4t + 10{,}145$

(b) $y'(t) = -206.973t^2 + 2416.68t - 5445.4$
   $y''(t) = -413.946t + 2416.68$

(c) $y'(t) > 0$ on $[5, 8]$

(d) 2005 ($t \approx 5.84$)

43. False. The Product Rule is
$[f(x)g(x)]' = f(x)g'(x) + g(x)f'(x).$

45. Answers will vary.

Section 8.2   *(page 642)*

Skills Warm Up   *(page 642)*

1. $y = x^2 - 2x$ **2.** $y = \dfrac{x - 3}{4}$

3. $y = 1, x \neq -6$ **4.** $y = -4, x \neq \pm\sqrt{3}$
5. $y = \pm\sqrt{5 - x^2}$ **6.** $y = \pm\sqrt{6 - x^2}$ **7.** $\frac{8}{3}$
8. $-\frac{1}{2}$ **9.** $\frac{5}{7}$

1. $-\dfrac{y}{x}$ **3.** $-\dfrac{x}{y}$ **5.** $\dfrac{1 - xy^2}{x^2 y}$ **7.** $\dfrac{y}{8y - x}$ **9.** 0

11. $-\dfrac{1}{10y - 2}$ **13.** 0   **15.** $-\dfrac{1}{4}$ **17.** $\dfrac{1}{2}$ **19.** -1

21. $-\frac{5}{4}$ **23.** $\frac{1}{4}$ **25.** $\frac{1}{3}$ **27.** 3   **29.** 0   **31.** 2

33. $\dfrac{1}{2y}, -\dfrac{1}{2}$

35. At $(8, 6)$: $y = -\frac{4}{3}x + \frac{50}{3}$
   At $(-6, 8)$: $y = \frac{3}{4}x + \frac{25}{2}$

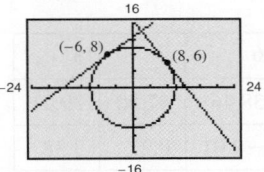

37. At $\left(1, \sqrt{5}\right)$: $15x - 2\sqrt{5}y - 5 = 0$
   At $\left(1, -\sqrt{5}\right)$: $15x + 2\sqrt{5}y - 5 = 0$

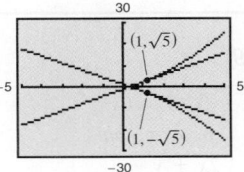

39. At $(0, 2)$: $y = 2$
   At $(2, 0)$: $x = 2$

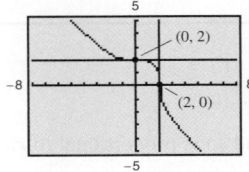

41. At $(-2, 1)$: $y = \frac{1}{2}x + 2$
At $\left(6, \frac{1}{5}\right)$: $y = -0.06x + 0.56$

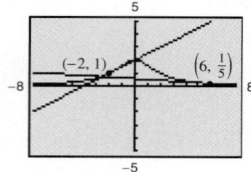

43. $-\dfrac{2}{p^2(0.00003x^2 + 0.1)}$　**45.** $-\dfrac{4xp}{2p^2 + 1}$

47. (a) -2

(b)

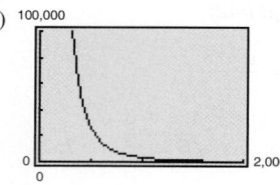

As more labor is used, less capital is available.
As more capital is used, less labor is available.

49. (a)

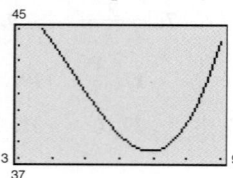

The number of cases of HIV/AIDS decreases from 2004 through 2007, then begins to increase.

(b) 2005

(c)

t	4	5	6	7	8
y	44.11	41.06	38.46	37.50	39.21
y'	-2.95	-3.00	-2.01	0.22	3.38

2005

Section 8.3　*(page 649)*

Skills Warm Up　*(page 649)*

1. $A = \pi r^2$　**2.** $V = \frac{4}{3}\pi r^3$　**3.** $S = 6s^2$
4. $V = s^3$　**5.** $V = \frac{1}{3}\pi r^2 h$　**6.** $A = \frac{1}{2}bh$
7. $-\dfrac{x}{y}$　**8.** $\dfrac{2x - 3y}{3x}$　**9.** $-\dfrac{2x + y}{x + 2}$
10. $-\dfrac{y^2 - y + 1}{2xy - 2y - x}$

1. (a) $\frac{3}{4}$　(b) 20　**3.** (a) $-\frac{5}{8}$　(b) $\frac{3}{2}$
5. (a) 36π in.^{2}/min　(b) 144π in.^{2}/min
7. If $\dfrac{dr}{dt}$ is constant, $\dfrac{dA}{dt} = 2\pi r \dfrac{dr}{dt}$ and so is proportional to r.
9. (a) $\dfrac{5}{2\pi}$ ft/min　(b) $\dfrac{5}{8\pi}$ ft/min
11. (a) 112.5 dollars/wk　(b) 7500 dollars/wk
(c) 7387.5 dollars/wk

13. (a) 9 cm^3/sec　(b) 900 cm^3/sec
15. (a) -18 in./sec　(b) 0 in./sec
(c) 6 in./sec　(d) 18 in./sec
17. -10.4 ft/sec; As $x \to 0$, $\dfrac{dx}{dt}$ increases.　**19.** 300 mi/h
21. (a) -750 mi/h　(b) 20 min
23. About 37.7 ft^3/min　**25.** 4 units/wk

Quiz Yourself　*(page 651)*

1. $6x - 2$　**2.** $\dfrac{4}{9\sqrt[3]{(x - 2)^7}}$　**3.** $6(x^2 + 1)(5x^2 + 1)$
4. $-\dfrac{60}{(2x + 5)^3}$　**5.** $-\dfrac{1}{32}$　**6.** -120　**7.** -96
8. 864 ft; 48 ft/sec; -32 ft/sec^2　**9.** $-\frac{2}{3}x + \frac{1}{3}$
10. $6x^2\sqrt{y}$　**11.** $-\dfrac{y - 1}{x - 1}$　**12.** $\dfrac{4xy}{3y^2 - 2x^2 + 1}$
13. $y = 2x + 1$　**14.** 2　**15.** $-\dfrac{4\sqrt{3}}{9}$

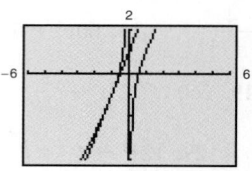

16. (a) \$190 per week　(b) \$20,000 per week
(c) \$19,810 per week

Section 8.4　*(page 659)*

Skills Warm Up　*(page 659)*

1. $x = 0, x = 8$　**2.** $x = 0, x = 24$　**3.** $x = \pm 5$
4. $x = 0$　**5.** $(-\infty, 3) \cup (3, \infty)$　**6.** $(-\infty, 1)$
7. $(-\infty, -2) \cup (-2, 5) \cup (5, \infty)$　**8.** $\left(-\sqrt{3}, \sqrt{3}\right)$
9. $x = -2$: -6　**10.** $x = -2$: 60
　　$x = 0$: 2　　　　$x = 0$: -4
　　$x = 2$: -6　　　$x = 2$: 60
11. $x = -2$: $-\frac{1}{3}$　**12.** $x = -2$: $\frac{1}{18}$
　　$x = 0$: 1　　　　$x = 0$: $-\frac{1}{8}$
　　$x = 2$: 5　　　　$x = 2$: $-\frac{3}{2}$

1. Increasing on $(-\infty, -1)$
Decreasing on $(-1, \infty)$
3. Increasing on $(-1, 0)$ and $(1, \infty)$
Decreasing on $(-\infty, -1)$ and $(0, 1)$
5. $x = \frac{3}{4}$　**7.** $x = 0, x = -3$　**9.** $x = \pm 2$
11. No critical numbers　**13.** Critical number: $x = 3$
Increasing on $(-\infty, \infty)$　Decreasing on $(-\infty, 3)$
Increasing on $(3, \infty)$

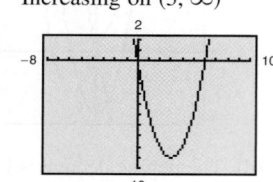

15. Critical number: $x = 1$
Increasing on $(-\infty, 1)$
Decreasing on $(1, \infty)$

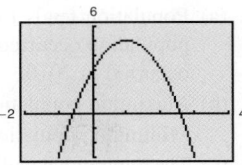

17. Critical numbers:
$x = -1, x = -\frac{5}{3}$
Increasing on $\left(-\infty, -\frac{5}{3}\right)$
and $(-1, \infty)$
Decreasing on $\left(-\frac{5}{3}, -1\right)$

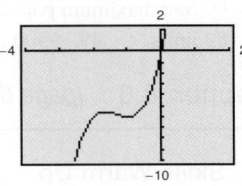

19. Critical numbers:
$x = 0, x = \frac{3}{2}$
Decreasing on $\left(-\infty, \frac{3}{2}\right)$
Increasing on $\left(\frac{3}{2}, \infty\right)$

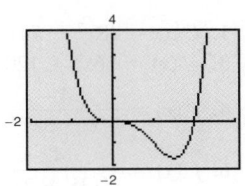

21. Critical number: $x = -2$
Decreasing on $(-\infty, -2)$
Increasing on $(-2, \infty)$

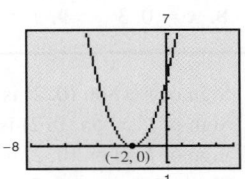

23. Critical number: $x = 1$
Increasing on $(-\infty, 1)$
Decreasing on $(1, \infty)$

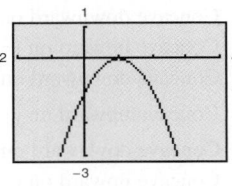

25. Critical number: $x = 0$
Increasing on $(-\infty, 0)$
and $(0, \infty)$

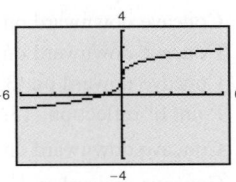

27. Critical numbers:
$x = -1, x = 1$
Decreasing on $(-\infty, -1)$
Increasing on $(1, \infty)$

29. Critical number: $x = -2$
Increasing on $(-\infty, -2)$
and $(-2, \infty)$

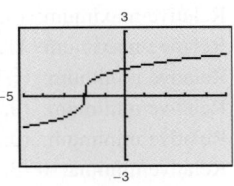

31. Critical numbers:
$x = -1, x = -\frac{2}{3}$
Decreasing on $\left(-1, -\frac{2}{3}\right)$
Increasing on $\left(-\frac{2}{3}, \infty\right)$

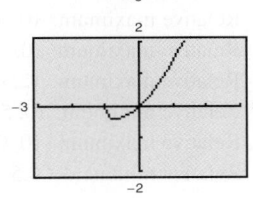

33. Critical numbers:
$x = -3, x = 3$
Decreasing on $(-\infty, -3)$
and $(3, \infty)$
Increasing on $(-3, 3)$

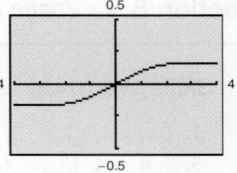

35. No critical numbers
Discontinuity: $x = 5$
Decreasing on $(-\infty, 5)$
and $(5, \infty)$

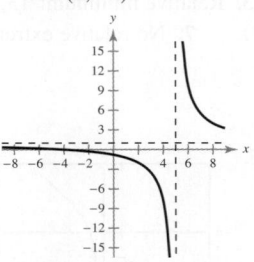

37. No critical numbers
Discontinuities: $x = \pm 4$
Increasing on $(-\infty, -4)$,
$(-4, 4)$, and $(4, \infty)$

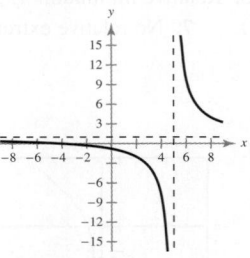

39. Critical number: $x = 0$
Discontinuity: $x = 0$
Increasing on $(-\infty, 0)$
Decreasing on $(0, \infty)$

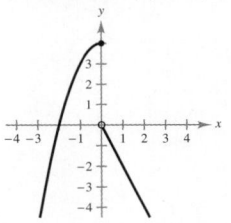

41. Critical numbers: $x = -1, 0$
Increasing on $(-\infty, -1)$
and $(0, \infty)$
Decreasing on $(-1, 0)$

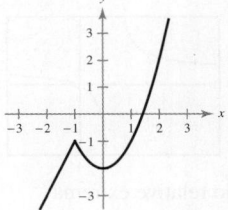

43. No critical numbers
$s'(t) = -3.196t + 45.61$
Increasing on $(3, 9)$

45. (a)

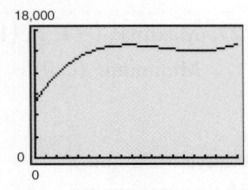

Increasing from 1970 to late 1988 and from late 2001 to 2008
Decreasing from late 1988 to late 2001

(b) $y' = 2.076t^2 - 100.22t + 1119.7$
Critical numbers: $t = 17.6, t = 30.7$
Therefore, the model is increasing from 1970 to late 1988 and from late 2001 to 2008 and decreasing from late 1988 to late 2001.

47. (a) Critical number: $x = 29,500$
Increasing on $(0, 29,500)$
Decreasing on $(29,500, 50,000)$

(b) You should charge the price that yields sales of $x = 29,500$ bags of popcorn. Because the function changes from increasing to decreasing at $x = 29,500$, the maximum profit occurs at this value.

CHAPTER 8

Section 8.5   *(page 668)*

Skills Warm Up   *(page 668)*

1. $0, \pm\frac{1}{2}$ **2.** $-2, 5$ **3.** 1 **4.** $0, 125$
5. $-4 \pm \sqrt{17}$ **6.** $1 \pm \sqrt{5}$
7. Negative   **8.** Positive   **9.** Positive
10. Negative   **11.** Increasing   **12.** Decreasing

1. Relative maximum: $(1, 5)$ **3.** Relative minimum: $(3, -9)$
5. Relative minimum: $(9, -2187)$ **7.** No relative extrema
9. Relative maximum: $(0, 15)$
   Relative minimum: $(4, -17)$
11. Relative maximum: $(-1, 2)$
13.

15.

Relative maximum: $(0, 0)$ Relative maximum: $\left(-1, -\frac{3}{2}\right)$
Relative minimum: $(8, -8)$

17.

No relative extrema
19. Minimum: $(2, 2)$ **21.** Maximum: $(0, 5)$
   Maximum: $(-1, 8)$ Minimum: $(3, -13)$
23. Minima: $(-1, -4), (2, -4)$
   Maxima: $(0, 0), (3, 0)$
25. Maximum: $(2, 1)$ **27.** Maxima: $\left(-1, \frac{1}{4}\right), \left(1, \frac{1}{4}\right)$
   Minimum: $\left(0, \frac{1}{3}\right)$ Minimum: $(0, 0)$
29. Maximum: $(-7, 4)$
   Minimum: $(1, 0)$
31. 2, absolute maximum (and relative maximum)
33. 1, absolute maximum (and relative maximum);
   2, absolute minimum (and relative minimum);
   3, absolute maximum (and relative maximum)
35. Maximum: $(5, 7)$ **37.** Maximum: $(2, 2.\overline{6})$
   Minimum: $(2.69, -5.55)$ Minima: $(0, 0), (3, 0)$
39. No relative maximum   **41.** Maximum: $\left(2, \frac{1}{2}\right)$
   Minimum: $(2, 12)$ Minimum: $(0, 0)$
43. Answers will vary. Sample answer:

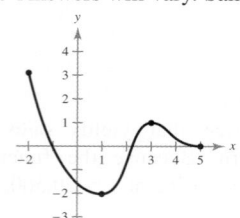

45. (a) Population tends to increase each year, so the minimum
   population occurred in 1790 and the maximum population
   occurred in 2010.
   (b) Maximum population: 310.07 million
   Minimum population: 3.69 million
   (c) The minimum population was about 3.69 million in 1790 and
   the maximum population was about 310.07 million in 2010.
47. 82 units   **49.** 3500 units, $2.25

Section 8.6   *(page 677)*

Skills Warm Up   *(page 677)*

1. $f''(x) = 48x^2 - 54x$ **2.** $g''(s) = 12s^2 - 18s + 2$
3. $g''(x) = 56x^6 + 120x^4 + 72x^2 + 8$
4. $f''(x) = \dfrac{4}{9(x-3)^{2/3}}$ **5.** $h''(x) = \dfrac{190}{(5x-1)^3}$
6. $f''(x) = -\dfrac{42}{(3x+2)^3}$ **7.** $x = \pm\dfrac{\sqrt{3}}{3}$
8. $x = 0, 3$ **9.** $t = \pm 4$ **10.** $x = 0, \pm 5$

1. Sign of $f'(x)$ on $(0, 2)$ is positive.
   Sign of $f''(x)$ on $(0, 2)$ is positive.
3. Sign of $f'(x)$ on $(0, 2)$ is negative.
   Sign of $f''(x)$ on $(0, 2)$ is negative.
5. Concave downward on $(-\infty, \infty)$
7. Concave upward on $(-\infty, 1)$
   Concave downward on $(1, \infty)$
9. Concave upward on $\left(-\infty, -\frac{1}{2}\right)$
   Concave downward on $\left(-\frac{1}{2}, \infty\right)$
11. Concave upward on $(-\infty, -2)$ and $(2, \infty)$
   Concave downward on $(-2, 2)$
13. Concave downward on $(-\infty, 3)$
   Concave upward on $(3, \infty)$
   Point of inflection: $(3, 0)$
15. Concave downward on $\left(-\infty, \frac{1}{2}\right)$
   Concave upward on $\left(\frac{1}{2}, \infty\right)$
   Point of inflection: $\left(\frac{1}{2}, -\frac{3}{2}\right)$
17. Concave upward on $(-\infty, \infty)$
   No inflection points
19. Concave downward on $(-\infty, 4)$
   Concave upward on $(4, \infty)$
   Inflection point: $(4, 16)$
21. Relative maximum: $(3, 9)$
23. Relative maximum: $(1, 3)$
   Relative minimum: $\left(\frac{7}{3}, \frac{49}{27}\right)$
25. Relative minimum: $(0, -3)$
27. Relative minimum: $(0, 1)$
29. Relative minima: $(-3, 0), (3, 0)$
   Relative maximum: $(0, 3)$
31. Relative maximum: $(0, 4)$ **33.** No relative extrema
35. Relative maximum: $(2, 9)$
   Relative minimum: $(0, 5)$
37. Relative maximum: $(0, 0)$
   Relative minima: $(-0.5, -0.052), (1, -0.\overline{3})$

39. Relative maximum: $(-2, 16)$
Relative minimum: $(2, -16)$
Point of inflection: $(0, 0)$

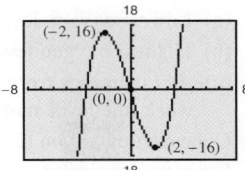

41. No relative maximum
Relative minimum: $(4, 4)$
Point of inflection:
$\left(12, \dfrac{8\sqrt{3}}{3}\right)$

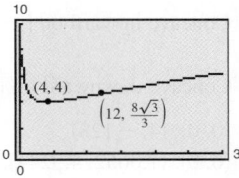

43. Relative maximum: $(0, 0)$
Relative minima: $(\pm 2, -4)$
Points of inflection:
$\left(\pm \dfrac{2\sqrt{3}}{3}, -\dfrac{20}{9}\right)$

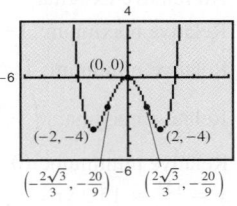

45. Relative maximum: $(-1, 0)$
Relative minimum: $(1, -4)$
Point of inflection: $(0, -2)$

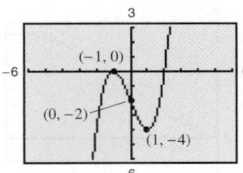

47. Relative minimum: $(-2, -2)$
No inflection points

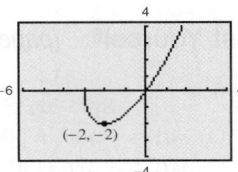

49. Relative maximum: $(0, 4)$
Points of inflection:
$\left(\pm \dfrac{\sqrt{3}}{3}, 3\right)$

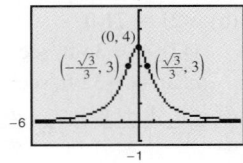

51.

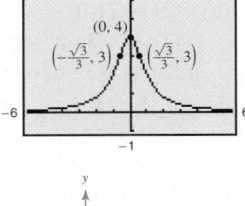

53.

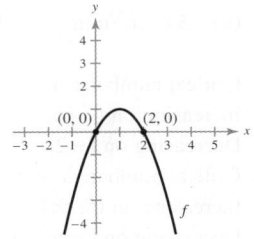

55.

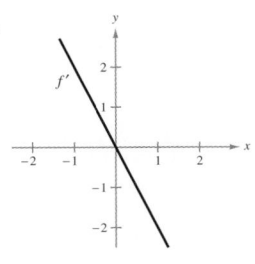

(a) f': Positive on $(-\infty, 0)$
f: Increasing on $(-\infty, 0)$
(b) f': Negative on $(0, \infty)$
f: Decreasing on $(0, \infty)$
(c) f': Not increasing
f: Not concave upward
(d) f': Decreasing on $(-\infty, \infty)$
f: Concave downward on $(-\infty, \infty)$

57. (a) f': Increasing on $(-\infty, \infty)$
(b) f: Concave upward on $(-\infty, \infty)$
(c) Relative minimum: $x = -2.5$
No inflection points

59. (a) f': Increasing on $(-\infty, 1)$
Decreasing on $(1, \infty)$
(b) f: Concave upward on $(-\infty, 1)$
Concave downward on $(1, \infty)$
(c) No relative extrema
Point of inflection: $x = 1$

61. $(200, 320)$ **63.** 8:30 P.M.

65.

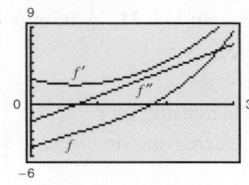

Relative minimum: $(0, -5)$
Relative maximum: $(3, 8.5)$
Point of inflection:
$\left(\dfrac{2}{3}, -3.2963\right)$

When f' is positive, f is increasing. When f' is negative, f is decreasing. When f'' is positive, f is concave upward. When f'' is negative, f is concave downward.

67.

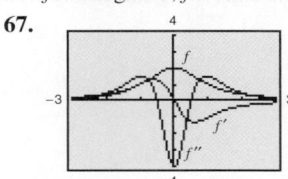

Relative maximum: $(0, 2)$
Points of inflection:
$(0.58, 1.5), (-0.58, 1.5)$

When f' is positive, f is increasing. When f' is negative, f is decreasing. When f'' is positive, f is concave upward. When f'' is negative, f is concave downward.

69. 120 units

71. (a)

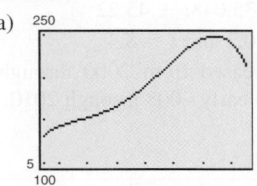

(b) 1995
(c) 2006
(d) Greatest: 2003
Least: 2009

73. (a)

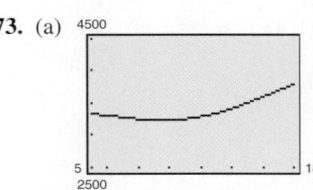

(b) Concave downward on $(5, 6.6517)$
Concave upward on $(6.6517, 16.4123)$
Concave downward on $(16.4123, 18)$
(c) Points of inflection: $(6.6517, 3291.0160)$ and
$(16.4123, 3638.4227)$
(d) The first inflection point is where the change in the number of veterans receiving benefits starts to increase after it has been decreasing. The second inflection point is where the change in the number of veterans receiving benefits starts to decrease again.

75. Answers will vary.

CHAPTER 8

Review Exercises   *(page 684)*

1. 6   **3.** $-\dfrac{120}{x^6}$ **5.** $\dfrac{35x^{3/2}}{2}$ **7.** $\dfrac{2}{x^{2/3}}$

9. 2   **11.** $\frac{512}{81}$

13. (a) $s(t) = -16t^2 + 5t + 30$ (b) About 1.534 sec
(c) About -44.09 ft/sec   (d) -32 ft/sec^2

15. $s'(t) = -\dfrac{2(t+1)}{(t^2+2t+1)^2};\ s''(t) = \dfrac{6}{(t^2+2t+1)^2}$

17. $-\dfrac{2x+3y}{3(x+y^2)}$ **19.** $\dfrac{2x-8}{2y-9}$ **21.** 5   **23.** 0

25. $y = \frac13 x + \frac13$ **27.** $y = \frac43 x + \frac23$

29. (a) 12π in.^{2}/min   (b) 40π in.^{2}/min   **31.** $\frac{1}{64}$ ft/min

33. (a) $\frac16$ ft/min   (b) $\frac34$ ft^3/min   **35.** $x = 1$

37. $x = -3, x = 3$ **39.** $x = 1, x = \frac73$

41. Critical number: $x = 1$ **43.** Increasing on $\left(-\frac12, \infty\right)$
    Increasing on $(1, \infty)$ Decreasing on $\left(-\infty, -\frac12\right)$
    Decreasing on $(-\infty, 1)$

45. Increasing on $(-\infty, 3)$ and $(3, \infty)$

47. The only critical number is $t \approx -10.85$. Any $t > -10.85$ produces a positive dR/dt, so the sales were increasing from 2004 through 2009.

49. (a)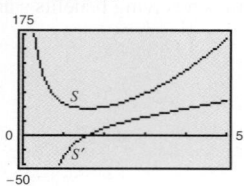

Revenues increased from 2000 through about 2007.
Revenues decreased from about 2008 through 2010.
(b) $R'(t) = -16.7334t^2 + 135.048t + 45.22$
Critical numbers: $t \approx -0.3$ and $t \approx 8.4$
Therefore, revenues increased from 2000 through early 2008 and decreased from early 2008 through 2010.

51. Relative maximum: $(0, -2)$
    Relative minimum: $(1, -4)$
53. Relative minimum: $(8, -52)$
55. Relative maxima: $(-1, 1), (1, 1)$
    Relative minimum: $(0, 0)$
57. Relative maximum: $(0, 6)$
59. Relative maximum: $(0, 0)$
    Relative minimum: $(4, 8)$
61. Maximum: $(0, 6)$
    Minimum: $\left(-\frac52, -\frac14\right)$
63. Maxima: $(-2, 17), (4, 17)$
    Minima: $(-4, -15), (2, -15)$
65. Maximum: $(1, 1)$
    Minimum: $(9, -3)$
67. Maximum: $(1, 1)$
    Minimum: $(-1, -1)$
69. $r \approx 1.58$ in.

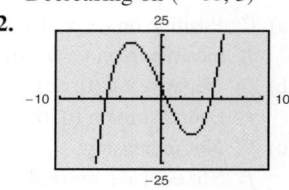

71. (a) Lowest oxygen level: $\frac12$ when $t = 1$
(b) Highest oxygen level: 1 when $t = 0$
(c) As t increases from 0 to 1, the level decreases. After $t = 1$ week, the level starts increasing toward 1.

73. Concave upward on $(2, \infty)$
    Concave downward on $(-\infty, 2)$

75. Concave upward on $\left(-\dfrac{2\sqrt3}{3}, \dfrac{2\sqrt3}{3}\right)$
    Concave downward on $\left(-\infty, -\dfrac{2\sqrt3}{3}\right)$ and $\left(\dfrac{2\sqrt3}{3}, \infty\right)$

77. $(0, 0), (4, -128)$

79. $(0, 0), (1.0652, 4.5244), (2.5348, 3.5246)$

81. No relative extrema

83. Relative maximum: $\left(-\sqrt3, 6\sqrt3\right)$
    Relative minimum: $\left(\sqrt3, -6\sqrt3\right)$

85. Relative maxima: $\left(-\dfrac{\sqrt2}{2}, \dfrac12\right), \left(\dfrac{\sqrt2}{2}, \dfrac12\right)$
    Relative minimum: $(0, 0)$

87. $x = \frac{137}{9} \approx 15.2$ yr   **89.** $\left(50, 166\frac23\right)$

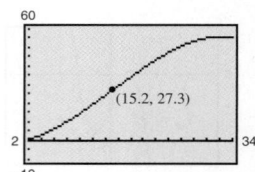

Test Yourself   *(page 688)*

1. 0   **2.** $-\dfrac{3}{8(3-x)^{5/2}}$ **3.** $-\dfrac{96}{(2x-1)^4}$

4. (a) $s(t) = -16t^2 + 30t + 75$
    $v(t) = -32t + 30$
    $a(t) = -32$
(b) $s(2) = 71$ ft
    $v(2) = -34$ ft/sec
    $a(2) = -32$ ft/sec^2

5. $\dfrac{dy}{dx} = -\dfrac{1+y}{x}$ **6.** $\dfrac{dy}{dx} = -\dfrac{1}{y-1}$ **7.** $\dfrac{dy}{dx} = \dfrac{x}{2y}$

8. (a) 3.75π cm^3/min   **9.** Critical number: $x = 0$
    (b) 15π cm^3/min   Increasing on $(0, \infty)$
                Decreasing on $(-\infty, 0)$

10. Critical numbers: $x = -2, x = 2$
    Increasing on $(-\infty, -2)$ and $(2, \infty)$
    Decreasing on $(-2, 2)$

11. Critical number: $x = 5$
    Increasing on $(5, \infty)$
    Decreasing on $(-\infty, 5)$

12.

Relative minimum: $(3, -14)$
Relative maximum: $(-3, 22)$

13.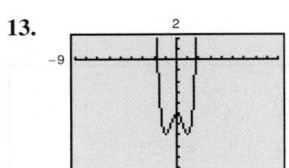

Relative minima:
$(-1, -7)$ and $(1, -7)$
Relative maximum: $(0, -5)$

14.

Relative maximum: $(0, 2.5)$

15. Minimum: $(-3, -1)$
Maximum: $(0, 8)$

16. Minimum: $(0, 0)$
Maximum: $(2.25, 9)$

17. Minimum: $(2\sqrt{3}, 2\sqrt{3})$
Maximum: $(1, 6.5)$

18. Concave upward: $(2, \infty)$
Concave downward: $(-\infty, 2)$

19. Concave upward: $\left(-\infty, -\frac{2\sqrt{2}}{3}\right)$ and $\left(\frac{2\sqrt{2}}{3}, \infty\right)$

Concave downward: $\left(-\frac{2\sqrt{2}}{3}, \frac{2\sqrt{2}}{3}\right)$

20. No point of inflection
The graph is concave upward on its entire domain.

21. Concave upward: $(-\infty, -3)$ and $(3, \infty)$
Concave downward: $(-3, 3)$
Points of inflection: $(-3, -175)$ and $(3, -175)$

22. Relative minimum: $(6, -166)$
Relative maximum: $(-2, 90)$

23. Relative minimum: $(3, -97.2)$
Relative maximum: $(-3, 97.2)$

Chapter 9

Section 9.1   *(page 695)*

1. $l = w = 25$ m   **3.** $l = w = 8$ ft
5. $x = 25$ ft, $y = \frac{100}{3}$ ft
7. (a) Proof
(b) $V_1 = 99$ in.³
$V_2 = 125$ in.³
$V_3 = 117$ in.³
(c) 5 in. × 5 in. × 5 in.
9. (a) $l = w = h = 20$ in.   (b) 2400 in.²
11. Width of rectangle: $\dfrac{100}{\pi} \approx 31.8$ m

Length of rectangle: 50 m
13. $l = w = 2\sqrt[3]{5} \approx 3.42$
$h = 4\sqrt[3]{5} \approx 6.84$

15. $V = 16$ in.³   **17.** 9 in. by 9 in.
19. Length: 3 units
Width: 1.5 units   **21.** Length: $5\sqrt{2}$ units
Width: $\dfrac{5\sqrt{2}}{2}$ units

23. Radius: about 1.51 in.
Height: about 3.02 in.

25. $(1, 1)$ **27.** $\left(3.5, \dfrac{\sqrt{14}}{2}\right)$ **29.** 18 in. × 18 in. × 36 in.

31. Radius: $\sqrt[3]{\dfrac{562.5}{\pi}} \approx 5.636$ ft

Height: about 22.545 ft

33. Radius of circle: $\dfrac{8}{\pi + 4}$

Side of square: $\dfrac{16}{\pi + 4}$

35. (a) $A(x) = \left(1 + \dfrac{4}{\pi}\right)x^2 - \dfrac{8}{\pi}x + \dfrac{4}{\pi}$
(b) $0 \le x \le 1$
(c)

(d) The total area is minimum when 2.24 feet is used for the square and 1.76 feet is used for the circle.
The total area is maximum when all 4 feet is used for the circle.

37. 4.75 weeks; 135 bushels; $3645

Section 9.2   *(page 705)*

1. 2000 units   **3.** 200 units   **5.** 200 units
7. 50 units   **9.** $60   **11.** $40
13. 3 units

$\overline{C}(3) = 17; \dfrac{dC}{dx} = 4x + 5$; when $x = 3$, $\dfrac{dC}{dx} = 17$

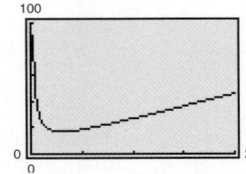

15. (a) $55   (b) $30.32
17. The maximum profit occurs when $s = 10$ (or $10,000).
The point of diminishing returns occurs at $s = \frac{35}{6}$ (or $5833.33).

19. 350 players   **21.** $50

23. C = cost under water + cost on land

$$= 25(5280)\sqrt{x^2 + 0.25} + 18(5280)(6 - x)$$

$$= 132{,}000\sqrt{x^2 + 0.25} + 570{,}240 - 95{,}040x$$

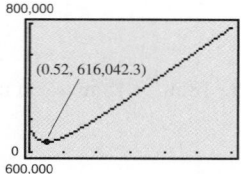

The line should run from the power station to a point across the river approximately 0.52 mile downstream.

$$\left(\text{Exact: } x = \frac{9\sqrt{301}}{301} \text{ mi}\right)$$

25. $v = 60$ mi/h

27. -1, unit elastic   **29.** $-\frac{2}{3}$, inelastic

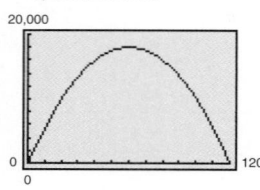   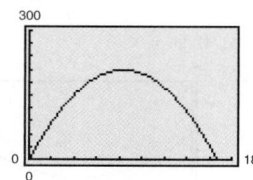

Elastic: $(0, 60)$ Elastic: $\left(0, 83\frac{1}{3}\right)$

Inelastic: $(60, 120)$ Inelastic: $\left(83\frac{1}{3}, 166\frac{2}{3}\right)$

31. $-\frac{25}{23}$, elastic

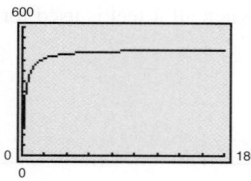

Elastic: $(0, \infty)$

33. (a) Elastic: $[0, 500)$

   Unit elastic: $x = 500$

   Inelastic: $(500, 1000]$

   (b) The revenue function increases on the interval $[0, 500)$, then is flat at 500, and decreases on the interval $(500, 1000]$.

35. 500 units $(x = 5)$

37. No; when $p = 8$, $x = 540$ and $\eta = -\frac{2}{3}$.

   Because $|\eta| = \frac{2}{3} < 1$, demand is inelastic.

39. (a) 2007

   (b) 2001

   (c) 2007: $237.55 million/yr

   2001: $17.78 million/yr

   (d)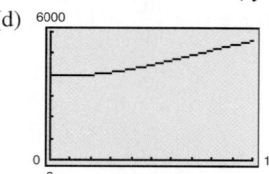

41. Proof   **43.** Answers will vary.

Section 9.3   *(page 716)*

Skills Warm Up   *(page 716)*

1. 3   **2.** 1   **3.** -11 **4.** 4   **5.** $-\frac{1}{4}$

6. -2 **7.** 0   **8.** 1

9. $\overline{C} = \dfrac{150}{x} + 3$ **10.** $\overline{C} = \dfrac{1900}{x} + 1.7 + 0.002x$

$\dfrac{dC}{dx} = 3$ $\dfrac{dC}{dx} = 1.7 + 0.004x$

11. $\overline{C} = 0.005x + 0.5 + \dfrac{1375}{x}$ **12.** $\overline{C} = \dfrac{760}{x} + 0.05$

$\dfrac{dC}{dx} = 0.01x + 0.5$ $\dfrac{dC}{dx} = 0.05$

1. Vertical asymptote: $x = 0$

   Horizontal asymptote: $y = 1$

3. Vertical asymptotes: $x = -1, x = 2$

   Horizontal asymptote: $y = 1$

5. Vertical asymptote: none

   Horizontal asymptote: $y = \frac{3}{2}$

7. Vertical asymptotes: $x = \pm 2$

   Horizontal asymptote: $y = \frac{1}{2}$

9. $x = 0, x = -3$ **11.** $x = -3$ **13.** $x = 4$ **15.** ∞

17. $-\infty$ **19.** ∞ **21.** 1   **23.** 7   **25.** $y = 2$

27. $y = 0$ **29.** No horizontal asymptote   **31.** $y = 5$

33. d   **34.** b   **35.** a   **36.** c

37. (a) ∞ (b) 5   (c) 0

39. (a) 0   (b) 1   (c) ∞

41.

x	10^0	10^1	10^2	10^3
$f(x)$	0.646	11.718	800.003	29,622.777

x	10^4	10^5	10^6
$f(x)$	980,000	31,422,776.6	998,000,000

$$\lim_{x \to \infty} \sqrt{x^3 + 6} - 2x = \infty$$

43.

x	10^0	10^1	10^2	10^3
$f(x)$	2.000	0.348	0.101	0.032

x	10^4	10^5	10^6
$f(x)$	0.010	0.003	0.001

$$\lim_{x \to \infty} \frac{x + 1}{x\sqrt{x}} = 0$$

45.

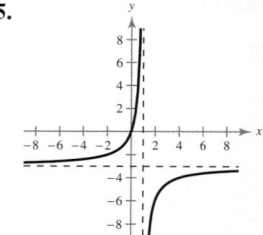

47.

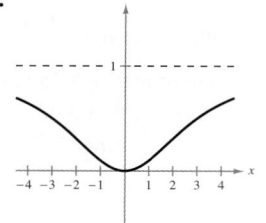

49.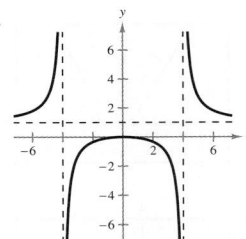

51.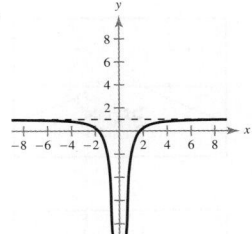

53.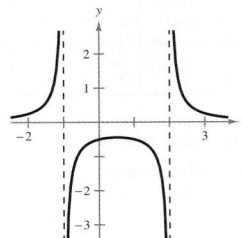

55.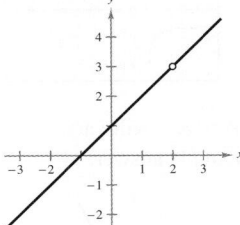

57.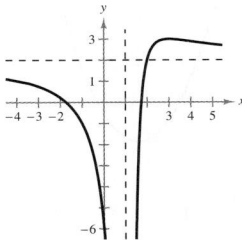

59.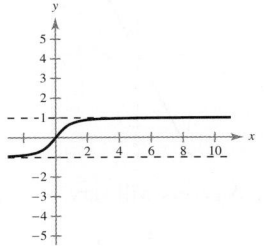

61. (a) $\overline{C} = 1.15 + \dfrac{6000}{x}$

(b) $\overline{C}(600) = 11.15$; $\overline{C}(6000) = 2.15$

(c) \$1.15; The cost approaches \$1.15 as the number of units produced increases.

63. (a) $\overline{P} = 35.4 - \dfrac{15,000}{x}$

(b) $\overline{P}(1000) = \$20.40$; $\overline{P}(10,000) = \$33.90$; $\overline{P}(100,000) = \$35.25$

(c) \$35.40; Explanations will vary.

65. (a) 25%: \$176 million; 50%: \$528 million; 75%: \$1584 million

(b) ∞; The limit does not exist, which means the cost increases without bound as the government approaches 100% seizure of the illegal drug entering the country.

67. (a)

n	1	2	3	4	5
P	0.5	0.74	0.82	0.86	0.89

n	6	7	8	9	10
P	0.91	0.92	0.93	0.94	0.95

(b) 1

(c)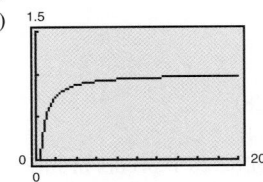

The percent of correct responses approaches 100% as the number of times the task is performed increases.

Quiz Yourself   *(page 719)*

1. (a) 100 ft by 50 ft   (b) 5000 ft^2 **2.** $7\frac{1}{2}$ in. by 10 in.

3. 400 units   **4.** 70 units   **5.** \$63   **6.** \$0.80

7. (a) 0.5   (b) Inelastic

(c)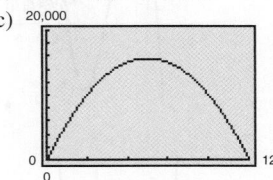

(d) Elastic: (0, 62.5)
Inelastic: (62.5, 125)

8. (a) 1   (b) Of unit elasticity

(c)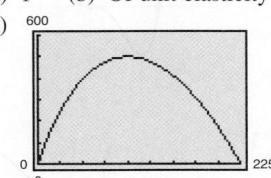

(d) Elastic: (0, 100)
Inelastic: (100, 225)

9. -1 **10.** $-\infty$ **11.** 10   **12.** $\frac{1}{3}$ **13.** -1 **14.** ∞

15. Vertical asymptote: $x = 1$
Horizontal asymptote: $y = 2$

16. Vertical asymptotes: $x = 0, x = 2$
Horizontal asymptote: $y = 0$

17. Vertical asymptote: $x = 3$
Horizontal asymptote: none

Section 9.4   *(page 727)*

Skills Warm Up   *(page 727)*

1. Vertical asymptote: $x = 0$
Horizontal asymptote: $y = 0$

2. Vertical asymptote: $x = 2$
Horizontal asymptote: $y = 0$

3. Vertical asymptote: $x = -3$
Horizontal asymptote: $y = 40$

4. Vertical asymptotes: $x = 1, x = 3$
Horizontal asymptote: $y = 1$

5. Decreasing on $(-\infty, -2)$
Increasing on $(-2, \infty)$

6. Increasing on $(-\infty, -4)$
Decreasing on $(-4, \infty)$

7. Increasing on $(-\infty, -1)$ and $(1, \infty)$
Decreasing on $(-1, 1)$

8. Decreasing on $(-\infty, 0)$ and $\left(\sqrt[3]{2}, \infty\right)$
Increasing on $\left(0, \sqrt[3]{2}\right)$

9. Increasing on $(-\infty, 1)$ and $(1, \infty)$

10. Decreasing on $(-\infty, -3)$ and $\left(\frac{1}{3}, \infty\right)$
Increasing on $\left(-3, \frac{1}{3}\right)$

CHAPTER 9

1.

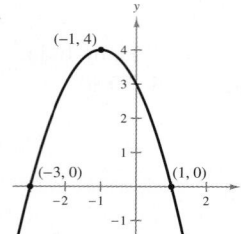

3.

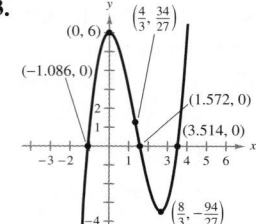

5.

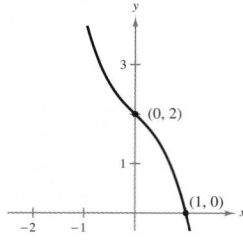

7.

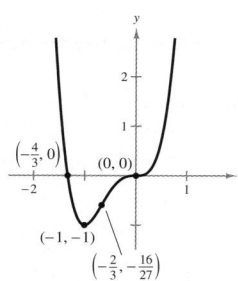

9.

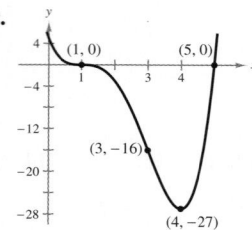

11.

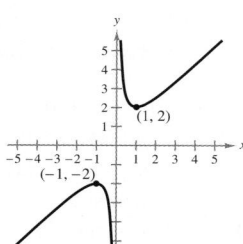

13.

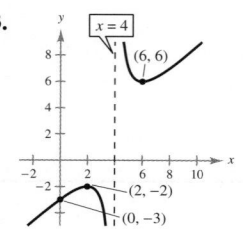

15.

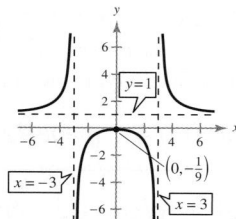

17.

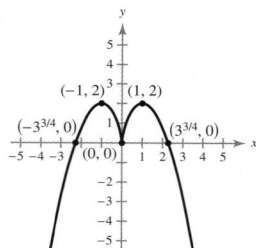

19.

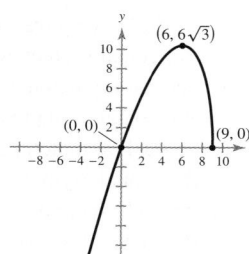

21.

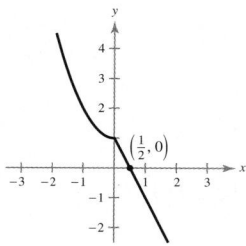

23.

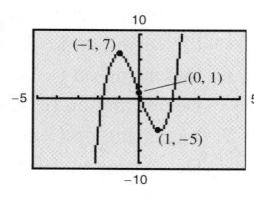

25.

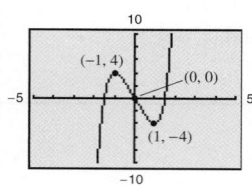

27.

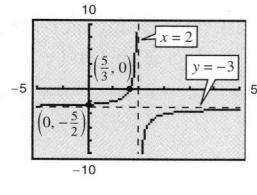

29.

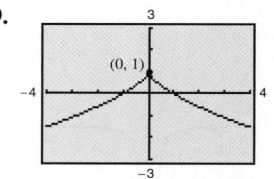

31.

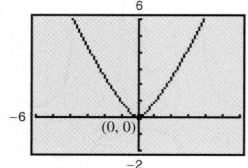

33.

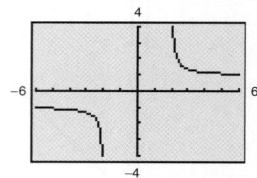

35.

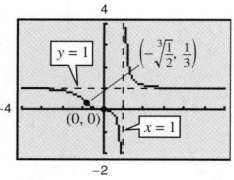

37. Answers will vary.
Sample answer:

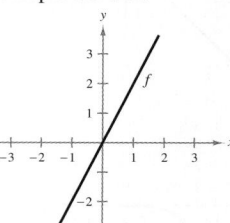

39. Answers will vary.
Sample answer:

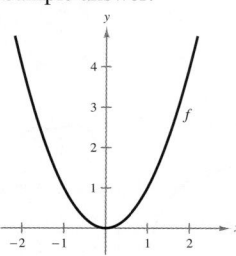

41. Answers will vary. Sample answer:

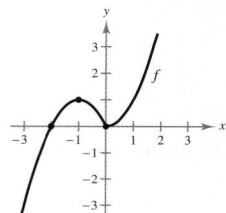

43. Answers will vary. Sample answer: $y = \dfrac{1}{x - 5}$

45. (a)

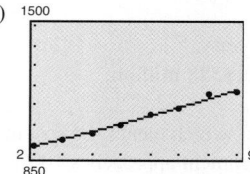

The model fits the data well.

(b) $1468.54

(c) No, because the benefits increase without bound as time approaches the year 2035 ($x = 35$), and the benefits are negative for the years past 2035.

47.

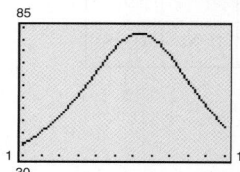

Absolute maximum: $(7, 82.28)$
Absolute minimum: $(1, 34.84)$
The maximum temperature of 82.28°F occurs in July.
The minimum temperature of 34.84°F occurs in January.

49.

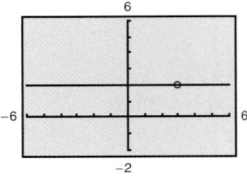

The rational function has the common factor $3 - x$ in the numerator and denominator. At $x = 3$, there is a hole in the graph, not a vertical asymptote.

51. (a) $f(x) = \dfrac{x^2 - 2x + 4}{x - 2} = \dfrac{x^2 - 2x}{x - 2} + \dfrac{4}{x - 2}$

$= \dfrac{x(x - 2)}{x - 2} + \dfrac{4}{x - 2} = x + \dfrac{4}{x - 2}$

(b)

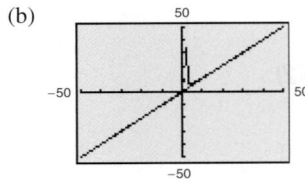

The graphs become almost identical as you zoom out.

(c) A slant asymptote is neither horizontal nor vertical. It is diagonal, following $y = x$.

Section 9.5 *(page 734)*

Skills Warm Up *(page 734)*

1. $\dfrac{dC}{dx} = 0.18x$ **2.** $\dfrac{dC}{dx} = 0.15$

3. $\dfrac{dR}{dx} = 1.25 + 0.03\sqrt{x}$ **4.** $\dfrac{dR}{dx} = 15.5 - 3.1x$

5. $\dfrac{dP}{dx} = -\dfrac{0.01}{\sqrt[3]{x^2}} + 1.4$ **6.** $\dfrac{dP}{dx} = -0.04x + 25$

7. $\dfrac{dA}{dx} = \dfrac{\sqrt{3}}{2}x$ **8.** $\dfrac{dA}{dx} = 12x$ **9.** $\dfrac{dC}{dr} = 2\pi$

10. $\dfrac{dP}{dw} = 4$ **11.** $\dfrac{dS}{dr} = 8\pi r$ **12.** $\dfrac{dP}{dx} = 2 + \sqrt{2}$

13. $A = \pi r^2$ **14.** $A = x^2$

15. $V = x^3$ **16.** $V = \frac{4}{3}\pi r^3$

1. $dy = 0.6$ **3.** $dy = -0.04$ **5.** $dy = 0.075$
$\Delta y = 0.6305$ $\Delta y \approx -0.0394$ $\Delta y \approx 0.0745$

7.

$dx = \triangle x$	dy	$\triangle y$	$\triangle y - dy$	$\dfrac{dy}{\triangle y}$
1.000	4.000	5.000	1.0000	0.8000
0.500	2.000	2.2500	0.2500	0.8889
0.100	0.400	0.4100	0.0100	0.9756
0.010	0.040	0.0401	0.0001	0.9975
0.001	0.004	0.0040	0.0000	1.0000

9.

$dx = \triangle x$	dy	$\triangle y$	$\triangle y - dy$	$\dfrac{dy}{\triangle y}$
1.000	-0.25000	-0.13889	0.11111	1.79999
0.500	-0.12500	-0.09000	0.03500	1.38889
0.100	-0.02500	-0.02324	0.00176	1.07573
0.010	-0.00250	-0.00248	0.00002	1.00806
0.001	-0.00025	-0.00025	0.00000	1.00000

11.

$dx = \triangle x$	dy	$\triangle y$	$\triangle y - dy$	$\dfrac{dy}{\triangle y}$
1.000	0.14865	0.12687	-0.02178	1.17167
0.500	0.07433	0.06823	-0.00610	1.08940
0.100	0.01487	0.01459	-0.00028	1.01919
0.010	0.00149	0.00148	-0.00001	1.00676
0.001	0.00015	0.00015	0.00000	1.00000

13. $5.20   **15.** $7.50   **17.** $-$1150   **19.** $dy = 24x^3\, dx$

21. $dy = 6x\, dx$ **23.** $dy = 12(4x - 1)^2\, dx$

25. $dy = -\dfrac{3}{(2x - 1)^2}\, dx$ **27.** $dy = \dfrac{-x}{\sqrt{9 - x^2}}\, dx$

29. $y = 28x + 37$

For $\Delta x = -0.01$, $f(x + \Delta x) = -19.281302$ and $y(x + \Delta x) = -19.28$.

For $\Delta x = 0.01$, $f(x + \Delta x) = -18.721298$ and $y(x + \Delta x) = -18.72$.

31. $y = x$

For $\Delta x = -0.01$, $f(x + \Delta x) = -0.009999$ and $y(x + \Delta x) = -0.01$.

For $\Delta x = 0.01$, $f(x + \Delta x) = 0.009999$ and $y(x + \Delta x) = 0.01$.

33. (a) $dP = $1160   (b) Actual: $1122.50

35. (a) $71.50   (b) $40.00   **37.** Approximately 19 deer

39. $R = -\frac{1}{3}x^2 + 100x$; $6

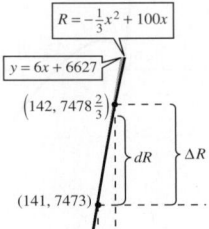

41. $\pm\frac{3}{4}$ in.^{2}; 2.08%   **43.** True

Review Exercises *(page 739)*

1. $l = w = 15$ m   **3.** 144 in.^{3}

5. 9 feet from the shorter post   **7.** $x = 900$

9. $x = 150$ **11.** (a) $24   (b) $8

13. (a) For $0 < x < 750$, $|\eta| > 1$ and the demand is elastic.
For $750 < x < 1500$, $|\eta| < 1$ and the demand is inelastic.
For $x = 750$, the demand has unit elasticity.
(b) From 0 to 750 units, revenue is increasing.
From 750 to 1500 units, revenue does not increase.

15. $x = -7, x = 0$ **17.** $x = -\frac{1}{2}$ **19.** $-\infty$ **21.** ∞

23. $\frac{2}{3}$ **25.** 0   **27.** $y = \frac{2}{3}$ **29.** $y = 0$ **31.** $y = -2$

33. (a) $\overline{C} = 0.75 + \dfrac{4000}{x}$

(b) $\overline{C}(100) = 40.75$
$\overline{C}(1000) = 4.75$

(c) The limit is 0.75. As more and more units are produced, the average cost per unit will approach \$0.75.

35. (a) 20%: \$62.5 million
50%: \$250 million
90%: \$2250 million

(b) The limit is ∞, meaning that as the percent gets very close to 100, the cost grows without bound.

37.
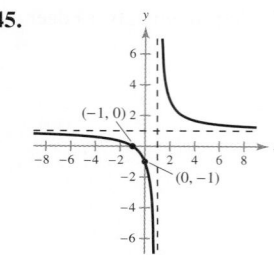

39.
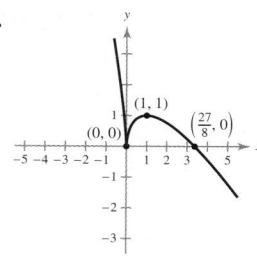

41.

43.

45.

47.

49. (a)
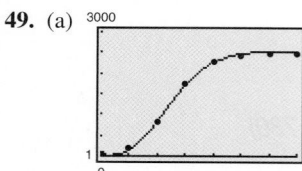

The model fits the data well.
(b) ≈ 2434 bacteria
(c) Answers will vary.

51. $dy = 0.08, \Delta y = 0.0802$
53. $dy = -2.1, \Delta y = -2.191$
55. \$800   **57.** \$15.25   **59.** $\approx$ \$4.52
61. $dy = 1.5x^2\, dx$ **63.** $dy = 18x(3x^2 - 2)^2\, dx$

65. $dy = -\dfrac{7}{(x+5)^2}\, dx$

67. (a) \$164   (b) \$163.2, a difference of \$0.80.

69. $B = 0.1\sqrt{5w}$

$\dfrac{dB}{dw} = \dfrac{0.05\sqrt{5}}{\sqrt{w}}$

$\Delta B \approx dB = \dfrac{0.05\sqrt{5}}{\sqrt{w}}\, dw$

$= \dfrac{0.05\sqrt{5}}{\sqrt{90}}(5)$

$\approx 0.059\ \text{m}^2$

Test Yourself   *(page 742)*

1. Vertical asymptote: $x = 1$
Horizontal asymptote: $y = 2$
2. Vertical asymptote: none
Horizontal asymptote: $y = 0$
3. Vertical asymptote: $x = 5$
Horizontal asymptote: $y = 3$
4. Vertical asymptotes: $x = 1$ and $x = -1$
Horizontal asymptote: $y = 1$
5. $-\infty$ **6.** ∞ **7.** ∞ **8.** 1   **9.** ∞ **10.** 3

11.
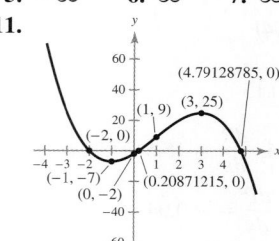

12.
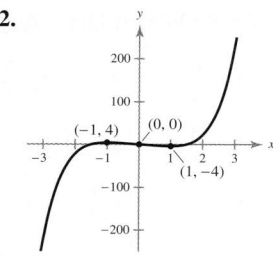

13.

14. $dy = 10x\, dx$ **15.** $dy = \dfrac{-4}{(x+3)^2}\, dx$

16. $dy = 3(x+4)^2\, dx$ **17.** 125 m by 125 m
18. (a) 20 in. by 20 in. by 20 in.
(b) 2400 in.^{2}
19. (312.5, 625)
20. $\Delta p = -0.05, dp = -0.05$
The values are the same.

Chapter 10

Section 10.1   *(page 748)*

Skills Warm Up   *(page 748)*

1. Horizontal shift to the left two units
2. Reflection about the x-axis
3. Vertical shift down one unit
4. Reflection about the y-axis
5. Horizontal shift to the right one unit
6. Vertical shift up two units
7. 125   **8.** 22.63   **9.** 9   **10.** $\frac{1}{125}$
11. $\frac{1}{2}$ **12.** $\frac{25}{64}$ **13.** 5   **14.** $\frac{4}{3}$ **15.** $-9, 1$
16. $2 \pm 2\sqrt{2}$ **17.** $1, -5$ **18.** $\frac{1}{2}, 1$

---

1. (a) 3125   (b) $\frac{1}{5}$ (c) 625   (d) $\frac{1}{125}$
3. (a) $\frac{1}{5}$ (b) 27   (c) 5   (d) 4096
5. 2 g
7.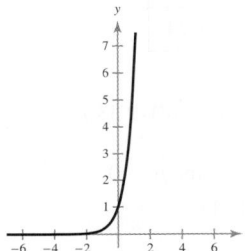
9.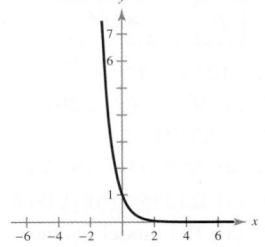
11.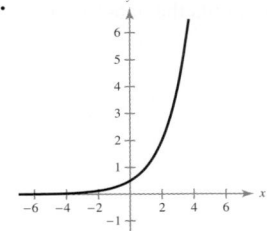
13.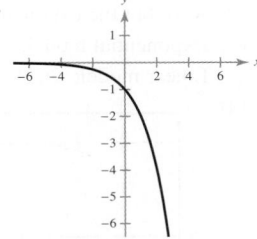
15.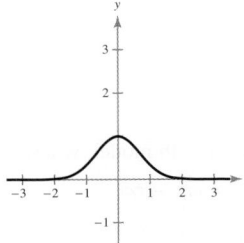
17.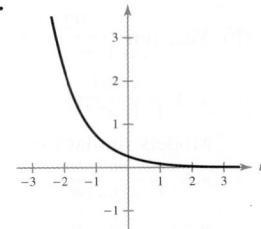

19. (a) $P(23) \approx 320.26$ million
    (b) $P(30) \approx 343.36$ million
21. (a) $V(5) \approx \$80,634.95$
    (b) $V(20) \approx \$161,269.89$
23. \$36.93

25. $V(t) = 28,000\left(\frac{3}{4}\right)^t$

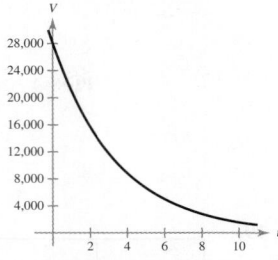

$V(4) = 28,000\left(\frac{3}{4}\right)^4 \approx 8859.38$

27. (a)

Year	2001	2002	2003	2004
Actual	37,188	38,221	39,165	40,201
Model	36,966	37,998	39,058	40,148

Year	2005	2006	2007	2008
Actual	40,520	41,746	43,277	46,025
Model	41,268	42,419	43,603	44,819

The model fits the data well. Explanations will vary.

(b)

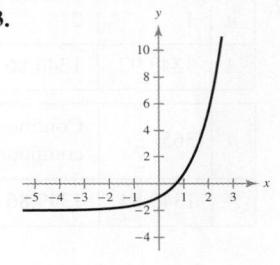

(c) 2014

Section 10.2   *(page 756)*

Skills Warm Up   *(page 756)*

1. Continuous on $(-\infty, \infty)$
2. Discontinuous at $x = \pm 2$
3. Discontinuous at $x = \pm\sqrt{3}$
4. Removable discontinuity at $x = 4$
5. $y = 0$ **6.** $y = 0$ **7.** $y = 4$ **8.** $y = \frac{1}{2}$
9. $y = \frac{3}{2}$ **10.** $y = 6$ **11.** $y = 0$ **12.** $y = 0$

---

1. (a) e^7 (b) e^{12} (c) $\dfrac{1}{e^6}$ (d) 1
3. (a) e^5 (b) $e^{5/2}$ (c) e^6 (d) e^7
5. f   **6.** e   **7.** d   **8.** b   **9.** c   **10.** a
11.
13.

15.

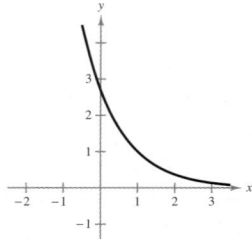

17.

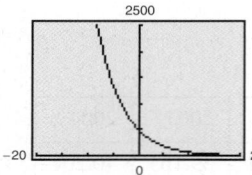

Horizontal asymptote: $N = 0$
Continuous on the entire
real number line

19.

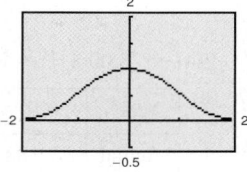

Horizontal asymptote: $g = 0$
Continuous on the entire
real number line

21.

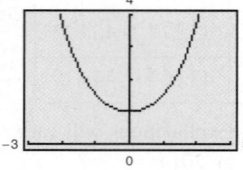

No horizontal asymptotes
Continuous on the entire
real number line

23.

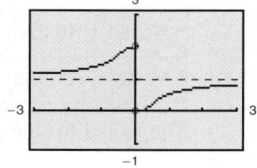

Horizontal asymptote: $y = 1$
Discontinuous at $x = 0$

25. (a)

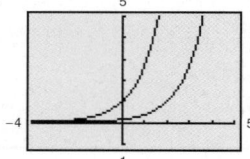

The graph of $g(x) = e^{x-2}$ is
shifted horizontally two units
to the right.

(b)

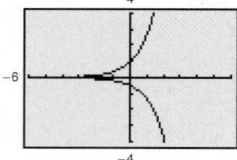

The graph of $h(x) = -\frac{1}{2}e^x$
decreases at a rate slower
than the rate at which the
graph of $f(x) = e^x$ increases.

(c)

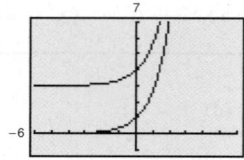

The graph of $q(x) = e^x + 3$
is shifted vertically three
units upward.

27.

n	1	2	4	12
A	1343.92	1346.86	1348.35	1349.35

n	365	Continuous compounding
A	1349.84	1349.86

29.

n	1	2	4	12
A	2191.12	2208.04	2216.72	2222.58

n	365	Continuous compounding
A	2225.44	2225.54

31.

t	1	10	20
P	96,078.94	67,032.00	44,932.90

t	30	40	50
P	30,119.42	20,189.65	13,533.53

33.

t	1	10	20
P	95,132.82	60,716.10	36,864.45

t	30	40	50
P	22,382.66	13,589.88	8251.24

35. \$107,311.12
37. (a) 9%   (b) 9.20%   (c) 9.31%   (d) 9.38%
39. \$6450.04
41. (a) \$849.53   (b) \$421.12   (c) $\lim\limits_{x\to\infty} p = 0$
43. (a) 0.1535   (b) 0.4866   (c) 0.8111
45. (a) The model fits the data well.
(b) $y = 637.11x + 5021.1$; The linear model fits the data
well, but the exponential model fits the data better.
(c) Exponential model: 2016
Linear model: 2021
47. (a)

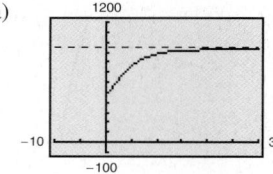

(b) Yes, $\lim\limits_{t\to\infty} \dfrac{925}{1 + e^{-0.3t}} = 925$

(c) $\lim\limits_{t\to\infty} \dfrac{1000}{1 + e^{-0.3t}} = 1000$

Models similar to this logistic growth model, where
$y = \dfrac{a}{1 + be^{-ct}}$ have a limit of a as $t \to \infty$.

49. (a) 0.536   (b) 0.666
(c)

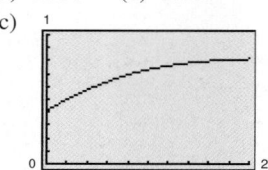

(d) Yes, $\lim\limits_{n\to\infty} \dfrac{0.83}{1 + e^{-0.2n}} = 0.83$

51. Amounts earned:

(a) \$5267.71   (b) \$5255.81   (c) \$5243.23

You should choose the certificate of deposit in part (a) because it earns more money than the others.

Section 10.3 *(page 765)*

Skills Warm Up   *(page 765)*

1. $\frac{1}{2}e^x(2x^2 - 1)$ **2.** $\frac{e^x(x + 1)}{x}$ **3.** $e^x(x - e^x)$

4. $e^{-x}(e^{2x} - x)$ **5.** $-\frac{6}{7x^3}$ **6.** $6x - \frac{1}{6}$

7. $6(2x^2 - x + 6)$ **8.** $\frac{t + 2}{2t^{3/2}}$

9. Relative maximum: $\left(-\frac{4\sqrt{3}}{3}, \frac{16\sqrt{3}}{9}\right)$

Relative minimum: $\left(\frac{4\sqrt{3}}{3}, -\frac{16\sqrt{3}}{9}\right)$

10. Relative maximum: $(0, 5)$

Relative minima: $(-1, 4), (1, 4)$

1. 0   **3.** $5e^{5x}$ **5.** $-2xe^{-x^2}$ **7.** $\frac{2}{x^3}e^{-1/x^2}$

9. $e^{4x}(4x^2 + 2x + 4)$ **11.** $-\frac{6(e^x - e^{-x})}{(e^x + e^{-x})^4}$

13. $-\frac{2e^x}{(e^x - 1)^2}$ **15.** $xe^x + e^x + 4e^{-x}$ **17.** 4   **19.** -3

21. $y = 2x - 3$ **23.** $y = \frac{4}{e^2}$ **25.** $y = 24x + 8$

27. $\frac{dy}{dx} = \frac{10 - e^y}{xe^y + 3}$ **29.** $\frac{dy}{dx} = \frac{e^{-x}(x^2 - 2x) + y}{4y - x}$

31. $6(3e^{3x} + 2e^{-2x})$ **33.** $32e^{4x}(x + 1)$

35.

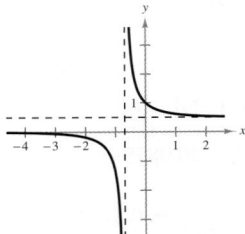

No relative extrema

No points of inflection

Horizontal asymptote to the right: $y = \frac{1}{2}$

Horizontal asymptote to the left: $y = 0$

Vertical asymptote: $x \approx -0.693$

37.

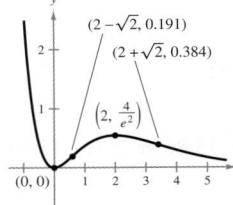

Relative minimum: $(0, 0)$

Relative maximum: $\left(2, \frac{4}{e^2}\right)$

Points of inflection:

$\left(2 - \sqrt{2}, 0.191\right),$

$\left(2 + \sqrt{2}, 0.384\right)$

Horizontal asymptote to the right: $y = 0$

39. $x = -\frac{1}{3}$ **41.** $x = 9$

43. (a)

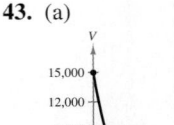

(b) $-\$5028.84/\text{yr}$

(c) $-\$406.89/\text{yr}$

(d) $V = -1497.2t + 15,000$

(e) In the exponential function, the initial rate of depreciation is greater than in the linear model. The linear model has a constant rate of depreciation.

45. (a)

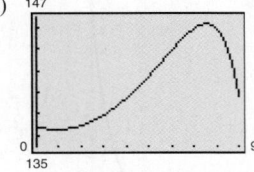

(b and c) 2000: -0.59 million people/yr

2004: 1.83 million people/yr

2009: -12.44 million people/yr

47. (a) \$433.31 per year

(b) \$890.22 per year

(c) \$21,839.26 per year

49. (a) $f(x) = \frac{1}{116\sqrt{2\pi}}e^{-(x-516)^2/26,912}$

(b)

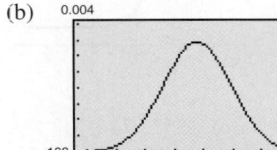

(c) $f'(x) = \frac{-1}{1,560,896\sqrt{2\pi}}(x - 516)e^{-(x-516)^2/26,912}$

(d) Answers will vary.

51.

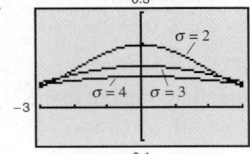

As σ increases, the graph becomes flatter.

53. Proof; maximum: $\left(0, \frac{1}{\sigma\sqrt{2\pi}}\right)$; answers will vary.

Sample answer:

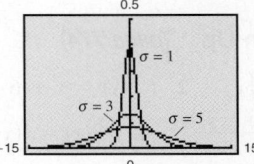

Quiz Yourself   *(page 767)*

1. 1024   **2.** 216   **3.** 27   **4.** $\sqrt{15}$

5. e^7 **6.** $e^{11/3}$ **7.** e^6 **8.** e^3

CHAPTER 10

9.
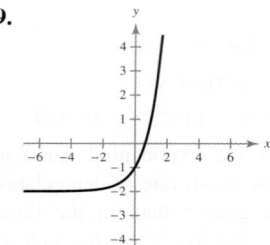

10.
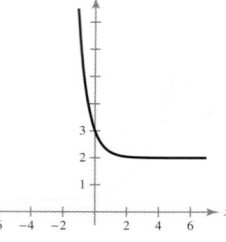

11. b   **12.** a

13.
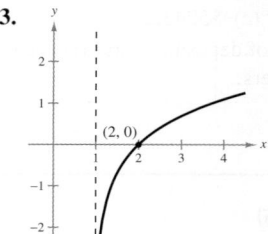

15.
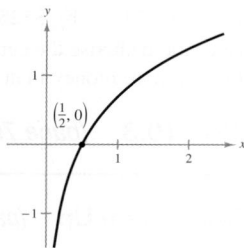

11.
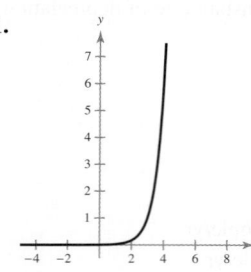

12.
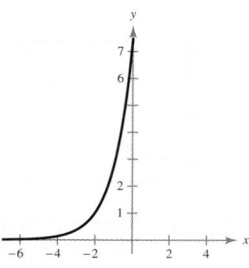

17.
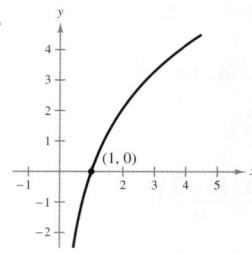

13.
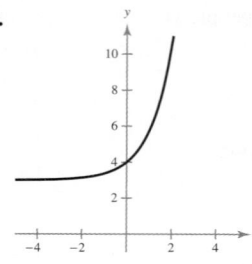

14.
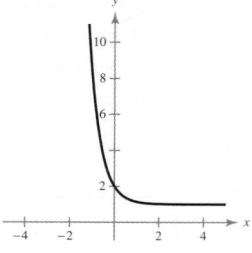

15. 22.69 grams   **16.** \$31.06

17. (a) \$3571.02   (b) \$3572.83   (c) \$3573.74

18. \$10,379.21   **19.** $5e^{5x}$ **20.** e^{x-4} **21.** $5e^{x+2}$

22. $e^x(2-x)$ **23.** $y = -2x + 1$

24.
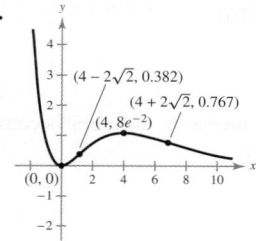

Relative maximum: $(4, 8e^{-2})$

Relative minimum: $(0, 0)$

Points of inflection:
$(4 - 2\sqrt{2}, 0.382)$,
$(4 + 2\sqrt{2}, 0.767)$

Horizontal asymptote to the right: $y = 0$

Section 10.4   *(page 774)*

Skills Warm Up   *(page 774)*

1. $f^{-1}(x) = \frac{1}{5}x$ **2.** $f^{-1}(x) = x + 6$

3. $f^{-1}(x) = \dfrac{x-2}{3}$ **4.** $f^{-1}(x) = \frac{4}{3}(x + 9)$

5. $x > -4$ **6.** Any real number x

7. $x < -1$ or $x > 1$ **8.** $x > 5$

9. \$3462.03   **10.** \$3374.65

1. $e^{0.6931.\,\cdots} = 2$ **3.** $e^{-1.6094\,\cdots} = 0.2$ **5.** $\ln 1 = 0$

7. $\ln(0.0498.\,.\,.) = -3$ **9.** c   **10.** d

19. x^2 **21.** $5x + 2$ **23.** $2x - 1$

25. $\ln 2 - \ln 3$ **27.** $\ln x + \ln y + \ln z$

29. $\frac{1}{3}\ln(2x + 7)$ **31.** $\ln z + 2\ln(z - 1)$

33. $\ln 3 + \ln x + \ln(x + 1) - 2\ln(2x + 1)$

35. Answers will vary.   **37.** Answers will vary.

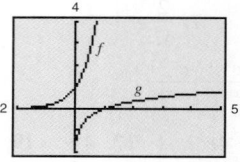

39. (a) 1.7917   (b) 0.4055   (c) 4.3944   (d) 0.5493

41. $\ln \dfrac{x-2}{x+2}$ **43.** $\ln \dfrac{x^3 y^2}{z^4}$ **45.** $\dfrac{\ln(x-6)^4}{\sqrt{3x+1}}$

47. $\ln \left[\dfrac{x(x+3)}{x+4}\right]^3$ **49.** $\ln \left[\dfrac{x(x^2+1)}{x+1}\right]^{3/2}$

51. $x = 4$ **53.** $x = \ln 4 - 1 \approx 0.3863$

55. $t = \dfrac{\ln 7 - \ln 3}{-0.2} \approx -4.2365$

57. $x = \frac{1}{2}\left(1 + \ln\frac{3}{2}\right) \approx 0.7027$ **59.** $x = 1$

61. $x = \dfrac{e^{2.4}}{2} \approx 5.5116$ **63.** $x = e^3 \approx 20.0855$

65. $x = \dfrac{6e^3}{e^3 - 1} \approx 6.314$ **67.** $x = \dfrac{\ln 15}{2 \ln 5} \approx 0.8413$

69. $t = \dfrac{\ln 2}{\ln 1.07} \approx 10.2448$

71. $t = \dfrac{\ln 3}{12 \ln[1 + (0.07/12)]} \approx 15.7402$

73. (a) 8.15 yr   (b) 12.92 yr

75. (a) 14.21 yr   (b) 13.89 yr   (c) 13.86 yr   (d) 13.86 yr

77. (a) $P(29) \approx 235{,}576$ (b) 2020

79. 9395 yr   **81.** 12,484 yr

83. (a) 80   (b) 57.5   (c) 10 mo

85. (a) ≈ 896 units   (b) ≈ 136 units

87.

x	y	$\dfrac{\ln x}{\ln y}$	$\ln \dfrac{x}{y}$	$\ln x - \ln y$
1	2	0	-0.6931	-0.6931
3	4	0.7925	-0.2877	-0.2877
10	5	1.4307	0.6931	0.6931
4	0.5	-2	2.0794	2.0794

89.

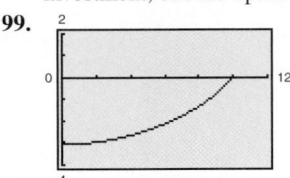

91. False. $f(x) = \ln x$ is undefined for $x \le 0$.

93. False. $f\left(\dfrac{x}{2}\right) = f(x) - f(2)$　　**95.** False. $u = v^2$

97. Options (b) and (c) will give you the same amount, but it makes more sense to double the rate, not the time. So option (b) is better than option (c). If you are looking for a long-term investment, choose option (a).

99.　　　　　　　　　　Answers will vary.

Section 10.5　(page 783)

Skills Warm Up　(page 783)

1. $2 \ln(x + 1)$　　**2.** $\ln x + \ln(x + 1)$

3. $\ln x - \ln(x + 1)$　　**4.** $3[\ln x - \ln(x - 3)]$

5. $\ln 4 + \ln x + \ln(x - 7) - 2 \ln x$

6. $3 \ln x + \ln(x + 1)$

7. $-\dfrac{y}{x + 2y}$　　**8.** $\dfrac{3 - 2xy + y^2}{x(x - 2y)}$

9. $-12x + 2$　　**10.** $-\dfrac{6}{x^4}$

---

1. $\dfrac{2}{x}$　　**3.** $\dfrac{2x}{x^2 + 3}$　　**5.** $\dfrac{1}{2(x - 4)}$　　**7.** $\dfrac{4}{x}(\ln x)^3$

9. $2 \ln x + 2$　　**11.** $\dfrac{2x^2 - 1}{x(x^2 - 1)}$　　**13.** $\dfrac{1}{x(x + 1)}$

15. $\dfrac{2}{3(x^2 - 1)}$　　**17.** $-\dfrac{4}{x(4 + x^2)}$　　**19.** $e^{-x}\left(\dfrac{1}{x} - \ln x\right)$

21. $\dfrac{e^x - e^{-x}}{e^x + e^{-x}}$　　**23.** 2　　**25.** -3　　**27.** 2　　**29.** 1.404

31. 5.585　　**33.** -0.631　　**35.** $(\ln 3)3^x$　　**37.** $\dfrac{1}{x \ln 2}$

39. $(2 \ln 4)4^{2x-3}$　　**41.** $\dfrac{2x + 6}{(x^2 + 6x) \ln 10}$　　**43.** $2^x(1 + x \ln 2)$

45. $y = 3x - 3 = 3(x - 1)$　　**47.** $y = 2x - e$

49. $y = -\dfrac{8}{5}x - 4$　　**51.** $y = \dfrac{1}{27 \ln 3}x - \dfrac{1}{\ln 3} + 3$

53. $\dfrac{2xy}{3 - 2y^2}$　　**55.** $\dfrac{y(1 - 6x^2)}{1 + y}$　　**57.** $y = x - 1$

59. $\dfrac{1}{2x}$　　**61.** $x(6 \ln x + 5)$　　**63.** $(\ln 5)^2 \, 5^x$

65. $\dfrac{d\beta}{dI} = \dfrac{10}{(\ln 10)I}$, so for $I = 10^{-4}$, the rate of change is about 43,429.4 db/W/cm².

67.　　　　　　　**69.**

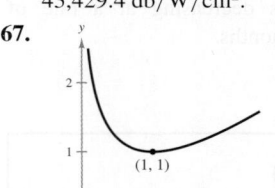　　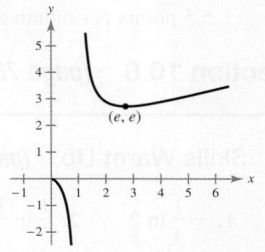

Relative minimum: $(1, 1)$　　Discontinuity: $x = 1$

Relative minimum: (e, e)

Point of inflection: $\left(e^2, \dfrac{e^2}{2}\right)$

71.

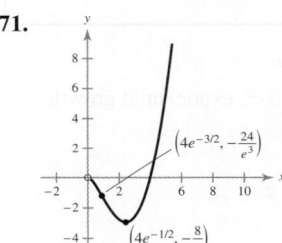

Relative minimum: $\left(4e^{-1/2}, \dfrac{-8}{e}\right)$

Point of inflection: $\left(4e^{-3/2}, \dfrac{-24}{e^3}\right)$

73. $-\dfrac{1}{p}, -\dfrac{1}{10}$

75. $p = 1000e^{-x}$

$\dfrac{dp}{dx} = -1000e^{-x}$

At $p = 10$, rate of change $= -10$.

$\dfrac{dp}{dx}$ and $\dfrac{dx}{dp}$ are reciprocals of each other.

77. (a) $\overline{C} = \dfrac{500 + 300x - 300 \ln x}{x}$

(b) Minimum of 279.15 at $e^{8/3}$

79. (a)　　　　　　　(b) 19.45 (thousand) per year

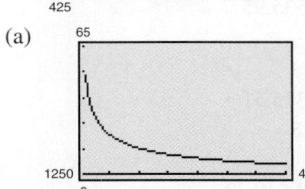

81. (a)

(b) $t \approx 30$; \$503,434.80　　(c) $t \approx 20$; \$386,685.60

(d) ≈ -0.081; ≈ -0.029

(e) For a higher monthly payment, the term is shorter, and the total amount paid is smaller.

CHAPTER 10

83. (a) $s(t) = 84.66 - 11.00 \ln t$

(b)
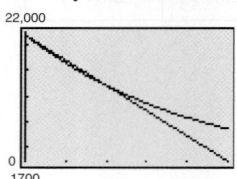

The model fits the data well.

(c) -5.5; The average score is decreasing at a rate of 5.5 points per month after 2 months.

Section 10.6   *(page 791)*

Skills Warm Up   *(page 791)*

1. $-\dfrac{1}{4} \ln 2$ **2.** $\dfrac{1}{5} \ln \dfrac{10}{3}$ **3.** $-\dfrac{\ln(25/16)}{0.01}$

4. $-\dfrac{\ln(11/16)}{0.02}$ **5.** $7.36e^{0.23t}$ **6.** $1.296e^{0.072t}$

7. $-33.6e^{-1.4t}$ **8.** $-0.025e^{-0.001t}$ **9.** 4

10. 12 **11.** $2x + 1$ **12.** $x^2 + 1$

1. $y = 2e^{0.1014t}$ **3.** $y = 4e^{-0.4159t}$

5. $y = 4\sqrt[3]{2}e^{((\ln 0.5)/3)t}$ **7.** $y = 10e^{2t}$, exponential growth

9. $y = 30e^{-4t}$, exponential decay

11. *Amount after 1000 years:* 6.48 g

Amount after 10,000 years: 0.13 g

13. *Initial quantity:* 6.73 g

Amount after 1000 years: 5.96 g

15. *Initial quantity:* 2.16 g

Amount after 10,000 years: 1.62 g

17. 68%   **19.** 15,642 yr

21. $k_1 = \dfrac{\ln 4}{12} \approx 0.1155$, so $y_1 = 5e^{0.1155t}$.

$k_2 = \dfrac{1}{6}$, so $y_2 = 5(2)^{t/6}$.

Explanations will vary.

23. (a) 1350   (b) $\dfrac{5 \ln 2}{\ln 3} \approx 3.15$ hr

(c) No. Answers will vary.

25. *Time to double:* 5.78 yr

Amount after 10 years: $3320.12

Amount after 25 years: $20,085.54

27. *Annual rate:* 8.66%

Amount after 10 years: $1783.04

Amount after 25 years: $6535.95

29. *Annual rate:* 9.50%

Time to double: 7.30 yr

Amount after 25 years: $5375.51

31. *Initial investment:* $6376.28

Time to double: 15.40 yr

Amount after 25 years: $19,640.33

33. $49,787.07

35. (a) Answers will vary.   (b) Answers will vary.

37. Answers will vary.

39. (a) $y = 21,500 - 3950x$

(b) $y = 21,500e^{-0.229x}$

(c) Linear Model:

after 1 year: $17,550

after 4 years: $5700

Exponential Model:

after 1 year: $17,099.56

after 4 years: $8602.50

(d)
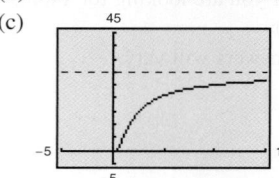

The exponential model depreciates slightly faster.

(e) After the second year, a buyer would gain an advantage by using the linear model, because it yields a lower value for the vehicle. A seller would want to use the exponential model, because it yields a higher value for the vehicle.

41. (a) $C = 30$

$k = \ln\left(\dfrac{1}{6}\right) \approx -1.7918$

(b) $30e^{-0.35836} = 20.9646$ thousand, or 20,965 units

(c)
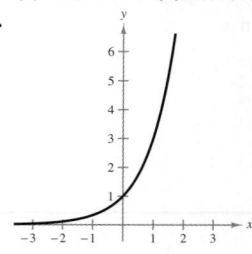

43. About 36 days

45. (a) $C \approx 81.090$, $k = \dfrac{\ln(45/40)}{-200} \approx -0.0005889$

(b) $x = 1/0.0005889 \approx 1698$ units, $p \approx \$29.83$

47. Answers will vary.

Review Exercises   *(page 798)*

1. (a) 16,384   (b) 117,649   (c) 0.0625   (d) 81

3.
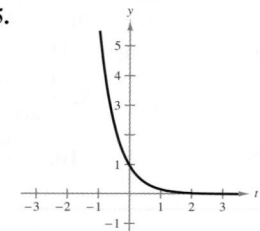

5.
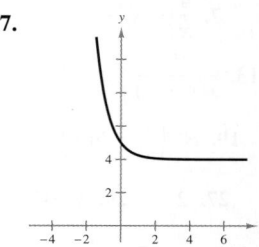

7.

9. (a) 5894.39 (thousand)   (b) 6203.76 (thousand)

11. (a) $69,295.66   (b) $233,081.88

13. (a) e^{10} (b) $\dfrac{1}{e^2}$ (c) $e^{11/2}$ (d) $\dfrac{1}{e^8}$

15.
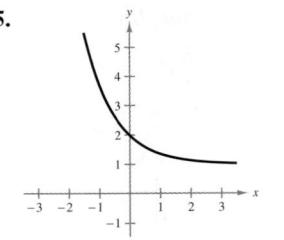
17.
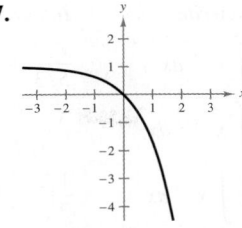

19.

n	1	2	4	12
A	\$1216.65	\$1218.99	\$1220.19	\$1221.00

n	365	Continuous compounding
A	\$1221.39	\$1221.40

21.

n	1	2	4	12
A	4231.80	4244.33	4250.73	4255.03

n	365	Continuous compounding
A	4257.13	4257.20

23. b   **25.** (a) 6%   (b) 6.09%   (c) 6.14%   (d) 6.17%
27. \$10,338.10
29. (a) \$8276.81   (b) \$7697.12   (c) \$7500
31. (a) The model fits the data very well.
   (b) $y = 116.85x + 111.1$
      The linear model fits the data moderately well.
      The exponential model is a better fit.
   (c) Exponential: \$4357.50 (million)
      Linear: \$1863.85 (million)
33. (a) $P \approx 1049$ fish
   (b)
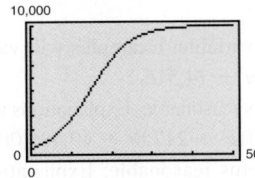
      13 months
   (c) Yes, P approaches 10,000 fish as t approaches ∞.
35. $8xe^{x^2}$ **37.** $\dfrac{1 - 2x}{e^{2x}}$ **39.** $-\dfrac{10e^{2x}}{(1 + e^{2x})^2}$
41. $y = 3 - x$ **43.** $y = \dfrac{x}{e}$
45.
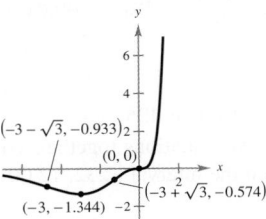
Relative minimum: $(-3, -1.344)$
Inflection points: $(0, 0)$, $(-3 + \sqrt{3}, -0.574)$,
      and $(-3 - \sqrt{3}, -0.933)$
Horizontal asymptote: $y = 0$

47.
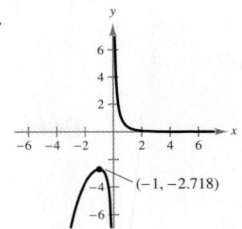
Relative maximum: $(-1, -2.718)$
Horizontal asymptote: $y = 0$
Vertical asymptote: $x = 0$

49. $e^{2.4849} \approx 12$ **51.** $\ln 4.4816 \approx 1.5$
53.
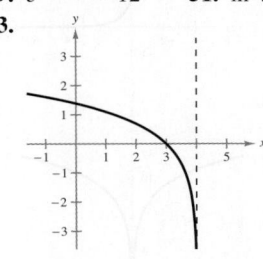
55.

57. $\ln x + \frac{1}{2}\ln(x - 1)$ **59.** $2 \ln x - 3 \ln(x + 1)$
61. $3[\ln(1 - x) - \ln 3 - \ln x]$
63. $\ln(2x^2 - x - 15)$ **65.** $4 \ln\left(\dfrac{x^5 - x^2}{x - 5}\right)$
67. 3   **69.** $e^3 \approx 20.09$ **71.** 1
73. $\dfrac{3 + \sqrt{13}}{2} \approx 3.3028$ **75.** $-\dfrac{\ln(0.25)}{1.386} \approx 1.0002$
77. $\frac{1}{2}(\ln 6 + 1) \approx 1.3959$ **79.** $\dfrac{\ln 1.1}{\ln 1.21} = 0.5$
81. (a) ≈ 28.07 years   (b) ≈ 27.75 years
   (c) ≈ 27.73 years   (d) ≈ 27.73 years
83. (a) 75   (b) 65.34   (c) ≈ 11 months
85. $\dfrac{2}{x}$ **87.** $\dfrac{1}{x} + \dfrac{1}{x - 1} - \dfrac{1}{x - 2} = \dfrac{x^2 - 4x + 2}{x(x - 2)(x - 1)}$
89. 2   **91.** $\dfrac{1 - 3 \ln x}{x^4}$ **93.** $\dfrac{4x}{3(x^2 - 2)}$
95. $\dfrac{2}{x} + \dfrac{1}{2(x + 1)}$ **97.** $\dfrac{1}{1 + e^x}$ **99.** 2   **101.** 0
103. 1.594   **105.** 1.500   **107.** $(2 \ln 5)5^{2x + 1}$
109. $\dfrac{2}{(2x - 1) \ln 3}$ **111.** $\dfrac{-1}{\ln 10} \cdot \dfrac{1}{x} = -\dfrac{1}{x \ln 10}$
113.
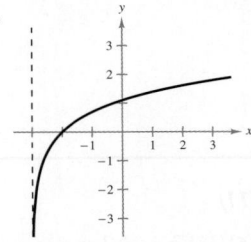
115.
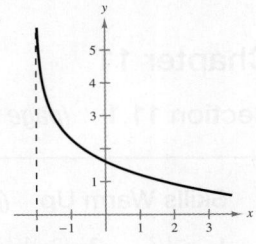
No relative extrema   No relative extrema
No points of inflection   No points of inflection
117. 2005: 256.4 (million)
      2008: 160.25 (million)
119. $y = 3e^{-0.27465t}$ **121.** 5.19 g; 0.10 g
123. 20.18 g; 17.88 g   **125.** 2.47 g; 1.85 g
127. 8.66 yr, \$1335.32, \$4433.43
129. 2%, 34.66 yr, \$24,730.82
131. (a) $D = 500e^{-0.38376t}$
   (b) 107.72 milligrams per milliliter

CHAPTER 10

Test Yourself   *(page 802)*

1. 1   **2.** $\frac{1}{256}$ **3.** $e^{9/2}$ **4.** e^{12}

5.
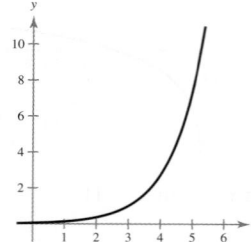

6.
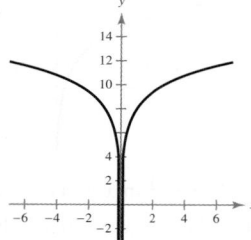

7.
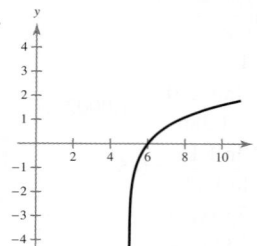

8.
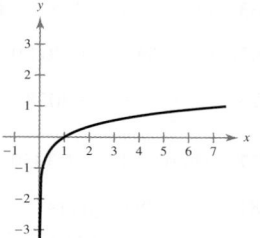

9.
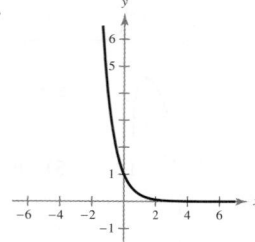

10.

11. $\ln 3 - \ln 2$ **12.** $\frac{1}{2}\ln(x + y)$ **13.** $\ln(x + 1) - \ln y$

14. $\ln[y(x + 1)]$ **15.** $\ln\dfrac{x^3}{(x - 1)^2}$ **16.** $\ln\dfrac{xy^4}{\sqrt{(z + 4)}}$

17. $x \approx 3.197$ **18.** $x \approx 1.750$ **19.** $x \approx 58.371$

20. (a) 17.67 yr   (b) 17.36 yr   (c) 17.33 yr   (d) 17.33 yr

21. $-3e^{-3x}$ **22.** $7e^{x+2} + 2$

23. $\dfrac{2x}{3 + x^2}$ **24.** $\dfrac{2}{x(x + 2)}$

25. (a) \$2241.54 million   (b) \$138.30 million/yr

26. 59.4%   **27.** 39.61 yr

Chapter 11

Section 11.1   *(page 811)*

Skills Warm Up   *(page 811)*

1. $x^{-1/2}$ **2.** $(2x)^{4/3}$ **3.** $5^{1/2}x^{3/2} + x^{5/2}$
4. $x^{-1/2} + x^{-2/3}$ **5.** $(x + 1)^{5/2}$ **6.** $x^{1/6}$
7. -12 **8.** -10 **9.** 14   **10.** 14

1–5. Answers will vary.   **7.** $u + C$ **9.** $6x + C$
11. $\frac{7}{2}x^2 + C$ **13.** $\frac{5}{3}t^3 + C$
15. $-\dfrac{5}{2x^2} + C$ **17.** $\frac{2}{5}y^{5/2} + C$

	Rewrite	Integrate	Simplify
19.	$\int x^{2/3}\,dx$	$\dfrac{x^{5/3}}{5/3} + C$	$\dfrac{3}{5}x^{5/3} + C$
21.	$\int x^{-3/2}\,dx$	$\dfrac{x^{-1/2}}{-1/2} + C$	$-\dfrac{2}{\sqrt{x}} + C$
23.	$\dfrac{1}{2}\int x^{-3}\,dx$	$\dfrac{1}{2}\left(\dfrac{x^{-2}}{-2}\right) + C$	$-\dfrac{1}{4x^2} + C$

25. $\dfrac{x^2}{2} + 3x + C$ **27.** $\dfrac{1}{4}x^4 + 2x + C$

29. $\dfrac{3}{4}x^4 - 2x^3 + 2x + C$ **31.** $\dfrac{1}{3}x^3 + \dfrac{5}{2}x^2 + x + C$

33. $2x + \dfrac{1}{2x^2} + C$ **35.** $3x^{2/3}(x + 2) + C$

37.
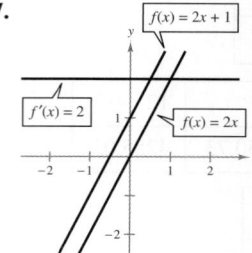

39.
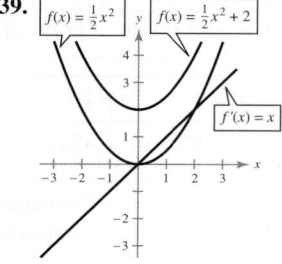

41. $f(x) = 2x^2 + 6$ **43.** $f(x) = x^2 + 4x + 7$

45. $f(x) = 5x^2 - 3x^4 + 200$ **47.** $f(x) = -\dfrac{1}{x^2} + \dfrac{1}{x} + \dfrac{1}{2}$

49. $f(x) = x^2 + x + 4$ **51.** $f(x) = \frac{9}{4}x^{4/3}$

53. $C = 85x + 5500$ **55.** $C = \frac{1}{10}\sqrt{x} + 4x + 750$

57. $R = 225x - \frac{3}{2}x^2$, $p = 225 - \frac{3}{2}x$

59. $P = -9x^2 + 1650x$ **61.** $P = -12x^2 + 805x + 68$

63. $s(t) = -16t^2 + 6000$; about 19.36 sec

65. $v_0 = 40\sqrt{22} \approx 187.62$ ft/sec

67. (a) $C = x^2 - 12x + 125$ (b) \$2025

$\overline{C} = x - 12 + \dfrac{125}{x}$

(c) \$125 is fixed. \$1900 is variable. Examples will vary.

69. (a) $P(t) = 79.4t^2 + 1758.6t + 74,515.2$

(b) 314,437.2; Yes, this seems reasonable. Explanations will vary.

71. (a) $M(t) = -0.035t^3 + 7.01t^2 + 217.8t + 49,486.005$

(b) 64,195.63; Yes, this seems reasonable. Explanations will vary.

73. (a)
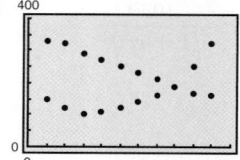

$\dfrac{dB}{dx} = -19.9x + 351$

$\dfrac{dC}{dx} = 5.38x^2 - 40.6x + 182$

(b) $B(x) = -9.95x^2 + 351x$;
$C(x) = 1.79x^3 - 20.3x^2 + 182x + 425$

(c) Graphing the benefit and cost equations together, you see that benefit exceeds cost on the interval (2.32, 12.00).

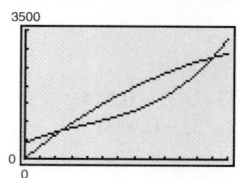

The company should produce from 3 to 11 units.

Section 11.2   *(page 821)*

Skills Warm Up   *(page 821)*

1. $\frac{1}{2}x^4 + x + C$ **2.** $\frac{3}{2}x^2 + \frac{2}{3}x^{3/2} - 4x + C$

3. $-\dfrac{1}{x} + C$ **4.** $-\dfrac{1}{6t^2} + C$ **5.** $\frac{4}{7}t^{7/2} + \frac{2}{5}t^{5/2} + C$

6. $\frac{4}{5}x^{5/2} - \frac{2}{3}x^{3/2} + C$ **7.** $\dfrac{5x^3 - 4}{2x} + C$

8. $\dfrac{-6x^2 + 5}{3x^3} + C$ **9.** $\frac{2}{5}\sqrt{x}(8x^2 + 15) + C$

$$\int u^n \frac{du}{dx}\, dx \qquad\qquad u \qquad\qquad \frac{du}{dx}$$

1. $\displaystyle\int (5x^2 + 1)^2(10x)\, dx$ $5x^2 + 1$ $10x$

3. $\displaystyle\int \sqrt{1 - x^2}\,(-2x)\, dx$ $1 - x^2$ $-2x$

5. $\displaystyle\int \left(4 + \frac{1}{x^2}\right)^5\!\left(\frac{-2}{x^3}\right) dx$ $4 + \dfrac{1}{x^2}$ $-\dfrac{2}{x^3}$

7. $\displaystyle\int \left(1 + \sqrt{x}\right)^3\!\left(\frac{1}{2\sqrt{x}}\right) dx$ $1 + \sqrt{x}$ $\dfrac{1}{2\sqrt{x}}$

9. $\frac{1}{5}(x - 1)^5 + C$ **11.** $\frac{1}{5}(1 + 2x)^5 + C$

13. $\frac{1}{2}(x^2 + 3x)^2 + C$ **15.** $\frac{2}{3}(4x^2 - 5)^{3/2} + C$

17. $\dfrac{1}{3(5 - 3x^2)^3} + C$ **19.** $\dfrac{(2x^3 - 1)^5}{30} + C$

21. $\frac{1}{3}(t^2 + 6)^{3/2} + C$ **23.** $\dfrac{1}{12(x^6 - 4)^2} + C$

25. $\frac{1}{10}(x^2 - 6x)^5 + C$ **27.** $-\dfrac{1}{2(x^2 + 2x - 3)} + C$

29. $-\dfrac{15}{8}(1 - x^2)^{4/3} + C$ **31.** $-\dfrac{3}{2(1 + x^2)} + C$

33. $-3\sqrt{2t + 3} + C$ **35.** $\frac{1}{4}(6x^2 - 1)^4 + C$

37. $\frac{3}{16}(4x + 3)^{4/3} + C$ **39.** $\sqrt{x^2 + 25} + C$

41. $\frac{2}{3}\sqrt{x^3 + 3x + 4} + C$

43. (a) $\frac{1}{3}x^3 - x^2 + x + C_1 = \frac{1}{3}(x - 1)^3 + C_2$

(b) Answers differ by a constant: $C_1 = C_2 - \frac{1}{3}$

(c) Answers will vary.

45. (a) $\frac{1}{6}x^6 - \frac{1}{2}x^4 + \frac{1}{2}x^2 + C_1 = \dfrac{(x^2 - 1)^3}{6} + C_2$

(b) Answers differ by a constant: $C_1 = C_2 - \frac{1}{6}$

(c) Answers will vary.

47. $f(x) = \frac{1}{12}(4x^2 - 10)^3 - 8$

49. (a) $C = 8\sqrt{x + 1} + 18$ (b) \$75.13

51. $x = \frac{1}{3}(p^2 - 25)^{3/2} + 24$ **53.** $x = \dfrac{6000}{\sqrt{p^2 - 16}} + 3000$

55. (a) $h = \sqrt{17.6t^2 + 1} + 5$ (b) 26 in.

57. (a) $Q = (x - 24{,}999)^{0.95} + 24{,}999$

(b)

x	25,000	50,000	100,000	150,000
Q	25,000	40,067.14	67,786.18	94,512.29
$x - Q$	0	9932.86	32,213.82	55,487.71

(c)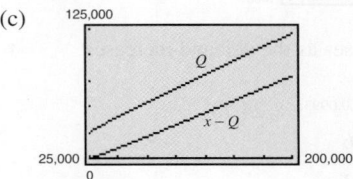

59. $-\frac{2}{3}x^{3/2} + \frac{2}{3}(x + 1)^{3/2} + C$

Section 11.3   *(page 828)*

Skills Warm Up   *(page 828)*

1. $x + 2 - \dfrac{2}{x + 2}$ **2.** $x - 2 + \dfrac{1}{x - 4}$

3. $x + 8 + \dfrac{2x - 4}{x^2 - 4x}$ **4.** $x^2 - x - 4 + \dfrac{20x + 22}{x^2 + 5}$

5. $\frac{1}{4}x^4 - \dfrac{1}{x} + C$ **6.** $\frac{1}{2}x^2 + 2x + C$

7. $\frac{1}{2}x^2 - \dfrac{4}{x} + C$ **8.** $-\dfrac{1}{x} - \dfrac{3}{2x^2} + C$

1. $e^{2x} + C$ **3.** $\frac{1}{4}e^{4x} + C$ **5.** $\frac{1}{5}e^{5x-3} + C$

7. $-\frac{9}{2}e^{-x^2} + C$ **9.** $\frac{5}{3}e^{x^3} + C$ **11.** $e^{x^2+x} + C$

13. $\ln|x + 1| + C$ **15.** $\ln|5x + 2| + C$

17. $-\frac{1}{2}\ln|3 - 2x| + C$ **19.** $\frac{2}{3}\ln|3x + 5| + C$

21. $\ln\sqrt{x^2 + 1} + C$ **23.** $\frac{1}{3}\ln|x^3 + 1| + C$

25. $\frac{1}{2}\ln|x^2 + 6x + 7| + C$ **27.** $\ln|\ln x| + C$

29. $\ln|1 - e^{-x}| + C$

31. $\frac{1}{4}x^2 - 4\ln|x| + C$; General Power Rule and Log Rule

33. $8x + 3\ln|x| - \dfrac{3}{x^2} + C$; General Power Rule and Log Rule

35. $e^x + 2x - e^{-x} + C$;
Exponential Rule and General Power Rule

37. $-\frac{2}{3}(1 - e^x)^{3/2} + C$; Exponential Rule

39. $\ln|e^x + x| + C$; Log Rule

41. $\frac{1}{7}\ln(7e^{5x} + 1) + C$; Log Rule

43. $\frac{1}{2}x^2 + 3x + 8\ln|x - 1| + C$;
General Power Rule and Log Rule

45. $x - 6\ln|x + 3| + C$;
General Power Rule, General Log Rule

47. $f(x) = \dfrac{-e^{2/x}}{2} + \dfrac{e^{1/2}}{2} + 6$

49. $f(x) = \frac{1}{2}x^2 + 5x + 8\ln|x - 1| - 8$

51. (a) $P(t) = 1000[1 + \ln(1 + 0.25t)^{12}]$

(b) $P(3) \approx 7715$ bacteria   (c) $t \approx 6$ days

53. (a) $p = -50e^{-x/500} + 45.06$

(b)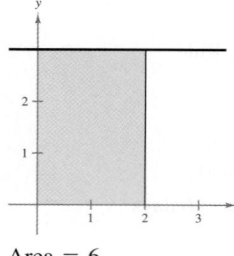

The price increases as the demand increases.

(c) 387

55. (a) $R(t) = 3223.56e^{0.0993t} + 24.78$

(b) $R(9) \approx \$7903.66$

57. False. $\ln x^{1/2} = \frac{1}{2}\ln x$

Quiz Yourself   *(page 830)*

1. $3x + C$ **2.** $5x^2 + C$ **3.** $-\dfrac{1}{4x^4} + C$

4. $\dfrac{x^3}{3} - x^2 + 15x + C$ **5.** $\dfrac{(6x+1)^4}{4} + C$

6. $\dfrac{1}{50}(5x^2 - 2)^5 + C$ **7.** $\dfrac{(x^2 - 5x)^2}{2} + C$

8. $-\dfrac{1}{2(x^3 + 3)^2} + C$ **9.** $\dfrac{2}{15}(5x + 2)^{3/2} + C$

10. $f(x) = 8x^2 + 1$ **11.** $f(x) = 3x^3 + 4x - 2$

12. (a) $C = -0.03x^2 + 16x + 9.03$ **(b)** \$9.03   **(c)** \$509.03

13. $f(x) = \dfrac{2}{3}x^3 + x + 1$

14. (a) 1000 bolts   **(b)** About 8612 bolts

15. $e^{5x+4} + C$ **16.** $e^{x^3} + C$

17. $\frac{1}{2}e^{(x^2 - 6x)} + C$ **18.** $\ln|2x - 1| + C$

19. $-\frac{1}{8}\ln|3 - 8x| + C$ **20.** $\frac{1}{6}\ln(3x^2 + 4) + C$

21. (a) $S(t) = 13.16t^2 + 848.99\ln(t) + 2504.44$

(b) \$5112.11 million

Section 11.4   *(page 840)*

Skills Warm Up   *(page 840)*

1. $\frac{3}{2}x^2 + 7x + C$ **2.** $\frac{2}{5}x^{5/2} + \frac{4}{3}x^{3/2} + C$

3. $\dfrac{1}{5}\ln|x| + C$ **4.** $-\dfrac{1}{6e^{6x}} + C$

5. $C = 0.008x^{5/2} + 29,500x + C$

6. $R = x^2 + 9000x + C$

7. $P = 25,000x - 0.005x^2 + C$

8. $C = 0.01x^3 + 4600x + C$

1.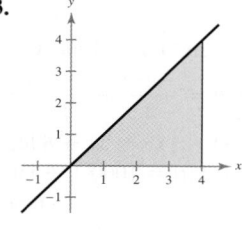

Area = 6

3.

Area = 8

5.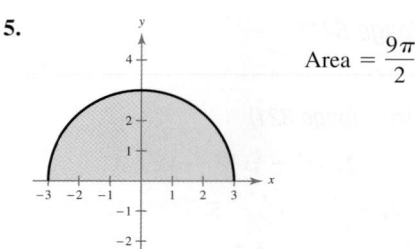

Area = $\dfrac{9\pi}{2}$

7. (a) 8   **(b)** 4   **(c)** -24 **(d)** 0

9. $\dfrac{1}{6}$ **11.** $\dfrac{1}{2}$ **13.** $6\left(1 - \dfrac{1}{e^2}\right)$ **15.** $8\ln 2 + \dfrac{15}{2}$

17. 1   **19.** $-\frac{5}{2}$ **21.** 38   **23.** $-\frac{15}{4}$ **25.** -4

27. $-\frac{27}{20}$ **29.** $6\ln 2 \approx 4.16$ **31.** 2

33. $-e^{-1} + 1 \approx 0.63$ **35.** $\frac{1}{3}\left[(e^2 + 1)^{3/2} - 2\sqrt{2}\right] \approx 7.157$

37. $\frac{1}{8}\ln 17 \approx 0.354$ **39.** 10   **41.** 39   **43.** 10

45. $4\ln 3 \approx 4.394$ **47.** \$6.75   **49.** \$22.50   **51.** \$3.97

53. Average = 12

$x = 2$

55. Average = $\dfrac{8}{3}$

$x = \pm\dfrac{2\sqrt{3}}{3} \approx \pm 1.155$

57. Average = $e - e^{-1} \approx 2.3504$

$x = \ln\left(\dfrac{e - e^{-1}}{2}\right) \approx 0.1614$

59. Average = $\frac{3}{4}\ln\frac{7}{3} \approx 0.6355$ **61.** $\frac{6}{5}$ **63.** 0

$x = 4/\ln(7/3) - 2 \approx 2.721$

65. (a) $\frac{1}{3}$ **(b)** $\frac{2}{3}$ **(c)** $-\frac{1}{3}$

Explanations will vary.

67. \$1925.23   **69.** \$16,605.21

71. \$2500   **73.** \$4565.65

75. (a) \$137,000   **(b)** \$214,720.93   **(c)** \$338,393.53

77. \$2623.94

79. (a) $M(t) = 273.78t^2 - 23.153t^3 - 331.258e^{-t} + 5438.258$

(b) \$8573.88 billion

Section 11.5   *(page 849)*

Skills Warm Up   *(page 849)*

1. $-x^2 + 3x + 2$ **2.** $-2x^2 + 4x + 4$

3. $-x^3 + 2x^2 + 4x - 5$ **4.** $x^3 - 6x - 1$

5. $(0, 4), (4, 4)$ **6.** $(1, -3), (2, -12)$

7. $(-3, 9), (2, 4)$ **8.** $(-2, -4), (0, 0), (2, 4)$

1. 36   **3.** 9   **5.** $e - 2$ **7.** $\dfrac{3}{2}$

9.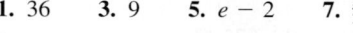

11.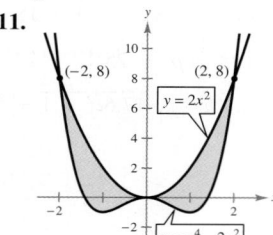

13. d

15.

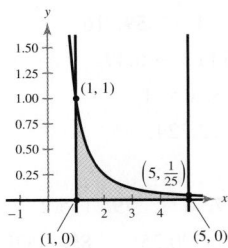

Area = $\frac{4}{5}$

17.

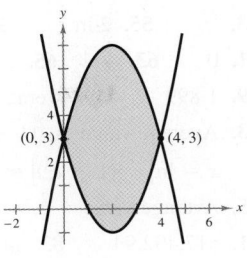

Area = $\frac{64}{3}$

19.

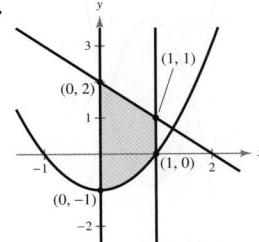

Area = $2\frac{1}{6}$

21.

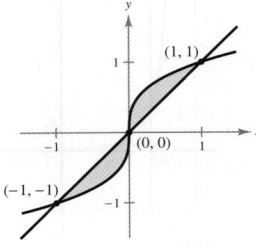

Area = $\frac{1}{2}$

23.

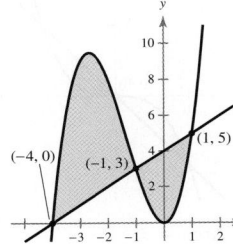

Area = $21\frac{1}{12}$

25.

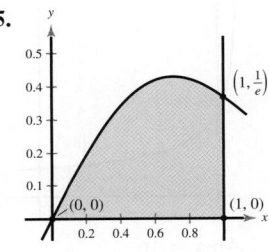

Area = $-\frac{1}{2}e^{-1} + \frac{1}{2}$

27.

Area = $(2e + \ln 2) - 2e^{1/2}$

29.

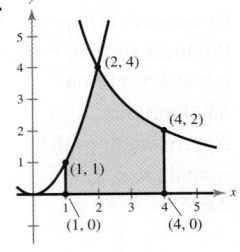

Area = $\frac{7}{3} + 8 \ln 2$

31.

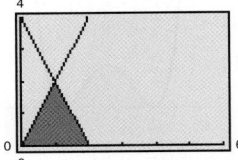

Area = $\displaystyle\int_0^1 2x\,dx + \int_1^2 (4 - 2x)\,dx$

33.

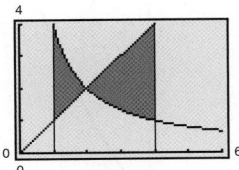

Area = $\displaystyle\int_1^2 \left(\frac{4}{x} - x\right) dx + \int_2^4 \left(x - \frac{4}{x}\right) dx$

35.

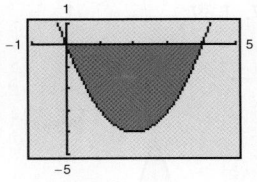

Area = $\frac{32}{3}$

37.

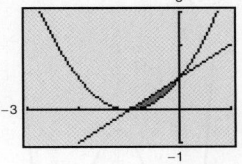

Area = $\frac{1}{6}$

39. 8

41. Consumer surplus = 1600   **43.** Consumer surplus = 500
Producer surplus = 400   Producer surplus = 2000

45. Consumer surplus = 640.00
Producer surplus ≈ 426.67

47. R_1; \$11.375 billion

49. \$573 million; Explanations will vary.

51. (a)

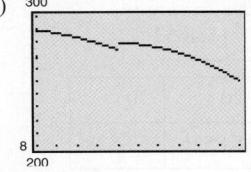

(b) 124.25 fewer pounds

53. \$333.33 million

55. CS = \$700,000
PS = \$1,375,000

57. 2077.10   **59.** Answers will vary.

Section 11.6   *(page 856)*

Skills Warm Up *(page 856)*

1. $\frac{1}{6}$ **2.** $\frac{3}{20}$ **3.** $\frac{7}{40}$ **4.** $\frac{13}{12}$ **5.** $\frac{61}{30}$ **6.** $\frac{53}{18}$
7. $\frac{2}{3}$ **8.** $\frac{4}{7}$ **9.** 0 **10.** 5

1. Approximation: 2   **3.** Approximation: 0.6730
Exact area: 2   Exact area: $\frac{2}{3} \approx 0.6667$

5. Approximation: 1.245
Exact area: $\frac{5}{4}$ = 1.25

7. 71.25   **9.** 1.079

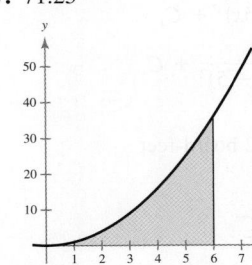

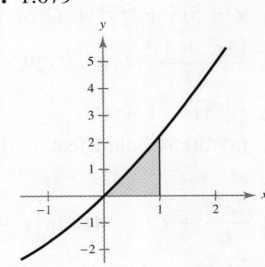

11. 24.28   **13.** 17.25

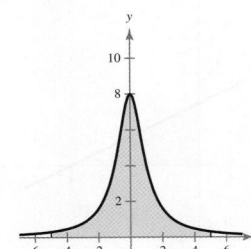

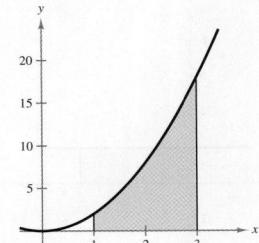

15. 34.25   **17.** 1.39

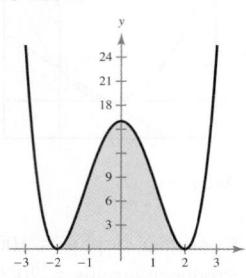

19.

n	10	20	30
Approximation	15.4543	15.4628	15.4644

n	40	50
Approximation	15.4650	15.4652

21.

n	10	20	30
Approximation	5.8520	5.8526	5.8528

n	40	50
Approximation	5.8528	5.8528

23. Area ≈ 54.6667,   **25.** Area ≈ 0.9163,
$n = 33$ $n = 3$

27. 9920 ft²   **29.** 381.6 mi²

31. Midpoint Rule: 3.1468
  Graphing utility: 3.141593

Review Exercises   *(page 862)*

1. $16x + C$ **3.** $\frac{3}{10}x^2 + C$ **5.** $x^3 + C$

7. $\frac{2}{3}x^3 + \frac{5}{2}x^2 + C$ **9.** $x^{2/3} + C$ **11.** $\frac{3}{7}x^{7/3} + \frac{3}{2}x^2 + C$

13. $\frac{4}{9}x^{9/2} - 2\sqrt{x} + C$ **15.** $6x^2 - 3$ **17.** $x^3 - 4x^2 + 15$

19. $s(t) = -16t^2 + 80t$
  5 seconds

21. $\frac{1}{4}(x + 4)^4 + C$ **23.** $\frac{1}{5}(5x + 1)^5 + C$

25. $x + 5x^2 + \frac{25}{3}x^3 + C$ or $\frac{1}{15}(1 + 5x)^3 + C_1$

27. $\dfrac{(3x^3 + 1)^3}{27} + C$ **29.** $\dfrac{-1}{12(2x^3 - 5)^2} + C$

31. $\frac{2}{5}\sqrt{5x - 1} + C$

33. (a) 30.54 board-feet   (b) 125.2 board-feet

35. $e^{4x} + C$ **37.** $-\frac{1}{5}e^{-5x} + C$

39. $\dfrac{7e^{3x^2}}{6} + C$ **41.** $\ln|x - 6| + C$

43. $\frac{2}{3}\ln|6x - 1| + C$ **45.** $-\frac{1}{3}\ln|1 - x^3| + C$

47.   **49.**

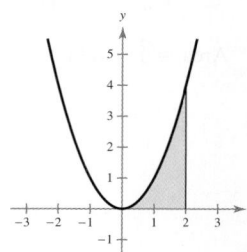 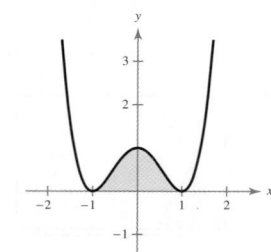

Area = 6   Area = 8

51. (a) 13   (b) 7   (c) 11   (d) 50

53. $\frac{32}{3}$ **55.** 2 ln 2   **57.** $4e^{1/2} - 4$ **59.** 16

61. 0   **63.** 4   **65.** 2   **67.** 5 ln 3 ≈ 5.49

69. 1.899   **71.** Average value = 3; $x = 1$

73. Average value = $\frac{2}{3}(1 - e^3) \approx -12.724$;
  $x = \ln\left[-\frac{1}{3}(1 - e^3)\right] \approx 1.850$

75. Average value = $\frac{2}{5}$; $x = \frac{25}{4}$ **77.** 0   **79.** 115.2

81. \$17,492.94   **83.** Increases by \$700.25   **85.** \$520.54

87.   **89.**

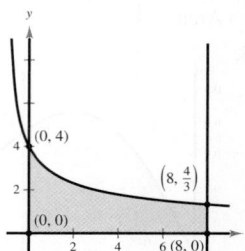 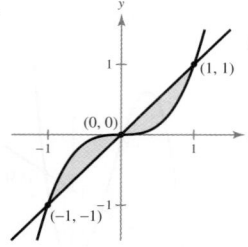

Area = $\frac{4}{9}$ Area = $\frac{64}{3}$

91.   **93.**

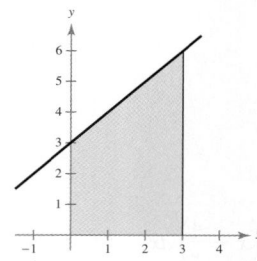 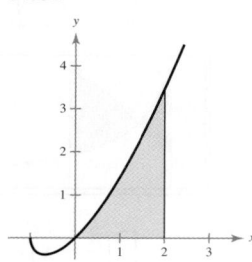

Area = 16   Area = $\frac{1}{2}$

95. Consumer surplus = 1417.5
  Producer surplus = 202.5

97. Consumer surplus = 1250
  Producer surplus = 1250

99. R_2; \$84.5 million

101. \$300 million

103. Approximation: 1.5
  Actual area: 1.5

105. 2.625   **107.** 1.070

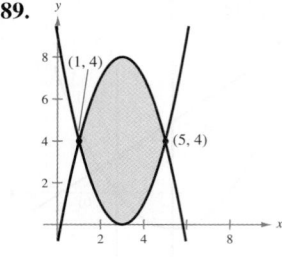

109. 13.5   **111.** 3.032

113. 9840 ft²

Test Yourself   *(page 866)*

1. $3x^3 - 2x^2 + 13x + C$ **2.** $\dfrac{(x+1)^3}{3} + C$

3. $\dfrac{2(x^4 - 7)^{3/2}}{3} + C$ **4.** $\dfrac{10x^{3/2}}{3} - 12x^{1/2} + C$

5. $5e^{3x} + C$ **6.** $\frac{3}{4}\ln|4x - 1| + C$

7. $f(x) = 3x^2 - 5x - 2$ **8.** $f(x) = e^x + x$

9. 8   **10.** 18   **11.** $\frac{2}{3}$

12. $2\sqrt{5} - 2\sqrt{2} \approx 1.644$ **13.** $\frac{1}{4}(e^{12} - 1) \approx 40{,}688.4$

14. $\ln 6 \approx 1.792$

15. (a) $S(t) = 2240e^{0.1013t} + 4.3906, \ 0 \le t \le 9$

(b) \$3661.68 million

16.

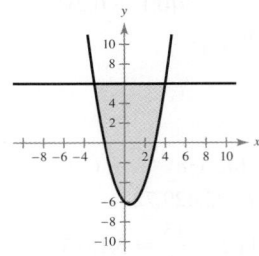

Area $= \dfrac{343}{6} \approx 57.167$

17.

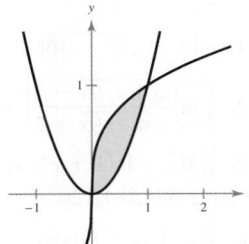

Area $= \dfrac{5}{12}$

18. Consumer surplus $= 20$ million

Producer surplus $= 8$ million

19. Midpoint Rule: $\frac{63}{64} \approx 0.9844$ **20.** Midpoint Rule: $\frac{21}{8} = 2.625$

Exact area: 1   Exact area: $\frac{8}{3} = 2.\overline{6}$

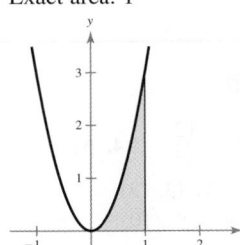

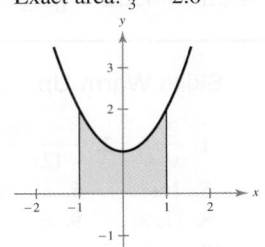

Chapter 12

Section 12.1   *(page 875)*

Skills Warm Up   *(page 875)*

1. $\dfrac{1}{x+1}$ **2.** $\dfrac{2x}{x^2 - 1}$ **3.** $3x^2e^{x^3}$

4. $-2xe^{-x^2}$ **5.** $e^x(x^2 + 2x)$ **6.** $e^{-2x}(1 - 2x)$

7. $\frac{64}{3}$ **8.** $\frac{4}{3}$ **9.** 36   **10.** 8

---

1. $u = x; \ dv = e^{3x}\,dx$ **3.** $u = \ln 2x; \ dv = x\,dx$

5. $\frac{1}{3}xe^{3x} - \frac{1}{9}e^{3x} + C$ **7.** $\dfrac{x^4}{16}(4\ln x - 1) + C$

9. $x\ln 2x - x + C$ **11.** $-x^2e^{-x} - 2xe^{-x} - 2e^{-x} + C$

13. $\frac{2}{3}x^{3/2}\left(\ln x - \frac{2}{3}\right) + C$ **15.** $2x^2e^x - 4e^xx + 4e^x + C$

17. $\frac{1}{4}e^{4x} + C$ **19.** $\frac{1}{4}xe^{4x} - \frac{1}{16}e^{4x} + C$

21. $-4e^{-x/4}(x + 4) + C$

23. $\frac{1}{2}t^2\ln(t+1) - \frac{1}{2}\ln(t+1) - \frac{1}{4}t^2 + \frac{1}{2}t + C$

25. $-e^{1/t} + C$ **27.** $\frac{1}{2}x^2(\ln x)^2 - \frac{1}{2}x^2\ln x + \frac{1}{4}x^2 + C$

29. $\dfrac{1}{3}(\ln x)^3 + C$ **31.** $-\dfrac{1}{x}(\ln x + 1) + C$

33. $\frac{2}{3}x(x-1)^{3/2} - \frac{4}{15}(x-1)^{5/2} + C$

35. $\dfrac{1}{4}x^4 + \dfrac{2}{3}x^3 + \dfrac{1}{2}x^2 + C$ **37.** $\dfrac{e^{2x}}{4(2x+1)} + C$

39. $\frac{5}{36}e^6 + \frac{1}{36} \approx 56.060$ **41.** $\frac{3}{2}\ln 3 - 1 \approx 0.648$

43. $\frac{1192}{15} \approx 79.467$ **45.** $e(2e - 1) \approx 12.060$

47. Area $= 2e^2 + 6 \approx 20.778$

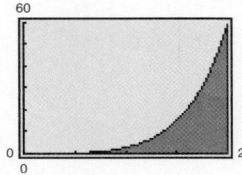

49. Area $= \dfrac{e - 2}{e}$ **51.** Area $= \dfrac{1}{9}(2e^3 + 1)$

≈ 0.2642 ≈ 4.575

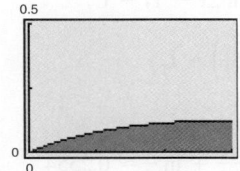   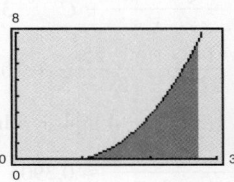

53. Proof   **55.** $\dfrac{e^{5x}}{125}(25x^2 - 10x + 2) + C$

57. $-\dfrac{1}{9x^3}(1 + 3\ln x) + C$ **59.** $\dfrac{3}{128} - \dfrac{379}{128}e^{-8} \approx 0.022$

61. $\dfrac{1{,}171{,}875}{256}\pi \approx 14{,}381.070$

63.

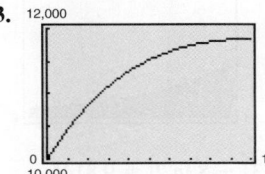

(a) Increase

(b) 113,212 units

(c) 11,321 units/yr

65. (a) $3.2\ln 2 - 0.2 \approx 2.018$

(b) $12.8\ln 4 - 7.2\ln 3 - 1.8 \approx 8.035$

67. \$18,482.03   **69.** \$931,265.10   **71.** \$4103.07

73. \$1,055,267   **75.** (a) \$1,200,000   (b) \$1,094,142.27

77. (a) \$18,000,000   (b) \$16,133,084   **79.** \$45,957.78

81. (a) \$17,378.62   (b) \$3681.26   **83.** ≈ 4.254

Section 12.2   *(page 883)*

Skills Warm Up   *(page 883)*

1. $x^2 + 8x + 16$ **2.** $x^2 - 2x + 1$

3. $x^2 + x + \frac{1}{4}$ **4.** $x^2 - \frac{2}{3}x + \frac{1}{9}$

5. $2e^x(x - 1) + C$ **6.** $x^3\ln x - \dfrac{x^3}{3} + C$

---

1. $\frac{1}{9}\left(\frac{2}{2+3x}+\ln|2+3x|\right)+C$

3. $\frac{2(3x-4)}{27}\sqrt{2+3x}+C$ **5.** $\ln\left(x^2+\sqrt{x^4-9}\right)+C$

7. $\frac{1}{2}(x^2-1)e^{x^2}+C$ **9.** $\ln\left|\frac{x}{1+x}\right|+C$

11. $-\frac{1}{3}\ln\left|\frac{3+\sqrt{x^2+9}}{x}\right|+C$

13. $-\frac{1}{2}\ln\left|\frac{2+\sqrt{4-x^2}}{x}\right|+C$ **15.** $\frac{3}{4}x^2[-1+2\ln(3x)]+C$

17. $3x^2-\ln(1+e^{3x^2})+C$

19. $\frac{2}{35}(x+3)^{3/2}(5x^2-12x+24)+C$

21. $\frac{1}{27}\left[\frac{4}{2+3t}-\frac{2}{(2+3t)^2}+\ln|2+3t|\right]+C$

23. $\frac{1}{\sqrt{3}}\ln\left|\frac{\sqrt{3+4x}-\sqrt{3}}{\sqrt{3+4x}+\sqrt{3}}\right|+C$

25. $-\frac{1}{2}x(2-x)+\ln|x+1|+C$

27. $\frac{1}{8}\left[\frac{-1}{2(3+2x)^2}+\frac{2}{(3+2x)^3}-\frac{9}{4(3+2x)^4}\right]+C$

29. $-\frac{\sqrt{1-x^2}}{x}+C$ **31.** $\frac{4}{9}x^3(3\ln 2x-1)+C$

33. $\frac{1}{27}\left(3x-\frac{25}{3x-5}+10\ln|3x-5|\right)+C$

35. $\frac{1}{9}(3\ln x-4\ln|4+3x|)+C$

37. $\frac{-2\sqrt{2}+4}{3}\approx 0.3905$ **39.** $-\frac{5}{9}+\ln\frac{9}{4}\approx 0.2554$

41. $12\left(2+\ln\left|\frac{2}{1+e^2}\right|\right)\approx 6.7946$ **43.** $8\ln 2-\frac{15}{8}\approx 3.6702$

45. Area $=\frac{1}{8\sqrt{3}}\approx 0.0722$ **47.** Area $=\frac{3\ln\frac{4}{5}+1}{36}\approx 0.0092$

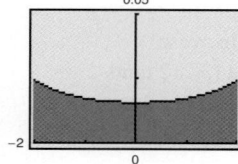
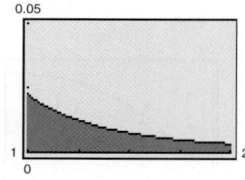

49. Area $=\frac{1}{4}\left[21\sqrt{5}-8\ln\left(\sqrt{5}+3\right)+8\ln 2\right]\approx 9.8145$

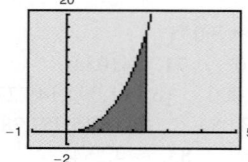

51. $x\left(\ln\frac{x}{3}-1\right)+C$ **53.** $\frac{2}{147}(7x+6)\sqrt{7x-3}+C$

55. (a) 0.483   (b) 0.283

57.

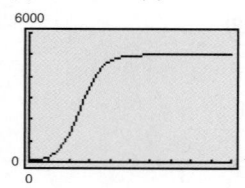

Average value: 401.40

59. $1138.43

61. Consumer surplus: ≈ 17.92
Producer surplus: 24

Quiz Yourself   *(page 885)*

1. $\frac{1}{5}xe^{5x}-\frac{1}{25}e^{5x}+C$ **2.** $3x\ln x-3x+C$

3. $\frac{1}{2}x^2\ln x+x\ln x-\frac{1}{4}x^2-x+C$

4. $\frac{2}{3}x(x+3)^{3/2}-\frac{4}{15}(x+3)^{5/2}+C$

5. $\frac{x^2}{4}\ln x-\frac{x^2}{8}+C$ **6.** $-\frac{1}{2}e^{-2x}\left(x^2+x+\frac{1}{2}\right)+C$

7. (a) 282,016 units   (b) 56,403 units

8. (a) \$784,000   (b) \$673,108.31

9. $\frac{1}{4}(2x-\ln|1+2x|)+C$ **10.** $10\ln\left|\frac{x}{0.1+0.2x}\right|+C$

11. $\ln\left|x+\sqrt{x^2-16}\right|-\frac{\sqrt{x^2-16}}{x}+C$

12. $\frac{1}{2}\ln\left|\frac{\sqrt{4+9x}-2}{\sqrt{4+9x}+2}\right|+C$

13. $\frac{1}{4}[4x^2-\ln(1+e^{4x^2})]+C$ **14.** $x^2e^{x^2+1}+C$

15. (a) \$84,281,126.52   (b) \$257,392,429.72

16. $\frac{8}{e}-4\approx -1.0570$ **17.** $10\ln 2-\frac{15}{4}\approx 3.1815$

18. $\frac{64}{3}\left(\sqrt{2}-1\right)\approx 8.8366$ **19.** $e-2\approx 0.7183$

20. $\frac{\sqrt{5}}{18}\approx 0.1242$ **21.** $\frac{1}{4}\left(\ln\frac{17}{19}-\ln\frac{7}{9}\right)\approx 0.0350$

Section 12.3   *(page 892)*

Skills Warm Up   *(page 892)*

1. $\frac{2}{x^3}$ **2.** $-\frac{96}{(2x+1)^4}$ **3.** $-\frac{12}{x^4}$ **4.** $6x-4$

5. $16e^{2x}$ **6.** $e^{x^2}(4x^2+2)$ **7.** $(3, 18)$

8. $(1, 8)$ **9.** $n<-5\sqrt{10},\ n>5\sqrt{10}$

10. $n<-5, n>5$

	Trapezoidal Rule	Simpson's Rule	Exact Value
1.	2.7500	2.6667	2.6667
3.	0.2704	0.2512	0.2499
5.	0.6941	0.6932	0.6931
7.	5.2650	5.3046	5.3333
9.	3.8643	3.3022	3.1809

11. (a) 0.783   (b) 0.785   **13.** (a) 2.540   (b) 2.541

15. (a) 3.283   (b) 3.240   **17.** (a) 1.470   (b) 1.463

19. (a) 1.879   (b) 1.888   **21.** \$21,831.20; \$21,836.98

23. \$678.36   **25.** 0.3413 = 34.13%

27. 0.5000 = 50.00%   **29.** 89,500 ft²

31. (a) $|E|\le\frac{1}{12}\approx 0.0833$ (b) $|E|\le 0$

33. (a) $|E|\le\frac{5e}{64}\approx 0.212$ (b) $|E|\le\frac{13e}{1024}\approx 0.035$

35. (a) $n=566$ (b) $n=16$

37. (a) $n=3280$ (b) $n=60$ **39.** 19.5215   **41.** 3.6558

43. (a) 36.2 years   (b) 36.2 years
(c) The results are the same.

45. 58.912 mg   **47.** 1878 subscribers

49. Answers will vary.

Section 12.4   *(page 902)*

Skills Warm Up   *(page 902)*

1. 9   **2.** 3   **3.** $-\frac{1}{8}$ **4.** Limit does not exist.

5. Limit does not exist.   **6.** -4

7. (a) $\frac{32}{3}b^3 - 16b^2 + 8b - \frac{4}{3}$ (b) $-\frac{4}{3}$

8. (a) $\dfrac{b^2 - b - 11}{(b-2)^2(b-5)}$ (b) $\dfrac{11}{20}$

9. (a) $\ln\left(\dfrac{5 - 3b^2}{b+1}\right)$ (b) $\ln 5 \approx 1.609$

10. (a) $e^{-3b^2}(e^{6b^2} + 1)$ (b) 2

1. Improper; The integrand has an infinite discontinuity when $x = \frac{2}{3}$, and $0 \le \frac{2}{3} \le 1$.

3. Not improper; continuous on $[0, 1]$

5. Not improper; continuous on $[0, 5]$

7. Converges; 1   **9.** Diverges   **11.** Diverges

13. Diverges   **15.** Diverges   **17.** Converges; 0

19. Converges; $\dfrac{1}{2\,(\ln 4)^2}$ **21.** 1   **23.** $\dfrac{1}{2}$ **25.** ∞

27. (a) 0.9026   (b) 0.0738   (c) 0.00235

29. (a) $\displaystyle\int_{35.5}^{\infty} f(x)\,dx \approx 0.9938$ (b) $\displaystyle\int_{35.9}^{\infty} f(x)\,dx \approx 0.6915$

31. $66,666.67   **33.** Yes, $360,000 < $400,000.

35. (a) $4,637,228   (b) $5,555,556

37. (a) $748,367.34   (b) $808,030.14   (c) $900,000.00

39. (a) $453,901.30   (b) $807,922.43   (c) $4,466,666.67

Review Exercises   *(page 908)*

1. $2\sqrt{x}\ln x - 4\sqrt{x} + C$ **3.** $xe^x + C$

5. $\frac{2}{15}(x-5)^{3/2}(3x+10) + C$ **7.** $x^2e^{2x} - xe^{2x} + \frac{1}{2}e^{2x} + C$

9. $3e^2 - \dfrac{3(e^2 - 1)}{2} \approx 12.584$ **11.** $16 - 20e^{-1/4} \approx 0.4240$

13. $90,634.62   **15.** $865,958.50

17. (a) $1,200,000   (b) $1,052,649.52

19. $\frac{1}{54}(9x^2 - 12x + 8\ln|3x+2|) + C$

21. $\dfrac{x}{2}\sqrt{x^2 - 16} - 8\ln\left(\sqrt{x^2 - 16} + x\right) + C$

23. $\dfrac{1}{9}\left(\dfrac{2}{2+3x} + \ln|2+3x|\right) + C$

25. $\sqrt{x^2 + 25} - 5\ln\left|\dfrac{5 + \sqrt{x^2 + 25}}{x}\right| + C$

27. $\dfrac{1}{4}\ln\left|\dfrac{x-2}{x+2}\right| + C$ **29.** $\dfrac{2}{3}(x-2)\sqrt{1+x} + C$

31. $2\sqrt{1+x} + \ln\left|\dfrac{\sqrt{1+x} - 1}{\sqrt{1+x} + 1}\right| + C$

33. (a) 0.675   (b) 0.290

35. Exact: $\frac{2}{3} \approx 0.6667$ **37.** Exact: $\frac{3}{8} = 0.375$
    Trapezoidal: 0.7050   Trapezoidal: 0.3786
    Simpson's: 0.6715   Simpson's: 0.3751

39. Exact: $2 - 2e^{-2} \approx 1.7293$ **41.** (a) 0.741   (b) 0.737
    Trapezoidal: 1.7652
    Simpson's: 1.7299

43. (a) 0.305   (b) 0.289   **45.** (a) 2.961   (b) 2.936

47. (a) $|E| \le \dfrac{e^4}{6} \approx 9.0997$ (b) $|E| \le \dfrac{e^4}{90} \approx 0.6066$

49. (a) $n = 214$ (b) $n = 2$

51. Converges, $-\frac{1}{4}$ **53.** Diverges   **55.** Diverges

57. $A \approx 4$ **59.** $A = 1$ **61.** $266,666.67   **63.** No

Test Yourself   *(page 910)*

1. $xe^{x+1} - e^{x+1} + C$ **2.** $3x^3\ln x - x^3 + C$

3. $-3x^2e^{-x/3} - 18xe^{-x/3} - 54e^{-x/3} + C$

4. (a) $\approx 6494.47 million   (b) $\approx 811.81 million

5. $\dfrac{1}{4}\left(\dfrac{7}{7+2x} + \ln|7+2x|\right) + C$ **6.** $x^3 - \ln(1 + e^{x^3}) + C$

7. $-\frac{2}{75}(2 - 5x^2)\sqrt{1 + 5x^2} + C$ **8.** $-1 + \frac{3}{2}\ln 3 \approx 0.6479$

9. $8\frac{2}{3}$ **10.** $4\ln\left[3\left(\sqrt{17} - 4\right)\right] + \sqrt{17} - 5 \approx -4.8613$

11. Exact: 18.0
    Trapezoidal: 18.28

12. Simpson's Rule: 41.3606; Exact: 41.1711

13. Converges; $\frac{1}{3}$ **14.** Converges; 12   **15.** Diverges

16. (a) $498.75   (b) Plan B, because $149 < $498.75.

Chapter 13

Section 13.1   *(page 917)*

Skills Warm Up   *(page 917)*

1. $2\sqrt{5}$ **2.** 5   **3.** 8   **4.** 8   **5.** $(4, 7)$

6. $(1, 0)$ **7.** $(0, 3)$ **8.** $(-1, 1)$

9. $(x-2)^2 + (y-3)^2 = 4$

10. $(x-1)^2 + (y-4)^2 = 25$

1.   **3.**

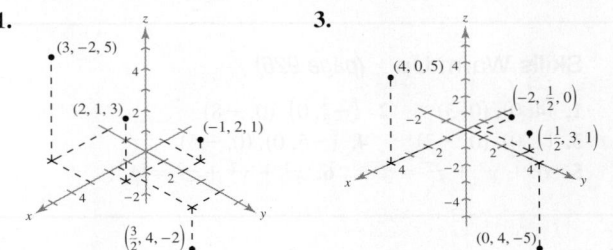

5. $(-3, 4, 5)$ **7.** $(10, 0, 0)$ **9.** 0   **11.** $3\sqrt{2}$

13. $\sqrt{206}$ **15.** $\left(2, -\frac{3}{2}, \frac{5}{2}\right)$ **17.** $\left(\frac{1}{2}, \frac{1}{2}, -1\right)$

19. $(6, -3, 5)$ **21.** $(1, 2, 1)$ **23.** $3, 3\sqrt{5}, 6$; right triangle

25. $\sqrt{41}, \sqrt{14}, \sqrt{41}$; isosceles triangle

27. $(0, 0, -5), (2, 2, -4), (2, -4, -1)$

29. $x^2 + (y-2)^2 + (z-2)^2 = 4$

31. $\left(x - \frac{3}{2}\right)^2 + (y-2)^2 + (z-1)^2 = \frac{21}{4}$

33. $(x-3)^2 + (y+2)^2 + (z+3)^2 = 16$

35. $(x-1)^2 + (y-3)^2 + z^2 = 10$

37. $(x+4)^2 + (y-3)^2 + (z-2)^2 = 4$

39. Center: $\left(\frac{5}{2}, 0, 0\right)$ **41.** Center: $(-2, 1, -4)$
    Radius: $\frac{5}{2}$ Radius: 5

43. Center: $(1, 3, 2)$
    Radius: $\dfrac{5\sqrt{2}}{2}$

45. 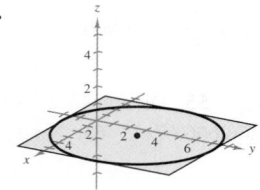   **47.**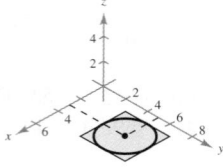

49. 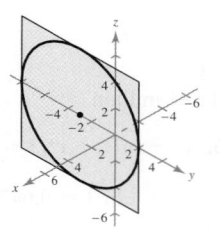   **51.**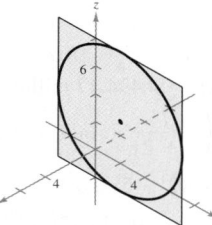

53. (a) 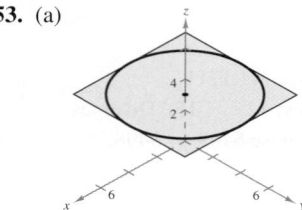   (b)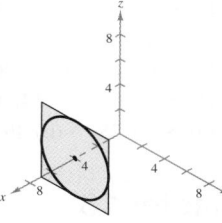

55. (a) 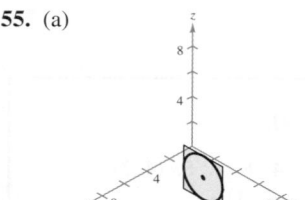   (b)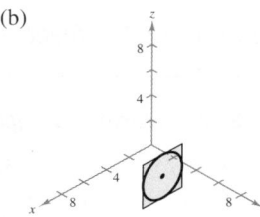

57. $x^2 + y^2 + z^2 = 6806.25$

Section 13.2   *(page 926)*

Skills Warm Up   *(page 926)*

1. $(4, 0), (0, 3)$ **2.** $\left(-\frac{4}{3}, 0\right), (0, -8)$

3. $(1, 0), (0, -2)$ **4.** $(-5, 0), (0, -5)$

5. $x^2 + y^2 + z^2 = \frac{1}{4}$ **6.** $x^2 + y^2 + z^2 = 4$

1. 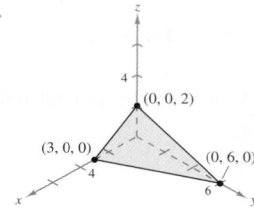   **3.**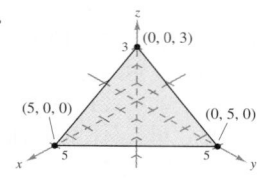

5. 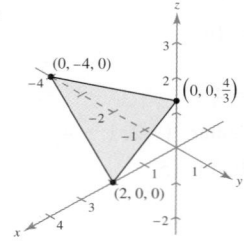   **7.**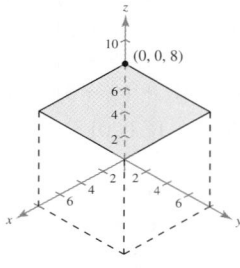

9. 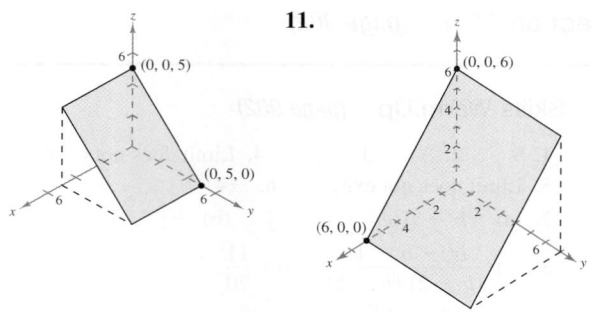   **11.**

13. Perpendicular   **15.** Parallel   **17.** Parallel
19. Neither parallel nor perpendicular   **21.** Perpendicular
23. c; Ellipsoid   **24.** e; Hyperboloid of two sheets
25. f; Hyperboloid of one sheet   **26.** b; Elliptic cone
27. d; Elliptic paraboloid   **28.** a; Hyperbolic paraboloid
29. (a) $x = \pm y$; Lines
  (b) $z = 9 - y^2$; Parabola
  (c) $z = x^2$; Parabola
  Hyperbolic parabola

31. (a) $\dfrac{x^2}{4} + y^2 = 1$; Ellipse   (b) $\dfrac{x^2}{4} + z^2 = 1$; Ellipse
  (c) $y^2 + z^2 = 1$; Circle
  Ellipsoid

33. (a) $z^2 - \dfrac{x^2}{9} = 1$; Hyperbola

  (b) $\dfrac{9}{13}z^2 - \dfrac{9}{208}y^2 = 1$; Hyperbola

  (c) $\dfrac{x^2}{135} + \dfrac{y^2}{240} = 1$; Ellipse

  Hyperboloid of two sheets
35. Ellipsoid   **37.** Hyperboloid of one sheet
39. Hyperbolic paraboloid   **41.** Elliptic paraboloid
43. Elliptic cone   **45.** Hyperboloid of two sheets
47. Hyperbolic paraboloid

49. $\dfrac{x^2}{3963^2} + \dfrac{y^2}{3963^2} + \dfrac{z^2}{3950^2} = 1$

51. (a)

Year	2004	2005	2006
x	33.1	34.9	37.4
y	13.2	13.8	14.9
z (actual)	15.5	16.3	17.8
z (approximated)	15.5	16.4	17.5

Year	2007	2008	2009
x	40.6	43.0	41.8
y	15.0	15.4	14.5
z (actual)	19.5	20.5	20.7
z (approximated)	19.4	20.7	20.4

The approximated values of z are very close to the actual values.
  (b) According to the model, increases in expenditures of recreation types y and z will correspond to an increase in expenditures of recreation type x.

Section 13.3   *(page 933)*

Skills Warm Up   *(page 933)*

1. 11   **2.** -16 **3.** 7   **4.** 4   **5.** $(-\infty, \infty)$
6. $(-\infty, -3) \cup (-3, 0) \cup (0, \infty)$
7. $[5, \infty)$ **8.** $\left(-\infty, -\sqrt{5}\right] \cup \left[\sqrt{5}, \infty\right)$
9. 55.0104   **10.** 6.9165

1. (a) $\dfrac{3}{2}$ (b) $-\dfrac{1}{4}$ (c) 6   (d) $\dfrac{5}{y}$ (e) $\dfrac{x}{2}$ (f) $\dfrac{5}{t}$
3. (a) 5   (b) $3e^2$ (c) $2e^{-1}$ (d) $5e^y$ (e) xe^2 (f) te^t
5. (a) $\frac{2}{3}$ (b) 0   **7.** (a) 90π (b) 50π
9. (a) \$20,655.20   (b) \$1,397,672.67   **11.** (a) 0   (b) 6
13. (a) $x^2 + 2x\,\Delta x + (\Delta x)^2 - 2y$ (b) $-2, \Delta y \ne 0$
15. Domain: all points (x, y) **17.** Domain: all points (x, y)
 inside and on the circle   Range: $[0, \infty)$
 $x^2 + y^2 = 16$
 Range: $[0, 4]$
19. Domain: all points (x, y) **21.** Domain: the half-plane
 such that $y \ne 0$ below the line $y = -x + 4$
 Range: $(0, \infty)$ Range: $(-\infty, \infty)$
23. Domain: all points (x, y) **25.** Domain: all points (x, y)
 inside and on the ellipse   such that $x \ne 0$
 $3x^2 + y^2 = 9$ Range: $(-\infty, \infty)$
 Range: $[0, 3]$
27. Domain: all points (x, y) **29.** Domain: all points (x, y)
 such that $x \ne 0$ and $y \ne 0$ such that $y \ge 0$
 Range: $(-\infty, 0)$ and $(0, \infty)$ Range: $(-\infty, \infty)$
31. b   **32.** d   **33.** a   **34.** c
35. The level curves are   **37.** The level curves are
 parallel lines.   circles.

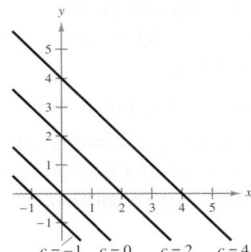

39. The level curves are   **41.** The level curves are
 hyperbolas.   circles.

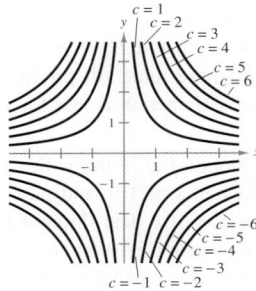

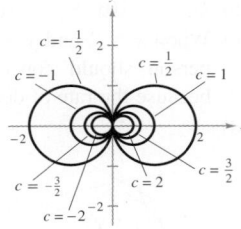

43. 135,540 units
45. (a) \$15,250   (b) \$18,425   (c) \$30,025

47.

R \ I	0	0.03	0.05
0	\$5187.48	\$3859.98	\$3184.67
0.28	\$4008.46	\$2982.67	\$2460.85
0.35	\$3754.27	\$2793.53	\$2304.80

49. (a) C   (b) A   (c) B
51. (a) \$2.78 earnings per share
 (b) x; Explanations will vary. Sample answer: The x-variable
 has a greater influence on the earnings per share because
 the absolute value of its coefficient is larger than the
 absolute value of the coefficient of the y-term.
53. Option (a): \$1003.73, \$240,895.20;
 Option (b): \$798.36, \$287,409.60;
 Option (c): \$1078.59, \$194,146.20;
 Answers will vary.

Section 13.4   *(page 944)*

Skills Warm Up   *(page 944)*

1. $\dfrac{x}{\sqrt{x^2 + 3}}$ **2.** $-6x(3 - x^2)^2$ **3.** $e^{2t+1}(2t + 1)$
4. $\dfrac{e^{2x}(2 - 3e^{2x})}{\sqrt{1 - e^{2x}}}$ **5.** $-\dfrac{2}{3 - 2x}$ **6.** $\dfrac{3(t^2 - 2)}{2t(t^2 - 6)}$
7. $-\dfrac{10x}{(4x - 1)^3}$ **8.** $-\dfrac{(x + 2)^2(x^2 + 8x + 27)}{(x^2 - 9)^3}$
9. $f'(2) = 8$ **10.** $g'(2) = \frac{7}{2}$

1. $\dfrac{\partial z}{\partial x} = 3; \dfrac{\partial z}{\partial y} = 5$ **3.** $f_x(x, y) = 3; f_y(x, y) = -12y$
5. $f_x(x, y) = \dfrac{1}{y}; f_y(x, y) = -\dfrac{x}{y^2}$
7. $f_x(x, y) = \dfrac{x}{\sqrt{x^2 + y^2}}; f_y(x, y) = \dfrac{y}{\sqrt{x^2 + y^2}}$
9. $\dfrac{\partial z}{\partial x} = 2xe^{2y}; \dfrac{\partial z}{\partial y} = 2x^2e^{2y}$
11. $h_x(x, y) = -2xe^{-(x^2 + y^2)}; h_y(x, y) = -2ye^{-(x^2 + y^2)}$
13. $\dfrac{\partial z}{\partial x} = -\dfrac{2y}{x^2 - y^2}; \dfrac{\partial z}{\partial y} = \dfrac{2x}{x^2 - y^2}$
15. $f_x(x, y) = 6x + y, 13; f_y(x, y) = x - 2y, 0$
17. $f_x(x, y) = 3ye^{3xy}, 12; f_y(x, y) = 3xe^{3xy}, 0$
19. $f_x(x, y) = -\dfrac{y^2}{(x - y)^2}, -\dfrac{1}{4}; f_y(x, y) = \dfrac{x^2}{(x - y)^2}, \dfrac{1}{4}$
21. $f_x(x, y) = \dfrac{3}{3x + 5y}, 1; f_y(x, y) = \dfrac{5}{3x + 5y}, \dfrac{5}{3}$
23. (a) 2   (b) 1   **25.** (a) -2 (b) -2
27. $w_x = y^2z^4$
 $w_y = 2xyz^4$
 $w_z = 4xy^2z^3$
29. $w_x = -\dfrac{2z}{(x + y)^2}$
 $w_y = -\dfrac{2z}{(x + y)^2}$
 $w_z = \dfrac{2}{x + y}$

31. $w_x = 2z^2 + 3yz, 2$

$w_y = 3xz - 12yz, 30$

$w_z = 4xz + 3xy - 6y^2, -1$

33. $w_x = \dfrac{x}{\sqrt{x^2 + y^2 + z^2}}, \dfrac{2}{3}$

$w_y = \dfrac{y}{\sqrt{x^2 + y^2 + z^2}}, -\dfrac{1}{3}$

$w_z = \dfrac{z}{\sqrt{x^2 + y^2 + z^2}}, \dfrac{2}{3}$

35. $w_x = 4xy^3z^2e^{2x^2}, -8\sqrt{e}$

$w_y = 3y^2z^2e^{2x^2}, 12\sqrt{e}$

$w_z = 2y^3ze^{2x^2}, -4\sqrt{e}$

37. $w_x = \dfrac{5}{5x + 2y^3 - 3z}, \dfrac{1}{5}$

$w_y = \dfrac{6y^2}{5x + 2y^3 - 3z}, \dfrac{6}{25}$

$w_z = -\dfrac{3}{5x + 2y^3 - 3z}, -\dfrac{3}{25}$

39. $(-6, 4)$ **41.** $(1, 1)$

43. $\dfrac{\partial^2 z}{\partial x^2} = 6x$

$\dfrac{\partial^2 z}{\partial y^2} = -8$

$\dfrac{\partial^2 z}{\partial y \partial x} = \dfrac{\partial^2 z}{\partial x \partial y} = 0$

45. $\dfrac{\partial^2 z}{\partial x^2} = 2$

$\dfrac{\partial^2 z}{\partial x \partial y} = \dfrac{\partial^2 z}{\partial y \partial x} = -2$

$\dfrac{\partial^2 z}{\partial y^2} = 6$

47. $\dfrac{\partial^2 z}{\partial x^2} = 108x^2(3x^4 - 2y^3)(11x^4 - 2y^3)$

$\dfrac{\partial^2 z}{\partial y^2} = 36y(2y^3 - 3x^4)(3x^4 - 8y^3)$

$\dfrac{\partial^2 z}{\partial x \partial y} = \dfrac{\partial^2 z}{\partial y \partial x} = 432x^3y^2(2y^3 - 3x^4)$

49. $\dfrac{\partial^2 z}{\partial x^2} = -\dfrac{y}{x^3}$

$\dfrac{\partial^2 z}{\partial x \partial y} = -\dfrac{x^2 - y^2}{2x^2y^2}$

$\dfrac{\partial^2 z}{\partial y \partial x} = -\dfrac{x^2 - y^2}{2x^2y^2}$

$\dfrac{\partial^2 z}{\partial y^2} = \dfrac{x}{y^3}$

51. $f_{xx}(x, y) = 12x^2 - 6y^2, 12$

$f_{xy}(x, y) = -12xy, 0$

$f_{yy}(x, y) = -6x^2 + 2, -4$

$f_{yx}(x, y) = -12xy, 0$

53. $f_{xx}(x, y) = e^{x^2}(4x^2y^3 + 2y^3), -6e$

$f_{xy}(x, y) = f_{yx}(x, y) = 6xy^2e^{x^2}, 6e$

$f_{yy}(x, y) = 6ye^{x^2}, -6e$

55. $f_{xx}(x, y, z) = 2$

$f_{xy}(x, y, z) = f_{yx}(x, y, z) = -3$

$f_{xz}(x, y, z) = f_{yy}(x, y, z) = f_{zx}(x, y, z) = 0$

$f_{yz}(x, y, z) = f_{zy}(x, y, z) = 4$

$f_{zz}(x, y, z) = 6z$

57. $f_{xx}(x, y, z) = -\dfrac{8yz}{(x + y)^3}$

$f_{xy}(x, y, z) = \dfrac{4z(x - y)}{(x + y)^3}$

$f_{xz}(x, y, z) = \dfrac{4y}{(x + y)^2}$

$f_{yy}(x, y, z) = \dfrac{8xz}{(x + y)^3}$

$f_{yx}(x, y, z) = \dfrac{4z(x - y)}{(x + y)^3}$

$f_{yz}(x, y, z) = -\dfrac{4x}{(x + y)^2}$

$f_{zz}(x, y, z) = 0$

$f_{zx}(x, y, z) = \dfrac{4y}{(x + y)^2}$

$f_{zy}(x, y, z) = -\dfrac{4x}{(x + y)^2}$

59. (a) At $(120, 160)$, $\dfrac{\partial C}{\partial x} \approx 154.77$; At $(120, 160)$, $\dfrac{\partial C}{\partial y} \approx 193.33$

(b) Racing bikes; Explanations will vary. Sample answer: The absolute value of dC/dy is greater than the absolute value of dC/dx at $(120, 160)$.

61. (a) About 113.72   (b) About 97.47

63. Complementary

65. (a) $\dfrac{\partial z}{\partial x} = 0.62$; $\dfrac{\partial z}{\partial y} = -0.41$

(b) For every increase of 1 billion dollars in expenditures on amusement parks and campgrounds, the expenditures for spectator sports will increase by 0.62 billion dollars. For every increase of 1 billion dollars in expenditures on live entertainment (excluding sports), the expenditures for spectator sports will decrease by 0.41 billion dollars.

67. $IQ_M(M, C) = \dfrac{100}{C}, IQ_M(12, 10) = 10$; For a child who has a current mental age of 12 years and a chronological age of 10 years, the IQ is increasing at a rate of 10 IQ points for every increase of 1 year in the child's mental age.

$IQ_C(M, C) = \dfrac{-100M}{C^2}, IQ_C(12, 10) = -12$; For a child who has a current mental age of 12 years and a chronological age of 10 years, the IQ is decreasing at a rate of 12 IQ points for every increase of 1 year in the child's chronological age.

69. $V_I(0.03, 0.28) \approx -14{,}478.99$

$V_R(0.03, 0.28) \approx -1391.17$

The rate of inflation has the greater negative influence on the growth of the investment because $|-14{,}478.99| > |-1391.17|$.

71. (a) $U_x = -10x + y$ (b) $U_y = x - 6y$

(c) When $x = 2$ and $y = 3$, $U_x = -17$ and $U_y = -16$. The person should consume one more unit of product y, because the rate of decrease of satisfaction is less for y.

(d)

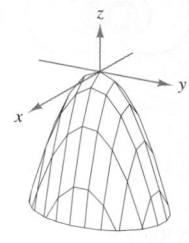

The slope of U in the x-direction is 0 when $y = 10x$ and negative when $y < 10x$. The slope of U in the y-direction is 0 when $x = 6y$ and negative when $x < 6y$.

Section 13.5   (page 953)

Skills Warm Up   (page 953)

1. $(3, 2)$ **2.** $(11, 6)$ **3.** $(1, 4)$ **4.** $(4, 4)$
5. $(5, 2)$ **6.** $(3, -2)$ **7.** $(0, 0), (-1, 0)$
8. $(-2, 0), (2, -2)$

9. $\dfrac{\partial z}{\partial x} = 12x^2$ $\dfrac{\partial^2 z}{\partial y^2} = -6$

$\dfrac{\partial z}{\partial y} = -6y$ $\dfrac{\partial^2 z}{\partial x \partial y} = 0$

$\dfrac{\partial^2 z}{\partial x^2} = 24x$ $\dfrac{\partial^2 z}{\partial y \partial x} = 0$

10. $\dfrac{\partial z}{\partial x} = 10x^4$ $\dfrac{\partial^2 z}{\partial y^2} = -6y$

$\dfrac{\partial z}{\partial y} = -3y^2$ $\dfrac{\partial^2 z}{\partial x \partial y} = 0$

$\dfrac{\partial^2 z}{\partial x^2} = 40x^3$ $\dfrac{\partial^2 z}{\partial y \partial x} = 0$

11. $\dfrac{\partial z}{\partial x} = 4x^3 - \dfrac{\sqrt{xy}}{2x}$ $\dfrac{\partial^2 z}{\partial y^2} = \dfrac{\sqrt{xy}}{4y^2}$

$\dfrac{\partial z}{\partial y} = -\dfrac{\sqrt{xy}}{2y} + 2$ $\dfrac{\partial^2 z}{\partial x \partial y} = -\dfrac{\sqrt{xy}}{4xy}$

$\dfrac{\partial^2 z}{\partial x^2} = 12x^2 + \dfrac{\sqrt{xy}}{4x^2}$ $\dfrac{\partial^2 z}{\partial y \partial x} = -\dfrac{\sqrt{xy}}{4xy}$

12. $\dfrac{\partial z}{\partial x} = 4x - 3y$ $\dfrac{\partial^2 z}{\partial y^2} = 2$

$\dfrac{\partial z}{\partial y} = 2y - 3x$ $\dfrac{\partial^2 z}{\partial x \partial y} = -3$

$\dfrac{\partial^2 z}{\partial x^2} = 4$ $\dfrac{\partial^2 z}{\partial y \partial x} = -3$

13. $\dfrac{\partial z}{\partial x} = y^3 e^{xy^2}$ $\dfrac{\partial^2 z}{\partial y^2} = 4x^2 y^3 e^{xy^2} + 6xy e^{xy^2}$

$\dfrac{\partial z}{\partial y} = 2xy^2 e^{xy^2} + e^{xy^2}$ $\dfrac{\partial^2 z}{\partial x \partial y} = 2xy^4 e^{xy^2} + 3y^2 e^{xy^2}$

$\dfrac{\partial^2 z}{\partial x^2} = y^5 e^{xy^2}$ $\dfrac{\partial^2 z}{\partial y \partial x} = 2xy^4 e^{xy^2} + 3y^2 e^{xy^2}$

14. $\dfrac{\partial z}{\partial x} = e^{xy}(xy + 1)$ $\dfrac{\partial^2 z}{\partial y^2} = x^3 e^{xy}$

$\dfrac{\partial z}{\partial y} = x^2 e^{xy}$ $\dfrac{\partial^2 z}{\partial x \partial y} = xe^{xy}(xy + 2)$

$\dfrac{\partial^2 z}{\partial x^2} = ye^{xy}(xy + 2)$ $\dfrac{\partial^2 z}{\partial y \partial x} = xe^{xy}(xy + 2)$

---

1. Critical point: $(-2, -4)$
No relative extrema
$(-2, -4, 1)$ is a saddle point.

3. Critical point: $(0, 0)$
Relative minimum: $(0, 0, 1)$

5. Critical point: $(1, 3)$
Relative minimum: $(1, 3, 0)$

7. Critical point: $(-1, 1)$
Relative minimum: $(-1, 1, -4)$

9. Critical point: $(8, 16)$
Relative maximum: $(8, 16, 74)$

11. Critical point: $(1, 1)$
Relative minimum: $(1, 1, 11)$

13. Critical points: $(0, 0), \left(\frac{4}{3}, \frac{4}{3}\right)$
Saddle point: $(0, 0, 1)$
Relative maximum: $\left(\frac{4}{3}, \frac{4}{3}, \frac{59}{27}\right)$

15. Critical point: $(0, 0)$
Saddle point: $(0, 0, 0)$

17. Critical points: $\left(\frac{1}{2}, \frac{1}{2}\right), \left(-\frac{1}{2}, -\frac{1}{2}\right)$
Relative maximum: $\left(\frac{1}{2}, \frac{1}{2}, e^{1/2}\right)$
Relative minimum: $\left(-\frac{1}{2}, -\frac{1}{2}, -e^{1/2}\right)$

19. Insufficient information   **21.** $f(x_0, y_0)$ is a saddle point.

23. $f(x_0, y_0)$ is a relative minimum.

25. Relative minima: $(a, 0, 0), (0, b, 0)$
Second-Partials Test fails at $(a, 0)$ and $(0, b)$.

27. Saddle point: $(0, 0, 0)$
Second-Partials Test fails at $(0, 0)$.

29. Relative minimum: $(0, 0, 0)$
Second-Partials Test fails at $(0, 0)$.

31. Relative minimum: $(1, -3, 0)$

33. 15, 15, 15   **35.** 20, 20, 20   **37.** $x_1 = 3, x_2 = 6$

39. $p_1 = 2500, p_2 = 3000$ **41.** $x_1 \approx 94, x_2 \approx 157$

43. 32 in. $\times$ 16 in. $\times$ 16 in.

45. Base dimensions: 3 ft $\times$ 3 ft
Height: 2 ft; Minimum cost: \$5.40

47. $x = 1.25, y = 2.5$; \$4.625 million

49. 500 smallmouth bass; 200 largemouth bass

51. Proof   **53.** True

Quiz Yourself   (page 956)

1. (a)

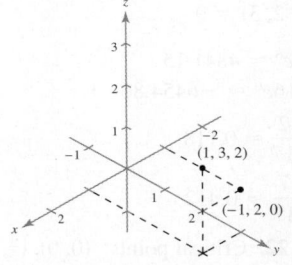

2. (a)

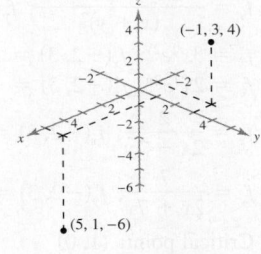

(b) 3   (c) $\left(0, \frac{5}{2}, 1\right)$

(b) $2\sqrt{35}$ (c) $(2, 2, -1)$

3. (a)

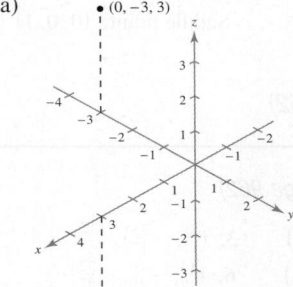

(b) $3\sqrt{6}$ (c) $\left(\frac{3}{2}, -\frac{3}{2}, 0\right)$

4. $(x - 2)^2 + (y + 1)^2 + (z - 3)^2 = 16$

5. $(x - 1)^2 + (y - 4)^2 + (z + 2)^2 = 11$

6. Center: $(4, 1, 3)$; radius: 7

7.

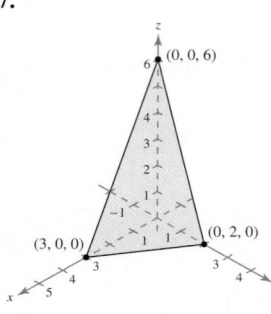

8.

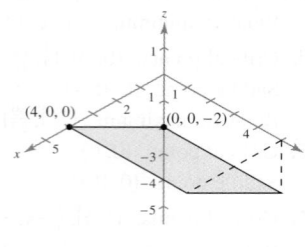

9.

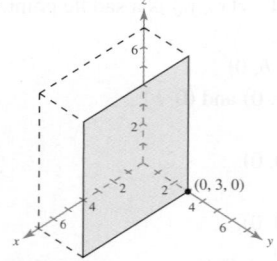

10. Ellipsoid

11. Hyperboloid of two sheets   **12.** Elliptic paraboloid

13. $f(1, 0) = 1$
   $f(4, -1) = -5$

14. $f(1, 0) = 2$
   $f(4, -1) = 3\sqrt{7}$

15. $f(1, 0) = 0$
   $f(4, -1) = \ln 6 \approx 1.79$

16. (a) Between 30° and 50°   (b) Between 40° and 80°
   (c) Between 70° and 90°

17. $f_x = 2x - 3$; $f_x(-2, 3) = -7$
   $f_y = 4y - 1$; $f_y(-2, 3) = 11$

18. $f_x = \dfrac{y(3 + y)}{(x + y)^2}$; $f_x(-2, 3) = 18$

   $f_y = \dfrac{-2xy - y^2 - 3x}{(x + y)^2}$; $f_y(-2, 3) = 9$

19. $f_x = 3x^2 e^{2y}$; $f_x(-2, 3) = 12e^6 \approx 4841.15$
   $f_y = 2x^3 e^{2y}$; $f_y(-2, 3) = -16e^6 \approx -6454.86$

20. $f_x = \dfrac{2}{2x + 7y}$; $f_x(-2, 3) = \dfrac{2}{17} \approx 0.118$

   $f_y = \dfrac{7}{2x + 7y}$; $f_y(-2, 3) = \dfrac{7}{17} \approx 0.412$

21. Critical point: $(1, 0)$
   Relative minimum:
   $(1, 0, -3)$

22. Critical points: $(0, 0)$, $\left(\frac{4}{3}, \frac{4}{3}\right)$
   Relative maximum:
   $\left(\frac{4}{3}, \frac{4}{3}, \frac{59}{27}\right)$
   Saddle point: $(0, 0, 1)$

23. $x = 80$, $y = 20$; $20,000

Section 13.6   *(page 962)*

Skills Warm Up   *(page 962)*

1. $\left(\frac{7}{8}, \frac{1}{12}\right)$ **2.** $\left(-\frac{1}{24}, -\frac{7}{8}\right)$ **3.** $\left(\frac{55}{12}, -\frac{25}{12}\right)$

4. $\left(\frac{22}{23}, -\frac{3}{23}\right)$ **5.** $\left(\frac{5}{3}, \frac{1}{3}, 0\right)$ **6.** $\left(\frac{14}{19}, -\frac{10}{19}, -\frac{32}{57}\right)$

7. $f_x = 2xy + y^2$ **8.** $f_x = 50y^2(x + y)$
   $f_y = x^2 + 2xy$ $f_y = 50y(x + y)(x + 2y)$

9. $f_x = 3x^2 - 4xy + yz$ **10.** $f_x = yz + z^2$
   $f_y = -2x^2 + xz$ $f_y = xz + z^2$
   $f_z = xy$ $f_z = xy + 2xz + 2yz$

1. $f(5, 5) = 25$ **3.** $f(4, 4) = 32$ **5.** $f(\sqrt{2}, 1) = 1$
7. $f(25, 50) = 2600$ **9.** $f(1, 1) = 2$
11. $f(2, 2) = e^4$ **13.** $f(9, 6, 9) = 432$

15. $f\left(\frac{1}{3}, \frac{1}{3}, \frac{1}{3}\right) = \frac{1}{3}$ **17.** $f\left(\frac{\sqrt{3}}{3}, \frac{\sqrt{3}}{3}, \frac{\sqrt{3}}{3}\right) = \sqrt{3}$

19. 20, 20, 20   **21.** 40, 40, 40   **23.** $3\sqrt{2}$ **25.** $\sqrt{3}$
27. 15 units × 10 units × 6 units   **29.** 12 ft × 12 ft × 18 ft
31. $x_1 = 145$ units, $x_2 = 855$ units

33. (a) $f\left(\frac{3125}{6}, \frac{6250}{3}\right) \approx 147,314$ (b) 1.473
   (c) 184,142 units   (d) 515,599 units

35. (a) $x \approx 317$ units, $y \approx 68$ units   (b) Answers will vary.

37. (a) 50 ft × 120 ft   (b) $2400

39. $x = \sqrt[3]{0.065} \approx 0.402$ L
   $y = \frac{1}{2}\sqrt[3]{0.065} \approx 0.201$ L
   $z = \frac{1}{3}\sqrt[3]{0.065} \approx 0.134$ L

41. About 190.7 g

43. (a) Cable television: $1200
      Newspaper: $600
      Radio: $900
   (b) About 3718 responses

Section 13.7   *(page 970)*

Skills Warm Up   *(page 970)*

1. 5.0225   **2.** 0.0189

3. $S_a = 2a - 4 - 4b$ **4.** $S_a = 8a - 6 - 2b$
   $S_b = 12b - 8 - 4a$ $S_b = 18b - 4 - 2a$

5. 15   **6.** 42   **7.** $\frac{25}{12}$ **8.** 14   **9.** 31   **10.** 95

1. $S = 1.6$; $S = 0.8259$ **3.** $S = 6.46$; $S = 0.125$
5. $y = x + \frac{2}{3}$ **7.** $y = -2.3x - 0.9$

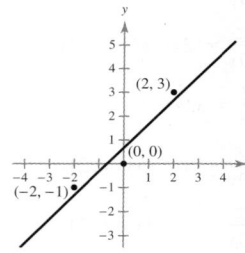

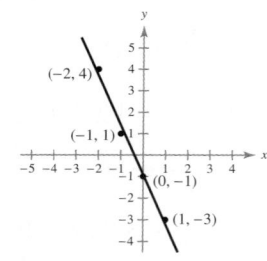

9. $y = 0.8x + 2$ **11.** $y = -1.1824x + 6.385$
13. (a) $y = 4.13t + 11.6$ (b) About $69.4 billion   (c) 2018
15. (a) $y = 0.138x + 22.1$ (b) 44.18 bushels/acre

17.

19.

Positive correlation,
$r \approx 0.9981$

No correlation, $r = 0$

21.

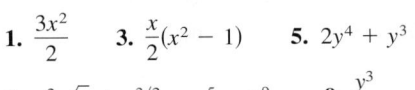

No correlation, $r \approx 0.0750$

23. False; The data modeled by $y = 3.29x - 4.17$ have a positive correlation.

25. True   **27.** True   **29.** Answers will vary.

Section 13.8   *(page 978)*

Skills Warm Up   *(page 978)*

1. 1   **2.** 6   **3.** 42   **4.** $\frac{1}{2}$ **5.** $\frac{19}{4}$

6. $\frac{16}{3}$ **7.** $\frac{1}{7}$ **8.** 4   **9.** ln 5   **10.** $\ln(e - 1)$

11. $\frac{e}{2}(e^4 - 1)$ **12.** $\frac{1}{2}\left(1 - \frac{1}{e^2}\right)$

13.   **14.**

15.   **16.**

---

1. $\dfrac{3x^2}{2}$ **3.** $\dfrac{x}{2}(x^2 - 1)$ **5.** $2y^4 + y^3 - 2y^2 - 16y$

7. $x^2\sqrt{x} + x^{3/2} - x^5 - x^9$ **9.** $\dfrac{y^3}{2}$ **11.** 3   **13.** 36

15. 64   **17.** $\frac{1}{2}$ **19.** $\frac{21}{2}$ **21.** $\frac{1}{3}$ **23.** 4   **25.** 24

27. 8   **29.** $\frac{16}{3}$ **31.** 36   **33.** 5   **35.** 2

37.

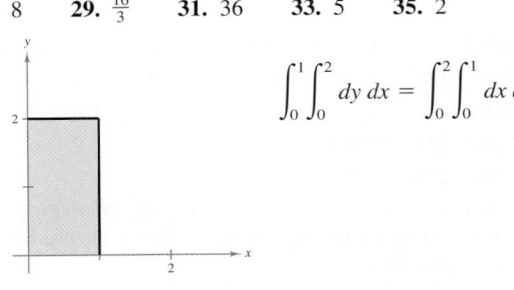

$$\int_0^1 \int_0^2 dy\,dx = \int_0^2 \int_0^1 dx\,dy = 2$$

39.

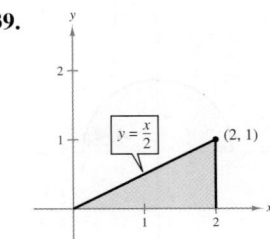

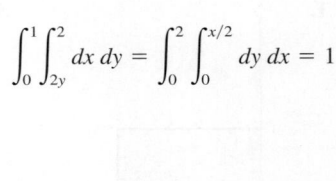

$$\int_0^1 \int_{2y}^2 dx\,dy = \int_0^2 \int_0^{x/2} dy\,dx = 1$$

41.

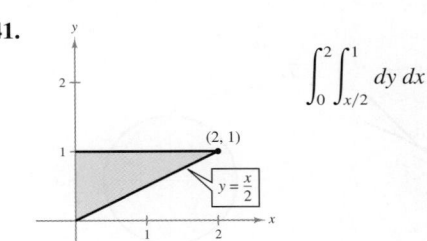

$$\int_0^2 \int_{x/2}^1 dy\,dx = \int_0^1 \int_0^{2y} dx\,dy = 1$$

43.

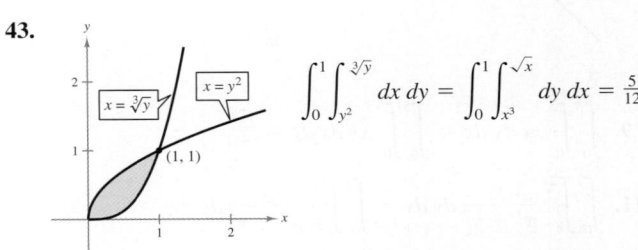

$$\int_0^1 \int_{y^2}^{\sqrt[3]{y}} dx\,dy = \int_0^1 \int_{x^3}^{\sqrt{x}} dy\,dx = \frac{5}{12}$$

45. (a) Answers will vary; $\frac{1}{2}(e^9 - 1) \approx 4051.042$
    (b) Answers will vary; $\frac{1}{2}(1 - e^{-4}) \approx 0.491$

47. 0.6588   **49.** 8.1747   **51.** 0.4521

53. 1.1190   **55.** True

Section 13.9   *(page 986)*

Skills Warm Up   *(page 986)*

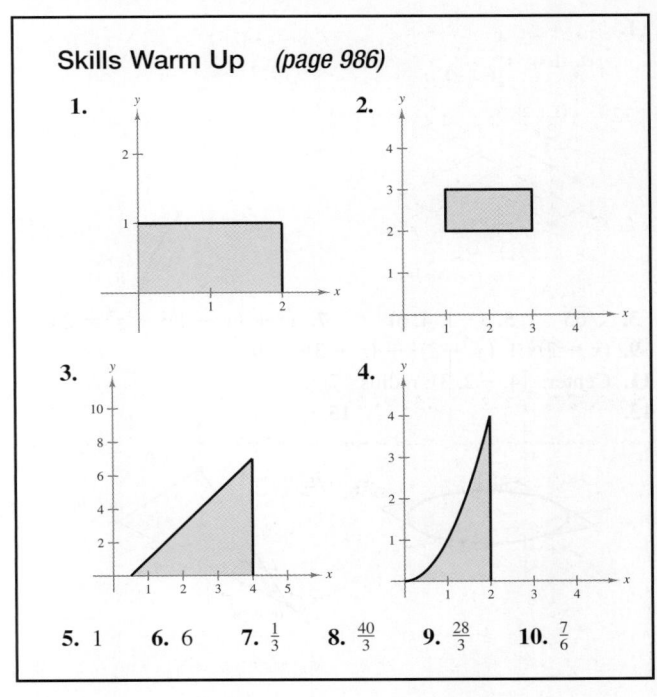

1.   **2.**

3.   **4.**

5. 1   **6.** 6   **7.** $\frac{1}{3}$ **8.** $\frac{40}{3}$ **9.** $\frac{28}{3}$ **10.** $\frac{7}{6}$

CHAPTER 13

1.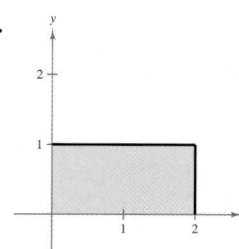

3.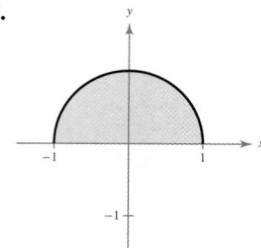

10

5.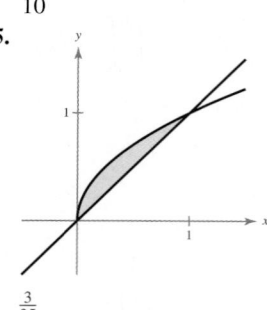

$\frac{3}{35}$

7.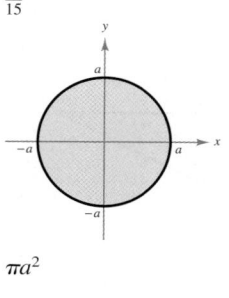

πa^2

$\frac{2}{15}$

9. $\displaystyle\int_0^3\int_0^5 xy\,dy\,dx = \int_0^5\int_0^3 xy\,dx\,dy = \frac{225}{4}$

11. $\displaystyle\int_0^2\int_x^{2x}\frac{y}{x^2+y^2}\,dy\,dx = \int_0^2\int_{y/2}^y \frac{y}{x^2+y^2}\,dx\,dy$

$+\displaystyle\int_2^4\int_{y/2}^2 \frac{y}{x^2+y^2}\,dx\,dy = \ln\frac{5}{2}$

13. 4   **15.** 12   **17.** 4   **19.** $\frac{40}{3}$ **21.** $\frac{81}{2}$

23. $\frac{134}{3}$ **25.** 10,000   **27.** $\frac{3}{2}$ **29.** $\frac{8}{3}$

31. $13,400   **33.** $75,125   **35.** 25,645.24

Review Exercises for Chapter 13   *(page 992)*

1.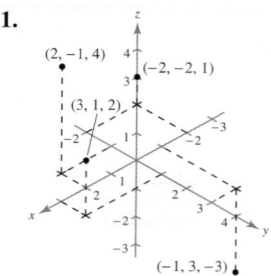

3. $\sqrt{65}$ **5.** $(-1,4,6)$ **7.** $x^2+(y-1)^2+z^2=25$

9. $(x-2)^2+(y+2)^2+(z+3)^2=9$

11. Center: $(4,-2,3)$; radius: 7

13.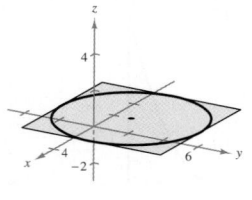

15.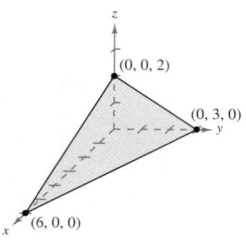

17.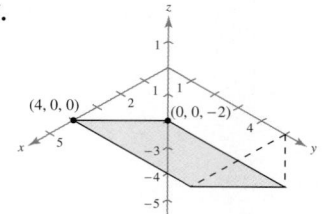

19. Sphere   **21.** Ellipsoid   **23.** Elliptic paraboloid

25. Top half of a circular cone

27. (a) 18   (b) 0   (c) -245 (d) -32

29. Domain: all points (x,y) inside or on the circle $x^2+y^2=1$

Range: $[0,1]$

31. Domain: all points (x,y)

Range: $(0,\infty)$

33. The level curves are lines of slope $-\frac{2}{5}$.

35. The level curves are hyperbolas.

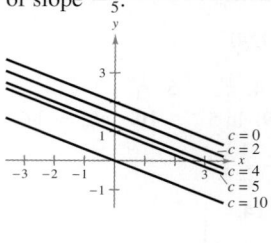

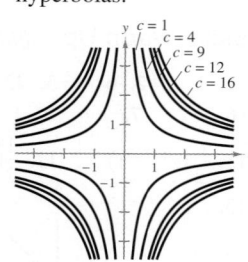

37. (a) No; the precipitation increments are 7.99 in., 9.99 in., 9.99 in., 9.99 in., and 19.99 in.

(b) Increase the number of level curves to correspond to smaller increments of precipitation.

39. (a) $2.49

(b) Because $\dfrac{\partial z}{\partial y}=0.060 > \dfrac{\partial z}{\partial x}=0.046$, y has the greater influence.

41. $f_x = 2xy+3y+2$

$f_y = x^2+3x-5$

43. $z_x = \dfrac{2x}{y^2}$

$z_y = \dfrac{-2x^2}{y^3}$

45. $f_x = \dfrac{5}{5x+4y}$

$f_y = \dfrac{4}{5x+4y}$

47. $f_x = ye^x + e^y$

$f_y = xe^y + e^x$

49. $w_x = yz^2$

$w_y = xz^2$

$w_z = 2xyz$

51. (a) -9 (b) -6 **53.** (a) -2 (b) -2

55. $f_{xx} = 6$

$f_{yy} = 12y$

$f_{xy} = f_{yx} = -1$

57. $f_{xx} = f_{yy} = f_{xy} = f_{yx} = \dfrac{-1}{4(1+x+y)^{3/2}}$

59. $f_{xx} = 10yz^3$ $f_{yx} = 1+10xz^3$ $f_{zx} = 30xyz^2$

$f_{xy} = 1+10xz^3$ $f_{yy} = -18yz$ $f_{zy} = 15x^2z^2 - 9y^2$

$f_{xz} = 30xyz^2$ $f_{yz} = 15x^2z^2 - 9y^2$ $f_{zz} = 30x^2yz$

61. (a) $C_x(500,250) = 99.50$

$C_y(500,250) = 140$

(b) Downhill skis; this is determined by comparing the marginal costs for the two models of skis at the production level $(500,250)$.

63. Critical point: $(0, 0)$

Relative minimum: $(0, 0, 0)$

65. Critical point: $(-2, 3)$

Saddle point: $(-2, 3, 1)$

67. Critical points: $(0, 0), \left(\frac{1}{6}, \frac{1}{12}\right)$

Relative minimum: $\left(\frac{1}{6}, \frac{1}{12}, -\frac{1}{432}\right)$

Saddle point: $(0, 0, 0)$

69. Critical points: $(1, 1), (-1, -1). (1, -1), (-1, 1)$

Relative minimum: $(1, 1, -2)$

Relative maximum: $(-1, -1, 6)$

Saddle points: $(1, -1, 2), (-1, 1, 2)$

71. $x_1 = 50, x_2 = 200$

73. At $(3, 6)$, the relative maximum is 36.

75. At $(2, 2)$, the relative minimum is 8.

77. At $\left(\frac{4}{3}, \frac{2}{3}, \frac{4}{3}\right)$, the relative maximum is $\frac{32}{27}$.

79. $x_1 = 378$ units; $x_2 = 623$ units

81. $y = \frac{60}{59}x - \frac{15}{59}$

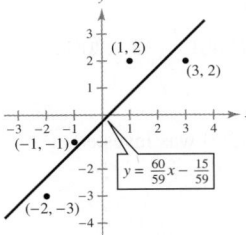

83. (a) $y = -2.6x + 347$

(b) 126 cameras

(c) About \$56.54

85. 1 **87.** $\frac{7}{4}$ **89.** $\frac{32}{3}$ **91.** $\frac{9}{2}$ **93.** 20 **95.** 8

97. $\frac{4096}{9}$ **99.** 3 **101.** \$5700 **103.** About \$155.69/ft²

Test Yourself *(page 504)*

1. (a)

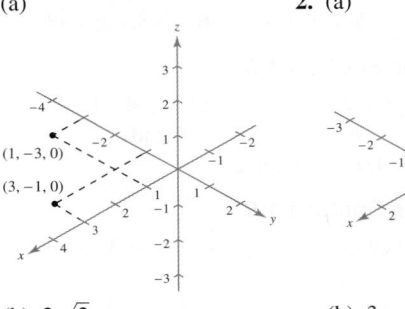

(b) $2\sqrt{2}$

(c) $(2, -2, 0)$

2. (a)

(b) 3

(c) $(-3, 1, 2.5)$

3. (a)

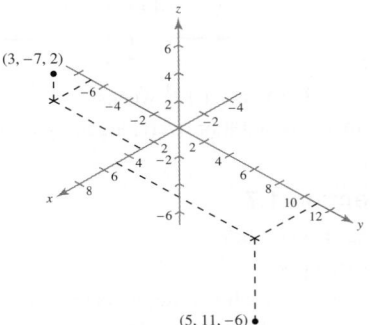

(b) $14\sqrt{2}$

(c) $(4, 2, -2)$

4. Center: $(10, -5, 5)$; radius: 5

5. Hyperboloid of one sheet

6. Elliptic cone **7.** Hyperbolic paraboloid

8. $f(3, 3) = 19$

$f(1, 4) = 6$

9. $f(3, 3) = \frac{3}{2}$

$f(1, 4) = -9$

10. $f(3, 3) = 0$

$f(1, 4) = 4 \ln \frac{1}{4} \approx -5.5$

11. $f_x = 6x + 9y^2;\ f_x(10, -1) = 69$

$f_y = 18xy;\ f_y(10, -1) = -180$

12. $f_x = (x + y)^{1/2} + \dfrac{x}{2(x + y)^{1/2}};\ f_x(10, -1) = \dfrac{14}{3}$

$f_y = \dfrac{x}{2(x + y)^{1/2}};\ f_y(10, -1) = \dfrac{5}{3}$

13. Critical point: $(1, -2)$; Relative minimum: $(1, -2, -23)$

14. Critical points: $(0, 0), (1, 1), (-1, -1)$

Saddle point: $(0, 0, 0)$

Relative maxima: $(1, 1, 2), (-1, -1, 2)$

15. About 128,613 units **16.** $y = 0.52x + 1.4$

17. $\frac{3}{2}$ **18.** 1 **19.** $\frac{4}{3}$ units² **20.** 48 **21.** $\frac{11}{6}$

CHAPTER 13

Answers to Checkpoints

Chapter 0

Checkpoints for Section 0.1

1 $x \geq 7$ denotes all real numbers greater than or equal to 7.
2 (a) $x \geq 5$ (b) $4 < y \leq 11$ (c) $-1 \leq z \leq 10$
3 12   **4** (a) $-|-6| = -|6|$ (b) $-|5| < |-5|$
5 8   **6** No

Checkpoints for Section 0.2

1 (a) $8, -15x$ (b) $4x^2, -3y, -7$ (c) $z^4, -10z^3, 1$
2 29   **3** (a) 33   (b) 1   (c) -21
4 (a) Multiplicative Identity Property
   (b) Commutative Property of Addition
5 $\dfrac{11x}{12}$ **6** $\dfrac{11}{12}$
7 Scientific calculator: $6\ \boxed{\times}\ \boxed{(}\ 8\ \boxed{y^x}\ 3\ \boxed{-}\ 481\ \boxed{)}\ \boxed{=}$
   Graphing calculator: $6\ \boxed{(}\ 8\ \boxed{\wedge}\ 3\ \boxed{-}\ 481\ \boxed{)}\ \boxed{\text{ENTER}}$
8 5.87

Checkpoints for Section 0.3

1 (a) 256   (b) 128   (c) $\dfrac{4}{3}$ **2** $\dfrac{z^{12}}{27x^6}$ **3** $\dfrac{7}{5}$
4 (a) 6.318×10^6 (b) 3.45×10^{-3}
5 (a) 428,000   (b) 0.0025   **6** 12.72 min
7 (a) -0.0545 (b) 5   **8** $7110.12   **9** 2.9%

Checkpoints for Section 0.4

1 (a) $\frac{1}{7}$ (b) 5   (c) 2   **2** $3x^2\sqrt{2x}$ **3** $3x\sqrt[3]{2x}$
4 $\dfrac{\sqrt[3]{2}}{2}$ **5** $8 + 2\sqrt{11}$ **6** 9   **7** $\sqrt{2}$ **8** $6\sqrt{x}$
9 2.621   **10** 4.164   **11** No

Checkpoints for Section 0.5

1 $-9x^2 + 3x + 7$; degree 2   **2** Not a polynomial
3 $2x^2 + 5x + 4$
4 (a) $x^2 + 14x + 45$ (b) $3x^2 - 2x - 1$
5 $x^4 - 2x^3 + 2x^2 - 11x + 4$ **6** $-x^2 + 9$
7 (a) $x^2 - 8x + 16$ (b) $x^3 - 9x^2 + 27x - 27$
8 $x^2 + 10x - y^2 + 25$ **9** $12,282.98
10 Volume $= 4x^3 - 44x^2 + 120x$
   $x = 2$ inches: $V = 96$ cubic inches
   $x = 3$ inches: $V = 72$ cubic inches

Checkpoints for Section 0.6

1 $(x + 1)(3x + 1)$ **2** $x(x + 1)(x - 1)$
3 $4(5 + y)(5 - y)$ **4** (a) $(2x - 1)^2$ (b) $(x - 6)^2$
5 (a) $(y - 1)(y^2 + y + 1)$ (b) $5(z - 2)(z^2 + 2z + 4)$
6 $(x + 3)(x - 2)$ **7** $(2x - 3)(x - 1)$
8 $(x + 1)(x^2 + 5)$ **9** $(2x - 3)(x + 4)$

Checkpoints for Section 0.7

1 (a) All real numbers except $x = 5$ (b) All real numbers
   (c) All nonnegative real numbers

2 $\dfrac{2(x + 1)}{3}, x \neq 1$ **3** $-\dfrac{3 + x}{2(x + 1)}, x \neq 1$
4 $\dfrac{1}{x + 1}, x \neq 2$ **5** $\dfrac{8}{5}, x \neq -y$ **6** $\dfrac{2(6 - x^2)}{3x}$
7 $\dfrac{9}{x}, x \neq 1$ **8** $\dfrac{1}{3}, x \neq 3$

Chapter 1

Checkpoints for Section 1.1

1 Identity   **2** (a) 2   (b) -1 **3** (a) -4 (b) $\frac{11}{3}$
4 Infinitely many solutions   **5** No solution   **6** 30
7 No solution   **8** 6   **9** 0.794   **10** 2003 $(t \approx 3)$

Checkpoints for Section 1.2

1 $950   **2** $p = 6800 + 0.05(6800)$ **3** 20%
4 1.5%   **5** 15 feet $\times$ 45 feet
6 1.1 hours   **7** 32 feet
8 $1200 was invested at 1% and $1300 was invested at 2%.
9 About 1.27 feet

Checkpoints for Section 1.3

1 (a) $-3, 4$ (b) $-6, -4$ **2** $0, 2$ **3** -2
4 ± 2 **5** $-3, 5$ **6** 6 feet $\times$ 14 feet
7 3 seconds   **8** ≈ 8.5 feet   **9** 2010 $(t \approx 10.80)$

Checkpoints for Section 1.4

1 One real solution   **2** $-1 \pm \sqrt{3}$ **3** $\frac{1}{3}$
4 $-0.831, 1.511$ **5** 2:00 P.M.   **6** ≈ 5.1 seconds

Checkpoints for Section 1.5

1 $0, \pm 1$ **2** $1, \pm \sqrt{2}$ **3** $\pm 2, \pm 1$ **4** 12
5 81   **6** $1, 3$ **7** $1, 5$ **8** 5 friends
9 $\approx 4.5\%$ **10** 4,010,025 copies

Checkpoints for Section 1.6

1 (a) $2 \leq x < 7$; bounded   (b) $-\infty < x < 3$; unbounded
2 $x < 1$
3 $x \geq -6$ **4** $-1 \leq x < 2$
5 $-9 \leq x \leq 5$ **6** $x < -4$ or $x > 2$
7 More than 3.5 miles   **8** At most $12\frac{1}{2}$ weeks
9 Undercharged by as much as $0.08 or overcharged by as much as $0.08

Checkpoints for Section 1.7

1 $(-2, 1)$ **2** $(-\infty, 1) \cup (2, \infty)$
3 (a) The solution set is empty.
   (b) The solution set consists of all real numbers except 1.
4 $(-\infty, 2] \cup (3, \infty)$ **5** At least $35 and at most $75
6 (a) $[-2, 2]$ (b) $(-\infty, \infty)$

Chapter 2

Checkpoints for Section 2.1

1

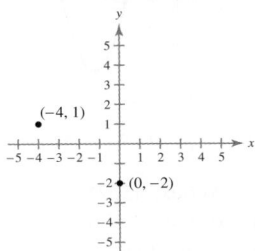

2 (a) 10   (b) $(-3, 2)$ **3** Yes

4 (a)   (b)

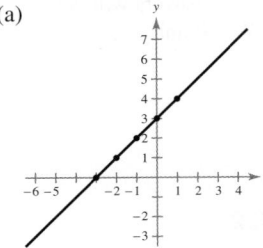 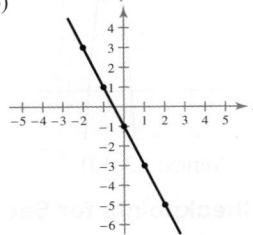

5 (a)   (b)

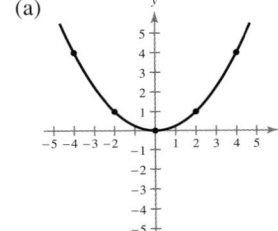 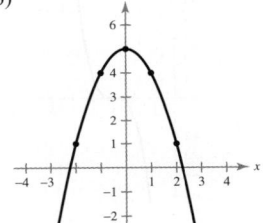

6 (a) $\left(-\frac{7}{4}, 0\right), (0, 7)$ (b) $(-1, 0), (1, 0), (0, -1)$

7 y-axis symmetry

8 (a)   (b)

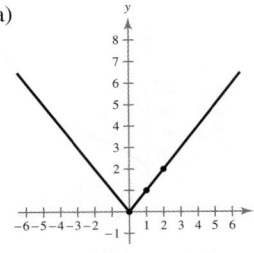 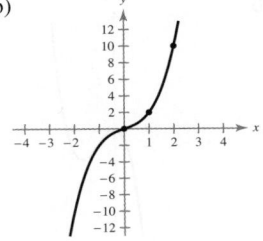

9 $x^2 + (y - 1)^2 = 25$

10

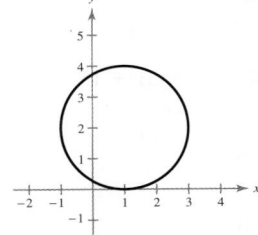

Checkpoints for Section 2.2

1 $\frac{1}{2}$

2 (a) $y = -2x + 8$ (b) $y = \frac{3}{2}x + 9$ (c) $y = 1$

3 $y = 0.2x + 1$

4 (a)   (b)

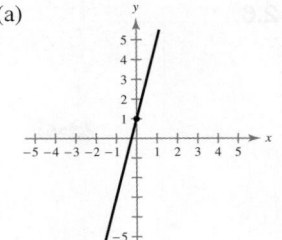 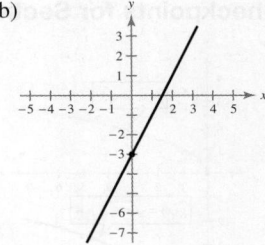

5 (a) $y = 3x - 2$ (b) $y = 3x + 6$ **6** $y = -4x + 4$

Checkpoints for Section 2.3

1 The model approximates the weight of the puppy best for $t = 2$ months and worst for $t = 10$ months.

2 $y = 0.06x$ **3** $y = 33.84615x$ **4** 6 P.M.

5 836,821 people   **6** $V = -195t + 2300$

7 (a) $y = 1.97x + 121.4$ (b)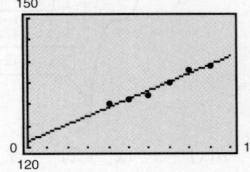

　　(c) $\approx 143,000,000$ tax returns

8

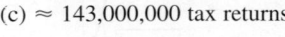

Answers will vary. Sample answer: $v = 50t - 150$

Checkpoints for Section 2.4

1 (a) No   (b) Yes   (c) No

2 (a) Yes   (b) No   **3** -3 **4** 10, 2

5 All real numbers   **6** $V = 4\pi h^3$ **7** No

8

t	9	10	11	12	13
p	160.2	168.9	177.6	186.2	194.9

t	14	15	16	17	18	19
p	221.9	239.3	247.3	246.0	235.3	215.3

Checkpoints for Section 2.5

1 Domain: $(-\infty, \infty)$ **2** (a) Yes   (b) Yes
　　Range: $[-3, \infty)$

3 Decreasing on $\left(-\infty, -\frac{3}{2}\right)$ and increasing on $\left(-\frac{3}{2}, \infty\right)$

4 $(2, 2)$ **5** (a)   (b) 8 days or less

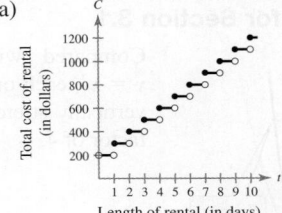

6 Neither

Checkpoints for Section 2.6

1

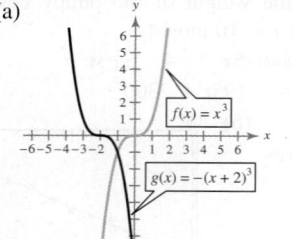

2 (a) The graph of g is a reflection of the graph of f in the x-axis.

(b) The graph of h is a reflection of the graph of f in the y-axis.

3 (a)   (b)

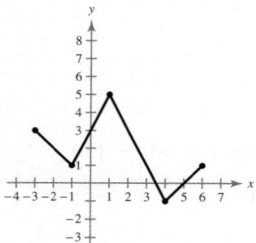

4 $h(x) = -(x - 2)^2 - 1$

5 (a) The graph of g is a vertical stretch of the graph of f by a factor of 4.

(b) The graph of h is a vertical shrink of the graph of f by a factor of $\frac{1}{4}$.

6

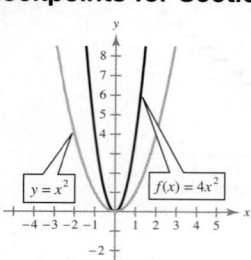

Checkpoints for Section 2.7

1 $2x^2 + x - 1$ **2** $-x - 7; -10$

3 Domain of f/h: All real numbers except $x = 3$

Domain of h/f: All real numbers except $x = 1$

4 (a) $x^2 + 2x - 1$ (b) $x^2 - 1$

5 $\sqrt{3 - x^2}; [-\sqrt{3}, \sqrt{3}]$

6 Answers will vary.

Sample answer: $f(x) = x^2 + 2$, $g(x) = x - 1$,

$h(x) = (x - 1)^2 + 2 = f(x - 1) = f(g(x))$

7 f represents the number of Independent senators.

8 About 1 hour and 18 minutes

Chapter 3

Checkpoints for Section 3.1

1

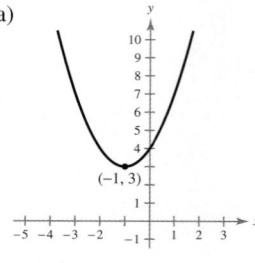

Compared with the graph of $y = x^2$, each output of $f(x) = 4x^2$ vertically stretches the graph by a factor of 4.

2 (a)   (b)

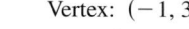

Vertex: $(-1, 3)$ Vertex: $(3, 2)$

3

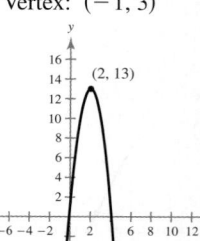

Vertex: $(2, 13)$

4 $f(x) = x^2 - 6x + 13$

5 ≈ 39.7 feet

6 30 moth sites;

Answers will vary.

7 30 units

Checkpoints for Section 3.2

1

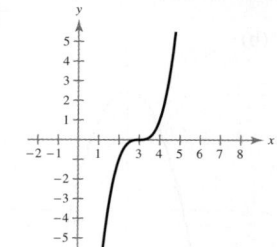

2 (a) Rises to the left; Rises to the right

(b) Falls to the left; Falls to the right

3 Rises to the left; Falls to the right   **4** ± 2 **5** $0, \pm 1$

6   **7**

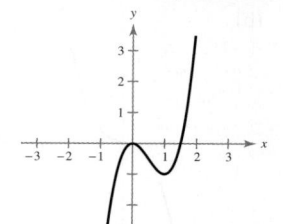   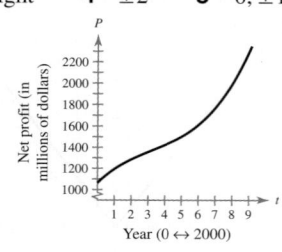

2006 $(t \approx 5.9)$

Checkpoints for Section 3.3

1 $(x - 4)(x - 3)(x + 1)$ **2** $x^2 - 2x + 4$

3 $5x^2 + 13 + \dfrac{2x + 13}{x^2 + 2x - 4}$ **4** $2x^2 + 3x + 15 - \dfrac{5}{x - 5}$

5 3

6

$$\begin{array}{r|rrrr} -2 & 1 & 6 & 7 & -6 & -8 \\ & & -2 & -8 & 2 & 8 \\ \hline & 1 & 4 & -1 & -4 & 0 \end{array}$$

$$\begin{array}{r|rrrr} -4 & 1 & 4 & -1 & -4 \\ & & -4 & 0 & 4 \\ \hline & 1 & 0 & -1 & 0 \end{array}$$

$f(x) = (x + 2)(x + 4)(x^2 - 1)$
$ = (x + 2)(x + 4)(x + 1)(x - 1)$

7 (a) $\approx 10.6\%$ (b) $\approx 15.9\%$

Checkpoints for Section 3.4

1 No rational zeros   **2** $-2, 1$ **3** $-2, -1, \frac{1}{2}$ **4** -2
5 The function has a zero between 1.3 and 1.4.   **6** -1.290
7 $-0.247, 1.445, 2.802$ **8** $x \approx 1.89$ **9** $\approx \$289,000$

Checkpoints for Section 3.5

1 (a) $5 + i$ (b) i **2** (a) $-8 + 12i$ (b) $16 - 30i$

3 $4 + i$ **4** $4 + 3i$ **5** $-\frac{3}{2} \pm \frac{\sqrt{7}}{2}i$

6
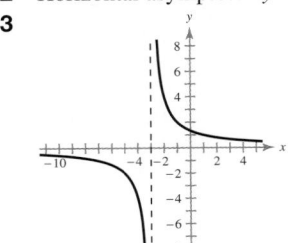

7 The complex number -3 is not in the Mandelbrot Set because for $c = -3$, the corresponding Mandelbrot sequence is $-3, 6, 33, 1086, 1,179,393, \ldots$, which is unbounded.

Checkpoints for Section 3.6

1 Four zeros: $\pm\sqrt{6}, \pm\sqrt{6}i$
2 (a) $\pm 1, \pm 3i$; $f(x) = (x - 1)(x + 1)(x - 3i)(x + 3i)$
   (b) $3, 3, -1, i, -i$;
     $g(x) = (x - 3)(x - 3)(x + 1)(x - i)(x + i)$
3 (a) Answers will vary. Sample answer: $f(x) = x^4 - 5x^2 - 36$
   (b) Answers will vary.
     Sample answer: $f(x) = x^3 - 11x^2 + 36x - 26$
4 (a) $(x^2 + 4)(x^2 - 3)$ (b) $(x^2 + 4)(x - \sqrt{3})(x + \sqrt{3})$
   (c) $(x + 2i)(x - 2i)(x - \sqrt{3})(x + \sqrt{3})$
   (d) Two irrational zeros and two imaginary zeros
5 $\frac{5}{3}, \pm 4i$

Checkpoints for Section 3.7

1 The domain of f is all real numbers except $x = 1$. As x approaches 1 from the left, $f(x)$ decreases without bound. As x approaches 1 from the right, $f(x)$ increases without bound.
2 Horizontal asymptote: $y = 1$; No vertical asymptotes
3
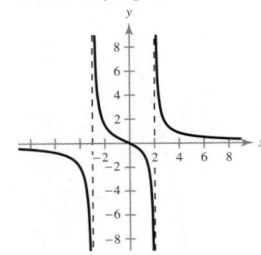
4
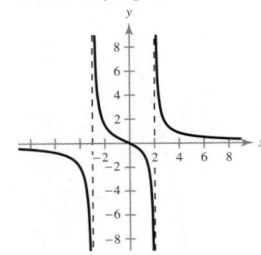
5
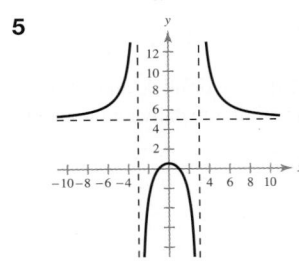
6
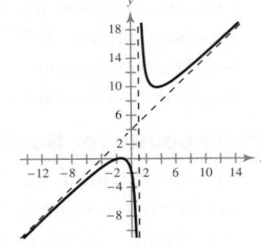
7 $\$1,066,667$ **8** ≈ 6.8 acres per person

Chapter 4

Checkpoints for Section 4.1

1 $f^{-1}(x) = 6x$ **2** $f^{-1}(x) = x - 10$
3 $\left(\sqrt[3]{x - 6}\right)^3 + 6 = x$; $\sqrt[3]{(x^3 + 6)} - 6 = x$
4 $g(x)$ **5** $f^{-1}(x) = \dfrac{x - 5}{4}$

6
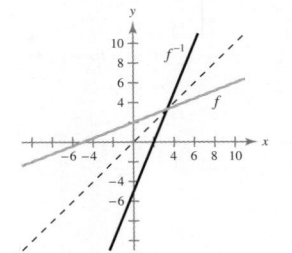
7
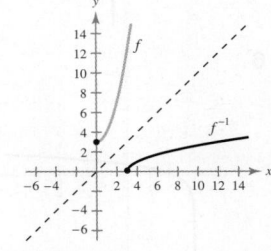

8 (a) f does not have an inverse function.
   (b) f has an inverse function.

Checkpoints for Section 4.2

1 3.918
2
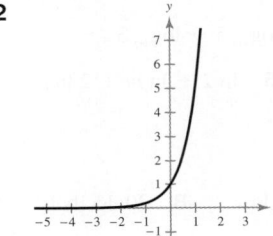
3
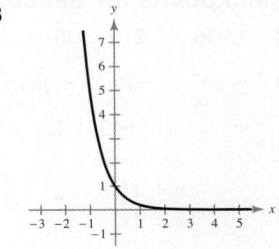
4
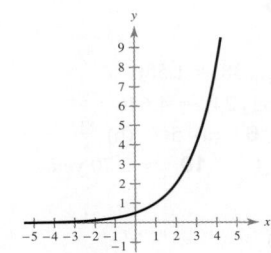
5 403.4287935; 0.0301973834
6
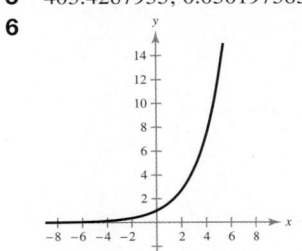
7 (a) $\$7927.75$ (b) $\$7935.08$ (c) $\$7938.78$
8 ≈ 9.971 pounds

Checkpoints for Section 4.3

1 (a) 2   (b) -2 **2** 2.301   **3** 0

4

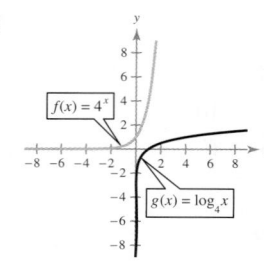

5

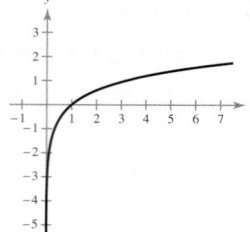

6

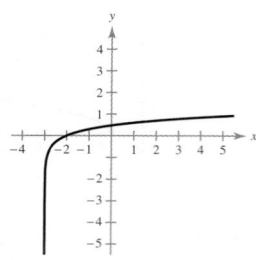

7 (a) 7   (b) 0   (c) $\frac{3}{4}$ **8** -2.303 **9** $(-5, \infty)$
10 103.0 in., 114.0 in., 128.2 in., 132.8 in., 141.7 in.

Checkpoints for Section 4.4

1 1.936   **2** 1.936   **3** $2 \log_{10} 5 - \log_{10} 3$

4 $-\ln \dfrac{2}{e} = -(\ln 2 - \ln e)$ **5** $\ln 2 + \ln m + 2 \ln n$

   $\qquad\qquad = -\ln 2 + \ln e$
   $\qquad\qquad = 1 - \ln 2$

6 $\log_{10} \dfrac{(x + 1)^2}{(x - 1)^3}$ **7** $\ln y = \dfrac{1}{2} \ln x$ **8** 30 decibels

Checkpoints for Section 4.5

1 (a) 4   (b) 216   (c) 5
2 (a) $\log_6 84 \approx 2.473$ (b) $\log_{10} 38 \approx 1.580$
3 $\ln 15 \approx 2.708$ **4** $\frac{1}{2}(7 + \log_4 24) \approx 4.646$
5 $\ln 3 \approx 1.099, \ln 4 \approx 1.386$ **6** (a) 5   (b) $\frac{81}{2}$
7 3   **8** $e^4 \approx 54.598$ **9** 4   **10** ≈ 2.70 years
11 2001 $(t \approx 11.01)$

Checkpoints for Section 4.6

1 2014 $(t \approx 23.89)$ **2** $y = 3e^{0.19617x}$ **3** ≈ 7681 years
4

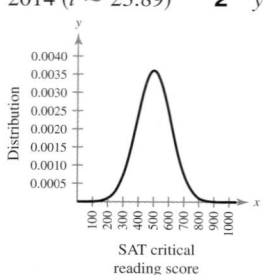

SAT critical
reading score

Average score: 501
5 9 days   **6** $\approx 31{,}622{,}777$; 100 times greater
7 1×10^{-5} mole of hydrogen per liter
8 Logistic growth model

Chapter 5

Checkpoints for Section 5.1

1 $(5, 1)$

2 \$4000 is invested at 3% and \$11,000 is invested at 5%.
3 $(-4, -7), (2, 5)$ **4** No solution   **5** $(4, 0)$
6 ≈ 4762 pairs of shoes   **7** 2006 $(t \approx 6.7)$

Checkpoints for Section 5.2

1 $\left(2, -\frac{1}{3}\right)$ **2** $(3, 2)$ **3** $(-3, -3)$
4 $(1, -4)$ **5** No solution   **6** $(a, -a - 5)$
7 Speed of plane: ≈ 471.18 miles per hour
   Speed of wind: ≈ 16.63 miles per hour
8 $(2{,}000{,}000, 290)$

Checkpoints for Section 5.3

1 $(-1, -3, 4)$ **2** $(1, 2, 3)$ **3** No solution
4 $\left(-\frac{3}{5}a + \frac{3}{10}, -\frac{9}{5}a + \frac{2}{5}, a\right)$ **5** $\left(\frac{1}{4}a, \frac{11}{4}a - 1, a\right)$
6 Answers will vary. Sample answer: \$650,000 at 5%, \$150,000
   at 7%, \$200,000 at 8%, \$200,000 at 10%
7 $y = 2x^2 - 2x + 3$

Checkpoints for Section 5.4

1

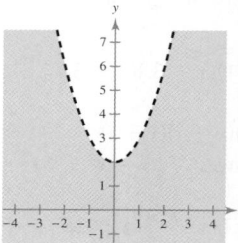

2

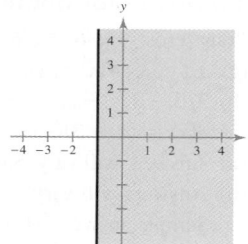

3

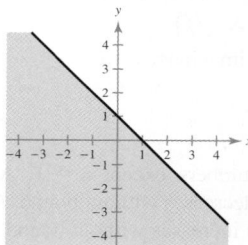

4

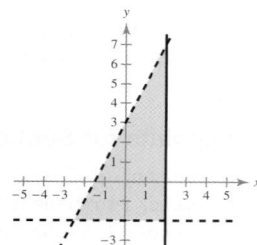

5

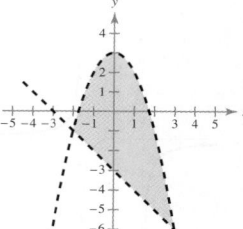

6 Consumer surplus: \$60,000,000
   Producer surplus: \$60,000,000
7 No. The combination of 4 cups of dietary drink X and 1 cup of
   dietary drink Y does not meet all the minimum daily requirements.

Checkpoints for Section 5.5

1 Maximum value at $(0, 3)$: 9
2 (a) Maximum value at $(27, 0)$: 135
   (b) Minimum value at $(0, 0)$: 0
3 No maximum value

4 The maximum profit would be $2925, and it would occur at monthly production levels of 1050 units of product I and 150 units of product II.

5 The minimum cost would be $0.56 per day, and it would occur when 1 cup of drink X and 4 cups of drink Y were consumed each day.

Chapter 6

Checkpoints for Section 6.1

1 2×3

2 $\begin{bmatrix} 4 & -3 & \vdots & -5 \\ -1 & 3 & \vdots & 12 \end{bmatrix}$ **3** Multiply the second row by $\frac{1}{3}$.
2×3

4 $\begin{cases} x - 2y + 5z = 3 \\ \quad\quad y + 4z = -3 \\ \quad\quad\quad\quad z = 2 \end{cases}$ **5** Row-echelon form

$(-29, -11, 2)$

6 $(-1, 0, -1, 3)$ **7** No solution   **8** $(6, 5, 5)$

9 $(-9a - 10, -5a - 2, a)$

Checkpoints for Section 6.2

1 $a_{11} = 5, a_{12} = 2, a_{21} = -1, a_{22} = 3$ **2** $\begin{bmatrix} 6 & -2 \\ -2 & -3 \end{bmatrix}$

3 (a) $\begin{bmatrix} 4 & 8 & -2 \\ 0 & 2 & 6 \\ -6 & 4 & 10 \end{bmatrix}$ (b) $\begin{bmatrix} 4 & 2 & -5 \\ -7 & 6 & 5 \\ -8 & 4 & 12 \end{bmatrix}$

4 $\begin{bmatrix} 1 & 1 \\ 12 & -4 \end{bmatrix}$ **5** $\begin{bmatrix} -24 & -4 \\ -12 & 32 \end{bmatrix}$

6 $\begin{bmatrix} 5 & \frac{1}{2} \\ \frac{1}{2} & 3 \end{bmatrix}$ **7** $\begin{bmatrix} 8 & -8 \\ 4 & -16 \\ -10 & 4 \end{bmatrix}$ **8** $\begin{bmatrix} -3 & -22 \\ 3 & 10 \\ -5 & 10 \end{bmatrix}$

9 Not possible   **10** $AB = [6], BA = \begin{bmatrix} 3 & -1 \\ -9 & 3 \end{bmatrix}$

11 $\begin{bmatrix} -2 & -3 \\ 6 & 1 \end{bmatrix}\begin{bmatrix} x_1 \\ x_2 \end{bmatrix} = \begin{bmatrix} -4 \\ -36 \end{bmatrix}, X = \begin{bmatrix} -7 \\ 6 \end{bmatrix}$

12 Women's team: $2490; Men's team: $2871

Checkpoints for Section 6.3

1 $AB = I$ and $BA = I$ **2** $\begin{bmatrix} 1 & 1 \\ -5 & -4 \end{bmatrix}$

3 $\begin{bmatrix} -4 & -2 & 5 \\ -2 & -1 & 2 \\ -1 & 0 & 1 \end{bmatrix}$ **4** $\begin{bmatrix} \frac{1}{10} & \frac{3}{10} \\ \frac{2}{5} & \frac{1}{5} \end{bmatrix}$ **5** $(-1, 2, 1)$

Checkpoints for Section 6.4

1 (a) -7 (b) 10 (c) 0

2 Minors: $M_{11} = -9, M_{12} = -10, M_{13} = 2, M_{21} = 5,$
$M_{22} = -2, M_{23} = -3, M_{31} = 13, M_{32} = 5,$
$M_{33} = -1$
Cofactors: $C_{11} = -9, C_{12} = 10, C_{13} = 2, C_{21} = -5,$
$C_{22} = -2, C_{23} = 3, C_{31} = 13, C_{32} = -5,$
$C_{33} = -1$

3 -32 **4** -133

Checkpoints for Section 6.5

1 14 **2** Not collinear   **3** $x - y + 2 = 0$

4 $\begin{bmatrix} 15 & 23 & 12 \end{bmatrix}\begin{bmatrix} 19 & 0 & 1 \end{bmatrix}\begin{bmatrix} 18 & 5 & 0 \end{bmatrix}$
$\begin{bmatrix} 14 & 15 & 3 \end{bmatrix}\begin{bmatrix} 20 & 21 & 18 \end{bmatrix}\begin{bmatrix} 14 & 1 & 12 \end{bmatrix}$

5 $110, -39, -59, 25, -21, -3, 23, -18, -5, 47, -20, -24,$
$149, -56, -75, 87, -38, -37$

6 OWLS ARE NOCTURNAL

Chapter 7

Checkpoints for Section 7.1

1 6 **2** (a) 4 (b) Does not exist   (c) 4

3 (a) 5 (b) 6 (c) 25 (d) -2 **4** 5 **5** 12

6 7 **7** $\frac{1}{4}$ **8** (a) -1 (b) 1 **9** 1

10 $\lim\limits_{x \to 1^-} f(x) = 18$ and $\lim\limits_{x \to 1^+} f(x) = 20$

$\lim\limits_{x \to 1^-} f(x) \neq \lim\limits_{x \to 1^+} f(x)$

11 Does not exist

Checkpoints for Section 7.2

1 (a) f is continuous on the entire real line.
(b) f is continuous on the entire real line.
(c) f is continuous on the entire real line.

2 (a) f is continuous on $(-\infty, 1)$ and $(1, \infty)$.
(b) f is continuous on $(-\infty, 2)$ and $(2, \infty)$.
(c) f is continuous on the entire real line.

3 f is continuous on $[2, \infty)$.   **4** f is continuous on $[-1, 5]$.

5

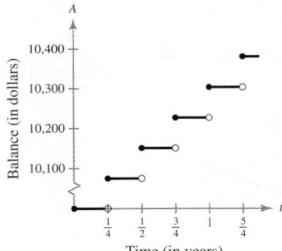

6 $A = 10,000(1 + 0.0075)^{[\![4t]\!]}$

Checkpoints for Section 7.3

1 3

2 For the months to the left of July on the graph, the tangent lines have positive slopes. For the months to the right of July, the tangent lines have negative slopes. The average daily temperature is increasing prior to July and decreasing after July.

3 4 **4** 2

5 $m = 8x$
At $(0, 1), m = 0.$
At $(1, 5), m = 8.$

6 $2x - 5$ **7** $-\dfrac{4}{t^2}$

Checkpoints for Section 7.4

1 (a) 0   (b) 0   (c) 0   (d) 0

2 (a) $4x^3$ (b) $-\dfrac{3}{x^4}$ (c) $2w$

3 $f'(x) = 3x^2$
$m = f'(-1) = 3;$
$m = f^{-1}(0) = 0;$
$m = f^{-1}(1) = 3$

4 (a) $8x$ (b) $\dfrac{8}{\sqrt{x}}$ **5** (a) $\dfrac{1}{4}$ (b) $-\dfrac{2}{5}$

6 (a) $-\dfrac{9}{2x^3}$ (b) $-\dfrac{9}{8x^3}$ **7** (a) $\dfrac{\sqrt{5}}{2\sqrt{x}}$ (b) $\dfrac{1}{4x^{3/4}}$

8 (a) $4x + 5$ (b) $4x^3 - 2$ **9** -1

10 $y = -x + 2$ **11** $0.34/\text{yr}$

Checkpoints for Section 7.5

1 (a) $0.5\overline{6}$ mg/ml/min   (b) 0 mg/ml/min
(c) -1.5 mg/ml/min

2 (a) -16 ft/sec   (b) -48 ft/sec   (c) -80 ft/sec

3 When $t = 1.75$, $h'(1.75) = -56$ ft/sec.
When $t = 2$, $h'(2) = -64$ ft/sec.

4 (a) $\frac{3}{2}$ sec   (b) -32 ft/sec

5 When $x = 100$, $\dfrac{dP}{dx} = \$16/\text{unit}$.

Actual gain = $16.06

6 $p = -0.057x + 23.82$, $R = -0.057x^2 + 23.82x$

7 Revenue: $R = 2000x - 4x^2$

Marginal revenue: $\dfrac{dR}{dx} = 2000 - 8x$; $0/\text{unit}$

8 $\dfrac{dP}{dx} = \$1.44/\text{unit}$

Actual increase in profit $\approx \$1.44$

Checkpoints for Section 7.6

1 $-27x^2 + 12x + 24$ **2** $\dfrac{2x^2 - 1}{x^2}$

3 (a) $18x^2 + 30x$ (b) $12x + 15$ **4** $-\dfrac{22}{(5x - 2)^2}$

5 $y = \frac{8}{25}x - \frac{4}{5}$;

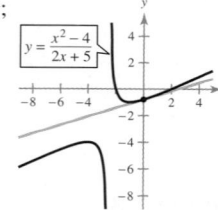

6 $\dfrac{-3x^2 + 4x + 8}{x^2(x + 4)^2}$ **7** (a) $\dfrac{2}{5}x + \dfrac{4}{5}$ (b) $3x^3$

8

t	0	1	2	3	4	5	6	7
$\dfrac{dP}{dt}$	0	-50	-16	-6	-2.77	-1.48	-0.88	-0.56

As t increases, the rate at which the blood pressure drops decreases.

Checkpoints for Section 7.7

1 (a) $u = g(x) = x + 1$, $y = f(u) = \dfrac{1}{\sqrt{u}}$
(b) $u = g(x) = x^2 + 2x + 5$, $y = f(u) = u^3$

2 $6x^2(x^3 + 1)$ **3** $4(2x + 3)(x^2 + 3x)^3$ **4** $y = \frac{1}{3}x + \frac{8}{3}$

5 $-\dfrac{8}{(2x + 1)^2}$ **6** $\dfrac{x(3x^2 + 2)}{\sqrt{x^2 + 1}}$ **7** $-\dfrac{12(x + 1)}{(x - 5)^3}$

8 About $3.48/\text{yr}$

Chapter 8

Checkpoints for Section 8.1

1 $f'(x) = 18x^2 - 4x$, $f''(x) = 36x - 4$, $f'''(x) = 36$, $f^{(4)}(x) = 0$

2 18   **3** $\dfrac{120}{x^6}$

4 Height = 144 ft   **5** -9.8 m/sec^2
Velocity = 0 ft/sec
Acceleration = -32 ft/sec^2

6

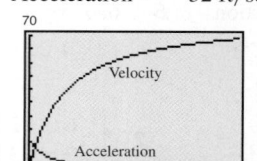

Acceleration approaches zero.

Checkpoints for Section 8.2

1 $-\dfrac{2}{x^3}$

2 (a) $12x^2$ (b) $6y\dfrac{dy}{dx}$ (c) $1 + 5\dfrac{dy}{dx}$ (d) $y^3 + 3xy^2\dfrac{dy}{dx}$

3 $\dfrac{dy}{dx} = -\dfrac{x - 2}{y - 1}$ **4** $\dfrac{3}{4}$ **5** $\dfrac{5}{9}$

6 $\dfrac{dx}{dp} = -\dfrac{2}{p^2(0.002x + 1)}$

Checkpoints for Section 8.3

1 9   **2** $12\pi \approx 37.7$ ft^2/sec
3 $1500/\text{day}$ **4** $28,400/\text{wk}$

Checkpoints for Section 8.4

1 $f'(x) = 4x^3$
$f'(x) < 0$ if $x < 0$; therefore, f is decreasing on $(-\infty, 0)$.
$f'(x) > 0$ if $x > 0$; therefore, f is increasing on $(0, \infty)$.

2 $\dfrac{dF}{dt} = -1.5348t + 2.872 < 0$ when $3 \le t \le 8$, which implies
that the consumption of fresh fruit was decreasing from 2003
through 2008.

3 $x = \frac{1}{2}$

4 Increasing on $(-\infty, -2)$ and $(2, \infty)$
Decreasing on $(-2, 2)$

5 Increasing on $(0, \infty)$
Decreasing on $(-\infty, 0)$

6 Increasing on $(-\infty, -1)$ and $(1, \infty)$
Decreasing on $(-1, 0)$ and $(0, 1)$

7 Because $f'(x) = -3x^2 = 0$ when $x = 0$ and because f is
decreasing on $(-\infty, 0) \cup (0, \infty)$, f is decreasing on $(-\infty, \infty)$.

8 $(0, 3000)$

Checkpoints for Section 8.5

1 Relative maximum at $(-1, 5)$
Relative minimum at $(1, -3)$

2 Relative minimum at $(3, -27)$

3 Relative maximum at $(1, 1)$
Relative minimum at $(0, 0)$

4 Absolute maximum at $(0, 10)$
Absolute minimum at $(4, -6)$

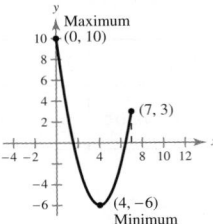

5

x (units)	24,000	24,200	24,300	24,400
P (profit)	$24,760	$24,766	$24,767.50	$24,768

x (units)	24,500	24,600	24,800
P (profit)	$24,767.50	$24,766	$24,760

Checkpoints for Section 8.6

1 (a) $f'' = -4$; because $f''(x) < 0$ for all x, f is concave downward for all x.

(b) $f''(x) = \dfrac{1}{2x^{3/2}}$; because $f''(x) > 0$ for all $x > 0$, f is concave upward for all $x > 0$.

2 Because $f''(x) > 0$ for $x < -\dfrac{2\sqrt{3}}{3}$ and $x > \dfrac{2\sqrt{3}}{3}$, f is concave upward on $\left(-\infty, -\dfrac{2\sqrt{3}}{3}\right)$ and $\left(\dfrac{2\sqrt{3}}{3}, \infty\right)$. Because $f''(x) < 0$ for $-\dfrac{2\sqrt{3}}{3} < x < \dfrac{2\sqrt{3}}{3}$, f is concave downward on $\left(-\dfrac{2\sqrt{3}}{3}, \dfrac{2\sqrt{3}}{3}\right)$.

3 f is concave upward on $(-\infty, 0)$.
f is concave downward on $(0, \infty)$.
Point of inflection: $(0, 0)$

4 f is concave upward on $(-\infty, 0)$ and $(1, \infty)$.
f is concave downward on $(0, 1)$.
Points of inflection: $(0, 1)$, $(1, 0)$

5 Relative minimum: $(3, -26)$

6 Point of diminishing returns: $x = \$150$ thousand

Chapter 9

Checkpoints for Section 9.1

1

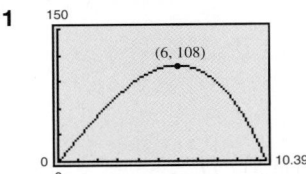

Maximum volume $= 108$ in.^{3}

2 $\left(\sqrt{\dfrac{1}{2}}, \dfrac{7}{2}\right)$ and $\left(-\sqrt{\dfrac{1}{2}}, \dfrac{7}{2}\right)$ **3** 8 in. by 12 in.

Checkpoints for Section 9.2

1 125 units yield a maximum revenue of $1,562,500.

2 400 units   **3** $6.25/unit   **4** $4.00

5 Demand is elastic when $0 < x < 144$.
Demand is inelastic when $144 < x < 324$.
Demand is of unit elasticity when $x = 144$.

Checkpoints for Section 9.3

1 (a) $\lim\limits_{x \to 2^-} \dfrac{1}{x - 2} = -\infty$ (b) $\lim\limits_{x \to 2^+} \dfrac{1}{x - 2} = \infty$

(c) $\lim\limits_{x \to -3^-} \dfrac{-1}{x + 3} = -\infty$ (d) $\lim\limits_{x \to -3^+} \dfrac{-1}{x + 3} = \infty$

2 $x = 0, x = 4$ **3** $x = 3$

4 $\lim\limits_{x \to 2^-} \dfrac{x^2 - 4x}{x - 2} = \infty$; $\lim\limits_{x \to 2^+} \dfrac{x^2 - 4x}{x - 2} = -\infty$

5 2

6 (a) $y = 0$ (b) $y = \dfrac{1}{2}$ (c) No horizontal asymptote

7 $C = 0.75x + 25,000$

$\overline{C} = 0.75 + \dfrac{25,000}{x}$

$\lim\limits_{x \to \infty} \overline{C} = \$0.75/\text{unit}$

8 No, the cost function is not defined at $p = 100$, which implies that it is not possible to remove 100% of the pollutants.

Checkpoints for Section 9.4

1

	$f(x)$	$f'(x)$	$f''(x)$	Shape of graph
x in $(-\infty, -1)$		$-$	$+$	Decreasing, concave upward
$x = -1$	-32	0	$+$	Relative minimum
x in $(-1, 1)$		$+$	$+$	Increasing, concave upward
$x = 1$	-16	$+$	0	Point of inflection
x in $(1, 3)$		$+$	$-$	Increasing, concave downward
$x = 3$	0	0	$-$	Relative maximum
x in $(3, \infty)$		$-$	$-$	Decreasing, concave downward

2

	$f(x)$	$f'(x)$	$f''(x)$	Shape of graph
x in $(-\infty, 0)$		$-$	$+$	Decreasing, concave upward
$x = 0$	5	0	0	Point of inflection
x in $(0, 2)$		$-$	$-$	Decreasing, concave downward
$x = 2$	-11	$-$	0	Point of inflection
x in $(2, 3)$		$-$	$+$	Decreasing, concave upward
$x = 3$	-22	0	$+$	Relative minimum
x in $(3, \infty)$		$+$	$+$	Increasing, concave upward

CHECKPOINTS

3

	$f(x)$	$f'(x)$	$f''(x)$	Shape of graph
x in $(-\infty, 0)$		+	−	Increasing, concave downward
$x = 0$	0	0	−	Relative maximum
x in $(0, 1)$		−	−	Decreasing, concave downward
$x = 1$	Undef.	Undef.	Undef.	Vertical asymptote
x in $(1, 2)$		−	+	Decreasing, concave upward
$x = 2$	4	0	+	Relative minimum
x in $(2, \infty)$		+	+	Increasing, concave upward

4

	$f(x)$	$f'(x)$	$f''(x)$	Shape of graph
x in $(-\infty, -1)$		+	+	Increasing, concave upward
$x = -1$	Undef.	Undef.	Undef.	Vertical asymptote
x in $(-1, 0)$		+	−	Increasing, concave downward
$x = 0$	−1	0	−	Relative maximum
x in $(0, 1)$		−	−	Decreasing, concave downward
$x = 1$	Undef.	Undef.	Undef.	Vertical asymptote
x in $(1, \infty)$		−	+	Decreasing, concave upward

5

	$f(x)$	$f'(x)$	$f''(x)$	Shape of graph
x in $(0, 1)$		−	+	Decreasing, concave upward
$x = 1$	−4	0	+	Relative minimum
x in $(1, \infty)$		+	+	Increasing, concave upward

Checkpoints for Section 9.5

1 $dy = 0.32$; $\Delta y = 0.32240801$ **2** $dR = \$22$; $\Delta R = \$21$
3 $dP = \$10.96$; $\Delta P = \$10.98$
4 (a) $dy = 12x^2\,dx$ (b) $dy = \frac{2}{3}\,dx$

(c) $dy = (6x - 2)\,dx$ (d) $dy = -\dfrac{2}{x^3}\,dx$

Chapter 10

Checkpoints for Section 10.1

1 (a) 243   (b) 3   (c) 64   (d) 8   (e) $\frac{1}{2}$ (f) $\sqrt{10}$
2 (a) 5.453×10^{-13} (b) 1.621×10^{-13}
(c) 2.629×10^{-14}

3 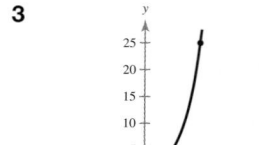   **4**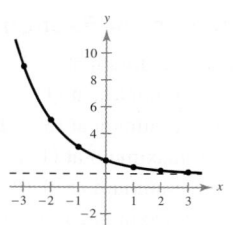

Checkpoints for Section 10.2

1

x	-2	-1	0	1	2
$g(x)$	$e^2 \approx 7.389$	$e \approx 2.718$	1	$\dfrac{1}{e} \approx 0.368$	$\dfrac{1}{e^2} \approx 0.135$

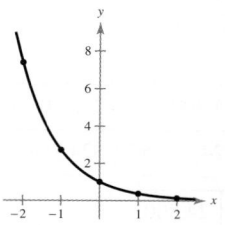

2 After 0 h, $y = 1.25$ g.
After 1 h, $y \approx 1.338$ g.
After 10 h, $y \approx 1.498$ g.
$\displaystyle\lim_{t \to \infty} \frac{1.50}{1 + 0.2e^{-0.5t}} = 1.50$ g

3 (a) \$4870.38   (b) \$4902.71
(c) \$4918.66   (d) \$4919.21
All else being equal, the more often interest is compounded, the greater the balance.
4 (a) 7%   (b) 7.12%   (c) 7.19%   (d) 7.23%
5 \$16,712.90

Checkpoints for Section 10.3

1 At $(0, 2)$, the slope is 2. At $(1, 2e)$, the slope is $2e$.
2 (a) $3e^{3x}$

(b) $-\dfrac{6x^2}{e^{2x^3}}$

(c) $8xe^{x^2}$

(d) $-\dfrac{2}{e^{2x}}$

3 (a) 0

(b) $3e^{3x+1}$

(c) $xe^x(x + 2)$

(d) $\frac{1}{2}(e^x - e^{-x})$

(e) $\dfrac{e^x(x - 2)}{x^3}$

(f) $e^x(x^2 + 2x - 1)$

4

![Graph with window from −30 to 30 horizontally and 0 to 75 vertically, showing a curve with minimum point labeled (0, 60).]

5 \$18.39/unit (80,000 units); \$1,471,517.77

6

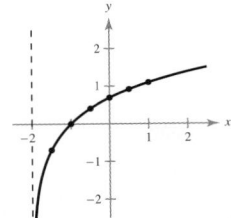

Points of inflection: $(-4, 0.060)$, $(4, 0.060)$

Checkpoints for Section 10.4

1

x	-1.5	-1	-0.5	0	0.5	1
$f(x)$	-0.693	0	0.405	0.693	0.916	1.099

2 (a) 3   (b) $x + 1$

3 (a) $\ln 2 - \ln 5$ (b) $\frac{1}{3} \ln(x + 2)$
  (c) $\ln x - \ln 5 - \ln y$ (d) $\ln x + 2 \ln(x + 1)$

4 (a) $\ln x^4 y^3$ (b) $\ln \frac{x + 1}{(x + 3)^2}$ **5** (a) $\ln 6$ (b) $5 \ln 5$

6 (a) e^4 (b) e^3 **7** 7.9 yr

Checkpoints for Section 10.5

1 $\frac{1}{x}$ **2** (a) $\frac{2x}{x^2 - 4}$ (b) $x(1 + 2 \ln x)$ (c) $\frac{2 \ln x - 1}{x^3}$

3 $\frac{1}{3(x + 1)}$ **4** $\frac{2}{x} + \frac{x}{x^2 + 1}$ **5** $y = 4x - 4$

6 Relative minimum: $(2, 2 - 2 \ln 2) \approx (2, 0.6137)$

7 $\frac{dp}{dt} = -1.3\%/\text{mo}$

The average score would decrease at a greater rate than the model in Example 7.

8 (a) 4   (b) -2 (c) -5 (d) 3

9 (a) 2.322   (b) 2.631   (c) 3.161   (d) -0.5

10

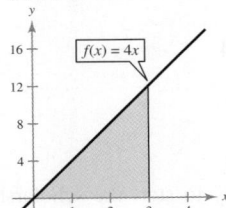

As time increases, the derivative approaches 0. The rate of change of the amount of carbon isotopes is proportional to the amount present.

Checkpoints for Section 10.6

1 About 2113.7 yr   **2** $y = 25e^{0.6931t}$

3 $r = \frac{1}{8} \ln 2 \approx 0.0866$ or 8.66%   **4** About 12.42 mo

Chapter 11

Checkpoints for Section 11.1

1 (a) $5x + C$ (b) $-r + C$ (c) $2t + C$ **2** $\frac{5}{2}x^2 + C$

3 (a) $-\frac{1}{x} + C$ (b) $\frac{3}{4}x^{4/3} + C$

4 (a) $\frac{1}{2}x^2 + 4x + C$ (b) $x^4 - \frac{5}{2}x^2 + 2x + C$

5 $\frac{2}{3}x^{3/2} + 4x^{1/2} + C$

6 General solution: $F(x) = 2x^2 + 2x + C$
  Particular solution: $F(x) = 2x^2 + 2x + 4$

7 $s(t) = -16t^2 + 32t + 48$. The ball hits the ground 3 seconds after it is thrown, with a velocity of -64 feet per second.

8 $C = -0.01x^2 + 28x + 12.01$
  $C(200) = \$5212.01$

Checkpoints for Section 11.2

1 (a) $\frac{(x^3 + 6x)^3}{3} + C$ (b) $\frac{2}{3}(x^2 - 2)^{3/2} + C$

2 $\frac{1}{36}(3x^4 + 1)^3 + C$ **3** $\frac{1}{9}(x^3 - 3x)^3 + C$

4 $2x^9 + \frac{12}{5}x^5 + 2x + C$ **5** $\frac{5}{3}(x^2 - 1)^{3/2} + C$

6 $-\frac{1}{3}(1 - 2x)^{3/2} + C$ **7** $\frac{1}{3}(x^2 + 4)^{3/2} + C$

8 About $34,068

Checkpoints for Section 11.3

1 (a) $3e^x + C$ (b) $e^{5x} + C$ (c) $e^x - \frac{x^2}{2} + C$

2 $\frac{1}{2}e^{2x+3} + C$ **3** $2e^{x^2} + C$

4 (a) $2 \ln|x| + C$ (b) $\ln|x^3| + C$ (c) $\ln|2x + 1| + C$

5 $\frac{1}{4} \ln|4x + 1| + C$ **6** $\frac{3}{2} \ln(x^2 + 4) + C$

7 (a) $4x - 3 \ln|x| - \frac{2}{x} + C$ (b) $2 \ln(1 + e^x) + C \, dx$

  (c) $\frac{x^2}{2} + x + 3 \ln|x + 1| + C$

Checkpoints for Section 11.4

1 $\frac{1}{2}(3)(12) = 18$

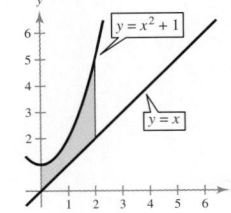

2 $\frac{22}{3}$ units^2 **3** 68

4 (a) $\frac{1}{4}(e^4 - 1) \approx 13.3995$ (b) $-\ln 5 + \ln 2 \approx -0.9163$

5 $\frac{13}{2}$ **6** (a) About $14.18   (b) $141.79

7 $13.70   **8** (a) $\frac{2}{5}$ (b) 0   **9** About $12,295.62

Checkpoints for Section 11.5

1 $\frac{8}{3}$ units^2

2 $\frac{32}{3}$ units^2 **3** $\frac{9}{2}$ units^2

4 $\frac{253}{12}$ units^2

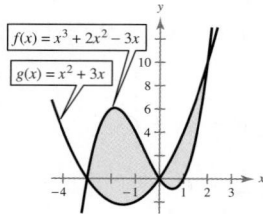

5 Consumer surplus: 40
Producer surplus: 20

6 The company can save $47.52 million.

Checkpoints for Section 11.6

1 $\frac{37}{8}$ units^2 **2** 0.436 unit^2 **3** 5.642 units^2

4 About 1.463

Chapter 12

Checkpoints for Section 12.1

1 $\frac{1}{2}xe^{2x} - \frac{1}{4}e^{2x} + C$ **2** $\frac{x^2}{2} \ln x - \frac{1}{4}x^2 + C$

3 $x \ln 2x - x + C$ **4** $e^x(x^3 - 3x^2 + 6x - 6) + C$

5 $e - 2$ **6** $538,145$ **7** $721,632.08$

Checkpoints for Section 12.2

1 $\frac{2}{3}(x - 4)\sqrt{2 + x} + C$ (Formula 19)

2 $\sqrt{x^2 + 16} - 4 \ln \left| \frac{4 + \sqrt{x^2 + 16}}{x} \right| + C$ (Formula 25)

3 $\frac{1}{4} \ln \left| \frac{x - 2}{x + 2} \right| + C$ (Formula 21)

4 $\frac{1}{3}[1 - \ln(1 + e) + \ln 2] \approx 0.12663$ (Formula 39)

5 $x(\ln x)^2 + 2x - 2x \ln x + C$ (Formula 44)

6 About 18.2%

Checkpoints for Section 12.3

1 3.2608   **2** 3.1956   **3** 1.154

Checkpoints for Section 12.4

1 (a) Converges; $\frac{1}{2}$ (b) Diverges

2 1   **3** $\frac{1}{2}$ **4** 0.0013 or 0.13%

5 No, you do not have enough money to start the scholarship
fund because you need $125,000.   ($125,000 > $120,000)

Chapter 13

Checkpoints for Section 13.1

1

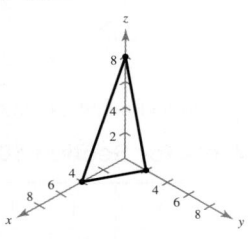

2 $2\sqrt{6}$ **3** $\left(-\frac{5}{2}, 2, -2\right)$

4 $(x - 4)^2 + (y - 3)^2 + (z - 2)^2 = 25$

5 $(x - 1)^2 + (y - 3)^2 + (z - 2)^2 = 38$

6 Center: $(-3, 4, -1)$; radius: 6

7 $(x + 1)^2 + (y - 2)^2 = 16$

Checkpoints for Section 13.2

1 x-intercept: $(4, 0, 0)$;
y-intercept: $(0, 2, 0)$;
z-intercept: $(0, 0, 8)$

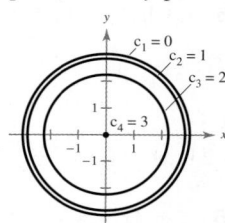

2 xy-trace: circle, $x^2 + y^2 = 1$
yz-trace: hyperbola, $y^2 - z^2 = 1$
xz-trace: hyperbola, $x^2 - z^2 = 1$
$z = 3$ trace: circle, $x^2 + y^2 = 10$
Hyperboloid of one sheet

3 (a) Elliptic paraboloid   (b) Elliptic cone

Checkpoints for Section 13.3

1 (a) 0   (b) $\frac{9}{4}$

2 Domain: $x^2 + y^2 \leq 9$
Range: $0 \leq z \leq 3$

3 For each value of c, the equation $f(x, y) = c$ is a circle (or
point) in the xy-plane.

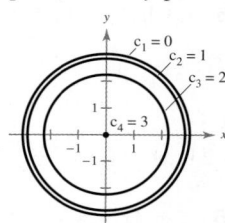

4 $f(1500, 1000) \approx 127{,}542$ units
$f(1000, 1500) \approx 117{,}608$ units
x, person-hours, has a greater effect on production.

5 (a) $M = 421.60/\text{mo}$
(b) Total paid $= (30 \times 12) \times 421.60$
$= 151{,}776$

Checkpoints for Section 13.4

1 $\dfrac{\partial z}{\partial x} = 4x - 8xy^3$

$\dfrac{\partial z}{\partial y} = -12x^2y^2 + 4y^3$

2 $f_x(x, y) = 2xy^3$; $f_x(1, 2) = 16$
$f_y(x, y) = 3x^2y^2$; $f_y(1, 2) = 12$

3 (a) $f_x(1, -1, 49) = 8$ (b) $f_y(1, -1, 49) = -18$

4 Substitute product relationship

5 $\dfrac{\partial w}{\partial x} = xy + 2xy \ln(xz)$

$\dfrac{\partial w}{\partial y} = x^2 \ln xz$

$\dfrac{\partial w}{\partial z} = \dfrac{x^2y}{z}$

6 $f_{xx} = 8y^2$
$f_{yy} = 8x^2 + 8$
$f_{xy} = 16xy$
$f_{yx} = 16xy$

7 $f_{xx} = 0 \qquad f_{xy} = e^y \qquad f_{xz} = 2$
$\quad f_{yx} = e^y \qquad f_{yy} = xe^y + 2 \qquad f_{yz} = 0$
$\quad f_{zx} = 2 \qquad f_{zy} = 0 \qquad f_{zz} = 0$

Checkpoints for Section 13.5

1 $f(-8, 2) = -64$: relative minimum
2 $f(0, 0) = 1$: relative maximum
3 $f(0, 0) = 0$: saddle point
4 $P(3.11, 3.81) = \$744.81$ maximum profit
5 $V\left(\frac{4}{3}, \frac{2}{3}, \frac{8}{3}\right) = \frac{64}{27}$ units^3

Checkpoints for Section 13.6

1 $V\left(\frac{4}{3}, \frac{2}{3}, \frac{8}{3}\right) = \frac{64}{27}$ units^3 **2** $f(187.5, 50) \approx 13,474$ units
3 About 26,740 units
4 $P(3.35, 4.26) = \$758.08$ maximum profit

Checkpoints for Section 13.7

1 For $f(x)$, $S = 10$. For $g(x)$, $S = 0.76$.
The quadratic model is a better fit.
2 $f(x) = \frac{6}{5}x + \frac{23}{10}$
3 $y = 16,194.4t + 132,405$
About 342,932,200 subscribers

Checkpoints for Section 13.8

1 (a) $\frac{1}{4}x^4 + 2x^3 - 2x - \frac{1}{4}$ (b) $\ln|y^2 + y| - \ln|2y|$

2 $\frac{25}{2}$ **3** $\displaystyle\int_2^4 \int_1^5 dx\,dy = 8$ **4** $\frac{4}{3}$

5 (a) 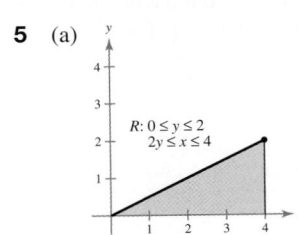 (b) $\displaystyle\int_0^4 \int_0^{x/2} dy\,dx$

$R: 0 \le y \le 2$
$2y \le x \le 4$

(c) $\displaystyle\int_0^2 \int_{2y}^4 dx\,dy = 4 = \int_0^4 \int_0^{x/2} dy\,dx$

6 $\displaystyle\int_{-1}^3 \int_{x^2}^{2x+3} dy\,dx = \frac{32}{3}$

Checkpoints for Section 13.9

1 $\frac{16}{3}$ **2** $e - 1$ **3** $\frac{176}{15}$
4 Integration by parts **5** 3

Answers to Tech Tutors

Section 0.2   *(page 11)*

11; 23

Section 1.2   *(page 85)*

$143.75

Section 1.4   *(page 106)*

1.854, −4.854

Section 2.2   *(page 175)*

$y = 2.2x + 11.3$

Section 2.6   *(page 221)*

(a) $g(x)$ is shifted four units to the right.

$h(x)$ is shifted four units to the right and three units upward.

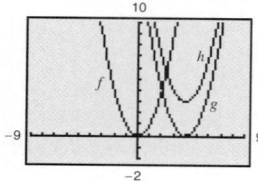

(b) $g(x)$ is shifted one unit to the left.

$h(x)$ is shifted one unit to the left and two units downward.

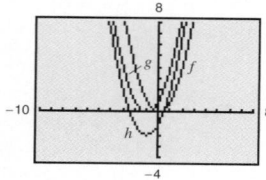

(c) $g(x)$ is shifted four units to the left.

$h(x)$ is shifted four units to the left and two units upward.

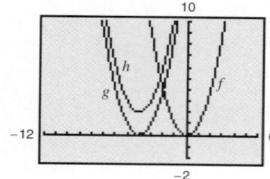

Section 3.6   *(page 307)*

(a) $x = 2$ (b) $x = 3$ (c) $x = 0$ (d) $x = \pm 1$

Section 4.6   *(page 388)*

$C = 138.78e^{0.167t}$; the models are the same.

Section 5.1   *(page 411)*

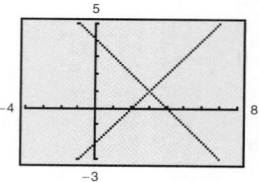

Solution: (3, 1)

Section 5.2   *(page 423)*

(16, 14)

Section 6.2   *(page 489)*

(a) $\begin{bmatrix} 6 & 6 & 12 \\ -9 & 0 & -3 \\ 6 & 3 & 6 \end{bmatrix}$

(b) $\begin{bmatrix} -2 & -2 & -4 \\ 3 & 0 & 1 \\ -2 & -1 & -2 \end{bmatrix}$

(c) $\begin{bmatrix} 4 & 6 & 12 \\ -10 & 4 & -6 \\ 7 & 0 & 4 \end{bmatrix}$

Section 7.2   *(page 554)*

Most calculators set in connected mode will join the two branches of the graph with a nearly vertical line near $x = 2$. This line is not part of the graph.

Section 10.5   *(page 779)*

Answers will vary.

Section 12.3   *(page 887)*

1.46265

Index

Applications (continued from front endsheets)